Forms Containing $u^2 \pm a^2$

38. $\displaystyle \int \frac{du}{a^2 - u^2} = \frac{1}{2a} \ln \left| \frac{a + u}{a - u} \right| + C$

39. $\displaystyle \int \frac{du}{u^2 - a^2} = \frac{1}{2a} \ln \left| \frac{u - a}{u + a} \right| + C$

40. $\displaystyle \int \frac{du}{\sqrt{u^2 \pm a^2}} = \ln|u + \sqrt{u^2 \pm a^2}| + C$

41. $\displaystyle \int \frac{du}{u\sqrt{a^2 \pm u^2}} = -\frac{1}{a} \ln \left| \frac{a + \sqrt{a^2 \pm u^2}}{u} \right| + C$

42. $\displaystyle \int \frac{du}{au^2 + c} = \frac{1}{\sqrt{ac}} \operatorname{Tan}^{-1} \left(u\sqrt{\frac{a}{c}} \right) + C, \quad ac > 0$

43. $\displaystyle \int \frac{du}{c^2 - u^2} = \frac{1}{2c} \ln \left| \frac{c + u}{c - u} \right| + C, \quad c^2 > u^2$

44. $\displaystyle \int \sqrt{u^2 \pm a^2}\, du = \frac{1}{2}[u\sqrt{u^2 \pm a^2} \pm a^2 \ln(u + \sqrt{u^2 \pm a^2})] + C$

45. $\displaystyle \int u\sqrt{u^2 \pm a^2}\, du = \frac{1}{3}(u^2 \pm a^2)^{3/2} + C$

46. $\displaystyle \int \frac{du}{\sqrt{u^2 \pm a^2}} = \ln(u + \sqrt{u^2 \pm a^2}) + C$

47. $\displaystyle \int \frac{du}{u\sqrt{u^2 - a^2}} = \frac{1}{a} \operatorname{Sec}^{-1} \left(\frac{u}{a} \right) + C$

48. $\displaystyle \int \frac{du}{u\sqrt{u^2 + a^2}} = -\frac{1}{a} \ln \left(\frac{a + \sqrt{u^2 + a^2}}{u} \right) + C$

49. $\displaystyle \int \frac{\sqrt{u^2 + a^2}}{u}\, du = \sqrt{u^2 + a^2} - a \ln \left(\frac{a + \sqrt{u^2 + a^2}}{u} \right) + C$

50. $\displaystyle \int \frac{\sqrt{u^2 - a^2}}{u}\, du = \sqrt{u^2 - a^2} - a \operatorname{Sec}^{-1} \left(\frac{u}{a} \right) + C, \quad a > 0$

51. $\displaystyle \int \frac{du}{\sqrt{(u^2 \pm a^2)^3}} = \frac{\pm u}{a^2 \sqrt{u^2 \pm a^2}} + C$

52. $\displaystyle \int \frac{\sqrt{u^2 \pm a^2}}{u^2}\, du = -\frac{\sqrt{u^2 \pm a^2}}{u} + \ln(u + \sqrt{u^2 \pm a^2}) + C$

53. $\displaystyle \int \frac{\sqrt{u^2 - a^2}}{u^3}\, du = -\frac{\sqrt{u^2 - a^2}}{2u^2} + \frac{1}{2a} \operatorname{Sec}^{-1} \left(\frac{u}{a} \right) + C, \quad a > 0$

54. $\displaystyle \int \sqrt{a^2 - u^2}\, du = \frac{1}{2} \left[u\sqrt{a^2 - u^2} + a^2 \operatorname{Sin}^{-1} \left(\frac{u}{a} \right) \right] + C, \quad a > 0$

55. $\displaystyle \int \frac{du}{\sqrt{a^2 - u^2}} = \operatorname{Sin}^{-1} \left(\frac{u}{a} \right) + C$

56. $\displaystyle \int \frac{du}{u\sqrt{a^2 - u^2}} = -\frac{1}{a} \ln \left(\frac{a + \sqrt{a^2 - u^2}}{u} \right) + C$

57. $\displaystyle \int \frac{du}{\sqrt{(a^2 - u^2)^3}} = \frac{u}{a^2 \sqrt{a^2 - u^2}} + C$

58. $\displaystyle \int u^2 \sqrt{a^2 - u^2}\, du = -\frac{u}{4} \sqrt{(a^2 - u^2)^3}$
$$+ \frac{a^2}{8} \left(u\sqrt{a^2 - u^2} + a^2 \operatorname{Sin}^{-1} \frac{u}{a} \right) + C, \quad a > 0$$

59. $\displaystyle \int \frac{u^2\, du}{\sqrt{a^2 - u^2}} = -\frac{u}{2} \sqrt{a^2 - u^2} + \frac{a^2}{2} \operatorname{Sin}^{-1} \left(\frac{u}{a} \right) + C, \quad a > 0$

60. $\displaystyle \int \frac{du}{u^2 \sqrt{a^2 - u^2}} = -\frac{\sqrt{a^2 - u^2}}{a^2 u} + C$

61. $\displaystyle \int \frac{\sqrt{a^2 - u^2}}{u^2}\, du = -\frac{\sqrt{a^2 - u^2}}{u} - \operatorname{Sin}^{-1} \left(\frac{u}{a} \right) + C, \quad a > 0$

Forms Containing $a + bu^2$

62. $\displaystyle \int \frac{du}{a + bu^2} = \frac{1}{\sqrt{ab}} \operatorname{Tan}^{-1} \frac{u\sqrt{ab}}{a} + C$

63. $\displaystyle \int \frac{du}{(u^2 + a^2)^2} = \frac{1}{2a^3} \operatorname{Tan}^{-1} \frac{u}{a} + \frac{u}{2a^2(u^2 + a^2)} + C$

64. $\displaystyle \int \frac{du}{u(a + bu^2)} = \frac{1}{2a} \ln \left| \frac{u^2}{a + bu^2} \right| + C$

Forms Involving $2au - u^2$

65. $\displaystyle \int \sqrt{2au - u^2}\, du = \frac{1}{2} \left[(u - a)\sqrt{2au - u^2} + a^2 \operatorname{Sin}^{-1} \left(\frac{u - a}{a} \right) \right] + C$

66. $\displaystyle \int \frac{du}{\sqrt{2au - u^2}} = 2 \operatorname{Sin}^{-1} \sqrt{\frac{u}{2a}} + C, \quad a > 0$

67. $\displaystyle \int \frac{du}{(2au - u^2)^{3/2}} = \frac{u - a}{a^2 \sqrt{2au - u^2}} + C$

68. $\displaystyle \int \frac{du}{\sqrt{2au + u^2}} = \ln|u + a + \sqrt{2au + u^2}| + C$

(Continued on back endsheet)

Calculus

Calculus

SECOND EDITION

DENNIS D. BERKEY
BOSTON UNIVERSITY

SAUNDERS COLLEGE PUBLISHING
New York   Chicago   San Francisco
Philadelphia   Montreal   Toronto
London   Sydney   Tokyo

Text Typeface: Times Roman
Compositor: York Graphic Services, Inc.
Acquisitions Editor: Robert Stern
Developmental Editors: Sandi Kiselica, Sarah Evans
Project Editor: Sally Kusch
Copy Editor: Will Eaton
Art Director: Carol Bleistine
Art Assistant: Doris Roessner
Text Designer: Emily Harste
Cover Designer: Lawrence R. Didona
Text Artwork: J&R Technical Services, Inc.
Production Managers: Tim Frelick, Harry Dean
Assistant Production Manager: JoAnn Melody

Cover Credit: A fractal—a set carried to infinity. Professor Larry Guseman and Robert Sparks, Texas A&M University.

Printed in the United States of America

CALCULUS, 2nd edition

0-03-008899-2

Library of Congress Catalog Card Number: 87-26531

789   032   987654321

PREFACE

This text is intended for use in a traditional three-semester or four-quarter sequence of courses on the calculus, populated principally by mathematics, science and engineering students. It reflects my own philosophy that freshman calculus should be taught so as to produce skilled practitioners who have some real feeling for the mathematical issues underlying the techniques they have acquired. It was written in order to provide students of widely varying interests and abilities with a highly readable exposition of the principal results, including ample motivation, numerous well-articulated examples, a rich discussion of applications, and a nontrivial description and use of numerical techniques. In particular, I have tried to indicate, in an unobtrusive manner, how computers can be used both to illustrate the theory and to provide approximate solutions to problems for which more elegant techniques break down.

LEVEL AND RIGOR: Nearly all topics in the traditional calculus curriculum are treated. The discussion is informal, and geometric arguments are used wherever possible. Limits are presented in Chapter 2, first informally, then followed by the formal $\delta-\epsilon$ definition. The formal definition, the examples of its use in proving limits, and the proofs of several limit theorems which appear in Chapter 2 may be omitted without loss of continuity by those instructors wishing to defer or minimize these formalities. When the topic of limits appears later, in the chapters on infinite series, it is treated more formally, reflecting the increased maturity of the reader by that point in the course. The definite integral is defined as a limit of Riemann sums, the natural logarithm is defined by integration, and the theorems of Green and Stokes, as well as the Divergence Theorem, are fully discussed.

STRATEGY/SOLUTION FORMAT IN EXAMPLES: An unusually large number of examples (more than 700) has been included. In many of these, especially in the early parts of the text, the solutions have been written in a two-column format, with one of the columns labelled ''Strategy.'' In this column the student will find, in very abbreviated form, a description of the principal steps involved in the fuller solution. Here I have attempted to help students identify the more general aspects of the particular solution, and to develop problem-solving strategies of their own.

ORGANIZATION AND CHANGES FROM THE FIRST EDITION: The order of topics is consistent with those of most popular texts. Many sections have been extensively rewritten from the first edition, and the treatments of several topics have been reorganized in response to user comments.

   Chapter 1 is a review of precalculus concepts which has been shortened significantly. The treatment of limits in Chapter 2 has been reorganized so that the formal definition and proofs of several limit theorems appear in the natural order of the chapter rather than in optional sections at the end. This change is in response to the increased mathematical abilities of high school graduates in recent years, although these formalities may be omitted without loss of continuity as noted above. The development of the derivative in Chapter 3 has been reorganized so that the derivatives of the trigonometric functions appear before the discussion of the Chain Rule, and the treatment of differentials has been rewritten and shortened significantly.

The treatment of applications of the derivative has been reorganized and condensed from the two chapters in the first edition into a single Chapter 4. This chapter begins with a discussion of the Mean Value and Extreme Value Theorems, moves next into the geometric applications of the derivative (increasing and decreasing functions, relative extrema, concavity and graphing) and concludes with discussions of extrema on closed intervals, max-min problems, and the derivative as a rate of change.

Chapter 5 is a brief chapter which addresses only the antiderivative and its application in solving separable differential equations. It is positioned so that antidifferentiation can be treated either as the final application of the derivative or as the first topic in a unit on integration. This ordering, which places work on antidifferentiation before the introduction of the definite integral, reflects a concern that I have held for some time that the definite integral is often presented almost simultaneously with the notion of antiderivative and the statement of the Fundamental Theorem of Calculus. As a result, students can leave their first calculus course thinking of a definite integral only as a difference between two values of an antiderivative that, coincidentally, might also be identified with the area of a certain planar region. By introducing antiderivatives first, together with their application to solving simple differential equations, the distinction between antiderivatives and limits of approximating Riemann sums can be clearly established before the important connection provided by the Fundamental Theorem is introduced. This approach should help students grasp more easily the notion of approximate integration as well as making the development of the various applications of the definite integral more accessible.

Chapter 6, on the definite integral, has been reorganized so that both parts of the Fundamental Theorem are presented at the same time, although a careful discussion of differentiating functions defined by integration appears in a separate section following the statement of the Fundamental Theorem. Applications of the definite integral, which appeared in two separate chapters in the first edition, have been consolidated and shortened into a single Chapter 7.

Chapters 8 through 10 treat the transcendental functions and techniques of integration in largely the same order as in the first edition. Chapter 11 is new and addresses the two topics of l'Hôpital's Rule (moved from its location in Chapter 5 of the first edition) and improper integrals.

The theory of infinite series has been reorganized from three chapters in the first edition into the two Chapters 12 and 13. Chapter 12 presents infinite sequences and series of constants while Chapter 13 discusses Taylor polynomials, Taylor series, and power series.

Material on geometry in the plane and in space appears in the same order as in the first edition, in Chapters 14 through 16, and the calculus in higher dimensions is treated in Chapters 17 through 20. This material has been extensively edited and considerably shortened. The treatment, however, is essentially that of the first edition.

Topics on differential equations, which appeared at various points throughout the first edition, have been consolidated into the concluding Chapter 21. This chapter contains new material on nonhomogeneous differential equations and numerical approximations to solutions of differential equations.

EXERCISES: More than 6,000 exercises are included, ranging from drill to challenging in type, and including many applied exercises from a broad range of disciplines. The extensive review exercises at the end of each chapter reflect the range of topics included in that chapter.

While many theorems, particularly those with instructive proofs, can and should be presented and proved, the time available to most of us for this task is not sufficient to allow a careful treatment of the least upper bound axiom, uniform continuity, differentiability of power series, or several other topics where statements of fact must simply be made. Honesty about these omissions, together with the right picture here and a good heuristic discussion there, can result in a presentation that is both factual and effective, and one that allows us to succeed in sharing with students the excitement of the triumphs of this classic subject.

COMPUTER-GENERATED COLOR FIGURES: New to this edition is the four-color insert in the multivariable calculus section. These computer-generated, four-color figures provide students with the ability to better view detailed, mathematically correct, three-dimensional surfaces. Some of the figures are rotated to show more than one perspective. These figures also appear within the text in line art form. The corresponding computer-drawn four-color figures are referenced in the line art captions.

ACKNOWLEDGMENTS: Many individuals played instrumental roles in the development of this text. It is my pleasure to acknowledge some of these here.

Twenty-nine teachers of the calculus scrutinized one or more drafts of the first edition, both on matters of content and to identify errors. These were

Paul Baum, Brown University
David Bellamy, University of Delaware
George Blakley, Texas A&M University
Jan List Boal, Georgia State University
Alan Candiotti, Drew University
George Feissner, SUNY at Cortland
Hebert Gindler, San Diego State University
Arthur Goodman, Queen's College, CUNY
Barry Granoff, Boston University
Douglas Hall, Michigan State University
Peter Herron, Suffolk Community College
Laurence Hoffman, Claremont Men's College
Adam Hulin, University of New Orleans
Frank Kocher, Pennsylvania State University
Carlon Krantz, Kean College
Paul Kumpel, SUNY at Stony Brook
Ed Landesman, University of California, Santa Cruz
Robert Lohman, Kent State University
Eldon Miller, University of Mississippi
Daniel Moran, Michigan State University
Richard Porter, Northeastern University
Thomas Rishel, Cornell University

Rainer Sachs, University of California, Berkeley
Philip Schaefer, University of Tennessee
Mark Schilling, University of Southern California
Donald Sherbert, University of Illinois
John Thorpe, SUNY at Stony Brook
William Wheeler, Indiana University
Robert Zink, Purdue University

The following mathematicians served as readers for the second edition:

Al Boggess, Texas A&M University
Phyllis Boutillier, Michigan Technological University
Lloyd Davis, College of San Mateo
Don Edmondson, University of Texas at Austin
Garrett Etgen, University of Houston
John Higgins, Brigham Young University
Ken Kramer, Queens College, CUNY
Steven Krantz, Washington University
Nicholas Krier, Colorado State University
Robert McFadden, Northern Illinois University
Robert Moreland, Texas Tech University
James Reeder, Honolulu Community College
Nathaniel Silver, University of Hartford
Richard Thompson, University of Arizona
Donald Wilken, SUNY, Albany
Stephen Willard, University of Alberta

Galley and page proofs were scrutinized for content and accuracy by Leon Gerber of St. John's University, Charles Stone of DeKalb College, Kathleen Hollowell of Newton High School (MA), Al Boggess, Steven Krantz, and Stephen Willard. Engineering applications were reviewed by Jeff Laible of the University of Vermont, Edgar Tacher of the University of Tulsa and Arthur Tiedmann of the University of Wisconsin. Kathleen Hollowell provided special assistance in this revision through thoughtful critique of the first edition and a careful review of all exercise sets.

Each of these individuals worked meticulously to ensure the accuracy of the text. Whatever errors might remain are, of course, the sole responsibility of the author. Any comments on correcting or improving the text will be gratefully acknowledged.

At Boston University I remain indebted to Tom Orowan for typing near-perfect drafts of the manuscript, and to Lisa Doherty for managing the flow of materials between Boston and Philadelphia. The BASIC programs included in the appendix were used by students on the University's time-sharing system, as well as on the author's personal computer.

Forewarned about hazards in dealings with publishers, I was delighted to experience a warm, professional, and highly supportive relationship with each of several key individuals at Saunders College Publishing: Mathematics Editor Robert Stern, Developmental Editors Sandi Kiselica and Sarah Evans, Project Editor Sally Kusch, and Publisher Don Jackson. Their commitment to excellence was a strong guiding force throughout the development of this edition.

The historical notes, which provide an important human contrast, were written by Professor Duane Deal of Ball State University.

The computer-drawn figures for the four-color insert were kindly provided by Larry Guseman and Robert Sparks of Texas A&M University. These figures were generated using a Silicon Graphics Iris 3130 computer.

On a more general and personal level, the writing of this text was supported by three very special groups of people. First, my students at Boston University, who have encouraged this project and helped sharpen my thinking about teaching and the calculus for the past decade. Second, my colleagues in the faculty and administration of Boston University, who believe deeply in the importance of effective teaching. And, most importantly, my family, who understood my need to write this book and shared fully and willingly in the sacrifices that were required. To all I am truly grateful.

Boston, Massachusetts   **Dennis D. Berkey**

CONTENTS OVERVIEW

CONTENTS

UNIT 1

PRELIMINARY NOTIONS

Isaac Newton

Gottfried Leibniz

NEWTON AND LEIBNIZ—THE UNIFICATION OF THE CALCULUS

More than once in history, a significant mathematical development has been independently discovered at about the same time by two mathematicians widely separated by geography and having no contact with one another. The greatest of these discoveries was that of the calculus. Isaac Newton (1642–1727) in England, and Gottfried Wilhelm Leibniz (1646–1717) in Germany, were completely unaware of each other's work. Newton developed his version of the calculus some ten years before Leibniz, but Newton was generally reluctant to publish his results, and Leibniz published his own version before Newton's publication. These time differences ultimately resulted in an extended controversy between English and Continental mathematicians concerning priority and the possibility of plagiarism. To their credit, however, Newton and Leibniz did not attack one another.

Isaac Newton was born on Christmas day in 1642 in a saddened household, as his father had died about two months earlier. He was a small and frail child, and not expected to live very long. He showed an early interest in mechanical contrivances. He constructed a clock, and a replica of a mill which was powered by a mouse, and he frightened the neighbors with a kite carrying a small fire. His mother wanted him to become a farmer, but he was not interested. His uncle suggested sending him to a university, and at the age of 19 he went to Trinity College at Cambridge, where he was an undistinguished undergraduate student.

In the first months of 1665 bubonic plague swept over England, and all public institutions were closed. Newton stayed on at Cambridge to finish his bachelor's degree that spring, and then returned to the family farm. There followed one of the most scientifically productive periods in human history. This 23-year-old youth, in the span of only 18 months, (a) proved that the binomial theorem was valid for fractional and negative exponents as well as for positive integral powers; (b) discovered, using prisms, that white light could be resolved into the colors of the rainbow, and then reconstituted into white light, leading to the theory of optics; (c) discovered the theory of universal gravitation and the inverse square law which was the key to understanding the structure of the solar system; and (d) discovered differential and integral calculus, and more importantly, the inverse relation of the two. The well-known incident of the falling apple which caused him to begin thinking of gravitation occurred during this period on the farm.

It was the infinite series generated by his fractional and negative exponents for binomials which caused young Newton to use these series as the basis for his calculus. He considered a curve as the result of a continuously moving point. The changing quantity he called a *fluent,* and its rate of change he named a *fluxion* (from the Latin *fluere,* to flow). The fluxion was thus what we shall refer to as a derivative, denoted $\dot{y}$ if y were the original fluent, or variable. Similarly, Newton saw that the original fluent y could be thought of as the fluxion of another function, which he designated $\overset{\square}{y}$ or $\overset{|}{y}$. We today call this an integral.

After the plague years, Isaac Newton returned to Cambridge to work on his master's degree, and at the age of only 26 was appointed to the prestigious endowed chair of mathematics, the Lucasian Professorship.

Newton was always reluctant to publish his mathematical and scientific discoveries, but friends eventually persuaded him to write and publish what was to become known as the calculus. This great work, entitled *Philosophiae naturalis principia mathematica,* or *Mathematical Principles of Natural Philosophy,* was published in 1687, about three years after Leibniz had published his own version of differential calculus in Germany.

(Photographs from the David Eugene Smith Papers, Rare Book and Manuscript Library, Columbia University.)

For about twenty years Newton made significant discoveries in many fields, but then the light of creative genius flickered. He suffered a long illness which affected his ability and, while he continued to work, the results were not of the quality of his earlier productivity. He was elected to Parliament, and then, in 1696, was appointed Warden of the Mint. In this capacity, and later as Master of the Mint, his great genius was spent in checking the quality of the metal in British coins. But also in this period, he was elected President of the Royal Society of London successively for 25 years, and presided over the study of the many scientific and mathematical accomplishments of the early eighteenth century.

Gottfried Wilhelm Leibniz was the son of a university professor who died when the boy was only six years old. He was a precocious student, and his teachers sought, unsuccessfully, to permit him to read only material they thought suitable for his age. After his father's death, he was taken from the school and permitted to read in his father's extensive library where he became essentially self-educated. He entered the University of Leipzig at age 15, received his bachelor's degree at 17, and his doctorate (in law) at 20. He became a diplomat by profession, but soon became interested in mathematics as a consuming avocation. By the time Leibniz was 30 in 1676, he had invented his calculus, which is in many ways the calculus as we use it today.

About 1673 he came to the realization that areas can be calculated by *summing* "infinitely thin" rectangles, and that tangents to curves involve *differences* in the x and y coordinates. This led him to suspect that the two processes are inverses of one another. In doing these calculations, he soon found himself immersed in infinite series representing functions, as had Newton several years before. Realizing the power of his discoveries, Leibniz set to work developing terminology and notation. On the night of October 29, 1675, he selected the integral sign $\int$ as representing the Latin word *summa,* or sum, and the derivative notation $\dfrac{dy}{dx}$.

Leibniz's, and the world's, first paper on differential calculus was published in 1684 with the title (translated) *A New Method for Maxima and Minima, and Also for Tangents. . . .* He gave formulas for differentials of powers, products, and quotients of functions—the same formulas we learn today. Two years later he published his integral calculus, in which the calculation of areas is shown to be the inverse operation of finding tangents.

Leibniz clearly published his calculus (1684–86) before Newton (1687), but Newton made his discoveries (1665–66) before Leibniz (1672–76). The supporters of each man commenced a great controversy concerning the original discoverer of the calculus. Leibniz and Newton corresponded, but it is a tribute to the character of both men that they did not personally enter into the quarrel. The unfortunate result, however, was that the mathematicians on the European continent lost the use of the ideas of Newton, especially concerning his theory of gravitation, and the English mathematicians lost ground due to their refusal to use the more powerful calculus notation of Leibniz. It was only about a hundred years ago that these attitudes faded.

Leibniz's life was primarily spent in serving a succession of petty German rulers. He became increasingly unhappy, and spent his later years writing about religious and philosophical questions. He never married. His later years were lonely, and he died an embittered man. His secretary was the only mourner at his funeral.

Newton, on the other hand, was respected and revered in his old age. He was knighted by Queen Anne in 1705, and was buried in Westminster Abbey with great honor.

Newton and Leibniz were both creative mathematicians of the first rank. The genius of neither is diminished by that of the other.

Chapter 1
Review of Precalculus Concepts

This chapter summarizes many of the concepts and techniques of precalculus mathematics used throughout the text. You will find little here which you have not seen before. However, there are several reasons why you should examine this chapter carefully.

First, you need to become familiar with the notation and style in which this text is written. Studying mathematics is similar to studying a foreign language in that a vocabulary must be mastered, as well as a set of concepts. In mathematics, the vocabulary is quite formal, consisting of symbols and terms that have very precise meanings. However, different authors may not use exactly the same notation. Let's agree on notation and basic concepts before discussing new ideas. This will make the material which follows seem less intimidating.

Second, as mathematics has come to be more widely used as a tool in the natural and social sciences, more students than ever before are studying the calculus. This means that the typical calculus class contains students with quite diverse backgrounds in mathematics. This chapter defines the level and scope of precalculus mathematics assumed throughout the text. It therefore provides a common base from which to proceed.

Finally, this chapter provides a gauge by which you can determine whether you are sufficiently well prepared to begin a course on the calculus. If, after carefully reviewing the chapter, you are unable to work most of the exercises successfully, you probably need to consider enrolling in a course on basic mathematical skills. On the other hand, many students will already have mastered the concepts presented here. If this is your situation, you should quickly survey the chapter to observe the notation, and then proceed into Chapter 2.

1.1 THE REAL NUMBER SYSTEM

Real Numbers

The **real numbers** may be represented by the set of all points lying on an infinite line. To do so we first select a point 0 on the line that we refer to as the **origin** (Figure 1.1). We then select a **unit** of measurement and locate the point on the line

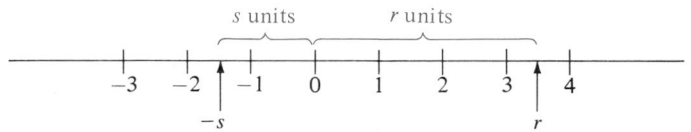

Figure 1.1 The real number line.

lying one unit to the right of 0. This point is labeled 1 (one). This procedure establishes the **scale** for the **number line** and also uniquely determines the location of each real number. Each positive real number r lies r units to the right of 0 and each negative real number, $-s$, lies s units to the left of 0.

There are three important special types of real numbers that can be identified by their decimal forms. The **integers** are the real numbers . . . $-2, -1, 0, 1, 2, ...$ Integers may be expressed in decimal form with only zeros to the right of the decimal point. The **rational** numbers are those real numbers that can be expressed as quotients of integers. In other words, if r is a rational number then $r = p/q$ for some integers p and q ($q \neq 0$). The decimal forms of rational numbers repeat eventually. For example, using a bar to identify the digits that repeat, we may write

$$\frac{1}{4} = 0.2500\overline{0} . . . ,$$

$$\frac{2}{9} = 0.222\overline{2} . . . ,$$

$$-\frac{25}{11} = -2.27\overline{27} . . . , \text{ and}$$

$$\frac{20}{13} = 1.538461\overline{538461} . . . .$$

(See Exercise 72 at the end of this section.)

Those real numbers whose decimal forms do not repeat are referred to as **irrational numbers.** The discovery nearly 2500 years ago that not every real number is also a rational number was both profound and traumatic, since the geometric mathematics of the early Greeks assumed that every length could be expressed as the ratio of two integral lengths. Most high school students today study the proof that the real number $\sqrt{2}$ is irrational, and the irrational number π plays a central role in Euclidean geometry.* Besides these two prominent examples, there are infinitely many irrational real numbers.

Sets of Real Numbers

There are several types of notation that we will use to specify sets of real numbers and operations on these sets. The most familiar notation to you is probably **set builder notation** in which one specifies a choice of variable and also a rule by which values of that variable are determined. For example, $\{x \mid 2 \leq x \leq 4\}$ is read, ''the set of all real numbers equal to or greater than 2 and equal to or less than 4.'' We will frequently refer to **intervals** of the real number line with the following notation:

$$[a, b] = \{x \mid a \leq x \leq b\},$$
$$[a, b) = \{x \mid a \leq x < b\},$$
$$(a, b] = \{x \mid a < x \leq b\},$$
$$(a, b) = \{x \mid a < x < b\}.$$

---

*The number $\pi = 3.14159 . . .$ is the ratio of the circumference of a circle to its diameter. Since this irrational number is often *approximated* by the fraction $\frac{22}{7}$, many students mistakenly believe that $\pi = \frac{22}{7}$ and, therefore, that π is rational. The proof that π is irrational is not simple.

We can denote intervals that extend infinitely in one direction by use of the infinity symbol, ∞, as follows:

$$[a, \infty) = \{x \mid a \le x\},$$
$$(-\infty, b) = \{x \mid x < b\}.$$

There are two other cases of intervals that extend infinitely in one direction, namely, (a, ∞) and $(-\infty, b]$; the interval $(-\infty, \infty)$ is the entire number line. It must be clearly understood that ∞ is not a number: it is shorthand for the phrase ''continuing infinitely through positive values'' (or negative values for $-\infty$). In sketching intervals, we use a solid dot to indicate an endpoint that is included and a hollow dot to indicate an endpoint that is excluded, as illustrated in Figures 1.2 and 1.3.

An interval of the form $[a, b]$ is called a **closed** interval because it includes both its endpoints. An interval of the form (a, b) is called an **open** interval because it includes neither of its endpoints. Intervals of the form $[a, b)$ or $(a, b]$ are referred to as either **half-open** or **half-closed.** The infinite intervals $[a, \infty)$ and $(-\infty, a]$ are called *closed,* and the infinite intervals (a, ∞) and $(-\infty, a)$ are called *open.*

If x is a number and A is a set of numbers (not necessarily an interval), the statement ''$x \in A$,'' read ''x is an element of set A,'' means that x is one of the numbers in set A. The negation of this statement is ''$x \notin A$,'' read ''x is not an element of set A.'' For example,

$$3 \in \{1, 3, 5\}, \qquad 2 \in [0, 7), \qquad \text{but} \qquad 6 \notin \{x \mid -2 \le x \le 2\}.$$

If A and B are sets of numbers, we say that A is a **subset** of B if every element of A is also an element of B. The notation for this is $A \subseteq B$. When both A and B are intervals, we say that A is a subinterval of B if $A \subseteq B$. For example

$$\{1, 3\} \subseteq \{1, 2, 3, 4\}, \qquad \text{and} \qquad [-1, 1] \subseteq (-\infty, 3).$$

The symbol for the empty set is $\varnothing$. It is the set with no elements.

There are two **operations** on sets that we shall use frequently. Given two sets, A and B, the **union** of the two sets is the set containing precisely those numbers which are elements of either set A or set B, or both. The symbol for the union of A and B is $A \cup B$. In other words,

$$A \cup B = \{x \mid x \in A \text{ or } x \in B\}.^*$$

For example,

$$\{1, 3\} \cup \{2, 3, 5\} = \{1, 2, 3, 5\}; \qquad [-3, 4] \cup (2, 6) = [-3, 6).$$

The **intersection** of two sets A and B is the set containing precisely those numbers which are elements of *both* set A and set B. The symbol for the intersection of A and B is $A \cap B$. Thus,

$$A \cap B = \{x \mid x \in A \text{ and } x \in B\}.$$

For example,

$$\{1, 3\} \cap \{2, 3, 5\} = \{3\}, \qquad \text{and} \qquad [-3, 4] \cap (2, 6) = (2, 4].$$

Figure 1.4 illustrates these two set operations for the interval examples above.

Figure 1.2 The interval $[a, b) = \{x \mid a \le x < b\}$.

Figure 1.3 The interval $(a, \infty) = \{x \mid a < x < \infty\}$.

---

*In mathematics, the usage of the word ''or'' is *inclusive.* This means that the statement ''U or V'' is interpreted as ''either U or V, or *both.*'' This is in contrast to the *exclusive* usage, in which ''U or V'' would be interpreted as ''either U or V, but *not* both.''

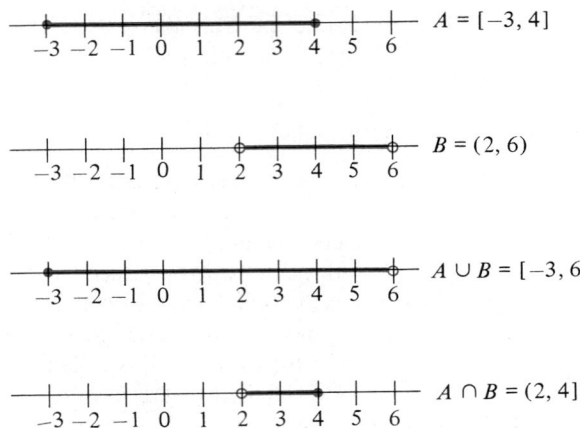

Figure 1.4 Set union and intersection.

Inequalities

There is a natural ordering for the real numbers. If b lies to the right of a on the number line, we say that b is greater than a, and we write $b > a$. Of course, this is the same as saying that a is less than b and writing $a < b$ (Figures 1.5 and 1.6). A purely algebraic way to say that b lies to the right of a is to say that the difference $(b - a)$ is positive. The formal definition of inequality is therefore the following.

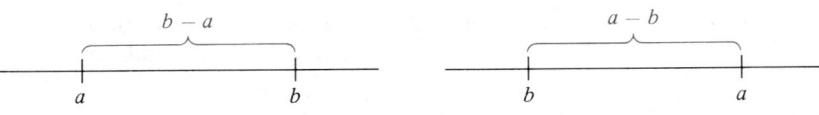

Figure 1.5 $a < b$, or $(b - a) > 0$. **Figure 1.6** $a > b$, or $(a - b) > 0$.

DEFINITION 1

The **inequality** $a < b$ means $b - a$ is positive. The inequality $a \leq b$ means that either $a = b$ or $a < b$.

We will often use the properties given by the following theorem. Their proofs are quite simple, and you are asked to provide them in Exercises 59, 60, and 63.

THEOREM 1
Properties of Inequality

Let x, y, z, and c be real numbers. Then

(i) Exactly one of the following holds: $x = y$, $x < y$, or $x > y$. (trichotomy)
(ii) If $x < y$ and $y < z$, then $x < z$. (transitivity)
(iii) If $x < y$, then $x + c < y + c$.
(iv) If $x < y$ and $c > 0$, then $cx < cy$.
(v) If $x < y$ and $c < 0$, then $cx > cy$.

Properties (iii)–(v) are frequently used in "solving" inequalities, that is, finding the numbers for which the inequalities are true statements. Property (iii) says that one can add a number to both sides of an inequality without changing the sense of the inequality. However, properties (iv) and (v) show that in multiplying both sides of an inequality by c, the sense of the inequality will be retained if $c > 0$ and reversed if $c < 0$. The incorrect use of property (v) is a frequent source of error.

Example 1

(a) To solve the inequality

$$x + 2 < 5$$

we add the constant -2 to both sides, using property (iii), to obtain the inequality

$$(x + 2) + (-2) < 5 + (-2),$$

or

$$x < 3.$$

The *solution set* is, therefore, $\{x \mid x < 3\} = (-\infty, 3)$.

(b) To solve the inequality

$$-\frac{3}{2}x \geq 9,$$

we multiply both sides by $-\frac{2}{3}$, using property (v). This gives the inequality

$$\left(-\frac{2}{3}\right)\left(-\frac{3}{2}x\right) \leq \left(-\frac{2}{3}\right) \cdot 9,$$

or

$$x \leq -6.$$

The solution set is $\{x \mid x \leq -6\} = (-\infty, -6]$. $\diamond$

The gist of "solving inequalities" is that we attempt to isolate the variable on one side of the inequality, just as we do in solving equations containing a variable. However, you must remember one critical difference—multiplying (or dividing) by a negative number reverses the sense of the inequality.

Example 2

Solve the inequality $7x - 4 \leq 3x + 8$.

Solution: Adding 4 to both sides of the inequality gives

$$7x \leq 3x + 12.$$

We next add $-3x$ to both sides (or, subtract $3x$ from both sides) to obtain

$$4x \leq 12.$$

Finally, we divide both sides by 4 $\left(\text{or, multiply both sides by } \frac{1}{4}\right)$ to conclude that

$$x \leq 3.$$

The solution set is therefore $\{x \mid x \leq 3\} = (-\infty, 3]$. $\diamond$

Quadratic inequalities are more difficult to solve than the simple linear inequalities of Examples 1 and 2, since they involve x^2 terms. The following example demonstrates how quadratic inequalities can be solved when we are fortunate enough to be able to "factor" the quadratic expression. Note the careful use of the notions of set unions and intersections.

Example 3

Solve the inequality $x^2 + x + 1 > 7$.

Strategy

Move all terms to the left side, leaving zero on the right.

Factor the left-hand side (if possible) and find its zeros.

These zeros divide the number line into intervals, on which $x^2 + x - 6$ will have constant sign.

Determine the sign of each factor on each interval.

From the signs of the factors, determine the sign of the quadratic expression on each interval.

Solution

Adding -7 to both sides of the inequality gives

$$x^2 + x - 6 > 0.$$

Since $x^2 + x - 6 = (x + 3)(x - 2)$, the inequality may be written as

$$(x + 3)(x - 2) > 0.$$

From this factored form we can see that $(x + 3)(x - 2) = 0$ if $x = -3$ or if $x = 2$. We may therefore determine the sign of the left-hand side on the three resulting intervals $(-\infty, -3)$, $(-3, 2)$, and $(2, \infty)$ by examining the signs of the individual factors:

(i) For $x \in (-\infty, -3)$, $x < -3$, so both

$$x + 3 < 0 \quad \text{and} \quad x - 2 < 0.$$

Thus, $(x + 3)(x - 2) > 0$ if $x \in (-\infty, -3)$.
(ii) For $x \in (-3, 2)$,

(a) $x > -3$, so $x + 3 > 0$.
(b) $x < 2$, so $x - 2 < 0$.

Thus, $(x + 3)(x - 2) < 0$ if $x \in (-3, 2)$.
(iii) For $x \in (2, \infty)$, $x > 2$, so both

$$x + 3 > 0 \quad \text{and} \quad x - 2 > 0.$$

Thus, $(x + 3)(x - 2) > 0$ if $x \in (2, \infty)$.

From cases (i)–(iii), we conclude that $(x + 3)(x - 2) > 0$ if $x \in (-\infty, -3) \cup (2, \infty)$. $\diamond$

REMARK: Figure 1.7 suggests a simple graphical device for visualizing the procedure in Example 3. Using one number line for each factor and one for the quadratic

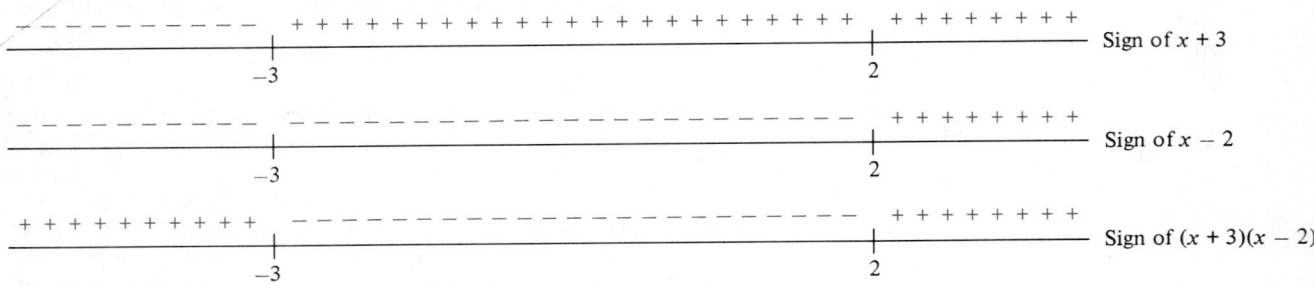

Figure 1.7 Graphical device for determining sign of a quadratic expression.

expression, mark the zeros, and then mark the signs of each factor on each interval. The signs for the quadratic on each interval are then easily recognized.

Absolute Value and Distance

In order to speak about the distance between two points on the number line, we need a way to refer to the magnitude of a number, regardless of its sign. We use the concept of the **absolute value** of a number x, denoted by $|x|$.

DEFINITION 2

$$|x| = \begin{cases} x & \text{if} \quad x \geq 0 \\ -x & \text{if} \quad x < 0. \end{cases}$$

For example, $|3| = 3$, $|-7| = -(-7) = 7$, and $|\pi - 4| = 4 - \pi$. (You should note that the absolute value of a number is *always* nonnegative. Even though the expression $-x$ appears in the second line of the definition of $|x|$, x itself is negative in this case, so $-x$ is positive.)

Example 4

Solve the equation $|2x - 3| = 6$.

Strategy
Write down the two cases from Definition 2.

Solve the two equations that result.

Solution
From Definition 2 either

$$2x - 3 = 6, \quad \text{or} \quad -(2x - 3) = 6,$$

so either

$$2x = 9, \quad \text{or} \quad -2x = 3.$$

These two equations produce the solutions

$$x_1 = \frac{9}{2}, \quad \text{and} \quad x_2 = -\frac{3}{2}. \qquad \diamond$$

The following theorem is useful in solving inequalities involving absolute values. Its proof is left to you as an exercise. (See Exercise 73.)

THEOREM 2

Let a be a positive number. The inequality

$$|x| < a$$

is equivalent to the double inequality

$$-a < x < a.$$

Example 5

Solve the inequality $|7 - 2x| \leq 5$.

Strategy
Use Theorem 2 to rewrite the inequality, with the goal of isolating x in the middle.

Solution
By Theorem 2 we can rewrite the inequality as

$$-5 \leq 7 - 2x \leq 5,$$

Subtract 7 from all parts.

so

$$-12 \le -2x \le -2,$$

Divide all terms by -2, remembering to reverse sense of all inequalities.

and

$$6 \ge x \ge 1.$$

The solution set is therefore $\{x \mid 1 \le x \le 6\}$, or $[1, 6]$. ◇

Using the concept of absolute value, we can define the distance between two numbers (points) on the number line.

DEFINITION 3	The **distance** between the numbers a and b is the number $|b - a|$.

Figure 1.8 illustrates the meaning of Definition 3. The distance between the numbers a and b is just the absolute value of their difference.

Figure 1.8 Distance on the number line.

The following example shows how we may use the notion of distance to interpret certain inequalities.

Example 6

Find all numbers x which satisfy the inequality

$$|x - 2| < 5.$$

Strategy
Interpret the statement geometrically, using the definition of distance.

Solution
Since $|x - 2|$ is the distance between the numbers x and 2, the inequality is describing precisely those numbers x which lie at a distance less than 5 from the number 2. These numbers are described by the inequality

$$-3 < x < 7.$$

Provide a sketch.

The solution set is $\{x \mid -3 < x < 7\} = (-3, 7)$.
See Figure 1.9. ◇

We shall make frequent use of the following properties of absolute value.

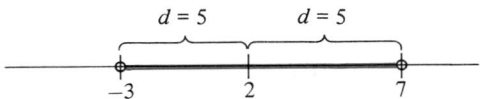

Figure 1.9 Solution of the inequality $|x - 2| < 5$.

THEOREM 3	For any real numbers x and y,
Properties of Absolute Value	

    (i) $|x| \geq 0$,

    (ii) $|xy| = |x| \cdot |y|$,

    (iii) $|x + y| \leq |x| + |y|$.

Statement (i) simply reminds us that absolute values are never negative. Statement (ii) says that the absolute value of a product is the same as the product of the absolute values. Statement (iii) is referred to as the **triangle inequality** (for reasons we shall see later) and states that the absolute value of the sum of two numbers can never exceed the sum of their absolute values. Notice that statement (iii) is an inequality rather than an equation. An example where equality fails is $|3 + (-5)| = |-2| = 2 < |3| + |-5|$.

Exercise Set 1.1

In each of Exercises 1–9, label the real number as either rational or irrational.

1. $\dfrac{22}{7}$ **2.** $3.141414\ldots$ **3.** $2.1010010001\ldots$

4. $\pi + 3$ **5.** $\dfrac{51}{17}$ **6.** $-6.163\overline{163}\ldots$

7. $\sqrt{256}$ **8.** $\sqrt{2}$ **9.** $1 + \sqrt{2}$

In Exercises 10–15, use the intervals $A = [-2, 5]$, $B = (-1, 6)$, and $C = (-\infty, 0)$ to find the indicated intervals.

10. $A \cup B$ **11.** $A \cap B$ **12.** $A \cup C$

13. $A \cap C$ **14.** $B \cup C$ **15.** $B \cap C$

16. Let $A = [2, 4)$. Find the interval B so that $A \cup B = [2, 6]$ and $A \cap B = (3, 4)$.

17. True or false? For a circle of positive radius, the radius and the circumference cannot both be rational numbers. What about area and radius? Area and circumference?

In each of Exercises 18–39, solve for x.

18. $2x + 3 < 9$ **19.** $x + 4 \leq 3x$

20. $2x + 7 > 4x - 5$ **21.** $6x - 6 \leq 8x + 8$

22. $-4 \leq 2(x + 2) < 10$ **23.** $(x - 3)(x + 1) < 0$

24. $(x + 7)(2x - 4) > 0$ **25.** $x(x + 6) > -8$

26. $x^2 + x < 0$ **27.** $x^3 + x^2 - 2x > 0$

28. $x^2 + x + 7 > 19$ **29.** $x^4 - 9x^2 < 0$

30. $|x - 7| = 2$ **31.** $|5x - 2| = 0$

32. $x + |x| = 0$ **33.** $2|3x - 1| = 22$

34. $x + |-3| = 7$ **35.** $(|x| + 6)^2 = 49$

36. $x - 6 = 2x - |\pi - 6|$ **37.** $x + 3|x| = 8$

38. $|(x + 2)^2 + 3| = 12$ **39.** $|x - 4| = |x - 7|$

40. True or false? $|a - b| = |b - a|$ for all real numbers a and b.

In Exercises 41–51, solve for x and sketch the solution set.

41. $|x - 3| \leq 2$ **42.** $|x + 5| < 4$

43. $|x + 2| > 1$ **44.** $3|x - 6| \geq 12$

45. $|2x - 7| \leq 3$ **46.** $|8 - 3x| \geq 5$

47. $|x - 1| = x - 1$ **48.** $x > |x|$

49. $|x + 3| + |x - 2| < 7$ **50.** $|x + 4| - |x - 1| < 4$

51. $|x + 2| + |x - 5| \geq 10$

In Exercises 52–55, solve the inequality geometrically by describing in words the meaning of the inequality in terms of distances.

52. $|x - 5| > 2$ **53.** $|x - 1| = |x + 3|$

54. $|x - 1| = 2|x - 3|$ **55.** $2 < |x - 4| < 5$

56. Write down an inequality whose solution is the set of all numbers lying at a distance greater than 4 from the number -5.

57. Write down an equation whose solution is the number whose distance from -2 is twice its distance from 12.

58. The terminology and operations described for sets of real numbers may be applied to more general kinds of sets. **Venn diagrams** may be used to illustrate these concepts, as shown below. The idea is simply that the shaded region represents the particular set.

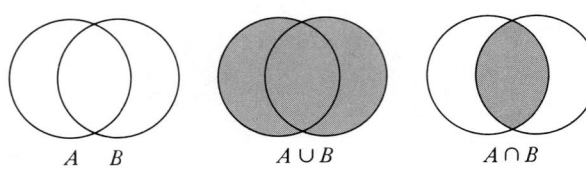

A B $\qquad$ $A \cup B$ $\qquad$ $A \cap B$

Draw Venn diagrams illustrating three sets A, B, and C, so that each pair shares a common area, for the following sets.

a. $A \cup B \cup C$ **b.** $A \cap B \cap C$ **c.** $A \cap (B \cup C)$

59. Prove Theorem 1, part (i), by examining the sign of $(y - x)$.

60. Prove Theorem 1, part (ii) by writing $(z - x) = (z - y) + (y - x)$ and applying Definition 1.

61. True or false? If $a^2 < b^2$ then $a < b$.

62. Prove that if $a^2 < b^2$ and $b > 0$ then $-b < a < b$. (*Hint:* Factor $b^2 - a^2$ as $(b - a) \cdot (b + a)$.)

63. Prove properties (iii)–(v) of Theorem 1.

64. Use Exercise 62 to solve the inequality $(x + 6)^2 < 49$.

65. Prove that $\sqrt{2}$ is not a rational number by filling in the details of the following argument:
a. Argue by contradiction. This means, assume that $\sqrt{2}$ *is* rational and seek a contradiction.
b. If $\sqrt{2}$ is rational, $\sqrt{2} = p/q$ where p and q have no common factors.
c. Thus $\sqrt{2}q = p$, so $2q^2 = p^2$.
d. This means p is divisible by 2 (why?), so $p = 2k$. Thus,
$$2q^2 = (2k)^2, \text{ so } q^2 = 2k^2.$$
e. This means q is divisible by 2.
f. Statements d and e contradict the last phrase of statement b. The assumption that $\sqrt{2}$ is rational is therefore false.

66. Prove that $|x| = \sqrt{x^2}$ for all real numbers x.

67. Prove the triangle inequality by noticing that
$$-|x| \le x \le |x| \quad \text{and} \quad -|y| \le y \le |y|$$
and adding to obtain $-(|x| + |y|) \le x + y \le |x| + |y|$.

68. True or false? Taking reciprocals reverses the sense of an inequality, i.e., if $a < b$ than $\dfrac{1}{a} > \dfrac{1}{b}$. If false, give a counterexample.

69. State and prove the correct version of the statement in Exercise 68.

70. Prove the inequality $|x| - |y| \le |x - y|$ by writing $x = y + (x - y)$ and applying the triangle inequality.

71. Does the same basic argument as Exercise 65 show that $\sqrt{3}$ is not rational?

72. The following demonstration shows how to recover the fractional form of a repeating decimal (rational number) $2.6\overline{36363}\ldots$.
$$x = 2.6\overline{36363}\ldots, \quad \text{so} \quad 100x = 263.6\overline{36363}\ldots,$$
so
$$99x = (100x - x) = 261.0\overline{000}\ldots.$$
Thus
$$x = \frac{261}{99}.$$
a. Use a similar procedure to find a fractional form for the rational number $1.341\overline{341341}\ldots$.
b. Generalize your findings to a statement about how to recover the fractional form for the repeating decimal $x = 0.\overline{a_1 a_2 \ldots a_n}$.

73. Prove Theorem 2 by considering the two cases $x \ge 0$ and $x < 0$ separately.

1.2 THE COORDINATE PLANE, DISTANCE, AND CIRCLES

The theory and techniques that we shall develop in the first two thirds of this text are designed for analyzing situations involving two variables. For example, we will consider the relationship between time and distance for particles moving along a line, and the relationship between pressure and temperature for a quantity of an ideal gas.

The Coordinate Plane

In order to illustrate these relationships graphically, we use a two-dimensional coordinate system constructed as follows. If x and y represent the two variables in question, we construct axes for x and for y (known as the **x-axis** and **y-axis,** respectively) so that the axes are perpendicular and so that the two origins lie at the same point. (Also, we follow the convention that the positive half of the horizontal axis extends to the right and the positive half of the vertical axis extends upward.) The plane determined by these two axes is called the **xy-coordinate plane.**

Each point P in the xy-plane is associated with a pair of real numbers, (x_0, y_0), called its coordinates. These coordinates are found by constructing lines through P parallel to both axes. The number x_0 on the x-axis where the vertical line crosses is called the **x-coordinate** of P. Similarly, the number y_0 where the horizontal line crosses the y-axis is called the **y-coordinate** of P. The coordinate axes divide the plane into four quadrants, numbered as in Figure 2.1. Figure 2.2 shows the coordinates of several points in the plane.

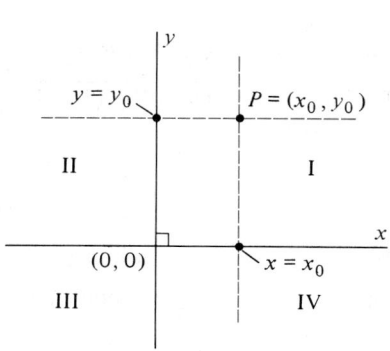

Figure 2.1 The xy-coordinate plane.

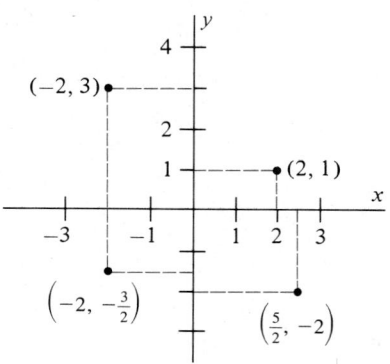

Figure 2.2 Points in the xy-plane.

To find the point Q in the xy-plane corresponding to the given coordinates (x_1, y_1), we reverse the procedure described above, constructing a vertical line through the point $x = x_1$ on the x-axis and a horizontal line through $y = y_1$ on the y-axis. The point where these lines cross is the desired point Q.

Distance

To calculate the distance between the points $P = (x_1, y_1)$ and $Q = (x_2, y_2)$ in the plane, we make use of the point $R = (x_2, y_1)$.

It then follows from the Pythagorean theorem, that

$$d^2 = |x_2 - x_1|^2 + |y_2 - y_1|^2$$
$$= (x_2 - x_1)^2 + (y_2 - y_1)^2.$$

(See Figure 2.3.) This explains our definition of distance in the plane.

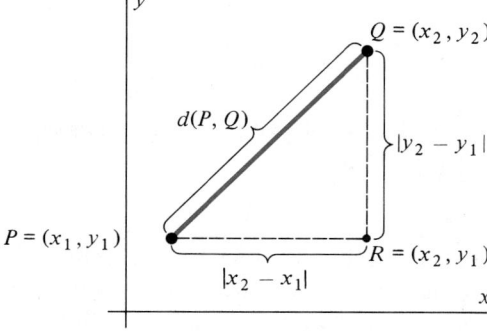

Figure 2.3 Distance between P and Q is $d(P, Q) = \sqrt{(x_2 - x_1)^2 + (y_2 - y_1)^2}$.

DEFINITION 4	The distance $d(P, Q)$ between the points $P = (x_1, y_1)$ and $Q = (x_2, y_2)$ in the xy-plane is the number $$d(P, Q) = \sqrt{(x_2 - x_1)^2 + (y_2 - y_1)^2}.$$ (Note that $d(P, Q)$ cannot be negative.)

Example 1

The distance between the points $P = (-4, 1)$ and $Q = (1, 3)$ is

$$\begin{aligned} d(P, Q) &= \sqrt{(1 - (-4))^2 + (3 - 1)^2} \\ &= \sqrt{5^2 + 2^2} \\ &= \sqrt{29}. \end{aligned}$$

◇

Example 2

Find the set of all points lying equidistant from the points $P = (-2, 1)$ and $Q = (4, -2)$.

Strategy

Select an arbitrary point (x, y) satisfying the condition. Try to find a relationship between x and y.

Use Definition 4 to convert the condition of equal distances to an equation in x and y.

Simplify the resulting equation as much as possible.

Solution

We let $T = (x, y)$ be any point lying equidistant from the points $P = (-2, 1)$ and $Q = (4, -2)$. This means that

$$d(P, T) = d(T, Q)$$

or

$$\sqrt{(x - (-2))^2 + (y - 1)^2} = \sqrt{(x - 4)^2 + (y - (-2))^2}.$$

Squaring both sides of this equation gives

$$(x + 2)^2 + (y - 1)^2 = (x - 4)^2 + (y + 2)^2,$$

so

$$(x^2 + 4x + 4) + (y^2 - 2y + 1) = (x^2 - 8x + 16) + (y^2 + 4y + 4).$$

This equation simplifies to the equation

$$12x = 6y + 15,$$

or

$$y = 2x - \frac{5}{2}.$$

The set of points satisfying this equation is the line sketched in Figure 2.4.

◇

Circles

Geometrically, a **circle** may be defined as the set of all points lying at a fixed distance (the **radius**) from a fixed point (the **center**). According to Definition 4, this means that the point (x, y) lies on the circle with radius r and center (h, k) if and only if

$$\sqrt{(x - h)^2 + (y - k)^2} = r, \qquad r > 0. \tag{1}$$

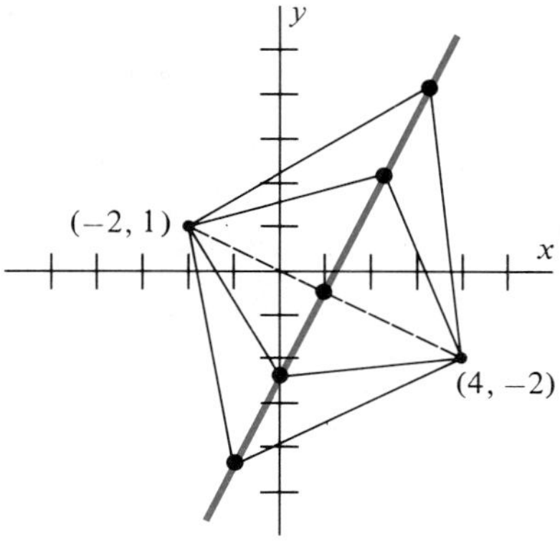

Figure 2.4 Points equidistant from two points lie along a line.

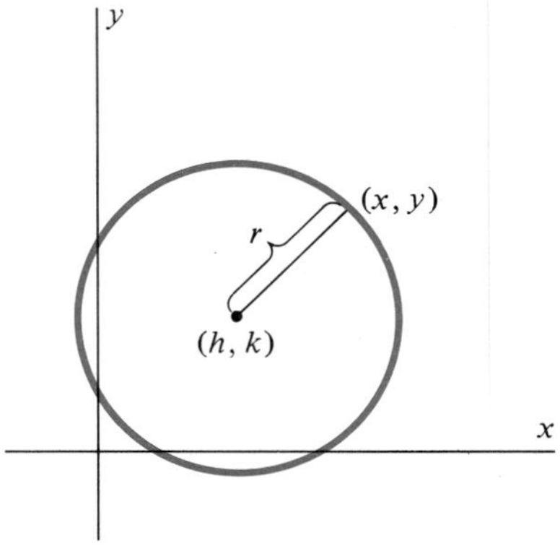

Figure 2.5 The circle $(x - h)^2 + (y - k)^2 = r^2$.

Squaring both sides of equation (1), we obtain the **standard** form for the equation of the circle with radius r and center (h, k):

$$(x - h)^2 + (y - k)^2 = r^2. \tag{2}$$

(See Figure 2.5.)

Example 3

An equation for the circle with center $(1, -2)$ and radius $r = 4$ is, according to equation (2),

$$(x - 1)^2 + (y - (-2))^2 = 4^2,$$

which simplifies to

$$(x - 1)^2 + (y + 2)^2 = 16,$$

or

$$x^2 - 2x + y^2 + 4y - 11 = 0. \qquad \diamondsuit$$

Example 3 had to do with finding an equation for a circle, given certain data. Example 4 involves the reverse problem—describing a circle from its equation. The procedure for doing so involves **completing the square.**

Example 4

Describe and sketch the graph of the equation

$$x^2 + y^2 + 6x - 2y + 6 = 0.$$

Strategy

Complete the square in x and y terms to try to bring equation into form of equation (1).

Solution

We write the given equation as

$$(x^2 + 6x) + (y^2 - 2y) + 6 = 0. \tag{3}$$

Completing the square in x.

To complete the square on the first term, we add the square of half the coefficient of x, that is, we add $\left(\dfrac{6}{2}\right)^2 = 9$. This gives

$$(x^2 + 6x) = (x^2 + 6x + 9) - 9 = (x + 3)^2 - 9. \tag{4}$$

Completing the square in y.

Similarly, we complete the square on the second term as

$$(y^2 - 2y) = (y^2 - 2y + 1) - 1 = (y - 1)^2 - 1. \tag{5}$$

Combining the results.

Substituting the expressions in equations (4) and (5) into equation (3) gives the equation

$$[(x + 3)^2 - 9] + [(y - 1)^2 - 1] + 6 = 0$$

Compare the equation obtained with equation (2).

or

$$(x + 3)^2 + (y - 1)^2 = 4.$$

The graph is therefore a circle with center $(-3, 1)$ and radius $r = \sqrt{4} = 2$ (see Figure 2.6). ◇

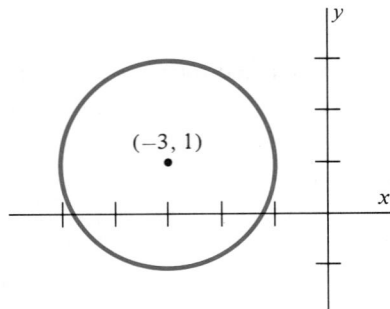

(−3, 1)

Figure 2.6 Circle $x^2 + y^2 + 6x - 2y = -6$.

Exercise Set 1.2

1. Find the distance between the following points.
   a. $(2, -1)$ and $(0, 2)$
   b. $(3, 1)$ and $(1, 3)$
   c. $(0, 2)$ and $(1, -9)$
   d. $(1, -3)$ and $(6, 6)$
   e. $(1, 1)$ and $(-1, -1)$
   f. $(-2, -2)$ and $(1, 2)$

2. Find two points (h, k) so that $h = 5$ and so that the distance between (h, k) and $(1, 3)$ equals 5.

3. Find two points in the xy-coordinate plane, each of which lies a distance of 4 units from both $(-2, 3)$ and $(2, 3)$.

4. True or false? *Every* point in the plane lying at a distance r from point P must lie on the circle with center P and radius r.

5. Find a so that the triangle with vertices $(-1, 0)$, $(2, 3)$, and $(a, 0)$ is isosceles. (*Hint:* There are 4 possible answers.)

6. Find b so that the triangle with vertices $(0, 0)$, $(2, 0)$, and $(1, b)$ is equilateral.

7. Find an equation for the set of all points lying equidistant from the points $(-1, -1)$ and $(3, 1)$.

8. Find an equation for the set of all points lying equidistant from the points $(1, 2)$ and $(5, -1)$.

9. Verify that the coordinates of the midpoint of the line segment joining the points (x_1, y_1) and (x_2, y_2) are $\left(\dfrac{x_1 + x_2}{2}, \dfrac{y_1 + y_2}{2}\right)$. (*Hint:* Use the distance formula.)

10. Find the midpoints of the line segments joining the following pairs of points. (See Exercise 9.)
    a. $(-1, 2)$ and $(6, 4)$
    b. $(1, 1)$ and $(-7, -3)$

In Exercises 11–14, sketch the set of points in the xy-plane satisfying the given property.

11. The set of all points lying more than $\sqrt{2}$ units from the point $(-3, 1)$.

12. The set of all points lying at least 3 units to the right of the y-axis and at least 3 units from the x-axis.

13. The set of all points lying no more than 2 units from either axis.

14. The set of all points lying more than 3 but less than 5 units from the point $(1, -2)$.

In Exercises 15–17, write the equation for the circle with the stated properties.

15. The circle with center $(0, 0)$ and radius 3.

16. The circle with center $(-2, 4)$ and radius 3.

17. The circle with center $(-6, -4)$ and radius 5.

18. Find an equation for the circle with center $(2, 3)$ that contains the point $(2, -1)$.

19. Find equations for the circles with radius 5 and containing the points $(-1, -4)$ and $(4, 1)$.

In each of Exercises 20–25, find the center and radius of the circle whose equation is given, and sketch the circle.

20. $x^2 - 2x + y^2 - 8 = 0$

21. $x^2 + y^2 + 4x + 2y - 11 = 0$

22. $x^2 + 14x + y^2 - 10y + 70 = 0$

23. $x^2 + y^2 - 2x - 6y + 3 = 0$

24. $x^2 - 2ax + y^2 + 4ay + 5a^2 - 1 = 0$

25. $x^2 + y^2 - 2by + b^2 - a^2 = 0$

26. Find an equation for the circle with center $(3, 2)$ which is tangent to the y-axis. (That is, which intersects the y-axis in a single point.)

27. Find the points of intersection of the line $x - y - 5 = 0$ with the circle $x^2 - 8x + y^2 - 4y + 11 = 0$.

28. True or false? Three points determine a circle. Why?

29. Show that the set of all points whose distance from the point $(-2, 1)$ is twice the distance from the point $(4, -2)$ is a circle. Find its radius and center.

30. Find an equation for the circle containing the points $(-2, 1)$, $(1, 4)$, and $(4, 1)$.

31. Find the family of *all* circles containing the two points $(1, 2)$ and $(-1, 2)$.

32. Show that the equation $(x - h)^2 + (y - k)^2 = d$ will have (1) infinitely many solutions (i.e., the points on a circle), (2) exactly one solution, or (3) no solutions, depending on whether d is positive, zero, or negative. (The last two cases are referred to as **degenerate** circles.)

1.3 LINEAR EQUATIONS

Equations whose graphs are straight lines are very important in the calculus. Indeed, one of the most fundamental problems in the entire subject is that of finding the equation of the line tangent to a given curve at a given point. More generally, lines are important because they represent the simplest relationship between two variables. For example, we shall see that for a freely falling body the relationship between velocity and time is a line. Even when the relationship between two variables is more complicated than simply a straight line, scientists often "linearize" their models. That is, they find the straight line which "best approximates" the true relationship, usually out of a desire to keep their models as simple as possible.

Slope

If $P_1 = (x_1, y_1)$ and $P_2 = (x_2, y_2)$ are two distinct points on a nonvertical line ℓ, the **slope** of ℓ is defined by the ratio

$$m = \frac{y_2 - y_1}{x_2 - x_1}, \qquad x_2 \neq x_1 \tag{1}$$

as illustrated in Figure 3.1.

For a given line, the slope m is independent of the particular choices for (x_1, y_1) and (x_2, y_2). (See Exercise 44.) Notice that expression (1) for slope is not defined if $x_1 = x_2$. For this reason, we say that a vertical line has no slope. For all other lines slope is a real number, and every real number can occur as the slope of some line. Some illustrations of lines with various slopes appear in Figure 3.2.

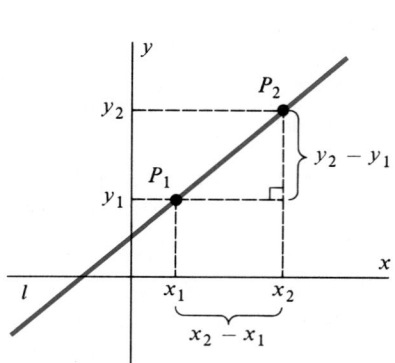

Figure 3.1 Slope of $\ell = \dfrac{y_2 - y_1}{x_2 - x_1}$.

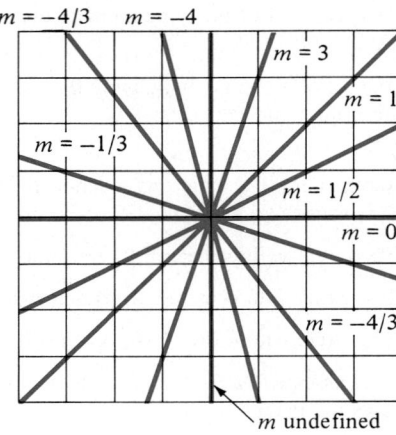

Figure 3.2 Some slopes.

Example 1

Find the slope of the line through the points

(a) $(-7, 2)$ and $(3, -3)$, and
(b) $(-1, -4)$ and $(7, 16)$.

Solution: From equation (1) we have

(a) $m = \dfrac{-3 - 2}{3 - (-7)} = \dfrac{-5}{10} = -\dfrac{1}{2}$, and

(b) $m = \dfrac{16 - (-4)}{7 - (-1)} = \dfrac{20}{8} = \dfrac{5}{2}$.   ◇

Equations for Lines

Suppose ℓ is a line with slope m which contains the point (x_1, y_1). To find an equation for ℓ we let $P = (x, y)$ be an arbitrary point on ℓ. Then, by equation (1), we obtain

$$m = \frac{y - y_1}{x - x_1}$$

so

$$y - y_1 = m(x - x_1),$$ (2a)

or

$$y = mx + b$$ (2b)

where $b = y_1 - mx_1$.

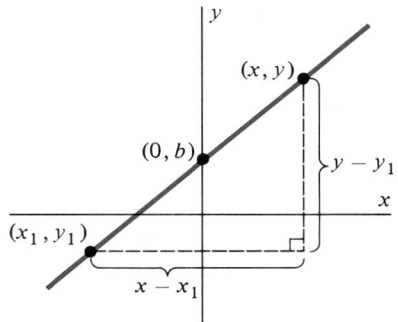

Figure 3.3 $y = mx + b$ has y intercept b.

Equation (2a) is called the **point-slope** equation for the line ℓ. Equation (2b) is called the **slope-intercept** equation for ℓ. This is because the constants m and b appearing in equation (2b) are the slope and y-intercept for ℓ, respectively. To see this, notice that if we set $x = 0$ in equation (2b), we obtain the statement $y = b$. This means that the point $(0, b)$ is on the graph of ℓ. (See Figure 3.3.)

Every nonvertical line can be described by an equation of the form (2b). Conversely, if (x_1, y_1) and (x_2, y_2) are any two distinct points whose coordinates satisfy equation (2b), then the slope of the line determined by these two points is

$$\text{Slope} = \frac{y_2 - y_1}{x_2 - x_1} = \frac{(mx_2 + b) - (mx_1 + b)}{x_2 - x_1} = \frac{m(x_2 - x_1)}{x_2 - x_1} = m.$$

It follows that any point (x, y) that satisfies equation (2b) lies on the line with slope m and y-intercept b. The following theorem summarizes our discussion.

THEOREM 4

The graph of the equation $y = mx + b$ is a line with slope m and y-intercept b. Conversely, the line with slope m and y-intercept b has this equation.

Example 2

Find an equation in slope-intercept form for the line through $(2, 5)$ with slope $m = 3$.

Solution: Substituting directly into equation (2a) gives

$$y - 5 = 3(x - 2),$$

so

$$y = 3x - 1$$

is the desired equation. ◇

Example 3

Find an equation for the line through $(-3, 4)$ and $(1, -2)$.

Strategy
First, find the slope using (1).

Solution
By equation (1), the slope of the line is

$$m = \frac{-2 - 4}{1 - (-3)} = -\frac{6}{4} = -\frac{3}{2}.$$

Then, use slope and one point to write an equation for the line.

Using the point $(-3, 4)$ and the slope $m = -\dfrac{3}{2}$ in equation (2a) then gives

$$y - 4 = \left(-\frac{3}{2}\right)(x + 3),$$

or

$$y = -\frac{3}{2}x - \frac{1}{2}.$$

(See Figure 3.4.) ◇

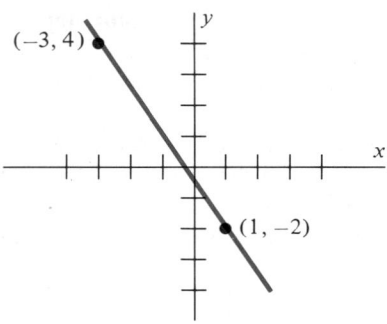

Figure 3.4 Line $y = -\dfrac{3}{2}x - \dfrac{1}{2}$.

Example 4

Find the slope and y-intercept for the line with equation

$$2x + 4y - 6 = 0.$$

Strategy

Bring equation to slope-intercept form of equation (2b).

Read off m and b.

Solution

We put the given equation in slope-intercept form by solving for y. We obtain

$$4y = -2x + 6,$$

so

$$y = -\frac{1}{2}x + \frac{3}{2}.$$

The line therefore has slope $-\dfrac{1}{2}$ and y-intercept $\dfrac{3}{2}$.

(See Figure 3.5.) ◇

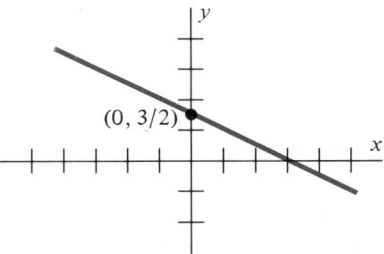

Figure 3.5 Line $y = -\dfrac{1}{2}x + \dfrac{3}{2}$.

Vertical and Horizontal Lines

A vertical line has the property that all x-coordinates of points on the line are the same, while the values of the y-coordinate(s) are unrestricted. The equation of a vertical line therefore has the form

$$x = a \qquad \text{(vertical line)}.$$

A horizontal line has the property that its slope is zero. Setting $m = 0$ in equation (2b) shows that a horizontal line has an equation of the form

$$y = b \qquad \text{(horizontal line)}.$$

Figure 3.6 shows the graphs of the vertical line $x = 2$ and the horizontal line $y = 1$.

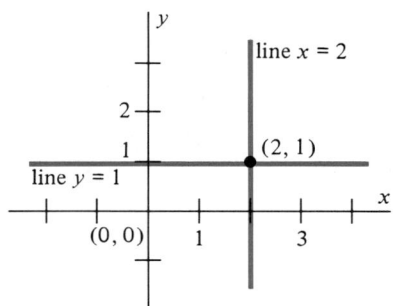

Figure 3.6 Vertical and horizontal lines.

Parallel and Perpendicular Lines

We will make use of the following theorem from elementary algebra.

THEOREM 5

Let ℓ_1 and ℓ_2 be two distinct lines with equations

$$\ell_1: y = m_1 x + b_1$$
$$\ell_2: y = m_2 x + b_2$$

where neither m_1 nor m_2 are zero. Then

(a) ℓ_1 and ℓ_2 are **parallel** if and only if $m_1 = m_2$.
(b) ℓ_1 and ℓ_2 are **perpendicular** if and only if $m_1 m_2 = -1$.

Theorem 5 states that parallel lines have the same slope, while the slopes of perpendicular lines are negative reciprocals of each other.

Example 5

For the line $\ell_1: 3y - 9x = 12$, find

(a) an equation of the line ℓ_2 which is parallel to ℓ_1 and contains the point (2, 0),
(b) an equation of the line ℓ_3 which is perpendicular to ℓ_1 and contains the point (3, −2).

Strategy

(a) Find the slope of ℓ_1 by putting equation in slope-intercept form.

Take $m_2 = m_1$.

Use point-slope form to find equation for ℓ_2.

Solution

(a) We first bring the equation for ℓ_1 into slope-intercept form by solving for y:

$$y = 3x + 4.$$

The slope of ℓ_1 is therefore $m_1 = 3$, so the slope of ℓ_2 must also be $m_2 = 3$. Since ℓ_2 contains the point (2, 0), we obtain

$$y - 0 = 3(x - 2)$$

or

$$\ell_2: y = 3(x - 2).$$

(b) Slope of ℓ_3 is negative reciprocal of m_1.

Use point-slope form to find equation for ℓ_3.

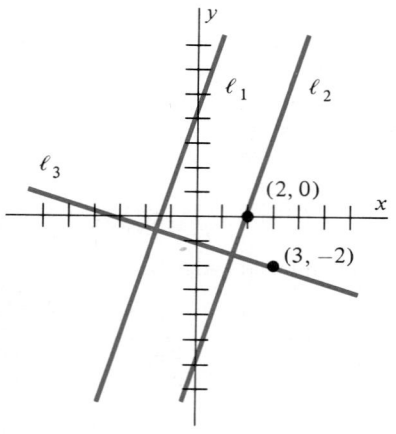

Figure 3.7 Lines in Example 5.

(b) Line ℓ_3 is perpendicular to ℓ_1, so $m_3 = -\dfrac{1}{m_1} = -\dfrac{1}{3}$. Since ℓ_3 contains the point $(3, -2)$ we obtain

$$y - (-2) = \left(-\frac{1}{3}\right)(x - 3),$$

so

$$y + 2 = -\frac{1}{3}(x - 3),$$

or

$$\ell_3: y = -\frac{1}{3}x - 1.$$

(See Figure 3.7.) ◇

Points of Intersection

Lines in the plane that are not parallel must intersect. If the lines are distinct (that is, if they are not the same line), they intersect at a single point. Since the coordinates of this point of intersection must satisfy the equations for both lines, we can find the point of intersection by solving both equations for y and then equating these two expressions. The result is a first degree equation in x that is easily solved.

Example 6

Find the point of intersection for the lines

$$\ell_1: 4x - 2y - 4 = 0,$$
$$\ell_2: x + y - 4 = 0.$$

Strategy

Solve both equations for y.

Equate expressions for y. Solve for x.

Substitute value for x into ℓ_1 or ℓ_2 to find y.

Solution

Solving both equations for y gives

$$\ell_1: y = 2x - 2,$$
$$\ell_2: y = -x + 4.$$

Equating expressions for y gives

$$2x - 2 = -x + 4$$

or

$$3x = 6.$$

Thus, $x = 2$. From the equation for ℓ_1 we find that $y = 2(2) - 2 = 2$. The point of intersection is therefore $(2, 2)$. ◇

Exercise Set 1.3

1. Find the slope of the line through the following pairs of points
  a. $(3, 1)$ and $(-1, 2)$ **c.** (a, b) and (b, a), $a \neq b$
  b. $(6, -2)$ and $(-1, -1)$ **d.** $(1, 1)$ and (a, a)

2. True or false? Every line has a slope.

3. True or false? Every slope determines one and only one line.

4. True or false? If ℓ_1 and ℓ_2 are perpendicular lines then ℓ_1 and ℓ_2 have precisely one point of intersection.

5. True or false? If ℓ_1 and ℓ_2 are parallel and distinct, they do not intersect.

6. Find a if the line through $(2, 4)$ and $(-2, a)$ has slope 3.

7. Find b if the line through $(b, 1)$ and $(1, -5)$ has slope 2.

In Exercises 8–22, find an equation for the line determined by the given information.

8. Slope 4 and y-intercept -2.

9. Slope -2 and y-intercept 5.

10. Slope zero and y-intercept -5.

11. Through $(-1, 6)$ and $(4, 12)$.

12. Through $(4, 1)$ and with slope 7.

13. Through $(1, 3)$ and with slope -3.

14. x-intercept -3 and y-intercept 6.

15. Through $(-3, 5)$ and vertical.

16. Through $(-3, 5)$ and horizontal.

17. Through $(-2, 4)$ and $(-6, 8)$.

18. Through $(0, 2)$ and $(-1, -4)$.

19. Through $(1, 4)$ and parallel to the line with equation $2x - 6y + 5 = 0$.

20. Through $(5, -2)$ and parallel to the line with equation $x - y = 2$.

21. Through $(1, 3)$ and perpendicular to the line with equation $3x + y = 7$.

22. Through $(4, -1)$ and perpendicular to the line through the points $(-2, 5)$ and $(-1, 9)$.

23. True or false? If x and y satisfy a linear equation, then y is proportional to x.

24. True or false? If x and y satisfy a linear equation, then y is proportional to $x + a$ for some constant a.

In Exercises 25–32, find the slope, x-intercept, and y-intercept of the line determined by the given equation. Graph the line.

25. $x = 7 - y$

26. $3x - 2y = 6$

27. $x + y + 3 = 0$

28. $2x = 10 - 3y$

29. $y = 5$

30. $y - 2x = 9$

31. $x = 4$

32. $y = x$

In each of Exercises 33–38, find the point(s) of intersection of the two lines, if any.

33. $3x - y - 1 = 0$
  $x + y - 3 = 0$

34. $x - 2y + 3 = 0$
  $x - 2y - 7 = 0$

35. $x - 3y + 3 = 0$
  $2x - 3y + 6 = 0$

36. $x - 2y + 4 = 0$
  $3x + 6y - 12 = 0$

37. $2x - y - 2 = 0$
  $8x - 4y + 2 = 0$

38. $3x - 2y + 2 = 0$
  $3x - y = 0$

39. Explain why any equation of the form $ax + by + c = 0$, where a and b are not both zero, must have a graph which is a line. (*Hint:* Show that if $b \neq 0$, the equation can be put in slope-intercept form. What about the remaining case, $b = 0$?)

40. Because of the result in Exercise 39, the equation $ax + by + c = 0$ is often referred to as the **general linear equation.** If only two points are required to determine a line, why does this equation have three constants?

41. By checking slopes, determine whether the following sets of points lie on a common line.
  a. $(1, 3)$, $(-2, 0)$, and $(4, 6)$
  b. $(2, -7)$, $(-2, -3)$, and $(-1, -4)$
  c. $(5, 15)$, $(0, 3)$, and $(-2, -7)$

42. Determine whether or not the points $(1, 3)$, $(3, 5)$, and $(4, 0)$ form the vertices of a right triangle.

43. The points $(-2, 2)$, $(4, 4)$, and $(0, a)$ are the vertices of a right triangle.
  a. For how many distinct values of a is this condition satisfied?
  b. Find these values of a.

44. Use information about similar triangles to explain why, for a given line, the calculation of slope is independent of the particular points chosen.

45. A market research firm determines that the demand (d) for a certain product, in terms of purchasers per thousand population, is 12 when the product is priced at $P = 20$ dollars, but that the demand is only 6 when the item is priced at $60. Write down a linear equation which models the relationship between demand (d) and price (P) according to these data. At what price level will demand reach zero?

46. A state has an income tax of 5% on all income over $5000.
  a. Write a linear equation which determines a person's tax, T, in terms of income, i, for a person earning more than $5000.

b. At what income level will a person owe $800 in state tax?

47. Find a linear equation which determines temperatures in degrees Fahrenheit from temperatures in degrees Celsius.

48. Find the distance from the point $(4, -2)$ to the line ℓ with equation $2x - y + 1 = 0$ as follows:
a. Find the slope m of the line ℓ.
b. The slope of the perpendicular from $(4, -2)$ to ℓ is then $-\dfrac{1}{m}$.
c. Write the equation for this line.
d. Find its point of intersection with ℓ.
e. Compute the desired distance.

49. As a beaker of water is heated the following temperatures are noted:

time (minutes)	0	10	20
temperature (°C)	22	42	62

a. Find a linear equation which relates time t to temperature T.
b. According to this equation what will the temperature of the water be after 25 minutes? 35 minutes? (Using an equation obtained from data to predict values of one variable from stated values of the other variable is called **extrapolation.**)

50. Temperatures T on the absolute, or Kelvin, scale are related to temperatures t on the Celsius scale by the equation $T = t + 273$.
At the Kelvin temperature 273 (0°C) the volume of a certain quantity of gas is 40 liters.
a. Find a linear relationship for volume in terms of temperature, according to Charles' Law: $V = kT = k(t + 273)$.
b. What is the volume of the gas at temperature 323 K?
c. What is the volume of the gas at temperature 100°C?

51. The length of a rod l is given by the linear equation $l = l_0(1 + at)$, where a is a constant called the coefficient of thermal expansion, l_0 is the length of the rod at 0°C, and t is the temperature of the rod. Find a if the rod is 100 cm long at temperature 0°C and 100.2 cm long at temperature 50°C.

52. Find an equation for the line through the points of intersection of the circles $x^2 + y^2 - 2y = 0$ and $x^2 - 2x + y^2 = 0$.

53. Find an equation for the line through the centers of the circles with equations $x^2 - 2x + y^2 - 4y + 1 = 0$ and $x^2 + y^2 + 2y = 0$.

1.4 FUNCTIONS

A linear equation $y = mx + b$ is but one type of rule that assigns a *unique* number y to each number x. More generally, a *function* is *any* rule that assigns a unique *value* to each element of a given set. For example,

(a) The equation $y = x^3$ assigns to each number x its (unique) cube. Thus, $y = x^3$ is a function.
(b) The rule, "measure his or her height," assigns a unique number (height) to each member of your calculus class. It is also a function.
(c) But, the equation $x^2 + y^2 = 1$ does *not* determine a function, since it does not determine a unique value of y for a given value of x. This is most readily seen by writing the equation in the form $y = \pm\sqrt{1 - x^2}$. It is then clear, for example, that both $y = 1$ and $y = -1$ are assigned to $x = 0$, violating the property of uniqueness.

Here is a more precise definition of what we are talking about.

DEFINITION 5

A **function** from set A to set B is a rule that assigns to each element $x \in A$ a **unique** element $y \in B$. We write $y = f(x)$ to indicate that the element y is the value assigned by the function f to the element x. The set A is called the **domain** of the function f. The set of all values $\{f(x) \mid x \in A\}$ is called the **range** of the function.

Thus, in order to specify a function, we must indicate both the domain of the function and the rule by which values of the function are determined. We will

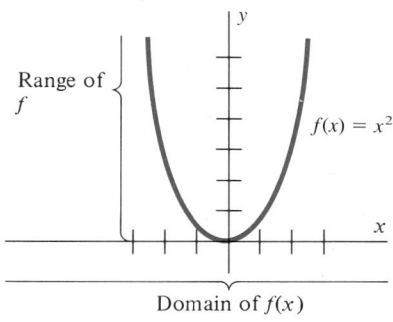

Figure 4.1 For the function $f = x^2$, domain of f is $(-\infty, \infty)$ and range of f is $[0, \infty)$.

almost always do this by writing an equation of the form $y = f(x)$, where $f(x)$ is an expression involving x's and constants. If we wish to specify a special subset of the real numbers as the domain of the function, we do so by writing a statement to this effect following the equation $y = f(x)$. Otherwise, the domain of the function consists of all numbers x for which the expression $f(x)$ is defined.

Example 1

For the function $f(x) = x^2$,

(a) the domain is the whole real line, since nothing has been indicated to restrict the domain;

(b) the range is the interval $[0, \infty)$, since all nonnegative numbers are squares (that is, have square roots), and all squares must be nonnegative.

(See Figure 4.1.) ◇

Example 2

For the function $f(x) = -x + 3, \; x \le 4$,

(a) the domain is specified to be the interval $(-\infty, 4]$ by the inequality $x \le 4$;

(b) the range is the interval $[-1, \infty)$, since $f(x) = -x + 3 \ge -1$ whenever $x \le 4$, and $y = -x + 3$ has a solution in $(-\infty, 4]$ whenever $y \ge -1$.

(See Figure 4.2.) ◇

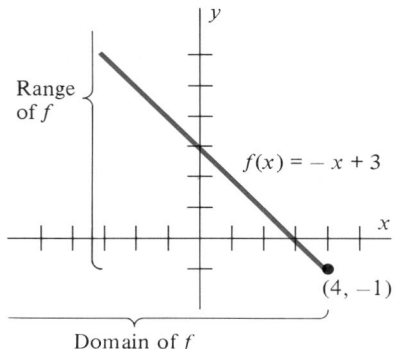

Figure 4.2 For $f(x) = -x + 3, \; x \le 4$, domain of f is $(-\infty, 4]$ and range of f is $[-1, \infty)$.

Example 3

For the function $f(x) = \dfrac{1}{x - 2}$,

(a) the domain is the set $(-\infty, 2) \cup (2, \infty)$. The number $x = 2$ is excluded from the domain, since the expression $\dfrac{1}{x - 2}$ is undefined for $x = 2$.

(b) The range is the set $(-\infty, 0) \cup (0, \infty)$, since every value $f(x)$ is a real number other than zero, and the equation $y = \dfrac{1}{x - 2}$ has a solution for every value y except $y = 0$ (see Figure 4.3). ◇

Example 3 points out one reason for which certain numbers must be excluded from the domain of a function, even though a domain is not explicitly stated: a denominator becomes zero. Example 4 presents another situation of this type, this time involving an even root, which exists only for nonnegative numbers.

Example 4

Find the domain and range of the function $f(x) = \sqrt{x + 3}$.

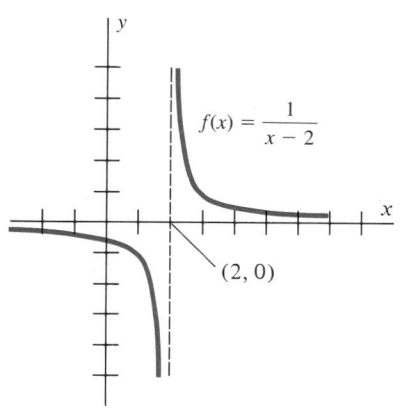

Figure 4.3 For $f(x) = \dfrac{1}{x - 2}$, domain is $(-\infty, 2) \cup (2, \infty)$ and range is $(-\infty, 0) \cup (0, \infty)$.

Solution: Because of the square root, $f(x)$ is defined only when $x + 3 \ge 0$, which gives $x \ge -3$. The domain is therefore $[-3, \infty)$. The range is $[0, \infty)$ since all nonnegative numbers y arise as square roots. (See Figure 4.4.) ◇

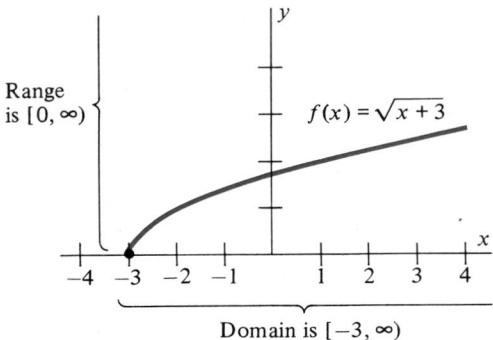

Figure 4.4 For $f(x) = \sqrt{x + 3}$, the domain is $[-3, \infty)$ and the range is $[0, \infty)$.

Example 5

The function $f(x) = \dfrac{x^2 - 1}{x - 1}$ is not defined for $x = 1$. For $x \neq 1$ we can write this function as

$$f(x) = \frac{x^2 - 1}{x - 1} = \frac{(x - 1)(x + 1)}{x - 1} = x + 1, \qquad x \neq 1.$$

That is,

$$f(x) = x + 1, \qquad x \neq 1.$$

This function is *not* the same as the function

$$g(x) = x + 1$$

which is defined for *all* x. Thus, even though the functions f and g have the same equations for $x \neq 1$, the functions are different because their domains are different. (See Figures 4.5 and 4.6.) ◇

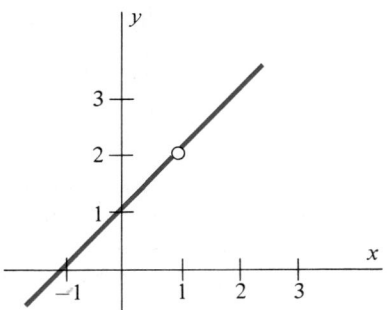

Figure 4.5 Graph of $f(x) = \dfrac{x^2 - 1}{x - 1}$. Here $f(1)$ is not defined.

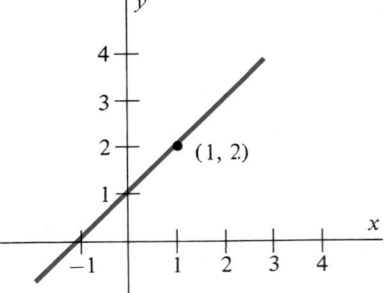

Figure 4.6 Graph of $g(x) = x + 1$. Here $g(1) = 2$.

Graphs of Functions: The Vertical Line Property

The **graph** of the function f is the set of all points (x, y) satisfying the equation $y = f(x)$. That is, the graph is the set

$$\{(x, y) \mid x \text{ is in the domain of } f, \text{ and } y = f(x)\}.$$

There is a simple but important property that distinguishes graphs of functions from graphs of equations that are not functions. It is called the **vertical line property,** and it means that *any vertical line* (that is, a line parallel to the *y*-axis) *can intersect the graph of a function at most once*. The reason for this is simple: a vertical line has equation $x = a$; but the graph of the *function f* has at most one point, $(a, f(a))$, whose *x*-coordinate is *a*. Figure 4.7 illustrates our previous conclusion, that $x^2 + y^2 = 1$ cannot be the graph of a function—the vertical line $x = a$ can intersect the graph twice. Figure 4.8 illustrates the vertical line property for an arbitrary function.

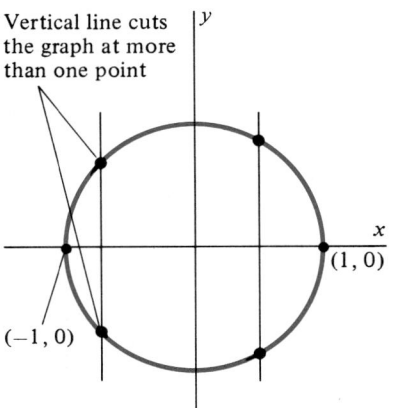

Figure 4.7 Graph of $x^2 + y^2 = 1$ does not have the vertical line property.

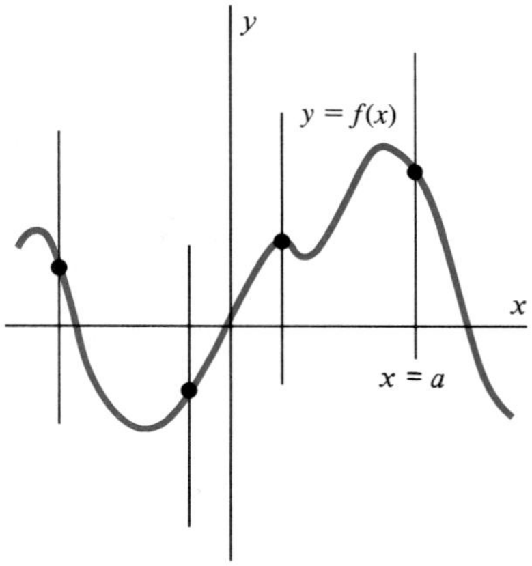

Figure 4.8 Vertical lines intersect the graph of a *function* at most once (Vertical Line Property).

Power and Quadratic Functions

Among the simplest types of functions are the power functions. These have the form

$$f(x) = x^n$$

where *n* is a positive integer. Graphs of several power functions appear in Figure 4.9.

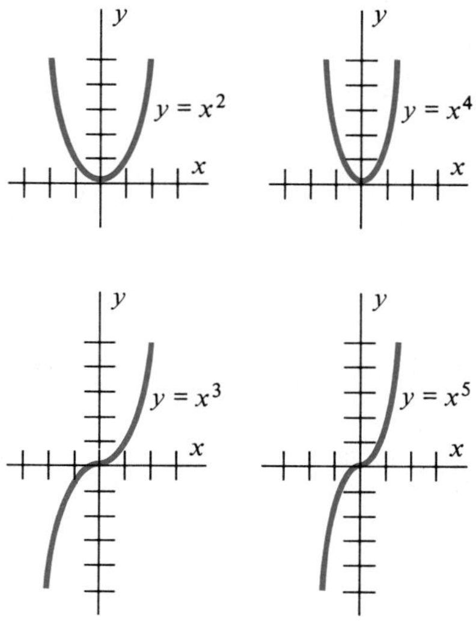

Figures 4.10 through 4.12 show the effect of introducing various constants into the equation $y = x^2$.

(a) The graph of $y = ax^2$, $a > 0$, opens upward, but does so more or less quickly depending on whether the constant *a* is large or small (Figure 4.10).
(b) The graph of $y = ax^2$, $a < 0$, opens downward (Figure 4.10).
(c) The graph of $y = ax^2 + b$ is like the graph of $y = ax^2$, except that it is "shifted" *b* units upward ($b > 0$) or $|b|$ units downward ($b < 0$, Figure 4.11).
(d) The graph of $y = a(x - c)^2 + b$ is like the graph of $y = ax^2 + b$, except that it is symmetric with respect to the vertical line $x = c$ rather than with respect to the *y*-axis (Figure 4.12).

A **quadratic function** is any function that can be written in the form

$$f(x) = Ax^2 + Bx + C \tag{1}$$

where *A*, *B*, and *C* are constants. By completing the square on the right-hand side of equation (1), we can bring the equation into the form $f(x) = a(x - c)^2 + b$. Thus, the graph of any quadratic function will be of the form illustrated in Figure 4.12. (Such curves are called **parabolas,** and they are discussed in detail in Chapter 13.)

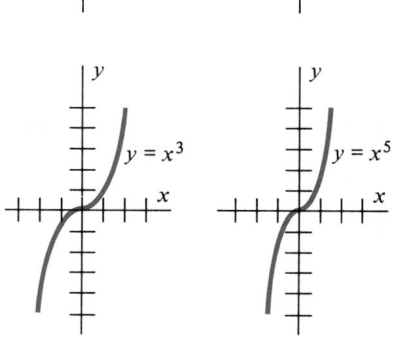

Figure 4.9 Some power functions.

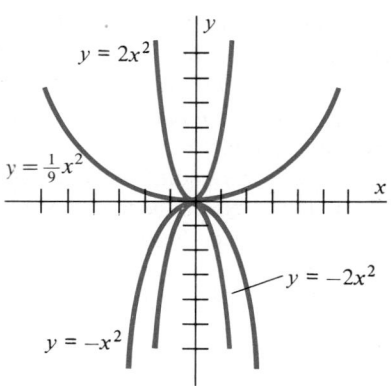

Figure 4.10 Functions of the form $f(x) = ax^2$.

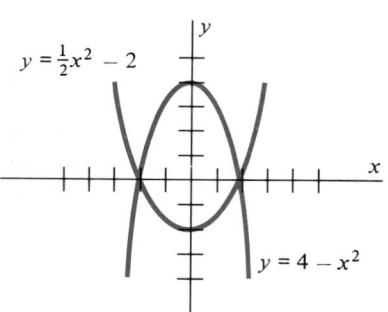

Figure 4.11 Functions of the form $h(x) = ax^2 + b$.

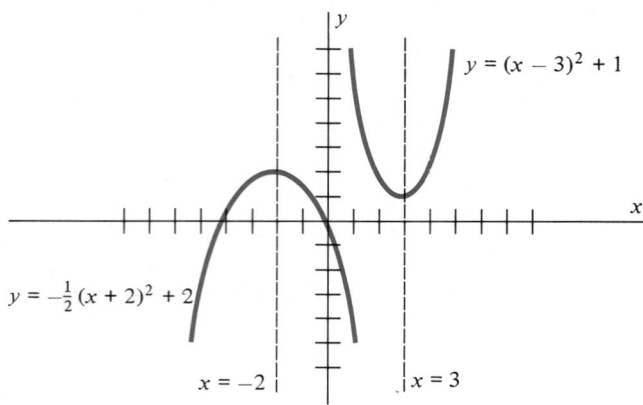

Figure 4.12 Functions of the form $f(x) = a(x - c)^2 + b$.

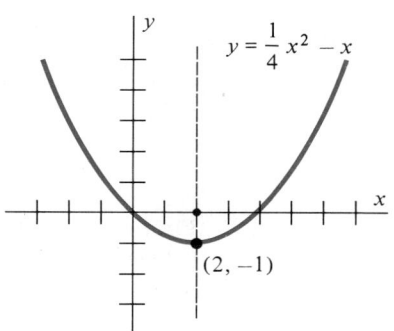

Figure 4.13 Graph of quadratic function $f(x) = \dfrac{1}{4}x^2 - x$.

Example 6

Sketch the graph of the quadratic function $f(x) = \dfrac{1}{4}x^2 - x$.

Strategy

Complete the square in x by

(1) factoring by $\dfrac{1}{4}$ and

(2) adding the square of half the coefficient of x.

Solution

We have

$$f(x) = \frac{1}{4}x^2 - x = \frac{1}{4}[x^2 - 4x]$$

$$= \frac{1}{4}[x^2 - 4x + 4] - \frac{1}{4}(4)$$

$$= \frac{1}{4}(x - 2)^2 - 1,$$

so

$$f(x) = \frac{1}{4}(x - 2)^2 - 1.$$

The graph appears in Figure 4.13.

◇

Polynomial Functions

A **polynomial function** is a function that can be written in the form

$$f(x) = a_n x^n + a_{n-1} x^{n-1} + \cdots + a_1 x + a_0$$

where $a_0, a_1, \ldots, a_n$ are constants and n is a positive integer. The integer n is called the **degree** of the polynomial, provided that $a_n \neq 0$. The following are examples of polynomial functions:

$$
\begin{array}{ll}
f(x) = x^3 - 7x + 2 & \text{(degree 3)} \\
f(x) = 1 - x^{10} & \text{(degree 10)} \\
y = (x - 3)^4 - 2(x + 5)^3 & \text{(degree 4)}.
\end{array}
$$

Rational Functions

A **rational function** is a quotient of two polynomial functions. A rational function is not defined for numbers x for which its denominator equals zero. For example, the rational function

$$f(x) = \frac{x^2 + 4}{x^3 + x^2 - 12x}$$

is not defined for $x = 0$, 3, or -4 since the denominator $x^3 + x^2 - 12x = x(x - 3)(x + 4)$ equals zero for these numbers x.

Absolute Value Function

The definition of the absolute value* of x, $|x|$, gives rise to the absolute value function $f(x) = |x|$. The graph of $f(x) = |x|$ appears in Figure 4.14.

Example 7

You can see the effect of the absolute value signs in the function $f(x) = |4 - x^2|$ by comparing Figure 4.15 with the graph of $g(x) = 4 - x^2$ in Figure 4.16. The "legs" of the graph of $4 - x^2$, which otherwise extend below the x-axis, have been turned upward. ◇

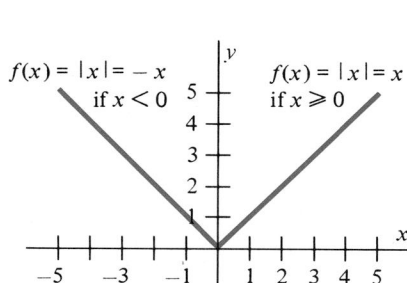

Figure 4.14 Graph of absolute value function.

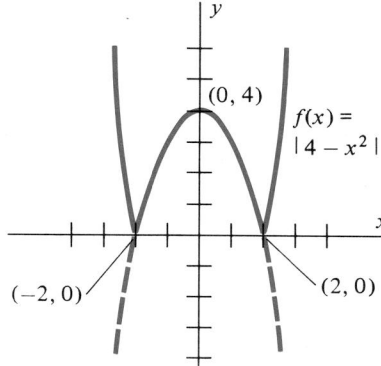

Figure 4.15 Graph of $f(x) = |4 - x^2|$. (Compare with Figure 4.16.)

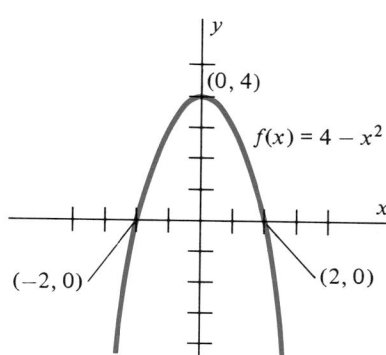

Figure 4.16 Graph of $f(x) = 4 - x^2$.

*See Section 1.1 for the definition of absolute value.

The Algebra of Functions

It is a simple matter to combine functions algebraically to form new functions that are sums, differences, multiples, products, or quotients of given functions. It is done by calculating the values of the individual terms in the combination separately and then forming the given combination using those values.

DEFINITION 6

Given the functions f and g and the number c, the functions $f + g, f - g, cf, fg$ and $\dfrac{f}{g}$ are defined by the following equations:

$$(f + g)(x) = f(x) + g(x)$$
$$(f - g)(x) = f(x) - g(x)$$
$$(cf)(x) = cf(x)$$
$$(fg)(x) = f(x)g(x)$$
$$(f/g)(x) = f(x)/g(x) \quad \text{(provided } g(x) \neq 0).$$

Of course, these combinations can be formed only for numbers x that are in the domains of both functions. While these definitions may seem obvious, we will need a clear understanding of these definitions in studying limits of functions in Chapter 2.

Example 8

Let $f(x) = x^3 + 2$ and $g(x) = \sqrt{x + 1}$. Then

(a) for $h_1 = f + g$, $h_1(x) = x^3 + 2 + \sqrt{x + 1}$, so
$$h_1(3) = f(3) + g(3) = [3^3 + 2] + \sqrt{3 + 1} = 29 + 2 = 31$$

and

$$h_1(1) = f(1) + g(1) = [1^3 + 2] + \sqrt{1 + 1} = 3 + \sqrt{2}$$

but

$h_1(-2)$ is undefined since $g(-2) = \sqrt{-2 + 1}$ is undefined.

(b) for $h_2 = \dfrac{f}{3g}$, $h_2(x) = \dfrac{x^3 + 2}{3\sqrt{x + 1}}$

so

$$h_2(3) = \frac{f(3)}{3g(3)} = \frac{3^3 + 2}{3\sqrt{3 + 1}} = \frac{27 + 2}{3\sqrt{4}} = \frac{29}{6}$$

and

$$h_2(0) = \frac{f(0)}{3g(0)} = \frac{0^3 + 2}{3\sqrt{0 + 1}} = \frac{2}{3}. \qquad \diamond$$

Composite Functions

Often, two functions are combined not by an algebraic operation such as addition, but by letting the second function act on values of the first function. The result is called a *composite function*.

For example, the function $h(x) = |1 - x^2|$ can be constructed out of the absolute value function and the quadratic function $1 - x^2$. Given a number x we first find $1 - x^2$, and then its absolute value $|1 - x^2|$. Here h is called a composite of the two functions, and we say that the absolute value function acts on values of the quadratic function.

DEFINITION 7

Given two functions f and g, the **composite function** $f \circ g$ is the result of the function f acting on values of the function g. That is,

$$(f \circ g)(x) = f(g(x)).$$

(See Figure 4.17.)

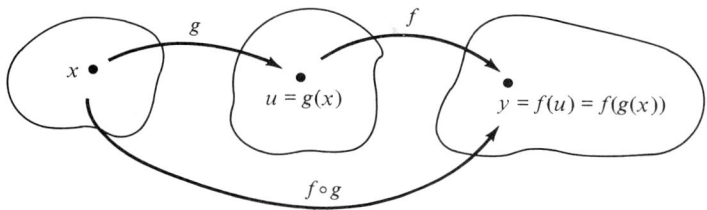

Figure 4.17 The composite function $f \circ g$ is the result of the function f acting on values of the function g.

The difference between composite functions, or functions formed *by composition,* and algebraic combinations of functions has to do with the order in which the operations are performed. In an algebraic combination the individual function values are found first and then combined. In the composite function $f \circ g$ we first find the value of the *inside* function $u = g(x)$, and then insert this number u into the *outside* function f to find $f(u) = f(g(x))$.

Thus, the *domain* of the composite function $f \circ g$ is the set of all x in the domain of g for which the number $u = g(x)$ lies in the domain of f.

Example 9

For the functions $f(x) = \sqrt{x}$ and $g(x) = x + 4$,

a. The composite function $f \circ g$ is

$$(f \circ g)(x) = f(g(x)) = f(x + 4) = \sqrt{x + 4}$$

with domain $[-4, \infty)$.

b. The composite function $g \circ f$ is

$$(g \circ f)(x) = g(f(x)) = g(\sqrt{x}) = \sqrt{x} + 4$$

with domain $[0, \infty)$. ◇

Example 10

The function $y = \dfrac{1}{x^2 + 4}$ may be thought of as a composite function $y = f(g(x))$ where the "inside" function is $u = g(x) = x^2 + 4$ and the "outside" function is $y = f(u) = \dfrac{1}{u}$. That is,

$$y = f(u) = \frac{1}{u} = \frac{1}{g(x)} = \frac{1}{x^2 + 4}.$$ ◇

Example 11

The following list shows how several additional functions may be regarded as composite functions of the form $(f \circ g)(x) = f(g(x))$.

Composite function $y = f(g(x))$	Outside function $y = f(u)$	Inside function $u = g(x)$
$y = (x^2 + 7)^{2/3}$	$f(u) = u^{2/3}$	$g(x) = x^2 + 7$
$y = 4 + \sqrt{2 + x^2}$	$f(u) = 4 + \sqrt{u}$	$g(x) = 2 + x^2$
$y = 4 + \sqrt{2 + x^2}$	$f(x) = 4 + u$	$g(x) = \sqrt{2 + x^2}$
$y = \dfrac{8}{\|3 + x^4\|}$	$f(u) = \dfrac{8}{\|u\|}$	$g(x) = 3 + x^4$
$y = \dfrac{8}{\|3 + x^4\|}$	$f(u) = \dfrac{8}{u}$	$g(x) = \|3 + x^4\|$

The last four entries illustrate that choices for the "inside" and "outside" functions are not necessarily unique. ◇

Exercise Set 1.4

1. Complete the following chart.

$f(x)$	$f(-2)$	$f(0)$	$f(4)$	$f(5)$	$\|f(3)\|$
$1 - 3x^2$	-11	1	-47	-74	26
$\dfrac{1}{x + 2}$	undefined	$\dfrac{1}{2}$			
$\dfrac{(x - 3)^2}{x^2 + 1}$	5				
$\sqrt{x + 4}$					
$\dfrac{1}{\sqrt{16 - x^2}}$					
$\begin{cases} 1 - x, & x < -3 \\ x - 1, & x > 1 \end{cases}$					

In Exercises 2–7, determine whether the given equation determines y as a function of x.

2. $2x - y = 7$

3. $xy = 5$

4. $xy^2 + 2x = 6$

5. $x^2 + 2x + y^2 = 8$

6. $x - y = x + y$

7. $y = 5$

8. True or false? The functions $f(x) = \dfrac{x^2 - 4}{x + 2}$ and $g(x) = x - 2$ are equal. (*Hint:* Do they have the same equations? Domains?)

In Exercises 9–20, state the domain of the given function.

9. $f(x) = x^5 - 5x^3$

10. $f(x) = \dfrac{x - 1}{x + 1}$, $x > 3$

11. $g(x) = \sqrt{x + 4}$

12. $h(x) = \sqrt{x(x - 1)}$

13. $f(t) = 1 + t^2$, $t \geq 0$

14. $f(x) = \dfrac{1}{1 - |x|}$

15. $h(s) = \sqrt{16 - s^2}$

16. $g(t) = \dfrac{1}{t^2 + 2t - 35}$

17. $f(x) = \begin{cases} \dfrac{1}{x}, & x > 0 \\ -\dfrac{1}{x}, & x < 0 \end{cases}$

18. $f(s) = s(s + 3)\sqrt{1 - s^2}$

19. $h(x) = \dfrac{\sqrt{x^2 + 2x}}{x - 2}$

20. $f(x) = \sqrt{6 - |x + 2|}$

21. True or false? For the rational function $f(x) = \dfrac{g(x)}{h(x)}$, if $h(x)$ is a polynomial of degree n then $f(x)$ will be undefined for n distinct values of x.

22. True or false? If a horizontal line intersects the graph of the equation $y = f(x)$ in more than one point, the equation $y = f(x)$ cannot determine y as a function of x. Explain.

For each of the equations in Exercises 23–28, y is determined as a quadratic function of x. Graph each by first completing the square.

23. $3x^2 - 2y - 8 = 0$

24. $3x^2 - y - 7 = 0$

25. $y - 3x^2 + x - \dfrac{1}{2} = 0$

26. $y - 2x^2 + 4x - 5 = 0$

27. $4x^2 + y - 24x + 34 = 0$

28. $y - \dfrac{1}{2}x^2 - \dfrac{1}{2}x - \dfrac{11}{24} = 0$

An equation of the form $x = ky^2$ determines x as a quadratic function of y. The graph will therefore be a parabola opening in either the positive x direction ($k > 0$) or the negative x direction ($k < 0$). In each of the Exercises 29–34, graph the given parabola.

29. $y^2 + x - 2 = 0$

30. $3y^2 - x - 4 = 0$

31. $3x + 2y^2 - 3 = 0$

32. $x - 3y^2 + 6y - 4 = 0$

33. $x - 2y^2 + 4y - 2 = 0$

34. $4y^2 - x - 24y + 34 = 0$

35. Find the point(s) of intersection of the line $x - y - 2 = 0$ and the parabola $y = 4 - x^2$.

36. Find the points of intersection of the parabolas $y = 6 - x^2$ and $y = x^2 - 2$.

In each of Exercises 37–42, graph the given function by plotting points.

37. $f(x) = \dfrac{1}{x - 1}$

38. $f(x) = \sqrt{2x + 1}$

39. $f(x) = \begin{cases} 7 - x, & x \leq 2 \\ 2x + 1, & x > 2 \end{cases}$

40. $f(x) = x(4 - 2x)$

41. $f(x) = \dfrac{x^2 - 1}{x - 1}$

42. $f(x) = \begin{cases} \sqrt{x + 2}, & -2 \leq x \leq 2 \\ 4 - x, & x > 2 \end{cases}$

In Exercises 43–48, sketch the graph by first noting where the expression inside the absolute value signs changes sign.

43. $f(x) = x + |x|$

44. $y = |1 - x^2|$

45. $y = \dfrac{x + 1}{|x - 1|}$

46. $f(x) = |x^2 - x + 2|$

47. $f(t) = \dfrac{1}{|t|}$

48. $g(s) = |s^2 - 1|$

49. A function $f(x)$ is called **even** if whenever x is in the domain of f, so is $-x$, and $f(-x) = f(x)$. It is called **odd** if whenever x is in the domain of f, so is $-x$, and $f(-x) = -f(x)$. Determine whether $f(x)$ is even, odd, or neither, and sketch the graph.

   a. $f(x) = x^2$ **e.** $f(x) = x^3 + x$
   b. $f(x) = 2 - x^2$ **f.** $f(x) = 2x^4 + x^2$
   c. $f(x) = x^3$ **g.** $f(x) = |x| + 2$
   d. $f(x) = 1 - x^3$ **h.** $f(x) = x^2 + x$

50. What symmetry properties does the graph of an even function possess? An odd function? (See Exercise 49.)

In Exercises 51–58, use the functions

$$f(x) = 3x + 1 \qquad\qquad g(x) = x^3 \qquad\qquad h(x) = \sqrt{x}$$

to form the indicated composite function.

51. $f \circ g$

52. $f \circ h$

53. $g \circ f$

54. $h \circ g$

55. $h(f(x))$

56. $h(f(g(x)))$

57. $f(g(h(x)))$

58. $g(h(f(x)))$

59. Given the function $f(x) = x^2$ with domain $\{x \mid -1 \leq x \leq 3\}$ and the function $g(x) = 2x + 6$ find
   a. the domain of $f \circ g$
   b. the range of $f \circ g$
   c. the domain of $g \circ f$
   d. the range of $g \circ f$.

60. Given the function $f(x) = x^2$ and the function $g(x) = 2x + 6$ with domain $\{x \mid -5 \leq x \leq 4\}$ find
   a. the domain of $f \circ g$
   b. the range of $f \circ g$
   c. the domain of $g \circ f$
   d. the range of $g \circ f$.

61. True or false? Every polynomial is a rational function.

62. A rectangle has area 36 cm^2. Find a function that expresses the width, w, in terms of the length, ℓ, of the rectangle. What is the domain of this function?

63. A water tank has the shape of a right circular cylinder. If the radius of the cylinder is fixed, express its volume as a function of its height.

64. Find two nonconstant functions f and g so that the composite function $f \circ g$ is constant.

65. Find a function f giving the volume of a cube in terms of the length of one of its edges.

66. Find a function giving the length of an edge of a cube in terms of its volume.

67. The frequency of x-rays emitted from a chemical element varies with the atomic number Z according to the equation $\gamma = a(Z - b)^2$, where a and b are constants. Discuss the nature of this graph.

68. For a moving automobile, air resistance is proportional to the square of the vehicle's speed. Write this statement in function form and discuss the nature of its graph.

69. Show that the quadratic function $f(x) = Ax^2 + Bx + C$ may be written as $f(x) = A(x - D)^2 + E$ with $D = \dfrac{-B}{2A}$ and $E = C - \dfrac{B^2}{4A}$. (*Hint:* Complete the square.)

70. What is the range of the quadratic function $f(x) = ax^2 + bx + c$ if $a > 0$? What if $a < 0$?

71. Find the range of the function:
   a. In Exercise 11
   b. In Exercise 13
   c. In Exercise 17
   d. In Exercise 20
   e. In Exercise 38

72. Use what you know about the graph of the quadratic function $f(x) = ax^2 + bx + c$, and the method of completing the square, to prove the quadratic formula for finding the zeros of the quadratic polynomial $ax^2 + bx + c$.

$$x = \frac{-b \pm \sqrt{b^2 - 4ac}}{2a}$$

1.5 TRIGONOMETRIC FUNCTIONS

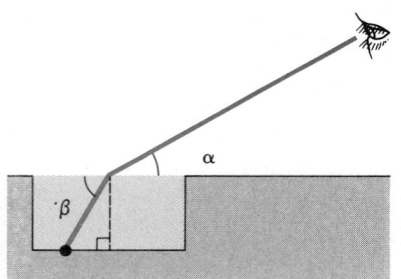

Many problems in mathematics, science, and engineering have to do with the measurement of angles. For example, in the study of optics a fundamental fact is that a ray of light will bend when passing from one medium into another (such as passing from water into air). This phenomenon is called **refraction.** Figure 5.1 is meant to illustrate the fact that the sizes of the angles associated with the refraction phenomenon are intimately related to the lengths of the sides of certain triangles. The subject of **trigonometry** addresses these kinds of questions, and the relationships that are obtained give rise to the six **trigonometric functions.**

Figure 5.1 Ray of light from coin in wishing well to observer's eye "bends" at water surface.

Radian Measure

You are already familiar with the measurement of angles in degrees, and you know that one complete revolution equals 360°. However, the degree system (which originated in ancient Babylonia) is poorly suited to the needs of calculus. We therefore define a more natural unit of angular measurement, the *radian.*

Consider a unit circle located in the xy-coordinate plane, with its center at the origin. Imagine a particle that travels counterclockwise along the circumference of the circle, starting at $(1, 0)$ and carrying with it one end of a line segment whose other end is pivoted at the origin (see Figure 5.2). The angle θ is defined by the positive x-axis and the rotating line segment. We say that the **radian measure** of the angle θ equals the distance travelled by the particle along the circumference. Since

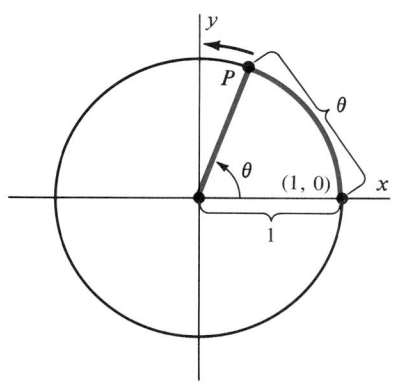

Figure 5.2 Unit circle. Angle measure in radians equals arc length in radii.

the circumference of the unit circle is 2π we have:

$$2\pi \text{ radians} = 360°.$$

The relation between any angle θ_r measured in radians and the same angle θ_d measured in degrees is therefore

$$\frac{\theta_r}{2\pi} = \frac{\theta_d}{360} \quad \text{or} \quad \theta_r = \frac{2\pi}{360}\theta_d.$$

Example 1

We have the following relationships.

θ_d	0°	30°	45°	60°	90°	120°	135°	180°	210°	270°	315°	360°
θ_r	0	$\dfrac{\pi}{6}$	$\dfrac{\pi}{4}$	$\dfrac{\pi}{3}$	$\dfrac{\pi}{2}$	$\dfrac{2\pi}{3}$	$\dfrac{3\pi}{4}$	π	$\dfrac{7\pi}{6}$	$\dfrac{3\pi}{2}$	$\dfrac{7\pi}{4}$	2π

We can imagine the point and line segment continuing to rotate for more than one complete revolution, defining angles greater than 360° or 2π.

This allows us to identify each real number t with a corresponding point P on the unit circle as follows. If $t > 0$, we allow our traveling point to go a distance t counterclockwise along the unit circle, starting from (1, 0). Let P denote the point on the circle where this motion ends. If $t < 0$, we use the same procedure, except that the point travels in the clockwise direction. This procedure of "wrapping the real number line around the unit circle" allows us to define the *trigonometric functions* for all real numbers.

Sine and Cosine Functions

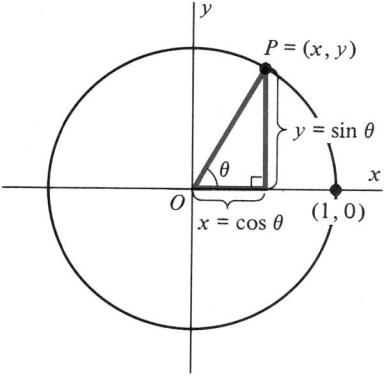

Figure 5.3 Definition of $\sin \theta$ and $\cos \theta$.

Let θ be any angle, measured in radians, and let P be the point on the unit circle associated with θ as described above. We define the **sine** of the angle θ, written $\sin \theta$, to be the y-coordinate of P. Similarly, the **cosine** of the angle, $\cos \theta$, is defined to be the x-coordinate of P. In other words,

$$\cos \theta = x$$
$$\sin \theta = y$$

(See Figure 5.3.)

In general, values of the sine and cosine are difficult to compute. (We will develop the ability to do this as part of our work on infinite series later in this text.) However, values of $\sin \theta$ and $\cos \theta$ have been tabulated for many angles (tables of these values appear at the end of this text). Moreover, most hand calculators are preprogrammed to give values of $\sin \theta$ and $\cos \theta$ with high degrees of precision. There are certain angles, however, for which values of $\sin \theta$ and $\cos \theta$ are easy to compute. These include multiples of $\pi/2$, and the angles involved in 30°-60°-90° and isosceles right triangles. This is because the ratios of sides in such triangles are easily found from elementary trigonometry (see Figure 5.4).

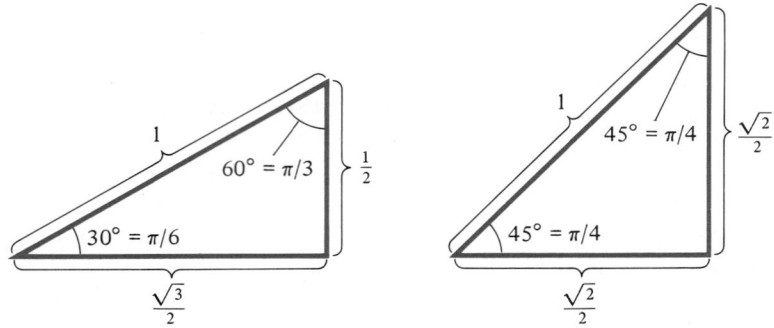

Figure 5.4 Ratios of sides in 30°-60°-90° and 45°-45°-90° triangles.

Example 2

Values of sin θ and cos θ for various angles θ are shown. Several such angles are sketched in Figure 5.5.

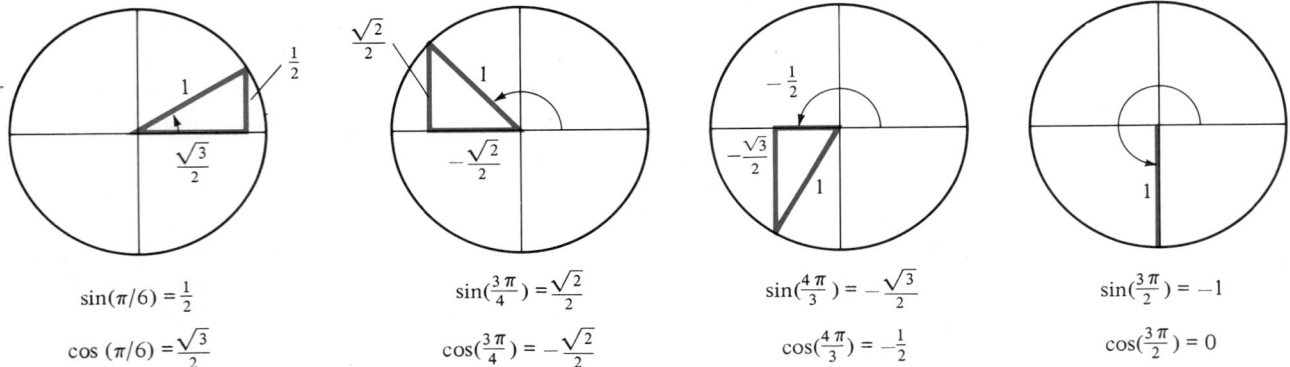

$$\sin(\pi/6) = \frac{1}{2}$$

$$\cos(\pi/6) = \frac{\sqrt{3}}{2}$$

$$\sin(\tfrac{3\pi}{4}) = \frac{\sqrt{2}}{2}$$

$$\cos(\tfrac{3\pi}{4}) = -\frac{\sqrt{2}}{2}$$

$$\sin(\tfrac{4\pi}{3}) = -\frac{\sqrt{3}}{2}$$

$$\cos(\tfrac{4\pi}{3}) = -\frac{1}{2}$$

$$\sin(\tfrac{3\pi}{2}) = -1$$

$$\cos(\tfrac{3\pi}{2}) = 0$$

Figure 5.5 Sin θ and cos θ for various angles.

θ	0	$\frac{\pi}{6}$	$\frac{\pi}{4}$	$\frac{\pi}{3}$	$\frac{\pi}{2}$	$\frac{2\pi}{3}$	$\frac{3\pi}{4}$	$\frac{5\pi}{6}$	π	$\frac{7\pi}{6}$	$\frac{5\pi}{4}$	$\frac{4\pi}{3}$	$\frac{3\pi}{2}$	$\frac{5\pi}{3}$	$\frac{7\pi}{4}$	$\frac{11\pi}{6}$
$\sin\theta$	0	$\frac{1}{2}$	$\frac{\sqrt{2}}{2}$	$\frac{\sqrt{3}}{2}$	1	$\frac{\sqrt{3}}{2}$	$\frac{\sqrt{2}}{2}$	$\frac{1}{2}$	0	$-\frac{1}{2}$	$-\frac{\sqrt{2}}{2}$	$-\frac{\sqrt{3}}{2}$	-1	$-\frac{\sqrt{3}}{2}$	$-\frac{\sqrt{2}}{2}$	$-\frac{1}{2}$
$\cos\theta$	1	$\frac{\sqrt{3}}{2}$	$\frac{\sqrt{2}}{2}$	$\frac{1}{2}$	0	$-\frac{1}{2}$	$-\frac{\sqrt{2}}{2}$	$-\frac{\sqrt{3}}{2}$	-1	$-\frac{\sqrt{3}}{2}$	$-\frac{\sqrt{2}}{2}$	$-\frac{1}{2}$	0	$\frac{1}{2}$	$\frac{\sqrt{2}}{2}$	$\frac{\sqrt{3}}{2}$

◇

Note that the definitions of sine and cosine apply to all real numbers (in radian units). For example, $\sin\left(\dfrac{9\pi}{2}\right) = \sin\left(\dfrac{\pi}{2}\right) = 1$, and $\cos\left(\dfrac{11\pi}{4}\right) = \cos\left(\dfrac{3\pi}{4}\right) = -\dfrac{\sqrt{2}}{2}$. Because angles with measure $t + 2n\pi$, $n = 0, \pm1, \pm2, \ldots$ are all associated with the same point P on the unit circle, we have the identities

$$\sin t = \sin(t + 2n\pi), \qquad n = \pm 1, \pm 2, \ldots .$$ (1a)
$$\cos t = \cos(t + 2n\pi), \qquad n = \pm 1, \pm 2, \ldots .$$ (1b)

Identities (1a) and (1b) say that the values of the sine and cosine are the same at any number located a multiple of 2π radians away from t as they are at the number t. For this reason, we say that these functions are **periodic,** with period $T = 2\pi$.

Graphs of $\sin t$ and $\cos t$ appear in Figure 5.6. Notice that, for both functions,

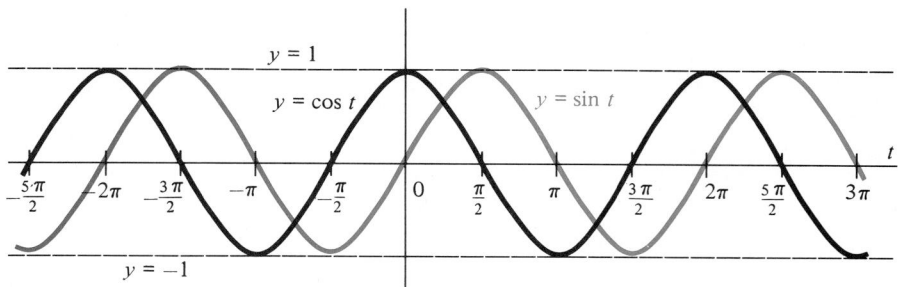

Figure 5.6 Graphs of functions $y = \sin t$ and $y = \cos t$.

the maximum value is 1 and the minimum value is -1. That is, $|\sin t| \le 1$ and $|\cos t| \le 1$.

The Other Trigonometric Functions

Four additional trigonometric functions are defined as the quotients and reciprocals of the sine and cosine functions. They are as follows:

(i) tangent: $\qquad \tan t = \dfrac{\sin t}{\cos t}, \qquad t \ne \dfrac{\pi}{2} + n\pi$

(ii) cotangent: $\qquad \cot t = \dfrac{\cos t}{\sin t}, \qquad t \ne n\pi$

(iii) secant: $\qquad \sec t = \dfrac{1}{\cos t}, \qquad t \ne \dfrac{\pi}{2} + n\pi$

(iv) cosecant: $\qquad \csc t = \dfrac{1}{\sin t}, \qquad t \ne n\pi$

Graphs of these functions appear in Figure 5.7. (Note that you can sketch these graphs geometrically, using Figure 5.6. For example, the y-coordinates of the points on the graph of $y = \tan t$ are just the quotients of the y-coordinates of the points on the sine and cosine curves.) You can see from the graphs that the tangent and cotangent functions are periodic with period $T = \pi$, whereas the other four trigonometric functions have a period of $T = 2\pi$. (See Exercise 27.)

Example 3

Find $\tan\left(\dfrac{\pi}{4}\right)$, $\cot\left(\dfrac{5\pi}{3}\right)$, $\sec\left(-\dfrac{2\pi}{3}\right)$, and $\csc(7\pi)$.

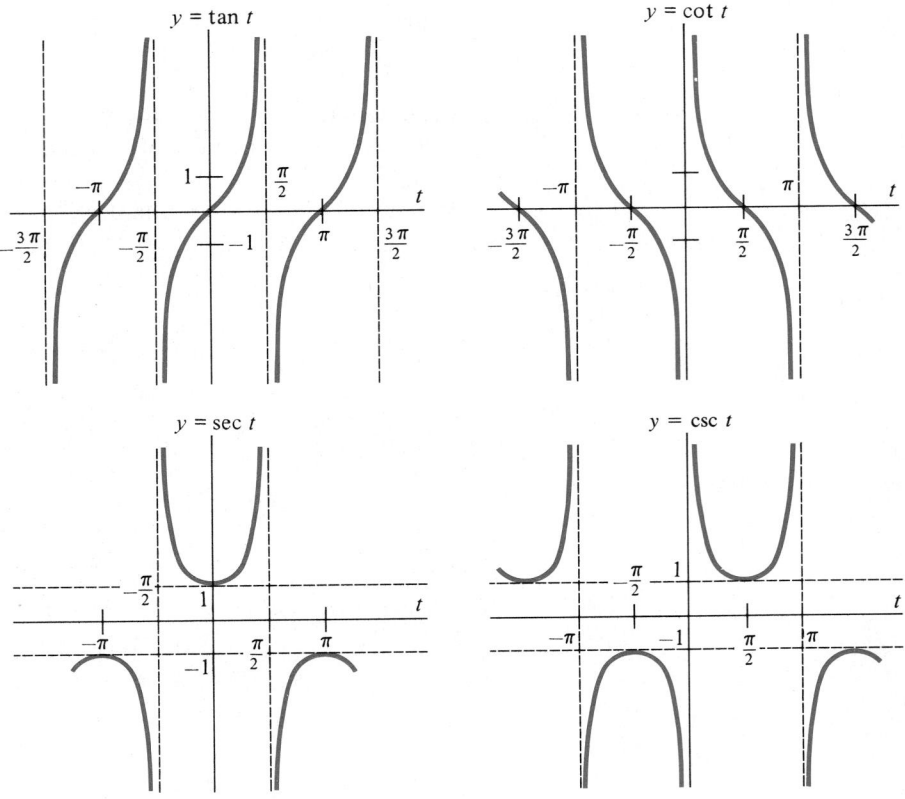

Figure 5.7 Graphs of tangent, cotangent, secant, and cosecant functions.

Solution

$$\tan\left(\frac{\pi}{4}\right) = \frac{\sin(\pi/4)}{\cos(\pi/4)} = \frac{\sqrt{2}/2}{\sqrt{2}/2} = 1$$

$$\cot\left(\frac{5\pi}{3}\right) = \frac{\cos(5\pi/3)}{\sin(5\pi/3)} = \frac{1/2}{-\sqrt{3}/2} = -\frac{1}{\sqrt{3}}$$

$$\sec\left(-\frac{2\pi}{3}\right) = \sec\left(\frac{4\pi}{3}\right) = \frac{1}{\cos(4\pi/3)} = \frac{1}{-1/2} = -2$$

$$\csc(7\pi) = \csc(\pi) = \frac{1}{\sin \pi}, \text{ which is } \textit{undefined} \text{ since } \sin \pi = 0. \qquad \diamondsuit$$

Trigonometric Identities

Because $\sin \theta$ and $\cos \theta$ are the y and x coordinates of a point on the unit circle, we have the identity

$$\sin^2 \theta + \cos^2 \theta = 1. \tag{2}$$

If $\cos \theta \neq 0$, we may divide each term of equation (2) by $\cos^2 \theta$ to obtain the identity

$$\tan^2 \theta + 1 = \sec^2 \theta.$$

Similarly, dividing (2) through by $\sin^2 \theta \neq 0$ gives

$$1 + \cot^2 \theta = \csc^2 \theta.$$

Here are some other useful identities:

$$\sin(\theta \pm \phi) = \sin \theta \cos \phi \pm \cos \theta \sin \phi$$
$$\cos(\theta \pm \phi) = \cos \theta \cos \phi \mp \sin \theta \sin \phi$$

$$\tan(\theta \pm \phi) = \frac{\tan \theta \pm \tan \phi}{1 \mp \tan \theta \tan \phi}, \quad 1 \mp \tan \theta \tan \phi \neq 0.$$

$$\sin^2 \theta = \frac{1}{2}(1 - \cos 2\theta)$$

$$\cos^2 \theta = \frac{1}{2}(1 + \cos 2\theta)$$

$$\sin 2\theta = 2 \sin \theta \cos \theta$$
$$\cos 2\theta = \cos^2 \theta - \sin^2 \theta$$
$$\sin \theta = \cos(\pi/2 - \theta)$$
$$\cos \theta = \sin(\pi/2 - \theta)$$
$$\sin(-\theta) = -\sin \theta$$
$$\cos(-\theta) = \cos \theta$$

Finally, for any triangle with interior angles α, β, and γ and corresponding opposite sides of length a, b, and c, we have the Law of Sines:

$$\frac{\sin \alpha}{a} = \frac{\sin \beta}{b} = \frac{\sin \gamma}{c}$$

and the Law of Cosines:

$$c^2 = a^2 + b^2 - 2ab \cos \gamma.$$

(See Figure 5.8.)

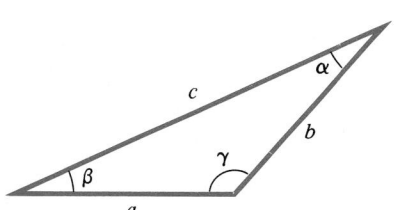

Figure 5.8

Exercise Set 1.5

1. Convert the following angles from degrees to radians.
   a. $30°$ **c.** $-15°$ **e.** $(x + 30)°$
   b. $75°$ **d.** $315°$ **f.** $2910°$

2. Convert the following angles from radians to degrees.
   a. π **c.** $9\pi/2$ **e.** $a + \pi$
   b. 1 **d.** $-21\pi/6$ **f.** $7\pi/16$

3. Find $\tan x$, $\cot x$, $\sec x$, and $\csc x$ for each of the following values of x.
   a. $x = \pi/6$ **c.** $x = -13\pi/3$ **e.** $x = -9\pi/4$
   b. $x = 5\pi/2$ **d.** $x = 9\pi$ **f.** $x = 7\pi/6$

4. Find all numbers x, with $0 \le x < 2\pi$, in radians, so that
   a. $\sin x = 0$
   b. $\sin x = \cos x$
   c. $\cos x = \sqrt{2}/2$
   d. $\tan x = -1$
   e. $\csc x = \sqrt{2}$
   f. $\sin 3x = 0$
   g. $\cos(\pi/2 + 2x) = 1$
   h. $\sin 2x = \cos x$
   i. $2 \cos^2 x - \cos x - 1 = 0$
   j. $3 \cos^2 x - \sin^2 x = 0$
   k. $\cos 2x + \cos x + 1 = 0$
   l. $\cos^2 x + \sin x = 0$
   m. $|\cos x| = \dfrac{\sqrt{3}}{2}$.

In Exercises 5–10, find all numbers t with $0 \le t < 2\pi$, in radians, satisfying all stated properties.

5. $\begin{cases} \sin t = \cos t \\ \tan t > 0 \end{cases}$

6. $\begin{cases} \sec t = 2 \\ \cot (t) < 0 \end{cases}$

7. $\begin{cases} \sin t > 0 \\ \tan t > 0 \end{cases}$

8. $\begin{cases} \sin t \ge \tan t \\ \sin t \ge 0 \end{cases}$

9. $\begin{cases} \cos t \ge 0 \\ \sec t \le 0 \end{cases}$

10. $\begin{cases} \cos t < 0 \\ \tan t < 0 \end{cases}$

In each of Exercises 11–16, graph the given function by comparing it with the function $y = \sin x$.

11. $y = 2 \sin x$

12. $f(x) = -2 \sin 3x$

13. $y = \pi \sin(\pi x)$

14. $f(x) = \sin(2x + \pi/2)$

15. $y = A \sin(\pi + x)$

16. $y = 4 \sin(2x + \pi)$

17. State the domains and ranges for each of the six trigonometric functions.

In Exercises 18–23, state whether the given function is even, odd, or neither. (See Exercise 49, Section 1.4.)

18. $y = \sin x$

19. $y = \cos x$

20. $y = \tan x$

21. $y = \dfrac{\sin x}{x}$

22. $y = 3 \sec x$

23. $y = \sin^2 x$

24. True or false? If the size of an angle is the same in both radians and degrees then that angle must be zero.

25. For acute angles θ (i.e., for $0 \leq \theta \leq 90°$), the six trigonometric functions may be defined as follows. First, construct a right triangle with θ as one of the three angles (Figure 5.9).

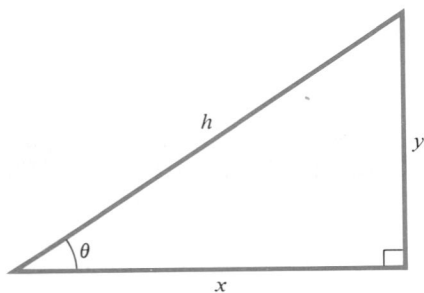

Figure 5.9 $x =$ side adjacent θ; $y =$ side opposite θ; $h =$ hypotenuse.

Label the side opposite θ as y, the side adjacent to θ as x, and the hypotenuse as h. Then,

$$\sin \theta = \frac{y}{h} \qquad \cos \theta = \frac{x}{h}$$

$$\tan \theta = \frac{y}{x} \qquad \cot \theta = \frac{x}{y}$$

$$\sec \theta = \frac{h}{x} \qquad \csc \theta = \frac{h}{y}$$

a. Show that these definitions are entirely equivalent to the definitions given in this section. (*Hint:* Divide all dimensions by h, construct coordinate axes.)

b. Explain how the above definitions can be extended to apply to angles of arbitrary size.

26. Use the results of Exercise 25 to explain why the following formula for the area of the triangle in Figure 5.9 is valid.

$$A = \frac{1}{2}xh \cdot \sin \theta$$

27. Use the definitions of the tangent and cotangent functions, together with the periodicity relations of the sine and cosine (equations (1)), to show that the period of the tangent and cotangent is $T = \pi$.

28. True or false? $\sin(x + y) = \sin x + \sin y$.

29. A pendulum consists of a bob hanging at the end of a cord 2 meters long. If the maximum angle of displacement for the pendulum is $\theta = \pi/6$, how far does the bob travel in one complete cycle? (Figure 5.10.)

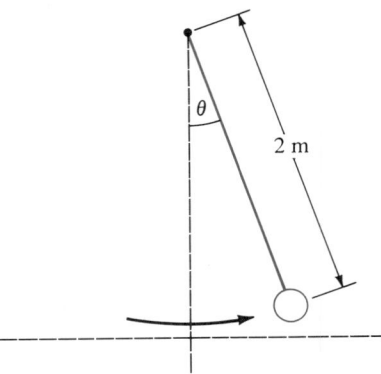

Figure 5.10 Pendulum in Exercise 29.

30. Points A and B are on opposite banks of a straight riverbed, as in Figure 5.11. Point A is directly opposite point B. What

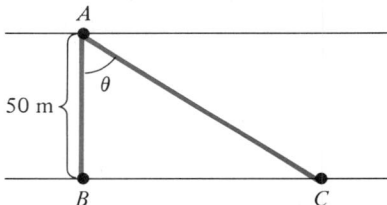

Figure 5.11 Points on opposite sides of a straight riverbed.

is the distance between points B and C if $\theta = \pi/6$?

31. What is the answer to the question in Exercise 30 if $\theta = \pi/3$?

32. A child stands 300 meters from a point directly below the location of an airplane. If the angle between ground level and the child's line of sight to the airplane is $\pi/6 = 30°$, what is the altitude of the airplane? (Neglect the height of the child.)

33. What is the altitude of the airplane in Exercise 32 if the angle is $\theta = \pi/4$?

34. Use the Law of Sines and Law of Cosines to find the missing sides and angles in the triangles shown in Figures 5.12 and 5.13.

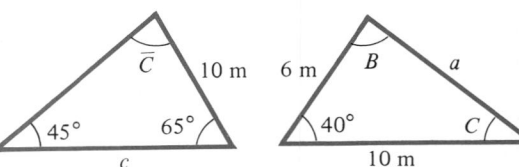

Figure 5.12　　　　**Figure 5.13**

SUMMARY OUTLINE OF CHAPTER 1

◆ The **real numbers** correspond to points on a line. The **integers** are the numbers . . . $-2, -1, 0, 1, 2, ...$ (page 3)
Rational numbers are quotients of integers. **Decimal expansions** for rational numbers terminate or repeat. Decimal expansions for **irrational** numbers neither repeat nor terminate.

◆ *Interval Notation:* (page 5)

$[a, b) = \{x \mid a \le x < b\}; \qquad (a, b) = \{x \mid a < x < b\}$
$[a, \infty) = \{x \mid a \le x\}; \qquad [a, b] = \{x \mid a \le x \le b\}$

◆ *Set Operations:* (page 5)

$A \cup B = \{x \mid x \in A \text{ or } x \in B\}$
$A \cap B = \{x \mid x \in A \text{ and } x \in B\}$

◆ The **inequality** $x < y$ means $y - x$ is positive. (page 6)

◆ *Theorem 1:* (Properties of Inequality) (page 6)

  (i) Either $x = y$, $x < y$, or $x > y$.
  (ii) If $x < y$ and $y < z$, then $x < z$.
  (iii) If $x < y$, then $x + c < y + c$ for any c.
  (iv) If $x < y$ and $c > 0$, then $cx < cy$.
  (v) If $x < y$ and $c < 0$, then $cx > cy$.

◆ *Absolute Value:* (page 9)

$$|x| = \begin{cases} x \text{ if } x \ge 0 \\ -x \text{ if } x < 0. \end{cases}$$

◆ *Theorem 2:* For $a > 0$, $|x| < a$ if and only if $-a < x < a$. (page 9)

◆ *Theorem 3:* (Properties of absolute value): For all x (page 11)

  (i) $|x| \ge 0$,
  (ii) $|xy| = |x| \cdot |y|$,
  (iii) $|x + y| \le |x| + |y|$.

◆ The **distance** D between (x_1, y_1) and (x_2, y_2) in the xy plane is the nonnegative number (page 14)
$D = \sqrt{(x_2 - x_1)^2 + (y_2 - y_1)^2}$.

◆ The **circle** with center (x_0, y_0) and radius r has equation $(x - x_0)^2 + (y - y_0)^2 = r^2$. (page 15)

◆ The **slope** of the **line** containing (x_1, y_1) and (x_2, y_2) is (page 16)

$m = \dfrac{y_2 - y_1}{x_2 - x_1}, \qquad x_2 \ne x_1.$

◆ An **equation** for the line through (x_1, y_1) with slope m is $(y - y_1) = m(x - x_1)$. (page 18)

◆ The **slope-intercept** form for this line is $y = mx + b$, $\qquad b = y$-intercept. (page 19)

◆ A **vertical** line has no slope, and equation $x = a$. A **horizontal** line has slope zero, and equation $y = b$. (page 20)

◆ Two lines with slope m_1 and m_2 are (a) **parallel** if $m_1 = m_2$, (b) **perpendicular** if $m_1 m_2 = -1$. (page 21)

◆ The equation $y = f(x)$ is a **function** if there is a **unique** y corresponding to each x in the **domain** of f. (page 24)

◆ The **domain** of f, if not explicitly stated, is the largest set for which $f(x)$ is defined. The **graph** of the function f (page 24)
is the graph of the equation $y = f(x)$.

◆ The graph of an equation of the form $y = a(x - b)^2 + c$ is a **parabola.** The graph of every **quadratic function** (page 27)
$f(x) = Ax^2 + Bx + C$ is a parabola. A quadratic function can be graphed by **completing the square** in x.

◆ **Radian** measure is defined by the equation 2π radians $= 360°$. If $P = (x, y)$ is a point on the unit circle and θ is the   (page 34) angle formed between the radius through P and the positive x-axis, then $\sin \theta = y$ and $\cos \theta = x$. Also,

$$\tan \theta = \frac{\sin \theta}{\cos \theta} = \frac{y}{x} \qquad \sec \theta = \frac{1}{\cos \theta} = \frac{1}{x}$$

$$\cot \theta = \frac{\cos \theta}{\sin \theta} = \frac{x}{y} \qquad \csc \theta = \frac{1}{\sin \theta} = \frac{1}{y}$$

when these expressions are defined. The functions $\sin \theta$, $\cos \theta$, $\sec \theta$, and $\csc \theta$ are **periodic** with period $T = 2\pi$. The functions $\tan \theta$ and $\cot \theta$ are periodic with period π.

REVIEW EXERCISES—CHAPTER 1

1. Give an example of a set with no least element.

2. For $A = [2, 7)$ and $B = (3, 9)$, find
   a. $A \cup B$ **b.** $A \cap B$

3. Which of the following sets have greatest elements?
   a. $\{1, 2, -3, 6\}$
   b. $[0, 5)$
   c. $\{x \mid 3 \le x \le 5\}$
   d. $\{x \mid x \text{ is a rational number}\}$
   e. $\{x \mid 0 \le x^2 \le 2 \text{ and } x \text{ is rational}\}$

4. List all subsets of the set $\{1, 7, 19, \pi\}$.

5. Find two sets of numbers A and B so that neither A nor B has a greatest element but so that $A \cap B$ has a greatest element.

6. Solve the inequality $2x - 7 \ge 9$.

7. Find $\{x \mid 6 \le x^2 - 3 \le 22\}$.

8. Let $A = \{x \mid x^2 \ge 4\}$, $B = \{x \mid -3 \le x \le 3\}$,   and $C = \{-4, -3, -2, -1, 0, 1, 2, 3, 4\}$: find
   a. $A \cap B$ **d.** $B \cap C$
   b. $A \cup B$ **e.** $A \cup (B \cap C)$
   c. $A \cap C$ **f.** $A \cap (B \cup C)$

9. Solve the inequality $4 \le (x + 2)^2 \le 36$.

10. Solve the inequality $|x - 7| \le 12$.

11. Find $\{x \mid 2 \le |x^2 - 2| \le 7\}$.

12. Let $A = \{x \mid |x| \le 9\}$ and $B = \{x \mid -2 < x < 2\}$: find $A \cup B$ and $A \cap B$.

Solve the following inequalities.

13. $|x - 3| \le 5$ **14.** $-3 \le |2x + 1| \le 11$

15. $|x + 2| \ge 6$ **16.** $|2x^2 + 1| \le 9$

17. $|9 - x^2| \ge 0$

18. $\cos x < \dfrac{\sqrt{3}}{2}$, $0 \le x \le 2\pi$

19. $|\sin x| > \dfrac{1}{2}$, $0 \le x \le 2\pi$

20. $\left(\sin x - \dfrac{1}{2}\right)\left(\cos x + \dfrac{1}{2}\right) > 0,$ $0 \le x \le 2\pi$

21. $\cos 2x + \sin x > 0,$ $0 \le x \le 2\pi$

22. Find the solution set for the equation $|x - 2| = |x + 2|$.

23. Determine which sets of points lie on a common line.
   a. $\{(0, 0), (-3, 6), (2, 4)\}$
   b. $\{(-4, 2), (-1, 5), (6, 10)\}$
   c. $\{(-3, 2), (-1, 1), (2, -3)\}$
   d. $\{(1, 3), (2, 4), (-2, 0)\}$

24. Find the area of the triangle with vertices $(0, 0)$, $(5, 0)$, and $(2, 3)$.

25. Find a if a right triangle has vertices $(0, 0)$, $(2, 4)$, and $(a, 0)$, and
   a. the right angle is at vertex $(a, 0)$.
   b. the right angle is at vertex $(2, 4)$.

26. Find an equation for the circle with center at $(2, -4)$ and diameter $d = 10$.

Find the center and radius of each of the following circles.

27. $x^2 + y^2 = 49$ **28.** $x^2 - 4x + y^2 = 0$

29. $x^2 - 2x + y^2 + 6y = -9$ **30.** $x^2 - 2x + y^2 + 2y = 14$

For each of the following lines find the slope, y-intercept, and x-intercept, if possible.

31. $x + y = 1$ **32.** $x - 3y + 4 = 0$

33. $y = 4$ **34.** $7x - 7y + 21 = 0$

35. $x - y = 5$ **36.** $3x = 6$

37. Find an equation for the line through the points $(-2, 1)$ and $(3, 3)$.

38. Find an equation for the line through the point $(2, -4)$ and parallel to the line with equation $2x - 4y = 14$.

39. The lines $y - ax = 1$ and $3y = 6x + 12$ are parallel. Find a.

40. Find an equation for the line perpendicular to the line with equation $3x - 6y = 8$ and passing through the origin.

41. Find an equation for the line through $(-1, 3)$ parallel to the line through the points $(1, 3)$ and $(6, -4)$.

42. Where does the line $ax + by + c = 0$ cross the x-axis?

43. The line $y = ax + b$ passes through the points $(2, 2)$ and $(4, -5)$. Find a and b.

44. Graph the region bounded by the graph of $y = |4 - x^2|$ and the x-axis.

45. Graph the region R where
$$R = \{(x, y) \mid y \le x + 4\} \cap \{(x, y) \mid y \le -2x + 8\}.$$

46. Sketch the region in the plane whose points satisfy all of the following inequalities.
   a. $x + y \ge 0$
   b. $2y \le 4x + 4$
   c. $x \le 4$

47. Sketch the region R consisting of all points satisfying both of the following inequalities
   a. $x^2 + y^2 \le 4$, and
   b. $x + y \ge 2$.

Convert the following decimals to quotients of integers.

48. $32.61\overline{61} \ldots$

49. $-6.214\overline{214} \ldots$

50. Draw Venn diagrams to illustrate the following statements.
   (a) $A \cap (B \cup C) = (A \cap B) \cup (A \cap C)$
   (b) $A \cup (B \cap C) = (A \cup B) \cap (A \cup C)$

51. Find the coordinates of the point of intersection of the lines with equations $x + 3y - 6 = 0$ and $x - y = 2$.

52. A beaker of water is being heated. The water is initially at $10°C$. After 3 minutes its temperature is $25°C$. Using this data, find a linear equation which will predict the temperature of the water at a later time.
   a. What is the predicted temperature after 5 minutes?
   b. What is wrong with extrapolating to predict the temperature after 30 minutes?

53. Find a function which converts temperatures in degrees Kelvin to temperature in Celsius degrees, given that $100°C = 373$ K and $0°C = 273$ K.

54. Find an equation which converts temperatures in Kelvin to temperatures in degrees Fahrenheit, given that $212°F = 373$ K and $32°F = 273$ K.

55. Find the point on the line $x + y = 2$ nearest the point $(4, 1)$.

56. When two chemicals combine to form a solution, the chemical used in greater quantity is called the *solvent*. *Raoult's law* states that in a dilute solution, the vapor pressure of the solvent, P_s, is proportional to the vapor pressure of the solvent when pure, P_p. Moreover, the constant of proportionality is the ratio of the number of moles of the solvent present, n_s, to the total number of moles in the solution, n_T.

   a. Write an equation expressing Raoult's law.
   b. In a certain solution, the number of moles of solvent is 85% of the total present. Write the equation for Raoult's law in this case, and find the vapor pressure of the solvent in solution if its vapor pressure when pure is 0.25 atmospheres.

In Exercises 57–62, state the domain of the given function.

57. $y = \sqrt{x^2 - 1}$

58. $y = \dfrac{1}{x + 2}$

59. $f(x) = \sin(1 - x^2)$

60. $f(x) = \dfrac{1}{1 + \cos x}$

61. $y = \dfrac{1}{\sqrt{1 - \sin^2 x}}$

62. $f(x) = \dfrac{1}{\sqrt{x(x + 2)}}$

In Exercises 63–66, state whether the given function is odd, even, or neither.

63. $f(x) = x^3 \sin x$

64. $f(x) = \dfrac{1 - \cos x}{x^2}$

65. $f(x) = (x - 1) \cdot (x + 1)$

66. $f(x) = \dfrac{x \tan x}{1 + x^3}$

In Exercises 67–70, use the technique of completing the square to graph the function.

67. $y - 2x^2 + 2x - \dfrac{7}{2} = 0$

68. $y^2 - 2x + 4y + 6 = 0$

69. $y^2 + 2x - y + \dfrac{3}{4} = 0$

70. $y - \dfrac{1}{6}x^2 - x - \dfrac{5}{2} = 0$

71. Find θ such that $0 \le \theta < 2\pi$ and
   a. $\sin \theta = \sqrt{3}/2$ and $\cos \theta < 0$
   b. $\sec \theta = \sqrt{2}$ and $\tan \theta = -1$
   c. $\sin 2\theta = \sin \theta$ and $\sin \theta \ne 0$.

72. Sketch a right triangle containing an angle θ for which
   a. $\sec \theta = 1$
   b. $\tan \theta = \sqrt{3}$
   c. $\csc \theta = -\sqrt{2}$.

In Exercises 73–78, let
$$f(x) = x^3 - x, \qquad g(x) = \frac{1}{1 - x}, \qquad \text{and} \qquad h(x) = \sin x.$$

Form the indicated composite function.

73. $g(h(x))$

74. $f(g(x))$

75. $h(g(x))$

76. $f(g(h(x)))$

77. $g(f(h(x)))$

78. $g(h(g(x)))$

79. Prove that the product of two odd functions is an even function.

80. Prove that the product of an odd function and an even function is an odd function.

81. Find an equation for the line tangent to the circle $(x - 6)^2 + (y - 4)^2 = 25$ at the point $(3, 8)$.

82. Find numbers a, b, and c so that the parabola $y = ax^2 + bx + c$ will have y-intercept $y = -6$, x-intercept $x = 3$, and vertex at $x = 2$.

83. Given the functions $f(x) = 4 - x^2$, with domain $(-\infty, \infty)$, and $g(x) = \sin x$, with domain $0 \le x \le 2\pi$, find the domain and range of the composite function $f \circ g$.

84. Given the functions $f(x) = \sqrt{x - \frac{1}{2}}$ and $g(x) = \sin x$, $0 \le x \le 2\pi$, find the domain and range of the composite function $f \circ g$.

85. Find the area of the triangle with vertices $(4, 3)$, $(9, 13)$, and $(20, 15)$.

Chapter 2
Limits of Functions

Here is where the calculus begins. So that you will have some sense of where we are headed, this chapter opens with a brief discussion of two of the major goals of the calculus. We turn quickly to practical matters, however, by developing in this chapter the concept of the *limit* of a function. This topic is so central to the calculus that it appears frequently throughout every chapter that follows.

2.1 TANGENTS, AREAS, AND LIMITS

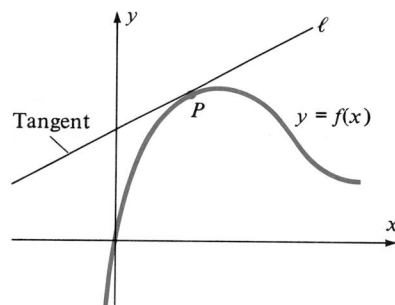

Figure 1.1 The Tangent Line Problem: Find the slope of ℓ.

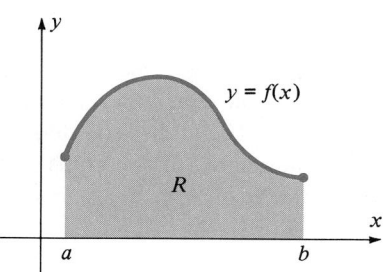

Figure 1.2 The Area Problem: Find the area of R.

Two general problems characterize much of what we shall do in this text:

The Tangent Line Problem: Find the slope of the line tangent to the graph of the function f at the point P (Figure 1.1).

The Area Problem: Find the area of the region R bounded by the graph of the function f and the x-axis for $a \leq x \leq b$ (Figure 1.2).

The **differential calculus** resolves the Tangent Line Problem, together with a host of related issues, all having to do with the rate of change of a function. The **integral calculus** addresses the Area Problem and other related issues.

In developing both the differential and integral calculus we will use the notation, theory, and techniques of elementary algebra and plane geometry, together with the powerful techniques of analytic geometry that synthesize these two subjects. (The development of analytic geometry is primarily due to René Descartes, 1596–1650, a French mathematician and philosopher.)

We shall see immediately, however, that something else is required. Whether one begins a study of the calculus with the Tangent Line Problem or the Area Problem, the notion of the **limit** of a function unavoidably surfaces as a fundamental issue. Because this concept underlies every aspect of the calculus, we need to develop it carefully at the outset. Our plan, therefore, is to begin the discussion of the Tangent Line Problem and then to devote the remaining part of this chapter to a careful study of the limit concept that arises in this first section. We will then be prepared for a rigorous study of the differential calculus beginning in Chapter 3.

The Tangent to a Curve

The first step in attacking the Tangent Line Problem is to define clearly what we mean by ''the line tangent to the graph of f at point P.'' From geometry we know that if the graph of f is an arc of a circle, then the tangent at point P may be defined

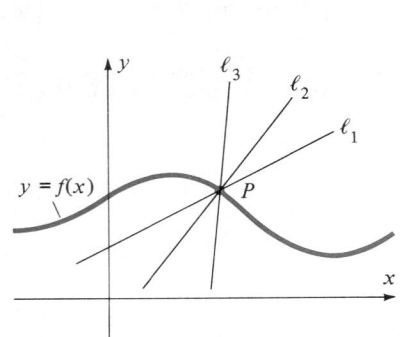

Figure 1.3 Lines intersecting graph of $y = f(x)$ "only at point P," which are not tangents.

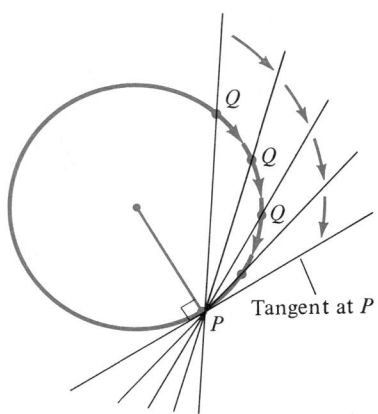

Figure 1.4 As Q "approaches P" along the circle, the secant through P and Q rotates into the tangent at P.

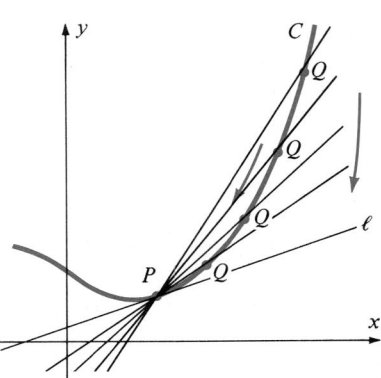

Figure 1.5 Tangent is limiting position of secant as $Q \to P$.

as the (unique) line that intersects the circle only at point P. This is a perfectly adequate definition for arcs of circles, but it fails for more general curves. For example, Figure 1.3 shows that several lines, none of which we would wish to call a tangent, can intersect the graph of f only at point P.

There is another way, however, of defining the tangent to a circle that does have a satisfactory generalization to more general curves. Figure 1.4 illustrates that a second point Q on the circle determines a *secant* through the points P and Q. As the point Q moves toward P along the circle, this secant rotates into the position of a fixed line through P. This fixed line is the tangent to the circle at P. This is the idea we shall use in defining the tangent to a more general curve.

DEFINITION 1

Let P and Q be points on a curve C. The line **tangent** to the curve C at the point P, if it exists, is the limiting position of the secant line through P and Q as Q approaches P along the curve C from either direction. (See Figure 1.5.)

Solving the Tangent Line Problem

We can now determine how to define the slope of the line tangent to the graph of a function f at a point P, if it exists, in a way that is consistent with our notion of a tangent to a curve. First, we let x_0 be the x-coordinate of P. Then P has coordinates $(x_0, f(x_0))$ since P is on the graph of f. To find a second point Q on the graph of f we choose any number $h \neq 0$ (positive or negative) and let $Q = (x_0 + h, f(x_0 + h))$. (See Figure 1.6.) Then

$$\left\{ \begin{matrix} \text{Slope of secant} \\ \text{through } P \text{ and } Q \end{matrix} \right\} = \frac{f(x_0 + h) - f(x_0)}{h}. \tag{1}$$

If the graph of f connects the points P and Q in an unbroken arc, as in Figure 1.6, as the number h approaches zero the point Q will approach P along the graph of f. Then the tangent to the graph of f at P is the "limiting position" of the secant through P and Q as h approaches zero. Thus, the slope of the tangent should be the

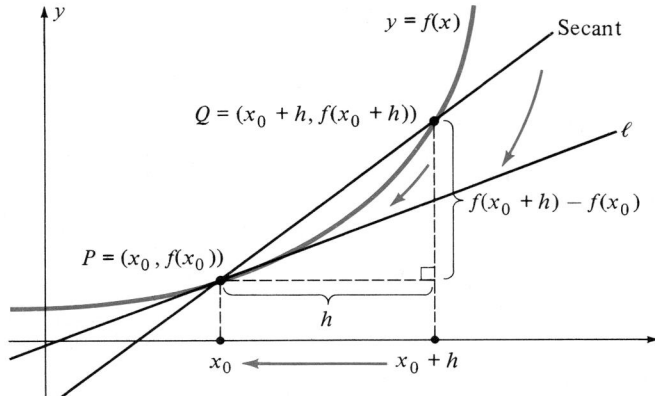

Figure 1.6 Slope of ℓ is limit of slope of secant as $h \to 0$.

"limiting value" of the slope of this secant as h approaches zero:

$$\begin{Bmatrix} \text{Slope of the} \\ \text{tangent at } P \end{Bmatrix} = \begin{Bmatrix} \text{limiting value of the slope of} \\ \text{the secant as } h \text{ approaches zero} \end{Bmatrix} \tag{2}$$

$$= \begin{Bmatrix} \text{limit as } h \text{ approaches zero of} \\ \dfrac{f(x_0 + h) - f(x_0)}{h} \end{Bmatrix}.$$

Using equations (1) and (2) we formally define the slope of the line tangent to the graph of a function at a point as follows.

DEFINITION 2	The **slope** of the line tangent to the graph of the function f at the point $(x_0, f(x_0))$, if it exists, is the number $$m = \lim_{h \to 0} \frac{f(x_0 + h) - f(x_0)}{h}. \tag{3}$$

The remaining sections of this chapter are devoted to making the *limit* concept, represented by the expression "$\lim_{h \to 0}$" in equation (3), precise. For now we shall interpret equation (3) as the result of allowing the number h in the quotient on the right-hand side to approach zero "by inspection." The following examples demonstrate how this may be done.

Example 1

Find the slope of the line tangent to the graph of $f(x) = x^2$ at the point $(2, 4)$.

Strategy

Identify x_0.

Find $f(x_0 + h)$ by substituting $x = x_0 + h$ in equation for $f(x)$.

Solution

Here $x_0 = 2$ and $f(x) = x^2$, so

$$f(x_0 + h) = f(2 + h) = (2 + h)^2$$
$$= 4 + 4h + h^2$$

and

$$f(x_0) = f(2) = 2^2 = 4.$$

Thus,

Set up the expression for the slope of the secant; substitute for $f(x_0 + h)$ and $f(x_0)$.

$$\frac{f(x_0 + h) - f(x_0)}{h} = \frac{(4 + 4h + h^2) - 4}{h}$$

$$= \frac{4h + h^2}{h}$$

$$= \frac{h(4 + h)}{h}$$

Simplify; factor h from the numerator. Divide by common factor of h.

$$= 4 + h \qquad (h \neq 0).$$

The slope of tangent is therefore

Find limit of resulting expression as h approaches zero.

$$m = \lim_{h \to 0}(4 + h) = 4.$$

(See Figure 1.7.)   ◇

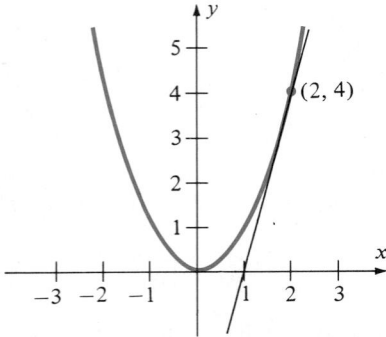

Figure 1.7 Slope of tangent to graph of $f(x) = x^2$ at $(1, 1)$ is $m = 2$.

Example 2

Find an equation for the line tangent to the graph of $f(x) = x^3 + 3x - 2$ at the point $(1, 2)$.

Solution: We begin by finding the slope of the tangent at $(1, 2)$, as in Example 1. With $x_0 = 1$ we have

$$f(1) = 1^3 + 3 \cdot 1 - 2 = 2$$

and

$$f(1 + h) = (1 + h)^3 + 3(1 + h) - 2$$
$$= (1 + 3h + 3h^2 + h^3) + (3 + 3h) - 2$$
$$= h^3 + 3h^2 + 6h + 2.$$

The slope of the secant through $(1, 2)$ and $(1 + h, f(1 + h))$ is therefore

$$\frac{f(1 + h) - f(1)}{h} = \frac{(h^3 + 3h^2 + 6h + 2) - 2}{h}$$

$$= \frac{h^3 + 3h^2 + 6h}{h}$$

$$= h^2 + 3h + 6$$

when $h \neq 0$. The slope of the tangent at $(1, 2)$ is therefore

$$m = \lim_{h \to 0}(h^2 + 3h + 6) = 6.$$

An equation for the tangent, which has slope $m = 6$ and passes through the point $(1, 2)$, is

$$y - 2 = 6(x - 1).$$ ◇

Exercise Set 2.1

In Exercises 1–15 use equation (3) to find the slope of the line tangent to the graph of the given function at the given point.

1. $f(x) = 3x - 2$, $P = (2, 4)$

2. $f(x) = 7 - 3x$, $P = (1, 4)$

3. $f(x) = 2x^2$, $P = (3, 18)$

4. $f(x) = 9x^2$, $P = (-1, 9)$

5. $f(x) = 2x^2 + 3$, $P = (1, 5)$

6. $f(x) = 5 - x^2$, $P = (-2, 1)$

7. $f(x) = 3x^2 + 4x + 2$, $P = (-2, 6)$

8. $f(x) = x^2 - 6x + 3$, $P = (2, -5)$

9. $f(x) = x^3 + 3$, $P = (2, 11)$

10. $f(x) = x^3 - 2x + 6$, $P = (1, 5)$

11. $f(x) = x^4$, $P = (-2, 16)$

12. $f(x) = ax^2 + bx + c$ where $x = 1$

13. $f(x) = ax^3 + bx^2 + cx + d$ where $x = 1$

14. $f(x) = \dfrac{1}{x}$ at $(1, 1)$

15. $f(x) = \dfrac{1}{x + 3}$ at $(-2, 1)$

16. Find an equation for the line tangent to the graph of the function $f(x) = 2x^2$ at the point $(2, 8)$.

17. Find an equation for the line tangent to the graph of $f(x) = ax^2$ at the point $(1, a)$.

18. Find an equation for the line tangent to the graph of $f(x) = 5 - x^2$ at the point $(-2, 1)$.

19. Find an equation for the line tangent to the graph of $f(x) = 2x^3 + x$ at the point where $x = 1$.

20. Find an equation for the line tangent to the graph of $f(x) = ax^3$ at the point $(1, a)$.

21. True or false? A tangent to the graph of a function cannot be horizontal. Why or why not?

22. Complete Table 1.1 for the function $f(x) = x^2 - 4x + 3$.

23. *(Calculator)* By using a calculator we can approximate the slope of the line tangent to the graph of f at $(x_0, f(x_0))$ by computing the slope of the secant, $\dfrac{f(x_0 + h) - f(x_0)}{h}$, for small numbers h. Approximate the slope of the line tangent to the graph of $f(x) = x^3 - 3x$ at the point $(2, 2)$ by completing Table 1.2.

Table 1.2

x	h	$\dfrac{[(2 + h)^3 - 3(2 + h)] - 2}{h}$
2	2	
2	1	
2	0.5	
2	0.2	
2	0.1	
2	0.05	
2	0.01	
2	0.005	

24. Use the method of Examples 1 and 2 to find the slope of the line tangent to the graph of $y = x^3 - 3x$ at the point $(2, 2)$. How does this compare with the results of Exercise 23?

25. Consider the graph of $f(x) = x^3 - 3x$ in Figure 1.8.
  a. Sketch tangents at the points on the graph with x-coordinates
$$-2, \ -\frac{3}{2}, \ -1, \ -\frac{1}{2}, \ 0, \ \frac{1}{2}, \ 1, \ \frac{3}{2}, \text{ and } 2.$$
  b. From your sketches estimate the slopes of each of these tangents.
  c. Use the method of Examples 1 and 2 to find the slopes of the tangents in part a. Compare with your slope estimates. How closely do they agree? Should they?

Table 1.1

x_0	$f(x_0)$	$\dfrac{f(x_0 + h) - f(x_0)}{h}$	Slope of tangent at $(x_0, f(x_0))$
-2			
-1			
0			
1			
2			
3			
x_0			

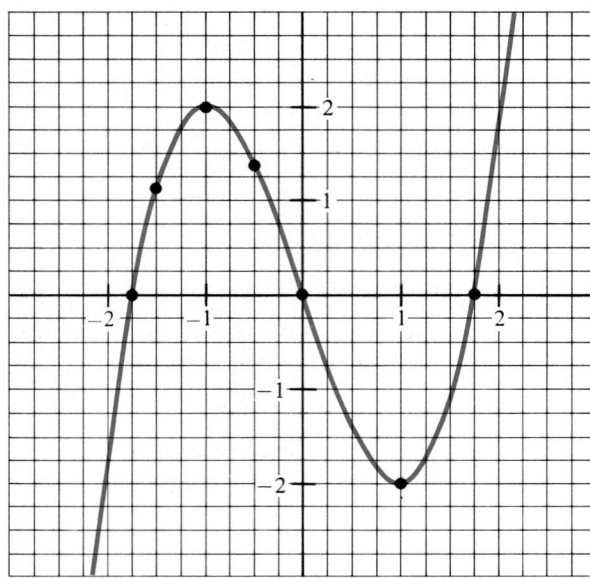

Figure 1.8

d. What is the relationship between these slopes and the high and low points on the graph for x between -2 and 2?

26. *(Calculator)* Use the method of Exercise 23 to approximate the slope of the line tangent to the graph of $y = \sin x$ at the point $(\pi/4, \sqrt{2}/2)$.

27. Show that the slope of the line tangent to the graph of $f(x) = x^2 + 6x + 1$ at the point where $x = a$ is $m = 2a + 6$. Use this information to find the number x where the slope of the tangent to the graph is 0.

28. Use the method of Exercise 27 to find the point(s) on the graph of $f(x) = x^2 - 3x + 1$ where the slope of the tangent at the point equals the y-coordinate of the point.

29. Given $f(x) = ax^2 + bx + c$.
  a. Use the method of Exercise 27 to show that the slope of the line tangent to the graph of f at $(x, f(x))$ is $m = 2ax + b$.

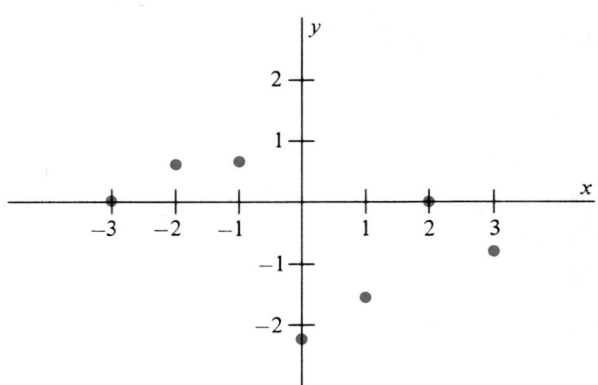

Figure 1.9

  b. The graph of this function is a parabola. Show that the slope of the tangent at the vertex of the parabola is 0.

30. Find the numbers a, b, and c such that the graph of $f(x) = ax^2 + bx + c$ has y-intercept $(0, 5)$, contains the point $(1, 2)$, and has a tangent with slope 3 when $x = 2$.

31. Prove that the line tangent to the graph of $f(x) = x^3$ at $(x_0, f(x_0))$ is always parallel to the line tangent to this graph at the point $(-x_0, f(-x_0))$. Is this true for any odd function? Why or why not?

32. Knowledge of the slopes of the tangents to the graph of a function often aids in understanding its behavior, even when the exact expression for the function is unknown. Use the indicated values of the slopes in Table 1.3 to sketch tangents at each of the points in Figure 1.9. Then sketch the graph of f as best you can. (Notice that the graph you sketch differs from what you might have sketched without the data of Table 1.3.)

Table 1.3

x	-3	-2	-1	0	1	2	3
slope	-2	4	0	-1	6	-2	2

2.2 LIMITS OF FUNCTIONS

The slope calculations of Section 2.1 involve special cases of the more general statement

$$L = \lim_{x \to a} f(x) \tag{1}$$

read, "The number L is the limit of the function f as x approaches a." A formal definition of limit, which allows us to actually prove statements in the form of equation (1), is given in the next section. The following intuitive explanation, however, enables us to calculate most limits that we shall encounter.

> Intuitively, the statement
>
> $$L = \lim_{x \to a} f(x)$$
>
> means that we can cause $f(x)$ to be as close to L as desired by choosing x sufficiently close to a.

Figure 2.1 illustrates this concept. As the number x is chosen close to the number a, the value $f(x)$ is close to the number L. Note that the number $x = a$ is not assumed to belong to the domain of f. (That is why Figure 2.1 shows a ''hole'' in the graph of

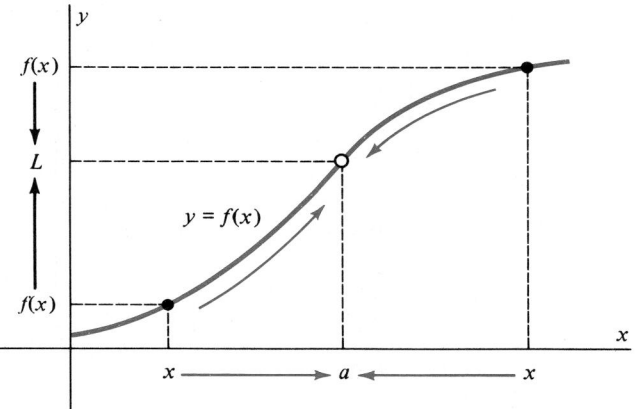

Figure 2.1 $L = \lim_{x \to a} f(x)$ means that $f(x)$ is near L when x is near a.

f at the point (a, L).) This emphasizes the fact that $\lim_{x \to a} f(x)$ is determined by the behavior of f for x near a *but not by the number $f(a)$ itself*. **Indeed, $f(a)$ need not even exist for $\lim_{x \to a} f(x)$ to be defined.**

Example 1

The function $f(x) = \dfrac{\sin x}{x}$ is undefined for $x = 0$. However, the limit

$$\lim_{x \to 0} \frac{\sin x}{x}$$

will be important in our work on the trigonometric functions in Chapter 3. Even though $\dfrac{\sin x}{x}$ is undefined for $x = 0$, the entries in Table 2.1 (obtained by use of a calculator) suggest that as x is chosen very close to 0, the value $\dfrac{\sin x}{x}$ is found very close to the number $L = 1$. Later in this chapter we shall actually prove that

$$\lim_{x \to 0} \frac{\sin x}{x} = 1.$$

◇

Table 2.1 $\lim\limits_{x\to 0} \dfrac{\sin x}{x} = 1$

x	$\dfrac{\sin x}{x}$
0.8	0.896695
0.5	0.958851
0.2	0.993347
0.08	0.998934
0.05	0.999583
0.02	0.999933
0.005	0.999996
0.002	0.999999
−0.002	0.999999
−0.005	0.999996
−0.02	0.999933
−0.05	0.999583
−0.08	0.998934
−0.2	0.993347
−0.5	0.958851
−0.8	0.896695

Table 2.2 $\lim\limits_{x\to 0} \dfrac{(x + 1)^3 - 1}{x} = 3$

x	$\dfrac{(x + 1)^3 - 1}{x}$
2.0	13.0000
1.5	9.7500
1.0	7.0000
0.5	4.7500
0.2	3.6400
0.1	3.3100
0.01	3.0301
0.001	3.0030
−0.001	2.9970
−0.01	2.9701
−0.1	2.7100
−0.2	2.4400
−0.5	1.7500
−1.0	1.0000 .
−1.5	0.7500
−2.0	1.0000

Example 2

Find $\lim\limits_{x\to 0} \dfrac{(x + 1)^3 - 1}{x}$.

Solution: Here the function $f(x) = \dfrac{(x + 1)^3 - 1}{x}$ is also undefined if $x = 0$. Experimenting with a calculator, however, gives results such as those in Table 2.2 which suggest that as x is chosen close to 0 the value $f(x)$ is close to $L = 3$. That is, the data suggest that

$$\lim_{x\to 0} \frac{(x + 1)^3 - 1}{x} = 3.$$

We can verify this limit using simple algebra. If $x \neq 0$ we can write the function f as

$$\frac{(x + 1)^3 - 1}{x} = \frac{(x^3 + 3x^2 + 3x + 1) - 1}{x}$$

$$= \frac{x^3 + 3x^2 + 3x}{x}$$

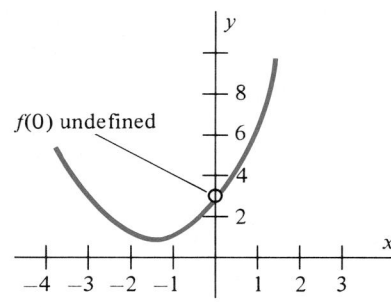

Figure 2.2 $f(x) = \dfrac{(x+1)^3 - 1}{x}$; $f(0)$ is undefined.

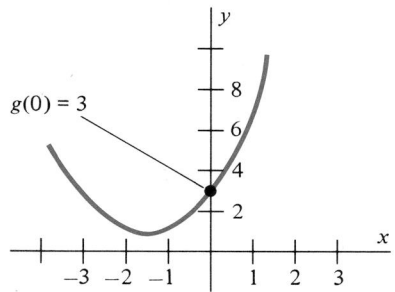

Figure 2.3 $g(x) = x^2 + 3x + 3$; $g(0) = 3$.

$$= \frac{x(x^2 + 3x + 3)}{x}$$

$$= x^2 + 3x + 3, \qquad x \neq 0.$$

We may therefore conclude that the functions

$$f(x) = \frac{(x+1)^3 - 1}{x} \qquad \text{and} \qquad g(x) = x^2 + 3x + 3$$

have the same values *except* at $x = 0$, where $f(0)$ is undefined. (See Figures 2.2 and 2.3.) Thus, the limits as x approaches zero of these two functions must be the same. (Remember, we care about only the values $f(x)$ for x **near** zero, not $f(0)$ itself.)

Our limit may therefore be calculated as follows:

$$\lim_{x \to 0} \frac{(x+1)^3 - 1}{x} = \lim_{x \to 0}(x^2 + 3x + 3) = 0 + 0 + 3 = 3. \tag{2}$$

Note that the technique used here to calculate the limit is the same as that of Section 2.1: Factor the term causing the zero from both the numerator and denominator and divide. ◇

REMARK: Some limits are quite obvious, and may be determined "by inspection." For example, in line (2) the limit

$$\lim_{x \to 0}(x^2 + 3x + 3) = 0 + 0 + 3 = 3$$

is obtained by noting that if x is close to zero so is x^2, and so is $3x$. The limit theorems established in Section 2.4 will justify such calculations. In the following example we use the limit

$$\lim_{x \to 2} \frac{x-1}{x+3} = \frac{2-1}{2+3} = \frac{1}{5},$$

also obtained "by inspection."

Example 3

Find $\lim\limits_{x \to 2} \dfrac{x^2 - 3x + 2}{x^2 + x - 6}$.

Strategy

Determine whether limit can be obtained "by inspection" by setting $x = 2$. This yields $\dfrac{0}{0}$.

Solution

First note that on setting $x = 2$ we obtain the quotient

$$\frac{2^2 - 3 \cdot 2 + 2}{2^2 + 2 - 6} = \frac{4 - 6 + 2}{4 + 2 - 6} = \frac{0}{0}$$

which is undefined. We therefore factor the numerator and denominator, obtaining, for $x \neq 2$,

Factor both numerator and denominator.

$$\frac{x^2 - 3x + 2}{x^2 + x - 6} = \frac{(x-2)(x-1)}{(x-2)(x+3)}$$

Divide both by common factor $x - 2$.

$$= \frac{x-1}{x+3} \qquad \left(\text{since } \frac{x-2}{x-2} = 1 \text{ if } x \neq 2\right).$$

The limit equals the limit of the resulting expression.

Thus,

$$\lim_{x \to 2} \frac{x^2 - 3x + 2}{x^2 + x - 6} = \lim_{x \to 2} \frac{x - 1}{x + 3} = \frac{2 - 1}{2 + 3} = \frac{1}{5}.$$ ◇

Example 4

Find $\lim\limits_{x \to \pi} \dfrac{\sin^2 x}{1 + \cos x}$.

Solution: First note that $\sin^2 \pi = (\sin \pi)^2 = 0$ and $1 + \cos \pi = 1 + (-1) = 0$ so both the numerator and denominator equal zero when x equals zero. To find an equivalent expression for $\dfrac{\sin^2 x}{1 + \cos x}$ we use the identity $\sin^2 x + \cos^2 x = 1$. Then $\sin^2 x = 1 - \cos^2 x$, and

$$\frac{\sin^2 x}{1 + \cos x} = \frac{1 - \cos^2 x}{1 + \cos x} = \frac{(1 - \cos x)(1 + \cos x)}{1 + \cos x} = 1 - \cos x$$

if $\cos x \neq -1$. Thus,

$$\lim_{x \to \pi} \frac{\sin^2 x}{1 + \cos x} = \lim_{x \to \pi} (1 - \cos x) = 1 - (-1) = 2.$$ ◇

Limits that Fail to Exist

The next two examples show ways in which limits can fail to exist.

Example 5

The limit $\lim\limits_{x \to 0} \dfrac{|x|}{x}$ does not exist. To see why, we use the definition of absolute value

$$|x| = \begin{cases} x & \text{if } x \geq 0 \\ -x & \text{if } x < 0 \end{cases}$$

to rewrite the function $\dfrac{|x|}{x}$ as

$$\frac{|x|}{x} = \begin{cases} \dfrac{x}{x} = 1 & \text{if } x > 0 \\ -\dfrac{x}{x} = -1 & \text{if } x < 0. \end{cases}$$

(See Figure 2.4.) This shows that when x is close to zero *and positive,* all values of the function $f(x) = \dfrac{|x|}{x}$ are $f(x) = 1$. But when x is close to zero *and negative,* values of the function f are $f(x) = -1$. For the limit to exist, values of f must lie close to a single number L when x is chosen close to 0 *on either side.* Since values of f are always 1 or -1, depending only on the sign of x, the limit does not exist. ◇

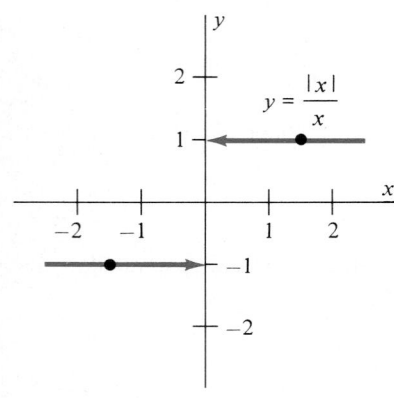

Figure 2.4 $\lim\limits_{x \to 0} \dfrac{|x|}{x}$ does not exist.

Example 6

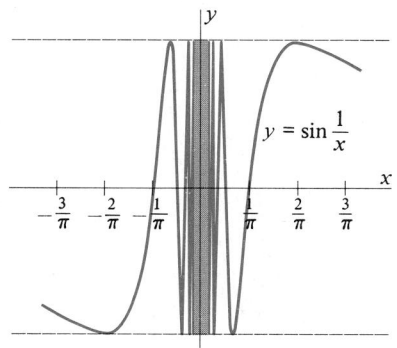

The limit $\lim\limits_{x \to 0} \sin\left(\dfrac{1}{x}\right)$ does not exist. Figure 2.5 shows that as x approaches zero from either direction the values $f(x) = \sin\left(\dfrac{1}{x}\right)$ oscillate between 1 and -1 at an increasingly rapid rate. This is because $\left|\dfrac{1}{x}\right|$ increases rapidly as x approaches zero from either direction. Thus, choosing x near zero will not force $f(x)$ to lie close to any particular number. ◇

Figure 2.5 $\lim\limits_{x \to 0} \sin\left(\dfrac{1}{x}\right)$ does not exist.

Exercise Set 2.2

For each of the functions whose graphs are sketched in Exercises 1–4, indicate whether

(i) $\lim\limits_{x \to a} f(x)$ exists and equals $f(a)$,

(ii) $\lim\limits_{x \to a} f(x)$ exists but does not equal $f(a)$, or

(iii) $\lim\limits_{x \to a} f(x)$ does not exist.

1. **2.**

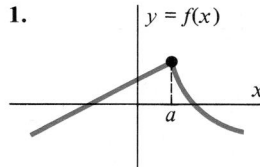

3. **4.**

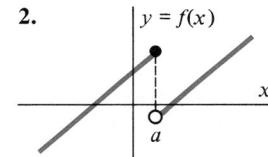
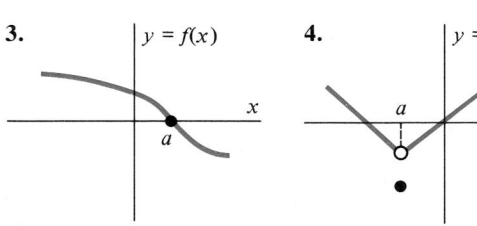

In Exercises 5–26 find the indicated limit, if it exists.

5. $\lim\limits_{x \to 2}(3 + 7x)$

6. $\lim\limits_{x \to 0}(x^3 - 7x + 5)$

7. $\lim\limits_{x \to 0} \dfrac{(3 + x)^2 - 9}{x}$

8. $\lim\limits_{x \to 3} \dfrac{x^2 - 9}{x - 3}$.

9. $\lim\limits_{h \to 0} \dfrac{h^2 - 1}{h - 1}$

10. $\lim\limits_{x \to 0} \dfrac{(x + 3)^3 - 27}{x}$

11. $\lim\limits_{x \to -2} \dfrac{x^2 - x - 6}{x + 2}$

12. $\lim\limits_{h \to 0} \dfrac{(4 + h)^2 - 16}{h}$

13. $\lim\limits_{x \to 0} \dfrac{1 - \cos^2 x}{\sin x \cos x}$

14. $\lim\limits_{x \to 0} \dfrac{\tan x}{\sin x}$

15. $\lim\limits_{h \to 0} \dfrac{\sin 2h}{\sin h}$ (*Hint:* Use identity $\sin 2h = 2 \sin h \cos h$)

16. $\lim\limits_{\theta \to 2} \dfrac{\theta^4 - 2^4}{\theta - 2}$

17. $\lim\limits_{x \to \pi/2} \sin 2x \csc x$

18. $\lim\limits_{x \to \pi/2} \dfrac{\sin 2x}{\cos x}$

19. $\lim\limits_{x \to -1} \dfrac{x^2 - 2x - 3}{x + 1}$

20. $\lim\limits_{x \to 4} \dfrac{x^2 - 2x - 8}{x - 4}$

21. $\lim\limits_{x \to \pi/2} \dfrac{\sec x \cos x}{x}$

22. $\lim\limits_{x \to 1} \dfrac{x^3 + 3x^2 - x - 3}{x^2 - 1}$

23. $\lim\limits_{x \to -1} \dfrac{x^3 + 3x^2 - x - 3}{x^2 - 1}$

24. $\lim\limits_{x \to 1} \dfrac{x^3 - 7x + 6}{x^2 + 2x - 3}$

25. $\lim\limits_{x \to -3} \dfrac{x^3 - 7x + 6}{x^2 + 2x - 3}$

26. $\lim\limits_{x \to 1} \dfrac{x^{5/2} - x^{1/2}}{x^{3/2} - x^{1/2}}$

In Exercises 27–32 find the slope of the line tangent to the graph of the given function at the given point, and find an equation for this tangent.

27. The graph of $y = 4 - 2x$ at $(3, -2)$.

28. The graph of $f(x) = 3x^2 + 4$ at $(1, 7)$.

29. The graph of $y = 9 - x^2$ at $(2, 5)$.

30. The graph of $y = \dfrac{1}{x}$ at $(2, \frac{1}{2})$.

31. The graph of $f(x) = \dfrac{3}{x + 2}$ at $(1, 1)$.

32. The graph of $f(x) = (x + 1)^4$ at $(1, 16)$.

33. True or false? If $f(a)$ does not exist then neither does $\lim_{x \to a} f(x)$.

34. According to our definition of limit, $\lim_{x \to 1} \sqrt{x - 1}$ does not exist. Why?

35. Figure 2.6 is a graph of the function $f(x) = x \sin\left(\dfrac{1}{x}\right)$.

   a. Does $\lim_{x \to 0} x \sin\left(\dfrac{1}{x}\right)$ exist? Why?

   b. What can you say about $\lim_{x \to 0} x^2 \sin\left(\dfrac{1}{x}\right)$?

(Compare these results with that of Example 6.)

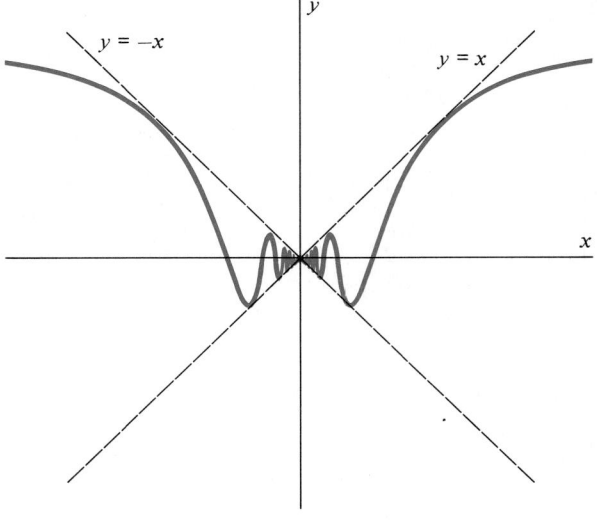

Figure 2.6 $f(x) = x \sin\left(\dfrac{1}{x}\right)$.

2.3 THE FORMAL DEFINITION OF LIMIT

In Section 2.2 we defined the limit of a function informally by saying that

$$L = \lim_{x \to a} f(x)$$

means that we can cause $f(x)$ to be as close to L as desired by choosing x sufficiently close to a.

From the viewpoint of precise mathematics this informal notion of limit is problematic. The difficulty lies in the use of the phrase "close to." A precise mathematical statement can involve constants, variables, equal signs, inequalities, arithmetic operations, and so forth, but not vague references to "closeness."

In order to capture our intuitive notion of limit with precise language, we do the following:

1. In place of the phrase "we can cause $f(x)$ to be as close to L as desired," we use the inequality

$$|f(x) - L| < \epsilon. \tag{1}$$

2. In place of the phrase "by choosing x sufficiently close to a," we use a similar inequality:

$$0 < |x - a| < \delta \tag{2}$$

where δ is a small positive number. (Remember, we do not want to allow $x = a$. That is the reason for the left part of this inequality.)

3. In order to connect these two phrases in the desired way, we say that, no matter what number ϵ is given, we can find a number δ so that if x satisfies inequality (2), then $f(x)$ will satisfy inequality (1). That is, we want to say that

$|x - a|$ small guarantees $|f(x) - L|$ small.

These conventions bring us to our formal definition of limit.

DEFINITION 3

Let $f(x)$ be defined for all x in an open interval containing a, except possibly at a. We say that the number L is the **limit** of the function f as x approaches a, written $L = \lim\limits_{x \to a} f(x)$, if and only if, given any number $\epsilon > 0$ there exists a corresponding number $\delta > 0$ so that

$$\text{if} \quad 0 < |x - a| < \delta, \quad \text{then} \quad |f(x) - L| < \epsilon.$$

In other words, $L = \lim\limits_{x \to a} f(x)$ means that we can cause $f(x)$ to be as close to L as desired (within ϵ units) by choosing x sufficiently close (within δ units) to a. (See Figure 3.1.)

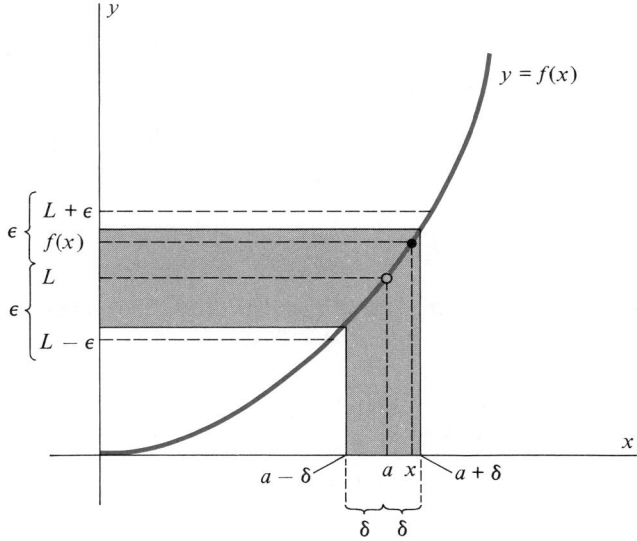

Figure 3.1 If $0 < |x - a| < \delta$, then $|f(x) - L| < \epsilon$.

The following examples show how Definition 3 can be used to prove statements of the form $L = \lim\limits_{x \to a} f(x)$. Before studying these examples you should note the limitations of Definition 3: *given* the number L, a criterion is available for proving that $L = \lim\limits_{x \to a} f(x)$. Definition 3 does not, however, provide any information as to how the number L is found. This is done largely by the techniques of the preceding examples.

Example 1

To illustrate Definition 3, we test the statement

$$\lim_{x \to 3}(2x + 1) = 7.$$

In this case, $f(x) = 2x + 1$, $L = 7$, and $a = 3$. If we are given a number $\epsilon > 0$, the inequality

$$|f(x) - L| < \epsilon$$

in Definition 3 is equivalent to the following inequalities:

$$|(2x + 1) - 7| < \epsilon \tag{3}$$
$$|2x - 6| < \epsilon$$
$$2|x - 3| < \epsilon$$

$$|x - 3| < \frac{\epsilon}{2} \tag{4}$$

Since $a = 3$, inequality (4) shows how small the number $|x - a| = |x - 3|$ must be made in order to "force" the equivalent inequality (3) to be true. For example,

(a) if $\epsilon = 0.5$ is given, we choose any x satisfying

$$0 < |x - 3| < \frac{\epsilon}{2} = \frac{0.5}{2} = 0.25.$$

Then inequality (4) holds. Since inequalities (4) and (3) are equivalent, this shows that

if $\quad 0 < |x - 3| < 0.25, \quad$ then $\quad |(2x + 1) - 7| < 0.5.$

(b) If $\epsilon = 0.1$ is given, we choose x "sufficiently close" to $a = 3$ so that

$$0 < |x - 3| < \frac{\epsilon}{2} = \frac{0.1}{2} = 0.05.$$

By the equivalence of inequalities (3) and (4), we conclude that if $0 < |x - 3| < 0.05$, then $|(2x + 1) - 7| < 0.1 (= \epsilon)$.   ◇

Example 2

Prove that $\lim\limits_{x \to 3}(2x + 1) = 7$.

Solution: In order to prove this limit we must use Definition 3, which requires us to show that, given any number $\epsilon > 0$, we can say how to find a number δ so that

if $\quad 0 < |x - 3| < \delta, \quad$ then $\quad |(2x + 1) - 7| < \epsilon.$

The calculations in Example 1 show that the inequality

$$|(2x + 1) - 7| < \epsilon \tag{5}$$

is equivalent to the inequality

$$|x - 3| < \frac{\epsilon}{2}. \tag{6}$$

This shows how to choose δ: with $\delta = \dfrac{\epsilon}{2}$, we will know that if $0 < |x - 3| < \delta$, then inequality (6) is true. Hence, so is inequality (5). Formally, the proof goes like this:

Let $\epsilon > 0$ be given. Choose $\delta = \epsilon/2$. It follows that if $0 < |x - 3| < \delta$ then

$$|(2x + 1) - 7| = |2x - 6|$$
$$= 2|x - 3|$$
$$< 2 \cdot \delta \quad \text{(since } |x - 3| < \delta)$$

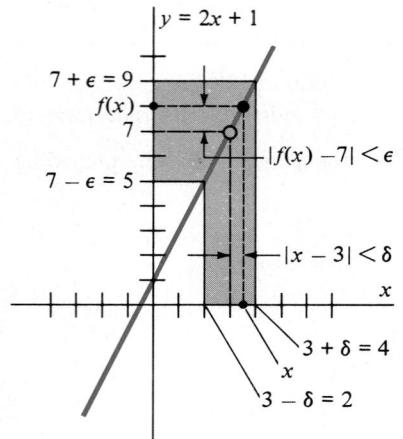

Figure 3.2 $\lim\limits_{x \to 3}(2x + 1) = 7$. Case $\epsilon = 2$, $\delta = \epsilon/2 = 1$.

$$= 2\left(\frac{\epsilon}{2}\right) \qquad \left(\text{since } \delta = \frac{\epsilon}{2}\right)$$
$$= \epsilon.$$

That is, if $0 < |x - 3| < \delta$, then $|(2x + 1) - 7| < \epsilon$ as required by Definition 3. (See Figure 3.2.) ◇

Example 3

Prove that $\lim\limits_{x \to 2}(4x + 3) = 11$.

Strategy

Identify $f(x)$, L, and a.

Beginning with the inequality

$|f(x) - L| < \epsilon$

try to find an equivalent inequality involving $|x - a|$.

Solution

Here $f(x) = 4x + 3$, $L = 11$, and $a = 2$. We have the following equivalent inequalities:

$$|f(x) - L| < \epsilon \tag{7}$$
$$|(4x + 3) - 11| < \epsilon$$
$$|4x - 8| < \epsilon$$
$$4|x - 2| < \epsilon$$
$$|x - 2| < \frac{\epsilon}{4}. \tag{8}$$

We will therefore want to take $\delta = \epsilon/4$, since the equivalence of inequalities (7) and (8) shows that if $|x - 2| < \delta = \dfrac{\epsilon}{4}$, then $|(4x + 3) - 11| < \epsilon$.

The formal proof:

(i) Take $\epsilon > 0$ arbitrary

(ii) Choose $\delta = \dfrac{\epsilon}{4}$

(iii) Show that

$0 < |x - a| < \delta$

guarantees that

$|f(x) - L| < \epsilon$.

Here is the proof:

Given $\epsilon > 0$, choose $\delta = \dfrac{\epsilon}{4}$. It follows that if

$$0 < |x - 2| < \delta$$

then

$$\begin{aligned} |(4x + 3) - 11| &= |4x - 8| \\ &= 4|x - 2| \\ &< 4 \cdot \delta \qquad (\text{since } |x - 2| < \delta) \\ &= 4\left(\frac{\epsilon}{4}\right) \\ &= \epsilon. \end{aligned}$$

Thus, with $\delta = \dfrac{\epsilon}{4}$, we have that

if $\quad 0 < |x - 2| < \delta, \quad$ then $\quad |(4x + 3) - 11| < \epsilon$. ◇

For linear functions, such as those in Examples 1–3, the relationship between δ and ϵ in Definition 3 is determined by the slope of the graph of the function. This is because the slope determines the rate at which $f(x)$ changes as x changes.

For nonlinear functions, proofs of particular limits are, in general, more complicated because the relationship between δ and ϵ also depends on the number a at which the limit is being evaluated. The next two examples concern nonlinear functions for which the particular limits in question can be proved in a straightforward manner. Beyond these special types of examples we shall not pursue further the topic of proving individual limits. Rather, we shall determine a set of *properties* of limits, from which individual limits can be deduced.

Example 4

Prove that $\lim\limits_{x \to 2}(x^2 - 4x + 7) = 3$.

Solution: In this limit we have $f(x) = x^2 - 4x + 7$, $a = 2$, and $L = 3$. Thus, if ϵ is a given positive number, the following inequalities are equivalent:

$$|f(x) - L| < \epsilon$$
$$|(x^2 - 4x + 7) - 3| < \epsilon \tag{9}$$
$$|x^2 - 4x + 4| < \epsilon$$
$$|(x - 2)^2| < \epsilon$$
$$|x - 2| < \sqrt{\epsilon}. \tag{10}$$

The equivalence of inequalities (9) and (10) shows that if we take $\delta = \sqrt{\epsilon}$, it follows that

$$\text{if} \quad 0 < |x - 2| < \delta, \qquad \text{then} \quad |(x^2 - 4x + 7) - 3| < \epsilon.$$

This proves that $\lim\limits_{x \to 2}(x^2 - 4x + 7) = 3$, according to Definition 3. ◇

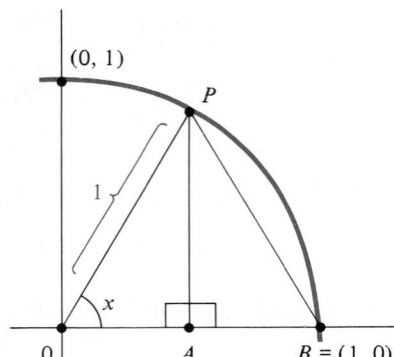

Figure 3.3 $|\sin x| = $ length of AP
$\leq$ length of BP
$\leq$ length of arc BP
$= |x|$, $\quad |x| < \pi/2$.

Example 5

Figure 3.3 illustrates the trigonometric identity

$$|\sin x| \leq |x|, \quad -\infty < x < \infty. \tag{11}$$

We may use this inequality to prove that

$$\lim_{x \to 0} \sin x = 0. \tag{12}$$

In this example we have $f(x) = \sin x$ and $a = L = 0$. Thus, if ϵ is a given positive number, the inequality

$$|f(x) - L| < \epsilon$$

is

$$|\sin x - 0| < \epsilon. \tag{13}$$

From inequalities (11) and (13) we may conclude that

$$\text{if} \quad |x| < \epsilon, \qquad \text{then} \qquad |\sin x| = |\sin x - 0| \leq |x| < \epsilon. \tag{14}$$

Since, in this example, $|x - a| = |x - 0| = |x|$, inequality (14) shows that if we take $\delta = \epsilon$, it follows that

$$\text{if} \quad 0 < |x - 0| < \delta, \qquad \text{then} \qquad |\sin x - 0| < \epsilon.$$

This proves equation (12) according to Definition 3. ◇

Exercise Set 2.3

Each of Exercises 1–6 present a limit of the form $L = \lim\limits_{x \to a} f(x)$.

For each given ϵ, find a number δ so that if $0 < |x - a| < \delta$ then $|f(x) - L| < \epsilon$. (These numbers δ are not unique.)

1. $\lim\limits_{x \to 2} (2x + 5) = 9$

   a. $\epsilon = 2$
   b. $\epsilon = 0.4$
   c. $\epsilon = 0.05$

2. $\lim\limits_{x \to -3} (1 - 4x) = 13$

   a. $\epsilon = 2$
   b. $\epsilon = 0.4$
   c. $\epsilon = 0.1$

3. $\lim\limits_{x \to 0} (x^2 + 3) = 3$

   a. $\epsilon = 2$
   b. $\epsilon = 1$
   c. $\epsilon = 0.3$

4. $\lim\limits_{x \to -2} \dfrac{1}{x - 2} = -\dfrac{1}{4}$

   a. $\epsilon = 2$
   b. $\epsilon = 1$
   c. $\epsilon = 0.5$

5. $\lim\limits_{x \to 1} \dfrac{x^2 - 1}{x - 1} = 2$

   a. $\epsilon = 2$
   b. $\epsilon = 0.8$
   c. $\epsilon = 0.05$

6. $\lim\limits_{x \to 4} \sqrt{2x + 1} = 3$

   a. $\epsilon = 0.5$
   b. $\epsilon = 0.2$
   c. $\epsilon = 0.05$

In Exercises 7–16 use Definition 3 to prove the stated limit.

7. $\lim\limits_{x \to 3} (x + 3) = 6$

8. $\lim\limits_{x \to 1} (2x - 3) = -1$

9. $\lim\limits_{x \to 4} (7 - 3x) = -5$

10. $\lim\limits_{x \to 2} (3x + 5) = 11$

11. $\lim\limits_{x \to 3} \dfrac{x^2 - 9}{x - 3} = 6$

12. $\lim\limits_{x \to 2} \dfrac{x^2 - 4}{x - 2} = 4$

13. $\lim\limits_{x \to 3} x^2 = 9$

14. $\lim\limits_{x \to 2} 3x^2 = 12$

15. $\lim\limits_{x \to 1} (x^2 - 2x + 4) = 3$

16. $\lim\limits_{x \to 3} (x^2 - 6x + 13) = 4$

17. Use the technique of Example 5 to prove that $\lim\limits_{x \to 0} \sin 2x = 0$.

18. Use Definition 3 to prove that $\lim\limits_{x \to 0} |x| = 0$.

19. Use the trigonometric inequality $|\sin x| \le |x|$ to prove that $\lim\limits_{x \to 0} x \sin x = 0$. (See Example 5.)

2.4 PROPERTIES OF LIMITS

The purpose of this section is to establish several properties of limits that will enable us to evaluate more complicated limits from two simple observations:

$$\lim_{x \to a} c = c, \qquad c \text{ constant} \tag{1}$$

and

$$\lim_{x \to a} x = a. \tag{2}$$

While each of these statements can be proved using the formal definition of limit (see Exercises 46 and 47), you may also regard them as sufficiently obvious to be accepted without proof. Statement (1) says that the limit of the constant function with values $f(x) = c$ for all x, is always the number c, regardless of the number a. Statement (2) says that the limit of the linear function $g(x) = x$ as x approaches a is the function value $a = g(a)$.

The following theorem establishes an ''algebra'' of limits, by which limits of sums, multiples, products, and quotients of functions can be calculated from the limits of the individual terms and factors. Its proof is given at the end of this section.

THEOREM 1

Assume that $\lim\limits_{x \to a} f(x) = L$ and $\lim\limits_{x \to a} g(x) = M$ both exist. Let c be any number. Then each of the following limits exists, with the value indicated:

(i) $\lim\limits_{x \to a} [f(x) + g(x)] = \lim\limits_{x \to a} f(x) + \lim\limits_{x \to a} g(x) = L + M$

(ii) $\lim\limits_{x \to a} [cf(x)] = c\left[\lim\limits_{x \to a} f(x)\right] = cL$

(iii) $\lim\limits_{x \to a}[f(x)g(x)] = \left[\lim\limits_{x \to a} f(x)\right] \cdot \left[\lim\limits_{x \to a} g(x)\right] = LM$

(iv) $\lim\limits_{x \to a} \dfrac{f(x)}{g(x)} = \dfrac{\lim\limits_{x \to a} f(x)}{\lim\limits_{x \to a} g(x)} = \dfrac{L}{M}$,   provided $M \neq 0$

Example 1

Find $\lim\limits_{x \to 2}(3x^2 + 6)$.

Solution:

$$\lim\limits_{x \to 2}(3x^2 + 6) = \lim\limits_{x \to 2}(3x^2) + \lim\limits_{x \to 2} 6 \qquad \text{(part i)}$$

$$= 3 \lim\limits_{x \to 2} x^2 + \lim\limits_{x \to 2} 6 \qquad \text{(part ii)}$$

$$= 3(\lim\limits_{x \to 2} x)(\lim\limits_{x \to 2} x) + \lim\limits_{x \to 2} 6 \qquad \text{(part iii)}$$

$$= 3 \cdot 2 \cdot 2 + 6 \qquad \text{(equations (1), (2))}$$

$$= 18.$$
◇

Example 2

Find $\lim\limits_{x \to -1}\left(\dfrac{2x + 3}{1 + x^2}\right)$.

Solution: Since the limit of the denominator is

$$\lim\limits_{x \to -1}(1 + x^2) = \lim\limits_{x \to -1} 1 + \lim\limits_{x \to -1} x^2 \qquad \text{(part i)}$$

$$= \lim\limits_{x \to -1} 1 + (\lim\limits_{x \to -1} x)(\lim\limits_{x \to -1} x) \qquad \text{(part iii)}$$

$$= 1 + (-1)(-1) \qquad \text{(equations (1), (2))}$$

$$= 2$$

which is nonzero, we may apply part (iv) to conclude that

$$\lim\limits_{x \to -1}\left(\dfrac{2x + 3}{1 + x^2}\right) = \dfrac{\lim\limits_{x \to -1}(2x + 3)}{\lim\limits_{x \to -1}(1 + x^2)} \qquad \text{(part iv)}$$

$$= \dfrac{2(\lim\limits_{x \to -1} x) + \lim\limits_{x \to -1} 3}{2} \qquad \text{(parts i, ii)}$$

$$= \dfrac{2(-1) + 3}{2}$$

$$= \dfrac{1}{2}.$$
◇

Equation (2), part (iii) of Theorem 1, and a straightforward application of mathematical induction give the following extension of Theorem 1 to integral powers of x and "power functions." (See Exercises 38 and 39.)

| THEOREM 2 | Assume that $\lim\limits_{x \to a} f(x) = L$ exists. For any positive integer $n = 1, 2, \ldots$ |

(i) $\lim\limits_{x \to a} x^n = a^n$

(ii) $\lim\limits_{x \to a} [f(x)]^n = [\lim\limits_{x \to a} f(x)]^n = L^n$.

The following examples show how Theorems 1 and 2 can be used to evaluate limits of polynomials, rational functions, and powers of these kinds of functions. We leave it as an exercise for you to verify which parts of Theorems 1 and 2 are used in each step.

Example 3

Find $\lim\limits_{x \to 2}(3x^4 + 7x^2 + 4x)$.

Solution:

$$\lim_{x \to 2}(3x^4 + 7x^2 + 4x) = 3(\lim_{x \to 2} x^4) + 7(\lim_{x \to 2} x^2) + 4(\lim_{x \to 2} x)$$
$$= 3(2)^4 + 7(2)^2 + 4(2)$$
$$= 3 \cdot 16 + 7 \cdot 4 + 4 \cdot 2$$
$$= 84.$$

$\diamond$

Example 4

Find $\lim\limits_{x \to -2}(x^{-3} - 3x^{-2} + 5x^3)$.

Solution:

$$\lim_{x \to -2}(x^{-3} - 3x^{-2} + 5x^3) = \lim_{x \to -2}\left(\frac{1}{x^3} + \frac{-3}{x^2} + 5x^3\right)$$
$$= \frac{1}{(-2)^3} + \frac{-3}{(-2)^2} + 5(-2)^3$$
$$= \frac{1}{-8} + \frac{-3}{4} + 5(-8)$$
$$= \frac{-1 - 6 - 320}{8}$$
$$= -\frac{327}{8}.$$

$\diamond$

Example 5

Find $\lim\limits_{x \to 2}(x^3 - 7x + 1)^3$.

Solution:

$$\lim_{x \to 2}(x^3 - 7x + 1)^3 = \left[\lim_{x \to 2}(x^3 - 7x + 1)\right]^3$$
$$= [2^3 - 7(2) + 1]^3$$
$$= (-5)^3$$
$$= -125.$$

◇

The next theorem enables us to calculate limits of fractional powers of x. Its proof is given in Appendix II.

THEOREM 3

Let m and n be positive integers. Then

(i) If m is even, $\displaystyle\lim_{x \to a} x^{n/m} = a^{n/m}$ for $0 < a < \infty$

(ii) If m is odd $\displaystyle\lim_{x \to a} x^{n/m} = a^{n/m}$ for $-\infty < a < \infty$.

Example 6

Find $\displaystyle\lim_{x \to 4}(3\sqrt{x} + x^{-3/2})$.

Solution: Using Theorems 1 and 3 we find that

$$\lim_{x \to 4}(3\sqrt{x} + x^{-3/2}) = \lim_{x \to 4}(3x^{1/2} + x^{-3/2})$$

$$= 3 \lim_{x \to 4} x^{1/2} + \frac{1}{\displaystyle\lim_{x \to 4} x^{3/2}} \qquad \text{(Theorem 1)}$$

$$= 3(4)^{1/2} + \frac{1}{(4)^{3/2}} \qquad \text{(Theorem 3)}$$

$$= 3 \cdot 2 + \frac{1}{8}$$

$$= \frac{49}{8}.$$

◇

Our final theorem is sometimes called the "Pinching Theorem." This name arises from its geometric interpretation as illustrated in Figure 4.1. Its proof is given in Appendix II.

THEOREM 4

Assume that $\displaystyle\lim_{x \to a} g(x)$ and $\displaystyle\lim_{x \to a} h(x)$ exist, and that $\displaystyle\lim_{x \to a} g(x) = L = \lim_{x \to a} h(x)$. If the function f satisfies the inequality

$$g(x) \le f(x) \le h(x)$$

for all x in an open interval containing a (except possibly at $x = a$), then $\displaystyle\lim_{x \to a} f(x) = L$ also.

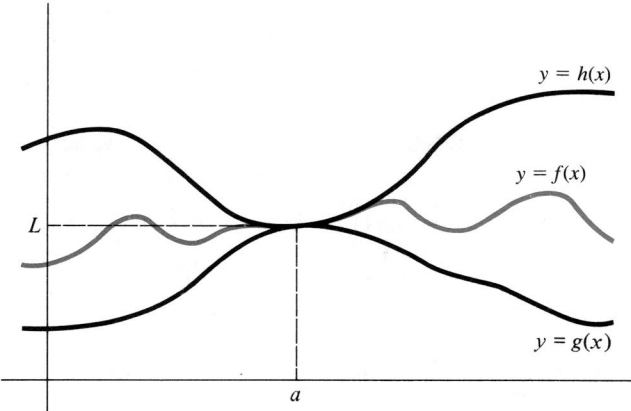

Figure 4.1 The Pinching Theorem.

Example 7

We may use the Pinching Theorem to show that

$$\lim_{x \to 0} x \sin \frac{1}{x} = 0. \tag{3}$$

To do so we first show that

$$\lim_{x \to 0} \left| x \sin \frac{1}{x} \right| = 0. \tag{4}$$

To prove limit (4) we note that since $|\sin u| \le 1$ for all u, we have the inequality

$$0 \le \left| x \sin \frac{1}{x} \right| = |x| \cdot \left| \sin \frac{1}{x} \right| \le |x| \cdot 1 = |x|.$$

Thus,

$$0 \le \left| x \sin \frac{1}{x} \right| \le |x|, \qquad -\infty < x < \infty. \tag{5}$$

Since $\lim_{x \to 0} |x| = 0$ (Exercise 18, Section 2.3), limit (4) follows from inequality (5) and the Pinching Theorem.

To prove limit (3) we use the inequality

$$-\left| x \sin \frac{1}{x} \right| \le x \sin \frac{1}{x} \le \left| x \sin \frac{1}{x} \right|.$$

Limit (3) now follows from this inequality, limit (4), and the Pinching Theorem. ◇

Example 8

Prove that $\lim_{x \to 0} \cos x = 1.$

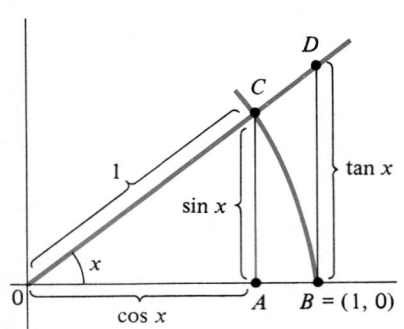

Figure 4.2

Solution: We use the Pinching Theorem and the fact that if $0 < a < 1$, then $0 < a < \sqrt{a} < 1$. On the interval $(-\pi/2, \pi/2)$, we have $\sin x < 1$ and $\cos x > 0$, so

$$1 \geq \cos x = \sqrt{1 - \sin^2 x} > 1 - \sin^2 x, \qquad -\pi/2 < x < \pi/2.$$

That is,

$$1 - \sin^2 x < \cos x \leq 1, \qquad -\pi/2 < x < \pi/2. \tag{6}$$

Since $\lim_{x \to 0}(1 - \sin^2 x) = 1 - (\lim_{x \to 0} \sin x)^2 = 1 - 0 = 1$, it follows from inequality (6) and the Pinching Theorem that $\lim_{x \to 0} \cos x = 1$. ◇

In our next example we return to the problem of calculating $\lim_{x \to 0} \dfrac{\sin x}{x}$. (See Example 1, Section 2.) The details of this example require an understanding of Figure 4.2, which concerns an arc of the unit circle subtended by an angle of size x radians, with $0 < x < \pi/2$. The following observations will help you with these details:

(a) The lengths of OC and OB are 1, since each segment is a radius for the unit circle.

(b) $\sin x = \dfrac{\text{length of } AC}{\text{length of } OC} = \text{length of } AC$.

(c) $\cos x = \dfrac{\text{length of } OA}{\text{length of } OC} = \text{length of } OA$.

(d) $\tan x = \dfrac{\text{length of } DB}{\text{length of } OB} = \text{length of } DB$.

Example 9

Show that $\lim_{x \to 0} \dfrac{\sin x}{x} = 1$.

Solution: We assume first that $0 < x < \pi/2$. From Figure 4.2 we observe that

$$\text{Area of triangle } OAC \leq \text{Area of sector } OBC \leq \text{Area of triangle } OBD. \tag{7}$$

Since

$$\text{Area of triangle } OAC = \frac{1}{2} \cos x \cdot \sin x$$

$$\text{Area of sector } OBC = \frac{1}{2}x \qquad \text{(it's } \frac{x}{2\pi} \cdot \pi r^2 = \frac{1}{2}xr^2 \text{ with } r = 1\text{)}$$

$$\text{Area of triangle } OBD = \frac{1}{2} \tan x$$

area of entire circle

fractional part of entire circle

inequality (7) becomes

$$\frac{1}{2} \cos x \cdot \sin x \leq \frac{1}{2}x \leq \frac{1}{2} \tan x.$$

Multiplying by 2 and dividing by $\sin x$ $\left(\text{which is nonzero since } 0 < x < \dfrac{\pi}{2}\right)$ gives

$$\cos x \leq \frac{x}{\sin x} \leq \frac{1}{\cos x}.$$

Inverting all terms then gives

$$\frac{1}{\cos x} \geq \frac{\sin x}{x} \geq \cos x \tag{8}$$

for $0 < x < \pi/2$. Similarly, for $-\pi/2 < x < 0$, inequality (8) also holds (see Exercise 49). Inequality (8) is therefore the situation described by the Pinching Theorem, with $g(x) = \dfrac{1}{\cos x}$, $f(x) = \dfrac{\sin x}{x}$, and $h(x) = \cos x$. Since, by Example 8,

$$\lim_{x \to 0} \frac{1}{\cos x} = 1 = \lim_{x \to 0} \cos x$$

it follows from the Pinching Theorem that

$$\lim_{x \to 0} \frac{\sin x}{x} = 1. \qquad \diamond$$

Example 10

Find $\displaystyle\lim_{x \to 0} \frac{\sin 2x}{x}$.

Strategy

Try to rewrite the function $\dfrac{\sin 2x}{x}$ with a factor $\dfrac{\sin x}{x}$.

Use Theorem 1 and facts that

$$\lim_{x \to 0} \frac{\sin x}{x} = 1$$

$$\lim_{x \to 0} \cos x = 1.$$

Solution

Using the identity $\sin 2x = 2 \sin x \cos x$, we obtain

$$\begin{aligned}
\lim_{x \to 0} \frac{\sin 2x}{x} &= \lim_{x \to 0} \frac{2 \sin x \cos x}{x} \\
&= \lim_{x \to 0} \left[2 \left(\frac{\sin x}{x} \right) \cos x \right] \\
&= 2 \left(\lim_{x \to 0} \frac{\sin x}{x} \right) \left(\lim_{x \to 0} \cos x \right) \\
&= 2 \cdot 1 \cdot 1 \\
&= 2. \qquad \diamond
\end{aligned}$$

Proof of Theorem 1

(Part i) To prove that $\displaystyle\lim_{x \to a} [f(x) + g(x)] = L + M$, using the formal definition of limit, we must show that we can make $|[f(x) + g(x)] - (L + M)|$ small by choosing x with $|x - a|$ sufficiently small. We do this by comparing

$$|[f(x) + g(x)] - (L + M)|$$

with the differences $|f(x) - L|$ and $|g(x) - M|$ as follows:

$$\begin{aligned}
|[f(x) + g(x)] - (L + M)| &= |[f(x) - L] + [g(x) - M]| \tag{9} \\
&\leq |f(x) - L| + |g(x) - M|.
\end{aligned}$$

Inequality (9) shows that the left side will be smaller than a given number ϵ whenever *each* of the two terms on the right is smaller than $\epsilon/2$. This can be accomplished by choosing $|x - a|$ sufficiently small because $\displaystyle\lim_{x \to a} f(x) = L$ and $\displaystyle\lim_{x \to a} g(x) = M$.

Here are the details:

Given $\epsilon > 0$, since $\lim_{x \to a} f(x) = L$ there exists a number δ_1 so that

$$\text{if} \quad 0 < |x - a| < \delta_1, \qquad \text{then} \quad |f(x) - L| < \frac{\epsilon}{2}. \tag{10}$$

Also, since $\lim_{x \to a} g(x) = M$, there exists a (possibly different) number δ_2 so that

$$\text{if} \quad 0 < |x - a| < \delta_2, \qquad \text{then} \quad |g(x) - M| < \frac{\epsilon}{2}. \tag{11}$$

Now let δ be the smaller of the numbers δ_1 and δ_2. Then,

if $\quad 0 < |x - a| < \delta,$
then $\quad both \quad 0 < |x - a| < \delta_1,$
and $\quad 0 < |x - a| < \delta_2,$

so by inequality (9) and statements (10) and (11) it follows that

$$\begin{aligned}
|[f(x) + g(x)] - (L + M)| &= |[f(x) - L] + [g(x) - M]| \\
&\leq |f(x) - L| + |g(x) - M| \\
&< \frac{\epsilon}{2} + \frac{\epsilon}{2} \\
&= \epsilon.
\end{aligned}$$

This proves that $\lim_{x \to a} [f(x) + g(x)] = L + M$. ◆

(Part ii) First note that if $c = 0$, the statement $\lim_{x \to a} cf(x) = cL$ is just $\lim_{x \to a} 0 = 0$, which is obviously true. We may therefore assume $c \neq 0$ is what follows. To prove that $\lim_{x \to a} cf(x) = cL$, we must show that we can make the difference $|cf(x) - cL|$ small by choosing x with $|x - a|$ small. Since

$$\begin{aligned}
|cf(x) - cL| &= |c[f(x) - L]| \\
&= |c| \cdot |f(x) - L|
\end{aligned} \tag{12}$$

we can make the left side of this equation smaller than a given number ϵ by requiring that the factor $|f(x) - L|$ on the right side be less than $\frac{\epsilon}{|c|}$. (Remember, $c \neq 0$.)

Here is the proof:

Given $\epsilon > 0$, since $\lim_{x \to a} f(x) = L$ there exists a number $\delta > 0$ so that

$$\text{if} \quad 0 < |x - a| < \delta, \qquad \text{then} \quad |f(x) - L| < \frac{\epsilon}{|c|}. \tag{13}$$

Thus, if $0 < |x - a| < \delta$, it follows from equation (12) and line (13) that

$$\begin{aligned}
|cf(x) - cL| &= |c[f(x) - L]| \\
&= |c| \cdot |f(x) - L| \\
&< |c| \left(\frac{\epsilon}{|c|} \right) \\
&= \epsilon.
\end{aligned}$$

This shows that $\lim_{x \to a} cf(x) = cL$. ◆

The proofs of parts (iii) and (iv) are similar, but slightly more complicated, and are found in Appendix II.

Exercise Set 2.4

In Exercises 1–22 use Theorems 1–3 to find the limit.

1. $\lim\limits_{x \to 3}(3x - 7)$

2. $\lim\limits_{x \to 2}(x^3 - 4x + 1)$

3. $\lim\limits_{x \to 2} \dfrac{x^2 + 3x}{x - 3}$

4. $\lim\limits_{x \to 5} \dfrac{x^2 - 5x}{x - 5}$

5. $\lim\limits_{x \to 4} \sqrt{x}(1 - x^2)$

6. $\lim\limits_{x \to \pi/4} \sin x \tan x$

7. $\lim\limits_{x \to \pi/4} \dfrac{\sin x}{\tan x}$

8. $\lim\limits_{x \to 2}(4\sqrt{x} - 5x^2)$

9. $\lim\limits_{x \to 1}(3x^9 - \sqrt[3]{x})$

10. $\lim\limits_{x \to 2} \dfrac{x^3 - 6x + 5}{x^2 + 2x + 2}$

11. $\lim\limits_{x \to 3} \dfrac{x^2 - 2x - 3}{x - 3}$

12. $\lim\limits_{x \to 3} \dfrac{x^2 - 10x + 21}{x - 3}$

13. $\lim\limits_{x \to 4} \dfrac{x^{3/2} + 2\sqrt{x}}{x^{5/2} + \sqrt{x}}$

14. $\lim\limits_{x \to 0} \dfrac{\sin 2x}{\sin x}$

15. $\lim\limits_{x \to 0} \dfrac{\tan x}{\sin 2x}$

16. $\lim\limits_{x \to 1} \dfrac{x^3 + 2x^2 + 2x - 5}{x^2 - 1}$

17. $\lim\limits_{x \to -8} \dfrac{x^{2/3} - x}{x^{5/3}}$

18. $\lim\limits_{x \to 4} \dfrac{x^{3/2} - x^{5/2}}{4 + \sqrt{x}}$

19. $\lim\limits_{x \to 4} \dfrac{x - 4}{\sqrt{x} - 2}$

20. $\lim\limits_{x \to -3} \dfrac{x^2 - 6x - 27}{x^2 + 3x}$

21. $\lim\limits_{x \to -1} (x^{7/3} - 2x^{2/3})^2$

22. $\lim\limits_{x \to -1} \dfrac{x^3 + x^2 - x - 1}{x^3 + 2x^2 + 2x + 1}$

In Exercises 23–26 use the facts that $\lim\limits_{x \to a} f(x) = 2$, $\lim\limits_{x \to a} g(x) = -3$, and $\lim\limits_{x \to a} h(x) = 5$ to find the limit.

23. $\lim\limits_{x \to a} 3 \cdot f(x) \cdot g(x)$

24. $\lim\limits_{x \to a} \dfrac{f(x) - g(x)}{[h(x)]^2}$

25. $\lim\limits_{x \to a} \dfrac{6f(x) - 4[g(x)]^2}{g(x) - 4f(x)}$

26. $\lim\limits_{x \to a}[f(x) + g(x)]^3$

In Exercises 27–32 use the fact that $\lim\limits_{x \to 0} \dfrac{\sin x}{x} = 1$ to find the limit.

27. $\lim\limits_{x \to 0} \dfrac{\sin x}{2x}$

28. $\lim\limits_{x \to 0} \dfrac{3 \sin x}{x}$

29. $\lim\limits_{x \to 0} \dfrac{\tan x}{4x}$

30. $\lim\limits_{x \to 0} \dfrac{\sec x \tan x}{x}$

31. $\lim\limits_{x \to 0} x^2 \cot x$

32. $\lim\limits_{x \to 0} \dfrac{\sin^2 x}{x}$

33. Suppose that $1 - x^2 \le f(x) \le 1 + x^2$ for all x. Find $\lim\limits_{x \to 0} f(x)$.

34. Suppose that $6x - x^2 \le f(x) \le x^2 - 6x + 18$ for all x. Find $\lim\limits_{x \to 3} f(x)$.

35. Suppose that $1 - x^4 \le f(x) \le \sec x$ for $x \in (-\pi/2, \pi/2)$. Find $\lim\limits_{x \to 0} f(x)$.

36. Use Theorem 1, part (i), to show that

$$\lim\limits_{x \to a}[f(x) + g(x) + h(x)] = \lim\limits_{x \to a} f(x) + \lim\limits_{x \to a} g(x) + \lim\limits_{x \to a} h(x)$$

provided each of the limits on the right side exists. (*Hint:* Apply Theorem 1 twice.)

37. Use Theorem 1, part (iii), to show that

$$\lim\limits_{x \to a}[f(x)g(x)h(x)] = (\lim\limits_{x \to a} f(x))(\lim\limits_{x \to a} g(x))(\lim\limits_{x \to a} h(x))$$

provided each of the limits on the right exists.

38. Use mathematical induction (see Appendix II) and Theorem 1 to prove that $\lim\limits_{x \to a} x^n = a^n$ as follows:

a. Explain why $\lim\limits_{x \to a} x^n = a^n$ when $n = 1$.

b. Assume that $\lim\limits_{x \to a} x^m = a^m$ for a positive integer m and explain why it follows that

$$\lim\limits_{x \to a} x^{m+1} = \lim\limits_{x \to a} x \cdot x^m = (\lim\limits_{x \to a} x)(\lim\limits_{x \to a} x^m)$$

$$= a \cdot a^m$$

$$= a^{m+1}.$$

c. Conclude that $\lim\limits_{x \to a} x^n = a^n$ for all positive integers n.

39. Use mathematical induction and Theorem 1 to prove that if $\lim\limits_{x \to a} f(x) = L$, then $\lim\limits_{x \to a}[f(x)]^n = L^n$ for each positive integer n. (*Hint:* See Exercise 38.)

40. Use Theorem 1, together with mathematical induction, to prove that for the polynomial $Q(x) = a_n x^n + a_{n-1}x^{n-1} + \cdots + a_1 x + a_0$,

$$\lim\limits_{x \to c} Q(x) = a_n c^n + a_{n-1}c^{n-1} + \cdots + a_1 c + a_0.$$

41. Give an example of a function f for which $\lim\limits_{x \to a} f(x) = L$ and $\lim\limits_{x \to b} f(x) = L$ with $a \ne b$. This shows that a function can have the same limit at more than one number x.

42. Let $f(x) = \dfrac{|x|}{x}$ and $g(x) = -\dfrac{|x|}{x}$, $x \neq 0$.

    a. Find the function $h = f + g$.

    b. Show that $\lim\limits_{x \to 0} h(x) = \lim\limits_{x \to 0}[f(x) + g(x)]$ exists, but that neither $\lim\limits_{x \to 0} f(x)$ nor $\lim\limits_{x \to 0} g(x)$ exists.

43. Find two functions f and g so that $\lim\limits_{x \to 0}[f(x)g(x)]$ exists, but either $\lim\limits_{x \to 0} f(x)$ or $\lim\limits_{x \to 0} g(x)$ fails to exist. (See Exercise 42.)

44. Is it possible to find a function f and a constant c so that $\lim\limits_{x \to a} cf(x)$ exists but $\lim\limits_{x \to a} f(x)$ does not exist? What if we require $c \neq 0$?

45. Find the number(s) a so that $\lim\limits_{x \to a}(x^2 - 2x - 5) = 10$.

46. Use the formal definition of limit to prove that $\lim\limits_{x \to a} c = c$ for any constant c. (*Hint:* Given $\epsilon > 0$, *any* number $\delta > 0$ will work.)

47. Use the formal definition of limit to prove that $\lim\limits_{x \to a} x = a$ for any number a. (*Hint:* Given $\epsilon > 0$, choose $\delta = \epsilon$.)

48. Show that $\lim\limits_{x \to 0} \dfrac{1 - \cos x}{x} = 0$ as follows:

    a. Write

$$\frac{1 - \cos x}{x} = \frac{1 - \cos x}{x}\left(\frac{1 + \cos x}{1 + \cos x}\right)$$

$$= \frac{1 - \cos^2 x}{x(1 + \cos x)}.$$

    b. Show that $\dfrac{1 - \cos x}{x} = \dfrac{\sin^2 x}{x(1 + \cos x)}$.

    c. Use the facts that $\lim\limits_{x \to 0} \sin x = 0$ and $\lim\limits_{x \to 0} \cos x = 1$ to obtain the desired limit.

49. Show that inequality (8) is true for $-\pi/2 < x < 0$ by using the fact that it is true for $0 < x < \pi/2$ and the identities $\sin(-x) = -\sin x$ and $\cos(-x) = \cos x$.

50. Use Theorems 1 and 3 to prove that, for $n/m < 0$,

    (i) if m is even, $\lim\limits_{x \to a} x^{n/m} = a^{n/m}$, $0 < a < \infty$

    (ii) if m is odd, $\lim\limits_{x \to a} x^{n/m} = a^{n/m}$, $a \neq 0$.

2.5 ONE-SIDED LIMITS

We have defined the statement $L = \lim\limits_{x \to a} f(x)$ to mean that $|f(x) - L|$ is small whenever $|x - a|$ is small (and nonzero) *regardless of whether x lies to the right of a or to the left*. Because of this "two-sided" property of limits, we say that $\lim\limits_{x \to a} f(x)$ does not exist for the function f whose graph is shown in Figure 5.1.

This graph, however, has the property that as x is chosen close to a *on the right*, the corresponding function values $f(x)$ are found close to the number L. This is the concept of **right-hand limit,** written $\lim\limits_{x \to a^+} f(x) = L$. Similarly, we write $\lim\limits_{x \to a^-} f(x) = M$ to mean that the values $f(x)$ are close to M when x is close to a *on the left*. The superscript $+$ or $-$ on the number a means that we consider only x's lying on the right $(+)$ or left $(-)$ side of a, respectively.

Intuitively, the statement

$$L = \lim\limits_{x \to a^+} f(x)$$

means that we can cause $f(x)$ to be as close to L as desired by choosing x *on the right side of a* and sufficiently close to a.

Similarly, the statement

$$M = \lim\limits_{x \to a^-} f(x)$$

means that we can cause $f(x)$ to be as close to M as desired by choosing x *on the left side of a* and sufficiently close to a. (See Figure 5.1.)

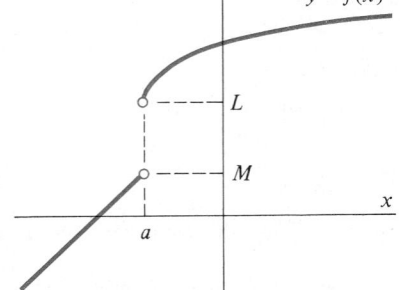

Figure 5.1 $f(x) \to L$ for $x > a$; $f(x) \to M$ for $x < a$.

Example 1

Because the square root function $f(x) = \sqrt{x}$ is not defined for $x < 0$, $\lim\limits_{x \to 0} \sqrt{x}$ does not exist. We may write, however, that

$$\lim_{x \to 0^+} \sqrt{x} = 0.$$ ◇

Example 2

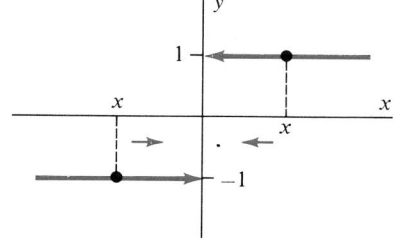

Figure 5.2 $\lim_{x \to 0^+} \dfrac{|x|}{x} = 1;$

$\lim_{x \to 0^-} \dfrac{|x|}{x} = -1.$

Figure 5.2 shows the graph of the function $f(x) = \dfrac{|x|}{x}$. In Section 2.2 we concluded

that $\lim_{x \to 0} \dfrac{|x|}{x}$ does not exist. The one-sided limits of this function at $a = 0$ do exist,

however. This is because

(a) For $x > 0$, $|x| = x$, so $\dfrac{|x|}{x} = \dfrac{x}{x} = 1$ and

$$\lim_{x \to 0^+} \frac{|x|}{x} = \lim_{x \to 0^+} 1 = 1, \quad \text{and}$$

(b) For $x < 0$, $|x| = -x$, so $\dfrac{|x|}{x} = \dfrac{-x}{x} = -1$, and

$$\lim_{x \to 0^-} \frac{|x|}{x} = \lim_{x \to 0^-} (-1) = -1.$$ ◇

Example 3

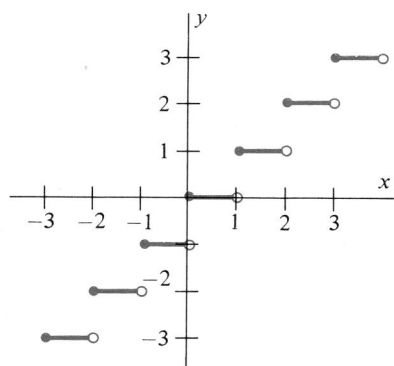

Figure 5.3 Graph of greatest integer function $f(x) = [x]$.

Figure 5.3 shows the graph of the **greatest integer** function which is defined by

$$[x] = \text{the largest integer } n \text{ with } n \le x.$$

Thus, $[2.5] = 2$, $[-1.2] = -2$, $[\pi] = 3$, $[7] = 7$, etc.

Although $\lim_{x \to n} [x]$ fails to exist for each integer n, the corresponding one-sided

limits do exist. In particular,

$$\lim_{x \to 2^+} [x] = 2; \quad \lim_{x \to 2^-} [x] = 1$$

$$\lim_{x \to -2^-} [x] = -3; \quad \lim_{x \to -2^+} [x] = -2.$$ ◇

Example 4

Figure 5.4 shows the graph of a function f giving the volume, as a function of temperature (in °C), occupied by a fixed mass (weight) of an antifreeze/water solution that freezes at temperature $-10°C$. If the function f is given on the interval $[-40, 40]$, except at $t = -10$, by

$$f(t) = \begin{cases} 20.2 + .02t, & -10 < t \le 40 \\ 25.1 + .01t, & -40 \le t < -10 \end{cases}$$

the limit as the temperature approaches the freezing point from the right is

$$\lim_{t \to -10^+} f(t) = \lim_{t \to -10^+} (20.2 + .02t)$$

$$= 20.2 + (.02)(-10)$$

$$= 20$$

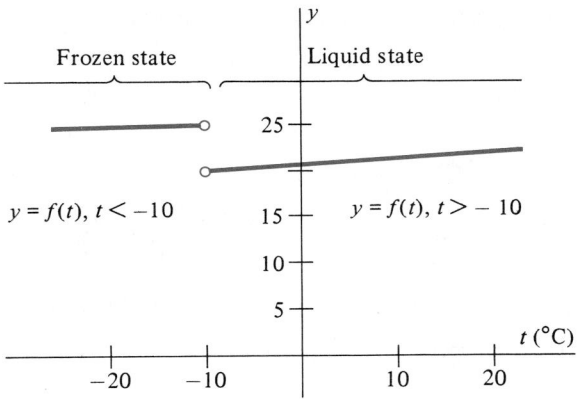

Figure 5.4 Graph of temperature versus volume function in Example 4.

while the limit as t approaches the melting point from the left is

$$\lim_{t \to -10^-} f(t) = \lim_{t \to -10^-} (25.1 + .01t)$$
$$= 25.1 + (.01)(-10)$$
$$= 25.$$

$\diamond$

Formal Definitions for One-Sided Limits

To formalize our intuitive notion of one-sided limit we modify the "two-sided" condition $0 < |x - a| < \delta$ in the definition of limit (Definition 3) to either the condition

$$a < x < a + \delta \qquad \text{(right-hand limit)}$$

or

$$a - \delta < x < a \qquad \text{(left-hand limit)}.$$

(See Figures 5.5 and 5.6.)

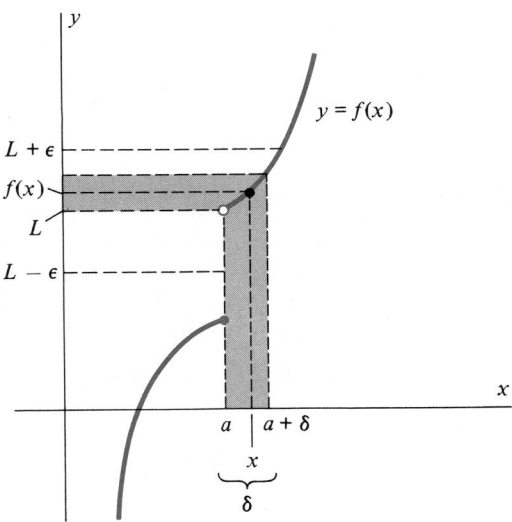

Figure 5.5 If $a < x < a + \delta$, then $|f(x) - L| < \epsilon$.

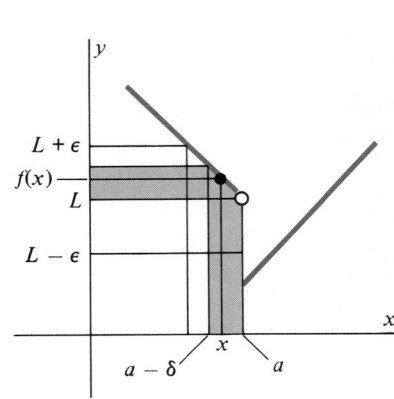

Figure 5.6 If $a - \delta < x < a$, then $|f(x) - L| < \epsilon$.

DEFINITION 4

We say that L is the limit of the function f as x approaches *a from the right,* written $L = \lim\limits_{x \to a^+} f(x)$ if and only if: for every $\epsilon > 0$ there exists a corresponding number $\delta > 0$ so that

$$\text{if}\quad a < x < a + \delta, \qquad \text{then}\quad |f(x) - L| < \epsilon.$$

Similarly, we say that M is the limit of f as x approaches *a from the left,* written $M = \lim\limits_{x \to a^-} f(x)$ if and only if: for every $\epsilon > 0$ there exists a corresponding $\delta > 0$ so that

$$\text{if}\quad a - \delta < x < a, \qquad \text{then}\quad |f(x) - M| < \epsilon.$$

Example 5

We may use Definition 4 to prove the one-sided limit $\lim\limits_{x \to 0^+} \sqrt{x} = 0$ as follows:

Strategy

Identify f, L, and a.

Examine the difference $|f(x) - L|$ to find a relationship with $x - a = x - 0 = x$.

Use this relationship to determine δ so that $a < x < a + \delta$ guarantees

$|f(x) - L| < \epsilon$.

Rewrite findings in the form given by Definition 4 to verify conclusions.

Solution

To prove that $\lim\limits_{x \to 0^+} \sqrt{x} = 0$ we begin by noting that $f(x) = \sqrt{x}$ and $L = 0$, so

$$\begin{aligned}
|f(x) - L| &= |\sqrt{x} - 0| \\
&= \sqrt{x}.
\end{aligned}$$

Thus, we will have $|f(x) - L| < \epsilon$ whenever $\sqrt{x} < \epsilon$. Squaring both sides of this last inequality shows that

$$\sqrt{x} < \epsilon \qquad \text{if} \qquad 0 < x < \epsilon^2. \tag{1}$$

Since $a = 0$, if we take $\delta = \epsilon^2$, the inequality on the right side of statement (1) is the required condition $a < x < a + \delta$. Thus,

Given $\epsilon > 0$, choose $\delta = \epsilon^2$. Then, if $0 < x < \delta$ we have $\sqrt{x} < \epsilon$, so $|\sqrt{x} - 0| < \epsilon$.

This shows that $\lim\limits_{x \to 0^+} \sqrt{x} = 0$, according to Definition 4. (See Figure 5.7.)

◇

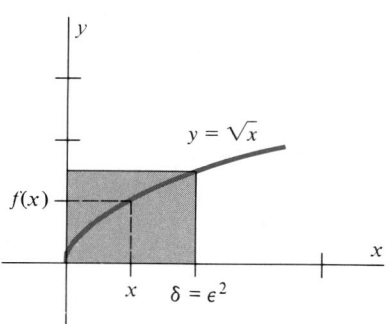

Figure 5.7 If $0 < x < \delta$, then $\sqrt{x} < \epsilon$.

Limits vs. One-Sided Limits

In each of Examples 1–5 $\lim\limits_{x \to a} f(x)$ fails to exist although one or both of the one-sided limits exists. You have no doubt noticed that when $\lim\limits_{x \to a} f(x)$ exists, so do both one-sided limits, although the latter convey no information beyond that given by $\lim_{x \to a} f(x)$. The following theorem, whose proof is sketched in Exercises 26 and 27, gives the relationship between "two-sided" and "one-sided" limits.

THEOREM 5

The limit $\lim_{x \to a} f(x)$ exists if and only if both $\lim_{x \to a^+} f(x)$ and $\lim_{x \to a^-} f(x)$ exist and are equal. That is,

$$\lim_{x \to a} f(x) = L \qquad \text{if and only if} \qquad \lim_{x \to a^+} f(x) = L = \lim_{x \to a^-} f(x).$$

Theorem 5 provides a way of proving the limit $L = \lim_{x \to a} f(x)$ that is sometimes simpler than applying the formal definition of limit directly: show that $\lim_{x \to a^+} f(x) = L = \lim_{x \to a^-} f(x)$.

Example 6

For the function

$$f(x) = \begin{cases} 4x - 3, & x \geq 1 \\ 2 - x^2, & x < 1 \end{cases}$$

prove that $\lim_{x \to 1} f(x) = 1$. (See Figure 5.8.)

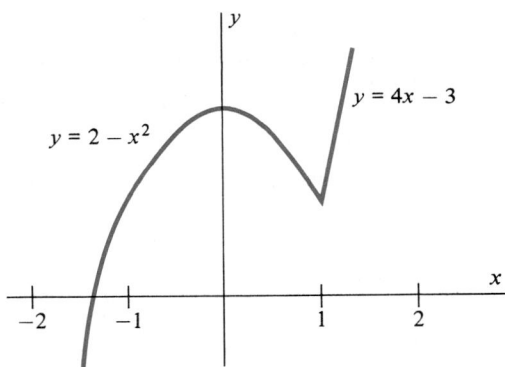

Figure 5.8 Graph of $f(x) = \begin{cases} 4x - 3, & x \geq 1 \\ 2 - x^2, & x < 1 \end{cases}$.

Strategy

Examine the one-sided limits individually. Show that $\lim_{x \to 1^+} f(x)$ is the same as $\lim_{x \to 1^+} g(x)$ for a function g for which $\lim_{x \to 1} g(x)$ is known, and conclude from Theorem 5 that these limits are equal.

Solution

To prove that $\lim_{x \to 1^+} f(x) = 1$, we note that if $x > 1$ then $f(x) = 4x - 3$.

It follows from Theorem 1 that the linear function $g(x) = 4x - 3$, defined for *all* x, has two-sided limit $\lim_{x \to 1} g(x) = 4(1) - 3 = 1$.

Thus, by Theorem 5, $\lim_{x \to 1^+} g(x) = \lim_{x \to 1^+}(4x - 3) = 1$ also. Thus,

$$\lim_{x \to 1^+} f(x) = \lim_{x \to 1^+}(4x - 3) = \lim_{x \to 1}(4x - 3) = 1.$$

Use the same strategy to establish the left-hand limit.

Use Theorem 5 to conclude that the limit equals the one-sided limits.

Similarly, if $x < 1$ then $f(x) = 2 - x^2$. We have already established that the quadratic polynomial $h(x) = 2 - x^2$ has limit $\lim\limits_{x \to 1} h(x) = 2 - 1^2 = 1$ so by Theorem 5 it has the one-sided limit $\lim\limits_{x \to 1^-} (2 - x^2) = 1$ also. Thus

$$\lim_{x \to 1^-} f(x) = \lim_{x \to 1^-} (2 - x^2) = \lim_{x \to 1} (2 - x^2) = 1.$$

Since $\lim\limits_{x \to 1^+} f(x) = 1 = \lim\limits_{x \to 1^-} f(x)$, Theorem 5 gives $\lim\limits_{x \to 1} f(x) = 1$. ◇

Exercise Set 2.5

1. For the function f in Figure 5.9 find
   a. $\lim\limits_{x \to 1^-} f(x)$

   b. $\lim\limits_{x \to 1} f(x)$

   c. $\lim\limits_{x \to 0} f(x)$

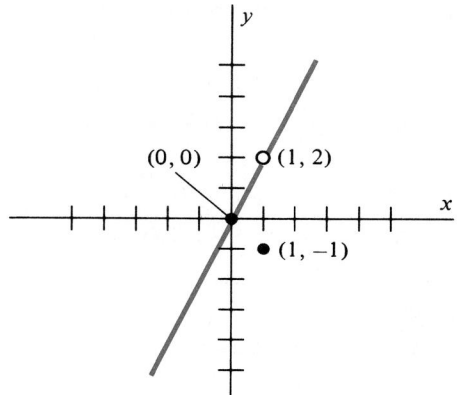

Figure 5.9

2. For the function f in Figure 5.10 find
   a. $\lim\limits_{x \to 1^-} f(x)$

   b. $\lim\limits_{x \to 1^+} f(x)$

   c. $\lim\limits_{x \to 1} f(x)$

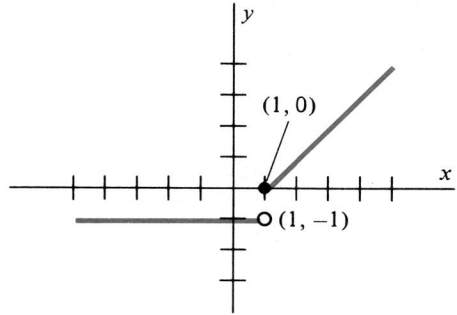

Figure 5.10

3. For the function f in Figure 5.11 find
   a. $\lim\limits_{x \to -2} f(x)$

   b. $\lim\limits_{x \to 2^-} f(x)$

   c. $\lim\limits_{x \to 2^+} f(x)$

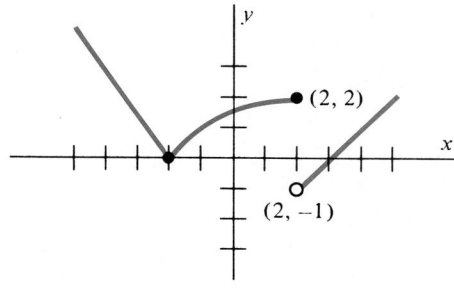

Figure 5.11

In Exercises 4–18 find the limit, if it exists.

4. $\lim\limits_{x \to 2^+} \sqrt{x - 2}$

5. $\lim\limits_{x \to 3^-} \dfrac{|x - 3|}{x - 3}$

6. $\lim\limits_{x \to 0^+} (x^{5/2} - 5x^{3/2})$

7. $\lim\limits_{x \to 1^-} \sqrt{1 - x}$

8. $\lim\limits_{x \to 2^-} \dfrac{x^2 - 4x + 4}{x - 2}$

9. $\lim\limits_{x \to 0^+} \dfrac{x - 1}{\sqrt{x} - 1}$

10. $\lim\limits_{x \to 0^+} \dfrac{\sqrt{x} + x^{4/3}}{6 - x^{3/4}}$

11. $\lim\limits_{x \to 3^+} [1 + x]$

12. $\lim\limits_{x \to 3^-} [1 + x]$

13. $\lim\limits_{x \to 0^+} \dfrac{3\sqrt{x} + 4x^2 - 6}{\sqrt{x} + \sqrt{x + 4}}$

14. $\lim\limits_{x \to 3^-} x|x|$

15. $\lim\limits_{x \to 0} [x^2 - 6x + 5]$

16. $\lim\limits_{x \to 4^-} \dfrac{\sqrt{x} - 4}{x + 2}$

17. $\lim\limits_{x \to 2^-} [x^2 - 6x + 5]$

18. $\lim\limits_{x \to 3^+} \dfrac{[x - 7]}{[x + 4]}$

19. Let

$$f(x) = \begin{cases} x + 2, & x \le -1 \\ -x, & x > -1. \end{cases}$$

a. Find $\lim\limits_{x \to -1^-} f(x)$.

b. Find $\lim\limits_{x \to -1^+} f(x)$.

c. Does $\lim\limits_{x \to -1} f(x)$ exist?

20. Let

$$f(x) = \begin{cases} \cos x, & x \le 0 \\ 1 - x, & x > 0. \end{cases}$$

a. Find $\lim\limits_{x \to 0^-} f(x)$.

b. Find $\lim\limits_{x \to 0^+} f(x)$.

c. Does $\lim\limits_{x \to 0} f(x)$ exist?

21. Let

$$f(x) = \begin{cases} x + 2, & x < 3 \\ 2x - 1, & x > 3. \end{cases}$$

Prove that $\lim\limits_{x \to 3} f(x) = 5$.

22. Let

$$f(x) = \begin{cases} x^2, & x \le 1 \\ x^5, & x > 1. \end{cases}$$

Prove that $\lim\limits_{x \to 1} f(x) = 1$.

23. Let

$$f(x) = \begin{cases} 2, & x \le -1 \\ -x, & -1 < x \le 1 \\ -x^2, & 1 < x. \end{cases}$$

a. Sketch the graph of f.

b. Does $\lim\limits_{x \to -1} f(x)$ exist? Why?

c. Does $\lim\limits_{x \to 1} f(x)$ exist? Why?

d. Prove any limit that exists in parts b and c.

24. Let

$$f(x) = \begin{cases} x + 2, & x \le -1 \\ cx^2, & x > -1. \end{cases}$$

Find c so that $\lim\limits_{x \to -1} f(x)$ exists.

25. Let

$$f(x) = \begin{cases} 3 - x^2, & x \le -2 \\ ax + b, & -2 < x < 2 \\ \dfrac{1}{2}x^2, & x \ge 2. \end{cases}$$

Find a and b so that $\lim\limits_{x \to -2} f(x)$ and $\lim\limits_{x \to 2} f(x)$ exist.

26. Prove the "if" part of Theorem 5 as follows.

a. Assume $\lim\limits_{x \to a^+} f(x) = L = \lim\limits_{x \to a^-} f(x)$.

b. Let ϵ be given. Explain why there exists a number δ_1, so that

if $\quad a < x < a + \delta_1 \quad$ then $\quad |f(x) - L| < \epsilon$.

c. For this same ϵ, explain why there exists a number δ_2 so that

if $\quad a - \delta_2 < x < a \quad$ then $\quad |f(x) - L| < \epsilon$.

d. Explain why, for this same ϵ and for $\delta = \min\{\delta_1, \delta_2\}$,

if $\quad 0 < |x - a| < \delta, \quad$ then $\quad |f(x) - L| < \epsilon$.

27. Prove the "only if" part of Theorem 5 as follows:

a. Assume that $\lim\limits_{x \to a} f(x) = L$ exists, according to Definition 1.

b. Show that $\lim\limits_{x \to a^+} f(x) = L$ and $\lim\limits_{x \to a^-} f(x) = L$, according to Definition 2.

2.6 CONTINUITY

In defining the limit $L = \lim\limits_{x \to a} f(x)$ we have repeatedly emphasized that $\lim\limits_{x \to a} f(x)$ and $f(a)$ need not be equal. (Indeed, $f(a)$ may not even be defined.) We now turn our attention to the case in which $f(a) = \lim\limits_{x \to a} f(x)$. When this happens we say that f is *continuous* at $x = a$.

DEFINITION 5

The function f is said to be **continuous** at $x = a$ if f is defined on an open interval containing a and $f(a) = \lim\limits_{x \to a} f(x)$.

It is important to note that the definition of continuity requires both *that the limit* $\lim\limits_{x \to a} f(x)$ *exists,* and *that the function f is defined at $x = a$*, in order that the condition

$$\lim\limits_{x \to a} f(x) = f(a)$$ be fulfilled.

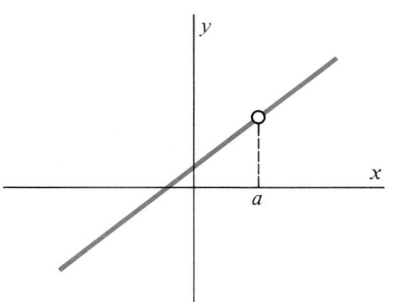

Figure 6.1 $f(a)$ fails to exist.

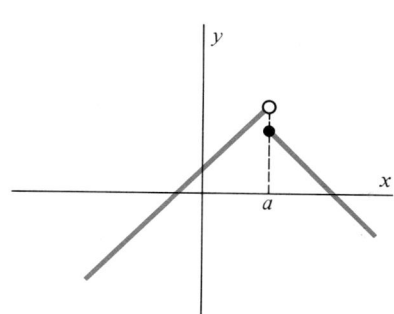

Figure 6.2 $\lim_{x \to a^-} f(x) \neq \lim_{x \to a^+} f(x)$;

$\lim_{x \to a} f(x)$ does not exist.

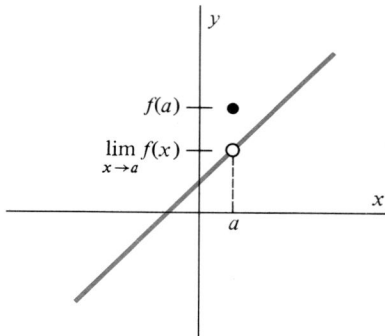

Figure 6.3 $f(a) \neq \lim_{x \to a} f(x)$.

If f is not continuous at $x = a$ we say that f is **discontinuous** at $x = a$. This is the case for each of the functions whose graphs appear in Figures 6.1–6.3. Geometrically, continuity is a property that assures that the graph of f will not have a hole or otherwise be "broken" at $(a, f(a))$ by ruling out each of the possibilities in Figures 6.1–6.3.

Example 1

Find the numbers x at which $f(x) = \dfrac{x^2 - 4}{x - 2}$ is continuous.

Strategy

Find the number(s) x for which the denominator equals zero. This is where f is undefined.

Show that

$\lim_{x \to a} f(x) = f(a)$

for all other x.

Solution

Since $x - 2 = 0$ when $x = 2$, the value $f(2)$ is not defined. The function is therefore *discontinuous* at $x = 2$.

For all numbers $a \neq 2$ we have

$$\lim_{x \to a} f(x) = \lim_{x \to a} \frac{x^2 - 4}{x - 2}$$
$$= \lim_{x \to a} \frac{(x - 2)(x + 2)}{x - 2}$$
$$= \lim_{x \to a} (x + 2)$$
$$= a + 2$$
$$= f(a).$$

Thus, f is continuous for all $x \neq 2$. (See Figure 6.4.) ◇

The discontinuity in Example 1 is called a **removable** discontinuity since we can eliminate (i.e., remove) the discontinuity at $x = 2$ by *defining* $f(2) = \lim_{x \to 2} f(x) = 4$. (See Figure 6.5.) In other words, we add $x = 2$ to the domain of f by defining the new function

$$\hat{f} = \begin{cases} \dfrac{x^2 - 4}{x - 2} & \text{if } x \neq 2 \\ 4 & \text{if } x = 2 \end{cases}$$

The function $\hat{f}$ is then continuous for all x, and agrees with f for $x \neq 2$.

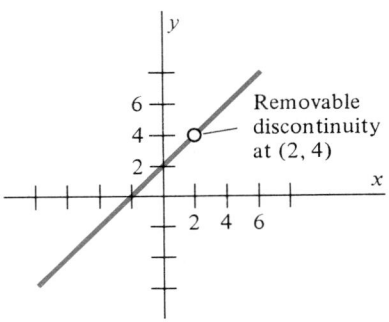

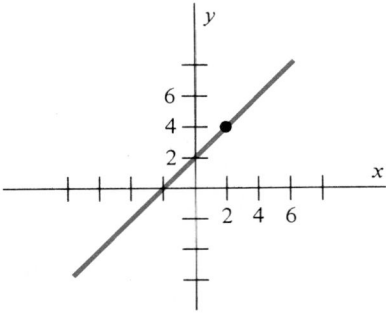

Figure 6.4 $f(x) = \dfrac{x^2 - 4}{x - 2}$ is discontinuous for $x = 2$.

Figure 6.5 Discontinuity in Figure 6.4 is *removed* by defining $f(2) = \lim\limits_{x \to 2} f(x) = 4$.

Example 2

Another function with a removable discontinuity is $f(x) = \dfrac{\sin x}{x}$. Although $f(0)$ is undefined, we have shown in Section 2.4 that $\lim\limits_{x \to 0} \dfrac{\sin x}{x} = 1$. We may therefore "remove" the discontinuity at $x = 0$ by defining

$$\hat{f}(x) = \begin{cases} \dfrac{\sin x}{x} & \text{for } x \neq 0 \\ 1 & \text{for } x = 0 \end{cases}$$

The function $\hat{f}$ is then continuous at $x = 0$. ◇

We leave as simple exercises the proofs that constant functions and the linear function $f(x) = x$ are continuous at each number x. Because continuity is defined in terms of limits, Theorems 1–3 on limits translate directly into the following theorems on continuity.

THEOREM 6

If the functions f and g are continuous at $x = a$ and if c is any real number, then the following functions are also continuous at $x = a$:

(i) $f + g$
(ii) cf
(iii) fg
(iv) $\dfrac{f}{g}$, *provided* $g(a) \neq 0$.

Proof: To prove that $f + g$ in (i) is continuous at $x = a$, we note first that $(f + g)(a) = f(a) + g(a)$ is defined, since both f and g are continuous at $x = a$. Moreover, $\lim\limits_{x \to a} f(x)$ and $\lim\limits_{x \to a} g(x)$ exist, and

$$\lim_{x \to a} (f + g)(x) = \lim_{x \to a} [f(x) + g(x)] \qquad \text{(Definition of } f + g\text{)}$$

$$= \lim_{x \to a} f(x) + \lim_{x \to a} g(x) \qquad \text{(Theorem 1)}$$

$$= f(a) + g(a) \qquad \text{(Continuity of } f, g \text{ at } a\text{)}$$
$$= (f + g)(a) \qquad \text{(Definition of } f + g\text{)}.$$

Thus $f + g$ satisfies the definition of continuity at $x = a$. The proofs of parts (ii)–(iv), which follow from the corresponding parts of Theorem 1, are left for you as exercises. ◆

Example 3

Later in this section we shall show that the function $f(x) = \sin x$ is continuous for all x. This fact, together with the continuity of the function $g(x) = x$ and part (iv) of Theorem 6 show that the function $h(x) = \dfrac{\sin x}{x}$ is continuous for all $x \neq 0$. Thus, the function $\hat{f}$ in Example 2 is continuous for all x. ◇

THEOREM 7

For each positive integer $n = 1, 2, \ldots$,

(i) the function $f(x) = x^n$ is continuous at all numbers x.
(ii) if the function g is continuous at $x = a$, the function $f(x) = (g(x))^n$ is continuous at $x = a$.

The proof follows from Theorem 2 and is left as an exercise.

Example 4

Theorems 6 and 7 combine to show that any polynomial is continuous at all numbers x, and any rational function is continuous wherever its denominator is not zero. In particular

(i) the polynomial $f(x) = x^3 - 2x^2 + 7$ is continuous at all x.
(ii) the rational function

$$f(x) = \frac{x^3 + x + 7}{x - 6}$$

is continuous at all x with $x \neq 6$.
(iii) the power function

$$g(x) = \left(\frac{x^3 + x + 7}{x - 6} \right)^4$$

is continuous at all x with $x \neq 6$. ◇

Continuity on Intervals

While Definition 5 concerns continuity at a particular number $x = a$, the following definition concerns continuity on an interval.

DEFINITION 6

(i) The function f is said to be continuous on the open interval (a, b) if it is continuous at each $x \in (a, b)$.
(ii) The function f is said to be continuous on the closed interval $[a, b]$ if it is continuous at each $x \in (a, b)$ and, in addition,

$$\lim_{x \to a^+} f(x) = f(a) \qquad \text{and} \qquad \lim_{x \to b^-} f(x) = f(b).$$

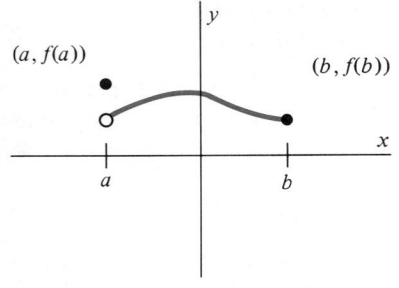

Figure 6.6 $y = f(x)$ is continuous on $(a, b]$ but not on $[a, b]$.

Continuity is defined in a similar way on half-open intervals, like $(a, b]$, and on infinite intervals, like $[a, \infty)$, by including the condition that the one-sided limit "from the inside" equals the function value at any included endpoint. Figure 6.6 illustrates why we must add this condition for included endpoints. *Note that the continuity of f on the closed interval $[a, b]$ does not guarantee the continuity of f at the endpoints $x = a$ or $x = b$ in the sense of Definition 5.*

Example 5

The function $f(x) = \sqrt{x}$ is continuous on the interval $[0, \infty)$. This is because $\lim_{x \to a} \sqrt{x} = \sqrt{a}$ for any $a \in (0, \infty)$, by Theorem 3, and $\lim_{x \to 0^+} \sqrt{x} = 0 = f(0)$ at the included endpoint by Example 5, Section 2.5 (see Figure 5.7). ◇

Example 6

The greatest integer function $f(x) = [x]$ is continuous on each interval of the form $[n, n + 1)$ where n is an integer. This is because $\lim_{x \to n^+} [x] = n = f(n)$. *This function, however, is discontinuous at each integer $x = n$. This is because* $\lim_{x \to n^+} [x]$ exists for every integer n, but $\lim_{x \to n} [x]$ does not. ◇

The notion of continuity on intervals is required to specify the continuity properties of rational power functions.

THEOREM 8

Let n and m be positive integers. The function $f(x) = x^{n/m}$ is

(i) continuous on $[0, \infty)$ if m is even;
(ii) continuous on $(-\infty, \infty)$ if m is odd.

Proof: To prove part (i) we recall from Theorem 3 that, for m even,

$$\lim_{x \to a} f(x) = \lim_{x \to a} x^{n/m} = a^{n/m} = f(a) \text{ if } 0 < a < \infty.$$

Thus, f is continuous on $(0, \infty)$. At $a = 0$, the fact that

$$\lim_{x \to 0^+} f(x) = \lim_{x \to 0^+} x^{n/m} = 0 = f(0)$$

follows by the same method of proof used to show

$$\lim_{x \to 0^+} \sqrt{x} = 0. \text{ (See Exercise 60.)}$$

To prove part (ii) we use Theorem 3 again. If m is odd, it guarantees that

$$\lim_{x \to a} f(x) = \lim_{x \to a} x^{n/m} = a^{n/m} = f(a)$$

for all numbers a. ◆

Example 7

Using Theorems 6–8 we may conclude that the following functions are continuous on the intervals stated.

(a) $f(x) = x^2 - 3\sqrt{x}$ on $[0, \infty)$

(b) $f(x) = \dfrac{x^2 - x^{2/3}}{1 - x}$ on $(-\infty, 1)$ and $(1, \infty)$

(c) $f(x) = \dfrac{6x^{2/3} + 5x^{3/2}}{x(x - 2)(x + 3)}$ on $(0, 2)$ and $(2, \infty)$ ◇

Alternate Definitions of Continuity

The following theorem gives two alternate definitions of continuity at $x = a$, both equivalent to Definition 5; both are more directly applicable than Definition 5 in certain situations.

THEOREM 9

Let the function f be defined on an open interval containing $x = a$. Then the following statements are equivalent.

(i) f is continuous at $x = a$ (Definition 5)

(ii) $\displaystyle\lim_{h \to 0} f(a + h) = f(a)$

(iii) Given any number $\epsilon > 0$ there exists a corresponding number $\delta > 0$ so that

$$\text{if } |x - a| < \delta, \quad \text{then} \quad |f(x) - f(a)| < \epsilon.$$

We shall not give a formal proof of Theorem 9, but here are the basic ideas. Writing $x = a + h$ we have $h = x - a$, so $\displaystyle\lim_{h \to 0} f(a + h) = \lim_{x \to a} f(x)$. Thus, statement (ii) is just the requirement of Definition 5 that $\displaystyle\lim_{x \to a} f(x)$ exists and equals $f(a)$.

Statement (iii) is just the formal definition of $\displaystyle\lim_{x \to a} f(x) = L$ and with $L = f(a)$, with the modification that we allow $|x - a| < \delta$ rather than the more restrictive condition that $0 < |x - a| < \delta$. That is, *we include the possibility that $x = a$.* This forces the conditions that $f(a)$ exist (not required in the definition of limit), and that $\displaystyle\lim_{x \to a} f(x) = f(a)$.

Example 8

Show that the function $f(x) = \sin x$ is continuous at all numbers x.

Strategy

Use Theorem 9 to establish continuity by showing

$\displaystyle\lim_{h \to 0} \sin(a + h) = \sin a.$

Rewrite $\sin(a + h)$ using the identity

$\sin(\alpha + \beta) = \sin \alpha \cos \beta + \cos \alpha \sin \beta.$

Use facts that

$\displaystyle\lim_{h \to 0} \cos h = 1$

$\displaystyle\lim_{h \to 0} \sin h = 0.$

Solution

The function $f(x) = \sin x$ is defined for all x. To establish continuity at $x = a$ we use the equivalent definition in part ii of Theorem 9:

$$\lim_{h \to 0} f(a + h) = \lim_{h \to 0} \sin(a + h)$$
$$= \lim_{h \to 0} [\sin a \cdot \cos h + \cos a \cdot \sin h]$$
$$= \sin a (\lim_{h \to 0} \cos h) + \cos a (\lim_{h \to 0} \sin h)$$
$$= \sin a \cdot (1) + \cos a \cdot (0)$$
$$= \sin a.$$
$$= f(a). \qquad ◇$$

Composite Functions

Recall that the composite function $f \circ g$ is defined by the equation $(f \circ g)(x) = f(g(x))$. For example, the function

$$y = \sin(\pi + x^2)$$

is a composite function, where the "inside" function

$$u = g(x) = \pi + x^2$$

is acted on by the "outside" function

$$f(u) = \sin u.$$

The following theorem provides a way of evaluating limits of composite functions when the "outside" function is continuous.

THEOREM 10

Let $\lim\limits_{x \to a} g(x) = L$ exist and assume that the function f is continuous at L. Then

$$\lim_{x \to a} f(g(x)) = f(L).$$

That is,

$$\lim_{x \to a} f(g(x)) = f(\lim_{x \to a} g(x)).$$

Theorem 10 says that when the "outside" function in the composition $f \circ g$ is continuous, we may "pass the limit inside the function f." The proof of Theorem 10 is given at the end of this section.

Example 9

Theorem 10 may be used as follows:

(a) $\lim\limits_{x \to 0} \sin(\pi + x^2) = \sin[\lim\limits_{x \to 0}(\pi + x^2)] = \sin(\pi + 0) = \sin \pi = 0.$

(b) $\lim\limits_{x \to 2} \sqrt{1 + x^3} = \sqrt{\lim\limits_{x \to 2}(1 + x^3)} = \sqrt{1 + 8} = 3.$ ◇

The next theorem asserts that "the composition of two continuous functions is a continuous function." The proof is given at the end of this section.

THEOREM 11

If g is continuous on an open interval containing $x = a$, and if f is continuous at $L = g(a)$, then the composite function $f \circ g$ is continuous at $x = a$.

Example 10

The composite function $y = \sin(\pi + x^2)$ is continuous on $(-\infty, \infty)$ since it has the form $y = f(g(x))$ where the "outside" function $f(u) = \sin u$ is continuous for all u and the "inside" function $g(x) = \pi + x^2$ is continuous for all x.   ◇

Example 11

The composite function $y = \sqrt{4 - x^2}$ is continuous on the interval $(-2, 2)$ since

(a) the "inside" function $g(x) = 4 - x^2$ is continuous on $(-2, 2)$ with corresponding values in the interval $(0, 4]$.

(b) the "outside" function $f(u) = \sqrt{u}$ is continuous on the interval $(0, \infty)$, which contains the interval $(0, 4]$.

By showing that $\lim\limits_{x \to -2^+} \sqrt{4 - x^2} = \sqrt{4 - 0} = 2$ and $\lim\limits_{x \to 2^-} \sqrt{4 - x^2} = \sqrt{4 - 0} = 2$, you can verify that $y = \sqrt{4 - x^2}$ is actually continuous on the *closed* interval $[-2, 2]$. ◇

The Intermediate Value Theorem

We shall make frequent use of the following property of continuous functions.

THEOREM 12 (Intermediate Value Theorem):	Let f be continuous on the closed interval $[a, b]$ with $f(a) \neq f(b)$. Let d be any number between $f(a)$ and $f(b)$. Then there exists at least one number $c \in (a, b)$ with $f(c) = d$.

The Intermediate Value Theorem states that a continuous function cannot "skip" any numbers in passing from any one of its values to another. Although a proof of this theorem is beyond the scope of this text, Figure 6.7 suggests a geometric

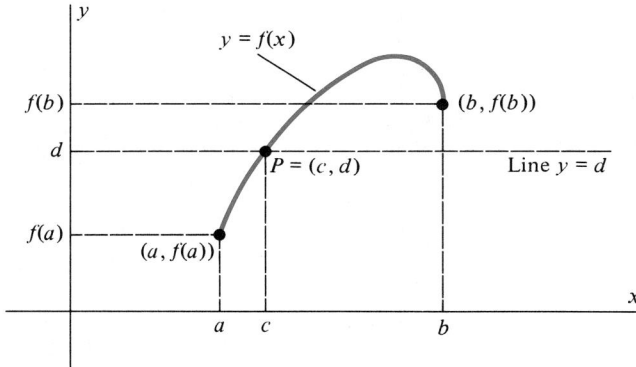

Figure 6.7 The Intermediate Value Theorem.

argument for its validity in the case $f(a) < f(b)$. If $f(a) < d < f(b)$ and f is continuous on $[a, b]$, the graph of f must intersect (not "jump over") the horizontal line $y = d$ at some point P. If P has coordinates $P = (c, d)$ we then have $f(c) = d$, as claimed.

The Intermediate Value Theorem is one of several **existence** theorems that we shall encounter. The reason for this terminology is that the theorem guarantees that at least one number c satisfying the stated condition (in this case, $f(c) = d$) exists, but it does not tell us how to find this number. The next example gives one case in which this number can actually be found.

Example 12

The function $f(x) = \sqrt{x^3 + 1}$ is continuous on the interval $[0, 2]$ as a result of Theorems 6, 7, 8, and 11. Since $f(0) = \sqrt{0 + 1} = 1$ and $f(2) = \sqrt{8 + 1} = 3$, the Intermediate Value Theorem guarantees that if d is any "intermediate value" with $1 < d < 3$, there exists a number $c \in (0, 2)$ with $f(c) = d$. In particular, if $d = 2$

then $f(c) = 2$ gives the equation $\sqrt{c^3 + 1} = 2$, which we can solve for c as follows:

$$\sqrt{c^3 + 1} = 2$$

gives

$$c^3 + 1 = 2^2 = 4$$

so

$$c^3 = 4 - 1 = 3$$

and

$$c = \sqrt[3]{3}. \qquad \Diamond$$

The Intermediate Value Theorem is useful in solving inequalities of the form $f(x) > 0$ or $f(x) < 0$. To do so, we first find the zeros of f. It then follows that $f(x)$ must be of constant sign on the resulting intervals, so the sign of f on any such interval can be determined by checking the sign of $f(x)$ at any number x in the interval. (If a and b are successive zeros of f, and if there were numbers x_1 and x_2 in (a, b) with $f(x_1) > 0$ and $f(x_2) < 0$, then the Intermediate Value Theorem would guarantee the existence of a number c in $(x_1, x_2) \subseteq (a, b)$ with $f(c) = 0$. But this cannot happen since a and b are *successive* zeros of f.)

Example 13

Use the Intermediate Value Theorem to solve the inequality $x^3 - 4x^2 - 5x > 0$.

Solution: The zeros of $f(x) = x^3 - 4x^2 - 5x$ are found by factoring:

$$f(x) = x^3 - 4x^2 - 5x = x(x^2 - 4x - 5) = x(x + 1)(x - 5)$$

so $f(x) = 0$ for $x = -1, 0$, and 5. The following table shows the result of checking the sign of f on the resulting intervals.

Interval	Test number x	$f(x)$	Conclusion
$(-\infty, -1)$	$x = -2$	$f(-2) = -14 < 0$	$f(x) < 0$ on $(-\infty, -1)$
$(-1, 0)$	$x = -\frac{1}{2}$	$f(-\frac{1}{2}) = \frac{11}{8} > 0$	$f(x) > 0$ on $(-1, 0)$
$(0, 5)$	$x = 1$	$f(1) = -8 < 0$	$f(x) < 0$ on $(0, 5)$
$(5, \infty)$	$x = 6$	$f(6) = 42 > 0$	$f(x) > 0$ on $(5, \infty)$

The solution of the inequality is therefore $(-1, 0) \cup (5, \infty)$. $\Diamond$

Proof of Theorem 10:   To prove $\lim_{x \to a} f(g(x)) = f(L)$ we must show that, given a number $\epsilon > 0$ there exists a corresponding number $\delta > 0$ so that

$$\text{if} \quad 0 < |x - a| < \delta, \qquad \text{then} \qquad |f(g(x)) - f(L)| < \epsilon. \tag{1}$$

Thus, we begin by assuming such a number $\epsilon > 0$ is given.

Since f is continuous at L, there exists, by part (iii) of Theorem 9, a number $\delta_1 > 0$ so that

$$\text{if} \quad |u - L| < \delta_1, \qquad \text{then} \qquad |f(u) - f(L)| < \epsilon. \tag{2}$$

Also, since $\lim_{x \to a} g(x) = L$ and δ_1 is a positive number, we may view δ_1 as another "given number ϵ" and conclude that there exists a number δ_2 so that

$$\text{if} \quad 0 < |x - a| < \delta_2, \quad \text{then} \quad |g(x) - L| < \delta_1. \tag{3}$$

Now let $u = g(x)$. Then, combining statements (2) and (3), we conclude that

$$\text{if} \quad 0 < |x - a| < \delta_2, \quad \text{then} \quad |g(x) - L| < \delta_1,$$

$$\text{so} \quad |f(g(x)) - f(L)| < \epsilon.$$

This is the required statement (1), with $\delta = \delta_2$. ◆

Proof of Theorem 11: Using the definition of continuity in part (iii) of Theorem 9, we let $\epsilon > 0$ be given and note, by the continuity of f at L, that there exists a number $\delta_1 > 0$ so that

$$\text{if} \quad |u - L| < \delta_1, \quad \text{then} \quad |f(u) - f(L)| < \epsilon. \tag{4}$$

Also, since g is continuous at a and δ_1 may be viewed as a given positive number, there is a number $\delta_2 > 0$ so that

$$\text{if} \quad |x - a| < \delta_2, \quad \text{then} \quad |g(x) - g(a)| < \delta_1. \tag{5}$$

Letting $u = g(x)$ and recalling that $L = g(a)$, we may combine statements (4) and (5) to conclude that

$$\text{if} \quad |x - a| < \delta_2, \quad \text{then} \quad |f(g(x)) - f(g(a))| < \epsilon.$$

This shows that the composite function $f \circ g$ is continuous at $x = a$. ◆

Exercise Set 2.6

In Exercises 1–14 find any numbers x at which the given function is discontinuous.

1. $f(x) = \dfrac{\sin x}{x}$

2. $f(x) = \sec x$

3. $f(x) = \dfrac{1}{4 - x^2}$

4. $f(x) = x \cot x$

5. $y = \dfrac{x^2 - 9}{x + 3}$

6. $f(x) = \dfrac{x}{\cos x}$

7. $f(x) = \dfrac{x + 2}{x^2 - x - 2}$

8. $y = \dfrac{x^2 + x + 1}{x^3 + 2x^2 - 3x}$

9. $y = x^{2/3} - x^{-2/3}$

10. $f(x) = \dfrac{1}{1 + x^{2/3}}$

11. $f(x) = \begin{cases} 1 - x, & x \le 2 \\ x - 1, & x > 2 \end{cases}$

12. $y = \begin{cases} x^2, & x < 0 \\ 3x, & x \ge 0 \end{cases}$

13. $f(x) = \begin{cases} -x, & x \le -1 \\ 4 - x^2, & -1 < x \le 2 \\ \frac{1}{2}x - 1, & x > 2 \end{cases}$

14. $y = \begin{cases} ax^2, & x \le 0 \\ bx^3, & 0 < x \le 1 \\ bx^4, & x > 1 \end{cases}$

In Exercise 15–26 state the intervals on which the function is continuous.

15. $y = \dfrac{x^2 - 36}{x - 6}$

16. $y = \dfrac{x^3 - 9x}{x^3 - 3x^2}$

17. $f(x) = \csc x$

18. $f(x) = \sqrt{x + 7}$

19. $f(x) = \dfrac{1}{\sqrt{x + 7}}$

20. $f(x) = \sqrt{x^3 - 4x}$

21. $y = x^{2/3} - x^{5/3}$

22. $f(x) = x^{5/2} + x^{-1/3}$

23. $f(x) = \sec x \tan x$

24. $y = \sqrt{\tan x}$

25. $f(x) = \dfrac{\tan x}{x^2 - x - 2}$

26. $f(x) = |9 - x^2|$

In Exercises 27–30 the function given has a removable discontinuity at $x = a$. Determine how to define $f(a)$ so that the function is continuous at a.

27. $f(x) = \dfrac{x^2 - 1}{x - 1}, \quad a = 1$

28. $f(x) = \begin{cases} x^2 + 1, & x < 1 \\ \sqrt{3 + x}, & x > 1 \end{cases}$ $a = 1$

29. $f(x) = \dfrac{\cos^2 x - 1}{\sin x}$, $a = 0$

30. $f(x) = \dfrac{x^2 + x - 2}{x^3 - x^2 - 6x}$, $a = -2$

In Exercises 31–33 find the constant k that makes the function continuous at $x = a$.

31. $y = \begin{cases} x^k, & x \le 2 \\ 10 - x, & x > 2 \end{cases}$ $a = 2$

32. $y = \begin{cases} k, & x \ge 1 \\ \dfrac{1}{\sqrt{kx^2 + k}}, & x < 1 \end{cases}$ $a = 1$

33. $h = \begin{cases} (x - k)(x + k), & x \le 2 \\ kx + 5, & x > 2 \end{cases}$ $a = 2$

In Exercises 34–41 use Theorem 10 to evaluate the limit.

34. $\lim\limits_{x \to 0} \sin\left(\pi - \dfrac{x}{2}\right)$

35. $\lim\limits_{x \to 8} (1 + \sqrt[3]{x})^5$

36. $\lim\limits_{x \to 0} \sqrt{\dfrac{1 - x}{1 + x}}$

37. $\lim\limits_{x \to \pi/2} \cos(\pi + x)$

38. $\lim\limits_{x \to 0} \left(\dfrac{3x + \sin x}{x}\right)^3$

39. $\lim\limits_{x \to 1} (3x^9 - \sqrt[3]{x})^6$

40. $\lim\limits_{x \to 2} \dfrac{\sqrt{4 - x}}{\sqrt{x + 2}}$

41. $\lim\limits_{x \to 0} \cos \pi(x + |x|)$

42. Prove the following equivalent statement of Theorem 10:

"If f is continuous on an open interval containing L, and if $\lim\limits_{x \to a} g(x) = L$, then, for $u = g(x)$,

$$\lim\limits_{x \to a} f(g(x)) = \lim\limits_{u \to L} f(u)."$$

Use the following steps:
a. The continuity of f at L gives $\lim\limits_{u \to L} f(u) = f(L)$.

b. Theorem 10 gives $\lim\limits_{x \to a} f(g(x)) = f(L)$.

c. Parts a and b give the desired result.

43. Use the result of Exercise 42 to show that $\lim\limits_{x \to 0} \dfrac{\sin ax}{x} = a$ as follows:
a. Define

$$f(u) = \begin{cases} \dfrac{\sin u}{u}, & u \ne 0 \\ 1, & u = 0 \end{cases}$$

and recall that f is continuous for all u (Example 3).

b. Use the fact that $\lim\limits_{u \to 0} \dfrac{\sin u}{u} = 1$ and the result of Exercise 42 to show that

$$\lim\limits_{x \to 0} \dfrac{\sin ax}{x} = \lim\limits_{x \to 0} \left(a \cdot \dfrac{\sin ax}{ax}\right)$$
$$= a \cdot \lim\limits_{x \to 0} \dfrac{\sin ax}{ax}$$
$$= a \cdot \lim\limits_{u \to 0} \dfrac{\sin u}{u}, \quad u = ax$$
$$= a.$$

In Exercises 44–49 use the fact that $\lim\limits_{x \to 0} \dfrac{\sin ax}{x} = a$ (Exercise 43) to evaluate the limit.

44. $\lim\limits_{x \to 0} \dfrac{\sin 6x}{x}$

45. $\lim\limits_{x \to 0} \dfrac{2x}{\sin x}$

46. $\lim\limits_{x \to 0} \dfrac{\tan 2x}{x}$

47. $\lim\limits_{x \to 0} \dfrac{\sin x}{\sin 3x}$

48. $\lim\limits_{x \to 0} \dfrac{\sin ax}{\sin bx}$

49. $\lim\limits_{x \to 0} x \csc 3x$

In Exercises 50–55 use the Intermediate Value Theorem to find the intervals on which the values of f are strictly positive or strictly negative.

50. $f(x) = x^2 - 4x - 5$

51. $f(x) = 9 - x^2$

52. $f(x) = x^3 - 3x - 2$

53. $f(x) = x^3 + 2x^2 - x - 2$

54. $f(x) = \sin \pi x$

55. $f(x) = \cos(x + \pi)$

In Exercises 56–63 solve the inequality $f(x) > 0$ or $f(x) < 0$ using the Intermediate Value Theorem.

56. $(x - 3)(x + 1) < 0$

57. $x(x + 6) > -8$

58. $x^2 + x < 0$

59. $x^3 + x^2 - 2x > 0$

60. $x^2 + x + 7 > 19$

61. $x^4 - 9x^2 > 0$

62. $x \sec x > 0$, $-2\pi < x < 2\pi$

63. $\sin x \cos x > 0$, $-2\pi < x < 2\pi$

64. Give an example to show that the composite function $f \circ g$ might be continuous at $x = a$ even though $\lim\limits_{x \to a} g(x) \ne g(a)$. Does this contradict Theorem 11?

65. Prove that the constant function with values $f(x) = c$ for all x is continuous at all x.

66. Prove that the linear function $f(x) = x$ is continuous at all x.

67. Prove parts (ii)–(iv) of Theorem 6.

68. Prove Theorem 7.

69. Prove that $\lim\limits_{x \to 0^+} x^{n/m} = 0$ if n and m are positive integers.

SUMMARY OUTLINE OF CHAPTER 2

◆ The line **tangent** to the curve C at point P is the limiting position of the secant through points P and Q as Q (page 46) approaches P along C.

◆ The **slope** of the line tangent to the graph of the function f at the point $(x_0, f(x_0))$, if it exists, is the limit (page 47)

$$m = \lim_{h \to 0} \frac{f(x_0 + h) - f(x_0)}{h}.$$

◆ Intuitively, $L = \lim_{x \to a} f(x)$ means that we can cause $f(x)$ to be as close to L as desired by choosing x sufficiently close (page 51) to a.

◆ Formally, we say that L is the **limit** of the function f as x approaches a, written $L = \lim_{x \to a} f(x)$, if and only if given any (page 57) number $\epsilon > 0$ there exists a corresponding number $\delta > 0$ so that

$$\text{if} \quad 0 < |x - a| < \delta, \qquad \text{then} \qquad |f(x) - L| < \epsilon.$$

◆ *Theorem:* If $\lim_{x \to a} f(x) = L$ and $\lim_{x \to a} g(x) = M$ and c is any constant, (page 61)

(i) $\lim_{x \to a}[f(x) + g(x)] = L + M$

(ii) $\lim_{x \to a}[cf(x)] = cL$

(iii) $\lim_{x \to a}[f(x)g(x)] = LM$

(iv) $\lim_{x \to a}\left(\dfrac{f(x)}{g(x)}\right) = \dfrac{L}{M}$, provided $M \neq 0$.

◆ *Theorem:* If $\lim_{x \to a} f(x) = L$ and n is any positive integer, (page 63)

a. $\lim_{x \to a} x^n = a^n$

b. $\lim_{x \to a} [f(x)]^n = [\lim_{x \to a} f(x)]^n = L^n$.

◆ *Theorem:* Let m and n be positive integers (page 64)

(i) If m is even, $\lim_{x \to a} x^{n/m} = a^{n/m}$ for $0 < a < \infty$.

(ii) If m is odd, $\lim_{x \to a} x^{n/m} = a^{n/m}$ for $-\infty < a < \infty$.

◆ *Theorem:* If $\lim_{x \to a} g(x) = \lim_{x \to a} h(x) = L$ exist and (page 64)

$g(x) \leq f(x) \leq h(x)$

for all x in an open interval containing a (except possibly at $x = a$), then $\lim_{x \to a} f(x) = L$ also.

◆ $L = \lim_{x \to a^+} f(x)$ if and only if, given $\epsilon > 0$ there exists a corresponding number $\delta > 0$ so that (page 73)

$$\text{if} \quad a < x < a + \delta, \qquad \text{then} \qquad |f(x) - L| < \epsilon.$$

◆ $M = \lim_{x \to a^-} f(x)$ if and only if, given $\epsilon > 0$ there exists a number $\delta > 0$ so that (page 73)

$$\text{if} \quad a - \delta < x < a, \qquad \text{then} \qquad |f(x) - M| < \epsilon.$$

◆ *Theorem:* The limit $L = \lim_{x \to a} f(x)$ exists if and only if $\lim_{x \to a^-} f(x)$ and $\lim_{x \to a^+} f(x)$ exist and $\lim_{x \to a^-} f(x) = L = \lim_{x \to a^+} f(x)$. (page 74)

◆ The function f is **continuous** at $x = a$ if f is defined on an open interval containing a and $\lim_{x \to a} f(x) = f(a)$. (page 76)

◆ **Theorem:** If f and g are continuous at $x = a$ and if c is any real number, then the following functions are also continuous at $x = a$:   (page 78)

(i) $f + g$
(ii) cf
(iii) fg
(iv) f/g, provided $g(a) \neq 0$.

◆ **Theorem:** For each positive integer $n = 1, 2, \ldots$ (page 79)

(i) The function $f(x) = x^n$ is continuous at all numbers x.
(ii) If the function g is continuous at $x = a$, so is the function $f(x) = [g(x)]^n$.

◆ The function f is **continuous on the open interval** (a, b) if f is continuous at each $x \in (a, b)$. The function f (page 79)
is **continuous on the closed interval** $[a, b]$ if f is continuous at each $x \in (a, b)$ and, in addition,

$$\lim_{x \to a^+} f(x) = f(a) \text{ and } \lim_{x \to b^-} f(x) = f(b).$$

◆ **Theorem:** Let m and n be positive integers. The function $f(x) = x^{n/m}$ is   (page 80)

(i) continuous on $[0, \infty)$ if m is even
(ii) continuous on $(-\infty, \infty)$ if m is odd.

◆ **Theorem:** The following conditions are equivalent for the function f defined on an open interval containing $x = a$:   (page 81)

(i) f is continuous at $x = a$.
(ii) $\lim_{h \to 0} f(a + h) = f(a)$.

(iii) Given any number $\epsilon > 0$ there exists a corresponding number $\delta > 0$ so that

$$\text{if } |x - a| < \delta \quad \text{then} \quad |f(x) - f(a)| < \epsilon.$$

◆ **Theorem:** Let $\lim_{x \to a} g(x) = L$ and assume that f is continuous at L. Then   (page 82)

$$\lim_{x \to a} f(g(x)) = f(L).$$

◆ **Theorem:** If g is continuous on an open interval containing $x = a$ and if f is continuous at $L = g(a)$, then the   (page 82)
composite function $f \circ g$ is continuous at $x = a$.

◆ **Theorem:** Let f be continuous on the interval $[a, b]$ with $f(a) \neq f(b)$ and let d lie between $f(a)$ and $f(b)$. Then there   (page 83)
exists at least one number $c \in (a, b)$ with $f(c) = d$.

REVIEW EXERCISES—CHAPTER 2

In Exercises 1–30 find the indicated limit.

1. $\lim_{x \to 2} (x^2 - x + 2)$

2. $\lim_{x \to 2} \dfrac{2x - 1}{x + 6}$

3. $\lim_{x \to 3} (x^4 - x - 1)$

4. $\lim_{x \to -5} \dfrac{x^2 - 25}{x + 5}$

5. $\lim_{x \to 3/2} \dfrac{4x^2 - 9}{2x - 3}$

6. $\lim_{x \to 3} \dfrac{x - 3}{x^2 - 9}$

7. $\lim_{x \to -2} \dfrac{x^2 + x - 2}{x + 2}$

8. $\lim_{x \to 1} \dfrac{x^2 + 6x - 7}{x - 1}$

9. $\lim_{x \to 1} \dfrac{x^2 + 2x - 3}{x^2 + x - 2}$

10. $\lim_{x \to -2} \dfrac{x^2 + 2x}{x^2 + x - 2}$

11. $\lim_{x \to 1^+} \sqrt{\dfrac{1 - x^2}{1 + x}}$

12. $\lim_{x \to 8} \sqrt{\dfrac{x - 7}{x + 2}}$

13. $\lim_{x \to 0} \dfrac{\tan x}{\sin x}$

14. $\lim_{x \to 0} \dfrac{3x}{\sin 2x}$

15. $\lim_{x \to 0} \dfrac{(2 + x)^2 - 4}{x}$

16. $\lim_{x \to 0} \dfrac{\sqrt{4 + x} - 2}{x}$

17. $\lim_{x \to 0} \dfrac{\sqrt[3]{x + 1} - 1}{x}$

18. $\lim_{x \to 2} \dfrac{3x^2 + x + 1}{1 - x^3}$

19. $\lim_{x \to 0} 3x \csc 4x$

20. $\lim_{x \to 0} \dfrac{\sqrt{4x^2 + 1}}{x^2 + 2}$

21. $\lim_{x \to 1} \dfrac{x^3 - 1}{x - 1}$

22. $\lim_{x \to 0} \dfrac{3x + 5x^2}{x}$

23. $\lim\limits_{x\to4^-}\sqrt{8-2x}$

24. $\lim\limits_{x\to0^+}\dfrac{1-x}{\sqrt{x}-1}$

25. $\lim\limits_{x\to1^+}\dfrac{|x-1|}{x-1}$

26. $\lim\limits_{x\to1^-}\dfrac{|x-1|}{x-1}$

27. $\lim\limits_{x\to0^+}\dfrac{2+\sqrt{x}}{2-\sqrt{x}}$

28. $\lim\limits_{x\to1^-}\sqrt{1-[x]}$

29. $\lim\limits_{x\to1^+}\sqrt{1-[x]}$

30. $\lim\limits_{x\to0^+}\dfrac{x-[x]}{x}$

In Exercises 31–42 find the intervals on which the given function is continuous.

31. $y=\dfrac{x^2-7}{x-2}$

32. $f(x)=\dfrac{x+2}{x^2-x-6}$

33. $f(x)=\dfrac{|x+2|}{x+2}$

34. $y=\sec x,\quad 0\le x\le2\pi$

35. $f(x)=\dfrac{\sin 3x}{2x}$

36. $f(x)=\sqrt{x^2-3x-4}$

37. $f(x)=\begin{cases}x-x^2, & x\le2\\ -x, & x>2\end{cases}$

38. $y=\begin{cases}\sin x, & x\le\pi/4\\ 1-\cos x, & x>\pi/4\end{cases}$

39. $f(x)=\begin{cases}\sqrt{x}, & x\le4\\ x-1, & x>4\end{cases}$

40. $f(x)=\begin{cases}1-t, & t<-2\\ -t, & -2\le t\le1\\ 1-2t^2, & 1<t\end{cases}$

41. $f(x)=\begin{cases}\dfrac{1}{1-x}, & x\ne1\\ 0, & x=1\end{cases}$

42. $f(x)=\begin{cases}\dfrac{x^2+3x-10}{x+5}, & x\ne-5\\ -7, & x=-5\end{cases}$

43. Use the Intermediate Value Theorem to prove that there exists a number x_0 so that $x_0^2-7x_0-2=0$.

44. Use the function

$$f(x)=\begin{cases}1, & x\le-1\\ \sqrt{x+2}, & -1<x<2\\ \cos\pi x, & 2\le x\end{cases}$$

to find:

a. $\lim\limits_{x\to-1^-}f(x)$
b. $\lim\limits_{x\to-1^+}f(x)$
c. $\lim\limits_{x\to-1}f(x)$
d. $\lim\limits_{x\to2^-}f(x)$
e. $\lim\limits_{x\to2^+}f(x)$
f. $\lim\limits_{x\to2}f(x)$

g. Is f continuous at $x=-1$?
h. Is f continuous at $x=2$?

45. Use the function

$$g(x)=\begin{cases}\tan x, & x<0\\ \sin x, & 0<x\le\dfrac{\pi}{2}\\ \dfrac{2x}{\pi}, & \dfrac{\pi}{2}<x\end{cases}$$

to find

a. $\lim\limits_{x\to0^-}g(x)$
b. $\lim\limits_{x\to0^+}g(x)$
c. $\lim\limits_{x\to0}g(x)$
d. $\lim\limits_{x\to\pi/2^-}g(x)$
e. $\lim\limits_{x\to\pi/2^+}g(x)$
f. $\lim\limits_{x\to\pi/2}g(x)$

g. Is g continuous at $x=0$?
h. Is g continuous at $x=\dfrac{\pi}{2}$?

46. The function

$$f(x)=\begin{cases}4-x+x^3, & x\le1\\ 9-ax^2, & x>1\end{cases}$$

is continuous at $x=1$. Find a.

47. Suppose that $1+4x-x^2\le f(x)\le x^2-4x+9$ for $x\ne2$. Find $\lim\limits_{x\to2}f(x)$.

48. Suppose $1-|x|\le f(x)\le\sec x$ for $-\frac12\le x\le\frac12$. Find $\lim\limits_{x\to0}f(x)$.

In Exercises 49–54 prove the stated limit.

49. $\lim\limits_{x\to3}(2x+3)=9$

50. $\lim\limits_{x\to1}(4x-3)=1$

51. $\lim\limits_{x\to2}(x^2+3)=7$

52. $\lim\limits_{x\to4}\dfrac{1}{x}=\dfrac14$

53. $\lim\limits_{x\to2^+}\sqrt{x-2}=0$

54. $\lim\limits_{x\to-3}|x+3|=0$

55. Given

$$f(x)=\begin{cases}2ax+b, & x\le3\\ ax+3b, & x>3\end{cases}$$

find a and b so that $\lim\limits_{x\to3}f(x)=10$.

56. For what real numbers a is $\lim\limits_{x\to a}(x^3-4x^2+3x-2)=-10$?

57. For

$$f(x)=\begin{cases}a^2x^2+bx-12, & x<1\\ ax+b, & x\ge1\end{cases}$$

find numbers a and b so that $\lim\limits_{x\to1}f(x)=2$.

58. True or false? If $f(x) \leq h(x) \leq g(x)$ for all x and both f and g are continuous at $x = a$, then h is continuous at $x = a$.

59. Prove that if $\lim_{x \to a} f(x)$ exists, then it is unique. That is, prove that if $\lim_{x \to a} f(x) = L$ and $\lim_{x \to a} f(x) = M$, then $L = M$.

60. Prove that if $f(x) \geq g(x)$ and if $\lim_{x \to a} g(x) = L$, then $\lim_{x \to a} f(x) \geq L$ if this limit exists.

61. If f is continuous at $x = a$, but $(f + g)$ is discontinuous at $x = a$ what can you say about the continuity of g at $x = a$?

UNIT 2

DIFFERENTIATION

Pierre de Fermat

Isaac Barrow

Augustin-Louis Cauchy

Isaac Newton wrote, ''If I seem to have seen further than others, it is because I stood on the shoulders of giants.'' He was referring to the several mathematicians who actually were performing differentiations and integrations in the years before Newton and Leibniz unified calculus into one subject.

One of the principal contributors was Pierre de Fermat (1601–1665). He was a lawyer by profession, and an amateur mathematician in the true sense of the term: a lover of mathematics.

Relatively little is known of his early life, and it is not even known for certain where he received all of his education. He was financially secure, and as a minor member of the nobility, was allowed to use the ''de'' in his name. His interests included studying the classics, and he was fluent in a half dozen languages.

Mathematically, he made significant contributions in a number of areas. He was a co-inventor with René Descartes of analytic geometry. He was particularly interested in the theory of numbers; the most famous theorem to which his name is attached is called Fermat's Last Theorem. It states that, for positive integral values of x, y, z, and n, there is no solution to the equation

$$x^n + y^n = z^n$$

for n larger than 2. After his death, the theorem was found written in the margin of one of his books, to which he added that he had found a marvelous proof, but that the margin was too small to contain it. Thousands of mathematician-hours have been spent on the theorem, but no one has ever been able to prove it.

A friend once wrote Fermat asking if he could determine whether the number 100,895,598,169 is prime. Fermat wrote back immediately to say that the factors are 898,423 and 112,303, and that each of these factors is itself prime. No one has ever been able to figure out how he was able to do it.

The French mathematician/astronomer Laplace called Fermat the discoverer of differential calculus. As early as 1629, before Newton was even born, Fermat was thinking of the differentiation process. Although he was hampered by a lack of the limit concept and of logical procedures, and although his notation was cumbersome, he essentially differentiated. He devised a method of finding maximum and minimum values which is still sometimes called Fermat's Method. He also found the tangent lines to curves of the form $y = f(x)$, noting that the process was similar to his maximum-minimum method.

Fermat determined volumes, centers of gravity, and the lengths of curves, but he did not realize the crucial relation: that integration and differentiation are inverse processes. For this reason, and his reluctance to publish his discoveries, he lost out as an acknowledged creator of the calculus.

Isaac Barrow (1630–1677) was a precocious and rebellious youngster, driving both his parents and his teachers to distraction. His father was heard to pray one time that if God had to take one of his children (child mortality was common in those times), that Isaac was the one who could most easily be spared. But Isaac Barrow survived, went to Trinity College at Cambridge, and stayed on as a scholar. At the relatively young age of 33 he became the first Lucasian Professor of Mathematics at Cambridge, an endowed chair with relatively few duties. After six years he resigned the chair. One frequently hears that he resigned so that his pupil, Isaac Newton, might fill the professorship, as Barrow recognized Newton's superior genius. Unfortunately, the story appears not to be true. There is no evidence that Barrow was ever Newton's tutor, though Newton may have attended public lectures given by Barrow. Actually, Barrow hoped for an appointment to a different position; he considered himself a theologian rather than a mathematician. Within a year he was appointed chaplain to the king.

Like Pierre de Fermat in France, Barrow performed differentiations much as we do today as he studied the drawing of lines tangent to curves. He also found areas between curved lines, which was close to the integration process. But he was without a theory of limits, and never realized the critically important inverse relationship between differentiation and integration.

Rolle's Theorem is named after Michel Rolle (1652–1719). His education was limited, and he learned most of his mathematics through his own efforts. He married young and had difficulty in supporting his family. But as a result of solving a difficult number theory problem at the age of 30, he was awarded a pension by the French government.

Rolle was at first a strong antagonist to the ''new'' calculus of Newton and Leibniz. He asserted that it was not based on logical grounds, and that it gave erroneous results. He called the calculus ''a collection of ingenious fallacies.'' Later he became convinced of the usefulness of the calculus. The theorem that bears his name, and which is used as a basis for proving the Mean Value Theorem, appears only incidentally in his writings about finding approximate solutions to equations. Rolle used it only in connection with polynomial functions, although we now know that it is applicable to a much wider variety of functions. His name was not applied to the theorem until 1846.

A nineteenth century French mathematician, Augustin-Louis Cauchy (1789–1857), is credited with bringing present-day standards of rigor to the calculus. In particular, he defined ''limit'' in a more precise way than had previously been done, and formally defined ''continuous function.'' Cauchy took the lead in defining the integral as the limit of a sum, rather than as the inverse of differentiation. He also first formally defined the derivative as the limit of the difference quotient, as it is done in today's textbooks.

Cauchy was born during the difficult years of the French Revolution. His early education was provided by his father, although he was fortunate to live not far from Pierre Laplace, who in turn introduced him to Joseph Lagrange; both were famous mathematicians. He became a civil engineer, although he was also interested in the ancient classics. Laplace and Lagrange persuaded him to pursue mathematics, and he became a teacher at the École Polytechnique in Paris. Reports on his teaching range from highest praise to one account from a student that Cauchy spent an entire hour extracting the square root of 17 by a method with which everyone in the class was already familiar. Others said he rambled incessantly.

His mathematical output was prodigious. In twenty years he published 589 papers in just one journal—and submitted many others which they were unable to publish. More theorems and concepts are named after Cauchy than after any other mathematician. His text *Cours d'analyse* (1821), which was based on his own lectures, became the model from which present day calculus texts are descended.

Death was sudden and abrupt. He was talking with the Archbishop of Paris and commented, ''Men pass away, but their deeds abide.'' With those words, he fell over dead.

Chapter 3
The Derivative

In this chapter we return to the problem of calculating the slope of the line tangent to the graph of a function. Using the properties of *limits,* developed in Chapter 2, we extend this idea to the more general concept of the *derivative* of a function, and we develop the properties of derivatives and the rules by which derivatives can be found. Applications of these ideas are presented in Chapter 4.

3.1 THE DERIVATIVE AS A FUNCTION

In Section 2.1 we determined that the slope of the line tangent to the graph of the function f at the point $(x_0, f(x_0))$, if it exists, is the limit

$$m = \lim_{h \to 0} \frac{f(x_0 + h) - f(x_0)}{h}. \tag{1}$$

The quotient $\dfrac{f(x_0 + h) - f(x_0)}{h}$ to which the limit is applied in equation (1) is called a **difference quotient** because it is the ratio of the change in the values of f and the change in x between the points $(x_0, f(x_0))$ and $(x_0 + h, f(x_0 + h))$ on the graph of f. Since we will find interpretations for the limit of this difference quotient other than just as the slope of a tangent, we refer to the limit in equation (1) more generally as the *derivative* of f at x_0, and we denote it by $f'(x_0)$.

DEFINITION 1

The **derivative of the function f at x_0** is the number

$$f'(x_0) = \lim_{h \to 0} \frac{f(x_0 + h) - f(x_0)}{h} \tag{2}$$

provided this limit exists.

Figure 1.1 shows the interpretation of $f'(x_0)$ as the slope of the line tangent to the graph of f at $(x_0, f(x_0))$. Figure 1.2 reminds us that the limit in equations (1) and (2) is "two-sided." The difference quotient $\dfrac{f(x_0 + h) - f(x_0)}{h}$ must be defined for $h < 0$ as well as for $h > 0$, and the limits in equations (1) and (2) must exist as h approaches zero from either the right or the left.

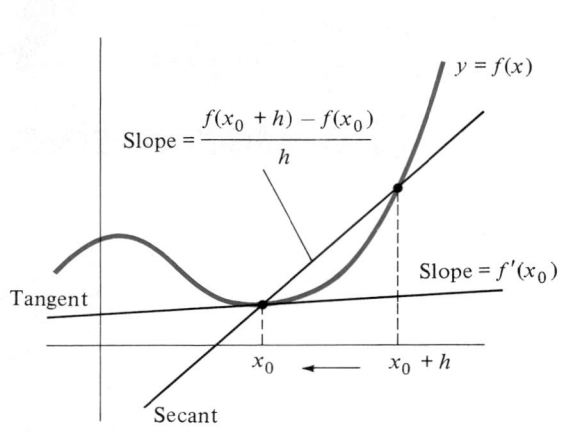

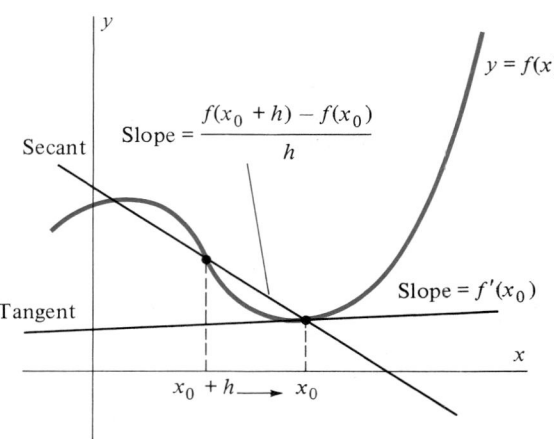

Figure 1.1 Difference quotient $\dfrac{f(x_0 + h) - f(x_0)}{h}$ is slope of the secant. Derivative $f'(x_0)$ is slope of the tangent at $(x_0, f(x_0))$.

Figure 1.2 Difference quotient $\dfrac{f(x_0 + h) - f(x_0)}{h}$ with $h < 0$ is also slope of secant.

We say that the function f is *differentiable* at x_0 if the limit $f'(x_0)$ in Definition 1 exists. We refer to the process of calculating the derivative $f'(x_0)$ as *differentiation* of the function f.

When $f'(x_0)$ exists for every x_0 in an interval I, the process of differentiation actually produces a new function f', defined on the interval I, with values $f'(x)$, $x \in I$.

DEFINITION 2

The derivative of the function f on the interval I, denoted by f', is the function with values

$$f'(x) = \lim_{h \to 0} \frac{f(x + h) - f(x)}{h}$$

provided this limit exists for all $x \in I$.

Since the derivative $f'(x)$ is the slope of the tangent to the graph of f at $(x, f(x))$, we refer to the derivative f' as the *slope function* associated with the function f. Figures 1.3 and 1.4 show the graph of a function f and the graph of the corresponding slope function f'.

You should note that the values $f'(x)$ of the derivative function are the slopes of the tangents to the graph of f. Mastery of the information in this chapter will enable you to determine and graph slope functions.

Example 1

For the linear function $f(x) = mx + b$, the derivative is

$$f'(x) = \lim_{h \to 0} \frac{f(x + h) - f(x)}{h}$$

$$= \lim_{h \to 0} \frac{[m(x + h) + b] - [mx + b]}{h}$$

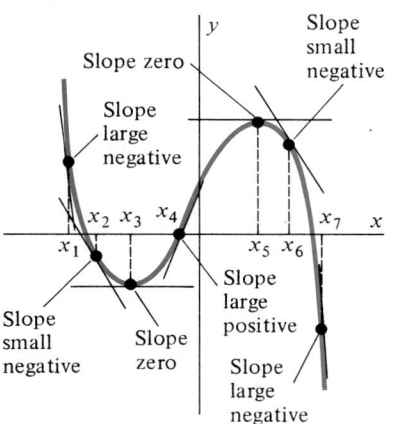

Figure 1.3 Graph of a function f, showing several tangents.

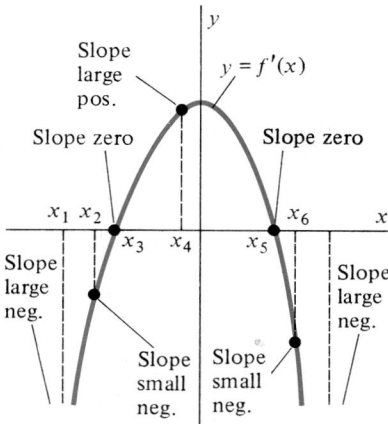

Figure 1.4 Graph of the "slope function" (derivative), f'.

$$= \lim_{h \to 0} \frac{(mx + mh + b) - (mx + b)}{h}$$

$$= \lim_{h \to 0} \frac{mh}{h} \qquad (h \neq 0)$$

$$= \lim_{h \to 0} m$$

$$= m.$$

See Figure 1.5. ◇

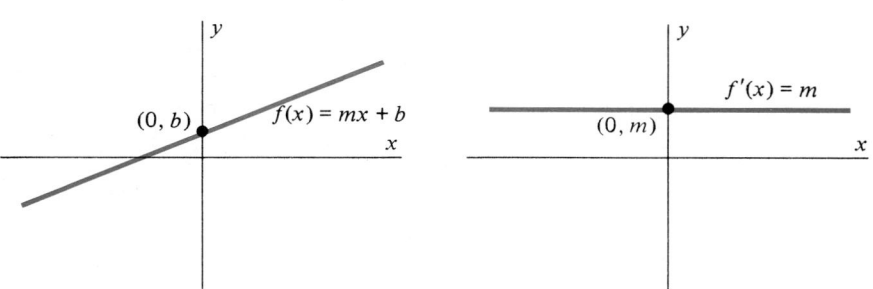

Figure 1.5 For the linear function $f(x) = mx + b$, the slope function (derivative) is $f'(x) = m$.

Example 2

The derivative of the function $f(x) = x^2$ is

$$f'(x) = \lim_{h \to 0} \frac{f(x + h) - f(x)}{h}$$

$$= \lim_{h \to 0} \frac{(x + h)^2 - x^2}{h}$$

$$= \lim_{h \to 0} \frac{(x^2 + 2xh + h^2) - x^2}{h}$$

$$= \lim_{h \to 0} \frac{2xh + h^2}{h}$$

$$= \lim_{h \to 0} (2x + h)$$

$$= 2x.$$

Thus, for $f(x) = x^2$, $f'(0) = 0$, $f'(-3) = -6$, $f'(2) = 4$, and $f'(10) = 20$ (see Figure 1.6). $\diamond$

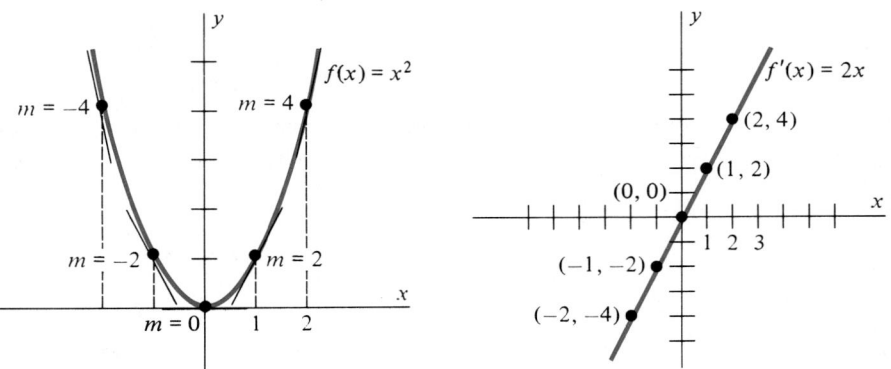

Figure 1.6 For the quadratic function $f(x) = x^2$, the slope function (derivative) is $f'(x) = 2x$.

Example 3

The derivative of the function $f(x) = \sqrt{x}$ is

$$f'(x) = \lim_{h \to 0} \frac{f(x + h) - f(x)}{h}$$

$$= \lim_{h \to 0} \frac{\sqrt{x + h} - \sqrt{x}}{h}$$

$$= \lim_{h \to 0} \left(\frac{\sqrt{x + h} - \sqrt{x}}{h} \right) \cdot \left(\frac{\sqrt{x + h} + \sqrt{x}}{\sqrt{x + h} + \sqrt{x}} \right)$$

$$= \lim_{h \to 0} \frac{(x + h) - x}{h(\sqrt{x + h} + \sqrt{x})}$$

$$= \lim_{h \to 0} \frac{h}{h(\sqrt{x + h} + \sqrt{x})}$$

$$= \lim_{h \to 0} \frac{1}{(\sqrt{x + h} + \sqrt{x})}$$

$$= \frac{1}{2\sqrt{x}}.$$

That is, for $f(x) = \sqrt{x}$, $f'(x) = \dfrac{1}{2\sqrt{x}}$. Thus, $f'(4) = \dfrac{1}{4}$, $f'(9) = \dfrac{1}{6}$, $f'(27) = \dfrac{1}{2\sqrt{27}}$, but $f'(0)$ is not defined (see Figure 1.7). $\diamond$

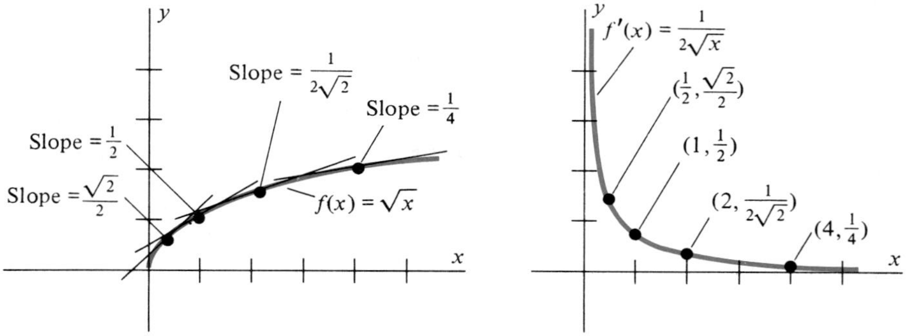

Figure 1.7   For the function $f(x) = \sqrt{x}$, the slope function (derivative) is $f'(x) = \dfrac{1}{2\sqrt{x}}$.

Properties of the Derivative

The goal of the remaining part of this chapter is to determine *properties* of the derivative and *rules* by which derivatives may be calculated. These results will free us from the somewhat cumbersome task of calculating derivatives directly from Definition 2.

Our first result enables us to calculate derivatives of functions that are "built up" from other functions by the processes of addition and multiplication by real numbers.

THEOREM 1

If the functions f and g are differentiable at x and c is any real number, then the functions $f + g$ and cf are also differentiable at x, and

(i) $(f + g)'(x) = f'(x) + g'(x),$ and
(ii) $(cf)'(x) = cf'(x)$.

Proof: The proof that $f + g$ is differentiable and equation (i) follow from the corresponding property of limits: the limit of the sum is the sum of the limits:

$$(f + g)'(x) = \lim_{h \to 0} \frac{(f + g)(x + h) - (f + g)(x)}{h} \qquad \text{(def. of derivative)}$$

$$= \lim_{h \to 0} \frac{[f(x + h) + g(x + h)] - [f(x) + g(x)]}{h} \qquad \text{(def. of } f + g\text{)}$$

$$= \lim_{h \to 0} \left\{ \frac{f(x + h) - f(x)}{h} + \frac{g(x + h) - g(x)}{h} \right\} \qquad \text{(rearranging)}$$

$$= \lim_{h \to 0} \frac{f(x + h) - f(x)}{h} + \lim_{h \to 0} \frac{g(x + h) - g(x)}{h} \qquad \text{(property of limits)}$$

$$= f'(x) + g'(x).$$

The proof of part (ii) is left for you as an exercise.   ◆

Example 4

For $h(x) = 3x^2 + 4\sqrt{x}$ find $h'(x)$.

Solution: This function has the form

$$h(x) = 3f(x) + 4g(x)$$

with

$$f(x) = x^2, \quad \text{and} \quad g(x) = \sqrt{x}.$$

Using Theorem 1 and the results of Examples 2 and 3 we find that

$$h'(x) = 3f'(x) + 4g'(x)$$

$$= 3(2x) + 4\left(\frac{1}{2\sqrt{x}}\right)$$

$$= 6x + \frac{2}{\sqrt{x}}. \qquad \diamond$$

Throughout this chapter we shall make frequent use of the relationship between differentiability and continuity, as described in the following theorem.

THEOREM 2   Let the function f be defined on an open interval containing x_0, If $f'(x_0)$ exists, then f is continuous at $x = x_0$.

Proof: To prove that f is continuous at x_0, we must show that $\lim_{h \to 0} f(x_0 + h) = f(x_0)$ (see Theorem 9, Chapter 2). Using the hypothesis that $f'(x_0)$ exists, and the "Product Rule" for limits (Theorem 1, part [iii], Chapter 2), we can write that

$$\lim_{h \to 0} f(x_0 + h) - f(x_0) = \lim_{h \to 0} [f(x_0 + h) - f(x_0)]$$

$$= \lim_{h \to 0} \left[\frac{f(x_0 + h) - f(x_0)}{h}\right] h$$

$$= \left(\lim_{h \to 0} \left[\frac{f(x_0 + h) - f(x_0)}{h}\right]\right) \cdot \left(\lim_{h \to 0} h\right)$$

$$= f'(x_0) \cdot 0$$

$$= 0,$$

which proves that $\lim_{h \to 0} f(x_0 + h) = f(x_0)$.   ◆

It is important to note that the converse of Theorem 2 is not true. That is, a function may be continuous at a number $x = a$ but not differentiable at a. One such example is the absolute value function $f(x) = |x|$ at $x = 0$. We leave it as an exercise for you to verify that f is continuous at $x = 0$, but that $f'(0)$ does not exist. (See Exercise 32.)

Exercise Set 3.1

1. Each of the Figures (i)–(v) is the graph of the slope function for one of the functions whose graph is one of Figures (a)–(e). Match the corresponding functions and derivatives.

a.

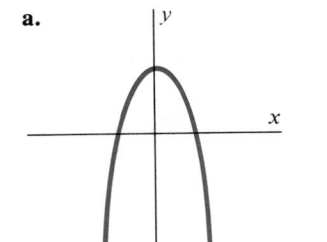

i.

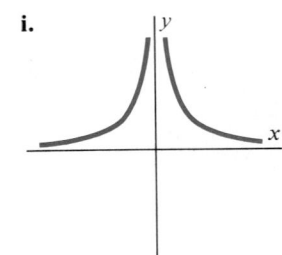

b.

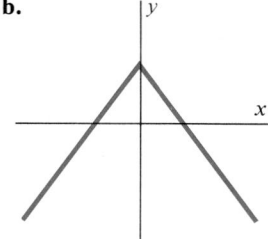

ii.

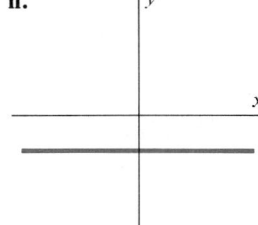

c.

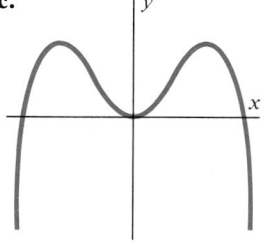

iii.

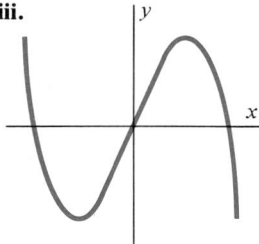

d.

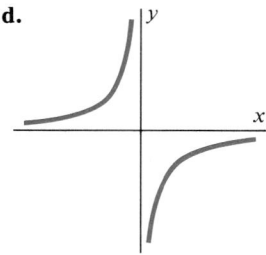

iv.

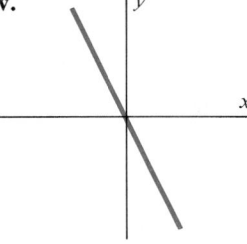

e.

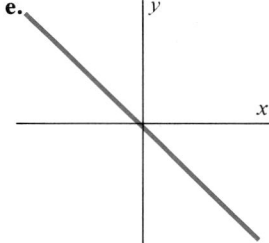

v.

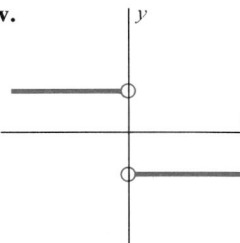

In Exercises 2–20, use Definition 2 to find f'.

2. $f(x) = x^2 - 7$

3. $f(x) = 2x^3 + 3$

4. $f(x) = 1 - x^3$

5. $f(x) = ax^3 + bx^2 + cx + d$

6. $f(x) = \dfrac{1}{x - 1}$

7. $f(x) = \dfrac{1}{2x + 3}$

8. $f(x) = \dfrac{1}{x^2 - 9}$

9. $f(x) = \sqrt{x + 1}$

10. $f(x) = \sqrt{2x + 3}$

11. $f(x) = \dfrac{1}{\sqrt{x + 1}}$

12. $f(x) = \dfrac{1}{ax + b}$

13. $f(x) = \dfrac{1}{x^2}$

14. $f(x) = x^4$

15. $f(x) = (x + 3)^3$

16. $f(x) = (x - 1)^2$

17. $f(x) = \dfrac{1}{\sqrt{x + 5}}$

18. $f(x) = \dfrac{1}{(x + 2)^2}$

19. $f(x) = \dfrac{3}{(x - 1)^2}$

20. $f(x) = \dfrac{-2}{\sqrt{x + 1}}$

21. Find an equation for the line tangent to the graph of $y = \dfrac{1}{2x + 3}$ at the point $(0, 1/3)$.

22. Find an equation for the line tangent to the graph of $f(x) = \sqrt{x + 1}$ at the point $(3, 2)$.

23. A **normal** to a curve at point P is a line through P perpendicular to the tangent. Find an equation for the normal to the graph of $f(x) = x^2 - 7$ at the point $P = (3, 2)$.

24. Find two points where tangents to the graph of $f(x) = x^2$ intersect on the y-axis and are perpendicular (see Figure 1.8).

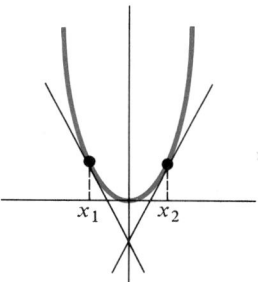

Figure 1.8

25. Find numbers x_1 and x_2 so that the triangle formed by the x-axis and the lines tangent to the graph of $f(x) = 1 - x^2$ at the points $(x_1, f(x_1))$ and $(x_2, f(x_2))$ is equilateral (see Figure 1.9).

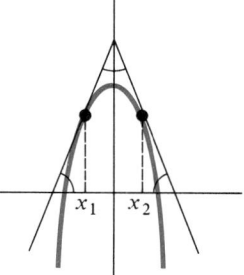

Figure 1.9

26. Find the point where the tangent to the graph of $y = x^2 + 4x + 4$ has slope 4.

27. Find a so that the graph of $y = 2 - ax^2$ has a tangent with slope 6 at $x = -1$.

28. Let $f(x) = ax^2 + bx + 3$. Find a and b so that the tangent to the graph of $y = f(x)$ at $(1, 5)$ has slope 1.

29. Find the constants a and b so that the graph of $y = ax^2 + b$ has tangent $y = 4x$ at the point $P = (1, 4)$.

30. *(Calculator)* Sketch the graph of $f(x) = \sin x$ for $0 \le x \le 2\pi$. At each of the values $x = 0, \pi/4, \pi/2, 3\pi/4, \ldots, 2\pi$ estimate the slope of the tangent by evaluating the difference quotient with $h = \pm 0.5, \pm 0.1, \pm 0.05$, and ± 0.01. Using these results sketch the graph of the "slope function" f'.

31. *(Calculator)* For $f(x) = \cos x$, estimate the value $f'(x)$ at $x = 0, \pi/4, \pi/2, 3\pi/4, \ldots, 2\pi$ by evaluating the difference quotient with $h = \pm 0.5, \pm 0.1, \pm 0.05$, and ± 0.01. Compare these estimates with the values of $g(x) = \sin x$ at the same angles. What relationship do you observe?

32. Show that the function $f(x) = |x|$ is continuous at $x = 0$, but that $f'(0)$ fails to exist. Explain why this shows that the converse of Theorem 2 is false.

33. Let $f(x) = x^3 + 2x^2 - 4x + 10$. For which numbers x is the slope of the line tangent to the graph of f greater than zero?

34. Let $f(x) = x^2 - 6x + 2$. For which numbers x is the slope of the line tangent to the graph of f less than zero?

35. Let

$$f(x) = \begin{cases} x^2 + 1, & x \ge 3 \\ 6x - 8, & x < 3. \end{cases}$$

a. Is f differentiable at $x = 3$? Why?
b. Is f continuous at $x = 3$? Why?

36. True or false: If the statement is true, explain why. If it is false, give a counterexample.
a. If $\lim_{x \to a} f(x)$ exists then f is continuous at a.

b. If f is continuous at a then $\lim_{x \to a} f(x)$ exists.

c. If $\lim_{x \to a} f(x)$ exists, then f is differentiable at a.

d. If f is differentiable at a, then $\lim_{x \to a} f(x)$ exists.

e. If f is continuous at a then f is differentiable at a.
f. If f is differentiable at a then f is continuous at a.

37. A function f has the following property:

$$f(x_1 + x_2) = f(x_1) + 2x_1x_2 + 3x_2 + x_2^2$$

for all numbers x_1, x_2. Use the definition of the derivative to find $f'(x)$.

3.2 RULES FOR CALCULATING DERIVATIVES

In this section we begin to develop a list of rules by which derivatives may be calculated. Before doing so, however, we introduce additional notation for the derivative:

> For the function $y = f(x)$,
>
> $$\frac{d}{dx} f(x) \quad \text{and} \quad \frac{dy}{dx} \quad \text{each mean } f'(x).$$

In other words, the symbol $\dfrac{d}{dx}$ means "the derivative with respect to x of." The notation $\dfrac{d}{dx}$ is usually credited to the mathematician Gottfried Leibniz (1646–1716).

Our first result is that the derivative of a constant function is zero.

THEOREM 3	Let f be the constant function with values $f(x) = c$ for all x. Then $f'(x) = 0$ for all x. That is,

$$\frac{d}{dx}(c) = 0.$$

Proof: By hypothesis, $f(x + h) = c = f(x)$, so

$$f'(x) = \lim_{h \to 0} \frac{f(x + h) - f(x)}{h} = \lim_{h \to 0} \frac{c - c}{h} = \lim_{h \to 0} \frac{0}{h} = \lim_{h \to 0} (0) = 0. \qquad \blacklozenge$$

The Power Rule

In Examples 1 and 2 of Section 3.1 we saw that

for $\quad f(x) = x, \qquad f'(x) = 1,$

and

for $\quad f(x) = x^2, \qquad f'(x) = 2x.$

Similarly, using the definition of the derivative you can show that

for $\quad f(x) = x^3, \qquad f'(x) = 3x^2,$
for $\quad f(x) = x^4, \qquad f'(x) = 4x^3,$
for $\quad f(x) = x^5, \qquad f'(x) = 5x^4,$

and so forth. Each of these results is a special case of the following rule.

THEOREM 4
Power Rule

Let n be any nonzero integer. The function $f(x) = x^n$ is differentiable for all x, and $f'(x) = nx^{n-1}$. That is,

$$\frac{d}{dx} x^n = nx^{n-1}, \qquad n = \pm 1, \pm 2, \dots .$$

Proof: We shall prove this result for the case $n > 1$ here. The proof for negative integers is given in Example 7. Using the Binomial Theorem we can write

$$(x + h)^n = x^n + nx^{n-1}h + \frac{n(n - 1)}{2} x^{n-2}h^2 + \cdots$$

$$+ \frac{n(n - 1)}{2} x^2 h^{n-2} + nxh^{n-1} + h^n.$$

We can therefore write the required difference quotient as follows:

$$\frac{f(x + h) - f(x)}{h} = \frac{(x + h)^n - x^n}{h}$$

$$= nx^{n-1} + \{\text{terms with factors of } h^P, \ P \geq 1\}.$$

We now observe that in the limit as $h \to 0$, all terms involving factors of h^P will approach zero. Thus, we conclude that

$$f'(x) = \lim_{h \to 0} \frac{(x + h)^n - x^n}{h} = nx^{n-1}.$$

(This proof must be regarded as intuitive rather than formal, since we have been vague about the nature of the "terms with factors of h" and their limits as $h \to 0$. A formal proof using the Principle of Mathematical Induction is outlined in Exercise 66.) $\qquad \blacklozenge$

Example 1

(a) For $f(x) = x^{27}$, $f'(x) = 27x^{26}$.

(b) For $y = x^{-4}$, $\dfrac{dy}{dx} = -4x^{-5}$.   ◇

Example 2

Using the Power Rule and Theorems 1 and 3 we find that

(a) $\dfrac{d}{dx}(3x^5 + 4) = \dfrac{d}{dx}(3x^5) + \dfrac{d}{dx}(4)$ (Theorem 1, part (i))

$\qquad\qquad\quad = 3\dfrac{d}{dx}(x^5) + \dfrac{d}{dx}(4)$ (Theorem 1, part (ii))

$\qquad\qquad\quad = 3 \cdot 5x^4 + 0$ (Power Rule and Theorem 3)

$\qquad\qquad\quad = 15x^4.$

That is, for $f(x) = 3x^5 + 4$, $f'(x) = 15x^4$.

(b) $\dfrac{d}{dx}(7x^4 - 5x^{-3}) = 7 \cdot \dfrac{d}{dx}(x^4) - 5\dfrac{d}{dx}(x^{-3})$ (Theorem 1, parts (i) and (ii))

$\qquad\qquad\quad = 7(4x^3) - 5(-3x^{-4})$ (Power Rule)

$\qquad\qquad\quad = 28x^3 + 15x^{-4}.$

That is, for $f(x) = 7x^4 - 5x^{-3}$, $f'(x) = 28x^3 + 15x^{-4}$.   ◇

As Example 2 suggests, the Power Rule may be combined with Theorem 1 to differentiate any polynomial.

Example 3

For $f(x) = 6x^5 - 3x^4 - 2x^3 + 4x^2 - 6x + 5$,

$\qquad f'(x) = 6(5x^4) - 3(4x^3) - 2(3x^2) + 4(2x) - 6(1)$

$\qquad\qquad = 30x^4 - 12x^3 - 6x^2 + 8x - 6.$ ◇

Differentiating Products

We can paraphrase Theorem 1, part (i), by saying that the derivative of the sum is the sum of the derivatives. This observation might lead you to suspect also that the derivative of a product of two functions would be the product of the individual derivatives. However, this is not true. A simple example is found by letting

$\qquad f(x) = x \qquad$ and $\qquad g(x) = x^3$.

Then,

$\qquad f'(x) = 1 \qquad$ and $\qquad g'(x) = 3x^2$.

Now the product is $h(x) = f(x)g(x) = x \cdot x^3 = x^4$, which has derivative $h'(x) = 4x^3$. However, the product of the individual derivatives is $f'(x)g'(x) = (1)(3x^2) = 3x^2$, not $4x^3$.

The correct procedure for differentiating a product is given by the following theorem.

| THEOREM 5 | Let f and g be differentiable at x. Then the product function fg is differentiable at x, and |
| Product Rule | |

$$(fg)'(x) = f'(x)g(x) + f(x)g'(x).$$

Example 4

For $f(x) = (2x + 7)(x - 9)$, find $f'(x)$.

Solution: One way to work this problem is to first multiply the two binomial terms,

$$f(x) = (2x + 7)(x - 9) = 2x^2 - 11x - 63,$$

and then differentiate this polynomial:

$$f'(x) = 4x - 11.$$

Another approach is to begin by applying the Product Rule:

$$f'(x) = \left[\frac{d}{dx}(2x + 7)\right](x - 9) + (2x + 7)\left[\frac{d}{dx}(x - 9)\right]$$
$$= 2(x - 9) + (2x + 7)(1)$$
$$= 4x - 11.$$

Of course, the result of either approach must be the same. ◇

Example 5

For the function $f(x) = (3x^3 - 6x)(9x^4 + 3x^3 + 3)$, it is easier to use the Product Rule to calculate the derivative:

$$f'(x) = \left[\frac{d}{dx}(3x^3 - 6x)\right](9x^4 + 3x^3 + 3) + (3x^3 - 6x)\left[\frac{d}{dx}(9x^4 + 3x^3 + 3)\right]$$
$$= (9x^2 - 6)(9x^4 + 3x^3 + 3) + (3x^3 - 6x)(36x^3 + 9x^2)$$
$$= 189x^6 + 54x^5 - 270x^4 - 72x^3 + 27x^2 - 18. \qquad ◇$$

Proof of Product Rule: By the definition of the derivative

$$(fg)'(x) = \lim_{h \to 0} \frac{f(x + h)g(x + h) - f(x)g(x)}{h}.$$

In order to factor the numerator, we subtract and add the quantity $f(x)g(x + h)$ and use the algebra of limits to obtain

$$(fg)'(x) = \lim_{h \to 0} \frac{[f(x + h)g(x + h) - f(x)g(x + h)] + [f(x)g(x + h) - f(x)g(x)]}{h}$$

$$= \lim_{h \to 0} \left[\left(\frac{f(x + h) - f(x)}{h}\right)g(x + h) + f(x)\left(\frac{g(x + h) - g(x)}{h}\right)\right]$$

$$= \left[\lim_{h \to 0} \frac{f(x + h) - f(x)}{h}\right][\lim_{h \to 0} g(x + h)] +$$

$$[\lim_{h \to 0} f(x)]\left[\lim_{h \to 0} \frac{g(x + h) - g(x)}{h}\right].$$

Now since the function g is differentiable at x it must be continuous at x, by Theorem 2. Thus $\lim\limits_{h \to 0} g(x + h) = g(x)$. Since x is fixed in this argument, $\lim\limits_{h \to 0} f(x) = f(x)$. We therefore conclude that

$$(fg)'(x) = f'(x)g(x) + f(x)g'(x)$$

and the proof is complete. ◆

Our final theorem is the rule for differentiating quotients. It is even more surprising than the rule for products.

THEOREM 6
Quotient Rule

Let f and g be differentiable at x with $g(x) \neq 0$. Then the quotient f/g is differentiable at x and

$$\left(\frac{f}{g}\right)'(x) = \frac{f'(x)g(x) - f(x)g'(x)}{[g(x)]^2}.$$

Example 6

Find $f'(x)$ for $f(x) = \dfrac{3x^2 + 7x + 1}{9 - x^3}$.

Solution: By the Quotient Rule we have

$$f'(x) = \frac{\left[\dfrac{d}{dx}(3x^2 + 7x + 1)\right](9 - x^3) - (3x^2 + 7x + 1)\left[\dfrac{d}{dx}(9 - x^3)\right]}{(9 - x^3)^2}$$

$$= \frac{(6x + 7)(9 - x^3) - (3x^2 + 7x + 1)(-3x^2)}{(9 - x^3)^2}$$

$$= \frac{3x^4 + 14x^3 + 3x^2 + 54x + 63}{x^6 - 18x^3 + 81}. \qquad \diamondsuit$$

Example 7

Prove the Power Rule for differentiating $f(x) = x^{-n}$ where n is a positive integer.

Solution: We write $f(x) = \dfrac{1}{x^n}$ and apply the Quotient Rule and the Power Rule for the case $n > 0$ to obtain

$$f'(x) = \frac{0 \cdot x^n - 1 \cdot nx^{n-1}}{x^{2n}}$$

$$= -nx^{n-1-2n}$$

$$= -nx^{-n-1}. \qquad \diamondsuit$$

Example 8

Find an equation for the line tangent to the graph of $y = x^2 + x^{-2}$ at the point $(1, 2)$.

Strategy

Find $f'(x)$ for $f(x) = x^2 + x^{-2}$ using the Power Rule in (1).

Slope is $f'(1)$.

Solution

For $f(x) = x^2 + x^{-2}$ we have

$$f'(x) = 2x - 2x^{-3}.$$

Then

$$f'(1) = 2 - 2 = 0$$

so the desired line is horizontal. Since this line has slope zero and contains $(1, 2)$ its equation is $y = 2$. ◇

Proof of Quotient Rule: By the definition of the derivative we have

$$\left(\frac{f}{g}\right)'(x) = \lim_{h \to 0} \frac{\dfrac{f(x + h)}{g(x + h)} - \dfrac{f(x)}{g(x)}}{h}$$

$$= \lim_{h \to 0} \frac{f(x + h)g(x) - f(x)g(x + h)}{hg(x)g(x + h)}.$$

We proceed to subtract and add the expression $f(x)g(x)$ in the numerator so as to be able to factor and again use the algebra of limits, to obtain

$$\left(\frac{f}{g}\right)'(x) = \lim_{h \to 0} \frac{[f(x + h)g(x) - f(x)g(x)] + [f(x)g(x) - f(x)g(x + h)]}{hg(x)g(x + h)}$$

$$= \lim_{h \to 0} \frac{\left[\dfrac{f(x + h) - f(x)}{h}\right]g(x) - f(x)\left[\dfrac{g(x + h) - g(x)}{h}\right]}{g(x)g(x + h)}$$

$$= \frac{\left[\lim_{h \to 0} \dfrac{f(x + h) - f(x)}{h}\right]g(x) - f(x)\left[\lim_{h \to 0} \dfrac{g(x + h) - g(x)}{h}\right]}{g(x)[\lim_{h \to 0} g(x + h)]}$$

$$= \frac{f'(x)g(x) - f(x)g'(x)}{[g(x)]^2}$$

(Notice that we have used the continuity of g in the denominator in the step above, just as we did in the proof of the Product Rule.) ◆

Exercise Set 3.2

In Exercises 1–34, find the derivative of the given function.

1. $f(x) = 8x^3 - x^2$

2. $f(x) = x - x^5$

3. $f(x) = ax^3 + bx$

4. $f(x) = a^5 + 3a^2x^2 + x^3$

5. $f(x) = \dfrac{x^2 + 5}{2}$

6. $f(x) = \dfrac{2}{3}x^3 + \dfrac{1}{2}x^2 + x$

7. $f(x) = (x - 1)(x + 2)$

8. $f(x) = (x^2 - 1)(2 - x)$

9. $f(x) = (3x^2 - 8x)(x^2 + 2)$

10. $f(x) = (x^2 + x + 1)(x + 1)$

11. $f(x) = (x^3 - x)^2$

12. $f(x) = \left(x^2 - \dfrac{3}{x^2}\right)^2$

13. $f(x) = 5x^2 + 2x + \dfrac{3}{x} - \dfrac{4}{x^2}$

14. $f(x) = \dfrac{x + 2}{x - 2}$

15. $f(x) = \dfrac{6}{3 - x}$

16. $f(x) = \dfrac{(8x + 2)(x + 1)}{x - 3}$

17. $f(x) = \dfrac{x^4 + 4x + 4}{1 - x^3}$

18. $f(x) = \dfrac{(2x + 1)(3x + 2)}{(x + 1)(x - 1)}$

19. $f(x) = 5x^{-3} - 2x^{-5}$

20. $f(x) = \dfrac{x^2 - 4}{x + 2}$

21. $f(x) = (x - 2)\left(x + \dfrac{1}{x}\right)$

22. $f(x) = \dfrac{1}{(x - 3)^2}$

23. $f(x) = \left(1 + \dfrac{3}{x}\right)^2$

24. $f(x) = \left(\dfrac{x - 1}{x}\right)^3$

25. $g(t) = \dfrac{t}{t^2 + t + 1}$

26. $f(s) = \dfrac{1 - s}{(1 + s)^2}$

27. $f(x) = \left(\dfrac{x + 1}{x - 1}\right)^2$

28. $f(t) = \dfrac{t - 4 + t^2}{t^3 + 3t^2 + 3}$

29. $f(x) = (x^5 + x^{-2})(x^3 - x^{-7})$

30. $f(x) = \dfrac{1}{(x - 6)^3}$

31. $f(x) = \dfrac{ax + b}{cx^2 + d}$

32. $f(u) = (u^2 + 4)^3$

33. $f(x) = \dfrac{x^{-3} - x^4}{x^5}$

34. $f(x) = ((x^2 + 1)^2 + 1)^2$

In each of Exercises 35–44, find f' in two ways—first by the Product Rule, then by first multiplying to eliminate the parentheses.

35. $f(x) = x(x + 1)$

36. $f(x) = (x + 2)(x - 1)^2$

37. $f(x) = (x^2 + 2)(x^2 - 2)$

38. $f(x) = \left(\dfrac{1}{x} + 1\right)\left(3 - \dfrac{2}{x^2}\right)$

39. $f(x) = (x^3 + 7)(3x^4 + x + 9)$

40. $f(x) = (1 - x^2)(1 + x^2)$

41. $f(x) = \left(\dfrac{1}{x + 1}\right)\left(\dfrac{2}{x + 2}\right)$

42. $f(s) = \left(s - \dfrac{3}{s^2}\right)\left(s + \dfrac{5}{s^2}\right)$

43. $f(u) = (u^2 + u + 1)(u^2 - u - 1)$

44. $f(x) = (x^{-2} + x^3)(x^2 - x^{-4})$

45. Use the Product Rule to establish the following formula for the derivative of the product of three functions:

$$(fgh)'(x) = f'(x)g(x)h(x) + f(x)g'(x)h(x) + f(x)g(x)h'(x).$$

In Exercises 46–51, use the result of Exercise 45 to find the derivative.

46. $f(x) = x(x + 1)(x + 2)$

47. $f(s) = (2s - 1)(s - 3)(s^2 + 4)$

48. $f(t) = (t^2 - 7)(3t^5 + t)(t^3 - 9)$

49. $f(x) = \left(\dfrac{1}{x}\right)\left(\dfrac{1}{x + 1}\right)\left(\dfrac{1}{x + 2}\right)$

50. $f(u) = (u^2 - 4)^3$

51. $f(x) = (2x^3 - 6x + 9)^3$

52. Use the Product Rule and the result of Exercise 45 to establish the following differentiation rules.
   a. $(f^2)'(x) = 2f(x)f'(x) \qquad (f^2(x) = [f(x)]^2)$.
   b. $(f^3)'(x) = 3f^2(x)f'(x) \qquad (f^3(x) = [f(x)]^3)$.

   Can you find a rule for differentiating f^4? What about f^n?

53. Find $f'(2)$ for $f(x)$ in Exercise 11.

54. Find $f'(-1)$ for $f(x)$ in Exercise 20.

55. Find $f'(3)$ for $f(x)$ in Exercise 24.

In each of Exercises 56–60, find an equation for the line tangent to the graph of the given function at the given point.

56. $f(x) = 3x^3 - 7$ at the point $(1, -4)$

57. $f(x) = \dfrac{x - 1}{x + 1}$ at the point $(1, 0)$

58. $f(x) = (x^2 + x)(1 - x^3)$ at the point $(2, -42)$

59. $f(x) = \left(1 - \dfrac{1}{x}\right)^2$ at the point $(1, 0)$

60. $f(x) = \dfrac{1}{x^3 + x}$ at the point $\left(2, \dfrac{1}{10}\right)$

61. Find the points at which the tangent to the graph of $f(x) = \dfrac{3x}{2x - 4}$ has slope $m = -3$.

62. Find the constant a if the graph of $y = \dfrac{1}{ax + 2}$ has tangent $4y + 3x - 2 = 0$ at $(0, 1/2)$.

63. Find the constant b if the graph of $y = \dfrac{b}{x^2}$ has tangent $4y - bx - 21 = 0$ when $x = -2$.

64. Use the Product Rule to find a formula for the derivative of $f(x) = \sqrt[3]{x}$. (*Hint:* Use the fact that $f(x)f(x)f(x) = x$.)

65. Use the method of Exercise 64 to find a formula for the derivative of $f(x) = x^{1/n}$, $x \neq 0$, where n is a positive integer.

66. Complete the steps in the following formal proof of Theorem 4 (the Power Rule):
   a. For $n = 1$, show that the function $f(x) = x$ is differentiable for all x and satisfies the equation $f'(x) = 1 \cdot x^{1-1} = x^0 = 1$. This shows that Theorem 4 holds for $n = 1$.

b. Assume that Theorem 4 holds for $n = k$ where k is a positive integer greater than 1.

c. Write $f(x) = x^{k+1} = x(x^k)$ and use assumption (b) and the Product Rule to show that f is differentiable for all x.

d. Use assumption (b) and the Product Rule to show that $f'(x) = (k + 1)x^k$ for all x.

e. Apply the Principle of Mathematical Induction to conclude that Theorem 4 holds for all integers $n \geq 1$.

3.3 DERIVATIVES OF THE TRIGONOMETRIC FUNCTIONS

The purpose of this section is to use the differentiation rules of Section 2, together with two limits established in Chapter 2, to find derivatives for all six of the trigonometric functions which were reviewed in Section 1.5.

The derivatives for $\sin x$ and $\cos x$ are given by the following theorem. The derivatives of the other four trigonometric functions follow from this theorem and the Quotient Rule.

THEOREM 7

The function $f(x) = \sin x$ is differentiable for all x, and $f'(x) = \cos x$. The function $g(x) = \cos x$ is differentiable for all x, and $g'(x) = -\sin x$. That is,

$$\frac{d}{dx} \sin x = \cos x \tag{1}$$

$$\frac{d}{dx} \cos x = -\sin x \tag{2}$$

Figure 3.1 shows the graphs of $f(x) = \sin x$ and $f'(x) = \cos x$ and reminds us of

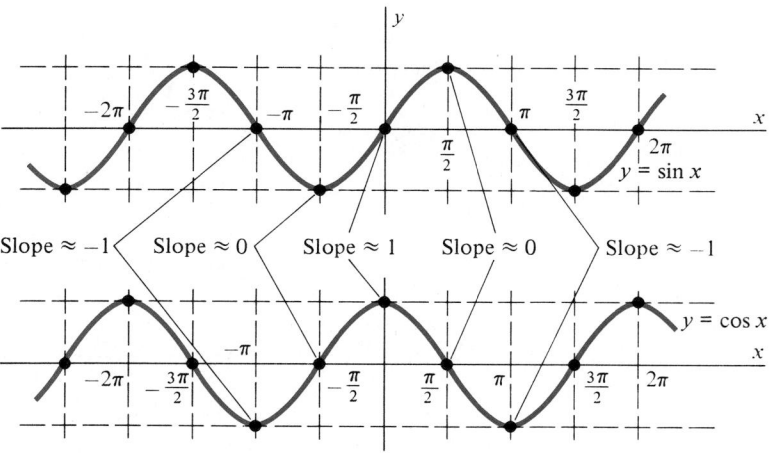

Figure 3.1 Graphs of $f(x) = \sin x$ and its derivative $f'(x) = \cos x$.

the "slope function" interpretation for f': the *slope* of the graph of $f(x) = \sin x$ at $x = x_0$ is given by the *value* $f'(x_0) = \cos x_0$ of the derivative at $x = x_0$. Since the function $f(x) = \sin x$ is *periodic,* with period 2π, this geometric interpretation of equation (1) means that the derivative $f'(x) = \cos x$ must also be periodic with period 2π, a fact that we have already noted in Section 1.5.

Proof of Theorem 7: The proof of this theorem uses the limit

$$\lim_{h \to 0} \frac{\sin h}{h} = 1, \tag{3}$$

established in Section 2.4, and the limit

$$\lim_{h \to 0} \frac{1 - \cos h}{h} = 0, \tag{4}$$

established in Exercise 48, Section 2.4.

To prove that $\dfrac{d}{dx} \sin x = \cos x$ we use the basic definition of the derivative and the addition rule for sines: $\sin(\alpha + \beta) = \sin \alpha \cos \beta + \cos \alpha \sin \beta$:

$$\begin{aligned}
\frac{d}{dx} \sin x &= \lim_{h \to 0} \frac{\sin(x + h) - \sin x}{h} \\
&= \lim_{h \to 0} \frac{(\sin x \cos h + \cos x \sin h) - \sin x}{h} \\
&= \lim_{h \to 0} \frac{\sin x(\cos h - 1) + \cos x \sin h}{h} \\
&= \lim_{h \to 0} \left\{ \sin x \left(\frac{\cos h - 1}{h} \right) + \cos x \left(\frac{\sin h}{h} \right) \right\} \\
&= \sin x \left(\lim_{h \to 0} \frac{\cos h - 1}{h} \right) + \cos x \left(\lim_{h \to 0} \frac{\sin h}{h} \right) \\
&= (\sin x)(0) + (\cos x)(1) \qquad \text{(limits (3) and (4))} \\
&= \cos x.
\end{aligned}$$

To prove that $\dfrac{d}{dx} \cos x = -\sin x$ we proceed as above using the addition rule for cosines: $\cos(\alpha + \beta) = \cos \alpha \cos \beta - \sin \alpha \sin \beta$:

$$\begin{aligned}
\frac{d}{dx} \cos x &= \lim_{h \to 0} \frac{\cos(x + h) - \cos x}{h} \\
&= \lim_{h \to 0} \frac{(\cos x \cos h - \sin x \sin h) - \cos x}{h} \\
&= \lim_{h \to 0} \frac{\cos x(\cos h - 1) - \sin x \sin h}{h} \\
&= \cos x \left(\lim_{h \to 0} \frac{\cos h - 1}{h} \right) - \sin x \left(\lim_{h \to 0} \frac{\sin h}{h} \right) \\
&= (\cos x)(0) - \sin x(1) \qquad \text{(limits (3) and (4))} \\
&= -\sin x.
\end{aligned}$$

Example 1

For $y = x^3 \sin x$ the derivative is

$$\begin{aligned}
\frac{d}{dx}(x^3 \sin x) &= \left(\frac{d}{dx} x^3 \right)(\sin x) + x^3 \left(\frac{d}{dx} \sin x \right) \qquad \text{(Product Rule)} \\
&= 3x^2 \sin x + x^3 \cos x. \qquad \text{(Equation (1))}
\end{aligned}$$

Example 2

Find $\dfrac{dy}{dx}$ for $y = \dfrac{2 - \cos x}{2 + \cos x}$.

Strategy

Apply Quotient Rule first.
Then apply equation (2).

Solution

$$\frac{dy}{dx} = \frac{(2 + \cos x)\left[\dfrac{d}{dx}(2 - \cos x)\right] - \left[\dfrac{d}{dx}(2 + \cos x)\right](2 - \cos x)}{(2 + \cos x)^2}$$

$$= \frac{(2 + \cos x)[-(-\sin x)] - (-\sin x)(2 - \cos x)}{(2 + \cos x)^2}$$

$$= \frac{4 \sin x}{(2 + \cos x)^2}. \qquad \diamondsuit$$

Example 3

Find the derivative of $y = \tan x$, where it is defined, and determine for which values of x the function $y = \tan x$ is differentiable.

Strategy

Write $\tan x = \dfrac{\sin x}{\cos x}$.

Apply Quotient Rule.

Use identity
$\cos^2 x + \sin^2 x = 1$.

$\dfrac{1}{\cos x} = \sec x$.

Solution

$$\frac{d}{dx} \tan x = \frac{d}{dx}\left(\frac{\sin x}{\cos x}\right)$$

$$= \frac{\cos x \cdot \dfrac{d}{dx}(\sin x) - \sin x \cdot \dfrac{d}{dx}(\cos x)}{\cos^2 x}$$

$$= \frac{\cos^2 x + \sin^2 x}{\cos^2 x}$$

$$= \left(\frac{1}{\cos x}\right)^2$$

$$= \sec^2 x,$$

that is,

$$\frac{d}{dx} \tan x = \sec^2 x.$$

Both $\tan x$ and $\sec x$ are defined for all x for which $\cos x \neq 0$. Since $\cos x$ is the only factor of the denominator in the above calculations, $\tan x$ is differentiable at all values of x with $\cos x \neq 0$ (i.e., except at odd multiples of $\pi/2$).

$\diamondsuit$

Since each of the remaining trigonometric functions is defined as a ratio involving $\sin x$, $\cos x$, or both, we may proceed as in Example 3 to determine their derivatives (see Exercise 29). The results are the following.

$$\frac{d}{dx} \tan x = \sec^2 x. \tag{5}$$

$$\frac{d}{dx} \cot x = -\csc^2 x. \tag{6}$$

$$\frac{d}{dx} \sec x = \sec x \tan x. \tag{7}$$

$$\frac{d}{dx} \csc x = -\csc x \cot x. \tag{8}$$

Example 4

Find $f'(x)$ for $f(x) = \sec x \cdot \tan x$.

Solution: Using the Product Rule and equations (5) and (7) we obtain

$$\frac{d}{dx}(\sec x \cdot \tan x) = \left(\frac{d}{dx} \sec x\right) \tan x + \sec x \left(\frac{d}{dx} \tan x\right)$$
$$= (\sec x \cdot \tan x) \tan x + \sec x(\sec^2 x)$$
$$= \sec x(\tan^2 x + \sec^2 x). \qquad \diamond$$

Example 5

Find $\dfrac{dy}{dx}$ for $y = \dfrac{\csc x}{1 + \cot x}$.

Solution: By the Quotient Rule and equations (6) and (8) we obtain

$$\frac{dy}{dx} = \frac{(1 + \cot x)\left(\dfrac{d}{dx} \csc x\right) - (\csc x)\left[\dfrac{d}{dx}(1 + \cot x)\right]}{(1 + \cot x)^2}$$
$$= \frac{(1 + \cot x)(-\csc x \cot x) - (\csc x)(-\csc^2 x)}{(1 + \cot x)^2}$$
$$= \frac{\csc^3 x - \csc x \cot x - \csc x \cot^2 x}{(1 + \cot x)^2}$$
$$= \frac{\csc x(\csc^2 x - \cot^2 x - \cot x)}{(1 + \cot x)^2}$$
$$= \frac{\csc x(1 - \cot x)}{(1 + \cot x)^2} \qquad (\csc^2 x - \cot^2 x = 1). \qquad \diamond$$

Example 6

Find an equation for the line tangent to the graph of $f(x) = \tan^2 x$ at the point $(\pi/4, 1)$.

Strategy	*Solution*
	The derivative is

Write $\tan^2 x$ as a product $(\tan x \cdot \tan x)$ and find $f'(x)$ using the Product Rule.

$$\frac{d}{dx}(\tan^2 x) = \frac{d}{dx}(\tan x \cdot \tan x)$$

$$= \left(\frac{d}{dx}\tan x\right)(\tan x) + (\tan x)\left(\frac{d}{dx}\tan x\right)$$

$$= \sec^2 x \cdot \tan x + \tan x \cdot \sec^2 x$$

$$= 2\sec^2 x \cdot \tan x.$$

Find the slope which is

$$m = f'\left(\frac{\pi}{4}\right).$$

So the slope of the tangent at $(\pi/4,\ 1)$ is

$$f'(\pi/4) = 2\left(\sec\frac{\pi}{4}\right)^2 \tan\frac{\pi}{4}$$

$$= 2(\sqrt{2})^2(1)$$

$$= 4.$$

Use slope and point $(\pi/4,\ 1)$ to write an equation for the tangent.

An equation for the tangent is therefore

$$y - 1 = 4\left(x - \frac{\pi}{4}\right)$$

or

$$4x - y = \pi - 1. \qquad \diamondsuit$$

Exercise Set 3.3

In Exercises 1–20 find $f'(x)$.

1. $f(x) = 4\cos x$

2. $f(x) = x\sin x$

3. $f(x) = x^3 \tan x$

4. $f(x) = \sin x \cdot \cos x$

5. $f(x) = (x^3 - 2)\cot x$

6. $f(x) = \cot x \cdot \csc x$

7. $f(x) = \sin x \cdot \sec x$

8. $f(x) = \sin^2 x$

9. $f(x) = x\cos x - x\sin x$

10. $f(x) = \cos x(x - \cot x)$

11. $f(x) = \sec x \cdot \tan x$

12. $f(x) = \csc^2 x \cot x$

13. $f(x) = \dfrac{x}{2 + \sin x}$

14. $f(x) = \dfrac{\tan x}{1 + x^2}$

15. $f(x) = \dfrac{\sin x - \cos x}{1 + \tan x}$

16. $f(x) = \dfrac{1 - \sin x}{1 + \sin x}$

17. $f(x) = \dfrac{x^2 + 4\cot x}{x + \tan x}$

18. $f(x) = \dfrac{x^2 + 4}{2 + \sec x}$

19. $f(x) = \dfrac{3\csc x}{4x^2 - 5\tan x}$

20. $f(x) = x\csc x - \dfrac{x}{\cot x}$

21. Find an equation for the line tangent to the graph of $y = x\sin x$ at the point $(\pi,\ 0)$.

22. Find an equation for the line tangent to the graph of $f(x) = \csc x \cdot \cot x$ at the point $\left(\dfrac{\pi}{4},\ \sqrt{2}\right)$.

23. For which number(s) x in the interval $[0,\ 4\pi]$ is the tangent to the graph of $y = \sec x$ horizontal?

24. Use the identity $\sin 2x = 2\sin x\cos x$ to find the derivative of the function $f(x) = \sin 2x$.

25. Use the identity $\sin^2 x = \frac{1}{2}(1 - \cos 2x)$ to find the derivative of $f(x) = \sin^2 x$.

26. Use the Product Rule to show that $\dfrac{d}{dx}\tan^2 x = \dfrac{d}{dx}\sec^2 x$.

27. Find the number(s) x in the interval $[0,\ 2\pi]$ for which the tangents to the graphs of $y = \sin x$ and $y = \cos x$ are
a. parallel
b. perpendicular.

28. Find all numbers x in the interval $[0,\ 2\pi]$ for which the graph of $y = f(x)$ has a horizontal tangent for
a. $f(x) = \cos x$ b. $f(x) = \sec x$ c. $f(x) = \csc x$.

29. Derive equations (6), (7), and (8).

30. Determine where the trigonometric functions $\cot x$, $\sec x$, and $\csc x$ are differentiable on the interval $[0,\ 2\pi]$.

3.4 THE DERIVATIVE AS VELOCITY

In addition to giving the slope of the line tangent to the graph of a function, the derivative may be used to define the *velocity* of a moving object.

Imagine an object moving along a line, such as a jogger on a footpath or an automobile on a highway. In physics, we define the **velocity** of such an object by the equation

$$\text{velocity} = \frac{\text{change in distance}}{\text{change in time}}. \tag{1}$$

Of course, what is meant by equation (1) is really an *average* velocity for the time period in question. For example, if a jogger wearing both a pedometer and a watch finds that she has traveled 12 kilometers in 45 minutes, the velocity associated with this time interval is

$$\text{velocity} = \frac{12 \text{ kilometers}}{3/4 \text{ hour}} = 16 \text{ kilometers per hour}.$$

However, at various times during the run the jogger will quite likely have been moving both faster and slower than 16 kilometers per hour.

There is a special setting in which we can use the theory of the derivative to define the velocity of an object *at each instant,* rather than having to settle for the average velocity over a finite time interval. First, the motion of the object must be along a line (we call this **rectilinear** motion), rather than along a general curve. Second, we must have available a **position function,** s, giving the location $s(t)$ of the object along the line at each time t (see Figure 4.1).

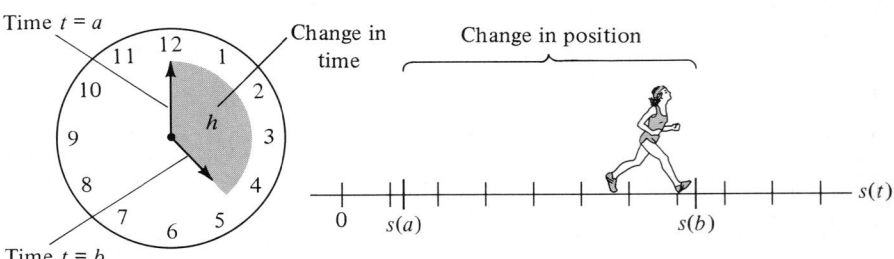

Figure 4.1 A *position function* $s(t)$ gives the location of an object along a (number) line at time t.

For example, in physics we are told that, near the surface of the earth, a freely falling body will have fallen $4.9t^2$ meters t seconds after its release. The motion of such an object is therefore along a (vertical) line, and its position function is $s(t) = -4.9t^2$. (The negative sign indicates position *below* the starting point. See Figure 4.2.)

To define the velocity of an object at time t_0 (sometimes referred to as the **instantaneous** velocity at time t_0), we begin by observing that if $h \neq 0$, then $s(t_0 + h) - s(t_0)$ is the change in the position of the object corresponding to the time interval with endpoints t_0 and $t_0 + h$. Thus, the expression

$$\left\{ \begin{matrix} \text{average} \\ \text{velocity} \end{matrix} \right\} = \frac{s(t_0 + h) - s(t_0)}{h} \tag{2}$$

is precisely the (average) velocity as defined by equation (1).

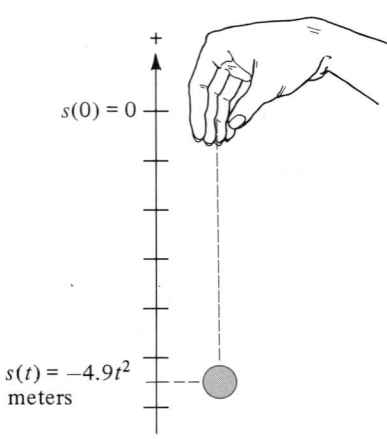

Figure 4.2 A body falling freely from rest falls $4.9t^2$ meters in t seconds.

We now argue that as $h \to 0$, the average velocity corresponding to the (shrinking) time interval with endpoints t_0 and $t_0 + h$ should provide an increasingly accurate measure of the velocity *at the instant* $t = t_0$. For this reason, we define the velocity at time $t = t_0$ to be the limiting value of these average velocities. That is, we *define* the velocity, $v(t_0)$, as

$$v(t_0) = \lim_{h \to 0} \frac{s(t_0 + h) - s(t_0)}{h} = s'(t_0) \tag{3}$$

whenever this limit exists. Equation (3) simply states that we have *defined* velocity as the derivative of the position function.

DEFINITION 3

If the differentiable function s gives the position at time t of an object moving along a line, then the velocity $v(t)$ at time t is the derivative

$$v(t) = s'(t).$$

That is,

$$v(t) = \frac{d}{dt} s(t).$$

Example 1

Starting at time $t = 0$, a particle moves along a line so that its position after t seconds is $s(t) = t^2 - 6t + 8$ meters.

(a) Find its velocity at time t.
(b) When is its velocity zero?

Solution: According to Definition 3, the velocity at time t is

$$v(t) = s'(t) = 2t - 6.$$

Setting $v(t) = 0$ gives $2t - 6 = 0$, so $t = 3$. The particle has zero velocity after 3 seconds (see Figures 4.3 and 4.4). ◇

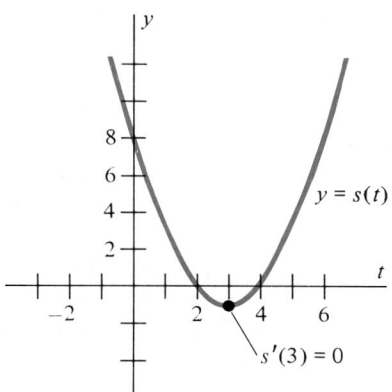

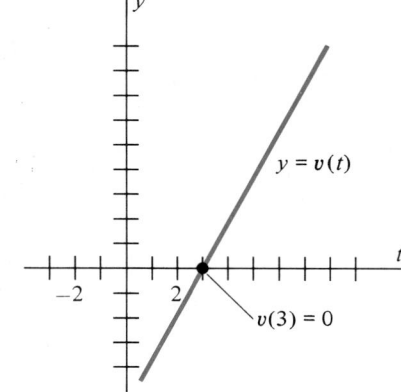

Figure 4.3 Position function $s(t) = t^2 - 6t + 8$.

Figure 4.4 Velocity function $v(t) = s'(t) = 2t - 6$.

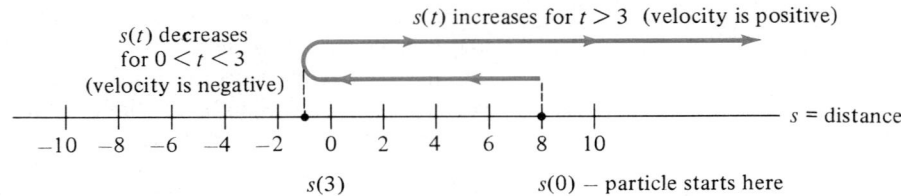

Figure 4.5 For $s(t) = t^2 - 6t + 8$ m/s, velocity is negative for $0 < t < 3$ and positive for $t > 3$.

Notice that the particle in Example 1 lies $s(0) = 8$ meters from the origin at time $t = 0$, but that as t increases from 0 to 3, $s(t)$ decreases from 8 meters to $s(3) = -1$ meter. If we think of the particle as moving along a number line, this says that the particle moves to the left for $0 \leq t < 3$ (Figure 4.5). This corresponds to the observation that $v(t)$ is negative for $t < 3$. For $t > 3$, the position $s(t)$ increases as t increases, so $v(t) > 0$ for $t > 3$.

Velocity versus Speed

In using velocity to describe motion along a line it is important that the sign (+ or −) associated with velocity agree with the choice of which direction along the line corresponds to positive values of position. We shall follow the usual conventions that:

(a) for horizontal motion, positive values of position lie to the *right* of the origin ($s = 0$), so motion to the right corresponds to positive velocity, while negative velocity indicates motion to the left;

(b) for vertical motion, positive values of position lie *above* the origin, so positive velocity indicates motion upward, and negative velocity is motion downward.

Speed, on the other hand, refers to the magnitude of velocity independent of its sign. Thus

$$\text{speed} = |v(t)|.$$

We may summarize the relationship between speed and velocity by saying that velocity indicates both a speed (its absolute value) and a direction (its sign).

Example 2

A pebble is dropped from an open window 100 meters above the ground. Find its velocity and its speed when it strikes the ground.

Solution: The motion of the pebble is along a vertical line. Following convention, we take the upward direction as positive. Also, we take the origin $s = 0$ to be at the height of the window (Figure 4.6). As previously stated, after t seconds the freely falling pebble will have position $s(t) = -4.9t^2$ meters. It will strike the ground when

$$s(t_0) = -4.9t_0^2 = -100 \text{ meters,}$$

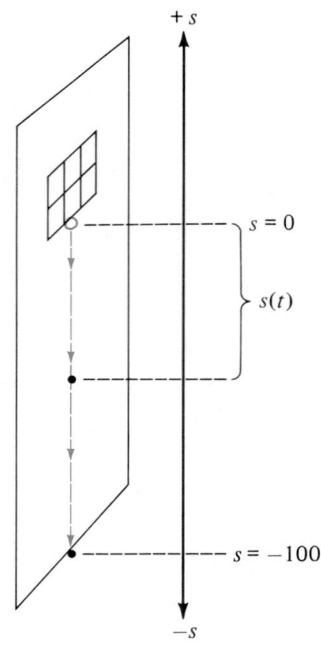

Figure 4.6

or when

$$t_0 = \sqrt{\frac{100}{4.9}} \approx 4.5 \text{ seconds.}$$

Since the corresponding velocity function is

$$v(t) = s'(t) = -2(4.9)t = -9.8t \text{ m/s,}$$

the velocity at the time of impact will be approximately

$$v(t_0) = v(4.5) = -(9.8)(4.5) = -44.1$$

meters per second (downward). The *speed* at impact will be

$$|v(4.5)| = |-44.1| = 44.1 \text{ m/s.} \qquad \diamond$$

Example 3

An object launched vertically upward from ground level with an initial velocity of 72 m/s will be located

$$s(t) = 72t - (4.9)t^2$$

meters above ground level after t seconds. According to this position function,

(a) when does the object stop rising?
(b) what is its maximum height?

Strategy

Establish an orientation for the line of motion and locate the origin.

Solution

From the statement of the problem we know that the motion is vertical, the origin is at ground level, and the positive direction is upward. The velocity function is

Calculate $v(t) = s'(t)$.

$$v(t) = \frac{d}{dt}(72t - 4.9t^2)$$
$$= 72 - 9.8t.$$

The object stops rising when $v(t_0) = 0$. Solve this equation for t_0.

Setting $v(t_0) = 0$ gives

$$72 - 9.8t_0 = 0$$

or

$$t_0 = \frac{72}{9.8} \approx 7.35 \text{ seconds.}$$

Maximum height is $s(t_0)$.

This is the time at which the object is at maximum height. The maximum height is approximately

$$s(t_0) = 72(7.35) - (4.9)(7.35)^2$$
$$= 264.5 \text{ meters.} \qquad \diamond$$

Example 4

Figure 4.7 shows a block attached to the end of a spring on a horizontal surface that we shall assume to be frictionless. If the block is oscillating back and forth along the

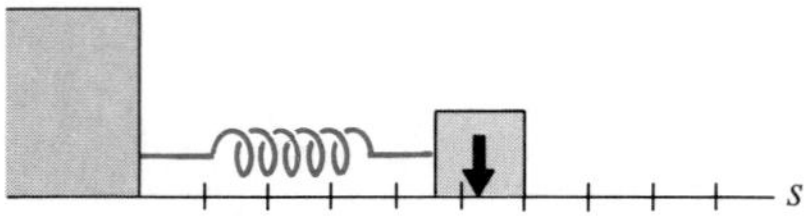

Figure 4.7 Block oscillating on a frictionless surface.

surface so that at time t (in seconds) the pointer is

$$s(t) = 20 + 6 \sin t$$

cm to the right of the fixed end of the spring, find

(a) its velocity at time t,
(b) its velocity at time $t = \pi$ seconds, and
(c) the times at which the block changes direction.

Strategy

(a) Find $v(t) = s'(t)$.

(b) Find $v(\pi)$.

(c) Set $v(t) = 0$ and solve.

Solution

(a) To find the velocity we find the derivative

$$v(t) = s'(t) = \frac{d}{dt}(20 + 6 \sin t)$$

$$= 6 \cos t.$$

(b) The velocity at time $t = \pi$ is

$$v(\pi) = 6 \cos \pi = 6(-1) = -6 \text{ cm/s}.$$

(c) The block changes direction at times t for which $v(t) = 0$.

The equation

$$v(t) = 6 \cos t = 0$$

has solutions $t = \dfrac{\pi}{2} + n\pi, \quad n = 0, 1, 2, \ldots$ ◇

Exercise Set 3.4

In Exercises 1–10, the function s gives the position of a particle moving along a line. Find **a.** the velocity function $v(t)$, **b.** the time(s) $t_0 \geq 0$ at which velocity is zero, and **c.** the intervals in $[0, \infty)$ on which velocity is positive.

1. $s(t) = 3t - 2$

2. $s(t) = t^2 - 6t + 4$

3. $s(t) = \dfrac{1}{1 + t}$

4. $s(t) = t^2 - 3t + 5$

5. $s(t) = t^3 - 9t^2 + 24t + 10$

6. $s(t) = t(t - 1)(t + 2)$

7. $s(t) = \sin t + \cos t$

8. $s(t) = 5 + 10 \cos t$

9. $s(t) = t^3 - 6t^2 + 9t + 7$

10. $s(t) = t^4 - 4t + 4$

11. A particle moves along a line so that its position at time t is
$$s(t) = \frac{t^2 + 2}{t + 1}$$
units. Find its velocity at time $t = 3$.

12. A particle moves along a line so that after t seconds its position is $s(t) = 6 + 5t - t^2$. Find its maximum distance from the origin during the time interval $[0, 6]$.

13. A particle moves along a line so that at time t its position is $s(t) = 6t - t^2$.
 a. What is its initial velocity, that is, v_0?
 b. When does it change direction?
 c. How fast is it moving when it reaches the origin the second time?

14. If a particle is projected vertically upward from ground level with an initial velocity v_0, its height after t seconds is $s(t) = v_0 t - 4.9t^2$ meters. Suppose $v_0 = 98$ meters per second.
 a. What is the velocity of the particle at time t?
 b. At what time does the particle reach its maximum height? (*Hint:* What is its velocity at maximum height?)

c. What is the maximum height?

d. How fast is it moving when it strikes the ground?

15. An object is dropped from rest and strikes the ground with velocity $v = -49$ m/s.

a. From what height was it dropped?

b. How long did it fall?

16. When an object is launched vertically from an initial height of s_0 meters with an initial velocity of v_0 m/s, its position after t seconds is

$$s(t) = s_0 + v_0 t - 4.9 t^2 \text{ meters.}$$

a. Find its velocity after 4 seconds.

b. Find its speed after 4 seconds.

17. Refer to Exercise 14. A flare is launched vertically from a tower 40 meters above ground level with an initial velocity of $v_0 = 400$ m/s.

a. Find its position function s.

b. Find its maximum height.

c. Find its speed when it strikes the ground (assuming it misses the tower on its way down).

18. A piece of tail section breaks loose from an airliner flying at an altitude of 10,000 meters. Neglecting air resistance,

a. when does the piece strike the ground?

b. with what velocity does the piece strike the ground?

19. A piston moves up and down in an engine block so that at time t it is $s(t) = 5 - 4 \sin t$ cm from the top of its cylinder. Find the time t at which the piston changes direction.

20. A mass at the end of a vibrating spring is located $s(t) = 30 + 10 \cos t$ cm from the fixed end of the spring t seconds after it is set in motion. If the motion is horizontal, find

a. the velocity at time t.

b. the times at which the mass changes direction.

21. The height of an object t seconds after it is thrown vertically into the air is given by the position function

$$s(t) = 800t - 16t^2.$$

a. Sketch the graph of the function $y = s(t)$.

b. Find an equation for its velocity, $v(t)$.

c. For which times t is the object rising?

d. For which times t is the object falling?

e. For which time(s) t is the object at rest before it strikes the ground?

f. What is the maximum height reached?

g. Determine the time required for the object to reach a height of 9600 feet.

h. When does the object strike the ground?

3.5 HIGHER ORDER DERIVATIVES

Since differentiation produces a new function f' from a given function f, it is entirely reasonable to ask what happens if we differentiate the function f'. The answer is simply that if f' itself is a differentiable function we will obtain its derivative, $(f')'$. We call this the **second derivative** of f, and we denote it by f''. That is,

$$f''(x) = (f')'(x).$$

The Leibniz notation for the second derivative of $y = f(x)$ is

$$\frac{d^2 y}{dx^2} = \frac{d}{dx}\left(\frac{dy}{dx}\right).$$

Similarly, beginning with f and differentiating three times produces the third derivative, which is denoted by

$$f'''(x), \quad \text{or} \quad \frac{d^3 y}{dx^3}$$

and so on for higher order derivatives. For $n \geq 4$, $f^{(n)}(x)$ denotes the value of the nth derivative of f at x.

Example 1

For the polynomial function

$$f(x) = x^4 - 4x^3 + 3x^2 - 5x + 3$$

we have

$$f'(x) = 4x^3 - 12x^2 + 6x - 5,$$
$$f''(x) = 12x^2 - 24x + 6,$$
$$f'''(x) = 24x - 24,$$
$$f^{(4)}(x) = 24,$$

and

$$f^{(n)}(x) = 0 \qquad \text{for} \qquad n \geq 5. \qquad \diamond$$

Example 2

Find the first three derivatives for the function

$$f(x) = x^3 + x \sin x.$$

Solution:
$$f'(x) = 3x^2 + \sin x + x \cos x$$
$$f''(x) = 6x + \cos x + \cos x - x \sin x$$
$$= 6x + 2 \cos x - x \sin x$$
$$f'''(x) = 6 - 2 \sin x - \sin x - x \cos x$$
$$= 6 - 3 \sin x - x \cos x \qquad \diamond$$

Example 3

It is an important property of the functions $f(x) = \sin x$ and $g(x) = \cos x$ that they reappear in the list of their own derivatives:

$$\begin{array}{ll}
f(x) = \sin x & g(x) = \cos x \\
f'(x) = \cos x & g'(x) = -\sin x \\
f''(x) = -\sin x & g''(x) = -\cos x \\
f'''(x) = -\cos x & g'''(x) = \sin x \\
f^{(4)}(x) = \sin x & g^{(4)}(x) = \cos x \\
f^{(5)}(x) = \cos x & g^{(5)}(x) = -\sin x \\
\text{etc.} & \text{etc.}
\end{array}$$

From these lists you can see that both $\sin x$ and $\cos x$ satisfy the *differential* (or, *derivative*) equations

(i) $f''(x) = -f(x)$
(ii) $f^{(4)}(x) = f(x)$.

Equation (i) is important in the modelling of oscillatory phenomena (such as problems in engineering involving springs), while equation (ii) occurs in problems involving the deflection of beams under heavy loading. We will have much more to say about differential equations later in the text. $\diamond$

Acceleration: For a particle moving along a line, acceleration is defined to be the (instantaneous) rate of change of velocity. Thus, if v is a differentiable velocity function, the acceleration $a(t)$ is defined as

$$a(t) = \lim_{h \to 0} \frac{v(t + h) - v(t)}{h}$$
$$= v'(t).$$

That is, acceleration is the first derivative of velocity. Recall that if $s(t)$ is the position of the particle, then $v(t) = s'(t)$, so acceleration is the second derivative of position:

$$a(t) = v'(t) = s''(t),$$

or

$$a = \frac{dv}{dt} = \frac{d^2s}{dt^2}.$$

Example 4

A particle moves along a line so that at time t seconds its position is $s(t) = t^3 - 6t^2 + 7t - 2$ meters.

(a) Find the acceleration, $a(t)$, at time t.
(b) Find the initial acceleration, $a(0)$.
(c) Find the time t_0 at which acceleration equals zero.

Solution: We have $v(t) = s'(t) = 3t^2 - 12t + 7$, so

$$a(t) = s''(t) = 6t - 12.$$

Thus the initial acceleration is $a(0) = -12$ m/s^2, and

$$a(t) = 0 \quad \text{when} \quad t = 2 \text{ seconds.} \qquad \diamondsuit$$

Example 5

A projectile is fired vertically upward from ground level with an initial velocity of 100 meters per second. In such a case the distance of the particle above ground level is given by the function $s(t) = 100t - 4.9t^2$ meters. Find the acceleration of the particle.

Solution: Here $v(t) = s'(t) = 100 - 9.8t$ m/s, so

$$a(t) = v'(t) = -9.8 \text{ m/s}^2.$$

(Note that the acceleration is both constant and downward. It is referred to as the **acceleration due to gravity.**) $\qquad \diamondsuit$

Exercise Set 3.5

In each of Exercises 1–14, find $f''(x)$.

1. $f(x) = 2x^3 - 6x$

2. $f(x) = a - bx - cx^2$

3. $f(x) = x^5 - 3x^{-2}$

4. $f(x) = \tan x$

5. $f(x) = \dfrac{1}{1 + x}$

6. $f(x) = \dfrac{ax + b}{cx + d}$

7. $f(x) = \sin^2 x$

8. $f(x) = \sin x \cdot \tan x$

9. $f(x) = \dfrac{1}{1 + \cos x}$

10. $f(x) = x \cos x$

11. $f(x) = x - x \sec x$

12. $f(x) = \dfrac{1 - \sin x}{\cos x}$

13. $f(x) = (x^3 + 1)^5$

14. $f(x) = \sec^2 x + \tan^2 x$

In each of Exercises 15–22, find $\dfrac{d^2y}{dx^2}$.

15. $y = x^{-2} - 2x^{-4}$

16. $y = x^2 + \dfrac{1}{x^2}$

17. $\dfrac{dy}{dx} = \sec x \cdot \tan x$

18. $\dfrac{dy}{dx} = \dfrac{x}{x + \cot x}$

19. $y = x^3 \cos x$

20. $\dfrac{dy}{dx} = x^4(1 - x^4)$

21. $y = \dfrac{1}{1 - x^2}$

22. $y = x^m + x^n$

In each of Exercises 23–26, find the indicated derivative.

23. $f''(x)$ for $f(x) = \dfrac{x}{x + 2}$

24. $f''(x)$ for $f(x) = x^3 \sin x$

25. $\dfrac{d^4y}{dx^4}$ for $y = (x^2 - 1)^4$

26. $\dfrac{d^3y}{dx^3}$ for $y = x^3 - 3x^{-2}$

27. Explain why a polynomial $p(x) = a_nx^n + \cdots + a_1x + a_0$ has derivatives of all orders (i.e., why $p(x)$ is **infinitely differentiable**).

28. A particle moves along a line so that at time t seconds its position is $s(t) = t^4 - 8t^2 + 2$.
  a. What is the velocity function?
  b. For what time intervals is velocity positive?

  c. What is the acceleration function?
  d. When is the acceleration positive?

29. A particle moves along a line with acceleration $a(t) = 2 - t$ for $0 \le t \le 4$.
  a. Sketch the graph of $a(t)$.
  b. If $v(t) =$ velocity, what can you say about $v(1)$ if $v(0) > 0$?
  c. What can you say about $v(4)$ if $v(3) < 0$?
  d. Must the particle necessarily change direction at some time t_0 where $0 \le t_0 \le 4$?
  e. Find an explicit position function $s(t)$ to support your answer to part (d).

30. Find a formula for $\dfrac{d^ny}{dx^n}$ if $y = (1 + x)^{-1}$.

31. Find a second degree polynomial $f(x)$ so that $f(2) = 2$, $f'(2) = 4$, and $f''(2) = 6$.

32. A particle moves with constant acceleration along a line. If $v(2) = 6$ meters per second and $v(5) = 15$ meters per second, find the acceleration.

33. Examine the first four derivatives of the trigonometric functions $\tan x$, $\cot x$, $\sec x$, and $\csc x$. Do you note any recurring pattern or otherwise see any pattern among these derivatives as happens for $\sin x$ and $\cos x$?

3.6 THE CHAIN RULE

The goal of this section is to extend our list of differentiation rules to include *composite functions* of the form $y = (f \circ g)(x) = f(g(x))$. The rule for differentiating the composite function $f \circ g$ is called the Chain Rule. Before discussing it in its most general form we consider the special case of a power g^n of a function g.

The Power Rule for Functions

If g is a differentiable function of x, the Product Rule may be applied to determine the derivative of its square:

$$(g^2)'(x) = (g \cdot g)'(x) = g'(x)g(x) + g(x)g'(x) = 2g(x)g'(x).$$

Similarly, for integers $n \ge 3$ repeated application of the Product Rule shows that

$$(g^n)'(x) = ng^{n-1}(x)g'(x). \tag{1}$$

Equation (1) is called the **Power Rule for Functions.** In Leibniz notation it is written as follows:
  If $u = g(x)$ and $y = u^n$, then

$$\frac{dy}{dx} = nu^{n-1} \cdot \frac{du}{dx}. \tag{2}$$

  The Power Rule says this: To find the derivative of the function $y = g^n(x) = [g(x)]^n$, let $u = g(x)$ and differentiate the function $y = u^n$ as if y were a function of the independent variable u. Then multiply the result by the derivative $\dfrac{du}{dx} = g'(x)$.

Example 1

Find $h'(x)$ for $h(x) = (x^4 - 6x^2)^3$.

Solution: This function has the form $h(x) = [g(x)]^3$ with "inside" function $g(x) = x^4 - 6x^2$. Since the derivative of this inside function is

$$g'(x) = 4x^3 - 12x$$

the Power Rule (equation 1) gives

$$h'(x) = 3[g(x)]^2 \cdot g'(x)$$
$$= 3(x^4 - 6x^2)^2(4x^3 - 12x).$$

Of course, we could also obtain h' by first writing h as the polynomial

$$h(x) = x^{12} - 18x^{10} + 108x^8 - 216x^6$$

and then differentiating:

$$h'(x) = 12x^{11} - 180x^9 + 864x^7 - 1296x^5.$$

We leave it for you to verify that these two results are the same. ◇

In Example 1 the advantage of the Power Rule solution is that the factors of $h'(x)$ are more easily recognized than in the Polynomial Rule solution, although both methods apply. In Example 2 the Power Rule is the only reasonable method available for finding the derivative.

Example 2

For $y = \left(\dfrac{x-3}{x^3+7}\right)^9$, find $\dfrac{dy}{dx}$.

Strategy

Identify the "inside" function u.

Solution

Here we define u to be the function

$$u = \frac{x-3}{x^3+7}.$$

Find $\dfrac{du}{dx}$.

Then, by the Quotient Rule,

$$\frac{du}{dx} = \frac{(x^3+7)(1) - (x-3)(3x^2)}{(x^3+7)^2}$$
$$= \frac{-2x^3 + 9x^2 + 7}{(x^3+7)^2}.$$

Apply the Power Rule (2) with $n = 9$.

Then, by the Power Rule for Functions,

$$\frac{dy}{dx} = 9u^8 \cdot \frac{du}{dx}$$

Substitute back for u in final answer.

$$= 9\left(\frac{x-3}{x^3+7}\right)^8\left[\frac{-2x^3 + 9x^2 + 7}{(x^3+7)^2}\right]$$
$$= \frac{9(-2x^3 + 9x^2 + 7)(x-3)^8}{(x^3+7)^{10}}.$$

◇

Composite Functions

$x \xrightarrow{\;g\;} u \xrightarrow{\;f\;} y$

$u = g(x)$

$y = f(u) = f(g(x))$

The power function $y = g^n(x)$ is a special case of the more general notion of a *composite function*. Composite functions have the form

$$(f \circ g)(x) = f(g(x)) \tag{3}$$

with an "outside" function f acting on values $u = g(x)$ of an "inside" function g. In the power function $y = g^n(x)$, the outside function is $f(u) = u^n$.

By way of review, here is a list of composite functions together with an indication of how they may be viewed as having the form (3).

$f(g(x))$	$f(u)$	$u = g(x)$
$\sqrt{x^2 + 3}$	$f(u) = \sqrt{u}$	$u = x^2 + 3$
$\dfrac{1}{7 - x}$	$f(u) = \dfrac{1}{u}$	$u = 7 - x$
$\dfrac{3}{(\sqrt{x} + 2)^4}$	$f(u) = \dfrac{3}{u^4}$	$u = \sqrt{x} + 2$
$\left(\dfrac{2x - 9}{2x + 9}\right)^3$	$f(u) = u^3$	$u = \dfrac{2x - 9}{2x + 9}$

Since the power function g^n is a special case of a composite function, we would expect the derivative of a composite function $f \circ g$ to be calculated in a manner consistent with the Power Rule for Functions. This is indeed the case, and the result is called the **Chain Rule.**

THEOREM 8
Chain Rule

If the function g is differentiable at x and the function f is differentiable at $u = g(x)$, then the composite function $f \circ g$ is differentiable at x, and

$$(f \circ g)'(x) = f'(g(x))g'(x).$$

That is,

$$\frac{d}{dx} f(g(x)) = f'(g(x))g'(x).$$

In Leibniz notation the rule is: If $y = f(u)$ and $u = g(x)$, then

$$\frac{dy}{dx} = \frac{dy}{du} \cdot \frac{du}{dx}.$$

The Chain Rule says this: To differentiate the composition $f \circ g$, first differentiate the "outside" function f as a function of $u = g(x)$. Then multiply the result by the derivative $\dfrac{du}{dx} = g'(x)$. Before concerning ourselves with justifying the Chain Rule we shall apply it in several examples.

Example 3

For $y = \dfrac{1}{(6x^3 - x)^4}$, find $\dfrac{dy}{dx}$.

Solution: Here we let u be the "inside" function $u = 6x^3 - x$. Then

$$y = \frac{1}{(6x^3 - x)^4} = \frac{1}{u^4} = u^{-4},$$

so

$$\frac{dy}{du} = -4u^{-5}, \quad \text{and} \quad \frac{du}{dx} = 18x^2 - 1.$$

Then, by the Chain Rule

$$\frac{dy}{dx} = \frac{dy}{du} \cdot \frac{du}{dx}$$
$$= -4u^{-5}(18x^2 - 1)$$
$$= -4(6x^3 - x)^{-5}(18x^2 - 1). \qquad \diamond$$

Example 4

Find $\dfrac{dy}{dx}$ for $y = \sin(6x^2 - x)$.

Solution: This function is composite. The outside function is

$$y = \sin u, \quad \text{with} \quad \frac{dy}{du} = \cos u.$$

The inside function is

$$u = 6x^2 - x, \quad \text{with} \quad \frac{du}{dx} = 12x - 1.$$

The Chain Rule therefore gives

$$\frac{dy}{dx} = \frac{dy}{du} \cdot \frac{du}{dx}$$
$$= (\cos u)(12x - 1)$$
$$= (12x - 1) \cos(6x^2 - x). \qquad \diamond$$

Example 5

Find $\dfrac{dy}{dx}$ for $y = \dfrac{1}{1 + \sin^2 x}$.

Solution: We may regard this as a composite function

$$y = \frac{1}{u} = u^{-1} \quad \text{with} \quad u = 1 + \sin^2 x.$$

Since

$$\frac{dy}{du} = -u^{-2} = \frac{-1}{u^2} \quad \text{and} \quad \frac{du}{dx} = 2 \sin x \cos x \qquad \text{(Power Rule)}$$

we have

$$\frac{dy}{dx} = \frac{-1}{u^2}(2 \sin x \cos x)$$

$$= \frac{-2 \sin x \cos x}{(1 + \sin^2 x)^2}.$$

◇

Justifying the Chain Rule

Here is an informal argument for the validity of the Chain Rule. We begin with the definition of the derivative of $f \circ g$:

$$(f \circ g)'(x) = \lim_{h \to 0} \frac{f(g(x + h)) - f(g(x))}{h}$$

and use some elementary algebra to obtain the statement

$$(f \circ g)'(x) = \lim_{h \to 0} \left[\frac{f(g(x + h)) - f(g(x))}{g(x + h) - g(x)} \right] \cdot \left[\frac{g(x + h) - g(x)}{h} \right]. \qquad (4)$$

We next let $u = g(x)$ and $v = g(x + h) - g(x)$. Then $g(x + h) = g(x) + v = u + v$, and we can write equation (4) as

$$(f \circ g)'(x) = \lim_{h \to 0} \left[\frac{f(u + v) - f(u)}{v} \right] \cdot \left[\frac{g(x + h) - g(x)}{h} \right]. \qquad (5)$$

Since g is differentiable at x, it is continuous at x by Theorem 2. Thus, by Theorem 9, Chapter 2

$$\lim_{h \to 0} [g(x + h) - g(x)] = \lim_{h \to 0} v = 0.$$

That is, $v \to 0$ as $h \to 0$. Using this observation and the algebra of limits we have from equation (5) that

$$(f \circ g)'(x) = \left[\lim_{v \to 0} \frac{f(u + v) - f(u)}{v} \right] \cdot \left[\lim_{h \to 0} \frac{g(x + h) - g(x)}{h} \right]$$

$$= f'(u) \cdot g'(x)$$

$$= f'(g(x)) \cdot g'(x)$$

as desired. This argument is not entirely rigorous, however, since closer attention needs to be paid to the question of whether the denominator v in equation (5) might equal zero for $h \neq 0$. A formal proof of the Chain Rule is given in Appendix II.

Exercise Set 3.6

In Exercises 1–10 form the composite function $y = f(g(x))$ with $u = g(x)$. Then find $\dfrac{dy}{dx} = \dfrac{dy}{du} \cdot \dfrac{du}{dx}$ using the Chain Rule.

1. $f(u) = u^3 + 1$
 $g(x) = 1 - x^2$

2. $f(u) = \dfrac{1}{1 + u}$
 $g(x) = 3x^2 - 7$

3. $f(u) = u(1 - u^2)$
 $g(x) = \dfrac{1}{x}$

4. $f(u) = \dfrac{u + 1}{u - 1}$
 $g(x) = \sin x$

5. $f(u) = \tan u$
 $g(x) = 1 + x^4$

6. $f(u) = 3u^2 + 5$
 $g(x) = \cos x$

7. $f(u) = \dfrac{1}{1 + u}$
 $g(x) = \sin^2 x$

8. $f(u) = u^2 - 1$
 $g(x) = 1 + \sec x$

9. $f(u) = \cos u$
 $g(x) = \sin x$

10. $f(u) = u^3 - 3u + \dfrac{1}{u}$
 $g(x) = \tan x$

In Exercises 11–40 find $\dfrac{dy}{dx}$ using the Chain Rule.

11. $y = (x^2 + 4)^3$

12. $y = (2 - 3x)^4$

13. $y = (\cos x - x)^6$

14. $y = (x^2 + 8x + 6)^5$

15. $y = x(x^4 - 5)^3$

16. $y = (x^2 - 7)^3(5 - x)^2$

17. $y = \dfrac{1}{(x^2 - 9)^3}$

18. $y = \dfrac{x + 3}{(x^2 - 6x + 2)^2}$

19. $y = (3 \tan x - 2)^4$

20. $y = (x^6 - x^2 + 2)^{-4}$

21. $y = \left(\dfrac{x - 3}{x + 3}\right)^4$

22. $y = \left(\dfrac{1 + \sin x}{1 - \sin x}\right)^{-6}$

23. $y = x \cos(1 - x^2)$

24. $y = (x \cos x - \sin x)^5$

25. $y = \dfrac{\tan^2 x + 1}{1 - x}$

26. $y = \dfrac{x + 2}{1 + \sec(1 + x^2)}$

27. $y = \dfrac{x}{(x^2 + x + 1)^6}$

28. $y = \csc^3(1 + x^3)$

29. $y = \left(\dfrac{ax + b}{cx + d}\right)^4$

30. $y = \left(\dfrac{a - bx}{c - dx}\right)^5$

31. $y = (\sec^3 x - 4x^2)^5$

31. $y = \dfrac{1}{1 + \cos^3 x}$

33. $y = \sin^3(x - \pi/2)$

34. $y = \tan^2(\pi - x^2)$

35. $y = \dfrac{x - \sin \pi x}{4 + \cos \pi x}$

36. $y = \dfrac{1}{(1 + x^4)^3}$

37. $y = x \cot\left(\dfrac{1 + x}{1 - x}\right)$

38. $y = \dfrac{x - \sin \pi x}{\tan^2(\pi x)}$

39. $y = \tan(6x) - 6 \tan x$

40. $y = (\sin \pi x - \cos \pi x)^4$

41. Apply the Chain Rule twice to show that the derivative of the composite function $y = f(g(h(x)))$ is

$$\frac{dy}{dx} = f'(g(h(x))) \cdot g'(h(x)) \cdot h'(x).$$

In Exercises 42–47 use the result of Exercise 41 to find $\dfrac{dy}{dx}$.

42. $y = \sin^4(x^3 + \pi x)$

43. $y = [1 + (x^2 - 3)^4]^6$

44. $y = \dfrac{1}{a + (bx + c)^4}$

45. $y = \cos^2\left(\dfrac{1 + x}{1 - x}\right)$

46. $y = [\sin \pi x - \cos \pi x]^6$

47. $y = 1 + (1 + (1 + x^2)^2)^2$

In each of Exercises 48–53 find an equation for the line tangent to the graph of $y = f(x)$ at point P:

48. $y = (1 - x^2)^3$, $\quad P = (1, 0)$

49. $y = \left(\dfrac{x}{x + 1}\right)^4$, $\quad P = (0, 0)$

50. $y = \sin(\pi + x^3)$, $\quad P = (0, 0)$

51. $y = x \cos(\pi + x^2)$, $\quad P = (0, 0)$

52. $y = \left(\dfrac{3x^2 + 1}{x + 3}\right)^2$, $\quad P = (-1, 4)$

53. $y = \tan\left(\dfrac{\dfrac{\pi}{4} - x}{1 + x}\right)$, $\quad P = (0, 1)$.

54. Let $h(x) = f(g(x))$. Find $h'(2)$ if $g(2) = 3$, $g'(2) = 7$ and $f'(3) = -2$.

55. Let $h(t) = [g^2(t) + 1]^3$. Find $h'(3)$ if $g(3) = -2$ and $g'(3) = -1$.

56. Let $h(t) = \sin(g(t))$. Find $h'(2)$ if $g(2) = 3\pi/4$ and $g'(t) = \sqrt{2}$.

57. Show that the function $f(t) = \sin kt$ satisfies the differential equation

$$f''(t) + k^2 f(t) = 0, \qquad -\infty < t < \infty.$$

58. Use the result of Exercise 57 to find a function of the form $f(t) = \sin kt$ satisfying the differential equation
   a. $f''(t) + 4f(t) = 0$
   b. $f''(t) + 25f(t) = 0$
   c. $\dfrac{d^2 y}{dt^2} + 5y = 0$
   d. $f''(t) = -9f(t)$.

59. Show that the function $f(t) = \cos kt$ satisfies the differential equation

$$f''(t) + k^2 f(t) = 0, \qquad -\infty < t < \infty.$$

60. Use the result of Exercise 59 to find functions of the form $f(t) = \cos kt$ satisfying the differential equations in Exercise 58, parts a–d.

61. Find an equation for the line tangent to the graph of the function $f(x) = \sin(x^2)$ at the point $(\sqrt{\pi/4}, \sqrt{2}/2)$.

62. Find an equation for the line tangent to the graph of $f(x) = \csc(\pi x)$ at the point $(\frac{1}{2}, 1)$.

63. A block oscillates at the end of a spring so that at time t (in seconds) it is $s(t) = 20 + 5 \sin(4t)$ cm from the fixed end of the spring. If the motion is along a line, find
   a. the velocity $v(t)$ of the block at time t.
   b. the acceleration $a(t)$ of the block at time t.

64. A particle moves along a line so that at time t (in seconds) it is located at position $s(t) = \left(\dfrac{t}{t + 1}\right)^2 + 10 \sin(\pi t)$. If distance is measured in meters, find
   a. the velocity of the particle at time t.
   b. the velocity $v(0)$ at time 0.
   c. the acceleration at time t.
   d. the acceleration $a(0)$ at time 0.

65. Verify equation (1) in the following way, using mathematical induction.

a. For $n = 1$, $(g^n)'(x) = ng^{n-1}(x)g'(x)$ since $g'(x) = 1[g(x)]^0 \cdot g'(x)$.

b. Assume that equation (1) holds for the integer $n > 1$.

c. Write $[g^{n+1}]'(x) = (g \cdot g^n)'(x)$ and use the Product Rule

to verify that

$$[g^{n+1}]'(x) = (n + 1)g^n(x) \cdot g'(x).$$

d. Conclude, from the principle of mathematical induction, that equation (1) holds for all integers $n \geq 1$.

3.7 IMPLICIT DIFFERENTIATION AND RATIONAL POWER FUNCTIONS

When a function is specified by an equation of the form $y = f(x)$ we say that we have y determined as an **explicit** function of x. This is because for each number x precisely one value of y is determined by substituting x into the right-hand side of the equation $y = f(x)$. However, some equations involving the two variables x and y determine interesting *relations* between these variables even though the equations can not be brought into the explicit form $y = f(x)$. For example, the equation

$$\sin(xy) = 1 \tag{1}$$

determines a seemingly complex relation between the variables x and y that, at the moment, we can study only by plotting pairs (x, y) where $\sin(xy) = 1$ (see Figure 7.1).

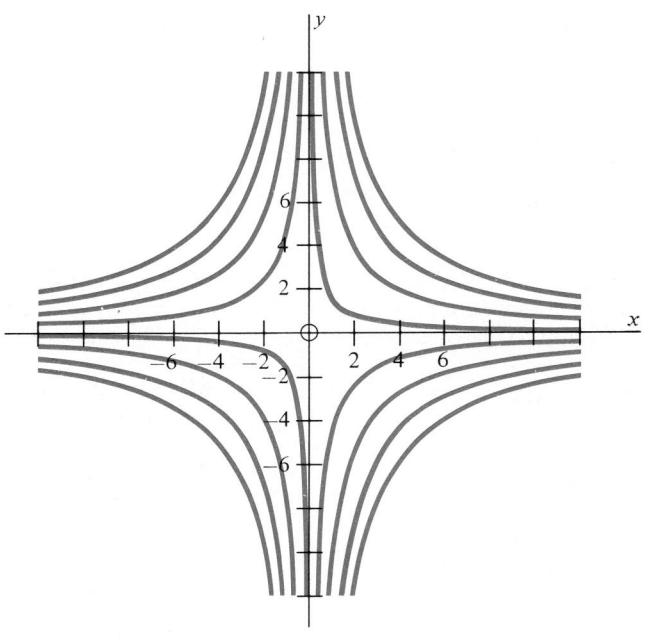

Figure 7.1 Some arcs of the graph of the relation $\sin(xy) = 1$.

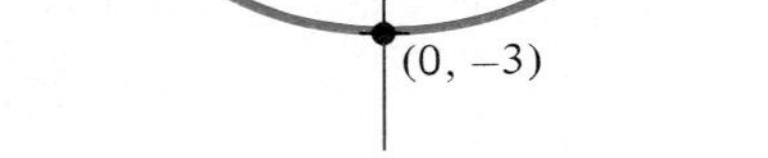

Figure 7.2 Elliptical graph of the equation $\dfrac{x^2}{16} + \dfrac{y^2}{9} = 1$.

Similarly, the equation

$$\frac{x^2}{16} + \frac{y^2}{9} = 1 \tag{2}$$

determines the familiar ellipse in the plane (see Figure 7.2) although it cannot be

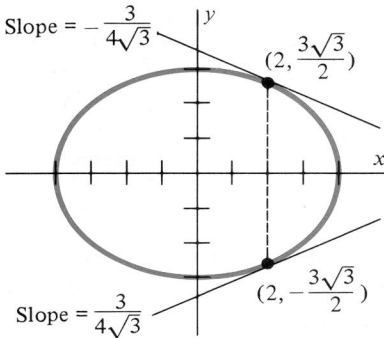

Figure 7.3 Although equation (2) does not define a function, tangents exist at all points on the graph.

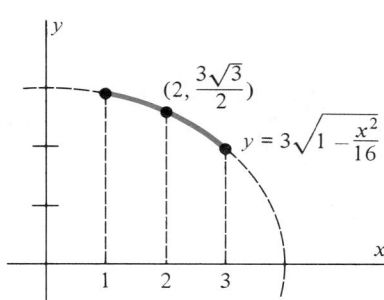

Figure 7.4 "Near" $\left(2, \dfrac{3\sqrt{3}}{2}\right)$ equation (2) does define a differentiable function of x.

brought into the form $y = f(x)$. The best we can do is solve for y as

$$y = \pm 3\sqrt{1 - x^2/16}.$$

(Because of the $\pm$ sign, the right-hand side does not determine a function of x since more than one value of y can correspond to a given number x.)

Regardless of whether an equation in x and y defines a function, near any particular point (x, y) on the graph of that equation we can ask about the slope of the tangent (see Figure 7.3). If the tangent at this point exists and is not vertical, its slope can often be obtained from the equation by application of the Chain Rule. The reason behind this observation is illustrated in Figure 7.4. When restricted to a small region near any particular point on its graph, the equation may indeed define a differentiable function whose derivative is the desired slope.

Example 1

Find the slope of the line tangent to the graph of the ellipse

$$\frac{x^2}{16} + \frac{y^2}{9} = 1$$

at the point $\left(2, \dfrac{3\sqrt{3}}{2}\right)$ (Figure 7.3).

Strategy

Force y to be a function of x by imposing the restriction that $y > 0$.

Differentiate both sides of equation (2), remembering that

$$\frac{d}{dx}\left(\frac{y^2}{9}\right) = \frac{2y}{9} \cdot \frac{dy}{dx},$$

according to the Chain Rule.

Solution

We see from Figure 7.3 that y can be considered a function of x if we impose the condition $y > 0$. Differentiating both sides of the equation, regarding y as an unspecified function of x, we obtain

$$\frac{2x}{16} + \frac{2y}{9} \cdot \frac{dy}{dx} = 0,$$

so

$$\frac{dy}{dx} = -\frac{9x}{16y}, \qquad y \neq 0.$$

Solve for $\dfrac{dy}{dx}$ and substitute values for x and y.

At the point $\left(2, \dfrac{3\sqrt{3}}{2}\right)$, we have $x = 2$ and $y = \dfrac{3\sqrt{3}}{2}$ so

$$\frac{dy}{dx} = -\frac{9(2)}{16\left(\dfrac{3\sqrt{3}}{2}\right)} = -\frac{\sqrt{3}}{4},$$

which is the desired slope. (Alternatively, we could have been asked for the slope of the tangent at the point $(2, -3\sqrt{3}/2)$. Then we would have used the restriction $y < 0$ to make y a function of x. The result would be $dy/dx = \sqrt{3}/4$ [see Figure 7.3].) ◇

The technique used in Example 1 is called **implicit differentiation.** It can be summarized as follows:

(i) Impose a condition on y so that the given equation implicitly defines y as an unspecified differentiable function of x.

(ii) Differentiate both sides of the given equation, remembering that the factor $\dfrac{dy}{dx}$ will occur in any differentiation of a term involving the function y, according to the Chain Rule.

(iii) Solve the resulting equation for $\dfrac{dy}{dx}$.

Although we can always carry out this procedure symbolically to obtain an expression for $\dfrac{dy}{dx}$ (assuming that the necessary differentiation formulas are known), there is a hazard with this procedure. The problem lies in the fact that the given equation need not define a differentiable function of x. Indeed, in Example 1 we could not obtain a value of $\dfrac{dy}{dx}$ at the point $(4, 0)$ since the expression $\dfrac{dy}{dx} = -\dfrac{9x}{16y}$ is undefined when $y = 0$. Geometrically, this corresponds to the observation that the tangent to the graph of equation (2) at the point $(4, 0)$ is vertical.

A careful resolution of this difficulty requires a theorem (called the Implicit Function Theorem) that gives precise conditions under which an equation determines y as a differentiable function of x near a particular point (x_0, y_0). Developing this theorem would take us far astray from our main objectives, so we simply refer you to more advanced texts for a discussion of this theorem. (See, for example, *Mathematical Analysis,* 2nd ed., by Tom Apostol, Addison-Wesley, 1974.) We will resolve this concern here by cautioning that you should always check to see that both the given equation and the resulting expression for $\dfrac{dy}{dx}$ are defined at the point (x_0, y_0) of interest.

Example 2

The equation

$$y^2 + x^2 y = 3x^2$$

determines y as a differentiable function of x near $(2, 2)$. (You may verify this by plotting a portion of the graph of the equation. We obtain y as a differentiable function of x in the region where $1 < x < 3$ and $1 < y < 3$.) Find an expression for $\dfrac{dy}{dx}$ near $(2, 2)$, and the slope of the line tangent to the graph of the equation at $(2, 2)$.

Strategy	*Solution*

Differentiate both sides with respect to x.

$$\underbrace{2y \cdot \frac{dy}{dx}}_{\frac{d}{dx}(y^2)} + \underbrace{2xy + x^2 \cdot \frac{dy}{dx}}_{\frac{d}{dx}(x^2y)} = 6x$$

Collect terms involving $\dfrac{dy}{dx}$.

Thus,

$$(x^2 + 2y)\frac{dy}{dx} = 6x - 2xy,$$

so

Solve for $\dfrac{dy}{dx}$.

$$\frac{dy}{dx} = \frac{6x - 2xy}{x^2 + 2y}.$$

Substitute given values for x and y to obtain the slope.

For $x = 2$, $y = 2$ we have

$$\frac{dy}{dx} = \frac{6 \cdot 2 - 2 \cdot 2 \cdot 2}{2^2 + 2 \cdot 2} = \frac{1}{2},$$

which is the desired slope.   ◇

Example 3

Find an equation of the line normal to the graph of

$$y^3 - x^2 = 7$$

at the point $(1, 2)$.

Solution: We begin by differentiating implicitly. We obtain

$$3y^2 \cdot \frac{dy}{dx} - 2x = 0$$

so

$$\frac{dy}{dx} = \frac{2x}{3y^2}.$$

At the point $(1, 2)$ the slope of the tangent is therefore

$$m_1 = \frac{2 \cdot 1}{3 \cdot 2^2} = \frac{1}{6},$$

so the slope of the normal, which is perpendicular to the tangent, is

$$m_2 = -\frac{1}{m_1} = -6.$$

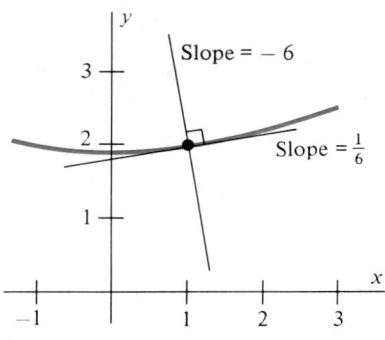

Figure 7.5 Normal to graph of $y^3 - x^2 = 7$ at $(1, 2)$ has slope -6.

An equation for the line is

$$y - 2 = -6(x - 1).$$

(See Figure 7.5.)

Given an equation in x and y we will sometimes need to compute $\dfrac{d^2y}{dx^2}$ as well as $\dfrac{dy}{dx}$. If the equation defines y implicitly as a twice differentiable function of x, we can often compute $\dfrac{d^2y}{dx^2}$ by applying the method of implicit differentiation twice.

Example 4

The equation $9x^2 + 4y^2 = 36$ defines y as a twice differentiable function of x near the point $(0, 3)$. Find $\dfrac{d^2y}{dx^2}$ for this function.

Strategy

Differentiate both sides of given equation and solve for $\dfrac{dy}{dx}$.

Differentiate resulting equation (using the Quotient Rule).

Solution

$$18x + 8y \cdot \frac{dy}{dx} = 0$$

$$\frac{dy}{dx} = -\frac{9x}{4y}.$$

Differentiating both sides of this equation gives

$$\frac{d^2y}{dx^2} = -\frac{(4y)9 - (9x) \cdot 4 \cdot \dfrac{dy}{dx}}{16y^2}$$

$$= -\frac{36y - 36x \cdot \dfrac{dy}{dx}}{16y^2}$$

Substitute for $\dfrac{dy}{dx}$ as found above.

$$= -\frac{36y - 36x\left(-\dfrac{9x}{4y}\right)}{16y^2}$$

$$= -\frac{9}{4y} - \frac{81x^2}{16y^3}. \qquad \diamond$$

Differentiating Rational Powers of x.

The following theorem tells us that the Power Rule extends to *all* power functions, $f(x) = x^r$, with r rational.

THEOREM 9
(Power Rule for the Case of Rational Exponents)

Let $f(x) = x^{p/q}$, where p and q are integers with $p \neq 0$ and $q \neq 0$. Then f is differentiable and

$$f'(x) = \frac{p}{q}x^{p/q - 1} \tag{3}$$

whenever $f(x)$ and the right side of equation (3) are defined. In Leibniz notation,

$$\frac{d}{dx}(x^{p/q}) = \left(\frac{p}{q}\right)x^{p/q - 1}. \tag{4}$$

Assuming $f(x) = x^{p/q}$ to be a differentiable function of x (a fact that we shall assume but not prove), we can establish equation (3) by implicit differentiation as follows. Let $x \neq 0$ be in the domain of the function

$$y = x^{p/q}$$

where p and $q \neq 0$ are integers. Then $y \neq 0$, and

$$y^q = (x^{p/q})^q = x^p. \tag{5}$$

If we assume y to be differentiable, then since q is an integer, y^q is differentiable. Applying the Power Rule (for integral powers) to both sides of equation (5) gives

$$qy^{q-1} \cdot \frac{dy}{dx} = px^{p-1}.$$

Recalling that $y = x^{p/q} \neq 0$, we can now solve for $\dfrac{dy}{dx}$ as

$$\frac{dy}{dx} = \left(\frac{p}{q}\right)x^{p-1}y^{1-q}$$

$$= \left(\frac{p}{q}\right)x^{p-1}(x^{p/q})^{1-q}$$

$$= \left(\frac{p}{q}\right)x^{p-1} \cdot x^{p/q - p}$$

$$= \left(\frac{p}{q}\right)x^{(p-1+p/q-p)}$$

$$= \left(\frac{p}{q}\right)x^{p/q - 1},$$

which is the formula in Theorem 9.

Example 5

For $f(x) = \sqrt[4]{x^3 - x^2 + 3}$, find $f'(x)$.

Strategy

Replace the radical sign with a fractional exponent.

Solution

Since $f(x) = (x^3 - x^2 + 3)^{1/4}$,

$$f'(x) = \frac{1}{4}(x^3 - x^2 + 3)^{-3/4} \cdot \frac{d}{dx}(x^3 - x^2 + 3)$$

$$= \frac{1}{4}(x^3 - x^2 + 3)^{-3/4}(3x^2 - 2x).$$

Apply Theorem 9 together with the Chain Rule.

◇

Example 6

Find the velocity of an object that moves along a line so that it is at location

$$s(t) = t^{2/3} \sin(1 + \sqrt{t})$$

at time t.

Solution: Using the Product Rule, Power Rule, and Chain Rule we obtain the velocity function

$$v(t) = s'(t) = \left(\frac{2}{3}t^{-1/3}\right) \sin(1 + t^{1/2}) + t^{2/3} \cos(1 + t^{1/2}) \cdot \frac{d}{dt}(1 + t^{1/2})$$

$$= \frac{2}{3}t^{-1/3} \sin(1 + t^{1/2}) + t^{2/3} \cos(1 + t^{1/2})\left(\frac{1}{2}t^{-1/2}\right)$$

$$= \frac{2}{3}t^{-1/3} \sin(1 + t^{1/2}) + \frac{1}{2}t^{1/6} \cos(1 + t^{1/2}).$$

(Note that $v(0)$ is undefined.)

◇

Example 7

Figure 7.6 shows the graph of $f(x) = x^{1/3}$, with several tangents sketched in. The derivative (slope function) for $f(x) = x^{1/3}$ is $f'(x) = \frac{1}{3}x^{-2/3}$.

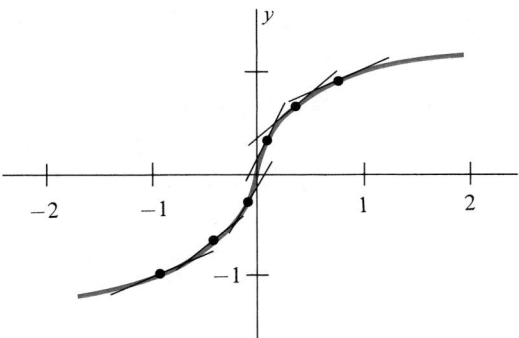

Figure 7.6 Graph of $f(x) = x^{1/3}$ and several tangents.

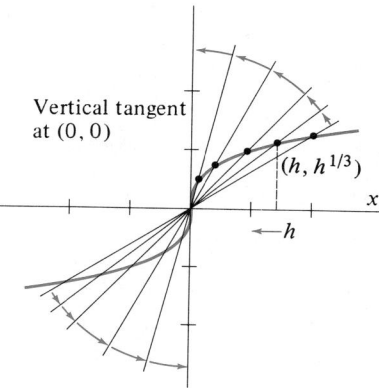

Figure 7.7 Secants through (0, 0) approach the y-axis as $h \to 0$. Tangent to $f(x) = x^{1/3}$ at (0, 0) is vertical.

Note that $f'(0)$ is undefined and that the slopes of the tangents approach $+\infty$ as x approaches zero from either direction.

However, the graph of $f(x) = x^{1/3}$ *does* have a tangent at $x = 0$ in the sense in which we defined tangents in Section 2.1. As Figure 7.7 illustrates, the slope of the secant line through (0, 0) and $(h, f(h)) = (h, h^{1/3})$ is

$$\frac{f(0 + h) - f(0)}{h} = \frac{h^{1/3} - 0}{h} = \frac{1}{h^{2/3}}.$$

Thus, the secant through (0, 0) and $(h, h^{1/3})$ rotates into the position of the vertical line through (0, 0) as h approaches zero. This shows that the graph of f may have a *vertical* tangent when $f'(a)$ fails to exist. ◇

Example 8

The graph of $f(x) = x^{2/3}$ appears in Figure 7.8. Since $f'(x) = \dfrac{2}{3}x^{-1/3}$, $f'(0)$ is undefined. Here the secants through (0, 0) have slope

$$\frac{(0 + h)^{2/3} - 0}{h} = \frac{1}{h^{1/3}}.$$

Again, the secants rotate into the y-axis as $h \to 0$. Thus, the graph of $f(x) = x^{2/3}$ has a vertical tangent at $x = 0$, even though $f'(0)$ is undefined.

Note, however, that the graph of $f(x) = x^{2/3}$ looks quite different from the graph of $g(x) = x^{1/3}$ near (0, 0). This is because $f(x) = x^{2/3} = (x^{1/3})^2$ is an *even* function, while $g(x) = x^{1/3}$ is an *odd* function. Because the tangent to the graph of $f(x) = x^{2/3}$ at (0, 0) is vertical, and because the branch of the graph to the left of $x = 0$ must be the mirror image of the branch to the right of $x = 0$, a "cusp" (sharp point) exists at (0, 0). ◇

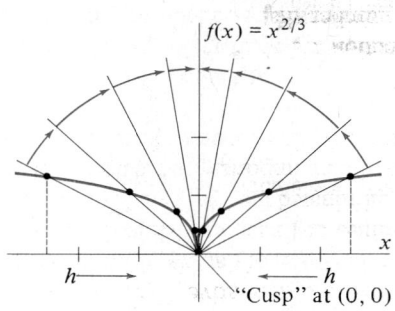

Figure 7.8 Tangent to graph of $f(x) = x^{2/3}$ at $(0, 0)$ is vertical. The graph has a "cusp" at $(0, 0)$.

Exercise Set 3.7

In Exercises 1–22, find $f'(x)$.

1. $f(x) = (x + 2)^{4/3}$

2. $f(x) = \sqrt[3]{x^2 + 1}$

3. $f(x) = \sqrt{x} + \dfrac{1}{\sqrt{x}}$

4. $f(x) = \dfrac{\sqrt[3]{x + 2}}{1 + \sqrt{x}}$

5. $f(x) = x^{2/3} + x^{-2/3}$

6. $f(x) = (3x^4 + 4x^3)^{-2}$

7. $f(x) = (x^2 - 1)^{-2}(x^2 + 1)^2$

8. $f(x) = (7 - x^2)^{2/3}$

9. $f(x) = \sqrt{1 + \sqrt[3]{x}}$

10. $f(x) = x^{5/2} + 3x^{2/3} + x^{-4/3}$

11. $f(x) = \dfrac{\sqrt[3]{x}}{\sqrt[3]{x} + x}$

12. $f(x) = \sqrt{x^3} \cdot (x^{-2} + 2x^{-1} + 1)^4$

13. $f(x) = \sqrt{x} + \sqrt[4]{x} + \sqrt[8]{x}$

14. $f(x) = \dfrac{x^{2/3}}{\sqrt[3]{x^2 + 1}}$

15. $f(x) = \dfrac{(x^2 + 1)^{2/3}}{(1 - x^2)^{4/3}}$

16. $f(x) = \dfrac{(x - 1)^{1/2} + (x + 1)^{1/3}}{(x + 2)^{1/4}}$

17. $f(x) = \sqrt{\dfrac{ax^2 + b}{cx + d}}$

18. $f(x) = \sqrt[3]{\sin x^2}$

19. $f(x) = \dfrac{x^{1/4}(x^{4/3} - x)}{(\sin^3 x - \sqrt{x})^{1/3}}$

20. $f(x) = \left[\dfrac{\tan(x^{3/2} - x^{1/2})}{1 + \sec(x^3 - \sqrt{x})} \right]^{4/3}$

21. $f(x) = \sin^2(x^{2/3} + x^{-4/3})^{5/2}$

22. $f(x) = (x^{-2/3} - x^{3/2})^{3/5} \cdot (6 - x^{5/3})^{-2/5}$

In Exercises 23–42, find $\dfrac{dy}{dx}$ by implicit differentiation.

23. $x^2 + y^2 = 25$

24. $x = \sin y$

25. $x^{1/2} + y^{1/2} = 4$

26. $x^2 + 2xy + y^2 = 8$

27. $x = \tan y$

28. $x^2 y + xy^2 = 0$

29. $x \sin y = y \cos x$

30. $x = y(y - 1)$

31. $(xy)^{1/2} = xy - x$

32. $y^2 = \sin^2 x - \cos^2 2x$

33. $x^3 + x^2 y + xy^2 + y^3 = 0$

34. $\cos(x + y) = y \sin x$

35. $\sqrt{x + y} = xy - x$

36. $y^2 = \dfrac{x + 1}{x^2 + 1}$

37. $\cot y = 3x^2 + \cot(x + y)$

38. $x^2 y^2 = 4$

39. $\sin(xy) = 1/2$

40. $\dfrac{1}{x} + \dfrac{1}{y} + \dfrac{1}{4} = 0$

41. $y^4 = x^5$

42. $\sin(x + y) + \cos(x - y) = 1$

In Exercises 43–46, find $\dfrac{dy}{dx}$ and $\dfrac{d^2y}{dx^2}$.

43. $y^2 = 4x$

44. $x^2 + y^2 = 1$

45. $y^2 - xy = 4$

46. $\sqrt{x} + \sqrt{y} = 1$

In each of Exercises 47–52, find an equation for the line tangent to the graph determined by the given equation at the given point.

47. $xy = 9$ $(3, 3)$ **48.** $x^2 + y^2 = 4$ $(\sqrt{2}, \sqrt{2})$

49. $x^3 + y^3 = 16$ $(2, 2)$

50. $x^2y^2 = 16$ $(-1, 4)$

51. $\dfrac{x + y}{x - y} = 4$ $(5, 3)$ **52.** $(y - x)^2 = x$ $(9, 12)$

53. Assume that an equation determines y as a differentiable function of x. Explain why $\dfrac{dy}{dx} = 0$ at points where the graph has a horizontal tangent.

54. Assume that an equation determines x as a differentiable function of y. Explain why $\dfrac{dx}{dy} = 0$ at points where the graph has a vertical tangent.

In Exercises 55–60, use the results of Exercises 53 and 54 to find all horizontal and vertical tangents.

55. $x^2 + y^2 = 2$

56. $x^2 + 4y^2 = 4$

57. $x^2 - y^2 = 1$

58. $(x - 1)^2 + (y + 2)^2 = 9$

59. $x^2 + y^2 + 2x + 4y = -4$

60. $(x + 1)(y - 1) = 1$

61. Find the value of a so that the circles with equations $(x - a)^2 + y^2 = 2$ and $(x + a)^2 + y^2 = 2$ intersect at points where their tangents are perpendicular.

62. Find an equation for the line normal to the curve $9x^2 + 16y^2 = 144$ at the point $(2, 3\sqrt{3}/2)$.

63. Prove that all normals to the curve $x^2 + y^2 = a^2$ pass through the origin.

64. Explain why the equation $4 + x^2 + y^2 = 2xy$ does not define a function $y = f(x)$ for *any* values of x and y. (*Hint:* Rewrite as $x^2 - 2xy + y^2 = -4$. What is the sign of the left-hand side?)

65. Some of the following equations do not determine y as a function of x. Others do, although we may not be able to solve them in practice. Which determine y as a function of x?

  a. $x = y^2 - 7$ **b.** $x = y^3$

  c. $x = y + y^3$ **d.** $x = y^3 + y^5 - 7$

  e. $x = y^3 + y^4 - y^5$ **f.** $x = 3 - y^4 + y$

66. Note that the graph of the equation $xy^2 + x^2y = 4x^2$ consists of the vertical line $x = 0$ (y-axis) in addition to the graph of the equation $y^2 + xy = 4x$. Why is this so? What does this say about the existence of a derivative $\dfrac{dy}{dx}$ ''near'' $x = 0$?

67. For the equation $x^2 + y^2 = 25$, find the slope of the line tangent to the graph at $(3, 4)$ by

  a. geometry;

  b. solving for y and then finding $\dfrac{dy}{dx}$;

  c. finding $\dfrac{dy}{dx}$ by implicit differentiation.

3.8 LINEAR APPROXIMATIONS AND DIFFERENTIALS

The fact that the derivative $f'(x_0)$ of a differentiable function f determines a line tangent to the graph of f at $(x_0, f(x_0))$ is the basis for several procedures for approximating values of f near x_0. While the availability of calculators and computers has virtually eliminated the ''by hand'' use of these procedures, the ideas on which they are based remain important in the design of computing algorithms and in generalizations that we shall encounter later in the text.

Increment Notation

In this section we shall make use of the increment notation Δx to denote a *fixed* nonzero number that is added to a given number x_0 to produce a second number $x = x_0 + \Delta x$. The reason for using increment notation, rather than writing $x =$

$x_0 + h$ as done previously, is that we are *not* interested here in studying a limit as $h \rightarrow 0$. Rather, we want to actually approximate the function value $f(x_0 + \Delta x)$ where both x_0 and Δx are fixed numbers. Writing $x = x_0 + \Delta x$ emphasizes the approach of drawing conclusions about $f(x)$ from what we know about $f(x_0)$, $f'(x_0)$ and the increment Δx.

Linear Approximations

Figure 8.1 illustrates a simple way to approximate $f(x_0 + \Delta x)$ when $f(x_0)$, $f'(x_0)$ and Δx are known. The idea is simply to approximate $f(x_0 + \Delta x)$ by the y-coordinate of the point *on the tangent* to the graph of f that has x-coordinate $x_0 + \Delta x$.

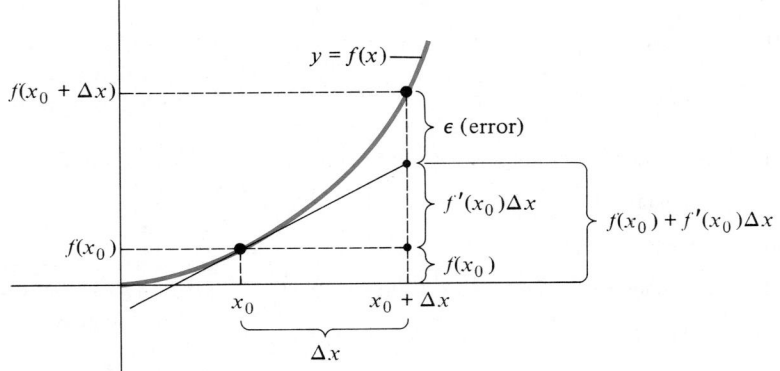

Figure 8.1 $f(x_0) + f'(x_0)\Delta x$ approximates $f(x_0 + \Delta x)$.

Since the slope of this tangent is $f'(x_0)$, the y-coordinate of the point on the tangent with x-coordinate $x_0 + \Delta x$ is $f(x_0) + f'(x_0)\Delta x$. We therefore have the **linear approximation**

$$f(x_0 + \Delta x) \approx f(x_0) + f'(x_0)\Delta x. \tag{1}$$

Of course, approximation (1) makes sense only when $f'(x_0)$ is defined.

Example 1

Use approximation (1) to obtain an approximation for the quantity $\sqrt{36 + \Delta x}$.

Solution: Here we take $x_0 = 36$ and $f(x) = \sqrt{x}$.
Then $f'(x) = \dfrac{1}{2\sqrt{x}}$, so approximation (1) gives

$$\sqrt{36 + \Delta x} = f(36 + \Delta x) \approx f(36) + f'(36)\Delta x$$

$$= \sqrt{36} + \frac{1}{2\sqrt{36}}\Delta x$$

$$= 6 + \frac{\Delta x}{12},$$

so

$$\sqrt{36 + \Delta x} \approx 6 + \frac{\Delta x}{12}.$$

Table 8.1 displays values of this approximation for various values of Δx and compares them with the actual values of $\sqrt{36 + \Delta x}$. Notice that the approximation improves as $\Delta x \to 0$. ◇

Absolute and Relative Errors

Table 8.1 also shows the absolute and relative errors in the approximations described in Example 1. In using the approximation

$$y_1 = f(x_0) + f'(x_0)\Delta x$$

to approximate the function value $y = f(x_0 + \Delta x)$,

(a) the **absolute error** is $|y_1 - y|$;

(b) the **relative error** is $\left|\dfrac{y_1 - y}{y}\right|$;

(c) the **percentage error** is the relative error multiplied by 100%.

Example 2

In the first row of Table 8.1 we find the approximation to $\sqrt{40}$ of $y_1 = 6.333333$, while the actual value (to 6 decimal places) is reported to be $y = 6.324555$. This gives an absolute error of

$$|y_1 - y| = |.008778| = .008778,$$

a relative error of

$$\left|\frac{y_1 - y}{y}\right| = \frac{.008778}{6.324555} = .001388,$$

and a percentage error of

$$.001388 \times 100\% = 0.1388\%. \qquad ◇$$

Table 8.1 Approximations to $\sqrt{36 + \Delta x}$

(a) Δx	(b) Actual value $\sqrt{36 + \Delta x}$	(c) Approximation $6 + \dfrac{\Delta x}{12}$	(d) Absolute error $|(b) - (c)|$	(e) Relative error $\left|\dfrac{(b) - (c)}{(b)}\right|$
4.0	6.324555	6.333333	.008778	.001388
2.0	6.164414	6.166667	.002253	.000365
1.5	6.123724	6.125000	.001276	.000208
1.0	6.082762	6.083333	.000571	.000094
0.5	6.041523	6.041667	.000144	.000024
0.1	6.008328	6.008333	.000006	.000001

Example 3

Find an approximation to sin 42°.

Strategy

Choose a convenient x_0 near 42°.

Solution

We take $x_0 = 45° = \pi/4$ since we are familiar with the fact that

$$\sin(\pi/4) = \frac{\sqrt{2}}{2}.$$

This means that $42° = x_0 + \Delta x$, so

Identify Δx.

$$\Delta x = 42° - 45° = -3°.$$

Convert Δx to radians.

We must convert Δx to radians, however:

$$\Delta x = \frac{-3°}{360°}(2\pi) = -.05236 \text{ radians}.$$

Calculate $f'(x)$.

Since $f(x) = \sin x$, $f'(x) = \cos x$. Applying approximation (1) we obtain

Apply (1).

$$\sin(42°) \approx \sin(\pi/4) + \cos(\pi/4)(-.05236)$$

$$= \frac{\sqrt{2}}{2} + \frac{\sqrt{2}}{2}(-.05236)$$

$$= .6701$$

The actual value is 0.6691.

Table 8.2 presents other approximations to $\sin(\frac{\pi}{4} + \Delta x)$ for various values of Δx (in radians). ◇

Table 8.2 Approximations to $\sin\left(\dfrac{\pi}{4} + \Delta x\right)$

(a) Δx (in radians)	(b) Actual value $\sin\left(\dfrac{\pi}{4} + \Delta x\right)$	(c) Approximation $\dfrac{\sqrt{2}}{2}(1 + \Delta x)$	(d) Absolute error $\|(b) - (c)\|$	(e) Relative error $\left\|\dfrac{(b) - (c)}{(b)}\right\|$
1.0	0.977062	1.41421	0.437151	0.447414
0.5	0.959549	1.06066	0.101111	0.105374
0.1	0.774165	0.777817	.003652	0.004718
0.05	0.741562	0.742462	.000901	0.000121
0.01	0.714140	0.714178	.000038	0.000053
0.005	0.710631	0.710642	.000011	0.000016

Geometrically, the error in using approximation (1) to calculate $f(x_0 + \Delta x)$ is represented by the vertical distance labelled ϵ in Figure 8.1. It is

$$\epsilon = f(x_0 + \Delta x) - [f(x_0) + f'(x_0)\Delta x]. \tag{2}$$

We can rewrite equation (2) in the form

$$\frac{f(x_0 + \Delta x) - f(x_0)}{\Delta x} = f'(x_0) + \frac{\epsilon}{\Delta x}, \qquad \Delta x \neq 0. \tag{3}$$

Now if f is differentiable at x_0, as $\Delta x \to 0$ the limit of the left-hand side of equation (3) is $f'(x_0)$. Thus we may conclude that not only does $\epsilon \to 0$ as $\Delta x \to 0$ but in fact

$$\lim_{\Delta x \to 0} \frac{\epsilon}{\Delta x} = 0. \tag{4}$$

We may interpret equation (4) by saying that as $\Delta x \to 0$ the error approaches zero "faster" than Δx.

The preceding discussion is so important in understanding the nature of a differentiable function that we summarize it as follows.

THEOREM 10

Let f be a differentiable function for all x in an open interval containing x_0. Then for all $\Delta x \neq 0$

$$f(x_0 + \Delta x) = f(x_0) + f'(x_0)\Delta x + \epsilon$$

where $\displaystyle \lim_{\Delta x \to 0} \frac{\epsilon}{\Delta x} = 0.$

Approximating Increments in y

Increment notation can also be used to indicate changes in the values of the function $y = f(x)$. If we let

$$\Delta y = f(x_0 + \Delta x) - f(x_0) \tag{5}$$

then Δy is the difference in the function values resulting from the increment Δx in the independent variable x from $x = x_0$ to $x = x_0 + \Delta x$. (A shortcoming of the Δy notation, however, is that while this increment depends on both x_0 and Δx, this is not reflected in the notation.)

Using the increment notation of equation (5) we can rewrite approximation (1), when f is differentiable at x_0, as

$$\Delta y \approx f'(x_0)\Delta x. \tag{6}$$

Figure 8.2 illustrates the interpretation of this form of our linear approximation, which says that *the change in the value of the function $y = f(x)$ resulting from an increment Δx in x from $x = x_0$ to $x = x_0 + \Delta x$ is approximated by the product of the derivative $f'(x_0)$ and the increment Δx.*

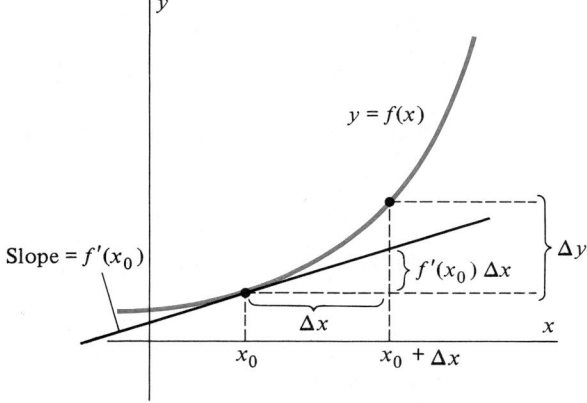

Figure 8.2 Δy is approximated by $f'(x_0)\Delta x$.

The form of the approximation in line (6) is useful in approximating the change in a quantity calculated from a measurement when small changes occur in the measurement. The following is a typical example.

Example 4

A ball bearing in the shape of a perfect sphere of radius 2 cm is to be machined from a metal alloy weighing 9 grams per cm³.

(a) Find the weight of a bearing meeting these specifications.

(b) Approximate the change in this weight resulting from an error in the radius of no more than ±.05 cm.

Solution: Since the volume of a sphere of radius r is $V = \dfrac{4}{3}\pi r^3$, the weight of a ball bearing of radius r made from this alloy is given by the function

$$W(r) = \left(\frac{4}{3}\pi r^3 \text{ cm}^3\right) \cdot \left(9\frac{\text{gm}}{\text{cm}^3}\right)$$
$$= 12\pi r^3 \text{ gm.}$$

(a) The weight of a bearing of radius $r_0 = 2$ cm is therefore

$$W(2) = 12\pi \cdot 2^3 = 96\pi \text{ gm.}$$

(b) If the actual measurement of the radius varies from $r_0 = 2$ cm by $\Delta r = \pm 0.05$, approximation (5) gives the approximation to the change ΔW in weight as

$$\Delta W \approx W'(r_0)\Delta r.$$

Since $W'(r) = 12\pi(3r^2) = 36\pi r^2$, this approximation, with $r_0 = 2$ and $\Delta r = \pm 0.05$, is

$$\Delta W \approx 36\pi(2^2)(\pm 0.05)$$
$$= \pm 7.2\pi \text{ grams.}$$

Thus, an error of $\Delta r = \pm 0.05$ cm in the radius will cause a fluctuation of *approximately* $\pm 7.2\pi$ gm in the weight.

To determine the accuracy of this approximation we compare, for example, the *actual* change ΔW in the weight corresponding to $\Delta r = 0.05$ with the *approximate* change, 7.2π. The *actual* change is

$$\Delta W = W(2.05) - W(2)$$
$$= 12\pi(2.05)^3 - 12\pi \cdot 2^3$$
$$= 7.3815\pi.$$

The relative error in this approximation is therefore

$$\frac{(7.3815\pi) - 7.2\pi}{7.3815\pi} = 0.0246,$$

a percentage error of 2.46%.   ◇

Differential Notation

Traditionally the symbol dx has been used to denote small changes in x, just as we have used the increment notation Δx, and the symbol dy has been used to represent *the approximation* to the resulting increment Δy given by the right side of approximation (6). That is,

$$dx = \Delta x \qquad (7)$$
$$dy = f'(x)dx \qquad (8)$$

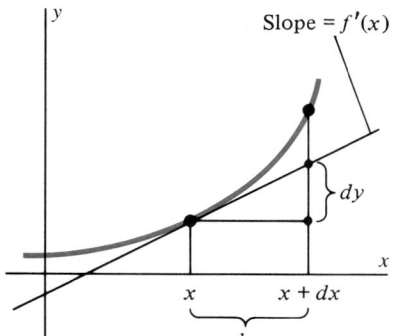

Figure 8.3 The differential $dy = f'(x)\, dx$.

The symbols dx and dy are referred to as **differentials.** Historically they have been used to argue that the derivative can be thought of as the ratio of *infinitesimals* giving the change dy in the function $y = f(x)$ resulting from a tiny change dx in x. The argument is based on the idea that the linear approximation in equation (6) becomes increasingly accurate as Δx approaches zero, so that the derivative

$$f'(x_0) = \lim_{\Delta x \to 0} \frac{f(x_0 + \Delta x) - f(x_0)}{\Delta x} = \lim_{\Delta x \to 0} \frac{\Delta y}{\Delta x}$$

may actually be regarded as the *quotient* of the infinitesimals dy and dx. (This is the origin of the Leibniz notation $\dfrac{dy}{dx}$ for the derivative $f'(x)$.)

We shall regard the differential dy, however, only as a notational device to represent the linear approximation $f'(x)\, dx$. (See Figure 8.3.) In using differential notation it is important to remember that the differential dy in equation (8) depends on (i) the value of the function f at the number $x = x_0$ in question, and (ii) the differential change $dx = \Delta x$ in x.

Example 5

Differential notation can be used to rewrite derivative formulas from the form $\dfrac{dy}{dx} = f'(x)$ in the *differential form* $dy = f'(x)\, dx$. While it appears that we do so simply by "multiplying through by dx" we are actually applying the definition of the symbol dy defined by equation (8). Here are several examples:

Function	Derivative	Differential form
$y = \sin x$	$\dfrac{dy}{dx} = \cos x$	$dy = \cos x\, dx$
$y = x^n$	$\dfrac{dy}{dx} = nx^{n-1}$	$dy = nx^{n-1}\, dx$
$y = \sqrt{x}$	$\dfrac{dy}{dx} = \dfrac{1}{2\sqrt{x}}$	$dy = \dfrac{dx}{2\sqrt{x}}$
$y = \dfrac{1}{x}$	$\dfrac{dy}{dx} = \dfrac{-1}{x^2}$	$dy = -\dfrac{dx}{x^2}$

$\diamond$

Exercise Set 3.8

In each of Exercises 1–10, use approximation (1) to approximate the given quantity.

1. $\sqrt{37}$

2. $\sqrt{48}$

3. $\dfrac{1}{\sqrt{17}}$

4. $\sqrt[3]{124}$

5. $\sin(2°)$

6. $\cos(88°)$

7. $\sqrt[3]{1004}$

8. $(31)^{3/5}$

9. $\dfrac{1}{\sqrt{26}}$

10. $\sin^2(46°)$

11. *(Calculator)* Use a calculator to find the exact value of the quantities in Exercises 1–10. For each problem compute the relative error by dividing the error by the actual value. Then find the percentage error by multiplying the relative error by 100%.

12. *(Calculator)* Find the relative and percentage errors for the approximation in Example 3.

In Exercises 13–18, find dy.

13. $y = \cos(1 - x^2)$

14. $y = \dfrac{x - 1}{x + 1}$

15. $y = x \sec(\pi - x)$

16. $y = \sqrt[3]{x^2 + x}$

17. $y = \dfrac{1 - x^3}{(2 + x)^2}$

18. $y = \dfrac{\sqrt{t}}{t^{2/3} - t}$

19. Find an approximation to $f(9)$ if $f(10) = 6$ and $f'(10) = -2$.

20. Find an approximation to $f(36)$ if $f(39) = 7$ and $f'(39) = 06.5$.

21. Water flows through a pipe at the rate of 3 liters per minute per square inch of cross-sectional area.
  a. Find the volume $V(r)$ per minute flowing through the pipe if the cross section is a circle of radius r.
  b. Find the volume per minute, $V(4)$, if the radius of the pipe is $r = 4$ inches.
  c. Approximate the change ΔV in the volume per minute resulting from an *increase* in the radius of $\Delta r = 0.5$ inches.

22. A roller bearing has the shape of a cylinder. If the radius of the base is to be $r = 2$ cm and the length is to be $\ell = 5$ cm, approximate the change ΔV in the volume if the length is precise but the radius varies by $\Delta r = 0.02$ cm.

23. A plot of ground is to be laid out in the shape of a square with sides of length $s = 200$ feet. Approximate the change ΔA in the area of the plot, in square feet, if each of the dimensions vary from the intended length by $\Delta s = 2$ feet.

24. The acceleration of a pendulum bob due to gravity is $a(\theta) = -32 \sin \theta$ where θ is the angle between the pendulum cord and the vertical (see Figure 8.4).

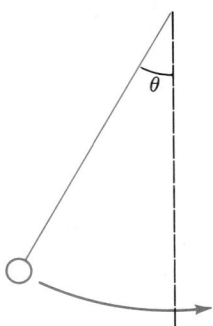

Figure 8.4

  a. Find a linear approximation to the acceleration $a(\theta)$ using $\theta_0 = 0$ degrees (in radians).
  b. Find the approximation to the acceleration when $\theta = \pi/6$ given by your answer to part a.
  c. What is the relative error in the approximation in part b?

25. The motion of a mass connected to the end of a spring is given by the position function

$$s(t) = 10 + 5 \sin(t + \pi/4)$$

where t is time in seconds.
  a. Find a linear approximation to $s(t)$ using $t_0 = \pi/4$.
  b. Find the approximation to $s(\pi/2)$ given by your answer to part a.
  c. What is the relative error in the approximation in part b?

3.9 NEWTON'S METHOD

This brief section notes an approximation procedure, due to Isaac Newton, for using the line tangent to the graph of f at $(x_0, f(x_0))$ to approximate not the values of $f(x)$ for x near x_0, but rather *the zeros* of f. Like the linear approximation procedure of Section 3.8, this method involves "following the tangent" to the desired approximation. But a new idea also enters, that of **iteration**—repeating the procedure over and over, until a desired degree of accuracy in the approximation is obtained. Although the method described here is now commonly implemented on computers rather than by hand, the mathematical basis for the method and the concept of iteration are important principles to be mastered for the effective use of this idea.

Let us consider a differentiable function f and the problem of approximating a zero c of f which is known to lie between the numbers $x = a$ and $x = b$ (Figure 9.1). Newton's method for approximating c calls for an initial guess, x_1, to be made. The line ℓ_1 tangent to the graph of $y = f(x)$ at $(x_1, f(x_1))$ is then constructed. By finding the x-intercept, x_2, of ℓ_1, we obtain a second approximation to c. Newton's famous observation is simply that for many functions the second approximation, x_2, is better

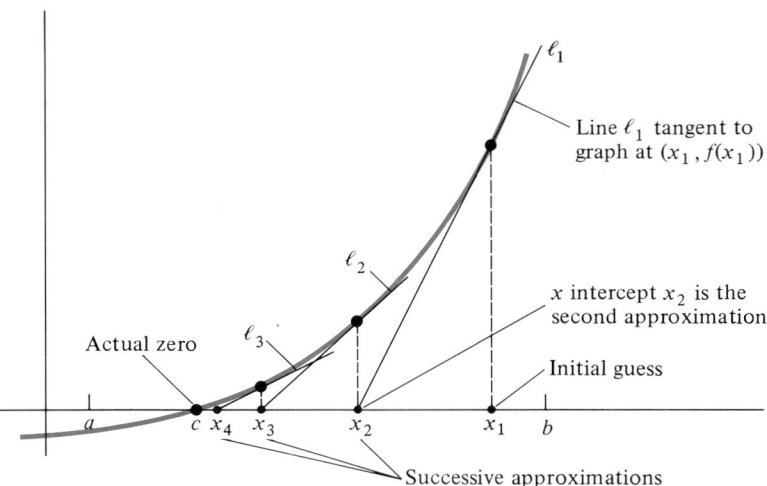

Figure 9.1 Newton's Method for approximating zero of $y = f(x)$.

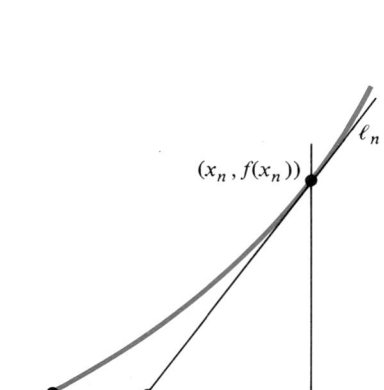

Figure 9.2 Obtaining the approximation x_{n+1} from x_n.

than the first, x_1. If the procedure is then repeated by finding the line ℓ_2 tangent to the graph of f at $(x_2, f(x_2))$, the x-intercept of ℓ_2 (say, x_3) provides an even closer approximation to c than x_2. The procedure is repeated again and again, until an approximation of sufficiently high accuracy is obtained.

We can obtain a simple equation for obtaining the $(n + 1)$st approximation, x_{n+1}, from the nth approximation, x_n, by considering Figure 9.2. Since the slope of line ℓ_n tangent to the graph of $y = f(x)$ at $(x_n, f(x_n))$ is given by the derivative $f'(x_n)$, the equation for ℓ_n can be written as

$$y - f(x_n) = f'(x_n)(x - x_n). \tag{1}$$

To find x_{n+1}, the x-intercept of ℓ_n, we set $y = 0$ and solve for x in equation (1). We obtain

$$x = x_n - \frac{f(x_n)}{f'(x_n)}, \qquad f'(x_n) \neq 0.$$

Since this is the desired approximation $x = x_{n+1}$, we have **the approximation scheme for Newton's Method:**

$$x_{n+1} = x_n - \frac{f(x_n)}{f'(x_n)}, \qquad f'(x_n) \neq 0. \tag{2}$$

Example 1

Use Newton's Method to approximate the zero of the function $f(x) = x^3 - 10$ lying between $x = 0$ and $x = 4$.

Strategy

First, verify that a zero exists in $[0, 4]$ by applying the Intermediate Value Theorem.

Solution

Before applying the method we should verify that the given function indeed has a zero between the given numbers. Since $f(0) = -10$ and $f(4) = 54$ are of opposite sign and f is continuous, the Intermediate Value Theorem guarantees that $f(x)$ must equal zero for some $x \in (0, 4)$.

Make an initial guess for the first approximation.

We begin by making an initial guess for c, say $x_1 = 3$. Then, from equation (2), with $n = 1$, we obtain the second approximation

Apply (2), using

$f'(x) = 3x^2$.

$$x_2 = 3 - \frac{f(3)}{f'(3)}$$

$$= 3 - \frac{(3^3 - 10)}{3(3)^2}$$

$$= 3 - \frac{17}{27} \approx 2.37.$$

Apply (2) again and so on.

The next approximation, x_3, is obtained from equation (2) by using $x_2 = 2.37$ and $n = 2$:

$$x_3 = 2.37 - \frac{f(2.37)}{f'(2.37)}$$

$$= 2.37 - \frac{(2.37)^3 - 10}{3(2.37)^2}$$

$$\approx 2.17,$$

and so on. $\diamondsuit$

Although the formula for Newton's Method is simple to state, the hand calculations quickly become tedious and the chance for error grows rapidly. However, this is precisely the sort of problem that is easy to implement on a hand calculator or on a computer. (A BASIC program for implementing Newton's Method is listed in Appendix I.) Table 9.1 contains the results obtained by using a computer to continue the calculations of this example through $n = 4$ iterations.

Table 9.1

n	x_n	$f(x_n)$	$f'(x_n)$	x_{n+1}
1	3.	17.	27.	2.370370
2	2.370370	3.318295	16.855967	2.173509
3	2.173509	0.267958	14.172419	2.154602
4	2.154602	0.002324	13.926924	2.154435

The advantage of listing partial calculations in a table such as this is that each approximation to the zero, x_n (column 2), can be compared directly to the function value (column 3) at each step. The results obtained in the second and fifth column of Table 9.1 suggest that $x_4 = 2.154602$ is a good approximation to the desired zero.

In Section 3.8 we used the method of linear approximation to approximate square roots of numbers lying close to perfect squares. Newton's Method can be used to approximate roots regardless of where the particular number lies.

Example 2

Use Newton's Method to approximate $\sqrt[3]{a}$.

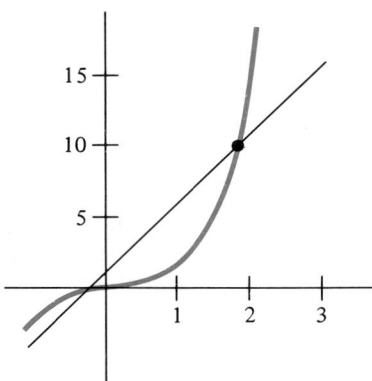

Figure 9.3

Solution: Finding the number x for which $\sqrt[3]{a} = x$ is equivalent to solving the equation $a = x^3$, which in turn is equivalent to finding a zero for the function $f(x) = x^3 - a$. This is the problem treated in Example 1 with $a = 10$. ◇

Example 3

Use Newton's Method to find the point in the right half-plane where the graphs of $f(x) = 2x^3$ and $g(x) = 5x + 1$ intersect (see Figure 9.3).

Solution: Finding a number x for which

$$2x^3 = 5x + 1$$

is equivalent to finding a zero of the function $h(x) = 2x^3 - 5x - 1$.
For this function the approximation scheme (2) becomes

$$x_{n+1} = x_n - \frac{2x_n^3 - 5x_n - 1}{6x_n^2 - 5}.$$

Since $h(0) = -1$ and $h(2) = 5$, we know by the Intermediate Value Theorem that a zero must lie in the interval $(0, 2)$. Using a first approximation $x_1 = 2$, we obtain the information contained in Table 9.2 for four iterations of Newton's Method.

Table 9.2

n	x_n	$f(x_n)$	$f'(x_n)$	x_{n+1}
1	2.0	5.0	19.0	1.736842
2	1.736842	0.794576	13.099723	1.676186
3	1.676186	0.037894	11.857599	1.672990
4	1.672990	0.000103	11.793380	1.672982

The desired point of intersection is approximately $(1.672982, 9.364914)$. ◇

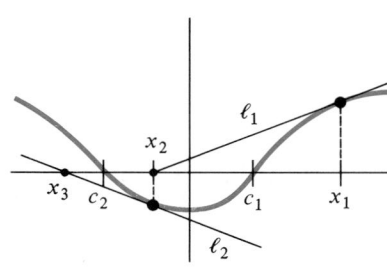

Figure 9.4 Initial approximation x_1 leads to zero c_2 where zero c_1 was desired.

When the approximations produced by Newton's Method approach the desired zero, we say that the method **converges** to that zero. Depending on the function and the initial approximation, Newton's Method may not converge to the desired zero. We comment on some of the reasons for this here, and pursue some of the details in the Exercise Set.

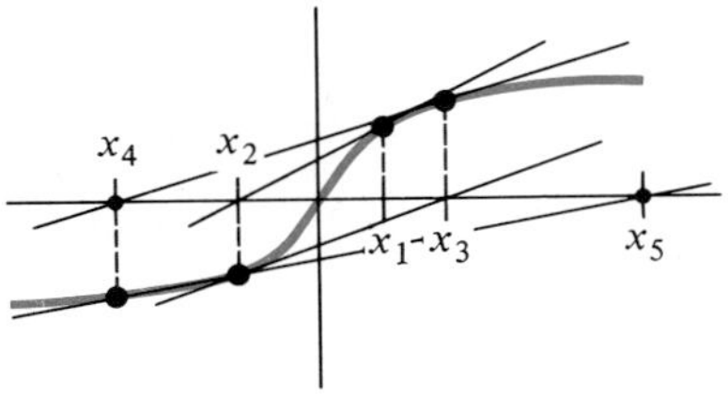

(1) The function may have more than one zero. Depending on the initial approximation x_1, the method may converge to a zero other than the one desired (see Figure 9.4).

(2) An approximation x_k may be obtained for which $f'(x_k) = 0$. In this case the denominator in the expression for x_{k+1} is zero, so the method fails.

(3) The slope of the graph of f may be such that the approximations do not converge, but instead move away from the desired zero (see Figure 9.5 and Exercise 14) or simply oscillate between two or more distinct approximations.

Figure 9.5 Iterates move away from zero rather than converging.

Exercise Set 3.9

In Exercises 1–8, use Newton's Method to approximate the zero of the given function lying between the numbers $x = a$ and $x = b$. In each case, first verify that such a zero indeed exists. If a calculator or computer is not available to you, only two or three iterations should be performed. If a calculator or computer is available, a table such as Table 9.1 should be constructed.

1. $f(x) = x^2 - \dfrac{7}{2}x + \dfrac{3}{4}$, $a = 0$, $b = 2$

2. $f(x) = 1 - x - x^2$, $a = -2$, $b = -1$

3. $f(x) = 1 - x - x^2$, $a = 0$, $b = 1$

4. $f(x) = x^3 + x - 3$, $a = 1$, $b = 2$

5. $f(x) = \sqrt{x + 3} - x$, $a = 1$, $b = 3$

6. $f(x) = x^3 + x^2 + 3$, $a = -3$, $b = -1$

7. $f(x) = x^4 - 5$, $a = -2$, $b = -1$

8. $f(x) = 5 - x^4$, $a = 1$, $b = 2$

9. Use the results of Exercise 5 to find a point of intersection between the graphs of $f(x) = \sqrt{x + 3}$ and $g(x) = x$.

10. Use Newton's Method to approximate

  a. $\sqrt{40}$ **b.** $\sqrt[3]{49}$

  c. $\sqrt[5]{18}$ **d.** $\sqrt{37}$

11. Show how the formula for the approximation scheme in Newton's Method can be obtained from the formula for linear approximation in Section 3.8. (Newton's Method can therefore be viewed as repeated application of the idea of linear approximation.)

12. What results from applying Newton's Method if the (lucky) first guess is precisely the desired zero?

13. Use a calculator or computer to find an approximation to the point where the graphs of $f(x) = 2x$ and $g(x) = \tan x$ intersect for $0 < x \le \pi/2$.

14. Attempt to use Newton's Method to find the zero of $f(x) = (x - 2)^{1/3}$ with initial guess $x_1 = 3$. What happens? Why?

15. The graphs of the functions f and g in Example 3 cross at two other points. Find them.

SUMMARY OUTLINE OF CHAPTER 3

◆ The **derivative** of f is defined by (page 94)

$$f'(x) = \lim_{h \to 0} \frac{f(x + h) - f(x)}{h}.$$

◆ The derivative of f at $x = a$ is the slope of the line tangent to the graph of $y = f(x)$ at $(a, f(a))$. The equation of this line is $y - f(a) = f'(a)(x - a)$. (page 94)

◆ **Theorem:** If f is differentiable at $x = a$, then f is continuous at $x = a$. (page 98)

◆ **Notation:** For $y = f(x)$, $\dfrac{dy}{dx}$ means $f'(x)$; $\dfrac{d^2y}{dx^2}$ means $f''(x)$. (page 100)

◆ If $s(t)$ denotes the location of a particle moving along a line, then the **velocity** of the particle is $v(t) = s'(t)$. (page 113)

◆ We have the following differentiation formulas:

If $h(x) = f(x) + g(x)$, $h'(x) = f'(x) + g'(x)$. (page 97)

If $h(x) = cf(x)$, $h'(x) = cf'(x)$. (page 97)

If $h(x) = f(x)g(x)$, $h'(x) = f'(x)g(x) + f(x)g'(x)$. (page 103)

If $h(x) = \dfrac{f(x)}{g(x)}$, $h'(x) = \dfrac{f'(x)g(x) - f(x)g'(x)}{g^2(x)}$. (page 104)

If $h(x) = \sin x$, $h'(x) = \cos x$. (page 107)

If $h(x) = \cos x$, $h'(x) = -\sin x$. (page 107)

If $h(x) = f(x)^n$, $h'(x) = nf(x)^{n-1}f'(x)$.   (page 120)

If $h(x) = f(g(x))$, $h'(x) = f'(g(x))g'(x)$.   (page 122)

◆ When y is not an explicit function of x we can often calculate $\dfrac{dy}{dx}$ by the technique of **implicit differentiation.**   (page 128)

◆ By using the tangent to $y = f(x)$ at $(x, f(x))$, we approximate $f(x + \Delta x)$ by the **linear approximation** $f(x + \Delta x) \approx$ (page 136)
$f(x) + f'(x)\Delta x$.

◆ **Differential Notation:** If $y = f(x)$, $dy = f'(x)dx$.   (page 140)

◆ **Newton's Method** uses the approximation   (page 143)

$$x_{n+1} = x_n - \frac{f(x_n)}{f'(x_n)}$$

to approximate solutions to $f(x) = 0$.

REVIEW EXERCISES—CHAPTER 3

In Exercises 1–28, find the derivative of the given function.

1. $f(x) = \dfrac{x}{3x - 7}$

2. $y = \dfrac{\sqrt{x}}{1 + \sqrt{x}}$

3. $y = \dfrac{6x^3 - x^2 + x}{3x^5 + x^3}$

4. $f(x) = \dfrac{x^3 - 1}{x^3 + 1}$

5. $f(t) = \sqrt{t}\,\sin t$

6. $y = (6x - 4)^{-3}$

7. $y = \dfrac{1}{x + x^2 + x^3}$

8. $f(s) = \left(s + \dfrac{1}{s}\right)^5$

9. $g(t) = (t^{-2} - t^{-3})^{-1}$

10. $y = \left(\dfrac{4x^3 + 3}{x + 2}\right)^4$

11. $y = x^2 \sin x^2$

12. $f(x) = \tan^2 x^2$

13. $y = \dfrac{1}{(x^4 + 4x^2)^3}$

14. $f(t) = \tan t \cdot \csc t$

15. $f(x) = \dfrac{2x - 7}{\sqrt{x^2 + 1}}$

16. $g(t) = 4t^2\sqrt{t} + t^3\sqrt{t}$

17. $y = ((x^2 + 1)^3 - 7)^5$

18. $y = [\sin(1 + x^2) + x]^3$

19. $f(x) = \sqrt{\dfrac{3x - 9}{x^2 + 3}}$

20. $y = \csc(\cot(3x))$

21. $f(s) = \tan^2 s \cdot \cot^5 s$

22. $y = \sqrt[3]{\sin x}$

23. $y = \dfrac{\sqrt{1 + \sin x}}{\cos x}$

24. $f(x) = \dfrac{x + \sqrt{x}}{\sqrt{1 + x^3}}$

25. $f(t) = t^{9/2} - 6t^{5/2}$

26. $y = \dfrac{\sqrt[3]{x}}{x + \sqrt[3]{x}}$

27. $y = \sqrt{x^3}(x^{-3} - 2x^{-1} + 2)^2$

28. $f(x) = x|x|$

Find $f'(x)$ and $f''(x)$ for

29. $f(x) = x \sin x$

30. $f(x) = \dfrac{x}{x + 1}$

31. $f(x) = \sec x \cdot \tan x$

32. $f(x) = \dfrac{1}{\sqrt{1 + x^2}}$

Find the velocity and acceleration of a particle moving along a line if its position along a line after t seconds is

33. $s(t) = t^2 - 1 + t^{1/2}$

34. $s(t) = \sqrt{t}\,\sin t$

35. $s(t) = \sqrt{2 + t^2}$

36. $s(t) = \dfrac{\sin t}{1 + \sqrt{t}}$

Use only the definition of the derivative to find $\dfrac{dy}{dx}$ for

37. $y = 6x - 2$

38. $y = \dfrac{1}{x + 1}$

39. $y = \sqrt{x + 1}$

40. $y = \dfrac{1}{\sqrt{2x + 1}}$

Find $\dfrac{dy}{dx}$ if

41. $6x^2 - xy - 4y^2 = 0$

42. $6x^3 - 2y^2 = x$

43. $x = y + y^2 + y^3$

44. $\sin \sqrt{x} + \sin \sqrt{y} = 1$

45. $xy = \cot xy$

46. $(2x^2y^3)^{1/3} = 1$

47. $x \sin y = y$

48. $\sqrt{x} + \sqrt{y} = K$

Find $\dfrac{d^2y}{dx^2}$ if

49. $x^2 + y^2 = 4$ **50.** $\sin xy = 1/2$

Find an equation for the line tangent to the graph at the indicated point.

51. $\dfrac{1}{x} + \dfrac{1}{y} = 4$, $\left(\dfrac{1}{2}, \dfrac{1}{2}\right)$

52. $y = \tan x - x$, $\left(\dfrac{\pi}{4}, 1 - \dfrac{\pi}{4}\right)$

53. $f(x) = \dfrac{\cos x}{\sin x}$, $\left(\dfrac{\pi}{4}, 1\right)$

54. $\sqrt{x} + \sqrt{y} = 1$, $\left(\dfrac{1}{4}, \dfrac{1}{4}\right)$

55. $f(x) = \sec x \tan x$, $\left(\dfrac{\pi}{4}, \sqrt{2}\right)$

Find the differential dy:

56. $y = ax + b$ **57.** $y = x \sin x$

58. $y = \dfrac{x + 1}{x + 2}$ **59.** $y = \sqrt{1 + x^2}$

60. $y = \dfrac{\sqrt{x + 1}}{\sqrt{x + 2}}$ **61.** $y = \cos \sqrt{x}$

Find a linear approximation to

62. $\sqrt{65}$ **63.** $(8.2)^{2/3}$

64. $\cos 44°$ **65.** $(0.96)^3$

Determine which of the following functions is differentiable at $x = 2$.

66. $f(x) = |x - 2|$

67. $f(x) = |x^2 - 6x + 8|$

68. $f(x) = \begin{cases} x^3 - 8, & x \le 2 \\ 6x - 12, & x > 2 \end{cases}$

69. $f(x) = \begin{cases} 4 - x^2, & x \le 2 \\ 2x - 4, & x > 2 \end{cases}$

70. A particle moves along a line so that its location at time t is given by a polynomial of degree 2. Prove that it can change direction at most once. Must it change direction?

71. The radius of a sphere is increased from 20 cm to 20.5 cm. Use differentials to approximate the change in volume. What is the relative error in this calculation? What is the percentage error?

72. Find $h'(3)$ if $h(x) = f(g(x))$, $g(3) = 4$, $g'(3) = 2$, and $f'(4) = 5$.

73. An isosceles triangle has base 10 cm and base angles $\pi/4$. Approximate the change in the area of the triangle, using differentials, if the base angles are increased 0.05 radian.

74. Prove that all normals to a circle meet at the center. What about ellipses?

75. The adiabatic law for the expansion of air is $PV^{1.4} = C$ where P = pressure, V = volume, and C is a constant. Approximate the percentage change in P caused by a 1% change in V.

76. Find the slope of the line tangent to the circle $(x + 2)^2 + (y - 3)^2 = 4$ at the point $(-1, 3 + \sqrt{3})$.

77. At what point(s) is the line tangent to the graph of $y = 4x^3 - 4x + 4$ parallel to the line with equation $y - 8x + 6 = 0$?

78. For $f(x) = \sin 2x$, find the numbers x in the interval $[-\pi, \pi]$ for which the slope of the line tangent to the graph of $y = f'(x)$ equals 2.

79. Give an example of a function f with $f'(x) = 2x^{1/4}$.

80. The graphs of $f(x) = x^2 - 4x + 4$ and its derivative are plotted on the same set of axes. Tangents to the graph of f are drawn at the points of intersection of these two graphs. Find the point of intersection of these two tangents.

81. The position of a particle moving along a horizontal line is given by the function $s(t) = \dfrac{1}{4}t^4 - 3t^3 + 12t^2 - 20t + 8$.

   a. Find the velocity function $v(t)$.
   b. Find the acceleration function $a(t)$.
   c. For which t is the particle moving to the right?
   d. For which t is the particle moving to the left?
   e. For which t is the particle at rest?
   f. At which t does the particle change direction?
   g. For which t is the acceleration zero?

82. For the function $y = A \sin x + B \tan x$ find A and B if
   a. the slope of the tangent to the graph at

      $x = 0$ is $m_1 = 4$, and

   b. the slope of the tangent to the curve at

      $x = \dfrac{\pi}{4}$ is $m_2 = 4 + \sqrt{2}$.

83. Prove that the derivative of an odd function is an even function.

84. A particle moves along a line so that after t seconds it is at $s(t) = t^3 + 3t + 3$ on the number line. Find the time at which its velocity and acceleration are of equal magnitude.

85. If the composite function $y = f(g(x))$ is differentiable at $x = a$ must both f and g be differentiable at $g(a)$ and a respectively?

Chapter 4
Applications of the Derivative

This chapter is devoted to the ways in which the theory of the derivative developed in Chapter 3 can be applied to determine properties of differentiable functions and to solve certain types of applied problems. The applications discussed here are organized around two principal interpretations of the derivative for the function f: as the slope of the tangent to the graph of $y = f(x)$, and as the rate of change of the function f.

In the first part of the chapter (Sections 4.1–4.6) we shall show how the derivative enables us to determine the shape of the graph of a function—the location of its high and low points, how it bends on various intervals, its behavior near discontinuities, and its behavior for large values of its independent variable. Knowing how to quickly determine the nature of the graph of a function is a powerful skill in solving applied problems involving quantities determined by differentiable functions.

In the latter part of the chapter we shall apply this information to problems involving the maximum or minimum value of a function and to problems concerning the rates of change among two or more functions.

4.1 EXTREME VALUES AND THE MEAN VALUE THEOREM

There are many situations in which one wishes to find the maximum or minimum value of a function. Here are but a few examples.

a. An aircraft engineer may wish to find the air speed at which an aircraft operates most efficiently.
b. A chemist may wish to find a temperature at which a certain chemical reaction proceeds most rapidly.
c. A manufacturing company may wish to determine the production rate at which its average cost per item is a minimum.

In order to solve applied problems of this type we must first determine how to use the derivative to analyze the behavior of a given function. This section presents two important tools which we shall use in this analysis: the Extreme Value Theorem and the Mean Value Theorem.

Extreme Values

By the extreme values of a function we simply mean its largest and smallest values. In order to be precise in using these terms we must also specify the set of numbers for which the function is defined.

DEFINITION 1
Extreme Values

Let the function f be defined for all numbers x in a set S.

(i) The **maximum value** of f on the set S is the value M for which

$f(c) = M$ for at least one number $c \in S$, and
$f(x) \leq M$ for *all* $x \in S$.

(ii) The **minimum value** of f on the set S is the value m for which

$f(d) = m$ for at least one number $d \in S$, and
$f(x) \geq m$ for all $x \in S$.

(iii) By the **extreme values** (or **extrema**) of f on S we mean both the maximum and minimum values of f on S.

Figure 1.1 illustrates this terminology for a function f defined on an interval $S = [a, b]$. Note that the extrema $M = f(c)$ and $m = f(d)$ are *values* of the function f, while c and d are the numbers in the interval $[a, b]$ at which these extrema occur.

It is not necessary that a given function have either a maximum or a minimum value on a given set S. For example, the function $f(x) = x^2$ does not have a maximum value on the set $S = (-\infty, \infty)$ of all real numbers (Figure 1.2), and the function $g(x) = x$ does not have a minimum value on the set $S = (-1, 1]$ (See Figure 1.3).

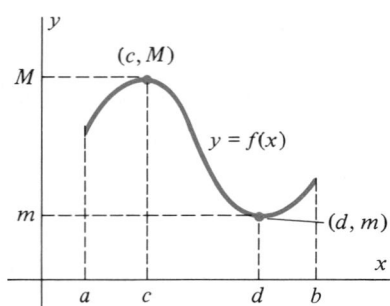

Figure 1.1 Maximum value $M = f(c)$ and minimum value $m = f(d)$ for f on $[a, b]$.

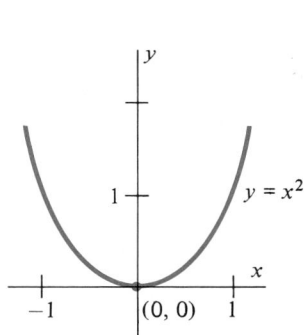

Figure 1.2 The function $f(x) = x^2$ has no maximum value on $S = (-\infty, \infty)$. (The minimum value is $f(0) = 0$.)

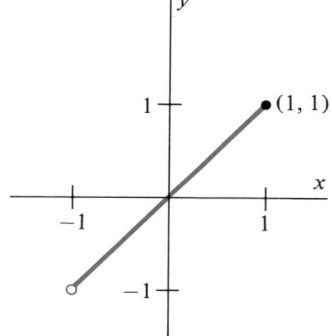

Figure 1.3 The function $g(x) = x$ has no minimum value on $S = (-1, 1]$. (The maximum value is $g(1) = 1$.)

An important property of continuous functions, however, is that *a continuous function will always have both a maximum and a minimum value on a closed finite interval $S = [a, b]$*. Although the proof of this fact is beyond the scope of this text, it is a theorem on which much of what follows depends because it guarantees the *existence* of the extrema which we shall seek.

THEOREM 1
Existence of Extreme Values

If the function f is continuous on the closed finite interval $[a, b]$ then f has both a maximum and a minimum value on $[a, b]$.

Extreme Values and the Derivative

Figure 1.4 suggests an important relationship between the derivative and extreme values for a differentiable function which occur at interior points of an interval $[a, b]$ on which f is defined. It is simply that $f'(c) = 0$ *if $f(c)$ is an extreme value for f and $a < c < b$.*

The reason for this fact is simple: if $f(c)$ is the maximum value for f, then $(c, f(c))$ is the "high point" on the graph of f, so the tangent to the graph of f at this point should be horizontal. Since $f'(c)$ is the slope of this tangent, this means that $f'(c) = 0$. The same reasoning leads to the conclusion that $f'(c) = 0$ if $f(c)$ is the minimum value of f on $[a, b]$ and $a < c < b$. Here is the formal statement of this observation.

THEOREM 2 **Extreme Value Theorem**	Let f be continuous on $[a, b]$ and let $c \in (a, b)$. If $f(c)$ is the maximum or minimum value of f on $[a, b]$ and if $f'(c)$ exists then $f'(c) = 0$.

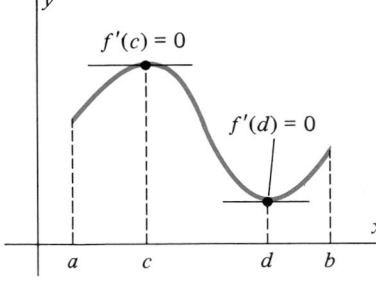

Figure 1.4 If $f(c)$ is the maximum or minimum value of f on $[a, b]$, if $c \in (a, b)$, and if $f'(c)$ exists then $f'(c) = 0$.

Proof: Let $c \in (a, b)$ and consider first the case where $f(c)$ is the maximum value of f on $[a, b]$. We shall show that both of the statements $f'(c) > 0$ and $f'(c) < 0$ lead to contradictions, leaving us with the conclusion that $f'(c) = 0$ as claimed.

If it were the case that $f'(c) > 0$, this would mean that

$$\lim_{h \to 0} \frac{f(c + h) - f(c)}{h} > 0. \tag{1}$$

From the definition of limit and from inequality (1) it follows that

$$\frac{f(c + h) - f(c)}{h} > 0 \tag{2}$$

for h sufficiently close to zero. Note also that since $a < c < b$, the number $c + h$ lies in the open interval (a, b) for h sufficiently small. For such h it follows from inequality (2) that if $h > 0$,

$$f(c + h) - f(c) > 0. \tag{3}$$

It now follows from inequality (3) that $f(c) < f(c + h)$, contradicting the fact that $f(c)$ is the maximum value of f on $[a, b]$.

This same argument shows that if we assume that $f'(c) < 0$ we are led to the inequality

$$\frac{f(c + h) - f(c)}{h} < 0$$

which, for $h < 0$, shows that $f(c + h) - f(c) > 0$ again giving the contradiction $f(c) < f(c + h)$. Thus, if $f(c)$ is the maximum value of f on $[a, b]$ and $c \in (a, b)$ we must have $f'(c) = 0$ when $f'(c)$ exists.

The proof for the case where $f(c)$ is the minimum value of f on $[a, b]$ follows by the same argument. ◆

REMARK: It is important to note that the Extreme Value Theorem applies only to the case in which the extreme value $f(c)$ occurs *at an interior point $c \in (a, b)$* of the closed interval $[a, b]$ (i.e., *not* at an endpoint) and *where the derivative $f'(c)$ exists.* In later sections we will consider situations in which extrema occur at endpoints or where $f'(c)$ fails to exist.

The following examples illustrate the Extreme Value Theorem.

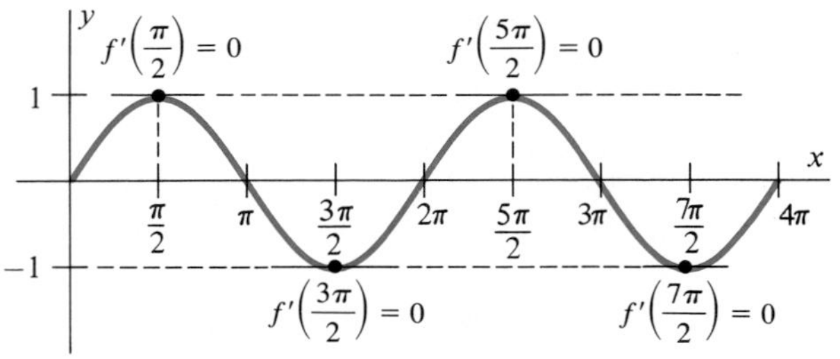

$$f'\left(\frac{\pi}{2}\right) = 0 \qquad f'\left(\frac{5\pi}{2}\right) = 0$$

$$f'\left(\frac{3\pi}{2}\right) = 0 \qquad f'\left(\frac{7\pi}{2}\right) = 0$$

Figure 1.5 Maximum and minimum values of $f(x) = \sin x$ on $[0, 4\pi]$ occur where $f'(x) = \cos x$ equals zero.

Example 1

Figure 1.5 shows the graph of the continuous function $f(x) = \sin x$ on the interval $[0, 4\pi]$. Note that

(a) The maximum value $f\left(\dfrac{\pi}{2}\right) = f\left(\dfrac{5\pi}{2}\right) = 1$ occurs where the derivative $f'(x) = \cos x$ equals zero:

$$f'\left(\frac{\pi}{2}\right) = \cos\left(\frac{\pi}{2}\right) = 0$$

$$f'\left(\frac{5\pi}{2}\right) = \cos\left(\frac{5\pi}{2}\right) = 0.$$

(b) The minimum value $f\left(\dfrac{3\pi}{2}\right) = f\left(\dfrac{7\pi}{2}\right) = -1$ also occurs where the derivative $f'(x) = \cos x$ equals zero:

$$f'\left(\frac{3\pi}{2}\right) = \cos\left(\frac{3\pi}{2}\right) = 0$$

$$f'\left(\frac{7\pi}{2}\right) = \cos\left(\frac{7\pi}{2}\right) = 0.$$

Note also that this example shows that *the maximum and minimum values of a function may occur at more than one number x.* ◇

Example 2

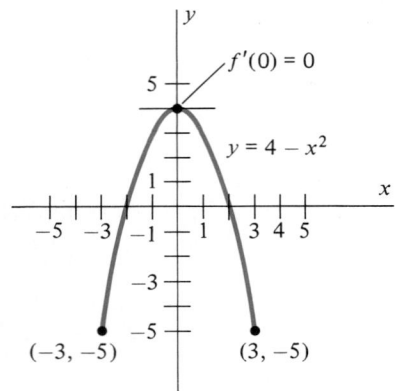

Figure 1.6 The maximum value of $f(x) = 4 - x^2$ on $[-3, 3]$ is $f(0) = 4$, where $f'(0) = 0$. The minimum value is $f(-3) = f(3) = -5$, which occurs at the endpoints of $[-3, 3]$.

Figure 1.6 shows the graph of the function $f(x) = 4 - x^2$ on the interval $[-3, 3]$. Note that the maximum value of f on this interval is

$$f(0) = 4 - 0 = 4$$

and that, since $f'(x) = -2x, f'(0) = 0$ according to the Extreme Value Theorem.

The minimum value of f on $[-3, 3]$, however, is $f(-3) = f(3) = -5$. Note that we do *not* have $f'(c) = 0$ for $c = -3$ or $c = 3$, since

$$f'(-3) = -2(-3) = 6 \text{ and } f'(3) = -2(3) = -6.$$

This is because the minimum value occurs at the *endpoints* of the interval rather than at an interior point. Thus, the Extreme Value Theorem does not apply at these points. ◇

The Extreme Value Theorem will be used frequently throughout this chapter to establish properties of continuous functions and procedures for finding various types of extrema on both finite and infinite intervals. It is also used to establish the *Mean Value Theorem,* the second principal tool that will be required in the analysis that follows. The key to proving the Mean Value Theorem is Rolle's Theorem, which follows as a direct consequence of the Extreme Value Theorem.

THEOREM 3
Rolle's Theorem

Let f be continuous on the interval $[a, b]$, let $f'(x)$ exist for each $x \in (a, b)$ and let $f(a) = f(b)$. Then there exists at least one number $c \in (a, b)$ for which $f'(c) = 0$.

Proof: Let K denote the common value $K = f(a) = f(b)$ and note first that if f is a *constant* function on $[a, b]$ we must have $f(x) = K$ for all $x \in [a, b]$. Thus, $f'(x) = 0$ for all $x \in (a, b)$ so any number $c \in (a, b)$ satisfies the conclusion $f'(c) = 0$.

If f is not a constant function, then since f is continuous on $[a, b]$ Theorem 1 guarantees that f must have a maximum value M on $[a, b]$ with $M > K$ or a minimum value m with $m < K$. This is because $f(a) = f(b) = K$. (See Figure 1.7.) In

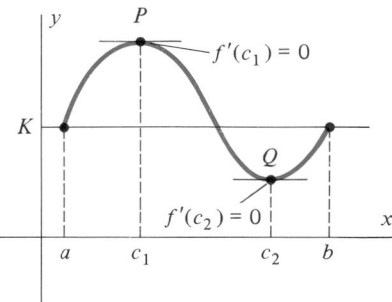

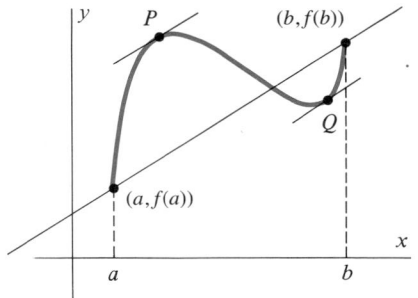

Figure 1.7 Rolle's Theorem: If f is continuous on $[a, b]$ and differentiable on (a, b) and if $f(a) = f(b)$, then $f'(c) = 0$ for at least one $c \in (a, b)$.

Figure 1.8 If the graph in Figure 1.7 is "tilted," the tangents at P and Q are still parallel to the line through $(a, f(a))$ and $(b, f(b))$.

either case we have the existence of an extreme value for f which must occur at a number $c \in (a, b)$. Since $f'(c)$ exists, the Extreme Value Theorem guarantees that $f'(c) = 0$. ◆

The Mean Value Theorem

We may paraphrase Rolle's Theorem by saying for a differentiable function f that "if the *net* change in the values $f(x)$ between $x = a$ and $x = b$ is zero, then the derivative $f'(x)$ must also equal zero somewhere between a and b." We may also interpret this result geometrically by saying that at some number $c \in (a, b)$ *the tangent to the graph of f must be parallel to the line through $(a, f(a))$ and $(b, f(b))$.*

The Mean Value Theorem generalizes Rolle's Theorem by extending these observations to functions for which $f(a) \neq f(b)$. In order to appreciate how simply this generalization is obtained, think of the graph of f shown in Figure 1.8 as the result of simply "tilting" the graph in Figure 1.7 so that $f(b) \neq f(a)$. If this were the case we would still expect the tangents at points P and Q to be parallel to the line through

$(a, f(a))$ and $(b, f(b))$. Since the line through $(a, f(a))$ and $(b, f(b))$ has slope $\dfrac{f(b) - f(a)}{b - a}$, and since the tangent at a point $(c, f(c))$ has slope $f'(c)$, we would expect to have

$$f'(c) = \frac{f(b) - f(a)}{b - a}$$

when c is the x-coordinate of point P or point Q. This is the statement of the Mean Value Theorem.

THEOREM 4
Mean Value Theorem

Let f be continuous on $[a, b]$ and let $f'(x)$ exist for each $x \in (a, b)$. Then there exists at least one number $c \in (a, b)$ for which

$$f'(c) = \frac{f(b) - f(a)}{b - a}. \tag{4}$$

(See Figure 1.9.)

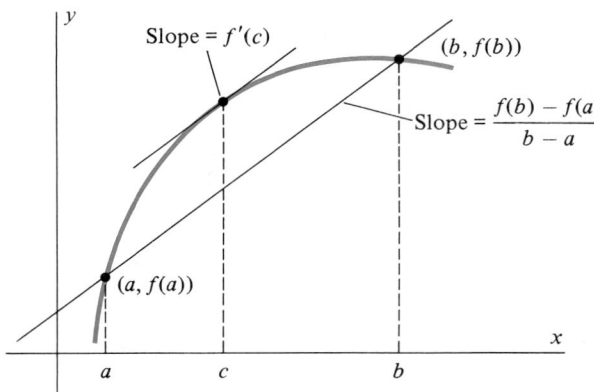

Figure 1.9 Mean Value Theorem: $f'(c) = \dfrac{f(b) - f(a)}{b - a}$.

Before giving a proof of the Mean Value Theorem we comment further on its interpretation and present several examples.

The number $\dfrac{f(b) - f(a)}{b - a}$ on the right side of equation (4), which is the *slope* of the line through $(a, f(a))$ and $(b, f(b))$, is referred to as the *average change* in f, per unit change in x, over the interval $[a, b]$. The Mean Value Theorem therefore says that *the average* change in f over $[a, b]$ is given by the derivative at one (or more) number c. This ability to represent the change in f *over an interval* by a value of its derivative *at a single number* is an important tool that will be used to analyze the behavior of functions in the next section.

The next example reminds you that we have encountered this concept of average change in f previously in our study of rectilinear motion.

Example 3

For an object travelling along a line (such as an automobile on a highway) if $s(t)$ represents the position of the object at time t we define the *average* velocity of the object between times $t = a$ and $t = b$ to be

$$\text{average velocity} = \frac{s(b) - s(a)}{b - a} \quad \left(= \frac{\text{change in distance}}{\text{change in time}} \right).$$

Recall also that we have defined the *instantaneous* velocity of the object at time t to be the derivative

$$v(t) = \lim_{h \to 0} \frac{s(t + h) - s(t)}{h} = s'(t).$$

Thus, for a differentiable position function $y = s(t)$ the Mean Value Theorem says that the *average* velocity from time $t = a$ to time $t = b$ equals the *instantaneous* velocity $v(c) = s'(c)$ for at least one time $t = c$ between a and b.

For example, if an automobile travels 90 miles in 2 hours and $s(t)$ represents the distance travelled after t hours, the *average* velocity during this time period is

$$\text{average velocity} = \frac{s(2) - s(0)}{2 - 0}$$

$$= \frac{90 - 0}{2}$$

$$= 45 \text{ mi/h.}$$

We may therefore conclude, by the Mean Value Theorem, that the automobile has velocity $v(c) = 45$ mi/h at at least one instant $t = c$ during this period. ◇

Example 4

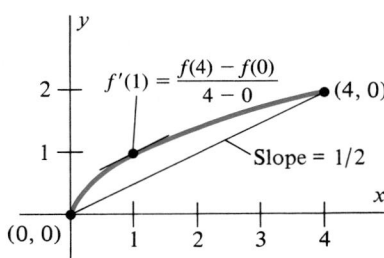

Figure 1.10 For $f(x) = \sqrt{x}$ on [0, 4] the number $c = 1$ satisfies the Mean Value Theorem.

Figure 1.10 shows the graph of the function $f(x) = \sqrt{x}$ on the interval [0, 4]. Since f is continuous on [0, 4] and $f'(x)$ exists for each $x \in (0, 4)$, the Mean Value Theorem guarantees the existence of a number c for which

$$f'(c) = \frac{f(4) - f(0)}{4 - 0} \tag{5}$$

$$= \frac{\sqrt{4} - \sqrt{0}}{4}$$

$$= \frac{1}{2}.$$

To find c we note that $f'(x) = \dfrac{1}{2\sqrt{x}}$, so equation (5) becomes

$$\frac{1}{2\sqrt{c}} = \frac{1}{2}.$$

Thus, $c = 1$.

Example 5

The function $f(x) = x^{2/3}$ does not satisfy the conditions of the Mean Value Theorem on the interval $[-1, 1]$ because $f'(x) = \dfrac{2}{3}x^{-1/3} = \dfrac{2}{3\sqrt[3]{x}}$ so $f'(0)$ is not defined. In this example we have

$$\frac{f(b) - f(a)}{b - a} = \frac{(1)^{2/3} - (-1)^{2/3}}{2 - (-2)} = 0$$

but $f'(x) \neq 0$ for all x with $-1 < x < 1$ and $x \neq 0$. (See Figure 1.11.) $\quad\diamond$

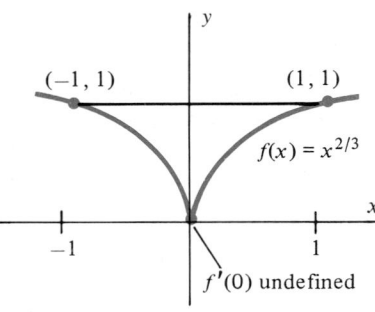

Figure 1.11 The function $f(x) = x^{2/3}$ does not satisfy the conditions of the Mean Value Theorem on $[-1, 1]$ since $f'(0)$ is undefined.

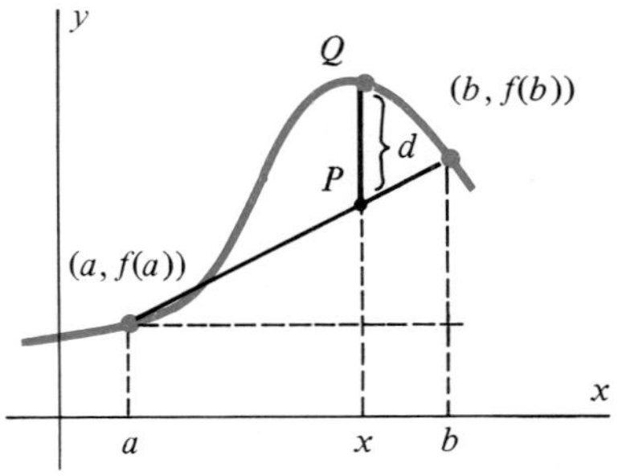

Figure 1.12 Proof of Mean Value Theorem involves finding the maximum value of the distance function d.

Proof of the Mean Value Theorem

The key to proving the Mean Value Theorem is to remember that the number c which satisfies Rolle's Theorem is the number for which $f(x)$ is a maximum on $[a, b]$. Since we have generalized Rolle's Theorem by "tilting" the line l through $(a, f(a))$ and $(b, f(b))$, we now seek the number $x = c$ for which the vertical distance $d(x)$ between the graph of f and the line l is a maximum. (See Figure 1.12.)

Since the line l contains the points $(a, f(a))$ and $(b, f(b))$, the coordinates of a point $P = (x, y)$ on l must satisfy the equation

$$\frac{y - f(a)}{x - a} = \frac{f(b) - f(a)}{b - a} \qquad (= \text{slope of } l)$$

so

$$y = f(a) + \left[\frac{f(b) - f(a)}{b - a}\right](x - a)$$

if $P = (x, y)$ is on l. The vertical distance (or its negative) between the point $P = (x, y)$ on l and the point $Q = (x, f(x))$ on the graph of f is therefore

$$d(x) = f(x) - \left\{f(a) + \left[\frac{f(b) - f(a)}{b - a}\right](x - a)\right\}. \tag{6}$$

We next note the following facts about the function $y = d(x)$:

(1) It is continuous on $[a, b]$ and differentiable on (a, b) because f has these properties.

(2) Letting $x = a$ in (6) shows that

$$d(a) = f(a) - \left\{ f(a) + \frac{f(b) - f(a)}{b - a}(0) \right\} = 0.$$

(3) Similarly, letting $x = b$ in (6) shows that

$$d(b) = f(b) - \left\{ f(a) + \frac{f(b) - f(a)}{b - a}(b - a) \right\} = 0.$$

Thus, the function d satisfies all hypotheses of Rolle's Theorem so, by Rolle's Theorem, there exists a number $c \in (a, b)$ for which

$$d'(c) = 0. \tag{7}$$

But from equation (6) we see that

$$d'(x) = f'(x) - \frac{f(b) - f(a)}{b - a} \tag{8}$$

so from (7) and (8) we conclude that

$$f'(c) - \frac{f(b) - f(a)}{b - a} = 0,$$

which gives the conclusion of the Mean Value Theorem. ◆

Exercise Set 4.1

In Exercises 1–10 the maximum or minimum value of f on the interval $[a, b]$ occurs at the number c. In each exercise, (a) verify that $f'(c) = 0$, (b) determine by inspection whether $f(c)$ is the maximum or minimum value of f on $[a, b]$, and (c) sketch the graph of $y = f(x)$ for $x \in [a, b]$.

1. $f(x) = x^2 - 4x + 3$ $\quad c = 2$ $\quad [0, 4]$

2. $f(x) = \sin\left(x + \dfrac{\pi}{4}\right)$ $\quad c = \dfrac{\pi}{4}$ $\quad [0, 2\pi]$

3. $f(x) = x^2 - 7$ $\quad c = 0$ $\quad [-3, 4]$

4. $f(x) = 6 - (x - 2)^2$ $\quad c = 2$ $\quad [0, 5]$

5. $f(x) = \cos(\pi x)$ $\quad c = 1$ $\quad [0, 2]$

6. $f(x) = 4 + 2x^2$ $\quad c = 0$ $\quad [-2, 2]$

7. $f(x) = \dfrac{1}{1 + x^2}$ $\quad c = 0$ $\quad [-2, 2]$

8. $f(x) = 1 - \sec x$ $\quad c = 0$ $\quad \left[-\dfrac{\pi}{2}, \dfrac{\pi}{2}\right]$

9. $f(x) = 2 - \sin x$ $\quad c = \dfrac{\pi}{2}$ $\quad [0, \pi]$

10. $f(x) = \csc x$ $\quad c = \dfrac{\pi}{2}$ $\quad \left[\dfrac{\pi}{4}, \dfrac{3\pi}{4}\right]$

11. Sketch the graph of $f(x) = |x|$ for $-1 \le x \le 1$. Does the Mean Value Theorem apply for f on $[-1, 1]$? Why or why not?

12. The function $f(x) = \dfrac{1}{x - 3}$ is differentiable on the open interval $(0, 3)$, yet the Mean Value Theorem does not apply on $[0, 3]$. Why?

In each of Exercises 13–20, verify that the given function satisfies the conditions of the Mean Value Theorem on the stated interval.

13. $f(x) = 2x - 7,$ $\quad x \in [-1, 5]$

14. $f(x) = 4 - x^2,$ $\quad x \in [0, 2]$

15. $f(x) = x^2 + 2x,$ $\quad x \in [0, 4]$

16. $f(x) = 3 \sin 2x,$ $\quad x \in [0, \pi/2]$

17. $f(x) = x^2 + 2x - 3,$ $\quad x \in [-3, 0]$

18. $f(x) = x^3 - 2x + 4,$ $\quad x \in [0, 2]$

19. $y = x^{2/3},$ $\quad x \in [0, 8]$

20. $y = \dfrac{1}{(x - 2)^2},$ $\quad x \in [3, 5]$

21. Show that if f satisfies the conditions of the Mean Value Theorem on $[a, b]$ and $|f'(x)| \leq M$ for all $x \in (a, b)$ then

$$|f(b) - f(a)| \leq M(b - a).$$

22. Show that if f satisfies the hypotheses of Exercise 21 then

$$f(x) \leq f(a) + M(b - a)$$

for all $x \in [a, b]$.

23. Use the results of Exercise 21 to show that

$$|\sin x - \sin y| \leq |x - y|$$

for any numbers x and y.

24. Extend the results of Exercise 21 to the case $m \leq f'(x) \leq M$, $x \in [a, b]$ and obtain the inequality

$$m(b - a) \leq |f(b) - f(a)| \leq M(b - a)$$

for differentiable functions f.

25. Suppose an automobile travels at speeds between 75 and 90 kilometers per hour for 6 hours. From Exercise 24 what can you conclude about the distance travelled during this time period?

26. Use the results of Exercise 22 to show that

$$6 < \sqrt{36.2} < 6.02$$

27. Rolle's Theorem shows that between any two zeros of a differentiable function there must always be a zero of the derivative. Is the following true? If f is differentiable for all x and if $f'(x_1) = f'(x_2) = 0$, f must have a zero in (x_1, x_2).

28. Use the Mean Value Theorem to show that if f and g are differentiable on (a, b) and continuous on $[a, b]$, if $f(a) = g(a)$, and if $f'(x) < g'(x)$ for all $x \in (a, b)$, then $f(b) < g(b)$. (*Hint:* Apply the Mean Value Theorem to the function $h = g - f$.)

(Calculator/Computer) In each of Exercises 29–31, use Newton's Method to approximate the number c satisfying the Mean Value Theorem for the given function on the given interval.

29. $f(x) = x^4 - 3x^2 + 2$, $\quad x \in [1, 4]$.

30. $f(x) = \tan x$, $\quad x \in \left[0, \dfrac{\pi}{4}\right]$.

31. $f(x) = \sin \sqrt{x}$, $\quad x \in \left[\dfrac{\pi^2}{16}, \dfrac{\pi^2}{4}\right]$.

32. Prove that if f is a quadratic function $f(x) = ax^2 + bx + c$ then the number c in the interval $[x_1, x_2]$ satisfying the Mean Value Theorem is just the midpoint $c = \dfrac{x_1 + x_2}{2}$.

4.2 INCREASING AND DECREASING FUNCTIONS

In this section we establish a relationship between the sign of the derivative and whether the graph of the function f is rising or falling. Figure 2.1 suggests that a rising graph corresponds to tangents with positive slope. Similarly, a falling graph suggests negative slopes. The following definition is useful in formalizing these observations.

DEFINITION 2

The function f is said to be **increasing** on an interval I if $f(x)$ is defined for all $x \in I$ and $f(x)$ increases as x increases. That is, f is increasing on I means that for any numbers x_1 and x_2 in I

if $\quad x_1 < x_2 \quad$ then $\quad f(x_1) < f(x_2)$.

Similarly, the statement that f is **decreasing** on I means that for any numbers x_1 and x_2 in I

if $\quad x_1 < x_2 \quad$ then $\quad f(x_1) > f(x_2)$.

(See Figures 2.2 and 2.3.)

Note that we refer to a function as being increasing or decreasing only *on intervals,* not at particular numbers. Also, note that we do not specify whether the interval I includes either or both its endpoints.

The following theorem shows how the sign of the first derivative determines whether a function is increasing or decreasing on a given interval.

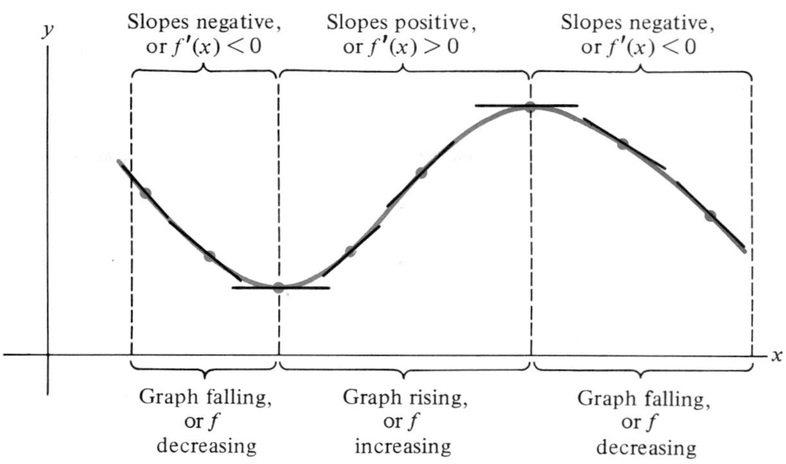

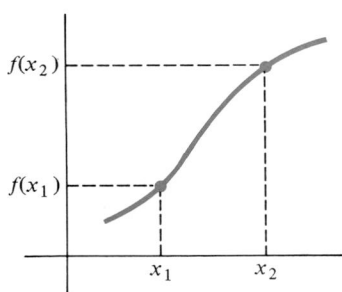

Figure 2.1 Relationship between graph rising and falling and sign of $f'(x)$.

Figure 2.2 f increasing: if $x_1 < x_2$, then $f(x_1) < f(x_2)$.

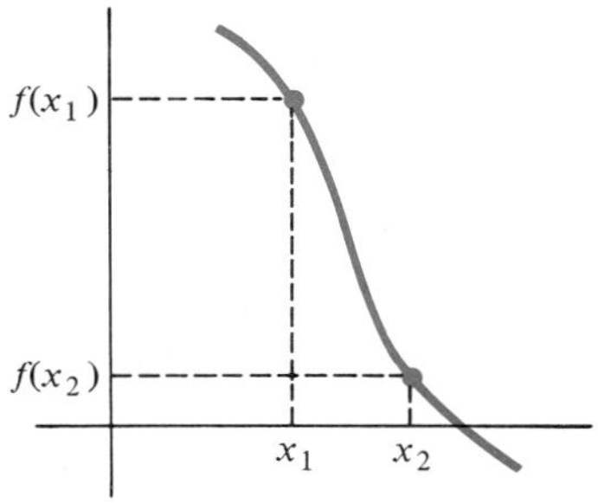

Figure 2.3 f decreasing: if $x_1 < x_2$, then $f(x_1) > f(x_2)$.

THEOREM 5

Let f be continuous on the interval I and let $f'(x)$ exist for all $x \in I$ except possibly at endpoints. Then

(i) If $f'(x) > 0$ for all x in I that are not endpoints, then f is increasing on I.
(ii) If $f'(x) < 0$ for all x in I that are not endpoints then f is decreasing on I.

Proof: To prove part (i) we use Definition 2 by assuming that x_1 and x_2 are any numbers in I with $x_1 < x_2$ and showing that $f(x_1) < f(x_2)$. For such numbers x_1 and x_2, our hypotheses guarantee that f is continuous on $[x_1, x_2]$ and that $f'(x)$ exists for all $x \in (x_1, x_2)$. Thus, the Mean Value Theorem guarantees the existence of a number $c \in (x_1, x_2)$ for which

$$\frac{f(x_2) - f(x_1)}{x_2 - x_1} = f'(c). \tag{1}$$

Since we are assuming that $f'(x) > 0$ for all $x \in (x_1, x_2)$, we know that $f'(c) > 0$. Thus, the left side of equation (1) is a positive number. Since $x_2 - x_1 > 0$, it follows that $f(x_2) - f(x_1) > 0$ also. That is, $f(x_1) < f(x_2)$. Since x_1 and x_2 were *any* numbers in I with $x_1 < x_2$, this shows that f is increasing on I. This proves statement (i).

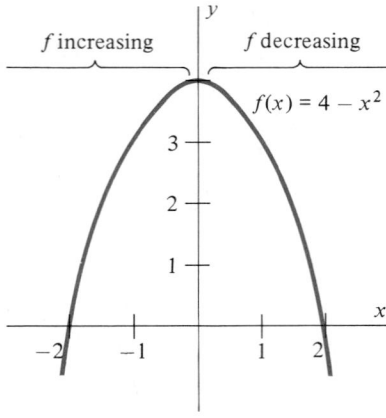

Figure 2.4 $f(x) = 4 - x^2$ is increasing on $(-\infty, 0]$ and decreasing on $[0, \infty)$.

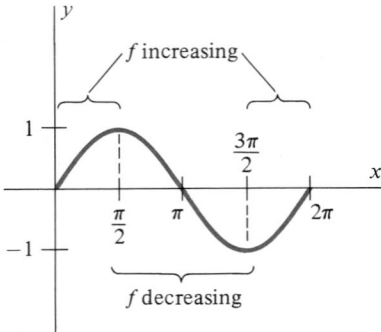

Figure 2.5 $f(x) = \sin x$ is increasing on $\left[0, \dfrac{\pi}{2}\right]$ and $\left[\dfrac{3\pi}{2}, 2\pi\right]$ and decreasing on $\left[\dfrac{\pi}{2}, \dfrac{3\pi}{2}\right]$.

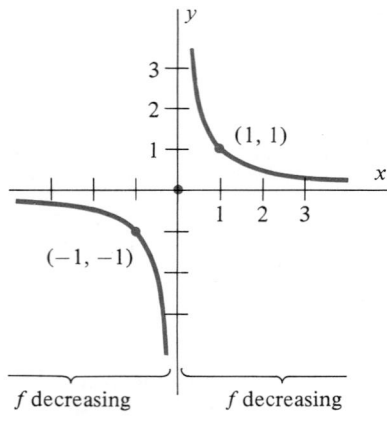

Figure 2.6 $f(x) = \dfrac{1}{x}$ is decreasing on $(-\infty, 0)$ and on $(0, \infty)$.

The proof for statement (ii) is the same except, since we are assuming $f'(x) < 0$ for $x \in (x_1, x_2)$, we have $f'(c) < 0$ in equation (1). It follows that $f(x_2) - f(x_1) < 0$, so $f(x_1) > f(x_2)$ and f is decreasing on I. ◆

Example 1

Use Theorem 5 to verify that the function $f(x) = 4 - x^2$ is

(a) increasing on the interval $(-\infty, 0]$, and
(b) decreasing on the interval $[0, \infty)$.

Solution: Since $f'(x) = -2x$ exists for all x, f is continuous and differentiable for all x, so Theorem 5 may be applied as follows:

(a) For $x \in (-\infty, 0)$, $f'(x) = -2x$ is *positive*. Thus, f is increasing on $(-\infty, 0]$. (We include the endpoint 0 because $f(0)$ is defined and f is continuous on the interval $(-\infty, 0]$.)
(b) For $x \in (0, \infty)$ $f'(x) = -2x$ is *negative* so f is decreasing on $[0, \infty)$. (See Figure 2.4.) ◇

Example 2

The function $f(x) = \sin x$ satisfies the conditions of Theorem 5 on the interval $[0, 2\pi]$. Theorem 5 shows that f is

(a) increasing on $\left[0, \dfrac{\pi}{2}\right]$, because $f'(x) = \cos x$ is positive for $x \in (0, \pi/2)$.

(b) decreasing on $[\pi/2, 3\pi/2]$ because $f'(x) = \cos x$ is negative on $\left(\dfrac{\pi}{2}, 3\pi/2\right)$.

(c) increasing on $\left[\dfrac{3\pi}{2}, 2\pi\right]$ because $f'(x) = \cos x$ is positive for $x \in \left(\dfrac{3\pi}{2}, 2\pi\right)$.

(See Figure 2.5.) ◇

Example 3

Figure 2.6 shows the graph of the function f defined as follows:

$$f(x) = \begin{cases} \dfrac{1}{x}, & x \neq 0 \\ 0, & x = 0. \end{cases}$$

Note that $f(x)$ is defined for all x, but that f is continuous only on the intervals $(-\infty, 0)$ and $(0, \infty)$. Since $f'(x) = \dfrac{-1}{x^2}$ is negative for all $x \neq 0$, we conclude that f is decreasing both on $(-\infty, 0)$ and on $(0, \infty)$.

Note that we may *not* conclude that f is decreasing on $(-\infty, 0]$ or on $[0, \infty)$ because f is not continuous on these intervals. In fact, f is *not* decreasing on $(-\infty, 0]$ or on $[0, \infty)$ as you are asked to show in Exercise 43. ◇

REMARK: While Theorem 5 gives conditions which *guarantee* (i.e., are *sufficient*) that f be increasing on an interval I, it is not *necessary* that an increasing function satisfy these conditions. The next two examples illustrate this point (which is valid also for decreasing functions).

Example 4

The function f defined on $[0, 2]$ by

$$f(x) = \begin{cases} x, & 0 \le x \le 1 \\ x + 1, & 1 < x \le 2 \end{cases}$$

is increasing on $[0, 2]$ as you can verify directly from Definition 2. (See Figure 2.7.) The function f does not satisfy the conditions of Theorem 5, however, since f is discontinuous at $x = 1$. ◇

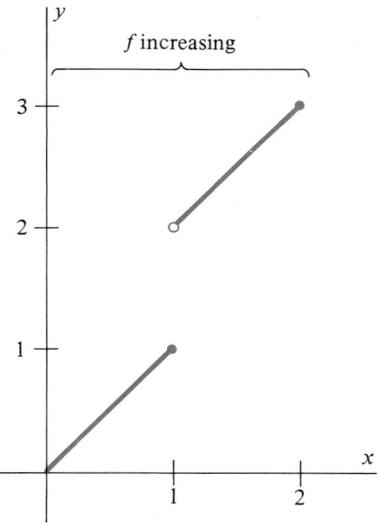

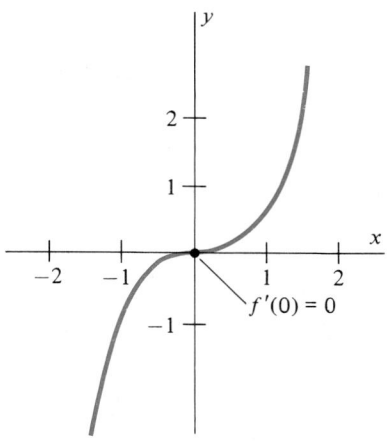

Figure 2.7 The function f in Example 4 is increasing on $[0, 2]$ although $f'(1)$ is undefined.

Figure 2.8 $f(x) = x^3$ is increasing on $(-\infty, \infty)$ even though $f'(0) = 0$.

Example 5

The function $f(x) = x^3$ is continuous and differentiable for all x. Since $x_2^3 > x_1^3$ whenever $x_2 > x_1$ this function is increasing on $(-\infty, \infty)$. We can use Theorem 5 to establish this conclusion in two steps as follows (see Figure 2.8):

(i) For $x < 0$ we have $f'(x) = 3x^2 > 0$. Since f is continuous at $x = 0$, we may conclude that f is increasing on $(-\infty, 0]$.

(ii) Similarly, since $f'(x) = 3x^2 > 0$ for $x > 0$, f is increasing on $[0, \infty)$.

Statements (i) and (ii) combine to tell us that f is increasing on $(-\infty, 0] \cup [0, \infty) = (-\infty, \infty)$. We cannot apply Theorem 5 directly to the interval $(-\infty, \infty)$, however, since $f'(0) = 0$. ◇

Critical Numbers

For functions that are differentiable except at (few) finitely many numbers x, Theorem 5 gives a procedure for finding the largest intervals on which f is either increasing or decreasing. A key concept in this procedure is that of a *critical number* for f, which we define as follows.

DEFINITION 3:

The **critical numbers** for a function f are those numbers c in the domain of f for which either $f'(c) = 0$ or $f'(c)$ fails to exist.

The reason for defining critical numbers is that they are useful in identifying the intervals on which Theorem 5 applies, as follows.

> **Procedure for Finding the Largest Intervals on which f is Either Increasing or Decreasing:**
>
> 1. Locate all critical numbers for f.
> 2. Find the intervals in the domain of f determined by the critical numbers and any numbers x for which $f(x)$ is undefined.
> 3. Apply Theorem 5 in each interval.

Example 6

Find the largest intervals on which the function

$$f(x) = \frac{1}{3}x^3 - x^2 - 3x + 4$$

is either increasing or decreasing.

Strategy

Find $f'(x)$.
Find the numbers x for which $f'(x) = 0$ or $f'(x)$ is undefined (critical numbers).
Apply Theorem 5 on each interval.

Solution

Here $f'(x) = x^2 - 2x - 3 = (x - 3)(x + 1)$.

Setting $f'(x) = (x - 3)(x + 1) = 0$ gives the critical numbers $x = -1$ and $x = 3$. There are no numbers x for which $f'(x)$ is undefined or for which $f(x)$ is undefined.

From the factored form of $f'(x)$, we can see (Figure 2.9) that

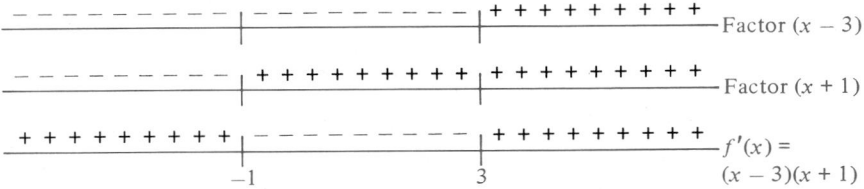

Figure 2.9 Sign analysis for $f'(x) = (x - 3)(x + 1)$ in Example 6.

(i) If $x < -1$, then both $(x - 3) < 0$ and $(x + 1) < 0$. Thus, $f'(x) > 0$ for $x \in (-\infty, -1)$, so f is increasing on $(-\infty, -1]$.
(ii) If $-1 < x < 3$, then $(x - 3) < 0$, but $(x + 1) > 0$. Thus $f'(x) < 0$ for $x \in (-1, 3)$, so f is decreasing on $[-1, 3]$.

(iii) If $x > 3$, then both $(x - 3) > 0$ and $(x + 1) > 0$. Thus, $f'(x) > 0$ for $x \in (3, \infty)$, so f is increasing on $[3, \infty)$.

(See Figure 2.10.) ◇

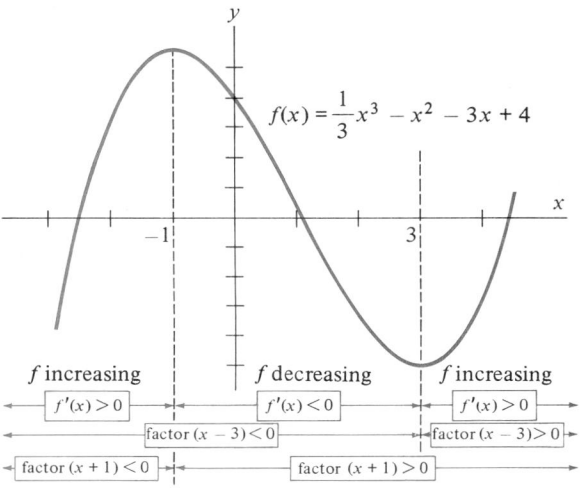

Figure 2.10 Graph of $f(x) = \dfrac{1}{3}x^3 - x^2 - 3x + 4$ and analysis of signs of factors of $f'(x) = x^2 - 2x - 3$.

Often it is difficult to determine the sign of $f'(x)$ on various intervals, especially when $f'(x)$ is not easily factored. When f' is a continuous function, a procedure based on the Intermediate Value Theorem may be used. The idea is simply that if a and b are successive critical numbers for f, then f' cannot change sign on the interval (a, b). (Otherwise the Intermediate Value Theorem would imply the existence of another zero for f' between a and b.) We can therefore determine the sign of f' on all of (a, b) by checking the sign of $f'(t)$ *at any particular* $t \in (a, b)$. We use this idea in the following example.

Example 7

Determine the largest intervals on which the function $f(x) = \dfrac{x^2}{x - 2}$ is increasing or decreasing.

Solution: The derivative is found using the Quotient Rule:

$$f'(x) = \frac{(x - 2)(2x) - (x^2)(1)}{(x - 2)^2} = \frac{x^2 - 4x}{(x - 2)^2} = \frac{x(x - 4)}{(x - 2)^2}.$$

Thus, $f'(x) = 0$ for $x = 0$ and $x = 4$, and $f'(2)$ is undefined, as is $f(2)$. The critical numbers are therefore $x = 0, 2,$ and 4, and we must inspect the sign of f' on each of the intervals $(-\infty, 0)$, $(0, 2)$, $(2, 4)$, and $(4, \infty)$. Table 2.1 shows the result of checking the sign of $f'(t)$ at an arbitrarily chosen test number t in each of these intervals, and the resulting conclusions from Theorem 5. (See Figure 2.11.) Note that our conclusions include the endpoints 0 and 4, where f is continuous, but not the endpoint 2 because $f(2)$ is undefined. ◇

Table 2.1 Analysis of $f(x) = \dfrac{x^2}{x-2}$ for increase and decrease

Interval I	Test number $t \in I$	Sign of $f'(t)$	Conclusion
$(-\infty, 0]$	$t = -1$	$f'(-1) = \dfrac{5}{9} > 0$	f increasing on $(-\infty, 0]$
$[0, 2)$	$t = 1$	$f'(1) = -3 < 0$	f decreasing on $[0, 2)$
$(2, 4]$	$t = 3$	$f'(3) = -3 < 0$	f decreasing on $(2, 4]$
$[4, \infty)$	$t = 5$	$f'(5) = \dfrac{5}{9} > 0$	f increasing on $[4, \infty)$

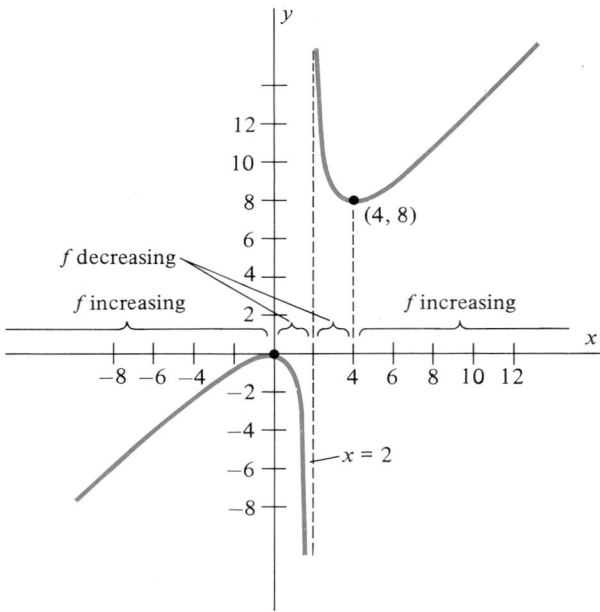

Figure 2.11 Graph of $f(x) = \dfrac{x^2}{x-2}$.

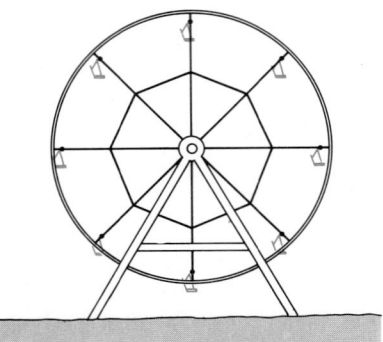

Figure 2.12 Ferris wheel in Example 8.

Example 8

Figure 2.12 shows a Ferris wheel of radius $r = 20'$ mounted $5'$ above ground level. When in motion the ferris wheel turns at a constant rate of $\dfrac{\pi}{4}$ radians per second. Figure 2.13 shows that after t seconds the point P that is originally located at position $\theta(0) = 0$ will be located at position $\theta(t) = \dfrac{\pi t}{4}$. For which times during the first 8 seconds of motion will the seat located at point P be (a) rising and (b) falling?

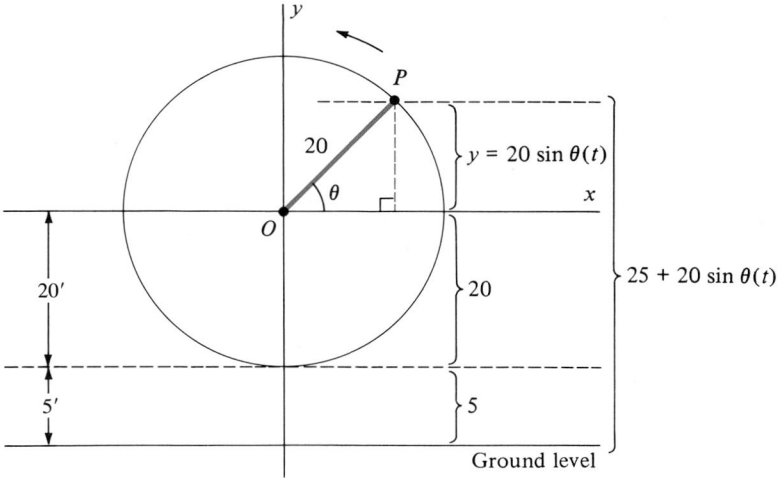

Figure 2.13   Height of point P is $h(t) = 25 + 20 \sin \theta(t)$.

Strategy

Find the function h giving the altitude of point P at time t.

Solution

With an xy-coordinate system superimposed as in Figure 2.13 the height of point P above ground level at time t is

$$h(t) = 25 + 20 \sin \theta(t).$$

Since $\theta(t) = \dfrac{\pi t}{4}$ radians, we have

$$h(t) = 25 + 20 \sin \frac{\pi t}{4}.$$

Then

Find $h'(t)$.

$$h'(t) = 20\left(\frac{\pi}{4}\right)\cos \frac{\pi t}{4}$$

$$= 5\pi \cos \frac{\pi t}{4}.$$

Thus, $h'(t) = 0$ if

Set $h'(t) = 0$ and solve to find the critical numbers for h in $[0, 8]$.

$$\cos\left(\frac{\pi t}{4}\right) = 0$$

which occurs when

$$\frac{\pi t}{4} = \frac{\pi}{2} + n\pi, \quad n = 0, 1, 2, \ldots$$

or

$$t = \frac{4}{\pi}\left(\frac{\pi}{2} + n\pi\right), \quad n = 0, 1, 2, \ldots$$

$$= 2 + 4n, \quad n = 0, 1, 2, \ldots.$$

Apply Theorem 5:
P is rising when h is increasing; P is falling when h is decreasing.

The critical numbers for h in the interval $[0, 8]$ are therefore $t = 2$ and $t = 6$. Applying Theorem 5 on the intervals $[0, 2]$, $[2, 6]$, and $[6, 8]$ gives the conclusions contained in Table 2.2. (See Figure 2.14.)

◇

Table 2.2

Time interval	Test time	$h'(t)$	Conclusion
$[0, 2]$	$t = 1$	$5\pi \cos \dfrac{\pi}{4} = \dfrac{5\pi\sqrt{2}}{2} > 0$	h increasing P rising
$[2, 6]$	$t = 4$	$5\pi \cos \pi = -5\pi < 0$	h decreasing P falling
$[6, 8]$	$t = 7$	$5\pi \cos \dfrac{7\pi}{4} = \dfrac{5\pi\sqrt{2}}{2} > 0$	h increasing P rising

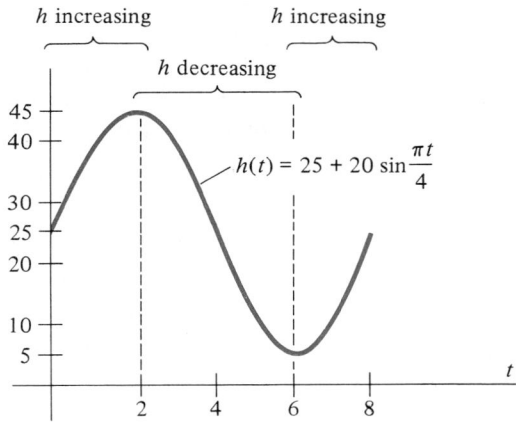

Figure 2.14 Altitude $h(t)$ of point P on ferris wheel in Example 8.

Exercise Set 4.2

In Exercises 1–14 find the largest intervals on which the function is (a) increasing and (b) decreasing. Use this information to sketch a rough graph of the function.

1. $f(x) = x^2 - 2x + 2$

2. $f(x) = x^2 - 6x + 7$

3. $f(x) = x^3 + 4$

4. $f(x) = 9 - x^2$

5. $f(x) = \cos x, \quad 0 \le x \le 2\pi$

6. $f(x) = \tan x, \quad -\pi \le x \le \pi$

7. $f(x) = x^4 - 1$

8. $f(x) = \dfrac{3}{x - 2}$

9. $f(x) = \dfrac{2}{x + 1}$

10. $f(x) = \dfrac{1}{x^2}$

11. $f(x) = \begin{cases} 4 - x^2, & -\infty < x \le 1 \\ x + 2, & 1 < x < \infty \end{cases}$

12. $f(x) = \begin{cases} \dfrac{x^2 - 1}{x - 1}, & x \ne 1 \\ 1, & x = 1 \end{cases}$

13. $f(x) = |x + 2|$

14. $f(x) = |4 - x^2|$

In Exercises 15–28 find the largest intervals on which f is (a) increasing or (b) decreasing.

15. $f(x) = \dfrac{x}{1 + x}$

16. $f(x) = \sqrt{x + 2}$

17. $f(x) = \dfrac{1}{1 + x^2}$

18. $f(x) = \dfrac{x}{1 + x^2}$

19. $f(x) = (x + 3)^{2/3}$

20. $f(x) = 1 - x^{2/3}$

21. $f(x) = \frac{1}{3}x^3 - 3x^2 - 7x + 5$

22. $f(x) = x^3 + 3x^2 + 10$

23. $f(x) = \dfrac{x^{1/3}}{x^{2/3} - 4}$

24. $f(x) = x^{2/3}(x + 8)^2$

25. $f(x) = \sec^2 x, \quad 0 \le x \le 2\pi$

26. $f(x) = \sin^2 x + \cos x, \quad 0 \le x \le 2\pi$

27. $f(x) = \frac{1}{4}x^4 - 2x^3 + \frac{3}{2}x^2 + 10x - 8$

28. $f(x) = \sqrt[3]{8 - x^3}$

29. The function $f(x) = 2x^3 - 3ax^2 + 6$ is decreasing only on the interval $[0, 3]$. Find a.

30. Find q so that the function $f(x) = 2x^2 + qx + 5$ is increasing on $[-3, \infty)$ and decreasing on $(-\infty, -3]$.

31. Find q so that the function $f(x) = \frac{1}{3}x^3 - x^2 + qx + 10$ is increasing on $(-\infty, -3]$ and $[5, \infty)$ and decreasing on $[-3, 5]$.

32. A function f has derivative $f'(x) = 2x - 6$. If f is defined for all x what is the largest interval on which f is increasing?

33. The function $f(x) = \sin ax$ has a critical number at $x = \dfrac{\pi}{6}$.

Describe the set of all numbers a for which this is true.

34. A rectangle of variable width and length has a fixed perimeter of 24 inches.
   a. Show that the area A of the rectangle is expressed as a function of its width x by the equation $A(x) = x(12 - x), \quad 0 \le x \le 12$.
   b. What is the largest subinterval of $[0, 12]$ for which $y = A(x)$ is an increasing function of x?
   c. What is the maximum value of the function $y = A(x)$ for $x \in [0, 12]$?

35. The position of a particle along a horizontal number line at time t is given by the function $s(t) = -t^2 + 6t - 8$.
   a. What is the largest time interval for which s is an increasing function? In which direction is the motion during this time?
   b. At what time does the particle change direction?

36. The sum of two positive numbers is 50. If x denotes one of these numbers,
   a. for which numbers x in the product of the two numbers an increasing function of x;

b. what is the maximum value of their product?

37. Let $f(x) = x + \dfrac{1}{x}, \quad x \ne 0$.
   a. Find the largest intervals on which f is an increasing function.
   b. Use the result of part a to conclude that

$$x + \frac{1}{x} \ge 2$$

   for all $x > 0$.
   c. Can you establish the inequality in part b without using calculus? (*Hint:* Begin with the inequality $(x - 1)^2 \ge 0$.) What about the case $x < 0$?

38. In physics the *total energy* of a particle is defined by the equation

$$E = \frac{mc^2}{\sqrt{1 - \dfrac{v^2}{c^2}}}, \quad 0 < v < c$$

where v = velocity, m = mass, and c = speed of light. Show that total energy is an increasing function of velocity.

39. In electronics, the ratio $R(\omega)$ of the output voltage to the input voltage in a certain "low-pass" RC circuit is given by the equation

$$r(\omega) = \frac{1/\omega C}{\sqrt{R^2 + \left(\dfrac{1}{\omega C}\right)^2}}$$

where the constants R and C denote the resistance and capacitance of the circuit, and ω is the frequency of the current. Show that r is a decreasing function of the independent variable ω for $0 < \omega < \infty$.

40. The speed of a satellite moving in a stable orbit about the earth is a function of the altitude h of the satellite above the earth's surface. It is given by the equation

$$S(h) = \sqrt{\frac{GM_e}{R_e + h}} = \sqrt{GM_e}(R_e + h)^{-1/2}$$

where the constants G, M_e, and R_e are defined as follows:

G = gravitational constant,
M_e = mass of the earth,
R_e = radius of the earth.

Show that the speed of the satellite decreases as altitude increases.

41. Show that the function f in Example 4 is increasing on $[0, 2]$.

42. Use Definition 2 to show that if f is increasing on $[a, b]$ and on $[b, c]$, then f is increasing on $[a, b] \cup [b, c] = [a, c]$.

43. Show that the function f in Example 3 is not decreasing on $(-\infty, 0]$ or on $[0, \infty)$.

4.3 RELATIVE EXTREMA

Recall from Section 4.1 that the maximum value of the function f on the interval $[a, b]$ occurs at $c \in [a, b]$ if $f(c) \geq f(x)$ for all $x \in [a, b]$. This definition is appropriate for finding the *single* largest value of the given function on a *closed, bounded* interval. However, our interest here is in using the theory of the derivative to study the graph of $y = f(x)$ for all x in the domain of f. Although the (absolute) maximum and minimum values for f remain of interest, we also want to know how to locate the *relative* maxima and minima as illustrated in Figure 3.1. We define such *relative extrema* as follows.

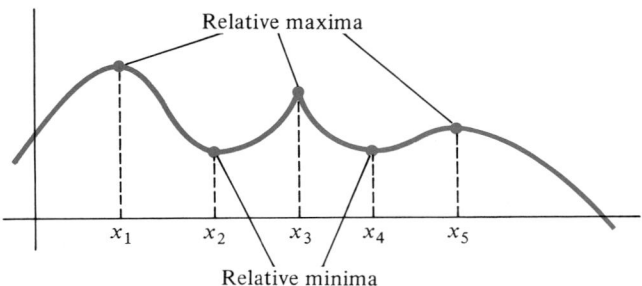

Figure 3.1 Relative extrema for the function f.

DEFINITION 4	The value $f(c)$ is a **relative maximum** for the function f if there exists a number $h > 0$ so that the interval $[c - h, c + h]$ is in the domain of f, and $f(c)$ is the maximum value for f on $[c - h, c + h]$, that is, so that

$$f(c) \geq f(x) \qquad \text{for all} \qquad x \in [c - h, c + h].$$

Similarly, the value $f(c)$ is a **relative minimum** if there exists a number $h > 0$ so that the interval $[c - h, c + h]$ is in the domain of f, and $f(c)$ is the minimum value for f on $[c - h, c + h]$, that is, so that

$$f(c) \leq f(x) \qquad \text{for all} \qquad x \in [c - h, c + h].$$

The term **relative extrema** refers to both relative maxima and relative minima.

Intuitively, this definition states that a relative maximum occurs when the number $f(c)$ is greater than or equal to $f(x)$ for all x "near" c. Figure 3.2 illustrates this concept for the relative maximum which occurs at $c = x_3$ in Figure 3.1. In Figure 3.3, the relative minimum $f(x_2)$ is shown.

Definition 4 is stated in a manner that emphasizes the relationship between *relative extrema* and *extrema on closed finite intervals*. A relative extremum $f(c)$ becomes an extremum if we restrict the interval on which the function is defined to a "small" interval containing $x = c$.

If a function f has a largest value on its domain, say $f(x) = M$, we will refer to M as the maximum or **absolute maximum,** for f, and similarly for **absolute minima.** However, functions whose domains are not restricted to finite closed intervals need not have **absolute extrema,** as Example 1 shows.

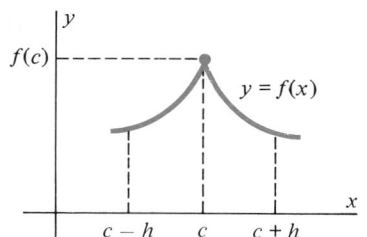

Figure 3.2 Relative maximum is absolute maximum on some interval about c.

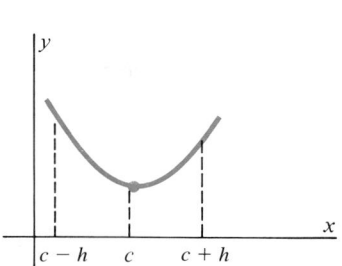

Figure 3.3 Relative minimum is absolute minimum on some interval about c.

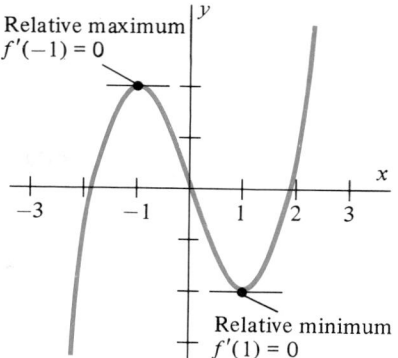

Figure 3.4 Graph of $f(x) = x^3 - 3x$ has relative maximum $f(-1) = 2$ and relative minimum $f(1) = -2$, but f has no *largest* or *smallest* values.

Example 1

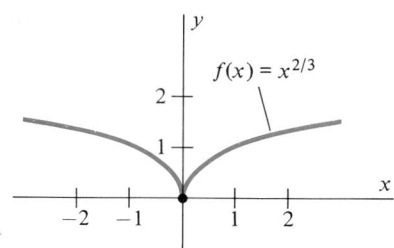

Figure 3.5 Function $f(x) = x^{2/3}$ has a relative minimum value $f(0) = 0$, but $f'(0)$ is undefined.

Figure 3.4 shows the graph of the function $f(x) = x^3 - 3x$. It has

(a) A relative maximum value $f(-1) = 2$ at $x_1 = -1$.
(b) A relative minimum value $f(1) = -2$ at $x_2 = 1$.

Note that the derivative is $f'(x) = 3x^2 - 3$, and that $f'(x) = 0$ for both $x = -1$ and $x = 1$. Note also that although $f(-1) = 2$ is a relative maximum, *the function f has no maximum value*. This is because $f(x)$ increases without bound as x increases beyond $x = 1$. Similarly, $f(1) = -2$ is a relative minimum, but the function f has no minimum value. ◇

Example 2

The function $f(x) = x^{2/3}$ has a relative minimum value of $f(0) = 0$ at $x = 0$. (See Figure 3.5.) This relative minimum is also the minimum value for f throughout its domain since $f(x) = x^{2/3} = (x^{1/3})^2$ is positive for all $x \neq 0$.

Note, however, that *the derivative* $f'(x) = \dfrac{2}{3}x^{-1/3} = \dfrac{2}{3\sqrt[3]{x}}$ *is undefined for* $x = 0$. This fact accounts for the cusp, or point, in the graph of f at $(0, 0)$. ◇

Finding Relative Extrema

Examples 1 and 2 show how relative extrema can occur at numbers x for which either $f'(x) = 0$ or else $f'(x)$ fails to exist. The following theorem guarantees that these are the only situations in which relative extrema occur.

THEOREM 6

Let f be a continuous function. If $f(c)$ is a relative extremum for f, then either $f'(c) = 0$ or else $f'(c)$ fails to exist.

Proof: If $f(c)$ is a relative extremum for f then, according to Definition 4, there is a number $h > 0$ so that f is defined on the closed bounded interval $[c - h, c + h]$

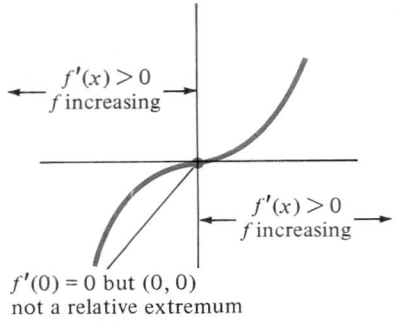

$f'(0) = 0$ but $(0, 0)$
not a relative extremum

Figure 3.6 Critical number for $f(x) = x^3$ does not yield a relative extremum.

and $f(c)$ is either the maximum or minimum value of f on this interval. Thus, by the Extreme Value Theorem, either $f'(c) = 0$ or else $f'(c)$ fails to exist. ◆

Recall that the number c is called a **critical number** for f if either $f'(c) = 0$ or $f'(c)$ fails to exist. Thus, Theorem 6 states that **relative extrema can occur only at critical numbers.**

However, not all critical numbers correspond to relative extrema. A typical example of this is found in the function $f(x) = x^3$. Here $f'(x) = 3x^2$, so $f'(0) = 0$, but $f(0)$ is neither a relative maximum nor a relative minimum (see Figure 3.6). In this example $f(x) = x^3$ is an increasing function for x on either side of the critical number $x = 0$. For a critical number to yield a relative extremum, the function must *change* from increasing to decreasing, or vice versa, at the critical number.

Since the sign of $f'(x)$ determines whether f is increasing or decreasing, we can use f' to identify relative maxima and minima, as well as critical numbers that yield neither. The procedure for doing this is called the *First Derivative Test*.

THEOREM 7
First Derivative Test

Let c be a critical number for f, let f be continuous at c, and let $f'(x)$ exist for all x in an interval $[c - h, c + h], h > 0$, except possibly at $x = c$. Then $f(c)$ is a relative extremum for f if and only if $f'(x)$ changes sign at $x = c$. Specifically, for $x \in [c - h, c + h]$,

(i) if $f'(x) > 0$ for $x < c$ and $f'(x) < 0$ for $x > c$, then $f(c)$ is a relative maximum;
(ii) if $f'(x) < 0$ for $x < c$ and $f'(x) > 0$ for $x > c$, then $f(c)$ is a relative minimum;
(iii) if $f'(x)$ does not change sign at $x = c$, then $f(c)$ is neither a relative maximum nor a relative minimum.

(See Figures 3.7 and 3.8.)

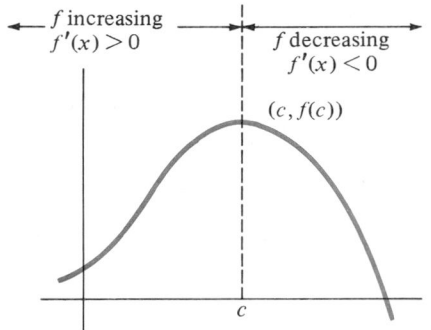

Figure 3.7 f changes from increasing to decreasing at a relative maximum.

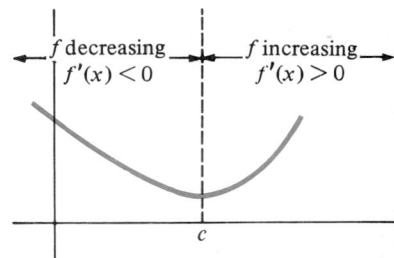

Figure 3.8 f changes from decreasing to increasing at a relative minimum.

Proof of Theorem 7: To prove statement (i) we note first that since $f'(x) > 0$ for $x \in (c - h, c)$ and f is continuous *at* c, it follows that f is *increasing* on $[c - h, c]$, so $f(c) \geq f(x)$ for all $x \in [c - h, c]$. Similarly, since $f'(x) < 0$ for $x \in (c, c + h)$ and f is continuous at c, it follows that f is *decreasing* on $[c, c + h]$, so $f(c) \geq f(x)$ for all $x \in [c, c + h]$. Together these two observations show that $f(c) \geq f(x)$ for *all* $x \in [c - h, c + h]$. Thus, $f(c)$ is a relative maximum.

The proofs of statements (ii) and (iii) follow from the same reasoning and are left as exercises. ◆

Taken together, Theorems 6 and 7 give us a complete procedure for finding and classifying relative extrema. The following examples illustrate this procedure.

Example 3

For $f(x) = 4x^2(1 - x^2)$, find and classify all relative extrema.

Strategy
Find $f'(x)$.

Solution:
We first find the relative extrema using Theorems 6 and 7. The derivative is

$$f'(x) = 8x(1 - x^2) + 4x^2(-2x)$$
$$= 8x - 16x^3$$
$$= 8x(1 - 2x^2).$$

Set $f'(x) = 0$ and solve to find critical numbers.

Identify intervals determined by the critical numbers.

The equation $f'(x) = 0$ yields the critical numbers $x = 0$, $x = -\sqrt{2}/2$, and $x = \sqrt{2}/2$. There are no other critical numbers since $f'(x)$ is defined for all x. We must therefore examine the sign of $f'(x)$ on each of the intervals $(-\infty, -\sqrt{2}/2)$, $(-\sqrt{2}/2, 0)$, $(0, \sqrt{2}/2)$, and $(\sqrt{2}/2, \infty)$. To do so we simply select a convenient number t in each interval and examine the sign of $f'(t)$. The results of doing so appear in Table 3.1.

Check the sign of f' on each interval.

Table 3.1

Interval I	Test number $t \in I$	Value $f'(t)$	Sign of $f'(t)$
$\left(-\infty, -\dfrac{\sqrt{2}}{2}\right)$	$t = -1$	$f'(-1) = 8$	$+$
$\left(-\dfrac{\sqrt{2}}{2}, 0\right)$	$t = -\dfrac{1}{2}$	$f'\left(-\dfrac{1}{2}\right) = -2$	$-$
$\left(0, \dfrac{\sqrt{2}}{2}\right)$	$t = \dfrac{1}{2}$	$f'\left(\dfrac{1}{2}\right) = 2$	$+$
$\left(\dfrac{\sqrt{2}}{2}, \infty\right)$	$t = 1$	$f'(1) = -8$	$-$

Apply First Derivative Test to identify relative extrema.

From the results of Table 3.1, as illustrated in Figure 3.9, and the First Derivative Test, we conclude that $f(-\sqrt{2}/2) = 1$ and $f(\sqrt{2}/2) = 1$ are relative maxima, and that $f(0) = 0$ is a relative minimum. The graph of f appears in Figure 3.10. Note that $f(-\sqrt{2}/2)$ and $f(\sqrt{2}/2)$ are indeed absolute maxima, but that no absolute minimum exists. ◇

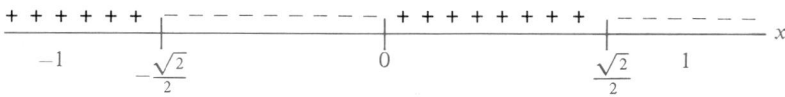

Figure 3.9 Sign of $f'(x)$ in Example 3.

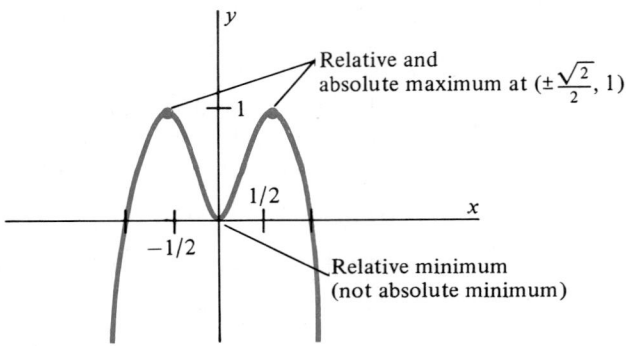

Figure 3.10 Relative extrema for $f(x) = 4x^2(1 - x^2)$.

Example 4

Find all relative extrema for the function

$$f(x) = \sin x + \cos x.$$

Solution: Although the domain of f is $(-\infty, \infty)$, f is periodic, of period 2π. This means that $f(x + 2n\pi) = f(x)$ for all integers $n = \pm 1, \pm 2, \ldots$. Thus, if we can identify the relative extrema in the interval $[0, 2\pi]$, we can find all others by simply adding $2n\pi$ to the ones found in $[0, 2\pi]$. Setting $f'(x) = 0$ gives

$$f'(x) = \cos x - \sin x = 0,$$

or

$$\sin x = \cos x.$$

The only solutions of this equation in $[0, 2\pi]$ are $x = \pi/4$ and $x = 5\pi/4$. Since $f'(x)$ is defined for all x this means that the only critical numbers in $[0, 2\pi]$ are $x = \pi/4$ and $x = 5\pi/4$. Table 3.2 shows the result of selecting one "test number" in each of the resulting intervals and checking the sign of $f'(t)$. ◇

Table 3.2

Interval I	Arbitrary test number $t \in I$	$f'(t) = \cos t - \sin t$	Sign of $f'(t)$
$\left(0; \dfrac{\pi}{4}\right)$	$t = \dfrac{\pi}{6}$	$f'\left(\dfrac{\pi}{6}\right) = \dfrac{\sqrt{3}}{2} - \dfrac{1}{2}$	$+$
$\left(\dfrac{\pi}{4}, \dfrac{5\pi}{4}\right)$	$t = \pi$	$f'(\pi) = -1$	$-$
$\left(\dfrac{5\pi}{4}, 2\pi\right)$	$t = \dfrac{3\pi}{2}$	$f'\left(\dfrac{3\pi}{2}\right) = 0 - (-1)$	$+$

From the results of Table 3.2, together with the First Derivative Test, we may conclude that the sign of f' is as illustrated in Figure 3.11 and that

(i) $f\left(\dfrac{\pi}{4} + 2n\pi\right) = \sqrt{2}$ is a relative maximum for $n = 0, \pm 1, \pm 2, \ldots,$

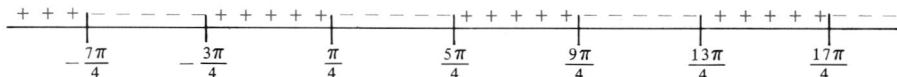

Figure 3.11 Sign of $f'(x)$ for $f(x) = \sin x + \cos x$.

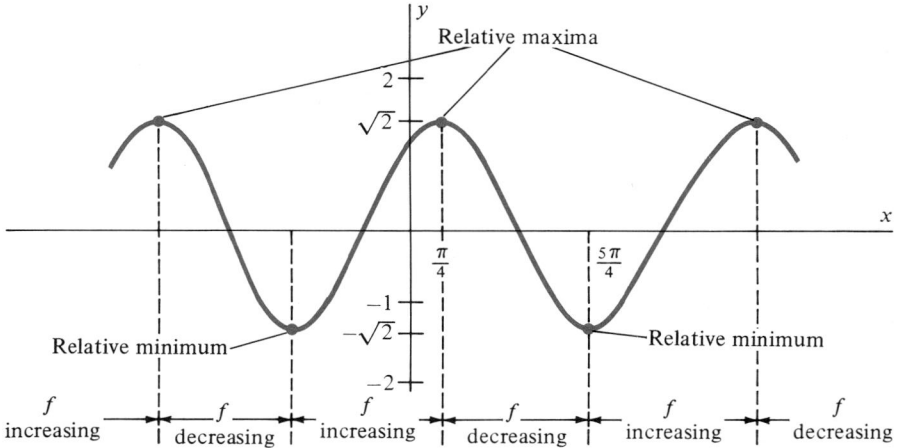

Figure 3.12 Relative extrema for $y = \sin x + \cos x$.

(ii) $f\left(\dfrac{5\pi}{4} + 2n\pi\right) = -\sqrt{2}$ is a relative minimum for $n = 0, \pm 1, \pm 2, \ldots$.

(See Figure 3.12.)

Note that the values $\sqrt{2}$ and $-\sqrt{2}$ are also the absolute maximum and minimum values of f, respectively. $\diamond$

Finding Absolute Extrema on Unbounded Intervals

For certain types of functions we can apply our techniques for finding relative extrema to the problem of finding the (absolute) maximum or minimum value of the function on its domain. This situation occurs when we know, from physical principles or from the context of a problem, that such a maximum or minimum exists and that it does not occur at an endpoint of the domain. That leaves us with the conclusion that the desired maximum or minimum corresponds to a relative extremum.

The following examples are typical of word problems that can be solved using this idea. Note in both examples that the function for which the extremum is sought must first be expressed as a function of a *single* variable. To do so we eliminate other variables using *auxiliary* equations derived from the given information.

Example 5

Find the rectangle with area 64 square inches for which the *perimeter* is a minimum.

Strategy

Assign variable names to the two dimensions, length and width. Express perimeter in terms of these variables.

Solution

If we let x and y denote the length and width of the rectangle, its perimeter is

$$P = 2x + 2y. \tag{1}$$

Find an *auxiliary equation* involving the two variables.

To express P as a function of a *single* independent variable we use the information that area equals 64 in^2 to obtain the *auxiliary* equation

$$xy = 64$$

Solve the auxiliary equation for one variable in terms of the other. Substitute for one of the variables in equation (1), thus *eliminating* the second variable, obtaining P as a function of *one* variable.

which we can solve for y as

$$y = \frac{64}{x}. \tag{2}$$

Substituting this expression for y in equation (1) gives

$$P = 2x + 2\left(\frac{64}{x}\right)$$

so

$$P(x) = 2x + \frac{128}{x} \tag{3}$$

is the function for which we seek the absolute minimum value.

Find the domain of the function P.

Before proceeding further we note that the *domain* of the function P in equation (3) is $(0, \infty)$. That is because a rectangle with area 64 in^2 can have as its length *any* positive number x, as long as its width is $y = \frac{64}{x}$. (See Figure 3.13.) *We are therefore seeking the minimum value of P on the interval* $(0, \infty)$.

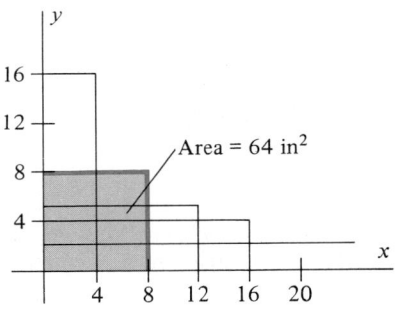

Figure 3.13 Rectangles with area $A = xy = 64$ in^2.

Find the relative extrema for P on its entire domain by setting $P'(x) = 0$ and solving for the critical numbers.

To find the relative extrema for P on $(0, \infty)$, we find the derivative

$$P'(x) = 2 - \frac{128}{x^2} \tag{4}$$

and note that the equation $P'(x) = 0$ gives

$$\frac{128}{x^2} = 2$$

or

$$x^2 = \frac{128}{2} = 64$$

which has solutions $x = \pm 8$. Thus, since $P'(x)$ is defined for all $x > 0$, *the only critical number for P in $(0, \infty)$ is $x = 8$*.

By checking the sign of $P'(x)$ on the intervals $(0, 8)$ and $(8, \infty)$, you can see that

Determine whether the critical number yields a relative minimum. If so, determine whether this relative minimum is the absolute minimum by examining the sign of $P'(x)$.

P is *decreasing* on $(0, 8]$, and
P is *increasing* on $[8, \infty)$.

(See Figure 3.14.)

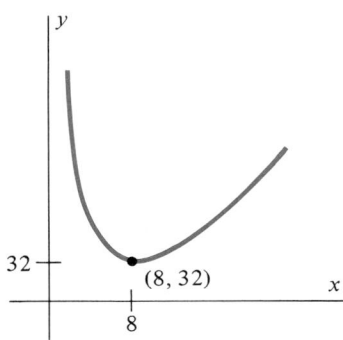

Figure 3.14 Minimum value of perimeter function $P(x) = 2x + \dfrac{128}{x}$ occurs at relative minimum $P(8) = 32$.

Figure 3.15 $D(x) = \sqrt{x^2 - 7x + 16}$ is the distance from the point $P = (x, y)$ to the point $(4, 0)$.

From these observations we may conclude that the minimum value of P on $(0, \infty)$ corresponds to the relative minimum

$$P(8) = 2(8) + \frac{128}{8}$$
$$= 32$$

and that this value of the perimeter occurs when the dimensions of the rectan-

State the conclusion in terms of the original question.

gle are $x = 8$ and $y = \dfrac{64}{8} = 8$. That is, the rectangle of area 64 in^2 with the smallest perimeter is a *square* of side $s = 8$. ◇

Example 6

Find the point on the graph of $y = \sqrt{x}$ nearest the point $(4, 0)$.

Solution: Following the steps outlined in Example 5, we first express the distance D between a point $P = (x, y)$ on the graph of $y = \sqrt{x}$ and the point $(4, 0)$ as

$$D = \sqrt{(x - 4)^2 + (y - 0)^2}. \tag{5}$$

(See Figure 3.15.) Since the point $P = (x, y)$ lies on the graph of $y = \sqrt{x}$, this is our auxiliary equation and we may substitute $y = \sqrt{x}$ in equation (5) to obtain the distance D as a function of x alone:

$$D(x) = \sqrt{(x - 4)^2 + (\sqrt{x})^2}$$
$$= \sqrt{x^2 - 7x + 16}.$$

Since the square root function $f(x) = \sqrt{x}$ is defined for all $x \geq 0$, we seek the minimum value of the distance function D on the interval $[0, \infty)$.

To find the relative extrema for D we first find

$$D'(x) = \frac{d}{dx}[(x^2 - 7x + 16)^{1/2}]$$

$$= \frac{1}{2}(x^2 - 7x + 16)^{-1/2}(2x - 7)$$

$$= \frac{2x - 7}{\sqrt{x^2 - 7x + 16}}.$$

Setting $D'(x) = 0$ gives $2x - 7 = 0$, or $x = \frac{7}{2}$. Since $D'(x)$ is defined for all $x \in [0, \infty)$, the number $x = \frac{7}{2}$ is the only critical number for D in $[0, \infty)$. Checking the sign of $D'(x)$ for x in the intervals $\left(0, \frac{7}{2}\right)$ and $\left(\frac{7}{2}, \infty\right)$ gives the conclusions that

D is decreasing on $\left[0, \frac{7}{2}\right]$, and

D is increasing on $\left[\frac{7}{2}, \infty\right)$.

Thus, $D\left(\frac{7}{2}\right) = \sqrt{\left(\frac{7}{2}\right)^2 - 7\left(\frac{7}{2}\right) + 16} = \frac{\sqrt{15}}{2}$ is a relative minimum for D on $[0, \infty)$ and also the *absolute* minimum for D on this interval.

The point $\left(\frac{7}{2}, \sqrt{\frac{7}{2}}\right)$ is therefore the point nearest the point $(4, 0)$ on the graph of $y = \sqrt{x}$. $\diamond$

Exercise Set 4.3

In Exercises 1–10 find the largest intervals on which f is increasing or decreasing. Find all relative extrema for f. Use this information in sketching the graph of f.

1. $f(x) = x^2 - 4x + 6$

2. $f(x) = x(x - 4)$

3. $f(x) = 9 - (x + 2)^2$

4. $f(x) = (x - 3)(x + 5)$

5. $f(x) = 4 + x^{2/3}$

6. $f(x) = 7 - 2x + x^2$

7. $f(x) = |4 - x^2|$

8. $f(x) = |x^2 - 9|$

9. $f(x) = \sin\left(x + \frac{\pi}{4}\right), \quad 0 \leq x \leq 2\pi$

10. $f(x) = \cos\left(\frac{\pi}{2} - 2x\right), \quad 0 \leq x \leq \pi$

In Exercises 11–22 find all relative extrema for f.

11. $f(x) = 4x^3 + 9x^2 - 12x + 7$ **12.** $f(x) = x^3 + x^2 - 8x + 8$

13. $f(x) = x + \sin x$ **14.** $f(x) = \tan^2 x$

15. $f(x) = 6x^{5/2} - 70x^{3/2} + 15$ **16.** $f(x) = \sqrt[3]{x}(x - 7)^2$

17. $f(x) = \frac{x}{1 + x^2}$ **18.** $f(x) = \frac{\sqrt{x}}{4 - x}$

19. $f(x) = |\sin x|$ **20.** $f(x) = |\cos x|$

21. $f(x) = x^{5/3} - 5x^{2/3} + 3$ **22.** $f(x) = \frac{1 - x}{1 + x^2}$

23. Explain why a function of the form $y = x^2 + bx + c$ always has precisely one relative minimum and no relative maximum.

24. What is the maximum number of local extrema that a polynomial of degree n can have? Could such a function have fewer?

25. The function $f(x) = x^2 - ax + b$ has a relative minimum at $x = 2$. Find a.

26. The function $f(x) = x^3 + ax^2 + bx + 7$ has relative extrema at $x = 1$ and $x = -3$.
  a. Find a and b.
  b. Classify the extrema (as relative maxima or minima).

27. The function $f(x) = a(x^2 - bx + 16)$ has a relative extremum at $x = 5$.
  a. Find b.
  b. Determine the sign of a if $(5, f(5))$ is a relative minimum.

28. The function $y = x^3 + ax + 7$ has a critical number at $x = 0$. Find a.

29. Two sides of a triangle are of length a. Find the length of the third side so that area is a maximum.

30. A salt container is to be made in the shape of a right circular cylinder and is to contain 500 cm^3 of salt. Find the dimensions for the container that requires the least amount of material.

31. Find the dimensions of the box with square top and bottom and rectangular sides of capacity V that can be constructed from a minimum amount of material.

32. Prove that the function $f(x) = ax^2 + bx + c$, $a \neq 0$, always has precisely one relative extremum which occurs at the number $x = -\dfrac{b}{2a}$.

33. The rate at which an outbreak of influenza spreads through a college dormitory is proportional to the product of the number of students infected and the number of students not yet infected. Show that the influenza is spreading most rapidly when precisely one half of the students are infected. (Assume that infected students remain so for the duration of the epidemic.)

34. Find the point on the parabola $y = x^2$ nearest the point $(-3, 0)$.

35. A storage bunker is to contain 96 cubic meters and is to have a square base and vertical sides. If the material for the base costs \$10 per square meter and the material for the sides and top costs \$6 per square meter, find the dimensions that minimize cost.

36. Show that of all right circular cylinders of fixed volume V the one with minimum surface area has height equal to the diameter of its base.

37. Prove that the distance from a point $P = (x_0, y_0)$ in the plane to the line with equation $ax + by + c = 0$ is the number
$$d = \frac{|ax_0 + by_0 + c|}{\sqrt{a^2 + b^2}}.$$

38. Prove statements (ii) and (iii) of Theorem 7.

4.4 SIGNIFICANCE OF THE SECOND DERIVATIVE: CONCAVITY

The information in Sections 4.1–4.3 provides several techniques that are useful in sketching the graph of a function f. Locating the relative extrema for f and determining whether f is increasing or decreasing between critical numbers enables us to find the "high and low" points, which we may then connect with "smooth" arcs if f is differentiable.

Figure 4.1 illustrates, however, that more must be known about f in order to correctly sketch its graph. In both figures the graphs have a relative maximum at

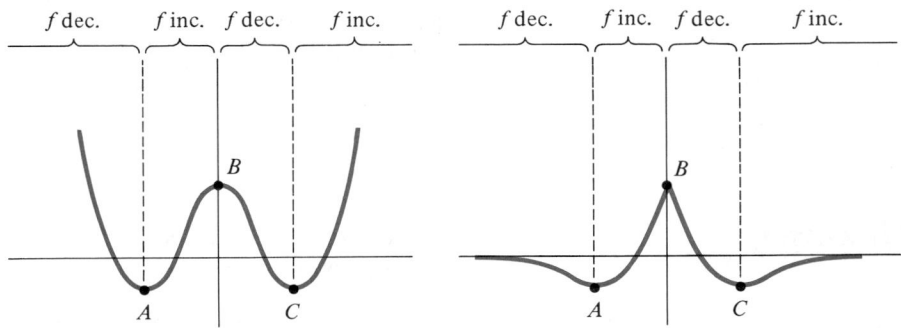

Figure 4.1 Two graphs with the same relative extrema and increasing/decreasing properties may have very different shapes depending on how they bend (concavity).

point B and relative minima at points A and C. What distinguishes these two graphs, however, is the way in which they "bend" between relative extrema, and how they "bend" as $|x|$ becomes large. These are the issues of *concavity,* which we discuss here, and *asymptotes* which we discuss in the next section.

Suppose that the function f is differentiable on the interval I. If the derivative f' is an increasing function on I, then the slopes of the lines tangent to the graph of f increase (causing a tangent to rotate counterclockwise) as x increases. This means that the graph will have a "cupped up" shape as illustrated in Figure 4.1.

For this reason we define the *concavity* of the graph of f in terms of whether its derivative f' is an increasing or decreasing function.

DEFINITION 5	Let f be continuous on an interval I and let $f'(x)$ exist for all $x \in I$ except possibly at endpoints. We say that the graph of f is (a) **concave up** on I if f' is an increasing function on I, where defined; (b) **concave down** on I if f' is a decreasing function on I, where defined. (See Figures 4.2 and 4.3.)

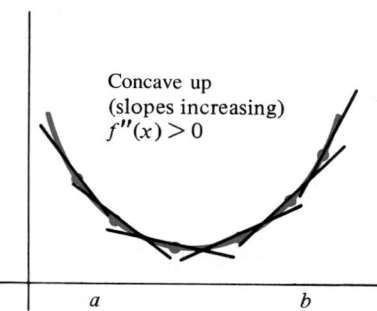

Figure 4.2 Graph of f is concave up if f' is increasing, that is, $f''(x) > 0$.

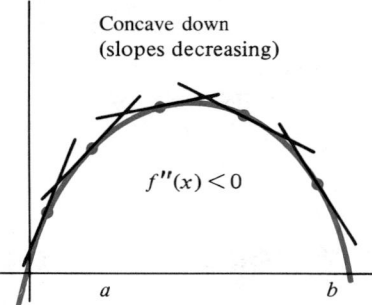

Figure 4.3 Graph of f is concave down if f' is decreasing, that is, $f''(x) < 0$.

REMARK: We are defining concavity in analytic terms (that is, by use of $f'(x)$) rather than geometrically, in order quickly to develop the relationship between concavity and the second derivative. A more geometric approach would be to say that the graph of f is concave up on I if, for each $c \in I$, the graph "lies above the tangent at $(c, f(c))$," as is the case in Figure 4.2. In Exercise 41 you are asked to show that this geometric condition implies that $f''(x) > 0$ on I, the condition we have chosen to use. A similar geometric definition can be given for concave down.

The following theorem shows that concavity is determined by the sign of the second derivative.

THEOREM 8	Let f be continuous on I and let $f''(x)$ exist for all $x \in I$ except possibly at endpoints. Then the graph of f is (a) concave up on I if $f''(x) > 0$ for all $x \in I$ that are not endpoints, and (b) concave down on I if $f''(x) < 0$ for all $x \in I$ that are not endpoints.

Proof: To prove part (a) we note that, since $f''(x) = \dfrac{d}{dx}[f'(x)]$, if $f''(x) > 0$ for all $x \in I$ that are not endpoints, it follows from Theorem 5 that f' is an increasing function on I, except possibly at endpoints. Thus, the graph of f is concave up on I according to Definition 5. Similar reasoning proves part (b). ◆

When the function f has a continuous derivative f', Theorem 8 provides *a procedure for finding the largest intervals on which the graph of f is concave up or concave down:*

(i) Find all numbers c for which $f''(x) = 0$ or $f''(c)$ fails to exist.
(ii) Determine the sign of $f''(x)$ for x in each of the resulting intervals and apply Theorem 8.

As in the procedure for finding intervals on which f is increasing or decreasing, we must be careful to also note numbers c at which $f(c)$ or $f'(c)$ fails to exist.

Points on the graph of f that separate arcs of opposite concavity, loosely speaking, are called *inflection points*. Here is a precise definition of this term.

DEFINITION 6 **Inflection Point**	The point $P = (c, f(c))$ is called an **inflection point** for the function f if f is continuous at $x = c$ and there exists a number $h > 0$ so that the graph of f is either (a) concave down on $[c - h, c]$ and concave up on $[c, c + h]$, or (b) concave up on $[c - h, c]$ and concave down on $[c, c + h]$.

Example 1

Figure 4.4 shows the graph of $f(x) = \sin x$ on the interval $I = [0, 2\pi]$. For this function we have

$$f'(x) = \cos x, \quad \text{and} \quad f''(x) = -\sin x$$

so $f''(x) = 0$ if $-\sin x = 0$, which occurs on I at $x = 0$, $x = \pi$, and $x = 2\pi$. Since f'' is continuous on I, we check the concavity of the graph of f on the intervals $[0, \pi]$ and $[\pi, 2\pi]$ by checking the sign of $f''(t)$ at a test number t in each of the intervals $(0, \pi)$ and $(\pi, 2\pi)$ and applying Theorem 8. The results of doing so are recorded in Table 4.1.

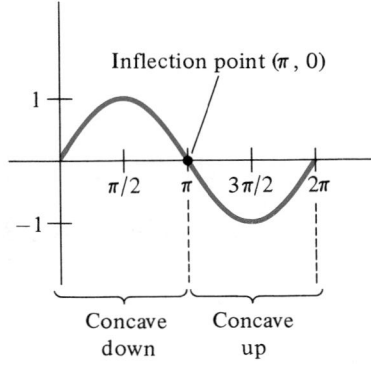

Figure 4.4 Graph of $y = \sin x$ is concave down on $[0, \pi]$ and concave up on $[\pi, 2\pi]$.

Table 4.1

Interval	Test number	$f''(t)$	Conclusion
$[0, \pi]$	$t = \pi/2$	$-\sin\left(\dfrac{\pi}{2}\right) = -1$	Concave down
$[\pi, 2\pi]$	$t = \dfrac{3\pi}{2}$	$-\sin\left(\dfrac{3\pi}{2}\right) = 1$	Concave up

Thus, the graph of f is concave down on $[0, \pi]$ and concave up on $[\pi, 2\pi]$. This means that the point $(\pi, f(\pi)) = (\pi, 0)$ is an inflection point, according to Definition 6. ◇

Example 2

Determine the concavity of the graph of the function $f(x) = x^4 - 3x^2 + 2$.

Strategy

Calculate f''.

Solution

Since $f(x) = x^4 - 3x^2 + 2$ we have

$$f'(x) = 4x^3 - 6x$$

and

$$f''(x) = 12x^2 - 6.$$

Find all x with $f''(x) = 0$ or $f''(x)$ undefined.

Setting $f''(x) = 0$ gives $12x^2 = 6$, or $x^2 = \frac{1}{2}$.

The zeros of the second derivative $f''(x)$ are, therefore,

$$x_1 = -\frac{\sqrt{2}}{2}, \quad \text{and} \quad x_2 = \frac{\sqrt{2}}{2}.$$

Determine the sign of f'' on each of the resulting intervals by checking $f''(t)$ at a single number. Then apply Theorem 8.

There are no numbers x for which $f''(x)$ is undefined. Since f'' is a continuous function (it's a polynomial), it must have constant sign on each of the intervals determined by its zeros. We may therefore determine the sign of f'' on each of these intervals by checking the sign of $f''(t)$ at a "test number" t. The results of doing so appear in Table 4.2.

Table 4.2

Interval I	Test number $t \in I$	Value of $f''(t)$	Sign of $f''(t)$	Concavity
$\left(-\infty, -\frac{\sqrt{2}}{2}\right)$	$t = -1$	$f''(-1) = 6$	$+$	up
$\left(-\frac{\sqrt{2}}{2}, \frac{\sqrt{2}}{2}\right)$	$t = 0$	$f''(0) = -6$	$-$	down
$\left(\frac{\sqrt{2}}{2}, \infty\right)$	$t = 1$	$f''(1) = 6$	$+$	up

From the results in Table 4.2 and by Theorem 8, we may conclude that the graph of f is

(i) concave up on $\left(-\infty, -\frac{\sqrt{2}}{2}\right]$,

(ii) concave down on $\left[-\frac{\sqrt{2}}{2}, \frac{\sqrt{2}}{2}\right]$,

(iii) concave up on $\left[\frac{\sqrt{2}}{2}, \infty\right)$.

Since the concavity of the graph of $y = f(x)$ changes at $(-\sqrt{2}/2, f(-\sqrt{2}/2)) = (-\sqrt{2}/2, 3/4)$ and at $(\sqrt{2}/2, f(\sqrt{2}/2)) = (\sqrt{2}/2, 3/4)$, both of these points are *inflection points* (see Figure 4.5). ◇

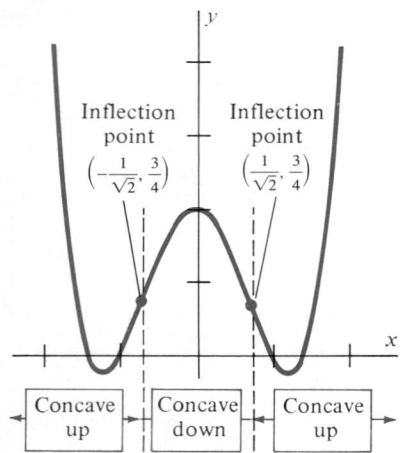

Figure 4.5 Concavity for graph of $f(x) = x^4 - 3x^2 + 2$.

REMARK: It is not necessary that $f''(c) = 0$ at an inflection point $(c, f(c))$. Figure 4.6 shows that the graph of $f(x) = |4 - x^2|$ has inflection points $(-2, 0)$ and $(2, 0)$, even though $f''(-2)$ and $f''(2)$ fail to exist.

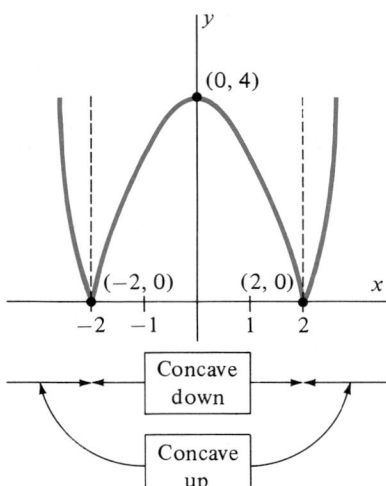

Figure 4.6 Graph of $f(x) = |4 - x^2|$ has inflection points at $(-2, 0)$ and $(2, 0)$ where $f''(x)$ fails to exist.

It is important to note that the point $(x_0, f(x_0))$ need not be an inflection point for the graph of f, even though $f''(x_0) = 0$ or $f''(x_0)$ fails to exist. The following example shows that the concavity can be the same on either side of such a point.

Example 3

Determine the concavity and find the inflection points for the graph of $f(x) = x^{2/3} - \dfrac{1}{5}x^{5/3}$.

Solution: We have

$$f'(x) = \frac{2}{3}x^{-1/3} - \frac{1}{3}x^{2/3},$$

so

$$f''(x) = -\frac{2}{9}x^{-4/3} - \frac{2}{9}x^{-1/3}$$

$$= -\frac{2}{9}x^{-4/3}(1 + x)$$

$$= \frac{-2(1 + x)}{9x^{4/3}}.$$

Thus $f''(x) = 0$ for $x = -1$, and $f''(x)$ is undefined for $x = 0$. The sign of f'' must therefore be checked on the intervals $(-\infty, -1)$, $(-1, 0)$, and $(0, \infty)$. This time, rather than substituting particular test numbers and constructing a table, we examine the signs of the numerator and denominator on each interval, as illustrated in Figure 4.7.

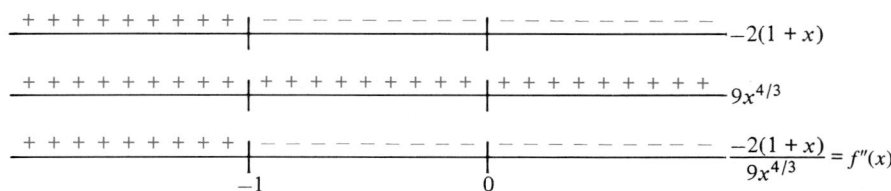

Figure 4.7 Sign analysis for $f''(x) = \dfrac{-2(1 + x)}{9x^{4/3}}$.

From Figure 4.7, we can see that f'' is positive on $(-\infty, -1)$ but negative on both $(-1, 0)$ and $(0, \infty)$. Thus, by Theorem 8, the graph is

(i) concave up on $(-\infty, -1]$,
(ii) concave down on $[-1, 0]$ *and* on $[0, \infty)$.

This means that the point $(-1, f(-1)) = \left(-1, \dfrac{6}{5}\right)$ is an inflection point, but that the point $(0, f(0)) = (0, 0)$ is *not* an inflection point (see Figure 4.8). ◇

The Second Derivative Test for Extrema

There is a very direct relationship between concavity and the nature of relative extrema as the following theorem shows.

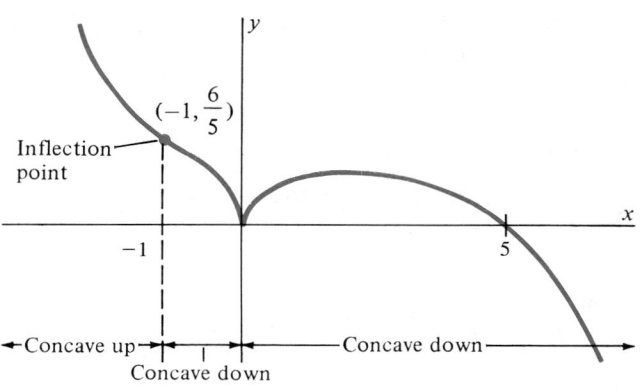

Figure 4.8 Graph of $f(x) = x^{2/3} - \frac{1}{5}x^{5/3}$.

THEOREM 9
Second Derivative Test

Let f be differentiable on an open interval containing the critical number $x = a$, with $f'(a) = 0$. Suppose also that $f''(a)$ exists.

(i) If $f''(a) < 0$, then $f(a)$ is a relative maximum.
(ii) If $f''(a) > 0$, then $f(a)$ is a relative minimum.
(iii) If $f''(a) = 0$, there is no conclusion.

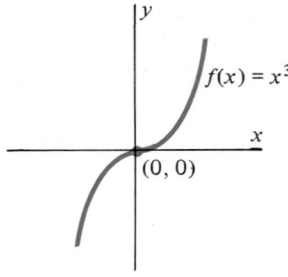

Figure 4.9 $f'(0) = f''(0) = 0$, but $(0, 0)$ is not an extremum.

Proof: A formal proof of Theorem 9 is outlined in Exercise 42. We prefer to give only an informal sketch of the argument here to avoid clouding the simple idea with details.

In case (i), the condition $f''(a) < 0$ may be interpreted as saying that the graph of f is concave down near $x = a$. This means that f' is a decreasing function near a. Since $f'(a) = 0$, it follows that $f'(x) > 0$ for $x < a$ and $f'(x) < 0$ for $x > a$. Thus, by the First Derivative Test, $f(a)$ is a relative maximum. A similar argument applies for statement (ii). ◆

In applying the Second Derivative Test be sure to note statement (iii)—if $f''(a) = 0$, we obtain no conclusion about whether $f(a)$ is a relative extremum. The following simple examples show why.

Example 4

For $f(x) = x^3$, $f'(x) = 3x^2$, and $f''(x) = 6x$. Thus $x = 0$ is a critical number, and $f''(0) = 0$. Here $(0, f(0)) = (0, 0)$ is neither a relative maximum nor a relative minimum (see Figure 4.9). ◇

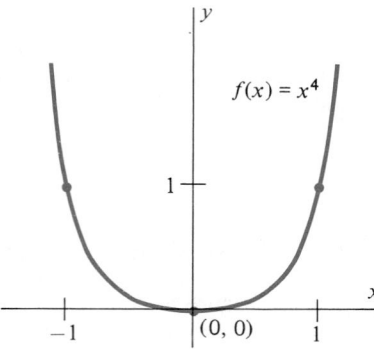

Figure 4.10 $f'(0) = f''(0) = 0$; $(0, 0)$ is a relative minimum.

Example 5

For $f(x) = x^4$, $f'(x) = 4x^3$ and $f''(x) = 12x^2$. Again, $x = 0$ is a critical number, and $f''(0) = 0$. For this function, $(0, f(0)) = (0, 0)$ is a relative minimum (see Figure 4.10). ◇

Example 6

The function $f(x) = -x^4$ has derivatives $f'(x) = -4x^3$ and $f''(x) = -12x^2$. Here also, $x = 0$ is a critical number, and $f''(0) = 0$. This time $(0, f(0)) = (0, 0)$ is a relative maximum (see Figure 4.11). ◇

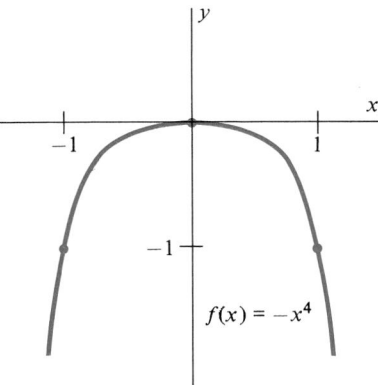

$f(x) = -x^4$

Figure 4.11 $f'(0) = f''(0) = 0$; $(0, 0)$ is a relative maximum.

Example 7

Find all relative extrema for the function $f(x) = x^4 - 3x^2 + 2$ in Example 2.

Strategy

Set $f'(x) = 0$ to find critical numbers.

Classify critical numbers using the Second Derivative Test.

Solution

Setting $f'(x) = 4x^3 - 6x = 0$, we obtain the equation

$$2x(2x^2 - 3) = 0$$

so $x = 0$, $x = -\sqrt{3/2}$, and $x = \sqrt{3/2}$ are critical numbers. We can classify these by the Second Derivative Test.

Since $f''(x) = 12x^2 - 6$ we find that

$f''(-\sqrt{3/2}) = 12 > 0$, so $(-\sqrt{3/2}, -1/4)$ is a relative minimum;
$f''(0) = -6 < 0$, so $(0, 2)$ is a relative maximum;
$f''(\sqrt{3/2}) = 12 > 0$, so $(\sqrt{3/2}, -1/4)$ is a relative minimum (see Figure 4.5). ◇

Example 8

Discuss the relative extrema and concavity for the function $y = x + \sin x$.

Strategy

Find $\dfrac{dy}{dx}$ and $\dfrac{d^2y}{dx^2}$.

Evaluate $\dfrac{d^2y}{dx^2}$ at critical numbers to check for extrema by Second Derivative Test. If inconclusive, use First Derivative Test.

Solution

Here $\dfrac{dy}{dx} = 1 + \cos x$, so $\dfrac{d^2y}{dx^2} = -\sin x$.

The equation $\dfrac{dy}{dx} = 0$ yields the critical numbers $x = \pm\pi$, $\pm 3\pi$, $\pm 5\pi$, ... At such points $\dfrac{d^2y}{dx^2} = 0$, so the Second Derivative Test is incon-

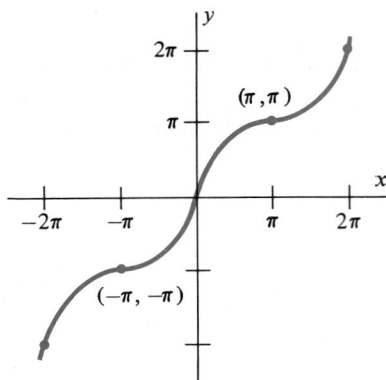

Figure 4.12 Graph of $f(x) = x + \sin x$.

clusive. However, since $\dfrac{dy}{dx} = 1 + \cos x \geq 0$ for all x, the First Derivative Test shows that there are no relative extrema.

The equation $\dfrac{d^2y}{dx^2} = -\sin x = 0$ has solutions $x = 0, \pm\pi, \pm 2\pi, \ldots$.

Examine sign of d^2y/dx^2 between zeros of $d^2y/dx^2 = 0$ to determine concavity by Theorem 8.

Inspection of the sign of $\dfrac{d^2y}{dx^2}$ on the resulting intervals shows that the graph of y is concave up on the intervals $[\pi, 2\pi], [3\pi, 4\pi], \ldots, [(2n-1)\pi, 2n\pi], \ldots$ and concave down on the intervals $[0, \pi], [2\pi, 3\pi], \ldots, [2n\pi, 2(n+1)\pi], \ldots$

Thus all points $(n\pi, n\pi), n = 0, \pm 1, \pm 2, \ldots$ are inflection points (see Figure 4.12). ◇

Exercise Set 4.4

In Exercises 1–10 determine the largest intervals on which the graph of f is concave up or concave down. Find any inflection points. Use this information to sketch the graph of f.

1. $f(x) = x^2 - 4x - 5$

2. $f(x) = 9 - x^3$

3. $f(x) = |1 - x^2|$

4. $f(x) = |x^2 - 4|$

5. $f(x) = \cos x, \quad 0 \leq x \leq 2\pi$

6. $f(x) = \tan x, \quad -\pi/2 < x < \pi/2$

7. $f(x) = (x + 3)^3$

8. $f(x) = \dfrac{4}{x - 2}$

9. $f(x) = \sqrt{x + 2}$

10. $f(x) = x^{2/3}$

In Exercises 11–20 find the largest intervals on which the graph of f is concave up or concave down. Find any inflection points.

11. $f(x) = \dfrac{x}{x + 1}$

12. $f(x) = (x + 2)^{1/3}$

13. $f(x) = (2x + 1)^3$

14. $f(x) = (1 - 4x)^3$

15. $f(x) = 2x^3 - 3x^2 + 18x - 12$

16. $f(x) = 2x^3 + 12x^2 + 18x + 12$

17. $f(x) = x^4 + 2x^3 - 36x^2 + 24x - 6$

18. $f(x) = \dfrac{1}{12}x^4 - \dfrac{2}{3}x^3 + 2x^2 + 5x - 8$

19. $f(x) = x^{5/3} - 5x^{2/3} + 3$

20. $f(x) = \dfrac{\sqrt[3]{x}}{1 - x}$

In each of Exercises 21–28, determine whether f has a relative extremum at the given value of x, using either the First or Second Derivative Test.

21. $f(x) = \sin^2 x, \quad x = \pi/2$

22. $f(x) = x^2 + \dfrac{2}{x}, \quad x = 1$

23. $f(x) = x^4 - 4x^3 - 48x^2 + 24x + 20, \quad x = 4$

24. $f(x) = 2x^3 - 3x^2, \quad x = 1$

25. $f'(x) = \dfrac{x - 1}{x + 1}, \quad x = 1$

26. $f'(x) = \dfrac{1 - \sin 2x}{\cos x}, \quad x = \pi/4$

27. $f'(x) = (x - 1)(x + 2), \quad x = -2$

28. $f'(x) = \sqrt{x^3 - 4x}, \quad x = 2$

In Exercises 29–32 find all relative extrema, determine the largest intervals on which the graph is concave up or concave down, and sketch the graph.

29. $f(x) = x^3 - 3x^2 + 6$ **30.** $y = x^4 + 4x^3 - 8x^2 - 48x + 9$

31. $y = |9 - x^2|$ **32.** $f(x) = (x - 8)^{2/3}$

33. What conditions must hold for the constants a, b, and c in order that the general quadratic function $f(x) = ax^2 + bx + c$ be concave up for all x?

34. Find an example of a function f and a number a so that $f(a)$ is a relative maximum but $f''(a) = 0$.

35. Find an example of a function f and a number b so that $f(b)$ is a relative minimum but $f''(b) = 0$.

36. Find an example of a function f and a number c for which $f'(c) = f''(c) = 0$ but $f(c)$ is not a relative extremum.

37. How many inflection points can a polynomial of degree $n = 2$ possess? $n = 3$? $n = k$?

38. Find an example of a function f which is concave up on $(-\infty, 2]$, concave down on $[2, \infty)$, increasing on $(-\infty, 1]$ and $[3, \infty)$, and decreasing on $[1, 3]$.

39. Find an example of a function which is concave down on $(-\infty, 0]$, concave up on $[0, \infty)$, and decreasing throughout its domain.

40. True or false? If $f(0) = 0$ and f is concave down for all x, then $f(x) \le 0$ for all x.

41. Let f be differentiable on an open interval I. Assume that throughout I "the graph of f is above the tangent." (That is, for each $c \in I$ if $y = g(x)$ is an equation for the tangent at $(c, f(c))$, then $g(x) < f(x)$ for all $x \in I$ with $x \neq c$.) Show that f' is increasing on I.

42. Prove statement (i) of the Second Derivative Test (Theorem 9) as follows:
   a. Assume that $f''(a) = L < 0$
   b. Show that $\lim\limits_{x \to a} \dfrac{f'(x)}{x - a} = L < 0$
   c. Conclude that there exists an interval $I = (a - h, a + h)$ so that $\dfrac{f'(x)}{x - a} < 0$ whenever $x \in I$.
   d. Show that $f'(x) > 0$ for $x \in (a - h, a)$
   e. Show that $f'(x) < 0$ for $x \in (a, a + h)$
   f. Apply the First Derivative Test to conclude that $(a, f(a))$ is a relative maximum.

4.5 CURVE SKETCHING I: ASYMPTOTES

The preceding discussions of relative extrema and concavity are useful in sketching graphs of functions on finite intervals when the graphs remain bounded. In order to sketch graphs of functions on infinite intervals or where function values become unbounded, we need to discuss briefly the notions of limits at infinity, infinite limits, and asymptotes.

Limits at Infinity

The function $f(x) = -\dfrac{2x + 1}{x + 2}$ is an example of a function with values approaching a constant as x increases without bound. In fact, both the entries of Table 5.1 and the graph of f in Figure 5.1 suggest that the values $f(x)$ approach the number $L = -2$ as x becomes very large.

In this case we write

$$\lim_{x \to \infty} -\frac{2x + 1}{x + 2} = -2$$

and we refer to the line $y = -2$ as a *horizontal asymptote*. More generally, we make the following informal definitions.

Table 5.1

x	$f(x) = -\dfrac{2x + 1}{x + 2}$
0	-0.50000
5	-1.57143
10	-1.75000
20	-1.86364
50	-1.94231
100	-1.97059
250	-1.98811
500	-1.99402
1,000	-1.99701
2,000	-1.99850
5,000	-1.99940
10,000	-1.99997

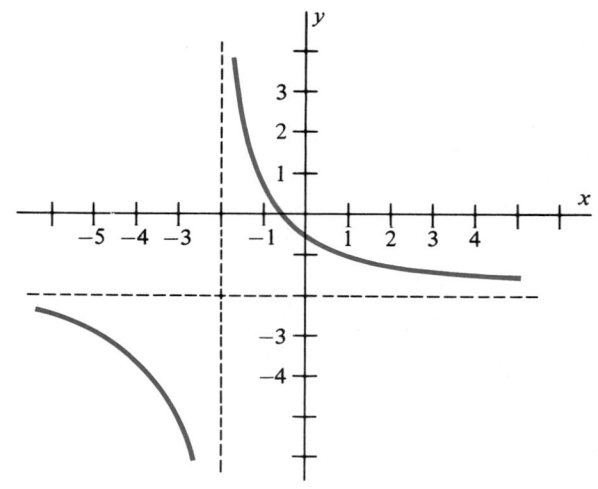

Figure 5.1 Graph of $f(x) = -\dfrac{2x + 1}{x + 2}$.

DEFINITION 7

The expression $\lim\limits_{x \to \infty} f(x) = L$ means that the values $f(x)$ approach the number L as x increases without bound. The expression $\lim\limits_{x \to -\infty} f(x) = M$ means that the values $f(x)$ approach the number M as x decreases without bound.

Definition 7 is an informal working definition. A formal definition is given in Exercise 47.

An *asymptote* for the graph of f is just any straight line "approached" by the graph of f. Using Definition 7 we may define a horizontal asymptote as follows.

DEFINITION 8

The line $y = L$ is a **horizontal asymptote** for the graph of the function f if either

$$L = \lim_{x \to \infty} f(x) \qquad \text{or} \qquad L = \lim_{x \to -\infty} f(x).$$

(See Figure 5.2.)

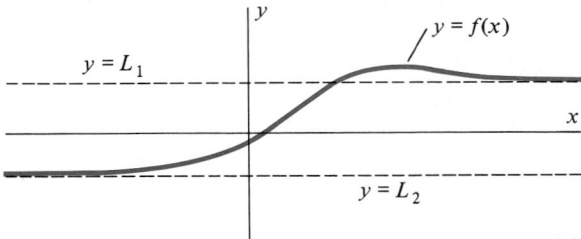

Figure 5.2 Horizontal asymptotes $y = L_1$ and $y = L_2$.

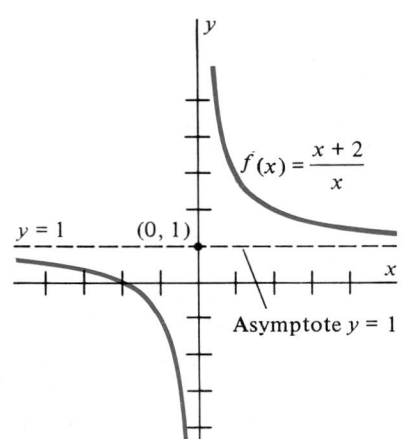

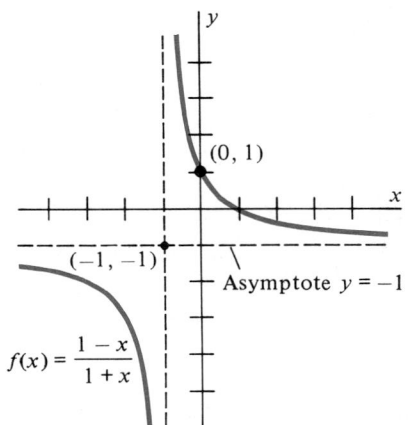

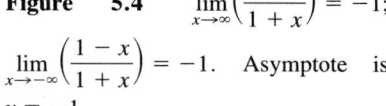

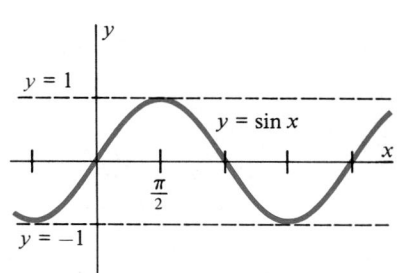

Figure 5.3 $\lim\limits_{x\to\infty}\left(\dfrac{x+2}{x}\right)=1;$

$\lim\limits_{x\to-\infty}\left(\dfrac{x+2}{x}\right)=1.$ Asymptote is $y=1.$

Figure 5.4 $\lim\limits_{x\to\infty}\left(\dfrac{1-x}{1+x}\right)=-1;$

$\lim\limits_{x\to-\infty}\left(\dfrac{1-x}{1+x}\right)=-1.$ Asymptote is $y=-1.$

Figure 5.5 $\lim\limits_{x\to\infty}\sin x$ does not exist.

Example 1

(a) $\lim\limits_{x\to\infty}\dfrac{x+2}{x}=\lim\limits_{x\to\infty}\left(1+\dfrac{2}{x}\right)=(1+0)=1.$

Thus, the line $y=1$ is a horizontal asymptote for the graph of $f(x)=\dfrac{x+2}{x}$. (Figure 5.3.)

(b) $\lim\limits_{x\to\infty}\dfrac{1-x}{1+x}=\lim\limits_{x\to\infty}\left(\dfrac{1-x}{1+x}\right)\left(\dfrac{1/x}{1/x}\right)=\lim\limits_{x\to\infty}\left(\dfrac{\dfrac{1}{x}-1}{\dfrac{1}{x}+1}\right)=\dfrac{0-1}{0+1}=-1.$

Thus, the line $y=-1$ is a horizontal asymptote for the graph of $f(x)=\dfrac{1-x}{1+x}$. (Figure 5.4.)

(c) $\lim\limits_{x\to\infty}\sin x$ does not exist because $\sin x$ oscillates between $y=-1$ and $y=1$ as $x\to\infty$. (See Figure 5.5.)

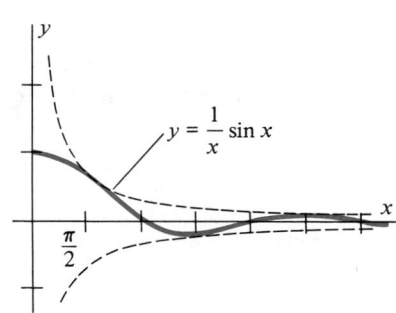

Figure 5.6 $\lim\limits_{x\to\infty}\dfrac{1}{x}\sin x=0.$ Asymptote is $y=0.$

(d) $\lim\limits_{x\to\infty}\dfrac{\sin x}{x}=0$ since $\left|\dfrac{\sin x}{x}\right|=\left|\dfrac{1}{x}\right|\cdot|\sin x|\le\dfrac{1}{x}\to0$ as $x\to\infty.$ Thus, the line $y=0$ is a horizontal asymptote for $f(x)=\dfrac{\sin x}{x}$. (Figure 5.6.) ◇

Evaluating Limits at Infinity

To handle nontrivial examples of limits as $x\to\infty$ or as $x\to-\infty$, we need to know the algebraic properties of such limits. The following theorem states that limits at infinity have the same algebraic properties as limits evaluated at real numbers. Its proof is left as Exercise 52.

THEOREM 10

Suppose that $\lim_{x \to \infty} f(x)$ and $\lim_{x \to \infty} g(x)$ both exist. Then

(i) $\lim_{x \to \infty} [f(x) + g(x)] = \lim_{x \to \infty} f(x) + \lim_{x \to \infty} g(x)$,

(ii) $\lim_{x \to \infty} [cf(x)] = c \lim_{x \to \infty} f(x)$, c constant,

(iii) $\lim_{x \to \infty} [f(x)g(x)] = [\lim_{x \to \infty} f(x)] \cdot [\lim_{x \to \infty} g(x)]$,

(iv) $\lim_{x \to \infty} \left[\dfrac{f(x)}{g(x)} \right] = \dfrac{\lim_{x \to \infty} f(x)}{\lim_{x \to \infty} g(x)}$,   provided $\lim_{x \to \infty} g(x) \neq 0$.

Theorem 10 also holds for limits as $x \to -\infty$. Theorem 10 will be used together with the following limits, which we state without proof, in evaluating limits at infinity.

Let $r = p/q$ be any positive rational number. Then

(i) $\lim_{x \to \infty} \dfrac{1}{x^r} = 0$ if q is even,   and   (1)

(ii) $\lim_{x \to \pm\infty} \dfrac{1}{x^r} = 0$ if q is odd.   (2)

Technique (Limits at Infinity):   To evaluate limits of the form

$$\lim_{x \to \infty} \frac{f(x)}{g(x)} \qquad \text{or} \qquad \lim_{x \to -\infty} \frac{f(x)}{g(x)}$$

where $f(x)$ and $g(x)$ are polynomials, divide both $f(x)$ and $g(x)$ by the highest power of x present. Then use limits (1) and (2).

Although the technique is stated for rational functions only, it works equally as well when $f(x)$ and $g(x)$ involve fractional powers of x. Also, note that there is no difficulty in assuring $x \neq 0$ in executing this technique, since we may assume that x is large.

Example 2

Using the above technique we obtain the following limits.

Strategy

Divide numerator and denominator by x^2.

Apply Theorem 10 (algebra of limits) and limit (1).

Solution

(a) $\lim_{x \to \infty} \dfrac{3x^2 + 7x - 4}{1 - x^2} = \lim_{x \to \infty} \dfrac{3 + \dfrac{7}{x} - \dfrac{4}{x^2}}{\dfrac{1}{x^2} - 1}$

$= \dfrac{3 + 7\left(\lim_{x \to \infty} \dfrac{1}{x} \right) - 4\left(\lim_{x \to \infty} \dfrac{1}{x^2} \right)}{\left(\lim_{x \to \infty} \dfrac{1}{x^2} \right) - 1}$

$$= \frac{3 + 7(0) - 4(0)}{0 - 1}$$

$$= -3.$$

Divide by x^3.

(b) $\displaystyle\lim_{x \to -\infty} \frac{x - 3}{2x + x^3} = \lim_{x \to -\infty} \frac{\dfrac{1}{x^2} - \dfrac{3}{x^3}}{\dfrac{2}{x^2} + 1}$

Apply Theorem 10 and limit (2).

$$= \frac{\left(\displaystyle\lim_{x \to -\infty} \frac{1}{x^2}\right) - 3\left(\displaystyle\lim_{x \to -\infty} \frac{1}{x^3}\right)}{2\left(\displaystyle\lim_{x \to -\infty} \frac{1}{x^2}\right) + 1}$$

$$= \frac{0 - 3(0)}{2(0) + 1}$$

$$= 0.$$

Divide by $x^{2/3}$. Apply Theorem 10 and limit (1).

(c) $\displaystyle\lim_{x \to \infty} \frac{\sqrt{x} + \sqrt[3]{x}}{x^{2/3}} = \lim_{x \to \infty} \frac{\dfrac{1}{x^{1/6}} + \dfrac{1}{x^{1/3}}}{1}$

$$= \frac{\left(\displaystyle\lim_{x \to \infty} \frac{1}{x^{1/6}}\right) + \left(\displaystyle\lim_{x \to \infty} \frac{1}{x^{1/3}}\right)}{1}$$

$$= \frac{0 + 0}{1}$$

$$= 0. \qquad \diamondsuit$$

Infinite Limits

Many functions do not have limits as $x \to \pm\infty$. For example, in the case of the parabola $y = x^2$, neither $\displaystyle\lim_{x \to \infty} x^2$, nor $\displaystyle\lim_{x \to -\infty} x^2$ exists in the sense of Definition 7.

We can write, however,

$$\lim_{x \to \infty} x^2 = \infty \qquad (3)$$

and

$$\lim_{x \to -\infty} x^2 = \infty, \qquad (4)$$

which mean that the values $f(x) = x^2$ *increase without bound* as $x \to \infty$ or as $x \to -\infty$ (Figure 5.7).

Since the symbol ∞ represents not a number, but rather a concept, equations (3) and (4) do *not* say that the corresponding limits exist. Rather, they give us information about why the limit fails to exist: $f(x) = x^2$ becomes infinitely large as $x \to \pm\infty$.

This situation is to be contrasted with the situation which occurs for the function $f(x) = \sin x$ (Figure 5.5). In this case $\displaystyle\lim_{x \to \infty} f(x)$ fails to exist, but we cannot write $\displaystyle\lim_{x \to \infty} \sin x = \infty$ since the values $f(x) = \sin x$ do not increase without bound. Rather, they simply fail to *converge* to a single number L as $x \to \infty$.

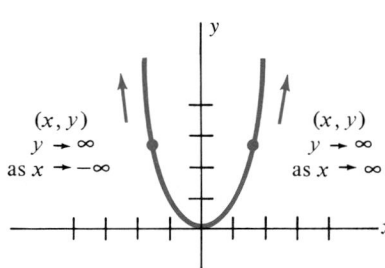

(x, y)
$y \to \infty$
as $x \to -\infty$

(x, y)
$y \to \infty$
as $x \to \infty$

Figure 5.7 For $f(x) = x^2$, both $\displaystyle\lim_{x \to \infty} f(x) = \infty$ and $\displaystyle\lim_{x \to -\infty} f(x) = \infty$.

We summarize all this by saying that for the expression $\lim_{x \to \infty} f(x)$, one of three situations occurs:

(i) $\lim_{x \to \infty} f(x)$ exists in the sense of Definition 7. That is, there is a *number L* so that
$$\lim_{x \to \infty} f(x) = L;$$

(ii) $\lim_{x \to \infty} f(x)$ fails to exist but $\lim_{x \to \infty} f(x) = \infty$ or $\lim_{x \to \infty} f(x) = -\infty$ (such as for $f(x) = x^2$);

(iii) $\lim_{x \to \infty} f(x)$ fails to exist, and neither $\lim_{x \to \infty} f(x) = \infty$ nor $\lim_{x \to \infty} f(x) = -\infty$ (such as for $f(x) = \sin x$).

The obvious corresponding statements hold for $\lim_{x \to -\infty} f(x)$.

Example 3

Find $\lim_{x \to \infty} \dfrac{4x^3 + 3x + 3}{2x^2 + x}$.

Strategy

Use technique of Example 2—divide both numerator and denominator by x^3.

Solution

$$\lim_{x \to \infty} \frac{4x^3 + 3x + 3}{2x^2 + x} = \lim_{x \to \infty} \frac{4 + \dfrac{3}{x^2} + \dfrac{3}{x^3}}{\dfrac{2}{x} + \dfrac{1}{x^2}}.$$

Examine limits of numerator and denominator separately, using Theorem 10 and limit (2).

In the numerator, we have
$$\lim_{x \to \infty} \left(4 + \frac{3}{x^2} + \frac{3}{x^3} \right) = 4 + 0 + 0 = 4$$

while in the denominator,
$$\lim_{x \to \infty} \left(\frac{2}{x} + \frac{1}{x^2} \right) = 0 + 0 = 0.$$

Since the denominator "shrinks" to zero through positive values while the numerator approaches 4, the quotient "blows up," i.e., increases without bound. Thus
$$\lim_{x \to \infty} \frac{4x^3 + 3x + 3}{2x^2 + x} = +\infty. \qquad \diamondsuit$$

Vertical Asymptotes

Infinite limits can also occur for $\lim_{x \to a} f(x)$, $\lim_{x \to a^+} f(x)$, or $\lim_{x \to a^-} f(x)$, that is, for one- or two-sided limits at $x = a$. For example, for $f(x) = \dfrac{1}{(x-2)^2}$, $\lim_{x \to 2} f(x) = +\infty$, since $(x-2)^2$ approaches zero through positive values as $x \to 2$. (Figure 5.8.) We formalize statements like $\lim_{x \to 2} f(x) = +\infty$ with the following working definition.

DEFINITION 9

The statement $\lim\limits_{x \to a} f(x) = \infty$ means that the values $f(x)$ increase without bound as x approaches a from either direction. The statement $\lim\limits_{x \to a^+} f(x) = \infty$ means that the values of $f(x)$ increase without bound as x approaches a from the right. Similar meanings are attached to the statements

$$\lim\limits_{x \to a} f(x) = \infty, \qquad \lim\limits_{x \to a} f(x) = -\infty, \qquad \lim\limits_{x \to a^+} f(x) = -\infty, \qquad \text{and}$$

$$\lim\limits_{x \to a^-} f(x) = -\infty.$$

DEFINITION 10

We say that the graph of $y = f(x)$ has a **vertical asymptote** at $x = a$ if any of the infinite limits in Definition 9 exists.

The technique for finding vertical asymptotes is to simply look for those numbers for which the denominator of $f(x)$ becomes zero.

Example 4

For $f(x) = \dfrac{1}{(x - 2)^3}$ note that if $x - 2$ is positive, so is $(x - 2)^3$. But if $(x - 2)$ is negative, $(x - 2)^3$ is also. Thus,

(a) if x approaches 2 from the right, $(x - 2)^3$ approaches zero through positive values, so

$$\lim\limits_{x \to 2^+} \frac{1}{(x - 2)^3} = \infty;$$

(b) if, however, x approaches 2 from the left, $(x - 2)^3$ is *negative*, so $(x - 2)^3$ approaches zero through negative values. Thus

$$\lim\limits_{x \to 2^-} \frac{1}{(x - 2)^3} = -\infty.$$

The graph of f has a vertical asymptote at $x = 2$. (Figure 5.9.) ◇

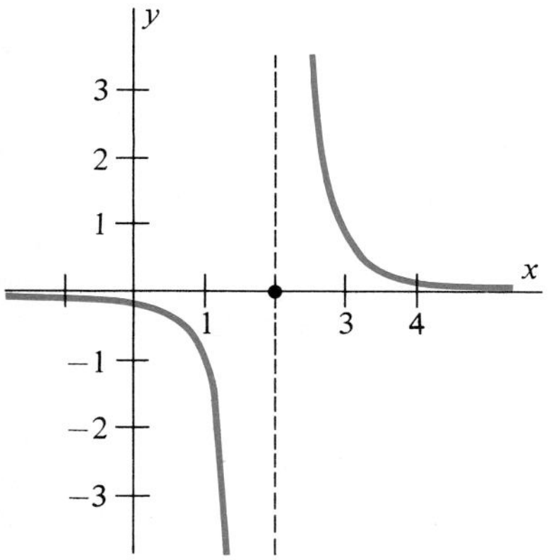

Example 5

Find all vertical asymptotes for the graph of $f(x) = \dfrac{x^2}{4 - x^2}$ and determine the corresponding limits.

Solution: The denominator can be factored as follows:

$$f(x) = \frac{x^2}{4 - x^2} = \frac{x^2}{(2 - x)(2 + x)}.$$

The denominator therefore equals zero for $x = -2$ and $x = 2$. These numbers are candidates for vertical asymptotes. To find the four corresponding one-sided limits, we must determine the signs of each of the factors of f. This information is easily obtained by use of a marked number line such as Figure 5.10.

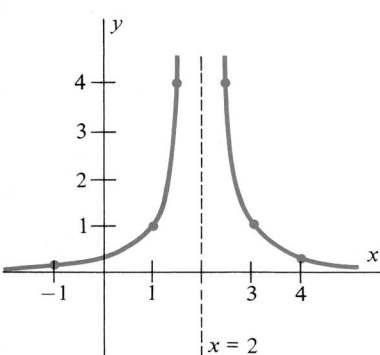

Figure 5.8 $\lim\limits_{x \to 2} \dfrac{1}{(x - 2)^2} = \infty.$

Figure 5.9 $\lim\limits_{x \to 2^+} \dfrac{1}{(x - 2)^3} = \infty;$ $\lim\limits_{x \to 2^-} \dfrac{1}{(x - 2)^3} = -\infty.$ Asymptote is $x = 2$.

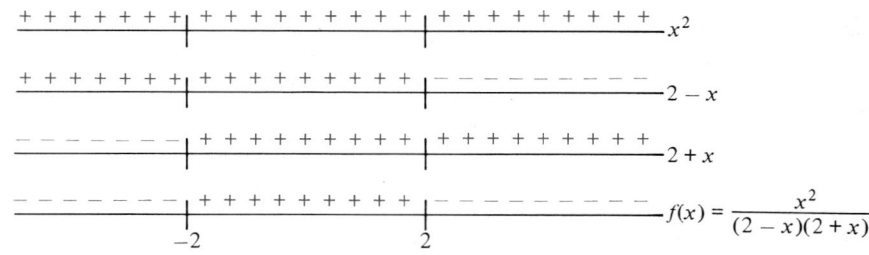

Figure 5.10 Sign analysis for $f(x)$.

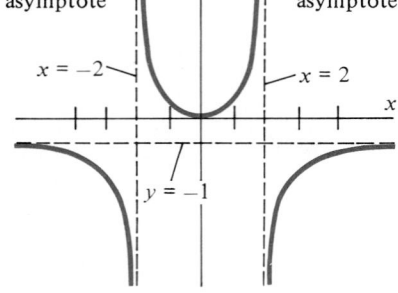

Figure 5.11 $f(x) = \dfrac{x^2}{4 - x^2}$.

Since the numerator of $f(x) = \dfrac{x^2}{4 - x^2}$ approaches 4 and the denominator approaches zero as $x \to \pm 2$, the one-sided limits at $x = \pm 2$ are all infinite. Figure 5.10 allows us to determine the correct signs. We have

$$\lim_{x \to -2^-} \frac{x^2}{4 - x^2} = -\infty,$$

$$\lim_{x \to -2^+} \frac{x^2}{4 - x^2} = \infty,$$

$$\lim_{x \to 2^-} \frac{x^2}{4 - x^2} = \infty, \quad \text{and}$$

$$\lim_{x \to 2^+} \frac{x^2}{4 - x^2} = -\infty.$$

The vertical asymptotes are $x = -2$ and $x = 2$. (See Figure 5.11. Note also that the graph has a *horizontal* asymptote $y = -1$.) ◇

Exercise Set 4.5

In Exercises 1–26, find the indicated limits.

1. $\displaystyle \lim_{x \to \infty} \frac{3x^2 + 2}{10x^2 - 3x}$

2. $\displaystyle \lim_{x \to \infty} \frac{6x^4 - 8x}{7 - 3x^4}$

3. $\displaystyle \lim_{x \to \infty} \frac{x(4 - x^3)}{3x^4 + 2x^2}$

4. $\displaystyle \lim_{x \to \infty} \left(\frac{1}{x}\right)^5$

5. $\displaystyle \lim_{x \to -\infty} x^{-1/3}$

6. $\displaystyle \lim_{x \to -\infty} \frac{2x + 6}{x^2 + 1}$

7. $\displaystyle \lim_{x \to \infty} \frac{3x^2 + 7x}{1 - x^4}$

8. $\displaystyle \lim_{x \to \infty} \frac{x^4 - 4}{x^3 + 7x^2}$

9. $\displaystyle \lim_{x \to \infty} \frac{\sin x}{x}$

10. $\displaystyle \lim_{x \to \infty} \frac{x^3 - 2x^2 + 5x - 4}{x - 3}$

11. $\displaystyle \lim_{x \to -\infty} \frac{x^4 - 3 \sin x}{3x + 5x^5}$

12. $\displaystyle \lim_{x \to \infty} \frac{x^2|2 + x^2|}{4 - x^4}$

13. $\displaystyle \lim_{x \to \infty} \frac{\sqrt{x - 1}}{x^2}$

14. $\displaystyle \lim_{x \to \infty} \frac{x^{2/3} + x^{4/3}}{x^2}$

15. $\displaystyle \lim_{x \to \infty} \frac{x^{2/3} + x}{1 + x^{3/4}}$

16. $\displaystyle \lim_{x \to \infty} \frac{\sqrt{x} + 7}{1 - \sqrt[3]{x}}$

17. $\displaystyle \lim_{x \to 2^+} \frac{1}{x - 2}$

18. $\displaystyle \lim_{x \to -2^-} \frac{1}{(x + 2)^2}$

19. $\displaystyle \lim_{x \to \pi/2^-} \tan x$

20. $\displaystyle \lim_{x \to \pi/2^+} \tan x$

21. $\displaystyle \lim_{x \to 0^-} \frac{|x|}{x}$

22. $\displaystyle \lim_{x \to 0^+} \frac{|x|}{x}$

23. $\displaystyle \lim_{x \to 1^+} \frac{1}{(x - 1)^{1/3}}$

24. $\displaystyle \lim_{x \to 1^-} \frac{1}{(x - 1)^{1/3}}$

25. $\displaystyle \lim_{x \to 1^+} \frac{1}{(x - 1)^{2/3}}$

26. $\displaystyle \lim_{x \to 1^-} \frac{1}{(x - 1)^{2/3}}$

In Exercises 27–34, find the horizontal asymptotes for the given functions.

27. $y = \dfrac{1}{1 + x}$

28. $y = \dfrac{1 + x}{3 - x}$

29. $y = \dfrac{2x^2}{x^2 + 1}$

30. $y = \dfrac{x^2}{1 - x}$

31. $y = \dfrac{2x^2}{(x^2 + 1)^2}$

32. $y = \dfrac{x^2 + 3}{1 - 3x^2}$

33. $y = 3 + \dfrac{\sin x}{x}$

34. $y = \dfrac{4x - \sqrt{x}}{x^{2/3} + x}$

35. The function $y = \dfrac{ax + 7}{4 - x}$ has a horizontal asymptote of $y = 3$. Find a.

Find all vertical asymptotes for the function in

36. Exercise 17.

37. Exercise 18.

38. Exercise 23.

39. Exercise 24.

40. Exercise 25.

41. Exercise 26.

42. The function $y = \dfrac{x^r + 3x}{7 - x^{4/3}}$ has a horizontal asymptote of $y = -1$. Find r.

43. The function $y = \dfrac{\pi + ax^r}{1 - 3x^{2/3}}$ has a horizontal asymptote of $y = -2$. Find a and r.

44. The function $y = \dfrac{3x^r + 2x^3}{rx^3}$ has a horizontal asymptote of $y = 5/3$. Find r.

45. The function $y = \dfrac{1}{x^2 + ax + b}$ has vertical asymptotes $x = 3$ and $x = 5$. Find a and b.

46. Show that $\displaystyle \lim_{x \to \infty} \dfrac{a_n x^n + a_{n-1}x^{n-1} + \cdots + a_1 x + a_0}{b_m x^m + b_{m-1}x^{m-1} + \cdots + b_1 x + b_0}$ (where $a_n \neq 0$, $b_m \neq 0$) equals
a. 0 if $n < m$,

b. ∞ if $n > m$ and a_n and b_m have like signs,

c. a_n/b_m if $n = m$.

47. A formal way of saying $\displaystyle \lim_{x \to \infty} f(x) = L$ is the following: "Given $\epsilon > 0$ there is an integer N so that if $x > N$ then $|f(x) - L| < \epsilon$." Use this definition to prove that

a. $\displaystyle \lim_{x \to \infty} \dfrac{1}{x} = 0$,

b. $\displaystyle \lim_{x \to \infty} \dfrac{3x + 1}{x} = 3$,

c. $\displaystyle \lim_{x \to \infty} \dfrac{\sin x}{x + 2} = 0$.

48. State a definition analogous to that of Exercise 47 for $\displaystyle \lim_{x \to -\infty} f(x) = L$ and use it to prove that:

a. $\displaystyle \lim_{x \to -\infty} \dfrac{1}{x} = 0$,

b. $\displaystyle \lim_{x \to -\infty} \dfrac{x}{2x + 1} = \dfrac{1}{2}$,

c. $\displaystyle \lim_{x \to -\infty} \dfrac{6x - 2}{x + 1} = 6$.

49. The formal way to say that $\displaystyle \lim_{x \to a} f(x) = \infty$ is: "Given $N > 0$ there exists a number $\delta > 0$ so that if $|x - a| < \delta$, then $f(x) > N$." Use this definition to prove that

a. $\displaystyle \lim_{x \to 0} \dfrac{1}{|x|} = \infty$,

b. $\displaystyle \lim_{x \to 0} \dfrac{1}{x^2} = \infty$,

c. $\displaystyle \lim_{x \to \pi/2} |\tan x| = \infty$.

50. State a formal definition for the statement $\displaystyle \lim_{x \to a} f(x) = \infty$ and use it to prove that $\displaystyle \lim_{x \to \pi/2^-} \tan x = \infty$.

51. State a formal definition for the statement $\displaystyle \lim_{x \to a} f(x) = -\infty$ and use it to prove that $\displaystyle \lim_{x \to 0^-} \dfrac{1}{x} = -\infty$.

52. Use the definition in Exercise 47 to prove Theorem 10.

4.6 CURVE SKETCHING II: SUMMARY OF TECHNIQUES

Sections 1–5 of this chapter have addressed ways in which information about limits and derivatives can be used to identify certain properties of the graph of a function. Our purpose here is to summarize these and other ideas that one should use in sketching the graph of a function. The goal in mastering these techniques is to become able to determine quickly a function's behavior without having to develop a precise graph.

To plot the graph of a function f:

(1) Determine the domain and, if possible, the range.
(2) If possible, locate the zeros by solving the equation $f(x) = 0$.
(3) Locate any vertical or horizontal asymptotes (Section 4.5).
(4) Find all critical numbers, classify the extrema, and determine whether f is increasing or decreasing on the resulting intervals (Sections 4.2 and 4.3).
(5) Determine the concavity and locate the inflection points (Section 4.4).
(6) Calculate the values of the function at a few convenient numbers and locate the corresponding points on the graph. Then sketch in the graph according to the above information.

Example 1

Sketch the graph of the function $f(x) = x^4 - 24x^2 + 20$.

Solution: Following the above outline we note the following.

(1) Since f is a polynomial, $f(x)$ is defined for all x.
(2) The range of the polynomial is not clear from its equation, so we skip this step.
(3) Since $\lim_{x \to +\infty} f(x) = +\infty$ there are no horizontal asymptotes. There are no vertical asymptotes because $f(x)$ is defined for all x.
(4) The first derivative is

$$f'(x) = 4x^3 - 48x.$$

Setting $f'(x) = 0$ gives

$$4x^3 - 48x = 4x(x^2 - 12) = 0$$

which has solutions $x = 0$ and $x = \pm\sqrt{12} = \pm 2\sqrt{3}$. These are the only critical numbers since $f'(x)$ is defined for all x.

The second derivative is

$$f''(x) = 12x^2 - 48$$

so

$$f''(-2\sqrt{3}) = 12(-2\sqrt{3})^2 - 48 = 96 > 0,$$
$$f''(0) = -48 < 0,$$
$$f''(2\sqrt{3}) = 12(2\sqrt{3})^3 - 48 = 96 > 0.$$

Thus, the second derivative test shows that f has a relative maximum at

$$(0, f(0)) = (0, 20)$$

and relative minima at

$$(-2\sqrt{3}, f(-2\sqrt{3})) = (-2\sqrt{3}, -124)$$

and

$$(2\sqrt{3}, f(2\sqrt{3})) = (2\sqrt{3}, -124).$$

As a result, f is decreasing on the intervals $(-\infty, -2\sqrt{3}]$ and $[0, 2\sqrt{3}]$ and increasing on the intervals $[-2\sqrt{3}, 0]$ and $[2\sqrt{3}, \infty)$ as you can verify.

(5) Setting $f''(x) = 12x^2 - 48 = 0$ gives $x = \pm2$. Checking the sign of $f''(x)$ on the resulting intervals shows that the graph of f is

concave up on $(-\infty, -2]$ and $[2, \infty)$,
concave down on $[-2, 2]$.

Thus, the points $(-2, f(-2)) = (-2, -60)$ and $(2, f(2)) = (2, -60)$ are inflection points.

(6) The graph of f is sketched in Figure 6.1.   ◇

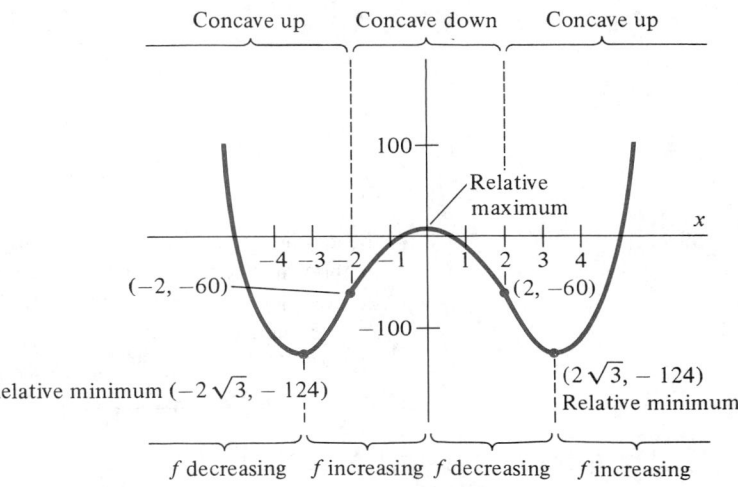

Figure 6.1   Graph of $f(x) = x^4 - 24x^2 + 20$.

Example 2

Sketch the graph of $f(x) = \dfrac{x}{1 - x^2}$.

Solution: Following the outline stated above, we find that:

(1) The *domain* of f is all numbers except $x = 1$ and $x = -1$. That is because the zeros of the denominator are $x = \pm1$. The *range* is all real numbers. To see this note that $f(0) = 0$, but that $f(x)$ becomes infinitely large and positive as x increases toward 1. Similarly, as x decreases from 0 toward -1, $f(x)$ becomes infinitely large and negative.

(2) The equation $f(x) = 0$ has the single solution $x = 0$.

(3) Since the denominator approaches zero as x approaches 1 or -1, but the numerator does not, there are vertical asymptotes at $x = \pm1$. The line $y = 0$ (i.e., the x-axis) is a horizontal asymptote, since

$$\lim_{x \to \pm\infty}\left(\frac{x}{1 - x^2}\right) = \lim_{x \to \pm\infty}\left(\frac{\dfrac{1}{x}}{\dfrac{1}{x^2} - 1}\right) = \frac{0}{0 - 1} = 0.$$

(4) To find the critical numbers we use the equation

$$f'(x) = \frac{d}{dx}\left(\frac{x}{1-x^2}\right) = \frac{(1)(1-x^2) - (x)(-2x)}{(1-x^2)^2} = \frac{1+x^2}{(1-x^2)^2} = 0.$$

This equation has no solutions, since $1 + x^2 \neq 0$ for all x. Although $f'(x)$ is undefined for $x = \pm 1$, these numbers are not in the domain of f. There are, therefore, no critical numbers; hence, no relative extrema. Since $1 + x^2 \geq 1$ for all x and $(1-x^2)^2 > 0$ for $x \neq \pm 1$, we can see that $f'(x) > 0$ if $x \neq \pm 1$. Thus, f is increasing on all intervals in its domain.

(5) Since

$$f''(x) = \frac{d}{dx}\left[\frac{1+x^2}{(1-x^2)^2}\right] = \frac{2x(1-x^2)^2 - (1+x^2)(2)(1-x^2)(-2x)}{(1-x^2)^4}$$

$$= \frac{2x^3 + 6x}{(1-x^2)^3}$$

$$= \frac{2x(x^2+3)}{(1-x^2)^3},$$

the candidate for an inflection point is $x = 0$ ($f''(0) = 0$). We note also that $f''(x)$ is undefined for $x = \pm 1$. The sign of $f''(x)$ on each of the resulting intervals can be obtained from Figure 6.2, which records the sign of each of the factors $2x$, $x^2 + 3$, and $(1-x^2)^3$.

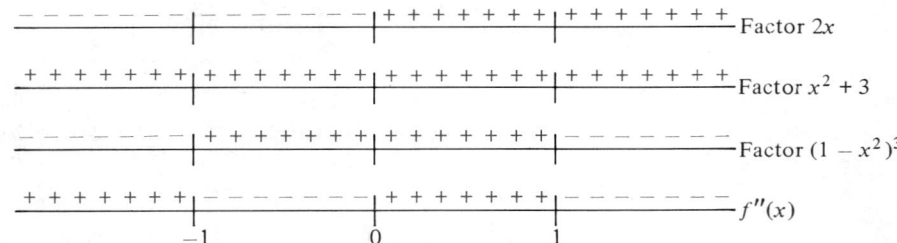

Figure 6.2 Sign analysis for $f''(x) = \dfrac{2x(x^2+3)}{(1-x^2)^3}$.

From the sign of $f''(x)$ we conclude that the graph of $y = f(x)$ is concave up on $(-\infty, -1)$ and $[0, 1)$ and concave down on $(-1, 0]$ and $(1, \infty)$.

The point $(0, f(0)) = (0, 0)$ is an inflection point.

(6) Since $f(-2) = 2/3$, $f(0) = 0$, and $f(2) = -2/3$, the points $(-2, 2/3)$, $(0, 0)$, and $(2, -2/3)$ are on the graph. The graph is sketched in Figure 6.3. ◇

Example 3

Sketch the graph of $f(x) = \dfrac{\sin x}{2 + \cos x}$ for $-\pi \leq x \leq \pi$.

Solution: Proceeding as in Example 2, we find that

(1) The domain of f is explicitly stated as the interval $[-\pi, \pi]$. Since $|\sin x| \leq 1$

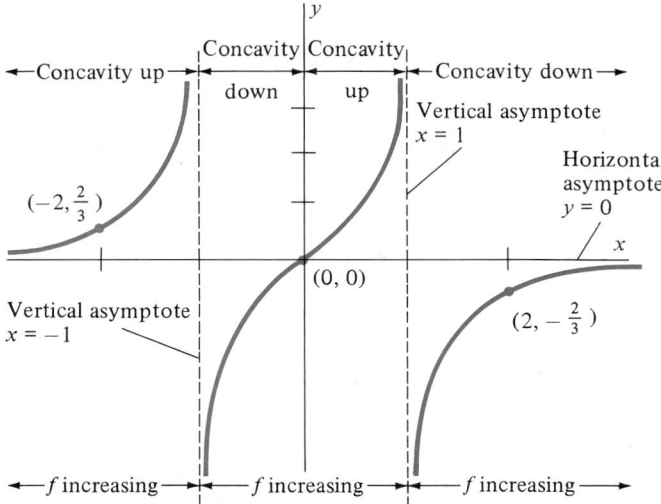

Figure 6.3 Graph of $f(x) = \dfrac{x}{1 - x^2}$.

and $|\cos x| \le 1$ for all x, the denominator can never be smaller than $2 + (-1) = 1$, so

$$|f(x)| = \left| \frac{\sin x}{2 + \cos x} \right| \le |\sin x| \le 1.$$

Thus, the range must lie within the interval $[-1, 1]$.

(2) The only solution of the equation $f(x) = \dfrac{\sin x}{2 + \cos x} = 0$ in $[-\pi, \pi]$ occurs when $\sin x = 0$, that is, at $x = -\pi, 0,$ or π.

(3) Since $f(x)$ is defined for all x, there are no vertical asymptotes. Since the domain of f is restricted to a finite interval there can be no horizontal asymptotes.

(4) The first derivative is

$$f'(x) = \frac{d}{dx}\left(\frac{\sin x}{2 + \cos x} \right) = \frac{(2 + \cos x)(\cos x) - (\sin x)(-\sin x)}{(2 + \cos x)^2}$$

$$= \frac{2 \cos x + 1}{(2 + \cos x)^2}.$$

Thus $f'(x) = 0$ if $2 \cos x + 1 = 0$, or $\cos x = -1/2$. The solutions of this equation in $[-\pi, \pi]$ are $x = \pm\dfrac{2\pi}{3}$. Since $f'(x)$ is defined for all x, these are the critical numbers.

By checking the sign of $f'(x)$ at one ''test number'' t in each of the resulting intervals, we obtain the information in Table 6.1 on the intervals on which f is increasing or decreasing.

Consequently $f\left(-\dfrac{2\pi}{3}\right) = -\dfrac{\sqrt{3}}{3} \approx -0.58$ is a relative minimum and $f\left(\dfrac{2\pi}{3}\right) = \dfrac{\sqrt{3}}{3}$ is a relative maximum, by the First Derivative Test.

Table 6.1 Analysis of sign of $f'(x)$.

Interval I	Test number $t \in I$	Sign of $f'(t)$	Conclusion
$\left(-\pi, -\dfrac{2\pi}{3}\right)$	$t = -\dfrac{5\pi}{6}$	$-$	f decreasing on $\left[-\pi, -\dfrac{2\pi}{3}\right]$
$\left(-\dfrac{2\pi}{3}, \dfrac{2\pi}{3}\right)$	$t = 0$	$+$	f increasing on $\left[-\dfrac{2\pi}{3}, \dfrac{2\pi}{3}\right]$
$\left(\dfrac{2\pi}{3}, \pi\right)$	$t = \dfrac{5\pi}{6}$	$-$	f decreasing on $\left[\dfrac{2\pi}{3}, \pi\right]$

(5) $f''(x) = \dfrac{(2 + \cos x)^2(-2 \sin x) - (2 \cos x + 1) \cdot 2(2 + \cos x)(-\sin x)}{(2 + \cos x)^4}$

$= \dfrac{2 \sin x(\cos x - 1)}{(2 + \cos x)^3}$

so $f''(x) = 0$ if $2 \sin x(\cos x - 1) = 0$. The solutions of this equation are $x = 0$ and the endpoints $x = -\pi$ and $x = \pi$. Since $\cos x - 1 \leq 0$ and $2 + \cos x \geq 0$ for all x, the sign of $f''(x)$ will be the opposite of the sign of the factor $\sin x$. Thus, $f''(x) > 0$ on $(-\pi, 0)$ and $f''(x) < 0$ on $(0, \pi)$. The graph is therefore concave up on $[-\pi, 0]$ and concave down on $[0, \pi]$, and the point $(0, 0)$ is inflection point. The graph appears in Figure 6.4. $\diamond$

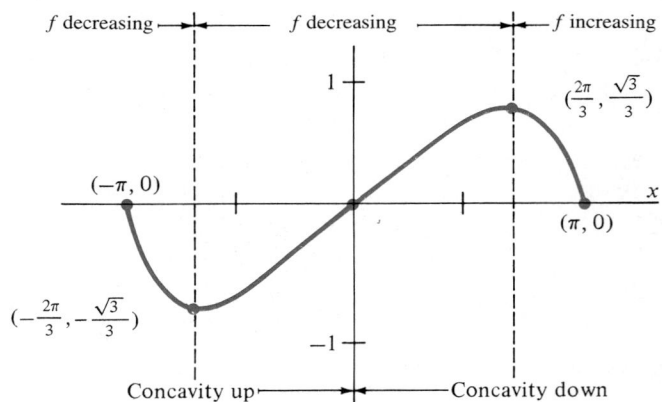

Figure 6.4 Graph of $f(x) = \dfrac{\sin x}{2 + \cos x}$.

Example 4

Sketch the graph of the function $f(x) = \dfrac{x^2 - x - 2}{x - 1}$.

Solution: Following the above outline, we begin by writing f in factored form as

$$f(x) = \dfrac{(x + 1)(x - 2)}{x - 1}.$$

From this equation for f we can observe that

(1) The *domain* of f is all real numbers except $x = 1$. The *range* is not obvious, so

we shall not dwell on this point. (Remember, we are after a *quick* sketch.)
(2) The zeros of f are $x = -1$ and $x = 2$.
(3) Since the denominator approaches zero as x approaches 1, but the numerator does not, the line $x = 1$ is a vertical asymptote. There are no horizontal asymptotes, since

$$\lim_{x \to \pm\infty} \frac{x^2 - x - 2}{x - 1} = \lim_{x \to \pm\infty} \frac{1 - \dfrac{1}{x} - \dfrac{2}{x^2}}{\dfrac{1}{x} - \dfrac{1}{x^2}} = \pm\infty.$$

(4) The derivative is

$$f'(x) = \frac{(x - 1)(2x - 1) - (x^2 - x - 2)}{(x - 1)^2}$$

$$= \frac{x^2 - 2x + 3}{(x - 1)^2}$$

$$= \frac{(x - 1)^2 + 2}{(x - 1)^2}$$

$$= 1 + \frac{2}{(x - 1)^2}.$$

Since $f'(x)$ is positive for all x except $x = 1$, where $f(1)$ and $f'(1)$ are undefined, we conclude that
(a) there are no critical numbers for f,
(b) f is increasing on $(-\infty, 1)$ and on $(1, \infty)$, and
(c) there are no relative extrema.
(5) The second derivative is $f''(x) = \dfrac{-4}{(x - 1)^3}$. Since $f''(x)$ is positive for $x < 1$, negative for $x > 1$ and since $f(1)$ is undefined, we conclude that
(a) there are no inflection points,
(b) the graph of f is concave up on $(-\infty, -1)$, and
(c) the graph of f is concave down on $(1, \infty)$.
(6) Noting that $f(-1) = 0$, $f(0) = 2$, and $f(2) = 0$ we obtain the sketch in Figure 6.5. $\diamondsuit$

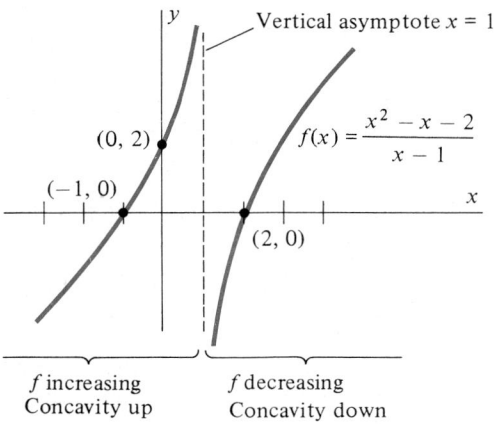

Figure 6.5 Graph of $f(x) = \dfrac{x^2 - x - 2}{x - 1}$.

Exercise Set 4.6

For each of Exercises 1–40, sketch the graph of f following the outline of this section.

1. $f(x) = x^2 - 2x - 8$

2. $f(x) = 9x - x^2$

3. $f(x) = 2x^3 - 3x^2$

4. $f(x) = 3x^4 - 4x^3$

5. $f(x) = x^3 + x^2 - 8x + 8$

6. $f(x) = x^3 + 2x^2 - x - 2$

7. $f(x) = x^3 - 7x + 6$

8. $f(x) = \dfrac{x - 1}{x + 1}$

9. $f(x) = \dfrac{x + 4}{x - 4}$

10. $f(x) = \sin x + \cos x$

11. $f(x) = 9x - x^{-1}$

12. $f(x) = x + 3x^{2/3}$

13. $f(x) = \tan^2 x + 1$

14. $f(x) = \sqrt{x + 4}$

15. $f(x) = |4 - x^2|$

16. $f(x) = x^2 + \dfrac{2}{x}$

17. $f(x) = \sin x \cos x$

18. $f(x) = (x + 1)^{5/3} - (x + 1)^{2/3}$

19. $f(x) = \dfrac{1}{x(x - 4)}$

20. $f(x) = x^2 - \dfrac{9}{x^2}$

21. $f(x) = \dfrac{x}{(2x + 1)^2}$

22. $f(x) = \sqrt{2x - x^2}$

23. $f(x) = (x - 3)^{2/3} + 1$

24. $f(x) = \sqrt{6x - x^2 - 8}$

25. $f(x) = \dfrac{1}{3}x^3 - x^2 - 3x + 4$

26. $f(x) = \dfrac{1}{12}x^4 - \dfrac{2}{3}x^3 + 2x^2 + 5x - 8$

27. $f(x) = \dfrac{x^2 - 4x + 5}{x - 2}$

28. $f(x) = \dfrac{3}{2}x^{2/3} - x$

29. $f(x) = 4x^2(1 - x^2)$

30. $f(x) = x^4 - 3x^2 + 2$

31. $f(x) = x^{2/3} - \dfrac{1}{5}x^{5/3}$

32. $f(x) = \dfrac{2x + 1}{x + 2}$

33. $f(x) = \dfrac{x^2}{9 - x^2}$

34. $f(x) = \dfrac{x^2}{x^2 - 16}$

35. $f(x) = 5x^3 - x^5$

36. $f(x) = x^4 - 2x^2 + 1$

37. $f(x) = 16 - 20x^3 + 3x^5$

38. $f(x) = x^4 - 18x^2 + 32$

39. $f(x) = \dfrac{1 - x^2}{x^3}$

40. $f(x) = 3x^{5/3} - 15x^{2/3} + 5$

4.7 EXTREMA ON CLOSED BOUNDED INTERVALS; MAX–MIN PROBLEMS

Many applications of the derivative involve finding the (absolute) maximum or minimum value of a continuous function f on a *closed bounded* interval $[a, b]$. We shall discuss several examples of this problem later in this section. We begin, however, by summarizing the relevant information that we have developed up to this point.

1. If f is *continuous* on $[a, b]$, its maximum and minimum values on $[a, b]$ both exist (Theorem 1). These are the values M and m, respectively, for which

   $f(x) \leq M$ for all $x \in [a, b]$, and
   $f(x) \geq m$ for all $x \in [a, b]$

   (Definition 1).

2. If the value $f(c)$ is the maximum or minimum value of f on $[a, b]$ and $c \in (a, b)$ is not an endpoint, then either

   $$f'(c) = 0, \quad \text{or else} \quad f'(c) \text{ fails to exist} \tag{1}$$

   (Theorem 2, Extreme Value Theorem). The numbers $c \in (a, b)$ satisfying one of the two conditions in line (1) are called the *critical numbers* for f on $[a, b]$ (Definition 3).

From these two observations we may conclude that *the maximum and minimum values of a continuous function f on a closed finite interval $[a, b]$ exist and occur either at critical numbers or at endpoints of the interval $[a, b]$.*

This observation gives the following procedure.

> **Procedure for Finding the Maximum and Minimum Values of a Continuous Function f on a Closed Finite Interval $[a, b]$:**
>
> 1. Find all numbers $c \in (a, b)$ for which either $f'(c) = 0$ or $f'(c)$ does not exist. (These are the critical numbers.)
> 2. Compute $f(a)$, $f(b)$, and all values $f(c)$ where c is a critical number for f in (a, b). (That is, check $f(x)$ at the *endpoints* and at all *critical numbers*.)
> 3. Select the largest and smallest of the values computed in Step 2. These are the maximum and minimum values, respectively.

Absolute Versus Relative Extrema

Before illustrating this technique we again note the distinction between finding the absolute extrema for f on $[a, b]$ and finding the *relative* extrema for f as done in Section 4.3.

If f is continuous on $[a, b]$ and $f(c)$ is an absolute extremum (that is, the absolute maximum or minimum) for f on $[a, b]$, then $f(c)$ will also be a relative extremum for f if $c \in (a, b)$. Note, however, that

(i) a *relative* extremum for f on $[a, b]$ need not be an *absolute* extremum for f on $[a, b]$;

(ii) an *absolute* extremum for f on $[a, b]$ need not be a *relative* extremum because the absolute extremum may occur at an *endpoint*. (Such extrema are called *endpoint extrema*.)

Figure 7.1 illustrates both point (i) and point (ii).

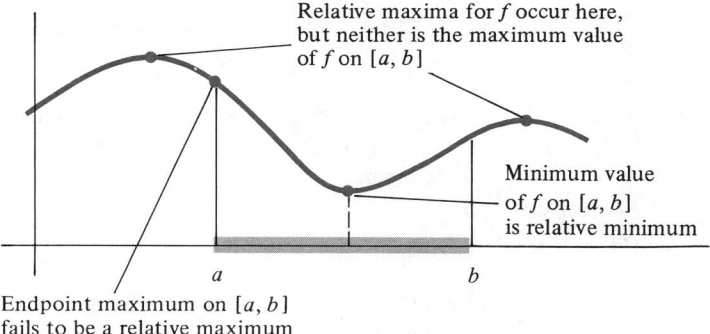

Figure 7.1 Relationships between extrema on $[a, b]$ and relative extrema.

Example 1

Find the maximum and minimum values of $f(x) = x^3 - 3x$ on the interval $[-2, 2]$ and the corresponding numbers x.

Strategy
Verify that f is continuous on $[-2, 2]$.

Solution
Since $f(x) = x^3 - 3x$ is a polynomial, we know that f is continuous on $[-2, 2]$. We have

$$f'(x) = 3x^2 - 3,$$

Set $f'(x) = 0$ to find critical numbers.

so setting $f'(x) = 0$ gives the equation

$$3x^2 - 3 = 0,$$

Find remaining critical numbers where $f'(x)$ is undefined. (In this case, there are none.)

which has solutions $x = \pm 1$. There are no numbers x for which $f'(x)$ is undefined. The critical numbers are, therefore, $x = -1$ and $x = 1$. The values of f at the critical numbers and endpoints are as follows:

Find $f(x)$ for all critical numbers and endpoints.

$$\begin{aligned}
f(-2) &= (-2)^3 - 3(-2) = -2 && \text{(endpoint)} \\
f(-1) &= (-1)^3 - 3(-1) = 2 && \text{(critical number)} \\
f(1) &= (1)^3 - 3(1) = -2 && \text{(critical number)} \\
f(2) &= 2^3 - 3(2) = 2 && \text{(endpoint)}
\end{aligned}$$

Find max and min by inspecting these values. Note the corresponding numbers x.

Thus, the maximum value of f on $[-2, 2]$ is 2, which occurs both at $x = -1$ and at $x = 2$. The minimum is -2, which occurs both at $x = -2$ and at $x = 1$. The graph of $y = f$ appears in Figure 7.2. ◇

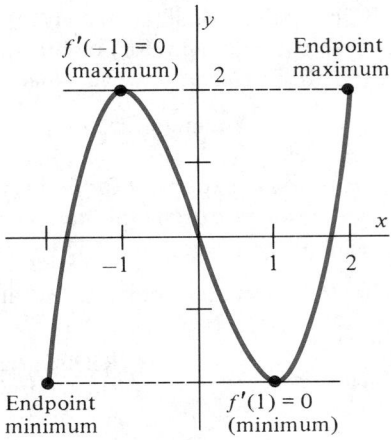

Figure 7.2 For $f(x) = x^3 - 3x$ on $[-2, 2]$, maximum value $= 2$ and minimum value $= -2$.

Example 2

Find the maximum and minimum values of the function $f(x) = 2 + 2x - 3x^{2/3}$ on the interval $[-1, 2]$.

Strategy
Verify f continuous.

Solution
The function f is continuous at $[-1, 2]$ since it is the sum of the polynomial $2 + 2x$ and a multiple of the rational power function $x^{2/3}$. Since $f'(x) = 2 - 2x^{-1/3}$, setting $f'(x) = 0$ gives the equation

Set $f'(x) = 0$ and solve to find critical number(s).

$$2 - 2x^{-1/3} = 0,$$

or

$$x^{1/3} = 1.$$

Determine where

$$f'(x) = 2 - \frac{2}{\sqrt[3]{x}}$$

The only solution of the equation $f'(x) = 0$ is therefore $x = 1$. Also, $f'(x)$ is undefined at $x = 0$. The critical numbers are therefore $x = 0$ and $x = 1$. Examining $f(x)$ at the critical numbers and endpoints we see that

$$f(-1) = 2 + 2(-1) - 3(-1)^{2/3} = -3 \qquad \text{(endpoint)}$$

is undefined to find remaining critical numbers.

Inspect $f(x)$ at critical numbers and endpoints to identify max and min.

$$f(0) = 2 + 2(0) - 3(0)^{2/3} = 2 \qquad \text{(critical number)}$$
$$f(1) = 2 + 2(1) - 3(1)^{2/3} = 1 \qquad \text{(critical number)}$$
$$f(2) = 2 + 2(2) - 3(2)^{2/3} = 6 - 3\sqrt[3]{4} \approx 1.2378. \quad \text{(endpoint)}$$

We conclude that the maximum value of $f(x)$ on $[-1, 2]$ is $f(0) = 2$ and that the minimum is $f(-1) = -3$. (Note that $f(1) = 1$ is a *relative* minimum for f, but not *the* minimum value of f on $[-1, 2]$.) The graph of f appears in Figure 7.3.   ◇

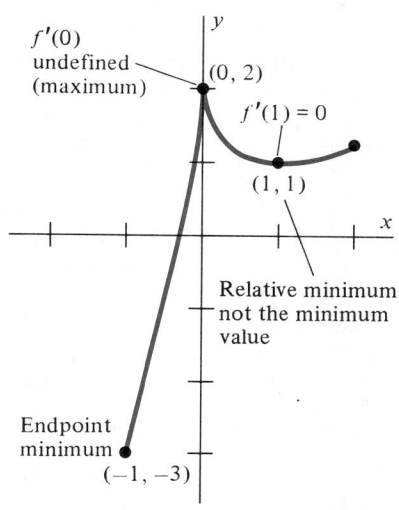

Figure 7.3 For $f(x) = 2 + 2x - 3x^{2/3}$ on $[-1, 2]$, maximum value $= 2$ and minimum value $= -3$.

Applied Maximum–Minimum Problems

One of the most important (and difficult) skills needed in becoming a successful practitioner of the calculus is the ability to formulate a given problem in precise mathematical terms. The remaining examples of this section involve finding the maximum or minimum value of a function on a closed finite interval, but these are "word problems" stated in prose rather than the "equation-type" problems of Examples 1 and 2.

The following diagram illustrates the major conceptual steps involved in applying mathematics to solve problems of this type.

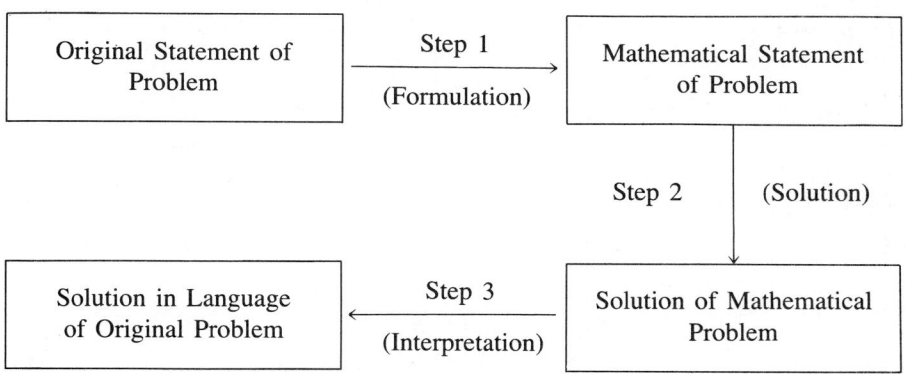

Diagram 7.1 Steps in solving applied problems.

The following procedure refines further the three principal steps in Diagram 7.1.

> ## Procedure for Solving Applied Max–Min Problems
> *(Formulation)*
>
> 1. Assign letter names to all variables and, if possible, draw a sketch showing all variables and constants in the problem.
> 2. Identify the variable for which the extremum is sought. Find an equation for this variable in terms of other variables and constants (principal equation).
> 3. Find any other equations involving the given variables and constants (auxiliary equations).
>
> *(Mathematical Solution)*
>
> 4. Use auxiliary equation(s) to substitute for variable(s) in the principal equation until it expresses the variable of interest as a *function* of a *single* independent variable.
> 5. Determine the interval $[a, b]$ on which the function in Step 4 is defined.
> 6. Find the maximum or minimum value for the function on $[a, b]$ by the technique described earlier in this section.
>
> *(Interpretation)*
>
> 7. Describe the solution found in Step 6 in the language of the original problem.

The remaining pages of this section consist of examples of problems solved by this procedure. While they are obviously contrived, these examples and the exercises at the end of the section provide a "laboratory" setting in which you can gain experience with the several steps in problem solving outlined above.

Example 3

Find two nonnegative numbers whose sum is 10 and the sum of whose squares is a maximum.

Strategy
Label variables.

Solution
Let x and y be the two numbers and let S be the desired sum. We want to maximize

Find equation for S by summing squares.

$$S = x^2 + y^2 \qquad \text{(principal equation)} \tag{2}$$

subject to the condition that

Auxiliary equation states that sum is 10.

$$x + y = 10 \qquad \text{(auxiliary equation)}. \tag{3}$$

Eliminate y in (2), using auxiliary equation.

To eliminate the variable y in equation (2), we solve equation (3) for y to obtain

$$y = 10 - x$$

and substitute into equation (2), which gives

$$S(x) = x^2 + (10 - x)^2.$$

Find domain of S in the form $[a, b]$ from auxiliary equation and other given information.

Equation (3) together with the fact that x and y are nonnegative imply that

$$0 \le x \le 10.$$

We therefore seek the maximum value of the function S on the interval $[0, 10]$.

Setting $S'(x) = 0$ gives the equation

Set $S'(x) = 0$ and solve to find critical numbers.

$$S'(x) = 2x - 2(10 - x) = 0,$$

so

$$x = 5.$$

Identify those critical numbers lying in the domain $[0, 10]$.

Since $S'(x)$ is defined for all $x \in [0, 10]$, the only critical number in $(0, 10)$ is $x = 5$. The maximum value of S will therefore occur among the following values:

Inspect $S(x)$ at critical number and endpoints to find maximum.

$$
\begin{aligned}
S(0) &= 0^2 + 10^2 = 100 && \text{(endpoint)} \\
S(5) &= 5^2 + 5^2 = 50 && \text{(critical number)} \\
S(10) &= 10^2 + 0^2 = 100 && \text{(endpoint).}
\end{aligned}
$$

The maximum occurs at the endpoints 0 and 10, both of which give the solution as the pair of numbers 0 and 10. ◇

Example 4

An open box is to be made from a square piece of cardboard measuring 12 inches on a side by cutting a square from each corner and folding up the sides as in Figure 7.4. Find the dimensions for which the volume of the resulting box is a maximum.

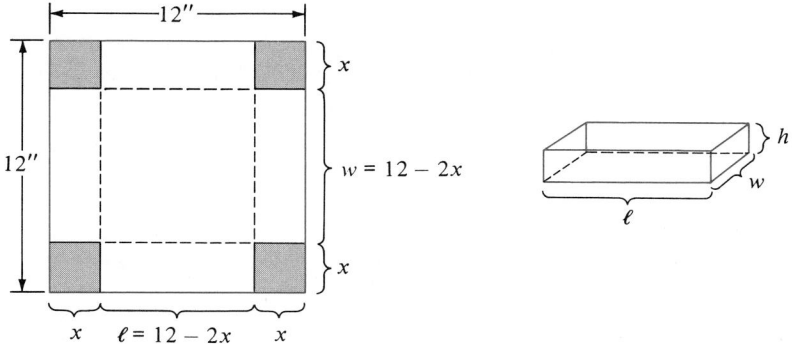

Figure 7.4 Box made from 12″ × 12″ square by cutting squares from each corner and folding up flaps.

Strategy

Find an equation for the volume V.

Solution

The volume of the box is

$$V = l \cdot w \cdot h$$

where l = length, w = width, and h = height. If we let x denote the length (and width) of the square cut from each corner, the dimensions of the box can be expressed in terms of x as

Express each dimension in terms of x, the side of the square cut out.

$$
\begin{aligned}
l &= 12 - 2x, \\
w &= 12 - 2x, \\
h &= x.
\end{aligned}
$$

The volume may now be expressed as a function of x:

Substitute in equation for V to express V as a function of x alone.

$$V(x) = (12 - 2x)(12 - 2x)(x)$$
$$= 4x^3 - 48x^2 + 144x.$$

Find the closed interval $[a, b]$ in which x lies.

Since the side of the piece cut from a corner can be no longer than $\dfrac{12}{2} = 6$ inches, the variable x is restricted to the closed interval $[0, 6]$.

Set $V'(x) = 0$ to find its critical numbers in (a, b).

The derivative for V is

$$V'(x) = 12x^2 - 96x + 144$$
$$= 12(x^2 - 8x + 12)$$
$$= 12(x - 2)(x - 6)$$

so the equation $V'(x) = 0$ gives the single critical number $c = 2$ in the interval $(0, 6)$. Since

Compute $V(x)$ at each endpoint and critical number.

$$V(0) = 0 \qquad \text{(endpoint)}$$
$$V(2) = 128, \qquad \text{(critical number)}$$

and

$$V(6) = 0, \qquad \text{(endpoint)}$$

Select the largest of these values.

the maximum volume of 128 in^3 occurs when $x = 2$, which corresponds to the dimensions $l = 12 - 2(2) = 8$ inches, $w = 8$ inches and $h = x = 2$ inches. ◇

Example 5

A water trough is to be constructed from three metal sheets 1 meter wide and 6 meters long plus end panels in the shape of trapezoids. Find the angle at which the long panels should be joined so as to provide a trough of maximum volume (see Figures 7.5 and 7.6).

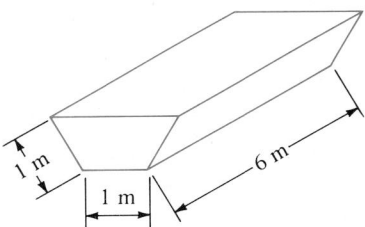

Figure 7.5 Water trough.

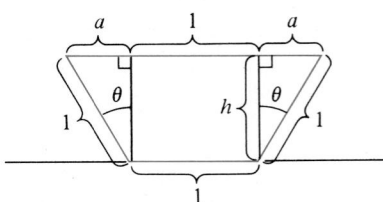

Figure 7.6 End view of the trough.

Strategy

Label variables.

Solution

Consider the end view as shown in Figure 7.6. Let θ be the angle between the side panel and the vertical, h the height, and a the length of the side of the triangle opposite the angle θ. Then

Use

$$h = \frac{h}{1} = \frac{\text{adjacent}}{\text{hypotenuse}}$$
$$= \cos \theta.$$

$$h = \cos \theta, \qquad \text{and} \qquad a = \sin \theta.$$

The end panel is a trapezoid with small base $b = 1$ and large base $B = 1 + 2a$. The volume of the trough is therefore

Find equation for V using formula for area of a trapezoid.

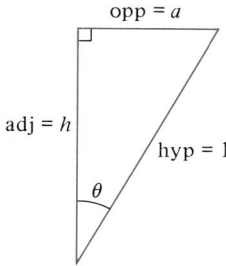

opp = a

adj = h

hyp = 1

θ

Domain of V is $[0, \pi/2]$, from sketch.

Set $V'(\theta) = 0$ to find critical numbers.

Remaining critical numbers occur when $V'(\theta) = 0$.

Determine which critical numbers lie in domain $[0, \pi/2]$.

Inspect $V(\theta)$ at critical numbers and endpoints to determine maximum.

$$V = \frac{1}{2}(B + b) \cdot h \cdot 6$$

$$= \frac{1}{2}[(1 + 2a) + 1] \cdot h \cdot 6$$

$$= 6(1 + \sin \theta)\cos \theta.$$

Clearly θ must lie within the interval $[0, \pi/2]$. Since V is expressed as a function of θ alone, we may proceed to find the critical numbers for θ:

$$\frac{dV}{d\theta} = -6 \sin \theta + 6 \cos^2 \theta - 6 \sin^2 \theta$$

$$= -6 \sin \theta + 6(1 - \sin^2 \theta) - 6 \sin^2 \theta$$
$$= -12 \sin^2 \theta - 6 \sin \theta + 6$$
$$= -6(2 \sin \theta - 1)(\sin \theta + 1).$$

The equation $dV/d\theta = 0$ therefore yields two critical numbers since

$$2 \sin \theta - 1 = 0 \text{ implies } \sin \theta = \frac{1}{2}, \quad \text{or} \quad \theta = \frac{\pi}{6},$$

and

$$\sin \theta + 1 = 0 \text{ implies } \sin \theta = -1, \quad \text{or} \quad \theta = -\frac{\pi}{2}.$$

Since $dV/d\theta$ is defined for all θ, and the critical number $\theta = -\pi/2$ does not lie within $(0, \pi/2)$, the only critical number for θ in $(0, \pi/2)$ is $\theta = \pi/6$. The maximum value for V must lie among the numbers

$$V(0) = 6(1 + 0)(1) = 6 \qquad \text{(endpoint)}$$

$$V(\pi/6) = 6\left(1 + \frac{1}{2}\right)\left(\frac{\sqrt{3}}{2}\right) = \frac{9\sqrt{3}}{2} \approx 7.79 \quad \text{(critical number)}$$

$$V(\pi/2) = 6(1 + 1)(0) = 0 \qquad \text{(endpoint)}$$

The angle corresponding to the maximum volume is therefore $\theta = \pi/6 = 30°$.

◇

Example 6

Find the right circular cylinder of maximum volume that can be inscribed in a sphere of radius 10 cm.

Solution: Figure 7.7 shows the cylinder of radius r and height h with edges touching the sphere. The equation for its volume is

$$V = \pi r^2 h. \tag{4}$$

Since the cross section of the sphere in the xy-plane is a circle of radius 10, it has equation

$$x^2 + y^2 = 10^2. \tag{5}$$

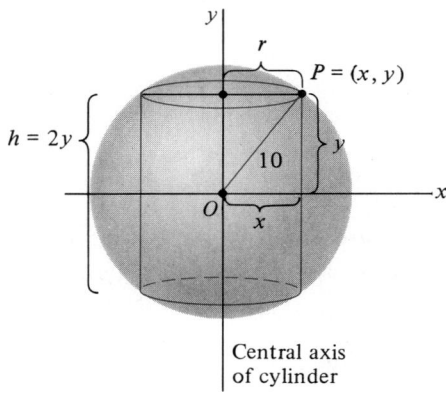

Figure 7.7 Cylinder within a sphere.

Finally, since the edges of the cylinder touch the sphere, we have

$$r = x \tag{6}$$

and

$$h = 2y. \tag{7}$$

Substituting in equation (4) for r and h from equations (6) and (7) gives

$$V = \pi x^2 (2y) = 2\pi x^2 y. \tag{8}$$

Writing equation (5) in the form $x^2 = 100 - y^2$ and substituting for x^2 in equation (8) then gives volume as a function of y alone:

$$V(y) = 2\pi(100 - y^2)y = 2\pi(100y - y^3), \qquad 0 \le y \le 10.$$

The derivative is

$$V'(y) = 2\pi(100 - 3y^2)$$

so the equation $V'(y) = 0$ gives $3y^2 = 100$, or $y = \dfrac{10}{\sqrt{3}}$. This is the only critical number for V. Since $0 \le y \le 10$, checking $V(y)$ at the critical number and endpoints gives

$$V(0) = 2\pi(100 \cdot 0 - 0^3) = 0 \qquad \text{(endpoint)}$$

$$V\left(\frac{10}{\sqrt{3}}\right) = 2\pi\left[100\left(\frac{10}{\sqrt{3}}\right) - \left(\frac{10}{\sqrt{3}}\right)^3\right] = \frac{4000\pi}{3\sqrt{3}} \qquad \text{(critical number)}$$

$$V(10) = 2\pi(100 \cdot 10 - 10^3) = 0. \qquad \text{(endpoint)}$$

The maximum volume of $\dfrac{4000\pi}{3\sqrt{3}} = \dfrac{4000\pi\sqrt{3}}{9}$ therefore corresponds to the cylinder of radius $r = x = \sqrt{100 - y^2} = \dfrac{10\sqrt{6}}{3}$ and height $h = 2y = \dfrac{20}{\sqrt{3}} = \dfrac{20\sqrt{3}}{3}$.

$\diamondsuit$

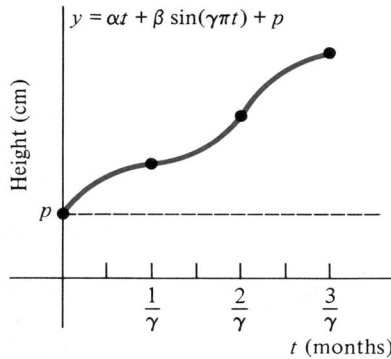

$y = \alpha t + \beta \sin(\gamma \pi t) + p$

Height (cm)

p

$\dfrac{1}{\gamma}$　$\dfrac{2}{\gamma}$　$\dfrac{3}{\gamma}$

t (months)

Figure 7.8 A model for animal growth.

Example 7

Researchers interested in modelling the rate at which animals (including humans) grow know that growth is not uniform. Periods of rapid growth often occur between periods of very slow growth.

A scientist interested in modelling the growth pattern for a particular species wishes to use an equation of the form

$$y = \alpha t + \beta \sin(\gamma \pi t) + p$$

as a growth model, since this function possesses the growth characteristics described above (see Figure 7.8). Here t represents time in months after birth and y represents height in centimeters. The parameters (constants) α, β, γ and p are positive numbers calculated by the scientist on the basis of observed data. (This process, called ''fitting'' the model to the data, is a subject in courses on mathematical statistics.) For a model of this type the growth rate is defined to be the derivative $\dfrac{dy}{dt}$.

If $\gamma = 1/4$ and $\beta = \alpha = 1$, find the maximum and minimum rates of growth and the times at which they occur during the first year of growth according to the proposed model.

Solution: The growth rate is

$$\frac{dy}{dt} = 1 + \frac{\pi}{4}\cos\left(\frac{\pi t}{4}\right).$$

The mathematical problem is to find the extreme values for this function on the interval $[0, 12]$.* To obtain the critical numbers for $\dfrac{dy}{dt}$ we differentiate, obtaining

$$\frac{d^2y}{dt^2} = -\frac{\pi^2}{16}\sin\left(\frac{\pi t}{4}\right).$$

This derivative equals zero when $\sin\left(\dfrac{\pi t}{4}\right) = 0$, that is, when

$$\frac{\pi t}{4} = 0,\ \pm\pi,\ \pm 2\pi,\ \ldots$$

or

$$t = 0,\ \pm 4,\ \pm 8,\ \pm 12,\ \ldots.$$

Now only the critical numbers $t = 4$ and 8 lie within the interval $(0, 12)$. Since $\dfrac{d^2y}{dt^2}$ is never undefined, there are no other critical numbers. Table 7.1 shows the growth rate at each critical number and endpoint.

_____

*Note here that we are seeking the extreme values of the *derivative* $\dfrac{dy}{dt}$ rather than the extreme values of y. We therefore examine $\dfrac{d}{dt}\left(\dfrac{dy}{dt}\right) = \dfrac{d^2y}{dt^2}$ for critical numbers.

Table 7.1

t	0	4	8	12
$\dfrac{dy}{dt}$	$1 + \dfrac{\pi}{4}$	$1 - \dfrac{\pi}{4}$	$1 + \dfrac{\pi}{4}$	$1 - \dfrac{\pi}{4}$

The maximum growth rate is observed to be $1 + \dfrac{\pi}{4} \approx 1.785$ cm/month, and this rate occurs at birth and at age 8 months. The minimum growth rate $1 - \dfrac{\pi}{4} \approx$ .215 cm/month occurs at ages 4 months and 12 months (see Figure 7.9). ◇

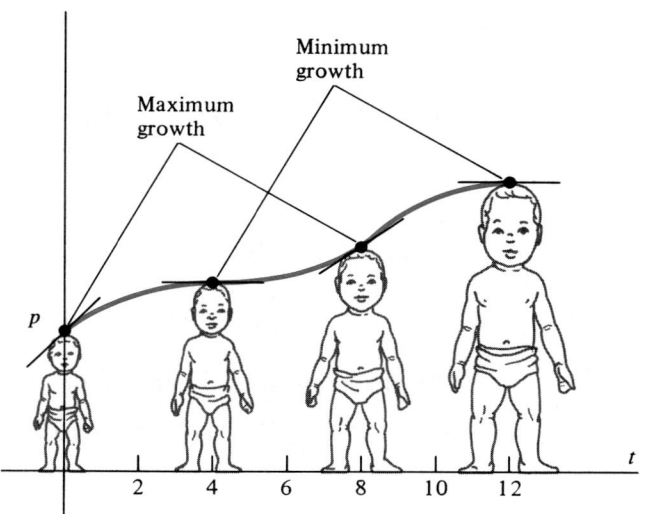

Figure 7.9 Times of minimum and maximum growth.

Example 8

Two laws in optics can be established using the calculus and a principle discovered by Pierre de Fermat, the seventeenth century French mathematician. Fermat's principle states that a ray of light traveling in a uniform medium will follow the path of minimum time.

The first law is the Law of Reflection, which states that a ray of light which strikes a flat reflecting surface at an angle α will be reflected away at that same angle, i.e., that the angle of incidence equals the angle of reflection (see Figure 7.10). You are asked to show this in Exercise 57.

The second law, first discovered by the seventeenth century Dutch mathematician Willebrord Snell, concerns a ray of light that passes from one medium, in which light travels at a velocity v_1, into a second medium, in which light travels at a velocity v_2. Within each medium the path of minimum time is a straight line. If $v_1 \neq v_2$, however, the path followed between a point A in medium 1 and a point B in medium 2 will consist of two line segments meeting at a point P on the surface

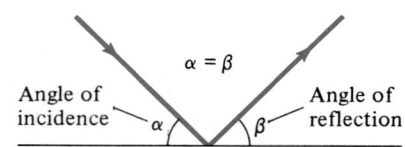

Angle of incidence α $\alpha = \beta$ Angle of reflection β

Figure 7.10 Law of Reflection.

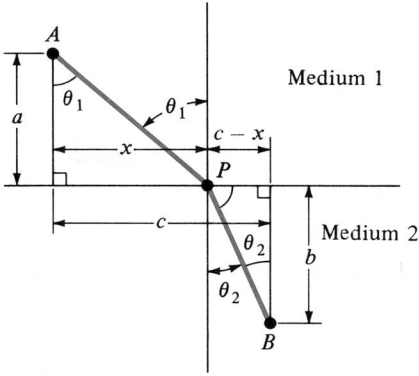

Figure 7.11 Snell's Law of Refraction.

separating the media (see Figure 7.11). This "bending" of the light ray is called refraction, and Snell's Law of Refraction states that

$$\frac{\sin \theta_1}{v_1} = \frac{\sin \theta_2}{v_2}$$

where θ_1 and θ_2 are the angles formed by the ray and the lines normal to the surface at P in each of the media.

To establish this law, we let a and b denote the lengths of the normals from points A and B to the surface, we let c be the distance between these normals, and we let x be the distance along the surface from the normal through A to point P.

Since the time required for the light to travel from point A to point P is

$$T_1 = \frac{\sqrt{a^2 + x^2}}{v_1} \qquad \left(\text{time} = \frac{\text{distance}}{\text{velocity}} \right)$$

and the time required for light to travel from point P to point B is

$$T_2 = \frac{\sqrt{(c - x)^2 + b^2}}{v_2},$$

we seek to minimize the total time

$$T(x) = T_1 + T_2 = \frac{\sqrt{a^2 + x^2}}{v_1} + \frac{\sqrt{(c - x)^2 + b^2}}{v_2}. \tag{9}$$

It is reasonable to assume that x lies within the closed interval $[0, c]$. Since equation (9) for T contains only one independent variable x, we may proceed to calculate $T'(x)$. We obtain

$$T'(x) = \frac{x}{v_1 \sqrt{a^2 + x^2}} - \frac{c - x}{v_2 \sqrt{(c - x)^2 + b^2}}.$$

This derivative is defined for all x, so the only way in which a critical number can arise is from the equation $T'(x) = 0$. This equation gives the condition that

$$\frac{x}{v_1 \sqrt{a^2 + x^2}} = \frac{c - x}{v_2 \sqrt{(c - x)^2 + b^2}}. \tag{10}$$

From Figure 7.11 we can see that

$$\frac{x}{\sqrt{a^2 + x^2}} = \sin \theta_1; \qquad \frac{c - x}{\sqrt{(c - x)^2 + b^2}} = \sin \theta_2.$$

Using these equations, we obtain from equation (10) that

$$\frac{\sin \theta_1}{v_1} = \frac{\sin \theta_2}{v_2},$$

which is Snell's Law. (The verification that this condition actually corresponds to a minimum rather than a maximum is less straightforward here than in the other examples. One reasons on physical grounds that a minimum must exist for $x \in (0, c)$. Since the condition of Snell's Law arises from the only critical number in $(0, c)$, it must correspond to the minimum travel time.) ◇

Exercise Set 4.7

In Exercises 1–22 find the maximum and minimum values for f on the given interval and the corresponding numbers x.

1. $f(x) = x^2(x - 1), \qquad x \in [0, 3]$

2. $f(x) = |x - 2|, \qquad x \in [0, 5]$

3. $f(x) = \dfrac{1}{x(x - 4)}, \qquad x \in [1, 3]$

4. $f(x) = x(x^2 - 2), \qquad x \in [-1, 2]$

5. $f(x) = x^2(x - 3), \qquad x \in [-1, 2]$

6. $f(x) = \sin\left(\dfrac{x}{2}\right), \qquad x \in [0, \pi]$

7. $f(x) = 1 - \tan^2 x, \qquad x \in [-\pi/4, \pi/4]$

8. $f(x) = \sin x \cos x, \qquad x \in [-\pi/2, \pi/2]$

9. $f(x) = \sin x + \cos x, \qquad x \in [0, \pi]$

10. $f(x) = (\sqrt{x} - x)^2, \qquad x \in [0, 4]$

11. $f(x) = x^{2/3} + 2, \qquad x \in [-2, 1]$

12. $f(x) = x + \dfrac{1}{x}, \qquad x \in [1/2, 2]$

13. $f(x) = \left(\dfrac{x + 1}{x - 1}\right)^2, \qquad x \in [-3, 0]$

14. $f(x) = \sec x, \qquad x \in [-\pi/4, \pi/4]$

15. $f(x) = \dfrac{x^2}{1 + x}, \qquad x \in [-1/2, 2]$

16. $f(x) = \dfrac{x^3}{1 + x}, \qquad x \in [0, 2]$

17. $f(x) = x \sin x, \qquad x \in [-\pi/2, \pi/2]$

18. $f(x) = 8x^{1/3} - 2x^{4/3}, \qquad x \in [-1, 8]$

19. $f(x) = \sqrt{x}(1 - x^2), \qquad x \in [0, 4]$

20. $f(x) = \dfrac{x^{2/3}}{2 + \sqrt[3]{x}}, \qquad x \in [-1, 8]$

21. $f(x) = \dfrac{\sqrt{x}}{1 + x}, \qquad x \in [0, 4]$

22. $f(x) = \dfrac{\sqrt{x}}{1 + \sqrt[3]{x}}, \qquad x \in [0, 8]$

23. Find the numbers b and c if the function $f(x) = x^2 + bx + c$ has minimum value $f(3) = -7$ on the interval $[0, 5]$.

24. Let $f(x) = ax^3 - bx$. Find a and b. if $f(2) = 4$ is the maximum value of $f(x)$ on $[0, 4]$.

25. The sum of two nonnegative numbers is 10. Find these numbers if
a. their product is as small as possible;
b. the sum of their squares is as small as possible;
c. the sum of their squares is as large as possible.

26. The sum of two nonnegative numbers is 36. Find these numbers if the first plus the square of the second is
a. a maximum;
b. a minimum.

27. A rectangular play yard is to be constructed along the side of a house by erecting a fence on three sides, using the house wall as the fourth wall of the play yard. Find the dimensions that produce the play yard of maximum area if 20 meters of fence is available for the project.

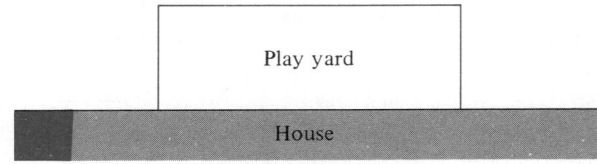

28. A model for the spread of disease assumes that the rate at which a disease spreads is proportional to the product of the number of people infected and the number not infected. Assume the size of the population to be a constant N. When is the disease spreading most rapidly?

29. A farmer has 120 meters of fencing with which he plans to make a rectangular pig pen. The pen is to have one internal fence running parallel to the end fences that divides the pen into two sections. Find the dimensions that produce the pen of maximum area if the length of the larger section is to be twice the length of the smaller section.

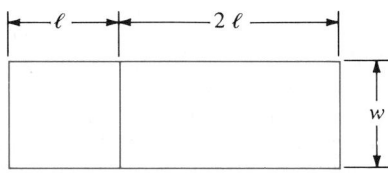

30. Find the minimum and maximum values of the slopes of the lines tangent to the graph

$$y = x^3 - 9x^2 + 7x - 6, \qquad 1 \le x \le 4,$$

and the points where these slopes occur.

31. An open box is to be made from a rectangular sheet of cardboard of dimensions 16 cm by 24 cm by cutting out squares of equal size from each of the four corners and bending up the flaps. Find the dimensions of the box of largest volume which can be made in this way.

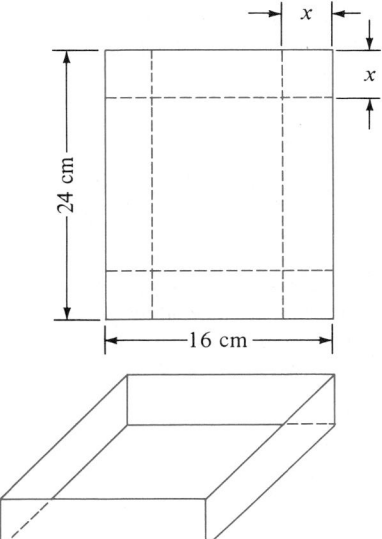

32. A rectangle is inscribed in a right triangle with sides of length 6 cm, 8 cm, and 10 cm, respectively. Find the dimensions of the rectangle of maximum area if two sides of the rectangle lie along two sides of the triangle.

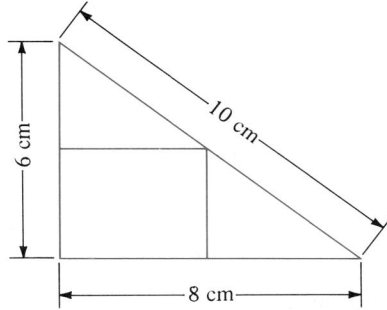

33. A pattern for a rectangular box with a top is to be cut from a sheet of cardboard measuring 10 cm by 16 cm. Find the dimensions of the box for which volume is a maximum.

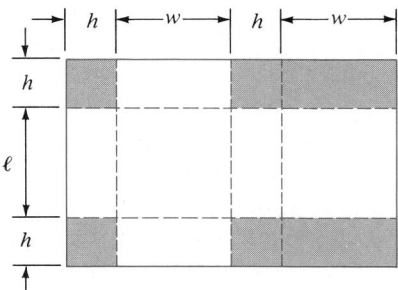

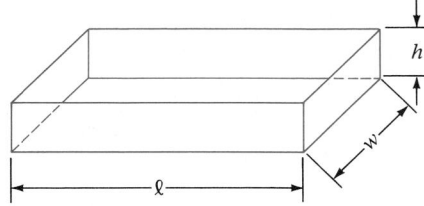

34. A window has the shape of a rectangle surmounted by a semicircle. Find the dimensions that provide maximum area if the perimeter of the window is 10 meters.

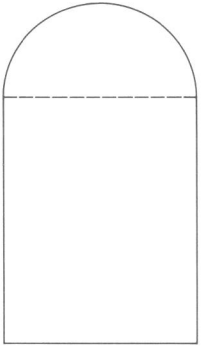

35. A rectangle is inscribed in an isosceles triangle with base 6 cm and height 4 cm. Find the dimensions of the rectangle of maximum area if one side of the rectangle lies along the base of the triangle.

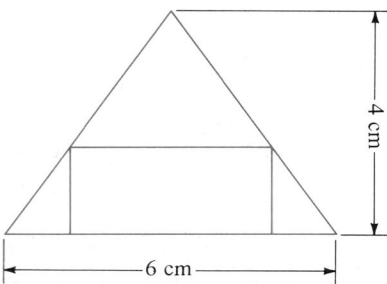

6 cm

36. A right triangle with hypotenuse 10 cm is rotated about one of its legs to sweep out a right circular cone. Of all such triangles, which generates the cone of maximum volume?

37. A rectangle has two vertices on the x-axis and the other two above the x-axis and on the graph of the equation $y = 4 - x^2$. Find the dimensions for which the area of such a rectangle is a maximum.

38. A triangle has two sides of length a and b. The angle at the vertex where these sides meet is θ. Find the value of θ for which the area of the triangle is a maximum.

39. A sector of a circle of radius r and angle θ is to have fixed perimeter P. Find the dimensions r and θ that maximize the area.

40. Find the dimensions of the right triangle with hypotenuse $h = 2$ and maximum area.

41. An orchard presently has 25 trees per acre. The average yield has been calculated to be 495 apples per tree. It is predicted that for each additional tree planted per acre the yield will be reduced by 15 apples per tree. According to this information, how many additional trees per acre should be planted in order to maximize yield?

42. Find the dimensions of the rectangle of largest area that can be inscribed in a semicircle of radius 8 cm.

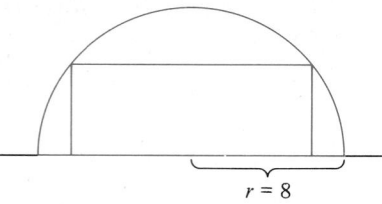

$r = 8$

43. Find the dimensions and the area for the rectangle of maximum area that can be inscribed in a circle of radius 4.

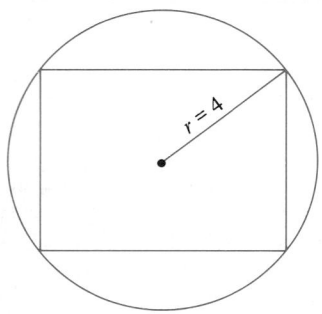

$r = 4$

44. A rectangular beam is to be cut from a round log 20 cm in diameter. If the strength of the beam is proportional to the product of its width and the square of its depth, find the dimensions of the cross section for the beam of maximum strength.

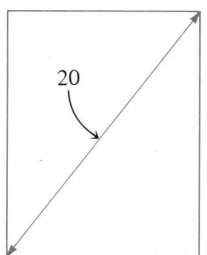

20

45. A format for textbook page layouts is to be chosen so that each printed page has a 4-cm margin at top and bottom and a 2-cm margin on the left and right sides. The rectangular region of printed matter is to have area 800 cm². Find the dimensions for the textbook pages that minimize their areas.

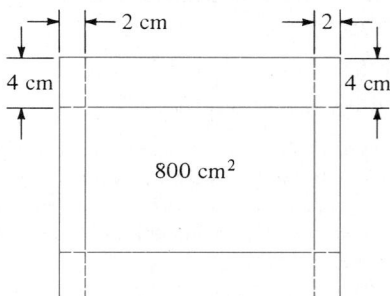

2 cm   2

4 cm   4 cm

800 cm²

46. Find the dimensions of the rectangle of maximum area that can be inscribed in the ellipse $x^2 + 4y^2 = 4$.

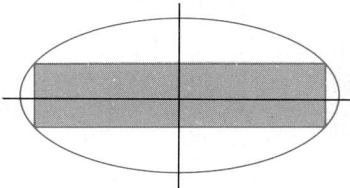

47. A right circular cylinder is inscribed in a right circular cone of radius 3 cm and height 5 cm. Find the dimensions of the cylinder of maximum volume.

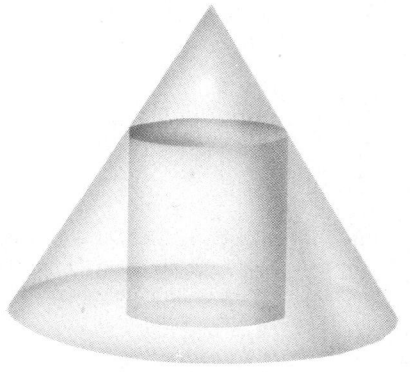

48. A right circular cone is inscribed in a sphere of radius 10 cm. Find the dimensions for which the volume of the cone is a maximum.

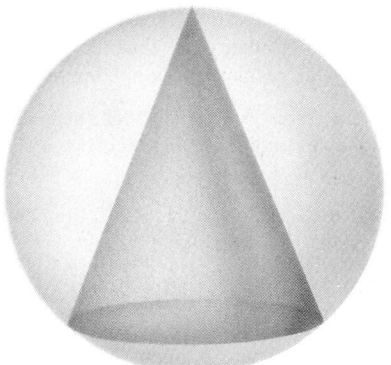

49. For the growth model of Example 7, find the time at which minimum and maximum growth occurs if $\alpha = 3$, $\beta = 2$, $\gamma = 1/3$, and $0 \leq t \leq 9$.

50. Prove that $\sin x \leq x$ for $x \geq 0$. (*Hint:* On the interval $[0, 1]$ find the minimum value of the function $f(x) = x - \sin x$. Then argue the case on $[1, \infty)$ from known properties of $\sin x$.)

51. Find the point on the graph of $y = \sqrt{x}$ nearest the point $(2, 0)$. (*Hint:* Assume x lies in an interval of the form $[0, a]$ where a is large. Then minimize the *square* of the distance.)

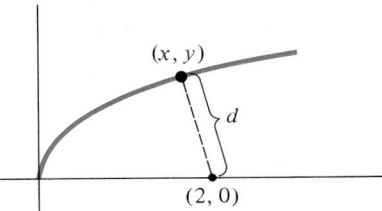

52. Find the point on the ellipse $x^2 + 4y^2 = 4$ nearest the point $(1, 0)$.

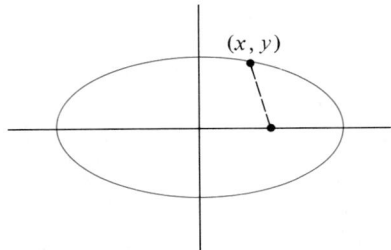

53. A wire 50 cm long is to be cut into two pieces, one of which is to be bent into the shape of a circle and the other of which is to be bent into the shape of a square. Find where the wire should be cut so that the area of the resulting figures is
a. a maximum;
b. a minimum.

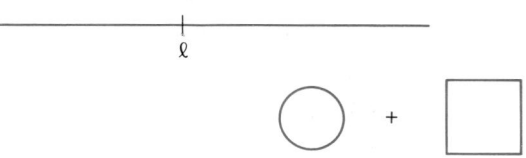

54. United States postal regulations limit the size of a parcel to be mailed parcel post at second class post offices, according to the rule "length plus girth not to exceed 100 inches." (Girth is the largest perimeter perpendicular to the length.) Find the dimensions of the rectangular box of maximum volume that can meet this restriction. (Neglect the thickness of the material. Then attack the problem in two steps: (1) For fixed girth, find the relationship between width and height that maximizes the area of the cross section. (2) Then find the relationship between girth and length that maximizes volume.)

55. Work Problem 53 if one of the two pieces of wire is bent into the shape of an equilateral triangle rather than a circle.

56. The illumination at point Q from a source of light at point P is proportional to the intensity of the light source at P and inversely proportional to the square of the distance from P to Q. Two light sources are 20 meters apart. Find the point on the line joining the two points of minimum illumination if one source is twice as strong as the other.

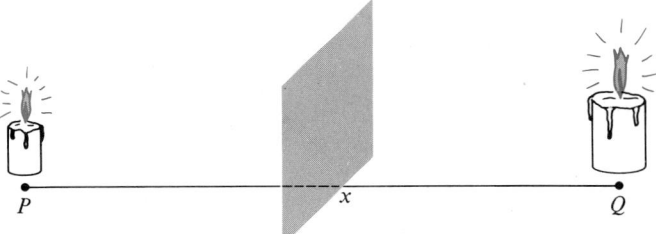

57. Prove the Law of Reflection: A ray of light striking a flat reflecting surface at an angle α will be reflected away at the same angle, i.e., the angle of incidence equals the angle of reflection (see Example 8).

58. A swimmer is in the ocean 100 meters from a straight shoreline. A person in distress is located on the shoreline 300 meters from the point on the shoreline closest to the swimmer. If the swimmer can swim 3 meters per second and run 5 meters per second, what path should the swimmer follow in order to reach the distressed person as quickly as possible?

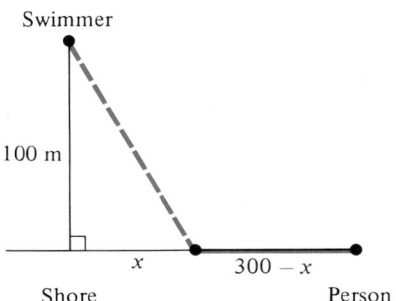

59. An underground telephone cable is to be laid between two boat docks on opposite banks of a straight river. One boathouse is 600 meters downstream from the other. The river is 200 meters wide. If the cost of laying the cable is $50 per meter under water and $30 per meter on land, how should the cable be laid to minimize cost?

60. Observations by plant biologists support the thesis that the survival rate for seedlings in the vicinity of a parent tree is proportional to the product of the density of seeds on the ground and their probability of survival against herbivores. (The density of herbivores tends to decrease as distance from the tree increases since the density of the food supply also decreases.) Let x denote the distance in meters from the trunk of the tree. The results of sampling indicate that the density of seeds on the ground for $0 \leq x \leq 10$ is given by

$$d(x) = \frac{1}{1 + (0.2x)^2}$$

and the probability of survival against herbivores is

$$p(x) = (0.1)x.$$

Find the distance, according to the model proposed above, at which the survival rate is a maximum.

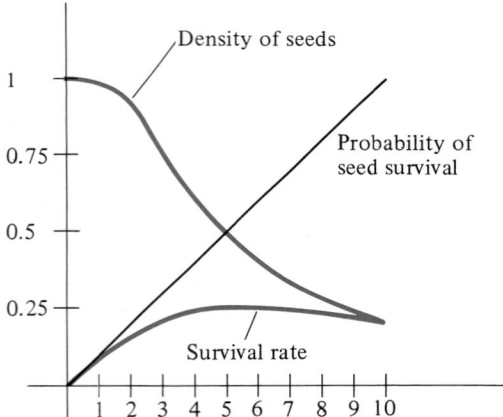

61. In Exercise 60, suppose that the density of seeds for $1 \leq x \leq 9$ is given by the function

$$d(x) = \frac{1}{x}$$

and that the probability of a seed surviving against herbivores is

$$p(x) = \frac{1}{(x - 10)^2}.$$

Find the distance from the parent tree at which seed survival will be a minimum.

62. Use the fact that the lateral surface area of a right circular cone is $S = \pi r \sqrt{r^2 + h^2}$ to find the dimensions of a drink-

ing cup in this shape if the volume of the cup is 36π cm³ and the surface area is a minimum.

63. A rectangle is inscribed inside the ellipse $\dfrac{x^2}{4} + \dfrac{y^2}{9} = 1$ with its sides parallel to the coordinate axes. Find the dimensions for which the area of the rectangle is a maximum.

64. Find the length of the longest rod which can be carried horizontally around a (square) corner from a corridor 4 ft wide into another corridor 4 ft wide.

65. Work Problem 64 if the corridors are each 8 ft high and the rod need not be carried horizontally.

66. Work Problem 64 for a 27 in wide corridor meeting a 64 in wide corridor, and a horizontal rod.

67. Assume that the velocity at which automobiles will travel along a certain section of highway is modeled by the function $v(\rho) = \dfrac{1}{4}(\rho - 2)^2$ where ρ is the density of automobiles (in units of automobiles per kilometer) and $\rho \in [0, 2]$.
   a. Find the density for which velocity will be a maximum.
   b. If the flow rate is defined to be $q = \rho v(\rho)$, find the density at which flow will be a maximum for the velocity function $v(\rho)$ given above.

68. To find the maximum and minimum values of the *periodic* function $f(x) = \sin x$ on the entire number line $(-\infty, \infty)$, we need only find these numbers on $[0, 2\pi]$ (or on *any* interval of length 2π). This is because the graph of $f(x) = \sin x$ on $(-\infty, \infty)$ consists of copies of the graph on the closed interval $[0, 2\pi]$. Generalize this observation to find the maximum and minimum values of the following periodic functions on $(-\infty, \infty)$.
   a. $f(x) = \sin x - \cos x$
   b. $f(x) = 2 \sin(\pi/2 - x)$
   c. $f(x) = 2 \sin x \cos x.$

69. Figure 7.12 shows the graph of $f(x) = |x^2 - 5x + 4|$, which has maximum value $f(0) = 4 = f(5)$ and minimum value $f(1) = f(4) = 0$ for $x \in [0, 5]$. By comparing the graph of f with the graph of $g(x) = x^2 - 5x + 4$ in Figure 7.13, you

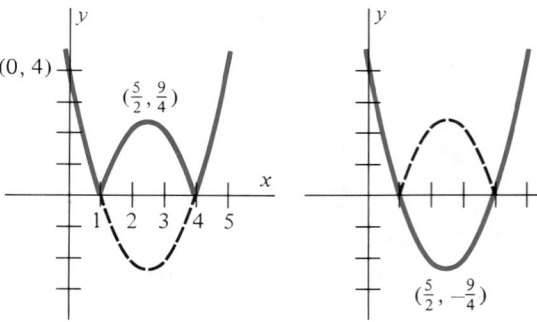

Figure 7.12 Graph of $f(x) = |x^2 - 5x + 4|$.

Figure 7.13 Graph of $g(x) = x^2 - 5x + 4$.

can see that the effect of the absolute value sign is to "turn the portion of the graph extending below the x-axis upward."

a. Show that f may be rewritten as

$$f(x) = \begin{cases} x^2 - 5x + 4, & x \leq 1 \\ -x^2 + 5x - 4, & 1 < x < 4 \\ x^2 - 5x + 4, & 4 \leq x. \end{cases}$$

b. Show that f is differentiable for $x \neq 1$ and $x \neq 4$, and that

$$f'(x) = \begin{cases} 2x - 5, & x < 1 \\ 5 - 2x, & 1 < x < 4 \\ 2x - 5, & 4 < x. \end{cases}$$

c. By examining the difference quotient for f, show that $f'(1)$ and $f'(4)$ do not exist.

d. Generalize your findings to a rule for finding the maximum and minimum values for the function $|h(x)|$, $x \in [a, b]$.

In Exercises 70–74, use the results of Exercise 69 to find the maximum and minimum values for the given function on the given interval and the corresponding numbers x.

70. $f(x) = |x - 3|$, $x \in [-2, 6]$

71. $f(x) = |3x - 5|$, $x \in [0, 4]$

72. $f(x) = |x^2 - x - 6|$, $x \in [-3, 5]$

73. $f(x) = |\cos x|$, $x \in [\pi/4, \pi]$

74. $f(x) = \dfrac{|x|}{4 + x}$, $x \in [-2, 4]$.

4.8 THE DERIVATIVE AS A RATE OF CHANGE

Often one encounters problems in which two or more variables are functions of time. For example, as an ice cube melts, its volume, its weight, and each of its dimensions change continuously as time passes. The objective of this section is to determine how to calculate and compare the rates at which the variables in such a problem are changing.

If f is a function of $t = $ time, then the **average rate of change** in f from time t to time $t + h$ is given by the quotient

$$\frac{f(t + h) - f(t)}{h} = \frac{\text{Change in } f}{\text{Change in time}}.$$

We define the **instantaneous rate of change** of f to be the limiting value of the average rate of change from time t to time $t + h$ as $h \to 0$, that is,

$$\begin{Bmatrix} \text{Rate of change of} \\ f \text{ at time } t \end{Bmatrix} = \lim_{h \to 0} \frac{f(t + h) - f(t)}{h}.$$

If this limit exists, the right side of the equation is the derivative, that is, we have

$$\begin{Bmatrix} \text{Rate of change of} \\ f \text{ at time } t \end{Bmatrix} = f'(t).$$

In other words, *the derivative $f'(t)$ is the (instantaneous) rate of change of the function f at time t.*

Example 1

For the function $f(t) = \sin \pi t$, the average rate of change over the interval $[1/4, 1/2]$ is

$$\frac{\Delta f}{\Delta t} = \frac{\sin(\pi/2) - \sin(\pi/4)}{1/2 - 1/4} = \frac{1 - \sqrt{2}/2}{1/4} = 4 - 2\sqrt{2} \approx 1.17,$$

while the instantaneous rate of change at $t = 1/4$ is

$$f'(\pi/4) = \pi \cos(\pi/4) = \frac{\pi\sqrt{2}}{2} \approx 2.22.$$

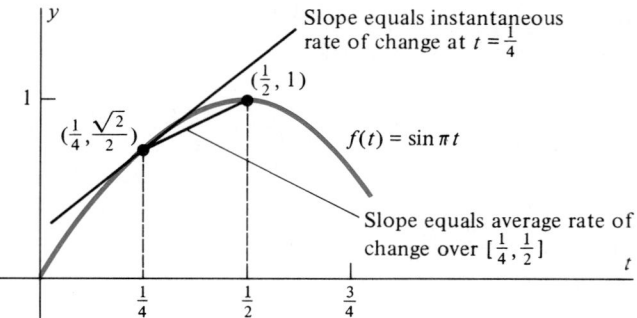

Figure 8.1 Average rate of change versus instantaneous rate of change.

(See Figure 8.1.)

We have already encountered such rates in the position/velocity problem. There $v(t)$ was defined as the rate of change of position, $s(t)$, per unit time. It was shown that $v(t) = s'(t)$ when s was differentiable.

Related Rate Problems

When two or more differentiable functions of time are related by a single equation, we can often obtain the relationship between their respective rates of change by differentiating both sides of that equation with respect to t. In so doing we are making use of the fact that the equation $f = g$ means that f and g *are the same function*, since the equation is assumed true for all values of t.* We may therefore conclude that $f' = g'$, since we have merely differentiated two different forms for the same function. Problems of this general type are referred to as **related rate problems.**

Example 2

The functions f and g are related by the equation

$$f(t) = 3[g(t)]^2 + 10, \qquad -\infty < t < \infty. \tag{1}$$

Find the rate at which the function f is changing at $t = 4$ if $g(4) = 2$ and $g'(4) = -5$.

Strategy

First find the rate $f'(t)$ by differentiating both sides of equation (1). Then substitute given data for t, $g(t)$ and $g'(t)$.

Solution

Using the Chain Rule we obtain

$$f'(t) = 3 \cdot 2g(t) \cdot g'(t)$$
$$= 6g(t)g'(t).$$

Substituting $t = 4$, $g(4) = 2$ and $g'(4) = -5$ then gives the rate

$$f'(4) = 6(2)(-5) = -60.$$

---

*For example, if $ax^2 + b = 3x^2 + 4$ we may conclude, by differentiating both sides, that $2ax = 6x$. But we cannot differentiate both sides of the equation $x^2 = 4$ to conclude that $2x = 0$, since the equation $x^2 = 4$ is true only for the two numbers $x = 2$ and $x = -2$. It is not an equation involving *functions*.

The next example uses this same basic idea. But it involves two geometric variables (volume and radius), each of which must be regarded as functions of time.

Example 3

A balloon in the shape of a sphere is being inflated so that the volume is increasing by 100 cubic centimeters per second. At what rate is the radius increasing when the radius is 9 cm?

Strategy Name all variables.	*Solution* We let V = volume, r = radius, and t = time. We know that
Find an equation relating these variables.	$$V = \frac{4}{3}\pi r^3.$$
Differentiate both sides of the equation, using the Chain Rule where necessary.	We assume that both V and r are differentiable functions of t. Then $$\frac{dV}{dt} = 4\pi r^2 \cdot \frac{dr}{dt},$$ so
Solve for the desired rate, $\dfrac{dr}{dt}$.	$$\frac{dr}{dt} = \frac{1}{4\pi r^2} \cdot \frac{dV}{dt}.$$
Substitute in given values of other rates and variables.	We are given that $\dfrac{dV}{dt} = 100$ cm³/s. When $r = 9$ cm, $$\frac{dr}{dt} = \frac{1}{4\pi(9 \text{ cm})^2} \cdot 100 \text{ cm}^3/\text{s}$$ $$= \frac{25}{81\pi} \approx .098 \text{ cm/s} \qquad \diamond$$

In Example 3, there was no difficulty in determining the basic equation relating the variables. Often, however, related rate problems do not clearly indicate what the basic relationship between the variables might be. We therefore suggest the following *procedure for solving related rate problems:*

(i) If the problem is geometric in nature (or can be interpreted geometrically), draw a sketch.

(ii) Label the important quantities as variables or constants, according to the statement of the problem.

(iii) From the sketch, together with known relationships among the variables (either given in the problem or known from geometry or trigonometry), write down an equation relating the relevant variables.

(iv) Differentiate both sides of the equation, using the Chain Rule where necessary, to obtain an equation relating the rates.

(v) Solve for the desired rate. After doing so, substitute given data into this expression to obtain the solution.

Example 4

A physics student is standing 30 meters from a straight section of railroad track in order to perform an experiment on the Doppler effect. A train is approaching, moving along the track at 90 kilometers per hour. How fast is the distance between the train and the student decreasing when the train is 50 meters from the student?

Strategy

Draw a sketch.

Solution

The situation is sketched in Figure 8.2. We let T denote the location of the train, S the location of the student, and P the point on the tracks nearest the student. Also, we let

$$x = \text{distance from } T \text{ to } P, \quad \text{and}$$
$$y = \text{distance from } T \text{ to } S.$$

The problem is, therefore, to find $\dfrac{dy}{dt}$ when $y = 50$ m.

Use Pythagorean Theorem to find an equation in x and y.

Since T, P, and S are vertices of a right triangle, we have the equation

$$y^2 = x^2 + 30^2. \tag{2}$$

Differentiate both sides of (2).

Assuming both x and y to be differentiable functions of t, we differentiate both sides of (2) to obtain

$$2y \cdot \frac{dy}{dt} = 2x \cdot \frac{dx}{dt},$$

so

Solve for $\dfrac{dy}{dt}$.

$$\frac{dy}{dt} = \frac{x}{y} \cdot \frac{dx}{dt}. \tag{3}$$

Find x when $y = 50$.

Now when $y = 50$ m, we obtain from equation (2)

$$x = \sqrt{50^2 - 30^2} = \sqrt{40^2} = 40 \text{ m}.$$

Insert given values for all variables and rates.

We are given that $\dfrac{dx}{dt} = -90$ km/h, so when $y = 50$ m we have from equation (3)

$$\frac{dy}{dt} = \left(\frac{40 \text{ m}}{50 \text{ m}}\right)(-90 \text{ km/h})$$

$$= -72 \text{ km/h}. \qquad \diamondsuit$$

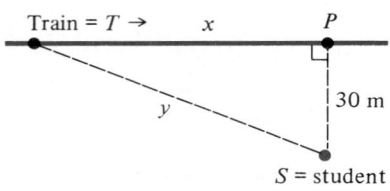

Train = $T \rightarrow$ $\quad x \quad$ P

30 m

y

S = student

Figure 8.2 Sketch for Doppler effect experiment.

Example 5

A coffee maker has the shape of a double cone 20 cm high. The radii at both the top and the base are 4 cm. Coffee is flowing from the top section into the bottom section at a rate of 4 cm³/s. At what rate is the level of coffee in the top section falling when the coffee in the top section is 4 cm deep?

Strategy

Draw a sketch.

Label variables.

Solution

The coffee maker is sketched in Figure 8.3. An idealized sketch of the top portion appears in Figure 8.4.

We let h denote the depth of the coffee in the top section and we let r denote the radius of its surface.

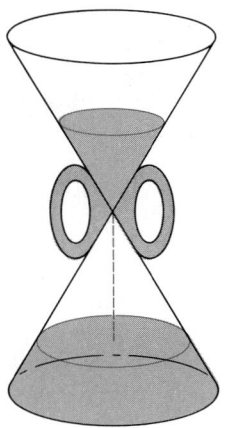

Figure 8.3 Coffee maker in shape of double cone.

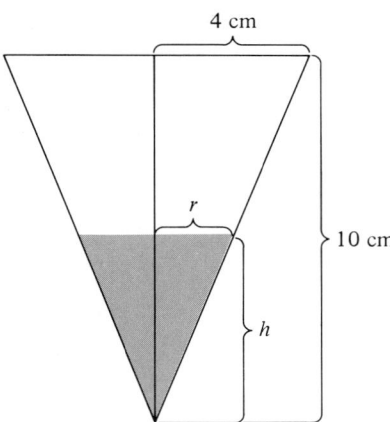

Figure 8.4 Top section of coffee maker.

Use equation for volume of a cone.

The formula for the volume of coffee in the upper cone is therefore

$$V = \frac{1}{3}\pi r^2 h. \tag{4}$$

No data given on $\dfrac{dr}{dt}$, so r must be eliminated from (4) before differentiation.

Now if we proceed directly to differentiate both sides of equation (4) with respect to t, we will obtain an equation involving dV/dt, dr/dt, and dh/dt, since each of the variables, V, h, and r is a function of time. However, we have no given information on $\dfrac{dr}{dt}$. We therefore attempt to find an equation expressing r in terms of V and/or h so that r can be eliminated from equation (4). To do so we observe from Figure 8.4 that, by similar triangles,

Use similar triangles to find equation relating r and h.

$$\frac{r}{h} = \frac{4}{10},$$

so

$$r = \frac{2}{5}h \tag{5}$$

Insert expression for r in (4).

is the desired equation. Substituting this expression for r in (4) gives

$$V = \frac{1}{3}\pi\left(\frac{2}{5}h\right)^2 h$$

$$= \frac{4\pi}{75}h^3. \tag{6}$$

Differentiating both sides of this equation with respect to t, we obtain

Differentiate both sides.

$$\frac{dV}{dt} = \frac{4\pi}{25}h^2 \cdot \frac{dh}{dt},$$

so

Solve for $\dfrac{dh}{dt}$.

$$\frac{dh}{dt} = \frac{25}{4\pi h^2} \cdot \frac{dV}{dt}.$$

Insert given information.

(Note that the sign of $\dfrac{dV}{dt}$ is negative.)

Substituting the given information $dV/dt = -4$ cm³/s and $h = 4$ cm then gives

$$\frac{dh}{dt} = \frac{25}{4\pi(4 \text{ cm})^2}(-4 \text{ cm}^3/\text{s})$$

$$= -\frac{25}{16\pi} \approx -0.497 \text{ cm/s.} \qquad \diamond$$

REMARK 1: Equation (5) is an example of an **auxiliary** equation. Its purpose is to allow one of the variables in the principal equation (4) to be eliminated. The strategy in the above example was to use the auxiliary equation to eliminate the variable r since no information about r was given. In general, you should look for auxiliary equations in problems of this type when you encounter more variables than can be evaluated with the given data.

REMARK 2: In the solution to Example 5, it was important that the two sides of equation (6) were differentiated *before* the particular values for the variables were substituted. Otherwise the desired rates would not have appeared. A common error in problems of this kind is to substitute the given values of the variables into the principal equation before differentiating. Be very careful not to make this mistake.

Example 6

An airplane is flying a level course due east at a speed of 3 kilometers per minute. A second airplane is flying a level course due south at a speed of 2 kilometers per minute at an altitude 4 kilometers below that of the first plane. At one instant the second airplane is directly beneath the first. At what rate is the distance between the two airplanes increasing one minute later?

Solution: The situation is sketched in Figure 8.5. In Figure 8.6, points P_1 and P_2 represent the first and second airplanes, respectively, and O_1 and O_2 are the locations of the planes when the first is directly above the second. We name variables as follows:

$$x = \text{distance between } O_1 \text{ and } P_1$$
$$y = \text{distance between } O_2 \text{ and } P_2$$
$$w = \text{distance between } O_2 \text{ and } P_1$$
$$s = \text{distance between } P_1 \text{ and } P_2.$$

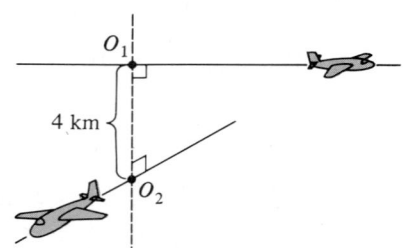

Figure 8.5 Two airplanes.

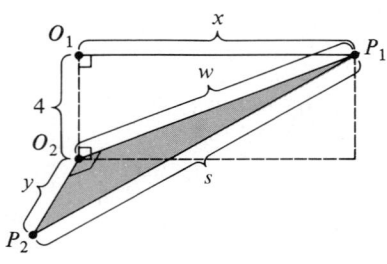

Figure 8.6 P_1 and P_2 denote the two airplanes.

The rate to be calculated is $\dfrac{ds}{dt}$. Since s is the length of the hypotenuse of the right triangle $P_1O_2P_2$, we have the equation

$$s^2 = w^2 + y^2. \qquad (7)$$

We are given information about $\dfrac{dy}{dt}$, the speed of the second airplane, but we do not have direct information about the variable w. We must therefore find an auxiliary equation relating w to one or more of the other variables. From Figure 8.6 we can see that w is the length of the hypotenuse of triangle $P_1O_1O_2$, so we obtain the auxiliary equation

$$w^2 = x^2 + 4^2.$$

Substituting this expression for w^2 into the principal equation (7), we obtain the equation

$$s^2 = x^2 + y^2 + 16.$$

We now differentiate implicitly with respect to t. This gives

$$2s \cdot \frac{ds}{dt} = 2x \cdot \frac{dx}{dt} + 2y \cdot \frac{dy}{dt} \qquad \text{so}$$

$$\frac{ds}{dt} = \frac{1}{s}\left(x \cdot \frac{dx}{dt} + y \cdot \frac{dy}{dt}\right). \qquad (8)$$

We are given

$$\frac{dx}{dt} = 3 \ \frac{\text{km}}{\text{min}} \qquad \text{and} \qquad \frac{dy}{dt} = 2 \ \frac{\text{km}}{\text{min}}.$$

Therefore, one minute after crossing, we will have $x = 3$ km and $y = 2$ km and $s = \sqrt{3^2 + 2^2 + 16} = \sqrt{29}$ km. Substituting these values into equation (8) gives

$$\frac{ds}{dt} = \left(\frac{1}{\sqrt{29} \text{ km}}\right)[(3 \text{ km})(3 \text{ km/min}) + (2 \text{ km})(2 \text{ km/min})]$$

$$= \frac{13}{\sqrt{29}} \approx 2.414 \text{ km/min}. \qquad \diamond$$

Exercise Set 4.8

In each of Exercises 1–8, find the average rate of change of the given function for the given time interval. Then find the instantaneous rate of change per unit time. Note any values of t where the instantaneous rate is undefined.

1. $s(t) = (1 + t)^2, \qquad t \in [2, 3]$

2. $s(t) = \dfrac{1}{1 - t}, \qquad t \in [2, 2.5]$

3. $f(t) = \sqrt{t}(1 - t^3), \qquad t \in [0, 1]$

4. $g(t) = \sin(\pi - t^2), \qquad t \in [-\pi/2, \pi/2]$

5. $r(t) = t^{5/2} - 2t^{3/2}, \qquad t \in [1, 4]$

6. $w(t) = \dfrac{\sqrt{t}}{1 + t^2}, \qquad t \in [0, 5]$

7. $k(t) = \tan(\pi t), \qquad t \in \left[\dfrac{1}{6}, \dfrac{1}{3}\right]$

8. $v(t) = (t^{-2} + 2)^3, \qquad t \in [1, 2]$

In Exercises 9–13 find the rate $f'(t)$ for the given values of g, g' and t.

9. $f(t) = 2[g(t)]^3 + 5, \qquad t = 1, \quad g(1) = 3, \quad g'(1) = -2$

10. $f(t) = \sqrt{2 + g(t)}, \qquad t = 0, \quad g(0) = 3, \quad g'(0) = 4$

11. $f(t) = \dfrac{1}{1 + g(t)}, \qquad t = 2, \quad g(2) = 3, \quad g'(2) = -2$

12. $[f(t)]^2 + [g(t)]^3 = 5, \qquad t = 1, \quad g(1) = 6, \quad g'(1) = -2, \quad f(1) = 7$

13. $\sin(f(t)) = [g(t)]^2, \qquad t = 0, \quad g(0) = 1, \quad g'(0) = -2, \quad f(0) = 0$

14. A spherical balloon is being inflated so that the radius is increasing at a rate of 3 cm/s. Find the rate at which the volume is increasing when $r = 10$ cm.

15. At what rate is the diagonal of a cube increasing if the edges are increasing at a rate of 2 cm/s?

16. When a pebble is tossed into a still pond, ripples move out from the point where the stone hits in the form of concentric circles. Find the rate at which the area of the disturbed water is increasing when the radius of the outermost circle equals 10 meters if this radius is increasing at a rate of 2 meters per second.

17. A snowball, in the shape of a sphere, is melting so that the radius is decreasing at a uniform rate of 1 centimeter per second. How fast is the volume decreasing when the radius equals 6 centimeters?

18. A point moves along the graph of $y = x^{5/2}$ so that its x-coordinate increases at the constant rate of 2 units per second. Find the rate at which its y-coordinate is increasing as it passes the point (4, 32).

19. A conical water tank with vertex down has a radius of 20 meters and a depth of 20 meters. Water is being pumped into the tank at the rate of 40 cubic meters per minute. How fast is the level of the water rising when the water is 8 meters deep?

20. A radio transmitter is located 3 kilometers from a fairly straight section of interstate highway. A truck is travelling away from the transmitter along the highway at a speed of 80 kilometers per hour. How fast is the distance between the truck and the transmitter increasing when they are 5 kilometers apart?

21. The area of a rectangle, whose length is twice its width, is increasing at the rate of 8 cm²/s. Find the rate at which the length is increasing when the width is 5 cm.

22. A ladder 5 meters long is leaning against a wall. The base of the ladder is sliding away from the wall at a rate of 1 meter per second. How fast is the top of the ladder sliding down the wall at the instant when the base is 3 meters from the wall?

23. A boat is being pulled to shore by a rope attached to a windlass atop a pier. The height of the windlass above the water is 6 meters, and the rope is being wound in at the rate of 5 meters per minute. How fast is the boat approaching the shore when it is 8 meters away?

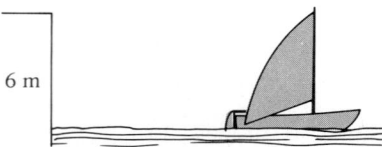

6 m

24. The law of cosines, relating the lengths of the three sides of a triangle, is

$$C^2 = A^2 + B^2 - 2AB \cos \theta$$

where θ is the angle opposite side C. Sides $A = 1$ and $B = 2$ are of fixed length, and the angle θ is increasing at the rate of 0.2 radians/minute. Find the rate at which C is increasing at the instant when $C = \sqrt{3}$.

25. A woman 1.6 meters tall is walking away from a lamp post 10 meters tall. If the woman is walking at a speed of 1.2 meters per second, how fast is her shadow increasing when she is 15 meters from the lamp post?

26. The lengths of the two equal sides in an isosceles triangle are increasing at the rate of 2 cm/s. The base of the triangle has fixed length 18 cm. Find the rate at which the area of the triangle is increasing when its height equals 12 cm.

27. Following the outbreak of an epidemic, a population of $N(t)$ people can be regarded as being made up of immunes, $I(t)$, and susceptibles, $S(t)$, that is,

$$N(t) = I(t) + S(t).$$

$I(t)$ includes both those who have contracted the disease and those who cannot contract the disease. If the rate of decrease

of susceptibles is 20 persons per day, and the rate of increase of immunes is 24 persons per day, how fast is the population growing?

28. Gravel is being poured onto a pile at the rate of 180 cubic meters per minute. The pile is in the shape of an inverted cone whose diameter is always three times its height. Find the rate at which the diameter of the base is increasing when the pile is 6 meters high.

29. At noon a truck leaves Columbus, Ohio, driving east at a speed of 40 kilometers per hour. An hour later a second truck leaves Columbus driving north at a speed of 60 kilometers per hour. At what rate is the distance between the two trucks increasing at 2:00 p.m. that day?

30. A water trough has end pieces in the shape of inverted isosceles triangles with bases 60 cm and heights 40 cm. The trough is 4 meters long. Water is being pumped into the trough at a rate of 9000 cm³ per second. How fast is the level of the water in the trough rising when the water is 10 cm deep?

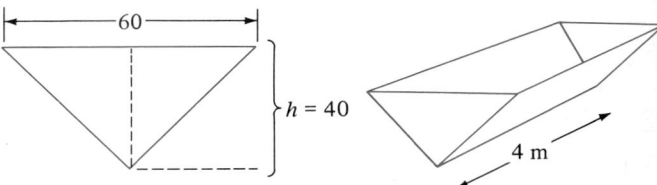

31. Work Exercise 30 if the ends of the trough have the shape of equilateral triangles whose sides have length 60 cm.

32. An observer stands 150 meters from a fireworks display rocket which is fired directly upward. When the rocket reaches a height of 200 meters, it is travelling at a speed of 12 meters per second. At what rate is the angle of elevation formed with the observer increasing at that instant?

33. A point moves along the unit circle at a constant speed. Find the points at which the x- and y-coordinates are
   a. changing at the same rates,
   b. changing at opposite rates.

34. For a body moving through air, **drag** is defined as the force opposing the motion of the body. For such a body, the drag D is jointly proportional to the squares of its velocity V and its surface area S, that is,

$$D = kV^2 S^2.$$

Find the rate at which drag is increasing if the body is undergoing an acceleration of 8 m/s² at the instant when the velocity equals 30 m/s.

35. On a particular autumn day, the sun moves across the sky at the rate of 15° per hour. How fast is the shadow cast by a building 30 meters high increasing at the moment when the evening sun is 30° above the horizon?

36. For a gas at temperature T confined within a container of volume V at a pressure P, the gas law states that the ratio $\dfrac{PV}{T}$ is a constant, that is, $PV = kT$. A 1000 cm³ tank of oxygen is being heated so that the temperature of the oxygen is increasing at a rate of 2°C per minute. Find the rate at which the pressure inside the tank is increasing.

37. How fast is the level of the coffee in the bottom of the coffee pot in Example 5 rising at the instant when the coffee in the bottom of the pot is 4 cm deep?

38. The gravitational force of attraction between two bodies of mass m_1 and m_2 is given by Newton's Law of Gravitation as

$$F = G\frac{m_1 m_2}{r^2}$$

where G is the gravitational constant and r is the distance between the two bodies. Find the rate at which the distance between the bodies is changing if the force is changing at the rate α.

39. A lighthouse lies 200 meters from a straight shore. The light rotates at a rate of 2 revolutions per minute. A piling marks the spot on the beach nearest the lighthouse. Find the speed at which the light beam is moving along the shore at a point 200 meters from the piling.

40. An *idealized heat engine*, designed to provide a model for studying the question of maximum possible efficiency, was proposed by the French engineer Sadi Carnot in 1824. For the Carnot engine the efficiency is defined by the equation

$$E = 1 - \frac{T_c}{T_h}$$

where T_h is the intake temperature and T_c is the exhaust temperature in degrees Kelvin.
   a. If T_c is increasing by 3 K per minute and the efficiency is not changing, what is the rate of change of T_h?
   b. If the efficiency is decreasing by 2% per hour and T_c is fixed, what is the rate of change of T_h?

41. If a ray of monochromatic light travelling in a vacuum makes an angle of incidence α with the normal to a surface of a substance a and an angle of refraction β in the substance, then **Snell's Law** states that

$$\frac{\sin \alpha}{\sin \beta} = C_a$$

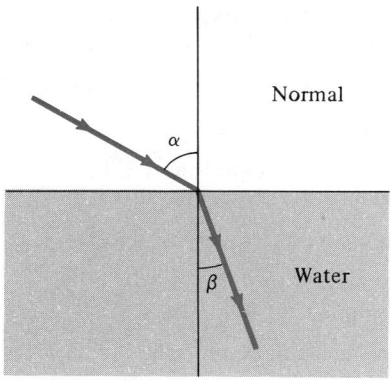

where C_a is a constant, called the index of refraction for the substance a. For water, the index of refraction is $C_w = 1.33$. If the angle of incidence of a light ray striking water is decreasing at the rate of 0.2 radians per second, find an expression for the rate at which the angle of refraction is decreasing.

42. A submarine running due north at a depth of 1 kilometer and a speed of 60 kilometers per hour passes directly beneath a ship sailing due west at 30 kilometers per hour. At what rate is the distance between the two vessels increasing 20 minutes later?

43. The velocity of viscous fluid flowing through a circular tube is not the same at all points of a cross section. Provided the velocity is not too great, the flow has maximum velocity at the center and decreases to zero at the walls. For a point at a radial distance r from the center of the tube, the velocity of the flow is given by

$$v = \frac{\alpha}{L}(R^2 - r^2)$$

where R is the radius of the tube, L is its length, and α is a constant. Find the acceleration of the fluid moving at the center of a tube if
   a. $L = 25$ cm is fixed and R is increasing at a rate of 0.2 cm/min at the instant when $R = 10$ cm.
   b. $R = 10$ cm is fixed and L is increasing at a rate of 0.5 cm/sec at the instant when $L = 25$ cm.

SUMMARY OUTLINE OF CHAPTER 4

◆ *Theorem* (**Mean Value Theorem**): If f is continuous on $[a, b]$ and differentiable on (a, b), there exists $c \in (a, b)$ (page 154) with

$$f'(c) = \frac{f(b) - f(a)}{b - a}$$

◆ The continuous function f is **increasing** on an interval I if $f'(x) > 0$ for all $x \in I$, and **decreasing** on I if $f'(x) < 0$ for all $x \in I$ except possibly at endpoints. (page 158)

◆ The number $f(c)$ is a relative **maximum** if $f(c) \geq f(x)$ for all x near c. The number $f(c)$ is a relative **minimum** if (page 168)
$f(c) \leq f(x)$ for all x near c. Relative maxima and minima are referred to as **relative extrema.**

◆ **Theorem:** If $f(c)$ is a relative extremum, then either $f'(c) = 0$ or $f'(c)$ fails to exist. (page 169)

◆ **Theorem:** The graph of $y = f(x)$ is (page 178)

(a) **concave up** on I if $f''(x) > 0$ for all $x \in I$,
(b) **concave down** on I if $f''(x) < 0$ for all $x \in I$.

◆ **Theorem:** If $f'(c) = 0$ and $f''(c)$ exists, then $f(c)$ is a (page 182)

(a) relative maximum if $f''(c) < 0$,
(b) relative minimum if $f''(c) > 0$.

◆ To find the maximum and minimum values of a differentiable function f on the interval $[a, b]$ we (page 201)

(i) find the critical numbers by finding all x for which $f'(x) = 0$ or $f'(x)$ is undefined, and
(ii) inspect $f(x)$ for all critical numbers and both endpoints.

The largest of the values observed in (ii) is the **maximum value of $f(x)$** on $[a, b]$, the smallest is the **minimum.**

◆ If f is a differentiable function of time, the (instantaneous) rate of change of f per unit time is the derivative $f'(t)$. (page 217)

◆ If two differentiable functions f and g are related by the equation $f = g$, we may differentiate both sides of the (page 218)
equation to obtain the equation $f' = g'$ relating their rates of change.

REVIEW EXERCISES—CHAPTER 4

In Exercises 1–20, find the largest intervals on which f is increasing and decreasing, find all relative extrema and points of inflection, determine the concavity, and sketch the graph.

1. $y = 4x - x^2$

2. $y = x(x - 1)(x + 3)$

3. $f(x) = x^2 - 2x + 3$

4. $y = \sin 4x$

5. $f(x) = 2 \sin(\pi x)$

6. $y = \dfrac{1 - x}{x}$

7. $f(t) = 2 \cos^2(2t)$

8. $f(x) = \sqrt{1 - \sin^2 x}$

9. $y = \dfrac{x - 3}{x + 3}$

10. $f(t) = \dfrac{t^2}{t^2 - 1}$

11. $f(x) = x\sqrt{16 - x^2}$

12. $y = \dfrac{\sqrt{x}}{1 + \sqrt{x}}$

13. $y = \dfrac{t^2}{t^2 + 9}$

14. $y = \dfrac{2}{\sqrt{x}} + \dfrac{\sqrt{x}}{2}$

15. $f(x) = x^4 - 2x^2$

16. $y = (x + 1)(x - 1)^2$

17. $f(x) = x + \cos x$

18. $y = t^3 - 3t^2 + 2$

19. $f(x) = x^2 + \dfrac{2}{x}$

20. $y = \tan x + \cot x$

In Exercises 21–36, find the maximum and minimum values of the function on the given interval and the corresponding numbers x.

21. $f(x) = \dfrac{1}{x^2 - 4}$, $x \in [-1, 1]$

22. $f(x) = \sin x + \cos x$, $x \in [-\pi, \pi]$

23. $y = x^3 - 3x^2 + 1$, $x \in [-1, 1]$

24. $f(x) = x\sqrt{1 - x}$, $x \in [-3, 0]$

25. $f(x) = x + x^{2/3}$, $x \in [-1, 1]$

26. $y = x - 2|x|$, $x \in [-3, 2]$

27. $y = x - \sqrt{1 - x^2}$, $x \in [-1, 1]$

28. $f(x) = \sqrt[3]{x} - x$, $x \in [-1, 1]$

29. $f(x) = x - 4x^{-2}$, $x \in [-3, -1]$

30. $y = \dfrac{x}{1 + x^2}$, $x \in [-2, 2]$

31. $y = x^4 - 2x^2$, $x \in [-2, 2]$

32. $f(x) = |x + \sin x|$, $x \in [-\pi/2, \pi/2]$

33. $f(x) = \dfrac{2 + \cos x}{2 - \cos x}$, $x \in [-\pi, \pi]$

34. $y = \dfrac{x - 1}{x^2 + 3}$, $x \in [-2, 4]$

35. $y = \dfrac{x + 2}{\sqrt{x^2 + 1}}$, $x \in [0, 2]$

36. $f(x) = |x^2 - 1|$, $x \in [-3, 3]$

37. The sum of two numbers is 10. If one is increasing, what can you say about the other?

38. The sum of two positive numbers is 10. For what value(s) will the difference of their squares be a maximum?

39. Sketch a possible graph for a function that is continuous for all x and for which $f(2) = 4$, $f'(2) = 1$, and $f''(x) < 0$ for all x.

40. Sketch a possible graph for a function that is continuous for all x and for which $f(-2) = -f(2) = -3$, $f'(-2) = f'(2) = 0$, $f''(-2) > 0$, and $f''(2) < 0$.

41. A fence 3 meters tall is parallel to the wall of a building and 1 meter from the building. What is the length of the shortest ladder which can extend from the ground over the fence to the building wall?

42. Use the Intermediate Value Theorem and Rolle's Theorem to prove that if f has a continuous derivative for $x \in [a, b]$, if $f(a)$ and $f(b)$ have opposite signs, and if $f'(x) \neq 0$ for all $x \in [a, b]$, then the equation $f(x) = 0$ has *precisely* one root in $[a, b]$.

43. Prove that for any quadratic function $f(x) = ax^2 + bx + c$ and any interval $[\alpha, \beta]$, the midpoint $\gamma = \frac{1}{2}(\alpha + \beta)$ satisfies the conclusion of the Mean Value Theorem.

44. Sketch examples of graphs of functions defined on an interval $[a, b]$ for which the conclusion of the Mean Value Theorem is satisfied:
  a. at precisely two points;
  b. at precisely three points.

45. Show that the function $f(t) = t^3 - 6t^2 + 14t + 5$ is increasing for all t. (*Hint:* Complete the square on $f'(t)$.)

46. For $f(x) = x^2 + bx + c$ find a condition involving b and c which will ensure that $f(x) > 0$ for all x. (*Hint:* Ensure that the minimum value is greater than zero.)

47. The function $f(x) = \dfrac{x}{1 + ax^2}$ has minimum values at $x = \pm 3$ for $x \in [-5, 5]$. Find a.

48. Find the number most exceeded by its square root.

49. The sides of a regular hexagon are increasing at a rate of 2 cm/s. At what rate is the area increasing when the length of a side is 10 cm?

50. Prove that the function $f(x) = x^n + ax + b$ cannot have more than 2 zeros when n is even. What if n is odd?

51. A peach orchard has 25 trees per acre, and the average yield is 300 peaches per tree. For each additional tree planted per acre, the average yield per tree will be reduced by approximately 10 peaches. How many trees per acre will give the largest peach crop?

52. The volume of a sphere is increasing at a rate of 500 cm³ per second at the instant when the radius is 25 cm. At what rate is the radius increasing?

53. A swimming pool is to be constructed in the shape of a sector of a circle with radius r and central angle θ. Find r and θ if the area of the water surface is the constant S and the perimeter is to be a minimum.

54. A volume V of oil is spilled at sea. The spill takes the shape of a disc whose radius is increasing at a rate of $a/\sqrt{t}$ m/s where a is constant. Find the rate at which the thickness of the layer of oil is decreasing after 9 seconds, if the radius at that time is $r(9) = 6a$.

55. A ladder 10 meters high leans against a wall. The bottom of the ladder is sliding away from the wall at a rate of 1 meter per second. At what rate is the area of the triangle formed by the ladder, wall, and ground changing when the base of the ladder is 6 meters from the wall?

56. A pebble dropped in still water sends out ripples in the shape of concentric circles. If the radius of the outer ripple increases at a rate of 2 meters per second, how fast is the area of the disturbance increasing after 5 seconds?

57. Find the rectangle of maximum area that can be inscribed within the equilateral triangle of side 10.

58. Find the points on the graph of the hyperbola $x^2 - x + \dfrac{5}{4} - y^2 = 0$ nearest the origin.

59. Find the dimensions of the rectangle of maximum area that can be inscribed in the ellipse $x^2 + \dfrac{y^2}{4} = 1$.

60. A truck traveling 80 kilometers per hour is heading north on highway X which crosses east-west highway Y at point P. A car traveling east on highway Y at 60 kilometers per hour passed point P when the truck was 20 kilometers south of point P. At what rate is the distance between the truck and the car increasing when the truck passes point P?

61. A spherical snowball melts at a rate proportional to its surface area. Show that its radius decreases at a constant rate.

62. The strength of a wooden beam cut from a log is proportional to the product of its width and the square of its depth. Find the ratio of depth to width of the strongest beam that can be cut from a circular log.

63. A student 1.6 meters tall walks directly away from a street light 8 meters above the ground at a rate of 1.2 meters per second. Find the rate at which the student's shadow is increasing when the student is 20 meters from the point directly beneath the light.

64. A propane storage tank is to have the shape of a cylinder capped at both ends by hemispheres. If the hemispherical caps are three times as expensive to construct, per square unit of surface area, as the cylindrical walls, find the dimensions that minimize construction costs for a given volume.

65. What is the maximum possible area of an isosceles triangle that can be inscribed in a circle of radius r?

66. Mathematical statisticians often "estimate" the true value x of a parameter in a large population by extracting a sample from the population and calculating the value $\bar{x}$ of the parameter in that sample. For example, we could estimate the average age of all citizens of Chicago by calculating the average age for a sample of 100 citizens, assuming the sample to be representative of the population. Given sample values $x_1, x_2, \ldots, x_n$ of the parameter of interest, the estimate $\bar{x}$ which statisticians prefer to use is the one which minimizes the sum of the squares of the errors $(\bar{x} - x_1)^2 + (\bar{x} - x_2)^2 + \cdots + (\bar{x} - x_n)^2$. Show that the number $\bar{x}$ which satisfies this criterion is the simple average

$$\bar{x} = \frac{x_1 + x_2 + \cdots + x_n}{n}$$

67. When a set of data $(x_1, y_1), (x_2, y_2), \ldots, (x_n, y_n)$ appears to be related by an equation of the form $y = mx$, the statistician defines the "best" straight line modeling this relationship as the line $y = mx$ that minimizes the sum of the

squares of the errors

$$(mx_1 - y_1)^2 + (mx_2 - y_2)^2 + \cdots + (mx_n - y_n)^2.$$

Show that the value of m producing this "best" straight line is

$$m = \frac{x_1 y_1 + x_2 y_2 + \cdots + x_n y_n}{x_1^2 + x_2^2 + \cdots + x_n^2}.$$

68. A rectangular box is to be constructed so as to hold 1024 cm³. Its base is to have length twice its width. Material for the top costs 12¢ per square centimeter, and material for the sides and bottom cost 6¢ per square centimeter. Find the dimensions that will minimize cost.

69. For $f(x) = x^2 - 2x - 15$:
   a. Find a closed interval on which Rolle's Theorem applies.
   b. Rolle's Theorem guarantees the existence of a number c in the interval found in part a. Find c.

70. When the Mean Value Theorem is applied to the function $f(x) = x^3 + qx^2 + 5x - 6$ on the interval $[0, 2]$, the number c determined by the theorem is $c = 2$. What is q?

Chapter 5
Antidifferentiation

This brief chapter is focused on a single question: Given a function f, what can we say about a function F for which $F' = f$? Does such a function F exist? If so, what are its properties? Is it unique? What can we say about its graph based on what we know about f?

This is the question of finding an *antiderivative* for a given function f. Because much of the answer has to do with simply reversing the process of differentiation, this topic is properly viewed as part of our discussion of the derivative. Yet as we begin our discussion of integration (Chapter 6), the second major topic of the calculus, we shall see that the antiderivative plays a central role.

It is because of its dual relationship to both differentiation and integration that we have distinguished the topic of antidifferentiation with its own brief chapter. Regardless of whether it is presented as a further application of the derivative or as the first part of the discussion on integration, the antiderivative emerges in Chapter 6, in the statement of the Fundamental Theorem of Calculus, as the seminal notion linking these two principal concepts.

5.1 ANTIDERIVATIVES

In Chapter 3 we developed a number of rules for finding derivatives. We now wish to discuss the "inverse" process for differentiation. That is, given a function f, how do we use what we know about differentiation to find a function F *whose derivative is the given function f*? This is referred to as *antidifferentiation* and the function F is called an antiderivative for f.

DEFINITION 1	Let the function f be defined on an interval I. A function F is called an **antiderivative** for f on the interval I if F is differentiable on I and

$$F'(x) = f(x) \qquad \text{for all} \qquad x \in I. \tag{1}$$

Obviously, antidifferentiation requires the ability to "think backwards through" (that is, invert) the process of differentiation. Here are several straightforward examples of antiderivatives.

Example 1

The function $F(x) = x^2$ is an antiderivative for the function $f(x) = 2x$ on the interval $(-\infty, \infty)$ since

$$F'(x) = 2x = f(x) \qquad \text{for all} \qquad x \in (-\infty, \infty). \qquad \diamond$$

Example 2

The function $F(x) = \sin x$ is an antiderivative for the function $f(x) = \cos x$ on $(-\infty, \infty)$ since

$$F'(x) = \cos x = f(x) \qquad \text{for all} \qquad x \in (-\infty, \infty). \qquad \Diamond$$

Example 3

The absolute value function $F(x) = |x|$ is an antiderivative for the constant function $f(x) \equiv 1$ on the interval $(0, \infty)$. That is because $|x| = x$ if $x > 0$, so

$$F'(x) = \frac{d}{dx}(|x|) = \frac{d}{dx}(x) = 1 = f(x) \qquad \text{if} \qquad x \in (0, \infty).$$

Note, however, that $F(x) = |x|$ is *not* an antiderivative for f on $(-\infty, \infty)$ since $F'(x) = -1$ for $x < 0$ and $F'(0)$ is not defined. $\qquad \Diamond$

Example 3 illustrates why we are being careful to specify the *interval* on which we seek an antiderivative F for f. (If the interval is not explicitly stated, we shall assume it to be the domain for f.)

Antiderivatives are Not Unique

Example 3 also suggests that antiderivatives are not unique, because it has no doubt occurred to you that the function $G(x) = x$ is also an antiderivative for the constant function $f(x) \equiv 1$, both on $(0, \infty)$ and on $(-\infty, \infty)$. Before pursuing this observation more generally we note several other antiderivatives for the functions in Examples 1 and 2.

Example 4

The function $f(x) = 2x$ in Example 1 also has antiderivatives

$$F_1(x) = x^2 + 1, \quad F_2(x) = x^2 - 10, \quad \text{and } F_3(x) = x^2 + \pi$$

on the interval $(-\infty, \infty)$. $\qquad \Diamond$

Example 5

The function $f(x) = \cos x$ in Example 2 also has antiderivatives

$$F_1(x) = \sin x + 5, \quad F_2(x) = \pi + \sin x, \quad \text{and } F_3(x) = \sin x - 6$$

on the interval $(-\infty, \infty)$. $\qquad \Diamond$

The Geometry of Antiderivatives

Obviously, we need a clearer understanding of what to expect as the antiderivative of a given function. We therefore return to our geometric interpretation of the derivative.

Recall, starting with the differentiable function F, the process of *differentiation* associates a *slope*, $F'(x)$, with each number x at which F is differentiable. (Figure 1.1).

Thus, if we begin with the function f and seek F with $F' = f$, we are beginning with a slope function and asking "what function F has slope $f(x) = F'(x)$ at each

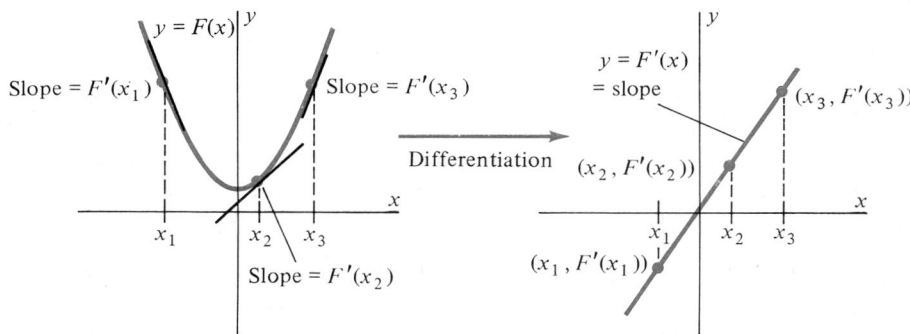

Figure 1.1 Differentiating F produces a slope, F'

x?'' Figure 1.2 illustrates the difficulty in asking such a question—since we do not know F, we do not know the particular points $(x, F(x))$ on the graph of F. We only know the slope at each point.

For example, suppose we ask for an antiderivative for the function $f(x) = \cos x$. That is, we seek F with $F'(x) = \cos x$. Figure 1.3 shows the required slopes $F'(x)$ for various values of x, and Figure 1.4 suggests several functions whose graphs have the required slopes. As Figure 1.4 suggests, we will see that the *graphs of all antiderivatives F have the same shape*. Two such graphs differ only in that one is a "vertical shift" of the other. In other words, if F_1 and F_2 are both antiderivatives for $f(x) = \cos x$, then $F_2(x) = F_1(x) + C$ for some constant C and all x.

Notice this difference between the process of differentiation and that of antidifferentiation: differentiating a particular function produces another (unique) function, while finding an antiderivative results in an entire family of functions. The following theorem confirms these observations.

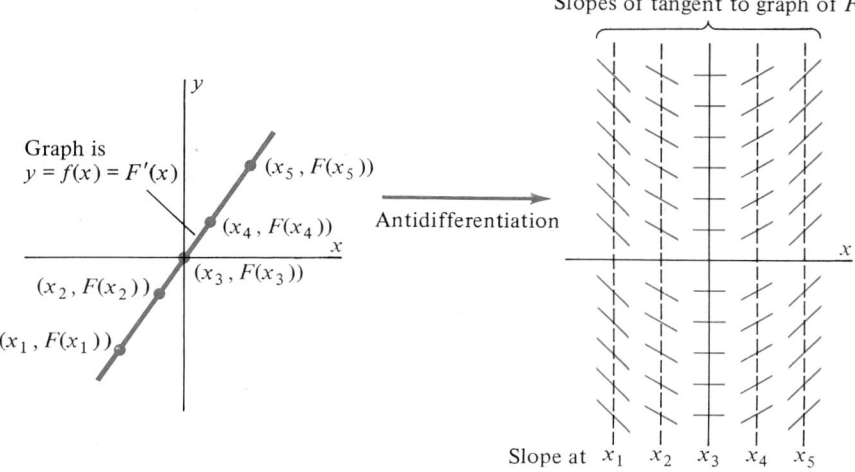

Figure 1.2 Geometric interpretation of antidifferentiation—values $f(x) = F'(x)$ interpreted as slopes of tangents to the graph of F.

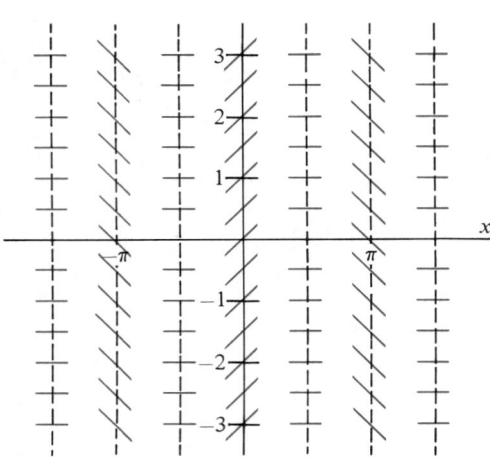

Figure 1.3 Slopes for $F(x)$ if $F'(x) = \cos x$.　　　**Figure 1.4** Antiderivatives for $F(x) = \cos x$.

THEOREM 1

Suppose the function F is an antiderivative for the function f on the interval I. That is, assume $F'(x) = f(x)$ for all $x \in I$. Then *any* antiderivative G for f on I must have the form

$$G(x) = F(x) + C, \qquad x \in I$$

for some constant C. Moreover, every function G of this form is an antiderivative for f.

In other words, *any two antiderivatives for the same function on the same interval differ at most by a constant.* Before proving Theorem 1 we establish a special case using the Mean Value Theorem.

THEOREM 2

The only antiderivatives of the zero function are the constant functions. That is, if $F'(x) = 0$ for all x in an interval I, then $F(x) = C$ for some constant C and all $x \in I$.

Proof: If $F'(x) = 0$ for all $x \in I$ then F satisfies the conditions of the Mean Value Theorem on I. Thus, if we let $a \in I$ be fixed, for any other number $x \in I$ the Mean Value Theorem guarantees that

$$\frac{F(x) - F(a)}{x - a} = F'(c) \tag{2}$$

for some number c between a and x. But $F'(x) = 0$ for *all* $x \in I$, so $F'(c) = 0$ regardless of the choice of x. Since $x - a \neq 0$, it now follows from equation (2) that $F(x) - F(a) = 0$ for all $x \in I$. Letting $F(a) = C$ then gives the conclusion that $F(x) = C$ for all $x \in I$. ◆

Proof of Theorem 1: Suppose that F and G are two antiderivatives for f on I. Then the function $h = G - F$ is differentiable on I, and

$$h'(x) = G'(x) - F'(x) = f(x) - f(x) = 0$$

for all $x \in I$. Thus, by Theorem 2 there exists a constant C so that $h(x) = C$ for all $x \in I$. That is, $G(x) - F(x) = C$ or $G(x) = F(x) + C$ for all $x \in I$. ◆

Notation and Terminology

Once we have found a particular antiderivative F for a function f, Theorem 1 guarantees that *all* antiderivatives for f have the form $F(x) + C$. We shall refer to the family of *all* antiderivatives for f as the *indefinite integral* of f, which we denote by $\int f(x)\, dx$.

DEFINITION 2

The **indefinite integral** of the function f on the interval I is the family of all antiderivatives for f on I and is denoted by $\int f(x)\, dx$. That is,

$$\int f(x)\, dx = F(x) + C \tag{3}$$

where $F'(x) = f(x)$ for all $x \in I$ and C denotes any constant.

The symbol $\int$ in Definition 2 is called an integral sign. It and the symbol dx enclose the function f for which F is an antiderivative. While the symbol dx suggests the differential discussed in Chapter 3, it should be regarded for now as simply part of the notation signifying the indefinite integral for f. The function f in equation (3) is referred to as the *integrand* of the indefinite integral. We refer to the process of finding the indefinite integral for f as *integrating* f or as *integration*.

Example 6

Using this notation and Theorem 1 we can give a complete description of the antiderivatives for the functions in Examples 1 and 2 by integration:

$$\int 2x\, dx = x^2 + C$$

and

$$\int \cos x\, dx = \sin x + C.$$ ◇

Table 1.1 shows several other indefinite integrals, each resulting from a differentiation fact established in Chapter 3. We refer to equations such as those in the second column as *integration formulas*.

Properties of Antiderivatives

Since antidifferentiation is, roughly speaking, the inverse process of differentiation, we have the following properties.

THEOREM 3

Suppose both f and g have antiderivatives on a common interval I. Then, on I,

(a) $\displaystyle\int [f(x) + g(x)]\, dx = \int f(x)\, dx + \int g(x)\, dx$, and

(b) $\displaystyle\int [cf(x)]\, dx = c \cdot \int f(x)\, dx$ for any constant c.

Table 1.1 Some differentiation and antidifferentiation results

$\dfrac{d}{dx}F(x) = f(x)$	$\displaystyle\int f(x)\,dx = F(x) + C$
$\dfrac{d}{dx}\left(\dfrac{1}{2}x^2\right) = x$	$\displaystyle\int x\,dx = \dfrac{1}{2}x^2 + C$
$\dfrac{d}{dx}\left(x^3 + \dfrac{1}{6}x^6\right) = 3x^2 + x^5$	$\displaystyle\int (3x^2 + x^5)\,dx = x^3 + \dfrac{1}{6}x^6 + C$
$\dfrac{d}{dt}(\cos t) = -\sin t$	$\displaystyle\int (-\sin t)\,dt = \cos t + C$
$\dfrac{d}{d\theta}(\tan \theta) = \sec^2 \theta$	$\displaystyle\int \sec^2 \theta\,d\theta = \tan \theta + C$
$\dfrac{d}{dx}(\sqrt{x}) = \dfrac{1}{2\sqrt{x}}$	$\displaystyle\int \dfrac{1}{2\sqrt{x}}\,dx = \sqrt{x} + C$
$\dfrac{d}{dt}\left(\dfrac{1}{2}\sin t^2\right) = t\cos t^2$	$\displaystyle\int t\cos t^2\,dt = \dfrac{1}{2}\sin t^2 + C$
$\dfrac{d}{dx}(\sqrt{ax^2 + bx + c}) = \dfrac{2ax + b}{2\sqrt{ax^2 + bx + c}}$	$\displaystyle\int \dfrac{2ax + b}{2\sqrt{ax^2 + bx + c}}\,dx = \sqrt{ax^2 + bx + c} + C$

Proof: Let F and G be particular antiderivatives for f and g, respectively. That is, assume that $F'(x) = f(x)$ and $G'(x) = g(x)$ for all $x \in I$. Then on the interval I,

(i) $\dfrac{d}{dx}[F(x) + G(x)] = F'(x) + G'(x) = f(x) + g(x)$,   and

(ii) $\dfrac{d}{dx}[cF(x)] = cF'(x) = cf(x)$.

Equation (i) shows that $F + G$ is an antiderivative for $f + g$ and equation (ii) shows that cF is an antiderivative for cf. ◆

Theorem 3 states that, in finding antiderivatives for sums or constant multiples of functions, we may first find antiderivatives for the individual functions and then form the appropriate sums or multiples, incorporating all arbitrary constants into a single constant.*

Example 7

Find $\displaystyle\int [7x^2 - 2x^3]\,dx$.

---

*Thus, if C_1 and C_2 denote arbitrary (and, therefore, unknown) constants, the sum $(C_1 + C_2)$ and the multiple kC_1 are again just unknown constants. We will therefore always combine arbitrary constants by equations of the form

$$C_1 + C_2 = C \quad\text{and}\quad kC_1 = C.$$

Strategy

Apply Theorem 3.

$$\int x^2\, dx = \frac{1}{3}x^3 + C_1$$

$$\int x^3\, dx = \frac{1}{4}x^4 + C_2$$

$$C = C_1 + C_2$$

Check result by differentiating.

Solution

$$\int [7x^2 - 2x^3]\, dx = 7\int x^2\, dx - 2\int x^3\, dx$$

$$= 7\left[\frac{x^3}{3} + C_1\right] - 2\left[\frac{1}{4}x^4 + C_2\right]$$

$$= \frac{7}{3}x^3 - \frac{1}{2}x^4 + C, \qquad C = 7C_1 - 2C_2.$$

To verify this result, we check

$$\frac{d}{dx}\left(\frac{7}{3}x^3 - \frac{1}{2}x^4 + C\right) = \frac{7}{3}(3)x^2 - \frac{1}{2}(4)x^3 + 0$$

$$= 7x^2 - 2x^3. \qquad \diamond$$

In Example 7, we have made use of the **Power Rule**

$$\int x^n\, dx = \frac{1}{n+1}x^{n+1} + C, \qquad n \ne -1,$$

which holds for all rational exponents n. The Power Rule may be verified by differentiation:

$$\frac{d}{dx}\left[\frac{1}{n+1}x^{n+1} + C\right] = \frac{1}{n+1}(n+1)x^n = x^n, \qquad n \ne -1.$$

The next example illustrates the fact that

$$\frac{d}{dx}\left\{\int f(x)\, dx\right\} = f(x) \tag{4}$$

when the function f has an antiderivative. The interpretation of this equation is simply that *the derivative of an antiderivative for f must be the function f itself.*

Example 8

Let $f(x) = x^3 + \sin x + \cos x$. Then,

$$\int f(x)\, dx = \int (x^3 + \sin x + \cos x)\, dx$$

$$= \int x^3\, dx + \int \sin x\, dx + \int \cos x\, dx \qquad \text{(Theorem 3)}$$

$$= \frac{1}{4}x^4 + (-\cos x) + \sin x + C \qquad \text{(Power Rule, Example 7, and Table 1.1.)}$$

$$= \frac{1}{4}x^4 - \cos x + \sin x + C,$$

so

$$\frac{d}{dx}\left\{ \int f(x) \, dx \right\} = \frac{d}{dx}\left\{ \frac{1}{4}x^4 - \cos x + \sin x + C \right\}$$

$$= \frac{1}{4}(4x^3) - (-\sin x) + \cos x + 0$$

$$= x^3 + \sin x + \cos x$$

$$= f(x)$$

as in equation (4). ◇

Position and Velocity as Antiderivatives

Returning now to the issues of position, velocity, and acceleration we can use the notation developed here to write the corresponding integration formulas:

Since $\quad s'(t) = v(t),\qquad \int v(t) \, dt = s(t) + C$ \hfill (5)

and

since $\quad v'(t) = a(t),\qquad \int a(t) \, dt = v(t) + C.$ \hfill (6)

The following examples illustrate how the constants in equations (5) and (6) are determined in particular situations.

Example 9

When a hockey player strikes a puck with a certain force, the puck moves along the ice with velocity $v(t) = 64 - \sqrt{t}$ meters per second at time t, where $0 \le t \le 8$ seconds. Let $s(t)$ denote the distance between the puck and the player after t seconds. Find $s(t)$.

Solution: Using (5) and the Power Rule we find the position function to be

$$s(t) = \int (64 - t^{1/2}) \, dt = 64t - \frac{2}{3}t^{3/2} + C.$$

To determine C we use the fact that $s(0) = 0$, which says that the distance between the player and the puck is zero at the initial time $t = 0$. This gives

$$0 = s(0) = 64(0) - \frac{2}{3}(0)^{3/2} + C = C,$$

so

$$C = 0.$$

Thus,

$$s(t) = 64t - \frac{2}{3}t^{3/2}.$$ ◇

The condition that $s(0) = 0$ in Example 9 is referred to as an **initial condition** since it provides the value $s(t)$ at a particular number t. (It is not necessary that this value be provided at $t = 0$. Information about $s(t)$ at *any* $t \in [0, 8]$ will allow us to determine C.) In problems involving antidifferentiation, we will need one initial condition to determine the arbitrary constant C for each antidifferentiation performed.

Example 10

When an object is moving freely in the atmosphere it is pulled toward the earth by the force of gravity. Near the surface of the earth the acceleration due to gravity is approximately 9.8 m/s². If a fireworks display rocket is fired vertically from ground level at an initial velocity of 45 m/s and fails to explode, when does it strike the ground?

Strategy

Find a from the problem statement.

Solution

Since the only force acting on the rocket is acceleration due to gravity, the acceleration function is the constant

$$a(t) = -9.8 \text{ m/s}^2.$$

(We take the positive direction to be upward, so the correct sign for $a(t)$ is $-$.)

Find v by using equation (6).

The general form of the velocity function is therefore

$$v(t) = \int (-9.8) \, dt$$

$$= -9.8t + C_1.$$

Apply initial condition $v(0) = 45$ to find C_1.

To find the constant C_1, we use the initial condition $v(0) = 45$. Setting $t = 0$ gives

$$v(0) = 45 = (-9.8)(0) + C_1$$

so

$$C_1 = 45.$$

The explicit velocity function is therefore

$$v(t) = -9.8t + 45 \text{ m/s}.$$

Find s from equation (5).

The position function is

$$s(t) = \int (-9.8t + 45) \, dt$$

$$= -4.9t^2 + 45t + C_2.$$

Apply initial condition $s(0) = 0$ to find C_2.

Since the rocket is fired from ground level, the initial condition for s is $s(0) = 0$. Thus

$$s(0) = 0 = -4.9(0)^2 + 45(0) + C_2,$$

so

$$C_2 = 0.$$

The explicit position function is therefore

$$s(t) = -4.9t^2 + 45t.$$

Set $s(t) = 0$ to find desired time.

The rocket strikes the ground when $s(t) = 0$ so we set

$$s(t) = -4.9t^2 + 45t = 0$$

and obtain $t = 0$ (launch), or

$$-4.9t + 45 = 0,$$

which gives

$$t = \frac{45}{4.9} \approx 9.18 \text{ seconds (after launching).}$$ ◇

Exercise Set 5.1

1. Figures a–d illustrate the graphs of four particular functions. Figures i–iv represent the associated slopes for the antideriva-tives. Match each function with the correct slope portrait for its antiderivative.

a.
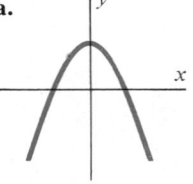

i.
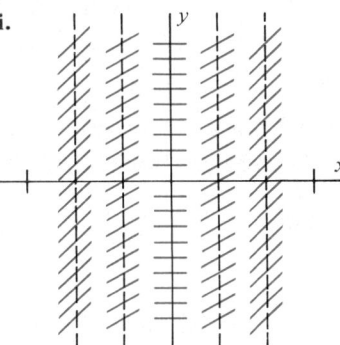

b.
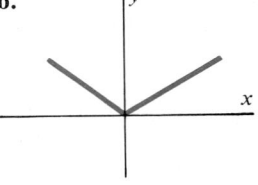

ii.
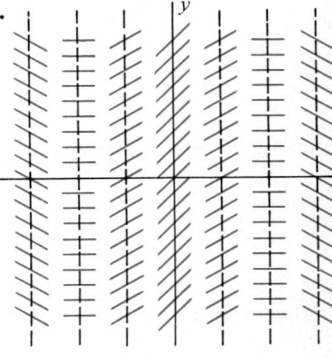

c.
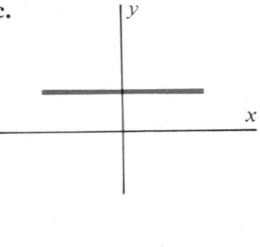

iii.
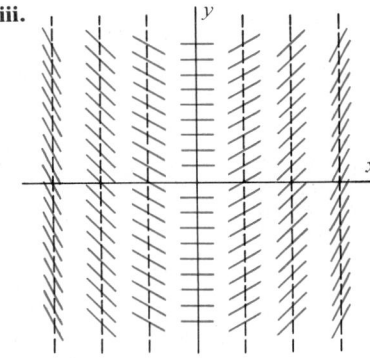

d.
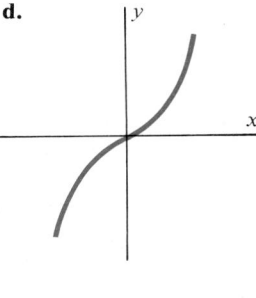

iv.
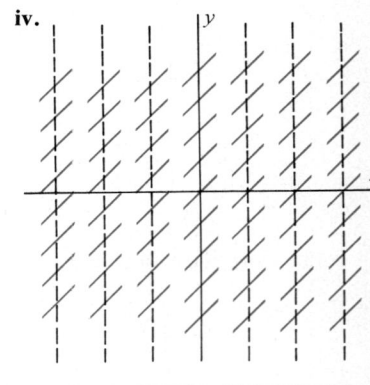

In each of Exercises 2–9, sketch the slope portrait for the antide-rivative of the given function.

2. $f(x) = x$
3. $f(x) = 1 - x$
4. $f(x) = \sqrt{x}$
5. $f(x) = \sin x$
6. $y = 1 - x^2$
7. $y = (1 - x)^2$
8. $y = 1/x$
9. $y = x^3 - 2$

In Exercises 10–25, find the indefinite integral.

10. $\displaystyle\int (2x^2 + 1)\, dx$

11. $\displaystyle\int (1 - x^3)\, dx$

12. $\displaystyle\int (x^{2/3} + x^{5/2})\, dx$

13. $\displaystyle\int \sqrt{x + 2}\, dx$

14. $\displaystyle\int \sin 4x\, dx$

15. $\displaystyle\int (2 \cos t - \sin t)\, dt$

16. $\displaystyle\int (\sin x + \sec^2 x)\, dx$

17. $\displaystyle\int \sqrt{t}(t^2 + 1)\, dt$

18. $\int (t - 1)(t + 1)\, dt$

19. $\int (t^2 + 1)(t + 2)\, dt$

20. $\int \dfrac{1}{\sqrt{x + 1}}\, dx$

21. $\int \dfrac{1}{\sqrt[3]{x}}\, dx$

22. $\int (\tan^2 x + 1)\, dx$

23. $\int (5\cos x + \sqrt{x})\, dx$

24. $\int (3x^2 - \sin x + 2\sec^2 x)\, dx$

25. $\int (6x + 5 - 3\cot x \csc x)\, dx$

In Exercises 26–32, find the position function s corresponding to the given velocity function and initial condition.

26. $v(t) = \cos t, \qquad s(0) = 2$

27. $v(t) = 1 + 2t, \qquad s(0) = 0$

28. $v(t) = 2t^2, \qquad s(0) = 4$

29. $v(t) = t(t + 2), \qquad s(0) = 2$

30. $v(t) = \sqrt{t}(t + 4), \qquad s(0) = 0$

31. $v(t) = \dfrac{(\sqrt{t} + 3)^3}{\sqrt{t}}, \qquad s(0) = 0$

32. $v(t) = (\sqrt{t} + 4)(\sqrt{t} - 4), \qquad s(0) = 5$

In Exercises 33–39, find the velocity function v and position function s corresponding to the given acceleration function and initial conditions.

33. $a(t) = 2, \qquad v(0) = 3, \qquad s(0) = 0$

34. $a(t) = -2, \qquad v(0) = 6, \qquad s(0) = 10$

35. $a(t) = 3t, \qquad v(0) = 0, \qquad s(0) = 20$

36. $a(t) = 200, \qquad v(0) = 100, \qquad s(0) = 200$

37. $a(t) = 4t + 4, \qquad v(0) = 8, \qquad s(0) = 12$

38. $a(t) = \cos t, \qquad v(0) = 1, \qquad s(0) = 9$

39. $a(t) = \sin t, \qquad v(0) = 0, \qquad s(0) = 2$

40. A stone is thrown vertically upward with an initial velocity 15 m/s. How long will it take for the stone to (a) stop rising, (b) strike the ground? (Use $a = -9.8$ m/s^2.)

41. How long will it take for the rocket in Example 10 to reach the highest point of its trajectory?

42. A particle moves along a line with constant acceleration $a = 3$ m/s^2.
   a. How fast is it moving after 6 seconds if its initial velocity is $v(0) = 10$ m/s?
   b. What is its initial velocity $v(0)$ if its speed after 3 seconds is 15 m/s?

43. A coin is dropped from the top of a building 150 meters high. Using $a = -9.8$ m/s^2,
   a. How long will it take for the coin to strike the ground?
   b. With what velocity will the coin strike the ground?

44. What constant acceleration will enable the driver of an automobile to increase its speed from 20 m/s to 25 m/s in 10 seconds?

45. What constant negative acceleration (deceleration) is required to bring an automobile travelling at a rate of 72 km/h to a full stop in 100 meters? (*Caution:* Be careful to work in common units.)

5.2 FINDING ANTIDERIVATIVES BY SUBSTITUTION

In Section 5.1 we noted that finding antiderivatives involves knowing how to "invert" the differentiation process. Since the Chain Rule is frequently used to differentiate functions, it is important to recognize when a given function f can be interpreted as the result of the Chain Rule having been applied to its antiderivative F.

Recall the statement of the Chain Rule for differentiating the composite function $F \circ g$:

$$(F \circ g)'(x) = F'(g(x)) \cdot g'(x). \tag{1}$$

Equation (1) tells us that an *antiderivative* for the function $F'(g(x))g'(x)$ on the right side is just the composite function $F \circ g$ which has been differentiated on the left side. We may therefore write

$$\int F'(g(x)) \cdot g'(x)\, dx = F(g(x)) + C$$

or, that

$$\int f(g(x)) \cdot g'(x) \, dx = F(g(x)) + C \qquad\qquad (2)$$

if $F' = f$.

Equation (2) is useful in finding the indefinite integral for a *product* $(f \circ g)g'$ when the first factor is a composite function $f \circ g$ and the second factor g' is the derivative of the "inside" function in the composition. In this case an antiderivative is the composition $F \circ g$ of an antiderivative F for the "outside" function with the "inside" function g.

Example 1

Find $\displaystyle\int (x^3 + 2x + 6)^4(3x^2 + 2) \, dx$.

Strategy

Identify the integrand as a composite function matching the integrand in equation (2).

Find an antiderivative F for the outside function f.

Apply equation (2).

Solution

The integrand has the form $f(g(x))g'(x)$ with

$$f(u) = u^4 \qquad \text{and} \qquad u = g(x) = x^3 + 2x + 6.$$

Since an antiderivative for $f(u) = u^4$ is

$$F(u) = \int u^4 \, du = \frac{1}{5}u^5 + C,$$

equation (2) tells us that

$$\int (x^3 + 2x + 6)^4(3x^2 + 2) \, dx = \frac{1}{5}(x^3 + 2x + 6)^5 + C,$$

which you can verify by differentiation. ◇

Example 2

Find $\displaystyle\int \frac{\cos x}{\sqrt{1 + \sin x}} \, dx$.

Solution: This integral can be written in the form

$$\int \frac{\cos x}{\sqrt{1 + \sin x}} \, dx = \int (1 + \sin x)^{-1/2} \cdot \cos x \, dx,$$

which we recognize as being of the form $\displaystyle\int f(g(x)) \cdot g'(x) \, dx$ with

$$f(u) = u^{-1/2} \qquad \text{and} \qquad u = g(x) = 1 + \sin x.$$

An antiderivative for f is $F(u) = \int u^{-1/2} \, du = 2u^{1/2} + C$, so it follows from equation (2) that

$$\int \frac{\cos x}{\sqrt{1 + \sin x}} \, dx = 2\sqrt{1 + \sin x} + C. \qquad\qquad ◇$$

The Method of Substitution

There is a way to formalize somewhat the procedure of identifying an integrand in the form $f(g(x))g'(x)$ and finding an antiderivative $F(g(x))$. It is based on the notation for the *differential du* of the function $u = g(x)$. Recall that definition:

$$\text{If}\quad u = g(x), \qquad \text{then} \qquad du = g'(x)\,dx \tag{3}$$

when g is differentiable at x. Using this notation we may write the integrand in equation (2) formally as

$$\underbrace{f(g(x))}_{f(u)}\,\underbrace{g'(x)\,dx}_{du} = f(u)\,du. \tag{4}$$

With the notation in line (4), equation (2) may be rewritten as

$$\int f(g(x))g'(x)\,dx = \int f(u)\,du = F(u) + C \tag{5}$$

$$\text{where} \qquad u = g(x) \qquad \text{and} \qquad F' = f.$$

Equation (5) should be regarded as a notational device for remembering equation (2). The advantage in using equation (5) is that *once we have made the substitution $u = g(x)$ and applied the differential notation $du = g'(x)\,dx$, the antiderivative F may be found by integrating the function f as a function of the independent variable u.* This is what is referred to as the **method of substitution.**

In using the method of substitution it is important to note that equation (5) results from the *notation $du = g'(x)\,dx$* for the differential du and *not* from its interpretation as an approximation.

Example 3

Find $\displaystyle\int \sin^3 x \cos x\,dx$ by the method of substitution.

Strategy

Since $(\sin x)' = \cos x$, the inside function is $u = \sin x$.

Substitute for u and du.

Find antiderivative with respect to u.

Substitute back in terms of x.

Solution

We use the substitution

$$u = \sin x; \qquad du = \cos x\,dx.$$

We obtain

$$\int \sin^3 x \cos x\,dx = \int u^3\,du$$

$$= \frac{1}{4}u^4 + C$$

$$= \frac{1}{4}\sin^4 x + C. \qquad \diamond$$

Example 4

Find $\displaystyle\int x^2(x^3 + 7)^4\,dx$.

Strategy

Take $u = x^3 + 7$ to be the inside function, since $du = 3x^2\ dx$ has a factor of x^2.

Since only $x^2\ dx$ appears in the integrand, solve for this term in terms of du.

Solution

We make the substitution

$$u = x^3 + 7, \qquad du = 3x^2\ dx.$$

Then $x^2\ dx = \dfrac{1}{3}\ du$, and we obtain

$$\int x^2(x^3 + 7)^4\ dx = \int (x^3 + 7)^4 x^2\ dx$$

Substitute for u and for du.

$$= \int u^4 \cdot \left(\frac{1}{3}\ du\right)$$

$$= \frac{1}{3}\int u^4\ du$$

$$= \frac{1}{3}\left(\frac{1}{5}u^5 + C_1\right)$$

Apply the Power Rule. Substitute back for u in terms of x.

$$= \frac{1}{15}(x^3 + 7)^5 + C, \qquad C = \frac{1}{3}C_1.$$

◇

Example 5

Find $\displaystyle\int \frac{x^2\ dx}{\sqrt{1 + x^3}}$.

Strategy

The only hope is in setting

$u = 1 + x^3$.

Solve for the factor $x^2\ dx$ in terms of du.

Solution

We make the substitution

$$u = 1 + x^3; \qquad du = 3x^2\ dx.$$

Then $x^2\ dx = \dfrac{1}{3}\ du$, so we obtain

Make the u-substitution.

$$\int \frac{x^2\ dx}{\sqrt{1 + x^3}} = \int \frac{\frac{1}{3}\ du}{\sqrt{u}}$$

Find the antiderivative with respect to u using Power Rule. Substitute back in terms of x.

$$= \frac{1}{3}\int u^{-1/2}\ du$$

$$= \left(\frac{1}{3}\right)[(2)u^{1/2} + C_1]$$

$$= \frac{2}{3}\sqrt{1 + x^3} + C, \qquad C = \frac{1}{3}C_1.$$

◇

Example 6

Find $\displaystyle\int \frac{\sin x \cos x\ dx}{\sqrt{1 + \sin^2 x}}$.

Solution: A first attempt might be to try the u-substitution

$$u = \sin x; \qquad du = \cos x\ dx.$$

Indeed, this works, since we obtain

$$\int \frac{\sin x \cos x \, dx}{\sqrt{1 + \sin^2 x}} = \int \frac{u \cdot du}{\sqrt{1 + u^2}}.$$

To handle the antiderivative on the right, we can make a second substitution

$$w = 1 + u^2; \qquad dw = 2u \, du,$$

so $u \, du = \dfrac{1}{2} dw$. With this substitution we have

$$\int \frac{\sin x \cos x \, dx}{\sqrt{1 + \sin^2 x}} = \int \frac{u \, du}{\sqrt{1 + u^2}} = \int \frac{\frac{1}{2} dw}{\sqrt{w}} = \int \frac{1}{2} w^{-1/2} \, dw$$
$$= \sqrt{w} + C$$
$$= \sqrt{1 + u^2} + C$$
$$= \sqrt{1 + \sin^2 x} + C.$$

A shorter solution to this problem is obtained if we simply take

$$u = 1 + \sin^2 x; \qquad du = 2 \sin x \cos x \, dx.$$

Then $\sin x \cos x \, dx = \dfrac{1}{2} du$, and we obtain

$$\int \frac{\sin x \cos x \, dx}{\sqrt{1 + \sin^2 x}} = \int \frac{\frac{1}{2} du}{\sqrt{u}} = \int \frac{1}{2} u^{-1/2} \, du = \sqrt{u} + C$$
$$= \sqrt{1 + \sin^2 x} + C.$$

Obviously, in some cases more than one substitution will work. ◇

Exercise Set 5.2

In Exercises 1–22 use the method of substitution to find the antiderivative.

1. $\displaystyle\int x\sqrt{x^2 + 1} \, dx$

2. $\displaystyle\int x \cos x^2 \, dx$

3. $\displaystyle\int \sec 2\theta \tan 2\theta \, d\theta$

4. $\displaystyle\int \csc(\pi x) \cot(\pi x) \, dx$

5. $\displaystyle\int x \csc^2(x^2) \, dx$

6. $\displaystyle\int \frac{\sin \sqrt{x} \cos \sqrt{x}}{\sqrt{x}} \, dx$

7. $\displaystyle\int (1 - t^2)\sqrt{3t^3 - 9t + 9} \, dt$

8. $\displaystyle\int (1 - x)^3 \, dx$

9. $\displaystyle\int \sec^2(x - \pi) \, dx$

10. $\displaystyle\int x \sec^2 x^2 \, dx$

11. $\displaystyle\int \frac{(2\sqrt{x} + 3)^2}{\sqrt{x}} \, dx$

12. $\displaystyle\int (t^3 - 6t + 7)^5 (2 - t^2) \, dt$

13. $\displaystyle\int x\sqrt{x - 1} \cdot \sqrt{x + 1} \, dx$

14. $\displaystyle\int \sqrt{1 - \sin x} \cos x \, dx$

15. $\displaystyle\int x \sec(\pi - x^2) \tan(\pi - x^2) \, dx$

16. $\displaystyle\int \frac{\sec^2 \sqrt{2x + 1}}{\sqrt{2x + 1}} \, dx$

17. $\displaystyle\int (x^5 - 2x^3)(x^6 - 3x^4)^{5/2} \, dx$

18. $\displaystyle\int \frac{x^3 + 2x}{\sqrt[3]{x^4 + 4x^2}} \, dx$

19. $\displaystyle\int \frac{t^2 + 2}{\sqrt{t^3 + 6t}} \, dt$

20. $\displaystyle\int \frac{\cos^3 \sqrt{x} \sin \sqrt{x}}{\sqrt{x}} \, dx$

21. $\displaystyle\int \frac{1}{(1 - 4x)^{2/3}} \, dx$

24. a. $\displaystyle\int \sin(3x) \, dx$

22. $\displaystyle\int \left(\frac{1 + \tan^2 \sqrt{x}}{\sqrt{x}} \right) dx$

b. $\displaystyle\int \sin(3x) \cos(3x) \, dx$

In Exercises 23–25 state a substitution appropriate to each part and find the antiderivative.

c. $\displaystyle\int \sin^4(3x) \cos(3x) \, dx$

23. a. $\displaystyle\int (3x + 4) \, dx$

25. a. $\displaystyle\int (x^2 + 6) \, dx$

b. $\displaystyle\int \sqrt{3x + 4} \, dx$

b. $\displaystyle\int \frac{x}{\sqrt{x^2 + 6}} \, dx$

c. $\displaystyle\int \sin(3x + 4) \, dx$

d. $\displaystyle\int \sin^2(3x + 4) \cos(3x + 4) \, dx$

c. $\displaystyle\int x \sec^2(x^2 + 6) \, dx$

5.3 DIFFERENTIAL EQUATIONS

Calculating functions from their slope functions and finding velocity and position from acceleration are not the only types of problems in which one attempts to determine a function from information about its derivative. For example, a classic experiment in elementary physics is that of sprinkling iron filings on a sheet of paper suspended over a horseshoe magnet (Figure 3.1). The iron filings assume the directions of the magnetic force field surrounding the magnet (Figure 3.2). Here we obtain information about the direction (slopes) of the field lines, and we can then ask for the equations of the field lines themselves.

The general theory for determining a function from certain types of information about its derivative(s) is referred to as the theory of **differential equations.** A differential equation is an equation involving one or more derivatives of an unknown function.

For example, since velocity is the derivative of position, we have already seen that the position function s satisfies the differential equation

$$s'(t) = v(t). \tag{1}$$

Figure 3.1 Iron filing–magnet experiment.

Later, we shall see that the function y representing the displacement of the end of an oscillating spring satisfies a differential equation of the form

$$\frac{d^2y}{dt^2} + b\frac{dy}{dt} + cy = 0. \tag{2}$$

Equation (1) is an example of a **first order** differential equation since it involves only a first derivative. Equation (2) is an example of a **second order** differential equation because it involves a second derivative.

A **solution** of a differential equation is a function that is differentiable as many times as the equation requires and that satisfies the equation. For example, we have already seen that the differential equation

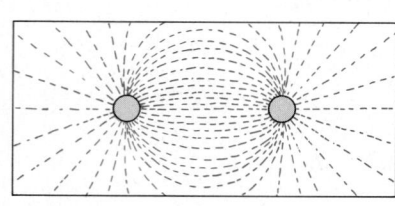

Figure 3.2 Filings point in direction of magnetic field lines.

$$\frac{dy}{dx} = \cos x$$

has the solution $y = \sin x$, while the differential equation

$$x\frac{d^2y}{dx^2} - \frac{dy}{dx} = 0$$

has solution $y = x^2$. (To see this note that $\dfrac{dy}{dx} = 2x$ and $\dfrac{d^2y}{dx^2} = 2$ for $y = x^2$. Thus,

$$x \cdot \frac{d^2y}{dx^2} - \frac{dy}{dx} = (x)(2) - 2 \cdot x = 0.)$$

Example 1

The function $y = \sin x$ satisfies the differential equation $\dfrac{d^2y}{dx^2} + y = 0$. To show this, we compute the second derivative:

$$\frac{dy}{dx} = \cos x \qquad \text{and} \qquad \frac{d^2y}{dx^2} = -\sin x.$$

Thus, substitution in the differential equation gives

$$\frac{d^2y}{dx^2} + y = (-\sin x) + \sin x$$

$$= 0$$

as required. Thus, $y = \sin x$ is a solution of the differential equation. $\diamond$

One of the simplest forms for a differential equation is

$$\frac{dy}{dx} = f(x). \tag{3}$$

Here, any antiderivative $y = F(x)$ for the function f is a solution, since $\dfrac{dy}{dx} = F'(x) = f(x)$. In fact, once we have identified a particular antiderivative F, Theorem 1 guarantees that all solutions of (3) have the form

$$y = F(x) + C, \qquad C \text{ constant.} \tag{4}$$

We refer to (4) as the **general solution** of differential equation (3) since all particular solutions of (3) may be obtained from (4) by appropriately specifying the constant C.

Frequently we will encounter differential equation (3) in the *differential* form

$$dy = f(x) \, dx. \tag{5}$$

To see that equations (3) and (5) are equivalent, recall that the differential dy is defined by the equation $dy = \dfrac{dy}{dx} \cdot dx$. Multiplying both sides of equation (3) by the differential dx gives equation (5).

Example 2

Find the general solution of the differential equation

$$3 \, dy - x \, dx = 0.$$

Strategy	*Solution*
Solve for $\dfrac{dy}{dx}$.	Here $\dfrac{dy}{dx} = \dfrac{x}{3}$.
Find y by antidifferentiation.	Thus

$$y = \int \frac{x}{3}\, dx$$

$$= \frac{x^2}{6} + C$$

is the general solution (see Figure 3.3).   ◇

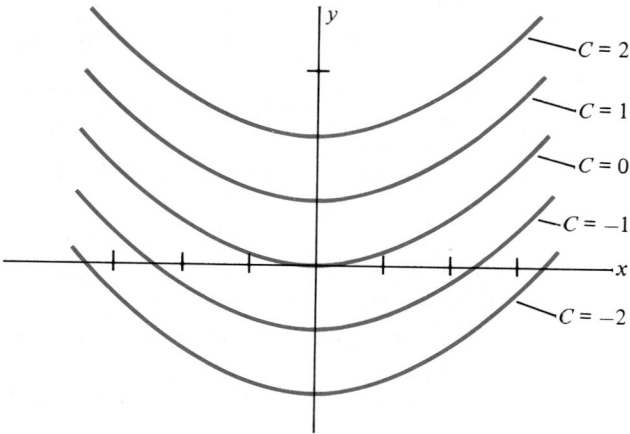

Figure 3.3 Solutions of $3\, dy - x\, dx = 0$ are $y = \dfrac{1}{6}x^2 + C$.

As Example 2 illustrates, the general solution of equation (3) is not a function but rather a *family* of functions of the form (4), one such function corresponding to each choice of the constant C. To obtain a *unique* solution to (3) we must also specify an *initial condition*. An **initial condition** is simply a point (x_0, y_0) through which the desired solution curve must pass. For example, if we specify the initial condition $y(1) = 3$ in Example 2, we obtain the equation

$$3 = y(1) = \frac{1}{6}(1)^2 + C = \frac{1}{6} + C,$$

so

$$C = 3 - \frac{1}{6} = \frac{17}{6}.$$

The particular solution satisfying this initial condition is therefore

$$y = \frac{1}{6}x^2 + \frac{17}{6}.$$

A problem consisting of a differential equation together with an initial condition is called an **initial value problem.**

Example 3

Solve the initial value problem

$$dy = x\sqrt{x^2 + 1} \, dx, \qquad y(0) = 1.$$

Strategy

Solution

Here $\dfrac{dy}{dx} = x\sqrt{x^2 + 1}$,

so

$$y = \int x\sqrt{x^2 + 1} \, dx$$

Find the indefinite integral by the substitution method with

$$u = x^2 + 1, \qquad du = 2x \, dx.$$

$$= \int u^{1/2} \left(\frac{1}{2} \, du\right)$$

$$= \frac{1}{2} \int u^{1/2} \, du$$

$$= \frac{1}{2} \left(\frac{2}{3} u^{3/2} + C_1\right)$$

$$= \frac{1}{3} u^{3/2} + C, \qquad C = \frac{1}{2} C_1$$

$$= \frac{1}{3} (x^2 + 1)^{3/2} + C.$$

Apply initial condition to find C.

Then $y(0) = 1$ gives the equation

$$1 = y(0) = \frac{1}{3} (0^2 + 1)^{3/2} + C$$

$$= \frac{1}{3} + C$$

so

$$C = \frac{2}{3}.$$

The desired solution is therefore

$$y = \frac{1}{3} (x^2 + 1)^{3/2} + \frac{2}{3}. \qquad \diamond$$

Separation of Variables

Another type of differential equation for which we can often find the general solution has the form

$$\frac{dy}{dx} = \frac{f(x)}{g(y)}, \qquad g(y) \neq 0. \tag{6}$$

Differential equations of this type are called **separable.** That is because if we multiply both sides of equation (6) by $g(y)$ we obtain the equation

$$g(y) \cdot \frac{dy}{dx} = f(x). \tag{7}$$

If we can find antiderivatives G for g and F for f, we will have succeeded in **separating the variables** in equation (6), since the left side of (7) is the derivative

$$\frac{d}{dx} G(y) = G'(y) \frac{dy}{dx} = g(y) \frac{dy}{dx}$$

of the composite function $G(y)$, and the right side of (7) is the derivative $F'(x) = f(x)$ of the function F. It then follows, from Theorem 1, that

$$G(y) = F(x) + C, \qquad C = \text{constant} \tag{8}$$

on the domain common to $F(x)$ and $G(y)$. Equation (8), in general, defines the solution y of equation (6) implicitly, but equation (8) can often be solved to produce the solution y as an explicit function of x.

REMARK: The differential notation $dy = \left(\dfrac{dy}{dx}\right) dx$ is often used to abbreviate the above discussion by saying that

$$\frac{dy}{dx} = \frac{f(x)}{g(y)} \qquad \text{implies that} \qquad g(y)\, dy = f(x)\, dx,$$

so

$$\int g(y)\, dy = \int f(x)\, dx,$$

and we obtain the solution y by "integrating both sides."

Example 4

Use the technique of separation of variables to find the general solution of the differential equation

$$\frac{dy}{dx} = \frac{x}{y}, \qquad y \neq 0.$$

Strategy

Separate variables.

Solution

Multiplying both sides by y and by dx gives

$$y\, dy = x\, dx$$

Integrate both sides.

so

$$\int y\, dy = \int x\, dx.$$

Thus

$$\frac{1}{2} y^2 + C_1 = \frac{1}{2} x^2 + C_2,$$

Solve for y^2, combining all arbitrary constants.

so

$$y^2 + 2C_1 = x^2 + 2C_2,$$

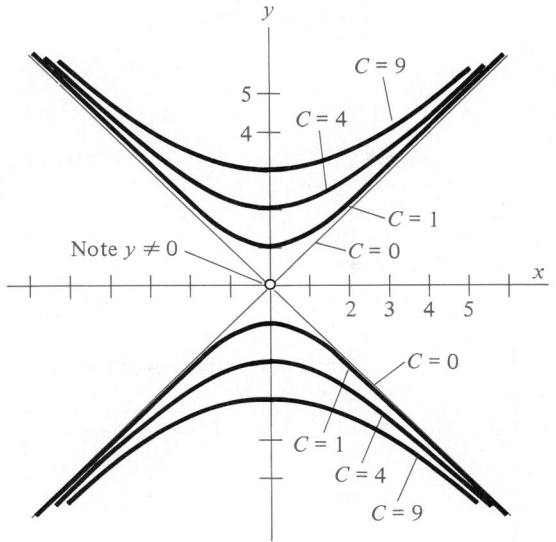

Figure 3.4 Graphs of solutions of the differential equation $\dfrac{dy}{dx} = \dfrac{x}{y}$.

or

$$y^2 = x^2 + C$$

where $C = 2C_2 - 2C_1$.

Graphs of this equation, for various values of C, appear in Figure 3.4. To verify this result we differentiate implicitly in the equation

$$y^2 = x^2 + C$$

to obtain

$$2y \cdot \frac{dy}{dx} = 2x$$

so

$$\frac{dy}{dx} = \frac{x}{y}, \qquad y \neq 0. \qquad\qquad \diamondsuit$$

Example 5

Find the solution of the initial value problem

$$dy = 2xy^2 \, dx, \qquad y(0) = 1/2.$$

Strategy

Separate variables.

Solution

We have

$$y^{-2} \, dy = 2x \, dx,$$

so

Integrate both sides.

$$\int y^{-2} \, dy = \int 2x \, dx$$

Solve for y.

and, therefore,

$$-y^{-1} + C_1 = x^2 + C_2.$$

Thus

$$\frac{1}{y} = -x^2 - C, \qquad C = C_2 - C_1$$

so

$$y = \frac{-1}{x^2 + C}.$$

From the initial condition $y(0) = 1/2$ we obtain

Apply initial condition to solve for C.

$$\frac{1}{2} = y(0) = \frac{-1}{0^2 + C} = \frac{-1}{C}$$

so

$$C = -2.$$

The solution is therefore

$$y = \frac{-1}{x^2 - 2}. \qquad \diamond$$

Example 6

A population of animals in an ecological niche grows according to the differential equation

$$\frac{dP}{dt} = \frac{100}{P}.$$

Here the function $y = P(t)$ represents the number of animals present after t years. Find the function P if initially $P(0) = 50$ animals are present. What is the size of the population after 12 years?

Solution: Separating variables in the differential equation

$$\frac{dP}{dt} = \frac{100}{P} \qquad\qquad (9)$$

gives

$$P \, dP = 100 \, dt.$$

Then

$$\int P \, dP = \int 100 \, dt$$

gives

$$\frac{1}{2}P^2 + C_1 = 100t + C_2$$

so

$$\frac{1}{2}P^2 = 100t + C_2 - C_1$$

or

$$P^2 = 200t + C, \qquad C = 2(C_2 - C_1).$$

Thus

$$P(t) = \sqrt{200t + C} \tag{10}$$

is the general solution of equation (9). (We take the sign of $P(t)$ to be positive since $P(t)$ is the size of a population which cannot be negative.)

To determine the constant C we set $t = 0$ in equation (10) and apply the initial condition $P(0) = 50$. This gives

$$P(0) = \sqrt{200(0) + C} = \sqrt{C} = 50.$$

Thus, $C = 50^2 = 2500$ and

$$P(t) = \sqrt{200t + 2500}$$

is the size of the population after t years.

After $t = 12$ years the population is of size

$$P(12) = \sqrt{2400 + 2500} = \sqrt{4900} = 70.$$

Differential equations are an important part of applied mathematics, physics, engineering mathematics, mathematical biology, and physical chemistry. In later chapters we shall encounter other types of differential equations, techniques for their solutions, and applications involving such equations.

Exercise Set 5.3

In Exercises 1–10, find the general solution of the differential equation.

1. $\dfrac{dy}{dx} = x - 1$

2. $\dfrac{dy}{dx} = 2 + \sec^2 x$

3. $\dfrac{dy}{dx} = x^2 - \dfrac{1}{x^2}$

4. $\dfrac{dy}{dx} = 4 \cos 2x$

5. $dy = (x - \sqrt{x})\, dx$

6. $dy + 3\, dx = 0$

7. $dy = \dfrac{x}{y}\, dx$

8. $\dfrac{dy}{dx} = \dfrac{x}{y^2}$

9. $dy = -4xy^2\, dx$

10. $dy = \dfrac{\sqrt{x}}{\sqrt{y}}\, dx$

In Exercises 11–18, find the solution of the initial value problem.

11. $\dfrac{dy}{dx} = \dfrac{x}{y}, \qquad y(1) = 4$

12. $2y\, dy = \sqrt{x}\, dx, \qquad y(0) = 2$

13. $dy = \dfrac{x^2}{(1 + x^3)^2}\, dx, \qquad y(0) = \dfrac{1}{2}$

14. $dy = \dfrac{x}{y\sqrt{1 + x^2}}\, dx, \qquad y(0) = 2$

15. $dy = y^2(1 + x^2)\, dx, \qquad y(0) = 1$

16. $\dfrac{dy}{dx} = 2x^2y^2, \qquad y(0) = -3$

17. $\dfrac{d^2y}{dx^2} = x^2, \qquad y(0) = 2, \qquad y'(0) = 4$

18. $\dfrac{d^2y}{dx^2} = \sin x, \qquad y(0) = \pi, \qquad y'(0) = 2$

19. The slope of the line tangent to the graph of $y = f(x)$ at $(x, f(x))$ is $x - \sqrt{x}, x > 0$. If the point $(1, 2)$ is on the curve, find f.

20. Find a curve in the xy-plane which contains the point $(2, 1)$ and whose normal at (x, y) has slope $y/x, \quad x > 0$.

21. The value of a bottle of a certain rare wine is increasing at a rate of $\sqrt{t+1}$ dollars per year when the wine is t years old. The wine sold for 5 dollars per bottle when new.

a. Write an initial value problem that determines the value $V(t)$ of the bottle of wine after t years.

b. Find the solution of this initial value problem.

22. A population of animals in an ecological niche is growing in time so that its rate of growth $\dfrac{dP}{dt}$ is related to its current size by the differential equation

$$\frac{dP}{dt} = \frac{900}{P^2}.$$

If time is measured in years and initially there are $P(0) = 10$

animals present, find the population function $P(t)$ giving the size of the population after t years.

23. An automobile manufacturer determines that the acceleration of one of its models under full throttle is related to its speed $v(t)$ by the differential equation

$$\frac{dv}{dt} = kv^{-1/2}$$

where k is a constant. In this equation v is in units of meters per second. Find the velocity $v(t)$ after t seconds if the velocity of the automobile at time $t = 0$ is $v(0) = 9$ m/s.

24. A population has size $P(t)$ at time t and grows according to the differential equation $\dfrac{dP}{dt} = 4\sqrt{P}$. Find $P(t)$, $t > 0$, if $P(0) = 10$.

SUMMARY OUTLINE OF CHAPTER 5

◆ The function F is an **antiderivative** for the function f on the interval I if $F'(x) = f(x)$ for all $x \in I$.   (page 229)

◆ The **indefinite integral** for f is $\displaystyle\int f(x)\, dx = F(x) + C$ where $F'(x) = f(x)$.   (page 233)

◆ We have the following **integration** formulas: (page 234)

$\int a\, dx = ax + C$

$\int x^n\, dx = \dfrac{1}{n+1}x^{n+1} + C, \quad n \neq -1 \quad$ (Power Rule)

$\int \cos x\, dx = \sin x + C$

$\int \sin x\, dx = -\cos x + C$

$\int \sec^2 x\, dx = \tan x + C$

$\int \csc^2 x\, dx = -\cot x + C$

$\int \sec x \tan x\, dx = \sec x + C$

$\int \csc x \cot x\, dx = -\csc x + C$

◆ **Theorem:** (Properties of antiderivatives)   (page 233)
(i) $\int [f(x) + g(x)]\, dx = \int f(x)\, dx + \int g(x)\, dx$.   (ii) $\int cf(x)\, dx = c \int f(x)\, dx$.

◆ The **method of substitution** is used to evaluate $\displaystyle\int f(g(x))g'(x)\, dx$ (page 241)

by the substitution $u = g(x)$, $du = g'(x)\, dx$ to obtain

$$\int f(g(x))g'(x)\, dx = \int f(u)\, du$$

$$= F(u) + C$$

$$= F(g(x)) + C$$

where $F' = f$.

◆ A **differential equation** of the form $\dfrac{dy}{dx} = \dfrac{f(x)}{g(y)}$ or $g(y)\,dy = f(x)\,dx$ may be solved by the method of   (page 247)

separation of variables:

$$\int g(y)\,dy = \int f(x)\,dx$$

so

$$G(y) = F(x) + C$$

where $G' = g$ and $F' = f$.

REVIEW EXERCISES—CHAPTER 5

In Exercises 1–14, find the indefinite integral for the given function.

1. $f(x) = 6x^2 - 2x + 1$

2. $f(x) = (x^2 - 6x)^2$

3. $y = \dfrac{\sin \sqrt{t}}{\sqrt{t}}$

4. $f(x) = x \cdot \sqrt{9x^4}$

5. $f(x) = 3\sqrt{x} + 3/\sqrt{x}$

6. $y = (t + \sqrt{t})^3$

7. $y = x \sec^2 x^2$

8. $f(s) = s\sqrt{9 - s^2}$

9. $f(x) = \dfrac{x^3 - 7x^2 + 6x}{x}$

10. $y = \dfrac{x^3 + x^2 - x + 2}{x + 2}$

11. $f(x) = x \cos(1 + x^2)$

12. $f(x) = \dfrac{\sec x}{1 + \tan^2 x}$

13. $y = (2x - 1)(2x + 1)$

14. $y = \dfrac{x}{4x^4 + 4x^2 + 1}$

In Exercises 15–18, find the particular function satisfying the stated conditions.

15. $f'(x) = 1 + \cos x$, $f(0) = 3$

16. $f'(x) = \dfrac{x}{\sqrt{1 + x^2}}$, $f(0) = 4$

17. $f''(x) = 3$, $f'(1) = 6$, $f(0) = 0$

18. $f''(x) = \sin x - \cos x$, $f'(0) = 3$, $f(0) = 0$

19. Find an equation for the graph whose slope at any point is twice the x-coordinate of that point, and that contains the point $(2, 9)$.

20. Find an equation for the graph containing the point $(0, 1)$ whose slope at any point is the quotient of its x-coordinate divided by its y-coordinate.

21. From what height must a ball be dropped in order to strike the ground with a velocity of -49 m/s? (The acceleration due to gravity is -9.8 m/s^2.)

In Exercises 22–25, find the solution of the initial value problem.

22. $\dfrac{dy}{dx} = \sec^2 x$, $y(0) = 1$

23. $\dfrac{dy}{dx} = \dfrac{x + 1}{y}$, $y(1) = 2$

24. $dy = 2xy^2\,dx$, $y(0) = -\dfrac{1}{2}$

25. $dy = \dfrac{\sqrt{x + 1}}{\sqrt{y}}\,dx$, $y(0) = 1$

26. A particle moves along a line with velocity $v(t) = 2t - (t + 1)^{-2}$.
   a. Find $s(t)$, its position at time t, if $s(0) = 0$.
   b. Find $a(t)$, its acceleration at time t.

27. A particle moves along a line so that after t seconds it is $s(t) = t^3 - 6t^2 + 9t - 4$ units on the positive side of the origin.
   a. Where does the particle lie at time $t = 0$?
   b. In which direction is it moving at time $t = 0$?
   c. When does it change direction?
   d. How many times does it change direction?

28. The value of a certain piece of real estate in San Diego has been increasing at a rate of $V'(t) = \sqrt{36 + t}$ thousand dollars per year since 1975. If its value in 1975 was $200,000, find its value $V(t)$, t years after 1975.

UNIT 3

INTEGRATION

Archimedes

Bonaventura Cavalieri

John Wallis

Georg Friedrich Bernhard Riemann

Carl Friedrich Gauss

THE ORIGINS OF INTEGRATION

Historically, the problem of finding the area of a region under a curve, which leads to the integral, was studied earlier and more extensively than the problem of finding the tangent to a curve, which leads to the derivative.

More than two millenia ago, the Greek mathematician Eudoxus (c. 408–355 B.C.) stated what is called the Method of Exhaustion. This principle says that if one successively subtracts from a quantity at least half of it, and from the remainder at least its half, and so on, eventually there will remain something smaller than any pre-assigned quantity. ("Quantity" may refer to a length, an area, or a volume.) As one example, Eudoxus applied this concept to achieve a difference between the area of a circle and that of an inscribed regular polygon. As the number of sides of the polygon becomes very large, the difference in areas becomes very small. Eudoxus eventually proved, from this result, that the ratio of the areas of two circles equals the ratio of the squares of their diameters. He did not actually determine a formula for the area of a circle, however. He also proved a similar theorem concerning the volumes of spheres. Archimedes wrote that Eudoxus was the first to prove that the volume of a cone is one-third the volume of a cylinder having the same base and altitude. The work of Eudoxus is contained in Book V of Euclid's *Elements*, and many applications are given in Book VI. Little is known of the life of Eudoxus, and none of his original work still exists. He was an authority in many fields, and wrote and lectured on astronomy, geography, music, medicine, and philosophy.

Archimedes of Syracuse (in Sicily)(287–212 B.C.) was the greatest mathematician of antiquity. He used a proof based on the Method of Exhaustion to determine an approximation to π, which he found to be between $3\frac{10}{71}$ and $3\frac{10}{70}$. He thought his greatest discovery was that if a right circular cylinder is circumscribed about a sphere, then the area of the sphere is exactly two-thirds the area of the cylinder, and the volume of the sphere is also exactly two-thirds the volume of the cylinder. His desire was that this theorem be inscribed on his tombstone. Three hundred years after his death, when his tomb was long lost, the Roman statesman Cicero hunted for the tomb, and actually found the tombstone with that geometrical figure engraved on it.

Archimedes, using the Method of Exhaustion, came very close to inventing the calculus two thousand years before Newton. It was not until 1906 that the discovery of a manuscript in a Constantinople library showed how he had thought. This manuscript, copied by a tenth-century scribe, contains several works of Archimedes; among them is a mathematical letter to his friend Eratosthenes, a librarian at Alexandria. The Archimedes material had been washed away in the thirteenth century so that the parchment might be reused for religious texts, but most of the earlier writing can be read. (Such a reused parchment is called a *palimpsest*.) Archimedes explained to his friend that he was not very certain of the validity of his method, though it seemed to give correct results. He would cut an area or volume to be measured into infinitely many parallel lines or sections, which (in his imagination) he would place at one end of a lever so as to balance an area or volume at the other end whose measure and center of gravity were known. Although the method was crude, it approximated the concept of the limit, and it got correct results.

In the letter to Eratosthenes mentioned above, Archimedes wrote, "I do believe that men of my time and of the future, and through this method, might find still other theorems which have not yet come to my mind." One of these men of the future was Bonaventura Cavalieri (1598–1647). In 1653 he published *Geometria indivisibilibus* in which is introduced his Method of Indivisibles. It was a precursor to integration, and an extension of Archimedes' work. Cavalieri did not define his terms well, but apparently an indivisible of a planar region is a line across that region, and an indivisible of a solid is a section by a plane. The region or solid is made up of infinitely many parallel chords or sections. If these parallel lines (or planes) are slid into other conformations, the total area (or volume) is unchanged. This principle can be used in ingenious ways to find areas and volumes of complex structures. Cavalieri was a brilliant thinker who wrote on astronomy and optics as well as mathematics, and is credited with introducing logarithms in Italy. As a teenager studying geometry, he was so excellent a student that he occasionally substituted for his teacher in mathematics classes at a nearby university.

John Wallis (1616–1703) was a bright child but he showed little aptitude for mathematics. Indeed, mathematics was not even a part of his grammar school curriculum, although he did learn Latin, Greek, and Hebrew. His acquaintance with mathematics came through reading books his brother had used when studying for a trade. At Cambridge University, mathematics ". . . was scarce looked upon . . . [as being] . . . in fashion." His interest in the subject came through his skill at deciphering captured letters in code written during the British civil war. He was appointed Professor of Geometry (i.e., of Mathematics) at Oxford, and continued in that post for an astonishing 54 years, until his death.

Wallis was one of the first to treat conics as equations of the second degree rather than as sections of a cone, and was the first to explain negative, fractional, and zero exponents in detail. He also created the symbol "∞" still used for infinity. Before Newton, he performed many definite integrations, essentially equivalent to $\int_0^1 x^m \, dx$, and his work with exponents allowed him to let m take on many values (except -1).

Wallis helped to found the Royal Society of London, and published for the first time since antiquity many editions of the writings of Greek mathematical and musical works. He was also one of the first to attempt to teach deaf mutes to speak. Wallis was very quarrelsome; modesty was not one of his virtues. He attacked violently (in print) some of the finest minds of his day.

During the eighteenth century, a vast amount of work was done on the development of the calculus. It was not, however, until the nineteenth century that the foundations of the subject were clearly stated. One of the most important contributors was the German mathematician Georg Friedrich Bernhard Riemann (1826–1866). It was he who finally put the integral onto a firm logical basis, treating it as the limit of upper and lower sums rather than solely as the inverse of differentiation. The modern presentation of the integral is mostly due to Riemann.

Riemann became a lecturer at the University of Göttingen at the age of 28. For his probationary lecture at the university, he had to present three possible titles; the first was almost always chosen. But Carl Friedrich Gauss selected the third title, and Riemann had to work frantically for two months to develop the material, about which he had previously done nothing. The result, however, was a triumph. Riemann's topic was the hypotheses that underlie the foundations of geometry; half a century later the concepts that he presented were used by Albert Einstein as the basis for the theory of general relativity. This lecture is perhaps the single most influential mathematics lecture ever given.

Riemann suffered ill health, and died at only 39 in Italy of tuberculosis. Though he published relatively little, his mathematical influence is still felt in analysis, number theory, and geometry.

Chapter 6
The Definite Integral

The principal goal of this chapter is to define and calculate the area of a region in the plane bounded by the graphs of one or more continuous functions. The solution of this problem will involve a new concept, the *definite integral* of a function over an interval, and we shall see that this concept is intimately related to the notion of the antiderivative introduced in Chapter 5.

Before beginning a careful treatment of the *area problem,* we review the use of sigma notation which will be used heavily in the next few sections.

6.1 REVIEW OF SIGMA NOTATION

The Greek letter Σ (capital sigma) is used to denote the sum of a finite number of similar terms when each is a function of the summation index j:

$$\sum_{j=1}^{n} f(j) \qquad \text{means} \qquad f(1) + f(2) + f(3) + \cdots + f(n).$$

Here are some examples of this notation.

Example 1

$$\sum_{j=1}^{4} (2j + 1) = [2(1) + 1] + [2(2) + 1] + [2(3) + 1] + [2(4) + 1]$$
$$= 3 + 5 + 7 + 9$$
$$= 24.$$

Example 2

$$\sum_{j=1}^{3} (9 - j^2) = (9 - 1^2) + (9 - 2^2) + (9 - 3^2)$$
$$= 8 + 5 + 0$$
$$= 13.$$

Example 3

$$\sum_{j=1}^{4} \sin\left(\frac{j\pi}{4}\right) = \sin\frac{\pi}{4} + \sin\frac{\pi}{2} + \sin\frac{3\pi}{4} + \sin\pi$$

$$= \frac{\sqrt{2}}{2} + 1 + \frac{\sqrt{2}}{2} + 0$$

$$= 1 + \sqrt{2}. \qquad\qquad\qquad \diamond$$

Sums indicated by sigma notation need not begin with the index $j = 1$. We may indicate any integer as the starting value of the index, with the understanding that the sum includes terms for each integer between and including the starting value and the terminal value. That is,

$$\sum_{j=k}^{k+p} f(j) \qquad \text{means} \qquad f(k) + f(k+1) + f(k+2) + \cdots + f(k+p).$$

Example 4

$$\sum_{j=3}^{5} \cos(j\pi) = \cos(3\pi) + \cos(4\pi) + \cos(5\pi)$$

$$= -1 + 1 + (-1)$$

$$= -1. \qquad\qquad\qquad \diamond$$

Example 5

$$\sum_{j=0}^{4} \left(\frac{j}{j+1}\right) = \frac{0}{0+1} + \frac{1}{1+1} + \frac{2}{2+1} + \frac{3}{3+1} + \frac{4}{4+1}$$

$$= 0 + \frac{1}{2} + \frac{2}{3} + \frac{3}{4} + \frac{4}{5}$$

$$= \frac{163}{60}. \qquad\qquad\qquad \diamond$$

Here are two formulas for finite sums which we shall use in Section 6.2. They are proved using mathematical induction. (See Appendix II.)

$$\sum_{j=1}^{k} j = 1 + 2 + 3 + \cdots + k = \frac{k(k+1)}{2}. \qquad\qquad (1)$$

$$\sum_{j=1}^{k} j^2 = 1 + 2^2 + 3^2 + \cdots + k^2 = \frac{k(k+1)(2k+1)}{6}. \qquad\qquad (2)$$

Example 6

Using equation (1) we find that

(a) $1 + 2 + 3 + \cdots + 6 = \displaystyle\sum_{j=1}^{6} j = \frac{6(6+1)}{2} = 21 \qquad (k = 6).$

(b) $1 + 2 + 3 + \cdots + 11 = \sum_{j=1}^{11} j = \dfrac{11(11 + 1)}{2} = 66 \qquad (k = 11)$.

(c) $4 + 5 + 6 + 7 + 8 = \sum_{j=1}^{8} j - \sum_{j=1}^{3} j = \dfrac{8(8 + 1)}{2} - \dfrac{3(3 + 1)}{2}$

$$= 36 - 6$$

$$= 30.$$ ◇

Example 7

Using equation (2) we find that

(a) $1 + 2^2 + 3^2 + \cdots + 9^2 = \sum_{j=1}^{9} j^2 = \dfrac{9(9 + 1)(2 \cdot 9 + 1)}{6}$

$$= 285.$$

(b) $7^2 + 8^2 + \cdots + 12^2 = \sum_{j=1}^{12} j^2 - \sum_{j=1}^{6} j^2$

$$= \dfrac{12(12 + 1)(2 \cdot 12 + 1)}{6} - \dfrac{6(6 + 1)(2 \cdot 6 + 1)}{6}$$

$$= 650 - 91$$

$$= 559.$$ ◇

Exercise Set 6.1

In Exercises 1–12 find the indicated sum.

1. $\displaystyle\sum_{j=1}^{6} (3j + 2)$

2. $\displaystyle\sum_{j=1}^{5} (2j^2 - 5)$

3. $\displaystyle\sum_{j=1}^{5} (2j - 5)^2$

4. $\displaystyle\sum_{j=1}^{5} \sin\left(\dfrac{j\pi}{2}\right)$

5. $\displaystyle\sum_{j=1}^{6} \cos\left(\dfrac{j\pi}{3}\right)$

6. $\displaystyle\sum_{j=1}^{5} \dfrac{\cos j\pi}{j + 1}$

7. $\displaystyle\sum_{j=3}^{6} (j^2 - 4j + 1)$

8. $\displaystyle\sum_{j=4}^{7} (j - 5)$

9. $\displaystyle\sum_{j=1}^{10} j$

10. $\displaystyle\sum_{j=1}^{8} (j + 2)$

11. $\displaystyle\sum_{j=1}^{8} j^2$

12. $\displaystyle\sum_{j=1}^{20} j^2$

13. Explain, using the properties of addition and multiplication for real numbers, why each of the following is true:

a. $\displaystyle\sum_{j=1}^{n} (x_j + y_j) = \sum_{j=1}^{n} x_j + \sum_{j=1}^{n} y_j$

b. $\displaystyle\sum_{j=1}^{n} cx_j = c\sum_{j=1}^{n} x_j$

c. $\displaystyle\sum_{j=1}^{n} (x_j - y_j) = \sum_{j=1}^{n} x_j - \sum_{j=1}^{n} y_j$

d. $\displaystyle\sum_{j=1}^{n} c = nc$.

In Exercises 14–19 use properties a–d in Exercise 13 together with equations (1) and (2) to find the sum.

14. $\displaystyle\sum_{j=1}^{10} 2j$

15. $\displaystyle\sum_{j=1}^{7} 3j^2$

16. $\displaystyle\sum_{j=1}^{8} (j^2 + j + 1)$

17. $\displaystyle\sum_{j=1}^{6} (2j^2 + 3j + 5)$

20. Find the integer n so that

$$\sum_{j=1}^{n} j = 55.$$

18. $\displaystyle\sum_{j=1}^{n} (j^2 + 2j + 3)$

19. $\displaystyle\sum_{j=1}^{n} (6j^2 - 4j)$

6.2 THE AREA PROBLEM: APPROXIMATING SUMS

In order to make use of the language of functions, we shall pursue the following formulation of the Area Problem.

> Area Problem: Let f be a continuous function for $x \in [a, b]$. Find the area of the region R bounded by the graph of f, the x-axis, and the lines $x = a$ and $x = b$.

Figure 2.1 represents a typical region R for the case of a nonnegative function f. Figure 2.2 shows our strategy for attacking the Area Problem. In order to define the

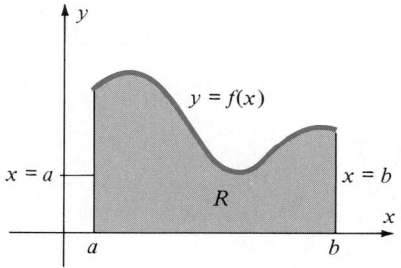

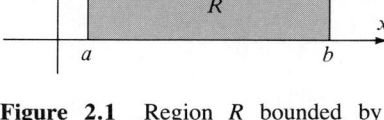

Figure 2.1 Region R bounded by graph of f and x-axis for $a \leq x \leq b$.

Figure 2.2 Approximating the region R by rectangles.

area A, we approximate the region R with rectangles, and we execute our approximation scheme in such a way that we will be able to calculate the area A as the limit of this sequence of approximations.

Figure 2.3 illustrates the basic idea behind the approximation of R by rectangles. As the number of rectangles increases, and the sizes of the individual rectangles

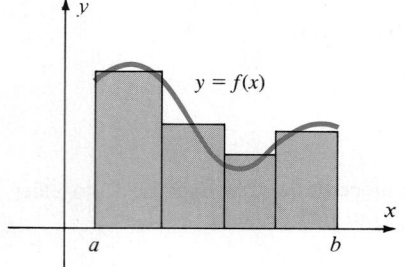
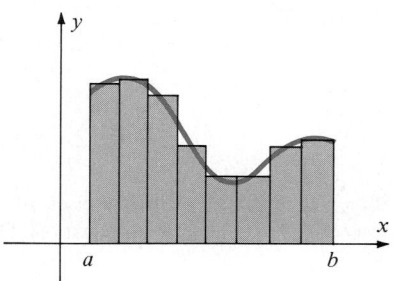
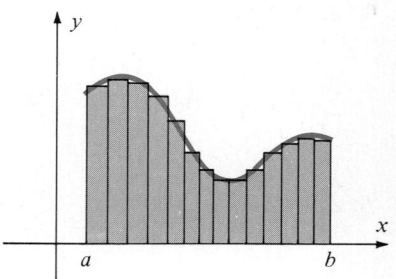

Figure 2.3 As the number n of subintervals increases, the approximation of R by rectangles becomes more precise.

decrease, the union of the set of approximating rectangles more accurately "fits" the region R. Our intention is to define the area A of R to be the limiting value of the areas associated with these approximations. Of course, we must first show that such a limit exists, and that is the purpose of this section.

Lower Approximating Sums

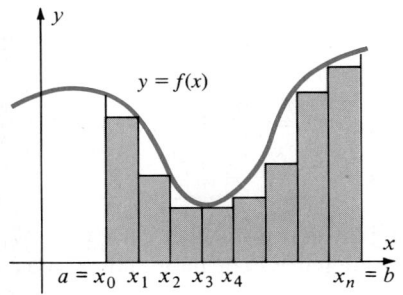

Figure 2.4 A lower approximating sum for R. Union of rectangles lies entirely within R.

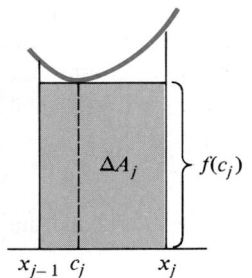

Figure 2.5 The jth approximating rectangle has area $\Delta A_j = f(c_j) \Delta x$.

A **lower approximating sum** $\underline{S}_n$ for the region R is an approximation of R by n rectangles of equal width *each of which is entirely contained within R* (see Figures 2.4 and 2.5). Since we use n rectangles of equal width, the width of each is $\Delta x = \dfrac{b - a}{n}$, and the endpoints of the resulting subintervals are

$$x_0 = a, \ x_1 = a + \Delta x, \ x_2 = a + 2\,\Delta x, \ \ldots, \ x_n = a + n\,\Delta x = b. \quad (1)$$

Obviously, we want the height of the rectangle constructed over the interval $[x_{j-1}, x_j]$ to be the minimum value of f on $[x_{j-1}, x_j]$. If f is continuous on $[x_{j-1}, x_j]$, there is at least one number $c_j \in [x_{j-1}, x_j]$ with

$$f(c_j) = \min \{f(x) \mid x_{j-1} \leq x \leq x_j\}$$

(Theorem 1, Chapter 4). Using this notation we can write the area ΔA_j of this jth *inscribed* rectangle as

$$\Delta A_j = f(c_j)\,\Delta x.$$

The lower approximating sum $\underline{S}_n$ is therefore

$$\underline{S}_n = \sum_{j=1}^{n} \Delta A_j$$

$$= \sum_{j=1}^{n} f(c_j)\,\Delta x.$$

Example 1

Find the lower approximating sum $\underline{S}_4$ for the area of the region R bounded by the graph of $f(x) = 4 - x^2$ and the x-axis between $x = 0$ and $x = 2$.

Solution: Since $n = 4$, the length of each subinterval is

$$\Delta x = \frac{2 - 0}{4} = \frac{1}{2}.$$

The endpoints of the subintervals are therefore

$$x_0 = 0, \qquad x_1 = \frac{1}{2}, \qquad x_2 = 1, \qquad x_3 = \frac{3}{2}, \qquad \text{and} \qquad x_4 = 2.$$

Since $f(x) = 4 - x^2$ is decreasing on $[0, 2]$, the minimum value of f on each subinterval occurs at the right endpoint. Thus,

$$c_1 = \frac{1}{2}, \qquad c_2 = 1, \qquad c_3 = \frac{3}{2}, \qquad \text{and} \qquad c_4 = 2.$$

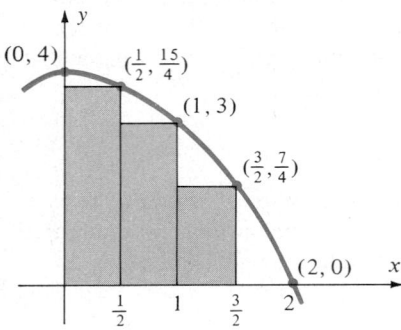

Figure 2.6 Lower approximating sum $\underline{S}_4$ in Example 1 for $f(x) = 4 - x^2$ on the interval [0, 2].

The lower approximating sum is, therefore,

$$
\begin{aligned}
\underline{S}_4 &= f\left(\frac{1}{2}\right) \cdot \left(\frac{1}{2}\right) + f(1)\left(\frac{1}{2}\right) + f\left(\frac{3}{2}\right)\left(\frac{1}{2}\right) + f(2)\left(\frac{1}{2}\right) \\
&= \left[4 - \left(\frac{1}{2}\right)^2\right]\left(\frac{1}{2}\right) + [4 - (1)^2]\left(\frac{1}{2}\right) + \left[4 - \left(\frac{3}{2}\right)^2\right]\left(\frac{1}{2}\right) + [4 - (2)^2]\left(\frac{1}{2}\right) \\
&= \left[\frac{15}{4} + 3 + \frac{7}{4} + 0\right]\left(\frac{1}{2}\right) \\
&= \frac{17}{4}.
\end{aligned}
$$

We will later show that the actual value of the area is $A = \dfrac{16}{3}$. Thus, $\underline{S}_4 < A$ (see Figure 2.6). ◇

A simple but important observation concerning lower sums is that, since each of the associated rectangles lies entirely within the region R, our definition of area should provide that

$$\underline{S}_n \le A \tag{2}$$

for all lower approximating sums $\underline{S}_n$.

Upper Approximating Sums

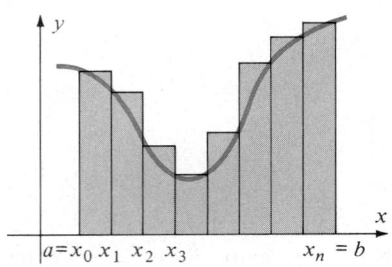

Figure 2.7 An upper approximating sum. The region R lies entirely within the union of the rectangles.

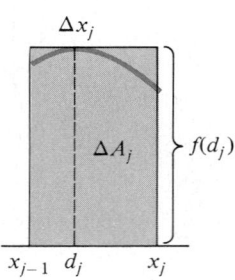

Figure 2.8 The jth approximating rectangle has area $\Delta A_j = f(d_j)\,\Delta x$.

If, instead of using the minimum value of f on each subinterval $[x_{j-1}, x_j]$, we use the maximum value, we obtain what is called an **upper approximating sum, $\overline{S}_n$**. That is, we take the height of the rectangle over the interval $[x_{j-1}, x_j]$ to be the value $f(d_j)$, where

$$f(d_j) = \max \{f(x) \mid x_{j-1} \le x \le x_j\}.$$

The area of the jth approximating rectangle is, therefore, $\Delta A_j = f(d_j)\Delta x$, and the upper approximating sum is

$$\overline{S}_n = \sum_{j=1}^{n} \Delta A_j = \sum_{j=1}^{n} f(d_j)\,\Delta x$$

(See Figures 2.7 and 2.8.)

Example 2

Find the upper approximating sum $\overline{S}_4$ for the region R in Example 1.

Solution: As in Example 1, we have subintervals of length $\Delta x = 1/2$ and with endpoints

$$x_0 = 0, \qquad x_1 = \frac{1}{2}, \qquad x_2 = 1, \qquad x_3 = \frac{3}{2}, \qquad \text{and} \qquad x_4 = 2.$$

However, since $f(x) = 4 - x^2$ is decreasing on [0, 2], the maximum value of f will occur at the *left* endpoint of each subinterval. We will therefore have

$$d_1 = 0, \qquad d_2 = \frac{1}{2}, \qquad d_3 = 1, \qquad \text{and} \qquad d_4 = \frac{3}{2}.$$

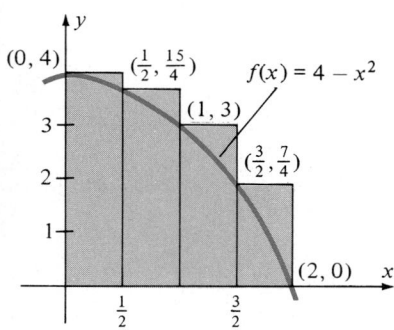

Figure 2.9 Upper approximating sum $\overline{S}_4$ in Example 2 for $f(x) = 4 - x^2$ on the interval $[0, 2]$.

The upper approximating sum $\overline{S}_4$ is

$$\overline{S}_4 = f(0)\left(\frac{1}{2}\right) + f\left(\frac{1}{2}\right)\left(\frac{1}{2}\right) + f(1)\left(\frac{1}{2}\right) + f\left(\frac{3}{2}\right)\left(\frac{1}{2}\right)$$

$$= [4 - (0)^2]\left(\frac{1}{2}\right) + \left[4 - \left(\frac{1}{2}\right)^2\right]\left(\frac{1}{2}\right) + [4 - (1)^2]\left(\frac{1}{2}\right) + \left[4 - \left(\frac{3}{2}\right)^2\right]\left(\frac{1}{2}\right)$$

$$= \left[4 + \frac{15}{4} + 3 + \frac{7}{4}\right]\left(\frac{1}{2}\right)$$

$$= \frac{25}{4}.$$

Since $A = 16/3$, we have $A < \overline{S}_4$. (See Figure 2.9.) ◇

Since, for upper approximating sums, the union of the approximating rectangles entirely contains the region R, our definition of area should provide that

$$A \le \overline{S}_n \tag{3}$$

for all upper approximating sums $\overline{S}_n$, as in Example 2.

Combining inequalities (2) and (3), we see that a definition of area should provide that

$$\underline{S}_n \le A \le \overline{S}_n \tag{4}$$

for all lower approximating sums $\underline{S}_n$ and upper approximating sums $\overline{S}_n$. Also, by the way $\underline{S}_n$ and $\overline{S}_n$ are defined, you can see that, in general, $\underline{S}_n$ increases and $\overline{S}_n$ decreases as $n \to \infty$. Now if it were the case that $\underline{S}_n$ and $\overline{S}_n$ had the same limit* S as $n \to \infty$, we could both satisfy inequality (4) and obtain an unambiguous definition of the area A of R by taking $A = S$ (see Figure 2.10). Indeed, this is precisely how things work out. Before taking up the general case, we look at a particular example.

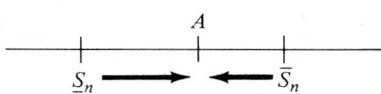

Figure 2.10 The area A of R is the unique number satisfying $\underline{S}_n \le A \le \overline{S}_n$ for all lower and upper approximating sums.

Example 3

Let R be the region bounded above by the graph of $f(x) = x^2$, below by the x-axis, on the left by $x = 0$, and on the right by $x = 1$. Show that the number $1/3$ satisfies the inequality

$$\underline{S}_n \le \frac{1}{3} \le \overline{S}_n$$

for all lower and upper approximating sums, and that

$$\lim_{n \to \infty} \underline{S}_n = \frac{1}{3} = \lim_{n \to \infty} \overline{S}_n.$$

---

*The limit as $n \to \infty$ of an approximating sum $\underline{S}_n$ or $\overline{S}_n$ means the same as the limit $\lim_{x \to \infty} f(x)$ of a function f, except that the expressions $\underline{S}_n$ and $\overline{S}_n$ are functions defined only for positive integers n. The techniques for evaluating $\lim_{n \to \infty} \underline{S}_n$ and $\lim_{n \to \infty} \overline{S}_n$ are the same as those for evaluating $\lim_{x \to \infty} f(x)$.

Solution: We shall make use of the formula

$$1 + 2^2 + 3^2 + \cdots + k^2 = \frac{k(k+1)(2k+1)}{6} \tag{5}$$

from Section 6.1.

To form the lower approximating sum for $f(x) = x^2$ on $[0, 1]$, we use n subintervals of equal size $\Delta x = \frac{1}{n}$, and endpoints

$$x_0 = 0, \quad x_1 = \frac{1}{n}, \quad x_2 = \frac{2}{n}, \quad \ldots, \quad x_n = \frac{n}{n} = 1.$$

Since $f(x) = x^2$ is *increasing* on $[0, 1]$, the minimum value of f on each subinterval $[x_{j-1}, x_j]$ will occur at the *left* endpoint, x_{j-1}. That is, $c_j = x_{j-1} = \frac{j-1}{n}$ for each $j = 1, 2, \ldots, n$. Thus,

$$\underline{S}_n = f(0) \cdot \frac{1}{n} + f\left(\frac{1}{n}\right) \cdot \frac{1}{n} + f\left(\frac{2}{n}\right) \cdot \frac{1}{n} + \cdots + f\left(\frac{n-1}{n}\right) \cdot \frac{1}{n}$$

$$= \left[0^2 + \left(\frac{1}{n}\right)^2 + \left(\frac{2}{n}\right)^2 + \cdots + \left(\frac{n-1}{n}\right)^2\right]\left(\frac{1}{n}\right)$$

$$= [1 + 2^2 + 3^2 + \cdots + (n-1)^2]\left(\frac{1}{n^3}\right).$$

Using formula (5) with $k = n - 1$, we can write this sum as

$$\underline{S}_n = \left\{\frac{(n-1)[(n-1)+1][2(n-1)+1]}{6}\right\}\left(\frac{1}{n^3}\right)$$

$$= \frac{(n-1)(n)(2n-1)}{6n^3}$$

so

$$\underline{S}_n = \frac{2n^2 - 3n + 1}{6n^2} = \frac{1}{3} - \left(\frac{3n-1}{6n^2}\right). \quad \text{(Figure 2.11.)} \tag{6}$$

To obtain the upper approximating sum we use $d_j = x_j$ on the interval $[x_{j-1}, x_j]$, $j = 1, 2, \ldots, n$. That is, d_j is the *right* endpoint of the jth interval. Thus,

$$\overline{S}_n = f\left(\frac{1}{n}\right) \cdot \frac{1}{n} + f\left(\frac{2}{n}\right) \cdot \frac{1}{n} + \cdots + f\left(\frac{n}{n}\right) \cdot \frac{1}{n}$$

$$= \left[\left(\frac{1}{n}\right)^2 + \left(\frac{2}{n}\right)^2 + \cdots + \left(\frac{n}{n}\right)^2\right]\left(\frac{1}{n}\right)$$

$$= [1 + 2^2 + 3^2 + \cdots + n^2] \cdot \left(\frac{1}{n^3}\right)$$

$$= \frac{n(n+1)(2n+1)}{6n^3} \quad \text{(using formula (5))}$$

$$= \frac{2n^2 + 3n + 1}{6n^2},$$

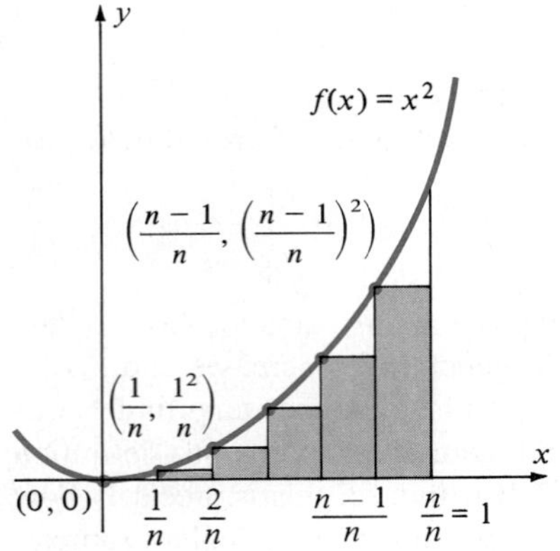

Figure 2.11 The lower approximating sum $\underline{S}_n$ for $f(x) = x^2$ on $[0, 1]$ is $\underline{S}_n = \frac{1}{3} - \left(\frac{3n-1}{6n^2}\right)$.

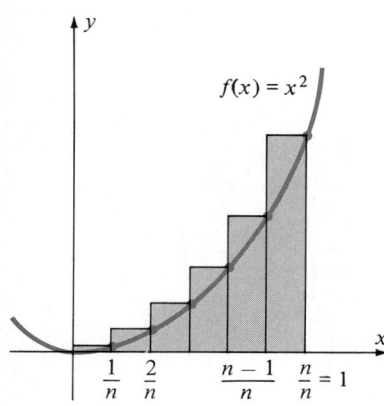

Figure 2.12 The upper approximating sum $\overline{S}_n$ for $f(x) = x^2$ on $[0, 1]$ is $\overline{S}_n = \dfrac{1}{3} + \left(\dfrac{3n + 1}{6n^2}\right)$.

so

$$\overline{S}_n = \frac{1}{3} + \left(\frac{3n + 1}{6n^2}\right). \qquad \text{(Figure 2.12.)} \qquad (7)$$

Now since the numbers $\dfrac{3n - 1}{6n^2}$ and $\dfrac{3n + 1}{6n^2}$ are positive for $n \geq 1$, we have from (6) and (7) that

$$\frac{1}{3} - \left(\frac{3n - 1}{6n^2}\right) = \underline{S}_n < \frac{1}{3} < \overline{S}_n = \frac{1}{3} + \left(\frac{3n + 1}{6n^2}\right) \qquad (8)$$

for all $n \geq 2$. Moreover, the number $1/3$ is the *only* number that can occupy the middle position in inequality (8), since

$$\lim_{n \to \infty} \underline{S}_n = \lim_{n \to \infty} \left[\frac{1}{3} - \left(\frac{3n - 1}{6n^2}\right)\right] = \left(\frac{1}{3} - 0\right) = \frac{1}{3},$$

and

$$\lim_{n \to \infty} \overline{S}_n = \lim_{n \to \infty} \left[\frac{1}{3} + \left(\frac{3n + 1}{6n^2}\right)\right] = \left(\frac{1}{3} + 0\right) = \frac{1}{3}.$$

We therefore define the area of the region R to be the number $A = 1/3$. ◇

The following theorem shows that the result of Example 3 is true for *any* continuous function.

THEOREM 1

Let f be continuous and nonnegative on the interval $[a, b]$. Let $\underline{S}_n$ and $\overline{S}_n$ denote the lower and upper approximating sums for f on $[a, b]$. Then $\lim_{n \to \infty} \underline{S}_n$ and $\lim_{n \to \infty} \overline{S}_n$ exist, and

$$\lim_{n \to \infty} \underline{S}_n = \lim_{n \to \infty} \overline{S}_n.$$

The proof of Theorem 1 is given in Appendix II. We shall focus here on the significance of this result.

Combining inequality (4) with the statement of Theorem 1, we conclude that the caption on Figure 2.10 is indeed correct: as $n \to \infty$, $\underline{S}_n$ increases, $\overline{S}_n$ decreases, and the two approximations meet at a single number $S = \lim_{n \to \infty} \underline{S}_n = \lim_{n \to \infty} \overline{S}_n$ "in the limit." This is the number that we shall define to be the area A of the region R. In doing so, however, we want to make use of one additional notion: an *arbitrary approximating sum* (or, just an "approximating sum").

Let's again divide the interval $[a, b]$ into n subintervals of equal length $\Delta x = \dfrac{b - a}{n}$, and with endpoints

$$a = x_0 < x_1 < x_2 < \cdots < x_n = b.$$

But this time, for the height of the rectangle over the jth interval $[x_{j-1}, x_j]$, we use $f(t_j)$, where t_j is *any* number in the interval $[x_{j-1}, x_j]$. (That is, $t_j \in [x_{j-1}, x_j]$ is *arbitrary*.) An (arbitrary) approximating sum is therefore

$$S_n = \sum_{j=1}^{n} f(t_j) \, \Delta x, \qquad t_j \in [x_{j-1}, x_j].$$

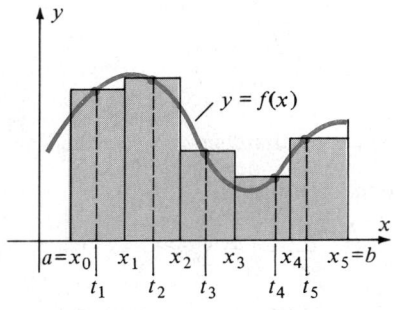

Figure 2.13 An arbitrary approximating sum S_5 for $f(x)$ on $[a, b]$. The numbers $t_j \in [x_{j-1}, x_j]$ are arbitrary.

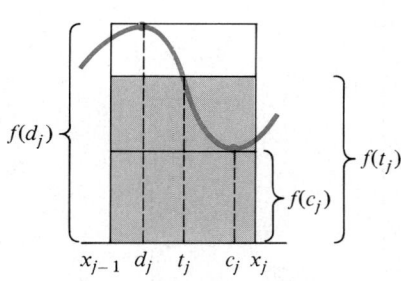

Figure 2.14 $f(c_j) \le f(t_j) \le f(d_j)$, so $\underline{S}_n \le S_n \le \overline{S}_n$.

Since $f(c_j) \le f(t_j) \le f(d_j)$ in each interval $[x_{j-1}, x_j]$, any approximating sum S_n must satisfy the inequality

$$\underline{S}_n \le S_n \le \overline{S}_n \tag{9}$$

for each $n = 1, 2, \ldots$ (see Figures 2.13 and 2.14). Since $\lim_{n \to \infty} \underline{S}_n = \lim_{n \to \infty} \overline{S}_n$, it follows from an argument analogous to the Pinching Theorem that

$$\lim_{n \to \infty} \underline{S}_n = \lim_{n \to \infty} S_n = \lim_{n \to \infty} \overline{S}_n.$$

This observation is so important that we state it as a corollary to Theorem 1.

COROLLARY 1

Let f be continuous and nonnegative on $[a, b]$. There exists a unique number S so that

$$S = \lim_{n \to \infty} S_n = \lim_{n \to \infty} \sum_{j=1}^{n} f(t_j)\, \Delta x, \qquad t_j \in [x_{j-1}, x_j]$$

regardless of how the numbers t_j are chosen in the intervals $[x_{j-1}, x_j]$.

The number S in Corollary 1 is how we define the area A of the region R in the statement of the Area Problem.

DEFINITION 1

Let R be the region bounded above by the graph of the continuous nonnegative function f, below by the x-axis, on the left by $x = a$, and on the right by $x = b$. The **area** of R is the number A defined by the equation

$$A = \lim_{n \to \infty} \sum_{j=1}^{n} f(t_j)\, \Delta x, \qquad t_j \in [x_{j-1}, x_j]$$

where $x_0, x_1, x_2, \ldots, x_n$ are the endpoints obtained by dividing $[a, b]$ into n subintervals of equal length $\Delta x = \dfrac{b - a}{n}$ and $t_j \in [x_{j-1}, x_j]$ is arbitrary.

Table 2.1 Lower and upper approximating sums for $f(x) = 3x^2 + 7$ on $[2, 3]$ obtained on a computer.

n	$\underline{S}_n$	$\overline{S}_n$
2	22.3749	29.8750
5	24.5199	27.5200
10	25.2549	26.7550
25	25.7008	26.3008
100	25.9250	26.0750
500	25.9850	26.0150

We can summarize the work of this section by saying that if f is continuous on $[a, b]$, all approximating sums (lower, upper, or arbitrary) have the same limit S as $n \to \infty$, and the area A of the region R is this limit. To find A for a particular region R, we may therefore construct *any* approximating sum S_n and calculate the limit

$$A = \lim_{n \to \infty} S_n.$$

Although the idea behind approximating sums is straightforward, the calculations involved in executing a particular problem are both repetitious and time-consuming. This is precisely the kind of situation in which computers can be very helpful. Program 2 in Appendix I is a BASIC program which computes a lower approximating sum for the function $f(x) = 3x^2 + 7$ on the interval $[2, 3]$. Program 3 is a BASIC program which computes upper approximating sums for this same function on $[2, 3]$. Using this program on a microcomputer we obtained the results in Table 2.1.

REMARK: The results of Table 2.1 illustrate vividly both the power and the limitations of computers in studying and applying the ideas of the calculus. Implementing Program 2 or Program 3 on a computer makes the calculation of particular approximating sums relatively effortless. (If you disagree with this statement, try producing the results of Table 2.1 by hand!) However, while it seems clear from Table 2.1 that the values of the approximating sums obtained are approaching a limiting value, it is not at all clear what the precise value of that limit is. By contrast, the method of approximating sums presented in Example 3, tedious as it may be, will provide a convincing demonstration that the limit in question is precisely 26 (see Exercise 15).

Exercise Set 6.2

In each of Exercises 1–6, find the lower approximating sum $\underline{S}_n$ for the area of the region bounded by the graph of f, the x-axis, the line $x = a$, and the line $x = b$.

1. $f(x) = x$, $\quad a = 0$, $\quad b = 4$, $\quad n = 4$

2. $f(x) = 2x + 5$, $\quad a = 1$, $\quad b = 3$, $\quad n = 4$

3. $f(x) = 2x^2 + 3$, $\quad a = 0$, $\quad b = 4$, $\quad n = 4$

4. $f(x) = 9 - x^2$, $\quad a = -3$, $\quad b = 0$, $\quad n = 3$

5. $f(x) = \sin x$, $\quad a = 0$, $\quad b = \pi$, $\quad n = 4$

6. $f(x) = x^3 + x + 1$, $\quad a = 0$, $\quad b = 6$, $\quad n = 3$

In each of Exercises 7–12, find the upper approximating sum $\overline{S}_n$ for the area of the region bounded by the graph of $f(x)$, the x-axis, the line $x = a$, and the line $x = b$.

7. $f(x) = 2x$, $\quad a = 0$, $\quad b = 3$, $\quad n = 3$

8. $f(x) = 6 - x$, $\quad a = 0$, $\quad b = 3$, $\quad n = 6$

9. $f(x) = 3x^2 + 10$, $\quad a = -1$, $\quad b = 3$, $\quad n = 4$

10. $f(x) = 4 - x^2$, $\quad a = 0$, $\quad b = 2$, $\quad n = 4$

11. $f(x) = \cos x$, $\quad a = -\pi/2$, $\quad b = \pi/2$, $\quad n = 4$

12. $f(x) = x^3 + x + 1$, $\quad a = 0$, $\quad b = 6$, $\quad n = 3$

13. Let $f(x) = 3x + 2$.
  a. Find the lower approximating sum $\underline{S}_4$ for $f(x)$ on $[0, 2]$ with $n = 4$ subintervals.
  b. Find an expression for the lower approximating sum $\underline{S}_n$ for $f(x)$ on $[0, 2]$ with n subintervals.
  c. Find $\lim_{n \to \infty} \underline{S}_n$ with $\underline{S}_n$ as in part (b).
  d. Find an expression for the upper approximating sum $\overline{S}_n$ for $f(x)$ on $[0, 2]$ with n subintervals.
  e. Find $\lim_{n \to \infty} \overline{S}_n$ and $\lim_{n \to \infty} (\overline{S}_n - \underline{S}_n)$.
  f. What do you conclude about the area bounded by the graph of $y = f(x)$, the x-axis, the line $x = 0$, and the line $x = 2$?

14. Let $f(x) = x^2$.
  a. Find the lower approximating sum $\underline{S}_n$ for $f(x)$ on the interval $[0, a]$.
  b. Find $\lim_{n \to \infty} \underline{S}_n$ with $\underline{S}_n$ as in part (a).
  c. Find the lower approximating sum $\underline{S}_n$ for the function $f(x) = x^2$ on the interval $[a, b]$ where $0 \le a \le b$.
  d. Find $\lim_{n \to \infty} \underline{S}_n$ with $\underline{S}_n$ as in part (c).
  e. How could you have answered part (d) from the answer to part (b) and area considerations alone?

f. What is the area of the region bounded by the graph of $y = x^2$ between $x = 1$ and $x = 5$?

15. Use the method of lower approximating sums to calculate the area of the region bounded above by the graph of $y = 3x^2 + 7$, on the left by $x = 2$, on the right by $x = 3$, and below by the x-axis. Compare your results with those of Table 2.1.

16. Rework Exercise 15 using upper approximating sums.

17. Rework Exercise 13 for the function $f(x) = 3x + 4$ and the interval $[1, 2]$.

18. Rework Exercise 13 for the function $f(x) = 8 - 2x$ and the interval $[0, 4]$.

19. Rework Exercise 15 for the function $f(x) = 3x^2 + 6$ and the interval $[0, 1]$.

20. Rework Exercise 15 for the function $f(x) = 2x^2 + x + 3$ and the interval $[0, 2]$.

21. Rework Exercise 15 for the function $f(x) = |1 - x^2|$ and the interval $[0, 2]$.

22. *(Calculator)* Use a hand calculator to approximate the area enclosed by the graph of the ellipse

$$\frac{x^2}{9} + \frac{y^2}{4} = 1$$

as follows.

a. Solve the equation for y to obtain

$$y = \pm 2\sqrt{1 - \frac{x^2}{9}}.$$

Choose the + sign to obtain the equation for the upper half of the figure.

b. Observe that the total area will be four times the area of the portion of the ellipse lying in the first quadrant.

c. Use your calculator and the function $f(x) = 2\sqrt{1 - x^2/9}$ to obtain the lower approximating sum $\underline{S}_3$. Multiply by four to estimate the total area.

d. Obtain an improved estimate by using $n = 6$.

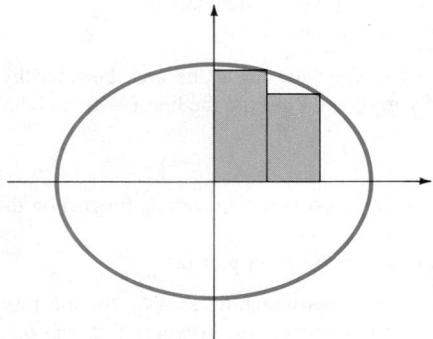

23. *(Calculator)* Consider the unit circle.

a. Write down the equation for the function $y = f(x)$ that describes the top half of the graph.

b. Use your calculator to obtain an estimate for the area of the portion of the circle lying in the first quadrant by calculating the lower approximating sum $\underline{S}_3$.

c. Calculate $\underline{S}_6$ for the same region.

d. Compare your estimates with what you know to be the exact value of that area.

24. *(Calculator)* Rework Exercise 22 using upper approximating sums.

25. *(Calculator)* Rework Exercise 23 using upper approximating sums.

26. *(Computer)*

a. Modify Program 2 to compute lower approximating sums for the function $f(x) = 3x^2 + 7$ on the interval $[A, B]$ instead of the interval $[2, 3]$.

b. Complete the following table using your modified program.

$[A, B]$	n	$\underline{S}_n$
$[1, 5]$	5	
$[1, 5]$	20	
$[1, 5]$	100	
$[5, 7]$	5	
$[5, 7]$	20	
$[5, 7]$	100	
$[1, 7]$	5	
$[1, 7]$	20	
$[1, 7]$	100	

c. What do you conjecture, based on your results in part (b) about the relationship between $\underline{S}_n$ on $[1, 7]$, $\underline{S}_n$ on $[1, 5]$, and $\underline{S}_n$ on $[5, 7]$?

27. *(Computer)* Make the same changes in Program 3 as indicated for Program 2 in Exercise 26. Then answer the same questions as in Exercise 26 for the corresponding upper sums $\overline{S}_n$.

28. *(Computer)* Modify Program 2 to compute lower approximating sums for the function $f(x) = \sqrt{1 + x^2}$ on the interval $[2, 3]$. Then use the program to approximate the area bounded by the graph of $f(x) = \sqrt{1 + x^2}$, the x-axis, the line $x = 2$, and the line $x = 3$.

29. *(Computer)* Modify the program you obtained in Exercise 28 so as to compute lower approximating sums for $f(x) = \sqrt{1 + x^2}$ on the interval $[a, b]$ (instead of $[2, 3]$).

30. *(Computer)* Write a program to approximate the area of the ellipse described in Exercise 22.

31. *(Computer)* Write a program to approximate the area bounded by the arc of $y = \sin x$ and the x-axis between $x = 0$ and $x = \pi/4$ using lower approximating sums.

6.3 RIEMANN SUMS: THE DEFINITE INTEGRAL

In this section, we generalize the notion of an approximating sum. We want to do this for two reasons. First, the approximating sum defined in Section 6.2 applies only to nonnegative functions and the problem of calculating areas of regions bounded by their graphs. We shall want to apply a similar approximation procedure in situations which can involve functions that have negative values or to issues other than area. Second, we want to free ourselves from the restriction of having to use subintervals of equal size, for reasons we shall see in this section.

Partitions and Norms

If $[a, b]$ is a closed interval, a **partition** P_n of $[a, b]$ is any set of $n + 1$ numbers $\{x_0, x_1, x_2, \ldots, x_n\}$ with

$$a = x_0 < x_1 < x_2 < \cdots < x_n = b.$$

As you can see in Figure 3.1, a partition P_n divides the interval $[a, b]$ into n nonoverlapping subintervals.

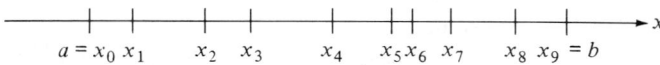

Figure 3.1 A partition P_9 for the interval $[a, b]$.

However, these subintervals are not necessarily of equal length. We define the **norm** of the partition P_n, denoted by $\|P_n\|$, to be the largest of these lengths. That is,

$$\|P_n\| = \max\{(x_1 - x_0), (x_2 - x_1), \ldots, (x_n - x_{n-1})\}.$$

Then

$$x_j - x_{j-1} \leq \|P_n\|, \qquad j = 1, 2, 3, \ldots, n. \tag{1}$$

Example 1

One partition of $[0, 4]$ is $P_7 = \left\{0, 1, \dfrac{3}{2}, \dfrac{7}{4}, 2, \dfrac{10}{3}, \dfrac{11}{3}, 4\right\}$. The norm of this partition is $\|P_7\| = \dfrac{10}{3} - 2 = \dfrac{4}{3}$. $\diamond$

If f is defined on $[a, b]$, we define a *Riemann sum** for f on $[a, b]$ in much the same way as we defined arbitrary approximating sums in Section 6.2, except that we do not require the subintervals to be of equal length. Nor do we require $f(x) \geq 0$.

DEFINITION 2

A **Riemann sum** for f on $[a, b]$ is any sum of the form

$$R_n = \sum_{j=1}^{n} f(t_j) \, \Delta x_j, \qquad t_j \in [x_{j-1}, x_j]$$

where $\{x_0, x_1, x_2, \ldots, x_n\}$ is a partition of $[a, b]$, $\Delta x_j = x_j - x_{j-1}$, and t_j is an element of $[x_{j-1}, x_j]$, $j = 1, 2, \ldots, n$.

*Riemann sums, and most of what is presented here as the theory of the definite integral, are originally the work of the German mathematician Bernhard Riemann, 1826–1866.

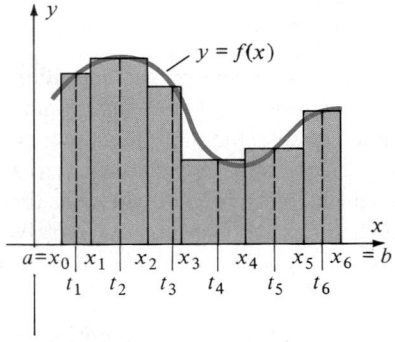

Figure 3.2 A Riemann sum R_6.

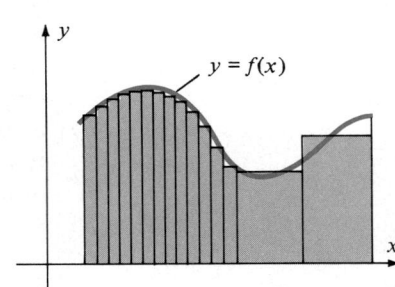

Figure 3.3 As $n \to \infty$, the widths of some rectangles may remain large.

Thus, every approximating sum S_n is also a Riemann sum, but the latter concept is more general. Figure 3.2 shows a geometric interpretation for a typical Riemann sum.

As in Section 6.2, we shall want to use Riemann sums to calculate areas (at least in principle). But in generalizing the notion of approximating sum we have introduced a minor difficulty. We must now be more careful about speaking of the limit of a sequence $\{R_n\}$ of Riemann sums as $n \to \infty$. The reason is illustrated by Figure 3.3. Since the lengths of the subintervals in the partition P_n are not necessarily equal, the lengths of all subintervals need not necessarily become small as n becomes large. If we want this to happen (as we do when approximating regions by rectangles) we must also specify that P_n is a partition for which $\|P_n\| \to 0$ as $n \to \infty$. By inequality (1) this guarantees that $\Delta x_j \to 0$ for all $j = 1, 2, \ldots, n$.

In order to extend and formalize the notion of the limit of a sequence of approximating sums for f on $[a, b]$ to more general Riemann sums, we must introduce the concept of the *definite integral* of f on $[a, b]$ and the property of f being *integrable* on $[a, b]$.

DEFINITION 3

The function f is said to be **integrable** on the interval $[a, b]$ with **definite integral** I if and only if for any given number $\epsilon > 0$ there exists a corresponding number $\delta > 0$ so that

$$\left| I - \sum_{j=1}^{n} f(t_j) \, \Delta x_j \right| < \epsilon \qquad (2)$$

for every Riemann sum $\sum_{j=1}^{n} f(t_j) \, \Delta x_j$ for f on $[a, b]$ with partition norm $\|P_n\| < \delta$.

We write

$$I = \lim_{n \to \infty} \sum_{j=1}^{n} f(t_j) \, \Delta x_j \qquad (3)$$

to mean that f is integrable on $[a, b]$ with definite integral I.

Definition (3) explains what we mean by saying that "the definite integral I is the limit of *any* sequence of Riemann sums for f on $[a, b]$ with the property that $\|P_n\| \to 0$ as $n \to \infty$." We shall abbreviate the formal statement of this concept using equation (3) because of its similarity with the corresponding statement for approximating sums with subintervals of equal length.

The following theorem allows us to dispense with much of the formality of Definition 3, however. It states that if f is continuous on $[a, b]$, all Riemann sums for f on $[a, b]$ with $\|P_n\| \to 0$ as $n \to \infty$ have the same "limit." That is, all continuous functions are integrable. Although a precise proof of this result is beyond the scope of this text, the basic ideas in the proof are the same as those given in Appendix II for the proof of Theorem 1.

THEOREM 2

If f is continuous on $[a, b]$ then f is integrable on $[a, b]$. That is, there exists a unique number I with

$$I = \lim_{n \to \infty} \sum_{j=1}^{n} f(t_j)\, \Delta x_j$$

for all Riemann sums for which $\|P_n\| \to 0$ as $n \to \infty$.

There do exist integrable functions that are not continuous. (See Exercises 39–40). We shall not need to consider such functions in this text, however, so we shall henceforth restrict our attention to definite integrals involving continuous functions.

Notation for the Definite Integral

When f is continuous on $[a, b]$ we use the richer notation

$$\int_a^b f(x)\, dx = I = \lim_{n \to \infty} \sum_{j=1}^{n} f(t_j)\, \Delta x_j$$

to denote the definite integral I in Definition 3. That is, $I = \int_a^b f(x)\, dx$. The advantage of this notation is that it indicates both the function f and the interval $[a, b]$ which determine the number $\int_a^b f(x)\, dx$.

The symbol $\int_a^b f(x)\, dx$ is read, "the definite integral from a to b of f with respect to x." As with the symbol $\int f(x)\, dx$ for antiderivatives, we refer to $\int$ as the **integral sign,** and we write the symbol dx following the **integrand** $f(x)$ to indicate that x is the independent variable for f. (We shall later see that the symbol dx has a meaning associated with the differential dx, as suggested by the Riemann sum. But for now simply regard dx as part of the notation identifying the definite integral.) The endpoints a and b are referred to as the **limits** of integration, and their presence at the extremities of the integral sign distinguishes the definite integral $\int_a^b f(x)\, dx$ from the antiderivative $\int f(x)\, dx$.

REMARK 1: It is very important to note that the definite integral $\int_a^b f(x)\, dx$ is a *number,* while the antiderivative $\int f(x)\, dx$ is a family of *functions.* That is,

$$\int_a^b f(x)\, dx \in (-\infty, \infty), \qquad \text{while} \qquad \int f(x)\, dx = F(x) + C$$

where $F'(x) = f(x)$. Although this is a point of possible confusion now, the results of Section 6.4 will reveal a strong relationship between definite integrals and antiderivatives, justifying this similarity of notation.

REMARK 2:   At this point, the only means available to us for evaluating the definite integral $\int_a^b f(x)\, dx$ results from Theorem 2: Find any sequence of Riemann sums S_n for $f(x)$ on $[a, b]$, with $\|P_n\| \to 0$ as $n \to \infty$. As defined in Definition 3, $\int_a^b f(x)\, dx$ will be the limit of this sequence.

Example 2

Since every approximating sum is also a Riemann sum, we may use the definite integral to write the result of Example 3, Section 6.2, as

$$\int_0^1 x^2\, dx = \frac{1}{3}.$$

$\diamond$

Example 3

We may use a sequence of Riemann sums to show that

$$\int_0^a \sqrt{x}\, dx = \frac{2}{3} a^{3/2}, \qquad a > 0$$

for any positive number a. To do so we shall partition the interval $[0, a]$ into n subintervals of *unequal* length, for reasons that will become clear in the calculation. We use the partition

$$P_n = \left\{0,\ \frac{a}{n^2},\ \frac{2^2 a}{n^2},\ \frac{3^2 a}{n^2},\ \dots,\ \frac{n^2 a}{n^2} = a\right\}$$

so that the jth subinterval is $\left[\dfrac{(j-1)^2 a}{n^2},\ \dfrac{j^2 a}{n^2}\right]$ with length

$$\Delta x = \frac{j^2 a}{n^2} - \frac{(j-1)^2 a}{n^2} = \frac{(2j-1)a}{n^2}. \tag{4}$$

Thus, the norm of our partition is

$$\|P_n\| = \max\left\{\frac{(2j-1)a}{n^2}\ \middle|\ 1 \le j \le n\right\}$$

$$= \frac{(2n-1)a}{n^2}$$

and $\|P_n\| \to 0$ as $n \to \infty$. Finally, in each subinterval we evaluate the function $f(x) = \sqrt{x}$ at the right endpoint $t_j = \dfrac{j^2 a}{n^2}$.

Thus

$$f(t_j) = \sqrt{\frac{j^2 a}{n^2}} = \frac{\sqrt{a}\, j}{n}, \qquad j = 1, 2, \dots, n, \tag{5}$$

which explains why we are using endpoints involving $\dfrac{j^2}{n^2}$ rather than $\dfrac{j}{n}$. Using equations (4) and (5) and summation formulas (1) and (2) from Section 6.1 we now

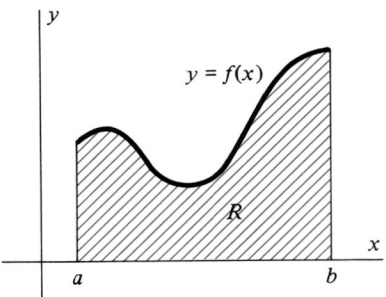

Figure 3.4 Area of $R = \int_a^b f(x)\ dx$ when $f(x) \geq 0$ for all $x \in [a, b]$.

obtain the desired integral as the limit of these Riemann sums:

$$\int_0^a \sqrt{x}\ dx = \lim_{n \to \infty} \sum_{j=1}^n f(t_j)\ \Delta x_j$$

$$= \lim_{n \to \infty} \sum_{j=1}^n \left(\frac{\sqrt{aj}}{n} \right) \left[\frac{(2j-1)a}{n^2} \right]$$

$$= \lim_{n \to \infty} \left\{ a^{3/2} \sum_{j=1}^n \left[\frac{j(2j-1)}{n^3} \right] \right\}$$

$$= a^{3/2} \lim_{n \to \infty} \left[\frac{1}{n^3} \left(2 \sum_{j=1}^n j^2 - \sum_{j=1}^n j \right) \right]$$

$$= a^{3/2} \lim_{n \to \infty} \left\{ \frac{1}{n^3} \left[\frac{2n(n+1)(2n+1)}{6} - \frac{n(n+1)}{2} \right] \right\}$$

$$= a^{3/2} \lim_{n \to \infty} \left(\frac{4n^3 + 3n^2 - n}{6n^3} \right)$$

$$= \frac{2}{3} a^{3/2}.$$

Note that we could *not* have obtained this integral as a limit of approximating sums using subintervals of equal length without access to a formula for summing function values of the form $f(t_j) = \sqrt{t_j}$, $j = 1, 2, \ldots, n$. ◇

The Definite Integral and Area

It is important to note at this point that while approximating sums for areas were defined in Section 6.2 only for nonnegative functions, *our definitions of Riemann sums and the definite integral $\int_a^b f(x)\ dx$ do not require that f be a nonnegative function, nor does Theorem 2*. Thus, the definite integral $\int_a^b f(x)\ dx$ exists whenever f is continuous on $[a, b]$, regardless of the signs of the function values $f(x)$.

The following theorem states, however, that *when f is nonnegative and continuous on $[a, b]$, the definite integral $\int_a^b f(x)\ dx$ is the area of the region bounded by the graph of f and the x-axis*. (See Figure 3.4.)

THEOREM 3

Let f be continuous on $[a, b]$ with $f(x) \geq 0$ for all $x \in [a, b]$. Then the area A of the region R bounded above by the graph of f, below by the x-axis, on the left by $x = a$, and on the right by $x = b$ is given by the definite integral

$$A = \int_a^b f(x)\ dx.$$

Proof: According to Definition 1, the area A is defined to be the limit

$$A = \lim_{n \to \infty} \sum_{j=1}^n f(t_j)\ \Delta x, \qquad t_j \in [x_{j-1}, x_j] \tag{6}$$

of a sequence of approximating sums for the region R. Since every approximating sum is also a Riemann sum, we may combine Definition 3 with equation (6) to

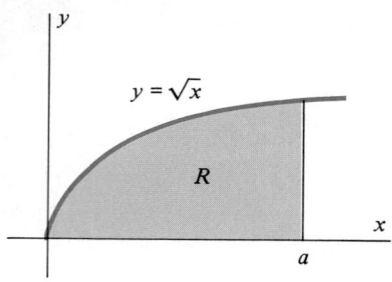

Figure 3.5 Area of R is $A =$
$$\int_0^a \sqrt{x}\, dx = \frac{2}{3}\, a^{3/2}, \quad a > 0.$$

conclude that

$$A = \lim_{n \to \infty} \sum_{j=1}^{n} f(t_j)\, \Delta x = \int_a^b f(x)\, dx. \qquad \blacklozenge$$

Example 4

According to Theorem 3 and the result of Example 3, the area of the region bounded by the graph of $f(x) = \sqrt{x}$ and the x-axis between $x = 0$ and $x = a$, $a > 0$, is

$$A = \int_0^a \sqrt{x}\, dx$$

$$= \frac{2}{3} a^{3/2}.$$

(See Figure 3.5.) $\qquad \Diamond$

Example 5

When the function f is nonnegative we may sometimes be able to evaluate $\int_a^b f(x)\, dx$ by identifying the integral with a region for which we already know the area.

(a) Figure 3.6 shows that the graph of the constant function $f(x) \equiv c$, $c > 0$, bounds a rectangle of area $c(b - a)$ over the interval $[a, b]$. Thus,

$$\int_a^b c\, dx = c(b - a), \qquad c > 0.$$

(b) Figure 3.7 shows that the graph of the function $f(x) = x$ bounds a trapezoid over the interval $[a, b]$ if $0 < a < b$. Since this trapezoid has bases of length $B_1 = f(a) = a$ and $B_2 = f(b) = b$, and altitude $h = b - a$, its area is $A = \frac{1}{2}(B_1 + B_2)h = \frac{1}{2}(a + b)(b - a) = \frac{1}{2}(b^2 - a^2)$. Thus,

$$\int_a^b x\, dx = \frac{1}{2}(b^2 - a^2), \qquad 0 < a < b.$$

(c) Figure 3.8 shows that the graph of the function $f(x) = \sqrt{a^2 - x^2}$ bounds a

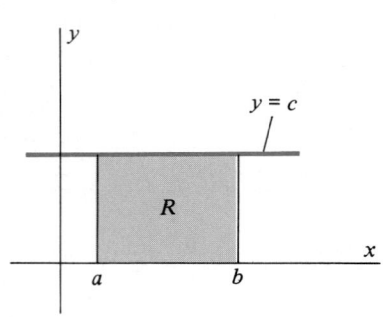

Figure 3.6 Area of R is $A =$
$$\int_a^b c\, dx = c(b - a).$$

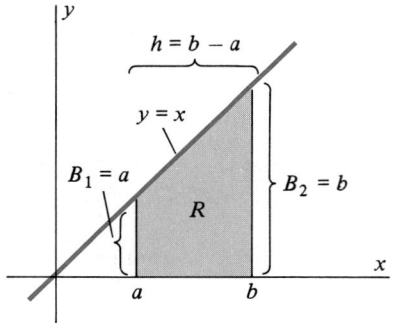

Figure 3.7 Area of R is $A =$
$$\int_a^b x\, dx = \frac{1}{2}(b^2 - a^2).$$

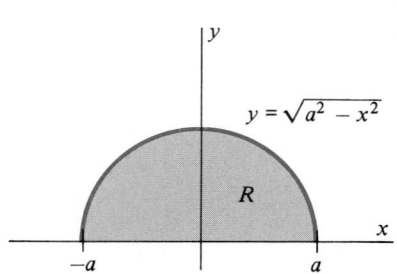

Figure 3.8 Area of R is
$$A = \int_{-a}^a \sqrt{a^2 - x^2}\, dx = \frac{1}{2}\pi a^2.$$

semicircle of radius $r = a$ over the interval $[-a, a]$ when $a > 0$. Since the area of this semicircle is $A = \dfrac{1}{2}\pi r^2 = \dfrac{1}{2}\pi a^2$, we have

$$\int_{-a}^{a} \sqrt{a^2 - x^2} \, dx = \frac{1}{2}\pi a^2.$$

◇

Extending the Definition of

$$\int_{a}^{b} f(x) \, dx$$

Since $\int_{a}^{b} f(x) \, dx$ has been defined for *intervals* $[a, b]$, it is implicit in Definition 3 that $a < b$. We will have need to work with definite integrals with $a \geq b$, which we now define.

DEFINITION 4

Let f be integrable on $[a, b]$. Then

(a) $\displaystyle\int_{a}^{a} f(x) \, dx = 0$

(b) $\displaystyle\int_{b}^{a} f(x) \, dx = -\int_{a}^{b} f(x) \, dx.$

These definitions have simple geometric motivation. Statement (a) says that a region of zero width and finite height has zero area, at least when $f(x) \geq 0$. The motivation for (b) is this. If $a < b$, the integral on the right is "from left to right." Since the integral on the left is then "from right to left," we are reversing the order in which we march along the number line adding up terms in the Riemann sums. The effect is to change the sign on each term $\Delta x_j = (x_j - x_{j-1})$, leaving $f(t_j)$ unchanged, and therefore, changing the sign on the Riemann sums associated with $\int_{a}^{b} f(x) \, dx$. We may paraphrase (b) by saying that "reversing the order of integration changes the sign of the integral."

Properties of Definite Integrals

We shall make frequent use of two theorems giving properties of definite integrals. The first states that definite integrals share two properties with derivatives and antiderivatives: the definite integral of a sum of two functions equals the sum of the definite integrals of the individual functions, and the definite integral of a multiple of a function equals that multiple of the definite integral of the function.

THEOREM 4

Let f and g be continuous on $[a, b]$ and let c be any constant. Then

(i) $\displaystyle\int_{a}^{b} [f(x) + g(x)] \, dx = \int_{a}^{b} f(x) \, dx + \int_{a}^{b} g(x) \, dx$

(ii) $\displaystyle\int_{a}^{b} cf(x) \, dx = c\int_{a}^{b} f(x) \, dx.$

We shall not give a rigorous proof of Theorem 4, but here is the basic idea. If $\sum_{j=1}^{n} f(t_j) \, \Delta x_j$ and $\sum_{j=1}^{n} g(t_j) \, \Delta x_j$ are Riemann sums for f and g, respectively, on the interval $[a, b]$ *with the same partition P,* and the same numbers t_j in each subinterval, then $\sum_{j=1}^{n}[f(t_j) + g(t_j)] \, \Delta x_j$, with the same partition P and numbers t_j, is a

Riemann sum for the function $f + g$ on $[a, b]$. Since

$$\sum_{j=1}^{n} [f(t_j) + g(t_j)] \, \Delta x_j = \sum_{j=1}^{n} f(t_j) \, \Delta x_j + \sum_{j=1}^{n} g(t_j) \, \Delta x_j$$

the limit as $n \to \infty$ of the sum on the left will exist and equation (i) will result from taking limits as $n \to \infty$ of each of these sums. The argument for equation (ii) is similar.

Example 6

From Examples 2 and 5 we know that

$$\int_0^1 c \, dx = c, \qquad c > 0,$$

$$\int_0^1 x \, dx = \frac{1}{2},$$

and

$$\int_0^1 x^2 \, dx = \frac{1}{3}.$$

From these facts and Theorem 4 we may conclude that

(a) $\displaystyle\int_0^1 (x + x^2) \, dx = \int_0^1 x \, dx + \int_0^1 x^2 \, dx \qquad$ (Property (i))

$$= \frac{1}{2} + \frac{1}{3}$$

$$= \frac{5}{6}.$$

(b) $\displaystyle\int_0^1 6x^2 \, dx = 6 \int_0^1 x^2 \, dx \qquad$ (Property (ii))

$$= 6\left(\frac{1}{3}\right)$$

$$= 2.$$

(c) $\displaystyle\int_0^1 [3x^2 - 6x + 7] \, dx = \int_0^1 3x^2 \, dx + \int_0^1 (-6x) \, dx + \int_0^1 7 \, dx$

$$= 3 \int_0^1 x^2 \, dx - 6 \int_0^1 x \, dx + \int_0^1 7 \, dx$$

$$= 3\left(\frac{1}{3}\right) - 6\left(\frac{1}{2}\right) + 7$$

$$= 5. \qquad\qquad\qquad\qquad\qquad\qquad \diamond$$

The following theorem states that a definite integral may be "split" into the sum of two (or more) definite integrals as follows.

THEOREM 5

Let f be continuous on an interval containing the numbers a, b, and c. Then

$$\int_a^b f(x)\,dx = \int_a^c f(x)\,dx + \int_c^b f(x)\,dx$$

regardless of how the numbers a, b, and c are ordered.

Figure 3.9 illustrates the statement of Theorem 5 when f is nonnegative on $[a, b]$ and $a < c < b$. In this case the result is obvious from area considerations. Theorem 5 is true, however, whether or not f is nonnegative and whether or not c lies between a and b.

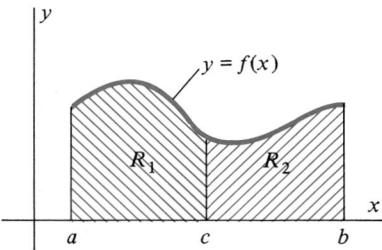

Figure 3.9 $\displaystyle\int_a^b f(x)\,dx$ = Area of $R_1 \cup R_2$

$$= \text{Area of } R_1 + \text{Area of } R_2$$

$$= \int_a^c f(x)\,dx + \int_c^b f(x)\,dx.$$

Proof of Theorem 5: When c lies between a and b the idea behind the proof of Theorem 5 is to require that c be an endpoint in all partitions in a sequence R_n of Riemann sums with limit $\int_a^b f(x)\,dx$. Write $n = m + p$ where m is the number of subintervals in $[a, c]$ and p is the number of subintervals in $[c, b]$. Use the notation $z_k = x_{m+k}$ for the x_j's to the right of c, and the notation $u_k = t_{m+k}$ for the t_j's to the right of c. Then

$$\int_a^b f(x)\,dx = \lim_{n\to\infty} \sum_{j=1}^{n} f(t_j)\,\Delta x_j$$

$$= \lim_{n\to\infty}\left[\sum_{j=1}^{m} f(t_j)\,\Delta x_j + \sum_{k=1}^{p} f(u_k)\,\Delta z_k\right] \qquad (n = m + p)$$

$$= \lim_{m\to\infty}\sum_{j=1}^{m} f(t_j)\,\Delta x_j + \lim_{p\to\infty}\sum_{k=1}^{p} f(u_k)\,\Delta z_k$$

$$= \int_a^c f(x)\,dx + \int_c^b f(x)\,dx.$$

This is messy, but it is important that you at least appreciate that this is why we need to allow partitions with unequal subinterval lengths. If all subintervals were of equal length, we could not guarantee that c would correspond to an endpoint, so we could not "break the sum into two pieces at $x = c$" as we have done here.   ◆

We will next consider the case $c < a < b$, leaving the other five possible orderings for you (Exercise 35).

Since $c < a < b$, we have already explained why

$$\int_c^b f(x)\ dx = \int_c^a f(x)\ dx + \int_a^b f(x)\ dx. \tag{7}$$

But, $\int_c^a f(x)\ dx = -\int_a^c f(x)\ dx$ by Definition 4. Solving equation (7) for $\int_a^b f(x)\ dx$ and using this fact gives

$$\int_a^b f(x)\ dx = -\int_c^a f(x)\ dx + \int_c^b f(x)\ dx$$

$$= \int_a^c f(x)\ dx + \int_c^b f(x)\ dx,$$

as claimed.

Example 7

Given that $\int_0^2 f(x)\ dx = 4$, $\int_2^3 f(x)\ dx = 2$, and $\int_2^5 f(x)\ dx = 10$ we may conclude from Definition 4 and Theorem 5 that

(a) $\displaystyle\int_0^5 f(x)\ dx = \int_0^2 f(x)\ dx + \int_2^5 f(x)\ dx = 4 + 10 = 14.$

(b) $\displaystyle\int_5^0 f(x)\ dx = -\int_0^5 f(x)\ dx = -14.$

(c) $\displaystyle\int_3^5 f(x)\ dx = \int_0^5 f(x)\ dx - \int_0^3 f(x)\ dx$

$$= \int_0^5 f(x)\ dx - \left[\int_0^2 f(x)\ dx + \int_2^3 f(x)\ dx\right]$$

$$= 10 - (4 + 2)$$

$$= 4. \qquad \diamond$$

Exercise Set 6.3

1. Given that $\displaystyle\int_0^2 f(x)\ dx = 3$, $\displaystyle\int_2^5 f(x)\ dx = -2$, and

$\displaystyle\int_5^8 f(x)\ dx = 5$, find

a. $\displaystyle\int_2^0 f(x)\ dx$ **b.** $\displaystyle\int_0^5 f(x)\ dx$

c. $\displaystyle\int_2^8 f(x)\ dx$ **d.** $\displaystyle\int_0^8 f(x)\ dx$

e. $\displaystyle\int_5^0 f(x)\ dx$ **f.** $\displaystyle\int_8^2 f(x)\ dx$

2. Given that $\displaystyle\int_1^3 f(x)\ dx = 5$ and $\displaystyle\int_1^3 g(x)\ dx = -2$, find

a. $\displaystyle\int_1^3 [f(x) + g(x)]\ dx$ **b.** $\displaystyle\int_1^3 [6g(x) - 2f(x)]\ dx$

c. $\displaystyle\int_3^1 2f(x)\ dx - \int_1^3 g(x)\ dx$ **d.** $\displaystyle\int_1^3 [2f(x) + 3]\ dx$

e. $\displaystyle\int_3^1 [g(x) - 4f(x) + 5]\ dx$

f. $\displaystyle\int_1^3 [1 - 4g(x) + 3f(x)]\ dx$

In Exercises 3–16, use the facts that

$$\int_a^b x^2\ dx = \frac{b^3}{3} - \frac{a^3}{3}; \quad \int_a^b x\ dx = \frac{b^2}{2} - \frac{a^2}{2};$$

$\displaystyle\int_a^b c\ dx = c(b - a)$ for *any* constants a, b, and c to find the value of the given integral (see Exercises 33, 34).

3. $\displaystyle\int_0^5 6\ dx$ **4.** $\displaystyle\int_0^3 (2x + 5)\ dx$

5. $\int_{-2}^{4} (3x - 2)\, dx$

6. $\int_{5}^{2} (5x + 7)\, dx$

7. $\int_{4}^{0} (9 - 2x)\, dx$

8. $\int_{2}^{5} 2x^2\, dx$

9. $\int_{3}^{1} (6 - x^2)\, dx$

10. $\int_{-4}^{-1} (3x^2 - 1)\, dx$

11. $\int_{0}^{1} (x^2 - x + 2)\, dx$

12. $\int_{-2}^{2} (2x^2 - 3x + 2)\, dx$

13. $\int_{0}^{4} x(2 + x)\, dx$

14. $\int_{3}^{-1} (x - 2)(x + 2)\, dx$

15. $\int_{-4}^{-1} 2x(1 - 2x)\, dx$

16. $\int_{-1}^{1} (ax^2 + bx + c)\, dx$

In Exercises 17–24, sketch a region in the plane whose area is given by the definite integral. Find the area of that region by geometry.

17. $\int_{-2}^{3} (x + 1)\, dx$

18. $\int_{0}^{3} \sqrt{9 - x^2}\, dx$

19. $\int_{-2}^{2} |x - 1|\, dx$

20. $\int_{-1}^{1} \sqrt{1 - x^2}\, dx$

21. $\int_{-4}^{0} (\sqrt{4 - (x + 2)^2} + 1)\, dx$

22. $\int_{1}^{4} |2x - 7|\, dx$

23. $\int_{-3}^{3} (4\sqrt{9 - x^2} + 3)\, dx$

24. $\int_{0}^{4} (5x - \sqrt{16 - x^2})\, dx$

(Calculator) The **midpoint rule** for approximating $\int_a^b f(x)\, dx$ uses a Riemann sum where $f(x)$ is evaluated in each subinterval at the midpoint, i.e., we take $\Delta x = (b - a)/n$, $x_0 = a$, $x_1 = a + \Delta x$, $x_2 = a + 2\,\Delta x$, ..., $x_n = a + n\,\Delta x = b$ and in each interval $[x_{j-1}, x_j]$ we choose $t_j = (x_{j-1} + x_j)/2$. The *midpoint rule* approximation is then

$$\int_a^b f(x)\, dx \approx \frac{b - a}{n}\{f(t_1) + f(t_2) + \cdots + f(t_n)\}.$$

(See Figure 3.10.)

In Exercises 25–30, use a calculator or computer to obtain the midpoint rule approximation for the given integral and the given value of n. Compare the results with those obtained using lower Riemann sums with the same values n to approximate the integrals.

25. $\int_{0}^{3} (x^3 + 1)\, dx$, $\quad n = 6$

26. $\int_{0}^{\pi/2} \sin x\, dx$, $\quad n = 6$

27. $\int_{0}^{\pi/4} \tan x\, dx$, $\quad n = 4$

28. $\int_{0}^{\pi} \cos x\, dx$, $\quad n = 6$

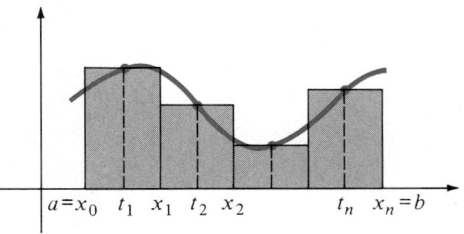

Figure 3.10 Approximation by midpoint rule: $t_j = \dfrac{x_{j-1} + x_j}{2}$.

29. $\int_{0}^{4} (x^4 + x^3)\, dx$, $\quad n = 8$

30. $\int_{0}^{2} \sqrt{4 - x^2}\, dx$, $\quad n = 8$

31. Use the definition of the integral to prove that $\int_{a}^{b} x\, dx = \dfrac{b^2}{2} - \dfrac{a^2}{2}$ for any constants a and b.

32. Use the definition of the integral to prove that $\int_{a}^{b} x^2\, dx = \dfrac{b^3}{3} - \dfrac{a^3}{3}$ for any constants a and b.

33. Prove the remaining cases of Theorem 5.

34. Give an argument for statement (ii) of Theorem 4 similar to that given for statement (i).

35. Prove that if $f(x)$ and $g(x)$ are continuous on $[a, b]$, and if $f(x) \le g(x)$ for all $x \in [a, b]$, then

$$\int_{a}^{b} f(x)\, dx \le \int_{a}^{b} g(x)\, dx.$$

(*Hint:* Write $g(x) = f(x) + h(x)$. What do you know about the sign of $h(x)$? Of $\int_a^b h(x)\, dx$?)

36. Prove that if f is continuous on $[a, b]$, then

$$\left| \int_{a}^{b} f(x)\, dx \right| \le \int_{a}^{b} |f(x)|\, dx.$$

37. Consider the functions

$$f(x) = x, \qquad 0 \le x \le 1$$
$$g(x) = \begin{cases} x, & 0 \le x < 1 \\ 2, & x = 1 \end{cases}$$

on the interval $[0, 1]$.

a. Note that f is continuous, and therefore integrable, on $[0, 1]$ but that g is *not* continuous on $[0, 1]$, because $g(1) \ne \lim\limits_{x \to 1^-} g(x)$. (See Figure 3.11.)

b. Explain why that if P_n is any partition of $[0, 1]$ then the

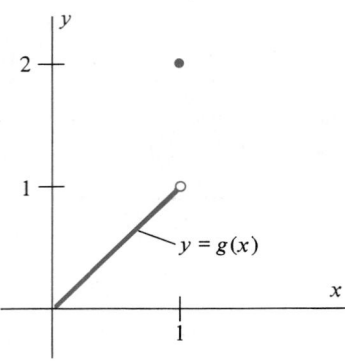

Figure 3.11

Riemann sums over this partition

$$\sum_{j=1}^{n} f(t_j)\, \Delta x_j \qquad \text{and} \qquad \sum_{j=1}^{n} g(t_j)\, \Delta x_j$$

are equal if the numbers t_j in both sums are the same and if we choose t_n with $t_n \neq 1$.

c. Conclude that

$$\lim_{n \to \infty} \sum_{j=1}^{n} g(t_j)\, \Delta x_j = \lim_{n \to \infty} \sum_{j=1}^{n} f(t_j)\, \Delta x_j$$

$$= \int_0^1 x\, dx$$

$$= \frac{1}{2},$$

so g is integrable on $[0, 1]$ with $\int_0^1 g(x)\, dx = 1$.

38. Generalize the result of Exercise 37 to conclude that if g is a function with a "jump discontinuity" as in

$$g(x) = \begin{cases} f_1(x), & a \le x \le b \\ f_2(x), & b < x \le c \end{cases}$$

with $f_1(b) \neq \lim_{x \to b^+} f_2(x)$, then if f_1 and f_2 are continuous, g is integrable on $[a, c]$ with

$$\int_a^c g(x)\, dx = \int_a^b f_1(x)\, dx + \int_b^c \hat{f}_2(x)\, dx$$

where $\hat{f}_2$ is the function

$$\hat{f}_2(x) = \begin{cases} \lim_{x \to b^+} f_2(x), & x = b \\ f_2(x), & b < x \le c \end{cases}$$

obtained from f_2 by "patching up" the discontinuity for g at $x = b$. (See Figure 3.12.)

39. Generalize the result of Exercise 38 to conclude that a function g is integrable on $[a, b]$ if it is continuous on $[a, b]$ except at finitely many numbers where it has "jump discontinuities" as in Figure 3.12. (We say that g has a jump discontinuity at $c \in (a, b)$ if the one-sided limits $\lim_{x \to c^-} g(x)$ and $\lim_{x \to c^+} g(x)$ both exist but are unequal; g has a jump discontinuity at the endpoint a if $g(a) \neq \lim_{x \to a^+} g(x)$ and similarly for the right endpoint b.)

40. Explain how the definite integral for the discontinuous function g in Exercise 39 would be calculated.

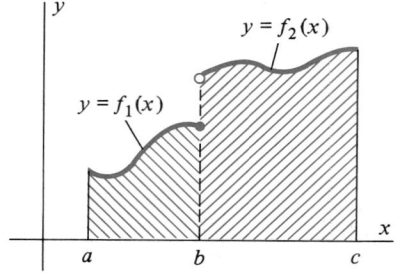

Figure 3.12

6.4 THE FUNDAMENTAL THEOREM OF CALCULUS

The purpose of this section is to present a remarkably simple procedure for evaluating definite integrals. The statement of this procedure is part of a very important result called the Fundamental Theorem of Calculus. We shall state the Fundamental Theorem here, sketching the proof only briefly, and then focus on how it is used to evaluate definite integrals. In Section 6.5 we shall provide the additional details of the proof, related results, and other applications of this theorem.

The Area Function

$$A(x) = \int_a^x f(t)\, dt$$

Let a be any (fixed) number. Then, for each $x \in (-\infty, \infty)$ we define

$$A(x) = \int_a^x f(t)\, dt. \tag{1}$$

That is, $A(x)$ is *the value of the definite integral* of f from a to x.

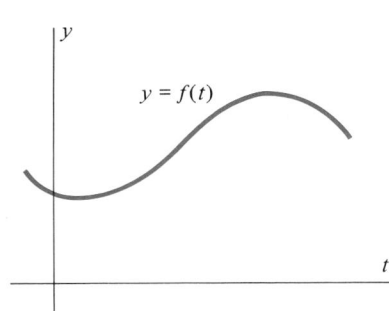

Figure 4.1 The continuous function $y = f(t)$.

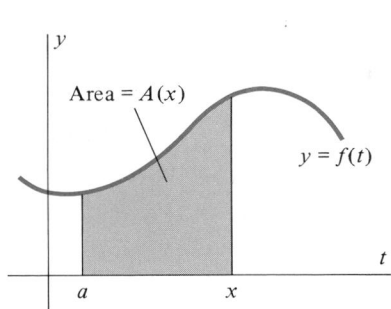

Figure 4.2 The area function $A(x) = \int_a^x f(t)\, dt$.

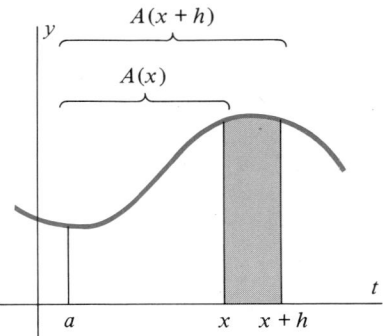

Figure 4.3 Area of shaded region is $A(x + h) - A(x)$.

Figure 4.1 shows the graph of a function $y = f(t)$ which, for simplicity, we assume to be defined and continuous for all $t \in (-\infty, \infty)$. Note that we are using the letter t to denote the independent variable for f. That is because we wish to use the letter x to denote the independent variable for another function, $y = A(x)$, which we define as follows.

Figure 4.2 shows that when $a < x$ and $f(t)$ is nonnegative for all $t \in [a, x]$ the value $A(x)$ is the area of the region bounded by the graph of f and the x-axis over the interval $[a, x]$. For this reason the function $y = A(x)$ is sometimes called an *area function* for f. But these restrictions are unnecessary. The value $A(x)$ in equation (1) exists for all x, regardless of the sign of $f(t)$.

Newton and Leibniz independently discovered a remarkable fact about the function $y = A(x)$: *it is differentiable, and its derivative is*

$$A'(x) = f(x). \tag{2}$$

That is, the derivative $A'(x)$ is just the value of the integrand $f(x)$ evaluated at $t = x$!

Here is a sketch of why equation (2) is true in the case where $x > a$ and f is nonnegative on $[a, x]$. (A complete proof is given in the next section.) If $h > 0$ is small, the difference $A(x + h) - A(x)$ is the area of a region that can be approximated by a rectangle of height $f(x)$ and width $(x + h) - x = h$. (See Figures 4.3 and 4.4.) That is,

$$A(x + h) - A(x) \approx f(x) \cdot h$$

or, since $h \neq 0$,

$$\frac{A(x + h) - A(x)}{h} \approx f(x).$$

We now argue that as $h \to 0^+$ the rectangle more precisely approximates the shaded region in Figure 4.4 so the above approximations yield

$$\lim_{h \to 0^+} \frac{A(x + h) - A(x)}{h} = f(x).$$

A similar argument applies for $h < 0$, and leads to the conclusion that

$$A'(x) = \lim_{h \to 0} \frac{A(x + h) - A(x)}{h} = f(x)$$

as stated in equation (2).

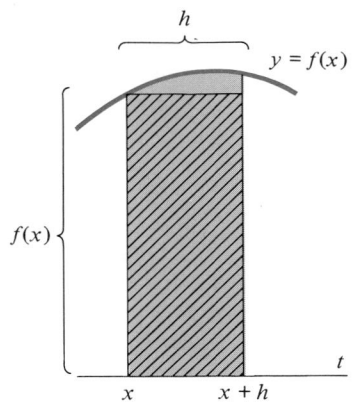

Figure 4.4 Rectangle of area $f(x) \cdot h$ approximates region of area $A(x + h) - A(x)$.

We can now apply equation (2) to find a formula for evaluating $\int_a^b f(t)\,dt$ when f is continuous on $[a, b]$. Since equation (2) states that the function $y = A(x)$ is an *antiderivative* for f on $[a, b]$, Theorem 1 of Section 5.1 guarantees that if F is any *other* antiderivative for f on $[a, b]$, then

$$F(x) = A(x) + C$$

for some constant C and all $x \in [a, b]$. Thus

$$
\begin{aligned}
F(b) - F(a) &= [A(b) + C] - [A(a) + C] \\
&= A(b) - A(a) \\
&= \int_a^b f(t)\,dt - \int_a^a f(t)\,dt \qquad \text{(equation (1))} \\
&= \int_a^b f(t)\,dt \qquad \left(\int_a^a f(t)\,dt = 0 \right).
\end{aligned}
$$

That is,

$$\int_a^b f(t)\,dt = F(b) - F(a) \tag{3}$$

where F is *any* antiderivative for f on $[a, b]$.

Equations (2) and (3) constitute the two parts of the Fundamental Theorem of Calculus, which we state formally as follows.

THEOREM 6
(Fundamental Theorem of Calculus)

(Part i) Let f be continuous on an open interval I containing the number a and let

$$A(x) = \int_a^x f(t)\,dt$$

for each $x \in I$. Then A is differentiable on I, and

$$A'(x) = f(x).$$

That is,

$$\frac{d}{dx}\left\{ \int_a^x f(t)\,dt \right\} = f(x), \qquad x \in I.$$

(Part ii) Let f be continuous on $[a, b]$ and let F be any antiderivative for f on $[a, b]$. Then

$$\int_a^b f(x)\,dx = F(b) - F(a). \tag{4}$$

(In writing $\int_a^b f(x)\,dx$ in equation (4) we have returned to the use of x as the name for the independent variable for f. It does not matter whether we call this "dummy variable" x or t, as in equation (3), because it is the only variable appearing in these equations.)

We shall defer further comment on part (i) of the Fundamental Theorem, and its proof, to the next section. Part (ii), which we have proved using part (i), states that a definite integral may be evaluated by finding any antiderivative F for the integrand f, if we can do so, and evaluating F at the limits of integration.

In using the Fundamental Theorem we shall make use of the notation

$$[F(x)]_a^b = F(b) - F(a).$$

For example,

$$[\sin x]_{\pi/4}^{\pi/2} = \sin \frac{\pi}{2} - \sin \frac{\pi}{4} = 1 - \frac{\sqrt{2}}{2}$$

$$[x^2 + 4]_2^3 = (3^2 + 4) - (2^2 + 4) = 5,$$

etc.

Example 1

$$\int_1^2 (4x + 6) \, dx = [2x^2 + 6x]_1^2 = (2 \cdot 2^2 + 6 \cdot 2) - (2 \cdot 1^2 + 6 \cdot 1)$$

$$= (8 + 12) - (2 + 6)$$

$$= 12$$

since an antiderivative for $f(x) = 4x + 6$ is $F(x) = 2x^2 + 6x$. ◇

Example 2

$$\int_{-2}^2 (x^3 - 3x^2 + 2x - 5) \, dx = \left[\frac{x^4}{4} - x^3 + x^2 - 5x \right]_{-2}^2$$

$$= \left(\frac{2^4}{4} - 2^3 + 2^2 - 5 \cdot 2 \right)$$

$$- \left(\frac{(-2)^4}{4} - (-2)^3 + (-2)^2 - 5(-2) \right)$$

$$= (4 - 8 + 4 - 10) - (4 + 8 + 4 + 10)$$

$$= -36$$

since an antiderivative for $f(x) = x^3 - 3x^2 + 2x - 5$ is

$$F(x) = \frac{x^4}{4} - x^3 + x^2 - 5x.$$ ◇

Example 3

$$\int_0^{3\pi/2} \cos x \, dx = [\sin x]_0^{3\pi/2} = \sin\left(\frac{3\pi}{2} \right) - \sin(0)$$

$$= -1 - 0$$

$$= -1$$

since an antiderivative for $f(x) = \cos x$ is $F(x) = \sin x$. ◇

Example 4

$$\int_1^4 \frac{x + 1}{\sqrt{x}} \, dx = \int_1^4 \left(\sqrt{x} + \frac{1}{\sqrt{x}} \right) dx$$

$$= \int_1^4 (x^{1/2} + x^{-1/2}) \, dx$$

$$= \left[\frac{2}{3}x^{3/2} + 2x^{1/2}\right]_1^4$$

$$= \left(\frac{2}{3} \cdot 4^{3/2} + 2 \cdot 4^{1/2}\right) - \left(\frac{2}{3} \cdot 1^{3/2} + 2 \cdot 1^{1/2}\right)$$

$$= \frac{2}{3} \cdot 8 + 2 \cdot 2 - \frac{2}{3} - 2$$

$$= \frac{20}{3}$$

since an antiderivative for $f(x) = x^{1/2} + x^{-1/2}$ is $F(x) = \frac{2}{3}x^{3/2} + 2x^{1/2}$.   ◇

REMARK 1:   In order to use the Fundamental Theorem, it is necessary to be able to find an antiderivative F for f. Here is a summary list of the integration formulas we have developed up to this point:

$$\int m \, dx = mx + C \qquad\qquad \int \sec^2 x \, dx = \tan x + C$$

$$\int x^n \, dx = \frac{x^{n+1}}{n+1} + C, \qquad n \neq -1 \qquad \int \csc^2 x \, dx = -\cot x + C$$

$$\int \sin x \, dx = -\cos x + C \qquad\qquad \int \sec x \tan x \, dx = \sec x + C$$

$$\int \cos x \, dx = \sin x + C \qquad\qquad \int \csc x \cot x \, dx = -\csc x + C$$

REMARK 2:   You may wonder why we are writing just F rather than $F + C$ for an antiderivative of f. The answer is that we need only find *any particular* antiderivative for f, since all antiderivatives yield the same value for the expression $F(b) - F(a)$. To see this note that

$$[F(x) + C]_a^b = [F(b) + C] - [F(a) + C]$$
$$= F(b) - F(a) + (C - C)$$
$$= F(b) - F(a).$$

Example 5

Find $\displaystyle\int_{-1}^5 |x - 2| \, dx$.

Solution: Applying the definition of absolute value we see that

$$|x - 2| = \begin{cases} x - 2 & \text{if} \quad x - 2 \geq 0 \\ -(x - 2) & \text{if} \quad x - 2 < 0. \end{cases}$$

That is,

$$|x - 2| = \begin{cases} x - 2 & \text{if} \quad x \geq 2 \\ 2 - x & \text{if} \quad x < 2. \end{cases}$$

(See Figure 4.5.)

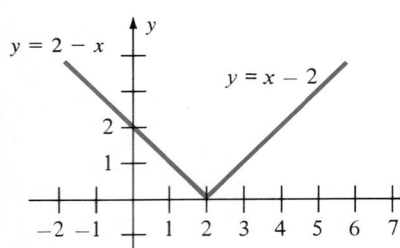

Figure 4.5   Graph of $f(x) = |x - 2|$ (Example 5).

We therefore evaluate the integral separately on $[-1, 2]$ and on $[2, 5]$, using the corresponding part of the definition of $|x - 2|$ on each interval:

$$\int_{-1}^{5} |x - 2| \, dx = \int_{-1}^{2} |x - 2| \, dx + \int_{2}^{5} |x - 2| \, dx$$

$$= \int_{-1}^{2} (2 - x) \, dx + \int_{2}^{5} (x - 2) \, dx$$

$$= \left[2x - \frac{x^2}{2} \right]_{-1}^{2} + \left[\frac{x^2}{2} - 2x \right]_{2}^{5}$$

$$= \left[\left(2 \cdot 2 - \frac{2^2}{2} \right) - \left(2(-1) - \frac{(-1)^2}{2} \right) \right]$$

$$+ \left[\left(\frac{5^2}{2} - 2 \cdot 5 \right) - \left(\frac{2^2}{2} - 2 \cdot 2 \right) \right]$$

$$= 9.$$

$\diamond$

Substitutions in Definite Integrals

In the next example, a substitution, first introduced in Section 5.2, is used to find the antiderivative. Note that we substitute back in terms of x's *before* the integral is evaluated.

Example 6

Find $\displaystyle\int_{-1}^{3} \frac{x}{\sqrt{7 + x^2}} \, dx$.

Solution: To find the antiderivative $\displaystyle\int \frac{x \, dx}{\sqrt{7 + x^2}}$, we use the substitution

$$u = 7 + x^2; \qquad du = 2x \, dx.$$

Then $x \, dx = \dfrac{1}{2} \, du$, and we have

$$\int \frac{x \, dx}{\sqrt{7 + x^2}} = \int \frac{\frac{1}{2} \, du}{\sqrt{u}} = \frac{1}{2} \int u^{-1/2} \, du = \frac{1}{2} \cdot (2u^{1/2} + C_1)$$

$$= \sqrt{7 + x^2} + C, \qquad C = \frac{1}{2} C_1.$$

We therefore use $F(x) = \sqrt{7 + x^2}$ and the Fundamental Theorem to find that

$$\int_{-1}^{3} \frac{x \, dx}{\sqrt{7 + x^2}} = [\sqrt{7 + x^2}]_{-1}^{3} = \sqrt{7 + 3^2} - \sqrt{7 + (-1)^2}$$

$$= \sqrt{16} - \sqrt{8}$$

$$= 4 - 2\sqrt{2}.$$

$\diamond$

In evaluating definite integrals involving substitutions, it is often simpler to change the limits of integration, as well as the integrand, from x's to u's, and then to evaluate the antiderivative directly as a function of u. The following theorem justifies doing this.

THEOREM 7

Let g' be continuous on $[a, b]$ and let f be continuous on an interval I containing the values $u = g(x)$, $x \in [a, b]$. If f has an antiderivative on I, then

$$\int_a^b f(g(x)) \cdot g'(x) \, dx = \int_{g(a)}^{g(b)} f(u) \, du.$$

Proof: If F is an antiderivative for f, then the composite function $F \circ g$ is an antiderivative for $(f \circ g)g'$. Thus, by the Fundamental Theorem,

$$\int_a^b f(g(x))g'(x) \, dx = [F(g(x))]_a^b = F(g(b)) - F(g(a)).$$

But, again using the Fundamental Theorem, we can write this last expression as

$$F(g(b)) - F(g(a)) = [F(u)]_{g(a)}^{g(b)} = \int_{g(a)}^{g(b)} f(u) \, du. \qquad \blacklozenge$$

Example 7

Using Theorem 7 we can evaluate the definite integral $\int_{-1}^{3} \dfrac{x \, dx}{\sqrt{7 + x^2}}$ in Example 6 as follows:

As before, we use the substitution

$$u = 7 + x^2; \qquad du = 2x \, dx,$$

so $x \, dx = \dfrac{1}{2} \, du$. Also, we change limits by noting that

(i) when $x = -1$, $\quad u = 7 + (-1)^2 = 8$, $\quad$ and
(ii) when $x = 3$, $\quad u = 7 + (3)^2 = 16$.

Thus, by Theorem 7,

$$\int_{-1}^{3} \frac{x \, dx}{\sqrt{7 + x^2}} = \frac{1}{2} \int_8^{16} u^{-1/2} \, du$$

$$= \frac{1}{2} \left[2u^{1/2} \right]_8^{16}$$

$$= \sqrt{16} - \sqrt{8}$$

$$= 4 - 2\sqrt{2}. \qquad \diamondsuit$$

Example 8

Find $\displaystyle\int_0^{\pi/4} \tan^2 x \, \sec^2 x \, dx$.

Solution: To find an antiderivative for $f(x) = \tan^2 x \, \sec^2 x$, we use the substitution

$$u = \tan x; \qquad du = \sec^2 x \, dx.$$

(The substitution $u = \sec x$ would not work, since we would have $du = \sec x \tan x \, dx$, leaving one factor of $\tan x$ not accounted for.)

To transform the limits of integration via the substitution, we note that

$$\text{if } x = 0, \quad \text{then} \quad u = \tan(0) = 0;$$

$$\text{if } x = \frac{\pi}{4}, \quad \text{then,} \quad u = \tan\left(\frac{\pi}{4}\right) = 1.$$

Thus,

$$\int_0^{\pi/4} \tan^2 x \sec^2 x \, dx = \int_0^1 u^2 \, du = \left[\frac{1}{3}u^3\right]_0^1 = \frac{1}{3} \cdot 1^2 - \frac{1}{3} \cdot 0^2 = \frac{1}{3}. \qquad \diamond$$

Example 9

Find $\displaystyle\int_{-1}^2 \frac{x^2 \, dx}{(x^3 + 4)^2}$.

Solution: Here we use the substitution

$$u = x^3 + 4; \quad du = 3x^2 \, dx.$$

Thus, $x^2 \, dx = \dfrac{1}{3} du$. Also,

$$\text{if } x = -1, \quad u = (-1)^3 + 4 = -1 + 4 = 3;$$
$$\text{if } x = 2, \quad u = 2^3 + 4 = 8 + 4 = 12.$$

Thus,

$$\int_{-1}^2 \frac{x^2 \, dx}{(x^3 + 4)^2} = \int_3^{12} \frac{\frac{1}{3} du}{u^2} = \frac{1}{3}\int_3^{12} u^{-2} \, du$$

$$= \left[-\frac{1}{3}u^{-1}\right]_3^{12}$$

$$= -\frac{1}{3}\left(\frac{1}{12}\right) - \left[-\frac{1}{3}\left(\frac{1}{3}\right)\right]$$

$$= \frac{1}{12}. \qquad \diamond$$

Exercise Set 6.4

In each of Exercises 1–38, evaluate the definite integral using the Fundamental Theorem of Calculus and, where necessary, a substitution.

1. $\displaystyle\int_{-2}^2 (x^3 - 1) \, dx$

2. $\displaystyle\int_0^4 (x^2 - 2x + 3) \, dx$

3. $\displaystyle\int_2^0 x(\sqrt{x} - 1) \, dx$

4. $\displaystyle\int_9^4 \left(\sqrt{x} + \frac{1}{\sqrt{x}}\right) dx$

5. $\displaystyle\int_0^{2\pi/3} \sin(2x) \, dx$

6. $\displaystyle\int_{-1}^2 \left(\frac{2}{x^3} + 5x\right) dx$

7. $\displaystyle\int_0^{\pi/4} (1 - \cos 2x) \, dx$

8. $\displaystyle\int_0^3 t(\sqrt[3]{t} - 2) \, dt$

9. $\displaystyle\int_0^1 2x(x^2 + 1)^2 \, dx$

10. $\displaystyle\int_0^1 x\sqrt{1 - x^2} \, dx$

11. $\displaystyle\int_0^4 |x - 3| \, dx$

12. $\displaystyle\int_4^9 \left(\sqrt{x} - \frac{1}{\sqrt{x}}\right) dx$

13. $\displaystyle\int_1^2 x\sqrt{4 - x^2} \, dx$

14. $\displaystyle\int_0^4 |x - \sqrt{x}| \, dx$

15. $\int_1^3 (t^2 + 2)^2 \, dt$

16. $\int_0^1 (x^{3/5} - x^{5/3}) \, dx$

17. $\int_1^2 \frac{1 - x}{x^3} \, dx$

18. $\int_0^4 \frac{dt}{\sqrt{2t + 1}}$

19. $\int_0^{\sqrt{\pi/2}} t \sin(\pi - t^2) \, dt$

20. $\int_0^{\pi/2} \cos^3 t \sin t \, dt$

21. $\int_{-\pi/4}^{\pi/4} \frac{\sin x}{\cos^2 x} \, dx$

22. $\int_0^{\pi/3} \cos t \sqrt{1 - \sin^2 t} \, dt$

23. $\int_0^1 x\sqrt{ax^2 + b} \, dx$

24. $\int_0^2 \frac{x}{\sqrt{16 + x^2}} \, dx$

25. $\int_{-1}^1 \frac{x}{(1 + x^2)^3} \cdot dx$

26. $\int_0^3 x\sqrt{9 - x^2} \, dx$

27. $\int_{\pi^2/2}^{\pi^2} \frac{\sin \sqrt{x}}{\sqrt{x}} \, dx$

28. $\int_0^{\pi/4} \sin x\sqrt{\cos x} \, dx$

29. $\int_1^2 \frac{x}{(2x^2 - 1)^3} \, dx$

30. $\int_4^9 x^{1/2}(1 - x^{3/2}) \, dx$

31. $\int_0^4 |9 - x^2| \, dx$

32. $\int_{-2}^2 |1 - x^2| \, dx$

33. $\int_0^1 \frac{1}{(4 - x)^2} \, dx$

34. $\int_{-1}^4 |x^2 - x - 2| \, dx$

35. $\int_0^{\pi/4} \sin^2 x \, dx$ (*Hint:* Use the identity
$$\sin^2 x = \frac{1}{2} - \frac{1}{2} \cos 2x.)$$

36. $\int_0^\pi \cos^2 x \, dx$ (*Hint:* Use the identity
$$\cos^2 x = \frac{1}{2} + \frac{1}{2} \cos 2x.)$$

37. $\int_{-2}^3 f(x) \, dx$, where $f(x) = \begin{cases} x^2 - 1 & \text{for} \quad x \leq 1 \\ x - 1 & \text{for} \quad x > 1 \end{cases}$

38. $\int_1^5 f(x) \, dx$, where $f(x) = \begin{cases} \sqrt{x} & \text{for} \quad 0 \leq x \leq 4 \\ 2x^2 - 7x & \text{for} \quad x > 4. \end{cases}$

In Exercises 39–45, find (a) the definite integral of the given function over the given interval and (b) the area bounded by the graph of the given function and the x-axis over the given interval.

39. $f(x) = 2x + 1$,   [0, 3]

40. $f(x) = 3 - x$,   [0, 2]

41. $f(x) = 3 - x^2$, $[0, \sqrt{3}]$

42. $f(x) = 2x^2 - 1$,   [1, 4]

43. $f(x) = 2x^2 + 6x + 1$,   [1, 2]

44. $f(x) = x^2 + 6$,   [0, 3]

45. $f(x) = x(x + 1)$,   [0, 3]

46. Show that

a. $\int_0^{\pi/2} \sin x \, dx = \int_0^{\pi/2} \cos x \, dx,$

b. $\int_0^{\pi/2} \sin^2 x \, dx = \int_0^{\pi/2} \cos^2 x \, dx.$

What do you conjecture about $\int_0^{\pi/2} \sin^n x \, dx$ and $\int_0^{\pi/2} \cos^n x \, dx$? Can you prove this?

6.5 THE FUNDAMENTAL THEOREM (CONTINUED)

The purpose of this section is to prove part (i) of the Fundamental Theorem of Calculus. A careful proof of this result uses another important theorem about definite integrals that we shall first develop.

The Mean Value Theorem for Integrals

We begin by recalling that if f is continuous on $[a, b]$ then f has a minimum value m and a maximum value M on the closed interval $[a, b]$. That is,

$$m \leq f(t) \leq M \quad \text{for all} \quad t \in [a, b]. \tag{1}$$

Now let $P = \{a = x_0, x_1, x_2, \ldots, x_n = b\}$ be any partition of $[a, b]$. As before, for each $j = 1, 2, \ldots, n$ we let $\Delta x_j = x_j - x_{j-1}$ and we let $t_j \in [x_{j-1}, x_j]$ be arbitrary. It then follows from inequality (1) that

$$\sum_{j=1}^n m \, \Delta x_j \leq \sum_{j=1}^n f(t_j) \Delta x_j \leq \sum_{j=1}^n M \, \Delta x_j. \tag{2}$$

But

$$\sum_{j=1}^{n} m\Delta x_j = m\sum_{j=1}^{n} \Delta x_j = m[(x_1 - x_0) + (x_2 - x_1) + \cdots + (x_n - x_{n-1})]$$

$$= m(x_n - x_0)$$

$$= m(b - a)$$

and, similarly,

$$\sum_{j=1}^{n} M\Delta x_j = M(b - a).$$

Thus, inequality (2) becomes

$$m(b - a) \le \sum_{j=1}^{n} f(t_j)\Delta x_j \le M(b - a). \tag{3}$$

Since we have shown that inequality (3) must hold for *any* Riemann sum for f on $[a, b]$, it follows from the definition of $\int_a^b f(x) \, dx$ as the limit of Riemann sums that

$$m(b - a) \le \int_a^b f(x) \, dx \le M(b - a)$$

or, that

$$m \le \frac{1}{b - a}\int_a^b f(x) \, dx \le M. \tag{4}$$

Now since f is continuous on $[a, b]$, inequality (4) and the Intermediate Value Theorem guarantee the existence of a number $c \in (a, b)$ for which

$$f(c) = \frac{1}{b - a}\int_a^b f(x) \, dx.$$

This is the statement of the Mean Value Theorem for Integrals.

THEOREM 8
Mean Value Theorem for Integrals

If f is continuous on $[a, b]$ there exists a number $c \in (a, b)$ for which

$$f(c) = \frac{1}{b - a}\int_a^b f(x) \, dx. \tag{5}$$

The number $\bar{f} = \dfrac{1}{b - a}\displaystyle\int_a^b f(x) \, dx$ is called the mean value, or *average value*, for the function f on $[a, b]$. The reason for this terminology is illustrated in Figure 5.1 in the case where f is nonnegative on $[a, b]$. Written in the form

$$\int_a^b f(x) \, dx = f(c)(b - a) = \bar{f} \cdot (b - a)$$

equation (5) states that the area of the region bounded by the graph of f and the x-axis over the interval $[a, b]$ is the same as the area of the rectangle with base $(b - a)$ and height $\bar{f}$. Thus, in this case the mean value $\bar{f}$ determines the constant function $y = \bar{f}$ whose graph bounds a region of the same area over $[a, b]$ as does the

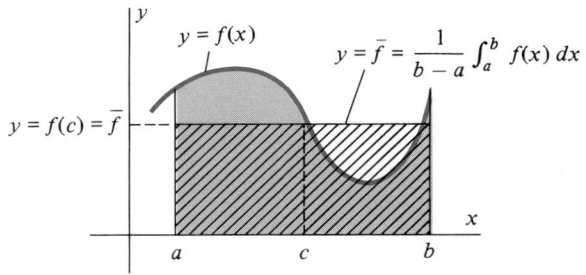

Figure 5.1 The mean value $\bar{f}$ satisfies the equation $\bar{f} \cdot (b - a) = \int_a^b f(x) \, dx.$

graph of f. Regardless of the sign of $f(t)$ on $[a, b]$, $\bar{f}$ is the number for which the product $\bar{f} \cdot (b - a)$ equals the value of the integral $\int_a^b f(x) \, dx$.

Example 1

Since we have shown in Example 5 of Section 6.3 that

$$\int_0^2 x \, dx = \frac{1}{2}(2^2 - 0^2) = 2$$

the mean value for the function $f(x) = x$ on $[0, 2]$ is

$$\bar{f} = \frac{1}{(2 - 0)} \int_0^2 x \, dx = \frac{1}{2}(2) = 1 \qquad \text{(Figure 5.2)}. \qquad \diamond$$

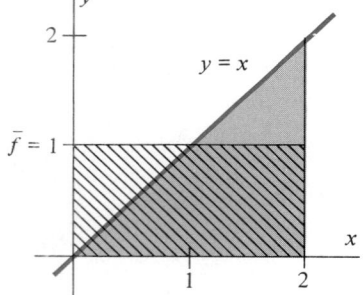

Figure 5.2 Mean value for $f(x) = x$ on $[0, 2]$ is $\bar{f} = 1$.

Example 2

Also in Example 5, Section 6.3, we showed that

$$\int_{-a}^a \sqrt{a^2 - x^2} \, dx = \frac{\pi a^2}{2}.$$

The mean value for the function $f(x) = \sqrt{a^2 - x^2}$ on $[-a, a]$ is therefore

$$\bar{f} = \frac{1}{a - (-a)} \int_{-a}^a \sqrt{a^2 - x^2} \, dx = \frac{1}{2a}\left(\frac{\pi a^2}{2}\right) = \frac{\pi a}{4}.$$

(See Figure 5.3.) $\qquad \diamond$

Several important inequalities concerning definite integrals may be established using the Mean Value Theorem for integrals.

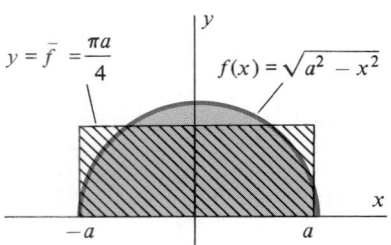

Figure 5.3 Mean value for $f(x) = \sqrt{a^2 - x^2}$ on $[-a, a]$ is $\bar{f} = \frac{\pi a}{4}$.

Example 3

Show that if f is continuous on $[a, b]$ and $f(x) \geq 0$ for all $x \in [a, b]$, then

$$\int_a^b f(x) \, dx \geq 0.$$

Solution: This follows directly from the Mean Value Theorem. Since $f(x) \geq 0$ for all $x \in [a, b]$, the number $f(c)$ in equation (5) is nonnegative. Since $(b - a) > 0$, equation (5) shows that the integral $\int_a^b f(x)\, dx$ is nonnegative. $\diamond$

Example 4

Show that if f and g are continuous on $[a, b]$ and if $f(x) \leq g(x)$ for all $x \in [a, b]$, then

$$\int_a^b f(x)\, dx \leq \int_a^b g(x)\, dx.$$

Solution: Since $f(x) \leq g(x)$ for all $x \in [a, b]$ we have $g(x) - f(x) \geq 0$ for all $x \in [a, b]$. It then follows from Example 3 that

$$\int_a^b [g(x) - f(x)]\, dx \geq 0$$

or

$$\int_a^b g(x)\, dx - \int_a^b f(x)\, dx \geq 0. \qquad\qquad \diamond$$

Proof of the Fundamental Theorem, Part (i)

We must show that if f is continuous on the open interval I, if $a \in I$, and if

$$A(x) = \int_a^x f(t)\, dt, \qquad x \in I$$

then A is differentiable on I and $A'(x) = f(x)$, $x \in I$.

To do so we let x be any number in I and we note that since the interval I is open, the number $x + h$ will lie in I for $|h|$ sufficiently small. Thus, the difference quotient

$$\frac{A(x + h) - A(x)}{h}$$

is defined for $|h|$ sufficiently small and its limit as $h \to 0$, if it exists, is the derivative $A'(x)$.

We assume first that $h > 0$ and consider the limit

$$\lim_{h \to 0^+} \frac{A(x + h) - A(x)}{h} = \lim_{h \to 0^+} \frac{1}{h} \left[\int_a^{x+h} f(t)\, dt - \int_a^x f(t)\, dt \right] \qquad (6)$$

$$= \lim_{h \to 0^+} \frac{1}{h} \int_x^{x+h} f(t)\, dt.$$

Since f is continuous on the interval $[x, x + h]$, the Mean Value Theorem for Integrals guarantees the existence of a number $c_h \in (x, x + h)$ for which

$$f(c_h) = \frac{1}{h} \int_x^{x+h} f(t)\, dt. \qquad (7)$$

Since the number c_h satisfies the inequality $x < c_h < x + h$ for $h > 0$, we have

$$\lim_{h \to 0^+} c_h = x. \qquad (8)$$

Also, since f is continuous on $[x, x + h]$ it follows from equation (8) that

$$\lim_{h \to 0^+} f(c_h) = f(x). \tag{9}$$

Combining equations (6), (7), and (9), we now conclude that

$$\lim_{h \to 0^+} \frac{A(x + h) - A(x)}{h} = \lim_{h \to 0^+} f(c_h) = f(x). \tag{10}$$

For $h < 0$ this same argument applied on the interval $[x + h, x]$ shows that

$$\lim_{h \to 0^-} \frac{A(x + h) - A(x)}{h} = f(x). \tag{11}$$

It now follows from equations (10) and (11) that

$$A'(x) = \lim_{h \to 0} \frac{A(x + h) - A(x)}{h} = f(x)$$

and, therefore, that A is differentiable at x. ◆

REMARK: It is important to note at this point that part (i) of the Fundamental Theorem guarantees that *every continuous function f has an antiderivative*, namely

$$A(x) = \int_a^x f(t) \, dt \tag{12}$$

as described in the Fundamental Theorem. The next example shows that the antiderivative A in equation (12) may sometimes be expressed in a more familiar form. Whether or not this is the case, however, equation (12) gives one way in which an antiderivative for a continuous function f may always be expressed.

Example 5

The function $f(x) = x^2$ is continuous for all x. Using equation (12) with $a = 0$ we may write one antiderivative for f as

$$A(x) = \int_0^x t^2 \, dt.$$

(See Figure 5.4.)

Since we already know that all antiderivatives for $f(x) = x^2$ are of the form $F(x) = \dfrac{1}{3}x^3 + C$, we may use part (ii) of the Fundamental Theorem to express $A(x)$ as

$$A(x) = \int_0^x t^2 \, dt = \left[\frac{1}{3}t^3 \right]_0^x = \frac{1}{3}x^3.$$

Differentiating both sides of this equation gives the result

$$A'(x) = x^2$$

as guaranteed by part (i) of the Fundamental Theorem. ◇

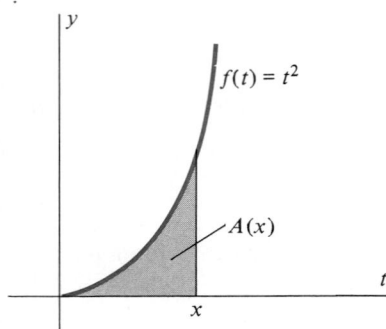

Figure 5.4 The antiderivative $A(x) = \int_0^x t^2 \, dt$ for $f(x) = x^2$ in Example 5.

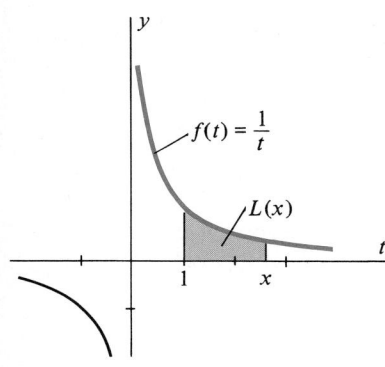

Figure 5.5 The antiderivative $L(x) = \int_1^x \frac{1}{t} dt$, $x > 0$ for $f(x) = \frac{1}{x}$ in Example 6.

Example 6

Figure 5.5 shows the function

$$L(x) = \int_1^x \frac{1}{t} dt \qquad x > 0 \qquad (13)$$

which, according to the Fundamental Theorem, is an antiderivative for the function

$$f(x) = L'(x) = \frac{1}{x}, \qquad x > 0.$$

Note that equation (13) is the *only* form of an antiderivative that we have available for the function $f(x) = \frac{1}{x}$ up to this point because the power rule $\int x^n \, dx = \frac{1}{n+1} x^{n+1} + C$ excludes the case $n = -1$. For this reason we shall study the function $y = L(x)$, called the natural logarithm function, extensively in Chapter 8. ◇

Other Functions Defined by Integration

Other types of functions involving integrals with the variable x in one or both limits of integration may be differentiated using the Fundamental Theorem, the Chain Rule, and properties of the definite integral. Here are several examples.

Example 7

For $F(x) = \int_2^x \sqrt{t} \sec \pi t \, dt$ find $F'(x)$.

Solution: The derivative follows immediately from the Fundamental Theorem, part (i):

$$F'(x) = \sqrt{x} \sin \pi x,$$

which is just the integrand evaluated at the upper limit x. ◇

Example 8

For $F(x) = \int_x^4 \sqrt{1 + t^2} \, dt$ find $F'(x)$.

Strategy

Change order of integration to obtain x in upper limit, using Theorem 4:

$$\int_a^b f(t) \, dt = -\int_b^a f(t) \, dt.$$

Apply Fundamental Theorem, part (i).

Solution

Since

$$F(x) = \int_x^4 \sqrt{1 + t^2} \, dt = -\int_4^x \sqrt{1 + t^2} \, dt,$$

we have

$$F'(x) = -\frac{d}{dx} \left\{ \int_4^x \sqrt{1 + t^2} \, dt \right\}$$

$$= -\sqrt{1 + x^2}.$$ ◇

Example 9

Find $\dfrac{d}{dx}\left\{\displaystyle\int_{1}^{x^3}\dfrac{1}{1+t^2}\,dt\right\}$.

Strategy

Express F as a composite function $y = F(u)$ with $u = x^3$ as upper limit of integration.

Solution

The function

$$F(x) = \int_{1}^{x^3}\frac{1}{1+t^2}\,dt$$

has the form

$$F(u) = \int_{1}^{u}\frac{1}{1+t^2}\,dt$$

with $u = x^3$. By the Chain Rule we have

Apply Chain Rule

and, in the first factor,

the Fundamental Theorem.

$$F'(x) = F'(u)\cdot u'(x)$$

$$= \frac{d}{du}\left\{\int_{1}^{u}\frac{1}{1+t^2}\,dt\right\}\cdot\frac{d}{dx}\,(x^3)$$

$$= \frac{1}{1+u^2}\cdot 3x^2$$

$$= \frac{3x^2}{1+(x^3)^2}$$

$$= \frac{3x^2}{1+x^6}.$$

$\diamondsuit$

Example 10

For $F(x) = \displaystyle\int_{\sqrt{x}}^{x^2+3}\cos^3 t\,dt,\qquad x > 0,\qquad$ find $F'(x)$.

Solution: We first use Theorem 5 to write

$$F(x) = \int_{\sqrt{x}}^{a}\cos^3 t\,dt + \int_{a}^{x^2+3}\cos^3 t\,dt$$

for any constant a. Then, proceeding as in Example 8 we reverse limits in the first integral obtaining

$$F(x) = -\int_{a}^{\sqrt{x}}\cos^3 t\,dt + \int_{a}^{x^2+3}\cos^3 t\,dt$$

and apply the Chain Rule and the Fundamental Theorem to obtain

$$F'(x) = (-\cos^3\sqrt{x})\left(\frac{d}{dx}\sqrt{x}\right) + [\cos^3(x^2+3)]\cdot\left[\frac{d}{dx}\,(x^2+3)\right]$$

$$= -\frac{\cos^3\sqrt{x}}{2\sqrt{x}} + 2x\cos^3(x^2+3).$$

$\diamondsuit$

Exercise Set 6.5

In Exercises 1–8 find the mean value $\bar{f} = \dfrac{1}{b-a}\displaystyle\int_a^b f(x)\,dx$ of the function f on the given interval.

1. $f(x) = x^2 + x + 1$, $[0, 1]$

2. $f(x) = \sqrt{x+2}$, $[2, 7]$

3. $f(x) = \sin x$, $[0, 2\pi]$

4. $f(x) = \sec^2 x$, $[0, \pi/4]$

5. $f(x) = \sec x \cdot \tan x$ $[0, \pi/4]$

6. $f(x) = x\sqrt{x^2 + 1}$ $[0, 2]$

7. $f(x) = (x + 4)^{2/3}$ $[-4, 4]$

8. $f(x) = x \cos x^2$ $[0, \sqrt{\pi}/2]$

In Exercises 9–14 first express F as a function not involving an integral sign by using part (ii) of the Fundamental Theorem (see Example 5). Then find $F'(x)$ by differentiation.

9. $F(x) = \displaystyle\int_1^x t\,dt$

10. $F(x) = \displaystyle\int_{-2}^x (t^2 + 2t)\,dt$

11. $F(x) = \displaystyle\int_0^x \cos t\,dt$

12. $F(x) = \displaystyle\int_2^x \sqrt{1 + t}\,dt$, $x \ge -1$

13. $F(x) = \displaystyle\int_0^x (at^2 + bt + c)\,dt$

14. $F(x) = \displaystyle\int_1^x \dfrac{t}{\sqrt{1 + t^2}}\,dt$

In Exercises 15–24 find $F'(x)$ using part (i) of the Fundamental Theorem: $\dfrac{d}{dx}\left\{\displaystyle\int_a^x f(t)\,dt\right\} = f(x)$.

15. $F(x) = \displaystyle\int_2^x t^2 \sin t^2\,dt$

16. $F(x) = \displaystyle\int_0^x \sqrt{t^4 + 1}\,dt$

17. $F(x) = \displaystyle\int_x^1 t^3 \cos^2 t\,dt$

18. $F(x) = \displaystyle\int_x^5 \sqrt{4 + t^3}\,dt$

19. $F(x) = \displaystyle\int_0^{3x} \sqrt{1 + \sin t}\,dt$

20. $F(x) = \displaystyle\int_2^{x^2} \dfrac{1}{1 + t^3}\,dt$

21. $F(x) = \displaystyle\int_{x^2}^x \cos^3(t + 1)\,dt$

22. $F(x) = \displaystyle\int_{-x}^x \sqrt{t^2 + 1}\,dt$

23. $F(x) = x\displaystyle\int_3^{x^2} (t^2 + 1)^{-3}\,dt$ **24.** $F(x) = \cos x\displaystyle\int_{\sin x}^\pi \sqrt{t^4 + 1}\,dt$

25. Use the Mean Value Theorem for Integrals to explain why the inequality
$$\frac{1}{2} \le \int_0^1 \frac{1}{x}\,dx \le 2$$
is true.

26. Find the number c that satisfies the Mean Value Theorem for Integrals for the given function and interval.

    a. $f(x) = 2x + 5$, $[4, 6]$
    b. $f(x) = 3x^2 - 4x + 1$, $[1, 4]$

27. Find the number a so that the average value of the function $f(x) = x^2 + ax + 3$ on the interval $[0, 6]$ is 27.

28. Find the number b so that the average value of the function $f(x) = 2x + 3$ on the interval $[1, b]$ is 11.

29. Show that for the function $f(x) = cx + d$ the average value of f on $[a, b]$ is $\dfrac{c}{2}(a + b) + d$ which occurs at $x = \dfrac{a + b}{2}$.

30. Use the Mean Value Theorem for Integrals to explain why the inequality
$$2 \le \int_0^2 \sqrt{1 + x^2}\,dx \le 2\sqrt{5}$$
is true.

31. Show that the average value for $f(x) = \sin x$ on $[0, 2\pi]$ is the same as the average value for $f(x) = \cos x$ on $[0, 2\pi]$. Do these average values occur at the same numbers $c \in [0, 2\pi]$?

32. Let f and g be continuous on $[a, b]$. Let $\bar{f}$ be the average value for f on $[a, b]$ and let $\bar{g}$ be the average value for g on $[a, b]$. Show that if $f(x) \le g(x)$ for all $x \in [a, b]$ then $\bar{f} \le \bar{g}$.

33. Show that the converse of the statement in Exercise 32 is false by finding an example of two continuous functions f and g on an interval $[a, b]$ for which $\bar{f} \le \bar{g}$ but for which $f(x) > g(x)$ for some $x \in [a, b]$.

34. Let f be continuous on $[a, b]$ with maximum value M and average value $\bar{f}$. Show that if $\bar{f} = M$, then f is constant on $[a, b]$ and $f(x) = M$ for all $x \in [a, b]$.

6.6 FINDING AREAS BY INTEGRATION

The purpose of this section is to study further the use of the definite integral in calculating areas of regions in the plane, extending its use to regions bounded by the

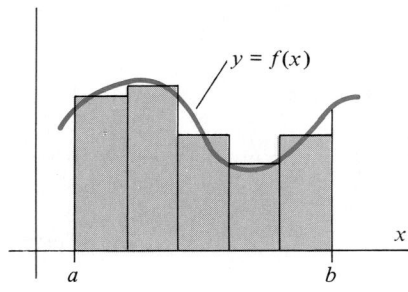

Figure 6.1 Area of $R = \int_a^b f(x)\,dx$

$$= \lim_{n \to \infty} \sum_{j=1}^n f(t_j)\,\Delta x.$$

graphs of two continuous functions, and to regions bounded by graphs of continuous functions with negative values. For the purposes of the area calculations of this section we shall restrict our considerations to Riemann sums with subintervals of equal length $\Delta x = x_j - x_{j-1}$, $j = 1, 2, \ldots, n$ as in Section 6.2.

Figure 6.1 summarizes the fact that when $f(x) \geq 0$ for all x in $[a, b]$ the area of the region R bounded by the graph of the continuous function f and the x-axis between $x = a$ and $x = b$ is given by the definite integral $\int_a^b f(x)\,dx$:

$$\text{Area of } R = \int_a^b f(x)\,dx \tag{1}$$

$$\text{if} \quad f(x) \geq 0 \text{ for all } x \text{ in } [a, b].$$

That is because the area of R is approximated by the areas of rectangles with heights $f(t_j)$ and widths Δx, as illustrated in Figure 6.2.

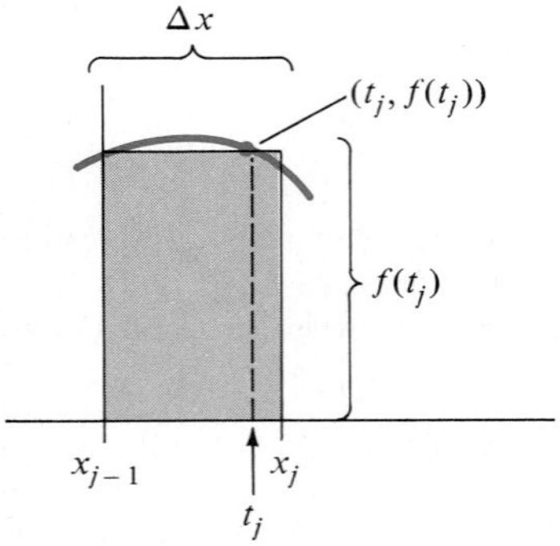

Figure 6.2 Area of jth approximating rectangle is $f(t_j)\,\Delta x$.

Example 1

Find the area of the region R bounded by the graph of $f(x) = \sqrt{x} + 2$ and the x-axis between $x = 1$ and $x = 4$. (See Figure 6.3.)

Solution: Here $f(x) > 0$ for all x in $[1, 4]$ so we apply equation (1) with $a = 1$ and $b = 4$:

$$\text{Area of } R = \int_1^4 (\sqrt{x} + 2)\,dx$$

$$= \int_1^4 (x^{1/2} + 2)\,dx$$

$$= \left[\frac{2}{3}x^{3/2} + 2x\right]_1^4$$

$$= \left(\frac{2}{3} \cdot 4^{3/2} + 2 \cdot 4\right) - \left(\frac{2}{3} \cdot 1^{3/2} + 2 \cdot 1\right)$$

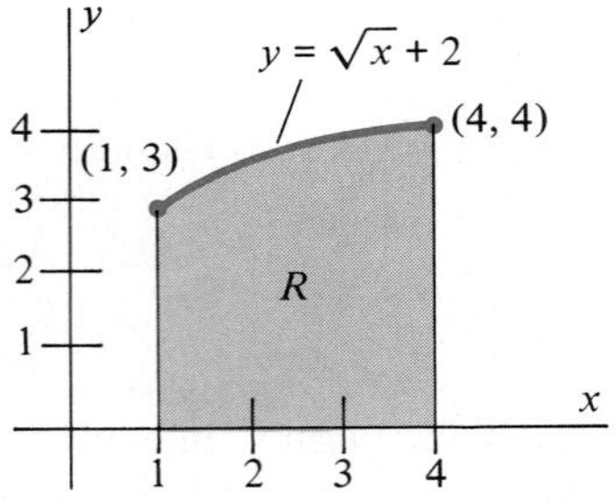

Figure 6.3 Region R in Example 1.

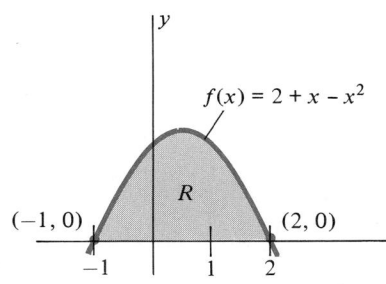

Figure 6.4 Region R in Example 2.

$$= \frac{2}{3}(8) + 8 - \frac{2}{3} - 2$$

$$= \frac{32}{3}.$$

In the next example the endpoints a and b defining the region R are not explicitly stated. They must be found by finding the *zeros* of the function f.

Example 2

Find the area of the region bounded by the graph of $f(x) = 2 + x - x^2$ and the x-axis.

Strategy

Sketch the graph to locate the region R.

Determine the interval $[a, b]$ defining the region by finding the zeros of f.

Use equation (1) to find the area of R.

Solution

The region bounded by the graph of $f(x) = 2 + x - x^2$ and the x-axis lies above the x-axis, as shown in Figure 6.4. To find the largest and smallest numbers x associated with this region we locate the zeros of f by factoring:

$$f(x) = 2 + x - x^2$$
$$= (2 - x)(1 + x).$$

Thus $x = a = -1$ and $x = b = 2$ are the zeros of f. The area of R is thus

$$\text{Area} = \int_{-1}^{2} (2 + x - x^2)\, dx$$

$$= \left[2x + \frac{1}{2}x^2 - \frac{1}{3}x^3 \right]_{-1}^{2}$$

$$= \left[2 \cdot 2 + \frac{1}{2} \cdot 2^2 - \frac{1}{3} \cdot 2^3 \right] - \left[2(-1) + \frac{1}{2}(-1)^2 - \frac{1}{3}(-1)^3 \right]$$

$$= \frac{9}{2}.$$

Example 3

Find the area of the region R bounded above by the graph of $y = |x - 3|$ and below by the x-axis for $1 \le x \le 6$.

Strategy

Use definition of absolute value to re-write $f(x) = |x - 3|$.

Calculate areas of regions over $[1, 3]$ and $[3, 6]$ separately, writing $f(x)$ as a simple linear function in each interval.

Solution

Since we do not have an antiderivative for $f(x) = |x - 3|$, we cannot proceed directly to apply equation (1). However, using the definition of absolute value we may rewrite f as

$$f(x) = |x - 3| = \begin{cases} x - 3 & \text{if } x \ge 3 \\ 3 - x & \text{if } x < 3. \end{cases}$$

This suggests that we calculate separately the areas of regions R_1 and R_2 as illustrated in Figure 6.5. We obtain

$$\text{Area of } R = \text{Area of } R_1 + \text{Area of } R_2$$

$$= \int_{1}^{3} (3 - x)\, dx + \int_{3}^{6} (x - 3)\, dx$$

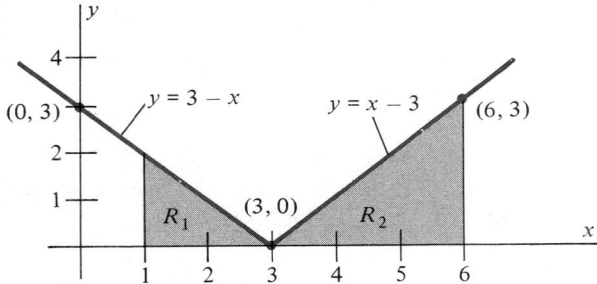

Figure 6.5 Region bounded by the graph of $f(x) = |x - 3|$ and the x-axis for $1 \leq x \leq 6$.

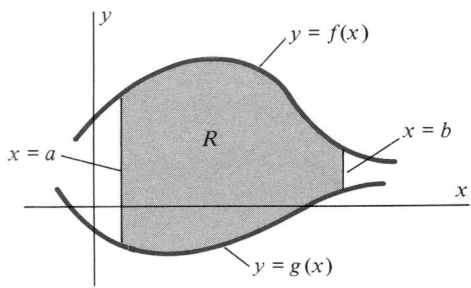

Figure 6.6 Region bounded by two graphs.

$$= \left[\left(3x - \frac{1}{2}x^2\right)\right]_1^3 + \left[\left(\frac{1}{2}x^2 - 3x\right)\right]_3^6$$

$$= \left\{\left(3 \cdot 3 - \frac{1}{2} \cdot 3^2\right) - \left(3 \cdot 1 - \frac{1}{2} \cdot 1^2\right)\right\}$$

$$+ \left\{\left(\frac{1}{2} \cdot 6^2 - 3 \cdot 6\right) - \left(\frac{1}{2} \cdot 3^2 - 3 \cdot 3\right)\right\}$$

$$= \frac{13}{2}. \qquad \diamond$$

The Area of a Region Bounded by Two Curves

Figure 6.6 shows a region R bounded by the graphs of two continuous functions— above by the graph of $y = f(x)$, and below by the graph of $y = g(x)$—for values of x between a and b.

We can use the definite integral to compute the area of such regions as follows. If we divide the interval $[a, b]$ into n subintervals of equal length $\Delta x = \dfrac{b - a}{n}$ and select one "test number" t_j in each subinterval, we can approximate the part of the region corresponding to that subinterval by the rectangle with width Δx and height $[f(t_j) - g(t_j)]$. (See Figures 6.7 and 6.8.) Summing the individual approximations

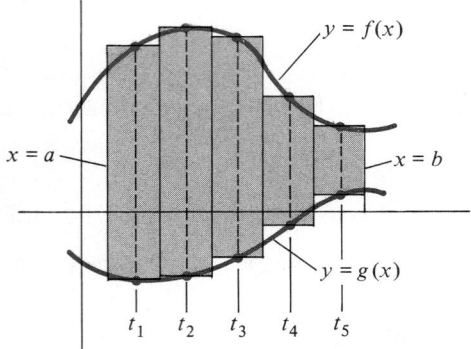

Figure 6.7 Approximating R by rectangles.

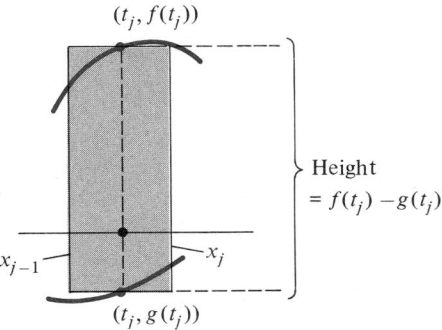

Figure 6.8 Area of jth rectangle is $[f(t_j) - g(t_j)]\Delta x$.

then gives the approximation to the area of R:

$$\text{Area of } R \approx \sum_{j=1}^{n} [f(t_j) - g(t_j)] \, \Delta x. \tag{2}$$

Now as n increases without bound the limit of the approximating sum on the right-hand side of equation (2) is the integral

$$\int_{a}^{b} [f(x) - g(x)] \, dx = \lim_{n \to \infty} \sum_{j=1}^{n} [f(t_j) - g(t_j)] \, \Delta x \tag{3}$$

which we define to be the area of R.

DEFINITION 5

The area of the region R bounded above by the graph of the continuous function f and below by the graph of the continuous function g, between $x = a$ and $x = b$, is

$$\text{Area of } R = \int_{a}^{b} [f(x) - g(x)] \, dx. \tag{4}$$

(Note that this requires $f(x) \geq g(x)$ for all x. See Figure 6.6.)

Note that equation (4) generalizes equation (1), since $g(x) = 0$ when the region R is bounded below by the x-axis. Notice also that there are no restrictions on the sign of $f(x)$ or $g(x)$ in equation (4)—either may be positive or negative. *It is essential that $f(x) \geq g(x)$, however,* so that the integrand $[f(x) - g(x)]$ is nonnegative.

Example 4

Find the area of the region R bounded by the graphs of the equations $y = -x^2 + 5x - 4$ and $y = -x - 4$ between $x = 0$ and $x = 6$. (See Figure 6.9.)

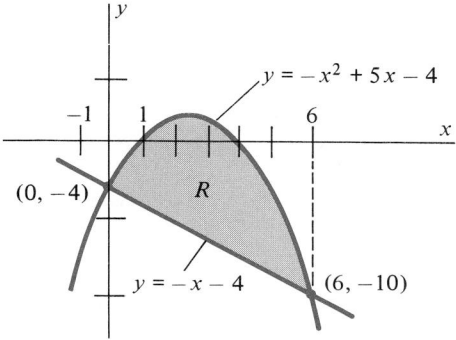

Figure 6.9 Region R in Example 4.

Strategy

Determine which graph is on top.

(To find area we must integrate top curve minus bottom.)

Apply equation (4).

Solution

By graphing the two equations (or by checking individual values) we find that the graph of $f(x) = -x^2 + 5x - 4$ forms the top boundary while the graph of $g(x) = -x - 4$ is the bottom boundary of the region R. Applying equation (4) gives

$$\text{Area of } R = \int_0^6 [(-x^2 + 5x - 4) - (-x - 4)] \, dx$$

$$= \int_0^6 (-x^2 + 6x) \, dx$$

$$= \left[-\frac{1}{3}x^3 + 3x^2 \right]_0^6$$

$$= \left(-\frac{1}{3} \right)6^3 + 3 \cdot 6^2$$

$$= 36. \qquad \diamondsuit$$

In Example 4 the region R was completely determined by the graphs of the two functions f and g, even though the numbers $x = 0$ and $x = 6$ were provided in the statement of the problem. If these numbers are not provided in such problems, it is necessary to first solve the equation $f(x) = g(x)$ in order to determine the horizontal extremities of the region R.

Example 5

Find the area of the region bounded by the graphs of the functions $y = x^4 + 1$ and $y = 2x^2$.

Strategy

Sketch region.

Find points of intersection.

Solve by factoring. (If the equation cannot be easily factored, use the quadratic formula.) Determine which curve is on top (see Figure 6.10).

Use Definition 5.

Solution

A rough sketch of the region shows that the graphs intersect at two points. To find these points we equate the two functions, obtaining

$$x^4 + 1 = 2x^2$$

or

$$x^4 - 2x^2 + 1 = (x^2 - 1)^2 = 0$$

so $x^2 - 1 = 0$ and $x = \pm 1$. The points of intersection are therefore $(-1, 2)$ and $(1, 2)$.

Checking any particular value of x in $(-1, 1)$ shows that $f(x) = x^4 + 1$ determines the upper boundary and $g(x) = 2x^2$ determines the lower boundary. Thus by Definition 5

$$A = \int_{-1}^1 [(x^4 + 1) - 2x^2] \, dx$$

$$= \left[\frac{1}{5}x^5 + x - \frac{2}{3}x^3 \right]_{x=-1}^{x=1}$$

$$= \left(\frac{1}{5} + 1 - \frac{2}{3} \right) - \left(-\frac{1}{5} - 1 + \frac{2}{3} \right)$$

$$= \frac{16}{15}. \qquad \diamondsuit$$

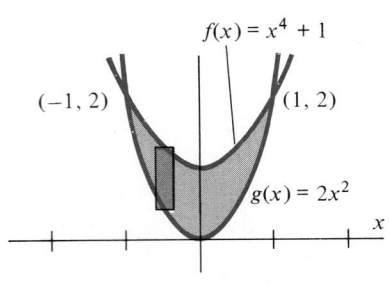

$f(x) = x^4 + 1$

$(-1, 2)$ $(1, 2)$

$g(x) = 2x^2$

Figure 6.10

Often graphs bounding the region in question cross several times. In such cases, we apply Definition 5 in each of the resulting subregions, being careful to note for each region the upper and lower boundaries.

Example 6

Find the area of the region bounded by the graphs of $f(x) = \sin x$ and $g(x) = 1/2$ for $0 \le x \le 2\pi$.

Solution: The two graphs cross where $\sin x = 1/2$. The solutions of this equation in $[0, 2\pi]$ are $x = \pi/6$ and $x = 5\pi/6$. By checking particular values of x we can see that, on the resulting intervals,

$$\sin x \le \frac{1}{2} \qquad \text{for} \qquad x \in \left[0, \frac{\pi}{6}\right],$$

$$\sin x \ge \frac{1}{2} \qquad \text{for} \qquad x \in \left[\frac{\pi}{6}, \frac{5\pi}{6}\right],$$

$$\sin x \le \frac{1}{2} \qquad \text{for} \qquad x \in \left[\frac{5\pi}{6}, 2\pi\right].$$

(See Figure 6.11.)

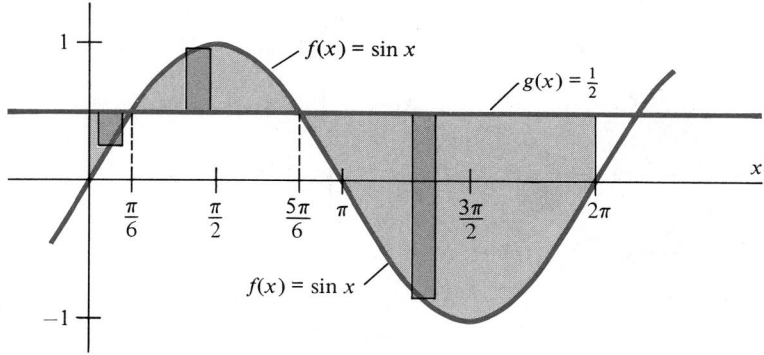

Figure 6.11 Area bounded by $f(x) = \sin x$ and $g(x) = 1/2$, $0 \le x \le 2\pi$.

Thus,

$$A = \int_0^{\pi/6}\left(\frac{1}{2} - \sin x\right) dx + \int_{\pi/6}^{5\pi/6}\left(\sin x - \frac{1}{2}\right) dx + \int_{5\pi/6}^{2\pi}\left(\frac{1}{2} - \sin x\right) dx$$

$$= \left[\frac{x}{2} + \cos x\right]_0^{\pi/6} + \left[-\cos x - \frac{x}{2}\right]_{\pi/6}^{5\pi/6} + \left[\frac{x}{2} + \cos x\right]_{5\pi/6}^{2\pi}$$

$$= \left[\left(\frac{\pi}{12} + \frac{\sqrt{3}}{2}\right) - (0 + 1)\right] + \left[\left(\frac{\sqrt{3}}{2} - \frac{5\pi}{12}\right) - \left(-\frac{\sqrt{3}}{2} - \frac{\pi}{12}\right)\right]$$

$$+ \left[(\pi + 1) - \left(\frac{5\pi}{12} - \frac{\sqrt{3}}{2}\right)\right]$$

$$= \frac{\pi}{3} + 2\sqrt{3} \approx 4.51. \qquad \diamond$$

Our next example illustrates that it may sometimes be easier to calculate the desired area by integrating with respect to y rather than with respect to x.

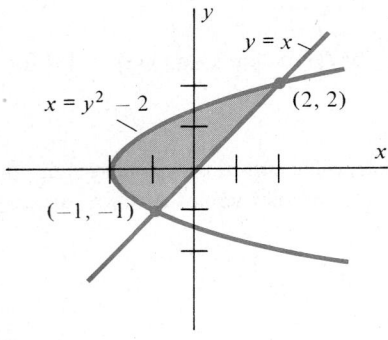

Figure 6.12

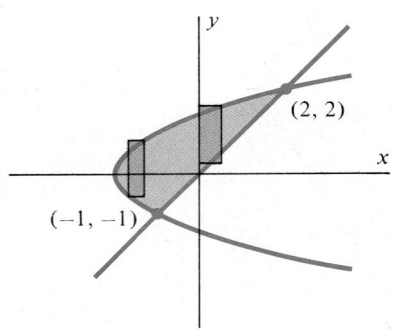

Figure 6.13

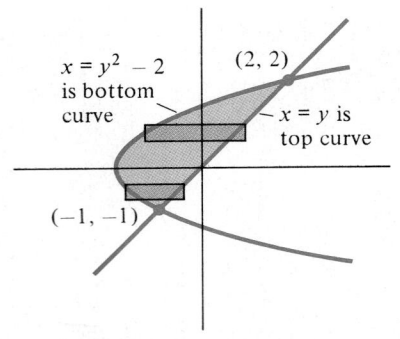

Figure 6.14

Example 7

Find the area bounded by the graphs of the equations $y = x$ and $x = y^2 - 2$.

Solution: The area in question is illustrated in Figure 6.12. The points of intersection are found by setting

$$y = y^2 - 2$$

or

$$y^2 - y - 2 = (y + 1) \cdot (y - 2) = 0,$$

so

$$y = -1 \quad \text{or} \quad y = 2.$$

The points are therefore $(-1, -1)$ and $(2, 2)$.

Figure 6.13 shows that by partitioning the x-axis (i.e., by using vertical rectangles) one encounters a difficulty—some approximating rectangles have both bottoms and tops on the graph of $x = y^2 - 2$, while others have tops on the graphs of $x = y^2 - 2$ and bottoms on the graph of $y = x$.

A simpler approach is to partition the y-axis, thus viewing y as the independent variable. Then all approximating rectangles run from the curve $x = y^2 - 2$ to the line $x = y$ (Figure 6.14). The area is thus calculated as

$$\text{Area} = \int_{y=-1}^{y=2} [y - (y^2 - 2)] \, dy$$

$$= \left[-\frac{1}{3}y^3 + \frac{1}{2}y^2 + 2y \right]_{-1}^{2}$$

$$= 9/2. \qquad \diamond$$

In general, when the region in question is bounded by the graphs of the functions $x = f(y)$ and $x = g(y)$ between $y = c$ and $y = d$ (see Figure 6.15), we have the area formula

$$\boxed{\begin{array}{l} \text{Area} = \displaystyle\int_c^d [f(y) - g(y)] \, dy \\[2mm] \text{if} \quad f(y) \geq g(y). \end{array}}$$

Example 8

Find the area of the region bounded by the graphs of the equations $3y - x = 6$, $x + y = -2$, and $x + y^2 = 4$ illustrated in Figure 6.16.

Solution: The graphs of the first two equations are lines. The graph of the third is a parabola.

As Figure 6.16 suggests, the simplest solution involves partitioning the y-axis, so that the approximating rectangles lie parallel to the x-axis. In this case, the approximating rectangles all will have their right edge lying on the parabola. Their left edge will lie on one of the two lines.

To pursue this approach, we express each equation as a function of y, obtaining

$$f(y) = 3y - 6, \qquad g(y) = -y - 2, \qquad \text{and} \qquad h(y) = 4 - y^2.$$

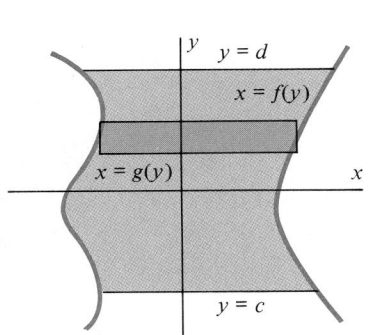

Figure 6.15 Region bounded by graphs of functions of y and horizontal lines.

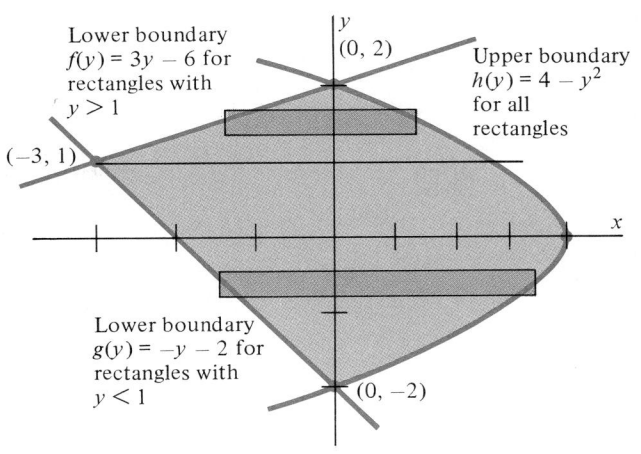

Figure 6.16 Region in Example 8: Partitioning the y-axis.

Equating these functions two at a time and solving gives the three points of intersection among these graphs: $(-3, 1)$, $(0, 2)$, and $(0, -2)$.

For $-2 \le y \le 1$, the left edges of the approximating rectangles are determined by the graph of $g(y) = -y - 2$, while for $1 \le y \le 2$ these edges intersect the graph of $f(y) = 3y - 6$. The area of the region is therefore

$$A = \int_{-2}^{1} [(4 - y^2) - (-y - 2)] \, dy + \int_{1}^{2} [(4 - y^2) - (3y - 6)] \, dy$$

$$= \int_{-2}^{1} (6 + y - y^2) \, dy + \int_{1}^{2} (10 - 3y - y^2) \, dy$$

$$= \left[6y + \frac{y^2}{2} - \frac{y^3}{3} \right]_{-2}^{1} + \left[10y - \frac{3y^2}{2} - \frac{y^3}{3} \right]_{1}^{2}$$

$$= \left[\left(6 + \frac{1}{2} - \frac{1}{3} \right) - \left(-12 + 2 + \frac{8}{3} \right) \right]$$

$$+ \left[\left(20 - 6 - \frac{8}{3} \right) - \left(10 - \frac{3}{2} - \frac{1}{3} \right) \right]$$

$$= \frac{50}{3} \approx 16.67. \qquad \diamond$$

Regions Bounded by Negative Functions

Definition 5 allows us to explain the relationship between the integral $\int_a^b g(x) \, dx$ and the area of the region R bounded by the graph of a continuous function g and the x-axis, for $a \le x \le b$, when $g(x) \le 0$ for all $x \in [a, b]$. (See Figure 6.17.) Since the x-axis is the graph of the constant function with values $f(x) = 0$ for all $x \in [a, b]$, the region R is bounded above by the graph of f and below by the graph of g, so

$$\text{Area of } R = \int_a^b [f(x) - g(x)] \, dx \qquad \text{(Definition 5)}$$

$$= \int_a^b [0 - g(x)] \, dx \qquad (f(x) = 0 \text{ for all } x)$$

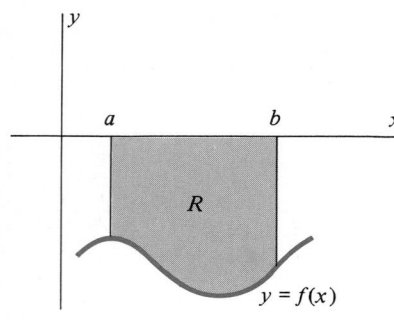

Figure 6.17 Area of $R = -\int_a^b f(x)\,dx$ if $f(x) \leq 0$ for $a \leq x \leq b$.

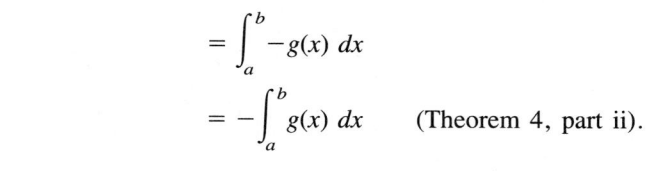

$$= \int_a^b -g(x)\,dx$$

$$= -\int_a^b g(x)\,dx \qquad \text{(Theorem 4, part ii).}$$

Thus

> if g is continuous on $[a, b]$ and $g(x) \leq 0$ for all $x \in [a, b]$, then the area of the region R bounded by the graph of g and the x-axis for $a \leq x \leq b$ is
>
> $$\text{Area of } R = -\int_a^b g(x)\,dx.$$

This is why we must be so careful to insure nonnegative integrands when calculating areas of regions using the definite integral.

Example 9

Figure 6.18 shows the graph of the function $f(x) = x^3$ for $-1 \leq x \leq 1$.

(a) To find the area of the region R bounded by the graph of $f(x) = x^3$ and the x-axis for $-1 \leq x \leq 1$, we must calculate separately the areas of the parts of the regions corresponding to the intervals $[-1, 0]$ and $[0, 1]$. That is because $f(x) \leq 0$ for $-1 \leq x \leq 0$. Thus,

$$\text{Area of } R = -\int_{-1}^0 x^3\,dx + \int_0^1 x^3\,dx$$

$$= -\left[\frac{1}{4}x^4\right]_{-1}^0 + \left[\frac{1}{4}x^4\right]_0^1$$

$$= -\left(0 - \frac{1}{4}\right) + \left(\frac{1}{4} - 0\right)$$

$$= \frac{1}{2}.$$

Figure 6.18 Graph of $f(x) = x^3$.

(b) The *value of the integral* $\int_{-1}^1 x^3\,dx$, however, is simply

$$\int_{-1}^1 x^3\,dx = \frac{1}{4}x^4\Big]_{-1}^1 = \frac{1}{4}(1)^4 - \frac{1}{4}(-1)^4 = 0.$$

Note the difference between parts (a) and (b)! In part (a) the integral was used to calculate the area of a region, and signs were chosen for $\int_a^b f(x)\,dx$ depending on whether the integrand was positive or negative. Part (b), however, is *not* a calculation of area, but rather simply the evaluation of a definite integral using the Fundamental Theorem. This example is intended to emphasize the difference between simply evaluating an integral and using the integral in the calculation of area. ◇

REMARK: Using our ability to evaluate definite integrals involving the absolute value function, we can restate the formula for the area A of the region bounded between the graphs of the functions f and g for $a \leq x \leq b$ as

$$A = \int_a^b |f(x) - g(x)| \, dx$$

regardless of the sign of $f(x) - g(x)$.

Exercise Set 6.6

In each of Exercises 1–8 find the area of the region bounded above by the graph of $y = f(x)$ and below by the x-axis for $a \le x \le b$.

1. $f(x) = 2x + 5$, $a = 0$, $b = 2$

2. $f(x) = 9 - x^2$, $a = -3$, $b = 3$

3. $f(x) = \dfrac{1}{\sqrt{x - 2}}$, $a = 3$, $b = 5$

4. $f(x) = \sin 2x$, $a = 0$, $b = \pi/2$

5. $f(x) = \dfrac{x + 1}{(x^2 + 2x)^2}$, $a = 1$, $b = 2$

6. $f(x) = \sqrt{1 + x^2}$, $a = 0$, $b = 1$

7. $f(x) = \sqrt{5 + x}$, $a = -4$, $b = 4$

8. $f(x) = \dfrac{x}{\sqrt{9 + x^2}}$, $a = 0$, $b = 4$

In Exercises 9–20, sketch the region bounded by the graphs of the given functions between the indicated values of x. Then calculate the area of the region.

9. $f(x) = x + 1$, $g(x) = -2x + 1$, $0 \le x \le 2$

10. $f(x) = 2x + 3$, $g(x) = x^2 - 4$, $-1 \le x \le 1$

11. $f(x) = \sqrt{x}$, $g(x) = -x^2$, $0 \le x \le 4$

12. $f(x) = \dfrac{1}{x^2}$, $g(x) = x^{2/3}$, $1 \le x \le 8$

13. $f(x) = \sin x$, $g(x) = \cos x$, $0 \le x \le 2\pi$

14. $f(x) = 1$, $g(x) = \cos x$, $0 \le x \le 2\pi$

15. $f(y) = y - 3$, $g(y) = y^2$, $-2 \le y \le 2$

16. $f(x) = x\sqrt{9 - x^2}$, $g(x) = -x$, $-3 \le x \le 3$

17. $f(x) = |4 - x^2|$, $g(x) = 5$, $-3 \le x \le 3$

18. $f(x) = \sin x$, $g(x) = x$, $0 \le x \le \pi$

19. $f(x) = \dfrac{x^2 - 1}{x^2}$, $g(x) = \dfrac{1 - x^2}{x^2}$, $1 \le x \le 2$

20. $f(x) = x^{2/3}$, $g(x) = x^{1/3}$, $-1 \le x \le 1$

In Exercises 21–29, sketch the region bounded by the graphs of the given equations. Then calculate the area of the region.

21. $y = 4 - x^2$, $y = x - 2$

22. $y = 9 - x^2$, $9y - x^2 + 9 = 0$

23. $y = x^2$, $y = x^3$

24. $y = x^3$, $y = x$

25. $x + y^2 = 4$, $y = x + 2$

26. $x = \sqrt{y}$, $x = \sqrt[3]{y}$

27. $2x = y^2$, $x + 2 = y^2$

28. $y = x^{2/3}$, $y = 2 - x^2$

29. $y = x^{2/3}$, $y = x^2$

30. Find the area of the region bounded by the graph of $y = x^{2/3}$ and the line $y = 1$ by partitioning the x-axis.

31. Rework Exercise 30, partitioning the y-axis.

32. Find the area of the region bounded by the graphs of $y = x^3$, $y = -x^3$, $y = 1$, and $y = -1$ by partitioning the x-axis.

33. Rework Exercise 32, partitioning the y-axis.

34. Find a number a so that the line $y = a$ divides the region bounded by the x-axis and the graph of the equation $y = 4 - x^2$ into two regions of equal area.

35. Verify that the method of this section produces the same value for the area of the region bounded by the following lines as does the appropriate formula from plane geometry.
a. $y = x$, $x = 3$, $y = -1$
b. $y = 2x + 2$, $y = -2x - 1$, $x = 1$, $x = 3$.

6.7 RULES FOR APPROXIMATING INTEGRALS

We have seen that many functions, such as $\sin x^2$ or $\sqrt{x^3 + 1}$, do not have antiderivatives that are easily found. The Fundamental Theorem of the Calculus is there-

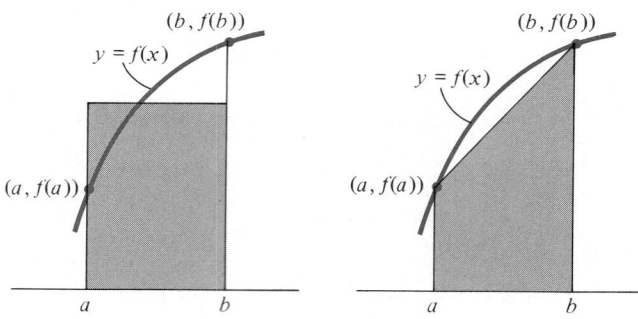

Figure 7.1 Approximations by trapezoids provide a "better fit," in general, than do approximations by rectangles.

fore not applicable to integrals involving such functions, and we can only approximate the values of these integrals.

In this section we present two techniques for approximating definite integrals, the Trapezoidal Rule and Simpson's Rule.

The Trapezoidal Rule

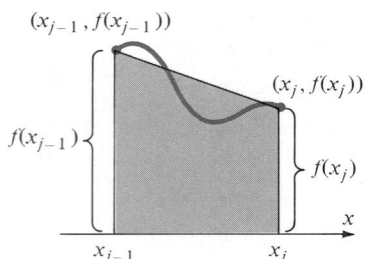

Figure 7.2 Trapezoidal approximation.

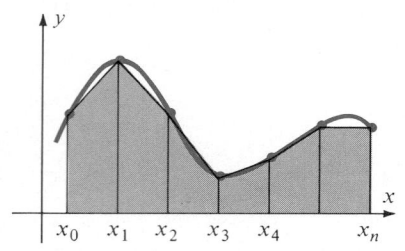

Figure 7.3 The Trapezoidal Rule.

Figure 7.1 illustrates the fact that, in intervals over which the graph of f is changing rapidly (the slope has relatively large positive or negative values), an approximating rectangle does not fit the curve well. There are large areas under the curve that are not included in the rectangle and vice versa. We can try to improve the estimate by drawing a line from $(a, f(a))$ to $(b, f(b))$, which is a secant line that approximates the curve between those points, and using the trapezoidal area below this line in the same way that we previously used approximating rectangles. From elementary geometry, the area of a trapezoid with parallel sides $f(a)$ and $f(b)$ and altitude $(b - a)$ is $\frac{1}{2}(b - a)[f(a) + f(b)]$.

To implement this idea, we divide the interval $[a, b]$ into n subintervals of equal length $\Delta x = \dfrac{b - a}{n}$, and we denote the endpoints of the subintervals by

$$a = x_0 < x_1 < x_2 < \cdots < x_n = b.$$

However, instead of using the area of an approximating rectangle in each subinterval we use the approximation

$$\int_{x_{j-1}}^{x_j} f(x)\, dx \approx \frac{1}{2}[f(x_{j-1}) + f(x_j)]\, \Delta x.$$

As illustrated in Figures 7.2 and 7.3, this quantity corresponds to the area of the trapezoid with vertices $(x_{j-1}, 0)$, $(x_j, 0)$, $(x_{j-1}, f(x_{j-1}))$, and $(x_j, f(x_j))$ when $f(x) \geq 0$ on the interval $[a, b]$.

Summing these individual approximations gives

$$\int_a^b f(x)\, dx \approx \frac{\Delta x}{2}\{[f(x_0) + f(x_1)] + [f(x_1) + f(x_2)] + [f(x_2) + f(x_3)]$$

$$+ \cdots + [f(x_{n-2}) + f(x_{n-1})] + [f(x_{n-1}) + f(x_n)]\}.$$

Notice that each function value $f(x_j)$, except $f(x_0)$ and $f(x_n)$, occurs twice in the above expression. That is because each such value is involved as a dimension in two trapezoids, once as a left vertical side and once as a right vertical side. We may therefore state our approximation as follows.

Trapezoidal Rule: Let $\Delta x = \dfrac{b - a}{n}$ and let

$$x_0 = a, \; x_1 = a + \Delta x, \; x_2 = a + 2\,\Delta x, \; \ldots, \; x_n = a + n\,\Delta x = b.$$

Then

$$\int_a^b f(x)\,dx \approx \frac{b - a}{2n}\{f(x_0) + 2f(x_1) + 2f(x_2) + \cdots + 2f(x_{n-1}) + f(x_n)\}.$$

Example 1

Approximate $\displaystyle\int_1^4 \frac{1}{x}\,dx$ using the Trapezoidal Rule with $n = 6$.

Solution: With $a = 1$, $b = 4$, and $n = 6$, we have $\Delta x = \dfrac{4 - 1}{6} = \dfrac{1}{2}$, so

$$x_0 = 1, \; x_1 = \frac{3}{2}, \; x_2 = 2, \; x_3 = \frac{5}{2}, \; x_4 = 3, \; x_5 = \frac{7}{2}, \; x_6 = 4.$$

(See Figure 7.4.)
The Trapezoidal Rule thus gives the approximation

$$\int_1^4 \frac{1}{x}\,dx \approx \frac{3}{12}\left\{1 + \frac{4}{3} + 1 + \frac{4}{5} + \frac{2}{3} + \frac{4}{7} + \frac{1}{4}\right\} = \frac{2361}{1680} = 1.4054. \qquad \diamond$$

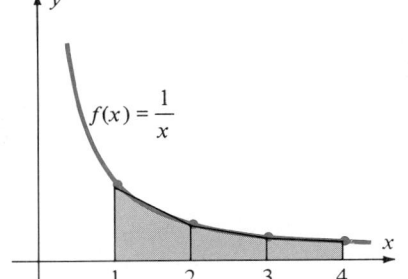

$f(x) = \dfrac{1}{x}$

Figure 7.4 Approximation in Example 1.

From the geometry of Figure 7.4, we can see that the approximation in Example 1 is actually *larger* than the actual value of the integral. This is because the graph of f is concave up on $[1, 4]$.

Error in Trapezoidal Approximations: In more advanced courses it is proved that the error in using the Trapezoidal Rule to approximate the integral $\int_a^b f(x)\,dx$ with n subdivisions has the following bound:

$$|\text{Error}| \le \frac{(b - a)^3 M}{12 n^2} \qquad (1)$$

where M is the maximum value of $|f''(x)|$ on the interval $[a, b]$.

Example 2

Find a bound on the error for the approximation in Example 1.

Solution: For $f(x) = \dfrac{1}{x}$, $f''(x) = \dfrac{2}{x^3}$ which has a maximum value of $M = 2$ on the interval $[1, 4]$. Thus, by inequality (1) we have

$$|\text{Error}| \le \frac{3^3}{12(6^2)}(2) = \frac{1}{8} = 0.125. \qquad \diamond$$

Table 7.1

[A, B]	n	S
[1, 4]	5	1.41348
[1, 4]	20	1.38805
[1, 4]	100	1.38636
[1, 4]	250	1.38631
[1, 20]	5	3.76955
[1, 20]	20	3.06566
[1, 20]	100	2.99872
[1, 20]	250	2.99621

Program 4 in Appendix I is a BASIC program implementing the Trapezoidal Rule. This program approximates the integral $\int_a^b \frac{1}{x}\, dx$.

Running this program on a computer, we obtained the results contained in Table 7.1. The student with access to a computer is encouraged to experiment with modifications of this program on integrals previously calculated using the Fundamental Theorem. Notice that the integral approximated in Table 7.1 *cannot* be evaluated by the Fundamental Theorem at this point as we do not yet have available an antiderivative for the integrand $1/x$.

Simpson's Rule

This approximation procedure uses parabolic arcs, rather than line segments, to approximate portions of the graph of $y = f(x)$. It uses the fact that if x_0, x_1, and x_2 are three values of x with

$$x_1 - x_0 = x_2 - x_1 = \Delta x$$

then the definite integral of the parabola passing through the three points (x_0, y_0), (x_1, y_1), and (x_2, y_2) equals

$$\frac{\Delta x}{3}[y_0 + 4y_1 + y_2]. \tag{2}$$

To see this let $x_0 = -\Delta x$, $x_1 = 0$, and $x_2 = \Delta x$ as in Figure 7.5. Let the parabola through (x_0, y_0), (x_1, y_1), and (x_2, y_2) be denoted by

$$y = ax^2 + bx + c.$$

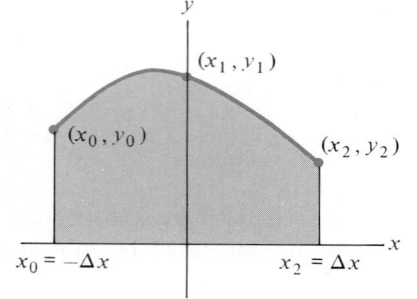

Figure 7.5

Then we have

$$y_0 = a(\Delta x)^2 - b(\Delta x) + c,$$
$$y_1 = c,$$

and

$$y_2 = a(\Delta x)^2 + b(\Delta x) + c.$$

Using these values we find that

$$\int_{-\Delta x}^{\Delta x} (ax^2 + bx + c)\, dx = \frac{1}{3}ax^3 + \frac{1}{2}bx^2 + cx \Big]_{-\Delta x}^{\Delta x}$$

$$= \frac{\Delta x}{3}[2a(\Delta x)^2 + 6c]$$

$$= \frac{\Delta x}{3}[y_0 + 4y_1 + y_2].$$

We apply this result to obtain Simpson's Rule. As before, we divide $[a, b]$ into n subintervals, but now we must require that n be an *even* integer. We then approximate the integral

$$\int_{x_{2j-2}}^{x_{2j}} f(x)\, dx$$

over each *pair* of subintervals by expression (2). Thus we are approximating the actual value of the integral of f over $[x_{2j-2}, x_{2j}]$ by the integral of the approximating parabola. We obtain the approximation

$$\int_a^b f(x)\, dx \approx \frac{\Delta x}{3}\{[f(x_0) + 4f(x_1) + f(x_2)] + [f(x_2) + 4f(x_3) + f(x_4)]$$

$$+ [f(x_4) + 4f(x_5) + f(x_6)] + \cdots$$
$$+ [f(x_{n-4}) + 4f(x_{n-3}) + f(x_{n-2})]$$
$$+ [f(x_{n-2}) + 4f(x_{n-1}) + f(x_n)]\}.$$

Again, notice that each $f(x_{2j})$ is counted twice, except for $f(x_0)$ and $f(x_n)$, for the same reason as in the Trapezoidal Rule. Letting $\Delta x = \dfrac{b-a}{n}$, we can state this approximation rule as follows:

Simpson's Rule: Let n be an even integer and let $\Delta x = \dfrac{b-a}{n}$.

Let

$$x_0 = a,\ x_1 = a + \Delta x,\ x_2 = a + 2\,\Delta x,\ \dots,\ x_n = a + n\,\Delta x = b.$$

Then

$$\int_a^b f(x)\, dx \approx \frac{b-a}{3n}\{f(x_0) + 4f(x_1) + 2f(x_2) + 4f(x_3)$$

$$+ 2f(x_4) + \cdots + 2f(x_{n-2}) + 4f(x_{n-1}) + f(x_n)\}.$$

(See Figure 7.6.)

Example 3

Use Simpson's Rule with $n = 6$ to approximate $\displaystyle\int_1^4 \frac{1}{x}\, dx$.

Solution: With $x_0, x_1, \dots, x_6$ as in Example 1, we obtain from the statement of Simpson's Rule the approximation

$$\int_1^4 \frac{1}{x}\, dx \approx \frac{4-1}{3\cdot 6}\left\{1 + 4\left(\frac{2}{3}\right) + 2\left(\frac{1}{2}\right) + 4\left(\frac{2}{5}\right) + 2\left(\frac{1}{3}\right) + 4\left(\frac{2}{7}\right) + \frac{1}{4}\right\}$$

$$= \frac{3497}{2520} = 1.3877. \qquad \diamond$$

Error in Simpson's Rule Approximations: For the approximation provided by Simpson's Rule it is elsewhere proved that

$$|\text{Error}| \le \frac{(b-a)^5 M}{180 n^4} \tag{3}$$

where M is the maximum value of $|f^{(4)}(x)|$, the fourth derivative, on the interval $[a, b]$.

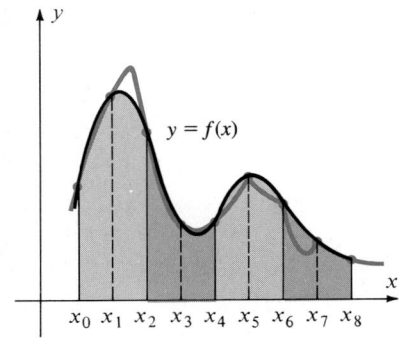

Figure 7.6 Approximation by Simpson's Rule.

y = f(x)

$x_0\ x_1\ x_2\ x_3\ x_4\ x_5\ x_6\ x_7\ x_8$

Example 4

Find how large n must be in order that the approximation of the integral in Example 3 by Simpson's Rule will be accurate to within 0.005.

Solution: For $f(x) = \dfrac{1}{x}$, $f^{(4)}(x) = 24x^{-5}$. The maximum value of this fourth derivative on $[1, 4]$ is $M = 24$. According to inequality (3) we must find n sufficiently large that

$$\frac{(4-1)^5}{180n^4}(24) \le 0.005.$$

Solving this inequality for n, we find that

$$n^4 \ge \frac{(24)(243)}{(.005)(180)} = 6480.$$

Here $n = 10$ will suffice since $10^4 = 10{,}000 > 6480$. ◇

Program 5 in Appendix I is a BASIC program which implements Simpson's Rule to approximate the integral $\displaystyle\int_a^b \frac{1}{x}\, dx$. Table 7.2 was obtained by use of this program. (Compare these results with those of Table 7.1 for the Trapezoidal Rule.)

In conclusion, several observations should be made about the two approximation procedures presented in this section.

(1) While the Trapezoidal Rule approximates curves by line segments, Simpson's Rule fits second degree curves to the given curve. Thus, we would expect Simpson's Rule to be more accurate for the same value of n. (Compare the results in Tables 7.1 and 7.2.)

(2) Since the error formula for the Trapezoidal Rule involves f'', the Trapezoidal Rule gives exact information for first degree polynomials, for which $f''(x) \equiv 0$.

(3) However, since the error formula for Simpson's Rule involves the fourth derivative of f, Simpson's Rule gives exact results for polynomials of degree three or less, for which $f^{(4)}(x) \equiv 0$.

Table 7.2

[A, B]	n	S
[1, 4]	6	1.38770
[1, 4]	20	1.38631
[1, 4]	100	1.38629
[1, 4]	250	1.38629
[1, 20]	6	3.21696
[1, 20]	20	3.00677
[1, 20]	100	2.99577
[1, 20]	250	2.99573

Exercise Set 6.7

In Exercises 1–5, use the Trapezoidal Rule with the given value of n to approximate the integral.

1. $\displaystyle\int_0^4 (x^3 - 7x + 4)\, dx;$ $n = 4$

2. $\displaystyle\int_0^2 \frac{1}{1+x^2}\, dx;$ $n = 4$

3. $\displaystyle\int_0^\pi \sin x\, dx;$ $n = 4$

4. $\displaystyle\int_0^\pi \sin x\, dx;$ $n = 6$

5. $\displaystyle\int_0^4 \frac{1}{1+x^2}\, dx;$ $n = 8$

In Exercises 6–10, use Simpson's Rule to approximate the integral.

6. In Exercise 1.

7. In Exercise 2.

8. In Exercise 3.

9. In Exercise 4.

10. In Exercise 5.

(*Calculator*) In Exercises 11–14, use the Trapezoidal Rule with $n = 6$ to approximate the given integral.

11. $\displaystyle\int_0^{\pi/4} \frac{1}{1+x^2}\, dx$

12. $\displaystyle\int_0^3 \sqrt{1+x^2}\, dx$

13. $\displaystyle\int_1^3 \sqrt{x^3 + 2}\, dx$

14. $\displaystyle\int_0^1 \sin x^2\, dx$

(Calculator) In Exercises 15–18, use Simpson's Rule with $n = 6$ to approximate the integral.

15. In Exercise 11.

16. In Exercise 12.

17. In Exercise 13.

18. In Exercise 14.

19. *(Computer)* Modify Program 4 in Appendix I to approximate the integral in Exercise 11 with $n = 5$, 20, 100, and 250.

20. *(Computer)* Modify Program 5 in Appendix I to approximate the integral in Exercise 11 with $n = 5$, 20, 100, and 250.

21. *(Calculator)* Use the Trapezoidal Rule with $n = 6$ to approximate

a. $\displaystyle\int_0^3 \sin \sqrt{x}\, dx$

b. $\displaystyle\int_0^{\pi/3} \frac{1}{\sqrt{\cos x}}\, dx.$

22. *(Calculator)* Repeat Exercise 21 using Simpson's Rule.

SUMMARY OUTLINE OF CHAPTER 6

◆ An **approximating sum** for the area of the region R bounded by the graph of $f(x) \geq 0$ and the x-axis for $a \leq x \leq b$ is (page 261) an expression of the form $S_n = \displaystyle\sum_{j=1}^{n} f(t_j)\, \Delta x$ where $\Delta x = \dfrac{b - a}{n}$ and $t_j \in [x_{j-1}, x_j]$.

◆ The **area** of R is $A = \displaystyle\lim_{n \to \infty} S_n = \lim_{n \to \infty} \sum_{j=1}^{n} f(t_j)\, \Delta x.$ (page 266)

◆ If f is continuous on $[a, b]$, the **definite integral** of f on $[a, b]$ is defined to be the limit of the approximating Riemann sums, that is, (page 270)

$$\int_a^b f(x)\, dx = \lim_{n \to \infty} \sum_{j=1}^{n} f(t_j)\, \Delta x_j, \qquad t_j \in [x_{j-1}, x_j].$$

◆ If $f(x) \geq 0$ on $[a, b]$ and f is continuous, then $\displaystyle\int_a^b f(x)\, dx$ is the **area** of the region bounded by the graph of f and the x-axis between $x = a$ and $x = b$. (page 273)

◆ *Properties of the Definite Integral* (page 275)

$$\int_a^b [f(x) + g(x)]\, dx = \int_a^b f(x)\, dx + \int_a^b g(x)\, dx$$

$$\int_a^b [cf(x)]\, dx = c \int_a^b f(x)\, dx$$

$$\int_a^b f(x)\, dx = \int_a^c f(x)\, dx + \int_c^b f(x)\, dx$$

$$\int_a^a f(x)\, dx = 0; \qquad \int_b^a f(x)\, dx = -\int_a^b f(x)\, dx$$

◆ The **Fundamental Theorem of Calculus** states that (page 282)

a. if f is continuous on an open interval I then $\dfrac{d}{dx}\left\{ \displaystyle\int_a^x f(t)\, dt \right\} = f(x),$ $a \in I$ for all $x \in I$, and

b. if $F'(x) = f(x)$ for all $x \in [a, b]$ then $\displaystyle\int_a^b f(x)\, dx = F(b) - F(a).$

◆ The *average value* of the continuous function f on $[a, b]$ is $\bar{f} = \dfrac{1}{b - a} \displaystyle\int_a^b f(x)\, dx.$ (page 289)

◆ *Theorem:* $\displaystyle\int_a^b f(x)\, dx = f(c)(b - a)$ for some number c in (a, b) if f is continuous on $[a, b]$. (page 289)

◆ **Trapezoidal Rule:** $\int_a^b f(x)\ dx \approx \dfrac{b-a}{2n}\{f(x_0) + 2f(x_1) + 2f(x_2) + \cdots + 2f(x_{n-1}) + f(x_n)\}.$ (page 307)

◆ **Simpson's Rule:** $\int_a^b f(x)\ dx \approx \dfrac{b-a}{3n}\{f(x_0) + 4f(x_1) + 2f(x_2) + 4f(x_3) + \cdots + 4f(x_{n-1}) + f(x_n)\},$ n even. (page 309)

REVIEW EXERCISES—CHAPTER 6

In Exercises 1–6, write **(a)** a lower approximating sum and **(b)** an upper approximating sum for the given function and interval, using n subintervals of equal size.

1. $f(x) = 3x - 1,$ $\quad x \in [0, 3],$ $\quad n = 6$

2. $f(x) = \dfrac{1}{x},$ $\quad x \in [1, 3],$ $\quad n = 4$

3. $f(x) = \dfrac{1}{1 + x^2},$ $\quad x \in [-1, 1],$ $\quad n = 6$

4. $f(x) = \sin \pi x,$ $\quad x \in [0, 1],$ $\quad n = 6$

5. $f(x) = \tan x,$ $\quad x \in [-\pi/3, \pi/3],$ $\quad n = 4$

6. $f(x) = \tan x,$ $\quad x \in [0, 1],$ $\quad n = 4$

7. Find the area of the region bounded above by the graph of $y = 2x - 2$ and below by the x-axis for $2 \le x \le 4$ by calculating the limit of lower approximating sums.

8. Rework Exercise 7 using upper approximating sums.

In Exercises 9–44, evaluate the definite integral using the Fundamental Theorem of Calculus.

9. $\int_0^9 \sqrt{x}\ dx$

10. $\int_0^3 3\sqrt{x + 1}\ dx$

11. $\int_0^1 (x^{2/3} - x^{1/2})\ dx$

12. $\int_1^2 \dfrac{1 - t}{t^3}\ dt$

13. $\int_0^1 x^3(x + 1)\ dx$

14. $\int_0^\pi \sin^2 x\ dx$ $\quad \left(\text{Hint: } \sin^2 \theta = \dfrac{1}{2} - \dfrac{1}{2}\cos 2\theta\right)$

15. $\int_0^1 (3x + 2)\ dx$

16. $\int_1^3 (7 + 3x)\ dx$

17. $\int_{-3}^5 (x^2 + 2)\ dx$

18. $\int_1^5 (3x^2 - 2)\ dx$

19. $\int_1^4 \sqrt{x}\ dx$

20. $\int_0^8 \sqrt{x + 1}\ dx$

21. $\int_1^4 \left(\sqrt{x} - \dfrac{1}{\sqrt{x}}\right)\ dx$

22. $\int_1^8 (x^{1/3} - 1)\ dx$

23. $\int_0^2 (x + 7)(2x + 2)\ dx$

24. $\int_1^4 (x^2 - 1)(x + 2)\ dx$

25. $\int_0^{\pi/4} \sin x\ dx$

26. $\int_{-\pi/4}^{-\pi/4} \cos x\ dx$

27. $\int_0^{\pi/4} \sin(2x)\ dx$

28. $\int_{\pi/4}^{\pi/2} \cos(\pi - 2x)\ dx$

29. $\int_1^2 (x + 4)^{10}\ dx$

30. $\int_0^4 (\sqrt{a} + \sqrt{x})^2\ dx$

31. $\int_2^4 \dfrac{t^2 - 2t}{t}\ dt$

32. $\int_1^2 \dfrac{1 + t}{t^3}\ dt$

33. $\int_0^3 \dfrac{dt}{(t + 1)^2}$

34. $\int_3^8 \dfrac{1}{\sqrt{x + 1}}\ dx$

35. $\int_0^{\pi/4} \dfrac{1}{\cos^2 x}\ dx$

36. $\int_0^{\pi/3} \dfrac{\sin x}{\cos^2 x}\ dx$

37. $\int_0^1 (x - \sqrt{x})^2\ dx$

38. $\int_1^4 x^{1/2}(1 + x^{3/2})^5\ dx$

39. $\int_1^8 t(\sqrt[3]{t} - 2t)\ dt$

40. $\int_1^4 \dfrac{x^2 + 2x + 4}{\sqrt{x}}\ dx$

41. $\int_1^2 \dfrac{1}{(1 - 2x)^3}\ dx$

42. $\int_{-\pi/4}^{\pi/4} \sec^2 t\ dt$

43. $\int_0^1 \dfrac{x^3 + 8}{x + 2}\ dx$

44. $\int_0^2 x^2\sqrt{x^3 + 1}\ dx$

In Exercises 45–51, find the area of the region bounded by the graph of the given function, the x-axis, the line $x = a$, and the line $x = b$.

45. $f(x) = \sqrt{x - 1},$ $\quad a = 1,$ $\quad b = 5$

46. $f(x) = \dfrac{1}{\sqrt{x - 1}},$ $\quad a = 2,$ $\quad b = 10$

47. $f(x) = (x^2 + 2)^2,$ $\quad a = 0,$ $\quad b = 1$

48. $f(x) = (x - 1)(x + 2),$ $\quad a = 0,$ $\quad b = 2$

49. $f(x) = 4x - x^2,$ $\quad a = 4,$ $\quad b = 5$

50. $f(x) = \sin \pi x,$ $\quad a = 0,$ $\quad b = 2$

51. $f(x) = 3 + 2x - x^2,$ $\quad a = 1,$ $\quad b = 4$

In each of Exercises 52–57, find the area of the region bounded above by the graph of $y = f(x)$ and below by the x-axis for the specified intervals.

52. $f(x) = x^3 - x,$ $a = 1,$ $b = 3$

53. $f(x) = x^2 - x - 2,$ $a = 2,$ $b = 4$

54. $f(x) = \sin(2x),$ $a = 0,$ $b = \dfrac{\pi}{4}$

55. $f(x) = \sqrt{1 - x},$ $a = 1,$ $b = 1$

56. $f(x) = \dfrac{1}{(x - 2)^2},$ $a = -2,$ $b = 1$

57. $f(x) = (ax^2 + 3)^2,$ $a = 0,$ $b = 3$

In each of Exercises 58–62, a region is described. Sketch the region, noting where the bounding function f is positive and where it is negative. Then calculate the area of the region using one or more integrals.

58. The region bounded by the graph $f(x) = 9 - x^2$ and the x-axis between $x = -3$ and $x = 3$.

59. The region bounded by the graph of $f(x) = x^2 + 2x - 3$ between $x = -3$ and $x = 1$.

60. The region bounded by the graph of $f(x) = (x - 1)^3$ between $x = 0$ and $x = 2$.

61. The region bounded by the graph of $f(x) = 2x - 4$ and the x-axis for $0 \le x \le 4$.

62. The region bounded by the graph of $f(x) = \sqrt{x} - 2$ and the x-axis for $0 \le x \le 9$.

In Exercises 63–74, find the area of the region bounded by the graphs of the given equations.

63. $y = x^3,$ $y = 0,$ $x = -2,$ $x = 2$

64. $y = 1 - x^2,$ $y = -x - 1,$ $x = -2,$ $x = 1$

65. $y = \sqrt{x},$ $y = -\sqrt{x},$ $x = 0,$ $x = 4$

66. $y = x^2,$ $y = 2 - x,$ $x = -2,$ $x = 1$

67. $y = x^2 - 4x + 2,$ $x + y = 6$

68. $y = 2 - 2x - x^2,$ $x = -y$

69. $y = \dfrac{x - 2}{\sqrt{x^2 - 4x + 8}},$ $x = 0,$ $y = 0$

70. $x - y^2 + 3 = 0,$ $x - 2y = 0$

71. $x + y^2 = 0,$ $x + y + 2 = 0$

72. $x = 2y^2 - 3,$ $x = y^2 + 1$

73. $x = y^{2/3},$ $x = y^2$

74. $y = 4x^2,$ $4x + y - 8 = 0$

75. Find the area of the region bounded by the parabola $y = x^2 + x - 2$ and the line through $(-1, -2)$ and $(1, 0)$.

76. Find the area of the region bounded by the parabola $y = 2 - x - x^2$ and the line through $(-1, 2)$ and $(1, 0)$.

77. Find the area of the region bounded by the graphs of the equations $3x + 5y = 23,$ $5x - 2y = 28,$ and $2x - 7y = -26$.

In Exercises 78–83, find the average value of the given function.

78. $y = x\sqrt{1 - x^2}$ on $[-1, 1]$

79. $y = \sin x \cos x$ on $[0, \pi/2]$

80. $y = x^3 \sin x^2$ on $[-\pi, \pi]$ (*Hint:* Use the fact that $y = f(x)$ is odd.)

81. $y = \dfrac{x}{\sqrt{1 + x^2}}$ on $[0, 2]$

82. $y = \sqrt{x} + \sqrt[3]{x}$ on $[0, 1]$

83. $y = \sin x$ on $[0, 10\pi]$

84. Show that the average value of $f(x) = \sin x$ on intervals of the form $[2n\pi, 2m\pi]$ is $\overline{f} = 0$, where n and m are integers. What about intervals of the form $[x_0 + 2n\pi, x_0 + 2m\pi]$?

85. Show that the average value of $f(x) = \sin^2 x$ on intervals of the form $[0, n\pi]$ is $\overline{f} = \frac{1}{2}$ where n is an integer.

86. Give an argument involving average values to show that
$$\lim_{L \to \infty} \frac{1}{L} \int_0^L \sin^2 x \, dx = \frac{1}{2}.$$
What about $\displaystyle\lim_{L \to \infty} \frac{1}{L} \int_0^L \cos^2 x \, dx$?

87. Let $F(x) = \displaystyle\int_0^x t^2 \sqrt{1 + t} \, dt$ for $x > -1$. Find

  a. $F(0)$ **b.** $F'(x)$

  c. $F'(3)$ **d.** $F'(2x)$

88. Let $F(x) = \displaystyle\int_0^{2x} \frac{1}{1 + t^2} \, dt$. Find

  a. $F'(x)$ **b.** $F'(1)$

  c. $F(x^2)$ **d.** $F'(x^2)$

89. True or false? $\dfrac{d}{dx}\left\{ \displaystyle\int_a^x f(t) \, dt \right\} = \displaystyle\int_a^x \frac{d}{dt}\left\{ f(t) \right\} dt.$

90. True or false? If $\displaystyle\int_a^b f(x) \, dx = 0$ then $f(x) \equiv 0$.

91. True or false? If $f(x)$ is continuous on $[a, b]$ and $\displaystyle\int_a^b f(x) \, dx = 0$, then $f(c) = 0$ for at least one $c \in [a, b]$.

92. True or false? If $\displaystyle\int_a^b |f(x)| \, dx = 0$ and $f(x)$ is continuous on $[a, b]$, then $f(x) = 0$ for all $x \in [a, b]$.

93. Find $\dfrac{d}{dx}\left\{ \displaystyle\int_{2x}^0 \sec t \, dt \right\}$.

94. Find $\dfrac{d}{dx}\left\{ \displaystyle\int_x^{x^2} \sqrt{1 + t^2} \, dt \right\}$.

In Exercises 95–100, sketch a region in the plane whose area corresponds to the given integral. Then find the value of the integral from area considerations.

95. $\int_{1}^{4} (2x + 1) \, dx$

96. $\int_{-1}^{1} \sqrt{1 - x^2} \, dx$

97. $\int_{0}^{3} (1 + \sqrt{9 - x^2}) \, dx$

98. $\int_{-3}^{3} |3 - x| \, dx$

99. $\int_{0}^{4} -\sqrt{16 - x^2} \, dx$

100. $\int_{-1}^{1} (3 + \sqrt{1 - x^2}) \, dx$

In Exercises 101–103, use the Trapezoidal Rule with n subdivisions to approximate the given integral.

101. $\int_{-2}^{2} \frac{1}{1 + x^2} \, dx, \qquad n = 4$

102. $\int_{0}^{4} \sqrt{1 + x^2} \, dx, \qquad n = 4$

103. $\int_{-1}^{2} \frac{1}{x + 2} \, dx, \qquad n = 6$

In Exercises 104–105, use Simpson's Rule with n subdivisions to approximate the given integral.

104. $\int_{0}^{2} \frac{1}{1 + x} \, dx, \qquad n = 6$

105. $\int_{0}^{\pi/3} \cos x \, dx, \qquad n = 6 \qquad$ (Use a cosine table or calculator.)

Chapter 7
Applications of the Definite Integral

In this chapter, you will encounter a variety of physical and mathematical problems, all of which share the property that their solutions arise in the form of definite integrals. In each instance, we will be led to the solution through a sequence of steps similar to those by which the definite integral arose as the solution to the area problem in Chapter 6.

Because of the strong parallels that exist among the problems and solutions of this chapter, we summarize here the essential concepts involved in the definition of the integral. You should refer to this introductory summary from time to time, as an aid to understanding the seminal idea of this chapter.

Summary Definition of the Integral $\int_a^b f(x)\,dx$: Let f be continuous on the interval $[a, b]$.

1. For each positive integer n, we partition the interval $[a, b]$ into subintervals of equal* length with endpoints

$$a = x_0 < x_1 < x_2 < \cdots < x_n = b$$

and with $(x_j - x_{j-1}) = \Delta x = \dfrac{b - a}{n}$ for $j = 1, 2, \ldots, n$.

2. A **Riemann sum** for f on $[a, b]$ is an expression of the form

$$\sum_{j=1}^{n} f(t_j)\,\Delta x$$

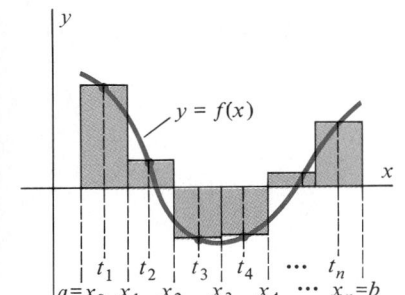

Figure 1 Approximating Riemann sum.

where t_j is an arbitrary number in the subinterval $[x_{j-1}, x_j]$ for each $j = 1, 2, \ldots, n$ (see Figures 1 and 2).

3. The limit of this Riemann sum as $n \to \infty$,

$$\int_a^b f(x)\,dx = \lim_{n \to \infty} \sum_{j=1}^{n} f(t_j)\,\Delta x,$$

exists independent of how the numbers $t_j \in [x_{j-1}, x_j]$ are chosen.

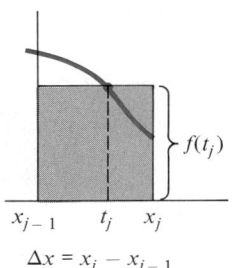

Figure 2 jth approximating rectangle.

*The most general definition of partition, as given in Chapter 6, allows for subintervals of varying length $\Delta x_j = (x_j - x_{j-1})$. Since all Riemann sums for f on $[a, b]$ have the limit $\int_a^b f(x)\,dx$ and since we will actually be constructing the partitions that are used here, we will work with **regular partitions**—those with subintervals of equal length $\Delta x = \dfrac{b - a}{n}$.

In each of the questions taken up in this chapter, the following problem-solving strategy will unfold:

(a) A continuous function will be identified that determines the quantity to be calculated.
(b) We will approximate the solution by assuming that this continuous function actually assumes only a finite number of distinct values which are the values $f(t_j)$ in the subintervals $[x_{j-1}, x_j]$. This will produce an approximation in the form of a Riemann sum.
(c) We will assume that as the number of subintervals becomes infinite, the approximations approach the actual value of the quantity to be calculated.
(d) The desired quantity will be defined as the limiting value of the approximating Riemann sum as $n \to \infty$. That is, the desired quantity will be obtained as a definite integral (point 3 above).

You are urged, in each case, to study carefully the procedure by which the solution is obtained rather than simply accepting the formula that results. Only by striving to understand the ways in which integrals arise from approximation schemes can you gain the insight you will need when encountering new problems whose solutions involve definite integrals.

7.1 CALCULATING VOLUMES BY SLICING

There are certain types of three-dimensional objects whose volumes can be calculated as definite integrals. For example, it is common for solid objects to be produced by a process of milling a rotating piece of stock. In using a lathe to produce a wooden table leg, a craftsman presses a chisel against a rapidly rotating block of wood (Figure 1.1). Similarly, a potter works a ball of clay into a vase by using a potter's wheel, which allows the clay to be rotated at a uniform speed about a central axis (Figure 1.2).

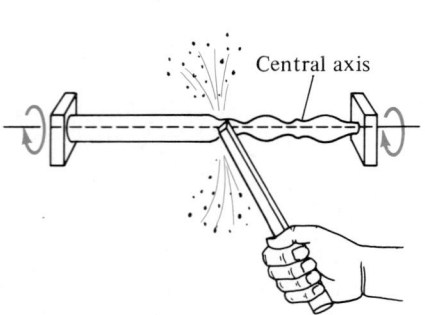

Figure 1.1 Table leg produced on a lathe by pressing chisel against rotating block of wood stock.

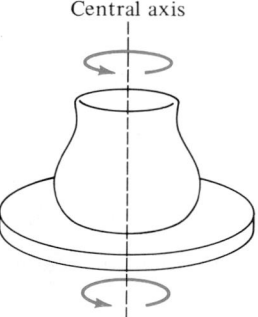

Figure 1.2 Pottery produced by shaping clay rotating on a wheel.

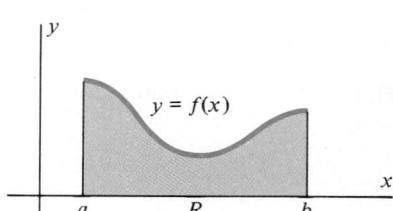

Figure 1.3 Region to be rotated.

The problem of calculating the volume of such solids is idealized mathematically as follows. Let f be a continuous nonnegative function for $a \le x \le b$. Let R denote the region bounded by the graph of f, the x-axis, and the lines $x = a$ and $x = b$

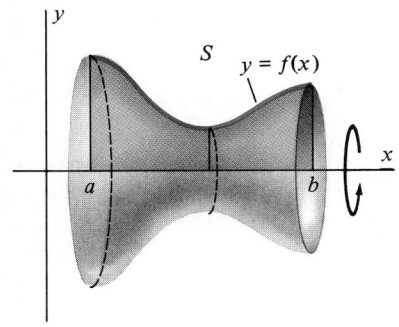

Figure 1.4 Solid obtained by rotating R about the x-axis.

(Figure 1.3). As the region R rotates about the x-axis (Figure 1.4), it sweeps out a **solid of revolution,** S. Just as for the lathe and pottery wheel illustrations, the cross sections for S taken perpendicular to the x-axis will be circles of radius $r = f(x)$. This is because the cross section taken at location x is described by rotating about the x-axis the line segment from $(x, 0)$ to $(x, f(x))$.

To find a formula for the volume of S, we begin by developing an approximation to the solid S. We do this by partitioning the interval $[a, b]$ into n equal subintervals of length $\Delta x = \dfrac{b - a}{n}$, and with endpoints $a = x_0, x_1, \ldots, x_n = b$. We arbitrarily select one "test number" t_j in each interval $[x_{j-1}, x_j]$, and we assume that the function f is constant throughout the interval $[x_{j-1}, x_j]$, with value $f(x) \equiv f(t_j)$.

In our approximation, corresponding to each interval $[x_{j-1}, x_j]$, we will be rotating a horizontal line segment $y = f(t_j)$ about the x-axis. This will generate a disc, of radius $r_j = f(t_j)$ and thickness Δx. The volume of this disc is, therefore,

$$\Delta V_j = \pi r_j^2 \, \Delta x = \pi f^2(t_j) \, \Delta x \qquad \text{(here } f^2(t) \text{ means } [f(t)]^2\text{).}$$

(See Figure 1.5.)

Summing the volumes of these individual discs from 1 to n gives the volume of our approximating solid as

$$\sum_{j=1}^{n} \Delta V_j = \sum_{j=1}^{n} \pi f^2(t_j) \, \Delta x.$$

(See Figure 1.6.) Next, we argue that as $n \to \infty$ and the thickness Δx of each individual disc becomes small, the volume of our approximating solid should approach the volume of S. That is, we want to *define* the volume V of S by the equation

$$V = \lim_{n \to \infty} \sum_{j=1}^{n} \Delta V_j = \lim_{n \to \infty} \sum_{j=1}^{n} \pi f^2(t_j) \, \Delta x.$$

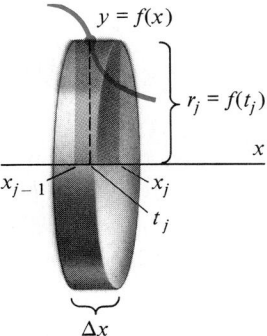

Figure 1.5 Rectangle of height $r_j = f(t_j)$ generates disc of volume $V_j = \pi f^2(t_j) \, \Delta x$.

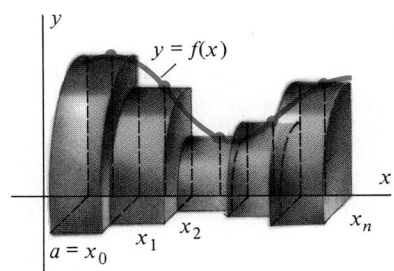

Figure 1.6 One quarter of the approximation to the volume of revolution S obtained by assuming f constant on subintervals.

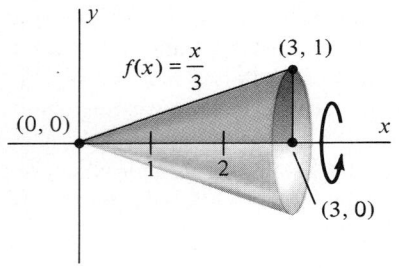

Figure 1.7 Cone as a volume of revolution.

Since the sum on the right is a Riemann sum for the function πf^2, we have arrived at the following definition:

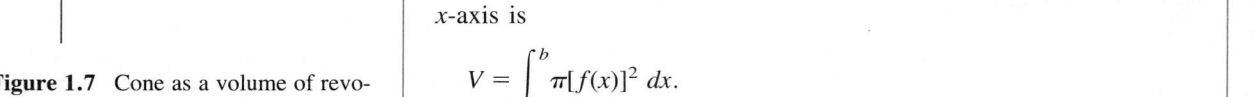

Let f be continuous and nonnegative for $a \leq x \leq b$ and let R denote the region bounded by the graph of $y = f(x)$, the x-axis, and the lines $x = a$ and $x = b$. The volume of the solid obtained by rotating R about the x-axis is

$$V = \int_a^b \pi[f(x)]^2 \, dx. \tag{1}$$

Example 1

Find the volume of the cone obtained by revolving about the x-axis the region bounded above by the graph of $f(x) = \dfrac{1}{3}x$ and below by the x-axis for $0 \leq x \leq 3$ (see Figure 1.7).

Solution: Using formula (1) with $f(x) = \dfrac{1}{3}x$ we obtain

$$V = \int_0^3 \pi\left(\frac{1}{3}x\right)^2 dx = \int_0^3 \frac{\pi}{9}x^2 \, dx = \frac{\pi}{27}x^3\bigg]_0^3 = \pi.$$

Since this cone has height $h = 3$ and base of radius $r = 1$, this result agrees with the formula from geometry for the volume of this cone:

$$V = \frac{1}{3}\pi r^2 h = \frac{1}{3}\pi(1)^2 \cdot 3 = \pi.$$

$\diamond$

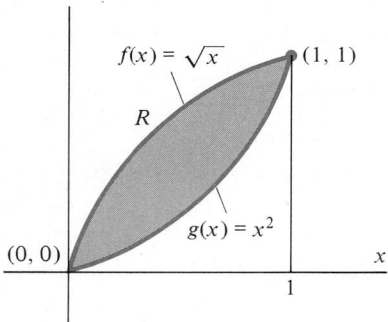

Figure 1.8 Region to be rotated about x-axis.

Example 2

Find the volume of the solid obtained by rotating the region bounded by the graphs of $f(x) = \sqrt{x}$ and $g(x) = x^2$ about the x-axis.

Strategy

Find points where the graphs cross.

View solid as difference of two solids of revolution.

Find the volume of each solid using formula (1).

Solution

The two graphs cross at $(0, 0)$ and $(1, 1)$ since the equation $\sqrt{x} = x^2$ implies $x = x^4$ or $x(1 - x^3) = 0$. Since $\sqrt{x} > x^2$ for $0 < x < 1$, the region is bounded above by the graph of $f(x) = \sqrt{x}$ and below by the graph of $g(x) = x^2$. As Figure 1.8 indicates, we may view the resulting solid as the solid obtained by rotation of $f(x) = \sqrt{x}$ from which the solid obtained by rotation of $g(x) = x^2$ is removed (see Figure 1.9). The calculation for volume, by equation (1), is therefore

$$V = \int_0^1 \pi(\sqrt{x})^2 \, dx - \int_0^1 \pi(x^2)^2 \, dx$$

$$= \frac{\pi}{2}x^2\bigg]_0^1 - \frac{\pi}{5}x^5\bigg]_0^1$$

$$= \frac{3\pi}{10}.$$

$\diamond$

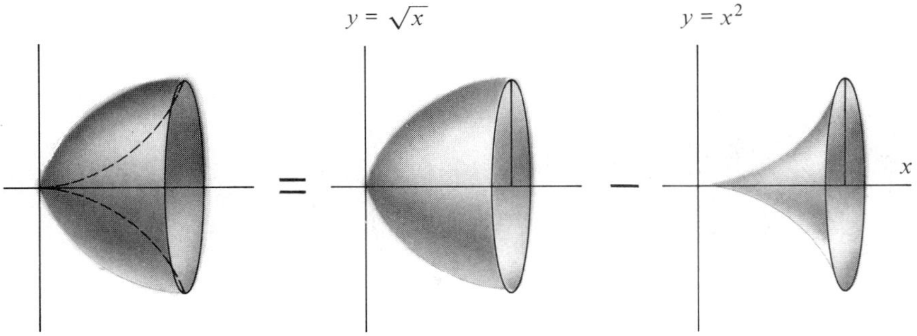

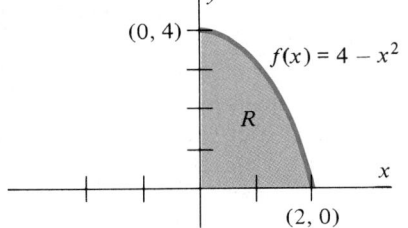

Figure 1.9 Volume obtained by expressing area between curves $\sqrt{x}$ and x^2 as the difference of volumes corresponding to upper curve $\sqrt{x}$ and lower curve x^2.

REMARK: As you may have observed, the volume in Example 2 could have been calculated by the single integral

$$V = \int_0^1 \pi[(\sqrt{x})^2 - (x^2)^2] \, dx.$$

In general, if the region R is bounded above by the graph of $y = f(x)$ and below by the graph of $y = g(x) \geq 0$ for $a \leq x \leq b$, then the formula for volume is

$$V = \int_a^b \pi([f(x)]^2 - [g(x)]^2) \, dx. \tag{2}$$

However, caution must be taken not to misinterpret the integrand as $[f(x) - g(x)]^2$.

Example 3

The region in the first quadrant bounded by the graph of $y = 4 - x^2$ and the coordinate axes is rotated about the y-axis. Find the volume of the resulting solid.

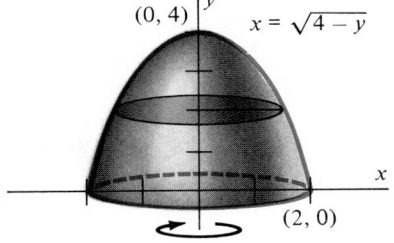

Figure 1.10 Region to be rotated about y-axis.

Solution: This problem is similar to that of Example 1 except that the roles of x and y are reversed. Solving the given equation for x as a function of y gives $f(y) = \sqrt{4 - y}$. Since the rotation is about the y-axis, the integration will be with respect to y, and the limits of integration are from $y = 0$ to $y = 4$ (see Figures 1.10 and 1.11). We obtain

$$V = \int_0^4 \pi[\sqrt{4 - y}]^2 \, dy$$

$$= \pi\left(4y - \frac{1}{2}y^2\right)\Big]_0^4$$

$$= 8\pi. \qquad \diamond$$

Figure 1.11 Cross sections perpendicular to y-axis are circles of radius $r = \sqrt{4 - y}$.

The result of Example 3 generalizes as follows:

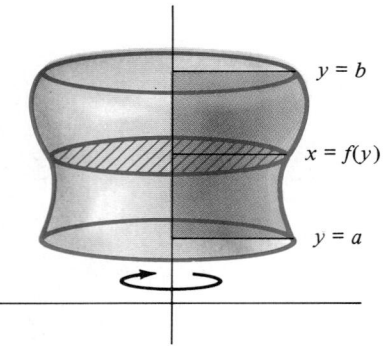

Figure 1.12 Volume of revolution about the y-axis.

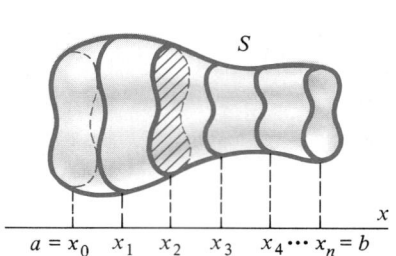

Figure 1.13 Cross sections of known area perpendicular to axis.

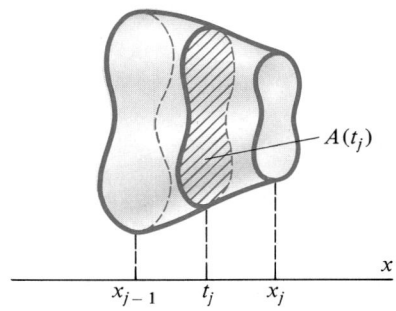

Figure 1.14 Slice over interval $[x_{j-1}, x_j]$.

> If the region R is bounded by the graph of the continuous function $x = f(y)$ and the y-axis from $y = a$ to $y = b$, then the volume of the solid obtained by rotating R about the y-axis is
>
> $$V = \int_a^b \pi[f(y)]^2 \, dy.$$
>
> (See Figure 1.12.)

Solids of Known Cross-Sectional Area

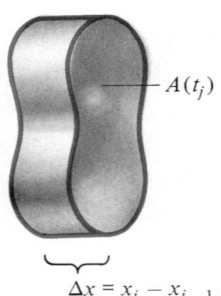

Figure 1.15 Volume element $A(t_j)\,\Delta x$ approximates volume of slice.

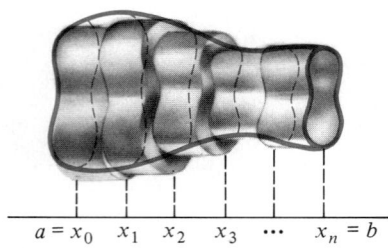

Figure 1.16 Approximation by volume elements.

For the solid of revolution S, the area of a cross section taken perpendicular to the x-axis at $x = x_0$ will be $A(x_0) = \pi f^2(x_0)$, since the radius of the disc-shaped cross section is $r = f(x_0)$ (see Figure 1.5). We can interpret formula (1) for the volume of S as the integral of the cross-sectional area $A(x) = \pi f^2(x)$ from $x = a$ to $x = b$. Our next objective is to show that the volume of *any* solid of known cross-sectional area (possibly not a solid of revolution) can be calculated in this same manner.

In particular, suppose that S is a solid for which the area of each cross section perpendicular to a given axis is known (Figure 1.13). Let $A(x)$ denote the area of the cross section taken at location x, and assume that the solid extends from $x = a$ to $x = b$. If we partition the interval $[a, b]$ into equal subintervals of length $\Delta x = \dfrac{b - a}{n}$ with endpoints $a = x_0 < x_1 < x_2 < \cdots < x_n = b$, the solid is partitioned by the cross sections taken at these endpoints into n slices of equal thickness Δx (see Figure 1.14). By selecting one number t_j in each subinterval $[x_{j-1}, x_j]$, we approximate the volume of the slice over the interval $[x_{j-1}, x_j]$ by the volume ΔV_j of the cylinder with base area $A(t_j)$ and thickness Δx (Figure 1.15), that is,

$$\Delta V_j = A(t_j)\,\Delta x.$$

Summing these approximations for $j = 1, 2, \ldots, n$ gives the approximation to the volume V of S

$$V \approx \sum_{j=1}^{n} \Delta V_j = \sum_{j=1}^{n} A(t_j)\,\Delta x, \tag{3}$$

which should approach the volume V as $n \to \infty$ and the lengths Δx approach zero (Figure 1.16). If the function A is continuous for $a \le x \le b$, the sum on the right-hand side of approximation (3) is a Riemann sum which therefore has a limit as $n \to \infty$. We define the volume V of S to be this limit.

If $A(x)$ denotes the area of the cross section of S for $a \le x \le b$ and, if the function A is continuous on $[a, b]$, the volume V of S is

$$V = \int_a^b A(x) \, dx. \tag{4}$$

Notice that equation (4) generalizes the familiar formula for the volume of a cylinder with base area A and height h: $V = Ah$. It also contains formula (1) for the volume of a solid of revolution as a special case, since for such solids, $A(x) = \pi f^2(x)$.

Example 4

The base of a solid is a circle of radius 4 cm. All cross sections perpendicular to a particular axis are squares. Find the volume of this solid (see Figure 1.17).

Strategy

Find an expression for the area of a cross section. Begin by finding an equation for the boundary of the base.

Solution

If we impose an xy-coordinate system on the circular base so that the x-axis corresponds to the given axis, then the equation for the boundary of the base is

$$x^2 + y^2 = 16 \qquad \text{(Figure 1.18)}.$$

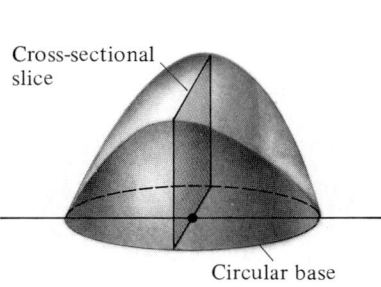

Cross-sectional slice

Circular base

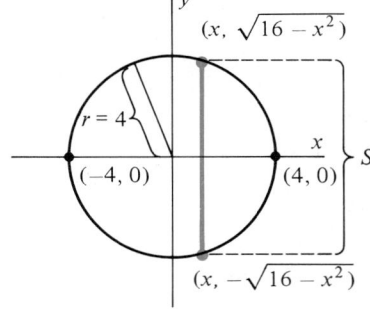

$(x, \sqrt{16 - x^2})$

$r = 4$

$(-4, 0)$ $\qquad$ $(4, 0)$ $\qquad$ S

x

$(x, -\sqrt{16 - x^2})$

Figure 1.17 Cross sections perpendicular to the x-axis.

Figure 1.18 Side $s = 2\sqrt{16 - x^2}$.

The equation for the upper semicircle is $y = \sqrt{16 - x^2}$, and the equation for the lower semicircle is $y = -\sqrt{16 - x^2}$. A cross section perpendicular to the x-axis will therefore intersect this circular base in a chord of length $s = 2\sqrt{16 - x^2}$. Since this chord is one side of the square cross section, the area of the cross section is

$$\begin{aligned} A(x) &= s^2 \\ &= (2\sqrt{16 - x^2})^2 \\ &= 64 - 4x^2. \end{aligned}$$

Find the limits of integration.

The smallest and largest values of x are, respectively, -4 and 4. The volume,

Apply (4).

by equation (4), is therefore

$$V = \int_{-4}^{4} A(x) \, dx = \int_{-4}^{4} (64 - 4x^2) \, dx$$

$$= 64x - \frac{4}{3}x^3 \Big]_{-4}^{4}$$

$$= \frac{1024}{3} \text{ cm}^3.$$

◇

Although the solid in Example 5 is a solid of revolution, the axis of rotation is neither the x- nor the y-axis. Note how using equation (4) helps clarify the issue.

Example 5

Find the volume of the solid generated by rotating the region bounded by the graph of $f(x) = \sqrt{4 - x}$ and the x-axis for $0 \le x \le 4$ about the line $y = -2$.

Solution: Since the region in question is being rotated about a line other than the x-axis, we cannot use formula (1). As illustrated in Figure 1.19, a cross section taken at location x will consist of a circle of radius $R = [\sqrt{4 - x} - (-2)]$ from which a smaller circle of radius $r = 2$ has been removed. The cross-sectional area is therefore

$$A(x) = \pi R^2 - \pi r^2$$
$$= \pi(\sqrt{4 - x} + 2)^2 - \pi \cdot 2^2$$
$$= \pi(4 - x + 4\sqrt{4 - x}).$$

We can now apply equation (4) to find that

$$V = \int_{0}^{4} \pi(4 - x + 4\sqrt{4 - x}) \, dx$$

$$= \pi\left(4x - \frac{1}{2}x^2 - \frac{8}{3}(4 - x)^{3/2}\right)\Big]_{0}^{4}$$

$$= \frac{88\pi}{3}.$$

◇

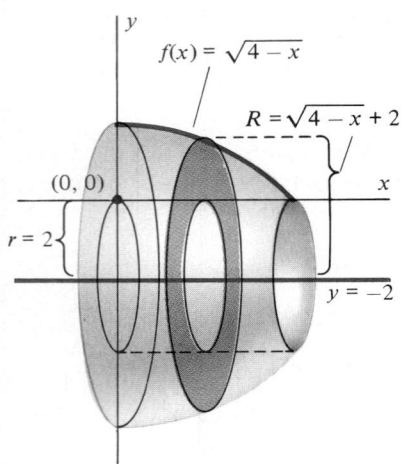

$f(x) = \sqrt{4 - x}$

$R = \sqrt{4 - x} + 2$

$(0, 0)$

$r = 2$

$y = -2$

Figure 1.19

Exercise Set 7.1

In Exercises 1–7, find the volume of the solid obtained by revolving the region bounded by the given curve and the x-axis, for $a \le x \le b$, about the x-axis.

1. $f(x) = 2x + 1, \qquad 1 \le x \le 4$

2. $f(x) = \sqrt{4x - 1}, \qquad 1 \le x \le 5$

3. $f(x) = \sin x, \qquad 0 \le x \le \pi$

$\left(Hint: \sin^2 x = \dfrac{1}{2} - \dfrac{1}{2}\cos 2x.\right)$

4. $f(x) = |x - 1|, \qquad 0 \le x \le 3$

5. $f(x) = \tan x, \qquad 0 \le x \le \pi/4$

6. $f(x) = \sqrt{4 - x^2}, \qquad 1 \le x \le 2$

7. $f(x) = \dfrac{\sqrt{1 + x}}{x^{3/2}}, \qquad 1 \le x \le 2$

In Exercises 8–11, find the volume of the solid obtained by revolving about the x-axis the region bounded by the given curves.

8. $f(x) = x^2, \qquad g(x) = x^3$

9. $f(x) = \dfrac{1}{4}x^2, \qquad g(x) = x$

10. $f(x) = \dfrac{1}{x}, \qquad g(x) = \sqrt{x}, \qquad 1 \le x \le 4$

11. $f(x) = \sin x,$ $g(x) = \cos x,$ $0 \le x \le \dfrac{\pi}{2}$

In each of Exercises 12–17, find the volume of the solid obtained by rotating the region bounded by the given curves about the *y*-axis.

12. $x + y = 4,$ $x = 0,$ $0 \le y \le 4$

13. $y = x^2,$ $y = 0,$ $0 \le x \le 2$

14. $y = 4,$ $y = x^2$

15. $y = x^3,$ $x = 2,$ $y = 0$

16. $y = x^2,$ $y = x^3$

17. $y = \dfrac{1}{x},$ $y = 0,$ $\dfrac{1}{4} \le x \le 1$

18. Find the volume of the right pyramid whose base is a square 10 cm on a side and whose altitude is 8 cm (Figure 1.20).

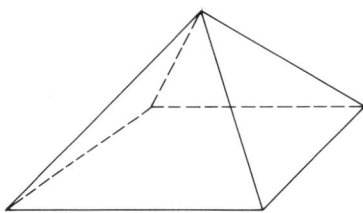

Figure 1.20

19. By integration, derive the formula for the volume of a sphere of radius *r*.

20. The base of a solid is a circle of radius 4. Find the volume of the solid if all cross sections perpendicular to a given axis are equilateral triangles (Figure 1.21).

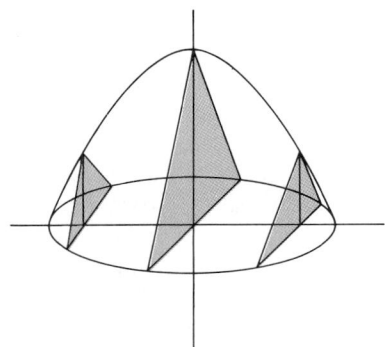

Figure 1.21

21. Find the volume of the solid in Exercise 20 if the cross sections perpendicular to the given axis are isosceles right triangles with bases lying along the base of the solid (Figure 1.22).

22. A hemispherical water tank of radius 10 meters contains water to a depth of 6 meters. How many cubic meters of water does the tank contain?

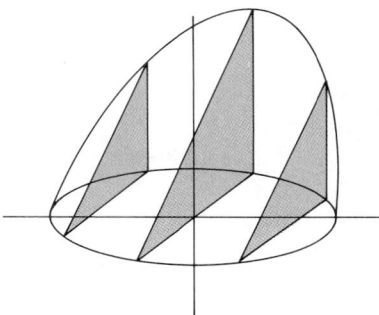

Figure 1.22

23. The base of a solid is the region bounded by the graphs of $f(x) = x^2$ and $g(x) = 8 - x^2$. Find the volume of the solid if all cross sections perpendicular to the *x*-axis are squares.

24. Find the volume of the solid in Exercise 23 if the cross sections perpendicular to the *x*-axis are semicircles.

25. Find the formula for the volume of the frustum of a cone with bases of radius *r* and *R*, and height *h* (see Figure 1.23).

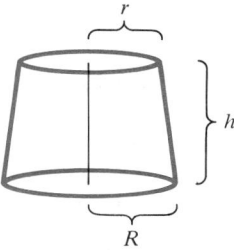

Figure 1.23

26. Find the volume of the solid generated by revolving the triangle with vertices $(0, 0)$, $(2, 5)$, and $(5, 0)$ about the *x*-axis.

27. Water is running into the tank in Exercise 22 at a rate of 10 cubic meters per minute. How fast is the water level rising when its depth is 4 meters?

28. A hole of radius 2 cm is drilled through the center of a spherical ball of radius 4 cm. What is the volume of the remaining solid?

29. Find the volume of the solid obtained by rotating the region bounded by the *x*-axis and the graph of $y = 1 - x^2$ about the line $y = -3$.

30. Find the volume of the solid obtained by rotating the region bounded by the graphs of $y = \sqrt{x}$ and $y = \dfrac{1}{2}x$ about the line $y = 4$.

31. Find the volume of the solid obtained when the region bounded by the graphs of $y = \sqrt{x}$, $y = 0$, and $x = 9$ is rotated about the line $y = -2$.

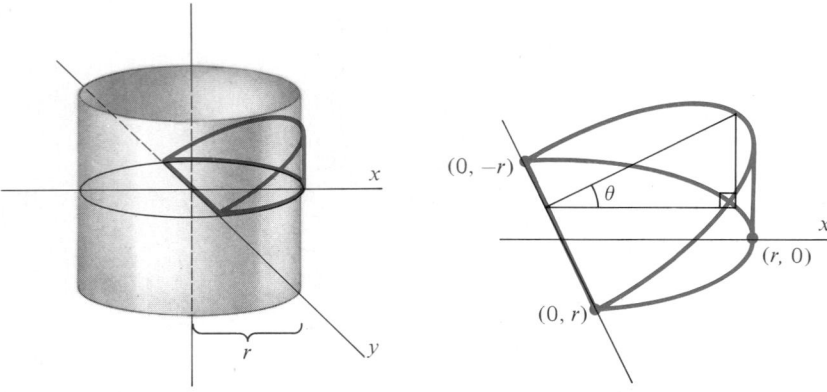

Figure 1.24

32. Find the volume of the solid obtained when the graph of the ellipse $\dfrac{x^2}{a^2} + \dfrac{y^2}{b^2} = 1$ is rotated about the x-axis.

33. Find the volume of the solid obtained when the region in Exercise 32 is rotated about the y-axis.

34. What formula is obtained from Exercises 32 and 33 when $a = b = r$?

35. When a right circular cylinder of radius r is sliced by two planes, one perpendicular to the central axis of the cylinder and the other at an angle θ with the first, as in Figure 1.24, a wedge results. Find its volume.

36. *(Calculator/Computer)* Use the Trapezoidal Rule with $n = 8$ to approximate the volume of the solid obtained by rotating

the region bounded by the graph of $f(x) = \dfrac{1}{\sqrt{1 + x^2}}$ and the line $y = 0$ for $0 \le x \le 4$ about the x-axis.

37. *(Calculator/Computer)* Use Simpson's rule with $n = 12$ to approximate the volume of the solid in Exercise 36.

38. *(Calculator/Computer)* Use an approximation scheme of your own choosing to approximate the volume of the solid obtained by rotating the region bounded by the graph of $y = x^{-1/2}$ and the x-axis for $1 \le x \le 4$ about the x-axis.

39. *(Calculator/Computer)* Approximate the volume of the solid obtained by revolving the region in Exercise 32 about the line $y = -2$ if $a = 3$ and $b = 1$.

7.2 CALCULATING VOLUMES BY THE METHOD OF CYLINDRICAL SHELLS

In addition to the slicing methods of Section 7.1, there is another approach to calculating volumes of solids of revolution. This method is often useful when a region is not adjacent to the axis about which it is to be rotated. While the slicing method is based on the idea of approximating cross sections taken perpendicular to the axis of rotation, the **method of cylindrical shells** uses approximating hollow cylinders centered about the axis of rotation.

To describe this method, we let R be the region bounded by the graph of the continuous nonnegative function f and the x-axis for $0 \le a \le x \le b$ (Figure 2.1). If the region R is rotated about the y-axis, a solid S is generated (Figure 2.2).

To approximate the volume of S, we partition the interval $[a, b]$ with numbers $a = x_0 < x_1 < x_2 < \cdots < x_n = b$ with $x_j - x_{j-1} = \Delta x = \dfrac{b - a}{n}$, $j = 1, 2, \ldots, n$. In each subinterval $[x_{j-1}, x_j]$, we arbitrarily select a number t_j. Then the area of the part of the region R over the interval $[x_{j-1}, x_j]$ is approximated by $f(t_j)\,\Delta x$, the area of the rectangle with base $[x_{j-1}, x_j]$ and height $f(t_j)$ (see Figure 2.3). As each rectangle rotates about the y-axis, it generates a cylindrical shell. The totality of all such shells (Figure 2.4) constitutes a solid whose volume we take as the approximation to the volume of S.

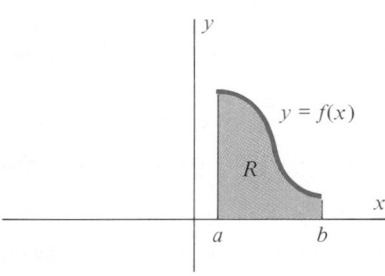

Figure 2.1 Region R to be rotated.

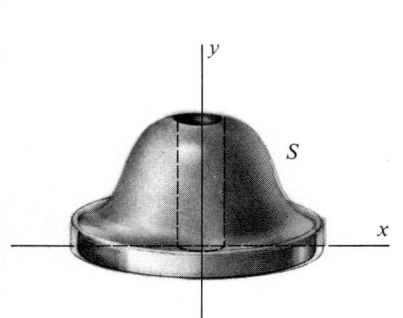

Figure 2.2 Solid obtained by rotating region R about the y-axis.

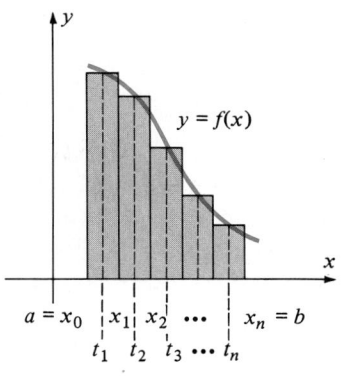

Figure 2.3 Approximating the area of R by rectangles.

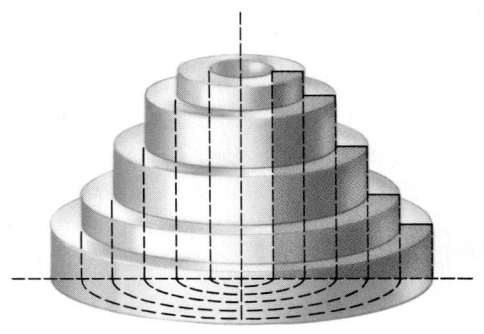

Figure 2.4 Rotating the approximating rectangle produces the approximation to the solid S.

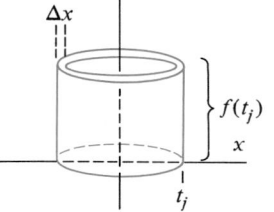

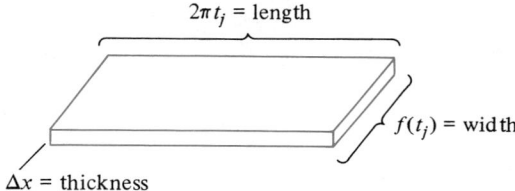

Figure 2.5 Volume of cylindrical shell is approximated by $\Delta V_j = 2\pi t_j f(t_j)\,\Delta x$.

Figure 2.5 shows how to think of the volume of one of these thin shells as the volume of the rectangular prism that results from ''cutting'' the shell along a vertical line and ''flattening it out.'' This volume is

$$\Delta V_j = 2\pi t_j f(t_j)\,\Delta x.$$

Summing these approximations for $j = 1$ to n gives an approximation to the volume of the solid

$$V \approx \sum_{j=1}^{n} 2\pi t_j f(t_j)\,\Delta x.$$

We therefore define the volume by the integral

$$\lim_{n\to\infty} \sum_{j=1}^{n} 2\pi t_j f(t_j)\,\Delta x = \int_{a}^{b} 2\pi x f(x)\,dx.$$

This is the number that we <u>define</u> to be the desired volume.

> If the region R bounded by the graph of the continuous nonnegative function f and the x-axis, for $0 \le a \le x \le b$, is rotated about the y-axis, the volume V of the resulting solid is
>
> $$V = \int_{a}^{b} 2\pi x f(x)\,dx. \tag{1}$$

Example 1

The region bounded by the graph of $y = -2x^2 + 8x - 6$ and the x-axis is rotated about the y-axis. Find the volume V of the resulting solid.

Strategy

Factor equation for y to find limits of integration.

Apply equation (1).

Solution

Since $y = -2x^2 + 8x - 6 = -2(x - 1)(x - 3)$, the region lies between the lines $x = 1$ and $x = 3$. Thus

$$V = \int_1^3 2\pi x[-2x^2 + 8x - 6] \, dx$$

$$= 2\pi \cdot \int_1^3 (-2x^3 + 8x^2 - 6x) \, dx$$

$$= 2\pi \left[-\frac{2}{4}x^4 + \frac{8}{3}x^3 - \frac{6}{2}x^2 \right]_1^3$$

$$= \frac{32\pi}{3}.$$

(See Figure 2.6.)

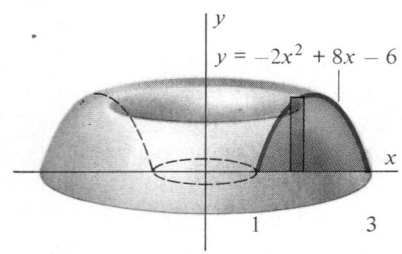

Figure 2.6 Volume of revolution about y-axis.

Equation (1) can be generalized to regions bounded below by curves other than the x-axis, as the following example shows.

Example 2

The region R is bounded by the graphs of $f(x) = -2x^2 + 8x - 6$ and $g(x) = 2x - 6$. Find the volume of the solid generated by revolving R about the y-axis.

Strategy

Set $f(x) = g(x)$ to find the points where the curves cross.

Note which curve bounds the top of the region and which curve bounds the bottom.

Apply equation (1) to both curves.

Solution

To find the points of intersection, we set $f(x) = g(x)$ and obtain the equation

$$-2x^2 + 8x - 6 = 2x - 6,$$

or

$$-2x^2 = -6x,$$

so $x = 0$ or $x = 3$. As Figure 2.7 illustrates, when the interval $[0, 3]$ is partitioned, the approximating rectangles are bounded above by $f(x) = -2x^2 + 8x - 6$ and below by $g(x) = 2x - 6$. Since the factor $f(x)$ in equation (1) represents the height of the approximating rectangles, in this case equation (1) is modified to the following:

$$V = \int_0^3 2\pi x[f(x) - g(x)] \, dx$$

$$= \int_0^3 2\pi x[-2x^2 + 8x - 6 - (2x - 6)] \, dx$$

$$= \int_0^3 2\pi x(-2x^2 + 6x) \, dx$$

$$= 2\pi \left[-\frac{2}{4}x^4 + \frac{6}{3}x^3 \right]_0^3$$

$$= 27\pi.$$

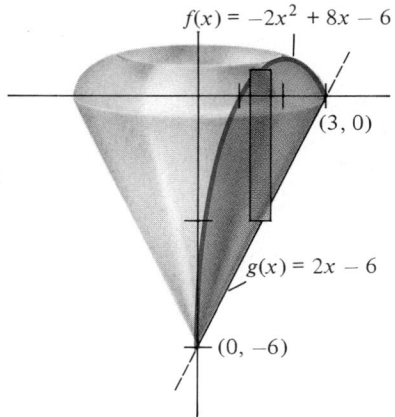

$f(x) = -2x^2 + 8x - 6$

$(3, 0)$

$g(x) = 2x - 6$

$(0, -6)$

Figure 2.7 Region R bounded above by $f(x)$ and below by $g(x)$.

REMARK 1: We may state the generalization of equation (1) observed in Example 2 as follows: If the region R bounded above by the graph of $y = f(x)$ and below by the graph of $y = g(x)$, for $0 \le a \le x \le b$, is rotated about the y-axis, the volume of the resulting solid is

$$V = \int_a^b 2\pi x[f(x) - g(x)]\ dx. \tag{2}$$

REMARK 2: You may have noticed that the integrands in equations (1) and (2) have simple geometric interpretations. If the region R to be rotated about the y-axis is sliced vertically at location x and the resulting line segment rotated about the y-axis, a band of height $[f(x) - g(x)]$ is generated (see Figure 2.8). Since the circumference of this band is $2\pi x$, the surface area of one "side" of the band is precisely the integrand $2\pi x[f(x) - g(x)]$ in equation (2) (Figure 2.9). We may therefore interpret equations (1) and (2) by saying that the volume V is found by integrating the "radial" cross section from $x = a$ to $x = b$.

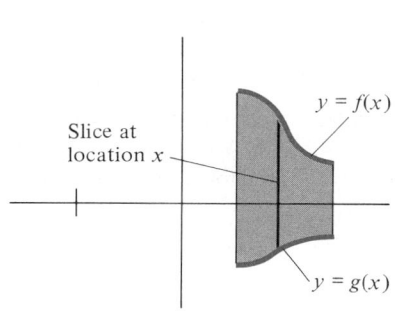

Figure 2.8 Vertical slice taken at location x.

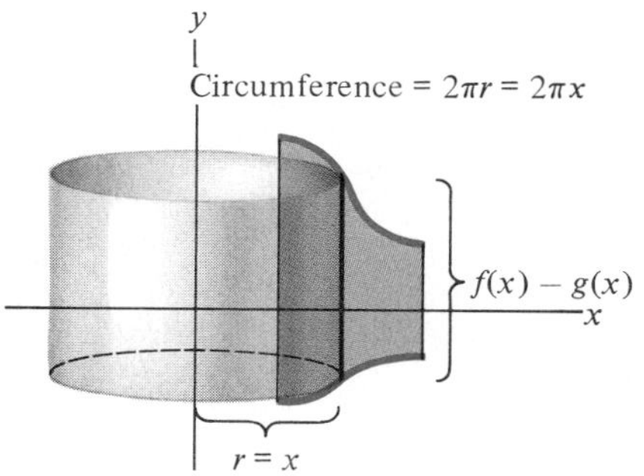

Figure 2.9 Slice generates band of surface area $2\pi x[f(x) - g(x)]$.

REMARK 3: Many of the problems on volumes of revolution in this section and the preceding section may be solved by using either the method of slicing by perpendicular cross sections or the method of cylindrical shells. Usually one method will be simpler to use, depending on the geometry of the region being rotated.

Example 3

Find the volume of the solid obtained by revolving about the line $x = -1$ the region R bounded by the graphs of $f(x) = x$ and $g(x) = (x - 2)^2$.

Solution: Figure 2.10 shows the region R which is bounded above by the graph of $f(x) = x$ and below by the graph of $g(x) = (x - 2)^2 = x^2 - 4x + 4$. To find the points of intersection of these two graphs we set $f(x) = g(x)$, which gives the equation

$$x^2 - 4x + 4 = x$$

or

$$x^2 - 5x + 4 = (x - 1)(x - 4) = 0.$$

The points of intersection are therefore $(1, 1)$ and $(4, 4)$.

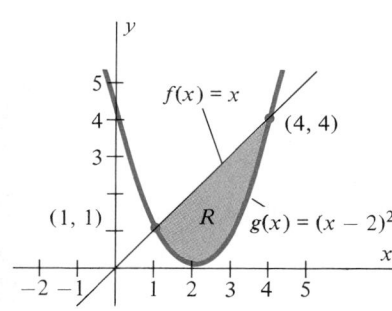

Figure 2.10 Region R in Example 3.

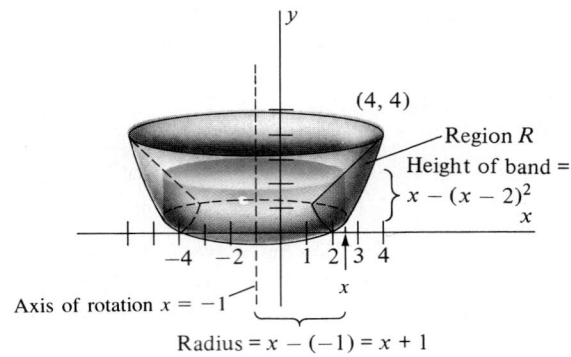

Figure 2.11 Vertical segment through R at x generates band of height $f(x) - g(x) = x - (x - 2)^2$ and radius $x + 1$.

Figure 2.11 shows that a vertical line segment through this region generates a circular band of height

$$f(x) - g(x) = x - (x^2 - 4x + 4) = -x^2 + 5x - 4.$$

Since the axis of rotation is the vertical line $x = -1$, the radius of this circular band is $x - (-1) = x + 1$. We must therefore replace the factor x in the integrand in formula (1) by the factor $x + 1$. The volume is therefore

$$V = \int_1^4 2\pi(x + 1)[f(x) - g(x)]\, dx$$

$$= \int_1^4 2\pi(x + 1)(-x^2 + 5x - 4)\, dx$$

$$= \int_1^4 2\pi(-x^3 + 4x^2 + x - 4)\, dx$$

$$= 2\pi\left(-\frac{x^4}{4} + \frac{4}{3}x^3 + \frac{x^2}{2} - 4x\right)\Bigg]_1^4$$

$$= \frac{63\pi}{2}.$$

Exercise Set 7.2

For each of Exercises 1–13, sketch the region bounded by the graphs of the given functions for the specified values of x or y. Then find the volume of the solid obtained by revolving the region about the y-axis.

1. $x + y = 1,\qquad x = 0,\qquad y = 0$

2. $y = \sqrt{x},\qquad y = 0,\qquad 1 \le x \le 4$

3. $y = x^3,\qquad y = 0,\qquad 1 \le x \le 3$

4. $y = \sqrt{1 + x^2},\qquad y = 0,\qquad 0 \le x \le 3$

5. $y = \sqrt{x},\qquad y = -x,\qquad x = 4$

6. $y = 1 + x + x^2,\qquad y = -2,\qquad 1 \le x \le 3$

7. $y = \dfrac{1}{x},\qquad y = 1,\qquad x = 4$

8. $y = 2,\qquad y = \sqrt[3]{x},\qquad x = 0$

9. $y = \sin x^2,\qquad y = 1,\qquad 0 \le x \le \sqrt{\dfrac{\pi}{2}}$

10. $y = \dfrac{1}{\sqrt{4 - x^2}},\qquad y = 0,\qquad 0 \le x \le 1$

11. $y = \sqrt{9 - x^2},\qquad y = 0,\qquad 0 \le x \le 3$

12. $y = \dfrac{1}{\sqrt{x}} + \sqrt{x},\qquad y = 0,\qquad 1 \le x \le 4$

13. $y = \dfrac{\sqrt{1 + x^{3/2}}}{\sqrt{x}}$, $y = 0$, $1 \le x \le 4$

In Exercises 14–21, use either the method of cylindrical shells or the method of slicing to find the volume of the solid described.

14. The region R, bounded by the graphs of $y = \sin x^2$ and $y = -x$, for $0 \le x \le \sqrt{\pi}$ is rotated about the y-axis.

15. The region bounded by the graph of $x = y^2$ and the line $x = 4$ is revolved about the line $x = -1$.

16. The region bounded by the graph of $x = \sqrt{4 + y}$, the line $x = 0$, and the line $y = 0$ is revolved about the line $x = -2$.

17. The region bounded by the graphs of $y = x^2$ and $y = \sqrt{x}$ is revolved about the line $y = -2$.

18. The region contained within the triangle with vertices $(1, 0)$, $(2, 4)$, and $(4, 0)$ is rotated about the y-axis.

19. The region bounded by the graphs of $y = x$ and $y = x^3$ is revolved about the y-axis.

20. The region bounded by the graphs of $y = 4$ and $y = x^2$ is revolved about the line $y = -2$.

21. The triangle with vertices $(1, 3)$, $(1, 7)$, and $(4, 7)$ is revolved about the line $y = 1$.

22. Use the method of cylindrical shells to find the volume of the solid obtained by rotating the region bounded by the ellipse $\dfrac{x^2}{a^2} + \dfrac{y^2}{b^2} = 1$ about the y-axis.

23. Figure 2.12 shows a solid called a *torus*. A **torus** is generated by revolving a disc about an axis to which it is not

Figure 2.12 Torus

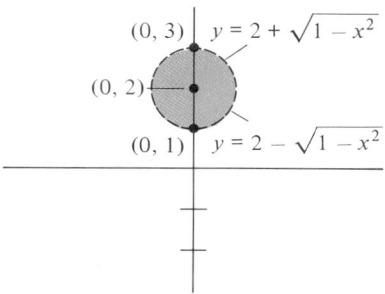

Figure 2.13 Torus generated by revolving this circle about the x-axis.

adjacent. Figure 2.13 shows a region R that is interior to the circle of radius $r = 1$ and center $(0, 2)$. Let T be the torus obtained by revolving R about the x-axis.

a. Show that the volume of T can be expressed as

$$V = \pi \int_{-1}^{1} [\sqrt{1 - x^2} + 2]^2 \, dx - \pi \int_{-1}^{1} [2 - \sqrt{1 - x^2}]^2 \, dx.$$

b. Show that V simplifies to the integral

$$V = 8\pi \int_{-1}^{1} \sqrt{1 - x^2} \, dx.$$

c. Evaluate the integral in (b) by interpreting it as the area of a semicircle.

d. Find the volume V of T.

24. *(Calculator/Computer)* Use the method of cylindrical shells and the Trapezoidal Rule to approximate the volume of the solid obtained by revolving the region bounded by the graph of $y = \sin x$ and the x-axis, for $0 \le x \le \pi$ about the y-axis.

25. *(Calculator/Computer)* Use Simpson's Rule to approximate the volume described in Exercise 24.

26. *(Calculator/Computer)* Approximate the volume of the solid obtained by revolving the region bounded by the graph of $y = \sqrt{1 + x^4}$, the line $y = 0$, and the lines $x = 0$ and $x = 2$ about the axis $y = -1$.

7.3 ARC LENGTH AND SURFACE AREA

The problems of computing the length of a curve in the plane and the surface area of a solid of revolution are closely related. We treat both problems in this section and then turn to more physical applications of the definite integral in the remaining sections of the chapter.

Arc Length

The arc length problem is to define and calculate the length of the arc of the graph of the function f from the point $(a, f(a))$ to the point $(b, f(b))$ (Figure 3.1). In the analysis that follows we will discover what specific conditions f need satisfy, but at the outset it seems necessary to assume at least that f be continuous for $a \le x \le b$.

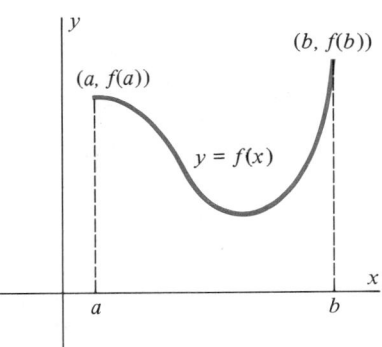

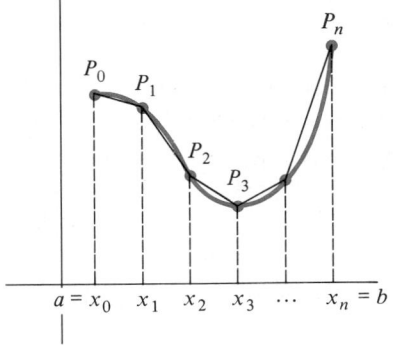

Figure 3.1 Arc whose length is to be calculated.

Figure 3.2 Approximating arc by polygonal path.

We approach this problem in a familiar way, by partitioning the interval $[a, b]$ into n subintervals of equal length. As before, we use partitions $a = x_0 < x_1 < x_2 < \cdots < x_n = b$, where $x_j - x_{j-1} = \Delta x = \dfrac{b - a}{n}$ for $j = 1, 2, \ldots, n$. On each subinterval $[x_{j-1}, x_j]$, we wish to approximate the length of the arc by an expression that we can easily calculate. We do this by connecting the endpoints of the arc, $P_{j-1} = (x_{j-1}, f(x_{j-1}))$ and $P_j = (x_j, f(x_j))$ with a line segment ℓ_j (see Figure 3.2).

We will approximate the length of the arc over the subinterval $[x_{j-1}, x_j]$ by the length of the line segment ℓ_j. The length of the entire arc is thus approximated by the length of the polygonal path through $P_0, P_1, P_2, \ldots, P_n$. If the limit of this approximation exists as $n \to \infty$, we will define this limit as the arc length in question.

An application of the distance formula gives the length of ℓ_j as

$$\text{Length } \ell_j = \sqrt{(x_j - x_{j-1})^2 + [f(x_j) - f(x_{j-1})]^2} \qquad (1)$$
$$= \sqrt{1 + \left[\frac{f(x_j) - f(x_{j-1})}{x_j - x_{j-1}}\right]^2} \cdot (x_j - x_{j-1}).$$

If f is differentiable on $[x_{j-1}, x_j]$, the Mean Value Theorem guarantees the existence of a number $t_j \in [x_{j-1}, x_j]$ so that

$$f'(t_j) = \frac{f(x_j) - f(x_{j-1})}{x_j - x_{j-1}}. \qquad \text{(See Figure 3.3.)}$$

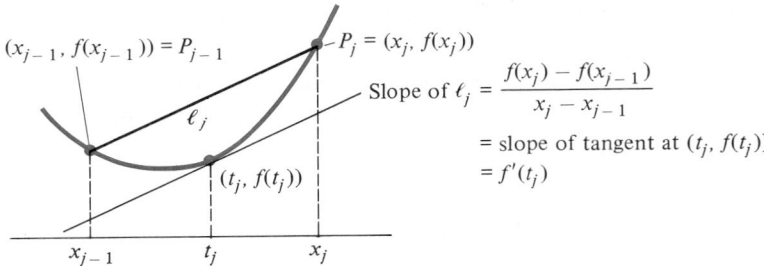

Figure 3.3 Slope of ℓ_j equals $f'(t_j)$ by Mean Value Theorem.

Substituting this expression into equation (1) gives

$$\text{Length of } \ell_j = \sqrt{1 + [f'(t_j)]^2} \cdot \Delta x.$$

Summing these expressions over all $j = 1, 2, \ldots, n$ gives the approximation

$$\text{Arc Length} \approx \sum_{j=1}^{n} \sqrt{1 + [f'(t_j)]^2} \, \Delta x. \tag{2}$$

Now the sum on the right side of approximation (2) will be a Riemann sum if the function f' is continuous. For this to be true, f' must be a continuous function on $[a, b]$. When this is the case, we *define*

$$L = \lim_{n \to \infty} \sum_{j=1}^{n} \sqrt{1 + [f'(t_j)]^2} \, \Delta x$$

$$= \int_a^b \sqrt{1 + [f'(x)]^2} \, dx.$$

We summarize our definition as follows:

If f has a continuous derivative on $[a, b]$, the length of the graph of $y = f(x)$ from $(a, f(a))$ to $(b, f(b))$ is given by

$$L = \int_a^b \sqrt{1 + [f'(x)]^2} \, dx. \tag{3}$$

or, in Leibniz notation,

$$L = \int_a^b \sqrt{1 + \left[\frac{dy}{dx}\right]^2} \, dx.$$

Example 1

Verify that the expression for arc length in equation (3) agrees with the distance formula for the case of a nonvertical line segment joining two points in the plane.

Strategy

First, determine the length according to the distance formula.

Solution

Let $P = (x_1, y_1)$ and $Q = (x_2, y_2)$ be two such points. Then the distance formula gives the length L of the line segment PQ as

$$L = \sqrt{(\Delta x)^2 + (\Delta y)^2} = \sqrt{(x_2 - x_1)^2 + (y_2 - y_1)^2}. \tag{4}$$

Find an equation for the line in question. Use form

$$y - y_1 = m(x - x_1)$$

where m = slope.

On the other hand, the equation for the line passing through P and Q is

$$y - y_1 = \frac{y_2 - y_1}{x_2 - x_1}(x - x_1),$$

so

Differentiate to find $\dfrac{dy}{dx}$.

$$\frac{dy}{dx} = \frac{y_2 - y_1}{x_2 - x_1}.$$

Equation (3) therefore gives the length of PQ as

Apply equation (3) to find length by integration.

$$L = \int_{x_1}^{x_2} \sqrt{1 + \left(\frac{y_2 - y_1}{x_2 - x_1}\right)^2}\, dx$$

Note that

$$= \left\{\sqrt{1 + \left(\frac{y_2 - y_1}{x_2 - x_1}\right)^2}\right\} \cdot x\Bigg]_{x_1}^{x_2}$$

$$\sqrt{1 + \frac{y_2 - y_1}{x_2 - x_1}}$$

$$= \sqrt{1 + \left(\frac{y_2 - y_1}{x_2 - x_1}\right)^2} \cdot (x_2 - x_1)$$

is a constant.

$$= \sqrt{(x_2 - x_1)^2 + (y_2 - y_1)^2},$$

Check that the two results agree.

which agrees with expression (4). ◇

Example 2

Find the length of the arc of the graph of $y = x^{3/2}$ between $(1, 1)$ and $(4, 8)$.

Strategy

Check that $\dfrac{dy}{dx}$ is continuous.

Solution

Here $\dfrac{dy}{dx} = \dfrac{3}{2}x^{1/2}$ is continuous on the interval $[1, 4]$, so equation (3) applies.

Apply equation (3).

We obtain

Use the substitution $u = 1 + \dfrac{9}{4}x$;

$$L = \int_1^4 \sqrt{1 + \left[\frac{3}{2}x^{1/2}\right]^2}\, dx$$

$du = \dfrac{9}{4}\, dx$. Then

$$= \int_1^4 \sqrt{1 + \frac{9}{4}x}\, dx$$

$$\int \sqrt{1 + \frac{9}{4}x}\, dx = \int \sqrt{u} \cdot \frac{4}{9}\, du$$

$$= \frac{4}{9} \cdot \frac{2}{3} u^{3/2} + C$$

$$= \frac{4}{9} \cdot \frac{2}{3}\left(1 + \frac{9}{4}x\right)^{3/2}\Bigg]_1^4$$

$$= \frac{4}{9} \cdot \frac{2}{3}\left(1 + \frac{9}{4}x\right)^{3/2} + C$$

$$= \frac{8}{27}\left[10^{3/2} - \left(\frac{13}{4}\right)^{3/2}\right] \approx 7.634.$$

is the antiderivative.

(Figure 3.4.) ◇

Example 3

Find the length of the arc of the graph of the equation

$$6xy - y^4 - 3 = 0 \tag{5}$$

from $(19/12, 2)$ to $(14/3, 3)$.

Strategy

Solve for x rather than y due to presence of y^4 term.

Solution

The form of the equation suggests that we solve for x as a function of y. Doing so we obtain

$$x = f(y) = \frac{1}{6}y^3 + \frac{1}{2y}.$$

Then

$$f'(y) = \frac{1}{2}y^2 - \frac{1}{2y^2}.$$

Rewrite equation (3) for functions of y.

The form of equation (3) for the arc of the graph of $x = f(y)$ connecting (a, c) and (b, d) is

$$L = \int_c^d \sqrt{1 + [f'(y)]^2} \; dy.$$

Apply equation obtained.

We therefore obtain

(By squaring the binomial term and factoring the result, the term under the radical can be brought into the form of a perfect square.)

$$L = \int_2^3 \sqrt{1 + \left[\frac{1}{2}y^2 - \frac{1}{2y^2}\right]^2} \; dy$$

$$= \int_2^3 \sqrt{1 + \left[\frac{1}{4}y^4 - \frac{1}{2} + \frac{1}{4y^4}\right]} \; dy$$

$$= \int_2^3 \sqrt{\frac{1}{4}y^4 + \frac{1}{2} + \frac{1}{4y^4}} \; dy$$

$$= \int_2^3 \sqrt{\left(\frac{1}{2}y^2 + \frac{1}{2y^2}\right)^2} \; dy$$

$$= \int_2^3 \left(\frac{1}{2}y^2 + \frac{1}{2y^2}\right) \; dy$$

$$= \frac{1}{6}y^3 - \frac{1}{2y}\Big]_2^3$$

$$= \frac{13}{3} - \frac{13}{12} = \frac{13}{4}.$$

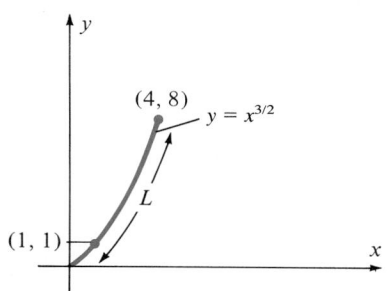

(4, 8)
$y = x^{3/2}$
L
(1, 1)

Figure 3.4

REMARK: You have no doubt observed that the integral in equation (3) is, in general, very difficult to evaluate by use of the Fundamental Theorem. Although we will study additional techniques of integration in Chapter 10, nevertheless we will remain unable to find convenient antiderivatives for most integrands arising from an application of equation (3). Thus, the procedures for approximating integrals will be needed in most practical applications of the formulas for arc length.

Surface Area

We saw in Section 7.1 that when the region bounded by the graph of a continuous function f and the x-axis for $a \leq x \leq b$ is revolved about the x-axis, a solid of revolution results. We wish now to determine how to calculate the lateral surface area of such a solid.

As for the arc length problem, we partition $[a, b]$ with numbers $a = x_0 < x_1 < x_2 < \cdots < x_n = b$ where $x_j - x_{j-1} = \Delta x = \dfrac{b - a}{n}$, $j = 1, 2, \ldots, n$. We then connect successive points $P_j = (x_j, f(x_j))$ with line segments to form a polygonal approximation to the graph of $y = f(x)$ from $P_0 = (a, f(a))$ to $P_n = (b, f(b))$ (see Figure 3.5). As this polygonal path is revolved about the x-axis, it sweeps out a solid whose surface area we will take as an approximation to the surface area S of the solid in question (Figure 3.6).

As the line segment joining P_{j-1} and P_j revolves about the x-axis (Figure 3.7), it sweeps out a frustum of a cone, with radii $f(x_{j-1})$ and $f(x_j)$, and slant height

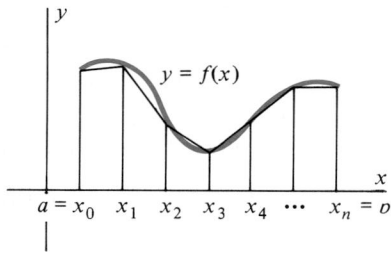

$y = f(x)$

$a = x_0 \quad x_1 \quad x_2 \quad x_3 \quad x_4 \quad \cdots \quad x_n = o$

Figure 3.5   Polygonal approximation to graph of $y = f(x)$.

$$P_{j-1}P_j = \sqrt{(x_j - x_{j-1})^2 + [f(x_j) - f(x_{j-1})]^2}.$$

Since the formula for the lateral surface area of a frustum of a cone with radii r and R and slant height ℓ is $s = \pi(r + R)\ell$ (Figure 3.8), the surface area of the

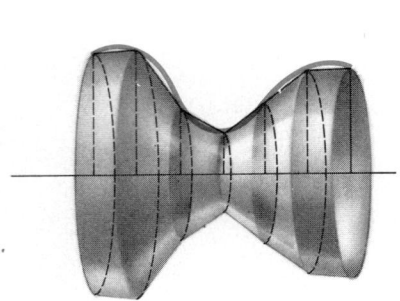

Figure 3.6 Figure approximating solid of revolution for surface area problem.

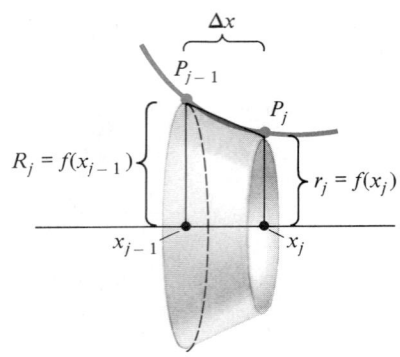

Figure 3.7 Element of surface area is frustum of cone with height Δx.

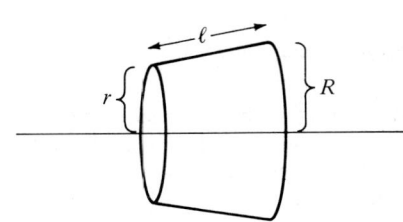

Figure 3.8 $s = \pi(r + R)\ell$.

frustum generated by $P_{j-1}P_j$ is

$$\Delta S_j = \pi(f(x_{j-1}) + f(x_j))\sqrt{(x_j - x_{j-1})^2 + [f(x_j) - f(x_{j-1})]^2}$$

$$= \pi(f(x_{j-1}) + f(x_j))\sqrt{1 + \left[\frac{f(x_j) - f(x_{j-1})}{x_j - x_{j-1}}\right]^2} \cdot \Delta x.$$

The sum of these approximations leads, as in the development of the formulas for volumes of solids of revolution and for arc length, to the following definition of surface area.

If the function f has a continuous derivative on the interval $[a, b]$, then the surface area S of the solid of revolution obtained by revolving about the x-axis the region bounded by the graph of $y = f(x) \geq 0$ and the x-axis, for $a \leq x \leq b$, is

$$S = \int_a^b 2\pi f(x)\sqrt{1 + [f'(x)]^2}\, dx \qquad (6)$$

or, in Leibniz notation,

$$S = \int_a^b 2\pi y\sqrt{1 + \left[\frac{dy}{dx}\right]^2}\, dx. \qquad (7)$$

REMARK: It is important to observe that we require $y = f(x) \geq 0$ in equations (6) and (7). If $f(x)$ is negative for some $x \in [a, b]$, we must replace $f(x)$ by $|f(x)|$ in (6) and y by $|y|$ in (7).

Example 4

Find the surface area of the band of the sphere generated by revolving the arc of the circle $x^2 + y^2 = r^2$ lying above the interval $[-a, a]$, $a < r$, about the x-axis (Figure 3.9).

Solution: Here $y = \sqrt{r^2 - x^2}$, so

$$\frac{dy}{dx} = \frac{-x}{\sqrt{r^2 - x^2}}.$$

Figure 3.9

Using equation (7), we obtain

$$S = \int_{-a}^{a} 2\pi\sqrt{r^2 - x^2} \sqrt{1 + \left[\frac{-x}{\sqrt{r^2 - x^2}}\right]^2} \, dx$$

$$= \int_{-a}^{a} 2\pi\sqrt{r^2 - x^2} \sqrt{\frac{(r^2 - x^2) + x^2}{r^2 - x^2}} \, dx$$

$$= \int_{-a}^{a} 2\pi\sqrt{r^2} \, dx$$

$$= 2\pi r x]_{-a}^{a}$$

$$= 4\pi ar.$$

◇

Example 5

Is it correct in Example 4 to set $a = r$ and conclude that the surface area of the sphere of radius r is $S = 4\pi r^2$?

Solution: Although the conclusion is correct, the reasoning is faulty. Applying formula (7) requires that the function f have a continuous derivative for all x in the interval in question. Since $\dfrac{dy}{dx} = \dfrac{-x}{\sqrt{r^2 - x^2}}$ is undefined for $x = r$, equation (7) cannot be applied on the interval $[-r, r]$. We will obtain a satisfactory solution to this problem in Chapter 15 when we study curves described by parametric equations. ◇

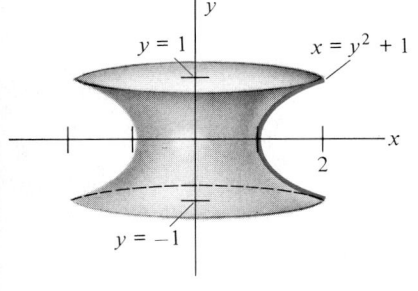

Figure 3.10

Example 6

Find an integral giving the surface area of the "spool" obtained by revolving the region bounded by the graph of $y^2 - x + 1 = 0$ and the y-axis, for $-1 \le y \le 1$, about the y-axis (Figure 3.10).

Solution: Solving for x as a function of y we obtain

$$x = y^2 + 1$$

so

$$\frac{dx}{dy} = 2y.$$

Since $\dfrac{dx}{dy}$ is continuous for $-1 \le y \le 1$, we may use the analogue of equation (7) for rotations about the y-axis. We obtain

$$S = \int_{-1}^{1} 2\pi x \sqrt{1 + \left[\frac{dx}{dy}\right]^2} \, dy, \qquad \text{or}$$

$$S = \int_{-1}^{1} 2\pi(y^2 + 1)\sqrt{1 + 4y^2} \, dy.$$

As will be the case for many functions to which equation (7) is applied, we cannot find a convenient antiderivative for the integrand in the above integral for surface area. Table 3.1, however, shows the results of using the Trapezoidal Rule to approximate the value of this integral. ◇

Table 3.1

n	Trapezoidal approximation
5	27.5483
10	26.5413
25	26.2583
50	26.2179
100	26.2078
200	26.2052

Exercise Set 7.3

In each of Exercises 1–11, find the length of the arc determined by the given equation.

1. $y = 2x + 3$, from $(-3, -3)$ to $(2, 7)$

2. $y = x^{3/2}$, from $(0, 0)$ to $(4, 8)$

3. $y - 2 = (x + 2)^{3/2}$ from $(-2, 2)$ to $(2, 10)$

4. $3y = (x^2 + 2)^{3/2}$ for $0 \le x \le 2$

5. $3y = 2(x + 1)^{3/2}$ from $(0, 2/3)$ to $(3, 16/3)$

6. $y = \dfrac{x^3}{3} + \dfrac{1}{4x}$ for $1 \le x \le 3$

7. $(3y - 6)^2 = (x^2 + 2)^3$ for $0 \le x \le 3$

8. $y = \dfrac{x^4}{4} + \dfrac{1}{8x^2}$ for $1 \le x \le 2$

9. $y = \dfrac{1}{6}x^3 + \dfrac{1}{2}x^{-1}$, from $\left(1, \dfrac{2}{3}\right)$ to $\left(4, \dfrac{259}{24}\right)$

10. $y = \dfrac{2}{3}x^{3/2} - \dfrac{1}{2}x^{1/2}$, from $\left(1, \dfrac{1}{6}\right)$ to $\left(3, \dfrac{3\sqrt{3}}{2}\right)$

11. $y = \dfrac{x^3}{6} + \dfrac{1}{2x}$, from $\left(1, \dfrac{2}{3}\right)$ to $\left(3, \dfrac{14}{3}\right)$

In each of Exercises 12–19, find the area of the lateral surface of the solid generated by revolving the stated region about the given axis.

12. The region bounded by the graph of $y = 2x + 3$ and the x-axis for $1 \le x \le 3$ is revolved about the x-axis.

13. The region bounded by the graph of $y = x^3$ and the x-axis for $0 \le x \le 3$ is revolved about the x-axis.

14. The region bounded by the graph of $y = x^{1/3}$ and the y-axis for $0 \le y \le 2$ is revolved about the y-axis.

15. The region bounded by the graph of $x = 2\sqrt{y}$ and the y-axis for $1 \le y \le 4$ is revolved about the y-axis.

16. The region bounded by the graph of $y = \sqrt{x}$ and the x-axis for $1 \le x \le 2$ is revolved about the x-axis.

17. The region bounded by the graph of $y = \sqrt{2x - 1}$ and the x-axis for $2 \le x \le 8$ is revolved about the x-axis.

18. The region bounded by the graph of $12xy - 3y^4 = 4$ and the y-axis for $1 \le y \le 3$ is revolved about the y-axis.

19. The region bounded by the graph of $y = \dfrac{x^3}{4} + \dfrac{1}{3x}$ and the x-axis for $1 \le x \le 3$ is revolved about the x-axis.

20. *(Calculator/Computer)* The small arc of the circle with equation $x^2 + (y - 1)^2 = 25$ from $(3, 5)$ to $(4, 4)$ is revolved about the x-axis. Use the Trapezoidal Rule to approximate the lateral surface area of the resulting solid.

21. *(Calculator/Computer)* A circle has center $(1, 1)$ and radius 5. The small arc joining the points $(4, 5)$ and $(5, 4)$ on the circle is revolved about the x-axis. Use the Trapezoidal Rule to approximate the lateral surface area of the resulting solid.

22. *(Calculator/Computer)* Write down an integral giving the length of the arc of the graph of $f(x) = \sin x$ between $(0, 0)$ and $(\pi/2, 1)$. Use the Trapezoidal Rule with $n = 10$ to approximate the value of this integral.

23. *(Calculator/Computer)* Write down an integral giving the length of the graph of the function $f(x) = x^2$ lying between $(-1, 1)$ and $(1, 1)$. Approximate this integral using Simpson's Rule with $n = 10$.

24. *(Calculator/Computer)* Write down an integral giving the lateral surface area of the figure obtained by revolving the upper half of the graph of the ellipse $\dfrac{x^2}{a^2} + \dfrac{y^2}{b^2} = 1$, for $-\dfrac{a}{2} \le x \le \dfrac{a}{2}$, about the x-axis. Approximate the value of this integral for $a = 4$ and $b = 3$ using Simpson's Rule with $n = 10$.

7.4 DISTANCE AND VELOCITY

In Chapter 3, we saw that the functions describing the position, velocity, and acceleration of an object moving along a line were related by the process of differentiation. (In particular, $v(t) = s'(t)$, $a(t) = v'(t)$.) In this section we shall show that the distance D travelled by such an object from time $t = a$ to time $t = b$ can be calculated as a definite integral involving the velocity function v. In particular,

$$D = \int_a^b |v(t)|\, dt. \tag{1}$$

We begin with the problem of an object moving along a line with velocity $v(t)$ at time t. The question we want to answer is this: Given that v is a known, continuous function, how can we calculate the distance travelled by the object from time $t = a$ to time $t = b$?

For the romantics among us, here is one centuries-old approach to solving this problem.

> As for distance sailed east or west, early mariners relied solely on dead reckoning, estimating their speed by peering over the side and noting the rate at which foam slipped past the planking. The results they came up with were little more than a compound of hunch, guesswork and intuition. Samuel Eliot Morison calculates that on Columbus' first Atlantic passage, he consistently overestimated the distance of his day's run by about 9 per cent, which scales up to a 90-mile error for every 1,000 miles traveled. However, the remarkable fact is not how far off these navigators were, but how amazingly close they came using such primitive methods.
>
> Progress in the following centuries came slowly. In the late 1500s, some ingenious mariner—name and nationality unknown—invented the common, or chip, log, a simple speed-measuring device. The chip log was a piece of wood shaped like a slice of pie. Attached to it was a light line, knotted at certain equal intervals. To check the ship's speed, a man threw the log overboard, turned a small 30-second sandglass, and began counting the knots that slipped through his fingers as the ship left the log astern. The number of knots he counted in 30 seconds was translated into the nautical miles per hour his ship was traveling. With this device began usage of the term "knots" to mean "nautical miles per hour."
>
> *Life Science Library/SHIPS* by Edward V. Lewis, Robert O'Brien, and the Editors of LIFE. Time-Life Books, Inc., Publisher. © 1965 Time Inc.

In our more mathematical setting, we begin by partitioning the time interval $[a, b]$ into n subintervals of equal length, $\Delta t = \dfrac{b - a}{n}$, and with endpoints

$$t_0 = a, \quad t_1 = a + \Delta t, \quad t_2 = a + 2\,\Delta t, \quad \ldots, \quad t_n = a + n\,\Delta t = b.$$

Our idea is to assume that $v(t)$ is constant on each subinterval of time, to approximate the distance travelled during each subinterval, and then to sum the individual distances to obtain an approximation to D, the distance travelled by the object.

To do so we let c_j denote any number (time) in the jth interval $[t_{j-1}, t_j]$, and we assume that $v(t) \equiv v(c_j)$ for all $t \in [t_{j-1}, t_j]$. Using the definition

$$\text{distance} = \text{speed} \times \text{time}$$

we approximate the distance ΔD_j travelled by the object during the time subinterval

$[t_{j-1}, t_j]$ as

$$\Delta D_j = |v(c_j)|(t_j - t_{j-1}) = |v(c_j)|\, \Delta t, \qquad j = 1, 2, \ldots, n.$$

(Recall, if $v(t)$ is velocity, speed = $|v(t)|$. We care only about the rate of the motion, not its direction.) Summing these individual approximations gives an approximation to D as

$$D \approx \sum_{j=1}^{n} \Delta D_j = \sum_{j=1}^{n} |v(c_j)|\, \Delta t, \qquad c_j \in [t_{j-1}, t_j]. \tag{2}$$

Now the sum on the right-hand side of approximation (2) is a Riemann sum for the function $|v|$ on the interval $[a, b]$. If v is continuous on $[a, b]$, so is $|v|$, so this Riemann sum will have as its limit the integral

$$\int_a^b |v(t)|\, dt = \lim_{n \to \infty} \sum_{j=1}^{n} |v(c_j)|\, \Delta t, \qquad c_j \in [t_{j-1}, t_j].$$

We therefore conclude that

$$D = \lim_{n \to \infty} \sum_{j=1}^{n} |v(c_j)|\, \Delta t = \int_a^b |v(t)|\, dt. \tag{3}$$

Example 1

Starting at time $t = 0$, a particle moves along a line with velocity $v(t) = 2 + \sqrt{t}$ m/s. How far does it travel during the first 4 seconds?

Solution: Here $v(t) = 2 + \sqrt{t} \geq 0$ for all $t \in [0, 4]$, so $|v(t)| = v(t)$ in (3). Thus, by (3),

$$D = \int_0^4 (2 + \sqrt{t})\, dt = 2t + \frac{2}{3}t^{3/2}\Big]_0^4 = \left(2 \cdot 4 + \frac{2}{3} \cdot 8\right) - (0)$$

$$= \frac{40}{3} \text{ meters.} \qquad \diamond$$

Example 2

A particle moves along a line with velocity $v(t) = 5 - t^2$ m/s. Find the distance travelled from time $t = 3$ s to time $t = 5$ s.

Solution: This time $v(t) = 5 - t^2 < 0$ for $t \in [3, 5]$, so $|v(t)| = -v(t)$ in equation (3). Thus,

$$D = \int_3^5 -(5 - t^2)\, dt = \frac{t^3}{3} - 5t\Big]_3^5 = \left(\frac{125}{3} - 25\right) - \left(\frac{27}{3} - 15\right)$$

$$= \frac{68}{3} \text{ meters.} \qquad \diamond$$

Example 3

A subway train accelerates at a rate of $a = 0.5$ m/s^2 until reaching its maximum speed of $v = 36$ kilometers per hour. How far does the train travel in the first two minutes, assuming maximum acceleration and cruising speed?

Solution: The velocity $v(t)$ of the train while undergoing acceleration $a = 0.5$ m/s^2 is

$$v(t) = \int a(t) \, dt$$

$$= \int 0.5t \, dt$$

$$= 0.5 + v_0$$

where the constant v_0 is the initial velocity $v_0 = v(0)$. Since the train begins from rest, we have $v_0 = v(0) = 0$. The velocity function is

$$v(t) = 0.5t, \qquad 0 \le t \le T$$

where T is the time at which the train reaches its maximum velocity $v_{max} = 36$ km/h. To find T we first convert this maximum velocity to units of m/s:

$$36 \text{ km/h} = 36{,}000 \text{ m/h} = \frac{36{,}000}{(60)^2} \text{ m/s} = 10 \text{ m/s}.$$

The train therefore reaches maximum speed when

$$v(T) = 0.5T = 10 \text{ m/s}$$

or at time

$$T = \frac{10}{0.5} = 20 \text{ s}.$$

After time $T = 20$ s the speed of the train remains constant at 10 m/s. The velocity function for the train during the first 2 minutes ($= 120$ s) is therefore

$$v(t) = \begin{cases} 0.5t, & 0 \le t \le 20 \\ 10, & 20 < t \le 120. \end{cases}$$

The distance travelled during this time is

$$D = \int_0^{120} v(t) \, dt$$

$$= \int_0^{20} (0.5t) \, dt + \int_{20}^{120} 10 \, dt$$

$$= 0.25t^2 \Big]_0^{20} + 10t \Big]_{20}^{120}$$

$$= 100 + 10(120 - 20)$$

$$= 1{,}100 \text{ m}$$

$$= 1.1 \text{ km}.$$

$\diamondsuit$

Obviously, a difficulty arises in applying (3) when v has both positive and negative values on $[a, b]$. In such cases, we must first find the zeros of v, and then apply formula (3) separately on each of the subintervals of $[a, b]$ determined by the zeros and the endpoints.

Example 4

If an object moves along a line with velocity $v(t) = t^3 - 3t^2 - t + 3$ m/s, how far does it travel from time $t = 0$ to time $t = 4$ seconds?

Solution: In order to determine the sign of $v(t)$ on $[0, 4]$, we must find the solutions of the equation $v(t) = 0$. Now $v(t)$ has factorization

$$v(t) = (t + 1)(t - 1)(t - 3),$$

so $v(t) = 0$ on $[0, 4]$ if $t = 1$ or $t = 3$. By evaluating $v(t)$ at numbers in each of the resulting intervals you can verify that

$$v(t) \geq 0 \quad \text{for} \quad t \in [0, 1]$$
$$v(t) \leq 0 \quad \text{for} \quad t \in [1, 3]$$
$$v(t) \geq 0 \quad \text{for} \quad t \in [3, 4].$$

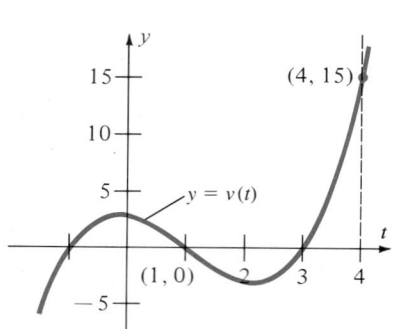

Figure 4.1 $v(t) = t^3 - 3t^2 - t + 3$ in Example 4.

(See Figure 4.1.)

Thus,

$$D = \int_0^1 (t^3 - 3t^2 - t + 3) \, dt + \int_1^3 -(t^3 - 3t^2 - t + 3) \, dt$$

$$+ \int_3^4 (t^3 - 3t^2 - t + 3) \, dt$$

$$= \left[\frac{t^4}{4} - t^3 - \frac{t^2}{2} + 3t \right]_0^1 + \left[-\frac{t^4}{4} + t^3 + \frac{t^2}{2} - 3t \right]_1^3$$

$$+ \left[\frac{t^4}{4} - t^3 - \frac{t^2}{2} + 3t \right]_3^4$$

$$= \left(\frac{7}{4} - 0 \right) + \left(\frac{9}{4} - \left(-\frac{7}{4} \right) \right) + \left(4 - \left(-\frac{9}{4} \right) \right).$$

$$= 12 \text{ meters.} \qquad \diamondsuit$$

Example 5

The position of the end of a vibrating spring is given by the function $s(t) = 6 \cos \pi t$, where t is in seconds and s is in centimeters. How far does the end of the spring travel between times $t = 0$ and $t = 3$?

Strategy

Differentiate s to find v.

Solution

Since $s(t) = 6 \cos \pi t$ is the *position* of the end of the spring, its velocity is

$$v(t) = \frac{d}{dt} (6 \cos \pi t) = -6\pi \sin \pi t.$$

Set $v(t) = 0$ to find zeros in $[0, 3]$.

Then $v(t) = 0$ when $\sin \pi t = 0$, which happens for

$$\pi t = n\pi, \quad \text{or} \quad t = n = 0, \pm 1, \pm 2, \ldots .$$

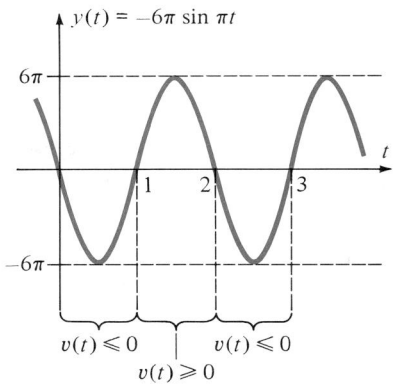

Figure 4.2 Velocity function in Example 5.

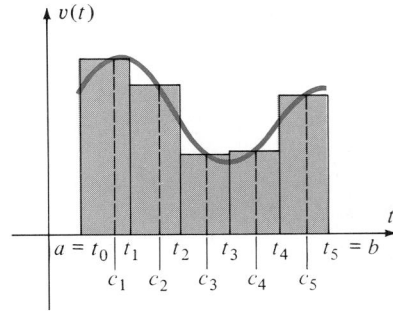

Figure 4.3 Riemann sum R_5 for $D = \int_a^b |v(t)|\, dt$.

The zeros for v in $[0, 3]$ are therefore $t = 0, 1, 2,$ and 3. By checking particular numbers, we find that

$$
\begin{array}{lll}
v(t) \leq 0 & \text{for} & t \in [0, 1] \\
v(t) \geq 0 & \text{for} & t \in [1, 2] \\
v(t) \leq 0 & \text{for} & t \in [2, 3].
\end{array}
$$

Determine the sign of $v(t)$ on each of the resulting intervals.

(See Figure 4.2.)
Thus, by (3),

Apply the definition (3) on each interval according to the sign of $v(t)$.

$$
\begin{aligned}
D &= \int_0^1 -v(t)\, dt + \int_1^2 v(t)\, dt + \int_2^3 -v(t)\, dt \\
&= \int_0^1 6\pi \sin \pi t\, dt + \int_1^2 (-6\pi \sin \pi t)\, dt + \int_2^3 6\pi \sin \pi t\, dt \\
&= -6 \cos \pi t]_0^1 + 6 \cos \pi t]_1^2 - 6 \cos \pi t]_2^3 \\
&= [(-6)(-1) - (-6)(1)] + [(6)(1) - (6)(-1)] \\
&\quad + [(-6)(-1) - (-6)(1)] \\
&= 36 \text{ cm.} \qquad \qquad \Diamond
\end{aligned}
$$

You have undoubtedly noted the similarity between the calculation of area and the calculation of distance. Both are nonnegative quantities, and both are calculated by integrating functions which themselves are nonnegative. In fact, a sketch of a Riemann sum for the integral $D = \int_a^b |v(t)|\, dt$ (Figure 4.3) shows that the sum could just as well be interpreted as an approximation to the area of the region bounded by the graph of $y = |v(t)|$ and the t-axis for $a \leq t \leq b$.

For the case of a nonnegative velocity function v, we know from Chapter 3 that the distance travelled from time $t = a$ to time $t = b$ by a particle travelling with velocity $v(t)$ is

$$
D = s(b) - s(a) \tag{4}
$$

where s is an antiderivative (a position function) for v. To verify that equations (3) and (4) define the same quantity, we use the Fundamental Theorem of Calculus. If

$s'(t) = v(t)$, we have

$$D = \int_a^b v(t) \, dt = s(t)]_a^b = s(b) - s(a).$$

The cases $v(t) \leq 0$ and $v(t)$ of mixed sign are similar and are left as Exercises 24 and 25.

Exercise Set 7.4

In Exercises 1–10, calculate the distance travelled by a particle moving along a line with the given velocity from time $t = a$ to time $t = b$.

1. $v(t) = 9 - t^2$, $\quad a = 0$, $\quad b = 3$

2. $v(t) = \sqrt{t + 1}$, $\quad a = 0$, $\quad b = 8$

3. $v(t) = \sin t$, $\quad a = 0$, $\quad b = \pi$

4. $v(t) = t - t^3$, $\quad a = 0$, $\quad b = 1$

5. $v(t) = 2 + t^2$, $\quad a = 0$, $\quad b = 5$

6. $v(t) = t^2 + t - 6$, $\quad a = 0$, $\quad b = 2$

7. $v(t) = \cos \pi t$, $\quad a = \dfrac{1}{2}$, $\quad b = \dfrac{3}{2}$

8. $v(t) = \dfrac{1 - t}{\sqrt{t}}$, $\quad a = 2$, $\quad b = 4$

9. $v(t) = \sin^2 t \cos t$, $\quad a = 0$, $\quad b = \dfrac{\pi}{2}$

10. $v(t) = \sqrt{t}(t + 2)$, $\quad a = 0$, $\quad b = 1$

In each of Exercises 11–16, the motion of a particle along a line is described by its velocity function. Taking care to note where $v(t)$ changes sign, compute the distance travelled by the particle during the time interval specified (see Example 4).

11. $v(t) = t - 4$, $\quad t = 0$ to $t = 6$

12. $v(t) = 2 - t^2$, $\quad t = 0$ to $t = 4$

13. $v(t) = \sin(2t)$, $\quad t = 0$ to $t = \pi/2$.

14. $v(t) = t^{1/3} - t^{2/3}$, $\quad t = 0$ to $t = 8$

15. $v(t) = (t - 1)^5$, $\quad t = 0$ to $t = 2$

16. $v(t) = \dfrac{t^2 - 9}{t + 3}$, $\quad t = 0$ to $t = 4$

17. The *displacement* of a particle from time $t = a$ to time $t = b$ is the difference $s(b) - s(a)$, where $s(t)$ is the position of the particle at time t and the motion is along a line. Find the displacement of the particle described in
   a. Exercise 11
   b. Exercise 12
   c. Exercise 13
   d. Exercise 14

18. The motion of a particle along a line is described by the position function $s(t) = t^2 - 7t + 10$.
   a. Find its velocity at time t.
   b. Find its acceleration at time t.
   c. For which times t is the particle at rest?
   d. For which times t is the particle moving to the right?
   e. For which times t is the particle moving to the left?
   f. Find the distance travelled by the particle from $t = 1$ to $t = 8$.
   g. Find the displacement of the particle between times $t = 1$ and $t = 8$.

19. The motion of a particle along a line is described by the velocity function $v(t) = -t^2 + 5t - 6$. The position of the particle at time $t = 0$ is $t(0) = 4$.
   a. Find its position $s(t)$ at time t.
   b. Find its acceleration $a(t)$ at time t.
   c. For which times t is the particle at rest?
   d. For which times t is the particle moving to the right?
   e. For which times t is the particle moving to the left?
   f. Find the distance travelled by the particle from time $t = 0$ to $t = 3$.
   g. Find the displacement of the particle between times $t = 0$ and $t = 3$.

20. The motion of a particle along a line is described by the following equations:

$$a(t) = 2t - 4$$
$$v(0) = 3$$
$$s(0) = 5$$

Here a is acceleration, v velocity, s position.
   a. Find the velocity $v(t)$ at time t.
   b. Find the position $s(t)$ at time t.
   c. For which time(s) t is the particle at rest?
   d. For which times t is the particle moving to the right?
   e. For which times t is the particle moving to the left?
   f. Find the distance travelled by the particle from $t = 0$ to $t = 3$.
   g. Find the displacement of the particle between times $t = 0$ and $t = 3$.

21. Find the distance travelled by the subway train in Example 3 if we add the conditions that the train also *decelerates* at the constant rate $a = -0.5$ m/s^2 and must come to a complete stop at time $t = 2$ minutes.

22. An ultratrain has a maximum cruising speed of 108 km/h and the capability of maximum acceleration of $a_1 = 1.0$ m/s^2 and maximum deceleration of $a_2 = -1.0$ m/s^2. How far does the train travel in 10 minutes if it begins from rest, travels under maximum acceleration until reaching its maximum cruising speed, and remains at this speed until time $t = 10$ minutes?

23. What is the maximum distance that the train in Exercise 22 can travel in 10 minutes if we add the condition that it must have come to a complete stop by time $t = 10$ minutes?

24. Verify that $\int_a^b |v(t)|\, dt = -[s(b) - s(a)] = s(a) - s(b)$ if $s'(t) = v(t)$ and $v(t) \leq 0$ for all $t \in [a, b]$.

25. What can you say about the relationship between $\int_a^b |v(t)|\, dt$ and the number $s(b) - s(a)$ when $v(t)$ has both positive and negative values on $[a, b]$?

7.5 HYDROSTATIC PRESSURE

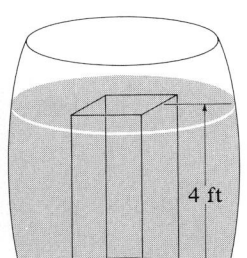

Figure 5.1

The word hydrostatic refers to properties of still fluids (as opposed to hydrodynamic, which refers to properties of moving fluids). Our interest here is in calculating pressures and forces exerted by water on the walls of containers.*

The **pressure** acting on an object is defined to be the force per unit area acting on that object. For example, if a rain barrel is filled to a depth $h = 4$ feet with water, then above each square foot of the bottom of the barrel extends a rectangular column of water of volume 1 ft $\times$ 1 ft $\times$ 4 ft = 4 ft^3. Since the density of water is 62.4 lb/ft^3, the pressure exerted on the bottom of the barrel by the column of water is

$$p = \text{density} \times \text{height}$$
$$= (62.4 \text{ lb/ft}^3) \times (4 \text{ ft})$$
$$= 249.6 \text{ lb/ft}^2.$$

(Figure 5.1.)

The total force F exerted on the (horizontal) bottom of the barrel is found by multiplying the area A of the base by the pressure p. If, in this example, the base is a circle of radius 2 feet, then

$$A = \pi r^2 = 4\pi \text{ ft}^2,$$

so

$$F = pA = (249.6 \text{ lb/ft}^2) \cdot (4\pi \text{ ft}^2)$$
$$= 998.4\pi \text{ lb}.$$

The problem of calculating the force acting on the side wall of the barrel is less straightforward. The preceding discussion shows that pressure depends on depth. However, **Pascal's Principle** states that the pressure exerted by a fluid on a submerged particle acts equally in all directions. Thus, the hydrostatic pressure at any point depends *only* on depth. However, the calculation of the total force acting on a side wall involves more than a simple multiplication, since the wall exists at various depths.

To pursue this question, we idealize it somewhat by considering a submerged object whose face consists of a region bounded above by the line $h = a$, below by the line $h = b$, and whose width at depth h is determined by the function $w(h)$. We take the surface of the water to lie at depth $h = 0$, we take the positive h-axis to

---

*The same principles apply to all fluids. For simplicity, we will concentrate on the most common fluid, water. For a fluid of density ρ lb/ft^3, the constant 62.4 is replaced by the constant ρ.

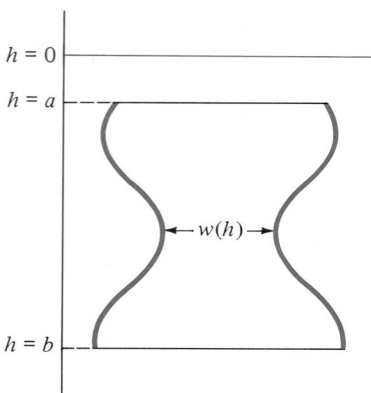

Figure 5.2 Face of submerged object; $w(h)$ is the width at depth h.

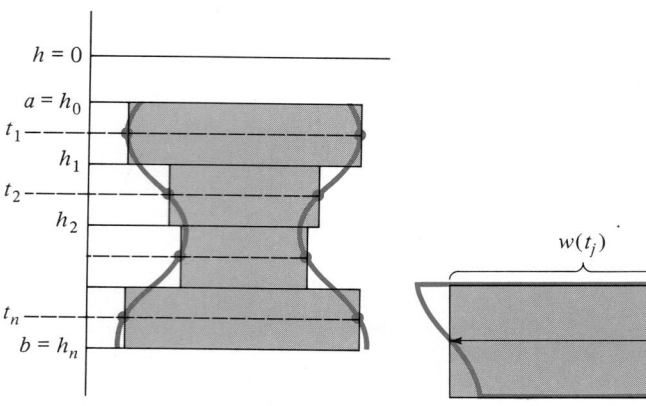

Figure 5.3 Approximating rectangles and test points t_j.

Figure 5.4 jth approximating rectangle.

extend vertically downward, and we assume that the face of the object is sitting vertically in the fluid (see Figure 5.2).

Since pressure varies continuously with the depth, we partition the interval $[a, b]$ into n subintervals of equal length with endpoints $a = h_0 < h_1 < h_2 < \cdots < h_n = b$, and we select one arbitrary number t_j in each interval $[h_{j-1}, h_j]$. We then make two types of approximations:

1. We approximate the slice of the face lying between depth h_{j-1} and depth h_j by the *rectangle* with width $w(t_j)$ and height $\Delta h = \dfrac{b - a}{n}$ (see Figures 5.3 and 5.4).

2. We approximate the depth of each point in the jth rectangle by the depth t_j.

That is, we assume the width $w(t)$ to be the constant $w(t_j)$ throughout the jth rectangle, and we assume the pressure throughout the jth rectangle to be constant and equal to $p(t_j) = (62.4)t_j$. Since the jth rectangle has area $\Delta A_j = w(t_j)\Delta h$, the force ΔF_j acting on the jth rectangle is

$$\Delta F_j = (62.4)t_j\Delta A_j = (62.4)t_jw(t_j)\Delta h, \qquad \Delta h = \frac{b - a}{n}.$$

Finally, we approximate the total force F acting on the face by the sum of these individual forces:

$$F \approx \sum_{j=1}^{n} \Delta F_j = \sum_{j=1}^{n} (62.4)t_jw(t_j)\,\Delta h. \tag{1}$$

If the width of the face is a continuous function of depth for $a \le h \le b$, then the function $(62.4)hw(h)$ is a continuous function of h. Consequently, the sum on the right-hand side of approximation (1) is a Riemann sum that must have a limit as $n \to \infty$. We *define* this limit as the total force acting on the face, that is, we define

$$F = \lim_{n \to \infty} \sum_{j=1}^{n} (62.4)t_jw(t_j)\,\Delta h = \int_{a}^{b} (62.4)hw(h)\,dh. \tag{2}$$

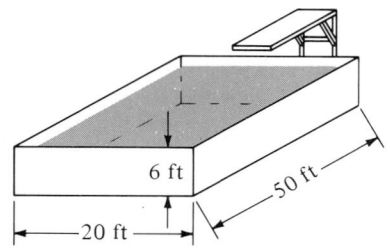

Figure 5.5 Swimming pool.

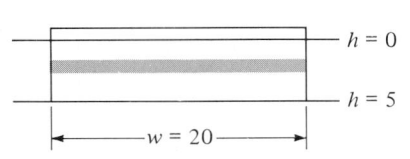

Figure 5.6 End wall of pool.

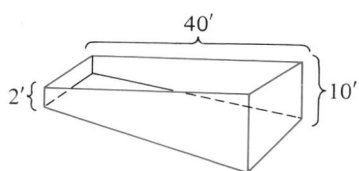

Figure 5.7 Swimming pool in Example 2.

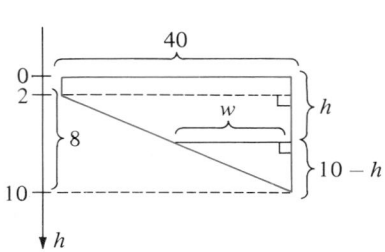

Figure 5.8 $\dfrac{w}{10-h} = \dfrac{40}{8}$, $2 \le h < 10$.

Equation (2) states that the total force (in pounds) acting on a submerged vertical face lying between depths $h = a$ and $h = b$ and whose width at depth h is $w(h)$ (all dimensions in feet) is found by integrating the product $(62.4)hw(h)$ over the interval $[a, b]$.

Example 1

An above-ground swimming pool has the shape of a rectangular box 50 feet long, 20 feet wide, and 6 feet high (Figure 5.5). Find the force exerted on an end wall when the pool is filled to a depth of 5 feet (Figure 5.6).

Solution: We take the water level to be depth $h = 0$. Then the submerged portion of the wall is a rectangle, extending from depth $h = 0$ to depth $h = 5$ feet. The width, $w = 20$ feet, is constant. An application of formula (2) gives

$$F = \int_0^5 (62.4)h(20)\, dh$$
$$= 624h^2\big]_0^5$$
$$= 15{,}600 \text{ pounds.} \qquad \diamond$$

Example 2

Figure 5.7 shows a 40-ft swimming pool 20 ft wide with a floor that slopes from a depth of 2 ft at one end of the pool to a depth of 10 ft at the other end. Find the hydrostatic force on one side wall of this pool when it is full of water.

Solution: The top 2-ft part of this wall is a rectangle of width $w(h) = 40$ ft., $0 \le h \le 2$. The force acting on this part is therefore

$$F_1 = \int_0^2 (62.4)h(40)\, dh = 2496 \cdot \frac{h^2}{2}\bigg]_0^2 = 4{,}992 \text{ lb.}$$

For depths between $h = 2$ ft and $h = 10$ ft the width of a horizontal strip of the wall depends upon its depth. By similar triangles (see Figure 5.8) we have

$$\frac{w}{10 - h} = \frac{40}{8},$$

so

$$w(h) = 5(10 - h), \qquad 2 \le h \le 10.$$

The force acting on this part of the wall is therefore

$$F_2 = \int_2^{10} (62.4)h[5(10 - h)]\, dh$$
$$= 312 \int_2^{10} h(10 - h)\, dh$$
$$= 312\left(5h^2 - \frac{1}{3}h^3\right)\bigg]_2^{10}$$
$$= 46{,}592 \text{ lb.}$$

The total force on the wall is therefore

$$F_1 + F_2 = 51{,}584 \text{ lb.} \qquad \diamond$$

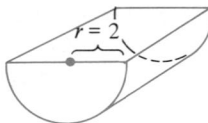

Figure 5.9 Water tank with semicircular end panels.

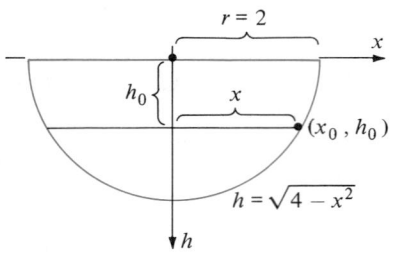

Figure 5.10 The width of a horizontal slice at depth h_0 is $w(h_0) = 2x_0 = 2\sqrt{4 - h_0^2}$.

Example 3

Figure 5.9 shows a water trough with end panels in the shape of a semicircle of radius 2 ft. Find the hydrostatic force on one of these end panels when the trough is full of water.

Solution: Figure 5.10 shows that if we impose an x-h coordinate system with origin at the center of the semicircle, then $x^2 + h^2 = 4$, and the width of a horizontal slice across the end panel at depth h is

$$w(h) = 2x = 2\sqrt{4 - h^2}.$$

According to equation (2) the force on this end panel is

$$F = \int_0^2 (62.4)h \cdot 2\sqrt{4 - h^2}\, dh.$$

To evaluate this integral we use the substitution

$$u = 4 - h^2; \qquad du = -2h\, dh,$$

and we change limits of integration according to the equations:

$$\text{if } h = 0, \qquad u = 4 - 0^2 = 4$$
$$\text{if } h = 2, \qquad u = 4 - 2^2 = 0.$$

The integral then becomes

$$F = \int_4^0 (62.4)(-1)\sqrt{u}\, du$$

$$= -\frac{2}{3}(62.4)u^{3/2}\Big]_4^0$$

$$= \frac{2}{3}(62.4)(8)$$

$$= 332.8 \text{ lb.} \qquad \Diamond$$

Exercise Set 7.5

1. The vertical ends of a water trough are rectangles 3 feet wide and 18 inches high. Find the hydrostatic force on the end panels when the trough is full of water.

2. A cardboard container for mineral water has a square base and rectangular sides 5 inches wide and 10 inches high. Find the force acting on a side when the carton is full of water.

3. The vertical ends of a water trough are isosceles right triangles with the right angles at the bottom. Find the hydrostatic force acting on an end when the trough is filled to a depth of 2 feet (Figure 5.11).

4. The vertical ends of a water trough are equilateral triangles whose sides are each 2 feet long. Find the hydrostatic force on an end panel when the trough is full of water.

5. The wall of a dam has the shape of the parabola $y = \frac{1}{100}x^2$.

   Find the hydrostatic force acting on the wall if the wall is 80

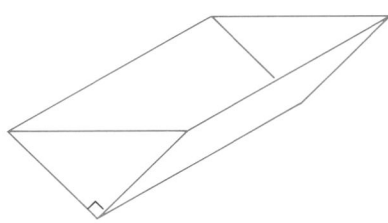

Figure 5.11

feet wide across its top and the dam is full of water. (*Hint:* The given equation assumes a coordinate system whose origin lies at the bottom of the wall. You have to find the depth h from the top of the wall.)

6. The end panels of a water trough are rectangles 2 feet wide and one foot high. To what depth must the trough be filled so that the force acting on an end panel is exactly half the force acting on the end panel when the trough is full?

7. Find the hydrostatic force on an end panel of a water trough in the shape of the lower half of an ellipse with equation

$$\frac{x^2}{16} + \frac{y^2}{4} = 1 \qquad (x \text{ and } y \text{ are in units of feet}).$$

8. A submarine has rectangular windows 10 inches wide and 5 inches high. Find the hydrostatic force on such a window when the top of the window lies 30 feet below water level.

9. Find the hydrostatic force on the submarine window in Exercise 8 if it has the shape of an equilateral triangle with side $s = 10$ inches, bottom side horizontal, and the top of the window 30 feet below water level.

10. Suppose an object is submerged in a fluid weighing c pounds per cubic foot. How should equation (2) be modified so as to give the force acting on a face of this object?

11. Use the result of Exercise 10 to find the force acting on one wall of a carton of heavy cream weighing 72 pounds per cubic foot if the walls of the carton are rectangles of width 4 inches and height 6 inches and the carton is full of cream.

12. A fuel oil truck has a tank in the shape of a cylinder whose base is described by the ellipse $\frac{x^2}{64} + \frac{y^2}{36} = 1$. The tank is mounted horizontally on the truck (i.e., the elliptical end panels are vertical). Find the force acting on one end panel

when the tank is half filled with fuel oil weighing 58 pounds per cubic foot (see Exercise 10).

13. (Calculator/Computer) Use Simpson's Rule to approximate the force acting on an end panel of the oil tank in Exercise 12 when the tank is full of oil.

14. If a rectangular plate is suspended in water, the theory of this section enables us to compute the hydrostatic force acting on one face of the plate. Why does the plate not move as a result of this force?

In each of Exercises 15–17, compute the hydrostatic force on a vertical plate of the given shape submerged to the given depth d below water level. In each case, compare your result with the product $F = (62.4)A \cdot h$ where A is the area of the figure and h is the depth of the center of the object below water level.

15. A rectangle of width 15 inches and height 10 inches lying $d = 2$ feet below water level.

16. A square of side 2 feet lying $d = 6$ feet below water level.

17. A rectangle of width 3 feet and height 2 feet lying 7 feet below water level.

18. What do you conjecture, based on the results of Exercises 15–17, about the relationship between the area, depth of center, and hydrostatic force acting on a submerged plate? Do the results of Exercise 13 support this conjecture? (These ideas are pursued in Section 7.7 on Centers of Gravity.)

7.6 WORK

When a constant force F moves an object through a distance d, the product

$$W = Fd \tag{1}$$

is defined as the **work** done on that object. For example, in lifting a 3-pound textbook 2 feet you do an amount of work equal to $W = (2 \text{ feet}) \times (3 \text{ lb}) = 6$ foot-pounds (Figure 6.1).

The choice of the unit foot-pounds reflects the definition of work as the product of force times distance. If force and distance are measured in units other than feet and pounds, the units for work are modified appropriately (inch-pounds or newton-meters, for example).

In many practical situations, the calculation of work cannot be done by a simple multiplication because the force F required to move an object is not constant. For example, the force required to raise a ship's anchor depends on the length of cable, ℓ, between the ship and the anchor (Figure 6.2). This is because the cable usually has substantial weight, and both the cable and the anchor must be raised.

When the motion of the object is along a line, the force required to move the object can sometimes be written as a function F of the object's position x along the line (as we will see in Example 1). In such cases we can generalize equation (1) by use of the definite integral, as follows.

Suppose that an object is to be moved from location $x = a$ to location $x = b$. We can approximate the work done on the object over the entire interval $[a, b]$ by partitioning $[a, b]$ into small subintervals, assuming $F(x)$ to be constant on each

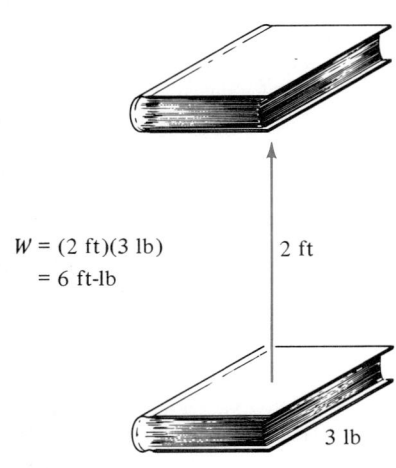

$W = (2 \text{ ft})(3 \text{ lb})$
$= 6 \text{ ft-lb}$

2 ft

3 lb

Figure 6.1

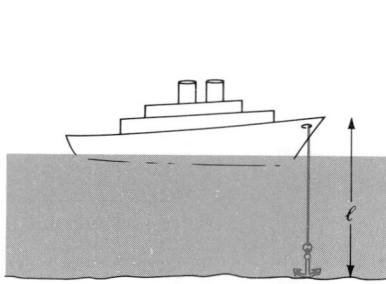

Figure 6.2 Force required to raise anchor depends upon ℓ.

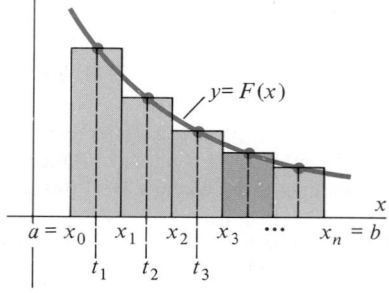

Figure 6.3 Approximating work done over $[a, b]$ by variable force $F(x)$.

subinterval, and then summing the resulting values of work required on each subinterval (Figure 6.3). To do so we choose a partition $a = x_0 < x_1 < x_2 < \cdots < x_n = b$ so that $x_j - x_{j-1} = \Delta x = \dfrac{b - a}{n}$ for each $j = 1, 2, \ldots, n$, and we select one number t_j in each interval $[x_{j-1}, x_j]$. If we approximate the force $F(x)$ throughout the interval $[x_{j-1}, x_j]$ by the constant $F(t_j)$, then by equation (1) the work ΔW_j required to move the object from x_{j-1} to x_j is the product $\Delta W_j = F(t_j)(x_j - x_{j-1}) = F(t_j)\,\Delta x$. By summing these approximations, we obtain the approximation for the work done in moving the object from $x = a$ to $x = b$ as

$$W \approx \sum_{j=1}^{n} \Delta W_j = \sum_{j=1}^{n} F(t_j)\,\Delta x. \tag{2}$$

If the force function F is continuous for $x \in [a, b]$, then the sum on the right-hand side of approximation (2) is a Riemann sum that has a limit as $n \to \infty$. Since our approximation was designed to provide an increasingly accurate measurement of the actual value of the work as $n \to \infty$ (and, hence, as the size of the intervals approaches zero), we *define* work to be the limiting value of this approximating sum, that is,

$$W = \lim_{n \to \infty} \sum_{j=1}^{n} F(t_j)\,\Delta x = \int_{a}^{b} F(x)\,dx. \tag{3}$$

Equation (3) states that the work done by a variable force in moving an object along a line from location $x = a$ to location $x = b$ is found by integrating the force function from $x = a$ to $x = b$. Figure 6.3 illustrates a typical force function and the geometric interpretation of our approximation scheme. Although the region bounded by the graph of F does not correspond to any physical aspect of the work problem, the area of that region is precisely the work being calculated.

Example 1

A principle from elementary physics known as **Hooke's Law** states that the force required to stretch or compress a spring a distance of x units from its natural length is proportional to that distance, that is,

$$F(x) = kx \qquad \text{(Hooke's Law)}. \tag{4}$$

Here k is a constant, called the **spring constant,** which depends on the particular spring. Since the force varies with distance, calculating the work done in stretching or compressing a spring requires use of equation (3) as the following examples illustrate.

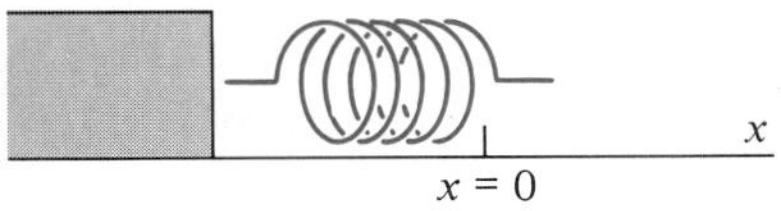

(a) A spring has spring constant $k = 20$ lb/ft. The work done in stretching the spring 6 inches beyond its natural length is found by use of equation (3) and (4):

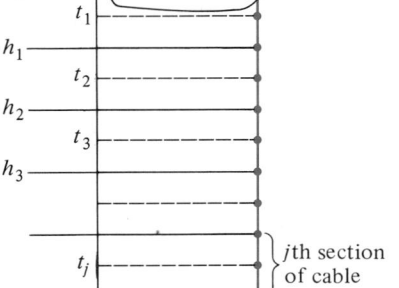

Figure 6.4 Spring in natural state.

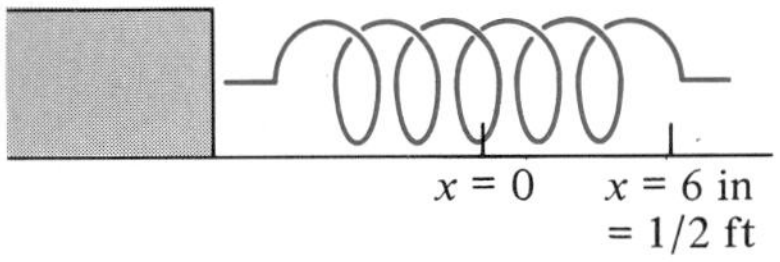

Figure 6.5 Spring stretched· 6 in $= \dfrac{1}{2}$ ft.

$$W = \int_0^{1/2} F(x)\, dx$$

$$= \int_0^{1/2} 20x\, dx$$

$$= 10x^2 \big]_0^{1/2}$$

$$= 10\left(\frac{1}{4} - 0\right)$$

$$= 2.5 \text{ foot-pounds.}$$

(Figures 6.4 and 6.5.)

(b) The work done in stretching the spring in part (a) from 3 inches $= \frac{1}{4}$ ft beyond its natural length to 6 inches $= \frac{1}{2}$ ft beyond its natural length is

$$W = \int_{1/4}^{1/2} 20x\, dx = 10x^2 \Big]_{1/4}^{1/2} = \frac{15}{8} \text{ foot-pounds.}$$

(c) Suppose that 4 foot-pounds of work are required to stretch a spring 1 foot beyond its natural length. We may calculate the spring constant from these data since

$$4 = \int_0^1 kx\, dx = \frac{k}{2} x^2 \Big]_0^1 = k/2.$$

Thus, $4 = k/2$, so $k = 8$ lb/ft. $\diamond$

Example 2

Suppose that a ship's anchor weighs 2 tons (4000 pounds) in water and that the anchor is hanging taut from 100 feet of cable. Find the work required to wind in the anchor if the cable weighs 20 pounds per foot in water.

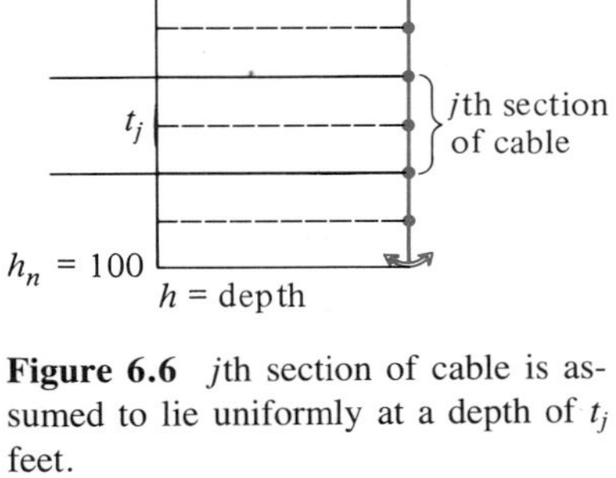

Solution: This example does not quite fit the general framework within which equation (3) was developed. Although the anchor is being moved from depth $h = 100$ to depth $h = 0$, the cable itself is not being lifted 100 feet. Rather, each "slice" of cable lying at depth h feet is being lifted only h feet. To approximate the work done on the entire cable, we can partition the interval [0, 100] as usual, with equally spaced numbers $h_0 = 0 < h_1 < h_2 < \cdots < h_n = 100$ (Figure 6.6). Then the force required to lift each section of cable, ΔF_j, is its weight, that is,

$$\Delta F_j = (20 \text{ lb/ft})(\Delta h \text{ feet}).$$

By selecting one number t_j in each interval $[h_{j-1}, h_j]$, we approximate the work required to lift the jth section of cable to the surface by the product of the force ΔF_j

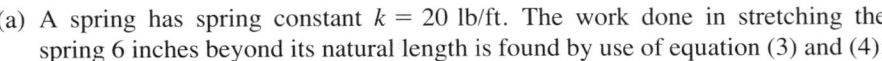

Figure 6.6 jth section of cable is assumed to lie uniformly at a depth of t_j feet.

and the distance t_j, that is,

$$\Delta W_j \approx \Delta F_j \cdot t_j$$
$$= 20t_j \, \Delta h \text{ ft-lb.}$$

The work W_c required to raise the cable is therefore approximated by the sum

$$W_c \approx \sum_{j=1}^{n} \Delta W_j = \sum_{j=1}^{n} 20t_j \, \Delta h. \tag{5}$$

Since the expression on the right side of equation (5) is a Riemann sum for the function $f(h) = 20h$, we conclude that

$$W_c = \lim_{n \to \infty} \sum_{j=1}^{n} 20t_j \, \Delta h \tag{6}$$

$$= \int_{0}^{100} 20h \, dh.$$

Since the work required to raise a 4000-lb anchor 100 feet is $W_a = (4000 \text{ lb}) \times (100 \text{ ft}) = 400{,}000$ ft-lb, we obtain the total work required as

$$W = W_a + W_c = 400{,}000 \text{ ft-lb} + \int_{0}^{100} 20h \, dh$$

$$= 400{,}000 \text{ ft-lb} + 10h^2 \big]_{0}^{100}$$
$$= 500{,}000 \text{ ft-lb.} \qquad \diamondsuit$$

REMARK: Notice that the integral obtained for the work W_c required to raise the cable in equation (6) cannot be obtained by applying equation (3). The reason is that equation (3) applies to a variable force acting to move an object from location $x = a$ to location $x = b$. However, equation (6) refers to moving a one-dimensional object of uniform weight through a *variable* distance. We could attempt to generalize equation (6), but the wording would be clumsy and, necessarily, vague. Instead, you should note the following cautions:

(i) *Equation (3) may be applied only when a variable force $F(x)$ moves the point at which it is applied from location $x = a$ to location $x = b$.* (Thus, equation (3) was correctly applied in Example 1, since the point of application corresponded to the end of the spring. However, in Example 2 there was no such "point of application.")
(ii) In all other problems involving the calculation of work, you should
    (a) partition the relevant interval,
    (b) approximate the work required in each interval,
    (c) write the resulting approximation to the total work as a Riemann sum, and
    (d) obtain the appropriate integral as the limit of the approximating Riemann sum.

Example 3

A swimming pool is 40 feet long and 20 feet wide. The floor of the pool has a constant slope from a depth of 2 feet at one end to a depth of 10 feet at the other. Find the work required to pump all the water out through a valve located at the top edge of the pool when the pool is full (Figure 6.7).

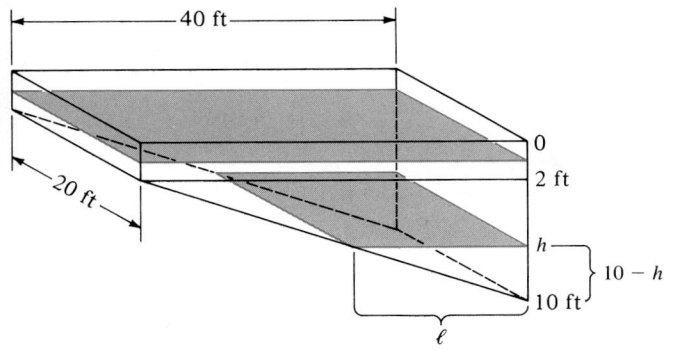

Figure 6.7 Swimming pool with sloping floor.

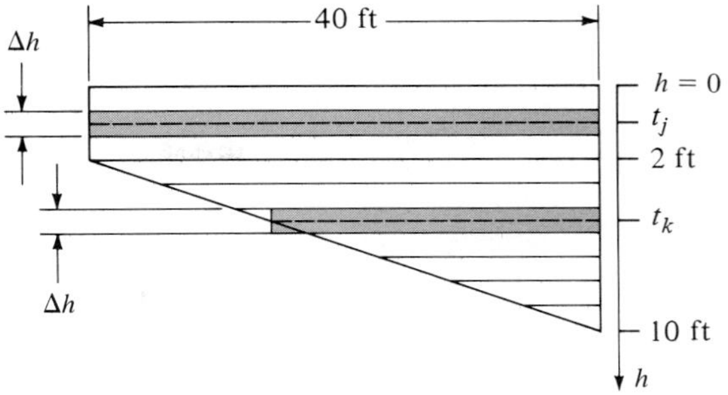

Figure 6.8 Partitioning the depth axis.

Strategy

Label the independent variable.

Treat the problem in two parts according to the geometry.

Partition the interval [0, 2] and approximate the volume of each slab.

$\Delta F = (62.4)\, \Delta V$

$\Delta W = t \cdot \Delta F.$

Sum the individual approximations. The result is a Riemann sum.

Obtain the integral as the limit of the Riemann sum.

Use similar triangles to find length as a function of depth.

Solution

Let the variable h denote depth. We imagine the result of partitioning the interval [0, 10] and slicing the water volume by horizontal planes into slabs of thickness Δh. As Figure 6.8 illustrates, the resulting slabs are of two types, those with fixed length 40 ft (lying above depth 2 ft) and those with variable length (lying below depth 2 ft). We treat the two types separately.

For depths t_j with $0 \le t_j \le 2$, the slab of thickness Δh has volume

$$\begin{aligned} \Delta V_j &= (20\ \text{ft})(40\ \text{ft})(\Delta h\ \text{ft}) \\ &= 800\ \Delta h\ \text{ft}^3. \end{aligned}$$

The force (weight) required to lift this slab is therefore

$$\begin{aligned} \Delta F_j &= (62.4\ \text{lb/ft}^3)(800\ \Delta h\ \text{ft}^3) \\ &= 49{,}920\ \Delta h\ \text{lb}. \end{aligned}$$

If we assume this slab to be concentrated at depth t_j, the work required to lift it to the top of the pool will be

$$\Delta W_j = t_j\, \Delta F_j = 49{,}920 t_j\, \Delta h\ \text{ft-lb}.$$

Assuming the interval [0, 2] to have been partitioned into n subintervals, the work W_T required to pump out the top 2 feet of water is approximated by

$$W_T \approx \sum_{j=1}^{n} \Delta W_j = \sum_{j=1}^{n} 49{,}920 t_j\, \Delta h,$$

so

$$W_T = \int_{0}^{2} 49{,}920 h\ dh.$$

For depths t_k with $2 \le t_k \le 10$, the length ℓ_k of the slab depends on its depth. By similar triangles (Figure 6.9), we have

$$\frac{\ell_k}{10 - t_k} = \frac{40}{8},$$

so

$$\ell_k = 5(10 - t_k).$$

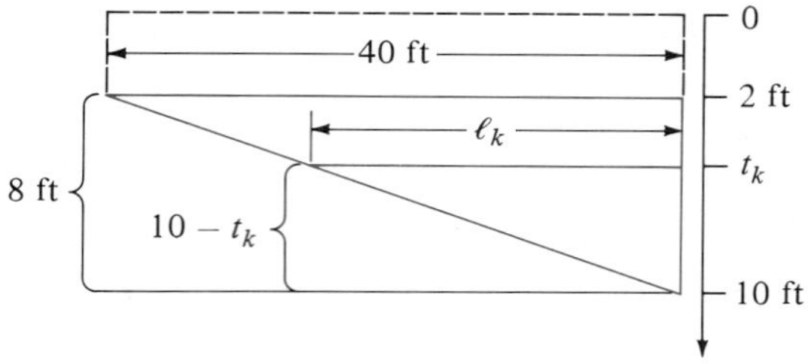

Figure 6.9 Similar triangles give length ℓ_k at depth t_k.

Proceed as in the first part.

A slab of water of thickness Δh assumed to be concentrated at depth t_k therefore has volume

$$\begin{aligned}\Delta V_k &= (20 \text{ ft})[5(10 - t_k)\text{ft}](\Delta h \text{ ft})\\ &= 100(10 - t_k)\ \Delta h \text{ ft}^3\end{aligned}$$

and requires a lifting force to overcome its weight of

$$\begin{aligned}\Delta F_k &= (62.4 \text{ lb/ft}^3)[100(10 - t_k)\ \Delta h \text{ ft}^3]\\ &= 6240(10 - t_k)\ \Delta h \text{ lb.}\end{aligned}$$

The work required to lift this slab to the top edge is therefore

$$\Delta W_k = 6240t_k(10 - t_k)\ \Delta h \text{ ft-lb.}$$

Summing over all slabs lying between depths 2 ft and 10 ft (assuming n such slabs) gives the approximation to W_B, the work required to pump out all water lying below 2 ft:

$$W_B \approx \sum_{k=1}^{n} \Delta W_k = \sum_{k=1}^{n} 6240t_k(10 - t_k)\ \Delta h.$$

Thus

$$W_B = \int_2^{10} 6240h(10 - h)\ dh.$$

Combine results.

Combining the expressions for W_T and W_B then gives the total work W required as

$$\begin{aligned}W &= W_T + W_B\\ &= \int_0^2 49{,}920h\ dh + \int_2^{10} 6240h(10 - h)\ dh\\ &= 24{,}960h^2]_0^2 + [31{,}200h^2 - 2080h^3]_2^{10}\\ &= 1{,}031{,}680 \text{ ft-lb}\\ &= 515.84 \text{ ft-tons.}\end{aligned}$$

1 ft-ton = 2000 ft-lb.

◇

Exercise Set 7.6

1. How much work is done in compressing a spring with spring constant $k = 40$ pounds per foot a distance of 6 inches?

2. A spring has spring constant $k = 5$ newtons per meter. How much work is done in stretching the spring 80 cm beyond its natural length? (Here work will be in units of newton · meters = joules.)

3. How much work is done in compressing the spring in Exercise 1 another 6 inches?

4. How much work is done in stretching the spring in Exercise 2 another 40 cm?

5. A force of 50 pounds stretches a spring 4 inches. Find the work required to stretch the spring an additional 4 inches.

6. If 40 foot-pounds of work are required to stretch a spring 6 inches beyond its natural length, what is its spring constant?

7. True or false? If W_1 is the work done on an object in moving it from point A to point B, and W_2 is the work done on the same object in moving it from point B to point C, then the work W done in moving the object from point A to point C satisfies the equation $W = W_1 + W_2$. What property of definite integrals is involved in this question?

8. The work required to stretch a spring 6 centimeters beyond its natural length is $W = 2$ ergs. What is the value of the spring constant for this spring? (The units for this spring constant will be dynes per centimeter, where 1 erg = 1 dyne · centimeter.)

9. A spring hangs from one end, which is attached to a supporting beam. When a 10-pound weight is attached to the free end of the spring it stretches a distance of 2 feet.

    a. Find the spring constant k.
    b. Find the additional weight that must be added so that the spring will be stretched 3 feet beyond its natural length.
    c. If both weights are removed from the end of the spring, what is the work required to again stretch the spring a distance of 3 feet beyond its natural length?

10. A children's wading pool is 6 feet wide, 10 feet long, and $\frac{1}{2}$ foot deep. How much work is required to pump all the water out through a valve at the top edge of the pool when the pool is full of water?

11. A conical water tank has radius $r = 10$ feet and height $h = 12$ feet and is mounted with its base horizontal and its tip pointing downward. How much work is required to pump the water out through a valve in the top of the tank when the tank is full?

12. A water tank has the shape of a right circular cylinder with radius $r = 10$ feet and depth $h = 20$ feet. It is mounted with its axis vertical. The tank contains 10 feet of water. Find the work done in pumping this water out through a valve in the top of the tank.

13. Find the work done in pumping the water out of the tank in Exercise 12 when the tank is full of water.

14. A water tank has the shape of a right circular cone of height $h = 20$ feet and radius $r = 6$ feet. It is mounted with its axis vertical and tip down. It contains 4 feet of water. Find the work done in pumping this water out through a valve in the top of the tank.

15. Find the work done in pumping the water out of the tank in Exercise 14 if the tank is full of water.

16. Find the work done in Exercise 12 if the water is pumped to a height 10 feet above the top of the tank.

17. Find the work done in Exercise 14 if the water is pumped to a height of 15 feet above the top of the tank.

18. A hemispherical water tank has radius $r = 5$ feet. The tank is mounted with its circular base on top, lying horizontally. How much work is required to pump the water out through a valve on the top edge of the tank when the tank is full?

19. A 20-foot section of rope weighing 2 pounds per foot is hanging from a windlass (crank). How much work is done in winding up the entire length of rope?

20. A chain weighs 40 newtons per meter. How much work is done in winding a 30-meter section of chain hanging from a windlass?

21. Coulomb's Law governing the force of attraction between two electrically charged bodies states that the force of attraction or repulsion between two point charges is directly proportional to the product of the charges and inversely proportional to the square of the distance between them, that is,

$$F = K\frac{q_1 q_2}{r^2}$$

where q_1 and q_2 are the magnitudes of the charges at points P_1 and P_2, respectively, r is the distance between P_1 and P_2, and K is the constant of proportionality. Suppose that charges of equal magnitude q exist at points P_1 and P_2 which are 4 meters apart, and that the force with which the two charges repel each other is known to be 10 newtons.

    a. What is the value of the constant K in terms of q?
    b. How much work is required to bring one charge from point P_2 to a distance of 2 meters from point P_1?

22. Suppose that an ion with charge $+1$ is stationary and located at $(5, 0)$. How much work is required to move a second ion with charge $+1$ from $(0, 0)$ to $(4, 0)$? (See Exercise 21.)

23. What is the answer to Exercise 22 if a third ion with charge $+1$ is located at $(-8, 0)$ and stationary?

24. The weight of an object varies inversely with the square of its distance from the center of the earth, that is,

$$w(r) = \frac{k}{r^2}$$

where $w(r)$ is the weight of the object when located at a distance r from the earth's center. A satellite weighs 5 tons at the earth's surface, when it is 4000 miles from the earth's center.

    a. Find the value of k for this satellite.
    b. Find the amount of work done in propelling this satellite to a height of 100 miles above the earth's surface.

7.7 MOMENTS AND CENTERS OF GRAVITY

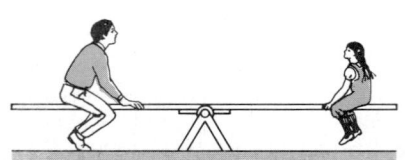

Figure 7.1 A larger weight at a smaller distance balances a smaller weight at a larger distance.

Most children can identify the two factors that determine whether they can "balance" another child seated opposite them on a seesaw: their weight and the distance at which they are seated from the pivot (Figure 7.1). Increasing either of these two quantities increases the tendency of the seesaw to rotate their end downward. We wish here to pursue the mathematical generalizations of these ideas.

We idealize the situation of the seesaw by imagining two objects of weight w_1 and w_2 placed on a flat weightless rod, which in turn is mounted on a fulcrum (pivot). The rod is assumed to be free to rotate about the fulcrum, and we assume the weights of the objects to be located at distances ℓ_1 and ℓ_2 from the fulcrum, respectively (Figure 7.2).

The tendency of the rod to rotate in the counterclockwise direction is measured by the **torque** $\ell_1 w_1$. Similarly, the torque $\ell_2 w_2$ measures the tendency of the rod to rotate in the clockwise direction. A physical principle called the **law of the lever** states that the rod will balance (equilibrium will occur) when the opposing torques are equal, that is, when

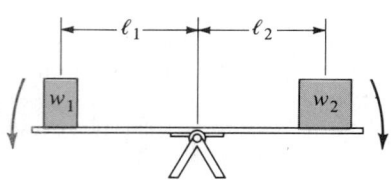

Figure 7.2 Weights balanced on a fulcrum.

$$\ell_1 w_1 = \ell_2 w_2. \tag{1}$$

We can further simplify the model by introducing a one-dimensional coordinate system along the rod (Figure 7.3). If we denote the point corresponding to the fulcrum as $x = 0$ and take the positive x-axis as the direction in which weights produce clockwise rotation, then for $x_1 = -\ell_1$ and $x_2 = \ell_2$ the equilibrium equation (1) becomes

$$x_1 w_1 + x_2 w_2 = 0. \tag{2}$$

Equation (2) can now be generalized to the condition for equilibrium for n weights $w_1, w_2, \ldots, w_n$ located at positions $x_1, x_2, \ldots, x_n$, respectively (Figure 7.4). The condition is

$$x_1 w_1 + x_2 w_2 + \cdots + x_n w_n = 0$$

or

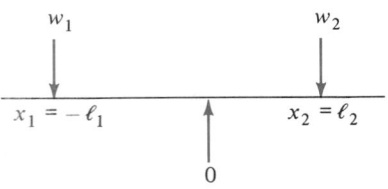

Figure 7.3 Assigning coordinates to the lever.

$$\sum_{j=1}^{n} x_j w_j = 0. \tag{3}$$

Finally, we observe that if the fulcrum is located at $x = \bar{x}$ rather than at $x = 0$ (which corresponds to moving the origin of the coordinate system, but *does not* change the physical situation), the (signed) distance of the weight w_j away from the fulcrum becomes $(x_j - \bar{x})$ rather than x_j (Figure 7.5). In this case the equilibrium condition (3) becomes

$$\sum_{j=1}^{n} (x_j - \bar{x}) w_j = 0. \tag{4}$$

Since we can write equation (4) as

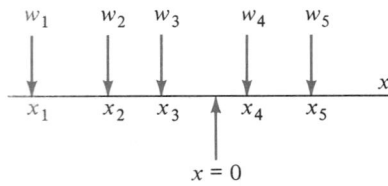

Figure 7.4 Equilibrium condition is $\Sigma x_j w_j = 0$ for fulcrum at $x = 0$.

$$\sum_{j=1}^{n} x_j w_j - \bar{x} \sum_{j=1}^{n} w_j = 0,$$

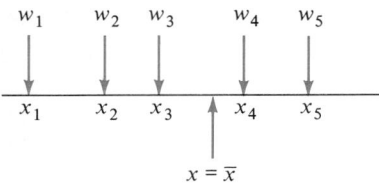

Figure 7.5 Fulcrum at $\bar{x}$. (Note that all we have done is change the label from 0 to $\bar{x}$.)

we may solve for $\bar{x}$ as

$$\bar{x} = \frac{\sum_{j=1}^{n} x_j w_j}{\sum_{j=1}^{n} w_j}. \tag{5}$$

Equation (5) specifies the location of the fulcrum in our idealized balance problem when equilibrium occurs. The advantage of this version of the equilibrium condition is that it enables us to calculate $\bar{x}$ from the weights $w_1, w_2, \ldots, w_n$ and the locations $x_1, x_2, \ldots, x_n$. Not surprisingly, $\bar{x}$ is referred to as the **center of gravity** of the system so described.

Example 1

A system of weights $w_1 = 10$ lb, $w_2 = 20$ lb, $w_3 = 10$ lb, $w_4 = 20$ lb, and $w_5 = 25$ lb is located along a line at points $x_1 = -6$ ft, $x_2 = -2$ ft, $x_3 = 1$ ft, $x_4 = 3$ ft, and $x_5 = 6$ ft, respectively. Find the center of gravity.

Solution:

$$\bar{x} = \frac{[(-6)(10) + (-2)(20) + (1)(10) + (3)(20) + (6)(25)] \text{ ft-lb}}{(10 + 20 + 10 + 20 + 25) \text{ lb}}$$

$$= \frac{24}{17} \text{ ft.} \qquad \diamond$$

REMARK 1: The physical definition of the weight w of an object is $w = mg$, where m is the mass of the object and g is the acceleration due to gravity. Since g is a constant, w and m are proportional. Thus, $\bar{x}$ in equation (5) is also the **center of mass** for the n particles whose masses are $m_1 = w_1/g$, $m_2 = w_2/g$, $\ldots$, $m_n = w_n/g$, since all factors of g in equation (5) will cancel. Henceforth we will refer exclusively to centers of mass.

REMARK 2: If we write the equilibrium equation (4) in terms of masses $m_j = w_j/g$, we obtain the equation

$$\sum_{j=1}^{n} (x_j - \bar{x})m_j = 0. \tag{6}$$

The sum on the left side of equation (6) is called the **first moment** of the mass system about the number $\bar{x}$. The first moment of a system about a number may be regarded as a measurement of the tendency of an idealized rod supported at $\bar{x}$ and with masses m_j at locations x_j to rotate, as in the earlier discussion involving torque. Since equation (5), written for masses $m_j = w_j/g$, becomes

$$\bar{x} = \frac{\sum_{j=1}^{n} x_j m_j}{\sum_{j=1}^{n} m_j}, \tag{7}$$

we may interpret $\bar{x}$ as *the first moment of the mass system about* $x = 0$ *divided by the total mass.*

The preceding ideas can be used *to calculate the center of mass of certain three-dimensional objects.*

The Center of Mass of a Rod

Imagine a cylindrical rod of length ℓ and uniform cross-sectional area A. Imagine also that the density ρ of the material is constant across any cross section of the rod, but that the density varies continuously along the rod.

Since the density of the rod is not uniform, the center of mass will not necessarily lie at the midpoint. To approximate the center of mass $\bar{x}$, we position the rod as in Figure 7.6 and we partition the interval $[0, \ell]$ into n subintervals of equal length

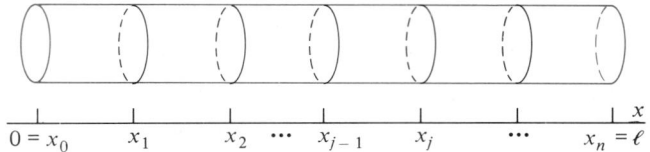

$$0 = x_0 \quad x_1 \quad x_2 \quad \cdots \quad x_{j-1} \quad x_j \quad \cdots \quad x_n = \ell$$

Figure 7.6 Rod of constant cross-sectional area and varying density.

with endpoints $0 = x_0 < x_1 < x_2 < \cdots < x_n = \ell$. By slicing the rod with planes perpendicular to the axis through each of the points $x_0, x_1, x_2, \ldots, x_n$, we divide the rod into n cylinders, each of length $\Delta x = x_j - x_{j-1} = \dfrac{\ell}{n}$ and of cross-sectional area A. The volume of each small cylinder is therefore $V_j = A\,\Delta x$.

Since the small cylinders are not of uniform density, we select one number t_j in each interval $[x_{j-1}, x_j]$ and *assume* that $\rho(x) \equiv \rho(t_j)$ throughout the interval $[x_{j-1}, x_j]$. This assumption allows us to approximate the mass m_j of the jth cylinder as

$$m_j \approx \rho(t_j)V_j = \rho(t_j)A\,\Delta x \qquad \text{(mass = density} \times \text{volume).} \tag{8}$$

Finally, we assume the mass m_j of each cylinder to be concentrated at the point $t_j \in [x_{j-1}, x_j]$. With these assumptions, we have described a system consisting of a finite number of masses, so we may apply the equilibrium condition of equation (6) for the center of mass $\bar{x}$. We obtain, from equation (6) and approximation (8), the condition

$$0 = \sum_{j=1}^{n}(t_j - \bar{x})m_j \approx \sum_{j=1}^{n}(t_j - \bar{x})\rho(t_j)A\,\Delta x. \tag{9}$$

If ρ is continuous on $[0, \ell]$, the sum in equation (9) is a Riemann sum for the function $f(x) = (x - \bar{x})\rho(x)A$. Thus,

$$0 = \lim_{n \to \infty} \sum_{j=1}^{n}(t_j - \bar{x})\rho(t_j)A\,\Delta x$$

$$= \int_0^\ell (x - \bar{x})\rho(x)A\,dx$$

$$= \int_0^\ell x\rho(x)A\,dx - \bar{x} \cdot \int_0^\ell \rho(x)A\,dx.$$

We can now solve for $\bar{x}$. We obtain

$$\bar{x} = \frac{\displaystyle\int_0^\ell x\rho(x)A \, dx}{\displaystyle\int_0^\ell \rho(x)A \, dx}. \tag{10}$$

Equation (10) gives the *center of mass for a rod of constant cross-sectional area A and density ρ(x)*. By analogy with the discrete case described by equation (7), the integral in the numerator is called the first moment of the rod about the endpoint $x = 0$, and the integral in the denominator is the total mass.

Example 2

A rod 10 cm long has uniform cross-sectional area $A = 4$ cm^2. Find the center of mass of the rod if

(a) the density ρ is uniform, or
(b) the density $\rho(x)$ varies linearly from 3 grams per cubic centimeter at one end to 6 grams per cubic centimeter at the other.

Solution: (a) We may apply equation (10) with $\rho(x) \equiv \rho$ (constant) to obtain

$$\bar{x} = \frac{\displaystyle\int_0^{10} \rho \cdot 4 \cdot x \, dx}{\displaystyle\int_0^{10} \rho \cdot 4 \, dx} = \frac{4\rho\left[\dfrac{1}{2}x^2\right]_0^{10}}{4\rho[x]_0^{10}} = \frac{4\rho \cdot 50}{4\rho \cdot 10} = 5,$$

which is the midpoint of the rod, as expected.

(b) For the variable density case, we must find an equation representing the density function. The statement that the density $\rho(x)$ varies linearly from $\rho(0) = 3$ to $\rho(10) = 6$ means that $\rho(x) = cx + d$ for some constants c and d. Since $\rho(0) = c \cdot 0 + d = 3$, we have $d = 3$. Then, from $\rho(10) = c \cdot 10 + 3 = 6$, we obtain $c = \dfrac{3}{10}$. The density function is therefore $\rho(x) = \dfrac{3}{10}x + 3$. Thus, from (10),

$$\bar{x} = \frac{\displaystyle\int_0^{10} 4x\left[\frac{3}{10}x + 3\right] dx}{\displaystyle\int_0^{10} 4\left[\frac{3}{10}x + 3\right] dx} = \frac{\displaystyle\int_0^{10}\left(\frac{6}{5}x^2 + 12x\right) dx}{\displaystyle\int_0^{10}\left(\frac{6}{5}x + 12\right) dx}$$

$$= \frac{\dfrac{2}{5}x^3 + 6x^2\Big]_0^{10}}{\dfrac{3}{5}x^2 + 12x\Big]_0^{10}}$$

$$= \frac{50}{9} \approx 5.56.$$

Because of the variation of the density, the center of mass is closer to the high-density end of the rod. ◇

Example 3

Find an equation for the center of mass $\bar{x}$ of a rod of length ℓ of constant density ρ whose cross-sectional area $A(x)$ is a continuous function for $0 \le x \le \ell$.

Solution: The differences between this question and the question answered by equation (10) are that (i) the density ρ is now assumed constant and (ii) the cross-sectional area A is now assumed to vary. We proceed with the same analysis as before, except that in approximation (8) we approximate the volume V_j by assuming the cross-sectional area $A(t_j)$ to be uniform throughout the jth cylinder. Thus, the analogue of approximation (8) is

$$m_j = \rho V_j \approx \rho A(t_j)\,\Delta x.$$

The analogue of equation (9) is

$$0 = \sum_{j=1}^{n}(t_j - \bar{x})m_j \approx \sum_{j=1}^{n}(t_j - \bar{x})\rho A(t_j)\,\Delta x,$$

which, in the limit as $n \to \infty$, gives

$$\int_0^\ell (x - \bar{x})\rho A(x)\,dx = 0.$$

Solving this equation for $\bar{x}$ produces the desired equation

$$\bar{x} = \frac{\displaystyle\int_0^\ell \rho x A(x)\,dx}{\displaystyle\int_0^\ell \rho A(x)\,dx} \tag{11}$$

for *the center of mass (or gravity) along the x-axis of a rod of uniform density and continuously varying cross-sectional area A(x)* (Figures 7.7 and 7.8). ◇

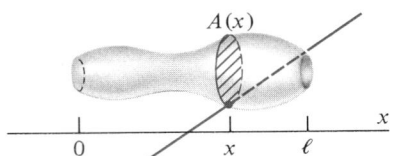

Figure 7.7 Rod of constant density and variable cross-sectional area.

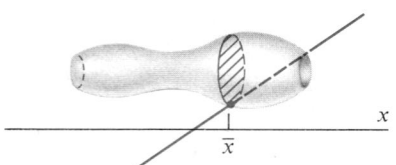

Figure 7.8 Rod "balances" at center of mass (gravity) $x = \bar{x}$.

Center of Mass of a Plate

We may apply the result of Example 3 to a thin plate of uniform density ρ and uniform thickness c. Suppose that the plate lies in the xy-plane and that the face is bounded above by the graph of $y = f(x)$ and below by the graph of $y = g(x)$ for $a \le x \le b$ (Figure 7.9). Then the area of a cross section taken perpendicular to the x-axis is $A(x) = c[f(x) - g(x)]$ (Figure 7.10). Regarding the plate as a rod of con-

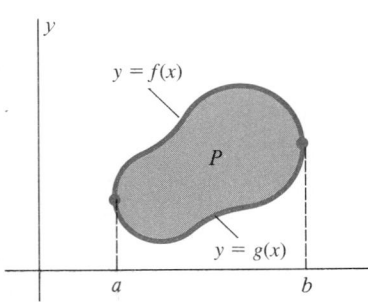

Figure 7.9 Face of thin plate bounded by graphs of $y = f(x)$ and $y = g(x)$.

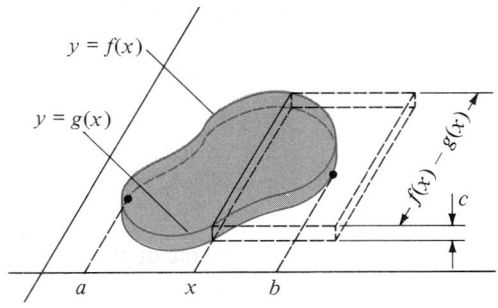

Figure 7.10 Cross section perpendicular to x-axis has area $A(x) = [f(x) - g(x)]c$.

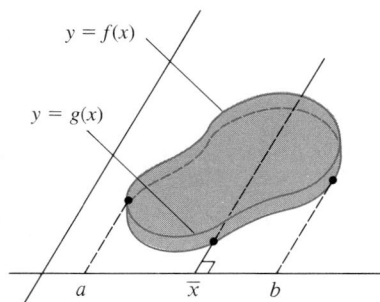

Figure 7.11 Plate ''balances'' at perpendicular to x-axis through center of mass $\bar{x}$.

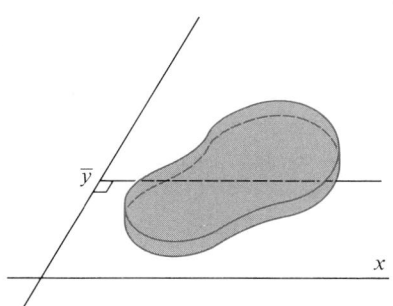

Figure 7.12 Plate ''balances'' at perpendicular to y-axis through center of mass $\bar{y}$.

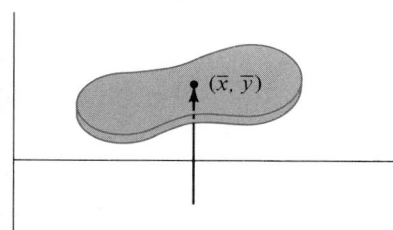

Figure 7.13 Plate balances at centroid $(\bar{x}, \bar{y})$.

stant density and variable cross-sectional area, we apply equation (11) to find the x-coordinate of the center of mass $\bar{x}$:

$$\bar{x} = \frac{\int_a^b c\rho x[f(x) - g(x)]\, dx}{\int_a^b c\rho[f(x) - g(x)]\, dx} = \frac{\int_a^b x[f(x) - g(x)]\, dx}{\int_a^b [f(x) - g(x)]\, dx}. \tag{12}$$

(See Figure 7.11.)

Similarly, a straightforward calculation outlined in Exercise 17 shows that the y-coordinate of the center of mass is

$$\bar{y} = \frac{1}{2} \frac{\int_a^b [f(x)^2 - g(x)^2]\, dx}{\int_a^b [f(x) - g(x)]\, dx}. \tag{13}$$

(See Figure 7.12.)

The point $(\bar{x}, \bar{y})$ whose coordinates are given by equations (12) and (13) is called the center of mass, or **centroid,** of the plate. Since the centroid lies on the lines representing the centers of gravity with respect to both the x- and y-axes, it is properly referred to as the (two-dimensional) center of gravity of the plate. It is the point where the plate ''balances'' (see Figure 7.13).

Note two things about equations (12) and (13). First, the constants c and ρ dropped out of both equations completely. This shows that the centroid is a purely geometric property of the face R of the plate, independent of both the uniform thickness and uniform density. Second, the denominators in both equations (12) and (13) are simply the area of the face. We may therefore restate our findings as follows:

If the region R is bounded above and below by the graphs of the continuous functions f and g for $a \le x \le b$, then the coordinates $(\bar{x}, \bar{y})$ of the centroid of R are

$$\bar{x} = \frac{1}{A}\int_a^b x[f(x) - g(x)]\, dx \tag{14}$$

$$\bar{y} = \frac{1}{2A}\int_a^b [f(x)^2 - g(x)^2]\, dx \tag{15}$$

where A is the area of R.

Example 4

Find the centroid of the region R bounded by the graphs of $y = \sqrt{x}$ and $y = x^2$.

Strategy

Find the upper and lower bounding functions.

Find

$$A = \int_a^b [f - g]\, dx.$$

Solution

For $0 \le x \le 1$, the region is bounded above by $f(x) = \sqrt{x}$ and below by $g(x) = x^2$ (Figure 7.14).

$$A = \int_0^1 (\sqrt{x} - x^2)\, dx = \frac{2}{3}x^{3/2} - \frac{1}{3}x^3 \Big]_0^1$$

$$= \frac{1}{3}.$$

Find $\bar{x}$ from (14).

We may apply equation (14) directly to obtain

$$\bar{x} = 3 \int_0^1 x[\sqrt{x} - x^2] \, dx$$

$$= 3 \left[\frac{2}{5} x^{5/2} - \frac{1}{4} x^4 \right]_0^1$$

$$= 3 \left(\frac{2}{5} - \frac{1}{4} \right) = \frac{9}{20}.$$

Apply (15) to find $\bar{y}$.

By equation (15)

$$\bar{y} = \frac{3}{2} \int_0^1 [(\sqrt{x})^2 - (x^2)^2] \, dx$$

$$= \frac{3}{2} \left(\frac{1}{2} x^2 - \frac{1}{5} x^5 \right) \Big]_0^1$$

$$= \frac{9}{20}.$$

The centroid of R is therefore

$$(\bar{x}, \bar{y}) = (9/20, 9/20). \qquad \diamond$$

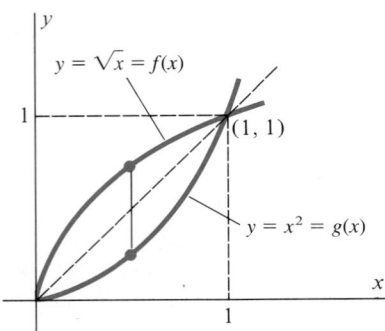

Figure 7.14

REMARK 3: The region R in Example 4 is symmetric with respect to the line $y = x$, meaning that the figure obtained by reflecting R in the line $y = x$ is congruent to R. Notice that the centroid $(9/20, 9/20)$ lies on this line. In fact, *if R is any region which is symmetric with respect to a line ℓ the centroid of R must lie on ℓ.* If this were not true, reflecting R in the line of symmetry would result in two congruent figures with different "centers of gravity," a physical impossibility. Happily, many of the figures for which one needs to find centroids in practice contain one or more lines of symmetry. In such cases, this observation can simplify the calculation of the centroid greatly.

Example 5

Find the centroid of the quarter circle R with center $(0, 0)$ and radius $r = 4$ lying in the first quadrant.

Strategy

Find A by geometry.

Solution

Since R is a quarter circle,

$$A = \frac{1}{4} \pi r^2 = \frac{16\pi}{4} = 4\pi.$$

Find upper and lower bounding functions and find $\bar{x}$ from (14). Centroid lies on line $x = \bar{x}$.

The region R is bounded above by $f(x) = \sqrt{16 - x^2}$ and below by $g(x) = 0$. Thus

$$\bar{x} = \frac{1}{4\pi} \int_0^4 x \sqrt{16 - x^2} \, dx$$

$$= \frac{1}{4\pi} \left(-\frac{1}{3} \right) (16 - x^2)^{3/2} \Big]_0^4 = \frac{16}{3\pi}.$$

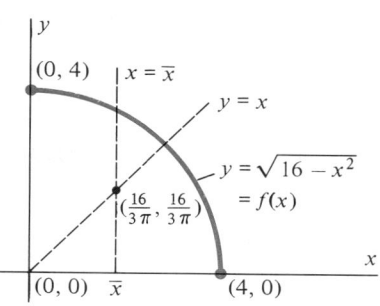

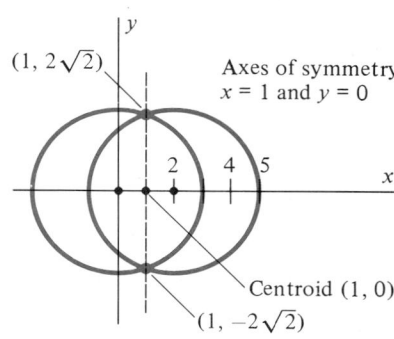

Figure 7.15 Centroid lies on intersection of $x = \bar{x}$ and $y = x$.

Figure 7.16 Centroid of region common to two circles.

| Use symmetry to obtain second line containing centroid. Centroid lies on the intersection. | Since R is symmetric about the line $y = x$, the centroid must lie on the line $y = x$. Thus $\bar{x} = \bar{y} = \dfrac{16}{3\pi}$, so the centroid is $\left(\dfrac{16}{3\pi}, \dfrac{16}{3\pi}\right)$ (see Figure 7.15). ◇ |

Example 6

Find the centroid of the region R common to both the circle $x^2 + y^2 = 9$ and the circle $(x - 2)^2 + y^2 = 9$ (Figure 7.16).

Strategy

Find points of intersection of the circles.

One line of symmetry is through the centers.

The other is through the points of intersection.

Solution

The circles intersect where $x^2 = (x - 2)^2$. Solving this equation gives $-4x + 4 = 0$ or $x = 1$. The points of intersection are therefore $(1, 2\sqrt{2})$ and $(1, -2\sqrt{2})$. Since both circles have centers on the x-axis, the region R is symmetric with respect to the x-axis. Since the circles have the same radius $r = 3$, the region is also symmetric with respect to the vertical line $x = 1$. The centroid must therefore lie on both lines $x = 1$ and $y = 0$, that is, $(\bar{x}, \bar{y}) = (1, 0)$. ◇

Exercise Set 7.7

1. Find the center of gravity for the system consisting of weights $w_1 = 1$ lb, $w_2 = 31$ lb, and $w_3 = 7$ lb located along a line at positions $x_1 = -4$, $x_2 = 2$, and $x_3 = 6$, respectively.

2. Find the center of mass for the system consisting of masses $m_1 = 2$ grams, $m_2 = 5$ grams, $m_3 = 1$ gram, and $m_4 = 10$ grams located along a line at positions $x_1 = -10$, $x_2 = -3$, $x_3 = 1$, and $x_4 = 5$, respectively.

3. An 80-pound child is sitting 3 feet from the pivot on a seesaw. How far from the pivot must a 50-pound child sit in order that the seesaw balance?

4. A manufacturing firm ships 100 items per year to city A and 300 items per year to city B. A straight section of interstate highway 200 miles long joins the two cities. The items are

shipped by truck along this highway. Where along this highway should the manufacturing plant be located if shipping costs are to be equal? What is the relationship between this problem and equation (1)?

5. A system of masses $m_1, m_2, \ldots, m_n$ are located in an xy-plane at points $(x_1, y_1), (x_2, y_2), \ldots, (x_n, y_n)$, respectively. By analogy with the discussion leading up to equations (4) and (6), explain why the center of mass with respect to the x-axis, $\bar{x}$, satisfies the equation

$$\sum_{j=1}^{n} (x_j - \bar{x}) m_j = 0.$$

(*Hint:* For the purposes of this calculation, we may regard the points as being distributed along a line parallel to the x-axis.)

6. Let $\bar{y}$ denote the center of mass for the system of Exercise 5. Explain why

$$\sum_{j=1}^{n} (y_j - \bar{y})m_j = 0.$$

7. Conclude from Exercise 5 and Exercise 6 that the center of mass $(\bar{x}, \bar{y})$ of the system of masses $m_1, m_2, \ldots, m_n$ located at points $(x_1, y_1), (x_2, y_2), \ldots, (x_n, y_n)$ is given by the coordinates

$$\bar{x} = \frac{\sum\limits_{j=1}^{n} x_j m_j}{\sum\limits_{j=1}^{n} m_j}, \qquad \bar{y} = \frac{\sum\limits_{j=1}^{n} y_j m_j}{\sum\limits_{j=1}^{n} m_j}.$$

8. Use the result of Exercise 7 to find the center of mass of the system consisting of masses $m_1 = 10$ grams, $m_2 = 15$ grams, $m_3 = 2$ grams, and $m_4 = 10$ grams located at points $(x_1, y_1) = (0, 0)$, $(x_2, y_2) = (2, -2)$, $(x_3, y_3) = (4, -2)$, and $(x_4, y_4) = (1, 1)$.

9. Find (x_3, y_3) if the mass system consisting of mass $m_1 = 2$ grams at point $(1, 1)$, mass $m_2 = 4$ grams at point $(-2, 3)$, and mass $m_3 = 3$ grams at point (x_3, y_3) has center of mass $(\bar{x}, \bar{y}) = (0, 1)$.

In each of Exercises 10–16, find the centroid of the region bounded by the given curves.

10. $y = x^3$ and $y = x^{1/3}$, for $0 \le x \le 1$

11. $y = x^2$ and $y = x^3$

12. $x = 4 - y^2$ and $x = 0$ (*Hint:* Obtain $\bar{y}$ by symmetry considerations. Use the substitution $u = 4 - x$ to find $\bar{x}$.)

13. $y = 4 - x$, $x = 0$, and $y = 0$

14. $y = x$, $y = 4 - x$, and $y = 0$

15. $y = \sqrt{4 - x^2}$ and $y = 0$

16. $x = y^2$ and $x = 4 - y^2$ (*Hint:* This can be done entirely by symmetry considerations.)

17. Assuming $f(x)$ and $g(x)$ to be continuous, derive the formula

$$\bar{y} = \frac{\dfrac{1}{2} \displaystyle\int_a^b [f(x)^2 - g(x)^2]\, dx}{\displaystyle\int_a^b [f(x) - g(x)]\, dx}$$

for the y-coordinate of the centroid of the region R bounded above by the graph of $y = f(x)$ and below by the graph of $y = g(x)$, for $a \le x \le b$, as follows:

a. Imagine the region R to be the face of a plate of uniform density ρ and uniform thickness c.

b. Partition the interval $[a, b]$ into n subintervals of equal

length $\Delta x = \dfrac{b - a}{n}$ and select one number t_j in each sub-interval $[x_{j-1}, x_j]$.

c. Approximate the "slice" of the plate over the subinterval $[x_{j-1}, x_j]$ to be the rectangular "slab" of height $[f(t_j) - g(t_j)]$, width Δx, thickness c, and density ρ. The mass of this slice is therefore approximated by $m_j \approx c\rho[f(t_j) - g(t_j)]\, \Delta x$.

d. Observe that the center of mass of this rectangular slab, with respect to the y-axis, must lie on the line

$$y = y_j = \frac{1}{2}[f(t_j) + g(t_j)].$$

e. Write down the analogue of equilibrium equation (6) as

$$0 = \sum_{j=1}^{n} (y_j - \bar{y})m_j \approx \sum_{j=1}^{n} (y_j - \bar{y})c\rho[f(t_j) - g(t_j)]\, \Delta x.$$

Substitute for y_j from (d) and apply $\lim\limits_{n\to\infty}$.

In Exercises 18–26, use equations (14) and (15) to find the centroid of the region bounded by the graphs of the given functions.

18. $f(x) = 1 - x^2$, $g(x) = 2$, for $-1 \le x \le 1$

19. $f(x) = x^3$, $g(x) = 0$, for $0 \le x \le 1$

20. $f(x) = \sqrt{x}$, $g(x) = 0$, for $0 \le x \le 4$

21. $f(x) = x$, $g(x) = -x$, for $0 \le x \le 4$

22. $f(x) = \sin x$, $g(x) = 0$, for $0 \le x \le \pi$

23. $f(x) = \sqrt{9 - x^2}$, $g(x) = 0$, $-3 \le x \le 3$

24. $f(x) = 1$, $g(x) = \cos x$, for $-\pi/2 \le x \le \pi/2$

25. $f(x) = x^2 + x + 1$, $g(x) = 0$ for $1 \le x \le 3$

26. $f(x) = 4$, $g(x) = 4 - x^2$

27. The density of a rod of length $\ell = 10$ cm is given by $\rho(x) = 1 + x$ grams/cm^3, where x represents distance from the lighter end. Find the center of mass of the rod.

28. Find the center of mass of a 10-cm rod if the density of the rod varies uniformly from 2 grams/cm^3 at one end to 12 grams/cm^3 at the other.

29. A right circular cone has radius $r = 4$ cm and height $h = 10$ cm. The density of the material from which the cone is made is uniform. Use equation (11) to find the center of mass of the cone.

30. Find the center of mass of the solid obtained by revolving about the x-axis the region bounded by the graph of $y = x^2$ for $0 \le x \le 2$. (Assume the solid to have uniform density.)

31. Give a symmetry argument to show that the centroid of a circle is its center.

32. Use a symmetry argument to find the centroid of the region bounded above by $y = \sin x$ and below by $y = -\sin x$ for $0 \le x \le \pi$.

33. Find the centroid of the region bounded above by the graph of $f(x) = 4 - x^2$ and below by the graph of the function

$$f(x) = \begin{cases} -x - 2 & \text{for} \quad -2 \le x \le 0 \\ x - 2 & \text{for} \quad 0 \le x \le 2 \end{cases}$$

34. From symmetry considerations, find the centroid of the region enclosed by the ellipse $x^2 + y^2 - xy = 6$. (*Hint:* Note that the equation is unchanged under the substitutions $y = x$ and $y = -x$.)

35. Show that the medians of a triangle intersect at the centroid.

7.8 THE THEOREM OF PAPPUS

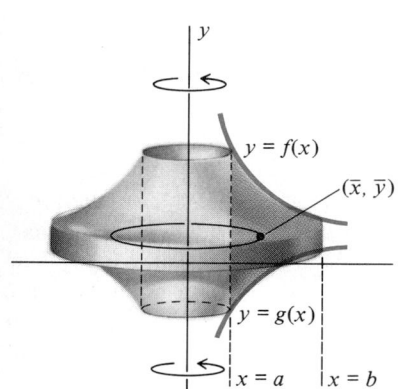

Figure 8.1 Centroid at $(\bar{x}, \bar{y})$ travels distance $2\pi\bar{x}$ as R is revolved about y-axis.

You may have noticed a similarity between equation (14) in Section 7.7 and the formula for finding the volume of a solid of revolution by the method of cylindrical shells. To make this observation more precise, suppose that a region R in the xy-plane is bounded above by the graph of $y = f(x)$ and below by the graph of $y = g(x)$, for $a \le x \le b$ (Figure 8.1).

If f and g are continuous, the volume V of the solid obtained by revolving R about the y-axis is

$$V = \int_a^b 2\pi x [f(x) - g(x)] \, dx. \tag{1}$$

Also, the x-coordinate of the centroid $(\bar{x}, \bar{y})$ of R is given by equation (14) of the preceding section as

$$\bar{x} = \frac{1}{A} \cdot \int_a^b x[f(x) - g(x)] \, dx. \tag{2}$$

By comparing equations (1) and (2), we see that

$$V = 2\pi A \bar{x}. \tag{3}$$

Now the quantity $2\pi\bar{x}$ is simply the circumference of the circle swept out by the centroid $(\bar{x}, \bar{y})$ as R is revolved about the y-axis. We have therefore proved a particular case of the following theorem, due to the Greek mathematician Pappus, who lived approximately 300 A.D.

THEOREM 1
The Theorem of Pappus

Let R be a region lying in a plane and let ℓ be a line not intersecting R. The volume V of the solid obtained by revolving the region R about the line ℓ is given by the equation $V = cA$ where A is the area of R and c is the circumference of the circle swept out by the centroid of R as R revolves about the line ℓ.

By using this theorem, we can calculate certain volumes for which the previous techniques of this chapter are inadequate.

Example 1

Find the volume of the torus obtained by revolving the circle $(x - 2)^2 + y^2 = 1$ about the y-axis. (See Figure 8.2.)

Solution: Let's first note why the method of shells will not handle this problem, at least with the integration techniques developed thus far. We can write the equation

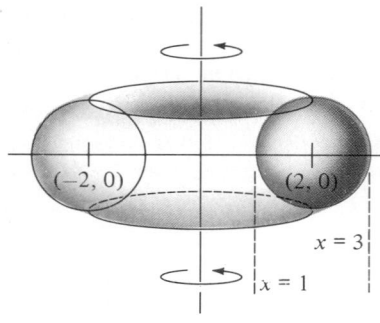

Figure 8.2 Torus obtained by revolving circle about *y*-axis.

for the upper semicircle as

$$y = \sqrt{1 - (x - 2)^2}$$

and the equation for the bottom semicircle as

$$y = -\sqrt{1 - (x - 2)^2}.$$

Applying the method of cylindrical shells gives the volume as

$$V = \int_1^3 2\pi x[2\sqrt{1 - (x - 2)^2}] \, dx,$$

and we have not yet developed a technique which will allow us to find an antiderivative for this integrand. (However, we *can* evaluate the definite integral by geometry—see Exercise 11.) But it is easy to see that

(i) the area of the circle is $A = \pi \cdot 1^2 = \pi$,
(ii) the centroid of the circle is $(\bar{x}, \bar{y}) = (2, 0)$, and
(iii) as the circle revolves around the *y*-axis, the centroid sweeps out a circle of radius $2\pi\bar{x} = 4\pi$.

Therefore, by Theorem 1, $V = (4\pi)\pi = 4\pi^2$. ◇

The Theorem of Pappus can also aid in certain calculations of hydrostatic pressure. Recall from Section 7.5 that if the flat vertical face of a submerged object lying between depths $h = a$ and $h = b$ has width $w(h)$, $a \le h \le b$ (all in feet), then the total hydrostatic force (in pounds) acting on that face is given by the integral

$$F = \int_a^b (62.4)h \cdot w(h) \, dh. \tag{4}$$

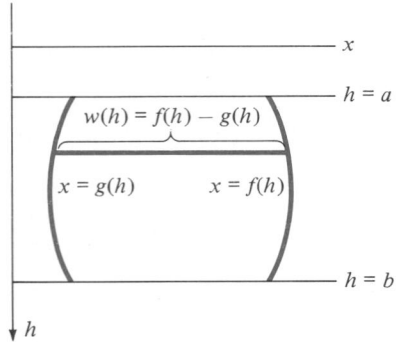

Figure 8.3 Submerged face.

If the right and left edges of the face are determined by the continuous functions f and g then we can write

$$w(h) = f(h) - g(h),$$

and equation (4) becomes

$$F = \int_a^b (62.4)h[f(h) - g(h)] \, dh. \tag{5}$$

(See Figure 8.3.)

Since the *h*-coordinate of the centroid of the face is given by

$$\bar{h} = \frac{1}{A} \int_a^b h[f(h) - g(h)] \, dh, \tag{6}$$

we conclude from equations (5) and (6) that

$$F = (62.4)A\bar{h}. \tag{7}$$

Equation (7) states that to calculate the hydrostatic force on a vertical face of a submerged object *we may regard the face as lying horizontally at the depth of its centroid.*

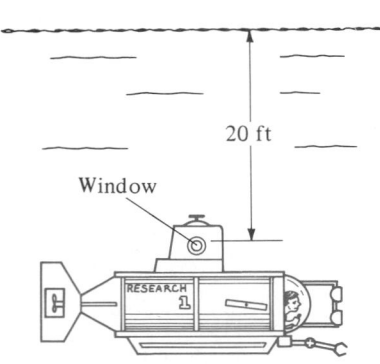

Figure 8.4 Circular submarine window of radius 6 inches.

Example 2

A submarine has a circular window of radius 6 inches. Find the hydrostatic force on the window when the submarine is submerged so that the top of the window lies 20 feet below the water level. Assume the window is mounted vertically. (See Figure 8.4.)

Solution: Since the window is positioned vertically with its top 20 feet below water level, the centroid lies at a depth of $\bar{h} = 20$ ft $+ 6$ in $= 246$ in, and the area of the window is $\pi(1/2)^2 = \pi/4$ ft$^2 = 36\pi$ in^2. Since we are working in units of inches rather than feet, we must convert 62.4 lb/ft^3 to $62.4/1728 = .03611$ lb/inch^3. From equation (7) we obtain

$$F = (.03611 \text{ lb/in}^3) \times 36\pi \text{ in}^2 \times 246 \text{ in}$$
$$\approx 1005 \text{ lb.} \qquad \diamond$$

Exercise Set 7.8

1. Find the volume of the torus obtained by revolving the region enclosed by the circle with the equation $x^2 + (y - 5)^2 = 9$ about the x-axis.

2. Find the volume of the torus obtained by revolving the region enclosed by the circle $(x - 2)^2 + (y + 3)^2 = 4$ about the line $y = 3$.

3. Use the Theorem of Pappus to find the volume of the solid obtained by revolving the region bounded by the graphs of $y = x^2$ and $y = \sqrt{x}$ about the y-axis (see Example 4, Section 7.7).

4. Find the volume of the solid obtained by revolving the region in the first quadrant bounded by the graph of the equation $x^2 + y^2 = 16$ about the line $y = -x$ (see Example 5, Section 7.7).

5. The region R common to the circles $x^2 + y^2 = 9$ and $(x - 2)^2 + y^2 = 9$ is revolved about the line $x = -3$. Find the volume of the resulting solid in terms of A, the area of R (see Example 6, Section 7.7).

6. Find the volume of the solid obtained by revolving the region bounded by the graph of $y = 2 + \sqrt{4 - x^2}$ and the x-axis for $-2 \leq x \leq 2$ about the x-axis.

7. Find the force exerted on the end wall of a swimming pool if the wall is a rectangle 10 feet wide and 8 feet deep. Assume the pool to be completely filled.

8. Find the force on a vertical window of a submarine if the window is in the shape of a circle of radius 4 inches and the top of the window is located 30 feet below the surface of the water.

9. Suppose the window in Exercise 8 has the shape of a rectangle 8 inches high and 6 inches long with a semicircle of radius 4 inches attached at either end. Find the force on the window (Figure 8.5).

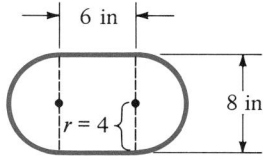

Figure 8.5

10. Find the volume of the solid obtained by revolving the graph of the semicircle $y = \sqrt{4 - x^2}$ about the line $y = 6$ (Figure 8.6).

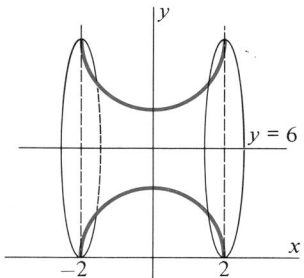

Figure 8.6

11. Show that the integral in Example 1 can be evaluated using the u-substitution $u = x - 2$ together with the observation that

$$\int_{-1}^{1} \sqrt{1 - u^2} \, du = \pi/2.$$

(This integral gives the area of the semicircle of radius 1.)

SUMMARY OUTLINE OF CHAPTER 7

◆ The volume of the solid whose cross-sectional area is $A(x)$ and that lies between $x = a$ and $x = b$ is   (page 321)

$$V = \int_{a}^{b} A(x) \, dx.$$

◆ The volume of the solid obtained by revolving about the x-axis the region bounded by the graph of $y = f(x) \geq 0$ and the x-axis for $a \leq x \leq b$ is (page 318)

$$V = \int_a^b \pi[f(x)]^2 \, dx.$$

◆ The volume of the solid obtained by rotating about the y-axis the region bounded by the graph of $y = f(x) \geq 0$ and the x-axis for $a \leq x \leq b$ is (page 325)

$$V = \int_a^b 2\pi x f(x) \, dx.$$

◆ The length of the graph of $y = f(x)$ from $(a, f(a))$ to $(b, f(b))$ is (page 331)

$$L = \int_a^b \sqrt{1 + [f'(x)]^2} \, dx.$$

◆ The lateral surface area of the solid obtained by rotating about the x-axis the region bounded by the graph of $y = f(x) \geq 0$, and the x-axis for $a \leq x \leq b$ is (page 334)

$$S = \int_a^b 2\pi f(x)\sqrt{1 + [f'(x)]^2} \, dx.$$

◆ The total distance travelled by an object moving along a line with velocity $v(t)$ at time t, between times $t = a$ and $t = b$, is (page 338)

$$D = \int_a^b |v(t)| \, dt.$$

◆ The **hydrostatic force** acting on a submerged face lying between depths $h = a$ and $h = b$ whose width at depth h is $w(h)$ is given by the integral (page 344)

$$F = \int_a^b (62.4) \cdot h \cdot w(h) \, dh.$$

◆ The **work** done by a variable force F in moving an object from point $x = a$ to point $x = b$ is given by the integral (page 348)

$$W = \int_a^b F(x) \, dx.$$

◆ The **center of mass** of a rod of constant cross-sectional area A and continuously varying density $\rho(x)$ is (page 357)

$$\bar{x} = \frac{\int_0^\ell x\rho(x)A \, dx}{\int_0^\ell \rho(x)A \, dx}.$$

◆ The **center of mass** of a rod of uniform density ρ and continuously varying cross-sectional area $A(x)$ is (page 358)

$$\bar{x} = \frac{\int_0^\ell \rho x A(x) \, dx}{\int_0^\ell \rho A(x) \, dx}.$$

◆ The **centroid** of the region R bounded above by the graph of $y = f(x)$ and below by the graph of $y = g(x)$ is $(\bar{x}, \bar{y})$, where (page 359)

$$\bar{x} = \frac{1}{A} \cdot \int_a^b x[f(x) - g(x)] \, dx \qquad (A = \text{area})$$

$$\bar{y} = \frac{1}{2A} \int_a^b [f(x)^2 - g(x)^2] \, dx \qquad (A = \text{area})$$

◆ **Theorem of Pappus:** The volume V of the solid obtained by revolving the region R about the line ℓ (not intersecting R) is $V = cA$, where A is the area of R and c is the circumference of the circle swept out by the centroid of R. (page 363)

REVIEW EXERCISES—CHAPTER 7

In Exercises 1–12, use integration to find the volume of the solid S described.

1. The sphere of radius r.

2. The solid obtained by rotating about the x-axis the region bounded by the line $x = 4$ and the parabola $y^2 = x$.

3. The solid obtained by rotating about the x-axis the region bounded by the graph of $y = 1/x$ and the x-axis for $1 \le x \le 4$.

4. The solid obtained by rotating about the x-axis the region bounded by the graph of $y = \sec\left(\dfrac{\pi x}{2}\right)$ and the x-axis for $-\dfrac{1}{2} \le x \le \dfrac{1}{2}$.

5. The solid obtained by rotating about the x-axis the region bounded by the graph of $y = \sin x$ and the x-axis for $0 \le x \le 3\pi/2$.

6. The solid obtained by rotating about the x-axis the region bounded by the graphs of $f(x) = \sqrt{1 - x^2}$ and $g(x) = 1/2$.

7. The solid obtained by rotating about the y-axis the region bounded by the graph of $f(x) = x^{2/3}$ and the x-axis for $0 \le x \le 2$.

8. The solid obtained by rotating about the y-axis the region bounded by the graph of $y = \sin x^2$ and the x-axis for $0 \le x \le \sqrt{\pi/2}$.

9. The solid obtained by rotating about the y-axis the region bounded by the graphs of $y_1 = \sin x^2$ and $y_2 = \cos x^2$ for $0 \le x \le \sqrt{\pi/2}$.

10. The solid obtained by rotating about the line $y = -2$ the region bounded by the graph of $y = 1/x^2$ and the x-axis for $1 \le x \le 4$.

11. The solid obtained by rotating about the line $y = -3$ the region bounded by the graph of $y = x^{2/3} + 1$ and the x-axis for $0 \le x \le 2$.

12. The solid obtained by rotating about the line $x = 2$ the region bounded by the graph of $y = 3 + x^3$ and the lines $x = 0$ and $y = 0$.

In Exercises 13–16, find the length of the graph of the given equation between the indicated points.

13. $y = 1 + x^{3/2}$ from $(0, 1)$ to $(4, 9)$

14. $x^2 = y^3$ from $(0, 0)$ to $(1, 1)$

15. $y^2 = (x + 1)^3$ from $(0, 1)$ to $(1, \sqrt{8})$

16. $y = \dfrac{1}{8}\left(x^4 + \dfrac{2}{x^2}\right)$ from $(1, 3/8)$ to $(2, 33/16)$

17. Find the volume of a triangular pyramid with base area A and height h.

18. The base of a solid is the region bounded by the ellipse $x^2 + \dfrac{y^2}{4} = 1$. Find the volume given that the cross sections perpendicular to the x-axis are
   a. squares,
   b. equilateral triangles.

19. A water tank has the shape of a hemisphere of radius r. To what percent capacity is it filled when the water has depth $\frac{1}{2}r$?

20. (Calculator/Computer) Approximate the length of the graph of $f(x) = \cos^2 x$ from $x = 0$ to $x = \pi$.

21. (Calculator/Computer) Approximate the lateral surface area of the solid obtained when one arc ($0 \le x \le \pi$, say) of the graph of $f(x) = \sin x$ is revolved about the x-axis.

22. A spring is stretched 10 inches by a force of 20 pounds. How much work is needed to stretch it 15 inches?

23. A cylindrical tank 10 feet in diameter and 20 feet high is full of water. How much work will be done in pumping all the water out through a valve in the top?

24. A tank in the shape of an inverted right circular cone has a radius of 8 feet and a depth of 12 feet. The tank is full of water. How much work will be done in pumping all the water to a height 10 feet above the top of the tank?

25. How much work is done in Problem 24 if only half the water is pumped to a height of 10 feet above the tank?

26. A 400-pound chain 20 feet long hangs from a windlass. How much work is done in winding in the chain?

27. How much work is done in Problem 26 if a 50-pound hook hangs at the end of the chain?

28. A water trough has end panels in the shape of trapezoids with lower bases 12 inches and upper bases 18 inches. The altitude of the trapezoid is 12 inches. Find the hydrostatic pressure on an end panel when the trough is full of water.

29. The vertical wall of a dam has the shape of the region enclosed by the parabola $y = \dfrac{1}{16}x^2$. Find the hydrostatic pressure on the wall when the water level behind it is 16 feet.

30. Show that the hydrostatic force against one face of an object submerged vertically in water is the product of the pressure at the centroid of the face times the area of the face.

31. A semicircular plate of radius 12 inches is submerged vertically in water with its diameter at the top edge and parallel to the surface of the water. Find the hydrostatic force on one face of the plate if the diameter lies 6 inches below water level.

32. A 20-cm rod has uniform cross-sectional area $A = 16\pi$ cm^2. Find its center of mass if its density varies uni-

formly from 2 grams per cm^3 at one end to 8 grams per cm^3 at the other.

33. A 10-cm rod has uniform density $\rho = 10$ grams per cm^3. Find the center of mass of the rod if its cross sectional area at a point x cm from one end is $(2 + .05x^2)$ cm^2.

34. Three particles of mass 2 grams, 5 grams, and 7 grams are located on the x-axis at points having coordinates -5, 1, and 4, respectively. Find the center of mass of the system.

35. Find the coordinates of the center of mass of the system consisting of four particles having coordinates $(-2, 4)$, $(-2, -2)$, $(0, 6)$, and $(1, -4)$ if the particles are of equal mass.

36. Find the center of mass of a rod of length 100 cm if the cross-sectional area is constant and the density is proportional to $1 + \sqrt{x}$ grams per cm^3 where x is the distance from one end.

37. Sketch an example of a region R in the plane whose centroid does not lie within R.

38. Find the centroid of the region bounded above by the graph of $y = x$, below by the x-axis, on the left by the line $x = 0$, and on the right by the line $x = 2$.

39. Find the centroid of the region bounded above by the graph of $y = 16 - x^2$ and below by the x-axis.

40. Find the centroid of the region bounded above by the graph of $y = x$ and below by the graph of $y = x^2$.

41. Find the centroid of the region bounded above by the graph of $y = 6x - x^2$ and below by the x-axis.

42. Find the centroid of the ellipse $9x^2 + 4y^2 = 36$.

43. Find the centroid of the region bounded by the graphs of $y = 6x - x^2$ and $y = 3 - |x - 3|$.

44. Use the Theorem of Pappus to find the volume of the cone obtained by rotating about the y-axis the triangle with vertices $(4, 0)$, $(0, 8)$, and $(0, 0)$.

45. Use the Theorem of Pappus to find the volume of the solid obtained by rotating the region bounded above by the graph of $y = \sqrt{4 - x^2}$ and below by the x-axis about the line
  a. $y = 0$,
  b. $y = -2$,
  c. $y = 6$.

UNIT 4

THE TRANSCENDENTAL FUNCTIONS

Leonhard Euler

Johann Bernoulli

Jakob Bernoulli

Marquis de l'Hôpital

Maria Gaetana Agnesi

At the beginning of the eighteenth century, calculus was still in its infancy. Logically precise foundations of the subject were yet to be developed, and only polynomials could be differentiated or integrated easily. The logarithmic and exponential functions were still to be invented, and the trigonometric functions were not yet really understood as functions.

One of the greatest innovators in mathematics then appeared: Leonhard Euler (1707–1783). Born in Switzerland, he spent many years in St. Petersburg (now Leningrad), and then taught in Berlin for 25 years. He was hired back to Russia by Catherine the Great and spent the last 17 years of his life there. He was perhaps the most prolific mathematician of all time—he published 520 books and papers and, after his death, the *Proceedings of the St. Petersburg Academy* continued to publish at least one new paper by Euler in every issue for the next 47 years! His output is seen to be all the more incredible by the fact that he was blind for virtually all of his second Russian period, dictating his work to one of his sons.

No one else did more to put calculus and analysis in its present form. Among the symbols Euler initiated are the sigma (Σ) for summation, e to represent the constant 2.71828 . . . , i for the imaginary $\sqrt{-1}$, and even a, b, and c for the sides of a triangle and A, B, and C for the opposite angles. Although William Jones had first used the symbol π for 3.14159 . . . in 1706, Euler made its use standard.

Euler wrote textbooks on differential and integral calculus; the general form of these texts is still in use today. His greatest work was the *Introductio in analysis infinitorum* (1748), in which Euler did for calculus what Euclid had done for geometry and the theory of numbers in ancient Greece. He systematized differentiation and the method of fluxions (integral calculus), creating a new system which is often called *analysis*—the study of infinite processes. The French physicist and astronomer Arago called Euler "analysis incarnate."

Euler transformed the trigonometric ratios into the trigonometric functions as we think of them today, and first used the abbreviations sin, cos, and tan. He treated logarithms and exponents as functions, whereas their creators (John Napier and Henry Briggs) had thought of them merely as tools to aid computation.

Leonhard Euler received both his bachelor's and his master's degrees at the age of 15, studying mathematics primarily under Johann Bernoulli. Bernoulli was one of a dozen mathematicians of that name, stretching over six generations, though the most famous are Johann (1667–1748) and his brother Jakob (1654–1705). The two brothers corresponded for many years with Leibniz about the calculus, and made many discoveries with which we are familiar today. Johann first introduced integration by partial fractions. He also discovered a relation between the trigonometric and logarithmic functions, thus paving the way to the realization that there are only two basic types of elementary functions: polynomial, rational, and algebraic functions on the one hand, and the transcendental (trigonometric, logarithmic, exponential, and hyperbolic) functions on the other. Johann and Jakob Bernoulli both studied arc length, the curvature of curves, and points of inflection; however, their ideas seem somewhat naive and inexact today. For example, one postulate states, "Each curved line consists of infinitely many straight lines, these themselves being infinitely small."

Jakob Bernoulli was the first to publish polar coordinates (although Newton had the general idea some years before). He studied the catenary or hanging chain curve, which is also known as the hyperbolic cosine curve. He suggested for the first time the term *calculus integralis,* or integral calculus as we call it today.

Many other mathematicians also contributed to the rapid growth of calculus in the eighteenth century. However, the greatest influence on the modern calculus classroom was that of the functions, symbols, and methods discovered and promoted by Leonhard Euler and the brothers Bernoulli.

A very useful theorem encountered in this unit is l'Hôpital's Rule, named after a French nobleman, Guillaume François Antoine de l'Hôpital (1661–1704). The Marquis de l'Hôpital wrote the first textbook on differential calculus in 1696. It was not what we would think of today as a text for teaching students, but was intended rather for presenting the new subject to mathematicians. The so-called "rule," which enables one to find the limit of a quotient whose numerator and denominator both tend to zero, appears in this book. However, the rule and much of the other material in the book was actually the work of Johann Bernoulli, l'Hôpital's teacher. They had an unusual agreement whereby Bernoulli agreed to turn over to l'Hôpital all of his mathematical discoveries, which l'Hôpital was entitled to claim as his own. L'Hôpital had plans for writing what would also have been the first text in integral calculus, but abandoned this project when Leibniz told him that he was planning such a work.

The first calculus textbook intended for young people was written by an Italian woman, Maria Gaetana Agnesi (1718–1799). She was a brilliant child who began learning foreign languages by the age of four, and by nine knew Latin, Greek, Hebrew, French, German, and Spanish fluently. When only ten she published her first book, on a printing press set up for her in her home. The book advocated education for girls and women.

Her father, a professor of mathematics, exploited her brilliance by establishing Sunday afternoon get-togethers at which learned men would sit in a circle around her while she lectured on some topic in areas that included mathematics, physics, logic, chemistry, and philosophy. She usually lectured in Latin, but would respond to questions in the language of the questioner. Her second book was a collection of 190 of these lectures, published when she was 21.

Maria Agnesi was the eldest of 21 children, and hence was expected to supervise the raising of the other 20. To explain mathematics to one of her brothers (and for other teenagers), she wrote a massive two-volume work of 1070 pages. She began with algebra, trigonometry, conic sections, and curve-sketching, but the main part was differential and integral calculus, continuing on into differential equations. It was immediately successful, and was translated from Italian into English. She was only 30 when it was published.

Agnesi was clearly much overworked, and though she had a brilliant mathematical mind, by the age of 40 she had completely lost interest in mathematics. She opened up her home to the hungry and homeless, and eventually she supervised an institution for indigent women. She died at the age of 80 and was buried in a common grave with fifteen other elderly women.

(Photograph of Leonhard Euler from the Library of Congress. Photographs of Johann and Jakob Bernoulli from the David Eugene Smith Papers, Rare Book and Manuscript Library, Columbia University. Photo of Maria Gaetana Agnesi from Culver Pictures.)

Chapter 8
Logarithmic and Exponential Functions

In previous courses on mathematics you encountered the *logarithm to the base b,* $\log_b x$, which was defined in terms of the *exponential* function $g(y) = b^y$ by the relation

$$y = \log_b x \quad \text{if and only if} \quad x = b^y. \tag{1}$$

Because *logarithm functions* are used frequently in science to describe the growth and decay of radioactive isotopes, animal populations, etc., we devote this chapter to a careful development of logarithm and exponential functions.

The principal difficulty in developing the concept of a logarithm function is that up to this point the exponential b^y in equation (1) has been defined only for exponents y that are *rational* numbers (that is, numbers that are quotients of integers). In order to apply the calculus to properly defined logarithm and exponential functions they must be continuous and differentiable, and therefore defined for *all* numbers at least on certain intervals.

In order to overcome this difficulty we shall use the Fundamental Theorem of Calculus to define the *natural logarithm* function as a certain integral, and then show that the exponential function can be obtained as the *inverse* of this integral. Along the way we shall also show that the functions defined in this way actually are extensions of the functions in line (1). Before beginning on this agenda, however, we review the concepts of $\log_b x$ and inverse function in Section 8.1.

8.1 REVIEW OF LOGARITHMS AND INVERSE FUNCTIONS

The **logarithm to the base b** of the number x is defined by the statement

$$y = \log_b x \quad \text{if and only if} \quad x = b^y. \tag{1}$$

In writing equation (1) we assume that b is a *positive* constant not equal to 1 and that $x > 0$. Thus, for example,

$$\log_{10} 100 = 2 \quad \text{because} \quad 100 = 10^2,$$
$$\log_2 8 = 3 \quad \text{because} \quad 8 = 2^3,$$
$$\log_{16} 4 = \frac{1}{2} \quad \text{because} \quad 4 = \sqrt{16} = 16^{1/2},$$

and

$$\log_2\left(\frac{1}{4}\right) = -2 \quad \text{because} \quad \frac{1}{4} = \frac{1}{2^2} = 2^{-2}.$$

The logarithm $\log_b x$ defined in statement (1) has the following properties:

(L1) $\log_b(uv) = \log_b u + \log_b v,$

(L2) $\log_b\left(\dfrac{u}{v}\right) = \log_b u - \log_b v,$

(L3) $\log_b(u^r) = r \log_b u, \qquad r$ rational.

These properties follow from statement (1) and the laws of exponents. (See Exercises 40–42.)

The Logarithm as an Inverse

It follows from statement (1) that if we calculate the logarithm of the number b^y we obtain simply the number y:

If $\quad x = b^y, \qquad$ then $\qquad \log_b(b^y) = \log_b x = y.$

That is,

$$\log_b(b^y) = y. \tag{2}$$

Equation (2) says that the logarithm reverses (or, *inverts*) the effect of the exponential function $g(y) = b^y$ acting on the number y. It is also the case that the exponential function "inverts" application of the logarithm:

If $\quad y = \log_b x, \qquad$ then $\quad b^{(\log_b x)} = b^y = x.$

That is,

$$b^{(\log_b x)} = x. \tag{3}$$

It is important to note that *equations (2) and (3) are valid, however, only when the logarithm and exponential functions have the same base b.*

Inverse Functions

Equations (2) and (3) are examples of identities that hold for pairs of functions f and g when one function is the *inverse* for the other.

DEFINITION 1

Let f be a function with domain D. The **inverse** function for f, written $g = f^{-1}$, is the function g defined by the equation

$$g(f(x)) = x \qquad \text{for all } x \in D.$$

That is, g is the inverse for the function f if the composite function $h = g \circ f$ is the identity function $h(x) = (g \circ f)(x) = x$ on the set D. (See Figure 1.1.) Thus, g is the inverse for f if g returns the value $y = f(x)$ to the number x for every x in the domain of f.

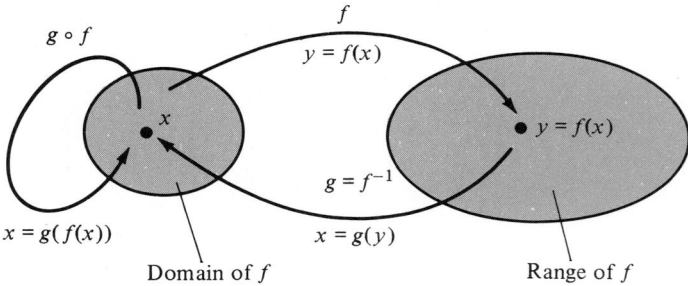

Figure 1.1 g is the inverse for f if $g(f(x)) = x$ for all $x \in D$.

REMARK 1: Be careful to note that we are using the notation f^{-1} to denote the inverse function for f, *not* the reciprocal function $1/f$. That is, $f^{-1}(f(x)) = x$ for all x in the domain of f.

REMARK 2: If g is the inverse function for f, then the *domain* of g is precisely the range of f (nothing more, nothing less) because g is *defined* by the statement

$$g(y) = x \quad \text{if and only if} \quad y = f(x).$$

Even though the *equation* for g may be defined on a larger set, the statement that $g = f^{-1}$ automatically restricts the domain of g to the range of f.

Example 1

The function $f(x) = 2x$ has domain $(-\infty, \infty)$. Its inverse is the function $g(x) = \dfrac{1}{2}x$.

That is because

$$\text{if} \quad y = f(x) = 2x, \quad \text{then} \quad g(y) = \frac{1}{2}y = \frac{1}{2}(2x) = x$$

for all $x \in (-\infty, \infty)$. That is,

$$g(f(x)) = \frac{1}{2}(f(x)) = \frac{1}{2}(2x) = x, \qquad x \in (-\infty, \infty). \qquad \diamond$$

Example 2

The function $f(x) = \sqrt{x}$ has domain $[0, \infty)$. Its inverse is the function $g(x) = x^2$, $x \geq 0$, since

$$\text{if} \quad y = f(x) = \sqrt{x}, \quad \text{then} \quad g(y) = y^2 = (\sqrt{x})^2 = x, \qquad x \in [0, \infty).$$

That is,

$$g(f(x)) = [f(x)]^2 = (\sqrt{x})^2 = x, \qquad x \in [0, \infty).$$

In this case the domain of the inverse function $f^{-1}(x) = g(x) = x^2$ is $[0, \infty)$ because this is the *range* of f. $\diamond$

Example 3

The exponential function $f(x) = 10^x$ is defined for all rational numbers. Its inverse function is the **common logarithm** function $g(x) = \log_{10} x$. This follows from equation (2): For x rational,

$$\text{if} \quad y = f(x) = 10^x \quad \text{then} \quad g(y) = \log_{10} y = \log_{10}(10^x) = x.$$

That is,

$$g(f(x)) = \log_{10}[f(x)] = \log_{10}(10^x) = x.$$

Thus, $f^{-1}(x) = g(x) = \log_{10} x$. $\diamond$

Example 4

The function $f(x) = x^2$ with domain $(-\infty, \infty)$ does *not* have an inverse. That is because each positive number in the range $R = [0, \infty)$ for f corresponds to *two* distinct numbers in the domain of f. For example, for $y = 4$ we have both $f(2) = 4$ and $f(-2) = 4$. It is therefore impossible to define $g(4)$ so that *both* $g(4) = 2$ and $g(4) = -2$. Thus, no function g can satisfy the requirement that $g(f(x)) = x$ for all $x \in (-\infty, \infty)$. ◇

Example 4 shows that *not every function has an inverse*. What is required for f to have an inverse is that each number y in the range of f be the function value for *only one* number x in the domain of f. This allows us to define the inverse function $g = f^{-1}$ by the rule: If $y = f(x)$, then $g(y) = f^{-1}(y) = x$. Functions that have this property are called *one-to-one* functions.

DEFINITION 2

The function f is said to be **one-to-one** if it satisfies the following property for all numbers x_1 and x_2 in its domain:

$$\text{if} \quad x_1 \neq x_2, \quad \text{then} \quad f(x_1) \neq f(x_2).$$

Figure 1.2 illustrates the geometric interpretation of the property of f being one-to-one. Since each number y in the range of f is the value $f(x)$ for only one number x, *any horizontal line* can intersect the graph of a one-to-one function at no more than one point. Figure 1.3 shows that this property fails for the function $f(x) = x^2$ which is not one-to-one. (Example 4.)

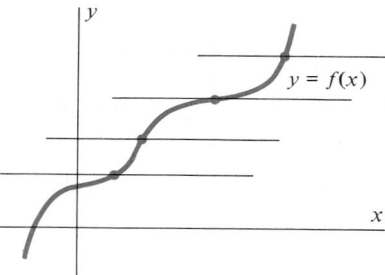

Figure 1.2 A horizontal line can intersect the graph of a one-to-one function at no more than one point.

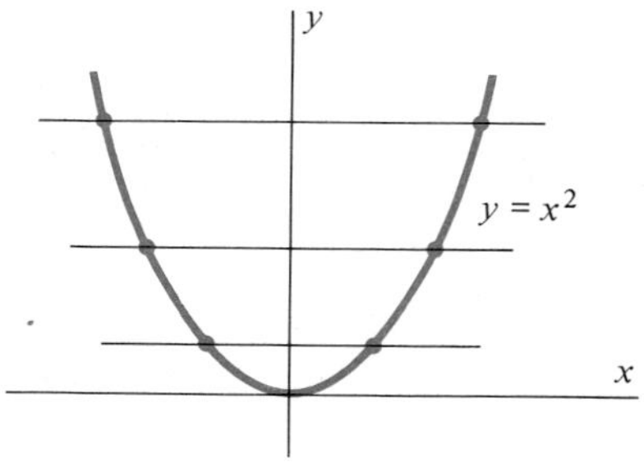

Figure 1.3 The function $f(x) = x^2$ is not one-to-one: Some horizontal lines intersect the graph at two points.

The following theorem summarizes our discussion about the existence of inverse functions.

THEOREM 1

If the function f is one-to-one, then

(i) the inverse function f^{-1} exists, and
(ii) the domain of f^{-1} is the range of f.

Proof: Let y be any number in the range of f. Then, since f is one-to-one, there is precisely one number x for which $f(x) = y$. Define $g(y) = x$. Then g is a function defined for all y in the range of f, and $g(f(x)) = x$ for all x in the domain of f. Thus, $g = f^{-1}$ according to Definition 1. ◆

The following theorem gives a useful criterion for f to have an inverse.

THEOREM 2

Let f be a function with domain D. Then f^{-1} exists if either

(i) f is increasing on D, or
(ii) f is decreasing on D.

Proof: To prove part (i) we assume that f is increasing on D. We shall show that this guarantees that f is one-to-one. To do so we let x_1 and x_2 be any two numbers in D with $x_1 \neq x_2$. Then, since f is increasing on D.

(a) if $x_1 < x_2$, then $f(x_1) < f(x_2)$;
(b) if $x_1 > x_2$, then $f(x_1) > f(x_2)$.

In either case it follows that $f(x_1) \neq f(x_2)$. Thus, f is one-to-one. It then follows from Theorem 1 that f^{-1} exists.

The proof for part (ii) is similar. ◆

Example 5

The function $f(x) = \tan x$ *restricted* to the domain $D = (-\pi/2, \pi/2)$ has an inverse (which we shall study in detail in Chapter 9). To see this we note that the derivative is

$$f'(x) = \frac{d}{dx}(\tan x) = \sec^2 x$$

which is positive for all $x \in (-\pi/2, \pi/2)$. Thus, f is increasing on $(-\pi/2, \pi/2)$, so f^{-1} exists by Theorem 2. ◇

Finding Inverse Functions

When a function f is specified by an equation of the form $y = f(x)$, we may sometimes succeed in finding an equation for the inverse function f^{-1} by solving the equation $y = f(x)$ for x in terms of y. This technique succeeds in the next example.

Example 6

Find the inverse function for $f(x) = 2x - 4$, $x \in (-\infty, \infty)$.

Solution: We solve the equation $y = 2x - 4$ for x:

$$y = 2x - 4$$
$$2x = y + 4$$
$$x = \frac{1}{2}y + 2$$

Thus, $f^{-1}(y) = \frac{1}{2}y + 2$ since

$$\text{if} \quad y = 2x - 4, \quad f^{-1}(y) = \frac{1}{2}y + 2 = \frac{1}{2}(2x - 4) + 2 = x$$

for all $x \in (-\infty, \infty)$. Written as a function of x, the inverse function is

$$f^{-1}(x) = \frac{1}{2}x + 2.$$

◇

Unfortunately, many functions have inverses that cannot be found by this method because the equation $y = f(x)$ cannot be solved algebraically for x. Examples include $y = x + x^3$, $-\infty < x < \infty$, and $y = \tan x$, $-\pi/2 < x < \pi/2$. We may always resort, however, to the defining equation $f^{-1}(f(x)) = x$, $x \in D$, when the inverse function f^{-1} is known to exist.

An Important Identity

Let f be a function with domain D and range R and assume that f^{-1} exists. Then, for each $y \in R$, we know that $x = f^{-1}(y)$ is in D. We may therefore apply f to both sides of this equation and conclude that

$$f(x) = f(f^{-1}(y)), \qquad y \in R. \tag{4}$$

Since $y = f(x)$, equation (4) gives the identity

$$\boxed{f(f^{-1}(y)) = y, \qquad y \in R.} \tag{5}$$

This equation shows that, if f has domain D, range R, and inverse f^{-1}, then *the function f is the inverse for f^{-1}*. (We shall use equation (5) in a critical way in later sections.)

Example 7

The function $f(x) = x^3$, with domain $(-\infty, \infty)$, has inverse function $f^{-1}(x) = x^{1/3}$. This is because,

$$\text{if} \quad y = x^3, \qquad \text{then} \quad f^{-1}(y) = y^{1/3} = (x^3)^{1/3} = x, \quad x \in (-\infty, \infty).$$

Note also that the *range* of f is $R = (-\infty, \infty)$ and that

$$f(f^{-1}(y)) = f(y^{1/3}) = (y^{1/3})^3 = y, \qquad y \in (-\infty, \infty)$$

as required by equation (5).

The Graph of the Inverse Function

Figure 1.4 shows the graphs of the functions considered in Examples 2, 6, and 7 with the corresponding inverse function graphed on the same set of axes. Note in each case that the graph of $y = f^{-1}(x)$ is the reflection across the line $y = x$ of the graph of $y = f(x)$. Indeed, this will always be the case, since if $b = f(a)$, then $a = f^{-1}(b)$. This means that the point (b, a) will lie on the graph of $y = f^{-1}(x)$ whenever (a, b) is on the graph of $y = f(x)$ and conversely.

Properties of Inverse Functions

The fact that the graph of f^{-1} is the reflection across the line $y = x$ of the graph of f suggests that a function and its inverse (when it exists) should have similar continuity and differentiability properties. The following two theorems address this issue. We shall not pursue formal proofs, although we shall assume these results in developing rules for differentiating inverses of the logarithm and trigonometric functions.

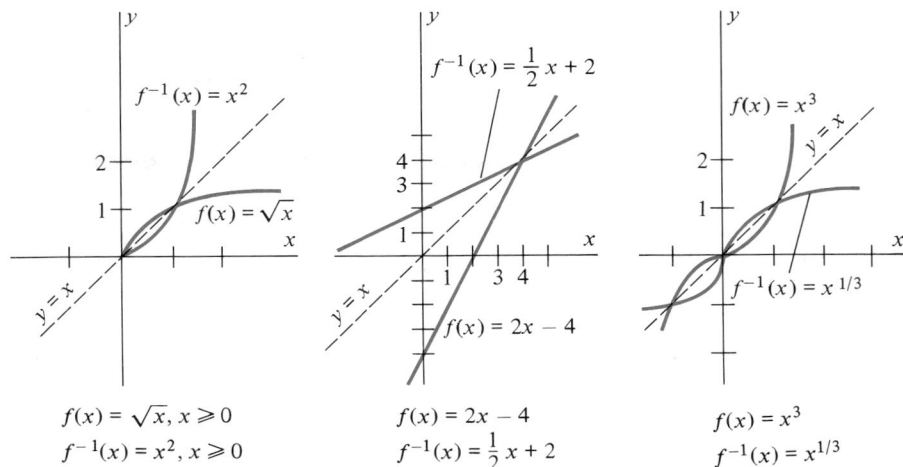

$f(x) = \sqrt{x}, \, x \geqslant 0$

$f^{-1}(x) = x^2, \, x \geqslant 0$

$f(x) = 2x - 4$

$f^{-1}(x) = \frac{1}{2}x + 2$

$f(x) = x^3$

$f^{-1}(x) = x^{1/3}$

Figure 1.4 The graph of f^{-1} is the reflection across the line $y = x$ of the graph of f. If (a, b) is on the graph of f, (b, a) is on the graph of f^{-1}.

THEOREM 3 **Continuity of Inverses**	If the function f has an inverse f^{-1} on an open interval I and if f is continuous at $a \in I$, then f^{-1} is continuous at $b = f(a)$.

In other words, the inverse of a continuous function is a continuous function where defined.

THEOREM 4	Let the function f be continuous on an open interval I and assume that f has an inverse function f^{-1}. Let $a \in I$. The inverse function f^{-1} is differentiable at $b = f(a)$ if (i) $f'(a)$ exists, and (ii) $f'(a) \neq 0$. In this case we have $$(f^{-1})'(b) = \frac{1}{f'(a)}.$$

Figure 1.5 suggests an intuitive justification for Theorem 4. The existence of the derivative $f'(a)$ means that the graph of f has a non-vertical tangent at (a, b). Reflecting this graph across the line $y = x$ produces a non-vertical tangent to the graph of f^{-1} at (b, a) with reciprocal slope *unless the tangent to the graph of f at (a, b) is horizontal* (in which case the tangent at (b, a) will be vertical with slope, and therefore $(f^{-1})'(b)$, undefined).

Figure 1.6 shows the graph of $f(x) = x^3$ and its inverse function $f^{-1}(x) = x^{1/3}$. Note that f is differentiable for all x, but that $f'(0) = 0$. Thus, the inverse function is *not* differentiable at $f(0) = 0$. This is evident from the equation for f^{-1} since

$$(f^{-1})'(x) = \frac{d}{dx}(x^{1/3}) = \frac{1}{3x^{2/3}}$$

which is undefined for $x = 0$.

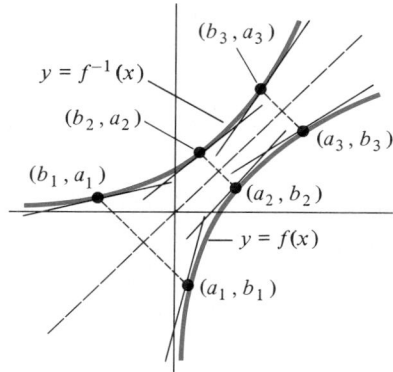

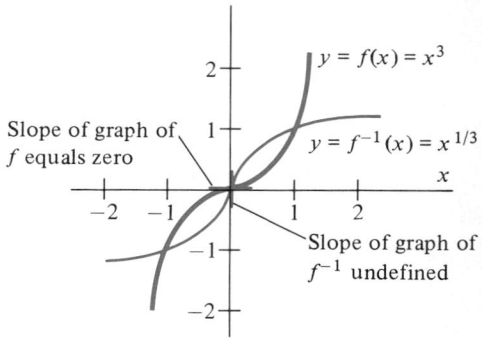

Figure 1.5 Slope of graph of f^{-1} at (b, a) is the reciprocal of slope of graph at (a, b).

Figure 1.6 For $f(x) = x^3$, inverse function is $f^{-1}(x) = x^{1/3}$, which is not differentiable at $x = 0$ because $f'(0) = 0$.

Exercise Set 8.1

In Exercises 1–10, find the logarithm.

1. $\log_{10} 1000$

2. $\log_7 343$

3. $\log_8 2$

4. $\log_4 (0.5)$

5. $\log_5 \left(\dfrac{1}{125} \right)$

6. $\log_2 \sqrt{2}$

7. $\log_4 8$

8. $\log_{343} 7$

9. $\log_{10} 1$

10. $\log_b 1$

In Exercises 11–18, solve for x.

11. $\log_x 16 = 4$

12. $\log_5 x = 125$

13. $\log_{1/2} x = 16$

14. $\log_{2/3} x = 125$

15. $\log_3 x = \dfrac{1}{6}$

16. $\log_x 243 = 5$

17. $\log_x 27 = \dfrac{3}{2}$

18. $\log_7 x = 1$

In Exercises 19–28, find the inverse function f^{-1}.

19. $f(x) = 3x - 2$

20. $f(x) = 7 - x$

21. $f(x) = \dfrac{1}{x + 2}$

22. $f(x) = \dfrac{2}{4 - x}$

23. $f(x) = \dfrac{x}{x + 3}$

24. $f(x) = \dfrac{x}{5 - x}$

25. $f(x) = 4 + \sqrt{x - 1}$

26. $f(x) = \sqrt[3]{x + 3}$

27. $f(x) = 3 + x^{5/3}$

28. $f(x) = \dfrac{1}{\sqrt{x + 1}}$

29. Show that the function $f(x) = \dfrac{1}{x}$ is its own inverse.

30. True or false? If f has an inverse, then no horizontal line can intersect the graph of f more than once. Why or why not?

31. Find the inverse of each of the following functions when restricted to the indicated domains:
  a. $f(x) = x^2 + 3$, $x \geq 0$
  b. $f(x) = (x + 1)^2$, $x \geq -1$
  c. $f(x) = x^2 + 4x + 4$, $x \geq -2$
  d. $f(x) = x^2 - 2x - 3$, $x \geq 1$

32. Explain why the function $f(x) = ax^2 + bx + c$, $a \neq 0$, has an inverse when restricted to the domain $\left[-\dfrac{b}{2a}, \infty \right)$. (*Hint:* Consider the quadratic formula, or find its relative maximum or minimum.)

33. For the function $f(x) = \sqrt[3]{x} + 5$
  a. Find the derivative $(f^{-1})'(2)$ using Theorem 4.
  b. Find the inverse function f^{-1}.
  c. Find $(f^{-1})'(x)$ from part b.
  d. Find $(f^{-1})'(2)$ from part c and verify that it agrees with your answer from part a.

34. Use Theorem 2 to verify that the function $f(x) = \sin x$ has an inverse when restricted to the domain $[-\pi/2, \pi/2]$.

35. Let g be the inverse function for $f(x) = \sin x$, $x \in [-\pi/2, \pi/2]$. (See Exercise 34.) Find, using Theorem 4,
  a. $g'(\pi/4)$
  b. $g'(0)$
  c. $g'(\pi/3)$
  d. $g'(-\pi/6)$

36. Verify that the function $f(x) = x + x^3$ has an inverse on $(-\infty, \infty)$. (*Hint:* Use Theorem 2.)

37. Let g be the inverse for the function $f(x) = x + x^3$. (See

Exercise 36.) Find:
 a. $g(0)$
 b. $g(2)$
 c. $g'(1)$
 d. $g'(2)$

38. The exponential function occurs naturally in formulas for the periodic compounding of interest. For example, if an annual rate of interest of r percent is applied to a principal amount P_0 placed in a savings account, the amount $P(1)$ on deposit at the end of one year is

$$P(1) = P_0 + rP_0 = (1 + r)P_0.$$

In the second year the interest rate r is applied to the new principal, $(1 + r)P_0$. Thus, the amount $P(2)$ on deposit at the end of two years is

$$P(2) = (1 + r)P_0 + r[(1 + r)P_0]$$
$$= (1 + r)^2 P_0.$$

Show that the amount $P(t)$ on deposit at the end of t years is given by the exponential function

$$P(t) = (1 + r)^t P_0.$$

39. Show that if interest is compounded n times per year at an annual percentage rate r, the amount $P(t)$ on deposit in a savings account after t years is

$$P(t) = \left(1 + \frac{r}{n}\right)^{nt} P_0$$

where P_0 is the amount of the original principal.

40. Prove property $(L1)$ of logarithms as follows:
 a. Let $x = \log_b m$ and $y = \log_b n$. Show that $m = b^x$ and $n = b^y$.
 b. By the laws of exponents, show that $mn = b^{x+y}$.
 c. Explain why $\log_b b^z = z$, z rational.
 d. Conclude that $\log_b(mn) = \log_b(b^{x+y}) = x + y = \log_b m + \log_b n$.

41. Prove property $(L2)$ (see Exercise 40).

42. Prove property $(L3)$ (see Exercise 40).

43. The Beer-Lambert law relates the absorption of light travelling through a material to the concentration and the thickness of the material. If I_0 and I denote the intensities of light of a particular wavelength before and after passing through the material, respectively, and if x denotes the length of the path followed by the beam of light passing through the material, then

$$\log_{10}\left(\frac{I}{I_0}\right) = kx$$

where k is a constant depending on the material. Express I as a function of x.

44. True or false? The values of $y = \log_5 x$, as defined in this section, must be rational numbers.

45. True or false? If $0 < x < y$ then $\log_a x < \log_a y$. Why?

8.2 THE NATURAL LOGARITHM FUNCTION

The purpose of this section is to introduce what we shall refer to as the *natural logarithm function*, written $y = \ln x$. We shall begin by answering what seems to be a completely unrelated question, and then show that the resulting function ($y = \ln x$) has the properties $L1$–$L3$ of logarithm functions as discussed in Section 8.1.

An Antiderivative for
$$f(x) = \frac{1}{x}$$

In discussing antiderivatives in Chapters 5 and 6, we were careful to note that the integration formula

$$\int x^n \, dx = \frac{1}{n + 1} x^{n+1} + C$$

does *not* apply when $n = -1$. The question of finding an antiderivative for $f(x) = \frac{1}{x}$ has therefore been addressed only by the Fundamental Theorem of Calculus. Recall, that theorem guarantees that, since the function $f(x) = \frac{1}{x}$ is continuous on $(0, \infty)$, the function

$$F(x) = \int_1^x \frac{1}{t} \, dt$$

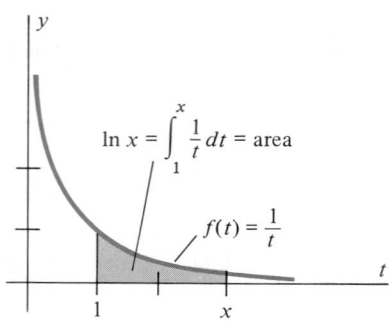

$$\ln x = \int_1^x \frac{1}{t}\,dt = \text{area}$$

$$f(t) = \frac{1}{t}$$

Figure 2.1 $\ln x > 0$ if $x > 1$.

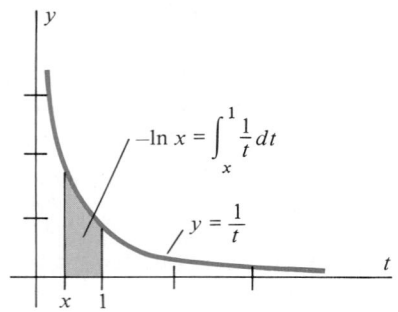

$$-\ln x = \int_x^1 \frac{1}{t}\,dt$$

$$y = \frac{1}{t}$$

Figure 2.2 $\ln x < 0$ if $x < 1$.

is differentiable on $(0, \infty)$, and $F'(x) = \dfrac{1}{x}$, $x > 0$. This function F, then, is an antiderivative for $f(x) = \dfrac{1}{x}$ on the interval $(0, \infty)$. Even though this function F is a *function defined by integration*, rather than a simple power of x, it is a legitimate answer to our question. This is the function that we shall henceforth denote by $\ln x$ and refer to as the *natural logarithm function*. That is,

$$\ln x = \int_1^x \frac{1}{t}\,dt, \qquad x > 0, \tag{1}$$

and $y = \ln x$ is a differentiable function for all x in the interval $(0, \infty)$ with

$$\frac{d}{dx}\ln x = \frac{1}{x}, \qquad x > 0. \tag{2}$$

Since a differentiable function is necessarily continuous, this shows that $f(x) = \ln x$ is continuous on $(0, \infty)$.

The geometric interpretation of the expression $y = \ln x$ is simple:

(a) If $x > 1$, $\ln x$ is the area of the region bounded by the graph of $f(t) = 1/t$ and the t-axis between $t = 1$ and $t = x$ (Figure 2.1).
(b) If $x < 1$, $\ln x$ is the *negative* of the area bounded by the graph of $f(t) = 1/t$ and the t-axis between $t = x$ and $t = 1$ (Figure 2.2).
(c) If $x = 1$, $\ln x = \displaystyle\int_1^1 \frac{1}{t}\,dt = 0$, so $\ln 1 = 0$.

REMARK: The natural logarithm function is defined with lower limit 1 in order to have $\ln 1 = 0$. Since $\log_a 1 = 0$ for all logarithm functions (because $a^0 = 1$ by definition), this agrees with the behavior we expect for $\ln x$.

For $x > 0$ the value $\ln x$ can be approximated to any desired accuracy by applying one of the procedures for approximating integrals (such as Simpson's Rule) to the integral in equation (1). A table of values for $\ln x$ appears in Appendix IV, and most hand calculators and computers are preprogrammed to provide values of $\ln x$. Note that $\ln x$ is not defined for $x \le 0$.

If u is a differentiable function of x, we may use the Chain Rule to generalize the differentiation formula of equation (2) to:

$$\frac{d}{dx}\ln u = \frac{1}{u}\cdot\frac{du}{dx}, \qquad u > 0. \tag{3}$$

Example 1

Find $f'(x)$ for (a) $f(x) = \ln 3x$, and (b) $f(x) = x\ln(1 + x^2)$.

Solution

(a) $f'(x) = \dfrac{d}{dx} \ln 3x = \dfrac{1}{3x} \cdot 3 = \dfrac{1}{x}$

(b) By the Product Rule and equation (3) we obtain

$$f'(x) = \frac{d}{dx}[x \ln(1 + x^2)] = (1)\ln(1 + x^2) + x \cdot \frac{d}{dx}[\ln(1 + x^2)]$$

$$= \ln(1 + x^2) + x\left(\frac{1}{1 + x^2}\right) \cdot \frac{d}{dx}(1 + x^2)$$

$$= \ln(1 + x^2) + x\left(\frac{1}{1 + x^2}\right)(2x)$$

$$= \ln(1 + x^2) + \frac{2x^2}{1 + x^2}.$$

$\diamondsuit$

But what does the function defined by equation (1) have to do with logarithms? The answer is provided by the following theorem which shows that *the function $y = \ln x$ satisfies properties (L1)–(L3) of logarithms.*

THEOREM 5

The natural logarithm function satisfies the following properties for all real numbers a, b, and r with $a > 0$, $b > 0$, and r rational:

(L1) $\ln(ab) = \ln a + \ln b$,

(L2) $\ln\left(\dfrac{a}{b}\right) = \ln a - \ln b$,

(L3) $\ln(a^r) = r \ln a$ (r rational).

The proof of Theorem 5 makes use of Theorem 1, Chapter 5, which states that *if two functions have the same derivative, they differ at most by a constant.* That is, if $f'(x) = g'(x)$, then $f(x) = g(x) + C$ for some constant C. We prove only property (L1), leaving (L2) and (L3) as exercises.

Proof of (L1): Let f and g be the functions

$$f(x) = \ln ax \quad \text{and} \quad g(x) = \ln x.$$

Then, by equation (3),

$$f'(x) = \frac{1}{ax} \cdot a = \frac{1}{x} = g'(x), \quad x > 0.$$

Thus, by Theorem 1 of Chapter 5

$$f(x) = g(x) + C, \quad x > 0$$

for some constant C, that is,

$$\ln ax = \ln x + C, \quad x > 0. \tag{4}$$

To determine the constant C, we use the fact that $\ln 1 = 0$. Setting $x = 1$ in equation

(4) gives

$$\ln a = 0 + C,$$

so $C = \ln a$. Equation (4), with $x = b$, now gives the desired result. ◆

The point of Theorem 5 is that the natural logarithm function satisfies all the properties of a logarithmic function. Moreover, $\ln x$ is defined for *all* real numbers $x > 0$, both rational and irrational. In Section 8.4 we will show that $\ln x$ can actually be written in the customary form, $\log_e x$, for a particular base number e, and we will justify the terminology *natural* logarithm. The remainder of this section concerns additional properties of $\ln x$.

Graph of $y = \ln x$

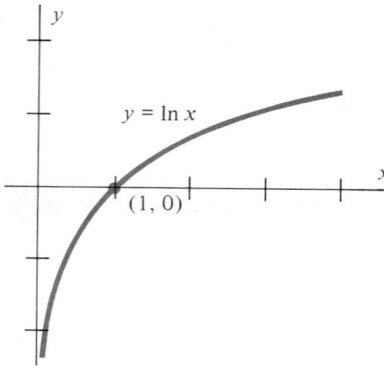

Figure 2.3 Graph of the natural logarithm function.

The graph of the natural logarithm function appears in Figure 2.3. It has the following properties:

(a) The domain of $y = \ln x$ is $(0, \infty)$.
(b) The function $y = \ln x$ is increasing for all x in its domain. (This follows from the fact that its derivative $1/x$ is positive for $x > 0$.)
(c) The graph of $y = \ln x$ is concave down on $(0, \infty)$. (The second derivative is

$$\frac{d^2}{dx^2} \ln x = \frac{d}{dx}\left(\frac{1}{x}\right) = -\frac{1}{x^2},$$

which is negative for all $x > 0$.)

(d) $\ln(1) = \int_1^1 \frac{1}{t}\, dt = 0.$

(e) $\lim_{x \to \infty} \ln x = +\infty$; $\quad \lim_{x \to 0^+} \ln x = -\infty$. Thus, the *range* of $f(x) = \ln x$ is $(-\infty, \infty)$.

(To see the first part of statement (e), we use the fact that $\ln x$ is an increasing function for all x. Thus $\ln 2 > \ln 1 = 0$. Since $\ln 2 > 0$,

$$\lim_{n \to \infty} \ln 2^n = \lim_{n \to \infty} n \cdot \ln 2 = +\infty$$

for any integer n. Since $\ln x$ is an increasing function, it follows that $\lim_{x \to \infty} \ln x = +\infty$.

The proof that $\lim_{x \to 0^+} \ln x = -\infty$ is similar).

The Number e

Since $y = \ln x$ is a continuous, increasing function whose range is the entire real line, the Intermediate Value Theorem guarantees that there is precisely one number x for which $\ln x = 1$.

We denote this number by the letter e, that is, we define the number e by the equation

$$\boxed{\ln e = 1.} \tag{5}$$

The number e has been shown to be an irrational number, and the decimal expansion for e correct to 12 decimal places is known to be

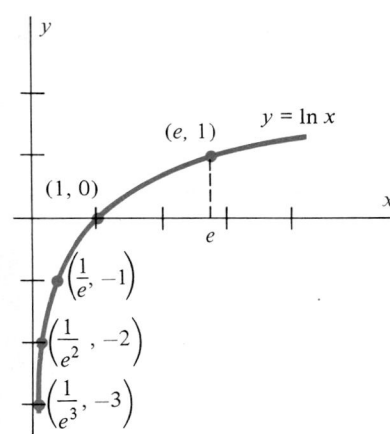

Figure 2.4 Points (e^n, n) on the graph of $y = \ln x$.

$$e \approx 2.718281828459.$$

We will see in the next two sections that the number e turns out to be the "natural" choice as a base for both the logarithmic and exponential functions. Note also that by property ($L3$) and equation (5) we have $\ln e^n = n \cdot \ln e = n \cdot 1 = n$ for all integers n:

$$\ln e^n = n.$$

This equation enables us to locate points on the graph of $y = \ln x$ of the form

$$(e^n, \ln e^n) = (e^n, n), \qquad n = 0, \pm 1, \pm 2, \ldots.$$

(See Table 2.1 and Figure 2.4.)

Table 2.1

n	$x = e^n$	$y = n \cdot \ln e$
-2	$\dfrac{1}{e^2} \approx 0.13534$	-2
-1	$\dfrac{1}{e} \approx 0.36788$	-1
0	$e^0 = 1$	0
1	$e \approx 2.71828$	1
2	$e^2 \approx 7.38906$	2

Natural Logarithms as Antiderivatives

From the differentiation formulas (2), we have the integration formula

$$\int \frac{1}{x}\, dx = \ln |x| + C, \qquad x \neq 0. \tag{6}$$

To see that equation (6) is valid for the case $x < 0$, recall that, in this case, $|x| = -x$. Thus

$$\frac{d}{dx} \ln |x| = \frac{d}{dx} \ln(-x) = \frac{1}{-x}(-1) = \frac{1}{x},$$

as required.

Example 2

Find $\displaystyle \int \frac{1}{2x + 3}\, dx.$

Solution: We use the substitution

$$u = 2x + 3; \qquad du = \left[\frac{d}{dx}(2x + 3)\right] dx = 2\, dx.$$

Then $dx = \dfrac{1}{2}\, du$, and we may apply equation (6) together with this substitution to obtain

$$\int \frac{1}{2x + 3}\, dx = \int \frac{1}{u}\left(\frac{1}{2}\right) du$$

$$= \frac{1}{2} \int \frac{1}{u}\, du$$

$$= \frac{1}{2} \ln |u| + C$$

$$= \frac{1}{2} \ln |2x + 3| + C. \qquad \diamond$$

Example 3

Find $\displaystyle\int \frac{1 + \cos x}{x + \sin x}\, dx, \qquad x + \sin x \neq 0.$

Solution: The numerator of the integrand appears to be the derivative of the denominator, so we try the substitution

$$u = x + \sin x; \qquad du = (1 + \cos x)\, dx.$$

Then

$$\int \frac{1 + \cos x}{x + \sin x}\, dx = \int \frac{1}{u}\, du$$

$$= \ln |u| + C$$
$$= \ln |x + \sin x| + C. \qquad \diamond$$

Example 4

Find $\displaystyle\int \frac{x + 1}{x^2 + 2x}\, dx.$

Solution: Note that the degree of the numerator is one less than the degree of the denominator. Accordingly, we make the substitution

$$u = x^2 + 2x.$$

Then

$$du = (2x + 2)\, dx = 2(x + 1)\, dx,$$

so

$$(x + 1)\, dx = \frac{1}{2}\, du.$$

With these substitutions we obtain

$$\int \frac{(x + 1)\, dx}{x^2 + 2x} = \int \frac{1}{u} \cdot \left(\frac{1}{2}\, du\right)$$

$$= \frac{1}{2} \int \frac{1}{u} \, du$$

$$= \frac{1}{2} \ln |u| + C$$

$$= \frac{1}{2} \ln |x^2 + 2x| + C. \qquad \diamond$$

Example 5

Find $\displaystyle\int \frac{x^2 + 3x}{x + 1} \, dx, \qquad x \neq -1.$

Strategy

The integrand is an improper fraction. Perform a division to express the integrand as the sum of a polynomial and a constant divided by $x + 1$.

Solution

We may rewrite the integrand as

$$\frac{x^2 + 3x}{x + 1} = x + 2 - \frac{2}{x + 1},$$

so

$$\int \frac{x^2 + 3x}{x + 1} \, dx = \int \left[x + 2 - \frac{2}{x + 1} \right] dx$$

Apply formula 6 in last term with

$u = x + 1, \qquad du = dx.$

$$= \int x \, dx + 2 \int dx - 2 \int \frac{1}{x + 1} \, dx$$

$$= \frac{x^2}{2} + 2x - 2 \ln |x + 1| + C. \qquad \diamond$$

Example 6

Find the area of the region bounded by the graph of $y = x - \dfrac{1}{x}$ and the x-axis between $x = 1$ and $x = e$.

Strategy

Determine where $f(x) = x - \dfrac{1}{x}$ is non-negative.

Use formula

$A = \displaystyle\int_a^b f(x) \, dx.$

Use facts that $\ln e = 1$ and $\ln 1 = 0$.

Solution

Since $x \geq \dfrac{1}{x}$ for $x \geq 1$, the function $y = x - \dfrac{1}{x}$ is nonnegative for $1 \leq x \leq e$.

The area A is therefore

$$A = \int_1^e \left(x - \frac{1}{x} \right) dx$$

$$= \left[\frac{1}{2} x^2 - \ln x \right]_1^e$$

$$= \left(\frac{1}{2} e^2 - \ln e \right) - \left(\frac{1}{2} - \ln 1 \right)$$

$$= \left(\frac{1}{2} e^2 - 1 \right) - \left(\frac{1}{2} - 0 \right)$$

Use approximation

$e \approx 2.71828.$

$$= \frac{1}{2}e^2 - \frac{3}{2} \approx 2.19.$$

(Figure 2.5.)

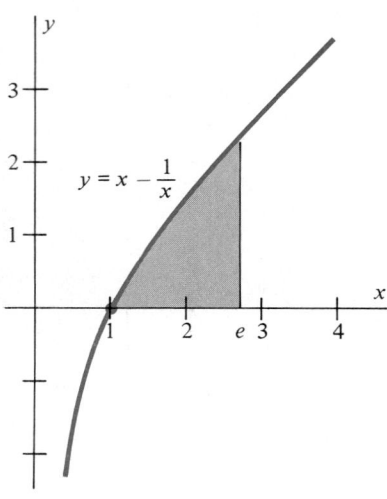

Figure 2.5 Graph of $y = x - 1/x$.

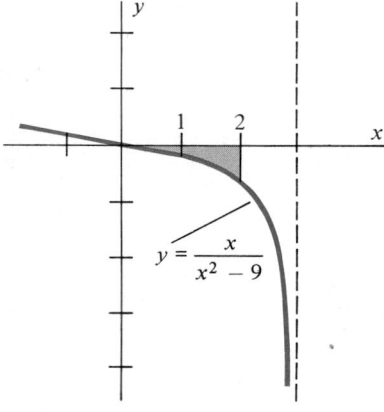

Figure 2.6 Graph of $y = \dfrac{x}{x^2 - 9}$.

Notice in the next example how we must take care to observe the absolute value signs in equation (6). The value of the integral is negative, since the region in the plane associated with the integral is bounded below by the nonpositive integrand (see Figure 2.6).

Example 7

Find $\displaystyle\int_0^2 \frac{x}{x^2 - 9}\, dx.$

Solution: We use the substitution

$$u = x^2 - 9; \qquad du = 2x\, dx.$$

Then $x\, dx = \dfrac{1}{2}\, du$, and the limits of integration change as follows:

$$\text{If } x = 0, \qquad u = 0^2 - 9 = -9$$
$$\text{If } x = 2, \qquad u = 2^2 - 9 = -5.$$

Then

$$\int_0^2 \frac{x}{x^2 - 9}\, dx = \int_{-9}^{-5} \frac{1}{u} \cdot \frac{1}{2}\, du = \frac{1}{2} \int_{-9}^{-5} \frac{1}{u}\, du$$

$$= \frac{1}{2} \ln |u|\,\Big]_{-9}^{-5}$$

$$= \frac{1}{2}[\ln |-5| - \ln |-9|]$$

$$= \frac{1}{2}(\ln 5 - \ln 9)$$

$$\approx -0.294.$$

The next example involves a natural logarithm, although this may not be obvious on first glance.

Example 8

Find $\displaystyle\int \frac{dx}{\sqrt{x}\,(1 + \sqrt{x})}.$

Solution: We try the substitution

$$u = 1 + \sqrt{x}; \qquad du = \frac{1}{2\sqrt{x}}\, dx.$$

Then $\dfrac{dx}{\sqrt{x}} = 2\ du$, and we obtain

$$\int \frac{dx}{\sqrt{x}\,(1 + \sqrt{x})} = \int \frac{1}{u} \cdot 2\ du = 2 \int \frac{1}{u}\ du$$

$$= 2\ \ln |u| + C$$
$$= 2\ \ln(1 + \sqrt{x}) + C.$$

$$(|1 + \sqrt{x}| = 1 + \sqrt{x} \quad \text{since} \quad \sqrt{x} \ge 0.)$$

Exercise Set 8.2

1. Use properties of ln x to solve for x. (You will need to use the table in Appendix IV or a hand calculator.)
a. $\ln 2x = 0$
b. $2 \ln x = \ln 2x$
c. $\ln(2/x) - \ln x = 0$
d. $\ln 3^{x^2} = 0$
e. $\sqrt{\ln \sqrt{x}} = 1$
f. $3 \ln x + x = 2 + \ln x^3$
g. $\displaystyle\int_1^x \frac{3}{t}\,dt = 6$
h. $\displaystyle\int_2^x \frac{1}{t}\,dt = 0$

2. Why may we conclude that the function $y = \ln x$ is continuous?

3. True or false? Every number is a natural logarithm.

In Exercises 4–19, find the derivative.

4. $y = \ln 2x$
5. $y = x \ln x$
6. $y = \ln(6 - x^2)$
7. $f(x) = \ln \sqrt{x^3 - x}, \quad x > 1$
8. $f(t) = \ln(\ln t), \quad t > 1$
9. $y = \sin(\ln x)$
10. $h(t) = \sqrt{1 + \ln t}$
11. $f(x) = \dfrac{x}{1 + \ln x}$
12. $f(x) = x^2 \ln^2 x$
13. $y = (3 \ln \sqrt{x})^4$
14. $f(x) = (\sin x) \ln(1 + \sqrt{x})$
15. $f(t) = \dfrac{\ln(a + bt)}{\ln(c + dt)}$
16. $y = \ln(\sin t - t \cos t)$
17. $f(x) = \ln \cos x$
18. $y = x \ln(x^2 - x - 3), \quad x > 3$
19. $u(t) = \dfrac{t}{1 + \ln^2 t}$

In Exercises 20–23, find $\dfrac{dy}{dx}$ by implicit differentiation.

20. $y = \ln(xy)$
21. $\ln(x + y) + \ln(x - y) = 1$
22. $x \ln y = 3$
23. $x \ln y + y \ln x = x$

In Exercises 24–26, find $\dfrac{dy}{dx}$.

24. $y = \displaystyle\int_5^x \frac{1}{t}\,dt$
25. $y = \displaystyle\int_x^1 \frac{1}{2t}\,dt, \quad x > 0$
26. $y = \displaystyle\int_1^{x^2} \ln 2t\,dt$

In each of Exercises 27–31, sketch the graph of the given function, noting all relative extrema.

27. $y = x \ln x, \quad x > 0$
28. $y = x - \ln x, \quad x > 0$
29. $y = \ln(2 + \sin x)$
30. $y = \ln(1 + x^2)$
31. $\ln(xy) = 1$

Evaluate the following integrals.

32. $\displaystyle\int \frac{dx}{2x + 1}$
33. $\displaystyle\int \frac{dx}{1 - x}$
34. $\displaystyle\int \frac{x\,dx}{x^2 + 1}$
35. $\displaystyle\int \frac{x - 1}{x^2 - 2x}\,dx$
36. $\displaystyle\int \frac{x}{1 - 3x^2}\,dx$
37. $\displaystyle\int \frac{x^2 + 3}{x^3 + 9x}\,dx$
38. $\displaystyle\int \cot x\,dx$
39. $\displaystyle\int \frac{dx}{x \ln x}$
40. $\displaystyle\int \frac{\sin t\,dt}{4 + 2 \cos t}$
41. $\displaystyle\int \frac{\ln^2 x}{x}\,dx$
42. $\displaystyle\int \frac{x^2}{x + 1}\,dx$
43. $\displaystyle\int \frac{1}{\sqrt{x}(1 - \sqrt{x})}\,dx$
44. $\displaystyle\int \frac{1 - 2t^2}{1 - t}\,dt$
45. $\displaystyle\int \frac{x^4 + 3x^2 + x + 1}{x + 1}\,dx$
46. $\displaystyle\int_1^e \frac{1}{x}\,dx$
47. $\displaystyle\int_e^{e^2} \frac{1}{x \ln x}\,dx$
48. $\displaystyle\int_1^e \frac{\ln x}{x}\,dx$
49. $\displaystyle\int_e^{e^2} \frac{1}{x \ln(x^2)}\,dx$

50. $\int_2^3 \frac{x}{x^2 + 1}\, dx$

51. $\int_1^2 \left(\frac{1}{1 + x} - \frac{1}{2 + x}\right) dx$

52. Find the equation of the line tangent to the graph of the equation $y = x(\ln x)^2 + \dfrac{x}{\ln x}$ at the point $(e, 2e)$.

53. Find the area of the region bounded by the graph of $y = \dfrac{1}{x - 2}$ and the x-axis for $3 \le x \le 4$.

54. Find the area of the region in the first quadrant bounded by the graphs of $x + y - 6 = 0$ and $xy = 8$.

55. Let ϵ be a small real number.
 a. Use differentials to obtain a formula for approximating the quantity $\ln(1 + \epsilon)$.
 b. Use the formula obtained in part (a) to approximate $\ln(0.8)$, $\ln(0.95)$, $\ln(1.05)$, and $\ln(1.2)$.
 c. Compare results obtained in part (b) with those obtained from a hand calculator or the table in Appendix IV.

56. True or false? For a given value of $x > 0$, the graphs of $y = \ln x$ and $y = \ln ax$ have the same slope.

57. Find the volume of the solid generated by rotating the region bounded by the graph of $y = \sqrt{\dfrac{2}{x - 1}}$ and the x-axis, for $3 \le x \le 5$, about the x-axis.

58. Verify that the function $F(x) = x \ln x - x + C$ is an antiderivative for $f(x) = \ln x$, $x > 0$, that is, show that

$$\int \ln x\, dx = x \ln x - x + C, \qquad x > 0.$$

59. Use the result of Exercise 58 to find the average value of the natural logarithm function on the interval $[1, e]$.

60. A particle moves along a line with acceleration $a(t) = \dfrac{1}{t + 2}$ m/s^2. Find the distance travelled by the particle during the time interval $[0, 4]$ if $v(0) = 0$. (*Hint:* Use Exercise 58.)

In Exercises 61–65, solve the differential equation.

61. $\dfrac{dy}{dx} = \dfrac{1}{x}, \qquad x > 0$

62. $(x^2 + 3)\, dy - 2x\, dx = 0$

63. $\cos x\, dy = \sin x\, dx$

64. $\dfrac{1}{y}\, dx = x\, dy$

65. $(2x + 1)\, dy = y^2\, dx$

66. Economists define the **growth of a function** $y = f(t)$ as the ratio

$$G = \frac{\dfrac{dy}{dt}}{y} = \frac{f'(t)}{f(t)} = \frac{y'}{y}.$$

For example, if $f(2) = 6$ and $f'(2) = 3$, then the growth of the function $y = f(t)$ at time $t = 2$ is $G = \dfrac{3}{6} = 0.5$.

 a. Show that the growth of a function may be calculated as the derivative of the natural logarithm of that function, that is, $G = \dfrac{d}{dt}\ln(f(t))$.
 b. Show that the growth of the product of two functions is the sum of the individual growths.
 c. An oil company determines that the price it obtains for heating oil is increasing at a rate of 15% per year but that the number of gallons of heating oil sold is decreasing by 10% per year. Find the rate at which revenues obtained from the sale of heating oil are increasing.

67. The vapor pressure P, in mm, of a certain fluid is related to its temperature T by the equation

$$\ln P = \frac{-2000}{T} + 5.5.$$

 a. Find $\dfrac{dP}{dT}$.
 b. Find $\dfrac{dT}{dP}$.

68. Prove statement $(L2)$ of Theorem 5.

69. Prove statement $(L3)$ of Theorem 5 for the case r rational. (The case r irrational must await the extension of the Power Rule $\dfrac{d}{dx} x^r = rx^{r-1}$ to irrational exponents. This is done in Section 8.4.)

70. Establish the inequality $2 \le e \le 4$ by explaining how each of the following statements leads to the next.
 a. For $1 \le x \le 2, \qquad \dfrac{1}{2} \le \dfrac{1}{x} \le 1$
 b. $\int_1^2 \dfrac{1}{2}\, dx \le \int_1^2 \dfrac{1}{x}\, dx \le \int_1^2 1\, dx$
 c. $\dfrac{1}{2} \le \ln 2 \le 1$
 d. $1 \le 2 \ln 2 \le 2$
 e. $\ln e \le \ln 4 \le \ln e^2$
 f. $e \le 4 \le e^2$
  (*Hint:* Use the fact that $f(x) = \ln x$ is an increasing function.)
 g. $2 \le e \le 4$

8.3 THE NATURAL EXPONENTIAL FUNCTION

In Section 8.2 we saw that the natural logarithm function is an increasing function for all $x \in (0, \infty)$ with range $(-\infty, \infty)$. These conditions are sufficient to guarantee the existence of the *inverse* of the natural logarithm function (see Theorem 2). We denote this function initially by exp(x). In other words, we *define* the exponential function as follows:

$$y = \exp(x) \qquad \text{if and only if} \qquad x = \ln y. \tag{1}$$

The definition embodied in statement (1) says that "the value of the exponential function exp(x) is the number y whose natural logarithm is the number x." Read the opposite way, the statement says that "the value of the natural logarithm function ln y is the number x for which the value of the exponential function is y."

Now recall from Section 8.1 that if the function g is the inverse of the function f then f is also the inverse of g on the range of f. Thus, $y = \ln x$ is the inverse of $y = \exp(x)$, and we have both the identities

$$\exp(\ln x) = x, \quad x > 0 \qquad \text{and} \qquad \ln(\exp(x)) = x, \quad -\infty < x < \infty.$$

The Graph of $y = \exp(x)$

Since $g(x) = \exp(x)$ is the inverse of $f(x) = \ln x$, the graph of $y = \exp(x)$ is the reflection across the line $y = x$ of the graph of $y = \ln x$. It therefore has the following properties (Figure 3.1):

1. The domain of $y = \exp(x)$ is $(-\infty, \infty)$ since this is the range of $y = \ln x$, that is, the function $y = \exp(x)$ is defined for all numbers.
2. The range of $y = \exp(x)$ is $(0, \infty)$. This is because the domain of $y = \ln x$ is $(0, \infty)$.
3. Since all points (e^n, n), $n = 0, \pm 1, \pm 2, \ldots$ lie on the graph of $y = \ln x$, the points (n, e^n), $n = 0, \pm 1, \pm 2, \ldots$ lie on the graph of the exponential function.
4. Since $y = \ln x$ is increasing for all $x \in (0, \infty)$, the function $y = \exp(x)$ is increasing for all $x \in (-\infty, \infty)$ (see Exercise 72).

Figures 3.2 and 3.3 show graphs of various functions of the form $y = \exp(kx)$.

Exp(x) is an Exponential Function

At this point we can show that the function exp(x) is really an exponential function, in the usual sense. In particular, for any rational number x,

$$\boxed{\exp(x) = e^x} \tag{2}$$

where e is the number defined by the equation ln $e = 1$. This follows from the fact that $\ln e^x = x \cdot \ln e = x \cdot 1 = x$ for any rational number x, according to Theorem 5. Since $\ln(e^x) = x$, $\exp(x) = e^x$ by statement (1).

From now on we will use the notation e^x rather than exp(x) to denote the natural exponential function. However, you should keep in mind the fact that statements (1) and (2) together define e^x for *all* real exponents x, both rational and irrational.

The following theorem shows that the function e^x satisfies the characteristic properties of exponential functions for *all* numbers x.

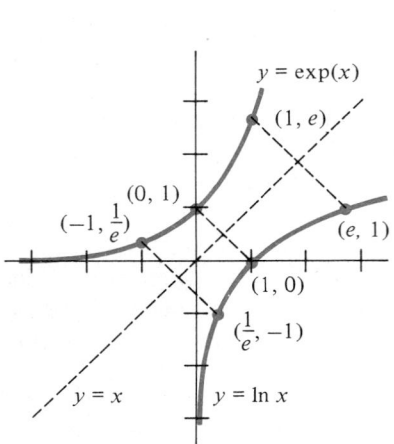

Figure 3.1 Function $y = \exp(x)$ is the inverse of the natural log function $y = \ln x$.

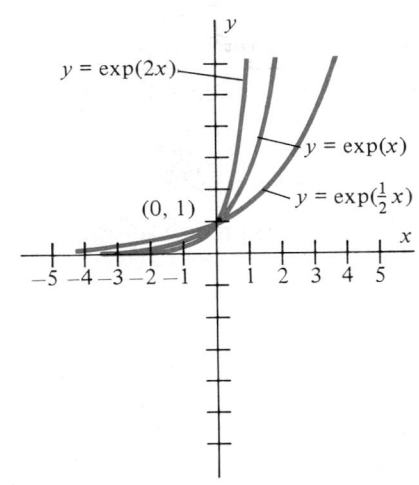

Figure 3.2 Graphs of $y = \exp(kx)$, $k > 0$.

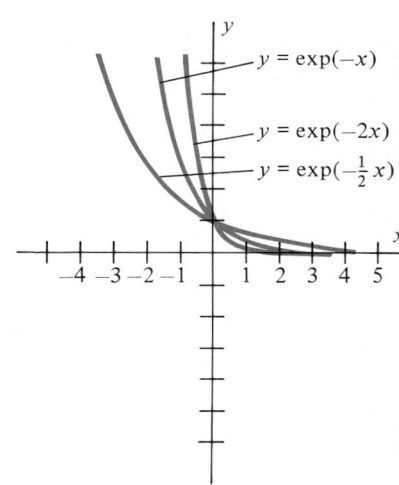

Figure 3.3 Graphs of $y = \exp(kx)$, $k < 0$.

THEOREM 6

Let x_1, x_2, and r be any real numbers with r rational. Then

(i) $e^{x_1} \cdot e^{x_2} = e^{x_1 + x_2}$,

(ii) $\dfrac{e^{x_1}}{e^{x_2}} = e^{x_1 - x_2}$,

(iii) $[e^{x_1}]^r = e^{rx_1}$, r rational.

Proof: Let $y_1 = e^{x_1}$ and $y_2 = e^{x_2}$.

Then

$$\ln y_1 = x_1 \quad \text{and} \quad \ln y_2 = x_2.$$

By property ($L1$), Section 8.2,

$$x_1 + x_2 = \ln y_1 + \ln y_2 = \ln(y_1 y_2).$$

Thus

$$e^{x_1 + x_2} = e^{\ln(y_1 y_2)}$$
$$= y_1 y_2$$
$$= e^{x_1} \cdot e^{x_2}$$

as required. This proves statement (i). The proofs of statements (ii) and (iii) are similar, and are left as Exercises 73 and 74. ◆

The Derivative of $y = e^x$

In asking for the derivative of the exponential function, we uncover one of the most remarkable facts of the calculus. Recall from Section 8.1 that if the function f has an inverse g, and if f is differentiable in an interval containing x_0, then g is also differentiable at $y_0 = f(x_0)$ if $f'(x_0) \neq 0$. Thus, since $y = \ln x$ is differentiable and increasing throughout its domain, its derivative exists for all $x \in (0, \infty)$ and is never zero. By Theorem 2 we may therefore conclude that its inverse, $y = e^x$, is differentiable for all x in $(-\infty, \infty)$, the range of $y = \ln x$.

To obtain the derivative of $y = e^x$, we apply the Chain Rule to the identity

$$\ln e^x = x, \qquad -\infty < x < \infty. \tag{3}$$

Differentiating both sides of equation (3) we obtain

$$\frac{1}{e^x} \cdot \frac{d}{dx} e^x = 1, \qquad -\infty < x < \infty. \tag{4}$$

Multiplying both sides of equation (4) by e^x gives the desired result:

$$\boxed{\frac{d}{dx} e^x = e^x.} \tag{5}$$

In other words, the exponential function $y = e^x$ is its own derivative! This function, and its multiples, are the only functions in the calculus with this property. (In fact, we could develop the entire theory of exponential and logarithm functions by seeking a function f for which $f'(x) = f(x)$.) Figure 3.4 illustrates the geometric interpretation of formula (5): The slope of the tangent to the graph of $y = e^x$ at (x, e^x) is precisely the y-coordinate e^x.

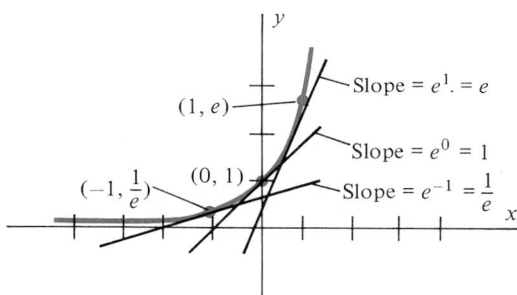

Figure 3.4 Slope of $y = e^x$ equals y-coordinate.

Example 1

Find $\dfrac{dy}{dx}$ for (a) $y = e^{6x}$, and (b) $y = e^{x \sin x}$.

Solution

(a) Equation (5) and the Chain Rule give

$$\frac{d}{dx}(e^{6x}) = e^{6x} \cdot \frac{d}{dx}(6x) = 6e^{6x}.$$

(b) Similarly,

$$\frac{d}{dx}(e^{x \sin x}) = e^{x \sin x} \cdot \frac{d}{dx}(x \sin x)$$

$$= e^{x \sin x}[\sin x + x \cos x].$$

Example 2

Find all relative extrema for the function $f(x) = x^2 e^{-x}$.

Strategy

Find f' using the product rule.

Solution

The first derivative is

$$f'(x) = \left(\frac{d}{dx} x^2\right) e^{-x} + x^2\left(\frac{d}{dx} e^{-x}\right)$$

$$= 2xe^{-x} - x^2 e^{-x}$$

$$= x(2 - x)e^{-x}.$$

Set $f'(x) = 0$ and solve to find the critical numbers.

Since $e^{-x} = \dfrac{1}{e^x}$ is nonzero for all x, the equation $f'(x) = 0$ gives $x = 0$ or $x = 2$. These are the critical numbers.

The second derivative is

Find f'' using the product rule.

$$f''(x) = \frac{d}{dx}(2xe^{-x} - x^2 e^{-x})$$

$$= (2e^{-x} - 2xe^{-x}) - (2xe^{-x} - x^2 e^{-x})$$

$$= (2 - 4x + x^2)e^{-x}.$$

Determine the sign of $f''(c)$ for each critical number c.

Using f'' to check the critical numbers for relative extrema gives

$$f''(0) = 2e^0 = 2 > 0$$

and

$$f''(2) = (2 - 8 + 4)e^{-2} = -\frac{2}{e^2} < 0.$$

Apply the Second Derivative Test for relative extrema.

Thus, the second derivative test shows that $f(0) = 0$ is a relative minimum value for f, and $f(2) = 4e^{-2}$ is a relative maximum.

The graph of f appears in Figure 3.5. ◇

Integrals involving e^x

The differentiation formula (5) gives the integration formula

$$\int e^x \, dx = e^x + C. \tag{6}$$

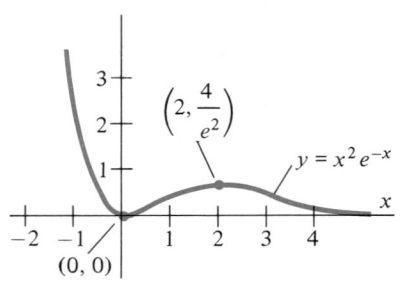

Figure 3.5 Graph of $f(x) = x^2 e^{-x}$ has relative minimum $f(0) = 0$ and relative maximum $f(2) = 4e^{-2}$.

Example 3

Find the following integrals:

(a) $\displaystyle\int e^{-3x} \, dx$

(b) $\displaystyle\int x^2 e^{x^3+1} \, dx.$

Strategy
Make a substitution with

$u = -3x$
$du = -3\ dx$.

Solution

(a) For $u = -3x$, $du = -3\ dx$, so $dx = -\dfrac{1}{3}\ du$. We may therefore write

$$\int e^{-3x}\ dx = \int e^{u}\left(-\frac{1}{3}\right)\ du$$

$$= -\frac{1}{3}\int e^{u}\ du$$

Use equation (6).

$$= -\frac{1}{3}e^{u} + C$$

$$= -\frac{1}{3}e^{-3x} + C.$$

Try to view the factor x^2 as a factor of du. This works for the substitution

$u = x^3 + 1$.

(b) If we use the substitution $u = x^3 + 1$, we have $du = 3x^2\ dx$, so $x^2\ dx = \dfrac{1}{3}\ du$. Thus

$$\int x^2 e^{x^3+1}\ dx = \int e^{u} \cdot \frac{1}{3} \cdot du$$

$$= \frac{1}{3}\int e^{u}\ du$$

Use equation (6).

$$= \frac{1}{3}e^{u} + C$$

$$= \frac{1}{3}e^{x^3+1} + C. \qquad \diamond$$

Example 4

Find the area of the region bounded by the graph of $y = xe^{1-x^2}$ and the x-axis for $-2 \le x \le 2$ (see Figure 3.6).

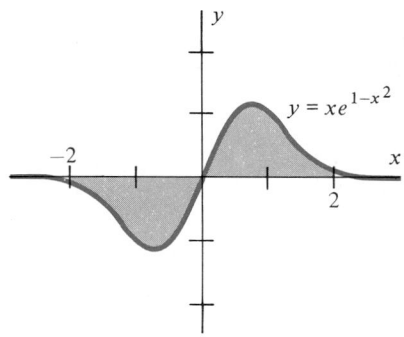

Figure 3.6 Graph of $y = xe^{1-x^2}$ showing region in Example 4.

Strategy
Determine where graph lies above x-axis.

Solution
Since $y = xe^{1-x^2}$ is an odd function, its graph is symmetric about the origin.

Area on $[-2, 2]$ equals twice the area on $[0, 2]$.

Since $xe^{1-x^2} \geq 0$ for $x > 0$, we may calculate the desired area A as

$$A = 2 \cdot \int_0^2 xe^{1-x^2}\, dx.$$

Find a substitution u so that x is a factor of du. The substitution $u = 1 - x^2$ works.

To evaluate this integral, we use the substitution

$$u = 1 - x^2, \qquad du = -2x\, dx.$$

Then

$$x\, dx = -\frac{1}{2}\, du$$

Change limits of integration according to u-substitution.

and the limits of integration become

$$u = 1 - 0^2 = 1 \qquad \text{if} \qquad x = 0$$
$$u = 1 - 2^2 = -3 \qquad \text{if} \qquad x = 2.$$

With these substitutions, we obtain

Apply equation (6).

$$A = 2 \int_0^2 xe^{1-x^2}\, dx$$

$$= 2 \int_1^{-3} e^u \left(-\frac{1}{2}\right) du$$

$$= 2\left(-\frac{1}{2}\right) \int_1^{-3} e^u\, du$$

Apply Fundamental Theorem of Calculus.

$$= -[e^u]_1^{-3}$$
$$= -[e^{-3} - e]$$

$$= e - \frac{1}{e^3}$$

$$\approx 2.67. \qquad\qquad \diamond$$

More on the Number e

The number e is difficult to approximate from the defining property $\ln e = \int_1^e \frac{1}{x}\, dx = 1$, since it appears as a limit of integration. This same number arises in a seemingly unrelated manner as the limit

$$e = \lim_{x \to \infty} \left(1 + \frac{1}{x}\right)^x. \tag{7}$$

It is much easier to approximate e as this limit, since we need only calculate values of the function $\left(1 + \frac{1}{x}\right)^x$ for x large. This limit is a special case of the more general statement

$$e^r = \lim_{x \to \infty} \left(1 + \frac{r}{x}\right)^x, \qquad r \text{ constant}, \tag{8}$$

which we shall prove in Chapter 11.

The results of approximating e by the expression $\left(1 + \dfrac{1}{n}\right)^n$ for various values of n appear in Table 3.1.

Table 3.1 Approximating $e \approx 2.7182818 \ldots$,
using $\left(1 + \dfrac{1}{n}\right)^n$

n	$\left(1 + \dfrac{1}{n}\right)^n$	n	$\left(1 + \dfrac{1}{n}\right)^n$
1	2.000000	500	2.715569
5	2.488320	1,000	2.716924
20	2.653298	2,500	2.717738
50	2.691588	5,000	2.718010
100	2.704814	10,000	2.718146
250	2.712865	100,000	2.718268

Example 5

According to equation (8) we have

(a) $\displaystyle\lim_{x \to \infty} \left(1 - \frac{1}{x}\right)^x = e^{-1} = \frac{1}{e}$

(b) $\displaystyle\lim_{x \to \infty} \left(1 + \frac{3}{x}\right)^x = e^3$

(c) $\displaystyle\lim_{x \to \infty} \left(1 + \frac{6}{x}\right)^{2x} = \lim_{x \to \infty} \left[\left(1 + \frac{6}{x}\right)^x\right]^2 = (e^6)^2 = e^{12}$

(d) $\displaystyle\lim_{x \to \infty} \left(1 - \frac{1}{x^2}\right)^x = \lim_{x \to \infty} \left[\left(1 - \frac{1}{x}\right)\left(1 + \frac{1}{x}\right)\right]^x$

$\displaystyle\qquad\qquad\qquad = \lim_{x \to \infty} \left[\left(1 - \frac{1}{x}\right)^x \cdot \left(1 + \frac{1}{x}\right)^x\right]$

$\qquad\qquad\qquad = e^{-1} \cdot e^1$

$\qquad\qquad\qquad = 1.$ ◇

Exercise Set 8.3

1. Simplify the following expressions

a. $e^{\ln 2}$
b. $e^{-\ln 4}$
c. $e^{(\ln x - \ln y)}$
d. $\ln e^{-x^2}$
e. $\ln x e^{\sqrt{x}} - \ln x$
f. $e^{x \ln 2}$
g. $e^{\ln(1/x)}$
h. $e^{4 \ln x}$
i. $\ln x e^{x^2}$
j. $e^{x - \ln x}$

2. Solve the following equations for x.

a. $\ln x = 2$
b. $\ln x^2 = 9$
c. $e^{x^2} = 5$
d. $e^{2x} - 2e^x + 1 = 0$

3. Find y if $e^{x-y} = x + 3$.

4. Find y if $e^{(y-1)^2} = x^4 + 1$. $y > 1$.

In Exercises 5–20, find the derivative of the given function.

5. $y = e^{3x}$
6. $f(x) = xe^{-x}$

7. $f(t) = e^{\sqrt{t}}$
8. $y = \dfrac{e^x + 1}{e^x}$

9. $y = e^{x^2 - x}$
10. $f(t) = e^{\sin t}$

11. $f(x) = e^x \sin x$

12. $y = xe^{-3\ln x}$

13. $f(x) = \ln \dfrac{e^x + 1}{x + 1}$

14. $f(x) = \dfrac{e^x - 1}{e^x + 1}$

15. $f(x) = (2 - e^{x^2})^3$

16. $y = x \ln(e^x + x)$

17. $y = \dfrac{1}{2}(e^x + e^{-x})$

18. $f(x) = (x^2 + x - 1)e^{x^2+3}$

19. $y = e^{\sqrt{x}} \cdot \ln \sqrt{x}$

20. $f(t) = te^{(1/t)^2}$

In Exercises 21–24, find $\dfrac{dy}{dx}$ by implicit differentiation.

21. $e^{xy} = x$

22. $e^{x-y} = ye^x$

23. $\ln(x + 2y) = e^y$

24. $y^2 e^x + y \ln x = 2$

25. True or false? For $y = e^{kx}$, y' is proportional to y.

26. True or false? The equation $y = e^x$ has a solution x for every $y \geq 0$.

27. True or false? You can *always* divide by e^x.

In Exercises 28–41, evaluate the given indefinite integrals.

28. $\displaystyle \int e^{2x}\, dx$

29. $\displaystyle \int e^{-x}\, dx$

30. $\displaystyle \int e^{2x+6}\, dx$

31. $\displaystyle \int xe^{x^2+3}\, dx$

32. $\displaystyle \int x^2 e^{1-x^3}\, dx$

33. $\displaystyle \int e^{2x}(1 + e^{2x})^3\, dx$

34. $\displaystyle \int \dfrac{e^{\sqrt{x}}}{\sqrt{x}}\, dx$

35. $\displaystyle \int \dfrac{e^{1/x}}{x^2}\, dx$

36. $\displaystyle \int \dfrac{e^x}{\sqrt{e^x + 1}}\, dx$

37. $\displaystyle \int \dfrac{e^x}{1 + e^x}\, dx$

38. $\displaystyle \int \cos x \cdot e^{\sin x}\, dx$

39. $\displaystyle \int \dfrac{e^x}{(3 + e^x)^2}\, dx$

40. $\displaystyle \int \dfrac{1 + e^{-ax}}{1 - e^{-ax}}\, dx$

41. $\displaystyle \int \dfrac{(1 + e^{\sqrt{x}})e^{\sqrt{x}}}{\sqrt{x}}\, dx$

In Exercises 42–49, evaluate the definite integrals.

42. $\displaystyle \int_0^1 e^{2x}\, dx$

43. $\displaystyle \int_1^{\ln 4} e^{-x}\, dx$

44. $\displaystyle \int_0^1 xe^{1-x^2}\, dx$

45. $\displaystyle \int_0^{\pi/2} \cos x e^{\sin x}\, dx$

46. $\displaystyle \int_0^{\ln 2} \dfrac{e^x}{2 + e^x}\, dx$

47. $\displaystyle \int_1^{e^2} \dfrac{\ln x}{x}\, dx$

48. $\displaystyle \int_0^2 \dfrac{e^x + e^{-x}}{2}\, dx$

49. $\displaystyle \int_0^2 e^x\, dx$

50. Find the maximum value of the function $y = (3 - x^2)e^x$.

51. Find all relative extrema for the function $y = x^2 e^{1-x^2}$.

52. Find the area of the region bounded by the graph of $y = \ln x$ and the lines $y = 0$, $y = 1$ and the y-axis. (*Hint:* Integrate with respect to x.)

53. For the function $f(x) = xe^{2x}$, find
   a. the largest interval(s) on which f is increasing
   b. the largest interval(s) on which f is decreasing
   c. all relative extrema
   d. the largest interval(s) on which the graph of f is concave up
   e. the largest interval(s) on which the graph of f is concave down
   f. the point(s) of inflection.
   Use this information to sketch the graph of f.

54. Apply the instructions of Exercise 53 to the function $f(x) = xe^{1-x^3}$.

55. Find the length of the graph of $y = \dfrac{1}{2}(e^x + e^{-x})$ from $(0, 1)$ to $(\ln 2, 5/4)$.

56. Find the lateral surface area of the solid obtained by revolving the region bounded by the arc of the graph in Exercise 55 and the x-axis about the x-axis.

57. Solve the differential equation $\dfrac{dy}{dx} = 2xy$.

58. Determine c, $c > 4$, so that $\displaystyle \int_4^c \dfrac{1}{x - 3}\, dx = 1$.

59. The line $y = -\dfrac{1}{e}$ is tangent to the graph of $y = xe^x$ at point P. Find P.

60. Find the area of the region bounded by the graphs of $y = e^x$ and $y = e^{-x}$ for $-1 \leq x \leq 1$.

61. Find the volume of the solid obtained by revolving about the x-axis the region bounded by the graph of $y = e^{2x}$ and the x-axis for $0 \leq x \leq 2$.

62. Show that the function $y = Ae^{kt}$ satisfies the differential equation $y' = ky$ and the initial condition $y(0) = A$. Use this information to solve the following *initial value problems:*
   a. $y' = y$, $\quad y(0) = 1$
   b. $y' = \pi y$, $\quad y(0) = -2$.
   c. $y' + 3y = 0$, $\quad y(0) = 2$.

63. Find the average value of the function $y = xe^{1-x^2}$ on the interval $[-2, 2]$.

64. Show that $f(x) = \dfrac{e^x + e^{-x}}{2}$ is an even function, and that
$$g(x) = \dfrac{e^x - e^{-x}}{2}$$ is an odd function.

65. Lambert's law states that the intensity of light, after passing through a thickness l of absorbing liquid, is $I = I_0 e^{-kl}$ where I_0 and k are constants. Find k if $\dfrac{dI}{dl} = 4I$.

66. The function $f(x) = \dfrac{1}{\sigma\sqrt{2\pi}} e^{-(x-\mu)^2/2\sigma^2}$ is called the *normal* probability density function with mean μ and variance σ^2.
  a. Show that the graph of f is symmetric about the line $x = \mu$.
  b. Show that inflection points exist for $x = \mu \pm \sigma$.
  c. Graph f, obtaining the familiar bell-shaped curve.

67. Human growth in height from age one year to adulthood typically looks like the graph in Figure 3.7. A function that

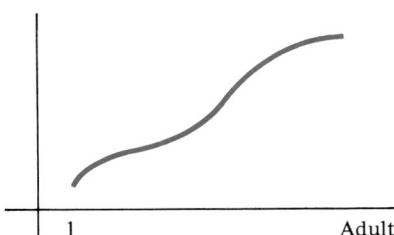

Figure 3.7 Typical double logistic.

has been used to model this kind of growth is the **double logistic**

$$y(t) = \frac{a}{1 + e^{-b_1(t-c_1)}} + \frac{f - a}{1 + e^{-b_2(t-c_2)}}$$

where the parameters a, b_1, b_2, c_1, c_2, and f are different for each individual.
  a. Show that the parameter f is adult height, that is, show that
  $$f = \lim_{t \to \infty} y(t).$$
  b. If $b_1 = b_2$ and $c_1 = c_2 = c$, show that the time of most rapid growth is $t = c$.

68. (*Computer*) An ion with n more electrons than protons carries a negative charge of $-np$ where p is the charge on one electron. An ion with m fewer electrons than protons carries a charge of mp. The attraction potential energy between two such ions is
$$E_+ = \frac{-mnp^2}{r}$$
where r is the distance separating the nuclei. However, since the electrons of the two ions repel each other there is a repulsion potential energy of
$$E_- = ae^{-r/b}$$
where a and b are constants. Assume $m = n = p = b = 1$ and $a = 10$, and let $E = E_+ + E_-$ be the total potential energy.
  a. Find the relative maximum and minimum values of the total potential energy on $(0, \infty)$ and the numbers r for which they occur. (*Hint:* You will need to use Newton's Method to solve the equation $\dfrac{dE}{dr} = 0$. There are two zeros of this equation.)
  b. Graph the function E.

69. (*Computer*) Solve the equation $e^{2x} - 8x + 1 = 0$. (*Hint:* Use Newton's Method.)

70. (*Computer*) Solve the equation $e^x + x = 5$, $x > 0$.

71. (*Computer*) Write a program to calculate entries for the following table, comparing actual values of e^x with those of the approximating polynomials
$$P_1(x) = 1 + x,$$
$$P_2(x) = 1 + x + \frac{x^2}{2!},$$
$$P_3(x) = 1 + x + \frac{x^2}{2!} + \frac{x^3}{3!},$$
$$P_4(x) = 1 + x + \frac{x^2}{2!} + \frac{x^3}{3!} + \frac{x^4}{4!}.$$
(Recall, $n! = n(n-1)(n-2) \cdot \cdots \cdot 2 \cdot 1$.)

x	e^x	$P_1(x)$	$P_2(x)$	$P_3(x)$	$P_4(x)$
1					
0.5					
2					
−2					

72. Prove that if the function f has an inverse f^{-1} defined for all x, then f^{-1} is increasing on the interval $[f(a), f(b)]$ if f is increasing on the interval $[a, b]$.

73. Prove statement (ii) of Theorem 6.

74. Prove statement (iii) of Theorem 6 for the case r rational.

8.4 EXPONENTIALS AND LOGS TO OTHER BASES

By use of the (natural) exponential function e^x, we can extend the domain of the exponential function $y = a^x$, $a > 0$, to *all* real numbers, both rational and irrational.

To do so we use the identity

$$a = e^{\ln a}, \qquad a > 0$$

to write

$$a^x = [e^{\ln a}]^x. \tag{1}$$

Now if x is rational, we can apply Theorem 6 to conclude that

$$[e^{\ln a}]^x = e^{x \ln a}. \tag{2}$$

Combining equations (1) and (2) we conclude that

$$\boxed{a^x = e^{x \ln a}} \tag{3}$$

for x rational. However, the right side of equation (3) is defined for all real numbers x and all real numbers $a > 0$. We therefore *define* the function $y = a^x$ by equation (3). Our demonstration shows that this definition agrees with the algebraic definition of a^x when x is rational, so we have now succeeded in extending this definition to all real numbers x. Using equation (3) we can extend the statements of Theorems 5 and 6 as follows:

> In Theorem 5, the statement
>
> $$\ln a^r = r \ln a \tag{4}$$
>
> holds for all real numbers $a > 0$ and r, whether or not r is rational.

Proof: Using equation (3) with $x = r$, we have

$$\ln a^r = \ln(e^{r \ln a}) = \ln e^{(r \ln a)} = r \ln a.$$

◆

> In Theorem 6, the statement
>
> $$[e^x]^r = e^{rx} \tag{5}$$
>
> holds for all real numbers x and r, whether or not r is rational.

Proof: Let $y = (e^x)^r$. Then, by equation (4),

$$\ln y = \ln [e^x]^r = r \ln e^x = rx.$$

Thus, $y = e^{rx}$, so $[e^x]^r = e^{rx}$.

◆

The following theorem shows that we have succeeded in preserving the usual laws of exponents in making this extension.

THEOREM 7

If $a > 0$ is any real number, the equations

(E1) $\qquad a^x a^y = a^{x+y}$

(E2) $\qquad a^x / a^y = a^{x-y}$

(E3) $\qquad (a^x)^y = a^{xy}$

hold for all real numbers x and y.

Proof: We prove only (E3), leaving (E1) and (E2) as exercises:

$$
\begin{aligned}
(a^x)^y &= (e^{x \ln a})^y && \text{(equation (3))} \\
&= e^{xy \ln a} && \text{(equation (5))} \\
&= a^{xy} && \text{(equation (3))}.
\end{aligned}
$$

(See Exercises 30 and 31.) ◆

Differentiating $y = a^x$

Since the exponential function $y = e^x$ and the linear function $y = (\ln a)x$ are both differentiable, equation (3) defines the function $y = a^x$ as the composition of two differentiable functions. As such, $y = a^x$ must be differentiable. To find its derivative we apply the Chain Rule to the right-hand side of equation (3):

$$
\begin{aligned}
\frac{d}{dx} a^x &= \frac{d}{dx} e^{x \ln a} \\
&= e^{x \ln a} \cdot \frac{d}{dx} (x \ln a) \\
&= e^{x \ln a} \cdot \ln a \\
&= a^x \ln a.
\end{aligned}
$$

That is,

$$
\boxed{\frac{d}{dx} a^x = a^x \ln a.}
\qquad (6)
$$

Example 1

Find $\dfrac{dI}{dx}$ for the Beer-Lambert law of Exercise 43, Section 8.1.

Solution: Recall that

$$
I = I_0 10^{kx}
$$

where I_0 and k are constants. Equation (6) gives

$$
\frac{dI}{dx} = I_0 10^{kx} \cdot \ln 10 \cdot \frac{d}{dx} (kx) = k I_0 10^{kx} \ln 10. \qquad \diamond
$$

Observe that equation (6) agrees with the differentiation formula for $y = e^x$ when $a = e$, since $\ln e = 1$. This fact explains why the number e is referred to as the *natural* choice for the base of an exponential function: among all bases, e is the one for which the derivative of the exponential function has the simplest form.

The integration formula corresponding to the differentiation formula (6) is

$$
\boxed{\int a^x \, dx = \frac{a^x}{\ln a} + C, \qquad a > 0, \qquad a \neq 1.}
\qquad (7)
$$

Example 2

Find

$$
\int \sin x \cos x \cdot 10^{\sin^2 x} \, dx.
$$

Solution: Since the exponent is $\sin^2 x$, we try the substitution

$$u = \sin^2 x; \qquad du = 2 \sin x \cos x \, dx.$$

Then $\sin x \cos x \, dx = \dfrac{1}{2} \, du$, and we obtain

$$\int \sin x \cos x \cdot 10^{\sin^2 x} \, dx = \int 10^u \cdot \frac{1}{2} \, du$$

$$= \frac{1}{2} \int 10^u \, du$$

$$= \frac{1}{2 \ln 10} 10^u + C$$

$$= \frac{1}{2 \ln 10} 10^{\sin^2 x} + C. \qquad \diamondsuit$$

The Function $y = \log_a x$

For any positive real number $a \neq 1$, equation (3) together with the algebraic definition of logarithm allows us to view the equation $y = \log_a x$ as defining a differentiable function of $x > 0$. To see this we write

$$y = \log_a x \qquad \text{if and only if} \qquad x = a^y = e^{y \ln a}. \tag{8}$$

Applying the natural logarithm to both sides of the equation

$$x = e^{y \ln a}$$

gives

$$\ln x = \ln[e^{y \ln a}] = y \ln a,$$

so

$$y = \frac{\ln x}{\ln a}, \qquad a \neq 1. \tag{9}$$

Thus, the function $\log_a x$ is simply a multiple of the function $\ln x$, and conversely. Statements (8) and (9) together give the equation

$$\log_a x = \frac{\ln x}{\ln a}, \qquad a \neq 1. \tag{10}$$

Example 3

Find formulas for converting common logs to natural logs and vice versa.

Strategy

Find ln 10 from a natural log table or calculator.

Solution

Common logs are logarithms to the base $a = 10$. Since

$$\ln(10) = 2.303,$$

equation (10) gives

Write equation (10) with $a = 10$.

$$\log_{10} x \approx \frac{1}{2.303} \ln x = .434 \ln x.$$

Solve equation (10) for ln x.

Solving for $\ln x$ we obtain

$$\ln x \approx 2.303 \log_{10} x.$$

$\diamond$

The differentiation formula for $y = \log_a x$ can be found by differentiating both sides of equation (10). We obtain

$$\frac{d}{dx} \log_a x = \frac{1}{x \ln a}, \qquad x > 0, \qquad a \neq 1. \tag{11}$$

REMARK: Equation (11) shows why $\ln x = \log_e x$ is called the *natural* logarithm function. Among all logarithm functions, $\ln x$ is the one for which the derivative has simplest form, since the factor $\ln a$ is just the integer 1.

Example 4

Find the minimum value of the function

$$y = \log_{10}(1 + x^2), \qquad -\infty < x < \infty.$$

Solution: To find the critical numbers we set

$$\frac{dy}{dx} = \frac{2x}{(1 + x^2) \ln 10} = 0 \tag{12}$$

and obtain the single critical number $x = 0$.

From the form of $\frac{dy}{dx}$ in equation (12) it is easy to see that $\frac{dy}{dx} < 0$ if $x < 0$, and $\frac{dy}{dx} > 0$ if $x > 0$. Thus, $y = \log_{10}(1 + x^2)$ decreases on $(-\infty, 0)$ and increases on $(0, \infty)$. The value $y = 0$ at the point $(0, 0)$ is therefore a relative (and absolute) minimum value of y on $(-\infty, \infty)$.

$\diamond$

Logarithmic Differentiation

There are two types of functions for which it is advantageous to calculate the derivative by differentiating the natural logarithm of the function rather than the function itself. The first concerns functions of the form

$$y = u(x)^{v(x)}, \qquad u(x) \geq 0. \tag{13}$$

Since the exponent $v(x)$ is not constant, the Power Rule is not applicable. Moreover, the base $u(x)$ is not constant, so the rule (6) for differentiating an exponential function does not apply. However, by applying the natural logarithm function to both sides of equation (13) we obtain

$$\ln y = \ln[u(x)^{v(x)}] = v(x) \ln u(x). \tag{14}$$

We may then obtain $\frac{dy}{dx}$ from equation (14) by the technique of implicit differentiation.

Example 5

Find $\frac{dy}{dx}$ for $y = x^{\sin x}$, $\qquad x > 0$

Strategy

Take natural logs of both sides.

Solution

Since $y = x^{\sin x}$ we use equation (4) to write

$$\ln y = \ln[x^{\sin x}] = \sin x \cdot \ln x.$$

Differentiate both sides, using Chain Rule on left, Product Rule on right.

Differentiating implicitly we find that

$$\frac{1}{y} \cdot \frac{dy}{dx} = \cos x \cdot \ln x + \frac{\sin x}{x},$$

so

Solve for $\dfrac{dy}{dx}$ and substitute for y.

$$\frac{dy}{dx} = y\left[\cos x \cdot \ln x + \frac{\sin x}{x}\right]$$

$$= x^{\sin x}\left[\cos x \cdot \ln x + \frac{\sin x}{x}\right].$$

◇

The second type of function for which this technique is useful is one that is complicated by a large number of algebraic operations. Properties (L1)–(L3) can often be applied to greatly simplify the form of $\ln y$. The derivative $\dfrac{dy}{dx}$ is then calculated implicitly as above.

Example 6

Find $\dfrac{dy}{dx}$ for $y = \dfrac{\sqrt{1 + x^2}(3x + 2)^3}{\sqrt[3]{x^2(x + 1)}}$.

Strategy

Take natural logs of both sides.

Use properties (L1)–(L3) of natural logs to simplify right-hand side.

Solution

We obtain

$$\ln y = \ln\left\{\frac{\sqrt{1 + x^2}(3x + 2)^3}{\sqrt[3]{x^2(x + 1)}}\right\}$$

$$= \ln[(1 + x^2)^{1/2}(3x + 2)^3] - \ln[x^2(x + 1)]^{1/3}$$

$$= \frac{1}{2}\ln(1 + x^2) + 3\ln(3x + 2)$$

$$- \frac{1}{3}[2 \cdot \ln x + \ln(x + 1)].$$

Differentiate both sides.

Differentiating both sides then gives

$$\frac{1}{y} \cdot \frac{dy}{dx} = \frac{1}{2} \cdot \frac{2x}{1 + x^2} + 3 \cdot \frac{3}{3x + 2} - \frac{1}{3}\left[\frac{2}{x} + \frac{1}{x + 1}\right]$$

$$= \frac{x}{1 + x^2} + \frac{9}{3x + 2} - \frac{2}{3x} - \frac{1}{3(x + 1)},$$

so

Solve for $\dfrac{dy}{dx}$.

$$\frac{dy}{dx} = y\left[\frac{x}{1 + x^2} + \frac{9}{3x + 2} - \frac{2}{3x} - \frac{1}{3(x + 1)}\right].$$

If necessary, you could substitute for y in terms of x.

◇

Exercise Set 8.4

1. Write each of the following in the form e^u.
  a. $2x$
  b. π^3
  c. $7^{\ln x}$
  d. $4^{\sqrt{2}}$
  e. $3^{\sin x}$
  f. $2^x \cdot 4^{1-x}$

2. Use a calculator with a natural exponential key or a table of values for the exponential function to find decimal expressions for the numbers in Exercise 1, parts (b) and (d).

3. Find a formula for converting powers of 10 to powers of e and vice versa.

4. True or false? $e^{x \ln a} = a^x$ for any a and x. Prove your answer.

In Exercises 5–20, find the derivative of the given function.

5. $y = 3^x$

6. $f(x) = \pi^{x^2 - 1}$

7. $f(x) = \log_{10}(2x - 1)$

8. $y = x10^{x-1}$

9. $y = \log_{10}(\ln x)$

10. $f(x) = 2^{1-x} \cdot \log_2 \sqrt{x}$

11. $g(t) = t^2 \log_2 t$

12. $y = \log_{10} e^{\sqrt{t^2+1}}$

13. $f(x) = \pi^x + x^\pi$

14. $y = \log_{10} 2^{x^2}$

15. $y = x^x$

16. $y = x^{\cos x}$

17. $y = x^{\sqrt{x}}$

18. $y = x^{\ln x}$

19. $y = (\cos x)^{\sin x}$, $\qquad \cos x \geq 0$

20. $y = [\ln x]^x$

In Exercises 21–28, evaluate the given integral.

21. $\displaystyle\int 5^x \, dx$

22. $\displaystyle\int x2^{1-x^2} \, dx$

23. $\displaystyle\int \frac{\pi^{\sqrt{x}}}{\sqrt{x}} \, dx$

24. $\displaystyle\int 2^{1-x}2^{1+x} \, dx$

25. $\displaystyle\int a^{2\ln x} \, dx$

26. $\displaystyle\int 7^{ax^2+bx}\left(ax + \frac{b}{2}\right) dx$

27. $\displaystyle\int_0^2 x3^{x^2} \, dx$

28. $\displaystyle\int_0^2 2^{3t-1} \, dt$

29. True or false? The number a^x is nonzero for all x if $a > 0$.

30. Prove statement $(E1)$ in Theorem 7.

31. Prove statement $(E2)$ in Theorem 7.

32. Sketch the graph of the function $y = x2^{1-x}$ by finding
  a. the relative extrema,
  b. the intervals on which y is increasing,
  c. the inflection points,
  d. the concavity.

33. Find the number $a > 0$ so that the function $y = a^x$ satisfies the differential equation $y' - 2y = 0$.

34. *(Calculator)* Use differentials to obtain a formula for approximating $\log_{10}(10^n + x)$ where n is an integer and x is a small real number. Use this formula to approximate
  a. $\log_{10}(10.2)$
  b. $\log_{10}(98)$
  c. $\log_{10}(995)$
  d. $\log_{10}(.09)$

35. Let R be the region bounded by the graphs of $y = 3^x$, $y = 0$, $x = 0$, and $x = 1$. Find the volume of the solid obtained by revolving this region about the x-axis.

36. Let R be the region bounded by the graph of $f(x) = 2^{x^2}$ and the line $y = 16$. Find the volume of the solid obtained by revolving this region about the y-axis.

37. Find an equation for the line tangent to the graph of $y = \log_{10} x$ at $x = 1$.

38. If P_0 dollars are invested at 10% interest, compounded annually, the amount on deposit after n years is $P(n) = (1.10)^n P_0$ (see Exercise 38, Section 8.1). Find the number r so that

$$P(n) = (1.10)^n P_0 = e^{nr} P_0, \qquad n = 1, 2, \ldots .$$

39. The owner of a valuable oil painting estimates that the value of the painting will be approximately $V(t) = (5000)2^{\sqrt{t}}$ dollars over the next few years. Here t represents time in years, and the present value of the painting is $V(0) = 5000$ dollars. How fast will the value of the painting be increasing in 4 years?

In Exercises 40–45, find $\dfrac{dy}{dx}$ by logarithmic differentiation.

40. $y = \dfrac{x(x + 1)(x + 2)}{(x + 3)(x + 4)(x + 5)}$

41. $y = \sqrt[3]{\dfrac{x + 2}{x + 2}}$

42. $y = \dfrac{\sqrt[3]{x^2 + 1}}{(x + 1) \cdot (x + 2)^2}$

43. $y = \dfrac{(x^2 + 2)^3 \cdot (x - 1)^5}{x \cdot \sqrt{x + 1} \cdot \sqrt{x + 2}}$

44. $y = \sqrt{x\sqrt{x\sqrt{x}}}$

45. $y = x^{\sqrt{x+1}}$

46. Prove that the Power Rule, $\dfrac{d}{dx} x^r = rx^{r-1}$, is valid for all r, both rational and irrational.

8.5 EXPONENTIAL GROWTH AND DECAY

Imagine the following experiment associated with a course in biology. On a certain day, a number n of fruit flies are placed in an enclosed environment such as a large bell jar. If the environment is supportive (e.g., sufficient food supply and proper temperature), the number of fruit flies will increase as time passes. The experiment is to record the number of fruit flies present after t days, $N(t)$, and to find a mathematical relationship between time and population size. In experiments of this kind, data such as those presented in Table 5.1 are often obtained. (See Figure 5.1.)

Table 5.1 Typical data on growth of fruit flies

t (days)	0	4	8	12	16	20	24
$N(t)$ (fruit flies)	10	18	35	72	107	208	361

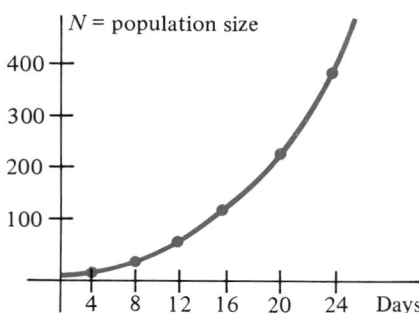

Figure 5.1 Plot of typical data on growth of fruit flies.

On examining these data, one observes that the population size, $N(t)$, increases more rapidly as this population size itself grows, i.e., the larger the population the more rapid the growth rate. Biologists refer to this phenomenon as the **Law of Natural Growth,** which is often stated as:

$$\{\text{Rate of growth of population}\} \propto \{\text{Present population size}\}. \tag{1}$$

The symbol "$\propto$" means "is proportional to." The statement of proportionality, $A \propto B$, is equivalent to the equation $A = kB$, where k is a constant. We can develop a mathematical model for populations that grow according to law (1) by assuming that the population size $N(t)$ may be approximated by a differentiable function of time. Since the derivative $N'(t) = \dfrac{dN}{dt}$ gives the rate of growth (increase or decrease) of the population size, the mathematical formulation of the growth law (1) is

$$\frac{dN}{dt} = kN. \tag{2}$$

The constant k in equation (2) is referred to as the **growth constant.** Since the mathematical model (2) consists of a *differential equation,* we must find a differentiable function $y = N(t)$ that satisfies the equation.

We have already determined (Section 8.3) that any exponential function of the form $y = Ce^{kt}$ satisfies equation (2), since

$$\frac{dy}{dt} = \frac{d}{dt}(Ce^{kt})$$

$$= C\left(\frac{d}{dt}e^{kt}\right)$$

$$= kCe^{kt}$$

$$= ky.$$

In Exercise 29 you will show that, other than the trivial solution $y = 0$, there are no other possible solutions of equation (2). In other words, all solutions of the differential equation (2) must have the form $y = Ce^{kt}$. Finally, we can be more specific about the constant C. Setting $t = 0$ in the equation $y = Ce^{kt}$ gives

$$y(0) = Ce^{k \cdot 0} = C,$$

so $C = y(0)$. We refer to the constant $C = y(0) = y_0$ as the **initial condition** associated with the differential equation (2). We summarize our findings as follows.

THEOREM 8

The unique solution of the differential equation

$$\frac{dy}{dt} = ky, \tag{3}$$

satisfying the initial condition $y(0) = y_0$, is the exponential function

$$y = y_0 e^{kt}. \tag{4}$$

Theorem 8 provides the strategy for solving problems involving populations or quantities that satisfy the Law of Natural Growth (1): If we can safely assume that the size or amount of the population or quantity y can be viewed as a differentiable function of t (which usually represents time), then y has the form given by equation (4). By substituting known data into equation (4) one obtains the growth constant k and the initial population size or amount, y_0. The function y is then completely determined.

Graphs of the growth function (4) are shown in Figure 5.2 for the two general cases $k > 0$ (growth) and $k < 0$ (decay).

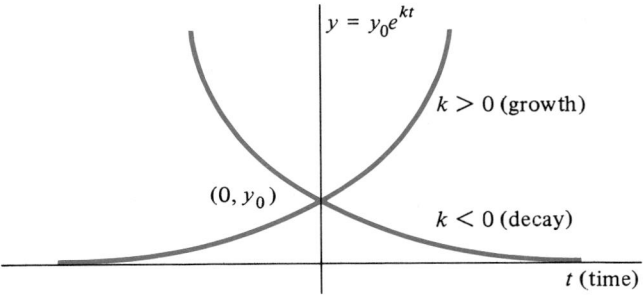

Figure 5.2 Graph of the natural growth function.

Example 1

Assume that the rate of growth of a population of fruit flies is proportional to the size of the population at each instant of time. If 100 fruit flies are present initially and 300 are present after 10 days, how many will be present after 15 days?

Strategy

Apply Theorem 8 to find the general form of the solution.

Find $N_0 = N(0)$ from given data.

Take natural logs of both sides and solve for k.

State explicit formulation of $N(t)$.

Find $N(15)$ using tables or calculator, rounding to nearest integer.

Solution

Since it is stated that the population size N satisfies the differential equation

$$\frac{dN}{dt} = kN,$$

the function N is given by equation (4):

$$N(t) = N_0 e^{kt}, \qquad N(0) = N_0. \tag{5}$$

We are given that $N = 100$ when $t = 0$, that is,

$$N_0 = 100.$$

Also, we are given that $N = 300$ when $t = 10$. Inserting these values in (5) gives

$$300 = 100 e^{10k},$$

so

$$e^{10k} = 3.$$

Thus

$$10k = \ln 3,$$

so

$$k = \frac{\ln 3}{10}.$$

The function $N(t)$ is therefore

$$N(t) = 100 e^{(\ln 3/10)t},$$

so

$$N(15) = 100 e^{15 \ln 3/10} = 100 e^{3/2 \ln 3} = 100 \cdot 3^{3/2}$$
$$\approx 520 \text{ fruit flies.}$$ ◇

Example 2

Certain radioactive isotopes, such as uranium, are known to decay at a rate proportional to the amount present. If a block of 50 grams of such material decays to 40 grams in 5 days, find the half-life of this material. (The half-life is the amount of time required for the material to decay to half its original amount.)

Strategy

Label the variables.

State form of A from Theorem 8.

Solution

Let $A(t)$ denote the amount present at time t. Since A is given as satisfying the differential equation (3), by Theorem 8, A has the form

$$A(t) = A_0 e^{kt}, \qquad t = \text{time.} \tag{6}$$

We are given

Apply initial condition
$A(0) = 50 = A_0$.

$$A_0 = A(0) = 50 \text{ grams.}$$

To find k we use the data $A(5) = 40$ together with equation (6). We obtain

Use equation $A(5) = 40$ to obtain equation for k.

$$40 = A_0 e^{5k} = 50 e^{5k},$$

so

$$e^{5k} = \frac{40}{50} = \frac{4}{5}.$$

Solve for k by taking ln's of both sides.

Thus

$$k = \frac{1}{5} \ln\left(\frac{4}{5}\right). \tag{7}$$

Write down condition for T: $A(T) = \frac{1}{2}A(0)$.

We seek a time T so that

$$A(T) = \frac{1}{2}A_0. \tag{8}$$

Using (6) we obtain

Combine (6) and (8).

$$A_0 e^{kT} = \frac{1}{2}A_0,$$

so

$$e^{kT} = \frac{1}{2}.$$

Solve for T, taking ln's of both sides and using (7).

Thus

$$T = \frac{\ln\left(\frac{1}{2}\right)}{k}.$$

Using the value for k in (7), we find

$$T = \frac{\ln\left(\frac{1}{2}\right)}{\frac{1}{5}\ln\left(\frac{4}{5}\right)} = \frac{5[\ln 1 - \ln 2]}{\ln 4 - \ln 5} \approx 15.53 \text{ days.} \qquad \diamond$$

COMPOUND INTEREST: In Exercise 39, Section 8.1, it was shown that if P_0 dollars are invested at annual interest rate r (in decimal form) compounded n times per year, the amount on deposit after t years is

$$P(t) = \left(1 + \frac{r}{n}\right)^{nt} P_0 \text{ dollars.} \tag{9}$$

Many banks advertise "continuous" compounding of interest. This means that the number of compoundings per year is regarded as having become infinitely large (i.e., interest is added to your account "each instant").

To obtain an expression for $P(t)$ under continuous compounding of interest we apply $\lim\limits_{n\to\infty}$ to expression (9):

$$P(t) = \lim_{n\to\infty}\left\{\left(1 + \frac{r}{n}\right)^{nt}P_0\right\} \tag{10}$$

$$= P_0\left[\lim_{n\to\infty}\left(1 + \frac{r}{n}\right)^n\right]^t.$$

We now use equation (8), Section 8.3,

$$e^r = \lim_{n\to\infty}\left(1 + \frac{r}{n}\right)^n$$

in equation (10) to obtain *the equation for continuous compounding of interest*

$$P(t) = P_0e^{rt}. \tag{11}$$

REMARK: Note that the function P in equation (11) is the solution of the differential equation

$$P'(t) = rP(t) \tag{12}$$

with $P(0) = P_0$. Equation (12) is another way of describing continuous compounding of interest since it says that the rate $P'(t)$ at which the amount $P(t)$ is growing at each instant is the interest rate r applied to the present size of the investment.

Example 3

Find the value of an initial deposit of $1,000 invested at 12% annual interest compounded continuously for 6 years.

Solution: We have $P_0 = 1000$, $r = 0.12$, and $t = 6$. From equation (11),

$$P(6) = 1000e^{6(.12)}$$
$$= 2504.43 \text{ dollars.} \qquad \diamondsuit$$

Example 4

Just as the amount of money available after compounding P_0 dollars continuously for t years can be found using equation (11), the *initial* amount P_0 needed to produce a given amount $P_t = P(t)$ after t years can be calculated. To do this we multiply both sides of equation (11) by e^{-rt} to obtain

$$P_0 = P_t e^{-rt}. \tag{13}$$

In this case P_0 is called the **present value** of an investment that will yield P_t dollars in t years if compounded continuously at r percent. Find the size of the deposit which, when compounded continuously for 10 years at 12% interest, will yield $20,000.

Solution: From equation (13) we obtain

$$P_0 = e^{-(0.12)\cdot 10}(20,000)$$
$$= 6,023.88 \text{ dollars.} \qquad \diamondsuit$$

Example 5

The owner of a stand of timber estimates that his timber is presently worth $5,000 and that the timber increases in value with time according to the formula

$$V(t) = 5000e^{\sqrt{t}/2} \text{ dollars}$$

where t is time in years. If the prevailing interest rates over the foreseeable future are expected to average 12%, when should he sell his timber so as to maximize profit?

Strategy

Convert the revenue received at a future date to present value, using (13).

Solution

Since dollars received at different times have different present values, we must discount the changing value of the revenue received for the sale of the timber by converting it to its *present value*. Using equation (13) we find the present value of the revenue received in selling the timber after t years to be

$$R(t) = V(t)e^{-.12t}$$
$$= 5000e^{(\sqrt{t}/2 - .12t)}$$

To maximize R we set $R'(t) = 0$:

Set $R'(t) = 0$.

$$R'(t) = 5000e^{(\sqrt{t}/2 - .12t)}\left[\frac{1}{4\sqrt{t}} - .12\right] = 0. \tag{14}$$

Since the exponential function is never zero we must have

$$\frac{1}{4\sqrt{t}} - .12 = 0$$

Solve for t.

or

$$\sqrt{t} = \left(\frac{1}{.48}\right),$$

so

$$t = \left(\frac{1}{.48}\right)^2 = 4.34 \text{ years},$$

Verify that the critical point obtained yields a maximum.

or approximately 4 years and 4 months. (We can verify that this value of t indeed produces a profit maximum by applying the first derivative test to R' in equation (14).) ◇

Exercise Set 8.5

In Exercises 1–6, find an exponential function of the form $y = Ae^{kt}$ which satisfies the given conditions.

1. $y' = 2y$ and $y(0) = 1$

2. $y' = -4y$ and $y(0) = 2$

3. $y' - 5y = 0$ and $y(1) = 1$

4. $y(0) = 2$ and $y(2) = 6$

5. $y(0) = 1$ and $y(1) = e^{-2}$

6. $y(0) = y(1) = 0$

7. True or false? The half-life of an isotope depends upon the amount present. Explain.

8. A radioactive isotope I decays exponentially with a half-life of 20 days. If 50 mg of the isotope remain after 10 days,
   a. How much of the isotope was present initially?
   b. How much will remain after 30 days?
   c. When will 90% of the initial amount have disintegrated?

9. Suppose an object travelling through a fluid is subject only to a force resisting its motion, and this resisting force is proportional to the velocity of the object. Let's write this resisting force as $f_r = cv$ where c is constant and v denotes velocity. By Newton's second law we have $f_r = ma$ where m is the mass of the object and a is its acceleration. Since $a = \dfrac{dv}{dt}$, we may combine our two equations to obtain

$$m\frac{dv}{dt} = -cv, \quad \text{or} \quad \frac{dv}{dt} = -\left(\frac{c}{m}\right)v.$$

a. Find a solution $v(t)$ of this differential equation for which $v(0) = v_0$.

b. Find a solution of the equation for $v_0 = 10$ m/sec, $c = 40$, and $m = 10$.

c. Does $v(t)$ approach a limit as $t \to \infty$? If so, what is this limit?

10. In a **first order chemical reaction** the rate of disappearance of reactant is proportional to the amount present. Chemists use the notation $[A]$ for the concentration of reactant A in moles per liter. Thus, in first order reactions, the concentration $[A]$ as a function of time is a solution of the differential equation

$$\frac{d[A]}{dt} = -k[A].$$

a. Find an expression for the concentration $[A(t)]$ after t seconds if $[A_0] = [A(0)]$.

b. If the reaction begins with a concentration $[A_0] = 10$ moles per liter, and 6 moles per liter of reactant A remain after 20 minutes, find the concentration of reactant A one hour after the reaction begins.

11. The number of bacteria in a certain culture grows from 50 to 400 in 12 hours. Assuming that the rate of increase is proportional to the number of bacteria present

a. How long does it take for the number of bacteria present to double?

b. How many bacteria will be present after 16 hours?

12. If we assume temperature to be constant at all altitudes, the rate of decrease in atmospheric pressure p, as a function of altitude, h, is proportional to p. That is,

$$\frac{dp}{dh} = -kp$$

where the constant k is approximately $k = 0.116$ km^{-1}. Find the pressure in atmospheres 8 km (5 mi) above sea level if the pressure at sea level is 1 atmosphere.

13. Land bought for speculation is expected to be worth $V(t) = 10{,}000(1.2)^{\sqrt{t}}$ dollars after t years. If the cost of money holds constant at 10%, when should the land be sold so as to maximize profits? Does the original price of the land affect this calculation?

14. In a simple model of the growth of a biological population, the rate of increase of N, the number of individuals, is proportional to the difference between the birth rate, b, and the death rate, d. That is, the simple birth-death model of population growth is

$$\frac{dN}{dt} = bN - dN.$$

a. Find an expression for population size, N, as a function of the birth rate r, death rate d, and initial population size N_0.

b. If in a population modelled by this process there are 20 births per year per 100 individuals and 12 deaths per year per 100 individuals, find the anticipated size of the population in 20 years if its present size is $N_0 = 200$ individuals.

15. When a foreign substance is introduced into the body, the body's defense mechanisms move to break down the substance and excrete it. The rate of excretion is usually proportional to the concentration present in the body, and the half-life of the resulting exponential decay is referred to as the **biological half-life** of the substance. If, after 12 hours, 30% of a massive dosage of a substance has been excreted by the body, what is the biological half-life of the substance?

16. All living matter contains two types of carbon, ^{14}C and ^{12}C, in its molecules. While the organism is alive the ratio of ^{14}C to ^{12}C is constant, as the ^{14}C is interchanged with fixed levels of $^{14}CO_2$ in the atmosphere. However, when a plant or animal dies this replenishment ceases, and the radioactive ^{14}C present decays exponentially with a half-life of 5760 years. By examining the $^{14}C/^{12}C$ ratio, archeologists can determine the percent of the original ^{14}C level remaining and therefore date a once living fossil.

a. If 80% of the original amount of ^{14}C remains, how old is the fossil?

b. A fossil contains 10% of its original amount of ^{14}C. What is its age?

17. The body concentrates iodine in the thyroid gland. This observation leads to the treatment of thyroid cancer by injecting radioactive iodine into the bloodstream. One isotope used has a half-life of approximately 8 days and decays exponentially in time. If 50 micrograms of this isotope are injected, what amount remains in the body after three weeks?

18. A chemical dissolves in water at a rate proportional to the amount still undissolved. If 20 grams of the chemical are placed in water and 10 grams remain undissolved 5 minutes later, when will 90% of the chemical be dissolved?

19. The value of a case of a certain French wine t years after being imported to the United States is $V(t) = 100(1.5)^{\sqrt{t}}$. If the cost of money under continuous compounding of interest is expected to remain constant at 12.5% per year, when

should the wine be sold so as to maximize profits? (Assume storage costs to be negligible.)

20. Answer the following questions to become familiar with the concept of **effective annual yield:**

a. Verify that under *continuous* compounding of interest at the rate $r_c = 10\%$, an initial deposit of 100 dollars will grow in one year to the amount

$$P(1) = 100e^{(.10)(1)} = 100(1.10517)$$
$$= \$110.52.$$

b. Note that under *a single annual* compounding of interest, an interest rate $r_a = 10.52\%$ would be required to produce the amount $P(1)$ in part a in one year.

c. The *effective annual yield* of an interest rate compounded continuously is the interest rate required, under a *single* annual compounding, to produce the same amount of interest in one year. What is the effective annual yield of the interest rate $r_c = 10\%$ compounded continuously?

d. What is the effective annual yield of the interest rate $r = 15\%$ compounded continuously?

21. What rate of interest, compounded continuously, will produce an effective annual yield of 8%?

22. An annuity pays 10% interest, compounded continuously. What amount of money, deposited today, will have grown to $2,500 in 8 years?

23. Find the effective annual yield (see Exercise 20) on
a. an interest rate of 10% compounded semiannually,

b. an interest rate of 10% compounded quarterly,
c. an interest rate of 10% compounded daily,
d. an interest rate of 10% compounded continuously.

24. How long does it take for a deposit of P_0 dollars to double at 10% interest compounded continuously?

25. Show that the effective annual rate of interest, i, for continuous compounding at rate r is $i = e^r - 1$.

26. What is the present value of $1000 five years from now if the prevailing rate of interest, when compounded continuously, is 10%? What is the present value if 10% is the prevailing effective annual yield? (See Exercise 20.)

27. True or false? If the population $P(t)$ of a city is increasing at a rate of 5% per year, then the function $P(t)$, if differentiable, satisfies the equation $P'(t) = .05P(t)$. If this is false, what is the correct equation?

28. The population of a certain city is increasing at a rate of 5% per year. If the population $P(t)$ is assumed differentiable, and if $P'(t) = kP(t)$, find k (see Exercise 27).

29. Prove that $y = Ce^{kt}$ is the only nontrivial ($y \neq 0$) solution of the differential equation $y' = ky$ as follows.
a. Assume that the differentiable function $y = f(t)$ is also a solution.
b. Show that the product $g(t) = f(t)e^{-kt}$ is differentiable, and that $g'(t) = 0$.
c. Conclude that $g(t) = C$ for some constant C, that is, necessarily $f(t) = Ce^{kt}$.

SUMMARY OUTLINE OF CHAPTER 8

◆ **Definition:** $\log_b x = y$ if and only if $b^y = x.$ (page 371)

◆ **Properties of logarithms:**   (page 372)

    (L1) $\log_b(mn) = \log_b m + \log_b n.$ (L2) $\log_b(m/n) = \log_b m - \log_b n.$ (L3) $\log_b(n^r) = r \log_b n.$

◆ **Definition:** $\ln x = \int_1^x \frac{1}{t} \, dt, \quad x > 0.$ (page 380)

◆ **Theorem:** $\frac{d}{dx} \ln x = \frac{1}{x}; \quad \int \frac{1}{x} \, dx = \ln |x| + C.$ (page 380)

◆ **Definition:** e is defined by the equation $\ln e = 1.$ $e \approx 2.718281828459.$ (page 382)

◆ **Definition:** $y = e^x$ if and only if $x = \ln y.$ (page 389)

◆ **Theorem:** $\frac{d}{dx} e^x = e^x; \quad \int e^x \, dx = e^x + C.$ (page 391)

◆ **Alternate definition of e^r:** $e^r = \lim_{k \to \infty} \left(1 + \frac{r}{k}\right)^k, \quad e = \lim_{k \to \infty} \left(1 + \frac{1}{k}\right)^k.$ (page 394)

◆ **Definition:** $a^x = e^{x \ln a}, \quad a > 0.$ (page 398)

◆ **Theorem:** $\frac{d}{dx} a^x = a^x \ln a; \quad \int a^x \, dx = \frac{a^x}{\ln a} + C.$ (page 399)

◆ **Definition:** $y = \log_a x$ if and only if $x = a^y = e^{y \ln a}$.   (page 400)

◆ **Conversion Formula:** $\log_a x = \dfrac{\ln x}{\ln a}, \qquad a \neq 1.$ (page 400)

◆ **Model for Natural Growth:** $\dfrac{dy}{dt} = ky.$ (page 404)

◆ **Solution of Model:** $y = Ce^{kt}, \qquad C = y(0).$ (page 405)

◆ **Formula for continuous compounding of interest:** $P(t) = e^{rt}P_0.$ (page 408)

◆ **Present value of P_t dollars in t years:** $P(0) = e^{-rt}P_t.$ (page 408)

REVIEW EXERCISES—CHAPTER 8

1. Simplify the following.
  a. $8^{2/3}$ **b.** $36^{-5/2}$
  c. $16^{5/4}$ **d.** $4^{-3/4}$

2. Find the following logarithms:
  a. $\log_2 16$ **b.** $\log_8 16$
  c. $\ln e^2$ **d.** $\log_2 3\sqrt{2}$

3. Find the inverses of the following functions:
  a. $f(x) = x + 2$ **b.** $g(x) = x^2, \quad x \geq 0$
  c. $y = 2x + 1$ **d.** $y = \sqrt{x}$

4. Solve for x:
  a. $3 \ln x - \ln 3x = 0$ **b.** $\ln x^3 = 3$
  c. $\displaystyle\int_1^x \frac{2}{t} \, dt = 4$ **d.** $\displaystyle\int_2^x \frac{1}{t} \, dt = 3 + \ln 2$

Find the derivative of each function with respect to its independent variable.

5. $y = x^2 \ln(x - a)$ **6.** $f(x) = xe^{1-x}$

7. $f(x) = \ln(\ln^2 x)$ **8.** $y = (2 \ln \sqrt{x})^3$

9. $y = e^{x^2} \tan 2x$ **10.** $f(t) = \sin^2 t \cos(\ln t)$

11. $f(t) = \dfrac{te^t}{1 + e^t}$ **12.** $y = x \ln(\sqrt{x} - e^{-x})$

13. $y = \dfrac{\ln(t - a)}{\ln(t^2 + b)}$ **14.** $f(t) = \ln \cos t$

15. $f(t) = e^{\sqrt{t} - \ln t}$ **16.** $y = \dfrac{1}{a} \ln\left(\dfrac{a + bx}{x}\right)$

17. $y = \dfrac{1}{3x} + \dfrac{1}{4} \ln\left(\dfrac{1 - 2x}{\sqrt{x}}\right)$

18. $f(t) = \sin(\ln t + te^{\sqrt{t}})$

19. $x \ln y^2 + y \ln x = 1$ $\left(\text{Find } \dfrac{dy}{dx}.\right)$

20. $y = e^{4x}(e^{-x} + \ln x)^2$

21. $x^2 + e^{xy} - y^2 = 2$ $\left(\text{Find } \dfrac{dy}{dx}.\right)$

22. $y = xe^{1 - \sqrt{x}\ln x}$

23. $e^{xy} = \sqrt{xy}$ $\left(\text{Find } \dfrac{dy}{dx}.\right)$ **24.** $y = \sqrt{e^{-x} + e^x}$

25. $f(x) = \ln \sqrt{e^{2x} + \sin \sqrt{x}}$ **26.** $y = \displaystyle\int_1^{\sin x} \ln(t + 1) \, dt$

Find the integral.

27. $\displaystyle\int \frac{x \, dx}{1 - x^2}$ **28.** $\displaystyle\int xe^{1-x^2} \, dx$

29. $\displaystyle\int \frac{x + 1}{4x + 2x^2} \, dx$ **30.** $\displaystyle\int \frac{dx}{(x + 1) \ln^3(x + 1)}$

31. $\displaystyle\int \sqrt{e^x} \, dx$ **32.** $\displaystyle\int (e^x + 1)^2 \, dx$

33. $\displaystyle\int_e^{e^2} \frac{1}{x\sqrt{\ln x}} \, dx$ **34.** $\displaystyle\int \frac{e^x - e^{-x}}{e^x + e^{-x}} \, dx$

35. $\displaystyle\int_0^1 \frac{x^3 - 1}{x + 1} \, dx$ **36.** $\displaystyle\int_1^4 \frac{e^{\sqrt{x}}}{\sqrt{x}} \, dx$

37. $\displaystyle\int_0^2 x(e^{x^2} + 1) \, dx$ **38.** $\displaystyle\int_4^9 \frac{1}{\sqrt{x}(1 + \sqrt{x})} \, dx$

39. $\displaystyle\int e^x(1 - e^{2x})^3 \, dx$ **40.** $\displaystyle\int \frac{\cot \sqrt{x}}{\sqrt{x}} \, dx$

41. $\displaystyle\int \frac{2x + 3x^2}{x^3 + x^2 - 7} \, dx$ **42.** $\displaystyle\int_1^{\ln 2} (e^x + 1)(e^x - 1) \, dx$

43. $\displaystyle\int_{-3}^{-2} \frac{x}{x^2 + 1} \, dx$

44. $\displaystyle\int_0^1 \left(\frac{1}{x + 1} - \frac{1}{x + 2}\right) dx$

Use logarithmic differentiation to find $\dfrac{dy}{dx}$ for

45. $y = x^{\sqrt{x}}$

46. $y = \sqrt{\dfrac{(x-1)^2(x+1)}{x(x+3)^3}}$

47. Find all relative extrema for the function $y = x^2 \ln x$.

48. Find the maximum and minimum values of the function $f(x) = e^{x^2/3}$ for x in the interval $[-1, \sqrt{8}]$.

49. On what intervals is the graph of the function $y = \ln(1 + x^2)$ concave up?

50. Find the area of the region bounded by the graph of $y = \dfrac{x}{x^2+1}$ and the x-axis for $-2 \le x \le 2$.

51. Find the area of the region bounded by the graphs of $y = \dfrac{1}{x}$ and $y = \dfrac{10x - 21}{4x - 10}$. (*Hint:* Divide first.)

52. Find all relative extrema for the function $y = x^2 - \ln x^2$ and sketch the graph.

53. Find the volume of the solid generated by rotating the portion of the graph of $y = \sqrt{\dfrac{2}{x-1}}$ from $x = 3$ to $x = 5$ about the x-axis.

54. The function

$$f(x) = \begin{cases} \lambda e^{-\lambda x}, & x > 0 \\ 0, & x \le 0 \end{cases}$$

is called the **exponential probability density function** with parameter λ. For any density function, the probability that x will fall in the interval $[a, b]$ is

$$P\{a \le x \le b\} = \int_a^b f(x)\,dx.$$

For $\lambda = 1$, find the following probabilities associated with the exponential distribution:
a. $P\{0 \le x \le 1\}$,
b. $P\{0 \le x \le \ln 2\}$,
c. $P\{0 \le x \le 4\}$.

55. Graph the exponential probability density function for $\lambda = 1$ (Exercise 54).

56. (*Calculator/Computer*) The *standard* normal probability density function is $f(x) = \dfrac{1}{\sqrt{2\pi}}e^{-x^2/2}$. Use a program for numerical integration to approximate

a. $P\{-1 \le x \le 1\} = \displaystyle\int_{-1}^{1} \dfrac{1}{\sqrt{2\pi}}e^{-x^2/2}\,dx$

b. $P\{0 \le x \le 2\}$.

57. Sketch the graph of the standard normal probability density function in Exercise 56.

58. Find a formula for approximating the function $y = x \ln x^2$ near $x = e$. Use this formula to approximate $y(2.70) = (2.70)\ln(2.70)^2$.

59. Approximate $e^{\sqrt{4.1}}$ by differentials.

60. Let $f(x) = \dfrac{e^x + e^{-x}}{2}$, $g(x) = \dfrac{e^x - e^{-x}}{2}$.
a. Show that $f'(x) = g(x)$ and that $g'(x) = f(x)$.
b. Show that both functions satisfy the differential equation $y'' - y = 0$.
c. Find two solutions of the differential equation $y'' - k^2 y = 0$.

61. A particle, starting from rest, moves along a line with acceleration $a(t) = \dfrac{1}{(t+2)^2}$ m/s². Find
a. the velocity function $v(t)$,
b. the distance travelled after 10 seconds.

62. (*Calculator*) Approximate $\displaystyle\int_0^5 e^{x^2}\,dx$ using a rectangular rule with $n = 10$.

63. Find r if $2^x = e^{rx}$ for all x.

64. Find the equation of the line tangent to the graph of $y = xe^{2x}$ at the point $(\ln 2, 4 \cdot \ln 2)$.

65. (*Calculator*) Use Simpson's rule with $n = 10$ to approximate
a. $\ln 4$ **b.** $\ln 5$ **c.** $\ln(1/2)$

66. (*Calculator/Computer*) Use Newton's Method to approximate the solution of the equation $2 + \ln x = x$.

Find the solution of the initial value problem:

67. $\dfrac{dy}{dx} = 2y$

   $y(0) = 1$

68. $\dfrac{dy}{dx} + y = 0$

   $y(\ln 2) = 2$

69. $2y' + 4y = 0$

   $y(0) = \pi$

70. $y' + y = 2$

   $y(0) = 1$

71. $4y' - 4y + 4 = 0$

   $y(0) = 1$

72. $(y' + y)^2 = 0$

   $y(0) = 1$

73. Find the average value of the function $y = xe^{x^2+1}$ on the interval $[0, \sqrt{\ln 2}]$.

74. (*Calculator/Computer*) Approximate the solution of the equation $e^x + 2x = 0$.

75. The population of the United States in 1970 was 203 million. By 1980 the population had grown to 227 million. Assuming exponential growth, what will the population be in 1990? 2000?

76. In the chemical reaction called the inversion of raw sugar, the inversion rate is proportional to the amount of raw sugar remaining. If 100 kilograms of raw sugar is reduced to 75 kilograms in 6 hours, how long will it be until
  a. half the raw sugar has been inverted?
  b. 90% of the raw sugar has been inverted?

77. A population of fruit flies grows exponentially. If initially there were 100 flies and if after 10 days there were 500 flies, how many flies were present after 4 days?

78. A snowball melts at a rate proportional to its surface area. If the snowball originally had a radius of 10 cm and, after 20 minutes, its radius is 8 cm, find
  a. a differential equation for the radius r,

b. a solution of this differential equation involving two constants,
c. an explicit solution of this equation, and
d. the radius of the snowball after one hour.

(*Hint:* In part (a) show that $\dfrac{dv}{dt} = s\dfrac{dr}{dt}$ where $v =$ volume and $s =$ surface area.)

79. 100 grams of a radioactive substance is reduced to 40 grams in 6 hours. If the decay is exponential, what is the half-life?

80. Show that the exponential function $f(x) = Ce^{kx}$ satisfies the differential equation $f'(x + a) = kf(x + a)$ for every real number a.

Chapter 9
Trigonometric and Inverse Trigonometric Functions

We have already encountered the six trigonometric functions and their derivatives. The objectives of this chapter are to

(i) obtain antiderivatives for all six trigonometric functions,
(ii) develop techniques for integrating more complicated expressions involving trigonometric functions,
(iii) define the inverses of the trigonometric functions and find their derivatives and antiderivatives,
(iv) define the hyperbolic functions, and
(v) study some models of physical phenomena involving these functions.

9.1 INTEGRALS OF THE TRIGONOMETRIC FUNCTIONS

We have already determined the following differentiation and integration formulas:

1. $\dfrac{d}{dx} \sin x = \cos x$ 1′. $\displaystyle\int \cos x \, dx = \sin x + C$

2. $\dfrac{d}{dx} \cos x = -\sin x$ 2′. $\displaystyle\int \sin x \, dx = -\cos x + C$

3. $\dfrac{d}{dx} \tan x = \sec^2 x$ 3′. $\displaystyle\int \sec^2 x \, dx = \tan x + C$

4. $\dfrac{d}{dx} \cot x = -\csc^2 x$ 4′. $\displaystyle\int \csc^2 x \, dx = -\cot x + C$

5. $\dfrac{d}{dx} \sec x = \sec x \tan x$ 5′. $\displaystyle\int \sec x \cdot \tan x \, dx = \sec x + C$

6. $\dfrac{d}{dx} \csc x = -\csc x \cot x$ 6′. $\displaystyle\int \csc x \cdot \cot x \, dx = -\csc x + C$

Identities involving the six trigonometric functions, together with graphs of these six functions, appear in Chapter 1. Remember, all angles are to be measured in *radian* measure when working with trigonometric functions in calculus.

415

Integrals of tan x and sec x

The problem of finding integrals for $\sin x$ and $\cos x$ was simple to solve, since these functions occur as derivatives of $-\cos x$ and $\sin x$, respectively. However, this is not the case for the remaining four trigonometric functions.

To find the integral for $y = \tan x$ we make use of the formula

$$\int \frac{1}{u}\, du = \ln |u| + C \tag{1}$$

from Chapter 8. Since $\tan x = \dfrac{\sin x}{\cos x}$, we let $u = \cos x$. Then $du = -\sin x\, dx$. Using equation (1) we obtain

$$\int \tan x\, dx = \int \frac{\sin x}{\cos x}\, dx$$

$$= -\int \frac{1}{u}\, du$$

$$= -\ln |u| + C$$
$$= -\ln |\cos x| + C.$$

Since $-\ln |\cos x| = \ln |\cos x|^{-1} = \ln |\sec x|$, we have the formula

$$\int \tan x\, dx = \ln |\sec x| + C, \qquad \cos x \neq 0. \tag{2}$$

To find the antiderivative for $y = \cot x$, we write $\cot x = \dfrac{\cos x}{\sin x}$ and let $u = \sin x$. Proceeding as above we obtain

$$\int \cot x\, dx = \ln |\sin x| + C, \qquad \sin x \neq 0. \tag{3}$$

Example 1

Find $\displaystyle\int x \cot x^2\, dx$.

Strategy
Find a substitution so that x is a factor of du.

Make substitution.

Solution
We make the substitution

$$u = x^2; \qquad du = 2x\, dx.$$

Then $x\, dx = \dfrac{1}{2}\, du$, and we obtain

$$\int x \cot x^2\, dx = \int \cot u \cdot \frac{1}{2}\, du$$

$$= \frac{1}{2} \int \cot u\, du$$

Apply equation (3).

$$= \frac{1}{2} \ln |\sin u| + C$$

Substitute back.

$$= \frac{1}{2} \ln |\sin x^2| + C. \qquad \diamond$$

Example 2

Find $\int \tan^3 x \, dx$.

Solution: We use the identity $\tan^2 x = \sec^2 x - 1$ to write

$$\int \tan^3 x \, dx = \int \tan x [\sec^2 x - 1] \, dx$$

$$= \int \tan x \sec^2 x \, dx - \int \tan x \, dx.$$

In the first integral we use the substitution

$$u = \tan x; \qquad du = \sec^2 x \, dx$$

to obtain

$$\int \tan x \sec^2 x \, dx = \int u \, du$$

$$= \frac{1}{2} u^2 + C$$

$$= \frac{1}{2} \tan^2 x + C.$$

The second integral is given by equation (2). Thus,

$$\int \tan^3 x \, dx = \frac{1}{2} \tan^2 x - \ln |\sec x| + C. \qquad \diamond$$

Integrals of sec x and csc x

The integrals for $\sec x$ and $\csc x$ also involve the natural logarithm function, although the derivations are not so straightforward. To handle $\sec x$ we multiply by $\dfrac{\sec x + \tan x}{\sec x + \tan x}$ to obtain

$$\int \sec x \, dx = \int \sec x \cdot \left(\frac{\sec x + \tan x}{\sec x + \tan x} \right) dx$$

$$= \int \frac{\sec x \tan x + \sec^2 x}{\sec x + \tan x} \, dx$$

$$= \int \frac{1}{\sec x + \tan x} \left[\frac{d}{dx} (\sec x + \tan x) \right] dx$$

$$= \ln |\sec x + \tan x| + C \qquad \text{(by equation (1)).}$$

Thus,

$$\int \sec x \, dx = \ln |\sec x + \tan x| + C. \tag{4}$$

Similarly, we can show that

$$\int \csc x \, dx = \ln |\csc x - \cot x| + C. \tag{5}$$

(See Exercise 36.)

Example 3

Find the length of the graph of the function

$$y = \ln \sec x \text{ from } (0, 0) \text{ to } (\pi/3, \ln 2).$$

Strategy

Find $\dfrac{dy}{dx}$.

Use formula for arc length

$$L = \int_a^b \sqrt{1 + \left[\dfrac{dy}{dx}\right]^2} \, dx.$$

Use identity

$1 + \tan^2 x = \sec^2 x$.

Use formula (4).

Solution

Since $\dfrac{dy}{dx} = \tan x$ we have

$$L = \int_0^{\pi/3} \sqrt{1 + \tan^2 x} \, dx$$

$$= \int_0^{\pi/3} \sqrt{\sec^2 x} \, dx$$

$$= \int_0^{\pi/3} \sec x \, dx$$

$$= \ln \left| \sec x + \tan x \right| \Big]_0^{\pi/3}$$

$$= \ln |2 + \sqrt{3}| - \ln |1 + 0| = \ln(2 + \sqrt{3})$$
$$\approx 1.317. \qquad \diamond$$

Example 4

Find $\displaystyle\int \frac{\csc \sqrt{x}}{\sqrt{x}} \, dx$.

Strategy

Find a substitution involving $x^{-1/2}$ as factor of du.

Solution

We make the substitution

$$u = \sqrt{x} = x^{1/2}; \qquad du = \frac{dx}{2\sqrt{x}} = \frac{1}{2} x^{-1/2} \, dx.$$

Then $x^{-1/2} \, dx = 2 \, du$, and

Make substitution.

$$\int \frac{\csc \sqrt{x}}{\sqrt{x}} \, dx = \int \csc u \cdot 2 \, du$$

Factor constant.

Apply formula (5).

Substitute back.

$$= 2 \int \csc u \cdot du$$

$$= 2 \ln |\csc u - \cot u| + C$$

$$= 2 \ln |\csc \sqrt{x} - \cot \sqrt{x}| + C. \qquad \diamond$$

The exercise set below provides a general review of integration of trigonometric functions as well as exercises concerning equations (3)–(5).

Exercise Set 9.1

In Exercises 1–30, find the integral.

1. $\int \cos 3x \, dx$

2. $\int \sec^2 2x \, dx$

3. $\int \sec 2x \tan 2x \, dx$

4. $\int x \csc x^2 \, dx$

5. $\int (\tan^2 x + 1) \, dx$

6. $\int \dfrac{\tan \sqrt{x}}{\sqrt{x}} \, dx$

7. $\int \dfrac{x}{\cos x^2} \, dx$

8. $\int \dfrac{1}{\sqrt{x} \sec \sqrt{x}} \, dx$

9. $\int x \tan(3x^2 - 1) \, dx$

10. $\int \sin x \sqrt{1 + \cos x} \, dx$

11. $\int \tan^3 x \sec^2 x \, dx$

12. $\int \cot x \cdot \ln(\sin x) \, dx, \qquad \sin x > 0$

13. $\int \sec^2 x e^{\tan x} \, dx$

14. $\int \dfrac{\cos^2(2x - 1)}{\sin(2x - 1)} \, dx$

15. $\int \dfrac{\sec^2 x + 1}{x + \tan x} \, dx$

16. $\int \dfrac{\csc^2(x + \pi/4)}{\cot^3(x + \pi/4)} \, dx$

17. $\int x \cot^2(1 - x^2) \, dx$

18. $\int \dfrac{\sec^2 x}{\sqrt{\tan x}} \, dx$

19. $\int (\sec x + \tan x) \, dx$

20. $\int \dfrac{\sin x + \cos x}{\sin x - \cos x} \, dx$

21. $\int (1 - x)\sec(x^2 - 2x) \, dx$

22. $\int \dfrac{\sec^5 \sqrt{x} \tan \sqrt{x}}{\sqrt{x}} \, dx$

23. $\int \sqrt{1 - \sin^2 x} \, e^{\sin x} \, dx$

24. $\int \dfrac{\sec \ln x}{x} \, dx$

25. $\int \dfrac{dx}{\sqrt{x}(1 + \cos \sqrt{x})}$

26. $\int \csc^4(2x) \, dx$

27. $\int \dfrac{\cos 4x}{\sqrt{1 + \sin 4x}} \, dx$

28. $\int \sin^3 2\theta \, d\theta$

29. $\int \dfrac{\sec^2 \theta \, d\theta}{\sqrt{\tan \theta + 1}}$

30. $\int \dfrac{\tan x}{1 + \tan^2 x} \, dx$

31. Find the area bounded by the graph of $y = x \sec x^2$ and the x-axis for $0 \le x \le \sqrt{\pi/3}$.

32. Find the area bounded by the graph of $y = \sec x \cdot \tan x$ and the x-axis for $-\pi/3 \le x \le \pi/3$.

33. Find the average value of the function $f(x) = \csc x$ for $\pi/6 \le x \le \pi/3$.

34. Find $\int \sin x \cdot \cos x \, dx$ in two ways. (*Hint:* $\sin 2x = 2 \sin x \cos x$.)

35. Find $\int \tan x \sec^2 x \, dx$ in two ways.

36. Derive formula (5).

37. Find the length of the curve $y = \ln \sin x$ from $(\pi/6, -\ln 2)$ to $(\pi/3, \ln(\sqrt{3}/2))$.

38. Find the volume of the solid obtained by revolving the graph of $y = \sec x \tan x$, $-\pi/4 \le x \le \pi/4$, about the x-axis.

39. Find the volume of the solid obtained by revolving about the x-axis the region bounded by the graphs of $y = \sec x \tan x$ and $y = \tan x$ for $-\pi/4 \le x \le \pi/4$.

40. Find the area of one of the regions bounded by the graph of $y = \sec x$ and the line $y = -2$.

41. Use Simpson's Rule with $n = 6$ to approximate the length of the graph of $y = \cos x$ for $0 \le x \le 2\pi$.

9.2 INTEGRALS INVOLVING PRODUCTS OF TRIGONOMETRIC FUNCTIONS

Integrals that involve products of the trigonometric functions can often be handled by use of identities and appropriate substitutions. For each of the various types of integrals discussed in this section we will present a typical example, followed by a general statement of strategy, followed by more challenging examples. You should be cautioned that integrating expressions of this kind is rarely straightforward. Indeed, the best recipe for success is knowledge of general strategies coupled with the experience of working many such problems.

Integrals Involving Odd Powers of Sine and Cosine

Example 1

Find $\int \sin^3 x \cos x \, dx$.

Solution: We make the substitution

$$u = \sin x; \qquad du = \cos x \, dx$$

so that $\cos x$ is a factor of du. Then

$$\int \sin^3 x \cos x \, dx = \int u^3 \, du$$

$$= \frac{1}{4} u^4 + C$$

$$= \frac{1}{4} \sin^4 x + C. \qquad \diamondsuit$$

General Strategy: In an integral of the form

$$\int \sin^n x \cdot \cos^m x \, dx, \qquad \text{with } n \text{ or } m \text{ odd:}$$

(i) *If m is odd,* make the substitution $u = \sin x$. Then $du = \cos x \, dx$ involves one factor of $\cos x$. The remaining even factors of $\cos x$ may be converted to a function of $\sin x$ by the identity $\cos^2 x = 1 - \sin^2 x$. The integral then has the form

$$\int \sin^n x [1 - \sin^2 x]^k \cdot \cos x \, dx = \int u^n (1 - u^2)^k \, du.$$

(ii) *If n is odd,* make the substitution $u = \cos x$. Then $du = -\sin x \, dx$ and the remaining even factors of $\sin x$ may be converted to a function of $\cos x$. Proceed as above.

Example 2

Find $\int \sin^2 x \cos^3 x \, dx$.

Strategy

Use one factor of $\cos x$ for du.

Solution

The exponent on $\cos x$ is odd, so we make the substitution

$$u = \sin x; \qquad du = \cos x \, dx.$$

Then, using the identity $\cos^2 x = 1 - \sin^2 x$, we obtain

Convert remaining factors of cos x to functions of sin x.

$$\int \sin^2 x \cos^3 x \, dx = \int \sin^2 x [1 - \sin^2 x] \cos x \, dx$$

$$= \int u^2 (1 - u^2) \, du$$

$$= \frac{1}{3} u^3 - \frac{1}{5} u^5 + C$$

$$= \frac{1}{3} \sin^3 x - \frac{1}{5} \sin^5 x + C. \qquad \diamond$$

Example 3

Find $\int \sin^5 2x \, dx$.

Strategy

Use one factor of $\sin 2x$ for du.

Solution

Even though no factor of $\cos x$ is present, we may make the substitution

$$u = \cos 2x; \qquad du = -2 \sin 2x \, dx.$$

We obtain

$$\int \sin^5 2x \, dx = \int \sin^4 2x \cdot \sin 2x \, dx$$

Convert remaining $\sin^4 2x$ to $[1 - \cos^2 2x]^2$.

$$= \int [1 - \cos^2 2x]^2 \cdot \sin 2x \, dx$$

Make substitution.

$$= \int [1 - u^2]^2 \left(-\frac{1}{2} \right) du$$

$$= -\frac{1}{2} \int (1 - 2u^2 + u^4) \, du$$

$$= -\frac{1}{2} u + \frac{1}{3} u^3 - \frac{1}{10} u^5 + C$$

$$= -\frac{1}{2} \cos 2x + \frac{1}{3} \cos^3 2x - \frac{1}{10} \cos^5 2x + C. \qquad \diamond$$

Integrals Involving Even Powers of Sine and Cosine

Example 4

Find $\int \sin^2 x \cos^2 x \, dx$.

Strategy

Use half angle formulas to reduce exponents by half.

Solution

We use the identities

$$\sin^2 x = \frac{1}{2} - \frac{1}{2} \cos 2x$$

$$\cos^2 x = \frac{1}{2} + \frac{1}{2} \cos 2x$$

to obtain

$$\int \sin^2 x \cos^2 x \, dx = \int \left(\frac{1}{2} - \frac{1}{2} \cos 2x\right)\left(\frac{1}{2} + \frac{1}{2} \cos 2x\right) dx$$

$$= \int \left(\frac{1}{4} - \frac{1}{4} \cos^2 2x\right) dx$$

$$= \int \left[\frac{1}{4} - \frac{1}{4}\left(\frac{1}{2} + \frac{1}{2} \cos 4x\right)\right] dx$$

$$= \int \left(\frac{1}{8} - \frac{1}{8} \cos 4x\right) dx$$

$$= \frac{x}{8} - \frac{1}{32} \sin 4x + C. \qquad \diamondsuit$$

Use identity

$$\cos^2 2x = \frac{1}{2} + \frac{1}{2} \cos 4x$$

(i.e., substitute for $\cos^2$ a second time).

General Strategy: In integrals of the form

$$\int \sin^n x \cos^m x \, dx$$

where both n and m are even, use the identities

$$\sin^2 \theta = \frac{1}{2} - \frac{1}{2} \cos 2\theta \qquad (1)$$

$$\cos^2 \theta = \frac{1}{2} + \frac{1}{2} \cos 2\theta \qquad (2)$$

repeatedly until an integrand involving only constants and cosine terms is obtained.

Example 5

Find $\int \cos^2\left(\frac{x}{2}\right)\sin^4\left(\frac{x}{2}\right) dx$.

Strategy

Both exponents are even so we use identities (1) and (2) with $\theta = x/2$.

Solution

$$\int \cos^2\left(\frac{x}{2}\right)\sin^4\left(\frac{x}{2}\right) dx = \int \cos^2\left(\frac{x}{2}\right)\left[\sin^2\left(\frac{x}{2}\right)\right]^2 dx$$

$$= \int \left[\frac{1}{2} + \frac{1}{2} \cos x\right]\left[\frac{1}{2} - \frac{1}{2} \cos x\right]^2 dx$$

$$= \frac{1}{8} \int [1 + \cos x][1 - 2 \cos x + \cos^2 x] \, dx$$

$$= \frac{1}{8} \int [1 - \cos x - \cos^2 x + \cos^3 x] \, dx$$

Use identity (2) again on $\cos^2 x$.

$$= \frac{1}{8} \int \left[1 - \cos x - \left(\frac{1}{2} + \frac{1}{2} \cos 2x\right) + \cos^3 x\right] dx$$

$$= \frac{1}{8} \int \left[\frac{1}{2} - \cos x - \frac{1}{2} \cos 2x + \cos^3 x \right] dx$$

Handle $\cos^3 x$ term as in Example 2.

$$= \frac{1}{8} \int \left[\frac{1}{2} - \cos x - \frac{1}{2} \cos 2x + (1 - \sin^2 x) \cos x \right] dx$$

$$= \frac{1}{8} \int \left[\frac{1}{2} - \frac{1}{2} \cos 2x - \sin^2 x \cos x \right] dx$$

$$= \frac{x}{16} - \frac{1}{32} \sin 2x - \frac{1}{24} \sin^3 x + C. \qquad \diamond$$

Example 6

Find $\int \sin^4 x \, dx$.

Strategy

Use identity (1) with $\theta = x$.

Solution

$$\int \sin^4 x \, dx = \int [\sin^2 x]^2 \, dx$$

$$= \int \left[\frac{1}{2} - \frac{1}{2} \cos 2x \right]^2 dx$$

Use identity (2) on $\cos^2 2x$ with $\theta = 2x$.

$$= \int \left[\frac{1}{4} - \frac{1}{2} \cos 2x + \frac{1}{4} \cos^2 2x \right] dx$$

$$= \int \left[\frac{1}{4} - \frac{1}{2} \cos 2x + \frac{1}{4} \left(\frac{1}{2} + \frac{1}{2} \cos 4x \right) \right] dx$$

$$= \int \left(\frac{3}{8} - \frac{1}{2} \cos 2x + \frac{1}{8} \cos 4x \right) dx$$

$$= \frac{3x}{8} - \frac{1}{4} \sin 2x + \frac{1}{32} \sin 4x + C. \qquad \diamond$$

Integrals Involving Powers of Secant and Tangent

Example 7

Find $\int \tan^2 x \sec^2 x \, dx$.

Solution: We make the substitution

$$u = \tan x; \qquad du = \sec^2 x \, dx.$$

Then

$$\int \tan^2 x \sec^2 x \, dx = \int u^2 \, du$$

$$= \frac{1}{3} u^3 + C$$

$$= \frac{1}{3} \tan^3 x + C. \qquad \diamond$$

Example 8

Find $\int \tan^3 x \sec x \, dx$.

Solution: We use the identity $\tan^2 x = \sec^2 x - 1$ to write

$$\int \tan^3 x \sec x \, dx = \int \tan^2 x \tan x \sec x \, dx$$

$$= \int (\sec^2 x - 1) \tan x \sec x \, dx.$$

We now make the substitution

$$u = \sec x; \quad du = \tan x \sec x \, dx.$$

With this substitution we have

$$\int \tan^3 x \sec x \, dx = \int (u^2 - 1) \, du$$

$$= \frac{1}{3} u^3 - u + C$$

$$= \frac{1}{3} \sec^3 x - \sec x + C.$$

$\diamondsuit$

General Strategy: In integrals of the form

$$\int \sec^n x \tan^m x \, dx:$$

(i) *If n is even,* write $\sec^{n-2} x$ as a function of $\tan x$ using the identity $\sec^2 x = \tan^2 x + 1$. Then make the substitution $u = \tan x$. The remaining factor $\sec^2 x \, dx$ becomes du.

(ii) *If m is odd,* write $\tan^{m-1} x$ as a function of $\sec x$ using the identity above. Make the substitution $u = \sec x$ using one factor of $\sec x$, and using the remaining factor of $\tan x$ as $du = \sec x \tan x \, dx$.

(iii) *If n is odd and m is even,* the technique of integration by parts is needed. (This will be discussed in Chapter 10.)

Example 9

Find $\int \tan^4 x \, dx$.

Strategy	*Solution*
Use identity $\tan^2 x = \sec^2 x - 1$.	$\int \tan^4 x \, dx = \int (\sec^2 x - 1) \tan^2 x \, dx$
Use identity again.	$= \int (\sec^2 x \tan^2 x - \tan^2 x) \, dx$

Let $u = \tan x$.

$du = \sec^2 x\, dx$.

in first term.

$$= \int (\sec^2 x \tan^2 x - \sec^2 x + 1)\, dx$$

$$= \frac{1}{3} \tan^3 x - \tan x + x + C.$$

◇

Integrals of the Forms

$\int \sin mx \cos nx\, dx$

$\int \sin mx \sin nx\, dx$

$\int \cos mx \cos nx\, dx$

In these integrals the general strategy is to use the identities

$$\sin mx \cdot \cos nx = \frac{1}{2}[\sin(m + n)\, x + \sin(m - n)\, x] \tag{3}$$

$$\sin mx \cdot \sin nx = \frac{1}{2}[\cos(m - n)\, x - \cos(m + n)\, x] \tag{4}$$

$$\cos mx \cdot \cos nx = \frac{1}{2}[\cos(m + n)\, x + \cos(m - n)\, x]. \tag{5}$$

(See Exercises 45–47.)

Example 10

Find $\int \sin 2x \cos 5x\, dx$.

Solution:

We use identity (3) to write

$$\int \sin 2x \cos 5x\, dx = \int \frac{1}{2}[\sin(2x + 5x) + \sin(2x - 5x)]\, dx$$

$$= \frac{1}{2} \int (\sin 7x + \sin(-3x))\, dx$$

$$= \frac{1}{2}\left[\left(\frac{1}{7}\right) \cdot (-\cos 7x) + \left(-\frac{1}{3}\right)(-\cos(-3x))\right] + C$$

$$= -\frac{1}{14} \cos 7x + \frac{1}{6} \cos(-3x) + C$$

$$= -\frac{1}{14} \cos 7x + \frac{1}{6} \cos 3x + C.$$

◇

Exercise Set 9.2

In Exercises 1–36, find the integral.

1. $\int \sin^2 2x\, dx$

2. $\int_0^{\pi/4} \sec^2 x\, dx$

3. $\int \sin x \cos^2 x\, dx$

4. $\int \sin x \cos^3 x\, dx$

5. $\int \sin^5 x\, dx$

6. $\int \tan x \sec^2 x\, dx$

7. $\int_0^{\pi/2} \sin^2 x \cos^3 x\, dx$

8. $\int_0^{\pi/4} \sin^2 x \cos^2 x\, dx$

9. $\int \sec^3 x \tan^3 x\, dx$

10. $\int \sec^4(2x - 1)\, dx$

11. $\int_0^{\pi} \sin^6 x\, dx$

12. $\int \tan^3 x \sec^4 x\, dx$

13. $\int \cot^3 x\, dx$

14. $\int \sin^3 x \sqrt{\cos x}\, dx$

15. $\displaystyle\int_0^{\pi/4} \sin x \sin 3x \, dx$

16. $\displaystyle\int \sin 5x \cos 3x \, dx$

17. $\displaystyle\int (\sin x + \cos x)^2 \, dx$

18. $\displaystyle\int \sin^2 x \cos^5 x \, dx$

19. $\displaystyle\int \tan^4 x \sec^4 x \, dx$

20. $\displaystyle\int \cot^5 x \, dx$

21. $\displaystyle\int \tan^5 x \sec^3 x \, dx$

22. $\displaystyle\int \frac{\sec^2 x}{1 + \tan x} \, dx$

23. $\displaystyle\int \sin x \sin 2x \, dx$

24. $\displaystyle\int \sin 4x \cos 3x \, dx$

25. $\displaystyle\int \frac{\cos^3 x}{\sqrt{\sin x}} \, dx$

26. $\displaystyle\int (\tan x + \cot x)^2 \, dx$

27. $\displaystyle\int \frac{\sec^2 x}{(1 + \tan x)^4} \, dx$

28. $\displaystyle\int \csc x \cdot \cot x \, dx$

29. $\displaystyle\int \cos 5x \cdot \cos 3x \, dx$

30. $\displaystyle\int \cos 7x \sin 2x \, dx$

31. $\displaystyle\int \frac{\sin^3 \theta}{\cos \theta} \, d\theta$

32. $\displaystyle\int \frac{\tan^3 \theta}{\sec \theta} \, d\theta$

33. $\displaystyle\int \sec^2 x\sqrt{\tan x} \, dx$

34. $\displaystyle\int \sin ax \cdot \sin bx \cdot \cos cx \, dx$

35. $\displaystyle\int \sin ax \cdot \cos bx \cdot \cos cx \, dx$

36. $\displaystyle\int \sin(ax + b) \cos(cx + d) \, dx$

Evaluate the following definite integrals which arise often in applied mathematics (n and m are integers).

37. $\displaystyle\frac{1}{\pi} \cdot \int_{-\pi}^{\pi} \cos n\theta \, d\theta, \qquad n \geq 1$

38. $\displaystyle\frac{2}{\pi} \cdot \int_0^{\pi} \sin n\theta \, d\theta, \qquad n \geq 1$

39. $\displaystyle\int_{-L}^{L} \cos^2\left(\frac{n\pi x}{L}\right) dx$

40. $\displaystyle\int_0^{2\pi} \sin^2 nx \, dx$

41. $\displaystyle\int_0^{2\pi} \sin nx \cos mx \, dx$

42. $\displaystyle\int_0^{2\pi} \cos nx \cos mx \, dx$

Show how the following reduction formulas are obtained.

43. $\displaystyle\int \tan^n x \, dx = \frac{\tan^{n-1} x}{n-1} - \int \tan^{n-2} x \, dx, \qquad n \neq 1$

44. $\displaystyle\int \cot^n x \, dx = -\frac{\cot^{n-1} x}{n-1} - \int \cot^{n-2} x \, dx, \qquad n \neq 1$

45. Prove identity (3) by adding the identities
$$\sin(\theta_1 + \theta_2) = \sin \theta_1 \cos \theta_2 + \cos \theta_1 \sin \theta_2,$$
$$\sin(\theta_1 - \theta_2) = \sin \theta_1 \cos \theta_2 - \cos \theta_1 \sin \theta_2.$$

46. Prove identity (4) by subtracting the identities
$$\cos(\theta_1 - \theta_2) = \cos \theta_1 \cos \theta_2 + \sin \theta_1 \sin \theta_2,$$
$$\cos(\theta_1 + \theta_2) = \cos \theta_1 \cos \theta_2 - \sin \theta_1 \sin \theta_2.$$

47. Prove identity (5).

48. Find the average value of the function $y = \tan x \sec^2 x$ for $-\pi/4 \leq x \leq \pi/4$.

49. Find the volume of the solid obtained by revolving about the x-axis the region bounded by the graph of $y = \tan x \sec x$ for $0 \leq x \leq \pi/4$.

50. Find the area of the region bounded by the graphs of $f(x) = \tan^2 x$ and $g(x) = -\tan^2 x$ for $0 \leq x \leq \pi/4$.

51. Find the area of the region bounded above by the graph of $y = \sec x$ and below by the x-axis for $0 \leq x \leq \pi/4$.

52. Find the average value of the function $f(x) = \sin^2 x$ on $[0, \pi]$.

53. A particle moves along a line with acceleration function $a(t) = \cos^2 t$. Find
 a. its velocity function $v(t)$, if $v(0) = 0$;
 b. its position function $s(t)$, if $s(0) = 4$.

54. Find the volume of the solid obtained by revolving about the x-axis the region bounded by the graph of $f(x) = \sec^2 x \tan x$ and the x-axis for $0 \leq x \leq \pi/4$.

9.3 THE INVERSE TRIGONOMETRIC FUNCTIONS

Consider the following problem. A television camera is located 3 kilometers from the launch pad for a rocket that will propel a new satellite into orbit. The problem is to determine the angle $\theta(t)$ at which the camera should be inclined at each instant t so as to track the rocket during the initial phase of its ascent, when its trajectory is nearly vertical.

If the acceleration $a(t)$ of the rocket is known, the height of the tip of the rocket above the ground, $s(t)$, may be calculated by integration. Thus, at each instant t we

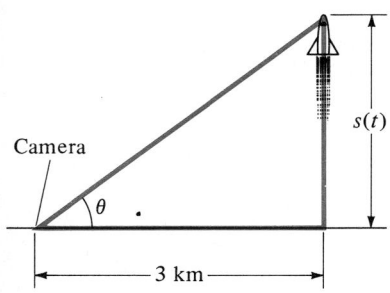

Figure 3.1 Elevation of camera determined by equation (1).

may compute $\tan \theta$ by the equation

$$\tan \theta = \frac{s}{3}. \tag{1}$$

(See Figure 3.1.)

The difficulty with equation (1) is that it describes the *tangent* of the desired angle rather than the angle θ itself. Thus, if we are to solve equation (1) for θ, we must find an *inverse* for the tangent function.

With this simple motivation in mind, we therefore set out to find inverses for each of the six trigonometric functions. As we succeed at doing so, we will also obtain some very useful integration formulas.

The Inverse Tangent Function

Consider the graph of $y = \tan x$ (Figure 3.2) and its reflection in the line $y = x$ (Figure 3.3). On observing these two graphs, one immediately concludes that there is no hope of defining an inverse for the function $y = \tan x$ for all x in the domain of this function. The difficulty is that $y = \tan x$ is *periodic*. In particular, $\tan(x + n\pi) = \tan x$ for all $n = \pm 1, +2, \ldots$, so each number y in the range of $y = \tan x$ corresponds to *an infinite number* of x's. Thus, the graph of the reflection (Figure 3.3) is not the graph of a *function*.

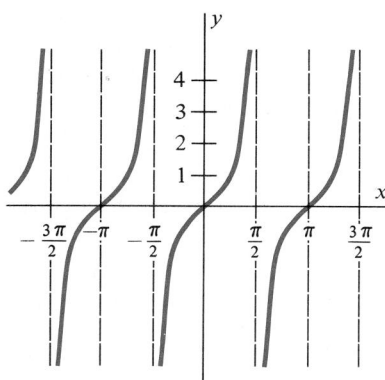

Figure 3.2 Graph of $y = \tan x$.

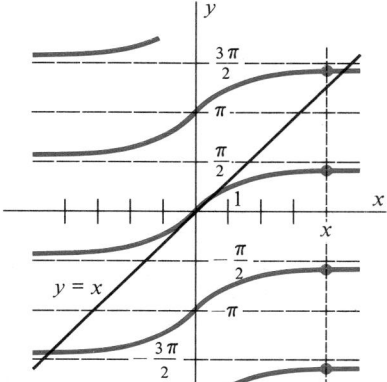

Figure 3.3 Reflection of $y = \tan x$ in $y = x$. (Many y's correspond to a single number x.)

However, all is not lost. Notice in Figure 3.2 that the arc of the graph of $y = \tan x$ corresponding to the interval $-\pi/2 < x < \pi/2$ has the property that each value y corresponds to a *unique* number x. This is because the function $y = \tan x$ is *increasing* throughout the interval $(-\pi/2, \pi/2)$. Our strategy is therefore to restrict the domain of the function $y = \tan x$ to the interval $(-\pi/2, \pi/2)$ and then define the *inverse* of the tangent function, denoted by Tan^{-1}, by the equation

$$\text{Tan}^{-1} y = x \quad \text{if and only if} \quad y = \tan x \quad \underline{\text{and}} \quad -\frac{\pi}{2} < x < \frac{\pi}{2}.$$

For example, $\text{Tan}^{-1}\left(\frac{\sqrt{3}}{2}\right) = \frac{\pi}{3}$ since $\tan \frac{\pi}{3} = \frac{\sqrt{3}}{2}$.

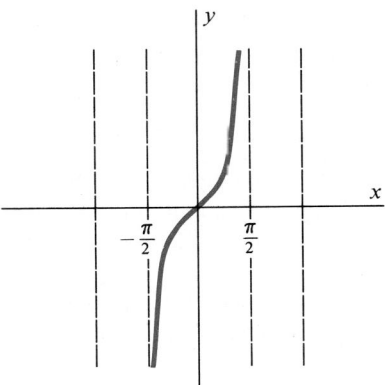

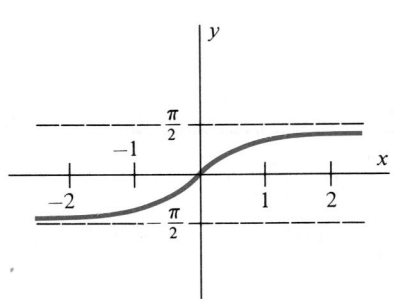

Figure 3.4 Graph of restricted function $y = \text{Tan } x$.

Figure 3.5 Reflection of graph of $y = \text{Tan } x$ is the graph of the inverse $y = \text{Tan}^{-1} x$.

We denote the restriction of the tangent function to the domain $(-\pi/2, \pi/2)$ by $y = \text{Tan } x$, and we refer to its graph as the **principal branch** of the graph of $y = \tan x$. Figures 3.4 and 3.5 show that the reflection of the graph of $y = \text{Tan } x$ in the line $y = x$ indeed represents the graph of a function. Figure 3.5 is therefore the graph of the *inverse* of the restricted tangent function $y = \text{Tan } x$.

Since the restricted tangent function $y = \text{Tan } x$ has domain $(-\pi/2, \pi/2)$ and range $(-\infty, \infty)$, the inverse will be defined on $(-\infty, \infty)$ with values in $(-\pi/2, \pi/2)$. The formal definition is the following.

DEFINITION 1

For $x \in (-\infty, \infty)$ the **inverse tangent function,** $\text{Tan}^{-1} x$, is defined by

$$y = \text{Tan}^{-1} x \quad \text{if and only if} \quad x = \tan y \quad \text{with} \quad -\frac{\pi}{2} < y < \frac{\pi}{2}.$$

Be careful to note that $\text{Tan}^{-1} x$ means the inverse of the function $y = \text{Tan } x$, *not* $(\tan x)^{-1}$. Other notation for the inverse tangent function common in textbooks is $y = \text{Arc tan } x$.

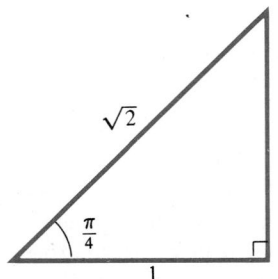

Figure 3.6 $\tan(\pi/4) = 1$; $\text{Tan}^{-1}(1) = \pi/4$.

Example 1

Find (a) $\text{Tan}^{-1}(1)$, (b) $\text{Tan}^{-1}(0)$, and (c) $\sin(\text{Tan}^{-1}(1/\sqrt{3}))$.

Solution:

(a) We seek an angle y so that $-\pi/2 < y < \pi/2$, and $\tan y = 1$. This angle is $y = \pi/4$. Thus,

$$\text{Tan}^{-1}(1) = \frac{\pi}{4}, \quad \text{since} \quad \tan\left(\frac{\pi}{4}\right) = 1 \quad \text{(Figure 3.6)}.$$

(b) $\text{Tan}^{-1}(0) = 0$, since $\tan(0) = 0$.

(c) Since $\tan(\pi/6) = 1/\sqrt{3}$, $\sin(\text{Tan}^{-1}(1/\sqrt{3})) = \sin(\pi/6) = 1/2$ (see Figure 3.7).

$\diamondsuit$

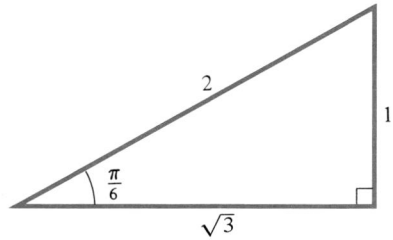

Figure 3.7 $\tan(\pi/6) = 1/\sqrt{3}$;
$\text{Tan}^{-1}(1/\sqrt{3}) = \pi/6$;
$\sin(\text{Tan}^{-1}(1/\sqrt{3})) = 1/2$.

Example 2

At what angle should the camera be inclined when the rocket in Figure 3.1 is 7.8 kilometers above the ground?

Solution: From equation (1), when $s = 7.8$ km we obtain

$$\tan \theta = \frac{7.8}{3} = 2.6.$$

If you have access to a hand calculator or computer with the ability to compute values of the inverse tangent, simply compute $\text{Tan}^{-1}(2.6)$. If not, use the table of trigonometric functions in Appendix IV to find an angle whose tangent is approximately 2.6. By either method the answer is

$$\theta = \text{Tan}^{-1}(2.6) \approx 1.20 \text{ radians}$$

$$\approx 69°. \qquad \diamond$$

Other Inverse Trigonometric Functions

To find inverses for the remaining five trigonometric functions, we use the same method followed to obtain $\text{Tan}^{-1} x$, namely:

(i) Restrict the domain of the trigonometric function $y = f(x)$ to an interval I on which each y in the range corresponds to exactly one x in the interval I.
(ii) Define the inverse function by the usual definition

$$y = f^{-1}(x) \quad \text{if and only if} \quad x = f(y) \quad \text{with} \quad y \in I.$$

We obtain the following inverse functions.

DEFINITION 2

The remaining five inverse trigonometric functions are defined as follows:

(a) For $-1 \le x \le 1$ and $-\pi/2 \le y \le \pi/2$ the **inverse sine function** is defined by $y = \text{Sin}^{-1} x$ if and only if $x = \sin y$.
(b) For $-1 \le x \le 1$ and $0 \le y \le \pi$ the **inverse cosine function** is defined by $y = \text{Cos}^{-1} x$ if and only if $x = \cos y$.
(c) For $|x| \ge 1$ and $y \in [0, \pi/2) \cup [\pi, 3\pi/2)$ the **inverse secant function** is defined by $y = \text{Sec}^{-1} x$ if and only if $x = \sec y$.
(d) For $|x| \ge 1$ and $y \in (0, \pi/2] \cup (\pi, 3\pi/2]$ the **inverse cosecant function** is defined by $y = \text{Csc}^{-1} x$ if and only if $x = \csc y$.
(e) For $-\infty < x < \infty$ and $0 < y < \pi$ the **inverse cotangent function** is defined by $y = \text{Cot}^{-1} x$ if and only if $x = \cot y$.

(Note that the domains of both $\text{Sec}^{-1} x$ and $\text{Csc}^{-1} x$ are unions of disjoint intervals. This is because the *ranges* of sec x and csc x consist of disjoint intervals. It is possible to choose different domains for $\text{Sec}^{-1} x$ and $\text{Csc}^{-1} x$, such as $[0, \pi/2) \cup (\pi/2, \pi]$ for $\sec^{-1} x$.)

Figures 3.8 through 3.11 illustrate the principal branches of the sine and secant functions, together with the graphs of the corresponding inverse functions. (The graphs of the other inverse functions are constructed in a similar manner. See Exercise 36.)

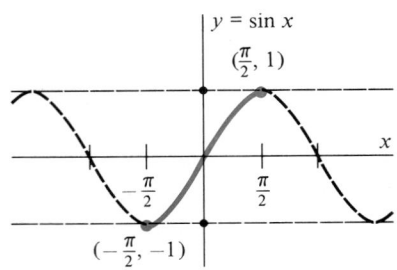

Figure 3.8 Principal branch of $y = \sin x$ is $-\pi/2 \le x \le \pi/2$.

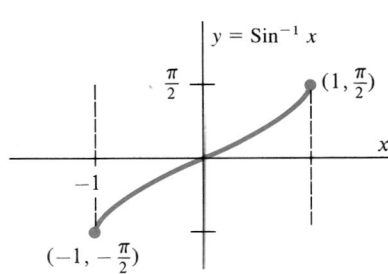

Figure 3.9 Graph of $y = \mathrm{Sin}^{-1} x$ is reflection of graph of principal branch of $y = \sin x$ in line $y = x$.

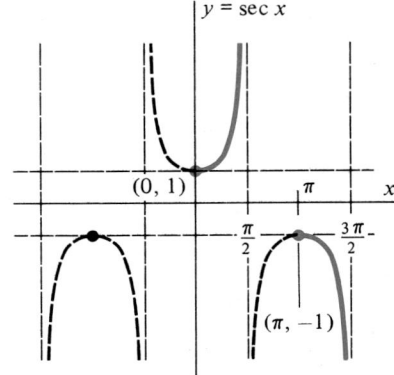

Figure 3.10 Principal branch of $y = \sec x$ is $x \in [0, \pi/2) \cup [\pi, 3\pi/2)$.

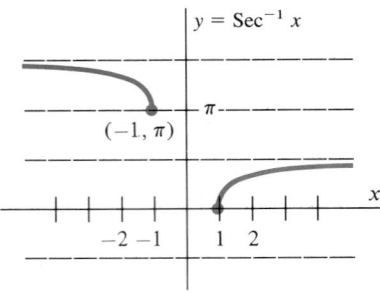

Figure 3.11 Graph of $y = \mathrm{Sec}^{-1} x$ is reflection of principal branch of $y = \sec x$ in line $y = x$.

Example 3

Find (a) $\mathrm{Sin}^{-1}(-\sqrt{3}/2)$, (b) $\mathrm{Csc}^{-1}(2)$, and (c) $\tan(\mathrm{Cos}^{-1}(-1/2))$.

Solution:

(a) An angle x that satisfies $\sin x = -\sqrt{3}/2$ and $-\pi/2 \le x \le \pi/2$ is $x = -\pi/3$. Thus, $\mathrm{Sin}^{-1}(-\sqrt{3}/2) = -\pi/3$ (Figure 3.12).

(b) An angle x with $\csc x = 2$ and $x \in (0, \pi/2] \cup (\pi, 3\pi/2]$ is $x = \pi/6$. Thus, $\mathrm{Csc}^{-1}(2) = \pi/6$ (Figure 3.13).

(c) An angle x with $\cos x = -1/2$ and $0 \le x \le \pi$ is $x = 2\pi/3$. Thus, $\tan(\mathrm{Cos}^{-1}(-1/2)) = \tan(2\pi/3) = -\sqrt{3}$ (Figure 3.14). ◇

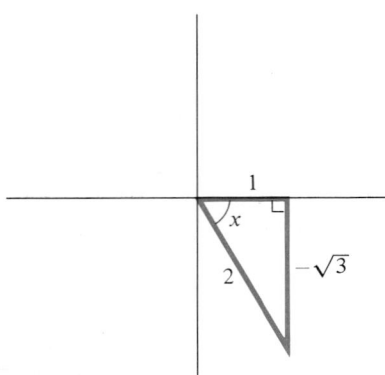

Figure 3.12 $\sin\left(-\dfrac{\pi}{3}\right) = -\dfrac{\sqrt{3}}{2}$ so $\mathrm{Sin}^{-1}\left(-\dfrac{\sqrt{3}}{2}\right) = -\dfrac{\pi}{3}$.

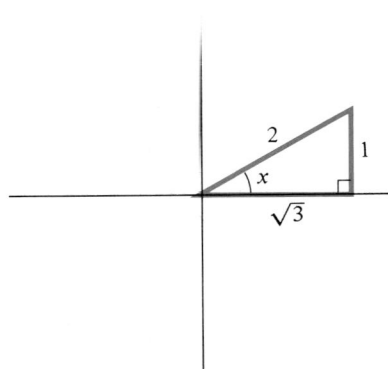

Figure 3.13 $\csc\left(\dfrac{\pi}{6}\right) = 2$ so $\mathrm{Csc}^{-1}(2) = \dfrac{\pi}{6}$.

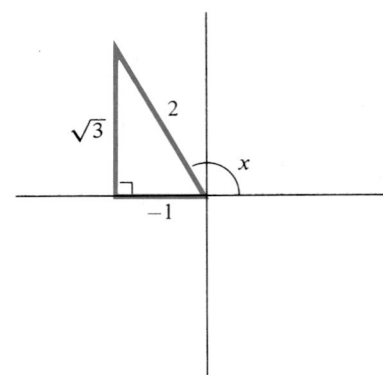

Figure 3.14 $\cos\left(\dfrac{2\pi}{3}\right) = -\dfrac{1}{2}$ so $\mathrm{Cos}^{-1}\left(-\dfrac{1}{2}\right) = \dfrac{2\pi}{3}$, $\tan\left(\mathrm{Cos}^{-1}\left(-\dfrac{1}{2}\right)\right) = -\sqrt{3}$.

The following example demonstrates the rather remarkable fact that a composition of a trigonometric function with an inverse trigonometric function produces an algebraic function.

Example 4

Let $-1 \le x \le 1$. Express $y = \sin(\text{Cos}^{-1} x)$ as an algebraic function of x.

Solution

We use the identity

$$\sin \theta = \sqrt{1 - \cos^2 \theta}.$$

Then

$$\sin(\text{Cos}^{-1} x) = \sqrt{1 - [\cos(\text{Cos}^{-1} x)]^2}$$
$$= \sqrt{1 - x^2}.$$

(See Figure 3.15.)

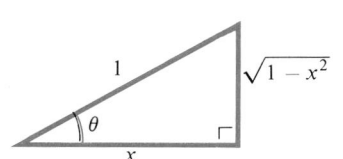

Figure 3.15 $\cos \theta = x$; $\sin \theta = \sqrt{1 - x^2}$.

Exercise Set 9.3

In Exercises 1–19, find the values indicated.

1. $\text{Sin}^{-1}(1/2)$

2. $\text{Cos}^{-1}(-\sqrt{3}/2)$

3. $\text{Tan}^{-1}(0)$

4. $\text{Sec}^{-1}(2)$

5. $\text{Cos}^{-1}(-1)$

6. $\text{Cot}^{-1}(-\sqrt{3})$

7. $\text{Sin}^{-1}(1) + \text{Cos}^{-1}(1)$

8. $\text{Tan}^{-1}\left(-\dfrac{1}{\sqrt{3}}\right)$

9. $\tan\left(\text{Sin}^{-1}\left(\dfrac{1}{2}\right)\right)$

10. $\text{Cos}^{-1}(\sin \pi/4)$

11. $\sec(\text{Tan}^{-1}(\sqrt{3}))$

12. $\text{Cot}^{-1}\left(\tan\left(\dfrac{2\pi}{3}\right)\right)$

13. $\cos[2\,\text{Tan}^{-1}(\sqrt{3})]$

14. $\sin\left[2\,\text{Cot}^{-1}\left(\dfrac{3}{4}\right)\right]$

15. $\tan^2\left(\pi - \text{Sin}^{-1}\left(\dfrac{\sqrt{2}}{2}\right)\right)$

16. $\text{Sec}^{-1}\left(2\tan\dfrac{\pi}{4}\right)$

17. $\text{Cos}^{-1}\left(2 - \sqrt{2}\sin\left(\dfrac{\pi}{4}\right)\right)$

18. $\sec(2\,\text{Tan}^{-1} 1)$

19. $\tan(\text{Sec}^{-1} 2)$

In Exercises 20–29, find an algebraic expression for the given function of x.

20. $y = \cos(\text{Sin}^{-1} x)$

21. $y = \tan(\text{Sin}^{-1} x)$

22. $y = \tan(\text{Tan}^{-1} x)$

23. $y = \sin[2\,\text{Sin}^{-1}(x)]$

24. $y = \sin(\text{Tan}^{-1} x)$

25. $y = \sin(\text{Sec}^{-1} x)$

26. $y = \cos(\text{Tan}^{-1} x)$

27. $y = \tan(\text{Cot}^{-1} x)$

28. $y = \tan(\text{Sec}^{-1} x)$

29. $y = \cos(\text{Csc}^{-1} x)$

30. Explain why the identity $\sin(\text{Sin}^{-1} x) = x$ is valid for $-1 \le x \le 1$. Are similar identities true for the other inverse trigonometric functions? If so, for what values of x do they hold?

31. True or false? $\text{Sec}^{-1}(\cos \pi/4) = \dfrac{2}{\sqrt{2}}$.

32. The selection of the principal branch of a trigonometric function is somewhat arbitrary.
   a. Show that the selection of the principal branch of the sine function as $\pi/2 \le x \le 3\pi/2$ leads to a legitimate inverse function for $y = \sin x$.
   b. What would be the disadvantage of choosing the principal branch of $y = \sin x$ to be $0 \le x \le \pi/2$?
   c. What properties should the principal branch of a trigonometric function possess?

33. A lighthouse lies 100 meters off a straight shoreline. A boat dock is located on the shoreline at the point nearest the lighthouse. Write an equation describing θ, the angle between the light beam and the line joining the lighthouse and the dock, as a function of the distance x between the dock and the point where the beam of light strikes the shore (see Figure 3.16).

34. Snell's Law of Refraction states that when a ray of light passes from one medium to another, the angles θ_1 and θ_2 formed between the rays and the normal (perpendicular line) to the boundary are related by the equation $\dfrac{\sin \theta_1}{n_1} = \dfrac{\sin \theta_2}{n_2}$.

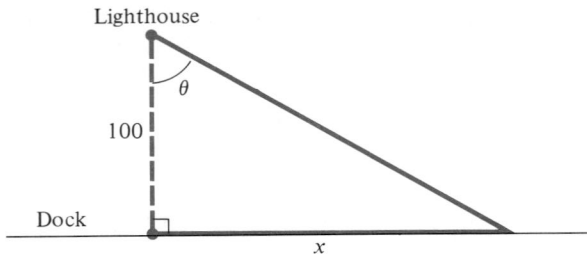

Figure 3.16 Lighthouse problem.

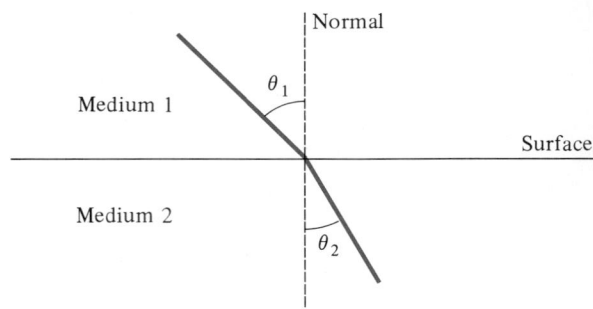

Figure 3.17 Snell's Law of Refraction.

where n_1 and n_2 are constants. Express θ_2 as a function of θ_1 (see Figure 3.17).

35. A patrol car is parked on an overpass 20 meters above a roadway, and the officer is using radar to time the cars moving along the roadway. Ignoring the heights of the cars, express the angle formed between the radar beam and the vertical when a car is x meters away from the point on the roadway directly beneath the patrol car.

36. Graph the inverse functions $y = \text{Cot}^{-1} x$, $y = \text{Cos}^{-1} x$, and $y = \text{Csc}^{-1} x$.

9.4 DERIVATIVES OF THE INVERSE TRIGONOMETRIC FUNCTIONS

Each inverse trigonometric function f^{-1} is differentiable throughout its domain D except at numbers $y = f(x)$ with $f'(x) = 0$, according to Theorem 4, Chapter 8. To obtain these derivatives we shall use the identity

$$f(f^{-1}(x)) = x, \qquad x \in D \tag{1}$$

and differentiate implicitly to find $\dfrac{dy}{dx} = \dfrac{d}{dx} f^{-1}(x)$.

For the function $y = \text{Tan}^{-1} x$, identity (1) becomes

$$\tan y = x, \qquad -\infty < x < \infty.$$

Differentiating implicitly with respect to x gives

$$\sec^2 y \cdot \frac{dy}{dx} = 1,$$

so

$$\frac{dy}{dx} = \frac{1}{\sec^2 y}$$

$$= \frac{1}{1 + (\tan y)^2} \qquad (\sec^2 y = 1 + \tan^2 y)$$

$$= \frac{1}{1 + x^2} \qquad (\tan y = x).$$

$$\boxed{\frac{d}{dx} \text{Tan}^{-1} x = \frac{1}{1 + x^2}, \qquad -\infty < x < \infty.} \tag{2}$$

Example 1

For the functions

(a) $f(x) = \text{Tan}^{-1}(3x)$

and

(b) $g(x) = x \, \text{Tan}^{-1} e^x$,

we have

(a) $f'(x) = \dfrac{1}{1 + (3x)^2} \cdot \dfrac{d}{dx}(3x)$

$\qquad = \dfrac{3}{1 + 9x^2}$,

and

(b) $g'(x) = \left[\dfrac{d}{dx}(x) \right] \text{Tan}^{-1} e^x + x\left(\dfrac{d}{dx} \text{Tan}^{-1} e^x \right)$

$\qquad = \text{Tan}^{-1} e^x + \dfrac{x}{1 + (e^x)^2}\left(\dfrac{d}{dx} e^x \right)$

$\qquad = \text{Tan}^{-1} e^x + \dfrac{xe^x}{1 + e^{2x}}$. $\qquad\qquad\qquad\qquad \diamond$

To find the derivative of the function $y = \text{Sin}^{-1} x$, we begin with the identity

$$\sin y = x, \qquad |x| \le 1, \qquad -\pi/2 \le y \le \pi/2.$$

Differentiating implicitly with respect to x gives

$$\cos y \frac{dy}{dx} = 1,$$

so

$$\frac{dy}{dx} = \frac{1}{\cos y}$$

$$= \frac{1}{\sqrt{1 - (\sin y)^2}} \qquad (\cos y = +\sqrt{1 - \sin^2 y} \quad \text{for} \quad -\pi/2 \le y \le \pi/2)$$

$$= \frac{1}{\sqrt{1 - x^2}} \qquad (\sin y = x)$$

where we must require that $1 - x^2 > 0$.

$$\boxed{\frac{d}{dx} \text{Sin}^{-1} x = \frac{1}{\sqrt{1 - x^2}}, \qquad |x| < 1.} \qquad\qquad (3)$$

We may establish differentiation formulas for each of the remaining inverse trigonometric functions by proceeding in exactly the same fashion. The results are the following:

$$\frac{d}{dx}\operatorname{Cos}^{-1}x = \frac{-1}{\sqrt{1-x^2}}, \qquad |x| < 1 \tag{4}$$

$$\frac{d}{dx}\operatorname{Cot}^{-1}x = \frac{-1}{1+x^2}, \qquad -\infty < x < \infty \tag{5}$$

$$\frac{d}{dx}\operatorname{Sec}^{-1}x = \frac{1}{x\sqrt{x^2-1}}, \qquad |x| > 1 \tag{6}$$

$$\frac{d}{dx}\operatorname{Csc}^{-1}x = \frac{-1}{x\sqrt{x^2-1}}, \qquad |x| > 1 \tag{7}$$

It is important to note the restrictions $|x| < 1$ in (4) and $|x| > 1$ in (6) and (7). If these inequalities do not hold the corresponding formulas are meaningless.

Example 2

Find $\dfrac{dy}{dx}$ for $y = x^3 \operatorname{Sin}^{-1}x + \operatorname{Cos}^{-1}\sqrt{x}$.

Solution: Using equations (3) and (4) we obtain

$$\frac{dy}{dx} = \left(\frac{d}{dx}x^3\right)\operatorname{Sin}^{-1}x + x^3\left(\frac{d}{dx}\operatorname{Sin}^{-1}x\right) + \frac{-1}{\sqrt{1-(\sqrt{x})^2}}\cdot\frac{d}{dx}(\sqrt{x})$$

$$= 3x^2\operatorname{Sin}^{-1}x + \frac{x^3}{\sqrt{1-x^2}} - \frac{1}{2\sqrt{x}\sqrt{1-x}}.$$

Note that this derivative exists only for $0 < x < 1$.   ◇

Example 3

Prove the identity $\operatorname{Tan}^{-1}x = \pi/2 - \operatorname{Tan}^{-1}(1/x)$ by differentiation.

Strategy
Show

$$\frac{d}{dx}\operatorname{Tan}^{-1}x = -\frac{d}{dx}\operatorname{Tan}^{-1}\frac{1}{x}$$

by computing $\dfrac{d}{dx}\operatorname{Tan}^{-1}\left(\dfrac{1}{x}\right)$.

Solution
By equation (2) we have

$$\frac{d}{dx}\operatorname{Tan}^{-1}\left(\frac{1}{x}\right) = \frac{1}{1+\left(\frac{1}{x}\right)^2}\cdot\left(-\frac{1}{x^2}\right)$$

$$= \frac{-1}{1+x^2}$$

$$= -\frac{d}{dx}\operatorname{Tan}^{-1}x.$$

Recall, if $f' = g'$ then $f = g(x) + C$.

Thus

$$\operatorname{Tan}^{-1}x = -\operatorname{Tan}^{-1}\left(\frac{1}{x}\right) + C$$

To find C, substitute particular value of x.

for some constant C. To find C we substitute $x = 1$ to obtain

$$\operatorname{Tan}^{-1}1 = -\operatorname{Tan}^{-1}1 + C$$

Use fact that $\text{Tan}^{-1}\, 1 = \dfrac{\pi}{4}$.

or

$$\frac{\pi}{4} = -\frac{\pi}{4} + C,$$

so

$$C = \frac{\pi}{2},$$

as required.   ◇

The differentiation formulas (2), (3), and (6) immediately give the following integration formulas:

$$\int \frac{dx}{\sqrt{1 - x^2}} = \text{Sin}^{-1}\, x + C \qquad (8)$$

$$\int \frac{dx}{1 + x^2} = \text{Tan}^{-1}\, x + C \qquad (9)$$

$$\int \frac{dx}{x\sqrt{x^2 - 1}} = \text{Sec}^{-1}\, x + C \qquad (10)$$

(We do not state the integration formulas corresponding to differentiation formulas (4), (5), and (7), since they involve only the negatives of the above integrals.)

The next two examples show how we can use substitutions to bring integrals into the form of equations (8–10).

Example 4

Find $\displaystyle\int \frac{1}{\sqrt{9 - x^2}}\, dx.$

Strategy

Factor out 9 to obtain form of equation (8).

Solution

$$\int \frac{dx}{\sqrt{9 - x^2}} = \int \frac{dx}{\sqrt{9\left(1 - \dfrac{x^2}{9}\right)}} = \int \frac{dx}{3\sqrt{1 - \left(\dfrac{x}{3}\right)^2}}$$

Make substitution.

We make the substitution

$$u = \frac{x}{3}; \qquad du = \frac{1}{3}\, dx.$$

Then $dx = 3\, du$, so our integral becomes

Apply equation (8).

$$\frac{1}{3} \int \frac{3 \cdot du}{\sqrt{1 - u^2}} = \int \frac{du}{\sqrt{1 - u^2}}$$

$$= \text{Sin}^{-1}\left(\frac{x}{3}\right) + C.$$ ◇

Example 5

Find $\displaystyle\int_1^3 \frac{dx}{\sqrt{x}(1+x)}$.

Strategy

Make a substitution so that $1/\sqrt{x}$ is a factor of du. Here $u = \sqrt{x}$ works.

Change limits of integration.

Use formula (9).

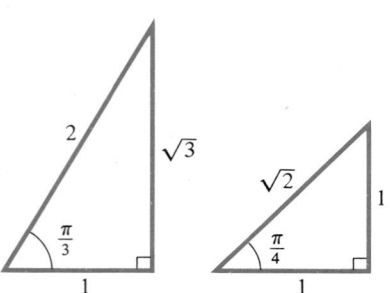

Solution

We make the substitution

$$u = \sqrt{x}; \qquad du = \frac{1}{2\sqrt{x}}\,dx.$$

Thus $\dfrac{1}{\sqrt{x}}\,dx = 2\,du$. We must also change limits of integration:

$$\begin{aligned} \text{If } x = 1, & \quad u = \sqrt{1} = 1.\\ \text{If } x = 3, & \quad u = \sqrt{3}. \end{aligned}$$

With these substitutions we obtain

$$\int_1^3 \frac{dx}{\sqrt{x}(1+x)} = \int_1^{\sqrt{3}} \frac{2\,du}{1+u^2}$$

$$= 2\,\text{Tan}^{-1}\,u\,\Big]_1^{\sqrt{3}}$$

$$= 2(\text{Tan}^{-1}\,\sqrt{3} - \text{Tan}^{-1}\,1)$$

$$= 2(\pi/3 - \pi/4)$$

$$= \pi/6.$$

$\diamondsuit$

The derivatives of inverse trigonometric functions are algebraic functions. Among other things, this remarkable fact allows us to obtain a simple procedure for approximating the irrational number π, as the following example shows.

Example 6

Use integration formula (9) to approximate π.

Strategy

The idea here is to select limits so as to obtain π as the value of the definite integral. We then approximate this integral.

Solution

Since

$$\int \frac{dx}{1+x^2} = \text{Tan}^{-1}\,x + C,$$

we have

$$\int_0^1 \frac{dx}{1+x^2} = \text{Tan}^{-1}\,x\,\Big]_0^1$$

$$= \text{Tan}^{-1}(1) - \text{Tan}^{-1}(0)$$

$$= \pi/4.$$

Thus

$$\pi = 4\int_0^1 \frac{dx}{1+x^2}.$$

We may therefore approximate π by approximating the definite integral. The results of using Simpson's Rule with various values of n to approximate this integral appear in Table 4.1.

$\diamondsuit$

Table 4.1

n	Approximation to π using Simpson's Rule
4	3.14156863
10	3.14159262
20	3.14159253
50	3.14159251

Exercise Set 9.4

In Exercises 1–14, find the derivative.

1. $y = \text{Sin}^{-1} 3x$

2. $f(x) = x \, \text{Tan}^{-1}(x - 1)$

3. $f(t) = \text{Sin}^{-1} \sqrt{t}$

4. $y = x^3 \, \text{Csc}^{-1}(1 + x)$

5. $f(x) = \text{Sin}^{-1} e^{-x}$

6. $y = \text{Tan}^{-1} \sqrt{x^2 - 1}$

7. $y = \sqrt{\text{Cos}^{-1} x}$

8. $f(x) = \text{Sin}^{-1} x^2 - x \, \text{Sec}^{-1}(x + 3)$

9. $f(x) = \ln \text{Tan}^{-1} x$

10. $y = \text{Tan}^{-1} \left(\dfrac{1 - x}{1 + x} \right)$

11. $y = \text{Sec}^{-1} \left(\dfrac{1}{x} \right)$

12. $y = e^{\text{Tan}^{-1} \sqrt{x}}$

13. $f(x) = \dfrac{\text{Tan}^{-1} x}{1 + x^2}$

14. $f(x) = x^2 \, \text{Sin}^{-1} 2x$

In Exercises 15–32, find the indicated integral.

15. $\displaystyle \int \frac{1}{4 + x^2} \, dx$

16. $\displaystyle \int_0^1 \frac{dx}{\sqrt{1 - x^2}}$

17. $\displaystyle \int \frac{1}{2x\sqrt{x^2 - 16}} \, dx$

18. $\displaystyle \int \frac{\cos x}{1 + \sin^2 x} \, dx$

19. $\displaystyle \int \frac{x}{\sqrt{1 - x^2}} \, dx$

20. $\displaystyle \int_1^2 \frac{1}{x\sqrt{x^2 - 1}} \, dx$

21. $\displaystyle \int \frac{1}{\sqrt{e^{2x} - 1}} \, dx$

22. $\displaystyle \int \frac{x}{\sqrt{1 - x^4}} \, dx$

23. $\displaystyle \int \frac{e^{\sqrt{x}}}{\sqrt{x}(1 + e^{2\sqrt{x}})} \, dx$

24. $\displaystyle \int \frac{x^2}{\sqrt{1 - x^6}} \, dx$

25. $\displaystyle \int \frac{x^2}{1 + x^6} \, dx$

26. $\displaystyle \int \frac{\sin x}{1 + \cos^2 x} \, dx$

27. $\displaystyle \int_0^1 \frac{\text{tan}^{-1} x}{1 + x^2} \, dx$

28. $\displaystyle \int \frac{\cos x}{\sqrt{4 - \sin^2 x}} \, dx$

29. $\displaystyle \int_1^2 \frac{dx}{x^2 + 2}$

30. $\displaystyle \int_2^3 \frac{dx}{x\sqrt{16x^2 - 25}}$

31. $\displaystyle \int \frac{dx}{2x\sqrt{4x^2 - 1}}$

32. $\displaystyle \int_1^e \frac{1}{x\sqrt{1 - \ln^2 x}} \, dx$

33. Use separation of variables (see Section 5.3) to solve the initial value problem

$$\frac{dy}{dx} = 4 + y^2, \qquad y(0) = 1.$$

34. Solve the initial value problem

$$\frac{dy}{dx} = \sqrt{1 - y^2}, \qquad y(0) = 1.$$

35. In the radar problem of Exercise 35, Section 9.3, how fast must the radar gun rotate to track an automobile 40 meters away from the point directly under the radar gun if the automobile is moving at a rate of 120 km/hr?

36. A tapestry 3 meters high hangs on a wall so that its lower edge is 1 meter above an observer's eye level. How far from the wall should the observer stand so as to maximize the angle subtended in the observer's eye by the tapestry?

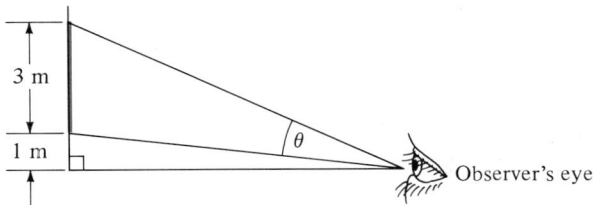

37. A ladder 10 meters long leans against a wall. If the base of the ladder is slipping away from the wall at a rate of 2 meters per second, how fast is the angle between the ladder and the wall increasing when the top of the ladder is 8 meters above the ground?

38. Analysis shows that the first minimum for the diffraction pattern of a circular aperture of diameter d, assuming certain conditions, is given by the equation

$$\sin \theta = 1.22 \left(\frac{\lambda}{d} \right), \qquad 0 < \theta < \frac{\pi}{2}$$

where λ is the wavelength of light (a constant). Assuming that θ is a differentiable function of d, show that θ is a decreasing function of d.

39. Find an equation for the line tangent to the graph of $y = \text{Sin}^{-1} x^2$ at the point $(\sqrt{2}/2, \pi/6)$.

40. Find the area of the region bounded by the graph of $f(x) = \dfrac{x^2}{1 + x^6}$ and the x-axis for $0 \le x \le 1$.

41. Find the volume of the solid generated by revolving the region bounded by the graph of $f(x) = \dfrac{1}{\sqrt{1 - x^4}}$ and the x-axis, for $0 \le x \le \sqrt{2}/2$, about the y-axis.

42. Find the average value of the function $f(x) = \dfrac{xe^{-x^2}}{\sqrt{1 - e^{-2x^2}}}$ on the interval $[\sqrt{\ln 2}, 1]$.

43. Derive formula (4).

44. Derive formula (5).

45. Derive formula (6).

46. Derive formula (7).

9.5 THE HYPERBOLIC FUNCTIONS

Recall from Chapter 8 that the function $y = Ce^x$ is the only nontrivial differentiable function satisfying the differential equation

$$\frac{dy}{dx} = y. \tag{1}$$

We have already seen that equation (1) has various practical interpretations corresponding to the various interpretations of the first derivative (e.g., velocity, rate of growth, and slope of tangent). Since the second derivative also has meaningful interpretations (acceleration, concavity, change in rate of growth) it is natural to ask which nonzero functions satisfy the second order differential equation.

$$\frac{d^2y}{dx^2} = y. \tag{2}$$

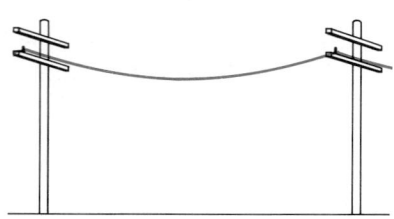

Figure 5.1 Cable suspended from two points hangs in the shape of a *catenary,* which is described by the hyperbolic cosine function.

A quick check shows that both $y_1 = e^x$ and $y_2 = e^{-x}$ are solutions of equation (2). Moreover, certain combinations of these two functions also satisfy equation (2) and, in addition, satisfy properties quite similar to the properties of the trigonometric functions. We are referring to the following functions.

DEFINITION 3

The hyperbolic sine and cosine functions, sinh x and cosh x, are defined as follows:

$$\sinh x = \frac{1}{2}(e^x - e^{-x}), \tag{3}$$

$$\cosh x = \frac{1}{2}(e^x + e^{-x}). \tag{4}$$

Roughly speaking, cosh x represents the average of exponential growth and exponential decay, while sinh x represents half the difference between these two phenomena.

The *hyperbolic functions* are important in applied problems in physics and engineering in which quantities of interest are described not by explicit functions but by *differential equations.* That is why we have come upon these functions by asking for the solutions of certain rather simple differential equations. For example, a cable suspended from two points hangs in the shape of a *catenary* (see Figure 5.1), which is described by the hyperbolic cosine function. (See Exercise 56.) We shall see in Chapter 14 why the term *hyperbolic* is used in conjunction with these functions.

Derivative Formulas

The differentiation formulas for the hyperbolic functions closely resemble those for the trigonometric functions.

For example, from equations (3) and (4) we obtain

$$\frac{d}{dx} \sinh x = \frac{d}{dx}\left\{\frac{1}{2}(e^x - e^{-x})\right\}$$

$$= \frac{1}{2}(e^x + e^{-x})$$

$$= \cosh x,$$

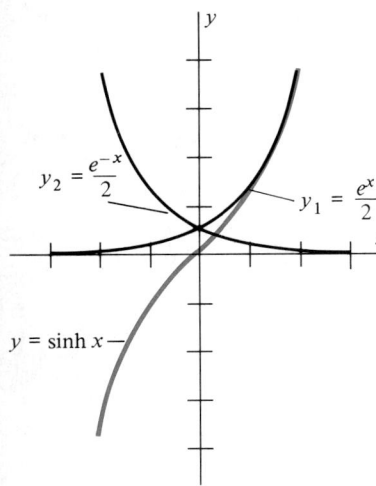

Figure 5.2 $\sinh x = \dfrac{e^x - e^{-x}}{2}$.

and

$$\frac{d}{dx}\cosh x = \frac{d}{dx}\left\{\frac{1}{2}(e^x + e^{-x})\right\}$$

$$= \frac{1}{2}(e^x - e^{-x})$$

$$= \sinh x.$$

(Note that the derivative of $\cosh x$ is $\sinh x$, not $-\sinh x$.)

As a consequence of equations (3) and (4), both $\sinh x$ and $\cosh x$ are their own second derivatives (just as e^x and e^{-x}):

$$\frac{d^2}{dx^2}(\sinh x) = \frac{d}{dx}(\cosh x) = \sinh x,$$

and

$$\frac{d^2}{dx^2}(\cosh x) = \frac{d}{dx}(\sinh x) = \cosh x.$$

Therefore, both $y_1 = \sinh x$ and $y_2 = \cosh x$ are solutions of differential equation (2).

REMARK: It is easy to show that, if f and g are solutions of differential equation (2), then $c_1 f + c_2 g$ is also a solution for any choice of constants c_1 and c_2. (This is called a **linear combination** of solutions.) The functions $\sinh x$ and $\cosh x$ are merely special linear combinations of the exponential solutions of (2).

Graphs of sinh x and cosh x

From the definition of $\sinh x$ we can see that

$$\sinh(-x) = \frac{1}{2}[e^{-x} - e^{-(-x)}] = \frac{1}{2}[e^{-x} - e^x] = -\frac{1}{2}[e^x - e^{-x}] = -\sinh x.$$

Thus, $\sinh x$ is an *odd* function, and its graph is symmetric with respect to the origin. Moreover, since

$$\frac{d}{dx}\sinh x = \cosh x = \frac{1}{2}[e^x + e^{-x}] > 0$$

for all x, $\sinh x$ is an increasing function on all intervals. Finally, since

$$\frac{d^2}{dx^2}\sinh x = \sinh x \qquad \text{is} \qquad \begin{cases} \text{positive, if } x > 0 \\ \text{negative, if } x < 0 \end{cases}$$

the graph of $\sinh x$ is concave down on $(-\infty, 0]$ and concave up on $[0, \infty)$ (see Figure 5.2).

A similar analysis shows that

(a) $\cosh x$ is an *even* function: $\cosh(-x) = \cosh x$ for all x. Its graph is therefore symmetric with respect to the y-axis.

(b) $\cosh x > 0$ for all x.

(c) $\cosh x$ is decreasing on $(-\infty, 0]$, increasing on $[0, \infty)$.

(d) The graph of $\cosh x$ is concave up on $(-\infty, \infty)$.

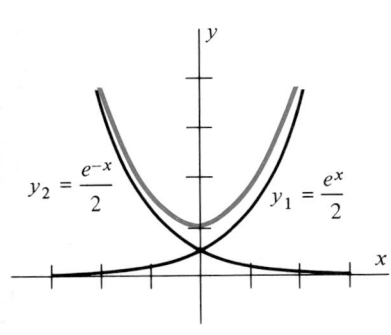

Figure 5.3 $\cosh x = \dfrac{e^x + e^{-x}}{2}$.

The graph of $\cosh x$ appears in Figure 5.3.

As with the trigonometric functions, we can define four other hyperbolic functions in terms of $\sinh x$ and $\cosh x$.

DEFINITION 4

The hyperbolic tangent, cotangent, secant, and cosecant functions are defined by

$$\tanh x = \frac{\sinh x}{\cosh x} = \frac{e^x - e^{-x}}{e^x + e^{-x}},$$

$$\coth x = \frac{\cosh x}{\sinh x} = \frac{e^x + e^{-x}}{e^x - e^{-x}},$$

$$\operatorname{sech} x = \frac{1}{\cosh x} = \frac{2}{e^x + e^{-x}},$$

$$\operatorname{csch} x = \frac{1}{\sinh x} = \frac{2}{e^x - e^{-x}}.$$

From Definition 4 we obtain the following differentiation formulas:

$$\frac{d}{dx} \tanh x = \operatorname{sech}^2 x \tag{5}$$

$$\frac{d}{dx} \coth x = -\operatorname{csch}^2 x \tag{6}$$

$$\frac{d}{dx} \operatorname{sech} x = -\operatorname{sech} x \cdot \tanh x \tag{7}$$

$$\frac{d}{dx} \operatorname{csch} x = -\operatorname{csch} x \cdot \coth x \tag{8}$$

Example 1

For $f(x) = x \operatorname{sech}^2 \sqrt{x}$,

$$f'(x) = 1 \cdot \operatorname{sech}^2 \sqrt{x} + x \cdot \frac{d}{dx} (\operatorname{sech}^2 \sqrt{x})$$

$$= \operatorname{sech}^2 \sqrt{x} + x \left\{ 2 \operatorname{sech} \sqrt{x} \cdot (-\operatorname{sech} \sqrt{x} \cdot \tanh \sqrt{x}) \cdot \frac{1}{2\sqrt{x}} \right\}$$

$$= \operatorname{sech}^2 \sqrt{x} - \sqrt{x} \operatorname{sech}^2 \sqrt{x} \tanh \sqrt{x}. \qquad \diamond$$

Identities

The hyperbolic functions satisfy identities similar to those for the trigonometric functions. For example,

$$\cosh^2 x = \left[\frac{1}{2}(e^x + e^{-x}) \right]^2$$

$$= \frac{1}{4}(e^{2x} + 2 + e^{-2x})$$

$$= 1 + \frac{1}{4}(e^{2x} - 2 + e^{-2x})$$

$$= 1 + \left[\frac{1}{2}(e^x - e^{-x})\right]^2$$

$$= 1 + \sinh^2 x.$$

Thus,

$$\cosh^2 x - \sinh^2 x = 1. \tag{9}$$

Similarly, we can show that

$$\text{sech}^2 x = 1 - \tanh^2 x \tag{10}$$

and

$$\text{csch}^2 x = \coth^2 x - 1. \tag{11}$$

(Other identities are presented in the exercise set.)

Example 2

Show that the function $y = \tanh(ax)$ is a solution of the differential equation

$$\frac{dy}{dx} = a(1 - y^2).$$

Solution: By formula (5), the derivative of the given function is

$$\frac{dy}{dx} = \frac{d}{dx}\tanh(ax) = a\,\text{sech}^2(ax).$$

Thus, with identity (10) we obtain

$$\frac{dy}{dx} = a\,\text{sech}^2(ax)$$

$$= a[1 - \tanh^2(ax)]$$
$$= a(1 - y^2). \qquad \diamond$$

Inverses

We have already verified that the hyperbolic sine function is increasing for all values of x. This condition guarantees that each value y corresponds to precisely one number x via the equation $y = \sinh x$. We may therefore define the **inverse** of the hyperbolic sine function $\sinh^{-1} x$ to be the function that reverses this correspondence. In other words,

$$y = \sinh^{-1} x \qquad \text{if and only if} \qquad x = \sinh y. \tag{12}$$

We can obtain an explicit formulation for the inverse hyperbolic sine function by combining equations (3) and (12): If

$$y = \sinh^{-1} x, \qquad \text{then} \qquad x = \sinh y = \frac{1}{2}(e^y - e^{-y}),$$

so

$$e^y - 2x - e^{-y} = 0. \tag{13}$$

Multiplying both sides of (13) by e^y gives

$$e^{2y} - 2xe^y - 1 = 0. \tag{14}$$

Viewing (14) as a quadratic expression in the variable e^y and applying the quadratic formula gives

$$e^y = \frac{2x \pm \sqrt{4x^2 + 4}}{2}$$

$$= x \pm \sqrt{x^2 + 1}.$$

Since e^y is never negative, and $\sqrt{x^2 + 1} > x$, the ambiguous sign $\pm$ must be $+$. Taking natural logs of both sides of the last equation now gives

$$y = \ln(x + \sqrt{x^2 + 1}),$$

so

$$\boxed{\sinh^{-1} x = \ln(x + \sqrt{x^2 + 1}), \qquad -\infty < x < \infty.} \tag{15}$$

By suitably restricting domains where necessary, we may similarly define inverse functions for each of the remaining hyperbolic functions. By rewriting each hyperbolic function in terms of exponential functions, we can then obtain explicit formulas for these inverse functions.

The results are as follows:

$$\cosh^{-1} x = \ln(x + \sqrt{x^2 - 1}), \qquad x \geq 1 \tag{16}$$

$$\operatorname{sech}^{-1} x = \ln\left(\frac{1 + \sqrt{1 - x^2}}{x}\right), \qquad 0 < x \leq 1 \tag{17}$$

$$\operatorname{csch}^{-1} x = \ln\left(\frac{1}{x} + \frac{\sqrt{1 + x^2}}{|x|}\right), \qquad x \neq 0 \tag{18}$$

$$\tanh^{-1} x = \frac{1}{2} \ln\left(\frac{1 + x}{1 - x}\right), \qquad |x| < 1 \tag{19}$$

$$\coth^{-1} x = \frac{1}{2} \ln\left(\frac{x + 1}{x - 1}\right), \qquad |x| > 1 \tag{20}$$

Derivative and integral formulas follow from equations (15)–(20) by applications of the corresponding rules for differentiating natural logarithms. For example, from (15) we have

$$\frac{d}{dx} \sinh^{-1} x = \frac{d}{dx} \ln(x + \sqrt{x^2 + 1})$$

$$= \frac{1 + \dfrac{1}{2\sqrt{x^2 + 1}} (2x)}{x + \sqrt{x^2 + 1}}$$

$$= \frac{\sqrt{x^2 + 1} + x}{\sqrt{x^2 + 1}(x + \sqrt{x^2 + 1})}$$

$$= \frac{1}{\sqrt{x^2 + 1}}.$$

Thus

$$\frac{d}{dx} \sinh^{-1} x = \frac{1}{\sqrt{x^2 + 1}}. \qquad (21)$$

We leave as exercises the formulas

$$\frac{d}{dx} \cosh^{-1} x = \frac{1}{\sqrt{x^2 - 1}}, \qquad x > 1 \qquad (22)$$

$$\frac{d}{dx} \tanh^{-1} x = \frac{1}{1 - x^2}, \qquad |x| < 1 \qquad (23)$$

$$\frac{d}{dx} \coth^{-1} x = \frac{1}{1 - x^2}, \qquad |x| > 1 \qquad (24)$$

$$\frac{d}{dx} \operatorname{sech}^{-1} x = \frac{-1}{x\sqrt{1 - x^2}}, \qquad 0 < x < 1 \qquad (25)$$

$$\frac{d}{dx} \operatorname{csch}^{-1} x = \frac{-1}{|x|\sqrt{1 + x^2}}, \qquad x \neq 0 \qquad (26)$$

By now you have no doubt realized that the hyperbolic functions and their inverses are really not "new" functions. Rather, they are simply combinations of exponential and logarithmic functions. However, these particular combinations occur often in certain applications. Because the differentiation and integration formulas for the hyperbolic functions are simpler than those for the corresponding exponential and logarithmic forms, it is convenient to use them as shorthand notation.

Exercise Set 9.5

In Exercises 1–10, find the value of the indicated expression.

1. $\sinh 0$

2. $\sinh 1$

3. $\cosh(\ln 2)$

4. $\sinh(\ln 4)$

5. $\tanh(\ln 2)$

6. $\operatorname{sech}(1)$

7. $\coth(\ln 4)$

8. $\operatorname{csch}(\ln \pi^3)$

9. $\sinh^{-1}(1)$

10. $\tanh^{-1}(1/2)$

In Exercises 11–32, find the derivative of the given function.

11. $y = \sinh 2x$

12. $f(x) = x \cosh x$

13. $f(x) = \sinh x \tanh x$

14. $y = \sinh^3(1 - x^2)$

15. $f(x) = \sqrt{\cosh(4x)}$

16. $y = \dfrac{1 - \cosh x}{1 + \cosh x}$

17. $y = \dfrac{1}{\cosh x}$

18. $f(x) = \ln(\tanh x^2)$

19. $f(x) = e^x \operatorname{csch} x^2$

20. $x + \cosh xy = y$

21. $y = \sinh^{-1} 2x$

22. $f(x) = \cosh^{-1}(1 + x^2)$

23. $f(s) = \tanh^{-1} s^2$

24. $y = \sqrt{\sinh^{-1} x}$

25. $f(x) = \ln \cosh^{-1} \pi x$

26. $y = \cosh^{-1} e^{x^2}$

27. $y = \cosh x \cosh x^2$

28. $f(x) = \ln(1 + \cosh \pi x)$

29. $y = \sinh \ln x^2$

30. $y = \sqrt{1 + \cosh^2 3x}$

31. $y = \tanh \sqrt{1 + x^2}$

32. $y = x \sinh^{-1}(1 + x^2)$

In Exercises 33–48, find the integrals using the differentiation formulas of this section.

33. $\displaystyle\int_1^2 \frac{1}{\sqrt{1 + x^2}} \, dx$

34. $\displaystyle\int_1^2 x \sinh x^2 \, dx$

35. $\displaystyle\int \tanh^2(2x) \, dx$

36. $\displaystyle\int e^x \sinh x \, dx$

37. $\displaystyle\int_0^2 \sinh x \cosh x \, dx$

38. $\displaystyle\int_0^1 \frac{1}{4 - x^2} \, dx$

39. $\displaystyle\int_2^6 \frac{dx}{\sqrt{x^2 - 4}}$

40. $\displaystyle\int_2^4 \frac{dx}{2 - x^2}$

41. $\displaystyle\int_0^1 \frac{1}{\sqrt{9x^2 + 25}} \, dx$

42. $\displaystyle\int_1^2 \frac{dx}{x\sqrt{4 + x^2}}$

43. $\displaystyle\int \frac{\sinh x}{\cosh x} \, dx$

44. $\displaystyle\int \frac{1}{\sinh^2 x} \, dx$

45. $\displaystyle\int \frac{\cosh x}{\sqrt{\sinh x}} \, dx$

46. $\displaystyle\int \frac{\sinh x}{1 + \cosh^2 x} \, dx$

47. $\displaystyle\int \tanh^2 x \, \text{sech}^2 x \, dx$

48. $\displaystyle\int \frac{1 - \tanh^2 x}{1 + \tanh^2 x} \, dx$

49. True or false? The hyperbolic sine function is a periodic function with period 2π.

Verify the following identities.

50. $1 - \tanh^2 x = \text{sech}^2 x$

51. $\cosh x \pm \sinh x = e^{\pm x}$

52. $\sinh 2x = 2 \sinh x \cosh x$

53. $\cosh 2x = 2 \sinh^2 x + 1$

Verify the addition formulas.

54. $\sinh(x + y) = \sinh x \cdot \cosh y + \sinh y \cdot \cosh x.$

55. $\cosh(x + y) = \cosh x \cdot \cosh y + \sinh x \cdot \sinh y.$

56. A flexible cable fastened at both ends hangs in the shape of the **catenary** $y = a \cosh(x/a)$. Find the length of the graph of the catenary $y = \cosh x$ from the point with x-coordinate 0 to the point with x-coordinate b.

57. Show that the functions $y_1 = \sinh(kx)$ and $y_2 = \cosh(kx)$ satisfy the differential equation $\dfrac{d^2y}{dx^2} = k^2 y$. Find two solutions of this equation that are not hyperbolic functions. How are these solutions related?

Find as many solutions as you can for the following differential equations.

58. $\dfrac{d^2y}{dx^2} - 4y = 0$

59. $y'' - k^2\pi^2 y = 0$

60. Verify that the function $y = A \sin x + B \cos x + C \sinh x + D \cosh x$ is a solution of the fourth order differential equation $\dfrac{d^4y}{dx^4} = y.$

61. Find the area of the region bounded by the graphs of $y = \coth x$, $y = \tanh x$, $x = \ln 2$ and $x = \ln 4$.

62. Find the volume of the solid obtained by rotating about the x-axis the region bounded above by the graph of $y = \cosh x$ and below by the x-axis, for $-\ln 2 \le x \le \ln 2$.

Find all relative extrema.

63. $f(x) = \cosh x - \sinh x$

64. $f(x) = 2 \cosh x + 5 \sinh x$

65. $f(x) = 5 \cosh x - 2 \sinh x$

In Exercises 66–68, find a second order differential equation satisfied by the function.

66. $f(x) = \cosh 2x$

67. $f(x) = \cosh 3x - 2 \sinh 3x$

68. $f(x) = A \cosh 6x + B \sinh 6x.$

69. Determine which of the hyperbolic trigonometric functions are odd and which are even.

70. Show that $e^x = \cosh x + \sinh x$. Find a similar expression for e^{-x}.

71. Verify differentiation formulas (5)–(8).

72. Verify identity (11).

73. According to the definition of $\cosh^{-1} x$ in equation (16), what is the *principal branch* of the graph of $y = \cosh x$?

74. Verify that the function $y = \cosh x$ is increasing on $[0, \infty)$ and therefore invertible on this interval. Sketch the graph of $y = \cosh^{-1} x$ by reflecting the principal branch of $y = \cosh x$ in the line $y = x$.

75. Find the principal branch for each of the functions $\tanh x$, $\coth x$, $\text{sech } x$, and $\text{csch } x$, according to equations (17)–(20), and verify that each function is invertible when restricted to its principal branch.

76. Sketch the graphs of the inverse functions defined by equations (17)–(20).

77. Verify differentiation formulas (22)–(26).

78. Consider the situation in which one edge of a rectangular metal plate is held at a constant high temperature, while the opposite edge is held at a constant low temperature; the remaining two edges are insulated. In determining an equation that gives the temperature at each point on the plate, part of the problem involves solving the *second order* differential equation

$$\frac{d^2g(y)}{dy^2} - k^2 g(y) = 0$$

where k is a nonzero constant. Verify that

$$g(y) = c_1 \sinh(ky) + c_2 \cosh(ky)$$

satisfies the differential equation for any choice of constants c_1 and c_2.

SUMMARY OUTLINE OF CHAPTER 9

◆ $\displaystyle\int \sin u \; du = -\cos u + C$ (page 415)

$\displaystyle\int \cos u \; du = \sin u + C$

$\displaystyle\int \tan u \; du = \ln |\sec u| + C, \qquad \cos u \neq 0$

$\displaystyle\int \cot u \; du = \ln |\sin u| + C, \qquad \sin u \neq 0$

$\displaystyle\int \sec u \; du = \ln |\sec u + \tan u| + C$

$\displaystyle\int \csc u \; du = \ln |\csc u - \cot u| + C$

◆ To integrate $\displaystyle\int \sin^m x \cos^n x \; dx$:   (page 420)

(i) If one of m or n is odd (say n), make the substitution $u = \sin x$, so $du = \cos x \; dx$. Change remaining even factors of $\cos x$ to factors of $\sin x$ using the identity $\cos^2 x = 1 - \sin^2 x$.

(ii) If both n and m are even, use the identities

$$\sin^2 x = \frac{1}{2} - \frac{1}{2} \cos 2x; \cos^2 x = \frac{1}{2} + \frac{1}{2} \cos 2x.$$

Use a similar strategy in integrals of the form $\displaystyle\int \sec^n x \tan^m x \; dx$.

◆ The **inverse trigonometric functions** are defined as follows: (page 427)

$$\begin{array}{llll}
y = \text{Tan}^{-1}\, x & \text{if and only if} & x = \tan y, & -\infty < x < \infty, \quad -\pi/2 < y < \pi/2 \\
y = \text{Sin}^{-1}\, x & \text{if and only if} & x = \sin y, & -1 \leq x \leq 1, \quad -\pi/2 \leq y \leq \pi/2 \\
y = \text{Cos}^{-1}\, x & \text{if and only if} & x = \cos y, & -1 \leq x \leq 1, \quad 0 \leq y \leq \pi \\
y = \text{Sec}^{-1}\, x & \text{if and only if} & x = \sec y, & |x| \geq 1, \quad y \in [0, \pi/2) \cup [\pi, 3\pi/2) \\
y = \text{Csc}^{-1}\, x & \text{if and only if} & x = \csc y, & |x| \geq 1, \quad y \in (0, \pi/2] \cup (\pi, 3\pi/2] \\
y = \text{Cot}^{-1}\, x & \text{if and only if} & x = \cot y, & -\infty < x < \infty, \quad 0 \leq y \leq \pi.
\end{array}$$

◆ *Differentiation and Integration Formulas for Inverse Trigonometric Functions:*   (page 433)

$$\frac{d}{dx} \text{Tan}^{-1}\, x = \frac{1}{1 + x^2} \qquad \int \frac{dx}{1 + x^2} = \text{Tan}^{-1}\, x + C$$

$$\frac{d}{dx} \text{Sin}^{-1}\, x = \frac{1}{\sqrt{1 - x^2}} \qquad \int \frac{dx}{\sqrt{1 - x^2}} = \text{Sin}^{-1}\, x + C, \quad |x| < 1$$

$$\frac{d}{dx} \text{Sec}^{-1}\, x = \frac{1}{x\sqrt{x^2 - 1}} \qquad \int \frac{dx}{x\sqrt{x^2 - 1}} = \text{Sec}^{-1}\, x + C, \quad |x| \geq 1$$

$$\frac{d}{dx} \text{Cot}^{-1}\, x = \frac{-1}{1 + x^2}$$

$$\frac{d}{dx} \text{Cos}^{-1}\, x = \frac{-1}{\sqrt{1 - x^2}}$$

$$\frac{d}{dx} \text{Csc}^{-1}\, x = \frac{-1}{x\sqrt{x^2 - 1}}$$

◆ The **hyperbolic functions** are as follows: (page 438)

$$\sinh x = \frac{1}{2}(e^x - e^{-x}) \qquad \coth x = \frac{\cosh x}{\sinh x}$$

$$\cosh x = \frac{1}{2}(e^x + e^{-x}) \qquad \text{sech } x = \frac{1}{\cosh x}$$

$$\tanh x = \frac{\sinh x}{\cosh x} \qquad \text{csch } x = \frac{1}{\sinh x}$$

◆ *Differentiation Formulas for Hyperbolic Functions:* (page 438)

$$\frac{d}{dx}\sinh x = \cosh x; \qquad \frac{d}{dx}\coth x = -\text{csch}^2 x;$$

$$\frac{d}{dx}\cosh x = \sinh x; \qquad \frac{d}{dx}\text{sech } x = -\text{sech } x \cdot \tanh x;$$

$$\frac{d}{dx}\tanh x = \text{sech}^2 x; \qquad \frac{d}{dx}\text{csch } x = -\text{csch } x \cdot \coth x$$

◆ The **inverse hyperbolic** functions satisfy the following identities: (page 442)

$$\sinh^{-1} x = \ln(x + \sqrt{x^2 + 1}), \qquad -\infty < x < \infty$$

$$\cosh^{-1} x = \ln(x + \sqrt{x^2 - 1}), \qquad x \geq 1$$

$$\text{sech}^{-1} x = \ln\left(\frac{1 + \sqrt{1 - x^2}}{x}\right), \qquad 0 < x \leq 1$$

$$\text{csch}^{-1} x = \ln\left(\frac{1}{x} + \frac{\sqrt{1 + x^2}}{|x|}\right), \qquad x \neq 0$$

$$\tanh^{-1} x = \frac{1}{2}\ln\left(\frac{1 + x}{1 - x}\right), \qquad |x| < 1$$

$$\coth^{-1} x = \frac{1}{2}\ln\left(\frac{x + 1}{x - 1}\right), \qquad |x| > 1$$

◆ Derivatives of the inverse hyperbolic functions: (page 443)

$$\frac{d}{dx}\sinh^{-1} x = \frac{1}{\sqrt{x^2 + 1}}$$

$$\frac{d}{dx}\cosh^{-1} x = \frac{1}{\sqrt{x^2 - 1}}, \qquad x > 0$$

$$\frac{d}{dx}\tanh^{-1} x = \frac{1}{1 - x^2}, \qquad |x| < 1$$

$$\frac{d}{dx}\coth^{-1} x = \frac{1}{1 - x^2}, \qquad |x| > 1$$

$$\frac{d}{dx}\text{sech}^{-1} x = \frac{-1}{x\sqrt{1 - x^2}}, \qquad 0 < |x| < 1$$

$$\frac{d}{dx}\text{csch}^{-1} x = \frac{-1}{|x|\sqrt{1 + x^2}}, \qquad x \neq 0$$

REVIEW EXERCISES—CHAPTER 9

In Exercises 1–9, evaluate the given expression.

1. $\text{Sin}^{-1}\left(\frac{\sqrt{3}}{2}\right)$

2. $\text{Cos}^{-1}(-1/2)$

3. $\text{Tan}^{-1}\sqrt{3}$

4. $\text{Sin}^{-1}(\sin \pi/4)$

5. $\text{Cos}^{-1}(\cos(-\pi/4))$

6. $\text{Tan}^{-1}(\sin \pi/2)$

7. $\cot\left(\text{Sin}^{-1}\frac{\sqrt{2}}{2}\right)$

8. $\tan(\text{Sec}^{-1}(-2))$

9. $\sinh(\ln 2)$

In Exercises 10–33, find the derivative of the given function.

10. $y = \text{Tan}^{-1} 3x$

11. $y = \text{Sin}^{-1} \sqrt{x}$

12. $f(x) = \text{Tan}^{-1}\left(\dfrac{2 - x}{2 + x}\right)$

13. $f(t) = \text{Cot}^{-1}(1 - t^2)$

14. $y = \cosh(\ln x)$

15. $f(x) = x^2 \sinh(1 - x)$

16. $y = \text{Tan}^{-1} \dfrac{\sqrt{x}}{2}$

17. $y = \csc(\cot 6x)$

18. $y = \ln^2(\cos^2 x^2)$

19. $f(x) = \text{Sin}^{-1}[\ln(2x + 1)]$

20. $y = \text{Sin}^{-1}(\cos e^{-x})$

21. $f(x) = \dfrac{\text{Tan}^{-1} x}{1 + x^2}$

22. $y = \dfrac{\text{Sin}^{-1} e^x}{1 + e^x}$

23. $f(x) = \text{Sec}^{-1} \sqrt{x^2 + 4}$

24. $y = \sqrt{\cosh x^2}$

25. $y = \dfrac{1}{\pi + \tanh x}$

26. $f(x) = \ln^2(\sinh x)$

27. $y = (\sinh x + \text{Cos}^{-1} 2x)^{1/5}$

28. $y = \ln |\tan x|$

29. $y = \ln(\text{Sin}^{-1} x)$

30. $f(x) = \dfrac{\tan 2x}{2 + \sec 2x}$

31. $f(x) = x \tanh^{-1}(\ln x)$

32. $y = x^2 \sinh^{-1}(e^x)$

33. $y = \ln \sqrt{\tanh^{-1}(x^2)}$

In Exercises 34–70, find the indicated integral.

34. $\displaystyle\int \dfrac{dx}{\sqrt{1 + 9x^2}}$

35. $\displaystyle\int \dfrac{dx}{\sqrt{4x^2 - 1}}$

36. $\displaystyle\int \dfrac{dx}{x\sqrt{4 + x^2}}$

37. $\displaystyle\int \sin^4 2x \cos 2x \, dx$

38. $\displaystyle\int \sec^5 x \tan x \, dx$

39. $\displaystyle\int_{\pi^2/4}^{\pi^2/16} \dfrac{\sin^2 \sqrt{x}}{\sqrt{x}} \, dx$

40. $\displaystyle\int_0^{\pi/2} \sin^{5/2} x \cos x \, dx$

41. $\displaystyle\int_{-1}^{1} \dfrac{dx}{\sqrt{2 - x^2}}$

42. $\displaystyle\int x \csc^2 x^2 \, dx$

43. $\displaystyle\int_0^{\pi/4} \sqrt{\tan x} \sec^2 x \, dx$

44. $\displaystyle\int \dfrac{\sec^2 \sqrt{x} \tan \sqrt{x}}{\sqrt{x}} \, dx$

45. $\displaystyle\int \dfrac{1 + \sin^2 2x}{\cos^2 2x} \, dx$

46. $\displaystyle\int \tan(\sec x) \sec x \tan x \, dx$ **47.** $\displaystyle\int_0^{\pi/8} \tan^2(2x) \sec^2(2x) \, dx$

48. $\displaystyle\int \dfrac{x}{x^4 + 1} \, dx$

49. $\displaystyle\int \dfrac{dx}{\sqrt{4 - x^2}}$

50. $\displaystyle\int \dfrac{e^{-x}}{2 + e^{-2x}} \, dx$

51. $\displaystyle\int \dfrac{\cos^{-1} x}{\sqrt{1 - x^2}} \, dx$

52. $\displaystyle\int \dfrac{\sqrt{x}}{x^3 + 4} \, dx$

53. $\displaystyle\int \dfrac{e^{\sqrt{x}}}{\sqrt{x}(1 + e^{2\sqrt{x}})} \, dx$

54. $\displaystyle\int_1^2 \dfrac{x^2}{x^6 + 9} \, dx$

55. $\displaystyle\int \dfrac{\sqrt{\sin^{-1} x}}{\sqrt{1 - x^2}} \, dx$

56. $\displaystyle\int \dfrac{\sin x}{\sqrt{2 - \cos^2 x}} \, dx$

57. $\displaystyle\int_0^{\pi/9} \sec^3 2x \tan 2x \, dx$

58. $\displaystyle\int_0^{\pi/2} \tan^3\left(\dfrac{x}{2}\right) dx$

59. $\displaystyle\int_0^{\pi/3} \sec^4 x \, dx$

60. $\displaystyle\int \sec^3 x \tan^3 x \, dx$

61. $\displaystyle\int_0^{\pi/3} \cos x \cos 5x \, dx$

62. $\displaystyle\int_0^{\pi/4} \sin x \cos 2x \, dx$

63. $\displaystyle\int \sqrt{\cos x} \sin^3 x \, dx$

64. $\displaystyle\int \cot^4 2x \, dx$

65. $\displaystyle\int \dfrac{\sin^3 x}{\cos^2 x} \, dx$

66. $\displaystyle\int_0^{\pi/2} \sin^2 x \cos^4 x \, dx$

67. $\displaystyle\int \dfrac{\cosh x}{1 + \sinh x} \, dx$

68. $\displaystyle\int x \coth x^2 \, dx$

69. $\displaystyle\int \sinh^3 x \, dx$

70. $\displaystyle\int \dfrac{e^{3x}}{\sqrt{9 + e^{6x}}} \, dx$

71. Find the area of the region bounded by the graph of $y = \sin^2 x$ and the x-axis between $x = 0$ and $x = \pi$.

72. Find the area of the region bounded by the graphs of $y = \sin^2 x$ and $y = \cos^2 x$ for x between $x = 0$ and $x = \pi/2$.

73. Find the volume of the solid generated by rotating about the x-axis the region bounded by the graph of $y = \tan x$ and the x-axis for $0 \le x \le \pi/4$.

74. Find the volume of the solid generated by rotating about the x-axis the region bounded by the graph of $y = \dfrac{1}{\sqrt{1 + x^2}}$ and the x-axis for $0 \le x \le 1$.

75. Find the volume of the solid generated by revolving about the y-axis the region bounded by the graph of $y = \dfrac{1}{1 + x^4}$ and the x-axis between $x = 0$ and $x = 2$.

76. Find the average value of the function $y = \cosh x$ on the interval $[-\ln 2, \ln 2]$.

77. Find $\dfrac{dy}{dx}$ if $y = \cosh^{-1} x^2 y$.

78. Find the area bounded by the graph of $y = \cosh x$ and the x-axis for $-1 \le x \le 1$.

79. For the function $y = \sinh x \cosh x$, find all relative extrema, determine the concavity, and sketch the graph.

80. A particle moves along a line with velocity $v(t) = \sin^2 \pi t$. Find the distance travelled by the particle between times $t = 0$ and $t = 4$.

81. Use differentials to approximate $\text{Sin}^{-1}(0.48)$.

82. Prove that $\cosh x > \sinh x$ for all x.

83. Find the equation of the line tangent to the graph of $y = \text{Sin}^{-1} x^2$ at the point with x-coordinate $\sqrt{2}/2$.

84. An airplane is flying directly away from a radar station at an altitude of 3 kilometers and a ground speed of 400 km/hr. How fast is the angle of elevation of the tracking antenna decreasing when the airplane is directly over a point 4 kilometers from the radar station? (Use inverse trigonometric functions.)

85. Find the points where the line tangent to the graph of $y = \text{Cot}^{-1} x$ is parallel to the line with equation $x + 5y - 10 = 0$.

86. A particle moves along a line with acceleration $a(t) = -1 - \dfrac{2t}{(1 + t^2)^2}$ m/s². Find the function $s(t)$ giving the distance travelled after t seconds if the particle starts at the origin with initial velocity 4 m/s.

87. A small boat is tied to a rope which is connected to a windlass (crank). The windlass is mounted on the edge of a dock 10 meters above the water level. The windlass is pulling the boat ashore by winding in the rope at the rate of 1 meter per second. Find the rate at which the angle between the rope and the horizontal is increasing at the instant when 20 meters of rope remain out.

88. Why is the hyperbolic sine function continuous throughout its domain?

Chapter 10
Techniques of Integration

The goal of this chapter is to develop several additional techniques for finding antiderivatives and evaluating definite integrals. Recall, we say that the function F is an *antiderivative* for the function f on the interval I if $F'(x) = f(x)$ for all $x \in I$. In this case we refer to the most general antiderivative

$$\int f(x)\ dx = F(x) + C \tag{1}$$

as the *indefinite integral* for f. The term *integration* refers to the process of finding the indefinite integral in equation (1).

One reason for developing additional techniques of integration is to expand the class of functions for which we can evaluate *definite integrals* using the Fundamental Theorem of Calculus:

$$\int_a^b f(x)\ dx = F(b) - F(a) \tag{2}$$

if $F'(x) = f(x)$ for $x \in [a, b]$.

This part of the Fundamental Theorem is not applicable to *all* continuous functions, however, because it requires us to find an antiderivative F for f. Some continuous functions simply do not have "closed form" antiderivatives.* For example, the function

$$f(x) = e^{x^2}$$

is continuous on $(-\infty, \infty)$, but there does not exist a "closed form" function F with $F'(x) = e^{x^2}$.

A second reason for developing additional techniques of integration is to better prepare you to solve the differential equations that arise in advanced courses in science, engineering, and applied mathematics. The techniques required to solve these differential equations involve much of what is discussed in this chapter.

Finally, it is important to note the major role computers now play in both evaluating definite integrals and finding antiderivatives of functions. We have seen in Chapter 6 that numerical methods such as the Trapezoidal Rule and Simpson's Rule, when implemented on computers, can be used to evaluate definite integrals to great accuracy without recourse to the Fundamental Theorem. A more recent develop-

---

*By a "closed form" antiderivative we mean a function not involving an integral sign. The Fundamental Theorem guarantees that $G(x) = \int_a^x f(t)\ dt$ is always an antiderivative for f when f is continuous on $[a, x]$, but G is not a "closed form" function.

ment is that various computer programs can now evaluate indefinite integrals symbolically for a wide class of functions. Because this technology is evolving so rapidly, we shall not cite specific products but note that they are becoming increasingly available at relatively modest cost. While the availability of these technological aids reduces the need to calculate integrals ''by hand,'' like the hand calculator they cannot eliminate entirely the need for familiarity with the underlying mathematical concepts.

10.1 INTEGRATION BY PARTS

This technique enables us to integrate certain functions that can be interpreted as products. Before discussing the technique generally we illustrate the basic idea with a typical example.

Example 1

Find $\int x \cos x \, dx$.

Solution: The simple substitution $u = \cos x$ is not helpful because it gives $du = -\sin x \, dx$, introducing a factor $(\sin x)$ which is not present in the integrand and leaving the factor x unaddressed.

The key to finding this integral is to note that the derivative of the *product* $h(x) = x \sin x$ involves the term $x \cos x$ which appears in the integrand:

$$\frac{d}{dx}(x \sin x) = \sin x + x \cos x.$$

Thus,

$$x \cos x = \frac{d}{dx}(x \sin x) - \sin x.$$

Integrating both sides of this equation and using the fact that $\int \left[\frac{d}{dx}(x \sin x) \right] dx = x \sin x + C$ gives

$$\int x \cos x \, dx = (x \sin x + C_1) - \int \sin x \, dx$$

$$= (x \sin x + C_1) + (\cos x + C_2)$$

$$= x \sin x + \cos x + C \qquad (C = C_1 + C_2). \qquad \diamondsuit$$

The key to finding the integral in Example 1 was to view the integrand $x \cos x$ as one of the two terms resulting from an application of the Product Rule to a function $h = fg$. To generalize this observation we recall that the Product Rule

$$(fg)'(x) = f'(x)g(x) + f(x)g'(x)$$

gives

$$f(x)g'(x) = (fg)'(x) - f'(x)g(x).$$

Integrating both sides of this equation gives

$$\int f(x)g'(x)\,dx = \int (fg)'(x)\,dx - \int f'(x)g(x)\,dx \tag{1}$$

$$= f(x)g(x) - \int f'(x)g(x)\,dx.$$

In writing equation (1) we have used the fact that $\int (fg)'(x)\,dx = (fg)(x) + C$ and we have not written a constant of integration with this term since one will appear when the integral on the right side of the equation is evaluated. Equation (1) is referred to as the **integration by parts formula:**

$$\int f(x)g'(x)\,dx = f(x)g(x) - \int f'(x)g(x)\,dx. \tag{2}$$

REMARK 1: The general strategy in using equation (2) is to pick the factors f and g' so that

(i) the antiderivative g can be found, and
(ii) the resulting integral $\int f'(x)g(x)\,dx$ is as simple to evaluate as possible.

REMARK 2: There is no need to introduce an arbitrary constant when finding an antiderivative g for g', since (2) holds for *any* antiderivative of g'. The simplest choice is to let the arbitrary constant be zero.

There is a somewhat simpler form by which the integration by parts formula can be remembered. If we make the substitutions

$$u = f(x), \qquad dv = g'(x)\,dx$$

then we can write

$$du = f'(x)\,dx, \qquad v = g(x)$$

and equation (2) becomes

$$\int u\,dv = uv - \int v\,du. \tag{3}$$

Equation (3) is the Leibniz notation form for the integration by parts formula.

Example 2

Find $\int x^2 \ln x\,dx$.

Strategy

The derivative of $\ln x$ is a power of x, so take $u = \ln x$, $dv = x^2\,dx$.

Solution

We let

$$u = \ln x, \qquad dv = x^2\,dx,$$

so

Find *du*, *v*.

$$du = \frac{1}{x} \, dx, \qquad v = \frac{1}{3}x^3.$$

Apply equation (3).

Integrating by parts then gives

$$\int x^2 \ln x \, dx = \frac{1}{3}x^3 \ln x - \int \frac{1}{3}x^3 \cdot \frac{1}{x} \, dx$$

$$= \frac{1}{3}x^3 \ln x - \frac{1}{3} \int x^2 \, dx$$

$$= \frac{1}{3}x^3 \ln x - \frac{1}{9}x^3 + C. \qquad \diamond$$

Often more than one application of the integration by parts formula is required, as the following example shows.

Example 3

Find $\displaystyle\int x^2 e^x \, dx$.

Strategy

Take $u = x^2$ since differentiation reduces the exponent by one.

Solution

We take

$$u = x^2, \qquad dv = e^x \, dx.$$

Then

$$du = 2x \, dx, \qquad v = e^x,$$

and integration by parts gives

Apply equation (3).

$$\int x^2 e^x \, dx = x^2 e^x - 2 \int x e^x \, dx. \qquad (4)$$

In $\displaystyle\int x e^x \, dx$ again take $u = x$, since differentiation will yield simply $du = 1 \cdot dx$.

The integral on the right requires a second application of the parts formula: In

$$\int x e^x \, dx,$$

we take

$$u = x, \qquad dv = e^x \, dx,$$

so

$$du = dx, \qquad v = e^x.$$

Apply equation (3) again.

Then, a second integration by parts gives

$$\int x e^x \, dx = x e^x - \int e^x \, dx \qquad (5)$$

$$= x e^x - e^x + C_1.$$

Combine results.

Combining lines (4) and (5) we now have

$$\int x^2 e^x \, dx = x^2 e^x - 2(xe^x - e^x + C_1)$$

$$= e^x(x^2 - 2x + 2) + C$$

where $C = -2C_1$. (Since C_1 is an arbitrary constant, C is also an arbitrary constant; we choose to write the answer in the simpler form.) ◇

In the exercise set you will encounter integrals requiring even more than two applications of the parts formula. A different type of problem requiring two applications of the parts formula is the following:

Example 4

Find $\int e^x \cos x \, dx$.

Solution: This time there is no clue as to which function to take as u. We arbitrarily choose

$$u = e^x, \qquad dv = \cos x \, dx.$$

Then

$$du = e^x \, dx, \qquad v = \sin x$$

and the integration by parts formula gives

$$\int e^x \cos x \, dx = e^x \sin x - \int e^x \sin x \, dx. \tag{6}$$

Now the integral on the right side of (6) seems no simpler than the integral on the left. However, let's try another application of the integration by parts formula in the integral $\int e^x \sin x \, dx$. Again taking $u = e^x$ we have

$$u = e^x, \qquad dv = \sin x \, dx$$

and

$$du = e^x \, dx, \qquad v = -\cos x.$$

Thus,

$$\int e^x \sin x \, dx = -e^x \cos x - \int (-\cos x) e^x \, dx \tag{7}$$

$$= -e^x \cos x + \int e^x \cos x \, dx.$$

Note that the desired integral has reappeared on the right side of equation (7)! This happens because the second derivatives of e^x and $\cos x$ (as well as $\sin x$, $\sinh x$, and $\cosh x$) are simply multiples of the original functions. Combining equations (6) and (7) gives

$$\int e^x \cos x \, dx = e^x \sin x - \left\{ -e^x \cos x + \int e^x \cos x \, dx \right\}$$

$$= e^x(\sin x + \cos x) - \int e^x \cos x \, dx.$$

We may now add $\int e^x \cos x \, dx$ to both sides to obtain

$$2 \int e^x \cos x \, dx = e^x(\sin x + \cos x) + C_1,$$

so

$$\int e^x \cos x \, dx = \frac{e^x}{2}(\sin x + \cos x) + C, \qquad C = \frac{C_1}{2}. \qquad \diamondsuit$$

We can apply the parts formula to any integral of the form $\int f(x) \, dx$ simply by taking $u = f(x)$ and $dv = dx$. The resulting integral may or may not be solvable.

Example 5

Find $\int \operatorname{Sin}^{-1} x \, dx$.

Strategy

Here our only choice is to take

$u = \operatorname{Sin}^{-1} x, \qquad dv = dx.$

Apply equation (3).

$\left(\text{Handle } \int \dfrac{x}{\sqrt{1 - x^2}} \, dx \text{ by a substitu-}\right.$

tion with $u = 1 - x^2.\Big)$

Solution

With $u = \operatorname{Sin}^{-1} x, \qquad dv = dx,$

$$du = \frac{dx}{\sqrt{1 - x^2}}, \qquad \text{and} \qquad v = x.$$

The parts formula gives

$$\int \operatorname{Sin}^{-1} x \, dx = x \operatorname{Sin}^{-1} x - \int \frac{x}{\sqrt{1 - x^2}} \, dx$$

$$= x \operatorname{Sin}^{-1} x + \sqrt{1 - x^2} + C. \qquad \diamondsuit$$

Example 6

Show that $\int \ln x \, dx = x \ln x - x + C.$

Solution: We do this by the method of Example 5. Since the only factor of the integrand is $\ln x$, we take

$$u = \ln x, \qquad dv = dx.$$

Then,

$$du = \frac{1}{x} \, dx, \qquad \text{and} \qquad v = x,$$

so an application of the integration by parts formula gives

$$\int \ln x \, dx = x \cdot \ln x - \int x \left(\frac{1}{x}\right) dx$$

$$= x \cdot \ln x - \int 1 \, dx$$

$$= x \ln x - x + C. \qquad \diamondsuit$$

The following is one of the trickier integrals involving integration by parts. It should be noted for future reference.

Example 7

Find $\int \sec^3 x \, dx$.

Strategy
Break $\sec^3 x$ into two factors.

Solution
We let

$$u = \sec x, \qquad dv = \sec^2 x \, dx.$$

Then

$$du = \sec x \tan x \, dx, \qquad v = \tan x,$$

so

Apply equation (3).

$$\int \sec^3 x \, dx = \sec x \tan x - \int \tan^2 x \sec x \, dx$$

Use identity $\tan^2 x = \sec^2 x - 1$.

$$= \sec x \tan x - \int [\sec^2 x - 1] \sec x \, dx$$

$$= \sec x \tan x - \int \sec^3 x \, dx + \int \sec x \, dx$$

$$= \sec x \tan x - \int \sec^3 x \, dx + \ln |\sec x + \tan x|.$$

Thus

Add $\int \sec^3 x \, dx$ to both sides and divide by 2.

$$\int \sec^3 x \, dx = \frac{1}{2} \sec x \tan x + \frac{1}{2} \ln |\sec x + \tan x| + C. \qquad \diamond$$

Finally, we note that the integration by parts formula can be applied to definite integrals as well as to indefinite integrals. The corresponding formulation of equation (2) is

$$\int_a^b f(x)g'(x) \, dx = f(x)g(x) \Big]_a^b - \int_a^b f'(x)g(x) \, dx. \tag{8}$$

Example 8

Find $\int_0^{\sqrt{\pi/2}} x^3 \cos x^2 \, dx$.

Strategy
We take $u = x^2$ so the remaining factor $dv = x \cos x^2 \, dx$ is a multiple of the derivative of $\sin x^2$.

Solution
Let

$$u = x^2, \qquad dv = x \cos x^2 \, dx.$$

Then

$$du = 2x \, dx, \qquad v = \frac{1}{2} \sin x^2,$$

so the parts formula gives

Apply (8).

$$\int_0^{\sqrt{\pi/2}} x^3 \cos x^2 \, dx = \frac{1}{2} x^2 \sin x^2 \Big]_0^{\sqrt{\pi/2}} - \int_0^{\sqrt{\pi/2}} x \sin x^2 \, dx$$

$$= \frac{1}{2}\left(\frac{\pi}{4}\right) \sin\left(\frac{\pi}{4}\right) - \left[-\frac{1}{2} \cos x^2\right]_0^{\sqrt{\pi/2}}$$

$$= \frac{\sqrt{2}\pi}{16} + \frac{1}{2}\left(\frac{\sqrt{2}}{2} - 1\right)$$

$$\approx .1312. \qquad \diamond$$

Exercise Set 10.1

In Exercises 1–36, evaluate the given integral.

1. $\displaystyle\int x \, e^x \, dx$

2. $\displaystyle\int x \ln x \, dx$

3. $\displaystyle\int x \cos x \, dx$

4. $\displaystyle\int x^2 \ln 2x \, dx$

5. $\displaystyle\int \mathrm{Tan}^{-1} x \, dx$

6. $\displaystyle\int x \sec x \tan x \, dx$

7. $\displaystyle\int x \sec^2 \pi x \, dx$

8. $\displaystyle\int e^{ax} \sin x \, dx$

9. $\displaystyle\int \sin(\ln x) \, dx$

10. $\displaystyle\int \sin x \sinh x \, dx$

11. $\displaystyle\int_0^1 x(x + 2)^8 \, dx$

12. $\displaystyle\int_0^4 x\sqrt{x + 1} \, dx$

13. $\displaystyle\int (\ln x)^2 \, dx$

14. $\displaystyle\int x \ln x^2 \, dx$

15. $\displaystyle\int xe^{2x} \, dx$

16. $\displaystyle\int_1^e x^3 \ln x \, dx$

17. $\displaystyle\int_0^1 x^3 e^{2x} \, dx$

18. $\displaystyle\int \sec^5 x \, dx$

19. $\displaystyle\int x \sinh x \, dx$

20. $\displaystyle\int_0^1 \frac{x^3}{\sqrt{x^2 + 1}} \, dx$

21. $\displaystyle\int e^{2x} \cos x \, dx$

22. $\displaystyle\int \mathrm{Cos}^{-1} 2x \, dx$

23. $\displaystyle\int x \, \mathrm{Sin}^{-1} x \, dx$

24. $\displaystyle\int (\mathrm{Cos}^{-1} x)^2 \, dx$

25. $\displaystyle\int \sqrt{x} \ln x \, dx$

26. $\displaystyle\int (\ln x)^2 \, dx$

27. $\displaystyle\int \frac{x^3}{\sqrt{1 + x^2}} \, dx$

28. $\displaystyle\int x^3\sqrt{9 - x^2} \, dx$

29. $\displaystyle\int x^3 e^{ax^2} \, dx$

30. $\displaystyle\int \tanh^{-1} x \, dx$

31. $\displaystyle\int \sec^3 x \tan^2 x \, dx$

32. $\displaystyle\int x \tanh^{-1} x \, dx$

33. $\displaystyle\int x \sinh^{-1} x \, dx$

34. $\displaystyle\int \sqrt{x} \, e^{-\sqrt{x}} \, dx$

35. $\displaystyle\int 2xe^{-\sqrt{x}} \, dx$

36. $\displaystyle\int \sinh^{-1} x \, dx$

37. Find the area of the region bounded by the graph of $y = x \sin \pi x$ and the x-axis between the lines $x = 0$ and $x = 1$.

38. Find the area of the region bounded by the graphs of $y = \ln(1 + x)$, $y = x$, and $x = 1$.

39. Find the volume of the solid obtained by revolving about the x-axis the region bounded by the graph of $y = \ln x$ and the x-axis for $1 \le x \le e$.

40. Find the volume of the solid obtained by revolving the region in Exercise 39 about the y-axis.

41. Find the average value of the function $f(x) = e^{-x} \sin \pi x$ for $x \in [0, 1]$.

42. Find the average value of the function $y = \mathrm{Sin}^{-1} x$ for $x \in [0, 1]$.

43. A particle moves along a line with velocity $v(t) = t \sin \pi t$ meters per second. Find the distance travelled by the particle between times $t = 0$ and $t = 2$ seconds.

44. A water tank has the shape of the solid obtained by revolving the region bounded by the graph of $y = \ln x$, the x-axis, and the line $x = 5$ about the x-axis. The tank is full of water. Find the work done in pumping all the water to the top of the tank.

45. A **reduction formula** is one that reduces the order of the integrand; by repeated applications the order can be reduced to zero or one. Establish the reduction formula

$$\int \sin^n x \, dx = -\frac{1}{n} \sin^{n-1} x \cos x + \frac{n-1}{n} \int \sin^{n-2} x \, dx$$

using integration by parts.

46. Use the formula in Exercise 45 to find $\int \sin^2 x \, dx$.

47. Use the formula in Exercise 45 to find $\int \sin^4 x \, dx$.

48. Establish the reduction formula

$$\int x^n e^x \, dx = x^n e^x - n \int x^{n-1} e^x \, dx$$

using integration by parts.

49. Use the formula in Exercise 48 to find $\int x^4 e^x \, dx$.

50. Establish the reduction formula

$$\int x^n \sin ax \, dx = -\frac{x^n}{a} \cos ax + \frac{n}{a} \int x^{n-1} \cos ax \, dx$$

using integration by parts.

51. Use the formula in Exercise 50 to find $\int x \sin 3x \, dx$.

52. Show that the integral $\int p(x)e^x \, dx$ can be handled by n applications of the integration by parts formula when $p(x)$ is a polynomial of degree n.

10.2 TRIGONOMETRIC SUBSTITUTIONS

Certain types of integrals may be evaluated by means of a trigonometric substitution of the form $x = a \sin \theta$, $x = a \tan \theta$, or $x = a \sec \theta$. The idea is based on the simple Pythagorean Theorem for right triangles, and it is used to handle integrands involving factors $\sqrt{x^2 + a^2}$, $\sqrt{a^2 - x^2}$, and $\sqrt{x^2 - a^2}$.

Integrals Involving $\sqrt{x^2 + a^2}$

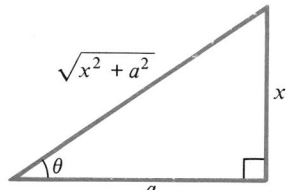

Figure 2.1 Triangle with hypotenuse $\sqrt{x^2 + a^2}$.

The idea is to visualize a right triangle for which the quantity $\sqrt{x^2 + a^2}$ is the length of one of the three legs. The simplest way to do this is to draw a right triangle whose two legs are labelled x and a, respectively (Figure 2.1). The Pythagorean Theorem then states that the length of the hypotenuse is $\sqrt{x^2 + a^2}$.

Next, let θ denote one of the acute angles of the triangle constructed above. Since a is constant, θ is a function of x, and vice versa. In fact, you can see that in Figure 2.1 x and θ are related by the equation

$$\frac{x}{a} = \tan \theta,$$

or

$$x = a \tan \theta. \tag{1}$$

If we make this substitution, the expression $\sqrt{x^2 + a^2}$ becomes

$$\sqrt{x^2 + a^2} = \sqrt{(a \tan \theta)^2 + a^2} = a\sqrt{\tan^2 \theta + 1} = a \sec \theta.$$

In other words, *the trigonometric substitution $x = a \tan \theta$ eliminates the radical.* This is the objective in using trigonometric substitutions. Of course, in using substitution (1) we must also substitute for the differential dx since we will now be integrating with respect to θ. From equation (1) and the definition of the differential we obtain the required equation

$$dx = \frac{d}{d\theta} (a \tan \theta) \, d\theta = a \sec^2 \theta \, d\theta.$$

Example 1

Find $\displaystyle\int \frac{dx}{\sqrt{x^2 + 1}}$.

Strategy

Solution

This integral has the form

$$\int \frac{dx}{\sqrt{x^2 + a^2}} \qquad \text{with} \qquad a = 1.$$

We therefore construct a right triangle with legs of length x and 1 and hypotenuse of length $\sqrt{x^2 + 1}$ (Figure 2.2). From the triangle, or from equation (1), we have the substitution equation

Write the equation for x as a function of θ. Differentiate this equation to obtain the expression for dx.

$$x = \tan\,\theta, \qquad -\frac{\pi}{2} < \theta < \frac{\pi}{2},\,{}^{*}$$

so

$$dx = \sec^2\,\theta\,d\theta.$$

With these substitutions we obtain

Substitute for x and for dx in the original integral.

$$\int \frac{dx}{\sqrt{x^2 + 1}} = \int \frac{\sec^2\,\theta\,d\theta}{\sqrt{\tan^2\,\theta + 1}}$$

Find the resulting integral in terms of θ.

$$= \int \frac{\sec^2\,\theta}{\sqrt{\sec^2\,\theta}}\,d\theta$$

$$= \int \sec\,\theta\,d\theta$$

$$= \ln\,|\sec\,\theta + \tan\,\theta| + C.$$

Substitute back in terms of x using the triangle.

From the triangle in Figure 2.2 we can read directly that

$$\sec\,\theta = \sqrt{x^2 + 1}; \qquad \tan\,\theta = x.$$

The integral is therefore

$$\int \frac{dx}{\sqrt{x^2 + 1}} = \ln\,|\sqrt{x^2 + 1} + x| + C.$$

(Since $\sqrt{x^2 + 1} + x > 0$ for all x, the absolute value signs may be dropped.) ◇

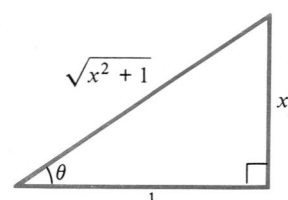

Figure 2.2

Strategy

Example 2

Find $\displaystyle\int \sqrt{x^2 + 4}\,dx$.

Solution

The integral has the form

$$\int \sqrt{x^2 + a^2}\,dx \qquad \text{with} \qquad a = 2.$$

---

*To guarantee that this substitution is reversible, we restrict the domain of $\tan\,\theta$ to its *principal values*.

We therefore construct a right triangle with legs of length x and 2 and hypotenuse $\sqrt{x^2 + 4}$ (Figure 2.3).

We make the substitution

Read off x as a function of θ from the triangle. Find dx by differentiation.

$$x = 2 \tan \theta, \qquad -\frac{\pi}{2} < \theta < \frac{\pi}{2}$$

$$dx = 2 \sec^2 \theta \, d\theta,$$

and the integral becomes

Substitute, simplify.

$$\int \sqrt{x^2 + 4} \, dx = \int \sqrt{4 \tan^2 \theta + 4} \cdot 2 \sec^2 \theta \, d\theta$$

$$= 4 \int \sqrt{\tan^2 \theta + 1} \cdot \sec^2 \theta \, d\theta$$

$$= 4 \int \sec^3 \theta \, d\theta$$

Use Example 7, Section 10.1:

$$\int \sec^3 x \, dx = \frac{1}{2} \sec x \tan x$$

$$+ \frac{1}{2} \ln |\sec x + \tan x| + C.$$

$$= 4\left\{ \frac{1}{2} \sec \theta \tan \theta + \frac{1}{2} \ln |\sec \theta + \tan \theta| \right\} + C$$

Substitute back in terms of x using the triangle.

$$= 2 \cdot \frac{\sqrt{x^2 + 4}}{2} \cdot \frac{x}{2} + 2 \ln \left| \frac{\sqrt{x^2 + 4}}{2} + \frac{x}{2} \right| + C$$

$$= \frac{1}{2} x \sqrt{x^2 + 4} + 2 \ln \left| \frac{\sqrt{x^2 + 4}}{2} + \frac{x}{2} \right| + C. \qquad \diamondsuit$$

Integrals Involving $\sqrt{a^2 - x^2}$

Here the idea is the same, except we must use a right triangle for which one of the sides has length $\sqrt{a^2 - x^2}$.

According to the Pythagorean Theorem, this can be achieved by labelling the hypotenuse a and one of the legs x. The other leg then has length $\sqrt{a^2 - x^2}$. As you can see from the triangle in Figure 2.4 this leads to the equation

$$\frac{x}{a} = \sin \theta, \qquad -\frac{\pi}{2} \le \theta \le \frac{\pi}{2}{}^*$$

and, therefore, to the substitution

$$x = a \sin \theta,$$
$$dx = a \cos \theta \, d\theta.$$

This gives

$$\sqrt{a^2 - x^2} = \sqrt{a^2 - a^2 \sin^2 \theta} = a\sqrt{1 - \sin^2 \theta} = a \cos \theta,$$

since $\cos \theta \ge 0$ for $-\pi/2 \le \theta \le \pi/2$.

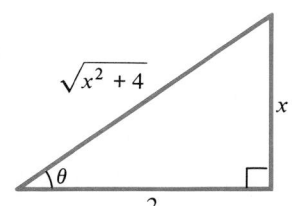

Figure 2.3

Figure 2.4 $x = a \sin \theta$.

Example 3

Find $\displaystyle \int \frac{dx}{x^2 \sqrt{9 - x^2}}$.

---

*Again, the domain is restricted to the principal values of $\sin \theta$.

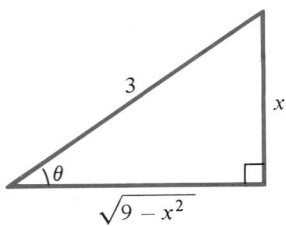

Figure 2.5

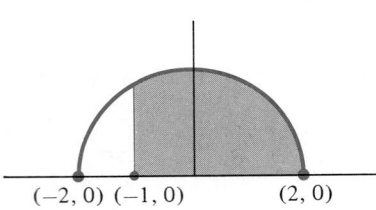

Figure 2.6 Region in Example 4.

Solution: Here $a = 3$ (Figure 2.5), so we use the substitution

$$x = 3 \sin \theta, \qquad -\pi/2 \le \theta \le \pi/2$$
$$dx = 3 \cos \theta \, d\theta.$$

We obtain

$$\int \frac{dx}{x^2\sqrt{9 - x^2}} = \int \frac{3 \cos \theta \, d\theta}{9 \sin^2 \theta \sqrt{9 - 9 \sin^2 \theta}} = \int \frac{3 \cos \theta \, d\theta}{9 \sin^2 \theta \cdot 3 \cdot \sqrt{1 - \sin^2 \theta}}$$

$$= \frac{1}{9} \int \frac{\cos \theta \, d\theta}{\sin^2 \theta \cdot \cos \theta}$$

$$= \frac{1}{9} \int \csc^2 \theta \, d\theta$$

$$= -\frac{1}{9} \cot \theta + C$$

$$= -\frac{1}{9} \cdot \frac{\sqrt{9 - x^2}}{x} + C. \qquad \diamond$$

Example 4

Find the area of the region bounded above by the semicircle $y = \sqrt{4 - x^2}$, below by the x-axis, and on the left by the line $x = -1$ (Figure 2.6).

Strategy

Write the integral giving the area.

Solution

The area is given by the integral

$$A = \int_{-1}^{2} \sqrt{4 - x^2} \, dx.$$

Use the substitution for form $\sqrt{a^2 - x^2}$.

To evaluate the integral, we use the trigonometric substitution

$$x = 2 \sin \theta, \qquad -\pi/2 \le \theta \le \pi/2,$$
$$dx = 2 \cos \theta \, d\theta \quad \text{(Figure 2.7)}.$$

Change limits of integration, using the function $\text{Sin}^{-1} \theta$.

To find the limits of integration with respect to θ, we note that

$$\sin \theta = -\frac{1}{2} \quad \text{when} \quad x = -1, \quad \text{and}$$

$$\sin \theta = 1 \quad \text{when} \quad x = 2.$$

To invert these relationships, we must recall the principal branch of the function $y = \text{Sin}^{-1} x$: $-\pi/2 \le y \le \pi/2$. Thus

$$\theta = \text{Sin}^{-1}(-1/2) = -\pi/6 \quad \text{when} \quad x = -1$$

and

$$\theta = \text{Sin}^{-1}(1) = \pi/2 \quad \text{when} \quad x = 2.$$

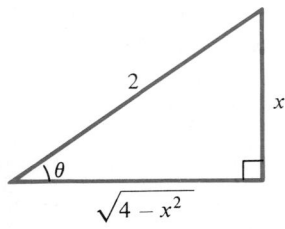

Figure 2.7

Substitute for x and dx, and use new limits of integration.

Our integral therefore becomes

$$A = \int_{-1}^{2} \sqrt{4 - x^2} \, dx = \int_{-\pi/6}^{\pi/2} \sqrt{4 - (2 \sin \theta)^2} \cdot 2 \cos \theta \, d\theta$$

$$= \int_{-\pi/6}^{\pi/2} 2\sqrt{1 - \sin^2 \theta} \cdot 2 \cos \theta \, d\theta$$

$$= 4 \int_{-\pi/6}^{\pi/2} \cos^2 \theta \, d\theta$$

Use identity

$$\cos^2 \theta = \frac{1}{2} + \frac{1}{2} \cos 2\theta.$$

$$= 4 \int_{-\pi/6}^{\pi/2} \left[\frac{1}{2} + \frac{1}{2} \cos 2\theta \right] d\theta$$

$$= 4 \left[\frac{\theta}{2} + \frac{1}{4} \sin 2\theta \right]_{-\pi/6}^{\pi/2}$$

$$= 4 \left[\pi/3 + \frac{\sqrt{3}}{8} \right]$$

$$\approx 5.055. \qquad \diamondsuit$$

REMARK: Notice that we did not have to substitute back in terms of x in the integral in Example 4. Once we have changed variables and limits of integration in a definite integral, we may proceed to evaluate the integral directly.

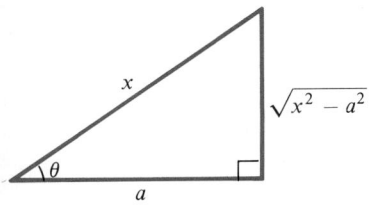

Figure 2.8 $x = a \sec \theta$.

Integrals Involving $\sqrt{x^2 - a^2}$

The situation for integrals involving $\sqrt{x^2 - a^2}$ is similar to those for the preceding two types except that we must label the hypotenuse x. Thus, one leg of the triangle is a, the other $\sqrt{x^2 - a^2}$. As suggested by Figure 2.8, we use the substitutions

$$x = a \sec \theta \qquad 0 \le \theta < \pi/2 \qquad \text{or} \qquad \pi \le \theta < 3\pi/2.$$
$$dx = a \sec \theta \tan \theta \, d\theta.$$

Then the radical $\sqrt{x^2 - a^2}$ simplifies to

$$\sqrt{x^2 - a^2} = \sqrt{a^2 \sec^2 \theta - a^2} = a\sqrt{\tan^2 \theta} = a \tan \theta,$$

since $\tan \theta \ge 0$ when $0 \le \theta < \pi/2$ or $\pi \le \theta < 3\pi/2$.

Example 5

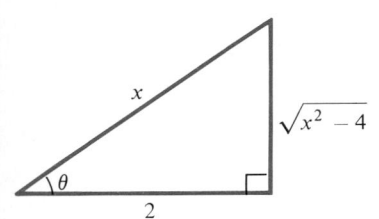

Figure 2.9 $x = 2 \sec \theta$.

Find $\displaystyle\int \frac{dx}{x^2\sqrt{x^2 - 4}}$.

Strategy

Solution

Because of the presence of the radical $\sqrt{x^2 - 4}$, we use the substitution (Figure 2.9)

$$x = 2 \sec \theta, \qquad 0 \le \theta < \pi/2 \qquad \text{or} \qquad \pi \le \theta < 3\pi/2,$$
$$dx = 2 \sec \theta \tan \theta \, d\theta.$$

The integral becomes

Make trigonometric substitution.

$$\int \frac{dx}{x^2\sqrt{x^2 - 4}} = \int \frac{2 \sec \theta \cdot \tan \theta \, d\theta}{(2 \sec \theta)^2\sqrt{4 \sec^2 \theta - 4}}$$

Use $\sec^2 \theta - 1 = \tan^2 \theta$.

$$= \int \frac{2 \sec \theta \tan \theta \, d\theta}{4 \sec^2 \theta \cdot 2 \tan \theta}$$

$$= \frac{1}{4} \int \frac{1}{\sec \theta} \, d\theta$$

$\sec \theta = \dfrac{1}{\cos \theta}$.

$$= \frac{1}{4} \int \cos \theta \, d\theta$$

From Figure 2.9,

$$\sin \theta = \frac{\sqrt{x^2 - 4}}{x}.$$

$$= \frac{1}{4} \cdot \sin \theta + C$$

$$= \frac{1}{4} \cdot \frac{\sqrt{x^2 - 4}}{x} + C. \qquad \diamond$$

In summary, the following trigonometric substitutions, together with the principal values indicated, are used to handle integrals involving the indicated radicals.

$\sqrt{x^2 + a^2}$	requires	$x = a \tan \theta,$	$-\pi/2 < \theta < \pi/2,$
$\sqrt{a^2 - x^2}$	requires	$x = a \sin \theta,$	$-\pi/2 \leq \theta \leq \pi/2,$
$\sqrt{x^2 - a^2}$	requires	$x = a \sec \theta,$	$0 \leq \theta < \pi/2$ or
			$\pi \leq \theta < 3\pi/2.$

Exercise Set 10.2

In Exercises 1–33, evaluate the given integral.

1. $\displaystyle\int \sqrt{4 - x^2}\, dx$

2. $\displaystyle\int \frac{1}{\sqrt{x^2 + 16}}\, dx$

3. $\displaystyle\int \frac{dx}{\sqrt{9 - x^2}}$

4. $\displaystyle\int \sqrt{x^2 - 1}\, dx$

5. $\displaystyle\int \frac{x}{\sqrt{x^2 - 4}}\, dx$

6. $\displaystyle\int \frac{x^2}{\sqrt{9 - x^2}}\, dx$

7. $\displaystyle\int \frac{\sqrt{1 - x^2}}{x}\, dx$

8. $\displaystyle\int \frac{1}{x^2\sqrt{x^2 + 16}}\, dx$

9. $\displaystyle\int \frac{1}{x^2\sqrt{x^2 - 4}}\, dx$

10. $\displaystyle\int \frac{x}{\sqrt{x^2 + 4}}\, dx$

11. $\displaystyle\int \frac{1}{(16 + x^2)^2}\, dx$

12. $\displaystyle\int \frac{dx}{(4 - x^2)^{3/2}}$

13. $\displaystyle\int \frac{x^2\, dx}{(1 - x^2)^{3/2}}$

14. $\displaystyle\int \frac{dx}{(x^2 - 4)^{3/2}}$

15. $\displaystyle\int x^3\sqrt{1 - x^2}\, dx$

16. $\displaystyle\int \frac{x^2}{\sqrt{4 + x^2}}\, dx$

17. $\displaystyle\int \frac{x^2\, dx}{(x^2 + 3)^{3/2}}$

18. $\displaystyle\int \frac{\sqrt{1 - x^2}}{x^4}\, dx$

19. $\displaystyle\int \frac{x^2}{\sqrt{x^2 + a^2}}\, dx$

20. $\displaystyle\int \frac{\sqrt{x^2 - a^2}}{x}\, dx$

21. $\displaystyle\int \frac{\sqrt{x^2 - a^2}}{x^2}\, dx$

22. $\displaystyle\int \frac{x^2}{\sqrt{x^2 - a^2}}\, dx$

23. $\displaystyle\int \frac{x\, dx}{\sqrt{(x^2 + a^2)^3}}$

24. $\displaystyle\int \frac{\sqrt{x^2 + a^2}}{x}\, dx$

25. $\displaystyle\int \frac{dx}{x\sqrt{x^2 - a^2}}$

26. $\displaystyle\int \frac{dx}{x\sqrt{x^2 + a^2}}$

27. $\displaystyle\int \frac{\sqrt{x^2 + a^2}}{x^2}\, dx$

28. $\displaystyle\int \frac{x^2\, dx}{(x^2 + a^2)^{3/2}}$

29. $\displaystyle\int \frac{x^2 + 3}{\sqrt{x^2 + 9}}\, dx$

30. $\displaystyle\int \frac{x - 5}{x^2 - 16}\, dx$

31. $\displaystyle\int \frac{x^2 - 3}{x\sqrt{x^2 + 4}}\, dx$

32. $\displaystyle\int \frac{1 - x}{x\sqrt{1 + x^2}}\, dx$

33. $\displaystyle\int \frac{x^2 - 4x + 5}{x\sqrt{9 - x^2}}\, dx$

34. Use trigonometric substitutions to obtain the formula

$$\int \frac{1}{1 + x^2}\, dx = \text{Tan}^{-1} x + C.$$

35. Find the area of the region bounded by the graph of $y = 5 - \sqrt{x^2 + 9}$ and the x-axis.

36. Find the volume of the solid generated by revolving about the y-axis the region bounded by the graph of $y = x\sqrt{9 - x^2}$ and the x-axis.

37. A water tank is in the shape of a cylinder with circular end panels of radius 5 meters. The tank is mounted with the circular ends vertical. To what percentage capacity is the tank filled when the depth of the water is 6 meters?

38. Find the area of the region enclosed by the ellipse

$$\frac{x^2}{4} + \frac{y^2}{3} = 1.$$

39. Find the length of the graph of $y = x^2$ from $(0, 0)$ to $(1/2, 1/4)$.

40. The region bounded above by the graph of $y = x(16 - x^2)^{1/4}$ and below by the x-axis is rotated about the x-axis. Find the volume of the resulting solid.

41. Find the length of the graph of $y = \ln x$ from $(1, 0)$ to $(3, \ln 3)$.

10.3 INTEGRALS INVOLVING QUADRATIC EXPRESSIONS

The method of substitution may be used together with the method of trigonometric substitutions to handle integrals involving quadratic expressions more general than those encountered in Section 10.2. In doing so two or more successive substitutions will sometimes be required. It is important to keep in mind a clear idea of which substitutions have been carried out and in what order so that you can correctly state the solution in terms of the original variable.

The type of integral we study here involves a square root of a general quadratic expression $ax^2 + bx + c$. Since the methods of trigonometric substitutions apply to many integrals involving perfect squares, we can handle integrals involving square roots of general quadratics by first completing the square and then applying a substitution (if necessary), followed by a trigonometric substitution. (Remember to keep the general strategy well in mind as there are several major steps involved in these solutions!)

Example 1

Find $\displaystyle\int \frac{dx}{\sqrt{x^2 - 2x + 5}}$.

Strategy

First complete the square under the radical.

Make the substitution

$u = x - 1$.

Write the original integral in terms of the substitution.

Solution

By completing the square, we find
$$\sqrt{x^2 - 2x + 5} = \sqrt{(x^2 - 2x + 1) + 4}$$
$$= \sqrt{(x - 1)^2 + 2^2}.$$

We therefore use the substitution
$$u = x - 1; \qquad du = dx$$

to write the integral as
$$\int \frac{dx}{\sqrt{x^2 - 2x + 5}} = \int \frac{dx}{\sqrt{(x - 1)^2 + 2^2}}$$
$$= \int \frac{du}{\sqrt{u^2 + 2^2}}.$$

The last integral requires the trigonometric substitution (Figure 3.1)
$$u = 2 \tan \theta, \qquad -\pi/2 < \theta < \pi/2,$$
$$du = 2 \sec^2 \theta \, d\theta.$$

With this substitution $\sqrt{u^2 + 2^2} = 2 \sec \theta$, so the integral becomes
$$\int \frac{du}{\sqrt{u^2 + 2^2}} = \int \frac{2 \sec^2 \theta \cdot d\theta}{2 \sec \theta}$$
$$= \int \sec \theta \, d\theta$$

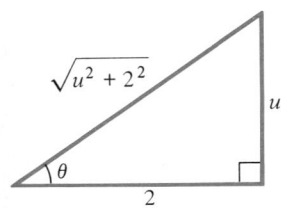

Figure 3.1 $u = 2 \tan \theta$
$du = 2 \sec^2 \theta \, d\theta.$

$$\int \sec \theta \, du = \ln |\sec \theta + \tan \theta| + C.$$

Substitute back in terms of u.

Substitute back in terms of x.

Use property: $\ln(z/2) = \ln z - \ln 2$.

Incorporate $-\ln 2$ in constant of integration.

$$= \ln |\sec \theta + \tan \theta| + C$$

$$= \ln \left| \frac{\sqrt{u^2 + 2^2}}{2} + \frac{u}{2} \right| + C$$

$$= \ln \left(\frac{\sqrt{(x-1)^2 + 2^2}}{2} + \frac{x-1}{2} \right) + C$$

$$= \ln(\sqrt{x^2 - 2x + 5} + x - 1) - \ln 2 + C$$

$$= \ln(\sqrt{x^2 - 2x + 5} + x - 1) + C. \qquad \diamond$$

Example 2

Find $\displaystyle \int \frac{x+3}{\sqrt{2x^2 - 8x}} \, dx$.

Strategy

Complete the square under the radical.

Solution

We have

$$\sqrt{2x^2 - 8x} = \sqrt{2(x^2 - 4x + 4) - 2(4)}$$
$$= \sqrt{2(x-2)^2 - 8}.$$

Make a substitution to simplify x term.

We therefore make the substitution

$$u = x - 2; \qquad du = dx.$$

Then

$$x = u + 2, \qquad \text{so} \qquad x + 3 = u + 5.$$

With these substitutions the integral becomes

$$\int \frac{x+3}{\sqrt{2x^2 - 8x}} \, dx = \int \frac{x+3}{\sqrt{2(x-2)^2 - 8}} \, dx$$

$$= \int \frac{u+5}{\sqrt{2u^2 - 8}} \, du$$

$$= \frac{1}{\sqrt{2}} \int \frac{u+5}{\sqrt{u^2 - 4}} \, du$$

Split integrand into two terms.

$$= \frac{1}{\sqrt{2}} \int \frac{u}{\sqrt{u^2 - 4}} \, du + \frac{1}{\sqrt{2}} \int \frac{5}{\sqrt{u^2 - 4}} \, du.$$

First integrand has form

$$\int w^{-1/2} \, dw \quad \text{where} \quad w = u^2 - 4.$$

The first of these two integrals is evaluated using a simple substitution and the Power Rule:

$$\frac{1}{\sqrt{2}} \int \frac{u}{\sqrt{u^2 - 4}} \, du = \frac{1}{2\sqrt{2}} \int \frac{2u}{\sqrt{u^2 - 4}} \, du$$

$$= \frac{1}{\sqrt{2}} \cdot \sqrt{u^2 - 4} + C_1.$$

The second integral requires the trigonometric substitution

$$u = 2 \sec \theta, \qquad 0 \le \theta < \pi/2 \qquad \text{or} \qquad \pi \le \theta < \frac{3\pi}{2},$$

$$du = 2 \sec \theta \tan \theta \, d\theta \qquad \text{(Figure 3.2)}.$$

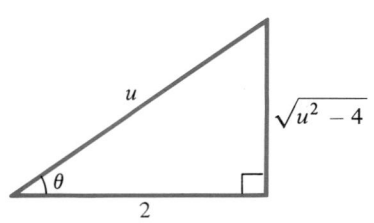

Figure 3.2 $u = 2 \sec \theta$.

Apply trigonometric substitution.

It becomes

$$\frac{1}{\sqrt{2}} \int \frac{5}{\sqrt{u^2 - 4}} \, du = \frac{1}{\sqrt{2}} \int \frac{5 \cdot 2 \cdot \sec \theta \tan \theta \, d\theta}{2 \tan \theta}$$

$$= \frac{5}{\sqrt{2}} \int \sec \theta \, d\theta$$

$$= \frac{5}{\sqrt{2}} \ln |\sec \theta + \tan \theta| + C_2$$

$$= \frac{5}{\sqrt{2}} \ln \left| \frac{u}{2} + \frac{\sqrt{u^2 - 4}}{2} \right| + C_2.$$

Combine results.

Let $C = C_1 + C_2$.

Incorporate factor $-\ln 2$ in C (see Example 1).

Combining these results and recalling that $u = x - 2$ we have

$$\int \frac{x + 3}{\sqrt{2x^2 - 8x}} \, dx = \frac{1}{\sqrt{2}} \sqrt{(x - 2)^2 - 4}$$

$$+ \frac{5}{\sqrt{2}} \ln \left| \frac{x - 2}{2} + \frac{\sqrt{(x - 2)^2 - 4}}{2} \right| + C$$

$$= \frac{1}{\sqrt{2}} \{ \sqrt{x^2 - 4x} + 5 \ln |x - 2 + \sqrt{x^2 - 4x}| \} + C.$$

◇

Exercise Set 10.3

Evaluate the following integrals.

1. $\int \dfrac{dx}{x^2 - 4x + 4}$

2. $\int \dfrac{dx}{x^2 - 6x + 12}$

3. $\int \dfrac{dx}{\sqrt{x^2 + 6x + 13}}$

4. $\int \dfrac{dx}{\sqrt{5 + 4x - x^2}}$

5. $\int \dfrac{dx}{\sqrt{x^2 - 6x}}$

6. $\int \dfrac{x \, dx}{x^2 + 6x + 15}$

7. $\int \dfrac{x \, dx}{\sqrt{6x - x^2}}$

8. $\int \dfrac{1 + x}{\sqrt{x^2 - 6x + 13}} \, dx$

9. $\int \dfrac{2x + 1}{\sqrt{18 + 6x + x^2}} \, dx$

10. $\int \dfrac{3x - 1}{\sqrt{x^2 + 8x}} \, dx$

11. $\int \dfrac{x + 2}{x^2 - 8x - 7} \, dx$

12. $\int \dfrac{2x - 1}{6x^2 + 12x} \, dx$

13. $\int \dfrac{x}{(x^2 + 2x + 2)^2} \, dx$

14. $\int (2x - 3)\sqrt{16 + 12x - 4x^2} \, dx$

15. $\int \dfrac{dx}{(9 - x)(88 - 18x + x^2)^{3/2}}$

16. $\int \dfrac{\sqrt{x^2 + 10x + 21}}{x^3 + 15x^2 + 75x + 125} \, dx$

17. $\int \dfrac{4x^2 + 20x + 25}{\sqrt{4x^2 + 20x + 29}} \, dx$

18. $\int \dfrac{dx}{(3 - x)\sqrt{18 - 6x + x^2}}$

10.4 THE METHOD OF PARTIAL FRACTIONS

In elementary algebra you learned the rule for adding two fractions with different denominators:

$$\frac{a}{b} + \frac{c}{d} = \frac{ad + bc}{bd}.$$

The idea was simply to find a common denominator. For the addition of rational functions of x, the rule is the same. For example,

$$\frac{1}{x+1} + \frac{3}{x-2} = \frac{1(x-2) + 3(x+1)}{(x+1)(x-2)} = \frac{4x+1}{x^2-x-2}. \tag{1}$$

By reversing this procedure, we obtain a technique for integrating rational functions of x such as that on the right side of equation (1). Using equation (1) we see that

$$\int \frac{4x+1}{x^2-x-2} \, dx = \int \left[\frac{1}{x+1} + \frac{3}{x-2} \right] dx \tag{2}$$

$$= \ln |x+1| + 3 \ln |x-2| + C.$$

Clearly the second integral in equation (2) is easier to handle than the first, since the terms in the integrand involve lower degree polynomials. The goal of the method of partial fractions is to decompose rational functions into sums of simpler terms, each of which can be integrated by methods we have already developed.

Before describing the method in its full detail, we will try to gain some insight into how one goes about finding such a decomposition. Obviously we need to keep two points in mind:

POINT 1: In searching for a decomposition of the rational function $\dfrac{p(x)}{q(x)}$ we must work only with denominators *which are factors of $q(x)$.* Otherwise, the terms obtained cannot possibly add up to $\dfrac{p(x)}{q(x)}$.

POINT 2: We will work only with proper fractions. That is, in all rational expressions $\dfrac{p(x)}{q(x)}$ we will require that the degree of $p(x)$ be less than the degree of $q(x)$. (If this is not the case we will first simplify by a polynomial long division.)

With these two points and our general goal in mind, we next take up several examples.

Example 1

Find a partial fraction decomposition for

$$f(x) = \frac{4x-7}{x^2-x-6}.$$

Strategy

Factor the denominator.

Solution

We first factor the denominator as

$$x^2 - x - 6 = (x+2)(x-3).$$

Allow one term for each factor of the denominator. (Assume constant numerators for first degree denominators.)

We therefore seek a decomposition of the form

$$\frac{4x-7}{x^2-x-6} = \frac{A}{x+2} + \frac{B}{x-3} \tag{3}$$

where A and B are constants. (Neither A nor B can be polynomials of degree higher than zero, or we would be introducing improper fractions, violating Point 2.)

Determine A and B by recombining fractions and equating numerators.

To determine A and B, we simply recombine the terms on the right of equation (3) according to the rule for adding fractions:

$$\frac{4x - 7}{x^2 - x - 6} = \frac{A}{x + 2} + \frac{B}{x - 3} \tag{4}$$

$$= \frac{A(x - 3) + B(x + 2)}{(x + 2) \cdot (x - 3)}$$

$$= \frac{(A + B)x + (-3A + 2B)}{(x + 2) \cdot (x - 3)}.$$

Collect like terms in x in numerator on right-hand side.

Now since the numerators of the two fractions on the left and right sides of equation (4) are equal, their coefficients of x must be the same and their constant terms must be the same. This gives the two equations

Use fact that if two polynomials are equal, coefficients of like powers of x must be the same (see Exercise 35).

Equate coefficients of like powers of x.

$$\begin{aligned} A + B &= 4 \quad &\text{(coefficients of } x\text{)}, \\ -3A + 2B &= -7 \quad &\text{(constants)}. \end{aligned}$$

Solve the resulting system either by substitution or by elimination.

The solution of this system of equations may be found by adding three times the top equation to the bottom equation. (Or, by substitution, if you prefer.) Doing so we obtain

$$5B = 5, \quad \text{so} \quad B = 1.$$

From either equation we then obtain $A = 3$.
We therefore have

$$\frac{4x - 7}{x^2 - x - 6} = \frac{3}{x + 2} + \frac{1}{x - 3}. \qquad \diamond$$

Example 2

Find $\displaystyle\int \frac{4x - 7}{x^2 - x - 6}\, dx$.

Solution: Using the results of Example 1 we find

$$\int \frac{4x - 7}{x^2 - x - 6}\, dx = \int \frac{3}{x + 2}\, dx + \int \frac{1}{x - 3}\, dx$$

$$= 3 \ln |x + 2| + \ln |x - 3| + C. \qquad \diamond$$

Example 3

Find a partial fraction decomposition for the function

$$f(x) = \frac{4x^2 - 3x + 1}{x^3 - x^2 + x}.$$

Strategy
Factor the denominator.

Solution
The denominator factors as

$$x^3 - x^2 + x = x(x^2 - x + 1).$$

Allow one term for each factor of the denominator. (Degree of numerator in each term is one less than the degree of the denominator.)

(This is the best we can do, since the term $x^2 - x + 1$ does not have linear factors.*) We therefore seek a partial fraction decomposition of the form

$$\frac{4x^2 - 3x + 1}{x^3 - x^2 + x} = \frac{A}{x} + \frac{Bx + C}{x^2 - x + 1}.$$

Here we must allow for a first degree numerator in the last term on the right since the denominator has degree two. To find the constants, A, B, and C we recombine terms to find that

$$\frac{4x^2 - 3x + 1}{x^3 - x^2 + x} = \frac{A}{x} + \frac{Bx + C}{x^2 - x + 1}$$

Recombine terms collecting like terms in x in numerator.

$$= \frac{A(x^2 - x + 1) + (Bx + C)(x)}{x(x^2 - x + 1)}$$

$$= \frac{(A + B)x^2 + (-A + C)x + A}{x(x^2 - x + 1)}.$$

Equate coefficients of like powers of x.

Equating coefficients of the various powers of x, we obtain the equations

Solve resulting system.

$$\begin{cases} A + B & = 4 & \text{(coefficients of } x^2) \\ -A & + C = -3 & \text{(coefficients of } x) \\ A & = 1 & \text{(constants)} \end{cases}$$

The system has solution $A = 1$, $B = 3$, $C = -2$. The partial fraction decomposition is therefore

$$\frac{4x^2 - 3x + 1}{x^3 - x^2 + x} = \frac{1}{x} + \frac{3x - 2}{x^2 - x + 1}.$$ ◇

Example 4

Find $\displaystyle\int \frac{2x^3 - 8x^2 + 9x + 1}{x^2 - 4x + 4}\, dx.$

Strategy

Integrand is an improper fraction. Reduce it to a proper fraction, plus a polynomial, by division.

Solution

Before seeking a partial fraction decomposition we note that the integrand is an improper fraction. We therefore perform a polynomial long division:

$$\begin{array}{r} 2x \\ x^2 - 4x + 4 \overline{)\,2x^3 - 8x^2 + 9x + 1} \\ \underline{2x^3 - 8x^2 + 8x} \\ x + 1 \end{array}$$

This calculation shows that

$$\frac{2x^3 - 8x + 9x + 1}{x^2 - 4x + 4} = 2x + \frac{x + 1}{x^2 - 4x + 4}.$$

---

*A quadratic polynomial $ax^2 + bx + c$ has linear factors $ax^2 + bx + c = a(x - A)(x - B)$ if and only if A and B are **roots** of the polynomial. Since the roots are given by the quadratic formula $x = \dfrac{-b \pm \sqrt{b^2 - 4ac}}{2a}$, the polynomial has linear factors only if the **discriminant** $b^2 - 4ac$ is nonnegative. (This statement is called the **Discriminant Test** for the existence of linear factors.) In this case $a = 1$, $b = -1$, and $c = 1$, so $b^2 - 4ac = (-1)^2 - (4)(1)(1) = 1 - 4 < 0$. Thus there can be no linear factors.

The first term presents no difficulties so we concentrate on finding a partial fraction decomposition for the second term. The denominator factors as

Factor denominator in second term.

$$x^2 - 4x + 4 = (x - 2)^2,$$

so we seek a decomposition of the form

$$\frac{x + 1}{x^2 - 4x + 4} = \frac{A}{x - 2} + \frac{B}{(x - 2)^2}.$$

Allow one term for each factor $(x - 2)$ and $(x - 2)^2$. (Only a constant numerator in the second term is required—see Exercise 36.)

Note that both $(x - 2)$ and $(x - 2)^2$ are factors of the denominator, so terms with both factors must be included. (However, we need not include an x term in the numerator of the last fraction since an x term will appear when we combine the fractions as written.) Combining terms we obtain

Combine terms.

$$\frac{x + 1}{x^2 - 4x + 4} = \frac{A}{x - 2} + \frac{B}{(x - 2)^2}$$

$$= \frac{A(x - 2) + B}{(x - 2)^2}$$

$$= \frac{Ax + (-2A + B)}{(x - 2)^2}.$$

Equate coefficients of like powers of x.

Equating like powers of x we obtain the equations

$$\begin{aligned} A &= 1 \quad \text{(coefficients of } x\text{)}, \\ -2A + B &= 1 \quad \text{(constants)}. \end{aligned}$$

This system has solution $A = 1$, $B = 3$, so

$$\frac{x + 1}{x^2 - 4x + 4} = \frac{1}{x - 2} + \frac{3}{(x - 2)^2}.$$

Combining our results we now have

$$\int \frac{2x^3 - 8x^2 + 9x + 1}{x^2 - 4x + 4} \, dx = \int 2x \, dx + \int \frac{1}{x - 2} \, dx + \int \frac{3}{(x - 2)^2} \, dx$$

$$= x^2 + \ln |x - 2| - \frac{3}{x - 2} + C. \qquad \diamondsuit$$

It is now time to state the formal procedure lurking behind Examples 1–4. The obvious first step in obtaining a partial fraction decomposition for the rational function $\dfrac{p(x)}{q(x)}$ is to factor the denominator. A theorem about polynomials tells us what to expect as the outcome: *every polynomial $q(x)$ with real coefficients can be factored into the product of only two types of factors:*

(i) powers of linear terms, that is, terms of the form $(x - a)^n$, and/or
(ii) powers of irreducible quadratic terms, that is, terms of the form $(x^2 + bx + c)^m$.

(Thus we are assured, for example, that a cubic polynomial such as $x^3 - 6x^2 + 5x - 9$ can be factored at least into a quadratic term and a linear term, if not three linear terms. Unfortunately, however, this theorem provides no indication of what these factors actually might be.)

We shall not prove this theorem here, but it is central to understanding why the following procedure encompasses all rational integrands.

Procedure for Finding the Partial Fraction Decomposition of the Rational function

$$f(x) = \frac{p(x)}{q(x)}$$

1. Check that the degree of $p(x)$ is smaller than the degree of $q(x)$. If not, perform a division.
2. Factor the denominator $q(x)$ into the product of powers of linear and irreducible quadratic terms. That is, obtain the factorization

$$q(x) = d(x - a_1)^{n_1} \cdots \cdots (x - a_k)^{n_k} \cdot (x^2 + b_1 x + c_1)^{m_1} \cdots$$
$$\cdot (x^2 + b_j x + c_j)^{m_j}.$$

3. For each linear factor $(x - a)^n$ allow the terms

$$\frac{A_1}{x - a} + \frac{A_2}{(x - a)^2} + \cdots + \frac{A_n}{(x - a)^n}$$

in the partial fraction expansion. For each factor $(x^2 + bx + c)^m$ allow the terms

$$\frac{B_1 x + C_1}{x^2 + bx + c} + \frac{B_2 x + C_2}{(x^2 + bx + c)^2} + \cdots + \frac{B_m x + C_m}{(x^2 + bx + c)^m}.$$

4. Combine all terms in the partial fraction expansion and collect coefficients of like powers of x.
5. Equate coefficients of like powers of x between the partial fraction expansion and the original function $\dfrac{p(x)}{q(x)}$.
6. Solve the resulting system for the unknown constants.

The procedure is stated in its fullest generality. Do not be intimidated by the notation, as usually only a small number of factors appears. However, in deciding what types of terms to allow, step 3 provides a helpful reference.

Example 5

The following are several examples of rational functions and the forms of their associated partial fraction expansions.

Function	*Partial Fraction Expansion*
$\dfrac{1}{(x - 1)(x + 2)}$	$\dfrac{A}{x - 1} + \dfrac{B}{x + 2}$
$\dfrac{x + 2}{(x - 1)^2}$	$\dfrac{A}{x - 1} + \dfrac{B}{(x - 1)^2}$
$\dfrac{x^2 - 6x + 1}{x(x^2 - x - 1)}$	$\dfrac{A}{x} + \dfrac{Bx + c}{x^2 - x - 1}$
$\dfrac{x + 7}{x^2(x - 2)^2}$	$\dfrac{A}{x} + \dfrac{B}{x^2} + \dfrac{C}{x - 2} + \dfrac{D}{(x - 2)^2}$
$\dfrac{x^3 + 6x + 1}{x(x^2 + x + 1)^2}$	$\dfrac{A}{x} + \dfrac{Bx + C}{x^2 + x + 1} + \dfrac{Dx + E}{(x^2 + x + 1)^2}$
$\dfrac{x^4 - x^2 + 1}{(x - 1)^3(x^2 + x + 2)^2}$	$\dfrac{A}{x - 1} + \dfrac{B}{(x - 1)^2} + \dfrac{C}{(x - 1)^3} + \dfrac{Dx + E}{x^2 + x + 2} + \dfrac{Fx + G}{(x^2 + x + 2)^2}$

◇

Example 6

Find $\displaystyle\int \frac{3x^4 + 9x^3 + 15x^2 + 10x + 4}{x(x^2 + 2x + 2)^2}\, dx$.

Solution: The numerator is fourth degree, and the denominator is fifth degree. Thus, division is not needed. Since $x^2 + 2x + 2$ is an irreducible quadratic, the denominator is in factored form and we proceed to seek the partial fraction decomposition:

$$\frac{3x^4 + 9x^3 + 15x^2 + 10x + 4}{x(x^2 + 2x + 2)^2}$$

$$= \frac{A}{x} + \frac{Bx + C}{x^2 + 2x + 2} + \frac{Dx + E}{(x^2 + 2x + 2)^2}$$

$$= \frac{A(x^2 + 2x + 2)^2 + x(Bx + C)(x^2 + 2x + 2) + x(Dx + E)}{x(x^2 + 2x + 2)^2}$$

$$= \frac{(A + B)x^4 + (4A + 2B + C)x^3 + (8A + 2B + 2C + D)x^2 + (8A + 2C + E)x + 4A}{x(x^2 + 2x + 2)^2}.$$

Equating coefficients of like powers of x gives the equations

$$
\begin{array}{lll}
A + B & = 3 & \text{(coefficients of } x^4) \\
4A + 2B + C & = 9 & \text{(coefficients of } x^3) \\
8A + 2B + 2C + D & = 15 & \text{(coefficients of } x^2) \\
8A + 2C + E & = 10 & \text{(coefficients of } x) \\
4A & = 4 & \text{(constants)}
\end{array}
$$

The solution of this system is $A = 1$, $B = 2$, $C = 1$, $D = 1$, and $E = 0$. We therefore conclude that

$$\int \frac{3x^4 + 9x^3 + 15x^2 + 10x + 4}{x(x^2 + 2x + 2)^2}\, dx$$

$$= \int \frac{1}{x}\, dx + \int \frac{2x + 1}{x^2 + 2x + 2}\, dx + \int \frac{x}{(x^2 + 2x + 2)^2}\, dx$$

$$= \int \frac{1}{x}\, dx + \int \frac{2x + 2}{x^2 + 2x + 2}\, dx - \int \frac{1}{(x - 1)^2 + 1}\, dx + \int \frac{x}{((x + 1)^2 + 1)^2}\, dx.$$

The first two of these integrals produce logarithms and the third yields an inverse tangent function. The fourth is handled by a trigonometric substitution. (See Exercise 13, Section 10.3.) The result is

$$\int \frac{3x^4 + 9x^3 + 15x^2 + 10x + 4}{x(x^2 + 2x + 2)^2}\, dx$$

$$= \ln |x| + \ln(x^2 + 2x + 2) - \frac{3}{2}\, \text{Tan}^{-1}(x + 1) - \frac{x + 2}{2[(x + 1)^2 + 1]} + C. \ \diamond$$

The Heaviside Method*

There is a procedure that often can speed the calculating of the unknown coefficients in an assumed partial fraction expansion. Suppose we wish to find constants

---

*This technique is due to Oliver Heaviside (1850–1925).

A, B, and C so that

$$\frac{p(x)}{(x - a)(x - b)(x - c)} = \frac{A}{x - a} + \frac{B}{x - b} + \frac{C}{x - c}. \tag{5}$$

To find A, we (i) multiply both sides of the equation by $(x - a)$, obtaining

$$\frac{p(x)}{(x - b)(x - c)} = A + B\left(\frac{x - a}{x - b}\right) + C\left(\frac{x - a}{x - c}\right). \tag{6}$$

Then we (ii) set $x = a$, obtaining from equation (6) the statement

$$A = \frac{p(a)}{(a - b)(a - c)}.$$

Similarly, we can find B and C by repeating steps (i) and (ii) with the factors $(x - b)$ and $(x - c)$, respectively. We find that

$$B = \frac{p(b)}{(b - a)(b - c)}; \qquad C = \frac{p(c)}{(c - a)(c - b)}.$$

This procedure, valid only for the case of distinct linear factors, is sometimes described by saying that one "strikes" the factor on the left side of equation (5) corresponding to the desired constant and then evaluates this expression at the corresponding value of x.

Example 7

Find $\displaystyle\int \frac{6x^2 + 6x - 6}{x^3 + 2x^2 - x - 2} \, dx$.

Strategy

Find the factors of $x^3 + 2x^2 - x - 2$.

Solution

To factor the denominator, we recall that $(x - a)$ can be a factor of $x^3 + 2x^2 - x - 2$ only if the integer a is a factor of the integer 2. Thus, the only possible linear factors are $(x \pm 1)$ and $(x \pm 2)$. In fact,

$$x^3 + 2x^2 - x - 2 = (x - 1)(x + 1)(x + 2).$$

Write the form of the partial fraction decomposition.

We therefore seek the partial fraction decomposition

$$\frac{6x^2 + 6x - 6}{(x - 1)(x + 1)(x + 2)} = \frac{A}{x - 1} + \frac{B}{x + 1} + \frac{C}{x + 2}. \tag{7}$$

Use the Heaviside method to find A, B, and C.

Using the Heaviside method to find A, we strike the factor $(x - 1)$ in the left side of (7) and set $x = 1$ to obtain

$$A = \frac{6(1)^2 + 6(1) - 6}{(1 + 1)(1 + 2)} = \frac{6}{6} = 1.$$

Similarly we find that

$$B = \frac{6(-1)^2 + 6(-1) - 6}{(-1 - 1)(-1 + 2)} = \frac{-6}{-2} = 3$$

and

$$C = \frac{6(-2)^2 + 6(-2) - 6}{(-2 - 1)(-2 + 1)} = \frac{6}{3} = 2.$$

Thus

$$\int \frac{6x^2 + 6x - 6}{x^3 + 2x^2 - x - 2} \, dx = \int \frac{1}{x - 1} \, dx + \int \frac{3}{x + 1} \, dx + \int \frac{2}{x + 2} \, dx$$

$$= \ln |x - 1| + 3 \ln |x + 1| + 2 \ln |x + 2| + C.$$

◇

Example 8

An important application of the method of partial fraction decomposition occurs in the logistic growth model. Recall from Chapter 8 that the Law of Natural Growth

$$\{\text{Rate of growth}\} \propto \{\text{Present size of population}\} \tag{8}$$

leads to the differential equation

$$\frac{dP}{dt} = kP, \tag{9}$$

which has solutions $P(t) = P_0 e^{kt}$. The constant k in equation (9) is referred to as the growth constant. A typical solution curve for equation (9) is depicted in Figure 4.1. One limitation of this growth model is that it assumes an unlimited environment. That is, when $k > 0$ the model assumes that the population can increase at an exponential rate forever.

The logistic growth model postulates a maximum possible population size K (sometimes called the **carrying capacity** of the system) by assuming that the growth rate is proportional to both present population size and the "unutilized capacity for growth" in the system (see Figure 4.2). That is,

$$\text{Rate of Growth} \propto (\text{Present size}) \cdot (\text{Unutilized capacity for growth}). \tag{10}$$

Since the fraction of the carrying capacity K unutilized at population size $P < K$ is $\left(\dfrac{K - P}{K}\right)$, the differential equation corresponding to the logistic growth law in (10) is

$$\frac{dP}{dt} = kP\left(\frac{K - P}{K}\right) \tag{11}$$

where k is again called the natural growth constant.

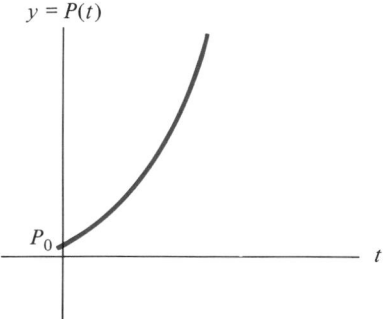

Figure 4.1 Model for natural growth assumes an unlimited environment.

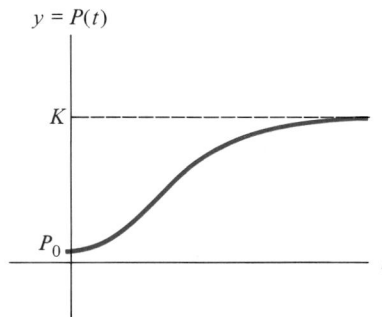

Figure 4.2 Logistic growth model assumes a maximum possible population size K.

To solve (11), we separate variables obtaining

$$\frac{dP}{P(K - P)} = \frac{k}{K} dt,$$

so

$$\int \frac{dP}{P(K - P)} = \int \frac{k}{K} dt = \left(\frac{k}{K}\right)t + C. \tag{12}$$

To find the integral on the left side, we use the method of partial fractions: Setting

$$\frac{1}{P(K - P)} = \frac{A}{P} + \frac{B}{K - P}$$

$$= \frac{A(K - P) + BP}{P(K - P)}$$

$$= \frac{(-A + B)P + AK}{P(K - P)}$$

gives the equations

$$\begin{cases} -A + B = 0 & \text{(coefficients of } P), \\ AK = 1 & \text{(constants).} \end{cases}$$

Thus, $A = B = \dfrac{1}{K}$. The integral on the left side of (12) is therefore

$$\int \frac{dP}{P(K - P)} = \frac{1}{K} \cdot \int \frac{1}{P} dP + \frac{1}{K} \cdot \int \frac{1}{K - P} dP$$

$$= \frac{1}{K} \ln P - \frac{1}{K} \ln(K - P)$$

$$= \frac{1}{K} \ln\left(\frac{P}{K - P}\right).$$

Thus equation (12) becomes

$$\frac{1}{K} \ln\left(\frac{P}{K - P}\right) = \frac{k}{K}t + C$$

or

$$\frac{P}{K - P} = e^{kt + KC}.$$

Solving for P now gives

$$P = \frac{K}{1 + Me^{-kt}} \qquad \text{where} \qquad M = e^{-KC}. \tag{13}$$

From the form of the solution in line (13) we can indeed see that

$$\lim_{t \to \infty} P(t) = \lim_{t \to \infty} \frac{K}{1 + Me^{-kt}} = K$$

for populations satisfying the logistic growth model. ◇

Exercise Set 10.4

In Exercises 1–30, find the indicated integral.

1. $\int \dfrac{1}{x(x+1)}\,dx$

2. $\int \dfrac{3x+2}{x(x+1)}\,dx$

3. $\int \dfrac{2x+4}{1-x^2}\,dx$

4. $\int \dfrac{7x+2}{x^2+x-2}\,dx$

5. $\int \dfrac{1}{1-x^2}\,dx$

6. $\int \dfrac{x^2+1}{x^2-1}\,dx$

7. $\int \dfrac{1}{x+x^3}\,dx$

8. $\int \dfrac{x^2+3x+3}{x(1+x)}\,dx$

9. $\int \dfrac{1}{(x+1)(x^2+1)}\,dx$

10. $\int \dfrac{ax^2+bx+a}{x+x^3}\,dx$

11. $\int \dfrac{4x^2+x+1}{(x+2)(x^2+1)}\,dx$

12. $\int \dfrac{1}{x(x+1)^2}\,dx$

13. $\int \dfrac{x^2+5x+1}{x(x+1)^2}\,dx$

14. $\int \dfrac{dx}{(x+1)^2}$

15. $\int \dfrac{2x^2+3x+2}{x^3+2x^2+x}\,dx$

16. $\int \dfrac{x^3+x^2+1}{x^4+2x^2+1}\,dx$

17. $\int \dfrac{1}{x^3+3x^2+2x}\,dx$

18. $\int \dfrac{3x^2+6x+2}{x(x+1)(x+2)}\,dx$

19. $\int \dfrac{5x^3-9x^2+3x-3}{x(x-3)(x^2+1)}\,dx$

20. $\int \dfrac{6x^3-4x^2+3x-3}{x(x-1)(1+x^2)}\,dx$

21. $\int \dfrac{4}{x^3+x}\,dx$

22. $\int \dfrac{dx}{x^3-6x^2+13x-10}$

23. $\int \dfrac{dx}{(a^2-x^2)^2}$

24. $\int \dfrac{2x^3-4x^2-7x-18}{x^2-2x-8}\,dx$

25. $\int \dfrac{5x^2+14x+11}{x^3+3x^2+4x-5}\,dx$

26. $\int \dfrac{2x^2-3x-2}{x^3+x^2-2x}\,dx$

27. $\int \dfrac{5x^2+2x+2}{x^3-1}\,dx$

28. $\int \dfrac{6x^2-21x-9}{x^3-6x^2+3x+10}\,dx$

29. $\int \dfrac{9x^2+26x-16}{x^3+2x^2-8x}\,dx$

30. $\int \dfrac{x^4-x^3+3x^2-10x+8}{x^3-x^2-4}\,dx$

31. Find the area of the region bounded above by the graph of $y = \dfrac{7x+3}{x^2+x}$ and below by the x-axis for $1 \le x \le 2$.

32. The region bounded above by the graph of $y = \dfrac{x}{x^2+1}$ and below by the x-axis for $0 \le x \le 1$ is revolved about the x-axis. Find the volume of the resulting solid.

33. The region bounded by the graph of $y = \dfrac{1}{(x+1)(x+3)}$ and the x-axis for $0 \le x \le 2$ is revolved about the y-axis. Find the volume of the resulting solid.

34. Find the average value of the function $f(x) = \dfrac{4x^2}{(1+x^2)^2}$ on the interval $[0, 1]$.

35. Prove that if

$$p(x) = a_nx^n + a_{n-1}x^{n-1} + \cdots + a_1x + a_0 \quad \text{and}$$
$$q(x) = b_nx^n + b_{n-1}x^{n-1} + \cdots + b_1x + b_0$$

are polynomials with $p(x) = q(x)$ for all x, then $a_n = b_n$, $a_{n-1} = b_{n-1}, \ldots, a_0 = b_0$. That is, if two polynomials are equal then corresponding coefficients of like powers of x must be equal. (*Hint:* Set $x = 0$ to conclude that $a_0 = b_0$. Differentiate.)

36. Show that the expression $\dfrac{A}{x-a} + \dfrac{B}{(x-a)^2}$ can always be brought to the form $\dfrac{Cx+D}{(x-a)^2}$ by correctly choosing A and B. (This shows that we need not allow for a first degree numerator in the term $\dfrac{B}{(x-a)^2}$ when seeking a partial fraction decomposition for $\dfrac{p(x)}{(x-a)^2}$.)

37. Find a solution of the logistic differential equation $\dfrac{dP}{dt} = 2P\left(\dfrac{100-P}{100}\right)$. What is the carrying capacity?

38. Let $P(t)$ be a solution of the differential equation $\dfrac{dP}{dt} = P(1-P)$ so that $P(0) > 0$. What is $\lim\limits_{t \to \infty} P(t)$?

39. Verify the evaluation of the integrals in Example 6.

10.5 MISCELLANEOUS SUBSTITUTIONS

With the exception of the integration by parts formula, all integration techniques encountered up to this point may be viewed as substitution methods of one sort or another. In this section we present several additional types of substitution techniques.

Radicals of Polynomial Expressions

In integrands involving fractional powers of polynomial expressions, such as $\sqrt[n]{p(x)}$, it is often useful to make a substitution of the form $u^n = p(x)$. This substitution will eliminate the radical sign since $\sqrt[n]{p(x)} = \sqrt[n]{u^n} = u$.

Example 1

Find $\displaystyle\int \frac{dx}{x\sqrt{x+1}}$, $\qquad x > 0$.

Strategy

To obtain a square under the square root we use

$u^2 = x + 1$.

Then solve for x and differentiate to find dx.

Substitute.

Form the partial fraction decomposition.

Integrate.

Solve $u^2 = x + 1$ for $u = \sqrt{x+1}$ and substitute back.

Solution

We make the substitution $u^2 = x + 1$. Then $x = u^2 - 1$, so $dx = 2\,u\,du$. The integral becomes

$$\int \frac{dx}{x\sqrt{x+1}} = \int \frac{2\,u\,du}{(u^2 - 1)\cdot u}$$

$$= \int \frac{2\,du}{u^2 - 1}$$

$$= \int \left\{ \frac{1}{u-1} - \frac{1}{u+1} \right\} du$$

$$= \ln\left| \frac{u-1}{u+1} \right| + C$$

$$= \ln\left(\frac{\sqrt{x+1} - 1}{\sqrt{x+1} + 1} \right) + C. \qquad \diamond$$

Example 2

Find $\displaystyle\int \frac{dx}{\sqrt{x} - \sqrt[3]{x}}$, $\qquad x > 0$.

Strategy

We need $x = u^n$ where both $\dfrac{n}{2}$ and $\dfrac{n}{3}$ are integers. Thus, we use $n = 6$.

Substitute and simplify.

Solution

We use the substitution $x = u^6$. Then $dx = 6u^5\,du$, so we obtain

$$\int \frac{dx}{\sqrt{x} - \sqrt[3]{x}} = \int \frac{6u^5\,du}{\sqrt{u^6} - \sqrt[3]{u^6}}$$

$$= \int \frac{6u^5\,du}{u^3 - u^2}$$

$$= 6\int \frac{u^3\,du}{u - 1}$$

Reduce integrand by division.

$$= 6 \int \left\{ u^2 + u + 1 + \frac{1}{u-1} \right\} du$$

Substitute back;

$$= 2u^3 + 3u^2 + 6u + 6 \ln |u - 1| + C$$

$u = x^{1/6}$.

$$= 2x^{1/2} + 3x^{1/3} + 6x^{1/6} + 6 \ln |x^{1/6} - 1| + C. \qquad \diamond$$

Example 3

Find $\int \sqrt{1 + e^x} \, dx$.

Strategy

Substitute a square for the quantity $1 + e^x$.

Solution

We use the substitution $u^2 = 1 + e^x$. Then

$$e^x = u^2 - 1,$$

so

Note that $u^2 > 1$ for all x, so these expressions are well defined.

$$x = \ln(u^2 - 1),$$

and

$$dx = \frac{2u \, du}{u^2 - 1}.$$

We obtain

Substitute and simplify.

$$\int \sqrt{1 + e^x} \, dx = \int \sqrt{u^2} \left(\frac{2u}{u^2 - 1} \right) du$$

$$= \int \frac{2u^2}{u^2 - 1} \, du$$

Reduce integrand by division.

$$= \int \left(2 + \frac{2}{u^2 - 1} \right) du$$

Form the partial fraction decomposition.

$$= \int \left(2 + \frac{1}{u - 1} - \frac{1}{u + 1} \right) du$$

Integrate.

$$= 2u + \ln \left| \frac{u-1}{u+1} \right| + C$$

Substitute back $u = \sqrt{1 + e^x}$.

$$= 2\sqrt{1 + e^x} + \ln \left(\frac{\sqrt{1 + e^x} - 1}{\sqrt{1 + e^x} + 1} \right) + C. \qquad \diamond$$

Rational Expressions in Sine and Cosine

There is a special substitution that allows us to handle a number of integrals involving rational functions of $\sin x$ and $\cos x$. While the technique is not elegant, it is the only technique available for integrating such expressions. The substitution is

$$u = \tan \left(\frac{x}{2} \right). \qquad (1)$$

To find the values for $\sin x$ and $\cos x$ corresponding to substitution (1) we observe that

$$\cos \left(\frac{x}{2} \right) = \frac{1}{\sec \left(\frac{x}{2} \right)} = \frac{1}{\sqrt{1 + \tan^2 \left(\frac{x}{2} \right)}} = \frac{1}{\sqrt{1 + u^2}}$$

and

$$\sin\left(\frac{x}{2}\right) = \tan\left(\frac{x}{2}\right)\cos\left(\frac{x}{2}\right) = \frac{u}{\sqrt{1+u^2}}.$$

Thus

$$\sin x = 2\sin\left(\frac{x}{2}\right)\cos\left(\frac{x}{2}\right) = \frac{2u}{1+u^2} \tag{2}$$

and

$$\cos x = \cos^2\left(\frac{x}{2}\right) - \sin^2\left(\frac{x}{2}\right) = \frac{1-u^2}{1+u^2}. \tag{3}$$

Finally, from (1) we have $x = 2\,\mathrm{Tan}^{-1}\,u$, so

$$dx = \frac{2}{1+u^2}\,du. \tag{4}$$

Example 4

Find $\displaystyle\int \frac{dx}{2 + \cos x}$.

Strategy

Use substitution (1) in the form of equations (3) and (4).

Solution

Using equations (3) and (4) we obtain

$$\int \frac{dx}{2 + \cos x} = \int \frac{\dfrac{2}{1+u^2}}{2 + \left(\dfrac{1-u^2}{1+u^2}\right)}\,du$$

Simplify.

$$= \int \frac{\dfrac{2}{1+u^2}}{\dfrac{2(1+u^2) + (1-u^2)}{1+u^2}}\,du$$

Factor into form
$\displaystyle\int \frac{dz}{1+z^2}$.

$$= \int \frac{2\,du}{3+u^2}$$

$$= \frac{2}{3}\int \frac{du}{1 + \left(\dfrac{u}{\sqrt{3}}\right)^2}$$

$$= \frac{2}{\sqrt{3}}\int \frac{\dfrac{1}{\sqrt{3}}\,du}{1 + \left(\dfrac{u}{\sqrt{3}}\right)^2}$$

$$= \frac{2}{\sqrt{3}}\,\mathrm{Tan}^{-1}\left(\frac{u}{\sqrt{3}}\right) + C$$

Substitute back using (1).

$$= \frac{2}{\sqrt{3}}\,\mathrm{Tan}^{-1}\left(\frac{1}{\sqrt{3}}\tan\frac{x}{2}\right) + C.$$

◇

Example 5

Find $\displaystyle\int \frac{1 + \sin x}{1 + \cos x}\, dx.$

Strategy

Use substitution (1).

Solution

Using equations (2)–(4) we obtain

$$\int \frac{1 + \sin x}{1 + \cos x}\, dx = \int \frac{\left(1 + \dfrac{2u}{1 + u^2}\right)}{\left(1 + \dfrac{1 - u^2}{1 + u^2}\right)} \cdot \left(\frac{2}{1 + u^2}\right) du$$

Simplify.

$$= \int \left(\frac{1 + 2u + u^2}{2}\right) \cdot \left(\frac{2}{1 + u^2}\right) du$$

Multiply factors in integrand.

$$= \int \frac{1 + 2u + u^2}{1 + u^2}\, du$$

Simplify by division.

$$= \int \left(1 + \frac{2u}{u^2 + 1}\right) du$$

$$= u + \ln(u^2 + 1) + C$$

$$= \tan\left(\frac{x}{2}\right) + \ln\left(\tan^2\left(\frac{x}{2}\right) + 1\right) + C.\qquad\diamond$$

Exercise Set 10.5

Find the following integrals.

1. $\displaystyle\int \frac{\sqrt{x}}{1 + \sqrt{x}}\, dx$

2. $\displaystyle\int \frac{\sqrt{x + 1}}{x}\, dx$

3. $\displaystyle\int \frac{dx}{\sqrt{x} + 2\sqrt[3]{x}}$

4. $\displaystyle\int \frac{\sqrt[3]{x} + 1}{\sqrt[3]{x} - 1}\, dx$

5. $\displaystyle\int \frac{x}{\sqrt[3]{x + 1}}\, dx$

6. $\displaystyle\int \frac{x + 2}{\sqrt{x - 2}}\, dx$

7. $\displaystyle\int \frac{dx}{2 + \sin x}$

8. $\displaystyle\int \frac{dx}{1 - \sin x}$

9. $\displaystyle\int \frac{dx}{1 - \cos x}$

10. $\displaystyle\int \frac{x^3}{\sqrt{1 + x^2}}\, dx$

11. $\displaystyle\int \frac{x^2}{\sqrt[3]{1 + x}}\, dx$

12. $\displaystyle\int \frac{1 + \sqrt{x}}{1 - \sqrt{x}}\, dx$

13. $\displaystyle\int \frac{dx}{\sin x + \cos x}$

14. $\displaystyle\int \frac{1 - \sin x}{1 + \cos x}\, dx$

15. $\displaystyle\int \frac{x^3}{\sqrt{1 - 2x}}\, dx$

16. $\displaystyle\int x^2(x^2 + 1)^{3/2}\, dx$

17. $\displaystyle\int \sqrt{\frac{1 + x}{1 - x}}\, dx$

18. $\displaystyle\int \frac{dx}{1 + e^x}$

19. $\displaystyle\int \frac{dx}{x\sqrt{a + bx}}$

20. $\displaystyle\int x\sqrt{1 + x}\, dx$

21. $\displaystyle\int \frac{x}{\sqrt{2 + x}}\, dx$

10.6 THE USE OF INTEGRAL TABLES

The endpapers of this text constitute what is referred to as a *table of integrals*—a list of antiderivatives for various types of functions. While the list presented here is relatively short, many tables of integrals have been published that contain hundreds of entries. Many of the formulas contained in such tables are established using the techniques of this chapter. Such tables are very useful when you are faced with a large number of integrals to evaluate or when you are unwilling to spend the time required to execute the techniques of this section.

Using a table of integrals is a relatively straightforward task. You must match the form of the integrand with those appearing in the table, select an integral matching yours in form, and identify the corresponding values of the constants. For example, the integral

$$\int \frac{dx}{x(3 + 5x^2)}$$

is evaluated using formula (64), which states that

$$\int \frac{du}{u(a + bu^2)} = \frac{1}{2a} \ln \left| \frac{u^2}{a + bu^2} \right| + C.$$

To do so we must take $a = 3$, $b = 5$, and $u = x$. We obtain

$$\int \frac{dx}{x(3 + 5x^2)} = \frac{1}{2 \cdot 3} \ln \left| \frac{x^2}{3 + 5x^2} \right| + C.$$

However, it is often necessary to first perform a substitution to bring an integrand into the form of one of the entries in a table. In doing so you must be especially careful to carry out the substitution for all variables, including the differential dx. The following examples demonstrate this technique.

Example 1

Use the table of integrals to evaluate

$$\int \frac{dx}{1 + e^{\pi x}}.$$

Solution: We use formula (102):

$$\int \frac{du}{1 + e^u} = \ln \left(\frac{e^u}{1 + e^u} \right) + C.$$

To do so we must use the substitution

$$u = \pi x; \qquad du = \pi \, dx.$$

Then $dx = \dfrac{1}{\pi} \, du$, and we obtain

$$\int \frac{dx}{1 + e^{\pi x}} = \int \frac{\frac{1}{\pi} \, du}{1 + e^u} = \frac{1}{\pi} \int \frac{du}{1 + e^u} = \frac{1}{\pi} \ln \left(\frac{e^{\pi x}}{1 + e^{\pi x}} \right) + C. \qquad \diamondsuit$$

Example 2

Use the table of integrals to evaluate

$$\int x \sqrt{x^2 - 4x + 1} \, dx.$$

Solution: This time it is not so clear which formula might correspond to the given integral. Since the table contains no entries involving the general quadratic $ax^2 + bx + c$, we begin by completing the square on the quadratic term:

$$x^2 - 4x + 1 = (x^2 - 4x + 4) - 3 \tag{1}$$
$$= (x - 2)^2 - 3.$$

Our next step is to use the substitution

$$u = x - 2; \quad du = dx.$$

Then $x = u + 2$, and using equation (1), we obtain

$$\int x\sqrt{x^2 - 4x + 1}\ dx = \int x\sqrt{(x - 2)^2 - 3}\ dx \tag{2}$$

$$= \int (u + 2)\sqrt{u^2 - 3}\ du$$

$$= \int u\sqrt{u^2 - 3}\ du + 2\int \sqrt{u^2 - 3}\ du.$$

To handle the first integral, we use formula 45:

$$\int u\sqrt{u^2 \pm a^2}\ du = \frac{1}{3}(u^2 \pm a^2)^{3/2} + C. \tag{3}$$

The second integral is evaluated using formula 44:

$$\int \sqrt{u^2 \pm a^2}\ du = \frac{1}{2}[u\sqrt{u^2 \pm a^2} \pm a^2\ \ln(u + \sqrt{u^2 \pm a^2})] + C. \tag{4}$$

Taking $a = \sqrt{3}$ and the sign choice $\pm = -$, we combine (2), (3), and (4) to obtain

$$\int x\sqrt{x^2 - 4x + 1}\ dx = \frac{1}{3}[(x - 2)^2 - 3]^{3/2} +$$

$$2 \cdot \frac{1}{2}[(x - 2)\sqrt{(x - 2)^2 - 3} - 3\ \ln(x - 2 + \sqrt{(x - 2)^2 - 3})] + C. \qquad \diamond$$

Exercise Set 10.6

In Exercises 1–30, use the table of integrals in the endpapers to evaluate the integral.

1. $\displaystyle\int \frac{dx}{1 + e^{5x}}$

2. $\displaystyle\int \frac{dx}{x(3 + x)}$

3. $\displaystyle\int \frac{dx}{x^2(9 + 2x)^2}$

4. $\displaystyle\int \frac{5\ dx}{\sqrt{14x - x^2}}$

5. $\displaystyle\int \frac{x + 3}{2 + 5x^2}\ dx$

6. $\displaystyle\int \frac{3\ dx}{\sqrt{6x + x^2}}$

7. $\displaystyle\int \tan^4 \pi x\ dx$

8. $\displaystyle\int x^2 \ln x\ dx$

9. $\displaystyle\int \sin 6x \sin 3x\ dx$

10. $\displaystyle\int \frac{dx}{8 + 3e^{5x}}$

11. $\displaystyle\int \frac{dx}{x^2(2 - 3x)}$

12. $\displaystyle\int \sqrt{12x - 4x^2}\ dx$

13. $\displaystyle\int \frac{9\ dx}{25 - 4x^2}$

14. $\displaystyle\int \frac{5\ dx}{\sqrt{10x + x^2}}$

15. $\displaystyle\int \tan^3 5x\ dx$

16. $\displaystyle\int e^x \sin 3x\ dx$

17. $\displaystyle\int \frac{6x^2\ dx}{\sqrt{6 + x}}$

18. $\displaystyle\int \frac{\sqrt{9 - x^2}}{x^2}\ dx$

19. $\displaystyle\int \frac{6\ dx}{(14 - x^2)^{3/2}}$

20. $\displaystyle\int x \cos^{-1} 3x\ dx$

21. $\displaystyle\int \frac{\pi}{4 + 4\sin 2x}\ dx$

22. $\displaystyle\int \frac{4\ dx}{x\sqrt{9 - x^2}}$

23. $\displaystyle\int \frac{3\,dx}{x\sqrt{x^2+16}}$

24. $\displaystyle\int \sin \ln \pi x\,dx$

27. $\displaystyle\int \frac{7\,dx}{(x+4)\sqrt{x^2+8x+20}}$

28. $\displaystyle\int \frac{4\,dx}{4x^2+20x+16}$

25. $\displaystyle\int \frac{dx}{9x^2+12x+5}$

26. $\displaystyle\int \frac{dx}{\sqrt{7-6x-x^2}}$

29. $\displaystyle\int \frac{(2x^2-7)\,dx}{\sqrt{16-x^2}}$

30. $\displaystyle\int \frac{dx}{(2x-x^2)^{3/2}}$

SUMMARY OUTLINE OF CHAPTER 10

◆ Integration by parts formula: $\displaystyle\int u\,dv = uv - \int v\,du.$ (page 451)

◆ Trigonometric substitutions: (page 462)
   (a) $\sqrt{x^2+a^2}$ requires $x=a\tan\theta$, $-\pi/2 < \theta < \pi/2$; $dx = a\sec^2\theta\,d\theta$
   (b) $\sqrt{a^2-x^2}$ requires $x=a\sin\theta$, $-\pi/2 \le \theta \le \pi/2$; $dx = a\cos\theta\,d\theta$
   (c) $\sqrt{x^2-a^2}$ requires $x=a\sec\theta$, $0\le\theta<\pi/2$ or $\pi\le\theta<3\pi/2$; $dx = a\sec\theta\tan\theta\,d\theta$

◆ To handle integrals involving $\sqrt{ax^2+bx+c}$, complete the square and use the corresponding trigonometric substitution. (page 463)

◆ The method of partial fractions decomposes rational functions into sums of simpler form: (page 464)

$$\frac{p(x)}{(x-a)(x-b)} = \frac{A}{x-a} + \frac{B}{x-b}$$

$$\frac{p(x)}{(x-a)(x^2+bx+c)} = \frac{A}{x-a} + \frac{Bx+C}{x^2+bx+c}.$$

REVIEW EXERCISES—CHAPTER 10

In Exercises 1–75, evaluate the given integral.

1. $\displaystyle\int x\sqrt{x^2+9}\,dx$

2. $\displaystyle\int xe^{3x}\,dx$

3. $\displaystyle\int \sqrt{x^2+a^2}\,dx$

4. $\displaystyle\int \frac{dx}{x^2-6x+9}$

5. $\displaystyle\int \frac{2x+3}{x(x+3)}\,dx$

6. $\displaystyle\int \frac{\sqrt{x+1}}{\sqrt{x}}\,dx$

7. $\displaystyle\int x\cos\pi x\,dx$

8. $\displaystyle\int x\,\mathrm{Tan}^{-1}\,x\,dx$

9. $\displaystyle\int x\ln^2 x\,dx$

10. $\displaystyle\int \frac{dx}{\sqrt{x^2+a^2}}$

11. $\displaystyle\int \frac{dx}{x^2-4x+9}$

12. $\displaystyle\int \frac{4x+5}{x^2-x+2}\,dx$

13. $\displaystyle\int \frac{\sqrt{x+4}}{x}\,dx$

14. $\displaystyle\int \frac{1}{\sqrt[3]{x+1}}\,dx$

15. $\displaystyle\int_0^2 x\sqrt{x+2}\,dx$

16. $\displaystyle\int x(x+3)^6\,dx$

17. $\displaystyle\int \frac{dx}{\sqrt{a^2-x^2}}$

18. $\displaystyle\int \frac{dx}{\sqrt{x^2+8x+25}}$

19. $\displaystyle\int \frac{3x^2+x+1}{x^2(x+1)}\,dx$

20. $\displaystyle\int \frac{dx}{2\sqrt{x}+3\sqrt[3]{x}}$

21. $\displaystyle\int_0^{\pi/4} \tan^2 x\,dx$

22. $\displaystyle\int_1^e x^2\ln x\,dx$

23. $\displaystyle\int \sqrt{a^2-x^2}\,dx$

24. $\displaystyle\int \frac{dx}{\sqrt{1+4x-x^2}}$

25. $\displaystyle\int \frac{4x^2+x+2}{(x-2)(x^2+1)}\,dx$

26. $\displaystyle\int \frac{2+\sqrt{x}}{2-\sqrt{x}}\,dx$

27. $\displaystyle\int \frac{x+1}{\sqrt[3]{x+1}}\,dx$

28. $\displaystyle\int x\csc^2\pi x\,dx$

29. $\displaystyle\int \sqrt{x^2-a^2}\,dx$

30. $\displaystyle\int \frac{dx}{\sqrt{x^2-8x}}$

31. $\displaystyle\int \frac{x^2+2x+3}{(x+2)(x^2+x+1)}\,dx$

32. $\displaystyle\int \frac{x-3}{\sqrt{x+2}}\,dx$

33. $\displaystyle\int (x^2-4)^{3/2}\,dx$

34. $\displaystyle\int \frac{x\,dx}{x^2+4x+8}$

35. $\displaystyle\int \frac{3x^3+x+2}{x(1+x^3)}\,dx$

36. $\displaystyle\int \frac{dx}{1+2\sin x}$

37. $\int \csc^3 x \, dx$

38. $\int \dfrac{e^{2x}}{\sqrt{1-e^x}} \, dx$

57. $\int \dfrac{x^5}{(x^2+2)^2} \, dx$

58. $\int \sqrt{1-\cos x} \, dx$

39. $\int \dfrac{dx}{x\sqrt{9+x^2}}$

40. $\int \dfrac{2+x}{\sqrt{x^2-6x+13}} \, dx$

59. $\int \dfrac{1}{\sqrt{4x^2-25}} \, dx$

60. $\int \sqrt{1+\sin x} \, dx$

41. $\int \dfrac{3x^2-x+10}{(x^2+2)(4-x)} \, dx$

42. $\int \dfrac{dx}{1-2\sin x}$

61. $\int \ln^2 x \, dx$

62. $\int \dfrac{7\,dx}{13+x^2}$

43. $\int \dfrac{\cos x}{2-\cos x} \, dx$

44. $\int \dfrac{x^2}{\sqrt{1+x^2}} \, dx$

63. $\int \dfrac{3\,dx}{x^2-5}, \quad x > \sqrt{5}$

64. $\int \dfrac{\pi\,dx}{x\sqrt{36-x^2}}$

45. $\int \dfrac{dx}{e^x-e^{-x}}$

46. $\int \sec^5 x \, dx$

65. $\int \dfrac{dx}{x\sqrt{9-\pi x}}$

66. $\int \dfrac{4\,dx}{(7-3x)^2}$

47. $\int \dfrac{x}{a^2+x^2} \, dx$

48. $\int \dfrac{2x-1}{\sqrt{17+8x+x^2}} \, dx$

67. $\int \dfrac{5\,dx}{x^2(7-2x)}$

68. $\int \dfrac{dx}{x(\pi+4x)^2}$

49. $\int \dfrac{5x^2-2x+12}{x^3-x^2+4x-4} \, dx$

50. $\int \dfrac{dx}{1+x^3}$

69. $\int \dfrac{3x\,dx}{\sqrt{9+2x}}$

70. $\int \dfrac{4x^2\,dx}{\sqrt{49-x^2}}$

51. $\int \dfrac{dx}{1+e^x}$

52. $\int \dfrac{x}{\sqrt{36+x^2}} \, dx$

71. $\int \dfrac{dx}{x^2\sqrt{8-x^2}}$

72. $\int \dfrac{dx}{\sqrt{5x-x^2}}$

53. $\int \dfrac{3x+1}{\sqrt{x^2+8x}} \, dx$

54. $\int \dfrac{6x^2+5x-2}{x^3+x^2-2x} \, dx$

73. $\int \dfrac{dx}{9+4\cos x}$

74. $\int x \operatorname{Sin}^{-1} 2x \, dx$

55. $\int \dfrac{x^5\,dx}{\sqrt{1+x^2}}$

56. $\int \dfrac{1}{x^2\sqrt{x^2-25}} \, dx$

75. $\int x^3 \ln x \, dx$

Chapter 11
l'Hôpital's Rule and Improper Integrals

With the exception of asymptotes, which involve limits of the form $\lim_{x \to \infty} f(x) = L$ or $\lim_{x \to a} f(x) = \pm\infty$, almost every topic we have considered up to this point has concerned a continuous function on a bounded interval.

In this chapter we shall develop a method (l'Hôpital's Rule) for evaluating limits that yield *indeterminate* forms such as $\dfrac{0}{0}$ or $\dfrac{\infty}{\infty}$. Using this method we shall extend the notion of the definite integral $\int_a^b f(x)\,dx$ of a continuous function to *improper integrals* of the form $\int_a^\infty f(x)\,dx$ or of the form $\int_a^b f(x)\,dx$ where f is not necessarily continuous on $[a, b]$. This method will also be used frequently in Chapters 12 and 13, which deal with the theory of infinite series.

11.1 INDETERMINATE FORMS: l'HÔPITAL'S RULE

On several occasions we have encountered the indeterminate form 0/0 in attempting to evaluate a limit of the form

$$\lim_{x \to a} \frac{f(x)}{g(x)}. \tag{1}$$

Specifically, this difficulty arises in limit (1) when $\lim_{x \to a} f(x) = \lim_{x \to a} g(x) = 0$. For example, recall the limit

$$\lim_{x \to 0} \frac{\sin x}{x} \qquad \left(\frac{0}{0} \text{ form}\right),$$

which required an application of the Pinching Theorem to resolve. More generally, we have seen that this difficulty will always arise in applying the definition of the derivative:

$$f'(x) = \lim_{h \to 0} \frac{f(x + h) - f(x)}{h} \qquad \left(\frac{0}{0} \text{ form}\right).$$

The theory of the derivative may be applied to establish the following simple procedure for dealing with many limits of the type (1) that yield the indeterminate forms 0/0 or ∞/∞. Although the result is credited to the Frenchman G.F.A. l'Hôpital (1661–1704), it was actually discovered by his teacher, Johann Bernoulli (1667–1748).

THEOREM 1 l'Hôpital's Rule	Let lim denote one of the symbols $\lim_{x \to a}$, $\lim_{x \to a^+}$, $\lim_{x \to a^-}$, $\lim_{x \to \infty}$, or $\lim_{x \to -\infty}$ and let the functions f and g be differentiable where defined, except possibly at $x = a$. If

$$\lim f(x) = \lim g(x) = 0$$

or

$$\lim f(x) = \lim g(x) = \infty,$$

then

$$\lim \frac{f(x)}{g(x)} = \lim \frac{f'(x)}{g'(x)},$$

provided the limit on the right exists or is infinite.

In other words, l'Hôpital's Rule states that if on attempting to evaluate limit (1) we obtain the indeterminate form 0/0 or $\dfrac{\infty}{\infty}$, we may separately differentiate both numerator and denominator and try again. If we then succeed in obtaining a limit it will also be the limit of the original quotient. (Caution: differentiate f and g *separately*—do *not* apply the Quotient Rule.)

We will establish Theorem 1 after considering several examples and extensions.

Example 1

Since both $\lim_{x \to 0} \sin x = 0$ and $\lim_{x \to 0} x = 0$, l'Hôpital's Rule gives

$$\lim_{x \to 0} \frac{\sin x}{x} = \lim_{x \to 0} \frac{\dfrac{d}{dx}(\sin x)}{\dfrac{d}{dx}(x)} = \lim_{x \to 0} \frac{\cos x}{1} = 1. \qquad \diamond$$

Example 2

Find $\lim_{x \to 8} \dfrac{\sqrt[3]{x} - 2}{x - 8}$.

Strategy
Check that both $\sqrt[3]{x} - 2 \to 0$ and $x - 8 \to 0$ as $x \to 8$.

Solution
Since both the numerator and denominator approach zero as $x \to 8$, we apply l'Hôpital's Rule:

$$\lim_{x \to 8} \frac{\sqrt[3]{x} - 2}{x - 8} \qquad \left(\frac{0}{0} \text{ form} \right)$$

Apply l'Hôpital's Rule.

$$= \lim_{x \to 8} \frac{\dfrac{1}{3} x^{-2/3}}{1}$$

$$= \frac{1}{3(8)^{2/3}} = \frac{1}{12}. \qquad \diamond$$

REMARK 1: Frequently, on applying l'Hôpital's Rule to $\lim\limits_{x \to a} \dfrac{f(x)}{g(x)}$, we again obtain an indeterminate form. In such cases we simply apply l'Hôpital's Rule again, provided the hypotheses are fulfilled for the functions f' and g'.

Example 3

Find $\lim\limits_{x \to 0} \dfrac{x - \tan x}{x - \sin x}$.

Strategy

Verify that 0/0 form results as $x \to 0$.

Solution

Since both $x - \tan x \to 0$ and $x - \sin x \to 0$ as $x \to 0$, we apply l'Hôpital's Rule to find that

$$\lim_{x \to 0} \frac{x - \tan x}{x - \sin x} \qquad \left(\frac{0}{0} \text{ form}\right)$$

Apply l'Hôpital's Rule.

$$= \lim_{x \to 0} \frac{1 - \sec^2 x}{1 - \cos x} \qquad \left(\frac{0}{0} \text{ form again}\right)$$

Apply l'Hôpital's Rule again!

$$= \lim_{x \to 0} \frac{-2 \sec^2 x \tan x}{\sin x}$$

Simplify using definition of $\sec x$ and $\tan x$.

$$= \lim_{x \to 0} \frac{-2\left(\dfrac{1}{\cos x}\right)^2\left(\dfrac{\sin x}{\cos x}\right)}{\sin x}$$

$$= \lim_{x \to 0} \frac{-2}{\cos^3 x}$$

Evaluate limit.

$$= -2. \qquad \qquad \diamondsuit$$

REMARK 2: Be careful not to apply l'Hôpital's Rule in situations where the hypotheses are not fulfilled. If the limit (1) is not an indeterminate form, an application of l'Hôpital's Rule is incorrect. For example,

$$\lim_{x \to 0} \frac{\sin x}{x - \sin x} \qquad \left(\frac{0}{0} \text{ form}\right)$$

$$= \lim_{x \to 0} \frac{\cos x}{1 - \cos x} \qquad (not \text{ an indeterminate form})$$

$$= +\infty$$

since $(1 - \cos x)$ approaches 0 *through positive values* and $\cos x \to 1$ as $x \to 0$. However, applying l'Hôpital's Rule to the quotient $\dfrac{\cos x}{1 - \cos x}$ would lead to the incorrect result

$$\lim_{x \to 0} \frac{-\sin x}{\sin x} = -1.$$

Example 4

$$\lim_{x \to \pi/2^-} \frac{\tan^2 x}{\sec^2 x} \qquad \left(\frac{\infty}{\infty} \text{ form}\right)$$

$$= \lim_{x \to \pi/2^-} \frac{2 \tan x \sec^2 x}{2 \sec^2 x \tan x}$$

$$= \lim_{x \to \pi/2^-} (1) = 1.$$

◇

Example 5

$$\lim_{x \to \infty} \frac{x^2 + 5}{x + e^x} \qquad \left(\frac{\infty}{\infty} \text{ form}\right)$$

$$= \lim_{x \to \infty} \frac{2x}{1 + e^x} \qquad \left(\text{still } \frac{\infty}{\infty} \text{ form}\right)$$

$$= \lim_{x \to \infty} \frac{2}{e^x}$$

$$= 0.$$

◇

Example 6

$$\lim_{x \to 0^+} \frac{\sin x}{x + x^{3/2}} \qquad \left(\frac{0}{0} \text{ form}\right)$$

$$= \lim_{x \to 0^+} \frac{\cos x}{1 + \frac{3}{2}x^{1/2}}$$

$$= \frac{1}{1 + 0}$$

$$= 1.$$

◇

The justification for l'Hôpital's Rule depends upon a generalization of the Mean Value Theorem due to Augustin L. Cauchy (1789–1857). Recall, the Mean Value Theorem guarantees the existence of a number $c \in (a, b)$ so that

$$f'(c) = \frac{f(b) - f(a)}{b - a} \tag{2}$$

when f is differentiable on (a, b) and continuous on $[a, b]$. Cauchy's generalization arises by viewing the denominator on the right side of equation (2) as $g(b) - g(a)$, where g is the identity function $g(x) = x$, and then asking what results by allowing g instead to be *any* differentiable function of x. Since $g'(c) = 1$ when g is the identity function, the following result is a direct generalization of the Mean Value Theorem.

THEOREM 2
Cauchy's Mean Value Theorem

Let the functions f and g be continuous on the closed interval $[a, b]$ and differentiable on the open interval (a, b). Also, assume $g'(x) \neq 0$ for all x in (a, b). Then there exists a number c in (a, b) so that

$$\frac{f'(c)}{g'(c)} = \frac{f(b) - f(a)}{g(b) - g(a)}.$$

Proof: The Mean Value Theorem was proved by considering the function

$$d(x) = f(x) - \left[f(a) + \frac{f(b) - f(a)}{b - a}(x - a) \right]. \tag{3}$$

Since Theorem 2 is obtained from the Mean Value Theorem by generalizing the identity function $g(x) = x$ to more general functions g, we consider the function D obtained from (3) by replacing $b - a$ with $g(b) - g(a)$, and $x - a$ with $g(x) - g(a)$:

$$D(x) = f(x) - \left[f(a) + \frac{f(b) - f(a)}{g(b) - g(a)} (g(x) - g(a)) \right]. \tag{4}$$

For this function D to make sense, we must know that $g(b) - g(a) \neq 0$. However since g satisfies the hypotheses of the Mean Value Theorem,

$$\frac{g(b) - g(a)}{b - a} = g'(w)$$

for some number w in (a, b). Since we are assuming $g'(w) \neq 0$ for all w in (a, b) this shows that $g(b) - g(a) \neq 0$. Thus, there is no difficulty in defining $D(x)$ by equation (4).

The function D is continuous on $[a, b]$ and differentiable on (a, b) by our assumptions on f and g. Moreover, $D(a) = D(b) = 0$. We may therefore apply Rolle's Theorem to conclude that for some number $c \in (a, b)$

$$D'(c) = 0. \tag{5}$$

From equation (4) we can see that

$$D'(x) = f'(x) - g'(x) \left(\frac{f(b) - f(a)}{g(b) - g(a)} \right). \tag{6}$$

Letting $x = c$ and combining statements (5) and (6) then gives the desired result. ◆

Proof of l'Hôpital's Rule: We shall prove only the case

$$\lim_{x \to a^+} \frac{f(x)}{g(x)} = \lim_{x \to a^+} \frac{f'(x)}{g'(x)} \tag{7}$$

where $\lim_{x \to a^+} f(x) = \lim_{x \to a^+} g(x) = 0$. Since we are assuming the limit on the right to exist or be infinite, the functions f and g are defined and differentiable on an open interval (a, β) with $\beta > a$. We define the functions F and G on $[a, \beta)$ by

$$F(x) = \begin{cases} f(x), & x \neq a \\ 0, & x = a \end{cases} \qquad G(x) = \begin{cases} g(x), & x \neq a \\ 0, & x = a \end{cases}.$$

Then F and G are differentiable on (a, β) and

$$\lim_{x \to a^+} F(x) = \lim_{x \to a^+} f(x) = 0, \tag{8}$$

$$\lim_{x \to a^+} G(x) = \lim_{x \to a^+} g(x) = 0, \tag{9}$$

$$F'(x) = f'(x), \qquad x \in (a, \beta), \tag{10}$$

and

$$G'(x) = g'(x), \qquad x \in (a, \beta). \tag{11}$$

We next let $x > a$ with $a < x < \beta$ and note that the functions F and G satisfy the hypotheses of Cauchy's Mean Value Theorem on the closed interval $[a, x]$. Thus, for each such x there exists a number $c_x \in (a, x)$, which depends on x, with

$$\frac{F(x) - F(a)}{G(x) - G(a)} = \frac{F'(c_x)}{G'(c_x)}. \tag{12}$$

Since $F(a) = G(a) = 0$, equation (12) is just the equation

$$\frac{F(x)}{G(x)} = \frac{F'(c_x)}{G'(c_x)}, \qquad a < c_x < x. \tag{13}$$

Since c_x lies between a and x we must have $c_x \to a^+$ as $x \to a^+$. Thus, equations (8)–(11) and (13) give

$$\lim_{x \to a^+} \frac{f(x)}{g(x)} = \lim_{x \to a^+} \frac{F(x)}{G(x)} = \lim_{x \to a^+} \frac{F'(c_x)}{G'(c_x)}$$

$$= \lim_{c_x \to a^+} \frac{F'(c_x)}{G'(c_x)}$$

$$= \lim_{x \to a^+} \frac{F'(x)}{G'(x)}$$

$$= \lim_{x \to a^+} \frac{f'(x)}{g'(x)}.$$

This proves equation (7). ◆

Exercise Set 11.1

In Exercises 1–37, find the indicated limit.

1. $\lim\limits_{x \to 1} \dfrac{1 - x}{e^x - e}$

2. $\lim\limits_{x \to 0} \dfrac{\sin 5x}{x}$

3. $\lim\limits_{x \to 0} \dfrac{\sin x^2}{x}$

4. $\lim\limits_{x \to 2} \dfrac{x^3 - x^2 - x - 2}{x - 2}$

5. $\lim\limits_{x \to 0} \dfrac{1 - \cos x}{x^2}$

6. $\lim\limits_{x \to \pi/2} \dfrac{1 - \sin x}{\cos x}$

7. $\lim\limits_{x \to 1^-} \dfrac{\sqrt{1 - x^2}}{x - 1}$

8. $\lim\limits_{x \to 0^+} \dfrac{1 - \cos x}{x^3}$

9. $\lim\limits_{\theta \to 0} \dfrac{\tan \theta - \theta}{\theta - \sin \theta}$

10. $\lim\limits_{x \to \infty} \dfrac{x^3 + 2x + 1}{4x^3 + 1}$

11. $\lim\limits_{x \to 1} \dfrac{\ln x}{x - 1}$

12. $\lim\limits_{x \to 1} \dfrac{\sqrt{x} - \sqrt[4]{x}}{x - 1}$

13. $\lim\limits_{x \to 0^+} \dfrac{1 + \cos \sqrt{x}}{\sin x}$

14. $\lim\limits_{x \to 0} \dfrac{\ln(1 + x)}{1 - e^x}$

15. $\lim\limits_{x \to 0^+} \dfrac{x^2}{x - \sin x}$

16. $\lim\limits_{x \to \infty} \dfrac{\sqrt{1 + x^2}}{x}$

17. $\lim\limits_{x \to 1} \dfrac{e^{x^2} - e^x}{x^2 - 1}$

18. $\lim\limits_{x \to 0} \dfrac{\sin x - x \cos x}{x}$

19. $\lim\limits_{x \to 0} \dfrac{\sin x}{\sqrt[3]{x}}$

20. $\lim\limits_{x \to 0^+} \dfrac{\cos x - x}{\sqrt{x}}$

21. $\lim\limits_{x \to \infty} \dfrac{e^{2x}}{1 + x^2}$

22. $\lim\limits_{x \to \pi^+} \dfrac{\sin x}{\sqrt{x - \pi}}$

23. $\lim\limits_{x \to 0} \dfrac{\sin x - x}{x - \tan x}$

24. $\lim\limits_{x \to 0^+} \dfrac{\cos x - x}{x - \tan x}$

25. $\lim\limits_{x \to \infty} \dfrac{9 - x^3}{xe^{\pi x}}$

26. $\lim\limits_{x \to 1^-} \dfrac{x^{5/2} - 1 + \sqrt{1 - x}}{\sqrt{1 - x^2}}$

27. $\lim\limits_{x \to 0} \dfrac{\operatorname{Sin}^{-1} x}{x}$

28. $\lim\limits_{x \to \pi/2} \dfrac{\cos 3x}{\cos x}$

29. $\lim\limits_{x \to \pi/2} \dfrac{\tan(x/2) - 1}{x - \pi/2}$

30. $\lim\limits_{x \to \infty} \dfrac{\tan(1/x)}{1/x}$

31. $\lim\limits_{x \to \infty} \dfrac{\sqrt{x} - 3x^2}{x(6 - x)}$

32. $\lim\limits_{x \to \infty} \dfrac{(ax + b)^3}{(x + c)^3}$

33. $\lim\limits_{x \to \infty} \dfrac{\sqrt{x} - \sqrt{a}}{\sqrt{x} + \sqrt{a}}$

34. $\lim\limits_{x \to \infty} \dfrac{x^3 - 7x^2 + 6x - 5}{(x - 3)(5 - x^2)}$

35. $\lim\limits_{x \to \infty} \dfrac{3x^2 - 4}{4x^2 + 3}$

36. $\lim\limits_{x \to \infty} \dfrac{x^2(x - 1)(x + 3)}{(x^3 - 6)(2x^2 + x + 1)}$

37. $\lim\limits_{x \to \infty} \dfrac{\sin x}{x^2 + \pi}$

38. Extend the proof given for l'Hôpital's Rule to the case

$$\lim_{x\to a^-} \frac{f(x)}{g(x)} = \lim_{x\to a^-} \frac{f'(x)}{g'(x)}$$

where $\lim\limits_{x\to a^-} f(x) = \lim\limits_{x\to a^-} g(x) = 0$.

39. Combine the proof given in this section for l'Hôpital's Rule in the case $\lim\limits_{x\to a^+}$ with the proof given in Exercise 38 for the case $\lim\limits_{x\to a^-}$ to prove the case

$$\lim_{x\to a} \frac{f(x)}{g(x)} = \lim_{x\to a} \frac{f'(x)}{g'(x)}$$

when $\lim\limits_{x\to a} f(x) = \lim\limits_{x\to a} g(x) = 0$.

40. Extend the proof given for l'Hôpital's Rule to the case

$$\lim_{x\to\infty} \frac{f(x)}{g(x)} = \lim_{x\to\infty} \frac{f'(x)}{g'(x)}$$

when $\lim\limits_{x\to\infty} f(x) = \lim\limits_{x\to\infty} g(x) = 0$ as follows:

a. Set $t = \dfrac{1}{x}$ and conclude that

$$\lim_{x\to\infty} \frac{f(x)}{g(x)} = \lim_{t\to 0^+} \frac{f\left(\dfrac{1}{t}\right)}{g\left(\dfrac{1}{t}\right)}.$$

b. Show that

$$\lim_{t\to 0^+} \frac{f(1/t)}{g(1/t)} = \lim_{t\to 0^+} \frac{\left(-\dfrac{1}{t^2}\right)f'\left(\dfrac{1}{t}\right)}{\left(-\dfrac{1}{t^2}\right)g'\left(\dfrac{1}{t}\right)}$$

$$= \lim_{t\to 0^+} \frac{f'(1/t)}{g'(1/t)} = \lim_{x\to\infty} \frac{f'(x)}{g'(x)}.$$

c. Combine a and b to obtain the result.

11.2 OTHER INDETERMINATE FORMS

There are five indeterminate forms in addition to $\dfrac{0}{0}$ and $\dfrac{\infty}{\infty}$ to which l'Hôpital's Rule may be applied. These are: $\infty - \infty$, $0 \cdot \infty$, 1^∞, 0^0, and ∞^0. Here is an example of each:

$$\lim_{x\to\frac{\pi}{2}^-} (\tan x - \sec x) \qquad (\infty - \infty \text{ form})$$

$$\lim_{x\to\infty} e^{-x} \ln x \qquad (0 \cdot \infty \text{ form})$$

$$\lim_{x\to\infty} \left(1 + \frac{1}{x}\right)^x \qquad (1^\infty \text{ form})$$

$$\lim_{x\to 0^+} (\sin x)^x \qquad (0^0 \text{ form})$$

$$\lim_{x\to\frac{\pi}{2}^-} (\tan x)^{\cos x} \qquad (\infty^0 \text{ form})$$

The technique for evaluating limits yielding any of these five forms is to first rewrite the function in question in a form that yields one of the indeterminate forms $\dfrac{0}{0}$ or $\dfrac{\infty}{\infty}$. L'Hôpital's Rule (Section 11.1) is then applied to yield the desired limit.

In discussing these terms, we shall use the symbol lim to represent one of the limits $\lim\limits_{x\to a}$, $\lim\limits_{x\to a^+}$, $\lim\limits_{x\to a^-}$, $\lim\limits_{x\to\infty}$, or $\lim\limits_{x\to-\infty}$ as in Section 11.1.

The Form $\infty - \infty$

The form $\infty - \infty$ occurs in $\lim(f(x) - g(x))$ when $\lim f(x) = \infty$ and $\lim g(x) = \infty$. In this case the strategy is to rewrite the difference $f - g$ as a quotient to which l'Hôpital's Rule may be applied.

Example 1

Find $\lim\limits_{x \to \frac{\pi}{2}^-} (\tan x - \sec x)$.

Strategy

Rewrite difference as quotient by first rewriting $\tan x$ and $\sec x$ in terms of $\sin x$ and $\cos x$. Then combine over common denominator. Apply l'Hôpital's Rule to the quotient.

Solution

$$\lim\limits_{x \to \frac{\pi}{2}^-} (\tan x - \sec x) \qquad (\infty - \infty \text{ form})$$

$$= \lim\limits_{x \to \frac{\pi}{2}^-} \left(\frac{\sin x}{\cos x} - \frac{1}{\cos x} \right)$$

$$= \lim\limits_{x \to \frac{\pi}{2}^-} \left(\frac{\sin x - 1}{\cos x} \right) \qquad \left(\frac{0}{0} \text{ form} \right)$$

$$= \lim\limits_{x \to \frac{\pi}{2}^-} \left(\frac{\cos x}{-\sin x} \right)$$

$$= 0$$

since $\cos x \to 0$ and $\sin x \to 1$ as $x \to \dfrac{\pi}{2}^-$. $\diamondsuit$

The Form $0 \cdot \infty$

The limit $\lim f(x)g(x)$ yields the indeterminate form $0 \cdot \infty$ when $\lim f(x) = 0$ and $\lim g(x) = \infty$, or vice-versa. Rewriting the product fg as one of the quotients $\dfrac{f}{1/g}$ or $\dfrac{g}{1/f}$ will yield one of the indeterminate forms $\dfrac{0}{0}$ or $\dfrac{\infty}{\infty}$.

Example 2

Find $\lim\limits_{x \to \infty} xe^{-x}$.

Strategy

Rewrite factor

e^{-x} as $\dfrac{1}{e^x}$.

Apply l'Hôpital's Rule.

Solution

$$\lim\limits_{x \to \infty} xe^{-x} \qquad (\infty \cdot 0 \text{ form})$$

$$= \lim\limits_{x \to \infty} \frac{x}{e^x} \qquad \left(\frac{\infty}{\infty} \text{ form} \right)$$

$$= \lim\limits_{x \to \infty} \frac{1}{e^x}$$

$$= 0. \qquad \diamondsuit$$

Example 3

Find $\lim\limits_{x \to \infty} x \ln\left(1 + \frac{1}{x} \right)$.

Strategy

Rewrite factor x as $1/(1/x)$ to obtain factor $1/x$ in denominator.

Solution

$$\lim_{x \to \infty} x \ln\left(1 + \frac{1}{x}\right) \qquad (\infty \cdot 0 \text{ form})$$

$$= \lim_{x \to \infty} \frac{\ln\left(1 + \frac{1}{x}\right)}{\frac{1}{x}} \qquad \left(\frac{0}{0} \text{ form}\right)$$

Apply l'Hôpital's Rule.

$$= \lim_{x \to \infty} \frac{\left[\frac{1}{\left(1 + \frac{1}{x}\right)}\right]\left(-\frac{1}{x^2}\right)}{-\frac{1}{x^2}}$$

Simplify and evaluate limit.

$$= \lim_{x \to \infty} \left(\frac{1}{1 + \frac{1}{x}}\right)$$

$$= 1. \qquad \qquad \qquad \Diamond$$

The Form 1^∞

This form occurs in the limit $\lim[f(x)^{g(x)}]$ when $\lim f(x) = 1$ and $\lim g(x) = \infty$. In this case *we first consider the limit of the natural logarithm of the function* $y = f(x)^{g(x)}$. If $L = \lim \ln y$ exists, the continuity of the exponential function guarantees that

$$\lim[f(x)^{g(x)}] = \lim[e^{\ln y}] = e^{\lim(\ln y)} = e^L.$$

That is,

$$\text{if} \quad L = \lim \ln[f(x)^{g(x)}] = \lim[g(x) \cdot \ln f(x)],$$

$$\text{then} \quad \lim[f(x)^{g(x)}] = e^L.$$

The following examples show how this procedure enables us to apply l'Hôpital's Rule.

Example 4

Find $\lim\limits_{x \to \frac{\pi}{2}^-} (\sin x)^{\sec x}$.

Strategy

Verify 1^∞ form.

Use property $\ln y^r = r \ln y$ of logarithms to evaluate limit $\lim[\ln(\sin x)^{\sec x}]$ $= \lim[\sec x \cdot \ln \sin x]$.

Solution

Since $\sin x \to 1$ and $\sec x \to +\infty$ as $x \to \frac{\pi}{2}^-$, this limit has the indeterminate form 1^∞. We therefore let $y = (\sin x)^{\sec x}$ and consider

$$L = \lim_{x \to \frac{\pi}{2}^-} \ln[(\sin x)^{\sec x}]$$

$$= \lim_{x \to \frac{\pi}{2}^-} [\sec x \cdot \ln \sin x] \qquad (\infty \cdot 0 \text{ form})$$

$$= \lim_{x \to \frac{\pi}{2}^-} \frac{\ln \sin x}{\cos x} \qquad \left(\frac{0}{0} \text{ form}\right)$$

Apply l'Hôpital's Rule.

$$= \lim_{x \to \frac{\pi}{2}^-} \frac{\left(\dfrac{\cos x}{\sin x}\right)}{-\sin x}$$

$$= \lim_{x \to \frac{\pi}{2}^-} \left(-\frac{\cos x}{\sin^2 x}\right)$$

$$= 0.$$

Desired limit is e^L.

Thus, the desired limit is

$$\lim_{x \to \frac{\pi}{2}^-} (\sin x)^{\sec x} = e^L = e^0 = 1.$$ ◇

Example 5

Verify that

$$\lim_{x \to \infty} \left(1 + \frac{r}{x}\right)^x = e^r, \qquad -\infty < r < \infty.$$

Solution: Since this limit has the indeterminate form 1^∞, we let $y = \left(1 + \dfrac{r}{x}\right)^x$ and first investigate the limit $\lim\limits_{x \to \infty} \ln y$. We obtain

$$\lim_{x \to \infty} \ln y = \lim_{x \to \infty} \ln\left[\left(1 + \frac{r}{x}\right)^x\right]$$

$$= \lim_{x \to \infty} \left[x \cdot \ln\left(1 + \frac{r}{x}\right)\right] \qquad \text{(property of ln's)}$$

$$= \lim_{x \to \infty} \frac{\ln\left(1 + \dfrac{r}{x}\right)}{\dfrac{1}{x}} \qquad \left(\frac{0}{0} \text{ form}\right)$$

$$= \lim_{x \to \infty} \frac{\left(\dfrac{1}{1 + r/x}\right)\left(-\dfrac{r}{x^2}\right)}{\left(-\dfrac{1}{x^2}\right)} \qquad \text{(l'Hôpital's Rule)}$$

$$= \lim_{x \to \infty} \left(\frac{r}{1 + r/x}\right)$$

$$= r.$$

The desired limit is therefore

$$\lim_{x \to \infty} \left(1 + \frac{r}{x}\right)^x = \lim_{x \to \infty} y = e^r.$$

Setting $r = 1$ we have the important special case

$$e = \lim_{x \to \infty} \left(1 + \frac{1}{x}\right)^x.$$ ◇

The Forms 0^0 and ∞^0

These forms occur in $\lim[f(x)^{g(x)}]$ when $\lim f(x) = \lim g(x) = 0$ or when $\lim f(x) = \infty$ and $\lim g(x) = 0$. We use the same technique as for evaluating limits with form 1^∞.

Example 6

Find $\displaystyle\lim_{x \to 0^+} (\sin x)^x$.

Solution: Since $\sin x \to 0$ as $x \to 0^+$, this limit yields the indeterminate form 0^0. For

$$y = (\sin x)^x$$

we have

$$\lim_{x \to 0^+} \ln y = \lim_{x \to 0^+} \ln[(\sin x)^x]$$

$$= \lim_{x \to 0^+} [x \cdot \ln \sin x].$$

Now $\sin x$ *decreases* to zero as $x \to 0^+$, so the factor $\ln(\sin x)$ *decreases toward* $-\infty$ as $x \to 0^+$. Thus, the term $-\ln(\sin x)$ increases toward $+\infty$, and to be careful about claiming we have $0 \cdot \infty$ form, we write

$$\lim_{x \to 0^+} \ln y = (-1) \lim_{x \to 0^+} [x(-\ln \sin x)] \qquad (0 \cdot \infty \text{ form})$$

$$= (-1) \lim_{x \to 0^+} \frac{-\ln \sin x}{1/x} \qquad \left(\frac{\infty}{\infty} \text{ form}\right)$$

$$= (-1) \lim_{x \to 0^+} \frac{-\left(\dfrac{\cos x}{\sin x}\right)}{(-1/x^2)} \qquad (\text{l'Hôpital's Rule})$$

$$= (-1) \lim_{x \to 0^+} \frac{x^2}{\tan x} \qquad \left(\frac{0}{0} \text{ form}\right)$$

$$= (-1) \lim_{x \to 0^+} \frac{2x}{\sec^2 x} \qquad (\text{l'Hôpital's Rule})$$

$$= (-1)(0)$$

$$= 0.$$

Thus, $\displaystyle\lim_{x \to 0^+} (\sin x)^x = \lim_{x \to 0^+} y = e^0 = 1.$ ◇

Example 7

Find $\displaystyle\lim_{x \to \frac{\pi}{2}^-} (\tan x)^{\cos x}$.

Solution: Since $\tan x \to +\infty$ and $\cos x \to 0$ as $x \to \dfrac{\pi}{2}^-$, this limit yields the indeterminate form ∞^0. Letting

$$y = (\tan x)^{\cos x}$$

we have

$$\lim_{x \to \frac{\pi}{2}^-} \ln y = \lim_{x \to \frac{\pi}{2}^-} \ln(\tan x)^{\cos x}$$

$$= \lim_{x \to \frac{\pi}{2}^-} (\cos x) \ln(\tan x)$$

$$= \lim_{x \to \frac{\pi}{2}^-} \frac{\ln(\tan x)}{\sec x} \qquad \left(\frac{\infty}{\infty} \text{ form}\right)$$

$$= \lim_{x \to \frac{\pi}{2}^-} \frac{\left(\dfrac{\sec^2 x}{\tan x}\right)}{\sec x \tan x} \qquad [\text{l'Hôpital's Rule}]$$

$$= \lim_{x \to \frac{\pi}{2}^-} \frac{\sec x}{\tan^2 x} \qquad (\text{simplifying})$$

$$= \lim_{x \to \frac{\pi}{2}^-} \frac{\left(\dfrac{1}{\cos x}\right)}{\left(\dfrac{\sin x}{\cos x}\right)^2}$$

$$= \lim_{x \to \frac{\pi}{2}^-} \frac{\cos x}{\sin^2 x}$$

$$= \frac{0}{1}$$

$$= 0.$$

The limit is therefore $\displaystyle\lim_{x \to \frac{\pi}{2}^-} (\tan x)^{\cos x} = e^0 = 1.$ ◇

Exercise Set 11.2

In Exercises 1–28 use l'Hôpital's Rule and the techniques of this section to evaluate the limit.

1. $\displaystyle\lim_{x \to 0^+} \left(\frac{1}{x} - \frac{1}{x^2}\right)$

2. $\displaystyle\lim_{x \to 0^+} (\csc x - \cot x)$

3. $\displaystyle\lim_{x \to 0} x \cot x$

4. $\displaystyle\lim_{x \to 0} x \sin \frac{1}{x}$

5. $\displaystyle\lim_{x \to 0^+} x \ln x$

6. $\displaystyle\lim_{x \to 0} \left(\frac{1}{x} - \csc x\right)$

7. $\displaystyle\lim_{x \to 1^+} \left(\frac{1}{x-1} - \frac{x}{\sqrt{x-1}}\right)$

8. $\displaystyle\lim_{x \to \infty} e^{-x} \ln x$

9. $\displaystyle\lim_{x \to 0^+} \left(\frac{1}{x}\right) \tan^{-1} x$

10. $\displaystyle\lim_{x \to \infty} \frac{1}{x} \ln \frac{1}{x}$

11. $\displaystyle\lim_{x \to 0^+} x^x$

12. $\displaystyle\lim_{x \to 0^+} x^{\cos x}$

13. $\displaystyle\lim_{x \to \infty} (1 + x^2)e^{-2x}$

14. $\displaystyle\lim_{x \to 0^+} \sqrt{x^2}$

15. $\displaystyle\lim_{x \to 0^+} \tan x \cdot \ln \tan x$

16. $\displaystyle\lim_{x \to 0^+} (1 + 2x)^{\cot x}$

17. $\displaystyle\lim_{x \to \infty} \left(1 + \frac{2}{x}\right)^{3x}$

18. $\displaystyle\lim_{x \to \infty} \left(1 - \frac{2}{x}\right)^{5x}$

19. $\displaystyle\lim_{x \to 0^+} \cot x \cdot \text{Tan}^{-1} x$

20. $\displaystyle\lim_{x \to 0^+} \sqrt{x} \ln x$

21. $\displaystyle\lim_{x \to 0} (2 - e^x)^{1/x}$

22. $\displaystyle\lim_{x \to \infty} [\ln(x^2 + 3) - \ln x]$

23. $\displaystyle\lim_{x \to \infty} [\ln \sqrt{4x + 2} - \ln\sqrt{x + 3}]$

24. $\lim\limits_{x\to\infty} \left(\dfrac{x+1}{x}\right)^{-2x}$

25. $\lim\limits_{x\to\infty} [\ln \sqrt{x^6 + 3x^2} - \ln(2x^3)]$

26. $\lim\limits_{x\to 0^+} x^2 \cot x^2$

27. $\lim\limits_{x\to\infty} \left(\dfrac{x}{x+1}\right)^{x+1}$

28. $\lim\limits_{x\to 0^+} x^{\sqrt{x}}$

29. For the function $f(x) = x^2 e^x$
    a. find all relative extrema,
    b. find the horizontal asymptote,
    c. sketch the graph.

30. For which real numbers r is $\lim\limits_{x\to\infty} x^r \ln x = 0$?

31. Explain why $\lim\limits_{n\to\infty} x^n e^{-x} = 0$ for all integers n.

11.3 IMPROPER INTEGRALS

In Chapter 6 we encountered functions defined by integration of the form

$$F(t) = \int_a^t f(x)\, dx. \tag{1}$$

When $f(x) \geq 0$ for $x \geq a$, we identified the value $F(t)$ with the area of the region bounded by the graph of $y = f(x)$ and the x-axis between $x = a$ and $x = t$ (Figure 3.1).

Now if f in equation (1) is continuous on $[a, \infty)$, the function F is a perfectly well defined function for all $t > a$. We may therefore ask about its limit as $t \to \infty$. To do so, we define the **improper integral** $\displaystyle\int_a^\infty f(x)\, dx$ as this limit, when it exists (Figure 3.2).

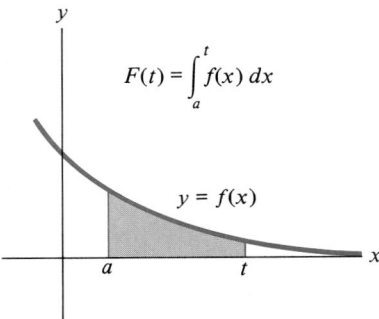

Figure 3.1 $F(t)$ defined by integration.

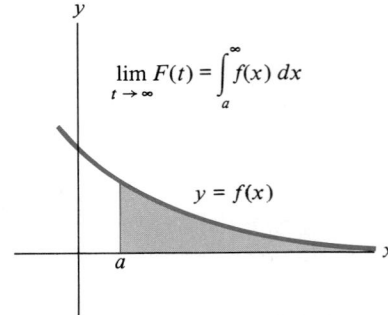

Figure 3.2 $\lim\limits_{t\to\infty} F(t)$ defines an improper integral.

DEFINITION 1

$$\int_a^\infty f(x)\, dx = \lim_{t\to\infty} \int_a^t f(x)\, dx.$$

If the limit in Definition (1) exists, we say that the improper integral **converges.** Otherwise is it said to **diverge.**

There are several important reasons for studying improper integrals at this point, including the following:

(i) It is not unreasonable to ask whether certain regions of infinite length might nevertheless have finite area. Since the definite integral determines area when

$f(x) \geq 0$, Definition 1 gives us a way to extend the notion of area to regions of infinite length.

(ii) In Chapter 8 we determined that the **present value** P_0 of an asset assumed to be worth P dollars after T years was $P_0 = e^{-rT}P$, assuming continuous compounding at the annual rate r. When the asset itself is income-producing, such as a forest continually harvested for lumber, P is a function of t, and the calculation becomes

$$P_0 = \int_0^T e^{-rt}P(t) \, dt.$$

(See Example 6.) The calculation of present value for an asset with an **infinite** life expectancy then leads to the improper integral

$$P_0 = \int_0^\infty e^{-rt}P(t) \, dt.$$

Calculation of improper integrals of the type given in Definition 1 is simply a matter of carefully evaluating the indicated limit.

Example 1

Find $\int_1^\infty \frac{1}{x^2} \, dx$.

Strategy   *Solution*

Apply Definition (1).

$$\int_1^\infty \frac{1}{x^2} \, dx = \lim_{t \to \infty} \int_1^t \frac{1}{x^2} \, dx$$

Evaluate the definite integral from $x = 1$ to $x = t$.

$$= \lim_{t \to \infty}\left\{ -\frac{1}{x} \Big]_{x=1}^{x=t} \right\}$$

$$= \lim_{t \to \infty}\left[\left(-\frac{1}{t}\right) - (-1) \right]$$

Evaluate the limit as $t \to \infty$ of this integral.

$$= 0 - (-1) = 1.$$

This improper integral converges.   ◇

Example 2

Find $\int_1^\infty \frac{1}{x} \, dx$.

Strategy   *Solution*

Apply Definition (1).

$$\int_1^\infty \frac{1}{x} \, dx = \lim_{t \to \infty} \int_1^t \frac{1}{x} \, dx$$

Evaluate the definite integral.

$$= \lim_{t \to \infty}\{\ln x]_{x=1}^{x=t}\}$$

$$= \lim_{t \to \infty}[\ln t - \ln 1]$$

Evaluate the limit using the fact that $\lim_{t \to \infty} \ln t = +\infty$.

$$= +\infty.$$

This improper integral diverges.   ◇

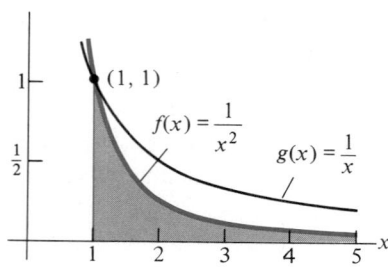

Figure 3.3 Graph of $f(x) = \dfrac{1}{x^2}$ approaches x-axis more quickly than graph of $g(x) = \dfrac{1}{x}$.

REMARK: It is instructive to note the difference between the improper integrals in Examples 1 and 2. Both are of the form $\displaystyle\int_1^\infty \frac{1}{x^p}\, dx$, but their behavior is quite different. As Figure 3.3 illustrates, the graph of $y = 1/x^2$ approaches the x-axis more quickly as x increases than does the graph of $y = 1/x$. In an intuitive sense, it is the rapidity with which the graph of such a function approaches the x-axis that determines whether its improper integral converges. However, this determination cannot be made simply by inspecting a graph. The next example gives a more precise criterion.

Example 3

For which numbers p does the improper integral $\displaystyle\int_1^\infty \frac{1}{x^p}\, dx$ converge?

Strategy

Evaluate the improper integral, carrying along the constant p.

Assume $p \ne 1$, so the integral is obtained via the Power Rule.

The existence of the limit depends upon the sign of the exponent $1 - p$. Examine each case.

Solution

We have

$$\int_1^\infty \frac{1}{x^p}\, dx = \lim_{t \to \infty} \int_1^t x^{-p}\, dx$$

$$= \lim_{t \to \infty} \left\{ \frac{x^{-p+1}}{-p+1} \right]_{x=1}^{x=t} \right\}, \qquad p \ne 1$$

$$= \lim_{t \to \infty} \left[\frac{t^{1-p}}{1-p} - \frac{1}{1-p} \right].$$

Three cases arise:

(a) If $1 - p < 0$, then $t^{1-p} \to 0$ as $t \to \infty$ and the integral converges.

(b) If $1 - p > 0$, then $t^{1-p} \to \infty$ as $t \to \infty$ and the integral diverges.

(c) In the case $1 - p = 0$, excluded above, the integral is $\displaystyle\int_1^\infty \frac{1}{x}\, dx$, which diverges (Example 2).

We therefore conclude that

$$\int_1^\infty \frac{1}{x^p}\, dx \text{ converges if and only if } p > 1.$$

$\diamond$

By use of the improper integral, we can extend the notion of area to regions of infinite length.

DEFINITION 2

Let f be continuous on $[a, \infty)$ with $f(x) \ge 0$ for all $x \ge a$. The area of the region bounded above by the graph of $y = f(x)$ and below by the x-axis and extending to the right of the line $x = a$ is defined to be

$$A = \int_a^\infty f(x)\, dx$$

if this integral converges.

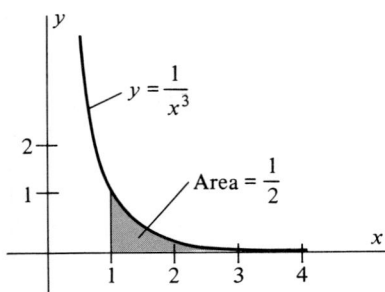

Figure 3.4 A region of infinite length with finite area.

REMARK: Our previous definition of area as

$$A = \int_a^b f(x)\, dx$$

involved A as an infinite limit of approximating sums. Definition 2 actually involves a "double limit," as both the number of approximating rectangles and the interval over which they are constructed become infinite. The notions of volume and surface area for solids of revolution may be extended to regions of infinite length in the same way.

Example 4

Find the area of the region bounded above by the graph of $y = 1/x^3$, below by the x-axis, and on the left by the line $x = 1$. (See Figure 3.4.)

Solution: By Definition 2 and Example 3 we have

$$A = \int_1^\infty \frac{1}{x^3}\, dx = \lim_{t \to \infty} \left\{ \int_1^t x^{-3}\, dx \right\} = \lim_{t \to \infty} \left\{ \frac{t^{-2}}{-2} - \frac{1^{-2}}{-2} \right\}$$

$$= \frac{1}{2}.$$ ◇

Example 5

Find the volume of the solid obtained by revolving about the x-axis the region R bounded above by the graph of $y = \dfrac{1}{x}$ and below by the x-axis for $1 \le x < \infty$ (Figure 3.5).

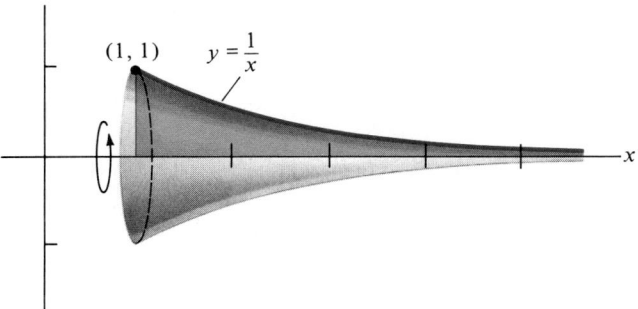

Figure 3.5 A region of infinite area generates a solid of finite volume.

Solution: This volume is $V = \int_1^\infty \pi \left[\dfrac{1}{x} \right]^2 dx = \pi$, by Example 1. Thus, although R has infinite area, it sweeps out a region of finite volume. (This peculiarity is a result only of the way in which we have extended these definitions to the infinite case. Such an intriguing object cannot be realized.)   ◇

Example 6

In Chapter 8 we saw that the **present value** of a sum P of money that will become available after t years is $P_0 = e^{-rt}P$, where r is the rate (in decimal form) of continuous compounding of interest. Closely related to this notion is that of a **revenue stream,** which is simply an anticipated flow of money over time. For example, consider the question of defining the present value of the profits from a small business over the next five years if profits are expected at a rate of $P(t)$ dollars per year, where $P(t)$ is a function of time.

If we partition the time interval $[0, 5]$ into n small intervals of length Δt, then the present value of the profits earned during the time interval $[t_{j-1}, t_j]$ is approximated by

$$\Delta P_j = e^{-rs_j}P(s_j) \, \Delta t \qquad 1 \le j \le n$$

where s_j is an arbitrary element of the interval $[t_{j-1}, t_j]$. An approximation to the total present value of these profits is then

$$P_0 \approx \sum_{j=1}^{n} e^{-rs_j}P(s_j) \, \Delta t.$$

As $n \to \infty$ this Riemann sum approaches the integral $\int_0^5 e^{-rt}P(t) \, dt$, so we define

$$P_0 = \int_0^5 e^{-rt}P(t) \, dt.$$

If we assume the business to have an infinite expected lifetime, the present value of all future profits is defined by the improper integral

$$P_0 = \int_0^\infty e^{-rt}P(t) \, dt.$$ ◇

Obviously, we can define improper integrals of the form $\int_{-\infty}^a f(x) \, dx$ by analogy with Definition 1. We can extend the notion of improper integral to integrals with two infinite limits as follows.

DEFINITION 3

The improper integral $\displaystyle\int_{-\infty}^{\infty} f(x) \, dx$ is defined by

$$\int_{-\infty}^{\infty} f(x) \, dx = \int_{-\infty}^{0} f(x) \, dx + \int_{0}^{\infty} f(x) \, dx.$$

The integral on the left is said to converge only when both integrals on the right converge. Otherwise it is said to diverge. (See Exercise 43.)

Example 7

Find $\displaystyle\int_{-\infty}^{\infty} \frac{1}{1 + x^2} \, dx$.

Solution: By Definition 3,

$$\int_{-\infty}^{\infty} \frac{1}{1+x^2}\, dx = \int_{-\infty}^{0} \frac{1}{1+x^2}\, dx + \int_{0}^{\infty} \frac{1}{1+x^2}\, dx$$

$$= \lim_{t\to-\infty}\{\mathrm{Tan}^{-1} x]_{x=t}^{x=0}\} + \lim_{t\to\infty}\{\mathrm{Tan}^{-1} x]_{x=0}^{x=t}\}$$

$$= \lim_{t\to-\infty}\{-\mathrm{Tan}^{-1}(t)\} + \lim_{t\to\infty}\{\mathrm{Tan}^{-1}(t)\}$$

$$= -\left(-\frac{\pi}{2}\right) + \frac{\pi}{2} = \pi. \qquad \diamond$$

A Comparison Test

We will sometimes need to determine whether an improper integral converges or diverges, even though we cannot find an antiderivative in closed form for its integrand. In such cases we can often resolve the issue of convergence for $\int_{a}^{\infty} f(x)\, dx$ by comparing the integrand f with another function g, and using information about $\int_{a}^{\infty} g(x)\, dx$.

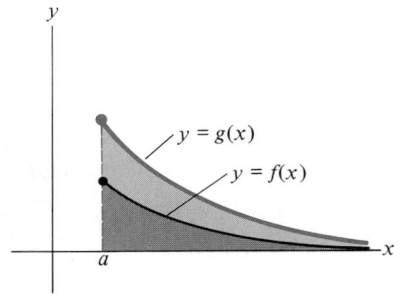

> **Comparison Test:** Suppose f and g are continuous on $[a, \infty)$ with $0 \le f(x) \le g(x)$ for all $x \ge a$. Then
>
> (i) If $\displaystyle\int_{a}^{\infty} g(x)\, dx$ converges, so does $\displaystyle\int_{a}^{\infty} f(x)\, dx$.
>
> (ii) If $\displaystyle\int_{a}^{\infty} f(x)\, dx$ diverges, so does $\displaystyle\int_{a}^{\infty} g(x)\, dx$.

Figure 3.6 Comparison Test: If $0 \le f(x) \le g(x)$ and $\int_{a}^{\infty} g(x)\, dx$ converges, then $\int_{a}^{\infty} f(x)\, dx$ converges.

Figure 3.6 illustrates statement (i), which we justify intuitively by saying that if the area of the region R_1 bounded by the graph of g and the x-axis, $a \le x < \infty$, is finite, then the area of the region R_2 bounded by the graph of f and the x-axis, $a \le x < \infty$, must also be finite, since $R_2 \subseteq R_1$. A similar argument applies to statement (ii): If R_2 is not finite, then R_1 cannot be finite. A more formal proof is given in Chapter 12 for an analogous statement concerning infinite series.

Example 8

Use the Comparison Test to determine whether the improper integral

$$\int_{1}^{\infty} e^{-(1+x^2)}\, dx$$

converges.

Solution: Since $1 + x^2 > x$ for $x \ge 1$ and e^{-x} is a *decreasing* function, we have the comparison

$$0 < e^{-(1+x^2)} < e^{-x}, \qquad x \ge 1.$$

Since

$$\int_{1}^{\infty} e^{-x}\, dx = \lim_{t\to\infty}(-e^{-x}]_{1}^{t} = \lim_{t\to\infty}\left(\frac{1}{e} - e^{-t}\right) = \frac{1}{e}$$

converges, so must

$$\int_1^\infty e^{-(1+x^2)} \, dx.$$

(Note, however, that we cannot determine the *value* of this integral by this method—but only whether or not it converges.) ◇

Improper Integrals—Infinite Integrands

A second general type of improper integral occurs when the integrand in a definite integral approaches $\pm\infty$ at one or both limits of integration. For example, suppose that f is defined and continuous on $[a, b)$ but that $\lim_{x \to b^-} f(x) = \pm\infty$ (Figure 3.7). In this case the Riemann integral $\int_a^b f(x) \, dx$ as defined in Chapter 6 does not exist. However, we define this **improper integral** as follows:

DEFINITION 4

If f is continuous on $[a, b)$ but $\lim_{x \to b^-} f(x) = \pm\infty$, we define

$$\int_a^b f(x) \, dx = \lim_{t \to b^-} \int_a^t f(x) \, dx.$$

If the limit on the right exists, the integral is said to converge. Otherwise it is said to diverge (Figure 3.8).

Example 9

Determine whether the improper integral

$$\int_0^1 \frac{x}{\sqrt{1 - x^2}} \, dx$$

converges or diverges.

Strategy

Determine where the integrand fails to exist.

Apply Definition 4.

Use a substitution with

$u = 1 - x^2, \qquad du = -2x \, dx.$

Solution

The integrand approaches $+\infty$ as $x \to 1^-$. We have

$$\int_0^1 \frac{x}{\sqrt{1 - x^2}} \, dx = \lim_{t \to 1^-} \left\{ \int_0^t \frac{x}{\sqrt{1 - x^2}} \, dx \right\}$$
$$= \lim_{t \to 1^-} \left\{ -\sqrt{1 - x^2} \Big|_{x=0}^{x=t} \right\}$$
$$= \lim_{t \to 1^-} \left\{ -\sqrt{1 - t^2} + 1 \right\}$$
$$= 1.$$

The integral converges. ◇

When the integrand approaches $\pm\infty$ at the left endpoint, the improper integral is defined by analogy with Definition 4 (see Figure 3.9). That is,

$$\int_a^b f(x) \, dx = \lim_{t \to a^+} \int_t^b f(x) \, dx.$$

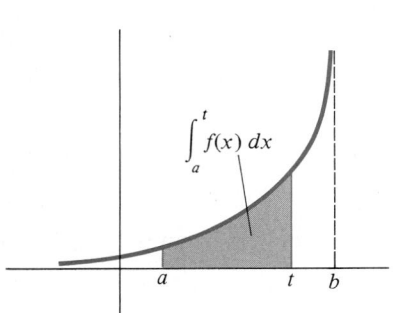

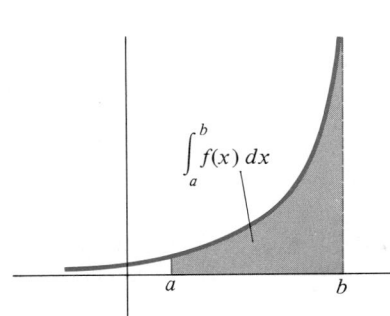

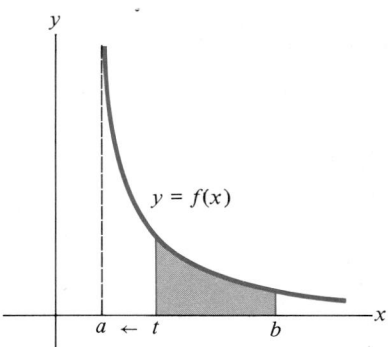

Figure 3.7 $\lim\limits_{x\to b^-} f(x) = \pm\infty.$

Figure 3.8 $\displaystyle\int_a^b f(x)\, dx =$

$\displaystyle\lim_{t\to b^-} \int_a^t f(x)\, dx.$

Figure 3.9 $\displaystyle\int_a^b f(x)\, dx =$

$\displaystyle\lim_{t\to a^+} \int_a^t f(x)\, dx.$

When an integrand approaches $\pm\infty$ at a point $c \in (a, b)$, we extend Definition 4 as follows:

$$\int_a^b f(x)\, dx = \int_a^c f(x)\, dx + \int_c^b f(x)\, dx \qquad (2)$$

$$= \lim_{t_1\to c^-} \int_a^{t_1} f(x)\, dx + \lim_{t_2\to c^+} \int_{t_2}^b f(x)\, dx.$$

(See Figure 3.10.) The improper integral on the left in equation (2) is said to converge only if both improper integrals on the right converge. Otherwise it is said to diverge.

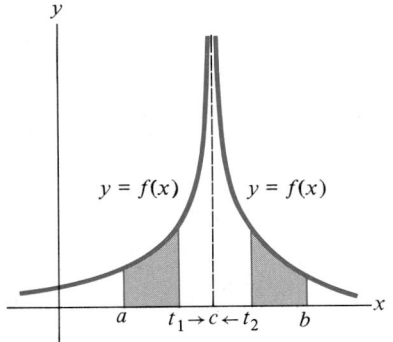

Figure 3.10 $\displaystyle\int_a^b f(x)\, dx = \int_a^c f(x)\, dx +$

$\displaystyle\int_c^b f(x)\, dx.$

Example 10

Determine whether the improper integral

$$\int_0^3 (x - 2)^{-4/3}\, dx$$

converges or diverges.

Solution: The integrand approaches $+\infty$ as $x \to 2$. Applying (2) we find that

$$\int_0^3 (x - 2)^{-4/3}\, dx = \lim_{t_1\to 2^-} \int_0^{t_1} (x - 2)^{-4/3}\, dx + \lim_{t_2\to 2^+} \int_{t_2}^3 (x - 2)^{-4/3}\, dx$$

$$= \lim_{t_1\to 2^-} \{-3(x - 2)^{-1/3}]_0^{t_1}\} + \lim_{t_2\to 2^+} \{-3(x - 2)^{-1/3}]_{t_2}^3\}$$

$$= \lim_{t_1\to 2^-} \left[\frac{-3}{\sqrt[3]{t_1 - 2}} + \frac{3}{\sqrt[3]{-2}}\right] + \lim_{t_2\to 2^+} \left[-3 + \frac{3}{\sqrt[3]{t_2 - 2}}\right]$$

$$= \infty + \infty = \infty.$$

Since both limits fail to exist, the improper integral diverges. ◇

Exercise Set 11.3

In Exercises 1–28, determine whether the improper integral converges. Evaluate those which converge.

1. $\int_3^\infty \frac{1}{x}\, dx$

2. $\int_2^\infty \frac{1}{(x+1)^2}\, dx$

3. $\int_1^\infty \frac{x}{\sqrt{1+x^2}}\, dx$

4. $\int_1^\infty \frac{dx}{4+x^2}$

5. $\int_0^\infty e^{-x}\, dx$

6. $\int_2^\infty \frac{x}{(4+x^2)^{3/2}}\, dx$

7. $\int_0^\infty e^{-x} \sin x\, dx$

8. $\int_0^\infty e^x \cos x\, dx$

9. $\int_{-\infty}^2 e^{2x}\, dx$

10. $\int_{-\infty}^0 \frac{1}{1+x^2}\, dx$

11. $\int_0^\infty xe^{-x}\, dx$

12. $\int_{-\infty}^\infty \frac{dx}{4+x^2}$

13. $\int_0^1 \frac{1}{x}\, dx$

14. $\int_0^1 \frac{1}{x^2}\, dx$

15. $\int_0^3 \sqrt{9-x^2}\, dx$

16. $\int_e^\infty \frac{1}{x \ln x}\, dx$

17. $\int_0^1 x \ln x\, dx$

18. $\int_0^1 \frac{dx}{\sqrt{1-x}}$

19. $\int_0^e x^2 \ln x\, dx$

20. $\int_0^1 x^{-2/3}\, dx$

21. $\int_{-1}^1 \frac{1}{x}\, dx$

22. $\int_0^3 \frac{dx}{x^2+2x-3}$

23. $\int_0^\infty e^{-2x} \cos 2x\, dx$

24. $\int_0^\infty e^{-3x} \sin 2x\, dx$

25. $\int_0^\infty \frac{e^{-2\sqrt{x}}}{\sqrt{x}}\, dx$

26. $\int_0^\infty \sqrt{x}\, e^x\, dx$

27. $\int_0^1 x \ln(1+x)\, dx$

28. $\int_0^1 \ln^2 x\, dx$

29. Show that the integral

$$\int_0^\infty \sin x\, dx$$

diverges. Thus, it is not necessary that

$$\lim_{t\to\infty} \int_a^t f(x)\, dx = \pm\infty$$

for an improper integral to diverge.

30. Give a geometric argument that $\int_1^\infty \frac{dx}{x^2} = \int_0^1 \frac{dx}{\sqrt{x}} - 1$.

31. For which numbers p does the integral $\int_0^1 x^p\, dx$ converge?

32. Find the numbers p for which the integral $\int_0^\infty x^p\, dx$ converges.

33. Give a geometric argument to show that if $0 \le f(x) \le g(x)$ for all $x > a$, then

$$\int_a^\infty g(x)\, dx \text{ diverges} \qquad \text{if} \qquad \int_a^\infty f(x)\, dx \text{ diverges.}$$

Use the Comparison Test to determine whether the following integrals converge or diverge.

34. $\int_1^\infty \frac{1}{1+x^3}\, dx$

35. $\int_1^\infty \frac{e^x}{\sqrt{1+x^2}}\, dx$

36. $\int_1^\infty \frac{1}{\sqrt{1+x^5}}\, dx$

37. $\int_1^\infty \frac{|\sin x|}{x^2}\, dx$

38. $\int_1^\infty e^{\sqrt{x}}\, dx$

39. $\int_1^\infty \sqrt{e^x + \sin x}\, dx$

40. $\int_1^\infty \frac{dx}{\sqrt{e^{x^2} + x + \cos x}}$

41. $\int_1^\infty e^{-x} \sin x\, dx$

42. Explain why

$$\int_{-\infty}^\infty f(x)\, dx = \lim_{t\to\infty} \int_{-t}^t f(x)\, dx$$

is wrong. (*Hint:* Consider the function $f(x) = x^3$, for example.)

43. Show that if $\int_{-\infty}^\infty f(x)\, dx$ converges then

$$\int_{-\infty}^\infty f(x)\, dx = \int_{-\infty}^a f(x)\, dx + \int_a^\infty f(x)\, dx$$

for all $a \in (-\infty, \infty)$.

44. Consider the region R bounded above by the graph of $y = 1/x$ and below by the x-axis for $1 \le x < \infty$.
 a. Show that the surface area of the solid obtained by revolving R about the x-axis is infinite.
 b. Compare the result of part (a) with that of Example 5. Comment on the observation that Figure 3.5 represents a horn of infinite length that (i) cannot be painted with a finite amount of paint but (ii) can be filled with a finite amount of paint.

45. Find the volume and the surface area of the solid obtained by revolving about the x-axis the region bounded above by the graph of $f(x) = e^{-x}$ and below by the x-axis for $1 \le x < \infty$.

46. Find the integers n for which the integral $\int_0^1 x^n \ln x \, dx$ converges.

47. A business expects to generate one million dollars profit per year indefinitely.

   a. Find the present value of its anticipated profits over the next five-year period, assuming an annual rate of interest of $r = 8\%$.

   b. Find the present value of all anticipated future profits.

48. What is the present value of an investment that will produce revenues at the rate of $R(t) = 1000 + 200t$ dollars per year, assuming an annual rate of continuous compounding of $r = 10\%$,

a. over the period of the next 5 years?

b. over an infinite future?

49. Show that

$$\int_0^1 [\ln x]^n \, dx = (-1)^n \cdot n!$$

$$(n! = n(n-1)(n-2) \cdots \cdots 2 \cdot 1).$$

50. Show that $\int_1^\infty \dfrac{dx}{1 + e^x} = \ln(1 + e) - 1$.

51. Use the notion of an improper integral to explain how you can conclude from the result of Example 4 in Section 7.3 that the surface area of the sphere is $4\pi r^2$.

SUMMARY OUTLINE OF CHAPTER 11

◆ **l'Hôpital's Rule** states that if f and g are differentiable (except possibly at $x = a$), if lim denotes one of $\lim_{x \to a}$, $\lim_{x \to a}$,   (page 485)
$\lim_{x \to a^+}$, $\lim_{x \to \infty}$, or $\lim_{x \to -\infty}$, and if either $\lim f(x) = 0 = \lim g(x)$ or $\lim f(x) = \infty = \lim g(x)$,

then

$$\lim \frac{f(x)}{g(x)} = \lim \frac{f'(x)}{g'(x)}$$

provided that the limit on the right exists or is infinite.

◆ Limits involving the **indeterminate forms** $\dfrac{0}{0}$, $\dfrac{\infty}{\infty}$, $\infty - \infty$, 1^∞, 0^0, and ∞^0 are evaluated, if possible, by l'Hôpital's   (page 490)
Rule.

◆ The **improper integral** $\int_a^\infty f(x) \, dx$ is defined by $\int_a^\infty f(x) \, dx = \lim_{t \to \infty} \int_a^t f(x) \, dx$.   (page 496)

◆ If $\lim_{x \to b^-} f(x) = \pm\infty$, the **improper integral** $\int_a^b f(x) \, dx$ is defined by $\int_a^b f(x) \, dx = \lim_{t \to b^-} \int_a^t f(x) \, dx$.   (page 502)

REVIEW EXERCISES—CHAPTER 11

In Exercises 1–16, evaluate the limit using l'Hôpital's Rule if necessary.

1. $\lim_{x \to 0} \dfrac{\sin 3x}{\sin 4x}$

2. $\lim_{x \to 0} \dfrac{\sin x}{1 - e^x}$

3. $\lim_{x \to 0^+} \dfrac{\tan x}{x^2}$

4. $\lim_{x \to 2} \dfrac{x - 2}{x^2 + x - 6}$

5. $\lim_{x \to 0} \dfrac{x - \sin x}{\tan x}$

6. $\lim_{x \to 0} \dfrac{\pi - \csc x}{\pi + \cot x}$

7. $\lim_{x \to 0^+} \dfrac{\sin \sqrt{x}}{\sqrt{x}}$

8. $\lim_{x \to \infty} \dfrac{\pi/2 - \mathrm{Tan}^{-1} x}{xe^{-x}}$

9. $\lim_{x \to \infty} \dfrac{\ln \sqrt{x}}{\sqrt{x}}$

10. $\lim_{x \to -\infty} \dfrac{e^{-x}}{x^2}$

11. $\lim_{x \to \infty} \dfrac{x^3 + e^x}{xe^{2x}}$

12. $\lim_{x \to \infty} x^2 e^{-x}$

13. $\lim_{x \to 0^+} x^x$

14. $\lim_{x \to 0^+} \tan x \ln x$

15. $\lim_{x \to 0^+} x^{\sin 2x}$

16. $\lim_{x \to \infty} (\ln x)^{e^{-x}}$

In Exercises 17–26, evaluate the improper integral.

17. $\int_1^\infty xe^{-x} \, dx$

18. $\int_{-\infty}^0 \dfrac{6x}{1 + x^2} \, dx$

19. $\displaystyle\int_0^2 \frac{2x+1}{x^2+x-6}\,dx$

20. $\displaystyle\int_0^4 \frac{\ln\sqrt{x}}{\sqrt{x}}\,dx$

23. $\displaystyle\int_0^{\pi/2} \tan x\,dx$

24. $\displaystyle\int_e^\infty \frac{1}{x\ln x}\,dx$

21. $\displaystyle\int_{-\infty}^\infty xe^{-x^2}\,dx$

22. $\displaystyle\int_1^\infty \frac{1}{x\sqrt{x^2-1}}\,dx$

25. $\displaystyle\int_0^4 \frac{1}{\sqrt{16-x^2}}\,dx$

26. $\displaystyle\int_{-\infty}^\infty \mathrm{Tan}^{-1}\,x\,dx$

UNIT 5

THE THEORY OF INFINITE SERIES

Brook Taylor

Colin Maclaurin

Joseph Fourier

Karl Weierstrass

THE THEORY OF INFINITE SERIES

It was noted in the introduction to Unit I that Newton and Leibniz both found themselves expressing functions as infinite sums. Later workers in analysis, particularly Leonhard Euler and Johann Bernoulli, also dealt with these infinite series. This early work was not on a firm mathematical footing, however, and operations were sometimes performed that just happened to work because of the circumstances in a particular problem. The development of a logical theory was yet to appear, a process that unfolded over a long period of time.

Brook Taylor (1685–1731) and Isaac Newton knew each other through their membership in the Royal Society of London. Newton was its president for many years, and Taylor served for several years as its secretary. Taylor was brought up in a well-to-do home, and music was an important part of his early life. In fact, Taylor and Newton jointly wrote a work entitled *On Musick,* although it was never completed or published. Taylor thought highly of the new Newtonian calculus and attempted to clarify the subject in his writings. His writing was rather murky, however, and his exposition not very successful.

In this unit we write functions in the form of a special type of infinite series called Taylor's Series. Taylor developed the series as a result of a chance remark made by a friend in a coffeehouse, and some have questioned whether Taylor should receive full credit. Although he published the concept in 1713, it was so badly written that it had little immediate impact: it took Euler's work 40 years later to make the series concept well known.

Also introduced in this unit is Maclaurin's series, a special case of Taylor's series which was used by Taylor (as was acknowledged by Maclaurin in 1742). Colin Maclaurin (1698–1746), a Scotsman, was the most outstanding British mathematician in the generation following Newton. He was a prodigy, matriculating at the University of Glasgow at the age of eleven. He received his M.A. degree at 15. By age 19 he was a college mathematics teacher, and at 21 he published his first mathematical work of importance.

In 1719 Maclaurin met Isaac Newton, and he quickly became a disciple of Newtonian calculus. When Bishop George Berkeley wrote a tract attacking Newton's fluxions, Maclaurin responded in 1742 with his *Treatise on Fluxions,* the first complete and systematic presentation of Newton's calculus. Although it was not a textbook and was still not the final answer to rigor in calculus, it stood as a standard for nearly a century. Maclaurin was not comfortable with the limit concept, and based calculus on geometry instead. An unfortunate consequence of this emphasis was that his writing failed to make clear the useful applications of the subject. Further, the English mathematicians continued to use Newton's inadequate symbolism instead of Leibniz's superior symbolism, and during the 18th century British mathematics lost its preeminence.

Among Maclaurin's mathematical discoveries was what is now known as Cramer's rule for evaluating determinants. There is some poetic justice to this, for Maclaurin didn't discover Maclaurin's series just as Cramer didn't discover Cramer's rule. (In fairness, it should be noted that Cramer's notation was better than Maclaurin's.)

Joseph Fourier (1768–1830) was the only mathematician ever to serve as Governor of Lower Egypt. He had supported the French Revolution, and was rewarded with an appointment to the École Polytechnique. However, he had always wanted to be an army officer, a career denied him because he was the son of a tailor. When the opportunity came to accompany Napoleon on a military campaign in Egypt, Fourier resigned his teaching position and went along, and was appointed Governor in 1798. When the British took Egypt in 1801, Fourier returned to France.

Fourier studied the flow of heat in metallic plates and rods. The theory that he developed now has applications in industry and in the study of the temperature of the earth's interior. He discovered that many functions could be expressed as infinite sums of sine and cosine terms, now called a trigonometric series, or Fourier series. A paper that he submitted to the Academy of Science in Paris in 1807 was studied by several eminent mathematicians and rejected because he failed to prove his claims. They suggested that he reconsider and refine his paper, and even made heat flow the topic for a prize to be awarded in 1812. Fourier won the prize, but the Academy still declined to publish his paper because of its lack of rigor. (When Fourier became the secretary of the Academy, in 1824, the 1812 paper was published without change.)

As Fourier grew older, he developed at least one peculiar notion. Whether influenced by his stay in the heat of Egypt or by his own studies of the flow of heat in metals, he became obsessed with the idea that extreme heat was the natural condition for the human body. He was always heavily bundled in woolen clothing, and kept his rooms at high temperatures. He died in his sixty-third year, as expressed by Howard Eves in *An Introduction to the History of Mathematics,* ''thoroughly cooked.''

Most creative mathematicians have shown their genius at an early age. A notable exception was Karl Weierstrass (1816–1897), who did not really become a mathematician until he was 40 years old. He had studied law at the University of Bonn at his father's insistence, but he spent much of his time fencing and drinking. He did not complete his studies, and did not get his degree even after four years. Instead he turned to mathematics, but did not complete his degree in that subject, either. Eventually Weierstrass became a *gymnasium* (high school) teacher in a variety of subjects including not only mathematics and physics, but also German, history, botany, geography, gymnastics, and calligraphy. The mathematician in him was struggling to break through, and he did some mathematical research, although he had no contact with other mathematicians. He did manage to publish a few minor papers that attracted some attention. After thirteen years of secondary level teaching, Weierstrass obtained a position at a technical school, and was then called to the University of Berlin. Here his mathematical and pedagogical abilities surfaced, and he is considered the greatest teacher of higher mathematics of the nineteenth century, as judged by the number of his students who became significant researchers. Judged on his own mathematical creativity, he is sometimes called the leading analyst of his time, and the father of modern analysis.

Weierstrass worked in several areas, but his principal contribution was the study of functions of complex numbers through power series, which will be mentioned in this unit. This was an extension of the work of such earlier mathematicians as Maclaurin, Taylor, and Euler, but with a new rigor.

With all this creativity, Weierstrass did not publish many papers, and much of what he accomplished is known only through notes taken in class by his students. Indeed, he was uninterested in publicity or fame, and seemingly did not mind that some of his students used material from his lectures as their own. Hence, the record of the mathematical achievements of Karl Weierstrass has never been completely sorted out.

Chapter 12
The Theory of Infinite Series

In the preceding eleven chapters we have seen several instances involving a *sequence* of values, one corresponding to each of the positive integers $n = 1, 2, 3, \ldots$. For example,

(1) The definite integral $\int_a^b f(x)\, dx$ has been interpreted as the limit of a sequence S_n of Riemann Sums

$$S_n = \sum_{j=1}^{n} f(t_j)\, \Delta x_j, \qquad n = 1, 2, 3, \ldots.$$

(2) The amount $P(n)$ on deposit in a savings account paying interest annually at the rate r (in decimal form) n years after an initial deposit of P_0 dollars is

$$P(n) = (1 + r)^n P_0, \qquad n = 1, 2, 3, \ldots.$$

(3) Approximations to a zero of a differentiable function f are given by the formula for Newton's Method:

$$x_{n+1} = x_n - \frac{f(x_n)}{f'(x_n)}, \qquad n = 1, 2, 3, \ldots$$

where x_1 is the initial approximation.

The purpose of this unit is to develop a formal theory for dealing with expressions such as these which are defined only for integer values of the independent variable n. Our principal interest is in knowing how to calculate limits as $n \to \infty$ of such expressions and in applying this knowledge to answer practical questions.

Chapter 12 concerns infinite *sequences* and infinite *series of constants* (i.e., numbers). Chapter 13 applies the results of Chapter 12 to the important problems of approximating differentiable functions by polynomials and representing functions by infinite series.

12.1 INFINITE SEQUENCES

Up to this point we have rather casually defined an infinite sequence to be an unending string of numbers of the form

$$a_1, a_2, a_3, \ldots, a_n, \ldots \tag{1}$$

It is important to note, however, that expression (1) indicates an *order* in which these numbers appear in the string (the subscripts) as well as the numbers themselves (the a_n's). A more precise notion of an infinite sequence is therefore the following.

DEFINITION 1   An infinite sequence is a function whose domain is the positive integers.

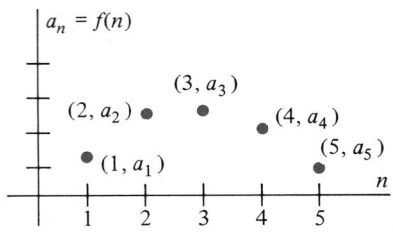

Figure 1.1   Plotting the sequence $\{a_1, a_2, a_3, \ldots\}$ where $a_n = f(n)$.

For example, the infinite sequence $\{1^2, 2^2, 3^2, 4^2, 5^2, \ldots, n^2, \ldots\}$ can be viewed as the set of values of the function $f(n) = n^2$, where $f(1) = 1^2$ is the first term, $f(2) = 2^2$ is the second term, and so on. Using the function concept, we can graph sequences on a coordinate plane. Figure 1.1 shows the graph of an arbitrary sequence $\{a_n\}$.

Usually, a sequence will be specified by a rule of the form $a_n = f(n)$ that determines the nth term of the sequence for each integer n, just as functions are usually specified by an equation of the form $y = f(n)$. For example, the rule $a_n = 2^n$ determines the sequence

$$\{2^n\} = \{2^1, 2^2, 2^3, 2^4, 2^5, \ldots\}$$
$$= \{2, 4, 8, 16, 32, \ldots\},$$

while the rule $a_n = (-1)^n$ determines the sequence

$$\{(-1)^n\} = \{-1, 1, -1, 1, -1, 1, -1, \ldots\}.$$

In such cases the term $a_n = f(n)$ is referred to as the **general term** of the sequence. Note that we use braces $\{\ \}$ to denote the entire sequence.

Example 1

Write out the first few terms of the sequences whose general terms are

(a) $a_n = 2n + 1$,

(b) $a_n = 2 + \dfrac{(-1)^n}{n}$.

Solution: In part (a) we have

$$a_1 = 2(1) + 1 = 3, \qquad a_2 = 2(2) + 1 = 5, \qquad a_3 = 2(3) + 1 = 7, \text{ and so on,}$$

so

$$\{2n + 1\} = \{3, 5, 7, 9, 11, 13, 15, \ldots\},$$

while in (b) we have

$$a_1 = 2 + \frac{(-1)}{1} = 1, \qquad a_2 = 2 + \frac{(-1)^2}{2} = \frac{5}{2},$$

$$a_3 = 2 + \frac{(-1)^3}{3} = \frac{5}{3}, \qquad \text{and so on,}$$

so

$$\left\{2 + \frac{(-1)^n}{n}\right\} = \left\{1, \frac{5}{2}, \frac{5}{3}, \frac{9}{4}, \frac{9}{5}, \frac{13}{6}, \frac{13}{7}, \frac{17}{8}, \frac{17}{9}, \ldots\right\}.$$

Graphs of these two sequences appear in Figures 1.2 and 1.3.   ◇

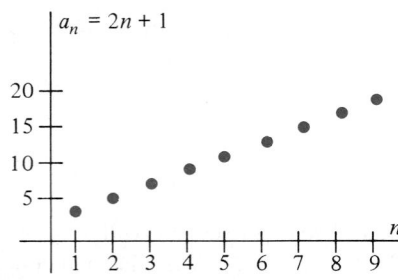

Figure 1.2   Graph of sequence $\{a_n\} = \{2n + 1\}$.

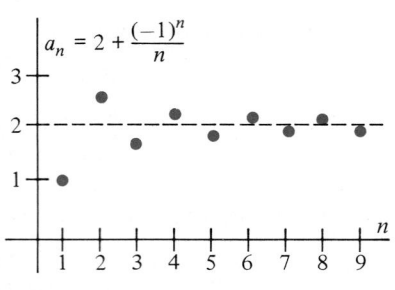

Figure 1.3   Graph of the sequence $\{a_n\} = \left\{2 + \dfrac{(-1)^n}{n}\right\}$.

There is an important difference between the sequence $\{2n + 1\}$ in Figure 1.2 and the sequence $\left\{2 + \dfrac{(-1)^n}{n}\right\}$ in Figure 1.3. In the first of these, the terms of the sequence increase uniformly, not approaching any particular number. In fact, in the language of Chapter 4, we would say that

$$\lim_{n \to \infty} \{2n + 1\} = +\infty$$

since the terms of the sequence increase without bound as $n \to \infty$.

However, the terms of the sequence $\left\{2 + \dfrac{(-1)^n}{n}\right\}$ "approach" the number $L = 2$ as $n \to \infty$, which we wish to write as

$$2 = \lim_{n \to \infty} \left\{2 + \frac{(-1)^n}{n}\right\}.$$

This notion of the *limit* of a sequence is just the notion of $\lim_{x \to \infty} f(x) = L$ developed for more general functions in Chapter 4. That is,

> $L = \lim_{n \to \infty} a_n$ means that the numbers a_n approach the number L as n increases without bound.

Using this working definition, together with some simple algebra, we can evaluate many types of limits by the same techniques used in Chapter 4 to evaluate $\lim_{x \to \infty} f(x)$.

Example 2

Find $\lim_{n \to \infty} \dfrac{6n^3 + 5n^2 + 7}{4n^3 - 2n + 2}$.

Strategy

Divide all terms by n^3 (the highest power of n in the denominator).

Use fact that

$$\frac{c}{n^k} \to 0 \quad \text{as} \quad n \to \infty$$

if $k > 0$.

Solution

$$\lim_{n \to \infty} \frac{6n^3 + 5n^2 + 7}{4n^3 - 2n + 2} = \lim_{n \to \infty} \frac{6 + \dfrac{5}{n} + \dfrac{7}{n^3}}{4 - \dfrac{2}{n^2} + \dfrac{2}{n^3}}$$

$$= \frac{6 + 0 + 0}{4 - 0 + 0} = \frac{3}{2}. \qquad \diamond$$

Example 3

Find $\lim_{n \to \infty} \left[\ln(n + 4) - \dfrac{1}{2}\ln(n)\right]$.

Strategy

Use properties of $\ln x$ to reduce expression to the logarithm of a single number.

Solution

$$\lim_{n \to \infty} [\ln(n + 4) - 1/2 \ln(n)] = \lim_{n \to \infty} [\ln(n + 4) - \ln(n^{1/2})]$$

$$= \lim_{n \to \infty} \ln\left(\frac{n + 4}{\sqrt{n}}\right)$$

Divide both terms in numerator by $\sqrt{n}$.

$$= \lim_{n \to \infty} \ln\left(\sqrt{n} + \frac{4}{\sqrt{n}}\right)$$

Use fact that $\ln x \to \infty$ as $x \to \infty$.

$$= \infty,$$

$$\text{since } \sqrt{n} \to \infty \text{ and } \frac{4}{\sqrt{n}} \to 0 \text{ as } n \to \infty. \qquad \diamond$$

Example 4

Find $\lim_{n \to \infty} (-1)^n\left(\frac{n+1}{n}\right)$.

Strategy

Because of the factor $(-1)^n$, examine even and odd terms separately.

Divide through by n.

Divide through by n.

Solution

For even integers n, $(-1)^n = 1$, so we find that

$$\lim_{n \to \infty} (-1)^n\left(\frac{n+1}{n}\right) \qquad \text{(even integers only)}$$

$$= \lim_{n \to \infty} (1)\left(1 + \frac{1}{n}\right) = 1.$$

However, for odd integers n, $(-1)^n = -1$, so

$$\lim_{n \to \infty} (-1)^n\left(\frac{n+1}{n}\right) \qquad \text{(odd integers only)}$$

$$= \lim_{n \to \infty} (-1)\left(1 + \frac{1}{n}\right) = -1.$$

Since this shows that the terms do not approach a *single* number as $n \to \infty$, this limit does not exist (see Figure 1.4). $\qquad \diamond$

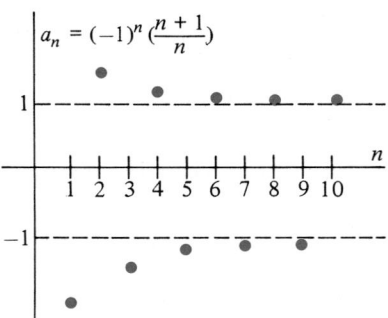

Figure 1.4 Terms of sequence $(-1)^n\left(\frac{n+1}{n}\right)$ approach both 1 and -1 as $n \to \infty$. The limit does not exist.

If $L = \lim_{n \to \infty} a_n$ exists, we say that the sequence $\{a_n\}$ **converges**. Otherwise the sequence is said to **diverge**. Note from Examples 3 and 4 that a sequence may diverge either because a_n becomes infinite as $n \to \infty$ or because a_n, remaining bounded, fails to approach a *single* number as $n \to \infty$.

We next state a formal definition of the limit of a sequence. Note the similarity with the formal definition for $L = \lim_{x \to \infty} f(x)$ given in Chapter 4.

DEFINITION 2
Formal Definition of Limit

We say that $L = \lim_{n \to \infty} a_n$ if and only if for each number $\epsilon > 0$ there exists an integer N so that

$$\text{if} \quad n > N, \quad \text{then} \quad |a_n - L| < \epsilon.$$

Definition 2 says this: If $L = \lim_{n \to \infty} a_n$, we will find all terms of the sequence $\{a_n\}$, beyond the Nth term, lying within ϵ units of the number L. Since the integer N, in general, depends upon the number ϵ, we expect to have to look further along the sequence to observe this "closeness" as ϵ decreases in size. Figures 1.5 through 1.7 illustrate choices of N corresponding to three different values of ϵ for a typical sequence $\{a_n\}$.

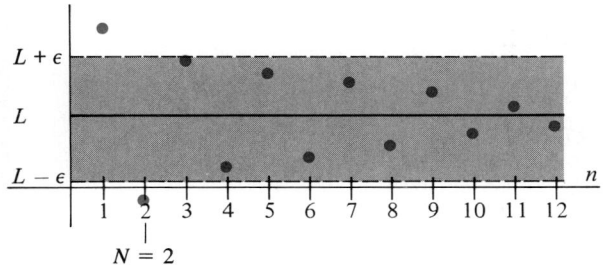

Figure 1.5 Large ϵ.

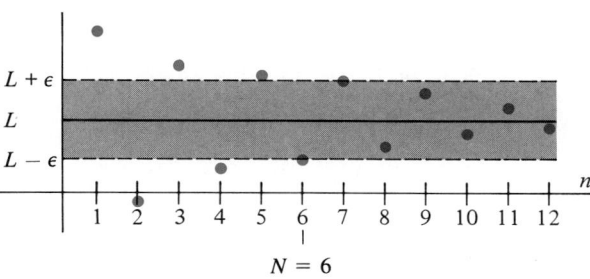

Figure 1.6 Medium ϵ.

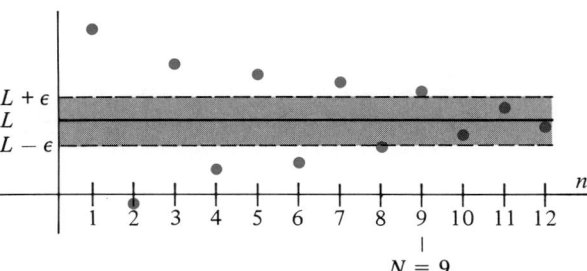

Figure 1.7 Small ϵ.

Definition 2 allows us to rigorously prove statements of the form $L = \lim_{n \to \infty} a_n$ once we have found L. It does not, however, tell us how L is determined from the general term for the sequence $\{a_n\}$. For this, the intuitive notion of limit and familiarity with examples such as Examples 2–4 above are essential.

The following examples illustrate how Definition 2 is used.

Example 5

Prove that $\lim_{n \to \infty} \dfrac{1}{n} = 0$.

Strategy

Set up the inequality

$$|a_n - L| < \epsilon$$

and solve for n to find a relationship between n and ϵ.

Solution

According to Definition 2, we allow $\epsilon > 0$ to be any fixed positive number. Since $a_n = 1/n$ and $L = 0$, we must determine how large to choose n to guarantee that

$$\left| \frac{1}{n} - 0 \right| = \frac{1}{n} < \epsilon. \tag{2}$$

Solving inequality (2) for n we see that it is equivalent to the inequality

$$n > \frac{1}{\epsilon}. \tag{3}$$

Choose N large enough that the inequality holds whenever $n > N$.

We therefore take N to be any integer larger* than $1/\epsilon$. Then, inequality (3) holds, whenever $n > N$, which guarantees that inequality (2) holds. In other words, if $N > 1/\epsilon$ then

$$\left| \frac{1}{n} - 0 \right| < \epsilon \qquad \text{whenever} \qquad n > N,$$

as required by Definition 2. ◇

Example 6

Prove that $\displaystyle \lim_{n \to \infty} \frac{2n - 1}{n + 2} = 2$.

Strategy

Set up the inequality

$$|a_n - L| < \epsilon.$$

Solution

We assume $\epsilon > 0$ to be an arbitrary fixed number. Since $a_n = \dfrac{2n - 1}{n + 2}$ and $L = 2$, we must determine how large to choose n to guarantee that

$$\left| \frac{2n - 1}{n + 2} - 2 \right| < \epsilon. \tag{4}$$

Solve this inequality for n.

Inequality (4) is equivalent to the inequality

$$\left| \frac{2n - 1 - 2(n + 2)}{n + 2} \right| < \epsilon,$$

or

$$\frac{5}{n + 2} < \epsilon$$

$$\Leftrightarrow 5 < \epsilon(n + 2)$$

$$\Leftrightarrow \frac{5}{\epsilon} < n + 2$$

$$\Leftrightarrow n > \frac{5}{\epsilon} - 2.$$

$$\frac{5}{n + 2} < \epsilon. \tag{5}$$

Solving inequality (5) for n we find that it is equivalent to the inequality

$$n > \frac{5}{\epsilon} - 2. \tag{6}$$

Choose n sufficiently large that the desired inequality holds.

We therefore take N to be any integer larger than $\dfrac{5}{\epsilon} - 2$. Then, inequality (6)

---

*Since ϵ is a positive number, so is $1/\epsilon$. It is a property of the real numbers that, given any positive number a, an integer N can be found with $N > a$.

holds whenever $n > N$. Since inequality (6) is equivalent to inequality (4), this shows that, for $N > \dfrac{5}{\epsilon} - 2$, we may conclude that

$$\text{if} \quad n > N \qquad \text{then} \qquad \left| \frac{2n - 1}{n + 2} - 2 \right| < \epsilon,$$

as required by Definition 2. ◇

Definition 2 can be used to prove several theorems giving rules by which limits of sequences may be calculated. Since the proofs of these theorems are similar to the proofs given in Chapters 2 and 4 for the corresponding theorems on limits of functions, we leave them as exercises.

THEOREM 1
Properties of Limits of Sequences

If $\lim\limits_{n \to \infty} a_n = L$, $\lim\limits_{n \to \infty} b_n = M$ and c is any real number, then

(i) $\lim\limits_{n \to \infty} (a_n + b_n) = L + M$,

(ii) $\lim\limits_{n \to \infty} (ca_n) = cL$,

(iii) $\lim\limits_{n \to \infty} (a_n b_n) = LM$,

(iv) $\lim\limits_{n \to \infty} \left(\dfrac{a_n}{b_n} \right) = \dfrac{L}{M}$, $\qquad b_n \neq 0, \qquad M \neq 0.$

Theorem 1 together with the proof in Example 5 makes legitimate the calculations in Example 2. The next theorem addresses a situation that occurred in Example 3.

THEOREM 2

Suppose that $\lim\limits_{n \to \infty} a_n = L$ and each number a_n lies in the domain of the function f. If f is continuous at $x = L$ then

$$\lim_{n \to \infty} f(a_n) = f(L).$$

Example 7

Since $f(x) = \tan x$ is continuous for $-\pi/2 < x < \pi/2$,

$$\lim_{n \to \infty} \tan\left(\frac{\pi n^2 + 1}{3 - 4n^2} \right) = \tan\left[\lim_{n \to \infty} \left(\frac{\pi n^2 + 1}{3 - 4n^2} \right) \right] = \tan\left(-\frac{\pi}{4} \right) = -1. \qquad ◇$$

Example 8

Since $f(x) = \sqrt{x}$ is continuous for $x \geq 0$,

$$\lim_{n \to \infty} \sqrt{\frac{4n + 1}{n}} = \left(\lim_{n \to \infty} \frac{4n + 1}{n} \right)^{1/2} = \sqrt{4} = 2. \qquad ◇$$

The following is the analogue for sequences of the Pinching Theorem for functions.

THEOREM 3
Pinching Theorem

Let $\{a_n\}$, $\{b_n\}$, and $\{c_n\}$ be sequences and let P be a positive integer. Let $a_n \leq b_n \leq c_n$ for all integers $n \geq P$. If

$$\lim_{n \to \infty} a_n = L = \lim_{n \to \infty} c_n,$$

then $\lim_{n \to \infty} b_n = L$ also.

Example 9

Show that $\lim_{n \to \infty} \dfrac{1}{n^p} = 0$ if $p \geq 1$.

Solution: If $p \geq 1$, $n^p \geq n$ for all $n = 1, 2, 3, \ldots$. Thus

$$0 \leq \frac{1}{n^p} \leq \frac{1}{n}, \qquad n \geq 1.$$

Since $\lim_{n \to \infty} \{0\} = 0 = \lim_{n \to \infty} \left\{\dfrac{1}{n}\right\}$, the conclusion follows by the Pinching Theorem. ◇

Example 10

Find $\lim_{n \to \infty} \dfrac{\sin n}{n}$.

Strategy
Find bounds on $\sin n$.

Solution
We have $|\sin n| \leq 1$ for all n, that is,

$$-1 \leq \sin n \leq 1, \qquad n \geq 1.$$

Thus

Divide by n to find bounds on $\dfrac{\sin n}{n}$.

$$-\frac{1}{n} \leq \frac{\sin n}{n} \leq \frac{1}{n}.$$

Apply Pinching Theorem.

Since $\lim_{n \to \infty} \left(-\dfrac{1}{n}\right) = 0 = \lim_{n \to \infty} \left(\dfrac{1}{n}\right)$,

$$\lim_{n \to \infty} \frac{\sin n}{n} = 0$$

by the Pinching Theorem. ◇

We summarize the ideas of this section by noting that finding the limit of sequence $\{a_n\}$ is similar to finding horizontal asymptotes for the function f with $f(n) = a_n$, $n = 1, 2, \ldots$. The principal difference is that $\{a_n\}$ is a function defined only for positive integers. Thus $L = \lim_{n \to \infty} a_n$ if $L = \lim_{x \to \infty} f(x)$, but the converse need not be true (see Exercise 35).

Exercise Set 12.1

In each of Exercises 1–28, write out the first four terms of the given sequence and determine whether the sequence converges or diverges. If the sequence converges, find its limit.

1. $\left\{\dfrac{n}{2n+1}\right\}$

2. $\left\{\dfrac{2n-1}{n+3}\right\}$

3. $\left\{\dfrac{n-4}{n^2+2}\right\}$

4. $\left\{\dfrac{n^2+1}{3n(n+2)}\right\}$

5. $\left\{\dfrac{1}{1+n^2}\right\}$

6. $\left\{\dfrac{1}{e^n}\right\}$

7. $\{\sqrt{5}\}$

8. $\left\{\dfrac{(n-1)(n+1)}{2n^2+2n+2}\right\}$

9. $\left\{\dfrac{20n}{1+\sqrt{n}}\right\}$

10. $\left\{\dfrac{6-n^{3/2}}{(\sqrt{n}+1)^2}\right\}$

11. $\left\{\dfrac{3+(-1)^n\sqrt{n}}{n+2}\right\}$

12. $\{(-1)^n\sin n\}$

13. $\left\{\sqrt{1+\dfrac{1}{n}}\right\}$

14. $\left\{1+\dfrac{(-1)^n}{2^n}\right\}$

15. $\left\{\cos\left(\dfrac{n-1}{n^2}\right)\right\}$

16. $\left\{\dfrac{n+1}{n}\right\}$

17. $\left\{\dfrac{n^{3/2}+2}{2n^{3/2}}\right\}$

18. $\left\{\dfrac{e^n-e^{-n}}{e^n+e^{-n}}\right\}$

19. $\left\{\dfrac{1}{n}-\dfrac{1}{n+1}\right\}$

20. $\left\{\dfrac{2^n}{5^{n+2}}\right\}$

21. $\{\sqrt{n+1}-\sqrt{n}\}$

22. $\left\{\dfrac{\cos^2 n}{n}\right\}$

23. $\left\{\dfrac{\sqrt{2n^2+1}}{n}\right\}$

24. $\left\{\tan^{-1}\left(\dfrac{n+2}{2}\right)\right\}$

25. $\left\{n\cdot\sin\dfrac{\pi}{2n}\right\}$

26. $\left\{\ln\dfrac{n^2+1}{(n+2)(n+3)}\right\}$

27. $\left\{\left(1+\dfrac{1}{n}\right)^n\right\}$

28. $\left\{\left(1-\dfrac{1}{n}\right)^n\right\}$

In Exercises 29–33, use the definition of limit to prove that $\lim\limits_{n\to\infty} a_n = L$.

29. $a_n = \dfrac{3}{n}$; $\quad L = 0$

30. $a_n = \dfrac{1}{2n+1}$; $\quad L = 0$

31. $a_n = \dfrac{n}{3n+1}$; $\quad L = \dfrac{1}{3}$

32. $a_n = \dfrac{3n-1}{n+1}$; $\quad L = 3$

33. $a_n = \dfrac{n^2+2n+3}{1+n^2}$; $\quad L = 1$

34. Use Theorem 2 to show that $\lim\limits_{n\to\infty}\sqrt[n]{a} = \lim\limits_{n\to\infty} e^{\ln a^{1/n}} = 1$ if $a > 0$.

35. Let $a_n = \sin \pi n$, $n = 1, 2, \ldots$, and let $f(x) = \sin \pi x$. Then $f(n) = a_n$. Show that $\lim\limits_{n\to\infty} a_n = 0$ but that $\lim\limits_{x\to\infty} f(x)$ does not exist.

36. Prove that if $\lim\limits_{n\to\infty} a_n$ exists, this limit is unique. (*Hint:* Assume that both $\lim\limits_{n\to\infty} a_n = L$ and $\lim\limits_{n\to\infty} a_n = M$. Then show $L = M$ by examining the inequality $|L - M| \le |L - a_n| + |a_n - M|$.)

37. A sequence is called **bounded** if there is a number M so that $|a_n| \le M$ for all terms a_n of the sequence. Prove that a convergent sequence must be bounded.

38. Give an example showing that a bounded sequence need *not* converge.

39. Prove that the sum of two bounded sequences is again bounded. What about the product of two bounded sequences? The quotient?

40. Prove that $\lim\limits_{n\to\infty} a_n = L$ if and only if $\lim\limits_{n\to\infty} |a_n - L| = 0$.

41. Prove that $\lim\limits_{n\to\infty} \dfrac{1}{a^n} = 0$ if $a > 1$, using the Pinching Theorem.

42. Prove part (i) of Theorem 1.

43. Prove part (ii) of Theorem 1.

44. Prove part (iii) of Theorem 1.

45. Prove part (iv) of Theorem 1.

46. Prove Theorem 2.

47. Prove Theorem 3.

12.2 MORE ON INFINITE SEQUENCES

If $\{a_n\}$ is a sequence and f is a function for which $f(n) = a_n$, $n = 1, 2, \ldots$, then we can conclude that $\lim_{n \to \infty} a_n = L$ whenever $\lim_{x \to \infty} f(x) = L$. (See Figure 2.1.)

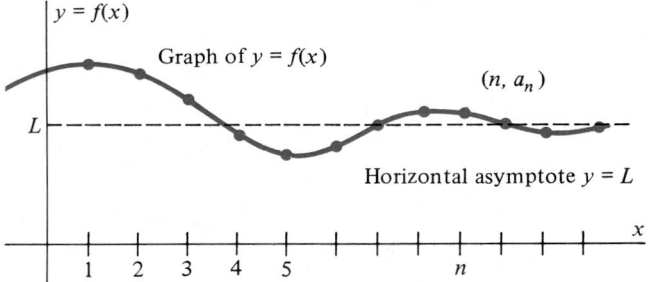

Figure 2.1 If $f(n) = a_n$ and $\lim_{x \to \infty} f(x) = L$, then $\lim_{n \to \infty} a_n = L$ also.

This observation allows us to reformulate the problem of finding $\lim_{n \to \infty} a_n$ as the problem of finding $\lim_{x \to \infty} f(x)$. In the latter problem, l'Hôpital's Rule may often be successfully applied to calculate an otherwise ambiguous limit.

Example 1

Find $\lim_{n \to \infty} \dfrac{n}{e^n}$.

Solution: The limit is not obvious because both numerator and denominator become infinite as $n \to \infty$. Using the above notion together with l'Hôpital's Rule gives

$$\lim_{n \to \infty} \frac{n}{e^n} = \lim_{x \to \infty} \frac{x}{e^x} \qquad \left(\frac{\infty}{\infty} \text{ form}\right)$$

$$= \lim_{x \to \infty} \frac{\dfrac{d}{dx}(x)}{\dfrac{d}{dx}(e^x)} \qquad \text{(l'Hôpital's Rule)}$$

$$= \lim_{x \to \infty} \frac{1}{e^x}$$

$$= 0. \qquad\qquad\qquad\qquad\qquad \diamond$$

We next establish several limits that we shall use frequently.

$$\lim_{n \to \infty} x^n = 0 \qquad \text{if} \qquad |x| < 1. \tag{1}$$

Proof: Let ϵ be an arbitrary positive number. Then, since

$$\lim_{n\to\infty} \frac{1}{n} = 0,$$

$\lim_{n\to\infty} \epsilon^{1/n} = \epsilon^0 = 1$. Thus, since $|x| < 1$, there exists an integer N, by Definition 2, such that $\epsilon^{1/n} > |x|$ whenever $n > N$. Thus

$$|x|^n < (\epsilon^{1/n})^n = \epsilon, \qquad n > N.$$

Equivalently, noting that $|x|^n = |x^n|$, we have

$$|x^n - 0| < \epsilon \qquad \text{whenever} \qquad n > N.$$

Thus,

$$\lim_{n\to\infty} x^n = 0 \qquad \text{by Definition 2.} \qquad \blacklozenge$$

$$\lim_{n\to\infty} \frac{x^n}{n!} = 0, \qquad -\infty < x < \infty. \tag{2}$$

Proof: To establish this limit we shall show that

$$\lim_{n\to\infty} \left|\frac{x^n}{n!}\right| = 0, \qquad -\infty < x < \infty. \tag{3}$$

(See Exercise 28.)

Let x be given and let N be an integer such that $N > |x|$. Write

$$J = \left|\left(\frac{x}{1}\right)\left(\frac{x}{2}\right)\left(\frac{x}{3}\right)\cdots\left(\frac{x}{N-1}\right)\right|$$

and note that J is constant since N and x are fixed. Then for $n > N$ we have

$$\left|\frac{x^n}{n!}\right| = \left|\left(\frac{x}{1}\right)\left(\frac{x}{2}\right)\left(\frac{x}{3}\right)\cdots\left(\frac{x}{N-1}\right)\left(\frac{x}{N}\right)\cdots\left(\frac{x}{n}\right)\right|$$

$$= J\left|\left(\frac{x}{N}\right)\left(\frac{x}{N+1}\right)\cdots\left(\frac{x}{n}\right)\right| \qquad (n-N+1) \text{ factors}$$

$$\leq J \cdot \left|\frac{x}{N}\right|^{(n-N+1)}.$$

Since $|x| < N$, we have $\left|\frac{x}{N}\right| < 1$, so $\lim_{n\to\infty} \left|\frac{x}{N}\right|^n = 0$. Thus,

$$\lim_{n\to\infty} J \cdot \left|\frac{x}{N}\right|^{(n-N+1)} = J \cdot \left|\frac{x}{N}\right|^{(-N+1)} \cdot \lim_{n\to\infty} \left|\frac{x}{N}\right|^n = 0.$$

This establishes (3) by the preceding inequality. $\qquad \blacklozenge$

$$\lim_{n\to\infty} \frac{\ln(n)}{n} = 0. \tag{4}$$

Proof: $\lim\limits_{n \to \infty} \dfrac{\ln(n)}{n} = \lim\limits_{x \to \infty} \dfrac{\ln x}{x} = \lim\limits_{x \to \infty} \dfrac{\frac{1}{x}}{1} = 0$, by l'Hôpital's Rule. ◆

$$\lim_{n \to \infty} \sqrt[n]{n} = 1.$$ (5)

Proof: $\lim\limits_{n \to \infty} \ln(n^{1/n}) = \lim\limits_{n \to \infty} \dfrac{\ln(n)}{n} = 0$, by (4).

Thus,

$$\lim_{n \to \infty} n^{1/n} = e^0 = 1.$$ ◆

Example 2

$$\lim_{n \to \infty} \left(\frac{5}{n}\right)^{1/n} = \lim_{n \to \infty} \frac{\sqrt[n]{5}}{\sqrt[n]{n}}$$

$$= \frac{\lim\limits_{n \to \infty} \sqrt[n]{5}}{\lim\limits_{n \to \infty} \sqrt[n]{n}}$$

$$= \frac{1}{1} = 1$$

by statement (5). ◇

We shall have need of one further theorem on sequences, which we develop here.

A set S of numbers is said to be *bounded* if there exists a number M so that $|x| \le M$ for every number $x \in S$. This number M is called an *upper bound* for S. A fundamental property of the real number system states that among all such upper bounds M there can always be found a smallest bound. This property is referred to as the **completeness axiom,** and it is formally stated as follows.

Completeness Axiom for Real Numbers: If S is any nonempty bounded set of real numbers there exists a least upper bound L for S. That is, there exists a number L for which

(i) $x \le L$ for every $x \in S$, and
(ii) if M is any upper bound for S, then $L \le M$.

For example, the set $S = \{x \mid -3 \le x \le 3\}$ is bounded since $|x| \le 3$ for any $x \in S$. The number 7 is an *upper bound* for S since $x < 7$ for all $x \in S$. The numbers 10, 37, and 1001 are also upper bounds for S. The number $L = 3$ is the *least* upper bound for S, because any *other* upper bound for S is greater than 3.

The completeness axiom and its role in the definition of the real number system are topics for more advanced courses. We shall use this axiom to establish a theorem about *increasing* sequences. As for functions, we define the sequence $\{a_n\}$ to be **increasing** if $a_n > a_m$ whenever $n > m$.

THEOREM 4   Every bounded increasing sequence converges.

Proof: We use the definition of the limit of a sequence and the completeness axiom. Let $\epsilon > 0$ be given, and let $\{a_n\}$ denote a bounded increasing sequence. Then the set of numbers $\{a_1, a_2, a_3, \ldots\}$ is bounded, so it has a least upper bound L according to the completeness axiom. We will complete the proof by showing that $\lim\limits_{n \to \infty} a_n = L$.

Since $\epsilon > 0$, we must have $L - \epsilon < L$, so $L - \epsilon$ cannot be an upper bound for the sequence. Thus, there must exist an integer N such that

$$a_N > L - \epsilon. \qquad \text{(Figure 2.2.)}$$

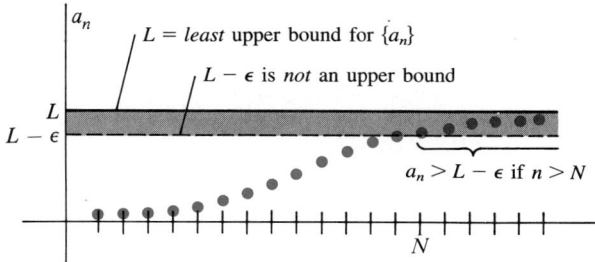

Figure 2.2 A bounded increasing sequence converges to its least upper bound.

But since $\{a_n\}$ is increasing we must have $a_n > a_N$ for all $n > N$. Thus we have the following five numbers in increasing order:

$$L - \epsilon < a_N < a_n \leq L < L + \epsilon, \qquad n > N.$$

It now follows from $L - \epsilon < a_n < L + \epsilon$ that

$$|a_n - L| < \epsilon \qquad \text{whenever} \qquad n > N.$$

Thus, $L = \lim\limits_{n \to \infty} a_n$ according to Definition 2. ◆

There is an obvious extension of Theorem 4 to bounded *decreasing* sequences. If $\{a_n\}$ is a bounded decreasing sequence, then the sequence $\{-a_n\}$ is a bounded *increasing* sequence, which, by Theorem 4, has a limit, say L. By Theorem 1 we then have

$$\lim_{n \to \infty} a_n = -\lim_{n \to \infty} \{-a_n\} = -L.$$

We summarize these remarks as follows.

COROLLARY 1   Every bounded decreasing sequence converges.

Recursively Defined Sequences

In many applications, particularly in computer science, sequences are defined *recursively* rather than as functions of the integer n. One of the most famous recursively defined sequences is the sequence of **Fibonacci numbers**

$$1, 1, 2, 3, 5, 8, 13, 21, 34, \ldots . \tag{6}$$

This sequence was first discovered by the Italian mathematician Leonardo of Pisa (who also went by the name Fibonacci) around the year 1200 A.D. (Fibonacci is regarded by many as the most brilliant of the pre-Renaissance mathematicians). The Fibonacci sequence $\{F_n\}$ in (6) is determined by the rules

$$F_1 = 1, \tag{7}$$

$$F_2 = 1, \quad \text{and} \tag{8}$$

$$F_{n+2} = F_n + F_{n+1}, \quad n = 1, 2, 3, 4, \ldots . \tag{9}$$

That is, every term in the sequence beyond the second is found by adding the two preceding terms. This is what we mean by saying that the sequence $\{a_n\}$ is **recursively defined**: the term a_n is a function of one or more preceding terms, such as a_{n-1}, a_{n-2}, etc. But a_n is *not* written as an explicit function of n.

It is often possible to rewrite a recursively defined sequence as an explicit function of n, and vice versa. For example, it has been shown that the nth term of the Fibonacci sequence can be written

$$F_n = \frac{1}{\sqrt{5}}\left(\frac{1 + \sqrt{5}}{2}\right)^{n+1} - \frac{1}{\sqrt{5}}\left(\frac{1 - \sqrt{5}}{2}\right)^{n+1}.$$

(See Exercise 20 for a biological interpretation of the Fibonacci sequence.) Another such example is the factorial sequence

$$\{a_n\} = \{n!\} = \{1!, 2!, 3!, \ldots\} = \{1, 2, 6, 24, \ldots\}.$$

It can be defined recursively as the sequence $\{f_n\} = \{f_1, f_2, f_3, \ldots\}$ where

$$f_1 = 1$$
$$f_n = nf_{n-1}, \quad n = 2, 3, 4, \ldots .$$

However, not every sequence $\{a_n\}$ can be defined recursively.

Recursively defined sequences occur frequently in the analysis of computer algorithms. In such situations one is concerned only with how to determine the next term in the sequence given the present term (and, possibly, several preceding terms), not with the correspondence between integers n and the terms a_n. Exercises 20–24 in this section concern recursively defined sequences.

Exercise Set 12.2

In each of Exercises 1–18, find the indicated limit, if it exists.

1. $\lim\limits_{n \to \infty} n \sin\left(\dfrac{2}{n}\right)$

2. $\lim\limits_{n \to \infty} \dfrac{\sin^3 n}{n}$

3. $\lim\limits_{n \to \infty} \sqrt[n]{4n}$

4. $\lim\limits_{n \to \infty} \dfrac{\ln(n)}{\sqrt{n}}$

5. $\lim\limits_{n \to \infty} (n + 1)e^{-n}$

6. $\lim\limits_{n \to \infty} \dfrac{n}{e^n}$

7. $\lim\limits_{n \to \infty} \dfrac{n^n}{n!}$

8. $\lim\limits_{n \to \infty} n^{3/n}$

9. $\lim\limits_{n \to \infty} (n + \pi)^{1/n}$

10. $\lim\limits_{n \to \infty} \left(1 - \dfrac{3}{n}\right)^n$

11. $\lim\limits_{n \to \infty} \dfrac{3^n}{(n + 3)!}$

12. $\lim\limits_{n \to \infty} \left(\dfrac{e}{n} \ln \dfrac{e}{n}\right)$

13. $\lim\limits_{n \to \infty} \dfrac{n^2 \ln(n)}{2^n}$

14. $\lim\limits_{n \to \infty} n^{\sin(\pi/n)}$

15. $\lim\limits_{n \to \infty} \left(\dfrac{n + 3}{n}\right)^n$

16. $\lim\limits_{n \to \infty} \left(1 + \dfrac{1}{n^2}\right)^n$

17. $\lim\limits_{n \to \infty} \sqrt[n]{n^3}$

18. $\lim\limits_{n \to \infty} \dfrac{n - \sin n}{n + \cos n}$

19. Prove that $\lim\limits_{n\to\infty} x^n$ does not exist if $|x| > 1$.

20. The Fibonacci sequence $\{1, 1, 2, 3, 5, 8, 13, 21, 34, \ldots\}$ arises as a mathematical model for the size of a population of rabbits under the following conditions. We assume that we begin with a single pair of rabbits, that this and each other pair of rabbits become fertile one month after birth, that each pair of fertile rabbits gives birth to one new pair of rabbits each month, and that no rabbits die. If F_n represents the number of pairs of rabbits in the population after n months, show that
   a. $F_1 = 1$,
   b. $F_2 = 1$, and
   c. $F_{n+2} = 2F_n + (F_{n+1} - F_n)$. (*Hint:* The term $2F_n$ is explained as follows. Every pair of rabbits that was present two months ago is still present along with one pair of offspring. The second term accounts for the fact that those rabbits which were fertile two months ago have produced *two* pairs of offspring since then, one of which is not counted in the first term.)
   d. Conclude from (c) that $F_{n+2} = F_n + F_{n+1}$.

21. Let $\{a_n\}$ be a sequence recursively defined by the equations

    $a_0 = 1$

    $a_n = 2a_{n-1}, \qquad n = 1, 2, \ldots .$

    Show that the general term for this sequence is $a_n = 2^n$.

22. Find the general term for the sequence $\{a_n\}$ recursively defined by the equations

    $a_0 = 4$

    $a_n = a_{n-1} + 1, \qquad n = 1, 2, \ldots .$

23. Find the general term for the sequence $\{a_n\}$ recursively defined by the equations

    $a_0 = -5$

    $a_n = a_{n-1} + 2, \qquad n = 1, 2, \ldots .$

24. For the Fibonacci sequence (Exercise 20) show that
    a. $F_{n+3} = 2F_{n+1} + F_n$
    b. $F_{n+4} = 3F_{n+1} + 2F_n$
    c. $F_{n+p} = F_p F_{n+1} + F_{p-1} F_n, \qquad p = 3, 4, \ldots .$

25. Give an example of a bounded sequence that does not converge.

26. Give an example of an increasing sequence that does not converge.

27. Must every convergent sequence be bounded and either increasing or decreasing?

28. Prove that if $\lim\limits_{n\to\infty} |a_n| = 0$, then $\lim\limits_{n\to\infty} a_n = 0$.

12.3 INFINITE SERIES

By an *infinite series* we mean an expression of the form

$$\sum_{k=1}^{\infty} a_k = a_1 + a_2 + a_3 + \cdots . \tag{1}$$

Unlike the situation for finite sums, we cannot associate a "sum" with an infinite series simply by "adding up" the terms $a_1, a_2, \ldots$ because this would require that we perform an infinite number of additions, something not even a high-speed computer can accomplish in a finite amount of time.

The method for evaluating improper integrals of the form $\int_a^\infty f(x)\, dx$ provides the idea by which we shall determine whether an infinite series has a sum—that is, we first find the *partial sums* $S_n = \sum_{k=1}^{n} a_k$ of the series in line (1) and then ask whether the limit $\lim\limits_{n\to\infty} S_n = \lim\limits_{n\to\infty} \sum_{k=1}^{n} a_k$ of these partial sums exists. If it does, this limit is what we shall call the sum of the infinite series.

DEFINITION 3

An **infinite series** is an expression of the form

$$\sum_{k=1}^{\infty} a_k = a_1 + a_2 + a_3 + \cdots.$$

The infinite series

$$\sum_{k=1}^{\infty} a_k$$

is said to **converge** to the **sum** S if

$$S = \lim_{n \to \infty} S_n,$$

where S_n denotes the nth **partial sum**

$$S_n = a_1 + a_2 + a_3 + \cdots + a_n = \sum_{k=1}^{n} a_k.$$

If the limit S does not exist, the series $\sum_{k=1}^{\infty} a_k$ is said to **diverge.**

Example 1

The repeating decimal 0.66666 may be interpreted as the infinite series

$$0.666\overline{66} = .6 + (.06) + (.006) + (.0006) + \cdots$$

$$= \frac{6}{10} + \frac{6}{10^2} + \frac{6}{10^3} + \frac{6}{10^4} + \cdots$$

$$= \sum_{k=1}^{\infty} \frac{6}{10^k}.$$

Let us verify that this interpretation is consistent with the ordinary notion that $0.6\overline{66} = 2/3$. The nth partial sum of this series is

$$S_n = \frac{6}{10} + \frac{6}{10^2} + \frac{6}{10^3} + \cdots + \frac{6}{10^n}. \tag{2}$$

Now note that each term of the sum is 10 times the following term. Multiplying both sides of equation (2) by $\frac{1}{10}$ gives

$$\frac{1}{10}S_n = \frac{6}{10^2} + \frac{6}{10^3} + \frac{6}{10^4} + \cdots + \frac{6}{10^{n+1}}. \tag{3}$$

Subtracting corresponding sides of equation (3) from those of equation (2) gives

$$S_n - \frac{1}{10}S_n = \frac{6}{10} - \frac{6}{10^{n+1}},$$

so

$$S_n = \frac{10}{9}\left(\frac{6}{10} - \frac{6}{10^{n+1}}\right) = \frac{2}{3}\left(1 - \frac{1}{10^n}\right).$$

According to the definition of an infinite series, the sum of the series is therefore

$$S = \lim_{n\to\infty} S_n = \lim_{n\to\infty} \frac{2}{3}\left(1 - \frac{1}{10^n}\right) = 2/3.$$

This shows that

$$\frac{2}{3} = \sum_{k=1}^{\infty} \frac{6}{10^k} = .66666\overline{6}.$$

Example 2

The infinite series

$$\sum_{k=1}^{\infty} (-1)^k = -1 + 1 - 1 + 1 - 1 + \cdots$$

does not converge. To see why, observe that the partial sums are

$$S_1 = -1$$
$$S_2 = -1 + 1 = 0$$
$$S_3 = -1 + 1 - 1 = -1$$
$$S_4 = -1 + 1 - 1 + 1 = 0$$
$$\cdot$$
$$\cdot$$
$$\cdot$$
$$S_{2n-1} = -1 + 1 - 1 + \cdots + 1 - 1 = -1 \qquad (2n - 1 \text{ terms})$$
$$S_{2n} = -1 + 1 - 1 + \cdots + 1 - 1 + 1 = 0 \qquad (2n \text{ terms}).$$

Thus, the terms of the sequence $\{S_n\}$ are alternately -1 or 0, so $\lim_{n\to\infty} S_n$ does not exist.

Example 3

Determine whether the infinite series

$$\sum_{k=1}^{\infty} \frac{1}{k(k + 1)}$$

converges. If it does, find its sum.

Solution: By the method of partial fractions we can show that

$$\frac{1}{k(k + 1)} = \frac{1}{k} - \frac{1}{k + 1}, \qquad k = 1, 2, 3, \ldots .$$

We can therefore write the partial sum S_n for this series as

$$S_n = \sum_{k=1}^{n} \frac{1}{k(k+1)} = \frac{1}{1 \cdot 2} + \frac{1}{2 \cdot 3} + \frac{1}{3 \cdot 4} + \cdots + \frac{1}{(n-1)n} + \frac{1}{n(n+1)}$$

$$= \left[\frac{1}{1} - \frac{1}{2} \right] + \left[\frac{1}{2} - \frac{1}{3} \right] + \left[\frac{1}{3} - \frac{1}{4} \right]$$

$$+ \cdots + \left[\frac{1}{n-1} - \frac{1}{n} \right] + \left[\frac{1}{n} - \frac{1}{n+1} \right]$$

$$= 1 - \frac{1}{n+1}$$

since all other terms in this "telescoping sum" cancel. This shows that

$$\sum_{k=1}^{\infty} \frac{1}{k(k+1)} = \lim_{n \to \infty} S_n = \lim_{n \to \infty} \left(1 - \frac{1}{n+1} \right) = 1.$$

Thus the series converges, and its sum is $S = 1$. $\diamondsuit$

REMARK: Up to this point we have been indexing all infinite series so that the first term has index $k = 1$. This is not necessary. For example, we could rewrite

$$\sum_{k=1}^{\infty} a_k \qquad \text{as} \qquad \sum_{k=2}^{\infty} b_k$$

where $b_k = a_{k-1}$, since in this case we would have

$$\sum_{k=2}^{\infty} b_k = b_2 + b_3 + b_4 + b_5 + \cdots$$

$$= a_{(2-1)} + a_{(3-1)} + a_{(4-1)} + a_{(5-1)} + \cdots$$
$$= a_1 + a_2 + a_3 + a_4 + \cdots.$$

Similarly, we can write

$$\sum_{k=1}^{\infty} a_k \qquad \text{as} \qquad \sum_{k=0}^{\infty} c_k, \qquad \text{where} \qquad c_k = a_{k+1}.$$

Note also that *the convergence of an infinite series does not depend on any finite number of terms*. For example, if $\sum_{k=1}^{\infty} a_k$ is a convergent series with sum S, then the

series $\sum_{k=0}^{\infty} a_k$ and $\sum_{k=2}^{\infty} a_k$ also converge, and their sums are

$$\sum_{k=0}^{\infty} a_k = a_0 + \sum_{k=1}^{\infty} a_k = a_0 + S$$

and

$$\sum_{k=2}^{\infty} a_k = \sum_{k=1}^{\infty} a_k - a_1 = S - a_1.$$

Another way to say this is that *modifying a convergent series by adding or subtracting a finite number of terms always gives another convergent series,* but with a (possibly) different sum.

Because of these observations we shall often express an infinite series as just Σa_k without making the starting value of the index k explicit.

Geometric Series

In elementary algebra one encounters the formula for the sum of a **geometric progression:**

$$1 + x + x^2 + x^3 + \cdots + x^{n-1} = \frac{1 - x^n}{1 - x}, \qquad x \neq 1. \tag{4}$$

(Equation (4) may be verified by multiplying both sides by $1 - x$. On the left side all terms cancel except $1 - x^n$.)

In equation (4) the variable x, referred to as the **ratio term,** can represent any number except $x = 1$. This formula leads directly to the definition of the **geometric series** with ratio term x:

$$\sum_{k=0}^{\infty} x^k = 1 + x + x^2 + x^3 + \cdots + x^k + \cdots.$$

The formula for the partial sum of this series is given by equation (4):

$$S_n = \frac{1 - x^n}{1 - x}, \qquad x \neq 1.$$

To determine the numbers x for which the geometric series converges, we apply Theorem 1 to conclude that

$$\lim_{n \to \infty} S_n = \lim_{n \to \infty} \frac{1 - x^n}{1 - x} \tag{5}$$

$$= \frac{1 - \lim_{n \to \infty} x^n}{1 - x}.$$

Now by equation (1), Section 12.2, $\lim_{n \to \infty} x^n = 0$ if $|x| < 1$ and, by Exercise 19, Section 12.2, $\lim_{n \to \infty} x^n$ does not exist if $|x| > 1$. Thus (5) shows that the geometric series $\Sigma_{k=0}^{\infty} x^k$

(i) converges to $S = \lim_{n \to \infty} S_n = \dfrac{1}{1 - x}$ if $|x| < 1$, and

(ii) diverges if $|x| > 1$.

The two remaining cases are $x = \pm 1$. If $x = 1$, the geometric series becomes the constant series

$$\sum_{k=0}^{\infty} 1^k = 1 + 1 + 1 + \cdots,$$

which diverges. If $x = -1$, the series is

$$\sum_{k=0}^{\infty} (-1)^k = 1 - 1 + 1 - 1 + \cdots,$$

which diverges, as in Example 2.

This completely determines the convergence properties of the geometric series, as summarized in the following theorem.

THEOREM 5
Convergence of Geometric Series

If $|x| < 1$, the geometric series converges to the sum

$$\sum_{k=0}^{\infty} x^k = \frac{1}{1 - x}.$$

If $|x| \geq 1$, the geometric series $\sum_{k=0}^{\infty} x^k$ diverges.

Example 4

The series

$$1 + \frac{2}{3} + \frac{4}{9} + \frac{8}{27} + \cdots + \frac{2^k}{3^k} + \cdots = \sum_{k=0}^{\infty} \frac{2^k}{3^k}.$$

is a geometric series with $x^k = \frac{2^k}{3^k} = \left(\frac{2}{3}\right)^k$ and $x = \frac{2}{3}$. Since $\left|\frac{2}{3}\right| < 1$, the series converges and its sum, by Theorem 5, is

$$S = \sum_{k=0}^{\infty} \left(\frac{2}{3}\right)^k = \frac{1}{1 - \frac{2}{3}} = 3. \qquad \diamond$$

Example 5

The series

$$\frac{1}{2} + \frac{1}{4} + \frac{1}{8} + \frac{1}{16} + \cdots + \frac{1}{2^k} + \cdots$$

has the form of the geometric series with $x^k = \frac{1}{2^k} = \left(\frac{1}{2}\right)^k$ except that the term $1 = \left(\frac{1}{2}\right)^0$ is missing. In order to use Theorem 5, we write

$$\sum_{k=1}^{\infty} \frac{1}{2^k} = \frac{1}{2} + \frac{1}{4} + \frac{1}{8} + \frac{1}{16} + \cdots + \frac{1}{2^k} + \cdots$$

$$= \left\{ 1 + \frac{1}{2} + \frac{1}{4} + \frac{1}{8} + \frac{1}{16} + \cdots + \frac{1}{2^k} + \cdots \right\} - 1$$

$$= \left(\sum_{k=0}^{\infty} \frac{1}{2^k} \right) - 1$$

$$= \left(\frac{1}{1 - \frac{1}{2}} \right) - 1$$

$$= 1.$$

The Algebra of Convergent Infinite Series

The following theorem shows that a limited amount of algebra may be performed on convergent infinite series.

THEOREM 6

Suppose that the series Σa_k and Σb_k both converge, with sums

$$S = \sum a_k \quad \text{and} \quad T = \sum b_k.$$

Then

(i) the series $\Sigma(a_k + b_k)$ converges, with sum

$$\sum (a_k + b_k) = S + T, \quad \text{and}$$

(ii) for any real number c the series $\Sigma c a_k$ converges, with sum

$$\sum c a_k = cS.$$

Proof: To prove part (i) we let S_n denote the nth partial sum for Σa_k and we let T_n denote the nth partial sum for Σb_k. Then $(S_n + T_n)$ is a partial sum for the series $\Sigma(a_k + b_k)$, since

$$\sum_{k=1}^{n} (a_k + b_k) = (a_1 + b_1) + (a_2 + b_2) + \cdots + (a_n + b_n)$$

$$= (a_1 + a_2 + \cdots + a_n) + (b_1 + b_2 + \cdots + b_n)$$
$$= S_n + T_n.$$

Applying Definition 3 and Theorem 1, part (i), we see that

$$\sum_{k=1}^{\infty} (a_k + b_k) = \lim_{n \to \infty} (S_n + T_n)$$

$$= \lim_{n \to \infty} S_n + \lim_{n \to \infty} T_n$$

$$= S + T.$$

The proof of part (ii) is similar and is left as an exercise. ◆

REMARK: We may paraphrase Theorem 6 by writing

$$\sum(a_k + b_k) = \sum a_k + \sum b_k \tag{6}$$

and

$$\sum ca_k = c\sum a_k. \tag{7}$$

Equations (6) and (7) must be read from right to left—if the sums on the right exist then so do the sums on the left, and equality holds.

Example 6

To find the sum of the series

$$\sum_{k=0}^{\infty} \frac{3 \cdot 2^k + 3^k}{5^k} = 4 + \frac{9}{5} + \frac{21}{25} + \frac{51}{125} + \cdots$$

we use equations (6) and (7) together with Theorem 5:

$$\sum_{k=0}^{\infty} \frac{3 \cdot 2^k + 3^k}{5^k} = \sum_{k=0}^{\infty} \left(\frac{3 \cdot 2^k}{5^k} + \frac{3^k}{5^k} \right)$$

$$= 3\sum_{k=0}^{\infty} \frac{2^k}{5^k} + \sum_{k=0}^{\infty} \frac{3^k}{5^k}$$

$$= 3\sum_{k=0}^{\infty} \left(\frac{2}{5}\right)^k + \sum_{k=0}^{\infty} \left(\frac{3}{5}\right)^k$$

$$= 3\left(\frac{1}{1 - \frac{2}{5}}\right) + \left(\frac{1}{1 - \frac{3}{5}}\right)$$

$$= 3\left(\frac{5}{3}\right) + \frac{5}{2}$$

$$= \frac{15}{2}. \qquad \diamond$$

Necessary Condition for Convergence

The following theorem gives a condition which must be satisfied by *any* convergent series.

THEOREM 7
Necessary Condition for
Convergence

If the infinite series $\sum a_k$ converges then $\lim\limits_{k \to \infty} a_k = 0$.

Proof: Let S_k denote the kth partial sum for Σa_k. If Σa_k converges there exists a sum S for the series. That is, $\lim_{k \to \infty} S_k = S$. Then $\lim_{k \to \infty} S_{k-1} = S$ also. But $a_k = S_k - S_{k-1}$, so

$$
\begin{aligned}
\lim_{k \to \infty} a_k &= \lim_{k \to \infty} (S_k - S_{k-1}) \\
&= \lim_{k \to \infty} S_k - \lim_{k \to \infty} S_{k-1} \\
&= S - S \\
&= 0.
\end{aligned}
$$

◆

REMARK: It is important to note that Theorem 7 is *not* useful in demonstrating that the series Σa_k converges. Although the condition $\lim_{k \to \infty} a_k = 0$ is **necessary** for convergence (meaning all convergent series have this property), it is **not sufficient** to guarantee convergence (meaning some divergent series also have this property). However, Theorem 7 does establish the divergence of a series Σb_k for which $\lim_{k \to \infty} b_k \neq 0$. The Venn diagram in Figure 3.1 illustrates the type of series to which Theorem 7 applies.

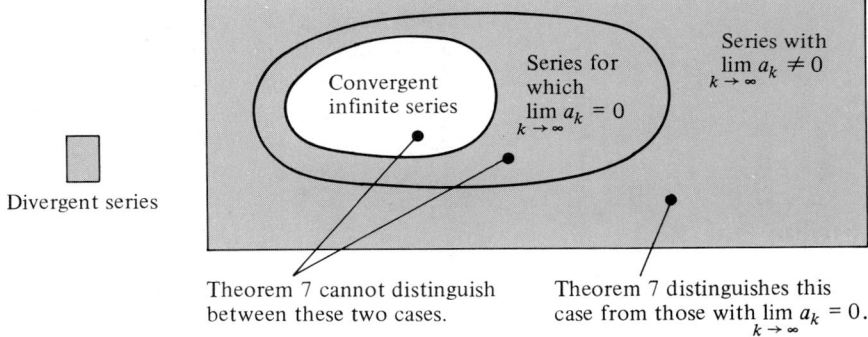

Figure 3.1 Diagram on the applicability of Theorem 7.

The following equivalent formulation of Theorem 7 summarizes these remarks and is the one you will generally find useful in your work.

COROLLARY 2

If $\lim_{k \to \infty} a_k \neq 0$, the series Σa_k diverges.

Example 7

$$
\sum_{k=1}^{\infty} \frac{k}{k + 2}
$$

diverges, by application of Corollary 2:

$$
\lim_{k \to \infty} a_k = \lim_{k \to \infty} \frac{k}{k + 2} = 1 \neq 0.
$$

◇

Example 8

$$\sum_{k=1}^{\infty} \cos(\pi k)$$

diverges, by Corollary 2, since $\lim_{k \to \infty} a_k = \lim_{k \to \infty} \cos(\pi k)$ does not exist. ◇

Example 9

Whether the series

$$\sum_{k=1}^{\infty} \frac{1}{k}$$

converges cannot be determined using Theorem 7 or Corollary 2, since $\lim_{k \to \infty} \frac{1}{k} = 0$.

However, the test discussed in the next section will show that this series diverges. ◇

Unfortunately, about the only types of infinite series yielding explicit formulas for their sums are the geometric series and those series whose partial sums telescope, as in Example 3. However, in most applications of the ideas developed in this chapter, we shall need only to determine whether a particular infinite series converges or diverges. The remaining sections of this chapter concern how this determination is made for various types of infinite series.

Exercise Set 12.3

In Exercises 1–6, write out the first four terms of the infinite series

1. $\sum_{k=1}^{\infty} \dfrac{\cos \pi k}{2^k}$

2. $\sum_{k=1}^{\infty} \dfrac{\sqrt{k} \sin \pi k}{k + 1}$

3. $\sum_{k=1}^{\infty} \dfrac{2^k + 1}{3^k + 2}$

4. $\sum_{k=0}^{\infty} \tan\left(\dfrac{\pi k}{3}\right)$

5. $\sum_{k=1}^{\infty} \ln\left(\dfrac{k}{k + 1}\right)$

6. $\sum_{k=1}^{\infty} k^k$

In Exercises 7–24, determine whether the given series converges or diverges. If it converges find its sum.

7. $\sum_{k=0}^{\infty} \dfrac{1}{7^k}$

8. $\sum_{k=1}^{\infty} \dfrac{1}{3^k}$

9. $\sum_{k=0}^{\infty} 4^k$

10. $\sum_{k=1}^{\infty} \dfrac{7^k + 3^k}{5^k}$

11. $\sum_{k=0}^{\infty} \dfrac{2^{2k}}{3^{3k}}$

12. $\sum_{k=2}^{\infty} \dfrac{-1}{3^k}$

13. $\sum_{k=0}^{\infty} \dfrac{1}{(2 + x)^k}, \quad |x| < 1$

14. $\sum_{k=2}^{\infty} \dfrac{1}{k(k + 1)}$

15. $\sum_{k=1}^{\infty} \left[\dfrac{1}{k + 2} - \dfrac{1}{k + 1}\right]$

16. $\sum_{k=2}^{\infty} \dfrac{2^{k+1} + 2 \cdot 7^k}{9^k}$

17. $\sum_{k=1}^{\infty} \cos \pi k$

18. $\sum_{k=2}^{\infty} \dfrac{1}{k^2 - 1}$

19. $\sum_{k=1}^{\infty} \dfrac{1}{k^2 + 5k + 6}$

20. $\sum_{k=4}^{\infty} \dfrac{1}{k^2 - 9}$

21. $\sum_{k=1}^{\infty} \ln\left(\dfrac{k}{k + 1}\right)$

22. $\sum_{k=1}^{\infty} \dfrac{2^{k-2} + 3^{k+1}}{5^k}$

23. $\sum_{k=0}^{\infty} \dfrac{2^{k/2}}{3^k}$

24. $\sum_{k=1}^{\infty} \dfrac{3^k}{3^{k/2}}$

In Exercises 25–30, write the given decimal fraction as **(a)** an infinite series, and **(b)** the quotient of two integers.

25. $0.3\overline{33} \ldots$

26. $0.7\overline{77} \ldots$

27. $0.929\overline{292}.\ .\ .$

28. $0.3215\overline{15}.\ .\ .$

29. $0.412\overline{412}.\ .\ .$

30. $0.0213434\overline{34}.\ .\ .$

In Exercises 31–34, use Theorem 5 on the convergence of geometric series to establish the stated fact.

31. $\displaystyle\sum_{k=0}^{\infty}(-1)^k x^k = \dfrac{1}{1+x}$ if $|x| < 1$

32. $\displaystyle\sum_{k=0}^{\infty} x^{2k} = \dfrac{1}{1-x^2}$ if $|x| < 1$

33. $\displaystyle\sum_{k=0}^{\infty}\dfrac{x^k}{y^k} = \dfrac{y}{y-x}$ if $|x| < |y|$

34. $\displaystyle\sum_{k=1}^{\infty} x^k = \dfrac{x}{1-x}$ if $|x| < 1$.

35. When dropped from a height h, a ball rebounds to a height $\frac{2}{3}h$. Write an infinite series expressing the total distance travelled by the ball as it bounces an infinite number of times. What is this distance?

36. Let Σa_k be the series $\displaystyle\sum_{k=0}^{\infty}\left(1 + \dfrac{1}{2^k}\right)$ and let Σb_k be the series $\displaystyle\sum_{k=0}^{\infty}(-1)$. Show that the statement $\Sigma(a_k + b_k) = \Sigma a_k + \Sigma b_k$ is false, but that $\Sigma(a_k + b_k)$ converges.

37. If Σca_k converges for a particular number c, must the series Σa_k converge? Why or why not?

38. Prove Theorem 6, part (ii).

39. Prove that if Σa_k diverges, so must Σca_k for any $c \neq 0$.

40. Prove that if $\displaystyle\sum_{k=1}^{\infty} a_k$ converges, then $\displaystyle\sum_{k=p}^{\infty} a_k$ converges for any $p \geq 1$. $\left(\textit{Hint:}\ \text{If } S_n \text{ is the } n\text{th partial sum for } \displaystyle\sum_{k=1}^{\infty} a_k, \text{ then } S_{p+n-1} - S_{p-1} \text{ is the } n\text{th partial sum for } \displaystyle\sum_{k=p}^{\infty} a_k.\right)$

41. Find a formula that defines the partial sum S_n for the series Σa_k recursively, in terms of the partial sum S_{n-1} and the nth term a_n.

42. (*Computer*) Program 6 in Appendix I is a BASIC program that computes partial sums for the geometric series $\displaystyle\sum_{k=p}^{\infty} ax^k$.

For example, partial sums for the series $\displaystyle\sum_{k=2}^{\infty} 7\cdot\left(\dfrac{4}{5}\right)^k$ are obtained by specifying $p = 2$, $a = 7$, and $x = \dfrac{4}{5}$. Results appear in Table 3.1.

Table 3.1

n	$S_n = \displaystyle\sum_{k=2}^{n} 7\left(\dfrac{4}{5}\right)^k$
5	13.224959
10	19.393521
25	22.294217
50	22.399598
100	22.399997
200	22.399998
500	22.399998

Show that $\displaystyle\lim_{n\to\infty} S_n = \sum_{k=2}^{\infty} 7\left(\dfrac{4}{5}\right)^k = 22.4$.

In Exercises 43–46, use Program 6 to find the partial sums S_5, S_{10}, S_{20}, S_{50}, and S_{100}. Then find $\displaystyle\lim_{n\to\infty} S_n$.

43. $\displaystyle\sum_{k=0}^{\infty}\left(\dfrac{2}{3}\right)^k$

44. $\displaystyle\sum_{k=3}^{\infty} 4\left(\dfrac{9}{10}\right)^k$

45. $\displaystyle\sum_{k=5}^{\infty} -2\left(\dfrac{5}{7}\right)^k$

46. $\displaystyle\sum_{k=1}^{\infty} 6\left(\dfrac{3}{2}\right)^k$

12.4 THE INTEGRAL TEST

In this section and in Sections 12.5 and 12.6 we present five tests for convergence for series *with positive terms*. It is especially important to note that these tests apply only to series Σa_k for which $a_k > 0$ for all k.

The Integral Test

This test exploits the relationship between an infinite series of positive terms and the improper integral of a positive continuous function.

THEOREM 8 **Integral Test**	Let $a_k > 0$ for all $k = 1, 2, \ldots$ and let f be a continuous decreasing function defined on $[1, \infty)$ so that $f(k) = a_k$ for each $k = 1, 2, \ldots$. Then $\displaystyle \sum a_k$ converges if and only if $\displaystyle \int_1^\infty f(x)\,dx$ converges. (See Figures 4.1 and 4.2.) That is, the series and the improper integral either both converge or both diverge.

Since $f(x) > 0$ for all $x \in [1, \infty)$, we may interpret these two conclusions geometrically by saying that

(i) If the area of the region R_f, bounded by the graph of f and the x-axis for $1 \leq x < \infty$, is finite, so is the area, R_s, of the region enclosed by the rectangles of area $a_1, a_2, a_3, \ldots$ (Figure 4.1).

(ii) If the area of the region R_f is infinite, so is the area of the region R_s (Figure 4.2).

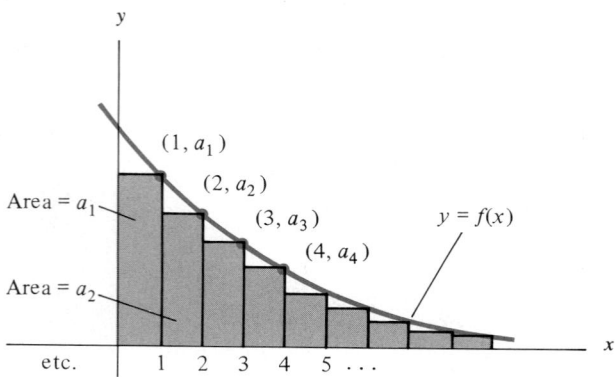

Figure 4.1 $\displaystyle \sum a_k$ converges if $\displaystyle \int_1^\infty f(x)\,dx < \infty$.

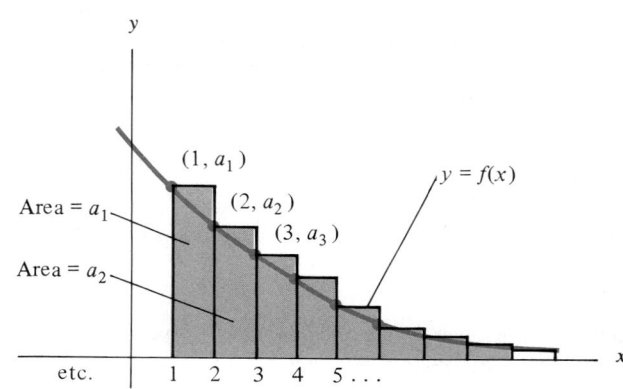

Figure 4.2 $\displaystyle \sum a_k$ diverges if $\displaystyle \int_1^\infty f(x)\,dx = \infty$.

Proof of Theorem 8: Since f is decreasing on $[1, \infty)$, for any integer $k \geq 2$,

$$f(x) \geq f(k) = a_k \qquad \text{for all } x \in [k-1, k], \qquad \text{and} \qquad (1)$$

$$f(x) \leq f(k) = a_k \qquad \text{for all } x \in [k, k+1]. \qquad (2)$$

(See Figure 4.3.)

Inequalities (1) and (2) imply that

$$\int_{k-1}^{k} f(x)\,dx \geq a_k, \qquad \text{and} \qquad (3)$$

$$\int_{k}^{k+1} f(x)\,dx \leq a_k, \qquad k = 2, 3, 4, \ldots \qquad \text{(Figure 4.4).} \qquad (4)$$

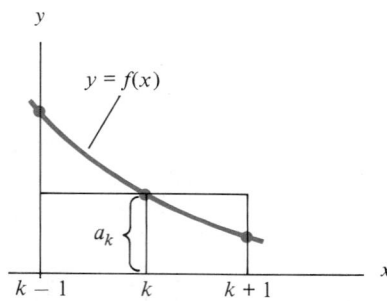

Figure 4.3 $f(x) \geq a_k$, $x \leq k$; $f(x) \leq a_k$, $x \geq k$.

For any integer $n > 2$ we may sum inequality (3) from $k = 2$ to n to conclude that

$$\int_1^n f(x)\,dx = \sum_{k=2}^{n} \int_{k-1}^{k} f(x)\,dx \geq \sum_{k=2}^{n} a_k = S_n - a_1 \qquad (5)$$

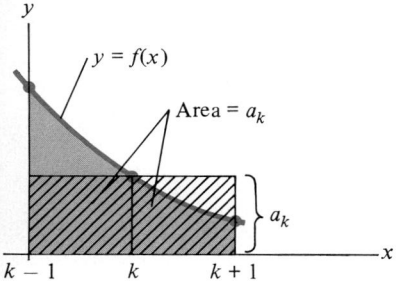

Figure 4.4

$$\int_{k-1}^{k} f(x)\, dx \geq a_k \geq \int_{k}^{k+1} f(x)\, dx.$$

The block of height a_k extending from $k - 1$ to k has area a_k; so does the block from k to $k + 1$.

where S_n denotes the nth partial sum $\sum_{k=1}^{n} a_k$. Thus,

$$S_n \leq a_1 + \int_{1}^{n} f(x)\, dx < a_1 + \int_{1}^{\infty} f(x)\, dx. \tag{6}$$

(The last inequality holds since $f(x) > 0$ on $[1, \infty)$.) Inequality (6) shows that if the improper integral $\int_{1}^{\infty} f(x)\, dx$ converges, the sequence $\{S_n\}$ of partial sums for the series Σa_k is a *bounded* sequence. Since $a_k > 0$ for each k, $\{S_k\}$ is also an *increasing* sequence. Since, by Theorem 4, every bounded increasing sequence converges, the sequence $\{S_n\}$ converges. That is, *if the improper integral $\int_{1}^{\infty} f(x)\, dx$ converges, so does the infinite series Σa_k.*

The converse conclusion is obtained from inequality (4). Summing both sides from $k = 1$ to $k = n$ shows that

$$\int_{1}^{n+1} f(x)\, dx \leq \sum_{k=1}^{n} a_k = S_n. \tag{7}$$

Now $\int_{1}^{n+1} f(x)\, dx$ is an increasing function of n since $f(x) > 0$ for $x \in [1, \infty)$. Thus, if the improper integral $\int_{1}^{\infty} f(x)\, dx$ diverges, we may conclude from inequality (7) that

$$\lim_{n \to \infty} S_n \geq \lim_{n \to \infty} \int_{1}^{n+1} f(x)\, dx = +\infty,$$

so the sequence $\{S_n\}$ diverges. This shows that *if the improper integral diverges, so does the series Σa_k.* This completes the proof. ◆

Example 1

The **harmonic series** is the series

$$\sum_{k=1}^{\infty} \frac{1}{k} = 1 + \frac{1}{2} + \frac{1}{3} + \frac{1}{4} + \cdots + \frac{1}{k} + \cdots.$$

To test this series for convergence, we use the function $f(x) = \dfrac{1}{x}$, which satisfies all hypotheses of the integral test.

Since

$$\int_{1}^{\infty} \frac{1}{x}\, dx = \lim_{t \to \infty} \int_{1}^{t} \frac{1}{x}\, dx = \lim_{t \to \infty} \ln t = +\infty$$

the harmonic series diverges. (Note that $\lim_{k \to \infty} a_k = \lim_{k \to \infty} \dfrac{1}{k} = 0$, although this series does not converge.) ◇

Example 2

To test the series

$$\sum_{k=1}^{\infty} k e^{-k^2}$$

for convergence, we use the function $f(x) = xe^{-x^2}$. Since

$$f'(x) = e^{-x^2} - 2x^2e^{-x^2} = (1 - 2x^2)e^{-x^2} < 0 \qquad \text{if} \qquad x > \frac{\sqrt{2}}{2},$$

we are assured that f is decreasing on $[1, \infty)$. Since

$$\int_1^\infty xe^{-x^2}\, dx = \lim_{t \to \infty} \int_1^t xe^{-x^2}\, dx$$

$$= \lim_{t \to \infty} \left. -\frac{1}{2}e^{-x^2} \right]_1^t$$

$$= \lim_{t \to \infty} \left(\frac{1}{2e} - \frac{1}{2e^{t^2}} \right) = \frac{1}{2e}$$

the infinite series $\sum_{k=1}^\infty ke^{-k^2}$ converges, by the Integral Test. $\qquad \diamond$

Example 3

A **p-series** is a series of the form

$$\sum_{k=1}^\infty \frac{1}{k^p} = 1 + \frac{1}{2^p} + \frac{1}{3^p} + \frac{1}{4^p} + \cdots, \qquad p > 0.$$

Determine for which values of p a p-series converges.

Strategy

Identify a function f with which to apply the Integral Test.

Verify that f is decreasing on $[1, \infty)$.

Set up the improper integral and evaluate, carrying along the unknown constant p.

Determine the values of p for which the improper integral converges.

Handle the remaining case $(p = 1)$ directly (see Example 1).

Solution

We use the function $f(x) = \dfrac{1}{x^p} = x^{-p}$.

Since

$$f'(x) = -px^{-p-1} = \frac{-p}{x^{p+1}}$$

is negative for $x > 0$, the function f is decreasing on $[1, \infty)$.
The improper integral for f is

$$\int_1^\infty x^{-p}\, dx = \lim_{t \to \infty} \int_1^t x^{-p}\, dx$$

$$= \lim_{t \to \infty} \left. \frac{x^{-p+1}}{1 - p} \right]_1^t \qquad (p \neq 1)$$

$$= \left\{ \lim_{t \to \infty} \frac{t^{-p+1}}{1 - p} \right\} - \frac{1}{1 - p}.$$

This limit will exist only if the exponent of t is negative; that is, if

$$-p + 1 < 0, \qquad \text{or} \qquad p > 1.$$

Thus, the improper integral converges if $p > 1$, and diverges if $0 < p < 1$. In the case $p = 1$, (excluded above), the p-series becomes simply the harmonic series, which diverges.

Apply the Integral Test.

The conclusion of Example 3 is this:

The p-series

$$\sum_{k=1}^{\infty} \frac{1}{k^p} = 1 + \frac{1}{2^p} + \frac{1}{3^p} + \frac{1}{4^p} + \cdots, \qquad p > 0$$

converges if $p > 1$ and diverges if $0 < p \leq 1$.

Example 4

The series

$$\sum_{k=1}^{\infty} \frac{1}{k^3} = 1 + \frac{1}{2^3} + \frac{1}{3^3} + \frac{1}{4^3} + \cdots$$

$$= 1 + \frac{1}{8} + \frac{1}{27} + \frac{1}{64} + \cdots$$

converges because it is a p-series with $p = 3 > 1$.

Example 5

The series

$$\sum_{k=1}^{\infty} \left(\frac{1}{k}\right)^{2/3} = 1 + \left(\frac{1}{2}\right)^{2/3} + \left(\frac{1}{3}\right)^{2/3} + \cdots$$

diverges because it is a p-series $\sum_{k=1}^{\infty} \left(\frac{1}{k}\right)^{2/3} = \sum_{k=1}^{\infty} \frac{1}{k^{2/3}}$ with $p = \frac{2}{3} < 1$.

Exercise Set 12.4

In Exercises 1–20, use the Integral Test to determine whether the given series converges or diverges.

1. $\sum \dfrac{1}{2k + 1}$

2. $\sum \dfrac{1}{(3k + 1)^2}$

3. $\sum \dfrac{1}{k \cdot \ln k}$

4. $\sum \dfrac{1}{k(\ln k)^2}$

5. $\sum \dfrac{1}{1 + k^2}$

6. $\sum k^2 e^{-k^3}$

7. $\sum \dfrac{1}{\sqrt{k + 1}}$

8. $\sum \dfrac{k + 1}{k}$

9. $\sum k^2 e^{-k^3}$

10. $\sum \dfrac{\text{Tan}^{-1} k}{1 + k^2}$

11. $\sum \dfrac{k}{\sqrt{1 + k^2}}$

12. $\sum \dfrac{k^2 - 2}{k^2 + 2}$

13. $\sum \dfrac{1}{\sqrt{k}}$

14. $\sum \dfrac{1}{k^2}$

15. $\sum k(1 + k^2)^{-3}$

16. $\sum k^2 e^{-k}$

17. $\sum \dfrac{\ln k}{k}$

18. $\sum \dfrac{k}{1 + k^2}$

19. $\sum \dfrac{k^2}{k^3 + 1}$

20. $\sum \dfrac{2k^2}{\sqrt{k^3 + 5}}$

21. Let Σa_k be a convergent infinite series that satisfies the hypotheses of the integral test. Show, using inequalities (5) and (7), that

$$\int_{n+1}^{\infty} f(x)\,dx \le \sum_{k=n+1}^{\infty} a_k \le \int_n^{\infty} f(x)\,dx \tag{8}$$

where $f(k) = a_k$ and f is continuous and decreasing on $[1, \infty)$. Inequality (8) gives an estimate for the **truncation error** $R_n = \sum_{n+1}^{\infty} a_k$ which occurs in using the partial sum

$$S_n = \sum_{k=1}^n a_k$$ to approximate the sum of a convergent series

$$S = \sum_{k=1}^{\infty} a_k$$ when $\{a_k\}$ is a decreasing sequence with

$$\lim_{k \to \infty} a_k = 0.$$

22. Use the result of Exercise 21 to estimate the magnitude of the error that occurs in approximating the sum of the p-series

$$\sum_{k=1}^{\infty} \frac{1}{k^2} \text{ by the partial sum } \sum_{k=1}^{10} \frac{1}{k^2}.$$

23. Use the result of Exercise 21 to determine how large n must be taken so that the partial sum S_n for the series $\sum_{k=2}^{\infty} \frac{1}{k \ln^2 k}$ differs from its sum S by less than .01.

24. Use the result of Exercise 21 to determine how large n must be taken so that the partial sum S_n for the series $\sum_{k=1}^{\infty} ke^{-k^2}$ differs from its sum S by less than .02.

25. Prove that the harmonic series $\sum_{k=1}^{\infty} \frac{1}{k}$ diverges without using the Integral Test, as follows.

a. Write the harmonic series as

$$\sum_{k=1}^{\infty} \frac{1}{k} = \{1\} + \left\{\frac{1}{2}\right\} + \underbrace{\left\{\frac{1}{3} + \frac{1}{4}\right\}}_{2^1 \text{ terms}} + \underbrace{\left\{\frac{1}{5} + \frac{1}{6} + \frac{1}{7} + \frac{1}{8}\right\}}_{2^2 \text{ terms}}$$

$$+ \left\{\frac{1}{9} + \cdots + \frac{1}{16}\right\} \quad (2^3 \text{ terms})$$

$$+ \left\{\frac{1}{17} + \cdots + \frac{1}{32}\right\} \quad (2^4 \text{ terms})$$

$$\cdots + \left\{\frac{1}{2^n + 1} + \cdots + \frac{1}{2^{n+1}}\right\} \quad (2^n \text{ terms}).$$

b. Show that the sums of the terms in each of the blocks above are greater than $\frac{1}{2}$.

c. Conclude that $S_{2^{n+1}} \ge 1 + \frac{(n+1)}{2}$.

d. From (c), conclude that $\lim_{n \to \infty} S_n = +\infty$.

26. Use the technique of Exercise 25 to prove the "rule of thumb" for the partial sum $S_{2^n} = \sum_{k=1}^{2^n} \frac{1}{k}$ of the harmonic series: $S_{2^n} \ge 1 + \frac{n}{2}$.

12.5 THE COMPARISON TESTS

Like the Integral Test, the next test is motivated by area considerations. However, instead of comparing terms of a given series with certain integrals, we compare them with terms of another series.

THEOREM 9	Let Σa_k and Σb_k be infinite series with $0 < a_k \le b_k$ for each $k = 1, 2, 3, \ldots$.
Basic Comparison Test	Then

(i) If Σb_k converges, so does Σa_k.
(ii) If Σa_k diverges, so does Σb_k.

Proof: We let

$$S_n = \sum_{k=1}^n a_k \quad \text{and} \quad T_n = \sum_{k=1}^n b_k$$

be the corresponding partial sums. Since $a_k \le b_k$ for all k, we have

$$S_n \le T_n \quad \text{for all} \quad n = 1, 2, \ldots \tag{1}$$

To prove (i), we note that if Σb_k converges, $\lim\limits_{n \to \infty} T_n$ exists. Thus the sequence $\{T_n\}$ is bounded and, by inequality (1), so is the sequence $\{S_n\}$. Since $a_k > 0$ for all k, this shows that $\{S_n\}$ is a bounded increasing sequence, so by Theorem 4, $\lim\limits_{n \to \infty} S_n$ exists. That is, Σa_k converges.

To prove (ii), we note that since $\{S_n\}$ is an increasing sequence, if Σa_k diverges, we must have $\lim\limits_{n \to \infty} S_n = +\infty$. Thus, by inequality (1)

$$\lim_{n \to \infty} T_n \geq \lim_{n \to \infty} S_n = +\infty,$$

so Σb_k diverges. ◆

Example 1

The series $\sum \dfrac{1}{1 + k^3}$ converges, by comparison with the convergent p-series $\sum \dfrac{1}{k^3}$, since $\dfrac{1}{1 + k^3} < \dfrac{1}{k^3}$. ◇

Example 2

To determine the convergence of the series $\sum \dfrac{\sqrt{k} - 1}{k^2 + 2}$, we make the comparison

$$\frac{\sqrt{k} - 1}{k^2 + 2} < \frac{\sqrt{k}}{k^2 + 2} < \frac{\sqrt{k}}{k^2} = \frac{1}{k^{3/2}}.$$

Since the series $\sum \dfrac{1}{k^{3/2}}$ is a p-series with $p = 3/2 > 1$, it converges. So, therefore, does the series $\sum \dfrac{\sqrt{k} - 1}{k^2 + 2}$. ◇

Example 3

The series

$$\sum_{k=1}^{\infty} \frac{1 + \ln k}{k}$$

diverges by comparison with the harmonic series $\sum\limits_{k=1}^{\infty} \dfrac{1}{k}$, since

$$\frac{1 + \ln k}{k} > \frac{1}{k} \qquad \text{if} \qquad k > 1.$$ ◇

REMARK 1: Successful application of the Comparison Test to the series Σa_k involves finding another series Σb_k with which you can make a useful comparison. If you suspect that Σa_k converges, you should look for a *convergent* series Σb_k with $a_k \leq b_k$. But if you suspect that Σa_k diverges, look for a divergent series Σb_k with $a_k \geq b_k$.

REMARK 2: We say that the series Σa_k *is dominated by* the series Σb_k (or, that Σb_k dominates Σa_k) if $a_k \le b_k$ for all k. Using this terminology we may paraphrase Theorem 9 by saying that a positive series that is dominated by a convergent series must converge, while a series that dominates a positive divergent series must diverge.

It is important to note that two cases remain (dominated by a divergent series and dominating a convergent series) in which no valid conclusions can be drawn. Table 5.1 may help you remember which comparisons yield valid conclusions.

Table 5.1 Conclusions concerning Σa_k when compared with Σb_k.

	Σb_k converges	Σb_k diverges
$0 < a_k \le b_k$	Σa_k converges	no valid conclusion
$a_k \ge b_k > 0$	no valid conclusion	Σa_k diverges

The Basic Comparison Test is used whenever an appropriate comparison series can be found. As the list of series with which you are familiar expands, your facility with the Basic Comparison Test will grow.

The following variation on the Basic Comparison Test is often easier to apply.

THEOREM 10
Limit Comparison Test

Let Σa_k and Σb_k be series with $a_k > 0$, $b_k > 0$ for all $k = 1, 2, \ldots$. If the limit

$$\rho = \lim_{k \to \infty} \frac{a_k}{b_k}$$

exists, and $\rho \ne 0$, then either both series converge or both series diverge.

The Limit Comparison Test states that if the ratio of the general terms of two positive series tends to a positive limit, then the two series have the same convergence property—either both converge or both diverge.

Before proving this theorem, we consider several examples of its use.

Example 4

The series

$$\sum_{k=1}^{\infty} \frac{1}{2k + 1}$$

diverges, as can be shown by Theorem 10. Comparing this series with the harmonic series

$$\Sigma b_k = \sum_{k=1}^{\infty} \frac{1}{k},$$

we find that

$$\rho = \lim_{k \to \infty} \left(\frac{\dfrac{1}{2k+1}}{\dfrac{1}{k}} \right) = \lim_{k \to \infty} \frac{k}{2k+1} = \frac{1}{2} > 0.$$

Since the limit $\rho > 0$ exists and since the harmonic series diverges, so must the given series. $\diamond$

Example 5

To apply the Limit Comparison Test to the series

$$\sum_{k=1}^{\infty} \frac{\sqrt{k}+2}{k^2+k+1},$$

we observe that the general term is a quotient containing a highest exponent of 1/2 in its numerator and a highest exponent of 2 in its denominator. This suggests a comparison with the series whose general term is $b_k = \dfrac{k^{1/2}}{k^2} = \dfrac{1}{k^{3/2}}$. We therefore take Σb_k to be the p-series

$$\sum_{k=1}^{\infty} \frac{1}{k^{3/2}}.$$

We obtain

$$\rho = \lim_{k \to \infty} \left(\frac{\dfrac{\sqrt{k}+2}{k^2+k+1}}{\dfrac{1}{k^{3/2}}} \right) = \lim_{k \to \infty} \left(\frac{k^2 + 2k^{3/2}}{k^2+k+1} \right) = 1 > 0.$$

We may now conclude that, since the p-series $\displaystyle\sum_{k=1}^{\infty} \frac{1}{k^{3/2}}$ converges, so does the series

$$\sum_{k=1}^{\infty} \frac{\sqrt{k}+2}{k^2+k+1}.$$ $\diamond$

Example 6

Determine whether the series

$$\sum_{k=1}^{\infty} \sin\left(\frac{\pi}{k}\right)$$

converges.

Solution: The Limit Comparison Test may be applied by recalling that we have previously established limit

$$\rho = \lim_{k\to\infty} \frac{\sin\left(\dfrac{\pi}{k}\right)}{\left(\dfrac{\pi}{k}\right)} = \lim_{x\to 0^+} \frac{\sin x}{x} = 1.$$

Thus, by Theorem 10, the series $\displaystyle\sum_{k=1}^{\infty} \sin\left(\frac{\pi}{k}\right)$ has the same convergence property as

does the series $\displaystyle\sum_{k=1}^{\infty} \frac{\pi}{k}$. Since $\displaystyle\sum_{k=1}^{\infty} \frac{\pi}{k} = \pi\sum_{k=1}^{\infty}\frac{1}{k}$ is a multiple of the (divergent) har-

monic series, the series $\displaystyle\sum_{k=1}^{\infty} \sin\left(\frac{\pi}{k}\right)$ diverges. $\diamond$

Proof of Theorem 10: Since all terms of Σa_k and Σb_k are positive, and $\rho \neq 0$, we must have

$$\rho = \lim_{k\to\infty} \frac{a_k}{b_k} > 0. \tag{2}$$

Now let $\epsilon = \dfrac{\rho}{2} > 0$. Then by (2) and the definition of the limit of a sequence there exists an integer N such that

$$\left|\frac{a_k}{b_k} - \rho\right| < \epsilon \qquad \text{whenever} \qquad k > N. \tag{3}$$

Statement (3) is equivalent to the statement

$$\rho - \epsilon < \frac{a_k}{b_k} < \rho + \epsilon, \qquad k > N. \tag{4}$$

Substituting $\epsilon = \dfrac{1}{2}\rho$ and multiplying through by $b_k > 0$ in (4) gives

$$\frac{1}{2}\rho \cdot b_k < a_k < \frac{3}{2}\rho \cdot b_k, \qquad k > N. \tag{5}$$

Using the left half of inequality (5), we conclude that Σa_k dominates a multiple of Σb_k, since $\Sigma\dfrac{1}{2}\rho b_k = \left(\dfrac{1}{2}\rho\right)\Sigma b_k$. Thus Σb_k converges if Σa_k converges (Theorem 9). The right half of inequality (5) shows that Σa_k is dominated by $\Sigma\dfrac{3}{2}\rho b_k = \left(\dfrac{3}{2}\rho\right)\Sigma b_k$, so Σb_k diverges if Σa_k diverges. These two conclusions taken together establish Theorem 10. $\blacklozenge$

Before concluding this section, we need to repeat an observation which applies to each of the tests embodied by Theorems 8, 9, and 10. We have previously noted that if the series $\Sigma_{k=N}^{\infty} a_k$ converges, then so does the series $\Sigma_{k=1}^{\infty} a_k$, and vice versa.

This observation means that we need not require that the conditions of Theorems 8 through 10 hold on the full series $\sum_{k=1}^{\infty} a_k$ but rather only on any "tail" of the series of the form $\sum_{k=N}^{\infty} a_k$.

For example, the series $\sum_{k=1}^{\infty} \dfrac{7+k}{1+k^3}$ may be successfully compared to the series

$$\sum_{k=1}^{\infty} \frac{2}{k^2} = 2\sum_{k=1}^{\infty} \frac{1}{k^2} \text{ for } k > 7 \text{ since}$$

$$\frac{7+k}{1+k^3} < \frac{k+k}{1+k^3} = \frac{2k}{1+k^3} < \frac{2k}{k^3} = \frac{2}{k^2} \,.$$

whenever $k > 7$. The Comparison Test then shows that the "tail" $\sum_{k=8}^{\infty} \dfrac{7+k}{1+k^3}$ con-

verges, since the p-series $\sum_{k=1}^{\infty} \dfrac{1}{k^2}$ converges. Thus the original series $\sum_{k=1}^{\infty} \dfrac{7+k}{1+k^3}$ also

converges.

Exercise Set 12.5

In Exercises 1–6, use the Basic Comparison Test to determine whether the given series converges or diverges.

1. $\sum \dfrac{1}{1+k^2}$

2. $\sum \dfrac{1}{k^{1/2} + k^{3/2}}$

3. $\sum \dfrac{\sqrt{k}}{1+k^3}$

4. $\sum \dfrac{\sqrt{k}}{1+k}$

5. $\sum \dfrac{3}{\sqrt{k}+2}$

6. $\sum (k-1)e^{-k}$

In Exercises 7–10, use the Limit Comparison Test to determine whether the given series converges or diverges.

7. $\sum \dfrac{k+3}{2k^2+1}$

8. $\sum \dfrac{k^2-4}{k^3+k+5}$

9. $\sum \dfrac{k+\sqrt{k}}{k+k^3}$

10. $\sum \dfrac{2k+2}{\sqrt{k^3+2}}$

In Exercises 11–30, determine whether the series converges or diverges and state which test you used.

11. $\sum \dfrac{\cos^2 \pi k}{k^2}$

12. $\sum \dfrac{k(k+1)}{(k+2)(k^2+1)}$

13. $\sum \cos\left(\dfrac{\pi k}{4}\right)$

14. $\sum \dfrac{2k^2+3k-1}{k^4-6k+10}$

15. $\sum \dfrac{\sin(\pi k)}{k}$

16. $\sum \dfrac{2^k}{3^k+1}$

17. $\sum \dfrac{1}{\sqrt{4k(k+1)}}$

18. $\sum \dfrac{2k+2}{\sqrt{k^3+2}}$

19. $\sum \dfrac{1}{\sqrt[3]{k^2+2k}}$

20. $\sum \dfrac{k+1}{2\ln k}$

21. $\sum \dfrac{\mathrm{Tan}^{-1} k}{k^2}$

22. $\sum \dfrac{k+1}{\sqrt{k^{3/2}+1}}$

23. $\sum \dfrac{\ln k}{1+\ln k}$

24. $\sum \dfrac{\ln(k+1)}{k+2}$

25. $\sum \dfrac{n^2+3}{2n^4+n-6}$

26. $\sum \dfrac{\ln n}{n^3}$

27. $\sum \dfrac{3^k}{k+7}$

28. $\sum \dfrac{k+3}{(k+2)2^k}$

29. $\sum \dfrac{\sqrt{k}}{\cos(2k-6)+k^2}$

30. $\sum \dfrac{\mathrm{Tan}^{-1}\sqrt{k}}{\pi+6k^2}$

31. Prove that if Σa_k and Σb_k are positive term series and $\lim_{k \to \infty} \dfrac{a_k}{b_k} = 0$, then Σa_k converges if Σb_k converges.

32. Prove that if Σa_k and Σb_k are positive term series and $\lim_{k \to \infty} \dfrac{a_k}{b_k} = +\infty$, then Σa_k diverges if Σb_k diverges.

33. Prove that if Σa_k converges, then $\lim_{N \to \infty} \sum_{k=N}^{\infty} a_k = 0$. That is, if Σa_k converges, its "tails" tend toward zero.

12.6 THE RATIO AND ROOT TESTS

In this section we discuss two additional tests for series with positive terms. The first involves the limit of the ratio of successive terms of the series.

THEOREM 11
Ratio Test

Let $a_k > 0$ for all $k = 1, 2, 3, \ldots$ and let

$$\rho = \lim_{k \to \infty} \frac{a_{k+1}}{a_k}. \tag{1}$$

Then, provided this limit exists,

(i) If $\rho < 1$, the series $\sum_{k=1}^{\infty} a_k$ converges.

(ii) If $\rho > 1$, the series $\sum_{k=1}^{\infty} a_k$ diverges.

(iii) If $\rho = 1$, no conclusion may be drawn.

Before proving Theorem 11, we discuss several examples in which the Ratio Test applies.

Example 1

For the series $\sum \frac{k^2}{3^k}$, the kth term is $a_k = \frac{k^2}{3^k}$, so $a_{k+1} = \frac{(k+1)^2}{3^{k+1}}$. The limit in (1) is therefore

$$\rho = \lim_{k \to \infty} \frac{\dfrac{(k+1)^2}{3^{k+1}}}{\dfrac{k^2}{3^k}} = \lim_{k \to \infty} \frac{3^k(k+1)^2}{3^{k+1}k^2} = \lim_{k \to \infty} \frac{1}{3}\left(\frac{k+1}{k}\right)^2 = \frac{1}{3}.$$

Since $\rho = \frac{1}{3} < 1$, the series $\sum \frac{k^2}{3^k}$ converges by the Ratio Test. $\diamond$

REMARK: The Ratio Test is often useful in dealing with series involving factorial expressions. Recall that

$$k! = k(k-1)(k-2) \cdots \cdot 3 \cdot 2 \cdot 1.$$

Thus, ratios of factorials can be simplified as follows:

$$\frac{(k+1)!}{k!} = \frac{(k+1)k!}{k!} = k + 1$$

$$\frac{(k+2)!}{k!} = \frac{(k+2)(k+1)k!}{k!} = (k+2)(k+1)$$

$$\frac{k!}{(k+p)!} = \frac{k!}{(k+p)(k+p-1) \cdot \ldots \cdot (k+1) \cdot k!}$$

$$= \frac{1}{(k+p)(k+p-1) \cdot \ldots \cdot (k+1)}.$$

Example 2

For the series $\sum \dfrac{k^2}{(k+1)!}$ the limit (1) is

$$\rho = \lim_{k \to \infty} \frac{\dfrac{(k+1)^2}{[(k+1)+1]!}}{\dfrac{k^2}{(k+1)!}} = \lim_{k \to \infty} \frac{(k+1)!(k+1)^2}{(k+2)!(k^2)}$$

$$= \lim_{k \to \infty} \left(\frac{1}{k+2}\right)\left(\frac{k+1}{k}\right)^2$$

$$= 0 \cdot 1^2 = 0.$$

Thus, the series $\sum \dfrac{k^2}{(k+1)!}$ converges, by the Ratio Test. ◇

Example 3

For the series $\sum \dfrac{k!}{k^k}$ the ratio limit is

$$\rho = \lim_{k \to \infty} \frac{\dfrac{(k+1)!}{(k+1)^{k+1}}}{\dfrac{k!}{k^k}} = \lim_{k \to \infty} \frac{(k+1)!k^k}{k!(k+1)^{k+1}}$$

$$= \lim_{k \to \infty} (k+1)\left[\frac{k^k}{(k+1)^{k+1}}\right]$$

$$= \lim_{k \to \infty} \left(\frac{k+1}{k+1}\right)\left[\frac{k^k}{(k+1)^k}\right]$$

$$= \lim_{k \to \infty} \left(\frac{k}{k+1}\right)^k$$

$$= \lim_{k \to \infty} \frac{1}{\left(1 + \dfrac{1}{k}\right)^k} = \frac{1}{e}.$$

The series $\sum \dfrac{k!}{k^k}$ therefore converges, by the Ratio Test. ◇

Example 4

If we attempt to apply the Ratio Test to the series $\sum \dfrac{k}{k^2+1}$ we find that

$$\rho = \lim_{k \to \infty} \frac{\left(\dfrac{k+1}{(k+1)^2+1}\right)}{\left(\dfrac{k}{k^2+1}\right)}$$

$$= \lim_{k \to \infty} \frac{(k+1)(k^2+1)}{k[(k+1)^2+1]} = \lim_{k \to \infty} \frac{k^3+k^2+k+1}{k^3+2k^2+2k} = 1,$$

so the Ratio Test is inconclusive. However, we can handle this series by a limit comparison with the harmonic series $\Sigma b_k = \Sigma\left(\dfrac{1}{k}\right)$ (Theorem 10). We find

$$\lim_{k\to\infty}\left(\frac{a_k}{b_k}\right) = \lim_{k\to\infty}\frac{\left(\dfrac{k}{k^2+1}\right)}{\left(\dfrac{1}{k}\right)} = \lim_{k\to\infty}\frac{k^2}{k^2+1} = 1.$$

Thus, since the harmonic series diverges, so does the series $\Sigma\dfrac{k}{k^2+1}$. ◇

In Chapter 13 we shall concern ourselves with series whose terms contain variables. For such series the Ratio Test is often helpful in determining those values of the variable for which the series converges. The following is one such example.

Example 5

For which numbers $x > 0$ does the infinite series $\Sigma x^k/k$ converge?

Solution: Applying the Ratio Test we find that

$$\rho = \lim_{k\to\infty}\left(\frac{\dfrac{x^{k+1}}{k+1}}{\dfrac{x^k}{k}}\right) = \lim_{k\to\infty}\frac{x^{k+1}\cdot k}{x^k(k+1)}$$

$$= \lim_{k\to\infty}x\left(\frac{k}{k+1}\right)$$

$$= x.$$

Since $\rho = x$, the series will converge for $\rho = x < 1$, by the Ratio Test, and diverge for $x > 1$. The case $\rho = x = 1$ is inconclusive by the Ratio Test, but if $x = 1$ the series becomes simply the harmonic series, which diverges. Thus, the series $\Sigma x^k/k$, $x > 0$, converges only if $0 < x < 1$. ◇

Proof of Theorem 11: To prove statement (i), ($\rho < 1$ implies convergence), we will compare the series $\Sigma_{k=1}^{\infty} a_k$ with a geometric series. Since $\rho < 1$, we can find a number γ so that $\rho < \gamma < 1$.
Since

$$\lim_{k\to\infty}\frac{a_{k+1}}{a_k} = \rho,$$

there exists an integer N so that

$$\frac{a_{k+1}}{a_k} \le \gamma \qquad \text{whenever} \qquad k > N. \tag{2}$$

(To see this take $\epsilon = \gamma - \rho > 0$ in Definition 2.)
Inequality (2) shows that

$$a_{N+1} \le \gamma a_N$$
$$a_{N+2} \le \gamma a_{N+1} \le \gamma^2 a_N$$

and, in general, that

$$a_{N+k} \le \gamma^k a_N, \qquad k = 1, 2, 3, \ldots . \tag{3}$$

Now let $b_k = \gamma^k a_N$ for $k = 1, 2, \ldots$. Then $\Sigma b_k = \Sigma \gamma^k a_N = a_N \Sigma \gamma^k$ is a convergent series since $\gamma < 1$ and the series $\Sigma \gamma^k$ is geometric. Thus, inequality (3) shows that the series $\Sigma_{k=1}^{\infty} a_{N+k}$ is dominated by the convergent series $\Sigma_{k=1}^{\infty} b_k$. The series $\Sigma_{k=1}^{\infty} a_{N+k} = \Sigma_{k=N+1}^{\infty} a_k$ therefore converges, by the Comparison Test. This shows that the series Σa_k converges.

The proof of statement (ii) is similar (except that $\rho > 1$, so the comparison is with a divergent geometric series) and is left as an exercise.

To prove statement (iii) note that

(a) the harmonic series $\sum \dfrac{1}{k}$ diverges, although

$$\rho = \lim_{k \to \infty} \frac{\dfrac{1}{k+1}}{\dfrac{1}{k}} = \lim_{k \to \infty} \frac{k}{k+1} = 1, \qquad \text{and}$$

(b) the p-series $\sum \dfrac{1}{k^2}$ converges, although

$$\rho = \lim_{k \to \infty} \frac{\dfrac{1}{(k+1)^2}}{\dfrac{1}{k^2}} = \lim_{k \to \infty} \frac{k^2}{(k+1)^2} = 1. \qquad \blacklozenge$$

The Root Test

The following test for convergence uses information about the kth root of the general term for Σa_k.

THEOREM 12 Root Test	Let $a_k > 0$ for all $k = 1, 2, 3, \ldots$ and let $\rho = \lim\limits_{k \to \infty} \sqrt[k]{a_k}$. Then (i) If $\rho < 1$, the series $\displaystyle\sum_{k=1}^{\infty} a_k$ converges. (ii) If $\rho > 1$, the series $\displaystyle\sum_{k=1}^{\infty} a_k$ diverges. (iii) If $\rho = 1$, no conclusions may be drawn.

The Root Test is primarily used in series involving kth powers.

Example 6

The series $\Sigma e^k / k^k$ converges. This can be shown using the Root Test since

$$\rho = \lim_{k \to \infty} \left(\frac{e^k}{k^k} \right)^{1/k} = \lim_{k \to \infty} \frac{e}{k} = 0 < 1. \qquad \diamondsuit$$

Example 7

For the series $\Sigma k^2/2^k$ we find that

$$\rho = \lim_{k \to \infty} \left(\frac{k^2}{2^k}\right)^{1/k} = \lim_{k \to \infty} \frac{(k^{1/k})^2}{2} = \frac{1}{2} < 1$$

so the series converges.

◇

Proof of Theorem 12: We proceed as in the proof of the Ratio Test. If $\rho < 1$, there exists a constant γ so that $\rho < \gamma < 1$. Then, since $\lim_{k \to \infty} \sqrt[k]{a_k} = \lim_{k \to \infty} a_k^{1/k} = \rho$, there exists a constant N so that

$$a_k^{1/k} < \gamma \quad \text{whenever} \quad k > N. \tag{4}$$

(To see this take $\epsilon = \gamma - \rho$ in Definition 2.) Inequality (4) may be written

$$a_k < \gamma^k, \quad k \geq N,$$

which shows that the series $\Sigma_{k=N}^{\infty} a_k$ is dominated by the geometric series $\Sigma_{k=N}^{\infty} \gamma^k$. Since $\gamma < 1$, this geometric series converges. Thus the series $\Sigma_{k=N}^{\infty} a_k$ converges by comparison with a geometric series. This proves statement (i).

The proof of statement (ii) is similar to that of (i) and is left as an exercise. The proof of statement (iii) proceeds as in the proof of Theorem 11. We note that

(a) $\rho = \lim_{k \to \infty} \left(\frac{1}{k}\right)^{1/k} = \lim_{k \to \infty} \frac{1}{\sqrt[k]{k}} = 1$, so the series $\sum \frac{1}{k}$ diverges.

(b) $\rho = \lim_{k \to \infty} \left(\frac{1}{k^2}\right)^{1/k} = \lim_{k \to \infty} \left(\frac{1}{\sqrt[k]{k}}\right)^2 = 1$, so the series $\sum \frac{1}{k^2}$ converges. ◆

Exercise Set 12.6

In each of Exercises 1–20, determine whether the given series converges or diverges and state the test used to verify your conclusion.

1. $\sum \dfrac{2^k}{k+2}$

2. $\sum \dfrac{k3^k}{(k+1)!}$

3. $\sum k^{10}e^{-k}$

4. $\sum \dfrac{k!}{2^{k+2}}$

5. $\sum \dfrac{\ln k}{e^k}$

6. $\sum \dfrac{(3k)!}{(k!)^3}$

7. $\sum \dfrac{k+2}{1+k^3}$

8. $\sum \left(\dfrac{k}{2k+1}\right)^k$

9. $\sum \dfrac{1}{(\ln k)^k}$

10. $\sum \dfrac{k!}{k^k}$

11. $\sum \left(1 + \dfrac{2}{k}\right)^k$

12. $\sum \left(\dfrac{k}{k+1}\right)^k$

13. $\sum \dfrac{3^k}{k^3 2^{k+2}}$

14. $\sum \dfrac{k^3 2^{k+3}}{2^{2k}}$

15. $\sum \dfrac{(k!)^2}{(3k)!}$

16. $\sum \left(\dfrac{k}{1+k^3}\right)^k$

17. $\sum \dfrac{(k+2)!}{4!k!2^k}$

18. $\sum \dfrac{1}{(k+1)!}$

19. $\sum \dfrac{k!}{e^{3k}}$

20. $\sum \dfrac{1}{k\sqrt{\ln k}}$

In Exercises 21–24, find those numbers $x > 0$ for which the given series converges.

21. $1 + x + \dfrac{x^2}{2!} + \dfrac{x^3}{3!} + \cdots = \sum_{k=0}^{\infty} \dfrac{x^k}{k!}$ $(0! = 1)$

22. $1 + x^2 + x^4 + \cdots = \sum_{k=0}^{\infty} x^{2k}$

23. $1 + \dfrac{x^2}{2} + \dfrac{x^4}{4} + \cdots = 1 + \displaystyle\sum_{k=1}^{\infty} \dfrac{x^{2k}}{2k}$

24. $1 + \dfrac{x^2}{2} + \dfrac{x^4}{3} + \cdots = \displaystyle\sum_{k=0}^{\infty} \dfrac{x^{2k}}{k+1}$

25. Prove Theorem 11, part (ii). (*Hint:* If $\rho > 1$, let γ be a constant so that $1 < \gamma < \rho$. Then, if $\dfrac{a_{k+1}}{a_k} > \gamma$, it follows that $a_k > \gamma^k$. Proceed as in the proof of part (i).)

26. Prove Theorem 12, part (ii). (*Hint:* If $\rho > 1$, let γ be a constant so that $1 < \gamma < \rho$. Then if $\sqrt[k]{a_k} > \gamma$, it follows that $a_k > \gamma^k$. Proceed as in the proof of part (i).)

27. For the *sequence* $\{a_k\}$ with $a_k > 0$, prove that if $\displaystyle\lim_{k \to \infty} \dfrac{a_{k+1}}{a_k} < 1$, then $\displaystyle\lim_{k \to \infty} a_k = 0$.

28. For the *sequence* $\{a_k\}$ with $a_k > 0$, if $\displaystyle\lim_{k \to \infty} \dfrac{a_{k+1}}{a_k} > 1$, what can you say about $\displaystyle\lim_{k \to \infty} a_k$?

12.7 ABSOLUTE AND CONDITIONAL CONVERGENCE

Each of the convergence tests discussed thus far applies only to series with strictly positive terms. In this section we discuss two criteria by which series containing both positive and negative terms may be tested for convergence.

Absolute Convergence

One straightforward procedure for testing the series Σa_k for convergence is simply to replace each term by its absolute value and then test the resulting series, $\Sigma |a_k|$, for convergence. This is the idea of *absolute convergence*.

DEFINITION 4

The series Σa_k is said to **converge absolutely** if the series of absolute values, $\Sigma |a_k|$, converges.

Example 1

The series

$$\sum_{k=1}^{\infty} \frac{(-1)^{n+1}}{k^2} = 1 - \frac{1}{4} + \frac{1}{9} - \frac{1}{16} + \frac{1}{25} - \cdots$$

converges absolutely since

$$\sum_{k=1}^{\infty} \left| \frac{(-1)^{n+1}}{k^2} \right| = \sum_{k=1}^{\infty} \frac{1}{k^2}$$

is a convergent *p*-series. ◇

Example 2

The series

$$\sum_{k=1}^{\infty} (-1)^{k+1} = 1 - 1 + 1 - 1 + \cdots$$

does not converge absolutely since $\displaystyle\sum_{k=1}^{\infty} |(-1)^{n+1}| = 1 + 1 + 1 + \cdots$ diverges. ◇

The following theorem shows that the terminology of Definition 4 is well chosen—absolutely convergent series indeed converge.

THEOREM 13

If the series $\Sigma|a_k|$ converges then so does the series Σa_k. That is, every absolutely convergent series converges.

Proof: Assume that the series $\Sigma|a_k|$ converges. Since the inequality

$$-|a_k| \leq a_k \leq |a_k|$$

holds for all terms a_k, we may add $|a_k|$ to all terms in this inequality, to get

$$0 \leq a_k + |a_k| \leq 2|a_k|, \qquad k = 1, 2, 3, \ldots \tag{1}$$

Since the series $\Sigma|a_k|$ converges, so does the series $2\Sigma|a_k| = \Sigma 2|a_k|$. Thus inequality (1) shows that the series $\Sigma(a_k + |a_k|)$ is a series with positive terms* that is dominated by a convergent series. The Comparison Test therefore guarantees that the series $\Sigma(a_k + |a_k|)$ converges. We may now apply Theorem 6 (Section 12.3) to conclude that the series

$$\Sigma a_k = \Sigma(a_k + |a_k|) - \Sigma|a_k|$$

converges. This completes the proof. ◆

Theorem 13 is simple to apply. If the series Σa_k contains negative terms, we test the positive series $\Sigma|a_k|$ for convergence. If $\Sigma|a_k|$ converges, so does Σa_k. However, if $\Sigma|a_k|$ diverges, we can draw no conclusions about the original series Σa_k.

Example 3

To test the series

$$\sum_{k=0}^{\infty} \frac{(-1)^k k^2}{(k+1)!}$$

for convergence, we apply the Ratio Test to the series

$$\sum_{k=0}^{\infty} \left| \frac{(-1)^k k^2}{(k+1)!} \right| = \sum_{k=0}^{\infty} \frac{k^2}{(k+1)!}.$$

We find that

$$\rho = \lim_{k \to \infty} \frac{a_{k+1}}{a_k} = \lim_{k \to \infty} \frac{\dfrac{(k+1)^2}{(k+2)!}}{\dfrac{k^2}{(k+1)!}} = \lim_{k \to \infty} \frac{(k+1)^2}{k^2(k+2)} = 0.$$

The series $\displaystyle\sum_{k=0}^{\infty} \frac{(-1)^k k^2}{(k+1)!}$ therefore converges absolutely. ◇

---

*If $a_k < 0$, then $a_k + |a_k| = 0$, so this term drops out of the series. This is why $\Sigma(a_k + |a_k|)$ may be regarded as a series with positive terms.

Example 4

The series

$$\sum_{k=0}^{\infty} \frac{\cos \pi k}{2^k} = 1 - \frac{1}{2} + \frac{1}{4} - \frac{1}{8} + \cdots$$

contains both positive and negative terms. To test for absolute convergence we consider the series (which turns out to be a geometric series)

$$\sum_{k=0}^{\infty} \left| \frac{\cos \pi k}{2^k} \right| = \sum_{k=0}^{\infty} \frac{|\cos \pi k|}{2^k} = \sum_{k=0}^{\infty} \frac{1}{2^k}$$

$$= \sum_{k=0}^{\infty} \left(\frac{1}{2} \right)^k = \frac{1}{1 - 1/2} = 2 \qquad \text{(geometric)}.$$

The series $\sum_{k=0}^{\infty} \dfrac{\cos \pi k}{2^k}$ therefore converges absolutely. ◇

Example 5

For the series

$$\sum_{k=1}^{\infty} \frac{(-1)^{k+1}}{k} = 1 - \frac{1}{2} + \frac{1}{3} - \frac{1}{4} + \cdots$$

the series of absolute values is the harmonic series $\sum_{k=1}^{\infty} 1/k$, which diverges. Theorem 13 therefore gives no conclusion concerning the convergence of the series $\sum_{k=1}^{\infty} \dfrac{(-1)^{k+1}}{k}$. This question will be resolved shortly. ◇

Example 6

If the variable x is unrestricted, the series

$$\sum_{k=1}^{\infty} \frac{x^k}{k} = x + \frac{x^2}{2} + \frac{x^3}{3} + \cdots$$

contains negative terms for $x < 0$. To determine the numbers x for which this series converges we consider the series of absolute values

$$\sum_{k=1}^{\infty} \left| \frac{x^k}{k} \right| = \sum_{k=1}^{\infty} \frac{|x|^k}{k}.$$

Applying the Ratio Test to this series, we find that, for $x \neq 0$,

$$\rho = \lim_{k \to \infty} \frac{|a_{k+1}|}{|a_k|} = \lim_{k \to \infty} \frac{\dfrac{|x|^{k+1}}{k+1}}{\dfrac{|x|^k}{k}} = \lim_{k \to \infty} |x| \left(\frac{k}{k+1} \right) = |x|.$$

Thus, $\rho < 1$ only if $|x| < 1$. This shows that $\displaystyle\sum_{k=1}^{\infty} x^k/k$ converges if $|x| < 1$. However, we obtain no conclusions for $|x| \geq 1$ by this procedure. $\diamond$

Alternating Series

Several types of infinite series containing negative terms can be shown to converge even though they do not converge absolutely. Among these are certain of the *alternating series*.

DEFINITION 5

An **alternating series** is a series of the form

$$\sum_{k=0}^{\infty} (-1)^k a_k = a_0 - a_1 + a_2 - a_3 + a_4 - \cdots$$

where $a_k > 0$ for each $k = 0, 1, 2, 3, \ldots$.

The following theorem indicates the conditions under which an alternating series converges. We will demonstrate its use before we give a proof.

THEOREM 14
Alternating Series Test

Let $a_k > 0$ for each $k = 0, 1, 2, \ldots$. The alternating series

$$\sum_{k=0}^{\infty} (-1)^k a_k$$

converges if both the following hold:

(i) The terms of the sequence $\{a_k\}_{k=0}^{\infty}$ are decreasing, and
(ii) $\displaystyle\lim_{k \to \infty} a_k = 0$.

Note that requirement (ii) in Theorem 14 is just a restatement of the necessary condition for convergence given by Theorem 7.

Example 7

For the series

$$\sum_{k=1}^{\infty} \frac{(-1)^{k+1}}{k} = 1 - \frac{1}{2} + \frac{1}{3} - \frac{1}{4} + \cdots$$

the general term is $(-1)^{k+1} a_k$, where $a_k = \dfrac{1}{k} > 0$. Since the sequence $\left\{\dfrac{1}{k}\right\}$ is decreasing and $\displaystyle\lim_{k \to \infty} \frac{1}{k} = 0$, the series converges by Theorem 14. $\diamond$

Combining the results of Examples 5 and 7, we see that the **alternating harmonic series** $\displaystyle\sum_{k=1}^{\infty} \frac{(-1)^{k+1}}{k} = 1 - \frac{1}{2} + \frac{1}{3} - \frac{1}{4} + \cdots$ converges, but does not converge absolutely. (This shows that the converse of Theorem 13 is not true.) Such series are said to *converge conditionally*.

DEFINITION 6	The series Σa_k is said to **converge conditionally** if Σa_k converges and $\Sigma|a_k|$ diverges.

Example 8

Determine whether the series

$$\sum_{k=1}^{\infty} \frac{(-1)^{k+1}k}{1+k^2}$$

converges absolutely, converges conditionally, or diverges.

Strategy

First, test for absolute convergence.

Solution

The series of absolute values is

$$\sum_{k=1}^{\infty} \frac{k}{1+k^2}.$$

Use the Integral Test on the series of absolute values.

We may test this series by the Integral Test. Since

$$\int_1^{\infty} \frac{x}{1+x^2} \, dx = \lim_{t\to\infty} \int_1^t \frac{x}{1+x^2} \, dx$$

$$= \lim_{t\to\infty} \frac{1}{2} \ln(1+t^2) - \frac{1}{2} \ln 2 = +\infty,$$

the series of absolute values diverges. Thus, the original series does not converge absolutely.

Since the series is alternating but does not converge absolutely, apply the alternating series test.

We next test for conditional convergence using Theorem 14. Since

$$\frac{d}{dx}\left(\frac{x}{1+x^2}\right) = \frac{1+x^2-x(2x)}{(1+x^2)^2} = \frac{1-x^2}{(1+x^2)^2}$$

is negative for $x>1$, the sequence $\left\{\dfrac{k}{1+k^2}\right\}$ is decreasing. Also, $\lim_{k\to\infty} \dfrac{k}{1+k^2} = 0$. Thus, both conditions of Theorem 14 are met, so the series converges conditionally. ◇

Example 9

For which numbers x does the series

$$\sum_{k=1}^{\infty} \frac{x^k}{k}$$

converge?

Solution: We have already determined that the series converges absolutely for $|x| < 1$ (Example 6). If $|x| > 1$, we may apply l'Hôpital's Rule (differentiating with respect to k) to conclude that

$$\lim_{k \to \infty} \frac{|x|^k}{k} \qquad \left(\frac{\infty}{\infty} \text{ form}\right)$$

$$= \lim_{k \to \infty} \frac{|x|^k \ln|x|}{1} = +\infty.$$

Thus, $\lim_{k \to \infty} \dfrac{x^k}{k}$ does not exist if $|x| > 1$, so $\displaystyle\sum_{k=1}^{\infty} \dfrac{x^k}{k}$ diverges for these values of x, by Theorem 7. The remaining cases are $x = 1$ and $x = -1$.

If $x = 1$, the series is the harmonic series $\sum_{k=1}^{\infty} 1/k$, which diverges. If $x = -1$, the series is

$$\sum_{k=1}^{\infty} \frac{(-1)^k}{k} = -1 + \frac{1}{2} - \frac{1}{3} + \frac{1}{4} - \cdots$$

This is the negative of the alternating harmonic series, which converges conditionally (Example 7).

We have therefore shown that the series $\sum_{k=1}^{\infty} x^k/k$ converges for $x \in [-1, 1)$ and diverges for all other numbers x. In particular, the series converges absolutely for $x \in (-1, 1)$ and converges conditionally for $x = -1$. $\diamondsuit$

Proof of Theorem 14: For the alternating series

$$\sum_{k=0}^{\infty} (-1)^k a_k,$$

the even partial sums have the form

$$S_{2n} = (a_0 - a_1) + (a_2 - a_3) + \cdots + (a_{2n-2} - a_{2n-1}) + a_{2n}.$$

Since the sequence $\{a_k\}$ is decreasing, each of the terms in parentheses is positive. This shows that all even partial sums are positive. Also, since $a_{2n+1} > a_{2n+2}$

$$\begin{aligned} S_{2n+2} &= S_{2n} - a_{2n+1} + a_{2n+2} \\ &= S_{2n} - (a_{2n+1} - a_{2n+2}) \\ &< S_{2n}. \end{aligned} \qquad (2)$$

Thus, the sequence of even partial sums $\{S_{2n}\}$ is decreasing. In particular, this shows that the sequence of positive terms $\{S_{2n}\}$ is bounded above by $S_0 = a_0$ and below by zero. Thus $\{S_{2n}\}$ is a bounded decreasing sequence which, by Corollary 1, must converge. We denote the limit of this sequence by $S = \lim_{n \to \infty} S_{2n}$.

Next, since $\lim_{n \to \infty} a_{2n+1} = 0$, we see that

$$\begin{aligned} \lim_{n \to \infty} S_{2n+1} &= \lim_{n \to \infty} (S_{2n} - a_{2n+1}) \\ &= S - 0 \\ &= S, \end{aligned}$$

so both even and odd partial sums converge to S. Thus $S = \lim_{n \to \infty} S_n$ and the series $\sum_{k=0}^{\infty} (-1)^k a_k$ converges. $\blacklozenge$

Estimating Remainders for Alternating Series

For convergent alternating series satisfying the hypotheses of Theorem 14, there is a simple way to obtain both an estimate of the sum and a bound on the error associated with that estimate.

As in line (2) in the proof of Theorem 14, we may show that the odd partial sums of an alternating series $\sum_{k=0}^{\infty} (-1)^k a_k$ form an *increasing* sequence:

$$S_{2n+1} = S_{2n-1} + (a_{2n} - a_{2n+1}) \tag{3}$$
$$> S_{2n-1}, \quad n = 1, 2, \ldots$$

since $a_{2n} > a_{2n+1}$.

If we write

$$S = \sum_{k=0}^{\infty} (-1)^k a_k,$$

inequalities (2) and (3) together imply that

$$S_{2p-1} < S < S_{2m} \tag{4}$$

for any positive integers p and m. That is, *any* even partial sum is larger than *any* odd partial sum, and the sum of the series lies between them (see Figure 7.1).

Now let

$$S = S_n + R_n \tag{5}$$

where

$$R_n = \sum_{k=n+1}^{\infty} (-1)^k a_k.$$

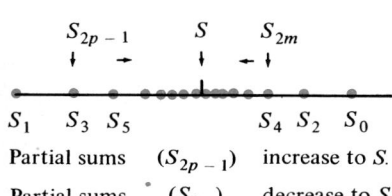

S_{2p-1} S S_{2m}

S_1 S_3 S_5 S_4 S_2 S_0

Partial sums (S_{2p-1}) increase to S.

Partial sums (S_{2m}) decrease to S.

Figure 7.1

Then R_n is the **truncation error,** or **remainder,** associated with the approximation of S by the partial sum S_n. From (4) and (5) it follows, by setting $m = n$ and $p = n + 1$, that

$$|R_{2n}| = |S - S_{2n}| < |S_{2n+1} - S_{2n}| = a_{2n+1}. \tag{6}$$

Similarly, by setting $m = p = n$ we find

$$|R_{2n-1}| = |S - S_{2n-1}| < |S_{2n} - S_{2n-1}| = a_{2n}. \tag{7}$$

Thus, regardless of whether we truncate an alternating series after an even or an odd number of terms, *the absolute value of the remainder (or tail) is less than the absolute value of the next term in the series*. This result is sufficiently important that we carefully summarize what we have proved in the form of a theorem.

THEOREM 15
Truncation Error

The absolute value of the error associated with approximating the sum of the convergent alternating series satisfying the hypotheses of Theorem 14

$$\sum_{k=0}^{\infty} (-1)^k a_k$$

by the partial sum

$$S_n = \sum_{k=0}^{n} (-1)^k a_k$$

is less than the first truncated term, a_{n+1}.

Example 10

Find an approximation to the sum of the alternating harmonic series

$$\sum_{k=0}^{\infty} (-1)^k \left(\frac{1}{k+1}\right) = 1 - \frac{1}{2} + \frac{1}{3} - \frac{1}{4} + \cdots$$

accurate to within .05.

Solution: If we approximate this sum by the partial sum

$$S_n = \sum_{k=0}^{n} (-1)^k \left(\frac{1}{k+1}\right),$$

the error in this approximation is less than $a_{n+1} = \dfrac{1}{n+2}$. We therefore choose n sufficiently large to satisfy the inequality

$$\frac{1}{n+2} \leq .05 = \frac{1}{20}.$$

This gives

$$n + 2 \geq 20, \quad \text{or} \quad n \geq 18.$$

The partial sum S_{18} therefore has the desired accuracy. It is

$$S_{18} = 1 - \frac{1}{2} + \frac{1}{3} - \frac{1}{4} + \cdots + \frac{1}{19} \approx 0.72.$$

$\diamondsuit$

Example 11

How many terms must be included in the partial sum for the alternating series

$$\sum_{k=0}^{\infty} \frac{(-1)^k}{1 + k^2} = 1 - \frac{1}{2} + \frac{1}{5} - \frac{1}{10} + \cdots$$

to ensure that the partial sum approximates the sum of the series accurately to within 0.01?

Solution: It is easier to work with S_{n-1}, so that the absolute value of the first truncated term is $a_n = \dfrac{1}{1 + n^2}$. We therefore need

$$\frac{1}{1 + n^2} \leq \frac{1}{100}, \quad \text{or} \quad n^2 \geq 99.$$

The choice $n = 10$ therefore will suffice, and the partial sum S_9 will have the desired accuracy.

$\diamondsuit$

Exercise Set 12.7

In each of Exercises 1–30, determine whether the series converges absolutely, converges conditionally, or diverges.

1. $\displaystyle\sum_{k=1}^{\infty} \frac{(-1)^k}{k^3}$

2. $\displaystyle\sum_{k=1}^{\infty} \frac{(-1)^k}{2k+1}$

3. $\displaystyle\sum_{k=1}^{\infty} \frac{(-1)^{k+1}k^2}{(k+2)!}$

4. $\displaystyle\sum_{k=2}^{\infty} (-1)^{k+1} \frac{k}{\ln k}$

5. $\displaystyle\sum_{k=1}^{\infty} (-1)^k \frac{k!}{(2k+1)!}$

6. $\displaystyle\sum_{k=1}^{\infty} \frac{(-k)^3}{3^k}$

7. $\displaystyle\sum_{k=1}^{\infty} \frac{(-1)^k}{1+\sqrt{k}}$

8. $\displaystyle\sum_{k=1}^{\infty} \frac{\cos \pi k}{k}$

9. $\displaystyle\sum_{k=1}^{\infty} \frac{\sin\left(\frac{\pi k}{2}\right)}{k}$

10. $\displaystyle\sum_{k=1}^{\infty} \frac{(-1)^k}{\sqrt{k(k+2)}}$

11. $\displaystyle\sum_{k=1}^{\infty} \frac{(-1)^k \cdot k^3}{2^{k+2}}$

12. $\displaystyle\sum_{k=2}^{\infty} \frac{k^2(-1)^{k+1}}{\ln k}$

13. $\displaystyle\sum_{k=2}^{\infty} \frac{k^2(-1)^{k+1}}{\ln k}$

14. $\displaystyle\sum_{k=1}^{\infty} \frac{(2k+1)(-1)^k}{6k+2}$

15. $\displaystyle\sum_{k=1}^{\infty} \frac{(-1)^k \sqrt{k}}{k+1}$

16. $\displaystyle\sum_{k=1}^{\infty} \frac{(-1)^{k+1} 2^k}{k^3 \cdot 3^{k+2}}$

17. $\displaystyle\sum_{k=2}^{\infty} \frac{(-1)^k}{k \ln^2 k}$

18. $\displaystyle\sum_{k=1}^{\infty} \frac{(-1)^{k+1} k!}{6^k}$

19. $\displaystyle\sum_{k=1}^{\infty} \frac{(-1)^k}{1+k^2}$

20. $\displaystyle\sum_{k=1}^{\infty} \sin\left(\frac{k\pi}{4}\right)$

21. $\displaystyle\sum_{k=2}^{\infty} \frac{(-1)^k k}{\ln \sqrt{k}}$

22. $\displaystyle\sum_{k=1}^{\infty} \frac{(-1)^k k^2}{(2k+1)(k+3)}$

23. $\displaystyle\sum_{k=1}^{\infty} \frac{(-1)^k \sinh k}{3e^{2k}}$

24. $\displaystyle\sum_{k=1}^{\infty} \frac{(-1)^k \operatorname{Tan}^{-1} k}{\sqrt{k}}$

25. $\displaystyle\sum_{k=1}^{\infty} \frac{\cos \pi k}{k+2}$

26. $\displaystyle\sum_{k=2}^{\infty} \frac{(-1)^k}{\sqrt[k]{\ln k}}$

27. $\displaystyle\sum_{k=2}^{\infty} \frac{k \sin\left[\frac{(2k+1)\pi}{2}\right]}{\sqrt{1+e^{\sqrt{k}}}}$

28. $\displaystyle\sum_{k=2}^{\infty} \frac{(-1)^k(k^{3/2}+3k)}{7-k^2+2k^{5/2}}$

29. $\displaystyle\sum_{k=1}^{\infty} \frac{\sinh k - \cosh k}{k}$

30. $\displaystyle\sum_{k=2}^{\infty} \frac{(-1)^k \cosh 2k}{ke^k}$

In Exercises 31–34, determine the numbers x for which the given series converges absolutely.

31. $\displaystyle\sum_{k=0}^{\infty} \frac{x^k}{k!} = 1 + x + \frac{x^2}{2!} + \frac{x^3}{3!} + \cdots$

32. $\displaystyle\sum_{k=0}^{\infty} \frac{x^k}{2k+1} = 1 + \frac{x}{3} + \frac{x^2}{5} + \cdots$

33. $\displaystyle\sum_{k=0}^{\infty} (-1)^k \frac{x^{2k+1}}{(2k+1)!} = x - \frac{x^3}{3!} + \frac{x^5}{5!} - \frac{x^7}{7!} + \cdots$

34. $\displaystyle\sum_{k=0}^{\infty} (-1)^k \frac{x^{2k}}{(2k)!} = 1 - \frac{x^2}{2!} + \frac{x^4}{4!} - \frac{x^6}{6!} + \cdots$

35. How large must n be taken so that the partial sum $S_n = \displaystyle\sum_{k=0}^{n} (-1)^k \left(\frac{1}{1+k}\right)$ approximates the sum of the alternating harmonic series accurate to within .01?

36. With what accuracy does the sum

$$\sum_{k=0}^{10} \frac{(-1)^k}{(k+1)^3}$$

approximate the sum of the series

$$\sum_{k=0}^{\infty} \frac{(-1)^k}{(k+1)^3}?$$

37. Prove that if Σa_k and Σb_k converge absolutely, then so does $\Sigma(a_k + b_k)$.

38. Prove that if Σa_k converges absolutely and c is any real number, the series Σca_k converges absolutely.

39. Prove for the (not necessarily positive) series Σa_k that if

$$\rho = \lim_{k \to \infty} \left| \frac{a_{k+1}}{a_k} \right|,$$

   (i) Σa_k converges absolutely if $0 \le \rho < 1$;
   (ii) Σa_k diverges if $\rho > 1$;
   (iii) no conclusions may be drawn if $\rho = 1$.

40. Prove that Σa_k^2 converges if Σa_k converges absolutely. Is the converse true?

SUMMARY OUTLINE OF CHAPTER 12

◆ An **infinite sequence** $\{a_n\}$ is a function whose domain is the set of positive integers. (page 510)

◆ $L = \lim\limits_{n \to \infty} a_n$ if, given $\epsilon > 0$, there exists an integer N so that $|a_n - L| < \epsilon$ whenever $n > N$. (page 513)

◆ The sequence $\{a_n\}$ is said to **converge** if $L = \lim\limits_{n \to \infty} a_n$ exists. (page 512)

◆ The nth **partial sum** of the infinite series $\sum\limits_{k=1}^{\infty} a_k$ is the sum of the first n terms: $S_n = \sum\limits_{k=1}^{n} a_k$. (page 524)

◆ S is called the **sum** of the infinite series if $S = \lim\limits_{n \to \infty} S_n = \lim\limits_{n \to \infty} \sum\limits_{k=1}^{n} a_k$. (page 524)

◆ The series Σa_k is said to **converge** if the sum S exists. (page 524)

◆ The **geometric series** $\sum\limits_{k=0}^{\infty} x^k = 1 + x + x^2 + x^3 + \cdots$ converges, if $|x| < 1$, to the sum $S = \dfrac{1}{1-x}$, and diverges otherwise. (page 527)

◆ The **harmonic series** $\sum\limits_{k=1}^{\infty} \dfrac{1}{k} = 1 + \dfrac{1}{2} + \dfrac{1}{3} + \dfrac{1}{4} + \cdots$ diverges. (page 535)

◆ The **p-series** $\sum\limits_{k=1}^{\infty} \dfrac{1}{k^p} = 1 + \dfrac{1}{2^p} + \dfrac{1}{3^p} + \dfrac{1}{4^p} + \cdots$ converges if $p > 1$ and diverges otherwise. (page 536)

◆ A **necessary condition** for Σa_k to converge: $\lim\limits_{k \to \infty} a_k = 0$ (however, this condition does not guarantee convergence). (page 530)

◆ *Tests for convergence of Σa_k when $a_k > 0$ for all k:*
  (1) (Comparison Test)   Let $0 < a_k \le b_k$ for all k. (page 538)
     a. If Σb_k converges, Σa_k converges.
     b. If Σa_k diverges, Σb_k diverges.

  (2) (Integral Test)   If $f(k) = a_k$ for all k and f is decreasing on $[1, \infty)$ then Σa_k converges if and only if (page 534)
     $\displaystyle\int_1^{\infty} f(x)\, dx$ converges.

  (3) (Ratio Test)   If $\rho = \lim\limits_{k \to \infty} \dfrac{a_{k+1}}{a_k}$ exists, then (page 544)
     a. Σa_k converges if $\rho < 1$.
     b. Σa_k diverges if $\rho > 1$.
     c. No conclusion may be drawn if $\rho = 1$.

  (4) (Root Test)   If $\rho = \lim\limits_{k \to \infty} \sqrt[k]{a_k}$ exists, same conclusions as for the Ratio Test. (page 547)

  (5) (Limit Comparison Test)   If $b_k > 0$ for all k and $\rho = \lim\limits_{k \to \infty} \dfrac{a_k}{b_k} > 0$ exists, then Σa_k and Σb_k either both converge or both diverge. (page 540)

◆ The series Σa_k **converges absolutely** if $\Sigma |a_k|$ converges. (page 549)

◆ *Theorem:* If $\Sigma |a_k|$ converges, then Σa_k converges. (page 550)

◆ An **alternating series** is a series of the form $\sum\limits_{k=0}^{\infty} (-1)^k a_k$ with all $a_k > 0$. (page 552)

◆ **Theorem:**   An alternating series converges if   (page 552)
  a. $a_k > a_{k+1}$ for all k,   and
  b. $\lim\limits_{k \to \infty} a_k = 0$.

◆ The absolute value of the **error** associated with the approximation of the sum of the alternating series $\sum\limits_{k=0}^{\infty} (-1)^k a_k$ by   (page 555)

the partial sum $S_n = \sum\limits_{k=0}^{n} (-1)^k a_k$ is less than the first discarded term, a_{n+1}.

REVIEW EXERCISES—CHAPTER 12

In Exercises 1–10, determine whether the given sequence converges. If it does, find its limit.

1. $\left\{ \sin \dfrac{n\pi}{2} \right\}$

2. $\left\{ \dfrac{1}{\sqrt{n} + 1} \right\}$

3. $\left\{ \dfrac{\sqrt{n}}{\ln(\sqrt{n} + 1)} \right\}$

4. $\left\{ \left(1 - \dfrac{2}{n}\right)^{2n} \right\}$

5. $\left\{ \dfrac{3^n}{n!} \right\}$

6. $\left\{ \dfrac{(-1)^{n+1}}{\sqrt{n^2 + 1}} \right\}$

7. $\left\{ \dfrac{(-1)^{n+1} \cdot n^3}{n(1 - n + n^2)} \right\}$

8. $\left\{ \dfrac{n^2 + 2n + 2}{n^{3/2}} \right\}$

9. $\{ e^{2 \ln (n)} \}$

10. $\{ \ln(n + 1) - \ln(n - 1) \}$

In Exercises 11–14, find the sum of the given geometric series.

11. $1 + \dfrac{1}{6} + \dfrac{1}{36} + \dfrac{1}{216} + \cdots$

12. $10 + \dfrac{10}{2} + \dfrac{10}{4} + \dfrac{10}{8} + \cdots$

13. $\dfrac{2}{3} + \dfrac{4}{9} + \dfrac{8}{27} + \cdots$

14. $1 - \dfrac{1}{5} + \dfrac{1}{25} - \dfrac{1}{125} + \cdots$

In Exercises 15 and 16, express the repeating decimal as an infinite series and find its rational form.

15. $.37\overline{37} \ldots$

16. $.0263263\overline{263} \ldots$

In Exercises 17–20, find the sum of the series.

17. $\sum\limits_{k=3}^{\infty} \left(\dfrac{1}{4}\right)^k$

18. $\sum\limits_{k=1}^{\infty} \dfrac{3^k + 2^k}{5^k}$

19. $\sum\limits_{k=2}^{\infty} \dfrac{20}{3^{k+1}}$

20. $\sum\limits_{k=1}^{\infty} \left[\dfrac{3}{2^k} + \dfrac{1}{(k + 1)(k + 2)} \right]$

In Exercises 21–58, determine whether the given series converges or diverges and state which test you used.

21. $\sum\limits_{k=1}^{\infty} \dfrac{1}{k(k + 1)}$

22. $\sum\limits_{k=1}^{\infty} \dfrac{3^k}{k^3}$

23. $\sum\limits_{k=3}^{\infty} \dfrac{2^k}{k^5}$

24. $\sum\limits_{k=1}^{\infty} \dfrac{(-1)^k}{2k + 1}$

25. $\sum\limits_{k=1}^{\infty} \dfrac{k^3}{k!}$

26. $\sum\limits_{k=0}^{\infty} \dfrac{1}{10k + 2}$

27. $\sum\limits_{k=2}^{\infty} \dfrac{\cos \pi k}{\sqrt{k}}$

28. $\sum\limits_{k=2}^{\infty} \dfrac{k^k}{k!}$

29. $\sum\limits_{k=3}^{\infty} \dfrac{\ln k}{k^2}$

30. $\sum\limits_{k=0}^{\infty} \dfrac{2^k + k}{(k + 1)!}$

31. $\sum\limits_{k=1}^{\infty} \dfrac{1}{9 + k^2}$

32. $\sum\limits_{k=2}^{\infty} \dfrac{(-1)^k k}{\ln k}$

33. $\sum\limits_{k=2}^{\infty} \dfrac{2}{k \ln(k + 1)}$

34. $\sum\limits_{k=1}^{\infty} \dfrac{\sqrt{k}}{2^k}$

35. $\sum\limits_{k=0}^{\infty} \dfrac{k^4 \cdot 2^k}{(k + 2)!}$

36. $\sum\limits_{k=1}^{\infty} (-1)^k \dfrac{\ln k}{k}$

37. $\sum\limits_{k=0}^{\infty} \dfrac{1}{\sqrt{4 + k^2}}$

38. $\sum\limits_{k=1}^{\infty} \left(\dfrac{k - 1}{k + 1}\right)^k$

39. $\sum\limits_{k=1}^{\infty} (-1)^{k+1} \dfrac{\sqrt{k}}{2k + 1}$

40. $\sum\limits_{k=1}^{\infty} k^2 e^{-k^3}$

41. $\displaystyle\sum_{k=1}^{\infty} \left(\frac{k!}{k^k}\right)^k$

42. $\displaystyle\sum_{k=2}^{\infty} (-1)^k \frac{k^2 + 2}{1 + k + k^2}$

43. $\displaystyle\sum_{k=1}^{\infty} 2ke^{-k}$

44. $\displaystyle\sum_{k=0}^{\infty} \left(\frac{k}{1+k}\right)^k$

45. $\displaystyle\sum_{k=0}^{\infty} (-1)^k \sin \pi k$

46. $\displaystyle\sum_{k=2}^{\infty} \frac{(2k+1)!}{k^2(k+1)!}$

47. $\displaystyle\sum_{k=1}^{\infty} \frac{k!}{e^{2k}}$

48. $\displaystyle\sum_{k=2}^{\infty} \frac{1}{(k+6)^{3/2}}$

49. $\displaystyle\sum_{k=1}^{\infty} (-1)^k \frac{3}{\sqrt{k}}$

50. $\displaystyle\sum_{k=2}^{\infty} \frac{1}{(\ln k)^k}$

51. $\displaystyle\sum_{k=1}^{\infty} \frac{2^{2k-1}}{(2k-1)!}$

52. $\displaystyle\sum_{k=2}^{\infty} \frac{1}{\sqrt{k}(\ln k)^3}$

53. $\displaystyle\sum_{k=0}^{\infty} \frac{(-1)^k}{1 + k^2}$

54. $\displaystyle\sum_{k=1}^{\infty} \mathrm{Tan}^{-1} k$

55. $\displaystyle\sum_{k=1}^{\infty} k\left(\frac{\pi}{k}\right)^k$

56. $\displaystyle\sum_{k=0}^{\infty} \frac{(-1)^k k(k+2)}{(k+1)!}$

57. $\displaystyle\sum_{k=1}^{\infty} \left(\frac{k}{2k+1}\right)^k$

58. $\displaystyle\sum_{k=0}^{\infty} \frac{(-1)^k \mathrm{Sin}^{-1}\left(\frac{1}{k+1}\right)}{k+1}$

In Exercises 59–68, determine whether the given series converges absolutely, converges conditionally, or diverges.

59. $\displaystyle\sum_{k=0}^{\infty} \frac{(-1)^{k+1}}{3k+2}$

60. $\displaystyle\sum_{k=2}^{\infty} \frac{(-1)^k}{k \ln k}$

61. $\displaystyle\sum_{k=1}^{\infty} (-1)^k \frac{k^k}{k!}$

62. $\displaystyle\sum_{k=0}^{\infty} \frac{\cos \pi k}{1 + k^2}$

63. $\displaystyle\sum_{k=1}^{\infty} \frac{(-1)^{k+1}}{k(k+1)}$

64. $\displaystyle\sum_{k=0}^{\infty} (-1)^k \left(\frac{2}{3}\right)^k$

65. $\displaystyle\sum_{k=1}^{\infty} \frac{k!}{(-2)^k}$

66. $\displaystyle\sum_{k=1}^{\infty} \frac{\cos \frac{\pi k}{3}}{1 + k^2}$

67. $\displaystyle\sum_{k=1}^{\infty} \frac{(-1)^k k^2}{(k+1)!}$

68. $\displaystyle\sum_{k=1}^{\infty} \frac{(-1)^{k+1}}{\sqrt{k(k+1)}}$

In Exercises 69–72, find a bound on the error associated with the approximation of the sum of the given series by the partial sum S_4 consisting of the sum of the first four terms.

69. $\displaystyle\sum_{k=1}^{\infty} \frac{2(-1)^k}{1 + k^2} = -1 + \frac{2}{5} - \frac{2}{10} + \cdots$

70. $\displaystyle\sum_{k=1}^{\infty} \frac{(-1)^k}{k^4} = -1 + \frac{1}{16} - \frac{1}{81} + \cdots$

71. $\displaystyle\sum_{k=1}^{\infty} \frac{(-1)^{k+1}}{2k+1} = \frac{1}{3} - \frac{1}{5} + \frac{1}{7} - \frac{1}{9} + \cdots$

72. $\displaystyle\sum_{k=1}^{\infty} \frac{(-1)^{k+1} \sin^2\left(\frac{\pi}{2k}\right)}{k+1} = \frac{1}{2} - \frac{1}{6} + \frac{1}{16} - \cdots$

73. Define the sequence $\{a_k\}$ by $a_1 = c$, $a_{k+1} = a_1 \cdot a_k$, $k \geq 1$. For what values of c does $\{a_k\}$ converge?

74. Prove that if the sequence $\{a_n\}$ is not bounded it cannot converge.

75. If Σa_k converges but $\Sigma(a_k + b_k)$ diverges, what can you conclude about Σb_k?

76. Explain how one could use an infinite series to determine at what time between 1 P.M. and 2 P.M. is the minute hand on a clock directly over the hour hand. (*Hint:* The minute hand moves 12 times as fast as the hour hand.)

77. A ball is dropped from a height of 10 meters. Each time it strikes the ground, it rebounds to 5/6 of the height from which it fell. Find the total distance travelled by the ball
   a. when it has struck the ground 10 times.
   b. when it has come to rest.

78. Prove that if Σa_k diverges then $\Sigma c a_k$ diverges for any number $c \neq 0$.

79. Let Σa_k and Σb_k be series with positive terms. If $\Sigma(a_k + b_k)^2$ converges, what can you conclude about
   a. Σa_k^2?
   b. Σb_k^2?
   c. $\Sigma a_k b_k$?

80. Let $p(k)$ and $q(k)$ be polynomials in k with positive coefficients. Let n be the degree of $p(k)$ and let m be the degree of $q(k)$. Show that
   a. $\displaystyle\sum \frac{p(k)}{q(k)}$ converges if $m > n + 1$
   b. $\displaystyle\sum \frac{p(k)}{q(k)}$ diverges if $m \leq n + 1$.

Chapter 13
Taylor Polynomials and Power Series

This chapter begins by posing the *approximation problem:* given a differentiable function *f*, how can we approximate values of *f* by values of certain polynomials (the latter being easier to calculate)?

We shall answer this question by developing the notion of *Taylor Polynomials.* In doing so we shall see that Taylor Polynomials may be regarded as partial sums for *power series,* which are infinite series involving powers of an independent variable. Applying the theory and techniques developed in Chapter 12 for infinite series of constants, we shall show that a convergent power series determines a differentiable function, and that many familiar functions, including the trigonometric, exponential, and logarithmic functions, may be represented as power series.

13.1 THE APPROXIMATION PROBLEM AND TAYLOR POLYNOMIALS

The **approximation problem** is this: *Given a function f and a number x = a, how can we approximate values f(x) for x near a using polynomials, and what is the accuracy of these approximations?*

Of course, if *f* is a polynomial the answer is easy. We simply use *f* to approximate itself, with complete accuracy. Our real interest is in knowing how to approximate transcendental functions, like sin *x* or e^x, or algebraic functions like $\sqrt[5]{x}$.

The reason for our interest in approximating such functions is that although we have developed a rich theory of differentiation and integration for these functions, we do not yet have available a means for calculating their values, except in special cases. For example, how does one calculate sin 53°, $e^{0.05}$, or $\sqrt[5]{216}$? The theory of differentials (Chapter 3) is a first step in this direction, although it is quite limited. Even if you feel secure in the knowledge that your calculator is equipped with a special key for each of these functions or that a five-place table for such functions will always be available somewhere, the question remains as to how that calculator was programmed or how the entries in such tables are determined.

As stated above, our objective is to approximate arbitrary functions using *polynomials.* The reason for using polynomials is that values of polynomial functions can be easily computed with complete accuracy by simple multiplications and additions. In other words, we want to know how to approximate troublesome functions with convenient ones.

Taylor Polynomials

Given the function *f*, suppose we ask for the polynomial of lowest degree that best approximates *f(x)* near *x = a*. Since the polynomials of lowest degree are constants,

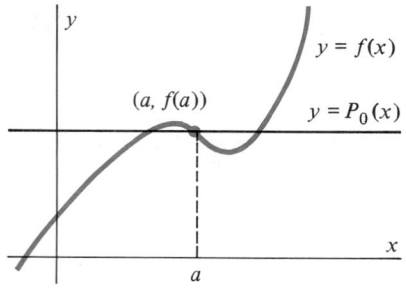

Figure 1.1 Constant polynomial $P_0 \equiv f(a)$ approximates f near a.

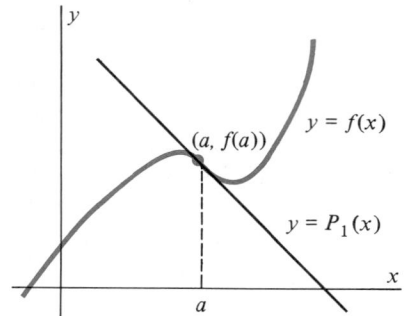

Figure 1.2 First degree polynomial $P_1(x) = f'(a)(x - a) + f(a)$.

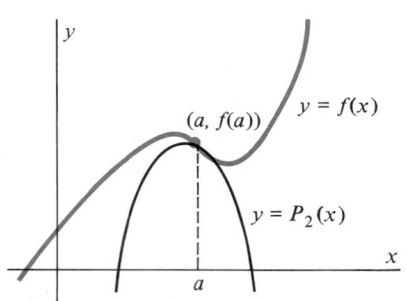

Figure 1.3 Second degree approximating polynomial P_2.

the obvious answer to this question is the constant polynomial

$$P_0(x) \equiv f(a) \qquad \text{(Figure 1.1)}. \tag{1}$$

Next, let us ask for the polynomial of degree one (that is, a linear function) that best approximates f near $x = a$. We shall write this polynomial in the form

$$P_1(x) = b + c(x - a) \tag{2}$$

and seek to determine the constants b and c. Since we again want $P_1(a) = f(a)$, we obtain $b = f(a)$. Now by the theory of the derivative, the line through $(a, f(a))$ that best approximates f has slope $f'(a)$, if $f'(a)$ exists. Thus, we also want $f'(a) = P_1'(a) = c$. P_1 is therefore

$$P_1(x) = f(a) + f'(a)(x - a). \qquad \text{(See Figure 1.2.)} \tag{3}$$

Continuing in this way, we write the second degree polynomial approximating $f(x)$ as

$$P_2(x) = b + c(x - a) + d(x - a)^2.$$

The condition that $P_2(a) = f(a)$ gives $b = f(a)$, and the condition that the polynomial and the function f have the same slope at $x = a$ gives $P_2'(a) = c = f'(a)$. To determine the remaining constant d we require that P_2 and f have the same measure of concavity at $x = a$; that is, we set $P_2''(a) = f''(a)$.

This gives the equation

$$P_2''(a) = 2d = f''(a).$$

Thus, $d = \dfrac{f''(a)}{2}$, and P_2 is therefore (see Figure 1.3)

$$P_2(x) = f(a) + f'(a)(x - a) + \frac{f''(a)}{2}(x - a)^2. \tag{4}$$

We must carry this analysis one step further for the general form of these polynomials to become clear. Although we do not have available a geometric interpretation for derivatives of order higher than two, the idea is that for an approximating polynomial P_n of degree n, each derivative of order 1 through n must equal the corresponding derivative of f at $x = a$.

Write the general third-degree polynomial $P_3(x)$ in the form

$$P_3(x) = b + c(x - a) + d(x - a)^2 + e(x - a)^3.$$

Then, requiring that P_3 and f, together with their first three derivatives, agree at $x = a$ gives

$$P_3(a) = b = f(a); \qquad\qquad b = f(a)$$
$$P_3'(a) = c = f'(a); \qquad\qquad c = f'(a)$$
$$P_3''(a) = 2d = f''(a); \qquad\qquad d = \frac{f''(a)}{2}$$
$$P_3'''(a) = 3 \cdot 2 \cdot e = f'''(a); \qquad e = \frac{f'''(a)}{3!}.$$

The polynomial P_3 is therefore (see Figure 1.4)

$$P_3(x) = f(a) + f'(a)(x - a) + \frac{f''(a)}{2}(x - a)^2 + \frac{f'''(a)}{3!}(x - a)^3. \tag{5}$$

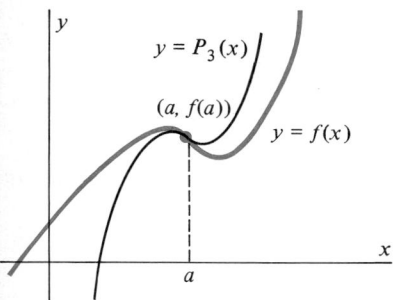

Figure 1.4 Third degree approximating polynomial P_3.

(The notation $n!$ in equation (5) is the **factorial symbol,** defined by

$$n! = n(n-1)(n-2)\cdots 3 \cdot 2 \cdot 1.$$

We also define $0! = 1$ and $1! = 1$.)

We can generalize equations (1)–(5) by stating that the approximating polynomial of degree n for the function f near $x = a$ is, for $n \geq 2$,

$$P_n(x) = f(a) + f'(a)(x-a) + \frac{f''(a)}{2}(x-a)^2 \qquad (6)$$
$$+ \frac{f'''(a)}{3!}(x-a)^3 + \cdots + \frac{f^{(n)}(a)}{n!}(x-a)^n.$$

The polynomial P_n in (6) is referred to as the nth **Taylor polynomial*** for f, expanded about $x = a$. The ideas we discuss here were developed by the English mathematician Brook Taylor (1686–1731) early in the eighteenth century.

Figures 1.1 through 1.4 suggest that the polynomials P_n more closely approximate the function f near $x = a$ as the degree of P_n increases. We will verify this observation for particular choices of the function f. In Section 13.2 we will take up the question of how well the polynomial P_n approximates f. The remaining part of this section concerns finding P_n for various choices of f and a.

Example 1

Find P_0, P_1, P_2, and P_3 for the function $f(x) = e^x$ at $a = 0$ and use these polynomials to approximate $e = f(1)$.

Solution: The values of $f(x) = e^x$ and the first three derivatives at $a = 0$ are as follows:

$$\begin{aligned}
f(x) &= e^x; & f(0) &= e^0 = 1 \\
f'(x) &= e^x; & f'(0) &= 1 \\
f''(x) &= e^x; & f''(0) &= 1 \\
f'''(x) &= e^x; & f'''(0) &= 1.
\end{aligned}$$

From this information and equation (6) we obtain

$$P_0(x) = 1$$
$$P_1(x) = 1 + x$$
$$P_2(x) = 1 + x + \frac{x^2}{2}$$
$$P_3(x) = 1 + x + \frac{x^2}{2} + \frac{x^3}{3!}.$$

Using these polynomials we may approximate the value $f(1) = e^1 = e$ by the numbers

$$P_0(1) = 1$$
$$P_1(1) = 1 + 1 = 2$$

---

*There is a slight abuse of terminology here. Since it may be the case that $f^{(n)}(a) = 0$, an nth Taylor polynomial as defined above may actually be a polynomial of lower degree (see Example 2).

$$P_2(1) = 1 + 1 + \frac{1^2}{2} = \frac{5}{2} = 2.5$$

$$P_3(1) = 1 + 1 + \frac{1^2}{2} + \frac{1^2}{6} = \frac{8}{3} = 2.6666\ \overline{\ }\ \ldots$$

Figure 1.5 shows graphs of these Taylor polynomials and the approximations $P_n(1)$ to the value $f(1) = e^1 = e$. Table 1.1 compares, to four decimal place accu-

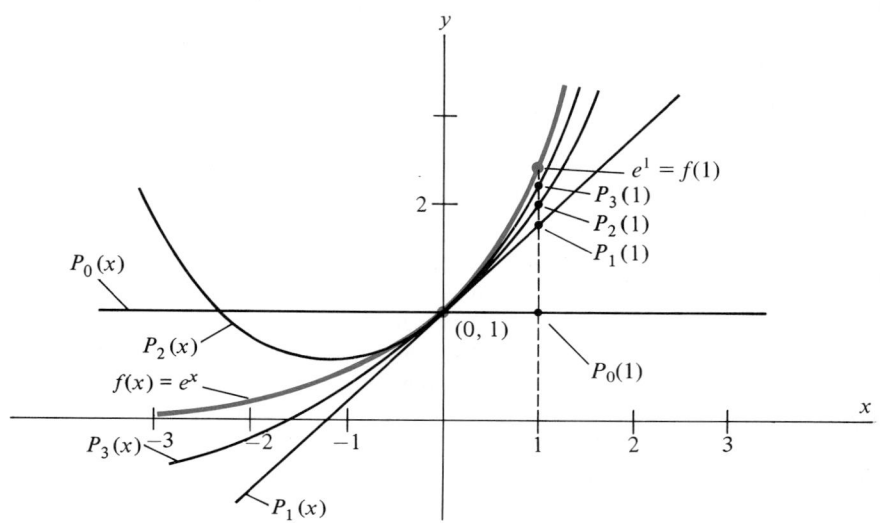

Figure 1.5 Taylor polynomials for $f(x) = e^x$ expanded about $a = 0$ and approximations $P_n(1)$ to $e^1 = e$.

racy, values of these Taylor polynomials and the function $f(x) = e^x$ for several different numbers x.

Table 1.1 Values of Taylor approximations to e^x near $a = 0$

x	$P_0(x)$	$P_1(x)$	$P_2(x)$	$P_3(x)$	e^x
-1.5	1.0	-0.5	0.6250	0.0625	0.2231
-1.0	1.0	0	0.5	0.3333	0.3679
0	1.0	1.0	1.0	1.0	1.0
0.5	1.0	1.5	1.625	1.6458	1.6487
1.0	1.0	2.0	2.5	2.6667	2.7183
1.5	1.0	2.5	3.6250	4.1875	4.4817

Example 2

Find the Taylor polynomials $P_0, P_1, P_2, \ldots, P_5$ for $f(x) = \sin x$ expanded about $a = 0$.

Solution: Here

$$f(x) = \sin x; \qquad f(0) = 0$$
$$f'(x) = \cos x; \qquad f'(0) = 1$$
$$f''(x) = -\sin x; \qquad f''(0) = 0$$
$$f'''(x) = -\cos x; \qquad f'''(0) = -1$$
$$f^{(4)}(x) = \sin x; \qquad f^{(4)}(0) = 0$$
$$f^{(5)}(x) = \cos x; \qquad f^{(5)}(0) = 1.$$

Thus, by (6),

$$P_0(x) = 0$$
$$P_1(x) = 0 + 1 \cdot x = x$$

$$P_2(x) = 0 + 1 \cdot x + \frac{0}{2} \cdot x^2 = x$$

$$P_3(x) = 0 + 1 \cdot x + \frac{0}{2}x^2 + \frac{-1}{3!}x^3 = x - \frac{x^3}{3!}$$

$$P_4(x) = 0 + 1 \cdot x + \frac{0}{2}x^2 + \frac{-1}{3!}x^3 + \frac{0}{4!}x^4 = x - \frac{x^3}{3!}$$

$$P_5(x) = 0 + 1 \cdot x + \frac{0}{2}x^2 + \frac{-1}{3!}x^3 + \frac{0}{4!}x^4 + \frac{1}{5!}x^5 = x - \frac{x^3}{3!} + \frac{x^5}{5!}.$$

(See Figure 1.6.)

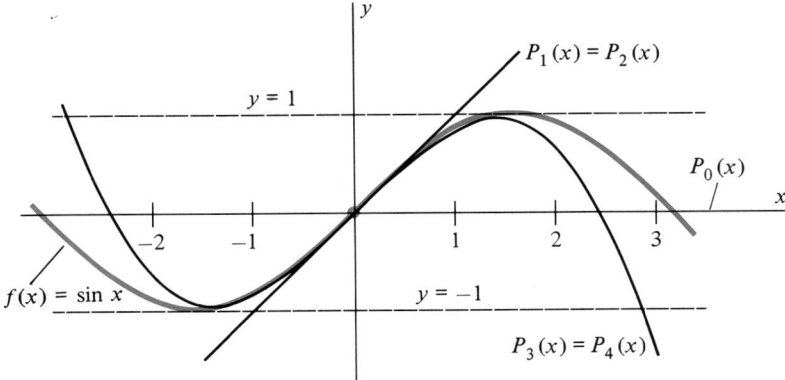

Figure 1.6 Five Taylor polynomials for $f(x) = \sin x$ at $a = 0$. Note that $P_1(x)$ and $P_2(x)$ coincide, as do $P_3(x)$ and $P_4(x)$.

Example 3

In Example 2 the even powers of x in the Taylor polynomials for $f(x) = \sin x$ at $a = 0$ dropped out, since even derivatives of $\sin x$ are zero at $a = 0$. For comparison, we next determine the fourth Taylor polynomial for $f(x) = \sin x$ expanded about $a = \pi/6$.

$$f(x) = \sin x; \qquad f(\pi/6) = \frac{1}{2}$$

$$f'(x) = \cos x; \qquad f'(\pi/6) = \frac{\sqrt{3}}{2}$$

$$f''(x) = -\sin x; \qquad f''(\pi/6) = -\frac{1}{2}$$

$$f'''(x) = -\cos x; \qquad f'''(\pi/6) = -\frac{\sqrt{3}}{2}$$

$$f^{(4)}(x) = \sin x; \qquad f^{(4)}(\pi/6) = \frac{1}{2}.$$

By (6),

$$P_4(x) = \frac{1}{2} + \frac{\sqrt{3}}{2}\left(x - \frac{\pi}{6}\right) + \left(\frac{1}{2}\right)\left(-\frac{1}{2}\right)\left(x - \frac{\pi}{6}\right)^2$$

$$+ \left(\frac{1}{3!}\right)\left(-\frac{\sqrt{3}}{2}\right)\left(x - \frac{\pi}{6}\right)^3 + \left(\frac{1}{4!}\right)\left(\frac{1}{2}\right)\left(x - \frac{\pi}{6}\right)^4$$

$$= \frac{1}{2} + \frac{\sqrt{3}}{2}\left(x - \frac{\pi}{6}\right) - \frac{1}{4}\left(x - \frac{\pi}{6}\right)^2 - \frac{\sqrt{3}}{12}\left(x - \frac{\pi}{6}\right)^3$$

$$+ \frac{1}{48}\left(x - \frac{\pi}{6}\right)^4.$$

Example 4

Find the Taylor polynomials $P_0, P_1, \ldots, P_4$ for the function $f(x) = \cos x$ expanded about $a = 0$.

Solution: The required derivatives are as follows:

$$f(x) = \cos x; \qquad f(0) = 1$$
$$f'(x) = -\sin x; \qquad f'(0) = 0$$
$$f''(x) = -\cos x; \qquad f''(0) = -1$$
$$f'''(x) = \sin x; \qquad f'''(0) = 0$$
$$f^{(4)}(x) = \cos x; \qquad f^{(4)}(0) = 1.$$

The Taylor polynomials are

$$P_0(x) = 1$$
$$P_1(x) = 1 + 0 \cdot x = 1$$

$$P_2(x) = 1 + 0 \cdot x + \frac{-1}{2}x^2 = 1 - \frac{x^2}{2}$$

$$P_3(x) = 1 + 0 \cdot x + \frac{-1}{2}x^2 + \frac{0}{3!}x^3 = 1 - \frac{x^2}{2}$$

$$P_4(x) = 1 + 0 \cdot x + \frac{-1}{2}x^2 + \frac{0}{3!}x^3 + \frac{1}{4!}x^4 = 1 - \frac{x^2}{2} + \frac{x^4}{4!}.$$

Example 5

Find the Taylor polynomials $P_0, P_1, \ldots, P_5$ expanded about $a = 1$ for the function $f(x) = \ln x$ and use these polynomials to approximate $\ln 2$.

Solution: We have

$$f(x) = \ln x; \qquad f(1) = \ln 1 = 0$$

$$f'(x) = \frac{1}{x} = x^{-1}; \qquad f'(1) = 1^{-1} = 1$$

$$f''(x) = -x^{-2}; \qquad f''(1) = -1^{-2} = -1$$
$$f'''(x) = 2x^{-3}; \qquad f'''(1) = 2 \cdot 1^{-3} = 2$$
$$f^{(4)}(x) = -6x^{-4}; \qquad f^{(4)}(1) = -6 \cdot 1^{-4} = -6$$
$$f^{(5)}(x) = 24x^{-5}; \qquad f^{(5)}(1) = 24 \cdot 1^{-5} = 24.$$

Thus,

$$P_0(x) = 0$$
$$P_1(x) = 0 + 1 \cdot (x - 1) = x - 1$$

$$P_2(x) = 0 + 1 \cdot (x - 1) + \frac{-1}{2}(x - 1)^2 = (x - 1) - \frac{1}{2}(x - 1)^2$$

$$P_3(x) = 0 + 1 \cdot (x - 1) + \frac{-1}{2}(x - 1)^2 + \frac{2}{3!}(x - 1)^3$$

$$= (x - 1) - \frac{1}{2}(x - 1)^2 + \frac{1}{3}(x - 1)^3$$

$$P_4(x) = 0 + 1 \cdot (x - 1) + \frac{-1}{2}(x - 1)^2 + \frac{2}{3!}(x - 1)^3 + \frac{-6}{4!}(x - 1)^4$$

$$= (x - 1) - \frac{1}{2}(x - 1)^2 + \frac{1}{3}(x - 1)^3 - \frac{1}{4}(x - 1)^4$$

and

$$P_5(x) = 0 + 1 \cdot (x - 1) + \frac{-1}{2}(x - 1)^2 + \frac{2}{3!}(x - 1)^3 + \frac{-6}{4!}(x - 1)^4$$

$$+ \frac{24}{5!}(x - 1)^5$$

$$= (x - 1) - \frac{1}{2}(x - 1)^2 + \frac{1}{3}(x - 1)^3 - \frac{1}{4}(x - 1)^4 + \frac{1}{5}(x - 1)^5.$$

Since $\ln 2 = f(2)$, we obtain approximations to $\ln 2$ from these polynomials by setting $x = 2$:

$$P_0(2) = 0$$
$$P_1(2) = (2 - 1) = 1$$

$$P_2(2) = (2 - 1) - \frac{1}{2}(2 - 1)^2 = 1 - \frac{1}{2} = \frac{1}{2} = 0.5$$

$$P_3(2) = (2 - 1) - \frac{1}{2}(2 - 1)^2 + \frac{1}{3}(2 - 1)^3 = 1 - \frac{1}{2} + \frac{1}{3} = \frac{5}{6} \approx 0.8333$$

$$P_4(2) = (2 - 1) - \frac{1}{2}(2 - 1)^2 + \frac{1}{3}(2 - 1)^3 - \frac{1}{4}(2 - 1)^4$$

$$= 1 - \frac{1}{2} + \frac{1}{3} - \frac{1}{4} = \frac{7}{12} \approx 0.5833$$

$$P_5(2) = (2-1) - \frac{1}{2}(2-1)^2 + \frac{1}{3}(2-1)^3 - \frac{1}{4}(2-1)^4 + \frac{1}{5}(2-1)^5$$

$$= 1 - \frac{1}{2} + \frac{1}{3} - \frac{1}{4} + \frac{1}{5}$$

$$= \frac{47}{60} \approx 0.7833.$$

By comparison, the exact value of ln 2, to four decimal places, is 0.6931. ◇

Exercise Set 13.1

In Exercises 1–16, find the nth Taylor polynomial for the function f expanded about $x = a$.

1. $f(x) = e^{-x}$, $a = 0$, $n = 4$

2. $f(x) = e^x$, $a = 1$, $n = 4$

3. $f(x) = \cos x$, $a = \pi/4$, $n = 6$

4. $f(x) = \sin x$, $a = \pi/3$, $n = 5$

5. $f(x) = \ln(1 + x)$, $a = 0$, $n = 4$

6. $f(x) = \dfrac{1}{1 + x}$, $a = 0$, $n = 4$

7. $f(x) = \tan x$, $a = 0$, $n = 3$

8. $f(x) = \dfrac{1}{1 + x^2}$, $a = 0$, $n = 2$

9. $f(x) = e^{x^2}$, $a = 0$, $n = 3$

10. $f(x) = \sqrt{x}$, $a = 2$, $n = 2$

11. $f(x) = \sqrt{1 - x}$, $a = 0$, $n = 3$

12. $f(x) = x \sin x$, $a = 0$, $n = 7$

13. $f(x) = \sec x$, $a = \pi/4$, $n = 3$

14. $f(x) = \text{Tan}^{-1} x$, $a = 0$, $n = 3$

15. $f(x) = \dfrac{1}{1 + e^x}$, $a = 0$, $n = 3$

16. $f(x) = x^4$, $a = 2$, $n = 4$

17. Compare the results of Exercises 5 and 6 and those of Exercises 8 and 14. Formulate a conjecture on the relationship between the Taylor polynomial for f and that for f' and test your conjecture on a few examples.

18. What is the relationship between the Taylor polynomial for f at $a = 0$ and that for $h(x) = xf(x)$? Can you prove this?

19. True or false? If f is an even function, the Taylor polynomial for f expanded about $a = 0$ will contain only even powers of x. Explain.

20. The Taylor polynomial P_n is defined for f if each of the first n derivatives of f exists at $x = a$. Does this imply that $f(x)$ is defined for all x for which $P_n(x)$ is defined? (*Hint:* Consider this question for the function in Example 5.)

13.2 TAYLOR'S THEOREM

If P_n denotes the nth Taylor polynomial for the function f at $x = a$, our claim is that $P_n(x)$ may be used to approximate $f(x)$ for x near a. In order to determine the accuracy of this approximation we write

$$f(x) = P_n(x) + R_n(x). \tag{1}$$

Since $R_n(x) = f(x) - P_n(x)$, the term $R_n(x)$ represents the *error* made in approximating $f(x)$ by $P_n(x)$ (see Figure 2.1). It is referred to as the **remainder** in a Taylor approximation.

Taylor's Theorem provides a way to estimate the size of $R_n(x)$ when making a Taylor approximation.

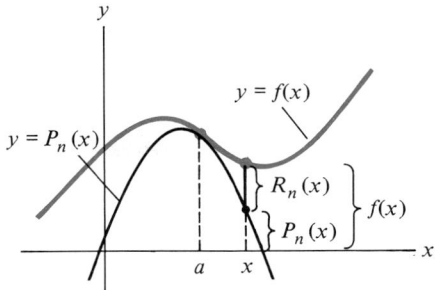

Figure 2.1 $R_n(x) = f(x) - P_n(x)$.

THEOREM 1
Taylor's Theorem

Let n be an integer and let f be a function for which $f^{(n+1)}(x)$ exists for each x in the closed interval $[\alpha, \beta]$. Let a be any number in the open interval (α, β). Then for each $x \in [\alpha, \beta]$ there exists a number c between a and x so that

$$f(x) = f(a) + f'(a)(x - a) + \frac{f''(a)}{2!}(x - a)^2 + \cdots$$

$$+ \frac{f^{(n)}(a)}{n!}(x - a)^n + \frac{f^{(n+1)}(c)}{(n+1)!}(x - a)^{n+1}.$$

(This is called Taylor's formula with remainder, or just Taylor's formula.) In other words,

$$f(x) = P_n(x) + R_n(x)$$

where $P_n(x)$ is the Taylor polynomial of degree n and

$$R_n(x) = \frac{f^{(n+1)}(c)}{(n+1)!}(x - a)^{n+1}. \tag{2}$$

We may paraphrase Taylor's Theorem by saying that the error involved in approximating a value of an $(n + 1)$ times differentiable function by the value of its Taylor polynomial $P_n(x)$ is just the term we would add to $P_n(x)$ to obtain $P_{n+1}(x)$, except that the coefficient $\dfrac{f^{(n+1)}(c)}{(n+1)!}$ is evaluated at a specific number *between* a and x rather than at $x = a$.

We will see in the next section how to determine a bound for $R_n(x)$.

Example 1

Write Taylor's formula using the third Taylor polynomial for $f(x) = e^x$ expanded about $a = 0$.

Solution: From Example 1 of Section 13.1 we have

$$P_3(x) = 1 + x + \frac{x^2}{2!} + \frac{x^3}{3!}.$$

For $f(x) = e^x$, $f^{(4)}(x) = e^x$, so $f^{(4)}(c) = e^c$. Thus, by Taylor's Theorem,

$$R_3(x) = \frac{e^c}{4!}(x - 0)^4$$

where c lies between 0 and x. Taylor's formula is therefore

$$e^x = 1 + x + \frac{x^2}{2} + \frac{x^3}{3!} + \frac{e^c}{4!}x^4. \qquad \diamondsuit$$

Since Taylor's Theorem is both important and useful, it is unfortunate that no easily motivated proof has been found. We present one of the standard proofs in the remainder of this section.

Proof of Theorem 1: In addition to the number a, choose a number $b \in [\alpha, \beta]$. Then we can define a number C (which depends on a, b, and n) by the equation

$$f(b) = f(a) + f'(a)(b - a) + \cdots + \frac{f^{(n)}(a)}{n!}(b - a)^n + C. \qquad (3)$$

That is, C is the difference $C = f(b) - P_n(b)$, where $P_n(b)$ is the Taylor polynomial of degree n expanded about a.

First note that if $b = a$, then (3) collapses to $f(a) = f(a)$, and $C = 0$. Thus, we consider $b \neq a$ (it makes no difference whether $a < b$ or $b < a$) and attempt to prove that C has the form of remainder $R_n(x)$ given in the theorem.

In order to do this, we construct a rather odd-looking function of x, whose purpose is to allow us to apply Rolle's Theorem on the interval between a and b:

$$F(x) = f(b) - \left\{ f(x) + f'(x)(b - x) + \cdots + \frac{f^{(n)}(x)}{n!}(b - x)^n \right.$$

$$\left. + C\frac{(b - x)^{n+1}}{(b - a)^{n+1}} \right\}. \qquad (4)$$

Now we show the following facts about F:

(i) If $x = a$, then

$$F(a) = f(b) - \left\{ f(a) + f'(a)(b - a) + \cdots \right.$$

$$\left. + \frac{f^{(n)}(a)}{n!}(b - a)^n + C\frac{(b - a)^{n+1}}{(b - a)^{n+1}} \right\}$$

$$= 0 \qquad \text{(by (3))}.$$

(ii) If $x = b$, then

$$F(b) = f(b) - \left\{ f(b) + f'(b)(b - b) + \cdots \right.$$

$$\left. + \frac{f^{(n)}(b)}{n!}(b - b)^n + C\frac{(b - b)^{n+1}}{(b - a)^{n+1}} \right\}$$

$$= f(b) - f(b) = 0.$$

(iii) Because $f^{(n+1)}(x)$ exists for each $x \in [\alpha, \beta]$, F is differentiable on the entire interval, and

$$F'(x) = -\frac{f^{(n+1)}(x)}{n!}(b - x)^n + (n + 1)C\frac{(b - x)^n}{(b - a)^{n+1}}.$$

To see that this is true, we simply differentiate F in equation (4), using the Product Rule, and note that many terms cancel:

$$F'(x) = 0 - \left\{ f'(x) + f''(x)(b - x) - f'(x) \right.$$

$$+ \frac{f'''(x)}{2}(b - x)^2 - f''(x)(b - x)$$

$$+ \frac{f^{(4)}(x)}{3!}(b - x)^3 - \frac{f'''(x)}{2}(b - x)^2$$

$$+ \cdots + \frac{f^{(n)}(x)}{(n-1)!}(b - x)^{n-1} - \frac{f^{(n-1)}(x)}{(n-2)!}(b - x)^{n-2}$$

$$+ \frac{f^{(n+1)}(x)}{n!}(b - x)^n - \frac{f^{(n)}(x)}{(n-1)!}(b - x)^{n-1} - (n + 1)C\frac{(b - x)^n}{(b - a)^{n+1}} \right\}$$

$$= -\frac{f^{(n+1)}(x)}{n!}(b - x)^n + (n + 1)C\frac{(b - x)^n}{(b - a)^{n+1}}.$$

From statements (i) and (ii) and Rolle's Theorem, it follows that there is a number c between a and b such that $F'(c) = 0$. Using statement (iii), we may write this condition as

$$\frac{f^{(n+1)}(c)}{n!}(b - c)^n = (n + 1)C\frac{(b - c)^n}{(b - a)^{n+1}}.$$

Solving for C now gives

$$C = \frac{f^{(n+1)}(c)}{(n + 1)!}(b - a)^{n+1}. \tag{5}$$

Since we have placed no restriction on b other than $b \neq a$ (and we already know that $C = 0$ if $b = a$), we can substitute x for b in (5), which yields exactly $C = R_n(x)$. This completes the proof. ◆

Exercise Set 13.2

In Exercises 1–14, write Taylor's formula (Theorem 1) for f using the nth Taylor polynomial expanded about $x = a$.

1. $f(x) = e^{-x}$, $n = 3$, $a = 0$

2. $f(x) = \sin x$, $n = 3$, $a = 0$

3. $f(x) = \cos x$, $n = 4$, $a = 0$

4. $f(x) = \sin x$, $n = 3$, $a = \dfrac{\pi}{4}$

5. $f(x) = \text{Tan}^{-1} x$, $n = 3$, $a = 0$

6. $f(x) = \sqrt{x}$, $n = 3$, $a = 4$

7. $f(x) = \dfrac{1}{1 + x^2}$, $n = 2$, $a = 1$

8. $f(x) = 3x^4 + 2x + 2$, $n = 3$, $a = 2$

9. $f(x) = \sec x$, $n = 2$, $a = \dfrac{\pi}{4}$

10. $f(x) = \ln(1 + x^2)$, $n = 3$, $a = 0$

11. $f(x) = \sinh x$, $n = 4$, $a = 0$

12. $f(x) = x \sinh x$, $n = 3$, $a = 0$

13. $f(x) = \cosh x$, $n = 3$, $a = \ln 2$

14. $f(x) = (1 + x)^{3/2}$, $n = 3$, $a = 0$

15. Let f be a polynomial of degree n. Use Theorem 1 to prove that $f = P_n$.

16. Refer to the precise hypotheses of Rolle's Theorem in Chapter 4. Then show that the hypothesis of Theorem 1 that $f^{(n+1)}(x)$ exists for each x in $[\alpha, \beta]$ can be relaxed to the condition that $f^{(n)}$ is continuous on $[\alpha, \beta]$ and $f^{n+1}(x)$ exists for each x in (α, β).

17. Use Taylor's Theorem to justify Newton's Method for approximating the root of a function.

18. Let P_n be the nth Taylor polynomial for $f(x) = \sin x$ expanded about $a = 0$. Show that $\lim_{n \to \infty} R_n(x) = 0$ if $|x| < 1$.

19. By multiplying both sides of the equation by $1 - x$, show that

$$\frac{1}{1 - x} = 1 + x + x^2 + x^3 + \cdots$$

$$+ x^n + \frac{x^{n+1}}{1 - x}, \qquad x \neq 1.$$

Conclude that $R_n(x) = \dfrac{x^{n+1}}{1 - x}$. Does this contradict Taylor's Theorem? Use Taylor's Theorem to verify that, for $f(x) = \dfrac{1}{1 - x}$ and $a = 0$, the nth Taylor polynomial for $f(x) = \dfrac{1}{1 - x}$ is

$$P_n(x) = 1 + x + x^2 + \cdots + x^n.$$

20. By substituting $-x$ for x in the equation in Exercise 19 conclude that

$$\frac{1}{1 + x} = 1 - x + x^2 - x^3 + \cdots$$

$$+ (-1)^n x^n + \frac{(-1)^{n+1} x^{n+1}}{1 + x}, \qquad x \neq -1.$$

Using Taylor's Theorem with $a = 0$, verify that the nth Taylor polynomial for $f(x) = \dfrac{1}{1 + x}$ is

$$P_n(x) = 1 - x + x^2 - x^3 + \cdots + (-1)^n x^n.$$

21. Replace x by x^2 in Exercise 20 to conclude that

$$\frac{1}{1 + x^2} = 1 - x^2 + x^4 - \cdots$$

$$+ (-1)^n x^{2n} + \frac{(-1)^{n+1} x^{2n+2}}{1 + x^2}.$$

What is the nth Taylor Polynomial P_n, expanded about $a = 0$, for $f(x) = \dfrac{1}{1 + x^2}$?

22. Replace x by t in Exercise 20 and integrate between 0 and x to conclude that

$$\int_0^x \frac{1}{1 + t} \, dt = \int_0^x \left(1 - t + t^2 - t^3 + \cdots \right.$$

$$\left. + (-1)^{n-1} t^{n-1} + \frac{(-1)^n t^n}{1 + t} \right) dt$$

$$= x - \frac{x^2}{2} + \frac{x^3}{3} - \frac{x^4}{4} + \cdots + \frac{(-1)^{n-1} x^n}{n}$$

$$+ (-1)^n \int_0^x \frac{t^n}{1 + t} \, dt.$$

What is the nth Taylor polynomial, expanded about $a = 0$, for $f(x) = \ln(1 + x)$?

23. Let P_n be as in Exercise 18. Show that $\lim_{n \to \infty} R_n(x) = 0$ for *all* x. $\left(\textit{Hint:} \text{ You must first show that } \lim_{n \to \infty} \dfrac{x^n}{n!} = 0. \right)$

24. Use Taylor's Theorem to prove the Binomial Theorem:

$$(x + a)^n = a^n + na^{n-1}x + \frac{n(n-1)}{2} a^{n-2} x^2 + \cdots$$

$$+ \binom{n}{r} a^{n-r} x^r + \cdots + nax^{n-1} + x^n$$

where n is a positive integer and $\binom{n}{r}$ is the **binomial coefficient**

$$\binom{n}{r} = \frac{n!}{r!(n - r)!}.$$

13.3 APPLICATIONS OF TAYLOR'S THEOREM

According to Taylor's Theorem, if we approximate $f(x)$ by the value of the nth Taylor polynomial $P_n(x)$ expanded about $x = a$, the error (remainder) in the calculation is

$$R_n(x) = \frac{f^{(n+1)}(c)}{(n + 1)!} (x - a)^{n+1} \tag{1}$$

where c lies between a and x. Since x is given, the choices that influence the size of the error involve the numbers a and n. Obviously, we will want to choose a close to

x so that the factors $(x - a)$ in $R_n(x)$ are small. Also, we will want to choose a so that the function value $f(a)$ and the various derivatives $f^{(k)}(a)$ are easy to compute.

Once a convenient value for a is chosen, the magnitude of the error depends only on the integer n. Since we seldom know the number c, we usually cannot determine $R_n(x)$ precisely. However, in most applications we are concerned only with knowing the approximate size of $R_n(x)$. For example, if we are to approximate $\ln(1.2)$ to within 0.01, we need only to be able to demonstrate that $|R_n(1.2)| < 0.01$.

In general, if we wish to achieve a Taylor approximation for $f(x)$ to within ϵ, the objective will be to find n sufficiently large to guarantee that the inequality

$$\left| \frac{f^{(n+1)}(c)}{(n + 1)!}(x - a)^{n+1} \right| < \epsilon$$

holds. By (1) this will assure the desired accuracy.

Example 1

Use the Taylor polynomial of degree 3 for $f(x) = \ln x$ expanded about $a = 1$ to approximate $\ln(1.5)$ and find the accuracy of this approximation.

Strategy	**Solution**
Find P_3 as in Section 13.2.	We have

$$f(x) = \ln x; \qquad f(1) = 0$$

$$f'(x) = \frac{1}{x}; \qquad f'(1) = 1$$

$$f''(x) = -\frac{1}{x^2}; \qquad f''(1) = -1$$

$$f'''(x) = \frac{2}{x^3}; \qquad f'''(1) = 2.$$

Thus

$$P_3(x) = 1(x - 1) - \frac{1}{2}(x - 1)^2 + \frac{2}{3!}(x - 1)^3.$$

Evaluate P_3 at $x = 1.5$ to obtain approximation $P_3(1.5)$.

The approximation is therefore

$$\ln(1.5) \approx P_3(1.5) = (.5) - \frac{1}{2}(.5)^2 + \frac{1}{3}(.5)^3$$

$$= .416\overline{6}.$$

Since $f^{(4)}(x) = -6x^{-4}$ for $f(x) = \ln x$, by (1) the error in the approximation is

Write expression for $|R_3(1.5)|$ using (1).

$$|R_3(1.5)| = \left| \frac{-6c^{-4}}{4!}(1.5 - 1)^4 \right| = \left| \frac{-6c^{-4}}{24}(.5)^4 \right|$$

$$= |c^{-4}| \cdot \frac{(.5)^4}{4}.$$

Since c is unknown, we must determine the largest possible value for $|f^{(4)}(c)|$ to obtain an upper bound on the error. We know that c is between $a = 1$ and $x = 1.5$.

Since $|c^{-4}| < 1$ for $1 < c < 1.5$, we have the inequality

$$|R_3(1.5)| < \frac{(.5)^4}{4} = .015625.$$

We may conclude only that the approximation is accurate to one decimal place. ◇

REMARK: When we say that a number is accurate to k decimal places, we mean that the error is less than $5 \times 10^{-(k+1)}$. That is, accuracy to one decimal place means an error less than $5 \times 10^{-2} = .05$, accuracy to two decimal places means an error less than .005, and so forth. Thus we could claim that the approximation in Example 1 was accurate to one decimal place, and the approximation in Example 2 (below) is accurate to three decimal places.

Example 2

Suppose we wish to use the approximation

$$\sin x \approx x$$

for numbers satisfying the inequality $|x| < \pi/45$. What is the maximum possible error associated with this approximation?

Strategy

Determine the Taylor polynomial $P_n(x)$ associated with the approximation.

Once $f(x)$, n, and a are known, use equation (1).

Estimate the maximum possible size of $|f''(c)|$. Use to obtain bound on $|R_2(x)|$.

Solution

For $f(x) = \sin x$, both $P_1(x) = x$ and $P_2(x) = x$ when $a = 0$. We may therefore take $n = 2$ as corresponding to the above approximation. Since $f'''(c) = -\cos(c)$, the error is

$$|R_2(x)| = \left| \frac{-\cos(c)}{3!}(x - 0)^3 \right|$$

$$= \frac{|\cos(c)|}{3!}|x|^3.$$

Since $|\cos(c)| \leq 1$ for all numbers c and since $|x| < \pi/45$, we have

$$|R_2(x)| \leq \frac{1}{3!}|x|^3 < \frac{1}{3!}\left(\frac{\pi}{45}\right)^3 < .00006.$$

The maximum error in the approximation is $.00006 = 6 \times 10^{-5}$. Note that if we had used $n = 1$ in these calculations, we would have had $|f''(c)| = |-\sin(c)| \leq 1$ and we could have claimed only that

$$|R_1(x)| \leq \frac{1}{2!}|x|^2 < \frac{1}{2}\left(\frac{\pi}{45}\right)^2 < .0025. \qquad ◇$$

Example 3

How large must n be chosen so that $\cos 48°$ is approximated with four decimal place accuracy using a Taylor polynomial for $f(x) = \cos x$ expanded about $a = 45° = \pi/4$?

Strategy

Set up the expression for $|R_n(x)|$ using

$$x = 48° = \frac{4\pi}{15}.$$

Find an upper bound for $|R_n(x)|$.

Solution

For $f(x) = \cos x$, $f^{(n+1)}(c)$ is either $\pm \sin c$ or $\pm \cos c$. In either case, $|f^{(n+1)}(c)| \leq 1$. Using this inequality we obtain

$$\left| R_n\left(\frac{4\pi}{15}\right) \right| = \left| \frac{f^{(n+1)}(c)}{(n + 1)!}\left(\frac{4\pi}{15} - \frac{\pi}{4}\right)^{n+1} \right| \qquad (2)$$

$$< \frac{1}{(n + 1)!} \cdot \left(\frac{\pi}{60}\right)^{n+1}.$$

Require that the upper bound be less than the desired degree of accuracy.

To ensure that $\left| R_n\left(\dfrac{4\pi}{15}\right) \right| < 5 \times 10^{-5}$, we need to find n large enough so that

$$\frac{1}{(n+1)!} \cdot \left(\frac{\pi}{60}\right)^{n+1} < 5 \times 10^{-5}. \tag{3}$$

We now proceed simply by trial and error to find n sufficiently large that (3) holds. Values of the left side of inequality (3) for various integers n are as follows

n	$\dfrac{1}{(n+1)!}\left(\dfrac{\pi}{60}\right)^{n+1}$
1	.0013708
2	.0000240
3	.0000003

Thus $n = 2$ is sufficient to give the desired accuracy. ◇

Example 4

Find a bound on the magnitude of $|x|$ so that the approximation

$$e^x \approx P_3(x) = 1 + x + \frac{x^2}{2!} + \frac{x^3}{3!}$$

is accurate to within .001.

Strategy

Solution

First note that the form of the given approximation is a Taylor polynomial expanded about $a = 0$, and that $f^{(n)}(a) = e^0 = 1$ for all n.

Set up the expression for $|R_3(x)|$ using (1).

Since $f^{(n)}(c) = e^c$ for all n, we have

$$|R_3(x)| = \left| \frac{e^c}{4!} x^4 \right| = \frac{e^c}{24} x^4.$$

Find an expression for the maximum size of the factor e^c.

Now since e^c is an increasing function and c lies between 0 and x, the maximum value of e^c for $-|x| < c < |x|$ is $e^{|x|}$. Thus

$$|R_3(x)| \leq \frac{e^{|x|}}{24} x^4. \tag{4}$$

Since e^x is what is being approximated, the factor $e^{|x|}$ must be replaced by a "safe" upper bound.

For $|R_3(x)|$ in (4) to be less than .001, we will clearly need to take $|x| < 1$. Thus we may safely bound the term $e^{|x|}$ by, say, 4, since $e^1 \approx 2.718 < 4.$[*] We therefore need to find the maximum value of x for which

Solve the resulting inequality for x^n.

$$\frac{4}{24} x^4 < .001, \quad \text{or} \quad x^4 < .006.$$

By trial and error (or by extracting the fourth root), find a value of x satisfying the inequality.

Since $(.25)^4 < .0039 < .006$, the bound $|x| < .25$ will assure the desired accuracy. ◇

---

[*] See Exercise 70, Section 8.2, where the inequality $e < 4$ was established.

The exercise set of this section deals with calculations similar to those of Examples 1–4. Before leaving this topic, however, we want to remark on two generalizations suggested by Example 3.

Since $\dfrac{\pi}{60} < 1$, we may write inequality (2) of Example 3 as

$$\left| R_n\left(\frac{4\pi}{15}\right) \right| < \frac{1}{(n+1)!}\left(\frac{\pi}{60}\right)^{n+1} < \frac{1}{(n+1)!}.$$

This shows that $R_n\left(\dfrac{4\pi}{15}\right) \to 0$ as $n \to \infty$. In other words, cos 48° can be approximated to any desired degree of accuracy by simply taking n sufficiently large. Since increasing n corresponds to taking polynomials of higher degree, this means that the *sequence* of estimates.

$$P_0\left(\frac{4\pi}{15}\right) = \frac{\sqrt{2}}{2}$$

$$P_1\left(\frac{4\pi}{15}\right) = \frac{\sqrt{2}}{2} - \frac{\sqrt{2}}{2}\left(\frac{\pi}{60}\right)$$

$$P_2\left(\frac{4\pi}{15}\right) = \frac{\sqrt{2}}{2} - \frac{\sqrt{2}}{2}\left(\frac{\pi}{60}\right) - \frac{\sqrt{2}}{4}\left(\frac{\pi}{60}\right)^2$$

$$P_3\left(\frac{4\pi}{15}\right) = \frac{\sqrt{2}}{2} - \frac{\sqrt{2}}{2}\left(\frac{\pi}{60}\right) - \frac{\sqrt{2}}{4}\left(\frac{\pi}{60}\right)^2 + \frac{\sqrt{2}}{12}\left(\frac{\pi}{60}\right)^3$$

$$\vdots$$

approaches the number $\cos\left(\dfrac{4\pi}{15}\right)$. In the language of Chapter 12 we will say that

(i) the **sequence** of approximations $\left\{ P_0\left(\dfrac{4\pi}{15}\right),\ P_1\left(\dfrac{4\pi}{15}\right),\ P_2\left(\dfrac{4\pi}{15}\right), \ldots \right\}$

approaches $\cos\left(\dfrac{4\pi}{15}\right)$ **in the limit,** that is,

$$\lim_{n\to\infty} P_n\left(\frac{4\pi}{15}\right) = \cos\left(\frac{4\pi}{15}\right);$$

(ii) the **series** of terms

$$\sum_{j=0}^{n} \frac{f^{(j)}(\pi/4)}{j!}\left(\frac{\pi}{60}\right)^j$$

$$= \frac{\sqrt{2}}{2} - \frac{\sqrt{2}}{2}\left(\frac{\pi}{60}\right) - \frac{\sqrt{2}}{4}\left(\frac{\pi}{60}\right)^2 + \cdots + \frac{f^{(n)}\left(\dfrac{\pi}{4}\right)}{n!}\left(\frac{\pi}{60}\right)^n$$

approaches $\cos\left(\dfrac{4\pi}{15}\right)$ **in the limit** as $n \to \infty$, that is,

$$\sum_{j=0}^{\infty} \frac{f^{(j)}\left(\dfrac{4\pi}{15}\right)}{j!}\left(\frac{\pi}{60}\right)^j = \cos\left(\frac{4\pi}{15}\right).$$

These observations give a glimpse of what is ahead: the limit

$$\lim_{n \to \infty} P_n(x) = \sum_{k=0}^{\infty} \frac{f^{(k)}(a)}{k!} (x - a)^k$$

will be viewed as an infinite series containing powers of the variable $x - a$ which, when it converges, will "represent" the function f. Before developing this notion of *Taylor series* we discuss the more general notion of *power series* in Section 13.4.

Exercise Set 13.3

In Exercises 1–8, use Taylor's Theorem to make the indicated approximation and estimate the accuracy using equation (1).

1. $\ln(1.5)$ $f(x) = \ln(x + 1)$, $a = 0$, $n = 3$

2. $\cos 36°$ $f(x) = \cos x$, $a = \pi/4$, $n = 2$

3. $\sqrt{3.91}$ $f(x) = \sqrt{x}$, $a = 4$, $n = 2$

4. $e^{0.2}$ $f(x) = e^x$, $a = 0$, $n = 3$

5. $\cos 1$ $f(x) = \cos x$, $a = \pi/3$, $n = 2$

6. $\mathrm{Sin}^{-1}(0.2)$ $f(x) = \mathrm{Sin}^{-1} x$, $a = 0$, $n = 1$

7. $\sqrt[3]{10}$ $f(x) = \sqrt[3]{x}$, $a = 8$, $n = 2$

8. $\mathrm{Tan}^{-1}\left(\dfrac{1}{2}\right)$ $f(x) = \mathrm{Tan}^{-1} x$, $a = 0$, $n = 2$

In Exercises 9–15, determine a bound on the accuracy of the given approximation for the indicated range of x.

9. $\sin x \approx x$, $|x| < .05$

10. $\sin x \approx x - \dfrac{x^3}{3!}$, $|x| < .15$

11. $\cos x \approx \dfrac{1}{2} - \dfrac{\sqrt{3}}{2}\left(x - \dfrac{\pi}{3}\right)$, $\left|x - \dfrac{\pi}{3}\right| < .05$

12. $\tan x \approx 1 + 2\left(x - \dfrac{\pi}{4}\right)$, $\left|x - \dfrac{\pi}{4}\right| < \dfrac{\pi}{36}$

13. $\sqrt[3]{1 + x} \approx 1 + \dfrac{x}{3}$, $|x| < .025$

14. $\ln x \approx (x - 1) - \dfrac{1}{2}(x - 1)^2 + \dfrac{1}{3}(x - 1)^3$, $|x - 1| < 0.1$

15. $\sqrt{1 + x} \approx 1 + \dfrac{x}{2}$, $0 < x < .02$

In Exercises 16–20, determine how large n must be taken to ensure accuracy to four decimal places in approximating

16. $\sqrt{38}$ using $f(x) = \sqrt{x}$, $a = 36$

17. $\ln 1.3$ using $f(x) = \ln(x + 1)$, $a = 0$

18. $\sin 9°$ using $f(x) = \sin x$, $a = 0$

19. $\cos 42°$ using $f(x) = \cos x$, $a = \pi/4$

20. $e^{0.3}$ using $f(x) = e^x$, $a = 0$.

21. Approximate e correct to four decimal places.

22. (Another way to approximate $\ln x$.)
   a. Find the third Taylor polynomial with $a = 0$ for $f(x) = \ln\left(\dfrac{1 + x}{1 - x}\right)$ including the remainder term.

   b. Find the number x for which $\dfrac{1 + x}{1 - x} = 1.5$.

   c. Find the accuracy in using the polynomial in part (a) to approximate $\ln(1.5)$.

   d. Compare this accuracy with that obtained in Example 1 using the third Taylor polynomial for $f(x) = \ln(1 + x)$.

23. A scientist needing to calculate $\cos x$ for small angles, say $|x| < 6°$, wonders how much accuracy is lost in simply using the approximation $\cos x \approx 1$. What is the answer?

24. Suppose that you needed to make many hand calculations of the function $f(x) = x^5 + 3x^3 + 2x + 6$ for numbers x between 0.8 and 1.0. Explain how you might obtain the approximation

$$f(x) \approx 12 + 16(x - 1).$$

What is the accuracy to be expected from such approximations? What if, instead, you use the approximation

$$f(x) \approx 12 + 16(x - 1) + 19(x - 1)^2?$$

25. Use Taylor's Theorem to prove the second derivative test for relative extrema.

26. Show that $\sin x = x - \dfrac{x^3}{3!} + R_4(x)$ where $|R_4(x)| \leq \dfrac{|x|^5}{5!}$. Use this to show, assuming that $R_4(x)$ is continuous, that

$$\int_0^1 \sin x \, dx = \int_0^1 \left(x - \frac{x^3}{3!} + R_4(x)\right) dx$$

$$= \frac{11}{24} + \int_0^1 R_4(x) \, dx.$$

Conclude that the number $\dfrac{11}{24}$ provides an approximation to the integral

$$\int_0^1 \sin x \, dx$$

with an error E no greater than

$$
\begin{aligned}
E = \left| \int_0^1 \sin x \, dx - \frac{11}{24} \right| &= \left| \int_0^1 R_4(x) \, dx \right| \\
&\leq \int_0^1 |R_4(x)| \, dx \\
&\leq \int_0^1 \frac{x^5}{5!} \, dx \\
&\leq \frac{1}{6!} \approx 0.0014.
\end{aligned}
$$

In Exercises 27–30, use the method of Exercise 26 to obtain an approximation accurate to two decimal places for the given integral.

27. $\displaystyle\int_0^1 \sin x^2 \, dx$

28. $\displaystyle\int_0^1 e^{x^2} \, dx$

29. $\displaystyle\int_0^1 \cos x^2 \, dx$

30. $\displaystyle\int_0^1 \frac{\sin x}{x} \, dx$

13.4 POWER SERIES

Because the geometric series $\sum_{k=0}^\infty x^k$ converges for $|x| < 1$, it actually represents a *function* with domain $(-1, 1)$. From the formula for its sum, we know that this function has equation $f(x) = \dfrac{1}{1 - x}$. That is,

$$f(x) = \frac{1}{1 - x} = \sum_{k=0}^\infty x^k, \qquad -1 < x < 1.$$

In the remaining sections of this chapter we study certain generalizations of the geometric series, called *power series,* to determine when they converge and the properties of the functions they represent. Among these, the Taylor Series will have the special form corresponding to the Taylor Polynomials discussed in Sections 13.1–13.3.

DEFINITION 1

A power series in powers of $x - a$ is an expression of the form

$$\sum_{k=0}^\infty a_k(x - a)^k = a_0 + a_1(x - a) + a_2(x - a)^2 + \cdots \tag{1}$$

$$+ a_k(x - a)^k + \cdots$$

where the coefficients $a_0, a_1, a_2, \ldots$ are constants and x is regarded as an independent variable.*

---

*Also it should be noted at the outset that any infinite series $\Sigma a_k(bx - c)^k$ involving powers of a *linear* function $f(x) = bx - c$ is a power series, since the general term may be written

$$a_k(bx - c)^k = a_k \left[b\left(x - \frac{c}{b}\right) \right]^k = b^k a_k \left(x - \frac{c}{b}\right)^k = A_k(x - C)^k$$

with $A_k = b^k a_k$ and $C = c/b$.

Since expression (1) may be simplified by the change of variable $u = x - a$, we will work almost exclusively with power series of the form

$$\sum_{k=0}^{\infty} a_k x^k = a_0 + a_1 x + a_2 x^2 + \cdots + a_k x^k + \cdots . \tag{2}$$

Here are several examples of power series:

$$\sum_{k=0}^{\infty} x^k = 1 + x + x^2 + x^3 + \cdots + x^k \cdots .$$

$$\sum_{k=0}^{\infty} \frac{x^k}{k!} = 1 + x + \frac{x^2}{2!} + \frac{x^3}{3!} + \cdots + \frac{x^k}{k!} + \cdots \qquad (0! = 1). \tag{3}$$

$$\sum_{k=0}^{\infty} \frac{(-1)^k}{1 + k} x^k = 1 - \frac{x}{2} + \frac{x^2}{3} + \cdots + \frac{(-1)^k}{1 + k} x^k + \cdots . \tag{4}$$

The techniques of Chapter 12 are available to us for determining the numbers x for which such series converge. For example, if we test the series in equation (3) for absolute convergence using the Ratio Test, we find that

$$\rho = \lim_{k \to \infty} \frac{\left| \dfrac{x^{k+1}}{(k+1)!} \right|}{\left| \dfrac{x^k}{k!} \right|} = \lim_{k \to \infty} \frac{|x|}{k + 1} = 0, \qquad x \neq 0,$$

for all $x \neq 0$, so the series in equation (3) converges for all x.

To determine the numbers x for which the series in equation (4) converges, we again apply the Ratio Test to test for absolute convergence: Since

$$\rho = \lim_{k \to \infty} \frac{\left| \dfrac{(-1)^{k+1} x^{k+1}}{k + 2} \right|}{\left| \dfrac{(-1)^k x^k}{k + 1} \right|} = \lim_{k \to \infty} \left(\frac{k + 1}{k + 2} \right) |x| = |x|, \qquad x \neq 0, \tag{5}$$

this power series converges absolutely if $|x| < 1$. Also, we can see that if $x = 1$ we obtain the alternating series $\sum_{k=0}^{\infty} \dfrac{(-1)^k}{1 + k}$, which converges, and when $x = -1$ we obtain the harmonic series $\sum_{k=0}^{\infty} \dfrac{1}{1 + k}$, which diverges. However, this leaves the convergence for $|x| > 1$ yet to be determined. Rather than pursue specific examples in this way, we now establish two theorems that tell us the convergence properties of power series in general.

THEOREM 2

(i) If the power series $\sum a_k x^k$ converges for $x = c \neq 0$, then it converges absolutely for all x with $|x| < |c|$.

(ii) If the power series $\sum a_k x^k$ diverges for $x = d$, then it diverges for all x with $|x| > |d|$.

Proof: To prove part (i) we assume that $\Sigma a_k c^k$ converges. It is a necessary condition (Theorem 7, Chapter 12) that $\lim_{k \to \infty} a_k c^k = 0$. Thus, there exists an integer N so that $|a_k c^k| < 1$ whenever $k > N$. Now let x be any number such that $|x| < |c|$, and let $\gamma = \dfrac{|x|}{|c|} < 1$. Then whenever $k > N$ we have

$$|a_k x^k| = \frac{|a_k c^k x^k|}{|c^k|} = |a_k c^k| \left(\frac{|x|}{|c|} \right)^k < \gamma^k.$$

This shows that the series $\displaystyle\sum_{k=N}^{\infty} |a_k x^k|$ is dominated by the convergent geometric series $\displaystyle\sum_{k=N}^{\infty} \gamma^k$. Thus, by the Comparison Test, the series $\displaystyle\sum_{k=N}^{\infty} |a_k x^k|$ converges. The series $\Sigma a_k x^k$ therefore converges absolutely if $|x| < |c|$.

To prove part (ii), we assume that the series $\Sigma a_k d^k$ diverges and we let x be any number so that $|x| > |d|$. Then $\Sigma a_k x^k$ cannot converge, since the convergence of $\Sigma a_k x^k$ would imply the convergence of $\Sigma a_k d^k$, by part (i). Thus $\Sigma a_k x^k$ diverges whenever $|x| > |d|$. This completes the proof. ◆

Theorem 2 gives insight into the geometry of the set of numbers x for which a given power series converges. Given the power series $\Sigma a_k x^k$, let S be this set.

Now suppose that there exists a number $d \notin S$. Then $\Sigma a_k d^k$ diverges, so by Theorem 2, $|x| \le |d|$ for every $x \in S$. This shows that the set S is bounded if it is not the entire real line, $(-\infty, \infty)$. By the Completeness Axiom there then exists a least upper bound r for the set S. Let us look at two cases:

(i) If $|x| < r$, then $|x|$ is not an upper bound for S, so there exists an element $c \in S$ with $|x| < c$. Since $c \in S$, $\Sigma a_k c^k$ converges. Thus $\Sigma a_k x^k$ converges absolutely, by Theorem 2.

(ii) If $|x| > r$, then $x \notin S$ so $\Sigma a_k x^k$ diverges.

Case (i) shows that S contains the interval $(-r, r)$, and that $\Sigma a_k x^k$ converges absolutely for every $x \in (-r, r)$. Case (ii) shows that the only other possible elements of S are the endpoints r and $-r$. For this reason the number r is called the **radius of convergence** of the power series. The interval $(-r, r)$, $[-r, r)$, $(-r, r]$, or $[-r, r]$ on which $\Sigma a_k x^k$ converges is called the **interval of convergence.** We summarize these findings as follows:

THEOREM 3

Given the power series $\Sigma a_k x^k$, precisely one of the following holds:

(a) The power series converges only for $x = 0$.

(b) There exists a positive number r so that the power series converges absolutely for $|x| < r$ and diverges for $|x| > r$. (The series may or may not converge for $x = \pm r$.)

(c) The power series converges for all x.

(See Figure 4.1).

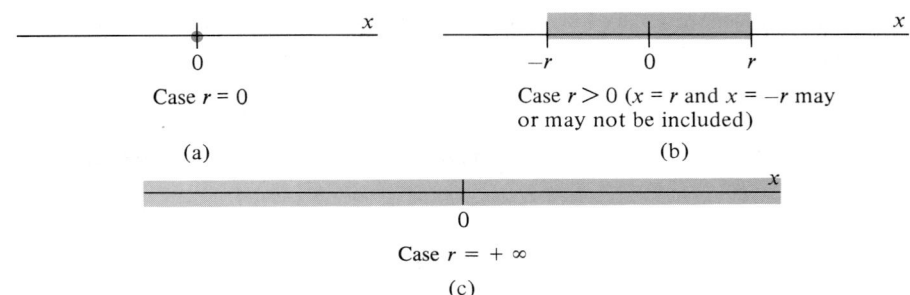

Figure 4.1 The geometry of the interval of convergence for the power series $\Sigma a_k x^k$.

Example 1

(a) The radius of convergence for the geometric power series Σx^k is $r = 1$, and the interval of convergence is $(-1, 1)$.

(b) The radius of convergence for the power series $\Sigma \dfrac{x^k}{k!}$ in equation (3) is $r = \infty$, and the interval of convergence is $(-\infty, \infty)$.

(c) To determine the radius of convergence of the power series $\Sigma \dfrac{(-1)^k}{1 + k} x^k$ in equation (4) we test for absolute convergence using the Ratio Test. We found in equation (5) that $\rho = |x|$. Thus, $\rho < 1$ if and only if $|x| < 1$, so $r = 1$ is the radius of convergence, and the power series converges absolutely for $|x| < 1$. We also checked the two values $x = r = 1$ and $x = -r = -1$, finding that the power series converges for $x = 1$ and diverges for $x = -1$. The interval of convergence for the power series $\displaystyle\sum_{k=0}^{\infty} \dfrac{(-1)^k}{1 + k} x^k$ is therefore $(-1, 1]$. ◇

Example 2

Find the radius and interval of convergence for the power series

$$\sum_{k=1}^{\infty} \frac{1}{k2^k} x^k.$$

Strategy

First, find the radius of convergence, using the Ratio Test. (The Root Test will also work well in this example.)

Solution

We test for absolute convergence, using the Ratio Test:

$$\rho = \lim_{k \to \infty} \frac{\left| \dfrac{1}{(k + 1)2^{k+1}} x^{k+1} \right|}{\left| \dfrac{1}{k2^k} \cdot x^k \right|}$$

$$= \lim_{k \to \infty} \left(\frac{k}{k + 1} \right) \cdot \left(\frac{1}{2} \right) \cdot |x| = \frac{1}{2}|x|.$$

Thus $\rho < 1$ if $|x| < 2$. The radius of convergence is therefore $r = 2$.

To determine the interval of convergence check the endpoints $x = r = 2$ and $x = -r = -2$.

If $x = 2$ we obtain the harmonic series

$$\sum_{k=1}^{\infty} \frac{1}{k \cdot 2^k}(2^k) = \sum_{k=1}^{\infty} \frac{1}{k} = 1 + \frac{1}{2} + \frac{1}{3} + \cdots ,$$

which diverges. If $x = -2$ we obtain the *alternating* harmonic series

$$\sum_{k=1}^{\infty} \frac{1}{k \cdot 2^k}(-2)^k = \sum_{k=1}^{\infty} \frac{(-1)^k}{k} = -1 + \frac{1}{2} - \frac{1}{3} + \cdots ,$$

which converges. The interval of convergence is therefore $[-2, 2)$. ◇

Example 3

Find the interval of convergence for the power series

$$\sum_{k=0}^{\infty} \frac{x^k}{3^k}.$$

Solution: Testing for absolute convergence using the Root Test we find that

$$\rho = \lim_{k \to \infty} \left\{ \left| \frac{x^k}{3^k} \right| \right\}^{1/k} = \lim_{k \to \infty} \frac{|x|}{3} = \frac{|x|}{3}.$$

Thus, $\rho < 1$ if $|x| < 3$. The radius of convergence is $r = 3$. (The Ratio Test will also work well in obtaining this conclusion.)

If $x = 3$, the series becomes

$$\sum_{k=0}^{\infty} \frac{3^k}{3^k} = \sum_{k=0}^{\infty} 1 = 1 + 1 + 1 + \cdots ,$$

which diverges. When $x = -3$, we obtain the divergent series

$$\sum_{k=0}^{\infty} \frac{(-3)^k}{3^k} = \sum_{k=0}^{\infty} (-1)^k = 1 - 1 + 1 - 1 + \cdots .$$

The interval of convergence is therefore $(-3, 3)$. ◇

Example 4

The radius of convergence for the power series

$$\sum_{k=0}^{\infty} k!x^k$$

is $r = 0$ since, by the Ratio Test,

$$\rho = \lim_{k \to \infty} \frac{|(k + 1)!x^{k+1}|}{|k!x^k|} = \lim_{k \to \infty} (k + 1)|x| = \begin{cases} 0, & x = 0 \\ \infty, & x \neq 0 \end{cases}.$$

The "interval" of convergence is simply $\{0\}$. ◇

The next example involves a power series of the form $\Sigma a_k(x - a)^k$. As we stated at the outset, we will use the change of variable $u = x - a$ to bring the power series into the form $\Sigma a_k u^k$. Note, however, that the interval of convergence will be centered at $x = a$ rather than at $x = 0$.

Example 5

Find the interval of convergence for the power series

$$\sum_{k=1}^{\infty} \frac{3^k}{k}(2x - 1)^k.$$

Strategy

Use the substitution $u = 2x - 1$ to bring power series to the form of equation (2). (The interval of convergence will therefore be centered about $x = 1/2$.) Apply Ratio Test to find radius of convergence in the variable u.

Solution

Letting $u = 2x - 1$ we obtain the series

$$\sum_{k=1}^{\infty} \frac{3^k}{k} u^k.$$

Testing this series for absolute convergence using the Ratio Test we find that

$$\rho = \lim_{k\to\infty} \frac{\left|\dfrac{3^{k+1} \cdot u^{k+1}}{(k+1)}\right|}{\left|\dfrac{3^k \cdot u^k}{k}\right|}$$

$$= \lim_{k\to\infty} 3\left(\frac{k}{k+1}\right)|u| = 3|u|.$$

Rewrite the inequality for $|u|$ in terms of the original variable x.

The series converges absolutely for $|u| < 1/3$ or for $|2x - 1| < 1/3$. This last inequality is equivalent to

$$-\frac{1}{3} < 2x - 1 < \frac{1}{3},$$

or

$$\frac{1}{3} < x < \frac{2}{3}.$$

Test the endpoints individually.

The radius of convergence (about $x = 1/2$) is therefore $r = 1/6$. At the endpoint $x = 2/3$ we obtain the series

$$\sum_{k=1}^{\infty} \frac{3^k}{k}\left(\frac{1}{3}\right)^k = \sum_{k=1}^{\infty} \frac{1}{k},$$

which diverges. At the endpoint $x = 1/3$ we obtain

$$\sum_{k=1}^{\infty} \frac{3^k}{k}\left(-\frac{1}{3}\right)^k = \sum_{k=1}^{\infty} \frac{(-1)^k}{k},$$

which converges. The interval of convergence is therefore $[1/3, 2/3)$.   ◇

Exercise Set 13.4

In Exercises 1–35, find the interval of convergence of the given power series.

1. $\displaystyle\sum_{k=0}^{\infty} \frac{x^k}{k+2}$

2. $\displaystyle\sum_{k=1}^{\infty} \frac{x^k}{2k}$

3. $\displaystyle\sum_{k=0}^{\infty} \frac{(-1)^{k+1}}{k!} x^k$

4. $\displaystyle\sum_{k=0}^{\infty} \frac{2^k x^k}{(k+1)!}$

5. $\displaystyle\sum_{k=1}^{\infty} \frac{(k^2+1)}{k!} x^k$

6. $\displaystyle\sum_{k=0}^{\infty} \frac{k \cdot x^k}{2^k}$

7. $\displaystyle\sum_{k=2}^{\infty} \frac{x^k}{\ln k}$

8. $\displaystyle\sum_{k=1}^{\infty} \frac{(-1)^k e^k}{k^2} x^k$

9. $\displaystyle\sum_{k=1}^{\infty} \frac{\cos \pi k}{1+k} x^k$

10. $\displaystyle\sum_{k=1}^{\infty} \frac{(2k+1)!}{2k!} x^{2k}$

11. $\displaystyle\sum_{k=1}^{\infty} \frac{(-1)^k}{k(k+1)} x^k$

12. $\displaystyle\sum_{k=1}^{\infty} k^2 c^k x^k$

13. $\displaystyle\sum_{k=1}^{\infty} \frac{(-1)^k}{k!} (x-3)^k$

14. $\displaystyle\sum_{k=1}^{\infty} \frac{k}{3^k} (x-\pi)^k$

15. $\displaystyle\sum_{k=0}^{\infty} k!(x-1)^k$

16. $\displaystyle\sum_{k=1}^{\infty} \frac{3^k}{k^2} (2x-1)^k$

17. $\displaystyle\sum_{k=2}^{\infty} \frac{(-1)^k}{\ln k} (3x-2)^k$

18. $\displaystyle\sum_{k=2}^{\infty} \frac{1}{(\ln k)^k} (x-1)^k$

19. $\displaystyle\sum_{k=1}^{\infty} \frac{1}{k+2} x^{2k+1}$

20. $\displaystyle\sum_{k=1}^{\infty} \frac{(x+2)^k}{(k+1)3^k}$

21. $\displaystyle\sum_{k=1}^{\infty} \frac{(x-3)^k}{k(k+1)}$

22. $\displaystyle\sum_{k=0}^{\infty} \frac{x^{2k+1}}{\pi^k}$

23. $\displaystyle\sum_{k=1}^{\infty} \frac{k(x-2)^k}{e^k}$

24. $\displaystyle\sum_{k=0}^{\infty} \frac{(2x+5)^k}{\sqrt{2k+8}}$

25. $\displaystyle\sum_{k=2}^{\infty} \frac{k(7x+1)^k}{2^k}$

26. $\displaystyle\sum_{k=1}^{\infty} \frac{(x-2)^k}{3^k \cdot k^2}$

27. $\displaystyle\sum_{k=3}^{\infty} kx^k$

28. $\displaystyle\sum_{k=1}^{\infty} \frac{x^k}{\ln(k+1)}$

29. $\displaystyle\sum_{k=1}^{\infty} \frac{(x-1)^k}{3^k \sqrt{k+1}}$

30. $\displaystyle\sum_{k=0}^{\infty} \frac{(2x-1)^k}{5^k}$

31. $\displaystyle\sum_{k=1}^{\infty} \frac{(k+4)x^k}{(k+1)(k+2)e^k}$

32. $\displaystyle\sum_{k=0}^{\infty} \frac{k^3(x-2)^k}{3^k}$

33. $\displaystyle\sum_{k=0}^{\infty} \frac{(-1)^k x^{2k+1}}{2(k+1)!}$

34. $\displaystyle\sum_{k=0}^{\infty} \frac{(-1)^k x^{2k}}{(2k)!}$

35. $\displaystyle\sum_{k=1}^{\infty} \frac{x^k}{k(k+1)}$

36. Show that the radius of convergence of the power series

$$\sum_{k=1}^{\infty} \frac{1}{k}(2x-3)^k \text{ is } r = 1/2.$$

37. Prove that if the interval of convergence of the power series $\Sigma a_k x^k$ is $[-r, r)$, then the power series is conditionally convergent, but not absolutely convergent, for $x = -r$.

38. Prove that if the power series $\Sigma a_k x^k$ has radius of convergence r, then the power series $\Sigma a_k x^{ck}$ has radius of convergence $r^{1/c}$, $c > 0$.

39. Prove that if the power series $\Sigma a_k x^k$ has radius of convergence r_a and if the power series $\Sigma b_k x^k$ has radius of convergence r_b, then the series $\Sigma(a_k + b_k)x^k$ converges absolutely for $|x| < c$ where c is the smaller of r_a and r_b.

40. Show that the power series $\Sigma a_k(x-a)^k$ always converges at $x = a$.

13.5 DIFFERENTIATION AND INTEGRATION OF POWER SERIES

Within its interval of convergence, a power series represents a perfectly legitimate function of x. Thus, if the power series $\Sigma a_k x^k$ converges on the interval I_a, we may write

$$f(x) = \sum a_k x^k, \qquad x \in I_a.$$

If $\Sigma b_k x^k$ is a second power series with interval of convergence I_b, then $g(x) = \Sigma b_k x^k$ is again a function of x, but g is defined on a (possibly) different interval than is f. However, if the intervals I_a and I_b overlap (that is, if $I = I_a \cap I_b$ is not empty), we may form sums and differences of f and g as follows:

$$(f + g)(x) = f(x) + g(x)$$
$$= \sum a_k x^k + \sum b_k x^k = \sum (a_k + b_k) x^k, \qquad x \in I_a \cap I_b$$
$$(f - g)(x) = f(x) - g(x)$$
$$= \sum a_k x^k - \sum b_k x^k = \sum (a_k - b_k) x^k, \qquad x \in I_a \cap I_b.$$

(The concepts of multiplication and division for power series are more complicated and will not be pursued here.)

Example 1

The formula for the sum of a geometric series shows that the function $f(x) = \dfrac{1}{1 - x}$ may be represented as a power series for $|x| < 1$:

$$\frac{1}{1 - x} = \sum_{k=0}^{\infty} x^k = 1 + x + x^2 + x^3 + \cdots. \tag{1}$$

Multiplying both sides of equation (1) by x shows that

$$\frac{x}{1 - x} = \sum_{k=0}^{\infty} x^{k+1} = x + x^2 + x^3 + x^4 + \cdots, \qquad |x| < 1. \tag{2}$$

Replacing x by $-x$ in (1) gives

$$\frac{1}{1 + x} = \sum_{k=0}^{\infty} (-x)^k = 1 - x + x^2 - x^3 + \cdots, \qquad |x| < 1. \tag{3}$$

Adding equations (1) and (2) (which is allowed because their intervals of convergence are identical) shows that

$$\frac{2}{(1 - x)(1 + x)} = \frac{1}{1 - x} + \frac{1}{1 + x}$$

$$= \sum_{k=0}^{\infty} x^k + \sum_{k=0}^{\infty} (-x)^k = 2 + 2x^2 + 2x^4 + 2x^6 + \cdots$$

$$= 2 \sum_{k=0}^{\infty} x^{2k}, \qquad |x| < 1. \qquad \Diamond$$

If you are beginning to suspect that within its interval of convergence the power series

$$f(x) = a_0 + a_1 x + a_2 x^2 + \cdots \tag{4}$$

behaves just like an "infinitely long polynomial," that is precisely the point we are trying to make. In fact, the following theorem shows that we may even differentiate expression (4) term by term, just as for polynomials, obtaining

$$f'(x) = a_1 + 2a_2 x + 3a_3 x^2 + \cdots,$$

which is the power series representation for the derivative of the function in equation (4). Its proof is given in more advanced courses.

THEOREM 4 **Differentiation of Power Series**	Suppose that the power series $\Sigma a_k x^k$ has a radius of convergence $r \neq 0$ and that the function f is defined to be its sum:

$$f(x) = \sum_{k=0}^{\infty} a_k x^k = a_0 + a_1 x + a_2 x^2 + a_3 x^3 + \cdots, \qquad |x| < r.$$

Then

(i) the function f is differentiable for $x \in (-r, r)$,

(ii) the power series $\displaystyle\sum_{k=0}^{\infty} k a_k x^{k-1}$ converges absolutely for each $x \in (-r, r)$, and

(iii) $f'(x) = \displaystyle\sum_{k=0}^{\infty} k a_k x^{k-1} = a_1 + 2a_2 x + 3a_3 x^2 + 4a_4 x^3 + \cdots, \qquad |x| < r.$

Example 2

Applying Theorem 4 to the geometric series

$$\frac{1}{1-x} = \sum_{k=0}^{\infty} x^k = 1 + x + x^2 + x^3 + \cdots, \qquad |x| < 1,$$

we conclude that

$$\frac{d}{dx}\left\{\frac{1}{1-x}\right\} = \frac{1}{(1-x)^2} = \sum_{k=0}^{\infty} k x^{k-1}$$

$$= 1 + 2x + 3x^2 + 4x^3 + \cdots, \qquad |x| < 1. \qquad \diamondsuit$$

Example 3

Find a power series representation for the function

$$f(x) = \frac{x}{(1-x^2)^2}.$$

Solution: Since f does not resemble any of the functions for which we already know the power series representations, we look for another way to attack the problem. Here f is easily integrated, leading to the result $f(x) = \dfrac{1}{2}\dfrac{d}{dx}\left(\dfrac{1}{1-x^2}\right)$. Now we can recognize the expression in parentheses as the sum of the geometric series in x^2, which converges for $|x^2| < 1$. Then we have

$$\frac{1}{1-x^2} = \sum_{k=0}^{\infty} (x^2)^k = \sum_{k=0}^{\infty} x^{2k} = 1 + x^2 + x^4 + x^6 + \cdots, \qquad |x| < 1.$$

We conclude from Theorem 4 that

$$\frac{x}{(1-x^2)^2} = \frac{1}{2}\frac{d}{dx}\left\{\sum_{k=0}^{\infty} x^{2k}\right\} = \frac{1}{2}\sum_{k=0}^{\infty} 2kx^{2k-1} = x + 2x^3 + 3x^5 + \cdots.$$

This series converges absolutely for $|x| < 1$. $\diamond$

Example 4

The power series

$$\sum_{k=0}^{\infty} \frac{x^k}{k!} = 1 + x + \frac{x^2}{2!} + \frac{x^3}{3!} + \cdots$$

converges for all x, as noted in Example 1, Section 13.4. On differentiating this series we find that

$$\frac{d}{dx}\left\{\sum_{k=0}^{\infty} \frac{x^k}{k!}\right\} = \frac{d}{dx}\left\{1 + x + \frac{x^2}{2!} + \frac{x^3}{3!} + \frac{x^4}{4!} + \cdots\right\}$$

$$= \left\{1 + \frac{2x}{2!} + \frac{3x^2}{3!} + \frac{4x^3}{4!} + \cdots\right\}$$

$$= \left\{1 + x + \frac{x^2}{2!} + \frac{x^3}{3!} + \cdots\right\}$$

$$= \sum_{k=0}^{\infty} \frac{x^k}{k!}.$$

That is, if $f(x) = \sum_{k=0}^{\infty} \frac{x^k}{k!}$, then $f'(x) = f(x)$ for all x. Recall that we have previously shown that the only function other than $f(x) \equiv 0$ that satisfies this differential equation is the function $f(x) = Ce^x$. This shows that, for some constant C,

$$Ce^x = \sum_{k=0}^{\infty} \frac{x^k}{k!} = 1 + x + \frac{x^2}{2!} + \frac{x^3}{3!} + \cdots.$$

Setting $x = 0$ shows that $C = 1$, so we conclude that the power series representation for the function e^x is

$$e^x = \sum_{k=0}^{\infty} \frac{x^k}{k!} = 1 + x + \frac{x^2}{2!} + \frac{x^3}{3!} + \cdots$$

and that this representation is valid for all x. $\diamond$

The statement of Theorem 4 raises an obvious question about integrals of functions represented by power series. The answer is just what you might suspect.

THEOREM 5
Integration of Power Series

Suppose that the power series $\Sigma a_k x^k$ has radius of convergence $r \neq 0$ and that the function f is defined as its sum:

$$f(x) = \sum_{k=0}^{\infty} a_k x^k = a_0 + a_1 x + a_2 x^2 + a_3 x^3 + \cdots, \qquad |x| < r.$$

Then

(i) the power series $\displaystyle\sum_{k=0}^{\infty} \left(\frac{a_k}{k+1}\right) x^{k+1}$ converges absolutely for each $x \in (-r, r)$,

and

(ii) $\displaystyle\int f(x)\, dx = \sum_{k=0}^{\infty} \left(\frac{a_k}{k+1}\right) x^{k+1} + C$

$$= \left\{ a_0 x + \frac{a_1}{2} x^2 + \frac{a_2}{3} x^3 + \frac{a_3}{4} x^4 + \cdots \right\} + C, \qquad |x| < r.$$

In other words, a power series may be integrated term by term within its radius of convergence. The proof of Theorem 5 is left for more advanced courses. Note that, rather than writing a constant of integration for each term, we collect all constants into a single number C.

Example 5

Since $\ln(1 + x) = \displaystyle\int \frac{1}{1+x}\, dx$, we integrate equation (3) according to Theorem 5.

$$\ln(1 + x) = \sum_{k=0}^{\infty} \left\{ \int (-x)^k\, dx \right\}$$

$$= \sum_{k=0}^{\infty} -\frac{(-x)^{k+1}}{k+1} + C$$

$$= \sum_{k=0}^{\infty} \frac{(-1)^k x^{k+1}}{k+1} + C$$

$$= \left\{ x - \frac{x^2}{2} + \frac{x^3}{3} - \frac{x^4}{4} + \cdots \right\} + C, \qquad |x| < 1.$$

Setting $x = 0$ gives $\ln 1 = 0 = C$, so

$$\ln(1 + x) = \sum_{k=0}^{\infty} -\frac{(-x)^{k+1}}{k+1}$$

$$= x - \frac{x^2}{2} + \frac{x^3}{3} - \frac{x^4}{4} + \cdots, \qquad |x| < 1. \tag{5}$$

REMARK: Equation (5) provides a practical means for calculating values of ln a for $0 < a < 2$. For example, to approximate ln(1.2) we set $x = 0.2$ and apply (5) to obtain

$$\ln(1.2) \approx .2 - \frac{(.2)^2}{2} + \frac{(.2)^3}{3} - \frac{(.2)^4}{4} = .182266$$

with an error of less than $\dfrac{.2^5}{5} = .000064$ (Theorem 15, Section 12.7).

Example 6

Replacing x by x^2 in equation (3) shows that

$$\frac{1}{1 + x^2} = \sum_{k=0}^{\infty} (-x^2)^k = 1 - x^2 + x^4 - x^6 + \cdots, \qquad |x| < 1.$$

Since $\displaystyle\int \frac{1}{1 + x^2}\, dx = \text{Tan}^{-1} x + C$, Theorem 5 gives

$$\text{Tan}^{-1} x = \sum_{k=0}^{\infty} \left\{ \int (-x^2)^k\, dx \right\} + C$$

$$= \sum_{k=0}^{\infty} \frac{(-1)^k}{2k + 1} \cdot x^{2k+1} + C$$

$$= \left\{ x - \frac{x^3}{3} + \frac{x^5}{5} - \cdots \right\} + C, \qquad |x| < 1.$$

Setting $x = 0$ gives $\text{Tan}^{-1}(0) = 0 = C$. Thus

$$\text{Tan}^{-1} x = \sum_{k=0}^{\infty} \frac{(-1)^k}{2k + 1} \cdot x^{2k+1} = x - \frac{x^3}{3} + \frac{x^5}{5} - \cdots, \qquad |x| < 1. \qquad \diamond$$

Exercise Set 13.5

In Exercises 1–10, find a power series representation for the given function using equation (1). State the radius of convergence for the power series obtained.

In Exercises 11–19, find a power series representation for the given function using Theorem 4. State the radius of convergence.

1. $\dfrac{1}{1 - 2x}$

2. $\dfrac{x^2}{1 - x}$

3. $\dfrac{1}{1 + 4x^2}$

4. $\dfrac{1}{1 - 9x^2}$

5. $\dfrac{x}{1 + x^2}$

6. $\dfrac{x}{1 - x^2}$

7. $\dfrac{x - 1}{x + 1}$

8. $\dfrac{x - 1}{1 + x^2}$

9. $\dfrac{1}{1 - x^4}$

10. $\dfrac{x}{4 + x^2}$

11. $f(x) = \dfrac{2}{(1 + x)^2}$ $\left(\text{Hint: } f(x) = -2\dfrac{d}{dx}\left(\dfrac{1}{1 + x}\right).\right)$

12. $f(x) = \dfrac{2}{(1 - x)^3}$ $\left(\text{Hint: } f(x) = \dfrac{d^2}{dx^2}\left(\dfrac{1}{1 - x}\right).\right)$

13. $f(x) = \dfrac{x}{(1 + x^2)^2}$

14. $f(x) = \dfrac{1 - x^2}{(1 + x^2)^2}$ $\left(\text{Hint: } f(x) = \dfrac{d}{dx}\left(\dfrac{x}{1 + x^2}\right).\right)$

15. $f(x) = \dfrac{1 + x^2}{(1 - x^2)^2}$

16. $f(x) = \dfrac{1}{(1 + 4x)^2}$

17. $f(x) = \dfrac{8x}{(1 + 4x^2)^2}$ (*Hint:* Use Exercise 3.)

18. $f(x) = \dfrac{1 + 2x - x^2}{(1 + x^2)^2}$ (*Hint:* Use Exercise 8.)

19. $f(x) = \dfrac{2}{(x + 1)^2}$ (*Hint:* Use Exercise 7.)

In Exercises 20–27, find a power series representation for the given function using Theorem 5. State the radius of convergence.

20. $f(x) = \ln(1 + x)$

21. $f(x) = \ln(1 - x)$

22. $f(x) = x \ln(1 + x)$

23. $f(x) = \text{Tan}^{-1}(2x)$

24. $f(x) = x \, \text{Tan}^{-1} x$

25. $f(x) = \ln(4 + x)$

26. $\displaystyle \int \frac{dx}{1 + x^4}$

27. $\ln(1 + x^2)$

28. Use the results of Exercises 20 and 21 to show that

$$\ln\left(\frac{1 + x}{1 - x}\right) = 2\left(x + \frac{x^3}{3} + \frac{x^5}{5} + \cdots\right), \qquad |x| < 1.$$

29. For the power series

$$\sum_{k=0}^{\infty} (-1)^k \frac{x^{2k}}{(2k)!} = 1 - \frac{x^2}{2!} + \frac{x^4}{4!} - \frac{x^6}{6!} + \cdots :$$

a. Use the Ratio Test to show that the series converges absolutely for all values of x.

b. Using Theorem 4, show that the function

$$f(x) = \sum_{k=0}^{\infty} (-1)^k \frac{x^{2k}}{(2k)!}$$

satisfies the differential equation $f''(x) = -f(x)$.

c. Show that $f(0) = 1$.

d. What function have you seen previously which satisfies both properties (b) and (c)?

30. For the power series

$$\sum_{k=0}^{\infty} (-1)^k \frac{x^{2k+1}}{(2k + 1)!} = x - \frac{x^3}{3!} + \frac{x^5}{5!} - \frac{x^7}{7!} + \cdots :$$

a. Use the Ratio Test to show that the series converges absolutely for all values of x.

b. Using Theorem 4, show that the function $g(x) = \displaystyle\sum_{k=0}^{\infty} (-1)^k \frac{x^{2k+1}}{(2k + 1)!}$ satisfies the differential equation $g''(x) = -g(x)$.

c. Show that $g(0) = 0$.

d. Show that $g'(x) = f(x)$ and $f'(x) = -g(x)$ where f is the function in Exercise 29. For what common function is $g(x)$ a power series representation?

31. Use the Chain Rule to show that if the power series

$$f(x) = \sum_{k=0}^{\infty} a_k(bx + c)^k$$

converges for $|bx + c| < r$, then

$$f'(x) = \sum_{k=0}^{\infty} kba_k(bx + c)^{k-1}, \qquad |bx + c| < r.$$

13.6 TAYLOR AND MACLAURIN SERIES

We are about to uncover a remarkable relationship between Taylor polynomials and power series, one that will provide a precise solution to the approximation problem posed in Section 13.1.

Recall that if the function f has $n + 1$ derivatives in an interval containing $x = a$, then the nth Taylor polynomial for f is

$$P_n(x) = f(a) + f'(a)(x - a)$$

$$+ \frac{f''(a)}{2!}(x - a)^2 + \cdots + \frac{f^{(n)}(a)}{n!}(x - a)^n \qquad (1)$$

$$= \sum_{k=0}^{n} \frac{f^{(k)}(a)}{k!}(x - a)^k.$$

We refer to the coefficient $\dfrac{f^{(k)}(a)}{k!}$ of $(x - a)^k$ as the **kth Taylor coefficient**. Also, if we write

$$f(x) = P_n(x) + R_n(x), \tag{2}$$

the remainder term $R_n(x)$ is bounded as follows (Taylor's Theorem):

$$|R_n(x)| \le \left| \frac{f^{(n+1)}(d)}{(n+1)!}(x-a)^{n+1} \right| \tag{3}$$

where d lies between a and x.

Now note that the right-hand side of equation (1) is precisely the nth partial sum for the power series

$$\sum_{k=0}^{\infty} \frac{f^{(k)}(a)}{k!}(x-a)^k = f(a) + f'(a)(x-a) +$$
$$\frac{f''(a)}{2!}(x-a)^2 + \frac{f'''(a)}{3!}(x-a)^3 + \cdots. \tag{4}$$

Series (4) is referred to as the **Taylor series** for f, expanded about $x = a$. It is simply the result of allowing the Taylor polynomial $P_n(x)$ to become "infinitely long." In other words,

$$\sum_{k=0}^{\infty} \frac{f^{(k)}(a)}{k!}(x-a)^k = \lim_{n \to \infty} \sum_{k=0}^{n} \frac{f^{(k)}(a)}{k!}(x-a)^k = \lim_{n \to \infty} P_n(x). \tag{5}$$

Of course, the expression in equation (4) makes sense only if the function f is infinitely differentiable. That is, the derivative $f^{(n)}(a)$ must exist for all integers $n = 1, 2, \ldots$ in order that all Taylor coefficients be defined.

The question of determining the numbers x for which a Taylor series converges to $f(x)$ may be handled by use of the remainder term, $R_n(x)$. Applying $\lim_{n \to \infty}$ to both sides of equation (2) and using (5) we see that

$$f(x) = \lim_{n \to \infty} P_n(x) + \lim_{n \to \infty} R_n(x)$$

$$= \sum_{k=0}^{\infty} \frac{f^{(k)}(a)}{k!}(x-a)^k + \lim_{n \to \infty} R_n(x).$$

This shows that *the Taylor series (4) converges to $f(x)$ if and only if $\lim_{n \to \infty} R_n(x) = 0$.*

Example 1

Find the Taylor series for the function $f(x) = \sin x$ expanded about $a = \pi/4$ and determine the numbers x for which the series converges to $f(x)$.

Strategy
Find the Taylor coefficients
$$\frac{f^{(k)}(a)}{k!}$$
with $a = \dfrac{\pi}{4}$.

Solution
We have

$$f(x) = \sin x; \qquad f\left(\frac{\pi}{4}\right) = \frac{\sqrt{2}}{2}$$

$$f'(x) = \cos x; \qquad f'\left(\frac{\pi}{4}\right) = \frac{\sqrt{2}}{2}$$

$$f''(x) = -\sin x; \qquad f''\left(\frac{\pi}{4}\right) = -\frac{\sqrt{2}}{2}$$

$$f'''(x) = -\cos x; \qquad f'''\left(\frac{\pi}{4}\right) = -\frac{\sqrt{2}}{2}$$

$$f^{(4)}(x) = \sin x; \qquad f^4\left(\frac{\pi}{4}\right) = \frac{\sqrt{2}}{2}$$

$$\vdots$$

$$f^{(2k)}(x) = (-1)^k \sin x; \qquad f^{(2k)}\left(\frac{\pi}{4}\right) = (-1)^k \frac{\sqrt{2}}{2}$$

$$f^{(2k+1)}(x) = (-1)^k \cos x; \qquad f^{(2k+1)}\left(\frac{\pi}{4}\right) = (-1)^k \frac{\sqrt{2}}{2}$$

$$\vdots$$

Insert the Taylor coefficients and $a = \pi/4$ in (4) to obtain the Taylor series.

The Taylor series is therefore

$$\sin x = \frac{\sqrt{2}}{2} + \frac{\sqrt{2}}{2}\left(x - \frac{\pi}{4}\right) - \frac{\sqrt{2}}{2 \cdot 2!}\left(x - \frac{\pi}{4}\right)^2$$
$$- \frac{\sqrt{2}}{2 \cdot 3!}\left(x - \frac{\pi}{4}\right)^3 + \frac{\sqrt{2}}{2 \cdot 4!}\left(x - \frac{\pi}{4}\right)^4 + \cdots.$$

Determine where $\lim_{n \to \infty} |R_n(x)| = 0$ to find the values of x for which the series converges to $f(x)$.

Since $f^{(n+1)}(c)$ is either $\pm \sin c$ or $\pm \cos c$, we know that $|f^{(n+1)}(c)| \le 1$ for all n. Thus

$$\lim_{n \to \infty} |R_n(x)| = \lim_{n \to \infty} \left| \frac{f^{(n+1)}(c)}{(n+1)!}\left(x - \frac{\pi}{4}\right)^{n+1} \right|$$

Use fact that $\lim_{n \to \infty} \dfrac{x^n}{n!} = 0$ (limit (2), Section 12.2).

$$\le \lim_{n \to \infty} \frac{\left| x - \frac{\pi}{4} \right|^{n+1}}{(n+1)!}$$

$$= 0$$

for all x. Thus the series converges to $\sin x$ for all x. $\diamond$

The special case of a Taylor series expanded about $a = 0$ is referred to as a **Maclaurin series:**

$$\sum_{k=0}^{\infty} \frac{f^{(k)}(0)}{k!} x^k = f(0) + f'(0)x + \frac{f''(0)}{2!}x^2 + \cdots + \frac{f^{(k)}(0)}{k!}x^k + \cdots. \tag{6}$$

Example 2

Find a Maclaurin series for $f(x) = e^x$ and determine the numbers x for which it converges to $f(x)$.

Solution: Every derivative of $f(x) = e^x$ is the same:

$$f(x) = e^x; \qquad f(0) = 1$$

$$f'(x) = e^x; \qquad f'(0) = 1$$

$$\vdots$$

$$f^{(n)}(x) = e^x; \qquad f^{(n)}(0) = 1$$

$$\vdots$$

Thus

$$e^x = 1 + x + \frac{x^2}{2!} + \frac{x^3}{3!} + \cdots + \frac{x^k}{k!} + \cdots$$

$$= \sum_{k=0}^{\infty} \frac{x^k}{k!}.$$

Here, for all n,

$$\left| f^{(n+1)}(c) \right| = \left| e^c \right| = e^c \le e^{|c|} \le e^{|x|}$$

since $|c| \le |x|$, so

$$\lim_{n \to \infty} |R_n(x)| = \lim_{n \to \infty} \left| \frac{e^c}{(n+1)!} x^{n+1} \right|$$

$$\le \lim_{n \to \infty} e^{|x|} \frac{|x|^{n+1}}{(n+1)!}$$

$$= 0$$

for all x. The Maclaurin series for e^x therefore converges to e^x for all x. (Note that this result was proved in Example 4, Section 13.5, by a different method.) ◇

Example 3

Find the Maclaurin series for $f(x) = \ln(1 + x)$.

Solution: Here

$$f(x) = \ln(1 + x); \qquad f(0) = 0$$
$$f'(x) = (1 + x)^{-1}; \qquad f'(0) = 1$$
$$f''(x) = -(1 + x)^{-2}; \qquad f''(0) = -1$$
$$f'''(x) = 2(1 + x)^{-3}; \qquad f'''(0) = 2$$

$$\vdots$$

$$f^{(k)}(x) = (-1)^{k+1}(k - 1)!(1 + x)^{-k}; \qquad f^{(k)}(0) = (-1)^{k+1}(k - 1)!$$

$$\vdots$$

Using (6) we obtain

$$\ln(1 + x) = 0 + 1 \cdot x - \frac{1}{2!}x^2 + \frac{2}{3!}x^3 - \cdots$$

$$+ \frac{(-1)^{k+1}(k - 1)!}{k!}x^k + \cdots$$

$$= x - \frac{x^2}{2} + \frac{x^3}{3} - \frac{x^4}{4} + \cdots + \frac{(-1)^{k+1}}{k}x^k + \cdots$$

$$= \sum_{k=1}^{\infty} \frac{(-1)^{k+1}}{k}x^k.$$

This is the power series that we showed to converge to $\ln(1 + x)$ for $|x| < 1$ in Example 5 of Section 13.5. ◇

Obviously, every Taylor or Maclaurin series is also a power series. We will now show that the converse is also true if f is represented as a power series with radius of convergence $r > 0$:

$$f(x) = a_0 + a_1x + a_2x^2 + \cdots + a_kx^k + \cdots, \qquad |x| < r \neq 0. \tag{7}$$

Using Theorem 4, we may repeatedly differentiate both sides of (7), obtaining

$$f'(x) = a_1 + 2a_2x + 3a_3x^2 + \cdots + ka_kx^{k-1} + \cdots$$
$$f''(x) = 2a_2 + 2 \cdot 3a_3x + \cdots + (k - 1)ka_kx^{k-2} + \cdots$$
$$\vdots$$

$$f^{(k)}(x) = k!a_k + (k + 1)!a_{k+1}x + \frac{(k + 2)!}{2!}a_{k+2}x^2 + \cdots. \tag{8}$$

Setting $x = 0$ in each equation causes all terms containing powers of x to vanish, so

$$f(0) = a_0, \qquad a_0 = \frac{f(0)}{0!}$$

$$f'(0) = a_1, \qquad a_1 = \frac{f'(0)}{1!}$$

$$f''(0) = 2a_2, \qquad a_2 = \frac{f''(0)}{2!}$$

$$\vdots \qquad\qquad \vdots$$

$$f^{(k)}(0) = k!a_k, \qquad a_k = \frac{f^{(k)}(0)}{k!}.$$

In other words, if f is represented by a convergent power series, as in equation (7), then the coefficients a_k must be precisely the Maclaurin coefficients

$$a_k = \frac{f^{(k)}(0)}{k!}.$$

If we replace x by $x - a$ in equation (7) and evaluate the derivatives at $x = a$, we prove the following theorem for Taylor series.

THEOREM 6

If the function f can be represented by a power series of the form

$$f(x) = a_0 + a_1(x - a) + a_2(x - a)^2 + \cdots + a_k(x - a)^k + \cdots$$

$$= \sum_{k=0}^{\infty} a_k(x - a)^k, \qquad |x| < r$$

with a radius of convergence $r > 0$, then the coefficients a_k in the power series expansion must be the Taylor coefficients

$$a_k = \frac{f^{(k)}(a)}{k!}, \qquad k = 0, 1, 2, \ldots.$$

REMARK: Be careful about what is being said. If we *know* that $f(x) = \Sigma a_k(x - a)^k$, then this series is the Taylor series for f. There are examples, however, for which the Taylor series $\Sigma a_k(x - a)^k$ for f converges, but $f(x) \neq \Sigma a_k(x - a)^k$. That is, *a convergent Taylor series for f need not converge to the value $f(x)$.* (See Exercise 37.)

Example 4

In Exercise 2 you are asked to obtain the Maclaurin series expansion

$$\sin x = x - \frac{x^3}{3!} + \frac{x^5}{5!} - \frac{x^7}{7!} + \cdots + \frac{(-1)^k x^{2k+1}}{(2k + 1)!} + \cdots. \tag{9}$$

To show that this series converges to $\sin x$ for all x, we must show that $\lim_{n \to \infty} |R_n(x)| = 0$ for all x.

We do so, using Taylor's theorem, by noting that the derivative $f^{(n+1)}$ for $f(x) = \sin x$ is one of $\pm \sin x$ or $\pm \cos x$. Thus, for any x, we know that

$$|f^{(n+1)}(c)| \leq 1$$

for all c between 0 and x. Thus, by Taylor's Theorem,

$$|R_n(x)| \leq \frac{1}{(n + 1)!} |x|^{n+1}, \qquad -\infty < x < \infty,$$

so

$$\lim_{n \to \infty} |R_n(x)| = \lim_{n \to \infty} \frac{|x|^{n+1}}{(n + 1)!} = 0, \qquad -\infty < x < \infty$$

according to equation (2), Section 12.2.

Since the Maclaurin series (9) is a power series that converges for all x, we may differentiate this series term by term, to obtain a power series for $\cos x = \dfrac{d}{dx} \sin x$:

$$\cos x = 1 - \frac{3x^2}{3!} + \frac{5x^4}{5!} - \frac{7x^6}{7!} + \cdots + \frac{(-1)^k (2k + 1)x^{2k}}{(2k + 1)!} + \cdots$$

$$= 1 - \frac{x^2}{2!} + \frac{x^4}{4!} - \frac{x^6}{6!} + \cdots + \frac{(-1)^k x^{2k}}{(2k)!} + \cdots,$$

which converges for all x. Theorem 6 then assures that this series is, in fact, the Maclaurin series for $f(x) = \cos x$. ◇

Example 5

Find a Maclaurin series for $x^3 e^{x^2}$.

Strategy

Do *not* calculate Taylor coefficients for $f(x) = x^3 e^{x^2}$. Work directly with the Maclaurin series for e^x and substitute x^2 for x.

Multiply by x^3.

Solution

Replacing x by x^2 in the series for e^x in Example 2 shows that

$$e^{x^2} = 1 + x^2 + \frac{x^4}{2!} + \frac{x^6}{3!} + \cdots = \sum_{k=0}^{\infty} \frac{x^{2k}}{k!},$$

which converges for all x. Thus

$$x^3 e^{x^2} = x^3 \left\{ 1 + x^2 + \frac{x^4}{2!} + \frac{x^6}{3!} + \cdots \right\}$$

$$= x^3 + x^5 + \frac{x^7}{2!} + \frac{x^9}{3!} + \cdots$$

$$= \sum_{k=0}^{\infty} \frac{x^{2k+3}}{k!}$$

converges for all x. By Theorem 6 this must be the Maclaurin series for $x^3 e^{x^2}$. ◇

The following examples show how Taylor and Maclaurin series may be used to calculate certain constants and integrals.

Example 6

Since the Maclaurin series

$$e^x = \sum_{k=0}^{\infty} \frac{x^k}{k!} = 1 + x + \frac{x^2}{2!} + \frac{x^3}{3!} + \cdots$$

converges to e^x for all x, we may use it to approximate e^x for any number x. Setting $x = 1$ gives

$$e = 1 + 1 + \frac{1}{2!} + \frac{1}{3!} + \cdots + \frac{1}{k!} + \cdots$$

$$= \sum_{k=0}^{\infty} \frac{1}{k!},$$

and setting $x = -1$, we obtain

$$\frac{1}{e} = 1 - 1 + \frac{1}{2!} - \frac{1}{3!} + \frac{1}{4!} - \cdots + \frac{(-1)^k}{k!} + \cdots$$

$$= \sum_{k=0}^{\infty} \frac{(-1)^k}{k!}.$$

Since this last series is an alternating series, we may approximate $\frac{1}{e}$ by, say,

$$\frac{1}{e} \approx \frac{1}{2!} - \frac{1}{3!} + \frac{1}{4!} - \frac{1}{5!} = .3667$$

with accuracy no worse than $\frac{1}{6!} = .0014$. ◇

Example 7

We may approximate the integral $\int_0^1 e^{-x^2}\, dx$ using Maclaurin series as follows. Since

$$e^x = 1 + x + \frac{x^2}{2!} + \frac{x^3}{3!} + \cdots + \frac{x^k}{k!} + \cdots$$

converges for all x, the series

$$e^{-x^2} = 1 - x^2 + \frac{x^4}{2!} - \frac{x^6}{3!} + \cdots + \frac{(-x^2)^k}{k!} + \cdots$$

also converges for all x. By Theorem 5 an antiderivative for e^{-x^2} is represented as

$$\int e^{-x^2}\, dx = \int 1\, dx - \int x^2\, dx + \int \frac{x^4}{2!}\, dx - \int \frac{x^6}{3!}\, dx + \cdots + C$$

$$= x - \frac{x^3}{3} + \frac{x^5}{5 \cdot 2!} - \frac{x^7}{7 \cdot 3!} + \cdots + \frac{(-1)^k x^{2k+1}}{(2k+1)k!} + \cdots + C.$$

Thus,

$$\int_0^1 e^{-x^2}\, dx = 1 - \frac{1}{3} + \frac{1}{5 \cdot 2!} - \frac{1}{7 \cdot 3!} + \cdots + \frac{(-1)^k}{(2k+1)k!} + \cdots.$$

Since this is an alternating series, we may approximate this series to any desired degree of accuracy by taking sufficiently many terms. For example, to obtain accuracy of .01, we find by trial and error that for $k = 4$,

$$\frac{1}{(2 \cdot 4 + 1)4!} = \frac{1}{9 \cdot 24} = \frac{1}{216} < \frac{1}{100}.$$

Thus, we use terms up through $k = 3$ to obtain

$$\int_0^1 e^{-x^2}\, dx \approx 1 - \frac{1}{3} + \frac{1}{5 \cdot 2!} - \frac{1}{7 \cdot 3!} \approx .74,$$

accurate to within .01. $\diamondsuit$

The Binomial Series

According to the Binomial Theorem, we know that

$$(1 + x)^n = 1 + nx + \frac{n(n-1)}{2} x^2 + \cdots$$

$$+ \binom{n}{k} x^k + \cdots + nx^{n-1} + x^n \tag{10}$$

when n is a positive integer. In writing (10) we have used the notation $\binom{n}{k}$ for the

binomial coefficient

$$\binom{n}{k} = \frac{n!}{k!(n-k)!} = \frac{n(n-1)(n-2) \cdots \cdot (n-k+1)}{k!}.$$

We can extend equation (10) to the case where n is not a positive integer. However, the result involves an infinite series, rather than a polynomial in x. The result is called the **Binomial Series:**

$$(1 + x)^r = 1 + rx + \frac{r(r - 1)}{2}x^2 + \frac{r(r - 1)(r - 2)}{3!}x^3 + \cdots$$

$$= 1 + \sum_{k=1}^{\infty} \frac{r(r - 1)(r - 2) \cdots (r - k + 1)}{k!}x^k, \qquad |x| < 1.$$

(11)

(Note that the infinite series in (11) reduces to the polynomial in (10) when $r = n$ is a positive integer, since the term $(r - k + 1)$ becomes zero when k reaches $r + 1$.)

Proving the validity of (11) involves showing that the right-hand side is precisely the Maclaurin series for $f(x) = (1 + x)^r$, and that the radius of convergence for this series is one. To do so we note that

$$f(x) = (1 + x)^r \qquad \text{gives} \qquad f(0) = 1$$
$$f'(x) = r(1 + x)^{r-1} \qquad \text{gives} \qquad f'(0) = r$$
$$f''(x) = r(r - 1)(1 + x)^{r-2} \qquad \text{gives} \qquad f''(0) = r(r - 1)$$

$$\vdots$$

$$f^{(k)}(x) = r(r - 1) \cdots (r - k + 1)(1 + x)^{r-k}$$
$$\text{gives} \qquad f^{(k)}(0) = r(r - 1) \cdots (r - k + 1).$$

Thus, the kth Maclaurin coefficient for $f(x) = (1 + x)^r$ is indeed

$$\frac{f^{(k)}(0)}{k!} = \frac{r(r - 1)(r - 2) \cdots (r - k + 1)}{k!}$$

as in the Binomial Series (11).

To verify that the series converges for $|x| < 1$, we let a_k denote the kth term

$$a_k = \frac{r(r - 1) \cdots (r - k + 1)}{k!}x^k.$$

Then

$$\rho = \lim_{k \to \infty} \left| \frac{a_{k+1}}{a_k} \right| = \lim_{k \to \infty} \left| \frac{r(r - 1) \cdots (r - k)}{r(r - 1) \cdots (r - k + 1)} \cdot \frac{k!}{(k + 1)!} \cdot \frac{x^{k+1}}{x^k} \right|$$

$$= \lim_{k \to \infty} \left| \frac{(r - k)}{1} \cdot \frac{1}{(k + 1)} \cdot x \right|$$

$$= \lim_{k \to \infty} \left| \frac{r - k}{k + 1} \right| \cdot |x|$$

$$= |x|.$$

Thus, if $|x| < 1$ we have $\rho < 1$, and the series converges absolutely by the Ratio Test. We assert, but do not prove, that this series converges *to* $(1 + x)^r$ for $|x| < 1$.

Example 8

The Binomial Series, with $r = \dfrac{1}{2}$, gives

$$\sqrt{1 + x} = (1 + x)^{1/2} = 1 + \frac{1}{2}x + \frac{\dfrac{1}{2}\left(\dfrac{1}{2} - 1\right)}{2}x^2 + \frac{\dfrac{1}{2}\left(\dfrac{1}{2} - 1\right)\left(\dfrac{1}{2} - 2\right)}{3!}x^3$$

$$+ \frac{\frac{1}{2}\left(\frac{1}{2} - 1\right)\left(\frac{1}{2} - 2\right)\left(\frac{1}{2} - 3\right)}{4!} x^4 + \cdots$$

$$= 1 + \frac{1}{2}x - \frac{1}{8}x^2 + \frac{1}{16}x^3 - \frac{5}{128}x^4 + \cdots, \qquad |x| < 1. \quad \diamond$$

Example 9

Replacing x by x^3 in Example 8 gives

$$\sqrt{1 + x^3} = (1 + x^3)^{1/2}$$

$$= 1 + \frac{1}{2}x^3 - \frac{1}{8}x^6 + \frac{1}{16}x^9 - \frac{5}{128}x^{12} + \cdots, \qquad |x| < 1. \qquad \diamond$$

Example 10

With $r = -\dfrac{1}{3}$ and $-x$ in place of x, (11) gives

$$\frac{1}{\sqrt[3]{1 - x}} = [1 + (-x)]^{-1/3}$$

$$= 1 + \left(-\frac{1}{3}\right)(-x) + \frac{\left(-\frac{1}{3}\right)\left(-\frac{1}{3} - 1\right)}{2}(-x)^2$$

$$+ \frac{\left(-\frac{1}{3}\right)\left(-\frac{1}{3} - 1\right)\left(-\frac{1}{3} - 2\right)}{3!}(-x)^3$$

$$+ \frac{\left(-\frac{1}{3}\right)\left(-\frac{1}{3} - 1\right)\left(-\frac{1}{3} - 2\right)\left(-\frac{1}{3} - 3\right)}{4!}(-x)^4 + \cdots$$

$$= 1 + \frac{1}{3}x + \frac{2}{9}x^2 + \frac{14}{81}x^3 + \frac{35}{243}x^4 + \cdots, \qquad |x| < 1. \qquad \diamond$$

Exercise Set 13.6

In Exercises 1–18, find the Taylor or Maclaurin series for the given function expanded about the given point and determine the numbers x for which the series converges.

1. $f(x) = e^{2x}, \quad a = 0$

2. $f(x) = \sin x, \quad a = 0$

3. $f(x) = \cos x, \quad a = \pi/4$

4. $f(x) = \sin x, \quad a = \pi/6$

5. $f(x) = 1 + x^2, \quad a = 2$

6. $f(x) = \dfrac{1}{1 + x}, \quad a = 0$

7. $f(x) = \ln(3 + x), \quad a = 0$

8. $f(x) = x \sin 2x, \quad a = 0$

9. $f(x) = 2^x, \quad a = 0$

10. $f(x) = (1 + x)^n, \quad a = 0$

11. $f(x) = (1 + x)^{3/2}, \quad a = 0$

12. $f(x) = \sqrt{x}, \quad a = 4$

13. $f(x) = \sqrt{x + 1}, \quad a = 0$

14. $f(x) = \dfrac{1}{x}, \quad a = 2$

15. $f(x) = \dfrac{\sin x}{x}$, $\quad a = 0$

16. $f(x) = x^2 e^{-x}$, $\quad a = 0$

17. $f(x) = x \sin x$, $\quad a = \pi/4$

18. $f(x) = x^2 \ln(1 + x)$, $\quad a = 0$

19. Find a Maclaurin series for $f(x) = \cos^2 x$ by use of the identity $\cos^2 x = \dfrac{1}{2}[1 + \cos 2x]$.

20. Find the first four terms of the Maclaurin series for $f(x) = \tan x$.

21. Find the first three terms of the Maclaurin series for $f(x) = \sec^2 x$, using the result of Exercise 20.

22. Find the Maclaurin series for $x \sin x^2$ using the Maclaurin series for $\sin x$.

23. Find a Maclaurin series for $\cos x^2$ using the technique of Exercise 22.

24. Find the Maclaurin series for $f(x) = \sinh x$ from the Maclaurin series for e^x and e^{-x}.

25. Find the Maclaurin series for $x \cosh x$.

26. Find the Maclaurin series for $f(x) = \sin^2 x$ (see Exercise 19).

In Exercises 27–32, use Taylor series to approximate the given quantity accurately to three decimal places.

27. $\sin 2°$

28. e^2

29. $\ln(1.1)$

30. $\displaystyle\int_0^{\pi/4} \sin x^2 \, dx$

31. $\displaystyle\int_0^{1/2} \dfrac{1}{1 + x^3} \, dx$

32. $\displaystyle\int_0^1 e^{-x^2} \, dx$

In Exercises 33–36, use the Binomial Series to find a power series for the given function and the radius of convergence.

33. $f(x) = \sqrt{1 + 2x}$

34. $f(x) = \sqrt[3]{27 + x}$

35. $f(x) = (9 + 3x)^{3/2}$

36. $f(x) = \dfrac{1}{\sqrt[3]{8 - x^2}}$

37. Define the function $f(x)$ by
$$f(x) = \begin{cases} e^{-(1/x^2)}, & x \neq 0 \\ 0, & x = 0. \end{cases}$$

a. Show that $f^{(n)}(0)$ exists and equals zero for every integer $n \geq 1$.

b. Show that the Maclaurin series for f is identically equal to zero for all values of x. Thus the function f does not equal its Maclaurin series in any interval containing zero. (*Remark:* This example shows that the existence of all derivatives is not a sufficient condition for a function to equal its Taylor series. The condition that $\lim_{n \to \infty} R_n(x) = 0$ is essential.)

38. Prove that power series representations are unique. That is, prove that if
$$f(x) = \sum_{k=0}^{\infty} a_k x^k \quad \text{and} \quad f(x) = \sum_{k=0}^{\infty} b_k x^k,$$
then $\quad a_0 = b_0, a_1 = b_1, \ldots, a_k = b_k, \ldots$

(*Hint:* Set $x = 0$ to obtain $a_0 = b_0$. Then differentiate and set $x = 0$ again, and so on.)

SUMMARY OUTLINE OF CHAPTER 13

◆ The nth Taylor polynomial for f, expanded about $x = a$, is $\qquad$ (page 563)

$$P_n(x) = f(a) + f'(a)(x - a) + \frac{f''(a)}{2!}(x - a)^2 + \cdots + \frac{f^{(n)}(a)}{n!}(x - a)^n = \sum_{j=0}^{n} \frac{f^{(j)}(a)}{j!}(x - a)^j.$$

◆ **Theorem:** If f is $(n + 1)$ times differentiable and $f(x) = P_n(x) + R_n(x)$, then $R_n(x) = \dfrac{f^{(n+1)}(c)}{(n + 1)!}(x - a)^{n+1}$ where c $\qquad$ (page 569) lies between a and x.

◆ A **power series** has the form $\displaystyle\sum_{k=0}^{\infty} a_k(x - a)^k$. $\qquad$ (page 578)

◆ **Theorem:** If the power series $\Sigma a_k x^k$ converges for $x = c$, it does so for all x with $|x| < |c|$. $\qquad$ (page 579)

◆ **Theorem:** The set of all x for which a power series converges is either (i) $\{a\}$, (ii) an interval with midpoint $x = a$ $\qquad$ (page 580) and radius $r \neq 0$, or (iii) $(-\infty, \infty)$. (The radius r is called the **radius of convergence.**)

◆ **Theorem:** If $f(x) = \Sigma a_k x^k$ with radius of convergence r, then $f'(x)$ exists, and $f'(x) = \Sigma k a_k x^{k-1}$ with radius of $\qquad$ (page 586) convergence r.

◆ **Theorem:** If $f(x) = \Sigma a_k x^k$ with radius of convergence r, then $\int f(x)\,dx = \Sigma\left(\dfrac{a_k}{k+1}\right)x^{k+1} + C$ with radius of convergence r. (page 588)

◆ A **Taylor series** is a power series of the form (page 591)

$$\sum_{k=0}^{\infty} \frac{f^{(k)}(a)}{k!}(x-a)^k = f(a) + f'(a)(x-a) + \frac{f''(a)}{2!}(x-a)^2 + \cdots + \frac{f^{(k)}(a)}{k!}(x-a)^k + \cdots.$$

If $a = 0$, this series is called a **Maclaurin series.**

◆ The **binomial series** is (page 598)

$$(1+x)^r = 1 + \sum_{k=1}^{\infty} \frac{r(r-1)(r-2)\cdots\cdots(r-k+1)}{k!}x^k,\ |x| < 1.$$

REVIEW EXERCISES—CHAPTER 13

In Exercises 1–10, find the Taylor polynomial of degree n for $f(x)$ expanded about $x = a$.

1. $f(x) = \sin 2x$, $\quad a = 0$, $\quad n = 5$

2. $f(x) = \ln(1 + x^2)$, $\quad a = 0$, $\quad n = 2$

3. $f(x) = x\cos x$, $\quad a = \pi/4$, $\quad n = 3$

4. $f(x) = \sqrt{1 + x^2}$, $\quad a = 0$, $\quad n = 3$

5. $f(x) = e^{x^2}$, $\quad a = 0$, $\quad n = 3$

6. $f(x) = \sqrt{2x + 3}$, $\quad a = 11$, $\quad n = 3$

7. $f(x) = \operatorname{Tan}^{-1} x$, $\quad a = 1$, $\quad n = 3$

8. $f(x) = x\ln(1 + x)$, $\quad a = 0$, $\quad n = 4$

9. $f(x) = 2^x$, $\quad a = 0$, $\quad n = 3$

10. $f(x) = \dfrac{1}{1 + x^3}$, $\quad a = 0$, $\quad n = 2$

In Exercises 11–15, find the accuracy of the approximation of the given quantity by the nth Taylor polynomial for the function $f(x)$ expanded about $x = a$.

11. $\sin 34°$, $\quad f(x) = \sin x$, $\quad a = \pi/6$, $\quad n = 3$

12. $\tan(\pi/12)$, $\quad f(x) = \tan x$, $\quad a = 0$, $\quad n = 2$

13. $\sqrt[5]{35}$, $\quad f(x) = \sqrt[5]{x}$, $\quad a = 32$, $\quad n = 2$

14. $\sec\left(\dfrac{3\pi}{16}\right)$, $\quad f(x) = \sec x$, $\quad a = \dfrac{\pi}{4}$, $\quad n = 2$

15. $e^{.25}$, $\quad f(x) = e^x$, $\quad a = 0$, $\quad n = 3$

16. Show that the fourth Taylor polynomial for $f(x) = \cosh x^2$ expanded about $a = 0$ is $P_4(x) = 1 + \dfrac{x^4}{2}$.

17. By integrating the $(n-1)$st Taylor polynomial for $f(x) = \dfrac{1}{1-x}$, show that

$$\ln\frac{1}{1-x} = x + \frac{x^2}{2} + \frac{x^3}{3} + \cdots + \frac{x^n}{n} + \int_0^x \frac{t^n\,dt}{1-t}.$$

18. Show that, if $P_n(x)$ is the nth Taylor polynomial for $f(x) = \sinh x$ expanded about $a = 0$, and if $Q_{n+1}(x)$ is the Taylor polynomial of degree $n + 1$ for $\cosh x$ expanded about $a = 0$, then

$$\frac{d}{dx}\,Q_{n+1}(x) = P_n(x).$$

What if $a \neq 0$?

19. Let $P_n(x)$ be the nth Taylor polynomial, expanded about $a = 0$, for $f(x) = x^{7/2}$. Show that $P_0(0) = P_1(0) = P_2(0) = P_3(0) = 0$, but that $P_n(0)$ is undefined for $n \geq 4$.

20. Let $P_n(x)$ be the nth Taylor polynomial for f expanded about $x = a$. If $P_n(x) = 0$ for all n for which $P_n(x)$ is defined, must f be the zero function? (*Hint:* See Exercise 19.)

In Exercises 21–26, find the interval of convergence for the given series.

21. $\Sigma k(x - 3)^k$

22. $\Sigma(-1)^k k^2 (x - 1)^k$

23. $\Sigma \dfrac{\sqrt{k}(x - 2)^k}{k + 3}$

24. $\Sigma \dfrac{(x + 2)^k}{2^k}$

25. $\Sigma \dfrac{(-1)^k x^k}{k\ln k}$

26. $\Sigma \dfrac{\ln k(x - 1)^k}{e^k}$

27. Show that $\displaystyle\sum_{k=0}^{\infty} a_k x^{k+1} = \sum_{k=1}^{\infty} a_{k-1} x^k.$

28. Show that $\displaystyle\sum_{k=1}^{\infty} k(k+1)x^k = \sum_{k=2}^{\infty} (k-1)kx^{k-1}$.

29. Find numbers $a_1, a_2, \ldots$ so that

$$\sum na_n x^{n-1} + c\sum a_n x^n = 0, \qquad c \neq 0.$$

What is the function f whose power series is $\sum a_k x^k$?

30. Find the numbers $a_2, a_3, a_4, \ldots$ so that

$$\sum_{n=2}^{\infty} n(n-1)a_n x^{n-2} + 4\sum_{n=0}^{\infty} a_n x^n = 0.$$

31. Find a power series representation for the function $f(x) = \dfrac{1}{1+x^4}$.

32. Find a Maclaurin series for the function $f(x) = \sin x \cos x$.

33. Find a Maclaurin series for $f(x) = \sqrt[3]{1+x^2}$.

34. Find a power series for the function $f(x) = \dfrac{1}{(1-x)(2-x)}$ using the geometric series. What is the radius of convergence?

35. Find the first 3 terms of the Taylor series for $f(x) = \sin \sqrt{x}$ expanded about $x = \pi^2/4$.

36. What is the radius of convergence of a power series for $e^{\sqrt{x}}$ expanded about $a = 0$?

37. Find a Maclaurin series for $f(x) = x^2 e^{x^2}$ using the series for e^x.

38. Find the Taylor series for $f(x) = \sqrt{x+4}$ expanded about $a = 2$. For what values of x does it converge?

39. Use a Taylor series to approximate $\sqrt{e}$ accurate to three decimal places.

UNIT 6

GEOMETRY IN THE PLANE AND IN SPACE

René Descartes

Pierre de Fermat

René Descartes (1596–1650) and Pierre de Fermat (1601–1665) had at least two things in common: both were French, and neither was a professional mathematician. They shared something else of far greater importance—each independently discovered analytic geometry.

Descartes left school at the age of 16 and went to Paris, where he occasionally studied mathematics. At 21 he became a soldier-for-pay, hiring out to the armies of various nations and minor political subdivisions. He alternated military service with study and travel throughout Europe, and met a number of mathematicians. His soldiering was apparently successful; he was offered a commission as lieutenant general, but declined it since he preferred having less responsibility and more time to think. After nine years of this life, his growing interest in matters of philosophy led him to the relative peace of the Netherlands, where he studied and wrote for 20 years.

On the night of November 10, 1619, Descartes had three dreams that profoundly affected his life and the development of mathematics. He claimed that in one of these dreams he was given the key to understanding nature. Although he never revealed precisely what this was, it is widely believed that he was referring to the relation between algebra and geometry. This may therefore be said to be the founding date of analytic geometry, although publication was not to occur for 18 years.

In 1637 Descartes published *Discours de la méthode* . . . , or *Discourse on the Method of Rightly Conducting the Reason and Seeking Truth in the Sciences*. This was a program for conducting philosophical research—an attempt to reach valid conclusions in many fields by systematic reasoning. The *Discourse* included three appendices. Each was a brilliant illustration of the application of the method to a different field. The first, *La dioptrique*, included the first published statement of the law of refraction (although the Dutch mathematician Willebrord Snell (1591–1626) had discovered it earlier). The second appendix, on meteorology, included an explanation of the colors of the rainbow. The third, and most famous, was *La géométrie*. Starting with an ancient problem from the Greek mathematician Pappus (third century A.D.), Descartes applied algebra to geometry and vice versa. This was not quite today's analytic geometry, however. Coordinate axes were implied but not used explicitly, and were thought of as oblique rather than the perpendicular axes that we now label Cartesian. The notions of distance, angle between lines, and slope were not included. In no case was a curve plotted from an equation. Descartes did not consider negative numbers or negative coordinates. Only one equation was considered in detail, the general second degree equation, and that was not discussed until Chapter 15. Descartes gave the conditions on the coefficients of this equation for the curve to be an ellipse, parabola, or hyperbola. The details were omitted, and indeed the entire work was difficult to read. Readers often found it difficult to see how the appendices were related to the philosophy set forth in the main text. In spite of these inadequacies, the concepts of analytic geometry were introduced in this little work, the only book about mathematics that Descartes ever produced. It was a remarkable and striking achievement.

Descartes was lured away from the Netherlands to a position with Queen Christina of Sweden, who was interested in philosophy and wanted him to establish an academy of sciences. Unfortunately, she wanted her tutoring in philosophy at 5:00 A.M. which seriously conflicted with Descartes' lifelong practice of late rising. He was unable to maintain his health in the cold, damp Swedish climate, and died of pneumonia at 54 years of age.

Pierre de Fermat has already been mentioned in the historical note preceding Unit 2, in which his early work on differentiation (before Newton and Leibniz) is discussed. As a lawyer, politician, and judge, he apparently had sufficient time to devote to the study of his real love, mathematics. His work on analytic geometry was actually done several years before that of Descartes, but he refused to publish his mathematical discoveries. What few items were printed during his lifetime were published by friends but without Fermat's name attached. Most of his work is known either through his notes, often written in the margins of books, or through the many letters he wrote to other mathematicians. The letters frequently contained original mathematics of such merit that copies were made and then recopied for circulation to other mathematicians. But Fermat was so casual about his work that he usually failed even to keep copies of his letters and essays for himself.

Fermat's work in analytic geometry was actually written several years before Descartes' discovery. It was not published, though, until 1679, fourteen years after Fermat's death and a half century after its completion. Descartes wrote in more modern notation than did Fermat, but the latter's ideas much more closely resemble modern mathematics. The 1679 volume, titled (in translation) *Introduction to Plane and Solid Loci,* presents in archaic terminology an astonishing amount of material which is familiar to us today. His linear equation written "*D in A aequitur B in E*" we would write as $Dx = By$, and Fermat sketched it as a line—or rather, as a half-line, since he did not accept negative coordinates. He dealt with circles, parabolas, ellipses, and hyperbolas. He even extended his analytic geometry to three-dimensional space, a generalization that had not occurred to Descartes.

Fermat considered the family of equations of the form $y = x^n$ where n is a positive or negative integer. In an attempt to find the maxima and minima of the associated curves, he found the derivatives of these special functions. He also found the tangents to these curves by using what was essentially the differentiation process, though without the idea of the limit. Later Fermat found the area between such a curve and a line, which corresponds to the integral of x^n. It is curious that he did not seem to have recognized the inverse nature of these two processes. Although Fermat is sometimes called a discoverer of the calculus, he worked with only a few polynomial functions, he lacked the concept of the limit, and he failed to note the Fundamental Theorem of Calculus. Thus, he cannot be said to have reached the intellectual summit that might have allowed him to be ranked with Newton and Leibniz.

Fermat's chief mathematical interest was neither the calculus nor analytic geometry. It involved such topics as prime numbers, perfect numbers, and magic squares, which fall under the heading of the theory of numbers. He further was one of the earliest to develop the theory of probability. He was a classical scholar and could read and write fluently in several languages: French, Spanish, Italian, Greek, and Latin. He even composed poetry in Latin. But his refusal to publish his work resulted in his brilliance not being recognized until long after his death.

(Photographs from the David Eugene Smith Papers, Rare Book and Manuscript Library, Columbia University.)

Chapter 14
The Conic Sections

The purpose of this chapter is to develop a broader familiarity with the types of planar curves that arise as graphs of equations of the form

$$Ax^2 + Bxy + Cy^2 + Dx + Ey + F = 0. \tag{1}$$

When at least one of A, B, and C is nonzero, equation (1) is referred to as the general second degree equation in the two variables x and y. We have previously examined one special case of this equation rather carefully.

If $A = C \neq 0$ and $B = 0$ the equation can be written

$$x^2 + y^2 + dx + ey + f = 0,$$

where $d = \dfrac{D}{A}$, $e = \dfrac{E}{A}$, $f = \dfrac{F}{A}$. By completing the squares in x and in y this equation can be brought to the standard form for the equation of a circle:

$$(x - h)^2 + (y - k)^2 = r^2.$$

(See Section 1.2.)

In the following three sections we shall see that if $B = 0$, the remaining possibilities for the graph of equation (1) are limited to the parabola, the ellipse, or the hyperbola. The case $B \neq 0$ is handled in Section 14.4, where we show that by

Circle.

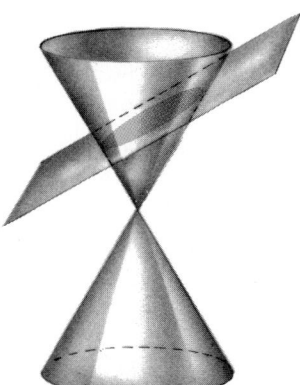

Ellipse.

Parabola.

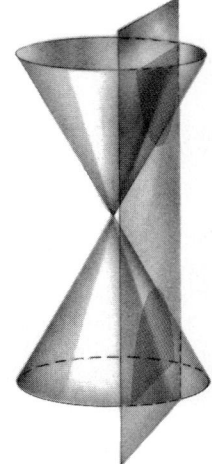

Hyperbola.

rotating the x and y axes we may find new coordinates for the plane in which the xy term in equation (1) is not present. Thus, we will show that the graph of any second degree equation of the form (1) must correspond to one of the four figures mentioned on page 605.

The circle, parabola, ellipse, and hyperbola are referred to as **conic sections** because each arises as the curve of intersection of an infinite cone and a plane, as illustrated on page 605.

We shall not pursue this three-dimensional interpretation of the conic sections further. Rather, we shall use the techniques of analytic geometry in the plane to analyze equation (1) in the cases corresponding to each of these figures.

14.1 PARABOLAS

We begin with the purely geometric definition of the parabola.

DEFINITION 1

A parabola is the set of all points P in the plane lying equidistant from a fixed line ℓ and a fixed point F.

In Definition 1 the fixed point F is called the **focus** of the parabola, and the fixed line ℓ is called the **directrix.** The line through the focus perpendicular to the directrix is called the **axis** of the parabola.

It is easy to see from Definition 1 that the point V on the axis midway between the focus and directrix lies on the parabola. This point is called the **vertex** of the parabola. A typical parabola, illustrating the requirement of Definition 1, is sketched in Figure 1.1.

To determine the equation satisfied by the coordinates of a point $P = (x, y)$ on the graph of a parabola, we position the parabola in the xy-plane so that the vertex V lies at the origin and so that the focus F lies at the point $(0, c)$ (see Figure 1.2). The directrix then has equation $\ell: y = -c$. The number c, representing the (signed) distance from the vertex to the focus, is referred to as the **focal length** of the parabola.

Now let $P = (x, y)$ be any point on the parabola, and let Q be the point on ℓ nearest P. Then $Q = (x, -c)$, and the distance from P to ℓ is the same as the distance from P to Q. Using the distance formula we can state the requirement

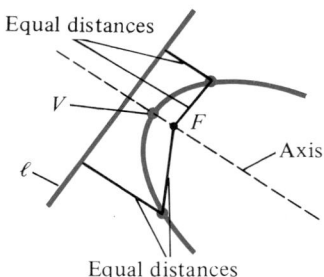

Figure 1.1 Parabola (ℓ = directrix, F = focus, V = vertex).

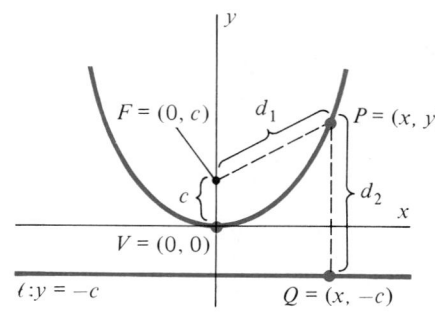

Figure 1.2 Parabola in standard position.

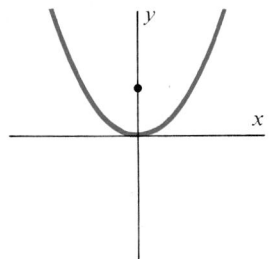

Figure 1.3 $x^2 = 4cy$, $c > 0$.

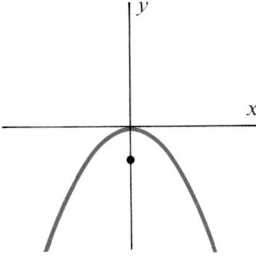

Figure 1.4 $x^2 = 4cy$, $c < 0$.

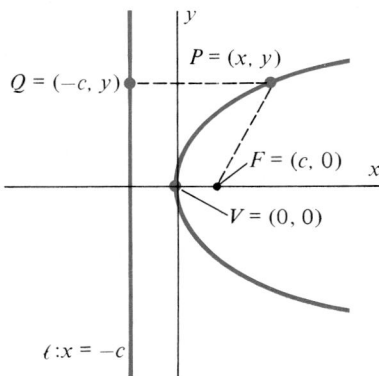

Figure 1.5 $y^2 = 4cx$, $c > 0$.

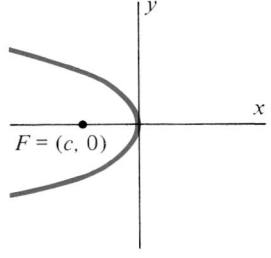

Figure 1.6 $y^2 = 4cx$, $c < 0$.

$d_1 = d_2$ of Definition 1 as

$$\sqrt{(x - 0)^2 + (y - c)^2} = \sqrt{(x - x)^2 + (y - (-c))^2},$$

or

$$\sqrt{x^2 + (y - c)^2} = |y + c|.$$

Squaring both sides gives

$$x^2 + (y - c)^2 = (y + c)^2$$

$$x^2 + (y^2 - 2cy + c^2) = y^2 + 2cy + c^2,$$

$$\boxed{x^2 = 4cy, \qquad \text{or} \qquad y = \frac{1}{4c}x^2.} \tag{1}$$

Equation (1) is the **standard form** for the parabola with vertex at the origin and focus on the y-axis. In Figure 1.2 we have assumed $c > 0$. However, the calculations above did not depend on this assumption, so equation (1) is also valid if $c < 0$. In the case $c < 0$ the parabola opens in the negative y-direction (see Figures 1.3 and 1.4).

REMARK: The above demonstration shows that any point (x, y) satisfying the *geometric* condition of Definition 1 satisfies the *algebraic* condition of equation (1). In Exercise 39 you are asked to show the converse—that any point (x, y) satisfying the algebraic condition (1) also satisfies the geometric condition of Definition 1.

For a parabola with vertex at $(0, 0)$, the axis may lie along the x-axis rather than the y-axis. In this case the focus is labelled $(c, 0)$, the directrix is $x = -c$, and the requirement of Definition 1 is that

$$\sqrt{(x - c)^2 + (y - 0)^2} = \sqrt{(x - (-c))^2 + (y - y)^2},$$

which simplifies to the equation

$$\boxed{y^2 = 4cx, \qquad \text{or} \qquad x = \frac{1}{4c}y^2.} \tag{2}$$

(See Figure 1.5.) Equation (2) is the standard form for the parabola with vertex at $(0, 0)$ and focus on the x-axis. When $c > 0$, the parabola opens in the positive x-direction. When $c < 0$ it opens in the negative x-direction (Figure 1.6).

Example 1

The equation $x^2 + 8y = 0$ may be written in the form

$$x^2 = -8y, \qquad \text{or} \qquad y = -\frac{1}{8}x^2$$

corresponding to equation (1). Its graph is therefore a parabola with vertex $(0, 0)$, with focal length $c = -2$, and opening in the negative y direction. Its focus is at $(0, -2)$ (see Figure 1.7). ◇

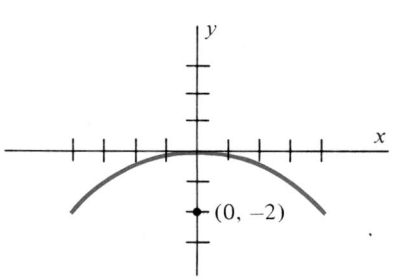

Figure 1.7 $x^2 = -8y$.

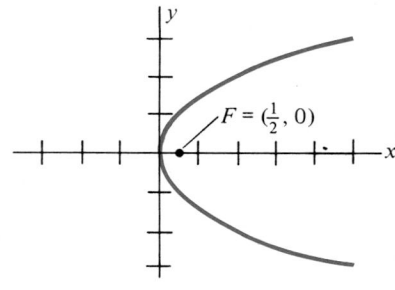

Figure 1.8 $y^2 = 2x$.

Example 2

The graph of the equation $\dfrac{1}{2}y^2 - x = 0$ can be written in the form of equation (2) as

$$y^2 = 2x, \quad \text{or} \quad x = \frac{1}{2}y^2.$$

Its graph is a parabola opening in the positive x direction with vertex $(0, 0)$, focal length $c = 1/2$, and focus $\left(\dfrac{1}{2}, 0\right)$ (Figure 1.8). $\qquad\qquad\Diamond$

Translating the Parabola

When the vertex of a parabola lies at a point (h, k) rather than at $(0, 0)$, we may apply the technique of **translation of axes** to analyze its graph. To do so we introduce new XY-coordinates, via the substitutions

$$X = x - h \quad \text{and} \quad Y = y - k. \tag{3}$$

These substitutions determine new X- and Y-axes, parallel to the x- and y-axes, but with the origin in XY-coordinates located at the vertex of the parabola (see Figure 1.9).

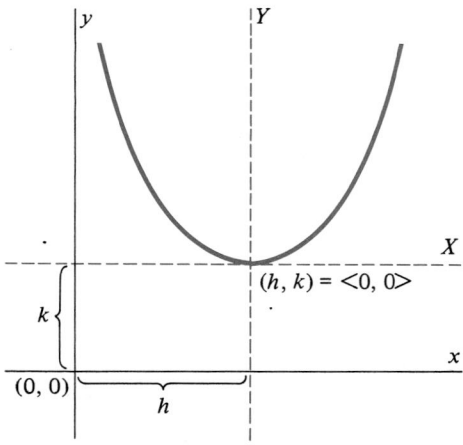

Figure 1.9 $(,) = xy$-coordinates $\langle \, , \, \rangle = XY$-coordinates.

If, after translating axes, the parabola opens along the Y-axis with focal length c, the equation for the parabola in XY-coordinates is

$$X^2 = 4cY, \tag{4}$$

according to equation (1). Combining equation (4) with substitutions (3) we conclude that

$$(x - h)^2 = 4c(y - k) \tag{5}$$

is the *standard form for the equation for the parabola with vertex at* (h, k), *focal length* c, *and axis parallel to the y-axis*. Similarly, from equation (2) and substitutions (3) it follows that

$$(y - k)^2 = 4c(x - h) \tag{6}$$

is the *standard form for the equation of the parabola with vertex at* (h, k), *focal length* c, *and axis parallel to the x-axis*.

As was the case for equations of circles in Section 1.2, the usual technique for analyzing the equation of a parabola is to first complete the square, if necessary, and then to bring the resulting equation into the form of equation (5) or equation (6).

Example 3

Describe the graph of the equation

$$(y - 1)^2 = 6(x + 2).$$

Strategy

Identify the form of the equation as that of equation (6).

Since the parabola opens in the x direction we have

    vertex $= (h, k)$,
    focal length $= c$,
    axis: $y = k$,
    focus: $(h + c, k)$.

Solution

We note that the equation has the form of equation (6) with

$$h = -2 \quad \text{and} \quad k = 1,$$

so the graph is a parabola with axis parallel to the x-axis. Since $6 = 4c$, we have $c = 3/2$.

The vertex is $(-2, 1)$, the focal length is $3/2$, the axis is $y = 1$, and the focus is $\left(-\dfrac{1}{2}, 1\right)$ (see Figure 1.10). $\diamond$

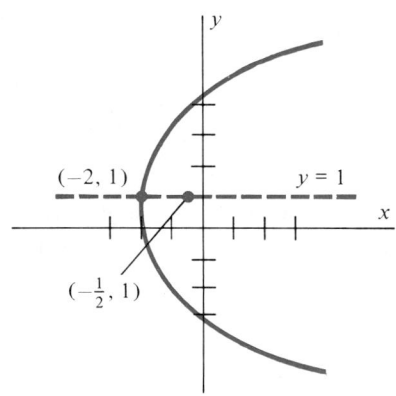

Figure 1.10 $(y - 1)^2 = 6(x + 2)$.

Example 4

Find the vertex, focus, axis, and focal length for the parabola with equation $x^2 - 6x + 6y - 3 = 0$.

Strategy

Solution

We first observe that

$$x^2 - 6x + 6y - 3 = 0$$

gives

Complete the square in x by adding 9 to both sides.

Move all terms except the squared term to the right side.

$$[x^2 - 6x + 9] + 6y - 3 = 9,$$
$$(x - 3)^2 + 6y = 12,$$
$$(x - 3)^2 = -6y + 12$$
$$= -6(y - 2)$$
$$= 4\left(-\frac{3}{2}\right)(y - 2).$$

Identify the form of the resulting equation.

This equation has the form of equation (5) with

$$h = 3, \qquad k = 2, \qquad c = -3/2.$$

Thus the vertex is (3, 2), the focal length is $c = -3/2$, the axis is $x = h = 3$, and the focus is $(h, k + c) = (3, 1/2)$. The graph appears in Figure 1.11. ◇

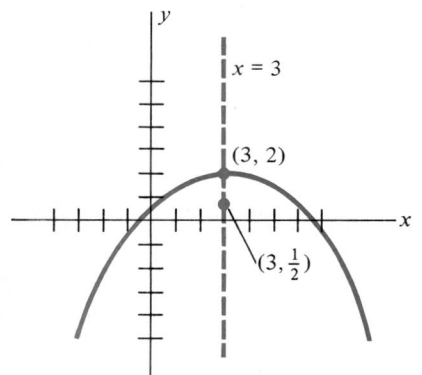

Figure 1.11 $x^2 - 6x + 6y - 3 = 0$.

Example 5

Find the equation of the parabola with directrix $y = -2$ and focus $F = (2, 5)$.

Solution: Since the axis lies perpendicular to the directrix, the axis is parallel to the y-axis. The standard equation for the parabola is therefore that of equation (5). Since the vertex lies midway between the directrix and the focus, the vertex has coordinates $\left(2, -2 + \frac{1}{2}(5 - (-2))\right) = (2, 3/2)$. Thus, $h = 2$ and $k = 3/2$. Since the distance from the focus to the vertex is $5 - 3/2 = 7/2$, the focal length is $c = 7/2$. The desired equation is therefore

$$(x - 2)^2 = 4\left(\frac{7}{2}\right)(y - 3/2),$$

or

$$(x - 2)^2 = 14(y - 3/2).$$ ◇

Reflecting Properties of Parabolas

An important application of parabolic arcs follows from the physical law that when a ray of light strikes a reflecting surface the angle of incidence equals the angle of reflection. The relationship between this law and parabolas lies in the fact that the line tangent to the graph of a parabola at a point P makes equal angles with the ray from P to the focus and the line through P parallel to the axis of the parabola (Figure 1.12 and Exercise 31).

Thus, rays of light parallel to the axis of a parabolic reflector will all be reflected precisely to the focus of the parabola (Figure 1.13). This principle is used in reflect-

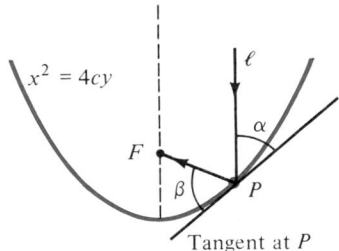

Figure 1.12 Reflecting property: $\alpha = \beta$ if ℓ is parallel to axis.

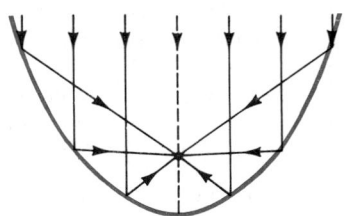

Figure 1.13 Rays parallel to the axis are all reflected through the focus of a parabola, and conversely.

ing telescopes, where the parallel rays of light from the stars are reflected off a parabolic mirror to the eyepiece located at the focus. The same principle applies to radar signals, which are gathered by means of parabolic reflecting ''dishes.''

The same principle in reverse explains the design of automobile headlamps. By placing the bulb at the focus of a parabolic mirror, one ensures that the beams of light will leave the headlamp parallel to the axis of the parabolic mirror.

Exercise Set 14.1

In Exercises 1–20, sketch the parabola whose equation is given, noting the vertex, focal length, axis, and focus.

1. $y = 2x^2$

2. $y = 4x^2 + 8$

3. $y = -2x^2 + 2$

4. $x = -y^2$

5. $x = 3y^2 + 1$

6. $x + 2y^2 = 4$

7. $8y + x^2 = 2$

8. $y - x^2 = 4$

9. $(x - 2)^2 = 8(y - 1)$

10. $x^2 = 12(y + 2)$

11. $(y - 2)^2 = 6(x + 3)$

12. $(y - 1)^2 = -7x$

13. $y^2 - 2y - 4x - 7 = 0$

14. $y^2 + 4y + x + 7 = 0$

15. $x^2 + 2x - 6y - 11 = 0$

16. $x^2 - 4x - 4y = 0$

17. $x^2 + 6x - 10y + 19 = 0$

18. $y^2 - 4y - 8x + 20 = 0$

19. $y^2 + 4y - 2x + 6 = 0$

20. $y^2 + 6y - 3x + 6 = 0.$

In Exercises 21–27, find an equation for the parabola with the stated properties.

21. Vertex $(0, 0)$ and focus $(0, 3)$.

22. Vertex $(-1, 2)$ and focus $(-1, -2)$.

23. Vertex $(1, 3)$ and directrix $y = -5$.

24. Focus $(-1, -1)$ and directrix $x = 3$.

25. Focus $(3, -6)$ and vertex $(-1, -6)$.

26. Vertex $(0, 4)$ and directrix $x = 2$.

27. Axis $x = 0$, directrix $y = -3$, and containing the point $(0, 1)$.

28. State a procedure for sketching the graph of a parabola according to Definition 1, assuming that you are given the equation for the directrix and the location of the focus and assuming that you can use only a ruler and compass.

29. The chord through the focus of a parabola perpendicular to its axis is called the **latus rectum** of the parabola (see Figure 1.14). Prove that the length of the latus rectum of a parabola is $|4c|$ where $|c|$ is the focal length.

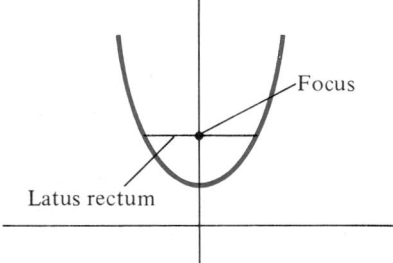

Figure 1.14 Latus rectum.

30. Prove that a circle that has a diameter corresponding to the latus rectum of a parabola must be tangent to the directrix of that parabola.

31. In Figure 1.12, α is the angle formed between the tangent at P and the line ℓ through P parallel to the axis of the parabola; β is the angle formed between the tangent at P and the line segment from P to the focus F. Prove that $\alpha = \beta$. (*Hint:* Construct a coordinate system with origin at the vertex of the parabola and extend ℓ through P toward the directrix. Also draw a line through F, perpendicular to the tangent at P, until it intersects ℓ. What can you show about the triangle formed in this way?)

32. Prove that the graph of the second degree equation

$$Ax^2 + Bxy + Cy^2 + Dx + Ey + F = 0$$

is a parabola with axis parallel to either the x- or y-axis whenever $B = 0$ and either $A = 0$ or $C = 0$ but $AC \neq 0$.

33. Find the area of the region bounded by the graph of $y = p(x)$ and the x-axis if the graph of $y = p(x)$ is a parabola with focus at the origin and directrix $y = -2$.

34. Find the volume of the solid obtained by revolving about the y-axis the area of the region bounded by the y-axis and the parabola with vertex $(-1, 0)$ and directrix $x = -3/2$.

35. Find the area of the region bounded by the x-axis and the parabola with vertex $(4, 2)$ and directrix $y = 4$.

36. Find the volume of the solid obtained by revolving about the y-axis the region bounded by the y-axis and the parabola with focus $(0, 4)$ and directrix $x = 4$.

37. Find the average value, on the interval $[0, 5]$, of the function whose graph is a parabola with focus $(2, 3)$ and vertex $(2, -1)$.

38. Find the area of the region bounded by the graph of the function in Exercise 34 and the line $x = 5$.

39. Show that if a point (x, y) satisfies the equation $x^2 = 4cy$ then it has the property that its distance from the directrix $y = -c$ equals its distance from the focus $(0, c)$. Thus, any point satisfying the algebraic condition $x^2 = 4cy$ for a parabola also satisfies the geometric condition of Definition 1.

14.2 THE ELLIPSE

The geometric definition of the ellipse is the following.

DEFINITION 2

An ellipse is the set of all points in the plane the sum of whose distances from two fixed points is a constant.

The two fixed points in Definition 2 are referred to as the **foci** of the ellipse. We may interpret Definition 2 by imagining a length of string tied at each end to one of two tacks driven into a flat surface. The tacks represent the foci. If a pencil is held against the surface as in Figure 2.1, so that the string is kept taut, the figure traced out by the possible locations for the pencil tip is an ellipse. This is because the sum of the distances of the pencil tip from the foci remains constant.

To determine the equation for the ellipse, we construct a coordinate system so that the foci for the ellipse are located at points $F_1 = (-c, 0)$ and $F_2 = (c, 0)$ (Figure 2.2). We let $P = (x, y)$ be an arbitrary point on the ellipse, and we let d_1 and d_2 be the distances from P to F_1 and from P to F_2, respectively. The condition that P be on the ellipse is that $d_1 + d_2$ equal a constant, which we take to be $2a$. (Note that we must have $a > c$, since the distance between the foci is $2c$.)

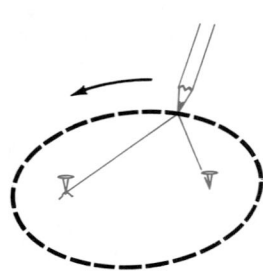

Figure 2.1 Tracing out an ellipse.

Using the distance formula, we can now write the requirement of Definition 2 as

$$d_1 + d_2 = 2a,$$

or

$$\sqrt{(x + c)^2 + y^2} + \sqrt{(x - c)^2 + y^2} = 2a.$$

To simplify this equation we move the second radical to the right-hand side and square both sides. We obtain

$$(x + c)^2 + y^2 = 4a^2 - 4a\sqrt{(x - c)^2 + y^2} + (x - c)^2 + y^2.$$

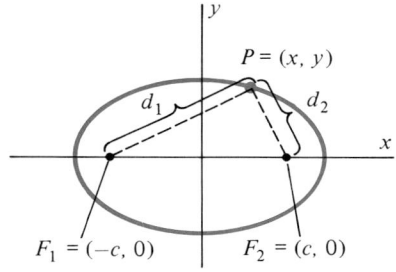

Figure 2.2 Ellipse in standard position.

After the squared terms are expanded, this simplifies to

$$2cx = 4a^2 - 4a\sqrt{(x - c)^2 + y^2} - 2cx,$$

so

$$a\sqrt{(x - c)^2 + y^2} = a^2 - cx.$$

Squaring both sides again gives

$$a^2[(x - c)^2 + y^2] = c^2x^2 - 2a^2cx + a^4,$$

or

$$(a^2 - c^2)x^2 + a^2y^2 = a^2(a^2 - c^2). \tag{1}$$

To simplify (1), we define the constant b by

$$b = \sqrt{a^2 - c^2}. \tag{2}$$

Substituting (2) into (1) then gives

$$b^2x^2 + a^2y^2 = a^2b^2,$$

or

$$\boxed{\dfrac{x^2}{a^2} + \dfrac{y^2}{b^2} = 1, \qquad a > b.} \tag{3}$$

Equation (3) is the *standard form for the equation of the ellipse with center at* (0, 0) *and foci on the x-axis*. (The **center** of an ellipse is the midpoint of the line segment joining the foci.) In Exercise 39 you are asked to show the converse—that any point satisfying equation (3) also satisfies the geometric condition of Definition 2.

The geometry associated with equation (3) is worth noting. Setting $y = 0$ in (3) shows that the x-intercepts of the ellipse are $x = \pm a$ (see Figure 2.3). The line segment joining the points $(-a, 0)$ and $(a, 0)$ is referred to as the *major* axis of the ellipse. Its length is $2a$. In general, the **major axis** of an ellipse is the chord passing through both foci.

Setting $x = 0$ in (3) shows that the y-intercepts are $y = \pm b$. The line segment joining $(0, b)$ and $(0, -b)$ is called the *minor* axis of the ellipse. Its length is $2b$. More generally, the **minor axis** of an ellipse is the chord through the center and perpendicular to the major axis.

Figure 2.4 illustrates the ellipse that results when the foci are located along the y-axis at $(0, c)$ and at $(0, -c)$. The equation describing this ellipse may be developed just as for the previous case. However, it is much simpler to note that in passing from Figure 2.3 to Figure 2.4 we have merely interchanged the roles of x and y. Thus,

$$\boxed{\dfrac{x^2}{b^2} + \dfrac{y^2}{a^2} = 1, \qquad a > b} \tag{4}$$

is *the standard form for the equation of the ellipse with center at* (0, 0) *and foci along the y-axis*.

Note in equation (4) that the constant a still represents half the length of the major axis and that b is again half the length of the minor axis. For either equation

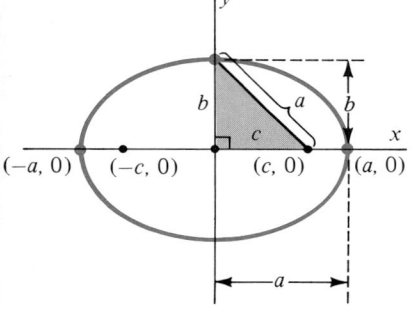

Figure 2.3 $\dfrac{x^2}{a^2} + \dfrac{y^2}{b^2} = 1.$

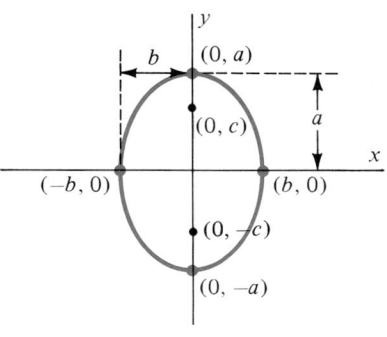

Figure 2.4 $\dfrac{x^2}{b^2} + \dfrac{y^2}{a^2} = 1.$

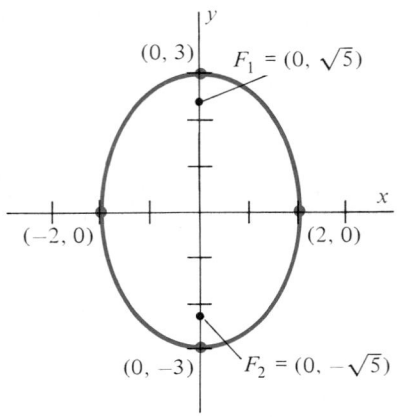

Figure 2.5 $\dfrac{x^2}{4} + \dfrac{y^2}{9} = 1.$

(3) or equation (4), the distance between the origin and a focus is found by solving equation (2) for c:

$$c = \sqrt{a^2 - b^2}. \tag{5}$$

Example 1

Describe the graph of the equation $\dfrac{x^2}{4} + \dfrac{y^2}{9} = 1.$

Solution: The equation has the form of equation (4) with $b^2 = 4$, $a^2 = 9$. Thus, $b = 2$ and $a = 3$. The figure is an ellipse with center at $(0, 0)$, with major axis vertical and of length $2a = 6$, and with minor axis of length $2b = 4$. From equation (5) we find that $c = \sqrt{9 - 4} = \sqrt{5}$, so the foci are located at $(0, \sqrt{5})$ and $(0, -\sqrt{5})$ (see Figure 2.5). $\diamond$

Example 2

Describe the graph of the equation $4x^2 + 16y^2 - 64 = 0$.

Solution: To bring the equation into standard form, we move the constant term to the right and divide both sides by this constant. We obtain

$$4x^2 + 16y^2 = 64,$$

so

$$\frac{x^2}{4^2} + \frac{y^2}{2^2} = 1.$$

By setting $y = 0$, we obtain the x-intercepts, $x = \pm 4$. The endpoints of the major axis are therefore $(-4, 0)$ and $(4, 0)$. Setting $x = 0$ gives the y-intercepts, $y = \pm 2$. The endpoints of the minor axis are $(0, 2)$ and $(0, -2)$. Since the equation is of the form of equation (3), with $a = 4$ and $b = 2$, we have from equation (5) that $c = \sqrt{4^2 - 2^2} = \sqrt{12}$. The foci are therefore $(\pm\sqrt{12}, 0)$ (Figure 2.6). $\diamond$

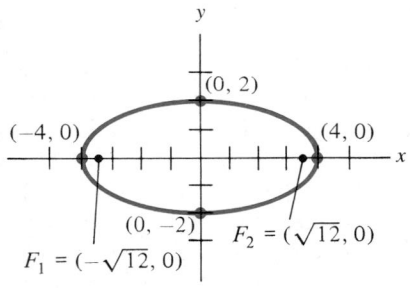

Figure 2.6 $4x^2 + 16y^2 - 64 = 0.$

Translating the Ellipse

When the center of an ellipse lies at the point (h, k) rather than at the origin, we can analyze its graph by first translating the coordinate axes by the substitutions

$$X = x - h \quad \text{and} \quad Y = y - k \quad \text{(Figure 2.7).} \tag{6}$$

In XY-coordinates, the center of the ellipse lies at the origin. If the major axis of the ellipse is parallel to the x-axis, the equation of the ellipse in XY-coordinates is, by equation (3),

$$\frac{X^2}{a^2} + \frac{Y^2}{b^2} = 1. \tag{7}$$

Using equation (6), we conclude that

$$\boxed{\frac{(x - h)^2}{a^2} + \frac{(y - k)^2}{b^2} = 1, \quad a > b} \tag{8}$$

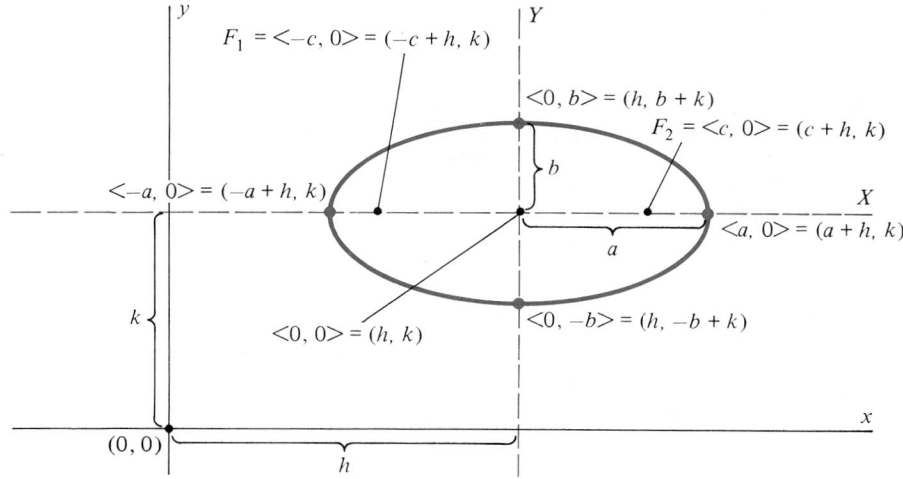

Figure 2.7   Ellipse $\dfrac{(x - h)^2}{a^2} + \dfrac{(y - k)^2}{b^2} = 1$, $a > b$. $\langle \, , \rangle$ = XY-coordinates, $(\, ,) = xy$-coordinates.

is *the standard form of the equation for the ellipse with center at (h, k) and major axis parallel to the x-axis*. In (8), the constant a represents half the length of the major axis while b is half the length of the minor axis. Similarly,

$$\frac{(x - h)^2}{b^2} + \frac{(y - k)^2}{a^2} = 1, \qquad a > b \tag{9}$$

is *the standard form of the equation for the ellipse with center at (h, k), with major axis parallel to the y-axis and of length 2a, and with minor axis of length 2b*.

For either equation (8) or equation (9), the distance between the center of the ellipse and either focus is given by c in equation (5).

Example 3

Describe the graph of the equation

$$4x^2 + 9y^2 - 16x - 54y + 61 = 0.$$

Strategy
Complete the square in both variables.

Solution
$$\begin{aligned} 4x^2 &+ 9y^2 - 16x - 54y + 61 \\ &= 4(x^2 - 4x + 4) + 9(y^2 - 6y + 9) + 61 - 16 - 81 \\ &= 4(x - 2)^2 + 9(y - 3)^2 - 36. \end{aligned}$$

Write the original equation in squared form.

The equation is therefore equivalent to

$$4(x - 2)^2 + 9(y - 3)^2 = 36,$$

Bring equation to form (8) or (9). Identify all constants.

or

$$\frac{(x - 2)^2}{3^2} + \frac{(y - 3)^2}{2^2} = 1.$$

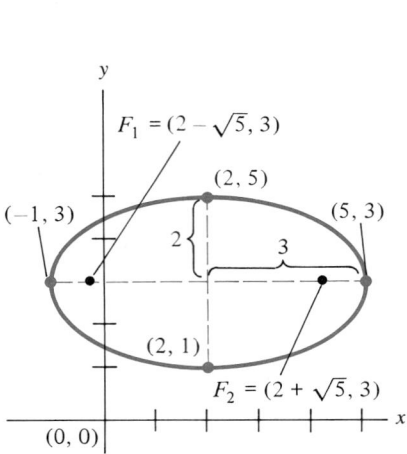

Figure 2.8 Graph of equation $4x^2 + 9y^2 - 16x - 54y + 61 = 0$.

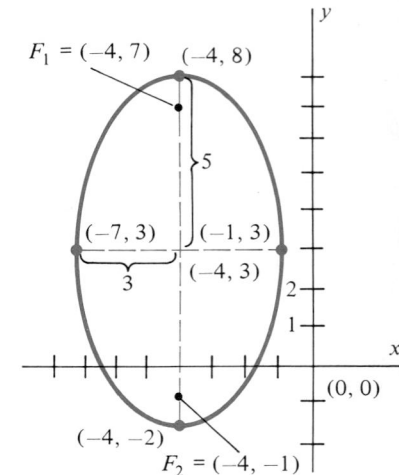

Figure 2.9

Center is (h, k).
Major axis has length $2a$.
Minor axis has length $2b$.
Foci are at $(h \pm c, k)$ where $c = \sqrt{a^2 - b^2}$.

This equation has the form of equation (8) with $h = 2$, $k = 3$, $a = 3$, $b = 2$, and $c = \sqrt{3^2 - 2^2} = \sqrt{5}$. The graph is an ellipse with center $(2, 3)$, major axis parallel to the x-axis and of length $2a = 6$, and minor axis of length $2b = 4$. Foci are at $(2 - \sqrt{5}, 3)$ and $(2 + \sqrt{5}, 3)$. The graph appears in Figure 2.8. ◇

Example 4

Find the equation for the ellipse with center $(-4, 3)$, with minor axis of length 6, and with foci at $(-4, 3 \pm 4)$.

Solution: The distance from the center of the ellipse to one of the foci is $c = 4$. The length of the minor axis is $2b = 6$, so $b = 3$. Thus, by equation (5),

$$a^2 = b^2 + c^2 = 3^2 + 4^2 = 5^2,$$

so $a = 5$. Also, $h = -4$ and $k = 3$ are the coordinates of the center. Since the major axis is vertical, we use equation (9) to write

$$\frac{(x + 4)^2}{3^2} + \frac{(y - 3)^2}{5^2} = 1.$$

(See Figure 2.9.) ◇

Reflecting Property of Ellipses

In Exercise 32 you are asked to show that the rays $\overline{PF}_1$ and $\overline{PF}_2$, from a point P on an ellipse to the foci, make equal angles with the line tangent to the ellipse at P. Applying the physical law that the angle of incidence equals the angle of reflection, we conclude that if a ray of light or a sound wave emanates from the focus F_1 of an elliptical reflecting surface, it will be reflected through the second focus F_2. Visitors to the United States Capitol Building hear tales of just such a phenomenon. In the days when congressmen were seated on the floor beneath the beautiful elliptical dome, the unfortunate politician seated at a focus risked losing whispered secrets to the statesman seated at the other focus (see Figures 2.10 and 2.11).

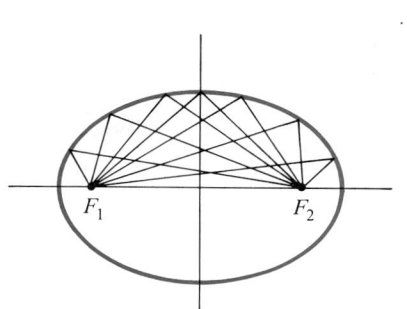

Figure 2.10 Reflecting property of ellipses.

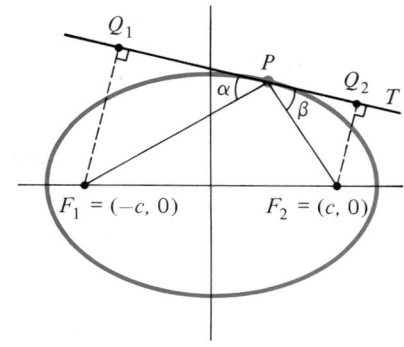

Figure 2.11 $\alpha = \beta$ if T is the tangent to the ellipse at P.

Exercise Set 14.2

In Exercises 1–16, find the center and foci for the given ellipse. State the length of the major and minor axes and sketch the graph.

1. $\dfrac{x^2}{9} + \dfrac{y^2}{1} = 1$

2. $\dfrac{x^2}{16} + \dfrac{y^2}{49} = 1$

3. $4x^2 + 16y^2 = 64$

4. $9x^2 + 4y^2 = 36$

5. $50x^2 + 18y^2 - 450 = 0$

6. $8x^2 + 18y^2 - 72 = 0$

7. $\dfrac{(x-2)^2}{3} + \dfrac{(y-1)^2}{4} = 1$

8. $\dfrac{(x-2)^2}{3} + \dfrac{(y+2)^2}{4} = 25$

9. $(x+3)^2 + 4(y-1)^2 = 16$

10. $5(x+2)^2 + 4(y-3)^2 = 25$

11. $9x^2 + 4y^2 - 18x + 8y = 23$

12. $x^2 + 4y^2 + 16y + 12 = 0$

13. $9x^2 + 25y^2 + 54x - 50y - 119 = 0$

14. $9x^2 + 4y^2 - 36x + 48y + 144 = 0$

15. $2x^2 + 3y^2 + 6y - 4x - 1 = 0$

16. $4x^2 + 2y^2 + 24x + 12y + 46 = 0$

In Exercises 17–23, find the equation for the ellipse with the stated properties.

17. Center at $(0, 0)$, major axis of length 8, minor axis of length 6, foci on x-axis.

18. Foci at $(\pm 4, 0)$, major axis of length 10.

19. Foci at $(\pm 4, 0)$, minor axis of length 6.

20. Foci at $(0, \pm\sqrt{3})$, minor axis of length 2.

21. Foci at $(4, 5)$ and $(4, -1)$, minor axis of length 8.

22. Center at $(2, -6)$, foci at $(2, -3)$, and $(2, -9)$, major axis of length 10.

23. Foci at $(-2, 3)$ and $(-2, 9)$, minor axis of length 8.

24. Find the area of the region enclosed by the ellipse
$$\frac{x^2}{4} + \frac{y^2}{9} = 1.$$

25. Find the volume of the solid obtained by revolving about the x-axis the region in the first quadrant bounded by the graph of the ellipse $25x^2 + 9y^2 = 225$.

26. Find the volume of the solid obtained by revolving the region described in Exercise 25 about the y-axis.

27. A chord through a focus of an ellipse and parallel to the minor axis is called a **latus rectum.** Find the length of a latus rectum for the ellipse
$$\frac{x^2}{a^2} + \frac{y^2}{b^2} = 1, \qquad a^2 > b^2.$$

28. The **eccentricity** of the ellipse $\dfrac{x^2}{a^2} + \dfrac{y^2}{b^2} = 1$ is defined to be the ratio
$$e = \frac{c}{a} = \frac{\sqrt{a^2 - b^2}}{a}.$$

   a. Show that $0 \le e < 1$ for every ellipse.
   b. Discuss the change in the nature of the ellipse as e approaches zero.
   c. Discuss the change in the nature of the ellipse as e approaches one.

29. True or false? A circle is a special case of an ellipse. Explain.

30. Show that for the general second degree equation

$$Ax^2 + Bxy + Cy^2 + Dx + Ey + F = 0$$

if $B = 0$, $A \neq 0$, and $C \neq 0$, the graph is an ellipse if A and B have the same sign.

31. Consider the equation for an ellipse: $\dfrac{x^2}{a^2} + \dfrac{y^2}{b^2} = 1$.

   a. Differentiate implicitly with respect to x to show that

   $$\frac{dy}{dx} = \frac{-b^2 x}{a^2 y}, \qquad y \neq 0.$$

   b. Use part (a) to show that the equation for the line tangent to the graph of this ellipse at (x_0, y_0) can be written

   $$(b^2 x_0)x + (a^2 y_0)y - a^2 b^2 = 0.$$

32. Use the result of Exercise 31 to prove that triangle $F_1 Q_1 P$ is similar to triangle $F_2 Q_2 P$ in Figure 2.11. Conclude that $\alpha = \beta$.

33. Find the equation for the set of all points the sum of whose distances from $(2, 3)$ and $(-4, 3)$ is 12.

34. Find the equation for the set of all points whose distances from the point $(4, 0)$ is half the distance from the line $y = 3$.

35. Find the volume of the solid obtained by revolving the region bounded by the ellipse $16x^2 + 9y^2 = 144$ about the y-axis.

36. A water tank has the shape of a cylinder with elliptical end panels. The end panels have major axes horizontal, of length 32 feet, and minor axes vertical, of length 18 feet. The tank is full. Find the work done in pumping the water out through a valve in the top of the tank.

37. Find the answer to Exercise 36 if the tank is only half full.

38. State the domain and range of the function $f(x) = \sqrt{b^2\left(1 - \dfrac{x^2}{a^2}\right)}$. What is the relationship between the graph of this function and the ellipse with equation $b^2 x^2 + a^2 y^2 = a^2 b^2$?

39. Show that if (x, y) is a point whose coordinates satisfy the equation

$$\frac{x^2}{a^2} + \frac{y^2}{b^2} = 1,$$

then (x, y) satisfies the geometric condition of Definition 2, that the sum of its distances from the foci $(-c, 0)$ and $(c, 0)$ is $2a$, where $c^2 = a^2 - b^2$.

14.3 THE HYPERBOLA

As for the parabola and the ellipse, we begin with the geometric definition of the hyperbola and then develop the corresponding equations.

DEFINITION 3

A hyperbola is the set of all points in the plane the difference of whose distances from two fixed points is a constant.

To interpret Definition 3, we locate two fixed points, F_1 and F_2 (called **foci**) in the plane. Let P be a point on the hyperbola, and let $k = d(P, F_1) - d(P, F_2)$. The hyperbola with foci F_1 and F_2 containing the point P is the set of all points Q such that either $d(Q, F_1) - d(Q, F_2) = k$ or else $d(Q, F_2) - d(Q, F_1) = k$ (see Figure 3.1).

To determine the equations for the hyperbola, we construct a pair of xy-coordinate axes so that the foci are located at $F_1 = (-c, 0)$ and $F_2 = (c, 0)$ (Figure 3.2). (We refer to the midpoint of the line segment $\overline{F_1 F_2}$ as the **center** of the hyperbola, so the center in this case is the origin.) Let $P = (x, y)$ be a point on the hyperbola, and let $2a$ be the constant referred to in Definition 3. Then P is on the hyperbola if and only if

$$|d(P, F_2) - d(P, F_1)| = 2a.$$

Using the distance formula and the coordinates for P, F_1, and F_2, we can rewrite this equation as

$$\left|\sqrt{(x - c)^2 + y^2} - \sqrt{(x + c)^2 + y^2}\right| = 2a, \qquad c > a. \tag{1}$$

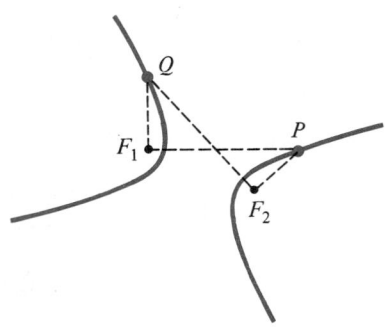

Figure 3.1 $d(P, F_1) - d(P, F_2) = d(Q, F_2) - d(Q, F_1)$.

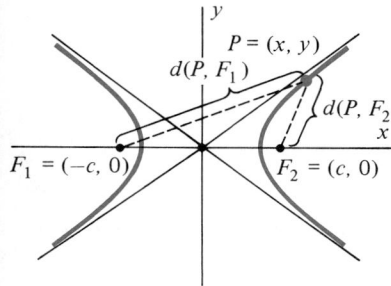

Figure 3.2 Hyperbola with center at (0, 0).

By squaring both sides of this equation and simplifying, just as we did in Section 14.2 for the ellipse, we can obtain the *standard form of the equation for the hyperbola with center at (0, 0) and foci on the x-axis:*

$$\frac{x^2}{a^2} - \frac{y^2}{b^2} = 1, \qquad b = \sqrt{c^2 - a^2}. \tag{2}$$

By setting $y = 0$ in equation (2), we obtain the x-intercepts of the hyperbola $x = \pm a$. The points $V_1 = (-a, 0)$ and $V_2 = (a, 0)$ are called the **vertices** of the hyperbola. (By setting $x = 0$ you will observe that the graph of equation (2) has no y-intercepts. This observation will help you in determining whether the foci of a given hyperbola lie along the x-axis or along the y-axis.)

In graphing equation (2) it is helpful to note that *the lines* $y = \pm\frac{b}{a}x$ *are asymptotes for the graph of equation (2)*. This means that the points on the graph of the hyperbola approach the lines $y = \pm\frac{b}{a}x$ as $|x| \to \infty$. To see this we solve (2) for y:

$$\frac{y^2}{b^2} = \frac{x^2}{a^2} - 1,$$

so

$$y = \pm b\sqrt{\frac{x^2}{a^2} - 1} = \pm\frac{b}{a}x\sqrt{1 - \frac{a^2}{x^2}}. \tag{3}$$

Since $\lim\limits_{|x|\to\infty} \sqrt{1 - \frac{a^2}{x^2}} = \sqrt{1 - 0} = 1$, equation (3) shows that

$$y = \pm\frac{b}{a}x \tag{4}$$

are the equations for asymptotes to the graph of equation (2) (see Figure 3.3).

It is worth noting the geometric significance of b in equation (2). If a rectangle is constructed with vertical sides passing through the vertices $(-a, 0)$ and $(a, 0)$ and with horizontal sides passing through the points $(0, b)$ and $(0, -b)$, the diagonals of this rectangle pass through $(0, 0)$ and have slope $\pm\frac{b}{a}$ (Figure 3.4).

Thus, the asymptotes (4) may be sketched by extending the diagonals of this rectangle. Also, the second equation in line (2) may be solved for c to give $c^2 = a^2 + b^2$. Thus, c may be interpreted as the length of the hypotenuse of one of the eight triangles determined by the diagonals of this rectangle.

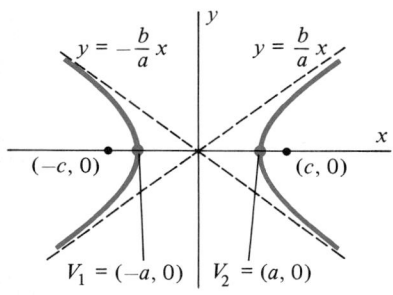

Figure 3.3 The hyperbola $\frac{x^2}{a^2} - \frac{y^2}{b^2} = 1.$

Example 1

Sketch the hyperbola $\dfrac{x^2}{9} - \dfrac{y^2}{4} = 1.$

Solution: This equation has the form of equation (2) with $a = 3$ and $b = 2$. The vertices are therefore $V_1 = (-3, 0)$ and $V_2 = (3, 0)$, and the asymptotes are $y = \pm\frac{2}{3}x$. The graph appears in Figure 3.5. ◇

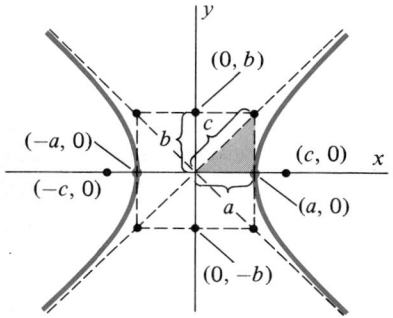

Figure 3.4 Relationships among the constants a, b, and c.

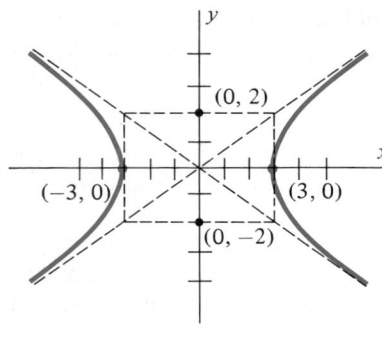

Figure 3.5 $\dfrac{x^2}{9} - \dfrac{y^2}{4} = 1.$

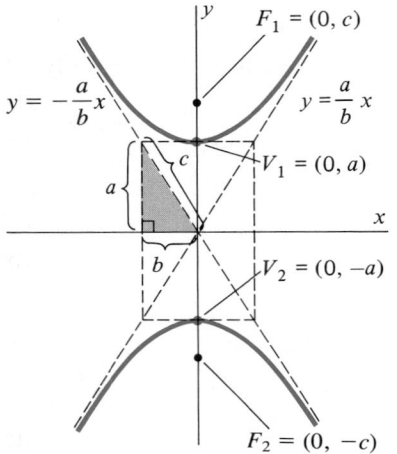

Figure 3.6 Hyperbola $\dfrac{y^2}{a^2} - \dfrac{x^2}{b^2} = 1.$

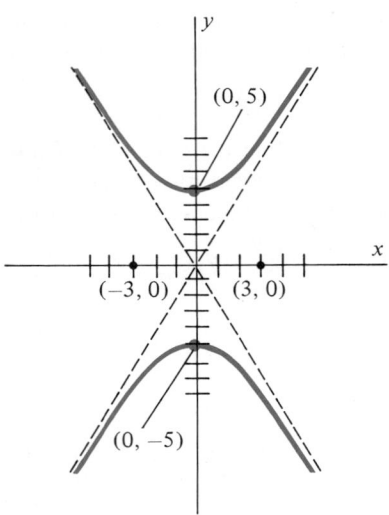

Figure 3.7 Graph of $25x^2 - 9y^2 + 225 = 0.$

If the foci of a hyperbola lie along the y-axis at $F_1 = (0, c)$ and $F_2 = (0, -c)$, the condition

$$|d(P, F_2) - d(P, F_1)| = 2a$$

becomes

$$\left|\sqrt{x^2 + (y + c)^2} - \sqrt{x^2 + (y - c)^2}\right| = 2a, \tag{5}$$

which simplifies to the *standard form of the equation for the hyperbola with center at (0, 0) and foci along the y-axis*

$$\frac{y^2}{a^2} - \frac{x^2}{b^2} = 1, \qquad b = \sqrt{c^2 - a^2}, \qquad c > a. \tag{6}$$

As in the previous case, we may solve for y to determine the equations for the asymptotes:

$$y^2 = a^2\left(\frac{x^2}{b^2} + 1\right),$$

so

$$y = \pm a\sqrt{\frac{x^2}{b^2} + 1} = \pm \frac{a}{b}x\sqrt{1 + \frac{b^2}{x^2}}. \tag{7}$$

Since $\displaystyle\lim_{x \to \pm\infty}\sqrt{1 + \frac{b^2}{x^2}} = 1$, equation (7) shows that *the asymptotes for the graph of equation (6) are given by*

$$y = \pm\frac{a}{b}x. \tag{8}$$

(See Figure 3.6.)

REMARK: A simple way to avoid confusion in working with equations (4) and (8) for the asymptotes associated with the standard form for a hyperbola is to note that in either case the equations are

$$y = \pm\left(\frac{\text{constant in denominator of } y \text{ term}}{\text{constant in denominator of } x \text{ term}}\right)^{1/2} \cdot x.$$

Example 2

Graph the hyperbola $25x^2 - 9y^2 + 225 = 0.$

Solution: Subtracting the constant term from both sides and dividing by -225 gives

$$\frac{y^2}{25} - \frac{x^2}{9} = 1, \qquad \text{or} \qquad \frac{y^2}{5^2} - \frac{x^2}{3^2} = 1.$$

This equation has the form of equation (6) with $a = 5$ and $b = 3$. By (8) the asymptotes are $y = \pm\dfrac{5}{3}x$. The vertices are $(0, a) = (0, 5)$ and $(0, -a) = (0, -5)$. The graph appears as Figure 3.7. $\diamond$

Translating the Hyperbola

If the center of a hyperbola is located at (h, k) rather than at $(0, 0)$, we translate axes using the equations

$$X = x - h \qquad \text{and} \qquad Y = y - k. \tag{9}$$

If the foci of the hyperbola lie on the X-axis, the equation in XY-coordinates, from (2), is

$$\frac{X^2}{a^2} - \frac{Y^2}{b^2} = 1, \tag{10}$$

and the equations for the asymptotes are

$$Y = \pm \frac{b}{a} X. \tag{11}$$

Using (9), we may write these equations in xy-coordinates. The equation

$$\frac{(x - h)^2}{a^2} - \frac{(y - k)^2}{b^2} = 1, \qquad b = \sqrt{c^2 - a^2}, \qquad a < c \tag{12}$$

is *the standard form of the equation for the hyperbola with center at (h, k) and foci on a line parallel to the x-axis*. The asymptotes for this hyperbola are

$$y - k = \pm \frac{b}{a}(x - h) \qquad \text{(see Figure 3.8).} \tag{13}$$

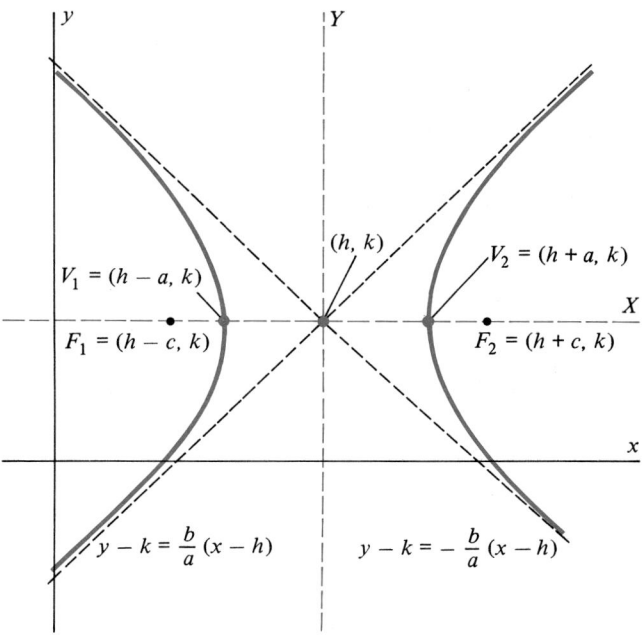

Figure 3.8 Hyperbola with center (h, k).

Similarly, *the standard form of the equation for the hyperbola with center at (h, k) and foci on a line parallel to the y-axis is*

$$\frac{(y - k)^2}{a^2} - \frac{(x - h)^2}{b^2} = 1, \qquad b = \sqrt{c^2 - a^2}, \qquad a < c, \tag{14}$$

and the asymptotes for this hyperbola are

$$y - k = \pm \frac{a}{b}(x - h). \tag{15}$$

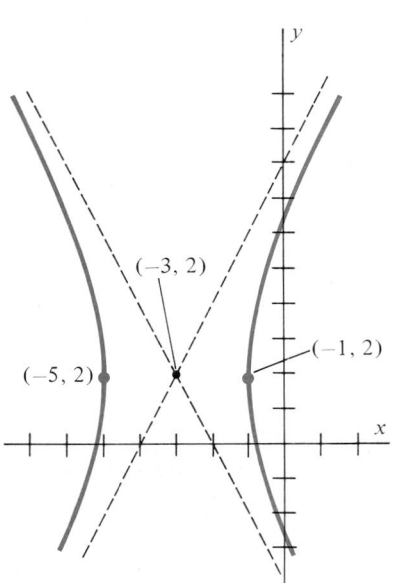

Figure 3.9 Graph of $16(x + 3)^2 - 4(y - 2)^2 = 64$.

Example 3

Describe the graph of the hyperbola

$$16(x + 3)^2 - 4(y - 2)^2 = 64.$$

Solution: Dividing both sides by 64 shows that the standard form for the hyperbola is

$$\frac{(x + 3)^2}{2^2} - \frac{(y - 2)^2}{4^2} = 1.$$

This equation has the form of equation (12) with $h = -3$, $k = 2$, $a = 2$, $b = 4$, and $c = \sqrt{a^2 + b^2} = \sqrt{20} = 2\sqrt{5}$. The hyperbola therefore has center $(-3, 2)$, vertices $(-5, 2)$ and $(-1, 2)$, foci $(-3 - 2\sqrt{5}, 2)$ and $(-3 + 2\sqrt{5}, 2)$, and asymptotes $y - 2 = \pm 2(x + 3)$. The graph appears as Figure 3.9. $\diamond$

Example 4

Describe the graph of the equation

$$25y^2 - 9x^2 - 50y - 54x - 281 = 0.$$

Solution: We complete the square in both variables on the left-hand side to find that

$$25(y^2 - 2y + 1) - 9(x^2 + 6x + 9) - 281 = 25 - 81,$$

so

$$25(y - 1)^2 - 9(x + 3)^2 = 225,$$

or

$$\frac{(y - 1)^2}{3^2} - \frac{(x + 3)^2}{5^2} = 1.$$

This equation has the form of equation (14) with $h = -3$, $k = 1$, $a = 3$, $b = 5$, and $c = \sqrt{a^2 + b^2} = \sqrt{34}$. Its graph is a hyperbola with center $(-3, 1)$, vertices $(-3, -2)$ and $(-3, 4)$, foci $(-3, 1 - \sqrt{34})$ and $(-3, 1 + \sqrt{34})$, and asymptotes $y - 1 = \pm \frac{3}{5}(x + 3)$ (see Figure 3.10). $\diamond$

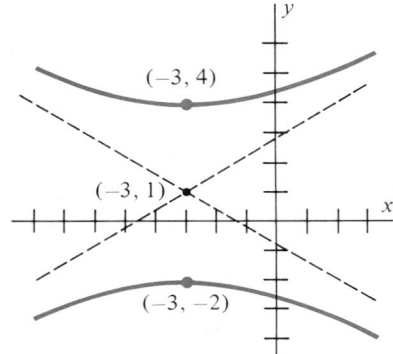

Figure 3.10 Graph of equation $25y^2 - 9x^2 - 50y - 54x - 281 = 0$.

Exercise Set 14.3

In Exercises 1–20, graph the hyperbola corresponding to the given equation. State the center and the equations for the asymptotes.

1. $\dfrac{x^2}{2^2} - \dfrac{y^2}{3^2} = 1$

2. $\dfrac{y^2}{4^2} - \dfrac{x^2}{2^2} = 1$

3. $\dfrac{x^2}{5} - \dfrac{y^2}{9} = 1$

4. $\dfrac{x^2}{9} - \dfrac{y^2}{2} = 1$

5. $16x^2 - 4y^2 - 64 = 0$

6. $x^2 - 9y^2 = 9$

7. $4y^2 - 7x^2 - 28 = 0$

8. $5x^2 - 9y^2 - 10x = 25$

9. $9x^2 - 4y^2 + 54x + 32y + 119 = 0$

10. $16y^2 - 4x^2 - 32y - 8x - 52 = 0$

11. $9x^2 - 4y^2 - 36x - 24y - 36 = 0$

12. $4y^2 - x^2 - 2x = 2$

13. $x^2 - 4y^2 = 4$

14. $4x^2 - y^2 + 24x + 4y + 28 = 0$

15. $2x^2 - y^2 - 4\sqrt{2}x + 2\sqrt{2}y = 6$

16. $3x^2 - 5y^2 + 6x - 20y = -8.$

17. $16x^2 - 25y^2 + 64x + 150y = 561$

18. $9x^2 - y^2 + 36x + 4y + 23 = 0$

19. $9y^2 - 16x^2 - 126y - 160x = 184$

20. $4y^2 - 16y - x^2 - 4x + 11 = 0$

In Exercises 21–27, find the equation for the hyperbola with the stated properties.

21. Foci at $(-2, 0)$ and $(2, 0)$, vertices at $(-1, 0)$ and $(1, 0)$.

22. Foci at $(2, 5)$ and $(2, -3)$, vertices at $(2, 3)$ and $(2, -1)$.

23. Vertices at $(0, 4)$ and $(0, 0)$, asymptotes $y = 2 \pm 2x$.

24. Foci at $(-3, 0)$ and $(3, 0)$, asymptotes $y = \pm\dfrac{3}{2}x$.

25. Foci at $(-5, 2)$ and $(3, 2)$, asymptotes $y - 2 = \pm 2(x + 1)$.

26. Center at the origin, one vertex $(0, 4)$, and containing the point $(2, 8)$.

27. Center at the origin, one vertex $(0, 4)$, and one focus $(0, -5)$.

28. Find an equation for the line tangent to the graph of the hyperbola $4x^2 - 16y^2 = 64$ at the point $(4\sqrt{2}, 2)$.

29. What properties do the graph of the hyperbola

$$\frac{x^2}{a^2} - \frac{y^2}{b^2} = 1$$

and its **conjugate** $\dfrac{y^2}{b^2} - \dfrac{x^2}{a^2} = 1$ have in common?

30. Show that the line tangent to the graph of the hyperbola $b^2x^2 - a^2y^2 = a^2b^2$ at the point (x_0, y_0) has slope $m = \dfrac{b^2x_0}{a^2y_0}$.

31. Using the result of Exercise 30, find the equation for the line tangent to the graph of $b^2x^2 - a^2y^2 = a^2b^2$ at the point (x_0, y_0).

32. Show that for the general second degree equation

$$Ax^2 + Bxy + Cy^2 + Dx + Ey + F = 0$$

if $B = 0$ and both $A \neq 0$ and $C \neq 0$, the graph is a hyperbola if A and C have opposite signs.

33. A chord of a hyperbola through a focus perpendicular to the principal axis is called a **latus rectum**. Show that the length of a latus rectum is $\dfrac{2b^2}{a}$.

34. Find an equation for the hyperbola with center at the origin, latus rectum of length 20, and one vertex at $(0, 4)$.

35. Show that any point (x, y) satisfying equation (2) satisfies the geometric condition of Definition 3.

14.4 ROTATION OF AXES

Up to this point, we have discussed all possible second degree equations of the form

$$Ax^2 + Bxy + Cy^2 + Dx + Ey + F = 0 \qquad (1)$$

except those involving an xy term. In this section we complete that discussion by considering the case $B \neq 0$. On encountering such an expression, our objective will be to find a suitable change of variables from (x, y)-coordinates to (x', y')-coordinates so that in the new coordinates equation (1) contains no $x'y'$ term.

The simplest change of variables that accomplishes this objective is a rotation of the xy-axes through an angle θ. After having constructed new x'- and y'-axes by

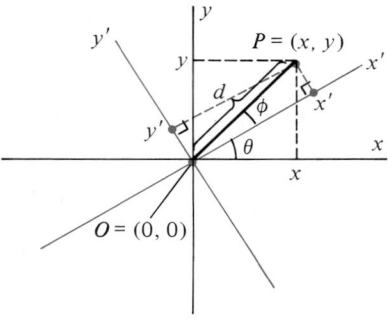

Figure 4.1 Rotation of axes through θ radians.

rotating the x- and y-axes counterclockwise through an angle θ, we observe that a point P with xy-coordinates (x, y) will have $x'y'$-coordinates given by the equations

$$x' = d \cos \phi, \tag{2a}$$

$$y' = d \sin \phi, \tag{2b}$$

where d is the distance from the origin, O, to P and ϕ is the angle between OP and the x'-axis (Figure 4.1).

Similarly, we can express the xy-coordinates of P as

$$x = d \cos(\theta + \phi), \tag{3a}$$

$$y = d \sin(\theta + \phi). \tag{3b}$$

Using the addition laws for sine and cosine we can combine equations (2) and (3) to obtain the equations

$$x = d(\cos \theta \cos \phi - \sin \theta \sin \phi) = x' \cos \theta - y' \sin \theta, \tag{4a}$$

$$y = d(\sin \theta \cos \phi + \cos \theta \sin \phi) = x' \sin \theta + y' \cos \theta. \tag{4b}$$

Solving this system for x' and y', we obtain

$$x' = x \cos \theta + y \sin \theta, \tag{5a}$$

$$y' = -x \sin \theta + y \cos \theta. \tag{5b}$$

By using equations (4) and (5), we can pass back and forth from xy- to $x'y'$-coordinates. In particular, if we substitute the right-hand sides of equations (4) for x and y into equation (1), we find that the coefficient of the $x'y'$ term is

$$2(C - A)\sin \theta \cos \theta + B(\cos^2 \theta - \sin^2 \theta)$$
$$= (C - A)\sin 2\theta + B \cos 2\theta.$$

If we wish this coefficient to be zero, we need only require that

$$\frac{\cos 2\theta}{\sin 2\theta} = \frac{A - C}{B}$$

or that*

$$\cot 2\theta = \frac{A - C}{B}, \qquad 0 < \theta < \frac{\pi}{2}. \tag{6}$$

Since θ is not yet specified, this shows that by choosing θ to satisfy equation (6) we can ensure that equation (1), when written in $x'y'$-coordinates, will contain no $x'y'$ term. Thus the graph of equation (1), in the rotated $x'y'$-coordinate system, will be a circle, a parabola, an ellipse, or a hyperbola.

Example 1

For the equation $5x^2 + 6xy + 5y^2 - 8 = 0$,

(a) find a rotation that eliminates the xy term, and
(b) sketch the graph.

---

*Notice that equation (6) is defined for all B except $B = 0$. In this case the original equation (1) contains no xy term, so a rotation is not needed.

Solution: We have $A = 5$, $B = 6$, and $C = 5$, so equation (6) becomes $\cot 2\theta = \dfrac{5 - 5}{6} = 0$. The equation $\cot 2\theta = 0$ has solution $2\theta = \pi/2$, or $\theta = \pi/4$. With this value of θ equations (4a) and (4b) become

$$x = \frac{\sqrt{2}}{2}x' - \frac{\sqrt{2}}{2}y',$$

$$y = \frac{\sqrt{2}}{2}x' + \frac{\sqrt{2}}{2}y'.$$

Inserting these expressions for x and y in the original equation gives

$$5\left(\frac{\sqrt{2}}{2}x' - \frac{\sqrt{2}}{2}y'\right)^2 + 6\left(\frac{\sqrt{2}}{2}x' - \frac{\sqrt{2}}{2}y'\right)\left(\frac{\sqrt{2}}{2}x' + \frac{\sqrt{2}}{2}y'\right)$$
$$+ 5\left(\frac{\sqrt{2}}{2}x' + \frac{\sqrt{2}}{2}y'\right)^2 - 8 = 0$$

or, on simplifying,

$$x'^2 + \frac{y'^2}{4} = 1.$$

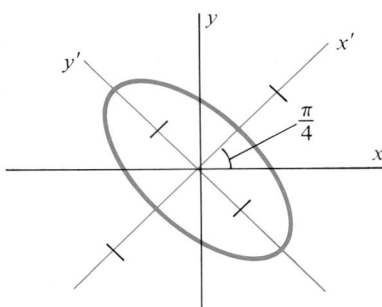

Figure 4.2 Graph of $5x^2 + 6xy + 5y^2 - 8 = 0$ is a rotated ellipse.

The graph of this ellipse is shown in Figure 4.2.   ◇

Example 2

Graph the equation $xy = 1$.

Strategy

Identify form of the equation.

Rotate axes to eliminate xy term.

Find the rotation angle from (6).

Express x and y as functions of x' and y' and insert these expressions into the given equation.

Bring resulting equation to standard form.

Solution

This is an equation of the form of the general second degree equation (1) with $B = 1$ and $A = C = D = E = 0$. Although we are already familiar with the general nature of its graph, we may use the method of rotation of axes to verify that the graph is, in fact, a hyperbola.

From equation (6) we find that

$$\cot 2\theta = \frac{A - C}{B} = 0,$$

so $\theta = \pi/4$, as in Example 1. From equations (4) we then obtain

$$x = \frac{\sqrt{2}}{2}x' - \frac{\sqrt{2}}{2}y',$$

$$y = \frac{\sqrt{2}}{2}x' + \frac{\sqrt{2}}{2}y'.$$

Inserting these expressions in $xy = 1$ gives

$$\left(\frac{\sqrt{2}}{2}x' - \frac{\sqrt{2}}{2}y'\right)\left(\frac{\sqrt{2}}{2}x' + \frac{\sqrt{2}}{2}y'\right) = 1,$$

or

$$\frac{x'^2}{2} - \frac{y'^2}{2} = 1.$$

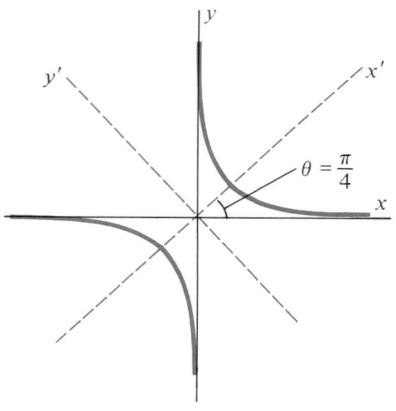

Figure 4.3 Graph of $xy = 1$ is a rotated hyperbola.

The graph is therefore a hyperbola in $x'y'$-coordinates, with center at $(0, 0)$ and asymptotes $y' = \pm x'$ (see Figure 4.3). ◇

We have now completed our analysis of the general second degree equation

$$Ax^2 + Bxy + Cy^2 + Dx + Ey + F = 0, \qquad (7)$$

where at least one of A and C is nonzero.

If $B \neq 0$, we can employ a rotation of axes, as described above, to obtain an equation in $x'y'$-coordinates of the form

$$A'x'^2 + C'y'^2 + D'x' + E'y' + F' = 0. \qquad (8)$$

The graph of equation (8) must then fall under one of the following cases:

(i) If $A' = C' \neq 0$, the graph of (8) is a circle (Section 1.2).
(ii) If $A' = 0$ or $C' = 0$ (but not both), the graph of (8) is a parabola (Exercise 32, Section 14.1).
(iii) If $A'C' \neq 0$ and $A' \neq C'$, the graph of (8) is
    (a) an ellipse, if A' and C' have the same sign (Exercise 30, Section 14.2),
    (b) a hyperbola, if A' and C' have opposite signs (Exercise 32, Section 14.3).

Since this list exhausts all possibilities for the coefficients in equation (7), the graph of any equation of the form (7) must be a circle, a parabola, an ellipse, or a hyperbola.

Exercise Set 14.4

In each of Exercises 1–6, find a rotation θ that eliminates the xy term.

1. $3x^2 + 2xy + y^2 + 6x + 9 = 0$

2. $6x^2 + 2xy + 2y^2 + 8y + 8 = 0$

3. $x^2 - 3xy + y^2 - x + y + 10 = 0$

4. $4x^2 + \sqrt{3}xy + 3y^2 + 2x + 7y - 8 = 0$

5. $5x^2 + 4xy + 2y^2 = 1$

6. $9x^2 - 24xy + 16y^2 - 40x - 30y + 100 = 0$

In each of Exercises 7–16, find a rotation θ that eliminates the xy term. Then use equations (4) to obtain an equation in $x'y'$-coordinates. Finally, graph the equation.

7. $5x^2 + 6xy + 5y^2 - 32 = 0$

8. $3x^2 + 10xy + 3y^2 - 32 = 0$

9. $3x^2 + 2\sqrt{3}xy + y^2 + 2x - 2\sqrt{3}y + 16 = 0$

10. $5x^2 + 4xy + 5y^2 = 21$

11. $9x^2 - 24xy + 16y^2 - 40x - 30y + 100 = 0$

12. $x^2 - 6xy + y^2 + 16 = 0$

13. $7x^2 - 2\sqrt{3}xy + 5y^2 - 16 = 0$

14. $23x^2 - 26\sqrt{3}xy - 3y^2 + 144 = 0$

15. $x^2 + 2xy + y^2 + \sqrt{2}x - \sqrt{2}y = 0$

16. $31x^2 + 10\sqrt{3}xy + 21y^2 = 144$

17. Find the endpoints of the major axis of the ellipse in Exercise 16.

18. Find the coordinates of the focus of the parabola in Exercise 15.

19. Find the vertices of the hyperbola in Exercise 14.

20. Find the endpoints of the minor axis of the ellipse in Exercise 13.

SUMMARY OUTLINE OF CHAPTER 14

◆ The standard forms of the equations for the **parabola** with vertex at (h, k) and focal length c are   (page 607)

$(x - h)^2 = 4c(y - k)$ (axis vertical),
$(y - k)^2 = 4c(x - h)$ (axis horizontal).

◆ The standard form of the equation for the **ellipse** with center at (h, k) is $\dfrac{(x-h)^2}{a^2} + \dfrac{(y-k)^2}{b^2} = 1$. (page 613)

◆ The standard forms of the equations for the **hyperbola** with center at (h, k) are (page 619)

$\dfrac{(x-h)^2}{a^2} - \dfrac{(y-k)^2}{b^2} = 1$ (foci on line parallel to x-axis),

$\dfrac{(y-k)^2}{a^2} - \dfrac{(x-h)^2}{b^2} = 1$ (foci on line parallel to y-axis).

◆ A **rotation of axes** to eliminate the xy term in $Ax^2 + Bxy + Cy^2 + Dx + Ey + F = 0$ is accomplished by the (page 624)
substitutions

$x = x' \cos\theta - y' \sin\theta, \qquad y = x' \sin\theta + y' \cos\theta$

where θ is determined by the equation $\cot 2\theta = \dfrac{A-C}{B}, \qquad 0 < \theta < \dfrac{\pi}{2}$.

REVIEW EXERCISES—CHAPTER 14

1. Write an equation for the circle with center $(3, -4)$ and radius 6.

2. Write an equation for the circle with center $(-2, -3)$ and radius 5.

3. Write an equation for the circle with center $(2, 3)$ that contains the point $(2, -1)$.

4. Write equations for the circles with radius 5 containing the points $(-1, -4)$ and $(4, 1)$. How many such circles exist?

5. Write an equation for the parabola with vertex $(0, 4)$ and directrix $x = -2$.

6. Write an equation for the parabola with axis horizontal, vertex on the y-axis, and containing the points $(2, 1)$ and $(18, 3)$.

7. Write an equation for the parabola with axis vertical and containing the points $(0, 5)$, $(2, 5)$, and $(3, 14)$.

8. Find an equation for the circle that passes through the focus and vertex of the parabola $y = 8x^2$ and that is tangent to the parabola at $(0, 0)$.

9. Find an equation of the ellipse with center $(-1, 1)$, focus $(1, 1)$, and vertex $(2, 1)$.

10. Find an equation for the ellipse with focus $(0, 1)$ and vertices $(0, 0)$ and $(0, 8)$.

11. Does the origin lie inside or outside the ellipse with equation $(x-4)^2 + 4(y-3)^2 = 4$?

Find the points of intersection of each of the following pairs of graphs.

12. The line $x + y - 4 = 0$ and the circle $x^2 + y^2 = 16$.

13. The line $x - y - 2 = 0$ and the parabola $y = 4 - x^2$.

14. The line $4y - 3x + 12 = 0$ and the ellipse $9x^2 + 16y^2 = 144$.

15. The parabola $y = 6 - x^2$ and the parabola $y = x^2 - 2$.

In Exercises 16–45, write the equation in standard form, identify the graph, and sketch.

16. $3y^2 - x - 4 = 0$

17. $x^2 - 2x + y^2 + 6y + 2 = 0$

18. $16x^2 + 7y^2 - 112 = 0$

19. $2x^2 - y^2 + 4 = 0$

20. $x^2 - 3y^2 + 3 = 0$

21. $16x^2 + 12y^2 - 192 = 0$

22. $3x^2 + 3y^2 - 3 = 0$

23. $9x^2 + 4y^2 - 36 = 0$

24. $16x^2 - 7y^2 + 112 = 0$

25. $4x^2 - 9y^2 - 36 = 0$

26. $4x^2 + 9y^2 - 36 = 0$

27. $y^2 + x - 2 = 0$

28. $7y^2 - 2x + 6 = 0$

29. $x^2 + 14x + y^2 - 10y + 70 = 0$

30. $9x^2 + 8y^2 - 72 = 0$

31. $9x^2 - 8y^2 - 72 = 0$

32. $4x^2 + y + 8 = 0$

33. $4x^2 + 25y^2 - 100 = 0$

34. $x^2 + 3y^2 - 3 = 0$

35. $3x^2 - 3y^2 + 12 = 0$

36. $x^2 - 6x + y^2 - 4y + 8 = 0$

37. $x^2 - 2ax + y^2 + 4ay + 5a^2 - 1 = 0$

38. $x^2 - 2x + y^2 - 8 = 0$

39. $y^2 - x^2 - 2 = 0$

40. $3x + 2y^2 - 3 = 0$

41. $x^2 + y^2 - 2x - 6y + 3 = 0$

42. $4x^2 - 25y^2 + 100 = 0$

43. $9y^2 - 25x^2 - 225 = 0$

44. $4x^2 - 9y^2 - 36 = 0$

45. $2x - 16y^2 - 12 = 0$.

Chapter 15
Polar Coordinates and Parametric Equations

Up to this point we have studied curves in the plane by regarding them as graphs of functions plotted in Cartesian (rectangular) coordinates. This chapter presents two additional ways in which we can represent curves in the plane: as graphs of functions of the form $r = f(\theta)$ plotted in *polar coordinates,* and as graphs determined by *parametric equations* for the x and y coordinates of points in Cartesian coordinates.

15.1 THE POLAR COORDINATE SYSTEM

Cartesian coordinates provide a convenient scheme to use when graphing functions whose variables represent naturally perpendicular quantities (length versus height, for example) or when attempting to approximate areas by rectangles. However, many plane curves can be more easily described in a coordinate system in which the coordinates of a point $P(r, \theta)$ measure its (radial) distance, r, along a ray from the origin that makes an angle θ with some fixed reference ray (usually the positive x-axis; see Figure 1.1).

These coordinates are especially convenient for describing curves that are symmetric with respect to the origin. An example of this situation occurs in the theory of electric potential fields. For a point charge, the lines representing equal electric potentials (called equipotentials) in any plane containing the point charge are circles with the point as center (Figure 1.2). These equipotential circles can most easily be described simply by stating their radius. Similarly, the lines of force emanating from such a point charge are rays, which are described completely by the angle they form with the positive x-axis.

Let O be any point in the plane, called the **pole,** and let ℓ be any ray emanating from O. The ray ℓ is called the **polar axis** for the plane. The choice of a pole and a polar axis determines a **polar coordinate system** for the plane in which any point P may be assigned **polar coordinates** $P = (r, \theta)$, where

(i) r, called the **radial variable,** is the distance from O to P, and
(ii) θ, called the **angular variable,** is the angle formed between the polar axis and the ray $\overline{OP}$, measured in the counterclockwise direction (Figure 1.1).

Figure 1.3 illustrates various points plotted in polar coordinates. As in most applications of trigonometry in the calculus, we shall work in radians.

In the xy-coordinate plane, the graphs of equations of the form $x =$ constant or $y =$ constant are lines. In the polar coordinate plane, graphs of equations of the form $r =$ constant are circles, while graphs of equations of the form $\theta =$ constant are rays (see Figure 1.4).

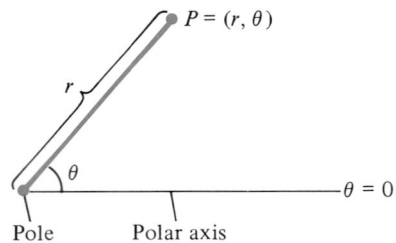

Figure 1.1 Polar coordinates for point P.

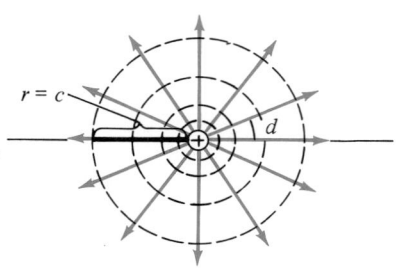

Figure 1.2 For a point charge (shown here as positive), equipotential lines are circles $r = c$ (=constant). Lines of force are rays $\theta = d$ (=constant).

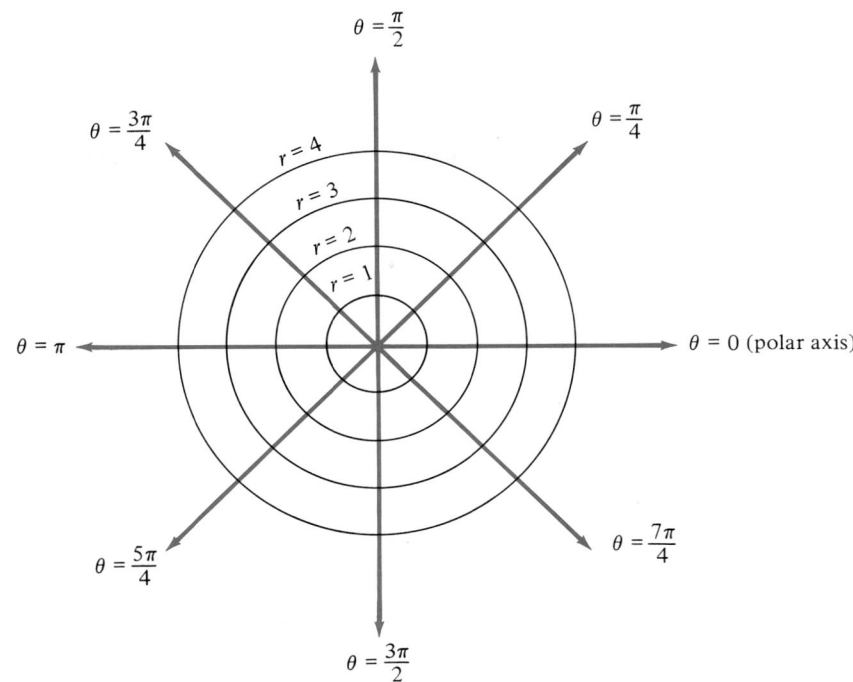

Figure 1.4 The polar coordinate plane.

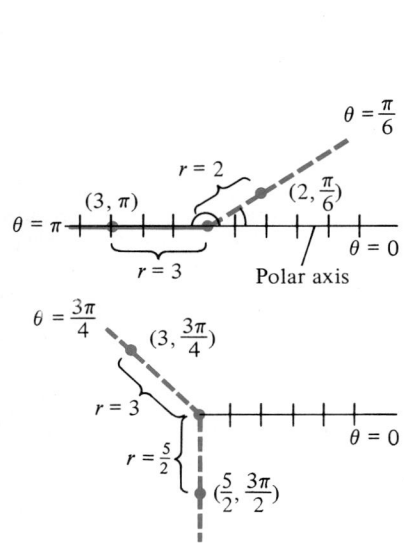

Figure 1.3 Polar coordinates for several points.

You may have already noticed that polar coordinates are not unique. Indeed, since the numbers $\theta = \theta_0 + 2n\pi$, $n = 0, \pm 1, \pm 2, \ldots$ all determine the same ray, we have

$$(2, \pi) = (2, 3\pi) = (2, 5\pi) = (2, -\pi) = \cdots,$$

$$\left(1, \frac{\pi}{4}\right) = \left(1, \frac{9\pi}{4}\right) = \left(1, \frac{17\pi}{4}\right) = \left(1, -\frac{7\pi}{4}\right) = \cdots,$$

and, in general,

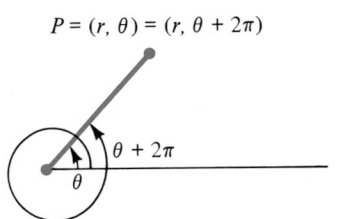

$$(r, \theta) = (r, \theta + 2n\pi), \qquad n = \pm 1, \pm 2, \pm 3, \ldots. \tag{1}$$

Figure 1.5 $(r, \theta) = (r, \theta + 2n\pi)$.

(See Figure 1.5.) Moreover, it is customary to extend the range of values for the radial variable r to include negative numbers via the identity

$$(-r, \theta) = (r, \theta + \pi). \tag{2}$$

(See Figure 1.6.) In other words, the point with polar coordinates $(-r, \theta)$ lies r units along the ray pointing in the direction opposite that specified by the angular variable θ.

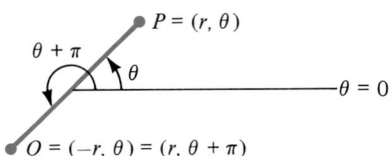

Figure 1.6 $(-r, \theta) = (r, \theta + \pi)$.

Example 1

By identities (1) and (2) we have

(a) $\left(3, \dfrac{\pi}{6}\right) = \left(3, \dfrac{13\pi}{6}\right) = \left(3, -\dfrac{11\pi}{6}\right),$

(b) $\left(-4, \dfrac{\pi}{4}\right) = \left(4, \dfrac{5\pi}{4}\right) = \left(4, -\dfrac{3\pi}{4}\right),$

(c) $\left(7, -\dfrac{\pi}{2}\right) = \left(-7, \dfrac{\pi}{2}\right) = \left(-7, \dfrac{5\pi}{2}\right) = \left(-7, -\dfrac{3\pi}{2}\right).$ ◇

The *pole O* in the polar coordinate plane has coordinates $(0, \theta)$ for every angle θ. It is the only point in the plane not associated with a unique angle θ in $[0, 2\pi)$.

Relationships Between Polar and Rectangular Coordinates

The relationships between polar and xy-coordinates are easily determined when the origin in xy-coordinates is identified with the pole, and the positive x-axis is taken to be the polar axis (see Figure 1.7). If the point P has rectangular coordinates (x, y) and polar coordinates (r, θ), and if $r > 0$, the definitions of $\sin \theta$ and $\cos \theta$ give

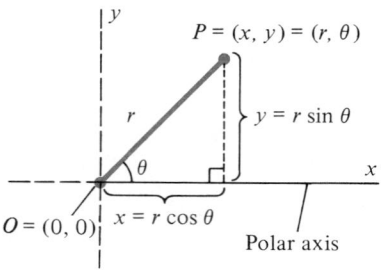

Figure 1.7 Relationships between rectangular and polar coordinates.

$$x = r \cos \theta \quad \text{and} \quad y = r \sin \theta. \tag{3}$$

In Exercise 54, you are asked to show that equations (3) hold for $r < 0$ as well. It follows from the equations in (3) (or from Figure 1.7) that

$$r = \sqrt{x^2 + y^2}; \quad \tan \theta = \dfrac{y}{x}, \quad x \neq 0. \tag{4}$$

The equations in line (3) are used in changing from polar to rectangular coordinates while the equations in line (4) allow us to change from rectangular to polar coordinates.

Example 2

To find the rectangular coordinates corresponding to the polar coordinates $(2, 2\pi/3)$, we use equations (3) with $r = 2$, $\theta = 2\pi/3$:

$$x = 2 \cos \dfrac{2\pi}{3} = 2\left(-\dfrac{1}{2}\right) = -1,$$

$$y = 2 \sin \dfrac{2\pi}{3} = 2\left(\dfrac{\sqrt{3}}{2}\right) = \sqrt{3}.$$

(See Figure 1.8.) ◇

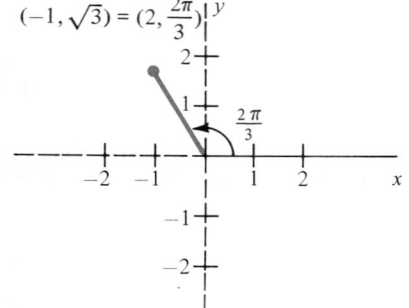

Figure 1.8

Example 3

To find the polar coordinates for the point with rectangular coordinates $(1, -1)$, we use the equations in (4):

$$r = \sqrt{x^2 + y^2} = \sqrt{1^2 + 1^2} = \sqrt{2}, \tag{5}$$

$$\tan \theta = \dfrac{y}{x} = \dfrac{-1}{1} = -1. \tag{6}$$

Now r is clearly determined by equation (5), but θ is *not* uniquely determined by equation (6), since both $\theta = 3\pi/4$ and $\theta = 7\pi/4$ satisfy $\tan \theta = -1$ (as do *all* an-

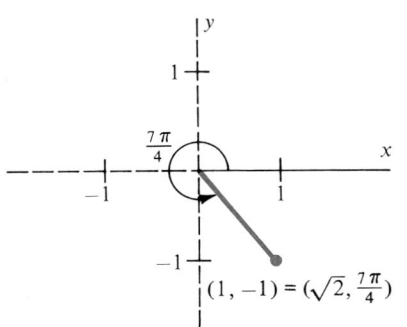

Figure 1.9

gles of the form $\theta = \dfrac{3\pi}{4} \pm n\pi$, $n = 1, 2, 3, \ldots$). The correct angle, $\theta = 7\pi/4$, is identified by noting that the point $(1, -1)$ lies in the fourth quadrant (Figure 1.9).

◇

REMARK: It is important to note, as in Example 3, that equations (4) do not uniquely determine θ for given values of x and y. This is because each value of $\tan \theta$ occurs twice for $\theta \in [0, 2\pi)$. This difficulty is overcome most simply by noting the quadrant in which the point (x, y) lies. An alternative approach is to first determine r from equation (4), and then determine θ as the (unique) simultaneous solution of equations (3): $x = r \cos \theta$; $y = r \sin \theta$.

Example 4

Graph the polar equation $r = 1 + \cos \theta$ and find the rectangular form for the equation.

Solution: One approach to graphing polar equations is to select several (convenient) values of θ (beginning with $\theta = 0$), calculate the corresponding values of r, and plot the points (r, θ) obtained. The graph is then completed by sketching a curve connecting these points. For the equation $r = 1 + \cos \theta$, we obtain the following points.

θ	0	$\dfrac{\pi}{6}$	$\dfrac{\pi}{4}$	$\dfrac{\pi}{3}$	$\dfrac{\pi}{2}$	$\dfrac{2\pi}{3}$	$\dfrac{3\pi}{4}$	$\dfrac{5\pi}{6}$	π
r	2	$\dfrac{2 + \sqrt{3}}{2}$	$\dfrac{2 + \sqrt{2}}{2}$	$\dfrac{3}{2}$	1	$\dfrac{1}{2}$	$\dfrac{2 - \sqrt{2}}{2}$	$\dfrac{2 - \sqrt{3}}{2}$	

θ	$\dfrac{7\pi}{6}$	$\dfrac{5\pi}{4}$	$\dfrac{4\pi}{3}$	$\dfrac{3\pi}{2}$	$\dfrac{5\pi}{3}$	$\dfrac{7\pi}{4}$	$\dfrac{11\pi}{6}$	2π
r	$\dfrac{2 - \sqrt{3}}{2}$	$\dfrac{2 - \sqrt{2}}{2}$	$\dfrac{1}{2}$	1	$\dfrac{3}{2}$	$\dfrac{2 + \sqrt{2}}{2}$	$\dfrac{2 + \sqrt{3}}{2}$	2

The graph is the **cardioid** in Figure 1.10. In Section 15.2 we shall discuss a technique for graphing polar equations which addresses the problem of ensuring that points obtained as above are connected in the correct manner.

To find the rectangular form of the equation $r = 1 + \cos \theta$, we multiply both sides by r to obtain

$$r^2 = r + r \cos \theta.$$

Using the equations in (3) and (4), we obtain

$$x^2 + y^2 = \sqrt{x^2 + y^2} + x.$$

Clearly the original equation is simpler to graph than its rectangular counterpart.

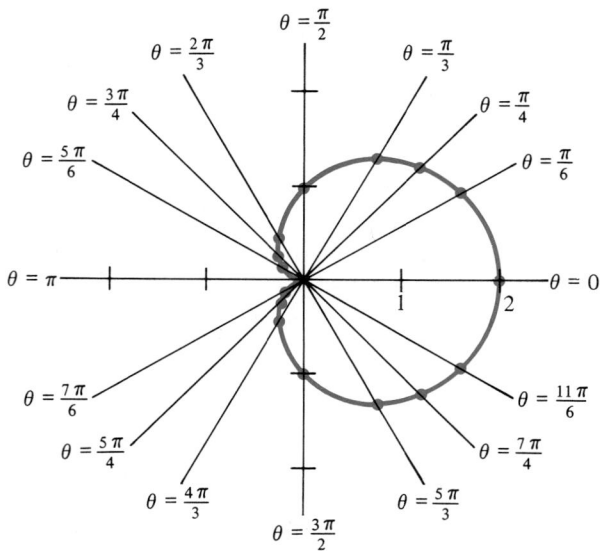

Figure 1.10 Cardioid $r = 1 + \cos \theta$.

Symmetry in Polar Coordinates

Perhaps you noticed in Example 4 that the graph of the equation $r = 1 + \cos \theta$ is symmetric about the x-axis. This is due to the identity $\cos(-\theta) = \cos \theta$, or $\cos(2\pi - \theta) = \cos \theta$, $0 \le \theta \le \pi$. More generally, graphs of polar equations $r = f(\theta)$ will be

  (i) *symmetric about the x-axis* if $(r, -\theta)$ lies on the graph whenever (r, θ) does (Figure 1.11);

  (ii) *symmetric about the y-axis* if $(r, \pi - \theta)$ lies on the graph whenever (r, θ) does (Figure 1.12);

  (iii) *symmetric about the origin* if $(-r, \theta) = (r, \theta + \pi)$ lies on the graph whenever (r, θ) does (Figure 1.13).

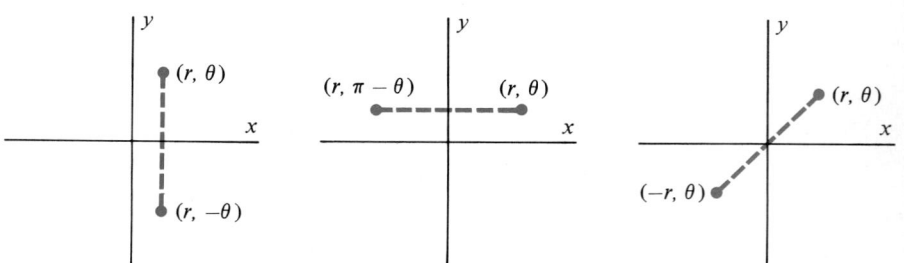

Figure 1.11 Symmetry about the x-axis.

Figure 1.12 Symmetry about the y-axis.

Figure 1.13 Symmetry about the origin.

Example 5

Graph the polar equation $r = \sin 2\theta$.

θ	0	$\dfrac{\pi}{6}$	$\dfrac{\pi}{4}$	$\dfrac{\pi}{3}$	$\dfrac{\pi}{2}$	$\dfrac{2\pi}{3}$	$\dfrac{3\pi}{4}$	$\dfrac{5\pi}{6}$	π
r	0	$\dfrac{\sqrt{3}}{2}$	1	$\dfrac{\sqrt{3}}{2}$	0	$\dfrac{-\sqrt{3}}{2}$	-1	$\dfrac{-\sqrt{3}}{2}$	0

Plotting these points gives the two leaves shown in Figure 1.14 (note that values of r are negative for $\pi/2 < \theta < \pi$).

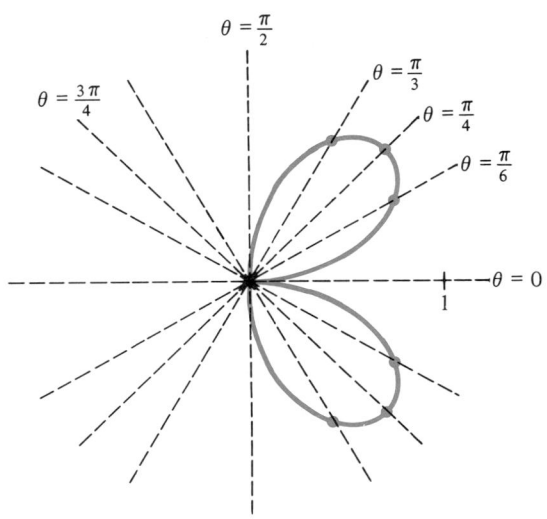

Figure 1.14 Graph of $r = \sin 2\theta$ for $0 \le \theta \le \pi$.

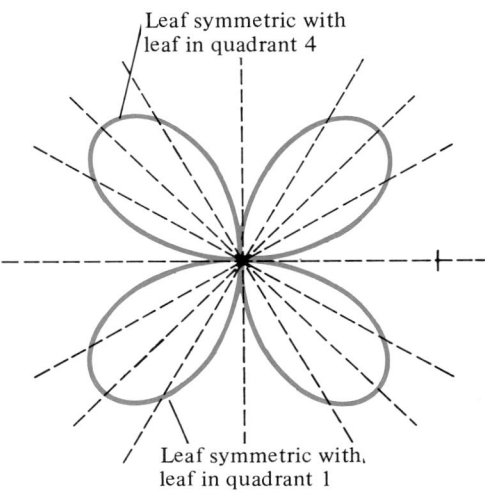

Figure 1.15 Graph of $r = \sin 2\theta$ for $0 \le \theta \le 2\pi$.

We can use symmetry considerations to obtain the remaining portion of the curve. To do so, we use the double angle identity to write

$$r(\theta) = \sin 2\theta = 2 \sin \theta \cos \theta.$$

Since $\sin(\theta + \pi) = -\sin \theta$ and $\cos(\theta + \pi) = -\cos \theta$, we obtain

$$r(\theta + \pi) = 2 \cdot \sin(\theta + \pi) \cos(\theta + \pi) = 2 \cdot \sin \theta \cos \theta = r(\theta).$$

This shows that $(r, \theta + \pi) = (-r, \theta)$ is on the graph whenever (r, θ) is. That is, we may sketch a leaf in the third quadrant symmetric with the leaf in the first quadrant, and a leaf in the second quadrant symmetric with the leaf in the fourth quadrant (Figure 1.15). (Actually we could obtain the entire graph by applying symmetry arguments just to the leaf obtained for $0 \le \theta \le \pi/2$. Can you see how this could be done?) The resulting figure is called a four-leaved rose. ◇

Example 6

For the polar equation $r = 2 \cos \theta$,

(a) sketch the graph;
(b) note any symmetries of the graph;
(c) find the rectangular form of the equation.

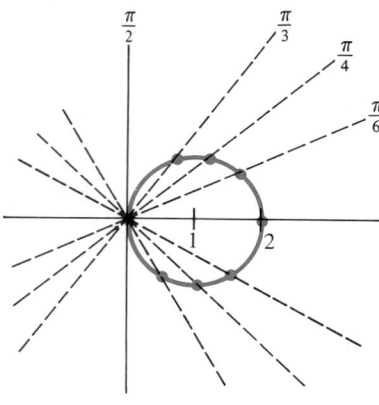

Figure 1.16 Graph of $r = 2 \cos \theta$.

Solution:

(a) For $0 \le \theta \le \pi$, we obtain the following points:

θ	0	$\dfrac{\pi}{6}$	$\dfrac{\pi}{4}$	$\dfrac{\pi}{3}$	$\dfrac{\pi}{2}$	$\dfrac{2\pi}{3}$	$\dfrac{3\pi}{4}$	$\dfrac{5\pi}{6}$	π
r	2	$\sqrt{3}$	$\sqrt{2}$	1	0	-1	$-\sqrt{2}$	$-\sqrt{3}$	-2

These points produce the curve shown in Figure 1.16.

For $\pi < \theta < 2\pi$, we note that $r(\theta_0 + \pi) = 2\cos(\theta_0 + \pi) = -2\cos(\theta_0) = -r(\theta_0)$ where $0 < \theta_0 < \pi$. Thus, these values simply trace over the same points a second time.

(b) Figure 1.16 suggests symmetry about the x-axis. This is indeed the case, since, for $r(\theta) = 2\cos\theta$,

$$r(-\theta) = 2\cos(-\theta) = 2\cos\theta = r(\theta).$$

Thus, $(r, -\theta)$ is on the graph whenever (r, θ) is.

(c) To find the rectangular form for $r = 2\cos\theta$, we multiply by r to obtain

$$r^2 = 2r\cos\theta.$$

Using equations (3) and (4) we obtain

$$x^2 + y^2 = 2x, \qquad \text{or} \qquad x^2 - 2x + y^2 = 0.$$

Completing the square in x gives

$$(x^2 - 2x + 1) + y^2 = 1$$

or

$$(x - 1)^2 + y^2 = 1.$$

The graph is therefore a circle of radius $r = 1$ with center $(1, 0)$. $\diamond$

Exercise Set 15.1

In Exercises 1–8, find rectangular coordinates for the point given in polar coordinates.

1. $(1, \pi/2)$

2. $(3, \pi/6)$

3. $(0, \pi)$

4. $(-2, \pi/4)$

5. $(\sqrt{2}, -\pi/4)$

6. $\left(-1, -\dfrac{3\pi}{2}\right)$

7. $(-3, \pi)$

8. $(\pi, -\pi)$

In Exercises 9–16, find polar coordinates, subject to the stated restrictions on θ, for the point given in rectangular coordinates.

9. $(1, 1)$, $0 < \theta < \pi$

10. $(1, 1)$, $\pi < \theta < 2\pi$

11. $(-3, 0)$, $-\pi/2 < \theta < \pi/2$

12. $(1, -\sqrt{3})$, $\pi < \theta < 2\pi$

13. $(1, -\sqrt{3})$, $0 < \theta < \pi$

14. $(-2, 2)$, $0 < \theta < \pi$

15. $(-2, 2)$, $3\pi < \theta < 4\pi$

16. $(-3, 4)$, $\pi < \theta < 2\pi$

In Exercises 17–24, identify all symmetries (about the x-axis, the y-axis, or the origin) possessed by the graph of the given equation.

17. $r = 4\cos\theta$

18. $r = 2\sin\theta$

19. $r = 1 + \sin\theta$

20. $r^2 = \cos\theta$

21. $r^2 = \cos 2\theta$

22. $r = \sin 3\theta$

23. $r = 2$

24. $r = 4\sin 2\theta$

In Exercises 25–32, find an equation in polar coordinates for the given equation.

25. $x^2 + y^2 = 4$

26. $y^2 = 16x$

27. $4x^2 + y^2 = 4$

28. $xy = 2$

29. $x = 6$

30. $y = 4$

31. $x^2 + y^2 + 2y = 0$

32. $y = x$

In Exercises 33–40, find an equation in rectangular coordinates for the given polar equation.

33. $r = 4 \sin \theta$

34. $r = 6$

35. $r = \tan \theta$

36. $r = \csc \theta$

37. $r = 4 \sec \theta$

38. $r = 1 + \sin \theta$

39. $r = \dfrac{1}{1 - \cos \theta}$

40. $r^2 = 4 \sec \theta$

In Exercises 41–50, sketch the graph of the given polar equation.

41. $r = \dfrac{1}{2}\theta$ (spiral of Archimedes)

42. $r = 2 \sin \theta$ (circle)

43. $r = 1 + \sin \theta$ (cardioid)

44. $r = \cos 2\theta$ (4-leaved rose)

45. $r = 1 + 2 \cos \theta$ (limaçon)

46. $r^2 = \cos 2\theta$ (lemniscate)

47. $r = e^{\theta}$, $\theta \geq 0$ (spiral)

48. $r = \dfrac{1}{e^{\theta}}$, $\theta \geq 0$

49. $r = 1 - 2 \cos \theta$ (limaçon)

50. $r = \sin 4\theta$ (8-leaved rose)

51. Prove that the graph of $r = 2 \sin \theta - 2 \cos \theta$ is a circle. Find its center and radius.

52. Prove that the graph of $r = a \sin \theta$ is a circle. Sketch the graph.

53. Prove that the graph of $r = a \cos \theta$ is a circle. Sketch the graph.

54. Show that equations (3) are valid for the case $r < 0$.

55. Show that the equation $y = ax$ of a nonvertical line through the origin is, in polar coordinates, $\tan \theta = a$.

56. The line $r = 4 \sec \theta$ is tangent to the graph of $r = a(1 + \cos \theta)$. Find a.

57. Find the area of the region enclosed by the graph of $r = 3 \cos \theta$. (*Hint:* See Exercise 53.)

58. Find the area of the region enclosed by the graph of $r^2 = 4 \cos^2 \theta$. (*Hint:* See Exercise 57.)

59. Find the vertices of the triangle determined by the lines

$$\tan \theta = 1 \quad \text{and} \quad r = 4 \sec \theta.$$
$$\theta = 0$$

60. Find the area of the triangle determined by the lines

$$\tan \theta = 2 \quad \text{and} \quad \tan \theta = -2.$$
$$r = 4 \csc \theta$$

15.2 GRAPHING TECHNIQUES FOR POLAR EQUATIONS

In Section 15.1 we graphed polar equations by plotting a few points (r_j, θ_j) and sketching in a curve. A slightly different approach is often simpler to use. That is simply to put the polar equation in the form $r = f(\theta)$, if possible, and then to "think dynamically," noting the behavior of r as θ sweeps out one or more revolutions about the pole.

For example, to graph the polar equation $r = a(1 + \cos \theta)$ we first note that the maximum value of $r = a(1 + 1) = 2a$ will occur when $\cos \theta = 1$, that is, when $\theta = 0$. The curve is therefore contained within the circle $r = 2a$ (Figure 2.1). To plot the curve we begin at the point $(2a, 0)$ and note that as θ increases from 0 to $\pi/2$, r *decreases* from length $r(0) = 2a$ to length $r(\pi/2) = a\left(1 + \cos \dfrac{\pi}{2}\right) = a$.

This observation enables us to sketch in the arc labelled 1 in Figure 2.2. We next note the behavior of r as θ increases from $\pi/2$ to π: r decreases from $r(\pi/2) = a$ to $r(\pi) = a(1 + (-1)) = 0$. This produces the arc labelled 2 in Figure 2.3. Similarly, arcs 3 and 4 are "swept out" as θ increases from π to 2π. Allowing θ to increase beyond 2π simply retraces the arcs already obtained, so the complete curve corresponding to $r = a(1 + \cos \theta)$ is obtained from $0 \leq \theta \leq 2\pi$ (Figure 2.3).

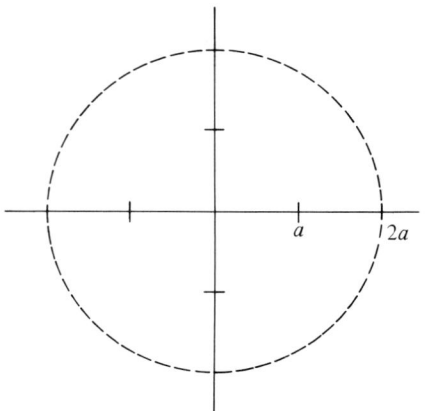

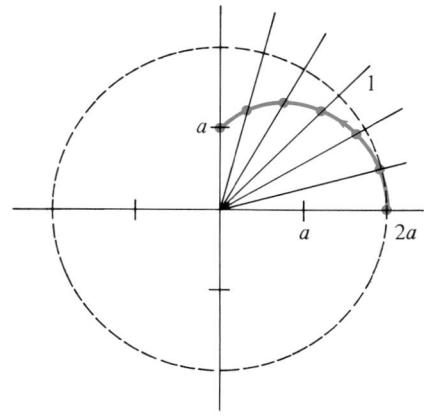

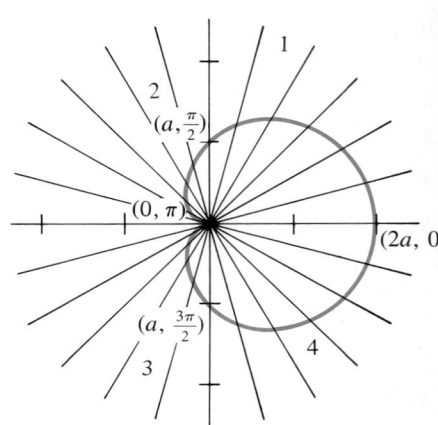

Figure 2.1 Circle $r = 2a$.

Figure 2.2 As θ increases toward $\pi/2$, r decreases from $2a$ to a.

Figure 2.3 Graph of $r = a(1 + \cos \theta)$.

Example 1

Sketch the graph of the polar equation $r = 4 \sin 3\theta$.

Solution: Since $|\sin 3\theta| \leq 1$ for all θ, the curve must lie within the circle $r = 4$. Beginning at the point with $\theta = 0$, we note the following:

(a) As θ turns from 0 to $\pi/6$, 3θ turns from 0 to $\pi/2$, so r *increases* from $r(0) = 4 \sin 0 = 0$ to $r(\pi/6) = 4 \sin 3(\pi/6) = 4$. This produces arc 1 in Figure 2.4.
(b) As θ turns from $\pi/6$ to $\pi/3$, 3θ turns from $\pi/2$ to π, so r *decreases* from $r(\pi/6) = 4$ to $r(\pi/3) = 0$ (arc 2 in Figure 2.4).
(c) As θ turns from $\pi/3$ to $\pi/2$, 3θ turns from π to $3\pi/2$, so r *decreases* from $r(\pi/3) = 0$ to $r(\pi/2) = 4 \sin 3(\pi/2) = -4$, producing arc 3 in Figure 2.5.
(d) As θ increases through the next quarter turn, from $\pi/2$ to π, 3θ increases from $3\pi/2$ to 3π. Thus, r increases from -4 to 4 for $\pi/2 \leq \theta \leq 5\pi/6$, producing arcs 4 and 5, and decreases from 4 to 0 for $5\pi/6 \leq \theta \leq \pi$, producing arc 6 (Figure 2.6).

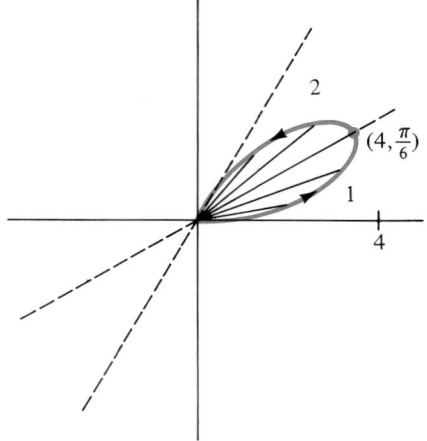

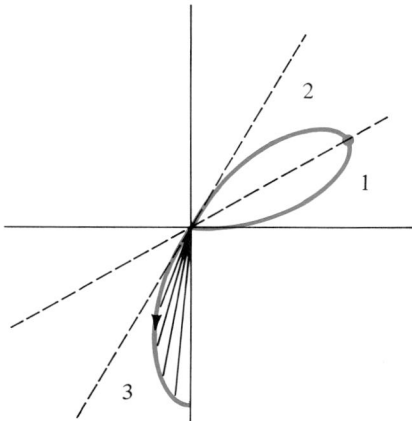

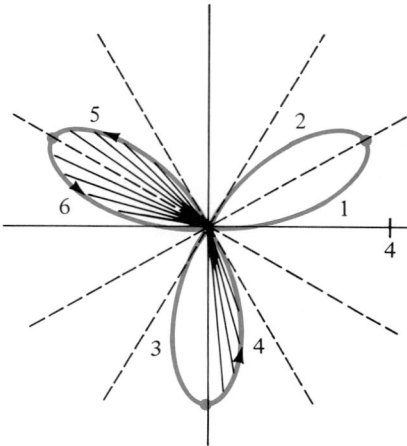

Figure 2.4 $r = 4 \sin 3\theta, 0 \leq \theta \leq \pi/3$.

Figure 2.5 $r = 4 \sin 3\theta, 0 \leq \theta \leq \pi/2$.

Figure 2.6 Three-leaved rose $r = 4 \sin 3\theta, 0 \leq \theta \leq \pi$.

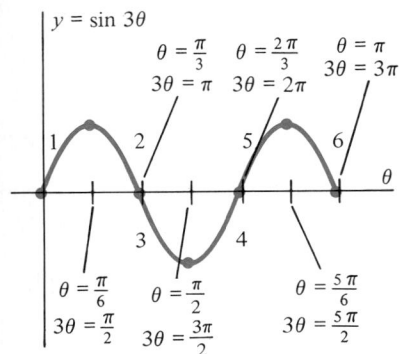

$y = \sin 3\theta$

$\theta = \frac{\pi}{3}$ $\theta = \frac{2\pi}{3}$ $\theta = \pi$

$3\theta = \pi$ $3\theta = 2\pi$ $3\theta = 3\pi$

1   2   5   6

3   4

$\theta = \frac{\pi}{6}$ $\theta = \frac{\pi}{2}$ $\theta = \frac{5\pi}{6}$

$3\theta = \frac{\pi}{2}$ $3\theta = \frac{3\pi}{2}$ $3\theta = \frac{5\pi}{2}$

Figure 2.7 Graph of $f(\theta) = \sin 3\theta$.

By noting the behavior of r for $\theta > \pi$, we can see that the entire curve (a three-leaved rose) has been obtained with only $0 \le \theta \le \pi$. (Figure 2.7 recalls the graph of the function $f(\theta) = \sin 3\theta$. Note how the arcs numbered 1–6 correspond to the arcs in Figures 2.4–2.6.) ◇

Example 2

Graph the lemniscate $r^2 = a \cos 2\theta$, $a > 0$.

Solution: Taking square roots of both sides shows that the given equation corresponds to the *pair* of equations

$$r_1 = \sqrt{a \cos 2\theta} \tag{1}$$

and

$$r_2 = -\sqrt{a \cos 2\theta}. \tag{2}$$

Thus, some values of θ will correspond to two values r, while others will not correspond to any real value r.

Beginning with $\theta = 0$, we observe that

(i) As θ increases from 0 to $\pi/4$, 2θ increases from 0 to $\pi/2$. Thus,

    (a) r_1 decreases from $r_1(0) = \sqrt{a}$ to $r_1(\pi/4) = 0$ (arc 1 in Figure 2.8),
    (b) r_2 increases from $r_2(0) = -\sqrt{a}$ to $r_2(\pi/4) = 0$ (arc 2).

(ii) As θ increases from $\pi/4$ to $3\pi/4$, 2θ increases from $\pi/2$ to $3\pi/2$. Since $\cos 2\theta < 0$ for $\pi/2 < 2\theta < 3\pi/2$, equations (1) and (2) have no solutions in this interval.

(iii) As θ increases from $3\pi/4$ to π, 2θ increases from $3\pi/2$ to 2π. Thus,

    (a) r_1 increases from $r_1(3\pi/2) = 0$ to $r_1(2\pi) = \sqrt{a}$ (arc 3 in Figure 2.9),
    (b) r_2 decreases from $r_2(3\pi/2) = 0$ to $r_2(2\pi) = -\sqrt{a}$ (arc 4).

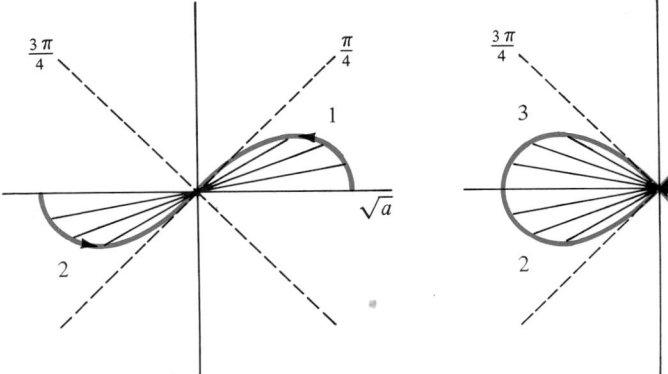

Figure 2.8 $r^2 = a \cos 2\theta$, $0 \le \theta \le \pi/4$.

Figure 2.9 $r^2 = a \cos 2\theta$.

Checking equations (1) and (2) for values of $\theta > \pi$ shows that all points on the graph are obtained by using only $0 \le \theta \le \pi/4$ and $3\pi/4 \le \theta \le \pi$. ◇

Intersections of Graphs in Polar Coordinates

In Section 15.3, we will need to be able to determine points of intersection for graphs of pairs of equations of the form

$$r_1 = f(\theta), \tag{3}$$

$$r_2 = g(\theta). \tag{4}$$

As for equations in rectangular coordinates, points of intersection may be found by equating the right-hand sides of equations (3) and (4) and solving the resulting equation for θ. However, unlike the situation for rectangular equations, *this method will not necessarily produce all points of intersection,* as the following examples show.

Example 3

Find all points of intersection for the graphs of the equations $r_1 = 1 + \cos \theta$ and $r_2 = 3 \cos \theta$.

Solution: Setting $r_1 = r_2$ gives the equation

$$1 + \cos \theta = 3 \cos \theta,$$

so

$$\cos \theta = 1/2.$$

The solutions of this equation for $-\pi < \theta < \pi$ are $\theta_1 = \pi/3$ and $\theta_2 = -\pi/3$. The corresponding points of intersection are $(3/2, \pi/3)$ and $(3/2, -\pi/3)$. However, as Figure 2.10 illustrates, the *pole* $(0, \theta)$ is also common to both graphs. Setting

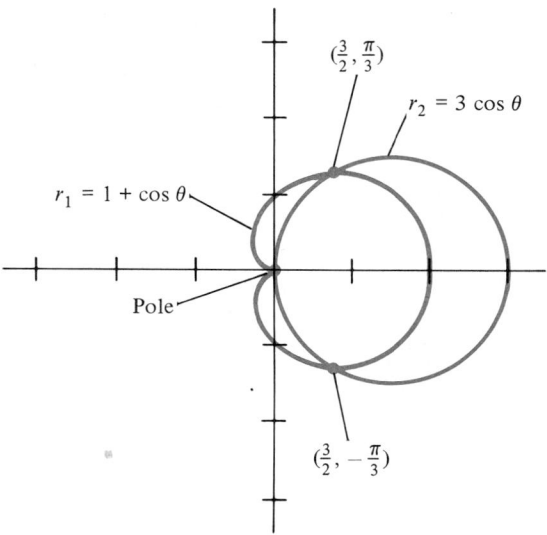

Figure 2.10 Graphs of r_1 and r_2 intersect in 3 points.

$r_1(\theta) = r_2(\theta)$ misses this point, since the pole corresponds to $\theta = \pi$ in the graph of r_1, but it corresponds to $\theta = \pi/2$ in the graph of r_2. ◇

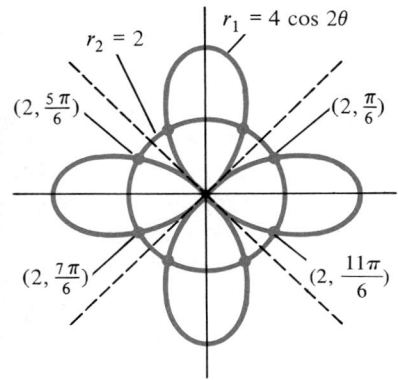

Figure 2.11 Graphs of r_1 and r_2 intersect in 8 points.

Example 4

Find all points of intersection of the four-leaved rose $r_1 = 4 \cos 2\theta$ and the circle $r_2 = 2$.

Solution: Setting $r_1 = r_2$ gives $4 \cos 2\theta = 2$, or $\cos 2\theta = 1/2$. Thus $2\theta = \pm\pi/3$ or $2\theta = \pm 7\pi/3$. The four distinct solutions of this equation in the interval $(0, 2\pi)$ are $\theta = \pi/6$, $5\pi/6$, $7\pi/6$, and $11\pi/6$. These give the four points of intersection $(2, \pi/6)$, $(2, 5\pi/6)$, $(2, 7\pi/6)$, and $(2, 11\pi/6)$. However, the method of setting $r_1 = r_2$ has failed to detect *four* additional points of intersection $(2, \pi/3)$, $(2, 2\pi/3)$, $(2, 4\pi/3)$, and $(2, 5\pi/3)$ (see Figure 2.11). The reason for this is that each of these last four points corresponds to *different* values of θ on the graph of r_1 than on the graph of r_2. ◇

Examples 3 and 4 show that simply setting $r_1 = r_2$ and solving the resulting equation $f(\theta) = g(\theta)$ for θ will not necessarily yield all points common to the graphs of equations (3) and (4). The reason for this difficulty is the nonuniqueness of polar coordinates. For example, the point $(2, \pi/3)$, common to both graphs $r_1 = 4 \cos 2\theta$ and $r_2 = 2$ in Example 4, actually occurs on the graph of r_1 in its equivalent form $(-2, 4\pi/3)$. Since these coordinates do not satisfy $r_2 = 2$, the point does not result from setting $r_1 = r_2$. Rather than trying to work out an algorithm allowing for all possible ways in which points of intersection of two polar equations can arise, *you should simply develop the habit of always sketching the two curves to ensure that all such points are found.*

Exercise Set 15.2

In Exercises 1–10, sketch the graph of the given polar equation using the method of Examples 1 and 2.

1. $r = a \cos \theta$

2. $r = a \sin \theta$

3. $r = a \sin 2\theta$

4. $r = a(1 + \cos \theta)$

5. $r = a(1 - \sin \theta)$

6. $r = 1 + 2 \sin \theta$

7. $r = 2\theta$

8. $r^2 = 4 \cos 2\theta$

9. $r = \sin \theta + \cos \theta$

10. $r = a(1 - \cos \theta)$

In Exercises 11–20, find all points of intersection of the graphs of the two given equations.

11. $r_1 = 2 \cos \theta$
$r_2 = 2 \sin \theta$

12. $r_1 = 2 \cos \theta$
$r_2 = 1$

13. $r_1 = 1 + \sin \theta$
$r_2 = 3 \sin \theta$

14. $r_1 = a \sin 3\theta$
$r_2 = a$

15. $r_1 = a(1 + \cos \theta)$
$r_2 = 3a \cos \theta$

16. $r_1 = 2a \cos 2\theta$
$r_2 = a$

17. $r_1 = a(1 + \cos \theta)$
$r_2 = a(1 - \cos \theta)$

18. $r_1 = \cos 2\theta$
$r_2 = \dfrac{\sqrt{2}}{2}$

19. $r = \dfrac{1}{\theta}$
$\theta = \pi/4$

20. $r_1 = a \sin 4\theta$
$r_2 = a$

21. Show that the graphs of the polar equations

$$r_1 = \frac{a}{1 \pm \cos \theta}; \qquad r_2 = \frac{a}{1 \pm \sin \theta}$$

are parabolas. (*Hint:* Convert to rectangular equations.)

22. Show that the graphs of the polar equations

$$r_1 = \frac{ab}{1 \pm b \cos \theta}; \qquad r_2 = \frac{ab}{1 \pm b \sin \theta}$$

are ellipses if $0 < b < 1$. (*Hint:* Convert to rectangular equations.)

In Exercises 23–28, find a rectangular form for the given polar equation, identify the graph as either a parabola or an ellipse, and sketch the graph. (See Exercises 21 and 22.)

23. $r = \dfrac{1}{1 + \cos \theta}$

24. $r = \dfrac{4}{1 - \sin \theta}$

25. $r = \dfrac{2}{1 + \dfrac{1}{2} \cos \theta}$

26. $2r = \dfrac{4}{2 - \sin \theta}$

27. $r - r \cos \theta = 2$

28. $3r + r \cos \theta - 6 = 0$

15.3 CALCULATING AREA IN POLAR COORDINATES

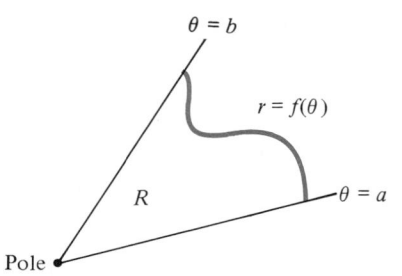

Figure 3.1 Region determined by $r = f(\theta)$, $a \le \theta \le b$.

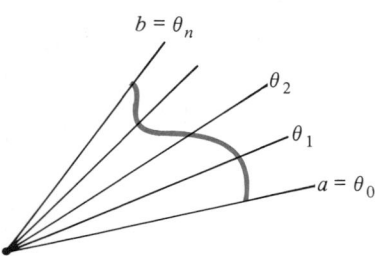

Figure 3.2 Partition $\theta_0 < \theta_1 < \cdots < \theta_n$ slices R into wedge-shaped regions.

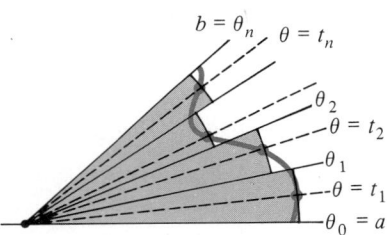

Figure 3.3 Area of R approximated by areas of sectors of circles.

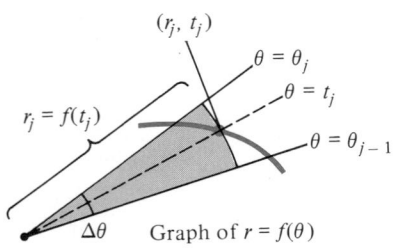

Figure 3.4 $\Delta A_j = \dfrac{1}{2} r_j^2 \, \Delta\theta$.

When a region R in the plane is determined by the graph of an equation written in the *polar* form $r = f(\theta)$, the area of R cannot be calculated by a direct application of the equations developed in Chapter 6. The difficulty lies in the fact that these formulas were obtained for functions $y = g(x)$ written in *rectangular* coordinates. To calculate areas of regions determined by the graphs of polar equations, we might try rewriting such equations in rectangular coordinates, using the substitutions $x = r \cos \theta$ and $y = r \sin \theta$, as in Section 15.1. However, this strategy often leads to one of two undesirable consequences. Either the resulting rectangular equations are much more complicated than the original polar equations, or, even worse, the resulting equations might not describe either variable as a *function* of the other.

A more satisfactory approach is to return to the basic theory of the definite integral and determine the formulas appropriate for calculating area in polar coordinates. The small amount of time spent in this effort will allow us to avoid the step of converting to rectangular coordinates for each area problem in polar coordinates.

Let us suppose that a region R is bounded by the graph of the function $r = f(\theta)$ for θ between the numbers $\theta = a$ and $\theta = b$ (Figure 3.1). Since θ is the independent variable, we partition the interval $[a, b]$ into n equal subintervals by using the increment $\Delta\theta = \dfrac{b - a}{n}$ and the endpoints

$$\theta_0 = a, \quad \theta_1 = a + \Delta\theta, \quad \theta_2 = a + 2\,\Delta\theta, \ldots, \quad \theta_n = a + n\,\Delta\theta = b.$$

The rays $\theta = \theta_j$, $j = 0, 1, 2, \ldots, n$ then "slice" the region R into n wedge-shaped subregions (Figure 3.2).

We now approximate the area of each of the subregions. To do so we choose one number t_j in each interval $[\theta_{j-1}, \theta_j]$, $j = 1, 2, \ldots, n$ (Figure 3.3). We then approximate the area of the jth subregion by the area of the circular **sector** of constant radius $r_j = f(t_j)$ and angle $\Delta\theta$ (see Figure 3.4).

Since the area of the circle of radius r_j is πr_j^2, the area of the circular sector comprising the fractional part $\dfrac{\Delta\theta}{2\pi}$ of the entire circle is

$$\Delta A_j = \pi r_j^2 \left(\frac{\Delta\theta}{2\pi} \right) = \frac{1}{2} r_j^2 \, \Delta\theta = \frac{1}{2}[f(t_j)]^2 \, \Delta\theta.$$

Summing these approximations, one for each subregion, gives an approximation to the area A of the whole region R of the form

$$A \approx \sum_{j=1}^{n} \Delta A_j = \sum_{j=1}^{n} \frac{1}{2}[f(t_j)]^2 \, \Delta\theta. \tag{1}$$

If the function $r = f(\theta)$ is continuous for $a \le \theta \le b$, the sum on the right-hand side of approximation (1) is a *Riemann* sum for the function $\dfrac{1}{2}[f(\theta)]^2$ which "converges" to the definite integral

$$\int_a^b \frac{1}{2} \cdot [f(\theta)]^2 \, d\theta = \lim_{n \to \infty} \sum_{j=1}^{n} \frac{1}{2}[f(t_j)]^2 \, \Delta\theta \tag{2}$$

as $n \to \infty$. We therefore argue that as $n \to \infty$, the circular sectors provide an increas-

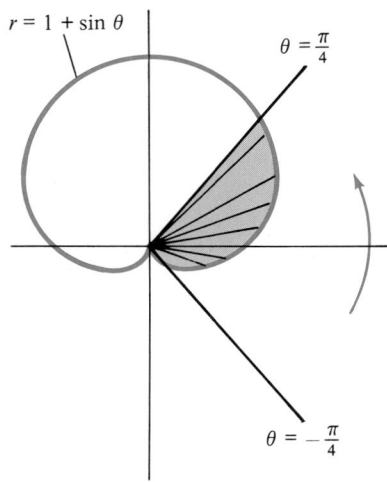

$r = 1 + \sin \theta$

$\theta = \dfrac{\pi}{4}$

$\theta = -\dfrac{\pi}{4}$

Figure 3.5 $-\pi/4 \leq \theta \leq \pi/4$.

ingly accurate approximation to the region R and that the area of R is correctly defined as follows.

> The area A of the region bounded by the graph of the continuous function $r = f(\theta)$ between the rays $\theta = a$ and $\theta = b$, $a < b$, is given by the definite integral
>
> $$A = \int_a^b \frac{1}{2}[f(\theta)]^2 \, d\theta. \tag{3}$$

The same cautions should be observed in using equation (3) as in calculating areas in rectangular coordinates. The region should first be carefully sketched, so that the proper limits of integration may be determined.

Example 1

Find the area of the region bounded by the rays $\theta = -\pi/4$ and $\theta = \pi/4$, and the graph of the equation $r = 1 + \sin \theta$.

Solution: The region is the portion of the cardioid swept out by the radius of length $r = 1 + \sin \theta$ as θ increases from $\theta = -\pi/4$ to $\theta = \pi/4$ (Figure 3.5). By equation (3) the area is

$$A = \int_{-\pi/4}^{\pi/4} \frac{1}{2}[1 + \sin \theta]^2 \, d\theta$$

$$= \int_{-\pi/4}^{\pi/4} \frac{1}{2}[1 + 2 \sin \theta + \sin^2 \theta] \, d\theta$$

$$= \int_{-\pi/4}^{\pi/4} \frac{1}{2}\left[1 + 2 \sin \theta + \left(\frac{1}{2} - \frac{1}{2} \cos 2\theta\right)\right] d\theta \quad \left(\text{using the identity}\right.$$

$$= \int_{-\pi/4}^{\pi/4} \frac{1}{2}\left[\frac{3}{2} + 2 \sin \theta - \frac{1}{2} \cos 2\theta\right] d\theta \qquad \left. \sin^2 \theta = \frac{1}{2} - \frac{1}{2} \cos 2\theta\right)$$

$$= \frac{1}{2}\left[\frac{3\theta}{2} - 2 \cos \theta - \frac{1}{4} \sin 2\theta\right]_{-\pi/4}^{\pi/4}$$

$$= \frac{1}{2}\left\{\left(\frac{3\pi}{8} - \sqrt{2} - \frac{1}{4}\right) - \left(-\frac{3\pi}{8} - \sqrt{2} + \frac{1}{4}\right)\right\}$$

$$= \frac{3\pi}{8} - \frac{1}{4} \approx 0.928. \qquad \diamondsuit$$

Example 2

Find the area of the region enclosed by the graph of the cardioid $r = 1 + \cos \theta$.

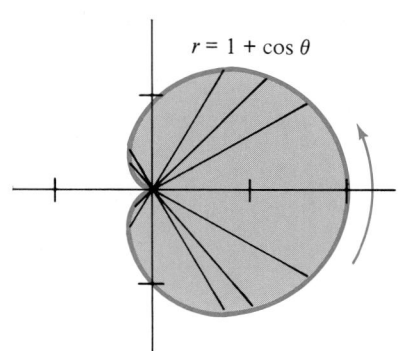

$r = 1 + \cos \theta$

Figure 3.6 $-\pi \leq \theta \leq \pi$.

Strategy

Sketch the region.

Find a pair of smallest and largest values of θ required to sweep out the region. These are the limits of integration. (Note that $0 \leq \theta \leq 2\pi$ works just as well as $-\pi \leq \theta \leq \pi$.)

Solution

The cardioid is sketched in Figure 3.6. Since the cardioid is the graph of the function $f(\theta) = 1 + \cos \theta$ as θ increases from $\theta = -\pi$ to $\theta = \pi$, the limits of integration are $a = -\pi$ and $b = \pi$. By equation (3) the area is

$$A = \int_{-\pi}^{\pi} \frac{1}{2}[1 + \cos \theta]^2 \, d\theta$$

Set up the integral given by equation (3).

$$= \frac{1}{2} \int_{-\pi}^{\pi} [1 + 2 \cos \theta + \cos^2 \theta] \, d\theta$$

Square $f(\theta)$ in the integrand and simplify, using the identity

$$\cos^2 \theta = \frac{1}{2} + \frac{1}{2} \cos 2\theta.$$

$$= \frac{1}{2} \int_{-\pi}^{\pi} \left[1 + 2 \cos \theta + \left(\frac{1}{2} + \frac{1}{2} \cos 2\theta \right) \right] d\theta$$

$$= \frac{1}{2} \int_{-\pi}^{\pi} \left[\frac{3}{2} + 2 \cos \theta + \frac{1}{2} \cos 2\theta \right] d\theta$$

Integrate.

$$= \frac{1}{2} \left[\frac{3\theta}{2} + 2 \sin \theta + \frac{1}{4} \sin 2\theta \right]_{-\pi}^{\pi}$$

$$= \frac{1}{2} \left\{ \left(\frac{3\pi}{2} + 2 \cdot 0 + \frac{1}{4} \cdot 0 \right) - \left(\frac{3}{2}(-\pi) + 2 \cdot 0 + \frac{1}{4} \cdot 0 \right) \right\}$$

$$= \frac{3\pi}{2}. \qquad \diamond$$

Often, symmetry considerations may be used to simplify the calculation of area, as the following example illustrates.

Example 3

Find the area of the region enclosed by the graph of the polar equation $r = a \sin 3\theta$.

Solution: The region is a three-leaved rose, as illustrated in Figure 3.7. Since the 3 leaves are congruent, and since the leaf determined by $0 \le \theta \le \pi/3$ is symmetric about the ray $\theta = \pi/6$, we may obtain the area of the entire figure by calculating the area of the half leaf determined by $0 \le \theta \le \pi/6$ and multiplying the result by 6. We obtain

$$A = 6 \cdot \int_0^{\pi/6} \frac{1}{2} [a \sin 3\theta]^2 \, d\theta$$

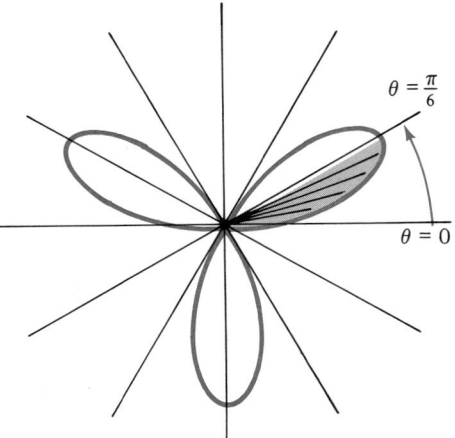

Figure 3.7 Area of three-leaved rose is 6 times area of region determined by $0 \le \theta \le \pi/6$.

$$= 3a^2 \int_0^{\pi/6} \sin^2 3\theta \, d\theta$$

$$= 3a^2 \int_0^{\pi/6} \left[\frac{1}{2} - \frac{1}{2} \cos 6\theta \right] d\theta$$

$$= 3a^2 \left[\frac{\theta}{2} - \frac{1}{12} \sin 6\theta \right]_0^{\pi/6}$$

$$= \frac{a^2 \pi}{4}.$$

$\diamondsuit$

As happens with equations in rectangular coordinates, an equation in polar coordinates may fail to determine r as a function of θ. The following example shows how symmetry may be used to overcome this difficulty.

Example 4

Find the area of the region enclosed by the lemniscate $r^2 = \cos 2\theta$.

Strategy

Sketch the figure. (Recall Example 2, Section 15.2.)

Take square roots to solve for r. Obtain a *function* $r = f(\theta)$.

Solution

The lemniscate is sketched in Figure 3.8. By extracting square roots on both sides of the equation $r^2 = \cos 2\theta$ we obtain the equation $r = \pm\sqrt{\cos 2\theta}$. By choosing the positive sign we obtain the function

$$f(\theta) = \sqrt{\cos 2\theta}.$$

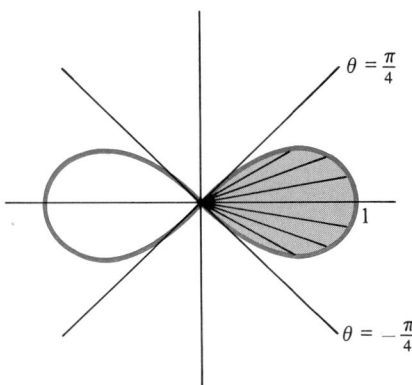

Figure 3.8 $-\pi/4 \le \theta \le \pi/4$ determines half the area of the lemniscate $r^2 = \cos 2\theta$.

Compare region determined by the function with that of the original equation.

This function is defined only for

$$-\frac{\pi}{4} \le \theta \le \frac{\pi}{4} \qquad \text{and} \qquad \frac{3\pi}{4} \le \theta \le \frac{5\pi}{4}.$$

Use symmetry to simplify calculation.

However, as θ ranges through these two intervals the two lobes of the lemniscate are traced out. We may therefore integrate over $-\pi/4 \le \theta \le \pi/4$ to obtain the area of one lobe and double the result. We obtain

Apply equation (3).

$$A = 2 \int_{-\pi/4}^{\pi/4} \frac{1}{2} [\sqrt{\cos 2\theta}]^2 \, d\theta$$

$$= \int_{-\pi/4}^{\pi/4} \cos 2\theta \, d\theta$$

$$= \frac{1}{2} \sin 2\theta \Big]_{-\pi/4}^{\pi/4}$$

$$= 1.$$

◇

Area Between Two Curves

When a region R lies between the graphs of two polar equations, as in Figure 3.9, we may calculate the area of R by subtracting the area enclosed by the inner curve from area enclosed by the outer curve. That is, if R lies inside the graph of $r = f(\theta)$ and outside the graph of $r = g(\theta)$, for $a \leq \theta \leq b$, the area A of R is given by the integral

$$A = \int_a^b \frac{1}{2} [f(\theta)]^2 \, d\theta - \int_a^b \frac{1}{2} [g(\theta)]^2 \, d\theta. \tag{4}$$

Example 5

Find the area of the region lying inside the circle $r = 3 \cos \theta$ and outside the cardioid $r = 1 + \cos \theta$.

Strategy

Sketch the region.
Determine the points of intersection.
This gives the limits of integration.

Solution

The graphs of the two equations are sketched in Figure 3.10. By solving the two equations simultaneously, we find the points of intersection to be

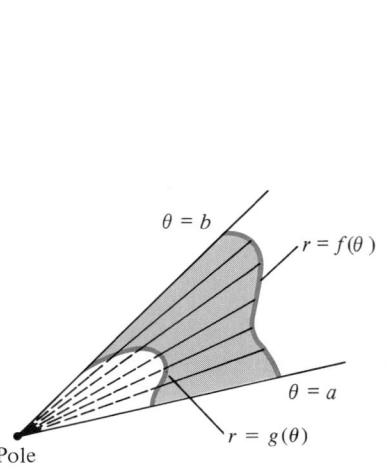

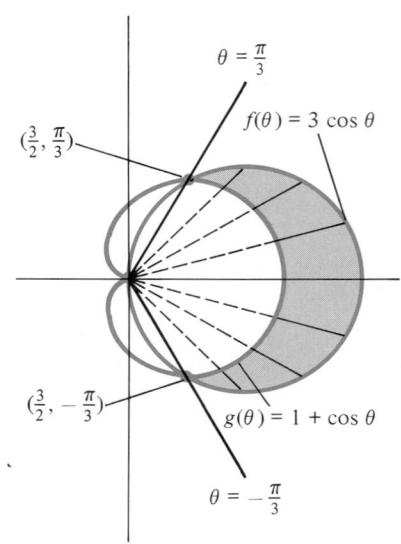

Figure 3.9 Region between graphs of $r = f(\theta)$ and $r = g(\theta)$ for $a \leq \theta \leq b$.

Figure 3.10

Determine which is the outer curve.

Apply equation (4).

Simplify integrand.

$\cos^2 \theta = \dfrac{1}{2} + \dfrac{1}{2} \cos 2\theta.$

(3/2, $\pm \pi/3$) (see Example 3, Section 15.2). The limits of integration are therefore $-\pi/3 \le \theta \le \pi/3$.

In this interval, the outer (greater) function is $f(\theta) = 3 \cos \theta$. The inner function is $g(\theta) = 1 + \cos \theta$. By equation (4) the area is

$$A = \int_{-\pi/3}^{\pi/3} \frac{1}{2}[3 \cos \theta]^2 \, d\theta - \int_{-\pi/3}^{\pi/3} \frac{1}{2}[1 + \cos \theta]^2 \, d\theta$$

$$= \int_{-\pi/3}^{\pi/3} \frac{1}{2}\{9 \cos^2 \theta - (1 + 2 \cos \theta + \cos^2 \theta)\} \, d\theta$$

$$= \int_{-\pi/3}^{\pi/3} \frac{1}{2}[8 \cos^2 \theta - 1 - 2 \cos \theta] \, d\theta$$

$$= \int_{-\pi/3}^{\pi/3} \frac{1}{2}\left[8\left(\frac{1}{2} + \frac{1}{2} \cos 2\theta\right) - 1 - 2 \cos \theta\right] d\theta$$

$$= \int_{-\pi/3}^{\pi/3} \left(\frac{3}{2} + 2 \cos 2\theta - \cos \theta\right) d\theta$$

$$= \frac{3\theta}{2} + \sin 2\theta - \sin \theta \Big]_{-\pi/3}^{\pi/3}$$

$$= \pi. \qquad \diamondsuit$$

Exercise Set 15.3

In Exercises 1–8, find the area of the region determined by the given equations and inequalities.

1. $r = 1 + \sin \theta,$ $\quad -\pi/2 < \theta < \pi/2$

2. $r = a \sin 2\theta,$ $\quad 0 \le \theta \le \pi/2$

3. $r = 2 \sin \theta,$ $\quad 0 \le \theta \le \pi$

4. $r = \theta,$ $\quad 0 \le \theta \le \pi$

5. $r = 2 + \sin \theta,$ $\quad 0 \le \theta \le \pi$

6. $r = a \cos 3\theta,$ $\quad \pi/2 < \theta < \dfrac{2\pi}{3}$

7. $r = 4 + \sin \theta,$ $\quad \pi/4 \le \theta \le \dfrac{3\pi}{4}$

8. $r = 3 \cos 2\theta,$ $\quad -\dfrac{\pi}{4} \le \theta \le \dfrac{\pi}{4}$

In Exercises 9–14, find the area of the region enclosed by the graph of the given equation.

9. $r = 2 \cos \theta$

10. $r = 1 + \cos \theta$

11. $r = a \sin 2\theta$

12. $r = a \sin 4\theta$

13. $r = a \cos 3\theta$

14. $r^2 = \cos \theta$

In Exercises 15–20, find the area of the region described.

15. The region inside the cardioid $r = 1 + \sin \theta$ and outside the circle $r = 2 \sin \theta$.

16. The region inside the cardioid $r = 1 + \cos \theta$ and outside the circle $r = 3 \cos \theta$. (*Hint:* You will need to treat the intervals $\left[\dfrac{\pi}{3}, \dfrac{\pi}{2}\right]$ and $\left[\dfrac{\pi}{2}, \pi\right]$ separately.)

17. The regions common to both cardioids $r = a(1 + \cos \theta)$ and $r = a(1 - \cos \theta)$.

18. The region outside the three-leaved rose $r = 4 \sin 3\theta$ and inside the circle $r = 4$.

19. The region common to the circles $r = 2 \cos \theta$ and $r = 2 \sin \theta$.

20. The region common to the circle $r = \dfrac{\sqrt{2}}{2}$ and the lemniscate $r^2 = \cos 2\theta$.

21. Find the area of the region bounded by the graphs of the spirals $r_1 = \theta$ and $r_2 = e^{\theta}$ for $0 \le \theta \le \pi$.

22. Find the area of the region inside the graph of $r = 3 + \sin \theta$ and outside the graph of $r = 4 \sin \theta$.

23. Find the area of the region outside the graph of $r = 1 + \cos \theta$ and inside the graph of $r = 2 - \cos \theta$.

24. Find the area of the region bounded by the graph of the equation $r - r \cos \theta = 2$ and the line $\theta = \dfrac{\pi}{2}$.

25. Show that the result of calculating the area of the region enclosed by the graph of $r = 2 \cos \theta$ in polar coordinates agrees with the result of calculating the area of the same region in rectangular coordinates.

26. Let $f(\theta) \geq g(\theta)$ for all θ. Sketch the region whose area is given by the integral $\int_a^b \frac{1}{2}[f(\theta) - g(\theta)]^2 \, d\theta$. Compare this region with the region whose area is given by the integral in equation (4). Show that, in general, the two integrals are not equal.

27. *(Computer)* Program 7 in Appendix I is a BASIC program

that approximates $\int_a^b \frac{1}{2}[f(\theta)]^2 \, d\theta$ for the function $f(\theta) = 1 + \sqrt{\sin \theta}$.

a. Use Program 7 to approximate the integral $\int_0^{\pi/2} \frac{1}{2}[1 + \sqrt{\sin \theta}]^2 \, d\theta$ with $n = 10$, 100, and 200.

b. Use Program 7 to approximate the area of the region enclosed by the graph of $f(\theta) = 1 + \sqrt{\sin \theta}$ for $0 \leq \theta \leq \pi$.

c. Modify Program 7 to approximate the area of the region enclosed by the graph of $f(\theta) = 1 - \sqrt{\cos \theta}$, $0 \leq \theta \leq \pi/2$.

15.4 PARAMETRIC EQUATIONS

Parametric equations provide a means of describing a curve in the plane by functions, even though the curve may not represent the graph of a function of the form $y = f(x)$. Rather than describing either coordinate as a function of the other, we allow both coordinates to be functions of a third (independent) variable, called a **parameter,*** which is usually denoted by t. That is, we describe the curve as the collection of points $P(t) = (x(t), y(t))$ where the parameter t ranges through some specified domain.

For example, the graph of the equation

$$x^2 + y^2 = 1 \tag{1}$$

is the familiar unit circle in the plane. However, equation (1) does not determine y as a function of x, since two distinct values of y correspond to each x with $-1 < x < 1$. We may use our knowledge of trigonometry to obtain parametric equations, or a **parameterization,** for the unit circle as follows. Let t denote the angle formed between the positive x-axis and the radius from $(0, 0)$ to (x, y), measured in the counterclockwise direction. Then

$$x = \cos t$$

$$y = \sin t$$

are the coordinate functions of the point (x, y). By inspection we can see that as t increases from $t = 0$ to $t = 2\pi$, the point $(x, y) = (\cos t, \sin t)$ traverses the unit circle once in the counterclockwise direction (Figure 4.1). We therefore say that a parameterization of the unit circle is given by the equations

$$\left. \begin{array}{l} x(t) = \cos t \\ y(t) = \sin t \end{array} \right\} \quad 0 \leq t < 2\pi. \tag{2a} \tag{2b}$$

It is important to note that parameterizations for curves are not unique. For example, each of the following pairs of equations also represents a parameterization

---

*The term *parameter* denotes a variable that often has no geometric or physical interpretation. Here the parameter t has no geometric interpretation with respect to the curve it helps describe. In some applications the parameter t represents time, while the variables x and y represent coordinates of points in the plane.

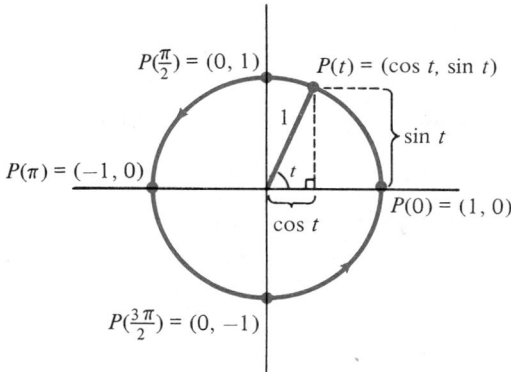

Figure 4.1 Parameterization of unit circle: $(x(t),$ $y(t)) = (\cos t, \sin t)$, $0 \leq t < 2\pi$.

of the unit circle

$$\left.\begin{aligned} x(t) &= \cos 2t \\ y(t) &= \sin 2t \end{aligned}\right\} \ 0 \leq t < \pi,$$

$$\left.\begin{aligned} x(t) &= \cos t \\ y(t) &= \sin(-t) \end{aligned}\right\} \ 0 \leq t < 2\pi.$$

Example 1

Find parametric equations for the line ℓ containing the point $P_0 = (2, 4)$ and with slope 3.

Solution: We first find the equation for ℓ in xy-coordinates, using the point-slope form. We obtain

$$y - 4 = 3(x - 2), \tag{3}$$

so

$$y = 3x - 2. \tag{4}$$

If we set $x = t$ in equation (4), we find that $y = 3t - 2$, and we obtain the parametric equations

$$\left.\begin{aligned} x(t) &= t \\ y(t) &= 3t - 2 \end{aligned}\right\} \ -\infty < t < \infty.$$

A different parameterization for ℓ may be obtained by setting $t = x - 2$ in equation (3). Then, $y - 4 = 3t$, so $y = 3t + 4$, and the parameterization is

$$\left.\begin{aligned} x(t) &= t + 2 \\ y(t) &= 3t + 4 \end{aligned}\right\} \ -\infty < t < \infty. \qquad \diamondsuit$$

Up to this point, we have considered only the problem of finding parameterizations for curves described by equations in xy-coordinates. Just as important is the problem of describing a curve that is presented in parametric form. One technique that is always available is to sketch the curve simply by selecting various values for t, calculating the corresponding coordinates of the point $P(t) = (x(t), y(t))$, and sketching in a curve passing through the various points obtained (in order!). However, through a familiarity with the xy-coordinate forms of various curves you can often convert the parametric equations directly into their xy counterparts.

Example 2

A point moves along a path with coordinates

$$(x(t), y(t)) = (a \cos t, b \sin t)$$

at time t. Describe the path.

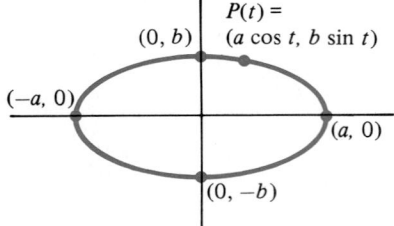

Figure 4.2 Parameterization for the ellipse $\dfrac{x^2}{a^2} + \dfrac{y^2}{b^2} = 1$.

Solution: By squaring both coordinate functions we observe that

$$[x(t)]^2 = a^2 \cos^2 t; \qquad [y(t)]^2 = b^2 \sin^2 t.$$

We obtain

$$\frac{[x(t)]^2}{a^2} + \frac{[y(t)]^2}{b^2} = \cos^2 t + \sin^2 t = 1.$$

The curve is, therefore, the ellipse $\dfrac{x^2}{a^2} + \dfrac{y^2}{b^2} = 1$ seen in Figure 4.2. ◇

Example 3

To convert the parametric equations

$$x(t) = t^2 - 2$$
$$y(t) = t^2 + 1$$

to an equation in x and y alone, we begin by noting that

$$y = t^2 + 1 = (t^2 - 2) + 3 = x + 3.$$

But the resulting equation is not simply $y = x + 3$. This is because the range of the function $x(t) = t^2 - 2$ is $[-2, \infty)$. (That is, we cannot have points (x, y) with $x < -2$, since $t^2 \geq 0$ for all t.) We must therefore restrict the domain by writing

$$y = x + 3, \qquad x \geq -2.$$

(See Figure 4.3.) ◇

Example 4

Parametric equations are useful in describing the motion of objects in space. For example, we may use them to determine the motion of a bullet fired horizontally from an altitude of 2 meters with an initial velocity of v_0 m/s.

If we let $(x(t), y(t))$ denote the position of the bullet t seconds after it is fired, as in Figure 4.4, we will show in Chapter 17 that (neglecting the effect of air resistance) the coordinate *functions* x and y are

$$x(t) = v_0 t,$$
$$y(t) = 2 - 4.9t^2.$$

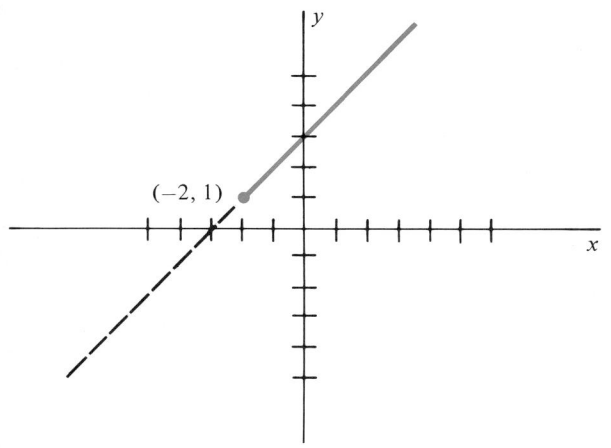

Figure 4.3 Graph of parametric equations $x(t) = t^2 - 2$, $y(t) = t^2 + 1$.

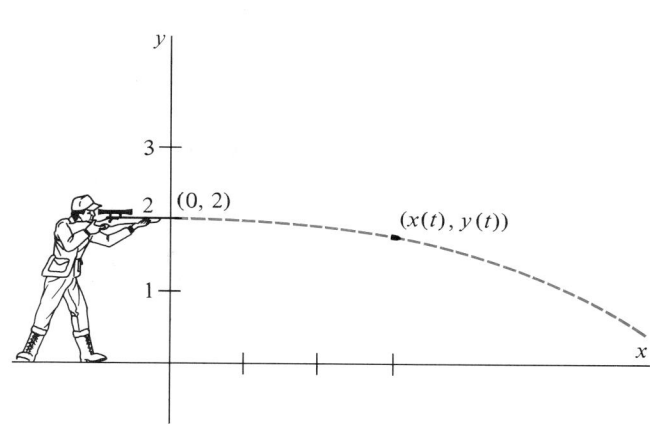

Figure 4.4 Bullet fired by a marksman actually follows a parabolic path.

To see that these equations, for $t \geq 0$, describe an arc of a parabola, we note that

$$y(t) = 2 - 4.9t^2$$

$$= 2 - \frac{4.9}{v_0^2}(v_0 t)^2$$

$$= 2 - \left(\frac{4.9}{v_0^2}\right)[x(t)]^2.$$

That is, $y = kx^2 + 2$ with $k = -\dfrac{4.9}{v_0^2}$. ◇

Having conveyed the general notion of parametric equations, we should pause at this point to sharpen our use of the terminology **curve.** In general, a curve C in the plane is the set of all points of the form

$$C = \{(x(t), y(t) \mid t \in I\} \tag{5}$$

where I is some interval, and where the coordinate functions x and y are **continuous** functions of $t \in I$.

The requirement that x and y be continuous assures that the curve will itself be continuous (unbroken) in the sense of the graph of a continuous function $y = f(x)$. The curve C in (5) is called **smooth** if the derivatives x' and y' are continuous functions of $t \in I$.

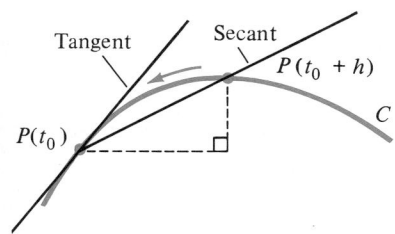

Figure 4.5 Slope of tangent is limit of slopes of secants.

Equations of Tangents

Suppose that the curve C is given in the parametric form of equation (5) and that we wish to find the slope of the line tangent to the curve at the point $P(t_0) = (x(t_0), y(t_0))$. (For example, the curve C might be the trajectory of a rocket, and we might wish to describe its angle of elevation as a function of time.) To do so, we begin by letting h be a small real number. Then $P(t_0 + h) = (x(t_0 + h), y(t_0 + h))$ is, in general, a second point on C (see Figure 4.5). If $x(t_0) \neq x(t_0 + h)$, we can write the slope of the secant through $P(t_0)$ and $P(t_0 + h)$ as

$$\text{Slope of secant} = \frac{y(t_0 + h) - y(t_0)}{x(t_0 + h) - x(t_0)}. \tag{6}$$

The limit of this slope as h tends to zero is the number m, the desired slope of the tangent at $P(t_0)$. By dividing both numerator and denominator in line (6) by $h \neq 0$, we find that

$$m = \lim_{h \to 0} \frac{\left(\dfrac{y(t_0 + h) - y(t_0)}{h} \right)}{\left(\dfrac{x(t_0 + h) - x(t_0)}{h} \right)} = \frac{y'(t_0)}{x'(t_0)}. \tag{7}$$

Obviously, m in equation (7) makes sense only if both $y'(t_0)$ and $x'(t_0)$ exist and $x'(t_0) \neq 0$. Also, recall that we needed to assume that $x(t_0 + h) \neq x(t_0)$. However, this assumption will be valid for all h near zero whenever $x'(t_0)$ exists and is not zero (see Exercise 42). Thus, we have shown that *if $x'(t_0)$ and $y'(t_0)$ exist, and if $x'(t_0) \neq 0$, the slope of the line tangent to the graph of the curve $C = \{(x(t), y(t)) \mid t \in I\}$ at $P(t_0) = (x(t_0), y(t_0))$ is*

$$m = \frac{y'(t_0)}{x'(t_0)}, \qquad x'(t_0) \neq 0. \tag{8}$$

The equation for this tangent line is therefore

$$y - y(t_0) = \frac{y'(t_0)}{x'(t_0)} (x - x(t_0))$$

or

$$x'(t_0)[y - y(t_0)] - y'(t_0)[x - x(t_0)] = 0. \tag{9}$$

If $x'(t_0) = 0$ and $y'(t_0) \neq 0$, equation (9) reduces to the equation $x = x(t_0)$ so the tangent is a vertical line through $(x(t_0), y(t_0))$. If both $x'(t_0) = 0$ and $y'(t_0) = 0$, no conclusion can be drawn about the tangent at $(x(t_0), y(t_0))$.

Example 5

Find the equation of the line tangent to the curve given by the parametric equations

$$\left. \begin{array}{l} x(t) = 2 \cos t \\ y(t) = 4 \sin t \end{array} \right\} \quad 0 \leq t < 2\pi$$

at the point $(\sqrt{2}, 2\sqrt{2})$.

Strategy

Find t_0.

Find $x'(t_0)$, $y'(t_0)$.

Solution

To find the number t_0 for which $P(t_0) = (\sqrt{2}, 2\sqrt{2})$, we set

$$x(t_0) = 2 \cos t_0 = \sqrt{2},$$

$$y(t_0) = 4 \sin t_0 = 2\sqrt{2}.$$

This gives $\cos t_0 = \sqrt{2}/2$, $\sin t_0 = \sqrt{2}/2$, so $t_0 = \pi/4$.

Then

$$x'(t_0) = -2 \sin(\pi/4) = -\sqrt{2}$$

and

$$y'(t_0) = 4 \cos(\pi/4) = 2\sqrt{2}.$$

Use equation (9).

Substituting into equation (9) then gives

$$-\sqrt{2}(y - 2\sqrt{2}) - 2\sqrt{2}(x - \sqrt{2}) = 0$$

or

$$y = -2x + 4\sqrt{2}. \qquad \text{(See Figure 4.6.)} \qquad \diamondsuit$$

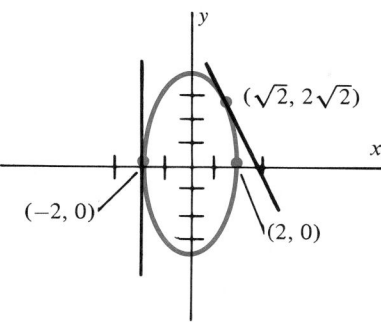

Figure 4.6 Tangents to curve in Example 5.

Example 6

Determine the points for which the curve in Example 5 has vertical tangents.

Strategy

Find t_0 so that $x'(t_0) = 0$.

Solution

Since $x'(t) = -2 \sin t$, the equation

$$x'(t) = -2 \sin t = 0$$

has solutions $t_0 = 0$, π for $0 \le t < 2\pi$.

Determine if $y'(t_0) \ne 0$ for each such t_0.

Since

$$y'(0) = 4 \cos 0 = 4 \ne 0 \qquad \text{and}$$
$$y'(\pi) = 4 \cos \pi = -4 \ne 0,$$

If so, the points

$$P(t_0) = (x(t_0), y(t_0))$$

yield vertical tangents.

the points

$$P(0) = (2 \cos 0, 4 \sin 0) = (2, 0)$$

and

$$P(\pi) = (2 \cos \pi, 4 \sin \pi) = (-2, 0)$$

yield vertical tangents (see Figure 4.6). $\qquad \diamondsuit$

Tangents to Polar Curves

If a curve in the plane is the graph of a polar equation of the form $r = f(\theta)$, we can use the equations

$$x = r \cos \theta \tag{10a}$$

$$y = r \sin \theta \tag{10b}$$

to obtain parametric equations for the curve. Multiplying the equation $r = f(\theta)$ on both sides by $\cos \theta$ gives the equation $r \cos \theta = f(\theta) \cos \theta$. Using (10a), we obtain

$$x(\theta) = f(\theta) \cos \theta. \tag{11a}$$

Similarly, multiplying both sides of $r = f(\theta)$ by $\sin \theta$ and using (10b), we obtain

$$y(\theta) = f(\theta) \sin \theta. \tag{11b}$$

If the derivative $f'(\theta) = \dfrac{dr}{d\theta}$ exists, we may differentiate both sides of equation (11a) and use equation (10b) to conclude that

$$x'(\theta) = f'(\theta) \cos \theta - f(\theta) \sin \theta \tag{12a}$$

$$= \frac{dr}{d\theta} \cos \theta - r \sin \theta.$$

Similarly, differentiating both sides of (11b) and using (10a), we obtain

$$y'(\theta) = f'(\theta) \sin \theta + f(\theta) \cos \theta \tag{12b}$$

$$= \frac{dr}{d\theta} \sin \theta + r \cos \theta.$$

Finally, we may apply equation (8) (with parameter θ rather than t) and equations (12a) and (12b) to conclude that *the slope of the line tangent to the graph of the polar equation $r = f(\theta)$ at the point (r, θ) is*

$$m = \frac{\dfrac{dr}{d\theta} \sin \theta + r \cos \theta}{\dfrac{dr}{d\theta} \cos \theta - r \sin \theta}. \tag{13}$$

As noted for equation (8), m in equation (13) is defined only if $x'(\theta) = \dfrac{dr}{d\theta} \cos \theta - r \sin \theta \neq 0$. If $x'(\theta) = 0$ and $y'(\theta) \neq 0$, the graph has a vertical tangent at (r, θ). If both $x'(\theta) = 0$ and $y'(\theta) = 0$, no conclusions may be drawn.

Example 7

Find the points where the four-leaved rose with equation

$$r = \sin 2\theta, \qquad 0 \leq \theta < 2\pi$$

has vertical tangents.

Strategy

Find a parameterization for the curve.

Solution

By equations (11a) and (11b), a parameterization for $f(\theta) = \sin 2\theta = 2 \sin \theta \cos \theta$ is

$$x(\theta) = 2 \sin \theta \cos^2 \theta$$

$$y(\theta) = 2 \sin^2 \theta \cos \theta.$$

Find the angles θ for which $x'(\theta) = 0$.

Thus,

$$x'(\theta) = 2 \cos^3 \theta - 4 \sin^2 \theta \cos \theta = 0$$

implies

$$\cos\theta(\cos^2\theta - 2\sin^2\theta) = 0.$$

Thus, $x'(\theta) = 0$ whenever

$$\cos\theta = 0 \qquad \text{or} \qquad \cos^2\theta = 2\sin^2\theta.$$

The equation $\cos\theta = 0$ has solutions $\theta = \pi/2$ and $\theta = 3\pi/2$ in the interval $[0, 2\pi)$. The equation $\cos^2\theta = 2\sin^2\theta$ gives $\tan^2\theta = \dfrac{1}{2}$ or $\tan\theta = \pm\dfrac{\sqrt{2}}{2}$, which has 4 solutions in the interval $[0, 2\pi)$, $\theta = \pm\text{Tan}^{-1}\left(\pm\dfrac{\sqrt{2}}{2}\right)$. The 6 solutions of $x'(\theta) = 0$ are therefore $\theta = \dfrac{\pi}{2}$, $\theta = \dfrac{3\pi}{2}$, and $\theta = \pm\text{Tan}^{-1}\left(\pm\dfrac{\sqrt{2}}{2}\right)$.

Find $y'(\theta)$.

To determine which of these yield vertical tangents we must examine

$$y'(\theta) = 4\sin\theta\cos^2\theta - 2\sin^3\theta$$
$$= 2\sin\theta[2\cos^2\theta - \sin^2\theta].$$

Determine if $y'(\theta) \neq 0$ for each θ with $x'(\theta) = 0$.

If $\theta = \pi/2$ or $3\pi/2$, then

$$y'(\theta) = -2\sin^3\theta \neq 0.$$

Similarly, if $\theta = \pm\text{Tan}^{-1}\left(\pm\dfrac{\sqrt{2}}{2}\right)$, then

$$y'(\theta) = 2\left(\pm\dfrac{\sqrt{3}}{3}\right)\left[2\cdot\dfrac{2}{3} - \dfrac{1}{3}\right] \neq 0.$$

If so, (r, θ) is a point that yields a vertical tangent. (See Figure 4.7.)

Thus, all 6 values of θ yield vertical tangents (note in Figure 4.7 that both $\theta = \pi/2$ and $\theta = 3\pi/2$ correspond to the origin). ◇

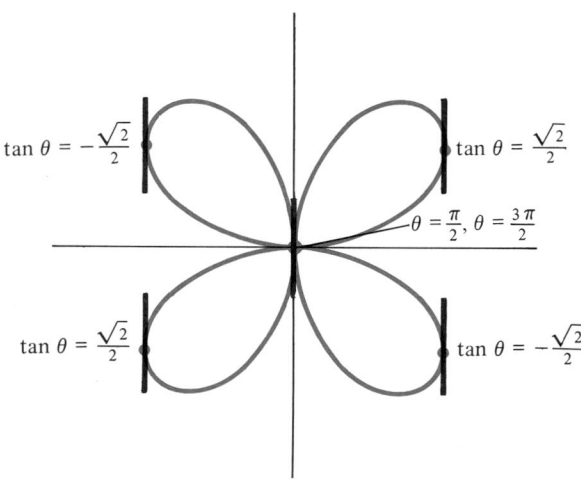

Figure 4.7 Points where graph of $r = \sin 2\theta$ has vertical tangents.

Exercise Set 15.4

In Exercises 1–12, sketch the curve described by the given parametric equations and find an equation in xy-coordinates whose graph contains the given curve.

1. $x(t) = t$
$y(t) = t + 6$

2. $x(t) = t + 2$
$y(t) = 3 - t$

3. $x(t) = 1 + 3t$
$y(t) = 2t + 2$

4. $x(t) = t$
$y(t) = t^2 + 2$

5. $x(t) = t^2 + 1$
$y(t) = t^2 - 1$

6. $x(t) = \sin t$
$y(t) = \cos^2 t$

7. $x(t) = e^t$
$y(t) = e^{3t}$

8. $x(t) = \sqrt{t}$
$y(t) = 3t - 2$

9. $x(t) = \sec t$
$y(t) = \tan t$

10. $x(t) = 3 \sin t$
$y(t) = 5 \cos t$

11. $x(t) = \sin t$
$y(t) = \sin 2t$

12. $x(t) = \cos t$
$y(t) = \sec t$

In Exercises 13–20, find the slope of the line tangent to the curve at the indicated point and the equation of the tangent.

13. $x(t) = t$
$y(t) = t^2 + 1$
$t = 2$

14. $x(t) = t^2$
$y(t) = 1 - t$
$t = 2$

15. $x(t) = \sin t$
$y(t) = \cos t$
$t = \pi/3$

16. $x(t) = \cos t$
$y(t) = \sin t$
$t = \pi/4$

17. $x(t) = \dfrac{1}{t}$
$y(t) = 3t^2 - 7$
$t = 2$

18. $x(t) = 3t^3$
$y(t) = \sin \pi t$
$t = 1$

19. $x(t) = 1 + \sqrt{t}$
$y(t) = 1 - \sqrt{t}$
$t = 4$

20. $x(t) = \sec t$
$y(t) = \tan t$
$t = \pi/4$

In Exercises 21–26, find the slope, if defined, of the line tangent to the graph of the polar equation at the point corresponding to the given value of θ.

21. $r = \cos \theta$
$\theta = \pi/4$

22. $r = 1 + \cos \theta$
$\theta = \pi/6$

23. $r = a \sin 2\theta$
$\theta = \pi/6$

24. $r = 2 \sin \theta$
$\theta = \pi/4$

25. $r = \theta$
$\theta = \pi/2$

26. $r = a \sin 3\theta$
$\theta = \pi/3$

In Exercises 27–32, find all points at which the curve described by the parametric equations has **(a)** a vertical tangent, **(b)** a horizontal tangent.

27. $x(t) = \cos t$
$y(t) = \sin t$

28. $x(t) = \sin 2t$
$y(t) = \sin t$

29. $x(t) = t^2 + 4$
$y(t) = 3t^2 - 6t + 2$

30. $x(t) = 5 + 2 \sin t$
$y(t) = 3 - \cos t$

31. $x(t) = 3t^2 + 6$
$y(t) = t - t^2$

32. $x(t) = t^{3/2}$
$y(t) = t + \cos t$

33. Find the point on the curve

$$C_1: \quad \begin{array}{l} x(t) = t \\ y(t) = 2t^2 + 3 \end{array}$$

where the tangent is parallel to the line

$$C_2: \quad \begin{array}{l} x(t) = t + 3 \\ y(t) = 4t - 10. \end{array}$$

34. Find parametric equations for the cardioid $r = 1 + \cos \theta$.

35. Find parametric equations for the three-leaved rose $r = a \sin 3\theta$.

36. Find parametric equations for the spiral $r = \theta$.

37. Find the points at which the cardioid $r = 1 + \sin \theta$ has vertical tangents.

38. Find the points at which the cardioid $r = 1 + \cos \theta$ has horizontal tangents.

39. Find the points at which the cardioid $r = 1 + \cos \theta$ has vertical tangents.

40. Show that the curve determined by the parametric equations

$$x(t) = a \cos t + h$$

$$y(t) = b \sin t + k$$

is an ellipse with center at (h, k).

41. A particle moves in the plane so that at time t it is at the point with coordinates $x(t) = t + 4$, $y(t) = 8 - t^2$. A second particle moves in the plane so that it is at the point with coordinates $x(t) = t + 4$, $y(t) = t + 6$, at time t.
   a. Find equations in xy-coordinates for each of the curves.
   b. Do the paths cross? If so, where?
   c. Do the particles collide? If so, where and at what time?

42. Prove that if $x(t)$ is differentiable at t_0 and if $x'(t_0) \neq 0$, then $x(t_0 + h) \neq x(t_0)$ for all sufficiently small h. (*Hint:* Use the differential approximation $x(t_0 + h) \approx x(t_0) + x'(t_0)h$. See Chapter 3.)

43. Show that the notion of curve is a more general concept than that of the graph of a continuous function $y = f(x)$. That is, show that every graph of a continuous function $y = f(x)$ can be expressed as a curve in parametric form, but that the converse is not true.

44. Prove that the graph of the function $y = f(x)$ is a *smooth* curve if $f'(x)$ is continuous for all x in the domain of f.

15.5 ARC LENGTH AND SURFACE AREA REVISITED

In Section 7.3 we developed formulas for arc length and surface area calculations associated with the graph of a differentiable function f. Since we have now seen that a curve is a more general notion than the graph of a function, we should ask whether the concepts of arc length and surface area can be extended to general curves. In both cases the answer is yes, and the development here parallels that of Section 7.3 very closely.

Arc Length

Suppose that C is a curve in the plane that is determined by the parametric equations

$$C: \quad \begin{matrix} x = x(t) \\ y = y(t) \end{matrix} \Big\} \; t \in I,$$

where I is some interval. Suppose further that a and b are numbers in I, with $a < b$, so that the arc of the curve from $P = (x(a), y(a))$ to $Q = (x(b), y(b))$ does not intersect itself, except possibly if $P = Q$. The problem is to define and calculate the length L of the arc of C connecting P and Q (see Figure 5.1).

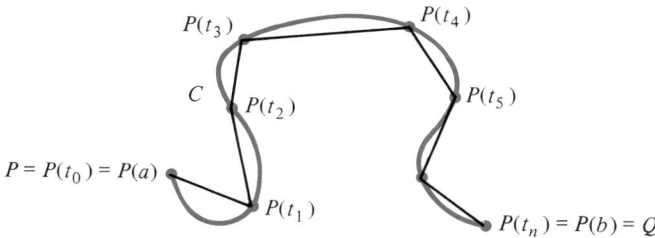

Figure 5.1 Polygonal path joining points $P(t_j) = (x(t_j), y(t_j))$ approximates C.

As in Section 2.3, we begin by partitioning the interval $[a, b]$ for the independent variable (parameter) t into n subintervals of equal length $\Delta t = \dfrac{b - a}{n}$ with endpoints

$$a = t_0 < t_1 < t_2 < \cdots < t_n = b.$$

For each integer $j = 0, 1, 2, \ldots, n$, we let $P(t_j) = (x(t_j), y(t_j))$. Then each $P(t_j)$ is a point on the arc of C joining P and Q. In each interval $[t_{j-1}, t_j]$, we use the length of the line segment from $P(t_{j-1})$ to $P(t_j)$ to approximate the length of the arc C_j connecting these two points. Using the distance formula, we can write this distance as

$$\Delta L_j = \sqrt{\Delta x_j^2 + \Delta y_j^2} \tag{1}$$
$$= \sqrt{[x(t_j) - x(t_{j-1})]^2 + [y(t_j) - y(t_{j-1})]^2}, \qquad j = 1, 2, \ldots, n.$$

(See Figure 5.2.)

If x is a differentiable function of t in each interval $[t_{j-1}, t_j]$, we may apply the Mean Value Theorem to conclude that there exists a number $c_j \in [t_{j-1}, t_j]$ so that

$$x'(c_j) = \frac{x(t_j) - x(t_{j-1})}{t_j - t_{j-1}} = \frac{x(t_j) - x(t_{j-1})}{\Delta t}.$$

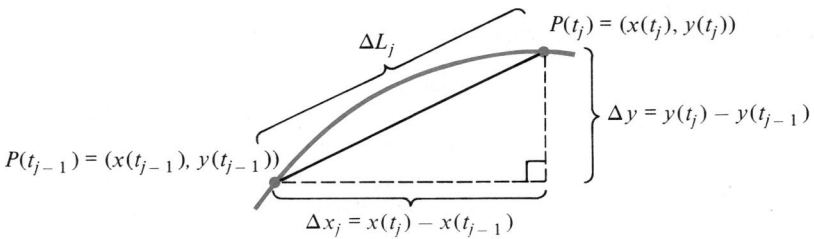

Figure 5.2 ΔL_j approximates length of arc from $P(t_{j-1})$ to $P(t_j)$.

Thus, we can write

$$x(t_j) - x(t_{j-1}) = x'(c_j)\,\Delta t, \qquad j = 1, 2, \ldots, n. \tag{2}$$

Similarly, if y is a differentiable function of t, there exist numbers $d_j \in [t_{j-1}, t_j]$ so that

$$y(t_j) - y(t_{j-1}) = y'(d_j)\,\Delta t, \qquad j = 1, 2, \ldots, n. \tag{3}$$

Combining equations (1)–(3), we obtain the length of the approximating line segment as

$$\Delta L_j = \sqrt{[x'(c_j)\,\Delta t]^2 + [y'(d_j)\,\Delta t]^2} \tag{4}$$
$$= \sqrt{[x'(c_j)]^2 + [y'(d_j)]^2}\,\Delta t, \qquad j = 1, 2, \ldots, n.$$

Finally, we approximate the length L of the arc of C from P to Q by the length of the polygonal path connecting the points $P(t_0), P(t_1), \ldots, P(t_n)$. The latter is simply the sum of the lengths ΔL_j in line (4). We obtain the approximation

$$L \approx \sum_{j=1}^{n} \Delta L_j = \sum_{j=1}^{n} \sqrt{[x'(c_j)]^2 + [y'(d_j)]^2}\,\Delta t. \tag{5}$$

The expression on the right-hand side of equation (5) is "almost" a Riemann sum. The difficulty is that the functions x' and y' are being evaluated at two (possibly) different points in each subinterval. However, in more advanced courses, it is shown that if both x' and y' are continuous on $[a, b]$ this difficulty can be overcome, and that as $n \to \infty$ the approximating sum converges to the integral

$$L = \int_a^b \sqrt{[x'(t)]^2 + [y'(t)]^2}\,dt. \tag{6}$$

As in Section 7.3, we argue that as $n \to \infty$ the polygonal paths more closely approximate the curve C, so that the limiting value of the length of the polygonal paths provides a reasonable definition of the length of the arc. It is important to keep in mind that *both x' and y' must be continuous on $[a, b]$ for formula (6) to be valid*. That is, C must be a *smooth* curve.

Example 1

Find the length of the arc of the curve

$$C: \quad \begin{aligned} x(t) &= \cos t + t \sin t \\ y(t) &= \sin t - t \cos t \end{aligned}$$

connecting the points $P = (x(0), y(0)) = (1, 0)$ and $Q = \left(x\left(\frac{\pi}{2}\right), y\left(\frac{\pi}{2}\right) \right) = \left(\frac{\pi}{2}, 1\right)$.

Solution: Here

$$x'(t) = -\sin t + \sin t + t \cos t = t \cos t$$

and

$$y'(t) = \cos t - \cos t + t \sin t = t \sin t.$$

By formula (6)

$$
\begin{aligned}
L &= \int_0^{\pi/2} \sqrt{(t \cos t)^2 + (t \sin t)^2} \, dt \\
&= \int_0^{\pi/2} \sqrt{t^2(\cos^2 t + \sin^2 t)} \, dt \\
&= \int_0^{\pi/2} t \, dt \\
&= \frac{t^2}{2} \Big]_0^{\pi/2} \\
&= \frac{\pi^2}{8}.
\end{aligned}
$$

◇

Arc Length in Polar Coordinates

In Section 15.4 we showed that the graph of the polar equation $r = f(\theta)$ can be parameterized, using θ as the parameter, as

$$x(\theta) = f(\theta) \cos \theta,$$
$$y(\theta) = f(\theta) \sin \theta.$$

If $f'(\theta)$ is continuous, so are the functions

$$x'(\theta) = f'(\theta) \cos \theta - f(\theta) \sin \theta$$

and

$$y'(\theta) = f'(\theta) \sin \theta + f(\theta) \cos \theta,$$

so we may apply equation (6) to determine a formula for arc length in polar coordinates. Before doing so we note that

$$[x'(\theta)]^2 = [f'(\theta)]^2 \cos^2 \theta - 2f'(\theta)f(\theta) \sin \theta \cos \theta + [f(\theta)]^2 \sin^2 \theta,$$

and

$$[y'(\theta)]^2 = [f'(\theta)]^2 \sin^2 \theta + 2f'(\theta)f(\theta) \sin \theta \cos \theta + [f(\theta)]^2 \cos^2 \theta,$$

so

$$[x'(\theta)]^2 + [y'(\theta)]^2 = [f'(\theta)]^2 + [f(\theta)]^2. \tag{7}$$

From equations (6) and (7) it follows that

$$L = \int_a^b \sqrt{[f'(\theta)]^2 + [f(\theta)]^2}\, d\theta \qquad (8)$$

gives the length of the arc of the graph of $r = f(\theta)$ from $\theta = a$ to $\theta = b$.

Example 2

Find the length of the cardioid $r = 1 + \cos \theta$.

Strategy

Determine the limits of integration. Verify that equation (8) applies.

Solution

Here $f(\theta) = r = 1 + \cos \theta$, and the cardioid is swept out as θ makes one complete revolution. Appropriate limits of integration are therefore $\theta = 0$ to $\theta = 2\pi$.

Since $f'(\theta) = -\sin \theta$ is continuous on $[0, 2\pi]$, we may apply formula (8):

Apply (8).

$$L = \int_0^{2\pi} \sqrt{(1 + \cos \theta)^2 + (-\sin \theta)^2}\, d\theta$$

$$= \int_0^{2\pi} \sqrt{(1 + 2 \cos \theta + \cos^2 \theta) + \sin^2 \theta}\, d\theta$$

$$= \int_0^{2\pi} \sqrt{2 + 2 \cos \theta}\, d\theta$$

Use identity

$$\cos^2 \phi = \frac{1}{2} + \frac{1}{2} \cos 2\phi$$

in reverse to simplify integrand.

$$= \int_0^{2\pi} \sqrt{4\left(\frac{1}{2} + \frac{1}{2} \cos \theta\right)}\, d\theta$$

$$= \int_0^{2\pi} \sqrt{4 \cos^2\left(\frac{\theta}{2}\right)}\, d\theta$$

$$= \int_0^{2\pi} \left|2 \cos\left(\frac{\theta}{2}\right)\right|\, d\theta$$

Use the symmetry of the cosine function to handle the absolute value signs. (Alternatively, we could have used the symmetry of the cardioid and integrated over $[0, \pi]$.)

$$= 2 \int_0^{\pi} 2 \cos\left(\frac{\theta}{2}\right)\, d\theta$$

$$= 8 \sin\left(\frac{\theta}{2}\right)\Big]_0^{\pi} = 8. \qquad \diamondsuit$$

Surface Area

Suppose that the region bounded by the smooth curve C with equations

$$C: \quad \left.\begin{array}{l} x = x(t) \\ y = y(t) \end{array}\right\} \; a \leq t \leq b$$

and the x-axis is to be revolved about the x-axis. In order to calculate the surface area of the resulting solid we shall need to require that $x'(t) \neq 0$ for all $t \in [a, b]$. (This last condition ensures that the curve will have at most one point with any given x-coordinate (see Exercise 35).)

We begin with the usual step of partitioning the interval $[a, b]$ into n equal subintervals of length $\Delta t = \dfrac{b - a}{n}$, and with endpoints

$$a = t_0 < t_1 < t_2 < \cdots < t_n = b.$$

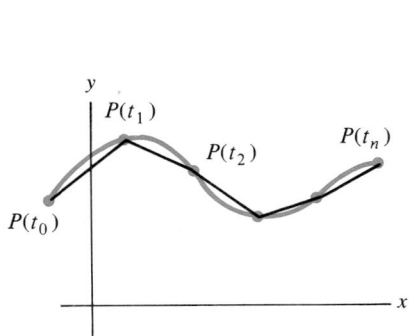

Figure 5.3 Curve to be revolved about the x-axis.

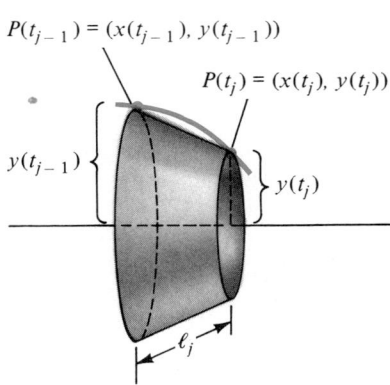

Figure 5.4 Frustum generated by $P(t_{j-1})P(t_j)$.

Let $P(t_j) = (x(t_j), y(t_j))$. We shall approximate C by the polygonal path joining these consecutive points on C (see Figure 5.3). As each of these line segments is revolved about the x-axis, it generates the frustum of a cone, as in Figure 5.4. The radii of the jth frustum are $y(t_{j-1})$ and $y(t_j)$, so the lateral surface area of the jth frustum is

$$\Delta S_j = \pi[y(t_{j-1}) + y(t_j)]\ell_j, \qquad j = 1, 2, \ldots, n$$

where

$$\ell_j = \sqrt{(\Delta x_j)^2 + (\Delta y_j)^2}$$
$$= \sqrt{[x(t_j) - x(t_{j-1})]^2 + [y(t_j) - y(t_{j-1})]^2}. \qquad \text{(See Figure 5.4.)}$$

As in our development of the formula for arc length, we obtain the following approximation to the lateral surface area S of the volume of revolution.

$$S \approx \sum_{j=1}^{n} \Delta S_j = \sum_{j=1}^{n} \pi[y(t_{j-1}) + y(t_j)]\sqrt{[x'(c_j)]^2 + [y'(d_j)]^2}\,\Delta t$$

which converges to the definite integral

$$S = \int_a^b 2\pi y(t)\sqrt{[x'(t)]^2 + [y'(t)]^2}\,dt. \qquad (9)$$

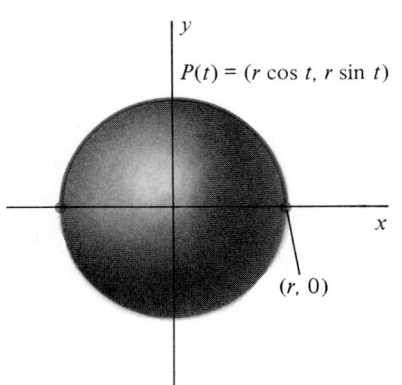

Figure 5.5 $S = 4\pi ar$.

Figure 5.6 $S = 4\pi r^2$.

Example 3

In Example 4, Section 7.3, it was shown that the surface area of the band of the sphere obtained by revolving the arc of the circle $x^2 + y^2 = r^2$ lying above the interval $[-a, a]$, $a < r$, about the x-axis is $S = 4\pi ar$ (Figure 5.5). Moreover, we noted in Example 5 of that section that one could not conclude from this calculation that the surface area of a sphere is $S = 4\pi r^2$.

We may now use equation (9) to demonstrate that the surface area of the sphere of radius r is indeed $S = 4\pi r^2$ (Figure 5.6). To do so we use the parameterization of the upper semicircle

$$\left.\begin{array}{l} x(t) = r\cos t \\ y(t) = r\sin t \end{array}\right\} \quad 0 \le t \le \pi.$$

Then

$$x'(t) = -r \sin t; \qquad y'(t) = r \cos t.$$

Both of these derivatives are continuous for all t, in contrast to $\dfrac{dy}{dx}$ (which is discontinuous at $x = r$ and $x = -r$). Thus,

$$S = \int_0^\pi 2\pi(r \sin t)\sqrt{(-r \sin t)^2 + (r \cos t)^2} \, dt$$

$$= \int_0^\pi 2\pi r^2 \sin t \, dt$$

$$= -2\pi r^2 \cos t]_0^\pi$$
$$= -2\pi r^2(-1 - 1)$$
$$= 4\pi r^2.$$

◇

Exercise Set 15.5

In Exercises 1–11, a curve is described by a pair of parametric equations. Find the length of the given curve.

1. $\begin{aligned} x(t) &= t \\ y(t) &= t^{3/2} \end{aligned} \qquad 0 \le t \le \dfrac{4}{9}$

2. $\begin{aligned} x(t) &= t + 1 \\ y(t) &= \dfrac{2}{3}t^{3/2} \end{aligned} \qquad 0 \le t \le 3$

3. $\begin{aligned} x(t) &= 3t^2 \\ y(t) &= 2t^3 \end{aligned} \qquad 0 \le t \le \sqrt{2}$

4. $\begin{aligned} x(t) &= \sin t \\ y(t) &= \cos t \end{aligned} \qquad 0 \le t \le 2\pi$

5. $\begin{aligned} x(t) &= 2(2t + 3)^{3/2} \\ y(t) &= 3(t + 1)^2 \end{aligned} \qquad 0 \le t \le 2$

6. $\begin{aligned} x(t) &= 2(1 - \cos t) \\ y(t) &= 2 \sin t \end{aligned} \qquad 0 \le t \le \pi$

7. $\begin{aligned} x(t) &= t^3 - 3t^2 \\ y(t) &= 3t^2 \end{aligned} \qquad 0 \le t \le 1$

8. $\begin{aligned} x(t) &= \ln \cos t \\ y(t) &= t \end{aligned} \qquad 0 \le t \le \pi/3$

9. $\begin{aligned} x(t) &= \cos^3 t \\ y(t) &= \sin^3 t \end{aligned} \qquad 0 \le t \le \pi$

10. $\begin{aligned} x(t) &= 3t^3 \\ y(t) &= 3t^2 \end{aligned} \qquad 0 \le t \le \sqrt{5}$

11. $\begin{aligned} x(t) &= e^t \cos t \\ y(t) &= e^t \sin t \end{aligned} \qquad 0 \le t \le \pi$

In Exercises 12–18, a curve is described by a pair of parametric equations. Find the surface area of the solid generated by revolving this curve about the x-axis.

12. $\begin{aligned} x(t) &= t \\ y(t) &= t^2 \end{aligned} \qquad 0 \le t \le 1$

13. $\begin{aligned} x(t) &= t^2 \\ y(t) &= 4t \end{aligned} \qquad 0 \le t \le 2$

14. $\begin{aligned} x(t) &= t \\ y(t) &= \dfrac{t^4}{4} + \dfrac{1}{8t^2} \end{aligned} \qquad 1 \le t \le 2$

15. $\begin{aligned} x(t) &= t \\ y(t) &= \sqrt{t} \end{aligned} \qquad 1 \le t \le 2$

16. $\begin{aligned} x(t) &= t \\ y(t) &= \dfrac{t^2 - 1}{2} \end{aligned} \qquad 0 \le t \le 1$

17. $\begin{aligned} x(t) &= 3t^2 \\ y(t) &= 2t^3 \end{aligned} \qquad 1 \le t \le 2$

18. $\begin{aligned} x(t) &= \cos^3 t \\ y(t) &= \sin^3 t \end{aligned} \qquad 0 \le t \le \pi$

19. Find the length of the cardioid $r = \cos^2\left(\dfrac{\theta}{2}\right)$.

20. Find the length of the cardioid $r = 1 + \cos \theta$.

21. Find the length of the spiral $r = e^{2\theta}$, $\quad 0 \le \theta \le \pi$.

22. Find the area of the surface generated when one arch of the cycloid $x(t) = t - \sin t$, $y(t) = 1 - \cos t$ is revolved about the x-axis.

23. Find the length of the circle $x(t) = 2 \cos t$, $y(t) = 2 \sin t$, $0 \le t \le 2\pi$.

24. Find the length of the spiral $r = 2\theta$, $\quad 0 \le \theta \le \pi$.

25. Show, using equation (8), that the circumference of the circle $r = 3 \cos \theta$ is 3π.

26. Find the length of the cardioid $r = a(1 + \sin \theta)$.

27. Find the length of the graph of $r = e^\theta$ for $0 \le \theta \le 2\pi$.

28. Find the length of the graph of the equation $r = 4 \sec \theta$ for

$$0 \le \theta \le \frac{\pi}{4}.$$

29. Find the surface area of the solid obtained by revolving the region bounded by the graph of $r = e^{\theta}$, $0 \le \theta \le \pi$, and the x-axis about the x-axis.

30. Find the area of the surface of the solid obtained by revolving the region bounded by the graph of $r = 4 \sin \theta$ about the line $\theta = 0$.

31. Determine the formula for the surface area of the volume of the solid generated when an arc of the curve determined by the parametric equations $x = x(t)$, $y = y(t)$ is revolved about the y-axis.

32. Use the result of Exercise 31 to find the surface area of the solid generated when the curve given in Exercise 13 is revolved about the y-axis.

33. Use the result of Exercise 31 to find the surface area of the solid obtained by revolving the curve in Exercise 15 about the y-axis.

34. *(Calculator/Computer)* Use Simpson's Rule to approximate the length of the ellipse

$$x(t) = 4 \cos t,$$
$$y(t) = 6 \sin t.$$

35. Show that the condition $x'(t) \ne 0$ is sufficient to guarantee that a differentiable curve C will have at most one point with any particular x-coordinate.

SUMMARY OUTLINE OF CHAPTER 15

◆ The **polar coordinates** (r, θ) for the point whose rectangular coordinates are (x, y) are determined by the equations   (page 630)
$r = \sqrt{x^2 + y^2}$, $\tan \theta = y/x$, while $x = r \cos \theta$, $y = r \sin \theta$.

◆ The following identities hold for polar coordinates: $(-r, \theta) = (r, \theta + \pi)$, $(r, \theta + 2n\pi) = (r, \theta)$,   (page 629)
$n = \pm 1, \pm 2, \ldots$.

◆ The graph of the polar equation $r = f(\theta)$ is   (page 632)

1. symmetric with respect to the pole if $f(\theta + \pi) = f(\theta)$;
2. symmetric with respect to the x-axis if $f(-\theta) = f(\theta)$;
3. symmetric with respect to the y-axis if $f(\theta) = f(\pi - \theta)$.

◆ The area A of the region bounded by the graph of the function $r = f(\theta)$ and the rays $\theta = a$ and $\theta = b$ is   (page 641)

$$A = \int_a^b \frac{1}{2} [f(\theta)]^2 \, d\theta.$$

◆ The slope of the line tangent to the curve determined by the **parametric equations** $x = x(t)$, $y = y(t)$, at the point   (page 650)
$(x(t_0), y(t_0))$ is

$$m = \frac{y'(t_0)}{x'(t_0)}$$

provided $x'(t_0)$ and $y'(t_0)$ exist and $x'(t_0) \ne 0$.

◆ The slope of the line tangent to the graph of the polar equation $r = f(\theta)$ at the point (r, θ) is   (page 652)

$$m = \frac{\dfrac{dr}{d\theta} \sin \theta + r \cos \theta}{\dfrac{dr}{d\theta} \cos \theta - r \sin \theta}.$$

◆ For the arc of the smooth curve C determined by the polar equations $x = x(t)$, $y = y(t)$, between the points   (page 656)
$(x(a), y(a))$ and $(x(b), y(b))$:

1. The length of the arc is

$$L = \int_a^b \sqrt{[x'(t)]^2 + [y'(t)]^2} \, dt.$$

2. The surface area of the volume obtained by revolving the arc about the x-axis is

$$S = \int_a^b 2\pi y(t) \sqrt{[x'(t)]^2 + [y'(t)]^2} \, dt.$$

◆ The length of the arc of the graph of the polar equation $r = f(\theta)$, from $\theta = a$ to $\theta = b$, is (page 658)

$$L = \int_a^b \sqrt{[f'(\theta)]^2 + [f(\theta)]^2}\, d\theta.$$

REVIEW EXERCISES—CHAPTER 15

In Exercises 1–6, sketch the curve described by the given parametric equations.

1. $\begin{aligned} x &= 5\cos\theta \\ y &= 5\sin\theta \end{aligned}$ $0 \le \theta \le 2\pi$

2. $\begin{aligned} x &= 3 + 2t \\ y &= 8 - 6t \end{aligned}$ $-\infty < t < \infty$

3. $\begin{aligned} x &= t\cos\pi t \\ y &= t\sin\pi t \end{aligned}$ $0 \le t \le 6$

4. $\begin{aligned} x &= t^2 + 1 \\ y &= t^4 - 4 \end{aligned}$ $-\infty < t < \infty$

5. $\begin{aligned} x &= \cos t \\ y &= \sin 2t \end{aligned}$ $0 \le t \le \pi$

6. $\begin{aligned} x &= 3\sqrt{t} + 1 \\ y &= 1 - \sqrt{t} \end{aligned}$ $0 \le t$

7. Eliminate the parameter in Exercise 3 to find an equation in x and y that represents the given curve.

8. Repeat Exercise 7 for the curve given in Exercise 4.

9. Repeat Exercise 7 for the curve given in Exercise 5.

In each of Exercises 10–13, find parametric equations for the given curves.

10. $x^2 + y^2 = 9$

11. $4x^2 + 9y^2 = 36$

12. $x + y = 7$

13. $x = 2y^2 + 4y + 5$

14. Show that the parametric equations $x(t) = t$, $y(t) = t^2$ describe the graph of the equation $y = x^2$ as well as the graph of the equation $y^3 = x^6$, but *not* the graph of $y^2 = x^4$.

15. Sketch the curve given parametrically in polar coordinates by the equations $r = 2t$, $\theta = \pi t^2$.

16. Sketch the curve given parametrically by the equations

$\theta = t$, $r = e^t$.

17. Show that the points on the curve determined by the parametric equations $x(t) = t^2 - 1$, $y(t) = 3t$ lie on a parabola. Find the equation for this parabola in xy-coordinates.

18. Find an equation in polar coordinates for the circle with center at the origin and radius 5.

In Exercises 19–28, sketch the graph of the given polar equation.

19. $r = \cos 2\theta$

20. $r = 3\theta$

21. $r = 5\cos 3\theta$

22. $r = |\sin 2\theta|$

23. $r = 6\sin\theta$

24. $r = a\tan\theta$

25. $r = 2 + \sin 2\theta$

26. $r = \dfrac{1}{2 - \cos\theta}$

27. $\theta = \pi/2$

28. $r = 1 - 2\sin\theta$

In Exercises 29–34, find the area of the region enclosed by the given curve.

29. $r = 2(1 + \sin 2\theta)$

30. $r = 2 - \cos\theta$

31. $r = a(1 + \sin\theta)$

32. $r = \sqrt{1 - \sin\theta}$

33. $r = 6\cos 3\theta$

34. $r = 2 + 4\sin\theta\cos\theta$

In Exercises 35–40, find the length of the curve given by the parametric equations.

35. $x = 1 + t$, $y = (1 + t)^{3/2}$, $0 \le t \le 1$.

36. $x = 2t + 1$, $y = t^2$, $0 \le t \le 1$

37. $x = 3t$, $y = t^3$, $0 \le t \le 1$

38. $x = \sin t - t$, $y = \cos t$, $0 \le t \le \pi$

39. $x = 1 - \cos t$, $y = \sin t$, $0 \le t \le \pi$

40. $x = 9t^2$, $y = 3t^3 - 9t^2$, $0 \le t \le 1$

In Exercises 41–44, find the length of the curve given by the polar equation.

41. $r = \cos\theta$, $-\pi/4 \le \theta \le \pi/4$

42. $r = \theta^2$, $1 \le \theta \le 2$

43. $r = 1 + \sin\theta$, $0 \le \theta \le \pi/2$

44. $r = 1 - \cos\theta$, $0 \le \theta \le \pi$

45. Find the area of the region common to the circles $r = \sin\theta$ and $r = \cos\theta$.

46. Find the area of the region enclosed by the lemniscate $r^2 = 4\cos 2\theta$.

47. Find the area of the region inside the circle $r = 6\cos\theta$ and outside the cardioid $r = 2(1 + \cos\theta)$.

48. Find the area of the region inside the circle $r = 4\cos\theta$ and outside the circle $r = 2$.

49. Find the area of the region common to the circle $r = 4\sin\theta$ and the circle $r = 2$.

50. Find the slope of the line tangent to the curve given by the parametric equations $x(t) = \sin t$, $y(t) = \sin 2t$ at the point where $t = \pi/4$.

Chapter 16
Vectors in the Plane and in Space

Up to this point we have used ordered pairs of numbers (a, b) to represent points in the plane. We shall now develop a richer interpretation for the ordered pair (a, b), that of a *vector* in the plane. We do this by defining operations of addition and multiplication by scalars (numbers) for such pairs. This will enable us to interpret the pair (a, b) as a displacement, a concept useful in physics and engineering, and to define the operations of addition and multiplication by scalars for functions that have vectors, rather than numbers, as values.

This chapter develops the concepts of vectors in the plane and in space. Chapters 17–20 use these concepts to study functions whose values are vectors, and functions of more than one variable.

16.1 VECTORS IN THE PLANE

Our discussion of parametric equations showed that it is useful to know about functions of the form $f(t) = (x(t), y(t))$, that is, functions whose values are points in the xy-plane rather than numbers. In order to be able to form sums and multiples of such functions, we shall need a way to add points in the plane, or multiply them by real numbers. The algebraic structure for doing this is provided by the concept of a *vector*.

DEFINITION 1

The set of all ordered pairs $\langle a, b \rangle$ of real numbers, together with the rules

(i) $\langle a_1, b_1 \rangle + \langle a_2, b_2 \rangle = \langle a_1 + a_2, b_1 + b_2 \rangle$
(ii) $c\langle a, b \rangle = \langle ca, cb \rangle$

for addition and for multiplication by real numbers c, is called the set of **vectors in the plane.** The numbers a and b are referred to as the **components** of the vector $\langle a, b \rangle$.

It is important to note that the vector $\langle a, b \rangle$ is an *ordered pair of numbers*, not a point (a, b) in the plane. We shall sometimes wish to associate the vector $\langle a, b \rangle$ with the point (a, b), but the former concept is more general.

The multiplication of the vector $\langle a, b \rangle$ by the real number c in equation (ii) is called **scalar multiplication,** and the number c is referred to as a **scalar.**

663

We shall observe the convention of denoting vectors by bold-faced letters. Thus if $v = \langle a_1, b_1 \rangle$ and $w = \langle a_2, b_2 \rangle$ are vectors, Definition 1 provides that

$$v + w = \langle a_1 + a_2, b_1 + b_2 \rangle,$$

etc.

Example 1

For $v = \langle 2, 5 \rangle$ and $w = \langle -3, 2 \rangle$, we have

(a) $v + w = \langle 2 + (-3), 5 + 2 \rangle = \langle -1, 7 \rangle$,
(b) $3v + w = \langle 3 \cdot 2, 3 \cdot 5 \rangle + \langle -3, 2 \rangle = \langle 3, 17 \rangle$,
(c) $4v + 2w = \langle 4 \cdot 2, 4 \cdot 5 \rangle + \langle 2(-3), 2 \cdot 2 \rangle = \langle 2, 24 \rangle$. ◇

We define the vector $-v$ as $(-1)v$. That is,

if $v = \langle a, b \rangle$, then $-v = \langle -a, -b \rangle$.

Also, we define $v - w$ as $v + (-w)$. That is, if

$$v = \langle a_1, b_1 \rangle, \qquad w = \langle a_2, b_2 \rangle,$$

then

$$v - w = v + (-w) = \langle a_1 - a_2, b_1 - b_2 \rangle.$$

The **zero vector** is defined to be the vector $\mathbf{0} = \langle 0, 0 \rangle$. Finally, we define two vectors to be **equal** if and only if they have the same components. That is,

$$\langle a_1, b_1 \rangle = \langle a_2, b_2 \rangle \qquad \text{if and only if} \qquad a_1 = a_2 \text{ and } b_1 = b_2.$$

Example 2

For $v = \langle 1, -3 \rangle$ and $w = \langle 2, 5 \rangle$,

(a) $v - w = \langle 1 - 2, -3 - 5 \rangle = \langle -1, -8 \rangle$.
(b) $2v - 3w = \langle 2, -6 \rangle - \langle 6, 15 \rangle = \langle -4, -21 \rangle$.
(c) $v - v = \langle 1 - 1, -3 + 3 \rangle = \langle 0, 0 \rangle = \mathbf{0}$. ◇

Vectors as Displacements

One interpretation of vectors is as displacements (meaning changes in location) in the plane. If P and Q are distinct points in the plane, the displacement $\overrightarrow{PQ}$ is represented by an arrow consisting of the line segment $\overline{PQ}$ to which a tip is added at Q. The tip distinguishes the *terminal point* Q from the *initial point* P. (See Figure 1.1.)

Thus, the displacement $\overrightarrow{PQ}$ has both a *length* (which we denote by $|\overrightarrow{PQ}|$ and which is the same as the length of $\overline{PQ}$) and a direction. Moreover, there is a natural way to form the "sum" of two displacements as long as the initial point of the second displacement coincides with the terminal point of the first one: $\overrightarrow{PQ} + \overrightarrow{QR} = \overrightarrow{PR}$. (See Figure 1.2.) Finally, the multiple $c\overrightarrow{PQ}$ may be defined to be the displacement that originates at P, that has either the same or the opposite direction as $\overrightarrow{PQ}$ depending on whether c is positive or negative, and that has length $|c|$ times the length of $\overrightarrow{PQ}$ (see Figure 1.3).

Here is the connection between displacements and vectors. Since a displacement indicates only a change in location (think of "walking three yards east," for exam-

Figure 1.1 The displacement $\overrightarrow{PQ}$.

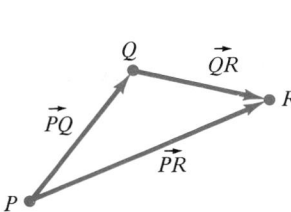

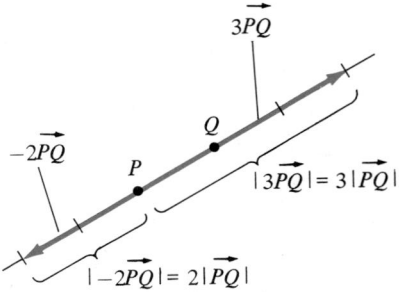

Figure 1.2 Adding two displacements: $\overrightarrow{PQ} + \overrightarrow{QR} = \overrightarrow{PR}$.

Figure 1.3 Multiples $c\overrightarrow{PQ}$ of the displacement $\overrightarrow{PQ}$.

ple), a displacement is determined only by the *changes* it causes in the x and y coordinates of the initial point, but not by the initial point itself. We therefore use vector notation to write

$$\overrightarrow{PQ} = \langle \Delta x, \Delta y \rangle, \tag{1}$$

which means that the displacement $\overrightarrow{PQ}$ consists of an increment Δx in the x-coordinate of the initial point and an increment Δy in the y-coordinate of the initial point. (See Figure 1.4.)

The point of equation (1) is that the vector $v = \langle \Delta x, \Delta y \rangle$ represents the displacement from *any* point $P = (x_1, y_1)$ to another point $Q = (x_2, y_2)$ as long as $x_2 - x_1 = \Delta x$ and $y_2 - y_1 = \Delta y$. We therefore regard two displacements as equal if they are represented by the same vector.

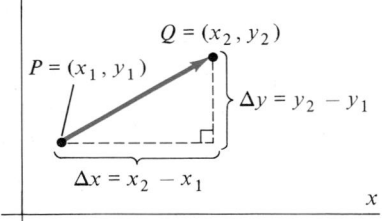

Figure 1.4 $\overrightarrow{PQ} = \langle \Delta x, \Delta y \rangle$
$= \langle x_2 - x_1, y_2 - y_1 \rangle$.

Example 3

Let $P = (0, 0)$, $Q = (2, 4)$, $R = (-3, 1)$, and $S = (-1, 5)$. Then

$$\overrightarrow{PQ} = \langle 2 - 0, 4 - 0 \rangle = \langle 2, 4 \rangle$$

and

$$\overrightarrow{RS} = \langle -1 - (-3), 5 - 1 \rangle = \langle 2, 4 \rangle,$$

so $\overrightarrow{PQ} = \overrightarrow{RS}$. (See Figure 1.5.) ◇

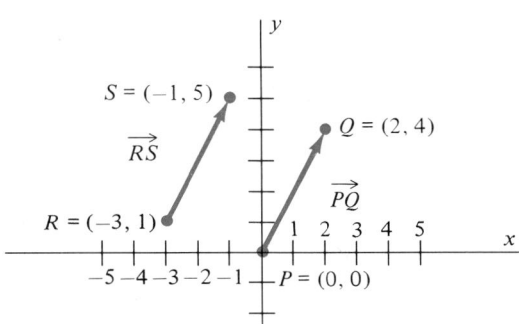

Figure 1.5 $\overrightarrow{PQ} = \overrightarrow{RS}$.

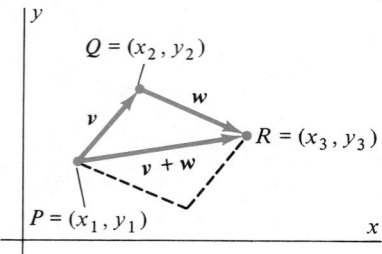

Figure 1.6 Head-to-tail addition for vectors as arrows in the plane.

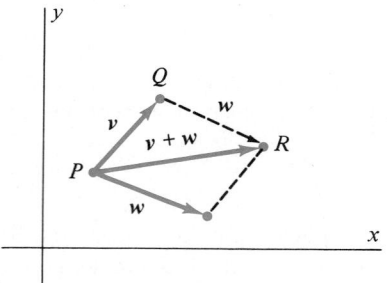

Figure 1.7 $v + w$ is the diagonal of the parallelogram determined by v and w.

Representing vectors by arrows (displacements) in the plane provides a geometric interpretation of the operations of addition and subtraction of vectors as follows.

Vector addition: If the vector $v = \langle a_1, b_1 \rangle$ is represented as an arrow originating at point $P = (x_1, y_1)$, this arrow has terminal point $Q = (x_2, y_2)$ with

$$x_2 = x_1 + a_1,$$
$$y_2 = y_1 + b_1.$$

If the vector $w = \langle a_2, b_2 \rangle$ is represented as an arrow originating at Q, this arrow has terminal point $R = (x_3, y_3)$ with

$$x_3 = x_2 + a_2 = x_1 + (a_1 + a_2),$$
$$y_3 = y_2 + b_2 = y_1 + (b_1 + b_2).$$

Since Definition 1 requires that

$$v + w = \langle a_1, b_1 \rangle + \langle a_2, b_2 \rangle = \langle a_1 + a_2, b_1 + b_2 \rangle,$$

this shows that $v + w$ is represented by the arrow originating at P and terminating at R. (See Figure 1.6.) Thus, the head-to-tail rule for adding arrows (displacements) corresponds to our formal definition of vector addition.

REMARK 1: Note in Figure 1.6 that the vector $v + w$ may also be interpreted as the diagonal of the parallelogram determined by the arrows for v and w, when positioned head-to-tail.

REMARK 2: Figure 1.7 shows that, if the arrow for w is translated (meaning moved so that its length and direction are preserved) so that it originates at P, the sum $v + w$ is still the diagonal of the parallelogram determined by v and w.

REMARK 3: More generally, the arrows for v and w may be translated to *any* initial point in the plane. As long as they originate at the same point, or are positioned head-to-tail, the sum $v + w$ is the diagonal of the parallelogram determined by v and w.

Vector Subtraction: The difference $v - w$ may be represented by an arrow which is the sum of the arrows for the vectors v and $-w$, as in Figures 1.8–1.10. That is because we have defined

$$v - w = v + (-w).$$

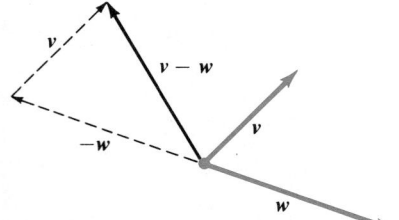

Figure 1.8 $v - w = v + (-w)$.

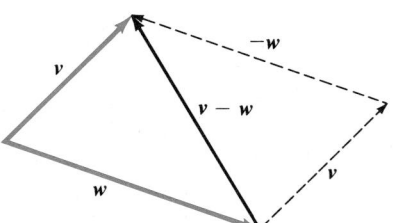

Figure 1.9 $v - w$.

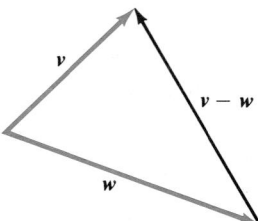

Figure 1.10 $v - w$.

Example 4

Let $P = (1, 2)$, $Q = (3, 6)$, and $R = (3, 0)$. Let $v = \overrightarrow{PQ}$ and $w = \overrightarrow{PR}$. Using the distance formula, we can see that

$$|v| = \text{length of } v = \text{length of } \overline{PQ} = \sqrt{(3-1)^2 + (6-2)^2} = 2\sqrt{5},$$
$$|w| = \text{length of } w = \text{length of } \overline{PR} = \sqrt{(3-1)^2 + (0-2)^2} = 2\sqrt{2}.$$

Also, the slope of v is $\dfrac{6-2}{3-1} = 2$, and the slope of w is $\dfrac{0-2}{3-1} = -1$. From this information it follows that

(a) v, when originating at $(3, 0)$, terminates at $(5, 4)$,
(b) w, when originating at $(3, 6)$, terminates at $(5, 4)$,
(c) $v + w$, when originating at $(1, 2)$, terminates at $(5, 4)$,
(d) $v - w$, when originating at $(3, 0)$, terminates at $(3, 6)$,
(e) $-w$, when originating at $(1, 2)$, terminates at $(-1, 4)$,
(f) $-2v$, when originating at $(1, 2)$, terminates at $(-3, -6)$,
(g) $-2v + w$, when originating at $(1, 2)$, terminates at $(-1, -8)$.

(See Figures 1.11 through 1.13.)

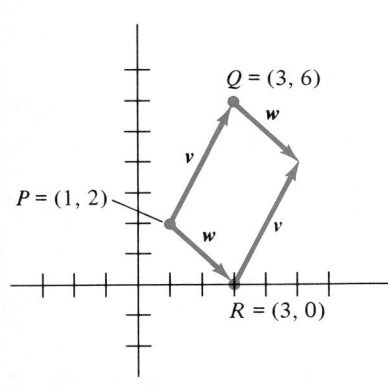

Figure 1.11

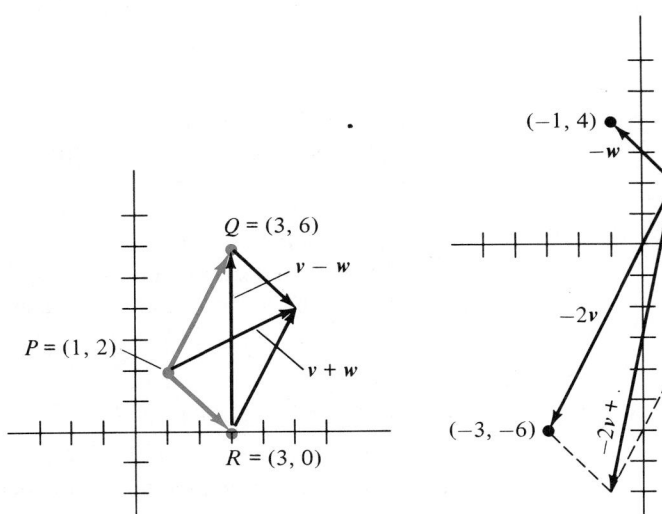

Figure 1.12   **Figure 1.13**

The next example shows how vector methods can be used in plane geometry.

Example 5

Use vectors to prove that the diagonals of a parallelogram bisect each other.

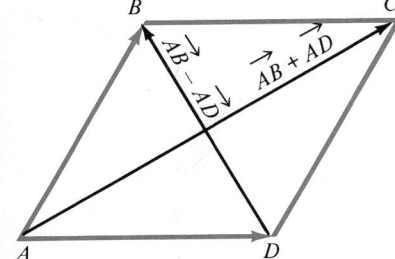

Figure 1.14 The diagonals of a parallelogram bisect each other.

Solution: Let the vertices of the parallelogram be A, B, C, and D, as in Figure 1.14.

Since $\overrightarrow{AC} = \overrightarrow{AB} + \overrightarrow{AD}$, the vector from A to the midpoint of diagonal $\overrightarrow{AC}$ is

$$v = \frac{1}{2}(\overrightarrow{AB} + \overrightarrow{AD}).$$

Since $\overrightarrow{DB} = \overrightarrow{AB} - \overrightarrow{AD}$, the vector from A to the midpoint of diagonal $\overrightarrow{DB}$ is

$$w = \overrightarrow{AD} + \frac{1}{2}(\overrightarrow{AB} - \overrightarrow{AD})$$

$$= \overrightarrow{AD} + \frac{1}{2}\overrightarrow{AB} - \frac{1}{2}\overrightarrow{AD}$$

$$= \frac{1}{2}(\overrightarrow{AB} + \overrightarrow{AD})$$

$$= v.$$

Since $v = w$, these midpoints are the same. ◇

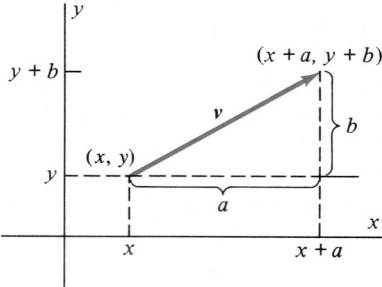

Figure 1.15 The vector $v = \langle a, b \rangle$.

The upshot of the preceding discussion is that a vector in the plane may be regarded as an ordered pair $\langle a, b \rangle$ of numbers, subject to the laws of addition and scalar multiplication given by Definition 1. Moreover, the vector $\langle a, b \rangle$ may be *represented* geometrically by an arrow originating at any point (x, y) and terminating at the point $(x + a, y + b)$ (Figure 1.15).

Properties of Vectors

Vectors in the plane satisfy many of the same properties as do real numbers. For example, vector addition is **commutative**. This means that $v + w = w + v$ for any two vectors v and w. However, not all properties of real numbers are shared by vectors. For example, we shall not encounter a useful way to define the *product* of two vectors as another vector in the plane.

The following theorem specifies the legitimate properties of vector addition and scalar multiplication for vectors in the plane.

THEOREM 1

Let u, v, and w be vectors in the plane and let a and b be real numbers. Then

(i) $v + w = w + v$ (commutativity),
(ii) $(u + v) + w = u + (v + w)$ (associativity),
(iii) $0 + v = v$ (identity for vector addition),
(iv) $1 \cdot v = v$ (identity for scalar multiplication),
(v) $a(v + w) = av + aw$ (scalar multiplication distributes over vector addition).
(vi) $(a + b)v = av + bv$.

To prove part (i), we write the vectors v and w in component form

$$v = \langle x_1, y_1 \rangle, \qquad w = \langle x_2, y_2 \rangle$$

and apply Definition 1:

$$v + w = \langle x_1, y_1 \rangle + \langle x_2, y_2 \rangle = \langle x_1 + x_2, y_1 + y_2 \rangle, \tag{2}$$

$$w + v = \langle x_2, y_2 \rangle + \langle x_1, y_1 \rangle = \langle x_2 + x_1, y_2 + y_1 \rangle. \tag{3}$$

Since addition for real numbers is commutative, $x_1 + x_2 = x_2 + x_1$, and $y_1 + y_2 = y_2 + y_1$. Thus, the right-hand sides of equations (2) and (3) represent the same components, so $v + w = w + v$.

You are asked to prove statements (ii) through (vi) in the exercise set. Each statement is proved by writing the vector(s) in component form and using the corresponding properties of real numbers.

Length of Vectors

When represented by an arrow originating at $(0, 0)$, the vector $v = \langle a, b \rangle$ terminates at the point (a, b). The obvious way to define the length of v is to use the distance formula in the plane (see Figure 1.16).

DEFINITION 2

The **length of the vector** $v = \langle a, b \rangle$ is $|v| = \sqrt{a^2 + b^2}$.

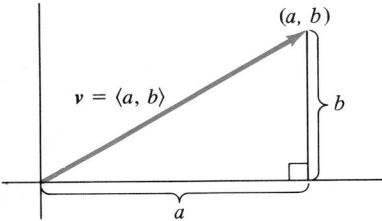

Figure 1.16 $|v| = \sqrt{a^2 + b^2}$.

Example 6

For $v = \langle 2, 1 \rangle$ and $w = \langle -1, 4 \rangle$,

(a) $|v| = \sqrt{2^2 + 1^2} = \sqrt{5}$,

(b) $|w| = \sqrt{(-1)^2 + 4^2} = \sqrt{17}$,

(c) $|2v - w| = \sqrt{(2 \cdot 2 - (-1))^2 + (2 \cdot 1 - 4)^2} = \sqrt{5^2 + 2^2} = \sqrt{29}$.   ◇

The following theorem summarizes the properties of length for vectors.

THEOREM 2

Let v and w be vectors and let c be a real number. Then

(i) $|v| \geq 0$; $|v| = 0$ if and only if $v = \mathbf{0}$,

(ii) $|cv| = |c| \cdot |v|$,

(iii) $|v + w| \leq |v| + |w|$.

Properties (i) and (ii) are obvious for our geometric concept of vectors. To prove them according to Definition 1, we write v in the component form $v = \langle a, b \rangle$. Then

$$|v| = \sqrt{a^2 + b^2} \geq 0 \qquad \text{for all} \qquad a, b,$$

which proves the first part of (i). The second part of (i) is proved by noting that

$$|v| = \sqrt{a^2 + b^2} = 0 \qquad \text{if and only if} \qquad a = b = 0,$$

in which case $v = \langle a, b \rangle = \langle 0, 0 \rangle = \mathbf{0}$. Statement (ii) is just as easy:

$$|cv| = |\langle ca, cb \rangle| = \sqrt{(ca)^2 + (cb)^2} = |c|\sqrt{a^2 + b^2} = |c| \cdot |v|.$$

Statement (iii) is referred to as the **triangle inequality.** It may be interpreted geometrically as saying that the length of one leg of a triangle cannot exceed the sum of the lengths of the other two legs (Figure 1.17). A proof of statement (iii) is given in Section 16.2.

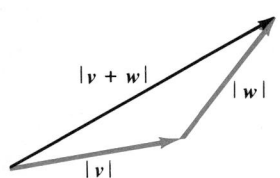

Figure 1.17 $|v + w| \leq |v| + |w|$.

Unit Vectors

A vector u with $|u| = 1$ is called a **unit vector.** In what follows we shall frequently need to find a unit vector u in the same direction as a given vector v. We may do so

using property (ii) of Theorem 2. Since

$$\left|\left(\frac{1}{|v|}\right)v\right| = \frac{1}{|v|} \cdot |v| = 1,$$

the vector

$$u = \frac{1}{|v|}v \tag{4}$$

is a unit vector pointing in the same direction as the vector v.

Example 7

Find a unit vector in the same direction as the vector $v = \langle 6, -4 \rangle$.

Solution:

$$|v| = \sqrt{6^2 + (-4)^2} = \sqrt{52} = 2\sqrt{13}.$$

Thus, by equation (4), the desired unit vector is

$$u = \frac{1}{2\sqrt{13}}\langle 6, -4 \rangle = \left\langle \frac{3}{\sqrt{13}}, \frac{-2}{\sqrt{13}} \right\rangle.$$

(Use Definition 2 to verify that $|u| = 1$.) ◇

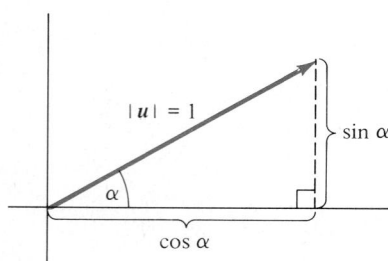

Figure 1.18 $u = \langle \cos \alpha, \sin \alpha \rangle$ is a unit vector.

A similar problem concerning unit vectors is that of finding a unit vector making an angle α with a given line or vector. For example, Figure 1.18 illustrates that *the unit vector forming an angle α with the positive x-axis is*

$$u = \langle \cos \alpha, \sin \alpha \rangle. \tag{5}$$

Equation (5) follows from the definition of the trigonometric functions $\sin \theta$ and $\cos \theta$.

Example 8

Find

(a) a unit vector u making an angle of $60°$ with the positive x-axis.
(b) a vector of length 5 making an angle of $45°$ with the positive x-axis.

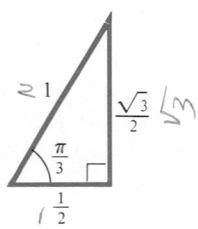

Strategy
(a) Use equation (5) and the facts

$$\cos 60° = \cos\left(\frac{\pi}{3}\right) = \frac{1}{2},$$

$$\sin 60° = \sin\left(\frac{\pi}{3}\right) = \frac{\sqrt{3}}{2}.$$

(b) First, use equation (5) to find a unit vector u in the given direction.

Solution
(a) Using equation (5) we obtain

$$u = \langle \cos 60°, \sin 60° \rangle$$

$$= \left\langle \frac{1}{2}, \frac{\sqrt{3}}{2} \right\rangle.$$

(b) By equation (5) a *unit* vector making an angle of $45°$ with the positive x-axis is

$$u = \langle \cos 45°, \sin 45° \rangle$$

$$= \left\langle \frac{\sqrt{2}}{2}, \frac{\sqrt{2}}{2} \right\rangle.$$

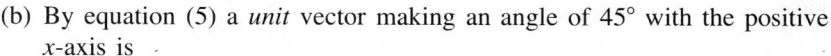

The solution is then

$$v = 5u.$$

The desired vector is therefore

$$v = 5u = \left\langle \frac{5\sqrt{2}}{2}, \frac{5\sqrt{2}}{2} \right\rangle.$$

◇

Unit Coordinate Vectors

Two special vectors enable us to develop yet another notation for representing vectors in the plane. They are the **unit coordinate vectors**

$$i = \langle 1, 0 \rangle \qquad \text{and} \qquad j = \langle 0, 1 \rangle.$$

(See Figure 1.19.) Using these two vectors, we may represent the vector $\langle a, b \rangle$ as

$$\langle a, b \rangle = a\langle 1, 0 \rangle + b\langle 0, 1 \rangle = ai + bj.$$

Geometrically, ai and bj represent the adjacent sides of the rectangle whose diagonal is $\langle a, b \rangle$ (Figure 1.20). When using this notation, we refer to the numbers a and b as the i and j components of the vector $ai + bj$, respectively.

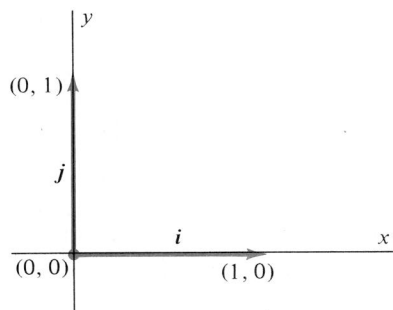

Figure 1.19 Unit coordinate vectors.

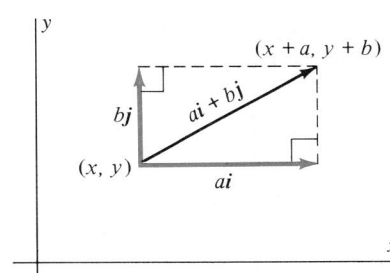

Figure 1.20 $\langle a, b \rangle = ai + bj$.

The principal advantage of the coordinate vector notation is that it helps avoid confusion between points and vectors. We shall make frequent use of this notation in what follows.

Example 9

For $v = \langle -2, 1 \rangle$ and $w = \langle 3, 4 \rangle$,

(a) $v = -2i + j;\qquad w = 3i + 4j,$
(b) $v + w = (-2i + j) + (3i + 4j) = (-2 + 3)i + (1 + 4)j = i + 5j,$
(c) $3v - 2w = (-6i + 3j) - (6i + 8j) = (-6 - 6)i + (3 - 8)j = -12i - 5j,$
(d) $|v| = |-2i + j| = \sqrt{(-2)^2 + 1^2} = \sqrt{5}.$

◇

Exercise Set 16.1

In Exercises 1–10, sketch the given vector originating at the given point P.

1. $\langle 1, 3 \rangle$ at point $P = (3, 2)$

2. $\langle -6, 2 \rangle$ at point $P = (-1, 1)$

3. $\langle -4, -2 \rangle$ at point $P = (2, 6)$

4. $\langle -2, 4 \rangle$ at point $P = (1, 5)$

5. $\langle 3, -5 \rangle$ at point $P = (1, 1)$

6. $i + j$ at point $P = (3, 5)$

7. $-i + 2j$ at point $P = (6, 0)$

8. $5i + 2j$ at point $P = (-3, 4)$

9. $-3i + 2j$ at point $P = (3, -2)$

10. $4j$ at point $P = (0, 2)$

In Exercises 11–16, let $P = (2, 5)$, $Q = (-1, 2)$, and $R = (2, -6)$. Find

11. $v = \overrightarrow{PQ}$

12. $v = \overrightarrow{PQ} + \overrightarrow{QR}$

13. $v = 2\overrightarrow{PQ} - \overrightarrow{RQ}$

14. $v = \overrightarrow{PR} + 4\overrightarrow{RQ}$

15. $v = 2\overrightarrow{RQ} - 2\overrightarrow{RP}$

16. $v = \overrightarrow{PQ} + 2\overrightarrow{QR} - 3\overrightarrow{RP}$

In Exercises 17–22, let $u = \langle 3, 1 \rangle$, $v = \langle -2, 4 \rangle$, and $w = \langle -4, -2 \rangle$. Find

17. $u + 2v$

18. $v - 2u$

19. $2u + 2v - 2w$

20. $-3u + 2v$

21. $v - u - w$

22. $7u + 3w - 6v$

In Exercises 23–30, let $u = 3i - j$, $v = 2i + 6j$, and $w = -i + j$. Find the indicated vector.

23. $u + 2v$

24. $u - 4w$

25. $u + v + w$

26. $u - v - w$

27. $3u + 4v - 2w$

28. $6u - 5w + v$

29. $3u + 3v + 3w$

30. $-u + v + 2w$

In Exercises 31–36, find the angle θ formed between the vector v and the positive x-axis.

31. j

32. $\langle \sqrt{3}, 1 \rangle$

33. $\langle -\sqrt{3}, 1 \rangle$

34. $3i - 3j$

35. $i - \sqrt{3}j$

36. $-i - j$

In Exercises 37–44, find the length of the given vector.

37. $\langle 6, -1 \rangle$

38. $\langle -3, 4 \rangle$

39. $\langle a, a^2 \rangle$

40. i

41. $i + j$

42. $3i + 4j$

43. $6i - 3j$

44. $4(i - 3j)$

45. Find a unit vector pointing in the direction opposite the positive x-axis.

46. Find a unit vector in the direction of $v = 3i + 4j$.

47. Find a unit vector making an angle of $120°$ with the positive x-axis.

48. Find a vector of length 3 in the direction opposite of $v = 2i - 3j$.

49. Find two unit vectors tangent to the graph of $y = x^3$ when positioned to originate at the point $(1, 1)$.

50. Show that the length of the vector originating at $P = (x, y)$ and terminating at $Q = (x + a, y + b)$ is $\sqrt{a^2 + b^2}$.

51. The points $A = (1, 1)$, $B = (5, 1)$, and $C = (6, 3)$ form 3 vertices of a parallelogram. Find the fourth vertex D if
   a. A and C lie on a diagonal,
   b. A and C lie on a common side.

52. Let $A = (0, 2)$, $B = (b, 5)$, $C = (5, 2)$, $D = (7, 5)$. Find b if $\overrightarrow{AB} + \overrightarrow{AC} = \overrightarrow{AD}$.

53. In the triangle with vertices A, B, and C, let D be the midpoint of side $\overline{AB}$ and let E be the midpoint of side $\overline{BC}$. Prove that $\overrightarrow{DE}$ is parallel to $\overrightarrow{AC}$ and half as long.

54. Prove that the midpoints of the sides of any quadrilateral form the vertices of a parallelogram.

55. Prove statements (ii) through (vi) of Theorem 1.

56. Let $v_1 = i + j$ and $v_2 = -i + j$. Show that for any vector w one can find constants c_1 and c_2 so that

$$w = c_1v_1 + c_2v_2.$$

(*Hint:* Express w in component form and obtain two linear equations for the unknowns c_1 and c_2.)

57. Generalize Exercise 56 by showing that if v_1 and v_2 are any nonzero and nonparallel vectors, then any vector w can be expressed as

$$w = c_1v_1 + c_2v_2$$

for appropriate c_1 and c_2.

58. Let P be a point in the plane and let a be the *position* vector $a = \overrightarrow{OP}$, where O is the origin. Let b be any nonzero vector in the plane. Show that the set of all terminal points of the vectors

$$r(t) = a + tb, \qquad -\infty < t < \infty$$

is a line, as follows.
   a. Show that the vector $r(t)$ originates at O for each number t. For each t, let $P(t)$ be the terminal point of the vector $r(t)$.
   b. For any $t_1 \neq t_2$, show that the vector $\overrightarrow{P(t_1)P(t_2)}$ is parallel to b. This shows that all points $P(t)$ lie on the same line ℓ.
   c. Show that for any point Q on ℓ there is a number t_0 so that $Q = P(t_0)$. This shows that every point on the line corresponds to a vector $r(t)$.

59. In Exercise 58, let $r(t) = x(t)i + y(t)j$, $a = a_1i + a_2j$, and $b = b_1i + b_2j$. Find parametric equations for the components $x(t)$ and $y(t)$. Use these equations to provide an alternate solution to Exercise 58.

60. Use Exercise 58 to show that if P and Q are distinct points in the plane and ℓ is the line through P and Q, the point R lies on ℓ if and only if

$$\overrightarrow{OR} = \overrightarrow{OP} + t\overrightarrow{PQ}$$

for some number t. (O denotes the origin.)

61. Under what geometric conditions does equality hold in the triangle inequality (statement (iii) of Theorem 2), that is, when does $|v + w| = |v| + |w|$?

16.2 THE DOT PRODUCT

Having defined addition and scalar multiplication for vectors in the plane, it is natural for us to ask whether one can find a useful way to define the product of two vectors. In this section we develop one such product, the *dot* product.

DEFINITION 3

The **dot product** of the vectors $v = \langle x_1, y_1 \rangle$ and $w = \langle x_2, y_2 \rangle$ is the number

$$v \cdot w = x_1 x_2 + y_1 y_2. \tag{1}$$

In other words, the dot product of two vectors is found by multiplying their corresponding components and adding the resulting products. It is important to note that the dot product $v \cdot w$ is a *number*, although each of the factors is a vector. For vectors written in unit coordinate notation, equation (1) becomes

$$(x_1 i + y_1 j) \cdot (x_2 i + y_2 j) = x_1 x_2 + y_1 y_2.$$

The dot product is also referred to as the **scalar product** or the **inner product.**

Example 1

For $v = \langle 2, 3 \rangle$ and $w = \langle -4, 5 \rangle$,

$$v \cdot w = 2(-4) + 3 \cdot 5 = 7. \qquad \diamondsuit$$

Example 2

For $v = i - 3j$ and $w = 3i + 3j$,

$$v \cdot w = 1 \cdot 3 + (-3)3 = -6. \qquad \diamondsuit$$

Example 3

$$i \cdot j = \langle 1, 0 \rangle \cdot \langle 0, 1 \rangle = 1 \cdot 0 + 0 \cdot 1 = 0. \qquad \diamondsuit$$

Example 4

$$(i + j) \cdot (i - j) = 1 \cdot 1 + 1(-1) = 0. \qquad \diamondsuit$$

We shall make frequent use of the properties of the dot product given by the following theorem.

THEOREM 3

Let u, v, and w be vectors in the plane and let c be a real number. Then

(i) $v \cdot v = |v|^2$,

(ii) $v \cdot w = w \cdot v$ ($\cdot$ is commutative),

(iii) $u \cdot (v + w) = u \cdot v + u \cdot w$ ($\cdot$ distributes over vector addition),

(iv) $(cv) \cdot w = c(v \cdot w)$ (scalars may be factored),

(v) $\mathbf{0} \cdot v = 0$,

(vi) $|v \cdot w| \le |v|\,|w|$ (Schwarz inequality).

Proof: The proof of the Schwarz inequality will be given later, using Theorem 4. The proofs of statements (i) through (v) follow directly from equation (1). For example, to prove statement (i) we write v in component form as $v = \langle x, y \rangle$. Then

$$v \cdot v = \langle x, y \rangle \cdot \langle x, y \rangle = x^2 + y^2 = (\sqrt{x^2 + y^2})^2 = |v|^2.$$

To prove (ii), we let $v = \langle x_1, y_1 \rangle$ and $w = \langle x_2, y_2 \rangle$. Then

$$v \cdot w = x_1 x_2 + y_1 y_2 = x_2 x_1 + y_2 y_1 = w \cdot v.$$

The proofs of statements (iii) through (v) are similar and are left as exercises. ◆

Angles Between Vectors

Examples 3 and 4 suggest that the dot product might have something to say about the angle between two vectors. In both cases the vectors are represented by perpendicular arrows, and in both cases the dot product is zero. In order to pursue this observation we need to define the concept of the angle between two vectors.

DEFINITION 4

Let the vectors $v = \langle x_1, y_1 \rangle$ and $w = \langle x_2, y_2 \rangle$ be represented by arrows in the plane, originating at the origin and terminating at the points (x_1, y_1) and (x_2, y_2), respectively. The **angle between the vectors** v and w is the smaller of the two angles formed between these arrows (Figure 2.1).

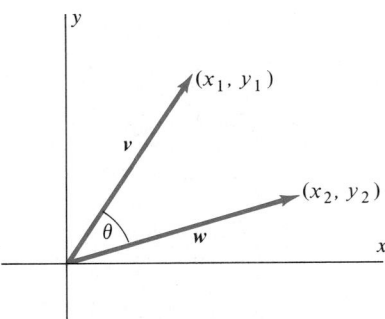

Figure 2.1 Angle between two vectors.

    In other words, the angle between two vectors is simply the angle determined by their representations as arrows.

    We say that two vectors are **perpendicular** if the angle between them is $\pi/2$. Perpendicular vectors are also called **orthogonal vectors.** We say that two vectors are **parallel** if the angle between them is 0 or π.

The relationship between the dot product $v \cdot w$ and the angle θ formed between v and w is given by the following theorem.

THEOREM 4

Let v and w be nonzero vectors in the plane, and let θ be the angle between v and w. Then

$$v \cdot w = |v|\,|w|\cos\theta. \tag{2}$$

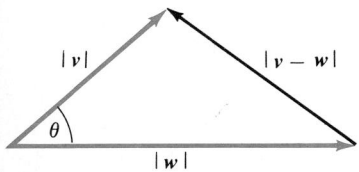

Figure 2.2 Law of Cosines: $|v - w|^2 = |v|^2 + |w|^2 - 2|v||w|\cos\theta.$

Proof: According to our geometric concept of vectors, the vectors v and w determine a triangle in the plane for which the side opposite angle θ has length $|v - w|$ (Figure 2.2). Thus, according to the law of cosines,

$$|v - w|^2 = |v|^2 + |w|^2 - 2|v|\,|w|\cos\theta. \tag{3}$$

Let's write v and w in component form as

$$v = \langle x_1, y_1 \rangle, \qquad w = \langle x_2, y_2 \rangle.$$

Then $v - w = \langle x_1 - x_2, y_1 - y_2 \rangle$, so equation (3) may be written in the form

$$(x_1 - x_2)^2 + (y_1 - y_2)^2 = (x_1^2 + y_1^2) + (x_2^2 + y_2^2) - 2|v|\,|w|\cos\theta$$

which simplifies to the equation

$$-2x_1x_2 - 2y_1y_2 = -2|v|\,|w|\cos\theta,$$

or

$$x_1x_2 + y_1y_2 = |v|\,|w|\cos\theta.$$

Since the left-hand side of this equation is $v \cdot w$, the proof is complete. ◆

When both $v \neq \mathbf{0}$ and $w \neq \mathbf{0}$, we may write equation (2) in the equivalent form

$$\cos\theta = \frac{v \cdot w}{|v|\,|w|}. \tag{4}$$

Equation (4) is useful in calculating angles between vectors, as the following example shows.

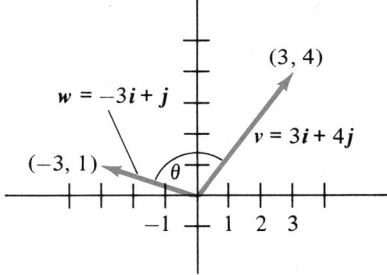

Figure 2.3

Example 5

Find the angle between the vectors $v = 3i + 4j$ and $w = -3i + j$ (Figure 2.3).

Solution: We use equation (4). Since

$$v \cdot w = 3(-3) + 4 \cdot 1 = -5,$$
$$|v| = \sqrt{3^2 + 4^2} = \sqrt{25} = 5,$$
$$|w| = \sqrt{3^2 + 1^2} = \sqrt{10},$$

we have

$$\cos\theta = \frac{-5}{5\sqrt{10}} \approx -0.3162.$$

Since we have implied in Definition 4 that $0 \le \theta \le \pi$, we have

$$\theta = \text{Cos}^{-1}\left(\frac{-5}{5\sqrt{10}}\right) \approx 0.602\pi \ (\approx 108.4°). \qquad \diamond$$

Theorem 4 provides a useful criterion for determining whether two nonzero vectors are orthogonal. If $|v| \ne 0$ and $|w| \ne 0$, equation (2) shows that $v \cdot w = 0$ if and only if $\cos \theta = 0$. Since θ is the angle between v and w, $0 \le \theta \le \pi$. Thus, $\cos \theta = 0$ if and only if $\theta = \pi/2$. Combining these observations gives the following result.

COROLLARY 1

The nonzero vectors v and w are orthogonal if and only if $v \cdot w = 0$.

Example 6

Show that the vectors $v = 4i - 6j$ and $w = 3i + 2j$ are orthogonal (Figure 2.4).

Solution: We use Corollary 1. Since

$$v \cdot w = 4 \cdot 3 + (-6) \cdot 2 = 0,$$

the vectors are orthogonal. $\qquad \diamond$

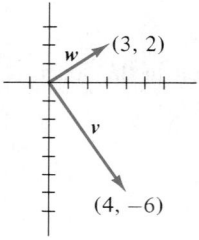

Figure 2.4

Example 7

Show that the vector $ai + bj$ is orthogonal to the line ℓ with equation

$$ax + by + c = 0.$$

(See Figure 2.5.)

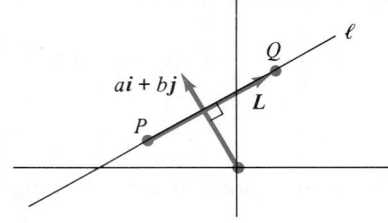

Figure 2.5

Solution: Our strategy is first to find a vector L parallel to the line ℓ and then to show that $(ai + bj) \cdot L = 0$. We begin by letting $P = (x_1, y_1)$ and $Q = (x_2, y_2)$ be distinct points on ℓ. Then $\overrightarrow{PQ}$ is a segment of ℓ, so $L = \overrightarrow{PQ}$ represents a vector parallel to ℓ, which has the component form

$$L = (x_2 - x_1)i + (y_2 - y_1)j.$$

Now, since P and Q lie on ℓ, both

$$ax_1 + by_1 + c = 0 \quad \text{and} \quad ax_2 + by_2 + c = 0.$$

Subtracting corresponding sides of these equations gives the equation

$$a(x_2 - x_1) + b(y_2 - y_1) = 0,$$

which we can rewrite as $(ai + bj) \cdot L = 0$. This shows that $ai + bj$ is orthogonal to L, and, hence, to ℓ. $\qquad \diamond$

We next use Theorem 4 to prove the Schwarz inequality of Theorem 3. Since this inequality is so important, we restate it here as a corollary of Theorem 4.

COROLLARY 2
Schwarz Inequality

For any vectors v and w,

$$|v \cdot w| \le |v| \, |w|.$$

Proof: If either $v = 0$ or $w = 0$, the inequality holds, since both sides are zero. Otherwise, an angle θ is determined between v and w. Since $|\cos \theta| \le 1$ for all θ, we may apply Theorem 4 to conclude that

$$|v \cdot w| = |v||w| \, |\cos \theta| \le |v| \, |w|. \qquad \blacklozenge$$

We can use the Schwarz inequality to prove the triangle inequality (Theorem 2, part (iii), Section 16.1):

$$|v + w| \le |v| + |w|. \tag{5}$$

The strategy will be to prove the inequality

$$|v + w|^2 \le (|v| + |w|)^2 \tag{6}$$

from which inequality (5) is obtained by taking square roots.

Applying Theorem 3, parts (i) through (iii), we find that

$$
\begin{aligned}
|v + w|^2 &= (v + w) \cdot (v + w) \\
&= v \cdot v + 2v \cdot w + w \cdot w \\
&= |v|^2 + 2v \cdot w + |w|^2.
\end{aligned}
\tag{7}
$$

Now the Schwarz inequality shows that

$$2v \cdot w \le 2|v \cdot w| \le 2|v| \, |w|. \tag{8}$$

Combining (7) and (8), we obtain the inequality

$$
\begin{aligned}
|v + w|^2 &\le |v|^2 + 2|v| \, |w| + |w|^2 \\
&= (|v| + |w|)^2,
\end{aligned}
$$

which proves inequality (6).

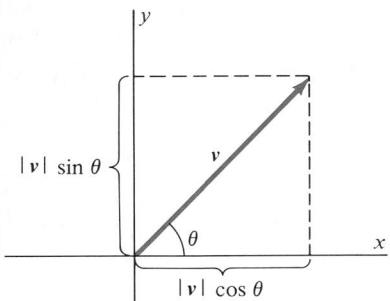

Figure 2.6 Component of v in x direction is $|v| \cos \theta = v \cdot i$.

Components

If v is a nonzero vector and θ is the angle formed between v and the unit coordinate vector i, then v can be written in the component form

$$v = |v|(\cos \theta)i + |v|(\sin \theta)\,j. \tag{9}$$

(See Figure 2.6.) Notice that we can write the component of v in the direction of the vector i as

$$|v| \cos \theta = |v| \, |i| \cos \theta = v \cdot i \qquad \text{(Theorem 4).} \tag{10}$$

Now suppose we are given a second vector w, with $w \ne 0$. Let θ be the angle formed between v and w. The number $|v| \cos \theta$ is called **the component of v with respect to w,** written $\operatorname{comp}_w v = |v| \cos \theta$ (Figure 2.7). The number $\operatorname{comp}_w v$ may be interpreted as the change, in the direction of w, represented by the vector v. Using the dot product we may write

$$|v| \cos \theta = \frac{|v| \, |w| \cos \theta}{|w|} = \frac{v \cdot w}{|w|}$$

so

$$\boxed{\operatorname{comp}_w v = \frac{v \cdot w}{|w|}.} \tag{11}$$

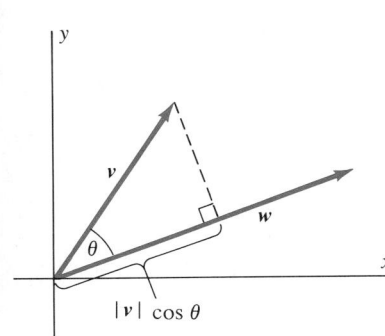

Figure 2.7 Component of v in the direction of w is $|v| \cos \theta = \dfrac{v \cdot w}{|w|}$.

Notice that $\operatorname{comp}_i v = \dfrac{v \cdot i}{|i|} = v \cdot i$, so $\operatorname{comp}_w v$ is a generalization of the concept of i-component in equation (10) (see Figure 2.7).

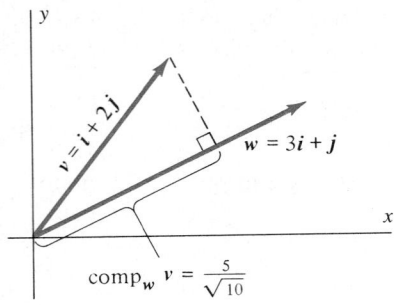

Figure 2.8

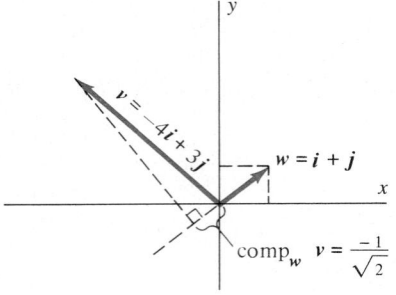

Figure 2.9

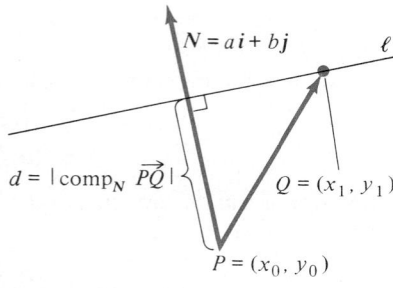

$d = |\text{comp}_N \, \overrightarrow{PQ}|$

Figure 2.10 Distance from (x_0, y_0) to ℓ: $ax + by + c = 0$ is $|\text{comp}_N \, \overrightarrow{PQ}|$.

Example 8

For $v = i + 2j$ and $w = 3i + j$,

$$\text{comp}_w \, v = \frac{1 \cdot 3 + 2 \cdot 1}{\sqrt{3^2 + 1^2}} = \frac{5}{\sqrt{10}} \qquad \text{(Figure 2.8).} \qquad \diamond$$

Example 9

For $v = -4i + 3j$ and $w = i + j$

$$\text{comp}_w \, v = \frac{(-4)(1) + 3 \cdot 1}{\sqrt{1^2 + 1^2}} = \frac{-1}{\sqrt{2}} \qquad \text{(Figure 2.9).} \qquad \diamond$$

The notion of component can be used to develop a formula for the distance from the point $P = (x_0, y_0)$ to the line ℓ with equation ℓ: $ax + by + c = 0$. The idea is summarized by Figure 2.10: If N is a vector orthogonal to ℓ, and if $Q = (x_1, y_1)$ is any point on ℓ, then $|\text{comp}_N \, \overrightarrow{PQ}|$ is the required distance. The details go like this:

If $Q = (x_1, y_1)$ is a point on ℓ, then

$$\overrightarrow{PQ} = (x_1 - x_0)i + (y_1 - y_0)j.$$

By Example 7, we know that a vector orthogonal to ℓ is $N = ai + bj$. The required distance is therefore

$$d = |\text{comp}_N \, \overrightarrow{PQ}| = \frac{|\overrightarrow{PQ} \cdot N|}{|N|}$$

$$= \frac{|a(x_1 - x_0) + b(y_1 - y_0)|}{\sqrt{a^2 + b^2}}$$

$$= \frac{|(ax_1 + by_1) - (ax_0 + by_0)|}{\sqrt{a^2 + b^2}}.$$

Since $Q = (x_1, y_1)$ is on ℓ, we have $ax_1 + by_1 + c = 0$, so $ax_1 + by_1 = -c$. We therefore find that

$$\boxed{d = \frac{|ax_0 + by_0 + c|}{\sqrt{a^2 + b^2}}} \qquad (12)$$

is the distance from the point $P = (x_0, y_0)$ to the line ℓ with equation $ax + by + c = 0$.

Example 10

The distance from the point $(4, -2)$ to the line with equation $3x - y + 4 = 0$ is, by equation (12),

$$d = \frac{|3(4) + (-1)(-2) + 4|}{\sqrt{3^2 + 1^2}} = \frac{18}{\sqrt{10}}. \qquad \diamond$$

Projections

Let v and w be nonzero vectors in the plane. The vector

$$\text{proj}_w \, v = \left(\frac{v \cdot w}{|w|^2}\right)w \qquad (13)$$

is called the **orthogonal projection** of v onto w. Since we can write

$$\text{proj}_w \ v = \left(\frac{v \cdot w}{|w|^2}\right)w = \left(\frac{v \cdot w}{|w|}\right)\frac{w}{|w|}, \tag{14}$$

we see that *proj$_w$ v is the product of the number comp$_w$ v and the unit vector* $\dfrac{w}{|w|}$ *in the direction of* w (see Figure 2.11).

Since $\dfrac{w}{|w|}$ is a unit vector, equation (14) shows that

$$|\text{proj}_w \ v| = \left|\frac{w \cdot v}{|w|}\right| = |\text{comp}_w \ v|. \tag{15}$$

That is, the length of the *vector* $\text{proj}_w \ v$ equals the absolute value of the *number* $\text{comp}_w \ v$.

The vector

$$\text{proj}_{\perp w} \ v = v - \text{proj}_w \ v \tag{16}$$

is called the **orthogonal projection of v perpendicular to w.**

It is obvious from equation (16) that

$$v = \text{proj}_w \ v + \text{proj}_{\perp w} \ v \qquad (\text{Figure } 2.12). \tag{17}$$

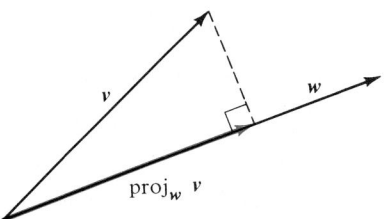

Figure 2.11 Projection of v onto w.

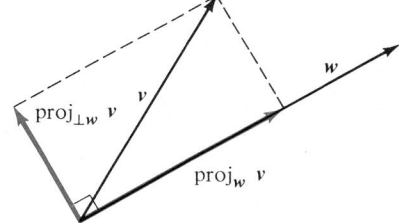

Figure 2.12 $v = \text{proj}_w \ v + \text{proj}_{\perp w} \ v$.

In other words, $\text{proj}_w \ v$ and $\text{proj}_{\perp w} \ v$ may be interpreted as adjacent sides of a parallelogram for which v is the diagonal. Indeed, we can say more: the parallelogram is actually a rectangle. That is,

$$(\text{proj}_w \ v) \cdot (\text{proj}_{\perp w} \ v) = 0. \tag{18}$$

The proof of (18) is simply a calculation:

$$(\text{proj}_w \ v) \cdot (\text{proj}_{\perp w} \ v) = \left[\left(\frac{v \cdot w}{|w|^2}\right)w\right] \cdot \left[v - \left(\frac{v \cdot w}{|w|^2}\right)w\right]$$

$$= \left(\frac{v \cdot w}{|w|^2}\right)w \cdot v - \left(\frac{v \cdot w}{|w|^2}\right)^2 w \cdot w$$

$$= \left(\frac{v \cdot w}{|w|}\right)^2 - \left(\frac{v \cdot w}{|w|^2}\right)^2 |w|^2$$

$$= 0.$$

Equation (18) explains the use of the terminology ''orthogonal'' projection in defining $\text{proj}_w \ v$. The projections $\text{proj}_w \ v$ and $\text{proj}_{\perp w} \ v$ allow us to express a given vector v as the sum of a vector parallel to a given vector or line, and a vector orthogonal to

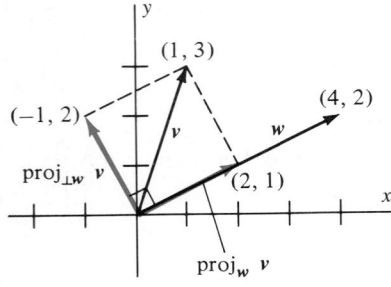

Figure 2.13 Projections of v onto and orthogonal to w.

the given vector or line. This is a particularly useful concept in physics, where one wishes to "resolve" a force or velocity vector into **component vectors** parallel and perpendicular to a given force or velocity vector.

Example 11

For $v = i + 3j$ and $w = 4i + 2j$,

(a) $\text{proj}_w v = \left[\dfrac{(i + 3j) \cdot (4i + 2j)}{|4i + 2j|^2}\right](4i + 2j)$

$= \left[\dfrac{1 \cdot 4 + 3 \cdot 2}{4^2 + 2^2}\right](4i + 2j)$

$= 2i + j,$

and

$\text{proj}_{\perp w} v = (i + 3j) - (2i + j) = -i + 2j$ (Figure 2.13).

(b) $\text{proj}_v w = \left[\dfrac{(4i + 2j) \cdot (i + 3j)}{|i + 3j|^2}\right](i + 3j)$

$= \left[\dfrac{4 \cdot 1 + 2 \cdot 3}{1^2 + 3^2}\right](i + 3j)$

$= i + 3j$ $(= v),$

and

$\text{proj}_{\perp v} w = (4i + 2j) - (i + 3j) = 3i - j$ (Figure 2.14). ◇

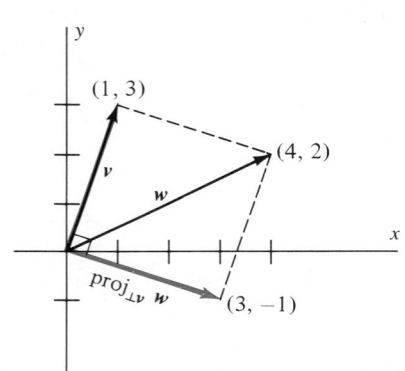

Figure 2.14 $v = \text{proj}_v w$.

Example 12

Find the force required to hold a 5000-kilogram automobile motionless on a 30° incline, assuming that the only force that must be overcome is that due to gravity.

Solution: We may represent the force due to gravity as a vector F pointing vertically downward. According to Newton's second law, the magnitude of this force vector is $|F| = mg = 5000g$, where $g = 9.8$ m/s² is the acceleration due to gravity. Thus

$$F = (-5000g)j.$$

By equation (5), Section 16.1, a unit vector u in the direction of the 30° incline is represented in Figure 2.15 and by the equation

$$u = \cos 30°i + \sin 30°j = \frac{\sqrt{3}}{2}i + \frac{1}{2}j.$$

The force due to F acting in the direction of the incline is therefore represented by the vector

$$F_1 = \text{proj}_u F = \frac{F \cdot u}{|u|^2}u = \frac{(-5000g)j \cdot \left(\dfrac{\sqrt{3}}{2}i + \dfrac{1}{2}j\right)}{1^2}\left(\frac{\sqrt{3}}{2}i + \frac{1}{2}j\right)$$

$$= -2500g\left(\frac{\sqrt{3}}{2}i + \frac{1}{2}j\right).$$

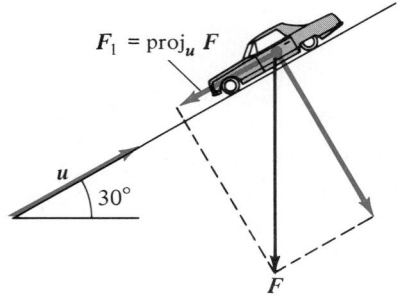

Figure 2.15 $|F_1| = |\text{proj}_u F|$ is the force required to prevent auto from rolling down the incline.

The magnitude of this force is

$$|F_1| = |\text{comp}_u \, F| = 2500g.$$

This is the force due to gravity which must be overcome to hold the automobile in place. ◇

Example 13

When a constant force of magnitude F moves the point at which it is applied a distance d *in the direction in which the force is applied,* the work done by the force is defined to be the product

$$W = Fd. \tag{19}$$

However, if the point of application is confined to move along a line in a direction different from that of the applied force, equation (19) does not apply. If the vector F represents the magnitude and direction of the applied force, and the vector d represents the magnitude and direction of the resulting displacement, the work done is defined to be the dot product

$$W = F \cdot d. \tag{20}$$

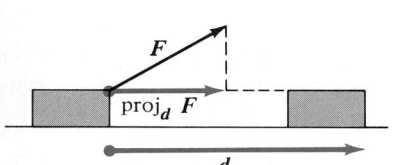

Figure 2.16 Work done by force F over the displacement d is $|\text{proj}_d \, F||d|$ $F \cdot d$.

Figure 2.16 illustrates this definition. The vector component of F acting in the direction of d is $\text{proj}_d \, F$, whose magnitude is $\text{comp}_d \, F = \dfrac{F \cdot d}{|d|}$. Since this is the force acting in the direction of motion, equation (19) gives

$$W = (\text{comp}_d \, F)(|d|) = \frac{F \cdot d}{|d|}(|d|) = F \cdot d. \qquad ◇$$

Exercise Set 16.2

In Exercises 1–6, find the dot product $v \cdot w$.

1. $v = i + 2j, \qquad w = 3i - j$

2. $v = \langle -3, 5 \rangle, \qquad w = \langle 1, 4 \rangle$

3. $v = \langle -3, 5 \rangle, \qquad w = \langle 6, -2 \rangle$

4. $v = ai + bj, \qquad w = ci + dj$

5. $v = i + 4j, \qquad w = i - 4j$

6. $v = 2j, \qquad w = i - 3j$

7. Find the cosine of the angle between the vectors $v = i - 3j$ and $w = -4i + j$.

8. Find the cosine of the angle between the vectors $v = \langle 3, 2 \rangle$ and $w = \langle -2, 2 \rangle$.

9. For the vectors $u = -2i + 4j$, $v = 3i + 5j$, and $w = 6i - 4j$ find

  a. $u \cdot v$ **b.** $u \cdot w$

  c. $u \cdot (v + w)$ **d.** $u \cdot (v - w)$

  e. $(u + v) \cdot (v - w)$ **f.** $u \cdot (2v + 3w)$

  g. $\text{comp}_v \, u$ **h.** $\text{comp}_w \, v$

  i. $\text{proj}_u \, v$ **j.** $\text{proj}_{\perp w} \, u$

10. Let $v = ai + j$ and $w = 3i + 4j$. Find a so that

  a. $v \cdot w = 0$ **b.** $v \cdot w = 10$

  c. $\text{proj}_w \, v = v$ **d.** $\text{proj}_w \, v = 0$

11. For $v = 2i - 3j$ and $w = 5i + j$, find

  a. $\text{comp}_w \, v$ **b.** $\text{comp}_i \, v$

  c. $\text{comp}_v \, w$ **d.** $\text{proj}_w \, v$

  e. $\text{proj}_{\perp w} \, v$ **f.** $\text{proj}_{\perp v} \, w$

12. Find the interior angles of the triangle with vertices $(0, 0)$, $(3, 0)$, and $(3, 4)$.

13. Determine which of the following pairs of vectors are orthogonal

  a. $v = 2i + 4j$ **b.** $v = 3i - 2j$
      $w = 5i - j$ $w = 10i + 15j$

  c. $v = i - 3j$ **d.** $v = i + j$
      $w = 6i + 2j$ $w = -i - j$

14. Find the number a so that the vectors $v = 3i - 5j$ and $w = ai + 3j$ are orthogonal.

15. Find two unit vectors orthogonal to $v = 2i + j$.

16. Find an example of three nonzero vectors u, v, and w for which $u \cdot v = u \cdot w$, but $v \neq w$. This shows that a "cancellation law" for dot products cannot hold.

17. Show that $v + w$ and $v - w$ are perpendicular if $|v| = |w| \neq 0$. Conclude that the diagonals of a rhombus are perpendicular. (A rhombus is a parallelogram with all sides of equal length.)

18. Express the vector $v = 3i + 4j$ as the sum of a vector parallel to $w = 3i + j$ and a vector orthogonal to w.

19. Express the unit coordinate vector i as the sum of a vector parallel to the vector $v = 2i - j$ and a vector perpendicular to v.

20. Find the distance from the point $(-4, 3)$ to the line with equation $2x - y - 3 = 0$.

21. Find the distance from the point $(3, 6)$ to the line with equation $x + y - 1 = 0$.

22. Show that the points $(1, 2)$, $(3, 4)$, and $(5, 2)$ are vertices of a right triangle. Which vertex corresponds to the right angle?

23. A child pulls a sled by exerting a 20-pound force on a rope which makes an angle of 45° with the horizontal. Find the work done by the child in pulling the sled a distance of 100 feet.

24. Prove statements (iii) through (v) of Theorem 3.

25. Assume $v \neq 0$ and $w \neq 0$. Under what conditions is $v \cdot w = |v| \cdot |w|$?

26. Find, to the nearest degree, the angle between $v = i + 4j$ and $w = 3i + 5j$.

27. Use the dot product to show that if $P = (a_1, b_1)$ and $Q = (a_2, b_2)$ are endpoints of the diameter of a circle and if $X = (x, y)$ is a point on the circle, then $\overrightarrow{XP}$ and $\overrightarrow{XQ}$ are orthogonal.

28. Use the result of Exercise 27 and the dot product to find an equation for the circle having (a_1, b_1) and (a_2, b_2) as endpoints of a diameter.

29. Prove the converse of Exercise 17: In a parallelogram, if the diagonals are perpendicular, then the parallelogram is a rhombus.

16.3 SPACE COORDINATES; VECTORS IN SPACE

Euclidean three-dimensional space provides a framework in which to discuss space curves, surfaces, and graphs of functions of two variables. By Euclidean space, we mean the coordinatizing of space by three mutually perpendicular coordinate axes, labelled as in Figure 3.1.

Scales for each of the axes are chosen so that the point of intersection of the three axes corresponds to zero on each axis. The usual convention in sketching the coordinate axes is to display the positive half of the x-axis as pointing toward the reader, the positive half of the y-axis as pointing to the right, and the positive half of the z-axis as pointing upward.

Using the scales on the three coordinate axes, we may identify each point P in space with a unique triple of numbers $P = (x_0, y_0, z_0)$ as follows: the plane through P perpendicular to the x-axis intersects the x-axis at x_0; the plane through P perpendicular to the y-axis intersects the y-axis at y_0; the plane through P perpendicular to the z-axis intersects the z-axis at z_0 (see Figure 3.2). Examples of several points in space are shown in Figure 3.3.

Figure 3.4 illustrates that the set of all points obtained by specifying two of the three coordinates (and allowing the third to range unrestricted) is a line parallel to the axis of the unrestricted variable. If one specifies only one variable, allowing all values of the other two, the figure obtained is a plane perpendicular to the axis of the restricted variable (Figure 3.5). Equations for lines and planes not parallel or perpendicular to coordinate axes are more complicated and are the subject of Section 16.5.

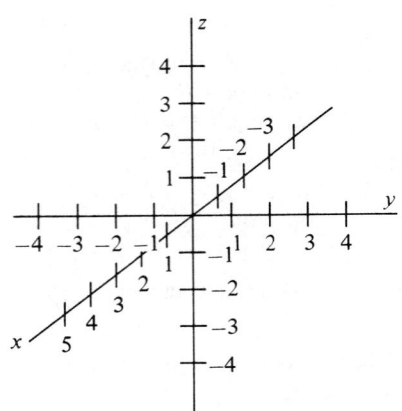

Figure 3.1 Coordinate axes for Euclidean space.

The Distance Formula

Let $P_1 = (x_1, y_1, z_1)$ and $P_2 = (x_2, y_2, z_2)$ be two points in space. To find the distance $d(P_1, P_2)$ between these points, we use the Pythagorean theorem twice.

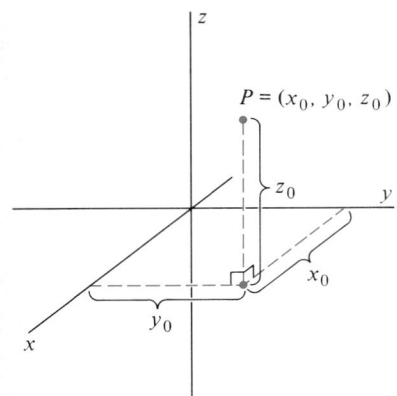

Figure 3.2 The coordinates of $P = (x_0, y_0, z_0)$.

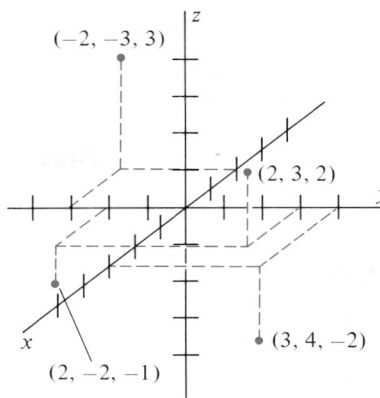

Figure 3.3 Points in space and their coordinates.

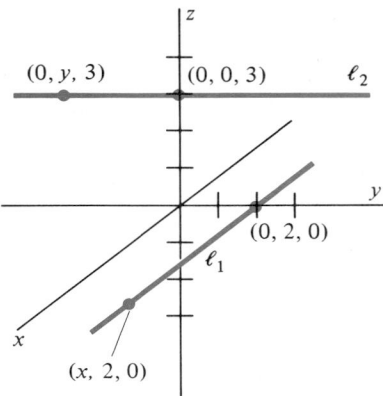

Figure 3.4 ℓ_1: The line $y = 2$, $z = 0$. ℓ_2: The line $x = 0$, $z = 3$.

Let $S = (x_1, y_2, z_1)$ and $R = (x_2, y_2, z_1)$, as in Figure 3.6. Then, since the segment $\overline{P_1 R}$ is the hypotenuse of the right triangle $\Delta P_1 RS$, the Pythagorean theorem gives

$$[d(P_1, R)]^2 = [d(S, R)]^2 + [d(P_1, S)]^2$$
$$= (x_2 - x_1)^2 + (y_2 - y_1)^2.$$

Similarly, the segment $\overrightarrow{P_1 P_2}$ is the hypotenuse of the right triangle $\Delta P_1 RP_2$. Thus, a second application of the Pythagorean theorem, together with the above equation, gives

$$[d(P_1, P_2)]^2 = [d(P_1, R)]^2 + [d(R, P_2)]^2$$
$$= (x_2 - x_1)^2 + (y_2 - y_1)^2 + (z_2 - z_1)^2.$$

Taking square roots of both sides of this equation gives the desired formula:

> The distance between $P_1 = (x_1, y_1, z_1)$ and $P_2 = (x_2, y_2, z_2)$ is
> $$d(P_1, P_2) = \sqrt{(x_2 - x_1)^2 + (y_2 - y_1)^2 + (z_2 - z_1)^2}.$$

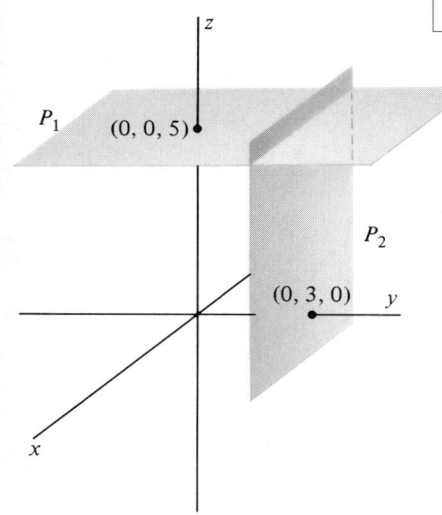

Figure 3.5 P_1: The plane $z = 5$. P_2: The plane $y = 3$.

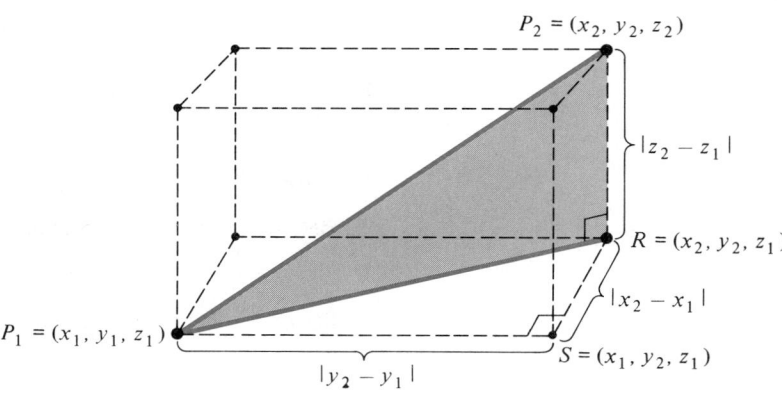

Figure 3.6 $d(P_1, P_2) = \sqrt{(x_2 - x_1)^2 + (y_2 - y_1)^2 + (z_2 - z_1)^2}.$

Example 1

The distance between $P = (-3, 0, 4)$ and $Q = (2, 5, -2)$ is

$$d = \sqrt{(2 - (-3))^2 + (5 - 0)^2 + (-2 - 4)^2}$$
$$= \sqrt{5^2 + 5^2 + (-6)^2} = \sqrt{86}.$$

$\diamondsuit$

Equations for Spheres

The sphere with center $C = (a, b, c)$ and radius $r > 0$ is the set of points $P = (x, y, z)$ for which

$$d(P, C) = r.$$

Using the distance formula, we may rewrite this equation as

$$\sqrt{(x - a)^2 + (y - b)^2 + (z - c)^2} = r,$$

or, in standard form,

$$\boxed{(x - a)^2 + (y - b)^2 + (z - c)^2 = r^2.}$$ (1)

Example 2

The sphere with center $C = (2, -3, 1)$ and radius $r = 3$ has equation

$$(x - 2)^2 + (y + 3)^2 + (z - 1)^2 = 9 \qquad \text{(Figure 3.7)}.$$

$\diamondsuit$

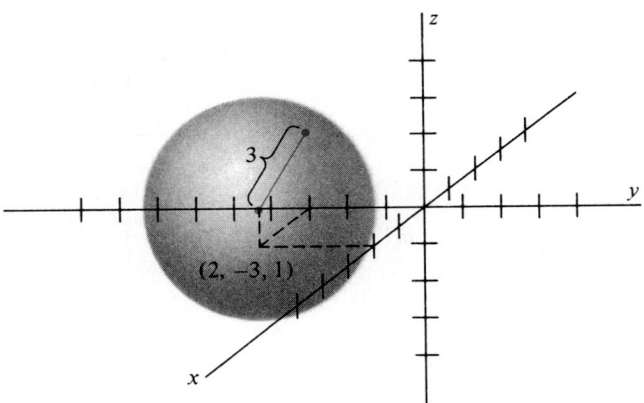

Figure 3.7 Sphere with center $(2, -3, 1)$ and radius 3.

Example 3

To determine whether the equation

$$x^2 + y^2 + z^2 - 4x + 6z - 3 = 0$$

is that of a sphere, we complete the square in each variable:

$$0 = x^2 + y^2 + z^2 - 4x + 6z - 3$$
$$= (x^2 - 4x) + y^2 + (z^2 + 6z) - 3$$
$$= (x^2 - 4x + 4) + y^2 + (z^2 + 6z + 9) - 3 - 4 - 9.$$

Thus,

$$(x - 2)^2 + y^2 + (z + 3)^2 = 16.$$

The graph is a sphere with center at $(2, 0, -3)$ and radius $r = 4$. ◇

Vectors in Space

To define vectors in space we must work with **ordered triples,** since points in space are determined by three coordinates.

DEFINITION 5

The set of all ordered triples $\langle x, y, z \rangle$ together with the laws

(i) $\langle x_1, y_1, z_1 \rangle + \langle x_2, y_2, z_2 \rangle = \langle x_1 + x_2, y_1 + y_2, z_1 + z_2 \rangle$

(ii) $c \langle x, y, z \rangle = \langle cx, cy, cz \rangle, \qquad c \in (-\infty, \infty)$

for addition and multiplication by real numbers is the **set of vectors in space.**

For the vector $v = \langle x_1, y_1, z_1 \rangle$, the numbers x_1, y_1, and z_1 are again referred to as the **components** of v. You have probably already noticed that Definition 5 prescribes that vector addition and scalar multiplication are to be carried out for vectors in space just as for vectors in the plane, except that now there is a third component to deal with. Accordingly, we define

(a) the **negative** of the vector $v = \langle x, y, z \rangle$ as

$$-v = \langle -x, -y, -z \rangle,$$

(b) the **difference** of two vectors $v = \langle x_1, y_1, z_1 \rangle$ and $w = \langle x_2, y_2, z_2 \rangle$ as

$$v - w = v + (-w) = \langle x_1 - x_2, y_1 - y_2, z_1 - z_2 \rangle,$$

(c) the **zero vector** as

$$0 = \langle 0, 0, 0 \rangle,$$

(d) **equality** of two vectors $v = \langle x_1, y_1, z_1 \rangle$ and $w = \langle x_2, y_2, z_2 \rangle$ by

$$v = w \qquad \text{if and only if} \qquad x_1 = x_2, \; y_1 = y_2 \text{ and } z_1 = z_2.$$

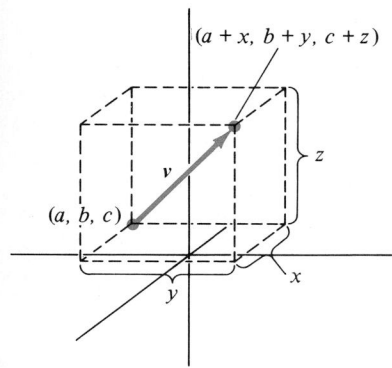

Figure 3.8 The vector $v = \langle x, y, z \rangle$.

Figure 3.9 The vector $v = \langle x, y, z \rangle$.

With these definitions, all properties of Theorem 1 for vectors in the plane carry over to vectors in space. The techniques of proof are the same as those used in the proof of Theorem 1, except that the third coordinate is included in all calculations (see Exercise 44).

The vector $v = \langle x_1, y_1, z_1 \rangle$ may be represented by an arrow originating at the origin and terminating at the point (x_1, y_1, z_1). More generally, if $P = (a, b, c)$ is any point in space, the vector $v = \langle x, y, z \rangle$ may be represented by an arrow originating at the point (a, b, c) and terminating at the point $(a + x, b + y, c + z)$ (see Figures 3.8 and 3.9).

As with vectors in the plane, we shall not concern ourselves with the distinction between vectors defined as ordered triples and their representations as arrows in space. For example, we say that the vector originating at the point $P = (x_1, y_1, z_1)$ and terminating at the point $Q = (x_2, y_2, z_2)$ may be written in component form as

$$\overrightarrow{PQ} = \langle x_2 - x_1, y_2 - y_1, z_2 - z_1 \rangle.$$

Finally, we define the **length of the vector** $v = \langle x, y, z \rangle$ by applying the distance formula in Figure 3.6:

$$
\text{For } v = \langle x, y, z \rangle,
$$
$$
|v| = \sqrt{x^2 + y^2 + z^2}.
$$

(2)

With this definition of length, all properties of Theorem 2 carry over to vectors in space. For example, to prove that

$$
|cv| = |c|\,|v|,
$$

(3)

we express v in component form as $v = \langle x, y, z \rangle$. Then

$$
\begin{aligned}
|cv| &= \sqrt{(cx)^2 + (cy)^2 + (cz)^2} \\
&= |c|\sqrt{x^2 + y^2 + z^2} \\
&= |c|\,|v|.
\end{aligned}
$$

You are asked to verify the remaining properties in Exercise 45.

Example 4

For the vectors $v = \langle 2, -1, 4 \rangle$ and $w = \langle 0, 3, -5 \rangle$,

(a) $v + w = \langle 2 + 0, -1 + 3, 4 + (-5) \rangle = \langle 2, 2, -1 \rangle$
(b) $3v = \langle 3 \cdot 2, 3(-1), 3 \cdot 4 \rangle = \langle 6, -3, 12 \rangle$
(c) $v - w = \langle 2 - 0, -1 - 3, 4 - (-5) \rangle = \langle 2, -4, 9 \rangle$
(d) $|2v + w| = |\langle 2 \cdot 2 + 0, 2(-1) + 3, 2 \cdot 4 + (-5) \rangle|$
    $= |\langle 4, 1, 3 \rangle| = \sqrt{4^2 + 1^2 + 3^2} = \sqrt{26}.$ ◇

Unit Coordinate Vectors

We define the three unit coordinate vectors in space by

$$
i = \langle 1, 0, 0 \rangle, \qquad j = \langle 0, 1, 0 \rangle, \qquad k = \langle 0, 0, 1 \rangle.
$$

(See Figure 3.10.) We may write the vector $v = \langle x, y, z \rangle$ in terms of i, j, and k as

$$
\begin{aligned}
v = \langle x, y, z \rangle &= x\langle 1, 0, 0 \rangle + y\langle 0, 1, 0 \rangle + z\langle 0, 0, 1 \rangle \\
&= xi + yj + zk \qquad \text{(Figure 3.11)}.
\end{aligned}
$$

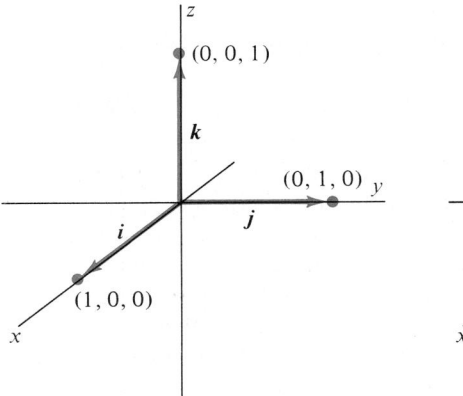

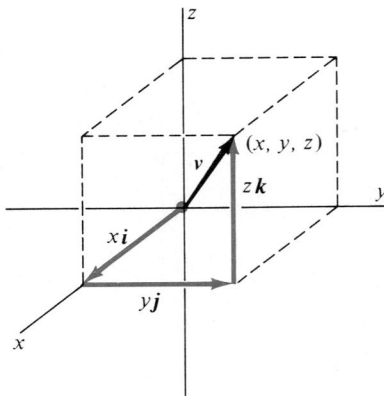

Figure 3.10 Unit coordinate vectors.   **Figure 3.11** $v = xi + yj + zk$.

The Dot Product

The idea of dot product carries over to vectors in space by the obvious generalization of Definition 3:

DEFINITION 6

The **dot product** of the vectors $v = \langle x_1, y_1, z_1 \rangle$ and $w = \langle x_2, y_2, z_2 \rangle$ is the number

$$v \cdot w = x_1 x_2 + y_1 y_2 + z_1 z_2.$$

All properties of Theorems 3 and 4 concerning the dot product remain true for vectors in space, including the Schwarz inequality

$$|v \cdot w| \le |v||w| \tag{4}$$

and the representation of the dot product in terms of the lengths of v and w and the angle θ between them:

$$v \cdot w = |v| \, |w| \cos \theta. \tag{5}$$

The proof of equation (5) is by direct analogy with that of Theorem 4, and the Schwarz inequality (4) follows from (5) as before since

$$|v \cdot w| = |v| \cdot |w||\cos \theta| \le |v| \, |w|.$$

(See Exercises 46 and 47.)

Example 5

For the vectors $v = 3i + 2j - k$ and $w = i + 4j + k$,

(a) $\begin{aligned}[t] v + 2w &= (3i + 2j - k) + 2(i + 4j + k) \\ &= (3 + 2 \cdot 1)i + (2 + 2 \cdot 4)j + (-1 + 2 \cdot 1)k \\ &= 5i + 10j + k. \end{aligned}$

(b) $v \cdot w = 3 \cdot 1 + 2 \cdot 4 + (-1)(1) = 3 + 8 - 1 = 10.$

(c) The cosine of the angle between v and w is

$$\cos \theta = \frac{v \cdot w}{|v| \cdot |w|} = \frac{10}{\sqrt{14} \cdot \sqrt{18}} = \frac{5}{3\sqrt{7}}.$$

(d) The **component** of v in the direction of w is the number

$$\text{comp}_w \, v = \frac{v \cdot w}{|w|} = \frac{10}{\sqrt{18}} = \frac{10}{3\sqrt{2}}.$$

(e) The **orthogonal projection** of v onto w is the vector

$$\begin{aligned} \text{proj}_w \, v = \frac{v \cdot w}{|w|^2} w &= \frac{10}{(\sqrt{18})^2}(i + 4j + k) \\ &= \frac{5}{9}i + \frac{20}{9}j + \frac{5}{9}k. \end{aligned}$$

(f) A **unit vector** in the direction of w is

$$\frac{1}{|w|}w = \frac{1}{\sqrt{18}}(i + 4j + k) = \frac{1}{\sqrt{18}}i + \frac{4}{\sqrt{18}}j + \frac{1}{\sqrt{18}}k. \qquad \diamondsuit$$

Direction Cosines

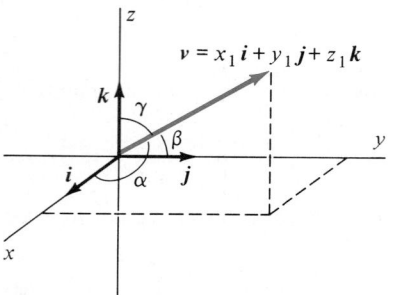

Figure 3.12 Direction angles for v.

Let $v = x_1 i + y_1 j + z_1 k$ be a nonzero vector, and let α, β, and γ denote the angles formed between v and the unit coordinate vectors i, j, and k, respectively (Figure 3.12).

We refer to α, β, and γ as the **direction angles** for v. Applying equation (5) we conclude that

$$\cos \alpha = \frac{v \cdot i}{|v| \, |i|} = \frac{v \cdot i}{|v|} = \frac{x_1}{|v|}; \qquad x_1 = |v| \cos \alpha \tag{6}$$

$$\cos \beta = \frac{v \cdot j}{|v| \, |j|} = \frac{v \cdot j}{|v|} = \frac{y_1}{|v|}; \qquad y_1 = |v| \cos \beta \tag{7}$$

and

$$\cos \gamma = \frac{v \cdot k}{|v| \, |k|} = \frac{v \cdot k}{|v|} = \frac{z_1}{|v|}; \qquad z_1 = |v| \cos \gamma. \tag{8}$$

Using (6) through (8), we may write the vector $v = x_1 i + y_1 j + z_1 k$ as

$$v = |v| \, [(\cos \alpha)i + (\cos \beta)j + (\cos \gamma)k]. \tag{9}$$

Equation (9) yields two important conclusions. First, calculating lengths on both sides of (9) shows that

$$|v| = |v| \, |\cos \alpha i + \cos \beta j + \cos \gamma k|$$
$$= |v| \sqrt{\cos^2 \alpha + \cos^2 \beta + \cos^2 \gamma}.$$

Cancelling factors of $|v| \neq 0$ and squaring both sides gives

$$\cos^2 \alpha + \cos^2 \beta + \cos^2 \gamma = 1. \tag{10}$$

Second, for a unit vector u, equation (9) becomes

$$u = (\cos \alpha)i + (\cos \beta)j + (\cos \gamma)k. \tag{11}$$

We refer to the cosines given in (6) through (8) as the **direction cosines** for the vector v. Note that these numbers are defined independent of the particular representation for v as an arrow in space. Equation (10) states that *the sum of the squares of the direction cosines is always one.* Equation (11) states that *the direction cosines of a unit vector are precisely its components.*

Example 6

Let $v = 2i + \sqrt{5}j + 4k$. The direction cosines for v are, by equations (6) through (8),

$$\cos \alpha = \frac{2}{|v|} = \frac{2}{\sqrt{4 + 5 + 16}} = \frac{2}{5};$$

$$\cos \beta = \frac{\sqrt{5}}{5}; \qquad \cos \gamma = \frac{4}{5}. \qquad\qquad \diamond$$

Example 7

A vector v makes angles of $45°$ with the vector i and $60°$ with the vector j.

(a) Find the angle between v and k.
(b) Find a unit vector in the direction of v.

Solution: We are given that

$$\cos \alpha = \cos 45° = \frac{\sqrt{2}}{2}; \qquad \cos \beta = \cos 60° = \frac{1}{2}.$$

We may therefore solve for $\cos \gamma$ using equation (10):

$$\cos^2 \gamma = 1 - (\cos^2 \alpha + \cos^2 \beta)$$

$$= 1 - \left(\frac{1}{2} + \frac{1}{4}\right)$$

$$= \frac{1}{4}.$$

Thus $\cos \gamma = \pm\frac{1}{2}$, so either $\gamma = 60°$ or $\gamma = 120°$.

If $\gamma = 60°$, a unit vector u in the direction of v is, by equation (11),

$$u = \frac{\sqrt{2}}{2}i + \frac{1}{2}j + \frac{1}{2}k.$$

If $\gamma = 120°$, then $\cos \gamma = -\frac{1}{2}$, so

$$u = \frac{\sqrt{2}}{2}i + \frac{1}{2}j - \frac{1}{2}k.$$

$\diamond$

Exercise Set 16.3

1. Plot the following points: $P_1 = (-2, 1, 5)$, $P_2 = (3, 0, 2)$, $P_3 = (1, 1, 5)$, $P_4 = (2, -3, 6)$, $P_5 = (-3, -4, -5)$.

2. Sketch the following planes in coordinate space.
 a. The xy-plane   **b.** The yz-plane
 c. The xz-plane   **d.** The plane $x = 2$
 e. The plane $y = 4$ **f.** The plane $z = -3$

3. Find the distance between the given pair of points.
 a. $P = (1, 0, 1)$, $Q = (3, 2, 1)$
 b. $P = (1, 2, -3)$, $Q = (-1, 4, 5)$
 c. $P = (-3, 6, 2)$, $Q = (0, 3, 0)$
 d. $P = (a, b, c)$, $Q = (2a, 2b, 2c)$

4. Find the midpoint of the line segment joining the points $P = (-4, 6, 1)$ and $Q = (-1, -3, -11)$.

5. Find the point one third of the distance from $P = (-4, 6, 1)$ to $Q = (-1, -3, -11)$.

6. Write an equation for the sphere with
 a. center $(0, 0, 0)$ and radius $r = 2$,
 b. center $(1, -1, 0)$ and radius $r = 1$,
 c. center $(-2, 3, -5)$ and radius $r = 3$,
 d. center $(4, 6, -2)$ and radius $r = 10$.

In Exercises 7–11, find the center and radius for the sphere with the given equation.

7. $x^2 + y^2 + z^2 - 4z = 5$

8. $x^2 + y^2 + z^2 - 2y - 4z = 4$

9. $x^2 + y^2 + z^2 - 4x + 2y - 6z = 2$

10. $x^2 + y^2 + z^2 - 4x + 4z = -4$

11. $x^2 + y^2 + z^2 - 6x + 2y + 4z = 11$

In Exercises 12–19, let $u = \langle 2, -1, 5 \rangle$, $v = \langle -3, 5, 0 \rangle$, and $w = \langle 3, 3, 1 \rangle$. Find the indicated vector or number.

12. $v + w$

13. $u + 6v$

14. $v - w$

15. $3u - 2v$

16. $\dfrac{v}{|v|}$

17. $|u + 4v|$

18. $v \cdot w$

19. $v \cdot (u + 2w)$

In Exercises 20–29, let $u = i + 2j - k$, $v = 3i - 2j + 2k$, and $w = 5i - j + 3k$. Find the indicated vector or number.

20. $u + 2v + w$

21. $v - 3w$

22. $|3v + w|$

23. $u \cdot (2v + 3w)$

24. $|v \cdot w + w \cdot u|$

25. $\dfrac{v}{|v|} + \dfrac{w}{|w|}$

26. $v \cdot w - |v|\,|w|$

27. $\operatorname{proj}_w v$

28. $\operatorname{comp}_u w$

29. $\operatorname{comp}_u v$

30. Find a unit vector in the direction of $w = i + 4j + 3k$.

31. Find the vector $v = \overrightarrow{PQ}$ for
  a. $P = (1, 0, 1)$, $Q = (3, 1, 3)$,
  b. $P = (-3, 2, 6)$, $Q = (1, 1, 1)$,
  c. $P = (1, -5, 3)$, $Q = (7, -2, 2)$.

32. Find the point D so that $\overrightarrow{AB} = \overrightarrow{CD}$ if $A = (3, 1, 6)$, $B = (-2, 1, 5)$, and $C = (6, -2, 2)$.

33. Find the direction cosines for the given vector.
  a. $v = 3i - j + 2k$
  b. $v = 6i - 2j + k$
  c. $\overrightarrow{PQ}$, where $P = (2, 1, 5)$ and $Q = (1, 3, 1)$

34. Which of the following sets of points are vertices of right triangles?
  a. $P = (1, 3, 2)$, $Q = (4, 1, 4)$, $R = (6, 5, 5)$
  b. $P = (0, 2, 5)$, $Q = (1, 3, 1)$, $R = (1, 4, 5)$
  c. $P = (-3, 1, 2)$, $Q = (1, -3, 2)$, $R = (2, -2, 2)$

35. We say that two vectors v and w in space are **parallel** if there is a constant c so that $v = cw$. The vectors are said to point in the same direction if $c > 0$, and they are said to point in opposite directions if $c < 0$.
  a. Find two vectors of length 2 parallel to the vector $v = 2i - 4j + 4k$.

b. Find the constant a so that the vectors $v = i - 3j + 4k$ and $w = ai + 9j - 12k$ are parallel.
  c. Find a vector of length 5 in the direction opposite that of $v = i - 2j + 3k$.
  d. Find a and b so that the vectors $3i - j + 4k$ and $ai + bj - 2k$ are parallel.

36. Describe the set of all points determined by the vector $r(t) = (i + 2j + k) + t(i + j + k)$ for $-\infty < t < \infty$, where $i + 2j + k$ originates at $(0, 0, 0)$.

37. Let $u = i - 2j + 3k$, $v = i + 2j + k$, and $w = i - 4j + k$. Find t so that the vector $u + tv$ is orthogonal to w.

38. Show that a nonzero vector is completely determined by its direction cosines and its length.

39. Which of the following triples can be direction angles of a single vector?
  a. $45°, 45°, 60°$
  b. $30°, 45°, 60°$
  c. $45°, 60°, 60°$

40. Use the dot product to prove the following statements.
  a. $|v + w| = |v - w|$ if and only if v and w are orthogonal.
  b. $|v + w|^2 = |v|^2 + |w|^2$ if and only if v and w are orthogonal.

41. Let v_1, v_2, and v_3 be mutually orthogonal vectors in space. Use the dot product to show that if c_1, c_2, and c_3 are scalars so that $c_1 v_1 + c_2 v_2 + c_3 v_3 = 0$, then $c_1 = c_2 = c_3 = 0$.

42. Give a geometric argument to show that three vectors v_1, v_2, and v_3 are coplanar if and only if there exist constants c_1 and c_2 so that $v_3 = c_1 v_1 + c_2 v_2$.

43. Let $v_1 = i + j$, $v_2 = j + k$, $v_3 = i + 2j + 3k$. Show that given any vector w there exist constants c_1, c_2, and c_3 so that $w = c_1 v_1 + c_2 v_2 + c_3 v_3$. (*Hint:* Express w in component form and obtain 3 linear equations for the constants c_1, c_2, and c_3.)

44. State and prove the analogue of Theorem 1 for vectors in space.

45. State and prove the analogue of Theorem 2 for vectors in space.

46. Prove the Schwarz inequality for vectors in space.

47. Prove equality (5) for vectors in space.

16.4 THE CROSS PRODUCT

In addition to the dot product, there is a second type of multiplication involving vectors. While the dot product of two vectors is a number, the *cross product* (or *vector* product) of two vectors is another vector. It is defined as follows.

DEFINITION 7

Let $v = x_1 i + y_1 j + z_1 k$ and $w = x_2 i + y_2 j + z_2 k$ be vectors in space. The **cross product**, $v \times w$, is the **vector**

$$v \times w = (y_1 z_2 - z_1 y_2)i + (z_1 x_2 - x_1 z_2)j + (x_1 y_2 - y_1 x_2)k. \tag{1}$$

There is a geometric interpretation of expression (1) similar to that developed for the dot product. Before taking up that issue, we look to several examples for insights.

Example 1

If $v = i$ and $w = j$, then $x_1 = y_2 = 1$ and $y_1 = z_1 = x_2 = z_2 = 0$ in Definition 7. Thus, by equation (1)

$$i \times j = (0 \cdot 0 - 0 \cdot 1)i + (0 \cdot 0 - 1 \cdot 0)j + (1 \cdot 1 - 0 \cdot 0)k.$$

That is,

$$i \times j = k \qquad \text{(Figure 4.1)}. \tag{2}$$

Similarly, you can verify that

$$j \times k = i \qquad \text{(Figure 4.2)} \tag{3}$$

and

$$k \times i = j \qquad \text{(Figure 4.3)}. \tag{4}$$

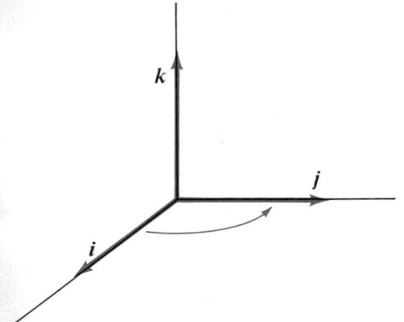

Figure 4.1 $i \times j = k$.

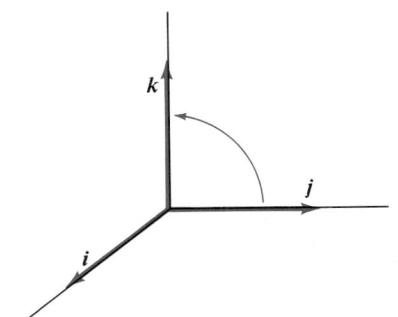

Figure 4.2 $j \times k = i$.

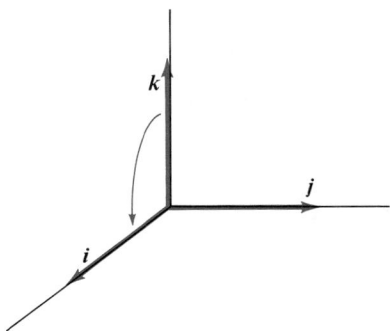

Figure 4.3 $k \times i = j$.

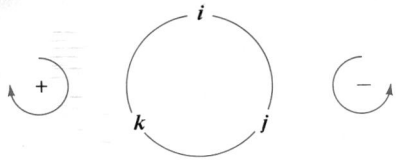

Figure 4.4 Device for remembering cross products of unit coordinate vectors.

The curved arrows in Figures 4.1 through 4.3 indicate the order of the vectors in the cross products.

However, you can also verify that the cross product is not commutative, since

$$j \times i = -k, \qquad k \times j = -i, \qquad \text{and} \qquad i \times k = -j. \tag{5}$$

Figure 4.4 presents a device for remembering these results. By beginning at the first factor and traversing the circle in the direction leading directly to the second factor, one determines the sign of the cross product: positive if the direction is clockwise (equations (2)–(4)); negative if the direction is counterclockwise (equations (5)). ◇

Example 2

For $v = 2i + j + 3k$ and $w = 4i - j + 2k$,

$$v \times w = [1 \cdot 2 - 3(-1)]i + [3 \cdot 4 - 2 \cdot 2]j + [2(-1) - 1 \cdot 4]k$$
$$= 5i + 8j - 6k.$$

$\diamondsuit$

Properties of the Cross Product

It is easy to see that in each of the cross products of Example 1 the vector $v \times w$ is orthogonal to both factors, v and w. The same is true for the vectors v and w of Example 2: $v \times w = 5i + 8j - 6k$ is orthogonal to $v = 2i + j + 3k$, since $(v \times w) \cdot v = 5 \cdot 2 + 8 \cdot 1 + (-6)(3) = 0$; $v \times w$ is orthogonal to w, since $(v \times w) \cdot w = 5 \cdot 4 + 8(-1) + 2(-6) = 0$.

These observations lead to the conjecture that $v \times w$ is *always* orthogonal both to v and to w. To prove this conjecture, we write v and w in the component forms $v = x_1 i + y_1 j + z_1 k$ and $w = x_2 i + y_2 j + z_2 k$. Then, using Definition 7, we find that

$$(v \times w) \cdot v = [(y_1 z_2 - z_1 y_2)i + (z_1 x_2 - x_1 z_2)j$$
$$+ (x_1 y_2 - y_1 x_2)k] \cdot [x_1 i + y_1 j + z_1 k]$$
$$= x_1(y_1 z_2 - z_1 y_2) + y_1(z_1 x_2 - x_1 z_2) + z_1(x_1 y_2 - y_1 x_2)$$
$$= 0.$$

Similarly, you can show that $(v \times w) \cdot w = 0$. Thus, it follows from Corollary 1 (Section 16.2) that $v \times w$ is orthogonal to both v and w.

This proves one part of the following theorem. The proofs of the other statements are carried out in the same way. Simply express all vectors in component form and apply Definition 7.

THEOREM 5
Properties of the Cross Product

Let u, v, and w be vectors and let c be a scalar. Then

(i) $v \times w = -w \times v$ (anticommutative)

(ii) $u \times (v + w) = u \times v + u \times w$ (multiplication distributes over addition)

(iii) $c(v \times w) = (cv) \times w = v \times (cw)$

(iv) $(v \times w) \perp v$; $(v \times w) \perp w$

(v) $v \times v = 0$ (self-annihilating)

The Determinant Notation

There is a useful formula for remembering equation (1) that involves the concept of **determinants**. The determinant of the 2×2 matrix

$$\begin{bmatrix} a_1 & b_1 \\ a_2 & b_2 \end{bmatrix}$$

is the number

$$\det \begin{bmatrix} a_1 & b_1 \\ a_2 & b_2 \end{bmatrix} = a_1 b_2 - b_1 a_2. \tag{6}$$

For example,

$$\det \begin{bmatrix} 3 & 2 \\ 4 & 1 \end{bmatrix} = 3 \cdot 1 - 2 \cdot 4 = 3 - 8 = -5.$$

The determinant of the 3×3 matrix

$$\begin{bmatrix} a_1 & b_1 & c_1 \\ a_2 & b_2 & c_2 \\ a_3 & b_3 & c_3 \end{bmatrix}$$

is the number

$$\det \begin{bmatrix} a_1 & b_1 & c_1 \\ a_2 & b_2 & c_2 \\ a_3 & b_3 & c_3 \end{bmatrix} = a_1 \cdot \det \begin{bmatrix} b_2 & c_2 \\ b_3 & c_3 \end{bmatrix} - b_1 \cdot \det \begin{bmatrix} a_2 & c_2 \\ a_3 & c_3 \end{bmatrix}$$

$$+ \, c_1 \cdot \det \begin{bmatrix} a_2 & b_2 \\ a_3 & b_3 \end{bmatrix} \quad (7)$$

$$= a_1(b_2c_3 - c_2b_3) - b_1(a_2c_3 - c_2a_3) + c_1(a_2b_3 - b_2a_3).$$

For example,

$$\det \begin{bmatrix} 1 & 2 & 3 \\ 4 & 5 & 6 \\ 7 & 8 & 9 \end{bmatrix} = 1 \cdot \det \begin{bmatrix} 5 & 6 \\ 8 & 9 \end{bmatrix} - 2 \cdot \det \begin{bmatrix} 4 & 6 \\ 7 & 9 \end{bmatrix} + 3 \cdot \det \begin{bmatrix} 4 & 5 \\ 7 & 8 \end{bmatrix} \quad (8)$$

$$= (5 \cdot 9 - 6 \cdot 8) - 2(4 \cdot 9 - 6 \cdot 7) + 3(4 \cdot 8 - 5 \cdot 7)$$

$$= 0.$$

The reason for bringing up the topic of determinants is that by comparing equations (1) and (7) you can see that the memory device

$$v \times w = \det \begin{bmatrix} i & j & k \\ x_1 & y_1 & z_1 \\ x_2 & y_2 & z_2 \end{bmatrix}$$

$$= (y_1z_2 - z_1y_2)i + (z_1x_2 - x_1z_2)j + (x_1y_2 - y_1x_2)k \quad (9)$$

provides a convenient way to remember equation (1). (Of course, equation (9) is not really a proper use of the determinant defined in equation (7), since the top row of the matrix consists of vectors rather than numbers.)

Determinants play an important role in the theory of systems of linear equations. Here we are making use of only a very limited aspect of determinants.

Example 3

For $v = 3i - j - 4k$ and $w = -2i + 2j + k$,

$$v \times w = \det \begin{bmatrix} i & j & k \\ 3 & -1 & -4 \\ -2 & 2 & 1 \end{bmatrix}$$

$$= ((-1)(1) - (-4)(2))i$$

$$\quad + ((-4)(-2) - (3 \cdot 1))j + (3 \cdot 2 - (-1)(-2))k$$

$$= 7i + 5j + 4k.$$

$\diamondsuit$

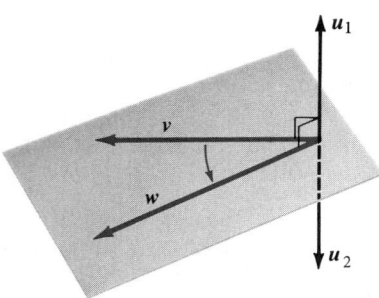

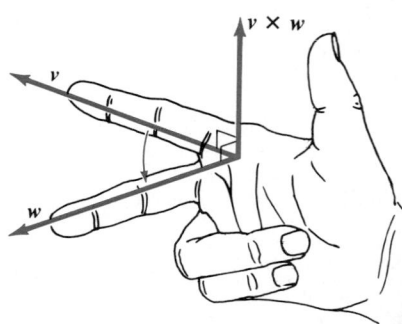

Figure 4.5 Both u_1 and u_2 are orthogonal to plane determined by v and w. Which one corresponds to $v \times w$?

Figure 4.6 Right-hand rule determines direction of $v \times w$.

The next item on our agenda is a theorem that allows us to interpret the cross product geometrically. Before stating this theorem we need to clarify an earlier point. We have already shown that $v \times w$ is perpendicular both to v and to w, so $v \times w$ is a vector perpendicular to the *plane* determined by v and w. This leaves two possible directions for $v \times w$, as Figure 4.5 illustrates.

However, from Figures 4.1 through 4.3 you can see that each of the cross products $i \times j, j \times k,$ and $k \times i$ satisfies the **right-hand rule:** *the thumb of a right hand points in the direction of $v \times w$ when the index finger points in the direction of v and the second finger points in the direction of w* (Figure 4.6).

In checking additional examples, you will observe that this rule holds in each case.

THEOREM 6

Let v and w be nonzero vectors forming an angle θ between them. Then

$$v \times w = (|v|\,|w|\,\sin\theta)N \tag{10}$$

where N is a unit vector orthogonal to both v and w, in the direction determined by the right-hand rule.

Proof: We have already noted the direction of $v \times w$. What remains to be shown is that

$$|v \times w| = |(|v|\,|w|\,\sin\theta)N| = |v|\,|w|\,|\sin\theta|. \tag{11}$$

To prove equation (11), we write v and w in component form as

$$v = x_1 i + y_1 j + z_1 k; \qquad w = x_2 i + y_2 j + z_2 k.$$

Then, using equation (1) and the definition of the dot product we find that

$$\begin{aligned}
|v \times w|^2 &= |(y_1 z_2 - z_1 y_2)i + (z_1 x_2 - x_1 z_2)j + (x_1 y_2 - y_1 x_2)k|^2 \\
&= (y_1 z_2 - z_1 y_2)^2 + (z_1 x_2 - x_1 z_2)^2 + (x_1 y_2 - y_1 x_2)^2 \\
&= (x_1^2 + y_1^2 + z_1^2)(x_2^2 + y_2^2 + z_2^2) - (x_1 x_2 + y_1 y_2 + z_1 z_2)^2 \\
&= |v|^2 |w|^2 - (v \cdot w)^2.
\end{aligned} \tag{12}$$

Now by Theorem 4,

$$(v \cdot w)^2 = |v|^2 |w|^2 \cos^2\theta. \tag{13}$$

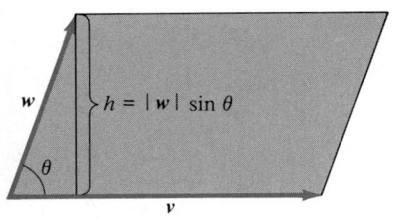

Figure 4.7 Area of parallelogram with base $|v|$ and adjacent side w is $|v \times w|$.

Combining equations (12) and (13) we have

$$
\begin{aligned}
|v \times w|^2 &= |v|^2|w|^2 - |v|^2|w|^2 \cos^2 \theta \\
&= |v|^2|w|^2 (1 - \cos^2 \theta) \\
&= |v|^2|w|^2 \sin^2 \theta.
\end{aligned}
$$

Taking square roots of both sides then gives equation (11). ◆

Figure 4.7 illustrates an immediate consequence of Theorem 6. If θ is the angle between v and w, the altitude of the parallelogram determined by the vectors v and w is $h = |w| \sin \theta$. Since the base has length $b = |v|$, the area A must be

$$
A = bh = |v|\,|w| \sin \theta = |v \times w|. \tag{14}
$$

Example 4

For $v = i + 2j - k$ and $w = -3i - j + 2k$,

$$
v \times w = \det \begin{bmatrix} i & j & k \\ 1 & 2 & -1 \\ -3 & -1 & 2 \end{bmatrix} = (4 - 1)i + (3 - 2)j + (-1 + 6)k
$$

$$
= 3i + j + 5k.
$$

The area of the parallelogram determined by v and w is, by equation (14),

$$
A = |3i + j + 5k| = \sqrt{9 + 1 + 25} = \sqrt{35} \qquad \text{(Figure 4.8)}. \qquad \diamond
$$

Example 5

Find the sine of the angle between the vectors v and w in Example 4.

Solution: Solving equation (14) for $\sin \theta$ gives

$$
\sin \theta = \frac{|v \times w|}{|v|\,|w|}.
$$

Since $|v \times w| = \sqrt{35}$, $|v| = \sqrt{1^2 + 2^2 + 1^2} = \sqrt{6}$, and $|w| = \sqrt{3^2 + 1^2 + 2^2} = \sqrt{14}$, we have

$$
\sin \theta = \frac{\sqrt{35}}{\sqrt{6} \cdot \sqrt{14}} = \frac{1}{2}\sqrt{\frac{5}{3}}. \qquad \diamond
$$

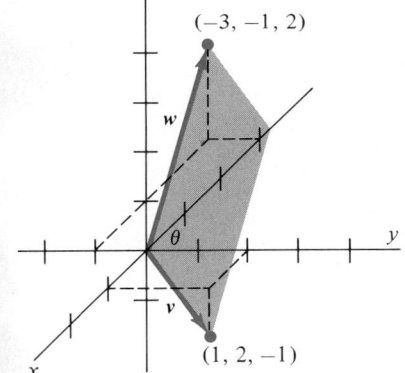

Figure 4.8

We can use equation (14) to learn something about 2×2 determinants as follows. Let $v = x_1 i + y_2 j$ and $w = x_2 i + y_2 j$. Then v and w are vectors in the plane $z = 0$, and the area of the parallelogram they determine is

$$
\begin{aligned}
A = |v \times w| &= \det \begin{bmatrix} i & j & k \\ x_1 & y_1 & 0 \\ x_2 & y_2 & 0 \end{bmatrix} \\
&= |(x_1 y_2 - y_1 x_2)k| \\
&= |x_1 y_2 - y_1 x_2| \\
&= \left| \det \begin{bmatrix} x_1 & y_1 \\ x_2 & y_2 \end{bmatrix} \right|.
\end{aligned}
$$

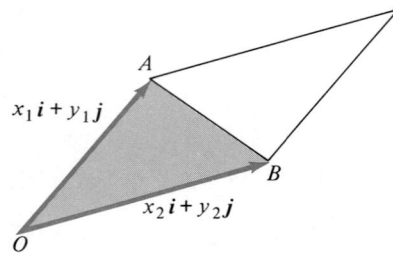

Figure 4.9 Area of $\triangle OAB$ is $\dfrac{1}{2}\left|\det\begin{bmatrix} x_1 & y_1 \\ x_2 & y_2 \end{bmatrix}\right|$

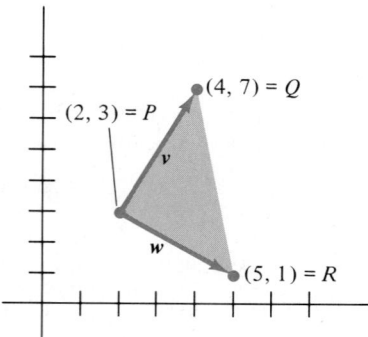

Figure 4.10

That is, the absolute value of the determinant of the matrix

$$\begin{bmatrix} x_1 & y_1 \\ x_2 & y_2 \end{bmatrix}$$

is the area of the parallelogram determined by the vectors $x_1 i + y_1 j$ and $x_2 i + y_2 j$. Since the diagonals of a parallelogram bisect the parallelogram into two congruent triangles, an even simpler interpretation of the 2×2 determinant is that *the absolute value of the determinant*

$$\det\begin{bmatrix} x_1 & y_1 \\ x_2 & y_2 \end{bmatrix}$$

is twice the area of the triangle determined by the vectors $x_1 i + y_1 j$ *and* $x_2 i + y_2 j$ (Figure 4.9).

Example 6

Find the area of the triangle with vertices $(2, 3)$, $(4, 7)$, and $(5, 1)$.

Solution: Let $P = (2, 3)$, $Q = (4, 7)$, and $R = (5, 1)$. Two of the sides may be interpreted as the vectors

$$v = \overrightarrow{PQ} = 2i + 4j$$

and

$$w = \overrightarrow{PR} = 3i - 2j \qquad \text{(Figure 4.10)}.$$

By the preceding observation, the area of the triangle is

$$A = \frac{1}{2}\left|\det\begin{bmatrix} 2 & 4 \\ 3 & -2 \end{bmatrix}\right| = \frac{1}{2}\cdot\left|2(-2) - 4(3)\right| = \frac{16}{2} = 8. \qquad \diamond$$

Clearly, the geometry associated with the dot and cross products is rich. We shall use these ideas in Section 16.5 to develop equations for lines and planes in space.

Exercise Set 16.4

In Exercises 1–6, let $u = i + j + k$, $v = 2i - j + 2k$, and $w = 3i + 4j - k$. Find the indicated vector or number.

1. $v \times w$

2. $u \times w$

3. $w \times u$

4. $u \times v$

5. $u \cdot (v \times w)$

6. $v \cdot (w \times u)$

In Exercises 7–14, let $u = 2i - j + 4k$, $v = i - 3k$, and $w = 2i + 3j + 4k$. Find the indicated vector or number.

7. $u \times v$

8. $v \times w$

9. $u \times w$

10. $3u \times 2w$

11. $u \times (v \times w)$

12. $(u \times v) \times w$

13. $u \cdot (v \times w)$

14. $w \cdot (v \times u)$

In Exercises 15–18, find the area of the parallelogram determined by the given vectors.

15. $v = 2i + j + k$, $\quad w = 3i - j - 2k$

16. $v = -3i + k$, $\quad w = 2i + 2j + k$

17. $v = 2i - j$, $\quad w = 4i + 2j$

18. $v = -6i + 4j - 3k$, $\quad w = i - j - k$

In Exercises 19–21, find the area of the triangle in space with given vertices.

19. $A = (3, 0, 1)$, $\quad B = (2, -1, 2)$, $\quad C = (1, 3, -2)$

20. $A = (5, 2, -2)$, $\quad B = (-1, 4, 2)$, $\quad C = (-4, 5, 3)$

21. $A = (0, 2, 1)$, $\quad B = (-4, 1, -2)$, $\quad C = (1, 1, -2)$

22. Find the area of the triangle with vertices $(2, 4)$, $(-3, 8)$, and $(-5, -2)$.

23. Find the area of the triangle with vertices $(1, 5)$, $(-3, -4)$, and $(4, 6)$.

24. Find a unit vector orthogonal to both $v = 2i + j - k$ and $w = i + j + 4k$.

25. Find a unit vector orthogonal to both $v = i + 3j$ and $w = 4i - j$.

26. Show that the cross product is not associative. That is, find vectors u, v, and w so that $u \times (v \times w) \neq (u \times v) \times w$.

27. Find two unit vectors orthogonal to $v = i + j$ and $w = 2i - j + 3k$.

28. Show that $|v \times w| = |v| \, |w|$ if v and w are orthogonal.

29. Show that if $u = x_1 i + y_1 j + z_1 k$, $v = x_2 i + y_2 j + z_2 k$, and $w = x_3 i + y_3 j + z_3 k$, then

$$u \cdot (v \times w) = \det \begin{bmatrix} x_1 & y_1 & z_1 \\ x_2 & y_2 & z_2 \\ x_3 & y_3 & z_3 \end{bmatrix}.$$

30. Show that the volume of the parallelepiped formed by u, v, and w is $|u \cdot (v \times w)|$. (*Hint:* $|v \times w|$ is the area of the base. Thus, $|u \cdot (v \times w)| = |v \times w| \cdot (|u| \, |\cos \theta|)$ is the area of the base times the altitude. Why?)

31. Use Exercises 29 and 30 to find the volume of the parallelepiped determined by the vectors $u = 2i + j - 4k$, $v = -i + 3j - 2k$, and $w = 4i - j + 3k$.

32. Find the volume of the parallelepiped determined by the vectors $u = -3i + j - k$, $v = 2i + 3j + 2k$, and $w = i + 4j + 2k$.

33. Prove that $u \times (v \times w) = (u \cdot w)v - (u \cdot v)w$.

34. Prove that $u \cdot (v \times w) = v \cdot (w \times u)$.

35. Prove that $u \cdot (v \times w) = w \cdot (u \times v)$.

36. Prove that $(u \times v) \times w = u \times (v \times w)$ only if $(u \times w) \times v = 0$.

37. Suppose that $u + v + w = 0$. Show that $u \times v = v \times w = w \times u$. What is the geometric interpretation of this result?

38. Why did we not try to define the cross product for vectors in the plane?

16.5 EQUATIONS FOR LINES AND PLANES

In the plane, a line ℓ is determined by its slope and one point on the line. However, a point and a slope do not provide sufficient information to determine a line in space. Indeed, even the term "slope" becomes ambiguous in space. For example, think about the problem of writing down an equation for the line determined by the beam from a searchlight as it sweeps across the sky. The beam is determined not only by a *point* (the location of the searchlight) and an *elevation* (perhaps the analogue of slope that you had in mind) but also by the angle at which the searchlight is *rotated* away from a fixed direction, say due north (Figure 5.1).

One of the simplest ways to think about how to find the general form for a line in space is to remember that two points determine a line. If P and Q are two points in space, then the vector $b = \overrightarrow{PQ}$ is parallel to the line ℓ containing P and Q. Moreover, any multiple tb is also parallel to ℓ (Figure 5.2).

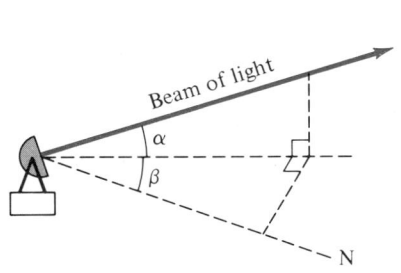

Figure 5.1 Searchlight positioned with elevation angle α and rotation angle β.

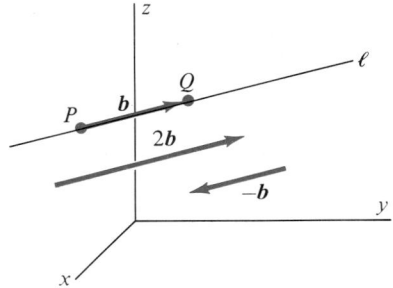

Figure 5.2 Multiples of $b = \overrightarrow{PQ}$ are parallel to ℓ.

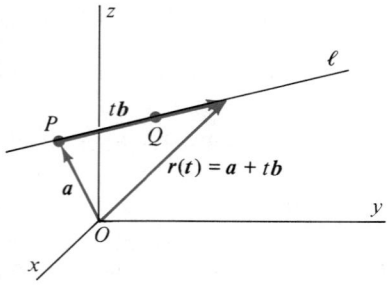

Figure 5.3 If $a = \overrightarrow{OP}$, the vector $r(t) = a + tb$ terminates on ℓ.

We can use the point P and the vector b to determine points on the line ℓ as follows. Let $a = \overrightarrow{OP}$ be the **position vector** originating at the origin O and terminating at the point P on ℓ. Then, since tb is parallel to ℓ, the vector $r(t) = a + tb$ also terminates on the line ℓ (Figure 5.3). As the values of t increase from 0 to ∞, the vector $a + tb$ terminates successively at all points on ℓ lying on one side of P, while as t decreases from 0 to $-\infty$ the vector $a + tb$ determines all points on ℓ lying on the opposite side of P (Figure 5.4). (See Exercises 58–60 of Section 16.1 for a more explicit demonstration of this concept.)

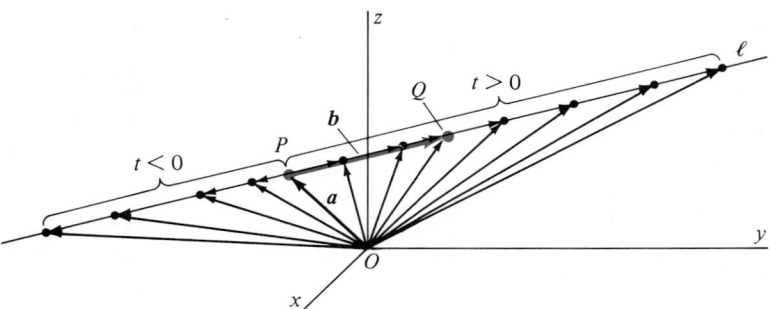

Figure 5.4 $\ell: \{r(t) = a + tb \mid -\infty < t < \infty\}$.

We can summarize this discussion by saying that a line ℓ in space is determined by a *position vector* a, originating at the origin, and a *direction vector* b as

$$\ell: r(t) = a + tb, \qquad -\infty < t < \infty. \tag{1}$$

Equation (1) is called the **vector form** of the equation for ℓ. If P and Q are distinct points on ℓ we can always take

$$a = \overrightarrow{OP} \tag{2}$$

and

$$b = \overrightarrow{PQ} \tag{3}$$

as in the preceding discussion. In this case equation (1) becomes

$$\ell: r(t) = \overrightarrow{OP} + t\overrightarrow{PQ}, \qquad -\infty < t < \infty.$$

More generally, equation (1) allows us to interpret the line ℓ as an infinite collection of vectors, each terminating at a point on ℓ.

Example 1

Find an equation for the line containing the point $P = (2, -1, 4)$ and parallel to the vector $v = i - j + 2k$.

Solution: The position vector is

$$a = \overrightarrow{OP} = 2i - j + 4k$$

and the direction vector is given as

$$b = v = i - j + 2k.$$

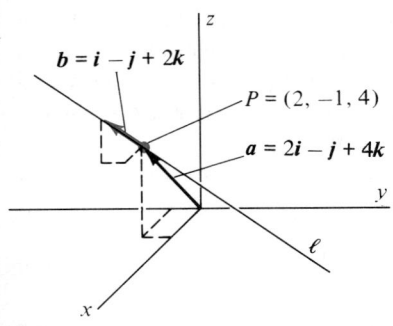

Figure 5.5

The line is, by equation (1), the set of all vectors of the form

$$r(t) = (2i - j + 4k) + t(i - j + 2k) \qquad \text{(Figure 5.5).} \qquad \diamond$$

Example 2

Find an equation for the line containing the points $P = (4, -1, 2)$ and $Q = (-3, 5, 1)$.

Solution: By equation (2) a position vector is

$$a = \overrightarrow{OP} = 4i - j + 2k$$

and, by (3), a direction vector is

$$b = \overrightarrow{PQ} = (-3 - 4)i + (5 - (-1))j + (1 - 2)k = -7i + 6j - k.$$

The line has vector equation

$$r(t) = (4i - j + 2k) + t(-7i + 6j - k) \qquad \text{(Figure 5.6).} \qquad \diamond$$

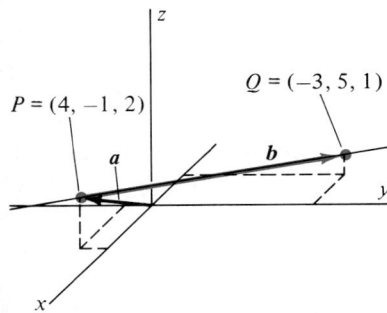

Figure 5.6

Example 3

Find the point, if it exists, where the lines

$$\ell_1: r_1(t) = (3i + 2j - k) + t(-6i + 4j + 3k)$$

$$\ell_2: r_2(t) = (5i + 4j + 7k) + t(14i - 6j + 2k)$$

intersect.

Solution: If ℓ_1 and ℓ_2 intersect at point P, it is not necessary that they do so for the same number t. The condition of intersection is therefore that

$$r_1(t_1) = r_2(t_2) \tag{4}$$

for some times t_1 and t_2. Equating i-components in equation (4) gives

$$3 - 6t_1 = 5 + 14t_2,$$

or

$$6t_1 + 14t_2 = -2. \tag{5}$$

Equating j-components in (4) gives the equation

$$2 + 4t_1 = 4 - 6t_2$$

or

$$4t_1 + 6t_2 = 2. \tag{6}$$

Solving equations (5) and (6) simultaneously (either by substitution or by elimination) gives $t_1 = 2$ and $t_2 = -1$. To ensure that equation (4) holds for these numbers t we must write out $r_1(t_1)$ and $r_2(t_2)$ and verify that the k-components are equal. Since

$$r_1(t_1) = (3i + 2j - k) + 2(-6i + 4j + 3k) = -9i + 10j + 5k$$

and

$$r_2(t_2) = (5i + 4j + 7k) + (-1)(14i - 6j + 2k) = -9i + 10j + 5k,$$

we can see that the lines indeed intersect at the point $(-9, 10, 5)$. $\diamond$

The Distance from a Point to a Line in Space

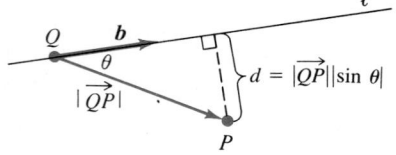

Figure 5.7 $d = |\overrightarrow{QP}||\sin \theta| = \dfrac{|\overrightarrow{PQ} \times b|}{|b|}$.

In space, a line does not determine a unique direction for vectors orthogonal to that line. (In fact, the family of vectors orthogonal to a given line determines an entire plane, as we shall see later in this section.) The problem of finding the distance d from a point $P = (x_0, y_0, z_0)$ to a line with equation $r(t) = a + tb$, must be handled differently from the corresponding problem in the plane.

Figure 5.7 illustrates that the desired distance d is given by the expression

$$d = |\overrightarrow{QP}||\sin \theta| \tag{7}$$

where Q is any point on ℓ and θ is the angle formed between $\overrightarrow{QP}$ and the direction vector b.

The right side of equation (7) is suggestive of the cross product. In fact, we may use Theorem 6 to rewrite equation (7) as

$$d = |\overrightarrow{QP}||\sin \theta| = \frac{|\overrightarrow{QP}||b||\sin \theta|}{|b|} = \frac{|\overrightarrow{PQ} \times b|}{|b|}.$$

Since $|b| \neq 0$ for all lines ℓ, we have the formula

$$\boxed{d = \frac{|\overrightarrow{PQ} \times b|}{|b|}} \tag{8}$$

for the distance from a point P to a line with direction vector b and containing the point Q.

Example 4

Find the distance from the point $P = (1, -1, 2)$ to the line with vector equation

$$\ell: r(t) = 3i - 2j + 4k + t(i - 2j + 2k).$$

Strategy

Find the direction vector b.

Find a point Q on the line ℓ.

Form the vector $\overrightarrow{QP}$.

Apply equation (8).

Solution

Since the equation for ℓ is in the form of equation (1) the direction vector is

$$b = i - 2j + 2k.$$

Setting $t = 0$ shows that the point

$$Q = (3, -2, 4)$$

lies on ℓ. Thus,

$$\begin{aligned} \overrightarrow{QP} &= (1 - 3)i + (-1 - (-2))j + (2 - 4)k \\ &= -2i + j - 2k. \end{aligned}$$

The distance from P to ℓ is therefore

$$d = \frac{1}{\sqrt{1^2 + (-2)^2 + 2^2}} \cdot \left| \det \begin{bmatrix} i & j & k \\ -2 & 1 & -2 \\ 1 & -2 & 2 \end{bmatrix} \right|$$

$$= \frac{1}{3} \left| -2i + 2j + 3k \right|$$

$$= \frac{\sqrt{17}}{3}.$$

◇

Parametric Equations for Lines

Suppose that the vectors $\boldsymbol{a}$ and $\boldsymbol{b}$ have components $\boldsymbol{a} = a_1\boldsymbol{i} + a_2\boldsymbol{j} + a_3\boldsymbol{k}$, and $\boldsymbol{b} = b_1\boldsymbol{i} + b_2\boldsymbol{j} + b_3\boldsymbol{k}$. Then if we write $\boldsymbol{r}(t) = x(t)\boldsymbol{i} + y(t)\boldsymbol{j} + z(t)\boldsymbol{k}$, the vector equation $\boldsymbol{r}(t) = \boldsymbol{a} + t\boldsymbol{b}$ for a line has component form

$$x(t)\boldsymbol{i} + y(t)\boldsymbol{j} + z(t)\boldsymbol{k} = (a_1\boldsymbol{i} + a_2\boldsymbol{j} + a_3\boldsymbol{k}) + t(b_1\boldsymbol{i} + b_2\boldsymbol{j} + b_3\boldsymbol{k})$$
$$= (a_1 + tb_1)\boldsymbol{i} + (a_2 + tb_2)\boldsymbol{j} + (a_3 + tb_3)\boldsymbol{k}.$$

Equating components in this equation gives the **parametric form** for the line ℓ with position vector $\boldsymbol{a} = a_1\boldsymbol{i} + a_2\boldsymbol{j} + a_3\boldsymbol{k}$ and direction vector $\boldsymbol{b} = b_1\boldsymbol{i} + b_2\boldsymbol{j} + b_3\boldsymbol{k}$:

$$\begin{cases} x(t) = a_1 + tb_1 \\ y(t) = a_2 + tb_2 \\ z(t) = a_3 + tb_3. \end{cases} \tag{9}$$

The point (x, y, z) is on ℓ if and only if the coordinates satisfy the equations in (9) for some single value of t.

Example 5

Find parametric equations for the line in Example 1.

Solution: From Example 1 we have

$$a_1 = 2, \qquad a_2 = -1, \qquad a_3 = 4, \qquad b_1 = 1, \qquad b_2 = -1, \qquad b_3 = 2.$$

Thus, by (9)

$$\begin{cases} x(t) = 2 + t \\ y(t) = -1 - t \\ z(t) = 4 + 2t \end{cases}$$

are the parametric equations for ℓ. ◇

Example 6

Find a vector parallel to the line with parametric equations

$$x = -4 + t, \qquad y = 3 - 6t, \qquad \text{and} \qquad z = 2 + 2t.$$

Solution: Comparing these equations with those in (9), we see that a direction vector for the line is

$$\boldsymbol{b} = b_1\boldsymbol{i} + b_2\boldsymbol{j} + b_3\boldsymbol{k} = \boldsymbol{i} - 6\boldsymbol{j} + 2\boldsymbol{k}.$$

This vector is parallel to the line. ◇

Symmetric Equations for Lines

Solving each of the equations in (9) for t gives the equations

$$t = \frac{x(t) - a_1}{b_1}; \qquad t = \frac{y(t) - a_2}{b_2}; \qquad t = \frac{z(t) - a_3}{b_3}.$$

Equating the right-hand sides of each of these equations gives the **symmetric** (or, **rectangular**) form of the equations for the line ℓ with position vector $\boldsymbol{a} = a_1\boldsymbol{i} +$

$a_2 \mathbf{j} + a_3 \mathbf{k}$ and direction vector $\mathbf{b} = b_1 \mathbf{i} + b_2 \mathbf{j} + b_3 \mathbf{k}$:

$$\boxed{\frac{x - a_1}{b_1} = \frac{y - a_2}{b_2} = \frac{z - a_3}{b_3}.} \tag{10}$$

Since the position vector $\mathbf{a}$ terminates at the point (a_1, a_2, a_3) on ℓ, we may also refer to equation (10) as the symmetric form for the equations of the line containing the point (a_1, a_2, a_3) and having direction vector $\mathbf{b} = b_1 \mathbf{i} + b_2 \mathbf{j} + b_3 \mathbf{k}$.

Example 7

Find symmetric equations for the line determined by parametric equations in Example 5.

Solution: Using the values of a_j and b_j, $j = 1, 2, 3$, in Example 5, together with equations in (10), we obtain

$$\frac{x - 2}{1} = \frac{y + 1}{-1} = \frac{z - 4}{2}. \qquad \diamond$$

Example 8

Find a unit vector orthogonal to the line ℓ with symmetric equations

$$\frac{x - 2}{3} = \frac{y - 4}{-2} = \frac{z + 1}{5}.$$

Strategy

Find a vector parallel to ℓ—its direction vector $\mathbf{b}$.

Find any vector $\mathbf{c}$ with $\mathbf{c} \cdot \mathbf{b} = 0$.

The desired unit vector is

$$\mathbf{u} = \left(\frac{1}{|\mathbf{c}|}\right)\mathbf{c}.$$

Solution

The direction vector for ℓ is, by equation (10),

$$\mathbf{b} = 3\mathbf{i} - 2\mathbf{j} + 5\mathbf{k}.$$

A vector $\mathbf{c} = c_1 \mathbf{i} + c_2 \mathbf{j} + c_3 \mathbf{k}$ is orthogonal to $\mathbf{b}$ if and only if

$$\mathbf{b} \cdot \mathbf{c} = 3c_1 - 2c_2 + 5c_3 = 0.$$

One solution (among many) of this equation is

$$c_1 = 2, \qquad c_2 = 3, \qquad c_3 = 0.$$

The vector $\mathbf{c} = 2\mathbf{i} + 3\mathbf{j}$ is therefore orthogonal to ℓ. A *unit* vector in the same direction is

$$\mathbf{u} = \frac{\mathbf{c}}{|\mathbf{c}|} = \frac{\mathbf{c}}{\sqrt{13}} = \frac{2}{\sqrt{13}}\mathbf{i} + \frac{3}{\sqrt{13}}\mathbf{j}. \qquad \diamond$$

Example 9

Find the symmetric equations for the line containing the points $P = (7, 3, -1)$ and $Q = (4, -5, -2)$.

Strategy

Find the direction vector $\mathbf{b} = \overrightarrow{PQ}$.

Solution

The direction vector for the line is

$$\begin{aligned}\mathbf{b} = \overrightarrow{PQ} &= (4 - 7)\mathbf{i} + (-5 - 3)\mathbf{j} + (-2 - (-1))\mathbf{k} \\ &= -3\mathbf{i} - 8\mathbf{j} - \mathbf{k}.\end{aligned}$$

Use $P = (a_1, a_2, a_3)$ and substitute into equations in (10).

Using the point $P = (7, 3, -1)$ on the line and the direction vector b we obtain from the equations in (10) that

$$\frac{x - 7}{-3} = \frac{y - 3}{-8} = \frac{z + 1}{-1}.$$

◇

Equations for Planes

In geometry a plane is characterized by the following two facts:

(i) Through any point P on a line ℓ, there is precisely one plane σ perpendicular to ℓ.
(ii) The point $Q \neq P$ lies on the plane σ in (i) if and only if the line through P and Q is perpendicular to ℓ.

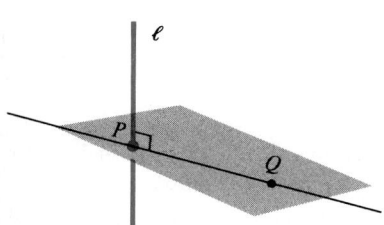

Figure 5.8 Plane σ determined by a point P and a line ℓ.

(Figure 5.8.)

We can use vectors to develop an analytic description of the plane σ as follows. If n is a direction vector for the line ℓ, the condition that the line through P and Q be perpendicular to the line ℓ is equivalent to the condition that the vectors $\overrightarrow{PQ}$ and n be orthogonal. By Corollary 1 of Section 16.2, this last condition is equivalent to the condition that $\overrightarrow{PQ} \cdot n = 0$. We may therefore write *the plane σ determined by the point P and the vector n* as

$$\sigma = \{Q \mid \overrightarrow{PQ} \cdot n = 0\}. \tag{11}$$

The vector n in equation (11) is referred to as a **normal vector** for the plane σ (Figure 5.9).

To develop an equation in xyz-coordinates for the plane σ, we express n in component form as $n = ai + bj + ck$, and we assume the point P to have coordinates $P = (x_0, y_0, z_0)$. Let the point Q have coordinates $Q = (x, y, z)$. Then the vector $\overrightarrow{PQ}$ is

$$\overrightarrow{PQ} = (x - x_0)i + (y - y_0)j + (z - z_0)k, \tag{12}$$

and the condition in line (11) is that

$$[(x - x_0)i + (y - y_0)j + (z - z_0)k] \cdot (ai + bj + ck) = 0. \tag{13}$$

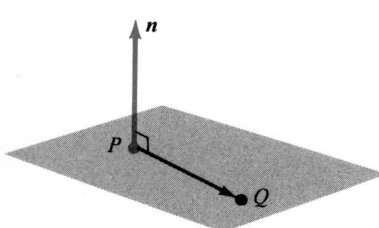

Figure 5.9 Plane σ determined by a point P and a normal vector n.

Equation (13) simplifies to the equation

$$a(x - x_0) + b(y - y_0) + c(z - z_0) = 0,$$

or

$$\boxed{ax + by + cz = ax_0 + by_0 + cz_0.} \tag{14}$$

Equation (14) is *the equation for the plane with normal vector $n = ai + bj + ck$ containing the point (x_0, y_0, z_0).* Notice that the coefficients of x, y, and z are just the respective components of the normal vector n, while the right-hand side is a *constant* determined from n and P.

Example 10

Find a vector normal to the plane with equation $3x + 5y - z = 9$.

Solution: The components of a normal vector are simply the coefficients of x, y, and z:

$$n = 3i + 5j - k.$$

◇

Example 11

Find an equation for the plane with normal vector $n = i + 4j + 2k$, and containing the point $P = (6, -3, 4)$.

Solution: Let $Q = (x, y, z)$ be a point in the desired plane. We are given

$$a = 1, \qquad b = 4, \qquad c = 2$$

and

$$x_0 = 6, \qquad y_0 = -3, \qquad z_0 = 4.$$

Substituting into equation (14) gives

$$x + 4y + 2z = 1 \cdot 6 + 4(-3) + 2 \cdot 4 = 2.$$

The desired equation is therefore

$$x + 4y + 2z = 2. \qquad \diamond$$

If a normal vector is not specified, we can sometimes determine a normal vector for a plane from other information about the plane. In particular, if we can find two vectors, v_1 and v_2, lying in the plane, then the cross product $n = v_1 \times v_2$ is a normal to the plane, since it is orthogonal both to v_1 and to v_2 (Figure 5.10).

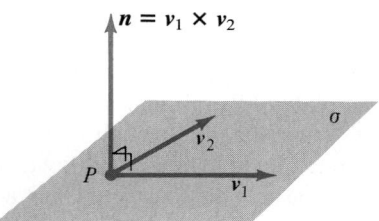

Figure 5.10 $n = v_1 \times v_2$ is orthogonal to the plane σ if v_1 and v_2 lie in σ.

Example 12

Find an equation for the plane determined by the three points $P = (-4, 0, 2)$, $Q = (1, -3, 1)$, and $R = (2, -2, 6)$.

Strategy

Find two vectors.

$$v_1 = \overrightarrow{PQ}, \qquad v_2 = \overrightarrow{PR}$$

in the plane.

Find a normal

$$n = v_1 \times v_2.$$

Substitute into equation (14), using n and P.

Solution

Two vectors in the plane are

$$v_1 = \overrightarrow{PQ} = 5i - 3j - k$$

and

$$v_2 = \overrightarrow{PR} = 6i - 2j + 4k.$$

A normal vector is therefore

$$n = v_1 \times v_2 = \det \begin{bmatrix} i & j & k \\ 5 & -3 & -1 \\ 6 & -2 & 4 \end{bmatrix}$$
$$= -14i - 26j + 8k.$$

Using n and the point P, we obtain from equation (14) that

$$-14x - 26y + 8z = -14(-4) + (-26)(0) + 8 \cdot 2$$
$$= 72.$$

The desired equation is

$$-14x - 26y + 8z = 72. \qquad \diamond$$

Example 13

Find the distance from the point $P = (3, -1, 2)$ to the plane with equation $2x - y + z = 4$.

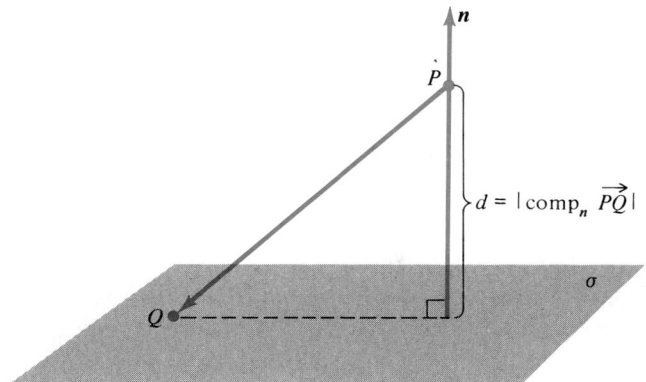

Figure 5.11 Distance from a point P to a plane σ.

Strategy	*Solution*
Find a normal n for the plane.	A normal vector for the plane is

$$n = 2i - j + k.$$

Set $x = y = 0$ and solve for z to find any convenient point Q in the plane.

A point Q in the plane is $Q = (0, 0, 4)$. A vector parallel to n originating at P and terminating in the plane is

$$v = \text{proj}_n \, \overrightarrow{PQ} \qquad \text{(see Figure 5.10)}.$$

Project $\overrightarrow{PQ}$ onto n. Recall that

$$|\text{proj}_n \, \overrightarrow{PQ}| = |\text{comp}_n \, \overrightarrow{PQ}|$$

(Section 16.2).

The desired distance is the length of this vector

$$|v| = |\text{comp}_n \, \overrightarrow{PQ}| = \frac{|\overrightarrow{PQ} \cdot n|}{|n|}$$

$$= \frac{|(-3i + j + 2k) \cdot (2i - j + k)|}{|2i - j + k|}$$

$$= \frac{5}{\sqrt{6}} \qquad \text{(Figure 5.11)}.$$

You may wish to try another convenient point Q, say $(2, 0, 0)$, and verify that the result is the same. ◇

Exercise Set 16.5

In Exercises 1–7, find a parametric form of the equations for the line described.

1. The line with direction vector $b = i + j - k$ and containing the point $P = (1, 2, 3)$.

2. The line with direction vector $b = 2i - j + 3k$ and containing the point $P = (3, -6, 2)$.

3. The line with position and direction vectors $a = b = 3i - j + 5k$.

4. The line containing the points $P = (1, 3, -1)$ and $Q = (7, -2, 5)$.

5. The line containing the points $P = (-4, 2, 1)$ and $Q = (-3, 5, 3)$.

6. The line with position vector $a = i + 2j - 6k$ and direction vector $b = 2i - j + 4k$.

7. The line through $(1, 2, 0)$ and orthogonal to *both* vectors $v = i + 3j + k$ and $w = j - 3k$.

In Exercises 8–11, find rectangular (symmetric) equations for the line determined by the given parametric equations.

8. $x(t) = 2t$
$y(t) = 3 - t$
$z(t) = t$

9. $x(t) = t + 7$
$y(t) = 4t - 6$
$z(t) = 3 - 2t$

10. $x(t) = 2t + 5$
$y(t) = t - 6$
$z(t) = 5t + 2$

11. $x(t) = t$
$y(t) = 8t - 6$
$z(t) = 4t + 4$

and

$$\ell_2: \frac{x + 4}{1} = \frac{y - 3}{2} = \frac{z + 1}{2}$$

as follows:

Find symmetric equations for

12. The line in Exercise 2.

13. The line in Exercise 3.

14. The line in Exercise 5.

In Exercises 15–18, find parametric equations for the line determined by the given symmetric equations.

15. $\dfrac{x}{3} = \dfrac{y}{2} = \dfrac{z}{5}$

16. $\dfrac{x - 2}{3} = \dfrac{y + 1}{-4} = \dfrac{z}{3}$

17. $\dfrac{x + 4}{4} = \dfrac{y - 2}{-2} = \dfrac{z + 3}{3}$

18. $x = \dfrac{y - 3}{2} = z + 1$

19. Find a vector parallel to the line in Exercise 9.

20. Find a vector parallel to the line in Exercise 16.

21. Find a direction vector for the line in Exercise 17.

22. Find a direction vector for the line in Exercise 18.

23. For the line ℓ with vector form $r(t) = a + tb$, $b \neq 0$, find the number t_0 for which $r(t_0) \perp b$. Conclude that $r(t_0) \perp \ell$.

24. Use Exercise 23 to show that the distance from the origin to the line ℓ with equation $r(t) = a + tb$ is $d = |r(t_0)|$, where t_0 is as in Exercise 23.

25. Use Exercise 24 to find the distance from the origin to the line with vector equation $r(t) = i + 2j + t(j - k)$.

26. Use Exercise 24 to find the distance from the origin to the line with symmetric equations

$$x - 1 = \frac{2 - y}{2} = \frac{z}{-2}.$$

27. Find the distance from the point $P = (2, -1, 4)$ to the line with equation $r(t) = (3i - j - k) + t(i + 2j + k)$.

28. Find the distance from the point $P = (1, 3, -2)$ to the line with equations $x(t) = 1 + t$, $y(t) = 3 - 2t$, $z(t) = 2t - 2$.

29. Find the distance from the point $P = (0, -2, 1)$ to the line with equations $\dfrac{x - 1}{4} = \dfrac{y + 3}{-2} = \dfrac{z + 1}{5}$.

30. For the line ℓ with vector form $r(t) = a + tb$, $b = b_1 i + b_2 j + b_3 k$, the numbers b_1, b_2, and b_3 are called the **direction numbers**. What condition must be satisfied by the direction numbers in order that the symmetric equations for ℓ exist?

31. Find the distance d between the lines

$$\ell_1: \frac{x - 2}{3} = \frac{y + 1}{2} = \frac{z - 3}{5}$$

a. Find direction vectors b_1 and b_2 for lines ℓ_1 and ℓ_2, respectively.

b. Observe that a vector perpendicular to both lines ℓ_1 and ℓ_2 is the cross product $n = b_1 \times b_2$.

c. Find a point P on ℓ_1 and a point Q on ℓ_2.

d. Obtain the distance d as

$$d = |\text{comp}_n \overrightarrow{PQ}|.$$

32. Use Exercise 31 to find the distance between the lines ℓ_1 and ℓ_2:

$$\ell_1: x(t) = 1 + t, \qquad y(t) = 5t, \qquad z(t) = 1 - t,$$
$$\ell_2: x(t) = 2 + t, \qquad y(t) = 2 - 3t, \qquad z(t) = 1 + 5t.$$

Two lines with vector equations $\ell_1: r_1(t) = a_1 + tb_1$ and $\ell_2: r_2(t) = a_2 + tb_2$ **intersect** if $r_1(t_1) = r_2(t_2)$ for some numbers t_1 and t_2. Find the point at which each of the following pairs of lines intersect.

33. $r_1(t) = (2i + j + 2k) + t(5i + j + 3k)$
$r_2(t) = (-4i + 7j + 10k) + t(3i - 3j - 4k)$

34. $\ell_1: x(t) = 1 + t, \qquad y(t) = 2 - 2t, \qquad z(t) = t + 5$
$\ell_2: x(t) = 2 + 2t, \qquad y(t) = 5 - 9t, \qquad z(t) = 2 + 6t$

In each of Exercises 35–39, find an equation for the plane with the given normal containing the given point.

35. $n = i + 2j - k$, $\quad P = (1, 2, -3)$

36. $n = 4i - j + k$, $\quad P = (-1, 3\ 5)$

37. $n = 2i - 2j + 3k$, $\quad P = (0, -2, 5)$

38. $n = i + j$ $\quad P = (4, 2, -3)$

39. $n = i + 2j + k$ $\quad P = (0, -1, 3)$

In Exercises 40–44, find an equation for the plane containing the given points.

40. $P = (0, 1, -2)$, $\quad Q = (1, 1, 1)$, $\quad R = (3, 5, 1)$

41. $P = (-2, 3, -4)$, $\quad Q = (1, 5, 1)$, $\quad R = (-2, -2, -2)$

42. $P = (7, 0, -2)$, $\quad Q = (1, -3, 4)$, $\quad R = (5, 2, -3)$

43. $P = (-1, 1, 1)$, $\quad Q = (0, 2, -1)$, $\quad R = (3, 5, -2)$

44. $P = (-2, 1, 5)$, $\quad Q = (2, 0, -2)$, $\quad R = (-1, -1, 3)$

45. Find a vector normal to the plane with equation $2x - 3y + z = 5$.

46. Find a vector normal to the plane containing $(-2, 1, 3)$, $(5, 1, 5)$, and $(0, 0, 2)$.

47. Find an equation for the plane perpendicular to the line with symmetric equation

$$\frac{x-2}{3} = \frac{1-y}{6} = \frac{z+2}{2}$$

containing the point $(3, -2, 5)$.

48. Find an equation for the plane containing the point $(3, -1, -1)$ and perpendicular to the line with parametric equations $x = 3 + t$, $y = -1 + 4t$, $z = 5 + 3t$.

49. Find parametric equations for the line of intersection of the planes with equations $x + y - 2z = 6$ and $2x - 4y + z = 3$.

50. The angle between two planes is equal to the angle between vectors normal to the two planes. Find the angle between the planes with equations $x + y + z = 6$ and $3x - y + 2z = 5$.

51. Find the angle between the planes with equations $x - y + z = 2$ and $x + 3y + 3z = -4$ (see Exercise 50).

52. Show that the distance from the point $P_0 = (x_0, y_0, z_0)$ to the plane with equation $ax + by + cz = d$ is

$$d = \frac{|ax_0 + by_0 + cz_0 - d|}{\sqrt{a^2 + b^2 + c^2}}.$$

(*Hint:* See Example 13.)

53. Find the distance from the point $P = (3, -6, 5)$ to the plane with equation $6x - y + z = 2$.

54. Show that the plane with equation $ax + by + cz = 1$ intersects the coordinate axes at the respective values $x = \frac{1}{a}$, $y = \frac{1}{b}$, $z = \frac{1}{c}$ if $a \neq 0$, $b \neq 0$, $c \neq 0$.

55. Find parametric equations for the line containing the point $(2, 4, -3)$ and perpendicular to the plane with equation $2x + 3y - 7z = 9$.

56. Determine which of the following sets of points are coplanar.
   a. $(2, 1, 0)$, $(3, 0, 5)$, $(1, 1, 1)$, $(2, 3, -12)$
   b. $(1, -6, 2)$, $(3, -5, 11)$, $(4, 0, 4)$, $(1, 5, -2)$
   c. $(1, 7, 2)$, $(3, 4, -2)$, $(1, 5, 1)$, $(1, 9, 3)$

57. Find an equation for the plane containing the point $(3, -1, 3)$ and the line with symmetric equations

$$\frac{x-2}{3} = \frac{y+1}{5} = \frac{z}{4}.$$

16.6 CYLINDRICAL AND SPHERICAL COORDINATES

We have seen that certain equations involving two variables are more appropriately graphed using polar coordinates than rectangular (Cartesian) coordinates for the plane. Similarly, certain surfaces in space are more easily described using *cylindrical* or *spherical* coordinates than by use of rectangular coordinates. In fact, some surfaces that are easily described in one of these coordinate systems are almost impossible to determine in rectangular coordinates.

Cylindrical Coordinates

In this coordinate system, the xy-plane is coordinatized by polar coordinates, and the third coordinate denotes the usual rectangular z-coordinate. That is, if the point P has cylindrical coordinates $P = (r, \theta, z)$ and rectangular coordinates $P = (x, y, z)$, we have the equations

$x = r \cos \theta,$	$r \geq 0$	(1a)
$y = r \sin \theta,$	$r \geq 0$	(1b)
$z = z$		(1c)

giving the rectangular coordinates in terms of the cylindrical coordinates, and the equations

$r = \sqrt{x^2 + y^2}$	(2a)
$\tan \theta = \dfrac{y}{x}$	(2b)
$z = z$	(2c)

for the cylindrical coordinates in terms of the rectangular coordinates (Figure 6.1). (Note that we now require $r \geq 0$, unlike the case for polar coordinates in the plane.)

Figure 6.2 shows why cylindrical coordinates are named as they are. The graph of the equation $r = r_0$, r_0 constant, is a cylinder in space. The graph of $\theta = \theta_0$, θ_0 constant, is a half plane, as in Figure 6.3. The graph of $z = z_0$, z_0 constant, is a horizontal plane, as in rectangular coordinates. Several points, plotted in both rectangular and cylindrical coordinates, appear in Figure 6.4.

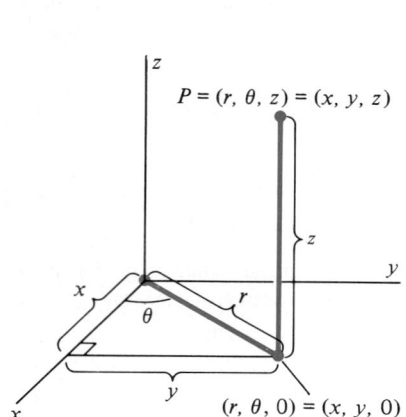

Figure 6.1 The cylindrical coordinate system.

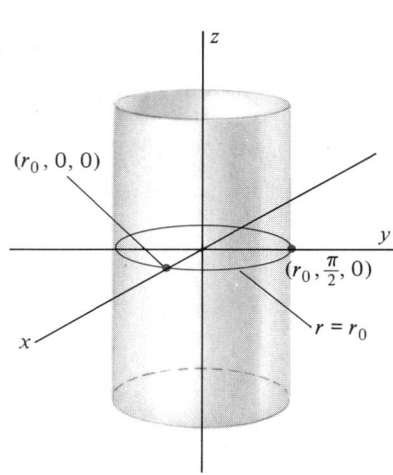

Figure 6.2 Graph of $r = r_0$ is a circular cylinder.

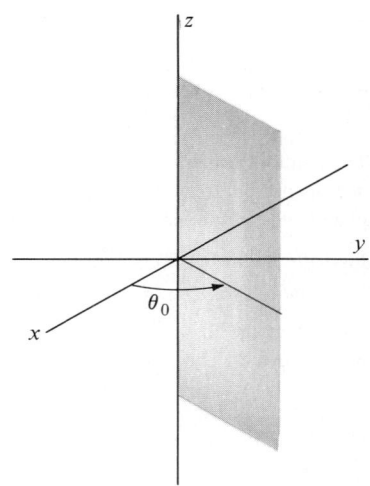

Figure 6.3 Graph of $\theta = \theta_0$ is a half plane.

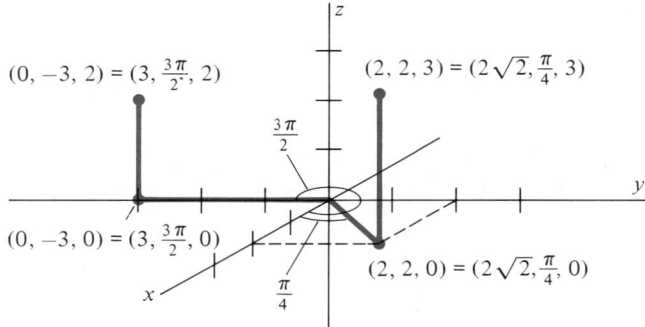

Figure 6.4 Four points in polar coordinates.

Example 1

Find cylindrical coordinates for the point with rectangular coordinates $(1, \sqrt{3}, 4)$.

Solution: Here $x = 1$, $y = \sqrt{3}$, $z = 4$. By equations (2a) through (2c),

$$r = \sqrt{1^2 + (\sqrt{3})^2} = 2,$$

$$\tan \theta = \sqrt{3} \quad \text{so} \quad \theta = \pi/3 \qquad \text{(since } x \text{ and } y \text{ place the point in the first quadrant),}$$

$$z = 4.$$

The cylindrical coordinates are $(2, \pi/3, 4)$. ◇

Example 2

Express the equation $z^2 = x^2 + y^2$ in cylindrical coordinates.

Solution: Using (1a) and (1b) the equation becomes

$$z^2 = (r \cos \theta)^2 + (r \sin \theta)^2$$
$$= r^2 \cdot (\cos^2 \theta + \sin^2 \theta) = r^2.$$

The equation therefore becomes

$$z = \pm r.$$

The graph of this equation is the cone in Figure 6.5. ◇

Example 3

Graph the equation $r^2 = a^2 \sin \theta$ in cylindrical coordinates.

Solution: Since the variable z is missing, the graph is a cylinder in the z direction whose trace in the xy-plane is the graph of the lemniscate $r^2 = a^2 \sin \theta$ (see Figure 6.6). ◇

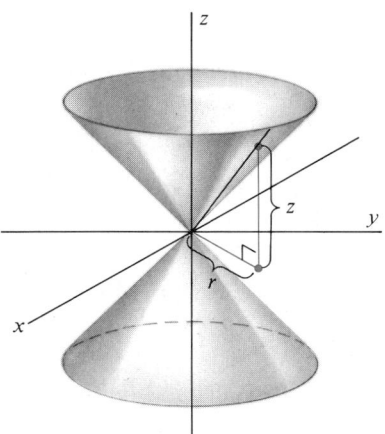
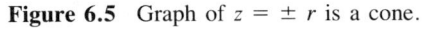

Figure 6.5 Graph of $z = \pm r$ is a cone.

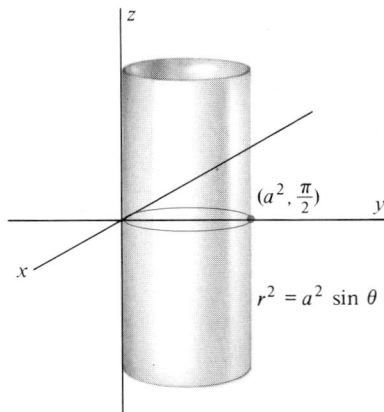

Figure 6.6 Graph of $r^2 = a^2 \sin \theta$ is a cylinder in cylindrical coordinates.

Spherical Coordinates

In spherical coordinates, a point is determined by the ordered triple $P = (\rho, \theta, \phi)$ where $\rho = |\overrightarrow{OP}|$ is the distance of the point P from the origin, θ is the polar angle associated with the vertical line through P, and ϕ is the (tilt) angle between the vector $\overrightarrow{OP}$ and the positive z-axis (Figure 6.7). By convention, we require $\rho \geq 0$, $0 \leq \theta < 2\pi$, and $0 \leq \phi \leq \pi$.

Figure 6.8 illustrates the reason for the terminology "spherical coordinates." The graph of the equation $\rho = \rho_0$, ρ_0 constant, is a sphere of radius ρ_0 centered at the origin. The graph of $\theta = \theta_0$, θ_0 constant, is a half plane, as in cylindrical coordinates. The graph of $\phi = \phi_0$, ϕ_0 constant, is a cone (Figure 6.9).

Figure 6.10 illustrates the relationships between the spherical coordinates (ρ, θ, ϕ) and the rectangular coordinates (x, y, z) for a point P. Since the vector $\overrightarrow{OP}$

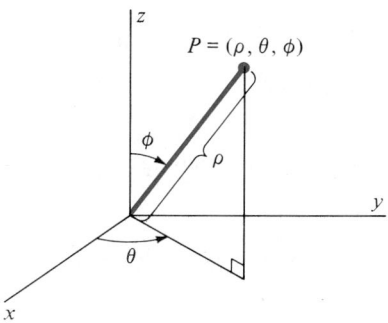

Figure 6.7 Spherical coordinates.

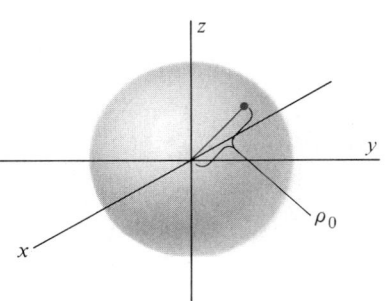

Figure 6.8 Graph of $\rho = \rho_0$ is a sphere.

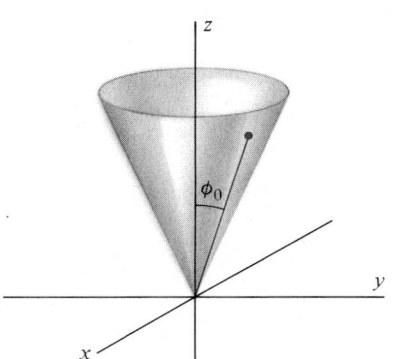

Figure 6.9 Graph of $\phi = \phi_0$ is a cone.

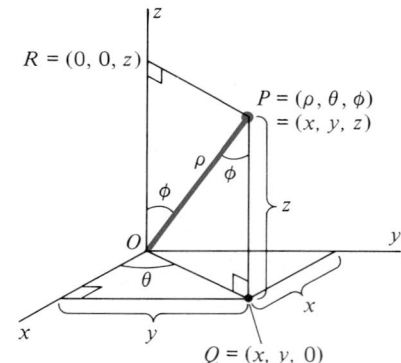

Figure 6.10 Relationships between rectangular and spherical coordinates.

is a diagonal of the rectangle $OQPR$,

$$|\overrightarrow{OQ}| = \rho \sin \phi, \qquad \text{and} \qquad z = \rho \cos \phi. \tag{3}$$

Also,

$$x = |\overrightarrow{OQ}| \cos \theta, \qquad \text{and} \qquad y = |\overrightarrow{OQ}| \sin \theta. \tag{4}$$

Combining equations (3) and (4) gives

$$
\begin{array}{lll}
x = \rho \sin \phi \cos \theta, & \rho \geq 0 & \text{(5a)} \\
y = \rho \sin \phi \sin \theta, & \rho \geq 0 & \text{(5b)} \\
z = \rho \cos \phi, & \rho \geq 0. & \text{(5c)}
\end{array}
$$

Also, from the distance formula we find

$$\rho = \sqrt{x^2 + y^2 + z^2}. \tag{6}$$

Equations for θ and ϕ may be easily derived from equations (3) and (4).

Example 4

The point P has spherical coordinates $P = (3, \pi/3, \pi/4)$. Find rectangular coordinates for P.

Solution: Here $\rho = 3$, $\theta = \pi/3$, and $\phi = \pi/4$. By equations (5a) through (5c),

$$x = 3 \sin\left(\frac{\pi}{4}\right) \cdot \cos\left(\frac{\pi}{3}\right) = 3\left(\frac{\sqrt{2}}{2}\right)\left(\frac{1}{2}\right) = \frac{3\sqrt{2}}{4},$$

$$y = 3 \sin\left(\frac{\pi}{4}\right) \cdot \sin\left(\frac{\pi}{3}\right) = 3\left(\frac{\sqrt{2}}{2}\right)\left(\frac{\sqrt{3}}{2}\right) = \frac{3\sqrt{6}}{4},$$

$$z = 3 \cos\left(\frac{\pi}{4}\right) = \frac{3\sqrt{2}}{2}.$$

The rectangular coordinates are $P = \left(\dfrac{3\sqrt{2}}{4}, \dfrac{3\sqrt{6}}{4}, \dfrac{3\sqrt{2}}{2}\right)$. ◇

Example 5

Express the equation $x^2 - y^2 + z^2 = 4$ in spherical coordinates.

Solution: Using equations (5a) through (5c) the equation becomes

$$(\rho \sin \phi \cos \theta)^2 - (\rho \sin \phi \sin \theta)^2 + (\rho \cos \phi)^2 = 4,$$
$$\rho^2 \sin^2 \phi[\cos^2 \theta - \sin^2 \theta] + \rho^2 \cos^2 \phi = 4,$$
$$\rho^2 \sin^2 \phi[1 - 2 \sin^2 \theta] + \rho^2 \cos^2 \phi = 4,$$
$$\rho^2 - 2\rho^2 \sin^2 \phi \sin^2 \theta = 4.$$

◇

Exercise Set 16.6

1. The following points are given in rectangular coordinates. Find their cylindrical coordinates.
a. $(1, 1, 0)$ b. $(\sqrt{3}, 1, 3)$
c. $(-1, 1, -2)$ d. $(-1, \sqrt{3}, 4)$
e. $(0, 3, -5)$ f. $(-\sqrt{2}, \sqrt{2}, \sqrt{2})$

2. The following points are given in cylindrical coordinates. Find their rectangular coordinates.
a. $(2, \pi/4, -3)$ b. $(1, \pi/6, 4)$
c. $(4, \pi/3, -5)$ d. $(2, 4\pi/3, 2)$
e. $(1, \pi, 1)$ f. $(5, 5\pi/3, 5)$

3. The following points are given in rectangular coordinates. Find their spherical coordinates.
a. $(1, 0, 0)$ b. $(1, 1, \sqrt{2})$
c. $(1, -1, \sqrt{2})$ d. $(1, \sqrt{3}, 2)$
e. $(-\sqrt{3}, 1, -2)$ f. $(-2, 2, 2\sqrt{2})$

4. The following points are given in spherical coordinates. Find their rectangular coordinates.
a. $(2, \pi/4, \pi/3)$ b. $(1, \pi/2, \pi)$
c. $(2, 3\pi/4, \pi/4)$ d. $(3, 3\pi/2, 2\pi/3)$
e. $(5, \pi/6, 5\pi/6)$ f. $(2, 5\pi/3, 3\pi/4)$

5. The following points are given in cylindrical coordinates. Find their spherical coordinates.
a. $(1, 0, 0)$ b. $(\sqrt{2}, -\pi/4, \sqrt{2})$
c. $(2, \pi/3, 2)$ d. $(2, \pi/4, 2)$
e. $(2, 5\pi/3, 0)$ f. $(2, \pi/6, 2)$

6. The following points are given in spherical coordinates. Find their cylindrical coordinates.
a. $(2, \pi/4, \pi/2)$ b. $(2, \pi/2, \pi/4)$
c. $(3, 2\pi/3, \pi/2)$ d. $(1, \pi/2, 2\pi/3)$
e. $(4, \pi/4, 0)$ f. $(2, \pi/3, \pi/2)$

In Exercises 7–14, an equation in cylindrical coordinates is given. Write the equation in rectangular coordinates and sketch the graph.

7. $z = 3$

8. $r = 2$

9. $z = 2r$

10. $z = r \sin \theta$

11. $r^2 + z^2 = 4$

12. $z^2 = r^2 \cos^2 \theta - 1$

13. $\cos^2 \theta - \sin^2 \theta = \dfrac{a^2}{r^2}$

14. $z = r^2 \cos^2 \theta$

In Exercises 15–22, an equation in rectangular coordinates is given. Write the equation in cylindrical coordinates.

15. $x^2 + y^2 = 9$

16. $x^2 + y^2 + z^2 = 9$

17. $x^2 + y^2 = 9z$

18. $y^2 + z^2 = 1$

19. $x^2 + z^2 = 4$

20. $x + y + z = 4$

21. $xy - ax = 4$

22. $x^2 - y^2 = 4$

In Exercises 23–30, write the equation in the exercise referred to in spherical coordinates.

23. Exercise 15

24. Exercise 16

25. Exercise 17

26. Exercise 18

27. Exercise 19

28. Exercise 20

29. Exercise 21

30. Exercise 22

16.7 QUADRIC SURFACES AND CYLINDERS

In Chapter 14 we showed that the graph of the general second degree equation in two variables

$$Ax^2 + Bxy + Cy^2 + Dx + Ey + F = 0$$

if not degenerate, is either a line, a parabola, a circle, an ellipse, or a hyperbola in the plane. The general second degree equation in three variables has the form

$$Ax^2 + By^2 + Cz^2 + Dxy + Exz + Fyz + Gx + Hy + Iz + J = 0. \qquad (1)$$

Equation (1) is called a **quadric equation.** Graphs of quadric equations are surfaces in space, called **quadric surfaces.** In this section we shall examine six general types of quadric surfaces and a seventh special case that we refer to as the **quadric cylinder.**

The six general quadric surfaces are

1. the ellipsoid,
2. the elliptic paraboloid,
3. the elliptic cone,
4. the hyperboloid of one sheet,
5. the hyperboloid of two sheets,
6. the hyperbolic paraboloid.

We will study each of these figures by finding its *traces* in the coordinate planes and in planes parallel to the coordinate planes. To understand what this means, let us first recall the equations for the three coordinate planes:

$$z = 0 \text{ is the equation for the } xy\text{-plane,} \qquad (2a)$$

$$y = 0 \text{ is the equation for the } xz\text{-plane,} \qquad (2b)$$

$$x = 0 \text{ is the equation for the } yz\text{-plane.} \qquad (2c)$$

(See Figure 7.1).

The **trace** of a surface in any plane is simply the intersection of the surface and the plane. The equation for this trace is obtained by substituting the constant value determining the plane for the corresponding variable in the equation for the surface. For example, we have already seen (Section 16.3) that the equation for the sphere

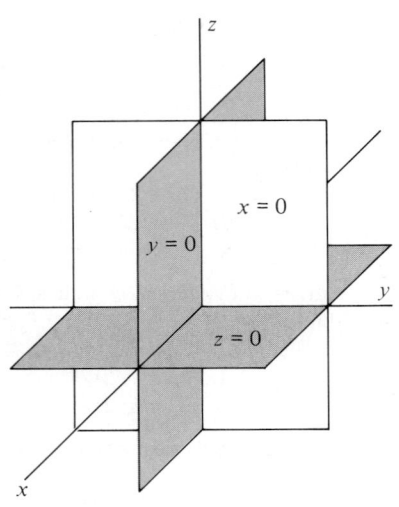

Figure 7.1 The three coordinate planes.

with center $(0, 0, 0)$ and radius r is

$$x^2 + y^2 + z^2 = r^2. \tag{3}$$

Combining equations (2a) and (3), we find that the equation for the trace of the sphere in the xy-plane is $x^2 + y^2 = r^2$. The trace is therefore the circle with center $(0, 0)$ and radius r in the xy-plane (Figure 7.2).

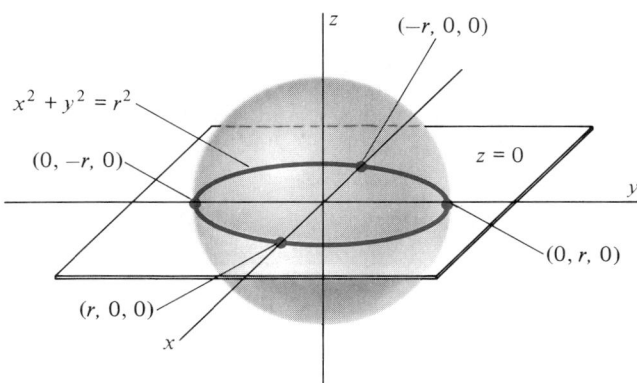

Figure 7.2 Trace of sphere $x^2 + y^2 + z^2 = r^2$ in plane $z = 0$ is circle $x^2 + y^2 = r^2$.

More generally, equations for planes *parallel* to the coordinate planes are

$$z = d, \quad \text{a plane parallel to the } xy\text{-plane}, \tag{4a}$$

$$y = d, \quad \text{a plane parallel to the } xz\text{-plane}, \tag{4b}$$

$$x = d, \quad \text{a plane parallel to the } yz\text{-plane}. \tag{4c}$$

Thus, to find the equation of the trace of the sphere in, say, the plane $z = d$, we combine equations (3) and (4a) to get $x^2 + y^2 + d^2 = r^2$, or $x^2 + y^2 = r^2 - d^2$. If $d^2 < r^2$, then the trace is again a circle (although smaller than the one in the xy-plane); if $d^2 > r^2$, there are no points of intersection (the plane is above or below the sphere).

The Ellipsoid: $\dfrac{x^2}{a^2} + \dfrac{y^2}{b^2} + \dfrac{z^2}{c^2} = 1$

Setting $z = 0$ shows that the trace in the xy-plane is the ellipse

$$\frac{x^2}{a^2} + \frac{y^2}{b^2} = 1.$$

Similarly, setting $y = 0$ and then $x = 0$ shows that the traces in the xz- and yz-plane are also ellipses. These three traces are shown in Figure 7.3.

Using Equations (4a) through (4c), we find that the traces in planes that are parallel to the coordinate planes and that intersect the ellipsoid are again ellipses. For example, setting $z = d$ gives the trace

$$\frac{x^2}{a^2} + \frac{y^2}{b^2} = 1 - \frac{d^2}{c^2}. \tag{5}$$

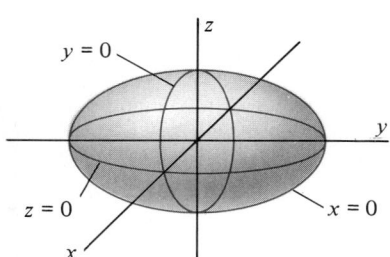

Figure 7.3 Traces of the ellipsoid $\dfrac{x^2}{a^2} + \dfrac{y^2}{b^2} + \dfrac{c^2}{d^2} = 1$ in the coordinate planes.

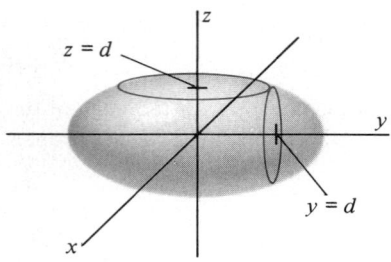

Figure 7.4 Traces of the ellipsoid in the planes $y = d$ and $z = d$.

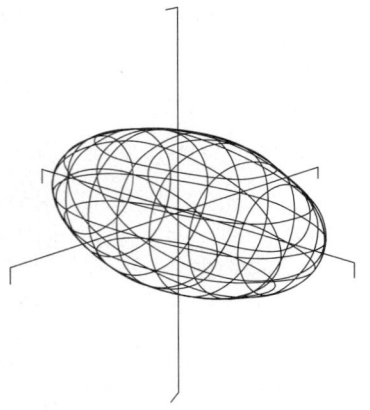

Figure 7.5 Various traces of an ellipsoid.

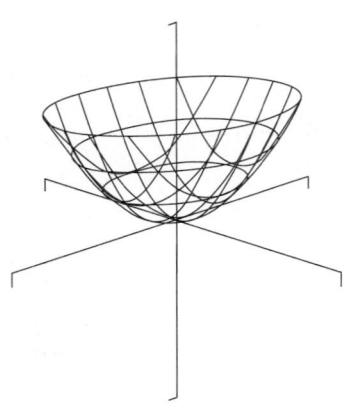

Figure 7.8 Various traces of an elliptic paraboloid.

The graph of equation (5) is an ellipse if $0 \le |d| < |c|$, it is a pair of points $(0, 0, \pm d)$ if $|d| = |c|$, and the equation has no solutions if $|d| > |c|$ (Figure 7.4). Figure 7.5 shows traces of the ellipsoid for various planes determined by equations (4a) through (4c).

The ellipsoid is symmetric with respect to each of the coordinate planes, since replacing x by $-x$, y by $-y$, or z by $-z$ does not change the equation. If $a = b = c$, the ellipsoid is, of course, a sphere.

The Elliptic Paraboloid: $\quad z = \dfrac{x^2}{a^2} + \dfrac{y^2}{b^2}$

Setting $x = 0$ shows that the trace of this figure in the yz-plane is the parabola $z = \dfrac{y^2}{b^2}$. Similarly, the trace in the xz-plane is the parabola $z = \dfrac{x^2}{a^2}$. The trace in the xy-plane ($z = 0$) is simply the origin $(0, 0)$ (Figure 7.6).

The traces in the planes $x = d_1$ and $y = d_2$ are again parabolas:

$$z = \frac{y^2}{b^2} + \frac{d_1^2}{a^2}; \qquad z = \frac{x^2}{a^2} + \frac{d_2^2}{b^2}.$$

However, traces in the plane $z = d$ are ellipses with equations

$$\frac{x^2}{a^2 d} + \frac{y^2}{b^2 d} = 1$$

if $d > 0$. If $d < 0$ there are no traces (Figure 7.7).

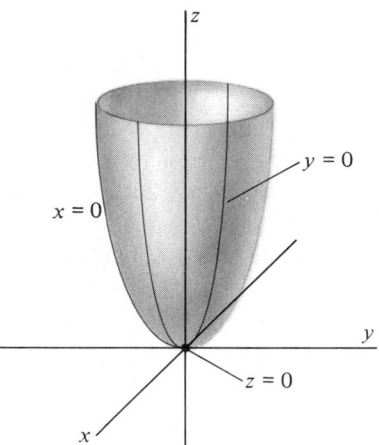

Figure 7.6 Traces of the elliptic paraboloid $z = \dfrac{x^2}{a^2} + \dfrac{y^2}{b^2}$ in the coordinate planes.

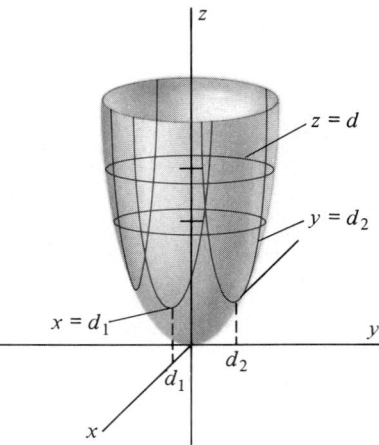

Figure 7.7 Traces of the elliptic paraboloid in planes $x = d_1$, $y = d_2$, $z = d$.

The elliptic paraboloid is symmetric with respect to the xz- and yz-planes, as you can see by replacing x by $-x$, or y by $-y$ in its equation (Figure 7.8).

The Elliptic Cone: $\quad z^2 = \dfrac{x^2}{a^2} + \dfrac{y^2}{b^2}.$

The difference between the equation of the elliptic cone and that of the elliptic paraboloid is the presence of z^2 rather than z. As a result, the traces in the xz-plane ($y = 0$) are the lines $z = \pm\dfrac{x}{a}$, and the traces in the yz-plane ($x = 0$) are the lines $z = \pm\dfrac{y}{b}$. The trace in the xy-plane ($z = 0$) is simply the origin $(0, 0, 0)$ (Figure 7.9).

In the planes $x = d$, the traces are the hyperbolas

$$\frac{a^2z^2}{d^2} - \frac{a^2y^2}{b^2d^2} = 1.$$

Similarly, the traces in the planes $y = d$ are hyperbolas (Figure 7.10). However, in

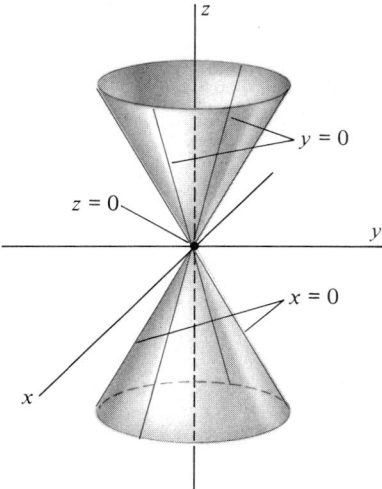

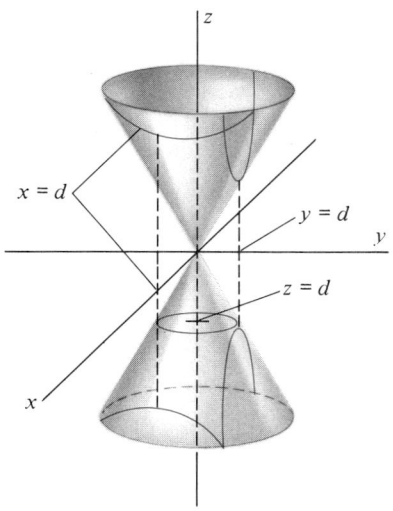

Figure 7.9 Traces of elliptic cone $z^2 = \dfrac{x^2}{a^2} + \dfrac{y^2}{b^2}$ in the coordinate planes.

Figure 7.10 Traces of elliptic cone in planes $x = d$, $y = d$, and $z = d$.

the planes $z = d$ and the traces are the ellipses with equations

$$\frac{x^2}{a^2d^2} + \frac{y^2}{b^2d^2} = 1.$$

The elliptic cone is symmetric with respect to each of the three coordinate planes. Also, it differs from the elliptic paraboloid in that it is unbounded in both the positive and negative z directions (Figure 7.11). If $a = b$, the cone is called a *right* or *circular* cone.

The Hyperboloid of One Sheet: $\dfrac{x^2}{a^2} + \dfrac{y^2}{b^2} - \dfrac{z^2}{c^2} = 1$

The trace in the xy-plane ($z = 0$) is the ellipse

$$\frac{x^2}{a^2} + \frac{y^2}{b^2} = 1,$$

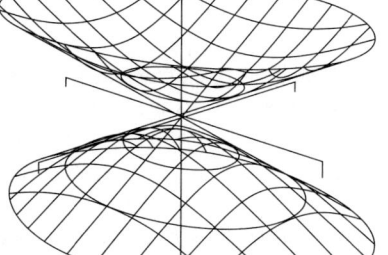

Figure 7.11 Various traces of an elliptic cone.

while traces in the other two coordinate planes are the hyperbolas

$$\frac{x^2}{a^2} - \frac{z^2}{c^2} = 1 \qquad \text{and} \qquad \frac{y^2}{b^2} - \frac{z^2}{c^2} = 1$$

(Figure 7.12). Similarly, setting $z = d$ shows that traces parallel to the xy-plane are ellipses, while the traces in planes parallel to the xz- or yz-planes are hyperbolas. The hyperboloid of one sheet is symmetric with respect to each of the coordinate axes (Figures 7.13 and 7.14).

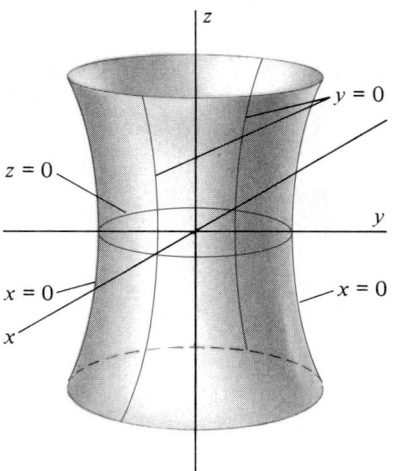

Figure 7.12 Traces of the hyperboloid of one sheet $\dfrac{x^2}{a^2} + \dfrac{y^2}{b^2} - \dfrac{z^2}{c^2} = 1$ in the coordinate planes.

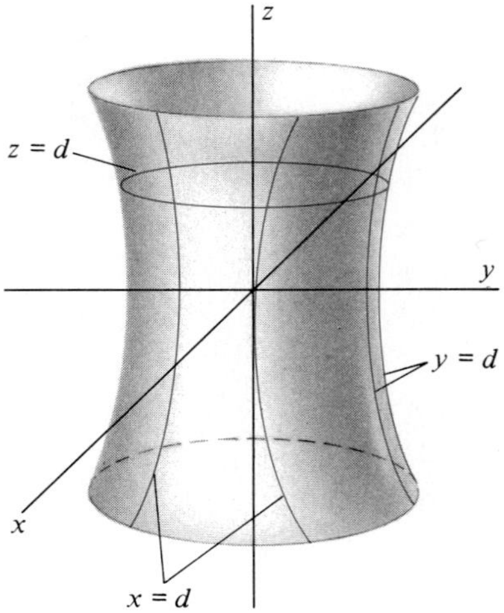

Figure 7.13 Traces of the hyperboloid of one sheet in the planes $x = d$, $y = d$, and $z = d$.

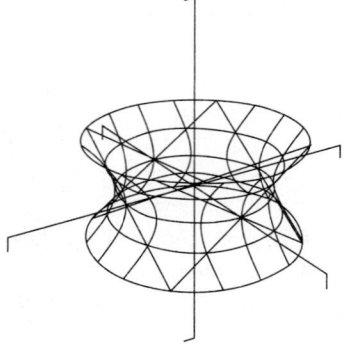

Figure 7.14 Various traces of a hyperboloid of one sheet.

The Hyperboloid of Two Sheets: $\quad \dfrac{x^2}{a^2} + \dfrac{y^2}{b^2} - \dfrac{z^2}{c^2} = -1$

Setting $z = 0$ shows that this surface has no trace in the xy-plane. The traces in the xz- and yz-planes are the hyperbolas

$$\frac{z^2}{c^2} - \frac{x^2}{a^2} = 1 \qquad \text{and} \qquad \frac{z^2}{c^2} - \frac{y^2}{b^2} = 1,$$

respectively. The traces in planes parallel to the xz- and yz-planes are also hyperbolas. However, traces in planes parallel to the xy-plane ($z = d$) are ellipses

$$\frac{x^2}{a^2} + \frac{y^2}{b^2} = \frac{d^2}{c^2} - 1$$

provided $|d| > |c|$. The hyperboloid of two sheets is symmetric with respect to each of the coordinate planes (Figures 7.15–7.17).

The Hyperbolic Paraboloid: $\quad z = \dfrac{y^2}{b^2} - \dfrac{x^2}{a^2}$

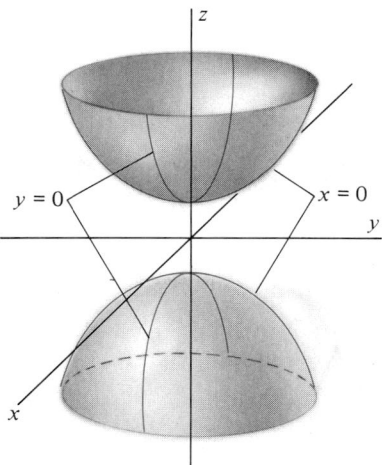

Figure 7.15 Traces of the hyperboloid of two sheets $\dfrac{x^2}{a^2} + \dfrac{y^2}{b^2} - \dfrac{z^2}{c^2} = -1$.

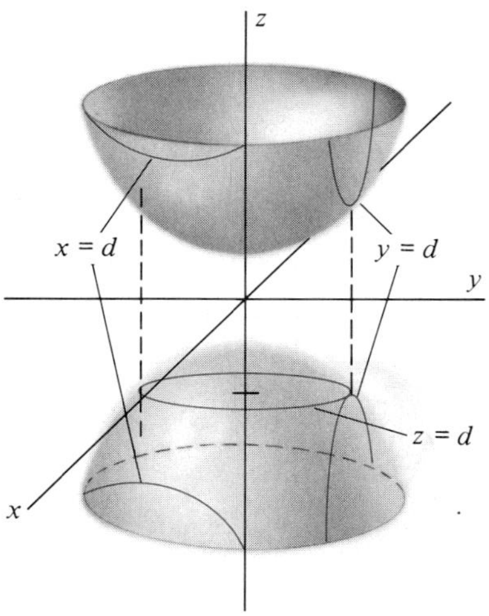

Figure 7.16 Traces of the hyperboloid of two sheets in planes parallel to the coordinate planes.

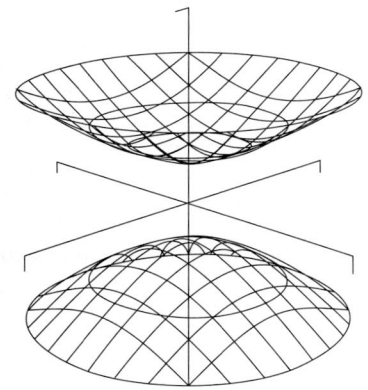

Figure 7.17 Traces of a hyperboloid of two sheets.

Setting $z = 0$ shows that the trace in the xy-plane is the pair of intersecting straight lines $y = \pm \left| \dfrac{b}{a} \right| x$. In the xz-plane ($y = 0$) the trace is the parabola $z = -\dfrac{x^2}{b^2}$ (Figure 7.18). The traces in planes parallel to the xz- and yz-planes are also parabolas. However, the traces in planes parallel to the xy-plane ($z = d$) are hyperbolas. The hyperbolic paraboloid is symmetric with respect to the xz- and yz-planes (Figure 7.19).

The six figures described here do not exhaust all possibilities for the graph of the general second degree equation (1). We shall not attempt to present an all encompassing discussion as we were able to do for the two variable case in Chapter 14. However, we do wish to point out that the technique of completing the square may often be applied to reduce a given second degree equation to one of the six types discussed above. In problems of this type it is often useful to use the **translation of**

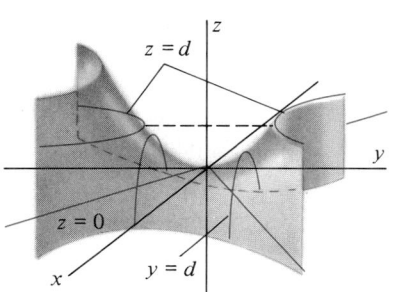

Figure 7.18 Traces of the hyperbolic paraboloid $z = \dfrac{y^2}{b^2} - \dfrac{x^2}{a^2}$.

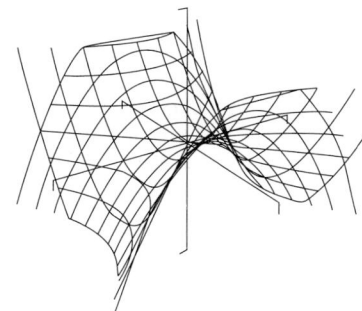

Figure 7.19 Various traces of a hyperbolic paraboloid.

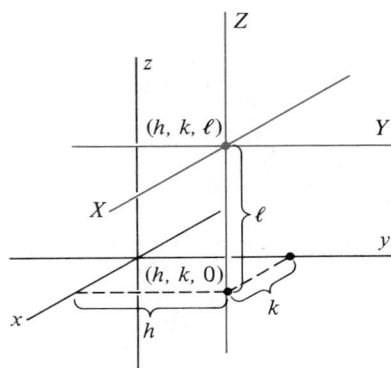

Figure 7.20 Translation of axes $X = x - h$, $Y = y - k$, $Z = z - \ell$.

axes defined by the following equations

$$X = x - h, \tag{6a}$$
$$Y = y - k, \tag{6b}$$
$$Z = z - \ell. \tag{6c}$$

As in the two variable case, these substitutions amount to a relocation of the coordinate axes so that the origin lies at (h, k, ℓ) and so that each of the coordinate axes lies parallel to its original position (see Figure 7.20).

Example 1

Describe the graph of the equation

$$x^2 + y^2 + 3z^2 - 2x + 4y - 4 = 0.$$

Strategy

Complete the square in x and in y.

Use translated variables (6a) through (6c) to simplify equation.

Identify the form of the equation obtained.

Solution

$$x^2 - 2x + y^2 + 4y + 3z^2 - 4 = 0$$
$$(x^2 - 2x + 1) + (y^2 + 4y + 4) + 3z^2 - 4 - 1 - 4 = 0$$
$$(x - 1)^2 + (y + 2)^2 + 3z^2 - 9 = 0$$

The given equation therefore has the form

$$X^2 + Y^2 + 3Z^2 = 9,$$

or

$$\frac{X^2}{3^2} + \frac{Y^2}{3^2} + \frac{Z^2}{(\sqrt{3})^2} = 1$$

with $X = x - 1$, $Y = y + 2$, and $Z = z$. The graph of this equation is an ellipsoid with center at $(1, -2, 0)$ (Figure 7.21). ◇

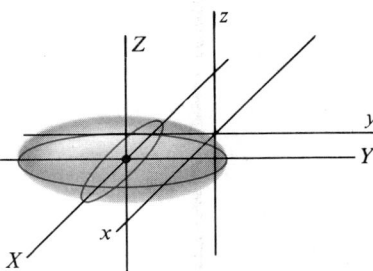

Figure 7.21 Ellipsoid $x^2 + y^2 + 3z^2 - 2x + 4y - 4 = 0$.

Cylinders

When an equation involves fewer variables than the number of axes on which it is graphed, the traces obtained by selecting various values of the missing variable must all be the same. The simplest example of this occurs when the equation $x = a$ is graphed in the xy-coordinate plane. Any trace obtained by setting $y = d$ is simply a point whose x-coordinate is a (Figure 7.22).

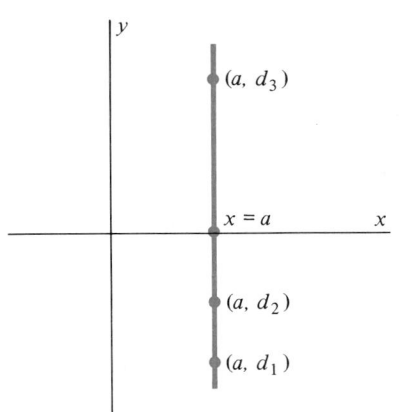

Figure 7.22 Graph of equation $x = a$ is a *cylinder* in the xy-plane.

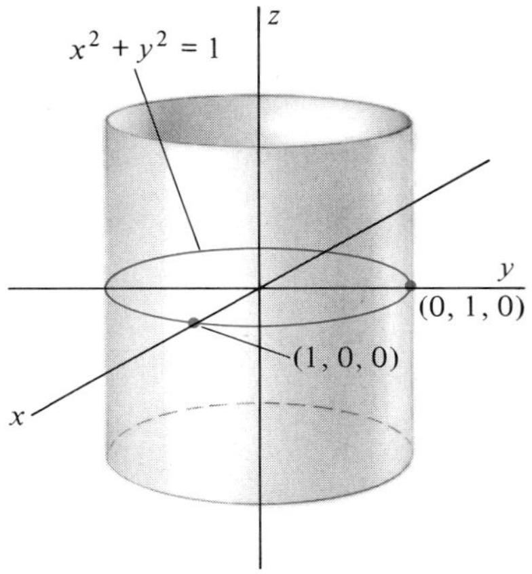

Figure 7.23 The equation $x^2 + y^2 = 1$ is a cylinder in xyz space.

A similar phenomenon occurs when an equation involving only two variables, say x and y, is graphed in xyz-space. Any trace corresponding to a fixed value of the missing variable, $z = d$, will be a "copy" of the graph of the equation in the xy-plane. The graph in xyz-space will therefore be a surface which can be thought of as an "infinite stack" of copies of the two dimensional figure. Graphs in xyz-space of equations with one or two variables missing are therefore called **cylinders.**

A typical cylinder is the graph of $x^2 + y^2 = 1$ in xyz-space (Figure 7.23). This is a special case of the **elliptic cylinder**

$$\frac{x^2}{a^2} + \frac{y^2}{b^2} = 1$$

in space. Three other examples of cylinders are

(a) a plane whose equation contains only two variables, such as $x + y = 2$ (Figure 7.24);

(b) the parabolic cylinder $y = x^2$ (Figure 7.25);

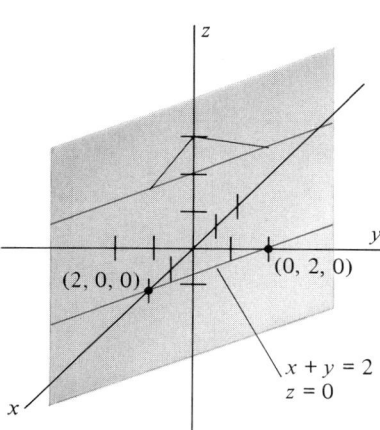

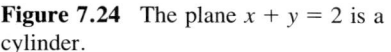

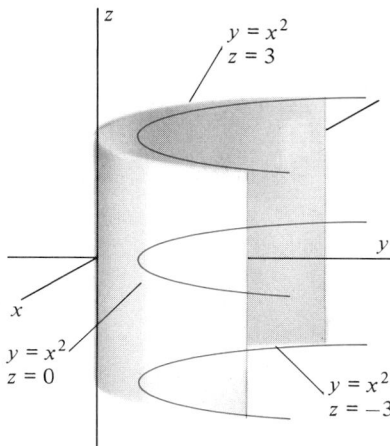

Figure 7.24 The plane $x + y = 2$ is a cylinder.

Figure 7.25 Parabolic cylinder $y = x^2$.

(c) the hyperbolic cylinder $\dfrac{y^2}{a^2} - \dfrac{x^2}{b^2} = 1$ (Figure 7.26).

Example 2

Sketch the cylinder $9x^2 + 4z^2 - 18x - 16z = 11$.

Solution: The first task is to complete the square in x and z:

$$9x^2 - 18x + 4z^2 - 16z = 11$$
$$9(x^2 - 2x) + 4(z^2 - 4z) = 11$$
$$9(x^2 - 2x + 1) + 4(z^2 - 4z + 4) = 11 + 9 \cdot 1 + 4 \cdot 4$$
$$9(x - 1)^2 + 4(z - 2)^2 = 36$$
$$\frac{(x - 1)^2}{2^2} + \frac{(z - 2)^2}{3^2} = 1.$$

This is the equation for an ellipse with center $(1, 2)$ in the xz-plane. Since the variable y is missing, the figure is an elliptic cylinder with central axis parallel to the y-axis (Figure 7.27). ◇

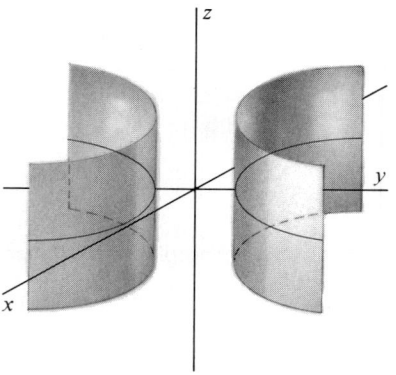

Figure 7.26 The hyperbolic cylinder $\dfrac{y^2}{a^2} - \dfrac{x^2}{b^2} = 1$.

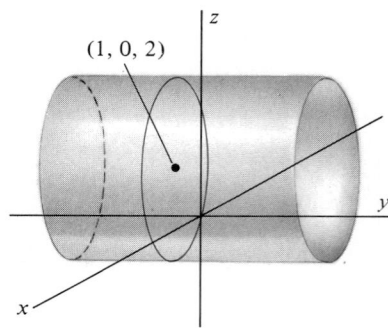

Figure 7.27 The elliptic cylinder $9x^2 - 18x + 4z^2 - 16z = 11$.

Exercise Set 16.7

In Exercises 1–20, describe the quadric surface that is the graph of the given equation.

1. $x^2 + y^2 + z^2 = 10$

2. $\dfrac{x^2}{4} + \dfrac{y^2}{9} + \dfrac{z^2}{16} = 1$

3. $\dfrac{x^2}{4} + \dfrac{y^2}{9} = z^2$

4. $6x^2 + 12y^2 - 8z^2 - 24 = 0$

5. $4x^2 + 9y^2 - 36z^2 + 36 = 0$

6. $9y^2 - 6x^2 = 54$

7. $2x^2 - 3y^2 + 8x + 6y - 6z + 5 = 0$

8. $8x^2 + 4y^2 - 2z^2 + 24y - 4z + 44 = 0$

9. $x^2 + 2y^2 - 2z^2 + 6x - 4y + 7 = 0$

10. $x^2 + 4y^2 - 4z^2 + 6x - 8y - 8z + 9 = 0$

11. $6x^2 + 9y^2 + 36y - 54z + 36 = 0$

12. $4x^2 + 3y^2 + 3z^2 - 12y + 18z + 27 = 0$

13. $9z^2 + 4y^2 - 36x = 0$

14. $x^2 - 4y^2 + 4z^2 = 0$

15. $z^2 + y^2 - x^2 - 1 = 0$

16. $6x^2 - 12y^2 + 8z^2 + 24 = 0$

17. $x^2 + y^2 + z^2 - 6x + 2y + 6z + 18 = 0$

18. $9x^2 + 4y^2 + 9z^2 - 36x + 8y - 18z + 13 = 0$

19. $x^2 + y^2 + 4x - 2y - 36z + 175 = 0$

20. $y^2 - x^2 - 6x - 2y - 4z - 16 = 0$

In Exercises 21–36, describe the given cylinder in xyz-space.

21. $x = 3$

22. $y = 6$

23. $x + y = 6$

24. $x^2 + y^2 = 5$

25. $\dfrac{x^2}{4} - \dfrac{y^2}{9} = 1$

26. $\dfrac{x^2}{2} + \dfrac{y^2}{4} = 1$

27. $xy = 1$

28.

29. $y = 4z^2$

30. x

31. $x^2 - z^2 = 1$

32. $z^2 -$

33. $z = 1 - y^2$

34. $(x - 1)^2$

35. $3x^2 + 12x - z + 16 = 0$

36. $9x^2 + 4z^2 + 18x - 16z - 11 = 0$

37. Write the equation for the surface obtained by revolving the graph of $y = 2x$ about the y-axis.

38. Describe the set of all points P for which the distance from P to the y-axis is twice the distance from P to the xz-plane.

39. Find an equation for the solid obtained by revolving the graph of $y = z^2$ about the y-axis.

SUMMARY OUTLINE OF CHAPTER 16

◆ A **vector** in the plane is an ordered pair of numbers $v = \langle a, b \rangle$ subject to the laws (page 663)

$$\langle a_1, b_1 \rangle + \langle a_2, b_2 \rangle = \langle a_1 + a_2, b_1 + b_2 \rangle,$$

$$c\langle a, b \rangle = \langle ca, cb \rangle, \qquad c \in (-\infty, \infty).$$

◆ A vector in space is an ordered triple $v = \langle a, b, c \rangle$ subject to the laws (page 685)

$$\langle a_1, b_1, c_1 \rangle + \langle a_2, b_2, c_2 \rangle = \langle a_1 + a_2, b_1 + b_2, c_1 + c_2 \rangle,$$

$$d\langle a, b, c \rangle = \langle da, db, dc \rangle, \qquad d \in (-\infty, \infty).$$

◆ The **unit coordinate vectors** are $i = \langle 1, 0, 0 \rangle$, $j = \langle 0, 1, 0 \rangle$, $k = \langle 0, 0, 1 \rangle$. (page 686)

◆ *Notation:* $v = \langle a, b, c \rangle = ai + bj + ck$. (page 686)

◆ The **length** of the vector $v = ai + bj + ck$ is $|v| = \sqrt{a^2 + b^2 + c^2}$. (page 686)

◆ Vector length has the following properties: (page 669)

(i) $|v| \geq 0$; $|v| = 0$ if and only if $v = 0$,
(ii) $|cv| = |c|\,|v|$,
(iii) $|v + w| \leq |v| + |w|$.

◆ A **unit vector** in the direction of v is $u = \dfrac{v}{|v|}$. (page 670)

◆ The **dot product** of the vectors $v = x_1 i + y_1 j + z_1 k$ and $w = x_2 i + y_2 j + z_2 k$ is the number (page 687)

$$v \cdot w = x_1 x_2 + y_1 y_2 + z_1 z_2.$$

◆ *Theorem:* $v \cdot w = |v|\,|w|\cos\theta$, where θ is the angle between v and w. (page 675)

◆ The vectors v and w are **orthogonal** (perpendicular) if and only if $v \cdot w = 0$. (page 676)

◆ The **component** of the vector v in the direction of the vector w is the **number** (page 677)

$$\mathrm{comp}_w\, v = \frac{v \cdot w}{|w|}.$$

◆ The **projection** of the vector v along the vector w is the **vector** (page 679)

$$\mathrm{proj}_w\, v = \left(\frac{v \cdot w}{|w|^2}\right) w.$$

distance from the point $P = (x_0, y_0)$ in the plane to the line with equation $ax + by + c = 0$ is
(page 678)

$$d = \frac{|ax_0 + by_0 + c|}{\sqrt{a^2 + b^2}}.$$

◆ The distance d between two points $P = (x_1, y_1, z_1)$ and $Q = (x_2, y_2, z_2)$ in space is
(page 683)

$$d = \sqrt{(x_2 - x_1)^2 + (y_2 - y_1)^2 + (z_2 - z_1)^2}.$$

◆ The **cross product** of the vectors $v = x_1 i + y_1 j + z_1 k$ and $w = x_2 i + y_2 j + z_2 k$ in space is the **vector**
(page 691)

$$v \times w = \det \begin{bmatrix} i & j & k \\ x_1 & y_1 & z_1 \\ x_2 & y_2 & z_2 \end{bmatrix} = (y_1 z_2 - z_1 y_2)i + (z_1 x_2 - x_1 z_2)j + (x_1 y_2 - y_1 x_2)k.$$

◆ **Theorem:** $v \times w = (|v| \, |w| \sin \theta)n$, where n is a unit vector orthogonal to both v and w in the direction determined by the right-hand rule.
(page 694)

◆ $|v \times w|$ is the area of the parallelogram determined by v and w.
(page 695)

◆ The **line** with position vector $a = a_1 i + a_2 j + a_3 k$ and direction vector $b = b_1 i + b_2 j + b_3 k$ has
(page 698)

   (i) vector equation $r(t) = a + tb, \quad -\infty < t < \infty$.
   (ii) parametric equations $x(t) = a_1 + tb_1, \; y(t) = a_2 + tb_2, \; z(t) = a_3 + tb_3$.
   (iii) symmetric equations $\dfrac{x - a_1}{b_1} = \dfrac{y - a_2}{b_2} = \dfrac{z - a_3}{b_3}$.

◆ The **distance** d from the point P to the line with direction vector b containing the point Q is
(page 700)

$$d = \frac{|\overrightarrow{PQ} \times b|}{|b|}.$$

◆ The **plane** with normal vector n containing the point P is the set of all points Q for which $\overrightarrow{PQ} \cdot n = 0$.
(page 703)

◆ If $n = ai + bj + ck$, $P = (x_0, y_0, z_0)$, and $Q = (x, y, z)$, the equation of the plane is
(page 703)

$$ax + by + cz = ax_0 + by_0 + cz_0.$$

◆ The cylindrical coordinates (r, θ, z) and the rectangular coordinates (x, y, z) for a point P are related via the equations
(page 707)

$$x = r \cos \theta, \quad r \geq 0, \qquad\qquad r = \sqrt{x^2 + y^2}$$
$$y = r \sin \theta, \quad r \geq 0, \quad \text{and} \quad \tan \theta = y/x,$$
$$z = z \qquad\qquad\qquad\qquad z = z.$$

◆ The spherical coordinates (ρ, θ, ϕ) and the rectangular coordinates (x, y, z) for a point P are related via the equations
(page 710)

$$x = \rho \sin \phi \cos \theta, \quad \rho \geq 0 \qquad\qquad \rho = \sqrt{x^2 + y^2 + z^2}$$
$$y = \rho \sin \phi \sin \theta, \quad \rho \geq 0 \quad \text{and} \quad \tan \theta = y/x, \quad x \neq 0$$
$$z = \rho \cos \phi, \quad \rho \geq 0 \qquad\qquad \tan \phi = \frac{\sqrt{x^2 + y^2}}{z}, \quad z \neq 0.$$

◆ **The Ellipsoid:** $\dfrac{x^2}{a^2} + \dfrac{y^2}{b^2} + \dfrac{z^2}{c^2} = 1$
(page 713)

◆ **The Elliptic Paraboloid:** $z = \dfrac{x^2}{a^2} + \dfrac{y^2}{b^2}$
(page 714)

◆ **The Elliptic Cone:** $z^2 = \dfrac{x^2}{a^2} + \dfrac{y^2}{b^2}$.
(page 714)

◆ **The Hyperboloid of One Sheet:** $\dfrac{x^2}{a^2} + \dfrac{y^2}{b^2} - \dfrac{z^2}{c^2} = 1$
(page 715)

◆ **The Hyperboloid of Two Sheets**: $\dfrac{x^2}{a^2} + \dfrac{y^2}{b^2} - \dfrac{z^2}{c^2} = -1$ (page 716)

◆ **The Hyperbolic Paraboloid**: $z = \dfrac{y^2}{b^2} - \dfrac{x^2}{a^2}$ (page 716)

REVIEW EXERCISES—CHAPTER 16

1. Find the distance between the points $P = (1, 2, -4)$ and $Q = (2, -5, 2)$.

2. Find an equation for the set of all points $P = (x, y, z)$ equidistant from the fixed points $P_1 = (1, -2, 1)$ and $Q = (3, 4, -3)$.

3. Describe the graph of the equation $x^2 + y^2 + z^2 - 6x + 4y - 2z + 10 = 0$.

4. Let $u = i + 2j - 3k$, $v = 2i + 2j + 6k$, $w = i - 4j + 3k$. Find
   a. $u + 2v$
   b. $u - v$
   c. $|3u + v|$
   d. $u \cdot v$
   e. $u \cdot (v \times w)$
   f. $|u + v - w|$

5. Find the cosine of the angle between the vectors $v = i - 3j + 2k$ and $w = 3i + 3j + 2k$.

6. Determine whether the following pairs of lines intersect and, if so, at what point.
   a. $x = t$, $y = t + 2$, $z = 2t - 4$;
      $x = 1 - t$, $y = 3 + t$, $z = 4t$
   b. $x = 1 + 2t$, $y = t - 2$, $z = 1 + t$;
      $x = 2t + 1$, $y = 4 - 2t$, $z = 5 - t$

7. Find parametric equations for the line containing the point $(2, 1, -3)$ that is perpendicular to both of these lines.
   ℓ_1: $x = 2t$, $y = 3 + t$, $z = 5t$
   ℓ_2: $x = t + 5$, $y = 6 - 4t$, $z = 3t + 4$

8. Find the direction cosines for the vector $v = 3i + 4j + 5k$.

9. Find an equation for the plane containing the points $(-1, 2, 1)$, $(2, 5, 3)$, and $(-4, 0, 2)$.

10. Find an equation for the plane containing the line
    ℓ_1: $r(t) = i - 3j + k + t(4i + 2j - k)$
    and the point $(1, 2, 1)$.

11. Find a vector normal to the plane with equation $8x + y - 2z = 5$.

12. Find the distance from the point $P = (1, 2, -4)$ to the plane with equation $3x + 2y - 5z = 5$.

13. Find a vector equation for the line of intersection of the planes with equations $2x + 3y - z = 4$ and $x - 3y + 5z = 2$.

14. Show that if $v \times w = 0$ and $w \neq 0$ then $v = cw$ for some constant c.

15. Show that $u \times (v \times w) = (u \cdot w)v - (u \cdot v)w$.

16. Show that $(u + v) \times (u - v) = 2v \times u$.

17. Sketch the graph of the equation $y^2 = 9 + z^2$.

18. Sketch the graph of the equation $x^2 + z^2 = 1 + y$.

19. Find rectangular coordinates for the point with spherical coordinates $(2, \pi/4, \pi/3)$.

20. Find parametric equations for the line containing the points $(3, 5, -6)$ and $(-2, 3, -1)$.

21. Find an equation for the plane containing the point $(2, -5, 3)$ and perpendicular to the line with parametric equations $x = 4 + 2t$, $y = 3 - t$, $z = 5 + 6t$.

22. Find an equation for the sphere with center on the z-axis and containing the points $(5, 0, 0)$ and $(0, 0, 4)$.

23. For $v = i + 2j + 4k$ and $w = 3j + 4k$, find
    a. $\text{comp}_w v$
    b. $\text{proj}_w v$

24. Find the distance from the point $(-4, 2, 5)$ to the line with equation $x = 3 + t$, $y = 4 - 2t$, $z = 5 + 5t$.

25. Find the area of the triangle with vertices $(1, 3, -2)$, $(1, 1, 1)$, and $(4, 0, 3)$.

26. Find a unit vector in the same direction as $v = i - 3j + 4k$.

27. Find an equation for the plane containing the points $(3, -2, 6)$ and $(4, -2, 2)$ that is perpendicular to the xz-plane.

28. Find cylindrical coordinates for the point with rectangular coordinates $(2, 2, 5)$.

29. Find symmetric equations for the line with parametric equations $x = 1 + 3t$, $y = 2 + 7t$, $z = 3 - t$.

In Exercises 30–33, find a rectangular equation for the given equation and sketch the graph.

30. $r = \cos \theta$

31. $\phi = \pi/4$

32. $\rho = 2 \sec \phi$

33. $\rho \sin \phi = 2 \cos \theta$

34. Find the vector of length 3 with direction cosines $\dfrac{\sqrt{2}}{2}$, 0, and $\dfrac{\sqrt{2}}{2}$.

35. For $v = 2i + 3j - k$ and $w = i + 3j + k$ find
    a. $\text{comp}_w v$
    b. $\text{proj}_v w$

36. Find the distance from the point $(1, 2, 1)$ to the plane containing the points $(1, 1, 1)$, $(2, -1, 5)$, and $(3, 1, -2)$.

37. Find the equation for the plane with intercepts $x = 3$, $y = 4$, $z = 5$.

38. Find an equation for the plane perpendicular to the line with symmetric equations

$$\frac{x - 2}{3} = \frac{y + 1}{2} = \frac{z - 3}{-2}$$

and containing the point $(3, -2, 3)$.

39. Find parametric equations for the line parallel to the line of intersection of the two planes $x + y + 2z = 6$ and $x - y - 2z = 4$ and containing the point $(5, 2, -3)$.

40. Find all vectors of length 2 orthogonal to both $v = 3i + 2j + k$ and $w = i + j - 2k$.

41. Find the area of the parallelogram determined by the vectors $v = 2i + 3j + k$ and $w = i - j + 2k$.

41. Prove that the planes $a_1x + b_1y + c_1z = d_1$ and $a_2x + b_2y + c_2z = d_2$ are perpendicular if and only if $a_1a_2 + b_1b_2 + c_1c_2 = 0$.

43. Show that the points $(1, 1, -1)$, $(0, 1, -1/2)$, $(-1, 1, 0)$, and $(0, 0, 1/4)$ all lie in the same plane.

In Exercises 44–50, describe the graph.

44. $9x^2 - 4y^2 + 36z^2 = -36$

45. $6x^2 + y^2 - 2z^2 = 6$

46. $36(x - 1)^2 + 18(y + 3)^2 + 8(z + 1)^2 = 72$

47. $-x^2 + y^2 = 9z$

48. $y^2 = 4x^2 + 2z$

49. $9x^2 + 9y^2 - 4z^2 + 18x - 16z - 43 = 0$

50. $(x + 1)(y - 3) = 1$

UNIT 7

CALCULUS IN HIGHER DIMENSIONS AND DIFFERENTIAL EQUATIONS

Josiah Willard Gibbs

George Gabriel Stokes

A host of mathematicians contributed to the several topics included in this unit: partial differentiation, multiple integration, and vector analysis. Several of these people have been discussed in other units.

Newton differentiated functions of two variables by means of formulas that we now obtain by partial differentiation. This work is recorded in his personal papers, but was not published. Leibniz also differentiated functions of two variables, but made little use of them. Jakob Bernoulli and his nephew Nicolaus used what were essentially partial derivatives around 1720, although they seemingly did not recognize that there is a difference between differentiation of a function of one variable and that for a function of two or more variables. The same symbol was used by many early writers for regular and partial derivatives, which of course led to much confusion. The "rounded d" symbol, ∂, was first used by Euler in 1776, but not in the way that we use it today. The first to use $\partial y/\partial x$ was the Frenchman Adrien-Marie Legendre in 1786, but it was more than a century before that notation came into general use.

Multiple integration was first used by Newton, but his arguments were geometrical and somewhat unclear. In the first half of the eighteenth century Euler used repeated integrations in order to integrate over a bounded domain. Joseph Louis Lagrange used a triple integral in a work on gravitation involving ellipsoids around 1775. By the nineteenth century the use of multiple integrals had become fairly common.

The concept of a vector was known in antiquity. Aristotle represented forces by vectors, and knew that two forces acting in different directions could be summed by what we call the parallelogram law. Representation of complex numbers in the plane was done by Wessel in Denmark, Argand in France, and Gauss in Germany between 1798 and 1806. It was fairly well known by 1830 that vectors could be nicely expressed as complex numbers. However, many physical situations involve more than two factors which are not all in the same plane. For many years mathematicians searched for a three-dimensional vector system which preserved all of the properties of real-number algebra, but their efforts were unsuccessful.

The problem was resolved by Ireland's greatest mathematician, William Rowan Hamilton (1805–1865). Hamilton was a largely self-taught youthful prodigy whose ability first manifested itself, as with several other mathematicians, in language. He read Greek, Latin, and Hebrew by the age of 5, and by the age of 14 he knew 14 languages including Sanskrit and Arabic. He became interested in mathematics through meeting an American calculating prodigy, and had soon read all four volumes of Newton's *Principia* and Laplace's *Celestial Mechanics,* in which he found a mathematical error. At the age of 22, still an undergraduate, he was appointed Professor of Astronomy at an Irish university and Royal Astronomer of Ireland.

While walking to Dublin, Hamilton suddenly realized the impossibility of a three-dimensional system with the desired properties, but that a *four*-dimensional system is possible. He stopped to carve the basic relations on the handrail of a bridge, and that carving can still be seen. He postulated a system with four mutually independent unit vectors, called *quaternions*. Quaternions do not obey the commutative law (that is, $A \cdot B \neq B \cdot A$); this was the first algebra in which such behavior was studied. The use of quaternions was advocated throughout the nineteenth century by Hamilton and others, and courses in the subject were taught in

British and American graduate schools well into the twentieth century. Eventually, however, it was superseded by the vector analysis pioneered by J. W. Gibbs. Hamilton did not lead a happy life; his marriage was unpleasant and his quaternions were not well-received on the Continent. He was knighted early, in 1835, but became an alcoholic in the last twenty years of his life.

Josiah Willard Gibbs (1839–1903) was an American mathematician and scientist in a day when American science was held in little esteem. Born in New Haven, Connecticut, he attended Yale College and obtained a Ph.D. in physics in 1863, one of the first doctorates to be granted in the United States. He then went to Europe for further study. Returning in 1871, he became Professor of Mathematical Physics at Yale, though he received no salary for the first nine years. He remained at Yale for the rest of his life, living in the house in which he had grown up, less than a block from the campus. In 1881 he had printed, at his own expense, a little pamphlet called *Elements of Vector Analysis* for the use of his students. In it, he created the subject much as it is known today. The pamphlet became well-known, and E. B. Wilson published a book called *Vector Analysis,* based on Gibbs' lectures, in 1901.

Oliver Heaviside (1850–1925) was another contributor to the development of vector analysis, and in fact he essentially created the subject independently of Gibbs. He had only an elementary school education in his native England, but embarked on a regimen of reading which resulted in his becoming a pragmatically successful mathematical physicist. He began his career as a telegrapher, but abandoned that work as a result of the deafness which bothered him all of his life. His creative work was disdained by many professional mathematicians particularly because of his lack of formal education and unorthodox methods. His work in electromagnetic theory was particularly brilliant when he generalized Maxwell's theory. Heaviside used ideas from Hamilton's theory of quaternions, and developed both scalar and vector multiplication. He never married, but lived with his parents until they died. He spent his last years in poverty and died in a nursing home.

Green's Theorem and Stokes' Theorem are two applications of vectors which are found in this unit. George Green (1793–1841), like Heaviside, had only an elementary school English education, and was largely self-taught. He was one of the first to treat static electricity and magnetism in mathematical terms. At the age of 36 Green decided to get a university education. He studied for four years, was admitted to Cambridge at the age of 40, and graduated four years later. His marks were not high, largely because he spent more time writing original mathematics than in studying for classes. In 1828 he had printed at his own expense his most important work, in which his mathematical treatment of electricity and magnetism appears. Here too is found what we now call Green's Theorem. It remained unknown until four years after his death when the British physicist Lord Kelvin became aware of the paper and had it reprinted in a German journal. By this time the Russian mathematician Michel Ostrogradski had also discovered the theorem; it is known by his name in the Soviet Union.

George Gabriel Stokes (1819–1903) was an Irishman who was Lucasian Professor of Mathematics at Cambridge, the same post held by Newton a century and a half earlier. Though prestigious, the position did not pay well, and Stokes had to teach at a School of Mines in order to make ends meet. Also like Newton, he served the Royal Society of London as an officer for many years—he was either president or secretary for 45 years. He studied various physical phenomena from a mathematical point of view, and through a brilliant paper established the foundations of hydrodynamics. The theorem named after him is a three-dimensional vector version of Green's Theorem.

(Photograph of J. W. Gibbs from the Library of Congress. Photograph of George G. Stokes from the David Eugene Smith Papers, Rare Book and Manuscript Library, Columbia University.)

Chapter 17
Vector-Valued Functions

In Chapter 1 we noted that a function need not have real numbers as its values. For example, the rule that assigns to each member of your immediate family the month of his or her birth is a function, and its values are months of the year.

In this chapter we shall discuss *vector-valued functions* of a real variable. These are functions defined on subsets of the real numbers, $(-\infty, \infty)$, but with vectors (either in the plane or in space) as values. In keeping with the vector notation of Chapter 16, we shall write the name f of a vector-valued function in boldface to remind you that its values $f(t)$ are vectors.

17.1 PROPERTIES OF VECTOR-VALUED FUNCTIONS

A **vector-valued function** is a function

$$f(t) = x(t)i + y(t)j + z(t)k \tag{1}$$

whose domain is a set of real numbers and whose values $f(t)$ are vectors.

In equation (1) we refer to the functions x, y, and z as the "component functions," to the values $x(t)$, $y(t)$, and $z(t)$ as the "scalar components," and to the vectors $x(t)i$, $y(t)j$, and $z(t)k$ as the "vector components." When we use the term "components" alone, it will always be clear from the context which type we mean.

By the **graph** of a vector function f, we shall mean the set of all points in space that are the terminal points of the arrows representing the vectors $f(t)$ *when $f(t)$ originates at the origin*. That is, we shall associate the vector $f(t) = x(t)i + y(t)j + z(t)k$ with the point $(x(t), y(t), z(t))$ in space.

We have already seen two examples of functions of this type:

(i) The **line** with vector equation

$$\begin{aligned} r(t) &= (a_1i + a_2j + a_3k) + t(b_1i + b_2j + b_3k) \\ &= (a_1 + tb_1)i + (a_2 + tb_2)j + (a_3 + tb_3)k \end{aligned}$$

has the form of equation (1) with $x(t) = (a_1 + tb_1)$, $y(t) = (a_2 + tb_2)$, and $z(t) = (a_3 + tb_3)$. The vector $r(t)$ is interpreted as extending from the origin to the point $P(t)$ whose coordinates are the components of the vector $r(t)$. If t denotes time, the vector function represents the motion of a particle moving in space.

(ii) The **curve** C in the plane determined by the parametric equations

$$C: \quad \begin{cases} x = x(t) \\ y = y(t) \end{cases} \quad a \le t \le b$$

may be interpreted as the graph of the vector function

$$f(t) = x(t)i + y(t)j, \qquad a \le t \le b$$

where the vector $f(t)$ extends from the origin to the point $(x(t), y(t))$ on C.

The concept of vector function allows us to pursue generalizations of both of these examples in a unified way. By interpreting $f(t)$ as a vector from the origin to a particle in space, we may study the motion of that particle, addressing the usual issues of position, speed, velocity, acceleration, distance, and elapsed time.

In studying vector-valued functions we shall make heavy use of the fact that the vector function $f(t)$ is the sum of three vector components, each of which involves a function of a single independent variable (Figure 1.1). The laws of vector algebra

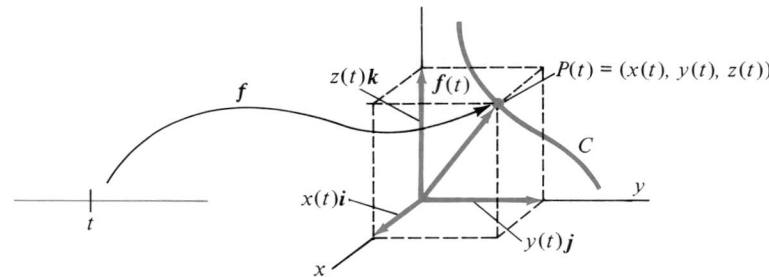

Figure 1.1 Graph of the vector-valued function $f(t) = x(t)i + y(t)j + z(t)k$.

allow us to perform algebraic operations with the components, while the theory of the calculus (Chapters 2–10) will enable us to address many of the issues associated with derivatives and integrals of vector functions.

Example 1

Sketch the graph of the vector function

$$f(t) = a \cos ti + a \sin tj + btk.$$

where a and b are constants.

Solution: We note that in any plane parallel to the xy-plane the distance from the point $(a \cos t, a \sin t, bt)$ to the point $(0, 0, bt)$ on the z-axis is

$$d = \sqrt{a^2 \cos^2 t + a^2 \sin^2 t} = |a|.$$

Thus, all points lie on the graph of the circular cylinder $x^2 + y^2 = a^2$. As t increases, the z-coordinate of the point determined by $f(t)$ increases uniformly. For every point $(x(t), y(t), z(t))$ on the graph, infinitely many points $(x(t), y(t), z(t) + 2\pi nb)$, $n = \pm 1, \pm 2, \ldots$ also lie on the graph. The graph is the **circular helix** in Figure 1.2. ◇

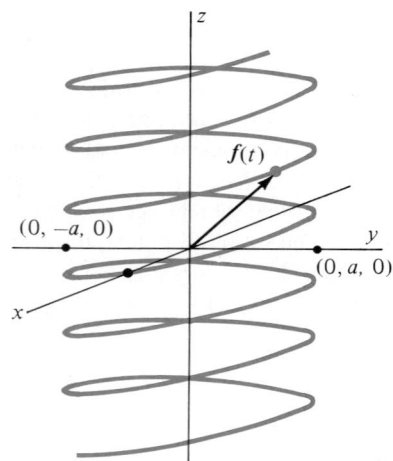

Figure 1.2 The circular helix $f(t) = a \cos ti + a \sin tj + btk$.

Algebra, Limits, and Continuity

Given two vector-valued functions

$$f(t) = x_1(t)i + y_1(t)j + z_1(t)k$$

and

$$g(t) = x_2(t)i + y_2(t)j + z_2(t)k,$$

we may form the following functions, as indicated.

(i) The **sum** of f and g is the vector-valued function $f + g$ where

$$(f + g)(t) = f(t) + g(t)$$
$$= [x_1(t) + x_2(t)]i + [y_1(t) + y_2(t)]j + [z_1(t) + z_2(t)]k.$$

In other words, we add vector functions component by component.

(ii) The **scalar multiple** of f by the real number c is the vector-valued function cf where

$$(cf)(t) = cf(t) = [cx_1(t)]i + [cy_1(t)]j + [cz_1(t)]k.$$

That is, multiplication of vector functions by scalars is done component by component.

(iii) The **dot product** of f and g is the **real-valued function** $f \cdot g$ where

$$(f \cdot g)(t) = f(t) \cdot g(t) = x_1(t)x_2(t) + y_1(t)y_2(t) + z_1(t)z_2(t).$$

The values of the dot product function are *numbers* rather than vectors.

(iv) The **cross product** of f and g is the vector-valued function $f \times g$ where

$$(f \times g)(t) = f(t) \times g(t)$$
$$= [y_1(t)z_2(t) - z_1(t)y_2(t)]i + [z_1(t)x_2(t) - x_1(t)z_2(t)]j$$
$$+ [x_1(t)y_2(t) - y_1(t)x_2(t)]k.$$

The values of the cross product function are *vectors*.

Example 2

From the vector-valued functions

$$f(t) = ti + t^2j + \sqrt{t}k, \qquad t \geq 0$$

and

$$g(t) = \cos t\,i + \sin t\,j + tk,$$

the following functions may be formed:

(a) $(f + g)(t) = (t + \cos t)i + (t^2 + \sin t)j + (\sqrt{t} + t)k, \qquad t \geq 0.$
(b) $3f(t) = 3ti + 3t^2j + 3\sqrt{t}\,k, \qquad t \geq 0.$
(c) $(f - 2g)(t) = (t - 2\cos t)i + (t^2 - 2\sin t)j + (\sqrt{t} - 2t)k, \qquad t \geq 0.$
(d) $(f \times g)(t) = \det \begin{bmatrix} i & j & k \\ t & t^2 & \sqrt{t} \\ \cos t & \sin t & t \end{bmatrix}$

$$= (t^3 - \sqrt{t}\sin t)i + (\sqrt{t}\cos t - t^2)j + (t\sin t - t^2\cos t)k. \qquad \diamond$$

Limits of Vector-Valued Functions

The limit of a vector-valued function is defined just as the limit of a real-valued function: $\lim_{t \to a} f(t) = L$ means that $|f(t) - L|$ is small whenever $|t - a|$ is small. (See Figure 1.3.) Note, however, that since $f(t)$ and L are vectors, the expression $|f(t) - L|$ is the *length* of the vector $f(t) - L$.

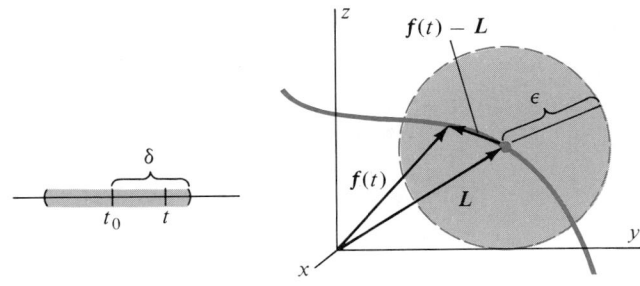

Figure 1.3 If $0 < |t - t_0| < \delta$, then $|f(t) - L| < \epsilon$.

DEFINITION 1

Let f be a vector-valued function that is defined on an open interval containing the number a, except possibly at $t = a$. We say that the vector L is the *limit* of the function f as t approaches a, written $L = \lim\limits_{t \to a} f(t)$, if, corresponding to each number $\epsilon > 0$ there is a number $\delta > 0$ so that

$$\text{if} \quad 0 < |t - a| < \delta, \quad \text{then} \quad |f(t) - L| < \epsilon.$$

The following theorem shows that Definition 1 is equivalent to the existence of limits for each of the component functions of f. This allows us to evaluate limits of vector-valued functions by using the techniques of Chapter 2 for limits of real-valued functions. The proof is given at the end of this section.

THEOREM 1

Let f be the vector-valued function

$$f(t) = x(t)i + y(t)j + z(t)k$$

and let

$$L = L_1 i + L_2 j + L_3 k$$

be a vector. Let f be defined on an open interval containing the number a except possibly at $t = a$. Then

$$L = \lim_{t \to a} f(t)$$

if and only if

$$L_1 = \lim_{t \to a} x(t), \qquad L_2 = \lim_{t \to a} y(t), \qquad \text{and} \qquad L_3 = \lim_{t \to a} z(t).$$

Theorem 1 states that the limit of a vector-valued function is the vector whose components are the limits of the individual components. If one or more of these component limits fails to exist we say that the limit of the vector-valued function fails to exist.

Example 3

Find $\lim\limits_{t \to \pi/3} (t^2 i + \sin tj + \cos tk)$.

Solution: The limits of the components are

$$L_1 = \lim_{t \to \pi/3} x(t) = \lim_{t \to \pi/3} t^2 = \frac{\pi^2}{9},$$

$$L_2 = \lim_{t \to \pi/3} y(t) = \lim_{t \to \pi/3} \sin t = \frac{\sqrt{3}}{2} \qquad \text{(Figure 1.4)},$$

$$L_3 = \lim_{t \to \pi/3} z(t) = \lim_{t \to \pi/3} \cos t = \frac{1}{2}.$$

Thus, by Theorem 1,

$$\lim_{t \to \pi/3} (t^2 i + \sin tj + \cos tk) = \frac{\pi^2}{9} i + \frac{\sqrt{3}}{2} j + \frac{1}{2} k.$$

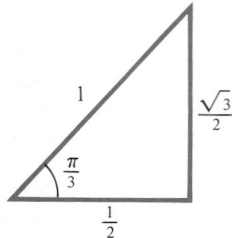

Figure 1.4

Continuity

Continuity for vector-valued functions is defined in terms of limits, just as for real-valued functions. Intuitively, the statement that f is continuous at $t = a$ means that $\lim_{t \to a} f(t) = f(a)$. The formal definition is the following.

DEFINITION 2

Let f be a vector-valued function that is defined on an open interval I containing the number $t = a$. We say that f is *continuous* at a if, corresponding to each number $\epsilon > 0$ there is a number $\delta > 0$ so that

$$\text{if} \quad |t - a| < \delta, \quad \text{then} \quad |f(t) - f(a)| < \epsilon.$$

We say that f is continuous on I if f is continuous at each number $a \in \boldsymbol{I}.$

Like the definition of limit, the definition of continuity for a vector-valued function is equivalent to the requirement that each of the component functions be continuous. We state this result as a corollary of Theorem 1 and we leave its proof, which is similar to the proof of Theorem 1, as an exercise.

COROLLARY 1

The vector-valued function $\boldsymbol{f}(t) = x(t)\boldsymbol{i} + y(t)\boldsymbol{j} + z(t)\boldsymbol{k}$ is **continuous** at $t = t_0$ if each of the component functions $x(t)$, $y(t)$, and $z(t)$ is continuous at t_0.

Note that each component of f must be continuous at t_0 if f is continuous at t_0. Thus, a vector-valued function is **discontinuous** at t_0 if one or more of its component functions is discontinuous at t_0.

Example 4

Find the intervals on which the vector-valued function

$$f(t) = \frac{1}{t}i + t^2 j + \frac{2}{t^2 - 4}k$$

is continuous.

Strategy
Determine the numbers t where one or more of the component functions is discontinuous.

Solution
The $\boldsymbol{i}$-component $x(t) = 1/t$ is discontinuous at $t = 0$.

The $\boldsymbol{j}$-component $y(t) = t^2$ is continuous for all t.

The $\boldsymbol{k}$-component $z(t) = \dfrac{2}{t^2 - 4} = \dfrac{2}{(t-2)(t+2)}$ is discontinuous for $t = -2, 2$.

The vector-valued function is continuous for all other values of t.

The vector-valued function f is therefore discontinuous at $t = -2, 0,$ and 2, so f is continuous on the intervals

$$(-\infty, -2), \ (-2, 0), \ (0, 2), \text{ and } (2, \infty). \qquad \diamond$$

Proof of Theorem 1: First, let us assume that

$$L_1 = \lim_{t \to t_0} x(t), \tag{2}$$

$$L_2 = \lim_{t \to t_0} y(t), \qquad \text{and} \tag{3}$$

$$L_3 = \lim_{t \to t_0} z(t). \tag{4}$$

Let $\epsilon > 0$ be given. According to the definition of the limit of a function of a single variable and equation (2), there exists a number δ_1 so that

$$\text{if} \quad 0 < |t - t_0| < \delta_1 \qquad \text{then} \qquad |x(t) - L_1| < \epsilon/3. \tag{5}$$

(You will see in a few lines why we use $\epsilon/3$ rather than just ϵ.) Similarly, it follows from equations (3) and (4) that there exist positive numbers δ_2 and δ_3 so that

$$\text{if} \quad 0 < |t - t_0| < \delta_2 \qquad \text{then} \qquad |y(t) - L_2| < \epsilon/3 \tag{6}$$

and

$$\text{if} \quad 0 < |t - t_0| < \delta_3 \qquad \text{then} \qquad |z(t) - L_3| < \epsilon/3. \tag{7}$$

If we take δ to be the smallest of the numbers δ_1, δ_2, and δ_3, then each of the inequalities on the left-hand sides of (5) through (7) is fulfilled when $0 < |t - t_0| < \delta$. In this case, using inequalities (5) through (7) and the triangle inequality, we find that

$$\begin{aligned}
|f(t) - L| &= |(x(t)i + y(t)j + z(t)k) - (L_1 i + L_2 j + L_3 k)| \\
&= |(x(t) - L_1)i + (y(t) - L_2)j + (z(t) - L_3)k| \\
&\le |x(t) - L_1| + |y(t) - L_2| + |z(t) - L_3| \\
&< \frac{\epsilon}{3} + \frac{\epsilon}{3} + \frac{\epsilon}{3} \\
&= \epsilon.
\end{aligned}$$

That is,

$$\text{if} \quad 0 < |t - t_0| < \delta, \qquad \text{then} \quad |f(t) - L| < \epsilon. \tag{8}$$

To prove the converse, we assume that whenever $\epsilon > 0$ is given we can find a number $\delta > 0$ so that statement (8) holds, and we show that this guarantees that statements (2)–(4) hold. To do so we observe that if t is any number for which $0 < |t - t_0| < \delta$ (with δ as in equation (8)) we have

$$\begin{aligned}
|x(t) - L_1| &= \sqrt{|x(t) - L_1|^2} \tag{8} \\
&\le \sqrt{|x(t) - L_1|^2 + |y(t) - L_2|^2 + |z(t) - L_3|^2} \\
&= |f(t) - L| \\
&< \epsilon.
\end{aligned}$$

That is, if $0 < |t - t_0| < \delta$, then $|x(t) - L_1| < \epsilon$. This shows that $\lim_{t \to t_0} x(t) = L_1$.

Similar comparisons show that $\lim_{t \to t_0} y(t) = L_2$ and that $\lim_{t \to t_0} z(t) = L_3$. This shows that if $L = \lim_{t \to t_0} f(t)$, then statements (2)–(4) hold. This completes the proof. ◆

Exercise Set 17.1

In Exercises 1–5, sketch the graph of the given vector function.

1. $f(t) = i + j + tk$

2. $f(t) = \cos ti + \sin tj + tk$

3. $f(t) = ti + \cos tj + \sin tk$

4. $f(t) = ti + t^2 j$

5. $f(t) = \sin ti + j + k$

The **implicit domain** of a vector-valued function f is the largest set of numbers t for which each of the component functions is defined. State the implicit domain for each of the following vector-valued functions.

6. $f(t) = ti + \sqrt{t}\,j + \sin tk$

7. $f(t) = t^2i - tj + \tan tk$

8. $f(t) = \dfrac{1}{\sqrt{1-t}}i + 2tj + \sin^2 tk$

9. $f(t) = \dfrac{1}{9-t^2}i + \dfrac{2}{1+t}j + \sqrt{t}\,k$

10. $f(t) = (t^2 - 1)i + \sec tj + t^3k$

11. $f(t) = t^2i + t^{3/2}j + t^{2/3}k$

12. $f(t) = \ln ti + \cos tj + \sqrt{9-t^2}k$

In Exercises 13–18, let $f(t) = ti + 3tj + t^2k$ and $g(t) = \sin ti + \cos tj + k$. Find the indicated functions.

13. $f + g$

14. $f - g$

15. $3f + 2g$

16. $4f - 3g$

17. $f \cdot g$

18. $f \times g$

In Exercises 19–24, let $f(t) = (1 + t)i + (1 - t^2)j + tk$, $g(t) = \sqrt{t}\,i + (1 - t)j + e^tk$, and $h(t) = \sin t$. Find the indicated functions.

19. $f + 3g$

20. hf

21. $h(f - g)$

22. $2f - 3hg$

23. $f \cdot g$

24. $3f \times 2g$

In Exercises 25–30, find the indicated limit.

25. $\lim\limits_{t \to \pi/4} (\sin ti + \tan tj + \cos tk)$

26. $\lim\limits_{t \to 2}(t^2i + (3t - 2)j + 6tk)$

27. $\lim\limits_{t \to 0}\left\{\left(\dfrac{\sin t}{t}\right)i + \left(\dfrac{1 - \cos t}{t}\right)j + e^{2t}k\right\}$

28. $\lim\limits_{t \to 0}\left\{\dfrac{\sin 3t}{t}i + \dfrac{\tan 2t}{3t}j + \ln(1 + t)k\right\}$

29. $\lim\limits_{t \to 3}\left\{\left(\dfrac{9 - t^2}{3 - t}\right)i + \left(\dfrac{t^2 + t - 12}{t - 3}\right)j + \left(\dfrac{t^3 - 13t + 12}{t - 3}\right)k\right\}$

30. $\lim\limits_{t \to 2}\left\{(t + 3)i + \left(\dfrac{t^2 - 4}{t - 2}\right)j + \left(\dfrac{t^3 + t^2 - 4t - 4}{t - 2}\right)k\right\}$

In Exercises 31–37, determine the intervals on which the given vector function is continuous.

31. $f(t) = 2ti + \cos tj + \tan tk$

32. $f(t) = \sqrt{9 - t^2}i + \dfrac{1}{t}j + \sin tk$

33. $f(t) = \dfrac{1}{1 + \sqrt{t}}i + \dfrac{1}{1 - \sqrt{t}}j + t^{3/2}k$

34. $f(t) = \left(\dfrac{1}{t^2 - t + 12}\right)i + \ln(2t)j + e^{-t}k$

35. $f(t) = \ln(1 + t^2)i + \ln(1 - t^2)j + \sqrt{t}\,k$

36. $f(t) = \begin{cases} ti + (3t - 2)j + (3 - t)k, & -\infty < t < 2 \\ 2i + (2 + t)j + tk, & 2 \le t < \infty \end{cases}$

37. $f(t) = \begin{cases} t^2i + (3 - t)j + 4tk, & -\infty < t < 1 \\ ti + (4 + t)j + (1 - t^2)k, & 1 \le t < \infty \end{cases}$

38. An automobile assembly robot turns a machine screw at a constant rate of 10π radians per second. The **pitch** of the screw (the number of threads per millimeter of length) is such that the screw advances 0.5 mm for each complete revolution.
  a. Write a vector function for the motion of a paint spot on one thread of the screw.
  b. At what rate, in mm/s, must the robot arm advance?

39. The **polarization** of a light wave is determined by the motion of the tip of the associated "electric vector" $E(t)$. If the motion follows a circular helix, the light is said to be circularly polarized. If a particular light wave is circularly polarized according to $E(t) = \cos(10^3t)i + \sin(10^3t)j + (3 \times 10^8)tk$, how far does the wave advance along the z-axis during one complete revolution of the electric vector?

40. Describe the graph of the vector function $f(t) = 3 \cos ti + 4 \sin tj + tk$.

41. Prove that if $\lim\limits_{t \to t_0} f(t) = L$ and $\lim\limits_{t \to t_0} g(t) = M$, then
$$\lim\limits_{t \to t_0} (f + g)(t) = L + M \text{ as follows.}$$
  a. Write $f(t)$ and $g(t)$ in component form.
  b. Obtain $(f + g)(t)$ in component form.

42. Prove that if f and g are continuous at $t = t_0$ then so is
  a. $f + g$,
  b. cf,
  c. $f \times g$.

43. Prove that if f and g are continuous vector-valued functions then so are the real-valued functions
  a. $f \cdot g$,
  b. $|f|$.

44. For each set of vector-valued functions, find the intervals on which f, g, h, and d are continuous.
  a. $f(t) = \sqrt{t}\,i - \dfrac{1}{t}j + tk$, $g(t) = -\sqrt{t}\,i + \dfrac{1}{t}j + 3tk$,
  $$h(t) = (f + g)(t)$$
  b. $f(t) = \dfrac{1}{t}i + \ln(1 + t)j + \sqrt{t}\,k$, $g(t) = ti + \sqrt{t}\,k$,
  $$d(t) = (f \cdot g)(t)$$

45. If $f + g$ is continuous, what can you say about the continuity of f and g? What conclusions can you draw from the continuity of the dot product $f \cdot g$?

46. Prove Corollary 1.

17.2 DERIVATIVES AND INTEGRALS OF VECTOR-VALUED FUNCTIONS

Let f be a vector-valued function and let t be a number in the domain of f. Also, assume that the numbers $t + h$ lie in the domain of f for h sufficiently small. The vector

$$\left(\frac{1}{h}\right)[f(t + h) - f(t)], \qquad h \neq 0 \tag{1}$$

is referred to as a **difference quotient** for the vector function $f(t)$. The vector in equation (1) may be interpreted as the average change in the vector $f(t)$ per unit change in t over an interval of length h. As for functions of a single variable, we are interested in knowing about the limiting value of this vector as $h \to 0$, which we refer to as the *derivative* of the vector-valued function f. The next three sections in part concern the geometric and physical interpretations of this derivative. Our interest here focuses on defining and calculating this particular limit.

DEFINITION 3

The vector-valued function f is said to be **differentiable** at the number t if the limit

$$f'(t) = \lim_{h \to 0} \frac{1}{h}[f(t + h) - f(t)] \tag{2}$$

exists. The vector $f'(t)$ is called the **derivative** of the vector-valued function f at t.

As for functions of a single variable, the vector-valued function f is said to be differentiable on an open interval I if $f'(t)$ in (2) exists for every $t \in I$. Thus, the derivative of a vector-valued function is again a vector-valued function, and the domain of f' is the subset of the domain of f for which the limit in line (2) exists.

The following theorem shows that the derivative f' may be calculated directly from the components of f.

THEOREM 2

Let $f(t) = x(t)i + y(t)j + z(t)k$. The vector-valued function f is differentiable if and only if each of the component functions x, y, and z is differentiable. In this case

$$f'(t) = x'(t)i + y'(t)j + z'(t)k. \tag{3}$$

Theorem 2 is proved by examining the difference quotient in Definition 3. According to Theorem 1,

$$f'(t) = \lim_{h \to 0} \frac{1}{h}[f(t + h) - f(t)]$$

$$= \lim_{h \to 0} \frac{1}{h}\{[x(t + h)i + y(t + h)j + z(t + h)k] - [x(t)i + y(t)j + z(t)k]\}$$

$$= \lim_{h \to 0} \left[\left(\frac{x(t + h) - x(t)}{h}\right)i + \left(\frac{y(t + h) - y(t)}{h}\right)j + \left(\frac{z(t + h) - z(t)}{h}\right)k\right]$$

$$= \left[\lim_{h \to 0} \left(\frac{x(t + h) - x(t)}{h}\right)\right]i + \left[\lim_{h \to 0} \left(\frac{y(t + h) - y(t)}{h}\right)\right]j$$

$$+ \left[\lim_{h \to 0} \left(\frac{z(t + h) - z(t)}{h}\right)\right]k.$$

This shows that f is differentiable if and only if each of the component functions is differentiable. Applying each of the indicated limits and using Definition 3 establishes equation (3).

Example 1

Let $f(t) = t^2 i + \cos tj + e^{3t}k$. Find $f'(t)$.

Solution: $f'(t) = \left[\dfrac{d}{dt}(t^2)\right]i + \left[\dfrac{d}{dt}(\cos t)\right]j + \left[\dfrac{d}{dt}(e^{3t})\right]k$

$\qquad\qquad = 2ti - \sin tj + 3e^{3t}k.$ ◇

If the derivative f' is itself differentiable, we define its derivative to be the *second* derivative of f. That is, $f'' = (f')'$. The second derivative of f is again a vector-valued function. Just as for real-valued functions, some or all derivatives of order higher than second may exist as well.

Example 2

Let $f(t) = \sin 2ti + \cos 2tj + \sqrt{t}k$. Find both f' and f''.

Solution: For $t > 0$ we have

$$f'(t) = 2 \cos 2ti - 2 \sin 2tj + \frac{1}{2}t^{-1/2}k,$$

and

$$f''(t) = -4 \sin 2ti - 4 \cos 2tj - \frac{1}{4}t^{-3/2}k.$$ ◇

Example 3

Show that the function $f(t) = A \sin \omega ti + B \cos \omega tj$ satisfies the **vector differential equation**

$$f''(t) + \omega^2 f(t) = 0. \tag{4}$$

Strategy
Calculate $f'(t)$ and $f''(t)$.

Solution
Here

$$f'(t) = \omega A \cos \omega ti - \omega B \sin \omega tj$$

and

$$f''(t) = -\omega^2 A \sin \omega ti - \omega^2 B \cos \omega tj.$$

Substitute f and f'' into the left-hand side of (4) and verify that 0 is obtained.

Substituting into equation (4) shows that

$$f''(t) + \omega^2 f(t) = [-\omega^2 A \sin \omega ti - \omega^2 B \cos \omega tj] + \omega^2[A \sin \omega ti + B \cos \omega tj]$$

$$= 0.$$ ◇

The following theorem shows how derivatives of various other vector-valued functions may be calculated.

THEOREM 3

Let the vector-valued functions f and g and the scalar-valued function $y = h(t)$ be differentiable on appropriate intervals and let c be a number. Then the vector-valued functions $f + g$, cf, hf, $f \times g$, and $f \circ h$, and the scalar-valued function $f \cdot g$ are differentiable, and

(i) $(f + g)'(t) = f'(t) + g'(t)$,
(ii) $(cf)'(t) = cf'(t)$,
(iii) $(hf)'(t) = h(t)f'(t) + h'(t)f(t)$,
(iv) $(f \times g)'(t) = [f(t) \times g'(t)] + [f'(t) \times g(t)]$,
(v) $(f \circ h)'(t) = h'(t)f'(h(t))$ (Chain Rule),
(vi) $(f \cdot g)'(t) = f(t) \cdot g'(t) + f'(t) \cdot g(t)$.

Before commenting on the proof of Theorem 3, we consider two examples of its use.

Example 4

Let $f(t) = t^2 i + e^t k$ and $g(t) = \sin t i + \cos t j + k$. Find the derivative of the function $r = f \times g$.

Solution: The derivatives of the given functions are

$$f'(t) = 2t i + e^t k \qquad \text{and} \qquad g'(t) = \cos t i - \sin t j.$$

According to part (iv) of Theorem 3 we have

$$r'(t) = [f(t) \times g'(t)] + [f'(t) \times g(t)]$$

$$= \det \begin{bmatrix} i & j & k \\ t^2 & 0 & e^t \\ \cos t & -\sin t & 0 \end{bmatrix} + \det \begin{bmatrix} i & j & k \\ 2t & 0 & e^t \\ \sin t & \cos t & 1 \end{bmatrix}$$

$$= [e^t \sin t i + e^t \cos t j - t^2 \sin t k]$$
$$\quad + [-e^t \cos t i + (e^t \sin t - 2t)j + 2t \cos t k]$$
$$= e^t(\sin t - \cos t)i + [e^t(\sin t + \cos t) - 2t]j$$
$$\quad + (2t \cos t - t^2 \sin t)k.$$

This same result may be obtained directly by noting that

$$r(t) = f(t) \times g(t)$$

$$= \det \begin{bmatrix} i & j & k \\ t^2 & 0 & e^t \\ \sin t & \cos t & 1 \end{bmatrix}$$

$$= -e^t \cos t i + (e^t \sin t - t^2)j + t^2 \cos t k$$

and differentiating component by component, according to Theorem 2.   ◇

Example 5

Let f and g be as in Example 4. Find the derivative of the scalar-valued function $s = f \cdot g$.

Solution: By Theorem 3, part (vi),

$$
\begin{aligned}
s'(t) &= f(t) \cdot g'(t) + f'(t) \cdot g(t) \\
&= [(t^2 i + e^t k) \cdot (\cos ti - \sin tj)] + [(2ti + e^t k) \cdot (\sin ti + \cos tj + k)] \\
&= t^2 \cos t + 2t \sin t + e^t.
\end{aligned}
$$

To calculate this derivative directly, we first compute

$$
\begin{aligned}
s(t) = f(t) \cdot g(t) &= t^2 \sin t + (0)(\cos t) + e^t \\
&= t^2 \sin t + e^t.
\end{aligned}
$$

Differentiating then gives the above result. ◇

Examples 4 and 5 suggest that Theorem 3 may be proved by writing the vector-valued functions f and g in component form and applying Theorem 2. This is indeed the case. We prove part (v) here and leave the other statements as exercises. To do so we write f in component form as

$$
f(t) = x(t)i + y(t)j + z(t)k.
$$

Then

$$
f(h(t)) = x(h(t))i + y(h(t))j + z(h(t))k.
$$

According to Theorem 2 and the Chain Rule of Chapter 3,

$$
\begin{aligned}
(f \circ h)'(t) &= [(x \circ h)'(t)]i + [(y \circ h)'(t)]j + [(z \circ h)'(t)]k \\
&= [x'(h(t)) \cdot h'(t)]i + [y'(h(t)) \cdot h'(t)]j + [z'(h(t)) \cdot h'(t)]k \\
&= h'(t)[x'(h(t))i + y'(h(t))j + z'(h(t))k] \\
&= h'(t) \cdot f'(h(t)).
\end{aligned}
$$

The result of the following example will be useful in later calculations.

Example 6

Let f be a vector-valued function and let c be a constant. Prove that if f is differentiable on some interval I and if $|f(t)| = c$ for all $t \in I$, then $f(t)$ and $f'(t)$ are orthogonal for all $t \in I$.

Strategy

Express $|f(t)|^2$ as a dot product using $|v|^2 = v \cdot v$.

The resulting expression equals c^2.

Differentiate both sides of this equation using part (vi) of Theorem 3.

Use fact that v is orthogonal to w if and only if $v \cdot w = 0$.

Solution

By Theorem 3, Chapter 16,

$$
|f(t)|^2 = f(t) \cdot f(t).
$$

Thus, since $|f(t)| = c$, we have

$$
f(t) \cdot f(t) = c^2, \qquad t \in I.
$$

Differentiating both sides of this equation with respect to t gives

$$
f(t) \cdot f'(t) + f'(t) \cdot f(t) = 0
$$

or,

$$
2f(t) \cdot f'(t) = 0.
$$

Thus $f'(t) \cdot f(t) = 0$, $t \in I$, so $f(t)$ and $f'(t)$ are orthogonal for all $t \in I$. ◇

Integrals of Vector-Valued Functions

Just as for real-valued functions, we say that the vector-valued function F is an *antiderivative* for the vector-valued function f on the interval I if $F'(t) = f(t)$ for all t in I.

If the vector-valued function

$$F(t) = X(t)i + Y(t)j + Z(t)k$$

is an antiderivative for

$$f(t) = x(t)i + y(t)j + z(t)k$$

on the interval I, it follows that any other antiderivative G for f on I must have the form

$$\begin{aligned} G(t) &= [X(t) + c_1]i + [Y(t) + c_2]j + [Z(t) + c_3]k \\ &= X(t)i + Y(t)j + Z(t)k + (c_1 i + c_2 j + c_3 k) \\ &= F(t) + C, \qquad C = (c_1 i + c_2 j + c_3 k). \end{aligned}$$

We refer to the family of *all* antiderivatives for f on I as the *indefinite integral* for f on I, denoted by $\int f(t)\, dt$, the component form of which is given by the following theorem.

THEOREM 4

The **indefinite integral** of the vector-valued function $f(t) = x(t)i + y(t)j + z(t)k$ on the interval I is

$$\int f(t)\, dt = \left[\int x(t)\, dt\right]i + \left[\int y(t)\, dt\right]j + \left[\int z(t)\, dt\right]k \tag{5}$$

provided that the integrals of the component functions exist on I.

REMARK: It is important to note in applying equation (5) that *three* constants of integration will arise—one for each of the antiderivatives of the respective components. That is, if $X'(t) = x(t)$, $Y'(t) = y(t)$, and $Z'(t) = z(t)$, we can write equation (5) as either

$$\int f(t)\, dt = [X(t) + c_1]i + [Y(t) + c_2]j + [Z(t) + c_3]k, \tag{6}$$

or,

$$\int f(t)\, dt = X(t)i + Y(t)j + Z(t)k + C \tag{7}$$

where

$$C = c_1 i + c_2 j + c_3 k.$$

A common mistake is to write the constant C in equation (7) as a number rather than as a vector.

Example 7

Find the indefinite integral $\int f(t)\, dt$ for

$$f(t) = 2ti + e^{-3t}j + \sec^2 tk.$$

Solution: By Theorem (4)

$$\int f(t)\,dt = \left[\int 2t\,dt\right]\!i + \left[\int e^{-3t}\,dt\right]\!j + \left[\int \sec^2 t\,dt\right]\!k$$

$$= (t^2 + c_1)i + \left(-\frac{1}{3}e^{-3t} + c_2\right)\!j + (\tan t + c_3)k$$

$$= t^2 i - \frac{1}{3}e^{-3t}j + \tan t k + C, \qquad C = c_1 i + c_2 j + c_3 k. \qquad \diamondsuit$$

Example 8

Find the vector-valued function f for which

$$f'(t) = i + t^3 j + \left(\frac{1}{1 + t^2}\right)\!k,$$

and $f(0) = 2i - j + 3k.$

Solution: By integration we obtain

$$f(t) = \int f'(t)\,dt$$

$$= \int \left[i + t^3 j + \left(\frac{1}{1 + t^2}\right)\!k\right]dt$$

$$= (t + c_1)i + \left(\frac{1}{4}t^4 + c_2\right)\!j + (\text{Tan}^{-1} t + c_3)k.$$

Setting $t = 0$ and using the given *initial condition* $f(0) = 2i - j + 3k$ gives the equation

$$f(0) = c_1 i + c_2 j + c_3 k = 2i - j + 3k.$$

Thus, $c_1 = 2$, $c_2 = -1$, and $c_3 = 3$, so

$$f(t) = (t + 2)i + \left(\frac{1}{4}t^4 - 1\right)\!j + (\text{Tan}^{-1} t + 3)k$$

is the desired solution. $\diamondsuit$

Finally, the definite integral of the vector-valued function f is defined as the vector whose components are the respective definite integrals of the components of f.

DEFINITION 4

Let $f(t) = x(t)i + y(t)j + z(t)k.$ If the definite integral of each of the components exists on the interval $[a, b]$, we define

$$\int_a^b f(t)\,dt = \left[\int_a^b x(t)\,dt\right]\!i + \left[\int_a^b y(t)\,dt\right]\!j + \left[\int_a^b z(t)\,dt\right]\!k.$$

Example 9

For $f(t) = \sin 3t\,i + \cos t\,j + k$, find

$$\int_0^{\pi/2} f(t)\,dt.$$

Solution: By Definition 4,

$$\int_0^{\pi/2} f(t)\,dt = \left[\int_0^{\pi/2} \sin 3t\,dt\right]i + \left[\int_0^{\pi/2} \cos t\,dt\right]j + \left[\int_0^{\pi/2} 1\,dt\right]k$$

$$= \left[-\frac{1}{3}\cos 3t\,\Big|_0^{\pi/2}\right]i + \left[\sin t\,\Big|_0^{\pi/2}\right]j + \left[t\,\Big|_0^{\pi/2}\right]k$$

$$= \left[0 - \left(-\frac{1}{3}\right)\right]i + [1 - 0]j + \left[\frac{\pi}{2} - 0\right]k$$

$$= \frac{1}{3}i + j + \frac{\pi}{2}k.$$

We may summarize the results of this section by saying that for the purposes of calculating limits, derivatives, or integrals, the vector-valued function $f(t) = x(t)i + y(t)j + z(t)k$ is regarded as a sum of three component functions of a single variable. In the remaining sections of this chapter, we shall see how these results may be combined with a broader interpretation of $f(t)$ as a vector in space to develop the concepts of tangent, normal, velocity, acceleration, and curvature associated with curves in space.

Exercise Set 17.2

In Exercises 1–4, state the intervals on which the given vector-valued function is differentiable.

1. $f(t) = ti + \cos tj + \sqrt{t}\,k$

2. $f(t) = t^{-2}i + |t - 3|j + t^3 k$

3. $f(t) = \ln(3 + t)i + \left(\dfrac{1}{t + 2}\right)j + e^t k$

4. $f(t) = \sqrt{1 - t^2}\,i + \ln tj + \dfrac{1}{1 - t}k$

In Exercises 5–10, find the derivative of the given vector-valued function.

5. $f(t) = ti + \sqrt{t}\,j$

6. $f(t) = i + \sin tj + \cos 2tk$

7. $f(t) = \sqrt{t}\,i + t^{-3/2}j + \ln(2t - 1)k$

8. $f(t) = \cos^2 ti + \sec tj + \mathrm{Tan}^{-1} tk$

9. $f(t) = \mathrm{Sin}^{-1} ti + \sqrt{1 + t^2}\,j + e^{-t^3}k$

10. $f(t) = e^{\sqrt{t}}i + 3j - \mathrm{Cos}^{-1} 2tk$

11. Find f'' for the function f in Exercise 7.

12. Find f'' for the function f in Exercise. 10.

13. Show that the function $f(t) = Ae^{\omega t}i + Be^{-\omega t}j$ satisfies the vector differential equation

$$f''(t) - \omega^2 f(t) = 0.$$

14. Show that the function

$$f(t) = C \sinh \omega ti + D \cosh \omega tj$$

satisfies the vector differential equation in Exercise 13.

In Exercises 15–20, let

$$f(t) = \sin ti + j + t^2 k,$$
$$g(t) = ti + \cos tk,$$
$$h(t) = e^{3t}.$$

Find the derivative of the given function.

15. $r(t) = (3f - 2g)(t)$

16. $r(t) = h(t)\,f(t)$

17. $r(t) = (f \times g)(t)$

18. $r(t) = (f \cdot g)(t)$

19. $r(t) = f(h(t))$

20. $r(t) = |g(t)|^2$

21. Verify that $f(t)$ and $f'(t)$ are orthogonal if f is the vector function $f(t) = \cos 2ti + \sin 2tj + 3k$.

In Exercises 22–26, find the indefinite integral for the given vector-valued function.

22. $f(t) = ti + \cos tj + \sin tk$

23. $f(t) = \sqrt{t}\, i + e^{2t}j + \dfrac{1}{t}k$

24. $f(t) = t \sin ti + \cos^2 tj + (1 - t)k$

25. $f(t) = \ln ti + \dfrac{1}{t \ln t}j$

26. $f(t) = t \sec^2(1 - t^2)i + \sqrt{1 - t}\, j + \sqrt{t}\, k$

27. Find f if $f'(t) = i + t^2j$ and $f(0) = 3i + 5j$

28. Find f if $f'(t) = \cos 2ti + \left(\dfrac{1}{1 + t^2}\right)j + \dfrac{t}{\sqrt{1 + t^2}}k$ and

$f(0) = -4i + k$.

29. Find f if $f'(t) = \cos^2 ti + \sin^2 tj + e^{-t}k$ and $f(0) = 3i - 6j + 2k$.

30. Find a solution of the vector differential equation $f'(t) = f(t)$ for which

$f(0) = i + 2j - k$.

31. Find a solution of the vector differential equation

$f'(t) + 4f(t) = 0$

satisfying the initial condition $f(0) = i + 4k$.

32. Find a family of solutions to the vector differential equation

$f''(t) - 2f(t) = 0$.

33. Find $\displaystyle\int_a^b (ti + t^2j - t^3k)\, dt$.

34. Find $\displaystyle\int_0^1 f(t)\, dt$ if $f(t) = t^2i - 2tj + \sqrt{t}\, k$.

35. Find $\displaystyle\int_0^{\pi/4} f(t)\, dt$ if $f(t) = \cos ti + \sin 2tj + \cos^2 tk$.

36. Find $\displaystyle\int_1^4 f(t)\, dt$ if $f(t) = e^ti + t^2j + \ln 2tk$.

37. Show that if $f'(t) = 0$ for all $t \in I$, then f is constant on the interval I.

38. Show that if $f'(t) = g'(t)$ for all $t \in I$ then $f(t) = g(t) + C$ for some vector C and all $t \in I$.

39. True or false? If f is differentiable on $[a, b]$, there exists a constant $c \in (a, b)$ so that

$$f'(c) = \dfrac{1}{b - a}[f(b) - f(a)].$$

40. Prove statements (i) through (iv) and (vi) of Theorem 3 by first writing the vector functions in component form.

41. Complete the following alternate proof of differentiation formula (iii) of Theorem 3:

$$(hf)'(t) = h'(t)\, f(t) + h(t)\, f'(t).$$

a. Show that we can write

$$(hf)'(t) = \lim_{h \to 0} \dfrac{1}{h}[h(t + h)\, f(t + h) - h(t)\, f(t)]$$

$$= \lim_{h \to 0} \dfrac{1}{h}[h(t + h) - h(t)]f(t + h)$$

$$+ \lim_{h \to 0} \dfrac{1}{h}[f(t + h) - f(t)]h(t).$$

b. Show that the limit in part (a) is

$$(hf)'(t) = h'(t)\, f(t) + h(t)\, f'(t).$$

42. Use a method similar to that of Exercise 41 to prove part (vi) of Theorem 3:

$$(f \cdot g)'(t) = f'(t) \cdot g(t) + f(t) \cdot g'(t).$$

17.3 TANGENT VECTORS AND ARC LENGTH

We have already seen that a curve C in the plane may be described by a pair of parametric equations. For example, the unit circle is parameterized by the equations

$$C: \begin{cases} x(t) = \cos t \\ y(t) = \sin t \end{cases} \quad 0 \le t < 2\pi.$$

In a similar way, we may regard the vector-valued function

$$C: \quad r(t) = x(t)i + y(t)j + z(t)k \qquad t \in I \tag{1}$$

as giving a parameterization of a curve C in space. That is, we interpret C as the set of all points $P(t) = (x(t), y(t), z(t))$ lying at the tips of the position vectors $r(t)$. (Recall that interpreting $r(t)$ as a **position vector** means that $r(t)$ originates at the origin.) As for curves in the plane, we shall say that the curve C is **differentiable** if the function r is differentiable on I, which in turn requires that each of the component functions x, y, and z is differentiable (Theorem 2). The curve C is called **smooth** if r' is continuous on I.

When the derivative $r'(t_0)$ exists, we refer to it as a *tangent vector*.

DEFINITION 5

For the differentiable curve C in equation (1), if $r'(t_0) \neq \mathbf{0}$, the vector

$$r'(t_0) = x'(t_0)i + y'(t_0)j + z'(t_0)k, \qquad t_0 \in I$$

is called a **tangent vector** for C at the tip of the vector $r(t)$ (Figure 3.1).

To see that Definition 5 agrees with our earlier concept of tangent, observe that the vector

$$r(t_0 + h) - r(t_0) \tag{2}$$

originates at the tip of $r(t_0)$ and terminates at the tip of $r(t_0 + h)$ (see Figure 3.1). This vector therefore determines a secant, joining the points $P(t_0)$ and $P(t_0 + h)$ on C. As $h \to 0$ the secant through these same points approaches the *tangent* to C at $r(t_0)$. However, we cannot obtain a direction vector for this line by simply applying $\lim_{h \to 0}$ to the vector in equation (1) since the length of this vector will approach zero as $h \to 0$. We therefore multiply the vector in (2) by the scalar $\dfrac{1}{h}$ (changing its length, but not its direction) before applying the limit. Thus, a direction vector for the tangent to C at $r(t_0)$ is

$$r'(t_0) = \lim_{h \to 0} \left\{ \frac{1}{h} [r(t_0 + h) - r(t_0)] \right\}. \tag{3}$$

(See Figure 3.2.) Since we assume $r'(t_0) \neq \mathbf{0}$, the tangent vector has nonzero length.

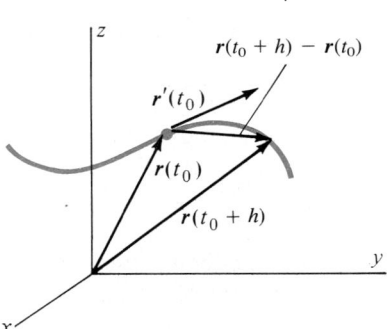

Figure 3.1 $r'(t_0)$ is tangent to C at tip of $r(t_0)$.

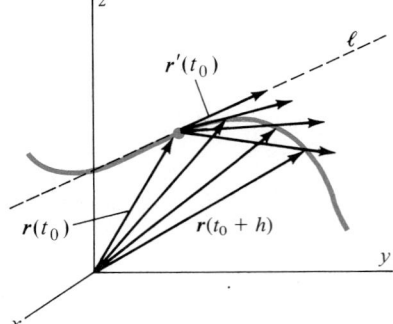

Figure 3.2 Line ℓ tangent to C at tip of $r(t_0)$ has direction vector $r'(t_0)$.

Example 1

The vector-valued function

$$r(t) = a \cos ti + a \sin tj, \qquad a > 0$$

has as its graph the circle in the *xy*-plane with center (0, 0) and radius *a*. For this function the tangent vector at $\mathbf{r}(t)$ is

$$\mathbf{r}'(t) = -a \sin t\mathbf{i} + a \cos t\mathbf{j}.$$

Note that

$$\mathbf{r}(t) \cdot \mathbf{r}'(t) = -a^2(\cos t)(\sin t) + a^2(\sin t)(\cos t)$$
$$= 0,$$

which confirms the familiar fact that the tangent to a circle at point *P* is perpendicular to the radius at that point. (See Figure 3.3.) ◇

The next example generalizes the observation of Example 1 to the circular helix of radius *a* in space.

Example 2

Show that a vector tangent to the circular helix

$$C: \quad \mathbf{r}(t) = a \cos t\mathbf{i} + a \sin t\mathbf{j} + b\mathbf{k}, \qquad 0 \le t < 2\pi$$

at $\mathbf{r}(t)$ is always orthogonal to the position vector $\mathbf{r}(t)$.

Solution: A vector tangent to *C* is, by Definition 5,

$$\mathbf{r}'(t) = -a \sin t\mathbf{i} + a \cos t\mathbf{j}.$$

Since $\mathbf{r}(t) \cdot \mathbf{r}'(t) = -a^2 \cos t \sin t + a^2 \sin t \cos t = 0$, the vectors $\mathbf{r}(t)$ and $\mathbf{r}'(t)$ are orthogonal. (Note that since $|\mathbf{r}(t)| = \sqrt{a^2 \cos^2 t + a^2 \sin^2 t + b^2} = \sqrt{a^2 + b^2}$ is constant, this result also follows by Example 6, Section 17.2.) (See Figure 3.4.) ◇

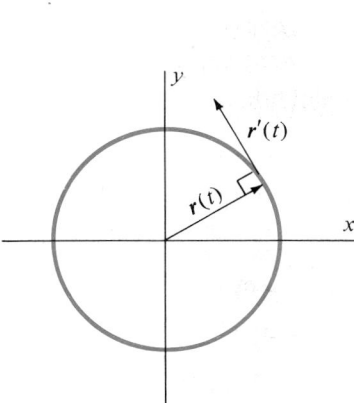

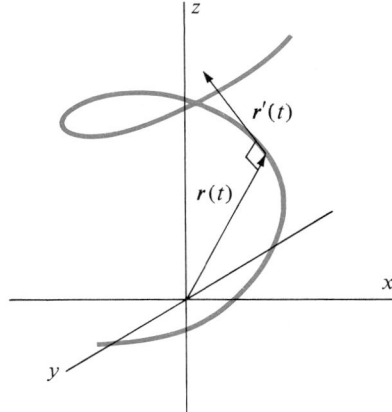

Figure 3.3 For the circle with parameterization

$$\mathbf{r}(t) = a \cos t\mathbf{i} + a \sin t\mathbf{j},$$

the tangent $\mathbf{r}'(t)$ is always orthogonal to the radius vector $\mathbf{r}(t)$.

Figure 3.4 For the circular helix

$$\mathbf{r}(t) = a \cos t\mathbf{i} + a \sin t\mathbf{j} + b\mathbf{k},$$

the tangent vector $\mathbf{r}'(t)$ is always orthogonal to the position vector $\mathbf{r}(t)$.

If $r'(t_0) \neq 0$, a unit vector T tangent to the curve C at $r(t_0)$ is given by

$$T = \frac{r'(t_0)}{|r'(t_0)|} \qquad \text{(unit tangent).} \tag{4}$$

Example 3

Find a unit vector tangent to the circular helix $r(t) = a \cos t i + a \sin t j + bt k$ at the point $(0, a, b\pi/2)$.

Strategy

Find t_0 for which $r(t_0)$ terminates at the point $\left(0, a, \dfrac{b\pi}{2}\right)$.

Solution

Setting $r(t_0) = 0i + aj + \dfrac{b\pi}{2}k$ gives the equations

$$a \cos t_0 = 0, \qquad a \sin t_0 = a, \qquad bt_0 = \frac{b\pi}{2},$$

to which the solution is $t_0 = \pi/2$.

Find $r'(t_0)$.

According to Definition 5 a tangent vector at $r(t_0) = r(\pi/2)$ is

$$r'\left(\frac{\pi}{2}\right) = -a \sin \left(\frac{\pi}{2}\right)i + a \cos \left(\frac{\pi}{2}\right)j + bk$$

$$= -ai + bk.$$

Find $|r'(t_0)|$.

This vector has length

$$\left|r'\left(\frac{\pi}{2}\right)\right| = \sqrt{a^2 + b^2}.$$

Apply (4).

The desired unit tangent is

$$T = \frac{r'\left(\dfrac{\pi}{2}\right)}{\left|r'\left(\dfrac{\pi}{2}\right)\right|} = \frac{-ai + bk}{\sqrt{a^2 + b^2}}. \qquad \diamondsuit$$

To find an equation for the *line* ℓ tangent to C at $r(t_0)$, we use the fact that $r(t_0)$ is a position vector for ℓ and $r'(t_0)$ is a direction vector for ℓ. Since t is the parameter for C, we should use a different letter to parameterize ℓ, say ω. We may then use equation (1), Section 16.5, to write a parameterization for ℓ as

$$\ell: \quad R(\omega) = r(t_0) + \omega r'(t_0). \tag{5}$$

(Be careful when using equation (5): $R(\omega)$ is the position vector of a point on ℓ, while $r(t_0)$ and $r'(t_0)$ are fixed vectors determined by the curve C.)

Example 4

Find vector, parametric, and symmetric equations for the line ℓ tangent to the curve

$$C: \quad r(t) = t^2 i + (3 - t)j + t^3 k$$

at the point $(4, 1, 8)$.

Strategy

Find t_0 for which $\boldsymbol{r}(t_0)$ terminates at the point $(4, 1, 8)$. $\boldsymbol{r}(t_0)$ is the position vector for ℓ.

Find $\boldsymbol{r}'(t_0)$. This is the direction vector for ℓ.

Use equation (5) to write the vector form of ℓ.

The parametric equations for x, y, and z are the components of $\boldsymbol{R}(\omega)$.

The symmetric equations are found from the parametric equations as in Section 16.5.

Solution

Setting $\boldsymbol{r}(t_0) = 4\boldsymbol{i} + \boldsymbol{j} + 8\boldsymbol{k}$ gives the equations

$$t^2 = 4, \qquad 3 - t = 1, \qquad \text{and} \qquad t^3 = 8$$

for which the solution is $t_0 = 2$.
Since $\boldsymbol{r}'(t) = 2t\boldsymbol{i} - \boldsymbol{j} + 3t^2\boldsymbol{k}$,

$$\boldsymbol{r}'(t_0) = \boldsymbol{r}'(2) = 4\boldsymbol{i} - \boldsymbol{j} + 12\boldsymbol{k}.$$

The vector form of ℓ is therefore

$$\ell: \quad \boldsymbol{R}(\omega) = (4\boldsymbol{i} + \boldsymbol{j} + 8\boldsymbol{k}) + \omega(4\boldsymbol{i} - \boldsymbol{j} + 12\boldsymbol{k}).$$

To find parametric equations for ℓ, we let $P(\omega) = (x(\omega), y(\omega), z(\omega))$ be a point on ℓ. Then

$$\begin{aligned} x(\omega) &= 4 + 4\omega && (\boldsymbol{i}\text{-components}) \\ y(\omega) &= 1 - \omega && (\boldsymbol{j}\text{-components}) \\ z(\omega) &= 8 + 12\omega && (\boldsymbol{k}\text{-components}) \end{aligned}$$

are the parametric equations for the coordinates of P. Solving each of these equations for ω and equating the results gives the symmetric equations

$$\frac{x - 4}{4} = 1 - y = \frac{z - 8}{12}$$

for ℓ. ◇

Arc Length

If C is a smooth curve in space parameterized by the vector function

$$\boldsymbol{r}(t) = x(t)\boldsymbol{i} + y(t)\boldsymbol{j} + z(t)\boldsymbol{k}$$

an expression for the length L of the arc of C from $\boldsymbol{r}(a)$ to $\boldsymbol{r}(b)$ is obtained in a manner similar to that for plane curves (see Section 7.3). First, the interval $[a, b]$ is partitioned into n subintervals of equal length $\Delta t = \dfrac{b - a}{n}$ and with endpoints $a = t_0 < t_1 < t_2 < \cdots < t_n = b$. This determines a polygonal path connecting the tips of the vectors $\boldsymbol{r}(t_0), \boldsymbol{r}(t_1), \ldots, \boldsymbol{r}(t_n)$ of total length

$$\sum_{j=1}^{n} \Delta L_j = \sum_{j=1}^{n} \sqrt{\Delta x_j^2 + \Delta y_j^2 + \Delta z_j^2} \tag{6}$$

where $\Delta x_j = x(t_j) - x(t_{j-1})$, $\Delta y_j = y(t_j) - y(t_{j-1})$, and $\Delta z_j = z(t_j) - z(t_{j-1})$. As before, applications of the Mean Value Theorem (once for each component) lead to an approximation of the form

$$L \approx \sum_{j=1}^{n} \sqrt{[x'(u_j)]^2 + [y'(v_j)]^2 + [z'(w_j)]^2} \; \Delta t_j. \tag{7}$$

As $n \to \infty$, $\Delta t \to 0$ and we obtain

$$L = \int_a^b \sqrt{[x'(t)]^2 + [y'(t)]^2 + [z'(t)]^2} \; dt \tag{8}$$

which we take as the definition of the arc length L. (Since this is familiar ground, we do not pursue the details of this development here.) Since the expression under the radical in equation (8) is equal to $r'(t) \cdot r'(t) = |r'(t)|^2$, the arc length may be expressed more compactly as

$$L = \int_a^b |r'(t)| \, dt. \tag{9}$$

Example 5

Find the length of the curve

$$C: \quad r(t) = \cos \pi t \mathbf{i} + \sin \pi t \mathbf{j} + t^{3/2} \mathbf{k}$$

from $r(0) = \mathbf{i}$ to $r(4) = \mathbf{i} + 8\mathbf{k}$.

Solution: Here

$$r'(t) = -\pi \sin \pi t \mathbf{i} + \pi \cos \pi t \mathbf{j} + \frac{3}{2}\sqrt{t}\, \mathbf{k}$$

so

$$|r'(t)| = \sqrt{\pi^2 \sin^2 \pi t + \pi^2 \cos^2 \pi t + \frac{9}{4}t}$$

$$= \sqrt{\pi^2 + \frac{9}{4}t}.$$

By formula (9),

$$L = \int_0^4 \sqrt{\pi^2 + \frac{9}{4}t} \, dt = \frac{8}{27}\left(\pi^2 + \frac{9}{4}t\right)^{3/2}\Bigg]_0^4$$

$$= \frac{8}{27}[(\pi^2 + 9)^{3/2} - \pi^3] \approx 15.1$$

(Figure 3.5). ◇

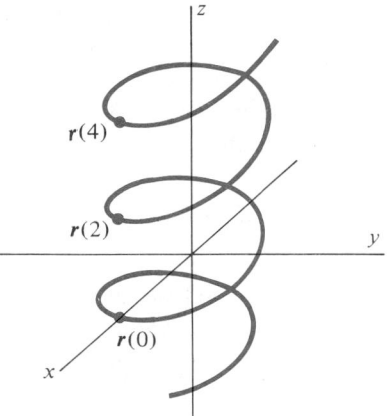

Figure 3.5

Parameterization by Arc Length

Up to this point, we have not concerned ourselves with the *rate* at which the vector-valued function f traces out a curve in space. In order to proceed further with our analysis of curves, we will now have to do so. A simple example of this type of concern is the observation that the functions $f(t) = a + tb$ and $g(t) = a + 2tb$ both trace out the same line in space, although a particle located at the tip of $g(t)$ moves along the line at a speed twice that of a particle located at the tip of $f(t)$.

A baseline for comparing the rate at which the vector function f traces out a curve in space is provided by the notion of *parameterization by arc length*. Intuitively, this notion says that as the parameter t varies through an interval of length t_0, an arc of this same length $s = t_0$ is traced out by the vector function f (Figure 3.6). A more precise formulation of this concept is the following.

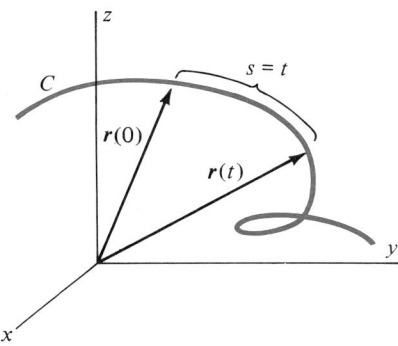

Figure 3.6 r is parameterized by arc length if $\Delta s = \Delta t$.

DEFINITION 6

Let the curve C be the graph of the differentiable function $r(t) = x(t)i + y(t)j + z(t)k$ defined on an interval I containing the number zero. The curve C (and the function r) is said to be **parameterized by arc length** if

$$\int_0^t |r'(\omega)| \, d\omega = t \tag{10}$$

for all t in the interval I.

Since the integral in equation (10) is the length of the arc of C from $r(0)$ to $r(t)$, Definition 6 agrees with the intuitive description of parameterization by arc length given above. Differentiating both sides of equation (10) with respect to t shows that the condition

$$\boxed{|r'(t)| = 1 \qquad \text{for all} \qquad t \in I} \tag{11}$$

is equivalent to C being parameterized by arc length. Thus, a curve is *parameterized by arc length if and only if its derivative has constant length equal to one*.

Example 6

The unit circle

$$r(t) = \cos t i + \sin t j, \qquad t \in [0, 2\pi)$$

is parameterized by arc length, since

$$|r'(t)| = |-\sin ti + \cos tj|$$
$$= \sqrt{\sin^2 t + \cos^2 t}$$
$$= 1$$

for all $t \in [0, 2\pi)$. (This should not be surprising. Both the circumference of the unit circle and the length of the interval $[0, 2\pi)$ are 2π.) ◇

Example 7

The circle of radius ρ centered at the origin is traced out by the vector function

$$r(t) = \rho \cos \alpha ti + \rho \sin \alpha tj, \qquad t \in \left[0, \frac{2\pi}{\alpha}\right), \qquad \alpha > 0.$$

Find the value of α for which this circle is parameterized by arc length.

Solution: We first find $|r'(t)|$:

$$r'(t) = -\alpha\rho \sin \alpha ti + \alpha\rho \cos \alpha tj.$$

Thus

$$|r'(t)| = \sqrt{(-\alpha\rho \sin \alpha t)^2 + (\alpha\rho \cos \alpha t)^2}$$
$$= |\alpha\rho|.$$

In order that equation (10) hold, we must have

$$|r'(t)| = |\alpha\rho| = 1.$$

We therefore take $\alpha = 1/\rho$. The circle of radius ρ is parameterized by arc length by the function

$$r(t) = \rho \cos\left(\frac{t}{\rho}\right)i + \rho \sin\left(\frac{t}{\rho}\right)j.$$ ◇

We conclude by highlighting the comment following equation (11). If the curve C determined by the vector function r is parameterized by arc length, the **unit tangent** $T(s)$ at $r(s)$ is

$$T(s) = r'(s). \tag{12}$$

Exercise Set 17.3

In Exercises 1–6, find a vector tangent to the given curve at the given point.

1. $r(t) = t^3i + \sqrt{t}\,j + 2k, \qquad t = 4$

2. $r(t) = \sin ti + \cos 2tj + e^{3t}k, \qquad t = 0$

3. $r(t) = \sec ti + \tan tj, \qquad t = \pi/4$

4. $r(t) = \sqrt{1 + t^2}\,i + \text{Tan}^{-1} tj + \ln(1 + t)k, \qquad t = 0$

5. $r(t) = (1 + t^2)i + (t - 4)j + 6tk$, at the point $(2, -5, -6)$

6. $r(t) = ai + btj + ct^2k$ at the point (a, b, c).

7. Find a unit vector tangent to the curve $r(t) = a \cos \omega ti + b \sin \omega tj$ at the point where $\omega t = \pi/4$.

In Exercises 8–10, let $r(t) = (t^2 + 2)i + 3tj$.

8. Find the number(s) t for which $r(t)$ and $r'(t)$ are orthogonal.

9. Find the number(s) t for which $r(t)$ and $r'(t)$ have the same direction.

10. Find the number(s) t for which $r(t)$ and $r'(t)$ have opposite direction.

11. Find an equation in vector form for the line tangent to the curve in Exercise 1 at the given point.

12. Find an equation in parametric form for the line tangent to the curve in Exercise 2 at the given point.

13. Find an equation in symmetric form for the line tangent to the curve in Exercise 5 at the given point.

14. Show that every tangent to the unit circle is orthogonal to the radius vector through the point of tangency.

15. Let $r(t) = e^t \cos t \, i + e^t \sin t \, j$. Find the angle between $r(t)$ and the tangent $r'(t)$ for
    a. $t = 0$,
    b. $t = \pi/4$.

In Exercises 16–20, find the length of the indicated arc.

16. $r(t) = \cos t \, i + \sin t \, j$, $\qquad 0 \le t \le \pi/4$

17. $r(t) = \cos t \, i + \dfrac{\sqrt{2}}{2} \sin t \, j + \dfrac{\sqrt{2}}{2} \sin t \, k$, $\quad 0 \le t \le \pi/2$

18. $r(t) = e^t \cos t \, i + e^t \sin t \, j$, $\qquad 0 \le t \le \pi/2$

19. $r(t) = t\,i + (t^2 + 1)j + t\,k$, $\qquad 0 \le t \le 1$

20. $r(t) = t^3 i + 3t^2 j + 6tk$, $\qquad 0 \le t \le 3$

21. Find a unit tangent $T(t)$ to the graph of $y = x^2 + 3$ at the point $(2, 7)$. (*Hint:* Let $x = t$. Then $y = t^2 + 3$, and a vector parameterization for the curve is $r(t) = t\,i + (t^2 + 3)j$.)

22. Use the method of Exercise 21 to find a unit tangent $T(t)$ to the graph of $y^3 = 1 - x^2$ at the point $(3, -2)$.

23. Determine the constant α so that the curve

$$r(t) = \cos \alpha t \, i + \sin \alpha t \, j + \alpha t \, k, \quad 0 \le t \le \frac{2\pi}{\alpha}, \quad \alpha > 0,$$

is parameterized by arc length.

24. Find a parameterization by arc length for the curve $y = 3x - 2$ so that $x = 0$ and $y = -2$ when $s = 0$.

25. Show that if $r(t) = x(t)i + y(t)j$ and $x'(t_0) \ne 0$, Definition 5 for the *tangent* at $r(t_0)$ agrees with our earlier definition of tangent for the curve C determined by the parametric equations

$$C: \begin{cases} x = x(t) \\ y = y(t). \end{cases}$$

26. Use Example 6, Section 17.2, to show that if $T(t) = \dfrac{r'(t)}{|r'(t)|}$ is differentiable on I, then $T'(t)$ is orthogonal to $T(t)$ for $t \in I$. If $T'(t) \ne 0$, we define $N(t) = \dfrac{T'(t)}{|T'(t)|}$ to be the **principal unit normal** to the curve $C = \{r(t) \mid t \in I\}$.

27. Find the principal unit normals to the curve with parameterization

$$r(t) = \sqrt{2} \cos t \, i + \sqrt{2} \sin t \, j, \qquad 0 \le t \le 2\pi.$$

(See Exercise 26.)

28. Show that the line with vector equation

$$R(\omega) = r(t_0) + \omega r''(t_0)$$

is orthogonal to the tangent vector $T(t_0)$ if $r(t)$ is a parameterization by arc length, $r(t)$ is twice differentiable at $r(t_0)$, $r'(t_0) \ne 0$, and $r''(t_0) \ne 0$. (We refer to this line as the line **normal** to the curve at $r(t_0)$.)

29. Find an equation for the line normal to the curve in Example 4 at the point $(4, 1, 8)$ (see Exercise 28).

30. Give an example to show that $r(t)$ is not always orthogonal to $r'(t)$.

17.4 VELOCITY AND ACCELERATION

If a particle moves in space so that its location at time t is the point $P(t) = (x(t), y(t), z(t))$, its motion may be described by the **position vector** function

$$r(t) = x(t)i + y(t)j + z(t)k. \tag{1}$$

That is, we think of the position of the particle as the tip of the vector $r(t)$ originating at the origin and terminating at the point $P(t)$.

In such cases the **average velocity** from time t to time $t + h$ is

$$\text{average velocity} = \frac{\text{change in position}}{\text{change in time}}$$

$$= \frac{1}{h} \Delta r$$

$$= \frac{1}{h} [r(t + h) - r(t)].$$

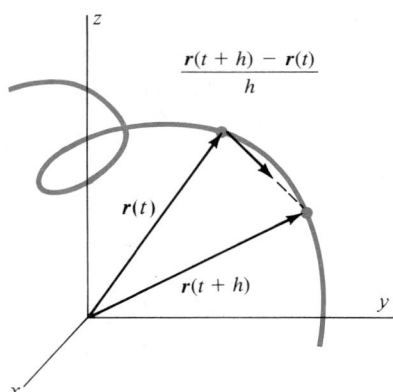

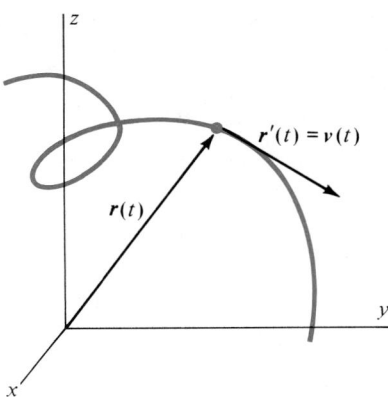

Figure 4.1 Average velocity is the vector $\dfrac{1}{h}[r(t + h) - r(t)]$.

Figure 4.2 Velocity is the vector $v(t) = r'(t) = \lim\limits_{h \to 0} \dfrac{1}{h}[r(t + h) - r(t)]$, and $|v(t_0)| =$ speed.

(See Figure 4.1.)

If our earlier concept of (instantaneous) velocity as the limit of average velocity as $h \to 0$ is to carry over to motion in space, we should define

$$\text{velocity at time } t = \lim_{h \to 0} \frac{1}{h}[r(t + h) - r(t)]. \tag{2}$$

(See Figure 4.2.)

We have already seen that the limit on the right-hand side of equation (2), when it exists, is the derivative $r'(t)$ of the vector-valued function r. According to Theorem 2, this limit exists precisely when each of the component functions x, y, and z is differentiable. The definition of the *velocity* of a particle moving in space is therefore the following.

DEFINITION 7

Let the vector function

$$r(t) = x(t)i + y(t)j + z(t)k$$

be the position function for a particle moving in space and let the parameter t denote time. If r is differentiable at time $t = t_0$, the **velocity** of the particle at $t = t_0$ is the vector

$$v(t_0) = r'(t_0) = x'(t_0)i + y'(t_0)j + z'(t_0)k. \tag{3}$$

If r is differentiable on an interval I, equation (3) determines the **velocity function** for r on the interval I.

Since velocity is just the derivative of the position function, we have already determined that $v(t_0)$ is *tangent* to the graph of r at $r(t_0)$, and points in the direction of increasing t, provided $v(t_0) \neq 0$ (Section 17.3).

As with real-valued functions, the length of the velocity vector $|v(t_0)| = |r'(t_0)|$ is defined to be the **speed** (rate of change of distance along the path) of the particle at time t_0.

Example 1

Show that a particle moving about the unit circle in the xy-plane with position function

$$\mathbf{r}(t) = \cos \alpha t \mathbf{i} + \sin \alpha t \mathbf{j}$$

moves at a constant speed.

Solution: By Definition 7, the velocity function is

$$\mathbf{v}(t) = -\alpha \sin \alpha t \mathbf{i} + \alpha \cos \alpha t \mathbf{j}.$$

The speed at time t is therefore

$$|\mathbf{v}(t)| = \sqrt{(\alpha \sin \alpha t)^2 + (\alpha \cos \alpha t)^2} = |\alpha|,$$

which is constant. For this reason, physicists call this **uniform circular motion.**

◇

As in the case of motion along a line, we define the *acceleration* of a particle moving in space to be the rate of change of velocity with respect to time.

DEFINITION 8

Let the vector function

$$\mathbf{r}(t) = x(t)\mathbf{i} + y(t)\mathbf{j} + z(t)\mathbf{k}$$

be the position function of a particle moving in space and let the parameter t denote time. If $\mathbf{r}$ is twice differentiable at time $t = t_0$, the **acceleration** of the particle at $t = t_0$ is the vector

$$\mathbf{a}(t_0) = \mathbf{r}''(t_0) = x''(t_0)\mathbf{i} + y''(t_0)\mathbf{j} + z''(t_0)\mathbf{k}.$$

Thus,

$$\mathbf{a}(t_0) = \mathbf{v}'(t_0). \tag{4}$$

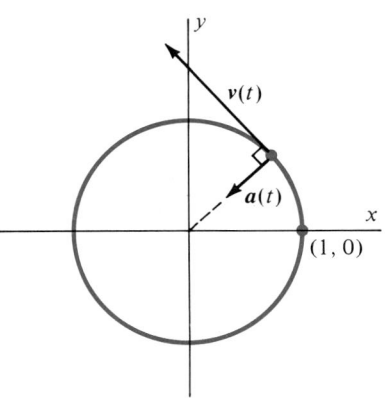

Figure 4.3 Velocity and acceleration vectors associated with uniform circular motion.

Example 2

Show that for the particle moving about the unit circle in Example 1, the acceleration vector always points toward the center of the circle and is of constant magnitude.

Solution: From the solution to Example 1 and equation (4) we have

$$\begin{aligned}
\mathbf{a}(t) = \mathbf{v}'(t) &= -\alpha^2 \cos \alpha t \mathbf{i} - \alpha^2 \sin \alpha t \mathbf{j} \\
&= -\alpha^2(\cos \alpha t \mathbf{i} + \sin \alpha t \mathbf{j}) \\
&= -\alpha^2 \mathbf{r}(t).
\end{aligned}$$

Thus, $\mathbf{a}(t)$ points in the direction opposite $\mathbf{r}(t)$ and has length

$$|\mathbf{a}(t)| = \sqrt{(\alpha^2 \cos \alpha t)^2 + (\alpha^2 \sin \alpha t)^2} = \alpha^2.$$

(See Figure 4.3.) Note that $\mathbf{v}(t)$ and $\mathbf{a}(t)$ are orthogonal, an observation that follows from Example 6, Section 17.2, since $|\mathbf{v}(t)| = |\alpha| = $ constant. ◇

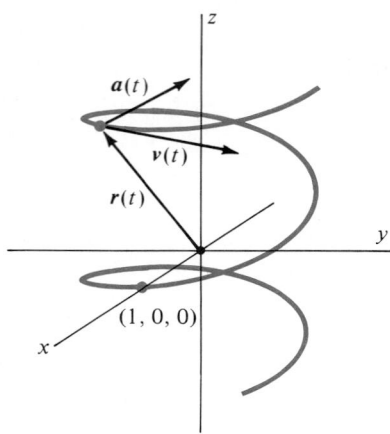

Figure 4.4 Velocity and acceleration vectors for the helix $r(t) = \cos ti + \sin tj + t^2 k$.

Example 3

Find the velocity and acceleration functions associated with the position function

$$r(t) = \cos ti + \sin tj + t^2 k$$

and show that $|a(t)|$ is constant.

Solution:

$$v(t) = r'(t) = -\sin ti + \cos tj + 2tk$$
$$a(t) = v'(t) = -\cos ti - \sin tj + 2k.$$

Since $|a(t)| = \sqrt{\cos^2 t + \sin^2 t + 4} = \sqrt{5}$, $|a(t)|$ is constant. The vectors $r(t)$, $v(t)$, and $a(t)$ are sketched in Figure 4.4.   ◇

Just as velocity and acceleration can be determined by differentiating the position function **r**, the functions for position and velocity can be obtained by integrating the acceleration function, provided antiderivatives of each component function can be found. More precisely, equation (4) gives

$$v(t) = \int a(t)\, dt, \tag{5}$$

and equation (3) shows that

$$r(t) = \int v(t)\, dt. \tag{6}$$

In using equations (5) and (6), you must remember that one constant of integration will appear in each component in each integration. These constants may be determined when *initial conditions* are specified for the functions **v** and **r**.

Example 4

A particle starts from rest at the point $P_0 = (2, 0, -3)$ and moves with an acceleration of $a(t) = 2i + 6tj$ m/s^2. Find

(a) the velocity function for the particle,
(b) the position function for the particle,
(c) the location of the particle after 3 seconds.

Strategy

Identify the initial conditions.

Obtain **v** by integrating **a**. (Don't forget that

$$\left[\int 0\, dt\right]k = (0 + c_3)k.)$$

Apply the initial condition $v_0 = \mathbf{0}$ to determine c_1, c_2, and c_3.

Solution

Since the particle starts at $P_0 = (2, 0, -3)$, we have

$$r_0 = r(0) = 2i - 3k.$$

Since the particle starts from rest, we have

$$v_0 = v(0) = \mathbf{0}.$$

Using equation (5), we have

$$v(t) = \int (2i + 6tj)\, dt$$
$$= (2t + c_1)i + (3t^2 + c_2)j + c_3 k.$$

Setting $t = 0$ and applying the initial condition $v(0) = \mathbf{0}$ gives

$$\mathbf{0} = c_1 i + c_2 j + c_3 k,$$

so
$$c_1 = c_2 = c_3 = 0.$$

The velocity function is therefore
$$v(t) = 2ti + 3t^2j.$$

Obtain r by integrating v.

Using equation (6), we find that
$$r(t) = \int (2ti + 3t^2j)\, dt$$
$$= (t^2 + d_1)i + (t^3 + d_2)j + d_3k.$$

Determine the constants d_1, d_2, and d_3 by applying the initial condition
$$r(0) = 2i - 3k.$$

Setting $t = 0$ and applying the initial condition $r(0) = 2i - 3k$ gives
$$2i - 3k = d_1i + d_2j + d_3k$$

so
$$d_1 = 2, \qquad d_2 = 0, \qquad \text{and} \qquad d_3 = -3.$$

Thus,
$$r(t) = (t^2 + 2)i + t^3j - 3k.$$

The location of the particle after 3 seconds is $r(3)$.

The location of the particle after 3 seconds is
$$r(3) = (3^2 + 2)i + 3^3j - 3k$$
$$= 11i + 27j - 3k,$$

or $(11, 27, -3)$. Note that, because there is no component of a in the z direction, the entire motion takes place in the plane $z = -3$. ◇

Projectile Motion

A frequent application of equations (5) and (6) concerns the trajectory followed by a projectile launched with a prescribed initial velocity, v_0. The term ''projectile'' refers to any object, such as a missile, a flare, or a baseball, which is not self-propelling. Thus, the trajectory followed by a projectile is completely determined by the speed and direction (i.e., the velocity) at which it is launched. In the discussion that follows, we shall assume that air resistance is negligible, so that the only force acting on the projectile after launch is the force due to gravity. Also, we shall take the y-axis as vertical and the x-axis as horizontal, with the positive x-axis pointing in the direction of the horizontal component of the initial velocity. The motion is then restricted entirely to the xy-plane.

The motion of projectiles (and, for that matter, of all moving objects) is governed by the vector form of Newton's second law of motion:
$$F = ma. \tag{7}$$

Here F is the vector sum of all forces acting on the projectile, m is its mass, and a is its acceleration.

According to our use of the term projectile, the only force acting on the object in flight is the force due to gravity,
$$F = -mgj \tag{8}$$

where $g = 9.81$ m/s^2 is the gravitational constant. (We use a negative sign since this force acts downward.)

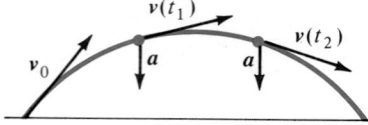

Figure 4.5 Motion of a projectile with initial velocity v_0.

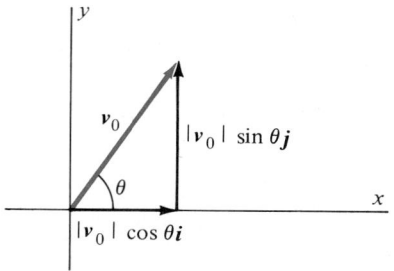

Figure 4.6 Components of initial velocity are

$$v_0 = |v_0| \cos \theta i + |v_0| \sin \theta j.$$

Combining equations (7) and (8) we find that a projectile in flight experiences the acceleration

$$a = -gj. \tag{9}$$

Figure 4.5 illustrates a typical trajectory for a projectile fired with initial velocity v_0.

Now consider the problem of determining the trajectory of a projectile fired from the origin at an angle θ with the horizontal and with an initial speed s_0. This means that the initial velocity v_0 has length $|v_0| = s_0$. We may therefore write v_0 in component form as

$$v_0 = s_0 \cos \theta i + s_0 \sin \theta j. \tag{10}$$

(See Figure 4.6.)

Applying equation (5) with $a(t)$ as in equation (9), we find

$$v(t) = \int a \, dt = \int -gj \, dt = c_1 i + (c_2 - gt)j.$$

Setting $t = 0$ and applying the initial condition given by equation (10), we obtain

$$v(0) = c_1 i + c_2 j = s_0 \cos \theta i + s_0 \sin \theta j.$$

Thus,

$$c_1 = s_0 \cos \theta; \qquad c_2 = s_0 \sin \theta,$$

and therefore

$$v(t) = s_0 \cos \theta i + (s_0 \sin \theta - gt)j.$$

According to equation (6), the position function is

$$r(t) = \int v(t) \, dt = \int [s_0(\cos \theta)i + (s_0 \sin \theta - gt)j] \, dt$$

$$= [s_0(\cos \theta)t + d_1]i + \left[s_0(\sin \theta)t - \frac{1}{2}gt^2 + d_2 \right]j.$$

Since the projectile is fired from the origin, we have the initial condition $r(0) = 0$. Setting $t = 0$ in the above expression for $r(t)$ and applying this initial condition gives

$$r(0) = d_1 i + d_2 j = 0,$$

so

$$d_1 = d_2 = 0.$$

The position function for the projectile motion is therefore

$$r(t) = [s_0(\cos \theta)t]i + \left[s_0(\sin \theta)t - \frac{1}{2}gt^2 \right]j.$$

We summarize our findings as follows:

> The trajectory of a projectile fired from the origin with initial speed s_0 and angle of elevation θ is parameterized by the vector-valued function
>
> $$r(t) = [s_0(\cos \theta)t]i + \left[s_0(\sin \theta)t - \frac{1}{2}gt^2 \right]j. \tag{11}$$

Example 5

A flare is fired from ground level at an elevation of 60° with an initial speed of 50 m/s. Find

(a) the maximum height of the trajectory,
(b) the length of time during which the flare is airborne, and
(c) the distance from the point of launch to the point of impact.

Strategy

Write the expression for $r(t) = x(t)i + y(t)j$ using (11).

Solution

With $\theta = 60° = \pi/3$ and $s_0 = 50$ m/s, $r(t)$ is

$$r(t) = 50\left(\cos\frac{\pi}{3}\right)ti + \left(50\left(\sin\frac{\pi}{3}\right)t - \frac{1}{2}gt^2\right)j$$

$$= 25ti + \left(25\sqrt{3}t - \frac{1}{2}gt^2\right)j.$$

Find the time, t_{max}, of maximum height by maximizing the *j*-component of $r(t)$.

The height of the flare is given by the *j*-component

$$y(t) = 25\sqrt{3}t - \frac{1}{2}gt^2.$$

The maximum height occurs when

$$y'(t) = 25\sqrt{3} - gt = 0,$$

or

$$t_{max} = \frac{25\sqrt{3}}{g} = \frac{25\sqrt{3}}{9.81} \approx 4.4 \text{ seconds.}$$

$y(t_{max})$ is the maximum height.

The maximum height is

$$y(t_{max}) = 25\sqrt{3}\left(\frac{25\sqrt{3}}{9.81}\right) - \frac{1}{2}(9.81)\left(\frac{25\sqrt{3}}{9.81}\right)^2$$

$$= \frac{\left(\frac{1}{2}\right)25^2 \cdot 3}{9.81} \approx 95.6 \text{ meters.}$$

Find the time, t_{end}, at which the projectile lands by setting $y(t) = 0$.

The projectile lands when the *j*-component reaches zero, that is, when

$$y(t) = 25\sqrt{3}t - \frac{1}{2}gt^2 = 0.$$

This equation has solutions $t = 0$ and $t = \dfrac{50\sqrt{3}}{g}$.

The time $t = 0$ corresponds to launch. The time at which the projectile returns to ground level is therefore

$$t_{end} = \frac{50\sqrt{3}}{g} = \frac{50\sqrt{3}}{9.81} \approx 8.8 \text{ seconds.}$$

Distance from launch point to impact point is $x(t_{end})$.

The distance from the point of launch to the point of impact is the *i*-component of $r(t_{end})$.

$$x(t_{end}) = 25(8.8) = 220 \text{ meters.}$$

◇

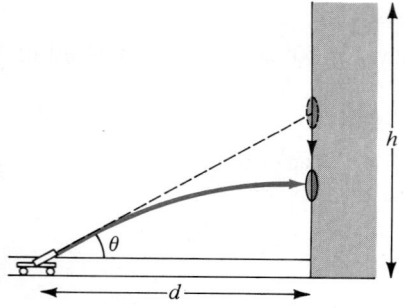

Figure 4.7 The cannon–target experiment.

Example 6

The following experiment is performed in an elementary physics class. A target is hung on a wall of a gymnasium at a height of h meters above the floor. A small cannon, which fires a lead slug, is located on the floor d meters from the wall. The muzzle velocity of the cannon is adjustable. The cannon is aimed at the bull's-eye on the target. By means of a trip mechanism, the target is released from its hanger at the precise instant at which the cannon is fired and falls toward the floor. The result of repeated trials of the experiment is that no matter what muzzle velocity is chosen, the slug always strikes the bull's-eye (assuming that d, θ, and the muzzle velocity are such that the slug does not strike the floor before reaching the wall). (See Figure 4.7.) How can this be explained?

Solution: Let θ be the angle of elevation of the barrel of the cannon when aimed at the bull's-eye, and let s_0 be the muzzle velocity of the cannon. We assume the cannon to be located at the origin of an xy-coordinate system. The trajectory of the slug is then given by $r(t)$ in equation (11).

The slug reaches the wall when the i-component of $r(t)$ satisfies the equation

$$x(t) = s_0(\cos \theta) \, t = d$$

which occurs at time $t_f = \dfrac{d}{s_0 \cos \theta}$. At this time the j-component of the slug is

$$y(t_f) = s_0 \sin \theta \left(\frac{d}{s_0 \cos \theta} \right) - \frac{1}{2} g \left(\frac{d}{s_0 \cos \theta} \right)^2$$

$$= d \tan \theta - \frac{1}{2} g t_f^2.$$

Since $\dfrac{h}{d} = \tan \theta$, $h = d \tan \theta$. Thus

$$y(t_f) = h - \frac{1}{2} g t_f^2$$

is the j-component of $r(t)$ when the slug reaches the wall.

Since the target is a freely falling body, after t_f seconds it has fallen a distance $\dfrac{1}{2} g t_f^2$, so the j-component of the bull's-eye's position vector after t_0 seconds is also $h - \dfrac{1}{2} g t_f^2$. Since the slug and the bull's-eye have the same position vector, $r(t_f) = d\mathbf{i} + (h - \dfrac{1}{2} g t_f^2)\mathbf{j}$, at time $t = t_f$, the slug must always score a direct hit on the target.   ◇

Exercise Set 17.4

In Exercises 1–7, find the velocity and acceleration functions corresponding to the given position functions.

1. $r(t) = i + 2tj$

2. $r(t) = 3ti + t^3j + (2t + 3)k$

3. $r(t) = \cos 3ti + \sin 3tj$

4. $r(t) = e^ti + e^{-t}j + \sqrt{t}k, \quad t \geq 1$

5. $r(t) = \ln t^2 i + \cos 2tj + \dfrac{1}{t}k, \quad t \geq 1$

6. $r(t) = t(t - 1)i + \text{Tan}^{-1} tj + te^t k$

7. $r(t) = e^t \cos ti + e^t \sin tj + \cos tk$

8. A particle moves with position function $r(t) = \cos \pi ti + \sin \pi tj$. Find its speed at time $t = 1/4$.

9. A particle moves with position vector $r(t) = t^2 i + 2tj + t^3 k$. Find its speed at time $t = 2$.

In Exercises 10–13, find the position function r from the given velocity function and the initial condition $r(0) = r_0$.

10. $v(t) = 2ti + t^2 j$, $r_0 = i + 4j$

11. $v(t) = \cos ti + \sin tj$, $r_0 = 3i + 2j$

12. $v(t) = e^t i + \sqrt{t}j + 2tk$, $r_0 = 6i + 4k$

13. $v(t) = \left(\dfrac{1}{1 + t^2}\right)i + te^{t^2}j$, $r_0 = -2i + j + 4k$

In Exercises 14–18, find the velocity function v and the position function r given the acceleration function a and the initial conditions $v_0 = v(0)$ and $r_0 = r(0)$.

14. $a = i + j$, $v_0 = i + j$, $r_0 = 0$

15. $a = 2i + k$, $v_0 = 0$, $r_0 = 3i - j + 4k$

16. $a = 6ti + e^t j$, $v_0 = 0$, $r_0 = 4i + j + 2k$

17. $a = \cos ti + \sin tj$, $v_0 = k$, $r_0 = 3i - j + 2k$

18. $a = e^t i + tj + e^{-t}k$, $v_0 = i + k$, $r_0 = i + 4j + k$

19. Verify that the vectors $v(t)$ and $a(t)$ of Example 2 are orthogonal for each t.

20. Prove that if a particle moves with constant speed its velocity and acceleration vectors are orthogonal.

21. A particle moves about a circle in the xy-plane with position function $r(t) = \rho \cos \alpha ti + \rho \sin \alpha tj$. Find a relationship between the radius of the circle and the speed of the particle.

22. A projectile is fired from ground level at an elevation angle of $\theta = 45°$ with an initial speed of 60 meters per second. Find

   a. the position function r of the projectile,
   b. the maximum altitude of the projectile,
   c. the flight time,
   d. the distance from the launch point to the impact point,
   e. the speed on impact.

23. Find each of the quantities in Exercise 22 for a projectile fired at an elevation angle of $\theta = 30°$ with an initial speed of 100 meters per second.

24. A bale of hay is dropped from an airplane flying at a ground speed of 200 km/h and an elevation of 1000 meters. How far from the drop point does the bale strike the ground?

25. A handgun used in a physics experiment simultaneously fires one slug through the barrel and drops a second slug mounted on the side of the barrel. Independent of the muzzle velocity, the two slugs are observed to strike the ground at the same instant when the gun is fired horizontally. Why?

26. How far along the path $r(t)$ does the particle of Example 4 travel during the first 3 seconds of its motion? What is its distance from the starting point P_0 at $t = 3$ seconds?

27. a. Write the integral that expresses the arc length of the motion of the flare in Example 5.
   b. *(Calculator/Computer)* Use Simpson's Rule with $n = 8$ to approximate the value of the integral in (a).

17.5 CURVATURE

Curvature is a measure of the rate at which a curve in space turns or twists. To make this idea more precise, we consider a curve C determined parametrically by the vector function

$$C: \quad r(s) = x(s)i + y(s)j + z(s)k, \qquad a \le s \le b. \tag{1}$$

We assume that C is parameterized by arc length, so that the tip of $r(s)$ moves along C uniformly. That is, we assume that $|r'(s)| = 1$, $a \le s \le b$.

Since $r'(s)$ is a unit tangent at each point $r(s)$, the second derivative r'' measures the rate at which the unit tangent changes (turns or twists) as s increases. This is analogous to the one variable case in which the second derivative f'' measures the rate at which the slope function f' is changing. However, the derivatives here are vectors rather than numbers. We therefore make the following definition.

| DEFINITION 9 | Let the curve C in (1) be parameterized by arc length, and let r in (1) be twice differentiable for $a \le s \le b$. The vector $r''(s)$ is called the **curvature vector** at $r(s)$. Its length, $|r''(s)|$, is called the **curvature** of C at $r(s)$. |

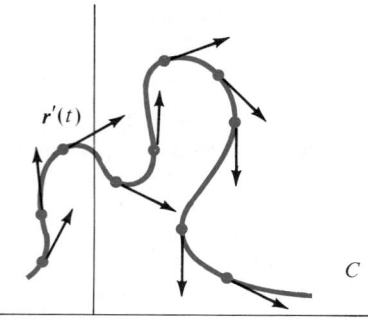

Figure 5.1 $\kappa(s)$ is large when C turns sharply.

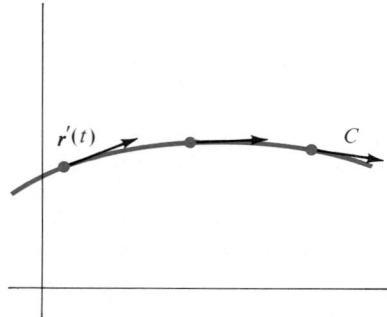

Figure 5.2 $\kappa(s)$ is small when C turns slowly.

REMARK: If C were not parameterized by arc length, then $|r''(s)|$ would depend upon both the direction and the magnitude of $r'(s)$. By insisting on parameterization by arc length, we have removed the dependence of the curvature on the "speed of the curve, $r'(s)$."

The Greek letter κ (kappa) is usually used to denote curvature. Thus, Definition 9 states that

$$\kappa(s) = |r''(s)|, \qquad a \leq s \leq b. \tag{2}$$

Figures 5.1 and 5.2 illustrate the essence of curvature: $\kappa(s)$ is large when C turns sharply (rapid change in $r'(s)$), and $\kappa(s)$ is small when C turns slowly (small change in $r'(s)$).

Example 1

In Example 7, Section 17.3, we determined that the circle of radius ρ centered at the origin is parameterized by arc length by the vector function

$$r(s) = \rho \cos\left(\frac{s}{\rho}\right)i + \rho \sin\left(\frac{s}{\rho}\right)j, \qquad 0 \leq s < 2\rho\pi.$$

The curvature vector is therefore

$$r''(s) = -\frac{1}{\rho} \cos\left(\frac{s}{\rho}\right)i - \frac{1}{\rho} \sin\left(\frac{s}{\rho}\right)j = -\frac{1}{\rho^2}r(s)$$

and the curvature at $r(s)$ is

$$\kappa(s) = |r''(s)| = \frac{1}{\rho}.$$

When the circle is allowed to have any center, a slight generalization of this argument shows that *the curvature of a circle of radius ρ is $\kappa = 1/\rho$ at each point on the circle* (see Exercise 33). $\diamond$

Figure 5.3 shows the curvature vectors for several circles of various radii.

For more general plane curves C parameterized by arc length, the expression $\rho(s) = \dfrac{1}{\kappa(s)}$ is called the **radius of curvature** at $r(s)$. Since $\rho(s)$ is the radius of the circle whose curvature is also $\kappa(s)$, the number $\rho(s)$ determines a circle that

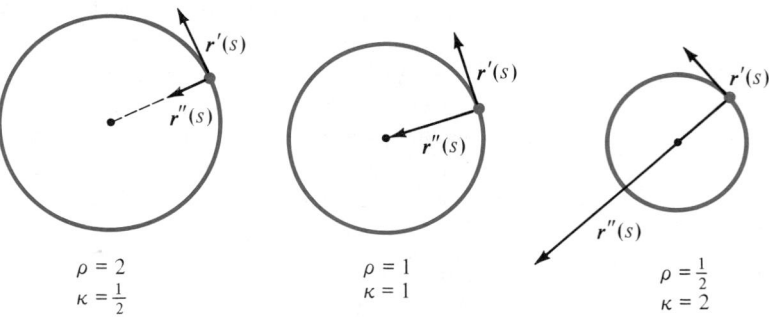

Figure 5.3 Tangent and curvature vectors for various circles.

(i) is tangent to C at $r(s)$,

(ii) has radius $\rho(s) = \dfrac{1}{\kappa(s)}$,

(iii) has the same curvature vector as C at $r(s)$.

This circle is called the **osculating circle,** or **circle of curvature,** at $r(s)$. The center of this circle is called the **center of curvature.** Figures 5.4 and 5.5 show the osculating circles at various points on a curve C.

For plane curves that are not parameterized by arc length, the following theorem indicates how curvature may be computed.

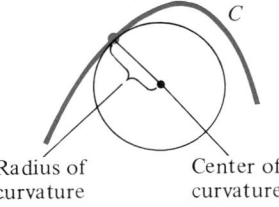

Radius of curvature · Center of curvature

Figure 5.4 Circle of curvature for a plane curve C.

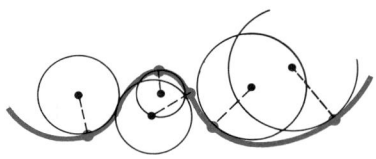

Figure 5.5 Osculating circles at various points on C.

THEOREM 5

Let the curve C be determined parametrically by the twice differentiable function

$$r(t) = x(t)i + y(t)j, \qquad a \le t \le b. \tag{3}$$

The curvature $\kappa(t)$ of C at the point $(x(t), y(t))$ is

$$\boxed{\kappa(t) = \frac{|x'(t)y''(t) - y'(t)x''(t)|}{[(x'(t))^2 + (y'(t))^2]^{3/2}}.} \tag{4}$$

Before indicating the proof of Theorem 5, we consider two examples.

Example 2

For the curve C given parametrically by the function

$$r(t) = (2t^2 + 1)i + (t^3 - 3)j, \qquad 1 \le t \le 3,$$

find

(a) the curvature function κ, and

(b) the radius of curvature at the point $(3, -2)$.

Solution: Here $x(t) = 2t^2 + 1$ and $y(t) = t^3 - 3$. Thus

$$x'(t) = 4t \qquad y'(t) = 3t^2,$$
$$x''(t) = 4 \qquad y''(t) = 6t.$$

According to Theorem 5, the curvature is

$$\kappa(t) = \frac{|4t \cdot 6t - 3t^2 \cdot 4|}{[(4t)^2 + (3t^2)^2]^{3/2}} = \frac{12t^2}{(16t^2 + 9t^4)^{3/2}}.$$

The point $(3, -2)$ corresponds to $t = 1$. The curvature at this point is therefore

$$\kappa(1) = \frac{12}{(16 + 9)^{3/2}} = \frac{12}{125}.$$

The radius of curvature at this point is

$$\rho(1) = \frac{1}{\kappa(1)} = \frac{125}{12}.$$

◇

Example 3

Find the curvature function κ for the graph of the function $f(t) = \sin t$.

Solution: We can write the function $f(t) = \sin t$ in parametric (vector) form as

$$r(t) = t\mathbf{i} + \sin t\mathbf{j}.$$

Thus, $x(t) = t$ and $y(t) = \sin t$, so

$$x'(t) = 1 \qquad y'(t) = \cos t,$$
$$x''(t) = 0 \qquad y''(t) = -\sin t.$$

By Theorem 5,

$$\kappa(t) = \frac{|1 \cdot (-\sin t) - 0 \cdot \cos t|}{[1^2 + (\cos t)^2]^{3/2}} = \frac{|\sin t|}{(1 + \cos^2 t)^{3/2}}.$$

Notice that $\kappa(t)$ is a minimum (zero) at $t = 0, \pm\pi, \pm2\pi, \ldots$ and that $\kappa(t)$ is a maximum (one) at $t = \pi/2 \pm n\pi, n = 1, 2, \ldots$.

◇

Example 3 shows how Theorem 5 may be used to calculate the curvature for the graph of a twice differentiable function $y = f(x)$. Such functions can always be written in vector form as

$$r(x) = x\mathbf{i} + f(x)\mathbf{j}.$$

Then $x' = 1$, $x'' = 0$, $y' = f'(x)$, and $y'' = f''(x)$, so the curvature $\kappa(x)$, according to Theorem 5, is

$$\kappa(x) = \frac{|0 \cdot f'(x) - f''(x) \cdot 1|}{[1 + (f'(x))^2]^{3/2}}.$$

We state this result as a corollary of Theorem 5.

COROLLARY 2

Let f be a twice differentiable function of x. The curvature $\kappa(x)$ at the point $(x, f(x))$ on the graph of $y = f(x)$ is

$$\kappa(x) = \frac{|f''(x)|}{[1 + (f'(x))^2]^{3/2}}.$$

Example 4

Show that the curvature of a nonvertical line in the plane is zero.

Solution: The equation for a nonvertical line in the plane can be written in the form $f(x) = mx + b$ where m and b are constants. Thus $f'(x) = m$ and $f''(x) = 0$. Thus, $\kappa(x) = 0$, according to Corollary 2. $\diamond$

Proof of Theorem 5: We will demonstrate the proof up to a point at which you can complete the argument with a calculation (Exercise 34).

The difficulty to be overcome is that the parameterization given by $r(t)$ in equation (3) is not necessarily a parameterization by arc length. We address this problem by letting ℓ denote the length of the arc of C from $r(a)$ to $r(b)$. Then the *arc length function*

$$s(t) = \int_a^t |r'(\omega)|\ d\omega$$

$$= \int_a^t \sqrt{[x'(\omega)]^2 + [y'(\omega)]^2 + [z'(\omega)]^2}\ d\omega \tag{5}$$

gives, as for curves in the plane, the length of the arc of C from $r(a)$ to $r(t)$. (See Section 17.3.) This function has the properties that

$$s(a) = 0,$$
$$s(b) = \ell,$$

and

$$s'(t) = |r'(t)|, \qquad a \le t \le b. \tag{6}$$

Now, suppose that the vector function

$$R(s) = X(s)i + Y(s)j, \qquad 0 \le s \le \ell$$

is an arc length parameterization for C. Then the composition of R with s in (5) is

$$R(s(t)) = X(s(t))i + Y(s(t))j, \qquad a \le t \le b.$$

It is important to note that $R(s(t)) = r(t)$ for each $t \in [a, b]$, since both vectors terminate at a point $s(t)$ units along C from $r(a)$ (see Figure 5.6).

Since $R(s)$ is a parameterization by arc length,

$$T(t) = R'(s) = \frac{d}{ds}\, R(s(t)) \tag{7}$$

is a unit tangent for C at $r(t) = R(s(t))$ (equation (12), Section 17.3). Differentiating both sides of equation (7) with respect to t, gives

$$T'(t) = \frac{d}{dt}\left(\frac{d}{ds}\, R\right)$$

$$= \frac{d}{ds}\left(\frac{d}{ds}\, R\right)\frac{ds}{dt} \qquad \text{(by Chain Rule)}$$

$$= R''(s)\, \frac{ds}{dt}$$

$$= R''(s)\, |r'(t)| \qquad \text{(by equation (6)),}$$

so

$$|R''(s)| = \frac{|T'(t)|}{|r'(t)|}. \tag{8}$$

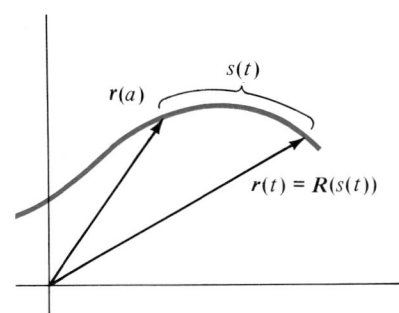

Figure 5.6 $R(s(t)) = r(t),\ a \le t \le b.$

$r(a)$

$s(t)$

$r(t) = R(s(t))$

The expression in equation (8) is the curvature of C at $r(t) = R(s(t))$. It is left for you to show that

$$\frac{|T'(t)|}{|r'(t)|} = \frac{|x''(t)y'(t) - x'(t)y''(t)|}{[(x'(t))^2 + (y'(t))^2]^{3/2}}, \tag{9}$$

which will complete the proof. ◆

Curvature for Space Curves

If C is a space curve parameterized by arc length by the vector function

$$C: \quad r(s) = x(s)i + y(s)j + z(s)k, \qquad a \le s \le b, \tag{10}$$

the vector $T(s) = r'(s)$ is a unit tangent to the curve C at $r(s)$. Since $|r'(s)| = 1$, the curvature vector $r''(s)$ is orthogonal to the tangent $T(s)$ (Example 6, Section 17.2). Thus, the curvature vector $r''(s)$ is **normal** to C at $r(s)$, and the **principal unit normal** to C at $r(s)$ is the vector

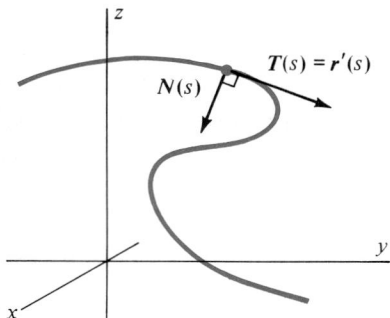

Figure 5.7 Principal unit normal is $N(s) = \dfrac{r''(s)}{|r''(s)|}$.

$$N(s) = \frac{1}{|r''(s)|} r''(s). \tag{11}$$

Of course, the principal unit normal in (11) is defined only when r is twice differentiable and $|r''(s)| \ne 0$ (see Figure 5.7). If C in (10) is a plane curve, the vector $N(s)$ points toward the center of curvature. Multiplying (11) through by $|r''(s)| = \kappa(s)$ and writing $r'(s) = T(s)$ gives the equation

$$T'(s) = \kappa(s)N(s), \tag{12}$$

an equation by which we could alternatively have defined curvature for space curves.

Normal and Tangential Components of Acceleration

We may apply these ideas to obtain an informative expression for the velocity of a particle moving in space with position function given by

$$r(t) = x(t)i + y(t)j + z(t)k. \tag{13}$$

We assume that r is twice differentiable (although not necessarily an arc length parameterization for the path).

Letting s denote arc length and $T = \dfrac{dr}{ds}$ the unit tangent, we apply the Chain Rule to obtain the velocity function v as

$$v(t) = \frac{dr}{dt} = \frac{dr}{ds}\frac{ds}{dt} = \frac{ds}{dt} T.$$

Similarly, the acceleration function is

$$a(t) = \frac{dv}{dt} = \frac{d}{dt}\left(\frac{dr}{ds}\frac{ds}{dt}\right) \tag{14}$$

$$= \frac{d^2s}{dt^2}\frac{dr}{ds} + \frac{ds}{dt}\left(\frac{d^2r}{ds^2}\frac{ds}{dt}\right)$$

$$= \frac{d^2s}{dt^2}\frac{dr}{ds} + \left(\frac{ds}{dt}\right)^2\left[\frac{d}{ds}\left(\frac{dr}{ds}\right)\right]$$

$$= \frac{d^2s}{dt^2} T(s) + \left(\frac{ds}{dt}\right)^2 \kappa(s)N(s) \qquad \text{(equation (12))}.$$

That is, we can express acceleration as the sum of a vector parallel to the unit tangent $T(s)$ and a vector parallel to the unit normal $N(s)$. It is customary to refer to the **tangential component of acceleration** as the coefficient $a_T = \dfrac{d^2 s}{dt^2}$, and to the **normal component of acceleration** as the coefficient $a_N = \left(\dfrac{ds}{dt}\right)^2 \kappa(s)$. We may then write

$$a = a_T T + a_N N. \tag{15}$$

The components a_T and a_N have straightforward interpretations. The tangential component is

$$a_T = \frac{d^2 s}{dt^2} = \frac{d}{dt}\left(\frac{ds}{dt}\right) = \frac{d}{dt}(|v(t)|),$$

so that a_T is just the rate at which speed is changing. For example, if the speed of the particle is constant, then $a_T = 0$. The normal component of acceleration can be written $a_N = \left(\dfrac{ds}{dt}\right)^2 \kappa(s) = |v|^2 \kappa$. Thus, a_N is influenced both by speed and by curvature. Since N is orthogonal to T, this explains why the force tending to pull an object (such as an automobile) from its path is influenced both by the speed of the object and by the curvature of the path.

Since $a_T = \dfrac{d}{dt}|v|$, computing a_T from r is usually simple. However, calculating a_N can be considerably more difficult. An aid to this calculation is obtained by using equation (15) and the fact that T and N are orthogonal unit vectors:

$$\begin{aligned}
|a|^2 = a \cdot a &= (a_T T + a_N N) \cdot (a_T T + a_N N) \\
&= (a_T)^2 |T|^2 + a_N |N|^2 \\
&= a_T^2 + a_N^2.
\end{aligned}$$

Thus

$$a_N = \sqrt{|a|^2 - a_T^2}. \tag{16}$$

Example 5

Find the tangential and normal components of acceleration for a particle moving along the helical path

$$r(t) = \cos t^2 i + \sin t^2 j + t k.$$

Strategy
Find $v(t) = r'(t)$.

Solution

Here

$$v(t) = -2t \sin t^2 i + 2t \cos t^2 j + k,$$

so

Find $|v(t)|$.

$$\begin{aligned}
|v(t)| &= \sqrt{4t^2 \sin^2 t^2 + 4t^2 \cos^2 t^2 + 1} \\
&= \sqrt{4t^2 + 1}.
\end{aligned}$$

Thus

$a_T = \dfrac{d}{dt}(|v(t)|).$

$$a_T = \frac{d}{dt}\left(\sqrt{4t^2 + 1}\right) = \frac{4t}{\sqrt{4t^2 + 1}}.$$

Find **a**.

To find **a** we differentiate **v**:

$$a(t) = (-2 \sin t^2 - 4t^2 \cos t^2)i + (2 \cos t^2 - 4t^2 \sin t^2)j,$$

so

Find $|a(t)|$.

$$|a(t)| = \{4 \sin^2 t^2 + 16t^2 \sin t^2 \cos t^2 + 16t^4 \cos^2 t^2 + 4 \cos^2 t^2 -$$
$$16t^2 \sin t^2 \cos t^2 + 16t^4 \sin^2 t^2\}^{1/2}$$
$$= \sqrt{4 + 16t^4}.$$

Apply equation (16).

Thus

$$a_N = \sqrt{|a|^2 - a_T^2} = \sqrt{(4 + 16t^4) - \frac{16t^2}{4t^2 + 1}}$$
$$= \sqrt{\frac{64t^6 + 16t^4 + 4}{4t^2 + 1}}.$$

Notice that $\lim_{t \to \infty} a_T = 2$, while $\lim_{t \to \infty} a_N = +\infty$. Can you explain why this is so?

◇

Vector Form of Curvature

A drawback of our work on curvature up to this point is that $\kappa(s)$ is difficult to compute, using Definition 9, when the curve C determined by r is not parameterized by arc length. We shall now develop a formula for κ that is easier to apply. We imagine the parameterization r for C as the position function of a particle moving in space. Then, writing

$$v(t) = \frac{dr}{dt} = \frac{dr}{ds} \cdot \frac{ds}{dt} = \left(\frac{ds}{dt}\right)T$$

and using equation (14) we find that

$$v \times a = \left(\frac{ds}{dt}\right)T \times \left[\left(\frac{d^2s}{dt^2}\right)T + \kappa\left(\frac{ds}{dt}\right)^2 N\right]$$
$$= \kappa\left(\frac{ds}{dt}\right)^3 (T \times N)$$

since $T \times T = 0$. Also, since T and N are orthogonal unit vectors,

$$|T \times N| = |T||N||\sin \theta| = 1.$$

Thus,

$$|v \times a| = \kappa\left(\frac{ds}{dt}\right)^3 = \kappa|v|^3,$$

so

$$\boxed{\kappa = \frac{|v \times a|}{|v|^3}.}$$
(17)

Example 6

Find the curvature of the helix

$$C: \quad r(t) = \cos t\,i + \sin t\,j + t\,k.$$

Solution: Here

$$v(t) = -\sin t\mathbf{i} + \cos t\mathbf{j} + \mathbf{k}$$

and

$$a(t) = -\cos t\mathbf{i} - \sin t\mathbf{j}.$$

Thus

$$v \times a = \det \begin{bmatrix} \mathbf{i} & \mathbf{j} & \mathbf{k} \\ -\sin t & \cos t & 1 \\ -\cos t & -\sin t & 0 \end{bmatrix}$$
$$= \sin t\mathbf{i} - \cos t\mathbf{j} + (\sin^2 t + \cos^2 t)\mathbf{k},$$

so

$$|v \times a| = \sqrt{\sin^2 t + \cos^2 t + 1} = \sqrt{2}.$$

Also,

$$|v| = \sqrt{\sin^2 t + \cos^2 t + 1} = \sqrt{2}.$$

Thus, by (17),

$$\kappa = \frac{\sqrt{2}}{(\sqrt{2})^3} = \frac{1}{2}.$$

$\diamondsuit$

Exercise Set 17.5

In Exercises 1–4, verify that the given curve is parameterized by arc length. Then find (a) the unit tangent $T(s)$, (b) the curvature vector $r''(s)$, and (c) the curvature $\kappa(s)$.

1. $r(s) = \dfrac{1}{2}s\mathbf{i} + \dfrac{\sqrt{3}}{2}s\mathbf{j}$

2. $r(s) = (4 + \cos s)\mathbf{i} + (2 + \sin s)\mathbf{j}$

3. $r(s) = \dfrac{\sqrt{2}}{2}\cos s\mathbf{i} + \dfrac{\sqrt{2}}{2}\sin s\mathbf{j} + \dfrac{\sqrt{2}}{2}s\mathbf{k}$

4. $r(s) = \sin\left(\dfrac{s}{2}\right)\mathbf{i} + \dfrac{\sqrt{3}}{2}s\mathbf{j} + \cos\left(\dfrac{s}{2}\right)\mathbf{k}$

In Exercises 5–10, find the curvature function $\kappa(t)$ for the given plane curves.

5. $r(t) = 3\mathbf{i} + t^2\mathbf{j}$

6. $r(t) = t\mathbf{i} + (t^2 + 3)\mathbf{j}$

7. $r(t) = t\mathbf{i} + e^t\mathbf{j}$

8. $r(t) = e^t\mathbf{i} + e^{-t}\mathbf{j}$

9. $r(t) = (3 + 2\sin t)\mathbf{i} + (5 + 2\cos t)\mathbf{j}$

10. $r(t) = (t - \sin t)\mathbf{i} + (1 - \cos t)\mathbf{j}$

In Exercises 11–16, find the curvature of the graph of the given function.

11. $f(x) = 3 + x^2$

12. $f(x) = \sqrt{x + 4}$

13. $f(x) = x^3$

14. $f(x) = \ln x$

15. $f(x) = \cos x$

16. $f(x) = e^{x^2}$

In Exercises 17–20, find the curvature $\kappa(t)$ for the space curve determined by $r(t)$.

17. $r(t) = \mathbf{i} + t\mathbf{j} + t^2\mathbf{k}$

18. $r(t) = \cos t\mathbf{i} + \sin t\mathbf{j} + \mathbf{k}$

19. $r(t) = e^t\cos t\mathbf{i} + e^t\sin t\mathbf{j} + t\mathbf{k}$

20. $r(t) = \ln t\mathbf{i} + t\mathbf{j} + t\mathbf{k}, \qquad t > 0$

21. Find the center of curvature for the graph of $y = \sin x$ at the point $(\pi/2, 1)$.

22. Find the point on the graph of $y = \ln x$ where curvature is a maximum.

23. For the graph of $y = \sqrt{x}$, find
  a. the curvature at $(1, 1)$,
  b. the center of curvature at $(1, 1)$,
  c. $\lim\limits_{x \to 0^+} \kappa(x)$.

24. Find the equation for the osculating circle (circle of curvature) for the graph of $y = e^{-x}$ at the point where $x = 1$.

25. Show that the curvature of a line in space is zero.

26. Show that if the curve C is the graph of the polar equation $r = f(\theta)$ and if $f''(\theta)$ exists, then the curvature $\kappa(\theta)$ is given by

$$\kappa(\theta) = \frac{|f(\theta)f''(\theta) - 2[f'(\theta)]^2 - [f(\theta)]^2|}{[(f'(\theta))^2 + (f(\theta))^2]^{3/2}}.$$

(*Hint: C* can be written as $r(\theta) = x(\theta)i + y(\theta)j,$ with $x(\theta) = f(\theta) \cos \theta, y(\theta) = f(\theta) \sin \theta.$)

In Exercises 27–30, use the result of Exercise 26 to find the curvature function $\kappa(\theta)$.

27. $r = 1 + \sin \theta$

28. $r = \theta$

29. $r = 1 - \cos \theta$

30. $r = e^\theta$

31. For a curve C in the plane, explain why the curvature vector $r''(s)$ always points in the direction of the concave side of the curve.

32. Find the curvature of the ellipse $x^2 + 2y^2 = 4$ at the point $(2, 0)$.

33. Prove that the curvature of a circle of radius ρ is $\kappa = \frac{1}{\rho}$.

34. Complete the proof of Theorem 5.

35. Find the tangential and normal components of acceleration for a particle moving along the helical path $2ti + \sin t^2 j + \cos t^2 k.$

SUMMARY OUTLINE OF CHAPTER 17

◆ A **vector-valued function** f has the form   (page 727)

$$f(t) = x(t)i + y(t)j, \quad \text{or} \quad f(t) = x(t)i + y(t)j + z(t)k$$

where $x(t)$, $y(t)$, and $z(t)$ are real-valued functions.

◆ For $f(t) = x(t)i + y(t)j + z(t)k,$ (page 730)

(i) $\lim_{t \to t_0} f(t) = \left[\lim_{t \to t_0} x(t) \right] i + \left[\lim_{t \to t_0} y(t) \right] j + \left[\lim_{t \to t_0} z(t) \right] k$

(ii) $f'(t) = x'(t)i + y'(t)j + z'(t)k$

(iii) $\int f(t)\, dt = \left[\int x(t)\, dt \right] i + \left[\int y(t)\, dt \right] j + \left[\int z(t)\, dt \right] k$

(iv) $\int_a^b f(t)\, dt = \left[\int_a^b x(t)\, dt \right] i + \left[\int_a^b y(t)\, dt \right] j + \left[\int_a^b z(t)\, dt \right] k.$

(v) f is continuous at t_0 if and only if $x(t)$, $y(t)$, and $z(t)$ are all continuous at t_0.

◆ Derivatives of vector-valued functions are calculated as follows: (page 734)

(i) $(f + g)'(t) = f'(t) + g'(t)$
(ii) $(cf)'(t) = cf'(t)$
(iii) $(hf)'(t) = h(t)f'(t) + h'(t)f(t)$
(iv) $(f \times g)'(t) = [f(t) \times g'(t)] + [f'(t) \times g(t)]$
(v) $(f \circ h)'(t) = h'(t)f'(h(t))$
(vi) $(f \cdot g)'(t) = f(t) \cdot g'(t) + f'(t) \cdot g(t).$

◆ If $r'(t_0) \neq 0$, the **tangent** to the graph of r at $r(t_0)$ is $r'(t_0) = x'(t_0)i + y'(t_0)j + z'(t_0)k.$ (page 742)

◆ The **unit tangent** to the graph of r at $r(t_0)$ is $T = \frac{1}{|r'(t_0)|} r'(t_0) \quad r'(t_0) \neq 0.$ (page 744)

◆ The curve C is **parameterized by arc length** on the interval I if   (page 747)

$$\int_0^t |r'(\omega)|\, d\omega = t$$

for all t in I. Equivalently, $|r'(t)| = 1, \quad t \in I.$

◆ For a particle moving in space with the twice differentiable **position function** r,   (page 750)

(i) the **velocity** function is $v = r'$

(ii) the **speed** is $|v(t)| = |r'(t)|$

(iii) the **acceleration** function is $a = v' = r''$.

◆ The **trajectory** of a projectile fired from the origin with initial speed s_0 and angle of elevation θ is   (page 754)

$$r(t) = s_0(\cos \theta)ti + \left(s_0 \sin \theta t - \frac{1}{2}gt^2\right)j.$$

◆ If C: $r(s)$ is parameterized by arc length, the **curvature** at $r(s)$ is $\kappa(s) = |r''(s)|$.   (page 757)

◆ The **radius of curvature** at $r(s)$ is   (page 758)

$$\rho(s) = \frac{1}{\kappa(s)}, \qquad \kappa(s) \neq 0.$$

◆ If the plane curve C is parameterized by the equations $x = x(t)$, $y = y(t)$, the **curvature** at $(x(t), y(t))$ is   (page 759)

$$\kappa(t) = \frac{|x'(t)y''(t) - y'(t)x''(t)|}{[(x'(t))^2 + (y'(t))^2]^{3/2}}.$$

◆ If the curve C is the graph of $y = f(x)$,   (page 760)

$$\kappa(x) = \frac{|f''(x)|}{[1 + (f'(x))^2]^{3/2}}.$$

◆ The **principal unit normal** to $r(s)$ is $N(s) = \dfrac{1}{|r''(s)|}r''(s)$ when $r(s)$ is parameterized by arc length.   (page 762)

◆ If r is a twice differentiable position function, its acceleration function can be written $a(t) = a_T T + a_N N$, where the   (page 763)
tangential component of acceleration is

$$a_T = \frac{d^2s}{dt^2}$$

and the **normal component of acceleration** is

$$a_N = \left(\frac{ds}{dt}\right)^2 \kappa(s) = |v|^2 \kappa.$$

◆ If the curve C is determined by the twice differentiable **position function** r, then   (page 764)

$$\kappa = \frac{|v \times a|}{|v|^3}$$

at each point on C.

REVIEW EXERCISES—CHAPTER 17

1. Find an equation in rectangular coordinates for the graph of the vector function $r(t) = a \cos ti + b \sin tj$.

2. Sketch the graph of the vector function $r(t) = i + \cos tj + k$.

3. Find the implicit domain of the vector function

$$r(t) = ti + \sqrt{1 - t^2}j + \frac{1}{t}k.$$

4. Show that the graph of the vector function $r(t) = \tan ti + \sec tj$, $-\pi/2 < t < \pi/2$, is one branch of the hyperbola with rectangular equation $y^2 - x^2 = 1$. What are the asymptotes for this curve?

5. Sketch the graph of the **cycloid** given by the vector function $r(t) = a(t - \sin t)i + a(1 - \cos t)j$, $t \geq 0$. Find the numbers t for which the tangent vector is zero. (The cycloid gives the path of a point on the rim of a wheel of radius a as the wheel rolls along a horizontal surface.)

In Exercises 6–8, find the indicated limit.

6. $\displaystyle\lim_{t \to 0^+}\left\{\cos 2ti + t \cos tj + \frac{t}{\sqrt{t} + 2}k\right\}$

7. $\displaystyle\lim_{t \to 0}\left\{\left(\frac{\sin 3t}{t}\right)i + \sqrt{t + 2}j + \frac{\tan 2t}{t}k\right\}$

8. $\lim\limits_{t \to 2^+} \left\{ \sqrt{2 + t} \, i + \ln \sqrt{t} \, j + \dfrac{t^3 - 8}{t - 2} k \right\}$

In Exercises 9–11, determine the numbers t for which the function is discontinuous.

9. $r(t) = \sin ti + \sec tj, \qquad 0 \le t \le 2\pi$

10. $r(t) = \dfrac{1}{\sqrt{t}} i + \tan tj + \ln(1 + t)k, \qquad 0 \le t \le \pi$

11. $r(t) = \dfrac{1}{t^2 - 9} i + \dfrac{t + 2}{t^2 - 4} j + tk, \qquad -4 \le t \le 4$

In Exercises 12–15, find r' and r''

12. $r(t) = e^{-2t} i + \cos\left(\dfrac{\pi t}{4}\right) j + k$

13. $r(t) = te^t i + te^{-t} j$

14. $r(t) = \text{Tan}^{-1} 2ti + \sqrt{t} j + tk$

15. $r(t) = \ln 3ti + e^{\sqrt{t}} j$

16. Find $\displaystyle\int_0^1 r(t) \, dt$ for r in Exercise 12.

17. Find $\displaystyle\int_0^1 r(t) \, dt$ for r in Exercise 13.

18. Find a vector function r for which $r'(t) = 4r(t)$ and $r(0) = 4i - j + \pi k$.

19. Find a vector function r for which $r''(t) = 9r(t)$ and $r(0) = 4i + 3j$.

20. Find the numbers α for which the graph of the vector function

$r(t) = \alpha ti + \cos(2\alpha t)j + \sin(2\alpha t)k, \qquad 0 \le t \le \pi$

is parameterized by arc length.

21. Find an arc length parameterization for the circle $x^2 + y^2 = 25$.

22. Find the unit tangent $T(t)$ for the ellipse $r(t) = a \cos ti + b \sin tj, \ 0 \le t < 2\pi$.

23. For the graph of the vector function $r(t) = ti + \cos tj + \sin tk$, find

a. the curvature, $\kappa(t)$,
b. the unit tangent $T(t)$.

24. A particle moves in space with position function $r(t) = a \cos ti + tbj + a \sin tk$. Show that
a. the velocity vector $v(t)$ and the acceleration vector $a(t)$ have constant lengths,
b. $v(t)$ and $a(t)$ are orthogonal for each t.

25. Find the length of the helix $r(t) = 3ti + \cos 2tj + \sin 2tk$ between the points $r(0)$ and $r(3)$.

26. Find the length of the curve determined by the position function $r(t) = \ln ti + tj$ between $r(1)$ and $r(2)$.

27. Find the curvature of the cubic curve $y = x^3$ at the point $(2, 8)$.

28. Find the curvature of the curve determined by the vector function $r(t) = 3i + \sqrt{t} j + t^2 k$ at the point $r(4)$.

29. Let C be a curve in space determined by the differentiable vector function r, $a \le t \le b$. Show that if $t_0 \ne a$, $t_0 \ne b$, and $r(t_0)$ is a point on C either nearest or farthest from the origin, then $r(t_0)$ and $r'(t_0)$ are orthogonal. (*Hint:* Consider $|r(t)|^2 = r(t) \cdot r(t)$.)

30. Use Newton's law $F = ma$ to show that if a particle moves in space subject to zero net force, then
a. the motion is along a line, and
b. both the tangential and normal components of acceleration are zero.

31. Prove that for a particle moving in space, the speed of the particle is constant if and only if its velocity and acceleration vectors are orthogonal at all points along its path.

32. Find the radius of curvature and an equation for the circle of curvature at the point $(0, 1)$ on the graph of $y = \cos x$.

33. Find the point on the graph of $y = e^x$ at which the radius of curvature is a minimum.

34. A particle moves along a curve with position function $r(t) = t \cos ti + t \sin tj$. Find its speed as a function of t.

35. A projectile is fired with an angle of elevation $\alpha = 30°$ and an initial speed $|v(0)| = 20$ m/s at a vertical wall 10 meters away. At what height does it strike the wall?

Chapter 18
Differentiation for Functions of Several Variables

The goal of this chapter is to extend the theory of differentiation to real-valued functions that involve more than one independent variable. Examples of such functions abound in nature and in the social sciences. Here are two.

(i) In chemistry, the ideal gas law relates the pressure, volume, and temperature of an ideal gas by the equation

$$PV = nRT$$

where n is the number of moles of the gas present and R is a constant. When written in the form

$$P = (nR)\left(\frac{T}{V}\right) = f(n, T, V),$$

this equation determines P as a function of n, T, and V.

(ii) In economics, one encounters the simple equation

$$R = Px$$

for the revenue obtained by selling x items at a price of P dollars per item. This equation has the form $R = f(P, x)$, meaning that one must know both the selling price and the number of sales in order to calculate revenue.

18.1 FUNCTIONS OF SEVERAL VARIABLES

By a **function of two variables,** we shall mean a rule that assigns a unique number to each *pair* (x, y) of numbers for which the rule is defined. Functions of two variables will generally be written in the form

$$z = f(x, y).$$

If the function f is defined for all ordered pairs (x, y), we may represent this situation symbolically as

$$f\colon \ \mathbb{R}^2 \to \mathbb{R}, \quad \text{or} \quad (x, y) \xrightarrow{\ f\ } z$$

where $\mathbb{R}^2$ represents the set of all pairs (x, y) with both $x \in \mathbb{R}$ and $y \in \mathbb{R}$. (You may think of $\mathbb{R}^2$ as simply the coordinate plane.) If the function f is defined only for (x, y) in a subset D of $\mathbb{R}^2$ (called the **domain** of the function), we write

$$f\colon \ D \to \mathbb{R}, \quad D \subset \mathbb{R}^2.$$

769

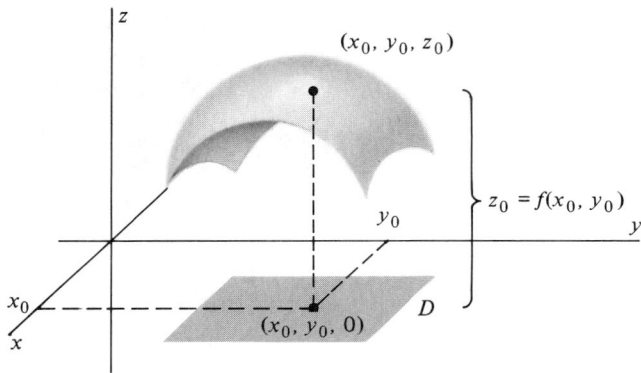

Figure 1.1 Graph of a function $z = f(x, y)$ of two variables with domain D. (See **Plates 1** and **2.**)

As for functions of a single variable, we will refer to the set of values $f(x, y)$, for $(x, y) \in D$, as the **range** of the function f.

Figure 1.1 shows how we may graph functions of two variables. We interpret the ordered pairs (x, y) as points in the xy-plane. We indicate the value of the function, $z = f(x, y)$, by plotting the point (x, y, z) in space. Then, the height of the point (x, y, z) above or below the point $(x, y, 0)$ represents the number (or **value**) z assigned by the function to the ordered pair (x, y). The set of all such points is called the **graph of the function** $z = f(x, y)$.

Notice that while graphs of functions of a single variable are *curves in the plane* (in general), graphs of functions of two variables are *surfaces in space*.

An example of a function of two variables is given by the equation

$$z = x^2 + y^2.$$

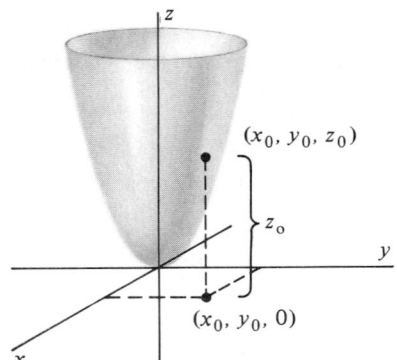

Figure 1.2 Paraboloid $z = x^2 + y^2$ is the graph of a function of two variables. (See **Plate 3.**)

The graph of this function is the paraboloid in Figure 1.2. However, the equation

$$z^2 = x^2 + y^2 \tag{1}$$

does *not* describe a function. The reason is that each pair of independent variables $(x, y) \neq (0, 0)$ corresponds to *two* values of z. This can be seen by writing equation (1) in the form

$$z = \pm\sqrt{x^2 + y^2}.$$

The graph of equation (1) is the cone in Figure 1.3.

Functions of three or more independent variables are defined in the analogous way. A function w of the three independent variables x, y, and z takes the form

$$w = f(x, y, z),$$

and we write

$$f: \ \mathbb{R}^3 \to \mathbb{R}, \qquad \text{or} \qquad (x, y, z) \xrightarrow{f} w.$$

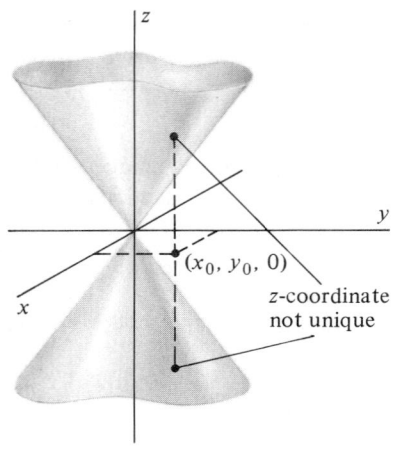

Figure 1.3 Cone $z^2 = x^2 + y^2$ is not the graph of a function of two variables. (See **Plate 4.**)

An example of a function of three variables is the temperature function for the room in which you are sitting. If you use a rectangular coordinate system to describe the space within the room, the number $w_0 = T(x_0, y_0, z_0)$ is the temperature at the point in the room with coordinates (x_0, y_0, z_0). Unfortunately, we cannot sketch a graph for a function of three variables, since all three axes are required just to plot the points in its domain.

More generally, a function of the n independent variables $x_1, x_2, \ldots, x_n$ has the form

$$w = f(x_1, x_2, \ldots, x_n).$$

Symbolically, we describe such functions by writing

$$f: \mathbb{R}^n \to \mathbb{R}, \quad \text{or} \quad (x_1, x_2, \ldots, x_n) \xrightarrow{f} w.$$

Graphing Techniques: Traces and Level Curves

Graphing a function of two variables is difficult, at best. Two techniques frequently give a general idea of the appearance of the graph of the function $z = f(x, y)$. The first is simply setting one of the two independent variables equal to a constant, so that we obtain a function of a single variable, whose graph can be sketched in the appropriate plane in space. Setting $x = c$ in $z = f(x, y)$ gives the equation $z = f(c, y) = g(y)$, whose graph is the intersection of the desired graph with the plane $x = c$; setting $y = c$ in $z = f(x, y)$ gives the equation $z = f(x, c) = h(x)$, whose graph is the intersection of the desired graph with the plane $y = c$.

We refer to the intersections of the graph of f with the planes $x = c$ or $y = c$ as the **traces** of f in the respective planes. By sketching traces of f in several planes, we can sometimes gain a fairly accurate picture of the graph of f.

Example 1

Sketch several traces for the graph of $z = x^2 + y^2$.

Solution: Tables 1.1 and 1.2 show the results of setting one of the independent variables equal to one of several constants.

Traces in each of these planes are indicated in Figures 1.4 and 1.5. They are combined in Figure 1.6. Figure 1.2 is a sketch of the corresponding paraboloid. ◇

Example 2

For the function $f(x, y) = xy$, the equations of various traces are given in Tables 1.3 and 1.4.

Table 1.1

Plane $y = c$	Function $z = f(x, c)$
0	$z = x^2$
1	$z = x^2 + 1$
2	$z = x^2 + 4$
-1	$z = x^2 + 1$
-2	$z = x^2 + 4$

Table 1.2

Plane $x = c$	Function $z = f(c, y)$
0	$z = y^2$
1	$z = y^2 + 1$
2	$z = y^2 + 4$
-1	$z = y^2 + 1$
-2	$z = y^2 + 4$

Table 1.3

Plane $y = c$	Equation $z = f(x, c)$
0	$z = 0$
1	$z = x$
2	$z = 2x$
-1	$z = -x$
-2	$z = -2x$

Table 1.4

Plane $x = c$	Equation $z = f(c, y)$
0	$z = 0$
1	$z = y$
2	$z = 2y$
-1	$z = -y$
-2	$z = -2y$

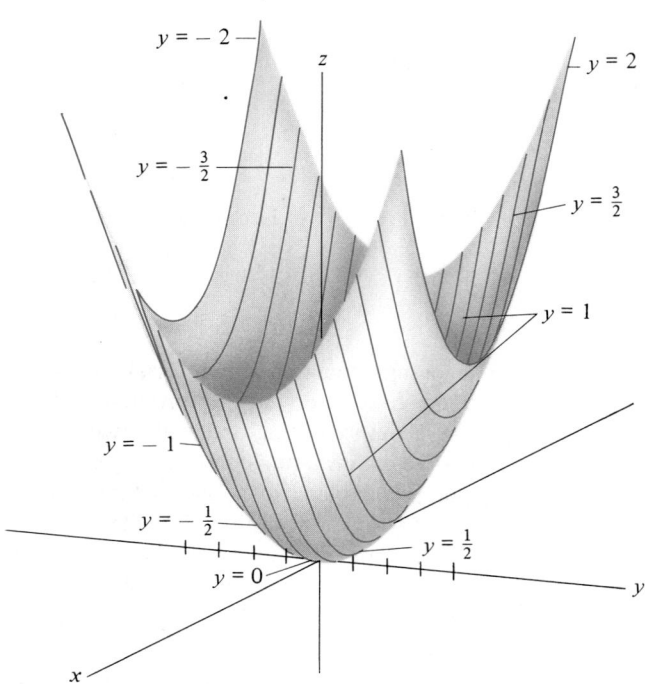

Figure 1.4 Traces of $z = x^2 + y^2$ in planes $y = c$.

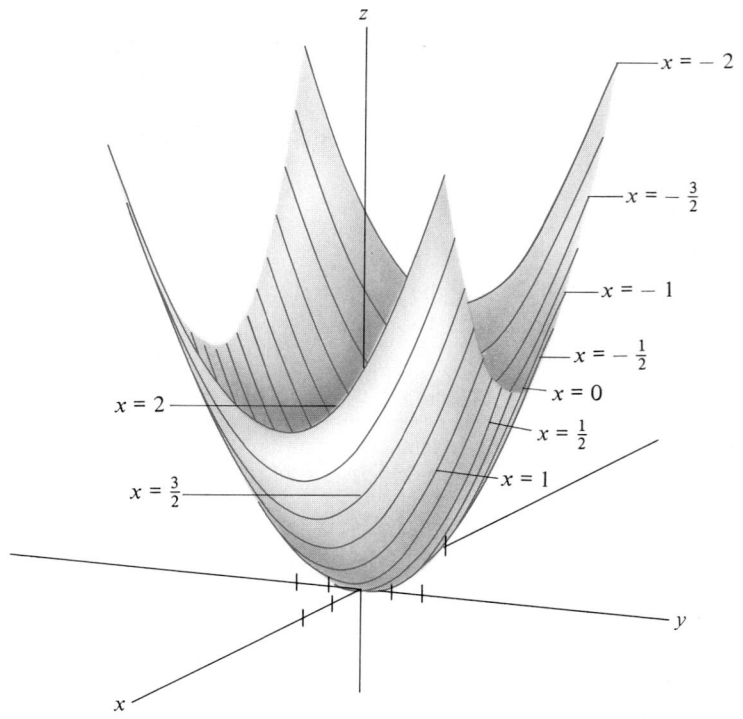

Figure 1.5 Traces of $z = x^2 + y^2$ in planes $x = c$.

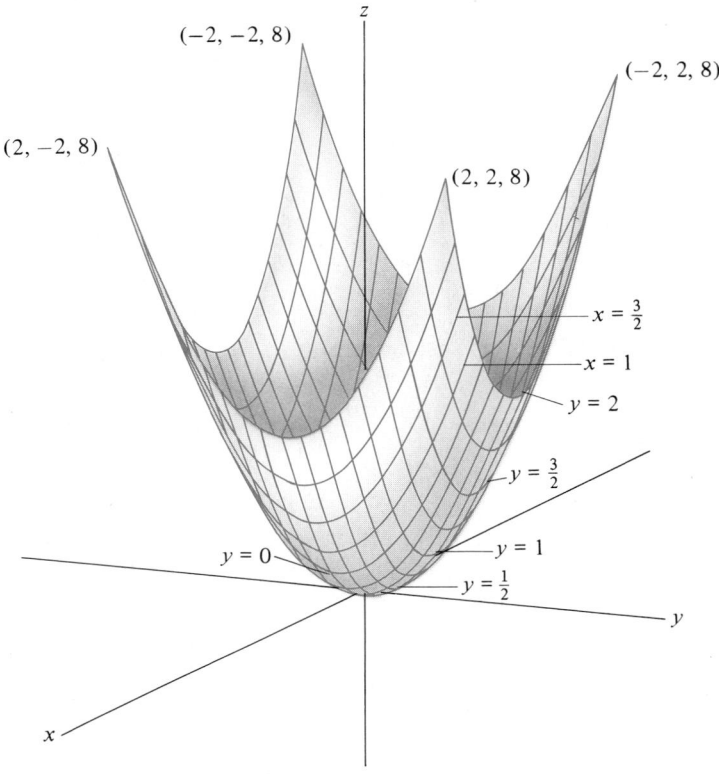

(−2, −2, 8)

(−2, 2, 8)

(2, −2, 8)

(2, 2, 8)

$x = \frac{3}{2}$

$x = 1$

$y = 2$

$y = \frac{3}{2}$

$y = 0$

$y = 1$

$y = \frac{1}{2}$

Figure 1.6 Traces of $z = x^2 + y^2$ in planes of both types.

These traces, and several others, are sketched in Figures 1.7 and 1.8. Figure 1.9 shows a portion of the graph of $f(x, y) = xy$. ◇

Example 3

For the function $f(x, y) = y^2 - x^2$, the equations of various traces are given in Tables 1.5 and 1.6. They are all parabolas.

Table 1.5

Plane $y = c$	Equation $z = f(x, c)$
0	$z = -x^2$
1	$z = 1 - x^2$
2	$z = 4 - x^2$
−1	$z = 1 - x^2$
−2	$z = 4 - x^2$

Table 1.6

Plane $x = c$	Equation $z = f(c, y)$
0	$z = y^2$
1	$z = y^2 - 1$
2	$z = y^2 - 4$
−1	$z = y^2 - 1$
−2	$z = y^2 - 4$

Figure 1.10 shows both types of traces. Figure 1.11 shows the graph of $f(x, y) = y^2 - x^2$. ◇

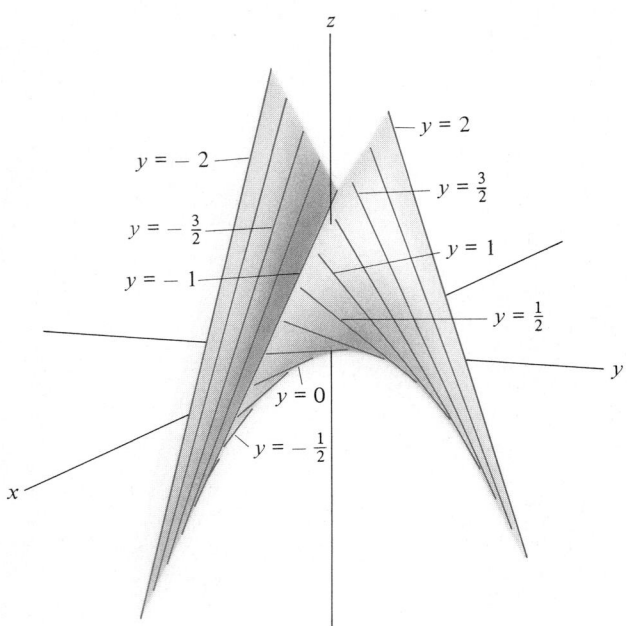

Figure 1.7 Traces of $z = xy$ in planes $y = c$.

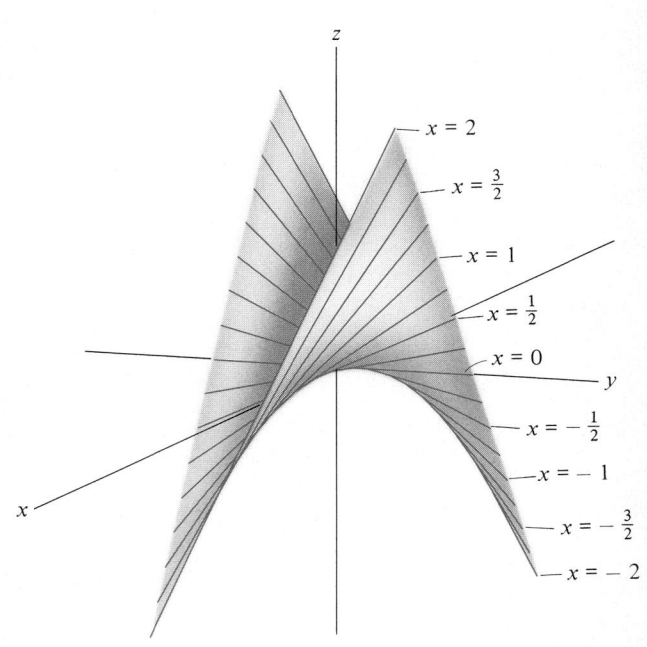

Figure 1.8 Traces of $z = xy$ in planes $x = c$.

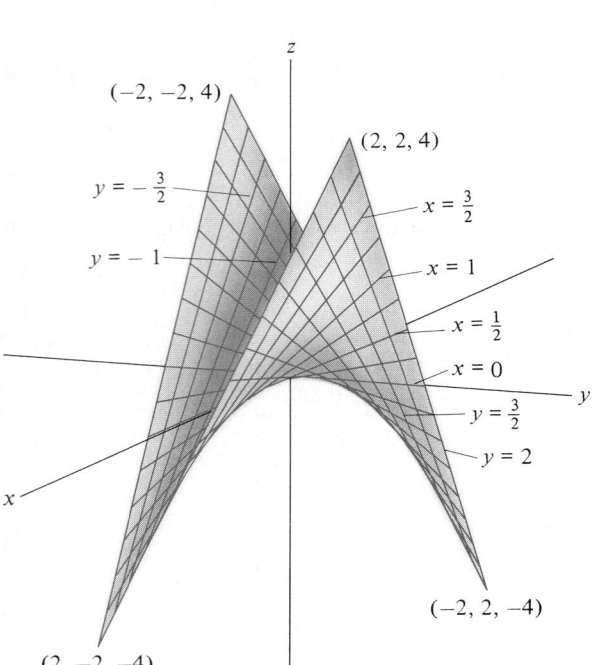

Figure 1.9 Graph of $f(x, y) = xy$.

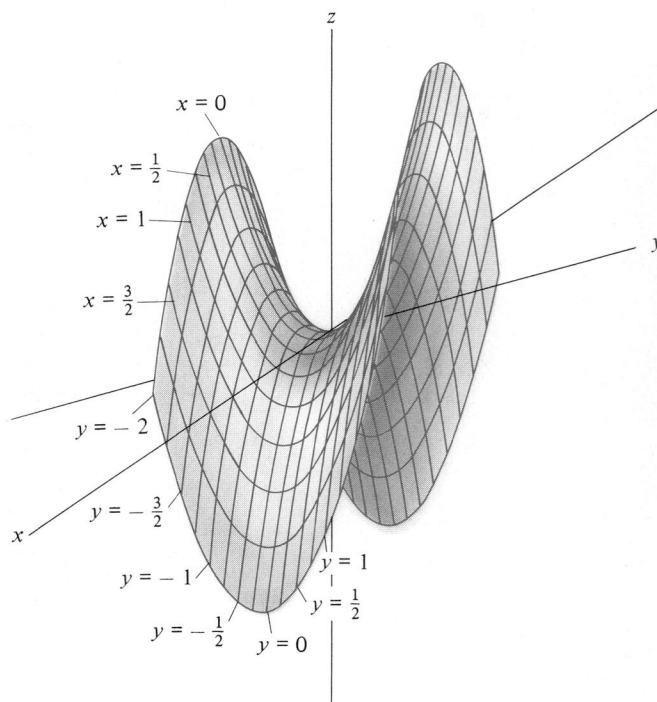

Figure 1.10 Traces of $z = y^2 - x^2$ in planes $y = c$ and $x = c$.

Level Curves and Surfaces

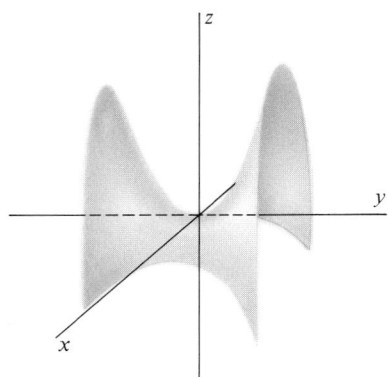

Figure 1.11 Graph of $z = y^2 - x^2$.

The idea behind using traces to describe the graph of $z = f(x, y)$ is that we actually try to sketch a three-dimensional graph. Another approach to representing a surface is to draw a purely two-dimensional sketch that conveys certain information about the surface. Here we want to know which points of the graph lie at various altitudes— that is, which points satisfy the equation $z = c$ for various values of c. This is the concept of **level curves.**

You have probably encountered level curves before. They are commonly used in topographical maps where one indicates the ''lay of the land'' by plotting curves of constant altitude (see Figures 1.12 and 1.13).

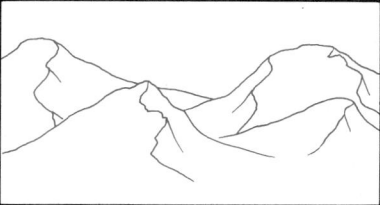

Figure 1.12 A picture of a mountainous region.

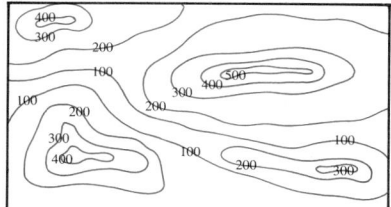

Figure 1.13 A topographical map of the region.

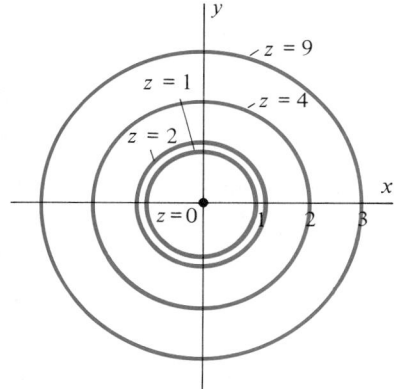

Figure 1.14 Level curves for $z = x^2 + y^2$ are circles $x^2 + y^2 = c$, $c \geq 0$.

To use the notion of level curves to describe the graph of $z = f(x, y)$, we set $z = c$ and graph the resulting equation $f(x, y) = c$ in the xy-plane. Doing so for several values of c conveys the same type of information about the graph of f as does a topographical map. Note that although setting $z = c$ produces a trace of f, we make no attempt to produce a three-dimensional sketch by this method. Also, you should be sure to label the value of z corresponding to each level curve sketched.

Example 4

Consider the function $z = x^2 + y^2$ in Example 1. Setting $z = c$ gives the equation $x^2 + y^2 = c$, valid for $c \geq 0$. The level curves are concentric circles, as sketched in Figure 1.14. ◇

Example 5

For the function $z = xy$ in Example 2, the level curves are $xy = c$, or $y = \dfrac{c}{x}$. These are the hyperbolas in Figure 1.15. ◇

Example 6

For the function $f(x, y) = y^2 - x^2$ in Example 3, the level curves have equation $y^2 - x^2 = c$ or $y = \pm\sqrt{x^2 + c}$. Graphs of several such curves appear in Figure 1.16. ◇

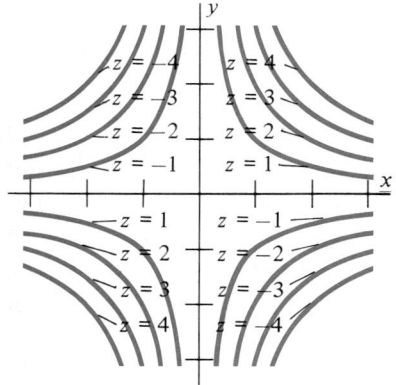

Figure 1.15 Level curves for $z = xy$ are curves $y = c/x$, $x \neq 0$. (See Figure 1.9.)

Although we cannot sketch graphs of functions of three variables, we can sketch **level surfaces** for functions of the form $w = f(x, y, z)$ by setting $w = c$ and sketching the graph of the equation $f(x, y, z) = c$. For example, level surfaces for the function $w = x^2 + y^2 + z^2$ are spheres $x^2 + y^2 + z^2 = c$, $c \geq 0$. More generally,

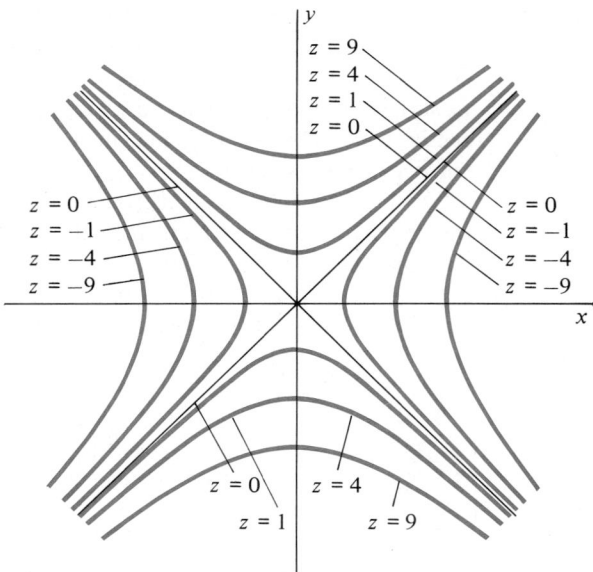

Figure 1.16 Level curves for $f(x, y) = y^2 - x^2$ are curves $y^2 - x^2 = c$. (See Figure 1.11.)

any function of two variables $z = f(x, y)$ may be regarded as a level surface for a function of three variables

$$g(x, y, z) = z - f(x, y) = 0.$$

We will exploit this observation later in finding planes tangent to graphs of functions of two variables.

A simple example of a level surface is the set of all points in your classroom at which the temperature $T(x, y, z)$ has a particular value. Such surfaces of constant temperature are called **isothermal surfaces.** Another example occurs in the theory of electricity and magnetism, where surfaces on which an electric potential is constant are called **equipotential surfaces.**

Neighborhoods

By a **neighborhood** of a point $P_0 = (x_0, y_0)$ in the plane, we shall mean an open **disc** N with center (x_0, y_0). That is,

$$N = \{(x, y) \mid \sqrt{(x - x_0)^2 + (y - y_0)^2} < r\} \tag{2}$$

where r is the radius of the disc N. (By analogy with open intervals on the real number line, the term "open" means that the boundary points are not included in the set.) Similarly, a neighborhood of a point $Q = (x_0, y_0, z_0)$ in space is an open **ball** N with center (x_0, y_0, z_0),

$$N = \{(x, y, z) \mid \sqrt{(x - x_0)^2 + (y - y_0)^2 + (z - z_0)^2} < r\}. \tag{3}$$

Using vector notation, we may generalize both (2) and (3) by saying that a neighborhood of the vector x_0 is a set of vectors

$$N = \{x \mid |x - x_0| < r\}. \tag{4}$$

In equation (4), r is called the **radius** of the neighborhood.

Finally, we shall define the term *deleted neighborhood* to mean all points in a neighborhood of x_0 except the vector x_0 itself. We shall use this terminology in defining $\lim_{x \to x_0} f(x)$ where we shall require $f(x)$ to be defined for all x in a neighborhood of x_0 except possibly at x_0 itself.

Limits

Just as we did for functions of a single variable, we want the statement

$$\lim_{x \to x_0} f(x) = L \tag{5}$$

to mean that the values $f(x)$ of the function f "approach" the number L as the vector (point) x approaches the fixed vector x_0. However, in formulating this definition we must realize that x may approach x_0 along many different paths. This observation points out a significant difference between functions of one variable and functions of several variables. In the one-variable case, a variable can approach a constant only from one of two directions along the number line. In writing statement (5) we shall mean that $f(x) \to L$ as $x \to x_0$ *regardless of the path* along which x approaches x_0. Figure 1.17 illustrates this concept. Figure 1.18 suggests a situation in which the limit in (5) fails to exist.

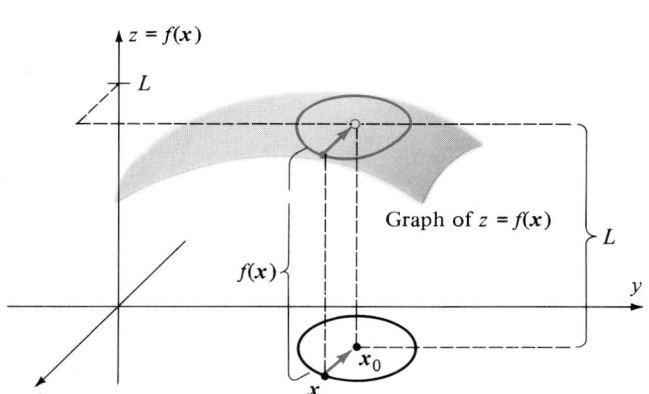

Figure 1.17 $\lim_{x \to x_0} f(x) = L$ if $f(x) \to L$ as x approaches x_0 from any direction.

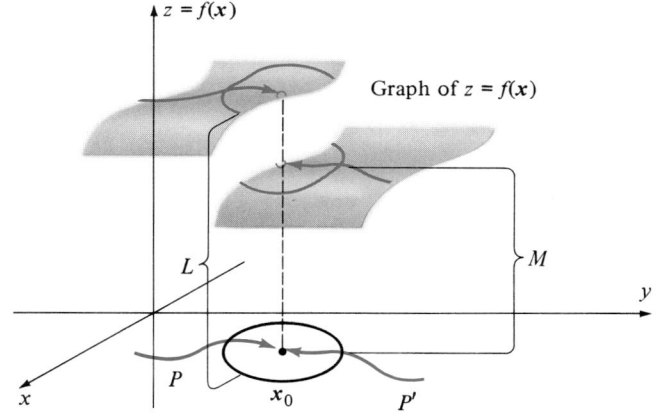

Figure 1.18 $f(x) \to L$ as $x \to x_0$ along path P, but $f(x) \to M$ as $x \to x_0$ along path P', so $\lim_{x \to x_0} f(x)$ does not exist.

Before formulating a precise definition for the limit of a function of several variables, we examine several examples.

Example 7

Consider the function $f(x, y) = 4 - x - y$. We claim that

$$\lim_{(x, y) \to (1, 1)} f(x, y) = 2. \tag{6}$$

Notice first that we have not used the vector notation of statement (5) in writing statement (6), since the number and names of the independent variables are known explicitly. From the form of $f(x, y)$ it is easy to see that $f(x, y) \to (4 - 1 - 1) = 2$

as $x \to 1$ and $y \to 1$. Furthermore, we can see that this result is independent of the path along which (x, y) approaches $(1, 1)$ by setting $z = f(x, y)$ and recalling that the graph of the equation $z = 4 - x - y$ is a plane (see Figure 1.19). Regardless of how (x, y) approaches $(1, 1)$, the point (x, y, z) on the plane approaches the point $(1, 1, 2)$. ◇

Example 8

Consider the function $f(x, y) = \dfrac{xy}{x^2 + y^2}$. We claim that

$$\lim_{(x, y) \to (0, 0)} \frac{xy}{x^2 + y^2} \tag{7}$$

does not exist. On first glance this conclusion may not be obvious. For example, we may allow (x, y) to approach $(0, 0)$ along the x-axis by setting $y = 0$. We obtain

$$\lim_{(x, 0) \to (0, 0)} \frac{xy}{x^2 + y^2} = \lim_{x \to 0} \frac{x \cdot 0}{x^2 + 0^2} = 0 \qquad \text{(see Figure 1.20).} \tag{8}$$

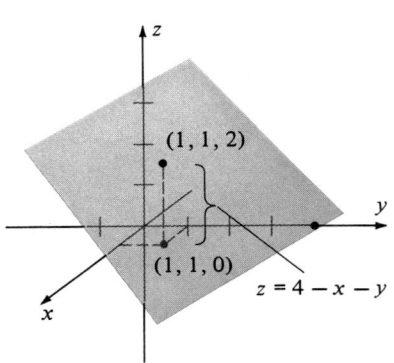

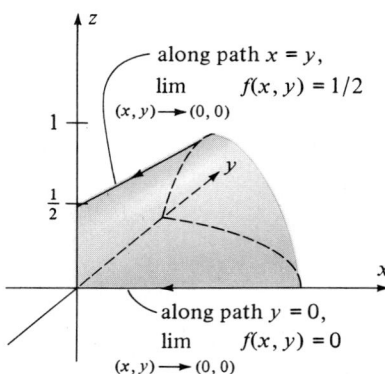

Figure 1.19 $\lim\limits_{(x,y) \to (1,1)} (4 - x - y) = 2.$

Figure 1.20 Portion of the graph of

$$f(x, y) = \frac{xy}{x^2 + y^2}.$$

$\lim\limits_{(x,y) \to (0,0)} f(x, y)$ does not exist. (xy-plane has been rotated 90° for clarity.)

Similarly, allowing (x, y) to approach $(0, 0)$ along the y-axis ($x = 0$) gives

$$\lim_{(0, y) \to (0, 0)} \frac{xy}{x^2 + y^2} = \lim_{y \to 0} \frac{0 \cdot y}{0^2 + y^2} = 0. \tag{9}$$

However, if we allow (x, y) to approach $(0, 0)$ along the line $y = x$, we find that

$$\lim_{(x, x) \to (0, 0)} \frac{xy}{x^2 + y^2} = \lim_{x \to 0} \frac{x^2}{x^2 + x^2} = \frac{1}{2}. \tag{10}$$

Since the result in (10) does not agree with that in (8) or in (9), we conclude that the limit in (7) cannot exist. (See Exercise 48 for a different look at this function.) ◇

Example 9

Find $\displaystyle\lim_{(x,\,y,\,z)\to(2,\,\pi/3,\,3)} x\cos yz.$

Solution: Since $f(x, y, z) = x \cos yz$ is a function of three variables, we cannot rely on a graph to determine this limit. We must, instead, rely on our intuition about both the cosine function and the operation of multiplication. As $y \to \pi/3$ and $z \to 3$, $yz \to (\pi/3)(3) = \pi$, so $\cos yz \to \cos \pi = -1$. Thus, as $x \to 2$, $x \cos yz \to 2(-1) = -2$. That is,

$$\lim_{(x,\,y,\,z)\to(2,\,\pi/3,\,3)} x\cos yz = -2. \qquad\qquad \diamond$$

Next, we give a formal definition for the limit in line (5).

DEFINITION 1

Let x be the position vector for the point $(x_1, x_2, \ldots, x_n)$ in $\mathbb{R}^n$. Let f be a function of n variables defined for all x in a deleted neighborhood of x_0. Let L be a real number. We say that L is the **limit of the function** f as x approaches x_0, written

$$\lim_{x\to x_0} f(x) = L,$$

if and only if, for every $\epsilon > 0$, there exists a number $\delta > 0$ so that

$$\text{if } 0 < |x - x_0| < \delta, \text{ then } |f(x) - L| < \epsilon. \qquad (11)$$

Note in statement (11) that the second inequality involves the *absolute value* of the *number* $f(x) - L$, while the first concerns the *length* of the *vector* $x - x_0$. When applying Definition 1 in a particular situation, you should rewrite the first inequality in (11) in the appropriate component form.

Example 10

Use Definition 1 to prove that

$$\lim_{(x,\,y)\to(0,\,0)} \sqrt{9 - x^2 - y^2} = 3.$$

Solution: Here $L = 3$, and $x_0 = \mathbf{0}$ is the position vector associated with the origin. We begin by rewriting the first inequality in (11) as

$$0 < \sqrt{(x - 0)^2 + (y - 0)^2} = \sqrt{x^2 + y^2} < \delta. \qquad (12)$$

Also, the second inequality in (11) becomes

$$\left|\sqrt{9 - x^2 - y^2} - 3\right| < \epsilon, \qquad (13)$$

or

$$3 - \sqrt{9 - x^2 - y^2} < \epsilon$$

since $\sqrt{9 - x^2 - y^2} \le 3$. Solving this inequality for the radical term, we obtain

$$\sqrt{9 - x^2 - y^2} > 3 - \epsilon,$$

which holds if and only if the following chain of equivalent inequalities holds:

$$9 - (x^2 + y^2) > (3 - \epsilon)^2 \qquad \text{(assuming } \epsilon < 3\text{)},$$
$$x^2 + y^2 < 6\epsilon - \epsilon^2,$$
$$\sqrt{x^2 + y^2} < \sqrt{6\epsilon - \epsilon^2}. \tag{14}$$

Now let's assume that $\epsilon > 0$ is a small ($\epsilon < 3$) positive given number. If we define the number δ to be $\delta = \sqrt{6\epsilon - \epsilon^2}$, then (12) holds whenever inequality (14) holds. Since inequality (14) is equivalent to inequality (13), this shows that

$$\text{if } 0 < \sqrt{x^2 + y^2} < \delta, \text{ then } |\sqrt{9 - x^2 - y^2} - 3| < \epsilon.$$

This proves the stated limit (see Figure 1.21). ◇

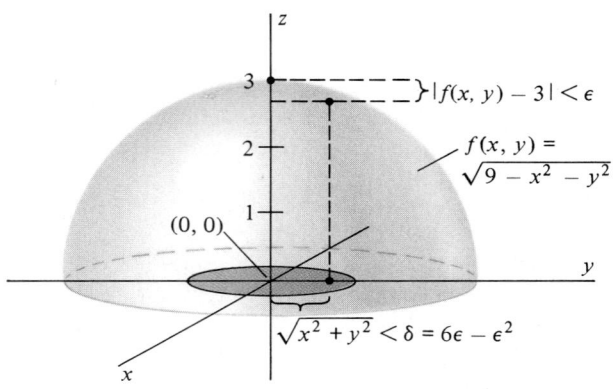

Figure 1.21 $\displaystyle\lim_{(x,y)\to(0,0)} \sqrt{9 - x^2 - y^2} = 3.$

The algebra of limits for functions of several variables is analogous to that for functions of a single variable. The following theorem may be proved using the same ideas used to prove Theorem 1, Chapter 2 (see Exercises 50 and 51).

THEOREM 1

Let f and g be functions of two or three variables defined in a deleted neighborhood of x_0. Suppose that both $\lim_{x\to x_0} f(x)$ and $\lim_{x\to x_0} g(x)$ exist. Let c be any constant. Then

(i) $\displaystyle\lim_{x\to x_0} [f(x) + g(x)] = \lim_{x\to x_0} f(x) + \lim_{x\to x_0} g(x),$

(ii) $\displaystyle\lim_{x\to x_0} cf(x) = c \cdot \lim_{x\to x_0} f(x),$

(iii) $\displaystyle\lim_{x\to x_0} f(x)g(x) = \left(\lim_{x\to x_0} f(x)\right)\left(\lim_{x\to x_0} g(x)\right),$

(iv) $\displaystyle\lim_{x\to x_0} \frac{f(x)}{g(x)} = \frac{\displaystyle\lim_{x\to x_0} f(x)}{\displaystyle\lim_{x\to x_0} g(x)}, \qquad \text{if } \lim_{x\to x_0} g(x) \neq 0.$

Continuity

Continuity for functions of several variables is defined just as for functions of a single variable.

DEFINITION 2

The function f is **continuous** at $x_0 \in \mathbb{R}^n$ if

(i) $f(x_0)$ is defined,

(ii) $\lim\limits_{x \to x_0} f(x)$ exists, and

(iii) $\lim\limits_{x \to x_0} f(x) = f(x_0)$.

In other words, f is continuous at x_0 if $f(x) \to f(x_0)$ as $x \to x_0$, regardless of the direction in which x approaches x_0. Here are some examples.

1. The function $f(x, y) = 4 - x - y$ in Example 7 is continuous at $(1, 1)$ since
$$\lim_{(x,\, y) \to (1,\, 1)} (4 - x - y) = 2 = f(1, 1).$$

2. The function in Figure 1.18 is discontinuous at x_0, since $\lim\limits_{x \to x_0} f(x)$ does not exist.

3. The function $f(x, y) = \dfrac{xy}{x^2 + y^2}$ in Example 8 is discontinuous at $(0, 0)$, since $f(0, 0)$ is undefined.

4. The function $f(x, y, z) = x \cos yz$ in Example 9 is continuous at $(2, \pi/3, 3)$ since
$$\lim_{(x,\, y,\, z) \to (2,\, \pi/3,\, 3)} x \cos yz = -2 = f(2, \pi/3, 3).$$

5. The function $f(x, y) = \sqrt{9 - x^2 - y^2}$ in Example 10 is continuous at $(0, 0)$ since
$$\lim_{(x,\, y) \to (0,\, 0)} \sqrt{9 - x^2 - y^2} = 3 = f(0, 0).$$

As is the case with limits of functions of several variables, continuity is a property that is often difficult to prove; however it can usually be determined by knowing how continuous functions combine to form other continuous functions. For example, since powers, products, and sums of continuous functions of a single variable are again continuous, we would expect a polynomial function such as

$$p(x, y) = x^3 y^2 + 6xy + xy^4$$

to be a continuous function of two variables. Indeed, using Definition 2 and the ideas of Chapter 2, we can prove that sums, multiples, products, powers, and quotients (where denominators are not zero) of continuous functions of several variables are continuous (see Exercises 52 and 53). Moreover, the composition of a continuous function of a single variable with a continuous function of n variables is a continuous function of n variables (see Exercise 54).

Points of discontinuity for functions of several variables occur for the usual reasons: a denominator becomes zero, an expression underneath a radical sign of even order becomes negative, and so forth.

Exercise Set 18.1

State the implicit domain for each of the following functions.

1. $f(x, y) = \dfrac{1}{x^2 + y^2}$

2. $f(x, y) = \dfrac{x - y}{x + y}$

3. $f(x, y) = \sqrt{y - x}$

4. $f(x, y) = \sqrt{1 - xy^2}$

5. $f(x, y) = \sin xy^2$

6. $f(x, y, z) = \sqrt{4 - x^2 - y^2 - z^2}$

7. $f(x, y, z) = xyz$

8. $f(x, y) = \dfrac{1}{x - 3} + \dfrac{2}{x - y}$

9. $f(x, y, z) = \dfrac{y}{xyz}$

10. Let $f(x, y) = x^2 + y^2$, and $g(z) = 2z + 3$. Write the composite function $h(x, y) = g(f(x, y))$ as an explicit function of x and y.

11. Let $f(x, y) = x + y^2$, and $g(z) = \sqrt{z}$.
  a. Write the composite function $h(x, y) = g(f(x, y))$ as an explicit function of x and y.
  b. Find the domain of the function h.

12. Let V be the volume of a right circular cone. Write $V = V(r, h)$ as a function of the radius r and height h.

13. Let S be the total exterior surface area of a right circular cylinder with top and bottom. Express S as a function $S(r, h)$ of the radius and the height of the cylinder.

14. An object is dropped from rest h meters above the ground. Express the time required for it to fall to the ground, $T(g, h)$ as a function of h and g, where g is the acceleration due to gravity (ignore air resistance).

In Exercises 15–22, evaluate the given limit.

15. $\lim\limits_{(x, y)\to(1, 3)} (x^2 - 2y)$

16. $\lim\limits_{(x, y)\to(1, -1)} \sqrt{x - y}$

17. $\lim\limits_{(x, y)\to(3, -1)} \dfrac{1}{\sqrt{x + y}}$

18. $\lim\limits_{(x, y)\to(1, -2)} \dfrac{x + y^3}{x^2 + 2xy + y^2}$

19. $\lim\limits_{(x, y)\to(\pi/2, 1)} x \cos xy$

20. $\lim\limits_{(x, y)\to(1, 0)} xe^{y-x}$

21. $\lim\limits_{(x, y)\to(\pi/2, 1)} \ln \sin xy$

22. $\lim\limits_{(x, y)\to(2, 2)} \dfrac{\mathrm{Tan}^{-1}(y/x)}{1 + xy}$

In Exercises 23–30, sketch several level curves for the given functions.

23. $f(x, y) = y - x^2$

24. $f(x, y) = x^2 + 4y^2$

25. $f(x, y) = 2xy$

26. $f(x, y) = x^2 - y^2$

27. $f(x, y) = xe^y$

28. $f(x, y) = x \cos y$

29. $f(x, y) = \sqrt{x + y}$

30. $f(x, y) = \sqrt{y^2 - x^2}$

In Exercises 31–38, sketch the graph of $z = f(x, y)$ by first sketching several traces of $z = f(x, y)$ in the planes $x = c$ and $y = c$.

31. $f(x, y) = x$

32. $f(x, y) = y^2$

33. $f(x, y) = x + y$

34. $f(x, y) = x \cos y$

35. $f(x, y) = x^2 + 4y^2$

36. $f(x, y) = y - x$

37. $f(x, y) = y \sin x$

38. $f(x, y) = y/x$

39. Find $\lim\limits_{(x, y)\to(1, 0)} \dfrac{xy - y}{x^2 + y^2 - 2x + 1}$, if it exists.

40. Is $f(x, y) = \dfrac{4xy}{x^2 + y^2}$ continuous at $(0, 0)$? What about
$$g(x, y) = \dfrac{4xy}{\sqrt{x^2 + y^2}} ?$$

41. Is the function
$$f(x, y) = \begin{cases} \dfrac{x^2 y}{x^3 + y^3}, & (x, y) \neq (0, 0) \\ 0, & (x, y) = (0, 0) \end{cases}$$
continuous at $(0, 0)$? (*Hint:* Consider the path $y = x$.)

42. Show that the function
$$f(x, y) = \begin{cases} \dfrac{x^3 y}{x^3 + y^5}, & (x, y) \neq (0, 0) \\ 0, & (x, y) = (0, 0) \end{cases}$$
is not continuous at $(0, 0)$.

43. Is the function
$$f(x, y) = \begin{cases} \dfrac{x^3}{x^2 + y^2}, & (x, y) \neq (0, 0) \\ 0, & (x, y) = (0, 0) \end{cases}$$
continuous at $(0, 0)$?

44. Let
$$f(x, y) = \begin{cases} \dfrac{\sin x \cos y}{x}, & (x, y) \neq (0, 0) \\ 0, & (x, y) = (0, 0). \end{cases}$$
Find $\lim\limits_{(x, y)\to(0, 0)} f(x, y)$. (*Hint:* Write $f(x, y) = g(x, y)h(x, y)$, $(x, y) \neq (0, 0)$, with $g(x, y) = \cos y$, $h(x, y) = \dfrac{\sin x}{x}$, and apply Theorem 1.)

45. Let
$$f(x, y) = \begin{cases} x \csc x \tan y, & (x, y) \neq (0, 0) \\ 0, & (x, y) = (0, 0). \end{cases}$$
Find $\lim\limits_{(x, y)\to(0, 0)} f(x, y)$ (see Exercise 44).

46. True or false? If $\lim_{(x, y) \to (a, b)} f(x, y) = L$, then $\lim_{x \to a} f(x, b) = L$. Explain.

47. True or false? If $\lim_{x \to a} f(x, b) = L$ and $\lim_{y \to b} f(a, y) = L$, then

$$\lim_{(x, y) \to (a, b)} f(x, y) = L.$$

48. Consider the function $f(x, y) = \dfrac{xy}{x^2 + y^2}$ whose graph appears in Figure 1.20.

    a. Letting $x = r \cos \theta$ and $y = r \sin \theta$, show that f may be expressed in polar coordinates as

$$f(r, \theta) = \frac{1}{2} \sin 2\theta.$$

    b. Conclude from part (a) that the values $f(r, \theta)$ are inde-pendent of r, depending only on θ. Explain why this shows that f is not continuous at the origin.

49. Prove that $\lim_{(x, y) \to (0, 0)} (x^2 + y^2) = 0$ using Definition 1.

50. Prove part (i) of Theorem 1.

51. Prove part (ii) of Theorem 1.

52. Prove that sums and constant multiples of continuous functions of several variables are continuous.

53. Prove that the product of two continuous functions of several variables is continuous.

54. Prove that if g is a continuous function of a single variable and if $y = f(x)$ is a continuous function of several variables, then the composite function $R(x) = g(f(x))$ is a continuous function of several variables.

18.2 PARTIAL DIFFERENTIATION

The *derivative* of the function f of one variable is the limit

$$f'(x) = \lim_{h \to 0} \frac{f(x + h) - f(x)}{h}, \tag{1}$$

and this limit measures the *rate of change* of $f(x)$ with respect to change in x. In the case of a function of two or more independent variables we have already seen that the question of calculating a rate of change for $z = f(x, y)$ at (x_0, y_0) by limits is a complicated issue, since the "nearby" point (x, y) may approach (x_0, y_0) along an infinite number of distinct paths.

    We begin by examining rates at which $f(x, y)$ changes along paths parallel to the coordinate axes. This is the concept of **partial differentiation.**

DEFINITION 3

Let $f(x, y)$ be defined in a neighborhood of (x_0, y_0). The **partial derivative of f with respect to x** at (x_0, y_0) is the number

$$\frac{\partial f}{\partial x}(x_0, y_0) = \lim_{h \to 0} \frac{f(x_0 + h, y_0) - f(x_0, y_0)}{h}, \tag{2}$$

if this limit exists. Similarly, the **partial derivative of f with respect to y** at (x_0, y_0) is the number

$$\frac{\partial f}{\partial y}(x_0, y_0) = \lim_{h \to 0} \frac{f(x_0, y_0 + h) - f(x_0, y_0)}{h}, \tag{3}$$

provided this limit exists.

Comparing equations (1) and (2), we see that the partial derivative $\dfrac{\partial f}{\partial x}(x_0, y_0)$ is simply the result of holding the variable y constant and differentiating the function $z = f(x, y_0)$ as a function of x alone. Similarly, the partial derivative $\dfrac{\partial f}{\partial y}(x_0, y_0)$

results from treating the variable x as the constant $x = x_0$ and differentiating $z = f(x_0, y)$ as a function of y alone. Thus, partial derivatives may be calculated by the rules developed earlier for differentiating functions of a single variable.

Example 1

Calculate the partial derivatives with respect to x and y for the function $f(x, y) = x^2 y^3 + e^x + \ln y$ and evaluate each at $(1, 4)$.

Solution: In computing the partial derivative with respect to x, we consider y (and any expression containing y alone) to be a constant. Thus,

$$\frac{\partial f}{\partial x}(x, y) = \left(\frac{d}{dx}x^2\right)y^3 + \frac{d}{dx}e^x + \frac{d}{dx}\ln y$$

$$= 2xy^3 + e^x + 0.$$

Similarly, when we consider x constant,

$$\frac{\partial f}{\partial y}(x, y) = x^2\left(\frac{d}{dy}y^3\right) + \frac{d}{dy}e^x + \frac{d}{dy}\ln y$$

$$= 3x^2y^2 + 0 + \frac{1}{y}, \qquad y \neq 0.$$

Thus,

$$\frac{\partial f}{\partial x}(1, 4) = 2 \cdot 1 \cdot 4^3 + e^1 = 128 + e$$

and

$$\frac{\partial f}{\partial y}(1, 4) = 3 \cdot 1^2 \cdot 4^2 + \frac{1}{4} = 48 + \frac{1}{4} = \frac{193}{4}. \qquad \diamondsuit$$

Example 2

For $f(x, y) = \sin xy$, we apply the rule $\dfrac{d}{dt}\sin at = a\cos at$, with one variable playing the role of a and the other playing the role of t:

$$\frac{\partial f}{\partial x}(x, y) = y\cos xy \qquad \text{and} \qquad \frac{\partial f}{\partial y}(x, y) = x\cos xy. \qquad \diamondsuit$$

Figures 2.1 and 2.2 illustrate the geometric interpretations of the partial derivatives in Definition 3. To interpret $\dfrac{\partial f}{\partial x}(x_0, y_0)$, we note that the set of points (x, y, z) in $\mathbb{R}^3$ with $y = y_0$ is a plane. The intersection of this plane with the graph of f is the *trace* of f in the plane $y = y_0$. This curve may be viewed as the graph of the function of one variable

$$h(x) = f(x, y_0), \qquad y_0 \text{ fixed.}$$

Since

$$h'(x_0) = \frac{\partial f}{\partial x}(x_0, y_0),$$

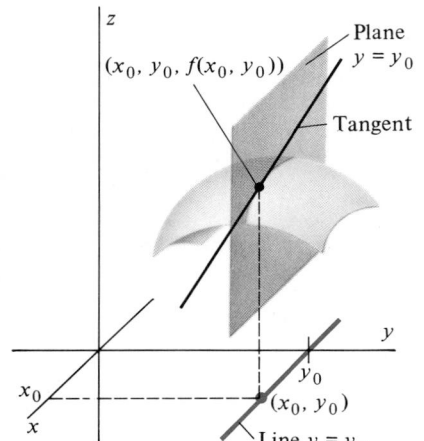

Figure 2.1 Partial derivative $\dfrac{\partial f}{\partial x}(x_0, y_0)$ is the slope of the line tangent to the trace of f in the plane $y = y_0$. (See **Plates 5** and **6**.)

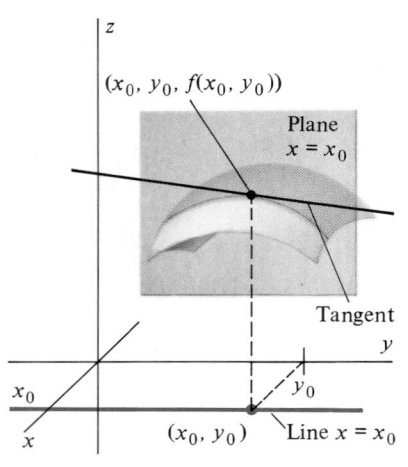

Figure 2.2 Partial derivative $\dfrac{\partial f}{\partial y}(x_0, y_0)$ is the slope of the line tangent to the trace of f in the plane $x = x_0$.

the partial derivative $\dfrac{\partial f}{\partial x}(x_0, y_0)$ gives the slope of the line tangent to this trace at the point $(x_0, y_0, f(x_0, y_0))$. A similar interpretation is valid for the partial derivative $\dfrac{\partial f}{\partial y}(x_0, y_0)$.

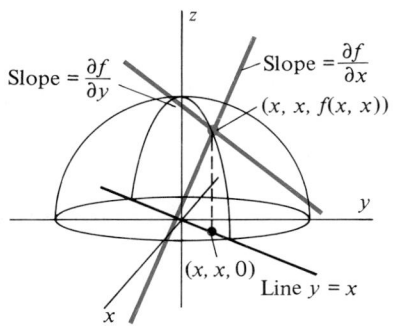

Slope $= \dfrac{\partial f}{\partial y}$

Slope $= \dfrac{\partial f}{\partial x}$

$(x, x, f(x, x))$

$(x, x, 0)$

Line $y = x$

Figure 2.3 $\dfrac{\partial f}{\partial x} = \dfrac{\partial f}{\partial y}$ if $y = x$.

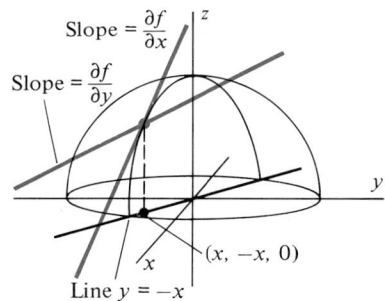

Slope $= \dfrac{\partial f}{\partial x}$

Slope $= \dfrac{\partial f}{\partial y}$

$(x, -x, 0)$

Line $y = -x$

Figure 2.4 $\dfrac{\partial f}{\partial x} = -\dfrac{\partial f}{\partial y}$ if $y = -x$.

Example 3

For the function $f(x, y) = \sqrt{1 - x^2 - y^2}$ show that

(i) $\dfrac{\partial f}{\partial x}(x, y) = \dfrac{\partial f}{\partial y}(x, y)$ if and only if $y = x$, and

(ii) $\dfrac{\partial f}{\partial x}(x, y) = -\dfrac{\partial f}{\partial y}(x, y)$ if and only if $y = -x$

and interpret this result geometrically.

Solution: Differentiating with respect to x, holding y constant, gives

$$\frac{\partial f}{\partial x}(x, y) = \frac{-x}{\sqrt{1 - x^2 - y^2}}.$$

Similarly, the partial derivative with respect to y is

$$\frac{\partial f}{\partial y}(x, y) = \frac{-y}{\sqrt{1 - x^2 - y^2}}.$$

Conclusions (i) and (ii) follow immediately.

Figure 2.3 shows the geometric interpretation of this result. The graph of $z = f(x, y) = \sqrt{1 - x^2 - y^2}$ is a hemisphere. The only points on the surface of this hemisphere for which the tangent in the plane parallel to the x-axis has the same slope as the tangent in the plane parallel to the y-axis are the points lying above the line $y = x$. Similarly, the points at which these tangents have opposite slope lie above the line $y = -x$ in the xy-plane (see Figure 2.4). ◇

Notation for Partial Derivatives

For a function of two variables $z = f(x, y)$, the symbol $\dfrac{\partial}{\partial x}$ denotes the partial derivative with respect to x just as the Leibniz notation $\dfrac{d}{dx}$ denotes the derivative of the function of a single variable. We also use subscripts to denote partial derivatives, such as $f_x(x, y)$ or z_x for $\dfrac{\partial f}{\partial x}(x, y)$.

We may summarize the various types of notation for partial derivatives as follows: If $z = f(x, y)$, then

$$\frac{\partial f}{\partial x}(x, y) = \frac{\partial}{\partial x}f(x, y) = z_x(x, y) = z_x,$$

and

$$\frac{\partial f}{\partial y}(x, y) = \frac{\partial}{\partial y}f(x, y) = z_y(x, y) = z_y.$$

Functions of More than Two Variables

Partial derivatives are defined for functions of more than two variables by the same idea used in Definition 3: Hold all variables constant except one, and differentiate the resulting function of a single variable as before. For example, if $w = f(x, y, z)$ is a function of the three independent variables x, y, and z, the three partial derivatives are defined as follows:

$$\frac{\partial f}{\partial x}(x, y, z) = \lim_{h \to 0} \frac{f(x + h, y, z) - f(x, y, z)}{h}$$

$$\frac{\partial f}{\partial y}(x, y, z) = \lim_{h \to 0} \frac{f(x, y + h, z) - f(x, y, z)}{h}$$

$$\frac{\partial f}{\partial z}(x, y, z) = \lim_{h \to 0} \frac{f(x, y, z + h) - f(x, y, z)}{h}$$

Partial derivatives for functions of more than three independent variables are defined and calculated in an analogous way.

Example 4

Let $f(x, y, z) = \sqrt{x}e^{y/z}$, $\quad z \neq 0$, $\quad x \geq 0$. Then

$$\frac{\partial f}{\partial x}(x, y, z) = \left\{\frac{d}{dx}\sqrt{x}\right\}e^{y/z} = \frac{1}{2\sqrt{x}} \cdot e^{y/z},$$

$$\frac{\partial f}{\partial y}(x, y, z) = \sqrt{x}\left\{\frac{\partial}{\partial y}(e^{y/z})\right\} = \sqrt{x}e^{y/z} \cdot \frac{1}{z} = \frac{\sqrt{x}}{z}e^{y/z},$$

$$\frac{\partial f}{\partial z}(x, y, z) = \sqrt{x}\left\{\frac{\partial}{\partial z}e^{y/z}\right\} = \sqrt{x}e^{y/z}\left(\frac{-y}{z^2}\right) = \frac{-y\sqrt{x}}{z^2}e^{y/z}. \qquad \diamond$$

Example 5

The radial rate of heat flow in a substance between two concentric spheres is given by the function

$$H(r, R, t, T) = \frac{(t - T)4\pi kRr}{R - r}$$

where k is a constant, and where the inner sphere has radius r and temperature t and the outer sphere has radius R and temperature T. Thus (using the Quotient Rule to differentiate with respect to r and R),

$$H_r = \frac{\partial}{\partial r}H(r, R, t, T) = \frac{(R - r)[(t - T)4\pi kR] - (-1)[(t - T)4\pi kRr]}{(R - r)^2}$$

$$= \frac{(t - T)4\pi kR^2}{(R - r)^2},$$

$$H_R = \frac{\partial}{\partial R}H(r, R, t, T) = \frac{(R - r)[(t - T)4\pi kr] - (1)[(t - T)4\pi kRr]}{(R - r)^2}$$

$$= \frac{-(t - T)4\pi kr^2}{(R - r)^2},$$

$$H_t = \frac{\partial}{\partial t} H(r, R, t, T) = \frac{\partial}{\partial t}\left[t\frac{4\pi kRr}{R-r} - T\frac{4\pi kRr}{R-r} \right] = \frac{4\pi kRr}{R-r},$$

$$H_T = \frac{\partial}{\partial T} H(r, R, t, T) = \frac{\partial}{\partial T}\left[t\frac{4\pi kRr}{R-r} - T\frac{4\pi kRr}{R-r} \right] = -\frac{4\pi kRr}{R-r}. \qquad \Diamond$$

Example 6

For the vectors $\boldsymbol{x} = x_1\boldsymbol{i} + x_2\boldsymbol{j} + x_3\boldsymbol{k}$ and $\boldsymbol{y} = y_1\boldsymbol{i} + y_2\boldsymbol{j} + y_3\boldsymbol{k}$, the dot product

$$\boldsymbol{x} \cdot \boldsymbol{y} = x_1y_1 + x_2y_2 + x_3y_3$$

may be viewed as a function of the six independent variables (components) x_1, x_2, x_3, y_1, y_2, y_3. Thus,

$$\frac{\partial}{\partial x_1}(\boldsymbol{x} \cdot \boldsymbol{y}) = y_1; \qquad \frac{\partial}{\partial y_1}(\boldsymbol{x} \cdot \boldsymbol{y}) = x_1,$$

$$\frac{\partial}{\partial x_2}(\boldsymbol{x} \cdot \boldsymbol{y}) = y_2; \qquad \frac{\partial}{\partial y_2}(\boldsymbol{x} \cdot \boldsymbol{y}) = x_2,$$

$$\frac{\partial}{\partial x_3}(\boldsymbol{x} \cdot \boldsymbol{y}) = y_3; \qquad \frac{\partial}{\partial y_3}(\boldsymbol{x} \cdot \boldsymbol{y}) = x_3.$$

Thus, the rate at which $\boldsymbol{x} \cdot \boldsymbol{y}$ changes with respect to change in the component x_1 is just the component y_1, and so forth. $\qquad \Diamond$

Limitations of Partial Derivatives

Before proceeding further, we should dispel a notion that students sometimes develop at this point. Although knowledge of the derivative $f'(x_0)$ completely determines the rate at which the function f changes at $x = x_0$, *knowledge of the partial derivatives $f_x(x_0, y_0)$ and $f_y(x_0, y_0)$ is not always sufficient to determine the rate at which the function of two variables $z = f(x, y)$ is changing at (x_0, y_0)*. The analogous statement for functions of more than two variables is also true.

To see this, let us consider the function

$$f(x, y) = \begin{cases} \dfrac{xy}{x^2 + y^2}, & (x, y) \neq (0, 0) \\ 0, & (x, y) = (0, 0) \end{cases}.$$

Using Definition 3, we find that

$$\frac{\partial f}{\partial x}(0, 0) = \lim_{h \to 0}\frac{f(0 + h, 0) - f(0, 0)}{h} = \lim_{h \to 0}\left(\frac{0}{h}\right) = 0$$

and

$$\frac{\partial f}{\partial y}(0, 0) = \lim_{h \to 0}\frac{f(0, 0 + h) - f(0, 0)}{h} = \lim_{h \to 0}\left(\frac{0}{h}\right) = 0.$$

Thus, both partial derivatives at $(0, 0)$ are zero. But it is false to conclude that the rate of change of $f(x, y)$ at $(0, 0)$ in *every direction* is zero, as can be seen from Figure 2.5. For example, a small change from the point $(0, 0)$ to the point (h, h) causes an abrupt change in the value $f(x, y)$ from $f(0, 0) = 0$ to $f(h, h) = \dfrac{1}{2}$, no matter how small h may be.

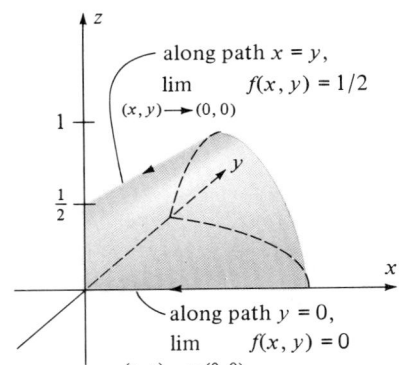

Figure 2.5 Portion of the graph of $f(x, y) = \dfrac{xy}{x^2 + y^2}$.

$\lim\limits_{(x,y) \to (0,0)} f(x, y)$ does not exist. (xy-plane has been rotated 90° for clarity.)

It is important to understand that partial derivatives give information about functions *only* in the directions of the coordinate axes. We must develop a richer theory of differentiation in order to obtain general conclusions about the rate of change of a function of several variables near a particular point.

Higher Order Partial Derivatives

Repeated applications of partial differentiation lead to **higher order partial derivatives.** There is nothing terribly complicated about this concept, except that we must be very careful about notation since we encounter **mixed partial derivatives,** in which one differentiation is performed with respect to a particular variable, followed by another differentiation with respect to a different variable.

We will use the following notation:

$$\frac{\partial^2 f}{\partial x^2}(x, y) = \frac{\partial^2}{\partial x^2}f(x, y) \qquad \text{means} \qquad \frac{\partial}{\partial x}\left(\frac{\partial f}{\partial x}(x, y)\right).$$

$$\frac{\partial^2 f}{\partial y \partial x}(x, y) = \frac{\partial^2}{\partial y \partial x}f(x, y) \qquad \text{means} \qquad \frac{\partial}{\partial y}\left(\frac{\partial f}{\partial x}(x, y)\right).$$

$$\frac{\partial^2 f}{\partial x \partial y}(x, y) = \frac{\partial^2}{\partial x \partial y}f(x, y) \qquad \text{means} \qquad \frac{\partial}{\partial x}\left(\frac{\partial f}{\partial y}(x, y)\right).$$

$$\frac{\partial^2 f}{\partial y^2}(x, y) = \frac{\partial^2}{\partial y^2}f(x, y) \qquad \text{means} \qquad \frac{\partial}{\partial y}\left(\frac{\partial f}{\partial y}(x, y)\right).$$

Note that the order in which the differentiations are performed is indicated by reading the "denominator" of the derivative notation from *right* to *left*. Similar definitions hold for third and higher order partial derivatives.

Example 7

For the function $f(x, y) = x^2 y^3 + \cos x \sin y$,

$$\frac{\partial^2 f}{\partial x^2}(x, y) = \frac{\partial}{\partial x}\left(\frac{\partial f}{\partial x}(x, y)\right) = \frac{\partial}{\partial x}(2xy^3 - \sin x \sin y) = 2y^3 - \cos x \sin y,$$

$$\frac{\partial^2 f}{\partial y \partial x}(x, y) = \frac{\partial}{\partial y}\left(\frac{\partial f}{\partial x}(x, y)\right) = \frac{\partial}{\partial y}(2xy^3 - \sin x \sin y) = 6xy^2 - \sin x \cos y,$$

$$\frac{\partial^2 f}{\partial x \partial y}(x, y) = \frac{\partial}{\partial x}\left(\frac{\partial f}{\partial y}(x, y)\right) = \frac{\partial}{\partial x}(3x^2 y^2 + \cos x \cos y) = 6xy^2 - \sin x \cos y$$

$$\frac{\partial^2 f}{\partial y^2}(x, y) = \frac{\partial}{\partial y}\left(\frac{\partial f}{\partial y}(x, y)\right) = \frac{\partial}{\partial y}(3x^2 y^2 + \cos x \cos y) = 6x^2 y - \cos x \sin y,$$

$$\frac{\partial^3 f}{\partial y^3}(x, y) = \frac{\partial}{\partial y}\left(\frac{\partial^2 f}{\partial y^2}(x, y)\right) = \frac{\partial}{\partial y}\left(\frac{\partial^2 f}{\partial y^2}\right) = \frac{\partial}{\partial y}(6x^2 y - \cos x \sin y)$$

$$= 6x^2 - \cos x \cos y. \qquad \diamond$$

We may also use subscript notation to indicate higher order partial derivatives as follows: If $z = f(x, y)$,

$$f_{xx}(x, y) \text{ or } z_{xx} \qquad \text{means} \qquad \frac{\partial^2 f}{\partial x^2}(x, y),$$

$$f_{xy}(x, y) \text{ or } z_{xy} \quad \text{means} \quad \frac{\partial^2 f}{\partial y \partial x}(x, y),$$

$$f_{yx}(x, y) \text{ or } z_{yx} \quad \text{means} \quad \frac{\partial^2 f}{\partial x \partial y}(x, y),$$

$$f_{yy}(x, y) \text{ or } z_{yy} \quad \text{means} \quad \frac{\partial^2 f}{\partial y^2}(x, y),$$

with similar statements holding for functions of more than two variables and for higher order derivatives.

It is important to note that, when subscripts are used, the differentiations are performed *in the order indicated by the subscripts, read from left to right*. This is just the opposite of the order in which the differentiations are indicated in Leibniz notation.

Example 8

For the function $f(x, y, z) = x^2 y^3 z^4$,

$$f_x = 2xy^3z^4; \qquad f_y = 3x^2y^2z^4; \qquad f_z = 4x^2y^3z^3$$

$$f_{xy} = \frac{\partial}{\partial y}(2xy^3z^4) = 6xy^2z^4; \qquad f_{yx} = \frac{\partial}{\partial x}(3x^2y^2z^4) = 6xy^2z^4$$

$$f_{yz} = \frac{\partial}{\partial z}(3x^2y^2z^4) = 12x^2y^2z^3; \qquad f_{zy} = \frac{\partial}{\partial y}(4x^2y^3z^3) = 12x^2y^2z^3$$

$$f_{xz} = \frac{\partial}{\partial z}(2xy^3z^4) = 8xy^3z^3; \qquad f_{zx} = \frac{\partial}{\partial x}(4x^2y^3z^3) = 8xy^3z^3$$

$$f_{xx} = 2y^3z^4; \qquad f_{yy} = 6x^2yz^4; \qquad f_{zz} = 12x^2y^3z^2$$

$$f_{xyz} = \frac{\partial}{\partial z}(6xy^2z^4) = 24xy^2z^3; \qquad f_{yzx} = \frac{\partial}{\partial x}(12x^2y^2z^3) = 24xy^2z^3$$

$$f_{xxx} = 0; \qquad f_{yyy} = 6x^2z^4; \qquad f_{zzz} = 24x^2y^3z. \qquad \Diamond$$

Equality of Mixed Partials

You have no doubt observed, in Example 7, that $f_{yx} = f_{xy}$ and, in Example 8, that $f_{xy} = f_{yx}$, $f_{yz} = f_{zy}$, and $f_{xz} = f_{zx}$. This is not true for all functions. However, when the function f and various of its partial derivatives are continuous, these mixed partials will be equal. The following theorem makes this precise. It is typically proved in courses on advanced calculus.

THEOREM 2
Equality of Mixed Partials

If the function $z = f(x, y)$ and the partial derivatives

$$\frac{\partial f}{\partial x}, \qquad \frac{\partial f}{\partial y}, \qquad \frac{\partial^2 f}{\partial x \partial y}, \qquad \text{and} \qquad \frac{\partial^2 f}{\partial y \partial x}$$

are all continuous in a neighborhood of the point (x_0, y_0), then

$$\frac{\partial^2 f}{\partial x \partial y}(x_0, y_0) = \frac{\partial^2 f}{\partial y \partial x}(x_0, y_0). \tag{4}$$

Equation (4) plays an important role in the theory of differentiation that follows.

Exercise Set 18.2

In Exercises 1–20, find all first order partial derivatives.

1. $f(x, y) = xy$

2. $z = \sqrt{x + y^2}$

3. $z = x \tan y^2$

4. $f(x, y) = xy^3 + \sqrt{y}$

5. $f(x, y) = e^{x^2 + y^2}$

6. $z = \text{Tan}^{-1}(y/x)$

7. $f(r, \theta) = r \sin(\pi/2 - \theta)$

8. $f(s, t) = \dfrac{s - t}{s + t}$

9. $z = \ln(xy^2 + x - y)$

10. $h(u, v) = e^{u-v} + e^{v-u}$

11. $f(x, y) = x^y$

12. $z = 2^x y^2$

13. $f(r, \theta) = r^2 \cos \theta$

14. $f(x, y, z) = x^3 e^y \ln z$

15. $f(x, y, z) = xy^3 - yz^2$

16. $w = \ln(x^2 + y^2 + z^2)$

17. $f(x, y, z) = \left(\dfrac{x - y}{x + y}\right)^z$

18. $f(u, v, w) = \dfrac{\sin u}{v^3 \, \text{Tan}^{-1} w}$

19. $f(r, s, t) = \dfrac{\sqrt{r} \, s \ln t}{\sqrt{s^2 - 2r + t}}$

20. $f(u, v, w) = \dfrac{ue^{vw}}{\sin^2 u + \tan^2 w}$

21. Find $\dfrac{\partial f}{\partial x}(2, 5)$ for $f(x, y) = xy^3 - y$.

22. Find $\dfrac{\partial f}{\partial y}(1, 0)$ for $f(x, y) = e^{x-y^2}$.

23. Find $z_x(2, 1)$ for $z = \sqrt{x + y^2}$.

24. Find f_{xx}, f_{xy}, f_{yx}, and f_{yy} for $f(x, y)$ in Exercise 4.

25. Find $\dfrac{\partial^2 f}{\partial r^2}$, $\dfrac{\partial^2 f}{\partial r \partial \theta}$, and $\dfrac{\partial^2 f}{\partial \theta^2}$ for $f(r, \theta)$ in Exercise 13.

26. Find f_{xx}, f_{xy}, f_{xz}, f_{yz}, f_{yy}, and f_{zz} for $f(x, y, z)$ in Exercise 15.

27. Find $w_{xx} + w_{yy} + w_{zz}$ for $w(x, y, z)$ in Exercise 16.

28. For a particle traveling in a circular orbit of radius r, the relation between angular speed ω and velocity v is $v = \omega r$. Find $\dfrac{\partial v}{\partial r}$ and $\dfrac{\partial v}{\partial \omega}$.

29. For the constant volume flow of an incompressible fluid through a tube of varying cross-sectional area, the equation $A_1 v_1 = A_2 v_2$ expresses the relationship between the respective cross-sectional areas and velocities at two points in the tube.
   a. Express v_2 as a function of A_2, A_1, and v_1.
   b. Find $\dfrac{\partial v_2}{\partial A_2}$, the rate at which v_2 changes with respect to change in A_2 alone.
   c. Suppose that $A_1 = 5$ cm^2, $A_2 = 3$ cm^2, and $v_1 = 20$ cm/s. Find the rate of change of v_2 with respect to A_2 if A_1 and v_1 are held constant.

d. With A_1, A_2, and v_1 as in (c), find the rate of change of A_2 with respect to change in v_1 if A_1 and v_2 are held constant.

In Exercises 30–36, use the equations

$$x = \rho \sin \phi \cos \theta$$
$$y = \rho \sin \phi \sin \theta$$
$$z = \rho \cos \phi$$

for changing from rectangular to spherical coordinates to find the indicated partial derivative.

30. $\dfrac{\partial x}{\partial \phi}$

31. $\dfrac{\partial y}{\partial \theta}$

32. $\dfrac{\partial z}{\partial \rho}$

33. $\dfrac{\partial x}{\partial \rho}$

34. $\dfrac{\partial y}{\partial \phi}$

35. $\dfrac{\partial z}{\partial \phi}$

36. $\dfrac{\partial \phi}{\partial z}$

37. For $f(x, y) = \displaystyle\int_x^{x+y} \cos t^2 \, dt$ find

   a. $\dfrac{\partial f}{\partial x}$ **b.** $\dfrac{\partial f}{\partial y}$

38. For $f(x, y, z) = \displaystyle\int_z^x \sqrt{t^3 + 1} \, dt - \int_y^z \sqrt{t^3 + 1} \, dt$ find

   a. $\dfrac{\partial f}{\partial x}$ **b.** $\dfrac{\partial f}{\partial y}$ **c.** $\dfrac{\partial f}{\partial z}$

39. Show that the function $f(x, y) = \text{Sin}^{-1}\left(\dfrac{x - y}{x + y}\right)$ is a solution of the differential equation

$$x \frac{\partial z}{\partial x} + y \frac{\partial z}{\partial y} = 0.$$

40. Show that the function $w = \ln(e^x + e^y + e^z)$ satisfies the partial differential equation

$$\frac{\partial w}{\partial x} + \frac{\partial w}{\partial y} + \frac{\partial w}{\partial z} = 1.$$

41. The **electric potential** at an axial point for a charged disc is given by

$$V = \frac{\sigma}{2\epsilon_0}(\sqrt{a^2 + r^2} - r)$$

where σ and ϵ_0 are constants, a is the radius of the disc, and r is the distance from the point to the disc.
   a. Find the rate of change of V with respect to a if r is held constant.
   b. Find the rate of change of V with respect to r if a is held constant.

42. The national unemployment rate u may be viewed as the dot product $u = y \cdot w$ where the components of the vector $v = \langle v_1, v_2, \ldots, v_n \rangle$ are the percentages of employable citizens in each of the n job categories and the components of the vector $w = \langle w_1, w_2, \ldots, w_n \rangle$ are the unemployment rates within each category.

a. Find $\dfrac{\partial u}{\partial v_j}$ and interpret this rate.

b. Find $\dfrac{\partial u}{\partial w_j}$ and interpret this rate.

43. Let $f(x, y, z) = e^{x+y+z}$. Show that all partial derivatives of all orders equal $f(x, y, z)$.

44. Show that a function of the form $f(x, y) = e^{kx}g(y)$ satisfies the differential equation $\dfrac{\partial f}{\partial x}(x, y) = kf(x, y)$.

45. Show that the function $f(x, y) = \sin xy$ satisfies the differential equation
$$x \frac{\partial f}{\partial x}(x, y) - y \frac{\partial f}{\partial y}(x, y) = 0.$$

46. Show that the function $f(x, y) = \sin xy$ satisfies the differential equation
$$x^2 \frac{\partial^2 f}{\partial x^2}(x, y) - y^2 \frac{\partial^2 f}{\partial y^2}(x, y) = 0.$$

47. Find a solution of the differential equation
$$x^n \frac{\partial^n f}{\partial x^n} = y^n \frac{\partial^n f}{\partial y^n}, \qquad n = 1, 2, 3, \ldots.$$
(See Exercises 45 and 46.)

48. Show that the function $y(x, t) = g(t - cx)$ satisfies the **one-dimensional wave equation**
$$\frac{\partial^2 y}{\partial t^2} = \frac{1}{c^2} \frac{\partial^2 y}{\partial x^2}.$$

49. **Laplace's equation** for the function $f(x, y)$ is
$$\frac{\partial^2 f}{\partial x^2} + \frac{\partial^2 f}{\partial y^2} = 0.$$
Show that the following functions satisfy Laplace's equation.
a. $f(x, y) = e^x \sin y$
b. $f(x, y) = e^{-x} \cos y$
c. $f(x, y) = \ln \sqrt{x^2 + y^2}$

50. Show that a polynomial $P(x, y)$ in x and y (like $P(x, y) = 3x^2y^3 + xy^6 - 5x^3y$) has the property that $\dfrac{\partial^2 P}{\partial x \partial y} = \dfrac{\partial^2 P}{\partial y \partial x}$.

18.3 TANGENT PLANES

For the function f of a single variable, knowledge of $f(a)$ and the derivative $f'(a)$ enables us to write an equation for the line tangent to the graph of f at the point $(a, f(a))$. The purpose of this brief section is to show how we can obtain an equation for a plane tangent to the graph of a function of two variables $z = f(x, y)$ from knowledge of its partial derivatives. A complete discussion of the conditions under which a tangent plane exists is beyond the scope of this text. The issue here is simply how one goes about using partial derivatives to find such a plane, assuming that it exists.

Suppose that f is a function of two variables that is defined in a neighborhood of the point (x_0, y_0). Assume also that the partial derivatives $\dfrac{\partial f}{\partial x}(x_0, y_0)$ and $\dfrac{\partial f}{\partial y}(x_0, y_0)$ both exist. From the discussion of the geometric interpretation of partial derivatives in Section 18.2, we may conclude the following (see Figure 3.1).

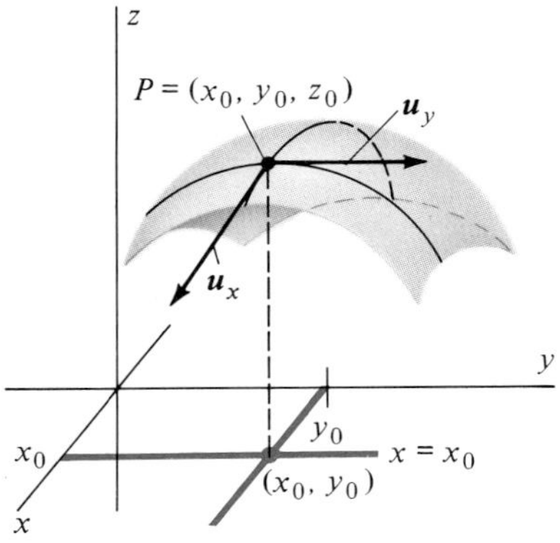

(i) The partial derivative $\dfrac{\partial f}{\partial x}(x_0, y_0)$ gives the slope of the line tangent to the trace of f in the plane $y = y_0$. Since the plane $y = y_0$ is parallel to the xz-coordinate plane, a vector parallel to this tangent line is
$$u_x = i + \frac{\partial f}{\partial x}(x_0, y_0)k.$$

Figure 3.1 Partial derivatives determine vectors $u_x = i + f_x(x_0, y_0)k$ and $u_y = j + f_y(x_0, y_0)k$ tangent to graph of $z = f(x, y)$ at point P.

(To check this statement, note that the slope of the vector, thought of as a line segment in the plane $y = y_0$, is $\dfrac{\partial f}{\partial x}(x_0, y_0)$, as required.)

(ii) The partial derivative $\dfrac{\partial f}{\partial y}(x_0, y_0)$ gives the slope of the line tangent to the trace of f in the plane $x = x_0$. Since the plane $x = x_0$ is parallel to the yz-coordinate plane, a vector parallel to this tangent line is

$$\boldsymbol{u}_y = \boldsymbol{j} + \frac{\partial f}{\partial y}(x_0, y_0)\boldsymbol{k}.$$

Since both vectors $\boldsymbol{u}_x$ and $\boldsymbol{u}_y$ must lie in the tangent plane, they together determine a normal to the plane as

$$\boldsymbol{N} = \boldsymbol{u}_y \times \boldsymbol{u}_x = \det\begin{bmatrix} \boldsymbol{i} & \boldsymbol{j} & \boldsymbol{k} \\ 0 & 1 & \dfrac{\partial f}{\partial y}(x_0, y_0) \\ 1 & 0 & \dfrac{\partial f}{\partial x}(x_0, y_0) \end{bmatrix} \tag{1}$$

$$= \frac{\partial f}{\partial x}(x_0, y_0)\boldsymbol{i} + \frac{\partial f}{\partial y}(x_0, y_0)\boldsymbol{j} - \boldsymbol{k}.$$

(See Figure 3.2.)

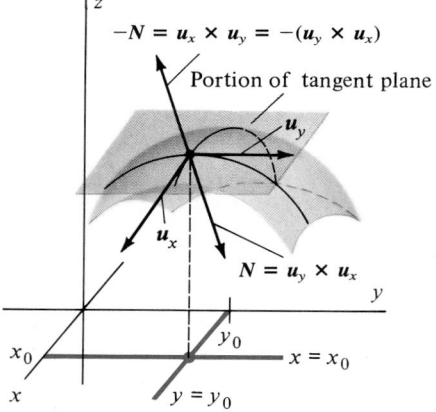

Figure 3.2 Vectors $\boldsymbol{u}_x$ and $\boldsymbol{u}_y$ determine normal $\boldsymbol{N} = \boldsymbol{u}_y \times \boldsymbol{u}_x$ for plane tangent to graph of $z = f(x, y)$ at point P.

According to equation (11), Section 16.5, an equation for the plane with normal vector $\boldsymbol{N}$, as above, and containing the point $(x_0, y_0, z_0) = (x_0, y_0, f(x_0, y_0))$ is

$$\frac{\partial f}{\partial x}(x_0, y_0)(x - x_0) + \frac{\partial f}{\partial y}(x_0, y_0)(y - y_0) - (z - z_0) = 0 \tag{2}$$

or

$$z = \frac{\partial f}{\partial x}(x_0, y_0)(x - x_0) + \frac{\partial f}{\partial y}(x_0, y_0)(y - y_0) + z_0. \tag{3}$$

Either equation (2) or equation (3) may be used to write the equation of the plane tangent to the graph of f at the point (x_0, y_0, z_0), where $z_0 = f(x_0, y_0)$. More generally, you may simply remember the idea used to develop these equations: the partial derivatives $\dfrac{\partial f}{\partial x}(x_0, y_0)$ and $\dfrac{\partial f}{\partial y}(x_0, y_0)$ determine **direction vectors, u_x and u_y,** whose cross product $N = u_y \times u_x$ is a normal to the desired plane.

Example 1

Find an equation for the plane tangent to the graph of $f(x, y) = x^2 + 4y^2$ at the point $(2, 1, 8)$.

Solution: The required partial derivatives are

$$\frac{\partial f}{\partial x}(2, 1) = 2x \left.\right|_{\substack{x=2 \\ y=1}} = 4,$$

$$\frac{\partial f}{\partial y}(2, 1) = 8y \left.\right|_{\substack{x=2 \\ y=1}} = 8,$$

and $x_0 = 2$, $y_0 = 1$, $z_0 = 8$. Thus, by equation (3), the equation is

$$z = 4(x - 2) + 8(y - 1) + 8$$

or

$$z = 4x + 8y - 8.\qquad\qquad\Diamond$$

Example 2

Find an equation for the line of intersection of the plane tangent to the graph of $z = 9 - x^2 - y^2$ at $(1, 2, 4)$ and the xy-coordinate plane.

Solution: Since

$$z_x(1, 2) = -2x \left.\right|_{\substack{x=1 \\ y=2}} = -2$$

and

$$z_y(1, 2) = -2y \left.\right|_{\substack{x=1 \\ y=2}} = -4,$$

the equation of the tangent plane, according to (3), is

$$z = -2(x - 1) - 4(y - 2) + 4$$

or

$$z = -2x - 4y + 14.$$

Setting $z = 0$ gives the line of intersection of this plane with the xy-plane as $2x + 4y - 14 = 0$ (see Figure 3.3). $\qquad\qquad\Diamond$

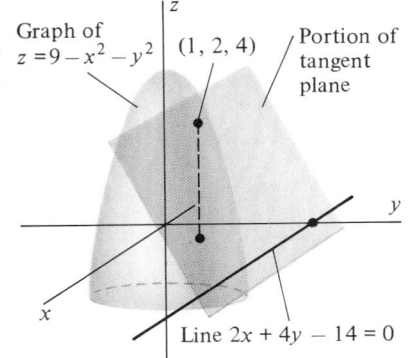

Figure 3.3 Plane tangent to graph of $y = 9 - x^2 - y^2$ at $(2, 1, 4)$ intersects xy-plane along line $2x + 4y - 14 = 0$. (See **Plate 7.**)

Normal Lines

If a surface has a tangent plane at point P, we say that a line through P is **normal** to the surface if it is normal to the tangent plane at P. Using N in (1) as a direction vector normal to the tangent plane for f at (x_0, y_0, z_0), we may write the normal line in vector form as

$$\ell: \mathbf{r}(t) = \mathbf{x}_0 + t\mathbf{N} \tag{4}$$

where $\mathbf{x}_0$ is the position vector $\mathbf{x}_0 = x_0\mathbf{i} + y_0\mathbf{j} + z_0\mathbf{k}$. Writing $\mathbf{r}(t)$ in component form as $\mathbf{r}(t) = x(t)\mathbf{i} + y(t)\mathbf{j} + z(t)\mathbf{k}$ and using (1), we may write (4) in the component form

$$x(t)\mathbf{i} + y(t)\mathbf{j} + z(t)\mathbf{k}$$

$$= (x_0\mathbf{i} + y_0\mathbf{j} + z_0\mathbf{k}) + t\left(\frac{\partial f}{\partial x}(x_0, y_0)\mathbf{i} + \frac{\partial f}{\partial y}(x_0, y_0)\mathbf{j} - \mathbf{k}\right). \tag{5}$$

Example 3

Find equations for the line normal to the graph of $z = e^{y-x^2}$ at the point $(1, \ln 2, 2/e)$.

Solution: For the function $f(x, y) = e^{y-x^2}$,

$$\frac{\partial f}{\partial x}(1, \ln 2) = -2xe^{y-x^2}\Big|_{\substack{x=1 \\ y=\ln 2}} = -2e^{\ln 2-1} = \frac{-4}{e}$$

and

$$\frac{\partial f}{\partial y}(1, \ln 2) = e^{y-x^2}\Big|_{\substack{x=1 \\ y=\ln 2}} = e^{\ln 2-1} = \frac{2}{e}.$$

Also, $x_0 = 1$, $y_0 = \ln 2$, and $z_0 = \dfrac{2}{e}$. The vector form for the normal line is therefore

$$\ell: \mathbf{r}(t) = \mathbf{i} + \ln 2\mathbf{j} + \frac{2}{e}\mathbf{k} + t\left(-\frac{4}{e}\mathbf{i} + \frac{2}{e}\mathbf{j} - \mathbf{k}\right). \qquad \diamond$$

Exercise Set 18.3

In each of Exercises 1–14, find an equation for the plane tangent to the graph of the given function at point P, assuming it exists.

1. $f(x, y) = x^2 + y^2$; $P = (1, 3, 10)$

2. $f(x, y) = x^2 + 2xy + y^2$; $P = (3, -1, 4)$

3. $z = x^2 + y^2 - xy - 4x - 2y$; $P = (1, -1, 1)$

4. $f(x, y) = 2x^2 - y^2$; $P = (2, \sqrt{5}, 3)$

5. $f(x, y) = \dfrac{x-2}{y+2}$; $P = (4, -1, 2)$

6. $f(x, y) = \sqrt{9 - x^2 - y^2}$; $P = (1, -2, 2)$

7. $f(x, y) = \ln xy$; $P = (1, 1, 0)$

8. $f(x, y) = e^{-x} \sin y$; $P = (0, \pi/6, 1/2)$

9. $z = \dfrac{x}{x^2 + y^2}$; $P = (1, 1, 1/2)$

10. $f(x, y) = \text{Tan}^{-1}(y/x)$; $P = (2, 2, \pi/4)$

11. $z = \ln y^x$, $P = (1, 1, 0)$

12. $z = \dfrac{y-6}{3x^2 - 1}$, $P = (1, 4, -1)$

13. $f(x, y) = \ln\left(\dfrac{y-x}{y+x}\right)$, $P = (0, e, 0)$

14. $f(s, t) = \dfrac{1 - \sqrt{s}}{t(s - \sqrt{t})}$, $P = \left(4, 1, -\dfrac{1}{3}\right)$

In Exercises 15–20, find a vector equation for the line normal to the graph of the given function at the point P, as described in the stated exercise.

15. Exercise 2.

16. Exercise 3.

17. Exercise 6.

18. Exercise 8.

19. Exercise 11.

20. Exercise 14.

21. Find the point on the graph of $z = -x^2 + xy + 2y^2$ where the tangent plane is parallel to the plane with equation $x - 14y + z = 4$.

22. Find a vector equation for the line of intersection of the plane tangent to the graph of $z = x^2 + 2y^2 - 4y + 2$ at $(2, 1, 4)$ and the xy-coordinate plane.

23. Show that all lines normal to the graph of $f(x, y) = x \sin y$ at points $(x, \pi/2)$ are parallel.

24. Show that at all points $(x, \pi/2)$, the planes tangent to the graph of $f(x, y) = x \sin y$ are the same. Find an equation for this plane.

25. Find an equation for the plane tangent to the paraboloid $z = 9x^2 + 4y^2$ at the point $(1, 2, 25)$.

26. Show that the volume of the tetrahedron formed by the planes $x = 0$, $y = 0$, and $z = 0$, and any tangent to the graph of the function $f(x, y) = \dfrac{c}{xy}$, $c > 0$ is $V = \dfrac{9c}{2}$.

27. Find an equation for the plane tangent to the sphere $x^2 + y^2 + z^2 = r^2$ at the point (x_0, y_0, z_0), $z_0 \neq 0$.

28. Find an equation for the plane tangent to the graph of the function $y = f(x, y)$ at the point $(x_0, y_0, z_0) = (x_0, f(x_0, z_0), z_0)$.

29. Use the result of Exercise 28 to find the equation of the plane tangent to the graph of $y = \text{Tan}^{-1}(z/x)$ at the point where $x = z = 1$.

30. Use the result of Exercise 28 to find a vector equation for the line of intersection of the plane tangent to the graph of $y = x^2 + 2xz - z^2$ at the point $(2, -8, -2)$ and the xz-coordinate plane.

31. Prove that every normal to a sphere passes through the center.

18.4 RELATIVE AND ABSOLUTE EXTREMA

One of the principal applications of the derivative for functions of a single variable is in finding relative and absolute extrema. The purpose of this section is to show how partial derivatives may be used to find relative extrema for functions of several variables. Although the ideas discussed here are not confined to functions of just two independent variables, we will deal mainly with this case because of the opportunity to interpret the results geometrically.

We begin by defining what we mean by relative extrema.

DEFINITION 4

The number $z_0 = f(x_0, y_0)$ is called a **relative maximum** for the function f if there exists a neighborhood N of (x_0, y_0) so that $f(x, y)$ is defined for all $(x, y) \in N$, and so that

$$f(x_0, y_0) \geq f(x, y)$$

for all $(x, y) \in N$.

The number $z_0 = f(x_0, y_0)$ is called a **relative minimum** for f if there exists a neighborhood N of (x_0, y_0) so that $f(x, y)$ is defined for all $(x, y) \in N$ and so that

$$f(x_0, y_0) \leq f(x, y)$$

for all $(x, y) \in N$.

The number $z_0 = f(x_0, y_0)$ is a **relative extremum** for the function $z = f(x, y)$ if it is either a relative maximum or a relative minimum.

Thus, a relative maximum for f is the precise analogue of a relative maximum for the function of a single variable: $z_0 = f(x_0, y_0)$ is the largest value $z = f(x, y)$ for all (x, y) "near" (x_0, y_0). A similar interpretation holds for relative minima.

Example 1

The function $f(x, y) = \sqrt{x^2 + y^2}$ has a relative minimum of $z_0 = 0$ at the point $(x_0, y_0) = (0, 0)$. This is easy to see, since

$$f(x, y) = \sqrt{x^2 + y^2} \geq 0 = f(0, 0)$$

for all points (x, y) (see Figure 4.1). ◇

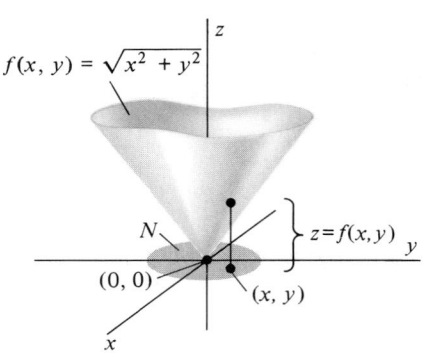

$f(x, y) = \sqrt{x^2 + y^2}$

N

$(0, 0)$

$z = f(x, y)$

(x, y)

Figure 4.1 $f(0, 0) \leq f(x, y)$ for all (x, y) near $(0, 0)$; $f(0, 0)$ is a relative minimum. (See **Plate 8.**)

One way to verify that the number $z_0 = f(x_0, y_0)$ is a relative extremum is to simply compare the number $f(x_0, y_0)$ with values of the function f for points (x, y) near (x_0, y_0). While such comparisons are sometimes difficult, if not impossible, to make, this method does handle a variety of polynomial functions in two variables. The idea is to write $x = x_0 + h$ and $y = y_0 + k$ and then to examine the sign of the difference:

$$f(x_0, y_0) - f(x_0 + h, y_0 + k). \tag{1}$$

If this difference is nonnegative for all small values of h and k, we conclude that $f(x_0, y_0)$ is a relative maximum. Similarly, if the difference in line (1) is nonpositive for all small values of h and k, we conclude that $f(x_0, y_0)$ is a relative minimum.

Example 2

Verify that $f(1, 2) = 4$ is a relative maximum for the function $f(x, y) = 2x + 4y - x^2 - y^2 - 1$.

Solution: Here $x_0 = 1$ and $y_0 = 2$, so the nearby point (x, y) is written $(x, y) = (x_0 + h, y_0 + k) = (1 + h, 2 + k)$. The difference in (1) is

$$\begin{aligned}
f(1, 2) - f(1 + h, 2 + k) \qquad\qquad\qquad\qquad &\tag{2}\\
= 4 - \{2(1 + h) + 4(2 + k) - (1 + h)^2 - (2 + k)^2 - 1\}&\\
= 4 - 2 - 2h - 8 - 4k + 1 + 2h + h^2 + 4 + 4k + k^2 + 1&\\
= h^2 + k^2.&
\end{aligned}$$

Since $h^2 + k^2 \geq 0$ for all h and k, the above calculation shows that the difference in (2) is nonnegative for all small h and k. Thus, $f(1, 2) = 4$ is a relative maximum for the function f (see Figure 4.2). ◇

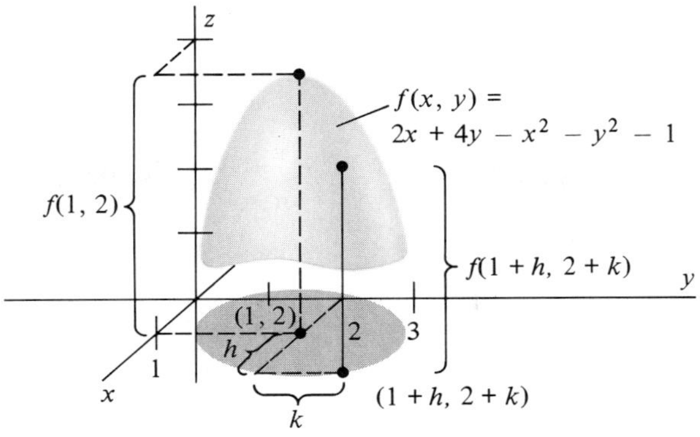

$$f(x, y) = 2x + 4y - x^2 - y^2 - 1$$

Figure 4.2 $f(1, 2) \geq f(1 + h, 2 + k)$ for small values of h and k; $f(1, 2)$ is a relative maximum. (See **Plate 9**.)

Finding Relative Extrema

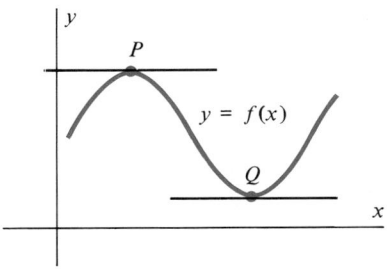

Figure 4.3 At relative extremum P or Q, $f'(x_0) = 0$ and tangent *line* is horizontal.

The preceding discussion addressed the issue of verifying that $f(x_0, y_0)$ is a relative extremum. But how do we find (x_0, y_0) to begin with? Recall the one-variable case again. If the function f has a relative extremum at $x = x_0$, then either $f'(x_0) = 0$ or else $f'(x_0)$ fails to exist. In the former case, the line tangent to the graph of f at $(x_0, f(x_0))$ is horizontal (see Figure 4.3). In the latter case, no tangent exists at $(x_0, f(x_0))$.

From a strictly geometric viewpoint, it seems that the same relationship should hold between the points (x_0, y_0), at which the function f has a relative extremum, and the *plane* tangent to the graph of f at $(x_0, y_0, f(x_0, y_0))$: At such points the tangent plane should either be horizontal or fail to exist (see Figure 4.4).

Since a tangent plane is determined at the point $(x_0, y_0, f(x_0, y_0))$ by the partial derivatives $\frac{\partial f}{\partial x}(x_0, y_0)$ and $\frac{\partial f}{\partial y}(x_0, y_0)$, these geometric observations lead to the following theorem.

THEOREM 3

If the number $z_0 = f(x_0, y_0)$ is a relative extremum for the function f at the point (x_0, y_0), one of the following two conditions must hold:

(i) $\frac{\partial f}{\partial x}(x_0, y_0) = \frac{\partial f}{\partial y}(x_0, y_0) = 0$, or

(ii) one or both of $\frac{\partial f}{\partial x}(x_0, y_0)$ and $\frac{\partial f}{\partial y}(x_0, y_0)$ fails to exist.

Proof: First consider the case where $f(x_0, y_0)$ is a relative maximum. We assume that both $\frac{\partial f}{\partial x}(x_0, y_0)$ and $\frac{\partial f}{\partial y}(x_0, y_0)$ exist, and we attempt to show that

$$\frac{\partial f}{\partial x}(x_0, y_0) = \frac{\partial f}{\partial y}(x_0, y_0) = 0.$$

Holding $y = y_0$ fixed, we define the function (of a single variable) g by

$$g(x) = f(x, y_0).$$

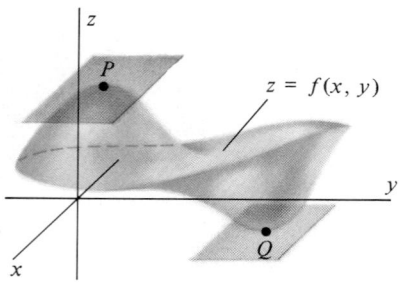

Figure 4.4 At relative extremum P or Q, $\dfrac{\partial f}{\partial x}(x_0, y_0) = \dfrac{\partial f}{\partial y}(x_0, y_0) = 0$ and tangent *plane* is horizontal.

Since $f(x_0, y_0)$ is a relative maximum, $f(x_0, y_0) \geq f(x, y)$ for all (x, y) near (x_0, y_0). Thus,

$$g(x_0) = f(x_0, y_0) \geq f(x, y_0) = g(x)$$

for all x near x_0. This shows that $g(x_0)$ is a relative maximum for the function g. Since the derivative of g is

$$g'(x) = \lim_{h \to 0} \frac{g(x + h) - g(x)}{h}$$

$$= \lim_{h \to 0} \frac{f(x + h, y_0) - f(x, y_0)}{h}$$

$$= \frac{\partial f}{\partial x}(x, y_0),$$

we know that $g'(x_0) = \dfrac{\partial f}{\partial x}(x_0, y_0)$ exists (Figure 4.5). But, if $g(x_0)$ is a relative maximum and $g'(x_0)$ exists, then $g'(x_0) = 0$. This shows that

$$\frac{\partial f}{\partial x}(x_0, y_0) = g'(x_0) = 0.$$

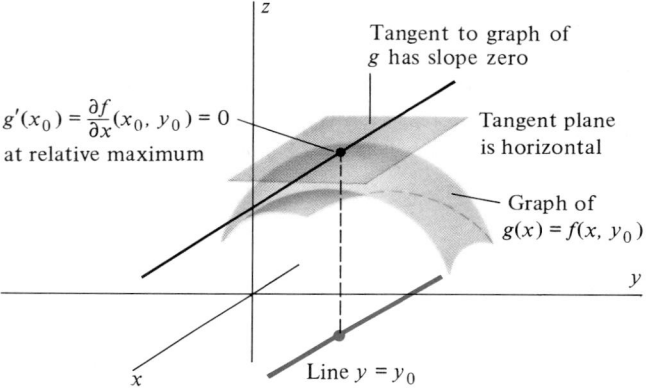

Figure 4.5 A relative maximum: If $\dfrac{\partial f}{\partial x}(x_0, y_0)$ exists, then $\dfrac{\partial f}{\partial x}(x_0, y_0) = 0$.

By repeating this argument with $x = x_0$ fixed, we can show that $\dfrac{\partial f}{\partial y}(x_0, y_0) = 0$ as well. Finally, we note that we have made no special use of the assumption that the extremum $f(x_0, y_0)$ was a relative maximum—the argument applies for relative minima as well. ◆

Theorem 3 determines a procedure for finding relative extrema for f: Find all points (x_0, y_0) where $\dfrac{\partial f}{\partial x}(x_0, y_0) = \dfrac{\partial f}{\partial y}(x_0, y_0) = 0$ or where either $\dfrac{\partial f}{\partial x}(x_0, y_0)$ or

$\frac{\partial f}{\partial y}(x_0, y_0)$ fails to exist. (We call these points **critical points.**) Then test each critical point to determine whether it yields a relative extremum.

Example 3

For the function $f(x, y) = 2x + 4y - x^2 - y^2 - 1$, setting both partial derivatives equal to zero gives the equations

$$\frac{\partial f}{\partial x}(x, y) = 2 - 2x = 0, \qquad \frac{\partial f}{\partial y}(x, y) = 4 - 2y = 0.$$

The (simultaneous) solution of this pair of equations is $x = 1$, $y = 2$. Thus, condition (i) in Theorem 3 yields the single critical point $(1, 2)$. We have verified that $f(1, 2)$ is a relative maximum in Example 2. Since the partial derivatives are defined for all (x, y), there are no points satisfying condition (ii) of Theorem 3. Thus, the only relative extremum for this function is the relative maximum at $(1, 2)$ (see Figure 4.2). ◇

Example 4

Find all relative extrema for the function

$$f(x, y) = \begin{cases} \sqrt{x^2 + y^2}, & (x, y) \neq (0, 0) \\ 0, & (x, y) = (0, 0). \end{cases}$$

Solution: The partial derivatives are

$$\frac{\partial f}{\partial x}(x, y) = \frac{x}{\sqrt{x^2 + y^2}}; \qquad \frac{\partial f}{\partial y}(x, y) = \frac{y}{\sqrt{x^2 + y^2}}.$$

Both partial derivatives are undefined at $(x, y) = (0, 0)$. For all other points at least one of the partial derivatives is nonzero. Since it is easy to see that $f(0, 0) = 0 < f(x, y)$ for all $(x, y) \neq (0, 0)$, $f(0, 0) = 0$ is a relative minimum. (Figure 4.1.) ◇

It is important to understand that the conditions of Theorem 3 do not *guarantee* that $f(x_0, y_0)$ is a relative extremum. Theorem 3 merely provides *necessary* conditions for an extremum. Without satisfying condition (i) or (ii) of Theorem 3, $f(x_0, y_0)$ cannot be a relative extremum. However, some critical points do not yield extrema. The following is a typical such example.

Example 5

For the function $f(x, y) = y^2 - x^2$, the partial derivatives are

$$\frac{\partial f}{\partial x} = -2x; \qquad \frac{\partial f}{\partial y} = 2y.$$

Thus, $\frac{\partial f}{\partial x}(0, 0) = \frac{\partial f}{\partial y}(0, 0) = 0$. However, the number $f(0, 0) = 0$ is neither a

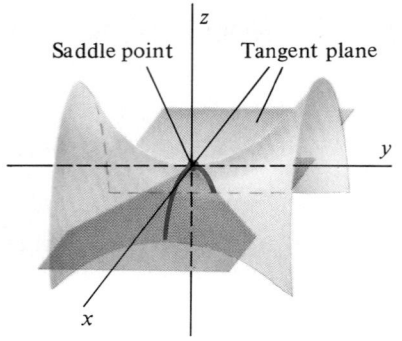

Figure 4.6 Graph of $z = y^2 - x^2$ has a *saddle* point, but no extremum, at $(0, 0)$. (See **Plate 10.**)

relative maximum nor a relative minimum. To see this, we compare $f(0, 0)$ with $f(h, k)$ where (h, k) is a point near $(0, 0)$. Since

$$f(0, 0) - f(h, k) = 0 - (k^2 - h^2) = h^2 - k^2,$$

we see that the sign of this difference depends only on the relative sizes of $|h|$ and $|k|$. Since this difference is not of constant sign, $f(0, 0)$ is neither a maximum nor a minimum.

Figure 4.6 illustrates why $f(x, y) = y^2 - x^2$ does not have an extremum at $(0, 0)$. Although both partial derivatives are zero, the function $g(x) = f(x, 0) = -x^2$ reaches a relative *maximum* while the function $h(y) = f(0, y) = y^2$ reaches a relative *minimum* at $(0, 0)$. ◇

In Example 5, the point $(0, 0)$ is called a *saddle point* for rather obvious reasons. The surface bows upward along one axis and downward along another. More generally, a point (x_0, y_0) in the domain of a function of two variables is called a **saddle point** if (x_0, y_0) is a critical point and if $f(x_0, y_0)$ is neither a relative maximum nor a relative minimum. Figure 4.7 shows the graph of another function that has a saddle point at $(0, 0)$ (see Exercise 20).

Second Derivative Test

There is a theorem that helps sort out actual extrema from saddle points. It may be regarded as the analogue of the Second Derivative Test for functions of a single variable. A proof of this theorem may be found in texts on advanced calculus.

THEOREM 4
Second Derivative Test

Let f be a function of two variables. Suppose that all second order partial derivatives of f are continuous in a neighborhood of (x_0, y_0) and that $\dfrac{\partial f}{\partial x}(x_0, y_0) = \dfrac{\partial f}{\partial y}(x_0, y_0) = 0$. Let

$$A = \frac{\partial^2 f}{\partial x^2}(x_0, y_0), \qquad B = \frac{\partial^2 f}{\partial y \partial x}(x_0, y_0), \qquad C = \frac{\partial^2 f}{\partial y^2}(x_0, y_0)$$

and

$$D = B^2 - AC.$$

Then

(i) If $D < 0$ and $A < 0$, $f(x_0, y_0)$ is a relative maximum.
(ii) If $D < 0$ and $A > 0$, $f(x_0, y_0)$ is a relative minimum.
(iii) If $D > 0$, (x_0, y_0) is a saddle point.
(iv) If $D = 0$, no conclusions may be drawn.

Theorem 4 verifies that the critical point $(0, 0)$ in Example 5 is a saddle point, since

$$A = \frac{\partial^2 f}{\partial x^2}(0, 0) = -2, \qquad B = \frac{\partial^2 f}{\partial y \partial x}(0, 0) = 0, \qquad C = \frac{\partial^2 f}{\partial y^2}(0, 0) = 2$$

and

$$D = B^2 - AC = 4 > 0.$$

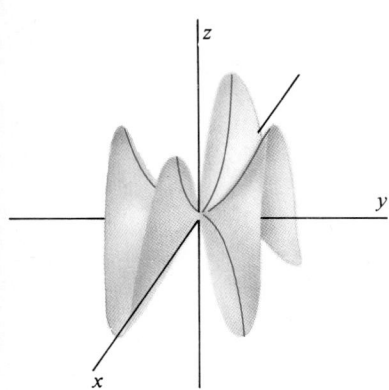

Figure 4.7 Graph of

$f(x, y) = x^4 - 3x^2y^2 + y^4$

has a saddle point at $(0, 0)$.

Similarly, for the function f in Example 3 and the critical point $(1, 2)$, we have

$$A = \frac{\partial^2 f}{\partial x^2}(1, 2) = -2, \qquad B = \frac{\partial^2 f}{\partial y \partial x}(1, 2) = 0, \qquad C = \frac{\partial^2 f}{\partial y^2}(1, 2) = -2$$

and

$$D = B^2 - AC = -4.$$

Thus, since $D < 0$ and $A < 0$, the Second Derivative Test agrees with the conclusion of Example 3, that $f(1, 2)$ is a relative maximum.

Example 6

Find and classify all relative extrema for the function $f(x, y) = x^4 + y^4 - 4xy$.

Solution: The partial derivatives are

$$\frac{\partial f}{\partial x} = 4x^3 - 4y; \qquad \frac{\partial f}{\partial y} = 4y^3 - 4x.$$

Since both partial derivatives are defined for all (x, y), the extrema can occur only at points where

$$\frac{\partial f}{\partial x} = 4x^3 - 4y = 0, \qquad \text{or} \qquad y = x^3 \tag{3}$$

and

$$\frac{\partial f}{\partial y} = 4y^3 - 4x = 0, \qquad \text{or} \qquad x = y^3. \tag{4}$$

Substituting for x in (3) using (4) gives the equation $y = y^9$. Thus, either $y = 0$, or else $y^8 = 1$, which gives $y = \pm 1$. Using (4) we find that $x = 0$ if $y = 0$, $x = 1$ if $y = 1$, and $x = -1$ if $y = -1$. The three critical points are therefore $(0, 0)$, $(1, 1)$, and $(-1, -1)$.

Next, we calculate the second order partials:

$$\frac{\partial^2 f}{\partial x^2} = 12x^2; \qquad \frac{\partial^2 f}{\partial y \partial x} = -4; \qquad \frac{\partial^2 f}{\partial y^2} = 12y^2.$$

At the critical point $(1, 1)$ we have

$$A = \frac{\partial^2 f}{\partial x^2}(1, 1) = 12, \qquad B = \frac{\partial^2 f}{\partial y \partial x}(1, 1) = -4, \qquad C = \frac{\partial^2 f}{\partial y^2}(1, 1) = 12,$$

and

$$D = B^2 - AC = 16 - 12 \cdot 12 = -128 < 0.$$

Thus, since $D < 0$ and $A > 0$, $f(1, 1) = -2$ is a relative minimum, according to Theorem 4.

At the critical point $(-1, -1)$ the values of A, B, C, and D are the same as for $(1, 1)$ so $f(-1, -1) = -2$ is also a relative minimum.

At the critical point $(0, 0)$, $A = C = 0$ and $B = -4$, so $D = B^2 - AC = 16 > 0$. Thus, $(0, 0)$ is a saddle point.   ◇

In each of Examples 7, 8, and 9, we obtain $D = 0$ at the critical point. These three examples show that any of the three possible outcomes (relative maximum, relative minimum, saddle point) may result when $D = 0$.

Example 7

For the function $f(x, y) = e^{-(x^4+y^4)}$, we have

$$\frac{\partial f}{\partial x} = -4x^3 e^{-(x^4+y^4)} \qquad \frac{\partial f}{\partial y} = -4y^3 e^{-(x^4+y^4)}$$

$$\frac{\partial^2 f}{\partial x^2} = (16x^6 - 12x^2)e^{-(x^4+y^4)} \qquad \frac{\partial^2 f}{\partial y^2} = (16y^6 - 12y^2)e^{-(x^4+y^4)}$$

$$\frac{\partial^2 f}{\partial y \partial x} = 16y^3 x^3 e^{-(x^4+y^4)}.$$

The point $(0, 0)$ is a critical point since $\dfrac{\partial f}{\partial x}(0, 0) = \dfrac{\partial f}{\partial y}(0, 0) = 0$. At this critical point

$$A = \frac{\partial^2 f}{\partial x^2}(0, 0) = 0; \qquad B = \frac{\partial^2 f}{\partial y \partial x}(0, 0) = 0; \qquad C = \frac{\partial^2 f}{\partial y^2}(0, 0) = 0.$$

Thus, the Second Derivative Test yields no conclusion about the critical point $(0, 0)$. However, it is easy to see that the expression $x^4 + y^4$ has a minimum at $(0, 0)$, so

$$f(x, y) = e^{-(x^4+y^4)} = \frac{1}{e^{x^4+y^4}}$$

has a relative maximum at $(0, 0)$ (see Figure 4.8). ◇

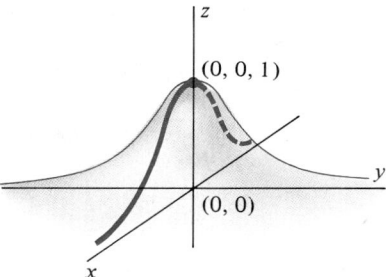

Figure 4.8 $f(x, y) = e^{-(x^4+y^4)}$ has a relative maximum at $(0, 0)$. (See **Plate 11.**)

Example 8

The function $f(x, y) = x^4 + y^4$ obviously has a relative minimum of $f(0, 0) = 0$, since $f(x, y) > 0$ at all other points. However, we find

$$A = \frac{\partial^2 f}{\partial x^2}(0, 0) = 12x^2 \bigg|_{x=0} = 0, \qquad B = \frac{\partial^2 f}{\partial y \partial x}(0, 0) = 0,$$

$$C = \frac{\partial^2 f}{\partial y^2}(0, 0) = 12y^2 \bigg|_{y=0} = 0, \qquad D = B^2 - AC = 0.$$

Thus, the Second Derivative Test fails to classify this critical point. ◇

Example 9

For the function $f(x, y) = x^3 - y^3$, the only simultaneous solution of the two equations

$$\frac{\partial f}{\partial x}(x, y) = 3x^2 = 0 \qquad \text{and} \qquad \frac{\partial f}{\partial y}(x, y) = -3y^2 = 0$$

is $x = y = 0$, so $(0, 0)$ is the only critical point. Since

$$\frac{\partial^2 f}{\partial x^2} = 6x, \qquad \frac{\partial^2 f}{\partial y \partial x} = 0, \qquad \frac{\partial^2 f}{\partial y^2} = -6y,$$

we have $A = B = C = D = 0$ at the critical point $(0, 0)$.

Thus, the test gives no information as to the nature of this critical point. However, since $f(x, y) = x^3 - y^3$ takes on both positive and negative values in every neighborhood of $(0, 0)$, the critical point $(0, 0)$ is a saddle point. ◇

Functions of More Than Two Variables

The discussion of this section has focused on the question of finding relative extrema for functions of two variables. For functions of more than two variables, the statement of Theorem 3 generalizes directly:

(a) If the function f of three variables has a relative extremum at (x_0, y_0, z_0), then either

$$\frac{\partial f}{\partial x} = \frac{\partial f}{\partial y} = \frac{\partial f}{\partial z} = 0$$

at this point, or else one or more of these partial derivatives fails to exist at (x_0, y_0, z_0).

(b) More generally, the function f of n variables can have a relative extremum at the point $(a_1, a_2, \ldots, a_n)$ only if one or more of the partial derivatives fails to exist, or if

$$\frac{\partial f}{\partial x_1} = \frac{\partial f}{\partial x_2} = \cdots = \frac{\partial f}{\partial x_n} = 0$$

at this point.

However, the Second Derivative Test does not generalize quite so easily. For functions of more than two variables, the classification of critical points is considerably more difficult and will not be pursued here.

Extrema on Closed Sets

Finally, we note that we have not addressed the question of *absolute* extrema, as we did for functions of a single variable in Chapter 4. In a manner analogous to the behavior of a function of one variable on a closed interval, a continuous function f of two variables will assume both an absolute maximum and an absolute minimum on any **closed bounded set** S in the plane. By a closed set in the plane we mean (roughly speaking) a set that includes its boundary, such as the disc $D = \{(x, y) \mid x^2 + y^2 \leq 1\}$ or the square $S = \{(x, y) \mid 0 \leq x \leq 1, 0 \leq y \leq 1\}$. As you might suspect, such absolute extrema will occur either at critical points or at points on the boundary of S. Checking for extrema among boundary points can be a quite complicated task, so we have chosen to defer this topic to more advanced courses where appropriate theory and techniques are developed. (However, this issue is addressed in simple settings in Exercises 29–32.)

Exercise Set 18.4

In Exercises 1–16, find all critical points. Classify each as a relative maximum, relative minimum, or saddle point.

1. $f(x, y) = x^2 + y^2 + 4y + 4$

2. $f(x, y) = x^2 + y^2 + 4x - 2y + 11$

3. $f(x, y) = x^2 - y^2 + 6x + 4y + 5$

4. $f(x, y) = 7 - 2x + 2y - x^2 - y^2$

5. $f(x, y) = xy + 9$

6. $f(x, y) = x^2 + y^4 - 2x - 4y^2 + 5$

7. $f(x, y) = 5x^2 + y^2 - 10x - 6y + 15$

8. $f(x, y) = x^2 + y^3 - 3y$

9. $f(x, y) = x^3 - y^3$

10. $f(x, y) = e^{x^2 - 2x + y^2 + 4}$

11. $f(x, y) = x^2 - xy$

12. $f(x, y) = e^x \cos y$

13. $f(x, y) = x^3 + y^3 + 4xy$

14. $f(x, y) = x^4 + y^4 - 4xy$

15. $f(x, y) = x \cos y$

16. $f(x, y) = \dfrac{y}{x} - \dfrac{x}{y}$

17. Show that the function $z = 4 - \sqrt{x^2 + y^2}$ has a relative maximum at $(0, 0)$. (Note that none of the partial derivatives exists at this point.)

18. Show that the function $z = x^2 + y^2 - 4x - 2y + 9$ has a relative minimum at $(2, 1)$ using the method of Example 2.

19. Find the highest point on the graph of $z = 2y^3 - 3x^2 - 3xy + 9x$ if the positive z-axis is upward.

20. Show that the function $f(x, y) = x^4 - 3x^2y^2 + y^4$ has a saddle point at $(0, 0)$ (see Figure 4.7).

21. Show that the function $f(x, y) = x^2 - y^2 + 2x + 4y - 3$ has a saddle point at $(-1, 2)$ by the method of Example 2.

22. Consider the function $f(x, y) = \dfrac{x^2 + y^2}{(x + y)^2}$.

a. Show that $\dfrac{\partial f}{\partial x}(x, y) = \dfrac{\partial f}{\partial y}(x, y) = 0$ if and only if $y = x$ and $x \neq 0$.

b. Show that f has a relative minimum at each point (x, x) with $x \neq 0$.

c. Show that the function $g(\lambda) = f(x, \lambda x) = \dfrac{1 + \lambda^2}{(1 + \lambda)^2}$, $\lambda \neq -1$, has a relative minimum at $\lambda = 1$.

d. Explain the geometric and algebraic relationships between the functions g and f.

23. A rectangular box, with a top, is to hold 16 cubic meters. Find the dimensions that produce the least expensive box if the material for the side walls is half as expensive as the material for the top and the bottom.

24. Find the point on the plane $2x + 3y + z - 14 = 0$ nearest the origin.

25. Find the point on the plane with equation $ax + by + cz = d$ nearest the origin.

26. Find the dimensions of the closed rectangular box of volume $V = 8000$ cm^3 and of minimum surface area.

27. Find the dimensions of the rectangular package of largest volume that can be mailed under the restrictions that length plus girth cannot exceed 84 inches. (Girth is the perimeter of the cross section taken perpendicular to the length.)

28. A manufacturer of tape recorders intends to market x recorders through retail outlets and y recorders through wholesale outlets. The manufacturer estimates that the revenue per recorder sold retail will be $r(x, y) = 250 - \dfrac{x}{20} - \dfrac{y}{5}$, while the revenue per recorder on the wholesale market is estimated to be $w(x, y) = 200 - \dfrac{x}{50} - \dfrac{y}{20}$. The cost of producing the recorders is known to be \$100 each. Using the model Profit = Revenue − Cost, find the number of recorders that should be marketed retail and the number that should be marketed wholesale so as to maximize profit.

29. Find the *absolute* maximum and minimum values of the function $f(x, y) = 2x + 2y + 4$ on the *closed* disc $D = \{(x, y) \mid x^2 + y^2 \le 1\}$. (*Hint:* To check for absolute extrema along the boundary $C = \{(x, y) \mid x^2 + y^2 = 1\}$ use the parameterization $x = \cos t$, $y = \sin t$, and optimize $f(x(t), y(t))$ as a function of t.)

30. Find the absolute maximum and minimum values of the function $f(x, y) = 3x - 2y^3 + 2$ on the (closed) square $S = \{(x, y) \mid 0 \le x \le 1, 0 \le y \le 1\}$ (see Exercise 29).

31. Find the absolute maximum and minimum values of the function $z = x + 2y + 6$ on the closed set consisting of the ellipse $4x^2 + 2y^2 = 4$ and its interior (see Exercise 29).

32. Find the absolute maximum and minimum values for the function $f(x, y) = x^2 + 2y^2 - 2xy - 6x + 4y$ on the rectangle $0 \le x \le 8$, $-2 \le y \le 2$.

33. Given a set of points $(x_1, y_1), (x_2, y_2), \ldots, (x_n, y_n)$, one way to define a line $y = mx + b$ that "best fits" these points is given by the Method of Least Squares. As illustrated by Figure 4.9, the distance from the data point (x_j, y_j) to the point on the line $y = mx + b$ with x-coordinate x_j is

$$|y_j - (mx_j + b)| = |y_j - mx_j - b|.$$

The Method of Least Squares defines the best fitting line $y = mx + b$ (called the **regression line**) to be the line that minimizes the sum of the squares of these individual distances:

$$S(m, b) = \sum_{j=1}^{n} (y_j - mx_j - b)^2.$$

Viewing $x_1, x_2, \ldots, x_n$ and $y_1, y_2, \ldots, y_n$ as fixed constants and m and b as independent variables, show that the values of m and b for which $S(m, b)$ is a minimum are given by the formulas

$$m = \frac{n \sum_{j=1}^{n} x_j y_j - \left(\sum_{j=1}^{n} x_j\right)\left(\sum_{j=1}^{n} y_j\right)}{n \sum_{j=1}^{n} x_j^2 - \left(\sum_{j=1}^{n} x_j\right)^2},$$

$$b = \frac{\left(\sum_{j=1}^{n} x_j^2\right)\left(\sum_{j=1}^{n} y_j\right) - \left(\sum_{j=1}^{n} x_j\right)\left(\sum_{j=1}^{n} x_j y_j\right)}{n \sum_{j=1}^{n} x_j^2 - \left(\sum_{j=1}^{n} x_j\right)^2}.$$

(*Hint:* Set $\dfrac{\partial S}{\partial m}(m, b) = 0$ and $\dfrac{\partial S}{\partial b}(m, b) = 0$ and solve the resulting system of equations.)

34. For a group of $n = 6$ calculus students, achievement test scores x_j and final course averages in calculus y_j were as follows:

j	1	2	3	4	5	6
x_j	52	46	69	54	61	48
y_j	74	66	94	91	84	80

The regression line for these data appears in Figure 4.10. Use the result of Exercise 33 to show that this regression line has equation $y = 0.95x + 29$.

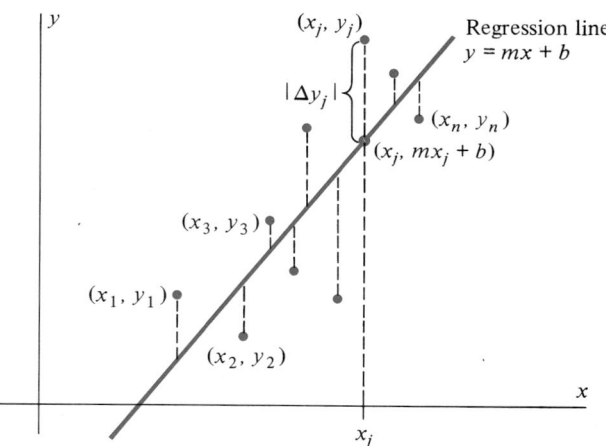

Figure 4.9 The Method of Least Squares determines the line that "best fits" the data points (x_1, y_1), $(x_2, y_2), \ldots, (x_n, y_n)$.

35. A regression line corresponding to a set of data points $(x_1, y_1), (x_2, y_2), \ldots, (x_n, y_n)$ provides a model for *predicting* a value of y corresponding to any particular value of x. Use the regression line obtained in Exercise 34 to predict the course average for a student whose achievement score is $x = 60$.

36. Given the five points whose coordinates are:

x	0	1	1	2	3
y	2	4	3	6	6

a. Find the regression line for the data using the Method of Least Squares.
b. Plot the data points and the regression line.
c. Find the predicted value for $x = 4$.

37. Repeat Exercise 36 for the data below.

x	6	8	9	10	12
y	2	5	5	7	9

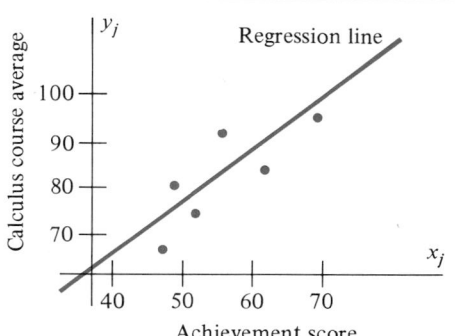

Figure 4.10 Achievement scores versus calculus course averages for six students.

38. For data consisting of ordered triples $(x_1, y_1, z_1), \ldots,$ (x_n, y_n, z_n), the analog of the regression line is the (regression) **plane** $z = ax + by + c$. This plane is defined as the plane that minimizes the sum of squares

$$S = \sum_{j=1}^{n} [z_j - (ax_j + by_j + c)]^2.$$

Show that the coefficients a, b, and c in this equation satisfy the equations

(i) $a \sum\limits_{j=1}^{n} x_j + b \sum\limits_{j=1}^{n} y_j + nc = \sum\limits_{j=1}^{n} z_j,$

(ii) $a \sum\limits_{j=1}^{n} x_j^2 + b \sum\limits_{j=1}^{n} x_j y_j + c \sum\limits_{j=1}^{n} x_j = \sum\limits_{j=1}^{n} x_j y_j,$

(iii) $a \sum\limits_{j=1}^{n} x_j y_j + b \sum\limits_{j=1}^{n} y_j^2 + c \sum\limits_{j=1}^{n} y_j = \sum\limits_{j=1}^{n} y_j z_j.$

39. Recall, from Section 7.7, that the *centroid* of the n data points $(x_1, y_1), (x_2, y_2), \ldots, (x_n, y_n)$ is the point $(\bar{x}, \bar{y})$ where

$$\bar{x} = \frac{1}{n} \sum_{j=1}^{n} x_j, \qquad \bar{y} = \frac{1}{n} \sum_{j=1}^{n} y_j.$$

Show that the centroid lies on the regression line determined by the Method of Least Squares.

40. Suppose that you suspected that the n data points (x_1, y_1), $(x_2, y_2), \ldots, (x_n, y_n)$ could be "fitted" well by a curve of the form $y = m \ln x + b$ (a common situation in biology and chemistry). Explain how the Method of Least Squares could be used to find the constants m and b.

41. Prove Theorem 4 in the case of the function $f(x, y) = ax^2 + byx + cy^2$.

18.5 APPROXIMATIONS AND DIFFERENTIALS

For the function of a single variable $y = f(x)$, the existence of the derivative $f'(x)$ leads to the approximation formula

$$f(x + \Delta x) \approx f(x) + f'(x)\, \Delta x \tag{1}$$

and to the definition of the differential

$$df = f'(x)\, dx. \tag{2}$$

In this section, we take up the analogous issues for functions of several variables. The key to the entire discussion (and to the discussions of the sections that follow) is the Approximation Theorem.

THEOREM 5
Approximation Theorem

Let f be a function of two variables. Let f and its first partial derivatives $\dfrac{\partial f}{\partial x}(x, y)$

and $\dfrac{\partial f}{\partial y}(x, y)$ be continuous in an open rectangle $R = \{(x, y) \mid a_1 < x < a_2, b_1 < y < b_2\}$ in the xy-plane. Let the point (x_0, y_0) and the point $(x_0 + \Delta x, y_0 + \Delta y)$ both lie in R. Then

$$f(x_0 + \Delta x, y_0 + \Delta y) = f(x_0, y_0) + \frac{\partial f}{\partial x}(x_0, y_0)\, \Delta x + \frac{\partial f}{\partial y}(x_0, y_0)\, \Delta y \tag{3}$$

$$+ \epsilon_1 \Delta x + \epsilon_2 \Delta y$$

where $\lim\limits_{\substack{\Delta x \to 0 \\ \Delta y \to 0}} \epsilon_1 = 0$ and $\lim\limits_{\substack{\Delta x \to 0 \\ \Delta y \to 0}} \epsilon_2 = 0$.

Theorem 5 provides the relationship between the value of the function f at (x_0, y_0) and the value of this function at the "nearby" point $(x_0 + \Delta x, y_0 + \Delta y)$.

Ignoring the error term $(\epsilon_1 \, \Delta x + \epsilon_2 \, \Delta y)$ in equation (3) leads to the approximation

$$f(x_0 + \Delta x, y_0 + \Delta y) \approx f(x_0, y_0) + \frac{\partial f}{\partial x}(x_0, y_0) \, \Delta x + \frac{\partial f}{\partial y}(x_0, y_0) \, \Delta y \qquad (4)$$

for functions of two variables.

Figure 5.1 provides a geometric interpretation of approximation (4). Since both $\dfrac{\partial f}{\partial x}(x_0, y_0)$ and $\dfrac{\partial f}{\partial y}(x_0, y_0)$ exist, they determine a "tangent" plane with equation

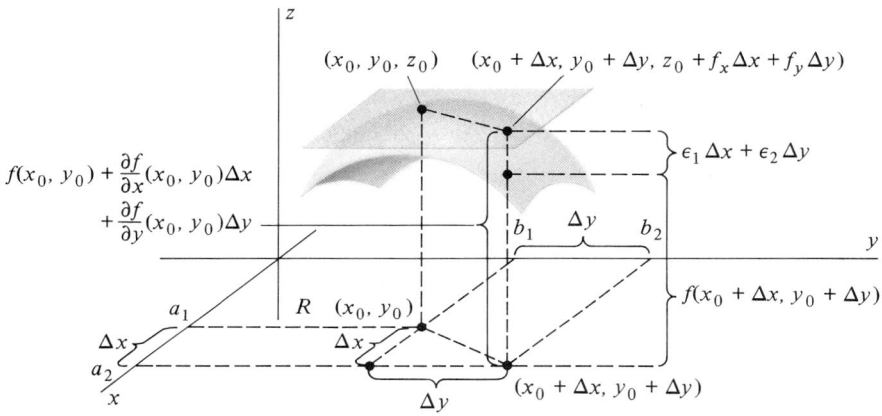

Figure 5.1 Point on tangent plane approximates point on graph of $z = f(x, y)$ above point $(x_0 + \Delta x, y_0 + \Delta y)$.

$$z = \frac{\partial f}{\partial x}(x_0, y_0)(x - x_0) + \frac{\partial f}{\partial y}(x_0, y_0)(y - y_0) + f(x_0, y_0) \qquad (5)$$

(equation (3), Section 18.3). To find the point on this plane above $(x, y) = (x_0 + \Delta x, y_0 + \Delta y)$, we set $x = x_0 + \Delta x$ and $y = y_0 + \Delta y$. Then $x - x_0 = \Delta x$ and $y - y_0 = \Delta y$, so (5) gives the z-coordinate as

$$z_T = \frac{\partial f}{\partial x}(x_0, y_0) \, \Delta x + \frac{\partial f}{\partial y}(x_0, y_0) \, \Delta y + f(x_0, y_0),$$

which is precisely the right-hand side of approximation (4). Thus, the left-hand side of approximation (4) is the z-coordinate of the point above $(x_0 + \Delta x, y_0 + \Delta y)$ *on the graph of f,* the right-hand side of (4) is the z-coordinate of the point above $(x_0 + \Delta x, y_0 + \Delta y)$ *on the tangent plane,* and the difference between these two numbers is the "error term," $\epsilon_1 \, \Delta x + \epsilon_2 \, \Delta y$.

The proof of Theorem 5 is given at the end of this section.

Example 1

To see how this all works out in a particular example let us consider the specific function $f(x, y) = 2x^2 + 4y^2$ and the problem of calculating $f(1 + \Delta x, 2 + \Delta y)$. Here

$$(x_0, y_0) = (1, 2)$$
$$f(x_0, y_0) = f(1, 2) = 2 \cdot 1^2 + 4 \cdot 2^2 = 18$$

$$\frac{\partial f}{\partial x}(x_0, y_0) = 4x \bigg|_{x=1} = 4$$

$$\frac{\partial f}{\partial y}(x_0, y_0) = 8y \bigg|_{y=2} = 16$$

and

$$
\begin{aligned}
f(1 + \Delta x, 2 + \Delta y) &= 2(1 + \Delta x)^2 + 4(2 + \Delta y)^2 \\
&= 2(1 + 2\,\Delta x + \Delta x^2) + 4(4 + 4\,\Delta y + \Delta y^2) \\
&= 18 + 4\,\Delta x + 16\,\Delta y + 2\,\Delta x^2 + 4\,\Delta y^2 \\
&= 18 + 4\,\Delta x + 16\,\Delta y + (2\,\Delta x)\Delta x + (4\,\Delta y)\Delta y.
\end{aligned}
\tag{6}
$$

$$
\underbrace{f(1, 2)} \quad \underbrace{\frac{\partial f}{\partial x}(1, 2)} \quad \underbrace{\frac{\partial f}{\partial y}(1, 2)} \quad \widetilde{\epsilon_1} \quad \widetilde{\epsilon_2}
$$

Thus, the approximation (4) is

$$f(1 + \Delta x, 2 + \Delta y) \approx 18 + 4\,\Delta x + 16\,\Delta y. \tag{7}$$

By comparing (6) and (7), we can see that the error in this approximation is

$$\epsilon_1\,\Delta x + \epsilon_2\,\Delta y = 2(\Delta x)^2 + 4(\Delta y)^2.$$

Thus, $\epsilon_1 = 2\,\Delta x$ and $\epsilon_2 = 4\,\Delta y$ in this example. $\diamond$

Example 2

Use approximation (4) to estimate the value of the expression $\sqrt{(3.04)^2 + (3.95)^2}$.

Solution: If we let $f(x, y) = \sqrt{x^2 + y^2}$, the problem is then to approximate $f(3.04, 3.95)$. To do so we note that for $x_0 = 3$ and $y_0 = 4$ it is easy to compute

$$f(x_0, y_0) = \sqrt{3^2 + 4^2} = \sqrt{25} = 5.$$

We therefore write

$$3.04 = x_0 + \Delta x \qquad \text{and} \qquad 3.95 = y_0 + \Delta y,$$

where

$$x_0 = 3, \qquad \Delta x = 0.04, \qquad y_0 = 4, \qquad \text{and} \qquad \Delta y = -0.05.$$

Also

$$\frac{\partial f}{\partial x}(x_0, y_0) = \frac{\partial}{\partial x}(\sqrt{x^2 + y^2}) \bigg|_{(3, 4)} = \frac{3}{\sqrt{3^2 + 4^2}} = 0.6$$

and

$$\frac{\partial f}{\partial y}(x_0, y_0) = \frac{\partial}{\partial y}(\sqrt{x^2 + y^2}) \bigg|_{(3, 4)} = \frac{4}{\sqrt{3^2 + 4^2}} = 0.8.$$

Thus, by (4),

$$
\begin{aligned}
\sqrt{(3.04)^2 + (3.95)^2} &\approx 5 + (0.6)(0.04) + (0.8)(-0.05) \\
&= 4.984.
\end{aligned}
$$

The actual value of this expression, to four decimal places, is 4.9844. The *relative error* in the approximation is therefore

$$\frac{4.9844 - 4.984}{4.9844} \approx .00008,$$

a *percentage error* of less than 0.01%. ◇

Approximation (4) is written in a form that is useful in approximating a particular value $f(x_0 + \Delta x, y_0 + \Delta y)$. By using the notation

$$\Delta f = f(x_0 + \Delta x, y_0 + \Delta y) - f(x_0, y_0),$$

we may rewrite approximation (4) as

$$\Delta f \approx \frac{\partial f}{\partial x}(x_0, y_0)\, \Delta x + \frac{\partial f}{\partial y}(x_0, y_0)\, \Delta y. \tag{8}$$

Approximation (8) is useful in approximating the *change* in the value of the function f due to changes Δx and Δy in the independent variables.

Example 3

The ideal gas law is $PV = nRT$, where P is pressure, V is volume, n is the number of moles of gas present, R is constant, and T is temperature. Approximate, according to the ideal gas law, the change in the volume of 1000 cc of gas at temperature 300 K and pressure 780 mm of mercury if the gas is heated by 10 K and the pressure is increased by 5 mm of mercury.

Solution: We first simplify the gas law by solving for the constant nR and using the given data. We find that

$$nR = \frac{PV}{T} = \frac{780 \cdot 1000}{300} = 2600.$$

Using this constant, we may solve the gas law for V as a function of T and P:

$$V(T, P) = \frac{nRT}{P} = \frac{2600T}{P}.$$

According to approximation (8),

$$\Delta V \approx \frac{\partial}{\partial T}\left(\frac{2600T}{P}\right)\Delta T + \frac{\partial}{\partial P}\left(\frac{2600T}{P}\right)\Delta P$$

$$= \frac{2600\, \Delta T}{P} - \frac{2600T\, \Delta P}{P^2}.$$

Using the data $T = 300$, $P = 780$, $\Delta T = 10$, and $\Delta P = 5$, we obtain

$$\Delta V \approx \frac{2600(10)}{780} - \frac{2600(300)(5)}{780^2} \tag{9}$$

$$= 26.92 \text{ cc.}$$

The actual value of V corresponding to the temperature $T = 310$ and pressure $P = 785$ mm of mercury is

$$V = \frac{2600(310)}{785} = 1026.75,$$

so the actual value of ΔV is 26.75. The approximation in (9) is therefore in error by $26.92 - 26.75 = 0.17$ cc, a percentage error of only $\left(\dfrac{0.17}{26.75}\right) \times 100\% = 0.64\%$.

$\diamond$

Differentials

The preceding ideas may be written a bit more compactly using the terminology of **differentials.** Recall that in the one-variable case the differential df for the differentiable function f is defined to be

$$df = f'(x)\, dx. \tag{10}$$

The differential in (10) provides an approximation to the change Δf in $f(x)$, corresponding to the change $\Delta x = dx$ in x, obtained by following the tangent at $(x, f(x))$ rather than the graph of f.

For precisely the same reason, we define the differential of the function f of two variables to be

$$df = \frac{\partial f}{\partial x}(x,\, y)\, dx + \frac{\partial f}{\partial y}(x,\, y)\, dy. \tag{11}$$

As Figure 5.2 illustrates, the differential df provides an approximation to the change Δf corresponding to the changes $dx = \Delta x$ in x and $dy = \Delta y$ in y, obtained by following the tangent plane at $(x, y, f(x, y))$ rather than the graph of f. The differential in equation (11) is sometimes referred to as the **total differential** for the function $z = f(x, y)$.

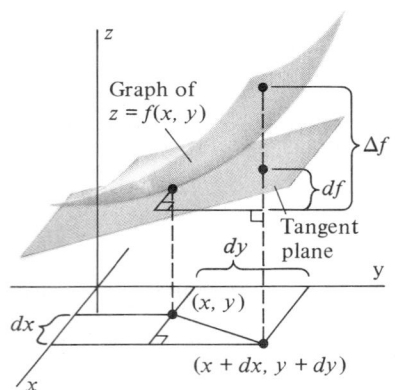

Figure 5.2 $df = f_x\, dx + f_y\, dy$ approximates change Δf by use of tangent plane.

Example 4

In economic theory, the Cobb-Douglas production function relating output production y to the input of labor L and capital K has the form

$$y = \lambda L^{\alpha} K^{1-\alpha}$$

where λ and α are positive constants. The change Δy in production resulting from a change dL in labor input and a change dK in capital input is therefore approximated by the total differential

$$dy = \frac{\partial}{\partial L}(\lambda L^{\alpha} K^{1-\alpha})\, dL + \frac{\partial}{\partial K}(\lambda L^{\alpha} K^{1-\alpha})\, dK$$

$$= \alpha \lambda L^{\alpha-1} K^{1-\alpha}\, dL + (1 - \alpha)\lambda L^{\alpha} K^{-\alpha}\, dK.$$

$\diamond$

Functions of Three Variables

The ideas of this section generalize directly to functions of three variables. The corresponding version of the Approximation Theorem (Theorem 5) is the following.

THEOREM 5'

Let f and its partial derivatives be continuous in the open rectangular box $B = \{(x, y, z) \mid a_1 < x < a_2, b_1 < y < b_2, c_1 < z < c_2\}$ in $\mathbb{R}^3$. If (x_0, y_0, z_0) and $(x_0 + \Delta x, y_0 + \Delta y, z_0 + \Delta z)$ both lie in B, then

$$f(x_0 + \Delta x, y_0 + \Delta y, z_0 + \Delta z) = f(x_0, y_0, z_0) + \frac{\partial f}{\partial x}(x_0, y_0, z_0)\Delta x$$

$$+ \frac{\partial f}{\partial y}(x_0, y_0, z_0)\Delta y + \frac{\partial f}{\partial z}(x_0, y_0, z_0)\Delta z$$

$$+ \epsilon_1 \Delta x + \epsilon_2 \Delta y + \epsilon_3 \Delta z$$

where $\epsilon_1 \to 0$, $\epsilon_2 \to 0$, and $\epsilon_3 \to 0$ as $\Delta x \to 0$, $\Delta y \to 0$, and $\Delta z \to 0$.

The proof of Theorem 5' is similar to that of Theorem 5 (see Exercise 31). Finally, the definition of the differential, suggested by Theorem 5', is

$$df = \frac{\partial f}{\partial x}(x, y, z) \, dx + \frac{\partial f}{\partial y}(x, y, z) \, dy + \frac{\partial f}{\partial z}(x, y, z) \, dz$$

or

$$dw = f_x \, dx + f_y \, dy + f_z \, dz$$

where $w = f(x, y, z)$.

Example 5

According to Newton's Law of Universal Gravitation, the force of attraction between two bodies of mass m_1 and m_2 is

$$F(m_1, m_2, r) = \frac{Gm_1m_2}{r^2},$$

where r is the distance between the bodies and G is the universal gravitation constant. The change ΔF in this force caused by changes dm_1 and dm_2 in the masses and dr in the distance is approximated by the differential

$$dF = \frac{\partial}{\partial m_1}\left(\frac{Gm_1m_2}{r^2}\right) dm_1 + \frac{\partial}{\partial m_2}\left(\frac{Gm_1m_2}{r^2}\right) dm_2 + \frac{\partial}{\partial r}\left(\frac{Gm_1m_2}{r^2}\right) dr$$

$$= \frac{Gm_2}{r^2} \, dm_1 + \frac{Gm_1}{r^2} \, dm_2 - \frac{2Gm_1m_2}{r^3} \, dr. \qquad \diamond$$

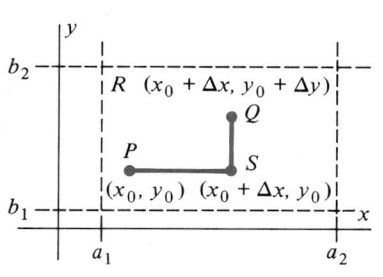

Figure 5.3 Diagram for proof of Theorem 5.

Proof of Theorem 5: Let $P = (x_0, y_0)$ and $Q = (x_0 + \Delta x, y_0 + \Delta y)$. Since R is a rectangle, the point $S = (x_0 + \Delta x, y_0)$ also lies in R, as do the line segments PS and SQ (see Figure 5.3).

Along the line segment PS the variable $y = y_0$ is fixed, so $g(x) = f(x, y_0)$ is a function of x alone, and $g'(x) = \frac{\partial f}{\partial x}(x, y_0)$ exists for all $x \in (x_0, x_0 + \Delta x)$. Thus, by the Mean Value Theorem, there exists a number $c \in (x_0, x_0 + \Delta x)$ so that

$$\frac{f(x_0 + \Delta x, y_0) - f(x_0, y_0)}{\Delta x} = \frac{\partial f}{\partial x}(c, y_0). \qquad (12)$$

Similarly, along the line segment SQ, the function $h(y) = f(x_0 + \Delta x, y)$ is a function of y alone, and $h'(y) = \dfrac{\partial f}{\partial y}(x_0 + \Delta x, y)$ for each $y \in (y_0, y_0 + \Delta y)$. Again by the Mean Value Theorem, there exists a number $d \in (y_0, y_0 + \Delta y)$ so that

$$\frac{f(x_0 + \Delta x, y_0 + \Delta y) - f(x_0 + \Delta x, y_0)}{\Delta y} = \frac{\partial f}{\partial y}(x_0 + \Delta x, d). \tag{13}$$

Next we multiply both sides of equation (12) by Δx and both sides of equation (13) by Δy. Adding the resulting equations gives

$$f(x_0 + \Delta x, y_0 + \Delta y) - f(x_0, y_0) = \frac{\partial f}{\partial x}(c, y_0)\, \Delta x \tag{14}$$

$$+ \frac{\partial f}{\partial y}(x_0 + \Delta x, d)\, \Delta y.$$

We now invoke the continuity of the partial derivatives. Since c lies between x_0 and $x_0 + \Delta x$, we have $c \to x_0$ as $\Delta x \to 0$. Thus, since $\dfrac{\partial f}{\partial x}(x, y)$ is continuous,

$$\lim_{\Delta x \to 0} \frac{\partial f}{\partial x}(c, y_0) = \frac{\partial f}{\partial x}(x_0, y_0). \tag{15}$$

Another way to write equation (15) is simply

$$\frac{\partial f}{\partial x}(c, y_0) = \frac{\partial f}{\partial x}(x_0, y_0) + \epsilon_1 \tag{16}$$

where $\epsilon_1 \to 0$ as $\Delta x \to 0$. Similarly, since d lies between y_0 and $y_0 + \Delta y$, $d \to y_0$ as $\Delta y \to 0$. Thus, since $\dfrac{\partial f}{\partial y}(x, y)$ is continuous, we conclude that

$$\lim_{\substack{\Delta x \to 0 \\ \Delta y \to 0}} \frac{\partial f}{\partial y}(x_0 + \Delta x, d) = \frac{\partial f}{\partial y}(x_0, y_0),$$

or, that

$$\frac{\partial f}{\partial y}(x_0 + \Delta x, d) = \frac{\partial f}{\partial y}(x_0, y_0) + \epsilon_2 \tag{17}$$

where $\epsilon_2 \to 0$ as $\Delta x \to 0$ and $\Delta y \to 0$.

Combining statements (14), (16), and (17) now gives

$$f(x_0 + \Delta x, y_0 + \Delta y) = f(x_0, y_0) + \frac{\partial f}{\partial x}(x_0, y_0)\, \Delta x + \frac{\partial f}{\partial y}(x_0, y_0)\, \Delta y$$

$$+ \epsilon_1\, \Delta x + \epsilon_2\, \Delta y$$

where $\epsilon_1 \to 0$ and $\epsilon_2 \to 0$ as $\Delta x \to 0$ and $\Delta y \to 0$ as desired. ◆

Exercise Set 18.5

In Exercises 1–12, find the total differential df.

1. $f(x, y) = x^2 y^4$

2. $f(x, y) = x^{2/3} y^{5/2}$

3. $f(x, y) = \sqrt{x^2 + y^4}$

4. $f(x, y) = x \sin y^2$

5. $f(x, y) = e^{\sqrt{x}} \cos y$

6. $f(x, y) = \text{Tan}^{-1}(y/x)$

7. $f(x, y, z) = x^2 y z^3$

8. $f(x, y, z) = \ln(x^2 + 3y + z^3)$

9. $f(x, y, z) = \dfrac{x}{y + 3^z}$

10. $f(x, y, z) = \sqrt{\dfrac{x}{y^2 + z^2}}$

11. $f(x, y, z) = \dfrac{x - y}{x^2 + y^2 + z^2}$

12. $f(x, y, z) = ze^{x^2 - y^3}$

13. Let $f(x, y) = 2xy^2 + x^2y$, $\quad x_0 = 2$, $\quad y_0 = 3$,
$\Delta x = 0.1$, and $\Delta y = 0.2$.
 a. Calculate $f(x_0, y_0)$ and $f(x_0 + \Delta x, y_0 + \Delta y)$.
 b. Calculate $\Delta f = f(x_0 + \Delta x, y_0 + \Delta y) - f(x_0, y_0)$.

14. *(Calculator)* Let $f(x, y) = \sin x \cos y$, $x_0 = \pi/2$, $y_0 = \pi/4$,
$\Delta x = \pi/16$, and $\Delta y = -\pi/16$. Answer the questions in Exercise 13.

In Exercises 15–22, use the total differential to approximate the given number.

15. $\sqrt{(3.02)^2 + (4.08)^2}$

16. $\sqrt{9.4} + \sqrt{15.6}$

17. $\sin 88° \cos 42°$

18. $(2.02)^3 + (3.96)^3$

19. $(5.03)^2(1.02)^3$

20. $(5.94) \ln(1.15)$

21. $(6.04)(3.1)(2.96)$

22. $\sqrt{(12.1)^2 + (4.05)^2 + (2.96)^2}$

23. A cylindrical bearing of radius 2 cm and length 6 cm is dipped in a molten brass solution to produce a brass coating. If the thickness of the coating varies from 0.02 cm to 0.06 cm, find an approximation for the volume of the coated bearing.

24. The money supply in the economy is defined by the equation $M = C + D$ where C is the currency outstanding in banks and D is the amount of money in demand and term deposits. Use differentials to approximate the percentage change in the money supply caused by a 10% increase in the outstanding currency and a 5% decrease in total deposits.

25. For the Cobb-Douglas production function $P = \lambda L^3 K^{3/2}$, use differentials to approximate the percentage change in P

caused by a 4% increase in labor L and a 10% decrease in capital K.

26. For the Cobb-Douglas production function $P = \lambda L^\alpha K^\beta$, what can you say about α and β if a percentage decrease in labor of γ%, combined with a percentage increase in capital of γ%, produces no change in P?

27. For small oscillations, the period of a simple pendulum is given by the equation

$$T = 2\pi \sqrt{\dfrac{s}{g}}$$

where s is the length of the pendulum, g is the acceleration due to gravity, and T is time. (If the units for s and g involve meters and seconds, the units for T are seconds.) What is the approximate error in the calculation of T resulting from its calculation using $s = 10$ m and $g = 9.8$ m/s^2 if the actual values of these quantities are $s = 10.2$ and $g = 9.75$?

28. Approximate $f(x_0, y_0, z_0)$ for $f(x, y, z) = \dfrac{\sqrt{x^2 - y^2}}{\tan z}$ and
$(x_0, y_0, z_0) = (5.04, 3.97, 0.78)$. (*Hint:* $\dfrac{\pi}{4} \approx 0.7853982$.)

29. When three resistors with resistances R_1, R_2, and R_3 are connected in parallel, the resistance of the resulting circuit is

$$\dfrac{1}{R} = \dfrac{1}{R_1} + \dfrac{1}{R_2} + \dfrac{1}{R_3}.$$

What is the maximum percentage change in R resulting from a change of no more than 10% in each of R_1, R_2, and R_3?

30. The work involved in an adiabatic (no heat lost to the surroundings) expansion of an ideal gas is given by the equation

$$w = nC_v(T_2 - T_1)$$

where C_v is the average heat capacity of the gas, and T_1 and T_2 are the initial and final temperatures. Show that if both T_1 and T_2 are increased by λ percent, the amount of work involved is also increased by λ percent.

31. Prove Theorem 5′, using the Mean Value Theorem three times.

18.6 CHAIN RULES

In this section, we use the Approximation Theorem to establish a generalization of the Chain Rule for functions of a single variable:

$$(f \circ g)'(x) = f'(g(x))g'(x).$$

Our generalization concerns the *composite* function

$$w(t) = f(x(t), y(t), z(t)), \qquad t \in (a, b).$$

It is important to note that this composite function is just a function of a single variable. Our interest is in knowing when the derivative $w'(t)$ exists and, if it does, how it may be calculated.

Before stating the theorem that answers this question, we describe a simple setting in which a composite function of this type occurs. Think of a region of airspace, such as that over the state of California, through which an airplane is about to fly. The region of airspace may be coordinatized with xyz-coordinates (which you may think of as representing latitude, longitude, and altitude, if you prefer), and the path of the airplane may be thought of as a curve parameterized by the vector function $r(t) = x(t)i + y(t)j + z(t)k$, $a \le t \le b$ giving the location of the airplane at time t. (See Figure 6.1.) Assuming a steady state temperature distribution through-

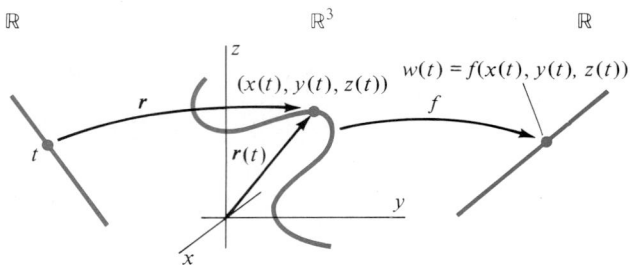

Figure 6.1 If $r: \mathbb{R} \to \mathbb{R}^3$ and $f: \mathbb{R}^3 \to \mathbb{R}$, the composite function $w(t) = f(r(t)) = f(x(t), y(t), z(t))$ maps $\mathbb{R}$ into $\mathbb{R}$.

out the region, we let $T(x, y, z)$ be the temperature at location (x, y, z) in the airspace. *The composite function*

$$w(t) = T(x(t), y(t), z(t))$$

gives the temperature outside the airplane at time t.

What factors determine the rate, $\dfrac{dw}{dt}$, at which the temperature outside the airplane changes? Obviously $\dfrac{dx}{dt}$, $\dfrac{dy}{dt}$, and $\dfrac{dz}{dt}$ are involved, since they are the components of the velocity of the airplane. But $\dfrac{dw}{dt}$ will also depend upon the partial derivatives $\dfrac{\partial T}{\partial x}$, $\dfrac{\partial T}{\partial y}$, and $\dfrac{\partial T}{\partial z}$, since these determine the rate of change of temperature with respect to change in position. The precise relationship between these rates is given by the following theorem.

THEOREM 6
Chain Rule

Let f be a continuous function with continuous partial derivatives for all (x, y, z) in an open "box"

$$Q = \{(x, y, z) \mid a_1 < x < b_1, a_2 < y < b_2, a_3 < z < b_3\}$$

in space. Assume that x, y, and z are functions of t so that $x'(t)$, $y'(t)$, and $z'(t)$ exist for each $t \in (a, b)$, and so that the point $(x(t), y(t), z(t))$ lies in Q for each $t \in (a, b)$. Then the composite function

$$w(t) = f(x(t), y(t), z(t))$$

Graphic Representations of
Mathematical Surfaces in Three Dimensions

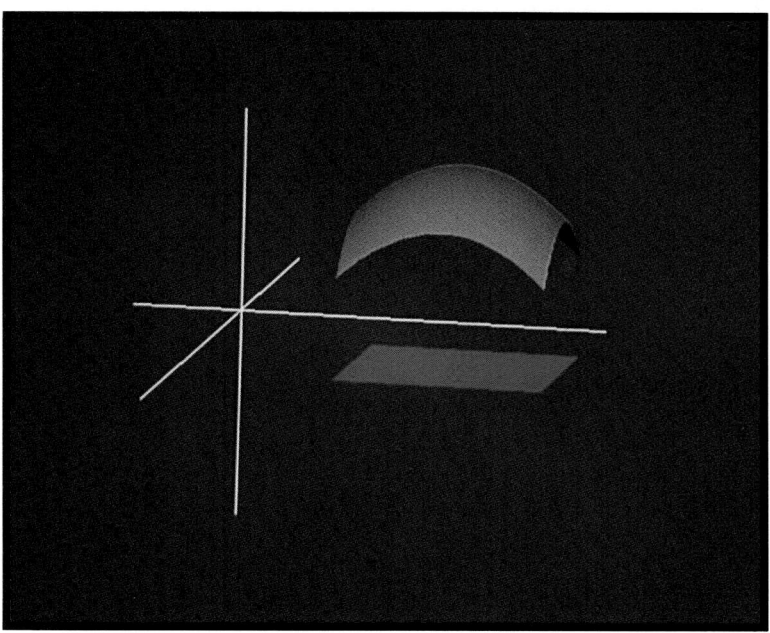

PLATE 1 Graph of a function $z = f(x, y)$ of two variables with domain D.

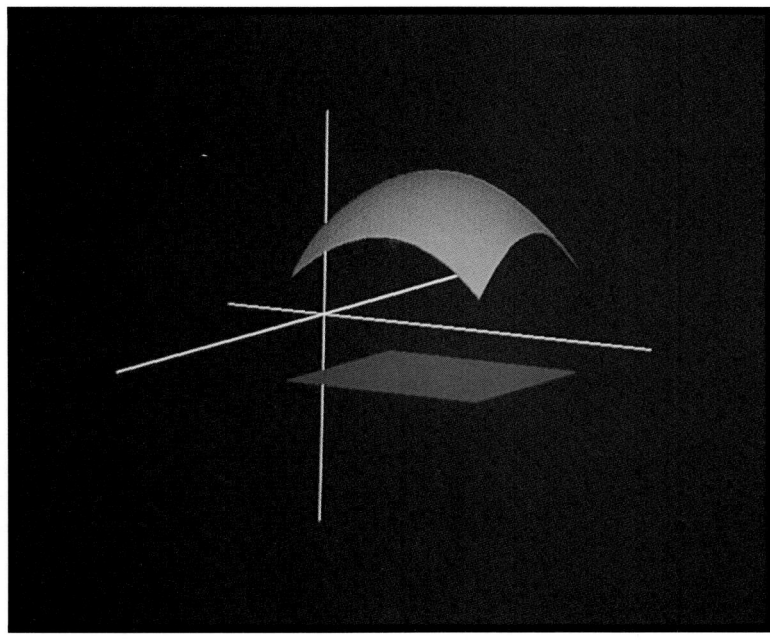

PLATE 2 Figure 18.1.1 (Plate 1) rotated.

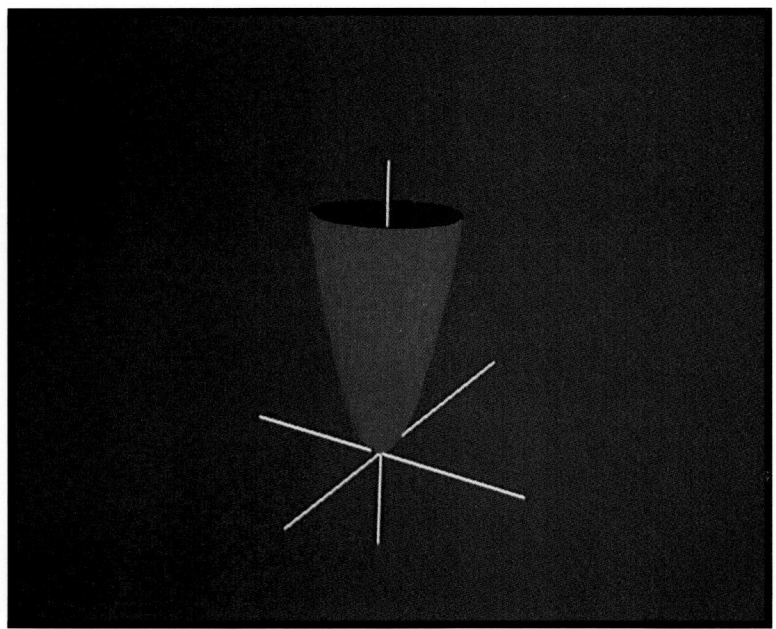

PLATE 3 Paraboloid $z = x^2 + y^2$ is the graph of a function of two variables.

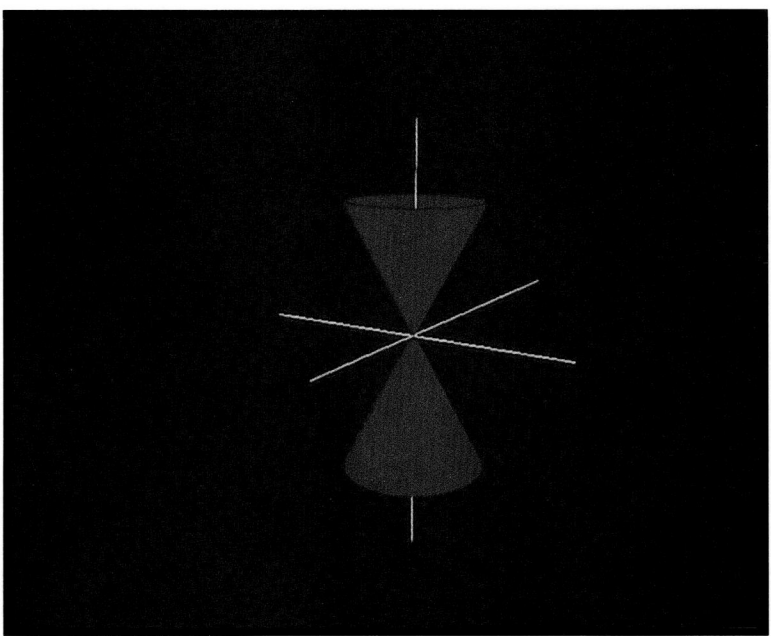

PLATE 4 A cone is the graph of the equation $z^2 = x^2 + y^2$.

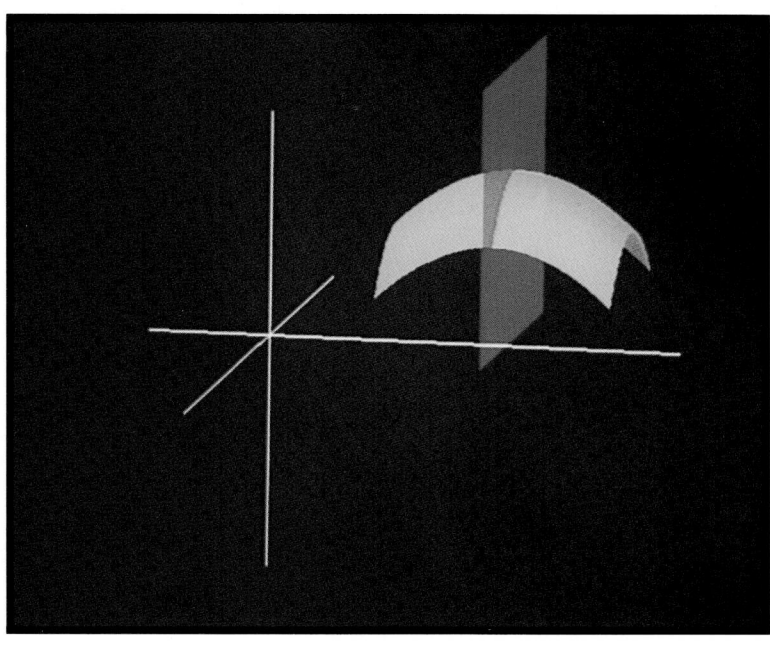

PLATE 5 Partial derivative $\dfrac{\partial f}{\partial x}(x_0, y_0)$ is the slope of the line tangent to the trace of $z = f(x, y)$ in the plane $y = y_0$.

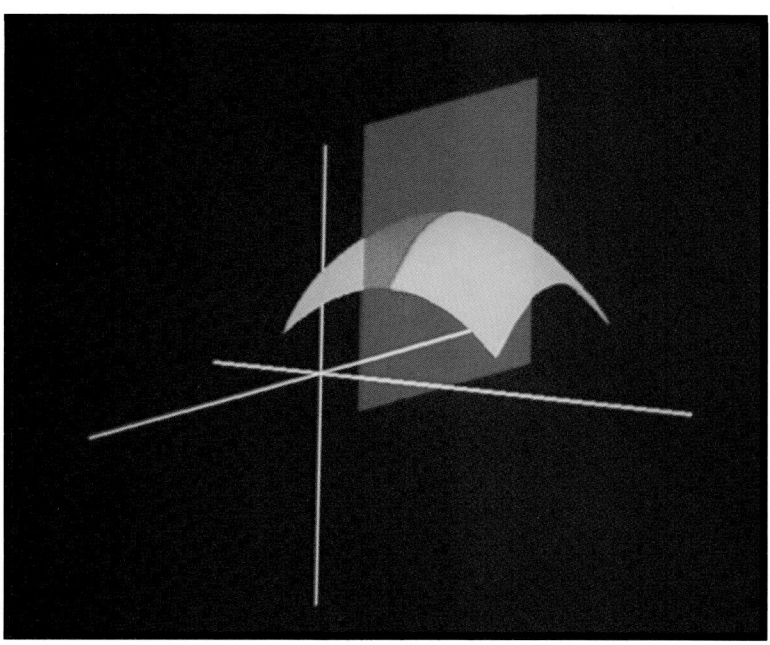

PLATE 6 Figure 18.2.1 (Plate 5) rotated.

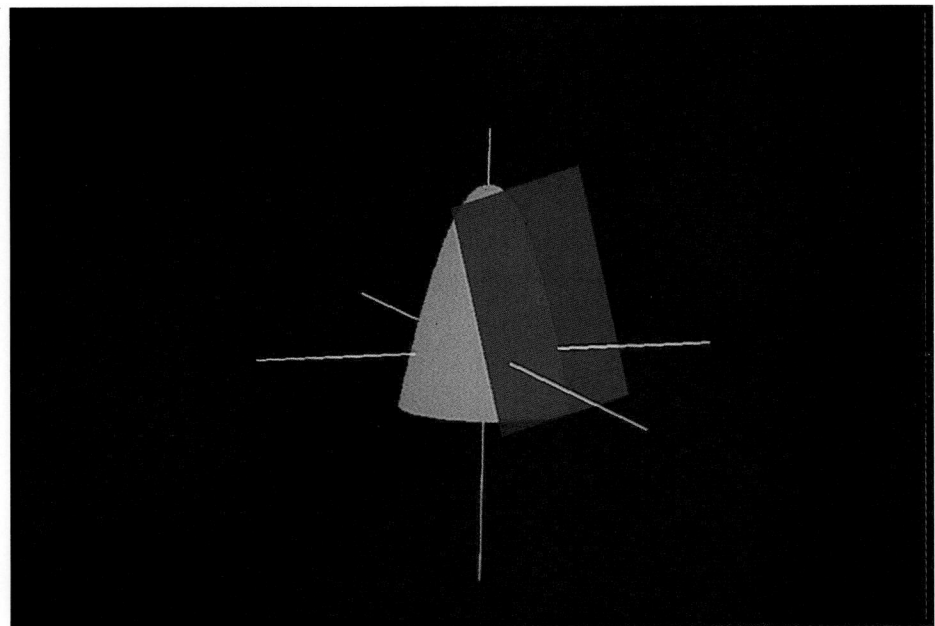

PLATE 7   Plane tangent to the graph of $y = 9 - x^2 - y^2$.

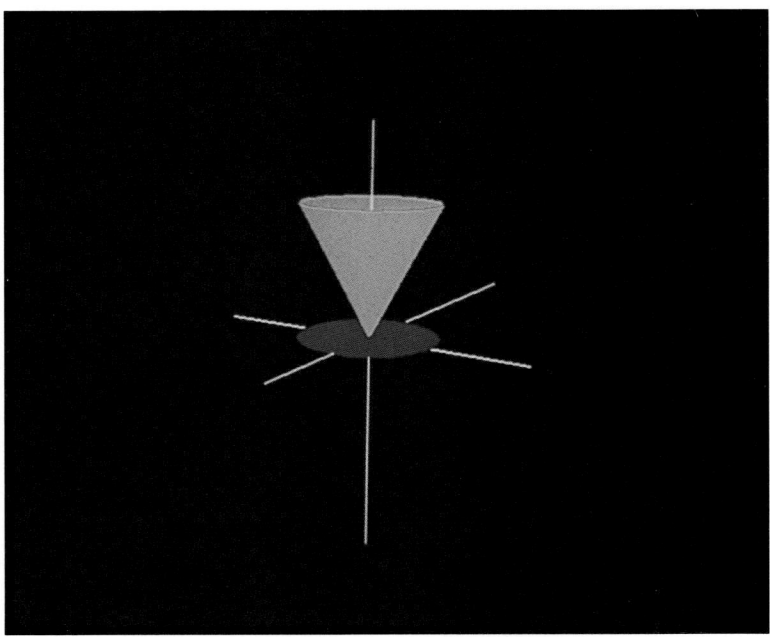

PLATE 8   Relative minimum for $z = \sqrt{x^2 + y^2}$ is at the vertex of a cone.

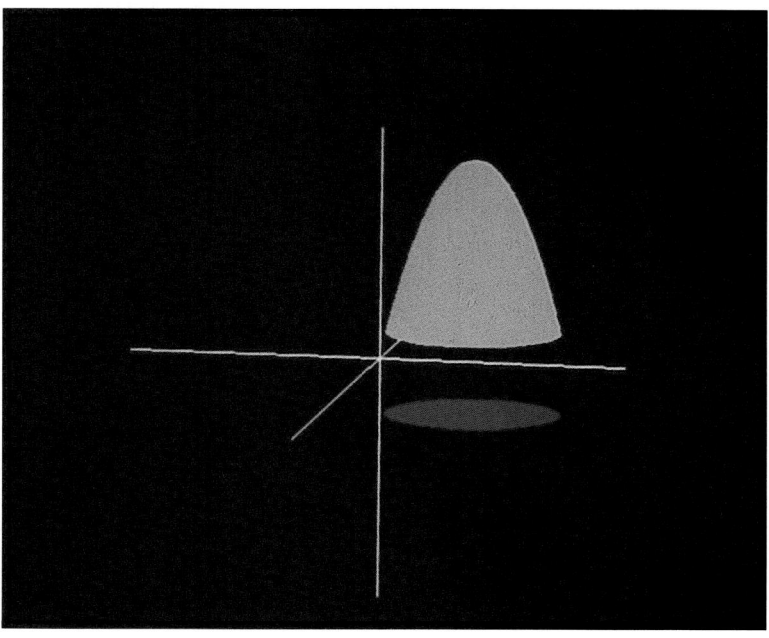

PLATE 9   Graph of a paraboloid $z = 2x + 4y - x^2 - y^2 - 1$ over a circle.

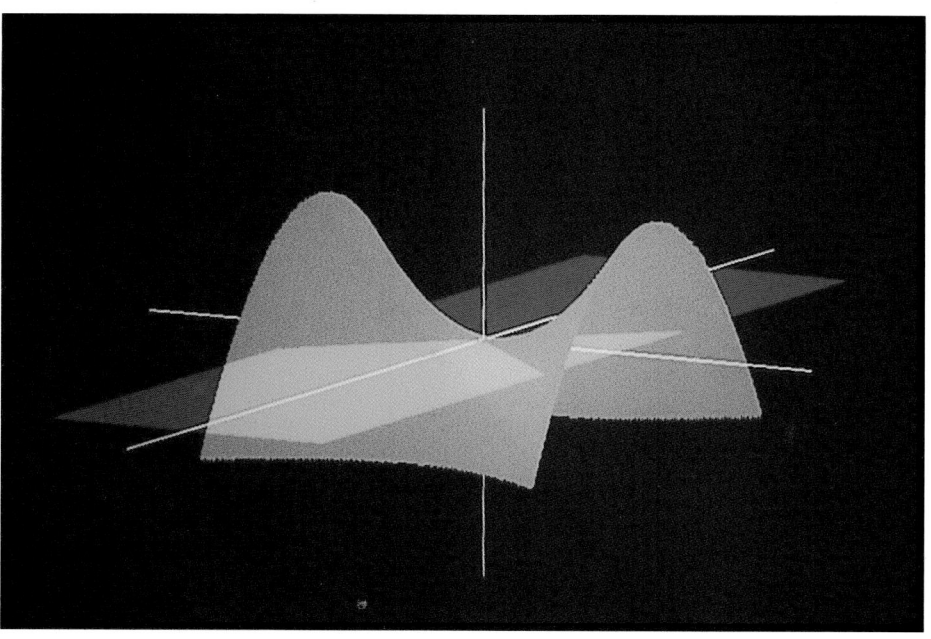

PLATE 10   Plane tangent to the graph of $z = y^2 - x^2$ at the saddle point.

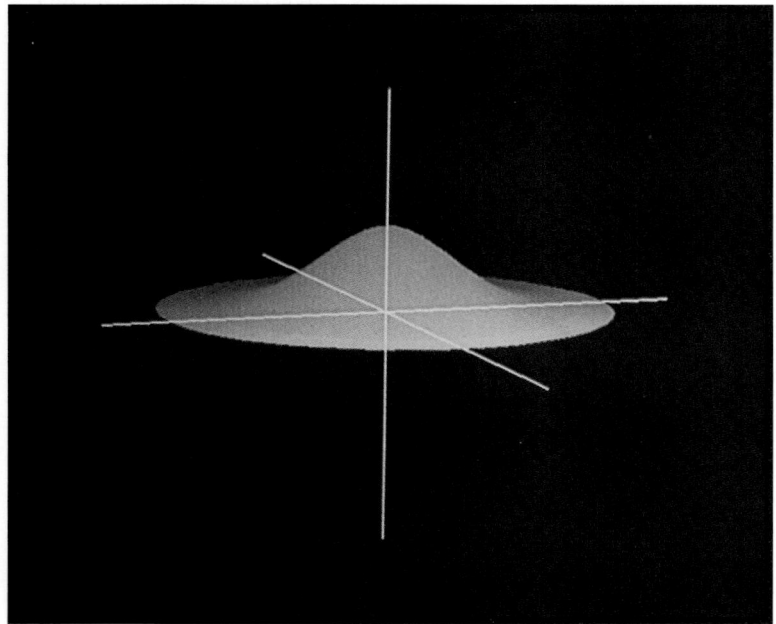

PLATE 11   Graph of $f(x, y) = e^{-(x^4 + y^4)}$.

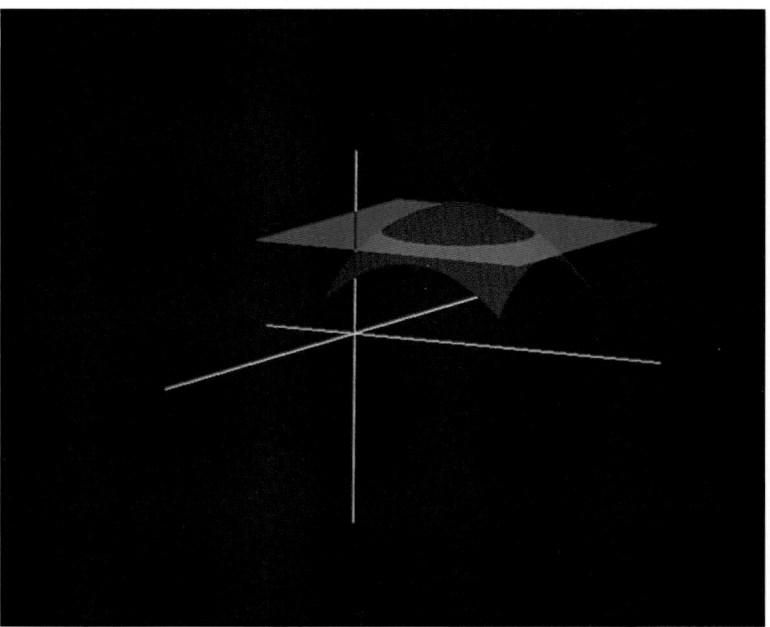

PLATE 12   Horizontal plane intersects the graph in the *level curve*.

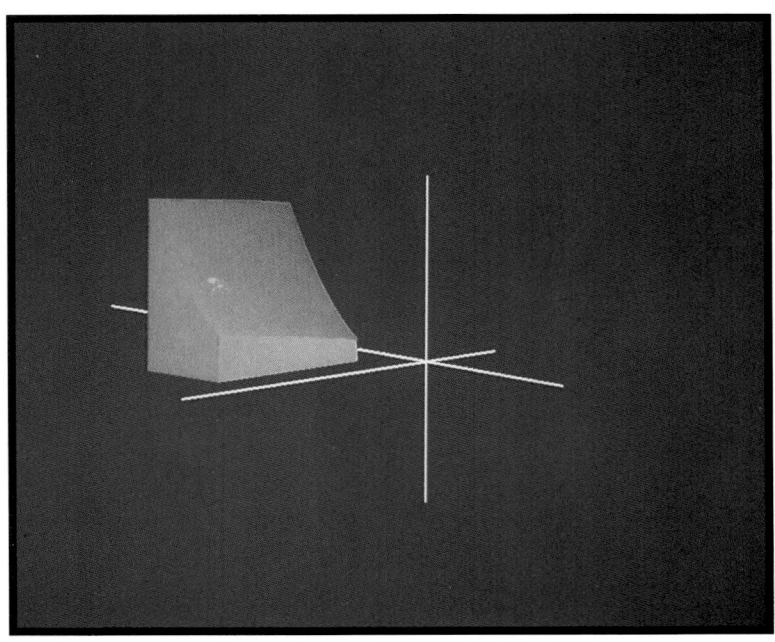

PLATE 13   Solid bounded by the graph of $f(x, y) = x^2 + 2y$ and xy plane.

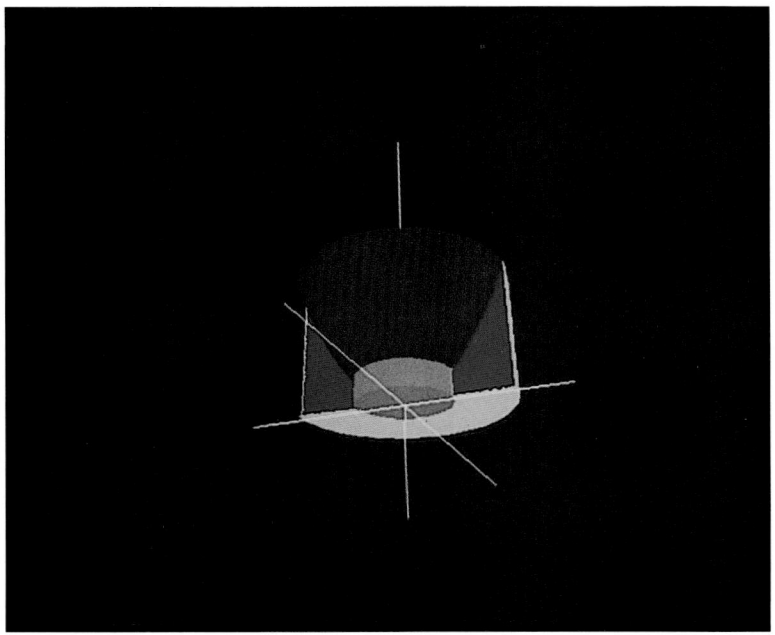

PLATE 14   One half of the bearing sleeve of Example 3, page 873.

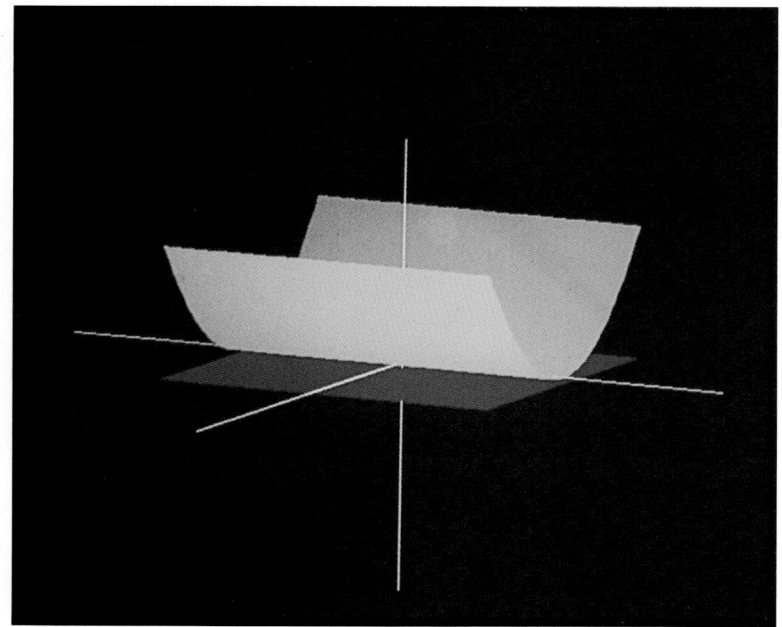

PLATE 15 Graph of $f(x, y) = x^2$ is a *cylinder*.

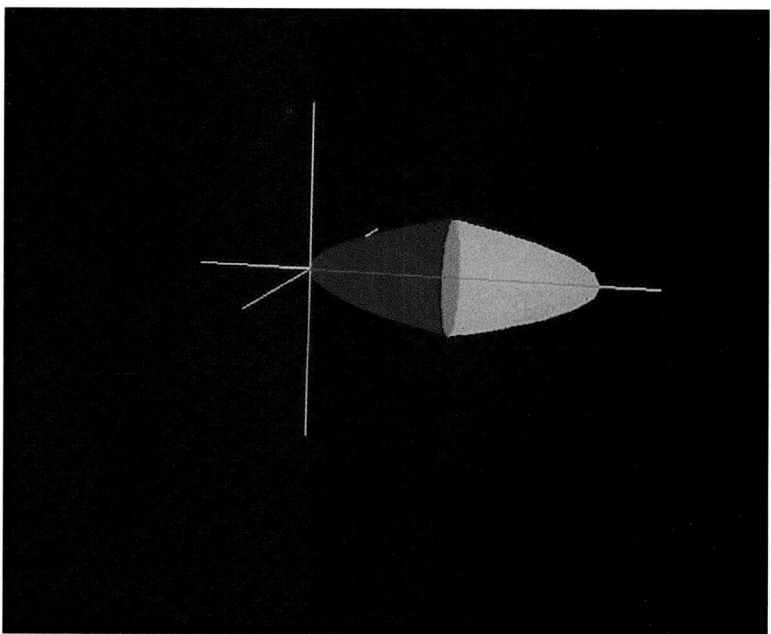

PLATE 16 Paraboloids intersecting in a circle.

is a differentiable function of $t \in (a, b)$, and

$$\frac{dw}{dt} = \frac{\partial f}{\partial x} \cdot \frac{dx}{dt} + \frac{\partial f}{\partial y} \cdot \frac{dy}{dt} + \frac{\partial f}{\partial z} \cdot \frac{dz}{dt}. \tag{1}$$

The proof of Theorem 6 is given at the end of this section.

Example 1

Let $f(x, y, z) = \sqrt{x}\, y^2 e^{2z}$, $x(t) = 3t^2 + 2$, $y(t) = 6t$, $z(t) = 1 - t^3$, and $w(t) = f(x(t), y(t), z(t))$. Find $w'(t)$.

Solution: We have

$$\frac{\partial f}{\partial x} = \frac{\partial}{\partial x}(\sqrt{x}\, y^2 e^{2z}) = \frac{y^2 e^{2z}}{2\sqrt{x}}$$

$$\frac{\partial f}{\partial y} = \frac{\partial}{\partial y}(\sqrt{x}\, y^2 e^{2z}) = 2\sqrt{x}\, y e^{2z}$$

and

$$\frac{\partial f}{\partial z} = \frac{\partial}{\partial z}(\sqrt{x}\, y^2 e^{2z}) = 2\sqrt{x}\, y^2 e^{2z}.$$

Also,

$$x'(t) = 6t, \qquad y'(t) = 6, \qquad \text{and} \qquad z'(t) = -3t^2.$$

According to Theorem 6 we must have

$$w'(t) = \frac{y^2 e^{2z}}{2\sqrt{x}}(6t) + 2\sqrt{x}\, y e^{2z}(6) + 2\sqrt{x}\, y^2 e^{2z}(-3t^2)$$

$$= \frac{108 t^3 e^{2(1-t^3)}}{\sqrt{3t^2 + 2}} + 72t\sqrt{3t^2 + 2}\; e^{2(1-t^3)} - 216t^4\sqrt{3t^2 + 2}\; e^{2(1-t^3)}$$

$$= \sqrt{3t^2 + 2}\; e^{2(1-t^3)}\left[\frac{108 t^3}{3t^2 + 2} + 72t - 216t^4\right]. \qquad \diamond$$

Example 2

A function giving the temperature $T(x, y, z)$ at a point (x, y, z) in space is

$$T(x, y, z) = \lambda\sqrt{x^2 + y^2 + z^2}$$

where λ is constant.

Find the rate of change of temperature with respect to t along the elliptical helix

$$r(t) = a \cos t\, i + b \sin t\, j + ct\, k.$$

Solution: The temperature function along the helix is the composite function $T(x(t), y(t), z(t))$ where

$$x(t) = a \cos t, \qquad y(t) = b \sin t, \qquad z(t) = ct.$$

Using equation (1), we obtain the desired rate as

$$\frac{dT}{dt} = \frac{\partial}{\partial x}(\lambda\sqrt{x^2 + y^2 + z^2}) \cdot \frac{d}{dt}(a \cos t)$$

$$+ \frac{\partial}{\partial y}(\lambda\sqrt{x^2 + y^2 + z^2}) \cdot \frac{d}{dt}(b \sin t) + \frac{\partial}{\partial z}(\lambda\sqrt{x^2 + y^2 + z^2}) \cdot \frac{d}{dt}(ct)$$

$$= \frac{\lambda x(-a \sin t)}{\sqrt{x^2 + y^2 + z^2}} + \frac{\lambda yb \cos t}{\sqrt{x^2 + y^2 + z^2}} + \frac{\lambda cz}{\sqrt{x^2 + y^2 + z^2}}$$

$$= \frac{\lambda(-a^2 + b^2) \sin t \cos t + \lambda c^2 t}{\sqrt{a^2 \cos^2 t + b^2 \sin^2 t + c^2 t^2}}.$$ ◇

Example 3

The **electric potential** at an axial point for a charged disc is given by the equation

$$V = \frac{\sigma}{2\epsilon_0}(\sqrt{a^2 + r^2} - r)$$

where a is the radius of the disc, r is the distance from the point to the disc, and σ and ϵ_0 are constants. If the radius of the disc is increasing at a rate of 2 cm/s and the point is moving away from the disc at a rate of 5 cm/s, find the rate at which the electric potential is changing when $a = 3$ cm and $r = 4$ cm.

Solution: Since V is a function of only two variables, a and r, we simply consider the third variable in the Chain Rule (1) to be $z = 0$, so $\frac{dz}{dt} = 0$. Applying the Chain Rule we find

$$\frac{dV}{dt} = \frac{\partial}{\partial a}\left[\frac{\sigma}{2\epsilon_0}(\sqrt{a^2 + r^2} - r)\right]\frac{da}{dt} \qquad (2)$$

$$+ \frac{\partial}{\partial r}\left[\frac{\sigma}{2\epsilon_0}(\sqrt{a^2 + r^2} - r)\right]\frac{dr}{dt}$$

$$= \frac{\sigma}{2\epsilon_0}\left(\frac{a}{\sqrt{a^2 + r^2}}\right)\frac{da}{dt} + \frac{\sigma}{2\epsilon_0}\left(\frac{r}{\sqrt{a^2 + r^2}} - 1\right)\frac{dr}{dt}.$$

We are given

$$\frac{da}{dt} = 2 \text{ cm/s}, \qquad \frac{dr}{dt} = 5 \text{ cm/s}, \qquad a = 3 \text{ cm, and } r = 4 \text{ cm.}$$

Substituting these values into (2) gives

$$\frac{dV}{dt} = \frac{\sigma}{2\epsilon_0}\left(\frac{3}{\sqrt{3^2 + 4^2}}\right)(2) + \frac{\sigma}{2\epsilon_0}\left(\frac{4}{\sqrt{3^2 + 4^2}} - 1\right)(5)$$

$$= \frac{\sigma}{10\epsilon_0}.$$ ◇

Differentiating Other Types of Composite Functions

Various other sorts of composite functions can be formed using functions of several variables, each of which may be differentiated using the Chain Rule. For example, if

$$x = x(s, t), \qquad y = y(s, t) \qquad \text{and} \qquad z = z(s, t)$$

are functions of two variables, and if

$$w = f(x, y, z)$$

is a function of three variables, the composite function

$$w(s, t) = f(x(s, t), y(s, t), z(s, t))$$

is again a function of two variables. To obtain $\dfrac{\partial w}{\partial s}$, we hold t constant and differentiate the resulting function of a single variable $g(s) = w(s, t)$ according to Theorem 6. Similarly, the partial derivative $\dfrac{\partial w}{\partial t}$ is obtained by holding s constant and differentiating with respect to t.

Example 4

Let $f(x, y) = x^2 y^3$ where the variables x and y are functions of the polar variables r and θ:

$$x(r, \theta) = r^2 \cos \theta, \qquad y(r, \theta) = r(1 - \sin \theta).$$

To find the partial derivative $\dfrac{\partial f}{\partial r}$, we treat θ as a constant, leaving x and y as functions of the single independent variable r, and apply Theorem 6:

$$\frac{\partial f}{\partial r} = \frac{\partial f}{\partial x} \cdot \frac{\partial x}{\partial r} + \frac{\partial f}{\partial y} \cdot \frac{\partial y}{\partial r}$$

$$= \left[\frac{\partial}{\partial x}(x^2 y^3) \right] \left[\frac{\partial}{\partial r}(r^2 \cos \theta) \right] + \left[\frac{\partial}{\partial y}(x^2 y^3) \right] \left[\frac{\partial}{\partial r}(r(1 - \sin \theta)) \right]$$

$$= 2xy^3 \cdot 2r \cos \theta + 3x^2 y^2 (1 - \sin \theta)$$

$$= 7r^6 \cos^2 \theta (1 - \sin \theta)^3.$$

Similarly,

$$\frac{\partial f}{\partial \theta} = \frac{\partial f}{\partial x} \cdot \frac{\partial x}{\partial \theta} + \frac{\partial f}{\partial y} \cdot \frac{\partial y}{\partial \theta}$$

$$= 2xy^3(-r^2 \sin \theta) + (3x^2 y^2)(-r \cos \theta)$$

$$= r^7 \cos \theta (1 - \sin \theta)^2 [2 \sin^2 \theta - 2 \sin \theta - 3 \cos^2 \theta]. \qquad \diamondsuit$$

Higher Order Derivatives

Using Theorem 6 we can work out formulas for higher order ordinary and partial derivatives of composite functions. It is usually simpler, however, to substitute directly for the independent variables after the first order derivatives are found, and then proceed to the second order derivatives by a second application of Theorem 6, as the following example illustrates.

Example 5

Let $f(x, y) = x^2 + 2xy$, $x = r \cos \theta$, and $y = r \sin \theta$. Find (a) $\dfrac{\partial^2 f}{\partial r^2}$ and (b) $\dfrac{\partial^2 f}{\partial \theta^2}$.

Solution: (a) Since $\dfrac{\partial^2 f}{\partial r^2} = \dfrac{\partial}{\partial r}\left(\dfrac{\partial f}{\partial r}\right)$, we begin by finding

$$\frac{\partial f}{\partial r} = \frac{\partial f}{\partial x}\cdot\frac{\partial x}{\partial r} + \frac{\partial f}{\partial y}\cdot\frac{\partial y}{\partial r}$$

$$= \frac{\partial}{\partial x}(x^2 + 2xy)\cdot\frac{\partial}{\partial r}(r\cos\theta) + \frac{\partial}{\partial y}(x^2 + 2xy)\cdot\frac{\partial}{\partial r}(r\sin\theta)$$

$$= (2x + 2y)(\cos\theta) + (2x)(\sin\theta)$$
$$= (2r\cos\theta + 2r\sin\theta)(\cos\theta) + (2r\cos\theta)(\sin\theta)$$
$$= 2r\cos^2\theta + 4r\sin\theta\cos\theta.$$

Thus

$$\frac{\partial^2 f}{\partial r^2} = \frac{\partial}{\partial r}(2r\cos^2\theta + 4r\cos\theta\sin\theta)$$

$$= 2\cos^2\theta + 4\sin\theta\cos\theta.$$

(b) Similarly, you can verify that

$$\frac{\partial^2 f}{\partial\theta^2} = 2r^2[\sin^2\theta - 4\sin\theta\cos\theta - \cos^2\theta]. \qquad \diamondsuit$$

Proof of Theorem 6: Let $t_0 \in (a, b)$ and let $\Delta t \neq 0$ be sufficiently small that $(t_0 + \Delta t)$ also lies in (a, b). Let

$$\begin{array}{lll}
x_0 = x(t_0) & & \Delta x = x(t_0 + \Delta t) - x(t_0) \\
y_0 = y(t_0) & & \Delta y = y(t_0 + \Delta t) - y(t_0) \\
z_0 = z(t_0), & \text{and} & \Delta z = z(t_0 + \Delta t) - z(t_0).
\end{array}$$

Then, according to the Approximation Theorem (Theorem 5'),

$$w(t_0 + \Delta t) = w(t_0) + \frac{\partial f}{\partial x}(x_0, y_0, z_0)\,\Delta x + \frac{\partial f}{\partial y}(x_0, y_0, z_0)\,\Delta y \qquad (3)$$

$$+ \frac{\partial f}{\partial z}(x_0, y_0, z_0)\,\Delta z + \epsilon_1\Delta x + \epsilon_2\Delta y + \epsilon_3\Delta z$$

where $\epsilon_1 \to 0$, $\epsilon_2 \to 0$, and $\epsilon_3 \to 0$ as $\Delta x \to 0$, $\Delta y \to 0$, and $\Delta z \to 0$.

Subtracting $w(t_0)$ from both sides of equation (3) and dividing by Δt gives the equation

$$\frac{w(t_0 + \Delta t) - w(t_0)}{\Delta t} = \frac{\partial f}{\partial x}(x_0, y_0, z_0)\frac{\Delta x}{\Delta t} + \frac{\partial f}{\partial y}(x_0, y_0, z_0)\frac{\Delta y}{\Delta t} \qquad (4)$$

$$+ \frac{\partial f}{\partial z}(x_0, y_0, z_0)\frac{\Delta z}{\Delta t} + \epsilon_1\frac{\Delta x}{\Delta t} + \epsilon_2\frac{\Delta y}{\Delta t} + \epsilon_3\frac{\Delta z}{\Delta t}.$$

Now according to our definitions, we have

$$\lim_{\Delta t \to 0}\frac{\Delta x}{\Delta t} = \lim_{\Delta t \to 0}\frac{x(t_0 + \Delta t) - x(t_0)}{\Delta t} = x'(t_0).$$

Thus,

$$\lim_{\Delta t \to 0}\left[\frac{\partial f}{\partial x}(x_0, y_0, z_0)\frac{\Delta x}{\Delta t}\right] \qquad (5)$$

$$= \frac{\partial f}{\partial x}(x_0, y_0, z_0)\cdot\lim_{\Delta t \to 0}\frac{\Delta x}{\Delta t} = \frac{\partial f}{\partial x}(x_0, y_0, z_0)x'(t_0)$$

and

$$\lim_{\Delta t \to 0} \epsilon_1 \cdot \frac{\Delta x}{\Delta t} = \left(\lim_{\Delta t \to 0} \epsilon_1 \right) \left(\lim_{\Delta t \to 0} \frac{\Delta x}{\Delta t} \right) = 0 \cdot x'(t_0) = 0. \tag{6}$$

Equations analogous to (5) and (6) hold for the variables y and z.

Finally, from equation (4) and the equations of the form (5) and (6) for each of the variables x, y, and z, we conclude that

$$w'(t_0) = \lim_{\Delta t \to 0} \frac{w(t_0 + \Delta t) - w(t_0)}{\Delta t}$$

$$= \frac{\partial f}{\partial x}(x_0, y_0, z_0)x'(t_0) + \frac{\partial f}{\partial y}(x_0, y_0, z_0)y'(t_0)$$

$$+ \frac{\partial f}{\partial z}(x_0, y_0, z_0)z'(t_0),$$

as required. ◆

Exercise Set 18.6

In Exercises 1–10, find the rate of change of f with respect to t, $\dfrac{df}{dt}$, along the given curves.

1. $f(x, y) = x^2 + y^2$, $\quad x(t) = 2t$, $\quad y(t) = 6 - t^2$

2. $f(x, y) = x - y^2$, $\quad x(t) = 3t^2 + 2$, $\quad y(t) = 4 + t$

3. $f(x, y) = xy^2$, $\quad r(t) = \cos t\mathbf{i} + \sin t\mathbf{j}$

4. $f(x, y) = x^2 - y^2$, $\quad r(t) = e^t\mathbf{i} + e^{-t}\mathbf{j}$

5. $f(x, y) = \sqrt{x + y}$, $\quad x(t) = \sqrt{t}$, $\quad y(t) = e^{t^2}$

6. $f(x, y) = y^2 - x^2$, $\quad x(t) = a \cos t$, $\quad y(t) = a \sin t$

7. $f(x, y, z) = xy - x + z^2$, $\quad r(t) = t^2\mathbf{i} - 2t\mathbf{j} + \sin t\mathbf{k}$

8. $f(x, y, z) = x^2 + y^2 + z^2$,
$\quad r(t) = a \cos \pi t\mathbf{i} + b \sin \pi t\mathbf{j} - t^2\mathbf{k}$

9. $f(x, y, z) = z(y^2 - x^2)$, $\quad x(t) = a \cosh t$,
$\quad y(t) = b \sinh t$, $\quad z(t) = e^{-2t}$

10. $f(x, y, z) = e^{xyz}$, $\quad x(t) = \ln \sqrt{t}$, $\quad y(t) = t \sin t$,
$\quad z(t) = 2^t$.

In Exercises 11–22, find the indicated derivative(s).

11. $f(x, y) = x^2y^3$, $\quad x(t) = \cos t$, $\quad y(t) = t \sin t$.
Find $\dfrac{df}{dt}$.

12. $f(x, y) = x^2y^3$, $\quad x(s, t) = st$, $\quad y(s, t) = s^2 - t^2$.
Find
a. $\dfrac{\partial f}{\partial s}$ **b.** $\dfrac{\partial f}{\partial t}$

13. $f(x, y, z) = x^2 + y^2 - z^2$, $\quad x(s, t) = e^{st}$,
$\quad y(s, t) = st$, $\quad z(s, t) = s - t$. Find
a. $\dfrac{\partial f}{\partial s}$ **b.** $\dfrac{\partial f}{\partial t}$

14. $f(x, y) = xy - y^2$, $\quad x(r, s, t) = rst$, $\quad y(r, s, t) = e^{rst}$.
Find
a. $\dfrac{\partial f}{\partial r}$ **b.** $\dfrac{\partial f}{\partial s}$ **c.** $\dfrac{\partial f}{\partial t}$

15. $f(x, y) = \sin(xy^2) - x^2y$, $\quad x(s, t) = s^2 - st$,
$\quad y(s, t) = t^2s^2$. Find
a. $\dfrac{\partial f}{\partial s}$ **b.** $\dfrac{\partial f}{\partial t}$

16. $f(x, y) = x^2e^{y-x}$, $\quad x(s, t) = s - t$, $\quad y(s, t) = \sqrt{s + t}$.
Find
a. $\dfrac{\partial f}{\partial s}$ **b.** $\dfrac{\partial f}{\partial t}$

17. $f(x, y, z) = xy^3z^2$, $\quad x(s, t) = s \sin t$,
$\quad y(s, t) = t \cos(s)$, $\quad z(s, t) = t^2 - s^2$. Find
a. $\dfrac{\partial f}{\partial s}$ **b.** $\dfrac{\partial f}{\partial t}$

18. $f(x, y) = e^{xy}$, $\quad x(r, s) = \sqrt{s^2 - r^2}$,
$\quad y(r, s) = \text{Tan}^{-1}\left(\dfrac{r}{s}\right)$. Find
a. $\dfrac{\partial f}{\partial r}$ **b.** $\dfrac{\partial f}{\partial s}$

19. $f(u, v) = e^{u^2} - e^{2v}$, $\quad u(x, y) = xy^2$, $\quad v(x, y) = x^2y$.
Find
a. $\dfrac{\partial f}{\partial x}$ **b.** $\dfrac{\partial f}{\partial y}$

20. $f(x, y, z) = x^2 - xy + yz$, $\quad x(r, \theta) = r \cos \theta$,
$\quad y(r, \theta) = 2r$, $\quad z(r, \theta) = r \sin \theta$. Find
a. $\dfrac{\partial f}{\partial r}$ **b.** $\dfrac{\partial f}{\partial \theta}$

21. $f(r, \theta) = r^2(1 - \cos \theta), \qquad r(t) = 1 + t^3,$

$\theta(t) = \sqrt{1 + t^2}$. Find $\dfrac{df}{dt}$.

22. $f(x, y, z) = \ln(x^2 + y^2 + z^2), \qquad x(u, v, w) = u \cos v,$

$y(u, v, w) = v \sin u, \qquad z(u, v, w) = uvw$. Find

a. $\dfrac{\partial f}{\partial u}$ **b.** $\dfrac{\partial f}{\partial v}$ **c.** $\dfrac{\partial f}{\partial w}$

23. The radius of the base of a cone is 6 cm and it is increasing at a rate of 2 cm/s. The height of the cone is 10 cm and it is increasing at a rate of 3 cm/s. At what rate is the volume increasing?

24. The radius of a cylinder is 8 cm and it is increasing at a rate of 2 cm/s. The height of the cylinder is 20 cm and it is increasing at a rate of 4 cm/s.
   a. Find the rate at which the volume is increasing.
   b. Find the rate at which the lateral surface area is increasing.

25. A particle moves along a helix with position function $r(t) = \cos t\, \mathbf{i} + \sin t\, \mathbf{j} + t\mathbf{k}$.
   a. Find the function $D(x, y, z)$ giving the distance from the particle to the origin.
   b. Find $\dfrac{dD}{dt}$, the rate at which the distance from the particle to the origin changes as a function of time.
   c. Show that $\lim\limits_{t \to \infty} D'(t) = 1$; $\lim\limits_{t \to -\infty} D'(t) = -1$.

26. In economic theory, the **rate of growth** (as opposed to rate of change) of a function f is defined to be the ratio

$$\frac{f'(t)}{f(t)} = \frac{d}{dt}[\ln(f(t))] = \frac{\text{marginal function}}{\text{total function}}.$$

Show, according to this definition, that if the rate of growth of consumption C is a, and if the rate of growth of population P is b, then the rate of growth of per capita consumption, $R = \dfrac{C}{p}$, is $a - b$.

27. The money supply in the economy is defined by the equation $M = C + D$ where C is the cash on deposit in banks and D is the total of all time and demand deposits. If m, c, and d are the respective rates of growth of M, C, and D, show that

$$m = \frac{cC}{C + D} + \frac{dD}{C + D}.$$

(See Exercise 26 for the definition of rate of growth.)

28. Let $f(x, y) = xe^{y^2}, \qquad x(r, \theta) = r \cos \theta,$

$y(r, \theta) = r \sin \theta$. Find

a. $\dfrac{\partial^2 f}{\partial r^2}$ **b.** $\dfrac{\partial^2 f}{\partial \theta^2}$

29. Let $f(x, y) = x^2 + y^4, \qquad x(s, t) = s^2 t,$

$y(s, t) = t^2 - s^2$. Find

a. $\dfrac{\partial^2 f}{\partial s^2}$ **b.** $\dfrac{\partial^2 f}{\partial t^2}$

30. Find $\dfrac{d^2 f}{dt^2}$ for the function $f(x, y)$ in Exercise 11.

31. Find $\dfrac{\partial^2 f}{\partial r^2}$ for the function $f(x, y)$ in Exercise 14.

32. Find an expression for the mixed second order partial derivative $\dfrac{\partial^2 f}{\partial s \partial t}$ for the composite function

$$f(x, y) = f(x(s, t), y(s, t)).$$

33. Use the result of Exercise 32 to find $\dfrac{\partial^2 f}{\partial r \partial \theta}$ for the function f in Exercise 28.

34. Use the result of Exercise 32 to find $\dfrac{\partial^2 f}{\partial s \partial t}$ for the function f in Exercise 29.

35. By use of the Chain Rule, show that Laplace's equation

$$\frac{\partial^2 v}{\partial x^2} + \frac{\partial^2 v}{\partial y^2} + \frac{\partial^2 v}{\partial z^2} = 0$$

in cylindrical coordinates is

$$\frac{\partial^2 v}{\partial r^2} + \frac{1}{r} \cdot \frac{\partial v}{\partial r} + \frac{1}{r^2} \cdot \frac{\partial^2 v}{\partial \theta^2} + \frac{\partial^2 v}{\partial z^2} = 0.$$

36. Use the Chain Rule to show that Laplace's equation

$$\frac{\partial^2 v}{\partial x^2} + \frac{\partial^2 v}{\partial y^2} + \frac{\partial^2 v}{\partial z^2} = 0$$

in spherical coordinates is

$$\frac{\partial^2 v}{\partial \rho^2} + \frac{2}{\rho} \cdot \frac{\partial v}{\partial \rho} + \frac{1}{\rho^2} \cdot \frac{\partial^2 v}{\partial \phi^2} + \frac{\cot \phi}{\rho^2} \cdot \frac{\partial v}{\partial \phi}$$

$$+ \frac{\csc^2 \phi}{\rho^2} \cdot \frac{\partial^2 v}{\partial \theta^2} = 0.$$

18.7 DIRECTIONAL DERIVATIVES; THE GRADIENT

If f is a function of two variables defined in a neighborhood of a point (x_0, y_0), the partial derivatives $\dfrac{\partial f}{\partial x}(x_0, y_0)$ and $\dfrac{\partial f}{\partial y}(x_0, y_0)$ measure the rate of change of f in the directions of the x- and y-coordinate axes, respectively. But how might we calculate the rate of change of f in an arbitrary direction? The answer to this question is provided by the *directional derivative*.

Suppose that we wish to calculate the rate of change of f at (x_0, y_0) in the direction determined by the unit vector $\boldsymbol{u} = u_1\boldsymbol{i} + u_2\boldsymbol{j}$ (Figure 7.1). If t is a number for which $0 < t < 1$, the point $(x_0 + tu_1, y_0 + tu_2)$ lies on the line between the two points (x_0, y_0) and $(x_0 + u_1, y_0 + u_2)$, so the difference

$$f(x_0 + tu_1, y_0 + tu_2) - f(x_0, y_0)$$

represents the change in $f(x, y)$ in the direction of $\boldsymbol{u}$ over an interval of length $\sqrt{(tu_1)^2 + (tu_2)^2} = t\sqrt{u_1^2 + u_2^2} = t$, since $|\boldsymbol{u}| = \sqrt{u_1^2 + u_2^2} = 1$ (see Figure 7.2). The definition of the directional derivative is therefore the following.

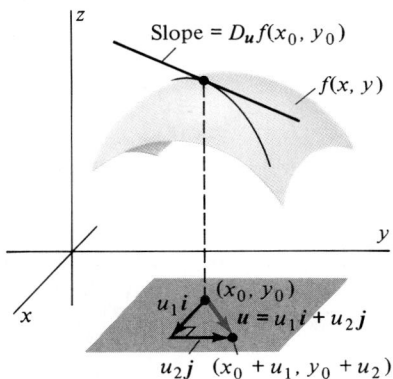

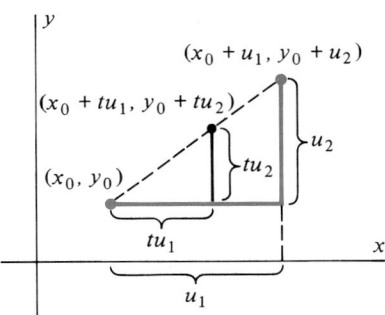

Figure 7.1 Directional derivative $D_{\boldsymbol{u}} f(x_0, y_0)$ is rate of change of f in the direction of the vector $\boldsymbol{u}$.

Figure 7.2 If $0 < t < 1$, the point $(x_0 + tu_1, y_0 + tu_2)$ lies between (x_0, y_0) and $(x_0 + u_1, y_0 + u_2)$.

DEFINITION 5

Let f be a function of two variables defined in a neighborhood of (x_0, y_0). Let $\boldsymbol{u} = u_1\boldsymbol{i} + u_2\boldsymbol{j}$ be a unit vector. The **directional derivative,** $D_{\boldsymbol{u}} f(x_0, y_0)$, in the direction of the unit vector $\boldsymbol{u},$ is the limit

$$D_{\boldsymbol{u}} f(x_0, y_0) = \lim_{t \to 0^+} \frac{f(x_0 + tu_1, y_0 + tu_2) - f(x_0, y_0)}{t}. \tag{1}$$

REMARK 1: Definition 5 has an obvious generalization to functions of more than two variables. If $\boldsymbol{u} = \langle u_1, u_2, \ldots, u_n \rangle$ is a unit vector in $\mathbb{R}^n$ and f is a function of n variables, the directional derivative of f at $(x_1, x_2, \ldots x_n)$ in the direction of $\boldsymbol{u}$ is the limit

$$D_{\boldsymbol{u}} f(x_1, x_2, \ldots, x_n) \tag{2}$$

$$= \lim_{t \to 0^+} \frac{f(x_1 + tu_1, x_2 + tu_2, \ldots, x_n + tu_n) - f(x_1, x_2, \ldots, x_n)}{t}.$$

Unfortunately, equations (1) and (2) fail to provide a simple procedure for actually calculating the directional derivative when it exists. When f satisfies the hypotheses of the Approximation Theorem (Theorem 5), we can use this result to establish such a procedure. First, we write the numerator in (1) as

$$f(x_0 + tu_1, y_0 + tu_2) - f(x_0, y_0) = \frac{\partial f}{\partial x}(x_0, y_0)(tu_1) + \frac{\partial f}{\partial y}(x_0, y_0)(tu_2)$$
$$+ \epsilon_1(tu_1) + \epsilon_2(tu_2)$$

where $\epsilon_1 \to 0$ and $\epsilon_2 \to 0$ as $t \to 0$. Dividing both sides by $t \neq 0$ and applying $\lim_{t \to 0^+}$ shows that

$$D_u f(x_0, y_0) = \lim_{t \to 0^+} \frac{\frac{\partial f}{\partial x}(x_0, y_0)(tu_1) + \frac{\partial f}{\partial y}(x_0, y_0)(tu_2) + \epsilon_1(tu_1) + \epsilon_2(tu_2)}{t}$$

$$= \lim_{t \to 0^+} \left\{ \frac{\partial f}{\partial x}(x_0, y_0)u_1 + \frac{\partial f}{\partial y}(x_0, y_0)u_2 + \epsilon_1 u_1 + \epsilon_2 u_2 \right\}$$

$$= \frac{\partial f}{\partial x}(x_0, y_0)u_1 + \frac{\partial f}{\partial y}(x_0, y_0)u_2 + \lim_{t \to 0^+} \epsilon_1 u_1 + \lim_{t \to 0^+} \epsilon_2 u_2.$$

Since $\epsilon_1 \to 0$ and $\epsilon_2 \to 0$ as $t \to 0$, the last two terms on the right side of the above equation are zero. We have therefore established the following theorem.

THEOREM 7

If the function f and its first partial derivatives are continuous in a neighborhood of (x_0, y_0), the directional derivative $D_u f(x_0, y_0)$ in the direction of the unit vector $u = u_1 i + u_2 j$ is given by

$$D_u f(x_0, y_0) = \frac{\partial f}{\partial x}(x_0, y_0)u_1 + \frac{\partial f}{\partial y}(x_0, y_0)u_2. \tag{3}$$

REMARK 2: The corresponding result for functions of n variables, $n \geq 3$, is that the directional derivative in (2) may be expressed as

$$D_u f(x_1, x_2, \ldots, x_n) = \frac{\partial f}{\partial x_1}(x_1, \ldots, x_n)u_1 \tag{4}$$
$$+ \frac{\partial f}{\partial x_2}(x_1, x_2, \ldots, x_n)u_2 + \cdots$$
$$+ \frac{\partial f}{\partial x_n}(x_1, x_2, \ldots, x_n)u_n$$

when f and its first order partial derivatives are continuous in a neighborhood of $(x_1, x_2, \ldots, x_n)$ in $\mathbb{R}^n$.

REMARK 3: It is important to note that the vector u in expressions (1) through (4) is a *unit* vector. If you wish to calculate the directional derivative of f in the direction of a vector w with $|w| \neq 1$, you must first obtain the unit vector $u = \frac{w}{|w|}$ in the direction of w.

REMARK 4: Note that if $u = i$ (so that $u_1 = 1$ and $u_2 = 0$), and if the partial derivative $\frac{\partial f}{\partial x}(x, y)$ exists, then the directional derivative $D_i f(x, y)$ is just $\frac{\partial f}{\partial x}(x, y)$.

Similarly, if $u = j$ and $\dfrac{\partial f}{\partial y}(x, y)$ exists, then $D_j f(x, y)$ is equal to $\dfrac{\partial f}{\partial y}(x, y)$. This is most easily seen from the form for $D_u f(x, y)$ in Theorem 7. However, $D_u f(x, y)$ may exist, as defined in Definition 5, for all u even when one or both of the partial derivatives do not exist (see Exercise 41).

Example 1

Find the directional derivative $D_u f(2, 1)$ where $f(x, y) = x^2 e^{3y}$ and $u = \dfrac{1}{\sqrt{5}} i + \dfrac{2}{\sqrt{5}} j$.

Solution: The partial derivatives are

$$\frac{\partial f}{\partial x}(2, 1) = 2xe^{3y}\bigg|_{(2,1)} = 4e^3$$

and

$$\frac{\partial f}{\partial y}(2, 1) = 3x^2 e^{3y}\bigg|_{(2,1)} = 12e^3.$$

Since

$$|u| = \sqrt{\frac{1}{5} + \frac{4}{5}} = 1,$$

we have from (3) that

$$D_u f(2, 1) = (4e^3)\left(\frac{1}{\sqrt{5}}\right) + 12e^3\left(\frac{2}{\sqrt{5}}\right)$$

$$= \frac{28e^3}{\sqrt{5}} \approx 251.5.$$

◇

Example 2

Find the directional derivative of the function $f = e^x \cos y + xz$ at the point $(1, \pi, -1)$ in the direction of the vector $w = i - 3j + 4k$.

Solution: The partial derivatives of f are

$$\frac{\partial f}{\partial x}(1, \pi, -1) = e^x \cos y + z\bigg|_{(1,\pi,-1)} = -e - 1,$$

$$\frac{\partial f}{\partial y}(1, \pi, -1) = -e^x \sin y\bigg|_{(1,\pi,-1)} = 0,$$

and

$$\frac{\partial f}{\partial z}(1, \pi, -1) = x\bigg|_{(1,\pi,-1)} = 1.$$

Since $|w| = \sqrt{1^2 + 3^2 + 4^2} = \sqrt{26}$, a *unit* vector in the direction of w is $u = \dfrac{1}{\sqrt{26}}(i - 3j + 4k)$, so $u_1 = \dfrac{1}{\sqrt{26}}$, $u_2 = \dfrac{-3}{\sqrt{26}}$, and $u_3 = \dfrac{4}{\sqrt{26}}$. According to (4),

$$D_u f(1, \pi, -1) = (-e - 1)\left(\frac{1}{\sqrt{26}}\right) + (0)\left(\frac{-3}{\sqrt{26}}\right) + (1)\left(\frac{4}{\sqrt{26}}\right)$$

$$= \frac{-e + 3}{\sqrt{26}}$$

$$\approx 0.055. \qquad \diamond$$

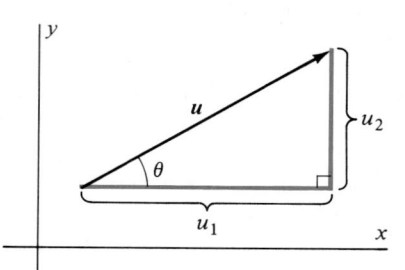

Figure 7.3 If $u = u_1 i + u_2 j$ with $|u| = 1$, then $u_1 = \cos \theta$, $u_2 = \sin \theta$.

If $u = u_1 i + u_2 j$ is a unit vector and θ is the angle formed between u and the positive x-axis, then

$$u_1 = \frac{u_1}{|u|} = \cos \theta, \qquad \text{and} \qquad u_2 = \frac{u_2}{|u|} = \sin \theta.$$

(See Figure 7.3.) Using these equations, we may write the directional derivative $D_u f(x_0, y_0)$ in (3) in the form

$$D_u f(x_0, y_0) = \frac{\partial f}{\partial x}(x_0, y_0) \cos \theta + \frac{\partial f}{\partial y}(x_0, y_0) \sin \theta. \qquad (5)$$

Equation (5) makes explicit the fact that when f and its first partial derivatives are continuous, the directional derivative depends only on the partial derivatives and the direction of the unit vector u.

Example 3

Let $f(x, y) = xy - y^3$. Find the unit vector u for which the directional derivative $D_u f(2, 1)$ is a maximum.

Solution: We begin by calculating the partial derivatives for f:

$$\frac{\partial f}{\partial x}(2, 1) = y\Big|_{(2,1)} = 1,$$

$$\frac{\partial f}{\partial y}(2, 1) = x - 3y^2\Big|_{(2,1)} = -1.$$

According to (5), the directional derivative $D_u f(2, 1)$ is

$$D_u f(2, 1) = (1)\cos \theta + (-1)\sin \theta$$
$$= \cos \theta - \sin \theta.$$

We must therefore find the value of θ for which the function

$$g(\theta) = \cos \theta - \sin \theta$$

is a maximum. To do so we set

$$g'(\theta) = -\sin \theta - \cos \theta = 0$$

and obtain the equation

$$\sin \theta = -\cos \theta, \qquad \text{or} \qquad \tan \theta = -1,$$

which has solutions $\theta = \dfrac{3\pi}{4}$ and $\theta = \dfrac{7\pi}{4}$ for $0 \le \theta \le 2\pi$.

Since

$$g''\left(\frac{3\pi}{4}\right) = -\cos\left(\frac{3\pi}{4}\right) + \sin\left(\frac{3\pi}{4}\right) = \sqrt{2} > 0$$

and

$$g''\left(\frac{7\pi}{4}\right) = -\cos\left(\frac{7\pi}{4}\right) + \sin\left(\frac{7\pi}{4}\right) = -\sqrt{2} < 0,$$

the angle $\theta = \dfrac{7\pi}{4}$ corresponds to the maximum. For this angle, $\boldsymbol{u} = (\cos \theta)\boldsymbol{i} + (\sin \theta)\boldsymbol{j} = \dfrac{\sqrt{2}}{2}\boldsymbol{i} - \dfrac{\sqrt{2}}{2}\boldsymbol{j}$ is the required unit vector. ◇

The Gradient

The form of the directional derivative given by Theorem 7 can be written as a dot product:

$$D_{\boldsymbol{u}}f(x_0, y_0) = \frac{\partial f}{\partial x}(x_0, y_0)u_1 + \frac{\partial f}{\partial y}(x_0, y_0)u_2 \tag{6}$$

$$= \left[\frac{\partial f}{\partial x}(x_0, y_0)\boldsymbol{i} + \frac{\partial f}{\partial y}(x_0, y_0)\boldsymbol{j}\right] \cdot [u_1\boldsymbol{i} + u_2\boldsymbol{j}].$$

The second factor in this dot product is just the unit vector $\boldsymbol{u} = u_1\boldsymbol{i} + u_2\boldsymbol{j}$. The first factor is called the *gradient* of f at (x_0, y_0). It is usually written as

$$\nabla f(x_0, y_0) = \frac{\partial f}{\partial x}(x_0, y_0)\boldsymbol{i} + \frac{\partial f}{\partial y}(x_0, y_0)\boldsymbol{j} \tag{7}$$

or just

$$\nabla f = \frac{\partial f}{\partial x}\boldsymbol{i} + \frac{\partial f}{\partial y}\boldsymbol{j}.$$

For functions of three variables, the gradient is

$$\nabla f = \frac{\partial f}{\partial x}\boldsymbol{i} + \frac{\partial f}{\partial y}\boldsymbol{j} + \frac{\partial f}{\partial z}\boldsymbol{k}.$$

Thus, the gradient ∇f is a *vector* whose components are the partial derivatives of f.

Example 4

For $f(x, y) = x \cos y$, the gradient is

$$\nabla f(x, y) = \left[\frac{\partial}{\partial x}(x \cos y)\right]\boldsymbol{i} + \left[\frac{\partial}{\partial y}(x \cos y)\right]\boldsymbol{j}$$

$$= \cos y\boldsymbol{i} - x \sin y\boldsymbol{j}$$

and the vector given by the gradient at $\left(2, \dfrac{\pi}{4}\right)$ is

$$\nabla f(2, \pi/4) = \frac{\sqrt{2}}{2}\boldsymbol{i} - \sqrt{2}\boldsymbol{j}. \qquad ◇$$

Example 5

For $f(x, y, z) = \sqrt{x} \, e^y \, \mathrm{Tan}^{-1} z$, the gradient is

$$\nabla f(x, y, z) = \frac{e^y \, \mathrm{Tan}^{-1} z}{2\sqrt{x}} \boldsymbol{i} + \sqrt{x} \, e^y \, \mathrm{Tan}^{-1} z \boldsymbol{j} + \frac{\sqrt{x} \, e^y}{1 + z^2} \boldsymbol{k},$$

and the vector given by the gradient at $(4, 0, 1)$ is

$$\nabla f(4, 0, 1) = \frac{\pi}{16} \boldsymbol{i} + \frac{\pi}{2} \boldsymbol{j} + \boldsymbol{k}. \qquad \diamond$$

Vector Notation and Gradients

Using vector notation, we may express the directional derivative as simply

$$D_{\boldsymbol{u}} f(\boldsymbol{x}) = \nabla f(\boldsymbol{x}) \cdot \boldsymbol{u}. \tag{8}$$

Equation (8) provides insight into the geometry associated with $\nabla f(\boldsymbol{x})$. From Theorem 4, Chapter 16, we know that

$$\begin{aligned}
\nabla f(\boldsymbol{x}) \cdot \boldsymbol{u} &= |\nabla f(\boldsymbol{x})| \cdot |\boldsymbol{u}| \cdot \cos \theta \\
&= |\nabla f(\boldsymbol{x})| \cos \theta \qquad (|\boldsymbol{u}| = 1)
\end{aligned} \tag{9}$$

where θ is the angle between the vectors $\nabla f(\boldsymbol{x})$ and $\boldsymbol{u}$. Combining equations (8) and (9) gives

$$D_{\boldsymbol{u}} f(\boldsymbol{x}) = |\nabla f(\boldsymbol{x})| \cos \theta. \tag{10}$$

Thus, since $-1 \le \cos \theta \le 1$ for all θ, equation (10) shows that

(i) $-|\nabla f(\boldsymbol{x})| \le D_{\boldsymbol{u}} f(\boldsymbol{x}) \le |\nabla f(\boldsymbol{x})|$, and
(ii) $D_{\boldsymbol{u}} f(\boldsymbol{x})$ assumes its maximum value when $\cos \theta = 1$, that is, when $\theta = 0$.

Since the case $\theta = 0$ occurs precisely when $\nabla f(\boldsymbol{x})$ and the direction vector $\boldsymbol{u}$ point in the same direction, conclusion (ii) says that $\nabla f(\boldsymbol{x})$ points in the direction of the maximum value of $D_{\boldsymbol{u}} f(\boldsymbol{x})$. That is,

$$\nabla f(\boldsymbol{x}) \text{ points in the direction of most rapid increase for } f \text{ at } \boldsymbol{x} \tag{11}$$

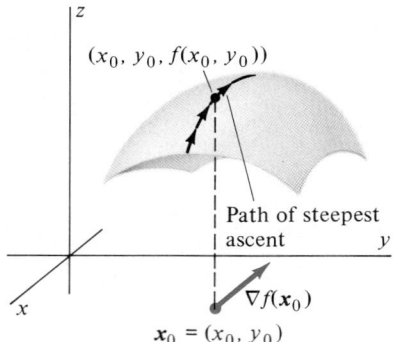

Figure 7.4 $\nabla f(\boldsymbol{x})$ points in the direction of most rapid increase of the function f at $\boldsymbol{x}$.

(assuming, of course, that the hypotheses of Theorem 7 hold).

This observation has several important applications. If $\boldsymbol{x} = (x, y)$ is a point in the domain of the function f, then $\nabla f(x, y)$ points in the direction in which a path through $(x, y, f(x, y))$ on the graph will rise most rapidly (see Figure 7.4). A path on a surface having the property that the tangent at each point of the path is parallel to the gradient of the function defining the surface is called a **path of steepest ascent.** Of obvious relevance to mountain climbers, this concept is also used to develop procedures for approximating extrema for complicated functions of several variables. (It should be geometrically obvious that proceeding in the opposite direction, $-\nabla f(x, y)$, leads to the **path of steepest descent.** In Exercise 33, you are asked to prove this analytically.)

A second application of observation (11) concerns particles moving so as to maximize an attribute of the medium through which they are moving. Examples of this are insects flying toward a light source (maximizing the intensity of light), heat-seeking missiles (maximizing temperature), and sharks in water (maximizing blood concentration). In such situations, if $A(\boldsymbol{x})$ is the value of the attribute at location x, the particle will move so that its tangent vector points in the direction of $\nabla A(x)$ (see Example 7).

Example 6

For the function $f(x, y) = 9 - \dfrac{x^2 + y^2}{4}$, the gradient at $x = (x, y)$ is

$$\nabla f(x) = -\frac{x}{2}i - \frac{y}{2}j$$

$$= -\frac{1}{2}(xi + yj)$$

$$= -\frac{1}{2}x.$$

Since $x = xi + yj$ is the position vector of the point (x, y), the vector $\nabla f(x) = -\dfrac{1}{2}x$ points toward the origin for all $(x, y) \neq (0, 0)$. In particular,

if $x = (2, 2)$, $\nabla f = -i - j$,

if $x = (1, 4)$, $\nabla f = -\dfrac{1}{2}i - 2j$,

if $x = (2, -1)$, $\nabla f = -i + \dfrac{1}{2}j$.

This should not be surprising, since the graph of f is a paraboloid. At any point on this surface, the z-coordinate is increased most rapidly by moving directly toward the z-axis (see Figure 7.5).   ◇

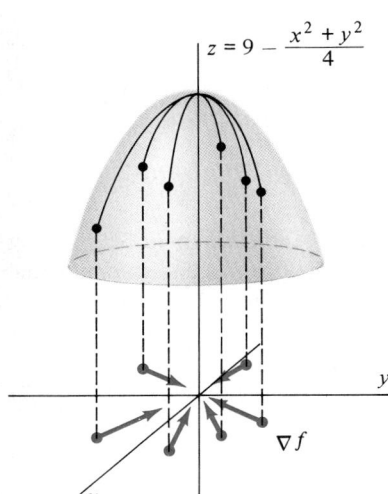

$z = 9 - \dfrac{x^2 + y^2}{4}$

∇f

Figure 7.5 $\nabla f(x) = -\dfrac{1}{2}x$ for

$f(x, y) = 9 - \dfrac{x^2 + y^2}{4}.$

Example 7

The temperature distribution across the surface of a rectangular plate is given by the function $T(x, y) = 100 - x^2 - 2y^2$. Find the path followed by a heat-seeking particle placed on the plate at the point $(4, 2)$.

Solution:

Let the path followed by the particle have parameterization

$$r(t) = x(t)i + y(t)j.$$

Assuming the components x and y to be differentiable functions of t, we recall from Chapter 17 that the tangent vector $T(t)$ at each point along the path is

$$T(t) = x'(t)i + y'(t)j.$$

Since the particle is heat seeking, this tangent should point in the same direction as the gradient

$$\nabla T(x, y) = -2xi - 4yj.$$

This will be the case when

$$x'(t) = -2x(t) \qquad \text{and} \qquad y'(t) = -4y(t).$$

These are the familiar differential equations for exponential decay. Their solutions are

$$x(t) = x(0)e^{-2t}, \quad \text{and} \quad y(t) = y(0)e^{-4t}.$$

Since the particle begins at $(4, 2)$, we have $x(0) = 4$ and $y(0) = 2$. Thus

$$x(t) = 4e^{-2t}, \quad y(t) = 2e^{-4t}$$

are the components of a parameterization of the path. We may eliminate the parameter t by noting that

$$y(t) = 2e^{-4t} = \frac{1}{8}(4e^{-2t})^2 = \frac{1}{8}[x(t)]^2.$$

The particle therefore approaches the origin along the parabola $y = \frac{1}{8}x^2$ (Figure 7.6).

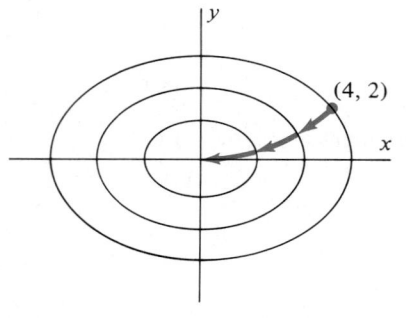

Figure 7.6 Heat-seeking particle approaches origin along parabolic path.

Level Curves and Surfaces

Let $r(t) = x(t)i + y(t)j + z(t)k$ be a parameterization for a curve in $\mathbb{R}^3$, and let $w = f(x, y, z)$ be a function of three variables. Using the gradient and assuming that all necessary derivatives exist, we may write the Chain Rule (Theorem 6)

$$\frac{d}{dt}f(x, y, z) = \frac{\partial f}{\partial x} \cdot \frac{dx}{dt} + \frac{\partial f}{\partial y} \cdot \frac{dy}{dt} + \frac{\partial f}{\partial z} \cdot \frac{dz}{dt}$$

as

$$\frac{d}{dt}f(r(t)) = \nabla f(r(t)) \cdot r'(t). \tag{12}$$

Equation (12) says that the derivative of the composite function $f(r(t))$ is the dot product of the gradient $\nabla f(r(t))$ with the tangent vector $r'(t) = x'(t)i + y'(t)j + z'(t)k$ for each t. Equation (12) also holds for curves $r(t) = x(t)i + y(t)j$ in the plane composed with functions $z = f(x, y)$ of two variables, as does Theorem 6. (Of course, the beauty of the vector notation is that we need not distinguish between these two cases in writing equation (12).)

Equation (12) reveals an important relationship between gradients and level curves. To see this, suppose that f is a function of two variables for which the equation

$$f(x, y) = k, \quad k \text{ constant} \tag{13}$$

determines a curve C in the xy-plane with parameterization

$$r(t) = x(t)i + y(t)j.$$

Then along this level curve the composite function

$$f(r(t)) = f(x(t), y(t)) \tag{14}$$

is constant, according to equation (13).

Now let $x_0 = r(t_0) = (x(t_0), y(t_0))$ be a point on C. If the functions f, x, and y satisfy the hypotheses of Theorem 6 in a neighborhood of x_0, we may differentiate

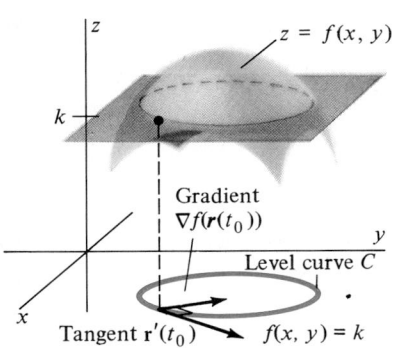

Figure 7.7 $\nabla f(x_0)$ is orthogonal to the level curve through x_0. (See **Plate 12.**)

both sides of equation (14), using equation (12), to conclude that

$$\frac{d}{dt}f(r(t))\bigg|_{t=t_0} = \nabla f(r(t_0)) \cdot r'(t_0) = 0. \tag{15}$$

Since $r'(t_0)$ is a vector tangent to the level curve C, equation (15) shows that *the gradient $\nabla f(r(t_0))$ is orthogonal to the level curve C at x_0* (see Figure 7.7). That is:

> Let f be a function of two variables. If f and its first partial derivatives are continuous, then at each point in the domain of f the gradient vector, if nonzero, is orthogonal to the level curve through that point. (16)

Example 8

The graph of the function $z = \sqrt{x^2 + y^2}$ is a cone. The level curves associated with this graph are the circles

$$\sqrt{x^2 + y^2} = k, \qquad \text{or} \qquad x^2 + y^2 = k^2.$$

The gradient for this function is

$$\nabla f(x, y) = \frac{x}{\sqrt{x^2 + y^2}}i + \frac{y}{\sqrt{x^2 + y^2}}j = \frac{xi + yj}{\sqrt{x^2 + y^2}}.$$

In vector notation, the gradient may be written as

$$\nabla f(x) = \frac{x}{|x|}.$$

This shows that the gradient vectors are parallel to radius vectors for the circular level curves and, therefore, orthogonal to the level curves (Figure 7.8). ◇

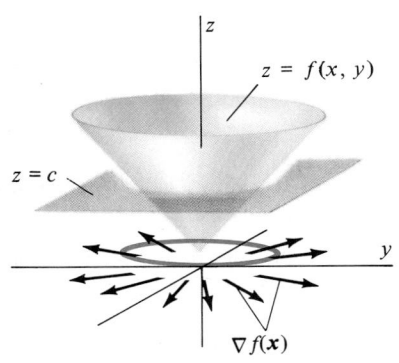

Figure 7.8 For $f(x, y) = \sqrt{x^2 + y^2}$ level curves are circles; $\nabla f(x)$ is parallel to a radius for the circle.

A similar relationship exists between gradients for functions of three variables and level surfaces. Suppose that

$$f(x, y, z) = k$$

is the equation of a level surface S for the function f, and suppose that the point x_0 lies on S. Choose a curve C on the surface S with parameterization

$$r(t) = x(t)i + y(t)j + z(t)k$$

for which $r(t_0) = x_0$ (see Figure 7.9). Then, along this curve

$$f(r(t)) = f(x(t), y(t), z(t)) = k, \tag{17}$$

that is, the composite function $f \circ r$ is constant. As before, if the function f and the component functions x, y, and z satisfy the hypotheses of Theorem 6 in a neighborhood of $x_0 = r(t_0)$, we may differentiate both sides of equation (17), using the Chain Rule, to conclude that

$$\frac{d}{dt}f(r(t))\bigg|_{t=t_0} = \nabla f(r(t_0)) \cdot r'(t_0) = 0. \tag{18}$$

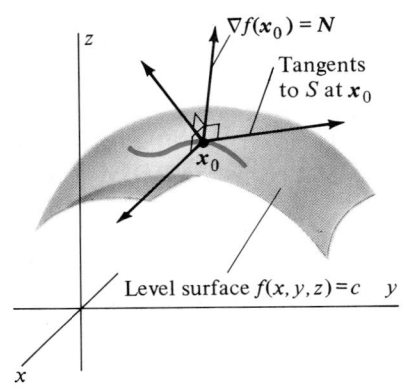

Figure 7.9 $\nabla f(x_0)$ is orthogonal to the level surface for f containing x_0.

Equation (18) shows that the gradient $\nabla f(x_0)$ is orthogonal to the tangent $r'(t_0)$ to the curve C through x_0 on the surface S. Since C was an *arbitrary* curve on S, it follows that $\nabla f(x_0)$ is orthogonal to *every* tangent to S at x_0. In particular, $\nabla f(x_0)$ is orthogonal to the tangent plane determined by $\dfrac{\partial f}{\partial x}(x_0)$ and $\dfrac{\partial f}{\partial y}(x_0)$ and therefore to the level surface S itself.

> Let f be a function of three variables. If f and its first partial derivatives are continuous, then at each point in the domain of f the gradient vector, if nonzero, is orthogonal to the level surface containing that point.　　(19)

Exercise Set 18.7

In Exercises 1–9, find the gradient of the given function at the point P.

1. $f(x, y) = x^2 y$,　　$P = (3, 1)$

2. $f(x, y) = xy^2 - ye^x$,　　$P = (0, 2)$

3. $f(x, y) = x \cos(y - x)$,　　$P = (\pi/2, \pi/4)$

4. $f(x, y) = \dfrac{2x}{y - x}$,　　$P = (2, 1)$

5. $f(x, y, z) = x^2 y + xz^2$,　　$P = (1, 1, 2)$

6. $f(x, y, z) = xe^{yz}$,　　$P = (2, 0, 1)$

7. $f(x, y, z) = x^2 y + xz^3 - y^2 z$,　　$P = (1, 2, 1)$

8. $f(x, y, z) = e^x \cos y - e^y \sin z$,　　$P = (0, \pi/4, \pi/3)$

9. $f(x, y, z) = x\sqrt{y} \cosh z$,　　$P = (2, 4, 1)$

In Exercises 10–19, find the directional derivative of the given function at the given point in the direction of the given vector.

10. $f(x, y) = 3x^2 - y^2$,　　$P = (1, 2)$,　　$w = i + j$

11. $f(x, y) = x^2 y^2$,　　$P = (-2, 3)$,　　$w = i - j$

12. $f(x, y) = \sin(y - x)$,　　$P = (\pi/2, \pi/4)$,　　$w = i + 2j$

13. $f(x, y) = \dfrac{x}{x + y}$,　　$P = (1, 2)$,　　$w = \sqrt{3}i + j$

14. $f(x, y) = xe^{y^2} - y$,　　$P = (1, 3)$,　　$w = \sqrt{2}i + \sqrt{2}j$

15. $f(x, y, z) = xy + xz + yz$,　　$P = (1, 2, 1)$,
$w = i + j - k$

16. $f(x, y, z) = \cosh x - \sinh y + \operatorname{Tan}^{-1} z$,
$P = (-1, 1, 3)$,　　$w = i - j - 2k$

17. $f(x, y, z) = xe^{yz}$,　　$P = (2, 0, 1)$,
$w = \sqrt{3}i + \sqrt{3}j - \sqrt{5}k$

18. $f(x, y, z) = xy^2 - x^2 z + (yz)^3$,　　$P = (1, 0, 1)$,
$w = i + 2j + k$

19. $f(x, y, z) = x \cos y - y \sinh z$,　　$P = (6, \pi/4, -1)$,
$w = i - 2j - 2k$

In Exercises 20–24, write vector equations for the normal and tangent lines to the given curves at the given points.

20. $2x^2 + 3y^2 = 11$,　　$P = (2, 1)$

21. $4x^2 - y^2 = 7$,　　$P = (2, 3)$

22. $3x^2 - y^4 + x = 13$,　　$P = (2, 1)$

23. $\sqrt{x} + \sqrt{y} = 4$,　　$P = (4, 4)$

24. $6x^2 - 4y^2 = 18$,　　$P = (3, -3)$

25. For $f(x, y) = xy^2 + ye^x$, find the directional derivative at $(0, 1)$ in the direction of most rapid increase of f.

26. For $f(x, y) = x^2 + 2y^3$, find the directional derivative at $(2, 3)$ in the direction toward the origin.

27. For $f(x, y)$ as in Exercise 25, find $D_u f(0, 1)$ in the direction of the origin.

28. For $f(x, y) = \dfrac{x}{x + y}$, find $D_u f(1, 1)$ in the direction of the point $(2, 3)$.

29. For f as in Exercise 28, find $D_u f(1, 1)$ in the direction of most rapid increase of f.

30. Use the gradient to find the point(s) on the graph of the hyperbola $3x^2 - 2y^2 - 6x + 8y = 3$ where the tangent is horizontal.

31. Use the gradient to find the point(s) on the graph of the ellipse $x^2 - 6x + 2y^2 - 4y = -7$ where the tangent is
a. horizontal,　　**b.** vertical.

32. Use equation (8) to show that the directional derivative of f in the direction of the unit vector u is the component of ∇f in the direction of u.

33. Use equation (8) to show that the negative gradient, $-\nabla f(x)$ points in the direction of most rapid decrease of the function f at x.

34. The temperature distribution in a room is given by the function $T(x, y, z) = 30 - (x^2 + 2y^2 + 3z^2)$. An insect flies so as to experience the most rapid decrease in temperature. In what direction does it move when located at point $(2, 1, 1)$?

35. Show that $|\nabla f(x_0)|$ is the maximum value of $D_u f(x_0)$ when f and its first partials are continuous at x_0.

36. Show that if $\cos \alpha$, $\cos \beta$, and $\cos \gamma$ are the direction cosines for the unit vector $\boldsymbol{u}$, then $D_u f = \dfrac{\partial f}{\partial x} \cos \alpha \boldsymbol{i} +$ $\dfrac{\partial f}{\partial y} \cos \beta \boldsymbol{j} + \dfrac{\partial f}{\partial z} \cos \gamma \boldsymbol{k}$.

37. Find the direction of steepest ascent at the point above $(4, 2)$ on the graph of $z = e^{x^2 + y^2}$.

38. Find the direction of steepest ascent at the point above $(1, 1)$ on the graph of $z = xe^y + ye^x$.

39. Find the direction of steepest *descent* at the point above $(1, 2)$ on the graph of $z = 4x^2 - 2y^2$.

40. The temperature distribution on a metal plate is $T(x, y) = 200 - (x^2 + 4y^2)$. Find the path followed by a heat-seeking particle placed at the point $(0, 2)$.

41. Show that if the partial derivative $\dfrac{\partial f}{\partial x}(x_0, y_0)$ exists, then $D_i f(x_0, y_0)$ exists and $D_i f(x_0, y_0) = \dfrac{\partial f}{\partial x}(x_0, y_0)$. Give an example to show that $D_u f(x_0, y_0)$ can exist for all $\boldsymbol{u}$ while $\dfrac{\partial f}{\partial x}(x_0, y_0)$ (and, therefore, $\nabla f(x_0, y_0)$) fails to exist.

18.8 CONSTRAINED EXTREMA: THE METHOD OF LAGRANGE MULTIPLIERS

As an application of the theory of the gradient, we discuss a method due to the French mathematician Joseph L. Lagrange (1736–1813) for finding extrema of functions of several variables subject to constraints. Examples of this kind of problem are the following:

Example 1

Find the maximum and minimum values of the function $f(x, y) = 2x^2 + 4y^2$, given that $x^2 + y^2 = 1$.

Example 2

Find the point(s) on the hyperboloid $z = (y + 1)^2 - (x - 2)^2 + 1$ nearest the point $(2, -1, 2)$.

Example 3

A cylindrical tin can, with a top and a bottom, is to be manufactured using 100 cm^2 of tin, ignoring waste. What dimensions produce the can of maximum volume?

Each of these examples involves finding the extreme values of a function

$$w = f(x), \qquad x \in \mathbb{R}^2 \text{ or } \mathbb{R}^3 \qquad \text{(function to be optimized)} \qquad (1)$$

subject to a **constraint** of the form

$$g(x) = 0 \qquad \text{(constraint equation)}. \qquad (2)$$

In Example 1 the function to be optimized is $f(x, y) = 2x^2 + 4y^2$, and the constraint is $x^2 + y^2 = 1$. The latter equation can be put in the form of equation (2) by writing it as

$$g(x, y) = x^2 + y^2 - 1 = 0.$$

In Example 2, we are to minimize the distance function $D(x, y, z) = \sqrt{(x - 2)^2 + (y + 1)^2 + (z - 2)^2}$ subject to the constraint that (x, y, z) lie on the hyperboloid. We can write this constraint in the form of equation (2) as

$$g(x, y, z) = z - (y + 1)^2 + (x - 2)^2 - 1 = 0.$$

In Example 3, the dimensions of the tin can are r (radius) and h (height), so we need to maximize the volume $V = \pi r^2 h$ subject to the constraint that total surface area equal 100 cm^2. This constraint may be expressed as the equation

$$g(r, h) = 2\pi r^2 + 2\pi rh - 100 = 0.$$

Each of these examples illustrates the important difference between finding *relative* extrema for functions f of several variables and finding *extrema in the presence of constraints,* or *constrained extrema.* The constraint equation $g(x) = 0$ restricts the points x that may be considered to just those satisfying this constraint. Constrained extrema in general will not correspond to relative extrema.

Figure 8.1 illustrates the geometry associated with the constrained extrema problem of Example 1. Notice how the constraint $x^2 + y^2 = 1$ restricts the graph of f to

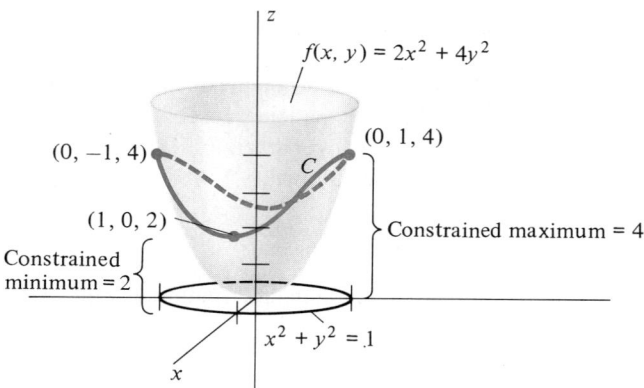

Figure 8.1 Constraint $g(x, y) = x^2 + y^2 - 1 = 0$ restricts graph of $f(x, y) = 2x^2 + 4y^2$ to the curve C.

just the curve C above the unit circle. Note also that neither the constrained maxima nor the constrained minima correspond to relative extrema for the function f.

It is sometimes possible to solve a problem involving a constraint by solving the constraint equation (2) for one of the independent variables and then substituting for this variable in equation (1). The method due to Lagrange is more general, in that it does not depend on our ability to solve the constraint equation for any particular variable. It is based on the following theorem.

THEOREM 8

Let x denote a point (vector) in either $\mathbb{R}^2$ or $\mathbb{R}^3$, and let f and g be functions of either two or three variables. Assume that both f and g have continuous partial derivatives in a neighborhood of the point x_0. If x_0 maximizes or minimizes the function

$$w = f(x) \tag{3a}$$

subject to the constraint

$$g(x) = 0, \tag{3b}$$

and if $\nabla g(x_0) \neq 0$, then

$$\nabla f(x_0) = \lambda \nabla g(x_0) \tag{4}$$

for some constant λ. In other words, $\nabla f(x_0)$ and $\nabla g(x_0)$ are parallel.

The method of Lagrange for finding relative extrema is now obvious—we simply check all points satisfying equation (4). Among the values $f(x_0)$ must lie the constrained extrema, if such extrema exist. We sketch a proof of Theorem 8 at the end of this section.

Theorem 8 establishes the **method of Lagrange multipliers** (for finding the extreme values of f subject to the constraint $g(x) = 0$):

1. Find all simultaneous solutions of the equations

$$\nabla f(x_0) = \lambda \nabla g(x_0) \tag{5}$$

and

$$g(x_0) = 0. \tag{6}$$

a. If $x = (x, y) \in \mathbb{R}^2$, equations (5) and (6) are equivalent to the three equations

$$
\begin{cases}
\dfrac{\partial f}{\partial x}(x_0, y_0) = \lambda \dfrac{\partial g}{\partial x}(x_0, y_0), \\[2mm]
\dfrac{\partial f}{\partial y}(x_0, y_0) = \lambda \dfrac{\partial g}{\partial y}(x_0, y_0), \\[2mm]
g(x_0, y_0) = 0.
\end{cases}
\tag{7}
$$

b. If $x = (x, y, z) \in \mathbb{R}^3$, equations (5) and (6) are equivalent to the four equations

$$
\begin{cases}
\dfrac{\partial f}{\partial x}(x_0, y_0, z_0) = \lambda \dfrac{\partial g}{\partial x}(x_0, y_0, z_0), \\[2mm]
\dfrac{\partial f}{\partial y}(x_0, y_0, z_0) = \lambda \dfrac{\partial g}{\partial y}(x_0, y_0, z_0), \\[2mm]
\dfrac{\partial f}{\partial z}(x_0, y_0, z_0) = \lambda \dfrac{\partial g}{\partial z}(x_0, y_0, z_0), \\[2mm]
g(x_0, y_0, z_0) = 0.
\end{cases}
\tag{8}
$$

2. Calculate $f(x_0)$ for all x_0 obtained in step 1.
3. On geometric, analytic, or physical grounds, determine which of the numbers obtained in step 2 correspond to constrained extrema.

Solution to Example 1: We wish to find the extreme values of

$$f(x, y) = 2x^2 + 4y^2$$

subject to

$$g(x, y) = x^2 + y^2 - 1 = 0.$$

The three equations corresponding to equations (7) are:

$$4x = 2\lambda x \qquad (f_x = \lambda g_x), \qquad (9)$$

$$8y = 2\lambda y \qquad (f_y = \lambda g_y), \qquad (10)$$

and

$$x^2 + y^2 - 1 = 0 \qquad (g = 0). \qquad (11)$$

To solve this system of three equations, we begin with equation (9). If $x = 0$ equation (9) is satisfied, equation (11) then becomes simply $y^2 - 1 = 0$, so $y = \pm 1$. We therefore obtain the two points $(0, 1)$ and $(0, -1)$ that must be checked for extrema.

On the other hand, if $x \neq 0$ in (9), we may divide both sides of (9) by x to obtain $4 = 2\lambda$, so $\lambda = 2$. Substituting this value of λ into (10) then gives $8y = 4y$, so y must equal zero. With $y = 0$, equation (11) becomes $x^2 - 1 = 0$, so $x = \pm 1$. We have therefore obtained two additional points, $(1, 0)$ and $(-1, 0)$.

Finally, we simply calculate the value of $f(x, y)$ for each of the four points $(0, 1)$, $(0, -1)$, $(1, 0)$, and $(-1, 0)$. We find

$$f(0, 1) = 4 \qquad \text{(maximum)},$$
$$f(0, -1) = 4 \qquad \text{(maximum)},$$
$$f(1, 0) = 2 \qquad \text{(minimum)},$$
$$f(-1, 0) = 2 \qquad \text{(minimum)}.$$

As Figure 8.1 illustrates, the constrained maximum is 4, occurring at $(0, 1)$ and $(0, -1)$, and the constrained minimum is $2 = f(1, 0) = f(-1, 0)$. $\diamondsuit$

REMARK 1: In reading these examples, you will notice that the method by which we solve the systems of equations (7) and (8) varies from problem to problem. The great versatility of the method of Lagrange multipliers is that it applies to a wide variety of functions and constraints. However, the resulting systems of equations will therefore be of various types, and often most or all of the resulting equations will be nonlinear. Your success in using this method will depend on your ability to solve these various sorts of systems of equations.

REMARK 2: Note that although we obtained the value $\lambda = 2$ at one point in the solution of Example 1, the value of λ did not appear in the solution. This will always be the case, since the Lagrange multiplier λ is an ''artificial'' variable that is introduced merely as a device by which we can solve for the ''real'' variables in the problem.

The solution of Example 2 is similar to that of Example 1, and we leave it to you as Exercise 15.

Solution to Example 3: Since the volume of the cylinder is $V(r, h) = \pi r^2 h$, the areas of the top and bottom are each πr^2, and the lateral surface area is $2\pi rh$, we must maximize the function

$$V(r, h) = \pi r^2 h$$

subject to the side condition

$$g(r, h) = 2\pi r^2 + 2\pi rh - 100 = 0.$$

The equations corresponding to equations (7) are therefore

$$2\pi rh = \lambda(4\pi r + 2\pi h) \qquad (V_r = \lambda g_r), \tag{12}$$

$$\pi r^2 = \lambda(2\pi r) \qquad (V_h = \lambda g_h), \tag{13}$$

and

$$2\pi r^2 + 2\pi rh - 100 = 0 \qquad (g = 0). \tag{14}$$

Solving equation (12) for λ gives

$$\lambda = \frac{rh}{2r + h}, \tag{15}$$

while solving equation (13) for λ gives

$$\lambda = \frac{r}{2}. \tag{16}$$

Equating the right sides of (15) and (16) we find

$$\frac{rh}{2r + h} = \frac{r}{2}, \qquad \text{so} \qquad 2r + h = 2h,$$

or $h = 2r$. Substituting $h = 2r$ for h in (14) then gives

$$2\pi r^2 + 2\pi r(2r) - 100 = 0,$$

which has solution $r = \dfrac{10}{\sqrt{6\pi}}$. Since it is clear from the geometry of the situation that at least one pair (r, h) must produce a maximum volume, we conclude that the maximum volume of

$$V = \pi \left(\frac{10}{\sqrt{6\pi}}\right)^2 \left(\frac{20}{\sqrt{6\pi}}\right) = \frac{1000}{3\sqrt{6\pi}} \text{ cm}^3$$

corresponds to the dimensions $r = \dfrac{10}{\sqrt{6\pi}}$ cm and $h = 2r = \dfrac{20}{\sqrt{6\pi}}$ cm. ◇

Example 4

A builder wishes to design a rectangular house containing V cubic meters of heated space, so as to minimize heating costs. One wall of the building is to face south. The annual heating costs are estimated to be $4 per square meter of floor space, $3 per square meter for all exterior wall not facing south, and $2 per square meter for exterior wall space facing south. What dimensions will produce the most energy-efficient building?

Solution: As in Figure 8.2, we let x denote the length of the wall facing south, and we refer to this dimension as the width of the house. We let y be the depth, and we let z be the height of the heated portion of the house. Since the area of the floor is xy, the area of each side wall is yz, and the area of each front and rear wall is xz, the

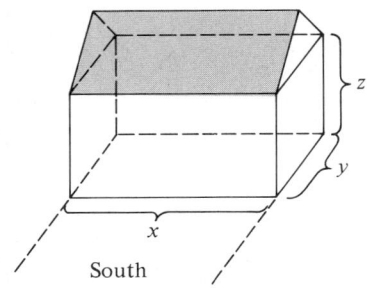

South

Figure 8.2

annual heating cost is

$$C(x, y, z) = 4xy + 3(2yz) + 3xz + 2xz$$

Cost for south wall

Cost for rear (north) wall

Costs for side walls (both)

Costs for roof (floor space)

$$= 4xy + 6yz + 5xz.$$

The constraint is that the volume of the building is to equal the constant V. That is,

$$g(x, y, z) = xyz - V = 0.$$

Applying equations (8), we find

$$4y + 5z = \lambda yz \qquad (C_x = \lambda g_x), \tag{17}$$

$$4x + 6z = \lambda xz \qquad (C_y = \lambda g_y), \tag{18}$$

$$5x + 6y = \lambda xy \qquad (C_z = \lambda g_z), \tag{19}$$

and

$$xyz - V = 0 \qquad (g = 0). \tag{20}$$

This time it is helpful to begin by multiplying both sides of equation (17) by x, both sides of (18) by y, and both sides of (19) by z. The result is the equations

$$4xy + 5xz = \lambda xyz, \tag{21}$$

$$4xy + 6yz = \lambda xyz, \tag{22}$$

and

$$5xz + 6yz = \lambda xyz. \tag{23}$$

Equating the left sides of (21) and (22) then gives $4xy + 5xz = 4xy + 6yz$, so $y = \dfrac{5}{6}x$. Similarly, equating the left sides of (22) and (23) gives $4xy + 6yz = 5xz + 6yz$, or $z = \dfrac{4y}{5} = \dfrac{2x}{3}$. With these substitutions (20) becomes

$$x\left(\frac{5}{6}x\right)\left(\frac{2}{3}x\right) = \frac{5}{9}x^3 = V.$$

so

$$x = \sqrt[3]{\frac{9V}{5}}, \qquad y = \frac{5}{6}\sqrt[3]{\frac{9V}{5}}, \qquad \text{and} \qquad z = \frac{2}{3}\sqrt[3]{\frac{9V}{5}}. \qquad \diamond$$

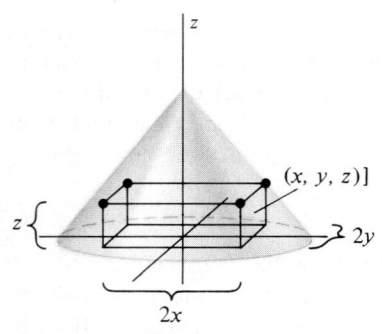

(x, y, z)]

2y

2x

Figure 8.3 Box inscribed within a cone.

Example 5

A rectangular box is to be inscribed in the cone $z = 9 - \sqrt{x^2 + y^2}$, $z \geq 0$ (see Figure 8.3). Find the dimensions for the box that maximize its volume.

Solution: It is clear that to achieve maximum volume we should position the box with one face lying in the xy-plane. If (x, y, z) denotes one corner of the box lying in the first octant, x, y, $z > 0$, the dimensions of the box are

$$\text{length} = 2x, \qquad \text{width} = 2y, \qquad \text{and} \qquad \text{height} = z.$$

The problem is therefore to maximize the function

$$V(x, y. z) = 4xyz$$

subject to the constraint

$$g(x, y, z) = z + \sqrt{x^2 + y^2} - 9 = 0.$$

Applying the Lagrange criterion (8), we find

$$4yz = \frac{\lambda x}{\sqrt{x^2 + y^2}} \qquad (f_x = \lambda g_x), \tag{24}$$

$$4xz = \frac{\lambda y}{\sqrt{x^2 + y^2}} \qquad (f_y = \lambda g_y), \tag{25}$$

$$4xy = \lambda, \tag{26}$$

and

$$z + \sqrt{x^2 + y^2} - 9 = 0. \tag{27}$$

Multiplying (24) through by x, (25) by y, and (26) by z gives

$$4xyz = \frac{\lambda x^2}{\sqrt{x^2 + y^2}}; \qquad 4xyz = \frac{\lambda y^2}{\sqrt{x^2 + y^2}}; \qquad 4xyz = \lambda z.$$

Equating the right-hand sides of these equations gives

$$\frac{\lambda x^2}{\sqrt{x^2 + y^2}} = \frac{\lambda y^2}{\sqrt{x^2 + y^2}} \tag{28}$$

and

$$\lambda z = \frac{\lambda x^2}{\sqrt{x^2 + y^2}}. \tag{29}$$

From (28) it follows that $y = x$, since neither x nor y can be negative. With $y = x$, (29) gives $z = \dfrac{x^2}{\sqrt{2x^2}} = \dfrac{x}{\sqrt{2}}$. Substituting these expressions for y and z in (27) then gives

$$\frac{x}{\sqrt{2}} + \sqrt{x^2 + x^2} - 9 = \left(\frac{1}{\sqrt{2}} + \sqrt{2}\right)x - 9 = 0,$$

so

$$x = \frac{9}{\dfrac{1}{\sqrt{2}} + \sqrt{2}} = \frac{9\sqrt{2}}{1 + 2} = 3\sqrt{2}.$$

Thus, $y = 3\sqrt{2}$ and $z = 3$. The maximum volume is therefore

$$V = 4(3\sqrt{2})(3\sqrt{2})(3) = 216.$$

◇

Proof of Theorem 8 (Sketch): Let's first consider the planar case $x = (x, y) \in \mathbb{R}^2$. Then the constraint equation $g(x) = g(x, y) = 0$ determines a curve C in the xy-plane on which the point x_0 must lie. This curve C is a *level curve* for the function g. Thus, by statement (16), Section 18.7, if $\nabla g(x_0) \neq 0$ then $\nabla g(x_0)$ is orthogonal to C at x_0. We will complete the proof for the planar case by showing that $\nabla f(x_0)$ is also orthogonal to C at x_0.

Let $r(t) = x(t)i + y(t)j$ be a parameterization for C so that $r(t_0) = x_0$ and so that $r'(t_0) \neq 0.$* Since $x_0 = r(t_0)$ maximizes $f(x)$ on C, the scalar function $f \circ r$ has a relative extremum at $t = t_0$. Thus, by the Chain Rule,

$$(f \circ r)'(t_0) = \nabla f(r(t_0)) \cdot r'(t_0) = 0. \tag{30}$$

This shows that $\nabla f(r(t_0)) = \nabla f(x_0)$ is orthogonal to $r'(t_0)$. Since $r'(t_0)$ is *tangent* to C at x_0, it follows that $\nabla f(x_0)$ is orthogonal to C at x_0.

Since $\nabla f(x_0)$, $\nabla g(x_0)$, and C all lie in the xy-plane, we may now conclude, since $\nabla f(x_0)$ and $\nabla g(x_0)$ are both orthogonal to C at x_0, that $\nabla f(x_0)$ and $\nabla g(x_0)$ are parallel. That is, $\nabla f(x_0) = \lambda \nabla g(x_0)$ for some constant λ.

The space case $x \in \mathbb{R}^3$ is similar to the planar case. However, the constraint equation $g(x) = 0$ now determines a surface S in $\mathbb{R}^3$ that is a level surface for the function $u = g(x)$. By statement (19), Section 18.7, if $\nabla g(x_0) \neq 0$ then $\nabla g(x_0)$ is orthogonal to S at x_0.

As before, we will show that $\nabla f(x_0)$ is also orthogonal to S at x_0. We begin by letting C be an arbitrary curve on S with parameterization $r(t) = x(t)i + y(t)j + z(t)k$ for which $r(t_0) = x_0$ and $r'(t_0) \neq 0$. Just as in the plane case, the scalar function $f \circ r$ has a relative extremum at t_0, so equation (30) again holds. Since $r'(t_0)$ is tangent to C at x_0, it is also tangent to S at x_0. This shows that $\nabla f(x_0)$ is orthogonal to a tangent at x_0. But since the curve C was arbitrary, so is its tangent $r'(t_0)$. It follows that $\nabla f(x_0)$ is orthogonal to *all* tangents to S at x_0, and therefore that $\nabla f(x_0)$ is orthogonal to S. Since two vectors orthogonal to a (smooth) surface at a common point must be parallel, it follows that $\nabla f(x_0) = \lambda \nabla g(x_0)$ for some constant λ. ◆

Exercise Set 18.8

In Exercises 1–14, find the maximum and minimum values of the given function subject to the given constraint.

1. $f(x, y) = 2x^2 + 4y^2$ subject to $x^2 + y^2 = 1$

2. $f(x, y) = x^2 + y$ subject to $x^2 + y^2 = 9$

3. $f(x, y) = x^3 - y^3$ subject to $x - y = 2$

4. $f(x, y) = xy$ subject to $x^2 + y^2 - 4y = 5$

5. $f(x, y) = xy$ subject to $x^2 + y^2 = 1$

6. $f(x, y) = y - x$ subject to $x^2 + y^2 = 2$

7. $f(x, y) = x^2 + 4x + 4y^2$ subject to $x^2 + 2y^2 = 4$

8. $f(x, y, z) = x + y + z$ subject to $x^2 + y^2 + z^2 = 4$

9. $f(x, y) = x^2 - 4x + 4y^2$ subject to $x^2 + y^2 = 1$

10. $f(x, y, z) = xyz$ subject to $2x^2 + y^2 + 4z^2 = 9$

11. $f(x, y, z) = x + 2y - z$ subject to $x^2 + y^2 + z^2 = 1$

12. $f(x, y, z) = \sqrt{xyz}$ subject to $x + y + z = 4$

13. $f(x, y, z) = x + y + z$ subject to $x^2 + y^2 + z^2 = 12$

---

*$r(t)$ can always be chosen so that $r'(t_0) \neq 0$ if $\nabla g(x_0) \neq 0$. This follows from the Implicit Function Theorem, a result usually discussed in courses on advanced calculus. Actually, the Implicit Function Theorem underlies several of the seemingly obvious statements being made here, which is why we refer to this argument only as a sketch of a proof.

14. $f(x, y, z) = x^2 + 2y^2 + 4z^2$ subject to $x^2 + y^2 + z^2 = 1$

15. Find the solution of Example 2.

16. Find the point on the ellipsoid $9x^2 + 36y^2 + 4z^2 = 36$ nearest the origin.

17. Find the point on the circle $(x - 3)^2 + (y + 2)^2 = 9$ nearest the origin.

18. Find the point on the sphere $x^2 + y^2 + z^2 = 1$ furthest from the point $(3, 2, 1)$.

19. Find the point on the ellipsoid $(x - 1)^2 + \dfrac{(y - 2)^2}{4} + \dfrac{(z - 1)^2}{9} = 1$ nearest the point $(1, -1, 1)$.

20. Find the point on the cone $z = \sqrt{x^2 + y^2}$ nearest the point $(3, 1, 0)$.

21. Find the point(s) on the hyperboloid $z = (y + 1)^2 - (x - 2)^2 + 1$ nearest the point $(2, -1, 1)$.

22. The plane $2y - 3z = 8$ intersects the cone $z^2 = 4x^2 + 4y^2$ in an ellipse. Find the highest and lowest points of intersection.

23. Find the dimensions of the rectangular box of maximum volume that can be inscribed in a sphere of radius r.

24. Find the dimensions of the rectangular box of maximum volume which can be inscribed in the ellipsoid $x^2 + 4y^2 + 2z^2 = 8$.

25. Find the dimensions for the cylindrical jar, with volume 2 liters, that has minimum exterior surface area. (Assume that the jar has a lid and that the thickness of its walls is negligible.)

26–31. Rework Exercises 23–28 of Section 18.4 using the method of Lagrange multipliers.

18.9 RECONSTRUCTING A FUNCTION FROM ITS GRADIENT

In several ways, the gradient has emerged as the analogue of the derivative for functions of a single variable. It is therefore reasonable to ask whether the concept of antidifferentiation makes sense for gradients and, if so, whether this concept is useful. The goal of this section is to define what we mean by the "antiderivative" of a gradient (we don't really use this particular terminology), to show how such functions can be found, and to prove a theorem indicating when such functions exist. We will restrict our discussion to functions of two variables, although analogous results may be developed in more general settings. The results of this section will be used extensively in Chapter 20.

Potential Functions

We begin by recalling that a gradient is a vector-valued function

$$\nabla f(x, y) = \frac{\partial f}{\partial x}(x, y)\mathbf{i} + \frac{\partial f}{\partial y}(x, y)\mathbf{j}.$$

The antidifferentiation question for gradients is therefore the following: Given a vector-valued function of the form

$$\mathbf{F}(x, y) = M(x, y)\mathbf{i} + N(x, y)\mathbf{j}$$

is there a function f of two variables for which

$$\mathbf{F}(x, y) = \nabla f(x, y)?$$

We prefer not to refer to f as the antiderivative for the vector-valued function $\mathbf{F}$. Instead, we call it a *potential function*, or simply a *potential*. (The reason for this particular terminology has to do with the roles played by such functions in mechanics and in the theory of electricity and magnetism.)

Given the vector-valued function $\mathbf{F}$, the task of finding a function f for which $\mathbf{F}(x, y) = \nabla f(x, y)$ is therefore referred to as finding a potential for $\mathbf{F}$. This is also what we mean by "reconstructing the function f from its gradient." As Example 1 shows, potentials are not unique.

DEFINITION 6

The function f of two variables is called a **potential** for the vector-valued function

$$F(x, y) = M(x, y)\mathbf{i} + N(x, y)\mathbf{j}$$

in a rectangle $Q \subseteq \mathbb{R}^2$ if $\dfrac{\partial f}{\partial x}(x, y)$ and $\dfrac{\partial f}{\partial y}(x, y)$ exist and if

$$F(x, y) = \nabla f(x, y)$$

for all (x, y) in Q.

Example 1

If $f(x, y) = xe^{2y}$, then

$$\nabla f(x, y) = e^{2y}\mathbf{i} + 2xe^{2y}\mathbf{j}.$$

Thus, the function

$$F(x, y) = e^{2y}\mathbf{i} + 2xe^{2y}\mathbf{j}$$

has the function $f(x, y) = xe^{2y}$ as a potential. However, the function $g(x, y) = xe^{2y} + C$ is also a potential for F, since

$$\nabla g(x, y) = e^{2y}\mathbf{i} + 2xe^{2y}\mathbf{j} = F(x, y).$$

Thus, if f is a potential for F, so is $f + C$ for any constant C. ◇

Finding Potentials

If a potential f for F exists, how is it found? According to Definition 6 the following must be true.

If $F(x, y) = M(x, y)\mathbf{i} + N(x, y)\mathbf{j} = \nabla f(x, y)$, then   (1)

$$M(x, y) = \frac{\partial f}{\partial x}(x, y)$$

and

$$N(x, y) = \frac{\partial f}{\partial y}(x, y). \qquad (2)$$

Equations (1) and (2) are the keys to finding f. Beginning with equation (1) and integrating "partially with respect to x" we find

$$f(x, y) = \int \left(\frac{\partial f}{\partial x}(x, y) \right) dx = \int M(x, y)\, dx = G_1(x, y) + h_1(y) + C_1. \quad (3)$$

By integrating "partially with respect to x" we mean treating y as a constant and integrating $M(x, y)$ as a function of x alone—just the reverse of partial differentiation. We must remember, however, that functions of y alone vanish entirely when differentiated partially with respect to x, so we must allow for the appearance of an entire function $h_1(y)$ of y alone when integrating partially with respect to x.

Applying the same idea to equation (2), we integrate partially with respect to y

to obtain an equation of the form

$$f(x, y) = \int \left(\frac{\partial f}{\partial y}(x, y) \right) dy = \int N(x, y)\, dy = G_2(x, y) + h_2(x) + C_2, \qquad (4)$$

where $h_2(x)$ is a function of x alone.

If we are fortunate, we can next equate the right-hand sides of equations (3) and (4) and determine the functions G_1, G_2, h_1, and h_2. The potential f can then be obtained from either equation (3) or equation (4).

Example 2

Let F be the vector-valued function

$$F(x, y) = (3x^2 + 2y^2)i + 4xyj.$$

Find a function f for which $F(x, y) = \nabla f(x, y)$ for all (x, y).

Strategy

Identify M and N.

Solution

Here $F(x, y) = M(x, y)i + N(x, y)j$ with

$$M(x, y) = 3x^2 + 2y^2; \qquad N(x, y) = 4xy.$$

Integrate M partially with respect to x to find an expression for $f(x, y)$.

Partially integrating M with respect to x according to equation (3) gives

$$f(x, y) = \int M(x, y)\, dx \qquad (5)$$

$$= \int (3x^2 + 2y^2)\, dx$$

$$= x^3 + 2xy^2 + h_1(y) + C_1$$

Integrate N partially with respect to y to obtain a second expression for $f(x, y)$.

where h_1 is a function of y alone. Then, partially integrating N with respect to y, we find

$$f(x, y) = \int N(x, y)\, dy \qquad (6)$$

$$= \int 4xy\, dy$$

$$= 2xy^2 + h_2(x) + C_2$$

where h_2 is a function of x alone.

Equate the two expressions for $f(x, y)$ and attempt to identify the unknown functions h_1 and h_2.

Equating these two expressions for $f(x, y)$ gives

$$x^3 + 2xy^2 + h_1(y) + C_1 = 2xy^2 + h_2(x) + C_2.$$

This equation is true if $h_1(y) = 0$, $h_2(x) = x^3$, and $C_1 = C_2 = C$. Both (5) and (6) then give the desired potential as

Read $f(x, y)$ from either expression.

$$f(x, y) = x^3 + 2xy^2 + C.$$

Partial differentiation verifies that

$$\nabla f(x, y) = F(x, y). \qquad \diamond$$

Example 3

Find a potential for the function

$$F(x, y) = (ye^{xy} - 2x \sin x^2)i + \left(\frac{1}{\sqrt{y}} + xe^{xy}\right)j.$$

Solution: Here $F(x, y) = M(x, y)i + N(x, y)j$ with

$$M(x, y) = ye^{xy} - 2x \sin x^2; \qquad N(x, y) = \frac{1}{\sqrt{y}} + xe^{xy}.$$

Integrating M partially with respect to x gives

$$f(x, y) = \int M(x, y)\, dx = \int (ye^{xy} - 2x \sin x^2)\, dx \qquad (7)$$
$$= e^{xy} + \cos x^2 + h_1(y) + C_1$$

where h_1 is a function of y alone. Integrating N partially with respect to y gives

$$f(x, y) = \int N(x, y)\, dy = \int \left(\frac{1}{\sqrt{y}} + xe^{xy}\right) dy \qquad (8)$$
$$= 2\sqrt{y} + e^{xy} + h_2(x) + C_2$$

where h_2 is a function of x alone. Equating the right-hand sides of equations (7) and (8) then shows that

$$h_1(y) = 2\sqrt{y}, \qquad h_2(x) = \cos x^2, \qquad \text{and} \qquad C_1 = C_2 = C. \qquad (9)$$

With the information in line (9), either of equations (7) and (8) gives that

$$f(x, y) = e^{xy} + \cos x^2 + 2\sqrt{y} + C.$$

Partial differentiation then verifies that $\nabla f(x, y) = F(x, y)$. ◇

Example 4

Show that the function

$$F(x, y) = yi - xj$$

has no potential. That is, F cannot be the gradient of a function f of two variables.

Solution: On the contrary, let us assume that there does exist a function f with $F(x, y) = \nabla f(x, y)$. Since

$$M(x, y) = y, \qquad \text{and} \qquad N(x, y) = -x,$$

we have

$$f(x, y) = \int M(x, y)\, dx = \int y\, dx = xy + h(y) + C_1 \qquad (10)$$

where h is a function of y alone. Furthermore, since we are assuming that $\dfrac{\partial f}{\partial y}(x, y)$ exists, h must be a differentiable function of y. We may therefore differentiate both sides of (10) with respect to y. Doing so, and using equation (2), gives

$$\frac{\partial f}{\partial y}(x, y) = x + h'(y) = -x = N(x, y).$$

Thus,

$$h'(y) = -2x. \tag{11}$$

But since h must be a function of y alone, so is h'. Thus, since h' cannot involve x, $\dfrac{\partial}{\partial x} h' = 0$. This shows that the result of differentiating both sides of equation (11) with respect to x is the contradictory statement $0 = -2$. The assumption that $F(x, y) = y\boldsymbol{i} - x\boldsymbol{j}$ has a potential therefore leads to a contradiction, and is therefore false. $F(x, y) = y\boldsymbol{i} - x\boldsymbol{j}$ has no potential. ◇

The result of Example 4 leaves us in need of information as to when a given vector function F has a potential. The information is supplied by the following theorem, whose proof employs the ideas of Examples 2 through 4.

THEOREM 9

Let M and N have continuous first partial derivatives in an open rectangle R in the xy-plane. Then the vector-valued function

$$F(x, y) = M(x, y)\boldsymbol{i} + N(x, y)\boldsymbol{j}$$

has a potential in R if and only if

$$\frac{\partial M}{\partial y}(x, y) = \frac{\partial N}{\partial x}(x, y) \tag{12}$$

for all (x, y) in R.

Before giving the proof, we note that condition (12) is satisfied for F in Example 2, since

$$\frac{\partial M}{\partial y}(x, y) = \frac{\partial}{\partial y}(3x^2 + 2y^2) = 4y = \frac{\partial N}{\partial x}(x, y).$$

Also, equation (12) holds for F in Example 3:

$$\begin{aligned}
\frac{\partial M}{\partial y}(x, y) &= \frac{\partial}{\partial y}(ye^{xy} - 2x \sin x^2) \\
&= e^{xy} + xye^{xy} \\
&= \frac{\partial}{\partial x}\left(\frac{1}{\sqrt{y}} + xe^{xy}\right) \\
&= \frac{\partial N}{\partial x}(x, y).
\end{aligned}$$

However, condition (12) fails for F in Example 4 since

$$\frac{\partial M}{\partial y} = \frac{\partial}{\partial y}(y) = 1 \neq -1 = \frac{\partial}{\partial x}(-x) = \frac{\partial N}{\partial x}.$$

Proof of Theorem 9: We first assume that F has a potential f and show that equation (12) holds. This means that

$$F(x, y) = \nabla f(x, y)$$

where $M(x, y) = \dfrac{\partial f}{\partial x}(x, y)$ and $N(x, y) = \dfrac{\partial f}{\partial y}(x, y)$. Since both M and N are assumed to have continuous first partial derivatives in R, it follows that

$$\frac{\partial^2 f}{\partial y \partial x} = \frac{\partial}{\partial y}\left(\frac{\partial f}{\partial x}\right) = \frac{\partial M}{\partial y}(x, y) \tag{13}$$

and

$$\frac{\partial^2 f}{\partial x \partial y} = \frac{\partial}{\partial x}\left(\frac{\partial f}{\partial y}\right) = \frac{\partial N}{\partial x}(x, y). \tag{14}$$

Moreover, by Theorem 2 the mixed partial derivatives on the left-hand sides of equations (13) and (14) are equal. Thus, the right-hand sides of these equations are equal, and condition (12) is obtained.

The remaining part of the proof involves assuming that condition (12) holds and actually finding (at least formally) a potential for F. Taking a cue from Examples 2 through 4 we begin by fixing a point (x_0, y_0) in R and defining the function f as

$$f(x, y) = \int_{x_0}^{x} M(t, y_0)\, dt + \int_{y_0}^{y} N(x, s)\, ds. \tag{15}$$

Figure 9.1 may help you understand how f has been ''pulled out of the hat.'' Since we want to construct f in such a way that $\dfrac{\partial f}{\partial x}(x, y) = M(x, y)$ and $\dfrac{\partial f}{\partial y}(x, y) =$

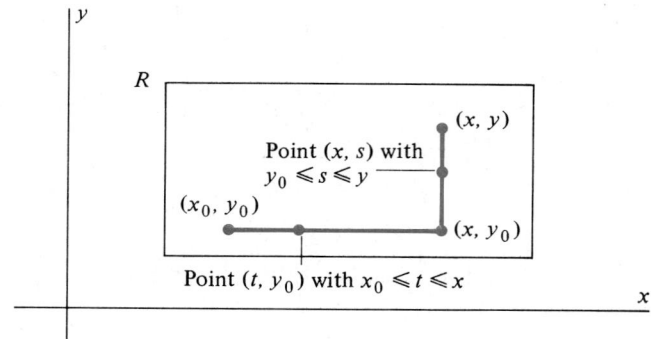

Figure 9.1 Domain of $M(x, y)$ and $N(x, y)$ in Theorem 9.

$N(x, y)$, the idea is to integrate M along a path involving only change in x and to integrate N along a path involving only change in y. The reason we require R to be a rectangle is simply to ensure that such a path lies entirely within R.

Since M and N are continuous in R, and since R is a rectangle, $f(x, y)$ is defined for all $(x, y) \in R$. Moreover, by the Fundamental Theorem of Calculus it follows that

$$\frac{\partial f}{\partial y}(x, y) = \frac{\partial}{\partial y}\left\{\int_{x_0}^{x} M(t, y_0)\, dt + \int_{y_0}^{y} N(x, s)\, ds\right\} \tag{16}$$

$$= \frac{\partial}{\partial y} \int_{y_0}^{y} N(x, s) \, ds$$

$$= N(x, y)$$

since the first integral inside the braces does not involve y. We also want to calculate $\frac{\partial f}{\partial x}(x, y)$, but this is a bit harder. We find that

$$\frac{\partial f}{\partial x}(x, y) = \frac{\partial}{\partial x} \left\{ \int_{x_0}^{x} M(t, y_0) \, dt + \int_{y_0}^{y} N(x, s) \, ds \right\} \tag{17}$$

$$= M(x, y_0) + \frac{\partial}{\partial x} \int_{y_0}^{y} N(x, s) \, ds.$$

In courses on advanced calculus it is proved that

$$\frac{\partial}{\partial x} \int_{y_0}^{y} N(x, s) \, ds = \int_{y_0}^{y} \frac{\partial}{\partial x} N(x, s) \, ds. \tag{18}$$

That is, we may pass the partial differentiation with respect to x under the y-integral sign in the integral on the right-hand side of equation (17). Using (18) we return to (17) and use condition (15) to find

$$\frac{\partial f}{\partial x}(x, y) = M(x, y_0) + \int_{y_0}^{y} \frac{\partial}{\partial x} N(x, s) \, ds \tag{19}$$

$$= M(x, y_0) + \int_{y_0}^{y} \frac{\partial}{\partial s} M(x, s) \, ds \qquad \left(\frac{\partial N}{\partial x}(x, s) = \frac{\partial M}{\partial s}(x, s) \right)$$

$$= M(x, y_0) + M(x, s) \Big]_{s=y_0}^{s=y}$$

$$= M(x, y_0) + [M(x, y) - M(x, y_0)]$$

$$= M(x, y).$$

Reviewing equations (16) and (19), we find that the function f defined in line (15) has the properties

$$\frac{\partial f}{\partial x}(x, y) = M(x, y); \qquad \frac{\partial f}{\partial y}(x, y) = N(x, y).$$

That is, $\nabla f = F$, so F indeed has a potential. ◆

Exercise Set 18.9

In Exercises 1–12, determine whether the given vector function has a potential. If so, find a potential for F.

1. $F(x, y) = i - j$

2. $F(x, y) = 6yi + 6xj$

3. $F(x, y) = \pi y i - \pi x j$

4. $F(x, y) = 3xi + 4yj$

5. $F(x, y) = (x^2 - y)i + (x + y^2)j$

6. $F(x, y) = \sin y i + x \cos y j$

7. $F(x, y) = (3x^2 \cos y - 1)i + (2y - x^3 \sin y)j$

8. $F(x, y) = (2xy^3 - 2x)i + (3x^2y^2 + 1)j$

9. $F(x, y) = \left(2xe^{xy} + x^2 \, ye^{xy} \right)i + \left(x^3 e^{xy} - \frac{1}{2\sqrt{y}} \right)j$

10. $F(x, y) = (x^3 - y \sin x)i + (y^3 - \cos x)j$

11. $F(x, y) = (2xy^2 - y \sin x)i + (2x^2y + \cos x)j$

12. $F(x, y) = 2x \, \text{Tan}^{-1} \, yi + \frac{x^2}{1 + y^2}j$

SUMMARY OUTLINE OF CHAPTER 18

◆ A **level curve** for the function $z = f(x, y)$ is the graph of the equation $f(x, y) = c$ in the xy-plane. A **level surface** for the function $w = f(x, y, z)$ is the graph of an equation $f(x, y, z) = c$ in space. (page 775)

◆ If $x \in \mathbb{R}^n$ and $f: \mathbb{R}^n \to \mathbb{R}$, $\lim\limits_{x \to x_0} f(x) = L$ means $|f(x) - L| \to 0$ as $x \to x_0$. (page 779)

◆ The function $f: \mathbb{R}^n \to \mathbb{R}$ is **continuous** at x_0 if $\lim\limits_{x \to x_0} f(x) = f(x_0)$. (page 781)

◆ The **partial derivatives** of f are the limits (page 783)

$$\frac{\partial f}{\partial x}(x, y, z) = \lim_{h \to 0} \frac{f(x + h, y, z) - f(x, y, z)}{h},$$

$$\frac{\partial f}{\partial y}(x, y, z) = \lim_{h \to 0} \frac{f(x, y + h, z) - f(x, y, z)}{h},$$

and

$$\frac{\partial f}{\partial z}(x, y, z) = \lim_{h \to 0} \frac{f(x, y, z + h) - f(x, y, z)}{h}.$$

◆ **Theorem:** $\dfrac{\partial^2 f}{\partial y \partial x}(x, y) = \dfrac{\partial^2 f}{\partial x \partial y}(x, y)$ if f, $\dfrac{\partial f}{\partial x}$, $\dfrac{\partial f}{\partial y}$, $\dfrac{\partial^2 f}{\partial x \partial y}$, and $\dfrac{\partial^2 f}{\partial y \partial x}$ are continuous. (page 789)

◆ The plane tangent to the graph of $z = f(x, y)$ at (x_0, y_0, z_0) has equation (page 792)

$$\frac{\partial f}{\partial x}(x_0, y_0)(x - x_0) + \frac{\partial f}{\partial y}(x_0, y_0)(y - y_0) - (z - z_0) = 0.$$

◆ **Theorem:** If f has a relative extremum at (x_0, y_0) then either (page 797)

(i) $\dfrac{\partial f}{\partial x}(x_0, y_0) = \dfrac{\partial f}{\partial y}(x_0, y_0) = 0$, or

(ii) one or both of $\dfrac{\partial f}{\partial x}(x_0, y_0)$ and $\dfrac{\partial f}{\partial y}(x_0, y_0)$ fail to exist.

◆ **Theorem:** If $\dfrac{\partial f}{\partial x}(x_0, y_0) = \dfrac{\partial f}{\partial y}(x_0, y_0) = 0$ and if $A = \dfrac{\partial^2 f}{\partial x^2}(x_0, y_0)$, $B = \dfrac{\partial^2 f}{\partial y \partial x}(x_0, y_0)$, $C = \dfrac{\partial^2 f}{\partial y^2}(x_0, y_0)$, and (page 800) $D = B^2 - AC$, then

(i) If $D < 0$, and $A < 0$, $f(x_0, y_0)$ is a relative maximum.
(ii) If $D < 0$ and $A > 0$, $f(x_0, y_0)$ is a relative minimum.
(iii) If $D > 0$, (x_0, y_0) is a saddle point.
(iv) If $D = 0$ there is no conclusion.

◆ **Theorem:** (Linear Approximation) If f and its first partials are continuous, then (page 806)

$$f(x_0 + \Delta x, y_0 + \Delta y) = f(x_0, y_0) + \frac{\partial f}{\partial x}(x_0, y_0)\,\Delta x + \frac{\partial f}{\partial y}(x_0, y_0)\,\Delta y + \epsilon_1 \Delta x + \epsilon_2 \Delta y$$

where $\epsilon_1 \to 0$ and $\epsilon_2 \to 0$ as $\Delta x \to 0$ and $\Delta y \to 0$.

◆ **Chain Rule:** If $w = f(x(t), y(t))$ then (page 814)

$$\frac{dw}{dt} = \frac{\partial f}{\partial x} \cdot \frac{dx}{dt} + \frac{\partial f}{\partial y} \cdot \frac{dy}{dt}.$$

◆ The **directional derivative** $D_u f(x_0, y_0)$ in the direction of the unit vector $u = u_1 i + u_2 j$ is (page 821)

$$D_u f(x_0, y_0) = \lim_{t \to 0^+} \frac{f(x_0 + tu_1, y_0 + tu_2) - f(x_0, y_0)}{t}.$$

◆ The **gradient** $\nabla f(x_0, y_0)$ of the function f at (x_0, y_0) is the **vector** (page 825)

$$\nabla f(x_0, y_0) = \frac{\partial f}{\partial x}(x_0, y_0)\mathbf{i} + \frac{\partial f}{\partial y}(x_0, y_0)\mathbf{j}$$

or

$$\nabla f(x_0, y_0, z_0) = \frac{\partial f}{\partial x}(x_0, y_0, z_0)\mathbf{i} + \frac{\partial f}{\partial y}(x_0, y_0, z_0)\mathbf{j} + \frac{\partial f}{\partial z}(x_0, y_0, z_0)\mathbf{k}.$$

◆ If $\dfrac{\partial f}{\partial x}(x, y)$ and $\dfrac{\partial f}{\partial y}(x, y)$ are continuous, then (page 826)

$$D_{\mathbf{u}} f(x_0, y_0) = \frac{\partial f}{\partial x}(x_0, y_0)u_1 + \frac{\partial f}{\partial x}(x_0, y_0)u_2$$

$$= \nabla f(x_0, y_0) \cdot \mathbf{u}.$$

◆ $\nabla f(x_0)$ points in the direction of most rapid increase of the function f at x_0. (page 826)

◆ $\nabla f(x_0)$ is orthogonal to the level curve for $z = f(x, y)$ at $x_0 = (x_0, y_0)$. (page 829)

◆ $\nabla f(x_0)$ is orthogonal to the level surface for $z = f(x, y, z)$ at $x_0 = (x_0, y_0, z_0)$. (page 830)

◆ The extreme values of the function $w = f(\mathbf{x})$ subject to the constraint $g(\mathbf{x}) = 0$ occur at points x_0 where $\nabla f(x_0) = \lambda \nabla g(x_0)$, where λ is constant. (page 832)

◆ The vector-valued function $\mathbf{F}(x, y) = M(x, y)\mathbf{i} + N(x, y)\mathbf{j}$ is said to have the **potential** f if $\mathbf{F}(x, y) = \nabla f(x, y)$. (page 839)

◆ **Theorem:** If M, N and their first partial derivatives are continuous in a rectangle R, then (page 843)

$$\mathbf{F}(x, y) = M(x, y)\mathbf{i} + N(x, y)\mathbf{j}$$

has a potential in R if and only if $\dfrac{\partial M}{\partial y} = \dfrac{\partial N}{\partial x}$ for all $(x, y) \in R$.

REVIEW EXERCISES—CHAPTER 18

1. Does $\displaystyle\lim_{(x,y)\to(0,0)} \frac{x^2 - y^2}{x^2 + y^2}$ exist? If so, find it.

2. Find $\displaystyle\lim_{(x,y)\to(0,0)} \frac{2x^2 y^3}{(x^2 + y^2)^2}$.

3. Sketch level curves for the function $f(x, y) = \dfrac{x}{x^2 + y^2}$ corresponding to levels $z = -2, -1, 1, 2, 4$.

4. Find $\dfrac{df}{dt}$ if $f(x, y) = x \operatorname{Tan}^{-1}\left(\dfrac{y}{x}\right)$, $x = 1 + t^2$, and $y = 1 - t$.

5. Find a vector normal to the curve $x^3 - 2xy^2 + y + 5 = 0$ at the point $(1, 2)$.

6. Find an equation for the plane tangent to the graph of the equation $x^3 + y^3 - 6xy + z = 0$ at the point $(2, 2, 8)$.

7. Find a vector normal to the surface $e^x \cos y - z = 4$ at the point $(0, \pi, -5)$.

8. What is the z-coordinate of the point $P = (1, 3, z)$ if P lies on the plane tangent to the ellipsoid $4x^2 + y^2 + 9z^2 = 17$ at the point $(1, 2, 1)$?

9. Find both first order partial derivatives for the function $f(x, y) = x \sin \sqrt{x^2 + y^2}$.

10. Find an equation for the plane tangent to the graph of $x^3 + y^3 + xz^2 + z^3 - 9 = 0$ at the point $(2, 1, -2)$.

11. Let $f(x, y, z) = xy^3 + x^2\sqrt{y^2 + z^2}$. Find the directional derivative of f in the direction of the vector $2\mathbf{i} + \mathbf{j} + 2\mathbf{k}$ at the point $(2, 3, 4)$.

12. Show that $f(x, y, z) = (ax + by + cz)^3$ satisfies the partial differential equation

$$x\frac{\partial f}{\partial x} + y\frac{\partial f}{\partial y} + z\frac{\partial f}{\partial z} = 3f.$$

13. Find $\dfrac{\partial f}{\partial r}$ and $\dfrac{\partial f}{\partial s}$ if $f(x, y) = 3x^3 + 2xy^2 - y^2$, $x = 2r + 5s$, and $y = r - 2s^2$.

14. Suppose that $w = f(x, y)$ has partial derivatives with re-

spect to both variables. Show that the function $z = f(x - y, y - x)$ satisfies the partial differential equation

$$\frac{\partial z}{\partial x} + \frac{\partial z}{\partial y} = 0.$$

15. Find an equation for the line tangent to the graph of $x \cos \pi y + x^2 e^y = 6$ at the point $(2, 0)$.

16. Find the directional derivative $D_u f$ of the function $f(x, y) = y^2 \operatorname{Sin}^{-1} x + y e^x$ at the point $(0, 1)$ in the direction of the vector $w = i + \sqrt{e}j$.

17. Let $f(x, y)$ and $g(x, y)$ have partial derivatives with respect to both variables. Show that $\nabla(\alpha f + \beta g) = \alpha \nabla f + \beta \nabla g$ where α and β are constants.

18. Find the maximum value of the directional derivative for the function $w = xy^3 + z \cos y - \ln(x^2 + y)$ at the point $(1, 0, 4)$.

19. Find a vector pointing in the direction of most rapid decrease for the function $f(x, y, z) = xyz - z \operatorname{Tan}^{-1}(y/x)$ at the point $(1, 2, 3)$.

20. Find an equation for the plane tangent to the surface $z^2 x - 2zy + e^{xy} = 9$ at the point $(2, 0, 2)$.

21. Show that the function $f(x, t) = e^{x+ct}$ is a solution of the wave equation

$$\frac{\partial^2 f}{\partial t^2} = c^2 \frac{\partial^2 f}{\partial x^2}.$$

22. Find the maximum and minimum values of the function $f(x, y) = 4x^2 + xy + 2y^2$ on the square $S = \{(x, y) \mid -1 \le x \le 1, -1 \le y \le 1\}$.

23. Use the method of Lagrange multipliers to find the rectangular box of largest volume that can be inscribed in the ellipsoid $x^2 + \dfrac{y^2}{9} + \dfrac{z^2}{4} = 1$.

Find and classify all relative extrema.

24. $f(x, y) = 4y^2 - 2x^2$

25. $f(x, y) = 3x^2 + xy - 6y^2$

26. $f(x, y) = x^2 y + xy^2 + 4x + 4y$

27. $f(x, y) = e^{x^2 - 4xy}$

28. $f(x, y) = \ln(1 + x^2 + y^2)$

29. $f(x, y) = e^{1+x^2-y^2}$

30. $f(x, y) = 6x^2 - 2x - 3xy + y^2 + 5y + 5$

31. The directional derivative of $w = f(x, y)$ at P in the direction of the vector $u_1 = j$ is $D_{u_1}f(P) = 3$, and the directional derivative of f at P in the direction of $u_2 = 3i + 4j$ is $D_{u_2}f(P) = 3$ also. Find

a. $\dfrac{\partial f}{\partial x}(P)$, **b.** $\dfrac{\partial f}{\partial y}(P)$, and **c.** $\nabla f(P)$.

32. The directional derivative of f at (x_0, y_0) in the direction of the unit vector u is 6. What is
a. $D_{2u}f(x_0, y_0)$?
b. $D_{-u}f(x_0, y_0)$?

33. Find a function f for which $\nabla f(x, y) = 2xe^y i + (x^2 e^y - \sin y)j$.

34. Find a potential for the vector function $F(x, y) = (ye^x + e^y)i + (1 + e^x + xe^y)j$.

35. When three resistors r_1, r_2, and r_3 are connected in parallel, the net resistance R is determined by the equation

$$\frac{1}{R} = \frac{1}{r_1} + \frac{1}{r_2} + \frac{1}{r_3}.$$

Suppose $r_1 = 10$ ohms, $r_2 = 20$ ohms, and $r_3 = 25$ ohms. Approximate the change in the value of R that results from each of r_1, r_2, and r_3 being increased by 10%.

36. Is the function

$$f(x, y) = \begin{cases} \dfrac{6xy}{x^2 + y^2}, & (x, y) \neq (0, 0) \\ 0, & (x, y) = (0, 0) \end{cases}$$

differentiable at $(0, 0)$? Why or why not?

37. Show that the function

$$w = e^{x-y} + \cos(y - z) + \sqrt{z - x}$$

satisfies the partial differential equation

$$\frac{\partial w}{\partial x} + \frac{\partial w}{\partial y} + \frac{\partial w}{\partial z} = 0.$$

38. An airplane flying due north at a speed of 200 km/h and an altitude of 2 km passes directly over an automobile travelling due east along a straight highway at a speed of 100 km/h. At what rate are the plane and the automobile moving apart after 6 minutes?

39. Find the maximum and minimum values of the function

$$f(x, y) = 3x - 2y + 5$$

on or inside the ellipse $\dfrac{x^2}{4} + \dfrac{y^2}{9} = 1$.

40. A closed rectangular box is to contain 1000 cm³. If the material for the top and bottom costs 2¢/cm² and the material for the sides costs 3¢/cm², find the dimensions which minimize cost.

41. Let u and v be distinct unit vectors. Is $D_{u+v}f(x_0, y_0) = D_u f(x_0, y_0) + D_v f(x_0, y_0)$? Why or why not?

42. Find an equation for the plane through the point $(1, 2, 1)$ with positive x-, y-, and z-intercepts that bounds the solid of least volume in the first octant.

43. Find the point on the surface

$$x^2 + y^2 + z^2 = 16$$

where the function $f(x, y, z) = x + 2y - 3z + 1$ is a maximum.

44. Suppose that the level surfaces $f(x, y, z) = K$ and $g(x, y, z) = L$ intersect in a curve C. Let x_0 maximize $h(x, y, z)$ on C, and assume that f, g, and h have continuous partial derivatives. Show that $\nabla f(x_0)$, $\nabla g(x_0)$, and $\nabla h(x_0)$ lie in a common plane when based at x_0. Conclude that $\nabla f(x_0) = \lambda \nabla g(x_0) + \mu \nabla h(x_0)$ for some constants λ and μ.

45. Prove that if f is differentiable for all (x, y) in a closed and bounded disc D, then f is bounded on D. (This means that there exists a number M with $|f(x, y)| \leq M$ for all $(x, y) \in D$.)

46. The function $f(x, y, z)$ is called homogeneous of degree n if $f(tx, ty, tz) = t^n f(x, y, z)$ for all x, y, z, and t. Show that if n is an integer and f has continuous partial derivatives, then

$$x\frac{\partial f}{\partial x} + y\frac{\partial f}{\partial y} + z\frac{\partial f}{\partial z} = nf$$

when f is homogeneous of degree n.

47. For the function $f(x, y) = \sqrt{x^2 + y^2}$, show that all directional derivatives exist at $(0, 0)$, but that neither partial derivative exists at $(0, 0)$.

48. Show that the spheres with equations $x^2 + (y - 1)^2 + z^2 = 2$ and $x^2 + (y + 1)^2 + z^2 = 2$ intersect in a circle. Then show that at each point on the circle of intersection the tangent planes to the two circles are orthogonal.

49. Use the method of Lagrange multipliers to find the point on the plane $x - 3y + 2z = 6$ nearest the origin.

50. Find a vector in the direction of most rapid increase of the function $f(x, y, z) = e^{x-z} \operatorname{Tan}^{-1}(y + z)$ at the point $(3, 1, 2)$.

51. Find an equation for the plane tangent to the surface $x^2 + 4y^2 + z^2 = 12$ at the point $(2, 1, 2)$.

52. Find $\dfrac{d^2 f}{dt^2}$ if $f(x, y) = e^{y^2 - x^2}$, $x(t) = 4 - t^2$, $y(t) = t \cos \pi t$.

53. Find the minimum value of the function $f(x, y) = 3y^2 - xy + x^2$ subject to the constraint that $x^2 + 3y^2 = 3$.

54. Find the maximum volume for a rectangular solid inscribed in the ellipsoid

$$\frac{x^2}{4} + \frac{y^2}{9} + \frac{z^2}{4} = 1.$$

55. Show that $u = f(xy)$ satisfies the partial differential equation

$$x\frac{\partial u}{\partial x} = y\frac{\partial u}{\partial y}.$$

55. Show that $u = f(x + 2y^2)$ satisfies the partial differential equation

$$\frac{\partial u}{\partial y} = 4y\frac{\partial u}{\partial x}.$$

57. Show that $u = f(t - x)$ is a solution of the partial differential equation

$$\frac{\partial^2 u}{\partial x^2} + 2\frac{\partial^2 u}{\partial x \partial t} + \frac{\partial^2 u}{\partial t^2} = 0.$$

58. Show that $u = f(x^2 + y^2)$ satisfies the partial differential equation

$$\frac{\partial u}{\partial y} = \frac{y}{x} \cdot \frac{\partial u}{\partial x}.$$

59. Show that if f is any differentiable function, the composite function $u(x, t) = f(bx - t)$ is a solution of the partial differential equation

$$\frac{\partial u}{\partial x} = -b\frac{\partial u}{\partial t}.$$

Chapter 19
Double and Triple Integrals

The goal of this chapter is to extend the theory of the Riemann integral to functions of two and three variables. You will find that most of what we do here is a straightforward generalization of the theory of integration for functions of a single variable.

19.1 THE DOUBLE INTEGRAL OVER A RECTANGLE

The original motivation for the definite integral was the problem of calculating the area of a region R bounded by the graph of the continuous nonnegative function $y = f(x)$ and the x-axis for $a \le x \le b$ (Figure 1.1). We did so by partitioning the

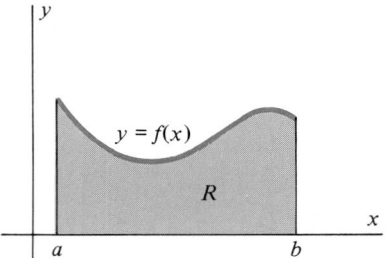

Figure 1.1 Region bounded by continuous nonnegative function $f(x)$.

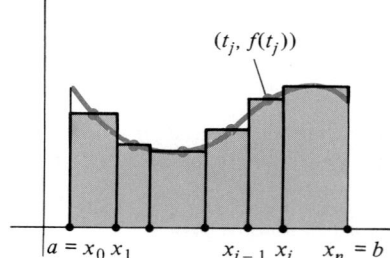

Figure 1.2 Riemann sum approximates region R by rectangles.

interval $[a, b]$ into subintervals of length $\Delta x_j = x_j - x_{j-1}$. After choosing one number t_j arbitrarily in each interval, we formed the approximating Riemann sum

$$S_n = \sum_{j=1}^{n} f(t_j)\, \Delta x_j$$

representing the sum of the areas of the rectangles illustrated in Figure 1.2. We then proved that the limit of this Riemann sum, as $n \to \infty$ and as the norm of the partition $\|P_n\| \to 0$, is the desired area. This led to the definition of the definite integral

$$\int_a^b f(x)\, dx = \lim_{n \to \infty} \sum_{j=1}^{n} f(t_j)\, \Delta x_j.$$

For functions of two variables, the primary motivation leading to the corresponding integral (called the *double* integral) will be the calculation of volume rather than area. This is because a region Q in the domain of $z = f(x, y)$ and the graph of f over

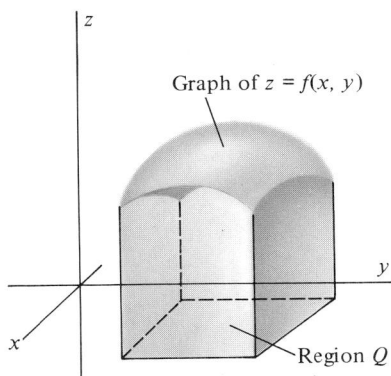

Figure 1.3 Graph of $z = f(x, y)$ over the region Q bounds a solid in xyz space.

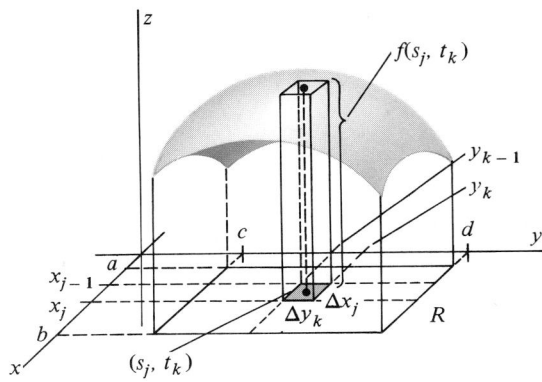

Figure 1.4 Volume of solid bounded by graph of f over R is approximated using rectangular prisms of volume $\Delta V_{jk} = f(s_j, t_k)\Delta x_j \Delta y_k$.

Q bound a solid in space (see Figure 1.3). We will approximate the volume of this solid by rectangular prisms (Figure 1.4), and the entire development will be very much analogous to the one-variable case.

Developing the Double Integral Over a Rectangle

We begin with the problem of calculating the volume V of a solid bounded above by the graph of the continuous nonnegative function $z = f(x, y)$, below by the rectangle $R = \{(x, y) \mid a \leq x \leq b, c \leq y \leq d\}$ in the xy-plane, and on four sides by the vertical planes $x = a$, $x = b$, $y = c$, and $y = d$ (see Figures 1.4 and 1.5).

Using the same terminology as in the one-variable case, we let $P_1 = \{a = x_0, x_1, x_2, \ldots, x_n = b\}$ be a partition of the interval $[a, b]$, and we let $P_2 = \{c = y_0, y_1, y_2, \ldots, y_m = d\}$ be a partition of the interval $[c, d]$. Also, we let

$$\Delta x_j = x_j - x_{j-1}, \quad j = 1, 2, \ldots, n$$

and

$$\Delta y_k = y_k - y_{k-1}, \quad k = 1, 2, \ldots, m.$$

As Figure 1.5 illustrates, these partitions determine a grid dividing the region R into rectangles R_{jk} of area $\Delta A_{jk} = \Delta x_j \Delta y_k$ for $j = 1, 2, \ldots, n$ and $k = 1, 2, \ldots, m$.

We refer to this grid as the *partition P* induced on R by the partitions P_1 and P_2, and we define the norm $\|P\|$ of this partition to be the larger of the norms $\|P_1\|$ and $\|P_2\|$ of the partitions P_1 and P_2. That is,

$$\|P\| = \max\{\|P_1\|, \|P_2\|\}$$
$$= \max\{\Delta x_1, \Delta x_2, \ldots, \Delta x_n, \Delta y_1, \Delta y_2, \ldots, \Delta y_m\}.$$

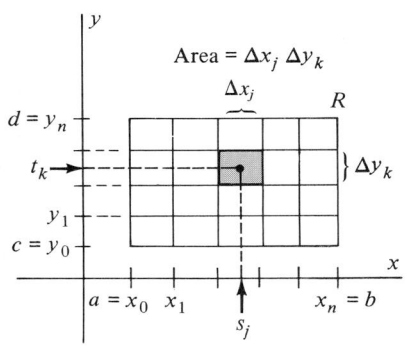

Figure 1.5 Rectangle R in xy-plane. Area of rectangle R_{jk} is $\Delta A = \Delta x_j \Delta y_k$.

Our idea is now to approximate the volume of the region above the rectangle R_{jk} and below the graph of f with the volume of the rectangular prism with base of area $\Delta A_{jk} = \Delta x_j \Delta y_k$. For the height of this prism we use the function value $f(s_j, t_k)$, where the "test point" (s_j, t_k) is chosen arbitrarily in the rectangle R_{jk}. This leads to the approximation

$$S_{n,\,m} = \sum_{j=1}^{n} \sum_{k=1}^{m} f(s_j, t_k)\,\Delta A_{jk}, \quad \Delta A_{jk} = \Delta x_j \Delta y_k, \tag{1}$$

which we refer to as a *Riemann double sum* for the function f on the rectangle R. Just as in the one-variable case, we now wish to ask whether the limit of such Riemann sums, as the sizes of the rectangles R_{jk} become very small, exists. Here is a formal definition of what we mean by the limit of the Riemann sums in equation (1).

DEFINITION 1

We say that the number S is the *limit* of the Riemann sum in equation (1) as $\|P\| \to 0$ (and, therefore, as $n \to \infty$ and $m \to \infty$), written

$$S = \lim_{\|P\| \to 0} \sum_{j=1}^{n} \sum_{k=1}^{m} f(s_j, t_k) \, \Delta A_{jk},$$

if, given any number $\epsilon > 0$, there exists a corresponding number $\delta > 0$ so that

$$\text{if } 0 < \|P\| < \delta, \text{ then } \left| S - \sum_{j=1}^{n} \sum_{k=1}^{m} f(s_j, t_k) \, \Delta A_{jk} \right| < \epsilon$$

regardless of how the test points (s_j, t_k) are chosen in R_{jk}.

Definition 1 simply says that the limit of the Riemann sums in equation (1) is the number S if these sums approach the number S as the number of rectangles R_{jk} in the partition P becomes infinite *and* as both dimensions (Δx_j and Δy_k) of each rectangle R_{jk} approach zero.

As in the one-variable case, the Riemann double sum in equation (1) will have a limit S when the function f is *continuous* on the rectangle R. We state this result without proof.

THEOREM 1

Let f be a function of two variables defined on the rectangle R. If f is continuous on R and P denotes a partition of R as described above, the limit

$$S = \lim_{\|P\| \to 0} \sum_{j=1}^{n} \sum_{k=1}^{m} f(s_j, t_k) \, \Delta A_{jk}$$

exists.

As before, we refer to the limit of the Riemann sum in equation (1), when it exists, as the Riemann integral of f on R. Because we are now in the two-variable case, we refer to this integral as a *double integral*.

DEFINITION 2

Let f be a continuous function of two variables on the rectangle $R = \{(x, y) \mid a \le x \le b, c \le y \le d\}$. The **double integral of f over the rectangle R** is the number

$$\iint_R f(x, y) \, dA = \lim_{\|P\| \to 0} \sum_{j=1}^{n} \sum_{k=1}^{m} f(s_j, t_k) \, \Delta A_{jk}, \qquad \Delta A_{jk} = \Delta x_j \Delta y_k$$

where P denotes a partition of the rectangle R and (s_j, t_k) is an arbitrary point in the rectangle $R_{jk} = \{(x, y) \mid x_{j-1} \le x \le x_j, y_{k-1} \le y \le y_k\}$.

In the symbol

$$\iint\limits_{R} f(x, y) \, dA,$$

we use two integral signs to indicate that it represents the result of a double limit process—the x-interval $[a, b]$ and the y-interval $[c, d]$ have both been partitioned into increasingly small subintervals. The subscript R denotes the rectangle over which the integral is evaluated. For now, the symbol dA (which may also be written $dx \, dy$) indicates that the Riemann sum has been obtained by partitioning R into rectangles of area $\Delta A_{jk} = \Delta x_j \Delta y_k$. As before, the function f is referred to as the **integrand.**

According to the development that led to the Riemann sum, we conclude that the volume V of the solid bounded above by the graph of the continuous nonnegative function f and below by the rectangle R in the xy-plane is

$$V = \iint\limits_{R} f(x, y) \, dA. \tag{2}$$

But how do we evaluate the integral in (2)? Ideally, there would be a result analogous to the Fundamental Theorem of Calculus that would enable us to do so.

Actually, the answer to this question is straightforward. Recall from Chapter 7 that the volume V is given by the definite integral

$$V = \int_a^b A(x) \, dx, \qquad a < b \tag{3}$$

when $A(x)$ is the area of the cross section taken perpendicular to the x-axis. Now if $x_0 \in [a, b]$ is fixed, the area of this cross section is just

$$A(x_0) = \int_c^d f(x_0, y) \, dy, \qquad c < d, \tag{4}$$

since the cross section is bounded above by the continuous function $g(y) = f(x_0, y)$ (see Figure 1.6).

Combining equations (3) and (4) we conclude that

$$V = \int_a^b \left\{ \int_c^d f(x, y) \, dy \right\} dx, \qquad a < b, \qquad c < d. \tag{5}$$

The meaning of equation (5) is that the volume V is calculated by first integrating f with respect to y (treating x as a constant) from c to d, and then integrating the resulting function of x from a to b. As Figure 1.7 illustrates, we could have begun by fixing y_0 and obtaining the area of the cross section perpendicular to the y-axis at $y = y_0$ as

$$A(y_0) = \int_a^b f(x, y_0) \, dx, \qquad a < b.$$

The resulting calculation for volume is

$$V = \int_c^d \left\{ \int_a^b f(x, y) \, dx \right\} dy, \qquad a < b, \quad c < d. \tag{6}$$

The integrals in (5) and (6) are called **iterated integrals,** because they involve the composition of two integrations, each with respect to a single variable. We

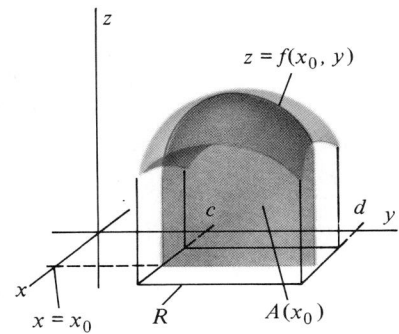

Figure 1.6 Area of cross section at x_0 is

$$A(x_0) = \int_c^d f(x_0, y) \, dy.$$

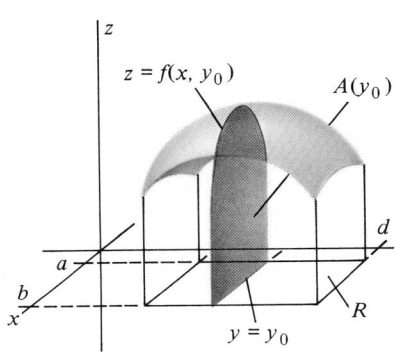

Figure 1.7 Area of cross section at y_0 is

$$A(y_0) = \int_a^b f(x, y_0) \, dx.$$

usually omit the braces and simply write

$$\int_a^b \int_c^d f(x, y)\, dy\, dx = \int_a^b \left\{ \int_c^d f(x, y)\, dy \right\} dx \qquad (7)$$

and

$$\int_c^d \int_a^b f(x, y)\, dx\, dy = \int_c^d \left\{ \int_a^b f(x, y)\, dx \right\} dy. \qquad (8)$$

It is important to note that iterated integrals are interpreted "from inside out," meaning just what is specified by equations (7) and (8).

Example 1

Evaluate the iterated integral

$$\int_1^2 \int_1^3 (x^2 + 2xy)\, dy\, dx$$

and interpret the result geometrically.

Solution: The integral is evaluated as in equation (7):

$$\int_1^2 \int_1^3 (x^2 + 2xy)\, dy\, dx = \int_1^2 \left\{ \int_1^3 (x^2 + 2xy)\, dy \right\} dx$$

$$= \int_1^2 \left\{ x^2 y + xy^2 \bigg]_{y=1}^{y=3} \right\} dx$$

$$= \int_1^2 [(3x^2 + 9x) - (x^2 + x)]\, dx$$

$$= \int_1^2 (2x^2 + 8x)\, dx$$

$$= \frac{2}{3} x^3 + 4x^2 \bigg]_1^2$$

$$= \left[\frac{2}{3}(8) + 4(4) \right] - \left[\frac{2}{3}(1) + 4(1) \right]$$

$$= \frac{50}{3}.$$

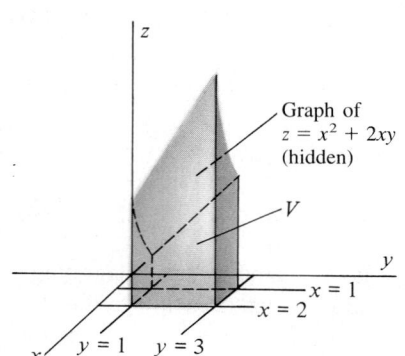

Figure 1.8 Volume of V is

$$\int_1^2 \int_1^3 (x^2 + 2xy)\, dy\, dx. \text{ (See } \textbf{Plate}$$
13.)

Since $f(x, y) = x^2 + 2xy$ is nonnegative for (x, y) in the rectangle $R = \{(x, y) \mid 1 \le x \le 2,\ 1 \le y \le 3\}$, the number $\frac{50}{3}$ is the volume of the solid bounded above by the graph of $f(x, y) = x^2 + 2xy$ and below by the xy-plane over the rectangle R (Figure 1.8). This volume may also be calculated using (8) as

$$\int_1^3 \int_1^2 (x^2 + 2xy)\, dx\, dy = \int_1^3 \left\{ \int_1^2 (x^2 + 2xy)\, dx \right\} dy$$

$$= \int_1^3 \left\{ \frac{x^3}{3} + x^2 y \bigg]_{x=1}^{x=2} \right\} dy$$

$$= \int_1^3 \left[\left(\frac{8}{3} + 4y \right) - \left(\frac{1}{3} + y \right) \right] dy$$

$$= \int_1^3 \left(\frac{7}{3} + 3y \right) dy$$

$$= \frac{7y}{3} + \frac{3y^2}{2} \Big]_1^3$$

$$= \frac{50}{3}. \qquad \diamond$$

Example 2

Calculate the volume of the solid over the square $R = \{(x, y) \mid -1 \le x \le 1, -1 \le y \le 1\}$ bounded above by the graph of $f(x, y) = 8 - x^2 - y^2$ and below by the xy-plane.

Solution: The desired volume may be calculated as

$$\iint_R (8 - x^2 - y^2) \, dA = \int_{-1}^1 \left\{ \int_{-1}^1 (8 - x^2 - y^2) \, dx \right\} dy$$

$$= \int_{-1}^1 \left\{ 8x - \frac{1}{3}x^3 - xy^2 \Big]_{x=-1}^{x=1} \right\} dy$$

$$= \int_{-1}^1 \left(\frac{46}{3} - 2y^2 \right) dy$$

$$= \frac{46y}{3} - \frac{2y^3}{3} \Big]_{y=-1}^{y=1}$$

$$= \frac{88}{3}.$$

(See Figure 1.9.) $\qquad \diamond$

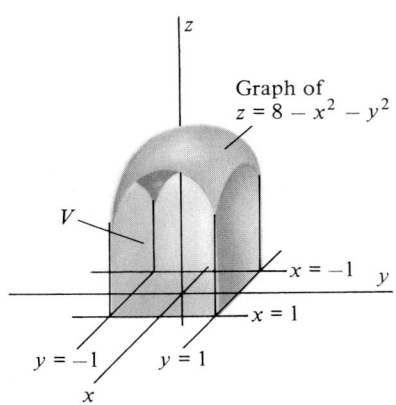

Graph of $z = 8 - x^2 - y^2$

V

$x = -1$

$x = 1$

$y = -1 \qquad y = 1$

Figure 1.9 Volume of V is
$\int_{-1}^1 \int_{-1}^1 (8 - x^2 - y^2) \, dx \, dy$.

Finally, we note that the definition of the double integral in Definition 2 does not depend on the function f being nonnegative. The double integral $\iint_R f(x, y) \, dA$ is therefore defined for any continuous function f, regardless of its sign on the rectangle R.

When $f(x, y) \ge 0$ for all (x, y) in R, we may combine statement (2) with statements (5) through (8) to obtain the equations

$$\iint_R f(x, y) \, dA = \int_a^b \int_c^d f(x, y) \, dy \, dx = \int_a^b \left\{ \int_c^d f(x, y) \, dy \right\} dx \qquad (9)$$

and

$$\iint_R f(x, y) \, dA = \int_c^d \int_a^b f(x, y) \, dx \, dy = \int_c^d \left\{ \int_a^b f(x, y) \, dx \right\} dy. \qquad (10)$$

Equations (9) and (10) are true regardless of the sign of $f(x, y)$, although we shall omit the details involved in proving the more general case (see Exercise 36 for a sketch of a proof of this statement). However, you must keep in mind that the integrals in (9) and (10) correspond to the volume of the region between the rectangle R and the graph of $z = f(x, y)$ *only when $f(x, y) \geq 0$* for all (x, y) in R, $a < b$, and $c < d$. We shall have more to say about properties of the double integral in the next section.

Exercise Set 19.1

In Exercises 1–14, evaluate the iterated integral.

1. $\displaystyle\int_0^1 \int_0^2 xy \, dx \, dy$

2. $\displaystyle\int_{-1}^1 \int_1^3 (y - x) \, dy \, dx$

3. $\displaystyle\int_1^3 \int_1^2 (4 + x - y) \, dx \, dy$

4. $\displaystyle\int_0^2 \int_{-1}^1 (x + y)^2 \, dx \, dy$

5. $\displaystyle\int_0^1 \int_0^{\pi/2} x \sin y \, dy \, dx$

6. $\displaystyle\int_0^1 \int_0^1 y e^{x-y^2} \, dy \, dx$

7. $\displaystyle\int_0^2 \int_1^e y^2 \ln x \, dx \, dy$

8. $\displaystyle\int_0^9 \int_1^4 \sqrt{\frac{y}{x}} \, dx \, dy$

9. $\displaystyle\int_0^1 \int_0^1 x \cosh y \, dx \, dy$

10. $\displaystyle\int_0^2 \int_0^2 xy e^{xy^2} \, dy \, dx$

11. $\displaystyle\int_0^1 \int_0^{\pi/2} xy \sin x \, dx \, dy$

12. $\displaystyle\int_0^{\pi} \int_0^1 \sinh x \cosh(\pi - y) \, dx \, dy$

13. $\displaystyle\int_0^1 \int_0^4 \frac{\sqrt{y}}{1 + x^2} \, dy \, dx$

14. $\displaystyle\int_0^{\pi/4} \int_0^{\pi/4} \tan x \sec^2 y \, dy \, dx$

In Exercises 15–21, evaluate the double integral over the rectangle R.

15. $\displaystyle\iint_R (x + y^2) \, dA$, $\quad R = \{(x, y) \mid 0 \leq x \leq 1, 0 \leq y \leq 1\}$

16. $\displaystyle\iint_R (x^2 + y^2) \, dA$, $\quad R = \{(x, y) \mid 0 \leq x \leq a, 0 \leq y \leq b\}$

17. $\displaystyle\iint_R x \cos y \, dA$, $\quad R = \{(x, y) \mid 0 \leq x \leq 4, 0 \leq y \leq \pi/2\}$

18. $\displaystyle\iint_R \frac{xy}{\sqrt{x^2 + y^2}} \, dA$, $\quad R = \{(x, y) \mid 1 \leq x \leq 2, 1 \leq y \leq 2\}$

19. $\displaystyle\iint_R xy \sec^2(xy^2) \, dA$, $\quad R = \{(x, y) \mid 0 \leq x \leq \pi/4, 0 \leq y \leq 1\}$

20. $\displaystyle\iint_R \frac{1}{\sqrt{x + y}} \, dA$, $\quad R = \{(x, y) \mid 4 \leq x \leq 8, 0 \leq y \leq 4\}$

21. $\displaystyle\iint_R y \cos(x + y) \, dA$, $\quad R = \{(x, y) \mid 0 \leq x \leq \pi/4, 0 \leq y \leq \pi/4\}$

In Exercises 22–27, use a double integral to calculate the volume of the solid bounded above by the graph of f and below by the rectangle R in the xy-plane.

22. $f(x, y) = 16 - 4x - 2y$, $\quad R = \{(x, y) \mid 0 \leq x \leq 2, 0 \leq y \leq 1\}$

23. $f(x, y) = x$, $\quad R = \{(x, y) \mid 0 \leq x \leq 2, 0 \leq y \leq 3\}$

24. $f(x, y) = 9 - x^2 - y^2$, $\quad R = \{(x, y) \mid -1 \leq x \leq 1, -1 \leq y \leq 1\}$

25. $f(x, y) = x \sin y$, $\quad R = \{(x, y) \mid 0 \leq x \leq 1, 0 \leq y \leq \pi/2\}$

26. $f(x, y) = \dfrac{\sqrt{x}}{1 + y^2}$, $\quad R = \{(x, y) \mid 0 \leq x \leq 4, 0 \leq y \leq 1\}$

27. $f(x, y) = \dfrac{e^{\sqrt{x}}}{\sqrt{xy}}$, $\quad R = \{(x, y) \mid 1 \leq x \leq 4, 1 \leq y \leq 9\}$

28. Show that if f is continuous on the rectangle R, then

$$\iint_R cf(x, y) \, dA = c \iint_R f(x, y) \, dA$$

for any constant c.

29. Show that if f and g are continuous on the rectangle R then

$$\iint_R [f(x, y) + g(x, y)] \, dA = \iint_R f(x, y) \, dA$$
$$+ \iint_R g(x, y) \, dA.$$

30. Show that if $f(x, y) \leq 0$ for all $(x, y) \in R = \{(x, y) \mid a \leq x \leq b, c \leq y \leq d\}$ and if V is the volume of the solid bounded by R and the graph of $z = f(x, y)$, then

$$V = -\iint_R f(x, y) \, dA.$$

31. Show that if f is continuous for $(x, y) \in R = \{(x, y) \mid a \le x \le b, c \le y \le d\}$ then

$$\int_a^b \int_c^d f(x, y) \, dy \, dx = -\int_b^a \int_c^d f(x, y) \, dy \, dx$$

$$= -\int_a^b \int_d^c f(x, y) \, dy \, dx$$

$$= \int_b^a \int_d^c f(x, y) \, dy \, dx.$$

32. *(Computer)* Program 8 in Appendix I is a BASIC computer program for calculating Riemann sums for double integrals. Use Program 8 to approximate the volume of the solid bounded above by the graph of $z = \sin \sqrt{xy}$ and below by the rectangle

$$R = \{(x, y) \mid 0 \le x \le 1, 0 \le y \le 1\}.$$

33. *(Computer)* Use Program 8 to approximate the double integral

$$\iint_R e^{x^2 + y^2} \, dA \qquad \text{where } R \text{ is the rectangle}$$

$$R = \{(x, y) \mid 0 \le x \le 1, 0 \le y \le 2\}.$$

34. *(Computer)* Use Program 8 to approximate the iterated integral

$$\int_0^1 \int_1^2 \sin(x^2 + y^2) \, dx \, dy.$$

35. *(Computer)* Use Program 8 to approximate the iterated integral

$$\int_1^3 \int_2^4 \sqrt{x^3 + y^3} \, dx \, dy.$$

36. The purpose of this exercise is to enable you to sketch a proof of the fact that

$$\iint_R f(x, y) \, dA = \int_a^b \int_c^d f(x, y) \, dy \, dx$$

where f is continuous and $R = \{(x, y) \mid a \le x \le b, c \le y \le d\}$.
a. Begin with the partitions $a = x_0 < x_1 < \cdots < x_n = b$

and $c = y_0 < y_1 < \cdots < y_m = d$ and the Riemann sum

$$S_{n, m} = \sum_{j=1}^n \sum_{k=1}^m f(s_j, t_k) \, \Delta y_k \Delta x_j$$

$$= \sum_{j=1}^n \left\{ \sum_{k=1}^m f(s_j, t_k) \, \Delta y_k \right\} \Delta x_j.$$

b. With j fixed, pick t_k in each interval $[y_{k-1}, y_k]$ to be the number for which

$$f(s_j, t_k) \, \Delta y_k = \int_{y_{k-1}}^{y_k} f(s_j, t) \, dt.$$

Why can this be done?
c. Conclude that

$$\sum_{k=1}^m f(s_j, t_k) \, \Delta y_k = \sum_{k=1}^m \int_{y_{k-1}}^{y_k} f(s_j, y) \, dy = \int_c^d f(s_j, y) \, dy.$$

Why is this true?
d. From (a) and (c) conclude that

$$S_{n, m} = \sum_{j=1}^n \left\{ \int_c^d f(s_j, y) \, dy \right\} \Delta x_j.$$

e. Note that $S_{n, m}$ in (d) has the form $S_{n, m} = \sum_{j=1}^n G(s_j) \, \Delta x_j$

where

$$G(s_j) = \int_c^d f(s_j, t) \, dt. \qquad \text{Conclude that}$$

$$\lim_{\|P\| \to 0} S_{n, m} = \lim_{n \to \infty} \sum_{j=1}^n G(s_j) \, \Delta x_j = \int_a^b G(x) \, dx$$

$$= \int_a^b \left\{ \int_c^d f(x, y) \, dy \right\} dx.$$

37. Show that if f is continuous on the interval $[a, b]$, and if g is continuous on the interval $[c, d]$, then

$$\iint_R f(x)g(y) \, dA = \left[\int_a^b f(x) \, dx \right] \cdot \left[\int_c^d g(y) \, dy \right]$$

where $R = \{(x, y) \mid a \le x \le b, c \le y \le d\}$.

19.2 DOUBLE INTEGRALS OVER MORE GENERAL REGIONS

Double integrals may be defined over regions in the plane more general than just rectangles. In particular, let Q be a region in the plane that is bounded (meaning it is contained in some finite rectangle) and whose boundary is a piecewise smooth curve that does not cross itself. A rectangle is certainly an example of such a region. But so is a pentagon, a triangle, an ellipse, or even a more general region such as that in Figure 2.1.

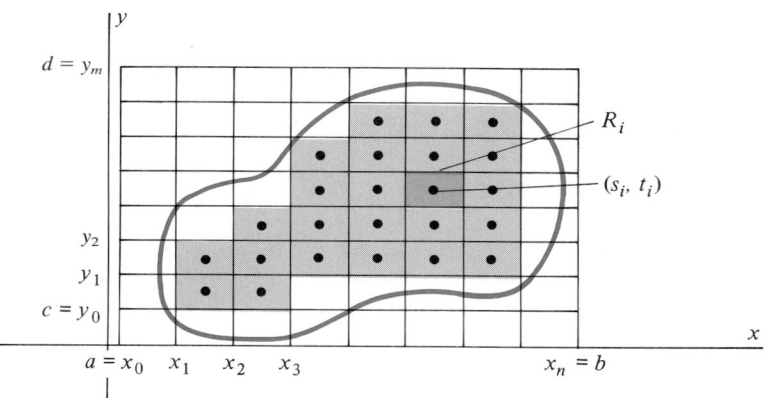

Figure 2.1 Partitioning a more general region Q.

Now suppose f is a continuous function of two variables defined on Q and that Q is entirely contained within a rectangle $R = \{(x, y) \mid a \leq x \leq b, c \leq y \leq d\}$. We construct a grid over R as before, which partitions R into smaller rectangles. We again denote the norm of this partition P by $\|P\|$.

Now some of these smaller rectangles will lie entirely within Q and some will not. Let $R_1, R_2, \ldots, R_m$ be a list of all such rectangles lying entirely within Q. For each such rectangle R_i, let (s_i, t_i) be a point in R_i. Finally, form the sum

$$S_m = \sum_{i=1}^{m} f(s_i, t_i) \, \Delta A_i, \qquad \Delta A_i = \text{area of } R_i. \tag{1}$$

Under the conditions stated on Q and on f, as $m \to \infty$ and as $\|P\| \to 0$ this Riemann sum will converge to the number we define to be the double integral of f over the region Q:

$$\iint_Q f(x, y) \, dA = \lim_{\|P\| \to 0} \sum_{i=1}^{m} f(s_i, t_i) \, \Delta A_i. \tag{2}$$

Statement (2) is actually a definition of the double integral in this more general setting, although we prefer not to state it formally as such since this would require repeating nearly all the preceding statements. Before looking at ways in which such integrals may be evaluated, we wish to emphasize two points concerning this definition.

(i) Although only one summation sign appears in (2), this definition agrees with Definition 2 when Q is a rectangle. The difference is only that we have used a two-dimensional counting procedure (rows by columns) in stating Definition 2. In statement (2) we have simply listed all included rectangles $R_1, R_2, \ldots, R_m$ using a single index.

(ii) When $f(x, y) \geq 0$ for all $(x, y) \in Q$, the double integral (2) gives the volume V of the solid bounded by the graph of $z = f(x, y)$ and the region Q. The reason for this is the same as for the simpler case of the double integral over a rectangle.

Regular Regions

There are two special types of regions for which the double integral in (2) can be evaluated as an iterated integral: *x*-simple and *y*-simple regions.

DEFINITION 3

A region Q in the *xy*-plane is called **y-simple** if there exist continuous functions g_1 and g_2 so that

$$Q = \{(x, y) \mid a \le x \le b, \ g_1(x) \le y \le g_2(x)\}.$$

The region Q is called **x-simple** if there exist continuous functions h_1 and h_2 so that

$$Q = \{(x, y) \mid c \le y \le d, \ h_1(y) \le x \le h_2(y)\}.$$

The region Q is called **regular** if it is both *x*-simple and *y*-simple.

Figure 2.2 gives two illustrations of *y*-simple regions. The condition that $g_1(x) \le y \le g_2(x)$ for all $x \in [a, b]$ simply means that *the vertical line segment*

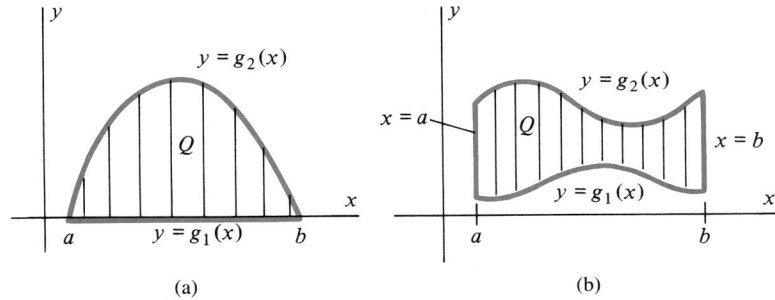

(a)   (b)

Figure 2.2 Two *y*-simple regions: Vertical lines intersect the boundary of Q at most twice.

connecting $(x, g_1(x))$ *and* $(x, g_2(x))$ *lies entirely within the region* Q. Another way to say this is that lines parallel to the *y*-axis intersect the boundary of Q at most twice.

Figure 2.3 shows two *x*-simple regions. This condition means that lines parallel to the *x*-axis intersect the boundary of Q at most twice. Note also that Figures 2.2(a)

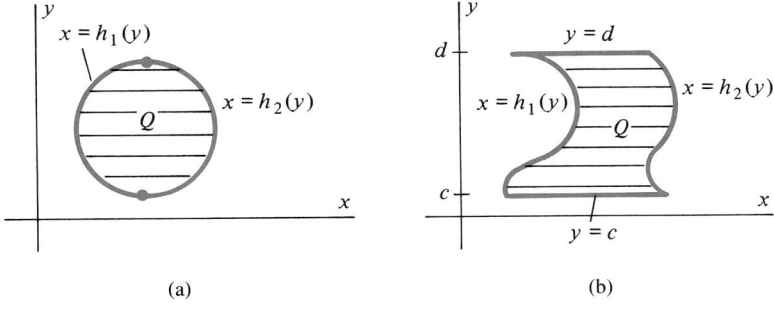

(a)   (b)

Figure 2.3 Two *x*-simple regions: Horizontal lines intersect the boundary of Q at most twice.

and 2.3(a) are *regular* (both x-simple and y-simple). However, Figure 2.2(b) is not x-simple, and Figure 2.3(b) is not y-simple.

The following theorem shows how double integrals over x-simple or y-simple regions may be evaluated as iterated integrals.

THEOREM 2

Let f be a continuous function of two variables on the region Q.

(i) If $Q = \{(x, y) \mid a \le x \le b, \ g_1(x) \le y \le g_2(x)\}$ is y-simple, then

$$\iint_Q f(x, y) \, dA = \int_a^b \int_{g_1(x)}^{g_2(x)} f(x, y) \, dy \, dx = \int_a^b \left\{ \int_{g_1(x)}^{g_2(x)} f(x, y) \, dy \right\} dx. \qquad (3)$$

(ii) If $Q = \{(x, y) \mid c \le y \le d, \ h_1(y) \le x \le h_2(y)\}$ is x-simple, then

$$\iint_Q f(x, y) \, dA = \int_c^d \int_{h_1(y)}^{h_2(y)} f(x, y) \, dx \, dy = \int_c^d \left\{ \int_{h_1(y)}^{h_2(y)} f(x, y) \, dx \right\} dy. \qquad (4)$$

Equation (3) says that if Q is y-simple, we first integrate f as a function of y alone, between the limits $g_1(x)$ and $g_2(x)$. The resulting function of x alone is then integrated over the constant limits $a \le x \le b$. Equation (4) says that if Q is x-simple, we first integrate f as a function of x alone, between the limits $h_1(y)$ and $h_2(y)$.

We shall not prove Theorem 2. However, Figure 2.4 embodies an argument for the validity of equation (4) when f is nonnegative on Q and Q is x-simple: For y-fixed, the integral $\displaystyle\int_{h_1(y)}^{h_2(y)} f(x, y) \, dx = A(y)$ gives the area of the cross section of the solid bounded by the graph of f and the region Q. The familiar formula for the volume V of a solid with known cross section then gives

$$V = \int_c^d A(y) \, dy = \int_c^d \int_{h_1(y)}^{h_2(y)} f(x, y) \, dx \, dy. \qquad (5)$$

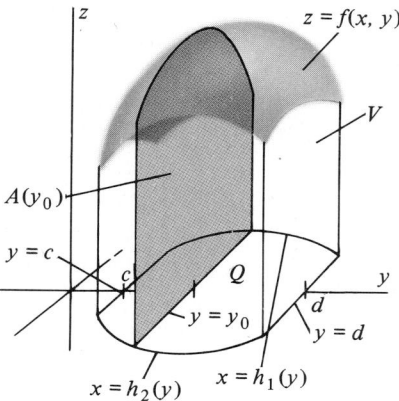

Figure 2.4 If Q is x-simple, area of cross section perpendicular to the y-axis at y_0 is

$$A(y_0) = \int_{h_1(y_0)}^{h_2(y_0)} f(x, y_0) \, dx.$$

Since the left side of equation (5) corresponds to the double integral $\iint\limits_{Q} f(x, y)\, dA$, we obtain equation (4).

In using Theorem 2 to evaluate double integrals, it is very important to first sketch the region Q to determine whether it is x-simple or y-simple. This determination will often dictate the order of integration in the iterated integral. (Of course, if Q is a regular region, you may proceed in either order. Just be careful that the limits of integration correspond to the chosen order of integration.)

Example 1

Evaluate the double integral $\iint\limits_{Q} (2xy + y^2)\, dA$ where Q is the triangle with vertices $(0, 0)$, $(1, 0)$, and $(1, 2)$.

Solution: The triangular region Q is sketched in Figure 2.5. Since Q is y-simple, we find the equation of the line segment joining $(0, 0)$ and $(1, 2)$. It is simply $y = 2x$. A vertical line segment through Q extends from $y = g_1(x) = 0$ to $y = g_2(x) = 2x$. That is, the limits of integration are determined by the inequalities

$$0 \le x \le 1 \qquad \text{and} \qquad 0 \le y \le 2x. \tag{6}$$

According to equation (3), the double integral is evaluated as

$$\iint\limits_{Q} (2xy + y^2)\, dA = \int_0^1 \int_0^{2x} (2xy + y^2)\, dy\, dx$$

$$= \int_0^1 \left\{ xy^2 + \frac{1}{3}y^3 \right]_{y=0}^{y=2x} \right\} dx$$

$$= \int_0^1 \frac{20x^3}{3}\, dx$$

$$= \frac{5}{3}x^4 \bigg]_0^1$$

$$= \frac{5}{3}.$$

Now the region Q is also x-simple. However, to use equation (4) we must first write the equation for the line segment joining $(0, 0)$ and $(1, 2)$ as a function of y. Solving $y = 2x$ for x gives $x = h_1(y) = y/2$ as the left boundary. As Figure 2.6

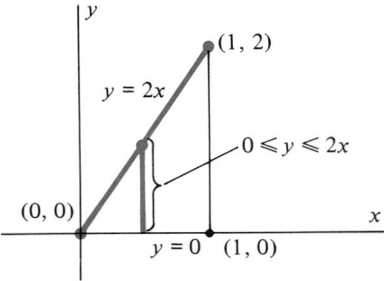

Figure 2.5 Treating the triangle Q as y-simple.

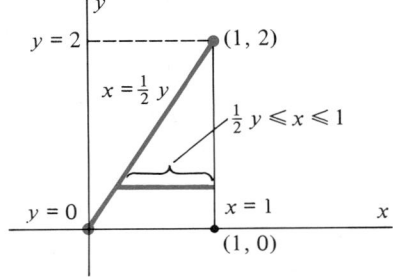

Figure 2.6 Treating the triangle Q as x-simple.

illustrates, $h_2(y) = 1$ is the right boundary of Q. The limits of integration are therefore

$$0 \le y \le 2 \quad \text{and} \quad \frac{y}{2} \le x \le 1. \tag{7}$$

Equation (4) then gives

$$\iint\limits_Q (2xy + y^2)\, dA = \int_0^2 \int_{y/2}^1 (2xy + y^2)\, dx\, dy$$

$$= \int_0^2 \left\{ x^2 y + xy^2 \Big]_{x=y/2}^{x=1} \right\} dy$$

$$= \int_0^2 \left(-\frac{3}{4}y^3 + y^2 + y \right) dy$$

$$= \left[-\frac{3}{16}y^4 + \frac{1}{3}y^3 + \frac{1}{2}y^2 \right]_0^2$$

$$= \frac{5}{3}. \qquad \diamond$$

REMARK: Notice that the order of integration is entirely determined by our choice of whether we describe Q as y-simple, using inequalities (6), or as x-simple, using inequalities (7). Whichever integration is performed last must involve only constant limits of integration, since your answer cannot contain variables.

Example 2

Evaluate the double integral

$$\iint\limits_Q 4xy\, dA$$

where Q is the region bounded by the graphs of the equations $y = x + 1$ and $x = 1 - y^2$.

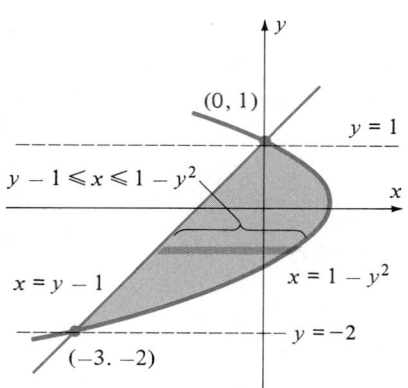

Figure 2.7 Region bounded by graphs of $y = x + 1$ and $x = 1 - y^2$ is x-simple.

Solution: The region Q is sketched in Figure 2.7. This region, like that in Example 1, is both x-simple and y-simple. However, we shall choose to work with Q as an x-simple region, since every horizontal line through Q originates on the line $y = x + 1$ and terminates on the parabola $x = 1 - y^2$. (Viewing Q as y-simple would be much messier—some vertical lines terminate on the line and others terminate on the parabola.)

Solving the equation $y = x + 1$ for x gives $x = h_1(y) = y - 1$ as the left boundary of Q. The right boundary is $h_2(y) = 1 - y^2$. The limits of integration are therefore

$$-2 \le y \le 1 \quad \text{and} \quad y - 1 \le x \le 1 - y^2.$$

The double integral is evaluated as

$$\iint\limits_Q 4xy\, dA = \int_{-2}^1 \int_{y-1}^{1-y^2} 4xy\, dx\, dy$$

$$= \int_{-2}^1 \left\{ 2x^2 y \Big]_{x=y-1}^{x=1-y^2} \right\} dy$$

$$= \int_{-2}^{1} (2y^5 - 6y^3 + 4y^2) \, dy$$

$$= \frac{1}{3}y^6 - \frac{3}{2}y^4 + \frac{4}{3}y^3 \bigg]_{-2}^{1}$$

$$= \frac{27}{2}.$$

◇

Example 3

Find the volume V of the solid bounded above by the graph of $f(x, y) = 2x^2y$ and below by the region Q in the xy plane bounded by the semicircle $y = \sqrt{1 - x^2}$ and the lines $y = 2$, $x = -1$, and $x = 1$.

Solution: The region Q is sketched in Figure 2.8. Since $f(x, y) \geq 0$ for all $(x, y) \in Q$, V is given by the double integral $\iint_{Q} 2x^2y \, dA$. As to the order of integration, we note that Q is y-simple, but not x-simple. We therefore integrate first with respect to y. The limits of integration are

$$-1 \leq x \leq 1 \qquad \text{and} \qquad \sqrt{1 - x^2} \leq y \leq 2.$$

The volume is

$$V = \iint_{Q} 2x^2y \, dA = \int_{-1}^{1} \int_{\sqrt{1-x^2}}^{2} 2x^2y \, dy \, dx$$

$$= \int_{-1}^{1} \left\{ x^2y^2 \bigg]_{y=\sqrt{1-x^2}}^{y=2} \right\} dx$$

$$= \int_{-1}^{1} [4x^2 - x^2(1 - x^2)] \, dx$$

$$= \int_{-1}^{1} (3x^2 + x^4) \, dx$$

$$= x^3 + \frac{1}{5}x^5 \bigg]_{-1}^{1}$$

$$= \frac{12}{5}.$$

◇

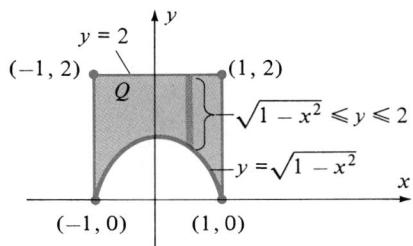

$y = 2$

$(-1, 2)$ $(1, 2)$

Q

$\sqrt{1 - x^2} \leq y \leq 2$

$y = \sqrt{1 - x^2}$

$(-1, 0)$ $(1, 0)$

Figure 2.8 Region bounded by $y = \sqrt{1 - x^2}$ and $y = 2$ is y-simple, but not x-simple.

Example 4

Find the volume V of the solid bounded by the graph of $f(x, y) = e^{x+y^2}$ and the region

$$Q = \{(x, y) \mid \ln y \leq x \leq \ln 2y, \, 1 \leq y \leq 2\}$$

in the xy plane.

Solution: The region Q is sketched in Figure 2.9. Although Q is both x-simple and y-simple, this time the order of integration is determined by the integrand $f(x, y) = e^{x+y^2}$. We must attempt to integrate first with respect to x, since there is no hope of finding an antiderivative with respect to y. Treating Q as an x-simple region, we use

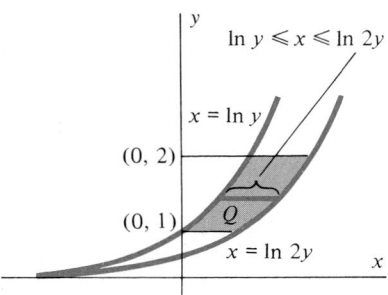

$\ln y \leq x \leq \ln 2y$

$x = \ln y$

$(0, 2)$

$(0, 1)$

Q

$x = \ln 2y$

Figure 2.9 Region Q in Example 4.

the limits

$$\ln y \le x \le \ln 2y \qquad \text{and} \qquad 1 \le y \le 2$$

and obtain

$$
\begin{aligned}
V = \iint_Q e^{x+y^2} \, dA &= \int_1^2 \int_{\ln y}^{\ln 2y} e^{x+y^2} \, dx \, dy \\
&= \int_1^2 \left\{ e^{x+y^2} \bigg]_{x=\ln y}^{x=\ln 2y} \right\} dy \\
&= \int_1^2 [(e^{\ln 2y})e^{y^2} - (e^{\ln y})e^{y^2}] \, dy \\
&= \int_1^2 (2ye^{y^2} - ye^{y^2}) \, dy \\
&= \int_1^2 ye^{y^2} \, dy \\
&= \frac{1}{2} e^{y^2} \bigg]_1^2 \\
&= \frac{e}{2}(e^3 - 1) \\
&\approx 25.94.
\end{aligned}
$$

◇

Finding Areas by Double Integration

The fact that volumes of solids may be calculated by use of double integrals means that areas of regions in the plane may also be calculated in this way. As Figure 2.10

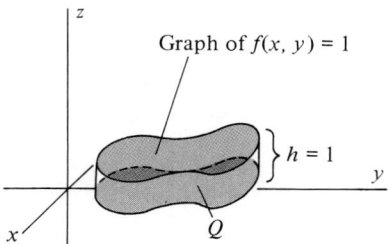

Figure 2.10 Area of $Q = \displaystyle\iint_Q 1 \, dA.$

illustrates, the area of Q is the same as the volume of the cylinder with base Q and uniform height $h = 1$. To calculate the area of Q, we simply evaluate the integral

$$
\text{Area of } Q = \iint_Q 1 \, dA. \tag{8}
$$

Example 5

Use a double integral to find the area of the region Q lying inside the circle $x^2 + y^2 = 4$ and above the line $y = 1$.

Solution: The points on the circle $x^2 + y^2 = 4$ with y-coordinate equal to 1 have x-coordinates $x = \pm\sqrt{4 - 1} = \pm\sqrt{3}$. The region may therefore be described by the inequalities

$$-\sqrt{3} \le x \le \sqrt{3}, \qquad 1 \le y \le \sqrt{4 - x^2}.$$

(See Figure 2.11.) According to (8),

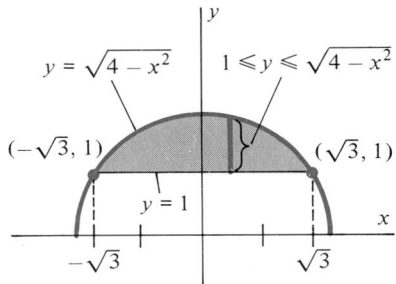

Figure 2.11 Region inside circle $x^2 + y^2 = 4$ and above $y = 1$.

$$\text{Area of } Q = \iint_Q 1 \ dA$$

$$= \int_{-\sqrt{3}}^{\sqrt{3}} \int_1^{\sqrt{4-x^2}} 1 \cdot dy \ dx$$

$$= \int_{-\sqrt{3}}^{\sqrt{3}} \left\{ \Big[y \Big]_{y=1}^{y=\sqrt{4-x^2}} \right\} dx$$

$$= \int_{-\sqrt{3}}^{\sqrt{3}} (\sqrt{4 - x^2} - 1) \ dx$$

$$= 2 \int_0^{\sqrt{3}} \sqrt{4 - x^2} \ dx - 2\sqrt{3}.$$

A straightforward application of the method of trigonometric substitutions shows that

$$\int \sqrt{4 - x^2} \ dx = 2 \ \text{Sin}^{-1} \left(\frac{x}{2} \right) + \frac{x\sqrt{4 - x^2}}{2} + C.$$

The desired area is therefore

$$2\left[2 \ \text{Sin}^{-1} \left(\frac{\sqrt{3}}{2} \right) + \frac{\sqrt{3}}{2} \right] - 2\sqrt{3} \approx 2.457. \qquad \diamondsuit$$

Interchanging the Order of Integration

There are occasions on which we need to interchange the order of integration in an iterated integral because an antiderivative cannot be found with respect to the "inside" variable. For example, in the iterated integral

$$\int_0^1 \int_{y^2}^1 ye^{x^2} \ dx \ dy \qquad\qquad (9)$$

we cannot find a (formal) antiderivative for the integrand ye^{x^2} with respect to x.

However, such integrals can often be evaluated by reversing the order of integration. In particular, the antiderivative of ye^{x^2} with respect to y is easily seen to be the function $\frac{1}{2}y^2 e^{x^2}$. But if we are going to reverse the order in which the integrations are performed, we must also determine correct limits of integration corresponding to this new order of integration. We do so as follows:

To reverse the order of integration in the iterated integral

$$\int_c^d \int_{h_1(y)}^{h_2(y)} f(x, y)\, dx\, dy:$$

(i) Identify the region Q for which the iterated integral can be written as the double integral

$$\int_c^d \int_{h_1(y)}^{h_2(y)} f(x, y)\, dx\, dy = \iint_Q f(x, y)\, dA.$$

(ii) Find constants a and b, and continuous functions g_1 and g_2, so that the region Q can be expressed as

$$Q = \{(x, y) \mid a \le x \le b,\ g_1(x) \le y \le g_2(x)\}.$$

(iii) Rewrite the iterated integral as

$$\int_c^d \int_{h_1(y)}^{h_2(y)} f(x, y)\, dx\, dy = \iint_Q f(x, y)\, dA$$

$$= \int_a^b \int_{g_1(x)}^{g_2(x)} f(x, y)\, dy\, dx.$$

Obviously, this procedure can be applied only when Q is a regular region. Also, though the procedure is stated for reversing the order of integration from $dx\, dy$ to $dy\, dx$, the procedure for changing from order $dy\, dx$ to order $dx\, dy$ is analogous. Finally, there is no guarantee that the resulting integral can be more easily evaluated than the original one.

Example 6

Use the procedure for reversing order of integration to evaluate the iterated integral

$$\int_0^1 \int_{y^2}^1 ye^{x^2}\, dx\, dy.$$

Solution: From the given limits of integration, the region Q is described by the inequalities

$$y^2 \le x \le 1, \qquad 0 \le y \le 1.$$

That is, Q is the region bounded between the graphs of $x = y^2$ and $x = 1$ for $0 \le y \le 1$. From Figure 2.12 (which must always be sketched when applying this method), we can see that Q is regular and can also be described by the inequalities

$$0 \le x \le 1, \qquad 0 \le y \le \sqrt{x} \qquad \text{(Figure 2.13)}.$$

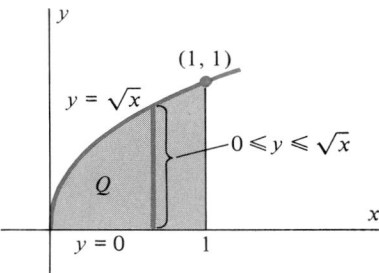

Figure 2.12 $Q = \{(x, y) \mid y^2 \leq x \leq 1, 0 \leq y \leq 1\}$.

Figure 2.13 $Q = \{(x, y) \mid 0 \leq x \leq 1, 0 \leq y \leq \sqrt{x}\}$.

Beginning with the given integral, we therefore reverse the order of integration as follows:

$$\int_0^1 \int_{y^2}^1 y e^{x^2}\, dx\, dy = \iint_Q y e^{x^2}\, dA = \int_0^1 \int_0^{\sqrt{x}} y e^{x^2}\, dy\, dx$$

$$= \int_0^1 \left\{ \left[\frac{1}{2} y^2 e^{x^2} \right]_{y=0}^{y=\sqrt{x}} \right\} dx$$

$$= \int_0^1 \frac{1}{2} x e^{x^2}\, dx$$

$$= \frac{1}{4} e^{x^2} \Big]_0^1$$

$$= \frac{1}{4}(e - 1)$$

$$\approx 0.43. \qquad \diamondsuit$$

REMARK: It is important to note that we cannot simply interchange limits of integration when we interchange the order of integration. There is no alternative to sketching the region Q and working out the new limits of integration from knowledge of the boundary of Q.

Example 7

Evaluate the iterated integral

$$\int_0^1 \int_0^{\sqrt{1-x}} x y^2\, dy\, dx$$

by first reversing the order of integration.

Solution: The given limits of integration are

$$0 \leq x \leq 1, \qquad 0 \leq y \leq \sqrt{1-x},$$

which describe the region Q bounded above by the graph of $y = \sqrt{1-x}$ and below by the x-axis for $0 \leq x \leq 1$ (see Figure 2.14). By solving the equation $y = \sqrt{1-x}$ for x, we find that this region may also be described by the inequalities

$$0 \leq x \leq 1 - y^2, \qquad 0 \leq y \leq 1.$$

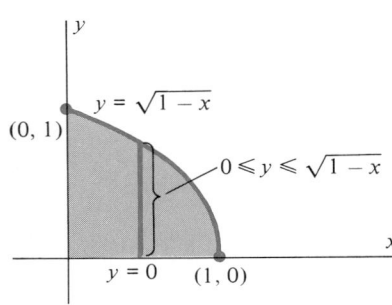

Figure 2.14 $Q = \{(x, y) \mid 0 \leq x \leq 1, 0 \leq y \leq \sqrt{1-x}\}$.

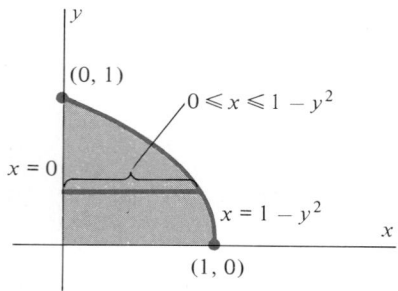

Figure 2.15 $Q = \{(x, y) \mid 0 \le x \le 1 - y^2,\ 0 \le y \le 1\}$.

(See Figure 2.15.) We may therefore evaluate the iterated integral as

$$\int_0^1 \int_0^{\sqrt{1-x}} xy^2\, dy\, dx = \int\int_Q xy^2\, dA = \int_0^1 \int_0^{1-y^2} xy^2\, dx\, dy$$

$$= \int_0^1 \left\{ \frac{1}{2}x^2 y^2 \right\}_{x=0}^{x=1-y^2} dy$$

$$= \int_0^1 \frac{1}{2}(1 - y^2)^2 y^2\, dy$$

$$= \int_0^1 \left(\frac{1}{2}y^2 - y^4 + \frac{1}{2}y^6 \right) dy$$

$$= \frac{1}{6}y^3 - \frac{1}{5}y^5 + \frac{1}{14}y^7 \Big]_0^1$$

$$= \frac{4}{105}. \qquad \diamond$$

Properties of Double Integrals

As defined in this section, the double integral satisfies the following properties:

(i) $\displaystyle\int\int_Q [f(x, y) + g(x, y)]\, dA = \int\int_Q f(x, y)\, dA + \int\int_Q g(x, y)\, dA,$

(ii) $\displaystyle\int\int_Q cf(x, y)\, dA = c\int\int_Q f(x, y)\, dA, \qquad c = \text{constant},$

(iii) $\displaystyle\int\int_{Q_1 \cup Q_2} f(x, y)\, dA = \int\int_{Q_1} f(x, y)\, dA + \int\int_{Q_2} f(x, y)\, dA.$

In (iii) we mean that $Q = Q_1 \cup Q_2$ is the union of two nonoverlapping regions Q_1 and Q_2, each satisfying the properties required of Q in this section. Property (iii) extends to unions of finitely many nonoverlapping regions $Q_1, Q_2, \ldots, Q_n$ of this type.

We shall not prove properties (i) through (iii). The proofs are analogous to those in the one variable case.

Exercise Set 19.2

In Exercises 1–12, sketch the region Q determined by the limits of integration and evaluate the iterated integral.

1. $\displaystyle\int_0^1 \int_0^x (y^2 - x^2)\, dy\, dx$

2. $\displaystyle\int_0^1 \int_0^{1-x} 2xy\, dy\, dx$

3. $\displaystyle\int_{-1}^0 \int_{-1}^{y+1} (xy - x)\, dx\, dy$

4. $\displaystyle\int_0^1 \int_0^y xye^{x^2}\, dx\, dy$

5. $\displaystyle\int_{-1}^1 \int_0^{1-x^2} xy\, dy\, dx$

6. $\displaystyle\int_0^1 \int_{-x}^{\sqrt{x}} \frac{y}{1+x}\, dy\, dx$

7. $\displaystyle\int_0^1 \int_{x^3}^{x^2} x\, dy\, dx$

8. $\displaystyle\int_0^{\sqrt{\pi/2}} \int_0^{\sqrt{y}} x \sin y^2\, dx\, dy$

9. $\displaystyle\int_0^{\pi/2} \int_0^{\sin x} 2y \cos x\, dy\, dx$

10. $\displaystyle\int_1^e \int_0^{\ln y} e^{x+y}\, dx\, dy$

11. $\displaystyle\int_0^1 \int_0^{y^2} e^{x/y}\, dx\, dy$

12. $\displaystyle\int_0^{\pi/2} \int_0^{\cos y} x \sin y\, dx\, dy$

In Exercises 13–22, evaluate the double integral.

13. $\displaystyle\int\int_Q \sqrt{x}\, y\, dA, \qquad Q = \{(x, y) \mid 0 \le x \le y^2,\ 0 \le y \le 1\}$

14. $\displaystyle\int\int_Q x \cos \pi y\, dA, \qquad Q = \{(x, y) \mid 0 \le x \le 1,\ 0 \le y \le x\}$

15. $\displaystyle\iint_Q (x + 2)\sqrt{1 + e^y} \, dA,$

$Q = \{(x, y) \mid 0 \le x \le e^{y/2}, \, 0 \le y \le 1\}$

16. $\displaystyle\iint_Q \frac{x^2}{1 + y} \, dA, \quad Q = \{(x, y) \mid 0 \le x \le 1, \, 0 \le y \le e^x - 1\}$

17. $\displaystyle\iint_Q xy \, dA, \qquad Q = \{(x, y) \mid y \le x \le \sqrt{y}, \, 0 \le y \le 1\}$

18. $\displaystyle\iint_Q (x^2 + y^2) \, dA,$

$Q = \{(x, y) \mid -2 \le x \le 2, \, -3 \le y \le 1 - x^2\}$

19. $\displaystyle\iint_Q y \, dA, \qquad Q = \{(x, y) \mid -1 \le x \le 1, \, e^x \le y \le e\}$

20. $\displaystyle\iint_Q ye^x \, dA, \, Q = \{(x, y) \mid -\ln y \le x \le \ln y, \, 1 \le y \le 2\}$

21. $\displaystyle\iint_Q x \, dA, \qquad Q = \{(x, y) \mid 0 \le x \le \sqrt{1 - y^2}, \, 0 \le y \le 1\}$

22. $\displaystyle\iint_Q xy \, dA, \qquad Q = \{(x, y) \mid 0 \le x^2 + y^2 \le 1\}$

In Exercises 23–28, sketch the region Q determined by the limits of integration, interchange the order of integration, and evaluate the given integral, where possible.

23. $\displaystyle\int_{-1}^{1} \int_{0}^{x+1} (x + y) \, dy \, dx$

24. $\displaystyle\int_{0}^{1} \int_{x^2}^{1} xe^{y^2} \, dy \, dx$

25. $\displaystyle\int_{0}^{1} \int_{0}^{y} xy^2 \, dx \, dy$

26. $\displaystyle\int_{-2}^{0} \int_{x^2}^{4} xe^{y^2} \, dy \, dx$

27. $\displaystyle\int_{1}^{e} \int_{0}^{\ln x} f(x, y) \, dy \, dx$

28. $\displaystyle\int_{0}^{1} \int_{0}^{\operatorname{Sin}^{-1} x} f(x, y) \, dy \, dx$

In Exercises 29–33, use a double integral to find the area of Q.

29. Q is the region bounded by the graphs of $y = 4 - x^2$ and the line $y = x + 2$.

30. Q is the region bounded by the graphs of $y = x^2$ and $y = x^3$.

31. Q is the region bounded by the graphs of $y = \sin x$ and $y = \cos x$ for $0 \le x \le \pi/4$.

32. Q is the region bounded by the graphs of $y = x^3$ and $y = 4x$.

33. Q is the region bounded by the graphs of $y = \sqrt{x}$ and $y = x^2$.

34. Use a double integral to find the volume of the tetrahedron with vertices $(1, 0, 0)$, $(0, 1, 0)$, $(0, 0, 0)$, and $(0, 0, 1)$.

35. Find the area of the ellipse $\dfrac{x^2}{4} + \dfrac{y^2}{9} = 1$ by use of a double integral.

36. Use a double integral to establish the formula for the volume of a right circular cone of radius r and height h.

37. Find the volume of the solid bounded by the paraboloid $z = \dfrac{1}{4}(x^2 + y^2)$ and the paraboloid $z = 5 - x^2 - y^2$.

38. Use a double integral to find the volume of the sphere $x^2 + y^2 + z^2 = 4$ lying above the plane $z = 1$.

39. Find the volume of the portion of the solid bounded by the plane $z = x$, the cylinder $x^2 + y^2 = 4$, and the plane $z = 0$.

40. Find the volume of the solid bounded by the coordinate planes and the plane $6x + 3y + 2z = 6$.

41. Sketch the region in the first octant common to the two cylinders $z^2 = 1 - x^2$ and $z^2 = 1 - y^2$. Find its volume.

19.3 DOUBLE INTEGRALS IN POLAR COORDINATES

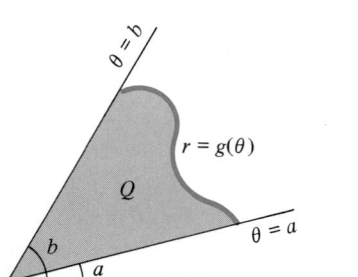

Figure 3.1 The region Q, bounded by the graph of $r = g(\theta)$ and the rays $\theta = a$ and $\theta = b$.

So far, we have defined double integrals only in the case where Q is a region in the xy- (Cartesian) coordinate plane and f is a function written in terms of Cartesian coordinates. We can also define the double integral when the region Q and the function f are described in polar coordinates.

Figure 3.1 shows a region Q bounded by the rays $\theta = a$ and $\theta = b$ and the graph of the function $r = g(\theta)$.

As Figure 3.2 illustrates, the circular arcs $r = r_j$ and the rays $\theta = \theta_k$ partition the region Q into wedge-shaped subregions R_{jk}. To determine the area of R_{jk} we use two facts (see Figure 3.3):

(i) The region R_{jk} lies between the concentric circles $r = r_{j-1}$ and $r = r_j$. The area of the entire annulus lying between these circles is therefore

$$\text{Area of annulus} = \pi r_j^2 - \pi r_{j-1}^2.$$

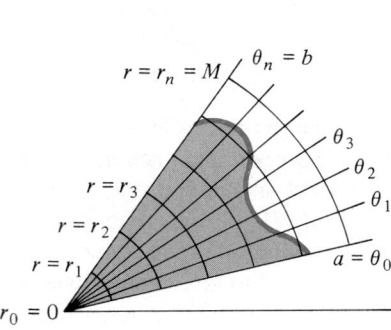
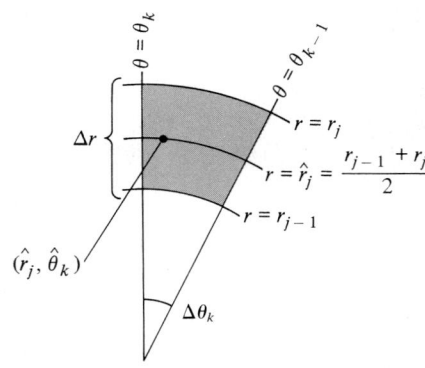

Figure 3.2 Q is partitioned into sub-regions by arcs $r = r_j$ and rays $\theta = \theta_k$.

Figure 3.3 Area of region R_{jk} is $\Delta A_{jk} = \hat{r}_j \, \Delta r_j \, \Delta \theta_k$.

(ii) The angle $\Delta \theta_k$, measured in radians, determines the proportion of the entire annulus lying in the region R_{jk}. That is,

$$\text{Proportion of annulus in } R_{jk} = \frac{\Delta \theta_k}{2\pi}.$$

From statements (i) and (ii), you can see that the area of the region R_{jk} is

$$\Delta A_{jk} = (\pi r_j^2 - \pi r_{j-1}^2)\left(\frac{\Delta \theta_k}{2\pi}\right) \tag{1}$$

$$= \frac{1}{2}(r_j^2 - r_{j-1}^2) \, \Delta \theta_k.$$

Since

$$r_j^2 - r_{j-1}^2 = (r_j + r_{j-1})(r_j - r_{j-1})$$

$$= 2\left(\frac{r_j + r_{j-1}}{2}\right)(r_j - r_{j-1}),$$

we may write

$$r_j^2 - r_{j-1}^2 = 2\hat{r}_j \, \Delta r_j \tag{2}$$

where $\hat{r}_j = \dfrac{r_j + r_{j-1}}{2}$ and $\Delta r_j = r_j - r_{j-1}$. It is important to note that the number $\hat{r}_j$ is simply the average of the radii r_{j-1} and r_j (see Figure 3.3). Combining (1) and (2), we may now write the area of the region R_{jk} as

$$\Delta A_{jk} = \hat{r}_j \, \Delta r_j \, \Delta \theta_k. \tag{3}$$

Now let f be a continuous function defined on the region Q, and let $\hat{\theta}_k$ be any number with $\theta_{k-1} \le \hat{\theta}_k \le \theta_k$. Then the number $f(\hat{r}_j, \hat{\theta}_k)$ is the value of the function f at the point $(\hat{r}_j, \hat{\theta}_k)$ in the region R_{jk}. If $f(r, \theta) \ge 0$ for all $(r, \theta) \in Q$, then the product $f(\hat{r}_j, \hat{\theta}_k) \, \Delta A_{jk}$ approximates the volume of the solid bounded by the graph of f over R_{jk}, and the sum

$$\sum_{j=1}^{n} \sum_{k=1}^{m} f(\hat{r}_j, \hat{\theta}_k) \, \Delta A_{jk} = \sum_{j=1}^{n} \sum_{k=1}^{m} f(\hat{r}_j, \hat{\theta}_k) \hat{r}_j \, \Delta r_j \, \Delta \theta_k \tag{4}$$

approximates the volume of the solid bounded by the graph of f over the entire region Q. (In writing the sums in line (4) we mean that the sums are taken only over those regions R_{jk} that lie entirely within Q.) We therefore define the double integral of $f(r, \theta)$ over Q as

$$\iint_Q f(r, \theta) \, dA = \lim_{\|P\| \to 0} \sum_{j=1}^{n} \sum_{k=1}^{m} f(\hat{r}_j, \hat{\theta}_k)\hat{r}_j \, \Delta r_j \, \Delta \theta_k \tag{5}$$

where $\|P\|$ is an appropriately defined norm for the partition P which we have described above.

The integral in (5) is defined whenever f is continuous on Q, regardless of the sign of $f(r, \theta)$. However, as for the double integral in Cartesian coordinates, if $f(r, \theta) \ge 0$ for all $(r, \theta) \in Q$, we have

$$\iint_Q f(r, \theta) \, dA = \text{volume of solid bounded above by the graph of} \tag{6}$$
$$f \text{ and below by } Q, \text{ and}$$

$$\iint_Q 1 \cdot dA = \text{area of the region } Q. \tag{7}$$

Of course, the integral defined by equation (5) is of little value unless we know how to evaluate the limit of the approximating sums. The following theorem, analogous to Theorem 2, shows how this may be done. The idea behind its proof is contained in equation (4).

THEOREM 3

Let Q be the region bounded by the graph of the continuous function $r = g(\theta)$ and the rays $\theta = a$ and $\theta = b$, as in Figure 3.1. Let f be a continuous function defined on the region Q. Then

$$\iint_Q f(r, \theta) \, dA = \int_a^b \int_0^{g(\theta)} f(r, \theta)r \, dr \, d\theta. \tag{8}$$

Theorem 3 says that the double integral in (5) may be evaluated as an iterated integral, where we integrate first with respect to r alone, for $0 \le r \le g(\theta)$, and then integrate the resulting function of θ over the limits $a \le \theta \le b$. However, *it is important to note that the integrand in the iterated integral is the product $f(r, \theta)r$, not just $f(r, \theta)$*. The reason for this can be seen from equation (3): The area of the region R_{jk} is $\Delta A_{jk} = \hat{r}_j \, \Delta r_j \, \Delta \theta_k$. In the Cartesian-coordinate case the rectangles R_{jk} used in approximating the area of Q had area $\Delta A_{jk} = \Delta x_j \, \Delta y_k$. The extra factor r appearing in the iterated polar double integral appears because of this formula for the area of the approximating region R_{jk}.

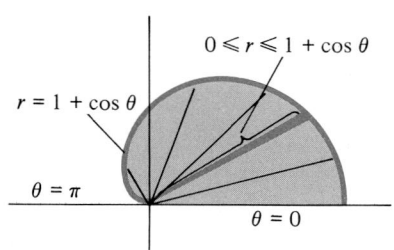

Figure 3.4   Region bounded by graph of $r = 1 + \cos \theta$ for $0 \le \theta \le \pi$.

Example 1

Evaluate the double integral $\iint_Q r \sin \theta \, dA$ where Q is the region inside the upper half of the cardioid $r = 1 + \cos \theta$ (see Figure 3.4).

Solution: Here $f(r, \theta) = r \sin \theta$, and the region Q may be described by the inequalities

$$0 \le r \le 1 + \cos \theta, \qquad 0 \le \theta \le \pi.$$

Thus, $g(\theta) = 1 + \cos \theta$ in equation (8), and

$$\iint_Q r \sin \theta \, dA = \int_0^\pi \int_0^{1+\cos \theta} r \sin \theta \cdot r \, dr \, d\theta \underset{\uparrow}{\hspace{1cm}} \text{note extra factor } r$$

$$= \int_0^\pi \int_0^{1+\cos \theta} r^2 \sin \theta \, dr \, d\theta$$

$$= \int_0^\pi \left\{ \frac{1}{3} r^3 \sin \theta \right]_{r=0}^{r=1+\cos \theta} \right\} d\theta$$

$$= \int_0^\pi \frac{1}{3} (1 + \cos \theta)^3 \sin \theta \, d\theta$$

$$= -\frac{1}{12} (1 + \cos \theta)^4 \Big]_{\theta=0}^{\theta=\pi}$$

$$= -\frac{1}{12} (0 - 2^4) = \frac{4}{3}. \qquad \diamond$$

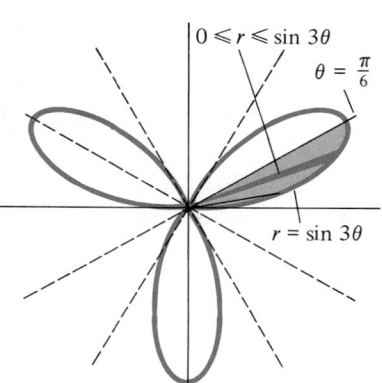

Figure 3.5 Region enclosed by three-leaved rose $r = \sin 3\theta$.

$0 \le r \le \sin 3\theta$

$\theta = \frac{\pi}{6}$

$r = \sin 3\theta$

Example 2

Find the area of the region enclosed by the three-leaved rose $r = \sin 3\theta$ (see Figure 3.5).

Solution: As illustrated in Figure 3.5, one sixth of the entire region Q lies between the rays $\theta = 0$ and $\theta = \pi/6$. Using the symmetry of the region, equation (7), and equation (8), we find that

$$\text{Area of } Q = \iint_Q 1 \, dA$$

$$= 6 \int_0^{\pi/6} \int_0^{\sin 3\theta} 1 \cdot r \, dr \, d\theta \underset{\uparrow}{\hspace{1cm}} \text{note extra factor } r$$

$$= 6 \int_0^{\pi/6} \left\{ \frac{r^2}{2} \right]_{r=0}^{r=\sin 3\theta} \right\} d\theta$$

$$= 3 \int_0^{\pi/6} \sin^2 3\theta \, d\theta$$

$$= \frac{3}{2} \int_0^{\pi/6} (1 - \cos 6\theta) \, d\theta \qquad \left(\sin^2 3\theta = \frac{1}{2} (1 - \cos 6\theta) \right)$$

$$= \frac{3}{2} \left[\theta - \frac{1}{6} \sin 6\theta \right]_{\theta=0}^{\theta=\pi/6}$$

$$= \frac{\pi}{4}. \qquad \diamond$$

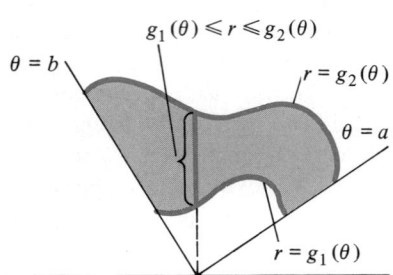

Figure 3.6 Region Q bounded by two polar curves, $r = g_1(\theta)$ and $r = g_2(\theta)$.

$\theta = b$

$g_1(\theta) \le r \le g_2(\theta)$

$r = g_2(\theta)$

$\theta = a$

$r = g_1(\theta)$

Figure 3.6 shows a region Q that is **r-simple:** it lies *between* the graphs of the polar equations $r = g_1(\theta)$ and $r = g_2(\theta)$ for $a \le \theta \le b$. That is,

$$Q = \{(r, \theta) \mid g_1(\theta) \le r \le g_2(\theta), \, a \le \theta \le b\}.$$

If we let
$$Q_1 = \{(r, \theta) \mid 0 \leq r \leq g_1(\theta), a \leq \theta \leq b\},$$
and
$$Q_2 = \{(r, \theta) \mid 0 \leq r \leq g_2(\theta), a \leq \theta \leq b\},$$
then $Q = Q_2 - Q_1$. Accordingly, we have

$$\iint_Q f(r, \theta) \, dA = \iint_{Q_2} f(r, \theta) \, dA - \iint_{Q_1} f(r, \theta) \, dA$$

$$= \int_a^b \int_0^{g_2(\theta)} f(r, \theta) \, r \, dr \, d\theta - \int_a^b \int_0^{g_1(\theta)} f(r, \theta) r \, dr \, d\theta$$

$$= \int_a^b \left\{ \int_0^{g_2(\theta)} f(r, \theta) r \, dr - \int_0^{g_1(\theta)} f(r, \theta) r \, dr \right\} d\theta$$

$$= \int_a^b \int_{g_1(\theta)}^{g_2(\theta)} f(r, \theta) r \, dr \, d\theta.$$

Thus, if Q is described by the polar inequalities
$$g_1(\theta) \leq r \leq g_2(\theta), \qquad a \leq \theta \leq b,$$
then

$$\iint_Q f(r, \theta) \, dA = \int_a^b \int_{g_1(\theta)}^{g_2(\theta)} f(r, \theta) r \, dr \, d\theta. \tag{9}$$

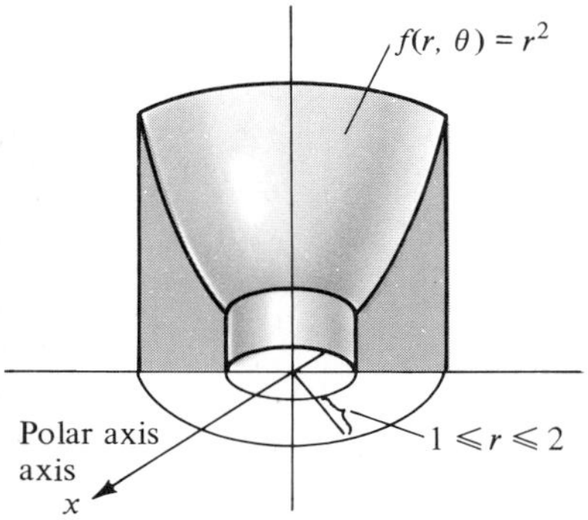

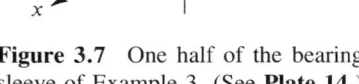

Figure 3.7 One half of the bearing sleeve of Example 3. (See **Plate 14.**)

Example 3

A bearing sleeve has the shape of the solid bounded above by the graph of $f(r, \theta) = r^2$ and below by the xy-plane for $1 \leq r \leq 2$. Find the volume of the sleeve.

Solution: The solid is illustrated in Figure 3.7. The region Q over which the solid is defined is described by the inequalities

$$1 \leq r \leq 2, \qquad 0 \leq \theta \leq 2\pi.$$

The volume is therefore

$$\iint_Q f(r, \theta) \, dA = \int_0^{2\pi} \int_1^2 r^2 \cdot r \, dr \, d\theta$$

$$= \int_0^{2\pi} \int_1^2 r^3 \, dr \, d\theta$$

$$= \int_0^{2\pi} \left\{ \frac{1}{4} r^4 \right]_{r=1}^{r=2} \right\} d\theta$$

$$= \int_0^{2\pi} \frac{15}{4} \, d\theta$$

$$= \frac{15\pi}{2}. \qquad \diamondsuit$$

Changing From Cartesian to Polar Coordinates

Often, a double integral in Cartesian coordinates is more easily evaluated by first changing to polar coordinates. For example, in Exercise 38 in the last section you were asked to find the volume of the sphere $x^2 + y^2 + z^2 = 4$ lying above the plane $z = 1$. This led to a double integral equivalent to

$$V = \int_{-\sqrt{3}}^{\sqrt{3}} \int_{-\sqrt{3-x^2}}^{\sqrt{3-x^2}} (\sqrt{4 - (x^2 + y^2)} - 1) \, dy \, dx. \tag{10}$$

(See Figure 3.8.)

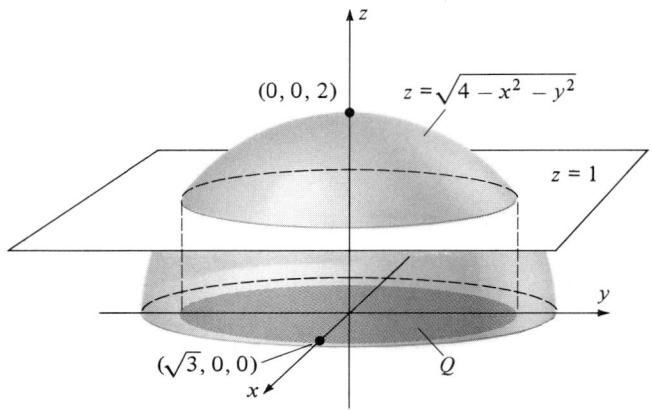

Figure 3.8 $Q = \{(x, y) \mid x^2 + y^2 \le 3\}$ may be described in polar coordinates as

$Q = \{(r, \theta) \mid 0 \le r \le \sqrt{3}, 0 \le \theta \le 2\pi\}.$

We can calculate this volume much more simply using polar coordinates, together with the equations

$$x = r \cos \theta \tag{11a}$$

$$y = r \sin \theta \tag{11b}$$

for changing from Cartesian coordinates to polar coordinates.

The region

$$Q = \{(x, y) \mid -\sqrt{3} \le x \le \sqrt{3}, -\sqrt{3 - x^2} \le y \le \sqrt{3 - x^2}\}$$

can be described in polar coordinates as

$$Q = \{(r, \theta) \mid 0 \le r \le \sqrt{3}, 0 \le \theta \le 2\pi\},$$

and the function $f(x, y) = \sqrt{4 - (x^2 + y^2)}$ in polar coordinates, obtained by using equations (11a) and (11b), is simply

$$f(r, \theta) = \sqrt{4 - (r^2 \cos^2 \theta + r^2 \sin^2 \theta)} = \sqrt{4 - r^2}.$$

The desired volume is then obtained using equations (6) and (8):

$$V = \iint_{Q} f(r \cos \theta, r \sin \theta) \, dA = \int_{0}^{2\pi} \int_{0}^{\sqrt{3}} (\sqrt{4 - r^2} - 1) r \, dr \, d\theta \quad \text{note extra } r$$

$$= \int_{0}^{2\pi} \left\{ -\frac{1}{3}(4 - r^2)^{3/2} - \frac{r^2}{2} \right]_{r=0}^{r=\sqrt{3}} \right\} d\theta$$

$$= \int_0^{2\pi} \left[-\frac{1}{3}(1 - 4^{3/2}) - \frac{3}{2} \right] d\theta$$

$$= \int_0^{2\pi} \frac{5}{6} d\theta = \frac{5\pi}{3}.$$

This example is typical of the following more general problem: Given an iterated integral in Cartesian coordinates, how can we evaluate the integral using an iterated integral in polar coordinates (without, of course, changing the value of the integral)? The answer is the following:

To express the iterated integral

$$\int_c^d \int_{h_1(y)}^{h_2(y)} f(x, y) \, dx \, dy \qquad \text{(or equivalent)}$$

in polar coordinates,

(i) express the region $Q = \{(x, y) \mid h_1(y) \leq x \leq h_2(y), c \leq y \leq d\}$ in polar coordinates as

$$Q = \{(r, \theta) \mid g_1(\theta) \leq r \leq g_2(\theta), a \leq \theta \leq b\}, \text{ and}$$

(ii) using the substitutions $x = r \cos \theta$ and $y = r \sin \theta$, replace the integrand

$$f(x, y) \qquad \text{by} \qquad f(r \cos \theta, r \sin \theta) \cdot r.$$

The result of steps (i) and (ii) is the iterated integral

$$\int_a^b \int_{g_1(\theta)}^{g_2(\theta)} f(r, \theta) r \, dr \, d\theta.$$

Another way to write this is simply that

$$\iint_Q f(x, y) \, dx \, dy = \iint_Q f(r \cos \theta, r \sin \theta) r \, dr \, d\theta. \qquad (12)$$

Equation (12) is referred to as a **change of variables** formula.

We shall not prove statement (12). However, it may be justified by comparing Theorem 2, which expresses $\iint_Q f \, dA$ as an iterated integral in Cartesian coordinates, with equation (9), which gives $\iint_Q f \, dA$ as an iterated integral in polar coordinates. We may paraphrase equation (12) by saying that in changing from Cartesian to polar coordinates we replace the *element of area*

$$dA = dx \, dy$$

in rectangular coordinates with the element of area

$$dA = r \, dr \, d\theta \qquad (13)$$

in polar coordinates. It is very important to note the extra factor r in (13), and to remember that, in the integral on the right side of equation (12), Q must be described using polar coordinates.

Example 4

Evaluate the integral

$$\int_{-2}^{2} \int_{-\sqrt{4-x^2}}^{\sqrt{4-x^2}} e^{x^2+y^2} \, dy \, dx$$

by first changing to polar coordinates.

Solution: From the limits of integration, we can see that the integral is being evaluated over the disc-shaped region

$$Q = \{(x, y) \mid -2 \le x \le 2, \ -\sqrt{4 - x^2} \le y \le \sqrt{4 - x^2}\}.$$

(See Figure 3.9.) This region can be described in polar coordinates by the inequalities

$$0 \le r \le 2, \qquad 0 \le \theta \le 2\pi \qquad \text{(Figure 3.10)}.$$

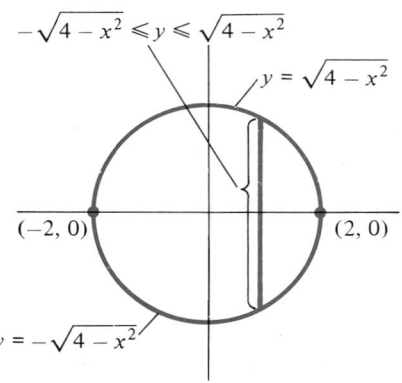

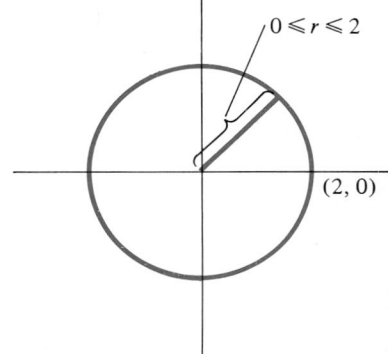

Figure 3.9 Q: $-2 \le x \le 2$
$-\sqrt{4 - x^2} \le y \le \sqrt{4 - x^2}$.

Figure 3.10 Q: $0 \le r \le 2$
$0 \le \theta \le 2\pi$.

With $x = r \cos \theta$ and $y = r \sin \theta$, the function $f(x, y) = e^{x^2+y^2}$ becomes

$$f(r \cos \theta, r \sin \theta) = e^{r^2 \cos^2 \theta + r^2 \sin^2 \theta}$$
$$= e^{r^2}.$$

Using (12), we obtain

$$\int_{-2}^{2} \int_{-\sqrt{4-x^2}}^{\sqrt{4-x^2}} e^{x^2+y^2} \, dy \, dx = \int_{0}^{2\pi} \int_{0}^{2} e^{r^2} \underbrace{r}_{} \, dr \, d\theta \qquad \text{note the extra factor } r$$

$$= \int_{0}^{2\pi} \left\{ \frac{1}{2} e^{r^2} \Big]_{r=0}^{r=2} \right\} d\theta$$

$$= \int_{0}^{2\pi} \frac{1}{2}(e^4 - 1) \, d\theta$$

$$= \pi(e^4 - 1). \qquad \diamond$$

REMARK: It is important to note that the iterated integral in Example 4 could not be evaluated in rectangular coordinates, since we would not be able to find an antiderivative for $e^{x^2+y^2}$ with respect to either x or y. Thus, it is sometimes essential that we change to polar coordinates, if possible. The next example involves a volume that could be calculated using Cartesian coordinates, but the calculation in polar coordinates is much easier. It is typical of the kind of problem we will encounter in Chapter 20.

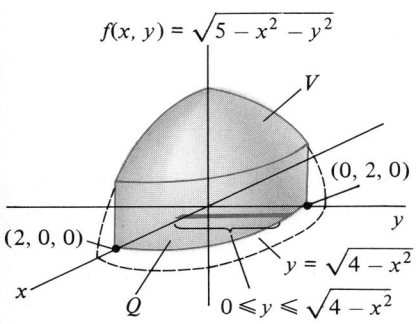

$f(x, y) = \sqrt{5 - x^2 - y^2}$

$(0, 2, 0)$

$(2, 0, 0)$

$y = \sqrt{4 - x^2}$

$0 \leqslant y \leqslant \sqrt{4 - x^2}$

Figure 3.11 Solid V over region Q in Example 5.

Example 5

Interpret the iterated integral

$$\int_0^2 \int_0^{\sqrt{4-x^2}} \sqrt{5 - x^2 - y^2} \; dy \; dx$$

geometrically, and evaluate the integral by first changing to polar coordinates.

Solution: Since the integrand is nonnegative, the integral may be interpreted as the volume of the solid bounded above by the graph of $f(x, y) = \sqrt{5 - x^2 - y^2}$ and below by the quarter disc $Q = \{(x, y) \mid 0 \leq x \leq 2, 0 \leq y \leq \sqrt{4 - x^2}\}$ of radius 2 (Figure 3.11). In polar coordinates, Q is determined by the inequalities

$$0 \leq r \leq 2, \qquad 0 \leq \theta \leq \frac{\pi}{2}.$$

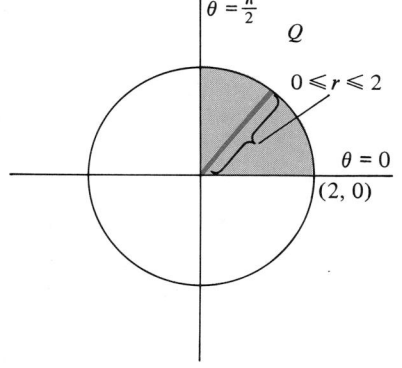

$\theta = \frac{\pi}{2}$

Q

$0 \leqslant r \leqslant 2$

$\theta = 0$

$(2, 0)$

Figure 3.12 Region Q in Example 5: $0 \leq r \leq 2, 0 \leq \theta \leq \pi/2$.

(See Figure 3.12.) Thus

$$\int_0^2 \int_0^{\sqrt{4-x^2}} \sqrt{5 - x^2 - y^2} \; dy \; dx$$

$$= \int_0^{\pi/2} \int_0^2 \sqrt{5 - (r \cos \theta)^2 - (r \sin \theta)^2} \cdot r \; dr \; d\theta \qquad \text{note the extra factor } r$$

$$= \int_0^{\pi/2} \int_0^2 \sqrt{5 - r^2} \; r \; dr \; d\theta$$

$$= \int_0^{\pi/2} \left\{ -\frac{1}{3}(5 - r^2)^{3/2} \Big]_{r=0}^{r=2} \right\} d\theta$$

$$= \int_0^{\pi/2} -\frac{1}{3}(1 - 5^{3/2}) \; d\theta$$

$$= \frac{\pi}{6}(5^{3/2} - 1).$$

$\diamond$

Exercise Set 19.3

In Exercises 1–7, sketch the region Q and evaluate the double integral $\iint_Q f(r, \theta) \; dA$ for the given function over the given region Q.

1. $f(r, \theta) = r$, $\quad Q = \{(r, \theta) \mid 0 \leq r \leq 1, 0 \leq \theta \leq 2\pi\}$

2. $f(r, \theta) = 1 - r$, $\quad Q = \{(r, \theta) \mid 0 \leq r \leq 2, 0 \leq \theta \leq \pi\}$

3. $f(x, y) = e^{x^2+y^2}$, $\quad Q = \{(r, \theta) \mid r \leq a\}$

4. $f(x, y) = e^{x^2+y^2}$, $\quad Q = \{(r, \theta) \mid 1 \leq r \leq 2, 0 \leq \theta \leq \pi/4\}$

5. $f(r, \theta) = 3r^2$, $\quad Q = \{(r, \theta) \mid 1 \leq r \leq 2, 0 \leq \theta \leq \pi/4\}$

6. $f(r, \theta) = 2 - r$, $\quad Q = \{(r, \theta) \mid 0 \leq r \leq (1 + \cos \theta), 0 \leq \theta \leq \pi/2\}$

7. $f(r, \theta) = r$, $\quad Q = \{(r, \theta) \mid 0 \le r \le \sin 3\theta,$
$\qquad\qquad\qquad\qquad 0 \le \theta \le \pi/3\}$

In Exercises 8–14, find the area of the region Q by use of a double integral.

8. Q is the region enclosed by the circle $r = 2 \cos \theta$.

9. Q is the region enclosed by the three-leaved rose $r = \sin 3\theta$.

10. Q is the region enclosed by the cardioid $r = a(1 + \cos \theta)$.

11. Q is the region inside the cardioid $r = 1 + \cos \theta$ and outside the circle $r = 1$.

12. Q is the region bounded by the graph of $r^2 = a^2 \sin 2\theta$.

13. Q is the region enclosed by the graph of $r = 2 - \sin \theta$.

14. Q is the region enclosed by the graph of the lemniscate $r^2 = 4 \cos^2 \theta$.

In Exercises 15–22, change the iterated integral from Cartesian to polar coordinates. Then evaluate the resulting integral.

15. $\displaystyle\int_0^1 \int_0^{\sqrt{1-x^2}} 2 \, dy \, dx$

16. $\displaystyle\int_{-2}^{\sqrt{2}} \int_x^{\sqrt{4-x^2}} 1 \, dy \, dx$

17. $\displaystyle\int_0^1 \int_0^{\sqrt{1-y^2}} e^{x^2+y^2} \, dx \, dy$

18. $\displaystyle\int_0^2 \int_{-\sqrt{2y-y^2}}^{\sqrt{2y-y^2}} \sqrt{x^2 + y^2} \, dx \, dy$

19. $\displaystyle\int_0^2 \int_0^{\sqrt{2x-x^2}} \frac{1}{\sqrt{x^2 + y^2}} \, dy \, dx$

20. $\displaystyle\int_0^1 \int_0^{\sqrt{4-x^2}} (x^2 + y^2)^{3/2} \, dy \, dx$

21. $\displaystyle\int_{-2}^2 \int_0^{\sqrt{4-x^2}} \frac{x}{\sqrt{x^2 + y^2}} \, dy \, dx$

22. $\displaystyle\int_0^1 \int_0^{\sqrt{1-y^2}} (x^2 + y^2)^{3/2} \, dx \, dy$

23. Find the volume of the portion of the cylinder $x^2 + (y - 1)^2 = 4$ bounded above by the plane $z = x + 4$ and below by the xy-plane.

24. Find the area of the region outside the spiral $r = \theta$ and inside the spiral $r = 2\theta$ for $0 \le \theta \le 2\pi$.

25. Find the volume of the region lying inside the sphere $x^2 + y^2 + z^2 = 4$ and outside the cylinder $x^2 + y^2 = 1$.

26. Find the volume of the region lying inside both the cone $z^2 = x^2 + y^2$ and the sphere $x^2 + y^2 + z^2 = 2$.

27. Find the volume of the solid bounded above by the plane $z = y + 2$ and below by the region inside the cardioid $r = 1 + \cos \theta$.

28. Use a double integral in polar coordinates to obtain the formula for the volume of a right circular cylinder of radius r and height h.

29. Find the volume of the solid bounded above by the graph of $z = 9 - x^2 - y^2$ and below by the graph of $z = 1 + x^2 + y^2$.

30. A hole 2 cm in diameter is drilled through the center of a spherical bearing of radius 3 cm. Find the volume of the remaining solid.

31. Change the order of integration in

$$\int_0^{\pi/2} \int_0^{\sin \theta} \sin \theta \, dr \, d\theta$$

and evaluate the integral.

32. Find the volume of the solid inside both the ellipsoid $z^2 + 4r^2 = 4$ and the cylinder $r = \sin \theta$.

33. Find the volume of the solid bounded by the cone $z = \sqrt{x^2 + y^2}$, the cylinder $x^2 + y^2 = 4$, and the xy-plane.

19.4 CALCULATING MASS AND CENTERS OF MASS

In this section we use the double integral to calculate mass and centroids for lamina (thin flat objects) lying in the plane. The difference between the discussion of Chapter 7 and what we do here is that we previously had assumed the density ρ of the material to be constant. Here we will treat the more general case of a variable density ρ. We assume the density function ρ to be continuous throughout the planar region Q that describes the lamina.

More specifically, let Q be a region in the plane (which we think of as the base of the lamina) and let $\rho(x, y)$ be the **mass per unit area** of the lamina at point (x, y). This is what we mean by density. (Thus, $\rho(x, y)$ is affected both by the thickness of the material and by its mass per unit volume. We will not be concerned about these two factors individually.) If R_j is a rectangle in Q of area ΔA, and if (s_j, t_j) is a point in R_j, then the product

$$\Delta M_j = \rho(s_j, t_j)\,\Delta A \qquad \text{(mass = mass per unit area times area)}$$

is an approximation to the mass of the lamina over the rectangle R_j. If, as in Section 19.2, the region Q is partitioned by a rectangular grid* and $R_1, R_2, \ldots, R_n$ is a list of all rectangles lying within Q, then

$$M \approx \sum_{j=1}^{n} \Delta M_j = \sum_{j=1}^{n} \rho(s_j, t_j)\,\Delta A \tag{1}$$

is an approximation to the mass of the lamina over Q. Since, as $n \to \infty$, the union of the rectangles R_j provides an increasingly accurate approximation to the region Q, we obtain M, the mass of Q, as the double integral

$$M = \iint_Q \rho(x, y)\,dA. \tag{2}$$

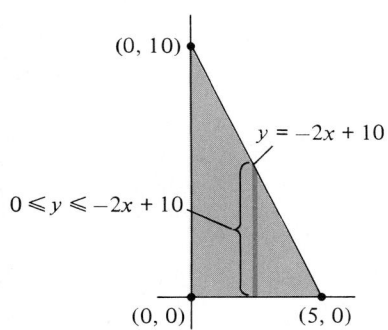

Figure 4.1 The triangular region in Example 1.

(0, 10)

$y = -2x + 10$

$0 \leq y \leq -2x + 10$

(0, 0) (5, 0)

Example 1

A thin plate has the shape of a right triangle with legs of length 5 cm and 10 cm. The density, in terms of mass per unit area, at each point on the plate is proportional to the square of the distance from the vertex corresponding to the right angle. What is the mass of the plate?

Solution: With the triangle positioned as in Figure 4.1, the region it occupies may be described by the inequalities

$$0 \leq x \leq 5, \qquad 0 \leq y \leq -2x + 10.$$

Since the right angle is at the origin, the density function is

$$\rho(x, y) = \lambda(x^2 + y^2)$$

where λ is constant. By (2), the mass is

$$M = \iint_Q \lambda(x^2 + y^2)\,dA = \int_0^5 \int_0^{-2x+10} \lambda(x^2 + y^2)\,dy\,dx$$

$$= \int_0^5 \lambda \left\{ x^2 y + \frac{1}{3}y^3 \right\}_{y=0}^{y=-2x+10} dx$$

$$= \int_0^5 \lambda \left(-\frac{14}{3}x^3 + 50x^2 - 200x + \frac{1000}{3} \right) dx$$

$$= \lambda \left[-\frac{7}{6}x^4 + \frac{50}{3}x^3 - 100x^2 + \frac{1000}{3}x \right]_0^5$$

$$= \frac{3125\lambda}{6}. \qquad \diamondsuit$$

*As we did in Chapter 7, we shall henceforth use *regular* grids to partition a region Q into rectangles of equal dimensions Δx and Δy, and areas $\Delta A = \Delta x\,\Delta y$.

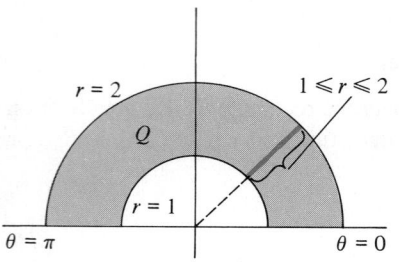

Figure 4.2 Annular region of Example 2.

Example 2

A machine part has the shape of a half annulus—the region between two concentric semicircles of radius $r = 1$ and $r = 2$ (see Figure 4.2). Find the mass of the part if the density at each point is proportional to the distance from that point to the common center of the two semicircles.

Solution: The density function can be written as

$$\rho(x, y) = \lambda\sqrt{x^2 + y^2}$$

where λ is constant. However, it is most convenient to describe the region Q in polar coordinates using the inequalities

$$0 \le \theta \le \pi, \qquad 1 \le r \le 2,$$

in which case the density function is $\rho(r, \theta) = \lambda r$.
The mass is therefore

$$M = \iint\limits_Q \lambda r \, dA = \int_0^\pi \int_1^2 \lambda r \cdot r \, dr \, d\theta$$

$$= \int_0^\pi \int_1^2 \lambda r^2 \, dr \, d\theta = \int_0^\pi \left\{ \frac{\lambda}{3} r^3 \right\}_{r=1}^{r=2} d\theta$$

$$= \int_0^\pi \frac{7\lambda}{3} \, d\theta = \frac{7\lambda\pi}{3}. \qquad \diamond$$

Centers of Mass

In Chapter 7 we determined that the x-coordinate of the *centroid* of the lamina with shape Q and constant density ρ is

$$\bar{x} = \frac{\displaystyle\int_a^b \rho x[f(x) - g(x)] \, dx}{\displaystyle\int_a^b \rho[f(x) - g(x)] \, dx}. \tag{3}$$

In writing equation (3), we assume that Q is a region that can be described by the inequalities

$$a \le x \le b, \qquad g(x) \le y \le f(x) \qquad \text{(Figure 4.3)}.$$

In other words, Q is y-simple.

To generalize to the case of a variable density function ρ, we begin with the integral in the numerator of $\bar{x}$ in (3), which we write as

$$\int_a^b \rho x[f(x) - g(x)] \, dx = \int_a^b \left\{ \rho xy\big]_{y=g(x)}^{y=f(x)} \right\} dx \tag{4}$$

$$= \int_a^b \int_{g(x)}^{f(x)} \rho x \, dy \, dx.$$

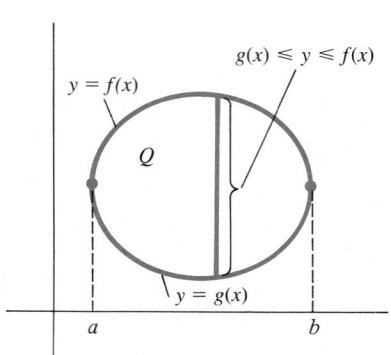

Figure 4.3

The right-hand side of (4) allows us to define the first moment of mass for Q about the y-axis when ρ is nonconstant. We simply replace ρ by $\rho(x, y)$ and define

$$M_y = \int_a^b \int_{g(x)}^{f(x)} x\rho(x, y) \, dy \, dx = \iint\limits_Q x\rho(x, y) \, dA.$$

Similarly, the first moment of mass for Q about the x-axis is defined to be

$$M_x = \iint\limits_Q y\rho(x, y)\, dA.$$

Finally, the denominator of $\bar{x}$ can be written

$$\int_a^b \rho[f(x) - g(x)]\, dx = \int_a^b \left\{ \rho y \right]_{y=g(x)}^{y=f(x)} \right\}\, dx$$

$$= \int_a^b \int_{g(x)}^{f(x)} \rho\, dy\, dx$$

$$= \iint\limits_Q \rho\, dy\, dx,$$

which is just the mass (area times density) of Q. We are therefore ready to generalize the concept of centroid to that of *center of mass*.

DEFINITION 4

Let Q be a region in the xy-plane that is either x-simple or y-simple, and let ρ be a continuous density function defined on Q. Then

(i) The **mass** of Q is the number

$$M = \iint\limits_Q \rho(x, y)\, dA.$$

(ii) The **first moment of mass for Q about the y-axis** is the number

$$M_y = \iint\limits_Q x\rho(x, y)\, dA.$$

(iii) The **first moment of mass for Q about the x-axis** is the number

$$M_x = \iint\limits_Q y\rho(x, y)\, dA.$$

(iv) The **center of mass** of Q is the point $(\bar{x}, \bar{y})$, where

$$\bar{x} = \frac{M_y}{M}, \qquad \bar{y} = \frac{M_x}{M}.$$

Because of the way the concept of centroid is generalized by Definition 4, the physical interpretation of the center of mass is the same as before—the lamina associated with Q will "balance" at the point $(\bar{x}, \bar{y})$.

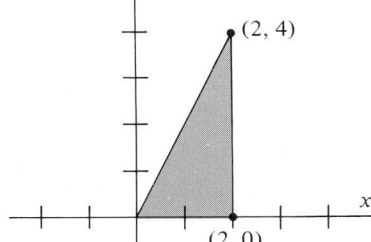

(2, 4)

(2, 0)

Figure 4.4

Example 3

A thin lamina has the shape of a triangle with vertices at $(0, 0)$, $(2, 0)$, and $(2, 4)$. The density function associated with the lamina has equation $\rho(x, y) = 4x + 2y + 2$. Find the mass and center of mass of the lamina (see Figure 4.4).

Solution: The region Q associated with the lamina may be described by the inequalities

$$0 \le x \le 2, \qquad 0 \le y \le 2x.$$

Thus, by Definition 4,

$$M = \iint\limits_{Q} (4x + 2y + 2)\, dA = \int_0^2 \int_0^{2x} (4x + 2y + 2)\, dy\, dx$$

$$= \int_0^2 \left\{ 4xy + y^2 + 2y]_{y=0}^{y=2x} \right\} dx$$

$$= \int_0^2 (12x^2 + 4x)\, dx$$

$$= 4x^3 + 2x^2]_0^2$$

$$= 40,$$

$$M_y = \iint\limits_{Q} x(4x + 2y + 2)\, dA = \int_0^2 \int_0^{2x} (4x^2 + 2xy + 2x)\, dy\, dx$$

$$= \int_0^2 \left\{ 4x^2 y + xy^2 + 2xy]_{y=0}^{y=2x} \right\} dx$$

$$= \int_0^2 (12x^3 + 4x^2)\, dx = 3x^4 + \frac{4}{3}x^3 \Big]_0^2$$

$$= \frac{176}{3},$$

and

$$M_x = \iint\limits_{Q} y(4x + 2y + 2)\, dA = \int_0^2 \int_0^{2x} (4xy + 2y^2 + 2y)\, dy\, dx$$

$$= \int_0^2 \left\{ 2xy^2 + \frac{2}{3}y^3 + y^2 \right]_{y=0}^{y=2x} \right\} dx$$

$$= \int_0^2 \left(\frac{40}{3}x^3 + 4x^2 \right) dx = \frac{10}{3}x^4 + \frac{4}{3}x^3 \Big]_0^2$$

$$= \frac{192}{3}.$$

Thus,

$$\bar{x} = \frac{M_y}{M} = \frac{\left(\dfrac{176}{3} \right)}{40} = \frac{22}{15}$$

and

$$\bar{y} = \frac{M_x}{M} = \frac{\left(\dfrac{192}{3} \right)}{40} = \frac{8}{5}.$$

The center of mass is therefore $(\bar{x}, \bar{y}) = \left(\dfrac{22}{15}, \dfrac{8}{5} \right)$. ◇

Example 4

Find the center of mass for the lamina described in Example 2.

Solution: The density function in Example 2 is

$$\rho(x, y) = \lambda\sqrt{x^2 + y^2}$$

and the region Q is described in polar coordinates as

$$0 \le \theta \le \pi, \qquad 1 \le r \le 2.$$

We evaluate the integral for M_y in polar coordinates as

$$M_y = \iint_Q x\rho(x, y)\, dA = \int_0^\pi \int_1^2 (r\cos\theta)\lambda r \cdot \underbrace{r\, dr\, d\theta}_{dA}$$

$$\rho = \lambda\sqrt{x^2 + y^2} = \lambda r$$
$$x = r\cos\theta$$

$$= \lambda \int_0^\pi \int_1^2 r^3 \cos\theta\, dr\, d\theta$$

$$= \lambda \int_0^\pi \left\{ \frac{1}{4}r^4 \cos\theta \Big]_{r=1}^{r=2} \right\} d\theta$$

$$= \lambda \int_0^\pi \frac{15}{4} \cos\theta\, d\theta$$

$$= \lambda\left(\frac{15}{4} \sin\theta \right)\Big]_{\theta=0}^{\theta=\pi}$$

$$= 0.$$

(This should not surprise you since both Q and ρ are symmetric with respect to the y-axis.) The integral for M_x is

$$M_x = \iint_Q y\rho(x, y)\, dA = \int_0^\pi \int_1^2 (r\sin\theta) \cdot \lambda r \cdot \underbrace{r\, dr\, d\theta}_{dA}$$

$$\rho = \lambda\sqrt{x^2 + y^2} = \lambda r$$
$$y = r\sin\theta$$

$$= \lambda \int_0^\pi \int_1^2 r^3 \sin\theta\, dr\, d\theta$$

$$= \lambda \int_0^\pi \left\{ \frac{1}{4}r^4 \sin\theta \Big]_{r=1}^{r=2} \right\} d\theta$$

$$= \lambda \int_0^\pi \frac{15}{4} \sin\theta\, d\theta$$

$$= \lambda\left(-\frac{15}{4} \cos\theta \right)\Big]_{\theta=0}^{\theta=\pi}$$

$$= \frac{15\lambda}{2}.$$

Since we found in Example 2 that $M = \dfrac{7\lambda\pi}{3}$, we have

$$\bar{x} = \frac{M_y}{M} = 0$$

and

$$\bar{y} = \frac{M_x}{M} = \frac{\left(\dfrac{15\lambda}{2}\right)}{\left(\dfrac{7\lambda\pi}{3}\right)} = \frac{45}{14\pi} \approx 1.02.$$

The center of mass is therefore $\left(0, \dfrac{45}{14\pi}\right)$. ◇

Exercise Set 19.4

In Exercises 1–10, find the mass of a lamina with shape given by the region Q and density function ρ.

1. $\rho(x, y) = x + y$,
$Q = \{(x, y) \mid 0 \le x \le 2, 0 \le y \le 1\}$

2. $\rho(x, y) = x^2 + y$,
$Q = \{(x, y) \mid 0 \le x \le 2, 0 \le y \le x\}$

3. $\rho(x, y) = 6 + x$,
$Q = \{(x, y) \mid -1 \le x \le 1, 0 \le y \le 1 - x^2\}$

4. $\rho(x, y) = xy$,
$Q = \{(x, y) \mid 0 \le x \le 1 - y^2, 0 \le y \le 1\}$

5. $\rho(x, y) = \sin(x + y)$,
$Q = \{(x, y) \mid 0 \le x \le \pi/4, 0 \le y \le \pi/4\}$

6. $\rho(x, y) = xy$, $Q = \{(x, y) \mid 0 \le x \le 1, 0 \le y \le 1\}$

7. $\rho(x, y) = x^2 + y^2$,
$Q = \{(x, y) \mid -1 \le x \le 1, 0 \le y \le \sqrt{1 - x^2}\}$

8. $\rho(x, y) = \sqrt{x^2 + y^2}$,
$Q = \{(r, \theta) \mid 0 \le r \le 2, 0 \le \theta \le \pi/2\}$

9. $\rho(x, y) = \dfrac{1}{\sqrt{x^2 + y^2}}$, Q is the annulus $1 \le r \le 2$

10. $\rho(x, y) = xy$,
$Q = \{(r, \theta) \mid 0 \le r \le 1, 0 \le \theta \le \pi/2\}$

11. Find the center of mass of the lamina in Exercise 1.

12. Find the center of mass of the lamina in Exercise 3.

13. Find the center of mass of the lamina in Exercise 8.

14. A lamina has the shape of a triangle with vertices $(0, 0)$, $(0, 4)$, and $(1, 0)$. The density at each point (x, y) is $\rho(x, y) = y - x + 8$. Find the mass of the lamina.

15. Find the center of mass of the lamina in Exercise 14.

16. Find the centroid of the region bounded by the graph of $r = \sin 2\theta$ for $0 \le \theta \le \pi/2$. (Assume $\rho(x, y) = 1$ throughout the region.)

17. Find the centroid of the planar region bounded by the parabola $y = 4 - x^2$ and the x-axis.

18. Find the centroid of the region bounded by the graph of the function $f(x) = 1/x$ and the line $2x + 2y = 5$.

19. Find the centroid of the region obtained by connecting the points $(0, 0)$, $(4, 0)$, $(4, 4)$, $(2, 4)$, $(2, 1)$, $(0, 1)$, and $(0, 0)$ in order by line segments.

20. Find the centroid of the region bounded by the graph of $y = x^2$ and $y = x^3$.

21. Show that the centroid of the region
$$R = \{(x, y) \mid 0 \le x \le a, 0 \le y \le \sqrt{a^2 - x^2}\}$$
is $(\bar{x}, \bar{y}) = \left(\dfrac{4a}{3\pi}, \dfrac{4a}{3\pi}\right)$.

22. Find the centroid of the region bounded by the graph of $f(x) = \sinh x$ and the x-axis for $0 \le x \le 1$.

23. Find the centroid of the region bounded by the graph of $x = y(4 - y)$ and the y-axis.

24. A lamina has the shape of a circle of radius R. The density at any point P is proportional to the distance from the center. What is the mass?

25. A lamina has the shape of a right triangle with legs of length 2 and 6. The density at any point is proportional to the distance from the longer leg. What is the mass?

26. Find the distance from the vertex at the right angle to the center of mass for the triangle in Exercise 25.

27. Assume that the region in Example 1 is oriented with the 10 cm leg of the triangle along the x-axis, again with the right angle at the origin. Show that the mass calculated in this way is the same as that found in Example 1.

28. Find the center of mass of a lamina with density function $\rho(r, \theta) = \lambda(1 + \cos^2 \theta)$ over the half annulus $Q = \{(r, \theta) \mid 1 \le r \le 2, 0 \le \theta \le \pi\}$. Locate the center of mass on a sketch of the region Q. State (in words) how the mass of the lamina is distributed, and what relation this has to the location of the center of mass.

19.5 SURFACE AREA

As our final application of double integrals, we consider the problem of calculating the area of a surface in space. We shall restrict our considerations to surfaces that are graphs of functions of two variables, although the ideas discussed here can be extended to more general types of surfaces.

Let Q be a region in the plane which is either x-simple or y-simple, and let the surface S be the graph of the continuous function $z = f(x, y)$ on Q. We partition the region Q with a rectangular grid, and we denote by $R_1, R_2, \ldots, R_n$ the rectangles lying entirely within Q. This grid is constructed so that each of the rectangles R_m has dimensions Δx and Δy (see Figure 5.1). The grid on Q partitions the surface S into

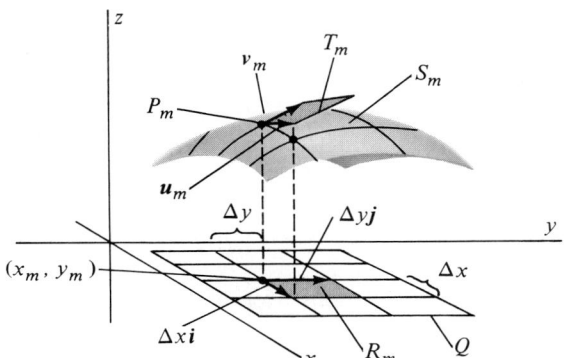

Figure 5.1 Surface is approximated by parallelograms pasted on above corners of approximating rectangles.

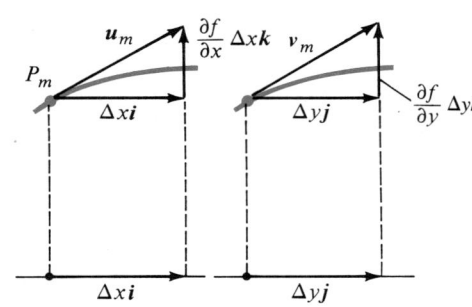

Figure 5.2 $u_m = \Delta x i + \dfrac{\partial f}{\partial x} \Delta x k$

$$v_m = \Delta y j + \dfrac{\partial f}{\partial y} \Delta y k.$$

patches $S_1, S_2, \ldots, S_n$. Specifically, by the patch S_m we mean the portion of the surface S lying above the rectangle R_m.

As Figure 5.1 suggests, the patch S_m is nearly a parallelogram, but not quite, since S_m is part of the (generally) curved surface S. We therefore construct a parallelogram T_m which approximates the patch S_m in the following way. Let P_m be the point on the corner of S_m nearest the z-axis and let (x_m, y_m) be the vertex of the rectangle R_m directly beneath the point P_m. Then the vector

$$u_m = \Delta x\, i + \frac{\partial f}{\partial x}(x_m, y_m)\, \Delta x\, k, \tag{1}$$

when originating at point P_m, terminates at some point above the vertex $(x_m + \Delta x, y_m)$ of R_m. Moreover, u_m is tangent to S at P_m, by definition of the partial derivative $\dfrac{\partial f}{\partial x}(x_m, y_m)$. Similarly, the vector

$$v_m = \Delta y\, j + \frac{\partial f}{\partial y}(x_m, y_m)\, \Delta y\, k, \tag{2}$$

when originating at P_m, terminates above the vertex $(x_m, y_m + \Delta y)$ of R_m and is tangent to S at P_m (see Figure 5.2).

From the properties noted for the vectors u_m and v_m, we conclude that u_m and v_m determine a parallelogram T_m that

(i) lies directly above the rectangle R_m (and, hence, the patch S_m), and
(ii) lies tangent to the surface S at point P_m.

The idea is therefore to use the area of the parallelogram T_m as an approximation to the area of the patch S_m. From Section 16.5, we recall that the area of the parallelogram T_m is equal to the length of the cross product of the vectors u_m and v_m, or

$$\text{Area of } T_m = |u_m \times v_m|$$

$$= \left| \left(\Delta x \, i + \frac{\partial f}{\partial x}(x_m, y_m) \, \Delta x \, k \right) \times \left(\Delta y \, j + \frac{\partial f}{\partial y}(x_m, y_m) \, \Delta y \, k \right) \right|$$

$$= (\Delta x \Delta y) \left| \det \begin{bmatrix} i & j & k \\ 1 & 0 & \dfrac{\partial f}{\partial x}(x_m, y_m) \\ 0 & 1 & \dfrac{\partial f}{\partial y}(x_m, y_m) \end{bmatrix} \right|$$

$$= (\Delta x \Delta y) \left| -\frac{\partial f}{\partial x}(x_m, y_m) i - \frac{\partial f}{\partial y}(x_m, y_m) j + k \right|$$

$$= \sqrt{\left[\frac{\partial f}{\partial x}(x_m, y_m) \right]^2 + \left[\frac{\partial f}{\partial y}(x_m, y_m) \right]^2 + 1} \; \Delta x \Delta y.$$

Summing these approximations over all patches $S_1, S_2, \ldots, S_n$ gives the approximation to the area of S as

$$\text{Area of } S \approx \sum_{m=1}^{n} \sqrt{\left[\frac{\partial f}{\partial x}(x_m, y_m) \right]^2 + \left[\frac{\partial f}{\partial y}(x_m, y_m) \right]^2 + 1} \; \Delta x \Delta y.$$

If the partial derivatives $\dfrac{\partial f}{\partial x}$ and $\dfrac{\partial f}{\partial y}$ are continuous on Q, this Riemann sum converges to an integral as $n \to \infty$ and as $\Delta x \to 0$ and $\Delta y \to 0$. This integral provides our definition of surface area.

DEFINITION 5

Let Q be a region in the plane that is either x-simple or y-simple, and let S be the graph of the function $z = f(x, y)$ for $(x, y) \in Q$. If $\dfrac{\partial f}{\partial x}(x, y)$ and $\dfrac{\partial f}{\partial y}(x, y)$ are continuous on Q, the **area of the surface S** is defined to be

$$A_s = \iint\limits_{Q} \sqrt{\left[\frac{\partial f}{\partial x}(x, y) \right]^2 + \left[\frac{\partial f}{\partial y}(x, y) \right]^2 + 1} \; dA. \tag{3}$$

Example 1

Find the area of the surface that is the graph of the equation $f(x, y) = x^2$ lying above the rectangle $Q = \{(x, y) \mid -1 \le x \le 1, -1 \le y \le 1\}$.

Solution: The surface is the "cylinder" sketched in Figure 5.3. Using the method of integration by trigonometric substitution, we may evaluate the surface area inte-

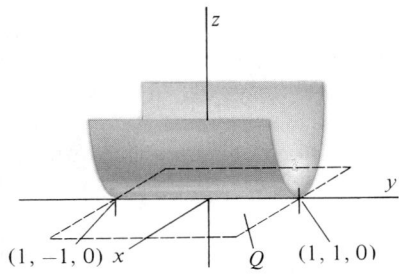

Figure 5.3 Cylinder $f(x, y) = x^2$ over the square

Q: $-1 \le x \le 1$, $-1 \le y \le 1$.

(See **Plate 15**.)

gral (3) as

$$A_s = \iint_Q \sqrt{(2x)^2 + 1} \, dA$$

$$= \int_{-1}^{1} \int_{-1}^{1} \sqrt{4x^2 + 1} \, dx \, dy$$

$$= \int_{-1}^{1} \left\{ \frac{1}{2} x \sqrt{4x^2 + 1} + \frac{1}{4} \ln |2x + \sqrt{4x^2 + 1}| \right]_{x=-1}^{x=1} \right\} dy$$

$$= \int_{-1}^{1} \left\{ \sqrt{5} + \frac{1}{4}[\ln(2 + \sqrt{5}) - \ln(\sqrt{5} - 2)] \right\} dy$$

$$= 2\sqrt{5} + \frac{1}{2}[\ln(2 + \sqrt{5}) - \ln(\sqrt{5} - 2)]$$

$$\approx 5.92. \qquad \diamond$$

Example 2

Find the surface area of the portion of the paraboloid

$$z = 4 - x^2 - y^2$$

lying above the xy-plane.

Solution: The paraboloid intersects the xy-plane in the circle $4 - x^2 - y^2 = 0$, or $x^2 + y^2 = 4$. (See Figure 5.4.) In xy-coordinates, this region may be described by the inequalities

$$-2 \le x \le 2, \qquad -\sqrt{4 - x^2} \le y \le \sqrt{4 - x^2}.$$

With $f(x, y) = 4 - x^2 - y^2$, we have

$$\frac{\partial f}{\partial x}(x, y) = -2x, \qquad \text{and} \qquad \frac{\partial f}{\partial y}(x, y) = -2y.$$

By (3) the integral giving the surface area is

$$A_S = \iint_Q \sqrt{4x^2 + 4y^2 + 1} \, dA = \int_{-2}^{2} \int_{-\sqrt{4-x^2}}^{\sqrt{4-x^2}} \sqrt{4x^2 + 4y^2 + 1} \, dy \, dx.$$

This integral is most easily evaluated in polar coordinates. With $x = r \cos \theta$ and $y = r \sin \theta$, and the region Q described as

$$0 \le \theta \le 2\pi, \qquad 0 \le r \le 2,$$

we have

$$A_s = \int_0^{2\pi} \int_0^2 \sqrt{4(r \cos \theta)^2 + 4(r \sin \theta)^2 + 1} \, r \, dr \, d\theta$$

$$= \int_0^{2\pi} \int_0^2 \sqrt{4r^2 + 1} \, r \, dr \, d\theta$$

$$= \int_0^{2\pi} \frac{1}{12}(4r^2 + 1)^{3/2} \Big]_{r=0}^{r=2} d\theta$$

$$= \frac{\pi}{6}(17^{3/2} - 1) \approx 11.5 \, \pi. \qquad \diamond$$

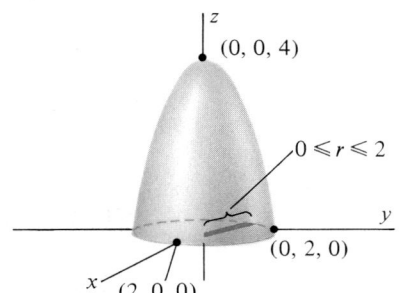

Figure 5.4 Paraboloid $z = 4 - x^2 - y^2$.

Example 3

Show that Definition 5 of surface area agrees with the usual definition of area when S is a flat surface over a rectangle in the plane.

Solution: If S is a flat surface in space, then S is a portion of a plane. Thus, we let

$$z = f(x, y) = Ax + By + C$$

be the equation for the plane, and we let

$$Q = \{(x, y) \mid a \le x \le b, c \le y \le d\}$$

be a rectangle in the plane.

As in Figure 5.1, the portion of the plane $Ax + By + C$ lying over the rectangle Q is a parallelogram. From (1) and (2) we see that the vectors

$$u = \Delta x\, i + \frac{\partial f}{\partial x}\Delta x\, k = (b - a)i + A(b - a)k$$

and

$$v = \Delta y\, j + \frac{\partial f}{\partial y}\Delta y\, k = (d - c)j + B(d - c)k,$$

when positioned at the vertex $(a, c, f(a, c))$, form two adjacent sides of the parallelogram. By a standard calculation involving the cross product, the area of the parallelogram is

$$A_s = |u \times v|$$

$$= \left| \det \begin{bmatrix} i & j & k \\ b - a & 0 & A(b - a) \\ 0 & d - c & B(d - c) \end{bmatrix} \right|$$

$$= |-A(b - a)(d - c)i - B(b - a)(d - c)j + (b - a)(d - c)k|$$

$$= (b - a)(d - c)\sqrt{A^2 + B^2 + 1}.$$

Using Definition 5 with $\dfrac{\partial f}{\partial x} = A$ and $\dfrac{\partial f}{\partial y} = B$ we find

$$A_s = \int_c^d \int_a^b \sqrt{A^2 + B^2 + 1}\; dx\, dy = (b - a)(d - c)\sqrt{A^2 + B^2 + 1}.$$

Thus, Definition 5 agrees with our usual concept of area for flat surfaces. ◇

The general notion of calculating surface area by double integration will be put to important use in the next chapter.

Exercise Set 19.5

In Exercises 1–10, find the surface area of the graph of $z = f(x, y)$ above the region Q in the plane.

1. $f(x, y) = x + y + 6$, $Q = \{(x, y) \mid 0 \le x \le 1, 0 \le y \le 1\}$

2. $f(x, y) = 9 - x + 2y$, $Q = \{(x, y) \mid 0 \le x^2 + y^2 \le 1\}$

3. $f(x, y) = 9 - x^2 - y^2$, $Q = \{(x, y) \mid 0 \le x^2 + y^2 \le 3\}$

4. $f(x, y) = 4 + y^2$, $Q = \{(x, y) \mid 0 \le x \le 1, 0 \le y \le 2\}$

5. $f(x, y) = 2 - x - y$, $Q = \{(x, y) \mid 0 \le x \le 2, 0 \le y \le 2 - x\}$

6. $f(x, y) = 3 + y^2$, $Q = \{(x, y) \mid 0 \le x \le 2, 0 \le y \le 2\}$

7. $f(x, y) = \sqrt{x^2 + y^2}$, $Q = \{(x, y) \mid 1 \le x^2 + y^2 \le 4\}$

8. $f(x, y) = x + y^2$, $Q = \{(x, y) \mid 0 \le x \le 1, \, 0 \le y \le 2\}$

9. $f(x, y) = \sqrt{3}\, y - x^2$, $Q = \{(x, y) \mid 0 \le x \le 1, \, 0 \le y \le 1\}$

10. $f(x, y) = x^2 + y$, $Q = \{(x, y) \mid 0 \le x \le 1, \, 0 \le y \le x\}$

11. Find the surface area of the portion of the graph of $z = y + 2x^2$ over the triangular region with vertices $(0, 0)$, $(0, 1)$, and $(1, 1)$.

12. Find the area of the part of the plane $x + y + z = 4$ bounded by the cylinder $x^2 + y^2 = 4$.

13. Find the surface area of the portion of the paraboloid $z = 16 - x^2 - y^2$ lying between the planes $z = 4$ and $z = 9$.

14. Find the surface area of the part of the sphere $x^2 + y^2 + z^2 = 4$ lying above the plane $z = 1$.

15. Find the surface area of the part of the hemisphere $z = \sqrt{4 - x^2 - y^2}$ lying inside the cylinder $x^2 + y^2 = 1$.

16. Find the surface area of the part of the paraboloid $z = 4 - x^2 - y^2$ lying inside the cylinder $x^2 + y^2 = 1$.

17. Use Simpson's Rule to approximate the surface area of the portion of the paraboloid $z = 3 - x^2 - y^2 + 2y$ lying above the plane $z = 2y + 2$.

18. Develop the formula for the surface area of a sphere using a double integral.

19. What relationship exists between the formula for the surface area of a solid of revolution and Definition 5?

20. (A coordinate-free formula for surface area)   Let u_m and v_m be the vectors in (1) and (2). Let ΔT_m be the area of the parallelogram tangent to S at P_m, over the rectangle R_m of area ΔA, as before.
   a. Show that $N_m = u_m \times v_m$ is normal to S at P_m.
   b. Show that $N \cdot k = \Delta x \Delta y = \Delta A$.
   c. Show that, also, $N \cdot k = |u_m \times v_m| \cos \theta$, where θ is the angle between N and k.
   d. Conclude from (b) and (c) that $\Delta T_m = |u_m \times v_m| = \dfrac{\Delta A}{\cos \theta} = \sec \theta\, \Delta A$.
   e. Conclude from (d) that $A_s = \displaystyle\iint_Q \sec \theta\, dA$.

19.6 TRIPLE INTEGRALS

In this section, we define the triple integral for a continuous function of three independent variables. As we did for double integrals, we shall first carry out this development for special types of regions Q (namely, boxes) and then indicate how the concept extends to more general regions.

The Triple Integral Over a Box

Let Q be the box-shaped region in $\mathbb{R}^3$ defined by the inequalities

$$a \le x \le b, \qquad c \le y \le d, \qquad p \le z \le q.$$

(See Figure 6.1.) Let f be a continuous function defined on Q. By constructing planes perpendicular to the x-axis at $x_0, x_1, \ldots, x_n$, planes perpendicular to the y-axis at $y_0, y_1, \ldots, y_m$, and planes perpendicular to the z-axis at $z_0, z_1, \ldots, z_\ell$, we partition the box Q into smaller rectangular boxes Q_{ijk}, each of which has volume $\Delta V_{ijk} = \Delta x_i \Delta y_j \Delta z_k$ (Figure 6.2).

Next, we select one point (s_i, t_j, u_k) in each box Q_{ijk}, and we form the approximating sum

$$S_n = \sum_{i=1}^{n} \sum_{j=1}^{m} \sum_{k=1}^{\ell} f(s_i, t_j, u_k)\, \Delta V_{ijk}. \tag{1}$$

By analogy with the one- and two-variable cases, this approximating sum is called a Riemann sum for f on Q and we say that the set of rectangular boxes constitutes a *partition* P of Q. As in the one- and two-variable cases, if f is continuous on Q, this sum approaches a limit as $\|P\| \to 0$. This limit is defined to be the **triple integral** of f on the box Q:

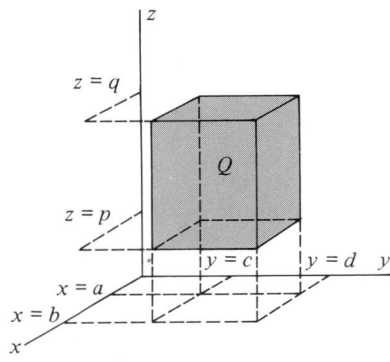

Figure 6.1   The rectangular box Q: $a \le x \le b$, $c \le y \le d$, $p \le z \le q$.

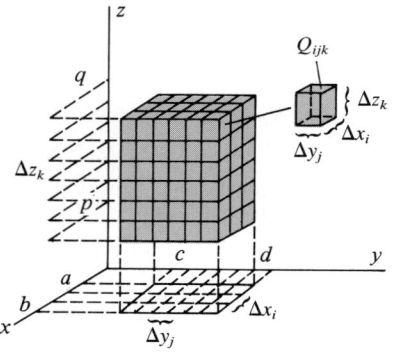

Figure 6.2 Partitioning the box into smaller boxes Q_{ijk} of volume $\Delta V_{ijk} = \Delta x_i \Delta y_j \Delta z_k$.

$$\iiint_Q f(x, y, z)\, dV = \lim_{\|P\| \to 0} \sum_{i=1}^{n} \sum_{j=1}^{m} \sum_{k=1}^{p} f(s_i, t_j, u_k)\, \Delta V_{ijk}. \tag{2}$$

In the special case $f(x, y, z) \equiv 1$, we can give a geometric interpretation of the triple integral in (1). The terms in the approximating Riemann sum (1) are just

$$f(s_i, t_j, u_k)\, \Delta V_{ijk} = 1 \cdot \Delta V_{ijk} = \text{volume of } Q_{ijk}.$$

Since the union of the boxes Q_{ijk} is the box Q, it follows that

$$\iiint_Q 1\, dV = \text{Volume of } Q. \tag{3}$$

That is, the triple integral of the function $f(x, y, z) \equiv 1$ over the box Q is just the volume of Q.

As for double integrals, triple integrals over boxes can be evaluated as iterated integrals. In particular, with Q as above, we have

$$\iiint_Q f(x, y, z)\, dV = \int_p^q \int_c^d \int_a^b f(x, y, z)\, dx\, dy\, dz. \tag{4}$$

In evaluating the iterated integral in (4), there is no reason why the first integration must be performed with respect to x. Since each of the variables ranges between constant limits, the order of integration can be any of the six possible orders xyz, xzy, yxz, yzx, zxy, or zyx.

We shall not prove equation (4), although it is easy to explain in the case $f(x, y, z) \equiv 1$. If v is any number in the z-interval $[p, q]$, the plane $z = v$ determines a rectangular cross section of Q of area

$$A(z) = \int_c^d \int_a^b 1 \cdot dx\, dy.$$

Thus,

$$\text{Volume of } Q = \int_p^q A(z)\, dz \tag{5}$$

$$= \int_p^q \left\{ \int_c^d \int_a^b 1 \cdot dx\, dy \right\} dz$$

$$= \int_p^q \int_c^d \int_a^b 1 \cdot dx\, dy\, dz.$$

Combining equations (3) and (5) results in equation (4) in this special case.

Example 1

Evaluate the triple integral $\iiint_Q xe^y \cos z\, dV$ where Q is the box $\{(x, y, z) \mid 0 \le x \le 2,\ 0 \le y \le \ln 2,\ 0 \le z \le \pi/2\}$.

Solution: Using equation (4), we find

$$\iiint\limits_{Q} xe^{y} \cos z \, dV = \int_{0}^{\pi/2} \int_{0}^{\ln 2} \int_{0}^{2} xe^{y} \cos z \, dx \, dy \, dz$$

$$= \int_{0}^{\pi/2} \int_{0}^{\ln 2} \left\{ \frac{x^2}{2} e^{y} \cos z \right]_{x=0}^{x=2} \right\} dy \, dz$$

$$= \int_{0}^{\pi/2} \int_{0}^{\ln 2} 2e^{y} \cos z \, dy \, dz$$

$$= \int_{0}^{\pi/2} \left\{ 2e^{y} \cos z \right]_{y=0}^{y=\ln 2} \right\} dz$$

$$= \int_{0}^{\pi/2} 2 \cos z \, dz$$

$$= 2 \sin z]_{z=0}^{z=\pi/2}$$

$$= 2. \qquad \diamond$$

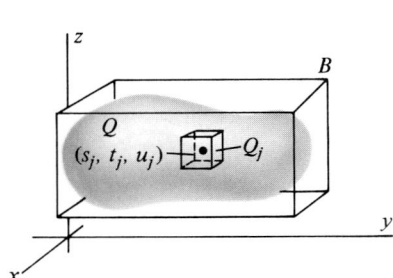

Figure 6.3 More general region Q enclosed by the box B.

Triple Integrals Over More General Regions

If Q is a bounded region in space (not necessarily a box), we define the triple integral of f over Q as follows. First, we find a box B containing the region Q. (See Figure 6.3.) Next, we partition the box B into smaller rectangular boxes, just as we did when Q itself was a box, and we let $Q_1, Q_2, \ldots, Q_n$ be a complete list of all such rectangular boxes *lying entirely within the region Q*. For each such box we select a point $(s_j, t_j, u_j) \in Q_j$ and we denote by ΔV_j the volume of Q_j. The limit of this sequence of approximating sums, when it exists, is called the **triple integral** of f over Q:

$$\iiint\limits_{Q} f(x, y, z) \, dV = \lim_{\|P\| \to 0} \sum_{j=1}^{n} f(s_j, t_j, u_j) \, \Delta V_j. \qquad (6)$$

The triple integral in (6) will exist whenever f is continuous on Q and Q is a "sufficiently nice" region in space. Rather than worry too much about the precise meaning of this last phrase, we shall state a theorem showing how triple integrals may be evaluated for regions of the type encountered in this text and in most applications. Before doing so, however, we need to make one observation concerning the approximating sum in equation (6). In the special case $f(x, y, z) \equiv 1$, the terms in the approximating sum are, as before, the volumes of the approximating boxes. The sum therefore approximates the volume of Q, and in the limit we obtain

$$\iiint\limits_{Q} 1 \cdot dV = \text{Volume of } Q \qquad (7)$$

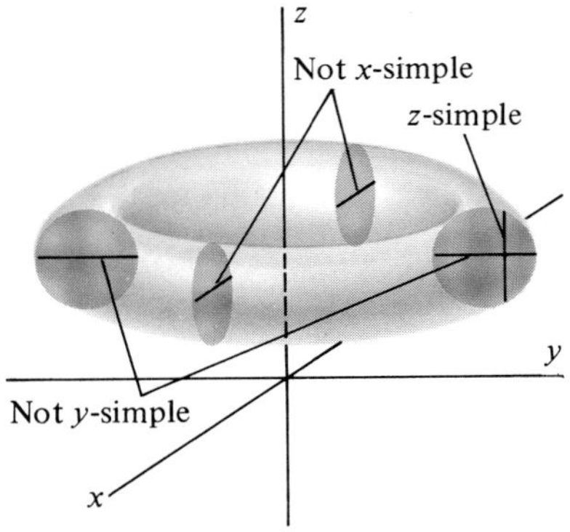

Figure 6.4 The torus is z-simple, but neither x-simple nor y-simple.

just as in the case when Q itself is a box.

Evaluating Triple Integrals

We say that a region Q in $\mathbb{R}^3$ is **z-simple** if every vertical line (that is, a line parallel to the z-axis) intersects the boundary of the region at most twice. The notions of **x-simple** and **y-simple** regions in $\mathbb{R}^3$ are defined accordingly. Figure 6.4 illustrates the fact that a torus is an example of a region that is z-simple, but neither x-simple nor y-simple. We shall describe a certain type of z-simple region Q for which the triple integral $\iiint\limits_Q f(x, y, z)\, dV$ can be evaluated as an iterated integral. The same result applies to regions that are x-simple or y-simple by interchanging the roles of x and z or y and z.

Specifically, let Q be a z-simple region in $\mathbb{R}^3$ that can be described by inequalities of the form

$$g_1(x, y) \le z \le g_2(x, y), \qquad h_1(x) \le y \le h_2(x), \qquad a \le x \le b. \tag{8}$$

In the inequalities in (8), we assume that g_1, g_2, h_1, and h_2 are continuous functions of their arguments. Figure 6.5 shows such a region. The following theorem shows how triple integrals over such regions may be evaluated.

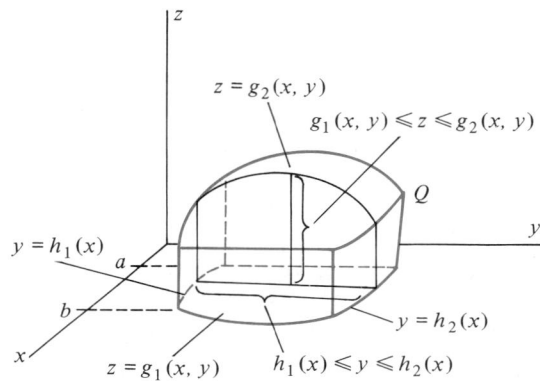

Figure 6.5 Region defined by inequalities (8).

THEOREM 4

Let Q be a region in $\mathbb{R}^3$ described by the inequalities

$$a \le x \le b, \qquad h_1(x) \le y \le h_2(x), \qquad g_1(x, y) \le z \le g_2(x, y),$$

where h_1, h_2, g_1, and g_2 are continuous functions. Let f be continuous on Q. Then

$$\iiint\limits_Q f(x, y, z)\, dV = \int_a^b \int_{h_1(x)}^{h_2(x)} \int_{g_1(x, y)}^{g_2(x, y)} f(x, y, z)\, dz\, dy\, dx.$$

A proof of Theorem 4 requires a deeper treatment of multiple integrals than we have given here and is left to more advanced courses. Before proceeding to apply this result, we need to emphasize two points.

REMARK 1: Theorem 4 remains true with the roles of the independent variables interchanged. For example, if the region Q is described by the inequalities

$$c \le y \le d, \qquad h_1(y) \le z \le h_2(y), \qquad g_1(y, z) \le x \le g_2(y, z),$$

then the iterated integral formula is

$$\iiint\limits_Q f(x, y, z)\, dV = \int_c^d \int_{h_1(y)}^{h_2(y)} \int_{g_1(y, z)}^{g_2(y, z)} f(x, y, z)\, dx\, dz\, dy.$$

REMARK 2:   Theorem 4 may be paraphrased this way. To evaluate the triple integral

$$\iiint_Q f(x, y, z)\ dV:$$

(i) Find the constants and/or functions that bound the region Q in each of the three directions corresponding to the coordinate axes. (Write these down!)

(ii) Evaluate $\iiint_Q f(x, y, z)\ dV$ as an iterated integral, integrating first with respect to a variable whose bounds depend on the other one or two variables, integrating second with respect to a variable whose limits involve the remaining variable, and integrating last with respect to a variable whose limits involve only constants.

It is important to note that we cannot integrate over limits involving a variable for which an integration has already been performed. For example, the expressions

$$\int_a^b \int_{g_1(x, y)}^{g_2(x, y)} \int_{h_1(y)}^{h_2(y)} f(x, y, z)\ dz\ dy\ dx$$

and

$$\int_{h_1(x)}^{h_2(x)} \int_{g_1(x, z)}^{g_2(x, z)} \int_a^b f(x, y, z)\ dx\ dy\ dz$$

are nonsense.

Example 2

Evaluate the triple integral

$$\iiint_Q 2xy\ dV,$$

where Q is the region inside the cylinder $x^2 + y^2 = 1$ bounded by the planes $x + y + z = 4$ and $z = -1$.

Solution: The region is sketched in Figure 6.6. Figure 6.7 shows the projection of Q into the xy-plane and illustrates how the inequalities involving x and y are obtained. The equation $x + y + z = 4$ gives $z = 4 - x - y$, so $-1 \le z \le 4 - x - y$. The region is therefore completely described by the inequalities

$$-1 \le x \le 1, \qquad -\sqrt{1 - x^2} \le y \le \sqrt{1 - x^2}, \qquad -1 \le z \le 4 - x - y.$$

Thus, according to Theorem 4,

$$\iiint_Q 2xy\ dV = \int_{-1}^1 \int_{-\sqrt{1-x^2}}^{\sqrt{1-x^2}} \int_{-1}^{4-x-y} 2xy\ dz\ dy\ dx$$

$$= \int_{-1}^1 \int_{-\sqrt{4-x^2}}^{\sqrt{4-x^2}} \left\{ 2xyz \Big]_{z=-1}^{z=4-x-y} \right\} dy\ dx$$

$$= \int_{-1}^1 \int_{-\sqrt{4-x^2}}^{\sqrt{4-x^2}} (10xy - 2x^2y - 2xy^2)\ dy\ dx$$

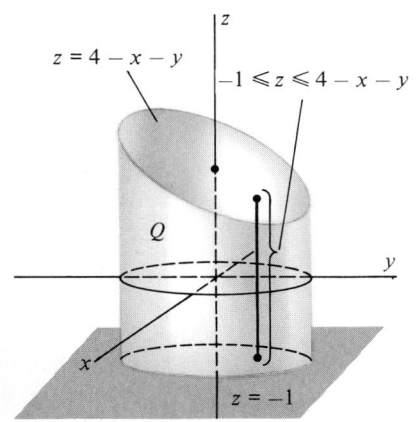

$z = 4 - x - y$

$-1 \le z \le 4 - x - y$

Q

$z = -1$

Figure 6.6 Region of Example 2.

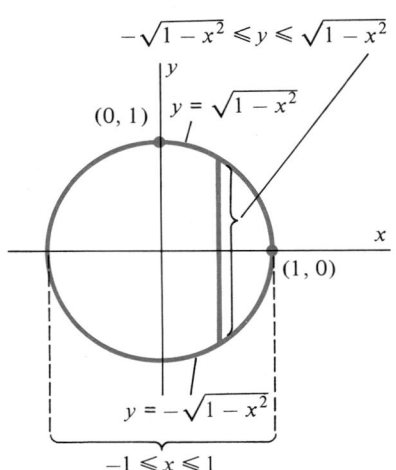

$-\sqrt{1 - x^2} \le y \le \sqrt{1 - x^2}$

$(0, 1)$ $y = \sqrt{1 - x^2}$

$(1, 0)$

$y = -\sqrt{1 - x^2}$

$-1 \le x \le 1$

Figure 6.7 Projection of Q into the xy-plane.

$$= \int_{-1}^{1} \left\{ 5xy^2 - x^2y^2 - \frac{2}{3}xy^3 \right]_{y=-\sqrt{4-x^2}}^{y=\sqrt{4-x^2}} \right\} dx$$

$$= \int_{-1}^{1} -\frac{4}{3}x(4 - x^2)^{3/2} \, dx$$

$$= \frac{4}{15}(4 - x^2)^{5/2} \Bigg]_{x=-1}^{x=1}$$

$$= 0.$$

(This result is explained by the fact that both the integrand and the region Q are symmetric with respect to the plane $y = x$.) ◇

REMARK: The integral in Example 2 could also have been evaluated as

$$\iiint_Q 2xy \, dV = \int_{-1}^{1} \int_{-\sqrt{1-y^2}}^{\sqrt{1-y^2}} \int_{-1}^{4-x-y} 2xy \, dz \, dx \, dy.$$

Example 3

Evaluate the triple integral

$$\iiint_Q (x - y + z) \, dV$$

where Q is the tetrahedron with vertices $(0, 0, 0)$, $(1, 0, 0)$, $(0, 2, 0)$, and $(0, 0, 4)$.

Solution: The tetrahedron is sketched in Figure 6.8. Substituting the given points into the equation $z = Ax + By + C$ shows that the plane bounding Q above has equation $z = 4 - 4x - 2y$. As Figure 6.9 illustrates, the base of Q is a triangle in the xy plane bounded by the lines $x = 0$, $y = 0$, and $y = -2x + 2$. The region Q

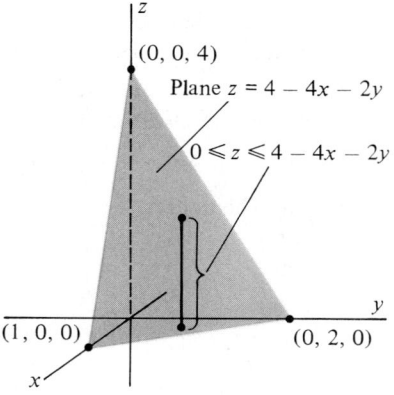

Figure 6.8 Tetrahedron of Example 3.

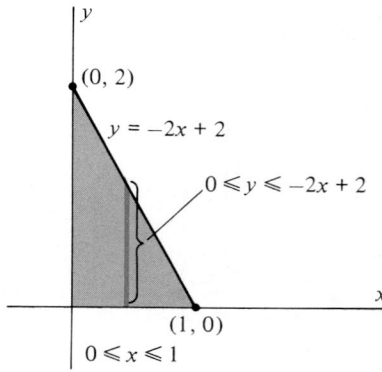

Figure 6.9 Projection of tetrahedron into the xy-plane.

may therefore be described by the inequalities

$$0 \le x \le 1, \qquad 0 \le y \le -2x + 2, \qquad 0 \le z \le 4 - 4x - 2y.$$

The integral is therefore

$$\iiint_Q (x - y + z)\, dV = \int_0^1 \int_0^{2-2x} \int_0^{4-4x-2y} (x - y + z)\, dz\, dy\, dx$$

$$= \int_0^1 \int_0^{2-2x} \left\{ (x - y)z + \frac{1}{2}z^2 \right\}_{z=0}^{z=4-4x-2y} dy\, dx$$

$$= \int_0^1 \int_0^{2-2x} (4x^2 - 12x + 10xy - 12y + 4y^2 + 8)\, dy\, dx$$

$$= \int_0^1 \left\{ (4x^2 - 12x + 8)y + 5xy^2 - 6y^2 + \frac{4}{3}y^3 \right\}_{y=0}^{y=2-2x} dx$$

$$= \int_0^1 \left(\frac{4}{3}x^3 - 4x + \frac{8}{3} \right) dx$$

$$= \frac{1}{3}x^4 - 2x^2 + \frac{8}{3}x \Big]_0^1$$

$$= 1. \qquad \diamond$$

REMARK: The integral in Example 3 could also have been evaluated as

$$\int_0^2 \int_0^{1-y/2} \int_0^{4-4x-2y} (x - y + z)\, dz\, dx\, dy,$$

or

$$\int_0^1 \int_0^{4-4x} \int_0^{2-2x-1/2z} (x - y + z)\, dy\, dz\, dx,$$

or as one of three other iterated integrals (see Exercise 29).

Example 4

Find the volume of the solid bounded by the paraboloids $y = 4 - x^2 - z^2$ and $y = x^2 + z^2$.

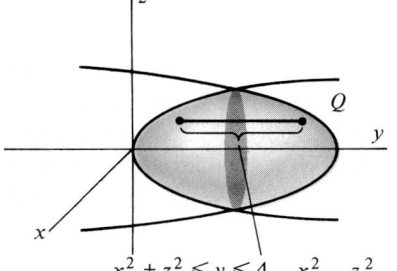

$x^2 + z^2 \leqslant y \leqslant 4 - x^2 - z^2$

Figure 6.10 Intersecting paraboloids of Example 4. (See **Plate 16**.)

Solution: The region is sketched in Figure 6.10. From the description of the region we can see that

$$x^2 + z^2 \le y \le 4 - x^2 - z^2. \tag{9}$$

We therefore seek inequalities for the variables x and z. Equating the two expressions for y gives

$$4 - x^2 - z^2 = x^2 + z^2$$

or

$$x^2 + z^2 = 2.$$

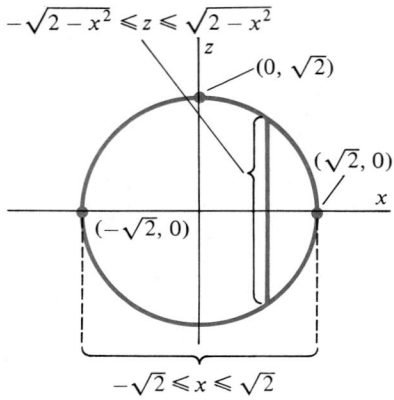

$-\sqrt{2-x^2} \leqslant z \leqslant \sqrt{2-x^2}$

$(0, \sqrt{2})$

$(\sqrt{2}, 0)$

$(-\sqrt{2}, 0)$

$-\sqrt{2} \leqslant x \leqslant \sqrt{2}$

Figure 6.11 Projection of Q into xz-plane.

This is the equation of the circle (in the plane $y = 2$) where the paraboloids intersect. The projection of this circle into the xz-plane is shown in Figure 6.11. From this sketch we see that the inequalities on x and on z are

$$-\sqrt{2} \leq x \leq \sqrt{2}, \qquad -\sqrt{2 - x^2} \leq z \leq \sqrt{2 - x^2}. \tag{10}$$

Using statement (7) and inequalities (9) and (10), we obtain the desired volume as

$$V = \iiint_Q 1 \, dV = \int_{-\sqrt{2}}^{\sqrt{2}} \int_{-\sqrt{2-x^2}}^{\sqrt{2-x^2}} \int_{x^2+z^2}^{4-x^2-z^2} 1 \, dy \, dz \, dx$$

$$= \int_{-\sqrt{2}}^{\sqrt{2}} \int_{-\sqrt{2-x^2}}^{\sqrt{2-x^2}} \left\{ \Big[y \Big]_{y=x^2+z^2}^{y=4-x^2-z^2} \right\} dz \, dx$$

$$= \int_{-\sqrt{2}}^{\sqrt{2}} \int_{-\sqrt{2-x^2}}^{\sqrt{2-x^2}} (4 - 2x^2 - 2z^2) \, dz \, dx$$

$$= \int_{-\sqrt{2}}^{\sqrt{2}} \left\{ (4 - 2x^2)z - \frac{2}{3}z^3 \Big]_{z=-\sqrt{2-x^2}}^{z=\sqrt{2-x^2}} \right\} dx$$

$$= \int_{-\sqrt{2}}^{\sqrt{2}} \frac{8}{3}(2 - x^2)^{3/2} \, dx \tag{11}$$

$$= 4\pi.$$

(The integral in line (11) is evaluated by means of a trigonometric substitution.) ◇

REMARK: The triple integral in Example 4 could also have been evaluated as

$$\int_{-\sqrt{2}}^{\sqrt{2}} \int_{-\sqrt{2-z^2}}^{\sqrt{2-z^2}} \int_{x^2+z^2}^{4-x^2-z^2} 1 \, dy \, dx \, dz.$$

Density

The triple integral may also be used to calculate the mass of a solid object, if we know the *density* of the material in units of mass per unit volume (such as gram/cm^3) as a continuous *density function* ρ. If Q is an object with a density function ρ of this type, we may approximate the mass of Q by partitioning Q into approximating boxes $Q_1, Q_2, \ldots, Q_n$, as before. If (s_j, t_j, u_j) is a point in Q_j, and if the volume ΔV_j of Q_j is small, then the quantity

$$M_j = \rho(s_j, t_j, u_j) \, \Delta V_j \qquad (\text{mass} = \text{density} \times \text{volume})$$

provides an approximation to the mass of the jth box Q_j. Summing these approximations over all boxes contained within Q gives the approximating sum

$$M \approx \sum_{j=1}^{n} M_j = \sum_{j=1}^{n} \rho(s_j, t_j, u_j) \, \Delta V_j.$$

Thus, the **mass** of Q is defined by the triple integral

$$\boxed{M = \iiint_Q \rho(x, y, z) \, dV.} \tag{12}$$

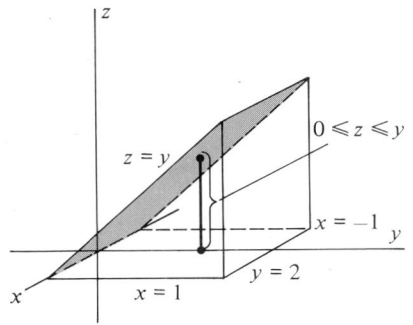

Figure 6.12 The wedge of Example 5.

That is, mass is the integral of the density function over the region Q, just as in the one- and two-variable cases.

Example 5

A small wedge has the shape of the region

$$Q = \{(x, y, z) \mid -1 \le x \le 1, \, 0 \le y \le 2, \, 0 \le z \le y\}.$$

Find its mass if the density at any point (x, y, z) is given by the density function $\rho(x, y, z) = 1 + x^2 + y^2$ gram/cm^3 and the dimensions for Q are in cm (Figure 6.12).

Solution: According to equation (12) and Theorem 4 the mass is

$$M = \iiint\limits_Q (1 + x^2 + y^2) \, dV = \int_{-1}^{1} \int_{0}^{2} \int_{0}^{y} (1 + x^2 + y^2) \, dz \, dy \, dx$$

$$= \int_{-1}^{1} \int_{0}^{2} \left\{ (1 + x^2 + y^2)z \, \right]_{z=0}^{z=y} \right\} dy \, dx$$

$$= \int_{-1}^{1} \int_{0}^{2} (y + x^2 y + y^3) \, dy \, dx$$

$$= \int_{-1}^{1} \left\{ \frac{1}{2} y^2 + \frac{1}{2} x^2 y^2 + \frac{1}{4} y^4 \, \right]_{y=0}^{y=2} \right\} dx$$

$$= \int_{-1}^{1} (6 + 2x^2) \, dx$$

$$= 6x + \frac{2}{3} x^3 \, \bigg]_{-1}^{1}$$

$$= \frac{40}{3} \text{ grams.} \qquad \diamondsuit$$

Moments and Center of Mass

Let Q denote a solid in $\mathbb{R}^3$, and let ρ be a continuous function giving the density $\rho(x, y, z)$ of Q at each point (x, y, z). As before, let Q be partitioned into rectangular boxes $Q_1, Q_2, \ldots, Q_n$ and let (s_j, t_j, u_j) be a point in Q_j, $j = 1, 2, \ldots, n$. Since s_j is the distance from the point (s_j, t_j, u_j) to the yz-plane, the product $s_j \rho(s_j, t_j, u_j) \, \Delta V_j$ may be interpreted as an approximation to the product of the mass of Q_j and the length of its "lever arm" extending from the yz-plane. By analogy with our previous discussions on moments (see Section 19.4), we define *the first moment, M_{yz}, of the solid Q with respect to the yz-plane* to be

$$M_{yz} = \iiint\limits_Q x\rho(x, y, z) \, dV = \lim_{n \to \infty} \sum_{j=1}^{n} s_j \rho(s_j, t_j, u_j) \, \Delta V_j.$$

Similarly, *the moments of Q with respect to the xz- and xy-planes* are

$$M_{xz} = \iiint\limits_Q y\rho(x, y, z) \, dV$$

and

$$M_{xy} = \iiint_Q z\rho(x, y, z)\, dV.$$

Finally, if M denotes the mass of Q, the **center of mass** of Q is the point $(\bar{x}, \bar{y}, \bar{z})$ where

$$\bar{x} = \frac{M_{yz}}{M}; \qquad \bar{y} = \frac{M_{xz}}{M}; \qquad \bar{z} = \frac{M_{xy}}{M}.$$

Example 6

Find the three first moments and the center of mass for the solid in Example 5.

Solution: With $\rho(x, y, z) = 1 + x^2 + y^2$ and Q as given in Example 5, we have

$$M_{yz} = \iiint_Q x(1 + x^2 + y^2)\, dV = \int_{-1}^{1} \int_0^2 \int_0^y (x + x^3 + xy^2)\, dz\, dy\, dx$$

$$= \int_{-1}^{1} \int_0^2 \left\{ (x + x^3 + xy^2)z \Big]_{z=0}^{z=y} \right\} dy\, dx$$

$$= \int_{-1}^{1} \int_0^2 [(x + x^3)y + xy^3]\, dy\, dx$$

$$= \int_{-1}^{1} \left\{ (x + x^3)\frac{y^2}{2} + \frac{1}{4}xy^4 \Big]_{y=0}^{y=2} \right\} dx$$

$$= \int_{-1}^{1} (6x + 2x^3)\, dx$$

$$= 3x^2 + \frac{1}{2}x^4 \Big]_{-1}^{1}$$

$$= 0.$$

Similar calculations show that

$$M_{xz} = \iiint_Q y(1 + x^2 + y^2)\, dV$$

$$= \int_{-1}^{1} \int_0^2 \int_0^y (y + x^2y + y^3)\, dz\, dy\, dx = \frac{896}{45}$$

and that

$$M_{xy} = \iiint_Q z(1 + x^2 + y^2)\, dV$$

$$= \int_{-1}^{1} \int_0^2 \int_0^y (z + x^2z + y^2z)\, dz\, dy\, dx = \frac{448}{45}.$$

Since we have previously calculated the mass to be $M = \dfrac{40}{3}$, the coordinates of the center of mass $(\bar{x}, \bar{y}, \bar{z})$ are

$$\bar{x} = \frac{M_{yz}}{M} = 0 \cdot \frac{3}{40} = 0,$$

$$\bar{y} = \frac{M_{xz}}{M} = \frac{896}{45} \cdot \frac{3}{40} = \frac{112}{75} \approx 1.49,$$

and

$$\bar{z} = \frac{M_{xy}}{M} = \frac{448}{45} \cdot \frac{3}{40} = \frac{56}{75} \approx 0.75. \qquad \diamondsuit$$

Exercise Set 19.6

In Exercises 1–8, evaluate the iterated integral.

1. $\displaystyle\int_0^1 \int_0^1 \int_0^1 xyz \, dx \, dy \, dz$

2. $\displaystyle\int_0^2 \int_{-\pi/2}^{\pi/2} \int_1^2 x \cos y e^z \, dx \, dy \, dz$

3. $\displaystyle\int_0^1 \int_0^y \int_0^x 3 \, dz \, dx \, dy$

4. $\displaystyle\int_1^3 \int_0^{\pi/4} \int_0^x \cos(x + y) \, dy \, dx \, dz$

5. $\displaystyle\int_0^2 \int_0^x \int_0^{x+y} z \, dz \, dy \, dx$

6. $\displaystyle\int_{-1}^1 \int_{-\sqrt{1-x^2}}^{\sqrt{1-x^2}} \int_0^{\sqrt{1-x^2-y^2}} 1 \, dz \, dy \, dx$ (*Hint*: Use geometry.)

7. $\displaystyle\int_{-1}^1 \int_0^y \int_0^x y e^{x^2+y^2} \, dz \, dx \, dy$ **8.** $\displaystyle\int_0^2 \int_0^x \int_{x+y}^{x^2+y^2} 1 \, dz \, dy \, dx$

9. Interchange the order of integration in

$$\int_0^1 \int_0^{2x} \int_0^{x+y} f(x, y, z) \, dz \, dy \, dx$$

from order $dz \, dy \, dx$ to order $dz \, dx \, dy$.

10. Use a triple integral to find the volume of the tetrahedron with vertices $(0, 0, 0)$, $(1, 0, 0)$, $(1, 1, 0)$, and $(1, 1, 1)$.

11. Sketch the solid whose volume is given by the iterated integral

$$\int_0^2 \int_0^{2x} \int_0^{x+y} dz \, dy \, dx.$$

12. Sketch the solid whose volume is given by the iterated integral

$$\int_{-1}^1 \int_{-\sqrt{1-x^2}}^{\sqrt{1-x^2}} \int_{\sqrt{x^2+y^2}}^{2-\sqrt{x^2+y^2}} dz \, dy \, dx.$$

13. Find the volume of the region in Exercise 12.

14. Find the volume of the tetrahedron bounded by the plane $x + y + z = 1$ and the coordinate planes $x = 0$, $y = 0$, and $z = 0$.

15. Find the volume of the region lying above the xy-plane, inside the cylinder $x^2 + y^2 = 9$, and below the plane $z = y + 3$.

16. Sketch the solid whose volume is given by the integral

$$V = \int_0^2 \int_0^{\sqrt{2x-x^2}} \int_0^{2-x} dz \, dy \, dx$$

and find the volume.

17. Evaluate the integral $\displaystyle\iiint_Q (3x + xz) \, dV$ where Q is the region bounded by the cylinder $x^2 + z^2 = 9$, the plane $y + z = 3$, and the plane $y = 0$.

18. Interchange the order of integration in the integral

$$\int_{-2}^2 \int_0^{\sqrt{4-x^2}} \int_{-\sqrt{4-x^2-y^2}}^{\sqrt{4-x^2-y^2}} f(x, y, z) \, dz \, dy \, dx$$

from $dz \, dy \, dx$ to $dy \, dz \, dx$.

19. Find the volume of the region bounded by the paraboloids $x = y^2 + z^2$ and $x = 2 - y^2 - z^2$.

20. Find the volume of the solid bounded above by the paraboloid $z = 2 - x^2 - y^2$ and below by the plane $z = 2 - 2x$.

21. Let Q be the solid bounded by the cylinder $x^2 + y^2 = 9$ and the planes $z = 0$ and $x + z = 3$. Find the mass of Q if the density at each point (x, y, z) is given by the function $\rho(x, y, z) = z$.

22. Find the volume of the region common to the cylinders $x^2 + z^2 = 1$ and $y^2 + z^2 = 1$.

23. The density at each point of the box $Q = \{(x, y, z) \mid 0 \le x \le 1, 0 \le y \le 2, 0 \le z \le 2\}$ is proportional to the square of the distance from the origin. Find the mass.

24. Find the center of mass of the region in Exercise 14 if the density is constant.

25. Find the center of mass of the solid in Exercise 15 if the density is constant.

26. Find the center of mass of the solid in Exercise 21.

27. Find the center of mass of the part of the region enclosed by the sphere $x^2 + y^2 + z^2 = 4$ lying in the first octant if the density is constant.

28. Find the center of mass of the cube $Q = \{(x, y, z) \mid 0 \le x \le 1, 0 \le y \le 1, 0 \le z \le 1\}$ if the density function is $\rho(x, y, z) = xyz$.

29. Find five other iterated integrals by which the triple integral of Example 3 may be evaluated.

30. Find the volume, mass, and center of gravity for the solid in Example 5 if the density is uniform $\rho(x, y, z) = 1$ gram/ cm^3. Compare your answers with those in Examples 5 and 6, and account for the differences in physical terms.

19.7 TRIPLE INTEGRALS IN CYLINDRICAL AND SPHERICAL COORDINATES

The purpose of this section is to define the triple integral for functions written in cylindrical or spherical coordinates. In part, we want to know how to calculate volumes and masses for solids described in these coordinate systems. Another reason for studying these topics is that certain triple integrals, originally expressed in Cartesian coordinates, are more easily evaluated by changing to either cylindrical or spherical coordinates.

Triple Integrals in Cylindrical Coordinates

Recall the relationship between cylindrical and Cartesian coordinates: If a point P has cylindrical coordinates $P = (r, \theta, z)$ and Cartesian coordinates $P = (x, y, z)$, then (r, θ) are the polar coordinates for the point (x, y) in the xy-plane (Figure 7.1). Now suppose that $\mathbb{R}^3$ is coordinatized by cylindrical coordinates, that Q is a region in $\mathbb{R}^3$, and that f is a continuous function defined on Q. We shall define the triple integral of f over Q in the usual way—by partitioning the region Q, forming an approximating sum, and obtaining the limit of the approximating sum.

Suppose that the region Q lies within the "cylindrical box" determined by the inequalities

$$r_a \le r \le r_b, \qquad \theta_a \le \theta \le \theta_b, \qquad z_a \le z \le z_b.$$

(See Figure 7.1.) We divide the region containing Q into small cylindrical blocks

$$Q_{ijk} = \{(r, \theta, z) \mid r_{i-1} \le r \le r_i, \theta_{j-1} \le \theta \le \theta_j, z_{k-1} \le z \le z_k\}$$

as illustrated in Figure 7.2. According to equation (3), Section 19.3, the area of the base of block Q_{ijk} is

$$\Delta A_{ijk} = \hat{r}_i \Delta r_i \Delta \theta_j$$

where $\hat{r}_i = \dfrac{1}{2}(r_{i-1} + r_i)$. Since the block Q_{ijk} has height Δz_k, the volume of the block Q_{ijk} is

$$\Delta V_{ijk} = \hat{r}_i \Delta r_i \Delta \theta_j \Delta z_k. \tag{1}$$

Now let θ_j^* be any number in the interval $[\theta_{j-1}, \theta_j]$ and let z_k^* be any number in the interval $[z_{k-1}, z_k]$. Then, since $\hat{r}_i \in [r_{i-1}, r_i]$, the point $(\hat{r}_i, \theta_j^*, z_k^*)$ lies in the

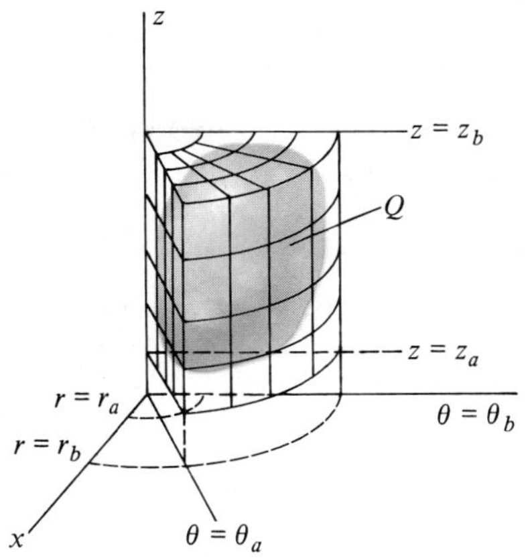

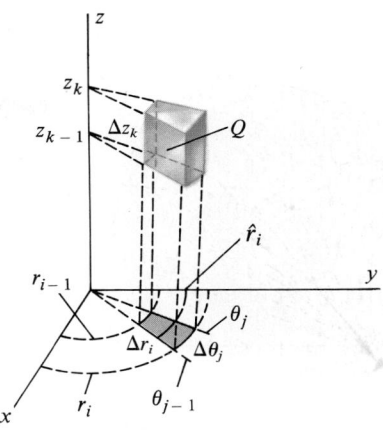

Figure 7.1 Partitioning the region Q using a grid specified in cylindrical coordinates.

Figure 7.2 Volume of circular wedge is

$$\Delta V_{ijk} = \hat{r}_i \, \Delta r_i \, \Delta \theta_j \, \Delta z_k$$

where $\hat{r}_i = \dfrac{r_{i-1} + r_i}{2}$.

block Q_{ijk}. The sum

$$\sum_i \sum_j \sum_k f(\hat{r}_i, \theta_j^*, z_k^*) \, \Delta V_{ijk} = \sum_i \sum_j \sum_k f(\hat{r}_i, \theta_j^*, z_k^*) \hat{r}_i \Delta r_i \Delta \theta_j \Delta z_k \qquad (2)$$

is the approximating Riemann sum for f over Q, where the sum is taken over all boxes Q_{ijk} lying entirely within Q. Its limit as the sizes of all blocks approach zero (written $\|P\| \to 0$ as before), which exists when f is continuous and Q is as described in the following theorem, is the triple integral of f over Q:

$$\iiint\limits_Q f(r, \theta, z) \, dV = \lim_{\|P\| \to 0} \sum_i \sum_j \sum_k f(\hat{r}_i, \theta_j^*, z_k^*) \hat{r}_i \Delta r_i \Delta \theta_j \Delta z_k. \qquad (3)$$

The following theorem shows how this triple integral may be evaluated for the types of regions encountered in this section and in most applications.

THEOREM 5

Let Q be a region in $\mathbb{R}^3$ of the form

$$Q = \{(r, \theta, z) \mid a \le \theta \le b, \, h_1(\theta) \le r \le h_2(\theta), \, g_1(r, \theta) \le z \le g_2(r, \theta)\}$$

where g_1, g_2, h_1, and h_2 are continuous functions. Let f be continuous on Q. Then

$$\iiint\limits_Q f(r, \theta, z) \, dV = \int_a^b \int_{h_1(\theta)}^{h_2(\theta)} \int_{g_1(r, \theta)}^{g_2(r, \theta)} f(r, \theta, z) \, r \, dz \, dr \, d\theta. \qquad (4)$$

REMARK 1: Note the extra factor r appearing in the iterated integral in (4). The reason for its appearance is the same as in Section 19.3: it can be considered a result of converting dV to cylindrical coordinates.

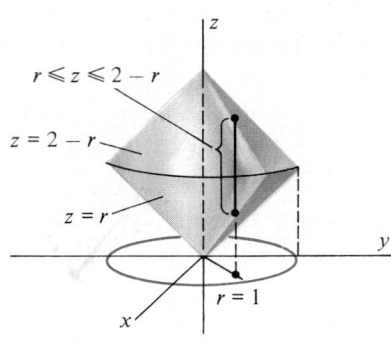

Figure 7.3 Solid Q in Example 1.

REMARK 2: When $f(r, \theta, z) \equiv 1$, it follows from equation (1) that the Riemann sum in (2) approximates the volume of Q. Thus

$$\text{Volume of } Q = \iiint\limits_{Q} dV \tag{5}$$

holds in cylindrical coordinates as well as in rectangular coordinates.

Example 1

A solid ''top'' has the shape of the region

$$Q = \{(r, \theta, z) \mid 0 \le r \le 1, 0 \le \theta \le 2\pi, r \le z \le 2 - r\}$$

as illustrated in Figure 7.3. Find the mass of the object if the density at any point is proportional to the distance from the z-axis.

Solution: The density function described here is $\rho(r, \theta, z) = \lambda r$, where λ is the constant of proportionality. By equation (12), Section 19.6, and Theorem 5, we have

$$\text{Mass} = \iiint\limits_{Q} \rho(r, \theta, z) \, dV = \iiint\limits_{Q} \lambda r \, dV$$

$$= \int_{0}^{2\pi} \int_{0}^{1} \int_{r}^{2-r} \lambda r^2 \, dz \, dr \, d\theta \qquad \text{note extra factor } r$$

$$= \int_{0}^{2\pi} \int_{0}^{1} \left\{ \lambda r^2 z \Big]_{z=r}^{z=2-r} \right\} dr \, d\theta$$

$$= \int_{0}^{2\pi} \int_{0}^{1} (2\lambda r^2 - 2\lambda r^3) \, dr \, d\theta$$

$$= \int_{0}^{2\pi} \left\{ \frac{2\lambda}{3} r^3 - \frac{\lambda}{2} r^4 \Big]_{r=0}^{r=1} \right\} d\theta$$

$$= \int_{0}^{2\pi} \left(\frac{\lambda}{6} \right) d\theta$$

$$= \frac{\pi\lambda}{3}. \qquad \diamond$$

Changing to Cylindrical Coordinates

You will sometimes find that a triple integral written in Cartesian coordinates is more easily evaluated by first changing to cylindrical coordinates. Doing so is analogous to changing a double integral from Cartesian to polar coordinates. Specifically, to write the iterated integral

$$\int_{a}^{b} \int_{h_1(x)}^{h_2(x)} \int_{g_1(x, y)}^{g_2(x, y)} f(x, y, z) \, dz \, dy \, dx$$

in cylindrical coordinates, we do the following:

(i) Express the region

$$Q = \{(x, y, z) \mid a \le x \le b, h_1(x) \le y \le h_2(x), g_1(x, y) \le z \le g_2(x, y)\}$$

in cylindrical coordinates as

$$Q = \{(r, \theta, z) \mid c \le \theta \le d, h_3(\theta) \le r \le h_4(\theta), g_3(r, \theta) \le z \le g_4(r, \theta)\}.$$

(ii) Using the substitutions $x = r \cos \theta$ and $y = r \sin \theta$, replace the integrand $f(x, y, z)$ by $f(r \cos \theta, r \sin \theta, z)r$. (Do not forget the extra factor r.)

(iii) Obtain the equation

$$\int_a^b \int_{h_1(x)}^{h_2(x)} \int_{g_1(x, y)}^{g_2(x, y)} f(x, y, z)\, dz\, dy\, dx \tag{6}$$
$$= \int_c^d \int_{h_3(\theta)}^{h_4(\theta)} \int_{g_3(r, \theta)}^{g_4(r, \theta)} f(r \cos \theta, r \sin \theta, z)\, r\, dz\, dr\, d\theta.$$

Equation (6) is verified by comparing Theorems 4 and 5, each of which expresses the triple integral $\iiint_Q f\, dV$ as one of the two iterated integrals in (6). One way to paraphrase equation (6) is to say that *in changing from Cartesian coordinates to cylindrical coordinates, the* **volume element**

$$dV = dz\, dy\, dx$$

in Cartesian coordinates is replaced by the volume element

$$dV = r\, dz\, dr\, d\theta \tag{7}$$

in cylindrical coordinates. It is very important to note the extra factor r that appears in the integrand on the right-hand side of (6) and in equation (7).

Example 2

Calculate the volume of the ellipsoid

$$4x^2 + 4y^2 + z^2 = 4.$$

Solution: The ellipsoid is sketched in Figure 7.4. Since the ellipsoid is symmetric with respect to the xy-plane, we may calculate the volume as twice the volume of the region Q lying above the xy-plane. Since this region is described by the inequalities

$$0 \le z \le 2\sqrt{1 - x^2 - y^2}, \qquad -\sqrt{1 - x^2} \le y \le \sqrt{1 - x^2}, \qquad -1 \le x \le 1,$$

the volume is given by the iterated integral

$$V = 2 \int_{-1}^1 \int_{-\sqrt{1-x^2}}^{\sqrt{1-x^2}} \int_0^{2\sqrt{1-x^2-y^2}} 1\, dz\, dy\, dx. \tag{8}$$

Clearly, this integral will be difficult to evaluate, so we try switching to cylindrical coordinates. In cylindrical coordinates the region Q is described by the inequalities

$$0 \le r \le 1, \qquad 0 \le \theta \le 2\pi, \qquad 0 \le z \le 2\sqrt{1 - r^2}$$

since $2\sqrt{1 - x^2 - y^2} = 2\sqrt{1 - (x^2 + y^2)} = 2\sqrt{1 - r^2}$. Using equation (6) we

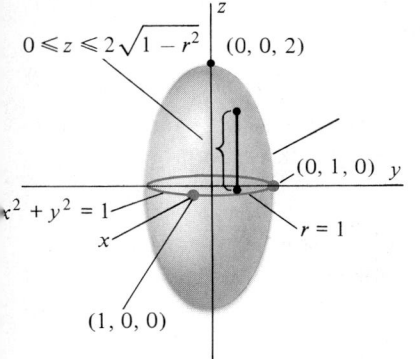

$0 \le z \le 2\sqrt{1 - r^2}$ $(0, 0, 2)$

$(0, 1, 0)$ y

$x^2 + y^2 = 1$

$r = 1$

$(1, 0, 0)$

Figure 7.4 Ellipsoid

$4x^2 + 4y^2 + z^2 = 4$

in Example 2.

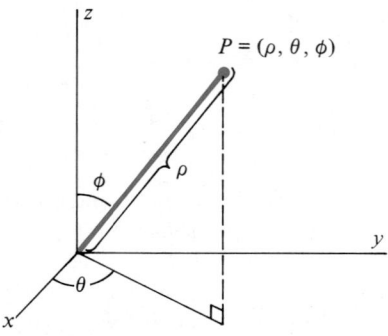

Figure 7.5 Spherical coordinates (ρ, θ, ϕ) for P.

may rewrite the integral (8) as

$$V = 2 \int_0^{2\pi} \int_0^1 \int_0^{2\sqrt{1-r^2}} r \, dz \, dr \, d\theta \qquad \text{note the extra factor } r$$

$$= 2 \int_0^{2\pi} \int_0^1 \left\{ zr \Big]_{z=0}^{z=2\sqrt{1-r^2}} \right\} dr \, d\theta$$

$$= 2 \int_0^{2\pi} \int_0^1 2r\sqrt{1-r^2} \, dr \, d\theta$$

$$= 2 \int_0^{2\pi} \left\{ -\frac{2}{3}(1-r^2)^{3/2} \Big]_{r=0}^{r=1} \right\} d\theta$$

$$= \frac{8\pi}{3}. \qquad \qquad \diamond$$

Spherical Coordinates

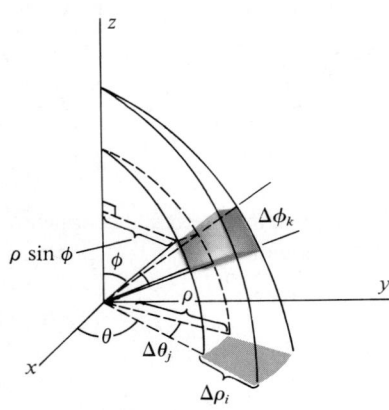

Figure 7.6 The spherical block.

Recall that the point P has spherical coordinates $P = (\rho, \theta, \phi)$ if $\rho \geq 0$ is its distance from the origin, θ is its rotation angle (as in cylindrical coordinates), and ϕ is its angle of inclination from the vertical (see Figure 7.5). Figure 7.6 illustrates the "spherical block" determined by the inequalities

$$\rho_{i-1} \leq \rho \leq \rho_i, \qquad \theta_{j-1} \leq \theta \leq \theta_j, \qquad \phi_{k-1} \leq \phi \leq \phi_k \qquad (9)$$

where $\Delta\rho_i = \rho_i - \rho_{i-1}$, $\Delta\theta_j = \theta_j - \theta_{j-1}$, and $\Delta\phi_k = \phi_k - \phi_{k-1}$. A fact, which we shall not prove, is that the volume of this spherical block is given by

$$\Delta V_{ijk} = \rho_i^{*2} \sin \phi_k^* \, \Delta\rho_i \Delta\theta_j \Delta\phi_k$$

for some $\rho_i^* \in [\rho_{i-1}, \rho_i]$ and $\phi_k^* \in [\phi_{k-1}, \phi_k]$ (see Exercise 31).

In order to define the integral of a continuous function f defined on Q, we select one point $(\rho_i^*, \theta_j^*, \phi_k^*)$ in each block Q_{ijk}, form the approximating sum, and evaluate the limit of this sum as the sizes of the blocks approach zero. When it exists, it is the triple integral of f over Q:

$$\iiint_Q f(\rho, \theta, \phi) \, dV$$
$$= \lim_{n\to\infty} \sum_i \sum_j \sum_k f(\rho_i^*, \theta_j^*, \phi_k^*)\rho_i^{*2} \sin \phi_k^* \, \Delta\rho_i \Delta\theta_j \Delta\phi_k.$$

As with other types of triple integrals, the triple integral in spherical coordinates may be evaluated as an iterated integral in certain cases.

| THEOREM 6 | Let Q be a region in $\mathbb{R}^3$ of the form |

$$Q = \{(\rho, \theta, \phi) \mid a \leq \theta \leq b, \, h_1(\theta) \leq \phi \leq h_2(\theta), \, g_1(\theta, \phi) \leq \rho \leq g_2(\theta, \phi)\}$$

where $h_1, h_2, g_1,$ and g_2 are continuous functions. Let f be continuous on Q. Then,

$$\iiint_Q f(\rho, \theta, \phi) \, dV = \int_a^b \int_{h_1(\theta)}^{h_2(\theta)} \int_{g_1(\theta, \phi)}^{g_2(\theta, \phi)} f(\rho, \theta, \phi) \, \rho^2 \sin \phi \, d\rho \, d\phi \, d\theta.$$

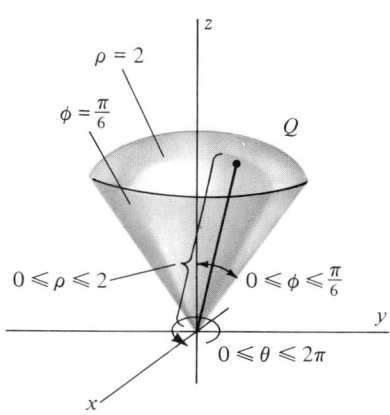

Figure 7.7 Region Q of Example 3.

REMARK: Note the extra factor $\rho^2 \sin \phi$ in the integrand in the iterated integral.

Example 3

Find the mass of the solid occupying the region

$$Q = \{(\rho, \theta, \phi) \mid 0 \le \rho \le 2, \ 0 \le \theta \le 2\pi, \ 0 \le \phi \le \pi/6\}$$

if the density at any point is proportional to its distance from the origin.

Solution: The region Q is the "ice cream cone" region sketched in Figure 7.7. As before, we obtain the mass of the solid by integrating the density function $D(\rho, \theta, \phi) = \lambda\rho$. (Note that we are now using ρ for radial distance, not density.) Thus, by Theorem 6

$$\text{Mass} = \iiint_Q \lambda\rho \, dV = \int_0^{2\pi} \int_0^{\pi/6} \int_0^2 (\lambda\rho)(\rho^2 \sin \phi) \, d\rho \, d\phi \, d\theta$$

note extra factor $\rho^2 \sin \phi$

$$= \int_0^{2\pi} \int_0^{\pi/6} \int_0^2 \lambda\rho^3 \sin \phi \, d\rho \, d\phi \, d\theta$$

$$= \int_0^{2\pi} \int_0^{\pi/6} \left\{ \frac{\lambda}{4}\rho^4 \sin \phi \right\}_{\rho=0}^{\rho=2} d\phi \, d\theta$$

$$= \int_0^{2\pi} \int_0^{\pi/6} 4\lambda \sin \phi \, d\phi \, d\theta$$

$$= \int_0^{2\pi} \left\{ -4\lambda \cos \phi \right\}_{\phi=0}^{\phi=\pi/6} d\theta$$

$$= \int_0^{2\pi} 4\lambda\left(1 - \frac{\sqrt{3}}{2}\right) d\theta$$

$$= 8\lambda\pi\left(1 - \frac{\sqrt{3}}{2}\right). \qquad \diamond$$

Example 4

Obtain the formula for the volume of a sphere of radius a using spherical coordinates.

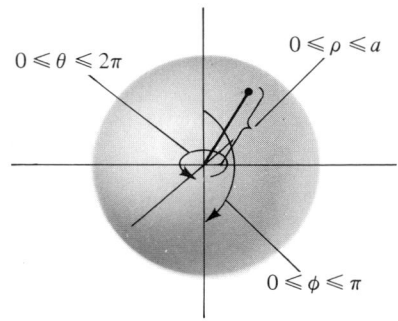

Figure 7.8 Sphere is determined by the inequalities

$0 \le \rho \le a$,
$0 \le \phi \le \pi, \ 0 \le \theta \le 2\pi$.

Solution: The sphere of radius a is described by the spherical inequalities

$$0 \le \rho \le a, \qquad 0 \le \phi \le \pi, \qquad 0 \le \theta \le 2\pi \qquad \text{(Figure 7.8)}.$$

Thus

$$\text{Volume} = \iiint_Q 1 \, dV = \int_0^{2\pi} \int_0^\pi \int_0^a \rho^2 \sin \phi \, d\rho \, d\phi \, d\theta$$

note extra factor $\rho^2 \sin \phi$

$$= \int_0^{2\pi} \int_0^\pi \left\{ \frac{1}{3}\rho^3 \sin \phi \right\}_{\rho=0}^{\rho=a} d\phi \, d\theta$$

$$= \int_0^{2\pi} \int_0^\pi \frac{a^3}{3} \sin \phi \, d\phi \, d\theta$$

$$= \int_0^{2\pi} \left\{ -\frac{a^3}{3} \cos \phi \right]_{\phi=0}^{\phi=\pi} \right\} d\theta$$

$$= \int_0^{2\pi} \frac{2a^3}{3} d\theta$$

$$= \frac{4}{3} \cdot \pi a^3.$$

$\diamond$

Changing From Cartesian to Spherical Coordinates

If we encounter a function f written in Cartesian coordinates for which we wish to calculate a triple integral over a spherical region, it is often useful to rewrite the function in spherical coordinates using the substitutions

$$x = \rho \sin \phi \cos \theta, \qquad y = \rho \sin \phi \sin \theta, \qquad z = \rho \cos \phi.$$

(See Section 16.6 for a review of how these equations are obtained.) We shall not write out a general formula for converting triple integrals from Cartesian coordinates to spherical coordinates, since the situations in which such changes are possible are almost exclusively limited to functions involving the expression

$$x^2 + y^2 + z^2 = \rho^2 \sin^2 \phi \cos^2 \theta + \rho^2 \sin^2 \phi \sin^2 \theta + \rho^2 \cos^2 \phi$$
$$= \rho^2 \sin^2 \phi (\cos^2 \theta + \sin^2 \theta) + \rho^2 \cos^2 \phi$$
$$= \rho^2.$$

Example 5

A hemispherical solid is described in Cartesian coordinates as

$$Q = \{(x, y, z) \mid 0 \le x^2 + y^2 \le 1, 0 \le z \le \sqrt{1 - x^2 - y^2}\}.$$

Find the mass of the object if the density at the point (x, y, z) is given by the function

$$D(x, y, z) = e^{-(x^2+y^2+z^2)^{3/2}}.$$

Solution: The hemisphere may be described by the inequalities

$$0 \le \rho \le 1, \qquad 0 \le \phi \le \pi/2, \qquad 0 \le \theta \le 2\pi,$$

and the density function, in spherical coordinates, is

$$D(\rho, \theta, \phi) = e^{-\rho^3}.$$

Thus,

$$\text{Mass} = \int_0^{2\pi} \int_0^{\pi/2} \int_0^1 e^{-\rho^3} \underbrace{\rho^2 \sin \phi}_{} \, d\rho \, d\phi \, d\theta \qquad \text{note extra factor } \rho^2 \sin \phi$$

$$= \int_0^{2\pi} \int_0^{\pi/2} \left\{ -\frac{1}{3} e^{-\rho^3} \sin \phi \right]_{\rho=0}^{\rho=1} \right\} d\phi \, d\theta$$

$$= \int_0^{2\pi} \int_0^{\pi/2} \frac{1}{3} \left(1 - \frac{1}{e} \right) \sin \phi \, d\phi \, d\theta$$

$$= \int_0^{2\pi} \left\{ \left(-\frac{e-1}{3e} \right) \cos \phi \Big]_{\phi=0}^{\phi=\pi/2} \right\} d\theta$$

$$= \int_0^{2\pi} \left(\frac{e-1}{3e} \right) d\theta$$

$$= \frac{2\pi(e-1)}{3e}.$$

◇

Exercise Set 19.7

In Exercises 1–8, use cylindrical coordinates.

1. Find the volume of the solid bounded by the graphs of $z = x^2 + y^2$ and $z = 9$.

2. Find the volume of the solid bounded by the graphs of the equations $z = 9 - x^2 - y^2$, $x^2 + y^2 = 3$, and $z = 9$.

3. Find the center of mass of a solid having the shape of the solid in Exercise 1 if the density is constant.

4. Find the volume of the region lying inside both the cylinder $x^2 - 2x + y^2 = 0$ and the sphere $x^2 + y^2 + z^2 = 4$.

5. Find the volume of the solid remaining when the region inside the cylinder $r = 3 \sin \theta$ is removed from the solid region bounded by the sphere $x^2 + y^2 + z^2 = 9$.

6. A solid is bounded by the cylinder $y^2 + z^2 = 4$ and the planes $x = 0$ and $x = 4$. Find its mass if the density at each point is proportional to the distance of that point from the central axis of the cylinder.

7. Find the volume of the solid bounded above and below by the cone $z^2 = 2x^2 + 2y^2$ and on the sides by the cylinder $x^2 + y^2 - 4y = 0$.

8. Find the volume of the solid bounded above by the plane $z = x$ and below by the paraboloid $z = x^2 + y^2$.

In Exercises 9–12, evaluate the iterated integral by first changing to cylindrical coordinates.

9. $\int_0^2 \int_{-\sqrt{2x-x^2}}^{\sqrt{2x-x^2}} \int_0^{\sqrt{x^2+y^2}} 1 \, dz \, dy \, dx$

10. $\int_{-1}^1 \int_{-\sqrt{1-x^2}}^{\sqrt{1-x^2}} \int_{\sqrt{x^2+y^2}}^1 x^2 \, dz \, dy \, dx$

11. $\int_{-2}^2 \int_{-\sqrt{4-x^2}}^{\sqrt{4-x^2}} \int_0^{y+2} xy \, dz \, dy \, dx$

12. $\int_{-1}^1 \int_0^{\sqrt{1-x^2}} \int_{x^2+y^2}^1 z \, dz \, dy \, dx$

In Exercises 13–18, use spherical coordinates.

13. Find the volume of the solid bounded above by the sphere $x^2 + y^2 + z^2 - 2z = 0$ and below by the cone $z = \sqrt{x^2 + y^2}$.

14. Find the mass of a sphere of radius $r = 2$ if the density at each point is proportional to the square of the distance from the center.

15. Find the center of mass of the quarter sphere

$$Q = \{(\rho, \theta, \phi) \mid 0 \le \rho \le 1, 0 \le \phi \le \pi/2, 0 \le \theta \le \pi\}$$

if the density is constant.

16. Find the mass of the solid lying between the spheres $x^2 + y^2 + z^2 = 1$ and $x^2 + y^2 + z^2 = 4$ if the density at each point is proportional to the reciprocal of the distance from the center of the spheres.

17. Find the volume of the solid remaining when the region lying inside the cone $z^2 = x^2 + y^2$ is removed from the region bounded by the sphere $x^2 + y^2 + z^2 = 4$.

18. Use spherical coordinates to obtain the formula for the volume of a right circular cone of radius r and height h.

In Exercises 19 and 20, evaluate the iterated integral by first changing to spherical coordinates.

19. $\int_{-1}^1 \int_{-\sqrt{1-x^2}}^{\sqrt{1-x^2}} \int_0^{\sqrt{1-x^2-y^2}} 3 \, dz \, dy \, dx$

20. $\int_{-2}^2 \int_{-\sqrt{4-x^2}}^{\sqrt{4-x^2}} \int_{-\sqrt{4-x^2-y^2}}^{\sqrt{4-x^2-y^2}} (x^2 + y^2) \, dz \, dy \, dx$

21. Two circular cylinders of radius R meet at right angles. Use triple integration to find the volume of the solid common to both cylinders.

22. Find the volume of the region lying inside the sphere $x^2 + y^2 + z^2 = 4$ and outside the cylinder $x^2 + z^2 = 1$.

23. Evaluate the triple integral

$$\iiint_Q \frac{z}{(x^2 + y^2)^{3/2}} \, dV$$

where Q is the region

$$Q = \{(x, y, z) \mid 1 \le x^2 + y^2 \le 3, 0 \le z \le 3\}.$$

24. Evaluate the triple integral $\iiint\limits_{Q} \cos \pi y \cdot \sqrt{x^2 + z^2} \, dV$

where Q is the region bounded by the cylinders $x^2 + z^2 = 1$ and $x^2 + z^2 = 4$, and the planes $y = -1$ and $y = 2$.

25. Evaluate the triple integral $\iiint\limits_{Q} \sqrt{\dfrac{x}{y^2 + z^2}} \, dV$ where Q is

the region bounded by the cone $y^2 + z^2 = x^2$, the cylinder $y^2 + z^2 = 4$, and the planes $x = 0$ and $x = 2$.

26. Find the volume of the solid that remains when the cone $3y^2 = x^2 + z^2$ is removed from the sphere $x^2 + y^2 + z^2 = 4$.

27. Find the volume of the smaller of the two parts of the sphere $\rho = 4$ determined by the plane $y = 2$.

28. Evaluate the integral $\iiint\limits_{Q} (x^2 + y^2) \, dV$ where Q is the

sphere $x^2 + y^2 + z^2 \le 4$.

29. Find the volume of the solid lying between the spheres $x^2 + y^2 + z^2 = 1$ and $x^2 + y^2 + z^2 = 9$ and inside the cone $y^2 = x^2 + z^2$.

30. Find the volume of the solid bounded below by the graph of $x^2 + y^2 + z^2 + 2z = 0$ and above by the graph of $z^2 = x^4 + 2x^2y^2 + y^4$.

31. Obtain the approximation $\Delta V_{ijk} \approx \rho_i^2 \sin \phi_k \, \Delta\rho_i \Delta\theta_j \Delta\phi_k$ for the spherical block in Figure 7.9 as follows.

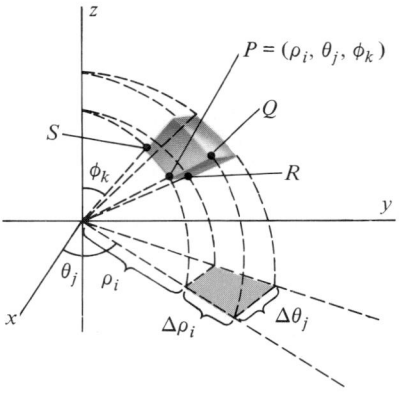

Figure 7.9 Spherical block of volume ΔV_{ijk}.

a. Observe that the spherical block is nearly a parallelepiped. We shall therefore approximate its volume by the product (length of PQ) · (length of PS) · (length of PR).

b. Show that the length of $PQ = \Delta\rho_i$.

c. Show that the length of the arc PS is $\rho_i \Delta\phi_k$. (*Hint:* It lies on a circle of radius ρ_i.)

d. Show that the length of the arc PR is $\rho_i \sin \phi_k \, \Delta\theta_j$. (*Hint:* It lies on a horizontal circle of radius $\rho_i \sin \phi_k$. Why?)

SUMMARY OUTLINE OF CHAPTER 19

◆ If f is continuous on the planar region $Q = \{(x, y) \mid a \le x \le b, \, g_1(x) \le y \le g_2(x)\}$ then

(page 860)

(i) $\displaystyle\iint\limits_{Q} f(x, y) \, dA = \int_a^b \int_{g_1(x)}^{g_2(x)} f(x, y) \, dy \, dx.$

(ii) Area of $Q = \displaystyle\iint\limits_{Q} 1 \, dA.$

◆ If f is continuous on the region Q described in polar coordinates as

(page 871)

$Q = \{(r, \theta) \mid g_1(\theta) \le r \le g_2(\theta), \, a \le \theta \le b\},$

then

(i) $\displaystyle\iint\limits_{Q} f(r, \theta) \, dA = \int_a^b \int_{g_1(\theta)}^{g_2(\theta)} f(r, \theta) \, r \, dr \, d\theta.$

(ii) Area of $Q = \displaystyle\int_a^b \int_{g_1(\theta)}^{g_2(\theta)} r \, dr \, d\theta$ if $g_1(\theta) \ge 0$ for $a \le \theta \le b$.

◆ In changing an iterated double integral from Cartesian to polar coordinates, the differential for area $dA = dx \, dy$ is replaced by the polar differential for area $dA = r \, dr \, d\theta$. (page 875)

◆ If a thin lamina with density ρ has the shape of the planar region Q, then (page 881)

(i) Mass $= M = \displaystyle\iint_Q \rho(x, y) \, dA.$

(ii) Center of Mass $= (\bar{x}, \bar{y})$ where $\bar{x} = \dfrac{M_y}{M}$, $\bar{y} = \dfrac{M_x}{M}$,

$$M_y = \iint_Q x\rho(x, y) \, dA,$$

and

$$M_x = \iint_Q y\rho(x, y) \, dA.$$

◆ The **surface area** of the graph of $z = f(x, y)$ over the region Q is   (page 886)

$$A_S = \iint_Q \sqrt{\left[\frac{\partial f}{\partial x}(x, y)\right]^2 + \left[\frac{\partial f}{\partial y}(x, y)\right]^2 + 1} \; dA.$$

◆ If the region Q in $\mathbb{R}^3$ is described by the inequalities $a \le x \le b$, $h_1(x) \le y \le h_2(x)$, $g_1(x, y) \le z \le g_2(x, y)$,   (page 890)
then

(i) $\displaystyle\iiint_Q f(x, y, z) \, dV = \int_a^b \int_{h_1(x)}^{h_2(x)} \int_{g_1(x, y)}^{g_2(x, y)} f(x, y, z) \, dz \, dy \, dx.$

(ii) Volume of $Q = \displaystyle\iiint_Q 1 \, dV.$

◆ If a solid has the shape of the region Q in $\mathbb{R}^3$ and density function ρ, then   (page 896)

(i) Mass $= M = \displaystyle\iiint_Q \rho(x, y, z) \, dV.$

(ii) Center of mass $= (\bar{x}, \bar{y}, \bar{z})$ where $\bar{x} = \dfrac{M_{yz}}{M}$, $\bar{y} = \dfrac{M_{xz}}{M}$, $\bar{z} = \dfrac{M_{xy}}{M}$,

$$M_{yz} = \iiint_Q x\rho(x, y, z) \, dV,$$

$$M_{xz} = \iiint_Q y\rho(x, y, z) \, dV,$$

$$M_{xy} = \iiint_Q z\rho(x, y, z) \, dV.$$

◆ If the region Q is described in cylindrical coordinates by the inequalities $a \le \theta \le b$, $h_1(\theta) \le r \le h_2(\theta)$,   (page 901)
$g_1(r, \theta) \le z \le g_2(r, \theta)$, then

$$\iiint_Q f(r, \theta, z) \, dV = \int_a^b \int_{h_1(\theta)}^{h_2(\theta)} \int_{g_1(r, \theta)}^{g_2(r, \theta)} f(r, \theta, z) r \, dz \, dr \, d\theta.$$

◆ In changing an iterated triple integral from Cartesian coordinates to cylindrical coordinates, the Cartesian differential   (page 903)
for volume $dV = dx \, dy \, dz$ is replaced by the cylindrical differential for volume $dV = r \, dz \, dr \, d\theta$.

◆ If the region Q in $\mathbb{R}^3$ is described in spherical coordinates by the inequalities $a \le \theta \le b$, $h_1(\theta) \le \phi \le h_2(\theta)$,   (page 904)
$g_1(\theta, \phi) \le \rho \le g_2(\theta, \phi)$, then

$$\iiint_Q f(\rho, \theta, \phi) \, dV = \int_a^b \int_{h_1(\theta)}^{h_2(\theta)} \int_{g_1(\theta, \phi)}^{g_2(\theta, \phi)} f(\rho, \theta, \phi)\rho^2 \sin \phi \, d\rho \, d\phi \, d\theta.$$

◆ In changing an iterated triple integral from Cartesian to spherical coordinates, the Cartesian differential for volume   (page 906)
$dV = dx \, dy \, dz$ is replaced by the spherical differential for volume $dV = \rho^2 \sin \phi \, d\rho \, d\phi \, d\theta$.

REVIEW EXERCISES—CHAPTER 19

1. Find the volume of the solid bounded above by the graph of $z = xy^2$ and below by the triangle with vertices $(0, 0)$, $(2, 0)$, and $(0, 1)$.

2. Evaluate the double integral $\iint\limits_{Q} \dfrac{x}{xy + 2} \, dA$ where Q is the rectangle

$$Q = \{(x, y) \mid 0 \le x \le 1, 0 \le y \le 1\}.$$

3. Evaluate $\iint\limits_{Q} e^{-x^2/2} \, dy \, dx$ where Q is the region

$$Q = \{(x, y) \mid 0 \le x \le 1, 0 \le y \le 2x\}.$$

4. Evaluate $\iint\limits_{Q} (x - 4y) \, dA$ where Q is the rectangle

$$Q = \{(x, y) \mid -1 \le x \le 1, 0 \le y \le 2\}.$$

5. Find the volume of the solid in the first octant bounded by the graphs of $z = y^2$, $y = x$, $z = 0$, and $y = 4$.

6. Change the order of integration:

$$\int_{-1}^{2} \int_{x^2-2}^{x} f(x, y) \, dy \, dx.$$

7. Find the surface area of that part of the sphere $x^2 + y^2 + z^2 = 4$ lying inside the cylinder $x^2 + y^2 - 2x = 0$.

8. Find the center of mass of a solid having the shape of the region bounded above by the paraboloid $4z = 4 - (x^2 + y^2)$ and below by the xy-plane if the density of the material is uniform.

9. Find the volume of the solid bounded by the cylinder $r = 2 \cos \theta$, the paraboloid $z = 2r^2$ and the plane $z = 0$ (use cylindrical coordinates).

10. Rewrite the integral

$$\int_{-\pi/2}^{\pi/2} \int_{0}^{2 \cos \theta} \int_{-\sqrt{4-r^2}}^{\sqrt{4-r^2}} r \, dz \, dr \, d\theta$$

in rectangular coordinates.

11. Find the volume of the solid bounded by the graph of $\dfrac{x}{a} + \dfrac{y}{b} + \dfrac{z}{c} = 1$ ($a > 0$, $b > 0$, $c > 0$) and the three coordinate planes.

12. Sketch the region Q corresponding to the iterated integral, reverse the order of integration, and evaluate the resulting integral:

$$\int_{-2}^{0} \int_{-\sqrt{x+2}}^{\sqrt{x+2}} y^2 \, dy \, dx.$$

13. Calculate the volume of the solid lying inside the cylinder $x^2 + y^2 = 4$ and between the planes $y + z = 9$ and $z = 0$.

14. Use a double integral to calculate the area of the region lying between the graph of $\sqrt{x} + \sqrt{y} = 1$ and the graph of $x + y = 1$.

15. Find the center of mass of a solid bounded by the cylinder $r = 2$, the cone $z = r$, and the plane $z = 0$ if the density of the material is uniform.

16. Calculate the volume of the ellipse

$$\frac{x^2}{4} + \frac{y^2}{9} + \frac{z^2}{4} = 1.$$

17. Find the volume of the region common to the sphere $r^2 + z^2 = a^2$ and the cylinder $r = a \cos \theta$.

18. Find the surface area of the paraboloid $z = x^2 + y^2$ lying between the planes $z = 1$ and $z = 9$.

19. Find the area of the region cut from the plane $z = 4y$ by the cylinder $x^2 + y^2 = 4$.

20. Find the volumes of the two regions cut from the sphere $\rho = 4$ by the plane $x = 2$.

21. Find the volume of the region common to the sphere $x^2 + y^2 + z^2 = 16$ and the cylinder $x^2 + z^2 = 4$.

22. Use a double integral to find the area enclosed by the lemniscate $r^2 = 2 \cos 2\theta$.

23. A solid is bounded below by the region bounded by the graphs of $y = x$ and $y = x^2 - 2$. It is bounded above by the plane $z - x + 2y = 10$. Find its volume.

24. Find $\iint\limits_{Q} \dfrac{\sin x}{x} \, dA$ where Q is the triangle with vertices $(0, 0)$, $(2, 0)$ and $(2, 2)$.

25. Find the centroid of the half disc $r = 4$, $0 \le \theta \le \pi$.

26. Find the mass of a right circular cone of radius r and height h if the density at each point is proportional to the distance of that point from the vertex.

27. Find $\iiint\limits_{Q} e^{(x^2+y^2+z^2)^{3/2}} \, dV$ where Q is the half sphere

$$Q = \{(r, \theta, \phi) \mid 0 \le r \le 1, 0 \le \phi \le \pi/2, 0 \le \theta \le 2\pi\}.$$

28. Show that if a thin lamina of constant density has the shape of the region Q described in polar coordinates, then the coordinates $(\bar{x}, \bar{y})$ of the center of mass may be calculated by the formulas

$$\bar{x} = \frac{1}{\text{area}} \iint\limits_{Q} r^2 \cos \theta \, dr \, d\theta;$$

$$\bar{y} = \frac{1}{\text{area}} \iint\limits_{Q} r^2 \sin \theta \, dr \, d\theta.$$

29. Use the result of Exercise 28 to calculate the center of mass

for a thin lamina of constant density whose shape is the cardioid $r = 1 + \cos \theta$.

30. Find the volume of the solid bounded above by the plane $z - x = 2$ and below by the paraboloid $z = x^2 + y^2$.

31. Find the volume of the region bounded by the paraboloids

$$z = 4 - x^2 + 2x - y^2 - 4y$$

and

$$z = x^2 - 2x + y^2 + 4y + 5.$$

32. Find the volume of the solid bounded above by the cylinder $z = 4 - x^2$ and below by the paraboloid $3x^2 + y^2 = z$.

33. A solid corresponds to the region bounded by the cylinder $x^2 + y^2 = 4$ and the planes $z = 0$ and $z = 4$. Calculate the mass of the solid if the density at each point is proportional to the distance from the xy-plane.

34. Evaluate the integral $\displaystyle\iint_Q \cos \sqrt{x^2 + y^2} \, dA$ where Q is the disc

$$Q = \{(x, y) \mid 0 \le x^2 + y^2 \le 4\}.$$

35. Evaluate the integral $\displaystyle\iint_Q \frac{1}{\sqrt{1 + x^2 + y^2}} \, dA$ where Q is the quarter circle

$$Q = \{(x, y) \mid 0 \le x \le 1, \, 0 \le y \le \sqrt{1 - x^2}\}.$$

36. Find the volume of the solid bounded above by the graph of $z = 4 - r$ and below by the region in the $r\theta$-plane bounded by the graph of $r = 3 \sin \theta$.

37. Find the volume of the solid bounded above by the cone $z^2 = x^2 + y^2$, on the sides by the cylinder $x^2 + y^2 - 4y = 0$, and below by the xy-plane.

38. Find the volume of the solid inside the cylinder $x^2 - 4x + y^2 = 0$ lying above the xy-plane and below the plane $z - x = 4$.

39. *(Computer)* Use a modification of Program 8 in Appendix I to approximate

$$\int_0^2 \int_1^3 \sin^2 \sqrt{x - y^2} \, dy \, dx.$$

40. Find the area of the region bounded by the graphs of $y = \sqrt{x + 2}$ and $x - 3y + 2 = 0$ by double integration.

41. Find the volume of the solid bounded by the cone $z^2 = x^2 + y^2$ and the cylinder $x^2 + y^2 = 9$.

42. Find the centroid of the region bounded by the graph of $r = \cos 2\theta$.

43. Find the surface area of the part of the sphere $x^2 + y^2 + z^2 = 4$ lying outside the cylinder $x^2 + y^2 = 1$.

Chapter 20
Topics in Vector Analysis

The goal of this chapter is to develop a theory of integral calculus for vector-valued functions defined in the plane or in space. These are functions of the form $w = F(r)$ where both w and r are vectors.

20.1 VECTOR FIELDS

To avoid confusion between functions of this type and vector-valued functions of a single real variable, we shall use the term *vector fields*.

Let Q be a subset of $\mathbb{R}^3$. A **vector field** on Q is a function

$$F(x, y, z) = M(x, y, z)i + N(x, y, z)j + P(x, y, z)k \qquad (1)$$

that assigns a vector $w = F(x, y, z)$ to each point (x, y, z) in Q.

If Q is a subset of $\mathbb{R}^2$, a vector field on Q is a function of the form

$$F(x, y) = M(x, y)i + N(x, y)j. \qquad (2)$$

The real-valued functions M, N, and P in equation (1) are referred to as the **component** functions of the vector field F. We say that a vector field is **continuous** if each of its component functions is continuous. Using the position vector

$$r = xi + yj + zk, \qquad (x, y, z) \in Q,$$

we may write the vector field in equation (1) as simply $w = F(r)$. Functions of the form $w = f(x, y, z)$, which assign real numbers to vectors, will now be referred to as **scalar functions** or scalar fields.

The following examples indicate the types of vector fields and scalar functions that we shall study in this chapter.

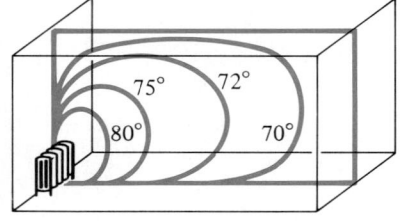

Figure 1.1 Isothermal curves in a cross section of a room with a single heat source.

Example 1

Figure 1.1 represents a room with a single radiator. The function T, which gives the temperature $T(x, y, z)$ at the point in the room with coordinates (x, y, z), is an example of a *scalar* function. The curves drawn in a cross section of the room are locations of points of equal temperature and are called **isothermal curves.**

Since warmer air rises and cooler air falls, the presence of a single heat source in a room causes air to flow about the room in **convection currents.** If the vector

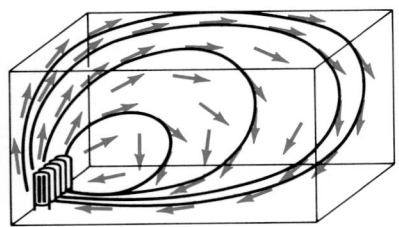

Figure 1.2 Tangents $F(x, y, z)$ to convection currents determine a vector field in the room.

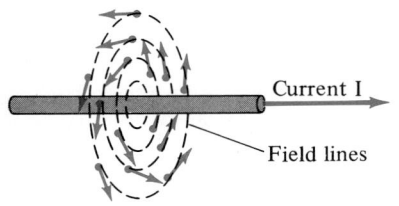

Figure 1.3 Tangents to magnetic field lines form a (magnetic) vector field in space.

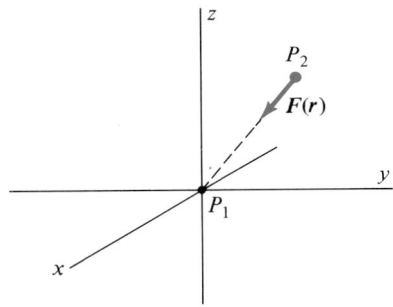

Figure 1.4 Gravitational force vector $F(r)$ acting on particle P_2.

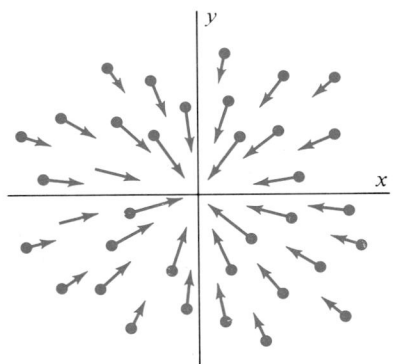

Figure 1.5 Gravitational force field (drawing represents xy-plane only).

$F(x, y, z)$ represents the velocity vector of the convection current passing through the point (x, y, z), the function $w = F(x, y, z)$ is an example of a vector field defined throughout the room (Figure 1.2). ◇

Example 2

An electric current of magnitude I flowing through a thin wire induces a **magnetic field** around the wire. (The statement of Ampère's Law, giving the relationship between the electric current I and its magnetic effect, involves the idea of a *line integral,* introduced in Section 20.2.) The set of vectors $F(x, y, z)$, giving the direction and intensity of the magnetic field at location (x, y, z), determines a vector field in the space surrounding the wire (Figure 1.3). ◇

Example 3

An example of a force field is a **gravitational field.** According to Newton's Law of Gravitation, the magnitude of the force of attraction exerted on a particle P_2 of mass m_2 by a particle P_1 of mass m_1 is given by the equation

$$F = \frac{Gm_1m_2}{r^2} \tag{3}$$

where r is the distance between the particles and G is a constant. If we assume the particle P_1 to be located at the origin in xyz-space, then the force in (3) is a function of the coordinates (x, y, z) giving the location of the particle P_2. That is,

$$F(x, y, z) = \frac{Gm_1m_2}{x^2 + y^2 + z^2}. \tag{4}$$

Since the force whose magnitude is given by (4) acts toward the origin, it acts in the direction opposite that of the position vector $r = xi + yj + zk$ of P_2. We may therefore write the force vector $F(x, y, z)$ as

$$F(x, y, z) = \left(\frac{Gm_1m_2}{(x^2 + y^2 + z^2)}\right)\left(-\frac{xi + yj + zk}{\sqrt{x^2 + y^2 + z^2}}\right) \tag{5}$$

$$= \left(\frac{-Gm_1m_2}{(x^2 + y^2 + z^2)^{3/2}}\right)(xi + yj + zk).$$

Using the position vector $r = xi + yj + zk$, we may write the force field in (5) as

$$F(r) = -\frac{Gm_1m_2}{|r|^3}r, \qquad r \neq 0 \qquad \text{(Figure 1.4)}. \tag{6}$$

Since the vector F in (5) and (6) is defined for all position vectors $r \neq 0$, either equation defines a vector field at all points of $\mathbb{R}^3$ except the origin (Figure 1.5). ◇

The gravitational field of Example 3 is an example of a **central force field,** since the force vector at each point points toward the origin. More generally, a central force field in space has the form

$$F(x, y, z) = f(x, y, z)(xi + yj + zk)$$

where f is a real-valued function of three variables.

Example 4

Another example of a central force field is obtained by letting $F(x, y, z)$ be the vector representing the force exerted on a particle P_2 with electric charge q_2 at location (x, y, z) by a particle P_1 with electric charge q_1 located at the origin. According to **Coulomb's Law,** this force vector is

$$F(r) = \frac{kq_1q_2}{|r|^3}r, \qquad r \neq 0 \tag{7}$$

where r is the position vector $r = xi + yj + zk$ of P_2, and k is a constant that depends on the choice of units for r, q_1, and q_2. Figure 1.6 represents the **electric force field** F (for positive charges P_2) due to a positive charge at the origin. Figure 1.7 represents the force field F (for positive charges P_2) due to a negative charge at

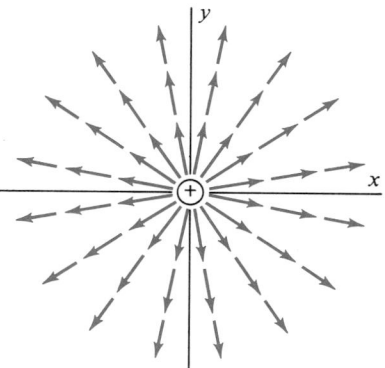

Figure 1.6 Electric force field due to a positive charge at the origin.

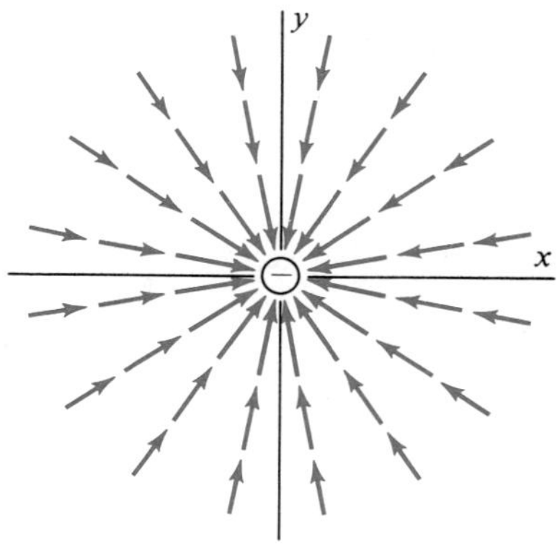

Figure 1.7 Electric field due to a negative charge at the origin.

the origin. (Although the force field is defined in space, we show only the force vectors lying in the xy-plane in these figures.) ◇

If charge is considered to be analogous to mass, it is clear from comparison of equations (6) and (7) that the gravitational and electric force fields have the same form; physicists say that both forces obey an *inverse square law*. The major difference is that the direction of $F(r)$ may be either toward or away from the origin for the electric field, but it can only be toward the origin for the gravitational field (since all masses are positive).

Electric force fields satisfy the **principle of superposition,** which states that the force acting on a charge at a point P due to two separate charges is the vector sum of the individual forces acting on the charge at P. Figure 1.8 represents the electric field, acting on a positive charge, that results from a positive charge at $(-1, 0)$ and a negative charge of equal magnitude at $(1, 0)$. Note that while the individual force fields in Figure 1.6 and 1.7 are central, their sum is not.

Example 5

If f is a differentiable scalar function of three variables, the gradient function

$$F(x, y, z) = \nabla f(x, y, z) = f_x(x, y, z)i + f_y(x, y, z)j + f_z(x, y, z)k \tag{8}$$

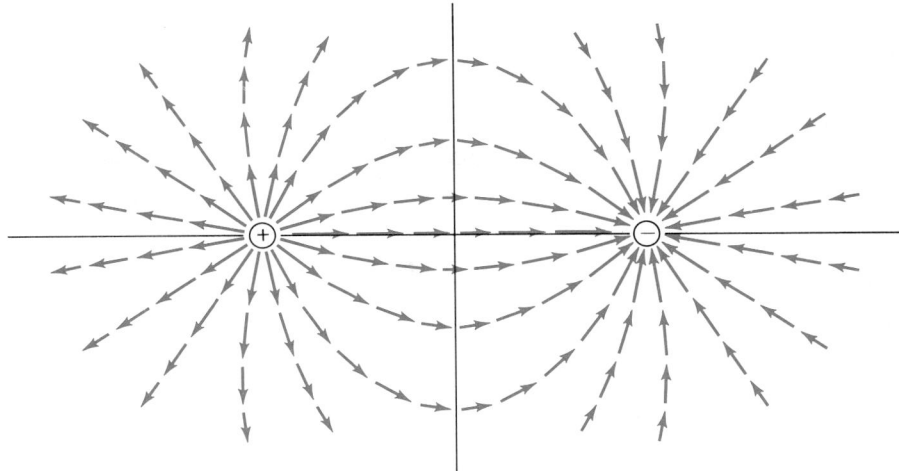

Figure 1.8 Electric force field due to two opposite charges is the vector sum of the individual central force fields.

defines a vector field on $\mathbb{R}^3$. If f is a differentiable scalar function of two variables, the corresponding gradient field on $\mathbb{R}^2$ is

$$F(x,\ y) = \nabla f(x,\ y) = f_x(x,\ y)\boldsymbol{i} + f_y(x,\ y)\boldsymbol{j}. \tag{9}$$

For example, the scalar function $f(x,\ y,\ z) = xe^{2y} \sin z$ yields the gradient vector field

$$F(x,\ y,\ z) = e^{2y} \sin z\boldsymbol{i} + 2xe^{2y} \sin z\boldsymbol{j} + xe^{2y} \cos z\boldsymbol{k}. \qquad \diamondsuit$$

As in Section 18.9, we will refer to the function f in equation (8) as a **scalar potential** for the vector field F. Although every differentiable scalar function yields a gradient vector field, a given vector field F need not be the gradient of a scalar potential. Those vector fields that can be written as gradients of scalar potentials are called **conservative vector fields.**

There are many other familiar examples of vector fields. The flow of air currents around a moving automobile or airplane determines a **velocity field.** The study of such vector fields is called *aerodynamics*. Similarly, the velocity field determined by the flow of water through a container, such as a pipe or a dam, is the subject of *hydrodynamics*.

Exercise Set 20.1

In Exercises 1–10, sketch enough vectors in the given vector field to get a sense of the nature of the vector field.

1. $F(x,\ y) = x\boldsymbol{i} - y\boldsymbol{j}$

2. $F(x,\ y) = -y\boldsymbol{i} + x\boldsymbol{j}$

3. $F(x,\ y) = 3\boldsymbol{i}$

4. $F(x,\ y) = -x\boldsymbol{i} - y\boldsymbol{j}$

5. $F(x,\ y) = \dfrac{x}{\sqrt{x^2 + y^2}}\boldsymbol{i} + \dfrac{y}{\sqrt{x^2 + y^2}}\boldsymbol{j}$

6. $F(x,\ y) = \dfrac{x}{\sqrt{x^2 + y^2}}\boldsymbol{i} - \dfrac{y}{\sqrt{x^2 + y^2}}\boldsymbol{j}$

7. $F(x,\ y,\ z) = \boldsymbol{j} + \boldsymbol{k}$

8. $F(x,\ y,\ z) = x\boldsymbol{i} + y\boldsymbol{j}$

9. $F(x,\ y,\ z) = 2x\boldsymbol{i} + 2y\boldsymbol{j} + 2z\boldsymbol{k}$

10. $F(x,\ y,\ z) = \boldsymbol{i} - \boldsymbol{j} + \boldsymbol{k}$

11. Which of the vector fields in Exercises 1–10 are central?

Find the gradient vector field $F = \nabla f$ for each of the functions in Exercises 12–17.

12. $f(x,\ y) = xy$

13. $f(x,\ y) = x^2 - y^2$

14. $f(x,\ y) = x \tan xy$

15. $f(x, y, z) = \sqrt{x^2 + y^2 + z^2}$

16. $f(x, y, z) = x \ln(y^2 + z^2)$

17. $f(x, y, z) = ze^{x-y}$

In Exercises 18–23, determine whether the given vector field is a gradient vector field. If so, find a potential ϕ with $F = \nabla \phi$ (see Section 18.9).

18. $F(x, y) = \sin y\mathbf{i} + \cos y\mathbf{j}$

19. $F(x, y) = (\tan xy - xy \sec^2 xy)\mathbf{i} + x^2 \sec^2 xy\mathbf{j}$

20. $F(x, y) = e^{\sqrt{xy}}\mathbf{i} - \sqrt{x}e^{\sqrt{xy}}\mathbf{j}$

21. $F(x, y) = \left[\ln(y - x) + \dfrac{x}{y - x}\right]\mathbf{i} + \left(\dfrac{x}{y - x}\right)\mathbf{j}$

22. $F(x, y, z) = x^2\mathbf{i} + x^2 \sin z\mathbf{j} - \cos yz\mathbf{k}$

23. $F(x, y, z) = yze^{xyz}\mathbf{i} + xze^{xyz}\mathbf{j} + xye^{xyz}\mathbf{k}$

24. The **flow lines** for a vector field F are curves $\boldsymbol{\alpha}(t)$ tangent to the vector field. That is, $\boldsymbol{\alpha}'(t) = F(\boldsymbol{\alpha}(t))$ for every t and every flow line. (An example of flow lines associated with a vector field in the plane is shown in Figure 1.9.) Sketch several flow lines for each of the vector fields in Exercises 1–10.

25. Imagine water flowing through a pipe as in Figure 1.10. At each point (x, y, z) within the pipe let $F(x, y, z)$ be the velocity vector for the water at that point. This determines a vector field within the pipe. Why are the vectors in the narrow part of the pipe larger in magnitude?

26. Show that the central force field $F(r) = \left(\dfrac{k}{|r|^3}\right)r$ is the gradient of the scalar field $f(r) = \dfrac{k}{|r|}$. (r denotes the position vector $r = x\mathbf{i} + y\mathbf{j} + z\mathbf{k}$ associated with the point (x, y, z).)

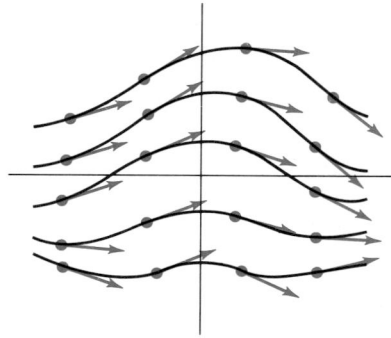

Figure 1.9 Flow lines for a vector field.

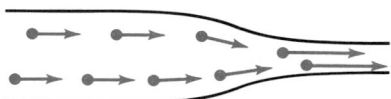

Figure 1.10 Velocity field in a water pipe.

27. Let $F(x, y, z)$ be the electric field describing the force on a particle with charge $q = +1$ at location (x, y, z) due to the combined effect of a charge $q_1 = +3$ at $(-1, 0, 0)$ and a charge $q_2 = -1$ at $(1, 0, 0)$.
  a. Find the force vector $F(0, 0, 0)$.
  b. Find the force vector $F(-2, 0, 0)$.
  c. Sketch enough vectors $F(x, y, 0)$ to get an idea of the force field in the xy-plane.
  d. Sketch several of the flow lines for the vector field (see Exercise 24).

20.2 WORK AND LINE INTEGRALS

When a force of magnitude F moves an object d units along a line, we define the **work** done on the object by the force as the product $W = Fd$. That is,

$$\text{Work} = (\text{force}) \times (\text{distance}) \tag{1}$$

for constant forces applied along a line.

   In Chapter 7 we applied the theory of the definite integral to calculate the work done by a continuously varying force f in moving an object from location $x = a$ to location $x = b$ along a line:

$$W = \lim_{n \to \infty} \sum_{j=1}^{n} f(t_j)\, \Delta x = \int_{a}^{b} f(x)\, dx,$$

which agrees with (1) when f is a constant force (Figures 2.1 and 2.2).

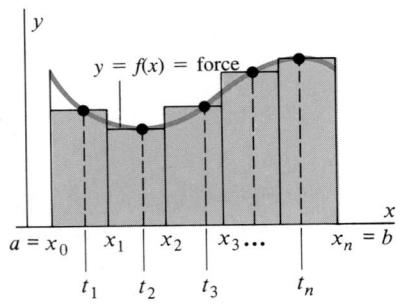

Figure 2.1 $W \approx \sum\limits_{j=1}^{n} f(t_j)\, \Delta x.$

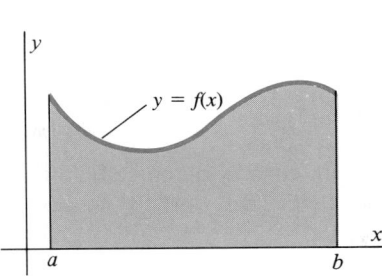

Figure 2.2 $W = \int_a^b f(x)\, dx.$

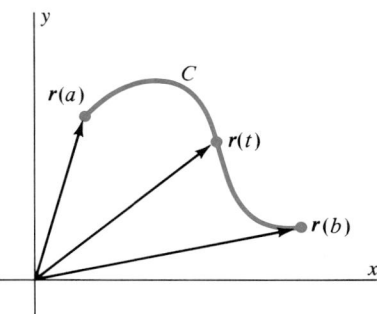

Figure 2.3 $C: \quad r(t) = x(t)i + y(t)j + z(t)k.$

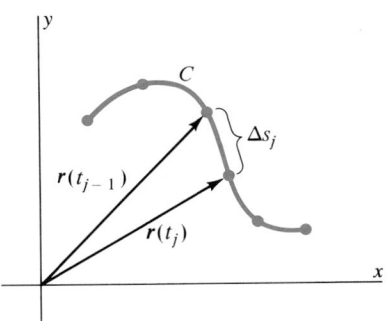

Figure 2.4 $\Delta s_j = \int_{t_{j-1}}^{t_j} |r'(t)|\, dt.$

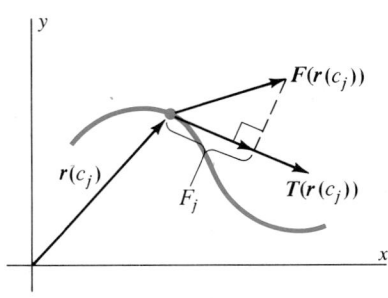

Figure 2.5 F_j is the component of the force field in the direction of the unit tangent at $r(c_j)$.

In this section we wish to generalize these concepts to calculate the work done on a particle by a force that moves the particle along a curve C lying in the plane or in space. Examples of this situation include small charged particles moving in electric fields, objects moving in gravitational fields, and others suggested by the discussion of Section 20.1.

To formulate the problem more precisely, assume that $\boldsymbol{F}$ is a vector field defined on an open box D of $\mathbb{R}^3$ that determines a force vector $\boldsymbol{F}(r)$ for each position vector $r = xi + yj + zk$ in D. Also, let C be a curve lying within D that is parameterized by the vector function

$$C: \quad r(t) = x(t)i + y(t)j + z(t)k, \qquad a \le t \le b \qquad \text{(see Figure 2.3)}. \qquad (2)$$

If we partition the parameter interval $[a, b]$ into n subintervals of equal length $\Delta t = \dfrac{b-a}{n}$ with endpoints $a = t_0 < t_1 < t_2 < \cdots < t_n = b$, the position vectors $r(t_0), r(t_1), \ldots, r(t_n)$ divide the curve C into n arcs. If the curve C is smooth (meaning that $r'(t)$ is continuous for $t \in [a, b]$), equation (9) of Section 17.3 gives the length of the jth arc as

$$\Delta s_j = \int_{t_{j-1}}^{t_j} |r'(t)|\, dt \qquad \text{(see Figure 2.4)}.$$

Under this assumption, the Mean Value Theorem for Integrals guarantees the existence of a number c_j in the interval $[t_{j-1}, t_j]$ for which

$$\Delta s_j = \int_{t_{j-1}}^{t_j} |r'(t)|\, dt = |r'(c_j)| \cdot (t_j - t_{j-1}) \qquad (3)$$

$$= |r'(c_j)|\, \Delta t.$$

To approximate the work ΔW_j done by the force field in moving a particle along the jth arc, we assume that the magnitude F_j of the force acting throughout the jth arc is the tangential component of $\boldsymbol{F}(r)$ at the point with position vector $r(c_j)$. That is, we use

$$F_j = \text{comp}_{T(r(c_j))}\, \boldsymbol{F}(r(c_j)) = \boldsymbol{F}(r(c_j)) \cdot T(r(c_j)) \qquad (4)$$

where $T(r(c_j))$ is the unit tangent to C at $r(c_j)$ (see Figure 2.5).

Multiplying (3) and (4) and summing over all n arcs, we obtain the approximation

$$W \approx \sum_{j=1}^{n} F_j \, \Delta s_j = \sum_{j=1}^{n} \boldsymbol{F}(\boldsymbol{r}(c_j)) \cdot \boldsymbol{T}(\boldsymbol{r}(c_j)) |\boldsymbol{r}'(c_j)| \, \Delta t$$

for the work done by the force field $\boldsymbol{F}$ in moving the particle over the curve C. We therefore define the work W as the limit as $n \to \infty$ of this approximating sum:

$$W = \lim_{n \to \infty} \sum_{j=1}^{n} \boldsymbol{F}(\boldsymbol{r}(c_j)) \cdot \boldsymbol{T}(\boldsymbol{r}(c_j)) |\boldsymbol{r}'(c_j)| \, \Delta t \qquad (5)$$

$$= \int_{a}^{b} \boldsymbol{F}(\boldsymbol{r}(t)) \cdot \boldsymbol{T}(\boldsymbol{r}(t)) |\boldsymbol{r}'(t)| \, dt.$$

Equation (5) may be simplified somewhat by recalling that, if $|\boldsymbol{r}'(t)| \neq 0,$* the unit tangent $\boldsymbol{T}(\boldsymbol{r}(t))$ to C at $\boldsymbol{r}(t)$ is given by

$$\boldsymbol{T}(\boldsymbol{r}(t)) = \frac{\boldsymbol{r}'(t)}{|\boldsymbol{r}'(t)|}. \qquad (6)$$

Combining equations (5) and (6) gives the desired definition of work done by a (continuous) force field in moving a particle along a smooth curve C.

DEFINITION 2

Let $\boldsymbol{F}$ be a continuous force field defined in some open box containing the smooth curve

$$C: \quad \boldsymbol{r}(t) = x(t)\boldsymbol{i} + y(t)\boldsymbol{j} + z(t)\boldsymbol{k}, \qquad a \leq t \leq b, \qquad |\boldsymbol{r}'(t)| \neq 0.$$

The **work done by the force field $\boldsymbol{F}$** in moving a particle along C from $\boldsymbol{r}(a)$ to $\boldsymbol{r}(b)$ is given by the definite integral

$$W = \int_{a}^{b} \boldsymbol{F}(\boldsymbol{r}(t)) \cdot \boldsymbol{r}'(t) \, dt. \qquad (7)$$

REMARK: For the curve C as in Definition 2,

$$\boldsymbol{r}'(t) = \frac{d}{dt}(\boldsymbol{r}(t)) = \frac{dx}{dt}\boldsymbol{i} + \frac{dy}{dt}\boldsymbol{j} + \frac{dz}{dt}\boldsymbol{k},$$

which suggests the notation

$$d\boldsymbol{r} = \boldsymbol{r}'(t) \, dt. \qquad (8)$$

Using (8), we sometimes abbreviate the integral in (7) as

$$W = \int_{C} \boldsymbol{F} \cdot d\boldsymbol{r}. \qquad (9)$$

In using equation (9), it is important to remember that one must first find a smooth

---

*The requirement that $|\boldsymbol{r}'(t)| \neq 0$ is not as restrictive as it might seem, since in most applications we are free to choose the particular parameterization used to describe the curve C.

parameterization $r(t)$ for C and then proceed as in equation (7). It is a fact, which we shall not prove, that the value of the integral in (9) does not depend on the particular parameterization chosen for C.

Example 1

Find the work done by the force field

$$F(x, y, z) = 2x\mathbf{i} + 3y\mathbf{j} + z\mathbf{k} \qquad \text{(see Figures 2.6 and 2.7)}$$

in moving a particle along the circular helix

$$C: \quad r(t) = \cos t\mathbf{i} + \sin t\mathbf{j} + t\mathbf{k}$$

from point $r(0) = \mathbf{i}$ to point $r(\pi) = -\mathbf{i} + \pi\mathbf{k}$.

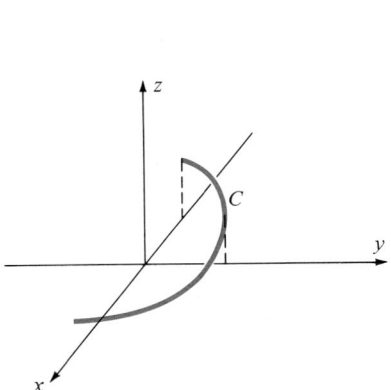

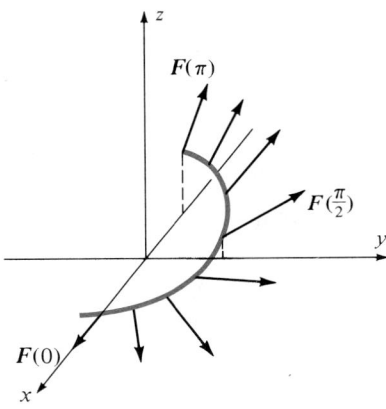

Figure 2.6 The helix C: $r(t) =$ $\cos t\mathbf{i} + \sin t\mathbf{j} + t\mathbf{k}$.

Figure 2.7 Selected vectors in the force field $F(x, y, z) = 2x\mathbf{i} + 3y\mathbf{j} + z\mathbf{k}$ for which (x, y, z) lies on C.

Solution: The parameterization given for C is

$$x(t) = \cos t; \qquad y(t) = \sin t; \qquad z(t) = t,$$

so

$$F(r(t)) = 2 \cos t\mathbf{i} + 3 \sin t\mathbf{j} + t\mathbf{k}.$$

Also,

$$r'(t) = -\sin t\mathbf{i} + \cos t\mathbf{j} + \mathbf{k}.$$

Thus, by Definition 2,

$$W = \int_0^\pi [2 \cos t\mathbf{i} + 3 \sin t\mathbf{j} + t\mathbf{k}] \cdot [-\sin t\mathbf{i} + \cos t\mathbf{j} + \mathbf{k}] \, dt$$

$$= \int_0^\pi (-2 \cos t \sin t + 3 \sin t \cos t + t) \, dt = \int_0^\pi (\sin t \cos t + t) \, dt$$

$$= \frac{1}{2} \sin^2 t + \frac{1}{2} t^2 \Big]_0^\pi = \frac{\pi^2}{2}. \qquad \diamond$$

Example 2

Find the work done by the force field

$$F(x, y) = 3y\mathbf{i} - x^2\mathbf{j}$$

in moving a particle along the plane curve $y = \sqrt{x}$ from the point $(1, 1)$ to the point $(4, 2)$.

Solution: Here we must first find a parameterization for the arc C of the graph of $y = \sqrt{x}$ from $(1, 1)$ to $(4, 2)$. We do so by setting $x(t) = t$, $y(t) = \sqrt{t}$. Then

$$\mathbf{r}(t) = t\mathbf{i} + \sqrt{t}\mathbf{j}, \qquad 1 \le t \le 4,$$

$$\mathbf{r}'(t) = \mathbf{i} + \frac{1}{2\sqrt{t}}\mathbf{j},$$

and

$$F(\mathbf{r}(t)) = 3\sqrt{t}\,\mathbf{i} - t^2\mathbf{j}.$$

Thus, by (7)

$$W = \int_1^4 [3\sqrt{t}\,\mathbf{i} - t^2\mathbf{j}] \cdot \left[\mathbf{i} + \frac{1}{2\sqrt{t}}\mathbf{j}\right] dt$$

$$= \int_1^4 \left(3t^{1/2} - \frac{1}{2}t^{3/2}\right) dt$$

$$= 2t^{3/2} - \frac{1}{5}t^{5/2}\Big]_1^4$$

$$= \frac{39}{5}.$$

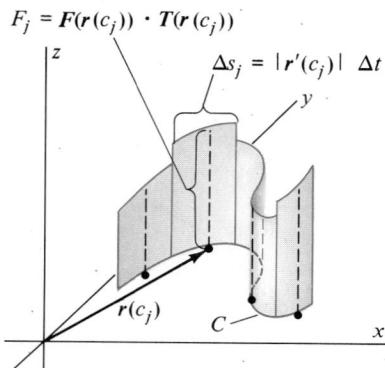

$F_j = F(\mathbf{r}(c_j)) \cdot T(\mathbf{r}(c_j))$

$\Delta s_j = |\mathbf{r}'(c_j)|\,\Delta t$

$\mathbf{r}(c_j)$

C

Figure 2.8 $F(\mathbf{r}(c_j)) \cdot T(\mathbf{r}(c_j))|\mathbf{r}'(c_j)|\,\Delta t$ in (5) is surface area of a curved slab over C.

Example 3

Show that the result of Example 2 remains unchanged if the parameterization $x(t) = t^2$, $y(t) = t$, $1 \le t \le 2$, is used for C.

Solution: Here $\mathbf{r}(t) = t^2\mathbf{i} + t\mathbf{j}$, $\mathbf{r}'(t) = 2t\mathbf{i} + \mathbf{j}$, and $F(\mathbf{r}(t)) = 3t\mathbf{i} - t^4\mathbf{j}$. Thus,

$$W = \int_1^2 [3t\mathbf{i} - t^4\mathbf{j}] \cdot [2t\mathbf{i} + \mathbf{j}]\,dt$$

$$= \int_1^2 (6t^2 - t^4)\,dt = 2t^3 - \frac{1}{5}t^5\Big]_1^2 = \frac{39}{5}.$$

This result illustrates our earlier remark that the value of the integral in (7) depends only on F and on C, not on the particular parameterization chosen for C.

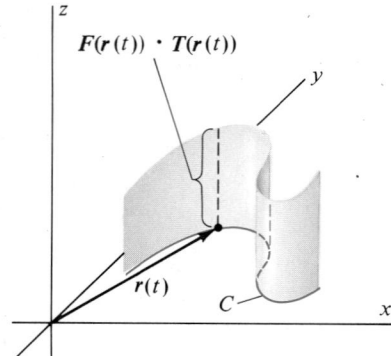

$F(\mathbf{r}(t)) \cdot T(\mathbf{r}(t))$

$\mathbf{r}(t)$

C

Figure 2.9 $\int_C F \cdot d\mathbf{r}$ is a surface area.

Figure 2.8 gives a geometric interpretation of the approximating sum for the integral in equation (5). As Figure 2.9 illustrates, the integral in equation (7) may be interpreted as surface areas of curved vertical "slabs" of variable height $F(\mathbf{r}(t)) \cdot T(\mathbf{r}(t))$.

Line Integrals

The existence of the integral in Definition 2 depends only on the mathematical properties of the force field F and the curve C. We may therefore generalize Definition 2 to apply to any vector field F, not necessarily a force field. The result is called a *line integral*.

DEFINITION 3

Let F be a continuous vector field in some open box D and let C be a smooth curve lying within D with parameterization $C = \{r(t) \mid a \le t \le b\}$. The **line integral** of F over C is defined as

$$\int_C F \cdot dr = \int_a^b F(r(t)) \cdot r'(t) \, dt. \tag{10}$$

Definitions 2 and 3 are the same except that Definition 2 refers to the more specific situation of calculating work when F is a force field.

The line integral in (10) may be written in various other ways. When the vector field F has the form

$$F(x, y, z) = M(x, y, z)i + N(x, y, z)j + P(x, y, z)k,$$

we use the notation

$$dr = dx\, i + dy\, j + dz\, k$$

to write $F \cdot dr$ as

$$F \cdot dr = M(x, y, z)\, dx + N(x, y, z)\, dy + P(x, y, z)\, dz.$$

If the parameterization for C has the form

$$r(t) = x(t)i + y(t)j + z(t)k,$$

then

$$r'(t) = x'(t)i + y'(t)j + z'(t)k.$$

With this notation, the line integral in (10) can be written

$$\int_C F \cdot dr = \int_a^b \{M(x(t), y(t), z(t))x'(t)$$
$$+ N(x(t), y(t), z(t))y'(t)$$
$$+ P(x(t), y(t), z(t))z'(t)\} \, dt. \tag{11}$$

Equation (11) may be abbreviated simply as

$$\int_C F \cdot dr = \int_C M(x, y, z)\, dx + N(x, y, z)\, dy + P(x, y, z)\, dz. \tag{12}$$

When the vector field F and the curve C are defined in the plane by

$$F(x, y) = M(x, y)i + N(x, y)j$$

and

$$r(t) = x(t)i + y(t)j, \qquad a \le t \le b,$$

the line integral in Definition 3 takes the form

$$\int_C F \cdot dr = \int_a^b [M(x(t), y(t))x'(t) + N(x(t), y(t))y'(t)] \, dt \tag{13}$$

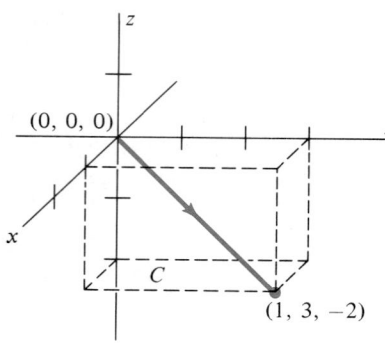

Figure 2.10 Path C from $(0, 0, 0)$ to $(1, 3, -2)$.

or

$$\int_C \mathbf{F} \cdot d\mathbf{r} = \int_C M(x, y)\, dx + N(x, y)\, dy. \tag{14}$$

Example 4

Evaluate the line integral $\displaystyle\int_C \mathbf{F} \cdot d\mathbf{r}$ where $\mathbf{F}$ is the vector field

$$\mathbf{F}(x, y, z) = x^2\mathbf{i} + xy\mathbf{j} + xz\mathbf{k}$$

and C is the line segment from $(0, 0, 0)$ to $(1, 3, -2)$ (see Figure 2.10).

Solution: Here $\mathbf{F} = M\mathbf{i} + N\mathbf{j} + P\mathbf{k}$ with

$$M(x, y, z) = x^2, \qquad N(x, y, z) = xy, \qquad P(x, y, z) = xz. \tag{15}$$

The line segment C from $(0, 0, 0)$ to $(1, 3, -2)$ may be parameterized with

$$x(t) = t, \qquad y(t) = 3t, \qquad z(t) = -2t \tag{16}$$
$$\text{for} \qquad 0 \le t \le 1 \qquad \text{(Figure 2.10)}.$$

Thus,

$$x'(t) = 1, \qquad y'(t) = 3, \qquad z'(t) = -2. \tag{17}$$

Combining equations (15) and (16) shows that

$$M(x(t), y(t), z(t)) = t^2; \qquad N(x(t), y(t), z(t)) = 3t^2; \tag{18}$$
$$P(x(t), y(t), z(t)) = -2t^2.$$

Using equations (11), (17), and (18), we find

$$\int_C \mathbf{F} \cdot d\mathbf{r} = \int_0^1 [(t^2)(1) + (3t^2)(3) + (-2t^2)(-2)]\, dt$$

$$= \int_0^1 14t^2\, dt = \frac{14}{3} t^3 \Big]_0^1 = \frac{14}{3}. \qquad \diamond$$

Example 5

Evaluate the line integral

$$\int_C xy\, dx - 2y^2\, dy$$

where C is the arc of the unit circle from $(1, 0)$ to $(0, 1)$ traversed counterclockwise (Figure 2.11).

Solution: The line integral has the form $\mathbf{F} = M\mathbf{i} + N\mathbf{j}$ where

$$M(x, y) = xy, \qquad N(x, y) = -2y^2.$$

A parameterization for the curve C is given by the equations

$$x(t) = \cos t, \qquad y(t) = \sin t, \qquad 0 \le t \le \pi/2 \qquad \text{(Figure 2.11)}.$$

Thus

$$x'(t) = -\sin t, \qquad y'(t) = \cos t.$$

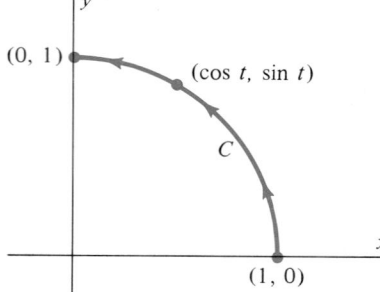

Figure 2.11 Path C from $(1, 0)$ to $(0, 1)$.

Equating the right sides of equations (13) and (14) and using the above functions we obtain

$$\int_C xy \, dx - 2y^2 \, dy = \int_0^{\pi/2} [(\cos t)(\sin t)(-\sin t) - 2(\sin^2 t)(\cos t)] \, dt.$$

$$= \int_0^{\pi/2} -3 \sin^2 t \cos t \, dt$$

$$= -\sin^3 t\big]_0^{\pi/2}$$

$$= -1. \qquad \diamondsuit$$

REMARK: It is important to note that the value of the line integral in Definition 3 depends on the *direction* in which the curve C is traced out by the parameterization $r(t)$. *Reversing the direction along C changes the sign of the line integral.* The reason for this can be most easily seen in line (5): Reversing the direction in which C is traversed changes the sign of the unit tangent vector $T(r(c_j))$.

To illustrate this remark, we evaluate the line integral in Example 5 over the same curve C, but traversed in the opposite direction, from $(0, 1)$ to $(1, 0)$. (We will refer to the curve C with its new orientation as $-C$). A parameterization for $-C$ is

$$x(t) = \cos\left(\frac{\pi}{2} - t\right), \qquad y(t) = \sin\left(\frac{\pi}{2} - t\right), \qquad 0 \le t \le \frac{\pi}{2},$$

so

$$x'(t) = \sin\left(\frac{\pi}{2} - t\right), \qquad y'(t) = -\cos\left(\frac{\pi}{2} - t\right),$$

and

$$\int_{-C} xy \, dx - 2y^2 \, dy = \int_0^{\pi/2} \left[\cos\left(\frac{\pi}{2} - t\right) \sin^2\left(\frac{\pi}{2} - t\right) \right.$$

$$\left. + 2 \sin^2\left(\frac{\pi}{2} - t\right) \cos\left(\frac{\pi}{2} - t\right) \right] dt$$

$$= \int_0^{\pi/2} 3 \sin^2\left(\frac{\pi}{2} - t\right) \cos\left(\frac{\pi}{2} - t\right) dt$$

$$= -\sin^3\left(\frac{\pi}{2} - t\right)\Big]_0^{\pi/2}$$

$$= 1.$$

This is the negative of the result obtained in Example 5. We summarize this remark by writing

$$\boxed{\int_{-C} F \cdot dr = -\int_C F \cdot dr.}$$

When the curve C is a line segment parallel to the x-axis, the line integral in (11) and (12) reduces to

$$\int_C F \cdot dr = \int_C M(x, y, z) \, dx = \int_a^b M(x(t), y(t), z(t))x'(t) \, dt.$$

Such integrals are called **line integrals with respect to x.** Line integrals with respect to y and with respect to z are defined similarly. Using these definitions we may write the line integral in (12) as

$$\int_C \mathbf{F} \cdot d\mathbf{r} = \int_C M(x, y, z)\, dx + \int_C N(x, y, z)\, dy + \int_C P(x, y, z)\, dz.$$

Piecewise Smooth Curves

Up to this point we have defined the line integral only for smooth curves C and continuous vector fields $\mathbf{F}$. (Recall that the curve $C = \{r(t) \mid a \le t \le b\}$ is *smooth* if $r'(t)$ exists and is continuous for all t in the interval $[a, b]$.) The reason why C has been taken to be smooth is that the integrand in the definite integral in line (10) involves the derivative, $r'(t)$. Thus, when $\mathbf{F}$ and r' are continuous, the theory developed in Chapter 6 guarantees that the definite integral in equation (10) exists.

However, this integral exists under slightly less restrictive conditions. In particular, it was stated in Chapter 6 that the definite integral exists when the integrand is merely piecewise continuous. Thus, we need only require that r' be piecewise continuous for the line integral in Definition 3 to exist. Curves of this type are called *piecewise smooth*. We shall refer to such curves as *paths* (see Figures 2.12 and 2.13).

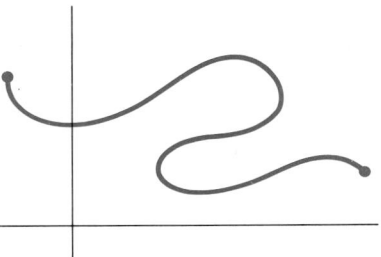

Figure 2.12 A smooth curve (r' continuous).

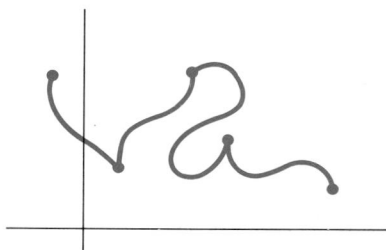

Figure 2.13 A piecewise smooth curve, or *path* (r' piecewise continuous).

DEFINITION 4

A curve $C = \{r(t) \mid a \le t \le b\}$ is **piecewise smooth** if there exist numbers $a = t_0 < t_1 < t_2 < \cdots < t_n = b$ so that r is continuous on $[a, b]$ and so that $r'(t)$ exists and is continuous on each of the subintervals $(t_0, t_1), (t_1, t_2), \ldots, (t_{n-1}, t_n)$. A piecewise smooth curve is called a **path.**

REMARK: For piecewise smooth arcs, we evaluate the line integral by first integrating over each of the subarcs on which r' is continuous and then summing the results:

$$\int_C \mathbf{F} \cdot d\mathbf{r} = \int_a^b \mathbf{F}(r(t)) \cdot r'(t)\, dt \tag{19}$$

$$= \int_{t_0}^{t_1} \mathbf{F}(r(t)) \cdot r'(t)\, dt + \cdots + \int_{t_{n-1}}^{t_n} \mathbf{F}(r(t)) \cdot r'(t)\, dt$$

$$= \int_{C_1} \mathbf{F} \cdot d\mathbf{r} + \int_{C_2} \mathbf{F} \cdot d\mathbf{r} + \cdots + \int_{C_n} \mathbf{F} \cdot d\mathbf{r}$$

where C_j denotes the arc $C_j = \{r(t) \mid t_{j-1} \le t \le t_j\}$.

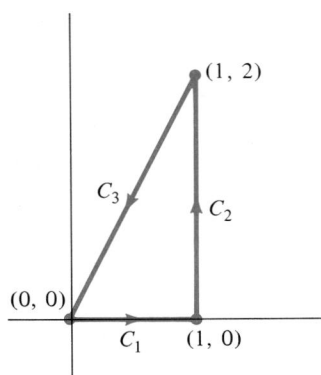

Figure 2.14

Example 6

Evaluate the line integral

$$\int_C (x + y)\, dx + xy\, dy$$

over the path C consisting of the line segments

C_1: from $(0, 0)$ to $(1, 0)$,

C_2: from $(1, 0)$ to $(1, 2)$,

C_3: from $(1, 2)$ to $(0, 0)$,

(See Figure 2.14.)

Solution: Here the vector field is $F(x, y) = M(x, y)\boldsymbol{i} + N(x, y)\boldsymbol{j}$ with

$$M(x, y) = x + y, \qquad N(x, y) = xy.$$

A parameterization for C_1 is

$$C_1: \quad \left. \begin{array}{l} x(t) = t \\ y(t) = 0 \end{array} \right\} 0 \le t \le 1.$$

On this arc, $x'(t) = 1$ and $y'(t) = 0$. The line integral over C_1 is therefore

$$\int_{C_1} (x + y)\, dx + xy\, dy = \int_0^1 [(t + 0)(1) + (t \cdot 0)(0)]\, dt$$

$$= \int_0^1 t\, dt = \frac{1}{2}.$$

A parameterization for C_2 is

$$C_2: \quad \left. \begin{array}{l} x(t) = 1 \\ y(t) = t \end{array} \right\} 0 \le t \le 2.$$

On this arc, $x'(t) = 0$ and $y'(t) = 1$. Thus,

$$\int_{C_2} (x + y)\, dx + xy\, dy = \int_0^2 [(1 + t)(0) + (1 \cdot t)(1)]\, dt$$

$$= \int_0^2 t\, dt = 2.$$

Finally, a parameterization for C_3 is

$$C_3: \quad \left. \begin{array}{l} x(t) = 1 - t \\ y(t) = 2 - 2t \end{array} \right\} 0 \le t \le 1.$$

On this arc, $x'(t) = -1$ and $y'(t) = -2$. Thus, by (13),

$$\int_{C_3} (x + y)\, dx + xy\, dy = \int_0^1 \{[(1 - t) + (2 - 2t)](-1)$$

$$+ (1 - t)(2 - 2t)(-2)\}\, dt$$

$$= \int_0^1 (-4t^2 + 11t - 7)\, dt = -\frac{17}{6}.$$

Applying equation (19), we conclude

$$\int_C (x + y)\, dx + xy\, dy = \int_{C_1} (x + y)\, dx + xy\, dy$$

$$+ \int_{C_2} (x + y)\, dx + xy\, dy$$

$$+ \int_{C_3} (x + y)\, dx + xy\, dy$$

$$= \frac{1}{2} + 2 - \frac{17}{6}$$

$$= -\frac{1}{3}. \qquad \diamond$$

Line Integrals with Respect to Arc Length

Thus far we have defined line integrals only for vector fields. Another type of line integral, called a **line integral with respect to arc length,** concerns a scalar valued function, say f on $\mathbb{R}^2$. Let C be a path with parameterization

$$C = \{r(t) = x(t)i + y(t)j \mid a \le t \le b\}.$$

The usual partitioning of the interval $[a, b]$ produces arcs $C_1, C_2, \ldots, C_n$ whose lengths Δs_j, as before, are

$$\Delta s_j = \sqrt{[x'(c_j)]^2 + [y'(c_j)]^2}\, \Delta t, \qquad j = 1, 2, \ldots, n,$$

where c_j is a number in the jth subinterval of $[a, b]$. If the function f is continuous on an open rectangle containing the path C, the Riemann sum

$$\sum_{j=1}^{n} f(x(c_j), y(c_j))\, \Delta s_j \qquad (20)$$

$$= \sum_{j=1}^{n} f(x(c_j), y(c_j))\sqrt{[x'(c_j)]^2 + [y'(c_j)]^2}\, \Delta t$$

converges to the definite integral

$$\int_C f(x, y)\, ds = \int_a^b f(x(t), y(t))\sqrt{[x'(t)]^2 + [y'(t)]^2}\, dt. \qquad (21)$$

The integral in equation (21) is called the *line integral of f over C with respect to arc length.* The difference between the integral in (21) and the line integral in Definition 3 is that (21) is an integral of a scalar function f with respect to change in arc length, while the integral in (10) is an integral of the dot product of a vector field and the unit tangent with respect to change in the parameter t. By using the differential for arc length

$$ds = \sqrt{[x'(t)]^2 + [y'(t)]^2}\, dt = |r'(t)|\, dt,$$

we may interpret the line integral in Definition 3 as a special case of the integral in (21) with $f(x(t), y(t)) = F(r(t)) \cdot T(r(t))$. That is,

$$\int_C F \cdot dr = \int_C F \cdot T\, ds. \qquad (22)$$

This can be seen by comparing the approximating sums in lines (5) and (20).

Although we shall work almost exclusively in what follows with line integrals as defined in Definition 3, equation (22) will be used in Section 20.4 to interpret certain statements about line integrals.

Example 7

Find $\int_C xy^3 \, ds$ where C is the quarter circle

$$C = \left\{ \cos t\mathbf{i} + \sin t\mathbf{j} \mid 0 \le t \le \frac{\pi}{2} \right\}.$$

Solution: Here $x(t) = \cos t$, $y(t) = \sin t$, $x'(t) = -\sin t$, and $y'(t) = \cos t$. Thus, by (21)

$$
\begin{aligned}
\int_C xy^3 \, ds &= \int_0^{\pi/2} (\cos t)(\sin^3 t)\sqrt{[-\sin t]^2 + [\cos t]^2} \, dt \\
&= \int_0^{\pi/2} \sin^3 t \cos t \, dt \\
&= \frac{1}{4} \sin^4 t \Big]_0^{\pi/2} \\
&= \frac{1}{4}.
\end{aligned}
$$

$\diamond$

It is important to note that, unlike line integrals of the form $\int_C \mathbf{F} \cdot d\mathbf{r}$, line integrals with respect to arc length are independent of the direction along which the curve C is traversed (see Exercise 28).

An application of line integrals with respect to arc length concerns finding the mass M of a thin wire of variable density whose shape is given by the curve $C = \{\mathbf{r}(t) = x(t)\mathbf{i} + y(t)\mathbf{j} \mid a \le t \le b\}$. If the mass density (in units of mass per unit length) is given by the continuous function f, the expression

$$\Delta m_j = f(x(c_j), y(c_j)) \, \Delta s_j$$

approximates the mass of a section of the wire of length Δs_j, one point of which has coordinates $(x(c_j), y(c_j))$. The sum in equation (20) therefore approximates the total mass of the wire, which is given precisely by the line integral with respect to arc length in equation (21).

For curves in space, line integrals with respect to arc length are defined in an entirely analogous manner. That is, if f is continuous in an open box containing the piecewise smooth curve

$$C = \{\mathbf{r}(t) = x(t)\mathbf{i} + y(t)\mathbf{j} + z(t)\mathbf{k} \mid a \le t \le b\},$$

then

$$
\begin{aligned}
&\int_C f(x, y, z) \, ds \\
&= \int_a^b f(x(t), y(t), z(t))\sqrt{[x'(t)]^2 + [y'(t)]^2 + [z'(t)]^2} \, dt.
\end{aligned}
$$

Exercise Set 20.2

In Exercises 1–8, calculate the work done by the force field F in moving a particle along the specified path C.

1. $F(x, y) = xi + yj$, C: $r(t) = t^2 i + (3 + t)j$, $0 \le t \le 2$

2. $F(x, y) = xyi + x^2 j$,
   C: $r(t) = \cos ti + \sin tj$, $0 \le t \le \pi$

3. $F(x, y) = x^2 i + y^2 j$,
   C: $r(t) = \sqrt{t}\, i + 3tj$, $1 \le t \le 4$

4. $F(x, y) = -yi + xj$,
   C: $r(t) = (t + 1)^2 i + (t - 1)^2 j$, $0 \le t \le 2$

5. $F(x, y) = (x^2 + y^2)i + xyj$,
   C: $r(t) + ti + t^2 j$, $0 \le t \le 2$

6. $F(x, y) = x^2 yi + xy^2 j$,
   C: $r(t) = ti + 2t^2 j$, $0 \le t \le 1$

7. $F(x, y, z) = xyi + xzj + yzk$,
   C: $r(t) = \cos ti + \sin tj + tk$, $0 \le t \le \pi$

8. $F(x, y, z) = x^2 i + y^2 j + z^2 k$,
   C: $r(t) = ti + 3t^2 j - 2tk$; $0 \le t \le 2$

In Exercises 9–15, evaluate the line integral of the vector field F over the indicated path C.

9. $F(x, y) = yi + 2xj$, C is the line segment from $(0, 0)$ to $(4, 2)$.

10. $F(x, y) = -x^2 i + y^2 j$, C is the upper unit semicircle from $(1, 0)$ to $(-1, 0)$.

11. $F(x, y) = (y - x)i + xyj$, C is the unit circle traversed counterclockwise.

12. $F(x, y) = xy^2 i + (x + y)j$, C is the triangular path from $(0, 0)$ to $(4, 0)$ to $(4, 2)$ to $(0, 0)$.

13. $F(x, y) = (x + y)i + (y^2 - x^2)j$, C is the triangle with vertices $(-1, 0)$, $(1, -4)$, $(0, 2)$ traversed counterclockwise.

14. $F(x, y, z) = xyi + yj + xzk$, C is the line segment from $(0, 0, 0)$ to $(2, 4, -6)$.

15. $F(x, y, z) = yi - xj + 2zk$, C is the arc of the circular helix $r(t) = \cos ti + \sin tj + tk$ from $(1, 0, 0)$ to $(0, 1, \pi/2)$.

16. Evaluate $\int_C xy\, dx$ where C is the arc of the unit circle from $(0, 1)$ to $(-1, 0)$ traversed counterclockwise.

17. Evaluate $\int_C (x^2 + y^2)\, dy$ where C is the path in Exercise 16.

18. Evaluate $\int_C F \cdot dr$ where $F(x, y) = (x^2 + y^2)i + 2xyj$ and C is the arc of the circle $x^2 + y^2 = 1$ from $(1, 0)$ to $(0, 1)$, followed by the line segment from $(0, 1)$ to $(-1, 1)$ (Figure 2.15).

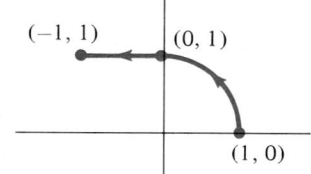

Figure 2.15

19. Evaluate $\int_C (x^2 + y^2 + z^2)\, dy$ along the line segment from $(0, 1, 0)$ to $(-1, 2, 1)$.

20. Find the work done on a particle by the force field $F(x, y) = (x - y)i + xyj$ in moving a particle counterclockwise around the ellipse $9x^2 + 4y^2 = 36$ from $(2, 0)$ to $(-2, 0)$.

21. Evaluate $\int_C xy\, dx + xy^2\, dy$ where C is the curve in Exercise 18.

22. Find the work done by the force field $F(x, y) = (xi - yj)$ in moving a particle along the path in Figure 2.16.

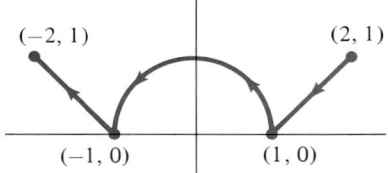

Figure 2.16

23. Evaluate $\int_C (x + y)\, dx + (x^2 - y^2)\, dy$ where C is the arc of the parabola $y = 2x^2$ from $(0, 0)$ to $(2, 8)$.

24. Evaluate $\int_C (x - y)\, dx + xy\, dy$ where C is the path consisting of the line segments from $(0, 1)$ to $(2, -1)$, from $(2, -1)$ to $(2, 5)$, and from $(2, 5)$ to $(4, 1)$.

25. Evaluate $\int_C x^2 y\, dx + xy^2\, dy$ where C is the path from $(0, 2)$ to $(4, 1)$ along the three line segments parallel to the x- and y-axes.

26. Let F be the central force field $F(r) = \dfrac{2}{|r|^3}r$, $r \ne 0$. Find the work done by F in moving a particle along the upper half of the unit circle from $(1, 0)$ to $(-1, 0)$.

27. Show that Definition 2 agrees with the definition of work when a force $f(x)$ is applied to a particle moving along the x-axis.

28. Show that the line integral with respect to arc length in line (21) is independent of the direction along which C is traversed.

Evaluate each of the following line integrals with respect to arc length.

29. $\displaystyle\int_C (x + 2y^2)\, ds$ where $C = \{r(t) = 2t i - 4t j \mid 0 \le t \le 1\}$

30. $\displaystyle\int_C x e^y\, ds$ where $C = \{r(t) = \cos t i + \sin t j \mid 0 \le t \le \pi\}$

31. $\displaystyle\int_C (8x + y - 3)\, ds$
where $C = \{r(t) = t i + 3j + t^2 k \mid 0 \le t \le 1\}$

32. Give an argument to support the conclusion that if the

curve C describes the shape of a thin wire of mass density $\rho(x, y, z)$ then the total mass of the wire is

$$M = \int_C \rho\, ds$$

$$= \int_a^b \rho(x(t), y(t), z(t))\sqrt{[x'(t)]^2 + [y'(t)]^2 + [z'(t)]^2}\, dt.$$

33. Refer to Exercise 32. Find the total mass of a wire with shape

$$C: \quad r(t) = \cos t i - \sin t j + 2k, \qquad 0 \le t \le \pi$$

if the density is $\rho(x, y, z) = x^2 + y^2 + z^2$.

34. Refer to Exercise 32. Find the mass of a wire of constant density $\rho(x, y, z) = k$ shaped like the helix $r(t) = 4\cos t i + 4\sin t j + 3t k$ for $0 \le t \le 2\pi$.

20.3 LINE INTEGRALS: INDEPENDENCE OF PATH

The Fundamental Theorem of Calculus allows us to evaluate definite integrals of functions of a single variable by means of an antiderivative. In this section, we develop a similar theorem for line integrals. To do so requires the concept of a *scalar potential* ϕ for F. Recall from Section 18.9 that the differentiable scalar function ϕ is a scalar potential for the vector function F if $\nabla\phi = F$.

Our Fundamental Theorem for line integrals is the following.

THEOREM 1

Let $C\{r(t) \mid a \le t \le b\}$ be a piecewise smooth curve in an open rectangle D. Let F be a vector field on D and let ϕ be a differentiable scalar function for which $F(r) = \nabla\phi(r)$ for all r in D. Then

$$\int_C F \cdot dr = \phi(r(b)) - \phi(r(a)).$$

Theorem 1 says this. If the vector field F is conservative, meaning that it is the gradient of some scalar function ϕ, then the line integral of F over C from $A = r(a)$ to $B = r(b)$ depends only on the values of the scalar potential ϕ at A and at B. Thus, if the vector field F is *conservative*, we can determine the value of the line integral by finding a scalar potential ϕ for F. Before proving Theorem 1, we present two examples.

Example 1

Find $\displaystyle\int_C F \cdot dr$ where F is the vector field $F(x, y) = 2xy i + x^2 j$ and C is the line segment from $(-1, 2)$ to $(4, -3)$.

Solution: Here $\phi(x, y) = x^2 y$ is a potential for F since

$$\nabla\phi(x, y) = 2xy i + x^2 j = F(x, y).$$

Thus, by Theorem 1,

$$\int_C \boldsymbol{F}(\boldsymbol{r}) \cdot d\boldsymbol{r} = \phi(4, -3) - \phi(-1, 2) = 4^2(-3) - (-1)^2 2 = -50. \qquad \diamond$$

Example 2

Evaluate the line integral

$$\int_C 2xye^{x^2} \, dx + e^{x^2} \, dy$$

over a path C from $(0, 1)$ to $(1, 3)$.

Solution: This integral has the form $\displaystyle\int_C \boldsymbol{F} \cdot d\boldsymbol{r}$ where

$$\boldsymbol{F}(x, y) = 2xye^{x^2}\boldsymbol{i} + e^{x^2}\boldsymbol{j}$$

and $d\boldsymbol{r} = dx\, \boldsymbol{i} + dy\, \boldsymbol{j}$. A potential for $\boldsymbol{F}$ is

$$\phi(x, y) = ye^{x^2}$$

since

$$\nabla\phi(x, y) = 2xye^{x^2}\boldsymbol{i} + e^{x^2}\boldsymbol{j} = \boldsymbol{F}(x, y).$$

Thus, by Theorem 1

$$\int_C 2xye^{x^2} \, dx + e^{x^2} \, dy = \phi(1, 3) - \phi(0, 1)$$

$$= 3e - e^0$$
$$= 3e - 1. \qquad \diamond$$

Proof of Theorem 1: We will need to make use of the Chain Rule for vector functions: If x and y are differentiable component functions, then

$$\frac{d}{dt}\phi(x(t), y(t)) = \phi_x(x(t), y(t))x'(t) + \phi_y(x(t), y(t))y'(t)$$

$$= \nabla\phi(x(t), y(t)) \cdot [x'(t)\boldsymbol{i} + y'(t)\boldsymbol{j}].$$

Using the position vector notation, this can be written

$$\frac{d}{dt}\phi(\boldsymbol{r}(t)) = \nabla\phi(\boldsymbol{r}(t)) \cdot \boldsymbol{r}'(t). \qquad (1)$$

Assume first that C is a smooth curve. Then, under the hypotheses of Theorem 1, equation (1) holds. Thus,

$$\int_C \boldsymbol{F} \cdot d\boldsymbol{r} = \int_a^b \boldsymbol{F}(\boldsymbol{r}(t)) \cdot \boldsymbol{r}'(t) \, dt$$

$$= \int_a^b \nabla\phi(\boldsymbol{r}(t)) \cdot \boldsymbol{r}'(t) \, dt$$

$$= \int_a^b \frac{d}{dt}[\phi(\boldsymbol{r}(t))] \, dt \qquad \text{(equation (1))}$$

$$= \phi(\boldsymbol{r}(t))]_a^b$$

$$= \phi(\boldsymbol{r}(b)) - \phi(\boldsymbol{r}(a)).$$

The extension to the case C piecewise smooth is easy and is left as an exercise.

◆

In trying to determine whether the vector field F is conservative, it is helpful to recall Theorem 9, Section 18.9: If the functions M and N are continuously differentiable in an open rectangle D, *the vector function*

$$F(x, y) = M(x, y)i + N(x, y)j$$

is conservative if and only if

$$M_y(x, y) = N_x(x, y) \tag{2}$$

for all (x, y) in D. Equation (2) provides a quick check of whether Theorem 1 applies for a vector field F in the plane. When equation (2) holds, the method of Section 18.9 can sometimes be applied to find the potential ϕ.

Example 3

Find the work done by the vector field

$$F(x, y) = (e^y + 3x^2y^2)i + (xe^y + 2x^3y + 2)j$$

in moving a particle from the point $(0, 0)$ to the point $(3, 2)$ in the plane.

Solution: Here $F(x, y) = M(x, y)i + N(x, y)j$ where

$$M(x, y) = e^y + 3x^2y^2, \qquad N(x, y) = xe^y + 2x^3y + 2.$$

Then

$$M_y(x, y) = e^y + 6x^2y = N_x(x, y),$$

so the vector field F is conservative. To find a potential ϕ with

$$\begin{aligned} \nabla\phi(x, y) &= \phi_x(x, y)i + \phi_y(x, y)j \\ &= M(x, y)i + N(x, y)j = F(x, y), \end{aligned} \tag{3}$$

we first equate i-components and integrate partially with respect to x to find that

$$\phi(x, y) = \int (e^y + 3x^2y^2)\, dx = xe^y + x^3y^2 + f(y) + C \tag{4}$$

where f is a function of y alone. Equating j-components in (3) and integrating partially with respect to y shows that

$$\phi(x, y) = \int (xe^y + 2x^3y + 2)\, dy = xe^y + x^3y^2 + 2y + g(x) + C \tag{5}$$

where g is a function of x alone.

Equating the expressions for ϕ in lines (4) and (5) shows that a scalar potential for F is

$$\phi(x, y) = xe^y + x^3y^2 + 2y.$$

We may therefore apply Theorem 1. The desired work is

$$\begin{aligned} W = \int_C F \cdot dr &= \phi(3, 2) - \phi(0, 0) \\ &= (3e^2 + 3^3 \cdot 2^2 + 2 \cdot 2) - 0 \\ &= 3e^2 + 112. \end{aligned}$$

◇

Whether or not an explicit potential ϕ can actually be found, Theorem 1 guarantees that the value of a line integral for a conservative vector field depends only on the endpoints of the path C and not on the path itself. (Physicists paraphrase this statement by saying that "work is a function of position, not of path.") The terminology we choose to use is that the line integral of a conservative force field is **independent of path.** The following theorem characterizes this notion.

THEOREM 2

Let C be a piecewise smooth curve in an open rectangle D. The line integral

$$\int_C \mathbf{F} \cdot d\mathbf{r}$$

is independent of path if and only if the vector field $\mathbf{F}$ is conservative in D.

The "if" part of this theorem follows directly from Theorem 1. The "only if" part is an important observation in more advanced courses in mathematics but will not be used here in a direct way. The proof is omitted.

The following corollary is a useful formulation of Theorem 2 and the statement concerning equation (2) for vector fields in the plane.

COROLLARY 1

Let D be an open rectangle in the plane and let C be any path from point A to point B lying within D. Let M and N be continuously differentiable in D. The line integral

$$\int_C M(x, y)\, dx + N(x, y)\, dy \tag{6}$$

is independent of path if and only if

$$M_y(x, y) = N_x(x, y) \tag{7}$$

for all (x, y) in D.

Example 4

Evaluate the line integral

$$\int_C (\cos xy - xy \sin xy)\, dx - x^2 \sin xy\, dy$$

where C is the upper unit semicircle from $(1, 0)$ to $(-1, 0)$.

Solution: This integral has the form (6) with

$$M(x, y) = \cos xy - xy \sin xy, \qquad N(x, y) = -x^2 \sin xy,$$

so

$$M_y(x, y) = -2x \sin xy - x^2 y \cos xy = N_x(x, y).$$

Thus, equation (7) holds, so the integral is independent of path. However, instead of applying the method of Example 3 to actually find the potential ϕ, we demonstrate how to apply Corollary 1 in evaluating this line integral.

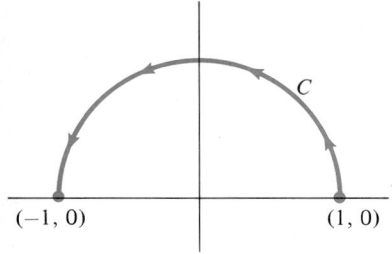

Figure 3.1 Path C from $(1, 0)$ to $(-1, 0)$.

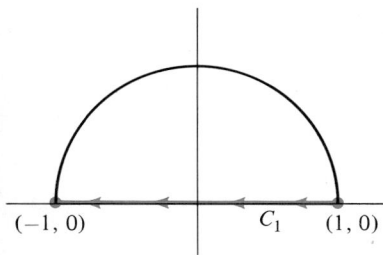

Figure 3.2 Path C_1 from $(1, 0)$ to $(-1, 0)$.

Since the integral is independent of path, we may choose a path that makes the line integral easier to compute than does C. In this case, we choose the path C_1 consisting of the (straight) line segment from $(1, 0)$ to $(-1, 0)$ (see Figures 3.1 and 3.2). A parameterization for C_1 is

$$C_1: \left.\begin{array}{l} x(t) = 1 - t \\ y(t) = 0 \end{array}\right\} \quad 0 \le t \le 2.$$

Thus, $x'(t) = -1$ and $y'(t) = 0$. Calculating the line integral according to Definition 2 gives

$$\int_C (\cos xy - xy \sin xy)\, dx - x^2 \sin xy\, dy$$

$$= \int_0^2 \{[\cos(0) - 0 \cdot \sin(0)](-1) - (1 - t)^2 \cdot \sin(0)(0)\}\, dt$$

$$= \int_0^2 (-1)\, dt$$

$$= -2.$$

Now we note that a scalar potential for this vector field is $\phi(x, y) = x \cos xy$. With this information the line integral could have been evaluated using Theorem 1 as

$$\phi(-1, 0) - \phi(1, 0) = (-1) \cos(0) - (1) \cos(0) = -2. \qquad \diamondsuit$$

REMARK: Since a line integral over a path from point A to point B for a conservative force field F depends only on the points A and B, the notation

$$\int_C F \cdot dr = \int_A^B F \cdot dr$$

is sometimes used. If $F = \nabla\phi$, Theorem 1 is stated in this notation as

$$\int_A^B F \cdot dr = \phi(B) - \phi(A).$$

In using this notation, one must exercise caution to ensure that the hypotheses of Theorem 1 are true. Otherwise, this notation is meaningless.

Example 5

In the above notation, the result of Example 4 can be stated as

$$\int_{(1, 0)}^{(-1, 0)} (\cos xy - xy \sin xy)\, dx - x^2 \sin xy\, dy = -2. \qquad \diamondsuit$$

Example 6

Although all of the preceding examples have dealt with paths in the plane, Theorems 1 and 2 apply also to paths and vector fields in space. The line integral

$$\int_C yz\, dx + xz\, dy + xy\, dz$$

over the helical path

$$C: \quad r(t) = \cos t\, i + \sin t\, j + t k, \qquad 0 \le t \le \pi/4$$

may be handled by means of Theorem 1 by noting that the vector field

$$F(x, y, z) = yz i + xz j + xy k$$

is the gradient of the scalar potential

$$\phi(x, y, z) = xyz.$$

Since the endpoints of C are $A = r(0) = (1, 0, 0)$ and $B = r(\pi/4) = (\sqrt{2}/2, \sqrt{2}/2, \pi/4)$, we have

$$\int_C yz\, dx + xz\, dy + xy\, dz = \int_{(1, 0, 0)}^{(\sqrt{2}/2,\, \sqrt{2}/2,\, \pi/4)} yz\, dx + xz\, dy + xy\, dz$$

$$= \phi\left(\frac{\sqrt{2}}{2}, \frac{\sqrt{2}}{2}, \frac{\pi}{4}\right) - \phi(1, 0, 0)$$

$$= \frac{\pi}{8}. \qquad \diamond$$

Line Integrals Over Closed Paths

The path (piecewise smooth curve) $C = \{r(t) \mid a \le t \le b\}$ is called **closed** if $r(a) = r(b)$. Thus, circles, ellipses, triangles, squares, and rectangles are all examples of closed paths in the plane. There is a very simple situation governing line integrals over closed paths. *If the vector field F is conservative, the line integral $\int_C F \cdot dr$ over a closed path is zero.* This statement follows directly from Theorem 1: Since a closed path $C = \{r(t) \mid a \le t \le b\}$ satisfies the equation $r(a) = r(b)$,

$$\int_C F \cdot dr = \phi(r(b)) - \phi(r(a)) = 0.$$

This fact is well known in physics. In terms of work, it says that the work done by a conservative force field (meaning that no energy is lost in the process) in moving a particle around a closed path is zero.

Conservation of Energy

Let's take the physics one step further. Let F be a conservative force field. If a particle of mass m moves in the vector field F with velocity $v(t) = r'(t)$, we define the **kinetic energy** of the particle at time t as

$$K(t) = \frac{1}{2} m |v(t)|^2. \tag{8}$$

Also, the change in **potential energy** U for the particle as it is moved from point A to point B is defined as the negative of the work done by the force field F in moving it from A to B:

$$U(B) - U(A) = -\int_A^B F \cdot dr. \tag{9}$$

Now if $a(t)$ denotes the acceleration of the particle at time t, Newton's second law of motion states that

$$F(r(t)) = ma(t) = mr''(t),$$

so we can rewrite the integral in Definition 2 as

$$F(r(t)) \cdot r'(t) = mr''(t) \cdot r'(t) \tag{10}$$

$$= \frac{1}{2}m[2r''(t) \cdot r'(t)]$$

$$= \frac{1}{2}m \frac{d}{dt}[r'(t) \cdot r'(t)]$$

$$= \frac{d}{dt}\left[\frac{1}{2}m|v(t)|^2\right].$$

Using (8) and (10) and the fact that F is conservative, we find that

$$\text{Work} = \int_A^B F \cdot dr = \int_a^b F(r(t)) \cdot r'(t) \, dt \tag{11}$$

$$= \int_a^b \frac{d}{dt}\left[\frac{1}{2}m|v(t)|^2\right] dt$$

$$= \frac{1}{2}m|v(b)|^2 - \frac{1}{2}m|v(a)|^2$$

$$= K(B) - K(A)$$

where $A = r(a)$, $B = r(b)$. That is,

$$\boxed{\text{Work} = K(B) - K(A).} \tag{12}$$

Equation (12) may be interpreted as the principle that *in a conservative system, the work done on an object (by a force) equals the change in its kinetic energy.* Finally, combining equations (9) and (11) gives the equation

$$U(A) - U(B) = K(B) - K(A)$$

or

$$\boxed{K(A) + U(A) = K(B) + U(B).} \tag{13}$$

Equation (13) is the famous **principle of conservation of energy:** In a conservative system, the sum of kinetic and potential energy of a moving particle remains constant from point to point.

Exercise Set 20.3

In Exercises 1–10, evaluate the line integral by verifying that the vector field is conservative and applying Theorem 1.

1. $\int_C F \cdot dr$, $F(x, y) = yi + xj$, C is a path from $(0, 0)$ to $(3, 1)$.

2. $\int_C F \cdot dr$, $F(x, y) = 3x^2y^2i + 2x^3yj$, C is a path from $(-3, 1)$ to $(2, 2)$.

3. $\int_C F \cdot dr$, $F(x, y) = ye^{xy}i + xe^{xy}j$, C is a path from $(0, 0)$ to $(1, 2)$.

4. $\int_C F \cdot dr$, $F(x, y, z) = yze^{xyz}i + xze^{xyz}j + xye^{xyz}k$, C is a path from $(0, 0, 0)$ to $(1, 0, 1)$.

5. $\int_C F \cdot dr$, $F(x, y) = 2xi + 2yj$, C is the upper unit semicircle traversed counterclockwise.

6. $\int_{(1, 1)}^{(2, 3)} (1 + 2xy^2) \, dx + 2x^2y \, dy$

7. $\int_{(0, 0)}^{(2, 1)} e^{y^2} \, dx + 2xye^{y^2} \, dy$

8. $\displaystyle\int_{(0,\,0)}^{(\pi/2,\,1)} y \sin xy\, dx + x \sin xy\, dy$

9. $\displaystyle\int_{(0,\,0)}^{(1,\,\pi/4)} e^x \sin y\, dx + e^x \cos y\, dy$

10. $\displaystyle\int_{(0,\,0,\,0)}^{(0,\,\pi/4,\,\pi/4)} e^x \sin y \cos z\, dx$
$\quad + e^x \cos y \cos z\, dy - e^x \sin y \sin z\, dz$

11. Let $F(x, y)$ define a conservative vector field throughout the plane. Let C_1 be the path along the upper semicircle from $(1, 0)$ to $(-1, 0)$. Let C_2 be the path along the x-axis from $(-1, 0)$ to $(1, 0)$. If $\displaystyle\int_{C_1} F \cdot dr = a$, what is $\displaystyle\int_{C_2} F \cdot dr$? Why?

12. Evaluate $\displaystyle\int_C \sin y\, dx - x \cos y\, dy$ where C is the curve with parameterization C: $\quad r(t) = \cos t i + \sin t j, \quad 0 \le t \le \pi$.

13. Use Theorem 2 to prove that the vector field F is conservative in the open set D if and only if $\displaystyle\int_C F \cdot dr = 0$ for every closed path in D.

14. Use Theorem 2 and Exercise 13 to prove that the line integral $\displaystyle\int_C F \cdot dr$ is independent of path in the open set D if and only if $\displaystyle\int_C F \cdot dr = 0$ for every closed path in D.

15. Let $F(x, y) = 2xye^{x^2y}i + x^2e^{x^2y}j$. Let C_1 be the path from $A = (-1, 0)$ to $B = (1, 0)$ clockwise along the unit circle, and let C_2 be the path consisting of the line segment from A to B. Use the method of Section 20.2 to evaluate each of the line integrals $\displaystyle\int_{C_1} F \cdot dr$ and $\displaystyle\int_{C_2} F \cdot dr$. Are they equal? Why?

16. Repeat Exercise 15 for the vector field $F(x, y) = x^3y^2i + x^3yj$.

17. Show that, "in a conservative force field, the force is equal to the negative gradient of the potential."

18. Find a force field F so that $\displaystyle\int_{(0,\,0,\,0)}^{(x,\,y,\,z)} F \cdot dr = xyz$ for all $(x, y, z) \in \mathbb{R}^3$, independent of path.

19. Are line integrals of $F(x, y) = (x - y)i + (x + y)j$ independent of path in $\mathbb{R}^2$? Why or why not?

20. **a.** Show that
$$\int_C \frac{1}{x^2 + y^2}(x\, dy - y\, dx) = 2\pi$$
where C is the unit circle oriented counterclockwise.
b. Show that the vector field
$$F(x, y) = \frac{x}{x^2 + y^2}i - \frac{y}{x^2 + y^2}j$$
is not conservative.
c. Show that, for
$$M(x, y) = \frac{x}{x^2 + y^2} \quad \text{and} \quad N(x, y) = \frac{-y}{x^2 + y^2},$$
$$M_y(x, y) = N_x(x, y), \text{ with } x \ne 0 \text{ and } y \ne 0.$$
d. Explain why parts (a) through (c) do not contradict the statement involving equation (2).

21. Prove Theorem 1 in the case C piecewise smooth.

20.4 GREEN'S THEOREM

The Fundamental Theorem of Calculus states that if $F'(x) = f(x)$ for all x in $[a, b]$, the definite integral $\displaystyle\int_a^b f(x)\, dx$ may be evaluated by means of the formula

$$\int_a^b f(x)\, dx = F(b) - F(a). \tag{1}$$

In other words, the value of the integral in (1) is completely determined by the associated antiderivative at the endpoints of the interval $[a, b]$. In this section, we establish a similar result for certain functions of two variables: The value of a double integral of such a function over a region R in the plane is the same as an associated line integral taken around the boundary of R. This remarkable result is due to George Green, an English mathematician and physicist (1793–1841).

The statement of Green's Theorem involves the concept of **simple closed curves**. A closed curve $C = \{r(t) \mid a \le t \le b\}$ is called simple if $r(t_1) \ne r(t_2)$ for all numbers t_1 and t_2 in (a, b) with $t_1 \ne t_2$. (Remember, we must have $r(a) = r(b)$ if C

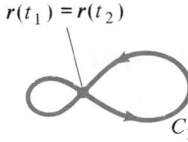

$r(t_1) = r(t_2)$

Figure 4.1 C_1 is not a simple closed curve.

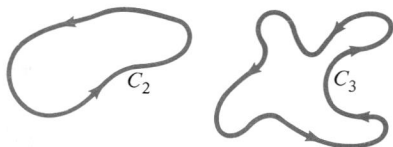

Figure 4.2 C_2 and C_3 are simple closed curves.

is closed.) In other words, *a simple closed curve does not cross itself* (see Figures 4.1 and 4.2). Be careful not to confuse the notion of simple closed *curves* with our earlier concepts of vertically simple or horizontally simple *regions*.

THEOREM 3
Green's Theorem

Let C be a piecewise smooth simple closed curve in the plane, oriented counterclockwise, and let Q be the region enclosed by C. If the functions $M(x, y)$ and $N(x, y)$ have continuous first partial derivatives in an open region containing Q, then

$$\int_C M(x, y)\, dx + N(x, y)\, dy = \iint_Q \left(\frac{\partial N}{\partial x} - \frac{\partial M}{\partial y} \right) dA. \tag{2}$$

Before discussing the proof of Green's Theorem, let's try to get a feel for what is going on by means of a few remarks and examples. We begin by expressing the line integral in (2) in vector form. If we write

$$\boldsymbol{F}(x, y) = M(x, y)\boldsymbol{i} + N(x, y)\boldsymbol{j}$$

and

$$C = \{\boldsymbol{r}(t) \mid a \leq t \leq b\},$$

then the conclusion (2) may be written as

$$\int_C \boldsymbol{F}(\boldsymbol{r}) \cdot d\boldsymbol{r} = \iint_Q \left(\frac{\partial N}{\partial x} - \frac{\partial M}{\partial y} \right) dA. \tag{3}$$

Now if $\boldsymbol{F}$ is a conservative vector field, equation (3) is not surprising, since both integrals are zero. The line integral is zero since C is closed, and the double integral is zero since $\dfrac{\partial N}{\partial x} - \dfrac{\partial M}{\partial y} = 0$ if $\boldsymbol{F}$ is conservative. Thus, Green's Theorem is telling us something about evaluating line integrals for *nonconservative* vector fields. Figure 4.3 illustrates the result geometrically: the surface area associated with the line integral is numerically equal to the volume associated with the double integral.

Example 1

Use Green's Theorem to evaluate the line integral

$$\int_C xy^2\, dx + 2x^3y\, dy$$

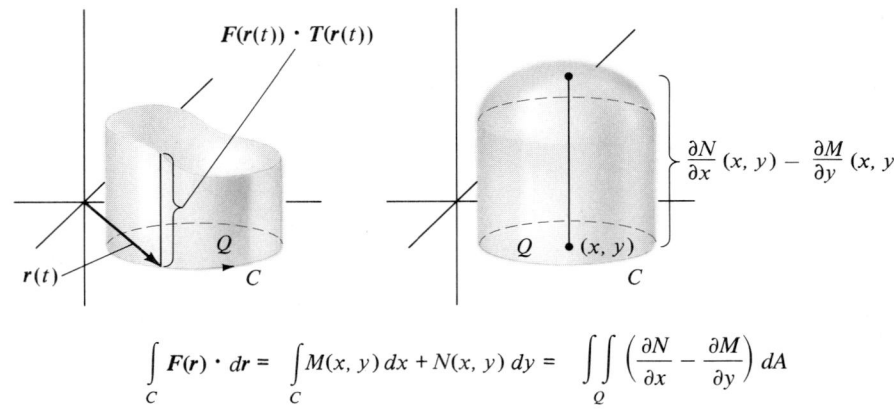

$$\int_C F(r) \cdot dr = \int_C M(x, y)\, dx + N(x, y)\, dy = \iint_Q \left(\frac{\partial N}{\partial x} - \frac{\partial M}{\partial y} \right) dA$$

Figure 4.3 Statement of Green's Theorem.

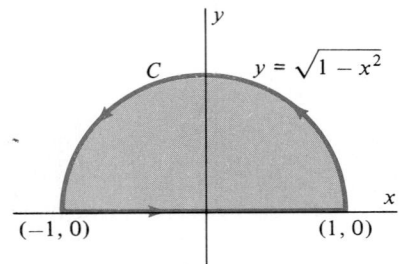

Figure 4.4 Region Q in Example 1.

where C is the path consisting of the upper unit semicircle traversed counterclockwise and the directed line segment from $(-1, 0)$ to $(1, 0)$ (see Figure 4.4).

Solution: Here $M(x, y) = xy^2$, $N(x, y) = 2x^3y$. The region Q enclosed by the path C may be described by the inequalities

$$Q: \quad -1 \le x \le 1; \quad 0 \le y \le \sqrt{1 - x^2}.$$

With $dA = dx\, dy$, Green's Theorem says that

$$\begin{aligned}
\int_C xy^2\, dx + 2x^3y\, dy &= \int_{-1}^{1} \int_0^{\sqrt{1-x^2}} \left[\frac{\partial}{\partial x}(2x^3y) - \frac{\partial}{\partial y}(xy^2) \right] dy\, dx \\
&= \int_{-1}^{1} \int_0^{\sqrt{1-x^2}} (6x^2y - 2xy)\, dy\, dx \\
&= \int_{-1}^{1} \{[(3x^2 - x)y^2]_0^{\sqrt{1-x^2}}\}\, dx \\
&= \int_{-1}^{1} (3x^2 - x)(1 - x^2)\, dx \\
&= \int_{-1}^{1} (-3x^4 + x^3 + 3x^2 - x)\, dx \\
&= \frac{4}{5}.
\end{aligned}$$

$\diamond$

Example 2

Use Green's Theorem to evaluate the line integral

$$\int_C xy\, dx + e^y\, dy$$

where C is the path from $(0, 0)$ to $(1, 1)$ along the graph of $y = x^2$ and from $(1, 1)$ to $(0, 0)$ along the graph of $y = \sqrt{x}$.

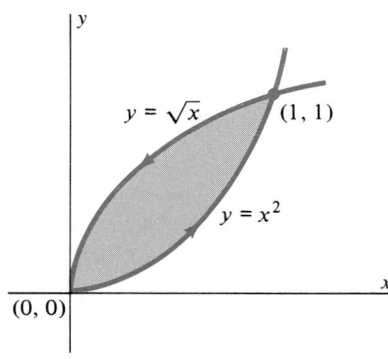

Figure 4.5 Region Q in Example 2.

Solution: Note in Figure 4.5 that C is oriented counterclockwise, as required. The region Q enclosed by C may be described by the inequalities

$$Q: \quad 0 \le x \le 1; \qquad x^2 \le y \le \sqrt{x}.$$

Since $M(x, y) = xy$ and $N(x, y) = e^y$, an application of Green's Theorem gives

$$\int_C xy \, dx + e^y \, dy = \int_0^1 \int_{x^2}^{\sqrt{x}} \left[\frac{\partial}{\partial x}(e^y) - \frac{\partial}{\partial y}(xy) \right] dy \, dx$$

$$= \int_0^1 \int_{x^2}^{\sqrt{x}} -x \, dy \, dx$$

$$= \int_0^1 \{-xy]_{x^2}^{\sqrt{x}}\} \, dx$$

$$= \int_0^1 (-x^{3/2} + x^3) \, dx$$

$$= -\frac{3}{20}. \qquad \diamond$$

Example 3

Use Green's Theorem to evaluate the line integral

$$\int_C xy \, dx + (x + y) \, dy$$

where C is the path indicated in Figure 4.6.

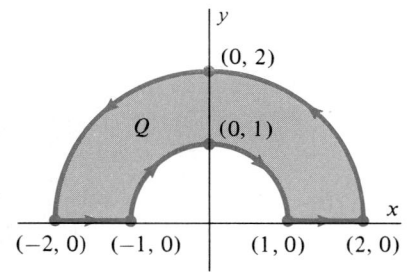

Figure 4.6

Solution: The region Q enclosed by C may be described in polar coordinates by the inequalities

$$Q: \quad 1 \le r \le 2; \qquad 0 \le \theta \le \pi.$$

Since $M(x, y) = xy$ and $N(x, y) = x + y$, we have

$$\frac{\partial N}{\partial x} - \frac{\partial M}{\partial y} = \frac{\partial}{\partial x}(x + y) - \frac{\partial}{\partial y}(xy) = 1 - x. \qquad (4)$$

Since we are using polar coordinates, we set

$$x = r \cos \theta, \qquad y = r \sin \theta, \qquad \text{and} \qquad dA = r \, dr \, d\theta. \qquad (5)$$

Combining (4) and (5) with the statement of Green's Theorem, we obtain

$$\int_C xy \, dx + (x + y) \, dy = \int_0^\pi \int_1^2 [1 - r \cos \theta] r \, dr \, d\theta$$

$$= \int_0^\pi \left[\frac{r^2}{2} - \frac{r^3}{3} \cos \theta \right]_{r=1}^{r=2} d\theta$$

$$= \int_0^\pi \left(\frac{3}{2} - \frac{7}{3} \cos \theta \right) d\theta$$

$$= \frac{3\theta}{2} - \frac{7}{3} \sin \theta \Big]_0^\pi$$

$$= \frac{3\pi}{2}. \qquad \diamond$$

There should be no doubt in your mind that the Green's Theorem solution to this problem is much easier than a direct evaluation of the line integral, since the latter would require four separate parameterizations, one for each arc of the curve C.

Calculating Area by Line Integrals

So far we have seen examples in which the double integral over Q has been easier to evaluate than the line integral over C in Green's Theorem. Sometimes it is the other way around. Since areas of certain regions in the plane may be calculated by means of a double integral, Green's Theorem enables us to find expressions for the area of Q (in Theorem 3) in terms of line integrals over C. For example, if we use the functions $M(x, y) = 0$ and $N(x, y) = x$, we obtain

$$\int_C x \, dy = \iint_Q \left[\frac{\partial}{\partial x}(x) - \frac{\partial}{\partial y}(0) \right] dA = \iint_Q 1 \, dA = \text{Area of } Q. \tag{6}$$

Similarly, by choosing $M(x, y) = y$ and $N(x, y) = 0$, we obtain

$$\int_C y \, dx = \iint_Q \left[\frac{\partial}{\partial x}(0) - \frac{\partial}{\partial y}(y) \right] dA = \iint_Q (-1) \, dA = -(\text{Area of } Q). \tag{7}$$

Subtracting the corresponding sides of equations (6) and (7) and dividing by 2 gives

$$\boxed{\text{Area of } Q = \frac{1}{2} \int_C x \, dy - y \, dx.} \tag{8}$$

Any of equations (6), (7), or (8) may be used in calculating the area of Q. In the following example, equation (8) provides an easy solution to a problem that otherwise requires the technique of integration by trigonometric substitution.

Example 4

Calculate the area A of the region enclosed by the ellipse

$$\frac{x^2}{a^2} + \frac{y^2}{b^2} = 1.$$

Solution: A parameterization for the ellipse with counterclockwise orientation is

$$x(t) = a \cos t, \qquad y(t) = b \sin t, \qquad 0 \le t \le 2\pi.$$

Thus

$$x'(t) = -a \sin t, \qquad \text{and} \qquad y'(t) = b \cos t.$$

By equation (8),

$$A = \frac{1}{2} \int_C x \, dy - y \, dx$$

$$= \frac{1}{2} \int_0^{2\pi} [(a \cos t)(b \cos t) - (b \sin t)(-a \sin t)] \, dt$$

$$= \frac{1}{2} \int_0^{2\pi} ab(\cos^2 t + \sin^2 t) \, dt$$

$$= \pi ab. \qquad \diamond$$

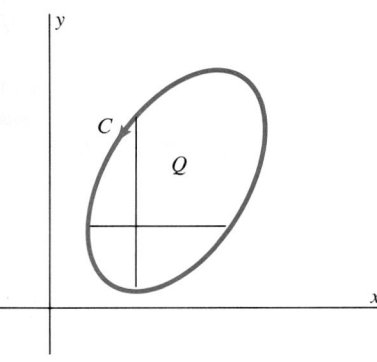

Figure 4.7 Q is both x-simple and y-simple.

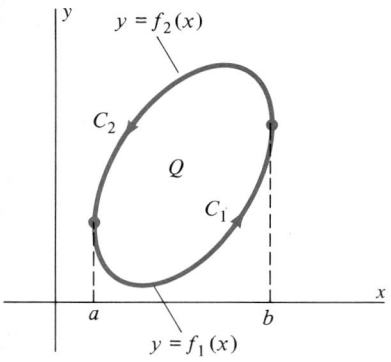

Figure 4.8 C_j is graph of $y = f_j(x)$, $j = 1, 2$.

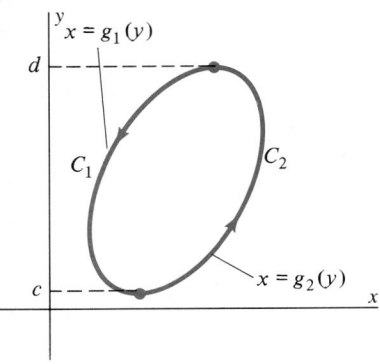

Figure 4.9 C_j is graph of $x = g_j(x)$, $j = 1, 2$.

Proof of Theorem 3: We will prove Theorem 3 only for regions which are both x-simple and y-simple and then comment on the idea by which the proof can be extended to more general regions.

Specifically, let us assume that Q is a region such as that in Figure 4.7 which can be described both as

$$Q = \{(x, y) \mid a \le x \le b, f_1(x) \le y \le f_2(x)\} \qquad \text{(Figure 4.8)} \qquad (9)$$

or as

$$Q = \{(x, y) \mid g_1(y) \le x \le g_2(y), c \le y \le d\} \qquad \text{(Figure 4.9)} \qquad (10)$$

where each of the functions $f_1, f_2, g_1,$ and g_2 is continuous and piecewise differentiable.

We will prove Theorem 3 by showing both that

$$\int_C M(x, y)\, dx = -\iint_Q \frac{\partial M}{\partial y}\, dA \qquad (11)$$

and that

$$\int_C N(x, y)\, dy = \iint_Q \frac{\partial N}{\partial x}\, dA. \qquad (12)$$

Adding the corresponding sides of equations (11) and (12) then establishes equation (2), which is what we are trying to prove.

To prove (11), we first use Figure 4.8 and equation (9) to find that

$$\int_C M(x, y)\, dx = \int_{C_1} M(x, y)\, dx + \int_{C_2} M(x, y)\, dx$$

$$= \int_a^b M(x, f_1(x))\, dx + \int_b^a M(x, f_2(x))\, dx$$

$$= \int_a^b M(x, f_1(x))\, dx - \int_a^b M(x, f_2(x))\, dx.$$

Thus,

$$\int_C M(x, y)\, dx = \int_a^b [M(x, f_1(x)) - M(x, f_2(x))]\, dx. \qquad (13)$$

On the other hand, we also have

$$\iint_Q \frac{\partial M}{\partial y}\, dA = \int_a^b \int_{f_1(x)}^{f_2(x)} \frac{\partial}{\partial y} M(x, y)\, dy\, dx \qquad (14)$$

$$= \int_a^b \{M(x, y)]_{y=f_1(x)}^{y=f_2(x)}\}\, dx$$

$$= \int_a^b [M(x, f_2(x)) - M(x, f_1(x))]\, dx.$$

Comparing equations (13) and (14) shows that equation (11) holds. Similarly, using Figure 4.9 and equation (10) we can show that equation (12) holds. (This step is left to you as Exercise 23.) This completes the proof for regions Q of the stated type. ◆

Extending Green's Theorem to More General Regions

Figure 4.10 shows a region that is neither x-simple nor y-simple, although it does satisfy the hypotheses of Theorem 3. To extend the preceding proof to this region, we make a "cut" (that is, we draw a line) from point P_1 to point P_2, chosen in such a way that the resulting subregions Q_1 and Q_2 are both x-simple and y-simple (See Figure 4.11.) Thus, if C_1 denotes the boundary of Q_1, and C_2 the boundary of Q_2,

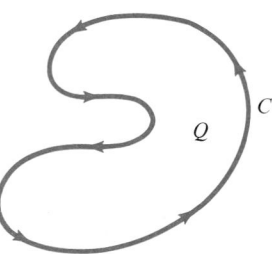

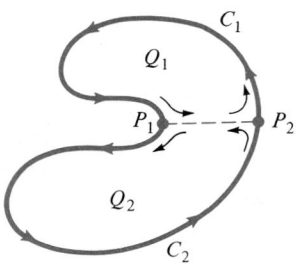

Figure 4.10 Q is neither x-simple nor y-simple.

Figure 4.11 Q_1 and Q_2 are both x-simple and y-simple.

Green's Theorem on the individual subregions shows that

$$\int_{C_1} M(x, y)\, dx + N(x, y)\, dy = \int\int_{Q_1} \left[\frac{\partial N}{\partial x} - \frac{\partial M}{\partial y}\right] dA \tag{15}$$

and

$$\int_{C_2} M(x, y)\, dx + N(x, y)\, dy = \int\int_{Q_2} \left[\frac{\partial N}{\partial x} - \frac{\partial M}{\partial y}\right] dA. \tag{16}$$

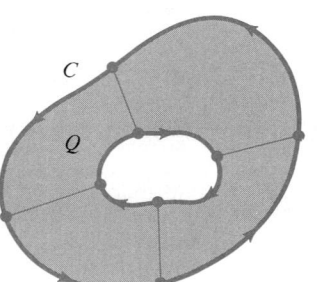

Figure 4.12 C consists of two arcs.

Adding the corresponding sides of these two equations then gives equation (2). The reason for this is that the contribution to the line integral in (15) resulting from the path from P_1 to P_2 is precisely the negative of the contribution of this same path to the line integral in (16), since the path is traversed in opposite directions by the two line integrals. Thus, the two line integrals taken over the path P_1P_2 "cancel."

Green's Theorem may be extended to regions even more general than those in our statement of Theorem 3. In particular, Green's Theorem holds for regions Q containing "holes," such as that in Figure 4.12. For such regions, the curve C is the entire boundary—both the "inner" and "outer" curves. The meaning of "counter-clockwise" for such regions is that each arc of C is traversed so as to keep the region Q on the left-hand side, as indicated in Figure 4.12. The idea by which Green's Theorem is extended to include such regions is again to make "cuts" (see Figure 4.13) dividing Q into subregions without holes, to which Theorem 3 applies. As before, the contributions to the line integral along each cut path sum to zero, so equation (2) results from adding the corresponding sides of the equations holding in each region.

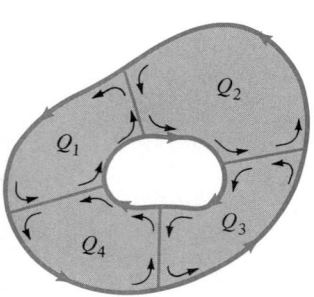

Figure 4.13 Cuts divide Q into regions without holes.

Other Formulations of Green's Theorem

There are two special forms in which Green's Theorem may be stated, each of which is a two-dimensional version of more general results yet to come. We will interpret each of these statements in terms of the flow of a thin layer of fluid, such as that very near the surface of the water in a swimming pool or in a cross section of a water pipe.

We begin by letting $F(x, y) = M(x, y)\mathbf{i} + N(x, y)\mathbf{j}$ be the vector field giving the velocity (speed and direction) of the fluid flow at each point (x, y) in the thin layer. Using equation (22), Section 20.2, we may write the vector form (3) of Green's Theorem as

$$\int_C \mathbf{F} \cdot \mathbf{T} \, ds = \iint_Q \left(\frac{\partial N}{\partial x} - \frac{\partial M}{\partial y} \right) dA \tag{17}$$

where C is any smooth closed curve in the fluid layer. The integrand $\mathbf{F} \cdot \mathbf{T}$ in the line integral in (17) is just the component of the vector field (i.e., the flow) in the direction of the unit tangent $\mathbf{T}$ to the curve C at each point (see Figure 4.14). If the fluid has a tendency to circulate around the curve C (such as what you see when you pull the plug in your bathtub), we would intuitively expect the line integral in (17) to be large, since the velocity vector $\mathbf{F}$ and the tangent vector $\mathbf{T}$ should point rather consistently either in approximately the same directions (when the fluid rotates counterclockwise) or in opposite directions (when the fluid rotates clockwise), as illustrated in Figure 4.15. When the fluid has little tendency to rotate, such as occurs in a constant flow field, we expect the line integral in (17) to be small since the component of $\mathbf{F}$ on $\mathbf{T}$ will be of opposite sign on opposite sides of C (Figure 4.16). For these reasons we say that the line integral in (17) is a measure of the

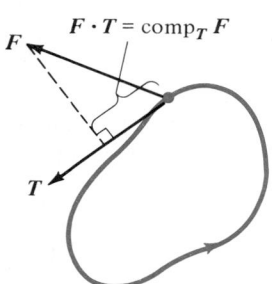

Figure 4.14 $\mathbf{F} \cdot \mathbf{T}$ is component of $\mathbf{F}$ in direction of $\mathbf{T}$.

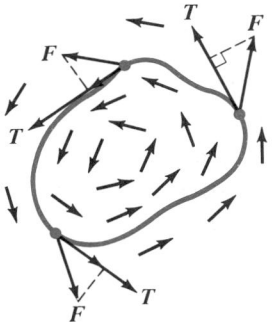

Figure 4.15 In a rotating fluid, $\left| \int_C \mathbf{F} \cdot \mathbf{T} \, ds \right|$ is large.

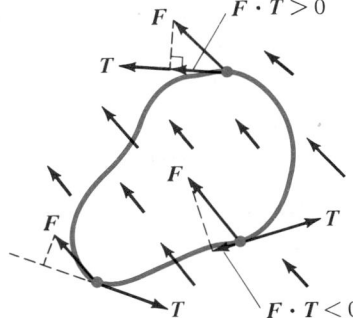

Figure 4.16 In a fluid with little rotation, $\left| \int_C \mathbf{F} \cdot \mathbf{T} \, ds \right|$ is small.

tendency of the fluid layer to circulate or to rotate. Vector fields in the plane for which this line integral is always zero are called **irrotational.**

Since the double integral in equation (17) has the same value as the line integral, it too measures the tendency of the fluid to rotate. However, it does so by integrating the scalar function $\dfrac{\partial N}{\partial x} - \dfrac{\partial M}{\partial x}$ over the region Q enclosed by C. For this reason we refer to this function as the **scalar curl** of the vector field $\mathbf{F}$. That is,

$$\text{curl } \mathbf{F} = \frac{\partial N}{\partial x} - \frac{\partial M}{\partial y}. \tag{18}$$

Using this definition, we may rewrite the formulation of Green's Theorem given

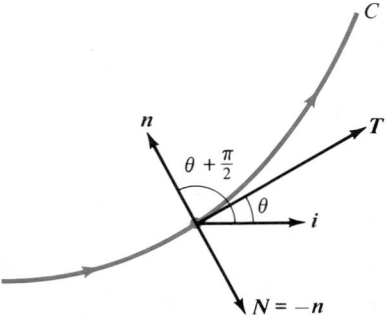

Figure 4.17 N is the outward unit normal.

in line (17) as

$$\int_C \boldsymbol{F} \cdot \boldsymbol{T} \, ds = \int\int_Q \text{curl } \boldsymbol{F} \, dA \qquad \text{(Circulation of } \boldsymbol{F} \text{ around } C\text{).} \quad (19)$$

Equation (19) is referred to as **Stokes' Theorem in the plane.** In Section 20.6 we shall study the generalizations of curl and of Stokes' Theorem to three dimensions.

A second vector formulation of Green's Theorem concerns the outward unit normal to the closed smooth simple curve C. To obtain this vector we let $\boldsymbol{T}$ be the unit tangent at a point on the curve C (oriented counterclockwise) and we let θ be the angle formed between $\boldsymbol{T}$ and the unit vector $\boldsymbol{i}$ (Figure 4.17). Then

$$\boldsymbol{T} = \cos \theta \boldsymbol{i} + \sin \theta \boldsymbol{j}. \qquad (20)$$

The unit vector $\boldsymbol{n}$ obtained by rotating $\boldsymbol{T}$ 90° in the counterclockwise direction is the **unit normal**

$$\begin{aligned} \boldsymbol{n} &= \cos(\theta + \pi/2)\boldsymbol{i} + \sin(\theta + \pi/2)\boldsymbol{j} \\ &= -\sin \theta \boldsymbol{i} + \cos \theta \boldsymbol{j}. \end{aligned}$$

However, this normal points inward for a closed curve C oriented counterclockwise, so the desired **outward unit normal N** is the vector

$$\boldsymbol{N} = -\boldsymbol{n} = \sin \theta \boldsymbol{i} - \cos \theta \boldsymbol{j}. \qquad (21)$$

Now recall from Section 17.3 that the unit tangent vector $\boldsymbol{T}$ may be written

$$\boldsymbol{T} = \frac{dx}{ds}\boldsymbol{i} + \frac{dy}{ds}\boldsymbol{j} \qquad (22)$$

where s is the arc length parameter and $\boldsymbol{r}(s) = x(s)\boldsymbol{i} + y(s)\boldsymbol{j}$ is a parameterization for C by arc length. From equations (20), (21), and (22) it follows that the outward unit normal $\boldsymbol{N}$ may be written

$$\boldsymbol{N} = \frac{dy}{ds}\boldsymbol{i} - \frac{dx}{ds}\boldsymbol{j}. \qquad (23)$$

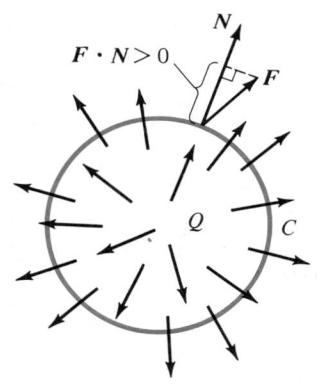

Figure 4.18 $\int_C \boldsymbol{F} \cdot \boldsymbol{N} \, ds > 0$ if net flow across C is outward.

For the vector field $\boldsymbol{F}(x, y) = M(x, y)\boldsymbol{i} + N(x, y)\boldsymbol{j}$, the expression (23) for $\boldsymbol{N}$ suggests that

$$\int_C \boldsymbol{F} \cdot \boldsymbol{N} \, ds = \int_C M(x, y) \, dy - N(x, y) \, dx, \qquad (24)$$

and Green's Theorem, applied to the right-hand side of equation (24) gives

$$\int_C M(x, y) \, dy - N(x, y) \, dx = \int\int_Q \left(\frac{\partial M}{\partial x} + \frac{\partial N}{\partial y} \right) dA. \qquad (25)$$

Finally, combining (24) and (25), we obtain

$$\int_C \boldsymbol{F} \cdot \boldsymbol{N} \, ds = \int\int_Q \left(\frac{\partial M}{\partial x} + \frac{\partial N}{\partial y} \right) dA. \qquad (26)$$

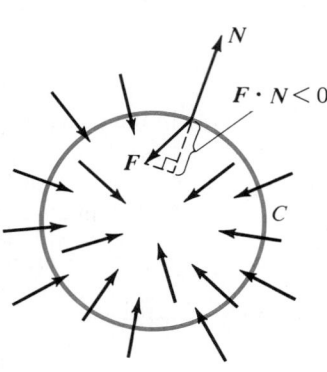

Figure 4.19 $\int_C \boldsymbol{F} \cdot \boldsymbol{N} \, ds < 0$ if net flow across C is inward.

Equation (26) is referred to as the *Divergence Theorem in the plane* for reasons that we will now explain. Since the scalar quantity $\boldsymbol{F} \cdot \boldsymbol{N}$ in the line integral is the

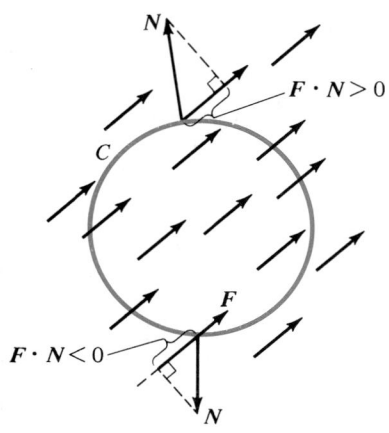

Figure 4.20 $\int_C F \cdot N \, ds = 0$ if net flow across C is zero.

component of the vector field (fluid flow) in the direction of the outward normal, the line integral in (26) measures the net flow of fluid outward across the boundary C of the region Q. The term **flux** is used to denote this net rate at which fluid flows across a boundary, and the line integral in (26) is called a **flux integral.** When the line integral is *positive,* more fluid is leaving Q than is entering. (Think of an open faucet somewhere inside Q.) When the line integral in (26) is *negative,* more fluid is flowing into Q than is leaving. (Think of an open drain somewhere inside Q.) When the line integral is small or zero, there is little or no net gain or loss of fluid in Q. (Think of fluid moving uniformly, or motionless fluid. See Figures 4.18 through 4.20.)

Since the double integral in (26) gives the same measure of net flow as does the line integral, the integrand $\dfrac{\partial M}{\partial x} + \dfrac{\partial N}{\partial y}$ is referred to as the **divergence** of the vector field F, written div F. That is

$$\text{div } F = \frac{\partial M}{\partial x} + \frac{\partial N}{\partial y}. \tag{27}$$

Using (27), we may rewrite the Divergence Theorem in the plane as

$$\int_C F \cdot N \, ds = \iint_Q \text{div } F \, dA. \qquad \text{(Flux of } F \text{ across } C\text{).} \tag{28}$$

In Section 20.7 we will encounter generalizations of both the concept of divergence of a vector field and the Divergence Theorem to three dimensions, the primary motivation being fluid flow in space.

Exercise Set 20.4

(In all exercises the closed curves C are oriented counterclockwise.) In Exercises 1–10, use Green's Theorem to evaluate the given line integral.

1. $\displaystyle\int_C xy \, dx + (x + y) \, dy$, C is the square with vertices $(0, 0)$, $(0, 1)$, $(1, 1)$, and $(1, 0)$.

2. $\displaystyle\int_C xy \, dx + (y - x) \, dy$, C is the unit circle.

3. $\displaystyle\int_C x^2 y^3 \, dx + (x^2 + 1) \, dy$, C is the square in Exercise 1.

4. $\displaystyle\int_C \sqrt{y} \, dx + \sqrt{x} \, dy$, C is the rectangle with vertices $(1, 1)$, $(1, 2)$, $(4, 1)$, and $(4, 2)$.

5. $\displaystyle\int_C xy^2 \, dy - x^2 y \, dx$, C is the quarter circular path from $(0, 0)$ to $(1, 0)$ to $(0, 1)$ to $(0, 0)$.

6. $\displaystyle\int_C e^x \tan y \, dx + e^x \sec^2 y \, dy$, C is the triangle with vertices $(-1, 2)$, $(3, 0)$, and $(1, 6)$.

7. $\displaystyle\int_C (x^3 + y) \, dx + (y - x^2) \, dy$, C is the boundary of the region bounded by the graphs of $y = x^2$ and $y = x^3$.

8. $\displaystyle\int_C (\text{Tan}^{-1} x + y^2) \, dx + (\ln^2 y - x^2) \, dy$, C is the circle $x^2 + y^2 = 4$.

9. $\displaystyle\int_C (\cos^5 x + \sqrt{x}) \, dx + \text{Tan}^{-1} y \, dy$, C is the ellipse $4x^2 + y^2 = 1$.

10. $\displaystyle\int_C \sqrt{x} \, dx + \ln(x^2 + y^2) \, dy$, C is the curve in Figure 4.6.

In Exercises 11–14, use Green's Thorem to find $\displaystyle\int_C F \cdot dr.$

11. $F(x, y) = 2yi - 3xj$, C is the unit circle.

12. $F(x, y) = xyi + x^2 j$, C is the square with vertices $(0, 0)$, $(0, 2)$, $(2, 2)$, and $(2, 0)$.

13. $F(x, y) = x^2(y^2 - x^2)i + \dfrac{2}{3}x^3 yj$, C is the triangle with vertices $(1, 1)$, $(4, 1)$, and $(3, 5)$.

14. $F(x, y) = e^x \sin y\mathbf{i} + e^x \cos y\mathbf{j}$, C is the ellipse $x^2 + 9y^2 = 1$.

15. Find the work done by the force field $F(x, y) = (e^{x^2} + 2y)\mathbf{i} + (ye^y - x)\mathbf{j}$ in moving a particle once around the path indicated in Figure 4.6.

16. Use equation (8) to find the area of the region bounded by the graphs of $y = x$ and $y = x^2$.

17. Use equations (8) to find the area of the region bounded by the graphs of $y = x$ and $y = \sqrt{x}$.

18. Find the area of the region enclosed by the graph of the parametric equations $x(t) = \sin t \cos t$, $y(t) = \sin t$ for $0 \le t \le \pi$.

19. Find the area of the region bounded by the graph of the parametric equations $x(t) = a \cos^3 t$, $y(t) = a \sin^3 t$, $0 \le t \le 2\pi$.

20. Let Q be a region in the plane whose boundary is a simple closed piecewise smooth curve C. Let A be the area of Q.

Show that the coordinates of the centroid of Q are

$$\bar{x} = \frac{1}{2A} \int_C x^2 \, dy, \qquad \bar{y} = -\frac{1}{2A} \int_C y^2 \, dx.$$

21. Use Green's Theorem to prove that if F is a conservative vector field in the plane, then any line integral $\int_C F \cdot dr$ over a piecewise smooth curve C is independent of path.

22. Use Green's Theorem to prove that if F is a vector field in the plane for which $\int_A^B F \cdot dr$ is independent of path for all A and B, then $\int_C F \cdot dr = 0$ for every simple closed path C.

23. Complete the proof of Theorem 3 by showing that equation (12) holds.

24. Let $F(x, y) = M(x, y)\mathbf{i} + N(x, y)\mathbf{j}$ where M and N have continuous first partial derivatives in some open rectangle D. Show that F is conservative in D if and only if curl $F = 0$ for all $(x, y) \in D$.

20.5 SURFACE INTEGRALS

The primary motivation for the remainder of this chapter will be to extend the concept of flux, introduced in Section 20.4, to the flow of fluids in and out of regions in space. Since regions in space are bounded by surfaces rather than curves, we must begin by developing the concept of integrals over surfaces, or *surface integrals*. Although we are primarily interested in defining the integral of a *vector field* F over a surface S (Figure 5.1), we shall begin by defining the integral of a scalar function f over a surface S (Figure 5.2). This is because the development of

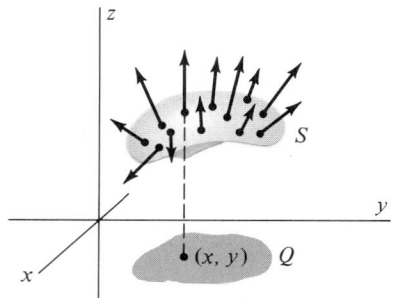

Figure 5.1 A vector field $F(x, y, z)$ assigns a vector to each point on the surface S.

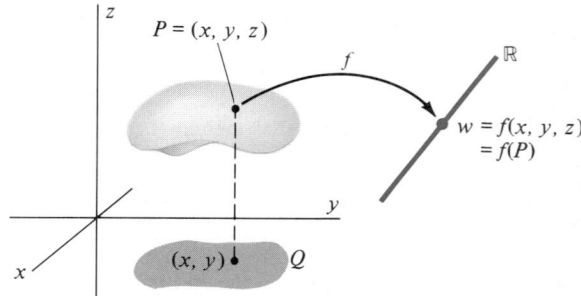

Figure 5.2 A scalar function $f(x, y, z)$ assigns a number to each point P on S.

the surface integral for scalar functions very closely parallels the discussion of surface area in Section 19.5.

By a surface S, we shall mean the graph of a continuous function of two variables, $z = g(x, y)$, over some regular region Q in the xy-plane. (Recall that the region Q is regular if it is both x-simple and y-simple.) Such a surface is illustrated in Figure 5.3. This is not the most general concept of surface, but techniques developed here will enable you to handle most surfaces encountered in practice.

Suppose that the continuous scalar function f is defined for all points $(x, y, g(x, y))$ on the surface S. Recall from Section 19.5 that a partition of the region Q into n rectangles of area $\Delta A_j = \Delta x\,\Delta y$ partitions "most" of the surface S into curvilinear "patches" S_j of area ΔS_j (Figure 5.4.) Proceeding in a purely

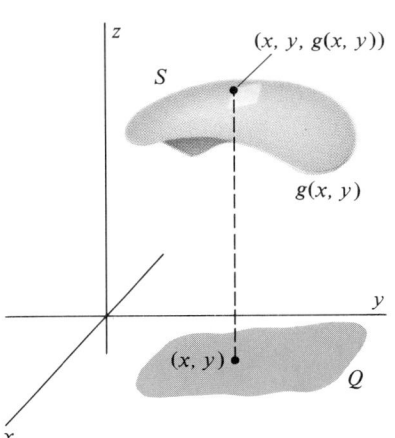

Figure 5.3 Surface S is the graph of a continuous function $z = g(x, y)$ over Q.

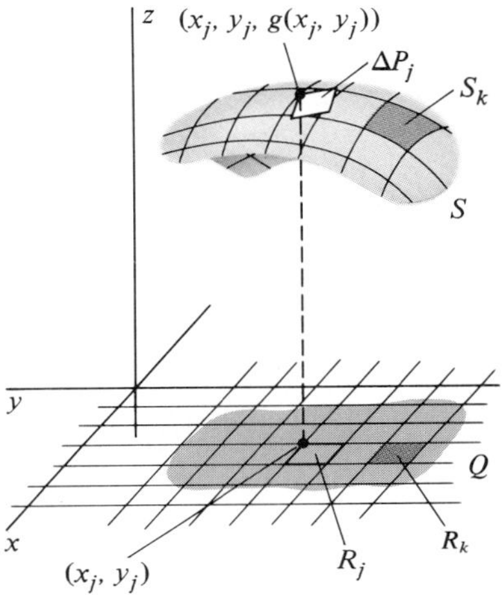

Figure 5.4 Parallelogram of area ΔP_j over rectangle R_j approximates the area ΔS_j of the "patch" S_j.

mathematical way, we form the product of the surface area ΔS_j of the patch S_j and the value $f(x_j, y_j, z_j)$ of the scalar function f at an arbitrary point (x_j, y_j, z_j) on the patch. Summing these products over all n patches gives the approximating sum

$$\sum_{j=1}^{n} f(x_j, y_j, z_j)\,\Delta S_j, \qquad (x_j, y_j, z_j) \in S_j.$$

The limit as the norm of the partition P approaches zero ($\|P\| \to 0$) of this approximating sum, when it exists, is defined as the surface integral of f over S:

$$\iint_S f(x, y, z)\,dS = \lim_{\|P\| \to 0} \sum_{j=1}^{n} f(x_j, y_j, z_j)\,\Delta S_j. \qquad (1)$$

We shall determine several ways in which this surface integral may be calculated. The first of these involves recalling that the area ΔS_j of the patch S_j may be approximated by the area ΔP_j of the parallelogram over R_j tangent to the surface S at one corner (Figure 5.4). That is, we have the approximation

$$\Delta S_j \approx \Delta P_j = \sqrt{\left(\frac{\partial g}{\partial x}\right)^2 + \left(\frac{\partial g}{\partial y}\right)^2 + 1}\, \Delta x\, \Delta y \tag{2}$$

where the partial derivatives $\dfrac{\partial g}{\partial x}$ and $\dfrac{\partial g}{\partial y}$ are evaluated at one corner, (x_j, y_j), of the rectangle R_j. Using approximation (2) in equation (1) leads to the following definition.

DEFINITION 5

Let Q be a regular region in the plane and let g be a continuous function defined on some open rectangle containing Q on which both $\dfrac{\partial g}{\partial x}$ and $\dfrac{\partial g}{\partial y}$ are continuous. Let S be the graph of g over Q, and let f be a continuous scalar function defined on S. The **surface integral** of f over S is

$$\iint_S f(x, y, z)\, dS = \iint_Q f(x, y, g(x, y))\sqrt{\left(\frac{\partial g}{\partial x}\right)^2 + \left(\frac{\partial g}{\partial y}\right)^2 + 1}\, dx\, dy. \tag{3}$$

Example 1

Evaluate the surface integral

$$\iint_S (x^2 + y + z)\, dS$$

where S is the graph of the function $g(x, y) = \sqrt{3}\, y - x^2$ over the unit square
$$Q = \{(x, y) \mid 0 \leq x \leq 1, 0 \leq y \leq 1\}.$$

Solution: The integrand is the function
$$f(x, y, z) = x^2 + y + z.$$

Also,
$$\frac{\partial g}{\partial x} = -2x, \qquad \text{and} \qquad \frac{\partial g}{\partial y} = \sqrt{3}.$$

Substituting into equation (3) gives

$$\iint_S (x^2 + y + z)\, dS = \iint_Q [x^2 + y + (\sqrt{3}\, y - x^2)]\sqrt{(-2x)^2 + (\sqrt{3})^2 + 1}\, dx\, dy$$

$$= \int_0^1 \int_0^1 (1 + \sqrt{3})y\sqrt{4x^2 + 4}\, dy\, dx$$

$$= (1 + \sqrt{3})\int_0^1 \{y^2\sqrt{x^2 + 1}\}_{y=0}^{y=1}\, dx$$

$$= (1 + \sqrt{3})\int_0^1 \sqrt{x^2 + 1}\, dx \qquad \text{(trigonometric substitution)}$$

$$= (1 + \sqrt{3})\left(\frac{\sqrt{2}}{2} + \frac{1}{2}\ln(1 + \sqrt{2})\right)$$

$$\approx 3.14. \qquad \diamond$$

Example 2

Evaluate the surface integral

$$\iint_S (x^2 + y^2 + z)\, dS$$

where S is the portion of the graph of the function $z = 4 - x^2 - y^2$ bounded below by the xy-plane.

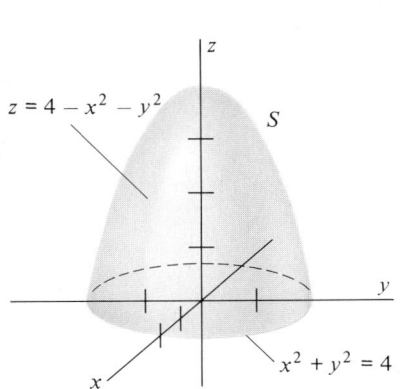

Figure 5.5 Surface S is graph of $z = 4 - x^2 - y^2$.

Solution: The integrand is $f(x, y, z) = x^2 + y^2 + z$, and the surface is the graph of the function $g(x, y) = 4 - x^2 - y^2$. Thus

$$\frac{\partial g}{\partial x} = -2x, \qquad \text{and} \qquad \frac{\partial g}{\partial y} = -2y.$$

The region Q in the xy-plane over which S is defined is

$$Q = \{(x, y) \mid x^2 + y^2 \le 4\} \qquad \text{(Figure 5.5)}.$$

According to Definition 5, the surface integral is

$$\iint_S (x^2 + y^2 + z)\, dS = \iint_Q [x^2 + y^2 + (4 - x^2 - y^2)]\sqrt{(-2x)^2 + (-2y)^2 + 1}\ dx\, dy$$

$$= \iint_Q 4\sqrt{4x^2 + 4y^2 + 1}\ dx\, dy.$$

Switching to polar coordinates allows us to evaluate this integral as

$$\iint_S (x^2 + y^2 + z)\, dS = \int_0^{2\pi} \int_0^2 4\sqrt{4r^2 + 1}\ r\, dr\, d\theta$$

$$= \int_0^{2\pi} \left\{ \left[\frac{1}{3}(4r^2 + 1)^{3/2} \right]_{r=0}^{r=2} \right\} d\theta$$

$$= \frac{2\pi}{3}(17^{3/2} - 1)$$

$$\approx 144.71. \qquad \diamondsuit$$

REMARK: For surfaces that are graphs of functions of the form $y = g(x, z)$ or $x = g(y, z)$, we may simply interchange roles among the variables x, y, and z as required. The following example presents one such situation. (Notice that the given surface cannot be described as the graph of a function of the form $z = g(x, y)$.)

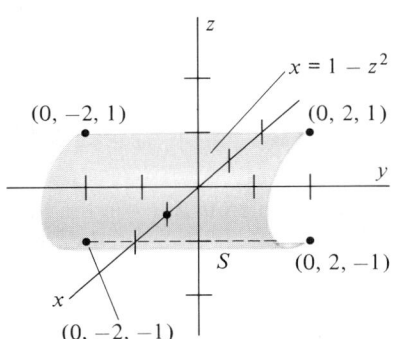

Figure 5.6 Surface S is graph of $x = 1 - z^2$.

Example 3

Evaluate the surface integral

$$\iint_S yz\, dS$$

where S is the portion of the graph of $x = 1 - z^2$ bounded by the yz-plane and the planes $y = -2$ and $y = 2$ (Figure 5.6).

Solution: In order to describe S as the graph of a function, we must take x as the dependent variable. Interchanging roles of x and z in Definition 5, we find that

$$f(x, y, z) = yz \qquad \text{and} \qquad g(y, z) = 1 - z^2.$$

Thus,

$$\frac{\partial g}{\partial z} = -2z \qquad \text{and} \qquad \frac{\partial g}{\partial y} = 0.$$

The region Q in the yz-plane over which S is defined is the rectangle

$$Q = \{(y, z) \mid -2 \leq y \leq 2,\ -1 \leq z \leq 1\}.$$

Definition 5 now gives

$$\iint_S yz\, dS = \iint_Q yz\sqrt{(-2z)^2 + 1}\ dz\, dy$$

$$= \int_{-2}^{2} \int_{-1}^{1} yz\sqrt{4z^2 + 1}\ dz\, dy$$

$$= \int_{-2}^{2} \left\{ \frac{y}{12}(4z^2 + 1)^{3/2} \Big]_{z=-1}^{z=1} \right\} dy$$

$$= \int_{-2}^{2} \frac{y}{12}(5^{3/2} - 5^{3/2})\ dy$$

$$= 0.$$

The result should not be surprising, since both f and the surface are symmetric with respect to both y and z. ◇

Calculating Mass

A straightforward application of surface integrals of the form (1) occurs in the calculation of the total mass of a hollow object (such as a basketball or a vase) whose mass density per unit surface area, $\rho(x, y, z)$, varies continuously over the surface of the object. That is, we assume that $\rho(x, y, z)$ gives the mass density (in grams/cm^2, say) of the object at location (x, y, z). Then if S_j is a small patch of area ΔS_j on the surface of the object, the total mass ΔM_j of S_j may be approximated by

$$\Delta M_j \approx \rho(x_j, y_j, z_j)\, \Delta S_j \qquad \left(\text{mass} = \frac{\text{mass}}{\text{unit area}} \times \text{area} \right) \qquad (4)$$

where (x_j, y_j, z_j) is a point on S_j. Summing (4) over all patches in a partition covering S, we arrive at the approximation to total mass:

$$M \approx \sum_{j=1}^{n} \rho(x_j, y_j, z_j)\, \Delta S_j. \qquad (5)$$

Comparing approximation (5) with equation (1), we conclude that in the limit as $\Delta S_j \to 0$, assuming appropriate conditions on the shape of the object and on the density function ρ,

$$M = \iint_S \rho(x, y, z)\, dS. \qquad (6)$$

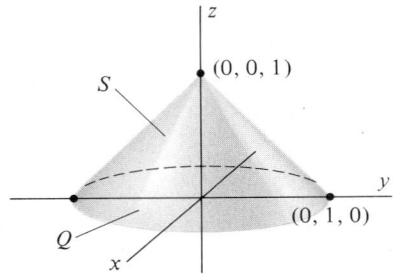

Figure 5.7

In other words, **the total mass of the object** is simply the integral, over the surface of the object, of the function giving the mass per unit area.

Example 4

Let S be the idealized cone described in cylindrical coordinates by the equation $z = 1 - r, 0 \leq r \leq 1$. Find the mass of S if the density at any point is proportional to the square of the distance between the point and the origin (see Figure 5.7).

Solution: Here $\rho(x, y, z) = \lambda(x^2 + y^2 + z^2)$ is the mass density, where λ is a constant. We will work with ρ in cylindrical coordinates:

$$\rho(r, \theta, z) = \lambda(r^2 + z^2).$$

Also, the surface S is described by the function

$$z = g(r, \theta) = 1 - r.$$

Using the result of Exercise 25 of this section (for evaluating a surface integral in cylindrical coordinates) together with equation (6), we find that

$$M = \iint_S \rho(r, \theta, z) \, dS$$

$$= \iint_Q \rho(r, \theta, g(r, \theta)) \sqrt{r^2 + r^2 \left(\frac{\partial g}{\partial r}\right)^2 + \left(\frac{\partial g}{\partial \theta}\right)^2} \, dr \, d\theta$$

$$= \int_0^{2\pi} \int_0^1 \lambda(r^2 + (1 - r)^2) \cdot \sqrt{r^2 + r^2(-1)^2 + 0^2} \, dr \, d\theta$$

$$= \int_0^{2\pi} \int_0^1 \sqrt{2}\lambda \cdot (2r^3 - 2r^2 + r) \, dr \, d\theta$$

$$= \frac{2\sqrt{2}\pi\lambda}{3}. \qquad \diamond$$

The concept of surface integral may be extended to various types of surfaces that do not fulfill the hypotheses of Definition 5. If a surface S is a union of a finite number of surfaces $S = S_1 \cup S_2 \cup \cdots \cup S_n$, each of which satisfies the hypotheses of Definition 5, we define the surface integral of $f(x, y, z)$ over S to be the sum of the individual surface integrals

$$\iint_S f(x, y, z) \, dS = \sum_{j=1}^n \iint_{S_j} f(x, y, z) \, dS_j, \qquad (7)$$

each of which is evaluated by use of Definition 5.

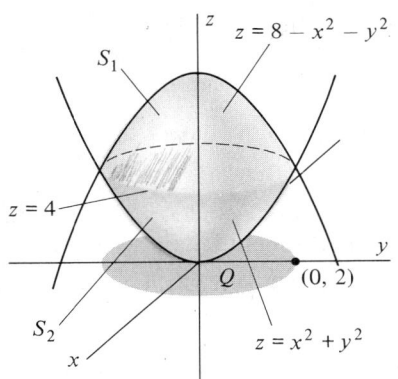

Figure 5.8 Surface S is not smooth along the curve $z = 4$.

Example 5

Evaluate $\iint_S (x^2 + y^2 + z) \, dS$ where the surface S is the boundary of the region bounded by the graphs of $g_1(x, y) = 8 - (x^2 + y^2)$ and $g_2(x, y) = x^2 + y^2$ (Figure 5.8).

Solution: The two graphs intersect above the circle $x^2 + y^2 = 4$. Let $Q = \{(x, y) \mid x^2 + y^2 \le 4\}$. We may therefore describe the surface S as $S = S_1 \cup S_2$ where

$$S_1 = \{(x, y, 8 - (x^2 + y^2)) \mid (x, y) \in Q\},$$
$$S_2 = \{(x, y, x^2 + y^2) \mid (x, y) \in Q\}.$$

On the surface S_1, with $f(x, y, z) = x^2 + y^2 + z$ and $g_1(x, y) = 8 - (x^2 + y^2)$, Definition 5 gives

$$\iint_{S_1} (x^2 + y^2 + z)\, dS = \iint_Q [x^2 + y^2 + (8 - (x^2 + y^2))]\sqrt{(-2x)^2 + (-2y)^2 + 1}\, dx\, dy$$

$$= \int_0^{2\pi} \int_0^2 8\sqrt{4r^2 + 1} \cdot r\, dr\, d\theta \qquad \text{(switching to polar coordinates)}$$

$$= \frac{4\pi}{3}(17^{3/2} - 1).$$

On the surface S_2, with $f(x, y, z) = x^2 + y^2 + z$ and $g_2(x, y) = x^2 + y^2$, we obtain

$$\iint_{S_2} (x^2 + y^2 + z)\, dS = \iint_Q [x^2 + y^2 + (x^2 + y^2)]\sqrt{(2x)^2 + (2y)^2 + 1}\, dx\, dy$$

$$= \int_0^{2\pi} \int_0^2 2r^2\sqrt{4r^2 + 1}\, r\, dr\, d\theta \qquad \text{(let } u = 4r^2 + 1\text{)}$$

$$= \frac{\pi}{20}(17^{5/2} - 1) - \frac{\pi}{12}(17^{3/2} - 1).$$

Combining these results according to equation (7) gives

$$\iint_S (x^2 + y^2 + z)\, dS = \frac{\pi}{20}(17^{5/2} - 1) + \frac{5\pi}{4}(17^{3/2} - 1)$$

$$\approx 458.3. \qquad \diamond$$

Definition 5 could not be applied directly to the entire surface S in Example 5 for two reasons. First, S is not the graph of a *function* $z = g(x, y)$, since there are two distinct points corresponding to some (x, y)-coordinates. Second, the surface S is not differentiable $\left(\text{meaning that } \dfrac{\partial g}{\partial x} \text{ and } \dfrac{\partial g}{\partial y} \text{ are not continuous}\right)$ along the "seam" where S_1 meets S_2.

Another difficulty that can arise in attempting to apply Definition 5 is that the integrand in the double integral has a singularity (i.e., becomes infinite) along the boundary of the region Q. In such cases we define $\iint_S f(x, y, z)\, dS$ to be the corresponding improper integral if this integral exists. The following example is typical. When combined with the notion of equation (7), this idea allows us to evaluate surface integrals over spheres, an important aspect of many applications of surface integrals.

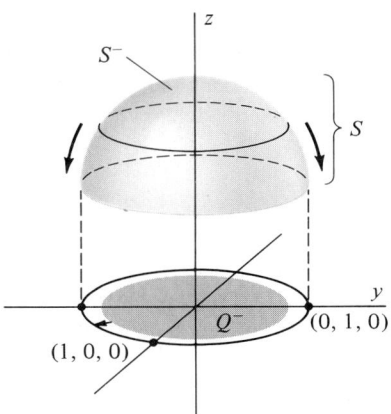

Figure 5.9 Surface integral over hemisphere $z = 1 + \sqrt{1 - x^2 - y^2}$ is evaluated as an improper integral.

Example 6

Evaluate the surface integral $\displaystyle\iint_S z \, dS$ where S is the hemisphere $z = 1 + \sqrt{1 - x^2 - y^2}$ (Figure 5.9).

Solution: Here $g(x, y) = 1 + \sqrt{1 - x^2 - y^2}$ and $f(x, y, z) = z$. Equation (3) becomes

$$\iint_S z \, dS = \iint_Q [1 + \sqrt{1 - x^2 - y^2}]$$

$$\cdot \sqrt{\left(\frac{-x}{\sqrt{1 - x^2 - y^2}}\right)^2 + \left(\frac{-y}{\sqrt{1 - x^2 - y^2}}\right)^2 + 1} \, dx \, dy$$

$$= \iint_Q \left[\frac{1}{\sqrt{1 - x^2 - y^2}} + 1\right] dx \, dy$$

where Q is the unit circle $x^2 + y^2 = 1$. However, this double integral is improper, since the denominator of the first term in the integrand vanishes as (x, y) approaches the boundary of Q. We evaluate this improper integral as suggested in Figure 5.9—by evaluating it over a disc Q^- of radius $\hat{r} < 1$ and then calculating the limit as $\hat{r}$ approaches 1. To do so we switch to polar coordinates and find that

$$\iint_S z \, dS = \iint_Q \left[\frac{1}{\sqrt{1 - r^2}} + 1\right] r \, dr \, d\theta$$

$$= \lim_{\hat{r} \to 1^-} \int_0^{\hat{r}} \int_0^{2\pi} \left[\frac{r}{\sqrt{1 - r^2}} + r\right] d\theta \, dr$$

$$= \lim_{\hat{r} \to 1^-} \int_0^{\hat{r}} 2\pi \left[\frac{r}{\sqrt{1 - r^2}} + r\right] dr$$

$$= \lim_{\hat{r} \to 1^-} 2\pi \left\{-\sqrt{1 - r^2} + \frac{1}{2}r^2\right\}\Big]_{r=0}^{r=\hat{r}}$$

$$= \lim_{\hat{r} \to 1^-} 2\pi \left\{-\sqrt{1 - \hat{r}^2} + \frac{1}{2}\hat{r}^2 - (-\sqrt{1})\right\}$$

$$= 3\pi. \qquad \diamond$$

Flux Integrals

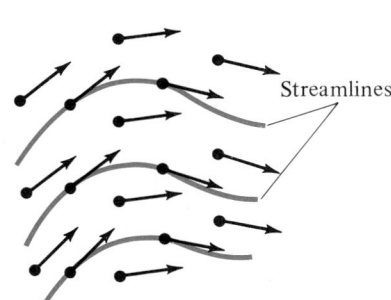

Figure 5.10 A velocity field and streamlines.

Returning now to our primary motivation concerning fluid flow, let us consider the motion of a fluid, such as water or a gas, through a region R in space. At each point (x, y, z) in the region, let $v(x, y, z)$ be the velocity vector for the motion of the fluid. That is, $v(x, y, z)$ is a vector pointing in the direction of motion whose length is the speed of the fluid at (x, y, z). We shall only consider the case of **steady state fluid motion.** This means that the velocity vector $v(x, y, z)$ does not change as time changes. Thus, $v(x, y, z)$ is a *vector field* defined on the region R. (The curves having the property of being tangent to the velocity field at each point are called the **streamlines** of the velocity field. See Figure 5.10.)

Now suppose that S is a permeable surface suspended in the fluid. (This means that the fluid can flow through the surface S, as coffee flows through a paper filter.)

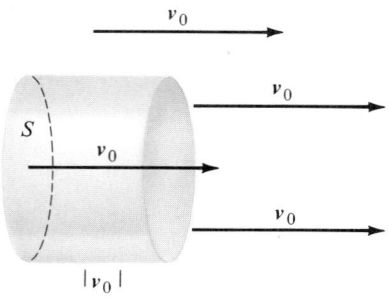

Figure 5.11 Volume of fluid flowing across S in unit time is $\Delta V = |v_0| \, \Delta S$.

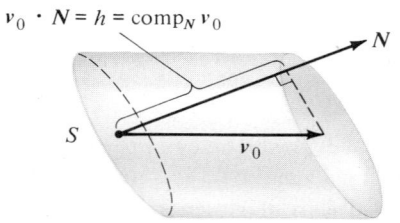

Figure 5.12 Volume of fluid crossing S in a unit of time is $\Delta V = (v_0 \cdot N) \, \Delta S$.

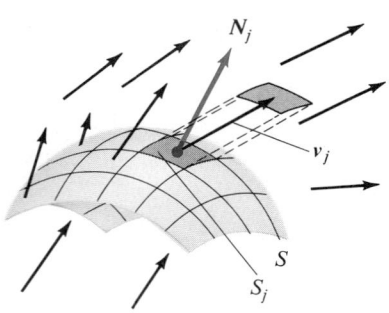

Figure 5.13 Volume of fluid crossing patch S_j is approximated by $\Delta V \approx (v_j \cdot N_j) \, \Delta S_j$.

For a steady state velocity field v, we define the **flux** of v across S as the rate at which mass is flowing across the surface S. Our interest here is in developing a formula by which flux can be calculated for a given surface and a given vector field.

We can simplify this problem somewhat by assuming that the density δ (mass per unit volume) is constant throughout the fluid. Since

$$\text{Mass} = (\text{density}) \times (\text{volume}) \tag{8}$$

we need only find the rate at which volume (i.e., fluid) is flowing across S. Multiplying by δ then gives the desired flux, according to equation (8).

The simplest case of this problem occurs when v is a constant vector field, $v(x, y, z) \equiv v_0$, and S is a flat surface perpendicular to v_0 (Figure 5.11). Since the speed at which fluid crosses S is $|v_0|$, in one unit of time the quantity

$$\Delta V = |v_0| \, \Delta S \tag{9}$$

crosses S, where ΔS is the area of S. If $v(x, y, z) = v_0$ is constant and S is flat, but not orthogonal to v_0, the situation is only slightly more complicated. Let N be a unit vector normal to S. (There are two possible choices for N. Take the one for which $0 \leq \theta \leq 90°$, where θ is the angle between N and v_0.) The volume of fluid that crosses S in a unit of time can be thought of as a prism with base of area ΔS and height $h = \text{comp}_N \, v_0 = v_0 \cdot N$, as illustrated in Figure 5.12. The desired volume is thus

$$\Delta V = (v_0 \cdot N) \, \Delta S. \tag{10}$$

The more general case of a curved surface S and a nonconstant velocity field v is handled in a familiar way. First, we assume that S has a normal vector $N(x, y, z)$ at each point and that $N(x, y, z)$ is continuous on S. Also, we assume that S can be recognized as having two sides and that $N(x, y, z)$ always remains on the same side of S. All these assumptions are met if S is the graph of a differentiable function $z = g(x, y)$ over a regular region Q in the plane.

By partitioning Q, as in Figure 5.4, we divide S into nonoverlapping patches $S_1, S_2, \ldots, S_n$. For each patch we choose one point $(x_j, y_j, z_j) \in S_j$, and we let ΔS_j denote the surface area of S_j. Also, we approximate $v(x, y, z)$ for every $(x, y, z) \in S_j$ by the vector $v_j = v(x_j, y_j, z_j)$ and we let $N_j = N(x_j, y_j, z_j)$. Using equation (10), we approximate the volume ΔV_j of fluid crossing the patch S_j in a unit of time as

$$\Delta V_j \approx (v_j \cdot N_j) \, \Delta S_j \qquad \text{(Figure 5.13).}$$

Summing these approximations over all patches gives the approximation

$$\Delta V \approx \sum_{j=1}^{n} (v_j \cdot N_j) \, \Delta S_j. \tag{11}$$

Finally, we argue that as the number of patches becomes large (and their individual sizes become small) the approximation in (11) should converge to the desired rate. Since the right-hand side of approximation (11) has the form of the approximating sum in equation (1), we conclude from equations (8) and (11) that the rate at which the mass of the fluid is flowing across S (that is, the flux) is given by the surface integral

$$\frac{dM}{dt} = \iint\limits_{S} \delta v(x, y, z) \cdot N(x, y, z) \, dS. \tag{12}$$

The integral in (12) is sometimes called the **rate of mass transport** across S in the direction of N.

Example 7

A fluid with mass density δ is emanating from the origin and flowing according to the central velocity field

$$v(x, y, z) = \lambda(x\boldsymbol{i} + y\boldsymbol{j} + z\boldsymbol{k})$$

where λ is constant. Find the rate of mass transport, $\dfrac{dM}{dt}$, across the square region $S = \{(x, y, z) \mid z = 2,\ -1 \le x \le 1,\ -1 \le y \le 1\}$.

Solution: The square S is horizontal, so we use the unit normal $N = \boldsymbol{k}$ at each point. Then $v(x, y, z) \cdot N = \lambda z$, so, by (12),

$$\frac{dM}{dt} = \iint\limits_{S} \delta \lambda z\ dS = \int_{-1}^{1} \int_{-1}^{1} 2\delta\lambda\ dx\ dy = 8\delta\lambda.$$

Note that the choice of $N = \boldsymbol{k}$ was arbitrary. If we had instead used $N = -\boldsymbol{k},$ the resulting integral would have yielded $\dfrac{dM}{dt} = -8\delta\lambda$. This simply says that the flow is in the direction *opposite* the vector $N = -\boldsymbol{k}$ (see Figures 5.14 and 5.15). ◇

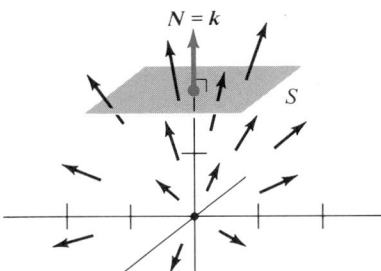

Figure 5.14 Flow across S in direction of $N = \boldsymbol{k}$.

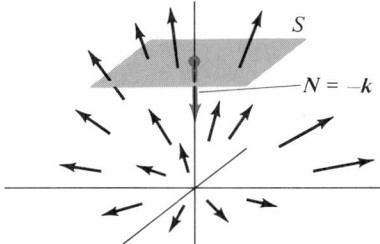

Figure 5.15 Flow in direction opposite to $N = -\boldsymbol{k}$.

A more general concept than that of equation (12) is obtained by letting F denote any vector field and $N = N(x, y, z)$ be a unit normal defined at each point on the surface S as described above. The flux of the vector field F across S in the direction of N is defined to be the surface integral

$$\text{Flux of } F \text{ across } S = \iint\limits_{S} F \cdot N\ dS. \tag{13}$$

Thus (12) is a special case of (13) with $F = \delta v.$ However, the vector field in (13) need not be a velocity field for a fluid. It may, for example, be an electric field induced by one or more point charges, or a gradient field for a temperature function.

In working with either equation (12) or equation (13) you will need to recall that a normal $N(x, y, z)$ to the surface S that is the graph of $z = g(x, y)$ can be found by writing S as a level surface

$$f(x, y, z) = z - g(x, y) = 0$$

and forming

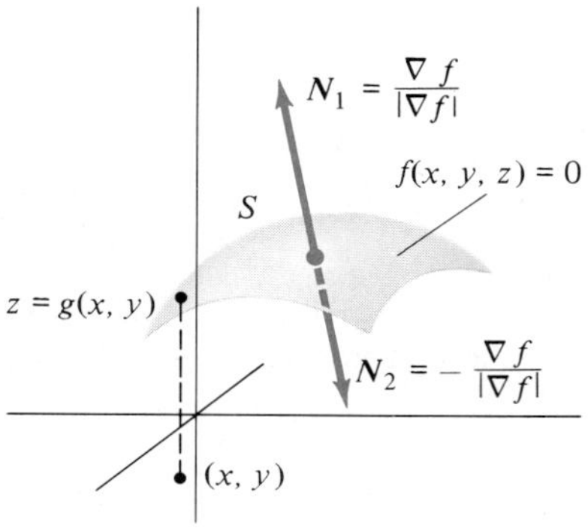

$$N_1 = \frac{\nabla f}{|\nabla f|} = \frac{-\dfrac{\partial g}{\partial x}i - \dfrac{\partial g}{\partial y}j + k}{\sqrt{\left(\dfrac{\partial g}{\partial x}\right)^2 + \left(\dfrac{\partial g}{\partial y}\right)^2 + 1}} \tag{14}$$

or

$$N_2 = -\frac{\nabla f}{|\nabla f|} = \frac{\dfrac{\partial g}{\partial x}i + \dfrac{\partial g}{\partial y}j - k}{\sqrt{\left(\dfrac{\partial g}{\partial x}\right)^2 + \left(\dfrac{\partial g}{\partial y}\right)^2 + 1}}. \tag{15}$$

Figure 5.16 Choosing a unit normal to graph of $f(x, y, z) = z - g(x, y) = 0$.

The choice between N_1 and N_2 is determined by whether you wish N to point upward (N_1) or downward (N_2) from the graph of $z = g(x, y)$ (Figure 5.16).

Before proceeding to our final example, we use equations (14) and (15) to express the flux integral (13) in a more manageable form. If the surface S is the graph of $z = g(x, y)$ and the vector field F has the component form

$$F(x, y, z) = M(x, y, z)i + N(x, y, z)j + P(x, y, z)k,$$

then the flux integral (13) corresponding to the upward unit normal N_1 in (14) is, according to Definition 5,

$$\iint_S F \cdot N\, dS = \iint_Q (Mi + Nj + Pk)$$

$$\cdot \left(\frac{-\dfrac{\partial g}{\partial x}i - \dfrac{\partial g}{\partial y}j + k}{\sqrt{\left(\dfrac{\partial g}{\partial x}\right)^2 + \left(\dfrac{\partial g}{\partial y}\right)^2 + 1}}\right)\left(\sqrt{\left(\dfrac{\partial g}{\partial x}\right)^2 + \left(\dfrac{\partial g}{\partial y}\right)^2 + 1}\right)dx\, dy$$

where Q is the region in the xy-plane over which S is defined. This simplifies to

$$\iint_S F \cdot N\, dS = \iint_Q \left[-M(x, y, g(x, y))\frac{\partial g}{\partial x} - N(x, y, g(x, y))\frac{\partial g}{\partial y}\right.$$
$$\left. + P(x, y, g(x, y))\right] dx\, dy. \tag{16}$$

Similarly, the flux integral (13) corresponding to the downward unit normal N_2 in (15) is

$$\iint_S F \cdot N\, dS = \iint_Q \left[M(x, y, g(x, y))\frac{\partial g}{\partial x} + N(x, y, g(x, y))\frac{\partial g}{\partial y}\right.$$
$$\left. - P(x, y, g(x, y))\right] dx\, dy. \tag{17}$$

Example 8

Calculate the flux over the sphere S with center $(0, 0, 0)$ and radius R associated with the (central) inverse square field

$$\mathbf{F}(\mathbf{r}) = \frac{\lambda \mathbf{r}}{|\mathbf{r}|^3}, \qquad \mathbf{r} \neq \mathbf{0}, \qquad \lambda \text{ constant.}$$

That is,

$$\mathbf{F}(x, y, z) = \lambda \cdot \frac{x\mathbf{i} + y\mathbf{j} + z\mathbf{k}}{(x^2 + y^2 + z^2)^{3/2}}, \qquad x^2 + y^2 + z^2 \neq 0.$$

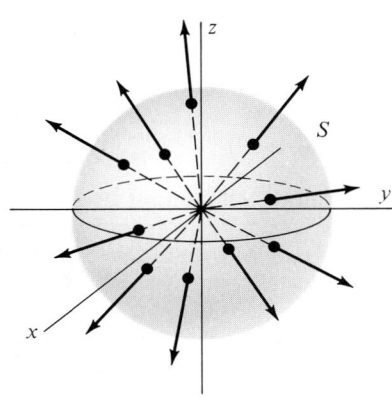

Figure 5.17 Flux of a central force field over a sphere is twice the flux over the upper hemisphere $z = \sqrt{R^2 - x^2}$.

Solution: Since the force field $\mathbf{F}$ has the same symmetry with respect to the origin as does the sphere S, we will calculate the flux over the upper hemisphere only and then double the result (see Figure 5.17). Note that we are dealing with an improper integral as in Example 6. Using equation (16) with

$$M(x, y, z) = \frac{\lambda x}{(x^2 + y^2 + z^2)^{3/2}} = \frac{\lambda x}{R^3},$$

$$N(x, y, z) = \frac{\lambda y}{(x^2 + y^2 + z^2)^{3/2}} = \frac{\lambda y}{R^3},$$

$$P(x, y, z) = \frac{\lambda z}{(x^2 + y^2 + z^2)^{3/2}} = \frac{\lambda z}{R^3},$$

and

$$g(x, y) = \sqrt{R^2 - x^2 - y^2} = z$$

gives

$$\iint_S \mathbf{F} \cdot \mathbf{N}\, dS = 2 \iint_Q \left[-\frac{\lambda x}{R^3} \left(\frac{-x}{\sqrt{R^2 - x^2 - y^2}} \right) \right.$$

$$\left. - \frac{\lambda y}{R^3} \left(\frac{-y}{\sqrt{R^2 - x^2 - y^2}} \right) + \frac{\lambda \sqrt{R^2 - x^2 - y^2}}{R^3} \right] dx\, dy$$

$$= 2 \iint_Q \frac{\lambda}{R\sqrt{R^2 - x^2 - y^2}}\, dx\, dy$$

$$= 2 \int_0^{2\pi} \int_0^R \frac{\lambda}{R\sqrt{R^2 - r^2}}\, r\, dr\, d\theta \qquad \text{(switch to polar coordinates)}$$

$$= 2 \int_0^{2\pi} -\left[\frac{\lambda}{R} \sqrt{R^2 - r^2} \right]_{r=0}^{r=R} d\theta$$

$$= 4\pi\lambda. \qquad\qquad\qquad \diamond$$

The surprising result of Example 8 is that *the flux of the inverse square field $\mathbf{F}$ over the sphere S is independent of the radius of S!* This result is true for any inverse square field, such as a gravitational field or an electric field. In the latter case the result of Example 8, together with another property of electric fields (Coulomb's

Law), gives **Gauss's Law:**

The flux of the electric field E through any closed surface S is

$$\iint\limits_{S} E \cdot N \, dS = 4\pi \sum_{j=1}^{n} q_j$$

where $q_1, q_2, \ldots, q_n$ are the point charges contained within S.

The Orientation of a Surface

When the surface S is the graph of a single function $z = g(x, y)$, there is no ambiguity about what is meant by the *upward* unit normal (as opposed to the downward unit normal), so a flux integral is easily determined to be of the form (16) or of the form (17). Also, in the cases where S is the graph of the function $y = g(x, z)$ or $x = g(y, z)$, the corresponding meaning of upward normal and the modifications of equations (16) and (17) should be clear. However, the situation becomes less clear as we think about surfaces of the form $S = S_1 \cup S_2 \cup \cdots \cup S_n$ where each S_j is as above. We need to devote just a few more lines to the meaning of N in the flux integral (13) for these more general cases.

According to our original motivation, flux is a measure of the *net* (positive minus negative) rate of flow across S. If more fluid is flowing upward across S: $z = g(x, y)$ than is flowing downward, flux is positive, although fluid may actually be flowing upward across some portions of S and downward across others. It is therefore crucial that N always point on the same "side" of S, since N provides the gauge with which the rate of flow is measured at each point. Thus, in order for the flux integral in (13) to make sense, the surface S must be **orientable,** that is, we must be able to identify two distinct sides of S and to assign a unique unit normal $N(x, y, z)$ that remains on one side or the other as (x, y, z) varies across S.

We have already seen that this is simple to do for surfaces of the form

$$S = \{(x, y, z) \mid z = g(x, y), \ (x, y) \in Q\}.$$

We just choose the unit normal with positive z-component (upward) or negative z-component (downward). For closed surfaces (such as a sphere or an ellipsoid) the two sides are the inside and the outside, and we must choose between the *inner* unit normal or the *outer* unit normal. (Figure 5.17 shows outer unit normals for a sphere.)

These two general situations (upper versus lower, or inner versus outer) encompass most of the surfaces with which you will need to deal, and we refer to any such surface as *orientable.* We shall not attempt to give a precise definition for this term (as is done in more advanced courses). However, you should know that not all surfaces are orientable. A classic example is the Möbius band, which is obtained by twisting one end of a long rectangular strip $180°$ and "gluing" it to the other end. As Figure 5.18 shows, a unit normal N beginning at point P and moving continuously once "around" the Möbius band arrives back at P pointing in the opposite direction.

In summary, we caution you to note that the surface integral in Definition 5 does not involve the notion of orientation for S. However, the flux integral in (13) does depend on an orientation for S. Reversing the orientation for S (and, hence, replacing N by $-N$) changes the sign of the integral in (13).

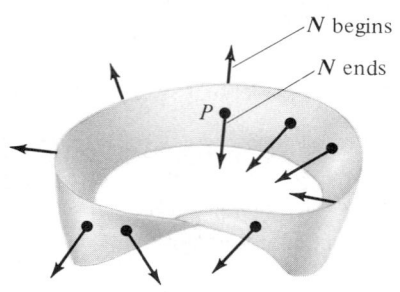

N begins

N ends

P

Figure 5.18 Möbius band is not orientable.

Exercise Set 20.5

1. Evaluate $\iint_S (x + 3y + z)\, dS$ over the portion of the plane

   $x - y + 2z = 4$ lying above the rectangle $Q = \{(x, y) \mid 0 \le x \le 1, 0 \le y \le 2\}$.

2. Evaluate the surface integral in Exercise 1 where S is the portion of the plane $x - y + 2z = 4$ lying above the unit circle.

3. Evaluate $\iint_S x^2 z\, dS$ where S is the portion of the plane

   $2x + 3y + z = 6$ lying in the first octant $(x \ge 0, y \ge 0, z \ge 0)$.

4. Evaluate $\iint_S y^2 z\, dS$ where S is the portion of the cone $z^2 = $

   $x^2 + y^2$ lying between the planes $z = 1$ and $z = 4$.

5. Evaluate the surface integral $\iint_S (x^2 + 5y - z)\, dS$ where S

   is the graph of the cylinder $z = x^2$ over the square $-1 \le x \le 1, -1 \le y \le 1$.

6. Evaluate $\iint_S (x^2 - y^2 + 1)\, dS$ where S is the part of the

   plane $y = x + 2$ inside the cylinder $x^2 + y^2 = 4$.

7. Find the mass of the part of a sphere $x^2 + y^2 + z^2 = 9$ lying inside the cylinder $x^2 + y^2 = 4$ if the density per unit area δ is constant.

8. Find $\iint_S F \cdot N\, dS$ where $F = 2xi - yj + 3zk$, S is the part

   of the plane $x - y + 2z = 4$ bounded by the planes $y = 0$, $y = -4$, $x = 0$, and $x = 4$, and N is the upward unit normal.

9. Find $\iint_S (x + y + z^2)\, dS$ where S is the hemisphere $z =$

   $\sqrt{4 - x^2 - y^2}$.

10. Evaluate $\iint_S (y - x)\, dS$ where S is the part of the plane

    $6x + 4y + 2z = 8$ lying inside the cylinder $x^2 + y^2 = 4$.

11. Find the flux $\iint_S F \cdot N\, dS$ where $F = xi + yj + zk$, S is the

    upper half sphere $z = \sqrt{1 - x^2 - y^2}$, and N is the upward unit normal.

12. Use the result of Exercise 11 to find $\iint_S F \cdot N\, dS$ where

$F = xi + yj + zk$, S is the entire unit sphere, and N is the outward unit normal.

13. Calculate the flux integral $\iint_S F \cdot N\, dS$ where $F = 2xi +$

    $yj + zk$, S is the surface of the paraboloid $z = x^2 + y^2$ bounded by the planes $z = 0$ and $z = 4$, and N is the outward unit normal.

14. Let $F = 2i - xj + yk$. Find the flux integral $\iint_S F \cdot N\, dS$

    where S is the portion of the cylinder $x^2 + y^2 = 1$ lying between the planes $z = 0$ and $z = 1$, and N is the outward unit normal.

15. Find the flux outward through the sphere $x^2 + y^2 + z^2 = a^2$ for the vector field $F(x, y, z) = zk$. What is the flux if $F(x, y, z) = yj$? Can you give a geometric argument for these results?

16. Find $\iint_S F \cdot N\, dS$ where S is the ellipsoid $4x^2 + y^2 +$

    $z^2 = 4$, $F = x^2 i + y^2 j + z^2 k$, and N is the outward unit normal. (This will require a geometric argument.)

17. Find the flux $\iint_S F \cdot N\, dS$ where $F = xi + yj + zk$, N is

    the outward unit normal, and S is the surface of the unit cube $\{(x, y, z) \mid 0 \le x \le 1, 0 \le y \le 1, 0 \le z \le 1\}$.

18. Suppose that $F(x, y, z) = axi + byj + czk$ and that S is part of the plane $ax + by + cz = d$. Show that the flux integral $\iint_S F \cdot N\, dS$ is $(a^2 + b^2 + c^2)A$ where A is the area of S, if

    N is the unit normal with $N \cdot ck \ge 0$.

19. Is the result of Example 8 the same if F is an inverse *cube*

    field $F(r) = \dfrac{r}{|r|^4}$? What is the result in this case?

20. Show that the *flux* of the vector field $F = Mi + Nj + Pk$ across a surface S can be written

$$\iint_S F \cdot N\, dS = \iint_S (P \cos \alpha + Q \cos \beta + R \cos \gamma)\, dS$$

where α, β, and γ are the direction numbers for the unit normal N.

For a surface S, the coordinates $(\bar{x}, \bar{y}, \bar{z})$ of the *centroid* are defined to be

$$\bar{x} = \frac{1}{A}\iint_S x\, dS, \quad \bar{y} = \frac{1}{A}\iint_S y\, dS, \quad \bar{z} = \frac{1}{A}\iint_S z\, dS$$

where A is the surface area of S. Use these formulas in Exercises 21–23.

21. Find the centroid of the hemisphere $z = \sqrt{a^2 - x^2 - y^2}$, $z \geq 0$. (*Hint:* Use symmetry as much as possible.)

22. Find the centroid of the cylinder $z = 1 - x^2$, $-2 \leq y \leq 2$, $-1 \leq x \leq 1$.

23. Find the centroid of the part of the spherical surface $x^2 + y^2 + z^2 = 1$ lying in the first octant.

24. Show that the surface integral, as defined by equation (1), can be written in the form

$$\iint_S f(x, y, z)\, dS = \iint_Q f(x, y, g(x, y)) \sec \gamma \, dx \, dy$$

where $\gamma = \gamma(x, y)$ is the angle between the upward normal to the graph of $z = g(x, y)$ at (x, y, z) and the vector $\mathbf{k}$. (*Hint:* Refer to Section 19.5, Exercise 20.)

25. Show that if the function $f(r, \theta, z)$ is written using cylindrical coordinates for $\mathbb{R}^3$ and if the surface S is the graph of the function $z = g(r, \theta)$ in polar coordinates, then the formula in Definition 5 becomes

$$\iint_S f(r, \theta, z)\, dS = \iint_Q f(r, \theta, g(r, \theta))$$

$$\cdot \sqrt{r^2 + r^2\left(\frac{\partial g}{\partial r}\right)^2 + \left(\frac{\partial g}{\partial \theta}\right)^2}\, dr \, d\theta.$$

20.6 STOKES' THEOREM

In this section we encounter a generalization of Green's Theorem, which we have earlier stated as

$$\int_C \mathbf{F} \cdot d\mathbf{r} = \iint_Q \left(\frac{\partial N}{\partial x} - \frac{\partial M}{\partial y}\right) dA = \iint_Q (\text{curl } \mathbf{F})\, dA. \tag{1}$$

Recall the setting of equation (1): $\mathbf{F}(x, y) = M(x, y)\mathbf{i} + N(x, y)\mathbf{j}$ is a vector field in the plane (with continuously differentiable components), C is a piecewise smooth simple closed curve enclosing the region Q, and curl $\mathbf{F}$ is the scalar function

$$\text{curl } \mathbf{F} = \frac{\partial N}{\partial x} - \frac{\partial M}{\partial y}. \tag{2}$$

All this takes place in the plane, $\mathbb{R}^2$.

The generalization referred to above is Stokes' Theorem, which differs from Green's Theorem in three ways. First, the setting for Stokes' Theorem is $\mathbb{R}^3$. C will now be a closed space curve, and the vector field $\mathbf{F}$ will have the form

$$\mathbf{F}(x, y, z) = M(x, y, z)\mathbf{i} + N(x, y, z)\mathbf{j} + P(x, y, z)\mathbf{k}. \tag{3}$$

Second, whereas the integral on the right-hand side of equation (1) is a double integral over the plane region Q, the right-hand side of Stokes' Theorem will involve a surface integral evaluated over a smooth surface S bounded by the closed space curve C. Figures 6.1 and 6.2 illustrate the geometry associated with these two theorems.

Stokes' Theorem (which we shall state in more precise terms later) says this:

$$\int_C \mathbf{F} \cdot d\mathbf{r} = \iint_S (\text{curl } \mathbf{F}) \cdot \mathbf{N}\, dS \tag{4}$$

where $\mathbf{N}$ is a unit normal for the surface S.

The strong similarity between equations (1) and (4) is obvious. However, the integrand in the surface integral in (4) has not yet been defined. This brings us to the third major difference between these two theorems.

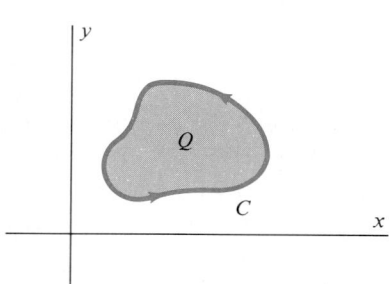

Figure 6.1 Curve C and region Q in Green's Theorem.

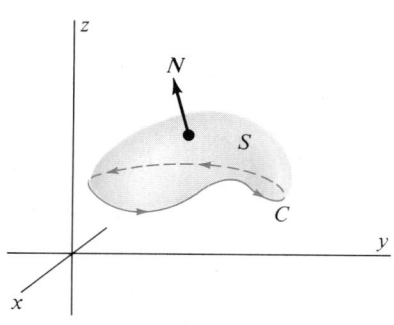

Figure 6.2 The surface S bounded by the space curve C in Stokes' Theorem.

Recall that the line integral in (1) was interpreted in Section 20.4 as the *circulation* around C of a fluid flowing with velocity field F. This interpretation issued from the equation

$$\int_C F \cdot dr = \int_C F \cdot T \, ds \tag{5}$$

discussed earlier. Now equation (5) holds for vector fields of the form (3) defined on space curves as well as in the planar case. Retaining the notion of fluid flow as our primary motivation, we would like to define curl F for F in (3) in such a way that equation (4) can be interpreted as a statement about fluid flow (and, of course, so that equation (4) is true!). The following definition of curl F does this. We shall defer further physical interpretation to the end of this section.

DEFINITION 6

Let F be a vector field of the form

$$F(x, y, z) = M(x, y, z)i + N(x, y, z)j + P(x, y, z)k.$$

If all first order partial derivatives for each component function M, N, and P exist at (x, y, z), **curl $F(x, y, z)$** is defined to be the vector

$$\text{curl } F(x, y, z) = \left(\frac{\partial P}{\partial y} - \frac{\partial N}{\partial z} \right)i + \left(\frac{\partial M}{\partial z} - \frac{\partial P}{\partial x} \right)j + \left(\frac{\partial N}{\partial x} - \frac{\partial M}{\partial y} \right)k \tag{6}$$

where each partial derivative is evaluated at (x, y, z).

It is important to note that curl $F(x, y, z)$ is a *vector* in $\mathbb{R}^3$ while curl $F(x, y)$ in (2) is a *number*. Now curl $F(x, y, z) \cdot N$ (the integrand in the surface integral in Stokes' Theorem) is a generalization of curl $F(x, y)$ (the integrand in the double integral in Green's Theorem) in the following sense. If we restrict F in (3) to the xy-plane by setting $z \equiv 0$ and $P(x, y, z) \equiv 0$, equation (6) gives

$$\text{curl } F(x, y, 0) = \left(\frac{\partial N}{\partial x} - \frac{\partial M}{\partial y} \right)k. \tag{7}$$

Since a unit normal to the xy-plane is $N = k$, we obtain

$$[\text{curl } F(x, y, 0)] \cdot N = \left[\left(\frac{\partial N}{\partial x} - \frac{\partial M}{\partial y} \right)k \right] \cdot k = \frac{\partial N}{\partial x} - \frac{\partial M}{\partial y}$$

$$= \text{curl } F(x, y), \text{ as claimed.}$$

You will often see the notation

$$\text{curl } F = \nabla \times F. \tag{8}$$

This results from thinking of the gradient ∇ as an "operator"

$$\nabla = \frac{\partial}{\partial x}i + \frac{\partial}{\partial y}j + \frac{\partial}{\partial z}k$$

and forming the cross product*

---

*As in the mnemonic for remembering how to compute cross products, the determinant here is only formal notation, since only the third row of the matrix actually contains numbers.

$$\nabla \times F = \det \begin{bmatrix} i & j & k \\ \dfrac{\partial}{\partial x} & \dfrac{\partial}{\partial y} & \dfrac{\partial}{\partial z} \\ M & N & P \end{bmatrix} \tag{9}$$

$$= \left(\frac{\partial P}{\partial y} - \frac{\partial N}{\partial z} \right) i + \left(\frac{\partial M}{\partial z} - \frac{\partial P}{\partial x} \right) j + \left(\frac{\partial N}{\partial x} - \frac{\partial M}{\partial y} \right) k.$$

Example 1

For the vector field $F(x, y, z) = yi - z^2 j + 2xk$, $M = y$, $N = -z^2$, and $P = 2x$. From equations (8) and (9) we find that

$$\text{curl } F = \det \begin{bmatrix} i & j & k \\ \dfrac{\partial}{\partial x} & \dfrac{\partial}{\partial y} & \dfrac{\partial}{\partial z} \\ y & -z^2 & 2x \end{bmatrix} = 2zi - 2j - k. \qquad \diamond$$

One final detail needs to be cleaned up before we state Stokes' Theorem. We shall need to require that the surface S be **simply connected.** Basically, this means that S contains no holes. Another way to say this is that, given any point P_0 on S, the boundary curve C can be "continuously contracted" across S to an arbitrarily small closed curve containing P_0. (If S had a hole you could not do this since the contraction of C would get "hung up" at the hole (see Figure 6.3).

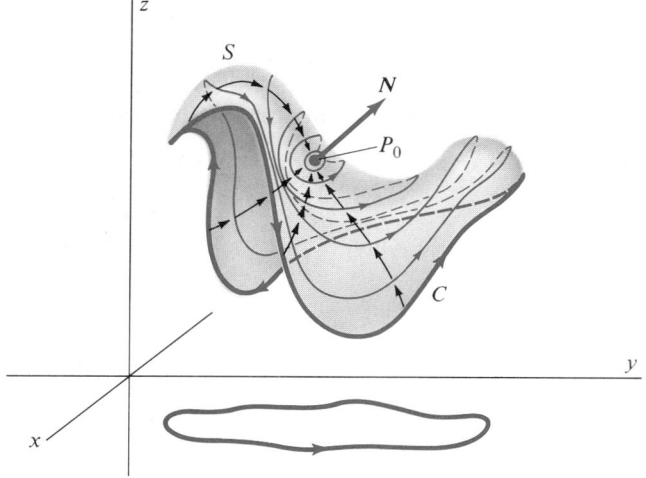

Figure 6.3 An oriented simply connected surface S. Contracting C continuously across S reveals the orientation of C induced by the orientation of S.

This notion of contracting C across S toward a point also allows us to clarify the issue of orientation. We shall require that the surface S be oriented. As defined in Section 20.5, this means that S comes equipped with a unit normal N that varies continuously across S, always remaining on the same side of S. In the statement of

Stokes' Theorem we must require that the orientation of the boundary curve C be the orientation *induced* on C by the orientation of S. Here's what we mean. Let N be the unit normal to S at the point P_0. If S is smooth, then very near P_0 the surface of S may be approximated by a plane tangent to S at P_0. Think of a very small circle C_S on this tangent plane with center P_0. This small circle approximates a small closed curve on S, and we can continuously contract C across S so that it takes the shape of this small closed curve. The point is that the orientation of all three curves (C, the small curve on S, and the small circle C_S on the tangent plane) must be the same. N determines an orientation for C_S by a variation of the right-hand rule: Think of wrapping the index finger of your right hand around C_S, keeping your thumb parallel to N. Your index finger points out the orientation of C_S (see Figure 6.4). This is the orientation induced by N on C.

We are now ready to state Stokes' Theorem.

THEOREM 4
Stokes' Theorem

Let S be a smooth, simply connected, oriented surface bounded by a piecewise smooth simple closed curve C with orientation induced by the orientation of S. Let

$$F(x, y, z) = M(x, y, z)\boldsymbol{i} + N(x, y, z)\boldsymbol{j} + P(x, y, z)\boldsymbol{k}$$

be a vector field defined on an open box D containing S and C so that each of the component functions M, N, and P is continuous and has continuous partial derivatives on D. Then

$$\int_C \boldsymbol{F} \cdot d\boldsymbol{r} = \iint_S (\text{curl } \boldsymbol{F}) \cdot N \, dS \tag{10}$$

where N is the unit normal to the oriented surface S.

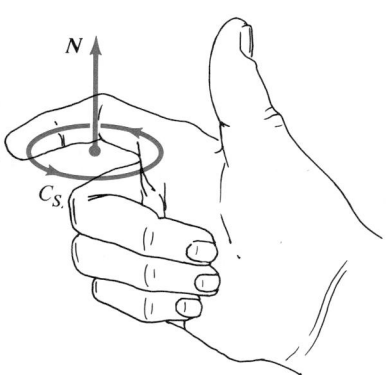

Figure 6.4 Right-hand rule determines orientation of C_S from direction of N.

Using equation (3), we may write equation (10) in the alternate form

$$\int_C M \, dx + N \, dy + P \, dz = \iint_S (\text{curl } \boldsymbol{F}) \cdot N \, dS.$$

We may paraphrase Stokes' Theorem by saying that "the line integral of the vector field around the boundary of S equals the integral of the normal component of curl $\boldsymbol{F}$ over the surface S."

Note that the particular shape of S is unimportant—the surface integral over S is entirely determined by the line integral along its boundary. For example, the surface integrals of $(\text{curl } \boldsymbol{F}) \cdot N$ over all of the surfaces in Figure 6.5 are the same, since each has the same boundary C.

A proof of Stokes' Theorem is beyond the scope of this text. (If you've noticed that the concepts of this chapter sometimes appear to be less precise than in earlier chapters, you are correct. We are now dabbling in ideas that require considerably more advanced mathematics to treat rigorously than we can hope to achieve in a first calculus course.) However, we can discuss why Stokes' Theorem should hold.

First consider the case of the tetrahedron sketched in Figure 6.6. We shall take the surface S to be the union of the three faces ABE, BDE, and DAE. Then S is bounded by the triangle ABD, which is the curve C. We orient S by taking the outward unit normal N at each point, which induces the orientation ABD on C.

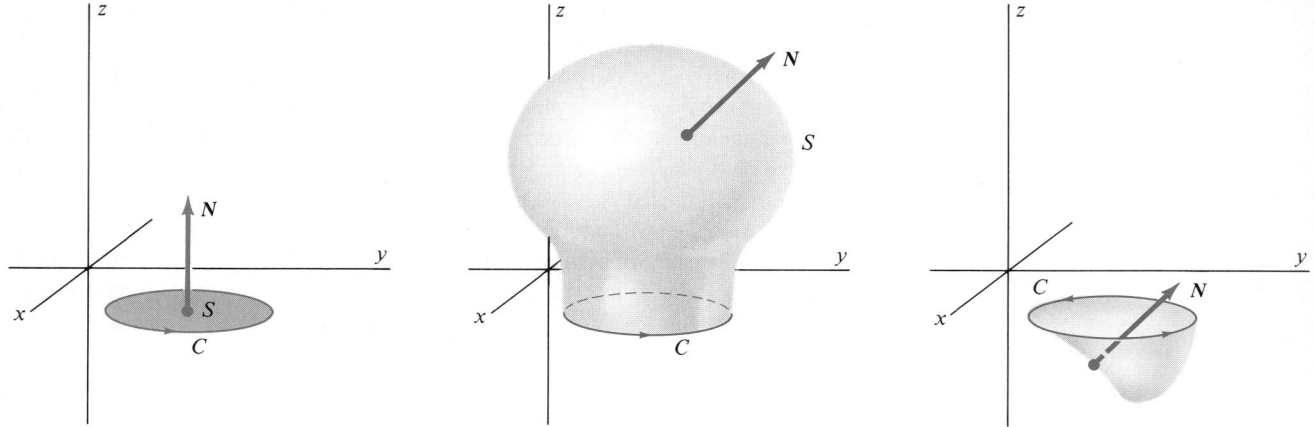

Figure 6.5 The integrals over all of these surfaces of (curl F) $\cdot$ N are the same: $\iint_S (\text{curl } F) \cdot N \, dS = \int_C F = dr$.

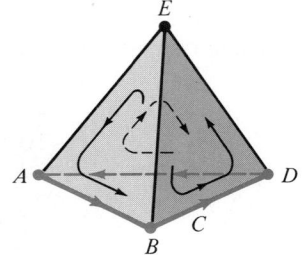

Figure 6.6 Stokes' Theorem holds for a tetrahedral surface.

Since each face of the surface is a plane region, we apply Green's Theorem to each face and obtain the three equations

$$\int_{ABE} F \cdot dr = \iint_{ABE} (\text{curl } F) \cdot N \, dS, \tag{11}$$

$$\int_{BDE} F \cdot dr = \iint_{BDE} (\text{curl } F) \cdot N \, dS, \tag{12}$$

$$\int_{DAE} F \cdot dr = \iint_{DAE} (\text{curl } F) \cdot N \, dS. \tag{13}$$

(In each of equations (11) through (13), we have used the remark concerning equation (7) to write the scalar curl in vector form.)

Next, we wish to add the corresponding sides in equations (11) through (13). When we do so, the contributions to the resulting line integral along the edges AE, BE, and DE will be zero, since each of these edges is traversed twice, once in each direction. Also, the three surface integrals should sum to the surface integral over S. The result, then, is

$$\int_{ABD} F \cdot dr = \iint_S (\text{curl } F) \cdot N \, dS$$

which is equation (10).

Now consider a more complex polyhedron S, such as that in Figure 6.7. The idea is the same. When we apply Green's Theorem to each face and sum the corresponding sides of the resulting equations, the only nonzero contributions to the line integral occur along the bottom edges. Along all other edges, the line integral will have been evaluated twice, once in each direction. Referring to the bottom edge as C, we again obtain equation (10).

The argument for a smooth surface S is to first approximate S by a sequence of

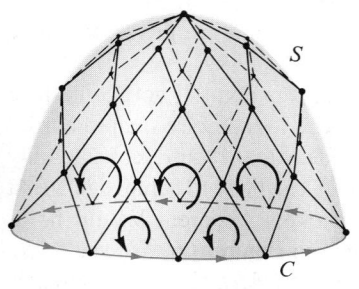

Figure 6.7 A polyhedral approximation to the surface S.

polyhedra, each of whose vertices lies on the surface S. As the number of vertices in the polyhedra becomes large, the polyhedra become increasingly accurate approximations to the surface. Since equation (10) holds for each polyhedron, it holds for the smooth surface "in the limit."

Example 2

Find $\displaystyle\int_C \boldsymbol{F} \cdot d\boldsymbol{r}$ where $\boldsymbol{F}$ is the vector field $\boldsymbol{F}(x, y, z) = y\boldsymbol{i} - z^2\boldsymbol{j} + 2x\boldsymbol{k}$ and C is the unit circle $C = \{(x, y, 0) \mid x^2 + y^2 = 1\}$ oriented in the counterclockwise direction.

Solution: We could evaluate the line integral directly (and, painfully) by parameterizing C by $x(t) = \cos t$, $y(t) = \sin t$, $0 \le t \le 2\pi$. Instead, we will apply Stokes' Theorem using any smooth surface S bounded by C. The simplest choice, of course, is the unit disc

$$S = \{(x, y, 0) \mid 0 \le x^2 + y^2 \le 1\}.$$

From Example 1 we know that

$$\operatorname{curl} \boldsymbol{F} = 2z\boldsymbol{i} - 2\boldsymbol{j} - \boldsymbol{k}$$

and a unit normal giving S the required orientation is $\boldsymbol{N} = \boldsymbol{k}$. Thus

$$(\operatorname{curl} \boldsymbol{F}) \cdot \boldsymbol{N} = (2z\boldsymbol{i} - 2\boldsymbol{j} - \boldsymbol{k}) \cdot \boldsymbol{k} = -1.$$

Thus, by Stokes' Theorem,

$$\int_C \boldsymbol{F} \cdot d\boldsymbol{r} = \iint_S (-1)\, dS = -\pi,$$

since π is the area of the unit disc. $\diamond$

Example 3

Verify Stokes' Theorem if $\boldsymbol{F}$ is the vector field

$$\boldsymbol{F}(x, y, z) = y\boldsymbol{i} - x^2\boldsymbol{j} + 2z^2\boldsymbol{k}$$

and S is the portion of the paraboloid $z = 4 - x^2 - y^2$ lying above the xy-plane. Take $\boldsymbol{N}$ to be the upward unit normal to S.

Solution: The curve C bounding S is the circle

$$C = \{(x, y, 0) \mid x^2 + y^2 = 4\}$$

which has parameterization $x(t) = 2 \cos t$, $y(t) = 2 \sin t$, $z(t) = 0$, $0 \le t \le 2\pi$. The line integral on the left side of equation (10) is therefore

$$\int_C \boldsymbol{F} \cdot d\boldsymbol{r} = \int_C y\, dx - x^2\, dy + 2z^2\, dz$$

$$= \int_0^{2\pi} (-4 \sin^2 t - 8 \cos^3 t)\, dt$$

$$= -4\pi.$$

To calculate the surface integral in (10), we first use (9) to find

$$\text{curl } \boldsymbol{F} = \nabla \times \boldsymbol{F} = \det \begin{bmatrix} \boldsymbol{i} & \boldsymbol{j} & \boldsymbol{k} \\ \dfrac{\partial}{\partial x} & \dfrac{\partial}{\partial y} & \dfrac{\partial}{\partial z} \\ y & -x^2 & 2z^2 \end{bmatrix} = -(2x + 1)\boldsymbol{k}.$$

Then, using equation (16), Section 20.5, we evaluate the surface integral as

$$\iint_S (\text{curl } \boldsymbol{F}) \cdot \boldsymbol{N} \, dS = \iint_S [-(2x + 1)\boldsymbol{k}] \cdot \boldsymbol{N} \, dS$$

$$= \iint_Q -(2x + 1) \, dx \, dy$$

$$= \int_0^{2\pi} \int_0^2 -(2r \cos \theta + 1) \, r \, dr \, d\theta \qquad \text{(switch to polar coordinates)}$$

$$= -4\pi,$$

which agrees with the value of the line integral. ◇

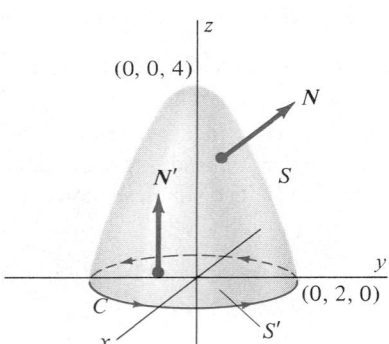

Figure 6.8 $\iint_S (\text{curl } \boldsymbol{F}) \cdot \boldsymbol{N} \, dS$ is the same as $\iint_{S'} (\text{curl } \boldsymbol{F}) \cdot \boldsymbol{N'} \, dS$, according to Stokes' Theorem.

REMARK: Notice that the calculation of the surface integral involved the same double integral as we would have obtained had the problem simply been to calculate $\iint_{S'} \text{curl } \boldsymbol{F} \cdot \boldsymbol{N} \, dS$ over the planar disc $S' = \{(x, y, 0) \mid 0 \le x^2 + y^2 \le 4\}$ (Figure 6.8). This is a direct illustration of the statement of Stokes Theorem—the value of the surface integral depends only on the line integral around the boundary.

Example 4

Find $\displaystyle\int_C \boldsymbol{F} \cdot d\boldsymbol{r}$ where the space curve C is the intersection of the plane $x - 2y + z = 5$ with the cylinder $x^2 + y^2 = 9$, oriented as in Figure 6.9, and $\boldsymbol{F}$ is the vector field $\boldsymbol{F}(x, y, z) = (x^2 - 3y^2)\boldsymbol{i} + (z^2 + y)\boldsymbol{j} + (x + 2z^2)\boldsymbol{k}$.

Solution: Attempting to evaluate the line integral directly would be difficult at best, so we resort to Stokes' Theorem, taking S to be the portion of the plane enclosed by C. From the equation of the plane and the orientation of C, we see that a unit normal to S is

$$\boldsymbol{N} = \frac{1}{\sqrt{6}}(\boldsymbol{i} - 2\boldsymbol{j} + \boldsymbol{k})$$

which is upward. Also,

$$\text{curl } \boldsymbol{F} = \det \begin{bmatrix} \boldsymbol{i} & \boldsymbol{j} & \boldsymbol{k} \\ \dfrac{\partial}{\partial x} & \dfrac{\partial}{\partial y} & \dfrac{\partial}{\partial z} \\ (x^2 - 3y^2) & (z^2 + y) & (x + 2z^2) \end{bmatrix} = -2z\boldsymbol{i} - \boldsymbol{j} + 6y\boldsymbol{k}.$$

The equation for S becomes $z = g(x, y) = 5 - x + 2y$. By Stokes' Theorem and

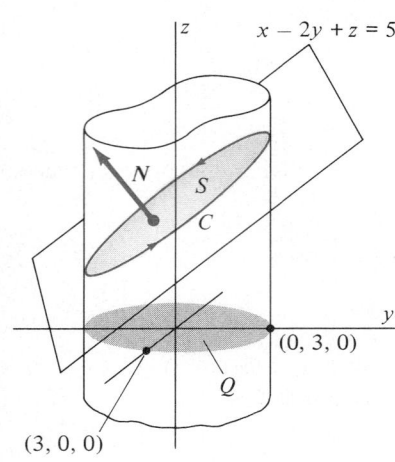

Figure 6.9 Line integral $\displaystyle\int_C \boldsymbol{F} \cdot d\boldsymbol{r}$ is evaluated as a surface integral $\displaystyle\iint_S (\text{curl } \boldsymbol{F}) \cdot \boldsymbol{N} \, dS$ over S.

equation (16), Section 20.5, we have

$$\int_C \boldsymbol{F} \cdot d\boldsymbol{r} = \iint_S (\text{curl } \boldsymbol{F}) \cdot \boldsymbol{N} \, dS$$

$$= \iint_Q \{-[-2(5 - x + 2y)](-1) - (-1)(2) + 6y\} \, dx \, dy$$

$$= \iint_Q (2x + 2y - 8) \, dx \, dy$$

$$= \int_0^{2\pi} \int_0^3 (2r \cos \theta + 2r \sin \theta - 8)r \, dr \, d\theta$$

$$= -72\pi.$$

(Note that we did not make use of the components of the unit normal $\boldsymbol{N}$—only the fact that it is oriented upward.) ◇

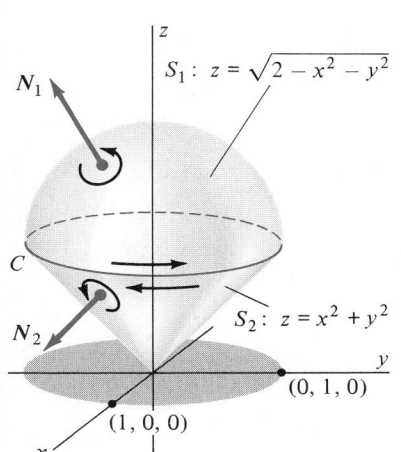

Figure 6.10 $\iint_S (\text{curl } \boldsymbol{F}) \cdot \boldsymbol{N} \, dS = 0$ for $S = S_1 \cup S_2$.

Example 5

Calculate $\displaystyle\iint_S (\text{curl } \boldsymbol{F}) \cdot \boldsymbol{N} \, dS$ where S is the closed surface shown in Figure 6.10 and $\boldsymbol{F}$ is a vector field satisfying the hypotheses of Stokes' Theorem.

Solution: The closed surface S may be written $S = S_1 \cup S_2$ where S_1 is the hemisphere $z = \sqrt{2 - x^2 - y^2}$ and S_2 is the graph of the cone $z = x^2 + y^2$ over the unit disc. These two surfaces share the common boundary curve

$$C = \{(x, y, 1) \mid x^2 + y^2 = 1\}.$$

However, the two surfaces induce opposite orientations on C. This is because the unit normal to S_1 is upward, inducing a counterclockwise orientation on C, while the unit normal to S_2 is downward, inducing a clockwise orientation on C. Thus, by Stokes' Theorem,

$$\iint_S (\text{curl } \boldsymbol{F}) \cdot \boldsymbol{N} \, dS = \iint_{S_1} (\text{curl } \boldsymbol{F}) \cdot \boldsymbol{N}_1 \, dS + \iint_{S_2} (\text{curl } \boldsymbol{F}) \cdot \boldsymbol{N}_2 \, dS$$

$$= \int_C \boldsymbol{F} \cdot d\boldsymbol{r} + \int_{-C} \boldsymbol{F} \cdot d\boldsymbol{r}$$

$$= \int_C \boldsymbol{F} \cdot d\boldsymbol{r} - \int_C \boldsymbol{F} \cdot d\boldsymbol{r}$$

$$= 0.$$ ◇

Curl F and Conservative Force Fields

For conservative planar vector fields $\boldsymbol{F}(x, y) = M\boldsymbol{i} + N\boldsymbol{j}$, we have previously seen that the condition that $\boldsymbol{F}$ be *conservative* in an open rectangle D is equivalent to the condition that curl $\boldsymbol{F}(x, y) = 0$ for all $(x, y) \in D$ (see Exercise 24, Section 20.4). We now sketch an argument for the same result in three dimensions.

First, suppose that curl $\boldsymbol{F}(x, y, z) = \boldsymbol{0}$ for all (x, y, z) in some simply connected region D in $\mathbb{R}^3$. Then if C is any closed path in D,

$$\int_C \boldsymbol{F} \cdot d\boldsymbol{r} = \iint_S \text{curl } \boldsymbol{F} \cdot \boldsymbol{N} \, dS = \iint_S \boldsymbol{O} \cdot \boldsymbol{N} \, dS = 0$$

where S is any smooth surface in D bounded by C. Thus, $\int_C \boldsymbol{F} \cdot d\boldsymbol{r} = 0$ for any closed path in D, so $\boldsymbol{F}$ is conservative (Exercise 13, Section 20.3).

Conversely, suppose that $\boldsymbol{F}$ is conservative in D. Then (again by Exercise 13, Section 20.3) $\int_C \boldsymbol{F} \cdot d\boldsymbol{r} = 0$ for any closed curve C in D. Now let (x_0, y_0, z_0) be any point in D, let C_r be a small circle of radius r and center (x_0, y_0, z_0), and let $\boldsymbol{N}$ be any unit normal to the plane containing C_r. Let S_r be the disc enclosed by C_r. Then, by Stokes' Theorem

$$\int_{C_r} \boldsymbol{F} \cdot d\boldsymbol{r} = \iint_{S_r} (\text{curl } \boldsymbol{F}) \cdot \boldsymbol{N} \, dS. \tag{14}$$

Now the line integral in (14) is zero, since $\boldsymbol{F}$ is conservative. Thus, $\iint_{S_r} (\text{curl } \boldsymbol{F}) \cdot \boldsymbol{N} \, dS = 0$, no matter how small r is chosen. From this statement we conclude that

$$(\text{curl } \boldsymbol{F}(x_0, y_0, z_0)) \cdot \boldsymbol{N} = 0. \tag{15}$$

(If the expression in (15) were either positive or negative, we could invoke the continuity of $\boldsymbol{F}$ to contract C_r to a small circle on which $(\text{curl } \boldsymbol{F}) \cdot \boldsymbol{N}$ is either always positive or always negative. The resulting surface integral in (14) would then be nonzero, a contradiction of equation (14).)

Since $\boldsymbol{N}$ was arbitrary, it follows from (15) that $\text{curl } \boldsymbol{F}(x_0, y_0, z_0) = \boldsymbol{0}$. Finally, since (x_0, y_0, z_0) was chosen arbitrarily in D, we conclude that $\text{curl } \boldsymbol{F}(x, y, z) = \boldsymbol{0}$ for all $(x, y, z) \in D$ when $\boldsymbol{F}$ is conservative.

The preceding argument is not a rigorous proof. However, it does convey the general flavor of the arguments linking the ideas of conservative vector fields, independence of path for line integrals, and the condition $\text{curl } \boldsymbol{F} = \boldsymbol{0}$. These relationships are summarized by the following theorem.

THEOREM 5

Let $\boldsymbol{F}(x, y, z) = M(x, y, z)\boldsymbol{i} + N(x, y, z)\boldsymbol{j} + P(x, y, z)\boldsymbol{k}$ be a vector field for which all of the component functions, together with all their first partial derivatives, are continuous in some open box D in $\mathbb{R}^3$. The following conditions are all equivalent:

(i) $\boldsymbol{F}$ is conservative in D.

(ii) $\int_C \boldsymbol{F} \cdot d\boldsymbol{r}$ is independent of path C in D.

(iii) $\int_C \boldsymbol{F} \cdot d\boldsymbol{r} = 0$ for every closed path C in D.

(iv) $\text{curl } \boldsymbol{F}(x, y, z) = \boldsymbol{0}$ for all $(x, y, z) \in D$.

Theorem 5 gives the following condition for determining when a vector field $\boldsymbol{F} = M\boldsymbol{i} + N\boldsymbol{j} + P\boldsymbol{k}$ is conservative.

COROLLARY 2

Let F and D be as in Theorem 5. Then F is conservative in D if and only if each of the following equations holds for all (x, y, z) in D:

(i) $\dfrac{\partial P}{\partial y}(x, y, z) = \dfrac{\partial N}{\partial z}(x, y, z),$

(ii) $\dfrac{\partial M}{\partial z}(x, y, z) = \dfrac{\partial P}{\partial x}(x, y, z),$

(iii) $\dfrac{\partial N}{\partial x}(x, y, z) = \dfrac{\partial M}{\partial y}(x, y, z).$

The proof of Corollary 2 consists of applying Theorem 5, parts (i) and (iv) together with the definition of curl F (Definition 6).

A Physical Interpretation of Curl

As a last look at Stokes' Theorem, we use equation (5) to write it in the form

$$\int_C F \cdot T \, ds = \iint_S (\text{curl } F) \cdot N \, dS. \tag{16}$$

Returning to our primary motivation of fluid flow, we recall from Section 20.4 that if F is the velocity field of a moving fluid in which the curve C is submerged, the line integral in (16), called the **circulation** of F around C, is a measure of the tendency of the fluid to rotate, or circulate, around the curve C. Now let $P_0(x_0, y_0, z_0)$ be a point in the fluid, and let C_r be a circle with radius r and center P_0 (see Figure 6.11). Applying the Mean Value Theorem for double integrals (an obvious generalization of the Mean Value Theorem for functions of a single variable which we have not previously stated) to equation (16), we find that for some point $P_r = (x_r, y_r, z_r)$ in S_r (the disc enclosed by C_r)

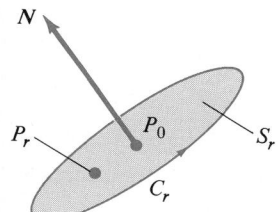

Figure 6.11

$$\int_{C_r} F \cdot T \, ds = \iint_{S_r} (\text{curl } F) \cdot N \, dS$$

$$= \pi r^2 [\text{curl } F(x_r, y_r, z_r)] \cdot N.$$

Letting $r \to 0$, we conclude that

$$[\text{curl } F(x_0, y_0, z_0)] \cdot N = \lim_{r \to 0} \frac{1}{\pi r^2} \int_{C_r} F \cdot T \, ds. \tag{17}$$

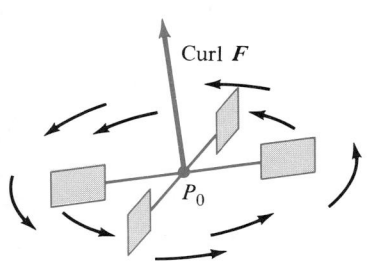

Figure 6.12 Paddle wheel interpretation of curl F. The wheel rotates most rapidly when the axis is parallel to curl F.

The interpretation of equation (17) is this. The right-hand side, as described earlier, is a measure of the tendency of the fluid to circulate, or rotate, about the small circle C_r. Since the normal N to the disc enclosed by this circle appears on the left-hand side of (17), this tendency to rotate will be the largest when C_r is positioned so that N is parallel to curl $F(x_0, y_0, z_0)$. Thus, curl $F(x_0, y_0, z_0)$ is a vector whose *magnitude* is a measure of the tendency of the fluid at (x_0, y_0, z_0) to rotate (as about the drain hole in a bathtub), and whose *direction* is along the axis about which the fluid has the maximal tendency to rotate.

Figure 6.12 shows an interpretation of equation (17) in terms of a paddle wheel. The motion of the fluid will cause the paddle wheel, based at P_0, to rotate most quickly when the axis points parallel to curl $F(x_0, y_0, z_0)$. For obvious reasons, fluid fields for which curl $F = 0$ (such as conservative fields) are called *irrotational*.

It is important to note that what we have achieved in these last few observations is an interpretation of curl $F(x, y, z)$ as a *local* property of the vector field—the tendency of the fluid *at location* (x, y, z) to rotate. In general, this tendency will vary from point to point within the fluid. What Stokes' Theorem says in this context is that the collective measure of this tendency taken over the entire surface S (that is, the value of the surface integral) is completely determined by, and equals, the tendency of the fluid to circulate around the boundary C (the value of the line integral).

Exercise Set 20.6

In Exercises 1–4, calculate curl F.

1. $F(x, y, z) = xi - yj + z^2k$

2. $F(x, y, z) = xy^2i + xz^2j + yz^2k$

3. $F(x, y, z) = xyzi - \cos(xy)j + \sin(yz)k$

4. $F(x, y, z) = e^{xy}i - x^2z^2j + \sqrt{xy}\,k$

In Exercises 5–8, determine whether the given vector field is conservative.

5. $F(x, y, z) = yzi + xzj + xyk$

6. $F(x, y, z) = 2xyz^2i + x^2z^2j + 2x^2yzk$

7. $F(x, y, z) = \sin yi + \cos xj + \sqrt{xy}\,k$

8. $F(x, y, z) = yze^{xyz}i + xze^{xyz}j + xye^{xyz}k$

9. Verify Stokes' Theorem for the vector field $F(x, y, z) = zi + xj + yk$ and the surface of the hemisphere $z = \sqrt{1 - x^2 - y^2}$.

10. Use Stokes' Theorem to calculate the line integral $\int_C F \cdot dr$ where $F(x, y, z) = x^2y^2i + x^2z^2j + y^2z^2k$ and C is the perimeter of the rectangle with vertices $(1, 1, 0)$, $(1, 5, 0)$, $(3, 1, 0)$, and $(3, 5, 0)$ traversed in this order.

11. Verify Stokes' Theorem for $F(x, y, z) = x^2yi + y^2zj + xzk$ where C is the boundary of the rectangle $Q = \{(x, y, 0) \mid 0 \le x \le 1, 0 \le y \le 2\}$ oriented counterclockwise.

12. Use Stokes' Theorem to calculate the line integral $\int_C F \cdot dr$ where $F(x, y, z) = x^2yi + y^2zj + xzk$ and C is the intersection of the plane $x + 3y + z = 4$ with the cylinder $x^2 + y^2 = 1$ with orientation induced by the upward unit normal to the plane.

13. Verify Stokes' Theorem for the portion of the plane $x + 2y + z = 2$ in the first octant, the vector field $F(x, y, z) = zi + xj + yk$, and the upward unit normal.

14. Calculate $\int_C F \cdot dr$ where $F(x, y, z) = xzi + 2zj - xyk$ and C is the intersection of the plane $y = z + 2$ and the cylinder $x^2 + y^2 = 4$. The orientation on C is that induced by the upward unit normal to the plane.

15. Evaluate $\int_C F \cdot dr$ around the unit circle in the xy-plane, counterclockwise, where

$$F(x, y, z) = (\sqrt{x} + y)i + (e^y - x)j + (\sin z + y)k.$$

16. Calculate $\int_C F \cdot dr$ where $F(x, y, z) = xyzi + xzk$ and where C is the intersection of the paraboloid $z = x^2 + y^2$ and the plane $z = 4$. The orientation on C is that induced by the upward unit normal on the paraboloid.

17. Verify Stokes' Theorem for the surface $z = 9 - x^2 - y^2$, $z \ge 0$ and the vector field $F(x, y, z) = x^2i + y^2j + z^2k$.

18. Show that curl $F = 0$ for $F(x, y, z) = x^2i + y^2j + z^2k$. Can you give a geometric interpretation for this result?

19. Let $F(x, y) = M(x, y)i + N(x, y)j$ be a conservative vector field. Let $G(x, y, z) = F(x, y) + \lambda zk$ where λ is constant. Show that $G(x, y, z)$ is conservative. Is the result true if zk is replaced by $g(z)k$?

20. Let $F(x, y, z)$ be a vector field satisfying the hypotheses of Stokes' Theorem. Show that $\iint_S (\text{curl } F) \cdot N \, dS = 0$ where S is the unit sphere.

21. Use Stokes' Theorem to show that a line integral is independent of parameterization if orientation is preserved.

20.7 THE DIVERGENCE THEOREM

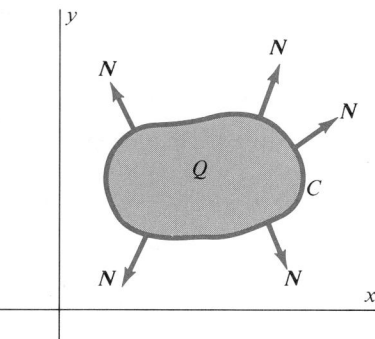

Figure 7.1 $\int_C F \cdot N \, dS$ is the flux of $F(x, y)$ across C in $\mathbb{R}^2$.

In this last section, we generalize equation (28) of Section 20.4,

$$\int_C F \cdot N \, ds = \iint_Q \text{div } F(x, y) \, dA \qquad \text{(flux of } F \text{ across } C), \tag{1}$$

to an appropriate equation for vector fields in $\mathbb{R}^3$. However, this time we begin with the physical interpretation of what we are after and see where the mathematics leads us.

Recall the interpretation for equation (1) in terms of fluid flow: F is a vector field in the plane giving the velocity vector $v(x, y) = F(x, y)$ for the steady state motion of a thin layer of fluid; C is a closed curve in the fluid layer; and N is the outward unit normal to C. In this setting the line integral $\int_C F \cdot N \, ds$ in (1) is interpreted as the **flux** of the fluid outward across C, that is, the net rate at which fluid is crossing the boundary C of Q (Figure 7.1). In Section 20.5 we have already encountered the notion of the flux of a vector field F across a closed surface S in $\mathbb{R}^3$.

$$\text{Flux} = \iint_S F \cdot N \, dS, \tag{2}$$

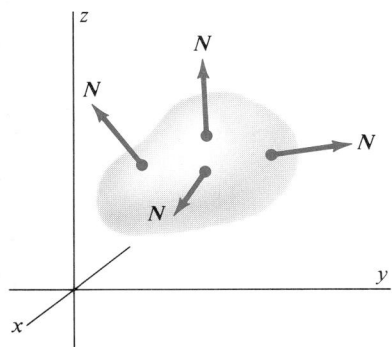

Figure 7.2 $\iint_S F \cdot N \, dS$ is the flux of $F(x, y, z)$ across S in $\mathbb{R}^3$.

where N is the outward unit normal to S at (x, y, z). That is, the natural generalization of the line integral in (1), at least in terms of fluid flow, is the surface integral in (2) (see Figure 7.2).

Now the right-hand side of equation (1) is a double integral evaluated over Q, the region bounded by C, of the scalar function div F, and we can paraphrase equation (1) by saying that "div F is a measure of the local behavior of F that, when integrated over the enclosed region Q, determines the flux of F across the boundary C."

Thus, to properly generalize equation (1) to a closed surface S in $\mathbb{R}^3$, we need to find a scalar function div F that is a measure of the local behavior of F and that, when integrated over the region R enclosed by S, gives the flux of F across the surface S. That is, div F must satisfy the equation

$$\iint_S F \cdot N \, dS = \iiint_R \text{div } F \, dV. \tag{3}$$

Although you can probably guess how we're going to define div F, let's proceed "experimentally" by investigating the surface integral in (3). For starters, we express F in component form as

$$F(x, y, z) = M(x, y, z)i + N(x, y, z)j + P(x, y, z)k. \tag{4}$$

Also, we assume that S is a closed surface that is the union $S = S_1 \cup S_2$ of two smooth surfaces, which are graphs of the functions

$$S_1: \quad z = g_1(x, y), \qquad (x, y) \in Q$$

$$S_2: \quad z = g_2(x, y), \qquad (x, y) \in Q$$

with $g_1(x, y) \le g_2(x, y)$ for all $(x, y) \in Q$. Let C be the boundary common to S_1 and S_2 (see Figure 7.3).

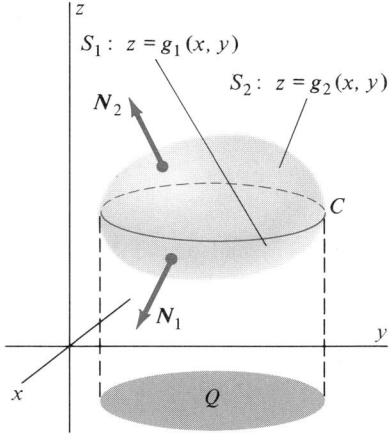

Figure 7.3 $S = S_1 \cup S_2$.

Using (4), we write the left side of equation (3) as

$$\iint_S \boldsymbol{F} \cdot \boldsymbol{N} \; dS = \iint_S (M\boldsymbol{i}) \cdot \boldsymbol{N} \; dS + \iint_S (N\boldsymbol{j}) \cdot \boldsymbol{N} \; dS + \iint_S (P\boldsymbol{k}) \cdot \boldsymbol{N} \; dS. \quad (5)$$

Let's now focus on the last integral on the right side of (5). Since the unit normal to S_2 is upward and the unit normal to S_1 is downward, we use both equation (16) and equation (17) of Section 20.5 to find that

$$\iint_S [P(x, y, z)\boldsymbol{k}] \cdot \boldsymbol{N} \; dS \quad (6)$$

$$= \iint_{S_2} [P(x, y, z)\boldsymbol{k}] \cdot \boldsymbol{N}_2 \; dS + \iint_{S_1} [P(x, y, z)\boldsymbol{k}] \cdot \boldsymbol{N}_1 \; dS$$

$$= \iint_Q P(x, y, g_2(x, y)) \; dx \; dy + \iint_Q -P(x, y, g_1(x, y)) \; dx \; dy$$

$$= \iint_Q \{P(x, y, z)]_{z=g_1(x, y)}^{z=g_2(x, y)}\} \; dx \; dy$$

$$= \iint_Q \left\{ \int_{g_1(x, y)}^{g_2(x, y)} \frac{\partial}{\partial z} P(x, y, z) \; dz \right\} dx \; dy$$

$$= \iiint_R \frac{\partial}{\partial z} P(x, y, z) \; dx \; dy \; dz.$$

(Note that we have changed the order of integration from $dz \; dx \; dy$ to $dx \; dy \; dz$.) That is,

$$\iint_S (P\boldsymbol{k}) \cdot \boldsymbol{N} \; dS = \iiint_R \frac{\partial P}{\partial z} \; dx \; dy \; dz. \quad (7)$$

Similarly, if S can be written as in Figure 7.4, we can show that

$$\iint_S (N\boldsymbol{j}) \cdot \boldsymbol{N} \; dS = \iiint_R \frac{\partial N}{\partial y} \; dx \; dy \; dz, \quad (8)$$

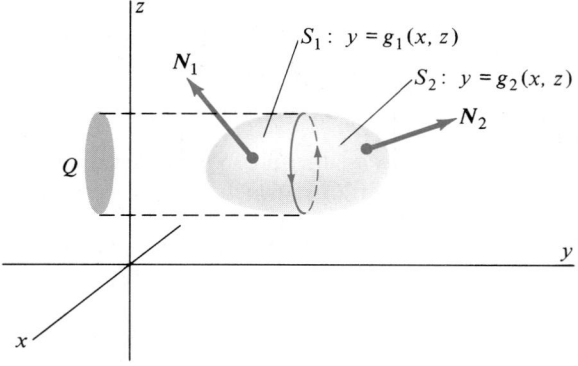

Figure 7.4 $S = S_1 \cup S_2$.

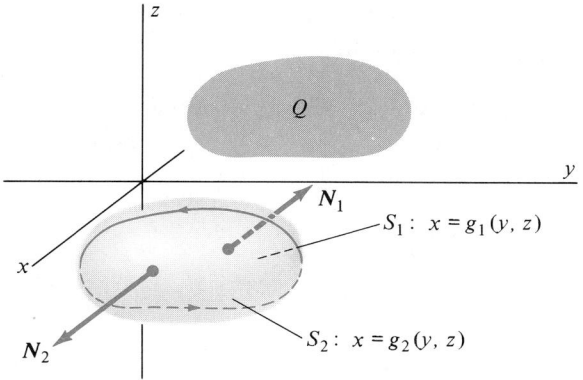

Figure 7.5 $S = S_1 \cup S_2$.

and if S can be written as in Figure 7.5, we obtain

$$\iint_S (M\boldsymbol{i}) \cdot \boldsymbol{N} \, dS = \iiint_R \frac{\partial M}{\partial x} \, dx \, dy \, dz. \tag{9}$$

Adding the corresponding sides of equations (7), (8), and (9), using equation (5), we obtain the equation

$$\iint_S \boldsymbol{F} \cdot \boldsymbol{N} \, dS = \iiint_R \left(\frac{\partial M}{\partial x} + \frac{\partial N}{\partial y} + \frac{\partial P}{\partial z} \right) dx \, dy \, dz. \tag{10}$$

Comparing equations (3) and (10) leads directly to the following definition.

DEFINITION 7

The **divergence of the vector field**

$$\boldsymbol{F}(x, y, z) = M(x, y, z)\boldsymbol{i} + N(x, y, z)\boldsymbol{j} + P(x, y, z)\boldsymbol{k}, \tag{11}$$

written div $\boldsymbol{F}$, is the scalar function

$$\text{div } \boldsymbol{F}(x, y, z) = \frac{\partial}{\partial x} M(x, y, z) + \frac{\partial}{\partial y} N(x, y, z) + \frac{\partial}{\partial z} P(x, y, z). \tag{12}$$

With this definition of div $\boldsymbol{F}$, the preceding discussion shows that equation (3) holds for smooth closed surfaces that are simultaneously x-simple, y-simple, and z-simple. This is our desired generalization of the planar equation (1). A more general and precisely stated version of equation (3) is our final result.

THEOREM 6
Divergence Theorem

Let S be a closed piecewise smooth surface enclosing the region R. Let $\boldsymbol{F}(x, y, z) = M(x, y, z)\boldsymbol{i} + N(x, y, z)\boldsymbol{j} + P(x, y, z)\boldsymbol{k}$ be a vector field for which all of the components, together with all their first partial derivatives, are continuous throughout R. Then

$$\iint_S \boldsymbol{F} \cdot \boldsymbol{N} \, dS = \iiint_R \text{div } \boldsymbol{F}(x, y, z) \, dx \, dy \, dz \tag{13}$$

where $\boldsymbol{N}$ is the outward unit normal to S.

REMARK 1: The Divergence Theorem was first discovered by the German mathematician Carl Friedrich Gauss and is often referred to as Gauss' Theorem.

REMARK 2: Using the operator notation

$$\boldsymbol{\nabla} = \frac{\partial}{\partial x}\boldsymbol{i} + \frac{\partial}{\partial y}\boldsymbol{j} + \frac{\partial}{\partial z}\boldsymbol{k}$$

we can write

$$\text{div } \boldsymbol{F} = \boldsymbol{\nabla} \cdot \boldsymbol{F} = \frac{\partial M}{\partial x} + \frac{\partial N}{\partial y} + \frac{\partial P}{\partial z}.$$

With this notation, the Divergence Theorem may be written either as

$$\iint_S \boldsymbol{F} \cdot \boldsymbol{N} \, dS = \iiint_R \boldsymbol{\nabla} \cdot \boldsymbol{F} \, dV, \qquad dV = dx \, dy \, dz \tag{14}$$

or as

$$\iint_S \boldsymbol{F} \cdot \boldsymbol{N} \, dS = \iiint_R \left(\frac{\partial M}{\partial x} + \frac{\partial N}{\partial y} + \frac{\partial P}{\partial z}\right) dx \, dy \, dz. \tag{15}$$

The appeal of equation (14) is that it expresses Theorem 6 in what is referred to as "coordinate free" form. On the other hand, equation (15) is the most explicit form for use when working in Cartesian (rectangular) coordinates.

REMARK 3: Our proof of the Divergence Theorem addressed only the case of a surface enclosing a region which was simultaneously x-simple, y-simple, and z-simple. Figures 7.6 and 7.7 illustrate how this proof may be extended to include

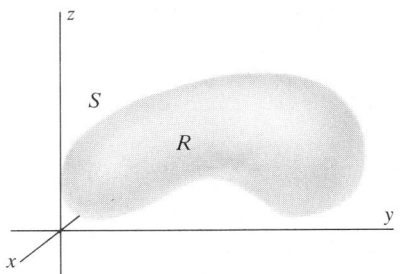

Figure 7.6 A region R that is not y-simple.

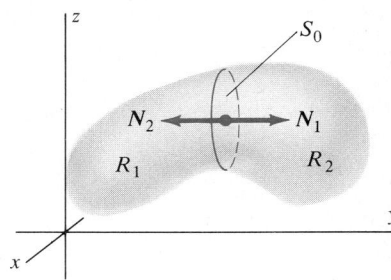

Figure 7.7 R_1 and R_2 are both y-simple. N_j is the outward unit normal for R_j, $j = 1, 2$.

more general regions. By constructing the surface S_0, we partition the region R into two regions, R_1 and R_2, both of which are simultaneously x-simple, y-simple, and z-simple. Letting S_1 denote the boundary of R_1 and S_2 the boundary of R_2, the previous proof shows that

$$\iint_{S_1} \boldsymbol{F} \cdot \boldsymbol{N} \, dS = \iiint_{R_1} \text{div } \boldsymbol{F} \, dx \, dy \, dz \tag{16}$$

and

$$\iint_{S_2} \boldsymbol{F} \cdot \boldsymbol{N} \, dS = \iiint_{R_2} \text{div } \boldsymbol{F} \, dx \, dy \, dz. \tag{17}$$

On the face S_0, the outward unit normal to S_1 is the *opposite* of the outward unit normal to S_2. Thus, the contributions over the face S_0 to the surface integrals in (16) and (17) cancel. Adding the corresponding sides of (16) and (17) then gives

$$\iint_{S} \boldsymbol{F} \cdot \boldsymbol{N} \, dS = \iint_{S_1} \boldsymbol{F} \cdot \boldsymbol{N} \, dS + \iint_{S_2} \boldsymbol{F} \cdot \boldsymbol{N} \, dS$$

$$= \iiint_{R_1} \text{div } \boldsymbol{F} \, dx \, dy \, dz + \iiint_{R_2} \text{div } \boldsymbol{F} \, dx \, dy \, dz$$

$$= \iiint_{R} \text{div } \boldsymbol{F} \, dx \, dy \, dz.$$

The same argument can be applied to more complicated surfaces.

Before proceeding to examples, let's nail down a bit more precisely what the Divergence Theorem says about fluid flow. Of course, the left-hand side of equation (13) is the **net flux** (rate out minus rate in) of the fluid outward across the surface S. However, the volume integral on the right-hand side of (13) can be thought of as a sum taken throughout R of the product div $\boldsymbol{F} \, dV$, where dV is an infinitesimal element of volume containing the point (x, y, z). It is this local property of div $\boldsymbol{F}$ that we wish to understand better.

To do so, let $\Delta V_\epsilon = \dfrac{4}{3} \pi \epsilon^3$ be the volume of a small sphere S_ϵ of radius ϵ and center (x_0, y_0, z_0) contained within S. According to Theorem 6

$$\text{Flux of } \boldsymbol{F} \text{ out of } S_\epsilon = \iiint_{R_\epsilon} \text{div } \boldsymbol{F} \, dx \, dy \, dz \tag{18}$$

where R_ϵ is the region enclosed by the sphere S_ϵ. We now proceed with an argument analogous to that of Section 20.6 for the interpretation of curl $\boldsymbol{F}$. The mean value property for triple integrals guarantees the existence of a point $(x_\epsilon, y_\epsilon, z_\epsilon)$ in R_ϵ for which

$$\iiint_{R_\epsilon} \text{div } \boldsymbol{F} \, dx \, dy \, dz = \text{div } \boldsymbol{F}(x_\epsilon, y_\epsilon, z_\epsilon) \, \Delta V_\epsilon. \tag{19}$$

Combining (18) and (19) we conclude that

$$\text{div } \boldsymbol{F}(x_\epsilon, y_\epsilon, z_\epsilon) = \frac{\text{Flux of } \boldsymbol{F} \text{ out of } S_\epsilon}{\Delta V_\epsilon}. \tag{20}$$

Since $(x_\epsilon, y_\epsilon, z_\epsilon) \to (x_0, y_0, z_0)$ as $\epsilon \to 0$, we conclude that

$$\boxed{\text{div } \boldsymbol{F}(x_0, y_0, z_0) = \lim_{\epsilon \to 0} \frac{\text{Flux of } \boldsymbol{F} \text{ out of } S_\epsilon}{\Delta V_\epsilon}.} \tag{21}$$

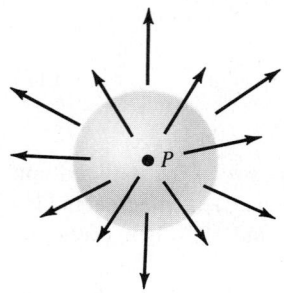

Figure 7.8 div $F > 0$; P is a source.

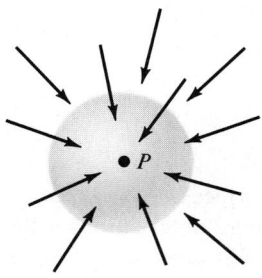

Figure 7.9 div $F < 0$; P is a sink.

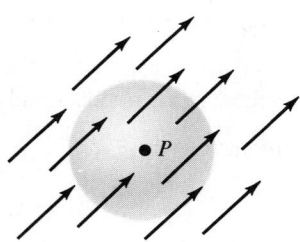

Figure 7.10 div $F = 0$; F is *incompressible* at P.

In other words, div $F(x_0, y_0, z_0)$ is the **flux per unit volume** at the point (x_0, y_0, z_0). Thus,

(i) If div $F(x_0, y_0, z_0) > 0$, more fluid is flowing out across a small sphere centered at (x_0, y_0, z_0) than is flowing in. In this case we say that the fluid is **expanding** at (x_0, y_0, z_0), or that (x_0, y_0, z_0) is a **source** for the vector field F (see Figure 7.8).

(ii) If div $F(x_0, y_0, z_0) < 0$, more fluid is flowing into a small sphere centered at (x_0, y_0, z_0) than is flowing out. The fluid is said, therefore, to be **contracting** at (x_0, y_0, z_0), and (x_0, y_0, z_0) is called a **sink** for F (see Figure 7.9).

(iii) If div $F(x_0, y_0, z_0) = 0$, the amount of fluid flowing out of the small sphere centered at (x_0, y_0, z_0) equals the amount flowing in. In this case the fluid is said to be **incompressible** at (x_0, y_0, z_0) (see Figure 7.10).

The powerful statement of the Divergence Theorem is that knowledge of the local property div $F(x, y, z)$, summed throughout the interior of S, determines the flux of F across the surface S, and conversely.

Example 1

Calculate the flux of the vector field

$$F(x, y, z) = (x - y^2)i + 3yj + (x^3 - z)k$$

outward over the sphere $S = \{(x, y, z) \mid x^2 + y^2 + z^2 = 9\}$.

Solution: Here

$$\text{div } F = \frac{\partial}{\partial x}(x - y^2) + \frac{\partial}{\partial y}(3y) + \frac{\partial}{\partial z}(x^3 - z)$$

$$= 1 + 3 - 1$$

$$= 3.$$

By the Divergence Theorem,

$$\text{Flux across } S = \iint_S F \cdot N \, dS = \iiint_R 3 \cdot dx \, dy \, dz = 3 \cdot \frac{4}{3}\pi \cdot 3^3 = 108\pi,$$

since the volume of the sphere of radius r is $\frac{4}{3}\pi r^3$. ◇

Example 2

Evaluate the surface integral

$$\iint_S F \cdot N \, dS$$

where $F(x, y, z) = xyi + xzj + yzk$ and N is the outward unit normal to the ellipsoid $S = \{(x, y, z) \mid x^2 + 4y^2 + z^2 = 1\}$.

Solution: We use the Divergence Theorem rather than trying to evaluate the surface integral directly. We have

$$\text{div } F = \frac{\partial}{\partial x}(xy) + \frac{\partial}{\partial y}(xz) + \frac{\partial}{\partial z}(yz) = 2y.$$

Letting R denote the region enclosed by the ellipsoid, we obtain

$$\iint_S \boldsymbol{F} \cdot \boldsymbol{N} \, dS = \iiint_R 2y \, dx \, dy \, dz$$

$$= \int_{-1}^{1} \int_{-(1/2)\sqrt{1-x^2}}^{(1/2)\sqrt{1-x^2}} \int_{-\sqrt{1-x^2-4y^2}}^{\sqrt{1-x^2-4y^2}} 2y \, dz \, dy \, dx.$$

Now this iterated integral can be evaluated directly. However, since the integrand satisfies the equation

$$\text{div } \boldsymbol{F}(x, -y, z) = -2y = -\text{div } \boldsymbol{F}(x, y, z),$$

and since the ellipsoid S is symmetric about the plane $y = 0$, the value of this integral is zero. ◇

Example 3

Calculate the value of the surface integral $\iint_S \boldsymbol{F} \cdot \boldsymbol{N} \, dS$ where $\boldsymbol{F}(x, y, z) = x^2 y \boldsymbol{i} + xy\boldsymbol{j} + y^2 z^3 \boldsymbol{k}$, S is the cube formed by the planes $x = \pm 1$, $y = \pm 1$, and $z = \pm 1$, and $\boldsymbol{N}$ is the outward unit normal on S.

Solution: Rather than calculate the surface integral directly for each of the six faces, we use the Divergence Theorem. Since

$$\text{div } \boldsymbol{F} = \frac{\partial}{\partial x}(x^2 y) + \frac{\partial}{\partial y}(xy) + \frac{\partial}{\partial z}(y^2 z^3) = 2xy + x + 3y^2 z^2$$

we have, from Theorem 6,

$$\iint_S \boldsymbol{F} \cdot \boldsymbol{N} \, dS = \int_{-1}^{1} \int_{-1}^{1} \int_{-1}^{1} (2xy + x + 3y^2 z^2) \, dx \, dy \, dz$$

$$= \int_{-1}^{1} \int_{-1}^{1} \left\{ \left[x^2 y + \frac{1}{2} x^2 + 3y^2 z^2 x \right]_{x=-1}^{x=1} \right\} dy \, dz$$

$$= \int_{-1}^{1} \int_{-1}^{1} 6y^2 z^2 \, dy \, dz$$

$$= \int_{-1}^{1} 2y^3 z^2 \Big]_{y=-1}^{y=1} dz$$

$$= \int_{-1}^{1} 4z^2 \, dz$$

$$= \frac{8}{3}. \qquad ◇$$

Example 4

Find the flux of the vector field

$$\boldsymbol{F}(x, y, z) = y^2 z^3 \boldsymbol{i} + 4x^2 yz^2 \boldsymbol{j} + x^2 z^3 \boldsymbol{k}$$

outward across the unit sphere $S = \{(x, y, z) \mid x^2 + y^2 + z^2 = 1\}$.

Solution: Let R be the unit ball $R = \{(x, y, z) \mid x^2 + y^2 + z^2 \leq 1\}$. We have

$$\text{div } F = \frac{\partial}{\partial x}(y^2 z^3) + \frac{\partial}{\partial y}(4x^2 yz^2) + \frac{\partial}{\partial z}(x^2 z^3) = 7x^2 z^2.$$

According to the Divergence Theorem,

$$\text{Flux across } S = \iint_S F \cdot N \, dS = \iiint_R 7x^2 z^2 \, dx \, dy \, dz.$$

To evaluate the triple integral, we use spherical coordinates:

$$x = \rho \cos \theta \sin \phi \qquad 0 \leq \rho \leq 1,$$
$$y = \rho \sin \theta \sin \phi \qquad 0 \leq \theta \leq 2\pi,$$
$$z = \rho \cos \phi \qquad 0 \leq \phi \leq \pi.$$

We obtain

$$\iiint_R 7x^2 z^2 \, dx \, dy \, dz = 7 \int_0^1 \int_0^{2\pi} \int_0^{\pi} [\rho \cos \theta \sin \phi]^2 [\rho \cos \phi]^2 \rho^2 \sin \phi \, d\phi \, d\theta \, d\rho$$

$$= 7 \int_0^1 \int_0^{2\pi} \int_0^{\pi} \rho^6 \cos^2 \theta \sin^3 \phi \cos^2 \phi \, d\phi \, d\theta \, d\rho$$

$$= 7 \int_0^1 \int_0^{2\pi} \int_0^{\pi} \rho^6 \cos^2 \theta [1 - \cos^2 \phi] \cos^2 \phi \sin \phi \, d\phi \, d\theta \, d\rho$$

$$= 7 \int_0^1 \int_0^{2\pi} \left\{ \rho^6 \cos^2 \theta \left[-\frac{1}{3} \cos^3 \phi + \frac{1}{5} \cos^5 \phi \right]_0^{\pi} \right\} d\theta \, d\rho$$

$$= 7 \cdot \frac{4}{15} \int_0^1 \int_0^{2\pi} \rho^6 \cos^2 \theta \, d\theta \, d\rho$$

$$= 7 \cdot \frac{4}{15} \cdot \pi \int_0^1 \rho^6 \, d\rho$$

$$= \frac{4\pi}{15}. \qquad \diamond$$

The Divergence Theorem finds wide application in physics, engineering, and applied mathematics. We have restricted our discussion to the example of fluid flow to keep the discussion straightforward and to allow you to focus primarily on the mathematics. The subject area of electricity and magnetism makes considerable use of each of the major theorems presented in this chapter (and involves another theorem due to Gauss).

Exercise Set 20.7

In Exercises 1–6, find div F.

1. $F(x, y, z) = x^2 i + y^2 j + z^2 k$

2. $F(x, y, z) = x^2 z i + y^2 x j + xz^2 k$

3. $F(x, y, z) = x i + xy j + xyz k$

4. $F(x, y, z) = (y - x) i + (y - z) j + (x - y) k$

5. $F(x, y, z) = \cos xy i + e^{xyz} j + y \sin(xz) k$

6. $F(x, y, z) = x \, \text{Tan}^{-1} yi - \sqrt{yz} \, j - z \sec y k$

7. Find the flux of the vector field $F(x, y, z) = z i + x j + y k$ out of the ellipsoid $x^2 + 9y^2 + 4z^2 = 1$.

8. Find the value of the surface integral $\iint_S F \cdot N \, dS$ where F

is the vector field $F(x, y, z) = -y^2i + xj + zk$, S is the sphere $x^2 + y^2 + z^2 = 5$ and N is the outward unit normal.

9. Find the flux of the vector field $F(x, y, z) = 3xi - z^2yj + 2zk$ outward across the rectangular box with vertices $(1, 0, 0)$, $(1, 3, 0)$, $(-2, 0, 0)$, $(-2, 3, 0)$, $(1, 0, 5)$, $(1, 3, 5)$, $(-2, 0, 5)$, and $(-2, 3, 5)$.

10. Find the value of the surface integral $\displaystyle\iint_S F \cdot N \, dS$ where F

    is the vector field $F(x, y, z) = 3xi - 4yj + 5zk$, and V is the volume of the solid enclosed by the smooth surface S.

11. Find the flux of the vector field $F(x, y, z) = 2xyi + z^2yj + xzk$ over the cube formed by the coordinate planes and the planes $x = 1$, $y = 1$, and $z = 1$.

12. Find the value of the surface integral $\displaystyle\iint_S F \cdot N \, dS$ where

    $F(x, y, z) = (x + e^y)i + (e^{xz} - y)j + (xy + z)k$, S is the cylinder $\{(x, y, z) \mid x^2 + y^2 = 4, 0 \le z \le 2\}$ and N is the outward unit normal.

13. Find the flux of the vector field $F(x, y, z) = x^2y^2i + xy^3j + xyk$ outward across the cylinder $\{(x, y, z) \mid x^2 + y^2 = 4, 0 \le z \le 2\}$. (*Hint:* Use cylindrical coordinates.)

14. Find the value of the surface integral $\displaystyle\iint_S F \cdot N \, dS$ where F

    is the vector field $F(x, y, z) = x^2yi + xy^2j + xyzk$, S is the surface of the quarter cylinder $C = \{(r, \theta, z) \mid 0 \le r \le 1, 0 \le \theta \le \pi/2, 0 \le z \le 1\}$.

15. Find the flux of the vector field $F(x, y, z) = x^3i + xz^2j + x^2zk$ outward across the sphere $S = \{(x, y, z) \mid x^2 + y^2 + z^2 = 4\}$. (Use spherical coordinates.)

16. Find the value of the surface integral $\displaystyle\iint_S F \cdot N \, dS$ where F

    is the vector field $F(x, y, z) = x^2yi + xy^2j + xyzk$, S is the unit sphere $x^2 + y^2 + z^2 = 1$ and N is the outward normal. (Use spherical coordinates.)

17. Find the flux of $F(x, y, z) = 6xi - yj + 4k$ across the ellipsoid $x^2 + 4y^2 + z^2 = 4$. (Calculate the volume of the ellipsoid as a volume of revolution.)

18. Verify the Divergence Theorem for the vector field $F(x, y, z) = xi + yj + zk$ and the closed surface S of the cylinder $\{(x, y, z) \mid x^2 + y^2 \le 1, 0 \le z \le 2\}$.

In Exercises 19–21, verify the stated identities for differentiable vector fields $F(x, y, z)$ and $G(x, y, z)$.

19. $\nabla \times (F + G) = \nabla \times F + \nabla \times G$

20. $\nabla \cdot (F + G) = \nabla \cdot F + \nabla \cdot G$

21. $\nabla \cdot (F \times G) = (\nabla \times F) \cdot G - (\nabla \times G) \cdot F$

The **Laplacian of the scalar field** $\phi = \phi(x, y, z)$ is defined by the equation

$$\nabla^2\phi = \nabla \cdot (\nabla\phi) = \frac{\partial^2\phi}{\partial x^2} + \frac{\partial^2\phi}{\partial y^2} + \frac{\partial^2\phi}{\partial z^2}.$$

The equation $\dfrac{\partial^2\phi}{\partial x^2} + \dfrac{\partial^2\phi}{\partial y^2} + \dfrac{\partial^2\phi}{\partial z^2} = 0$ is called **Laplace's equation.** Functions which satisfy Laplace's equation are called **harmonic functions.**

22. Show that if $F = \nabla\phi$, then ϕ is harmonic if and only if div $F = 0$.

23. Determine which of the following vector fields are gradients of harmonic scalar functions.
    a. $F(x, y, z) = yzi + xzj + xyk$
    b. $F(x, y, z) = 2xye^zi + x^2e^zj + x^2ye^zk$
    c. $F(x, y, z) = x \sin(yz)i + z \cos yzj + y \cos yzk$

24. For the inverse square field $F(r) = \dfrac{r}{|r|^3}$, $|r| \ne 0$, show that div $F = 0$.

25. For the inverse square field of Exercise 24 show that
    $$\iint_S F \cdot N \, dS = 0$$ if S is a surface containing a region R, as in the statement of the Divergence Theorem, so that $(0, 0, 0) \notin S \cup R$.

26. Let S be a sphere with center $(0, 0, 0)$. Show that $\displaystyle\iint_S F \cdot$

    $N \, dS = 4\pi$ where F is the inverse square field of Exercise 24, and S is the outward unit normal. Why does this not contradict Theorem 6?

27. Show that $\displaystyle\iint_S F \cdot N \, dS = 0$ if the vector field F is incom-

    pressible for all $(x, y, z) \in S$. (S, F, and N are as in Theorem 6.)

28. Show that if $\displaystyle\iint_S F \cdot N \, dS > 0$, then S must contain at least

    one source in its interior. (S, F, and N are as in Theorem 6.)

    What if $\displaystyle\iint_S F \cdot N \, dS < 0$?

29. True or false? If $\displaystyle\iint_S F \cdot N \, dS = 0$, S may contain neither

    sources nor sinks. Explain.

30. Let F, S, N, and R satisfy the hypotheses of Theorem 6 in a region Ω. If $\displaystyle\iint_S F \cdot N \, dS = 0$ for all closed surfaces S

    within Ω must $F \equiv 0$ for all $(x, y, z) \in \Omega$? Why or why not?

SUMMARY OUTLINE OF CHAPTER 20

◆ A **vector field** on $\mathbb{R}^3$ is a function   (page 912)

$$F(x, y, z) = M(x, y, z)\boldsymbol{i} + N(x, y, z)\boldsymbol{j} + P(x, y, z)\boldsymbol{k}.$$

◆ The **work** done by the force field F in moving an object along a curve $C = \{r(t) \mid a \leq t \leq b\}$ is   (page 918)

$$W = \int_C \boldsymbol{F} \cdot d\boldsymbol{r} = \int_a^b \boldsymbol{F}(\boldsymbol{r}(t)) \cdot \boldsymbol{r}'(t)\, dt.$$

◆ The **line integral** of F over $C = \{r(t) = x(t)\boldsymbol{i} + y(t)\boldsymbol{j} + z(t)\boldsymbol{k} \mid a \leq t \leq b\}$ is   (page 921)

$$\int_C \boldsymbol{F} \cdot d\boldsymbol{r} = \int_a^b \boldsymbol{F}(\boldsymbol{r}(t)) \cdot \boldsymbol{r}'(t)\, dt$$

$$= \int_a^b [M(x(t), y(t), z(t))x'(t) + N(x(t), y(t), z(t))y'(t) + P(x(t), y(t), z(t))z'(t)]\, dt$$

$$= \int_C M\, dx + N\, dy + P\, dz.$$

◆ The **line integral** of the scalar function f **with respect to arc length** over the path C (above) is   (page 926)

$$\int_C f(x, y, z)\, ds = \int_a^b f(x(t), y(t), z(t))\sqrt{[x'(t)]^2 + [y'(t)]^2 + [z'(t)]^2}\, dt.$$

◆ **Theorem 1:** Under appropriate hypotheses, if $F = \nabla\phi$ and $C = \{r(t) \mid a \leq t \leq b\}$, then   (page 929)

$$\int_C \boldsymbol{F} \cdot d\boldsymbol{r} = \phi(\boldsymbol{r}(b)) - \phi(\boldsymbol{r}(a)).$$

◆ **Theorem 2:** $\int_C \boldsymbol{F} \cdot d\boldsymbol{r}$ is independent of path if and only if F is **conservative** (i.e., $F = \nabla\phi$ for some scalar potential ϕ).   (page 932)

◆ **Green's Theorem:** Let C be a simple closed path, oriented counterclockwise, that encloses Q. Then   (page 937)

$$\int_C M\, dx + N\, dy = \iint_Q \left(\frac{\partial N}{\partial x} - \frac{\partial M}{\partial y}\right) dA.$$

◆ The **surface integral** of the scalar function f over the surface $S = \{(x, y, g(x, y)) \mid (x, y) \in Q\}$ is   (page 947)

$$\iint_S f(x, y, z)\, dS = \iint_Q f(x, y, g(x, y))\sqrt{\left(\frac{\partial g}{\partial x}\right)^2 + \left(\frac{\partial g}{\partial y}\right)^2 + 1}\, dx\, dy.$$

◆ The **flux** of the vector field F across the surface S in the direction indicated by the unit normal N is   (page 955)

$$\text{Flux} = \iint_S \boldsymbol{F} \cdot \boldsymbol{N}\, dS.$$

◆ The **curl** of the vector field $F = Mi + Nj + Pk$ is the **vector** curl   (page 961)

$$\boldsymbol{F} = \nabla \times \boldsymbol{F} = \left(\frac{\partial P}{\partial y} - \frac{\partial N}{\partial z}\right)\boldsymbol{i} + \left(\frac{\partial M}{\partial z} - \frac{\partial P}{\partial x}\right)\boldsymbol{j} + \left(\frac{\partial N}{\partial x} - \frac{\partial M}{\partial y}\right)\boldsymbol{k}.$$

◆ **Stokes' Theorem:** If the surface S is bounded by the curve C, then   (page 963)

$$\int_C \boldsymbol{F} \cdot d\boldsymbol{r} = \iint_S (\text{curl } \boldsymbol{F}) \cdot \boldsymbol{N}\, dS.$$

◆ The **divergence** of the vector field $F = Mi + Nj + Pk$ is the **scalar**   (page 973)

$$\text{div } \boldsymbol{F} = \frac{\partial M}{\partial x} + \frac{\partial N}{\partial y} + \frac{\partial P}{\partial z}.$$

◆ **Divergence Theorem (Gauss):** If the surface S encloses the region R, (page 973)

$$\iint_S \mathbf{F} \cdot \mathbf{N} \, dS = \iiint_R \operatorname{div} \mathbf{F} \, dV.$$

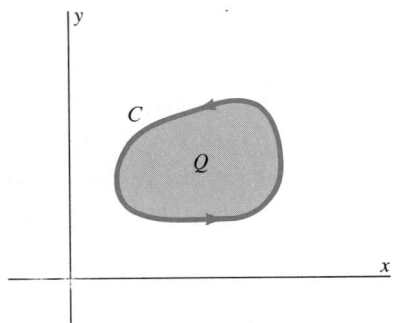

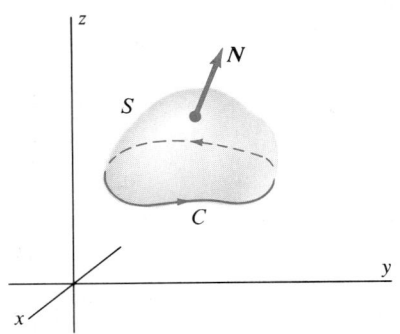

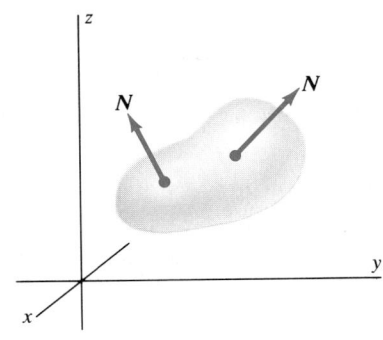

Green's Theorem

$$\int_C M \, dx + N \, dy = \iint_Q \left(\frac{\partial N}{\partial x} - \frac{\partial M}{\partial y} \right) dA$$

Stokes' Theorem

$$\int_C \mathbf{F} \cdot d\mathbf{r} = \iint_S (\operatorname{curl} \mathbf{F}) \cdot \mathbf{N} \, dS$$

Divergence Theorem

$$\iint_S \mathbf{F} \cdot \mathbf{N} \, dS = \iiint_R \operatorname{div} \mathbf{F} \, dV$$

REVIEW EXERCISES—CHAPTER 20

1. Compute the gradient vector field of the given function.
   a. $f(x, y, z) = y(\cos x)\ln z$
   b. $f(x, y, z) = x^3yz^2 + \sin xy$

2. Calculate the work done by the force field $\mathbf{F}(x, y) = x^2\mathbf{i} + 2xy\mathbf{j}$ when a particle is moved once around the triangle with vertices $(-1, 2)$, $(1, 0)$, and $(3, 2)$.

3. Show that the force field

   $$\mathbf{F}(x, y) = \frac{y}{x^2 + y^2}\mathbf{i} - \frac{x}{x^2 + y^2}\mathbf{j} \text{ is conservative.}$$

4. Show that curl $\mathbf{F} = \mathbf{0}$ where $\mathbf{F}$ is the force field in Exercise 3.

5. Let C be the unit circle. Show that $\displaystyle\iint_C \mathbf{F} \cdot d\mathbf{r} \neq 0$ where $\mathbf{F}$ is the vector field in Exercise 3. Does this result together with the result of Exercise 4 contradict Stokes' Theorem? Why or why not?

6. For $\mathbf{F}(x, y, z) = xy\mathbf{i} + (y^2 - x^2)\mathbf{j} + x^2z^2\mathbf{k}$ find
   a. curl $\mathbf{F}$ **b.** div $\mathbf{F}$

7. Is the vector field $\mathbf{F}(x, y, z) = y^2z\mathbf{i} + 2xyz\mathbf{j} + xy^2\mathbf{k}$ conservative? Why or why not?

8. Show that $\displaystyle\iint_S \mathbf{F} \cdot \mathbf{N} \, dS = 0$ if S is the unit sphere and $\mathbf{F}$ is the vector field $\mathbf{F}(\mathbf{r}) = \mathbf{r}_0$ where $\mathbf{r}_0$ is constant.

9. Find the flux of the vector field $\mathbf{F}(x, y, z) = zx\mathbf{i} + xy\mathbf{j} - x^2z\mathbf{k}$ outward over the tetrahedron with vertices $(0, 0, 0)$, $(0, 1, 0)$, $(1, 0, 0)$, and $(0, 0, 1)$.

10. Show that the vector field $\mathbf{F}(x, y, z) = x\mathbf{i} + y^2\mathbf{j} + z^3\mathbf{k}$ is irrotational.

11. Let $f(x, y, z)$ be a differentiable scalar function. Show that curl $(\nabla f) = \mathbf{0}$.

12. Let $\mathbf{F}(x, y, z)$ be a differentiable vector field. Show that div(curl $\mathbf{F}$) = 0.

13. True or false? If curl $\mathbf{F} = \mathbf{0}$ then $\mathbf{F} = \nabla\phi$ for some differentiable scalar function $\phi(x, y, z)$. Explain.

14. Find the value of the surface integral $\displaystyle\iint_S \mathbf{F} \cdot \mathbf{N} \, dS$ where $\mathbf{F}(x, y, z) = xy\mathbf{i} + x^2z\mathbf{j} + xz\mathbf{k}$, S is the unit sphere and N is the outward unit normal.

15. Let $\mathbf{F}(x, y, z) = ax\mathbf{i} + by\mathbf{j} + cz\mathbf{k}$. Find the flux of $\mathbf{F}$ over the closed surface S in terms of the volume V of the region enclosed by S.

16. Find $\displaystyle\int_C \mathbf{F} \cdot d\mathbf{r}$ where $\mathbf{F}(x, y) = x\mathbf{i} - 3xy\mathbf{j}$ and C is the square with vertices $(0, 0)$, $(0, 2)$, $(2, 2)$, $(0, 2)$ oriented counterclockwise.

17. Find $\int_C F \cdot dr$ where $F(x, y) = x^2 i - xy j$ and C is the upper unit semicircle from $(1, 0)$ to $(-1, 0)$.

18. Find $\int_{(1, 2)}^{(3, 2)} F \cdot dr$ where $F(x, y) = 4x^3 y^2 i + 2x^4 y j$.

In Exercises 19–23, evaluate the given line integral.

19. $\int_C xz \, dx + yz \, dy + z \, dz$ where C is parameterized by $r(t) = t i + j + \sin t k$, $0 \le t \le \pi$.

20. $\int_C y \, dx + x \, dy$, C is the triangle with vertices $(0, 0)$, $(4, 0)$, and $(2, 4)$, taken in that order.

21. $\int_C F \cdot dr$, $F(x, y, z) = 2y i - x j + 3z^2 k$, $r(t) = t i - 3t j + k$, $0 \le t \le 1$.

22. $\int_C F \cdot dr$, $F(x, y) = e^{2y} i + \cos \pi x j$, C is the triangle with vertices $(0, 0)$, $(1, 0)$, and $(0, 1)$ taken in that order.

23. $\int_C F \cdot dr$, $F(x, y) = xy^2 i + x^3 y j$, C is the upper unit semicircle taken counterclockwise.

In Exercises 24–28, evaluate the given surface integral.

24. $\iint_S (x^2 + 2) \, dS$, S is the portion of the plane $x + y + z = 6$ lying inside the cylinder $x^2 + y^2 = 9$.

25. $\iint_S (x + y) \, dS$, S is the portion of the plane $x - 2y + z = 8$ lying above the square with vertices $(0, 0, 0)$, $(1, 0, 0)$, $(1, 1, 0)$, $(0, 1, 0)$.

26. $\iint_S (x^2 + y^2 + 1) \, dS$, S is the part of the plane $z = x + 4$ inside the cylinder $x^2 + y^2 = 4$.

27. $\iint_S x^2 z \, dS$, S is the portion of the cone $z^2 = x^2 + y^2$ lying between the planes $z = 1$ and $z = 4$.

28. $\iint_S \sqrt{x^2 + y^2} \, dS$, S is the part of the graph of $z = 2xy$ lying inside the cylinder $x^2 + y^2 = 1$.

In Exercises 29–33, find the flux $\iint F \cdot N \, dS$ of the given vector field outward over the given surface.

29. $F(x, y, z) = xy i + y j + 2z k$, S is the graph of the paraboloid $z = x^2 + y^2$ for $1 \le z \le 4$, N is the downward unit normal.

30. $F(x, y, z) = yi + xj$, S is the unit sphere $x^2 + y^2 + z^2 = 1$.

31. $F(x, y, z) = yi + xj$, S is the cylinder $x^2 + y^2 = 1$, $0 \le z \le 1$, including the top and bottom discs.

32. $F(x, y, z) = (x + y^3)i + (z^2 - y)j + (x^3 - y^2)k$, S is the unit sphere $x^2 + y^2 + z^2 = 1$.

33. $F(x, y, z) = axi + byj + czk$, S is the graph of $z = 4 - x^2 - y^2$, $z \ge 0$, N is the upward unit normal.

In Exercises 34–37, evaluate the given line integral.

34. $\int_C F \cdot dr$, $F(x, y, z) = (x + y)i + (z - y)j + zk$, C is the circle $x^2 + y^2 = 4$ oriented counterclockwise.

35. $\int_C F \cdot dr$, $F(x, y, z) = 2y^2 i - 3z j + 2x k$, C is the triangle with vertices $(0, 0, 0)$, $(2, 4, 2)$, and $(0, 4, 0)$, taken in that order.

36. $\int_C F \cdot dr$, $F(x, y, z) = 3yi - zj + 4xk$, C is the boundary of the part of the unit sphere $x^2 + y^2 + z^2 = 1$ lying in the first quadrant with orientation induced by the outward unit normal to the sphere.

37. $\int_C F \cdot dr$, $F(x, y, z) = (x^3 - y)i + (\cos y - x)j + \sin z k$, C is the triangle with vertices $(0, 0, 0)$, $(4, 2, 0)$, and $(2, 4, 2)$, taken in that order.

38. Let $F(x, y, z) = x^3 i + z j + y k$. Find the work done by F on an object that moves from $(1, 0, 0)$ to $(-1, 0, \pi)$ along a straight line.

39. Find the work done by F in Exercise 38 if the motion is along the helix $r(t) = \cos t i + \sin t j + t k$.

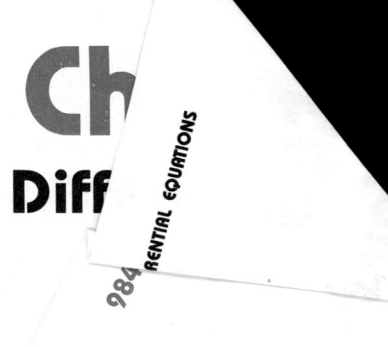

Ch

Diff

The purpose of this chapter is to indicate brie...
the calculus can be used to solve certain types of u.
(See Section 5.3, where the basic definitions associated wi...
and the method of separation of variables, were first introduceu.,

For a comprehensive treatment of the subject of differential equatio...
referred to courses and texts with this title. The treatment here, however, s...
enable you to appreciate the use made of differential equations in elementary
courses in science and engineering.

21.1 FIRST ORDER LINEAR DIFFERENTIAL EQUATIONS

Recall that a differential equation is an equation involving one or more derivatives
of an unknown function. Two examples of differential equations are

$$\frac{dy}{dt} = ky \tag{1}$$

and

$$f''(t) + 4f'(t) + f(t) = \cos t. \tag{2}$$

By a *solution* of a differential equation on an interval I we mean a function that is
differentiable on I as many times as the equation requires, and that satisfies the
equation. For example, we have already seen in Chapter 8 that the function $y = e^{kt}$
is a solution of equation (1) on $I = (-\infty, \infty)$ because $y = e^{kt}$ is differentiable for all
$t \in (-\infty, \infty)$ and

$$\frac{dy}{dt} = \frac{d}{dt}(e^{kt}) = k(e^{kt}) = ky, \qquad -\infty < t < \infty.$$

The *general solution* of a differential equation is a function involving one or
more constants that is a solution of the differential equation and from which all
particular solutions can be obtained by properly specifying the choices for these
constants. For example, the general solution of equation (1) is $y = Ce^{kt}$, since we
showed in Chapter 8 that all solutions of equation (1) are of this form.

The *order* of a differential equation is the order of the highest derivative in the
equation. Thus, equation (1) is a first order equation and equation (2) is a second
order equation.

In Section 5.3 we showed that a first order differential equation of the form

$$\frac{dy}{dt} = \frac{f(t)}{g(y)}$$

may be solved by the method of *separation of variables*. This involves rewriting the equation as

$$g(y) \cdot \frac{dy}{dt} = f(t)$$

and "integrating both sides:"

$$\int g(y)\left(\frac{dy}{dt}\right) dt = \int f(t)\, dt$$

or

$$\int g(y)\, dy = \int f(t)\, dt.$$

In the remaining part of this section we show how this technique, and the method of integrating factors, may be used to solve first order linear equations of the general form

$$\frac{dy}{dt} + p(t)y = g(t)$$

where the functions p and g (which may be constants, including zero) are continuous on the interval of interest.

Separation of Variables

First, suppose we arbitrarily add a constant b to the right side of equation (1). We obtain the differential equation

$$\frac{dy}{dt} = ky + b.$$

By letting $a = -\dfrac{b}{k}$ we can write this equation as

$$\frac{dy}{dt} = k(y - a). \tag{3}$$

The interpretation of equation (3) is that the rate of increase or decrease of y is proportional to the *difference* between y and the constant a. Before turning to practical examples of this type, we observe that equation (3) may be solved by the method of *separation of variables*. The following example illustrates the use of this method.

Example 1

Find a solution of the differential equation

$$\frac{dy}{dt} = 2 - ay, \qquad a \neq 0. \tag{4}$$

Strategy
Separate variables.

Solution
Recall that **separating variables** means isolating y terms on one side of the equation and t terms on the other. To do so we must assume $2 - ay \neq 0$. We

then divide both sides by $(2 - ay)$ and multiply both sides by dt to obtain

$$\frac{dy}{2 - ay} = dt.$$

Integrate both sides.

Integrating both sides then gives

$$\int \frac{dy}{2 - ay} = \int dt,$$

so

(Only one constant of integration need be displayed.)

$$-\frac{1}{a} \ln |2 - ay| = t + C \qquad (C \text{ arbitrary})$$

or

$$\ln |2 - ay| = -at - aC.$$

Thus

Solve for y by applying the exponential function to both sides.

$$|2 - ay| = e^{(-at - aC)}$$
$$2 - ay = \pm e^{-at - aC}$$
$$-ay = -2 \pm e^{-at - aC}$$
$$y = \frac{2}{a} + Ae^{-at}, \qquad A = \pm \frac{1}{a} e^{-aC} \tag{5}$$

A is determined by initial conditions.

where the constant A is to be determined from an initial condition. For example, if $y(0) = 1$ we obtain

$$1 = \frac{2}{a} + Ae^0, \qquad \text{so} \qquad A = \left(1 - \frac{2}{a}\right).$$

Verify that assumption on $2 - ay$ holds for solutions obtained.

To check that our assumption $2 - ay \neq 0$ holds, notice that from (5)

$$2 - ay = 2 - a\left[\frac{2}{a} + Ae^{-at}\right]$$
$$= -aAe^{-at},$$

which is not zero unless $A = 0$. However, if $A = 0$ equation (5) still gives the valid solution $y = \frac{2}{a}$. Thus, in all cases equation (5) represents a solution of (4).   ◇

Mixing Problems

Examples 2 and 3 illustrate how differential equations of the above type arise as models in certain "mixing" problems.

Example 2

A tank in which chocolate milk is being mixed contains a mixture of 460 liters of milk and 40 liters of chocolate syrup initially. Syrup and milk are then added to the tank at the rates of 2 liters per minute of syrup and 8 liters per minute of milk. Simultaneously, the mixture is withdrawn at the rate of 10 liters per minute. Assuming perfect mixing of the milk and syrup, find the function giving the amount of syrup in the tank at time t.

Strategy

Label variables.

Use fact that $\dfrac{dy}{dt}$ is a *rate* to find a differential equation satisfied by y.

Find y by solving this differential equation.

Determine the constant A by applying *initial conditions*.

State the solution.

Verify that restrictions on y are met.

Solution

If y represents the amount of syrup in the tank, $\dfrac{dy}{dt}$ is the rate at which the amount of syrup is changing. This rate is the rate at which syrup enters the tank (2 liters per minute) minus the rate at which syrup leaves the tank $\left(\dfrac{y}{500} \times 10 \text{ liters per minute}\right)$. The function y therefore satisfies the differential equation

$$\frac{dy}{dt} = 2 - .02y.$$

This is a particular form of equation (4) treated in Example 1, with $a = .02$. From (5) the solution is

$$y = Ae^{-.02t} + 100.$$

To determine A we apply the initial condition that $y = 40$ when $t = 0$. We obtain the equation

$$40 = A + 100,$$

so

$$A = -60.$$

The desired solution is therefore

$$y = 100 - 60e^{-.02t} \text{ liters.} \tag{6}$$

Note that the requirement of Example 1 that

$$y \neq \frac{2}{a} = \frac{2}{.02} = 100$$

is met since $60e^{-.02t} > 0$ for all t. The graph of the solution appears in Figure 1.1 and values for $y(t)$ are listed in Table 1.1. ◇

Table 1.1 Calculated values of $y(t)$ in Figure 1.1

t	0	1	2	3	5	10	20	100	200
$y = 100 - 60e^{-.02t}$	40	41.2	42.4	43.5	45.7	50.9	59.8	91.9	98.9

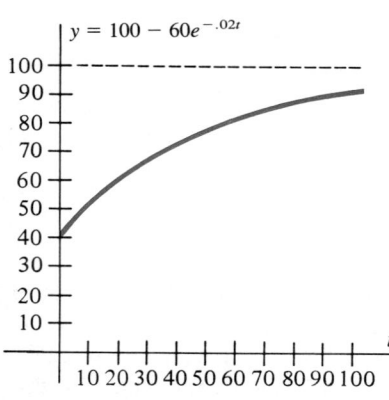

Figure 1.1 Solution of mixing problem in Example 2.

REMARK: Figure 1.1 illustrates an important property of solutions of differential equations of the form of equation (3), namely, that the solution approaches a horizontal asymptote (constant limit) as $t \to \infty$. For the solution y of Example 2 it is easy to see that

$$\lim_{t \to \infty} y = \lim_{t \to \infty} (100 - 60e^{-.02t}) = 100,$$

as Figure 1.1 suggests. This asymptote $y = 100$ is called the **steady state** part of the solution, while the term $60e^{-.02t}$ is referred to as the **transient** part of the solution. The result of Example 2 is precisely what intuition suggests—regardless of the initial concentration of syrup, the steady state concentration will be $100/500 = .2$, since 2 liters of syrup enter for every 8 liters of milk.

Example 3

Newton's law of cooling states that the rate at which an object gains or loses heat is proportional to the difference in temperature between the object and its surroundings. A bottle of soda at temperature 6°C is removed from a refrigerator and placed in a room at temperature 22°C. If, after 10 minutes, the temperature of the soda is 14°C, find its temperature after 20 minutes according to Newton's law.

Strategy	*Solution*
Label variables.	Let $y(t)$ denote the temperature of the soda t minutes after it is removed from the refrigerator. Newton's law of cooling states that
Interpret Newton's law as a differential equation for y.	$$\frac{dy}{dt} = k \cdot (22 - y).$$
	Thus
Separate variables.	$$\frac{dy}{22 - y} = k\,dt,$$
	so
	$$\int \frac{dy}{22 - y} = \int k\,dt.$$
Integrate.	Integrating on both sides we obtain
	$$-\ln|22 - y| = kt + C$$
Apply exponential function to both sides.	$$|22 - y| = e^{-kt-C}$$
	$$22 - y = \pm e^{-kt-C}.$$
	Thus
Solve for y.	$$y = 22 - Ae^{-kt}, \qquad A = \pm e^{-C}. \qquad (7)$$
Use initial data to find A.	To find A we apply the initial data* that $y_0 = 6°$ when $t = 0$. Inserting this information in (7) gives
	$$6 = 22 - Ae^0,$$
	so
	$$A = 16. \qquad (8)$$
Use additional data to find k.	To find k we use (7), (8), and the data $y(10) = 14$. We obtain the equation
	$$14 = 22 - 16e^{-10k}.$$
	Thus
	$$e^{-10k} = \frac{-8}{-16} = \frac{1}{2},$$
	so
$\ln(e^{-10k}) = -10k = \ln\left(\frac{1}{2}\right)$	$$k = \frac{\ln 2}{10}.$$
$\qquad = -\ln 2,$	

---

*Initial data refer to information about a function at *any* value of the independent variable, not necessarily $t = 0$.

so

$$k = \frac{-\ln 2}{-10} = \frac{\ln 2}{10}.$$

Find $y(20)$

$$e^{\ln 2^{-2}} = 2^{-2} = \frac{1}{4}.$$

Table 1.2

t	y in (9)
0	6.00
4	9.87
8	12.81
12	15.04
16	16.72
20	18.00
24	18.97
28	19.70
32	20.26
36	20.68

The function y is now completely determined as

$$y = 22 - 16e^{-\left(\frac{\ln 2}{10}\right)t}. \tag{9}$$

The temperature after 20 minutes is therefore

$$\begin{aligned}
y(20) &= 22 - 16e^{-\left(\frac{\ln 2}{10}\right)20} \\
&= 22 - 16e^{-2\ln 2} \\
&= 22 - 16e^{\ln 2^{-2}} \\
&= 22 - 16 \cdot \frac{1}{4} \\
&= 18°.
\end{aligned}$$

(See Table 1.2 and Figure 1.2.)

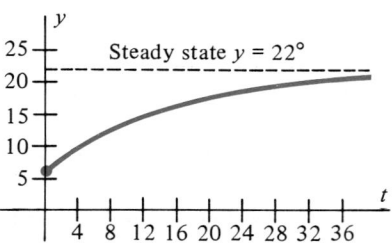

Figure 1.2 Graph of $y = 22 - 16e^{-\left(\frac{\ln 2}{10}\right)t}.$

Integrating Factors

A second generalization of equation (1) is to linear first order differential equations of the form

$$\frac{dy}{dt} + p(t)y = g(t) \tag{10}$$

where p and g are continuous functions of t. The differences between equation (10) and equation (1) involve these two functions. In equation (1) we have $p(t) \equiv -k$, and $g(t) \equiv 0$.

We may find solutions to equations of the form (10) if we let $u(t) = \displaystyle\int p(t)\, dt$ and note that *if we multiply both sides of* (10) *by the integrating factor*

$$e^{u(t)} = e^{\int p(t)\, dt} \tag{11}$$

it is easy to find the integral of the left side of equation (10). (In (11) the expression $\int p(t)\, dt$ denotes any *particular* antiderivative for p.) To demonstrate our remark we note that

$$\left[\frac{dy}{dt} + p(t)y\right] \cdot e^{u(t)} = \frac{dy}{dt}\, e^{\int p(t)\, dt} + yp(t)e^{\int p(t)\, dt} \tag{12}$$

$$= \frac{d}{dt}[ye^{\int p(t)\, dt}]$$

$$= \frac{d}{dt}[y \cdot e^{u(t)}].$$

To find a solution of the differential equation (10), we begin by multiplying both sides by an integrating factor e^u in (11) to obtain

$$\left[\frac{dy}{dt} + p(t)y\right]e^{u(t)} = g(t)e^{u(t)}.$$

Using (12) we can write this equation as

$$\frac{d}{dt}[ye^{u(t)}] = g(t)e^{u(t)},$$

so

$$ye^{u(t)} = \int g(t)e^{u(t)}\, dt + C$$

or

$$y = e^{-u(t)}\left\{\int g(t)e^{u(t)}\, dt + C\right\}. \tag{13}$$

The function y in (13) is the **general solution** of equation (10) because, as is shown in more advanced courses, every solution of (10) has the form (13) for some choice of the constant C.

Rather than remembering formula (13), you should simply remember that solving a differential equation of the form (10) requires the integrating factor $e^{u(t)} = e^{\int p(t)\, dt}$.

Example 4

Find the general solution of the differential equation

$$\frac{dy}{dt} + 2ty = t. \tag{14}$$

Solution: This equation has the form (10) with $p(t) = 2t$, $g(t) = t$. We therefore use the integrating factor

$$e^{u(t)} = e^{\int 2t\, dt} = e^{t^2}.$$

Multiplying both sides of (14) by e^{t^2} gives

$$e^{t^2} \cdot \frac{dy}{dt} + e^{t^2} \cdot 2t \cdot y = te^{t^2}.$$

We next rewrite the left side as a derivative

$$\frac{d}{dt}[e^{t^2} \cdot y] = e^{t^2} \cdot \frac{dy}{dt} + e^{t^2} \cdot 2t \cdot y$$

and integrate both sides:

$$e^{t^2}y = \int te^{t^2}\,dt$$

$$= \frac{1}{2}e^{t^2} + C.$$

Finally, we multiply both sides by e^{-t^2} to obtain

$$y = e^{-t^2}\left[\frac{1}{2}e^{t^2} + C\right]$$

$$= Ce^{-t^2} + \frac{1}{2}.$$

◇

Example 5

Find the solution of $\dfrac{dy}{dt} - \dfrac{2}{t}y = t^3$, $t > 0$ satisfying the initial condition $y(1) = 2$.

Strategy

Find an integrating factor.

Solution

We use the integrating factor

$$e^{u(t)} = e^{\int -2/t\,dt} = e^{-2\ln t} = e^{\ln t^{-2}} = \frac{1}{t^2}.$$

Multiply both sides of the equation by the integrating factor.

We then obtain

$$\frac{1}{t^2}\cdot\frac{dy}{dt} - \frac{2}{t^3}\cdot y = t^3\left(\frac{1}{t^2}\right).$$

We can write this equation as

Write the left side as a derivative.

$$\frac{d}{dt}\left[\frac{1}{t^2}y\right] = t,$$

so

Integrate both sides.

$$\frac{1}{t^2}y = \int t\,dt = \frac{1}{2}t^2 + C.$$

Solve for y.

Thus

$$y = t^2\left[\frac{1}{2}t^2 + C\right]$$

$$= \frac{1}{2}t^4 + Ct^2.$$

Apply initial conditions to solve for C.

The initial condition $y(1) = 2$ gives

$$2 = \frac{1}{2} + C,$$

so

$$C = 2 - \frac{1}{2} = \frac{3}{2}.$$

The desired solution is

$$y(t) = \frac{1}{2}t^4 + \frac{3}{2}t^2.$$

◇

Exercise Set 21.2

In Exercises 1–4, find a particular solution of the initial value problem. Identify the steady-state and transient parts of the solution.

1. $\dfrac{dy}{dt} = y + 1 \qquad y(0) = 1$

2. $\dfrac{dy}{dt} = 2y + 2 \qquad y(0) = -1$

3. $\dfrac{dy}{dt} = -y + \pi \qquad y(0) = \pi/2$

4. $\dfrac{dy}{dx} + y = 2 \qquad y(0) = 2$

In Exercises 5–14 find the general solution of the differential equation.

5. $y' + 2y = 4$

6. $y' + ay = b$

7. $y' - \dfrac{2}{t}y = t^4$

8. $y' = \dfrac{1}{t^2}y$

9. $y' - 2y = e^{2t} \sin t$

10. $y' = 2ty + 3t$

11. $y' + y = e^t$

12. $ty' - 2y = -2t$

13. $ty' + y = t$

14. $y' = y - 2e^{-t}$

15. Use the method of separation of variables to find a solution of the differential equation $\dfrac{dy}{dx} = axy$ where a is constant and $y > 0$.

16. A curve in the xy-plane has the property that at each point (x, y) the slope of the tangent equals xy, the product of the coordinates. Find an equation for such a curve containing the point $(0, 1)$ and sketch its graph. (*Hint:* Use the result of Exercise 15.)

17. A tank contains 200 liters of brine. The initial concentration of salt is 0.5 kilogram per liter of brine. Fresh water is then added at a rate of 3 liters per minute and the brine is drawn off from the bottom of the tank at the same rate. Assume perfect mixing.
   a. Find a differential equation for the number of kilograms of salt, $p(t)$, in the solution at time t.
   b. How much salt is present after 20 minutes?
   c. What is the steady state concentration of salt?

18. A bottle of water whose temperature is 26°C is placed in a room at temperature 12°C. If the temperature of the water after 30 minutes is 20°C, find its temperature after 2 hours.

19. A person initially places $100 in a savings account that pays interest at a rate of 10% per year compounded continuously. If the person makes additional deposits of $100 per year continuously throughout each year, find:
   a. a differential equation for $P(t)$, the amount on deposit after t years, and
   b. the amount on deposit after 7 years.

20. An electrical circuit contains a battery supplying E volts, a resistance R, and an inductance L. When the battery is connected, the resulting current I satisfies the differential equation

$$L\frac{dI}{dt} + RI = E, \qquad I(0) = 0.$$

   a. Find the solution I as a function of t.
   b. Obtain Ohm's Law, $I = E/R$, as the steady state part of the solution.

21. If, in the circuit described in Exercise 20, the inductance is replaced by a capacitance C, the equation becomes

$$R\frac{dI}{dt} + \frac{I}{C} = \frac{dE}{dt}.$$

   Find the solution $I(t)$, if the battery is switched off (i.e., $dE/dt = 0$ at the time $t = 0$.)

22. In the chemical reaction called inversion of raw sugar, the inversion rate is proportional to the amount of raw sugar remaining. If 100 kilograms of raw sugar is reduced to 75 kilograms in 6 hours, how long will it be until
   a. half the raw sugar has been inverted?
   b. 90% of the raw sugar has been inverted?

23. A pan of boiling water is removed from a burner and cools to 80°C in 3 minutes. Find its temperature after 10 minutes, according to Newton's law of cooling, if the surrounding temperature is 30°C.

24. A room containing 1000 cu ft of air is initially free of pollutants. Air containing 100 parts of pollutants per cu ft is pumped into the room at a rate of 50 cu ft per minute. The air in the room is kept perfectly mixed, and air is removed from the room at the same rate of 50 cu ft per minute. Find the function $P(t)$ giving the number of parts of pollutants per cubic foot in the room air after t minutes.

25. A tank contains a mixture of 500 gallons of salt brine (salt and water) initially containing 2 pounds of salt per gallon. A second mixture, containing 6 pounds of salt per gallon, is then pumped in at a rate of 10 gallons per minute. The resulting mixture is drawn off at the same rate. Assuming perfect mixing, find the function that describes the concentration of salt (pounds per gallon) in the mixture in the tank t minutes after mixing begins.

26. Influenza spreads through a university community at a rate proportional to the product of the proportion of those already infected and the proportion of those not yet infected. Assuming that those infected remain infected, find the proportion $P(t)$ infected after t days if initially 10% were infected and after 3 days 30% were infected.

27. An annuity is initially set up containing $10,000 to which interest is compounded continuously at a rate of 10% per year. The owner of the annuity withdraws $2000 per year in a continuous manner. When does the value of the annuity reach zero?

28. Certain biological populations grow in size according to the Gompertz growth equation

$$\frac{dy}{dt} = -ky \ln y$$

where k is a positive constant. Find the general solution of this differential equation.

29. The Weber-Fechner law in psychology is a model for the rate of change of the reaction y to a stimulus of strength s that is given by the differential equation

$$\frac{dy}{ds} = k\frac{y}{s}$$

where k is a positive constant. Find the general solution of this differential equation.

21.2 EXACT EQUATIONS

Certain types of first order differential equations may be solved by a technique discussed in Chapter 20 for vector functions. These are *exact* differential equations.

DEFINITION 1

Let the functions M and N be defined in an open rectangle R in $\mathbb{R}^2$. The differential equation

$$M(x, y) + N(x, y)\frac{dy}{dx} = 0, \qquad (x, y) \in R \tag{1}$$

is called *exact* if there exists a function f of two variables for which

$$M(x, y) = \frac{\partial f}{\partial x}(x, y) \text{ and } N(x, y) = \frac{\partial f}{\partial y}(x, y)$$

for all $(x, y) \in R$.

Example 1

The differential equation

$$e^{2y} + 2xe^{2y}\frac{dy}{dx} = 0$$

is exact since, for the function $f(x, y) = xe^{2y}$,

$$M(x, y) = e^{2y} = \frac{\partial}{\partial x}(xe^{2y}) = \frac{\partial f}{\partial x}$$

and

$$N(x, y) = 2xe^{2y} = \frac{\partial}{\partial y}(xe^{2y}) = \frac{\partial f}{\partial y}.$$

The general solution is therefore determined by the level curves for the function $f(x, y) = xe^{2y}$:

$$xe^{2y} = C.$$ ◇

Example 2

The differential equation

$$(3x^2 + 2y^2)\, dx + 4xy\, dy = 0$$

is exact, since the function $f(x, y) = x^3 + 2xy^2$ has the property that

$$M(x, y) = 3x^2 + 2y^2 = \frac{\partial}{\partial x}(x^3 + 2xy^2) = \frac{\partial f}{\partial x}$$

and

$$N(x, y) = 4xy = \frac{\partial}{\partial y}(x^3 + 2xy^2) = \frac{\partial f}{\partial y}.$$

The general solution is therefore

$$x^3 + 2xy^2 = C,$$

or

$$y = \pm\left(\frac{C - x^3}{2x}\right)^{1/2}.$$ ◇

Since Definition 1 may be paraphrased by saying that equation (1) is exact if the function f is a potential for the vector function $\mathbf{F}(x, y) = M(x, y)\mathbf{i} + N(x, y)\mathbf{j}$, Theorem 10 of Chapter 18 provides the following criterion for equation (1) to be exact.

THEOREM 1

Let the functions M and N have continuous partial derivatives in an open rectangle R in $\mathbb{R}^2$. The differential equation

$$M(x, y) + N(x, y)\frac{dy}{dx} = 0$$

is exact in R if and only if

$$\frac{\partial M}{\partial y}(x, y) = \frac{\partial N}{\partial x}(x, y)$$

for all (x, y) in R.

Example 3

The differential equation

$$xy^3\, dx + x^3y\, dy = 0$$

is not exact, since

$$\frac{\partial M}{\partial y}(x, y) = \frac{\partial}{\partial y}(xy^3) = 3xy^2 \neq 3x^2y = \frac{\partial}{\partial x}(x^3y) = \frac{\partial N}{\partial x}(x, y).$$ ◇

Example 4

Test the differential equation

$$(3x^2y + \cos x)\, dx + x^3\, dy = 0$$

for exactness. If it is exact, find a solution passing through the point $(1, \pi)$.

Solution: Here

$$M(x, y) = 3x^2y + \cos x; \qquad N(x, y) = x^3.$$

Since both M and N have continuous partial derivatives throughout $\mathbb{R}^2$, we may apply Theorem 1 to test for exactness. Since

$$\frac{\partial M}{\partial y}(x, y) = 3x^2 = \frac{\partial N}{\partial x}(x, y),$$

the equation is exact.

To solve the differential equation, we must find a potential f. To do so we integrate $M(x, y)$ partially with respect to x to find

$$f(x, y) = \int M(x, y)\, dx = \int (3x^2y + \cos x)\, dx$$

$$= x^3y + \sin x + h_1(y) + K_1.$$

Also,

$$f(x, y) = \int N(x, y)\, dy = \int x^3\, dy = x^3y + h_2(x) + K_2.$$

Equating these two expressions for f then gives

$$h_1(y) \equiv 0, \qquad h_2(x) = \sin x, \qquad \text{and} \qquad K_1 = K_2 = K.$$

Thus

$$f(x, y) = x^3y + \sin x + K$$

is the desired potential for any constant K. We therefore choose $K = 0$ and obtain the family of level curves

$$x^3y + \sin x = C \tag{2}$$

as the general solution for the differential equation. To find the particular curve passing through the point $(x_0, y_0) = (1, \pi)$, we set $x = 1$ and $y = \pi$ in (2) to find

$$1^3(\pi) + \sin(\pi) = \pi = C,$$

so $C = \pi$. A solution passing through $(1, \pi)$ is therefore determined by the equation

$$x^3y + \sin x = \pi$$

as

$$y = \frac{\pi - \sin x}{x^3}, \qquad x^3 \neq 0. \qquad\qquad \diamond$$

Example 4 demonstrates the general technique for solving exact equations. Once the equation is shown to be exact by use of Theorem 1, the solution $f(x, y) = C$ is found by integrating $M = \dfrac{\partial f}{\partial x}$ partially with respect to x, integrating $N = \dfrac{\partial f}{\partial y}$ partially with respect to y, and equating the resulting expressions for f. This is the same technique that was used in Chapter 18 to reconstruct a potential from its gradient.

Example 5

Solve the differential equation

$$\cos y \, dx + (3y^2 - x \sin y) \, dy = 0.$$

Solution: Writing the equation in the form

$$\cos y + (3y^2 - x \sin y) \frac{dy}{dx} = 0$$

and letting

$$M(x, y) = \cos y; \qquad N(x, y) = 3y^2 - x \sin y,$$

we verify that the equation is exact by noting that

$$\frac{\partial M}{\partial y} = -\sin y = \frac{\partial N}{\partial x}.$$

To find the solution $f(x, y) = C$, we first integrate M partially with respect to x to find that

$$f(x, y) = \int M(x, y) \, dx = \int \cos y \, dx = x \cos y + g(y).$$

Next, integrating N partially with respect to y gives

$$f(x, y) = \int N(x, y) \, dy = \int (3y^2 - x \sin y) \, dy$$
$$= y^3 + x \cos y + h(x).$$

Equating these two expressions for $f(x, y)$ gives

$$x \cos y + g(y) = y^3 + x \cos y + h(x),$$

so $g(y) = y^3$ and $h(x) \equiv 0$. Thus, $f(x, y) = y^3 + x \cos y$ and the general solution is

$$y^3 + x \cos y = C.$$

$\diamond$

Exercise Set 21.2

In Exercises 1–10, determine whether the given equation is exact. If it is, find a family of solution (level) curves of the form $f(x, y) = C$.

1. $2xy \, dy + x^2 \, dy = 0$

2. $y^2 + xy \dfrac{dy}{dx} = 0$

3. $\dfrac{x \, dx}{\sqrt{x^2 + y^2}} + \dfrac{y \, dy}{\sqrt{x^2 + y^2}} = 0$

4. $\cos y^2 - 2xy \sin y^2 \dfrac{dy}{dx} = 0$

5. $\dfrac{x}{y^2} \, dx + \dfrac{y}{x^2} \, dy = 0$

6. $\dfrac{y}{1 + x^2 y^2} + \dfrac{x}{1 + x^2 y^2} \dfrac{dy}{dx} = 0$

7. $(x - 2xy^2) \, dx + (y + x^2 y) \, dy = 0$

8. $(y^2 - 2xy) \, dx + (2xy - x^2) \, dy = 0$

9. $\dfrac{x \, dx}{\sqrt{x + y^2}} + \dfrac{y \, dy}{\sqrt{x + y^2}} = 0$

10. $\dfrac{1}{\sqrt{x - y^2}} - \dfrac{2y}{\sqrt{x - y^2}} \dfrac{dy}{dx} = 0$

11. Find a solution of the differential equation in Exercise 1 passing through the point $(1, 3)$.

12. Find a solution of the differential equation in Exercise 4 passing through the point $(2, 0)$.

13. Find a solution of the differential equation in Exercise 8 passing through the point $(1, 2)$.

14. Show that a separable differential equation $f(x) \, dx + g(y) \, dy = 0$ is always exact. What is the ''solution'' function $f(x, y)$?

15. Show that the differential equation $(1 + x^2y^2 + y) \, dx + x \, dy = 0$ is not exact. Then show that if this equation is multiplied through by the integrating factor $p(x, y) = \dfrac{1}{1 + x^2y^2}$ an exact differential equation results. Is a solution of this exact equation also a solution of the original equation?

21.3 SECOND ORDER LINEAR EQUATIONS

By adding a term involving the second derivative to the first order linear differential equation $y' = ky$ for natural growth, we obtain one type of *second order linear differential equation*,*

$$\frac{d^2y}{dt^2} + a \frac{dy}{dt} + by = 0 \tag{1}$$

where a and b are constants. From our experience with first order equations we might suspect that equation (1) has a solution of the form $y = e^{rt}$ where r is a constant. If this were true, the derivative of this proposed solution would be

$$\frac{dy}{dt} = re^{rt}, \qquad \text{and} \qquad \frac{d^2y}{dt^2} = r^2e^{rt}.$$

Substituting into equation (1) would therefore give

$$r^2e^{rt} + a(re^{rt}) + b(e^{rt}) = 0,$$

or

$$(r^2 + ar + b)e^{rt} = 0. \tag{2}$$

Since values of the exponential function $y = e^{rt}$ are always nonzero, the only way in which equation (2) can be true is if the *characteristic polynomial* $r^2 + ar + b$ equals zero. That is, r must be a (real) *root* of the quadratic polynomial $r^2 + ar + b$.

> Let r be a real number. The function $y = e^{rt}$ is a solution of the differential equation
>
> $$\frac{d^2y}{dt^2} + a \frac{dy}{dt} + by = 0$$
>
> if and only if the constant r is a real root of the characteristic polynomial
>
> $$r^2 + ar + b = 0.$$

---

* More precisely, equation (1) is referred to as a second order, constant-coefficient, linear, homogeneous equation. The term homogeneous refers to the fact that the right-hand side is zero. This is but one type of second order equation. There are many others, but we shall not pursue them here.

Thus, we expect to find two, one, or zero solutions of the differential equation (1) of the form $y = e^{rt}$, depending on the number of real roots r of the associated characteristic polynomial.

Example 1

For the differential equation

$$\frac{d^2y}{dt^2} - \frac{dy}{dt} - 2y = 0$$

the constant coefficients are $a = -1$ and $b = -2$, so the characteristic polynomial is

$$r^2 - r - 2 = (r - 2)(r + 1).$$

This characteristic polynomial has real roots $r_1 = 2$ and $r_2 = -1$. Two solutions of the differential equation are therefore

$$y_1 = e^{r_1 t} = e^{2t} \qquad \text{and} \qquad y_2 = e^{r_2 t} = e^{-t},$$

as you can verify. (We shall say more about the *general* solution of this equation later.) ◇

Example 2

The differential equation

$$\frac{d^2y}{dt^2} + 4\frac{dy}{dt} + 4y = 0$$

has characteristic polynomial $r^2 + 4r + 4 = (r + 2)^2$, which has only the single real root $r = -2$. Thus, the function $y_1 = e^{-2t}$ is a solution. ◇

REMARK 1: When the characteristic polynomial for equation (1) has only the single real root r, equation (1) will have both the functions $y_1 = e^{rt}$ and $y_2 = te^{rt}$ as solutions. Thus, the function $y_2 = te^{-2t}$ is also a solution of the equation in Example 2, as you can verify.

REMARK 2: When the characteristic polynomial $r^2 + ar + b$ has the *complex** root $r_1 = \alpha + i\beta$, $\beta \neq 0$, then the complex number $r_2 = \alpha - i\beta$ is also a root, and the differential equation (1) has the two solutions

$$y_1 = e^{\alpha t} \sin \beta t \qquad \text{and} \qquad y_2 = e^{\alpha t} \cos \beta t.$$

Example 3

For the differential equation

$$\frac{d^2y}{dt^2} - 2\frac{dy}{dt} + 5y = 0$$

---

* An introduction to complex numbers is presented in Appendix III. The ideas contained in that appendix are assumed in this discussion.

the characteristic polynomial is

$$r^2 - 2r + 5.$$

The roots of this polynomial, according to the quadratic formula, are

$$r = \frac{2 \pm \sqrt{(-2)^2 - 4(1)(5)}}{2}$$

$$= 1 \pm \frac{1}{2}\sqrt{-16}$$

$$= 1 \pm 2i, \qquad i^2 = -1.$$

Thus, the roots are $r_1 = \alpha + i\beta = 1 + 2i$ and $r_2 = \alpha - i\beta = 1 - 2i$ with $\alpha = 1$ and $\beta = 2$, so two solutions are

$$y_1 = e^t \sin 2t \qquad \text{and} \qquad y_2 = e^t \cos 2t,$$

which we leave for you to verify. ◇

The General Solution of Equation (1)

If y_1 and y_2 are any two solutions of equation (1), then so is any *linear combination* of these solutions of the form

$$y(t) = Ay_1(t) + By_2(t) \tag{3}$$

where A and B are constants. (See Exercise 36.) If the functions y_1 and y_2 are not multiples of each other, the function y is called the *general solution* of equation (1) since it can be shown that any solution can be written in this form for an appropriate choice of the constants A and B.

Example 4

(i) The general solution of the differential equation $\dfrac{d^2y}{dt^2} - \dfrac{dy}{dt} - 2y = 0$ in Example 1 is

$$y = Ae^{2t} + Be^{-t}.$$

(ii) The general solution of the differential equation $\dfrac{d^2y}{dt^2} + 4\dfrac{dy}{dt} + 4y = 0$ in Example 2 is

$$y = Ae^{-2t} + Bte^{-2t}.$$

(iii) The general solution of the differential equation $\dfrac{d^2y}{dt^2} - 2\dfrac{dy}{dt} + 5y = 0$ in Example 3 is

$$y = Ae^t \sin 2t + Be^t \cos 2t$$
$$= e^t(A \sin 2t + B \cos 2t). \qquad ◇$$

Initial Conditions

In general, two initial conditions are required to determine a unique solution of a second order differential equation, one for the solution y and one for its derivative $\dfrac{dy}{dt}$, *specified at the same number t_0*. Since the general solution of equation (1) has the form given by equation (3), two initial conditions will produce two linear equations for the constants A and B, from which these numbers may be determined.

Example 5

Find a solution of the initial value problem consisting of the differential equation

$$y'' - 2y' - 3y = 0 \tag{4}$$

and the initial conditions

$$y(0) = 0$$
$$y'(0) = 4.$$

Strategy

First, find the general solution by

a. finding the characteristic polynomial

b. identifying its roots r_1 and r_2

c. combining the solutions $e^{r_1 t}$ and $e^{r_2 t}$ as in equation (3).

Next, substitute initial conditions into general solution y, and its derivative y', to obtain two equations in A and B.

Solve the two equations for A and B by substitution or by elimination.

Find particular solution by substituting for A and B in general solution.

Solution

The characteristic polynomial for equation (4) is

$$r^2 - 2r - 3 = (r - 3)(r + 1),$$

which has roots $r_1 = 3$ and $r_2 = -1$. The general solution of equation (4) is therefore

$$y(t) = Ae^{3t} + Be^{-t}, \tag{5}$$

so the derivative y' has the form

$$y'(t) = 3Ae^{3t} - Be^{-t}. \tag{6}$$

Substituting the initial condition $y(0) = 0$ into equation (5) gives the equation

$$A + B = 0. \tag{7}$$

Substituting the initial condition $y'(0) = 4$ into equation (6) gives the equation

$$3A - B = 4. \tag{8}$$

Substituting $B = -A$ from equation (7) into equation (8) gives

$$3A - (-A) = 4$$

or

$$4A = 4.$$

Thus, $A = 1$ and $B = -A = -1$.

Equation (5) now becomes

$$y(t) = e^{3t} - e^{-t},$$

the desired solution. ◇

Modelling Oscillatory Motion

A version of equation (1) that occurs frequently in applications is

$$\frac{d^2 y}{dt^2} + by = 0, \qquad b > 0. \tag{9}$$

The characteristic polynomial for equation (9) is $r^2 + b$, which has the complex roots $r_1 = \sqrt{b}i$ and $r_2 = -\sqrt{b}i$. Since two solutions of this equation are $e^{\alpha t} \sin \beta t$

and $e^{\alpha t} \cos \beta t$ with $\alpha = 0$ and $\beta = \sqrt{b}$, we have:

The differential equation

$$\frac{d^2y}{dt^2} + by = 0, \qquad b > 0 \tag{10}$$

has general solution

$$y = A \sin \sqrt{b}\, t + B \cos \sqrt{b}\, t. \tag{11}$$

Example 6

Find the general solution of the differential equation

$$y'' + 9y = 0.$$

Solution: Since the differential equation has the form of equation (10), with $b = 9$ the general solution is

$$y = A \sin 3t + B \cos 3t. \qquad \diamond$$

Because solutions of equation (9) are made up of sine and cosine functions, versions of this equation occur in models for the motion of vibrating physical systems involving springs, pendulums, etc. Here is a typical situation.

The Harmonic Oscillator

Imagine a mass m lying on a frictionless surface and attached to a spring (see Figure 3.1). When the mass is moved x units from its natural rest position, the spring will exert a restoring force, F_s, proportional to the displacement x (Hooke's Law). Since the restoring force acts in the direction opposite to the motion of the mass, we write

$$F_s = -kx \tag{12}$$

where the constant k is called the **spring constant** and depends on the particular spring.

Newton's second law of motion states that the sum of all forces acting on the mass must equal the product of the mass m and acceleration a. Since the net force* acting on the mass after it is displaced and released is F_s, we can write Newton's law as

$$F_s = ma, \qquad a = \text{acceleration}.$$

Since the variable x represents the displacement of the mass from its rest position, the time derivative $\dfrac{dx}{dt}$ is the velocity of the mass, and the second derivative $\dfrac{d^2x}{dt^2}$ is its acceleration. Assuming the displacement x to be a twice differentiable function of t, we can rewrite Newton's law as

$$F_s = m \cdot \frac{d^2x}{dt^2}. \tag{13}$$

---

*The force of gravity is exactly counteracted by the supporting force exerted by the surface, so the net *vertical* force is zero.

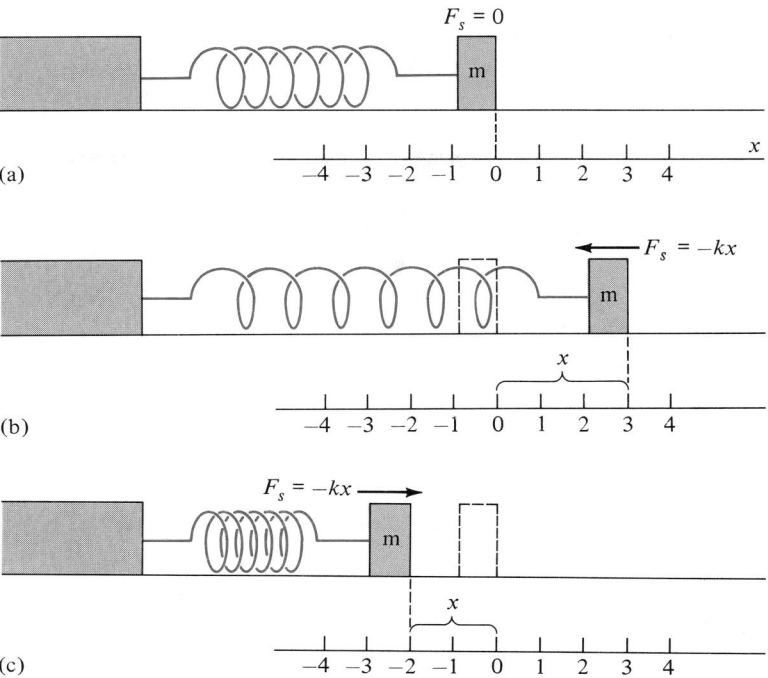

Figure 3.1 Mass-spring system (a) at equilibrium, (b) stretched, and (c) compressed. Displacement, x, is measured with respect to a particular point on the block.

Combining equations (12) and (13) now gives the differential equation

$$m \cdot \frac{d^2x}{dt^2} = -kx,$$

or

$$\frac{d^2x}{dt^2} + \frac{k}{m} \cdot x = 0. \qquad (14)$$

Equation (14) is called the **harmonic oscillator equation,** and it provides the desired mathematical model for the motion of the mass for the situation described above.

The solution of equation (14) is given by equation (11):

$$x(t) = A \cos \sqrt{\frac{k}{m}}\, t + B \sin \sqrt{\frac{k}{m}}\, t. \qquad (15)$$

Example 7

An 8-kg mass is attached to a spring as in Figure 3.1. The spring constant is $k = 2$ N/m. Find the resulting motion if the mass is displaced 10 cm from its equilibrium position and released from rest (zero initial velocity).

Strategy

Write differential equation (14), using given values of k and m.

Solution

Here $\dfrac{k}{m} = \dfrac{2}{8} = \dfrac{1}{4}$, so the differential equation (14) is

$$\frac{d^2x}{dt^2} + \frac{1}{4}x = 0.$$

Write general solution from (15).

From (15), with $\sqrt{k/m} = 1/2$, the general solution is

$$x(t) = A\,\cos\frac{t}{2} + B\,\sin\frac{t}{2}. \tag{16}$$

Set $x(0) = 10$ cm $= \dfrac{1}{10}$ m to find one equation in A and B.

To find A and B, we apply the given *initial conditions*

(i) $x(0) = \dfrac{1}{10}$ $\left(\text{initial displacement } 10 \text{ cm} = \dfrac{1}{10} \text{ m}\right)$, and

(ii) $x'(0) = 0$ (initial velocity zero).

From equation (16) and condition (i) we have

$$\frac{1}{10} = A\cos 0 + B\sin 0 = A,$$

so $A = \dfrac{1}{10}$. Since, from (16),

Differentiate $x(t)$ to obtain equation for $x'(t)$.

$$\frac{dx}{dt} = -\frac{A}{2}\sin\frac{t}{2} + \frac{B}{2}\cos\frac{t}{2}, \tag{17}$$

condition (ii) gives

Set $x'(0) = 0$ to obtain second equation in A and B.

$$0 = -\frac{A}{2}\sin 0 + \frac{B}{2}\cos 0 = \frac{B}{2}.$$

Thus, $B = 0$. The solution is therefore

$$x(t) = \frac{1}{10}\cos\frac{t}{2} \text{ (meters)}. \qquad \diamond$$

Example 8

Repeat Example 7, except that the mass now is released with an initial velocity of 3 m/s.

Strategy

Same solution as for Example 7, except that the condition $x'(0) = 0$ is replaced by the condition $x'(0) = 3$.

Solution

As in Example 7 we have

$$x(t) = A\,\cos\frac{t}{2} + B\,\sin\frac{t}{2}.$$

Also, we still have $x(0) = \dfrac{1}{10}$, so $A = \dfrac{1}{10}$.

However, the condition that the initial velocity is 3 m/s means that $x'(0) = 3$. Equation (17) therefore gives

Set $x'(0) = 3$ and solve for B.

$$3 = -\frac{A}{2}\sin 0 + \frac{B}{2}\cos 0 = \frac{B}{2},$$

so $B = 6$. In this case the solution is therefore

$$x(t) = \frac{1}{10}\cos\frac{t}{2} + 6\sin\frac{t}{2}.$$

◇

Exercise Set 21.3

In Exercises 1–6 find the general solution of the differential equation in the form $y = Ae^{r_1 t} + Be^{r_2 t}$ by finding two distinct real roots of the characteristic polynomial.

1. $\dfrac{d^2y}{dt^2} + 5\dfrac{dy}{dt} + 6y = 0$

2. $y'' + 3y' - 4y = 0$

3. $y'' - y = 0$

4. $\dfrac{d^2y}{dt^2} - \dfrac{dy}{dt} - 6y = 0$

5. $\dfrac{d^2y}{dt^2} + 3\dfrac{dy}{dt} - 10y = 0$

6. $y'' - 4y = 0$

In Exercises 7–10 find the general solution of the differential equation in the form $y = Ae^{rt} + Bte^{rt}$ by finding the single root r of the characteristic polynomial.

7. $y'' + 2y' + 1y = 0$

8. $y'' - 4y' + 4y = 0$

9. $\dfrac{d^2y}{dt^2} + 6\dfrac{dy}{dt} + 9y = 0$

10. $y'' - 2y' + 1y = 0$

In Exercises 11–14 find the general solution of the differential equation in the form $y = A\sin\sqrt{b}\,t + B\cos\sqrt{b}\,t$.

11. $\dfrac{d^2y}{dt^2} + 4y = 0$

12. $y'' + 16y = 0$

13. $y'' = -5y$

14. $\dfrac{d^2y}{dt^2} = -25y$

In Exercises 15–24 find the general solution of the differential equation.

15. $y'' - 3y' - 10y = 0$

16. $y'' + 8y' + 16y = 0$

17. $y'' - 4y' + y = 0$

18. $y'' + 6y' + 5y = 0$

19. $y'' - 2y' + 6y = 0$

20. $y'' - 6y' + 9y = 0$

21. $y'' - y' - 12y = 0$

22. $y'' - 2y' + 3y = 0$

23. $y'' + 2y' + 4y = 0$

24. $y'' + 4y' + 4y = 0$

In Exercises 25–30 find the solution of the initial value problem.

25. $y'' - y' - 6y = 0$
$y(0) = 0$
$y'(0) = 5$

26. $y'' - 5y' + 6y = 0$
$y(0) = 2$
$y'(0) = 0$

27. $y'' + 2y' + y = 0$
$y(0) = 3$
$y'(0) = 1$

28. $y'' - 4y' + 4y = 0$
$y(0) = 2$
$y'(0) = 5$

29. $y'' + 9y = 0$
$y(0) = 3$
$y'(0) = -3$

30. $y'' + y = 0$
$y(0) = 0$
$y'(0) = 2$

In Exercises 31–32 find the general solution by first converting the system of two first order equations into a second order equation.

31. $\dfrac{dy}{dt} = 4x$

$\dfrac{dx}{dt} = -y$

32. $\dfrac{dy}{dt} = 2(x - 3)$

$\dfrac{dx}{dt} = 8(y + 5)$

33. A 0.5 kg mass is attached to a spring with spring constant $k = 2$ N/m, as in Figure 3.1. The mass is pulled 10 cm from the equilibrium position and released with zero velocity. Find an equation describing the resulting motion.

34. A 0.5 kg mass is attached to a spring with spring constant $k = 100$ N/m, as in Figure 3.1. The mass is tapped while in its equilibrium position so as to give it an initial velocity of 10 cm/s.
   a. Find a function describing the resulting motion.
   b. When will the mass return to its equilibrium point for the first time?

35. Sketch the graph of the solution of the initial value problem

$$\frac{d^2x}{dt^2} + kx = 0$$

$x(0) = 1$
$x'(0) = 0$

for $k = 0, \pm 1, \pm 4, \pm 9$. What can you conclude about this solution as
   a. $k \to \infty$?
   b. $k \to -\infty$?
   c. $k \to 0^+$?
   d. $k \to 0^-$?

36. Verify that if y_1 and y_2 are solutions of the differential equation

$$y'' + ay' + by = 0,$$

then so is any function of the form

$$y(t) = Ay_1(t) + By_2(t).$$

21.4 NONHOMOGENEOUS LINEAR EQUATIONS

In Section 21.3 we noted that if y_1 and y_2 are solutions of the linear differential equation

$$\frac{d^2y}{dt^2} + a\,\frac{dy}{dt} + by = 0, \tag{1}$$

then so is any function y of the form

$$y(t) = C_1 y_1(t) + C_2 y_2(t) \tag{2}$$

where C_1 and C_2 are constants.

We say that the solutions y_1 and y_2 are *linearly independent* if they are not multiples of each other. It can be shown that when y_1 and y_2 are linearly independent solutions of equation (1), the general solution of equation (1) is given by the function y in line (2).

Equation (1) is called *homogeneous* because the term on the right-hand side is simply zero.

We next wish to determine the general solution of the *nonhomogeneous* differential equation of the form

$$\frac{d^2y}{dt^2} + a\,\frac{dy}{dt} + by = f(t) \tag{3}$$

for certain types of nonzero functions f.

We first note that the nonhomogeneous equation (3) does *not* have the property that the sum of two solutions of (3) is again a solution. (See Exercise 21.) The following theorem shows how the general solution of equation (3) is related to the general solution of the associated homogeneous equation (1). Its proof is sketched in Exercise 19.

THEOREM 2

Let y_p be any particular solution of the nonhomogeneous linear differential equation (3) and let y_g be the general solution of the associated homogeneous linear equation (1). Then the general solution of the nonhomogeneous equation (3) is

$$y(t) = y_g(t) + y_p(t). \tag{4}$$

Theorem 2 says that to find the general solution of the nonhomogeneous equation (3) we must do two things. First, find *any particular* solution y_p of equation (3). Then, find the *general* solution y_g of the "homogeneous part" given by equation (1). The general solution of (3) is the sum of these two solutions.

Example 1

The nonhomogeneous equation

$$y'' + y' - 2y = -4t \tag{5}$$

has the particular solution $y_p = 2t + 1$. To verify this note that

$$y_p' = 2 \qquad \text{and} \qquad y_p'' = 0,$$

so

$$y_p'' + y_p' - 2y_p = 0 + 2 - 2(2t + 1)$$
$$= -4t$$

as required.

To find the *general* solution we next note that the associated homogeneous equation

$$y'' + y' - 2y = 0 \tag{6}$$

has characteristic polynomial $r^2 + r - 2 = (r + 2)(r - 1)$, with roots $r_1 = -2$ and $r_2 = 1$, so the general solution of the homogeneous equation (6) is

$$y_g = C_1 e^{-2t} + C_2 e^t.$$

The general solution of equation (5) is therefore

$$y = C_1 e^{-2t} + C_2 e^t + 2t + 1$$

according to Theorem 2. ◇

The Method of Undetermined Coefficients

Although Theorem 2 specifies the form of the general solution of equation (3), it does not tell us how to find the particular solution y_p required to form the general solution.

For certain types of functions f, a simple observation enables us to find y_p. The idea is simply to think about which types of functions, when differentiated and combined according to the left side of equation (3), produce the function f on the right side of equation (3). We then substitute the "general forms" of such functions into the equation and see what happens.

For example, in the nonhomogeneous equation

$$y'' + 3y' - 4y = e^{2t} \tag{7}$$

we suggest that only functions of the form $y = Ae^{2t}$ are candidates for a solution since derivatives of Ae^{2t} are again of the form Ae^{2t}. With $y = Ae^{2t}$ we have

$$y' = 2Ae^{2t} \quad \text{and} \quad y''(t) = 4Ae^{2t},$$

so substituting into the left side of equation (7) gives

$$y'' + 3y' - 4y = 4Ae^{2t} + 3(2Ae^{2t}) - 4Ae^t \tag{8}$$
$$= 6Ae^{2t}.$$

Comparing the right sides of equations (7) and (8), we conclude that $y = Ae^{2t}$ is a solution of equation (7) if $e^{2t} = 6Ae^{2t}$. Thus, $6A = 1$, so $A = \frac{1}{6}$. A particular solution of equation (7) is therefore $y_p(t) = \frac{1}{6}e^{2t}$, as you can verify.

By the "general form" for a function f, we mean the most general form of a function y which, when combined with its derivatives $\dfrac{dy}{dt}$ and $\dfrac{d^2y}{dt^2}$ according to the left side of a linear second order differential equation

$$\frac{d^2y}{dt^2} + a\,\frac{dy}{dt} + by = f(t), \tag{9}$$

can produce the function f on the right side. Here is a list of the general forms associated with the functions f for which the method we are describing works best.

Function f	General Form
$f(t) = t$	$y = At + B$
$f(t) = t^2$	$y = At^2 + Bt + C$
$f(t) = e^{kt}$	$y = Ae^{kt}$
$f(t) = \sin kt$	$y = A \sin kt + B \cos kt$
$f(t) = \cos kt$	$y = A \sin kt + B \cos kt$

The *method of undetermined coefficients* for finding a particular solution of equation (9) is to simply insert the general form for the function f into equation (9) and attempt to solve for the "undetermined coefficients" A, B, etc.

Example 2

In the differential equation

$$y'' + 3y' + 2y = t^2 \tag{10}$$

the general form for the function $f(t) = t^2$ is

$$y = At^2 + Bt + C. \tag{11}$$

Then

$$y' = 2At + B$$

and

$$y'' = 2A.$$

Inserting these expressions in equation (10) gives

$$2A + 3(2At + B) + 2(At^2 + Bt + C) = t^2$$

or

$$2At^2 + (6A + 2B)t + (2A + 3B + 2C) = t^2.$$

We therefore obtain the system of equations

$$\begin{aligned} 2A &= 1 & \text{(coefficients of } t^2) \\ 6A + 2B &= 0 & \text{(coefficients of } t) \\ 2A + 3B + 2C &= 0 & \text{(constants)} \end{aligned}$$

which has solution $A = \frac{1}{2}$, $B = -\frac{3}{2}$, $C = \frac{7}{4}$. A particular solution of equation (10), given by equation (11), is therefore

$$y_p(t) = \frac{1}{2}t^2 - \frac{3}{2}t + \frac{7}{4}.$$

Since the homogeneous equation

$$y'' + 3y' + 2y = 0 \tag{12}$$

has characteristic polynomial $r^2 + 3r + 2 = (r + 2)(r + 1)$ with roots $r_1 = -2$ and $r_2 = -1$, the general solution of equation (12) is

$$y_g(t) = C_1 e^{-2t} + C_2 e^{-t},$$

and the general solution of equation (10) is

$$y(t) = C_1 e^{-2t} + C_2 e^{-t} + \frac{1}{2}t^2 - \frac{3}{2}t + \frac{7}{4}.$$

◇

Example 3

In the differential equation

$$y'' - 2y' + y = \cos t \tag{13}$$

the general form for the function $f(t) = \cos t$ is

$$y = A \sin t + B \cos t.$$

Then

$$y' = A \cos t - B \sin t$$

and

$$y'' = -A \sin t - B \cos t.$$

Inserting these expressions into equation (13) gives

$$(-A \sin t - B \cos t) - 2(A \cos t - B \sin t) + (A \sin t + B \cos t) = \cos t$$

or

$$2B \sin t - 2A \cos t = \cos t.$$

Thus, we obtain the equations

$$
\begin{aligned}
2B &= 0 && \text{(coefficients of } \sin t) \\
-2A &= 1 && \text{(coefficients of } \cos t)
\end{aligned}
$$

with solution $A = -\frac{1}{2}$ and $B = 0$. A particular solution of equation (13) is therefore

$$y_p(t) = -\frac{1}{2} \sin t.$$

Since the corresponding homogeneous equation

$$y'' - 2y' + y = 0 \tag{14}$$

has characteristic polynomial $r^2 - 2r + 1 = (r - 1)^2$ with the repeated root $r = 1$, the general solution of equation (14) is

$$y_g(t) = C_1 e^t + C_2 t e^t,$$

and the general solution of equation (13) is

$$y = C_1 e^t + C_2 t e^t - \frac{1}{2} \sin t.$$

◇

Superposition of Solutions

When the function f on the right side of the nonhomogeneous equation

$$\frac{d^2 y}{dt^2} + a \frac{dy}{dt} + by = f(t)$$

is a *linear combination* of two or more functions of the form t^n, e^{kt}, $\sin kt$ or $\cos kt$

the following theorem enables us to obtain a particular solution as a linear combination of solutions of equations involving only one of these functions on their right-hand sides.

THEOREM 3

Let y_1 be a solution of the equation

$$\frac{d^2y}{dt^2} + a\,\frac{dy}{dt} + by = f_1(t), \tag{15}$$

and let y_2 be a solution of the equation

$$\frac{d^2y}{dt^2} + a\,\frac{dy}{dt} + by = f_2(t). \tag{16}$$

Then the function

$$y(t) = \alpha y_1(t) + \beta y_2(t) \tag{17}$$

is a solution of the equation

$$\frac{d^2y}{dt^2} + a\,\frac{dy}{dt} + by = \alpha f_1(t) + \beta f_2(t). \tag{18}$$

Theorem 3 says that we may find a solution of equation (18) by first finding solutions y_1 and y_2 of equations (15) and (16), and then forming the linear combination $y(t) = \alpha y_1(t) + \beta y_2(t)$ of these solutions. The proof of this theorem is given in Exercise 20.

Example 4

We leave it as an exercise for you to verify that the equation

$$\frac{d^2y}{dt^2} + 4y = e^t$$

has the particular solution $y_1(t) = \frac{1}{5}e^t$ and that the equation

$$\frac{d^2y}{dt^2} + 4y = t^2$$

has the particular solution $y_2(t) = \frac{1}{4}t^2 - \frac{1}{8}$.
   Thus, according to Theorem 3,

   (i) the equation

$$\frac{d^2y}{dt^2} + 4y = 6e^t$$

   has solution

$$y(t) = 6y_1(t) = \frac{6}{5}e^t;$$

   (ii) the equation

$$\frac{d^2y}{dt^2} + 4y = -4t^2$$

has solution

$$y(t) = -4y_2(t) = -4\left(\frac{1}{4}t^2 - \frac{1}{8}\right) = -t^2 + \frac{1}{2};$$

(iii) the equation

$$\frac{d^2y}{dt^2} + 4y = 10e^t + 8t^2 \tag{19}$$

has solution

$$y = 10y_1(t) + 8y_2(t)$$
$$= 10\left(\frac{1}{5}e^t\right) + 8\left(\frac{1}{4}t^2 - \frac{1}{8}\right)$$
$$= 2e^t + 2t^2 - 1.$$

We leave it as a further exercise for you to verify that we could have proceeded to find a particular solution of equation (19) directly by beginning with the general form

$$y = Ae^t + Bt^2 + Ct + D$$

and applying the method of undetermined coefficients. ◇

A Complicating Issue

There is a situation in which the method of undetermined coefficients, as it has been described thus far, fails to produce a particular solution of the nonhomogeneous equation

$$\frac{d^2y}{dt^2} + a\frac{dy}{dt} + by = f(t). \tag{20}$$

This occurs when the function $y = f(t)$ is a solution of the *homogeneous* equation

$$\frac{d^2y}{dt^2} + a\frac{dy}{dt} + by = 0. \tag{21}$$

In this case, inserting the general form for the function $y = f(t)$ in the left side of equation (20) will simply yield the zero function, because of equation (21). In this case we modify the method of undetermined coefficients as follows:

> If the function $y = f(t)$ in equation (20) is a solution of the homogeneous equation (21), we seek a particular solution of equation (20) in the form $y = tg(t)$ where $g(t)$ is the general form for the function f.

We shall not "prove" this statement, because we have described the method of undetermined coefficients only as a way to *seek* particular solutions of equation (20). The following examples give you some indication, however, as to why this technique "works."

Example 5

For the nonhomogeneous equation

$$y'' + 2y' - 3y = e^t \tag{22}$$

the corresponding homogeneous equation is

$$y'' + 2y' - 3y = 0. \tag{23}$$

The characteristic polynomial for equation (23) is

$$r^2 + 2r - 3 = (r + 3)(r - 1),$$

which has roots $r_1 = -3$ and $r_2 = 1$, so the general solution of equation (23) is

$$y_g(t) = C_1 e^{-3t} + C_2 e^t. \tag{24}$$

Thus, the function $f(t) = e^t$ on the right side of equation (22) is a solution of equation (23) (with $C_1 = 0$ and $C_2 = 1$ in equation (24)).

If we were to seek a particular solution of equation (22) using the general form

$$y = Ae^t,$$

we would have $y' = Ae^t$ and $y'' = Ae^t$, and inserting these functions in equation (22) would give the "equation"

$$Ae^t + 2Ae^t - 3Ae^t = e^t. \tag{25}$$

Since "equation" (25) simplifies to the false statement $0 = e^t$, there is no number A for which $y = Ae^t$ is a solution of equation (22).

If, however, we begin with the general form

$$y = Ate^t \tag{26}$$

we obtain

$$y' = A(1 + t)e^t \quad \text{and} \quad y'' = A(2 + t)e^t.$$

Substituting these functions in equation (22) then gives

$$A(2 + t)e^t + 2A(1 + t)e^t - 3Ate^t = e^t,$$

so

$$A(2 + 2)e^t + A(1 + 2 - 3)te^t = e^t$$

or

$$4Ae^t = e^t.$$

Thus $A = \frac{1}{4}$, and a particular solution of equation (22) is

$$y_p(t) = \frac{1}{4}te^t$$

in equation (26). From equations (24) and (26) we may conclude that the *general* solution of equation (22) is

$$y(t) = C_1 e^{-3t} + C_2 e^t + \frac{1}{4}te^t. \qquad \diamond$$

Example 6

For the nonhomogeneous equation

$$y'' + y = \sin t \tag{27}$$

the associated homogeneous equation $y'' + y = 0$ has general solution

$$y_g(t) = C_1 \sin t + C_2 \cos t.$$

To find a particular solution of the nonhomogeneous equation (27) we must therefore use the general form

$$y = t(A \sin t + B \cos t)$$
$$= At \sin t + Bt \cos t.$$

We leave it as an exercise for you to verify that substituting this general form into equation (27) leads to the conclusion $A = 0$ and $B = -\frac{1}{2}$, so a particular solution of (27) is

$$y_p(t) = -\frac{1}{2} t \cos t,$$

and the general solution of (27) is

$$y(t) = C_1 \sin t + C_2 \cos t - \frac{1}{2} t \cos t. \qquad \diamond$$

Example 7

In the nonhomogeneous equation

$$y'' - 4y' + 4y = e^{2t} \qquad (28)$$

the associated homogeneous equation

$$y'' - 4y' + 4y = 0 \qquad (29)$$

has characteristic polynomial $r^2 - 4r + 4 = (r - 2)^2$. Since $r = 2$ is a *repeated* root, the general solution of (29) is

$$y_g(t) = C_1 e^{2t} + C_2 t e^{2t}.$$

This means that not only is the function $f(t) = e^{2t}$ a solution of (29), *but so is the function $tf(t) = te^t$*. To find a particular solution of (28) we must therefore begin with the general form obtained by multiplying by yet another factor of t:

$$y = t(Ate^{2t}) = At^2 e^{2t} \qquad (30)$$

gives

$$y' = 2Ate^{2t} + 2At^2 e^{2t}$$

and

$$y'' = 2Ae^{2t} + 8Ate^{2t} + 4At^2 e^{2t},$$

so

$$y'' - 4y' + 4y = (4A - 8A + 4A)t^2 e^{2t} + (8A - 8A)te^{2t} + 2Ae^{2t}$$

and we obtain the equation $2Ae^{2t} = e^{2t}$. Thus, $A = \frac{1}{2}$ and the particular solution of (28) of the form in line (30) is

$$y_p(t) = \frac{1}{2} t^2 e^{2t},$$

and the general solution of (28) is

$$y(t) = C_1 e^{2t} + C_2 t e^{2t} + \frac{1}{2} t^2 e^{2t}. \qquad \diamond$$

Exercise Set 21.4

In Exercises 1–12 find the general solution y of the given second order differential equation.

1. $y'' + 2y' - y = \sin 2t$

2. $y'' - 3y' + 4y = t$

3. $y'' + y = e^{-t}$

4. $y'' - 9y = t^2$

5. $y'' - 4y = 4t$

6. $y'' + 7y' + 12y = e^{2t}$

7. $y'' - 3y' - 10y = 5e^{2t}$

8. $y'' + y = t^2$

9. $y'' + 9y = \cos 3t$

10. $y'' + 2y' + y = 4e^{-t}$

11. $y'' - 3y' - 4y = 2e^{-t}$

12. $y'' + 5y' + 6y = 3e^{-3t}$

In Exercises 13–17 use Theorem 3 and the method of undetermined coefficients to find the general solution of the differential equation.

13. $y'' + 2y' + y = 3e^t + t^2$

14. $y'' - 2y' - 3y = 2e^{2t} + 3 \sin t$

15. $y'' - 2y' + y = 2 \cos 2t + \sin 2t + 2e^{-t}$

16. $y'' + y' = t^3 + 2t^2$

17. $y'' + y' + y = t^3 + 2t^2$

18. Solve the initial value problem

$$y'' - 2y' - 3y = 2e^t - 10 \sin t$$
$$y(0) = 2$$
$$y'(0) = 4.$$

19. Prove Theorem 2 by proving the following statements.

a. If y_p and z_p are any two particular solutions of the nonhomogeneous equation

$$y'' + ay' + by = f(t),$$

then the function $y = y_p - z_p$ is a solution of the *homogeneous* equation

$$y'' + ay' + by = 0.$$

(*Hint:* Simply substitute the function $y = y_p - z_p$ into the left side of the homogeneous equation.)

b. Let $y_g(t) = C_1 y_1(t) + C_2 y_2(t)$ be the general solution for the homogeneous equation. Conclude from part **a** that for some choice of the constants C_1 and C_2,

$$z_p(t) = C_1 y(t) + C_2 y_2(t) + y_p(t).$$

c. Conclude from parts **a** and **b** that *any* particular solution z_p of the nonhomogeneous equation must have the form

$$z_p = y_g + y_p$$

where y_g is the general solution of the homogeneous equation.

20. Prove Theorem 3 by simply substituting $y(t) = \alpha y_1(t) + \beta y_2(t)$ into equation (18) and verifying the equality.

21. Demonstrate that the sum $y = y_1 + y_2$ of two distinct solutions of the nonhomogeneous equation

$$y'' + ay' + by = f(t), \qquad f(t) \neq 0$$

is not again a solution of this equation. For what differential equation *is* y a solution?

22. The method of undetermined coefficients may also be applied to certain *first* order nonhomogeneous equations. Verify the following facts required in its application.

a. If a is a constant, the general solution of the first order homogeneous equation

$$y' + ay = 0$$

is

$$y_g(t) = Ce^{-at}.$$

b. If y_p is any particular solution of the nonhomogeneous equation

$$y' + ay = f(t),$$

the general solution of the nonhomogeneous equation is

$$y = Ce^{-at} + y_p.$$

(*Hint:* As in Exercise 19, verify that the difference of any two particular solutions is a solution of the homogeneous equation in part **a**.)

c. Conclude that if $f(t)$ is a function of the form t^n, e^{kt}, $\sin kt$, or $\cos kt$, and the method of undetermined coefficients can be used to find a particular solution of the equation

$$y' + ay = f(t),$$

then the general solution of this equation is as in part **b**.

Use the method of undetermined coefficients and the results of Exercise 22 to find the general solution of the differential equations in Exercises 23–30.

23. $y' - 3y = 6t - 14$

24. $y' + 2y = 3t$

25. $y' - 4y = 1 - 4t^2$

26. $y' + 5y = 10$

27. $y' - 4y = 2e^{3t}$

28. $y' + 2y = 2 \sin t + 11 \cos t$

29. $y' - 3y = 4e^t + 6t - 14$

30. $y' - 2y = -4t + 8 + 5 \sin t$

In Exercises 31–34 find the solution of the initial value problem.

31. $y' + 4y = 0$
$y(0) = 3$

32. $y' - 2y = 12 - 4t$
$y(0) = 8$

33. $y' + 3y = 5e^{2t}$
$y(0) = 7$

34. $y' - 2y = 2 \sin 4t + 16 \cos 4t$
$y(0) = 8$

21.5 POWER SERIES SOLUTIONS OF DIFFERENTIAL EQUATIONS

The theory of power series provides a method for solving certain types of differential equations. For differential equations of the form

$$f(x, y, y') = 0 \tag{1}$$

the idea is to express the (unknown) solution y as a power series in the independent variable x:

$$y = \sum_{k=0}^{\infty} a_k x^k. \tag{2}$$

Then, assuming (2) has a nonzero radius of convergence, we apply the theorem on differentiating power series to conclude that

$$y' = \sum_{k=0}^{\infty} k a_k x^{k-1}. \tag{3}$$

We next insert representations (2) and (3) into the differential equation (1). For each integer k, coefficients of x^k will appear in various places in equation (1). By equating coefficients of x^k on either side of equation (1) we can often find a set of equations which allow the coefficients $a_0, a_1, a_2, a_3, \ldots$ to be determined completely. In such cases the power series representation for the solution is obtained.

Example 1

Use the power series method to solve the differential equation

$$y' = y. \tag{4}$$

Solution: We assume that $y = \sum_{k=0}^{\infty} a_k x^k$, so that $y' = \sum_{k=0}^{\infty} k a_k x^{k-1}$, and both of these series have the same radius of convergence. Inserting these expansions in the differential equation (4) gives

$$\sum_{k=0}^{\infty} k a_k x^{k-1} = \sum_{k=0}^{\infty} a_k x^k$$

or

$$a_1 + 2a_2 x + 3a_3 x^2 + \cdots + k a_k x^{k-1} + \cdots$$
$$= a_0 + a_1 x + a_2 x^2 + \cdots + a_{k-1} x^{k-1} + \cdots. \tag{5}$$

Since the two series in equation (5) are equal, the coefficients of like powers of x must be the same. Thus,

$$\begin{array}{ll} a_1 = a_0 & \text{(constant terms)} \\ 2a_2 = a_1 & \text{(x terms)} \\ 3a_3 = a_2 & \text{(x^2 terms)} \end{array}$$

$$\cdot$$
$$\cdot$$
$$\cdot$$

$$ka_k = a_{k-1} \qquad (x^{k-1} \text{ terms})$$
$$\cdot$$
$$\cdot$$
$$\cdot$$

Using these equations, we may solve for each coefficient a_2, a_3, a_4, . . . in terms of the one preceding and, therefore, in terms of a_0:

$$a_1 = a_0$$

$$a_2 = \frac{1}{2}a_1 = \frac{1}{2}a_0 = \frac{1}{2!}a_0$$

$$a_3 = \frac{1}{3}a_2 = \frac{1}{3 \cdot 2}a_0 = \frac{1}{3!}a_0$$
$$\cdot$$
$$\cdot$$
$$\cdot$$

$$a_k = \frac{1}{k}a_{k-1} = \frac{1}{k(k-1) \cdot \ldots \cdot 3 \cdot 2}a_0 = \frac{1}{k!}a_0. \tag{6}$$

Returning to the expansion for y, we can now write

$$y = a_0 + a_0 x + \frac{1}{2!}a_0 x^2 + \frac{1}{3!}a_0 x^3 + \cdots + \frac{1}{k!}a_0 x^k + \cdots \tag{7}$$

$$= a_0 \left\{ 1 + x + \frac{x^2}{2!} + \frac{x^3}{3!} + \cdots + \frac{x^k}{k!} + \cdots \right\},$$

which we recognize as the power series representation for the function $y = a_0 e^x$. Since we have previously verified that this power series converges for all values of x, the solution $y = a_0 e^x$ is valid for all x. The constant a_0 is determined by an initial condition for the equation (4). ◇

REMARK: The result of Example 1 should not have been unexpected, since we determined in Chapter 8 that the solution of the differential equation $y' = y$ is $y = Ce^x$. However, we see here that the use of power series provides an entirely different approach to solving differential equations.

The equation

$$a_k = \frac{1}{k}a_{k-1}$$

on the left side of equation (6) warrants special attention. It is referred to as the **recurrence relation** for the differential equation because it gives the general formula by which each coefficient in the series expansion for the solution (except the first) may be determined from its predecessor. If you can succeed in finding a recurrence relation, you will be able to generate all coefficients in the expansion for the solution of a differential equation. However, two issues then remain to be addressed:

(1) Does the power series obtained from the recurrence relation actually converge? If so, for which values of x? You will have succeeded in finding a legitimate solution of the differential equation only if the power series has a nonzero radius of convergence.

(2) Can a closed form expression* for the power series solution be recognized, as in Example 1 where the solution was recognized as the power series for a_0e^x? The answer to this question is not critical if you are willing to accept solutions in the form of power series. In fact, many important functions in mathematical physics arise as series solutions of differential equations and do not have closed form expressions (see Example 3).

Example 2

Use the power series method to solve the initial value problem

$$\begin{cases} y' = xy \\ y(0) = 1. \end{cases}$$

Strategy

Assume a power series form for the solution y. Differentiate to find series form for y'.

Solution

We assume that

$$y = \sum_{k=0}^{\infty} a_k x^k, \qquad \text{so} \qquad y' = \sum_{k=0}^{\infty} k a_k x^{k-1}.$$

Insert the expansions for y and y' into the differential equation.

This gives

$$\sum_{k=0}^{\infty} k a_k x^{k-1} = x \sum_{k=0}^{\infty} a_k x^k$$

$$= \sum_{k=0}^{\infty} a_k x^{k+1}$$

Write out the first few terms, including the general terms on both sides *for the same power of x*. (Here we arbitrarily picked x^{k-1}.)

Equate coefficients of like powers of x.

or

$$a_1 + 2a_2 x + 3a_3 x^2 + \cdots + k a_k x^{k-1} + \cdots$$
$$= a_0 x + a_1 x^2 + a_2 x^3 + \cdots + a_{k-2} x^{k-1} + \cdots,$$

so

$$a_1 = 0 \qquad \text{(constant terms)}$$

$$a_2 = \frac{1}{2} a_0 \qquad \text{(x terms)}$$

$$a_3 = \frac{1}{3} a_1 = 0 \qquad \text{(x}^2 \text{ terms)}$$

$$\vdots$$

The recurrence relation is obtained by equating coefficients of x^{k-1}.

$$a_k = \frac{1}{k} a_{k-2} \qquad \text{(x}^{k-1} \text{ terms)}$$

$$\vdots$$

---

*A closed form expression is one which does not involve an infinite process, such as summation. $f(x) = \dfrac{1}{1-x}$ is expressed in closed form, while $f(x) = \sum_{k=0}^{\infty} x^k$ is not.

Determine the form of all coefficients using the recurrence relation.

From the recurrence relation $a_k = \dfrac{1}{k} a_{k-2}$ and the fact that $a_1 = 0$, we see that all odd coefficients are zero: $0 = a_1 = a_3 = a_5 = \cdots$.

Also we can see that the even coefficients are

$$a_2 = \frac{1}{2} a_0$$

$$a_4 = \frac{1}{4} a_2 = \frac{1}{4 \cdot 2} a_0 = \frac{1}{2^2}\left(\frac{1}{2 \cdot 1}\right)a_0 = \frac{1}{2^2} \cdot \frac{1}{2!} a_0$$

$$a_6 = \frac{1}{6} a_4 = \frac{1}{6 \cdot 4 \cdot 2} a_0 = \frac{1}{2^3}\left(\frac{1}{3 \cdot 2 \cdot 1}\right)a_0 = \frac{1}{2^3} \cdot \frac{1}{3!} a_0$$

$$\vdots$$

$$a_{2k} = \frac{1}{2k} a_{2k-2} = \frac{1}{2k(2k-2) \cdot \ldots \cdot 2} a_0 = \frac{1}{2^k} \cdot \frac{1}{k!} a_0.$$

Write the series for y using coefficients found above.

The power series for the solution y is therefore

$$y = a_0 + a_1 x + a_2 x^2 + a_3 x^3 + a_4 x^4 + \cdots$$

Try to bring the series for y into the form of a known power series. (Here we use the fact that

$$e^u = \sum_{k=0}^{\infty} \frac{u^k}{k!}$$

for all u.)

$$= a_0 + \frac{1}{2} a_0 x^2 + \frac{1}{2^2} \cdot \frac{1}{2!} a_0 x^4 + \frac{1}{2^3} \cdot \frac{1}{3!} a_0 x^6 + \cdots + \frac{1}{2^k} \cdot \frac{1}{k!} a_0 x^{2k} + \cdots$$

$$= a_0\left\{1 + \frac{x^2}{2} + \frac{\left(\dfrac{x^2}{2}\right)^2}{2!} + \frac{\left(\dfrac{x^2}{2}\right)^3}{3!} + \cdots + \frac{\left(\dfrac{x^2}{2}\right)^k}{k!} + \cdots\right\}$$

$$= a_0 \sum_{k=0}^{\infty} \frac{(x^2/2)^k}{k!}.$$

This is the power series expansion for $y = a_0 e^{x^2/2}$, which converges for all values of x. The general solution for the differential equation $y' - xy = 0$ is therefore $y = a_0 e^{x^2/2}$. The constant a_0 is determined from the initial condition:

Apply initial condition to determine a_0.

$$1 = y(0) = a_0 e^0 = a_0.$$

The solution of the initial value problem is $y = e^{x^2/2}$. ◇

Power series methods may be used in higher order differential equations as well. Note in the following example that the power series for y must be differentiated twice since the differential equation is of order 2.

Example 3

Find a power series solution for the differential equation

$$xy'' - y = 0.$$

Solution: We shall work with the equation in the form

$$xy'' = y. \tag{8}$$

We assume the solution y to have the form

$$y = \sum_{k=0}^{\infty} a_k x^k.$$

Then

$$y' = \sum_{k=0}^{\infty} k a_k x^{k-1}, \quad \text{and} \quad y'' = \sum_{k=0}^{\infty} k(k-1) a_k x^{k-2}.$$

Equation (8) becomes

$$x \sum_{k=0}^{\infty} k(k-1) a_k x^{k-2} = \sum_{k=0}^{\infty} a_k x^k,$$

or

$$2a_2 x + 3 \cdot 2a_3 x^2 + 4 \cdot 3 \cdot a_4 x^3 + \cdots + k(k-1) a_k x^{k-1} + \cdots$$
$$= a_0 + a_1 x + a_2 x^2 + a_3 x^3 + \cdots + a_{k-1} x^{k-1} + \cdots.$$

Thus

$$a_0 = 0$$

$$2a_2 = a_1 \text{ gives } a_2 = \frac{1}{2} a_1$$

$$3 \cdot 2 \cdot a_3 = a_2 \text{ gives } a_3 = \frac{1}{3 \cdot 2} a_2 = \frac{1}{3 \cdot 2^2} a_1 = \frac{1}{3(2!)^2} a_1$$

$$4 \cdot 3 \cdot a_4 = a_3 \text{ gives } a_4 = \frac{1}{4 \cdot 3} a_3 = \frac{1}{4 \cdot 3^2 \cdot 2^2} a_1 = \frac{1}{4(3!)^2} \cdot a_1$$

$$\vdots$$

$$k(k-1) a_k = a_{k-1} \text{ gives } a_k = \frac{1}{k(k-1)} a_{k-1} = \cdots = \frac{1}{k[(k-1)!]^2} \cdot a_1$$

The solution y is therefore

$$y = a_1 \left[x + \frac{1}{2} x^2 + \frac{1}{3 \cdot 2^2} x^3 + \frac{1}{4 \cdot (3!)^2} x^4 + \cdots + \frac{1}{k[(k-1)!]^2} x^k + \cdots \right]. \qquad (9)$$

To determine the radius of convergence for (9) we apply the Ratio Test.

$$\rho = \lim_{k \to \infty} \left| \frac{\dfrac{a_1}{(k+1)(k!)^2} x^{k+1}}{\dfrac{a_1}{k[(k-1)!]^2} x^k} \right| = \lim_{k \to \infty} \left(\frac{k}{k+1} \right) \left(\frac{1}{k^2} \right) |x| = 0$$

for all values of x, so the series in (9) converges for all x. We have therefore obtained a legitimate solution for the differential equation (8), although the solution is expressed in power series form and is not immediately recognizable as the power series for a known function in closed form. ◇

As you might suspect at this point, the theory associated with the use of power series in solving differential equations is not simple. In fact, the examples we have presented here were carefully chosen to convey the basic idea while avoiding the difficulties surrounding this method. These difficulties include the following:

(i) It is often possible only to obtain the first few terms of the series solution rather than the general term as we have succeeded in doing in these examples. This may leave the question of convergence unanswerable, although in many cases one can obtain at least an approximation to the solution.

(ii) The assumption that the solution y has a power series representation contains the implicit assumption that y has derivatives of all orders. Since the solution to a differential equation of degree n need have only n derivatives, the power series approach will necessarily fail to identify solutions to certain equations since it assumes too much.

Further work on power series methods in differential equations is left to more specialized courses. Should you take such a course you will find that the material in Chapter 13 constitutes an important foundation on which much of the work of that course depends.

Exercise Set 21.5

In Exercises 1–8, use the method of this section to find a power series form of the solution of the differential equation or initial value problem. Check your work by solving the equation by the method of separation of variables.

1. $y' + y = 0$

2. $y' + 2y = 0$

3. $y' - 6y = 0$

4. $y' + xy = 0$

5. $y' - 2xy = 0$

6. $y' + ax = 0$

7. $\begin{cases} y' + 4y = 0 \\ y(0) = 1 \end{cases}$

8. $\begin{cases} y' + xy = 0 \\ y(0) = 2 \end{cases}$

9. Find a power series solution for the second order differential equation $xy'' + y = 0$.

10. Find a power series solution for the initial value problem

$$\begin{cases} y'' + y = 0 \\ y(0) = 0 \\ y'(0) = 1. \end{cases}$$

(*Hint:* Let $y = \displaystyle\sum_{k=0}^{\infty} a_k x^k$. Since $y(0) = 0$, $a_0 = 0$. This observation simplifies the recurrence relation.)

11. Find a power series solution for the initial value problem

$$\begin{cases} y'' + y = 0 \\ y(0) = 1 \\ y'(0) = 0. \end{cases}$$

12. Conclude from Exercises 10 and 11 that the function $y = A \sin x + B \cos x$ is a solution of the differential equation $y'' + y = 0$.

21.6 APPROXIMATING SOLUTIONS OF DIFFERENTIAL EQUATIONS

There are many differential equations for which simple solutions cannot be found. In such cases, however, we can resort to certain approximation procedures to obtain information about the solution, just as we can use the Trapezoidal Rule or Simpson's Rule to approximate a definite integral when the corresponding antiderivative cannot be found.

To give you an idea of how such approximation procedures work, we discuss one such procedure, called Euler's method, in this section. It is named for the Swiss mathematician Leonhard Euler (1707–1783).

Euler's Method

Euler's method may be used to approximate the solution to an initial value problem of the form

$$\frac{dy}{dt} = f(t, y) \qquad (1)$$

$$y(a) = y_0$$

on some interval $[a, b]$. (We shall assume that f and its partial derivatives are continuous for all t and y, although less restrictive conditions can be given.)

Figure 6.1 illustrates the basic idea to be pursued. Assume that $y = f(t)$ is a particular solution of the given differential equation. We divide the interval $[a, b]$ into n subintervals of equal length $h = \dfrac{b - a}{n}$, obtaining the $(n + 1)$ endpoints

$$a = t_0 < t_1 < t_2 < \cdots < t_n = b.$$

Then, beginning at the left endpoint $t_0 = a$ we use the given information about the solution y (namely, the initial value $y(a) = y_0$) and its derivative to approximate the value $y(t_1)$, and we call the approximation y_1. (Note, as suggested by Figure 6.1,

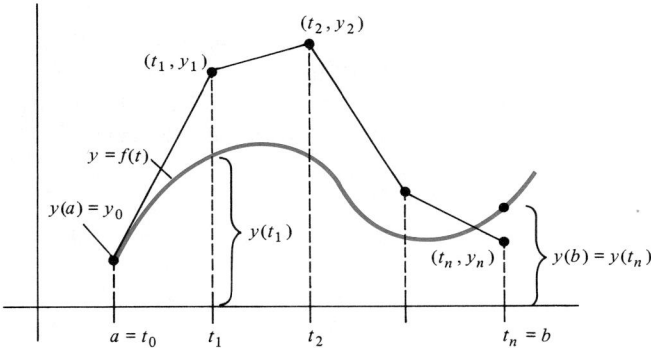

Figure 6.1 Approximation by Euler's method to the solution y of the initial value problem

$$y' = f(t, y), \quad y(a) = y_0.$$

that, in general, the approximation y_1 will not equal the function value $y(t_1)$, which we do not know.) Next, we use the assumed value y_1 and the number $t_1 = a + h$ to approximate $y(t_2)$, and we call this approximation y_2. Continuing in this way we finally approximate $y(t_n) = y(b)$ from the approximate value y_{n-1} and the number $t_{n-1} = a + (n - 1)h$. We therefore obtain *approximations* $y_1, y_2, \ldots, y_n$ to the (unknown) values of the solution $y(t_1), y(t_2), \ldots, y(t_n)$ at finitely many numbers t_j in $[a, b]$. The choice of the endpoint b and the number of intervals, n, depend upon which values of the solution you wish to approximate and with what accuracy.

Figure 6.2 illustrates how we get started in applying Euler's method. To find the approximation y_1 to the value $y(t_1)$, we follow the line tangent to the graph of y at the point $P_0 = (t_0, y_0)$ until we reach the point $P_1 = (t_1, y_1)$ with t-coordinate t_1. Since the slope of this line is $y'(t_0) = f(t_0, y_0)$, it follows that

$$\frac{y_1 - y_0}{t_1 - t_0} = f(t_0, y_0),$$

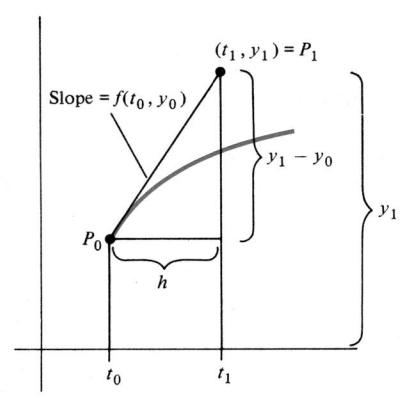

Figure 6.2 $y_1 = y_0 + f(t_0, y_0)h.$

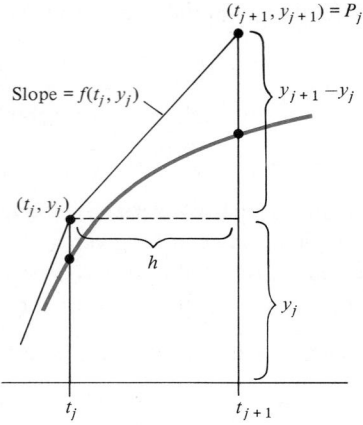

Figure 6.3 $y_{j+1} = y_j + f(t_j, y_j)h.$

and since $h = t_1 - t_0$,

$$y_1 = y_0 + f(t_0, y_0)h. \tag{2}$$

Once we have obtained the approximation y_1 from equation (2), we *pretend* that y_1 is the actual value of the solution y at $t = t_1$. Since we don't know the actual value $y(t_1)$, the approximation y_1 is the "next best thing." This "pretending" is necessary in order that we may carry out the same procedure to obtain the approximation y_2 for $y(t_2)$. Starting with the point (t_1, y_1) and the slope $f(t_1, y_1)$,* we obtain the y-coordinate of the point $P_2 = (t_2, y_2)$ from the equation

$$\frac{y_2 - y_1}{t_2 - t_1} = f(t_1, y_1).$$

That is, since $h = t_2 - t_1$,

$$y_2 = y_1 + f(t_1, y_1)h.$$

Continuing in this way (see Figure 6.3), we obtain each succeeding approximation y_{j+1} from the preceding y_j and t_j by the equation

$$\boxed{y_{j+1} = y_j + f(t_j, y_j)h \qquad j = 0, 1, \ldots, n - 1} \tag{3}$$

which is the summary statement of Euler's method.

The following examples show how a first order differential equation may be put in the form of equation (1), and how the solution of an associated initial value problem may be approximated using Euler's method.

Example 1

Use Euler's method to approximate the solution of the initial value problem

$$\begin{aligned} y' &= y(4 - y) \\ y(0) &= 1 \end{aligned} \tag{4}$$

on the interval $[0, 1]$ using 4 subintervals of equal length.

Solution: With $[0, 1]$ divided into 4 subintervals of length $h = \dfrac{1 - 0}{4} = \dfrac{1}{4}$, we have endpoints

$$t_0 = 0, \qquad t_1 = \frac{1}{4}, \qquad t_2 = \frac{1}{2}, \qquad t_3 = \frac{3}{4}, \qquad \text{and} \qquad t_4 = 1.$$

With $y_0 = y(0) = 1$ and $f(t, y) = y(4 - y)$, equation (3) gives

$$\begin{aligned} y_1 &= y_0 + y_0(4 - y_0)h \\ &= 1 + 1(4 - 1)\left(\frac{1}{4}\right) \\ &= 1 + \frac{3}{4} \\ &= 1.75 \end{aligned}$$

---

*The pretending occurs here in two places. Since we do not actually know $y(t_j)$, we do not know the actual slope $f(t_j, y(t_j))$, which we approximate by $f(t_j, y_j)$. Neither do we know the point of tangency, $(t_j, y(t_j))$, so we use (t_j, y_j) instead.

$$y_2 = y_1 + y_1(4 - y_1)h$$
$$= 1.75 + 1.75(4 - 1.75)(0.25)$$
$$\cong 2.7344,$$

etc. Table 6.1 shows the approximations y_1, y_2, y_3, y_4 together with the actual values $y(t_1)$, $y(t_2)$, $y(t_3)$, $y(t_4)$ of the solution $y = \dfrac{4}{1 + 3e^{-4t}}$ of the initial value problem to four decimal place accuracy. (This solution is obtained by the method of Section 21.1.) ◇

Table 6.1 Approximate and actual values of the solution to the initial value problem in Example 1. ($h = 0.25$)

t_j	0	.25	.50	.75	1.0
y_j	1	1.75	2.7344	3.5995	3.9599
$y(t_j)$	1	1.9015	2.8449	3.4802	3.7917

Example 2

Approximate the solution to the initial value problem in Example 1 on the interval $[0, 1]$ using $n = 10$ subintervals of equal size.

Solution: This time we have $h = \dfrac{1 - 0}{10} = \dfrac{1}{10}$, so Euler's equation becomes

$$y_{j+1} = y_j + y_j(4 - y_j)(0.10), \qquad j = 0, 1, 2, \ldots, 9.$$

Table 6.2 shows the resulting approximations, together with the actual values of the solution, to four decimal place accuracy. (Note the improvement in accuracy compared with Example 1.) ◇

Table 6.2 Values of the approximations and actual solution to the initial value problem in Example 2. ($h = 0.10$)

t_j	0	.1	.2	.3	.4	.5	.6	.7	.8	.9	1.0
y_j	1	1.3	1.651	2.0388	2.4387	2.8194	3.1523	3.4195	3.6180	3.7562	3.8478
$y(t_j)$	1	1.3285	1.7036	2.1013	2.4911	2.8449	3.1443	3.3829	3.5642	3.6970	3.7917

Example 3

Use Euler's method to approximate the solution of the initial value problem

$$y' + 2y = e^t \tag{5}$$
$$y(0) = 1$$

on the interval $[0, 1]$.

Solution: We begin by moving all terms in equation (5) except the derivative y' to the right-hand side, obtaining

$$\frac{dy}{dt} = e^t - 2y.$$

Thus, $f(t, y) = e^t - 2y$ in equation (3) for Euler's method. We shall obtain two approximations.

(a) Using $n = 4$ subdivisions, we have $h = \dfrac{1 - 0}{4} = 0.25$, and equation (3) becomes

$$y_{j+1} = y_j + (e^{t_j} - 2y_j)(0.25), \qquad j = 0, 1, 2, 3.$$

Thus since $y_0 = y(0) = 1$ is given,

$$\begin{aligned}
y_1 &= y_0 + (e^{t_0} - 2 \cdot y_0)(0.25) \\
&= 1 + (e^0 - 2 \cdot 1)(0.25) \\
&= 1 - 0.25 \\
&= 0.75, \\
y_2 &= y_1 + (e^{t_1} - 2y_1)(0.25) \\
&= 0.75 + [e^{.25} - 2(0.75)](0.25) \\
&= 0.6960,
\end{aligned}$$

etc. Table 6.3 shows the same approximations y_1, y_2, y_3, and y_4 to four decimal place accuracy, as well as the actual values $y(t_1)$, $y(t_2)$, $y(t_3)$, and $y(t_4)$ for the actual solution $y = \dfrac{1}{3}(2e^{-2t} + e^t)$.

Table 6.3 Approximate and actual values for the initial value problem in Example 3. ($h = 0.25$)

t_j	0	.25	.5	.75	1.0
y_j	1	.75	.6960	.7602	.9093
$y(t_j)$	1	.8324	.7948	.8544	.9963

(b) The same solution, approximated using 10 subintervals of $[0, 1]$ of length $h = \dfrac{1 - 0}{10} = 0.1$, is obtained using the Euler equation

$$y_{j+1} = y_j + (e^{t_j} - 2y_j)(0.1), \qquad j = 0, 1, \ldots, 9.$$

Table 6.4 shows the results of this approximation. ◇

Table 6.4 Approximate and actual values of the solution of the initial value problem in Example 3. ($h = 0.1$)

t_j	0	.1	.2	.3	.4	.5	.6	.7	.8	.9	1.0
y_j	1	.9	.8305	.7866	.7642	.7606	.7733	.8009	.8421	.8962	.9629
$y(t_j)$	1	.9142	.8540	.8158	.7968	.7948	.8082	.8356	.8764	.9301	.9963

Euler's method is one procedure for approximating the solution of an initial value problem involving a first order differential equation. While we have used examples in which an actual solution may be obtained for purposes of comparison, Euler's method may be applied to differential equations whose solutions are not known.

Obviously there is a fair amount of tedium in applying Euler's method. Calculators and especially computers can reduce this tedium to a minimum. Program 9 in

Appendix I is a BASIC Computer program which was used to perform the calculations in Tables 1–4. You may find it useful in working the exercises.

Several sources of error enter into the calculations that are involved in Euler's method. Since we obtain only approximations y_j instead of the actual value $y(t_j)$ at each step, the slope calculation $\dfrac{dy}{dt} = f(t_j, y_j)$ contains an error due to the difference $|y_j - y(t_j)|$, and the point of assumed tangency, (t_j, y_j) contains an error due to the same factor. Finally, roundoff error in such calculations is inescapable. In general, however, we can say that accuracy decreases as we attempt to approximate values of the solution further away from the number t_0 where the initial condition is prescribed, and that accuracy increases as the stepsize h decreases. These important issues of accuracy are pursued in more advanced courses.

Exercise Set 21.6

In each of Exercises 1–10 use Euler's method to approximate the value of the solution to the initial value problem at the endpoints of 4 subintervals of $[0, 1]$ of size $h = 0.25$.

1. $y' = 2y$

   $y(0) = 1$

2. $y' = 9y(1 - y)$

   $y(0) = \dfrac{1}{2}$

3. $y' + 2y = 4$

   $y(0) = 1$

4. $\dfrac{dy}{dx} = \dfrac{y + 1}{x + 1}$

   $y(0) = 0$

5. $\dfrac{dy}{dx} = x(y + 1)$

   $y(0) = 1$

6. $\dfrac{dy}{dx} = 2 - 4y$

   $y(0) = 1$

7. $\dfrac{dy}{dt} + 2ty = t$

   $y(0) = \dfrac{3}{2}$

8. $e^y \cdot \dfrac{dy}{dt} - t^2 = 0$

   $y(0) = 0$

9. $\dfrac{dy}{dt} = 2y - 2t$

   $y(0) = \dfrac{3}{2}$

10. $\dfrac{dy}{dt} - 2y = e^{2t} \sin t$

   $y(0) = 0$

11. Find the actual solution of the initial value problem in Exercise 3. Compute the actual value of the solution at each endpoint and compare it with the value from Euler's method.

12. An investor opens a savings account paying 10% interest compounded continuously with an initial deposit of $1000. The investor makes deposits of $1000 per year in a continuous manner.
   a. Show that the function P, giving the amount on deposit after t years, satisfies the differential equation $P'(t) = 0.10P(t) + 1000$.
   b. Use Euler's method to approximate the values $P(1)$, $P(2)$, $P(3)$, and $P(4)$.

SUMMARY OUTLINE OF CHAPTER 21

◆ First order differential equations of the form (page 984)

$$\frac{dy}{dt} = ky + b$$

may be solved by the method of *separation of variables*.

◆ First order differential equations of the form (page 988)

$$\frac{dy}{dt} + p(t)y = g(t)$$

may be solved using the *integrating factor* $e^{\int p(t)\, dt}$

◆ The differential equation (page 992)

$$M(x, y) + N(x, y)\frac{dy}{dx} = 0, \qquad (x, y) \in R \qquad (1)$$

is *exact* if there exists a function f for which

$$M(x, y) = \frac{\partial f}{\partial x}(x, y), \qquad N(x, y) = \frac{\partial f}{\partial y}(x, y).$$

◆ **Theorem:** Equation (1) is exact if and only if   (page 993)

$$\frac{\partial M}{\partial y}(x, y) = \frac{\partial N}{\partial x}(x, y), \qquad (x, y) \in R.$$

◆ The second order homogeneous linear differential equation $y'' + ay' + by = 0$ has characteristic polynomial   (page 996)
$r^2 + ar + b$ and general solution

    (i) $y = C_1 e^{r_1 t} + C_2 e^{r_2 t}$ if $r^2 + ar + b$ has two distinct real roots r_1 and r_2;

    (ii) $y = C_1 e^{rt} + C_2 t e^{rt}$ if $r^2 + ar + b$ has a repeated real root r;

    (iii) $y = e^{\alpha t}[C_1 \sin \beta t + C_2 \cos \beta t]$ if $r^2 + ar + b$ has the complex roots $r = \alpha \pm \beta i$; $\beta \neq 0$.

◆ The second order linear homogeneous equation $y'' + by = 0$, $b > 0$ has general solution $y =$ (page 1000)
$C_1 \sin \sqrt{b}t + C_2 \cos \sqrt{b}t$.

◆ The nonhomogeneous second order linear equation $y'' + ay' + by = f(t)$ may be solved by the *method of undeter-*   (page 1005)
mined coefficients when $f(t)$ is of the form t^n, e^{kt}, $\sin kt$ or $\cos kt$.

◆ A differential equation may sometimes be solved by assuming a *power series solution* of the form $y = \Sigma a_k x^k$,   (page 1013)
substituting into the differential equation, and solving the resulting equations for the coefficients a_k.

◆ *Euler's Method* for approximating solutions of the differential equation $y' = f(t, y)$ with $y(a) = y_0$ is based on the   (page 1019)
formula $y_{j+1} = y_j + f(t_j, y_j)h$, which gives the approximation y_{j+1} to the value $y(t_{j+1})$.

REVIEW EXERCISES—CHAPTER 21

In Exercises 1–30 find the general solution of the given differential equation.

1. $\dfrac{dy}{dx} = x - 3$

2. $\dfrac{dy}{dt} = 4 - \sqrt{t}$

3. $\dfrac{dy}{dt} = \dfrac{\sec \sqrt{t} \tan \sqrt{t}}{\sqrt{t}}$

4. $\dfrac{d^2 y}{dx^2} = \sqrt{1 + x}$

5. $\dfrac{dy}{dt} = 3ty^2$

6. $\dfrac{dy}{dt} = \dfrac{\sqrt{t + 1}}{y}$

7. $\dfrac{dy}{dx} = (1 + x)(2 + y)$

8. $ty' = \dfrac{t^2 + 3}{y}$

9. $y' + 2y = 4$

10. $y' = y(3 - y)$

11. $y' = y(4 + y)$

12. $ty' = y \ln t$

13. $y' + y \cos t = 0$

14. $y' = 4y \ln y$

15. $y' = 4y + 8$

16. $y' = 4y + 4t + 8$

17. $y'' + 4y' + 4y = 0$

18. $y'' - 3y' - 10y = 0$

19. $y'' - 2y' - 15y = \cos 2t$

20. $y'' + 9y = 5 + e^{2t}$

21. $y'' - 9y = 5 + e^{2t}$

22. $y' - 3y = -9 \sin 2t$

23. $t^2 y' + y = 0$

24. $\dfrac{dy}{dx} = y(2 - y)$

25. $\dfrac{d^2 y}{dt^2} - 5\dfrac{dy}{dt} - 14y = 2 + t$

26. $y'' + y' - 6y = 6 - e^{2t}$

27. $2xy^3 + 3x^2 y^2 \dfrac{dy}{dx} = 0$

28. $\cos y - x \sin y \dfrac{dy}{dx} = 0$

29. $2y - y^2 - 2x\dfrac{dy}{dx} = 0$

30. $(e^{xy} + xye^{xy})\dfrac{dy}{dx} + y^2 e^{xy} = 0$

In Exercises 31–38 find the solution of the initial value problem.

31. $\dfrac{dy}{dx} = \dfrac{x}{\sqrt{1 + x^2}}$

$y(0) = 3$

32. $y' = 6y$

$y(0) = 2$

33. $\dfrac{dy}{dt} = ty$

$y(0) = 3$

34. $y' = 4(1 - y)$

$y(0) = 1$

35. $\dfrac{d^2 y}{dt^2} = -9y$

$y(0) = 3$

$y'(0) = 9$

36. $y'' - 16y = 0$

$y(0) = 2$

$y'(0) = 8$

37. $y'' - 6y' + 9y = 0$

$y(0) = 0$

$y'(0) = 6$

38. $y' - 4y = -2e^{3x}$

$y(0) = 3$

$y'(0) = 10$

39. Find a differential equation satisfied by the function

a. $y = Ce^{2t}$

b. $y = C_1e^t + C_2e^{-t}$.

40. A tank contains 500 liters of brine with an initial concentration of salt of 1 kilogram of salt per liter. A second mixture of brine containing 2 kilograms of salt per liter is then added at a rate of 50 liters per minute. The tank is kept well mixed, and brine is drawn off the bottom at the same rate, 50 liters per minute. Find the concentration of salt t minutes after the second mixture begins entering the first.

41. An automobile radiator contains five gallons of pure anti-freeze. The owner begins adding fresh water at the rate of 1 gallon per minute, with the engine running to insure complete mixing, and draining the radiator at the same rate of 1 gallon per minute. Find the concentration $y(t)$ of antifreeze per gallon t minutes after the owner begins this process.

42. An investor places $5000 in a savings account paying 10% interest compounded continuously and pledges to make additional deposits of $2000 per year in a continual manner.

a. Find a differential equation for $P(t)$, the amount on deposit in this account t years after it is opened.

b. Use Euler's method to approximate $P(1)$, $P(2)$, $P(3)$, and $P(4)$.

43. A cold drink is removed from a refrigerator at temperature 40°F and is placed on a sunporch where the surrounding temperature is 90°F. After 5 minutes the temperature of the drink is 50°F. Find its temperature after 10 minutes.

44. Use a power series to find a solution to the initial value problem

$$\begin{cases} y' - xy = 0 \\ y(0) = 2. \end{cases}$$

45. Find a power series solution for the second order differential equation $xy'' - y = 0$.

46. Find a power series solution for the initial value problem

$$\begin{cases} xy'' + y = 0 \\ y(0) = 0 \\ y'(0) = 1. \end{cases}$$

Why is the condition $y(0) = 0$ *necessary*?

47. Find the general form of solutions to the differential equation

$$y'' - 4y' + 6y = 0.$$

48. Find the solution for the initial value problem

$$\begin{cases} y'' - 4y' + 8y = 0 \\ y(0) = 0 \\ y'(0) = 2. \end{cases}$$

Appendi

Basic Computer Progra...s

Included here are nine BASIC computer programs that are referred to in various examples and exercises throughout the text. They are presented as "bare-bones" prototypes, which those with access to computing facilities (personal computers, programmable calculators, or large computers) can use in designing programs that actually operate on particular machines. Notation appearing in these programs includes the following:

1. $a * b$ means the product ab.
2. $a \uparrow b$ means the exponentiation a^b.

3. a/b means the quotient $\dfrac{a}{b}$.

   Proper development of computer software requires full documentation, as well as the inclusion of checks to insure that the user does not attempt to supply inappropriate values to the program. (For example, in asking the user to specify the endpoints of an interval $[a, b]$, one should check to insure that $a < b$.) We have made no attempt to do either, since we wish to highlight only the algorithm involved in the program.

Program 1: Newton's Method for $f(x) = x^3 - 7$

```
10 DEF FNF (T) = T↑3 - 7
20 DEF FND (T) = 3*(T↑2)
30 PRINT "how many iterations?"
40 INPUT N
50 PRINT "what is your first guess?"
60 INPUT X
70 FOR I = 1 TO N
80   LET Z1 = FNF(X)
90   LET Z2 = FND(X)
100  LET W = X - (Z1/Z2)
110  PRINT I,W
120  LET X = W
130 NEXT I
140 END
```

Comment: This program implements Newton's Method to locate a zero of the function $f(x) = x^3 - 7$. The value of the function is computed in line 10, the derivative $f'(x) = 3x^2$ is computed in line 20, and the formula for Newton's Method is implemented in line 100. The user supplies the number of iterations and a first guess at the root.

Program 2: Lower Riemann Sums for $f(x) = 3x^2 + 7$

```
 10 DEF FNF (T) = 3*(T↑2) + 7
 20 PRINT "enter interval endpoints a,b"
 30 INPUT A,B
 40 PRINT "how many subintervals?"
 50 INPUT N
 60 LET D = (B - A)/N
 70 LET S = 0
 80 FOR I = 1 TO N
 90   LET X = A + (I - 1)*D
100   LET S = S + FNF(X)*D
110 NEXT I
120 PRINT S
130 END
```

Comment: This program computes a lower Riemann sum for the function $f(x) = 3x^2 + 7$ on the interval $[a, b]$ using n equal subintervals. The numbers a, b, and n are supplied by the user. The program computes the function value at left endpoints of the resulting subintervals, since $f(x)$ is an increasing function.

Program 3: Upper Riemann Sums for $f(x) = 3x^2 + 7$

```
 10 DEF FNF(T) = 3*(T↑2) + 7
 20 PRINT "enter interval endpoints a,b"
 30 INPUT A,B
 40 PRINT "how many subintervals?"
 50 INPUT N
 60 LET D = (B - A)/N
 70 LET S = 0
 80 FOR I = 1 TO N
 90   LET X = A + I*D
100   LET S = S + FNF(X)*D
110 NEXT I
120 PRINT S
130 END
```

Comment: This program computes an upper Riemann sum for the function $f(x) = 3x^2 + 7$ on the interval $[a, b]$, if $a > 0$, using n equal subintervals. The numbers a, b, and n are supplied by the user. The program computes the function values at right endpoints of the resulting subintervals, since f is an increasing function for $x \geq 0$.

Program 4: Trapezoidal Rule Applied to $f(x) = \dfrac{1}{x}$

```
 10 DEF FNF(T) = 1/T
 20 PRINT "enter interval endpoints a,b"
```

```
 30 INPUT A,B
 40 PRINT "how many subintervals?"
 50 INPUT N
 60 LET D = (B - A)/N
 70 LET S = FNF(A)
 80 FOR I = 1 TO (N - 1)
 90   LET X = A + I*D
100   LET S = S + 2*FNF(X)
110 NEXT I
120 LET S = S + FNF(B)
130 LET S = S*(D/2)
140 PRINT S
150 END
```

Comment: This program implements the Trapezoidal Rule for approximate integration of the function $f(x) = \dfrac{1}{x}$ on the interval $[a, b]$ with $a > 0$. The numbers a, b, and n (the number of subintervals) are supplied by the user.

Program 5: Simpson's Rule Applied to the function $f(x) = \dfrac{1}{x}$

```
 10 DEF FNF(T) = 1/T
 20 PRINT "enter interval endpoints a,b"
 30 INPUT A,B
 40 PRINT "how many subintervals (an even number)?"
 50 INPUT N
 60 LET C = (B - A)/N
 70 LET D = C/3
 80 LET S = FNF(A)
 90 FOR I = 1 TO (N - 1) STEP 2
100   LET S = S + 4*FNF(A + I*C)
110 NEXT I
120 FOR I = 2 TO (N - 2) STEP 2
130   LET S = S + 2*FNF(A + I*C)
140 NEXT I
150 LET S = S + FNF(B)
160 LET S = S*D
170 PRINT S
180 END
```

Comment: This program implements Simpson's Rule for approximate integration for the function $f(x) = 1/x$ on the interval $[a, b]$ with $a > 0$. The numbers a, b, and n (the number of subintervals, *an even number*) are supplied by the user.

Program 6: Partial Sums of the Geometric Series

```
 10 PRINT "enter P,a,x,n"
 20 INPUT P,A,X,N
 30 LET S = 0
```

```
40 FOR K = P TO N
50   LET T = A*X↑K
60   LET S = S + T
70 NEXT K
80 PRINT S
90 END
```

Comment: This program computes the partial sum

$$\sum_{k=p}^{n} ax^k$$

where the constants p, a, x, and n are supplied by the user.

Program 7: A Riemann Sum in Polar Coordinates for $f(\theta) = 1 + \sqrt{\sin \theta}$

```
10 DEF FNF(T) = 1 + SQR(SIN(T))
20 DEF FNG(T) = 0.5 * (T↑2)
30 PRINT "enter interval endpoints a,b"
40 INPUT A,B
50 PRINT "how many subintervals?"
60 INPUT N
70 LET D = (B - A)/N
80 LET S = 0
90 FOR I = 1 TO N
100   LET Y = FNF(A + I*D)
110   LET S = S + FNG(Y)*D
120 NEXT I
130 PRINT S
140 END
```

Comment: This program computes a Riemann sum, using n subintervals of equal length, for the integral

$$\int_{a}^{b} \frac{1}{2}(1 + \sqrt{\sin \theta})^2 \, d\theta$$

which approximates the area bounded by $f(\theta) = 1 + \sqrt{\sin \theta}$. The parameters a, b, and n are supplied by the user.

Program 8: Riemann Sum for a Double Integral

```
10 PRINT "enter a,b,c,d"
20 INPUT A,B,C,D
30 PRINT "how many subintervals?"
40 INPUT N
50 LET D1 = (B - A)/N
60 LET D2 = (D - C)/N
70 LET D3 = D1*D2
80 LET S = 0
90 FOR J = 1 TO N
100   FOR K = 1 TO N
```

```
110    LET S = S + (8 - 2*(A + J*D1) - 4*(C + K*D2))
120    NEXT K
130 NEXT J
140 LET S = S*D3
150 PRINT S
160 END
```

Comment: This program computes a Riemann sum for the double integral

$$\int_c^d \int_a^b (8 - 2x - 4y) \, dx \, dy$$

using a grid consisting of n^2 rectangles of dimension $(b - a)(d - c)$. The parameters a, b, c, d, and n are supplied by the user.

Program 9: Euler's Method Applied to the Initial Value Problem

$$\frac{dy}{dt} = e^t - 2y$$

$$y(0) = 1$$

```
10 LET Y = 1
20 LET T = 0
30 FOR I = 1 TO 4
40 T = T + 0.25
50 Y = Y + (EXP(T) - 2*Y)*(0.25)
60 PRINT T,Y
70 NEXT I
80 END
```

Comment: This program gives an approximation to the stated initial value problem on the interval [0, 1] using $n = 4$ subintervals. The equation for Euler's method is implemented in step 50. Note that the right side of this equation involves current values of both T and Y.

Appendix II
Some Additional Proofs

Mathematical Induction

The *principle of mathematical induction* is a property of the real number system (we will think of it as an axiom) that allows us to prove statements about positive integers. Represented generally as $P(n)$, such statements are usually formulas involving some or all of the integers $1, 2, 3, \ldots, n$. Two particular formulas that we wish to prove here are equations (1) and (2) of Section 6.1:

$$1 + 2 + 3 + \cdots + n = \frac{n(n + 1)}{2} \tag{1a}$$

and

$$1^2 + 2^2 + 3^2 + \cdots + n^2 = \frac{n(n + 1)(2n + 1)}{6}. \tag{1b}$$

The axiom that we shall use to prove these formulas is the following.

PRINCIPLE OF MATHEMATICAL INDUCTION (PMI)

If $P(n)$ represents a mathematical statement involving the positive integer n, then $P(n)$ is true for all positive integers n if both of the following conditions hold:

(i) $P(1)$ is true.

(ii) Whenever $P(n)$ is true, $P(n + 1)$ is also true.

The PMI is not surprising. If a statement $P(1)$ is true, then setting $n = 1$ shows that $P(2) = P(1 + 1)$ is true by statement (ii). Since $P(2)$ is then known to be true, we next set $n = 2$ and conclude, again by (ii), that $P(3) = P(2 + 1)$ is true. Continuing in this way we verify that $P(n)$ is true for each of the positive integers $1, 2, 3, \ldots$.

Proof of Formula (1a): We prove formula (1a) by applying the PMI to the formula

$$P(n): 1 + 2 + 3 + \cdots + n = \frac{n(n + 1)}{2}. \tag{2}$$

First, setting $n = 1$ gives the statement

$$P(1): 1 = \frac{1(1 + 1)}{2},$$

which is true, since $\dfrac{1(1 + 1)}{2} = \dfrac{1 \cdot 2}{2} = 1$. Thus, statement (i) of the PMI holds.

To show that statement (ii) of the PMI holds, we assume that $P(n)$ in (2) is true, and we attempt to show that

$$P(n + 1): 1 + 2 + 3 + \cdots + (n + 1) = \frac{(n + 1)[(n + 1) + 1]}{2} \tag{3}$$

is true. To do this we use statement (2) to write

$$1 + 2 + 3 + \cdots + n + (n + 1) = \{1 + 2 + 3 + \cdots + n\} + n + 1$$
$$= \frac{n(n + 1)}{2} + n + 1$$
$$= \frac{n(n + 1) + 2(n + 1)}{2}$$
$$= \frac{(n + 1)(n + 2)}{2}$$
$$= \frac{(n + 1)[(n + 1) + 1]}{2},$$

which establishes (3). Thus, by the PMI, statement (2) holds for all positive integers. This proves formula (1a). ◆

Proof of Formula (1b): This is done just as the proof of formula (1a), using the PMI. The statement to be proved is

$$P(n): 1^2 + 2^2 + 3^2 + \cdots + n^2 = \frac{n(n + 1)(2n + 1)}{6}. \tag{4}$$

The statement corresponding to $n = 1$ is

$$P(1): 1^2 = \frac{1(1 + 1)(2 \cdot 1 + 1)}{6},$$

which is true, so (i) of the PMI holds.

Next, we assume that $P(n)$ in (4) holds, and we attempt to prove the statement

$$P(n + 1): \tag{5}$$
$$1^2 + 2^2 + 3^2 + \cdots + (n + 1)^2 = \frac{(n + 1)[(n + 1) + 1][2(n + 1) + 1]}{6}.$$

We do this using statement (4):

$$1^2 + 2^2 + 3^2 + \cdots + (n + 1)^2 = \{1^2 + 2^2 + 3^2 + \cdots + n^2\} + (n + 1)^2$$
$$= \frac{n(n + 1)(2n + 1)}{6} + (n + 1)^2$$
$$= \frac{n(n + 1)(2n + 1) + 6(n + 1)^2}{6}$$
$$= \frac{(n + 1)[n(2n + 1) + 6(n + 1)]}{6}$$
$$= \frac{(n + 1)[2n^2 + 7n + 6]}{6}$$

$$= \frac{(n+1)[(n+2)(2n+3)]}{6}$$

$$= \frac{(n+1)[(n+1)+1][2(n+1)+1]}{6}.$$

Thus, $P(n+1)$ in (5) is true whenever $P(n)$ is true. Thus, statement (ii) of the PMI holds, so formula (1b) is proved by the PMI. ◆

**THEOREM 3,
CHAPTER 2**

Let m and n be positive integers.

(i) If n is even, $\lim_{x \to a} x^{m/n} = a^{m/n}$ whenever $0 < a < \infty$.

(ii) If n is odd, $\lim_{x \to a} x^{m/n} = a^{m/n}$ for all $-\infty < a < \infty$.

Proof: We first prove case (i). Actually, we need only prove that

$$\lim_{x \to a} \sqrt[n]{x} = \sqrt[n]{a}$$

from which statement (i) follows by Theorem 2. According to Definition 3, Chapter 2, we must show that if $\epsilon > 0$ is given, there can be found a corresponding number δ so that

$$|\sqrt[n]{x} - \sqrt[n]{a}| < \epsilon \qquad \text{whenever} \qquad 0 < |x - a| < \delta. \tag{6}$$

The first of these two inequalities holds the key to determining how to choose δ, so we begin by examining the inequality

$$|\sqrt[n]{x} - \sqrt[n]{a}| < \epsilon \tag{7}$$

that we can rewrite as

$$-\epsilon < \sqrt[n]{x} - \sqrt[n]{a} < \epsilon.$$

The strategy is now to isolate the expression $x - a$ in the middle position, which we do as follows: From the above we have

$$\sqrt[n]{a} - \epsilon < \sqrt[n]{x} < \sqrt[n]{a} + \epsilon, \tag{8}$$

so

$$(\sqrt[n]{a} - \epsilon)^n < x < (\sqrt[n]{a} + \epsilon)^n, \tag{9}$$

so

$$(\sqrt[n]{a} - \epsilon)^n - a < x - a < (\sqrt[n]{a} + \epsilon)^n - a.$$

(There is a potential difficulty here. Going from step (8) to step (9) is valid only if $\sqrt[n]{a} - \epsilon$ is nonnegative. We can insure that this is true by requiring that $\epsilon < \sqrt[n]{a}$ to begin with. There is no problem in making this assumption, since if statement (6) holds for a smaller value of ϵ than that originally given, it also must hold for the given value.)

Finally, we rewrite (9) as

$$-[a - (\sqrt[n]{a} - \epsilon)^n] < x - a < [(\sqrt[n]{a} + \epsilon)^n - a] \tag{10}$$

in which both expressions inside brackets are positive. Then, subject to the remark that $\epsilon < \sqrt[n]{a}$, inequality (10) is equivalent to inequality (7). Now suppose we take

$\delta > 0$ to be so small that both

$$\delta < [a - (\sqrt[n]{a} - \epsilon)^n] \qquad \text{and} \qquad \delta < [(\sqrt[n]{a} + \epsilon)^n - a]. \tag{11}$$

Then the inequality

$$0 < |x - a| < \delta \tag{12}$$

can be rewritten as

$$-\delta < x - a < \delta. \tag{13}$$

From (11) it follows that if (13) holds, so does (10). Since (13) is equivalent to (12), and (10) is equivalent to (7), this shows that

$$|\sqrt[n]{x} - \sqrt[n]{a}| < \epsilon \qquad \text{whenever} \qquad 0 < |x - a| < \delta,$$

which completes the proof of statement (i).

To prove (ii), we assume that n is odd. As in part (i) we show only that $\lim\limits_{x \to a} \sqrt[n]{x} = \sqrt[n]{a}$. If $a > 0$ the proof for part (i) applies. Thus, we must only concern ourselves with the case $a < 0$. But in this case $-a > 0$, so we may apply the proof of statement (i) to conclude that

$$\lim_{x \to a} \sqrt[n]{-x} = \sqrt[n]{-a}. \tag{14}$$

Of course, we now wish to simply "factor out" the -1's in (14), but we must do it rigorously. By Definition 3, Chapter 2, equation (14) means that, given $\epsilon > 0$, there exists a $\delta > 0$ so that

$$|\sqrt[n]{-x} - \sqrt[n]{-a}| < \epsilon \qquad \text{whenever} \qquad 0 < |-x - (-a)| < \delta. \tag{15}$$

Since, for odd n,

$$|\sqrt[n]{-x} - \sqrt[n]{-a}| = |(-1)^{\frac{1}{n}}\sqrt[n]{x} - (-1)^{\frac{1}{n}}\sqrt[n]{a}| = |(-1)(\sqrt[n]{x} - \sqrt[n]{a})|$$
$$= |\sqrt[n]{x} - \sqrt[n]{a}|$$

and

$$|-x - (-a)| = |(-1)(x - a)| = |x - a|,$$

inequality (15) is equivalent to the statement

$$|\sqrt[n]{x} - \sqrt[n]{a}| < \epsilon \qquad \text{whenever} \qquad 0 < |x - a| < \delta,$$

which shows that $\lim\limits_{x \to a} \sqrt[n]{x} = \sqrt[n]{a}$ in the present case. This completes the proof. ◆

THEOREM 1, CHAPTER 2,
(parts (iii) and (iv))

If $\lim\limits_{x \to a} f(x) = L$ and $\lim\limits_{x \to a} g(x) = M$, then

(iii) $\lim\limits_{x \to a} f(x)g(x) = LM,$ and

(iv) $\lim\limits_{x \to a} \dfrac{f(x)}{g(x)} = \dfrac{L}{M},$ $M \neq 0.$

Proof: To prove (iii), we must show that, given $\epsilon > 0$, there exists a $\delta > 0$ so that

$$|f(x)g(x) - LM| < \epsilon \qquad \text{whenever} \qquad 0 < |x - a| < \delta.$$

We do this by using the triangle inequality to write

$$|f(x)g(x) - LM| = |f(x)g(x) - f(x)M + f(x)M - LM| \tag{16}$$
$$\leq |f(x)g(x) - f(x)M| + |f(x)M - LM|$$
$$= |f(x)||g(x) - M| + |M||f(x) - L|.$$

We need to make three observations about the terms on the right side of inequality (16).

(a) Since $\lim_{x \to a} f(x) = L$, there exists a number $\delta_1 > 0$ so that

$$|f(x) - L| < 1 \qquad \text{whenever} \qquad 0 < |x - a| < \delta_1.$$

Thus,

$$|f(x)| < |L| + 1 \qquad \text{whenever} \qquad 0 < |x - a| < \delta_1. \tag{17}$$

(b) Also, since $\lim_{x \to a} f(x) = L$, there exists a number $\delta_2 > 0$ so that

$$|f(x) - L| < \frac{\epsilon}{2(|M| + 1)} \qquad \text{whenever} \qquad 0 < |x - a| < \delta_2. \tag{18}$$

(c) Finally, since $\lim_{x \to a} g(x) = M$, there exists a number δ_3 so that

$$|g(x) - M| < \frac{\epsilon}{2(|L| + 1)} \qquad \text{whenever} \qquad 0 < |x - a| < \delta_3. \tag{19}$$

Now, we take δ to be the smallest of the numbers δ_1, δ_2, and δ_3. Then each of the conditions in statements (17), (18), and (19) is fulfilled whenever $0 < |x - a| < \delta$. Combining inequality (16) with (17) through (19) then shows that

$$|f(x)g(x) - LM| \leq |f(x)| \cdot |g(x) - M| + |M| \cdot |f(x) - L|$$
$$< (|L| + 1)\left[\frac{\epsilon}{2(|L| + 1)}\right] + |M|\left[\frac{\epsilon}{2(|M| + 1)}\right]$$
$$< \frac{\epsilon}{2} + \frac{\epsilon}{2}$$
$$= \epsilon$$

whenever $0 < |x - a| < \delta$. This proves part (iii).

To prove (iv), we first show that

$$\lim_{x \to a} \frac{1}{g(x)} = \frac{1}{M}, \qquad M \neq 0. \tag{20}$$

Statement (iv) then follows from part (iii) and statement (20). To prove (20), we must show that, given $\epsilon > 0$, there exists a $\delta > 0$ so that

$$\left|\frac{1}{g(x)} - \frac{1}{M}\right| < \epsilon \qquad \text{whenever} \qquad 0 < |x - a| < \delta.$$

Since

$$\left|\frac{1}{g(x)} - \frac{1}{M}\right| = \left|\frac{M - g(x)}{Mg(x)}\right| = \left(\frac{1}{|M||g(x)|}\right)|M - g(x)|, \tag{21}$$

we note first that since $\lim_{x \to a} g(x) = M$, there exists a number δ_1 so that

$$|g(x) - M| < \frac{|M|}{2} \qquad \text{whenever} \qquad 0 < |x - a| < \delta_1. \tag{22}$$

From (22) it follows that $|g(x)| > \dfrac{|M|}{2}$, or

$$\frac{1}{|g(x)|} < \frac{2}{|M|} \qquad \text{whenever} \qquad 0 < |x - a| < \delta_1. \tag{23}$$

Next, again since $\lim_{x \to a} g(x) = M$, there exists a number δ_2 so that

$$|g(x) - M| < \frac{|M|^2 \epsilon}{2} \qquad \text{whenever} \qquad 0 < |x - a| < \delta_2. \tag{24}$$

Finally, we take δ to be the smaller of δ_1 and δ_2. Then from statements (21), (23), and (24) it follows that

$$\left| \frac{1}{g(x)} - \frac{1}{M} \right| = \left(\frac{1}{|M||g(x)|} \right) |M - g(x)|$$
$$< \left(\frac{1}{|M|} \right) \left(\frac{2}{|M|} \right) \left(\frac{|M|^2 \epsilon}{2} \right)$$
$$= \epsilon$$

whenever $0 < |x - a| < \delta$. This proves statement (iv). ◆

THEOREM 4,
CHAPTER 2

Assume that $\lim_{x \to a} g(x)$ and $\lim_{x \to a} h(x)$ exist, and that $\lim_{x \to a} g(x) = L = \lim_{x \to a} h(x)$. If the function f satisfies the inequality

$$g(x) \le f(x) \le h(x)$$

for all x in an open interval containing a (except possibly at $x = a$), then $\lim_{x \to a} f(x) = L$ also.

Proof of Theorem 4: Since

$$\lim_{x \to a} g(x) = L,$$

given $\epsilon > 0$, there is a number δ_1 so that when $0 < |x - a| < \delta_1$,

$$-\epsilon < g(x) - L < \epsilon$$

or, adding L to each term, so that

$$L - \epsilon < g(x) < L + \epsilon. \tag{25}$$

Similarly, since

$$\lim_{x \to a} h(x) = L,$$

there is a (possibly different) number δ_2 so that when $0 < |x - a| < \delta_2$,

$$L - \epsilon < h(x) < L + \epsilon. \tag{26}$$

Now since $g(x) \le f(x) \le h(x)$, we may use the left hand of inequality (25) and the right half of inequality (26) to conclude that

$$L - \epsilon < g(x) \le f(x) \le h(x) < L + \epsilon$$

when both $0 < |x - a| < \delta_1$ and $0 < |x - a| < \delta_2$.

This shows that if we take $\delta = \min \{\delta_1, \delta_2\}$, we will have

$$L - \epsilon < f(x) < L + \epsilon$$

or

$$-\epsilon < f(x) - L < \epsilon,$$

whenever $0 < |x - a| < \delta$, as required.   ◆

Proof of Theorem 8, Chapter 3 (Chain Rule): Let x_0 be in the domain of g. To prove that

$$(f \circ g)'(x) = f'(g(x_0))g'(x_0),$$

we begin by defining a new function F for t in the domain of f:

$$F(t) = \begin{cases} \dfrac{f(t) - f(g(x_0))}{t - g(x_0)} & \text{if} \quad t \ne g(x_0) \\ f'(g(x_0)) & \text{if} \quad t = g(x_0) \end{cases}$$

The function F is continuous at $t_0 = g(x_0)$, since

$$\lim_{t \to g(x_0)} F(t) = \lim_{t \to g(x_0)} \frac{f(t) - f(g(x_0))}{t - g(x_0)} = f'(g(x_0)) = F(g(x_0)).$$

The function F is also continuous at numbers $t = g(x) \ne g(x_0)$ since we have, for $s \ne g(x_0)$,

$$\lim_{s \to t} F(s) = \lim_{s \to t} \frac{f(s) - f(g(x_0))}{s - g(x_0)} = \frac{f(t) - f(g(x_0))}{t - g(x_0)} = F(t).$$

We wish now to work with the composite function $F \circ g$. Since F is continuous and g is differentiable (hence, continuous), the composite function $F \circ g$ is continuous on the domain of g. Thus, according to the definition of F,

$$\lim_{x \to x_0} F(g(x)) = F(g(x_0)) = f'(g(x_0)). \tag{27}$$

Next, we note that, if $x \ne x_0$, we can write

$$\frac{f(g(x)) - f(g(x_0))}{x - x_0} = F(g(x)) \left[\frac{g(x) - g(x_0)}{x - x_0} \right]. \tag{28}$$

To see this, note that if $t = g(x) \ne g(x_0)$, then

$$F(g(x)) = \frac{f(g(x)) - f(g(x_0))}{g(x) - g(x_0)}$$

$$F(g(x))[g(x) - g(x_0)] = f(g(x)) - f(g(x_0))$$

$$F(g(x)) \left[\frac{g(x) - g(x_0)}{x - x_0} \right] = \frac{f(g(x)) - f(g(x_0))}{x - x_0}$$

while if $g(x) = g(x_0)$ for $x \neq x_0$, both sides of (28) are zero. Using equations (27) and (28) together with the continuity of F and g, we may now conclude that

$$
\begin{aligned}
(f \circ g)'(x_0) &= \lim_{x \to x_0} \frac{f(g(x)) - f(g(x_0))}{x - x_0} \\
&= \lim_{x \to x_0} F(g(x)) \left[\frac{g(x) - g(x_0)}{x - x_0} \right] \\
&= f'(g(x_0)) g'(x_0).
\end{aligned}
$$

◆

Proof of Theorem 1, Chapter 6: This should really be regarded as a sketch of the proof, rather than a rigorous argument, since several facts must be assumed that we cannot prove here. However, the basic notion is a straightforward calculation.

To show that $\lim_{n \to \infty} \underline{S}_n$ exists, we note that $\underline{S}_{n+1} \geq \underline{S}_n$, since the subinterval size decreases as n increases. (Thus, the minimum values $f(c_j)$ can only increase, not decrease. Thus, $\{\underline{S}_n\}$ is an increasing sequence of numbers which is bounded above (by an upper sum). Such sequences always have limits, as we shall prove in Chapter 12. The proof that $\lim_{n \to \infty} \overline{S}_n$ exists is similar.

To prove $\lim_{n \to \infty} \underline{S}_n = \lim_{n \to \infty} \overline{S}_n$, it is sufficient to prove that

$$
\lim_{n \to \infty} (\overline{S}_n - \underline{S}_n) = 0. \tag{29}
$$

According to the definitions of $\underline{S}_n$ and $\overline{S}_n$ we have

$$
\lim_{n \to \infty} (\overline{S}_n - \underline{S}_n) = \lim_{n \to \infty} \left\{ \sum_{j=1}^{n} f(d_j)\, \Delta x - \sum_{j=1}^{n} f(c_j)\, \Delta x \right\} \tag{30}
$$

$$
= \lim_{n \to \infty} \sum_{j=1}^{n} [f(d_j) - f(c_j)]\, \Delta x.
$$

Now recall that $\Delta x = (x_j - x_{j-1}) = \dfrac{b - a}{n}$, so as $n \to \infty$ the width of each of the subintervals approaches zero. Since f is a continuous function, the difference between the maximum value $f(d_j)$ and the minimum value $f(c_j)$ must approach zero as the width of the interval $[x_{j-1}, x_j]$ approaches zero.* Thus, if we let

$$
M_n = \max \{ f(d_j) - f(c_j) \mid 1 \leq j \leq n \},
$$

we can state in mathematical notation both that

$$
f(d_j) - f(c_j) \leq M_n \qquad \text{for all } j = 1, 2, \ldots, n, \tag{31}
$$

(that is, M_n is the largest difference) and that

$$
\lim_{n \to \infty} M_n = 0 \tag{32}
$$

(that is, the differences approach zero as the interval width approaches zero).

---

*An entirely rigorous proof of this statement involves the concept of *uniform continuity,* a topic for more advanced courses on analysis.

Now, returning to equation (30) and using inequality (31) we obtain the statement

$$\lim_{n \to \infty} (\overline{S}_n - \underline{S}_n) = \lim_{n \to \infty} \sum_{j=1}^{n} [f(d_j) - f(c_j)] \, \Delta x \tag{33}$$

$$\leq \lim_{n \to \infty} \sum_{j=1}^{n} M_n \, \Delta x$$

$$= \lim_{n \to \infty} M_n \cdot \sum_{j=1}^{n} \Delta x.$$

Since

$$\sum_{j=1}^{n} \Delta x = \sum_{j=1}^{n} \left(\frac{b-a}{n} \right) = \frac{n(b-a)}{n} = b - a,$$

equation (32) and inequality (33) now give

$$\lim_{n \to \infty} (\overline{S}_n - \underline{S}_n) = (b - a) \lim_{n \to \infty} M_n = 0,$$

which establishes the desired equation (29).   ◆

Appendix III
Complex Numbers

In the set of real numbers, the quadratic equation

$$x^2 = -1 \tag{1}$$

has no solution, since the square of a real number is never negative. The *complex numbers* constitute a number system containing the real number system (just as the real numbers contain the integers), and in this system equations such as equation (1) have solutions.

To form the complex number system, we first define the complex number i by the equation

$$i^2 = -1. \tag{2}$$

Another way to write (2) is to say that $i = \sqrt{-1}$. Thus, the complex number i is a solution to equation (1), as is the number $-i = -\sqrt{-1}$. However, i and $-i$ are not the only complex numbers. The following definition fully determines the complex number system.

DEFINITION 1

The **complex numbers** are all numbers of the form

$$z = a + bi,$$

where a and b are real numbers, together with the following operations:

(i)   Scalar multiplication: $\lambda(a + bi) = \lambda a + (\lambda b)i, \quad \lambda \in \mathbb{R}$
(ii) Addition: $(a_1 + b_1 i) + (a_2 + b_2 i) = (a_1 + a_2) + (b_1 + b_2)i$
(iii) Multiplication: $(a_1 + b_1 i) \cdot (a_2 + b_2 i) = (a_1 a_2 - b_1 b_2) + (a_1 b_2 + a_2 b_1)i$

Notice the following about Definition 1:

(a) When $b = 0$, the complex number $z = a + 0i = a$ is a real number. Moreover, the definitions of addition and multiplication agree with the definitions of addition and multiplication of real numbers when $b_1 = b_2 = 0$. Thus, the set of complex numbers of the form $\{a + bi \mid a \in \mathbb{R}, b = 0\}$ is just the set $\mathbb{R}$ of real numbers.

(b) The definition of multiplication is obtained by assuming that the usual distributive and commutative laws hold for all expressions involving a_1, a_2, b_1, b_2, and i:

$$(a_1 + b_1 i) \cdot (a_2 + b_2 i) = a_1(a_2 + b_2 i) + (b_1 i)(a_2 + b_2 i)$$
$$= a_1 a_2 + a_1 b_2 i + a_2 b_1 i + b_1 b_2 i^2$$
$$= a_1 a_2 - b_1 b_2 + (a_1 b_2 + a_2 b_1)i.$$

For $z = a + bi$, we say that z is **pure imaginary** if $a = 0$.

Example 1

If $z_1 = 3 + 2i$ and $z_2 = 4 - i$, then

(i) $4z_1 = 4(3 + 2i) = 4 \cdot 3 + (4 \cdot 2)i = 12 + 8i.$

(ii) $z_1 + z_2 = (3 + 2i) + (4 - i) = (3 + 4) + (2 + (-1))i = 7 + i.$

(iii) $z_1 - z_2 = z_1 + (-z_2) = (3 + 2i) + (-1)(4 - i)$
$$= (3 + 2i) + (-4 + i)$$
$$= (3 + (-4)) + (2 + 1)i$$
$$= -1 + 3i.$$

(iv) $z_1 z_2 = (3 + 2i)(4 - i) = 3 \cdot 4 - (2)(-1) + [3(-1) + 2 \cdot 4]i$
$$= 14 + 5i.$$ $\diamond$

Geometric and Vector Interpretations

Since the complex number $z = a + bi$ is determined by a pair of real numbers, we refer to a as the **real part** (or real component) of z, and we refer to b as the **imaginary part** (imaginary component) of z. (Remember that b is a real number.) Plotting the real part of z on the x-axis (called the real axis) and the imaginary part of z on the y-axis (called the imaginary axis), we can plot the complex number $z = a + bi$ as the ordered pair (a, b) in the xy-plane (see Figure III.1). Note that the real numbers (considered as a subset of the complex numbers) correspond to points on the real axis, while pure imaginary numbers correspond to points on the imaginary axis.

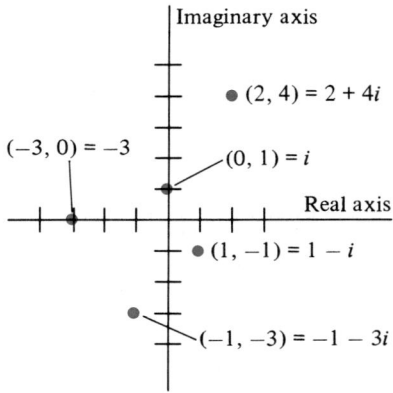

Figure III.1 Complex numbers $a + bi$ may be represented as points (a, b) in the complex plane.

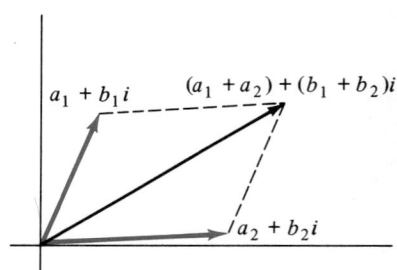

Figure III.2 Addition of complex numbers corresponds to vector addition.

Since points (a, b) in the plane may be regarded as position vectors $\langle a, b \rangle$ originating at the origin $(0, 0)$, we may interpret the complex number $a + bi$ as the position vector $\langle a, b \rangle$. As Figure III.2 illustrates, the definition of addition for complex numbers agrees with the definition of vector addition. You can verify that the corresponding definitions of scalar multiplication agree as well. The vector interpretation leads directly to the concept of the *modulus* (also called the length, or absolute value) of a complex number.

DEFINITION 2

The **modulus of the complex number** $z = a + bi$ is the real number $|z| = \sqrt{a^2 + b^2}$.

Note that the modulus of a complex number is just its absolute value when the number is real. A concept related to the modulus is the *conjugate* of a complex number.

DEFINITION 3

The **conjugate of the complex number** $z = a + bi$ is the complex number $\bar{z} = a - bi$.

The relationship between the modulus and the conjugate is that

$$z\bar{z} = (a + bi)(a - bi) = a^2 + b^2 = |z|^2.$$

Figure III.3 shows that the conjugate $\bar{z}$ is just the reflection in the real axis of the complex number z.

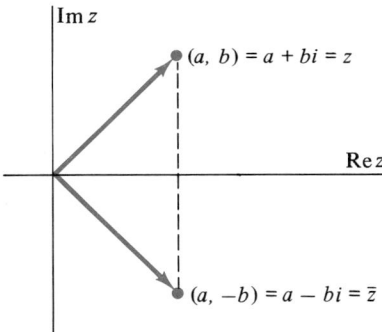

Im z

$(a, b) = a + bi = z$

Re z

$(a, -b) = a - bi = \bar{z}$

Figure III.3 The conjugate $\bar{z}$ of z is the reflection of z in the real axis.

Division for Complex Numbers

The concepts of modulus and conjugate allow us to define the operation of division. Since

$$z\bar{z} = |z|^2$$

it follows that

$$\left(\frac{1}{|z|^2}\right)(z\bar{z}) = 1,$$

so the reciprocal $\dfrac{1}{z}$, $z \neq 0$, is defined as

$$\frac{1}{z} = \frac{\bar{z}}{|z|^2}, \qquad z \neq 0.$$

Division is then defined as multiplication by a reciprocal.

DEFINITION 4

Let z_1 and z_2 be complex numbers with $z_2 \neq 0$. The quotient $\dfrac{z_1}{z_2}$ is defined to be

$$\frac{z_1}{z_2} = z_1\left(\frac{1}{z_2}\right) = \frac{z_1\bar{z}_2}{|z_2|^2}.$$

Again, this definition agrees with the definition of division in $\mathbb{R}$ when z_1 and z_2 are real.

Example 2

Let $z_1 = 1 + 3i$ and $z_2 = 2 - 4i$. Then

(i) $\bar{z}_1 = 1 - 3i$.

(ii) $z_1\bar{z}_1 = (1 + 3i)(1 - 3i) = (1 + 9) + (3 - 3)i = 10 = |z_1|^2$.

(iii) $\dfrac{1}{z_1} = \dfrac{\bar{z}_1}{|z_1|^2} = \dfrac{1}{10}(1 - 3i) = \dfrac{1}{10} - \dfrac{3}{10}i$.

(iv) $\dfrac{z_1}{z_2} = z_1\left(\dfrac{1}{z_2}\right) = \dfrac{z_1\bar{z}_2}{|z_2|^2} = \dfrac{1}{20}(1 + 3i)(2 + 4i)$

$$= \frac{1}{20}[(2 - 12) + (6 + 4)i]$$

$$= -\frac{1}{2} + \frac{1}{2}i. \qquad \diamond$$

Solutions of Quadratic Equations

We conclude by recalling that the solutions of the quadratic equation

$$ax^2 + bx + c = 0$$

are given by the quadratic formula as

$$x = -\frac{b}{2a} \pm \frac{\sqrt{b^2 - 4ac}}{2a}$$

when $b^2 - 4ac \geq 0$. Using complex numbers we may now write the roots as

$$x = -\frac{b}{2a} \pm \left(\frac{\sqrt{4ac - b^2}}{2a}\right)i$$

when $b^2 - 4ac < 0$.

Example 3

For the quadratic equation

$$x^2 - 2x + 3 = 0$$

the solutions are

$$x = -\frac{(-2)}{2} \pm \frac{\sqrt{(-2)^2 - 4(1)(3)}}{2}$$

$$= 1 \pm \frac{\sqrt{-8}}{2}$$

$$= 1 \pm \sqrt{2}\,i. \qquad \diamond$$

Appendix IV
Tables of Transcendental Functions

Trigonometric functions (x in radians)

x	$\sin x$	$\cos x$	$\tan x$
0.0	.00000	1.00000	0.00000
0.1	.09983	.99500	.10033
0.2	.19867	.98007	.20271
0.3	.29552	.95534	.30934
0.4	.38942	.92106	.42279
0.5	.47943	.87758	.54630
0.6	.56464	.82534	.68414
0.7	.64422	.76484	.84229
0.8	.71736	.69671	1.02964
0.9	.78333	.62161	1.26016
1.0	.84147	.54030	1.55741
1.1	.89121	.45360	1.96476
1.2	.93204	.36236	2.57215
1.3	.96356	.26750	3.60210
1.4	.98545	.16997	5.79788
1.5	.99749	.07074	14.10142
$\pi/2$	1.00000	.00000	∞
1.6	.99957	−.02920	−34.23254
1.7	.99166	−.12884	−7.69660
1.8	.97385	−.22720	−4.28626
1.9	.94630	−.32329	−2.92710
2.0	.90930	−.41615	−2.18504
2.1	.86321	−.50485	−1.70985
2.2	.80850	−.58850	−1.37382
2.3	.74571	−.66628	−1.11921
2.4	.67546	−.73739	−.91601
2.5	.59847	−.80114	−.74702
2.6	.51550	−.85689	−.60160
2.7	.42738	−.90407	−.47273
2.8	.33499	−.94222	−.35553
2.9	.23925	−.97096	−.24641
3.0	.14112	−.98999	−.14255
3.1	.04158	−.99914	−.04162
π	.00000	−1.00000	.00000
3.2	−.05837	−.99829	.05847
3.3	−.15775	−.98748	.15975

x	$\sin x$	$\cos x$	$\tan x$
3.4	−.25554	−.96680	.26432
3.5	−.35078	−.93646	.37459
3.6	−.44252	−.89676	.49347
3.7	−.52984	−.84810	.62473
3.8	−.61186	−.79097	.77356
3.9	−.68777	−.72593	.94742
4.0	−.75680	−.65364	1.15782
4.1	−.81828	−.57482	1.42353
4.2	−.87158	−.49026	1.77778
4.3	−.91617	−.40080	2.28585
4.4	−.95160	−.30733	3.09632
4.5	−.97753	−.21080	4.63733
4.6	−.99369	−.11215	8.86017
4.7	−.99992	−.01239	80.71271
$3\pi/2$	−1.00000	−.00000	$-\infty$
4.8	−.99616	.08750	−11.38487
4.9	−.98245	.18651	−5.26749
5.0	−.95892	.28366	−3.38051
5.1	−.92581	.37798	−2.44939
5.2	−.88345	.46852	−1.88564
5.3	−.83227	.55437	−1.50127
5.4	−.77276	.63469	−1.21754
5.5	−.70554	.70867	−.99558
5.6	−.63127	.77557	−.81394
5.7	−.55069	.83471	−.65973
5.8	−.46460	.88552	−.52467
5.9	−.37388	.92748	−.40311
6.0	−.27942	.96017	−.29101
6.1	−.18216	.98327	−.18526
6.2	−.08309	.99654	−.08338
2π	.00000	1.00000	.00000
6.3	.01681	.99986	.01682
6.4	.11655	.99318	.11735
6.5	.21512	.97659	.22028

Exponential functions

x	e^x	e^{-x}
0.00	1.00000	1.00000
.05	1.05127	.95123
.10	1.10517	.90484
.15	1.16183	.86071
.20	1.22140	.81873
.25	1.28403	.77880
.30	1.34986	.74082
.35	1.41907	.70469
.40	1.49182	.67032
.45	1.56831	.63763
.50	1.64872	.60653
.55	1.73325	.57695
.60	1.82212	.54881
.65	1.91554	.52205
.70	2.01375	.49659

x	e^x	e^{-x}
.75	2.11700	.47237
.80	2.22554	.44933
.85	2.33965	.42741
.90	2.45960	.40657
.95	2.58571	.38674
1.00	2.71828	.36788
2.00	7.38906	.13534
3.00	20.08554	.04979
4.00	54.59815	.01832
5.00	148.41316	.00674
6.00	403.42879	.00248
7.00	1096.63316	.00091
8.00	2980.95799	.00034
9.00	8103.08393	.00012
10.00	22026.46579	.00005

Note: $e^{a+x} = e^a e^x$

Natural logarithms

x	$\ln x$	x	$\ln x$
.1	−2.30258	2.6	.95551
.2	−1.60943	2.7	.99325
.3	−1.20396	2.8	1.02962
.4	−.91628	2.9	1.06471
.5	−.69314	3.0	1.09861
.6	−.51082	3.1	1.13140
.7	−.35666	3.2	1.16315
.8	−.22313	3.3	1.19392
.9	−.10535	3.4	1.22378
1.0	0.00000	3.5	1.25276
1.1	.09531	3.6	1.28093
1.2	.18232	3.7	1.30833
1.3	.26236	3.8	1.33500
1.4	.33647	3.9	1.36098
1.5	.40547	4.0	1.38629
1.6	.47000	4.1	1.41099
1.7	.53063	4.2	1.43508
1.8	.58779	4.3	1.45862
1.9	.64185	4.4	1.48160
2.0	.69315	4.5	1.50408
2.1	.74194	4.6	1.52606
2.2	.78846	4.7	1.54756
2.3	.83291	4.8	1.56862
2.4	.87547	4.9	1.58924
2.5	.91629	5.0	1.60944

x	$\ln x$	x	$\ln x$
5.1	1.62924	7.6	2.02815
5.2	1.64866	7.7	2.04122
5.3	1.66771	7.8	2.05412
5.4	1.68640	7.9	2.06686
5.5	1.70475	8.0	2.07944
5.6	1.72277	8.1	2.09186
5.7	1.74047	8.2	2.10413
5.8	1.75786	8.3	2.11626
5.9	1.77495	8.4	2.12823
6.0	1.79176	8.5	2.14007
6.1	1.80829	8.6	2.15176
6.2	1.82455	8.7	2.16332
6.3	1.84055	8.8	2.17475
6.4	1.85630	8.9	2.18605
6.5	1.87180	9.0	2.19722
6.6	1.88707	9.1	2.20827
6.7	1.90211	9.2	2.21920
6.8	1.91692	9.3	2.23001
6.9	1.93152	9.4	2.24071
7.0	1.94591	9.5	2.25129
7.1	1.96009	9.6	2.26176
7.2	1.97408	9.7	2.27213
7.3	1.98787	9.8	2.28238
7.4	2.00148	9.9	2.29253
7.5	2.01490	10.0	2.30259

Note: $\ln 10x = \ln x + \ln 10$

Appendix V
Geometry Formulas

(A = area, C = circumference, V = volume, S = surface area, r = radius, b = base, h = height)

Triangle

$$A = \frac{1}{2}bh$$

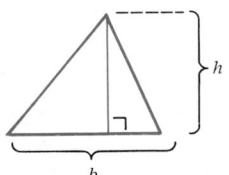

Rectangle

$$A = bh$$

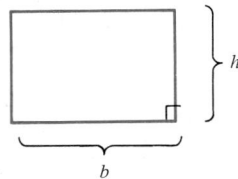

Circle

$$A = \pi r^2$$

$$C = \pi d = 2\pi r$$

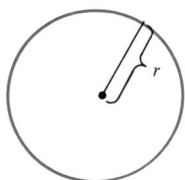

Parallelogram

$$A = bh$$

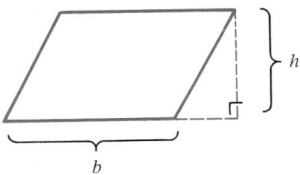

Trapezoid

$$A = \frac{1}{2}(b_1 + b_2)h$$

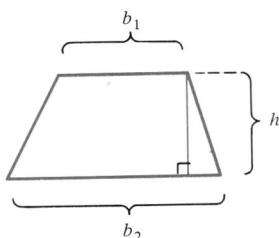

Cylinder with Parallel Bases

$$V = Bh$$

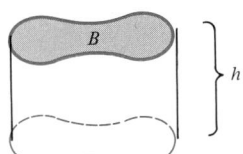

Right Circular Cylinder

$$V = \pi r^2 h$$

$$S = 2\pi rh + 2\pi r^2$$

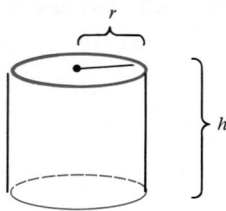

Sphere

$$V = \frac{4}{3}\pi r^3$$

$$S = 4\pi r^2$$

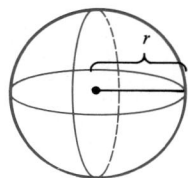

Right Circular Cone

$$V = \frac{1}{3}\pi r^2 h$$

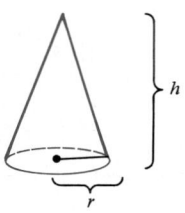

ANSWERS TO ODD-NUMBERED EXERCISES

CHAPTER 1

Exercise Set 1.1

1. Rational

3. Irrational

5. Rational

7. Rational

9. Irrational

11. $(-1, 5]$

13. $[-2, 0)$

15. $(-1, 0)$

17. True, true, true

19. $x \geq 2$

21. $x \geq -7$

23. $-1 < x < 3$

25. $x < -4$ or $x > -2$

27. $-2 < x < 0$ or $x > 1$

29. $-3 < x < 3$ and $x \neq 0$
Equivalently $-3 < x < 0$ or $0 < x < 3$

31. $x = 2/5$

33. $x = 4$ or $x = -10/3$

35. $x = 1$ or $x = -1$

37. $x = 2$ or $x = -4$

39. $x = 11/2$

41. $1 \leq x \leq 5$

43. $x > -1$ or $x < -3$

45. $2 \leq x \leq 5$

47. $x \geq 1$

49. $-4 < x < 3$

51. $x \leq -\dfrac{7}{2}$ or $x \geq \dfrac{13}{2}$

53. $x = -1$

55. x is more than 2 units, but less than 5 units, from 4.

57. $|x + 2| = 2|x - 12|$

61. False   **71.** Yes

Exercise Set 1.2

1. a. $\sqrt{13}$ **b.** $2\sqrt{2}$ **c.** $\sqrt{122}$
   d. $\sqrt{106}$ **e.** $2\sqrt{2}$ **f.** 5

3. $(0, 3 + 2\sqrt{3})$ and $(0, 3 - 2\sqrt{3})$

5. $a \in \{-1 + 3\sqrt{2}, -1 - 3\sqrt{2}, 5, 2\}$

7. $2x + y - 2 = 0$

11. (x, y) s.t. $(x + 3)^2 + (y - 1)^2 > 2$

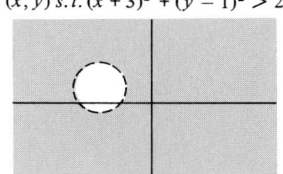

13. $\{(x, y \text{ s.t. } |x| \leq 2 \text{ and } |y| \leq 2\}$

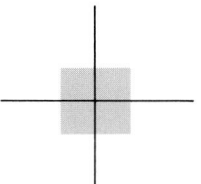

15. $x^2 + y^2 = 9$

17. $(x + 6)^2 + (y + 4)^2 = 25$

19. $(x - 4)^2 + (y + 4)^2 = 25$
   $(x + 1)^2 + (y - 1)^2 = 25$

21. Center $(-2, -1)$, radius 4

23. Center $(1, 3)$, radius $\sqrt{7}$

25. Center $(0, b)$, radius a

27. $(7, 2)$ and $(4, -1)$

31. $\{x^2 + (y - a)^2 = 1 + (a - 2)^2 \mid a \in \mathbb{R}\}$

Exercise Set 1.3

1. a. $-1/4$ **b.** $-1/7$ **c.** -1 **d.** 1

3. False

5. True

7. $b = 4$

9. $y = -2x + 5$

11. $6x - 5y = -36$

13. $y = -3x + 6$

15. $x = -3$

17. $y = -x + 2$

19. $x - 3y + 11 = 0$

21. $x - 3y = -8$

23. False

25. $m = -1$, x-intercept $= 7$, y-intercept $= 7$

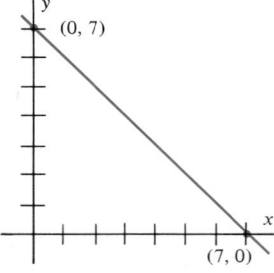

27. $m = -1$, x-intercept $= -3$, y-intercept $= -3$

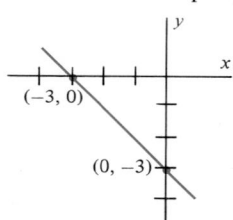

29. $m = 0$, no x-intercept, y-intercept $= 5$

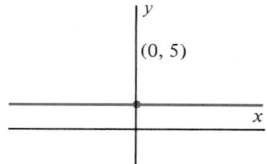

31. No slope, x-intercept $= 4$, no y-intercept

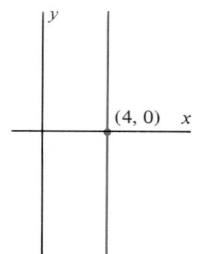

33. $(1, 2)$ **35.** $(-3, 0)$ **37.** No intersection

39. $y = -\dfrac{a}{b}x - \dfrac{c}{b}$; if $b = 0$, $x = -\dfrac{c}{a}$

41. a. Yes   **b.** Yes   **c.** No

43. a. 4 **b.** $-4, 6, 16, 0$

45. $P = 100 - \dfrac{20}{3}d$, $P = \$100$

47. $F = \dfrac{9}{5}C + 32$

49. a. $T = 2t + 22$ **b.** $72°C$, $92°C$

51. a. $1/25{,}000$

53. $y = 3x - 1$

Exercise Set 1.4

1.

$f(x)$	$f(-2)$	$f(0)$	$f(4)$	$f(5)$	$|f(3)|$
$1 - 3x^2$	-11	1	-47	-74	26
$\dfrac{1}{x+2}$	undefined	$\dfrac{1}{2}$	$\dfrac{1}{6}$	$\dfrac{1}{7}$	$\dfrac{1}{5}$
$\dfrac{(x-3)^2}{x^2+1}$	5	9	$\dfrac{1}{17}$	$\dfrac{2}{13}$	0
$\sqrt{x+4}$	$\sqrt{2}$	2	$2\sqrt{2}$	3	$\sqrt{7}$
$\dfrac{1}{\sqrt{16-x^2}}$	$\dfrac{1}{2\sqrt{3}}$	$\dfrac{1}{4}$	undefined	undefined	$\dfrac{1}{\sqrt{7}}$
$\begin{cases} 1-x, & x < -3 \\ x-1, & x > 1 \end{cases}$	undefined	undefined	3	4	2

3. Yes

5. No

7. Yes

9. $x \in (-\infty, \infty)$

11. $x \geq -4$

13. $t \in [0, \infty)$

15. $-4 \leq s \leq 4$

17. $x \neq 0$

19. $x \leq -2$ or $0 \leq x < 2$ or $2 < x$

21. False

23. $y = \dfrac{3}{2}x^2 - 4$

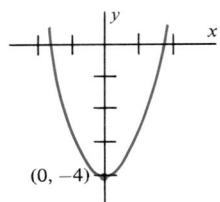

25. $y = 3\left(x - \dfrac{1}{6}\right)^2 + \dfrac{5}{12}$

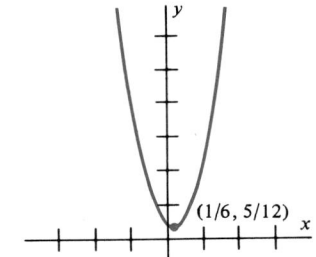

27. $y = -4(x - 3)^2 + 2$

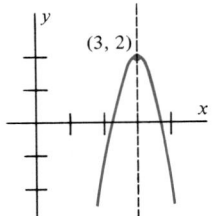

29. $x = -y^2 + 2$

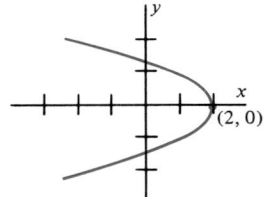

31. $x = -\dfrac{2}{3}y^2 + 1$

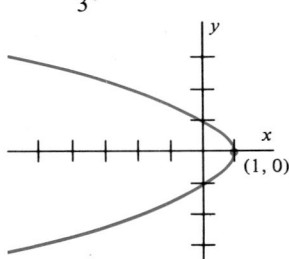

33. $x = 2(y - 1)^2$

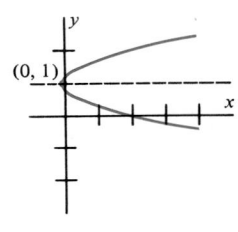

35. $(-3, -5), (2, 0)$

37.

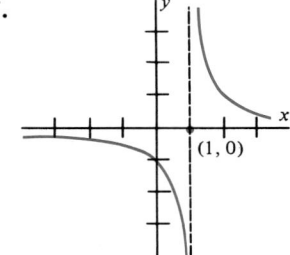

39.

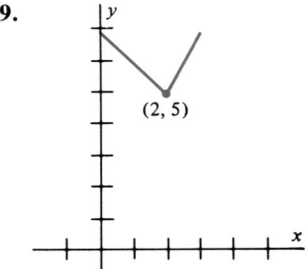

41.

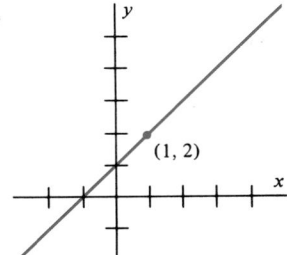

43.

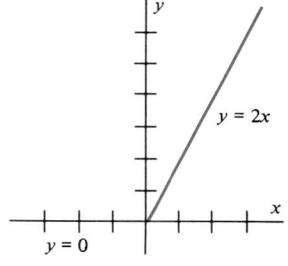

45.

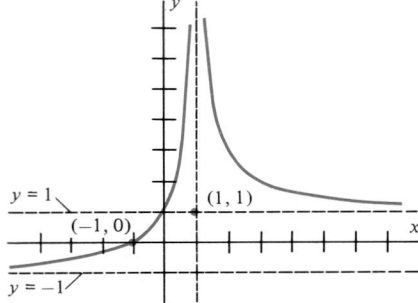

$y = 1$ $(1, 1)$
$(-1, 0)$
$y = -1$

47.

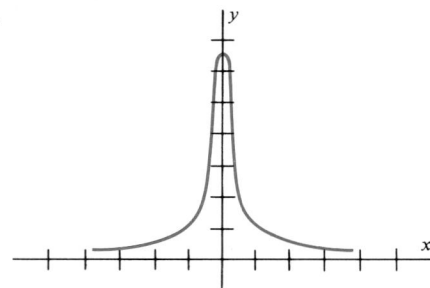

49. a. Even   **b.** Even
   c. Odd   **d.** Neither
   e. Odd   **f.** Even
   g. Even   **h.** Neither

51. $3x^3 + 1$ **53.** $(3x + 1)^3$

55. $\sqrt{3x + 1}$ **57.** $3x^{3/2} + 1$

59. a. $\left[-\dfrac{7}{2}, -\dfrac{3}{2}\right]$ **b.** $[0, 9]$
   c. $[-1, 3]$ **d.** $[6, 24]$

61. True   **63.** $f(h) = V = \pi r^2 h$

65. $f(e) = V = e^3$

67. $a(Z - b)^2$ is a parabola with vertex $(b, 0)$.

71. a. $R = [0, \infty)$ **b.** $R = [1, \infty)$
   c. $R = (0, \infty)$ **d.** $R = [0, \sqrt{6}]$
   e. $R = [0, \infty)$

Exercise Set 1.5

1. a. $\pi/6$ **b.** $5\pi/12$ **c.** $-\pi/12$

   d. $7\pi/4$ **e.** $\dfrac{2\pi}{360} \cdot (x + 30)$ **f.** $97\pi/6$

3.

	$\tan x$	$\cot x$	$\sec x$	$\csc x$
a.	$\sqrt{3}/3$	$\sqrt{3}$	$2/\sqrt{3}$	2
b.	undefined	0	undefined	1
c.	$-\sqrt{3}$	$-\sqrt{3}/3$	2	$-2/\sqrt{3}$
d.	0	undefined	-1	undefined
e.	-1	-1	$\sqrt{2}$	$-\sqrt{2}$
f.	$\sqrt{3}/3$	$\sqrt{3}$	$-2\sqrt{3}/3$	-2

5. $t = \pi/4,\ 5\pi/4$

7. $0 < t < \pi/2$

9. no t

11.

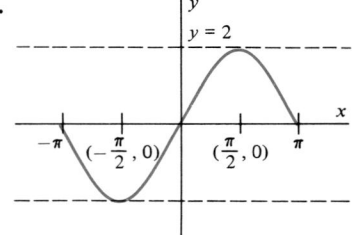

$y = 2$
$-\pi$ $\left(-\dfrac{\pi}{2}, 0\right)$ $\left(\dfrac{\pi}{2}, 0\right)$ π

13.

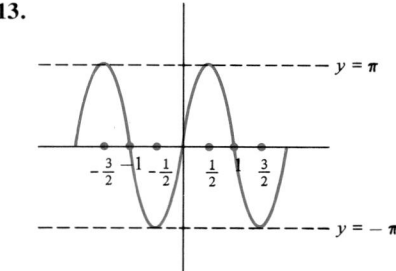

$y = \pi$
$-\dfrac{3}{2}$ -1 $-\dfrac{1}{2}$ $\dfrac{1}{2}$ 1 $\dfrac{3}{2}$
$y = -\pi$

15.

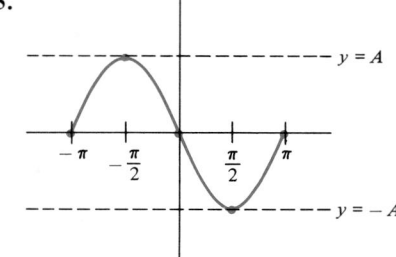

$y = A$
$-\pi$ $-\dfrac{\pi}{2}$ $\dfrac{\pi}{2}$ π
$y = -A$

17.

	DOMAIN	RANGE
$\sin x$	all x	$[-1, 1]$
$\cos x$	all x	$[-1, 1]$
$\tan x$	$x \neq \dfrac{\pi}{2} + n\pi$	$(-\infty, \infty)$
$\sec x$	$x \neq \dfrac{\pi}{2} + n\pi$	$(-\infty, -1] \cup [1, \infty)$
$\csc x$	$x \neq n\pi$	$(-\infty, -1] \cup [1, \infty)$
$\cot x$	$x \neq n\pi$	$(-\infty, \infty)$

19. Even   **21.** Even   **23.** Even

25. a)

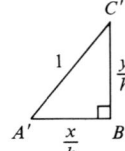

$\triangle ABC - \triangle A'B'C'$
so $\angle C'A'B' = \theta$

29. $4\pi/3$ meters

31. $50\sqrt{3}$ meters

33. 300 meters

Review Exercises—Chapter 1

1. The interval $(0, 1)$

3. **a, c.**

5. $A = (0, 2)$
$B = (0, 1] \cup (3, 4)$

7. $3 \le x \le 5$ or $-5 \le x \le -3$

9. $0 \le x \le 4$ or $-8 \le x \le -4$

11. $x = 0$ or $2 \le x \le 3$ or $-3 \le x \le -2$

13. $-2 \le x \le 8$

15. $x \ge 4$ or $x \le -8$

17. all x

19. $\dfrac{1}{6}\pi < x < \dfrac{5}{6}\pi$ or $\dfrac{7}{6}\pi < x < \dfrac{11}{6}\pi$

21. $0 < x < \dfrac{1}{2}\pi$ or $\dfrac{1}{2}\pi < x < \dfrac{7}{6}\pi$ or $\dfrac{11}{6}\pi < x \le 2\pi$

23. **d**

25. **a.** $a = 2$ **b.** $a = 10$

27. $(0, 0)$; 7

29. $(1, -3)$; 1

	Slope	y-intercept	x-intercept
31.	-1	1	1
33.	0	4	none
35.	1	-5	5

37. $5y - 2x = 9$

39. $a = 2$

41. $7x + 5y - 8 = 0$

43. $a = -\dfrac{7}{2}$, $b = 9$

45.

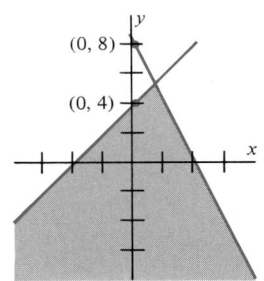

47.

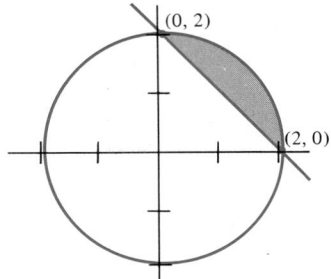

49. $-6208/999$

51. $(3, 1)$

53. $C = K - 273$

55. $\left(\dfrac{5}{2}, \dfrac{-1}{2}\right)$

57. $x \le -1$ or $x \ge 1$

59. All x

61. $x \ne \dfrac{\pi}{2} + n\pi$

63. Even

65. Even

67. $y = 2\left(x - \dfrac{1}{2}\right)^2 + 3$

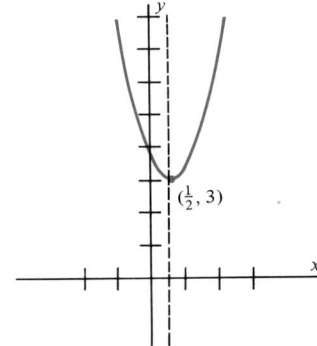

69. $x = -\dfrac{1}{2}\left(y - \dfrac{1}{2}\right)^2 - \dfrac{1}{4}$

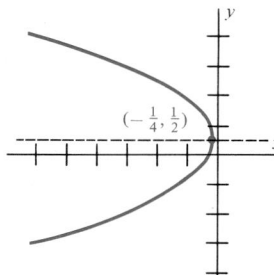

71. **a.** $2\pi/3$ **b.** $7\pi/4$ **c.** $\pi/3$ or $5\pi/3$

73. $\dfrac{1}{1 - \sin x}$

75. $\sin\left(\dfrac{1}{1 - x}\right)$

77. $\dfrac{1}{1 - \sin^3 x + \sin x}$

81. $3x - 4y + 23 = 0$

83. domain $= [0, 2\pi]$, range $= [3, 4]$

85. 50

CHAPTER 2

Exercise Set 2.1

1. 3   **3.** 12   **5.** 4

7. -8 **9.** 12   **11.** -32

13. $3a + 2b + c$ **15.** -1 **17.** $y = 2ax - a$

19. $y = 7x - 4$

21. False. The tangent to $y = x^2$ at $(0, 0)$ is horizontal.

23.

x	h	$\dfrac{(2 + h)^3 - 3(2 + h) - 2}{h}$
2	2	25
2	1	16
2	0.5	12.25
2	0.2	10.24
2	0.1	9.61
2	0.05	9.3025
2	0.01	9.0601
2	0.005	9.030025

25.

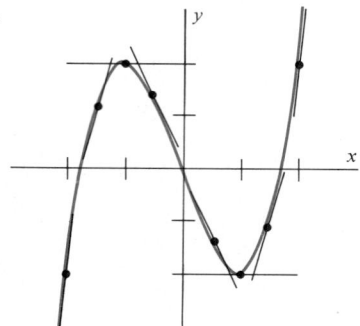

25. b–c.

x_0	-2	$-\dfrac{3}{2}$	-1	$-\dfrac{1}{2}$	0	$\dfrac{1}{2}$	1	$\dfrac{3}{2}$	2
slope	9	$\dfrac{15}{4}$	0	$-\dfrac{9}{4}$	-3	$-\dfrac{9}{4}$	0	$\dfrac{15}{4}$	9

d. The slope is 0 at the high and low points.

27. -3

Exercise Set 2.2

1. (i)   **3.** (i)   **5.** 17   **7.** 6   **9.** 1

11. -5 **13.** 0   **15.** 2   **17.** 0   **19.** -4

21. $\dfrac{2}{\pi}$ **23.** 2   **25.** -5

27. $m = -2$, $y = 4 - 2x$ **29.** $m = -4$, $y = 13 - 4x$

31. $m = -\dfrac{1}{3}$, $x + 3y = 4$ **33.** False

35. Both limits are 0.

Exercise Set 2.3

1. $\delta = \dfrac{1}{2}\epsilon$ **a.** 1   **b.** 0.2   **c.** 0.025

3. $\delta = \dfrac{1}{5}\epsilon$ **a.** 0.4   **b.** 0.2   **c.** 0.06

5. $\delta = \epsilon$ **a.** 2   **b.** 0.8   **c.** 0.05

Exercise Set 2.4

1. 2   **3.** -10 **5.** -30 **7.** $\dfrac{1}{2}\sqrt{2}$ **9.** 2   **11.** 4

13. $\dfrac{6}{17}$ **15.** $\dfrac{1}{2}$ **17.** $-\dfrac{3}{8}$ **19.** 4   **21.** 9   **23.** -18

25. $\dfrac{24}{11}$ **27.** $\dfrac{1}{2}$ **29.** $\dfrac{1}{4}$ **31.** 0   **33.** 1   **35.** 1

41. $\lim\limits_{x \to 0} (x^2 - x) = \lim\limits_{x \to 1} (x^2 - x) = 0$

43. $f(x) = g(x) = \dfrac{|x|}{x}$

45. $a = 5, -3$

Exercise Set 2.5

1. a. 2   **b.** 2   **c.** 0

3. a. 0   **b.** 2   **c.** -1

5. -1 **7.** 0   **9.** 1   **11.** 4   **13.** -3 **15.** 5   **17.** -3

19. a. 1   **b.** 1   **c.** 1

23. a. 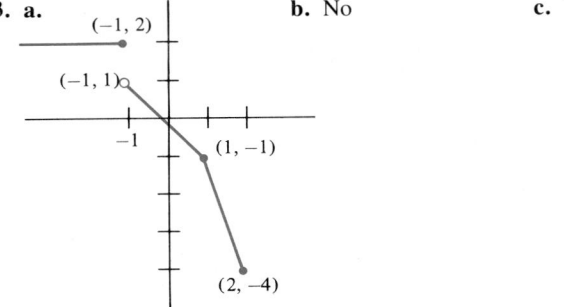 **b.** No   **c.** Yes

25. $a = \dfrac{3}{4}$, $b = \dfrac{1}{2}$

Exercise Set 2.6

1. 0   **3.** ± 2 **5.** -3 **7.** $-1, 2$

9. 0   **11.** 2   **13.** -1

15. $(-\infty, 6]$ and $[6, \infty)$ **17.** $(k\pi, (k+1)\pi)$

19. $(-7, \infty)$ **21.** $(-\infty, \infty)$

23. $\left(\left(k - \dfrac{1}{2}\right)\pi, \left(k + \dfrac{1}{2}\right)\pi\right)$

25. $\cdots, \left(-\dfrac{3}{2}\pi, -\dfrac{1}{2}\pi\right), \left(-\dfrac{1}{2}\pi, -1\right), \left(-1, \dfrac{1}{2}\pi\right),$
$\left(\dfrac{1}{2}\pi, 2\right), \left(2, \dfrac{3}{2}\pi\right), \left(\dfrac{3}{2}\pi, \dfrac{5}{2}\pi\right), \cdots$

27. $f(1) = 2$ **29.** $f(0) = 0$ **31.** $k = 3$ **33.** $k = -1$

35. 243   **37.** -1 **39.** 64   **41.** 1

45. 2   **47.** $\dfrac{1}{3}$ **49.** $\dfrac{1}{3}$

51. Positive: $(-3, 3)$; negative: $(-\infty, 3), (3, \infty)$

53. Positive: $(-2, -1), (1, \infty)$; negative: $(-\infty, -2), (-1, 1)$

55. Positive: $\left(\left(2k + \dfrac{1}{2}\right)\pi, \left(2k + \dfrac{3}{2}\right)\pi\right)$;
negative: $\left(\left(2k - \dfrac{1}{2}\right)\pi, \left(2k + \dfrac{1}{2}\right)\pi\right)$

57. $x < -4$ or $x > -2$

59. $-2 < x < 0$ or $x > 1$

61. $x < -3$ or $x > 3$

63. $-2\pi < x < -\dfrac{3}{2}\pi$ or $-\pi < x < -\dfrac{1}{2}\pi$ or
$0 < x < \dfrac{1}{2}\pi$ or $\pi < x < \dfrac{3}{2}\pi$

Review Exercises—Chapter 2

1. 4   **3.** 77   **5.** 6   **7.** -3 **9.** $\dfrac{4}{3}$

11. Does not exist.

13. 1   **15.** 4   **17.** $\dfrac{1}{3}$ **19.** $\dfrac{3}{4}$ **21.** 3

23. 0   **25.** 1   **27.** 1   **29.** 0

31. $(-\infty, 2), (2, \infty)$ **33.** $(-\infty, -2), (-2, \infty)$

35. $(-\infty, 0), (0, \infty)$ **37.** $(-\infty, \infty)$

39. $[0, 4], (4, \infty)$ **41.** $(-\infty, 1), (1, \infty)$

45. a. 0   **b.** 0   **c.** 0   **d.** 1
e. 1   **f.** 1   **g.** Yes   **h.** Yes

47. 5

55. $a = \dfrac{4}{3}$, $b = 2$

57. $a = 4, b = -2$ or $a = -3, b = 5$

61. g is discontinuous at a.

CHAPTER 3

Exercise Set 3.1

1. a. (iv)   **b.** (v)   **c.** (iii)   **d.** (i)   **e.** (ii)

3. $6x^2$ **5.** $3ax^2 + 2bx + c$

7. $-2/(2x + 3)^2$ **9.** $\dfrac{1}{2\sqrt{x + 1}}$

11. $-\dfrac{1}{2}(x + 1)^{-3/2}$ **13.** $-2x^{-3}$

15. $3(x + 3)^2$ **17.** $-\dfrac{1}{2}(x + 5)^{-3/2}$

19. $-6(x - 1)^{-3}$ **21.** $2x + 9y - 3 = 0$

23. $x + 6y = 15$ **25.** $x_1 = -\dfrac{1}{2}\sqrt{3}, x_2 = \dfrac{1}{2}\sqrt{3}$

27. $a = 3$ **29.** $a = 2, b = 2$

31. If $f(x) = \cos x$ then $f'(x) = -\sin x$

33. $x < -2$ or $x > \dfrac{2}{3}$

35. a. Yes   **b.** Yes

37. $2x + 3$

Exercise Set 3.2

1. $24x^2 - 2x$ **3.** $3ax^2 + b$

5. x

7. $2x + 1$

9. $12x^3 - 24x^2 + 12x - 16$

11. $6x^5 - 8x^3 + 2x$

13. $10x + 2 - 3x^{-2} + 8x^{-3}$

15. $6(3 - x)^{-2}$

17. $(1 - x^3)^{-2}(-x^6 + 12x^3 + 12x^2 + 4)$

19. $-15x^{-4} + 10x^{-6}$

21. $2x - 2 + 2x^{-2}$

23. $-6x^{-2} - 18x^{-3}$

25. $(t^2 + t + 1)^{-2}(1 - t^2)$

27. $(x - 1)^{-3}(-4x - 4)$

29. $8x^7 + 1 + 2x^{-3} + 9x^{-10}$

31. $(cx^2 + d)^{-2}(-acx^2 - 2bcx + ad)$

33. $-8x^{-9} + x^{-2}$

35. $2x + 1$

37. $4x^3$

39. $21x^6 + 88x^3 + 27x^2 + 7$

41. $-2(x^2 + 3x + 2)^{-2}(2x + 3)$

43. $4u^3 - 2u - 2$

47. $8s^3 - 21s^2 + 22s - 28$

49. $(x^3 + 3x^2 + 2x)^{-2}(-3x^2 - 6x - 2)$

51. $(18x^2 - 18)(2x^3 - 6x + 9)^2$

53. 132

55. $4/27$

57. $x - 2y - 1 = 0$

59. $y = 0$

61. $\left(3, \dfrac{9}{2}\right), \left(1, -\dfrac{3}{2}\right)$

63. $b = 7$

65. $\dfrac{1}{n}x^{(1/n)-1}$

Exercise Set 3.3

1. $-4 \sin x$

3. $3x^2 \tan x + x^3 \sec^2 x$

5. $3x^2 \cot x - (x^3 - 2) \csc^2 x$

7. $\sec^2 x$

9. $(1 - x) \cos x - (1 + x) \sin x$

11. $\sec x \tan^2 x + \sec^3 x$

13. $\dfrac{2 + \sin(-x) \cos x}{(2 + \sin x)^2}$

15. $\dfrac{(\cos x + \sin x)(1 + \tan x) - (\sin x - \cos x) \sec^2 x}{(1 + \tan x)^2}$

17. $\dfrac{(2x - 4 \csc^2 x)(x + \tan x) - (x^2 + 4 \cot x)(1 + \sec^2 x)}{(x + \tan x)^2}$

19. $\dfrac{(-3 \csc x \cot x)(4x^2 - 5 \tan x) - 3 \csc x (8x - 5 \sec^2 x)}{(4x^2 - 5 \tan x)^2}$

21. $y = -\pi x + \pi^2$

23. $x = 0, \pi, 2\pi, 3\pi, 4\pi$

25. $-2 \sin 2x$

27. a. $x = \dfrac{3}{4}\pi, \dfrac{7}{4}\pi.$ **b.** No x.

Exercise Set 3.4

1. a. 3 **b.** Never **c.** $[0, \infty)$

3. a. $-(1 + t)^{-2}$ **b.** Never **c.** Never

5. a. $3(t - 2)(t - 4)$ **b.** $2, 4$ **c.** $[0, 2) \cup (4, \infty)$

7. a. $\cos t - \sin t$

    b. $t = n + \dfrac{1}{4}\pi, \quad n = 0, 1, 2, \ldots$

    c. $\left(2n - \dfrac{3}{4}\pi, 2n + \dfrac{1}{4}\pi\right), \quad n = 1, 2, 3, \ldots$

9. a. $3(t - 3)(t - 1)$

    b. $1, 3$

    c. $[0, 1) \cup (3, \infty)$

11. $13/16$

13. a. 6 **b.** 3 **c.** -6

15. a. 122.5 m **b.** 5 s

17. a. $40 + 400t - 4.9t^2$ **b.** 8200 m **c.** 401 m/s

19. $\left(n + \dfrac{1}{2}\right)\pi, \quad n = 0, 1, 2, \ldots$

21. a.

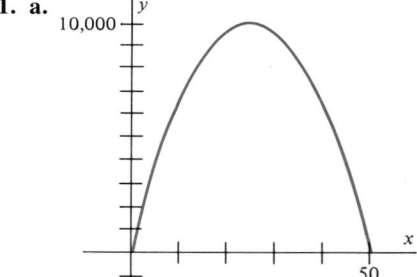

    b. $32(25 - t)$ **c.** $0 \le t \le 25$ **d.** $25 < t \le 50$
    e. $t = 25$ **f.** $10,000$ **g.** $t = 20, 30$
    f. $t = 50$

Exercise Set 3.5

1. $12x$

3. $20x^3 - 18x^{-4}$

5. $2(1 + x)^{-3}$

7. $2 \cos(2x)$

9. $(2 - \cos x)(1 + \cos x)^{-2}$

11. $-2 \sec x \tan x - x \sec x \tan^2 x - x \sec^3 x$

13. $30x(x^3 + 1)^3(7x^3 + 1)$

15. $6x^{-4} - 40x^{-6}$

17. $2 \sec^3 x - \sec x$

19. $6x \cos x - 6x^2 \sin x - x^3 \cos x$

21. $(2 + 6x^2)(1 - x^2)^{-3}$

23. $-4(x + 2)^{-3}$

25. $48(35x^4 - 30x^2 + 3)$

29. a.

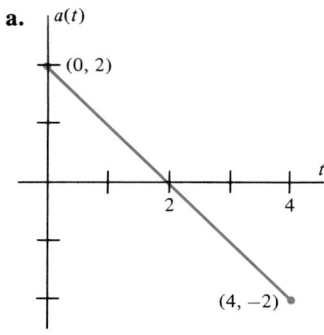

b. $v(1) > \dfrac{3}{2}$ c. $v(4) < -\dfrac{3}{2}$ d. No.

e. Let $s(t) = t^2 - \dfrac{1}{6}t^3$

31. $6 - 8x + 3x^2$

33. There is no simple pattern.

Exercise Set 3.6

1. $-6x(1 - x^2)^2$

3. $-x^{-2} + 3x^{-4}$

5. $4x^3 \sec^2(1 + x^4)$

7. $-(2 \sin x \cos x)(1 + \sin^2 x)^{-2}$

9. $-\sin(\sin x) \cos x$

11. $6x(x^2 + 4)^2$

13. $-6(\sin x + 1)(\cos x - x)^5$

15. $(x^4 - 5)^2(13x^4 - 5)$

17. $-6x(x^2 - 9)^{-4}$

19. $12 \sec^2 x(3 \tan x - 2)^3$

21. $24(x - 3)^3(x + 3)^{-5}$

23. $\cos(1 - x^2) + 2x^2 \sin(1 - x^2)$

25. $[2 \tan x \cdot \sec^2 x(1 - x) + (\tan^2 x + 1)](1 - x)^{-2}$

27. $(-11x^2 - 5x + 1)(x^2 + x + 1)^{-7}$

29. $4(ad - bc)(ax + b)^3(cx + d)^{-5}$

31. $5(\sec^3 x - 4x^2)^4(3 \sec^3 x \tan x - 8x)$

33. $3 \sin^2\!\left(x - \dfrac{\pi}{2}\right) \cos\!\left(x - \dfrac{\pi}{2}\right)$

35. $[4 - \pi + (1 - 4\pi) \cos \pi x + x \sin \pi x](4 + \cos \pi x)^{-2}$

37. $\cot\!\left(\dfrac{1 + x}{1 - x}\right) - \csc^2\!\left(\dfrac{1 + x}{1 - x}\right)\dfrac{2x}{(1 - x)^2}$

39. $6 \sec^2(6x) - 6 \sec^2 x$

43. $48x(x^2 - 3)^3[1 + (x^2 - 3)^4]^5$

45. $\dfrac{-4}{(1 - x)^2} \cos\!\left(\dfrac{1 + x}{1 - x}\right) \sin\!\left(\dfrac{1 + x}{1 - x}\right)$

47. $8x(1 + x^2)[1 + (1 + x^2)^2]$

49. Tangent: $y = 0$

51. Tangent: $y = -x$

53. Tangent: $y = 1 - \left(2 + \dfrac{\pi}{2}\right)x$

55. 300

61. Tangent: $y = \sqrt{\dfrac{1}{2}\pi}\left(x - \dfrac{1}{2}\sqrt{\pi}\right) + \dfrac{1}{2}\sqrt{2}$

63. a. $v = 20 \cos 4t$ **b.** $a = -80 \sin 4t$

Exercise Set 3.7

1. $\dfrac{4}{3}(x + 2)^{1/3}$

3. $\dfrac{1}{2}x^{-1/2} - \dfrac{1}{2}x^{-3/2}$

5. $\dfrac{2}{3}x^{-1/3} - \dfrac{2}{3}x^{-5/3}$

7. $-8x(x^2 + 1)(x^2 - 1)^{-3}$

9. $1/(6x^{2/3}\sqrt{1 + x^{1/3}})$

11. $\dfrac{-2x^{1/3}}{3(x^{1/3} + x)^2}$

13. $\dfrac{1}{2}x^{-1/2} + \dfrac{1}{4}x^{-3/4} + \dfrac{1}{8}x^{-7/8}$

15. $\dfrac{4x(x^2 + 3)}{3(x^2 + 1)^{1/3}(1 - x^2)^{7/3}}$

17. $\dfrac{1}{2}\left(\dfrac{ax^2 + b}{cx + d}\right)^{-1/2}\left\{\dfrac{acx^2 + 2adx - bc}{(cx + d)^2}\right\}$

19. $(\sin^3 x - \sqrt[3]{x})^{-4/3}\left\{(\sin^3 x - \sqrt[3]{x})\left(\dfrac{19}{12}x^{7/12} - \dfrac{5}{4}x^{1/4}\right)\right.$
$\left. -(x^{19/12} - x^{5/4})\dfrac{1}{3}\left[3 \sin^2 x \cos x - \dfrac{1}{2}x^{-1/2}\right]\right\}$

21. $\dfrac{10}{3} \sin(x^{2/3} + x^{-4/3})^{5/2} \cos(x^{2/3} + x^{-4/3})^{5/2}(x^{2/3} + x^{-4/3})^{3/2}$
$\cdot (x^{-1/3} - 2x^{-7/3})$

23. $-x/y$

25. $\sqrt{y}/\sqrt{x}$

27. $\cos^2 y$

29. $\dfrac{\sin y + y \sin x}{\cos x - x \cos y}$

31. $\dfrac{2\sqrt{xy}(y-1)-y}{x(1+2\sqrt{xy})}$

33. $\dfrac{-3x^2-2xy-y^2}{x^2+2xy+3y^2}$

35. $\dfrac{2(y-1)\sqrt{x+y}-1}{1-2x\sqrt{x+y}}$

37. $\dfrac{6x-\csc^2(x+y)}{\csc^2(x+y)-\csc^2 y}$

39. $-y/x$

41. $5x^4/4y^3$

43. $-4y^{-3}$

45. $8(2y-x)^{-3}$

47. $x+y=6$

49. $x+y-4=0$

51. $3x-5y=0$

55. $y=\sqrt{2}$
$y=-\sqrt{2}$
$x=\sqrt{2}$
$x=-\sqrt{2}$

57. $x=1$
$x=-1$

59. $y=-1$
$y=-3$
$x=0$
$x=-2$

61. $a=\pm 1$

65. b., c., d.

67. $-\dfrac{3}{4}$

23. $800\ \text{ft}^2$

25. a. 15 **b.** ≈ 15 **c.** 0.1

Exercise Set 3.9

1. 0.229309

3. 0.618034

5. 2.30278

7. -1.49535

9. 2.30278

13. 1.16556

15. $-1.46962,\ -0.203364$

Review Exercises—Chapter 3

1. $-7(3x-7)^{-2}$

3. $(3x^4+x^2)^{-2}(-36x^5+9x^4-12x^3+x^2-2x)$

5. $\dfrac{1}{2\sqrt{t}}\sin t+\sqrt{t}\cos t$

7. $(x^3+x^2+x)^{-2}(-3x^2-2x-1)$

9. $(t^{-2}-t^{-3})^{-2}(2t^{-3}-3t^{-4})$

11. $2x\sin x^2+2x^3\cos x^2$

13. $-3(x^4+4x^2)^{-4}(4x^3+8x)$

15. $(2+7x)(x^2+1)^{-3/2}$

17. $30x(x^2+1)^2[(x^2+1)^3-7]^4$

19. $(-3x^2+18x+9)/[2\sqrt{3x-9}(x^2+3)^{3/2}]$

21. $-3\cot^2 s\csc^2 s$

23. $(\cos^2 x\sqrt{1+\sin x})^{-1}\left(\dfrac{1}{2}\cos^2 x+\sin x+\sin^2 x\right)$

25. $\dfrac{9}{2}t^{7/2}-15t^{3/2}$

27. $\dfrac{1}{2}x^{-11/2}(12x^6-8x^5-4x^4-12x^3+20x^2-9)$

29. $f'=\sin x+x\cos x,\quad f''=2\cos x-x\sin x$

31. $f'=2\sec^3 x-\sec x,\quad f''=(6\sec^2 x-1)\sec x\tan x$

33. $v=2t+\dfrac{1}{2}t^{-1/2},\quad a=2-\dfrac{1}{4}t^{-3/2}$

35. $v=t/\sqrt{2+t^2},\quad a=2(2+t^2)^{-3/2}$

37. 6

39. $1/(2\sqrt{x+1})$

41. $(12x-y)/(x+8y)$

43. $1/(1+2y+3y^2)$

45. $-y/x$

47. $\sin y/(1-x\cos y)$

49. $-(y^2+x^2)/y^3=4/y^3$

51. $x+y-1=0$

53. $2x+y-1-\dfrac{\pi}{2}=0$

55. $y=3\sqrt{2}\left(x-\dfrac{\pi}{4}\right)+\sqrt{2}$

Exercise Set 3.8

1. $73/12\approx 6.0833333$

3. $31/128=0.24218750$

5. $\pi/90\approx 0.034906585$

7. $3004/300\approx 10.01333333$

9. $49/250=0.1960$

11.

Exercise	Calculator value	Relative error	% error
1	6.0827625	9.38×10^{-5}	9.38×10^{-3}
2	6.9282032	5.31×10^{-5}	5.31×10^{-3}
3	0.2425356	1.44×10^{-3}	1.44×10^{-1}
4	4.9866310	7.16×10^{-6}	7.16×10^{-4}
5	0.0348995	2.01×10^{-4}	2.01×10^{-2}
6	0.0348995	2.01×10^{-4}	2.01×10^{-2}
7	10.013316	1.70×10^{-6}	1.70×10^{-4}
8	7.8490486	1.21×10^{-4}	1.21×10^{-2}
9	0.1961161	5.92×10^{-4}	5.92×10^{-2}
10	0.5164497	6.96×10^{-6}	6.96×10^{-4}

13. $2x\sin(1-x^2)\,dx$

15. $[\sec(\pi-x)-x\sec(\pi-x)\tan(\pi-x)]\,dx$

17. $(2+x)^{-3}(-x^3-6x^2-2)\,dx$

19. 8

21. $12\pi=37.7\ \text{in}^3/\text{min}$

57. $(\sin x + x \cos x)\, dx$

59. $x(1 + x^2)^{-1/2}\, dx$

61. $\dfrac{-\sin \sqrt{x}}{2\sqrt{x}}\, dx$

63. 4.0667

65. 0.88

67. Not differentiable

69. Not differentiable

71. $0.075 = 7.5\%$

73. 2.5

75. -1.4%

77. $(1, 4),\ (-1, 4)$

79. $f = \dfrac{8}{5}x^{5/4}$

81. a. $(t - 2)^2(t - 5)$ **b.** $3(t - 2)(t - 4)$
c. $t > 5$ **d.** $t < 2$ or $2 < t < 5$
e. $t = 2, 5$ **f.** $t = 5$
g. $t = 2, 4$

85. No. $g(x) = x^2,\quad f(x) = |x|$

CHAPTER 4

Exercise Set 4.1

1. Minimum

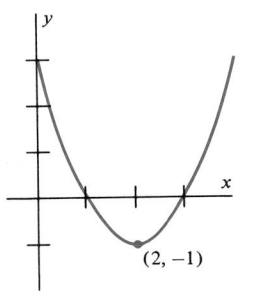

3. Minimum

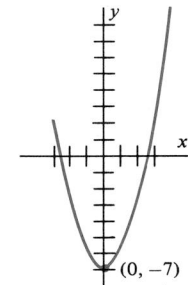

5. Minimum

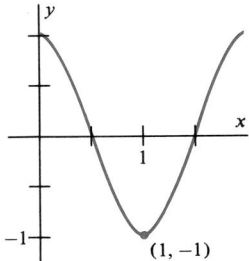

7. Maximum

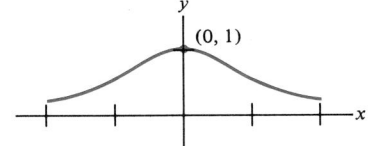

9. Minimum

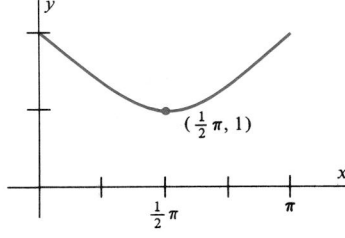

11. Does not apply since $f'(0)$ does not exist.

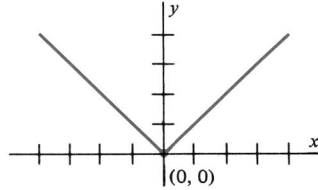

13–17. The given functions are differentiable, and hence continuous, for all x.

19. y is continuous on $[0, 8]$ and y' exists on $(0, 8)$.

25. Between 450 and 540 km.

27. No

29. 2.78851

31. 1.40649

Exercise Set 4.2

	Increasing	Decreasing
1.	$[1, \infty)$	$(-\infty, 1]$
3.	$(-\infty, \infty)$	
5.	$[\pi, 2\pi]$	$[0, \pi]$
7.	$[0, \infty)$	$(-\infty, 0]$
9.		$(-\infty, 1),\ (-1, \infty)$
11.	$(\infty, 0],\ [1, \infty)$	$[0, 1]$
13.	$[-2, \infty)$	$(-\infty, -2]$

1.

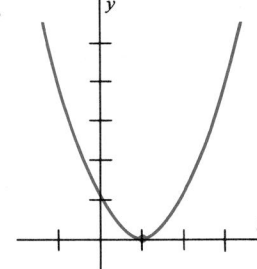

3.

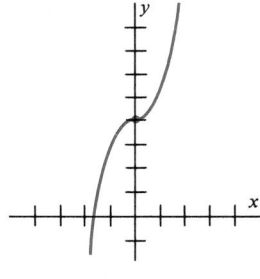

5.

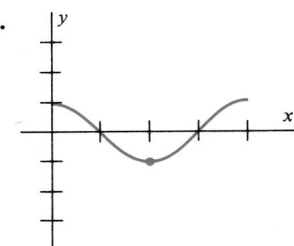

7.

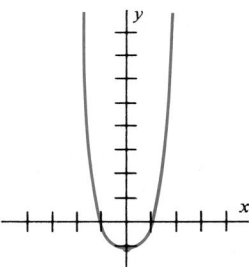

9.

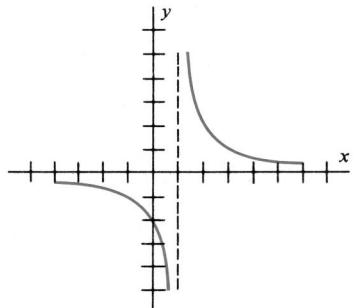

11.

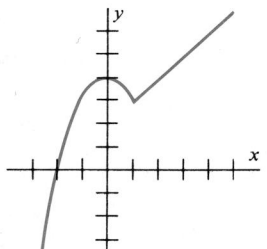

Increasing

13.

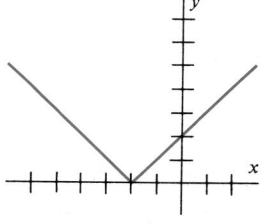

Decreasing

15. $(-\infty, -1), (-1, \infty)$

17. $(-\infty, 0]$ $\qquad$ $[0, \infty)$

19. $[-3, \infty)$ $\qquad$ $(-\infty, -3]$

21. $(-\infty, -1], [7, \infty)$ $\qquad$ $[-1, 7]$

23. $\qquad$ $(-\infty, -8), (-8, 0), (0, \infty)$

25. $\left[0, \frac{1}{2}\pi\right), \left(\pi, \frac{3}{2}\pi\right)$ $\qquad$ $\left(\frac{1}{2}\pi, \pi\right), \left(\frac{3}{2}\pi, 2\pi\right]$

27. $[-1, 2], [5, \infty),$ $\qquad$ $(-\infty, -1], [2, 5]$

29. 3 $\qquad$ **31.** -15

33. $a = 6k + 3,\quad k = 0, \pm1, \pm2, \ldots$

35. a. $[0, 3)$ to the right
$\quad$ **b.** $t = 3$

37. a. $(-\infty, -1], [1, \infty)$

Exercise Set 4.3

1. dec: $x < 2$
min: $(2, 2)$
inc: $x > 2$

3. inc: $x < -2$
max: $(-2, 9)$
dec: $x > -2$

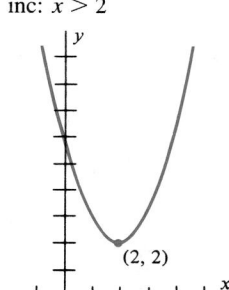

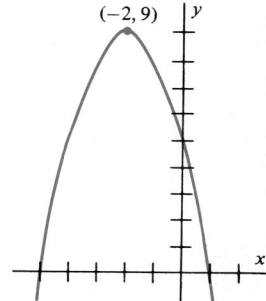

5. dec: $x < 0$
min: $(0, 4)$
inc: $x > 0$

7. dec: $x < -2$
min: $(2, 0)$
inc: $-2 < x < 0$
max: $(0, 4)$
dec: $0 > x > 2$
min: $(2, 0)$
inc: $x > 2$

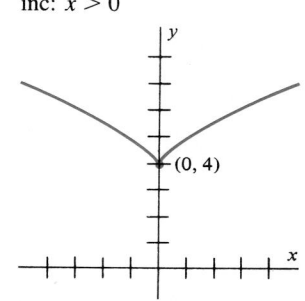

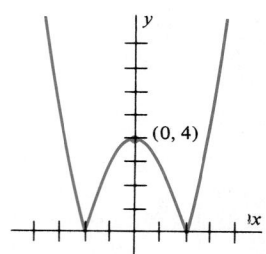

9. inc: $0 \le x \le \frac{1}{4}\pi$

max: $\left(\frac{1}{4}\pi, 1\right)$

dec: $\frac{1}{4}\pi < x < \frac{5}{4}\pi$

min: $\left(\frac{5}{4}\pi, -1\right)$

inc: $\frac{5}{4}\pi < x \le 2\pi$

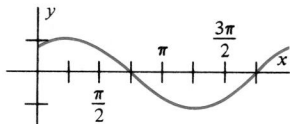

11. max: $(-2, 35)$ $\qquad$ min: $\left(\frac{1}{2}, \frac{15}{4}\right)$

13. No relative extrema

15. max: $(0, 15)$ $\qquad$ min: $(7, -504)$

17. max: $\left(1, \dfrac{1}{2}\right)$ min: $\left(-1, -\dfrac{1}{2}\right)$

19. max: $\left(\left(n + \dfrac{1}{2}\right)\pi, 1\right)$ min: $(n\pi, 0)$

21. max: $(0, 3)$ min: $(2, -1.76)$

25. 4

27. a. 10   **b.** $a > 0$

29. $\sqrt{2a}$

31. Length = width = height = $V^{1/3}$

35. Side of base = $\sqrt[3]{72} \approx 4.16$ m
Height = $96(72)^{-2/3} \approx 5.55$ m

Exercise Set 4.4

1. Concave up on $(-\infty, \infty)$

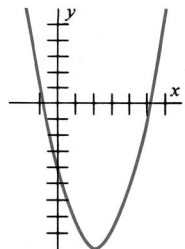

3. Concave up on $(-\infty, -1]$ and $[1, \infty)$, concave down on $[-1, 1]$, infl. pts. $(-1, 0)$ and $(1, 0)$

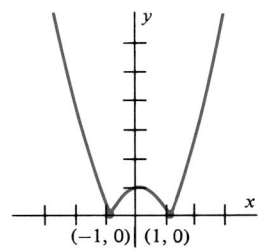

5. Concave up on $\left[\dfrac{1}{2}\pi, \dfrac{3}{2}\pi\right]$, concave down on $\left[0, \dfrac{1}{2}\pi\right]$ and $\left[\dfrac{3}{2}\pi, 2\pi\right]$, infl. pts. $\left(\dfrac{1}{2}\pi, 0\right)$ and $\left(\dfrac{3}{2}\pi, 0\right)$

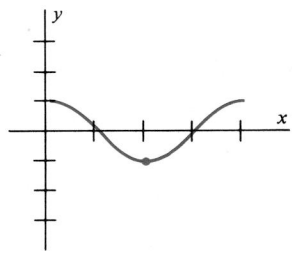

7. Concave up on $[-3, \infty)$, concave down on $(-\infty, -3]$, infl. pt. $(-3, 0)$

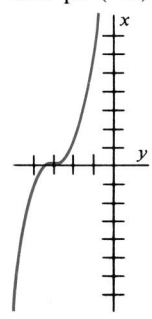

9. Concave down on $[-2, \infty)$

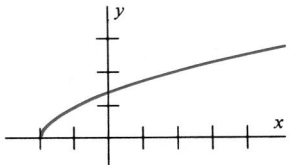

11. Concave up on $(-\infty, -1)$, concave down on $(-1, \infty)$

13. Concave up on $\left[-\dfrac{1}{2}, \infty\right)$, concave down on $\left(-\infty, -\dfrac{1}{2}\right]$, infl. pt. $\left(-\dfrac{1}{2}, 0\right)$

15. Concave up on $\left[\dfrac{1}{2}, \infty\right)$, concave down on $\left(-\infty, \dfrac{1}{2}\right]$, infl. pt. $\left(\dfrac{1}{2}, -\dfrac{7}{2}\right)$

17. Concave up on $(-\infty, -3]$ and $[2, \infty)$, concave down on $[-3, 2]$, infl. pts. $(-3, -375)$ and $(2, -70)$

19. Concave up on $[-1, 0]$ and $[0, \infty)$, concave down on $(-\infty, -1)$, infl. pt. $(-1, -3)$

21. rel. max.

23. No rel. extremum

25. rel. min.

27. rel. max.

29. max: $(0, 6)$
min: $(2, 2)$
up: $[1, \infty)$
down: $(-\infty, 1]$

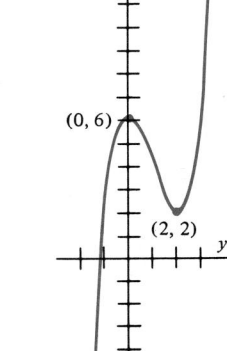

31. max: $(0, 9)$
min: $(-3, 0)$, $(3, 0)$
up: $(-\infty, -3]$, $[3, \infty)$
down: $[-3, 3]$

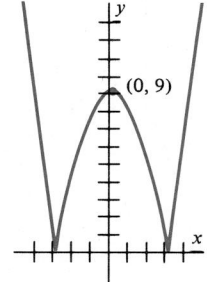

33. $a > 0$

35. Any constant function, $f(x) = c$, for any b.

37. None; one; $k - 2$.

39. $f(x) = -x^{1/3}$ is an example.

Exercise Set 4.5

1. 3/10 **3.** $-1/3$ **5.** 0 **7.** 0

9. 0 **11.** 0 **13.** 0 **15.** $+\infty$

17. $+\infty$ **19.** $+\infty$ **21.** -1 **23.** $+\infty$

25. $+\infty$ **27.** $y = 0$ **29.** $y = 2$ **31.** $y = 0$

33. $y = 3$ **35.** $a = -3$ **37.** $x = -2$ **39.** $x = 1$

41. $x = 1$

43. $r = 2/3$, $a = b$ **45.** $a = -8$, $b = 15$

Exercise Set 4.6

1.

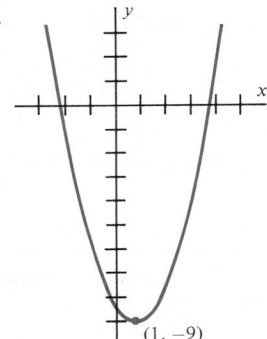

$(1, -9)$

3.

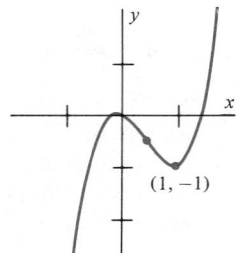

$(1, -1)$

5.

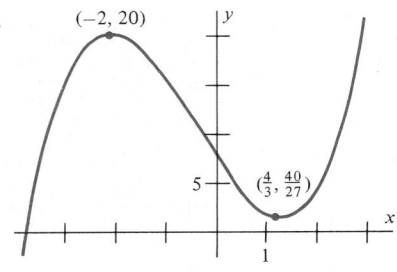

$(-2, 20)$

5 $\left(\frac{4}{3}, \frac{40}{27}\right)$

1

7.

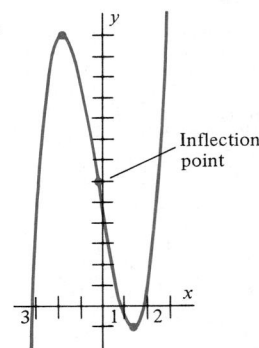

Inflection point

3 1 2

9.

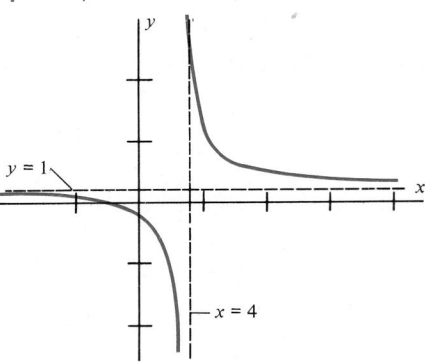

$y = 1$

$x = 4$

11.

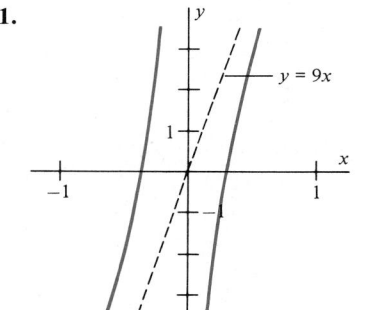

$y = 9x$

1

-1 1

13.

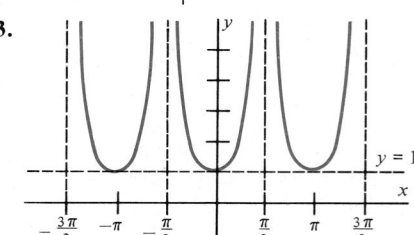

$y = 1$

$-\frac{3\pi}{2}$ $-\pi$ $-\frac{\pi}{2}$ $\frac{\pi}{2}$ π $\frac{3\pi}{2}$

15.

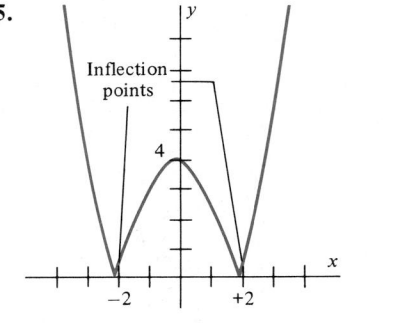

Inflection points

4

-2 $+2$

17.

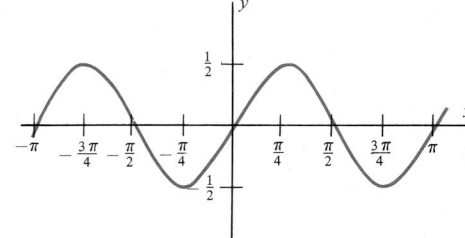

19.

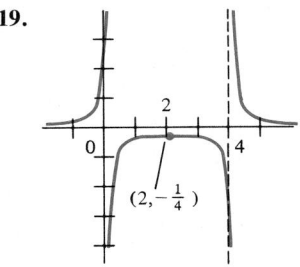

21.

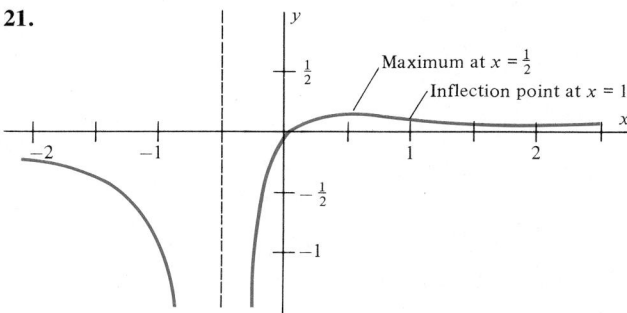

23.

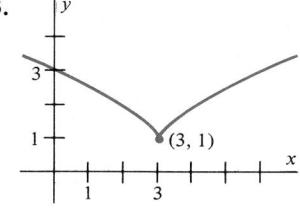

25.

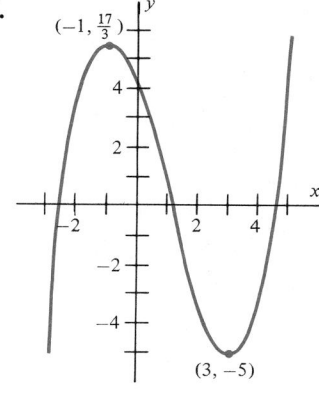

27.

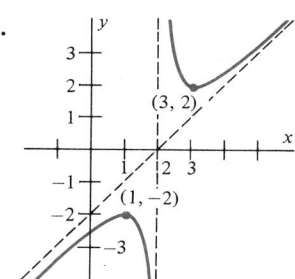

29.

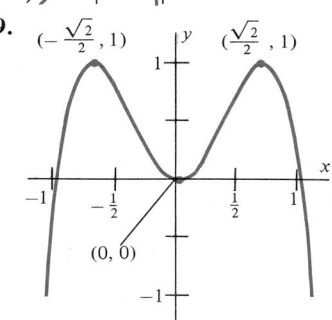

31.

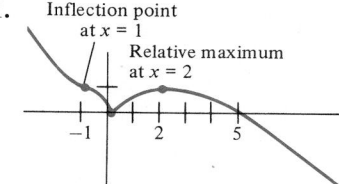

33.

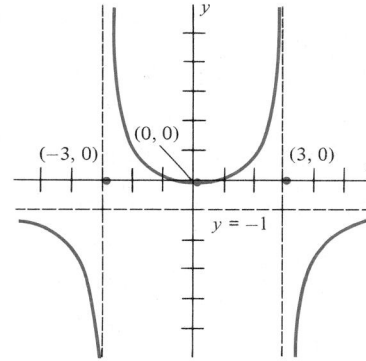

35.

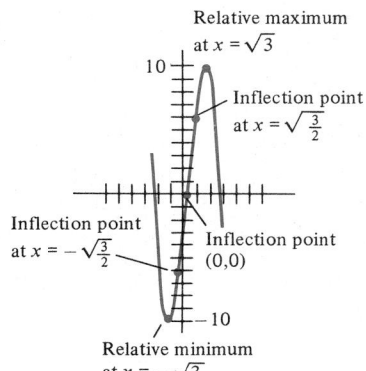

37.

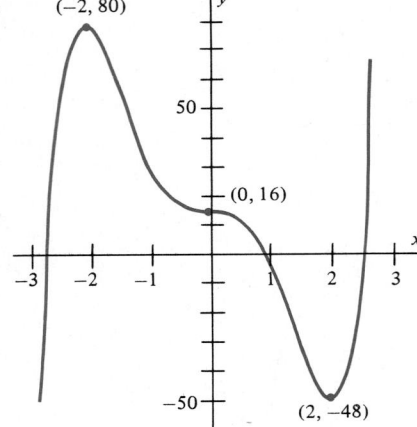

(−2, 80)

(0, 16)

(2, −48)

39.

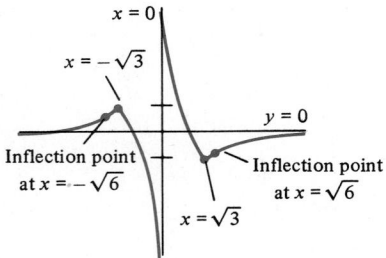

$x = 0$

$x = -\sqrt{3}$

$y = 0$

Inflection point at $x = -\sqrt{6}$

Inflection point at $x = \sqrt{6}$

$x = \sqrt{3}$

Exercise Set 4.7

	Critical numbers	Minimum	Maximum
1.	$0, \dfrac{2}{3}$	$f\left(\dfrac{2}{3}\right) = -\dfrac{4}{27}$	$f(3) = 18$
3.	2	$f(1) = f(3) = -\dfrac{1}{3}$	$f(2) = -\dfrac{1}{4}$
5.	$0, 2$	$f(-1) = f(2) = -4$	$f(0) = 0$
7.	0	$f\left(-\dfrac{\pi}{4}\right) = f\left(\dfrac{\pi}{4}\right) = 0$	$f(0) = 1$
9.	$\dfrac{\pi}{4}$	$f(\pi) = -1$	$f\left(\dfrac{\pi}{4}\right) = \sqrt{2}$
11.	0	$f(0) = 2$	$f(-2) \approx 3.59$
13.	-1	$f(-1) = 0$	$f(0) = 1$
15.	0	$f(0) = 0$	$f(2) = \dfrac{4}{3}$
17.	0	$f(0) = 0$	$f\left(\dfrac{\pi}{2}\right) = f\left(-\dfrac{\pi}{2}\right) = \dfrac{\pi}{2}$
19.	$0, \dfrac{1}{\sqrt{5}}$	$f(4) = -30$	$f\left(\dfrac{1}{\sqrt{5}}\right) = 4(5)^{-5/4} \approx 0.53$
21.	$0, 1$	$f(0) = 0$	$f(1) = \dfrac{1}{2}$

23. $b = -6$
$c = 2$

25. a. 0, 10 **b.** 5, 5 **c.** 0, 10

27. 5 m × 10 m

29. $\ell = 10$ m, $w = 20$ m

31. $\ell = 17.72$ cm
$w = 9.72$ cm
$h = 3.14$ cm

33. 6 cm × 6 cm × 2 cm

35. 3 cm wide × 2 cm high

37. $\dfrac{4}{\sqrt{3}}$ units wide × $\dfrac{8}{3}$ units high

39. $r = \dfrac{P}{4}$, $\theta = 2$ rad

41. 4 trees

43. $4\sqrt{2} \times 4\sqrt{2} = 32$

45. 24 cm × 48 cm

47. $r = 2$, $h = \dfrac{5}{3}$

49. min at $t = 3, 9$; max at $t = 0, 6$

51. $\left(\dfrac{3}{2}, \sqrt{\dfrac{3}{2}}\right)$

53. min: cut at 22.00 cm
max: cut at 50 cm

55. Cut at 28.25 cm

59. 450 m along land

61. Minimum 9/4000 at $x = 10/3$ m

63. $2\sqrt{2} \times 3\sqrt{2}$

65. $8\sqrt{3}$ ft

67. a. 0 **b.** $\dfrac{2}{3}$

71. max: (4, 7)
min: $\left(\dfrac{5}{3}, 0\right)$

73. max: $(\pi, 1)$
min: $\left(\dfrac{1}{2}\pi, 0\right)$

Exercise Set 4.8

1. $7; 2 + 2t$

3. $0; (1 - 7t^3)/2\sqrt{t}; f'(0)$ not defined

5. $\dfrac{17}{3}; \dfrac{5}{2}t^{3/2} - t^{1/2}$

7. $4\sqrt{3}; \pi \sec^2 \pi t$

9. -108

11. $\dfrac{1}{8}$

13. 4

15. $2\sqrt{3}$ cm/s

17. 144π cm^3/s

19. $40/64\pi \approx 0.199$ m/min

21. 0.8 cm/s

23. 6.25 m/min

25. 0.23 m/s

27. 4 persons per day

29. 68 km/hr

31. $\dfrac{9}{8}\sqrt{3}$ cm/s

33. a. $\left(-\dfrac{\sqrt{2}}{2}, \dfrac{\sqrt{2}}{2}\right), \left(\dfrac{\sqrt{2}}{2}, -\dfrac{\sqrt{2}}{2}\right)$

   b. $\left(\dfrac{\sqrt{2}}{2}, \dfrac{\sqrt{2}}{2}\right), \left(-\dfrac{\sqrt{2}}{2}, -\dfrac{\sqrt{2}}{2}\right)$

35. $10\pi \approx 31.416$

37. $\dfrac{25}{36\pi} \approx 0.22$ cm/s

39. $\dfrac{80\pi}{3} \approx 83.8$ m/s

41. $0.15 \dfrac{\cos \alpha}{\cos \beta}$ rad/s

43. a. $\dfrac{4\alpha}{25}$ cm/min^2

   b. -0.08α cm/s^2

Review Exercises—Chapter 4

1.

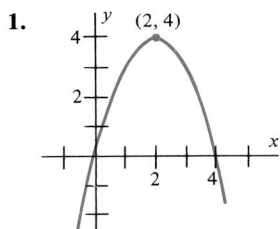

3.

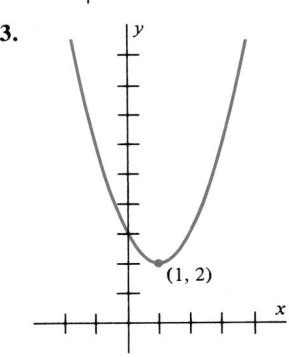

5.

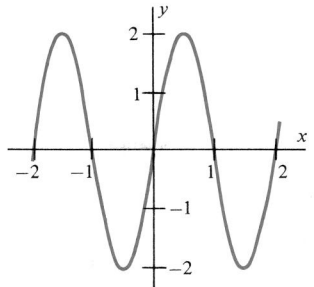

7.

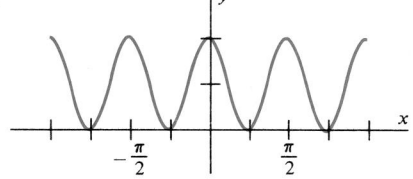

9.

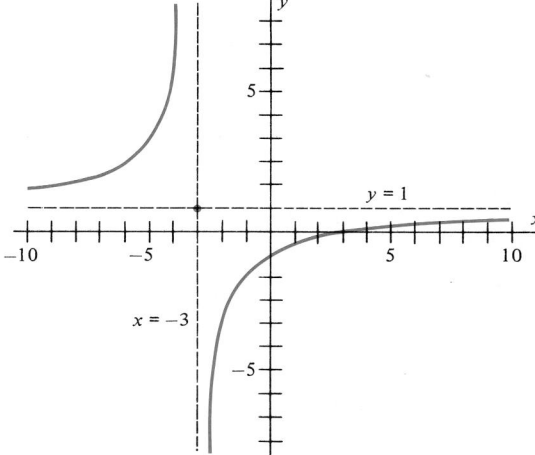

11.

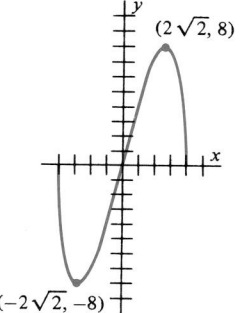

13.

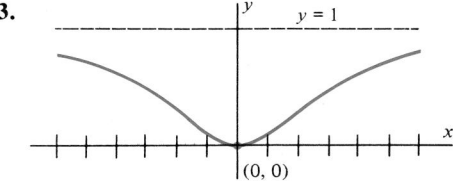

15.

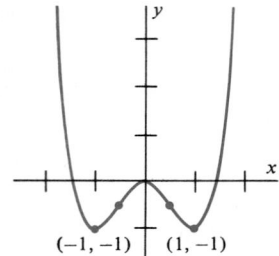

$(-1, -1)$ $(1, -1)$

17.

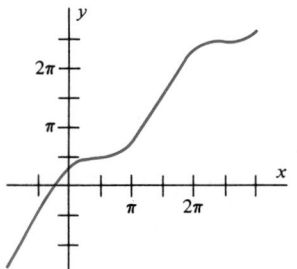

19.

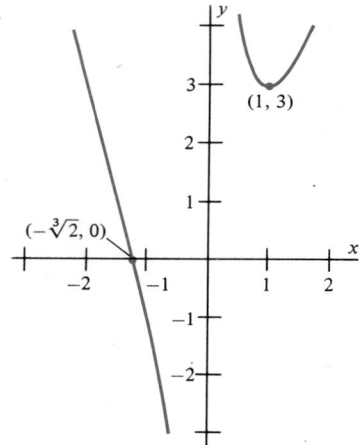

Minimum	Maximum
21. $f(-1) = f(1) = -\dfrac{1}{3}$	$f(0) = -\dfrac{1}{4}$
23. $y(-1) = -3$	$y(0) = 1$
25. $f(-1) = f(0) = 0$	$f(1) = 2$
27. $y\left(-\dfrac{1}{\sqrt{2}}\right) = -\sqrt{2}$	$y(1) = 1$
29. $f(-1) = -5$	$f(-2) = -3$
31. $y(-1) = y(1) = -1$	$y(-2) = y(2) = 8$
33. $f(-\pi) = f(\pi) = \dfrac{1}{3}$	$f(0) = 3$
35. $y(2) = \dfrac{4}{5}\sqrt{5}$	$y\left(\dfrac{1}{2}\right) = \sqrt{5}$

37. Decreasing

39.

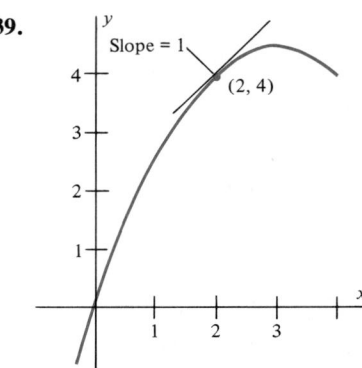

Slope = 1

$(2, 4)$

41. $(1 + 3^{2/3})^{3/2} \approx 5.41$ m

45. $f'(t) = 3(t - 2)^2 + 2 > 0$

47. $a = \dfrac{1}{9}$

49. $60\sqrt{3}$ cm²/s

51. 27.5 trees/acre

53. $\theta = 2$ rad, $r = \sqrt{S}$

55. $\dfrac{7}{4}$ m/s

57. Base = 5, height = $\dfrac{5}{2}\sqrt{3}$

59. Base = $\sqrt{2}$, height = $2\sqrt{2}$

63. 0.3 m/s

65. $\dfrac{3}{4}\sqrt{3}\ r^2$

69. a. $[-3, 5]$ **b.** Midpoint = 1

CHAPTER 5

Exercise Set 5.1

1. a. ii   **b.** i   **c.** iv   **d.** iii

3.

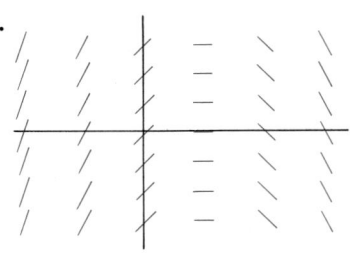

5.

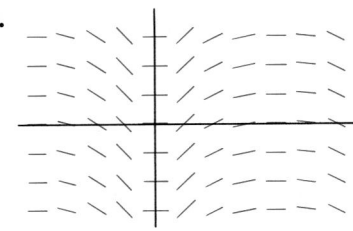

7.

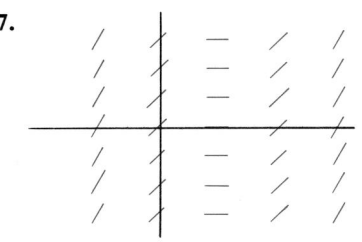

9.

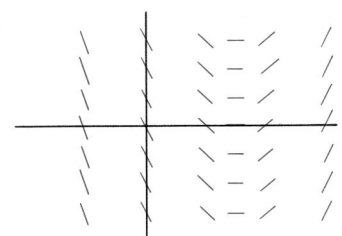

11. $x - \dfrac{1}{4}x^4 + C$

13. $\dfrac{2}{3}(x + 2)^{3/2} + C$

15. $2 \sin t + \cos t + C$

17. $\dfrac{2}{7}t^{7/2} + \dfrac{2}{3}t^{3/2} + C$

19. $\dfrac{t^4}{4} + \dfrac{2t^3}{3} + \dfrac{t^2}{2} + 2t + C$

21. $\dfrac{3}{2}x^{2/3} + C$

23. $5 \sin x + \dfrac{2}{3}x^{3/2} + C$

25. $3x^2 + 5x + 3 \csc x + C$

27. $s(t) = t + t^2$

29. $s(t) = \dfrac{1}{3}t^3 + t^2 + 2$

31. $s(t) = \dfrac{1}{2}(\sqrt{t} + 3)^4 - \dfrac{81}{2}$

33. $v(t) = 2t + 3$, $s(t) = t^2 + 3t$

35. $v(t) = \dfrac{3}{2}t^2$, $s(t) = \dfrac{1}{2}t^3 + 20$

37. $v(t) = 2t^2 + 4t + 8$, $s(t) = \dfrac{2}{3}t^3 + 2t^2 + 8t + 12$

39. $v(t) = -\cos t + 1$, $s(t) = -\sin t + t + 2$

41. $45/9.8 \approx 4.6$ s

43. a. $\sqrt{150/4.9} \approx 5.5$ s
  b. $9.8\sqrt{150/4.9} \approx 54$ m/s (downward)

45. $a = -2$ m/s^2

Exercise Set 5.2

1. $\dfrac{1}{3}(x^2 + 1)^{3/2} + C$

3. $\dfrac{1}{2}\sec(2\theta) + C$

5. $-\dfrac{1}{2}\cot(x^2) + C$

7. $-\dfrac{2}{27}(3t^3 - 9t + 9)^{3/2} + C$

9. $\tan(x - \pi) + C$

11. $\dfrac{1}{3}(2\sqrt{x} + 3)^3 + C$

13. $\dfrac{1}{3}(x^2 - 1)^{3/2} + C$

15. $-\dfrac{1}{2}\sec(\pi - x^2) + C$

17. $\dfrac{1}{21}(x^6 - 3x^4)^{7/2} + C$

19. $\dfrac{2}{3}\sqrt{t^3 + 6t} + C$

21. $-\dfrac{3}{4}(1 - 4x)^{1/3} + C$

23. $u = 3x + 4$
  a. $\dfrac{1}{6}(3x + 4)^2 + C$
  b. $\dfrac{2}{9}(3x + 4)^{3/2} + C$
  c. $-\dfrac{1}{3}\cos(3x + 4) + C$
  d. $\dfrac{1}{9}\sin^3(3x + 4) + C$

25. a. $\dfrac{1}{3}x^3 + 6x + C$

b. $u = x^2 + 6,\ \sqrt{x^2 + 6} + C$

c. $u = x^2 + 6,\ \dfrac{1}{2}\tan(x^2 + 6) + C$

Exercise Set 5.3

1. $y = \dfrac{1}{2}x^2 - x + C$

3. $y = \dfrac{1}{3}x^3 + x^{-1} + C$

5. $y = \dfrac{1}{2}x^2 - \dfrac{2}{3}x^{3/2} + C$

7. $y^2 = x^2 + C$

9. $y = \dfrac{1}{2x^2 + C}$

11. $y = \sqrt{x^2 + 15}$

13. $y = \dfrac{-1}{3(1 + x^3)} + \dfrac{5}{6}$

15. $y = \dfrac{-1}{x + \dfrac{x^3}{3} - 1}$

17. $y = \dfrac{1}{12}x^4 + 4x + 2$

19. $f(x) = \dfrac{1}{2}x^2 - \dfrac{2}{3}x^{3/2} + \dfrac{13}{6}$

21. a. $V'(t) = (t + 1)^{1/2},\ V(0) = 5$

b. $V(t) = \dfrac{2}{3}(t + 1)^{3/2} + \dfrac{13}{3}$

23. $v = \left(\dfrac{3}{2}kt + 27\right)^{2/3}$

Review Exercises—Chapter 5

1. $2x^3 - x^2 + x + C$

3. $-2\cos\sqrt{t} + C$

5. $2x^{3/2} + 6x^{1/2} + C$

7. $\dfrac{1}{2}\tan(x^2) + C$

9. $\dfrac{x^3}{3} - \dfrac{7}{2}x^2 + 6x + C$

11. $\dfrac{1}{2}\sin(1 + x^2) + C$

13. $\dfrac{4}{3}x^3 - x + C$

15. $x + \sin x + 3$

17. $\dfrac{3}{2}x^2 + 3x + 4$

19. $y = x^2 + 5$

21. 122.5 m

23. $y = x + 1$

25. $y = x + 1$

27. a. -4 **b.** to the right   **c.** $t = 1$ and $t = 3$ **d.** twice

CHAPTER 6

Exercise Set 6.1

1. 75   **3.** 45   **5.** 0   **7.** 18

9. 55   **11.** 204   **15.** 420   **17.** 275

19. $2n^3 + n^2 - n$

Exercise Set 6.2

1. 6   **3.** 40   **5.** $\dfrac{\sqrt{2}}{4}\pi$

7. 12   **9.** 85   **11.** $(2 + \sqrt{2})\dfrac{\pi}{4}$

13. a. $\dfrac{17}{2}$ **b.** $10 - \dfrac{6}{n}$

c. 10   **d.** $10 + \dfrac{6}{n}$

e. 10, 0   **f.** 10

15. 26

17. a. $\dfrac{65}{8}$ **b.** $\dfrac{17}{2} - \dfrac{3}{2n}$

c. $\dfrac{17}{2}$ **d.** $\dfrac{17}{2} + \dfrac{3}{2n}$

e. $\dfrac{17}{2},\ 0$ **f.** $\dfrac{17}{2}$

19. 7   **21.** 2

23. a. $y = \sqrt{1 - x^2}$

b. 0.56

c. 0.68

d. $\pi/4 \approx 0.785$

25. b. 0.896   **c.** 0.85

Exercise Set 6.3

1. a. -3 **b.** 1   **c.** 3

d. 6   **e.** -1 **f.** -3

3. 30   **5.** 6

7. -20 **9.** $\dfrac{-10}{3}$

11. $\dfrac{11}{6}$ **13.** $\dfrac{112}{3}$

15. -99

17. $A = 8\tfrac{1}{2}$

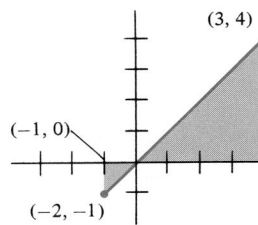

(3, 4)

(−1, 0)

(−2, −1)

19. $A = 5$

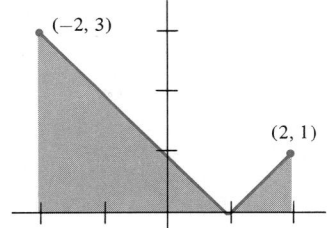

21. $2\pi + 4$ **23.** $18\pi + 18$

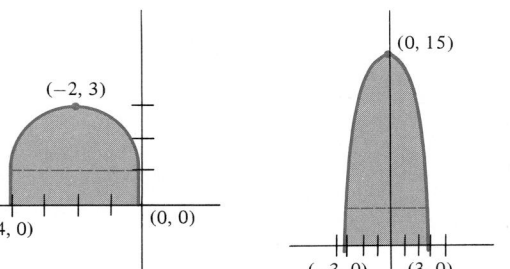

25. 22.97 **27.** 0.345 **29.** 265.64

Exercise Set 6.4

1. -4

3. $2 - \dfrac{8}{5}\sqrt{2}$

5. $\dfrac{3}{4}$

7. $\dfrac{\pi}{4} - \dfrac{1}{2}$

9. $\dfrac{7}{3}$

11. 5

13. $\sqrt{3}$

15. $\dfrac{1366}{15}$

17. $\dfrac{-1}{8}$

19. $\dfrac{1}{2} - \dfrac{\sqrt{2}}{4}$

21. 0

23. $\dfrac{1}{3a}[(a + b)^{3/2} - b^{3/2}]$

25. 0

27. $2 + 2\cos\dfrac{\pi}{\sqrt{2}}$

29. $\dfrac{6}{49}$

31. $\dfrac{64}{3}$

33. $\dfrac{1}{12}$

35. $\dfrac{\pi}{8} - \dfrac{1}{4}$

37. 2

39. 12

41. $2\sqrt{3}$

43. $\dfrac{44}{3}$

45. $\dfrac{27}{2}$

Exercise Set 6.5

1. $\dfrac{11}{6}$

3. 0

5. $\dfrac{4}{\pi}(\sqrt{2} - 1)$

7. $\dfrac{12}{5}$

9. $\dfrac{1}{2}x^2 - \dfrac{1}{2}$; x

11. $\sin x$; $\cos x$

13. $\dfrac{1}{3}ax^3 + \dfrac{1}{2}bx^2 + cx$; $ax^2 + bx + c$

15. $x^2 \sin x^2$

17. $-x^3 \cos^2 x$

19. $3\sqrt{1 + \sin 3x}$

21. $\cos^3(x + 1) - 2x\cos^3(x^2 + 1)$

23. $\displaystyle\int_3^{x^2} (t^2 + 1)^{-3}\, dt + 2x^2(x^4 + 1)^{-3}$

27. 4

31. No

33. $f(x) = x$, $g(x) \equiv \dfrac{2}{3}$ on $[0, 1]$

Exercise Set 6.6

1. 14

3. $2\sqrt{3} - 2$

5. $\dfrac{5}{48}$

7. $\dfrac{52}{3}$

9. 6

11. $\dfrac{80}{3}$

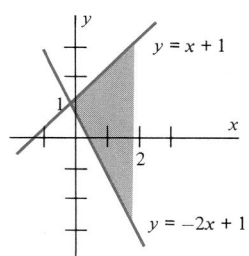

13. $4\sqrt{2}$

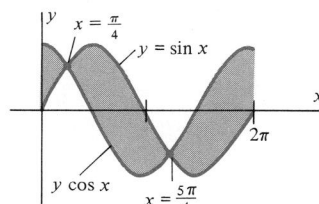

15. $\dfrac{52}{3}$

17. $\dfrac{44}{3}$

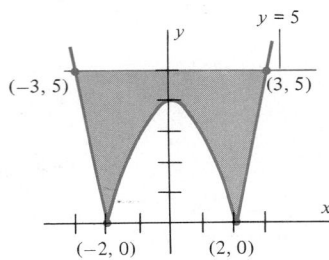

19. 1

23. $\dfrac{1}{12}$

27. $\dfrac{16}{3}$

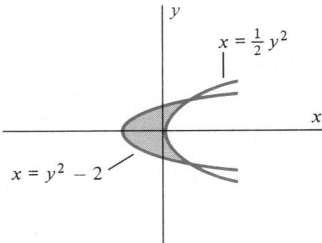

29. $\dfrac{8}{15}$

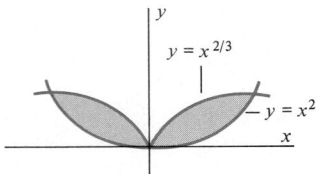

31. $\dfrac{4}{5}$

35. a. 8   **b.** 22

Exercise Set 6.7

1. 28

5. 1.3253

9. $\dfrac{\pi}{9}(4 + \sqrt{3})$

13. 6.5998

21. $\dfrac{125}{6}$

25. $\dfrac{125}{6}$

33. 3

3. $\dfrac{\pi + \sqrt{2}\pi}{4}$

7. 1.1051

11. 0.7842

15. 0.7853983

17. 6.5846

19.

n	s
5	0.6645375
20	0.6656965
100	0.6657707
250	0.6657731

21. a. 2.4565
b. 1.1713

Review Exercises—Chapter 6

1. a. $\dfrac{33}{4}$

b. $\dfrac{51}{4}$

5. a. -0.91
b. 0.91

9. 18

13. $\dfrac{9}{20}$

17. $\dfrac{200}{3}$

21. $\dfrac{8}{3}$

25. $1 - \dfrac{\sqrt{2}}{2}$

29. 28,542,630

33. $\dfrac{3}{4}$

37. $\dfrac{1}{30}$

41. $\dfrac{-2}{9}$

45. $\dfrac{16}{3}$

49. $\dfrac{7}{3}$

53. $\dfrac{26}{3}$

57. $\dfrac{243}{5}a^2 + 54a + 27$

3. a. 1.3949

b. 1.7282

7. 8

11. $\dfrac{-1}{15}$

15. $\dfrac{7}{2}$

19. $\dfrac{14}{3}$

23. $\dfrac{196}{3}$

27. $\dfrac{1}{2}$

31. 2

35. 1

39. -286.2381

43. $\dfrac{10}{3}$

47. $\dfrac{83}{15}$

51. $\dfrac{23}{3}$

55. $\dfrac{4}{3}\sqrt{2}$

59. $\dfrac{32}{3}$

61. 8

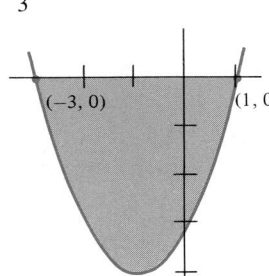

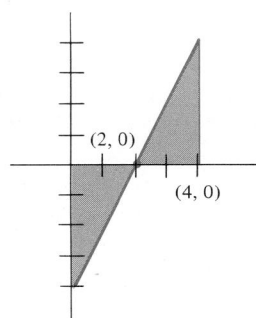

63. 8

65. $\dfrac{32}{3}$

67. $\dfrac{125}{6}$

69. $2\sqrt{2} - 2$

71. $\dfrac{9}{2}$

73. $\dfrac{8}{15}$

75. $\dfrac{4}{3}$

77. 31/2

79. $1/\pi$

81. $\dfrac{\sqrt{5}}{2} - \dfrac{1}{2}$

83. 0

85. 1/2

87. a. 0   **b.** $x^2\sqrt{1 + x}$
c. 18   **d.** $4x^2\sqrt{1 + 2x}$

89. False   **91.** True   **93.** $-2\sec(2x)$

95. 18

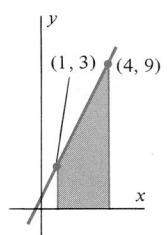

97. $\dfrac{9\pi}{4} + 3$

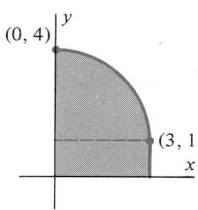

99. -4π

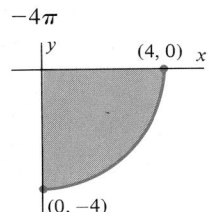

101. 2.2   **103.** 1.4054   **105.** 0.8660

CHAPTER 7

Exercise Set 7.1

1. 117π

3. $\dfrac{\pi^2}{2}$

5. $\pi\left(1 - \dfrac{\pi}{4}\right)$

7. $\dfrac{7\pi}{8}$

9. $\dfrac{128\pi}{15}$

11. π

13. 8π

15. $\dfrac{64\pi}{5}$

17. $\dfrac{3\pi}{2}$

19. $\dfrac{4\pi r^3}{3}$

21. $\dfrac{512}{3}$

23. $\dfrac{2048}{15}$

25. $\dfrac{1}{3}\pi h(R^2 + rR + r^2)$

27. $\dfrac{5}{32\pi}$ m/min

29. $\dfrac{136\pi}{15}$

31. $\dfrac{225}{2}\pi$

33. $V = \dfrac{4}{3}\pi a^2 b$

35. $\dfrac{2}{3}r^3 \tan\theta$

37. ≈ 4.16

39. ≈ 118.24

Exercise Set 7.2

1. $\dfrac{\pi}{3}$

3. $\dfrac{484}{5}\pi$

5. $\dfrac{1024}{15}\pi$

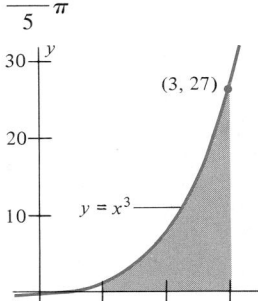

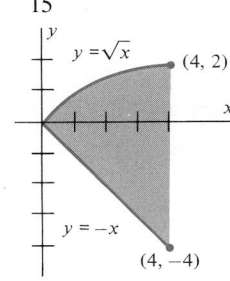

7. 9π

9. $\dfrac{\pi^2}{2} - \pi$

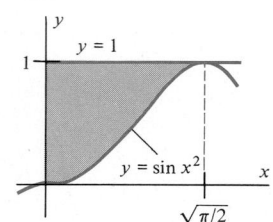

11. 18π

13. $24\pi - \dfrac{16\sqrt{2}}{9}\pi$

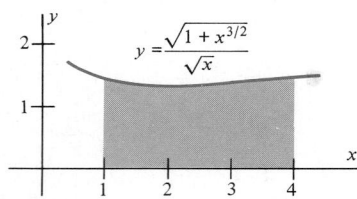

15. $\dfrac{1088}{15}\pi$

17. $\dfrac{49}{30}\pi$

19. $\dfrac{8}{15}\pi$

21. 56π

23. $4\pi^2$

25. 19.74

Exercise Set 7.3

1. $5\sqrt{5}$

3. $\dfrac{8}{27}(10\sqrt{10} - 1)$

5. $\dfrac{2}{3}(5\sqrt{5} - 2\sqrt{2})$

7. 12

9. $\dfrac{87}{8}$

11. $\dfrac{14}{3}$

13. $\dfrac{\pi}{27}(730\sqrt{730} - 1)$

15. $\dfrac{8\pi}{3}(5\sqrt{5} - 2\sqrt{2})$

17. $\dfrac{112}{3}\pi$

19. 151.63

21. $n = 4;\ S \approx 40.35$

23. 2.9579

Exercise Set 7.4

1. 18

3. 2

5. $\dfrac{155}{3}$

7. $\dfrac{2}{\pi}$

9. $\dfrac{1}{3}$

11. 10

13. 1

15. $\dfrac{1}{3}$

17. a. -6 **b.** $\dfrac{32}{3}$ **c.** 1   **d.** $-\dfrac{36}{5}$

19. a. $-\dfrac{1}{3}t^3 + \dfrac{5}{2}t^2 - 6t + 4$ **b.** $-2t + 5$

   c. $t = 2, 3$ **d.** $2 < t < 3$

   e. $0 \le t < 2$ or $t > 3$ **f.** $\dfrac{29}{6}$

   g. $\dfrac{-9}{2}$

21. 800 m

23. 17.1 km

25. distance $>$ displacement

Exercise Set 7.5

1. 421.2 lb

3. 117.66 lb

5. 341,000 lb

7. 665.6 lb

9. 571.9 lb

11. 3.3 lb

13. Equal force acting on opposite side of rectangular plate

15. 157.083 lb

17. 2995.2 lb

Exercise Set 7.6

1. 5 ft-lb

3. 15 ft-lb

5. 25 ft-lb

7. True

9. a. 5 lb/ft
   b. 5 lb
   c. 22.5 ft-lb

11. $74,880\pi$ ft-lb

13. $1,248,000\pi$ ft-lb

15. $74,880\pi$ ft-lb

17. 3833.9π ft-lb

19. 400 ft-lb

21. a. $\dfrac{160}{q^2}$
   b. 40 N-m

23. 0.758 K

Exercise Set 7.7

1. $\dfrac{100}{39}$

3. 4.8 ft

9. $x_3 = 2,\ y_3 = \dfrac{-5}{3}$

11. $\left(\dfrac{3}{5}, \dfrac{12}{35}\right)$

13. $\left(\dfrac{4}{3}, \dfrac{4}{3}\right)$

15. $\left(0, \dfrac{8}{3\pi}\right)$

19. $\left(\dfrac{4}{5}, \dfrac{2}{7}\right)$

21. $\left(\dfrac{8}{3}, 0\right)$

23. $\left(0, \dfrac{4}{\pi}\right)$

25. $\left(\dfrac{49}{22}, \dfrac{933}{220}\right)$

45. a. $\dfrac{32\pi}{3}$ **b.** $8\pi\left(\dfrac{4}{3} + \pi\right)$ **c.** $8\pi\left(3\pi - \dfrac{4}{3}\right)$

27. $\dfrac{115}{18}$

29. On the axis, 2.5 cm from the base

33. $\left(0, \dfrac{54}{55}\right)$

CHAPTER 8

Exercise Set 8.1

1. 3   **3.** 1/3   **5.** −3

7. 3/2   **9.** 0   **11.** 2

13. 1/65,536   **15.** $3^{1/6}$ **17.** 9

Exercise Set 7.8

1. $90\pi^2$

3. $\dfrac{3\pi}{10}$

19. $(y + 2)/3$ **21.** $\dfrac{1}{y} - 2$ **23.** $\dfrac{3y}{y - 1}$

5. $8\pi A$

7. 19,968 lb

25. $y^2 - 8y + 17$ **27.** $(y - 3)^{3/5}$

9. 1291.65 lb

11. $4\pi^2$

31. a. $\sqrt{3 - y}$ **b.** $\sqrt{y} - 1$ **c.** $\sqrt{y} - 2$ **d.** $\sqrt{y + 4} + 1$

33. a. 12   **b.** $y^3 - 5$ **c.** $3y^2$ **d.** 12

Review Exercises—Chapter 7

1. $\dfrac{4}{3}\pi r^3$

3. $\dfrac{3\pi}{4}$

35. a. $\dfrac{1}{4}\pi$ **b.** 1   **c.** 2   **d.** $\dfrac{2}{3}\sqrt{3}$

5. $\dfrac{3\pi^2}{4}$

7. $3\pi\sqrt[3]{4}$

37. a. 0   **b.** 1   **c.** 1   **d.** $\dfrac{1}{4}$

43. $I = I_0 10^{kx}$ **45.** False if $0 < a < 1$.

9. $\pi(\sqrt{2} - 1)$

11. $\pi\left(\dfrac{12}{7}\sqrt[3]{2} + \dfrac{48}{5}\sqrt[3]{4} + 14\right)$

13. $\dfrac{8}{27}(10\sqrt{10} - 1)$

15. $\dfrac{22\sqrt{22} - 13\sqrt{13}}{27}$

Exercise Set 8.2

1. a. 1/2   **b.** 2   **c.** $\sqrt{2}$ **d.** 0
   e. e^2 **f.** 2   **g.** e^2 **h.** 2

17. $\dfrac{1}{3}Ah$

19. 31.25%

3. True   **5.** $\ln x + 1$

21. The exact answer is 14.424 to 3 places.

7. $\dfrac{1}{2} \cdot \dfrac{3x^2 - 1}{x^3 - x}$ **9.** $\dfrac{\cos(\ln x)}{x}$

23. 980,177 ft-lb   **25.** 542,626 ft-lb

27. 5000 ft-lb   **29.** 204,472.32 lb

11. $\dfrac{\ln x}{(1 + \ln x)^2}$ **13.** $\dfrac{162(\ln \sqrt{x})^3}{x}$

31. 90.61 lb   **33.** 6.136 cm

35. $\left(-\dfrac{3}{4}, 1\right)$

15. $\dfrac{\dfrac{b}{a + bt}\ln(c + dt) - \dfrac{d}{c + dt}\ln(a + bt)}{\ln^2(c + dt)}$

37.

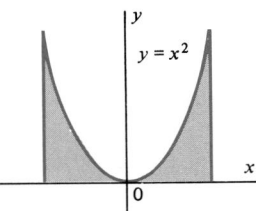

17. $-\tan x$ **19.** $\dfrac{(\ln t - 1)^2}{(1 + \ln^2 t)^2}$

21. $\dfrac{x}{y}$

39. (0, 6.4)

41. $\left(3, \dfrac{18}{5}\right)$

23. $\dfrac{xy - xy \ln y - y^2}{x^2 + xy \ln x}$

43. $\left(3, \dfrac{67}{15}\right)$

25. $-\dfrac{1}{2x}$

27.

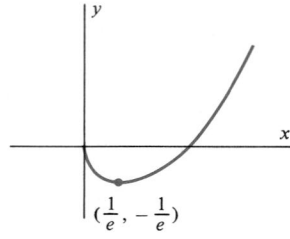

$\left(\frac{1}{e}, -\frac{1}{e}\right)$

29.

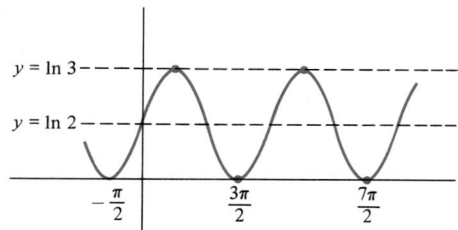

$y = \ln 3$

$y = \ln 2$

$-\frac{\pi}{2}$ $\frac{3\pi}{2}$ $\frac{7\pi}{2}$

31.

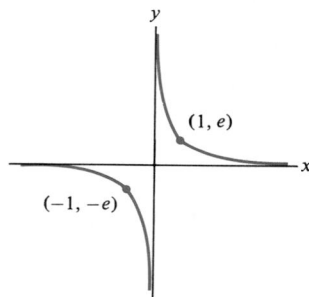

$(1, e)$

$(-1, -e)$

33. $-\ln|1 - x| + C$

35. $-\dfrac{1}{2}\ln|x^2 - 2x| + C$

37. $\dfrac{1}{3}\ln|x^3 + 9x| + C$

45. $\dfrac{x^4}{4} - \dfrac{x^3}{3} + 2x^2 - 3x + 4\ln|x + 1| + C$

47. $\ln 2$

49. $\left[\dfrac{1}{2}\ln|\ln x^2|\right]_e^{e^2} = \dfrac{1}{2}\ln 2$

51. $2\ln 3 - 3\ln 2$

55. a. $\ln(1 + \epsilon) \approx \ln 1 + \dfrac{\epsilon}{1} = \epsilon$

b. $\ln(0.8) = \ln(1 + (-0.2)) \approx -0.2$
$\ln(0.95) = \ln(1 - 0.05) \approx -0.05$
$\ln(1.05) = \ln(1 + 0.05) \approx 0.05$
$\ln(1.2) = \ln(1 + 0.2) \approx 0.2$

57. $\pi \displaystyle\int_3^5 \dfrac{2}{x - 1}\,dx = 2\pi\ln 2$

39. $\ln|\ln x| + C$

41. $\dfrac{1}{3}(\ln x)^3 + C$

43. $-2\ln|1 - \sqrt{x}| + C$

53. $\displaystyle\int_3^4 \dfrac{dx}{x - 2} = \ln 2$

59. $\dfrac{1}{e - 1}$

61. $y = \ln Cx$

63. $y = \ln|\sec x| + C$

65. $y = \dfrac{1}{C - \dfrac{1}{2}\ln(2x + 1)}$

67. a. $\dfrac{dP}{dT} = \dfrac{2000P}{T^2}$ **b.** $\dfrac{dT}{dP} = \dfrac{T^2}{2000P}$

Exercise Set 8.3

1. a. 2   **b.** $\dfrac{1}{4}$ **c.** $\dfrac{x}{y}$ **d.** $-x^2$

e. $\sqrt{x}$ **f.** 2^x **g.** $\dfrac{1}{x}$ **h.** x^4

i. $x^2 + \ln x$ **j.** $\dfrac{1}{x} \cdot e^x$

3. $y = x - \ln(x + 3)$ **11.** $e^x(\sin x + \cos x)$

5. $3e^{3x}$ **13.** $\dfrac{e^x}{e^x + 1} - \dfrac{1}{x + 1}$

7. $\dfrac{e^{\sqrt{t}}}{2\sqrt{t}}$ **15.** $-6x \cdot e^{x^2} \cdot (2 - e^{x^2})^2$

9. $(2x - 1)e^{x^2 - x}$ **17.** $\dfrac{1}{2} \cdot (e^x - e^{-x})$

19. $\dfrac{1}{2}\dfrac{e^{\sqrt{x}}}{\sqrt{x}} \cdot \left(\ln\sqrt{x} + \dfrac{1}{\sqrt{x}}\right)$ **21.** $\dfrac{1 - xy}{x^2}$

23. $\dfrac{1}{e^y(x + 2y) - 2}$ **25.** True

27. True   **29.** $-e^{-x} + C$

31. $\dfrac{1}{2}e^{x^2 + 3} + C$ **33.** $\dfrac{1}{8}(1 + e^{2x})^4 + C$

35. $-e^{1/x} + C$ **37.** $\ln(1 + e^x) + C$

39. $\dfrac{-1}{3 + e^x} + C$ **41.** $2e^{\sqrt{x}} + e^{2\sqrt{x}} + C$

43. $\dfrac{1}{e} - \dfrac{1}{4}$ **45.** $e - 1$

47. 2   **49.** $e^2 - 1$

51. max: $(-1, 1)$, $(1, 1)$ min: $(0, 0)$

53. a. $\left(-\dfrac{1}{2}, \infty\right)$ **b.** $\left(-\infty, -\dfrac{1}{2}\right)$

c. min: $\left(-\dfrac{1}{2}, -\dfrac{1}{2}e^{-1}\right) \approx \left(-\dfrac{1}{2}, -0.184\right)$

d. $(-1, \infty)$ **e.** $(-\infty, -1)$
f. $(-1, -e^{-2}) \approx (-1, -0.135)$

55. $\dfrac{3}{4}$

57. $y = Ce^{x^2}$

59. $(-1, -e^{-1})$

61. $\displaystyle\int_0^2 \pi(e^{2x})^2 \, dx = \dfrac{\pi}{4}(e^8 - 1)$

63. 0

65. -4

69. $0.407263,\ 0.933326$

71.

x	e^x	$P_1(x)$	$P_2(x)$	$P_3(x)$	$P_4(x)$
1	2.71828	2	2.5	2.66667	2.70833
0.5	1.64872	1.5	1.625	1.64583	1.64844
2	7.38906	3	5	6.3333	7
-2	0.135335	-1	1	-0.3333	0.3333

Exercise Set 8.4

1. a. $e^{\ln 2x}$ **b.** $e^{3\ln \pi}$ **c.** $e^{\ln 7 \cdot \ln x}$
d. $e^{\sqrt{2}\ln 4}$ **e.** $e^{\ln 3 \cdot \sin x}$ **f.** $e^{(2-x)\ln 2}$

3. $10^a = e^{a \ln(10)},\ e^b = 10^{b\,\log_{10}(e)}$

5. $y' = (\ln 3) \cdot 3^x$

7. $\dfrac{2}{(2x-1)\ln 10}$

9. $\dfrac{1}{x \cdot \ln 10 \cdot \ln x}$

11. $2t \log_2 t + \dfrac{t}{\ln 2}$

13. $\ln \pi \cdot \pi^x + \pi x^{\pi - 1}$

15. $x^x \cdot (\ln x + 1)$

17. $x^{\sqrt{x}-1/2}\left(1 + \dfrac{\ln x}{2}\right)$

19. $(\cos x)^{\sin x}\left[\cos x \cdot \ln(\cos x) + \dfrac{\sin^2 x}{\cos x}\right]$

21. $\dfrac{5^x}{\ln 5} + C$

23. $\dfrac{2}{\ln \pi}\pi^{\sqrt{x}} + C$

25. $\dfrac{1}{1 + 2\ln a} \cdot x^{(2\cdot\ln a + 1)} + C$

27. $\dfrac{40}{\ln 3}$

29. True

33. e^2

35. $\dfrac{4\pi}{\ln 3}$

37. $y = \dfrac{x-1}{\ln 10}$

39. $5000(\ln 2)$ dollars/year

41. $\dfrac{1}{3}\left(\dfrac{1}{x+2} - \dfrac{1}{x+3}\right)\sqrt[3]{\dfrac{x+2}{x+3}}$

43. $\left[\dfrac{6x}{x^2+2} + \dfrac{5}{x-1} - \dfrac{1}{x} - \dfrac{1}{2x+2} - \dfrac{1}{2x+4}\right] \cdot$
$\dfrac{(x^2+2)^3 \cdot (x-1)^5}{x\sqrt{x+1}\sqrt{x+2}}$

45. $\dfrac{2(x+1) + x\ln x}{2x\sqrt{x+1}} \cdot x^{\sqrt{x+1}}$

Exercise Set 8.5

1. e^{2t} **3.** $e^{5(t-1)}$ **5.** e^{-2t}

7. False

9. a. $v(t) = v_0 e^{(-c/m)t}$
b. $v(t) = 10e^{-4t}$
c. 0

11. a. 4 h   **b.** 800

13. Sell after 0.83 years. The original price does not affect this answer.

15. 23.3 h   **17.** 8.1 μg   **19.** 2.63 years

21. $\ln(1.08) \approx 0.070 = 7.70\%$

23. a. 10.25%   **b.** 10.38%
c. 10.516%   **d.** 10.517%

27. False. $P'(t) = \ln(1.05)P(t)$.

Review Exercises—Chapter 8

1. a. 4   **b.** $\dfrac{1}{7776}$ **c.** 32   **d.** $\dfrac{\sqrt{2}}{4}$

3. a. $y = x - 2$ **b.** $y = \sqrt{x}$
c. $y = \dfrac{1}{2}(x-1)$ **d.** $y = x^2$

5. $2x\ln(x-a) + \dfrac{x^2}{x-a}$ **7.** $\dfrac{2}{x\ln x}$

9. $2e^{x^2}(x \tan 2x + \sec^2 2x)$

11. $\dfrac{(1 + t + e^t)e^t}{(1 + e^t)^2}$

13. $\dfrac{(t^2 + b) \ln(t^2 + b) - 2t(t - a) \ln(t - a)}{(t - a)(t^2 + b)[\ln(t^2 + b)]^2}$

15. $\left(\dfrac{\sqrt{t}-2}{2t^2}\right)e^{\sqrt{t}}$

17. $\dfrac{-1}{3x^2} - \dfrac{1}{2(1 - 2x)} - \dfrac{1}{8x}$

19. $\dfrac{-(2xy \ln y + y^2)}{(2x^2 + xy \ln x)}$

21. $\dfrac{2x + ye^{xy}}{2y - xe^{xy}}$

23. $\dfrac{-y}{x}$

25. $\dfrac{4\sqrt{x}e^{2x} + \cos \sqrt{x}}{4\sqrt{x}(e^{2x} + \sin \sqrt{x})}$

27. $-\dfrac{1}{2} \ln |1 - x^2| + C$

29. $\dfrac{1}{4} \ln |4x + 2x^2| + C$

31. $2e^{x/2} + C$

33. $2(\sqrt{2} - 1)$

35. $\dfrac{5}{6} - 2 \ln 2$

37. $\dfrac{e^4 + 3}{2}$

39. $e^x - e^{3x} + \dfrac{3}{5}e^{5x} - \dfrac{1}{7}e^{7x} + C$

41. $\ln |x^3 + x^2 - 7| + C$

43. $-\dfrac{1}{2} \ln 2$

45. $\dfrac{1}{2}(\ln x + 2)x^{\sqrt{x}-(1/2)}$

47. $\left(\dfrac{1}{\sqrt{e}}, -\dfrac{1}{2e}\right)$

49. $(-1, 1)$

51. $\dfrac{15}{4} - \ln 16$

53. $2\pi \ln 2$

55.

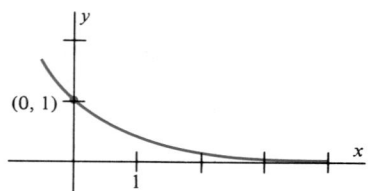

57.

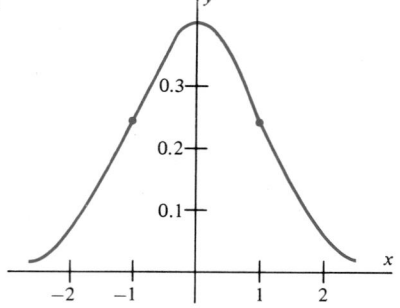

59. $\dfrac{41}{40}e^2$

61. a. $\dfrac{1}{2} - \dfrac{1}{t + 2}$ **b.** $5 - \ln 6$

63. $r = \ln 2$

65. a. 1.3856   **b.** 1.6101   **c.** -0.693150

67. $y = e^{2x}$ **69.** $y = \pi e^{-2x}$

71. $y \equiv 1$ **73.** $\dfrac{e}{2\sqrt{\ln 2}}$

75. a. 253.8 million   **b.** 283.8 million

77. $100(5)^{4/10}$ **79.** 4.5 hours

CHAPTER 9
Exercise Set 9.1

1. $\dfrac{1}{3} \sin 3x + C$

3. $\dfrac{1}{2} \sec 2x + C$

5. $\tan x + C$

7. $\dfrac{1}{2} \ln |\sec x^2 + \tan x^2| + C$

9. $\dfrac{1}{6} \ln |\sec(3x^2 - 1)| + C$

11. $\dfrac{1}{4} \tan^4 x + C$

13. $e^{\tan x} + C$

15. $\ln |x + \tan x| + C$

17. $\dfrac{1}{2} \cot(1 - x^2) + \dfrac{1}{2}(1 - x^2) + C$

19. $\ln |\sec x + \tan x| + \ln |\sec x| + C$

21. $\dfrac{1}{-2} \ln |\sec(x^2 - 2x) + \tan(x^2 - 2x)| + C$

23. $\begin{cases} e^{\sin x} + C, & \text{if } \cos x \geq 0 \\ -e^{\sin x} + C, & \text{if } \cos x < 0 \end{cases}$

25. $-2 \cot \sqrt{x} + 2 \csc \sqrt{x} + C$

27. $\dfrac{1}{2}\sqrt{1 + \sin 4x} + C$

29. $2\sqrt{\tan \theta + 1} + C$

31. $\dfrac{1}{2} \ln (2 + \sqrt{3})$

33. $-\dfrac{6}{\pi} \ln(2\sqrt{3} - 3)$

35. $\dfrac{1}{2} \tan^2 x + C$

37. $-\ln(2\sqrt{3} - 3)$

39. $\dfrac{\pi^2}{2} - \dfrac{4}{3}\pi$

41. $\dfrac{2\pi}{3}(1 + \sqrt{7}) \approx 7.6356$

Exercise Set 9.2

1. $\dfrac{1}{2}x - \dfrac{1}{8}\sin 4x + C$

3. $-\dfrac{1}{3}\cos^3 x + C$

5. $-\cos x + \dfrac{2}{3}\cos^3 x - \dfrac{1}{5}\cos^5 x + C$

7. $\dfrac{2}{15}$

9. $\dfrac{1}{5}\sec^5 x - \dfrac{1}{3}\sec^3 x + C$

11. $\dfrac{5\pi}{16}$

13. $\dfrac{-1}{2}\cot^2 x - \ln|\sin x| + C$

15. $\dfrac{1}{4}$

17. $x - \dfrac{1}{2}\cos 2x + C$

19. $\dfrac{1}{5}\tan^5 x + \dfrac{1}{7}\tan^7 x + C$

21. $\dfrac{1}{7}\sec^7 x - \dfrac{2}{5}\sec^5 x + \dfrac{1}{3}\sec^3 x + C$

23. $\dfrac{1}{2}\sin x - \dfrac{1}{6}\sin 3x + C$

25. $2(\sin x)^{1/2} - \dfrac{2}{5}(\sin x)^{5/2} + C$

27. $\dfrac{-1}{3(1 + \tan x)^3} + C$

29. $\dfrac{1}{16}\sin 8x + \dfrac{1}{4}\sin 2x + C$

31. $\dfrac{1}{2}\cos^2 \theta - \ln|\cos \theta| + C$

33. $\dfrac{2}{3}(\tan x)^{3/2} + C$

35. $\dfrac{-1}{4}\dfrac{\cos(a + b + c)x}{a + b + c} - \dfrac{1}{4}\dfrac{\cos(a + b - c)x}{(a + b - c)}$
$\quad - \dfrac{1}{4}\dfrac{\cos(a - b + c)x}{(a - b + c)} - \dfrac{1}{4}\dfrac{\cos(a - b - c)x}{(a - b - c)} + C$

37. 0   **39.** L

41. 0   **49.** $\dfrac{\pi}{3}$

51. $\ln(\sqrt{2} + 1)$

53. a. $\dfrac{1}{2}t + \dfrac{1}{4}\sin 2t$ **b.** $\dfrac{t^2}{4} - \dfrac{1}{8}\cos 2t + \dfrac{33}{8}$

55. $\ln(2 + \sqrt{3})$

Exercise Set 9.3

1. $\dfrac{\pi}{6}$ **3.** 0   **5.** π

7. $\dfrac{\pi}{2}$ **9.** $\dfrac{1}{\sqrt{3}}$ **11.** 2

13. $\dfrac{-1}{2}$ **15.** 1   **17.** 0

19. $\sqrt{3}$ **21.** $\dfrac{x}{\sqrt{1 - x^2}}$ **23.** $2x\sqrt{1 - x^2}$

25. $\dfrac{\sqrt{x^2 - 1}}{x}$ **27.** $\dfrac{1}{x}$ **29.** $\dfrac{\sqrt{x^2 - 1}}{x}$

31. False

33. $\theta = \text{Tan}^{-1}\dfrac{x}{100}$

35. $\theta = \text{Tan}^{-1}\dfrac{x}{20}$

Exercise Set 9.4

1. $\dfrac{3}{\sqrt{1 - 9x^2}}$ **3.** $\dfrac{1}{2\sqrt{t - t^2}}$ **5.** $\dfrac{-e^{-x}}{\sqrt{1 - e^{-2x}}}$

7. $\dfrac{-1}{2\sqrt{(\text{Cos}^{-1} x)(1 - x^2)}}$ **17.** $\dfrac{1}{8}\text{Sec}^{-1}\left(\dfrac{x}{4}\right) + C$

9. $\dfrac{1}{(1 + x^2)\,\text{Tan}^{-1} x}$ **19.** $-\sqrt{1 - x^2} + C$

11. $\dfrac{-1}{\sqrt{1 - x^2}}$ **21.** $\text{Sec}^{-1} e^x + C$

23. $2\,\text{Tan}^{-1} e^{\sqrt{x}} + C$

13. $\dfrac{1 - 2x\,\text{Tan}^{-1} x}{(1 + x^2)^2}$ **25.** $\dfrac{1}{3}\text{Tan}^{-1} x^3 + C$

15. $\dfrac{1}{2}\text{Tan}^{-1}\left(\dfrac{x}{2}\right) + C$ **27.** $\dfrac{\pi^2}{32}$

29. $\dfrac{1}{\sqrt{2}}\left(\text{Tan}^{-1}(\sqrt{2}) - \text{Tan}^{-1}\left(\dfrac{1}{\sqrt{2}}\right)\right) + C$

31. $\dfrac{1}{2}\text{Sec}^{-1} 2x + C$ **33.** $2\tan\left(2x + \text{Tan}^{-1}\dfrac{1}{2}\right)$

35. $\dfrac{1}{3}$ rad/s   **37.** $\dfrac{1}{4}$ rad/s

39. $\dfrac{2\sqrt{6}}{3}x + \left(\dfrac{\pi}{6} - \dfrac{2}{3}\sqrt{3}\right)$

41. $\dfrac{\pi^2}{6}$

Exercise Set 9.5

1. 0

3. 5/4

5. 3/5

7. 17/15

9. $\ln(1 + \sqrt{2})$

11. $2 \cosh 2x$

13. $\cos x \tanh x + \sin x \, \text{sech}^2 x$

15. $2(\cosh(4x))^{-1/2} \sinh(4x)$

17. $\dfrac{-\sinh x}{\cosh^2 x}$

19. $-2xe^x \, \text{csch } x^2 \coth x^2 + e^x \, \text{csch } x^2$

21. $\dfrac{2}{\sqrt{4x^2 + 1}}$

23. $\dfrac{2s}{1 - s^4}$

25. $\dfrac{\pi}{\cosh^{-1} \pi x \cdot \sqrt{\pi^2 x^2 - 1}}$

27. $\sinh x \cdot \cosh x^2 + 2x \cosh x \cdot \sinh x^2$

29. $\dfrac{2 \cosh(\ln x^2)}{x} = x + x^{-3}$

31. $\dfrac{x \, \text{sech}^2(\sqrt{1 + x^2})}{\sqrt{1 + x^2}}$

33. $\ln\left(\dfrac{2 + \sqrt{5}}{1 + \sqrt{2}}\right)$

35. $x - \dfrac{1}{2} \tanh(2x) + C$

37. $\dfrac{(e^2 - e^{-2})^2}{8}$

39. $\ln(3 + \sqrt{8})$

41. $\dfrac{1}{3} \ln\left(\dfrac{3}{5} + \dfrac{1}{5} \cdot \sqrt{34}\right)$

43. $\ln(\cosh x) + C$

45. $2(\sinh x)^{1/2} + C$

47. $\dfrac{1}{3} \tanh^3 x + C$

49. False

57. $z_1 = e^{kz}, \; z_2 = e^{-kz}$

$y_1 = \dfrac{z_1 - z_2}{2}, \; y_2 = \dfrac{z_1 + z_2}{2}$

59. $y_1 = C_1 \sinh k\pi x + C_2 \cosh k\pi x$
$y_2 = C_1 e^{k\pi x} + C_2 e^{-k\pi x}$

61. $\ln \dfrac{25}{17}$

63. No relative extrema.

65. min: $\left(\dfrac{1}{2} \ln \dfrac{7}{3}, \sqrt{21}\right)$

67. $y'' - 9y = 0$

69. Even: $\cosh x$ Odd: $\sinh x$
       $\text{sech } x$ $\tanh x$
                 $\coth x$
                 $\text{csch } x$

73. $[0, +\infty)$

75. $\tanh x$: $-\infty < x < \infty$
    $\coth x$: $x \in (-\infty, 0) \cup (0, \infty)$
    $\text{sech } x$: $x \geq 0$
    $\text{csch } x$: $x \in (-\infty, 0) \cup (0, \infty)$

Review Exercises—Chapter 9

1. $\dfrac{\pi}{3}$ **3.** $\dfrac{\pi}{3}$ **5.** $\dfrac{\pi}{4}$ **7.** 1

9. $\dfrac{3}{4}$ **11.** $\dfrac{1}{2\sqrt{x}(1 - x)}$

13. $\dfrac{2t}{1 + (1 - t^2)^2}$

15. $2x \sinh(1 - x) - x^2 \cosh(1 - x)$

17. $6 \csc(\cot 6x) \cot(\cot 6x) \cdot (\csc^2 6x)$

19. $\dfrac{2}{(2x + 1)\sqrt{1 - (\ln(2x + 1))^2}}$

21. $\dfrac{1 - 2x \, \text{Tan}^{-1} x}{(1 + x^2)^2}$

23. $\dfrac{x}{(x^2 + 4)(\sqrt{x^2 + 3})}$

25. $-\dfrac{\text{sech}^2 x}{(\pi + \tanh x)^2}$

27. $\dfrac{1}{5}(\sinh x + \text{Cos}^{-1} 2x)^{-4/5} \cdot \left(\cosh x - \dfrac{2}{\sqrt{1 - 4x^2}}\right)$

29. $\dfrac{1}{\text{Sin}^{-1} x\sqrt{1 - x^2}}$

31. $\tanh^{-1}(\ln x) + \dfrac{1}{1 - (\ln x)^2}$

33. $\dfrac{x}{\tanh^{-1}(x^2) \cdot (1 - x^4)}$

35. $\dfrac{1}{2} \cosh^{-1}(2x) + C = \dfrac{1}{2} \ln |2x + \sqrt{4x^2 - 1}| + C$

37. $\dfrac{1}{10} \sin^5 2x + C$

39. $\dfrac{-1}{2} - \dfrac{\pi}{4}$ **41.** $\dfrac{\pi}{2}$ **43.** $\dfrac{2}{3}$

45. $\tan 2x - x + C$ **47.** $\dfrac{1}{6}$

49. $\mathrm{Sin}^{-1}\left(\dfrac{x}{2}\right) + C$

51. $\dfrac{-1}{2}(\mathrm{Cos}^{-1} x)^2 + C$

53. $2\,\mathrm{Tan}^{-1}\,(e^{\sqrt{x}}) + C$

55. $\dfrac{2}{3}(\mathrm{Sin}^{-1} x)^{3/2} + C$

57. $\dfrac{1}{6}\left(\sec^3 \dfrac{2\pi}{9} - 1\right)$

59. $2\sqrt{3}$

61. $\dfrac{-\sqrt{3}}{16}$

63. $\dfrac{2}{7}\cos^{7/2} x - \dfrac{2}{3}\cos^{3/2} x + C$

65. $\cos x + \dfrac{1}{\cos x} + C$

67. $\ln|1 + \sinh x| + C$

69. $\dfrac{1}{3}\cosh^3 x - \cosh x + C$

71. $\dfrac{\pi}{2}$

73. $\pi\left(1 - \dfrac{\pi}{4}\right)$

75. $\pi\,\mathrm{Tan}^{-1}\,4$

77. $\dfrac{2xy}{\sqrt{x^4 y^2 - 1} - x^2}$

79. No relative extrema
Point of inflection at $x = 0$
$y'' < 0$ on $(-\infty, 0)$, concave down
$y'' > 0$ on $(0, \infty)$, concave up

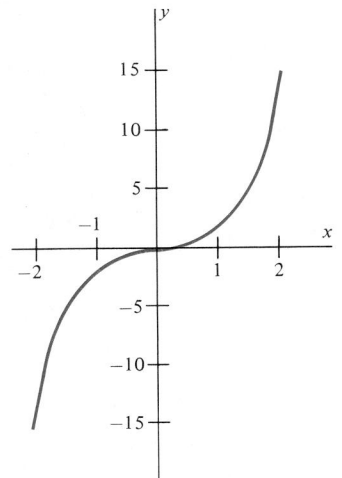

81. $28.677°$

83. $\dfrac{2\sqrt{6}}{3}x + \left(\dfrac{\pi}{6} - \dfrac{2\sqrt{3}}{3}\right)$

85. $x = 2,\quad y = \mathrm{Cot}^{-1}\,2$
$x = -2,\quad y = \mathrm{Cot}^{-1}(-2)$

87. $\dfrac{\sqrt{3}}{60}$ radians/s

CHAPTER 10

Exercise Set 10.1

1. $xe^x - e^x + C$

3. $x \sin x + \cos x + C$

5. $x\,\mathrm{Tan}^{-1}\,x - \dfrac{1}{2}\ln|1 + x^2| + C$

7. $\dfrac{x}{\pi}\tan \pi x - \dfrac{1}{\pi^2}\ln|\sec \pi x| + C$

9. $\dfrac{x}{2}\sin(\ln x) - \dfrac{x}{2}\cos(\ln x) + C$

11. $27761/18 \approx 1542.28$

13. $x(\ln x)^2 - 2x \ln x + 2x + C$

15. $\dfrac{x}{2}e^{2x} - \dfrac{1}{4}e^{2x} + C$

17. $\dfrac{e^2 + 3}{8}$

19. $x \cosh x - \sinh x + C$

21. $\dfrac{2}{5}e^{2x}\cos x + \dfrac{1}{5}e^{2x}\sin x + C$

23. $\left(\dfrac{x^2}{2} - \dfrac{1}{4}\right)\mathrm{Sin}^{-1}\,x - \dfrac{x}{4}\sqrt{1 - x^2} + C$

25. $\dfrac{2}{3}x^{3/2}\ln x - \dfrac{4}{9}x^{3/2} + C$

27. $\dfrac{1}{3}(x^2 - 2)\sqrt{x^2 + 1}$

29. $\dfrac{x^2}{2a}e^{ax^2} - \dfrac{1}{2a^2}e^{ax^2} + C$

31. $\dfrac{1}{4}\sec^3 x \tan x - \dfrac{1}{8}\sec x \tan x - \dfrac{1}{8}\ln|\sec x + \tan x| + C$

33. $\dfrac{2x^2 + 1}{4}\sinh^{-1} x - \dfrac{x}{4}\sqrt{1 + x^2} + C$

35. $-4x^{3/2}\,e^{-\sqrt{x}} - 12xe^{-\sqrt{x}} - 24\sqrt{x}\,e^{-\sqrt{x}} - 24\,e^{-\sqrt{x}} + C$

37. $\dfrac{1}{\pi}$

39. $\pi(e - 2)$

41. $\dfrac{1}{1 + \pi^2}\left(\dfrac{\pi}{e} + \pi\right)$

43. $\dfrac{4}{\pi}$ meters

47. $-\dfrac{1}{4}\sin^3 x \cos x - \dfrac{3}{8}\sin x \cos x + \dfrac{3}{8}x + C$

49. $(x^4 - 4x^3 + 12x^2 - 24x + 24)e^x + C$

51. $-\dfrac{x}{3}\cos 3x + \dfrac{1}{9}\sin 3x + C$

Exercise Set 10.2

1. $\frac{1}{2}x\sqrt{4-x^2} + 2\,\text{Sin}^{-1}\frac{x}{2} + C$

3. $\text{Sin}^{-1}\left(\frac{x}{3}\right) + C$

5. $\frac{1}{3}(x^2+8)\sqrt{x^2-4} + C$

7. $\ln\left|\frac{1-\sqrt{1-x^2}}{x}\right| + \sqrt{1-x^2} + C$

9. $-\frac{1}{16}\text{Sec}^{-1}\left(\frac{x}{2}\right) + \frac{1}{8}\frac{\sqrt{x^2-4}}{x^2} + C$

11. $\frac{1}{128}\text{Tan}^{-1}\left(\frac{x}{4}\right) + \frac{x}{32(16+x^2)} + C$

13. $\frac{x}{\sqrt{1-x^2}} - \text{Sin}^{-1}x + C$

15. $-\frac{1}{15}(3x^2+2)(1-x^2)^{3/2} + C$

17. $\ln\left|\frac{\sqrt{x^2+3}+x}{\sqrt{3}}\right| - \frac{x}{\sqrt{x^2+3}} + C$

19. $\sqrt{x^2-a^2} - a\,\text{Sec}^{-1}\left(\frac{x}{a}\right) + C$

21. $\ln\left|\frac{x+\sqrt{x^2-a^2}}{a}\right| - \frac{\sqrt{x^2-a^2}}{x} + C$

23. $\frac{-1}{\sqrt{x^2+a^2}} + C$

25. $\frac{1}{a}\text{Sec}^{-1}\left(\frac{x}{a}\right) + C$

27. $\frac{-\sqrt{x^2+a^2}}{x} + \ln(\sqrt{x^2+a^2}+x) + C$

29. $\frac{x\sqrt{x^2+9}}{2} - \frac{3}{2}\ln\left|\frac{\sqrt{x^2+9}+x}{3}\right| + C$

31. $\sqrt{x^2+4} - \frac{3}{2}\ln\left|\frac{\sqrt{x^2+4}-2}{x}\right| + C$

33. $-\sqrt{9-x^2} - 4\,\text{Sin}^{-1}\left(\frac{x}{3}\right) + \frac{5}{3}\ln\left|\frac{3-\sqrt{9-x^2}}{x}\right| + C$

35. $20 - 9\ln 3$

37. 62.64%

39. $\frac{1}{4}\sqrt{2} + \frac{1}{4}\ln(\sqrt{2}+1)$

41. $\sqrt{10} - \sqrt{2} + \ln\left|\frac{(\sqrt{10}-1)}{3(\sqrt{2}-1)}\right|$

Exercise Set 10.3

1. $\frac{-1}{x-2} + C$

3. $\ln\left|\frac{\sqrt{x^2+6x+13}+x+3}{2}\right| + C$

5. $\ln\left|\frac{(x-3)+\sqrt{x^2-6x}}{3}\right| + C$

7. $-\sqrt{6x-x^2} + 3\,\text{Sin}^{-1}\left(\frac{x-3}{3}\right) + C$

9. $2\sqrt{x^2+6x+18} - 5\ln\left|\frac{\sqrt{x^2+6x+18}-(x+3)}{3}\right| + C$

11. $\frac{1}{2}\ln|x^2-8x-7| + \frac{3}{\sqrt{23}}\ln\left|\frac{x-4-\sqrt{23}}{x-4+\sqrt{23}}\right| + C$

13. $-\frac{x+2}{2(x^2+2x+2)} - \frac{1}{2}\text{Tan}^{-1}(x+1) + C$

15. $\frac{-\sqrt{7}}{49}\ln\left|\frac{\sqrt{88-18x+x^2}-\sqrt{7}}{x-9}\right|$

$-\frac{1}{7\sqrt{88-18x+x^2}} + C$

17. $\frac{1}{2}\left(x+\frac{5}{2}\right)\sqrt{4x^2+20x+29}$

$-\ln\left|\frac{\sqrt{4x^2+20x+29}}{2} + \left(x+\frac{5}{2}\right)\right| + C$

Exercise Set 10.4

1. $\ln\left|\frac{x}{x+1}\right| + C$

3. $-3\ln|1-x| + \ln|1+x| + C$

5. $\frac{1}{2}\ln\left|\frac{1+x}{1-x}\right| + C$

7. $\ln|x| - \frac{1}{2}\ln|x^2+1| + C$

9. $\frac{1}{2}\ln|x+1| - \frac{1}{4}\ln(x^2+1) + \frac{1}{2}\text{Tan}^{-1}x + C$

11. $3\ln|x+2| + \frac{1}{2}\ln|x^2+1| - \text{Tan}^{-1}x + C$

13. $\ln|x| - \frac{3}{x+1} + C$

15. $2\ln|x| + \frac{1}{x+1} + C$

17. $\frac{1}{2}\ln|x| - \ln|x+1| + \frac{1}{2}\ln|x+2| + C$

19. $\ln |x| + 2 \ln |x - 3| + \ln |x^2 + 1| + C$

21. $4 \ln |x| - 2 \ln |x^2 + 1| + C$

23. $\dfrac{1}{4a^3} \ln \left| \dfrac{a + x}{a - x} \right| + \dfrac{1}{2a^2} \dfrac{x}{a^2 - x^2} + C$

25. $9 \ln |x + 3| + \dfrac{1}{2} \ln (x^2 + 2x + 6) + C$

27. $3 \ln |x - 1| + \ln |x^2 + x + 1| + C$

29. $2 \ln |x| + 6 \ln |x - 2| + \ln |x + 4| + C$

31. $4 \ln 3 - \ln 2$

33. $3\pi \ln 5 - 4\pi \ln 3$

37. $P = \dfrac{100}{1 + Me^{-2t}}, \ k = 100$

Exercise Set 10.5

1. $x - 2\sqrt{x} - 2 \ln (1 + \sqrt{x}) + C$

3. $2\sqrt{x} - 6\sqrt[3]{x} + 24\sqrt[6]{x} - 48 \ln |\sqrt[6]{x} + 2| + C$

5. $\dfrac{3}{5}(x + 1)^{5/3} - \dfrac{3}{2}(x + 1)^{2/3} + C$

7. $\dfrac{2\sqrt{3}}{3} \operatorname{Tan}^{-1}\left(\dfrac{2 \tan \left(\dfrac{x}{2} \right) + 1}{\sqrt{3}} \right) + C$

9. $\dfrac{-1}{\tan \left(\dfrac{x}{2} \right)} + C$

11. $\dfrac{3}{8}(1 + x)^{8/3} - \dfrac{6}{5}(1 + x)^{5/3} + \dfrac{3}{2}(1 + x)^{2/3} + C$

13. $\dfrac{\sqrt{2}}{2} \ln \left| \dfrac{\tan \left(\dfrac{x}{2} \right) - 1 + \sqrt{2}}{\tan \left(\dfrac{x}{2} \right) - 1 - \sqrt{2}} \right| + C$

15. $\dfrac{1}{56}(1 - 2x)^{7/2} - \dfrac{3}{40}(1 - 2x)^{5/2}$

$\qquad + \dfrac{1}{8}(1 - 2x)^{3/2} - \dfrac{1}{8}(1 - 2x)^{1/2} + C$

17. $\operatorname{Sin}^{-1} x - \sqrt{1 - x^2} + C$

19. $\begin{cases} \dfrac{1}{\sqrt{a}} \ln \left| \dfrac{\sqrt{a + bx} - \sqrt{a}}{\sqrt{a + bx} + \sqrt{a}} \right| + C, & \text{if } a > 0 \\ \dfrac{2}{\sqrt{-a}} \operatorname{Tan}^{-1} \sqrt{\dfrac{a + bx}{-a}} + C, & \text{if } a < 0 \end{cases}$

21. $\dfrac{2}{3}(2 + x)^{3/2} - 4(2 + x)^{1/2} + C$

Exercise Set 10.6

1. $\dfrac{1}{5} \ln \left(\dfrac{e^{5x}}{1 + e^{5x}} \right) + C$

3. $\dfrac{4}{729} \ln \left(\dfrac{9 + 2x}{x} \right) - \dfrac{(9 + 2x) + 2x}{81x(9 + 2x)} + C$

5. $\dfrac{1}{10} \ln(2 + 5x^2) + \dfrac{3}{\sqrt{10}} \operatorname{Tan}^{-1}\left(\dfrac{\sqrt{5}x}{\sqrt{2}} \right) + C$

7. $\dfrac{1}{\pi} \left\{ \dfrac{1}{3} \tan^3 \pi x - \tan \pi x + \pi x \right\} + C$

9. $\dfrac{\sin 3x}{6} - \dfrac{\sin 9x}{18} + C$

11. $\dfrac{-1}{2x} - \dfrac{3}{4} \ln \left| \dfrac{2 - 3x}{x} \right| + C$

13. $\dfrac{9}{20} \ln \left| \dfrac{2x + 5}{2x - 5} \right| + C$

15. $\dfrac{1}{10} \tan^2 5x - \dfrac{1}{5} \ln |\sec 5x| + C$

17. $\dfrac{4}{5}(3x^2 - 24x + 288)\sqrt{6 + x} + C$

19. $\dfrac{6x}{14\sqrt{14 - x^2}} + C$

21. $\dfrac{\pi}{8} \tan \left(\dfrac{\pi}{4} - x \right) + C$

23. $\dfrac{-3}{4} \ln \left| \dfrac{4 + \sqrt{x^2 + 16}}{x} \right| + C$

25. $\dfrac{1}{3} \operatorname{Tan}^{-1}(3x + 2) + C$

27. $\dfrac{-7}{2} \ln \left| \dfrac{2 + \sqrt{x^2 + 8x + 20}}{x + 4} \right| + C$

29. $9 \operatorname{Sin}^{-1}\left(\dfrac{x}{4} \right) - x\sqrt{16 - x^2} + C$

Review Exercises—Chapter 10

1. $\dfrac{1}{3}(x^2 + 9)^{3/2} + C$

3. $\dfrac{x}{2}\sqrt{x^2 + a^2} + \dfrac{a^2}{2} \ln \left| \dfrac{x + \sqrt{x^2 + a^2}}{a} \right| + C$

5. $\ln |x(x + 3)| + C$

7. $\dfrac{x}{\pi} \sin \pi x + \dfrac{1}{\pi^2} \cos \pi x + C$

9. $\dfrac{x^2}{2} \ln^2 x - \dfrac{x^2}{2} \ln x + \dfrac{x^2}{4} + C$

11. $\dfrac{\sqrt{5}}{5} \operatorname{Tan}^{-1}\left(\dfrac{x-2}{\sqrt{5}}\right) + C$

13. $2\sqrt{x+4} + 2\ln\left|\dfrac{\sqrt{x+4}-2}{\sqrt{x+4}+2}\right| + C$

15. $\dfrac{32 + 16\sqrt{2}}{15}$

17. $\operatorname{Sin}^{-1}\left(\dfrac{x}{a}\right) + C$

19. $\dfrac{-1}{x} + 3\ln|x+1| + C$

21. $1 - \dfrac{\pi}{4}$

23. $\dfrac{a^2}{2}\operatorname{Sin}^{-1}\left(\dfrac{x}{a}\right) + \dfrac{x}{2}\sqrt{a^2 - x^2} + C$

25. $4\ln|x-2| + \operatorname{Tan}^{-1}x + C$

27. $\dfrac{3}{5}(x+1)^{5/3} + C$

29. $\dfrac{x}{2}\sqrt{x^2 - a^2} - \dfrac{a^2}{2}\ln\left|\dfrac{x + \sqrt{x^2 - a^2}}{a}\right| + C$

31. $\ln|x+2| + \dfrac{2\sqrt{3}}{3}\operatorname{Tan}^{-1}\left(\dfrac{2x+1}{\sqrt{3}}\right) + C$

33. $\dfrac{1}{4}(x^3 - 10x)\sqrt{x^2 - 4} + 6\ln|x + \sqrt{x^2 - 4}| + C$

35. $2\ln|x| + \dfrac{2}{3}\ln|x+1|$
$\qquad\qquad + \dfrac{1}{6}\ln(x^2 - x + 1) + \dfrac{1}{\sqrt{3}}\operatorname{Tan}^{-1}\dfrac{2x-1}{\sqrt{3}} + C$

37. $-\dfrac{1}{2}\csc x \cot x + \dfrac{1}{2}\ln|\csc x - \cot x| + C$

39. $\dfrac{1}{6}\ln\left|\dfrac{\sqrt{9+x^2}-3}{\sqrt{9+x^2}+3}\right| + C$

41. $3\ln|x-4| - \dfrac{1}{\sqrt{2}}\operatorname{Tan}^{-1}\dfrac{x}{\sqrt{2}} + C$

43. $\dfrac{4}{\sqrt{3}}\operatorname{Tan}^{-1}\left[\sqrt{3}\tan\left(\dfrac{x}{2}\right)\right] - x + C$

45. $\dfrac{1}{2}\ln\left|\dfrac{e^x - 1}{e^x + 1}\right| + C$

47. $\dfrac{1}{2}\ln(a^2 + x^2) + C$

49. $3\ln|x-1| + \ln(x^2 + 4) + C$

51. $-\ln(e^{-x} + 1) + C$

53. $3\sqrt{x^2 + 8x} - 11\ln|(x+4) + \sqrt{x^2 + 8x}| + C$

55. $\dfrac{1}{5}(x^2 + 1)^{5/2} - \dfrac{2}{3}(x^2 + 1)^{3/2} + \sqrt{x^2 + 1} + C$

57. $\dfrac{x^2}{2} - 2\ln(x^2 + 2) - \dfrac{2}{x^2 + 2} + C$

59. $\dfrac{1}{2}\ln\left|\dfrac{2x + \sqrt{4x^2 - 25}}{5}\right| + C$

61. $x\ln^2 x - 2x\ln x + 2x + C$

63. $\dfrac{3}{2\sqrt{5}}\ln\dfrac{x - \sqrt{5}}{x + \sqrt{5}} + C$

65. $\dfrac{1}{3}\ln\left|\dfrac{\sqrt{9 - \pi x} - 3}{\sqrt{9 - \pi x} + 3}\right| + C$

67. $\dfrac{10}{49}\ln\left|\dfrac{x}{7 - 2x}\right| - \dfrac{5}{7x} + C$

69. $\dfrac{1}{2}(9 + 2x)^{3/2} - \dfrac{27}{2}(9 + 2x)^{1/2} + C$

71. $-\dfrac{\sqrt{8 - x^2}}{8x} + C$

73. $\dfrac{2\sqrt{65}}{65}\operatorname{Tan}^{-1}\left[\sqrt{\dfrac{5}{13}}\tan\left(\dfrac{x}{2}\right)\right] + C$

75. $\dfrac{x^4}{4}\ln x - \dfrac{x^4}{16} + C$

CHAPTER 11

Exercise Set 11.1

1. $-\dfrac{1}{e}$ **3.** 0 **5.** $\dfrac{1}{2}$

7. $-\infty$ **9.** 2 **11.** 1

13. $+\infty$ **15.** $+\infty$ **17.** $\dfrac{e}{2}$

19. 0 **21.** $+\infty$ **23.** $\dfrac{1}{2}$

25. 0 **27.** 1 **29.** 1

31. 3 **33.** 1 **35.** $\dfrac{3}{4}$ **37.** 0

Exercise Set 11.2

1. $-\infty$ **3.** 1 **5.** 0 **7.** $+\infty$

9. 1 **11.** 1 **13.** 0 **15.** 0

17. e^6 **19.** 1 **21.** e^{-1} **23.** $\ln 2$

25. $\ln\dfrac{1}{2}$ **27.** e^{-1}

29. a. max: $(-2, 4e^{-2})$, min: $(0, 0)$ **b.** $y = 0$

c.

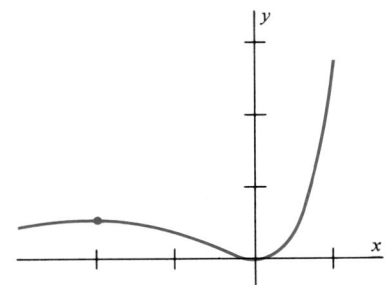

Exercise Set 11.3

1. Diverges

3. Diverges

5. 1

7. $\dfrac{1}{2}$

9. $\dfrac{e^4}{2}$

11. 1

13. Diverges

15. $\dfrac{9\pi}{4}$

17. $\dfrac{-1}{4}$

19. $\dfrac{2e^3}{9}$

21. Divergent

23. $\dfrac{1}{4}$

25. 1

27. $\dfrac{1}{4}$

31. $p > -1$

35. Diverges

37. Converges

39. Diverges

41. Converges

45. $\dfrac{\pi}{2}$, $\pi[\sqrt{2} + \ln(\sqrt{2} + 1)]$

47. a. 4,121,000 dollars
b. 12.5 million dollars

Review Exercises—Chapter 11

1. $\dfrac{3}{4}$

3. $+\infty$

5. 0

7. 1

9. 0

11. 0

13. 1

15. 1

17. $2e^{-1}$

19. Diverges

21. 0

23. Diverges

25. $\dfrac{\pi}{2}$

CHAPTER 12

Exercise Set 12.1

1. $\left\{\dfrac{1}{3}, \dfrac{2}{5}, \dfrac{3}{7}, \dfrac{4}{9}, \ldots\right\}$. Converges.

$$\lim_{n\to\infty} \frac{n}{2n+1} = \lim \frac{1}{2 + \dfrac{1}{n}} = \frac{1}{2}.$$

3. $\left\{-1, -\dfrac{1}{3}, -\dfrac{1}{11}, 0, \ldots\right\}$. Converges to 0.

5. $\left\{\dfrac{1}{2}, \dfrac{1}{5}, \dfrac{1}{10}, \dfrac{1}{17}, \ldots\right\}$. Converges to 0.

7. $\{\sqrt{5}, \sqrt{5}, \sqrt{5}, \sqrt{5}, \ldots\}$. Converges to $\sqrt{5}$.

9. $\left\{10, \dfrac{40}{1 + \sqrt{2}}, \dfrac{60}{1 + \sqrt{3}}, \dfrac{80}{3}, \ldots\right\}$. Diverges.

11. $\left\{\dfrac{2}{3}, \dfrac{3 + \sqrt{2}}{4}, \dfrac{3 - \sqrt{3}}{5}, \dfrac{5}{6}, \ldots\right\}$. Converges to 0.

13. $\left\{\sqrt{2}, \dfrac{1}{2}\sqrt{6}, \dfrac{2}{3}\sqrt{3}, \dfrac{1}{2}\sqrt{5}, \ldots\right\}$. Converges to 1.

15. $\left\{1, \cos\dfrac{1}{4}, \cos\dfrac{2}{9}, \cos\dfrac{3}{16}, \ldots\right\}$. Converges to $\cos 0 = 1$.

17. $\left\{\dfrac{3}{2}, \dfrac{1 + \sqrt{2}}{2\sqrt{2}}, \dfrac{2 + 3\sqrt{3}}{6\sqrt{3}}, \dfrac{5}{8}, \ldots\right\}$. Converges to $\dfrac{1}{2}$.

19. $\left\{\dfrac{1}{2}, \dfrac{1}{6}, \dfrac{1}{12}, \dfrac{1}{20}, \ldots\right\}$. Converges.

$$\lim\left(\frac{1}{n} - \frac{1}{n+1}\right) = \lim\left(\frac{1}{n(n+1)}\right) = 0.$$

21. $\{\sqrt{2} - 1, \sqrt{3} - 2, 2 - \sqrt{3}, \sqrt{5} - 2, \ldots\}$. Converges to 0.

23. $\left\{\sqrt{3}, \dfrac{3}{2}, \dfrac{\sqrt{19}}{3}, \dfrac{\sqrt{33}}{4}, \ldots\right\}$. Converges to $\sqrt{2}$.

25. $\left\{1, \sqrt{2}, \dfrac{3}{2}, 4\sin\dfrac{\pi}{8}, \ldots\right\}$. Converges to $\dfrac{\pi}{2}$.

27. $\left\{2, \dfrac{9}{4}, \dfrac{64}{27}, \dfrac{625}{256}, \ldots\right\}$. Converges to e.

39. Product is bounded, quotient need not be bounded.

Exercise Set 12.2

1. 2

3. 1

5. 0

7. Diverges

9. 1

11. 0

13. 0

15. e^3

17. 1

23. $a_n = 2n - 5$.

25. $\{1, -1, 1, -1, 1, -1, \ldots\}$

27. No. An example is $\left\{1, -\dfrac{1}{2}, \dfrac{1}{3}, -\dfrac{1}{4}, \dfrac{1}{5}, -\dfrac{1}{6}, \ldots\right\}$.

Exercise Set 12.3

1. $-\dfrac{1}{2} + \dfrac{1}{4} - \dfrac{1}{8} + \dfrac{1}{16} - \cdots$

3. $\dfrac{3}{5} + \dfrac{5}{11} + \dfrac{9}{29} + \dfrac{17}{83} + \cdots$

5. $\ln \dfrac{1}{2} + \ln \dfrac{2}{3} + \ln \dfrac{3}{4} + \ln \dfrac{4}{5} + \cdots$

7. Converges to $\dfrac{7}{6}$.　　　　　**9.** Diverges.

11. Converges; $\dfrac{27}{23}$.

13. Converges to $\dfrac{1}{1 - \dfrac{1}{2 + x}} = \dfrac{x + 2}{x + 1}$.

15. Converges to $-\dfrac{1}{2}$.　　　**17.** Diverges.

19. Converges to $\dfrac{1}{3}$.　　　　**21.** Diverges.

23. Converges to $\dfrac{3}{3 - \sqrt{2}}$.

25. a. $\displaystyle\sum_{k=1}^{\infty} \dfrac{3}{10^k}$　　　**b.** $\dfrac{1}{3}$

27. a. $\displaystyle\sum_{k=1}^{\infty} \dfrac{92}{100^k}$　　**b.** $\dfrac{92}{99}$

29. a. $\displaystyle\sum_{k=1}^{\infty} \dfrac{412}{1000^k}$　**b.** $\dfrac{412}{999}$

35. $5h$　　　　　　　　　**37.** No, if $c = 0$.

41. $S_n = S_{n-1} + a_n$

43.

	S_5	S_{10}	S_{20}	S_{50}	S_{100}
	2.736625514	2.96531694	2.999398543	2.999999997	3.

The actual sum is $\dfrac{1}{1 - \dfrac{2}{3}} = 3$.

45.

	S_5	S_{10}	S_{20}	S_{50}	S_{100}
	$-.3718688642$	-1.128682959	-1.295565042	-1.301540778	-1.301541025

The actual sum ≈ -1.301541025.

Exercise Set 12.4

1. Diverges　**3.** Diverges　　**5.** Converges

7. Diverges　**9.** Converges　　**11.** Diverges

13. Diverges　**15.** Converges　**17.** Diverges

19. Diverges　**23.** $n > \sqrt{\ln 25} \approx 1.794$

Exercise Set 12.5

1. Converges　**3.** Converges　**5.** Diverges

7. Diverges　**9.** Converges　**11.** Converges

13. Diverges　**15.** Converges　**17.** Diverges

19. Diverges　**21.** Converges　**23.** Diverges

25. Converges　**27.** Diverges　**29.** Converges

Exercise Set 12.6

1. Diverges; ratio test.

3. Converges; ratio test.

5. Converges; ratio test.

7. Converges; limit comparison test using $\displaystyle\sum \dfrac{1}{k^2}$.

9. Converges; root test.

11. Diverges, by Theorem 7.

13. Diverges; ratio test.

15. Converges; ratio test.

17. Converges; ratio test.

19. Diverges; ratio test.

21. All $x > 0$.

23. Converges for $x < 1$ and diverges for $x \geq 1$.

Exercise Set 12.7

1. Absolutely　　　　**3.** Absolutely

5. Absolutely　　　　**7.** Conditionally

9. Conditionally　　　**11.** Absolutely

13. Diverges　　　　　**15.** Conditionally

17. Absolutely　　　　**19.** Absolutely

21. Diverges　　　　　**23.** Absolutely

25. Conditionally　　　**27.** Absolutely

29. Absolutely　　　　**31.** All x

33. All x　　　　　　**35.** $n \geq 98$

Review Exercises—Chapter 12

1. Diverges　　　　　**3.** Diverges

5. Converges to 0　　　**7.** Diverges

9. Diverges

11. $\dfrac{6}{5}$

13. 2

15. $37\sum\limits_{k=1}^{\infty}\dfrac{1}{100^k}=\dfrac{37}{99}$

17. $\dfrac{1}{48}$

19. $\dfrac{10}{9}$

21. Converges **23.** Diverges

25. Converges **27.** Converges

29. Converges **31.** Converges

33. Diverges **35.** Converges

37. Diverges **39.** Converges

41. Converges **43.** Converges

45. Converges **47.** Diverges

49. Converges **51.** Converges

53. Converges **55.** Converges

57. Converges **59.** Converges conditionally

61. Diverges **63.** Converges absolutely

65. Diverges **67.** Converges absolutely

69. $\dfrac{1}{13}$ **71.** $\dfrac{1}{11}$

73. $-1<c\le 1$ **75.** Σb_k must diverge

77. a. 90.6192 meters
b. 110 meters

79. a. Converges
b. Converges
c. Converges

CHAPTER 13
Exercise Set 13.1

1. $1-x+\dfrac{1}{2}x^2-\dfrac{1}{6}x^3+\dfrac{1}{24}x^4$

3. $\dfrac{\sqrt2}{2}-\dfrac{\sqrt2}{2}\left(x-\dfrac{\pi}{4}\right)-\dfrac{\sqrt2}{4}\left(x-\dfrac{\pi}{4}\right)^2$
$+\dfrac{\sqrt2}{12}\left(x-\dfrac{\pi}{4}\right)^3+\dfrac{\sqrt2}{48}\left(x-\dfrac{\pi}{4}\right)^4$
$-\dfrac{\sqrt2}{240}\left(x-\dfrac{\pi}{4}\right)^5-\dfrac{\sqrt2}{1440}\left(x-\dfrac{\pi}{4}\right)^6$

5. $x-\dfrac{1}{2}x^2+\dfrac{1}{3}x^3-\dfrac{1}{4}x^4$

7. $x+\dfrac{1}{3}x^3$

9. $1+x^2$

11. $1-\dfrac{1}{2}x-\dfrac{1}{8}x^2-\dfrac{1}{16}x^3$

13. $\sqrt2+\sqrt2\left(x-\dfrac{\pi}{4}\right)+\dfrac{3\sqrt2}{2}\left(x-\dfrac{\pi}{4}\right)^2$
$+\dfrac{11\sqrt2}{6}\left(x-\dfrac{\pi}{4}\right)^3$

15. $\dfrac{1}{2}-\dfrac{1}{4}x+\dfrac{1}{48}x^3$

19. True

Exercise Set 13.2

1. $e^{-x}=1-x+\dfrac{1}{2}x^2-\dfrac{1}{6}x^3+\dfrac{e^{-c}}{24}x^4;$
c is between 0 and x

3. $\cos x=1-\dfrac{1}{2}x^2+\dfrac{1}{24}x^4-\dfrac{\sin c}{120}x^5;$
c is between 0 and x

5. $\text{Tan}^{-1}x=x-\dfrac{1}{3}x^3+\dfrac{c-c^3}{(1+c^2)^4}x^4;$
c is between 0 and x

7. $\dfrac{1}{1+x^2}=\dfrac{1}{2}-\dfrac{1}{2}(x-1)+\dfrac{1}{4}(x-1)^2$
$+4\dfrac{c-c^3}{(1+c^2)^4}(x-1)^3;$ c is between 1 and x

9. $\sec x=\sqrt2+\sqrt2\left(x-\dfrac{\pi}{4}\right)+\dfrac{3\sqrt2}{2}\left(x-\dfrac{\pi}{4}\right)^2$
$+\dfrac{(\sin c)(5+\sin^2 c)}{6\cos^4 c}\left(x-\dfrac{\pi}{4}\right)^3;$ c is between $\dfrac{\pi}{4}$ and x

11. $\sinh x=x+\dfrac{1}{6}x^3+\dfrac{\cosh c}{5!}x^5;$ c is between 0 and x

13. $\cosh x=\dfrac{5}{4}+\dfrac{3}{4}(x-\ln 2)+\dfrac{5}{8}(x-\ln 2)^2+\dfrac{1}{8}(x-\ln 2)^3$
$+\dfrac{1}{24}(\cosh c)(x-\ln 2)^4;$ c is between $\ln 2$ and x

21. $P_n(x)=1-x^2+x^4-x^6+\cdots,$ using all terms of degree $\le n$.

Exercise Set 13.3

Note: In Problems 1–20, there are many ways to handle the R_n and make accuracy estimates.

1. $\ln(1.5)\approx 0.416\overline{6}$ **3.** $\sqrt{3.91}\approx 1.99737$

5. $\cos 1\approx 0.540317$ **7.** $\sqrt[3]{10}\approx 2.152\overline{7}$

9. $|R_2(x)|\le 0.0000208$ **11.** $|R_1(x)|\le 0.00125$

13. $|R_1(x)| < 0.000118$

15. $|R_1(x)| < 0.00005$

17. $n = 6$

19. n is 2.

21. 2.7183

23. $|R_1(x)| < 0.005483$

27. 0.31

29. 0.9

Exercise Set 13.4

1. $(-1, 1)$

3. $(-\infty, \infty)$

5. $(-\infty, \infty)$

7. $[-1, 1)$

9. $(-1, 1]$

11. $[-1, 1]$

13. $(-\infty, \infty)$

15. Converges only for $x = 1$.

17. $\left(\dfrac{1}{3}, 1\right]$

19. $(-1, 1)$

21. $[2, 4]$

23. $(2 - e, 2 + e)$

25. $\left(-\dfrac{3}{7}, \dfrac{1}{7}\right)$

27. $(-1, 1)$

29. $[-2, 4)$

31. $[-e, e]$

33. $(-\infty, \infty)$

35. $[-1, 1]$

Exercise Set 13.5

1. $\sum_{k=0}^{\infty} 2^k x^k; \quad r = \dfrac{1}{2}$

3. $\sum_{k=0}^{\infty}(-1)^k 4^k x^{2k}; \quad r = \dfrac{1}{2}$

5. $\sum_{k=0}^{\infty}(-1)^k x^{2k+1}; \quad r = 1$

7. $-1 + \sum_{k=1}^{\infty}(-1)^{k+1} 2x^k; \quad r = 1$

9. $\sum_{k=0}^{\infty} x^{4k}; \quad r = 1$

11. $-2 \sum_{k=1}^{\infty}(-1)^k k x^{k-1}; \quad r = 1$

13. $\sum_{k=1}^{\infty} k(-1)^{k+1} x^{2k-1}; \quad r = 1$

15. $\sum_{k=0}^{\infty}(2k + 1)x^{2k}; \quad r = 1$

17. $\sum_{k=1}^{\infty} 2k(-1)^{k+1} 4^k x^{2k-1}; \quad r = \dfrac{1}{2}$

19. $\sum_{k=1}^{\infty} 2k(-1)^{k-1} x^{k-1}; \quad r = 1$

21. $-\sum_{k=0}^{\infty} \dfrac{x^{k+1}}{k + 1}; \quad r = 1$

23. $2 \sum_{k=0}^{\infty} \dfrac{(-1)^k 4^k x^{2k+1}}{2k + 1}; \quad r = \dfrac{1}{2}$

25. $\sum_{k=0}^{\infty} \dfrac{(-1)^k x^{k+1}}{4^{k+1} \cdot k + 1} + \ln(4); \quad r = 4$

27. $\sum_{k=0}^{\infty}(-1)^k \dfrac{x^{2k+2}}{k + 1}; \quad r = 1$

Exercise Set 13.6

1. $\sum_{k=0}^{\infty} \dfrac{2^k x^k}{k!}; \quad$ all x

3. $\sum_{k=0}^{\infty} \dfrac{\sqrt{2}(-1)^{k(k+1)/2}\left(x - \dfrac{\pi}{4}\right)^k}{2(k!)}; \quad$ all x

5. $5 + 4(x - 2) + (x - 2)^2; \quad$ all x

7. $\sum_{k=1}^{\infty}(-1)^{k-1}\dfrac{x^k}{k3^k} + \ln(3); \quad -3 < x \le 3$

9. $\sum_{k=0}^{\infty} \dfrac{x^k(\ln 2)^k}{k!}; \quad$ all x

11. $1 + \sum_{k=1}^{\infty} \dfrac{\dfrac{3}{2}\left(\dfrac{3}{2} - 1\right) \cdots \left(\dfrac{3}{2} - k + 1\right)x^k}{k!}; \quad |x| < 1$

13. See Example 8.

15. $\sum_{k=0}^{\infty} \dfrac{(-1)^k x^{2k}}{(2k + 1)!}; \quad$ all x

17. $\sum_{k=1}^{\infty} \dfrac{\sqrt{2}}{2}\left(x - \dfrac{\pi}{4}\right)^k\left[\dfrac{(-1)^{(k-1)(k-2)/2}}{(k - 1)!} + \dfrac{\pi}{4}\dfrac{(-1)^{(k-1)k/2}}{k!}\right]$
$$+ \dfrac{\pi\sqrt{2}}{8}$$

19. $1 + \sum_{k=1}^{\infty} \dfrac{2^{2k-1}x^{2k}(-1)^k}{(2k)!}; \quad$ all x

21. $1 + x^2. \quad$ Use $\dfrac{d}{dx}(\tan x) = \sec^2 x.$

23. $\sum_{k=0}^{\infty} \dfrac{(-1)^k x^{4k}}{(2k)!}$

25. $\sum_{k=0}^{\infty} \dfrac{x^{2k+1}}{(2k)!}$

27. 0.035   **29.** 0.095   **31.** 0.485

33. $1 + x - \sum_{k=2}^{\infty}(-1)^k \dfrac{1 \cdot 3 \cdots (2k - 3)}{k!}x^k;$
$$|x| < \dfrac{1}{2}$$

35. $27 + \sum_{k=1}^{\infty} \dfrac{\dfrac{3}{2}\left(\dfrac{3}{2} - 1\right) \cdots \left(\dfrac{3}{2} - k + 1\right)x^k}{3^{k-3}\,k!}; \quad |x| < 3$

Review Exercises—Chapter 13

1. $2x - \dfrac{4}{3}x^3 + \dfrac{4}{15}x^5$

3. $\dfrac{\pi\sqrt{2}}{8} + \dfrac{\sqrt{2}}{2}\left(1 - \dfrac{\pi}{4}\right)\left(x - \dfrac{\pi}{4}\right)$

$\quad - \dfrac{\sqrt{2}}{4}\left(2 + \dfrac{\pi}{4}\right)\left(x - \dfrac{\pi}{4}\right)^2 + \dfrac{\sqrt{2}}{12}\left(\dfrac{\pi}{4} - 3\right)\left(x - \dfrac{\pi}{4}\right)^3$

5. $1 + x^2$

7. $\dfrac{\pi}{4} + \dfrac{1}{2}(x - 1) - \dfrac{1}{4}(x - 1)^2 + \dfrac{1}{12}(x - 1)^3$

9. $1 + (\log 2)x + \dfrac{(\log 2)^2}{2}x^2 + \dfrac{(\log 2)^3}{6}x^3$

11. $\left|R_3\left(\dfrac{17\pi}{90}\right)\right| \le 9.89775 \cdot 10^{-7}$

13. $|R_2(35)| \le 0.000079$

15. $\left|R_3\left(\dfrac{1}{4}\right)\right| < 0.000488$

21. $(2, 4)$ **23.** $[1, 3)$ **25.** $(-1, 1]$

27. Replace $k - 1$ by k on the right.

29. $a_n = \dfrac{c^n a_0}{n!}(-1)^n;\ a_0 e^{-cx}$ **31.** $\sum_{k=0}^{\infty}(-1)^k x^{4k};\ |x| < 1$

33. $1 + \displaystyle\sum_{k=1}^{\infty} \dfrac{\dfrac{1}{3}\left(\dfrac{1}{3} - 1\right)\cdots\left(\dfrac{1}{3} - k + 1\right)x^{2k}}{k!}$

35. $1 - \dfrac{1}{2\pi^2}\left(x - \dfrac{\pi^2}{4}\right)^2 + \dfrac{1}{\pi^2}\left(x - \dfrac{\pi^2}{4}\right)^3$

37. $\displaystyle\sum_{k=0}^{\infty} \dfrac{x^{2k+2}}{k!}$ **39.** $e^{1/2} \approx 1.649$

CHAPTER 14

Exercise Set 14.1

	Vertex	Focal Length	Axis	Focus
1.	$(0, 0)$	$\dfrac{1}{8}$	$x = 0$	$\left(0, \dfrac{1}{8}\right)$
3.	$(0, 2)$	$\dfrac{1}{8}$	$x = 0$	$\left(0, \dfrac{15}{8}\right)$
5.	$(1, 0)$	$\dfrac{1}{12}$	$y = 0$	$\left(\dfrac{13}{12}, 0\right)$
7.	$\left(0, \dfrac{1}{4}\right)$	-2	$x = 0$	$\left(0, -\dfrac{7}{4}\right)$
9.	$(2, 1)$	2	$x = 2$	$(2, 3)$

	Vertex	Focal Length	Axis	Focus
11.	$(-3, 2)$	$\dfrac{3}{2}$	$y = 2$	$\left(-\dfrac{3}{2}, 2\right)$
13.	$(-2, 1)$	1	$y = 1$	$(-1, 1)$
15.	$(-1, -2)$	$\dfrac{3}{2}$	$x = -1$	$\left(-1, -\dfrac{1}{2}\right)$
17.	$(-3, 1)$	$\dfrac{5}{2}$	$x = -3$	$\left(-3, \dfrac{7}{2}\right)$
19.	$(1, -2)$	$\dfrac{1}{2}$	$y = -2$	$\left(\dfrac{3}{2}, -2\right)$

1.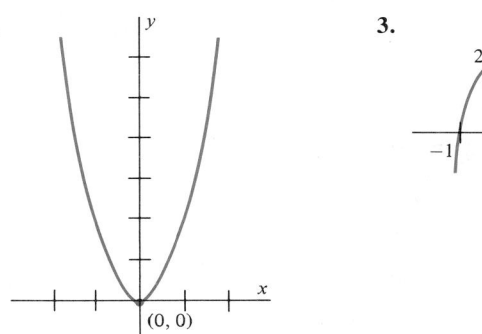

3.

5.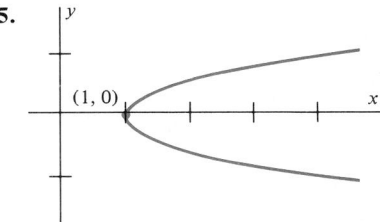

7.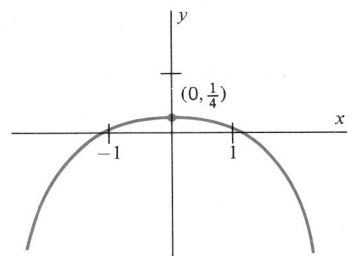

9.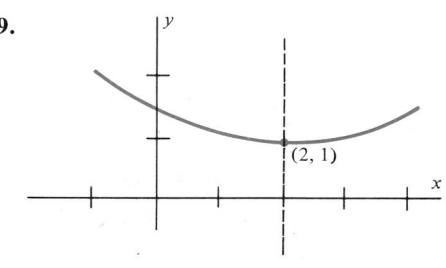

11.

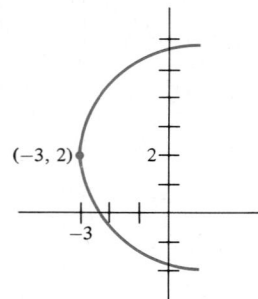

(−3, 2)

−3

13.

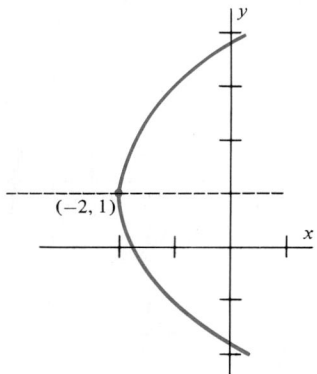

(−2, 1)

15.

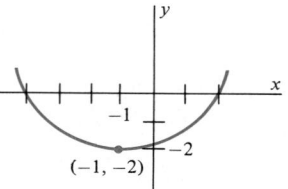

−1

−2

(−1, −2)

17.

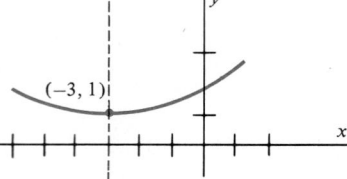

(−3, 1)

19.

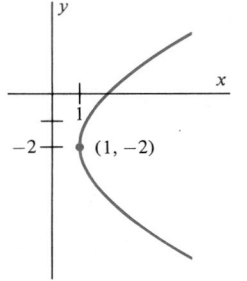

1

−2 (1, −2)

21. $x^2 = 12y$

23. $(x - 1)^2 = 32(y - 3)$

25. $(y + 6)^2 = 16(x + 1)$

27. $(y - 1)^2 = 16x$

33. $\dfrac{8}{3}$

35. $\dfrac{32}{2}$

37. -0.85

Exercise Set 14.2

	Center	Major Axis	Minor Axis	Foci
1.	$(0, 0)$	3	1	$(\pm 2\sqrt{2}, 0)$
3.	$(0, 0)$	4	2	$(\pm 2\sqrt{3}, 0)$
5.	$(0, 0)$	5	3	$(0, \pm 4)$
7.	$(2, 1)$	2	$\sqrt{3}$	$(2, 0), (2, 2)$
9.	$(-3, 1)$	4	2	$(-3 \pm 2\sqrt{3}, 1)$
11.	$(1, -1)$	3	2	$(1, -1 \pm \sqrt{5})$
13.	$(-3, 1)$	5	3	$(1, 1), (-7, 1)$
15.	$(1, -1)$	$\sqrt{3}$	$\sqrt{2}$	$(0, -1), (2, -1)$

1.

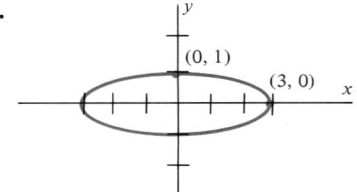

(0, 1)

(3, 0)

3.

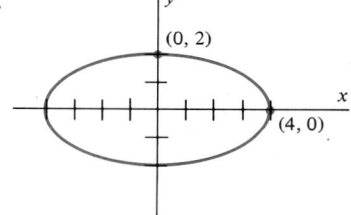

(0, 2)

(4, 0)

5.

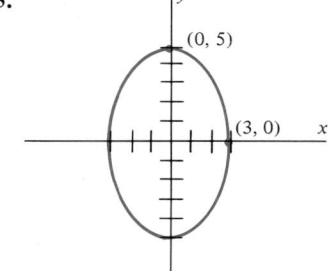

(0, 5)

(3, 0)

7.

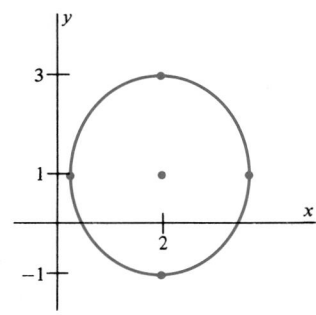

9.

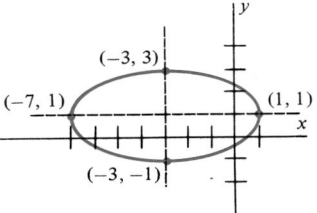

11.

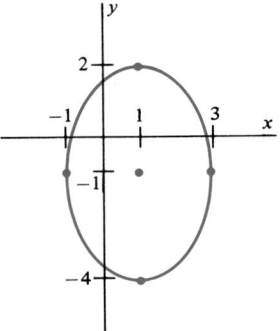

13.

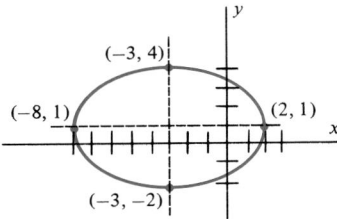

15.

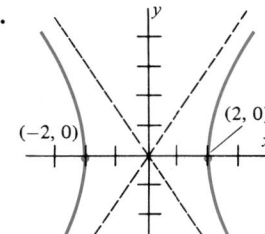

17. $\dfrac{x^2}{16} + \dfrac{y^2}{9} = 1$

19. $\dfrac{x^2}{25} + \dfrac{y^2}{9} = 1$

21. $\dfrac{(x-4)^2}{16} + \dfrac{(y-2)^2}{25} = 1$

23. $\dfrac{(x+2)^2}{16} + \dfrac{(y-6)^2}{25} = 1$

25. 50π **27.** $\dfrac{2b^2}{a}$ **29.** True

33. $\dfrac{(x+1)^2}{36} + \dfrac{(y-3)^2}{27} = 1$

35. 48π **37.** 180,945 ft-lb

Exercise Set 14.3

	Center	Asymptotes
1.	$(0, 0)$	$y = \pm\dfrac{3}{2}x$
3.	$(0, 0)$	$y = \pm\dfrac{3}{\sqrt{5}}x$
5.	$(0, 0)$	$y = \pm 2x$
7.	$(0, 0)$	$y = \pm\dfrac{\sqrt{7}}{2}x$
9.	$(-3, 4)$	$y - 4 = \pm\dfrac{3}{2}(x + 3)$
11.	$(2, -3)$	$y + 3 = \pm\dfrac{3}{2}(x - 2)$
13.	$(0, 0)$	$y = \pm\dfrac{1}{2}x$
15.	$(\sqrt{2}, \sqrt{2})$	$y - \sqrt{2} = \pm\sqrt{2}(x - \sqrt{2})$
17.	$(-2, 2)$	$y - 2 = \pm 3(x + 2)$
19.	$(-5, 7)$	$y - 7 = \pm\dfrac{5}{3}(x + 5)$

1.

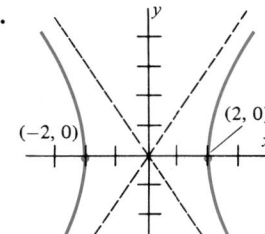

3.

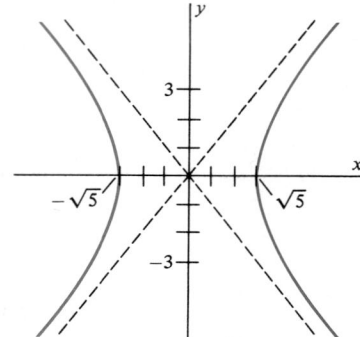

5.

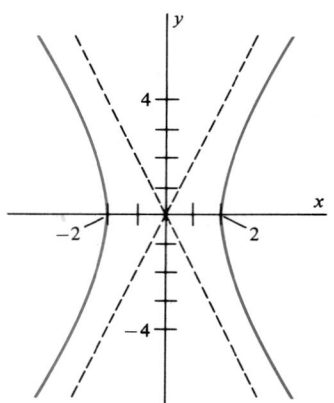

7.

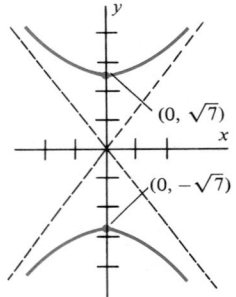

9.

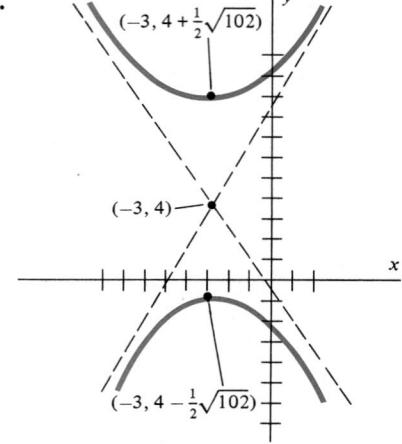

11.

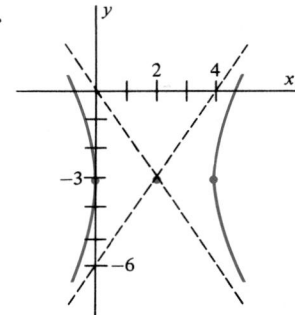

13.

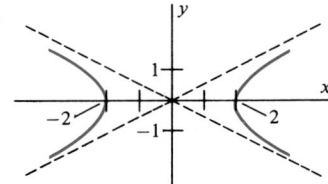

15.

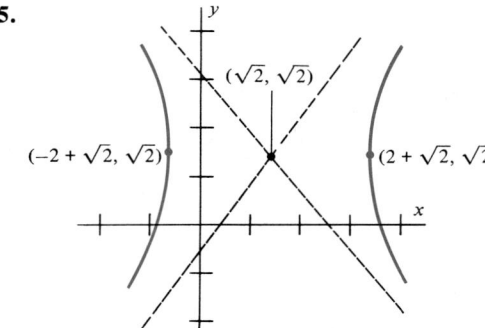

17.

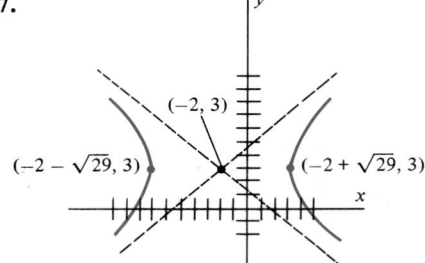

19.

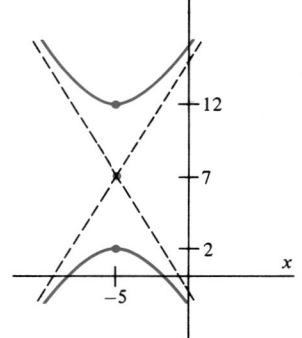

21. $x^2 - \dfrac{y^2}{3} = 1$

23. $\dfrac{(y-2)^2}{4} - x^2 = 1$

25. $\dfrac{(x+1)^2}{16} - \dfrac{(y-2)^2}{64} = \dfrac{1}{5}$

27. $\dfrac{y^2}{16} - \dfrac{x^2}{9} = 1$

29. They have the same asymptotes: $y = \pm\dfrac{b}{a}x$.

31. $b^2 x_0 x - a^2 y_0 y = a^2 b^2$

Exercise Set 14.4

1. $\theta = \dfrac{\pi}{8}$ **3.** $\theta = \dfrac{\pi}{4}$

5. $\theta = \dfrac{1}{2}\,\mathrm{Tan}^{-1}\dfrac{4}{3}$

7. $\theta = \dfrac{\pi}{4}, \quad \dfrac{x'^2}{4} + \dfrac{y'^2}{16} = 1$

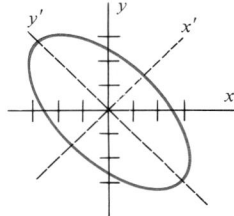

9. $\theta = \dfrac{\pi}{6}, \quad x'^2 = -(y'+4)$

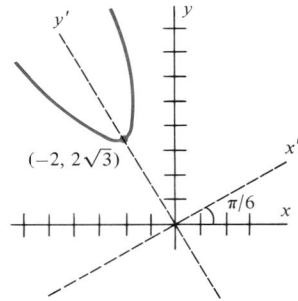

$(-2, 2\sqrt{3})$ $\pi/6$

11. $\theta = \mathrm{Sin}^{-1}\dfrac{3}{5}, \quad y'^2 = 2(x'-2)$

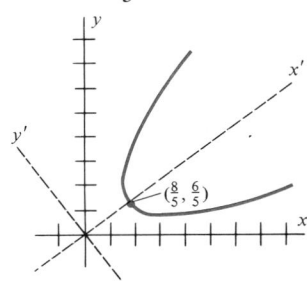

$\left(\dfrac{8}{5}, \dfrac{6}{5}\right)$

13. $\theta = -\dfrac{\pi}{6}, \quad \dfrac{x'^2}{2} + \dfrac{y'^2}{4} = 1$

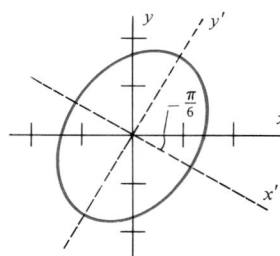

$-\dfrac{\pi}{6}$

15. $\theta = \dfrac{\pi}{4}, \quad x'^2 = y'$

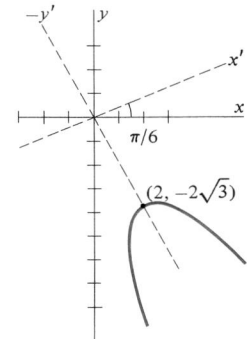

$\pi/6$ $(2, -2\sqrt{3})$

17. $\left(\mp\dfrac{3}{2}, \pm\dfrac{3\sqrt{3}}{2}\right)$ **19.** $\left(\pm\dfrac{3}{2}, \pm\dfrac{3\sqrt{3}}{2}\right)$

Review Exercises—Chapter 14

1. $(x-3)^2 + (y+4)^2 = 36$

3. $(x-2)^2 + (y-3)^2 = 16$

5. $(y-4)^2 = 8x$

7. $y = 3x^2 - 6x + 5$

9. $\dfrac{(x+1)^2}{9} + \dfrac{(y-1)^2}{5} = 1$

11. Outside

13. $(-3, -5)$ and $(2, 0)$

15. $(2, 2)$ and $(-2, 2)$

17. $(x-1)^2 + (y+3)^2 = 8,$ circle

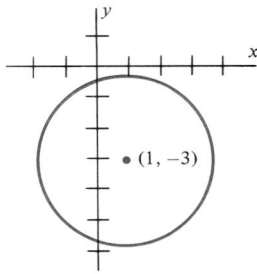

$\bullet\ (1, -3)$

19. $\dfrac{y^2}{4} - \dfrac{x^2}{2} = 1$ hyperbola

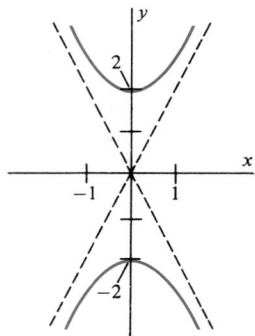

21. $\dfrac{x^2}{12} + \dfrac{y^2}{16} = 1$,   ellipse

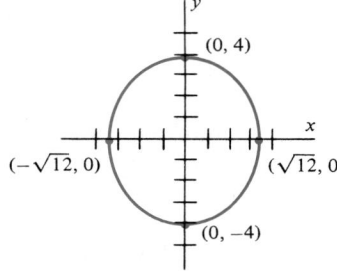

23. $\dfrac{x^2}{4} + \dfrac{y^2}{9} = 1$,   ellipse

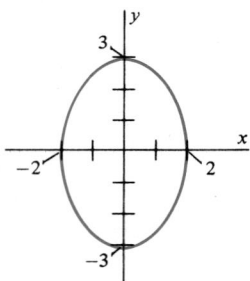

25. $\dfrac{x^2}{9} - \dfrac{y^2}{4} = 1$,   hyperbola

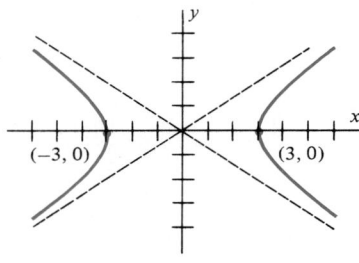

27. $y^2 = -(x - 2)$,   parabola

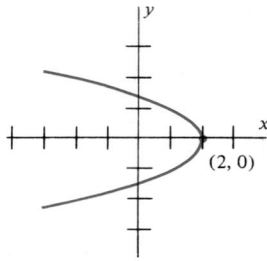

29. $(x + 7)^2 + (y - 5)^2 = 4$,   circle

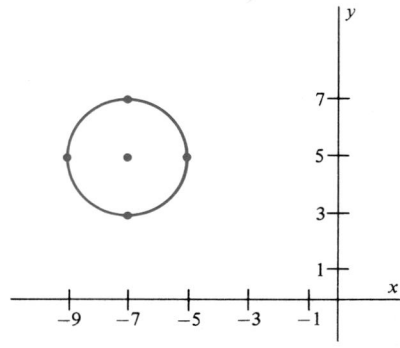

31. $\dfrac{x^2}{8} - \dfrac{y^2}{9} = 1$,   hyperbola

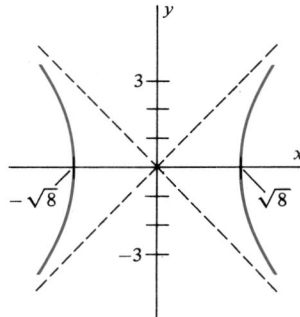

33. $\dfrac{x^2}{25} + \dfrac{y^2}{4} = 1$,   ellipse

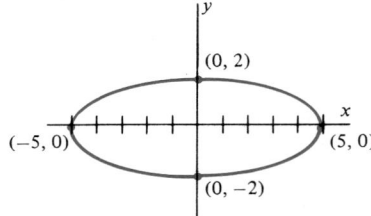

35. $\dfrac{y^2}{4} - \dfrac{x^2}{4} = 1$,   hyperbola

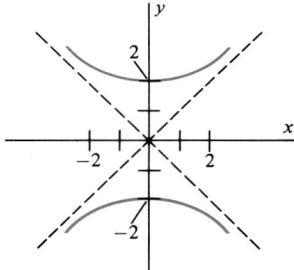

37. $(x - a)^2 + (y + 2a)^2 = 1$,   circle (Assume $a > 0$.)

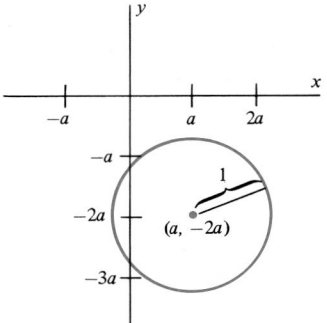

39. $y^2 - x^2 = 2$,   hyperbola

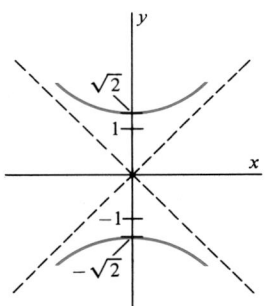

41. $(x - 1)^2 + (y - 3)^2 = 7$,   circle

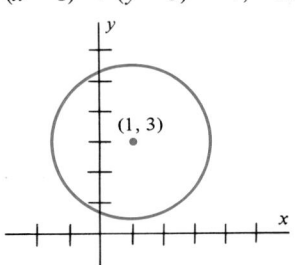

43. $\dfrac{y^2}{25} - \dfrac{x^2}{9} = 1$,   hyperbola

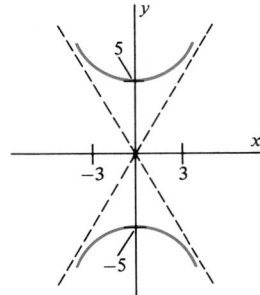

45. $y^2 = \dfrac{1}{8}(x - 6)$,   parabola

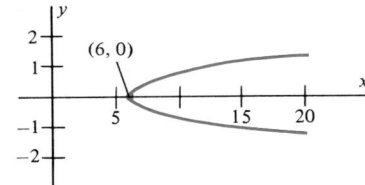

CHAPTER 15

Exercise Set 15.1

1. $P = (0, 1)$
3. $P = (0, 0)$
5. $P = (1, -1)$
7. $P = (3, 0)$
9. $P = \left(\sqrt{2}, \dfrac{\pi}{4}\right)$
11. $P = (-3, 0)$
13. $P = \left(-2, \dfrac{2\pi}{3}\right)$
15. $P = \left(-2\sqrt{2}, \dfrac{15\pi}{4}\right)$
17. Symmetric about the x-axis
19. Symmetric about the y-axis
21. x-axis, origin, y-axis
23. x-axis, y-axis, origin
25. $r = 2$
27. $r^2 = \dfrac{4}{1 + 3\cos^2\theta}$
29. $r = \dfrac{6}{\cos\theta} = 6\sec\theta$
31. $r = -2\sin\theta$
33. $x^2 + y^2 - 4y = 0$
35. $x^4 + x^2y^2 - y^2 = 0$
37. $x = 4$
39. $y^2 = 1 + 2x$

41.

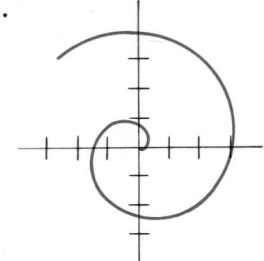

43.

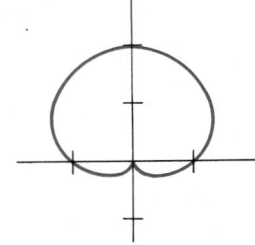

Exercise Set 15.2

1.

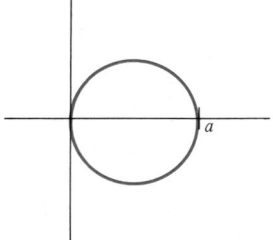

45.

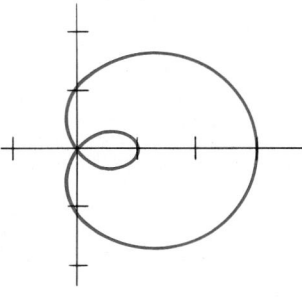

3.

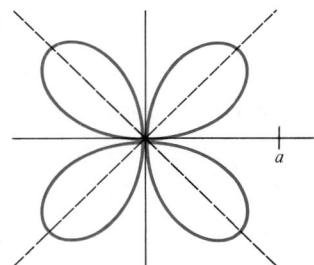

5.

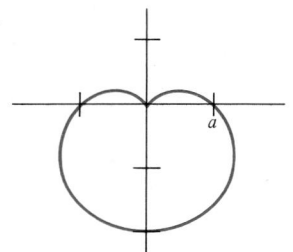

47.

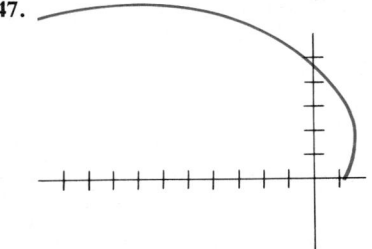

7.

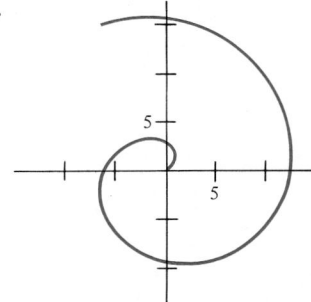

49.

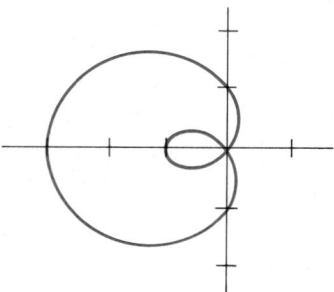

9.

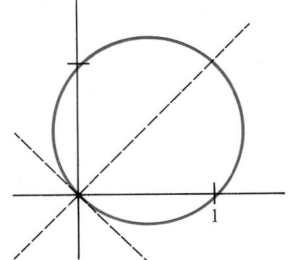

51. Center $\left(\sqrt{2}, \dfrac{3\pi}{4}\right)$ and radius $\sqrt{2}$

53. Center $\left(\dfrac{a}{2}, 0\right)$ and radius $\dfrac{a}{2}$

57. $\dfrac{9\pi}{4}$

59. $(0, 0)$, $(4, 0)$, $\left(4\sqrt{2}, \dfrac{\pi}{2}\right)$

11. $\left(\sqrt{2}, \dfrac{\pi}{4}\right)$, $\left(-\sqrt{2}, \dfrac{5\pi}{4}\right)$, $(0, \theta)$

13. $\left(\dfrac{3}{2}, \dfrac{\pi}{6}\right), \left(\dfrac{3}{2}, \dfrac{5\pi}{6}\right), (0, \theta)$

15. $\left(\dfrac{3a}{2}, \dfrac{\pi}{3}\right), \left(\dfrac{3a}{2}, \dfrac{5\pi}{3}\right), (0, \theta)$

17. $\left(a, \dfrac{\pi}{2}\right), \left(a, \dfrac{3\pi}{2}\right), (0, \theta)$

19. $\left(\dfrac{1}{\pi/4 + k\pi}\right), k = 0, \pm 1, \pm 2, \ldots$ **21.** $x^2 = a^2 \mp 2ay$

23. $y^2 = 1 - 2x$

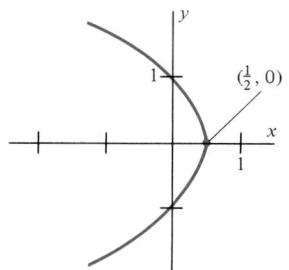

25. $\dfrac{\left(x + \dfrac{4}{3}\right)^2}{\left(\dfrac{8}{3}\right)^2} + \dfrac{y^2}{\left(\dfrac{4}{\sqrt{3}}\right)^2} = 1$ (ellipse)

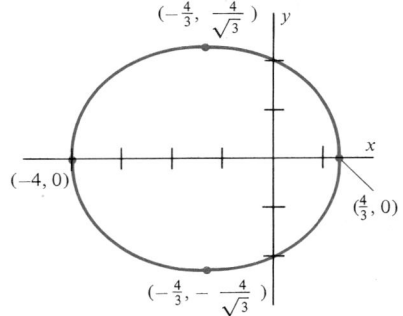

27. $y^2 = 4x + 4$ (parabola)

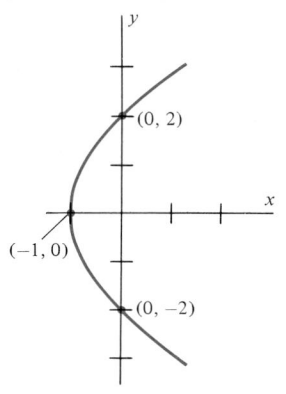

Exercise Set 15.3

1. $\dfrac{3\pi}{4}$

3. π

5. $\dfrac{9}{4}\pi + 4$

7. $\dfrac{33\pi}{8} + 4\sqrt{2} + \dfrac{1}{4}$

9. π

11. $\dfrac{\pi a^2}{2}$

13. $\dfrac{\pi a^2}{4}$

15. $\dfrac{\pi}{2}$

17. $\dfrac{3a^2\pi}{2} - 4a^2$

19. $\dfrac{\pi - 2}{2}$

21. $\dfrac{1}{4}e^{2\pi} - \dfrac{\pi^3}{6} - \dfrac{1}{4}$

23. $2\pi + 3\sqrt{3}$

25. π

27.

	$n = 10$	$n = 100$	$n = 200$
a.	2.587376	2.494899	2.489281
b.	4.885851	4.964704	4.966249
c.	0.138355	0.091575	0.089359

Exercise Set 15.4

1. $y = x + 6$

3. $y = \dfrac{2}{3}x + \dfrac{4}{3}$

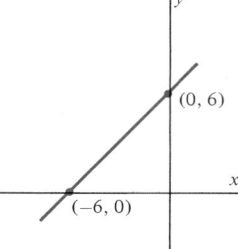

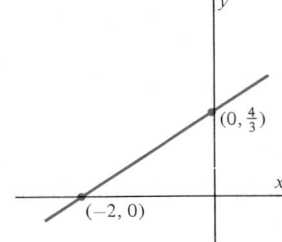

5. $y = x - 2;$ $x \geq 1$

7. $y = x^3$

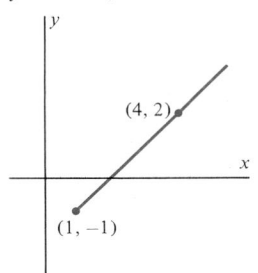

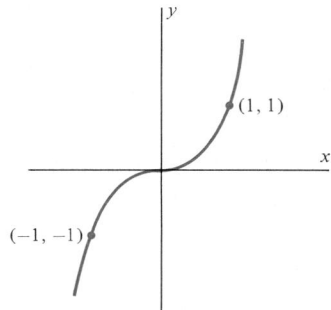

9. $x^2 - y^2 = 1$

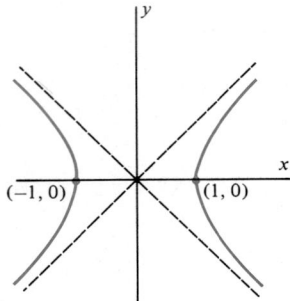

11. $y^2 = 4x^2(1 - x^2)$

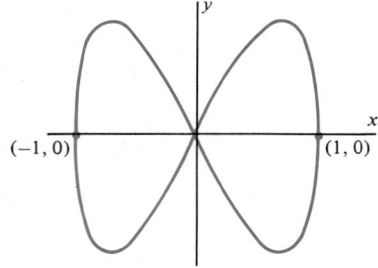

13. Slope $= 4$, $y = 4x - 3$

15. Slope $= -\sqrt{3}$, $y = -\sqrt{3}x + 2$

17. Slope $= -48$, $y = -48x + 29$

19. Slope $= -1$, $y = -x + 2$

21. 0   **23.** $\dfrac{5}{\sqrt{3}}$ **25.** $\dfrac{-2}{\pi}$

27. Vertical tangent: $(1, 0)$, $(-1, 0)$
Horizontal tangent: $(0, 1)$, $(0, -1)$

29. Vertical tangent at $(4, 2)$
Horizontal tangent at $(5, -1)$

31. Vertical tangent at $(6, 0)$
Horizontal tangent at $\left(\dfrac{27}{4}, \dfrac{1}{4}\right)$

33. $(1, 5)$

35. $x(\theta) = f(\theta)\cos\theta = a\sin 3\theta \cdot \cos\theta$
$y(\theta) = f(\theta)\sin\theta = a\sin 3\theta \cdot \sin\theta$

37. $(0, 0)$, $\left(\dfrac{3}{2}, \dfrac{\pi}{6}\right)$, $\left(\dfrac{3}{2}, \dfrac{5\pi}{6}\right)$

39. Vertical tangent at $(2, 0)$, $\left(\dfrac{1}{2}, \dfrac{2\pi}{3}\right)$, $\left(\dfrac{1}{2}, \dfrac{4\pi}{3}\right)$

41. a. $y = -(x^2 - 8x + 8)$
$y = x + 2$
b. Yes. At $(2, 4)$ and $(5, 7)$.
c. The particles collide at $t = -2$ and at $t = 1$ at the points
$(2, 4)$ and $(5, 7)$.

Exercise Set 15.5

1. $\dfrac{8}{27}(2^{3/2} - 1)$ **3.** $2(3^{3/2} - 1)$ **5.** 36

7. $(5^{3/2} - 8^{3/2}) + (12\sqrt{2} - 3\sqrt{5} + 12\ln(\sqrt{5} - 1)$
$\qquad\qquad\qquad\qquad\qquad - 12\ln(\sqrt{8} - 2))$

9. 3   **11.** $\sqrt{2}(e^{\pi} - 1)$

13. $\dfrac{16\pi}{3}(8^{3/2} - 8)$ **15.** $\dfrac{\pi}{6}[27 - 5^{3/2}]$

17. $\pi\left[133\sqrt{5} - \dfrac{7}{2}\sqrt{2} + \dfrac{3}{2}\ln(\sqrt{5} + 2) - \dfrac{3}{2}\ln(\sqrt{2} + 1)\right]$

19. 4   **21.** $\dfrac{\sqrt{5}}{2}[e^{2\pi} - 1]$ **23.** 4π

25. 3π **27.** $\sqrt{2}(e^{2\pi} - 1)$ **29.** $\dfrac{2}{5}\sqrt{2}\pi(e^{2\pi} - 1)$

31. $S = \displaystyle\int_a^b 2\pi x(t)\sqrt{(x'(t))^2 + (y'(t))^2}\, dt,\quad a \le t \le b$

33. $\dfrac{\pi}{64}[204\sqrt{2} - 36\sqrt{5} - \ln(17 + 12\sqrt{2}) + \ln(9 + 4\sqrt{5})]$

Review Exercises—Chapter 15

1. $x^2 + y^2 = 25$

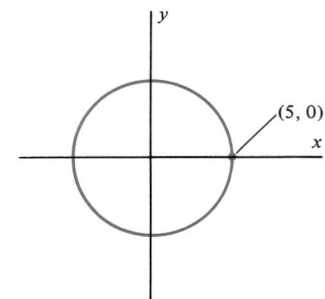

3. $r = \theta/\pi$

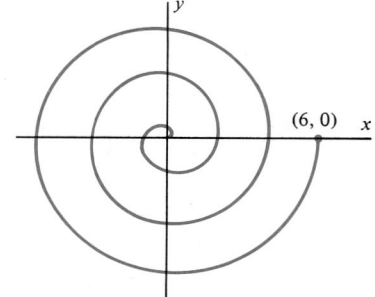

5. $y = 2x\sqrt{1 - x^2}$

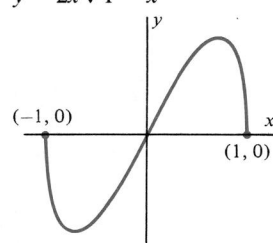

7. $y = x \tan \pi\sqrt{x^2 + y^2}$ **9.** $y^2 = 4x^2(1 - x^2)$

11. $x(t) = 3 \cos t$
$y(t) = 2 \sin t$

13. $x(t) = 2t^2 + 4t + 5$
$y(t) = t$

15.

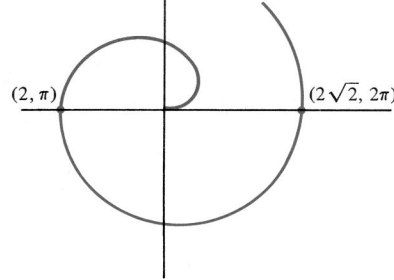

17. $x = \dfrac{y^2}{9} - 1$ (parabola)

19.

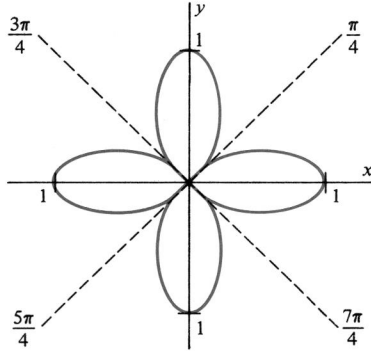

21.

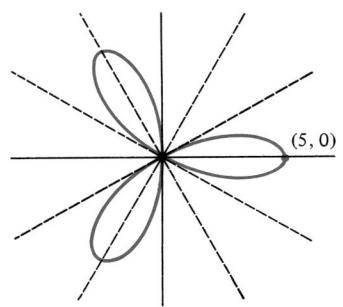

23.

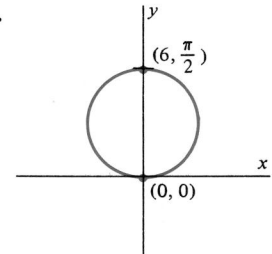

25.

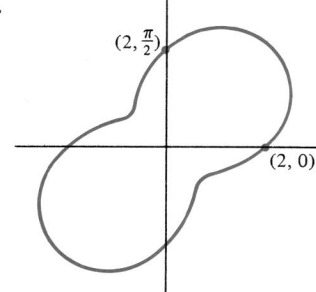

27. The y-axis.   **29.** 6π

31. $\dfrac{3a^2\pi}{2}$ **33.** 9π

35. $\dfrac{1}{27}(22^{3/2} - 13^{3/2})$ **37.** $\displaystyle\int_0^1 3\sqrt{1 + t^4}\,dt \approx 3.2683$

39. π **41.** $\dfrac{\pi}{2}$

43. $2\sqrt{2}$ **45.** $\dfrac{\pi}{8} - \dfrac{1}{4}$

47. 4π **49.** $\dfrac{8\pi}{3} - 2\sqrt{3}$

CHAPTER 16

Exercise Set 16.1

1.

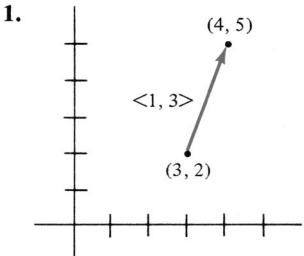

3.

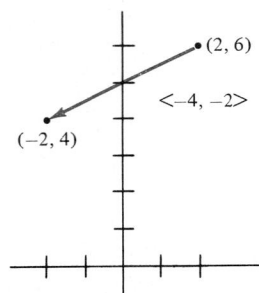

5.

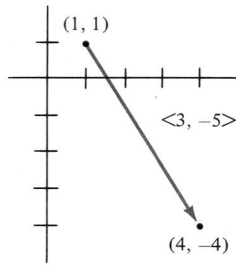

7. $\langle -1, 2 \rangle$

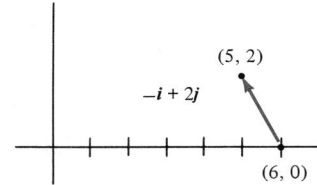

9. $\langle -3, 2 \rangle$

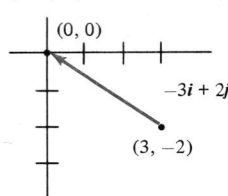

11. $\langle -3, -3 \rangle$

13. $\langle -3, -14 \rangle$

15. $\langle -6, -6 \rangle$

17. $\langle -1, 9 \rangle$

19. $\langle 10, 14 \rangle$

21. $\langle -1, 5 \rangle$

23. $\langle 7, 11 \rangle$

25. $\langle 4, 6 \rangle$

27. $\langle 19, 19 \rangle$

29. $\langle 12, 18 \rangle$

31. $\dfrac{\pi}{2}$

33. $\dfrac{5\pi}{6}$

35. $\dfrac{5\pi}{3}$

37. $\sqrt{37}$

39. $|\alpha|\sqrt{1 + \alpha^2}$

41. $\sqrt{2}$

43. $3\sqrt{5}$

45. $-i$

47. $\left\langle -\dfrac{1}{2}, \dfrac{\sqrt{3}}{2} \right\rangle$

49. $t\left\langle \dfrac{1}{\sqrt{10}}, \dfrac{3}{10} \right\rangle$

51. a. $D = (2, 3)$
   b. $D = (10, 3)$ or $(0, -1)$

61. When v and w have the same direction.

Exercise Set 16.2

1. 1

3. -28

5. -15

7. $-\dfrac{7}{\sqrt{170}}$

9. a. 14

b. -28

c. -14

d. 42

e. 78

f. -56

g. $\dfrac{14}{\sqrt{34}}$

h. $-\dfrac{1}{\sqrt{13}}$

i. $\left\langle -\dfrac{7}{5}, \dfrac{14}{5} \right\rangle$

j. $\left\langle \dfrac{16}{13}, \dfrac{24}{13} \right\rangle$

11. a. $\dfrac{7}{\sqrt{26}}$

b. 2

c. $\dfrac{7}{\sqrt{13}}$

d. $\left\langle \dfrac{35}{26}, \dfrac{7}{26} \right\rangle$

e. $\left\langle \dfrac{17}{26}, -\dfrac{85}{26} \right\rangle$

f. $\left\langle \dfrac{51}{13}, \dfrac{35}{13} \right\rangle$

13. b, and **c.**

15. $\left\langle \dfrac{1}{\sqrt{5}}, -\dfrac{2}{\sqrt{5}} \right\rangle$ and $\left\langle -\dfrac{1}{\sqrt{5}}, \dfrac{2}{\sqrt{5}} \right\rangle$

19. $\left\langle \dfrac{4}{5}, -\dfrac{2}{5} \right\rangle + \left\langle \dfrac{1}{5}, \dfrac{2}{5} \right\rangle$

21. $4\sqrt{2}$

23. $1000\sqrt{2}$ ft-lb

25. v and w being parallel and of same direction.

Exercise Set 16.3

1.

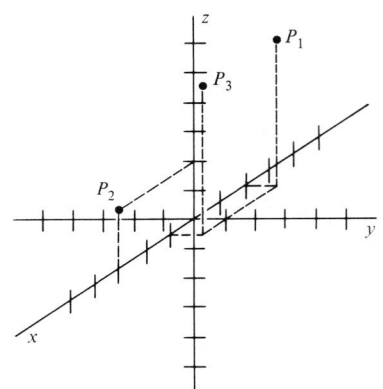

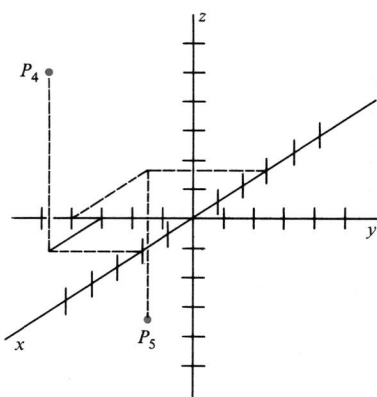

3. a. $2\sqrt{2}$ **b.** $6\sqrt{2}$ **c.** $\sqrt{22}$ **d.** $\sqrt{a^2 + b^2 + c^2}$

5. $(-3, 3, -3)$

7. Center $(0, 0, 2)$ and radius 3

9. Center $(2, -1, 3)$ and radius 4

11. Center $(3, -1, -2)$ and radius 5

13. $\langle -16, 29, 5 \rangle$ **19.** 1

15. $\langle 12, -13, 15 \rangle$ **21.** $\langle -12, 1, -7 \rangle$

17. $9\sqrt{6}$ **23.** -6

25. $\left\langle \dfrac{3}{\sqrt{17}} + \dfrac{5}{\sqrt{35}}, \dfrac{-2}{\sqrt{17}} - \dfrac{1}{\sqrt{35}}, \dfrac{2}{\sqrt{17}} + \dfrac{3}{\sqrt{35}} \right\rangle$

27. $\left\langle \dfrac{23}{7}, \dfrac{-23}{35}, \dfrac{69}{35} \right\rangle$ **29.** $\dfrac{-3}{\sqrt{6}}$

31. a. $\langle 2, 1, 2 \rangle$ **b.** $\langle 4, -1, -5 \rangle$ **c.** $\langle 6, 3, -1 \rangle$

33. a. $\cos \alpha = \dfrac{3}{\sqrt{14}}$, $\cos \beta = \dfrac{-1}{\sqrt{14}}$, $\cos \gamma = \dfrac{2}{\sqrt{14}}$

b. $\cos \alpha = \dfrac{6}{\sqrt{41}}$, $\cos \beta = -\dfrac{2}{\sqrt{41}}$, $\cos \gamma = \dfrac{1}{\sqrt{41}}$

c. $\langle -1, 2, -4 \rangle$

35. a. $\left\langle \dfrac{2}{3}, -\dfrac{4}{3}, \dfrac{4}{3} \right\rangle$ and $\left\langle -\dfrac{2}{3}, \dfrac{4}{3}, -\dfrac{4}{3} \right\rangle$

b. $a = -3$

c. $-\dfrac{5}{\sqrt{14}}i + \dfrac{10}{\sqrt{14}}j - \dfrac{15}{\sqrt{14}}k$

d. $a = -\dfrac{3}{2}$, $b = \dfrac{1}{2}$

37. $t = 2$ **39. c.**

Exercise Set 16.4

1. $-7i + 8j + 11k$ **3.** $5i - 4j - k$

5. 12

7. $3i + 10j + k$

9. $-16i + 8k$

11. $37i + 30j - 11k$

13. 40

15. $5\sqrt{5}$

17. 8

19. $\dfrac{5}{2}\sqrt{2}$

21. $\dfrac{5\sqrt{10}}{2}$

23. $\dfrac{23}{2}$

25. k or $-k$

27. $\dfrac{i}{\sqrt{3}} - \dfrac{j}{\sqrt{3}} - \dfrac{k}{\sqrt{3}}$ and $-\dfrac{i}{\sqrt{3}} + \dfrac{j}{\sqrt{3}} + \dfrac{k}{\sqrt{3}}$

31. 43

Exercise Set 16.5

1. $x(t) = 1 + t$, $y(t) = 2 + t$, $z(t) = 3 - t$

3. $x(t) = 3t$, $y(t) = -t$, $z(t) = 5t$

5. $x(t) = -4 + t$, $y(t) = 2 + 3t$, $z(t) = 1 + 2t$

7. $x(t) = 1 - 10t$, $y(t) = 2 + 3t$, $z(t) = t$

9. $\dfrac{x - 7}{1} = \dfrac{y + 6}{4} = \dfrac{z - 3}{-2}$

11. $\dfrac{x - 0}{1} = \dfrac{y + 6}{8} = \dfrac{z - 4}{4}$

13. $\dfrac{x - 3}{3} = \dfrac{y + 1}{-1} = \dfrac{z - 5}{5}$

15. $x(t) = 3t$, $y(t) = 2t$, $z(t) = 5t$

17. $x(t) = 4t - 4$, $y(t) = -2t + 2$, $z(t) = 3t - 3$

19. $i + 4j - 2k$ **21.** $4i - 2j + 3k$

23. $-\dfrac{a \cdot b}{|b|^2}$ **25.** $\sqrt{3}$

27. $\dfrac{\sqrt{210}}{3}$ **29.** $\dfrac{1}{15}\sqrt{1270}$

31. d. $\dfrac{16}{\sqrt{53}}$ **33.** $(2, 1, 2)$

35. $x + 2y - z = 8$ **37.** $2x - 2y + 3z = 19$

39. $x + 2y + z = 1$ **41.** $29x - 6y - 15z = -16$

43. $x - y = -2$ **45.** $2i - 3j + k$

47. $3x - 6y + 2z = 31$

49. $x(t) = \dfrac{27}{6} + \dfrac{7}{6}t$, $y(t) = \dfrac{9}{6} + \dfrac{5}{6}t$, $z(t) = t$

51. $\theta = \text{Cos}^{-1} \dfrac{1}{\sqrt{57}}$

53. $\dfrac{27}{\sqrt{38}}$

55. $x(t) = 2 + 2t, \; y(t) = 4 + 3t, \; z(t) = -3 - 7t$

57. $3x - y - z = 7$

Exercise Set 16.6

1. a. $\left(\sqrt{2}, \dfrac{\pi}{4}, 0\right)$

b. $\left(2, \dfrac{\pi}{6}, 3\right)$

c. $\left(\sqrt{2}, \dfrac{3\pi}{4}, -2\right)$

d. $\left(2, \dfrac{2\pi}{3}, 4\right)$

e. $\left(3, \dfrac{\pi}{2}, -5\right)$

f. $\left(2, \dfrac{3\pi}{4}, \sqrt{2}\right)$

3. a. $\left(1, 0, \dfrac{\pi}{2}\right)$

b. $\left(2, \dfrac{\pi}{4}, \dfrac{\pi}{4}\right)$

c. $\left(2, \dfrac{7\pi}{4}, \dfrac{\pi}{4}\right)$

d. $\left(2\sqrt{2}, \dfrac{\pi}{3}, \dfrac{\pi}{4}\right)$

e. $\left(2\sqrt{2}, \dfrac{5\pi}{6}, \dfrac{3\pi}{4}\right)$

f. $\left(4, \dfrac{3\pi}{4}, \dfrac{\pi}{4}\right)$

5. a. $\left(1, 0, \dfrac{\pi}{2}\right)$

b. $\left(2, -\dfrac{\pi}{4}, \dfrac{\pi}{4}\right)$

c. $\left(2\sqrt{2}, \dfrac{\pi}{3}, \dfrac{\pi}{4}\right)$

d. $\left(2\sqrt{2}, \dfrac{\pi}{4}, \dfrac{\pi}{4}\right)$

e. $\left(2, \dfrac{5\pi}{3}, \dfrac{\pi}{2}\right)$

f. $\left(2\sqrt{2}, \dfrac{\pi}{6}, \dfrac{\pi}{4}\right)$

7. $z = 3$

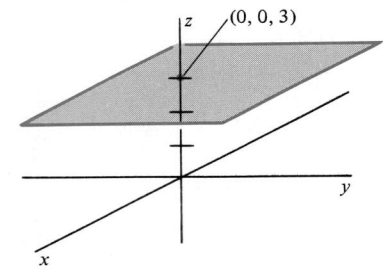

9. $z^2 = 4(x^2 + y^2)$

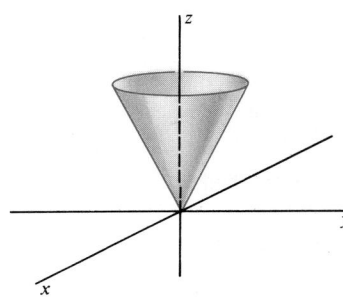

11. $x^2 + y^2 + z^2 = 4$

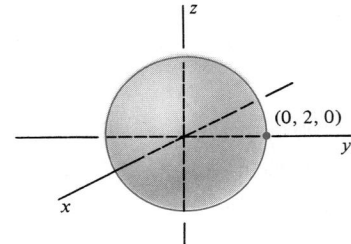

13. $x^2 - y^2 = a^2$

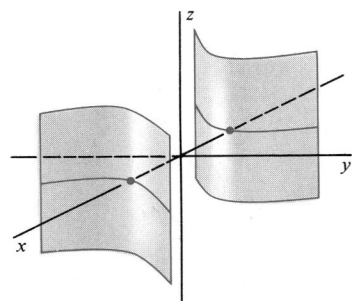

15. $r^2 = 9$ **17.** $r^2 = 9z$

19. $r^2 \cos^2 \theta + z^2 = 4$

21. $r^2 \cos \theta \sin \theta = 4 + ar \cos \theta$

23. $\rho^2 \sin^2 \phi = 9$

25. $\rho \sin^2 \phi = 9 \cos \phi$

27. $\rho^2 = \dfrac{4}{1 - \sin^2 \phi \sin^2 \theta}$

29. $\rho^2 \sin^2 \phi \cos \theta \sin \theta - a\rho \sin \theta \cos \theta = 4$

Exercise Set 16.7

Equation	Name	Center	yz-sections	xz-sections	xy-sections
1. $x^2 + y^2 + z^2 = 10$	sphere	$(0, 0, 0)$	circle*	circle*	circle*
3. $\dfrac{x^2}{4} + \dfrac{y^2}{9} = z^2$	elliptic cone	$(0, 0, 0)$	hyperbola**	hyperbola**	ellipse*

5. $-\dfrac{x^2}{9} - \dfrac{y^2}{4} + z^2 = 1$ hyperboloid of two sheets   (0, 0, 0)   hyperbola   hyperbola   ellipse*

7. $\dfrac{(x + 1)^2}{2} - \dfrac{(y - 1)^2}{3} = z$ hyperbolic paraboloid   none   parabola   parabola   hyperbola**

9. $\dfrac{(x + 3)^2}{4} + \dfrac{(y - 1)^2}{2} - \dfrac{z^2}{2} = 1$ hyperboloid of one sheet   (−3, 1, 0)   hyperbola   hyperbola   ellipse

11. $\dfrac{x^2}{9} + \dfrac{(y + 2)^2}{6} = z$ elliptic paraboloid   none   parabola   parabola   ellipse*

13. $\dfrac{z^2}{4} + \dfrac{y^2}{9} = x$ elliptic paraboloid   none   ellipse*   parabola   parabola

15. $-x^2 + y^2 + z^2 = 1$ hyperboloid of one sheet   (0, 0, 0)   circle   hyperbola   hyperbola

17. $(x - 3)^2 + (y + 1)^2 + (z + 3)^2 = 1$ sphere   (3, −1, 3)   circle*   circle*   circle*

19. $(x + 1)^2 + (y - 1)^2 = 36\left(z - \dfrac{85}{18}\right)$ paraboloid of revolution   none   parabola   parabola   circle*

21. Plane parallel to yz-plane

	Direction of Ruling	Cross-section
23.	z-axis	line
25.	z-axis	hyperbola
27.	z-axis	hyperbola
29.	x-axis	parabola
31.	y-axis	hyperbola
33.	x-axis	parabola
35.	y-axis	parabola

37. $y^2 = 4(x^2 + z^2)$

39. $y = x^2 + z^2$

Review Exercises—Chapter 16

1. $\sqrt{86}$

3. Center (3, −2, 1), radius 2

5. $\dfrac{-1}{\sqrt{77}}$

7. $x(t) = 23t + 2,\; y(t) = -t + 1,\; z(t) = -9t - 3$

9. $7x - 9y + 3z = -22$

11. $8\mathbf{i} + \mathbf{j} - 2\mathbf{k}$

13. $\mathbf{r}(t) = (2 - 12t)\mathbf{i} + 11t\mathbf{j} + 9t\mathbf{k}$

17.

(0, −3, 0)
(0, 3, 0)

19. $\left(\dfrac{\sqrt{6}}{2}, \dfrac{\sqrt{6}}{2}, 1\right)$

21. $2x - y + 6z = 27$

23. **a.** $\dfrac{22}{5}$ **b.** $\dfrac{66}{25}\mathbf{j} + \dfrac{88}{25}\mathbf{k}$

25. $\dfrac{1}{2}\sqrt{118}$ **27.** $4x + z = 18$

29. $\dfrac{x - 1}{3} = \dfrac{y - 2}{7} = \dfrac{z - 3}{-1}$

31. $z^2 = x^2 + y^2,\; z \geq 0$

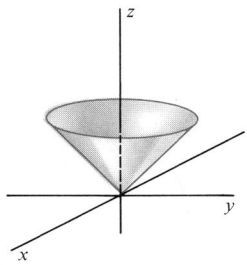

---

*Section may degenerate to point or empty set.
**Section may degenerate to two lines.

33. $(x - 1)^2 + y^2 = 1$

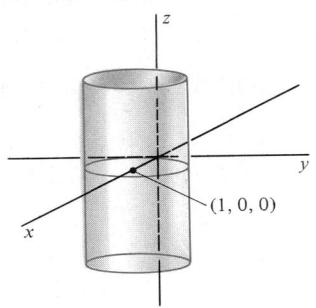

$(1, 0, 0)$

35. a. $\dfrac{10}{\sqrt{11}}$ **b.** $\dfrac{10}{7}i + \dfrac{15}{7}j - \dfrac{5}{7}k$

37. $\dfrac{x}{3} + \dfrac{y}{4} + \dfrac{z}{5} = 1$

39. $x(t) = 5,\ y(t) = -t + 2,\ z(y) = \dfrac{1}{2}t - 3$

41. $\sqrt{83}$

Equation	Name	Center	yz-section	xz-section	xy-section
45. $x^2 + \dfrac{y^2}{6} - \dfrac{z^2}{3} = 1$	hyperboloid of one sheet	$(0, 0, 0)$	hyperbola	hyperbola	ellipse
47. $\dfrac{y^2}{9} - \dfrac{x^2}{9} = z$	hyperbolic paraboloid	none	parabola	parabola	hyperbola*
49. $\dfrac{(x + 1)^2}{4} + \dfrac{y^2}{4} - \dfrac{(z - 2)^2}{9} = 1$	hyperboloid of one sheet	$(-1, 0, 0)$	hyperbola	hyperbola	circle

CHAPTER 17

Exercise Set 17.1

1.

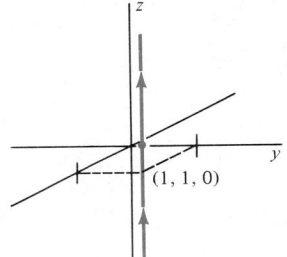

$(1, 1, 0)$

3.

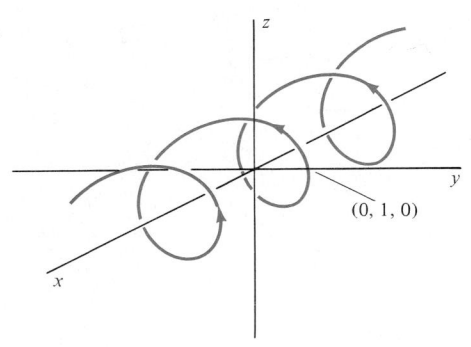

$(0, 1, 0)$

---

*Section may degenerate to two lines.

5.

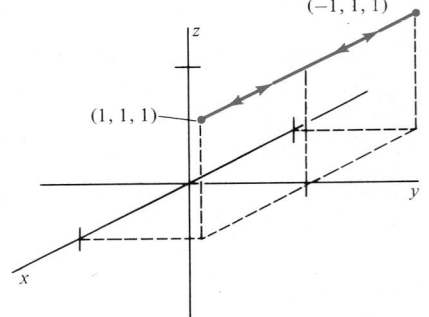

$(-1, 1, 1)$

$(1, 1, 1)$

7. $t \neq (2k + 1)\dfrac{\pi}{2},\quad k = 0,\ \pm 1,\ \pm 2, \ldots$

9. $t \geq 0$ and $t \neq 3$ **11.** $t \geq 0$

13. $(t + \sin t)i + (3t + \cos t)j + (t^2 + 1)k$

15. $(3t + 2 \sin t)i + (9t + 2 \cos t)j + (3t^2 + 2)k$

17. $t \sin t + 3t \cos t + t^2$

19. $(1 + t + 3\sqrt{t})i + (4 - 3t - t^2)j + (t + 3e^t)k$

21. $(1 + t - \sqrt{t}) \sin ti + (t - t^2) \sin tj + (t - e^t) \sin tk$

23. $\sqrt{t}(1 + t) + (1 - t)(1 - t^2) + te^t$

25. $\dfrac{\sqrt{2}}{2}i + j + \dfrac{\sqrt{2}}{2}k$ **27.** $i + k$

29. $6i + 7j + 14k$

31. $\left(-\dfrac{\pi}{2} + n\pi, \dfrac{\pi}{2} + n\pi\right), \ n = 0, \pm 1, \ldots$

33. $[0, 1)$ and $(1, \infty)$

35. $[0, 1)$

37. $(-\infty, 1)$ and $(1, \infty)$

39. $600{,}000\pi$ One complete revolution takes $t = \dfrac{\pi}{500}$ seconds.

45. No conclusions in either case.

Exercise Set 17.2

1. $(0, \infty)$

3. $(-3, -2)$ and $(-2, \infty)$

5. $i + \dfrac{1}{2\sqrt{t}}j$

7. $\dfrac{1}{2\sqrt{t}}i - \dfrac{3}{2}t^{-5/2}j + \dfrac{2}{2t - 1}k$

9. $\dfrac{1}{\sqrt{1 - t^2}}i + \dfrac{t}{\sqrt{1 + t^2}}j - 3t^2e^{-t^3}k$

11. $-\dfrac{1}{4}t^{-3/2}i + \dfrac{15}{4}t^{-7/2}j - \dfrac{4}{(2t - 1)^2}k$

15. $(3\cos t - 2)i + (6t + 2\sin t)k$

17. $-\sin ti + (3t^2 + \sin^2 t - \cos^2 t)j - k$

19. $3e^{3t}\cos e^{3t}i + 6e^{6t}k$

23. $\dfrac{2}{3}t^{3/2}i + \dfrac{1}{2}e^{2t}j + \ln|t|k + C$

25. $(t\ln t - t)i + \ln|\ln t|j + C$

27. $(3 + t)i + \left(5 + \dfrac{t^3}{3}\right)j$

29. $\left(3 + \dfrac{t}{2} + \dfrac{1}{4}\sin 2t\right)i + \left(\dfrac{t}{2} - \dfrac{1}{4}\sin 2t - 6\right)j$
$\qquad\qquad\qquad\qquad\qquad + (3 - e^{-t})k$

31. $e^{-4t}i + 4e^{-4t}k$

33. $\left(\dfrac{b^2 - a^2}{2}\right)i + \left(\dfrac{b^3 - a^3}{3}\right)j - \left(\dfrac{b^4 - a^4}{4}\right)k$

35. $\dfrac{\sqrt{2}}{2}i + \dfrac{1}{2}j + \left(\dfrac{\pi}{8} + \dfrac{1}{4}\right)k$

39. False

Exercise Set 17.3

1. $48i + \dfrac{1}{4}j$

3. $\sqrt{2}i + 2j$

5. $-2i + j + 6k$

7. $\dfrac{1}{\sqrt{a^2 + b^2}}(-ai + bj)$

9. $t = \sqrt{2}$

11. $(64i + 2j + 2k) + t\left(48i + \dfrac{1}{4}j\right)$

13. $\dfrac{2 - x}{2} = y + 5 = \dfrac{z + 6}{6}$

15. a. $\dfrac{\pi}{4}$

17. $\dfrac{\pi}{2}$

 b. $\dfrac{\pi}{4}$

19. $\dfrac{1}{2}\sqrt{6} + \dfrac{1}{2}\ln(2 + \sqrt{6}) - \dfrac{1}{4}\ln 2$

21. $\dfrac{1}{\sqrt{17}}(i + 4j)$

23. $\alpha = \dfrac{\sqrt{2}}{2}$

27. $\cos ti - \sin tj$

29. $(4i + j + 8k) + t(2i + 12k)$

Exercise Set 17.4

1. $v = 2j, \quad a = 0$

3. $v = -3\sin 3ti + 3\cos 3tj, \quad a = -9\cos 3ti - 9\sin 3tj$

5. $v = \dfrac{2}{t}i - 2\sin 2tj - \dfrac{1}{t^2}k, \quad a = \dfrac{-2}{t^2}i - 4\cos 2tj + \dfrac{2}{t^3}k$

7. $v = e^t(\cos t - \sin t)i + e^t(\sin t + \cos t)j - \sin tk,$
$\quad a = -2e^t\sin ti + 2e^t\cos tj - \cos tk$

9. $|v| = 2\sqrt{41}$

11. $(3 + \sin t)i + (3 - \cos t)j$

13. $(\text{Tan}^{-1} t - 2)i + \left(\dfrac{1}{2}e^{t^2} + \dfrac{1}{2}\right)j + 4k$

15. $v = 2ti + tk, \quad r = (t^2 + 3)i - j + \left(\dfrac{t^2}{2} + 4\right)k$

17. $v = \sin ti + (1 - \cos t)j + k,$
$\quad r = (4 - \cos t)i + (-1 + t - \sin t)j + (t + 2)k$

21. $\text{Speed} = |\alpha| \cdot \text{radius}$

23. a. $50\sqrt{3}\, ti + \left(50t - \dfrac{1}{2}gt^2\right)j$

 b. $\dfrac{1250}{9.8}$ m

 c. $\dfrac{100}{9.8}$ s

 d. $\dfrac{5000\sqrt{3}}{9.8}$ m

 e. 100 m/s

27. a. $\displaystyle\int_0^{50\sqrt{3}/g} \sqrt{2500 - 50\sqrt{3}gt + g^2t^2}\; dt$

   b. 304.6

Exercise Set 17.5

1. $T(s) = \dfrac{1}{2}i + \dfrac{\sqrt{3}}{2}j, \quad r''(s) = 0, \quad \kappa(s) = 0$

3. $T(s) = \dfrac{-\sqrt{2}}{2}\sin(s)i + \dfrac{\sqrt{2}}{2}\cos(s)j + \dfrac{\sqrt{2}}{2}k,$

   $r''(s) = \dfrac{-\sqrt{2}}{2}\cos(s)i - \dfrac{\sqrt{2}}{2}\sin(s)j, \quad \kappa(s) = \dfrac{\sqrt{2}}{2}$

5. 0

7. $\dfrac{e^t}{(1 + e^{2t})^{3/2}}$

9. $\dfrac{1}{2}$

11. $\dfrac{2}{(1 + 4x^2)^{3/2}}$

13. $\dfrac{|6x|}{(1 + 9x^4)^{3/2}}$

15. $\dfrac{|\cos x|}{(1 + \sin^2 x)^{3/2}}$

17. $\dfrac{2}{(1 + 4t^2)^{3/2}}$

19. $\dfrac{2e^t(1 + e^{2t})^{1/2}}{(2e^{2t} + 1)^{3/2}}$

21. $\left(\dfrac{\pi}{2}, 0\right)$

23. a. $\dfrac{2}{5^{3/2}}$ **b.** $\left(\dfrac{7}{2}, -4\right)$ **c.** 2

27. $\dfrac{3}{2^{3/2}} \dfrac{1}{(1 + \sin\theta)^{1/2}}$

29. $\dfrac{3}{2^{3/2}} \dfrac{1}{(1 - \cos\theta)^{1/2}}$

35. $a_T = \dfrac{2t}{\sqrt{t + 1}}, \quad a_N\sqrt{\dfrac{16t^6 + 16t^4 + 4}{t^2 + 1}}$

Review Exercises—Chapter 17

1. $\dfrac{x^2}{a^2} + \dfrac{x^2}{b^2} = 1$

3. $[-1, 0) \cup (0, 1]$

5. $t = 2n\pi, \quad n = 0, 1, 2, \dots$

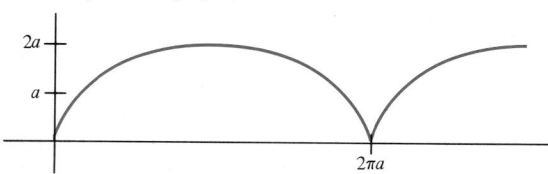

7. $3i + \sqrt{2}j + 2k$

9. $t = \dfrac{\pi}{2}, \quad \dfrac{3\pi}{2}$

11. $t = \pm 3, \pm 2$

13. $r'(t) = e^t(t + 1)i + e^{-t}(1 - t)j$
   $r''(t) = e^t(t + 2)i + e^{-t}(t - 2)j$

15. $r'(t) = t^{-1}i + \dfrac{1}{2}t^{-1/2}e^{\sqrt{t}}j$

   $r''(t) = -t^{-2}i + \dfrac{1}{4}(t^{-1} - t^{3/2})e^{\sqrt{t}}j$

17. $i + (1 - 2e^{-1})j$

19. $(4i + 3j)\cosh 3t + C\sinh 3t$ for any constant vector C

21. $r(t) = 5\cos\left(\dfrac{1}{5}t\right)i + 5\sin\left(\dfrac{1}{5}t\right)j$

23. a. $\kappa(t) = \dfrac{1}{2}$

   b. $T(t) = \dfrac{\sqrt{2}}{2}i - \dfrac{\sqrt{2}}{2}\sin tj + \dfrac{\sqrt{2}}{2}\cos tk$

25. $3\sqrt{13}$

27. $\dfrac{12}{(145)^{3/2}}$

33. $x = \dfrac{-\ln 2}{2}$

35. Approximately 4.14 m

CHAPTER 18

Exercise Set 18.1

1. $\{(x, y) \mid (x, y) \neq (0, 0)\}$

3. $\{(x, y) \mid y \geq x\}$

5. All $(x, y) \in \mathbb{R}^2$

7. All $(x, y, z) \in \mathbb{R}^3$

9. $\{(x, y, z) \mid xyz \neq 0\}$

11. a. $h(x, y) = \sqrt{x + y^2}$
   b. $\{(x, y) \mid x + y^2 \geq 0\}$

13. $S(r, h) = 2\pi r^2 + 2\pi rh$

15. -5

17. $\dfrac{1}{\sqrt{2}}$

19. 0

21. 0

23.

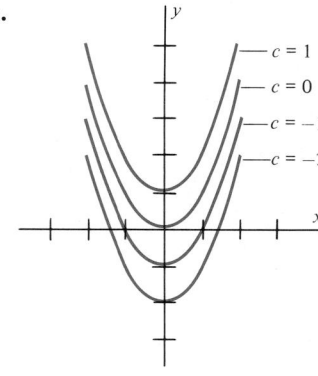

25.

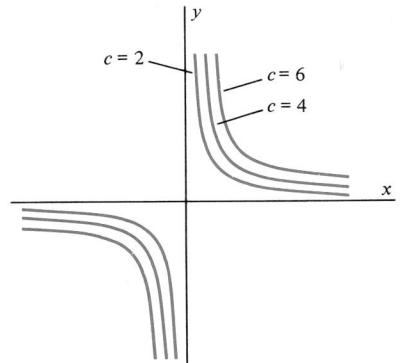

27.

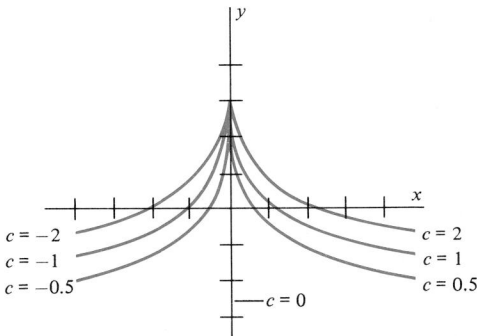

29.

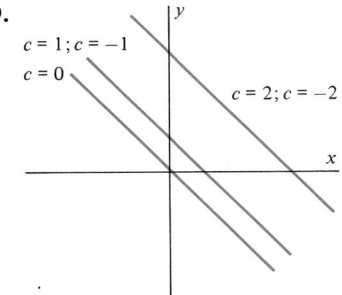

31.

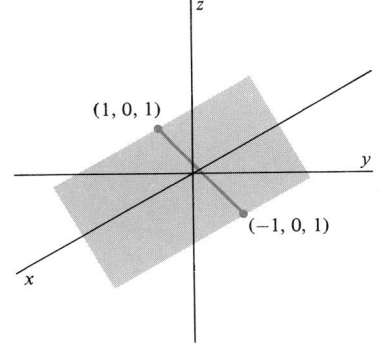

33.

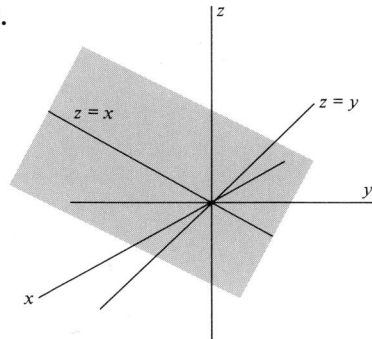

35.

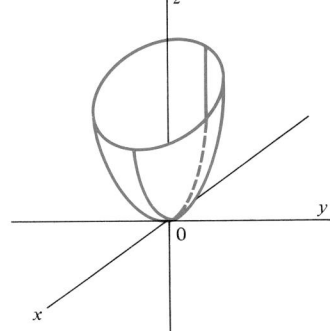

37.

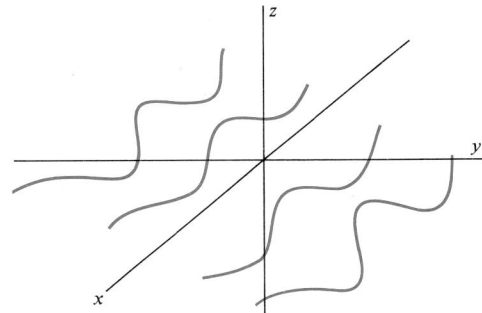

39. Does not exist   **41.** No

43. Yes   **45.** 0

47. False

Exercise Set 18.2

1. $\dfrac{\partial f}{\partial x} = y$, $\dfrac{\partial f}{\partial y} = x$

3. $\dfrac{\partial f}{\partial x} = \tan y^2$, $\dfrac{\partial f}{\partial y} = 2xy \sec^2 y^2$

5. $\dfrac{\partial f}{\partial x} = 2xe^{x^2+y^2}$, $\dfrac{\partial f}{\partial y} = 2ye^{x^2+y^2}$

7. $\dfrac{\partial f}{\partial r} = \sin\left(\dfrac{\pi}{2} - \theta\right)$, $\dfrac{\partial f}{\partial \theta} = -r \cos\left(\dfrac{\pi}{2} - \theta\right)$

9. $\dfrac{\partial f}{\partial x} = \dfrac{y^2 + 1}{xy^2 + x - y}, \; \dfrac{\partial f}{\partial y} = \dfrac{2xy - 1}{xy^2 + x - y}$

11. $\dfrac{\partial f}{\partial x} = yx^{y-1}, \; \dfrac{\partial f}{\partial y} = x^y \ln x$

13. $\dfrac{\partial f}{\partial r} = 2r \cos \theta, \; \dfrac{\partial f}{\partial \theta} = -r^2 \sin \theta$

15. $\dfrac{\partial f}{\partial x} = y^3, \; \dfrac{\partial f}{\partial y} = 3xy^2 - z^2, \; \dfrac{\partial f}{\partial z} = -2yz$

17. $\dfrac{\partial f}{\partial x} = \dfrac{2yz}{(x + y)^2}\left(\dfrac{x - y}{x + y}\right)^{z-1},$

$\dfrac{\partial f}{\partial y} = \dfrac{-2xz}{(x + y)^2}\left(\dfrac{x - y}{x + y}\right)^{z-1}$

$\dfrac{\partial f}{\partial z} = \left(\dfrac{x - y}{x + y}\right)^z \ln\left(\dfrac{x - y}{x + y}\right)$

19. $\dfrac{\partial f}{\partial r} = \dfrac{s \ln t \, (s^2 + t)}{2r^{1/2} \, (s - 2r + t)^{3/2}}$

$\dfrac{\partial f}{\partial s} = \dfrac{r^{1/2} \ln t \, (-2r + t)}{(s^2 - 2r + t)^{3/2}}$

$\dfrac{\partial f}{\partial t} = \dfrac{-r^{1/2} \, s \, (s^2 - 2r + t - \tfrac{1}{2}t \ln t)}{t \, (s^2 - 2r + t)^{3/2}}$

21. 125

23. $\dfrac{1}{2\sqrt{3}}$

25. $\dfrac{\partial^2 f}{\partial r^2} = 2 \cos \theta, \; \dfrac{\partial^2 f}{\partial r \, \partial \theta} = -2r \sin \theta, \; \dfrac{\partial^2 f}{\partial \theta^2} = -r^2 \cos \theta$

27. $\dfrac{2}{x^2 + y^2 + z^2}$

29. a. $v_2 = \dfrac{A_1 v_1}{A_2}$

b. $\dfrac{\partial v_2}{\partial A_2} = \dfrac{-A_1 v_1}{(A_2)^2}$

c. $\dfrac{-100 \text{ cm/s}}{9 \text{ cm}^2}$

d. $\dfrac{3 \text{ cm}^2}{20 \text{ cm/s}}$

31. $\rho \sin \phi \cos \theta$ **33.** $\sin \phi \cos \theta$ **35.** $-\rho \sin \phi$

37. a. $\cos(x + y)^2 - \cos x^2$
b. $\cos(x + y)^2$

41. a. $\dfrac{\partial V}{\partial a} = \dfrac{a\sigma}{2\epsilon_0 \sqrt{a^2 + r^2}}$

b. $\dfrac{\partial V}{\partial r} = \dfrac{\sigma}{2\epsilon_0}\left(\dfrac{r}{\sqrt{a^2 + r^2}} - 1\right)$

47. $f(x, y) = \sin(x, y)$

Exercise Set 18.3

1. $2x + 6y - z = 10$

3. $x - 5y - z = 5$

5. $x - 2y - z = 4$

7. $x + y - z = 2$

9. $y + 2z = 2$

11. $y - z = 1$

13. $2x + ez = 0$

15. $r(t) = 3i - j + 4k + t(4i + 4j - k)$

17. $r(t) = i - 2j + 2k + t(-i + 2j - 2k)$

19. $r(t) = i + j + t(j - k)$

21. $(2, 3, 20)$

25. $18x + 16y - z = 25$

27. $x_0 x + y_0 y + z_0 z = r^2$

29. $-x - 2y + z + \dfrac{\pi}{2} = 0$

Exercise Set 18.4

1. Relative minimum: $(0, -2, 0)$

3. Saddle point $(-3, 2, 0)$

5. Saddle point $(0, 0, 9)$

7. Relative minimum: $(1, 3, 1)$

9. Saddle point $(0, 0, 0)$

11. Saddle point $(0, 0, 0)$

13. Saddle point $(0, 0, 0)$
Relative maximum: $\left(-\dfrac{4}{3}, -\dfrac{4}{3}, \dfrac{64}{27}\right)$

15. Saddle points $\left(0, \dfrac{\pi}{2} + n\pi, 0\right)$

19. The graph is unbounded.

23. $\ell = 2$ m, $w = 2$ m, $h = 2$ m

25. $\left(\dfrac{ad}{a^2 + b^2 + c^2}, \dfrac{bd}{a^2 + b^2 + c^2}, \dfrac{cd}{a^2 + b^2 + c^2}\right)$

27. $\ell = 25''$, $w = h = 14''$

29. Absolute minimum: $-2\sqrt{2} + 4$
Absolute maximum: $2\sqrt{2} + 4$

31. Absolute minimum: 3
Absolute maximum: 9

35. 86

37. a. $y = \dfrac{23}{20}x - \dfrac{19}{4}$

b.

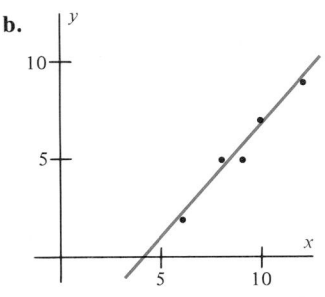

c. $y(4) = \dfrac{-3}{20}$

Exercise Set 18.5

1. $2xy^4\,dx + 4x^2y^3\,dy$

3. $\dfrac{x}{\sqrt{x^2 + y^4}}\,dx + \dfrac{2y^3}{\sqrt{x^2 + y^4}}\,dy$

5. $\dfrac{e^{\sqrt{x}}}{2\sqrt{x}}\cos y\,dx - e^{\sqrt{x}}\sin y\,dy$

7. $2xyz^3\,dx + x^2z^3\,dy + 3x^2yz^2\,dz$

9. $\dfrac{1}{y + 3^z}\,dx - \dfrac{x}{(y + 3^z)^2}\,dy - \dfrac{x3^z\,\ln 3}{(y + 3^z)^2}\,dz$

11. $\dfrac{y^2 + z^2 - x^2 + 2xy}{(x^2 + y^2 + z^2)^2}\,dx + \dfrac{y^2 - x^2 - z^2 - 2xy}{(x^2 + y^2 + z^2)^2}\,dy$
$$+ \dfrac{2zy - 2zx}{(x^2 + y^2 + z^2)^2}\,dz$$

13. a. $f(2, 3) = 48$
$f(2.1, 3.2) = 57.12$
b. $\Delta f = 9.12$

15. 5.076   **17.** 0.7441

19. 26.8   **21.** 55.44

23. $77.41 \le v \le 81.43$ **25.** -3%

27. 0.048 seconds   **29.** 10%

Exercise Set 18.6

1. $-16t + 4t^3$

3. $-\sin^3 t + 2\sin t\cos^2 t$

5. $\dfrac{1}{4\sqrt{t}\sqrt{\sqrt{t} + e^{t^2}}} + \dfrac{te^{t^2}}{\sqrt{\sqrt{t} + e^{t^2}}}$

7. $-6t^2 - 2t + 2\sin t\cos t$

9. $2e^{-3t}[a^2\cosh t + b^2\sinh t]$

11. $-2t^3\sin^4 t\cos t + 3t^2\sin^3 t\cos^2 t + 3t^3\sin^2 t\cos^3 t$

13. a. $2te^{2st} + 2st^2 - 2s + 2t$
b. $2se^{2st} + 2s^2t + 2s - 2t$

15. a. $[y^2\cos xy^2 - 2xy](2s - t) + [2xy\cos xy^2 - x^2](2t^2s)$
b. $[y^2\cos xy^2 - 2xy](-s) + [2xy\cos xy^2 - x^2](2s^2t)$

17. a. $y^3z^2\sin t - 3xy^2z^2t\sin s - 4sxy^3z$
b. $sy^3z^2\cos t + 3xy^2z^2\cos s + 4txy^3z$

19. a. $2ue^{u^2}y^2 - 4xye^{2v}$
b. $4xyue^{u^2} - 2x^2e^{2v}$

21. $6rt^2(1 - \cos\theta) + \dfrac{r^2t\sin\theta}{\sqrt{1 + t^2}}$

23. 116π cm³/s

25. a. $D(t) = \sqrt{1 + t^2}$

b. $D'(t) = \dfrac{t}{\sqrt{1 + t^2}}$

29. a. $12s^2t^2 - 8y^3 + 48s^2y^2$
b. $2s^4 + 8y^3 + 48t^2y^2$

31. $2s^2t^2e^{rst} - 4s^2t^2e^{2rst} + rs^3t^3e^{rst}$

33. $e^{y^2}[-2yr\sin^2\theta - \sin\theta + 2yr\cos^2\theta$
$$+ x(2 + 4y^2)r\sin\theta\cos\theta + 2xy\cos\theta]$$

Exercise Set 18.7

1. $6i + 9j$

3. $\dfrac{\sqrt{2}}{2}\left(1 - \dfrac{\pi}{2}\right)i + \dfrac{\sqrt{2}}{4}\pi j$

5. $6i + j + 4k$ **7.** $5i - 3j - k$

9. $\left(e + \dfrac{1}{e}\right)i + \dfrac{1}{4}\left(e + \dfrac{1}{e}\right)j + 2\left(e - \dfrac{1}{e}\right)k$

11. $-30\sqrt{2}$ **13.** $\dfrac{2\sqrt{3} - 1}{18}$

15. $\dfrac{2}{\sqrt{3}}$ **17.** $\dfrac{3\sqrt{3}}{\sqrt{11}}$

19. $\dfrac{13\sqrt{2}}{6} + \dfrac{1}{3}\left(\dfrac{1}{e} - e\right) + \dfrac{\pi}{12}\left(\dfrac{1}{e} + e\right)$

21. N: $2i + 3j + t(16i - 6j)$
T: $2i + 3j + t(6i + 16j)$

23. N: $4i + 4j + t(i + j)$
T: $4i + 4j + t(i - j)$

25. $\sqrt{5}$ **27.** -1 **29.** $\dfrac{\sqrt{2}}{4}$

31. a. $(3, 1 + \sqrt{2}), (3, 1 - \sqrt{2})$
b. $(1, 1), (5, 1)$

37. $2i + j$

39. $-i + j$

41. One example is $f(x, y) = \sqrt{x^2 + y^2}$ at $(0, 0)$.

Exercise Set 18.8

Maximum

1. $f(0, \pm 1) = 4$

3. None

5. $f\left(\dfrac{1}{\sqrt{2}}, \dfrac{1}{\sqrt{2}}\right)$

$= f\left(\dfrac{-1}{\sqrt{2}}, \dfrac{-1}{\sqrt{2}}\right) = \dfrac{1}{2}$

7. $f(2, 0) = 12$

9. $f\left(-\dfrac{2}{3}, \pm\dfrac{\sqrt{5}}{3}\right) = \dfrac{48}{9}$

11. $f\left(\dfrac{1}{\sqrt{6}}, \dfrac{2}{\sqrt{6}}, -\dfrac{1}{\sqrt{6}}\right)$

$= \sqrt{6}$

13. $f(2, 2, 2) = 6$

15. $f\left(2, -1 + \dfrac{\sqrt{2}}{2}, \dfrac{3}{2}\right) = f\left(2, -1 - \dfrac{\sqrt{2}}{2}, \dfrac{3}{2}\right) = \dfrac{3}{4}$

17. $\left(3 - 9\dfrac{\sqrt{13}}{13}, -2 + 6\dfrac{\sqrt{13}}{13}\right)$

19. $(1, 0, 1)$

21. $(2, -1, 1)$

23. $\dfrac{2r}{\sqrt{3}} \times \dfrac{2r}{\sqrt{3}} \times \dfrac{2r}{\sqrt{3}}$

25. $r = \sqrt[3]{\dfrac{1}{\pi}}, h = 2\sqrt[3]{\dfrac{1}{\pi}}$

27. $x = 2, y = 3, z = 1$

29. $x = y$ and $y = z$ $x = 20$

31. $x = 2500, y = 0$ yields max. revenue

Exercise Set 18.9

1. $x - y + C$

5. None

9. $x^2 e^{xy} - \sqrt{y} + C$

Review Exercises—Chapter 18

1. No

Minimum

1. $f(\pm 1, 0) = 2$

3. $f(1, -1) = 2$

5. $f\left(\dfrac{1}{\sqrt{2}}, \dfrac{-1}{\sqrt{2}}\right)$

$= f\left(\dfrac{-1}{\sqrt{2}}, \dfrac{1}{\sqrt{2}}\right) = -\dfrac{1}{2}$

7. $f(-2, 0) = -4$

9. $f(1, 0) = -3$

11. $f\left(-\dfrac{1}{\sqrt{6}}, -\dfrac{2}{\sqrt{6}}, \dfrac{1}{\sqrt{6}}\right)$

$= -\sqrt{6}$

13. $f(-2, -2, -2) = -6$

3. None

7. $x^3 \cos y - x + y^2 + C$

11. $x^2 y^2 + y \cos x + C$

3.

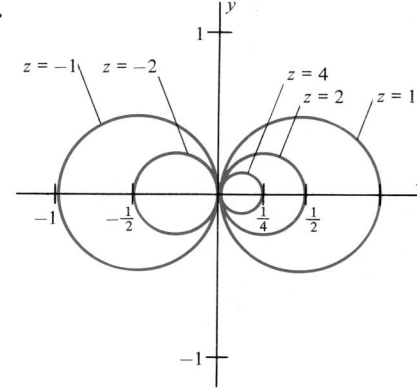

5. $5i - 7j$

7. $-i - k$

9. $\dfrac{\partial f}{\partial x}(x, y) = \sin \sqrt{x^2 + y^2} + \dfrac{x^2 \cos \sqrt{x^2 + y^2}}{\sqrt{x^2 + y^2}},$

$\dfrac{\partial f}{\partial y}(x, y) = \dfrac{xy \cos \sqrt{x^2 + y^2}}{\sqrt{x^2 + y^2}}$

11. $\dfrac{784}{15}$

13. $\dfrac{\partial f}{\partial r} = 18x^2 + 4y^2 + 4xy - 2y,$

$\dfrac{\partial f}{\partial s} = 45x^2 + 10y^2 - 16sxy + 8sy$

15. $5x + 4y = 10$

19. $-\dfrac{36}{5}i - \dfrac{12}{5}j - (2 - \text{Tan}^{-1} 2)k$

23. $\dfrac{2\sqrt{3}}{3} \times 2\sqrt{3} \times \dfrac{4\sqrt{3}}{3}$

25. $(0, 0, 0)$: saddle point

27. $(0, 0, 1)$: saddle point

29. $(0, 0, e)$: saddle point

31. a. 1
b. 3
c. $i + 3j$

33. $f(x, y) = x^2 e^y + \cos y + C$

35. R increases by 10%

39. Maximum: $5 + 6\sqrt{2}$
Minimum: $5 - 6\sqrt{2}$

41. No

43. $\left(\dfrac{4}{\sqrt{14}}, \dfrac{8}{\sqrt{14}}, \dfrac{-12}{\sqrt{14}}\right)$

49. $\left(\dfrac{3}{7}, \dfrac{-9}{7}, \dfrac{6}{7}\right)$

51. $x + 2y + z = 6$

53. $3 - \dfrac{1}{2}\sqrt{3}$

CHAPTER 19

Exercise Set 19.1

1. 1

3. 7

5. $\dfrac{1}{2}$

7. $\dfrac{8}{3}$

9. $\dfrac{1}{3}$

9. $\dfrac{1}{4}\left(e - \dfrac{1}{e}\right)$

11. $\dfrac{1}{2}$

13. $\dfrac{4\pi}{3}$

15. $\dfrac{5}{6}$

17. 8

19. $\dfrac{1}{2}\ln\sqrt{2}$

21. $\dfrac{\pi\sqrt{2}}{8} + 1 - \sqrt{2}$

23. 6

11. $\dfrac{1}{2}$

25. $\dfrac{1}{2}$

27. $8e(e - 1)$

33. The integral is 24.064 to 3 places.

35. The integral is 24.636 to 3 places.

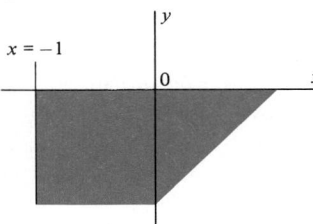

Exercise Set 19.2

1. $-\dfrac{1}{6}$

3. $\dfrac{13}{24}$

13. $\dfrac{2}{15}$

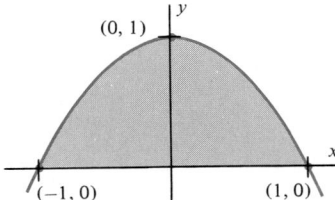

15. $\dfrac{1}{3}(1 + e)^{3/2} - \dfrac{4}{3}2^{3/2} + 2\sqrt{e}\sqrt{1 + e}$
$$+ 2\ln\dfrac{\sqrt{e} + \sqrt{1 + e}}{1 + \sqrt{2}}$$

17. $\dfrac{1}{24}$

5. 0

19. $\dfrac{3e^2 + e^{-2}}{4}$

21. $\dfrac{1}{3}$

23. 2

7. $\dfrac{1}{20}$

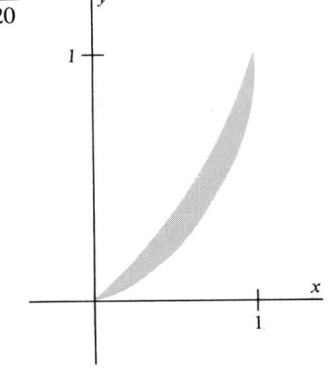

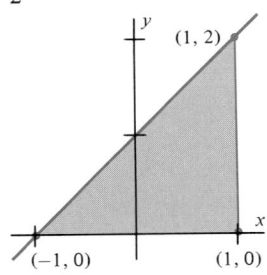

25. $\dfrac{1}{10}$

27. $\int_0^1 \int_{e^y}^e f(x, y)\, dx\, dy$

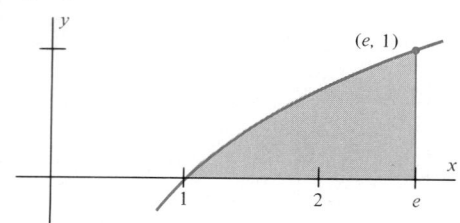

29. $\dfrac{9}{2}$

31. $\sqrt{2} - 1$

33. $\dfrac{1}{3}$

35. 6π

37. 10π

39. $\dfrac{16}{3}$

41. $\dfrac{2}{3}$

7. $\dfrac{4}{27}$

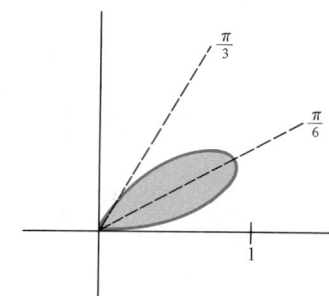

9. $\dfrac{\pi}{4}$

11. $2 + \dfrac{\pi}{4}$

13. $\dfrac{9\pi}{2}$

15. $\dfrac{\pi}{2}$

17. $\dfrac{\pi}{4}(e - 1)$

19. 2

21. 0

23. 16π

25. $2\sqrt{3}\,\pi$

27. 3π

29. 16π

31. $\dfrac{\pi}{4}$

33. $\dfrac{16\pi}{3}$

Exercise Set 19.3

1. $\dfrac{2\pi}{3}$

3. $\pi(e^{a^2} - 1)$

5. $\dfrac{45\pi}{16}$

Exercise Set 19.4

1. 3

3. 8

5. $\sqrt{2} - 1$

7. $\dfrac{\pi}{4}$

9. 2π

11. $(\bar{x}, \bar{y}) = \left(\dfrac{11}{9}, \dfrac{5}{9}\right)$

13. $(\bar{x}, \bar{y}) = \left(\dfrac{3}{\pi}, \dfrac{3}{\pi}\right)$

15. $(\bar{x}, \bar{y}) = \left(\dfrac{17}{54}, \dfrac{13}{9}\right)$

17. $(\bar{x}, \bar{y}) = \left(0, \dfrac{8}{5}\right)$

19. $(\bar{x}, \bar{y}) = (2.6, 1.7)$

21. $(\bar{x}, \bar{y}) = \left(\dfrac{4a}{3\pi}, \dfrac{4a}{3\pi}\right)$

23. $(\bar{x}, \bar{y}) = \left(\dfrac{8}{5}, 2\right)$

25. 4λ

Exercise Set 19.5

1. $\sqrt{3}$

3. $\dfrac{\pi}{6}(13^{3/2} - 1)$

5. $2\sqrt{3}$

7. $3\sqrt{2}\pi$

9. $\sqrt{2} + \ln(1 + \sqrt{2})$

11. $\dfrac{5\sqrt{2}}{12} + \dfrac{1}{4}\ln(3 + 2\sqrt{2})$

13. $\dfrac{\pi}{6}(7^3 - 29^{3/2})$

15. $4\pi(2 - \sqrt{3})$

17. $A \approx 7.904$

Exercise Set 19.6

1. $\dfrac{1}{8}$

3. $\dfrac{1}{2}$

5. $\dfrac{14}{3}$

7. 0

9. $\displaystyle\int_0^2 \int_{y/2}^1 \int_0^{x+y} f(x, y, z)\, dz\, dx\, dy$

11.

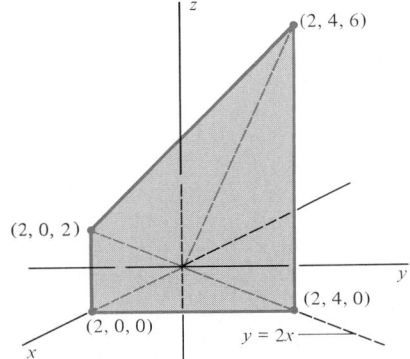

13. $\dfrac{2\pi}{3}$

15. 27π

17. 0

19. π

21. $\dfrac{405\pi}{8}$

23. 12

25. $(\bar{x}, \bar{y}, \bar{z}) = \left(0, \dfrac{3}{4}, \dfrac{15}{8}\right)$

27. $(\bar{x}, \bar{y}, \bar{z}) = \left(\dfrac{3}{4}, \dfrac{3}{4}, \dfrac{3}{4}\right)$

29. $\displaystyle\int_0^2 \int_0^{1-y/2} \int_0^{4-4x-2y} (x - y + z)\, dz\, dx\, dy,$

$\displaystyle\int_0^1 \int_0^{4-4x} \int_0^{2-2x-z/2} (x - y + z)\, dy\, dz\, dx,$

$\displaystyle\int_0^4 \int_0^{2-z/2} \int_0^{1-y/2-z/4} (x - y + z)\, dx\, dy\, dz,$

$\displaystyle\int_0^2 \int_0^{4-2y} \int_0^{1-y/2-z/4} (x - y + z)\, dx\, dz\, dy,$

$\displaystyle\int_0^4 \int_0^{1-z/4} \int_0^{2-2x-z/2} (x - y + z)\, dy\, dx\, dz$

Exercise Set 19.7

1. $\dfrac{81\pi}{2}$

3. $(\bar{x}, \bar{y}, \bar{z}) = (0, 0, 6)$

5. $18\pi + 24$

7. $\dfrac{2\sqrt{2}\, 4^4}{3^2} \approx 80$

9. $\dfrac{32}{9}$

11. 0

13. π

15. $(\bar{x}, \bar{y}, \bar{z}) = \left(0, \dfrac{3}{8}, \dfrac{3}{8}\right)$

17. $\dfrac{16\sqrt{2}\,\pi}{3}$

19. 2π

21. $V = \dfrac{16}{3}R^3$

23. $9\pi\left(1 - \dfrac{\sqrt{3}}{3}\right)$

25. $\dfrac{32\sqrt{2}\,\pi}{5}$

27. $\dfrac{40\pi}{3}$

29. $\dfrac{104\pi}{3}\left(1 - \dfrac{\sqrt{2}}{2}\right)$

Review Exercises—Chapter 19

1. $\dfrac{1}{15}$

3. $2\left(1 - \dfrac{1}{\sqrt{e}}\right)$

5. 64

7. $16\left(\dfrac{\pi}{2} - 1\right)$

9. 3π

11. $\dfrac{abc}{6}$

13. 36π

15. $(\bar{x}, \bar{y}, \bar{z}) = \left(0, 0, \dfrac{3}{4}\right)$

17. $\dfrac{2a^3}{9}(3\pi - 4)$

19. $4\sqrt{17}\,\pi$

21. $\dfrac{32\pi}{3}(8 - 3\sqrt{3})$

23. $\dfrac{977}{1017}$

25. $(\bar{x}, \bar{y}) = \left(0, \dfrac{16}{3\pi}\right)$

27. $\dfrac{2\pi}{3}(e - 1)$

29. $(\bar{x}, \bar{y}) = \left(\dfrac{5}{6}, 0\right)$

31. $\dfrac{81\pi}{4}$

33. $32\pi\rho$

35. $\dfrac{\pi}{2}(\sqrt{2} - 1)$

37. $\dfrac{256}{9}$

39. The integral is 2.2228 to 4 places.

41. 36π

43. $8\sqrt{3}\,\pi$

CHAPTER 20

Exercise Set 20.1

1.

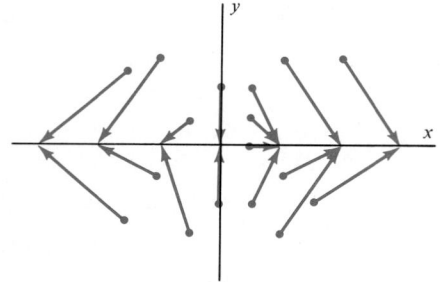

3.

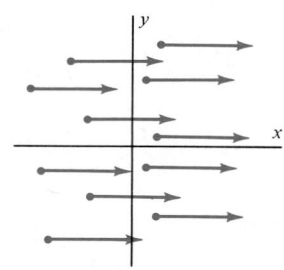

5.

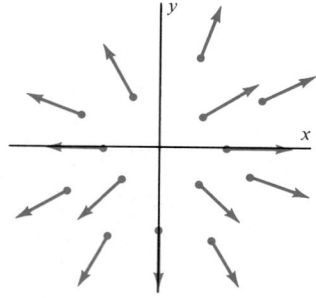

7.

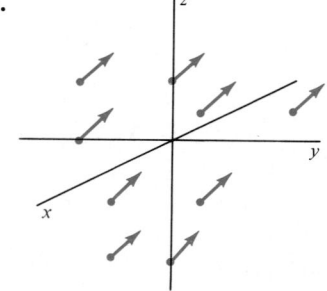

9.

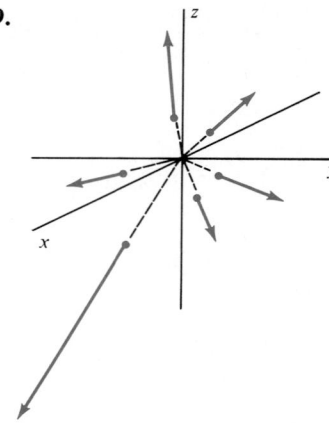

11. 4, 5, 9

13. $F = 2xi - 2yj$

15. $\dfrac{1}{\sqrt{x^2 + y^2 + z^2}}(xi + yj + zk)$

17. $e^{x-y}(zi - zj + k)$

19. Not a gradient

21. Not a gradient

23. $F = \nabla(e^{xyz})$

27. a. $F(0, 0, 0) = 4i$

  b. $F(-2, 0, 0) = -\dfrac{26}{9}i$

  c., d.

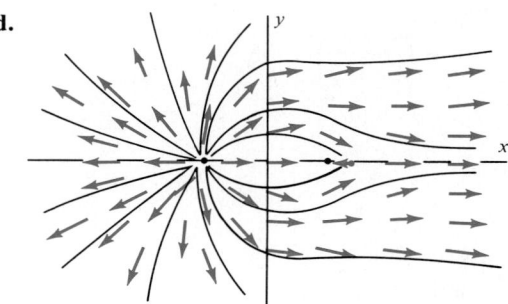

Exercise Set 20.2

1. 16

3. $\dfrac{1708}{3}$

5. $\dfrac{328}{15}$

7. $\dfrac{\pi^2}{4} + \pi$

9. 12

11. $-\pi$

13. -4

15. $\dfrac{\pi}{2}\left(\dfrac{\pi}{2} - 1\right)$

17. -1

19. 3

21. $\dfrac{1}{6} + \dfrac{\pi}{16}$

23. $\dfrac{-442}{3}$

25. $\dfrac{4}{3}$

31. $\dfrac{2}{3}(5^{3/2} - 1)$

29. $\dfrac{70\sqrt{5}}{3}$

33. 5π

21. $(\bar{x}, \bar{y}, \bar{z}) = \left(0, 0, \dfrac{a}{2}\right)$

23. $(\bar{x}, \bar{y}, \bar{z}) = \left(\dfrac{1}{2}, \dfrac{1}{2}, \dfrac{1}{2}\right)$

Exercise Set 20.3

For problems 1 through 10 $F = \nabla f$ with

f	$\int_c F \cdot dr$
1. xy	3
3. e^{xy}	$e^2 - 1$
5. $x^2 + y^2$,	0
7. xe^{y^2},	$2e$
9. $e^x \sin y$,	$e\dfrac{\sqrt{2}}{2}$

11. $-a$

19. No, the field is not conservative.

Exercise Set 20.4

1. $\dfrac{1}{2}$

5. $\dfrac{\pi}{8}$

9. 0

13. 0

17. $\dfrac{1}{6}$

3. $\dfrac{2}{3}$

7. $-\dfrac{11}{60}$

11. -5π

15. $-\dfrac{9\pi}{2}$

19. $\dfrac{3a^2\pi}{8}$

Exercise Set 20.5

1. $\dfrac{23\sqrt{6}}{4}$

5. 0

9. $\dfrac{32\pi}{3}$

13. 16π

17. 3

19. No, not the same: $\dfrac{4\pi}{r}$.

3. $\dfrac{27\sqrt{14}}{5}$

7. $48\pi(9 - 3\sqrt{5})$

11. 2π

15. $\dfrac{4\pi a^3}{3}$

Exercise Set 20.6

1. 0

3. $z \cos(yz)i + xyj + (y \sin xy - xz)k$

5. 0

7. Not conservative

9. π

11. $-\dfrac{2}{3}$

13. 4

15. -2π

17. 0

19. 0

Exercise Set 20.7

1. $2x + 2y + 2z$

3. $1 + x + yx$

5. $-y \sin(xy) + xze^{xyz} + xy \cos(xz)$

7. 0

9. -150

11. $\dfrac{11}{6}$

13. 0

15. $\dfrac{512\pi}{15}$

17. $\dfrac{80\pi}{3}$

23. a.

Review Exercises—Chapter 20

1. **a.** $-y \sin x \ln zi + \cos x \ln zj + \dfrac{y \cos x}{z}k$

  b. $[3x^2yz^2 + y \cos(xy)]i + [x^3z^2 + x \cos(xy)]j + 2x^3yzk$

3. $F = \nabla\left(-\text{Tan}^{-1}\left(\dfrac{y}{x}\right)\right)$

5. -2π

7. $F = \nabla(xy^2z)$

9. $\dfrac{1}{15}$

13. True

15. $(a + b + c)V$

17. $-\dfrac{4}{3}$

19. π

21. $-\dfrac{3}{2}$

23. $\dfrac{2}{5}$

25. $\sqrt{6}$

27. $\dfrac{1023\sqrt{2}\,\pi}{5}$

29. $-\dfrac{15\pi}{2}$

31. 0

33. $8\pi(a + b + c)$

35. $-\dfrac{164}{3}$

37. 0

39. 0

CHAPTER 21

Exercise Set 21.1

1. $y = 2e^t - 1$

3. $y = \dfrac{-\pi}{2}e^{-t} + \pi$

5. $y = Ce^{-2t} + 2$

7. $y = \dfrac{1}{3}t^5 + Ct^2$

9. $y = e^{2t}(-\cos t + C)$

11. $y = \dfrac{1}{2}e^t + Ce^{-t}$

13. $y = \dfrac{1}{2}t + Ct^{-1}$

15. $y = ke^{ax^2/2}$

17. a. $\dfrac{dp}{dt} = -\dfrac{3}{200}p$

  b. 74.08 kg

  c. $p = 0$

19. a. $\dfrac{dp}{dt} = 0.10p + 100$

  b. \$1215.13

21. $I = I_0e^{-t/RC}$

23. 52.8° C

25. $p = 6 - 4e^{-.02t}$

27. 6.93 years

29. $y = Cs^k$

Exercise Set 21.2

1. $x^2y = C$

3. $\sqrt{x^2 + y^2} = C$

5. Not exact

7. Not exact

9. $x^2 + y^2 = C$

11. $x^2y = 3$

13. $xy^2 - x^2y = 2$

Exercise Set 21.3

1. $y = Ae^{-3t} + Be^{-2t}$

3. $y = Ae^t + Be^{-t}$

5. $y = Ae^{2t} + Be^{-5t}$

7. $y = Ae^{-t} + Bte^{-t}$

9. $y = Ae^{-3t} + Bte^{-3t}$

11. $y = A \sin 2t + B \cos 2t$

13. $y = A \sin \sqrt{5}t + B \cos \sqrt{5}t$

15. $y = Ae^{5x} + Be^{-2x}$

17. $y = Ae^{(2+\sqrt{3})x} + Be^{(2-\sqrt{3})x}$

19. $y = e^x(A \sin \sqrt{5}x + B \cos \sqrt{5}x)$

21. $y = Ae^{4x} + Be^{-3x}$

23. $y = e^{-x}(A \sin \sqrt{3}x + B \cos \sqrt{3}x)$

25. $y = e^{3t} - e^{-2t}$

27. $y = 3e^{-t} + 4te^{-t}$

29. $y = -\sin 3t + 3 \cos 3t$

31. $x = A \sin 2x + B \cos 2x$

  $y = -2A \cos 2x + 2B \sin 2x$

33. $x(t) = \dfrac{1}{10} \cos 2t$

35. a. Oscillating with increasing frequency

  b. Exponential function with increasingly large exponent

  c and **d** approach constant solution $x(t) = 1$

Exercise Set 21.4

1. $y = C_1e^{(-1+\sqrt{2})t} + C_2e^{(-1-\sqrt{2})t} - \dfrac{5}{41} \sin 2t - \dfrac{4}{41} \cos 2t$

3. $y = C_1 \sin t + C_2 \cos t + \dfrac{1}{2}e^{-t}$

5. $y = C_1e^{2t} + C_2e^{-2t} - t$

7. $y = C_1e^{5t} + C_2e^{-2t} - \dfrac{5}{12}e^{2t}$

9. $y = \left(C_1 + \dfrac{1}{6}t\right) \sin 3t + C_2 \cos 3t$

11. $y = C_1e^{4t} + \left(C_2 - \dfrac{2}{5}t\right)e^{-t}$

13. $y = (C_1 + C_2t)e^{-t} + \dfrac{3}{4}e^t + t^2 - 4t + 6$

15. $y = (C_1 + C_2t)e^t - \dfrac{11}{25} \sin 2t - \dfrac{2}{25} \cos 2t + \dfrac{1}{2}e^{-t}$

17. $y = e^{-t/2}\left(C_1 \sin \frac{1}{2}\sqrt{3}t + C_2 \cos \frac{1}{2}\sqrt{3}t\right)$
$$+ t^3 - t^2 - 4t + 6$$

23. $y = Ce^{3t} - 2t + 4$

25. $y = Ce^{4t} + t^2 + \frac{1}{2}t - \frac{1}{8}$

27. $y = Ce^{4t} - 2e^{3t}$

29. $y = Ce^{3t} - 2e^t - 2t + 4$

31. $y = 3e^{-4t}$

33. $y = 5e^{-3t} + e^{2t}$

Exercise Set 21.5

1. $y = a_0 \sum\limits_{k=0}^{\infty} \dfrac{(-1)^k x^k}{k!} = a_0 e^{-x}$

3. $y = a_0 \sum\limits_{k=0}^{\infty} \dfrac{6^k x^k}{k!} = a_0 e^{6x}$

5. $y = a_0 \sum\limits_{k=0}^{\infty} \dfrac{x^{2k}}{k!} = a_0 e^{x^2}$

7. $y = \sum\limits_{k=0}^{\infty} \dfrac{(-1)^k 4^k x^k}{k!} = e^{-4x}$

9. $y = a_1 \sum\limits_{k=1}^{\infty} \dfrac{(-1)^k x^k}{k[(k-1)!]^2}$

11. $y = \sum\limits_{k=0}^{\infty} \dfrac{x^{2k}(-1)^k}{(2k)!} = \cos x$

Exercise Set 21.6

1.

t_j	0	.25	.5	.75	1.0
y_j	0	1.5	2.25	3.375	5.0625
$y(t_j)$	1	1.6487	2.7183	4.4817	7.3891

$$y = e^{2t}$$

3.

t_j	0	.25	.50	.75	1.0
y_j	1	1.5	1.75	1.875	1.9875
$y(t_j)$	1	1.3934	1.6321	1.7769	1.8647

$$y = 2 - e^{-2t}$$

5.

t_j	0	.25	.5	.75	1.0
y_j	1	1	1.125	1.3906	1.8389
$y(t_j)$	1	1.0635	1.2663	1.6496	2.2974

$$y = 2e^{\frac{1}{2}x^2} - 1$$

7.

t_j	0	.25	.50	.75	1.0
y_j	1.5	1.5	1.375	1.1563	0.9102
$y(t_j)$	1.5	1.4394	1.2788	1.0698	0.8679

$$y = e^{-t^2} + \frac{1}{2}$$

9.

t_j	0	.25	.5	.75	1.0
y_j	1.5	2.25	3.25	4.625	6.5625
$y(t_j)$	1.5	2.3987	3.7183	5.7317	8.8890

$$y = t + \frac{1}{2} + e^{2t}$$

11. See answer to Exercise 3.

Review Exercises—Chapter 21

1. $y = \dfrac{1}{2}x^2 - 3x + C$

3. $y = 2 \sec \sqrt{t} + C$

5. $y = \dfrac{-2}{3t^2 + C}$

7. $y = Ce^{(1+x)^2/2} - 2$

9. $y = Ce^{-2t} + 2$

11. $y = \dfrac{4}{Ce^{-4t} - 1}$

13. $y = Ce^{-\sin t}$

15. $y = Ce^{4t} - 2$

17. $y = Ce^{-4t} - 1$

19. $y = C_1 e^{5t} + C_2 e^{-3t} - \dfrac{4}{377} \sin 2t - \dfrac{19}{377} \cos 2t$

21. $y = C_1 e^{3t} + C_2 e^{-3t} - \dfrac{5}{9} - \dfrac{1}{5}e^{2t}$

23. $y = Ce^{1/t}$

25. $y = C_1 e^{7t} + C_2 e^{-2t} - \dfrac{1}{14}t - \dfrac{23}{196}$

27. $y = Cx^{-2/3}$

29. $y = \dfrac{2x}{C + x}$

31. $y = \sqrt{x^2 + 1} + 2$

33. $y = 3e^{t^2/2}$

35. $y = 3 \sin 3t + 3 \cos 3t$

37. $y = 6te^{3t}$

39. a. $y' = 2y$
    b. $y'' - y = 0$

41. $y = e^{-t/5}$

43. $58°$ F

45. $y = C_1\left(x + \dfrac{x^2}{2} + \dfrac{x^3}{3!2!} + \dfrac{x^4}{4!3!} + \dfrac{x^5}{5!4!} + \cdots\right)$

47. $y = e^{2t}\left(A \sin \sqrt{2}t + B \cos \sqrt{2}t\right)$

INDEX